REE

NAUTICAL ALMANAC

CARIBBEAN
1998

This publication contains tide and/or tidal current predictions and associated information produced by and obtained from the Department of Commerce, National Oceanic and Atmospheric Administration, National Ocean Service. This is not a National Ocean Service publication. The National Ocean Service is not responsible for any reproduction errors. These predictions satisfy all U.S. Coast Guard requirements including: 33 CFR Ch. I (7-1-91 Edition), 164.33 Charts and Publications.

REED'S

NAUTICAL ALMANAC

CARIBBEAN

1998

Editor

Ben Ellison

5th Annual Edition

REED'S tide and tidal current predictions use original data files from the National Oceanic and Atmospheric Administration (NOAA) for all stations included in the Nation Ocean Service (NOS) database.

For Caribbean locations not in the NOS database, REED'S uses the same prediction algorithms as the United Kingdom Hydrographic Service.

13A Lewis Street
Boston MA 02113
U.S.A.
Tel.: 617-248-0700 Fax: 617-248-5855
Web Site: www.treed.com
E-mail publisher: jdk@treed.com
E-mail editor: editor@treed.com

ISBN 1-884666-25-6

Printed in the U.S.A. by:

Courier Corp.
Pleasant Street
Westford, MA 01886

Contents

PREFACE 1998

Welcome to the 5th edition of *Reed's Nautical Almanac, Caribbean 1998* edition. As we go to press, we are weary but proud. We think we've produced our best edition yet, a remarkably well organized and comprehensive reference volume for mariners.

If you are familiar with *Reed's Almanacs*, you will notice a number of changes and additions to this edition:

- ➢ A new pagination system and simplified chapter plan—both designed to make it easier to find what you want.
- ➢ Tide tables now include sunrise and sunset times.
- ➢ 3 methods for determining the height of tide at any time.
- ➢ Numerous new or improved chartlets.
- ➢ Expanded planet, star, and eclipse information in the ephemeris (which is not just for celestial navigators).
- ➢ Amplitude tables for checking your compass at sea.
- ➢ A bibliography of cruising guides for the Caribbean region.

Many of you will notice that these are improvements that you suggested. And if you sent in a comment on a particular port or other subject, it is probably in the book. Thank you. We truly value your feedback and the information you find on the water. We have a couple of new ideas on how to collect your local knowledge and publish it for the benefit of all readers—see the Tides and Cruising Guides sections for more information.

If you are new to *Reed's*, you have some exploring to do. This book is literally several books in one; it is packed with useful information. We suggest that you scan the contents pages of each chapter and look through a country section of the Pilot to get a sense of what's here. As usual, all the information about countries and ports has been thoroughly checked, the navigation aids are up-to-date right to press time, and thc tide and current predictions are the best available.

1998 marks the 25th year that *Reed's* has published North American Almanacs, and more than 200 years as a trusted name in marine publishing. We feel happily poised between the past and the future. We are using the latest desktop publishing technology and numerous wonderful nautical software packages to produce these, reliable books. The possibilities are limitless. Please, use *Reed's*; and help us to make it even better.

Safe Passages,

Jerald Knopf, Publisher
Ben Ellison, Editor

LEGAL NOTES

Acknowledgements

The publisher and editor would like to thank the many people, government agencies, and businesses that have helped in the preparation of this new edition of Reed's Caribbean Almanac. Without their assistance, this publication would not have been possible.

We would particularly like to credit Nancy Hauswald for researching the Pilot chapter, Elbert Maloney for technical editing the Resources chapter, and Pamela Benner for her copy editing and proof reading throughout.

We thank Aaron Bowman and Michael Neal of Nautical Software; Jim Sexton and Dennis Mills of Nautical Technologies; Todd Clift of Chart Kit/BBA; and Catherine Hohenkerk of the Royal Greenwich Observatory.

Richard T. Stofer deserves special mention for an extrordinary volunteer job of checking the ephemeris, working the examples, and helping to expand the chapter.

Also, we acknowledge the dozens of tremendously helpful and courteous people at foreign consulates in the U.S., at U.S. government offices around the country, and at government and customs offices, tourist bureaus, and marinas throughout the Caribbean basin.

And finally, we thank those readers of the 1997 edition—John Stufflebeem, Geoff Kuenning, Dan Hagen, and Daniel Feinstein, to name a few—who responded to our request for corrections and suggestions. Thanks, and keep those comments and ideas coming.

Government Publications

Government publications for the waters covered in this almanac include:

U.S. Government Printing Office: Light List Volume III: Atlantic and Gulf Coasts

Defense Mapping Agency: Pub. 110, List of Lights, Radio Aids and Fog Signals; Pub. 117, Radio Navigational Aids; Pub. 140, Sailing Directions North Atlantic Ocean; Pub. 147, Sailing Directions for the Caribbean Sea, Volume I; Pub. 148, Sailing Directions for the Caribbean Sea, Volume II; Pub. 150, World Port Index; Pub. 151, Distances Between Ports; Notices to Mariners; U.S. Coast Guard: Local Notices to Mariners, District 7

National Oceanic and Atmospheric Administration: United States Coast Pilot, Volumes 4 and 5; Selected Worldwide Marine Weather Broadcasts

Data presented in the ephemeris has been supplied by H.M. Nautical Almanac Office, the Particle Physics and Astronomy Research Council, Royal Greenwich Observatory, England, and is reproduced with permission.

YACHTSMEN
FOOD & BEVERAGE PROVISIONING IN
THE BRITISH VIRGIN ISLANDS
WITH OVER 21 YEARS EXPERIENCE
WE SET THE STANDARD - ALL OTHERS TRY TO FOLLOW
DO YOURSELF A FAVOUR
CALL, FAX OR WRITE FOR OUR
25 PAGE CUSTOM ORDER FORM
ENSURE YOU GET ONLY THE FOOD YOU WANT
THE AMPLE HAMPER
P.O. BOX 73, ROAD TOWN, TORTOLA, BVI
EXPRESS MAIL: P.O. BOX 8309, CRUZ BAY, ST. JOHN, USVI 00831
WEST END
TEL. (809) 495-4684
FAX (809) 495-4957
ROAD TOWN
TEL. (809) 494-2494
FAX (809) 494-5349
E-mail ahamper@caribsurf.com
Internet Home Page
http://www.amplehamper.com

ASK 1,000 EXPERIENCED CRUISERS WHY THEY OWN A GRAND BANKS, AND YOU'LL HEAR 1,000 GOOD REASONS. HERE ARE 15 OF THEM.

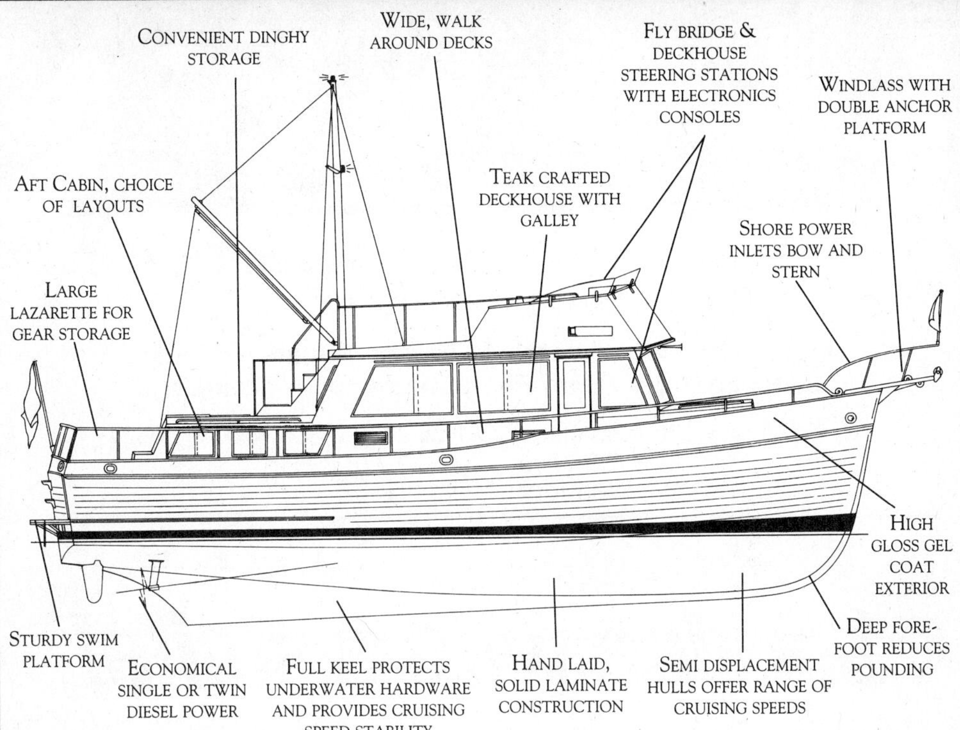

Profile of the Grand Banks 42 Classic.

WE ALL LEARN FROM EXPERIENCE. THAT'S WHY OVER THE YEARS THOUSANDS OF EXPERIENCED CRUISERS THROUGHOUT THE WORLD HAVE STEPPED UP TO A GRAND BANKS AFTER OWNING OTHER BOATS. IN THE VAST MAJORITY OF CASES, THESE OWNERS KEEP THEIR GRAND BANKS FOR YEARS OR TRADE UP TO NEWER, BIGGER MODELS.

YOU CAN TAKE ADVANTAGE OF ALL THIS KNOWLEDGE BY CONTACTING YOUR GRAND BANKS DEALER ABOUT A NEW OR PREVIOUSLY OWNED GRAND BANKS YACHT. AFTER ALL, THERE'S NO REASON TO GO THROUGH THE EXPERIENCE OF OWNING OTHER BOATS BEFORE FINDING HAPPINESS ON A GRAND BANKS. NEW OR USED, THINK OF ALL THE TIME (AND MONEY) YOU COULD SAVE BY BUYING A GRAND BANKS IN THE FIRST PLACE.

GRAND BANKS®

DEPENDABLE DIESEL CRUISERS

FOR MORE INFORMATION CONTACT A GRAND BANKS DEALER OR WRITE GRAND BANKS, 136 JOHN ST., SOUTHPORT, CT 06490 http://wwwgrandbanks.com

GB36 GB42 GB46 GB49 GB58

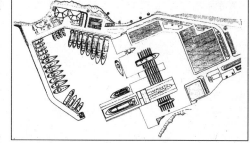

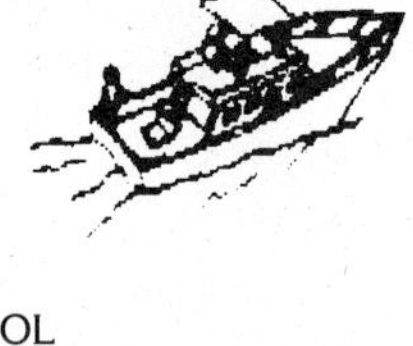

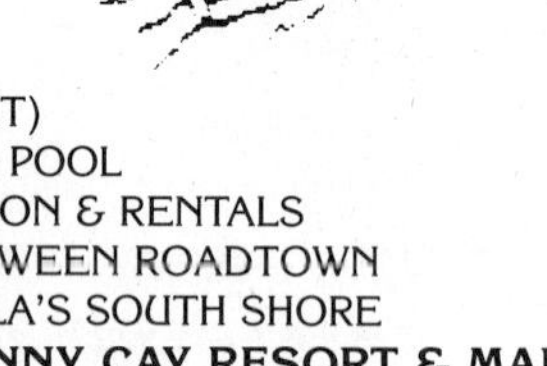

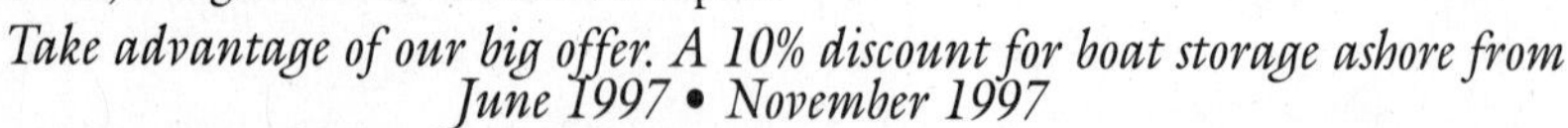

Coast Pilot

USING THE COAST PILOT

INTRODUCTION

The primary goal of *Reed's* Coast Pilot is to provide an organized list of the more important aids to navigation. These are useful in several ways:

- The positions of the aids provide the navigator with thousands of potential waypoints for use with GPS and Loran systems (please see caution below).
- The descriptions we transcribe from the government light lists give the navigator much more detailed information about the various lights and buoys than is presented on charts. This additional information can be critical, particularly when navigating at night or in reduced visibility.
- The prudent navigator can use this list to check the accuracy of charts. Even newly purchased charts often do not show changes to buoys and lights made in recent years.

Reed's also includes important cautions to navigators and brief descriptions of key harbors. This information is minimal, and we encourage navigators to accompany *Reed's* with official coast pilots, sailing directions, and/or privately published cruising guides.

ORGANIZATION

The Coast Pilot chapter is organized geographically, beginning with Bermuda and then proceeding clockwise around the Caribbean from southeast Florida, the Bahamas, and the Turks and Caicos through the Greater Antilles and Lesser Antilles to the north coast of South America and the Caribbean coasts of Central America and Mexico. Refer to the chapter table of contents on the preceding page or the Index in the back of the almanac to find a particular country or island section. Each section opens with information on entry procedures and other general items and resources of interest to mariners. This information is followed by lists of selected major lights, buoys, and daymarks; general piloting information; descriptions of selected ports; and reference chartlets of significant harbors.

WAYPOINT CAUTION

Whenever available, *Reed's* has listed the latitude and longitude of major aids to navigation. The listings are in degrees, minutes, and tenths of a minute. These positions have been obtained from government light lists and are for reference only. *Reed's* does not guarantee the accuracy of any latitude or longitude position. Navigators using these positions as waypoints in Loran or GPS units should plot the listed position on a large-scale chart of the area to ascertain its relationship to navigational hazards in the vicinity. Because floating aids to navigation, such as buoys, move about on their rodes, the positions of buoys should always be regarded as approximate.

Prudent navigators will use these waypoints with caution and will not rely upon them as their sole means of position-fixing.

CHARTLET CAUTION

Reed's has included in this almanac over 200 chartlets. Some are wide-area chartlets meant for general reference. Others are detailed harbor chartlets meant to illustrate the aids. Although most are actual reproductions of government charts, these chartlets are not suitable for navigation. They are resized, gray-scaled scanned images of color charts, and important details may have been lost in the process. Additionally they do not incorporate corrections to navigational aids, depths, or hazards reported since the chart was published. *(Note: The chartlets may be corrected from the aid-to-navigation listings, which are current through summer 1997, subject to the accuracy of government light lists.)*

Prudent mariners would not use these chartlets for navigation. Vessels should always be equipped with up-to-date charts along their intended route. Mariners should also keep abreast of chart changes as posted in the Notices to Mariners.

LIGHT LIST NOTES

The information in *Reed's* Coast Pilot chapter regarding aids to navigation is current through U.S. NIMA Notice to Mariners No. 33/97 (August 16, 1997). Virtually all aids that are listed in NIMA Pub. 110, List of Lights, for the Caribbean rim are contained in the *Pilot*, as well as some significant aids from USCG Volume III.

This year *Reed's* has initiated a new policy of indicating changes in aids that have taken place over the last year. Within an aid listing, underlined text indicates data that has changed or is new since September, 1996; ~~Strike-out~~ text indicates deleted data. This should highlight recent changes for mariners, many of which may not be included in even the most recently published charts.

USING THE AIDS TO NAVIGATION LISTINGS

Information for all the aids to navigation is presented in a set order. The first item is the *name of the aid* and *its designation,* followed by its *latitude* and *longitude,* usually given to the nearest tenth of a minute. Sometimes there is a brief description of the aid's location relative to fixed identifiable objects (i.e., piers, shoals, etc.).

For those aids that are lighted, the descriptive information that follows the location includes *light characteristics,* the *height of the light* (as measured above high water and generally given in feet [ft]), and its *visible range* in nautical miles (abbreviated M). The abbreviations used are described below and in the table of light characteristics that follows on page P 6.

A brief description of the aid's *physical characteristics,* or structure, generally follows the light characteristics information. Abbreviations are described below.

Any *additional notes,* such as limits of visibility or the presence of radar reflectors, racons, or other potentially confusing lights, follow the structure description. When bearings are given (usually for ranges), they are true bearings, not adjusted for magnetic variation. Limits of light sectors and arcs of visibility *as observed from a vessel* are given in clockwise order. Note that not all information is known for all aids to navigation.

Here is an example of a typical listing:

SMITH SHOAL LIGHT, on NE end of shoal, 24 43.1N, 81 55.3W. Fl W 6s, 54ft, 9M. On hexagonal pyramidal skeleton tower on piles.

The name of the aid to navigation is SMITH SHOAL and the aid is a light. Its position is 24 degrees, 43.1 minutes north latitude, 81 degrees, 55.3 minutes west longitude, at the northeast end of Smith Shoal. The light is white and flashes every 6 seconds and is located 54 feet above the high-water level. It has a range of 9 nautical miles. The light is located on a hexagonal pyramidal skeleton tower set on piles. There are no special remarks.

LIGHTS, BUOYS, AND DAYMARKS

Lights may vary from simple poles to substantial stone towers more than 100 feet high. In general, a "light" is a fixed structure that is lighted at night. Some major lights remain lit 24 hours a day; this information is generally noted. In addition to giving light characteristics, *Reed's* has, when available, included a brief physical description of each light to assist with identification during daylight hours also.

Buoys are floating aids to navigation. Within Canada and the United States, red buoys show even numbers (i.e., 2, 4, 6, etc.) and green buoys, odd numbers (i.e., 1, 3, 5, etc.). In a few areas, you may still encounter black buoys that have not yet been given a coat of green paint. The color of the light shown by the buoy will usually match the color of its hull. Center-of-channel buoys are often red and white. Buoys marking hazards are often yellow. An unlit red buoy is usually a nun; an unlit green buoy is a can. Other variations are generally indicated.

Daymarks are colored signs that can be seen during the day—and sometimes at night if illuminated by a spotlight—that are fixed to posts, lighthouses, buildings, and other structures. There are many types of daymarks, which can be very confusing to even the most informed navigators. The list of abbreviations that follows should help alleviate some of the confusion.

ABBREVIATIONS USED IN THE COAST PILOT

The following are standard abbreviations used to describe aids to navigation listed in the Coast Pilot chapter of this almanac. See page P 6 for light characteristic abbreviations.

N	north	**S**	south
W	west	**E**	east
R	red	**G**	green
W	white	**B**	black
Y	yellow	**Or**	orange
Bu	Blue		

Dir	directional
kHz	kilohertz
km	kilometer
LIB	lighted buoy
LNB	large navigational buoy
M	nautical mile
m	meters
ODAS	anchored oceangraphic data buoy
s	seconds
vert	vertical

P Pilot

DAYMARK ABBREVIATIONS

Following are some of the standard abbreviations used to describe the purpose and appearance of dayboards, according to the U.S. system. The first letter of the daymark listing indicates its basic purpose or shape as follows:

S=Square. Used to mark the port (left) side of channels when entering from seaward.

T=Triangle. Used to mark the starboard (right) side of channels when entering from seaward.

J=Junction. May be a square or a triangle. Used to mark channel junctions or bifurcations in the channel. May be used to mark wrecks or other obstructions that may be passed on either side. The color of the top band has lateral significance for the preferred channel.

M=Safe water. Octagonal. Used to mark the fairway or middle of the channel.

K=Range. Rectangular. When the front and rear daymarks are aligned on the same bearing, the vessel is on the azimuth of the range, which usually marks the center of the channel.

N=No lateral significance. Diamond or rectangular shaped. Used for special purposes as a warning, distance, or location marker.

The second letter in a daymark abbreviation indicates its color (G=green, etc.). When there is a third letter with no intervening hyphen, it refers to the color of the center stripe. This designation is used for range daymarks only.

Finally, a hyphen (-) indicates a mark on an intracoastal waterway, as follows:

-I= Intracoastal Waterway. Daymark with a yellow horizontal reflective strip.

-SY=Intracoastal Waterway. Daymark with a yellow reflective square. Indicates a port-hand mark for vessels crossing the waterway. May appear on a triangular daymark when the intracoastal coincides with a waterway having opposite conventional direction of buoyage.

-TY=Intracoastal Waterway. Daymark with a yellow reflective triangle. Indicates a starboard-hand mark for vessels crossing the waterway. May appear on a square daymark where the intracoastal coincides with a waterway having opposite conventional direction of buoyage.

The following abbreviations cover many of the major designations:

SG=Square green daymark with a green reflective border.

SG-I=Square green daymark with a green reflective border and a yellow reflective horizontal strip.

SG-SY=Square green daymark with a green reflective border and a yellow reflective square.

SG-TY=Square green daymark with a green reflective border and a yellow reflective triangle.

TR=Triangular red daymark with a red reflective border.

TR-I=Triangular red daymark with a red reflective border and a yellow reflective horizontal strip.

TR-SY=Triangular red daymark with a red reflective border and a yellow reflective square.

TR-TY=Triangular red daymark with a red reflective border and a yellow reflective triangle.

JG=Daymark with horizontal bands of green and red, green band topmost, with a green reflective border.

JG-I=Daymark with horizontal bands of green and red, green band topmost, with a green reflective border and a yellow reflective horizontal strip.

JG-SY=Daymark with horizontal bands of green and red, green band topmost, with a green reflective border and a yellow reflective square.

JG-TY=Daymark with horizontal bands of green and red, green band topmost, a green reflective border, and a yellow reflective triangle.

JR=Daymark with horizontal bands of green and red, red band topmost, with a red reflective border.

JR-I=Daymark with horizontal bands of green and red, red band topmost, a red reflective border and a yellow horizontal strip.

JR-SY=Triangular daymark with horizontal bands of green and red, red band topmost, with a red reflective border and a yellow reflective square.

JR-TY=Triangular daymark with horizontal bands of green and red, red band topmost, with a red reflective border and a yellow reflective triangle.

MR=Octagonal daymark with stripes of white and red with a white reflective border.

MR-I=Octagonal daymark with stripes of white and red, with a white reflective border and a yellow reflective horizontal strip.

CG=Diamond-shaped green daymark bearing small green diamond-shaped reflectors at each corner.

CR=Diamond-shaped red daymark bearing small red diamond-shaped reflectors at each corner.

KBG=Rectangular black daymark bearing a central green stripe.

KBG-I=Rectangular black daymark bearing a central green stripe and a yellow reflective horizontal strip.

KBR=Rectangular black daymark bearing a central red stripe.

KBR-I=Rectangular black daymark bearing a central red stripe and a yellow reflective horizontal strip.

KBW=Rectangular black daymark bearing a central white stripe.

KBW-I=Rectangular black daymark bearing a central white stripe and a yellow reflective horizontal strip.

KGB=Rectangular green daymark bearing a central black stripe.

KGB-I=Rectangular green daymark bearing a central black stripe and a yellow reflective horizontal strip.

KGR=Rectangular green daymark bearing a central red stripe.

KGR-I=Rectangular green daymark bearing a central red stripe and a yellow reflective horizontal strip.

KGW=Rectangular green daymark bearing a central white stripe.

KGW-I=Rectangular green daymark bearing a central white stripe and a yellow reflective horizontal strip.

KRB=Rectangular red daymark bearing a central black stripe.

KRB-I=Rectangular red daymark bearing a central black stripe and a yellow reflective horizontal strip.

KRG=Rectangular red daymark bearing a central green stripe.

KRG-I=Rectangular red daymark bearing a central green stripe and a yellow reflective horizontal strip.

KRW=Rectangular red daymark bearing a central white stripe.

KRW-I=Rectangular red daymark bearing a central white stripe and a yellow reflective horizontal strip.

KWB=Rectangular white daymark bearing a central black stripe.

KWB-I=Rectangular white daymark bearing a central black stripe and a yellow reflective horizontal strip.

KWG=Rectangular white daymark bearing a central green stripe.

KWG-I=Rectangular white daymark bearing a central green stripe and a yellow reflective horizontal strip.

KWR=Rectangular white daymark bearing a central red stripe.

KWR-I=Rectangular white daymark bearing a central red stripe and a yellow reflective horizontal strip.

NB=Diamond-shaped daymark divided into four diamond-shaped colored sectors with the sectors at the side corners white and the sectors at the top and bottom corners black, with a white reflective border.

NG=Diamond-shaped daymark divided into four diamond-shaped colored sectors with the sectors at the side corners white and the sectors at the top and bottom corners green, with a white reflective border.

NR=Diamond-shaped daymark divided into four diamond-shaped colored sectors with the sectors at the side corners white and the sectors at the top and bottom corners red, with a white reflective border.

NW=Diamond-shaped white daymark with an orange reflective border and black letters describing the information, or regulatory nature, of the mark.

NL=Rectangular white location marker with an orange reflective border and black letters indicating the location.

NY=Diamond-shaped yellow daymark with a yellow reflective border

P
Pilot

LIGHT CHARACTERISTICS

Abbrev	Example	Explanation	Graphic representation = 30 seconds
F	F W	FIXED: a continuous steady light	
	OCCULTING: total light greater than dark; eclipses equal and repeated regularly		
Oc	Oc W 10s	SINGLE OCCULTING: steady light with eclipse regularly repeated	10s
Oc(*n*)	Oc(2) W 15s	GROUP OCCULTING: a group of eclipses, number specified, regularly repeated	15s
Oc(*n*+*n*)	Oc(2+1) 15s	COMPOSITE GROUP OCCULTING: successive groups in a period have different numbers of eclipses	15s
Iso	Iso W 15s	ISOPHASE: a light where the duration of light and darkness are clearly equal	15s
	FLASHING: single light at regular intervals; duration of light less than dark		
Fl	Fl W 4s	SINGLE FLASHING: flash is regularly repeated at less than 50 flashes per minute	4s
LFl	LFl W 10s	LONG FLASHING: a flash of 2 or more seconds, regularly repeated	10s
Fl(*n*)	Fl(2) W 10s	GROUP FLASHING: successive groups of flashes, specified in number, regularly repeated	10s
Fl(*n*+*n*)	Fl(2+1) W 10s	COMPOSITE GROUP FLASHING: successive groups in a period have a different number of flashes	10s
	QUICK: rapid regularly repeated flashes, between 50 and 80 per minute (usually 60)		
Q	Q W	CONTINUOUS QUICK: light in which a flash is regularly repeated	
Q(*n*)	Q(3) W 10s	GROUP QUICK: quick light in which a specified group of flashes is regularly repeated	10s
VQ + LFl	VQ(6)+LFl 10s	QUICK plus LONG FLASH: quick light in which a sequence of flashes is interrupted by regular eclipses of constant and long duration	10s
	VERY QUICK: regularly repeated flashes, between 80 and 160 per minute (usually 120)		
VQ	VQ W	CONTINUOUS VERY QUICK: very quick light in which flash is regularly repeated	
VQ(n)	VQ(3) W 5s	GROUP VERY QUICK: specified group of flashes regularly repeated	5s
IVQ	IVQ W 10s	INTERRUPTED VERY QUICK: sequence of very quick flashes interrupted by regularly repeated eclipses of constant and long duration	10s
	ULTRA QUICK: flashes repeated at rate of not less than 160 per minute (usually 180)		
UQ	UQ W	CONTINUOUS ULTRA QUICK: ultra quick light in which flash is regularly repeated	
IUQ	IUQ W 10s	INTERRUPTED ULTRA QUICK: flashes interrupted by eclipses of long duration	10s
Mo(a)	Mo(a) W 6s	MORSE CODE: in which appearances of light of two clearly different durations are grouped to represent a character(s) in the Morse code	6s
F Fl	F W & Fl W 2.5s	FIXED AND FLASHING: steady light combined with one brilliant flash at regular intervals	
Alt	Alt RW 5s	ALTERNATING: a light that alters in color in successive flashing	R W R W R W R W R W R W 5s

U.S. CUSTOMS PROCEDURES

Foreign commercial vessels and some classes of U.S. commercial vessels are required to make their first U.S. landfall at a designated customs port of entry. If an emergency necessitates a stop at another port first, customs should be notified by phone immediately.

Other vessels, foreign and U.S., may proceed to any port, but must report to customs by phone at the first landfall and await instructions. Actual procedures within different districts vary; vessels may be asked to proceed to the nearest customs port of entry. (Note: Contacting customs by mobile phone while approaching but before actually arriving at the landfall does not fulfill customs requirements; however, many customs offices appreciate 24 hours advance notice of a yacht's arrival, particularly if the customs office is not located in the intended port of arrival or the presence of foreign nationals on board requires action by immigration officials.) Nothing should be allowed ashore from the vessel, and no person should come ashore other than the master until customs has been notified.

To report by phone from anywhere in the U.S., the vessel's master should first call the customs toll-free 24-hour communications center in Orlando, Florida: 800-XSECTOR (800-973-2867). Personnel at this center will refer the caller to the relevant phone numbers for the customs office or after-hours duty officer nearest the vessel's port of arrival.

The Orlando communications center does not service Puerto Rico or the U.S. Virgin Islands. The offices on those islands that clear in non-commercial vessels are listed below.

Puerto Rico: *Mayagüez,* 787-831-3342, -3343
San Juan, 787-253-4533, -4534, -4535, -4536
Ponce, 787-841-3130, -3131, -3132
Fajardo, 787-863-0950, -0102, -4075, -0811
Vieques, 787-741-8366
Culebra, 787-742-3531
Note: In Puerto Rico, Vieques, and Culebra, all vessels arriving after hours, or on Saturdays, Sundays, and holidays, must report to the San Juan station, which is at the airport, via telephone. San Juan will then issue the vessel a customs number or instruct the master to await officials. Normal working hours for most offices are 0800 to 1700 Monday through Friday.

Because the U.S. Virgins and Puerto Rico are in different customs jurisdictions, all vessels, U.S. and foreign, arriving in Puerto Rico from the U.S. Virgin Islands must report to customs. U.S. vessels traveling from Puerto Rico to the U.S. Virgin Islands do not need to clear out on departure. For additional information concerning Puerto Rico customs, contact the Customs Carrier Control Office (787-729-6797, -6778), which serves both commercial and private vessels.

U.S. Virgin Islands: *St. Croix,* 809-773-1011; Saturdays, 809-773-1011 or -778-2257; Sundays and holidays, 809-778-0216
St. John (Cruz Bay), 809-776-6741
St. Thomas, 809-774-6755

Failure to report to customs as required may result in a fine of up to $5,000 for the first violation and $10,000 for subsequent violations.

For U.S. vessels, there is no charge for a customs inspection. (The after-hours service charge was discontinued effective January 1995). However, pleasure craft 30 feet long or longer are required to purchase from the U.S. Customs Service an annual $25 Customs User Fee Decal in order to clear in. If the decal is not obtained before the vessel returns to U.S. waters, the vessel operator will be directed upon arrival to the nearest customs office to purchase one.

U.S.-documented or -registered vessels do not need to report their departure from the United States for foreign destinations.

Foreign vessels may be charged fees for formal entry, a permit to proceed, and clearing out. Necessary documents include ship's registration and clearance from the last port of call, plus any required immigration documents (which vary according to nationality), such as passports, visas, or other forms of identification. Pleasure boats from certain countries may obtain a cruising license for U.S. waters. After their initial entry, this license exempts them from formal reporting and clearance procedures at every subsequent port of call. (Foreign vessels will still need to report by phone their arrivals in new ports.) The license can be obtained from the District Director of Customs at the first port of arrival in the United States. The licenses are issued for no longer than a year. They do not exempt a vessel from applicable duties. Vessels from the following countries are eligible to obtain a cruising license:

Argentina, Australia, Austria, Bahamas, Belgium, Bermuda, Canada, Denmark, Finland, France, Germany, Greece, Honduras, Ireland, Italy, Jamaica, Liberia, Netherlands, New Zealand, Norway, Sweden, Switzerland, Turkey, United Kingdom (includes Turks and Caicos, St. Vincent, Northern Grenadine Islands, Cayman Islands, British Virgin Islands, and St. Christopher, Nevis, and Anguilla).

In the absence of a cruising license, foreign vessels must report at each U.S. port and must obtain clearance to proceed to the next port. Foreign vessels must clear customs before departing the U.S.

AREA DESCRIPTIONS

BERMUDA

This archipelago is a favorite offshore destination for U.S. mariners, as either a stopover on the way to the Caribbean or a cruising ground of its own. It lies just under 700 miles from both Newport, Rhode Island, and Norfolk, Virginia, and is approximately 900 miles north of St. Thomas in the Virgin Islands.

The major consideration voyaging to Bermuda is always the weather. Routes to the islands cross the major hurricane tracks. In addition, boats sailing from the U.S. mainland must cross the Gulf Stream.

The approaches to Bermuda are generally well marked. You will be beyond the range of reliable Loran coverage, but the island has a powerful radiobeacon and several long-range lights. Bermuda Harbour Radio is always standing by to aid those in trouble.

It is wise to stand well off Bermuda in heavy weather. Many vessels have encountered severe seas when deep ocean waves began tumbling on the Bermuda shelf waters. It is often hard to spot the edge of the reefs surrounding Bermuda. Keep in mind that the major beacons are situated on the reef itself; they should be given a wide berth in bad weather.

Bermuda offers a warm welcome to the visiting cruiser. The island is beautiful and safe. Most boating supplies are readily available. This is a great place to break up an offshore trip and recoup for the next leg.

FLORIDA, EAST COAST

From Fort Pierce Inlet to the Miami harbor entrance, a distance of just over 100 nautical miles, Florida's east coast generally trends south. Here, a low and sandy strip of vegetated land punctuated by conspicuous highrise buildings that become more dense close to Miami is separated from the mainland by a series of sounds and rivers interspersed with short stretches of canal, all part of the Intracoastal Waterway. There are several inlets, some of which are navigable only by small craft and with local knowledge. There are major deepwater ports at Fort Pierce, Palm Beach (Lake Worth Inlet), Port Everglades, and Miami. The coast is fairly bold, and the northward-setting Gulf Stream not far offshore.

Along the inside Intracoastal route, sailors can often raise sail and enjoy a good run as far as St. Lucie. Anchorages are not frequent, but they are well spaced. South of St. Lucie the character of the waterway changes. Highrise buildings begin to intrude upon the natural landscape. Anchorages are infrequent and often restricted by local laws. Most boats will have to proceed under power. The many opening bridges have restricted schedules. However, here the numerous marine facilities are among the finest in the country.

For vessels crossing the Gulf Stream to the Bahamas, Palm Beach is a favorite spot to depart for West End and the Abacos. Fort Lauderdale and Miami are good departure points for Bimini, Cat Cay, or Nassau.

FLORIDA, THE KEYS

The Keys are an archipelago of tropical islets surrounded by beautiful, but shallow, water. For boats drawing less than 4 or 5 feet, the inside route is possible. For deeper-draft boats, the often-boisterous Hawk Channel route is preferred. Passes between the two are restricted to a few good openings. Channel Five near Long Key and the Moser Channel near Marathon are the only choices for sailors.

The John Pennekamp Park, near Key Largo, is an undersea coral reef park. Everywhere, the coral is protected and should not be anchored upon or even touched. Those who go aground in coral are liable for fines.

Key West is the end of the line for most boaters in the Keys. The adventurous make the 60-mile passage to the Dry Tortugas to find a true "out island." Fort Jefferson guards a beautiful sandy lagoon sheltered by coral reefs. The nesting ground of the frigate bird is on aptly named Bird Key. There are no facilities for boaters—you must even remove your own trash.

THE BAHAMAS

Cruising in the Bahamas is very different from cruising U.S. waters. The waters are shallow and filled with shifting sands and coral reefs. Aids to navigation are few and far between, and none should be trusted. Loran is also unreliable.

The tools of Bahama navigation are simple—a good depthsounder and your eyes. Most boaters quickly learn to read the water depth ahead, utilizing the beautiful colors visible through the crystal clear water. Experienced Bahamas cruisers travel when the sun is high to prevent glare on the water from impeding visibility.

The VHF radio can be a handy tool. If you leave it on channel 16 while entering an unfamiliar harbor, advice will often be radioed out to you from a boater inside. During storms boaters relay information on location, wind speeds, and wind direction. In the evenings an informal chat-hour invariably commences around 1700.

Working your way from island to island you will encounter wild places with few people. The island chain is about 500 miles long, and most of the islands are uninhabited. Despite a few, often-repeated horror stories, most cruisers are much safer here than in their hometowns. Bahamians are friendly and helpful people.

THE CARIBBEAN

The islands of the Caribbean are many and varied. From the gorgeous reef diving of the Caymans to the old-world charm of San Juan to the crowded harbors of the Virgins and the boisterous sailing of the Grenadines, this is an area of contrasts with many cultures. Some people speak Spanish and others French, but most know at least a smattering of English. You must be more self-sufficient to cruise here. You may have to travel 200 miles to get fuel (Bahamas to Dominican Republic), and you may have to jug your water from an inland spigot. You learn to stock up when the food is plentiful and low in price (Puerto Rico, the Dominican Republic, and Venezuela). You will want to carry lots of heavy ground tackle, with plenty of anchor chain, to resist the chafe on coral bottoms.

Many of the harbors are not geared strictly to transient boaters, although charter areas do have marinas with stateside-type facilities. But many of the islands cater to boaters only as a courtesy; the harbors are designed for the handling of commercial ships and cruise liners. You may have to tie to a crumbling concrete wharf while clearing customs, or you may have to anchor out for weeks at a time. Of course, this is part of the great charm. Despite decades of tourism, Caribbean people have often maintained their own cultures in the face of mounting pressures to change.

Navigation in the Caribbean often involves simply sailing to the next island. Visibility is usually excellent. There is often plenty of wind—in the winter, 20 to 30 knots almost every day. The trades blow from the east day after day, week after week, month after month. When the wind switches, many harbors become uncomfortable for a few days. There are many harbors that are simply coves on the leeward side of an island.

The currents of the Caribbean tend to be the result of the constant northeast trade winds. However, they do vary with the strength of the wind, its direction, and the season. See the two current charts at the end of the Currents chapter for an indication of the general pattern to be expected.

Hurricanes are the scourge of the Caribbean. Experienced boaters plan to be south in the Grenadines or Venezuela from August through October. If you must stay in the islands during the hurricane season, plan your strategy well ahead of time. Find your hurricane hole and retreat to it early if hurricane winds threaten.

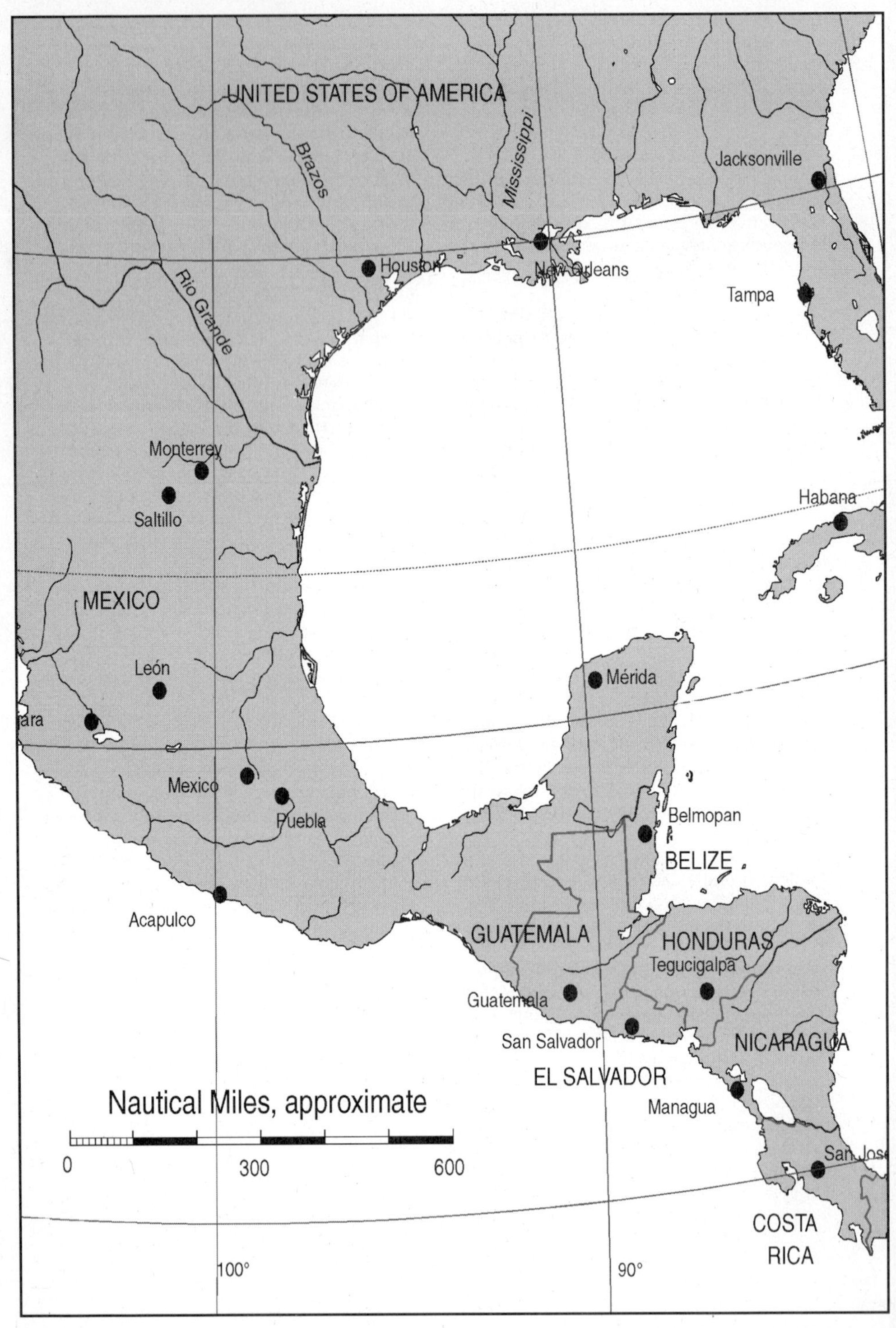

UNITED STATES OF AMERICA
Brazos
Mississippi
Jacksonville
Houston
New Orleans
Tampa
Rio Grande
Monterrey
Saltillo
Habana
MEXICO
León
Mérida
ara
Mexico
Puebla
Belmopan
BELIZE
Acapulco
GUATEMALA
HONDURAS
Tegucigalpa
Guatemala
San Salvador
NICARAGUA
EL SALVADOR
Managua
Nautical Miles, approximate
0
300
600
San Jose
COSTA RICA
100°
90°

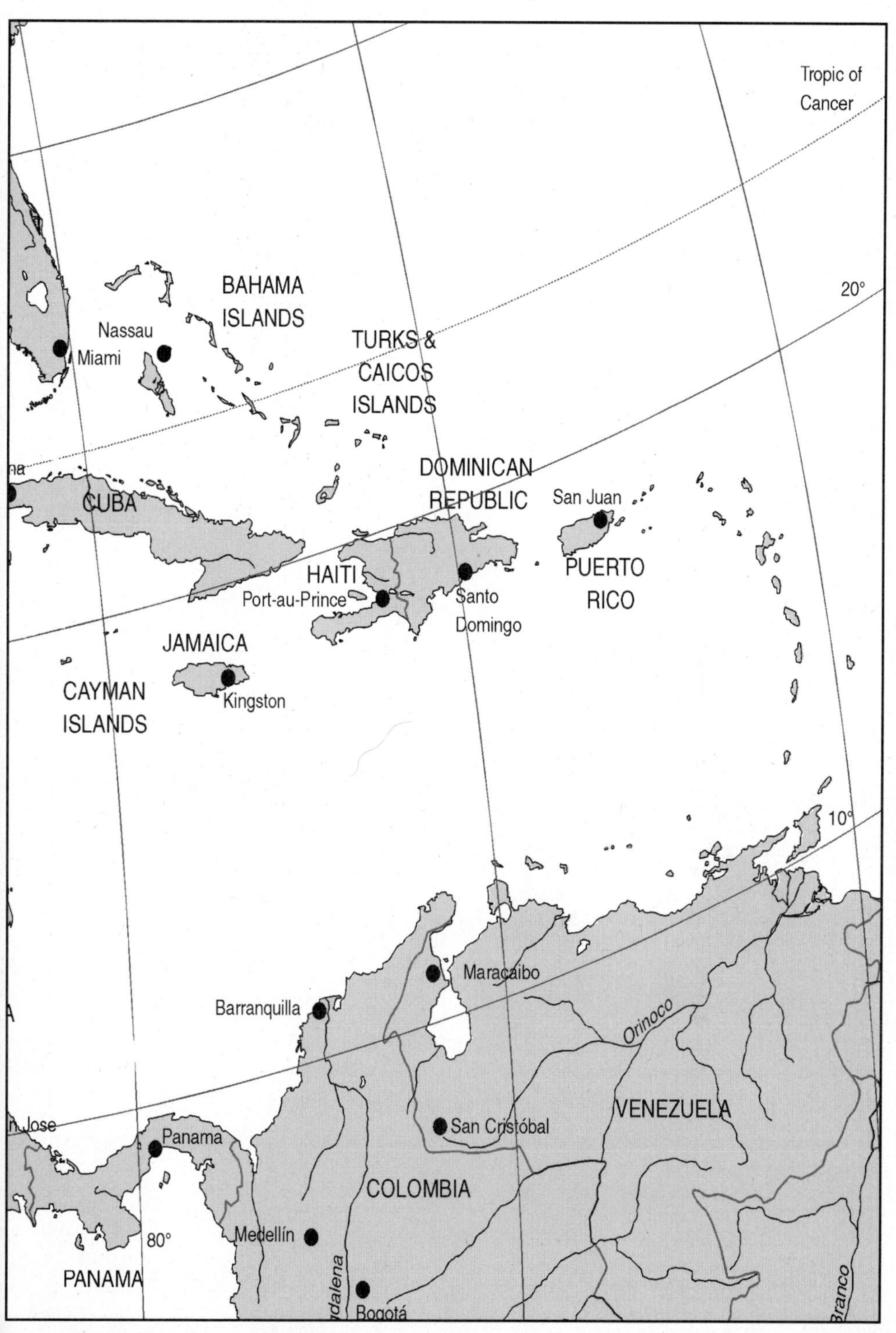
Tropic of Cancer
20°
BAHAMA ISLANDS
Nassau
Miami
TURKS & CAICOS ISLANDS
DOMINICAN REPUBLIC
CUBA
San Juan
HAITI
Port-au-Prince
Santo Domingo
PUERTO RICO
JAMAICA
Kingston
CAYMAN ISLANDS
10°
Maracaibo
Barranquilla
Orinoco
VENEZUELA
San Cristóbal
Panama
COLOMBIA
Medellín
80°
PANAMA
Bogotá
Branco

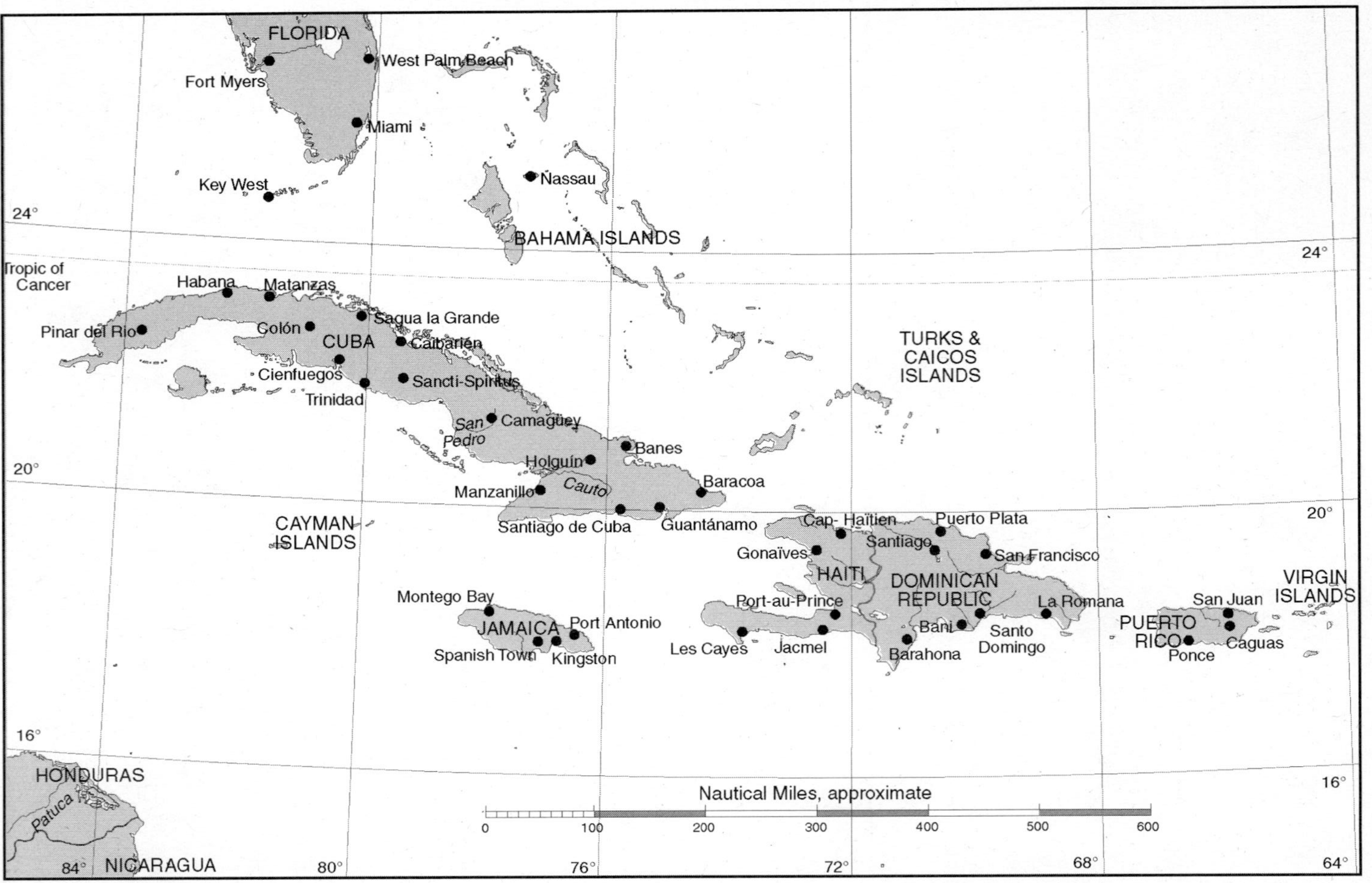
FLORIDA
Fort Myers
West Palm Beach
Miami
Key West
BAHAMA ISLANDS
Nassau
TURKS & CAICOS ISLANDS
CUBA
Habana
Matanzas
Colón
Cienfuegos
Trinidad
Sagua la Grande
Caibarien
Sancti-Spiritus
Camagüey
San Pedro
Holguin
Cauto
Banes
Manzanillo
Santiago de Cuba
Guantánamo
Baracoa
Pinar del Rio
Tropic of Cancer
CAYMAN ISLANDS
JAMAICA
Montego Bay
Spanish Town
Kingston
Port Antonio
HAITI
Cap-Haïtien
Gonaïves
Port-au-Prince
Jacmel
Les Cayes
DOMINICAN REPUBLIC
Santiago
Puerto Plata
San Francisco
Bani
Barahona
Santo Domingo
La Romana
PUERTO RICO
San Juan
Caguas
Ponce
VIRGIN ISLANDS
HONDURAS
NICARAGUA
Patuca
Nautical Miles, approximate
0
100
200
300
400
500
600
24°
20°
16°
84°
80°
76°
72°
68°
64°

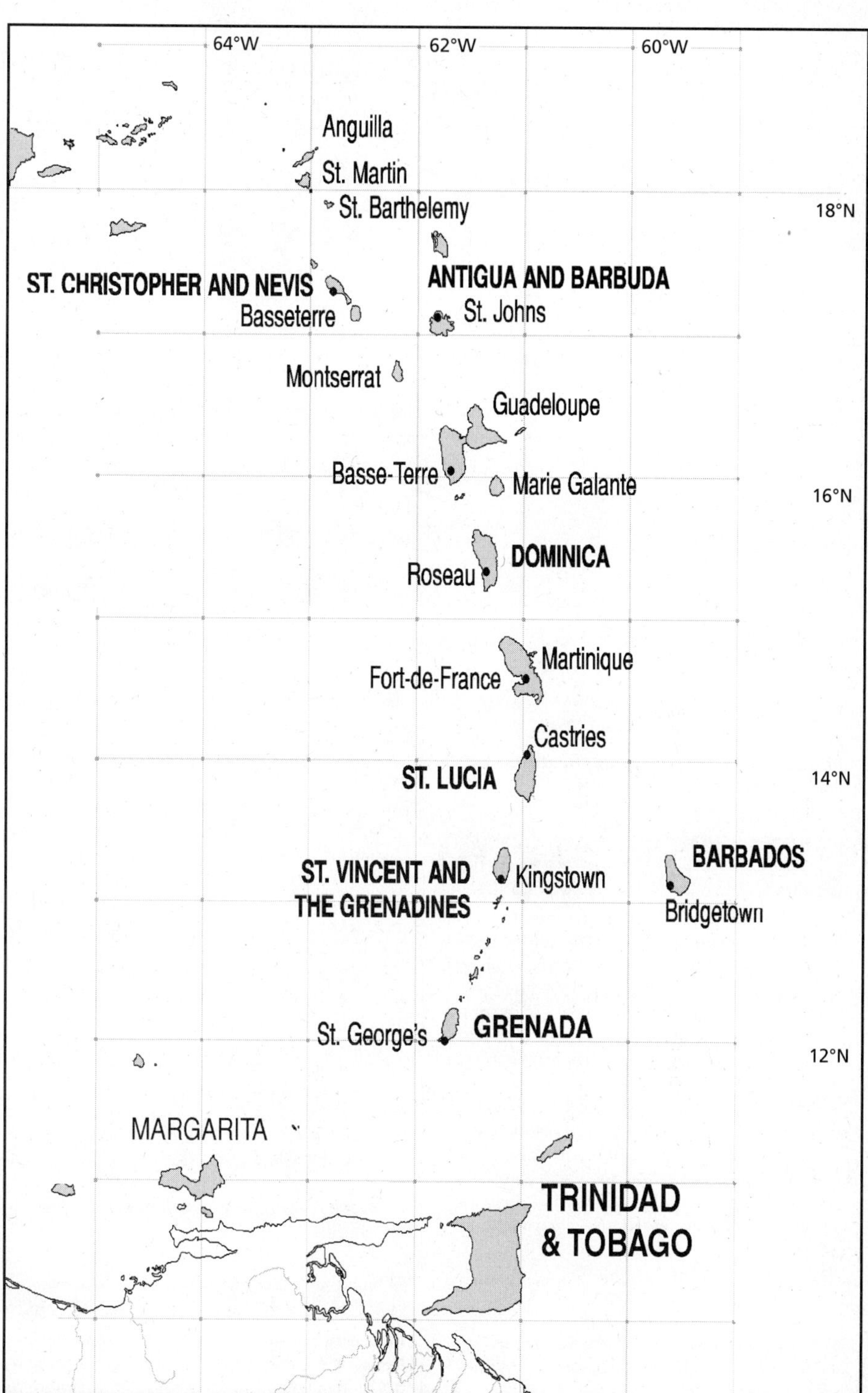
64°W
62°W
60°W
18°N
16°N
14°N
12°N
Anguilla
St. Martin
St. Barthelemy
ST. CHRISTOPHER AND NEVIS
Basseterre
ANTIGUA AND BARBUDA
St. Johns
Montserrat
Guadeloupe
Basse-Terre
Marie Galante
DOMINICA
Roseau
Martinique
Fort-de-France
Castries
ST. LUCIA
ST. VINCENT AND
THE GRENADINES
Kingstown
BARBADOS
Bridgetown
St. George's
GRENADA
MARGARITA
TRINIDAD
& TOBAGO

Bermuda

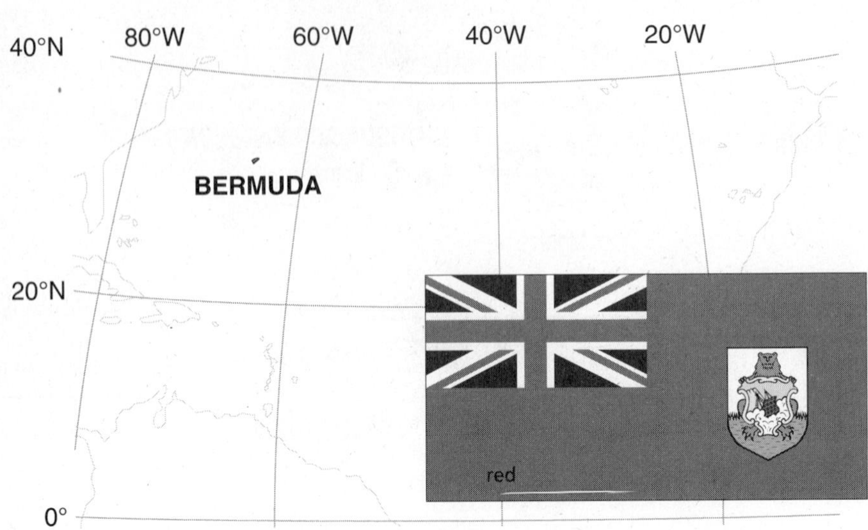

Please see cautions about chartlets and waypoints on page P 2.

ENTRY PROCEDURES

All visiting pleasure craft must contact Bermuda Harbour Radio (BHM), callsign ZBM, on VHF channel 16 (156.8MHz) prior to arrival. Bermuda Harbour Radio is operated by the Bermuda government and is the rescue coordination center for the Bermuda area and is in constant contact with U.S. and Caribbean rescue centers. You should attempt to contact BHM at 30 miles from the island. BHM will help you obtain customs, immigration, and health clearance forms when you clear in at St. George's Harbour. All visiting yachts are required to obtain these forms in St. George before proceeding elsewhere in Bermuda. The 24-hour clearance facility is located at the eastern end of Ordnance Island. This also applies to clearance on departure. All vessels should enter St. George's Harbour flying code flag "Q." At *Reed's* press time, BHM was establishing a Web site, to be on line by the end of 1997.

Visitors should have proof of citizenship and identification; passports are preferred. Visas are not required of most nationals to enter Bermuda. The vessel's master should have two copies of the crew list and stores list. All firearms must be declared and will be held or sealed by officials until the vessel leaves Bermuda. This regulation includes flare guns. A fee of $15 is levied for each person aboard (the fee is waived for sail-training vessels and yachts entered in official races). Shipboard pets arriving without proper documents are refused entry and will be returned to their point of origin or be confined to the vessel at a deepwater anchorage; address inquiries to: Department of Agriculture, Fisheries and Parks, PO Box HM 834, Hamilton HM BX; tel., 441-236-4201; fax, 441-236-7582; telex, 3246 CWAGY-BA; e-mail: agfish@ibl.bm. Importing fruits and vegetables from other countries is prohibited.

USEFUL INFORMATION

Language: English.

Currency: Bermuda dollar (BDA$), on par with U.S. dollar. U.S. currency is accepted everywhere,

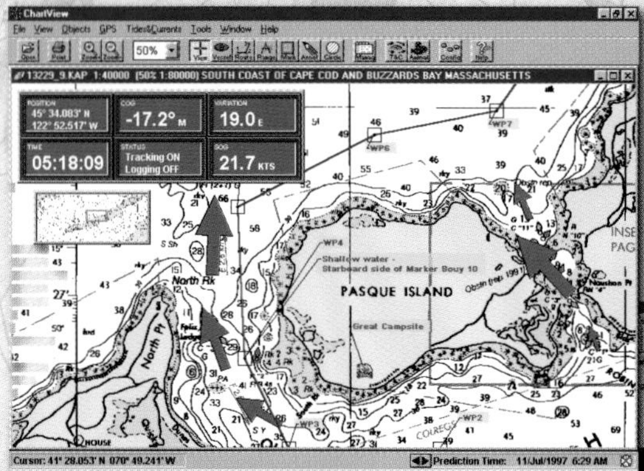

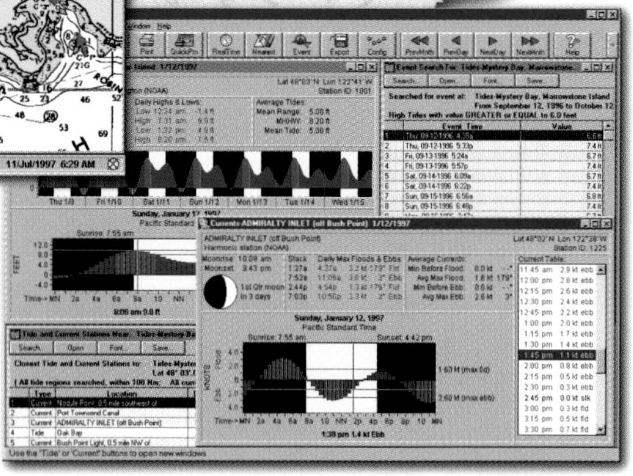

and credit cards are widely accepted. The Bank of Bermuda operates numerous ATM machines, many available 24 hours a day.

Telephones: Area code, 441; local numbers, seven digits.

Sewage and trash disposal: All boats should be fitted with either holding tanks or have U.S. Coast Guard–approved marine sanitation devises type 1 or 2. Whenever possible, however, boaters should use onshore sanitation facilities. Sewage disposal is strictly prohibited in St. George's and Hamilton Harbours and the marina basin at Dockyard. Arrangements for trash pickup may be made through the Corporation of St. George's (297-1532), the Corporation of Hamilton (292-1234), or through the marina or club where berthed.

Public holidays (1998): New Year's Day, January 1; Good Friday, April 10; Bermuda Day, May 25; Queen's Official Birthday, June 15; Cup Match Day, July 30; Somers Day, July 31; Labour Day, September 7; Remembrance Day, November 11; Christmas Day, December 25 ; Boxing Day, December 26.

Note: Although Monday, December 28 is not a public holiday, government offices will be closed that day.

Tourism offices: *In Bermuda:* Bermuda Department of Tourism, Global House, 43 Church St., Hamilton, HM 12 (PO Box HM 465, Hamilton HM BX); 441-292-0023. Visitors Service Bureau offices are located in King's Square, St. George's; on Front Street, Hamilton; at the airport; and in the Dockyard. *In the U.S.:* Bermuda Department of Tourism, Suite 201, 310 Madison Avenue, New York, NY 10017; 800-223-6106, 212-818-9800. Offices also in Atlanta, Boston, and Chicago. *In Canada:* Bermuda Department of Tourism, Suite 1004, 1200 Bay St., Toronto, Ont. M5R 2A5; 416-923-9600 or 800-387-1304.

Additional notes: The Bermuda Department of Tourism publishes an "information sheet" geared especially to private yachts. Actually many pages of highly useful information, it is generally updated late each spring. Once you have cleared at St. George's and obtained permission to cruise elsewhere, local navigation should present little difficulty during daylight hours in clear weather with up-to-date charts.

APPROACHES

NOTE: In Bermuda, offshore light towers are referred to as beacons.

NORTH ROCK BEACON, 32 28.5N, 64 46.0W. Fl (4) W 20s, 70ft, 12M. North cardinal mark; black over yellow tower. Worded NORTH ROCK in white letters on black background near top of tower. Radar reflector.

NORTH EAST BREAKER BEACON, 32 28.7N, 64 40.9W. Fl W 2.5s, 45ft,12M. Red tower on red tripod base, worded NORTH EAST in red letters on a white background near base of tower. Radar reflector. **RACON N** (– ·).

KITCHEN SHOAL BEACON, 32 26.0N, 64 37.6W. Fl (3) W 15s, 45ft, 12M. East cardinal mark; red and white horizontal striped tower on a tripod, worded KITCHEN in red letters on a white background at base of tower. Radar reflector.

Mills Breaker Buoy, 32 23.9N, 64 36.8W. V Q (3) 5s. East cardinal mark, black with a single horizontal yellow band; topmark, two black cones base to base; worded MILLS in black on a yellow background.

Sea Buoy, 32 22.9N, 64 36.9W. Mo (A) W 6s. Safe water mark, pillar with red and white vertical stripes with single red ball topmark; worded SB in white on side.

Spit Buoy, 32 22.6N, 64 38.4W. Q (3) W 10s. Eastern cardinal mark, black with a single horizontal yellow band; topmark, two black cones base to base; worded SPIT in black on a yellow background.

ST. DAVIDS ISLAND LIGHTHOUSE, on Mount Hill, 32 21.8N, 64 39.0W. Fl (2) W 20s, 213ft, 15M; F R G, 20M. White octagonal tower with red band, 72ft. W partially obscured 044°–135°; R partially obscured 044°–135°, R 135°–221°, G 221°–276°, R 276°–044°. Radiobeacon 031°, 380 meters. F R lights 0.95 mile SSW, 0.63 mile SW, 0.75 mile and 1.12 mile WNW.

KINDLEY FIELD AVIATION LIGHT, on control tower at airport, 32 21.9N, 64 40.5W. Al W W G 10s, 140ft, 15M. F R on tank 0.5 mile WNW, two F R at Ferry Reach swing bridge 0.8 mile WNW.

GIBBS HILL LIGHTHOUSE, summit, 32 15.1N, 64 50.0W. Fl W 10s, 354ft, 26M. White round iron tower, 133ft. Obscured 223°–228° and 229°–237°. F R obstruction light shown on top of lantern.

CHUB HEADS BEACON, 32 17.2N, 64 58.7W. Q (9) W 15s, 60ft, 12M. West cardinal mark, yellow and black beacon, name on side. Radar reflector. **RACON C** (– · – ·).

EASTERN BLUE CUT BEACON, NW reef, 32 24.0N, 64 52.6W. Mo (U) W 10s, 60ft, 12M. White fiberglass tower, black bands, black concrete base marked EASTERN BLUE CUT. Radar reflector.

P Pilot BERMUDA

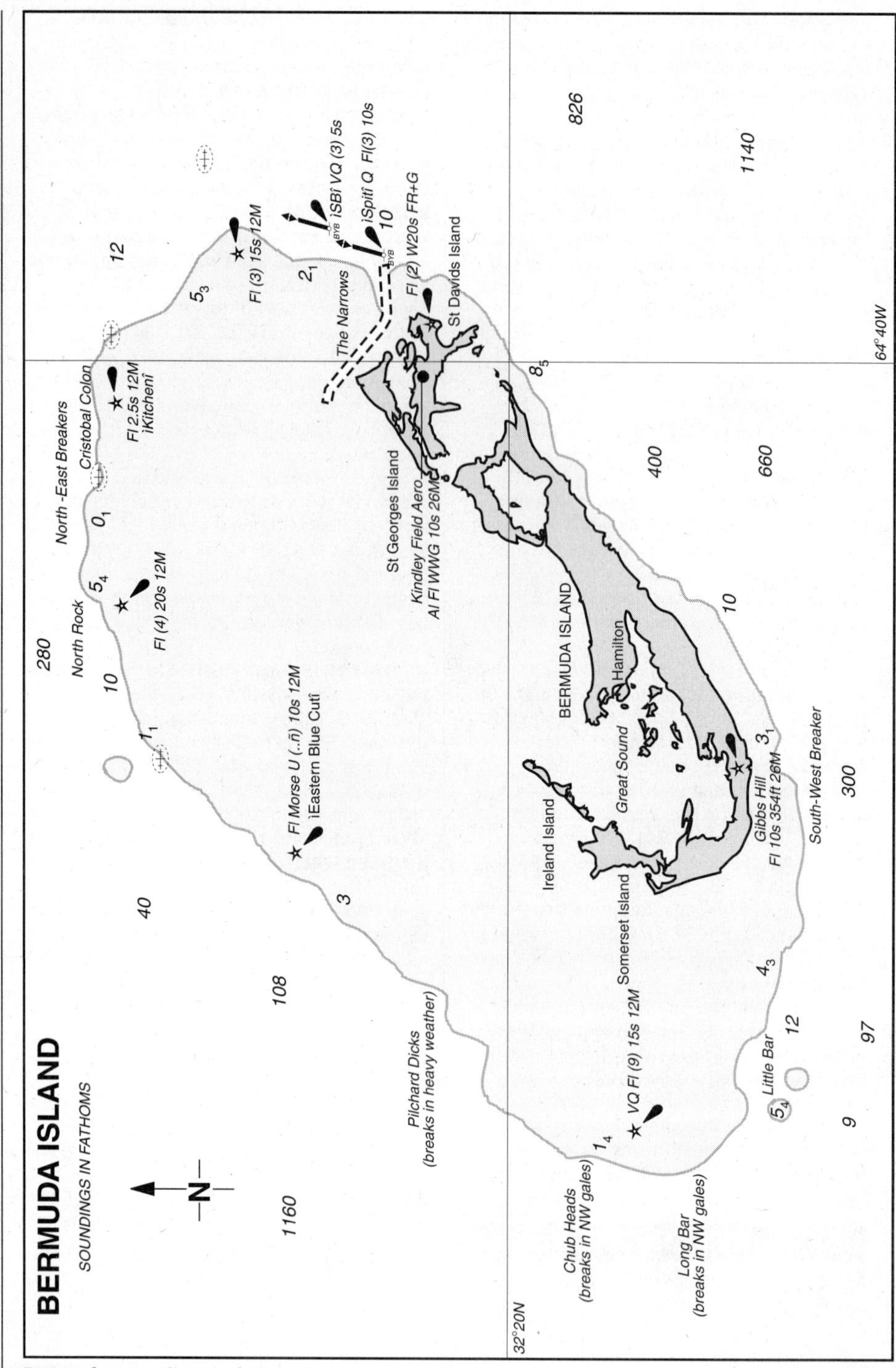

Bermuda, soundings in fathoms

TOWN CUT CHANNEL

GATES FORT LIGHT, N side, outer, 32 22.7N, 64 39.7W. F R, 46ft, 8M. White metal framework tower with black and white checkered daymark. Visible 250°–080°.

HIGGS ISLAND LIGHT, NE corner, 32 22.6N, 64 39.7W. F G, 48ft, 8M. Red and white checkered square on red metal framework tower, white bands. F R lights shown from Fort George flagstaff 1 mile W.

CHALK WHARF LIGHT, on NW side of channel, 32 22.7N, 64 39.9W. F R, 52ft, 8M. White metal framework tower with black and white checkered daymark. Visible 250°–095°.

HORSESHOE ISLAND LIGHT, on W corner of island, 32 22.6N, 64 39.8W. F G, 8M.

ST. GEORGE'S HARBOUR

THREE SISTERS SHOAL BEACON, 32 22.6N, 64 40.0W. V Q G, 4M. White beacon with green band.

HEN ISLAND BEACON, NW part, 32 22.5N, 64 40.5W. Fl G 1.5s, 16ft, 4M. White metal framework tower with green band, red and white checkered daymark.

ST. GEORGE'S

Port of entry

Bermuda Harbour Radio (ZBM): Yachts approaching Bermuda should contact Bermuda Harbour Radio on VHF channel 16 (working 27) or 2182 kHz (working 2582 kHz) when 30 miles out and when moving within Bermuda. BHR also listens on DSC channels VHF 70 and 2187.5 kHz; ITU HF channels 410, 603, 817, 1220, and 1618 are the station's working frequencies; a listening watch is not maintained on these channels. Telephone, 441-297-1010; fax, 441-297-1530; e-mail, rccbda@ibl.bm; telex, 3208 RCC BA.

Weather reports: Available from the Bermuda Yacht Reporting Service on Ordnance Island in St. George's 24 hours a day. Services include: display of the latest weather and wind forecast charts, marine advisories, Gulf Stream temperature analysis, and weather satellite photos. Recorded reports are available from the Bermuda Weather Service via dedicated phone lines: from overseas, dial 441-297-7977; from within Bermuda, for public forecast, dial 9770; for current weather, 9771; for the marine forecast, 9772; and for weather warnings, 9773. Bermuda Weather Service also offers free yacht departure weather briefing packets on request: tel., 441-293-6659; fax, 441-293-6658; Internet address, www.weather.bm. For Bermuda weather information from overseas, call 441-297-7977. Bermuda Harbour Radio transmits local and high-seas weather on 2582 kHz and VHF channel 27 at 1235 and 2035 GMT; on 518 kHz (Navtex station B) at 0010Z, 0410Z, 0810Z, 1210Z, 1610Z, and 2010Z; and continuously on 162.4 MHz (WX-02).

Bermuda Radio VRT: Use VHF channels 26 or 28 for marine phone calls.

Emergencies in port: For fire, ambulance, or police, call Bermuda Harbour Radio or dial 911 (no charge) on the telephone.

Dockage: The concrete wharves on the north side of Ordnance Island provide side-to tie-up. Rafting boats several deep is common. More dockage is available at bulkheads west of the island. Dockage is prohibited on the south side of Ordnance Island and at all commercial docks except in an emergency. St. George's Dinghy and Sports Club may have space at its docks near Town Cut Channel.

Anchorage: Designated yacht anchorages lie east of Ordnance Island and west of Hen Island.

Fuel: Fuel is available at the bulkhead west of Ordnance Island and at the west end of the harbor. Vessels must register in advance to obtain duty-free fuel upon departure.

Repairs: Good services are available. There are haulout facilities near the west end of the harbor and elsewhere in Bermuda. Several sail repair agents, a rigger, electronics specialists, and welding and machining services are also available. Most marine supplies are available.

Supplies: A small but well-stocked grocery is located in St. George's. Larger supermarkets are a short drive away.

Vessels in doubt about entering St. George's at night are advised to lie off until daylight or to anchor outside Town Cut Channel in Five Fathom Hole (32°23'N, 64°37'W). Keep in mind the possibility of encountering a large cruise ship or commercial vessel when approaching or transiting Town Cut Channel. The anchorages in St. George's are secure in most conditions, but other sheltered harbors are nearby if severe weather threatens. Most hurricanes pass to the west of Bermuda, with the peak season being August 15 to October 15.

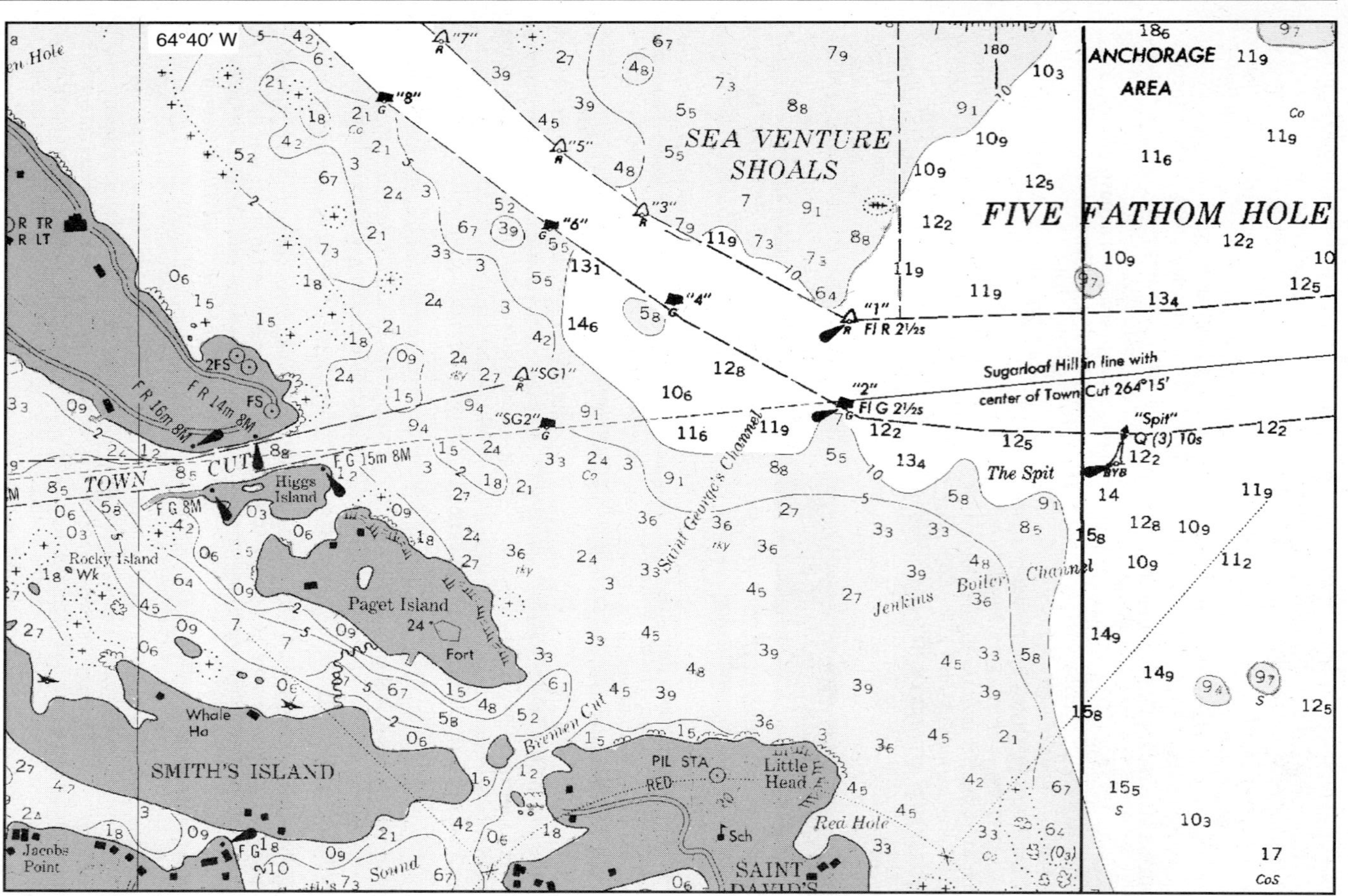

Town Cut, soundings in meters *(from DMA 26342)*

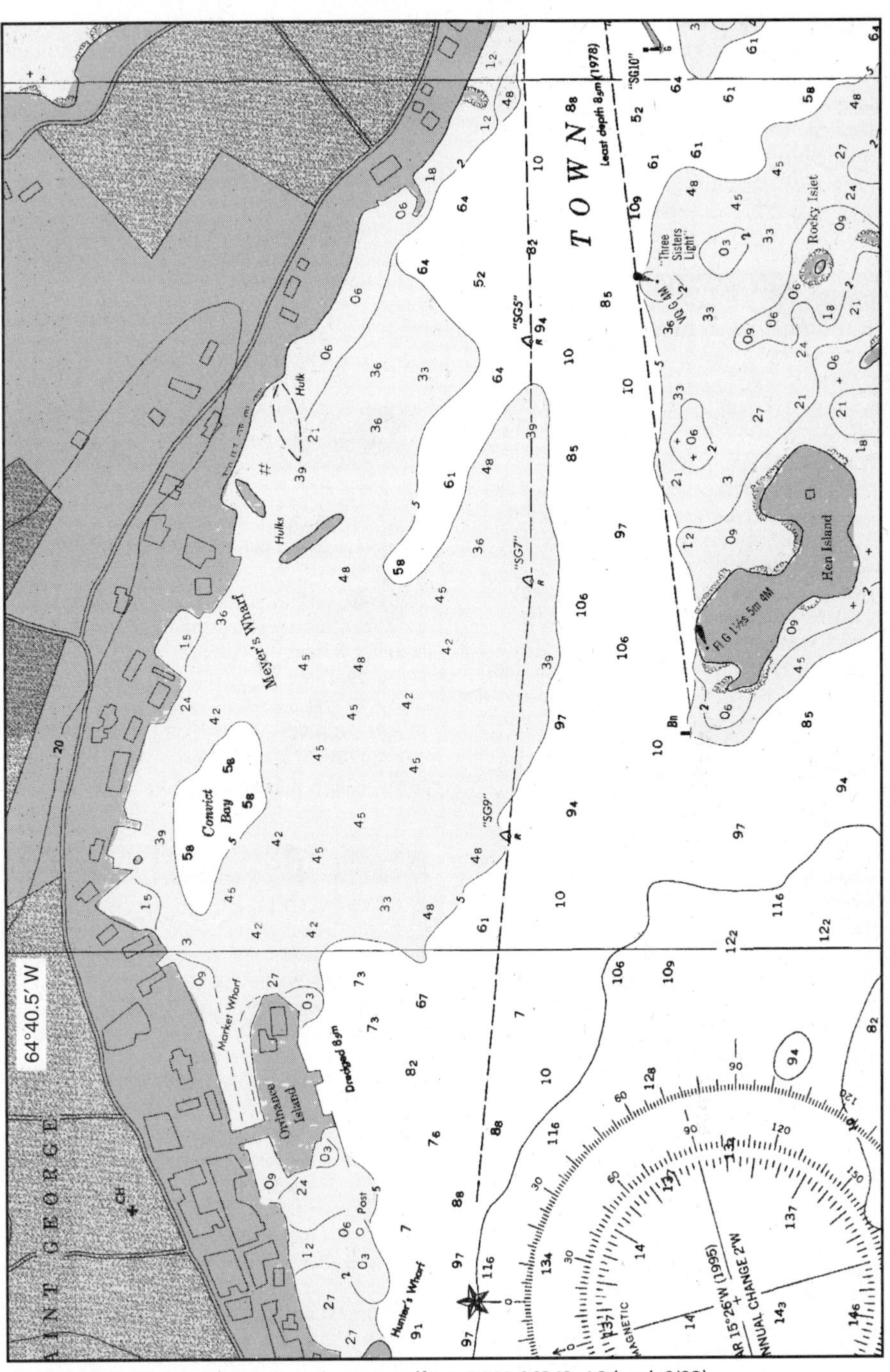

St. George's Harbour, soundings in meters *(from DMA 26342, 16th ed, 9/92)*

P
Pilot
BERMUDA

THE NARROWS

Buoy 1, Fl R 2.5s. Red.

Buoy 2, Fl G 2.5s. Green.

BEACON 10, 32 23.4N, 64 39.9W. Fl G 4s. Green port-hand beacon with topmark.

Buoy 15, Fl R 4s. Red.

Buoy 16, Fl (2) G 7.5s. Green.

SOUTH CHANNEL

BEACON 20, 32 20.9N, 64 44.4W. Fl G 2.5s. PORT (B), green beacon.

Buoy 21, Fl R 2.5s. Red, to be replaced with a beacon.

BEACON 22, Fl G 4s. Green.

SHELLY BAY SHOAL BEACON, Fl (2) W 7.5s. Middle ground mark, black beacon with one broad horizontal red band; topmark, two black spheres.

GIBBET ISLAND LIGHT, 32 19.3N, 64 44.6W. Fl R 4s, 24ft, 2M. Wood column, white base.

DEVONSHIRE DOCK LIGHT, W side of entrance, 32 18.4N, 64 46.3W. F G, 27ft, 2M. White post. F R on radio mast 0.7 mile and 1.2 mile ENE and 0.55 mile and 0.7 mile SE; two F R mark chimney 460 meters ENE.

BEACON 26, 32 18.7N, 64 47.5W. Fl G 2.5s. PORT (B), green beacon.

BEACON 29, 32 19.3N, 64 48.7W. Fl R 4s. STARBOARD (B), red beacon, topmark.

Elbow Buoy, Q (6) W + L Fl 10s. Yellow and black pillar, two cones point down; worded ELBOW.

GRASSY HILL BEACON 30, 32 19.1N, 64 48.7W. Fl G 4s, 16ft. PORT (B), green beacon.

HOGFISH BEACON, 32 18.6N, 64 49.3. Fl (2) Y 10s, 16ft, 5M. White masonry structure with black band around top. F R on radio mast 1.07 mile 290°.

Note: This year *Reed's* has initiated a new policy of indicating changes in aids that have taken place over the last year. Within an aid listing, underlined text indicates data that has changed or is new since September, 1996; ~~Strike-out~~ text indicates deleted data. This should highlight recent changes for mariners, many of which may not be included in even the most recently published charts.

THE DOCKYARD

NORTH BREAKWATER LIGHT, 32 19.3N, 64 50.0W. Fl R 4s, 12ft, 2M. Red structure on white bollard, 7ft.

SOUTH BREAKWATER LIGHT, 32 19.2N, 64 49.9W. Fl G 4s, 12ft, 3M. Green structure on white bollard, 7ft.

WEST CORNER OF SOUTH BASIN LIGHT, Dockyard gate, 32 19.2N, 64 50.3W. F G, 27ft, 2M. Visible 322°–350°.

IRELAND ISLAND WHARF LIGHT, on pier in front of captain-in-charge residence, 32 19.1N, 64 50.4W. F R, 11ft, 2M.

DUNDONALD CHANNEL

Buoy 33, Fl R 2.5s. Red.

BEACON 35, 32 18.8N, 64 49.7W. Fl R 4s. White column with red band.

Buoy 38, Fl G 2.5s. Green.

Buoy 40, V Q (9) W 15s. West cardinal mark; yellow with broad black band; topmark, two black cones point to point.

Buoy 99, Fl R 1.5s. Red.

Buoy 102, Fl G 4s. Green.

PEARL ISLAND LIGHT, 32 17.5N, 64 50.2W. Fl Y 4s, 20ft, 5M. White concrete beacon.

Buoy 103, Fl R 4s. Red.

GREAT SOUND

PLAICES POINT LIGHT, Somerset Island, 32 17.8N, 64 51.5W. F R, 20ft, 2M. Iron column. A F R light shows from center of Watford Bridge.

TWO ROCKS PASSAGE LIGHT, on N side of channel on islet off SW point of Agars Island, 32 17.5N, 64 48.6W. V Q G, 21ft, 5M. White structure with a green band.

TWO ROCKS PASSAGE LIGHT, S side, 32 17.4N, 64 48.6W. V Q R, 21ft, 5M. White structure with red band.

HEAD OF THE LANE CHANNEL, N side, 32 17.3N, 64 48.8W. Fl W 4s, 15ft, 1M. Black circular stone beacon with white base.

TEE ROCK LIGHT, SW of Royal Bermuda Yacht Club dock, 32 17.3N, 64 47.2W. Fl G 4s, 8ft, 3M. White circular concrete tower.

HINSON ISLAND LIGHT, NW point, Timlins Narrows, 32 17.0N, 64 48.3W. F R, 12ft, 5M. White structure with red band.

DAGGER ROCK LIGHT, 32 16.5N, 64 48.8W. Fl R 4s, 13ft, 4M. White column with red band.
RICKETTS ISLAND LIGHT, 32 16.5N, 64 49.7W. Fl R 2.5s, 18ft, 4M. White column with red top.
RIDDELLS BAY LIGHT, N side of entrance, 32 15.7N, 64 49.8W. Fl G 4s, 12ft, 1M. White round stone tower, red top.
PEROTS ISLAND LIGHT, SW end, 32 15.5N, 64 49.9W. F W. Occasional.

HOGFISH CUT CHANNEL

NOTE: This channel is not recommended. Mariners should not attempt it without local knowledge. All arriving yachts must first report to customs in St. George's.

POMPANO BEACON, 32 15.0N, 64 52.6W. V Q G. Green square on beacon. Turning mark for entrance into Hogfish Channel.
HOGFISH TRIPOD, 32 15.3N, 64 52.7W. Fl G 4s.
HOGFISH CUT BEACON, 32 15.5N, 64 52.9W. Fl R 4s, 13ft, 5M. Post.
WRECK HILL BEACON, 32 16.8N, 64 53.3W. V Q R. Red triangular daymark on beacon.

SEND FOR YOUR FREE 1998 SUPPLEMENT

Just return the post-paid reply card and questionnaire to receive your free list of navigation updates, changes, and special notices. Publishing date — Spring 1998.

FLORIDA

Note: This chapter only covers lights and harbors in the southeastern section of Florida. Refer to Reed's Nautical Almanac, East Coast edition for complete coverage.

Please see cautions about chartlets and waypoints on page P 2.

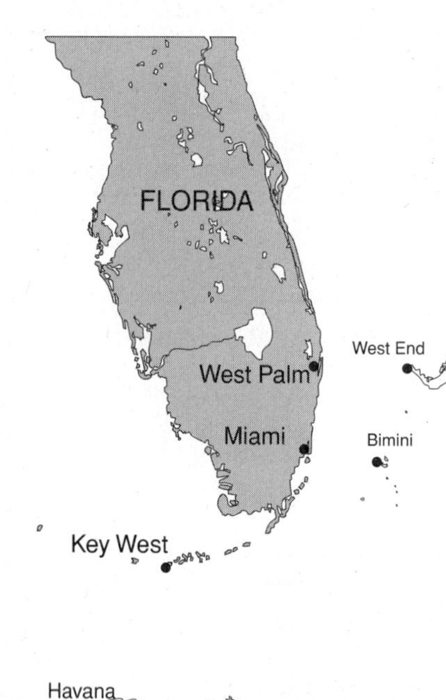

ENTRY PROCEDURES

For information on U.S. customs and immigration procedures refer to page P 7 in the introduction to this chapter.

Ports of entry in Florida are Fernandina Beach, Jacksonville, Port Canaveral, West Palm Beach, Port Everglades, Miami, Key West, Tampa, St. Petersburg, Apalachicola, Boca Grande, Carrabelle, Panama City, Pensacola, and Port St. Joe. Green Cove Springs and Fort Pierce are customs stations supervised by Jacksonville and West Palm Beach respectively.

USEFUL INFORMATION

Language: English.
Currency: U.S. dollar ($). Credit cards are widely accepted. ATM machines are numerous, and many are available 24 hours a day.
Telephones: U.S. country code, 1; area codes, three digits; local numbers, seven digits. To call outside the immediate local calling area but within the U.S., Canada, and much of the Caribbean, dial 1, the area code, and the local number. To reach areas with a country code other than 1, dial 011 and then the country code and number.
National public holidays (1998): New Year's Day, January 1; Martin Luther King, Jr.'s Birthday, January 19; Presidents' Day, February 16; Memorial Day, May 26; Independence Day, July 4; Labor Day, September 7; Columbus Day, October 12; Veteran's Day, November 11; Thanksgiving, November 26; Christmas Day, December 25.
Emergencies, U.S. Coast Guard: VHF channel 16 or 2182 kHz. Fort Pierce station (407-464-6101, 646-6135, 489-0489, 489-0789), Lake Worth Inlet station (407-844-5030), Fort Lauderdale station (954-927-1611, 927-1612), Miami Beach station (305-535-4370), Islamorada station (305-664-4404), Marathon station (305-743-6778), Key West station (305-292-8856). There is also a Coast Guard air station at Miami and a group base in Miami.
Emergencies, Florida Marine Patrol: tel., 800-DIALFMP (800-342-5367); cellular tel., *FMP.

FORT PIERCE INLET

Fort Pierce Fish Haven North Buoy, 27 31.2N, 80 16.9W. Orange and white bands, can. Private aid.
Fort Pierce Fish Haven South Buoy, 27 29.5N, 80 16.9W. Orange and white bands, can. Private aid.
Fort Pierce Inlet Lighted Buoy 2, 27 28.5N, 80 16.2W. Fl R 4s, 4M. Red.
ENTRANCE RANGE (Front Light), 27 28.1N, 80 18.2W. Q R, 15ft. KRW on skeleton tower on piles. Visible 2° each side of rangeline.
(Rear Light), 1,296 yards, 259.6° from front. Iso R 6s, 50ft. KRW on skeleton tower on piles. Visible 2° each side of rangeline.
Buoy 3, green can.
Buoy 4, red nun.
Buoy 4A, red nun.
Buoy 5, green can.
Buoy 6, red nun.
Lighted Buoy 7, Fl G 4s, 4M. Green.
INNER RANGE (Front Light), 27 28.3N, 80 17.8W. Q W, 20ft. KRW on skeleton tower on piles. Obscured 129°–355°, higher intensity on rangeline. **(Rear Light),** 265 yards, 061.7° from front. Iso W 6s, 32ft. KRW on skeleton tower on piles. Visible all around, higher intensity on rangeline.
CENTER RANGE (Front Light), 27 27.5N, 80 19.4W. F R, 53ft. On warehouse. Visible 2° each side of rangeline. **(Rear Light),** 630 yards, 242° from front. F R, 90ft. On telephone pole. Visible 2° each side of rangeline.
Buoy 8, red nun.
Buoy 9, green can.
Buoy 10, red nun.
Buoy 11, green can.
Buoy 12, red nun with yellow square.
LIGHT 13, 27 27.6N, 80 19.1W. Fl G 4s, 16ft, 4M. SG-SY on pile.

FORT PIERCE

27 28N, 80 19W
Customs station (supervised by West Palm Beach port of entry).
U.S. Coast Guard: Station located just inside Fort Pierce Inlet on the south side.
Dockage: Extensive selection of marinas both north and south on the ICW. Care should be used when approaching the marinas because of strong currents on the beam.
Anchorage: Pleasure vessels anchor at Faber Cove and in the turning basin.
Fuel: Available at marinas.
Repairs: Extensive. Haulout facilities.
Supplies: Extensive.

Fort Pierce Inlet is considered navigable by most craft under ordinary conditions and provides easy access to the Gulf Stream. Caution should be used when an ebb current meets heavy seas outside. Contact the Coast Guard if unsure of conditions. The port has some commercial and fishing facilities. The boatyards and marinas in this area can provide most services required by yachts. The city is a major sportfishing center.

FORT PIERCE TO ST. LUCIE

Capron Shoal Buoy 10A, E side of 18-foot shoal, 27 26.6N, 80 13.3W. Red nun.
Fort Pierce Yacht Club Race Course Buoy A, 27 26.2N, 80 16.0W. Yellow can. Private aid.
Fort Pierce Yacht Club Race Course Buoy B, 27 24.3N, 80 15.4W. Yellow can. Private aid.
St. Lucie Shoal Lighted Whistle Buoy 12, off N end of shoal, 27 23.3N, 80 08.0W. Fl R 6s, 4M. Red.
St. Lucie Power Plant Ocean Discharge Pipeline Obstruction Lighted Buoy, 27 21.3N, 80 13.8W. Q Y. Yellow, worded DANGER SUBMERGED DISCHARGE STRUCTURE. Private aid.
St. Lucie Power Plant Intake Pipe North Obstruction Lighted Buoy, 27 20.9N, 80 14.0W. Q Y. Yellow, worded DANGER SUBMERGED INTAKE STRUCTURE. Private aid.
St. Lucie Shoal Buoy 14, near SE end of shoal, 27 18.6N, 80 08.7W. Red nun.
North Point Groin Obstruction Daybeacon, 27 10.1N, 80 09.0W. NW on pile, worded DANGER GROIN. Private aid.

ST. LUCIE INLET

St. Lucie Entrance Lighted Whistle Buoy 2, 27 10.0N, 80 08.4W. Fl R 4s, 3M. Red.
~~Buoy 3~~, green can. Removed 4/96.
Buoy 4, red nun.
Lighted Buoy 5, Fl G 2.5s, 3M. Green.
Buoy 6, 27 10.1N, 80 09.6W. Red nun.
Buoy 8, red nun.
Buoy 9, green can.
Daybeacon 10, TR on pile.
Daybeacon 12, TR on pile.
Daybeacon 14, TR on pile.
Daybeacon 16, TR on pile.
Daybeacon 17, SG on pile.

ST. LUCIE

27 10N, 80 10W

Dockage: Many marinas in Manatee Pocket.

Anchorage: Excellent, well-sheltered anchorage in Manatee Pocket, which the Coast Pilot considers a hurricane hole.

Fuel: At marinas in Manatee Pocket or farther upriver at Stuart.

Repairs: Extensive. Haulout facilities. Additional repair facilities at Stuart.

Supplies: Good.

St. Lucie Inlet, the mouth of the St. Lucie River and southern end of the Indian River, is a "local knowledge"-only entrance. Sportfishermen use the inlet constantly; it may be possible to follow one to locate the best channel. Buoys are shifted frequently to match the changing conditions, and currents are strong. For information on local conditions, contact the Fort Pierce Coast Guard station and ask for the St. Lucie Coast Guard Auxiliary telephone number. Use care when negotiating the ICW in the vicinity of the inlet. Strong currents hit boats on the beam, and the area is prone to shoaling. Manatee Pocket is a pleasant anchorage well away from the traffic of the ICW. The St. Lucie River is the eastern end of the cross-Florida Okeechobee Waterway.

STUART

27 12N, 80 14W

Dockage: Marinas on both North and South Forks of the St. Lucie River plus the city dock.

Anchorage: Good anchorage on both the North and South Forks.

Fuel: Several fuel docks.

Repairs: Extensive.

Supplies: Good.

The Stuart area is well serviced by a variety of marinas and repair facilities. There are good anchoring opportunities, which are rare from here south to Miami. The beginning of the Okeechobee Waterway passes down the South Fork. If you are headed for Florida's west coast and can clear the 49-foot lift bridge at Mile 38, this waterway is an excellent alternative to the route south around the Florida Keys.

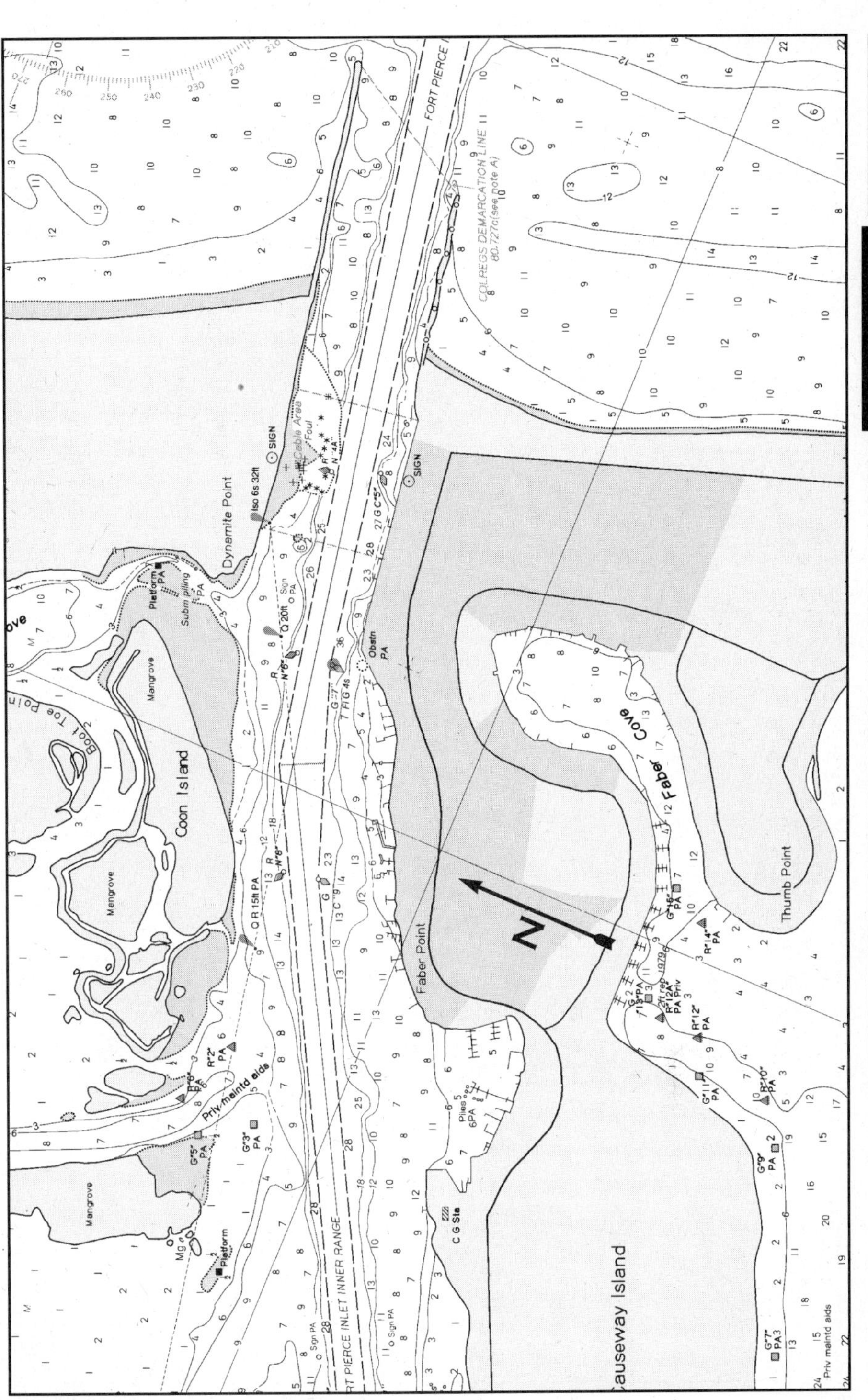

Fort Pierce, FL, soundings in feet *(from NOAA 11472)*

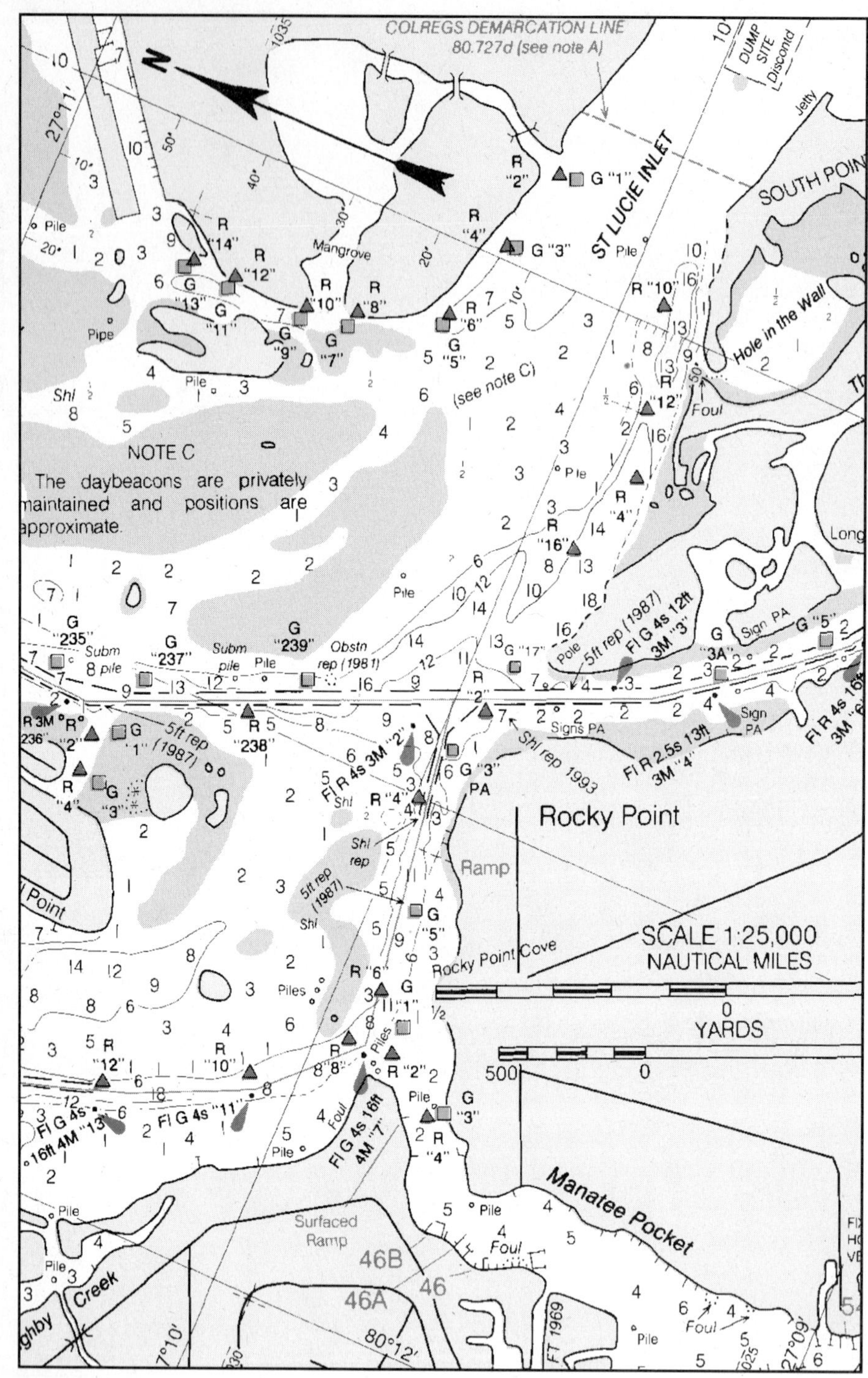

St. Lucie, FL, soundings in feet *(from NOAA11472)*

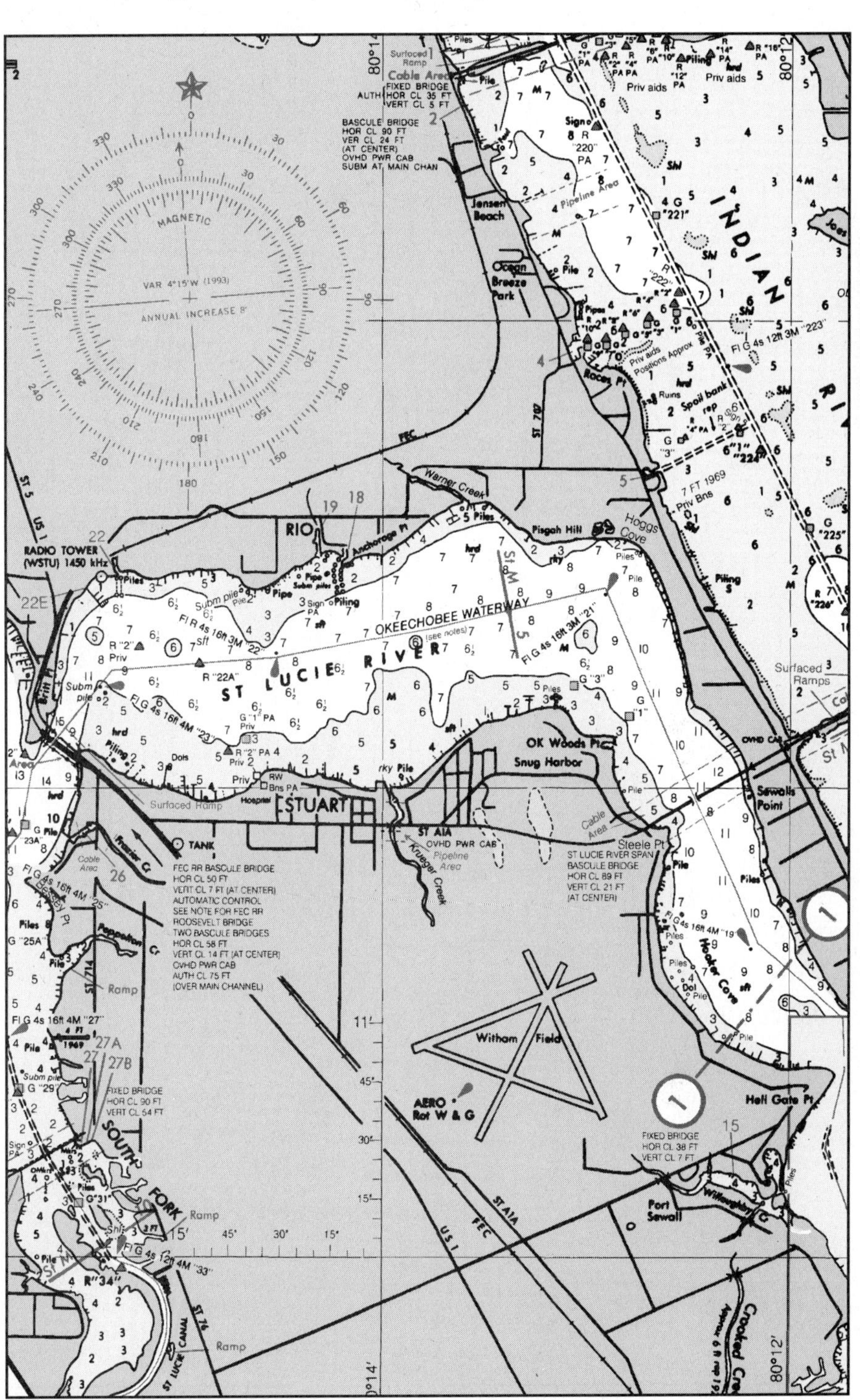

Stuart, FL, soundings in feet *(from NOAA 11428)*

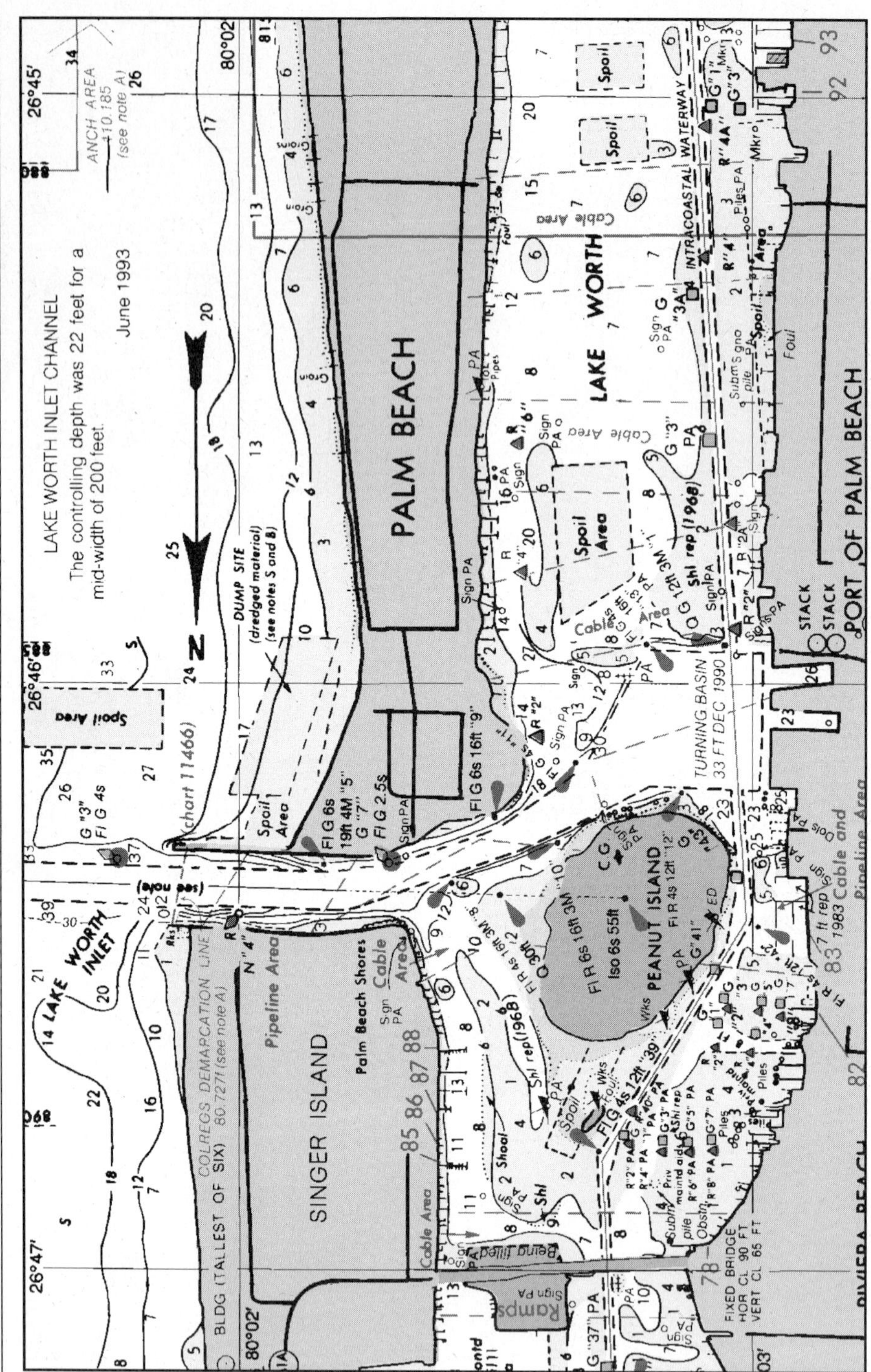

Lake Worth Inlet and Palm Beach, FL, soundings in feet *(from NOAA 11472)*

JUPITER INLET TO LAKE WORTH INLET

JUPITER INLET LIGHT, 26 56.9N, 80 04.9W. Fl (2) W 30s, 146ft, 25M. Red brick tower. Obscured 231°–234° when within 5.5 miles.
Jupiter Inlet South Jetty Daybeacon 1, 26 56.6N, 80 04.3W. SG on pile on jetty. Private aid.
Jupiter Inlet North Jetty Daybeacon 2, TR on pile on jetty. Private aid.
Palm Beach Artificial Reef North Buoy, 26 42.6N, 80 02.0W. White and orange bands; nun. Private aid.
Palm Beach Artificial Reef Middle Buoy, 26 42.3N, 80 02.0W. White and orange bands; nun. Private aid.
Palm Beach Artificial Reef South Buoy, 26 42.0N, 80 02.0W. White and orange bands; nun. Private aid.
Palm Beach Sailing Club Race Course Buoy B, 26 49.6N, 80 00.3W. Orange and white bands, nun. Private aid.
Palm Beach Sailing Club Race Course Buoy D, 26 42.0N, 80 01.7W. Orange and white bands, nun. Private aid.

LAKE WORTH INLET

Lake Worth Lighted Whistle Buoy LW, 26 46.3N, 80 00.6W. Mo (A) W, 6M. Red and white stripes with red spherical topmark.
Lighted Buoy 2, 26 46.4N, 80 01.5W. Fl R 2.5s, 4M. Red.
ENTRANCE RANGE (Front Light), 26 46.3N, 80 02.5W. Q W, 30ft. KRW on skeleton tower on dolphin. Visible 4° each side of rangeline.
(Rear Light), 440 yards, 271.5° from front. Iso W 6s, 55ft. KRW on skeleton tower. Visible 4° each side of rangeline.
Lighted Buoy 3, Fl G 4s, 3M. Green.
Buoy 4, red nun.
LIGHT 5, Fl G 6s, 19ft, 4M. SG on dolphin.
Lighted Buoy 7, Fl G 2.5s, 3M. Green.

SINGER ISLAND CHANNEL

Daybeacon 1, 26 46.5N, 80 02.4. SG on pile.
Daybeacon 3, SG on pile.
LIGHT 5, Fl G 4s, 16ft, 3M. SG on pile.
Daybeacon 7, SG on pile.
Daybeacon 8, TR on pile.
Daybeacon 9, SG on pile.

LAKE WORTH (CONT.)

LIGHT 8, Fl R 4s, 16ft, 3M. TR on pile.
Palm Beach Channel Daybeacon 2, TR on pile.
Palm Beach Channel Daybeacon 4, TR on pile.
Palm Beach Channel Daybeacon 6, TR on pile.
LIGHT 9, Fl G 6s, 16ft, 3M. SG on dolphin.
LIGHT 10, Fl R 6s, 16ft, 3M. TR on dolphin.
LIGHT 11, Fl G 4s, 16ft, 3M. SG on pile.
LIGHT 12, Fl R 4s, 12ft, 3M. TR on pile.
LIGHT 13, Fl G 2.5s, 16ft, 3M. SG on pile.
LAKE WORTH SOUTH LIGHT 1, 26 45.9N, 80 02.9W. Q G, 12ft, 4M. SG-SY on pile.

PALM BEACH

26 46N, 80 03W
Port of entry
U.S. Coast Guard: Lake Worth Inlet station, on the north side of the entrance channel, on the south side of Peanut Island.
Dockage: Many luxurious marinas.
Anchorage: Good anchorage in the northernmost portion of Lake Worth. Possible to anchor just east of Peanut Island.
Fuel: Many fuel docks.
Repairs: Extensive. Haulout facilities.
Supplies: Extensive.

Lake Worth Inlet, one of the best openings on Florida's coast, leads to Lake Worth, the ICW, and the Port of Palm Beach, a deepwater development with significant commercial activity. The communities of Palm Beach and West Palm Beach are among the wealthiest in the United States. Every conceivable marine service is available for yachts of any size, and the shopping on the island resort of Palm Beach is absolutely top-notch in both quality and price. The inlet is a favorite jumping-off spot for yachts heading to West End in the Bahamas.

BOCA RATON INLET

Red Reef Park Artificial Reef Buoys (6), 26 21.6N, 80 04.1W. White and orange bands on nun, worded ROCK. Private aids.
NORTH JETTY LIGHT 2, 26 20.1N, 80 04.2W. Q R, 14ft. TR on pile. Private aid.
SOUTH JETTY LIGHT 1, Q G, 14ft. SG on pile. Private aid.
DEERFIELD FISHING PIER NORTH LIGHT, 26 19.0N, 80 04.4W. F R, 20ft. On pier. Private aid.
DEERFIELD FISHING PIER SOUTH LIGHT, F R, 20ft. On pier. Private aid.

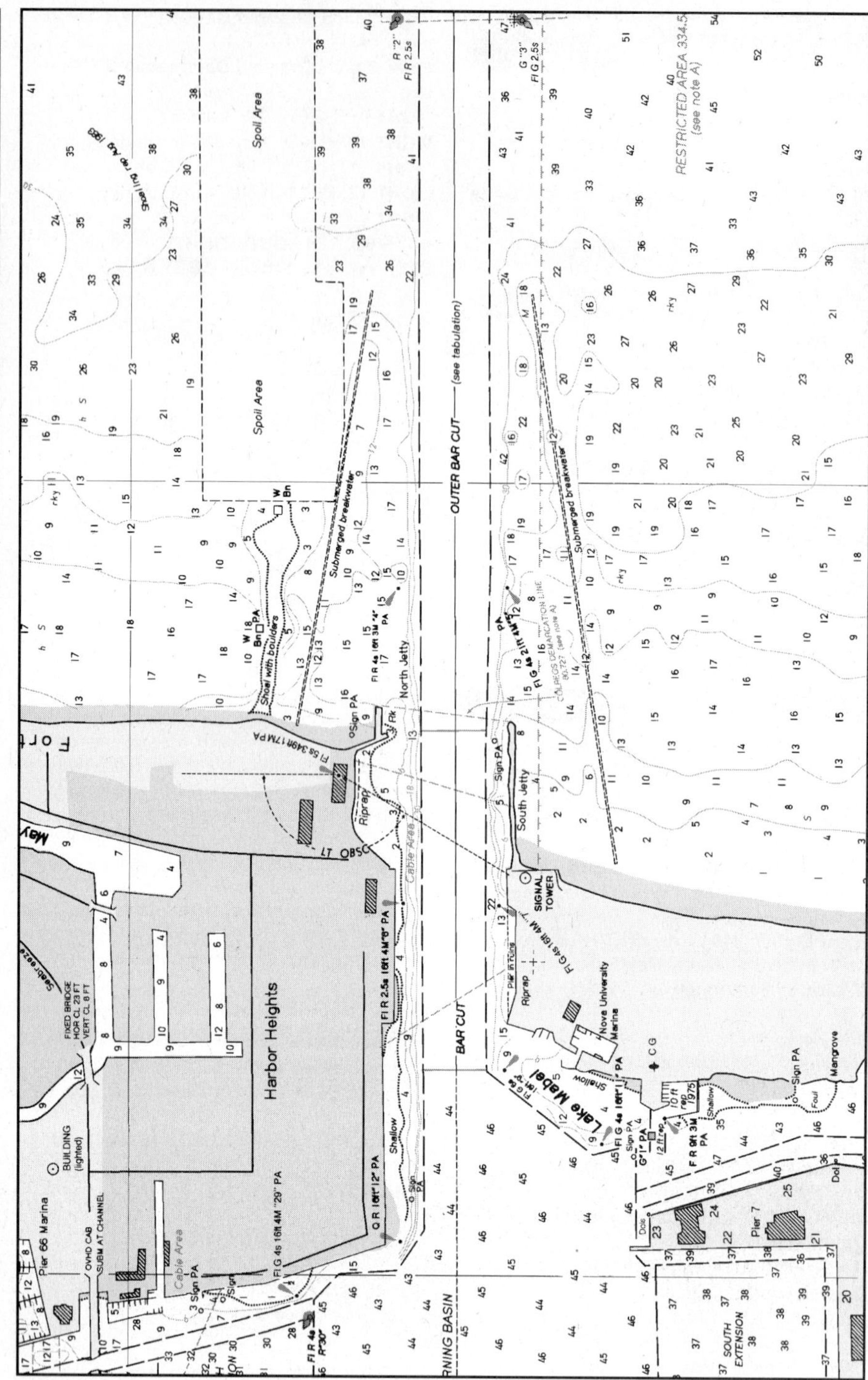

Port Everglades and Fort Lauderdale, FL, soundings in feet *(from NOAA 11470)*

HILLSBORO INLET

HILLSBORO INLET LIGHT, 26 15.6N, 80 04.9W. Fl W 20s, 136ft, 28M. Octagonal pyramidal iron skeleton tower with central stair cylinder, lower third of structure white, upper two-thirds black. Obscured 015°–186°.
Hillsboro Inlet Entrance Lighted Buoy HI, 26 15.1N, 80 04.5W. Mo (A) W, 5M. Red and white stripes with red spherical topmark.
ENTRANCE LIGHT 1, 26 15.3N, 80 04.8W. Fl G 4s, 8ft, 4M. SG on dolphin.
ENTRANCE LIGHT 2, Fl R 4s, 16ft, 4M. TR on dolphin.
Entrance Daybeacon 3, SG on pile.

PORT EVERGLADES

Gulfstream Sail Club Race Buoy B, 26 08.3N, 80 06.0W. Yellow can. Private aid.
Gulfstream Sail Club Race Buoy A, 26 06.8N, 80 06.1W. Yellow can. Private aid.
APPROACH LIGHT, 26 05.7N, 80 06.4W. Fl W 5s, 349ft, 17M. On building. Obscured 030°–180°.
Lighted Buoy PE, 26 05.5N, 80 04.8W. Mo (A) W, 7M. Red and white stripes with red spherical topmark. **RACON T** (–).
Spoil Bank Daybeacon, 26 05.8N, 80 06.0W. NW on pile, worded DANGER SHOAL.
Spoil Bank West Daybeacon, 26 05.8N, 80 06.2W. NW on pile, worded DANGER SHOAL.
Lighted Buoy 2, Fl R 2.5s, 3M. Red.
Lighted Buoy 3, Fl G 2.5s, 3M. Green.
ENTRANCE RANGE (Front Light), 26 05.6N, 80 07.5W. F G, 85ft. KRW on skeleton tower. Visible 2° each side of rangeline. **(Rear Light),** 924 yards, 269.5° from front. F G, 135ft. KRW on skeleton tower. Visible 2° each side of rangeline.
LIGHT 4, 26 05.7N, 80 06.1W. Fl R 4s, 16ft, 3M. TR on dolphin.
LIGHT 5, Fl G 4s, 21ft, 4M. SG on dolphin.
LIGHT 6, Fl R 2.5s, 16ft, 4M. TR on dolphin.
LIGHT 7, 26 05.6N, 80 06.7W. Fl G 4s, 16ft, 4M. SG on dolphin.
LIGHT 9, Fl G 6s, 16ft, 4M. SG on dolphin.
TURNING BASIN LIGHT 11, Fl G 4s, 16ft, 4M. SG-SY on dolphin.
TURNING BASIN LIGHT 12, Q R, 16ft, 3M. TR-SY on dolphin.
PIER 7 LIGHT, 26 05.6N, 80 07.0W. F W, 15ft. On pier. Private aid.
Stranahan River Lighted Buoy 30, Fl R 4s, 3M. Red with yellow triangle.
Coast Guard Station Daybeacon 1, 26 05.4N, 80 06.8W. SG on pile.
COAST GUARD STATION BREAKWATER LIGHT, 26 05.4N, 80 06.8W. F R, 9ft, 3M. On breakwater.
PORT EVERGLADES SOUTHPORT TURNING DOLPHIN LIGHT, 26 04.5N, 80 06.9W. F R 6ft. On pile. Private aid.
DANIA SOUND DANGER LIGHT, 26 03.9N, 80 06.8W. Q W, 16ft, 4M. NW on pile worded DANGER ROCKS.
DANIA BEACH FISHING PIER OBSTRUCTION LIGHT, 26 03.5N, 80 06.5W. F R, 18ft. On pier. Private aid.

FORT LAUDERDALE

26 06N, 80 07W
Port of entry: Port Everglades.
U.S. Coast Guard: Fort Lauderdale station on the east side of the ICW at Mile 1066.8, southeast of the turning basin.
City dockmaster: The dockmaster is in charge of the mooring area near the Las Olas Boulevard bridge and the city-owned berths on the New River, where the dockmaster's office is located (VHF channels 09 and 16; tel., 954-761-5423).
Dockage: Extensive private marinas line the waterways of Fort Lauderdale. There are three city dock facilities—Cooley's Landing on the New River, the Fort Lauderdale Municipal Marina, and the Las Olas Boulevard docks—but the Las Olas Boulevard docks are closed for renovation until approximately May 1998.
Moorings: Near Las Olas Boulevard bridge at the city-regulated moorings ($10/night). A dinghy dock and fresh water are provided.
Anchorage: If there is room, vessels may anchor for a maximum of 24 hours in the designated anchorage area north of the moorings by the Las Olas Boulevard bridge; a city ordinance limits both anchoring and mooring to a maximum of 30 nights in a calendar year. Vessels may not anchor if a mooring is available.
Fuel & supplies: Readily available.
Repairs: Extensive. Major haulout facilities.

Fort Lauderdale, just north of commercial Port Everglades, provides the most complete marine facilities on the U.S. east coast. Marinas accommodate the largest luxury yachts alongside the ordinary. Despite the huge number of berths, reservations are always recommended. Ashore are luxury hotels, restaurants, and endless shops. Port Everglades, a deepwater port with significant cruise ship activity as well as some commercial shipping, is a safe and well-marked inlet. This is an excellent departure port for trips to the Bahamas.

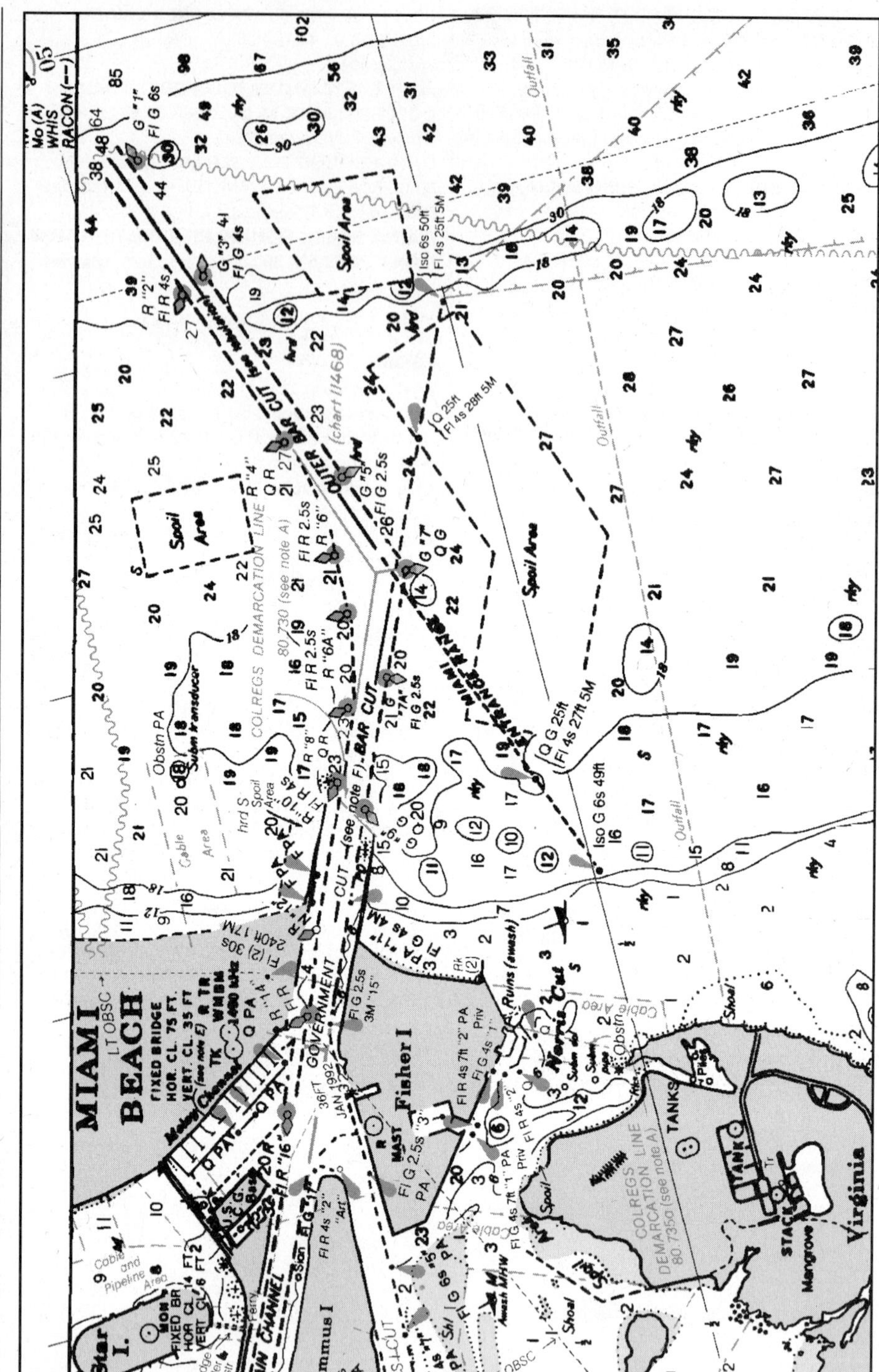

Miami, FL, soundings in feet *(from NOAA 11467)*

PORT EVERGLADES TO MIAMI

Parker Dorado Association Obstruction Daybeacon, 25 58.5N, 80 07.1W. NW on pile, worded DANGER SUBMERGED OUTFALL. Private aid.
BAKERS HAULOVER INLET JETTY LIGHT, 25 53.9N, 80 07.2W. Q W, 13ft. NW on pile, worded DANGER JETTY. Private aid.

MIAMI ENTRANCE

Miami Anchorage Buoy A, 25 48.3N, 80 05.7W. Yellow nun. Marks anchorage area E of Miami Beach.
Miami Anchorage Buoy B, 25 46.4N, 80 06.2W. Yellow nun. Marks anchorage area E of Miami Beach.
MIAMI APPROACH LIGHT, 25 46.0N, 80 08.0W. Fl (2) W 30s, 240ft, 17M. On building. Obscured 065°–195°.
Miami Lighted ~~Whistle~~ Buoy M, 25 46.1N, 80 05.0W. Mo (A) W, 7M. Red and white stripes with red spherical topmark. **RACON M** (– –).
ENTRANCE RANGE (Front Light), 25 45.1N, 80 07.7W. Q G, 25ft; Fl W 4s, 27ft, 5M. KRB on skeleton tower on piles. Visible 4° each side of rangeline. Passing light visible around horizon. **(Rear Light),** 670 yards, 249.5° from front. Iso G 6s, 49ft. KRB on skeleton tower on piles. Visible 4° each side of rangeline.
Lighted Buoy 1, Fl G 6s, 4M. Green.
Lighted Buoy 2, Fl R 4s, 4M. Red.
Lighted Buoy 3, Fl G 4s, 4M. Green.
Lighted Buoy 4, Q R, 4M. Red.
Lighted Buoy 5, Fl G 2.5s, 4M. Green.
Lighted Buoy 6, Fl R 2.5s, 4M. Red.
Lighted Buoy 7, Q G, 4M. Green.
GOVERNMENT CUT RANGE (Front Light), 25 45.2N, 80 06.5W. Q W, 25ft; Fl W 4s, 28ft, 5M. KRW on tower on piles. Visible 2° each side of rangeline. Passing light visible around horizon. **(Rear Light),** 785 yards, 114.8° from front. Iso W 6s, 50ft; Fl W 4s, 25ft, 5M. KRW on tower on piles. Visible 2° each side of rangeline. Passing light visible around horizon.
Lighted Buoy 6A, Fl R 2.5s, 4M. Red.
Lighted Buoy 9, Fl G 2.5s, 4M. Green.
Lighted Buoy 8, Q R, 4M. Red.
Lighted Buoy 11, Q G, 4M. Green.
Lighted Buoy 10, Fl R 4s, 4M. Red.
LIGHT 13, Fl G 4s, 16ft, 4M. SG on dolphin.
Buoy 12, red nun.
Lighted Buoy 14, Fl R 4s, 4M. Red.
LIGHT 15, Fl G 2.5s, 16ft, 3M. SG on dolphin.
Lighted Buoy 16, 25 46.0N, 80 08.5W. Fl R 4s, 4M. Red.
MAIN CHANNEL LIGHT 17, 25 46.0N, 80 08.7W. Fl G 4s, 16ft, 4M. SG on dolphin.
MAIN CHANNEL LIGHT 22, Fl R 4s, 16ft, 3M. TR on pile. Higher intensity beam toward Pier No. 1.

MELOY CHANNEL, MIAMI BEACH MARINA

PIER OBSTRUCTION SOUTH LIGHT, 25 46.0N, 80 08.4W. Q W, 12ft. On pile. Private aid.
BREAKWATER SOUTH LIGHT, Q W, 12ft. On pile. Private aid.
BREAKWATER NORTH LIGHT, Q W, 12ft. On pile. Private aid.

FISHERMANS CHANNEL

ARTICULATED LIGHT 2, 25 45.9N, 80 08.7W. Fl R 4s, 16ft, 3M. TR on mast.
LIGHT 3, 25 45.8N, 80 08.7W. Fl G 2.5s, 16ft, 3M. SG on pile.

MIAMI

25 46N, 80 08W
Port of entry
U.S. Coast Guard: Miami Beach station on Causeway Island, on the north side of the main channel.
Dockage: Extensive dockage is available at Miami Beach just inside Government Cut and in Miami at Bay Front Park just south of the Dodge Island Causeway. There are many additional marinas and repair facilities along the ICW and in the Miami River.
Anchorage: Good anchorage possibilities exist in Biscayne Bay north of the MacArthur Causeway. Beware of cable and pipeline areas marked on the charts.
Fuel: Many fuel docks. Beware of strong currents in the marina just inside Government Cut.
Repairs: Extensive. Haulout facilities.
Supplies: Extensive.

Miami is the largest city in Florida and has become the de facto capital of the Caribbean in recent years. Its deepwater port facilities inside Government Cut handle a large volume of cruise ship and commercial traffic day and night. The channel is safe and almost too well marked. The profusion of buoyage and lights can be confusing at night. From the marina at Miami Beach you can walk to museums, restaurants, and shops. The downtown marina complex on the mainland is in the heart of Miami's revitalized waterfront, where shops and restaurants overlook the harbor.

MIAMI TO FOWEY ROCKS

Bear Cut Daybeacon 2, 25 43.5N, 80 08.0W. TR on pile.

Cape Florida Daybeacon 2, TR on pile.

CAPE FLORIDA LIGHT 4, 25 41.1N, 80 11.0W. Fl R 2.5s, 16ft, 4M. TR on pile.

Biscayne National Park North Lighted Buoy N, 25 38.7N, 80 05.4W. Fl Y 2.5s, 4M. Yellow.

FOWEY ROCKS LIGHT, on outer line of reefs, 25 35.4N, 80 05.8W. Fl W 10s (2 R sectors), 110ft, W 15M, R 10M. Brown octagonal pyramidal skeleton tower enclosing white stair cylinder and octagonal dwelling, pile foundation. W 188°–359° and 022°–180°, R in intervening sectors.

BISCAYNE CHANNEL

LIGHT, 25 38.3N, 80 07.9W. Fl W 4s, 37ft, 5M. NG on skeleton tower on piles.

LIGHT 1, 25 38.7N, 80 08.0W. Fl G 2.5s, 16ft, 4M. SG on dolphin.

LIGHT 2, Fl R 2.5s, 16ft, 4M. TR on dolphin.

LIGHT 3, Fl G 4s, 12ft, 3M. SG on pile.

Daybeacon 4, TR on pile.

LIGHT 6, Fl R 4s, 12ft, 3M. TR on dolphin.

Daybeacon 7, SG on pile.

Daybeacon 8, TR on pile.

Daybeacon 10, TR on pile.

Daybeacon 12, TR on pile.

Daybeacon 14, TR on pile.

LIGHT 15, 25 39.1N, 80 10.5W. Fl G 4s, 16ft, 4M. SG on pile.

Daybeacon 16, TR on pile.

Daybeacon 18, TR on pile.

Daybeacon 19, SG on pile.

Daybeacon 20, TR on pile.

LIGHT 21, Fl G 2.5s, 16ft, 3M. SG on pile.

FOWEY ROCKS TO ALLIGATOR REEF

TRIUMPH REEF LIGHT 2TR, 25 28.6N, 80 06.7W. Fl R 2.5s, 19ft, 6M. TR on dolphin.

PACIFIC REEF LIGHT, 25 22.2N, 80 08.5W. Fl W 4s, 44ft, 9M. Black skeleton tower on piles.

Turtle Harbor Lighted Whistle Buoy 4, marks entrance between shoal patches, 25 17.5N, 80 09.9W. Fl R 6s, 5M. Red.

CARYSFORT REEF LIGHT, on outer line of reefs, 25 13.3N, 80 12.7W. Fl (3) W 60s (3 R sectors), 100ft, W 15M, R 13M. Brown octagonal pyramidal skeleton tower enclosing stair cylinder and conical dwelling, pile foundation. W 211°–018°, 049°–087°, and 145°–184°, R in intervening sectors.

ELBOW REEF LIGHT 6, 25 08.7N, 80 15.5W. Fl R 2.5s, 36ft, 6M. TR on skeleton tower on piles.

Dixie Shoal Buoy 8, on seaward edge of reef, 25 04.6N, 80 18.7W. Red nun.

MOLASSES REEF BOUNDARY

~~***Northwest Buoy,*** 25 00.7N, 80 22.4W. Yellow nun, worded NO ANCHOR ZONE. Private aid.~~

~~***Northeast Buoy,*** 25 00.6N, 80 22.4W. Yellow nun, worded NO ANCHOR ZONE. Private aid.~~

~~***Southwest Buoy,*** 25 00.6N, 80 22.4W. Yellow nun, worded NO ANCHOR ZONE. Private aid.~~

~~***Southeast Buoy,*** 25 00.6N, 80 22.4W. Yellow nun, worded NO ANCHOR ZONE. Private aid.~~

MOLASSES REEF LIGHT 10, 25 00.7N, 80 22.6W. Fl R 10s, 45ft, 13M. TR on square brown pyramidal skeleton tower, pile foundation.

Conch Reef Buoy 12, near seaward edge of reef, 24 57.0N, 80 27.5W. Red nun.

DAVIS REEF LIGHT 14, 24 55.5N, 80 30.2W. Fl R 4s, 22ft, 6M. TR on dolphin.

Crocker Reef Buoy 16, 24 54.5N, 80 31.5W. Red nun.

ALLIGATOR REEF LIGHT, on outer line of reef, 24 51.1N, 80 37.1W. Fl (4) W 60s (2 R sectors), 136ft, W 16M, R 13M. White octagonal pyramid skeleton tower enclosing stair cylinder and square dwelling, black pile foundation. R 223°–249° and 047°–068°.

HAWK CHANNEL: SOLDIER KEY TO CHANNEL FIVE

Soldier Key Daybeacon 2, 25 35.8N, 80 07.6W. TR on pile.

Fowey Rocks Daybeacon 3, 25 35.7N, 80 06.9W. SG on pile.

Soldier Key Daybeacon 4, 25 34.2N, 80 07.9W. TR on pile.

Ragged Key Daybeacon 7, 25 31.0N, 80 08.3W. SG on pile.

BOWLES BANK LIGHT 8, 25 30.4N, 80 08.7W. Fl R 4s, 16ft, 3M. TR on pile.

Bache Shoal Daybeacon 9, SG on pile.

BACHE SHOAL LIGHT 11BS, on SW edge of shoal, 25 29.2N, 80 08.9W. Fl G 6s, 18ft, 5M. SG on dolphin.

Bache Shoal Daybeacon 13, 25 28.7N, 80 09.2W. SG on pile.

ELLIOTT KEY

Daybeacon 14, TR on pile.
Daybeacon 15, SG on pile.
Daybeacon 16, 25 26.8N, 80 10.2W. TR on pile.
Daybeacon 17, SG on pile.
Daybeacon 18, TR on pile.
Daybeacon 19, SG on pile.
Pacific Reef Daybeacon 2, TR on pile.
Pacific Reef Daybeacon 3, 25 22.4N, 80 09.6W. SG on pile.
Pacific Reef Daybeacon 4, TR on pile.
LIGHT 20, 25 23.1N, 80 11.5W. Fl R 4s, 16ft, 3M. TR on pile.

CAESAR CREEK

Caesar Creek Daybeacon 1, 25 23.1N, 80 11.6W. SG on pile. Maintained by National Park Service.

NOTE: A series of twenty-five more red triangular and green square daybeacons on piles mark Caesar Creek to Junction Daybeacon A, a red and green banded daymark, red band topmost. Daybeacons 27 through 30, alternating red triangles and green squares on piles, mark the remainder of the channel. Caesar Creek Northeast Channel is marked by Daybeacons 2 through 6, also alternating red triangles and green squares. All Caesar Creek daybeacons are maintained by the National Park Service.

OLD RHODES KEY

Daybeacon 21, SG on pile.
LIGHT 22, Fl R 2.5s, 12ft, 4M. TR on pile.
Daybeacon 23, SG on pile.
Daybeacon 24, TR on pile.

ANGELFISH CREEK

LIGHT 2, 25 19.7N, 80 15.0W. Fl R 4s, 12ft, 3M. TR on pile.
Daybeacon 1, SG on pile.
Daybeacon 2A, TR on pile.
Daybeacon 3, SG on pile.
Daybeacon 3A, SG on pile.
Daybeacon 4, TR on pile.
Daybeacon 5, SG on pile.
LIGHT 6, Fl R 4s, 12ft, 3M. TR on pile.
Daybeacon 8, TR on pile.
Daybeacon 10, TR on pile.
Daybeacon 12, TR on pile.
LIGHT 14, 25 20.1N, 80 16.8W. Fl R 4s, 12ft, 3M. TR on pile.

VILLA CHANNEL

Daybeacon 1, 25 18.8N, 80 15.9W. SG on pile. Private aid.

NOTE: A series of twenty more red triangular and green square daybeacons mark Villa Channel. They are all private aids.

OCEAN REEF HARBOR ENTRANCE LIGHT 2, 25 18.5N, 80 16.1W. Fl R 4s, 16ft, 3M. TR on pile.

DISPATCH CREEK CHANNEL

Daybeacon 3, 25 18.3N, 80 16.2W. SG on pile. Private aid.

NOTE: A series of seventeen more red triangular and green square daybeacons on piles mark Dispatch Creek Channel. All are private aids.

HARBOR COURSE CREEK

Daybeacon 1, 25 18.6N, 80 17.0W. SG on pile. Private aid.
Daybeacon 2, TR on pile. Private aid.

TURTLE HARBOR

West Shoal Daybeacon 2, 25 19.4N, 80 12.7W. TR on pile.
WEST SHOAL PREFERRED CHANNEL LIGHT, 25 18.3N, 80 12.8W. Fl (2+1) G 6s, 16ft, 3M. JG on pile.
Daybeacon 1, SG on pile.
Daybeacon 3, SG on pile.
Daybeacon 4, TR on pile.
Daybeacon 5, SG on pile.
Daybeacon 6, TR on pile.

Key Largo Daybeacon 25, SG on pile.
Key Largo Daybeacon 27, SG on pile.
Key Largo Daybeacon 29, 25 14.3N, 80 17.0W. SG on pile.

BASIN HILLS LIGHT 31BH, Fl G 4s, 27ft, 4M. SG on skeleton structure on piles.
KEY LARGO LIGHT 32, 25 10.6N, 80 20.3W. Fl R 2.5s, 16ft, 3M. TR on pile.

BASIN HILLS CHANNEL

Daybeacon 2, 12ft. TR on pile. Private aid.
Daybeacon 3, 12ft. SG on pile. Private aid.

NOTE: Four more green square daymarks on piles mark Basin Hills Channel. All are private aids.

GARDEN COVE

Daybeacon 1, 25 10.6N, 80 21.1W. SG on pile. Private aid.

NOTE: A series of 24 more red triangular and green square daybeacons on pilings mark the Garden Cove and North Sound Creek channels. All are private aids.

Key Largo Daybeacon 32A, 25 08.8N, 80 21.2W. TR on pile.
Key Largo Daybeacon 33, SG on pile.

JOHN PENNEKAMP CORAL REEF STATE PARK

Buoy D, 25 07.4N, 80 18.0W. Orange and white bands, barrel. Private aid.
Buoy E, 25 06.7N, 80 18.5W. Orange and white bands, barrel. Private aid.
Buoy F, 25 06.6N, 80 20.7W. Orange and white bands, barrel. Private aid.
Buoy G, 25 03.2N, 80 20.2W. Orange and white bands, barrel. Private aid.

Carysfort North Reef North Obstruction Daybeacon, 25 13.5N, 80 12.5W. NW on pile, worded DANGER SHOAL. Private aid.
Carysfort North Reef South Obstruction Daybeacon, 25 13.2N, 80 12.7W. NW on pile, worded DANGER SHOAL. Private aid.
Carysfort South Reef North Obstruction Daybeacon, 25 12.8N, 80 13.1W. NW on pile, worded DANGER SHOAL. Private aid.
Carysfort South Reef South Obstruction Daybeacon, 25 12.4N, 80 13.3W. NW on pile, worded DANGER SHOAL. Private aid.
Key Largo Dry Rocks Obstruction Daybeacon, NW on pile, worded DANGER SHOAL. Private aid.
Grecian Rocks Obstruction Daybeacon, NW on pile, worded DANGER SHOAL. Private aid.
Cannon Patch Obstruction Daybeacon, NW on pile, worded DANGER SHOAL. Private aid.
Mosquito Bank Obstruction Daybeacon, NW on pile, worded DANGER SHOAL. Private aid.
White Banks Obstruction Daybeacon, 25 02.4N, 80 22.3W. NW on pile, worded DANGER SHOAL. Private aid.
French Reef Obstruction Daybeacon, 25 02.1N, 80 21.1W. NW on pile, worded DANGER SHOAL. Private aid.
Sand Island Obstruction Daybeacon, 25 01.1N, 80 22.0W. NW on pile, worded DANGER SHOAL. Private aid.
Molasses Reef North Obstruction Daybeacon, 25 00.8N, 80 22.4W. NW on pile, worded DANGER SHOAL. Private aid.
Molasses Reef South Obstruction Daybeacon, 25 00.6N, 80 22.7W. NW on pile, worded DANGER SHOAL. Private aid.

LARGO SOUND CHANNEL

LIGHT 2, 25 05.8N, 80 23.8W. Fl R 4s, 16ft, 4M. TR on dolphin.
Daybeacon 3, SG on pile.
Daybeacon 4, TR on pile.
Daybeacon 5, SG on pile.
Daybeacon 6, TR on pile.
Daybeacon 6A, TR on pile.
Daybeacon 7, SG on pile.
Daybeacon 7A, SG on pile.
Daybeacon 8, TR on pile.
LIGHT 9, Fl G 4s, 16ft, 3M. SG on pile.
Daybeacon 10, TR on pile.
Daybeacon 11, SG on pile.
Daybeacon 12, TR on pile.
Daybeacon 13, SG on pile.
Daybeacon 14, TR on pile.
Daybeacon 16, TR on pile.
Daybeacon 17, SG on pile.
Daybeacon 19, SG on pile.
Daybeacon 20, TR on pile.
Daybeacon 21, SG on pile.
Daybeacon 22, TR on pile.
Daybeacon 23, SG on pile.
MARVIN D. ADAMS WATERWAY LIGHT A, 25 08.2N, 80 23.7W. Fl W 6s, 12ft, 4M. NR on pile.
MARVIN D. ADAMS WATERWAY LIGHT B, I Fl W 6s, 12ft, 4M. NR on pile.

MOSQUITO BANK LIGHT 35, 25 04.3N, 80 23.6W. Fl G 4s, 37ft, 5M. SG on triangular pyramidal skeleton structure on piles.
MOSQUITO BANK LIGHT 2, Fl R 2.5s, 16ft, 4M. TR on pile.
Hidden Bay Daybeacon 2, 25 04.8N, 80 26.2W. TR on pile. Private aid.
Hidden Bay Daybeacon 4, TR on pile. Private aid.
KAWAMA MARINA AND YACHT CLUB LIGHT, 25 04.6N, 80 26.4W. Q R, 25ft. On roof of thatched hut. Private aid.
HARBORAGE YACHT CLUB ENTRANCE LIGHT 2, 25 04.1N, 80 27.7W. Fl R 4s, 12ft. TR on steel tower. Private aid.
Key Largo Daybeacon 37, SG on pile.
Key Largo Daybeacon 39, NW of shoal. SG on pile.

KEY LARGO

25 06N, 80 26W
Dockage: The marinas are in a series of canals northwest of Mosquito Bank on Key Largo. Approach depths are 4 to 5 feet.
Anchorage: Good anchorage in the lee of Rodriguez Key.
Fuel: At marina up the canal.
Repairs: Good. Haulout facilities available.
Supplies: Good.

The big attraction in the Key Largo area is the John Pennekamp Coral Reef State Park. The Key Largo National Marine Sanctuary extends seaward of the state park, further protecting the coral reef and associated marine life. Dive boats take visitors out to the area—anchoring is forbidden on the reefs. Ashore many restaurants and shops are strung out along U.S. 1. The marinas are small and low key, but provide most services. When approaching any docks do so at slow speed—shoaling occurs frequently.

MOLASSES REEF

Daybeacon 1, on S side of passage between shoals, 25 01.6N, 80 23.7W. SG on pile.
Daybeacon 3, SG on pile.
Daybeacon 5, SG on pile.

KEY LARGO OCEAN RESORT LIGHT 2, 25 02.5N, 80 29.3W. Fl R 4s. TR on pile. Private aid.

HENRY HARRIS PARK

Daybeacon 1, 25 01.3N, 80 29.6W. SG on pile. Private aid.
NOTE: A series of five more alternating red and green daybeacons on piles, numbered 2 through 6, mark the channel to Henry Harris Park. All are private aids.

OCEAN POINT

Daybeacon 1, 25 01.0N, 80 29.9W. SG on pile. Private aid.
NOTE: A series of five more daybeacons, alternately red and green, numbered 2 through 6, mark Ocean Point channel. All are private aids.

BLUE WATERS TRAILER VILLAGE

LIGHT 1, 25 00.8N, 80 30.3W. Fl G 6s, 12ft. SG on pile. Private aid.
NOTE: A series of five more alternating red and green daybeacons on piles, numbered 2 through 6, mark the Blue Waters Trailer Village channel. All are private aids.

TAVERNIER KEY LIGHT 2, 25 00.2N, 80 29.0W. Fl R 4s, 16ft, 3M. TR on pile.
Tavernier Key Daybeacon 3, SG on pile.
Tavernier Key Daybeacon 4, TR on pile.
Tavernier Ocean Shores Daybeacon 2, TR on pile. Private aid.
Tavernier Ocean Shores Daybeacon 4, TR on pile. Private aid.

TAVERNIER CREEK

LIGHT, 24 59.2N, 80 31.4W. Fl W 6s, 16ft, 5M. NR on pile.
NOTE: A series of seventeen red triangular and green square daybeacons, numbered 1 through 18, mark Tavernier Creek. Numbers 3 through 18 are private aids.

Coral Harbor Club Daybeacon 1, 24 58.3N, 80 33.1W. SG on pile. Private aid.
CORAL HARBOR CLUB LIGHT 2, Fl R 4s.TR on pile. Private aid.

OCEAN HARBOR

Ocean Harbor Daybeacon 1, 24 58.0N, 80 33.5W. SG on pile.
NOTE: A series of six red triangular and green square daybeacons, numbered 1 through 6, mark Ocean Harbor.

HEN AND CHICKENS SHOAL LIGHT 40, SE of shoal, 24 55.9N, 80 32.9W. Fl R 2.5s, 35ft, 5M. TR on triangular pyramidal structure on piles.

SNAKE CREEK

LIGHT 2, 24 56.5N, 80 34.9W. Fl R 4s, 16ft, 4M. TR on pile.
Daybeacon 1, SG on pile.
Daybeacon 3, SG on pile.
Daybeacon 4, TR on pile.
Daybeacon 4A, TR on pile.
Daybeacon 5, SG on pile.
Daybeacon 6, TR on pile.
Daybeacon 6A, TR on pile.
Daybeacon 7, SG on pile.
Daybeacon 7A, SG on pile.
Daybeacon 8, TR on pile.
Daybeacon 9, SG on pile.
Daybeacon 10, TR on pile.
LIGHT 12, Fl R 2.5s, 16ft, 3M. TR on pile.

WHALE HARBOR AND WINDLEY PASS

LIGHT 1, 24 55.8N, 80 35.7W. Fl G 2.5s, 14ft, 3M. SG on pile.

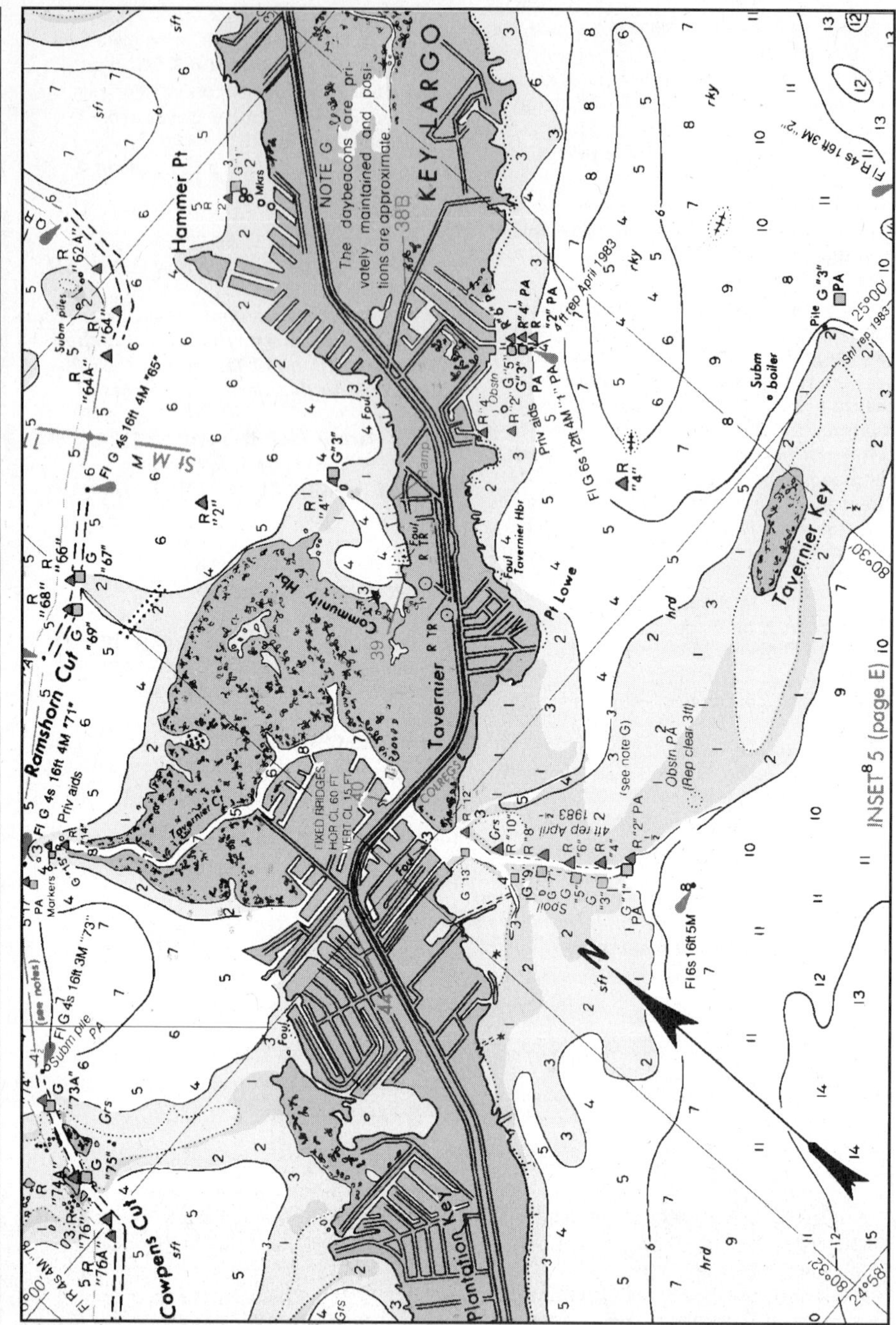

Key Largo, FL, soundings in feet *(from NOAA 11451)*

NOTE: A series of nine more red triangular and green square daybeacons on piles, numbered from 2 to 9, mark Whale Harbor channel to Windley Pass, which is marked by five more alternating red and green daybeacons, numbered 11 through 16. The Windley Pass beacons are private aids.

HOLIDAY ISLE MARINA

Daybeacon 1, 24 56.3N, 80 36.5W. SG on pile. Private aid.

NOTE: A series of five more red triangular and green square daybeacons, numbered 2 through 6, mark the Holiday Isle Marina channel. All are private aids.

BEACON REEF PIER LIGHT, 24 55.8N, 80 37.0W. Q W, 10ft. On pile. Private aid.

Upper Matecumbe Daybeacon 41, SG on pile.

LA SIESTA RESORT

Daybeacon 1, SG on pile. Private aid.

NOTE: A series of three more alternating red and green daybeacons on piles, numbered 2 through 4, mark the channel to La Siesta Resort. All are private aids.

BUD AND MARY'S FISHING MARINA

Daybeacon 1, 24 53.5N, 80 39.2W. SG on pile. Private aid.

NOTE: A series of five more red triangular and green square daybeacons on piles, beginning with number 5 and ending with 11, mark this channel. All are private aids.

YELLOW SHARK CHANNEL

Buoy 15, 24 53.9N, 80 39.7W. Green can. Private aid.

NOTE: A series of eleven more alternating green cans and red nuns, beginning with number 15 and ending with 27, mark this channel. All are private aids.

Tea Table Key Daybeacon 42, TR on pile.

TEA TABLE, PLANT, RACE, AND SHELL KEY CHANNELS

Buoy 1, 24 52.8N, 80 39.6. Green can. Private aid.

NOTE: A series of alternating green and red aids, all private, mark Tea Table Channel, Plant Channel, Shell Key Channel, and Race Channel. The aids in Tea Table Channel are daybeacons on piles. Aids in Plant, Shell Key, and Race channels are all buoys.

Alligator Reef Daybeacon 43, SG on pile.

INDIAN KEY CHANNEL

Daybeacon 1, 25 52.5N, 80 40.3W. SG on pile. Private aid.

NOTE: A series of seventeen more red triangular and green square daybeacons on piles, numbered consecutively from 2 through 18, mark Indian Key Channel. All are private aids.

LIGNUMVITAE AND WHEEL DITCH CHANNELS

NOTE: A series of alternating green and red private aids mark Lignumvitae Channel and Wheel Ditch Channel. The Lignumvitae Channel aids are daybeacons; the Wheel Ditch Channel aids are buoys. Between daybeacons 8 and 9 in Lignumvitae Channel are Shoal Buoys A (24 53.2N, 80 41.3W) and B (24 53.2N, 80 41.5W), which display orange and white bands with an orange diamond. These two are also private aids.

CALOOSA COVE CHANNEL

LIGHT 2, 24 50.1N, 80 44.8W. Fl R 6s, 10ft. TR and NL on dolphin, NL worded CALOOSA COVE. Private aid.

NOTE: A series of sixteen more red triangular and green square daybeacons, beginning with number 4 and ending with 21, mark the Caloosa Cove channel. All are private aids.

CHANNEL FIVE

Daybeacon 1, SG on pile.

LIGHT 2, 24 49.6N, 80 46.4W. Fl R 4s, 16ft, 4M, TR on pile.

OCEANSIDE ISLE CHANNEL

Buoy 1, 20 40.0N, 80 47.9W. Green can. Private aid.

NOTE: Five more alternating red nuns and green cans, all private aids, mark the Oceanside Isle Channel.

LAYTON CANAL CHANNEL

Daybeacon 1, SG on pile. Private aid.

Daybeacon 2, TR on pile. Private aid.

Daybeacon 3, SG on pile. Private aid.

ALLIGATOR REEF TO SOMBRERO KEY

Tennessee Reef East Lighted Buoy 18, 24 47.5N, 80 41.6W. Fl R 2.5s, 5M. Red.

TENNESSEE REEF LIGHT, on W side of shoal, 24 44.7N, 80 46.9W. Fl W 4s, 49ft, 8M. Small black house on hexagonal pyramidal skeleton tower on piles.

COFFINS PATCH LIGHT 20, 24 40.5N, 80 57.4W. Fl R 6s, 24ft, 6M. TR on dolphin.

SOMBRERO KEY LIGHT, 24 37.6N, 81 06.6W. Fl (5) W 60s (3 R sectors), 142ft, W 15M, R 12M. Brown octagonal pyramidal skeleton tower enclosing stair cylinder and square dwelling, pile foundation. W 222°–238°, 264°–066°, and 094°–163°, R in intervening sectors.

HAWK CHANNEL: LONG KEY TO MARATHON

LONG KEY LIGHT 44, FL R 6s, 4M. TR on pile.

DUCK KEY INLET CHANNEL

LIGHT 1, Fl W 4s. SG on pile. Private aid.

NOTE: A series of six more red triangular and green square daybeacons, numbered 2 through 8, mark the Duck Key Inlet channel. All are private aids.

Duck Key Channel Daybeacon 2, TR on pile. Private aid.

Duck Key Channel Daybeacon 4, TR on pile. Private aid.

LITTLE CRAWL KEY CHANNEL

Daybeacon 1, 24 44.4N, 80 58.2W. SG on pile. Private aid.

NOTE: A series of twelve more red triangular and green square daybeacons, beginning with number 2 and ending with 15, all private aids, mark Little Crawl Key Channel.

EAST TURTLE SHOAL LIGHT 45, 24 43.5N, 80 56.0W. Fl G 4s, 27ft, 5M. SG on tower.

West Turtle Shoal Daybeacon 47, SG on pile.

COCO PLUM CHANNEL

Daybeacon 1, 24 43.5N, 80 59.6W. SG on pile. Private aid.

NOTE: Nine more alternating red and green daybeacons on piles mark the Coco Plum Channel. All are private aids.

KEY COLONY BEACH

East Channel Daybeacon 2A, TR on pile. Private aid.

East Channel Daybeacon 3A, SG on pile. Private aid.

West Channel Daybeacon 1, SG on pile. Private aid.

West Channel Daybeacon 3, SG on pile. Private aid.

West Channel Daybeacon 6, TR on pile. Private aid.

Fat Deer Key Daybeacon 48, 24 41.5N, 81 01.5W. TR on pile.

EAST WASHERWOMAN SHOAL LIGHT 49, on N side of shoal, 24 40.0N, 81 04.0W. Fl G 4s, 36ft, 5M. SG on black triangular pyramidal structure on piles.

SISTER CREEK

LIGHT 2, 24 41.2N, 81 05.2W. Fl R 4s, 16ft, 3M. TR on pile.

Daybeacon 3, SG on pile.

Daybeacon 4, TR on pile.

Daybeacon 6, TR on pile.

Daybeacon 8, TR on pile.

BOOT KEY HARBOR

LIGHT 1, 24 42.0N, 81 07.2W. Fl G 4s, 16ft, 4M. SG on pile.

NOTE: A series of thirteen red triangular and green square daybeacons, beginning with number 2 and ending with 21, mark Boot Key Harbor. A series of seven private daybeacons, beginning with number 1 and ending with 9, mark the channel in the eastern part of the harbor.

MARATHON

24 42N, 81 06W

U.S. Coast Guard: Station on the Florida Bay (north) side of Vaca Key, east of Knight Key Channel.

Dockage: Many marinas line the shores of Boot Key Harbor.

Anchorage: Good anchorage in Boot Key Harbor.

Fuel: Many fuel docks.

Repairs: Extensive.

Supplies: Extensive. Haulout facilities.

Marathon is one of the few sheltered harbors on the Hawk Channel side of the Keys. Elaborate marine facilities provide most services. Nearby Moser Channel is one of only two channels through the Keys where high-level bridges allow masted vessels access to Florida Bay. The channel is a possible route for vessels drawing less than 8 feet headed north to the west coast of Florida. Ashore are many stores and restaurants. Marathon is the best town in the Keys to reprovision.

Marathon, FL, soundings in feet *(from NOAA 11451)*

SOMBRERO KEY TO KEY WEST

BIG PINE SHOAL LIGHT 22, on seaward edge of reef, 24 34.1N, 81 19.5W. Fl R 2.5s, 16ft, 6M. TR on dolphin.

LOOE KEY LIGHT 24, 24 32.8N, 81 24.2W. Fl R 4s, 20ft, 6M. TR on dolphin.

AMERICAN SHOAL LIGHT, 24 31.5N, 81 31.2W. Fl (3) W 15s (2 R sectors), 109ft, W 13M, R 10M. Brown octagonal pyramidal skeleton tower enclosing white stair cylinder and brown octagonal dwelling, pile foundation. W 270°–067°, R 067°–090°, obscured 090°–125°, W 125°–242°, R 242°–270°.

PELICAN SHOAL LIGHT 26, on seaward edge of reef, 24 30.3N, 81 36.0W. Fl R 6s, 22ft, 7M. TR on dolphin.

Eastern Sambo Daybeacon 28, 24 29.5N, 81 39.8W. TR on pile.

Western Sambo Daybeacon 30, 24 28.9N, 81 42.2W. TR on pile.

STOCK ISLAND APPROACH CHANNEL LIGHT 32, 24 28.4N, 81 44.5W. Fl R 2.5s, 19ft, 6M. TR on dolphin.

Key West Entrance Lighted Whistle Buoy KW, 24 27.7N, 81 48.1W. Mo (A) W, 6M. Red and white stripes with red spherical topmark.

HAWK CHANNEL: MOSER CHANNEL TO KEY WEST

MOSER CHANNEL

South Daybeacon 2, 24 41.1N, 81 10.0W. TR on pile.

South Daybeacon 4, TR on pile.

South Daybeacon 5, SG on pile.

South Daybeacon 6, TR on pile.

BAHIA HONDA KEY LIGHT 49A, 24 37.5N, 81 14.2W. Fl G 6s, 20ft, 6M. SG on piles.

NEWFOUND HARBOR KEYS LIGHT 50, 24 36.8N, 81 23.6W. Fl R 2.5s, 16ft, 4M. TR on pile.

NEWFOUND HARBOR CHANNEL

ENTRANCE LIGHT 2, 24 37.1N, 81 24.4W. Fl R 4s, 16ft, 3M. TR on dolphin.

Daybeacon 3, SG on pile.

Daybeacon 4, TR on pile.

Daybeacon 5, SG on pile.

Daybeacon 6, TR on pile.

Daybeacon 8, TR on pile.

RAMROD KEY

Approach Daybeacon 2, TR on pile. Private aid.

Shoal Daybeacon, 24 31.5N, 81 23.7W. NW on pile, worded SHOAL. Private aid.

Channel Daybeacon 1, SG on pile. Private aid.

Channel Daybeacon 2, TR on pile. Private aid.

Channel Daybeacon 3, SG on pile. Private aid.

Channel Daybeacon 5, SG on pile. Private aid.

NILES CHANNEL

Daybeacon 4, 24 37.4N, 81 24.9W. TR on pile.

Daybeacon 5, SG on pile.

Daybeacon 6, TR on pile.

PURITA DOCK OBSTRUCTION LIGHTS (2), 24 39.0N, 81 26.2W. Fl W 4s, 13ft. On pier. Private aids.

North Shoal Daybeacon, NW on pile.

South Shoal Daybeacon, 24 39.7N, 81 25.9W. NW on pile, worded SHOAL.

Shoal Buoy, 24 42.0N, 81 25.9W. Orange and white bands, worded DANGER SHOAL, can. Private aid.

LOGGERHEAD KEY LIGHT 50A, 24 35.8N, 81 27.3W. Fl R 6s, 16ft, 4M. TR on pile.

KEMP CHANNEL AND SUMMERLAND KEY

NOTE: This channel is marked by a series of twenty red triangular and green square daymarks, beginning with number 1 and ending with 25. Between daybeacons 17 and 18, a privately marked channel leads to Summerland Key (Daybeacon 1, 24 36.5N, 81 30.7W). Near the end of Kemp Channel is a white diamond-shaped Obstruction Daybeacon, displaying the words DANGER SUBMERGED PILES.

TROPICAL MARINA

Daybeacon 2, TR on pile. Private aid.

Daybeacon 4, TR on pile. Private aid.

Daybeacon 6, TR on pile. Private aid.

Daybeacon 8, TR on pile. Private aid.

PIRATES COVE,
UPPER SUGARLOAF SOUND,
SUGARLOAF KEY SHORES, AND CUDJOE KEY

NOTE: The Pirates Cove channel, marked by a series of ten red triangular and green square daymarks beginning with number 1, leads to the privately marked Upper Sugarloaf Sound, Sugarloaf Key Shores, and Cudjoe Key channels.

NINEFOOT SHOAL LIGHT, 24 34.1N, 81 33.1W. Fl W 2.5s, 18ft, 6M. NG on dolphin.

West Washerwoman Daybeacon 51, 24 33.3N, 81 33.8W. SG on pile.

West Washerwoman Daybeacon 53, SG on pile.

Pelican Key Daybeacon 55, SG on pile.

SADDLEBUNCH HARBOR

Daybeacon 1, 24 34.5N, 81 37.6W. SG on pile. Private aid.

NOTE: This channel is marked by eight more red triangular and green square daymarks, beginning with 2 and ending with 18. All are private aids.

TAMARAC PARK CHANNEL

Obstruction East Daybeacon, NW on pile. Private aid.

Obstruction West Daybeacon, NW on pile. Private aid.

NOTE: This channel is marked by five more red triangular and green square daymarks, beginning with 1 and ending with 6. All are private aids.

BOCA CHICA LIGHT 56, 24 33.1N, 81 41.2W. Fl R 4s, 14ft, 3M. TR on pile.

STOCK ISLAND
APPROACH CHANNEL

LIGHT 32, 24 28.4N, 81 44.5W. Fl R 2.5s, 19ft, 6M. TR on dolphin.

Daybeacon 2, TR on pile.

HAWK CHANNEL LIGHT 57, 24 31.9N, 81 45.5W. Fl G 4s, 16ft, 4M. SG on dolphin.

BOCA CHICA CHANNEL

LIGHT 1, 24 32.8N, 81 43.6W. Fl G 4s, 16ft, 4M. SG on pile.

Daybeacon 2, TR on pile.

Daybeacon 3, SG on pile.

Daybeacon 4, TR on pile.
Daybeacon 5, SG on pile.
Daybeacon 6, TR on pile.
Daybeacon 7, SG on pile.
LIGHT 8, Fl R 4s, 16ft, 3M. TR on pile.
Daybeacon 9, SG on pile.
Daybeacon 10, TR on pile.
Daybeacon 11, SG on pile.
Daybeacon 12, TR on pile.
Daybeacon 13, SG on pile.
Daybeacon 14, TR on pile.
Daybeacon 15, SG on pile.
LIGHT 16, Fl R 4s, 16ft, 3M. TR on pile.
Daybeacon 17, SG on pile.

STOCK ISLAND EAST CHANNEL

A Daybeacon 1A, 24 33.3N, 81 43.4W. SG on pile. Private aid.
A Daybeacon 3A, SG on pile. Private aid.
A Daybeacon 4A, TR on pile. Private aid.
A Daybeacon 5A, SG on pile. Private aid.
B Daybeacon 1B, 24 33.7N, 81 43.1W. SG on pile. Private aid.
B Daybeacon 2B, TR on pile. Private aid.
B Daybeacon 3B, SG on pile. Private aid.
B Daybeacon 4B, TR on pile. Private aid.
B Daybeacon 5B, SG on pile. Private aid.
B Daybeacon 6B, TR on pile. Private aid.
B Daybeacon 7B, SG on pile. Private aid.

SAFE HARBOR CHANNEL

LIGHT 2, 24 32.5N, 81 43.9W. Fl R 4s, 16ft, 3M. TR on dolphin.
LIGHT 3, Fl G 6s, 16ft, 4M. SG on dolphin.
LIGHT 4, Fl R 2.5s, 16ft, 3M. TR on pile.
Daybeacon 5, SG on pile.

KEY WEST
OCEANSIDE MARINA

NOTE: This channel is marked by ten alternating red triangular and green square daybeacons numbered from 1 through 10. All are private aids.

COW KEY CHANNEL

Daybeacon 1, 24 33.3N, 81 44.7W. SG on pile. Private aid.
NOTE: This channel is marked by eleven more red triangular and green square daybeacons, beginning with number 2 and ending with 20. All are private aids.

COW KEY WEST CHANNEL

Daybeacon 1A, 24 33.9N, 81 45.0W. SG on pile. Private aid.
Daybeacon 3A, SG on pile. Private aid.

CASA MARINA

Daybeacon 1, 24 32.4N, 81 47.4W. SG on pile. Private aid.
NOTE: This channel is marked by three more alternating red triangular and green square daybeacons numbered from 2 through 4. All are private aids.

KEY WEST HARBOR
MAIN CHANNEL

Entrance Lighted Whistle Buoy KW, 24 27.7N, 81 48.1W. Mo (A) W, 6M. Red and white stripes with red spherical topmark.
Lighted Bell Buoy 2, Fl R 4s, 3M. Red.
Lighted Buoy 3, Fl G 4s, 4M. Green.
RANGE (Front Light), 24 32.2N, 81 48.4W. Q G; Fl R 4s, 32ft, R 3M. KRW on skeleton tower on piles. Visible 2° each side of rangeline. Passing light obscured 326°–026°. **(Rear Light),** 1,310 yards, 356° from front. Iso G 6s, 75ft. KRW on skeleton tower on piles. Visible 2° each side of rangeline.
Lighted Bell Buoy 3A, 24 29.8N, 81 48.3W. Fl G 6s, 4M. Green.
Western Triangle Lighted Bell Buoy 5, at edge of shoal. Fl G 4s, 4M. Green.
Lighted Buoy 6, Fl R 2.5s, 4M. Red.
Lighted Buoy 7, Q G, 4M. Green.
CUT A RANGE (Front Light), 24 33.2N, 81 50.0W. Q G, 44ft. KRW on tower on piles. Visible 2° each side of rangeline. **(Rear Light),** 1,000 yards, 325° from front. Iso G 6s, 70ft. KRW on tower on piles. Visible 2° each side of rangeline.
Lighted Buoy 8, Q R, 3M. Red.
North Spoil Bank Buoy A, yellow can.
Lighted Buoy 9, Fl G 4s, 4M. Green.
CUT B RANGE (Front Light), 24 33.6N, 81 48.9W. Q R, 25ft. KRW on tower, Visible 4° each side of rangeline. **(Rear Light),** 300 yards, 003° from front. Iso R 6s, 40ft. KRW on tower. Visible 4° each side of rangeline.
Lighted Buoy 12, 24 32.0N, 81 48.9W. Q R, 3M. Red.
Lighted Buoy 13, Fl G 4s, 4M. Green.
Lighted Buoy 14, Q R, 3M. Red.
Lighted Buoy 15, Fl G 4s, 4M. Green.
KEY WEST HARBOR RANGE (Front Light), 24 34.7N, 81 48.0W. Q W, 19ft. KRW on dolphin. Visible all around, higher intensity on rangeline. **(Rear Light),** 781 yards, 024.3° from front. Iso W 6s, 36ft. KRW on piles. Visible 4° each side of rangeline.
Buoy 17, green can.

Buoy 19, green can.
TRUMAN ANNEX BARRIER PIER LIGHT, 24 33.4N, 81 48.5W. F W, 5ft. On pier. Private aid.
Daybeacon 21, SG on pile.
Lighted Buoy 23, Fl G 2.5s, 3M. Green.
Lighted Buoy 24, Fl R 4s, 3M. Red.
Lighted Buoy 25, Fl G 4s, 3M. Red.
TURNING BASIN LIGHT 27, 24 34.0N, 81 48.5W. Fl G 4s, 16ft, 4M. SG on dolphin.
Turning Basin Lighted Buoy 29, Fl G 4s, 4M. Green.
Turning Basin Daybeacon 31, SG on pile.

KEY WEST

24 34N, 81 48W
Port of entry
U.S. Coast Guard: Station located at Pier D2, north of main waterfront.
Dockage: Available at marinas in the former Truman Annex basin, Key West Bight, and Garrison Bight. Other marinas in the Key West area are on Stock Island. The marinas in the former Truman Annex basin and on Stock Island can accommodate deeper draft boats.
Anchorage: The best anchorage is just east of Wisteria Island. There may be some room near Light 2, but the holding is poor and the area crowded with moorings. A public dinghy landing is southwest of Light 2, in Key West Bight.
Fuel: Available in Key West Bight, Garrison Bight, and Stock Island.
Repairs: Good. Haulout facilities.
Supplies: Good. Walking distance to most shops.

Key West is the final stop for boats cruising the Keys. It has several well-marked approaches. The southern channel is the easiest. The Northwest Channel is shallower and has several hazards, including a submerged breakwater. Study the chart well if using the Northwest Channel after dark. The anchorage off Wisteria Island offers poor shelter and holding, but is within dinghy distance of town. If bad weather threatens, you are better off in a marina or some other harbor. This is no place to be in a hurricane. Ashore is a collage of everything touristy and entertaining. Live music, tie-dye shirts, open-air bars, and the Hemingway house all vie for your attention.

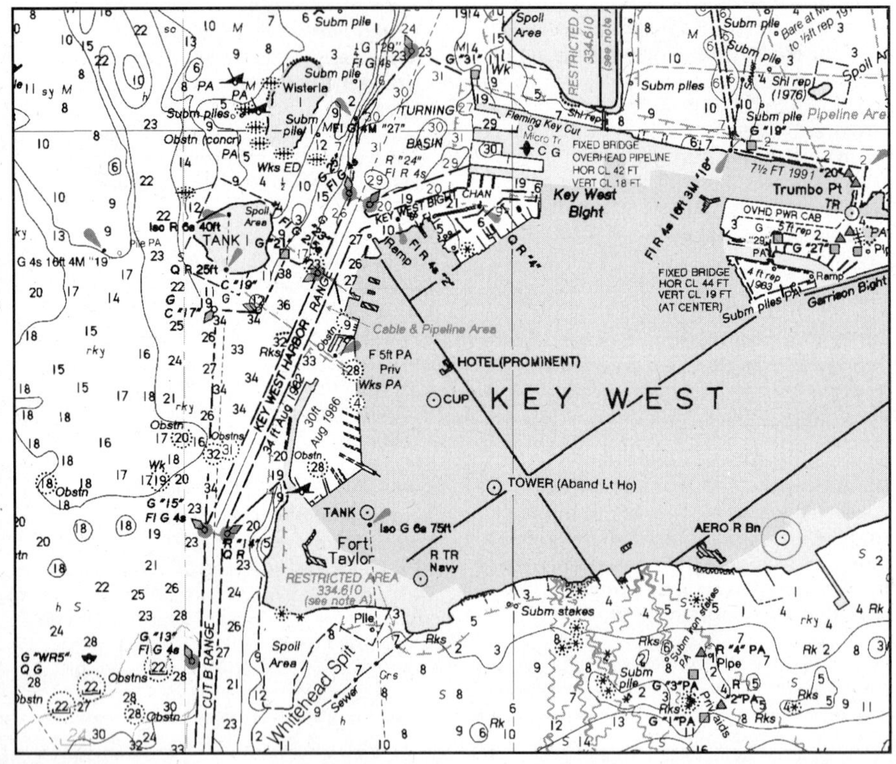

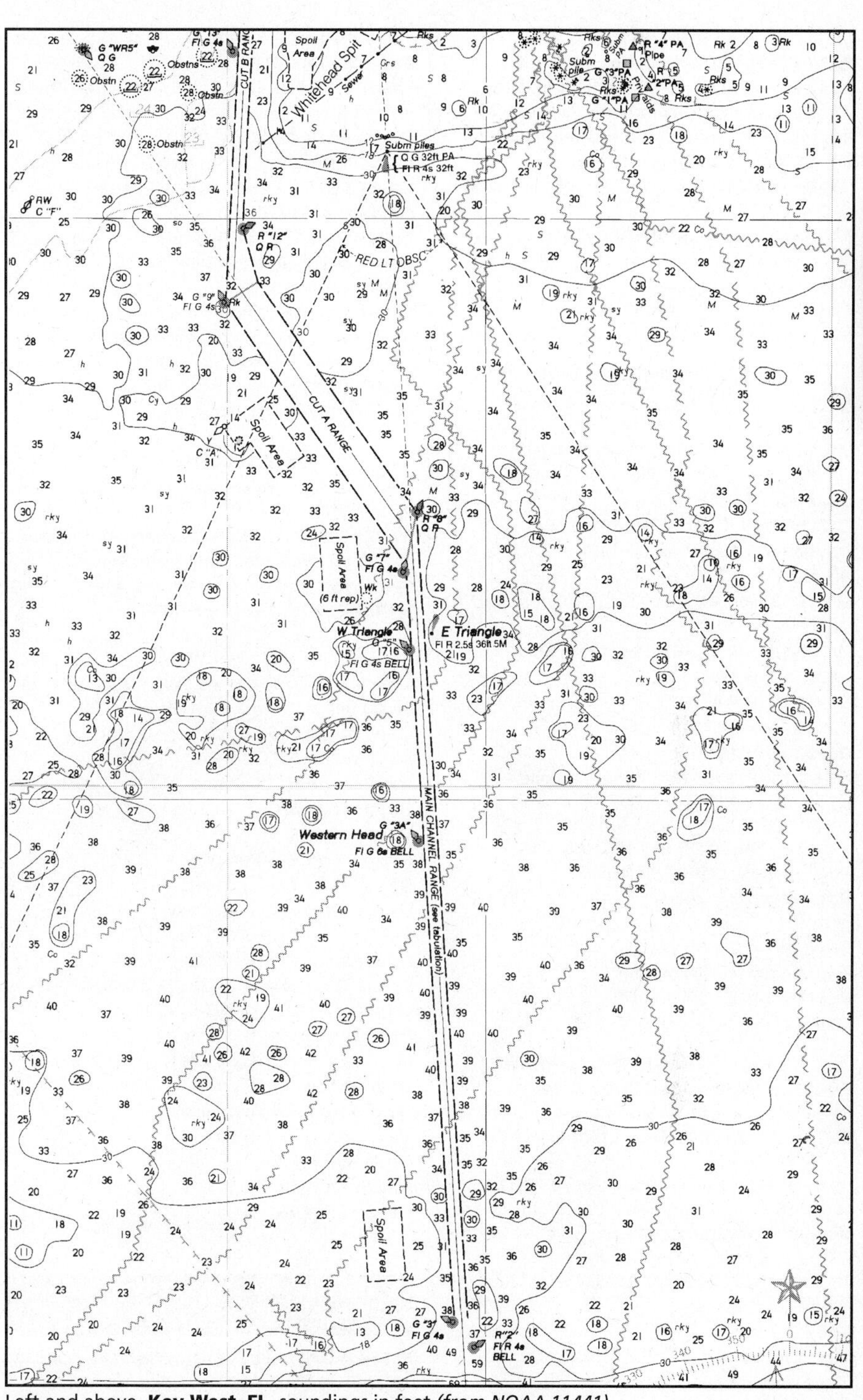

Left and above, **Key West, FL,** soundings in feet *(from NOAA 11441)*

P
Pilot
FLORIDA

FLEMING KEY

RANGE (Front Daybeacon), 24 34.5N, 81 47.8W. KRW on dolphin. Maintained by U.S. Navy. **(Rear Daybeacon),** 45 yards, 256° from front. KRW on dolphin. Maintained by U.S. Navy.
Daybeacon 2, TR on pile. Maintained by U.S. Navy.
KEY WEST BIGHT CHANNEL LIGHT 2, 24 33.7N, 81 48.3W. Fl R 4s, 16ft, 4M. TR on dolphin.
KEY WEST BIGHT CHANNEL LIGHT 4, Q R, 16ft, 4M. TR on dolphin.

GARRISON BIGHT CHANNEL

APPROACH LIGHT 2, 24 35.1N, 81 48.3W. Fl R 4s, 16ft, 3M. TR on pile.
LIGHT 3, Fl G 4s, 16ft, 4M. SG on pile.
Daybeacon 4, TR on pile.
Daybeacon 6, TR on pile.
LIGHT 8, Fl R 4s, 12ft, 3M. TR on pile.
Daybeacon 10, TR on pile.
Daybeacon 12, TR on pile.
LIGHT 13, Fl G 4s, 16ft, 4M. SG on dolphin.
Daybeacon 14, TR on pile.
LIGHT 16, Fl R 4s, 16ft, 3M. TR on pile.
LIGHT 17, Fl G 4s, 13ft, 4M. SG on pile.
LIGHT 18, Fl R 4s, 16ft, 3M. TR on pile.
Daybeacon 19, SG on pile.
Daybeacon 20, TR on pile.
LIGHT 21, Fl G 4s, 16ft, 4M. SG on pile.
Daybeacon 22, TR on pile.

GARRISON BIGHT CHANNEL TURNING BASIN

Daybeacon 24, 24 33.7N, 81 47.0W. TR on pile.
Daybeacon 25, SG on pile.
Daybeacon 26, TR on pile.
Daybeacon 27, SG on pile.
Daybeacon 29, SG on pile.

KEY WEST NORTHWEST CHANNEL

Entrance Lighted Bell Buoy 1, 24 38.8N, 81 54.0W. Fl G 2.5s, 4M. Green.
ENTRANCE RANGE (Front Light 6), 24 37.9N, 81 53.8W. Q R, 22ft, 4M. TR and KRW on piles. Visible all around. **(Rear Light),** 787 yards, 166° from front. Iso R 6s, 36ft. KRW on triangular pyramidal structure on piles. Visible all around, higher intensity on rangeline.
JETTY LIGHT A, 24 38.4N, 81 53.6W. Q Y, 16ft, 7M. NY on dolphin.
Buoy 2, red nun.
Lighted Buoy 3, Q G, 5M. Green.
Buoy 4, red nun.
Lighted Buoy 5, Fl G 2.5s, 4M. Green.
Daybeacon 8, TR on pile.
Buoy 9, green can.
LIGHT 10, 24 36.9N, 81 52.6W. Fl R 4s, 12ft, 4M. TR on pile.
Daybeacon 11, SG on pile.
LIGHT 12, Fl R 2.5s, 16ft, 4M. TR on pile.
LIGHT 14, Fl R 6s, 16ft, 4M. TR on pile.
Daybeacon 15, SG on pile.
LIGHT 15A, Fl G 4s, 14ft, 4M. SG on pile.
Daybeacon 16, TR on pile.
LIGHT 17, Fl G 2.5s, 16ft, 4M. SG on pile.
LIGHT 18, on E point of shoals, 24 33.4N, 81 49.7W. Fl R 2.5s, 36ft, 4M. TR on red triangular pyramidal skeleton structure on piles.
LIGHT 19, on SE end of middle ground. Fl G 4s, 16ft, 4M. SG on pile.

LAKE PASSAGE CHANNEL

Daybeacon 1, 24 34.2N, 81 50.5W. SG on pile. Private aid.
NOTE: This channel is marked by thirteen more red triangular and green square daymarks, beginning with number 2 and ending with 18. All are private aids.

CALDA CHANNEL

LIGHT 1, 24 37.8N, 81 49.6W. Fl G 4s, 16ft, 4M. SG on pile.
Daybeacon 3, SG on pile.
Daybeacon 5, SG on pile.
Daybeacon 6, TR on pile.
Daybeacon 6A, TR on pile.
Daybeacon 8, TR on pile.
Daybeacon 9, SG on pile.
Daybeacon 11, SG on pile.
Daybeacon 12, TR on pile.
Daybeacon 13, SG on pile.
Daybeacon 14, TR on pile.
Shoal Daybeacon, 24 36.7N, 81 48.3W. NW on pile, worded SHOAL.
Daybeacon 16, TR on pile.
Daybeacon 17, SG on pile.
Daybeacon 18, TR on pile.
Daybeacon 20, TR on pile.
Daybeacon 21, SG on pile.
Daybeacon 22, TR on pile.
Daybeacon 24, TR on pile.
Daybeacon 25, SG on pile.

KEY WEST TO THE DRY TORTUGAS

Eastern Dry Rocks Obstruction Daybeacon, 24 27.7N, 81 50.6W. NW on pile, worded DANGER CORAL REEF. Private aid.

Get Your Free
REED'S Supplement

Dear REED'S Customer,
Please take a moment to complete this questionnaire so we can continue to make REED'S your most important nautical resource. Thank you!

[ ] Please send me my **free** annual supplement to REED'S 1998 Almanac.

Name ______________________________
Address ______________________________
City ______________ State ________ Zip ________

Type of boat owned? ______________________________

[ ] Power [ ] Sail Builder ______________ Length ________

In what type of boating activity do you participate? [ ] Daysailing [ ] Fishing
[ ] Racing [ ] Coastal Cruising [ ] Offshore [ ] Commercial [ ] Other ________

How many days of the year do you spend on the water?
[ ] <20 [ ] 20 - 39 [ ] 40 - 59 [ ] 60 - 89 [ ] > 90 [ ] All Year

Where do you go? ______________________________

Topics or features in REED'S that are of particular interest or use to you?
[ ] Coast Pilot [ ] Harbor Charts [ ] Tides/Currents [ ] Communications
[ ] Electronics Navigation [ ] Nautical Ephemeris [ ] Other ________

How many years have you been buying REED'S? [ ] 1 - 3 [ ] 4 - 6 [ ] 7 - 10 [ ] >10

Fold along dotted line

How many times throughout the year do you refer to REED'S?
[ ] <20 [ ] 20 - 50 [ ] 51 - 100 [ ] >100

Would you find it valuable for REED'S to issue supplements more frequently than once a year? [ ] Yes [ ] No How often? [ ] Semi-Annually [ ] Quarterly

What other areas would you like REED'S to cover in future editions?
[ ] North California [ ] South California [ ] Baja [ ] Mexico (Pacific Coast)
[ ] Hawaii [ ] South America [ ] Canada [ ] Other ________

Do you have a copy of REED'S Nautical Companion? [ ] Yes [ ] No

Would you like more information on REED'S Companion? [ ] Yes [ ] No

What improvements can you suggest for REED'S Almanacs? ________
____
____

Other comments ______________________________
____
____

Please see reverse side

Are you planning a charter? [ ] Yes [ ] No [ ] Bareboat [ ] Crewed

Average days spent on a charter per year? [ ] 3 - 7 [ ] 8 - 14 [ ] >14

Where do you intend to charter? [ ] Caribbean [ ] East Coast [ ] Europe
[ ] Pacific Northwest [ ] West Coast [ ] Mexico [ ] Other ________________

In the next 12 months what do you expect to spend on all boating expenses?
[ ] <$1,000 [ ] $1,000 - $3,000 [ ] $3,001 - $5,000 [ ] over $5,000

What equipment do you have/intend to buy in the next 12 months?

	Have	Intend to Buy		Have	Intend to Buy
Navigation	[ ]	[ ]	Hardware	[ ]	[ ]
Marine Radio	[ ]	[ ]	Rigging	[ ]	[ ]
Sails	[ ]	[ ]	Weather	[ ]	[ ]
Canvas	[ ]	[ ]	Inflatable	[ ]	[ ]
Fishing Gear	[ ]	[ ]	Outboard	[ ]	[ ]

Thank you for taking the time to help us improve REED'S Nautical Almanac!

Fold along dotted line and seal

NO POSTAGE
NECESSARY
IF MAILED IN THE
UNITED STATES

BUSINESS REPLY MAIL

FIRST CLASS MAIL PERMIT NO. 8237 BOSTON MA

POSTAGE WILL BE PAID BY ADDRESSEE

THOMAS REED PUBLICATIONS, INC.
13A LEWIS STREET
BOSTON MA 02113-9838

Rock Key Obstruction Daybeacon, 24 27.6N, 81 51.5W. NW on pile, worded DANGER CORAL REEF. Private aid.

SAND KEY LIGHT, 24 27.2N, 81 52.6W. Fl (2) W 15s, 40ft, 13M. NR on square skeleton tower.

Western Dry Rocks Daybeacon K, 24 26.8N, 81 55.6W. NB on pile.

SAND KEY CHANNEL

Daybeacon 2, 24 27.6N, 81 52.8W. TR on pile.

Middle Ground Daybeacon 3, 24 29.0N, 81 52.9W. SG on pile.

Preferred Channel Buoy, red and green bands, nun.

KEY WEST SOUTHWEST CHANNEL

Lighted Buoy SW, 24 26.6N, 81 58.8W. Mo (A) W, 5M. Red and white stripes with red spherical topmark.

Buoy A, red and white stripes, spherical topmark, can.

Buoy B, red and white stripes, spherical topmark, can.

Buoy C, red and white stripes, spherical topmark, can.

Buoy 2, red nun.

Buoy 3, green can.

Buoy 4, red nun.

Buoy D, red and white stripes, spherical topmark, can.

Buoy E, red and white stripes, spherical topmark, can.

Buoy F, red and white stripes, spherical topmark, can.

Buoy G, red and white stripes, spherical topmark, can.

Wreck Lighted Buoy WR5, 24 32.6N, 81 49.6W. Q G, 3M. Green.

Boca Grande Channel Daybeacon 1, 24 37.3N, 82 04.4W. SG on pile.

Boca Grande Channel Daybeacon 2, TR on pile.

Coalbin Rock Buoy CB, on S side of shoal, 24 27.0N, 82 05.3W. Red and green bands, nun.

COSGROVE SHOAL LIGHT, 24 27.5N, 82 11.1W. Fl W 6s, 54ft, 9M. On hexagonal pyramidal skeleton tower on piles.

Marquesas Rock Buoy MR, on S side of reef. Red and green bands, nun.

TWENTY-EIGHT FOOT SHOAL LIGHT, 24 25.8N, 82 25.3W. Fl W 4s, 53ft, 9M. On hexagonal pyramidal skeleton tower on piles.

HALFMOON SHOAL LIGHT WR2, 24 33.5N, 82 26.4W. Fl R 6s, 19ft, 4M. TR on dolphin.

REBECCA SHOAL LIGHT, 24 34.7N, 82 35.1W. Fl W 6s (R sector), 66ft, W 9M, R 6M. Square skeleton tower on brown pile foundation. R 254°–302°.

DRY TORTUGAS LIGHT, Loggerhead Key, 24 38.0N, 82 55.2W. Fl W 20s, 151ft, 20M. Conical tower, lower half white, upper half black. Emergency light of reduced intensity when main light is extinguished.

NORTHWEST CHANNEL TO THE DRY TORTUGAS

SMITH SHOAL LIGHT, on NE end of shoal, 24 43.1N, 81 55.3W. Fl W 6s, 54ft, 9M. On hexagonal pyramidal skeleton tower on piles.

ELLIS ROCK LIGHT, 24 38.9N, 82 11.0W. Fl W 2.5s, 16ft, 3M. NB on dolphin.

NEW GROUND ROCKS LIGHT, 24 40.0N, 82 26.6W. Fl W 4s, 19ft, 7M. NB on pile.

DRY TORTUGAS

Lighted Buoy A, 24 34.0N, 82 54.0W. Fl Y 2.5s, 4M. Yellow.

Buoy B, yellow can.

Lighted Buoy C, 24 34.0N, 82 58.0W. Fl Y 4s, 4M. Yellow.

Buoy D, yellow can.

Lighted Buoy E, 24 39.0N, 82 58.0W. Fl Y 6s, 5M. Yellow.

Buoy F, yellow can.

Buoy G, yellow can.

Lighted Buoy H, 24 43.0N, 82 54.0W. Fl Y 2.5s, 4M. Yellow.

Lighted Buoy I, 24 43.5N, 82 52.0W. Fl Y 4s, 4M. Yellow.

Buoy J, yellow can.

Lighted Buoy K, 24 43.5N, 82 48.0W. Fl Y 6s, 5M. Yellow.

Lighted Buoy L, 24 42.0N, 82 46.0W. Fl Y 2.5s, 4M. Yellow.

Lighted Buoy M, 24 40.0N, 82 46.0W. Fl Y 4s, 4M. Yellow.

Buoy N, yellow can.

Lighted Buoy O, 24 37.0N, 82 48.0W. Fl Y 6s, 5M. Yellow.

Buoy P, yellow can.

Buoy Q, yellow can.

PULASKI SHOAL LIGHT, on E side of shoal, 24 41.6N, 82 46.4W. Fl W 6s, 56ft, 9M. On hexagonal pyramidal skeleton tower on piles.

SOUTHEAST CHANNEL

LIGHT 1, 24 35.6N, 82 52.4W. Fl G, 2.5s, 19ft, 3M. SG on dolphin.

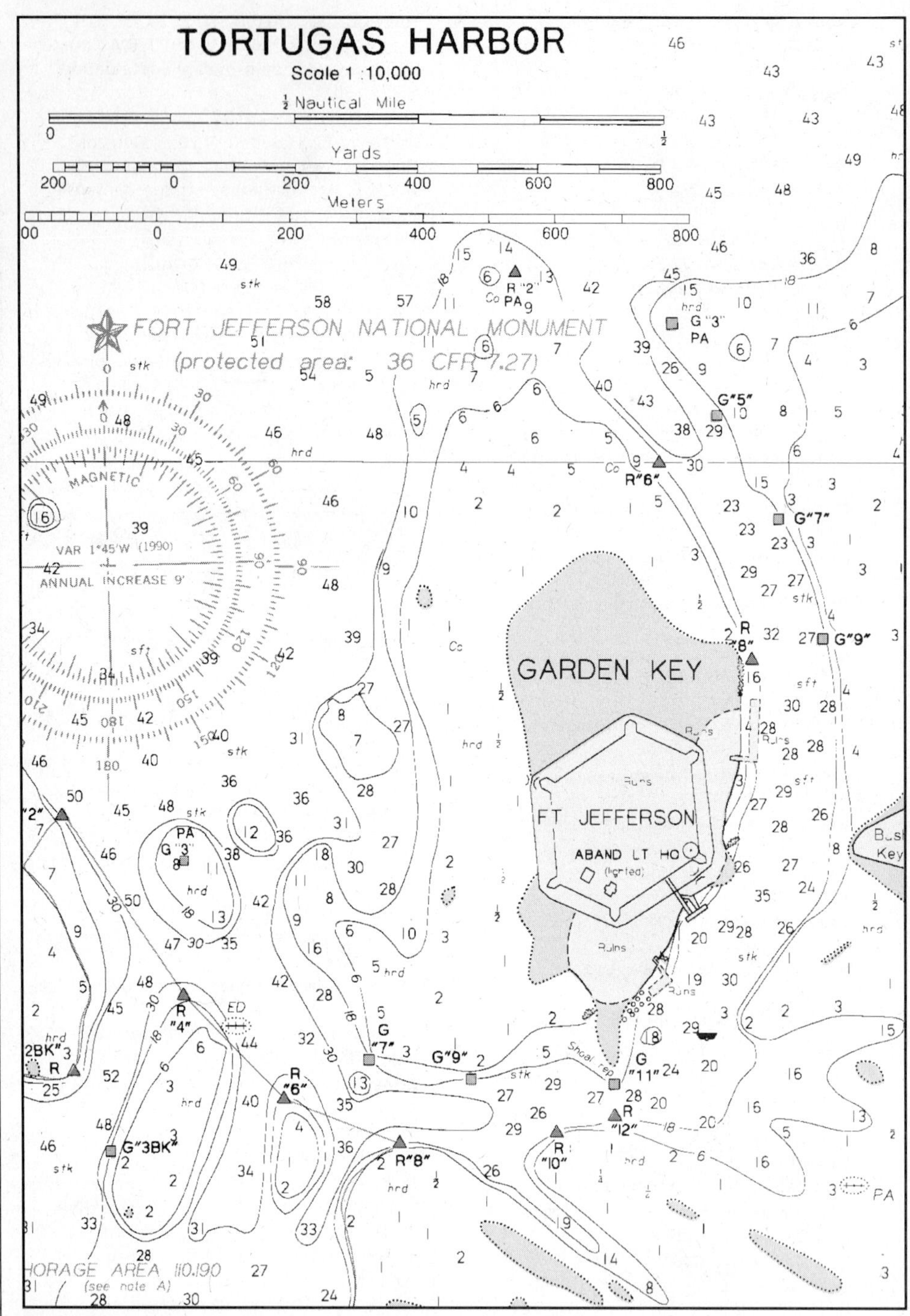

Dry Tortugas, FL, soundings in feet *(from NOAA 11438)*

Lighted Buoy 2, Fl R 4s, 3M. Red.
LIGHT 3, 24 38.3N, 82 51.7W. Fl G 4s, 16ft, 4M. SG on dolphin.
Daybeacon 4, TR on pile.
Middle Ground Daybeacon, on E side of shoal, 24 36.8N, 82 52.2W. JR on pile.

FORT JEFFERSON EAST CHANNEL

Daybeacon 2, 24 38.1N, 82 52.5W. TR on pile.
Daybeacon 3, SG on pile.
Daybeacon 5, SG on pile.
Daybeacon 6, TR on pile.
Daybeacon 7, SG on pile.
Daybeacon 8, TR on pile.
Daybeacon 9, SG on pile.

SOUTHWEST CHANNEL

Buoy 1, green can.
Daybeacon 3, 24 37.6N, 82 54.5W. SG on pile.
Daybeacon 4, TR on pile.
Daybeacon 6, TR on pile.
White Shoal North Daybeacon 7, SG on pile.

FORT JEFFERSON WEST CHANNEL

Daybeacon 2, 24 37.7N, 82 52.9W. TR on pile.
Daybeacon 3, SG on pile.
Daybeacon 4, TR on pile.
Daybeacon 6, TR on pile.
Daybeacon 7, SG on pile.
Daybeacon 8, TR on pile.
Daybeacon 9, SG on pile.
Daybeacon 10, TR on pile.
Daybeacon 11, SG on pile.
Daybeacon 12, TR on pile.

BIRD KEY HARBOR

Daybeacon 2BK, TR on pile.
Daybeacon 3BK, SG on pile.
Daybeacon 5BK, SG on pile.

DRY TORTUGAS

24 38N, 82 52W

Dry Tortugas National Park: Park headquarters and a Visitor Center are located in historic Fort Jefferson on Garden Key. Call on VHF channel 16. The NPS supervises, manages, and provides interpretive services for park islands, Fort Jefferson, and surrounding waters.
Dockage: Temporary landings may be made at the National Park Service dock at Fort Jefferson. There is no overnight dockage.
Anchorage: The principal anchorage is in Bird Key Harbor, southwest of Garden Key/Fort Jefferson, where the holding is good. Although there is no protection from high ground, the shoals break up most seas. The channel is marked by daybeacons. Other limited anchorages, depending on conditions, are northwest and southeast of Garden Key. The surrounding shallows are not marked, but may be easily seen when the sun is high. The dinghy beach on Garden Key is adjacent to the NPS dock. Anchoring in or damaging coral is prohibited. Vessels and dinghies in the vicinity of the fort should be alert for seaplanes, which use the adjacent waters for takeoffs and landings.
Fuel, Repairs, Supplies: None. No fresh water is available. There is no trash service; visitors are responsible for removing their own refuse.

Beautiful sandy beaches, crystal-clear water, and bountiful marine and birdlife reward visitors who make the often-rough 68-mile passage to the Dry Tortugas from Key West. The seven sand and coral islands of the Dry Tortugas ("dry" because there is no fresh water) are fully encompassed by the 100-square-mile Dry Tortugas National Park, and park regulations apply. An information packet is available from: National Park Service, Dry Tortugas National Park, P.O. Box 6208, Key West, FL 33041 (305-242-7700). The approaches are best made in daylight because of the many surrounding shoals that may be easily seen. Large areas of coral reef should be carefully avoided. The Southeast and Southwest channels are the principal approaches.

BAHAMAS

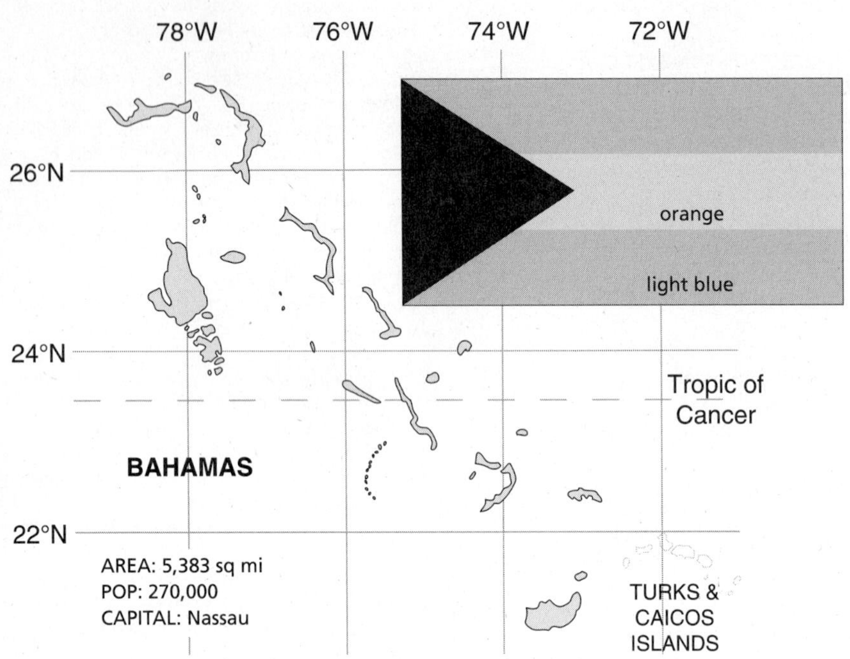

Please see cautions about chartlets and waypoints on page P 2.

ENTRY PROCEDURES

On arrival in the Bahamas, vessels must clear in with customs at one of the following ports of entry: on New Providence Island, Nassau; in the Abacos, Walkers Cay, Spanish Cay, Green Turtle Cay, Treasure Cay, Marsh Harbour, and Sandy Point; on Grand Bahama, Freeport Harbour, Bell Channel marinas, and West End; on Great Bahama Bank, Alice Town (North Bimini) and Cat Cay; in the Berry Islands, Great Harbour Cay and Chub Cay; on Andros, Fresh Creek (Andros Town), San Andros, and Congo Town; on Eleuthera, Harbour Island, Spanish Wells, Hatchet Bay, Rock Sound, and Governor's Harbour; in the Exumas, George Town; on Cat Island, Smith's Bay; on San Salvador, Cockburn Town; on Long Island, Stella Maris; and on Great Inagua Island, Matthew Town.

You should have proof of ownership of your boat. If you are planning additional stops in the Bahamas, request a cruising permit, for which there is a $10 charge. Vessels may remain in the Bahamas for up to one year duty free; duty exemptions may be granted for an additional two years upon application; application fees are $500 a year. Vessels with a cruising permit may bring in equipment and repair materials duty free. All firearms must be declared. Clearing-in costs during normal working hours at designated ports of entry are limited to the officer's transportation costs; outside office hours, overtime charges are also assessed. For more information, contact customs headquarters in Nassau (242-325-6551).

For U.S. and Canadian citizens, immigration requires either a passport or an original birth certificate or voter's registration card along with a photo identification for each person (Canadians without passports are allowed a three-week stay only). Other nationals should have passports; citizens of some countries may need visas. The immigration officer may use discretion in determining the period of stay, but in any event, no stays may exceed eight months. For more information, call immigration headquarters in Nassau (242-322-7530).

Fishing licenses for the boat are obtained when you clear customs. Licenses are $10 per vessel and are good for 180 days. There are stiff fines for fishing without a license. For more information on fishing rules and regulations, as well as a fishing guide, contact the Bahamas Tourism Office in Aventura, Florida (800-327-7678).

USEFUL INFORMATION

Language: English.
Currency: Bahamian dollar (B$), on par with the U.S. dollar.
Telephones: Area code, 242, became effective on January 1, 1997. Local numbers are seven digits.
Public holidays (1998): New Year's Day, January 1; Good Friday, April 10; Easter Monday, April 13; Whit Monday, June 1; Labour Day, June 7; Independence Day, July 10; Emancipation Day, August 4; Discovery Day, October 13; Christmas Day, December 25; Boxing Day, December 26.
Tourism offices: *In the Bahamas,* Ministry of Tourism, P.O. Box N 3701, Nassau (tel., 242-322-7500; fax, 242-328-0945). *In the U.S.:* Bahamas Tourism Office, 150 East 52nd St., 28th Floor N, New York, NY 10022 (tel., 800-422-4262, 212-758-2777; fax, 212-753-6531). Offices also in Atlanta, Charlotte (NC), Chicago, Dallas, Los Angeles, and Miami (Aventura, 800-327-7678). *In Canada:* Bahamas Tourism Office, 121 Bloor St. East, Ste. 1101, Toronto, Ont. M4W 3M5; (tel., 800-667-3777, 416-968-2999; fax, 416-968-6711).
Emergencies, Bahamas Air-Sea Rescue Association (BASRA): Maritime emergencies in the Bahamas are coordinated by BASRA, an association of volunteers throughout the main islands who stand by on VHF channel 16, on 2182 kHz, and on 4125 kHz. The Nassau headquarters, accessible by way of its dinghy dock just west of Paradise Island Bridge and Potter's Cay, can be reached by phone seven days a week from 9 am to 5 pm at 242-325-8864 (for after-hours emergencies, dial the police answering service at 242-322-3877). BASRA has a close working relationship with the U.S. Coast Guard; if you can't raise BASRA directly, try hailing the U.S. Coast Guard on VHF 16 or 2182 kHz. For more information, contact: BASRA, P.O. Box SS-6247, Nassau, Bahamas. Membership in BASRA costs $30/year.

WEST END AND LITTLE BAHAMA BANK

MEMORY ROCK LIGHT, 26 56.8N, 79 06.8W. Fl W 3s, 37ft, 11M. Metal tower, 23ft.
INDIAN CAY LIGHT, 26 43.0N, 79 00.1W. Fl W 6s, 40ft, 8M. Aluminum tower, 36ft.
SETTLEMENT POINT LIGHT, W end of island, 26 41.5N, 78 59.9W. Fl W 4s, 44ft, 6M. White steel tower, 32ft.
LITTLE SALE CAY LIGHT, 27 02.5N, 78 10.5W. Fl W 3s, 47ft, 9M. Metal tower, 19ft. Unreliable.

WEST END

26 42N, 78 59W
Port of entry: Customs is located in the yacht basin, near the marina docks. You may tie up in the marina and walk to the office.
Dockage: Available at the Jack Tar Marina (VHF channel 16; tel., 242-346-6211; fax, 242-346-6546). Entering the marina after dark may be difficult. Reservations are not accepted; slips are available on a first-come, first-served basis.
Anchorage: Possible between North Point and Settlement Point, but the holding is poor. This is a good anchorage if arriving after dark. Most vessels anchor outside the channel into the marina.
Services: Fuel, water, electricity, ice, and showers are available. Some groceries are available in town. Repair services and more extensive supplies are available in Freeport.

This is a convenient port of entry for many vessels traveling to the northern Bahamas. At night the tall television tower makes a good leading light. Approaching the marina, the first channel to the south leads to the commercial harbor. The second channel leads into the yacht facilities. The passages onto the Little Bahama Bank are very shoal here. Depths of less than 5 feet have been reported in the channel north of Indian Cay. Deeper-draft boats should enter the bank near Memory Rock.

WALKER'S CAY

27 16N, 78 24W
Port of entry: The marina can contact customs officials, who may be at the adjacent airstrip.

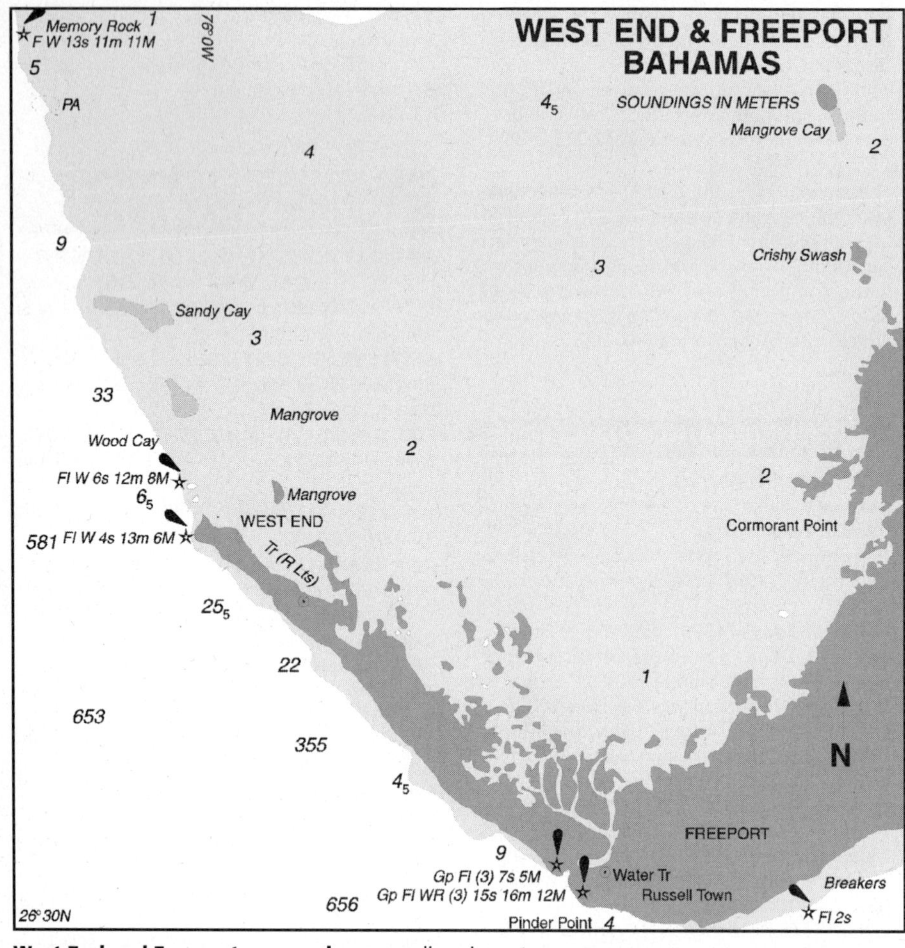

West End and Freeport approaches, soundings in meters

Dockage: Walker's Cay Marina (reservations, from U.S. locations, 800-432-2092; or locally, VHF channels 16 and 68; tel., 242-352-5252) encompasses the protected basin on the south side of the island. Dock space fills quickly during fishing tournaments, and reservations are advised.

Anchorage: There may be some room to anchor west of the western breakwater forming the marina. This area is shallow.

Services: Fuel (diesel and gas), water, ice, fishing supplies, and some limited groceries are available. The resort has accommodations, a restaurant, and full dive operations.

This island and resort are dedicated to sport fishing. The entry channel is very shoal and shifting. Stakes are moved to mark the deep water, but an approach in good light is recommended. Approach depths have been reported as less than 5 feet from the banks' side.

GREAT ABACO

CRAB CAY LIGHT, Angel Fish Point, 26 55.6N, 77 36.3W. Fl W 5s, 33ft, 8M. Metal tower, 23ft.

SPANISH CAY

26 58 N, 77 33W

Port of entry: Currently (summer 1997) open only on weekends. Call ahead to the Spanish Cay Club dockmaster (VHF channel 16; tel., 242-365-0083), who will alert customs.

Dockage: Slips are available.

Anchorage: Discouraged by club management.

Services: Fuel, ice, water, a full-service dive

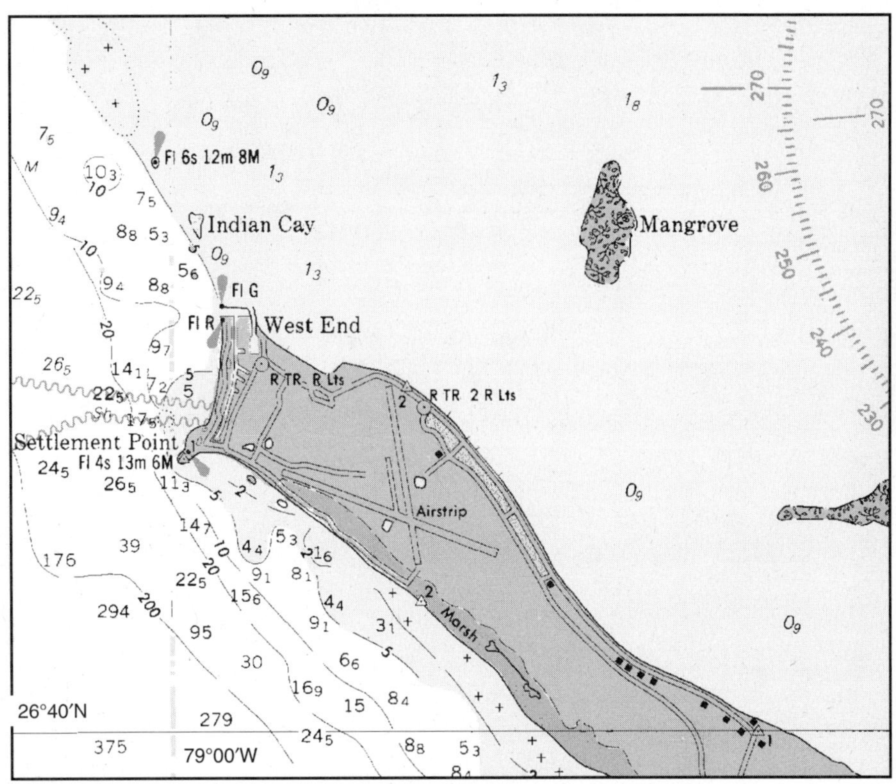

West End, soundings in meters *(from DMA 26323, 3rd ed, 8/86)*

shop, a ship's store with limited marine supplies, and resort services, including two restaurants, are available.

GREEN TURTLE CAY

26 46N, 77 20W

Port of entry: Proceed to the town dock at the village of New Plymouth and call Bahamas Customs (242-365-4077; airport office, 242-365-8602) 9 to 5 Monday through Friday; any work outside of those hours is overtime.

Dockage: Available in White Sound at the Bluff House Marina (tel., 242-365-4247; fax, 242-365-4248) and the Green Turtle Club (tel., 242-365-4271; fax, 242-365-4272) and in Black Sound at The Other Shore Club (242-365-4195) and Black Sound Marina (242-365-4531). All marinas monitor VHF channel 16. *Note: Space at Bluff House Marina will be limited until the end of 1997 while docks are torn out and replaced with upgraded services.*

Anchorage: Good anchorage is available in both Black Sound and White Sound. Some boats anchor west of the settlement of New Plymouth or in shallow Settlement Creek to the east of town.

Repairs: Haulout facilities and dry storage are available in Black Sound at Abaco Yacht Services (VHF channel 16; tel., 242-365-4033; fax, 242-365-4216).

Services: There is fuel at Bluff House Marina, Green Turtle Club, and The Other Shore Club. Water, electricity, ice, and showers are available at all marinas. The water is desalinated at the White Sound marinas. Black Sound Marina has some space set aside for dry and wet storage. New Plymouth settlement has a good selection of groceries, shops, and restaurants, plus banking services and marine hardware supplies.

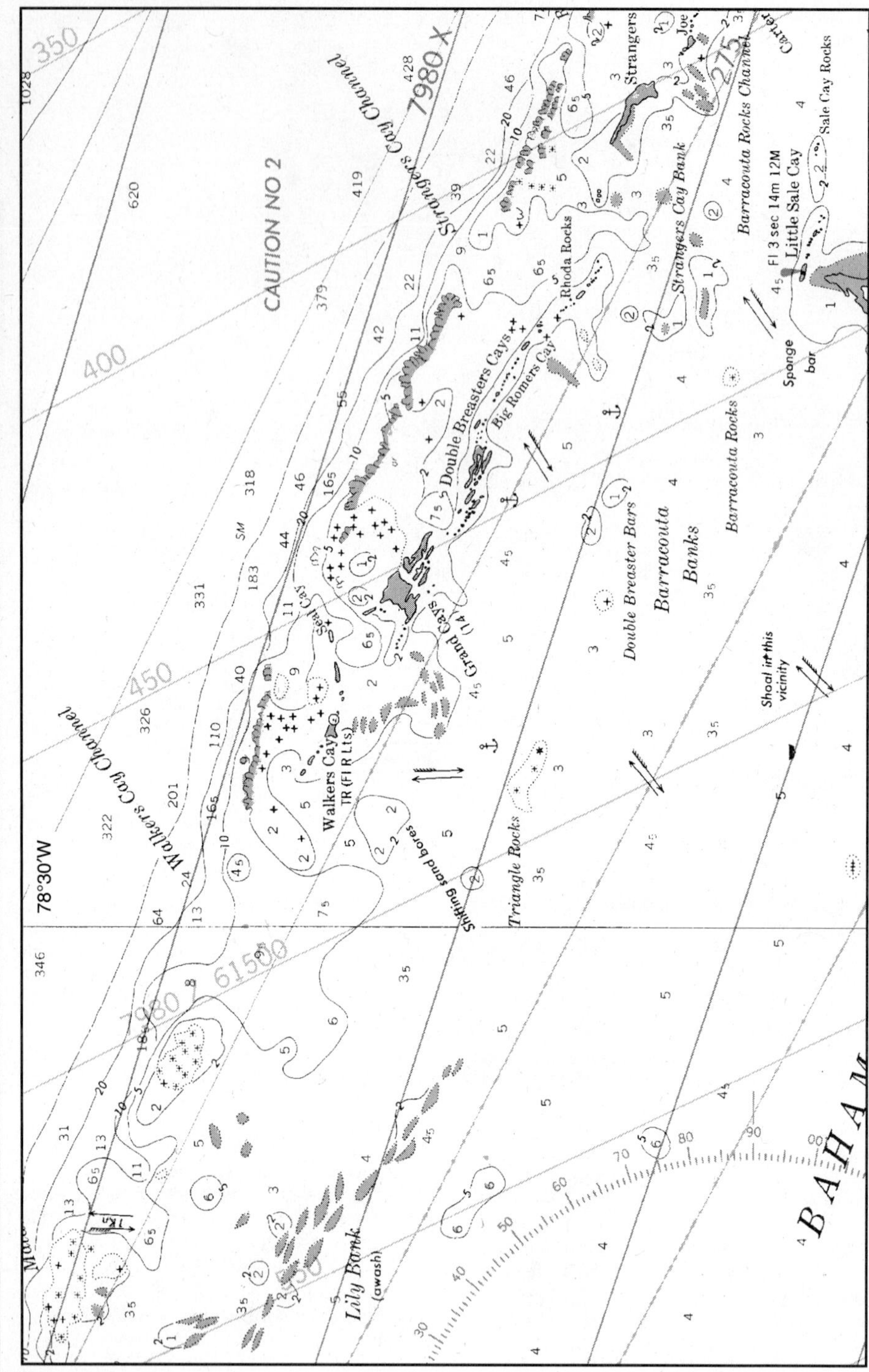

Walker's Cay, soundings in meters *(from DMA 26300, 6th ed, 7/81)*

GAMEFISH OF THE BAHAMAS

The following table will give you a general idea of the fishing prospects in the Bahamas. Check with Bahamian officials for the latest bag limits and other restrictions.

E = Excellent **G** = Good **F** = Fair **O** = Occasional **N** = None

	Jan	Feb	Mar	Apr	May	June	July	Aug	Sept	Oct	Nov	Dec
Blue Marlin	O	O	G	G	E	E	E	G	F	O	O	O
White Marlin	O	F	G	E	E	G	F	F	F	F	F	O
Sailfish	F	F	G	E	G	F	F	O	O	F	F	F
Swordfish	F	F	F	F	G	E	E	E	E	F	F	F
Dolphin	F	F	G	E	G	F	F	G	G	G	F	F
Wahoo	E	E	E	F	O	O	O	O	O	F	E	E
Kingfish	F	F	F	F	F	G	G	F	F	F	F	F
Mackerel	G	G	E	E	F	F	F	F	F	F	F	F
Allison Tuna	O	O	E	E	E	F	F	O	O	O	O	O
Blackfin Tuna	F	F	F	F	G	E	E	F	O	F	F	F
Bluefin Tuna	N	N	O	G	E	G	N	N	O	N	N	N
Bonito	F	F	F	F	G	E	E	F	N	F	F	F
Bonefish	G/E	G/E	E	E	G/E	G/E	G/E	G/E	G/E	G/E	G/E	G/E
Permit	F	F	G	E	E	E	E	G	F	F	G	G
Tarpon	O	O	O	F	G	G	F	F	O	O	O	O
Amberjack	F	F	E	E	E	E	E	E	F	F	F	F
Grouper	E	E	G	G	G	G	G	G	G	G	G	G
Snapper	G	G	G	E	E	E	E	E	E	G	G	G
Barracuda	G	G	G	G	G	E	E	E	G	G	G	G
Shark	G	G	G	E	E	E	G	G	G	G	G	G

BAHAMAS Pilot P

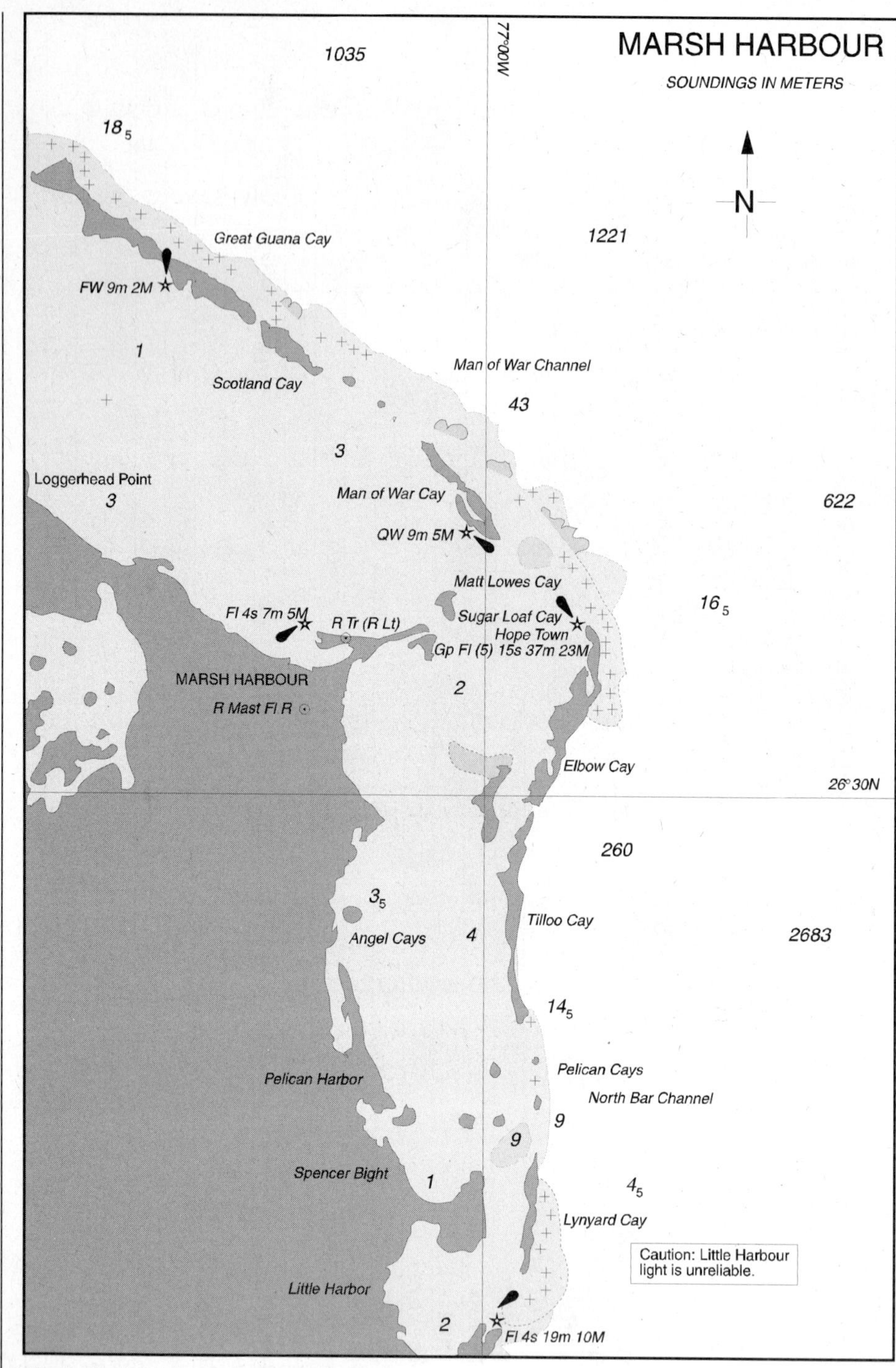

Marsh Harbour area, soundings in meters

WHALE CAY LIGHT, W point, 26 42.9N, 77 14.8W. Fl W 5s, 40ft, 8M. Black and aluminum structure.

WHALE CAY PASS

26 42N, 77 14W

CAUTION: This passage is dangerous during periods of heavy onshore seas and winds – conditions known locally as a "rage." **Large vessels have capsized while crossing Whale Cay Pass during a rage.**

A now-haphazardly marked ship channel formerly used by cruise ships leads in the direction of Guana Cay. To its south are several large spoil banks created from dredged material. Their size and location is changing, so careful navigation is necessary. The shallow "Don't Rock Passage" leading across the banks inside of Whale Cay to Great Abaco is very poorly defined and constantly changing; depths are reported to be less than 3 feet.

TREASURE CAY

26 40N, 77 17W
Port of entry: By special request only, with advance notice through the marina.
Dockage: Treasure Cay Marina (VHF channel 16; 242-365-8250; or in the U.S., 800-327-1584) offers luxurious dockage with all amenities.
Anchorage: It is possible to anchor outside the entrance channel to the marina; the better anchorage inside the entrance channel costs $8/night but permits use of marina facilities. Rental moorings are also available.
Services: Fuel, water, electricity, laundry, showers, and basic supplies are available. The resort attached to the marina has a restaurant and a few shops.

GREAT GUANA CAY LIGHT, 26 39.8N, 77 06.8W. F W, 30ft, 2M. White pole, 20ft.
MARSH HARBOUR, N side, Great Abaco Island, 26 33.1N, 77 03.4W. Fl W 4s, 23ft, 5M. White mast, 14ft.
MAN OF WAR CAY LIGHT, 26 35.2N, 77 00.0W. Q W, 30ft, 5M. White pole.

MARSH HARBOUR

26 33N, 77 03W
Port of entry: Clear in at the government wharf at the western end of the harbor (customs, 242-367-2525). A separate channel leads to the wharf. Alternatively, any of the marinas can arrange for customs to meet you at their docks.
Dockage: Several marinas offer dockage in the harbor: Conch Inn Marina (VHF channels 82, 16; tel., 242-367-4000; fax, 242-367-4004), Harbour View Marina (242-367-2182), Mangoes Marina (242-367-4255); Marsh Harbour Marina (tel., 242-367-2700; fax, 242-367-2033), and Triple J Marine (dock, 242-367-2287). Triple J is doubling its dock space to 32 slips and expects to have the work finished by late summer of 1997. On the Abaco Sound side is Boat Harbour Marina (242-367-2736; fax, 242-367-2819). All marinas monitor VHF channel 16.
Anchorage: Anchor anywhere depths and permanent moorings permit, without blocking the marked channels.
VHF radio weather: A new FM radio station (93.5FM), "Radio Abaco," broadcasts weather for cruisers in the Abacos at 9 a.m., 1 p.m., and toward the end of the 6 p.m. evening news (at approximately 6:25 p.m.). "Radio Abaco" also broadcasts local weather (Green Turtle Cay to Cherokee Sound) over VHF channel 68 at 0815 daily.
Services: All services are available except haulouts, which are undertaken in nearby Man of War Cay. This is the commercial center of the Abacos and is the best port for major stocking of groceries, doing laundry, obtaining parts, receiving mail, and arranging travel. Fresh water is available in large quantities.

Marsh Harbour is one of the largest towns in the Bahamas and the base for major and smaller charter operations. There is also a good selection of shops, restaurants, and nightlife ashore.

MAN OF WAR CAY

26 35N, 77 00W
Dockage: Available at Man-O-War Marina (VHF channel 16; tel., 242-365-6008) in the northern arm of the harbor. Approach depths at low tide are reported to be about 5 feet.
Anchorage: The southern harbor is a popular anchorage, where moorings may be available from private parties or Edwin's Boatyard. It may also be possible to anchor among the moored boats in the main (northern) harbor. Beware of shoal spots.
Repairs: This is one of the best places for major repairs in the Bahamas. There are haulout facilities and services for major repairs of all sorts. Edwin's Boatyard 1 and 2 are expert in wood repairs, and there is a sail loft and a well-stocked hardware store.
Services: All services are available. Most groceries are available, although the selection is not as great as in Marsh Harbour.

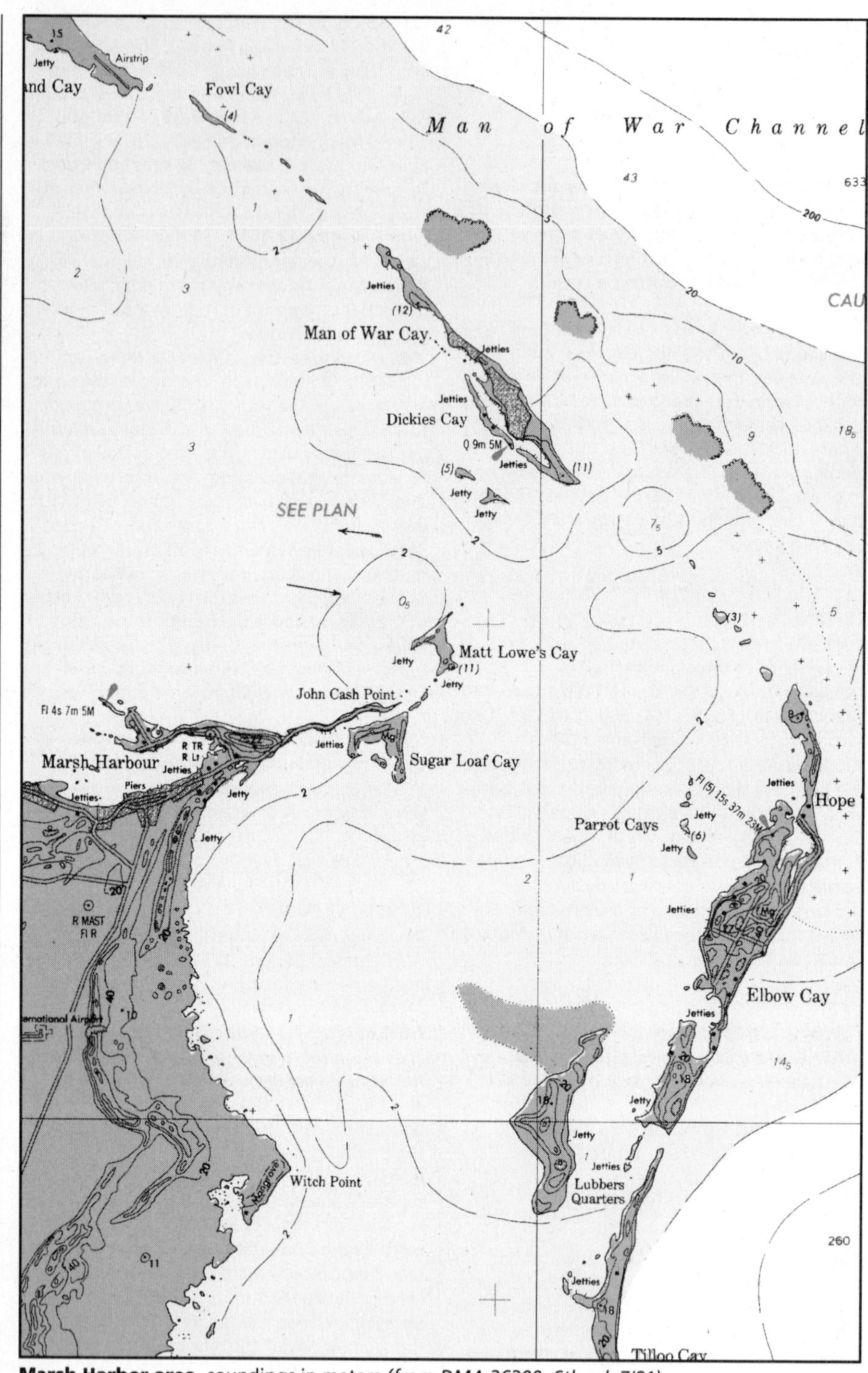

Marsh Harbor area, soundings in meters *(from DMA 26300, 6th ed, 7/81)*

ELBOW CAY LIGHT (Hope Town), 26 32.4N, 76 58.0W. Fl (5) W 15s, 120ft, 23M. White tower, red bands, white lantern, 89ft.

HOPE TOWN

26 33N, 76 58W

Dockage: Available at Club Soleil/Hope Town Marina (242-366-0003; fax, 242-366-0254) or Lighthouse Marina (tel., 242-366-0154; fax, 242-366-0171). The marinas monitor VHF channel 16. Approach depths have been reported as about 5 feet at low tide.

Moorings: A profusion of moorings now fill much of the harbor. Hope Town Marina/Club Soleil (VHF channel 16) controls most of the rental moorings and is able to take boats up to 48 feet. Abaco Yacht Charters (VHF channel 10; tel., 242-366-0151; fax, 242-366-0238) also rents out its moorings when its boats are out on charter.

Anchorage: It may be possible to find room among the moored boats, but be prepared for poor holding and limited swinging room.

Services: Fuel and marine repair services are available at Lighthouse Marina; water, ice, and electricity are at both marinas. There are several restaurants, a good selection of groceries, and a bakery.

SANDY CAY BEACON, 26 24.0N, 76 59.1W. Fl G 3s. Beacon.

LITTLE HARBOUR LIGHT, S side of entrance, Great Abaco Island, 26 19.7N, 76 59.6W. Fl W 4s, 61ft, 10M. White building. F W is shown when bar is dangerous.

LITTLE HARBOUR

26 20N, 77 00W

Moorings: For rental moorings, call "Pete's Pub" on VHF channel 16; the channel is usually monitored only in the late afternoons. The channel depths have been reported to be 3 to 4 feet.

Anchorage: Anchor off the docks or wherever depths allow.

Services: Meals and drinks are available at Pete's Pub. No other services are available.

Privately owned Little Harbour is the location of the famous Johnston Studios, where the late Randolph Johnston created lost-wax-process bronzes. Bronzes and jewelry are sold at the studios. This harbor is well sheltered in most weather, but the entrance channel is quite shoal. The pass from the ocean is tricky and should be attempted only in good light. If conditions are not ideal, enter the harbor by way of North Bar Channel, north of Lynyard Cay.

CHEROKEE SOUND LIGHT, on Duck Cay, 26 16N, 77 04W. F R, 29ft, 6M. White square stone building, 22ft. Visible 229°–094° except where obscured to the E by the high land of Cherokee Point.

ABACO LIGHT, HOLE IN THE WALL, S point of Great Abaco Island, 25 51.5N, 77 11.2W. Fl W 10s, 168ft, 23M. White tower, red top, white lantern, 93ft.

ROCKY POINT LIGHT, 25 59.9N, 77 24.3W. Fl W 6s, 33ft, 10M. Black metal framework, 27ft.

SANDY POINT LIGHT, near W extremity of Great Abaco Island, 26 01.6N, 77 24.0W. F W, 25ft, 5M. Pole, 20ft.

CHANNEL CAY LIGHT, 26 15.0N, 77 37.8W. Fl W 2.5s, 38ft, 7M. Black steel lattice structure. Reported extinguished (June 1981).

GRAND BAHAMA ISLAND, SOUTH SIDE

SWEETINGS CAY LIGHT, 26 36.7N, 77 54.0W. Fl W 6s, 23ft, 8M. Black tower.

RIDING POINT AVIATION LIGHT, 26 42.7N, 78 09.5W. Oc R 3s, 269ft. Red and white tower.

SOUTH RIDING POINT HARBOUR RANGE (Front Light), 26 37.5N, 78 13.1W. Oc G 3s, 39ft. Orange triangle on framework tower. **(Rear Light),** 60 meters, 340° from front.Oc G 3s, 52ft. Orange diamond on framework tower.

COMMUNICATIONS TOWER LIGHTS, 26 37.7N, 78 14.3W. 2 F R, 239ft. Red and white tower; lights in vertical line on tower.

PLATFORM LIGHT, 26 36.8N, 78 13.7W. Fl R 3s. Control building. Two Q R lights on east and west dolphins.

HIGH ROCK LIGHT, S side of Grand Bahama, 26 37.3N, 78 16.1W. F W, 25ft, 6M. White pole.

IONOSPHERIC TOWER AVIATION LIGHT, 26 37.1N, 78 18.8W. F R, Oc R 8s, 210ft.

BASSETT COVE TOWER AVIATION LIGHT, 26 36.8N, 78 19.4W. Q R, 2 F R, 407ft. Red and white tower; lights in vertical line.

BORE SITE TOWER AVIATION LIGHT, 26 36.7N, 78 20.9W. F W, 2 F R, Oc R 4s, 174ft. Red and white tower; lights in vertical line.

GRAND LUCAYAN WATERWAY LIGHT, W breakwater head, 26 32.4N, 78 33.4W. Fl (3) G 10s, 13ft, 3M. Concrete column, 7ft.

GRAND LUCAYAN WATERWAY LIGHT, E breakwater head, 26 32.4N, 78 33.3W. Fl (3) R 10s, 13ft, 3M. Concrete column, 7ft.
BAHAMIA MARINA LIGHT, Xanadu Marina, E breakwater head, 26 29.3N, 78 42.1W. Q R, 3M. Concrete pedestal.
BAHAMIA MARINA LIGHT, Xanadu Marina, W breakwater head, 26 29.3N, 78 42.2W. Q G, 3M. Concrete pedestal.

MARINAS EAST OF FREEPORT

26 29N 78 42W
Ports of entry: Pleasure boats are encouraged to clear at the Xanadu, Lucayan Marina Village, or Port Lucaya Marina rather than in the commercial port of Freeport. Marina staff notify customs officials of your arrival.
Dockage: Available at Xanadu Marina (242-352-6782), Running Mon Hotel and Marina (242-352-6834), Ocean Reef Resort (242-373-4662), Port Lucaya Marina Village (242-373-9090), and Lucayan Marina (242-373-8888). All monitor VHF channel 16. Ocean Reef Resort is a private club with limited room for nonmembers.
Services: All services are available in this area, which is a short bus or taxi ride from Freeport proper. Many tourist shops and restaurants are located near the marinas.

The first marked channel east of Freeport leads to Xanadu Marina (listed as Bahamia Marina on the charts). A little farther east is the entrance to Running Mon Marina; the entrance to the Ocean Reef Club is about 2.2 miles farther east at about 78° 39.6'W. Neither the Running Mon Marina nor the Ocean Reef Club is a customs port of entry. Off the waterway labeled Bell Channel are the Lucayan Marina Village and Port Lucaya marinas. About 4.5 miles east of Bell Channel is the entrance to the Grand Lucayan Waterway, a canal through Grand Bahama Island that provides a protected inside route to Little Bahama Bank for those who can clear the 27-foot fixed bridge.

FREEPORT INTERNATIONAL AIRPORT AVIATION LIGHT, 26 32.9N, 78 42.3W. Al W G 10s, 98ft, 40M.
FREEPORT LIGHT, Pinder Point, 26 30.4N, 78 45.9W. Fl (3) W R 15s, 54ft, 12M. White structure, black bands, 40ft. W 301°–113°, R 113°–301°. Occasional. Reported obscured to seaward by jetties and ships.

BORCO OIL TERMINAL NO. 1 JETTY LIGHT, SE end, 26 30.0N, 78 46.2W. Q W.
BORCO OIL TERMINAL NO. 1 JETTY LIGHT, NW end, 26 30.3N, 78 46.7W. Fl (3) W 7s, 5M. Dolphin.
BORCO OIL TERMINAL NO. 2 JETTY LIGHT, NW end, 26 30.5N, 78 46.7W. Q W. Dolphin.
FREEPORT WEST BREAKWATER LIGHT, 26 31.1N, 78 46.7W. Fl W 4s, 23ft, 2M. Metal structure, 10ft.
FREEPORT CHANNEL ENTRANCE, W side, 26 31.2N, 78 46.6W. Fl G 4s, 12ft, 2M. Metal structure.
FREEPORT CHANNEL ENTRANCE, E side, 26 31.2N, 78 46.5W. Fl R 4s, 12ft, 2M. Metal structure.

FREEPORT

26 31N 78 47W
Port of entry: Vessels may clear customs here, although the harbor is oriented to large commercial vessels. Pleasure boats usually clear at one of the marinas east of Freeport (see *Marinas East of Freeport*).
Freeport Harbour Control: Call on VHF channel 16 when approaching the harbor.
Dockage: Contact Freeport Harbour Control for details; pleasure boats may tie up free for a limited time only to clear customs. Most pleasure boats dock at one of the marinas east of Freeport.
Anchorage: None in the harbor.
Services: Fuel and water are available at the marinas to the east. Freeport has a good selection of supplies, including a variety of tourist oriented shops. The town is located well to the northeast of the harbor, a cab ride away.

Freeport is one of the leading commercial harbors in the Bahamas. Large cruise ships, cargo vessels, and oil tankers call here.

PINDER POINT RANGE (Front Light), 26 31.5N, 78 46.4W. F G, 26ft, 3M. Orange and white diamond-shaped daymark on beacon.
(Rear Light), 255 meters, 021°47′ from front. 26 31.6N, 78 46.4W. F G, 45ft, 3M. Orange and white diamond-shaped daymark on beacon.
SETTLEMENT POINT LIGHT, W end of island, 26 41.5N, 78 59.9W. Fl W 4s, 44ft, 6M. White steel tower, 32ft.

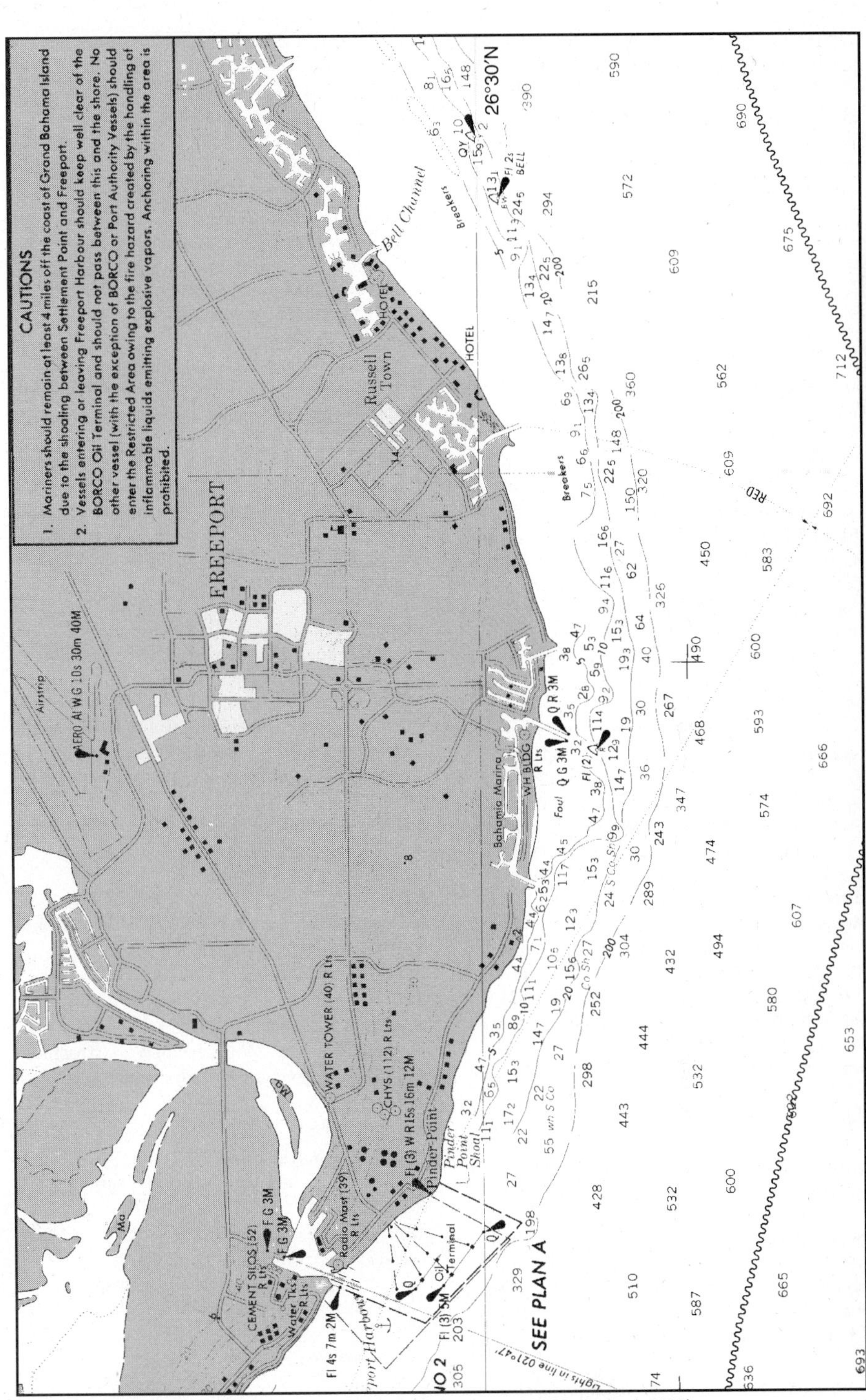

Freeport and Xanadu, soundings in meters *(from DMA 26323, 3rd ed, 8/86)*

BIMINI AND CAT CAY APPROACHES

GREAT ISAAC LIGHT, center of island, 26 01.8N, 79 05.4W. Fl W 15s, 152ft, 23M. White circular tower, 137ft.
NORTH ROCK LIGHT, 1.9 km N of North Bimini, 25 48.2N, 79 16.7W. Fl W 3s, 40ft, 8M. White skeleton steel tower, 29ft. Visible 022°–343°.
NORTH BIMINI ISLAND LIGHT, head of wharf, 25 43.7N, 79 19.1W. F W, 20ft, 5M. Gray steel framework tower. Two F R range lights indicate channel through reef on South Bimini. These lights are unreliable.
BIMINI AVIATION LIGHT, 25 42.5N, 79 16.3W. Mo (B) R 20s, 284ft, 23M. Steel framework tower.

BIMINI

25 43N, 79 18W
Port of entry: You may clear customs after securing your boat in a marina. Check with the dockmaster for customs forms, which you should take to the customs and immigration office near the ferry dock.
Dockage: Available at several marinas, among them Bimini Blue Water (VHF channel 68; tel., 242-347-3166; fax, 242-347-3293), Bimini Big Game Fishing Club and Hotel (VHF channel 16; tel., 242-347-3391; fax, 242-347-3392), and Weech's Bimini Dock (VHF channel 18; tel., 242-347-3028; fax, 242-347-3508). Services at the Bimini Big Game Fishing Club docks have been upgraded. Care should be taken when approaching any slip because of the strong currents running through the harbor.
Anchorage: Good anchorage is available off the marina docks toward the harbor's north end. Be careful not to block access to the marinas or the seaplane ramp. There are strong currents, and the bottom shoals rapidly to the east and north.
Services: Fuel, water, groceries, liquor, hardware, and some marine supplies are available.

Bimini is only 45 miles from Miami and makes a convenient first port of call for boaters. The entrance channel is frequently shifted to follow the deepest water. A privately maintained range on the South Bimini beach directs boats over the bar. The channel then parallels the beach running north into the harbor. If in any doubt, call one of the marinas on the VHF radio for directions. It is best to arrive with good light, when the shoals will be most visible.

GUN CAY LIGHT, near S point, 25 34.5N, 79 18.8W. Fl W 10s, 80ft, 23M. Conical stone tower, upper part red, lower part white. Obscured by Bimini between 176° and 198° when 8 miles distant. Unofficially reported unlit and much faded (1995.)

GUN CAY AND CAT CAY

25 34N, 79 18W
Port of entry: Clear customs at the docks on the east side of North Cat Cay. The Cay Cay Yacht Club dockmaster (VHF channel 16) will contact customs for you; a marina fee is charged to clear in here.
Dockage: Contact the Cat Cay Yacht Club on VHF channel 16 or at 242-347-3565—let it ring!
Anchorage: The best anchorages are on either side of Gun Cay, depending upon the weather. Honeymoon Harbor, on the north end of Gun Cay, is another favorite spot, but be careful not to get caught in a north wind.
Services: Fuel, water, electricity, a commissary, liquor, and a restaurant are available at the Cat Cay Yacht Club.

Transients may use the yacht club marina and restaurant for a 24-hour period, but otherwise are not allowed on the island. This is a good place to anchor before crossing the Great Bahama Bank. The channel between Gun Cay and North Cat Cay is easy when the sun is high. Hug the shore of Gun Cay closely until past the bar extending north from North Cat Cay. You can receive NOAA weather radio from Florida; it is advisable to cross the banks with a favorable forecast.

GREAT BAHAMA BANK

NORTH CAT CAY, head of breakwater, 25 34.0N, 79 18.3W. Fl W 2s, 10ft, 5M. Beacon, 12ft. Not visible W or SW of Cat Cay.

SOUTH RIDING ROCK LIGHT, 25 13.8N, 79 10.0W. Fl W 5s, 35ft, 11M. White framework structure.

SYLVIA BEACON, 25 28N, 79 01W. Fl W 5s, 33ft, 8M. Beacon on piles. (Unofficially reported missing, 1995.)

MACKIE SHOAL LIGHT, 25 41.2N, 78 39.2W. Fl W 2s, 15ft, 10M. White tripod, yellow top.

RUSSELL BEACON, 25 28.6N, 78 25.5W. Beacon destroyed; use caution as remains are dangerous.

Russell Lighted Buoy, 25 28.6N, 78 25.5W. Fl W 4s. Replaces Russell Beacon. Vessels anchor here at night. Use caution when approaching after dark.

NORTHWEST CHANNEL LIGHT, S side of E entrance, 25 27.2N, 78 09.8W. Fl W 5s, 33ft, 8M. White skeleton tower, 36ft. (Unofficially

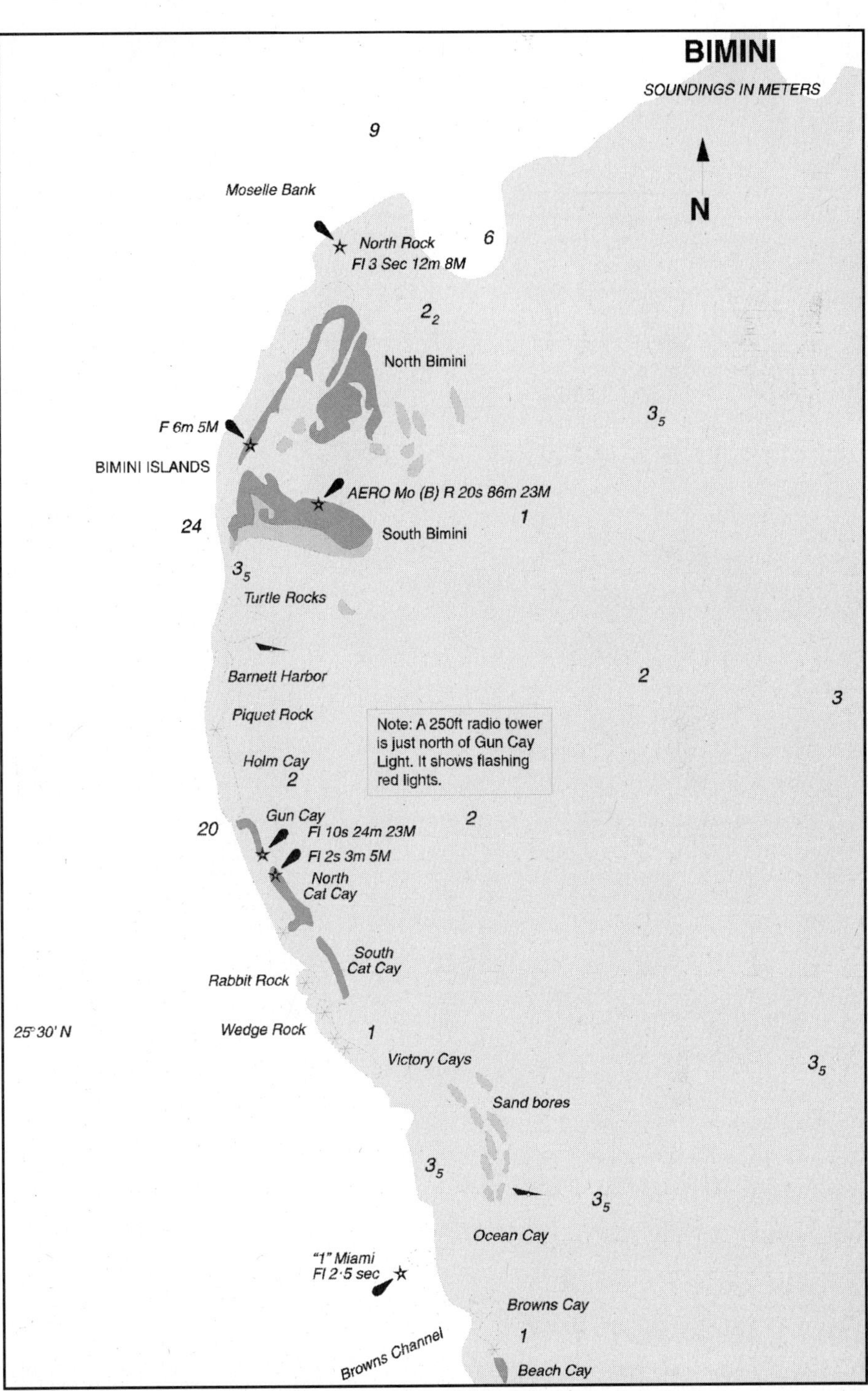

Bimini and Cat Cay approaches, soundings in meters *(after DMA 26324)*

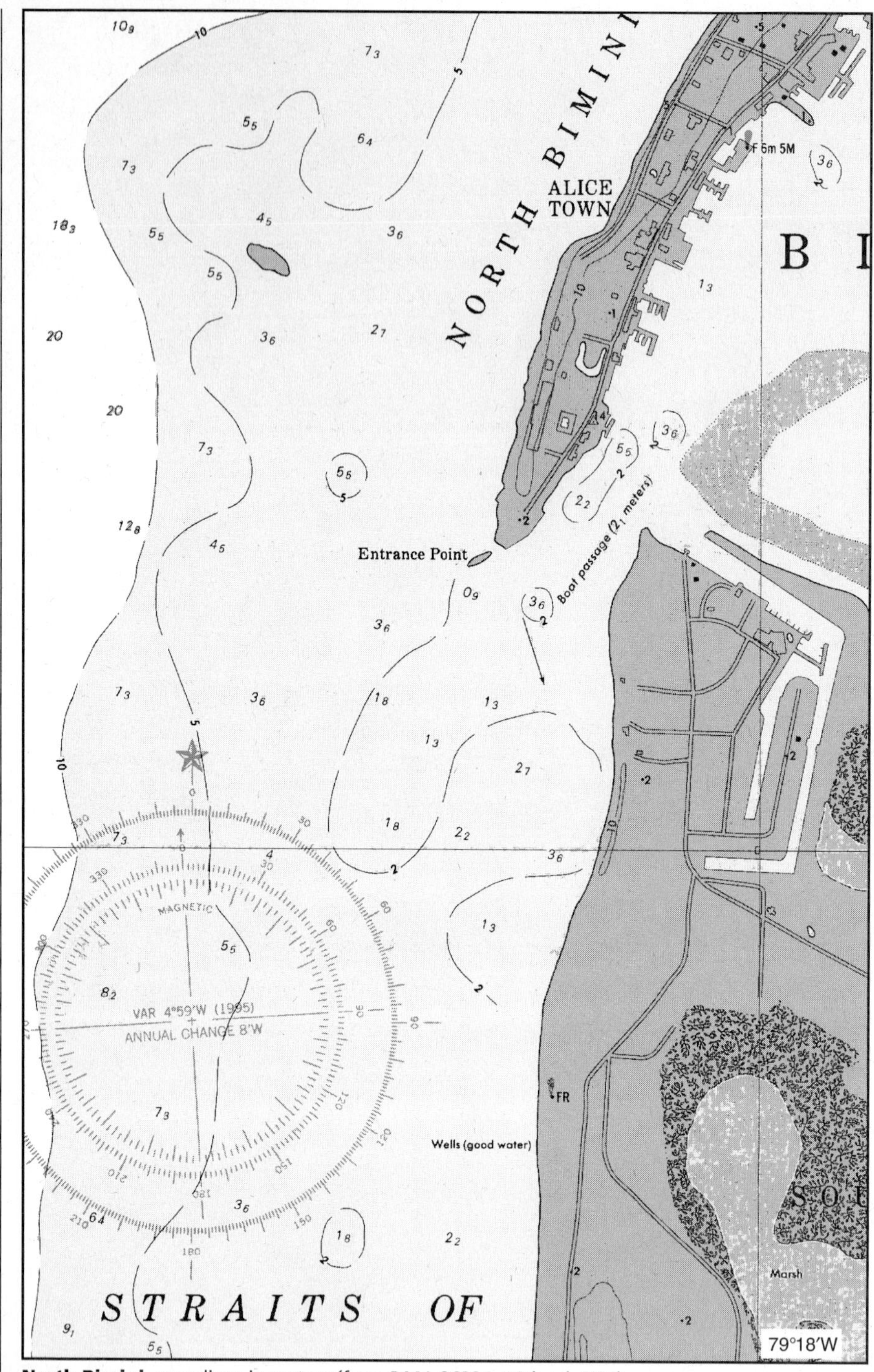

North Bimini, soundings in meters *(from DMA 26324, 16th ed, 1/95)*

reported missing and tower remains dangerous; replaced by temporary, unreliable lighted buoy just north of remains, 1995.)

Northwest Shoal Buoy, Fl W 2s. Red buoy. (Unofficially reported unreliable, 1995.)

GREAT BAHAMA BANK

Pilotage: Many vessels drawing less than 5 feet cross the banks from Gun Cay Light (25°34.5'N, 79°18.8'W) to the remains of Northwest Channel Light (25°27.2'N, 78°09.8'W), a distance of about 62 miles. Mariners are cautioned that the aids to navigation on the banks cannot be considered reliable. They may be missing, off station, or showing incorrect characteristics due to hurricane damage.

Leaving Gun Cay, you encounter a series of sand ridges within the first 10 miles. The controlling depth of this route has been reported as 5 feet. The water depths improve approximately 10 miles east of Gun Cay. Deeper-draft vessels may want to travel from North Rock Light (25°48.2'N, 79°16.7'W) to Northwest Channel, taking care to avoid Mackie Shoal in approximate position 25 41.2'N, 78 39.2'W. Vessels anchoring on the banks are advised to show an anchor light.

Good depths have been reported on the route from South Riding Rock Light (25°13.8'N, 79°10.0'W) to the Russell Beacon remains and buoy, then on to Northwest Channel. Vessels may get some protection anchoring between South Riding Rock and Castle Rock.

THE BERRY ISLANDS

CHUB POINT LIGHT, 25 23.8N, 77 54.2W. Fl W R 10s, 44ft, 7M. Aluminum-colored lattice structure, 32ft. W 320°–054°, R 054°–320°. Fl W and Fl G on water tower 0.8M to the NE.

FRAZER'S HOG CAY AVIATION LIGHT, 25 25.0N, 77 53.7W. Fl W G. Radiobeacon.

CHUB CAY

25 24N, 77 54W

Port of entry: Tie up in Chub Cay Club Marina to clear customs. The officials may have to travel from the airstrip. The club charges a clearance fee of $25 to vessels clearing in but not docking for the night.

Dockage: Contact Chub Cay Club Marina (VHF channels 68, dock, and 71, office; tel., 242-325-1490; fax, 242-322-5199). The boat basin is protected, and after recent dredging, 8 feet of water is reported in the entrance channel. Frazer's Cay Club and Marina (VHF channel 16; information in the U.S., 305-743-2420) on the channel east of Frazer's Hog Cay also has slips.

Anchorage: Boats crossing the banks and headed to Nassau often anchor in the bight just southwest of the marina entrance channel. This area is subject to a surge, and depths run less than 6 feet in some spots. Shallow-draft boats can get good protection between Mama Rhoda Rock and Chub Cay. Anchorage is also found in the channel along the eastern shore of Frazer's Hog Cay.

Services: Fuel, water, electricity, ice, a restaurant and bar, limited groceries, and minor repairs are available at Chub Cay. Water, ice, limited electricity, showers, moorings, and a bar are currently available at Frazer's Cay; limited restaurant service is available.

This is a good spot to break up the long journey from Nassau across the banks. From here it is about 50 miles to Nassau and 76 miles to Gun Cay. The main anchorage may become untenable in strong westerlies.

WHALE POINT LIGHT, SW point of Whale Cay, 25 23.7N, 77 48.1W. Fl W 4s, 70ft, 7M. Stone tower, white steel superstructure, 43ft.

LITTLE HARBOUR CAY LIGHT, 25 33.6N, 77 42.4W. Fl W 2.2s, 94ft, 9M. Black and white banded pipe, 23ft.

LITTLE HARBOUR CAY

25 34N, 77 43W

Anchorage only: Anchor between Little Harbour and Cabbage cays. A shallow channel with a reported depth of less than 3 feet leads north to a private dock on Little Harbour.

BULLOCK HARBOUR LIGHT, 25 45.8N, 77 52.1W. Fl R 6s, 36ft, 7M. Iron pyramidal structure, upper half white, lower half black.

GREAT HARBOUR CAY

25 45N, 77 52W

Pilotage: From the western tip of Little Stirrup Cay run on a course of 180° for about 2.8 miles to an unlit piled mark with a radar reflector. From here a course of 110° leads through a well-marked channel to the marina.

Port of entry: Tie up in Great Harbour Cay Marina. The dockmaster will give you customs and immigration forms and call customs officals for you (242-367-8204). Officials must come in from the airstrip.

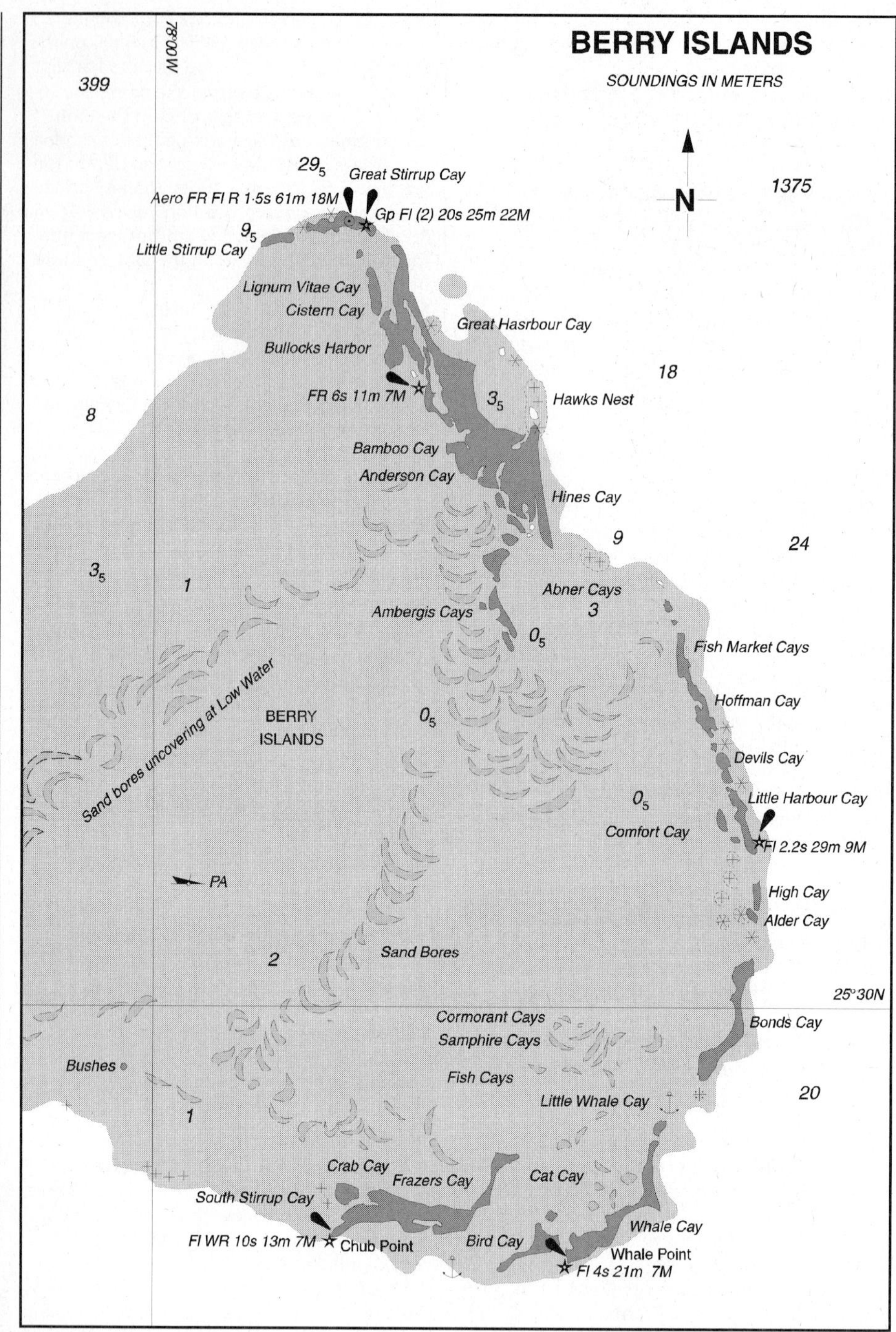

Berry Islands, soundings in meters

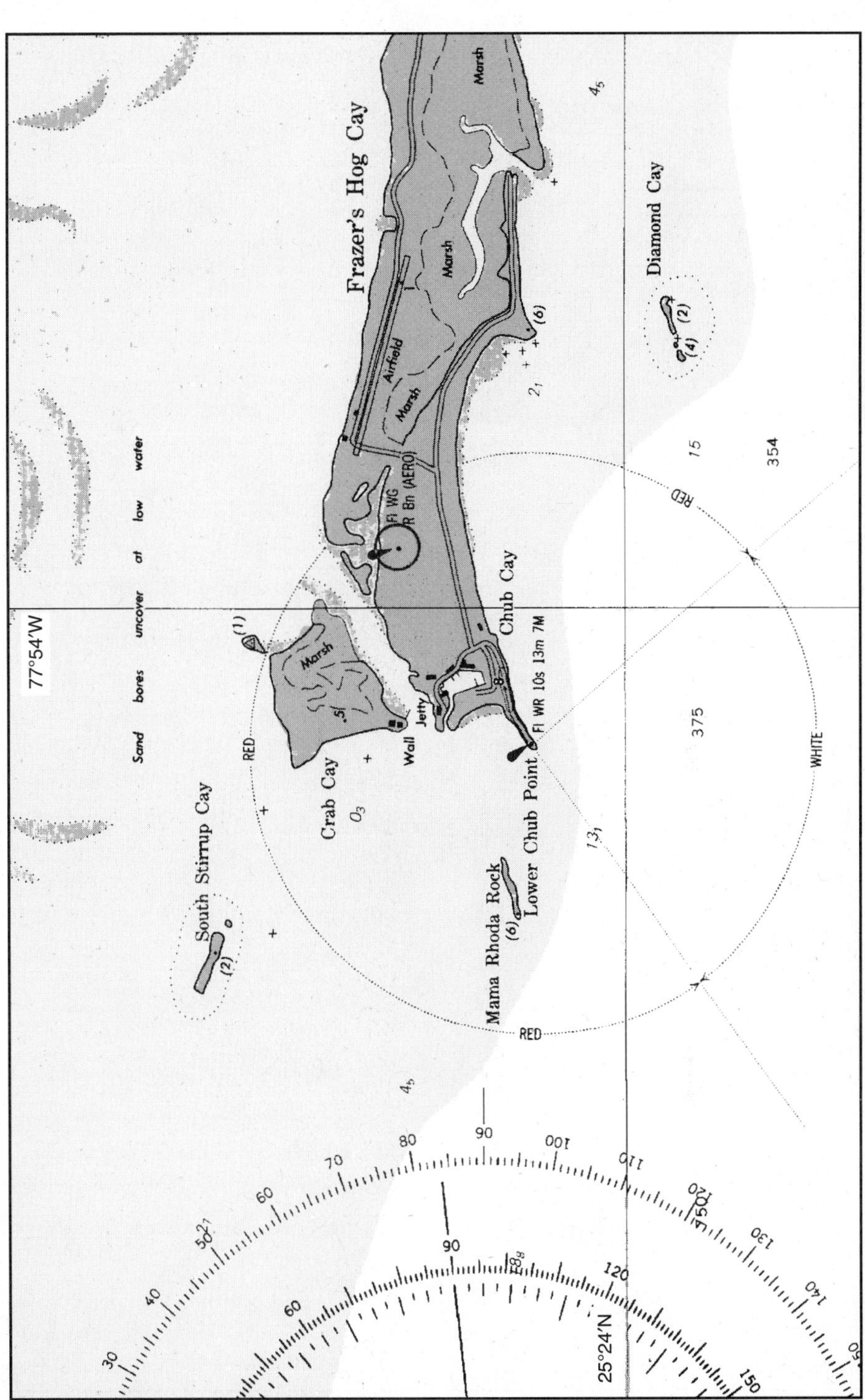

Chub Cay, soundings in meters *(from DMA 26328, 2nd ed, 11/94)*

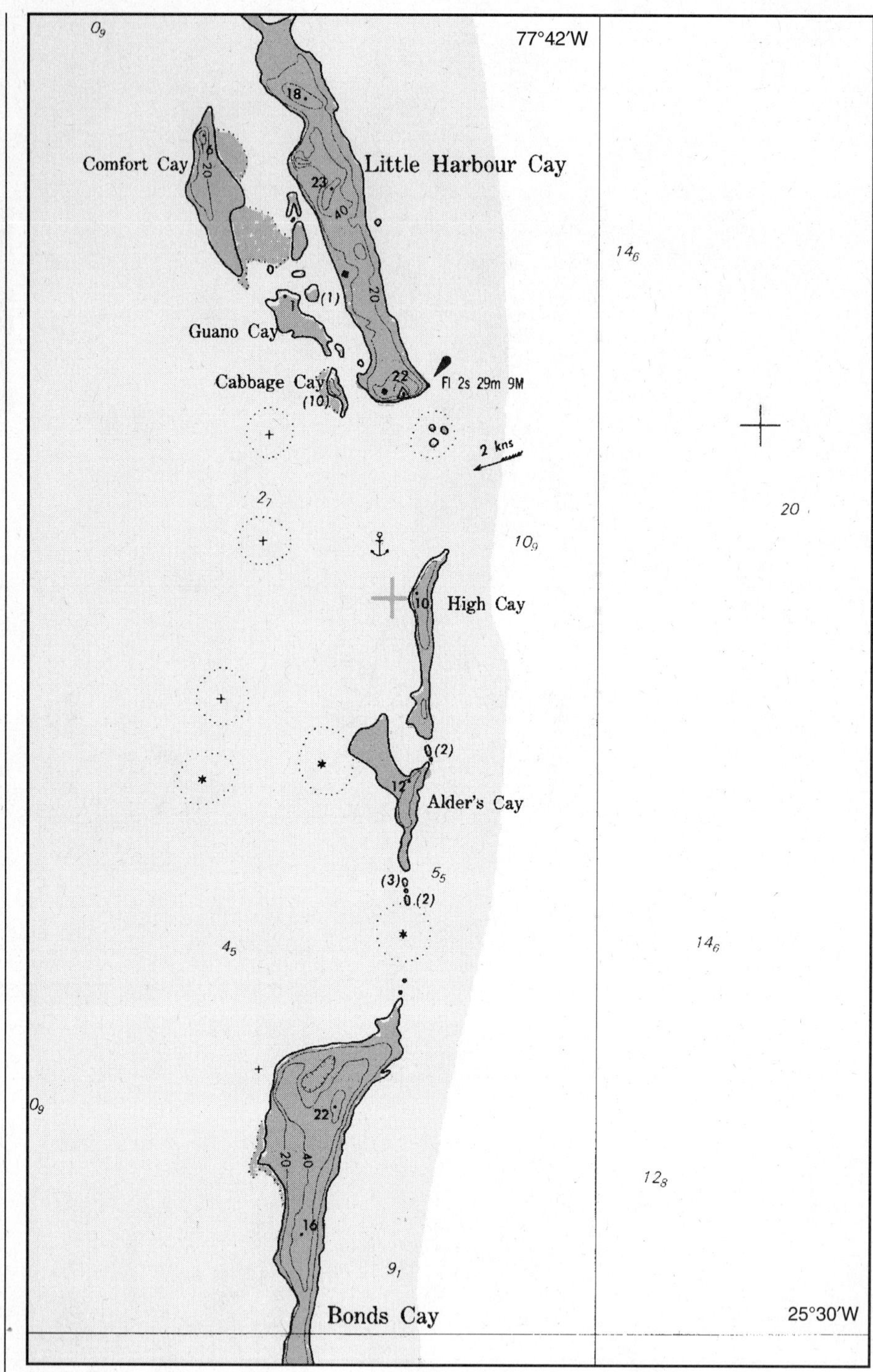

Little Harbour Cay, soundings in meters *(from DMA 26328, 2nd ed, 11/94)*

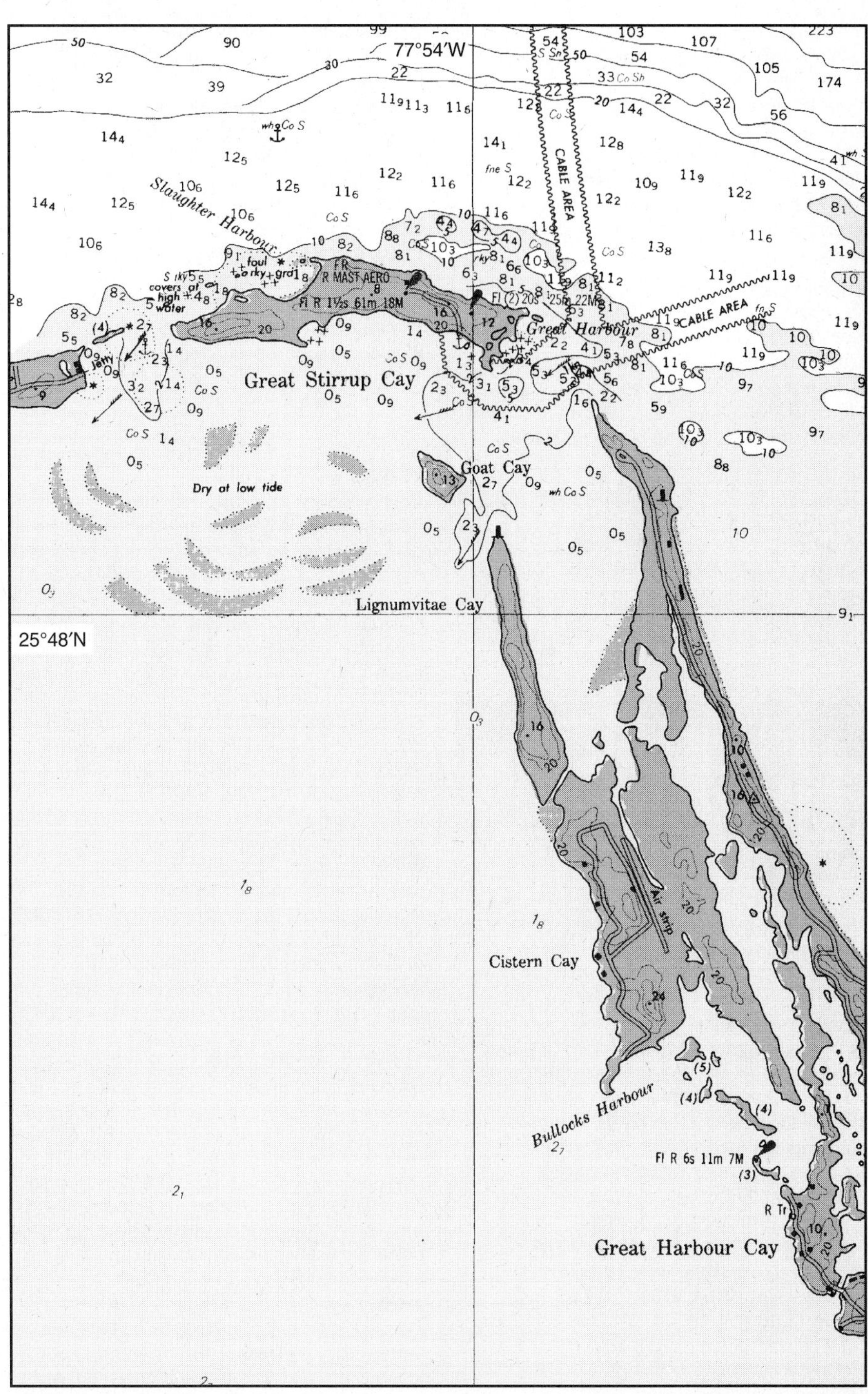

Great Harbour Cay and Great Stirrup Cay, soundings in meters *(from DMA 26328, 2nd ed, 11/94)*

P Pilot BAHAMAS

Dockage: Contact Great Harbour Cay Marina (VHF channels 16, 14, 68, and 72; tel., 242-367-8123; in the U.S., 561-585-0027). Call the U.S. number for reservations.
Anchorage: Anchor west of Bullock's Harbour or south of the marina channel or off the beach club on the east side of the island.
Services: Fuel, water, ice, limited groceries, a few restaurants, and accommodations are available.

GREAT STIRRUP CAY LIGHT, 640 meters from E end, 25 49.7N, 77 54.0W. Fl (2) W 20s, 82ft, 22M. White circular tower, 57ft.
GREAT STIRRUP CAY AVIATION LIGHT, 1,189 meters from E end, 25 49.8N, 77 54.4W. F R, 200ft, 18M. Radio mast. Obstruction light. (Reported unofficially to be missing, 1996).

GREAT STIRRUP CAY

25 50N, 77 54W
Anchorage only: Enter the anchorage at the eastern end of the cay. Cruise ships discharge passengers in the coves on the north side of the island. *Caution: The light has been reported as unreliable.*

NASSAU AND APPROACHES

CLIFTON TERMINAL LIGHT, commercial harbor on SW end of New Providence Island, 25 00.4N, 77 32.5W. 2 Q R (horiz), 121ft, 13M. Orange mast with white bands. 2 F R (horiz), 59ft, 10M. Same structure.
GOULDING CAY LIGHT, off W point of New Providence Island, 25 01.6N, 77 35.8W. Fl W 2s, 36ft, 8M. Gray structure, 49ft.
FORT FINCASTLE AVIATION LIGHT, 25 04.4N, 77 20.3W. Fl W 5s, 219ft, 28M. Gray concrete tower, 130ft. Signal light.
PARADISE (HOG) ISLAND LIGHT, W point, 25 05.2N, 77 21.1W. Fl W or R 5s, 68ft, W 13M. White conical stone tower, red bands, 63ft. Changed to Fl R 5s when bar is dangerous; obscured 334°–025°.
NASSAU EAST BREAKWATER LIGHT, head, 25 05.3N, 77 21.2W. Fl G 5s, 29ft. Reported destroyed (February 1991).
NASSAU WEST BREAKWATER LIGHT, head, 25 05.1N, 77 21.3W. Fl R 5s, 29ft. Tower.
NASSAU HARBOUR RANGE, W of town **(Front Light),** 25 04.7N, 77 21.0W. F G, 37ft, 7M. Red framework tower, 38ft. **(Rear Light),** 260 meters, 151.6° from front. F G, 61ft, 7M. Red framework tower, 36ft.
GOVERNMENT HOUSE LIGHT, 25 04.5N, 77 20.7W. Fl R 3s, 122ft, 10M. Green cupola on building, 68ft.
THE NARROWS LIGHT, between Paradise Island and Athol Island, 25 05.0N, 77 18.5W. Fl R 5s, 12ft, 2M. Red post, 12ft.
PORGEE ROCKS LIGHT, SE rock, 25 04.3N, 77 15.8W. Fl W 3s, 25ft, 5M. Gray structure, 23ft.
CHUB ROCKS LIGHT, 25 06.9N, 77 15.6W. Fl W 5s, 32ft, 4M. White framework tower, 26ft.
EAST END POINT LIGHT, on point S of East End Point, 25 02.3N, 77 16.8W. Fl W 6s, 57ft, 8M. White square stone building, 27ft. Visible 180°–056°.

NASSAU

25 05N, 77 21W
Port of entry: The customs office is located on Prince George Wharf, where the cruise ships dock. You may tie up here (use lots of fenders) or have customs called from a marina once you are dockside.
Nassau Harbour Control: All vessels must report to Nassau Harbour Control, which stands by on VHF channel 16 around the clock, when entering or moving within the harbor.
Radio communications and weather: Harbour Control uses VHF channel 09 as a working channel; large vessels (eg., cruise liners) routinely use VHF channel 14. The Nassau marine operator broadcasts traffic lists and the latest Nassau weather report on the even hour on VHF channel 27. Amateur weather broadcasts are also generally announced on VHF channel 16 at 0715, switching to VHF channel 72, and on 3696 kHz at 0720. The Waterway net comes on at 0745 on 7268 kHz LSB.
Dockage: Available west of the Paradise Island bridge at East Bay Marina (242-394-1816) and east of the bridge (low-water clearance, 72 feet) on Paradise Island at Hurricane Hole (242-363-3600) and Paradise Harbour Marina (242-363-2992) and on New Providence Island at Nassau Yacht Haven (242-393-8173), Nassau Harbour Club (242-393-0771), and Harbour Marina View (242-393-4182). All monitor VHF channel 16. Boats heading for East Bay Marina should call ahead for directions to the approach channel west of Potter's Cay used by the sand barges and not shown on the charts.
Anchorage: The holding ground in most of Nassau Harbour is reported to be very poor. Boats must stay clear of the cruise ship docks, the marinas, sand barge operations, and the seaplane landing area. There is limited anchorage

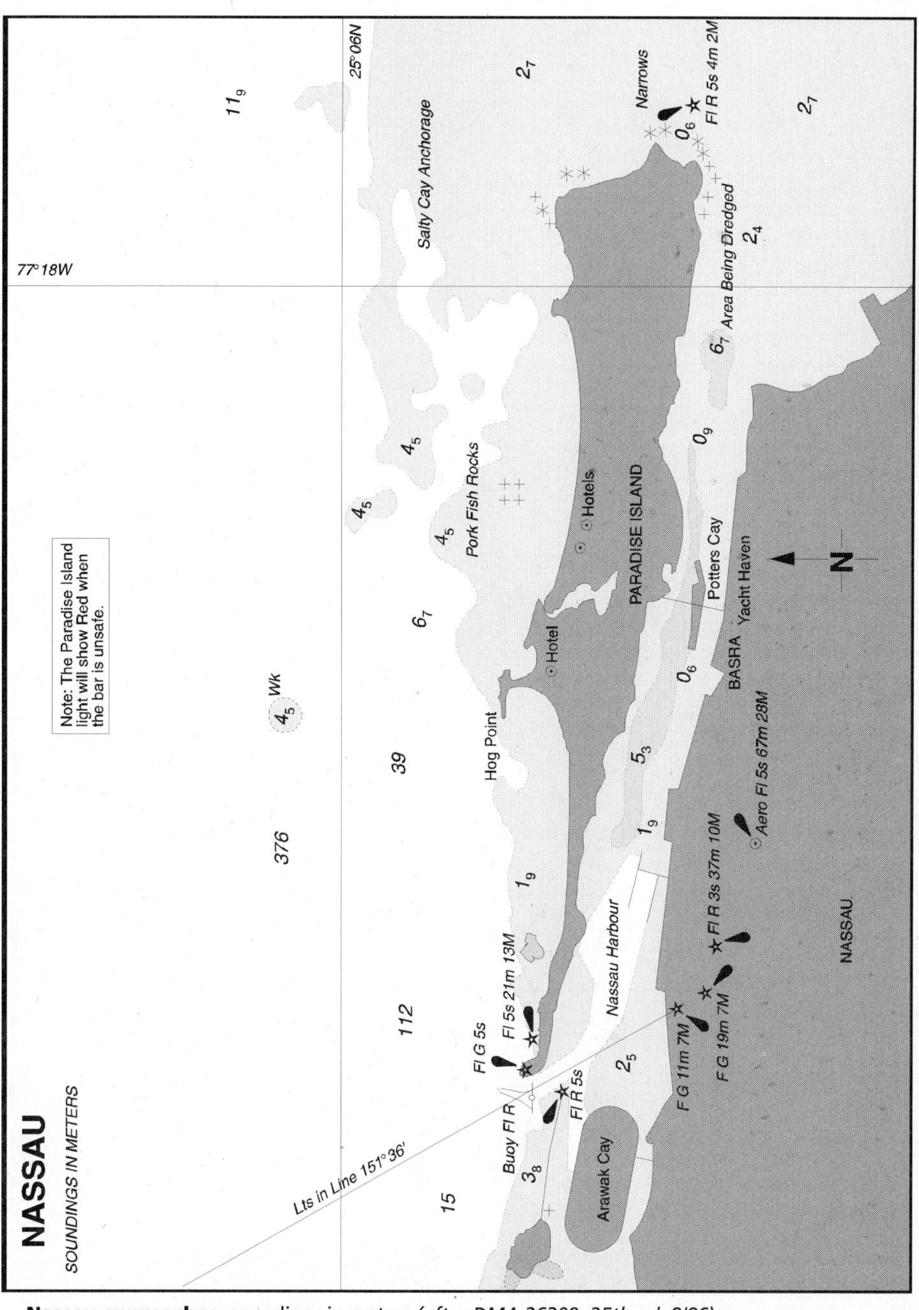

Nassau approaches, soundings in meters *(after DMA 26309, 25th ed, 8/86)*

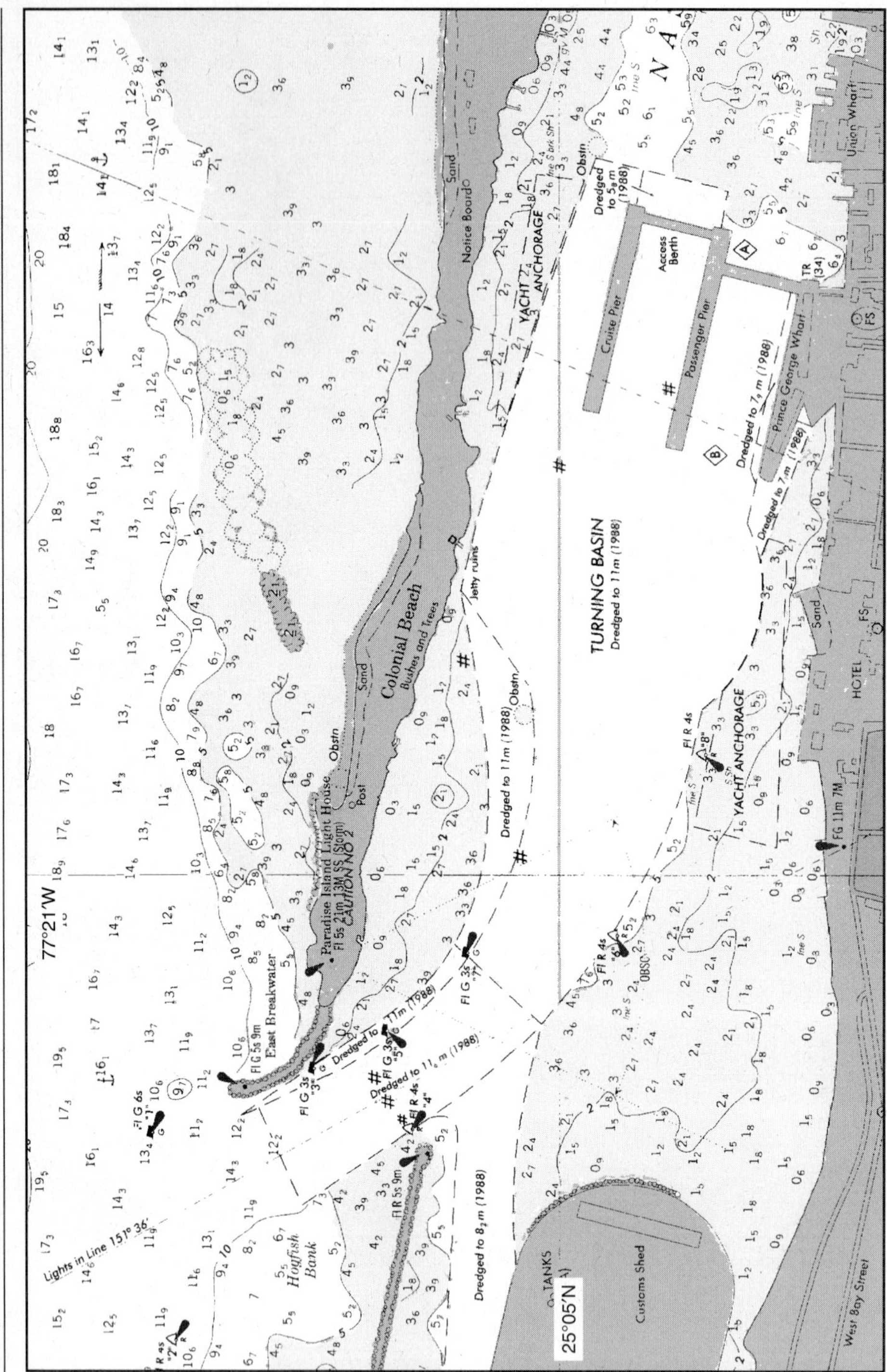

Nassau Harbour, western part, soundings in meters *(from DMA 26309)*

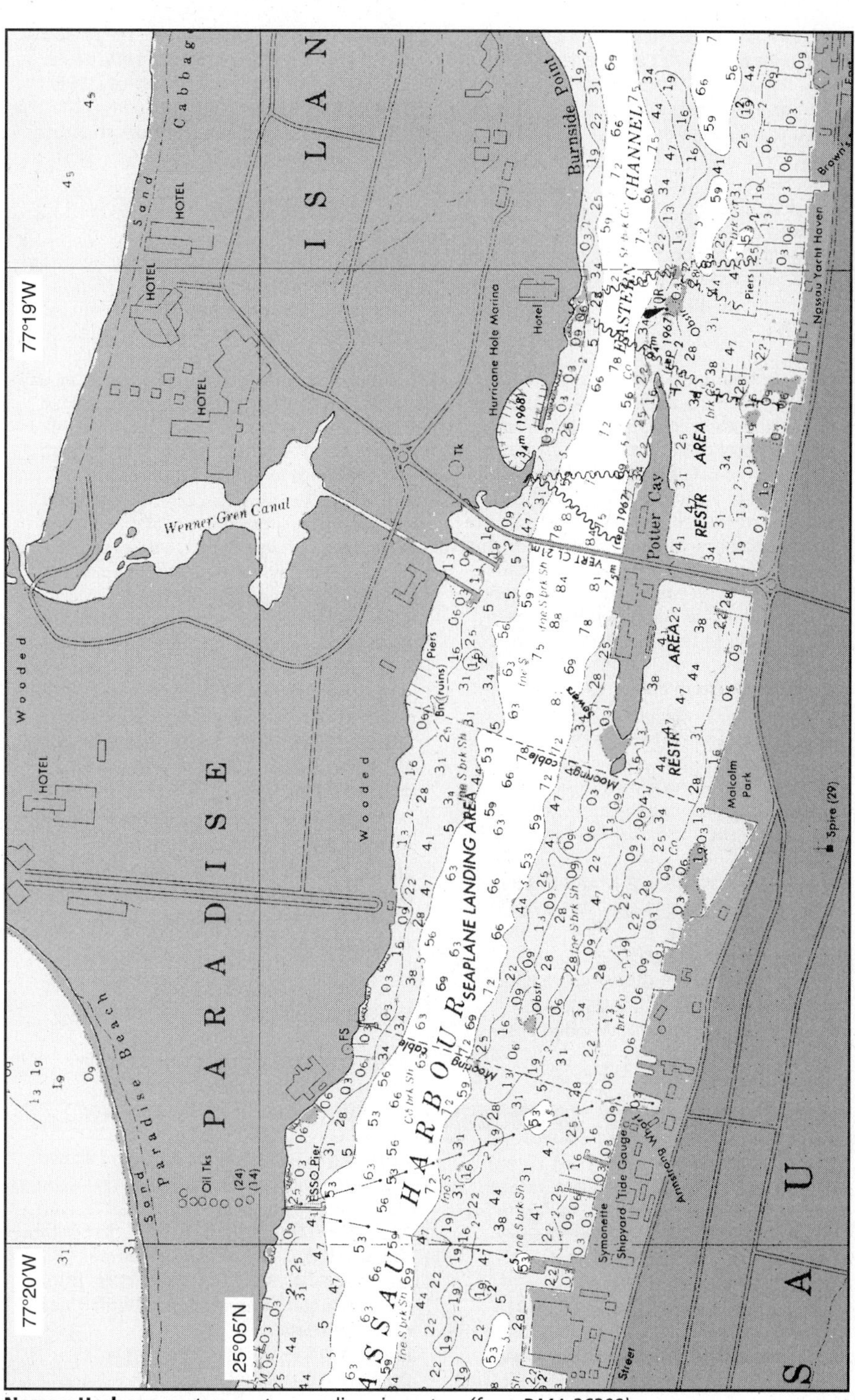

Nassau Harbour, eastern part, soundings in meters *(from DMA 26309)*

south of Paradise Island near the Club Med docks, west of Potter's Cay off the BASRA dock, and in the eastern harbor. All areas are subject to harbor traffic wakes. Avoid the charted shoal areas and hurricane chain and cable areas.
Repairs: Hauling facilities and repair services available.
Services: All services are available. Major repairs and haulouts may be done here. There are several large grocery stores, hardware stores, marine supply stores, and many restaurants, bars, casinos, shops, and modern services of all types.

Nassau Harbour has a bar across the entrance channel. If the Paradise Island bar light shows red, the channel is considered dangerous. In general, approaching the harbor after dark is not considered safe because of the difficulty of spotting navigation aids against the background of city lights. Mariners should be prepared to encounter large cruise ships, freighters, and barges in the harbor. Nassau Harbour Control will advise you of any traffic problems. Vessels heading for marinas in the eastern harbor unable to clear the Paradise Island bridge should approach via Chub or Porgee rocks off the eastern end of Paradise Island.

Nassau is the capital of the Bahamas and is an exciting port of call. There are many historic sights and interesting attractions ashore. The city is also a good place to conduct banking, place phone calls, and make travel arrangements.

PORT NEW PROVIDENCE MARINA LIGHT, entrance channel, 25 00.2N, 77 15.7W. Fl W 4s, 7ft, 4M. Wooden pile.
MARINA CHANNEL NO. 1, 25 00.3N, 77 15.7W. Fl W 4s, 7ft, 4M. Wooden pile.
MARINA CHANNEL NO. 2, 25 00.4N, 77 15.8W. Fl W 4s, 7ft, 4M. Wooden pile.

ANDROS

MORGANS BLUFF RANGE (Front Light), 25 10.5N, 78 02.1W. Q W, 29ft, 6M. Intensified on rangeline. **(Rear Light),** 223°44′ from front. Oc W 4s, 60ft, 6M. Intensified on rangeline.
MORGANS BLUFF DOCK LIGHT, 25 10.8N, 78 01.6W. Fl W 4s, 20ft, 4M.
NICOLLS TOWN RANGE (Front Light), 25 07.8N, 78 00.5W. Fl W 5s, 60ft, 8M. Red and aluminum colored structure. **(Rear Light),** 24 meters, 247° from front. Fl W 5s, 65ft, 10M. Red and aluminum colored structure.
STANIARD ROCK LIGHT, 24 51.1N, 77 52.5W. Fl W 4s, 18ft, 6M. Gray structure, 15ft. Reported unofficially out of service (1996).
STANIARD CREEK LIGHT, 24 50.8N, 77 54.0W. F R, 26ft, 6M. Wooden framework structure, 20ft. Reported unofficially missing (1996).

STANIARD CREEK

24 51N, 77 54W
Pilotage: The entrance to the channel to Staniard Creek and the Long Bay Cay Marina under development is privately marked by temporary stakes. The entrance is barely navigable at low tide; wait for high tide. For directions or assistance, call the new Long Cay Bay marina on VHF channel 68. A white-roofed house by the entrance to Settlement Creek is prominent. The Batelco tower (Fl R 3s, 200ft), some distance inland, is clearly visible.
Dockage: Two slips are currently available at the new marina development (VHF channels 16 and 68, call sign, Gunshot Base) under way on Long Bay Cay. Additional slips were planned for completion in 1997.
Services: Services at present are limited (fuel delivery can be arranged), but the development plans include fuel, restaurant/bar, accommodations, and other facilities.

FRESH CREEK ISLET LIGHT, S side, 24 44.3N, 77 46.9W. Fl W 6s, 10ft, 5M. Metal mast.
ANDROS TOWN LIGHT, S side of entrance, 24 43.7N, 77 47.8W. F W, 27ft, 6M. White quadrangular stone building, 24ft.

NOTE: AUTEC (ATLANTIC UNDERSEA TEST AND EVALUATION CENTER) AREAS ARE OFF LIMITS TO TRANSITING MARINERS, EXCEPT IN EMERGENCIES.

AUTEC SITE 1

NORTH ITT TOWER, 24 43.9N, 77 46.3W. Fl Y 4s, 5M. Steel skeleton tower.
SOUTH ITT TOWER, 24 42.4N, 77 45.3W. Fl Y 4s, 5M. Steel skeleton tower.
AVIATION LIGHT, 24 42.3N, 77 46.4W. Al Fl W G Y 2s, 136ft. Tower.
RANGE (Front Light), 24 42.3N, 77 45.9W. Q W, 25ft, 10M. White tower, red daymark with white stripe. **(Rear Light),** 335 meters, 223.8° from front. Iso W 6s, 33ft, 10M. White tower, red daymark with white stripe.
LIGHT 5, 24 42.8N, 77 45.3W. Fl G 4s, 18ft, 4M. Dolphin, green square daymark with green reflective border.
LIGHT 6, 24 42.9N, 77 45.3W. Fl r 4s, 19ft, 3M. Dolphin, red triangular daymark with red reflective border.

LIGHT 7, 24 42.7N, 77 45.6W. Fl G 4s, 16ft, 4M. Dolphin, green square daymark with green reflective border.
LIGHT 9, 24 42.5N, 77 45.7W. Fl G 6s, 16ft, 4M.
LIGHT 10, 24 42.5N, 77 45.8W. Fl R 4s, 16ft, 3M. Dolphin, red triangular daymark with red reflective border.
LIGHT 15, 24 42.3N, 77 45.8W. Fl G 6s, 16ft, 4M.

HIGH CAY LIGHT, 24 39.3N, 77 42.6W. Fl W 4s, 70ft, 6M. Metal mast.

AUTEC SITE 2

RANGE (Front Light), 24 29.9N, 77 43.1W. Q W, 22ft, 10M. White skeletal tower, red daymark with white stripes. **(Rear Light),** 745 meters, 269.9° from front. Iso W 6s, 42ft, 10M. White skeleton tower, red daymark with white stripes.
LIGHT 3, 24 29.9N, 77 42.1W. Fl G 4s, 19ft, 4M. Dolphin, green square daymark with green reflective border.
LIGHT 4, 24 29.9N, 77 43.0W. Fl R 4s, 19ft, 3M.
LIGHT 6, 24 29.9N, 77 42.2W. Fl R 4s, 19ft, 3M.
LIGHT 9, 24 29.9N, 77 42.6W. Fl G 4s, 16ft, 4M.
LIGHT 10, 24 29.9N, 77 42.6W. Fl R 4s, 16ft, 3M.
LIGHT 13, 24 29.9N, 77 43.0W. Fl G 4s, 17ft, 4M.

AUTEC SITE 3

ITT TOWER, 24 20.9N, 77 40.4W. Fl Y 4s, 33ft, 5M.
LIGHT 3, 24 20.1N, 77 40.4W. Fl G 4s, 19ft, 4M. Dolphin, green square daymark with green reflective border.
LIGHT 4, 24 20.1N, 77 40.1W. Fl R 4s, 20ft, 3M.
LIGHT 5, 24 20.2N, 77 40.7W. Fl G 4s, 19ft, 4M.
LIGHT 6, 24 20.2N, 77 40.9W. Fl G 4s, 16ft, 3M.
LIGHT 3, 24 20.2N, 77 41.0W. Fl G 4s, 16ft, 4M.

MIDDLE BIGHT LIGHT, on rock S side of channel, 24 18.8N, 77 40.3W. Fl W 5s, 17ft, 7M. White steel framework tower.
MANGROVE CAY, PEATS WHARF LIGHT, 24 14.5N, 77 38.4W. F R, 20ft, 7M.
SIRIOUS ROCK LIGHT, 24 13.0N, 77 36.0W. Fl W 3s, 29ft, 7M. Black and white mast on white cylindrical structure.

AUTEC SITE 4

ITT TOWER, 24 13.3N, 77 36.0W. Fl Y 4s, 33ft, 5M.
LIGHT ~~1~~ 3, 24 13.3N, 77 36.3W. Fl G 4s, 23ft, 4M.
LIGHT 2, 24 13.3N, 77 36.2W. Fl R 4s, 22ft, 3M.
LIGHT 4, 24 13.3,N, 77 36.3W. Fl R 4s, 20ft, 3M.

GREEN CAY LIGHT, W end, 24 02.2N, 77 11.2W. Fl W 3s, 33ft, 7M. Black structure, white house, 23ft.

AUTEC SITE 6

RANGE (Front Light), 24 00.4N, 77 31.7W. Q W, 34ft, 10M. Skeleton tower. **(Rear Light),** 760 meters, 257° from front. Iso W 6s, 68ft. Skeleton tower.
LIGHT 4, 24 00.5N, 77 30.1W. Fl R 4s, 16ft, 3M.
LIGHT 5, 24 00.5N, 77 30.1W. Fl G 4s, 22ft, 4M.
LIGHT 6, 24 00.5N, 77 30.2W. Fl R 4s, 21ft, 3M.
LIGHT 7, 24 00.5N, 77 30.7W. Fl G 4s, 21ft, 4M.
LIGHT 8, 24 00.5N, 77 30.4W. Fl R 4s, 21ft, 3M.
LIGHT 9, 24 00.5N, 77 30.4W. Fl G 4s, 20ft, 3M.
LIGHT 10, 24 00.6N, 77 30.5W. Fl R 4s, 20ft, 3M.
LIGHT 11, 24 00.6N, 77 30.7W. Fl G 4s, 18ft, 4M.
LIGHT 12, 24 00.6N, 77 30.7W. Fl R 4s, 17ft, 3M.
LIGHT 13, 24 00.5N, 77 30.9W. Fl G 4s, 20ft, 3M.
LIGHT 16, 24 00.5N, 77 31.3W. Fl R 4s, 16ft, 3M.
LIGHT 17, 24 00.4N, 77 31.6W. Fl G 4s, 16ft, 4M.

TINKER ROCKS LIGHT, 23 59.0N, 77 29.6W. Fl W 4.2, 32ft, 8M. Black mast on white cylindrical housing.
HIGH POINT CAY LIGHT, 23 55.6N, 77 29.0W. Fl W 5s. White building on piles.

AUTEC SITE 7

ITT TOWER, 23 54.5N, 77 28.5W. Fl Y 4s, 33ft, 5M.
~~**LIGHT 4,** 23 53.9N, 77 28.7W. Fl R 4s, 20ft, 3M.~~
LIGHT 5, 23 53.9N, 77 28.7W. Fl G 4s, 20ft, 4M.
LIGHT 7, 23 53.9N, 77 28.8W. Fl G 4s, 17ft, 4M.
LIGHT 10, 23 53.9N, 77 29.0W. Fl R 4s, 16ft, 3M.
LIGHT 11, 23 53.9N, 77 29.0W. Fl G 4s, 18ft, 4M.
LIGHT 12, 23 53.9N, 77 29.2W. Fl G 4s, 19ft, 3M.
LIGHT 14, 23 53.9N, 77 29.3W. Fl R 4s, 17ft, 3M.

BILLY ISLAND LIGHT, N end of Williams Island, 24 39.3N, 78 28.7W. F W, 22ft, 5M. Mast. Fishing light.

NASSAU TO ELEUTHERA

SIX SHILLING CHANNEL LIGHT, 4.3 km 235.5° from Six Shilling Cays Light, 25 15.1N, 76 56.4W. Fl R 4s, 16ft, 8M. Red and yellow structure.
SIX SHILLING CAYS LIGHT, 25 16.7N, 76 55.1W. Fl W 8s, 32ft, 10M. Gray steel framework tower, 33ft.
CURRENT ROCK LIGHT, 25 25N, 76 52W. Fl W 8s, 41ft, 7M. Framework structure, 33ft.
EGG ISLAND LIGHT, on summit of island, Northeast Providence Channel, 25 30N, 76 53W. Fl W 3s, 112ft, 12M. White metal tower, 59ft.

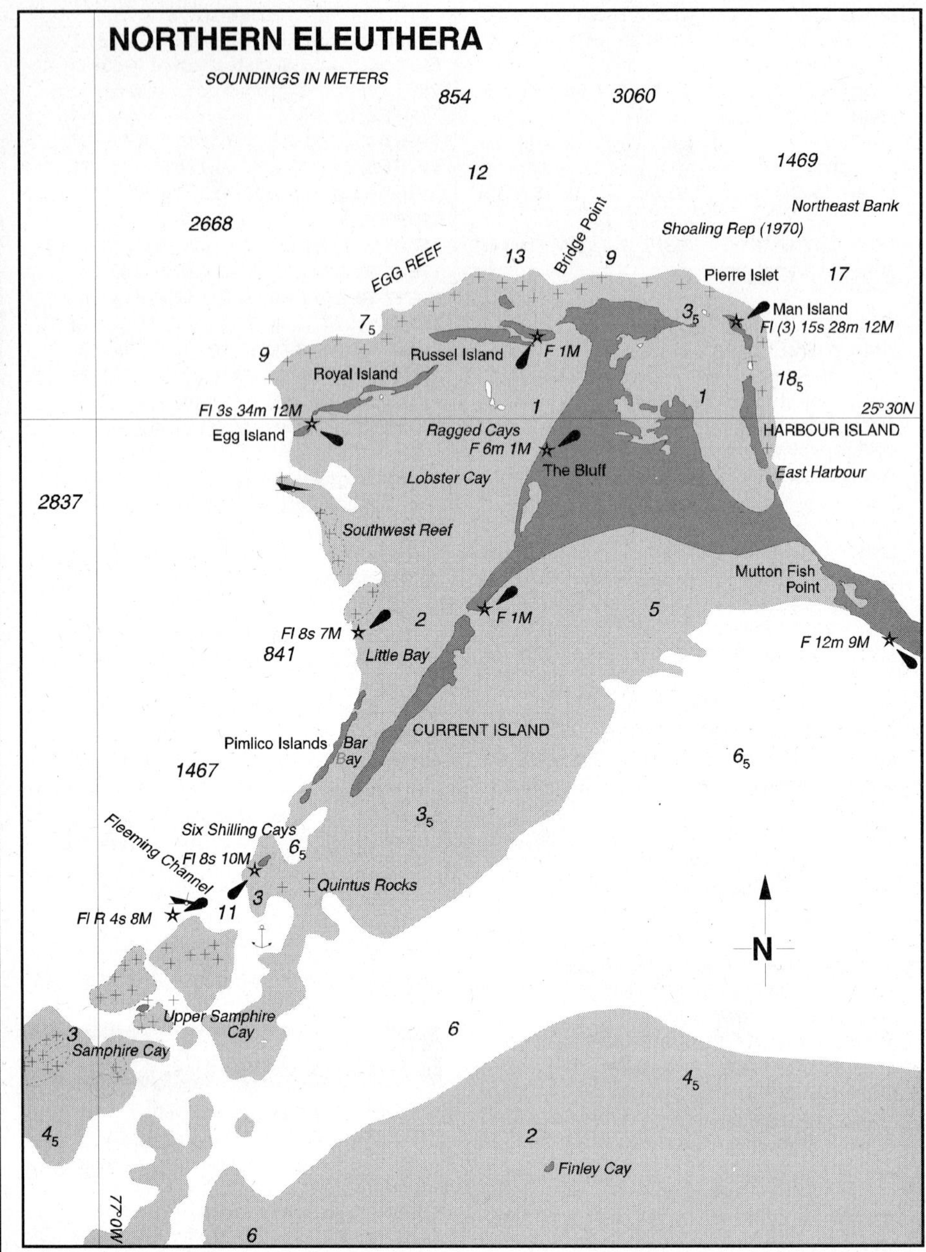

Northern Eleuthera, soundings in meters

ELEUTHERA

MAN ISLAND LIGHT, 25 32.8N, 76 38.5W. Fl (3) W 15s, 93ft, 12M. Aluminum framework tower, 60ft. Reported extinguished (March 1987).

HARBOUR ISLAND

25 30N, 76 38W
Port of entry: The marinas will alert customs to your arrival.
Dockage: You may tie up temporarily to the government dock at Dunmore Town. South of the dock is Valentine's Yacht Club and Marina (VHF channel 16; tel., 242-333-2141). About 0.5 mile south is Harbour Island Club and Marina (VHF channel 16; tel., 242-333-2427), which has a new lunchtime restaurant.
Anchorage: There is good holding in a soft and muddy bottom along the harbor within 0.5 mile of facilities ashore.
Services: There is fuel, water, electricity, and ice at the marinas. There are several good grocery stores and, in general, a very good selection of supplies. The town's resorts have a variety of bars, restaurants, and nightclubs.

The approaches to Harbour Island from Spanish Wells through Devil's Backbone are notoriously difficult. See the section on Spanish Wells for information on obtaining a local pilot to guide you. This route should not be attempted in strong north winds.

SPANISH WELLS LIGHT, 25 53N, 76 46W. F W, 6ft, 1M. Concrete column.

SPANISH WELLS

25 33N, 76 45W
Port of entry: The marina will alert customs to your arrival.
Dockage: Contact Spanish Wells Yacht Haven (VHF channel 16; tel., 242-333-4255; fax, 242-333-4649).
Moorings: Edsel Roberts on "Dolphin," controls a few moorings in East Basin. Others may be available through one of the Spanish Wells pilots (see below).
Anchorage: A few boats squeeze into the small pocket of deep water near the east end of town. Do not block the channel. Good anchorage is found in nearby Royal Harbour.
Services: Fuel, water, groceries, showers, hardware, and restaurants are ashore. Dockage is available at the Yacht Haven for those leaving their boats for extended periods. Pilots may be hired to assist vessels headed to Harbour Island; contact Cinnabar, or call "Spanish Wells pilot" on VHF channel 16.

THE BLUFF LIGHT, 25 31N, 76 45W. F W, 20ft, 1M. Mast.
CURRENT ISLAND LIGHT, 25 23N, 76 49W. F W, 20ft, 1M. Mast.
CURRENT LIGHT, at town of Current, 25 25N, 76 48W. F W, 12ft, 1M. Mast.
STAFFORD LIGHT, Gregory Town, 25 23.5N, 76 34.5W. F W, 41ft, 9M. Mast, 13ft.
HATCHET BAY LIGHT, W side of entrance, 25 20.5N, 76 29.8W. Fl W 15s, 57ft, 8M. Tapered cast iron pipe mast, 23ft. Two F R range lights in line on 022° are shown from the E side of the bay.

HATCHET BAY

25 21N, 76 30W
Port of entry: The marina will alert customs (242-332-2714) to your arrival.
Dockage: Marine Services of Eleuthera (VHF channel 16; tel., 242-335-0186) has a concrete bulkhead available for transient dockage.
Moorings: Contact Marine Services of Eleuthera. Short- and long-term rentals are available.
Anchorage: Near shore, the bottom is mud, so boats often anchor farther out. There is good shelter from the wind and seas. Stay clear of the commercial docks near the settlement of Alice Town. There is a dinghy dock at the end of the bulkhead.
Services: Diesel, water, electricity, marine hardware, and liquor are all available. There is a new grocery and an ice plant. Arrangements can be made to accommodate minor repairs.

AVIATION LIGHT, 25 16.1N, 76 19.1W. Iso R 3s. Radio mast.
CUPID CAY LIGHT, NW end of Governor's Harbour, 25 12N, 76 16W. Fl W 4s, 40ft, 8M. Aluminum colored steel tower, 32ft.

GOVERNOR'S HARBOUR

25 12N, 76 15W
Port of entry
Anchorage only: The holding is reported to be poor here. Moorings are reported to be hazardous.
Services: Groceries, stores, and restaurants are ashore.

NORTH PALMETTO POINT LIGHT, 25 10.8N, 76 11.4W. Iso W 4s, 73ft, 12M. White tower with black top and dwelling.
TARPUM BAY LIGHT, S end, 25 00.0N, 76 13.1W. F W, 35ft, 7M. Mast, 20ft.

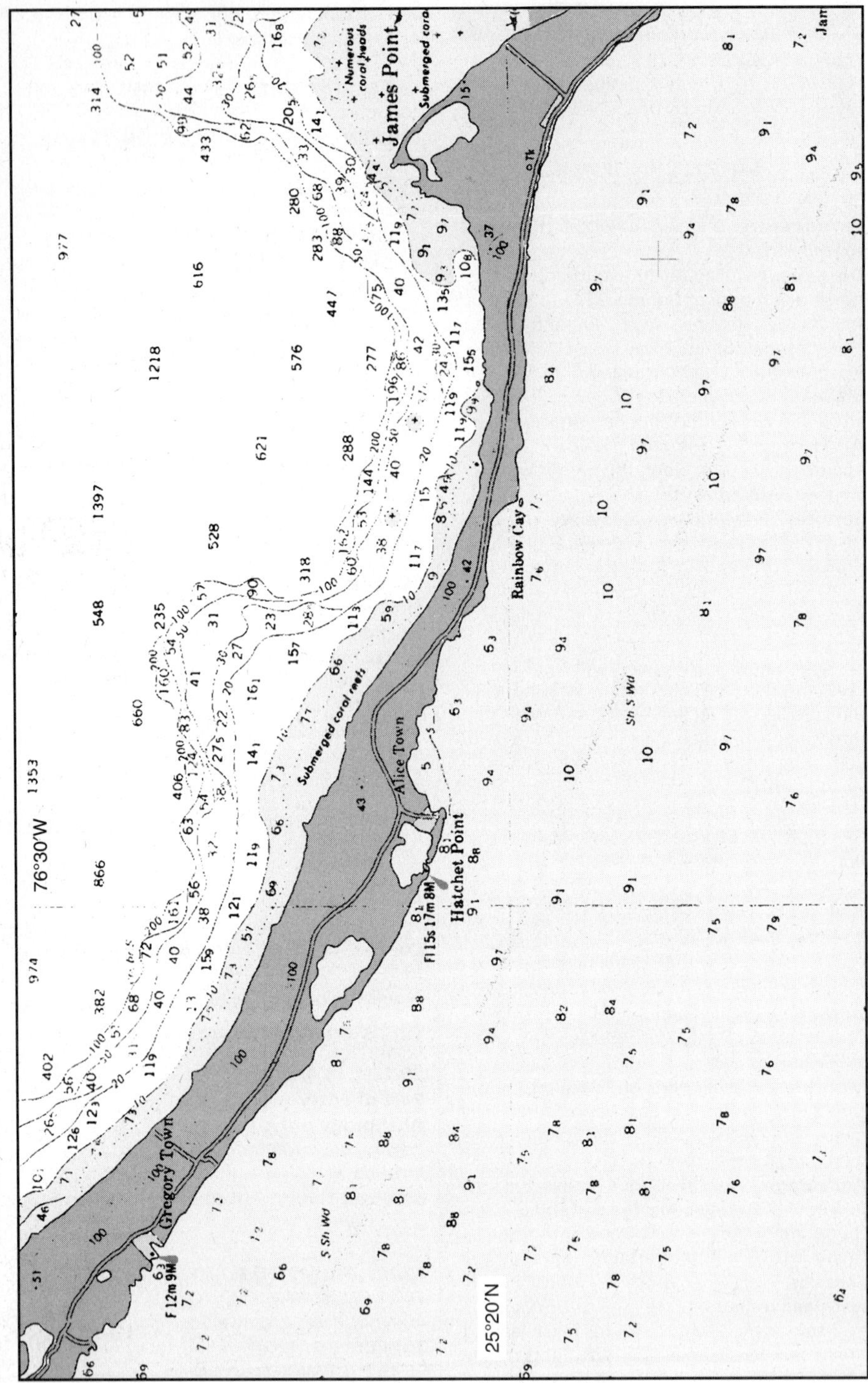

Hatchet Bay, soundings in meters *(from DMA 26307, 15th ed, 8/86)*

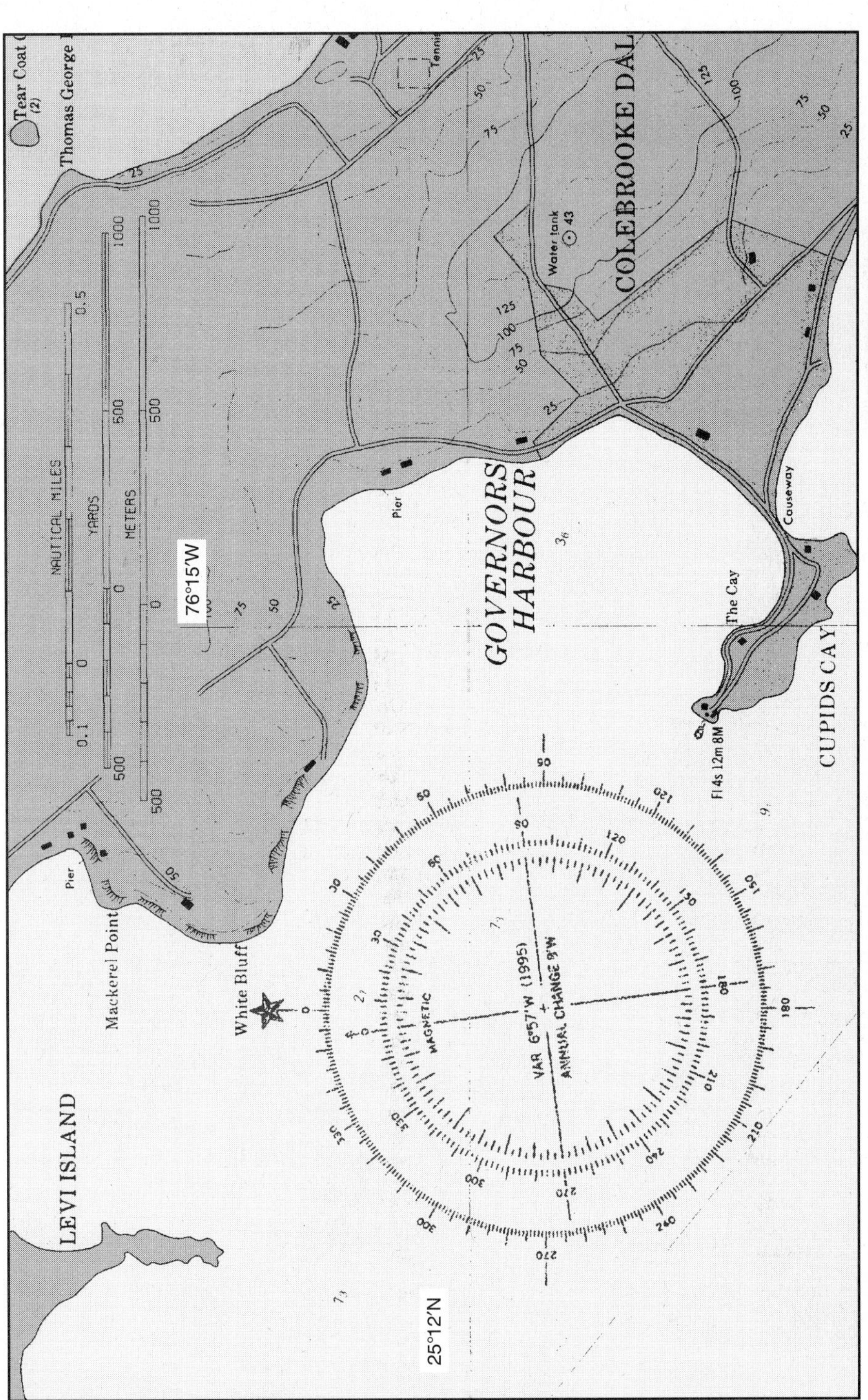

Governor's Harbour, soundings in meters *(from DMA 26307, 15th ed, 8/86)*

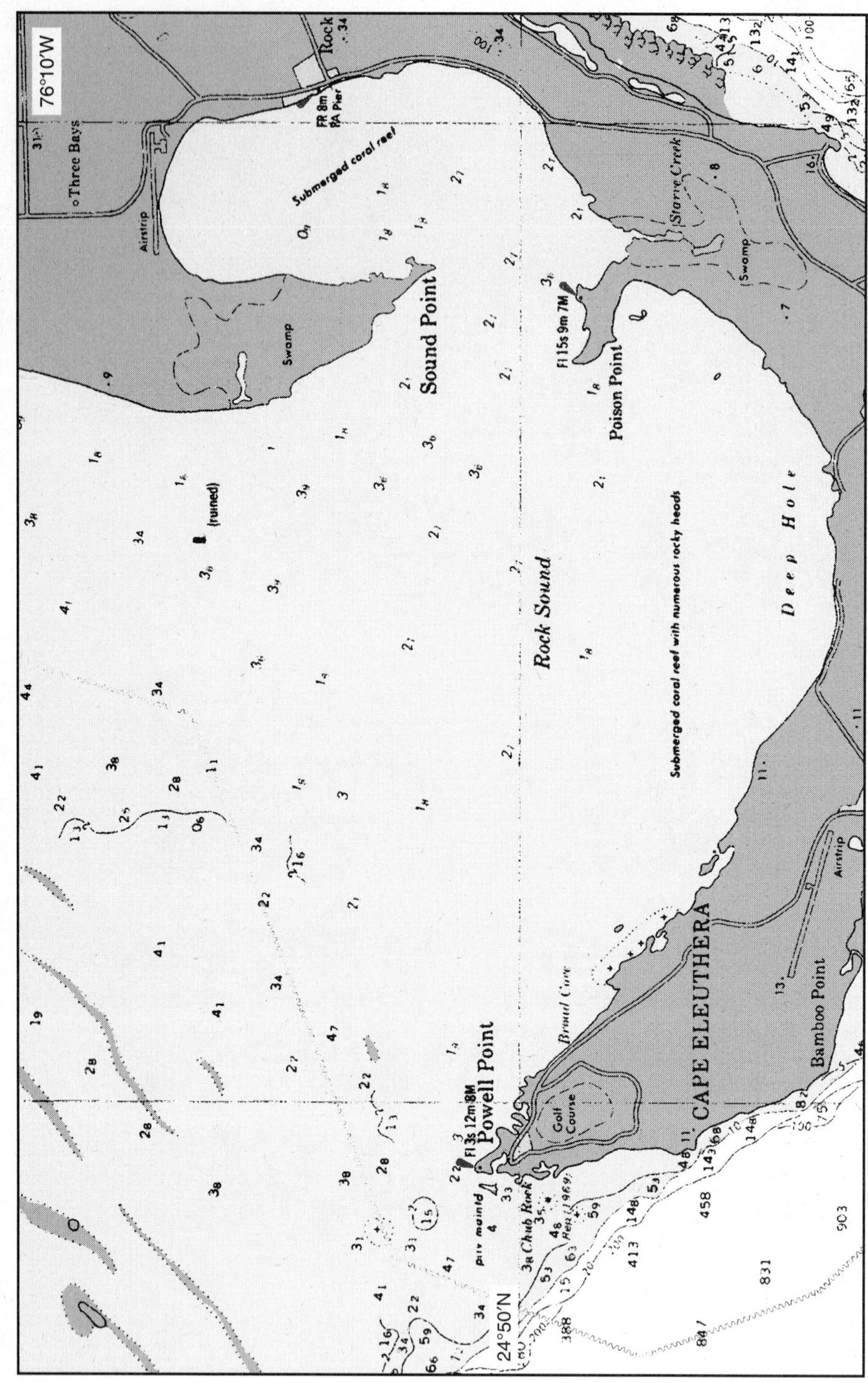

Rock Sound, soundings in meters *(from DMA 26307, 15th ed, 8/86)*

POISON POINT LIGHT, 24 52.4N, 76 13.7W. Fl W 15s, 29ft, 7M. White mast, 12ft.
ROCK SOUND SETTLEMENT LIGHT, 24 54.4N, 76 12.3W. F R, 25ft. Mast, 20ft.

ROCK SOUND

24 52N, 76 10W
Port of entry
Dockage: You may be able to tie up temporarily to the government wharf. Check your depths carefully as you approach—depths are 4 feet to 5 feet alongside.
Anchorage: Anchor wherever depths permit. There are many shallow spots throughout the harbor, some of which are hard coral.
Services: There are groceries and shops ashore. You may be able to obtain water at the government wharf via jugs or hose. Fuel may be jugged from a service station.

POWELL POINT LIGHT, 24 52.4N, 76 21.9W. Fl W 3s, 38ft, 8M. Aluminum-colored steel framework tower, 32ft.
FREE TOWN LIGHT, 24 48.1N, 76 18.2W. F W, 19ft, 7M. White pole.
WEMYSS BIGHT LIGHT, 24 45.0N, 76 14.1W. F W, 27ft, 2M. White pole, 18ft.

DAVIS HARBOUR

24 44N, 76 14W
Dockage: A privately marked channel leads to Davis Harbour Marina (VHF channel 16; tel., 242-334-6303) in a small dredged basin, where slips are available. It is reported to have fuel, water, a restaurant, and limited supplies.
Services: Fuel, water, and some ice is available. Restaurants are close by. Services may also be available in nearby Cape Eleuthera Marine.

ELEUTHERA POINT, SE extremity of island, 24 36.8N, 76 08.8W. Fl W 4.6s, 61ft, 6M. Light colored framework structure on white house, 25ft.

LITTLE SAN SALVADOR

LITTLE SAN SALVADOR LIGHT, 24 34.0N, 75 56.0W. Fl W 2.4s, 69ft, 13M. Gray metal framework structure, 30ft. Visible 240°–110°, obscured 110°–130°, visible 130°–140°, obscured 140°–170°, visible 170°–190°, obscured 190°–200°, visible 200°–220°, obscured 220°–240°.

LITTLE SAN SALVADOR

24 34N, 75 56W
Anchorage only: The anchorage in West Bay, by a spectacular crescent-shaped beach, is popular. As of April 1997, construction has begun on a cruise ship destination resort at this beach.

EXUMAS

BEACON CAY (NORTH ROCK) LIGHT, Ship Channel, 24 52.4N, 76 49.8W. Fl W R 3s, 58ft, 8M. Gray structure, 30ft. R 292°–303°, W 303°–292°; partially obscured by neighboring islands 319°–006°.

ALLAN'S CAY

24 45N, 76 50W
Anchorage only: This group of cays is a popular stop at the northern end of the Exumas chain of islands. The easiest approach is from the west. As there are no aids to navigation, the approach should be made in good light. Once inside, most boats anchor on either side of the shallow sandbar in the middle of the harbor. It is possible to anchor in the bight on the north end of Southwest Allan's Cay. Leaf Cay, the small cay to the east of Allan's, is famous for the iguanas that wander its beaches.

HIGHBORNE CAY

24 42N, 76 49W (Charted as Highburn Cay)
Dockage: Some dockage is available at the small marina (VHF channel 16) in the basin at the south end of this private island. A range on the hill may lead you clear of the coral shoals to the west of the marina. An approach in good light is recommended.
Anchorage: A few moorings are available for overnight rental in the small basin off the docks. There is no room to anchor. The small anchorage south of the basin is prone to surge.
Services: Fuel, water, public telephones, and some groceries are available.

ELBOW CAY LIGHT, W end, 24 31N, 76 49W. Fl W 2s, 46ft, 11M. Gray structure, 25ft.

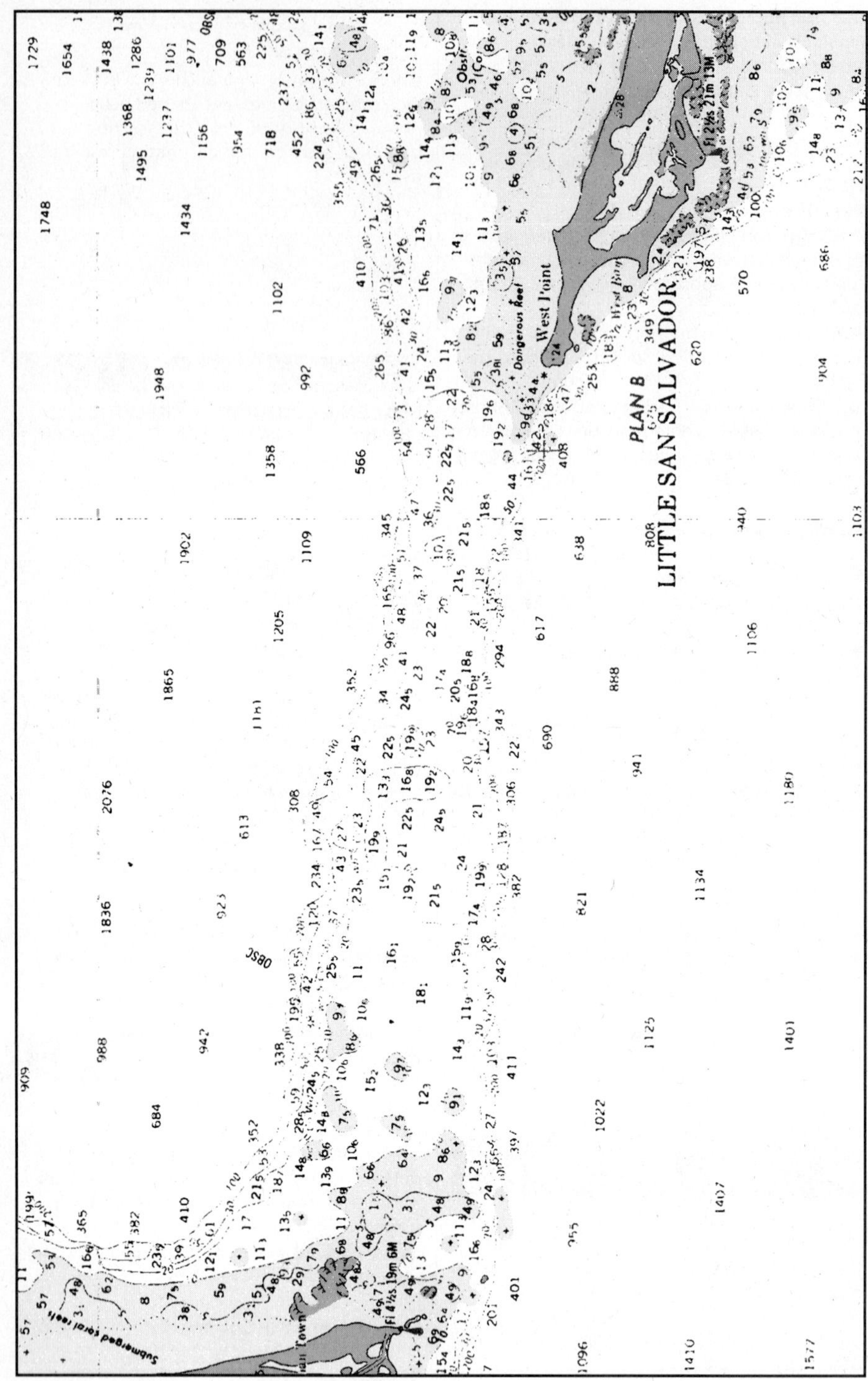

Little San Salvador, soundings in meters *(from DMA 26307, 15th ed, 8/86)*

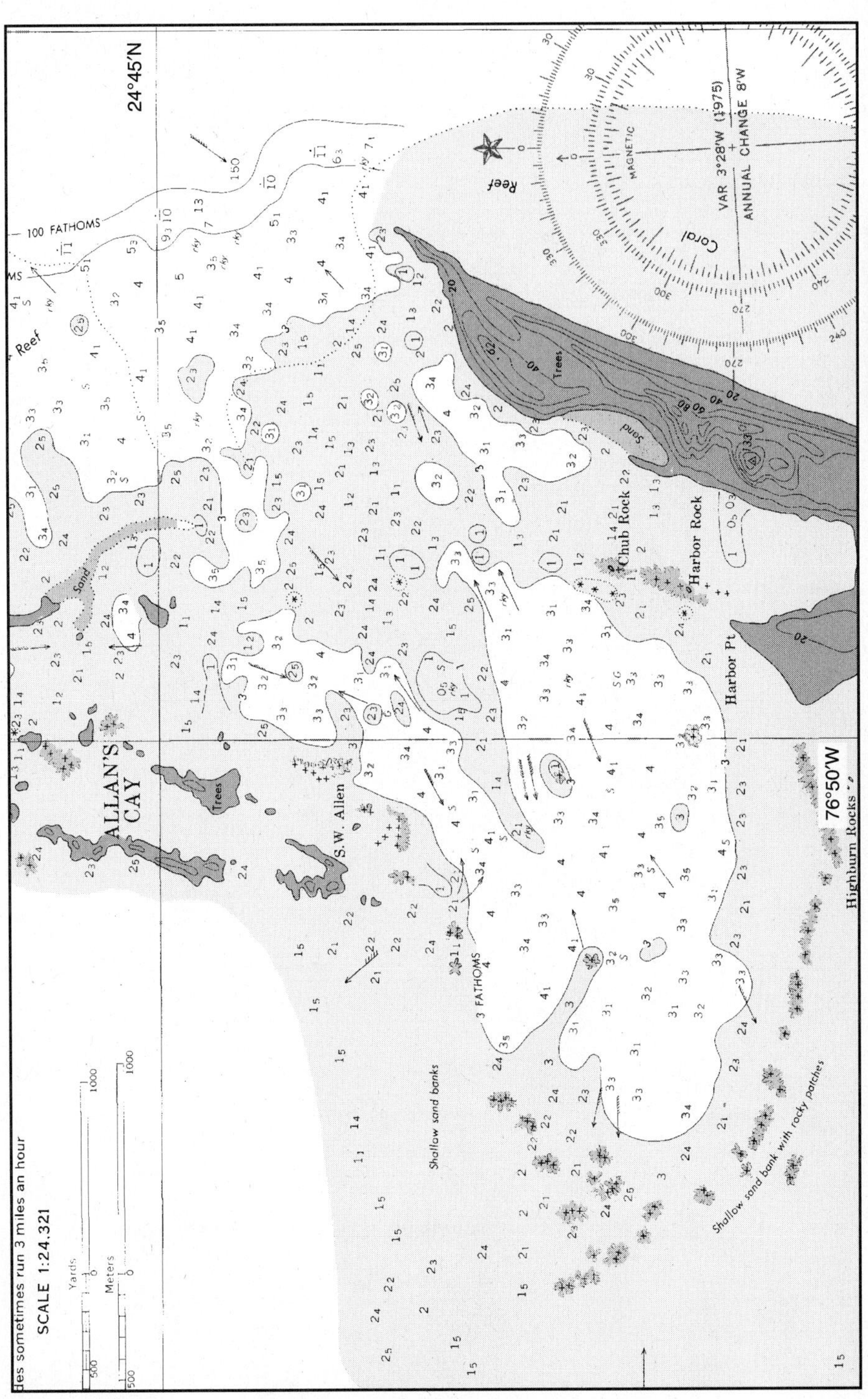

Highborne Cut, soundings in fathoms and feet *(from DMA 26257, 14th ed, 1/79)*

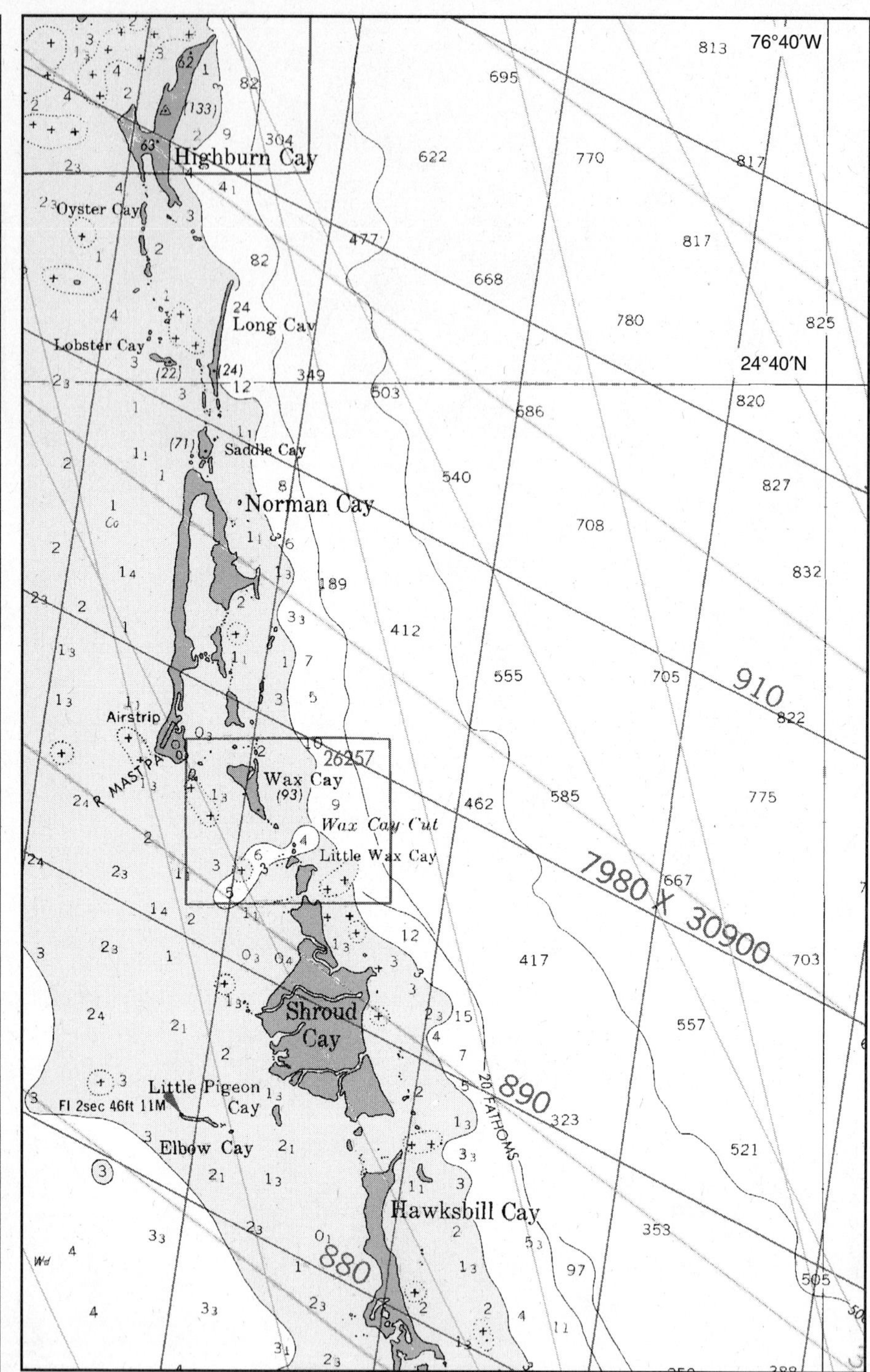

Highborne to Hawksbill, soundings in fathoms *(from DMA 26305, 4th ed, 3/80)*

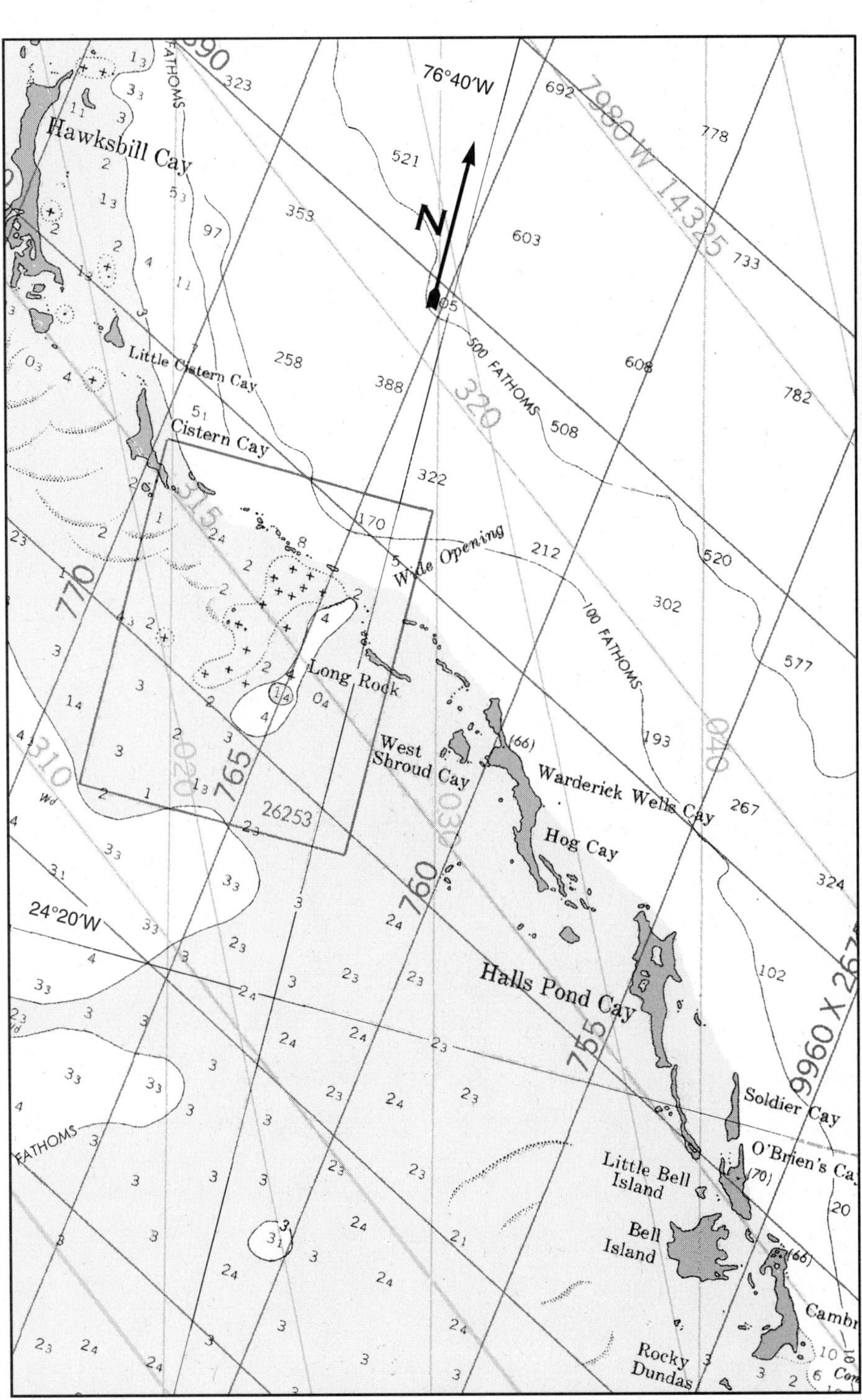

Warderick Wells, soundings in fathoms *(from DMA 26305, 4th ed, 3/80)*

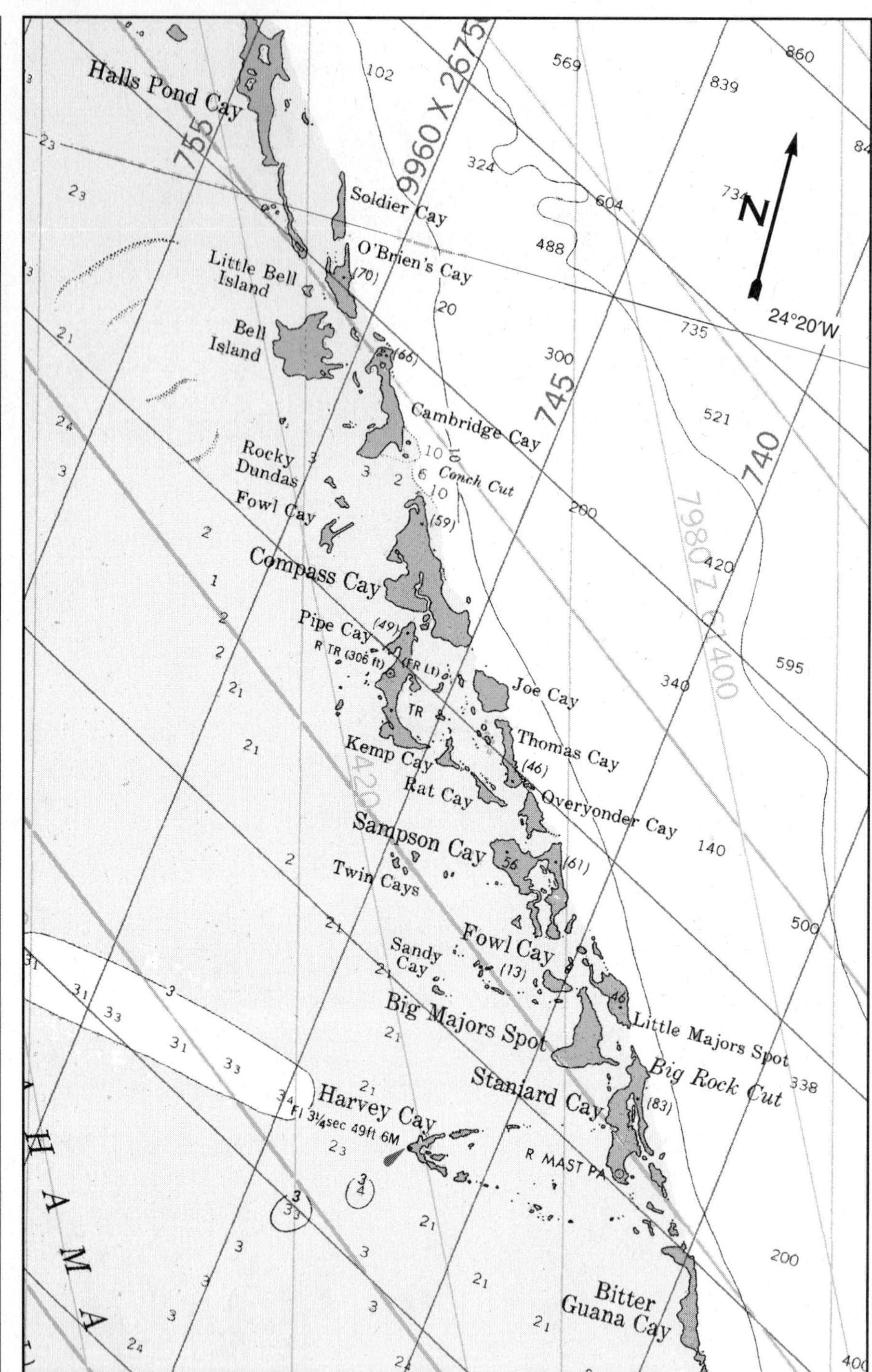

Sampson Cay and Staniel Cay, soundings in fathoms *(from DMA 26305, 4th ed, 3/80)*

WARDERICK WELLS

24 24N, 76 38W
Exuma Cays Land and Sea Park: The Bahamas National Trust administers the park, which has its headquarters at the north end of Warderick Wells Cay. Park rangers monitor VHF channel 16; call ahead before you approach the mooring area. All fishing and collection of plants, wildlife, and marine life, living or dead, are prohibited within the park. Visitors are urged to join the Exuma Cays Park Support Fleet for $30 a year. For more information contact: The Bahamas National Trust, P.O. Box N-4105, Nassau, Bahamas (tel., 242-393-1317; fax, 242-393-4978).
Moorings: There are 24 park moorings at north Warderick Wells and 5 in the south anchorage for the reasonable fee of $15 for two nights. Members of the Support Fleet moor for free (membership is available at the park headquarters or through the Nassau office).
Anchorage: Anchoring is no longer permitted in the mooring area by the north end of Warderick Wells Cay; there is an anchorage by Hog Cay at the south end of Warderick Wells Cay.
Services: The park headquarters provides maps of the trails ashore and descriptive information on the flora and fauna.

SAMPSON CAY

24 13N, 77 29W
Dockage: Contact Sampson Cay Club (VHF channel 16; tel., 242-355-2034), where very protected dockage is available in the lagoon for those leaving boats for extended periods.
Anchorage: Anchor west of the docks in good holding.
Services: The club has fuel, water, ice, some groceries, liquor, air refills, and a restaurant and bar. Reservations by 1600 are recommended for the restaurant (VHF channel 16). A seaplane is available for chartered flights and emergencies. This is also the headquarters of a salvage operation complete with a mobile crane.

STANIEL CAY

24 10N, 76 27W (Charted as Staniard Cay)
Dockage: Contact the Staniel Cay Yacht Club (242-355-2024), the Happy People Marina (242-355-2008), or the Thunderball Club for dockage. All three monitor VHF channel 16. Strong currents run parallel to the yacht club docks.
Moorings: Sampson Cay Club has recently added 10 moorings west of their docks and around to the southwest side of the cay, and Club Thunderball has approximately 12 moorings in the basin between the grotto and the club.
Anchorage: Boats anchor in several places here. A favorite bad-weather spot is in the narrow channel east of Big Majors Spot. Some boats anchor east of the yacht club docks or in the shallow basin south of the docks. Be wary of old moorings and coral south of the docks. The approach depths to all areas are reported to be around 6 feet.
Services: Fuel, water, ice, limited groceries, marine supplies, a restaurant, bars, and accommodations are available. General repair services are available.

HARVEY CAY LIGHT, 24 09N, 76 28W. Fl W 3.3s, 49ft, 6M. Gray beacon, 12ft.
BITTER GUANA CAY LIGHT, N end of Dotham Cut, 24 07N, 76 23W. Fl W 5s, 33ft.

LITTLE FARMER'S CAY

23 57N, 76 19W
Moorings and dockage: Contact Farmer's Cay Yacht Club on the VHF radio.
Anchorage: Make your approach in good light as this area is loaded with shoals and coral areas. There is a narrow strip of deep water along the shore of Great Guana Cay and one along the east side of Little Farmer's.
The mouth of the cove on the end of Big Farmer's is a good anchorage. Farmer's Cay Cut is a good passage to the deeper waters of Exuma Sound. The passage on the banks side gets quite shallow south of Little Farmers.
Services: Fuel, water, ice, telephone, electricity, and some groceries are available. There are a couple of small restaurants and bars.

GALLIOT CUT, N end of Cave Cay, 23 55.2N, 76 16.8W. Fl W 4s, 49ft, 7M. Aluminum colored steel framework tower.
CONCH CAY LIGHT, approach to George Town, 23 33.5N, 75 48.0W. Fl W 5s, 40ft, 8M. Gray metal framework tower, 30ft.
SIMON POINT LIGHT, approach to George Town, 23 32N, 75 48W. F W, 40ft, 5M. Mast, 15ft. (Reported destroyed; replaced by daymark.)

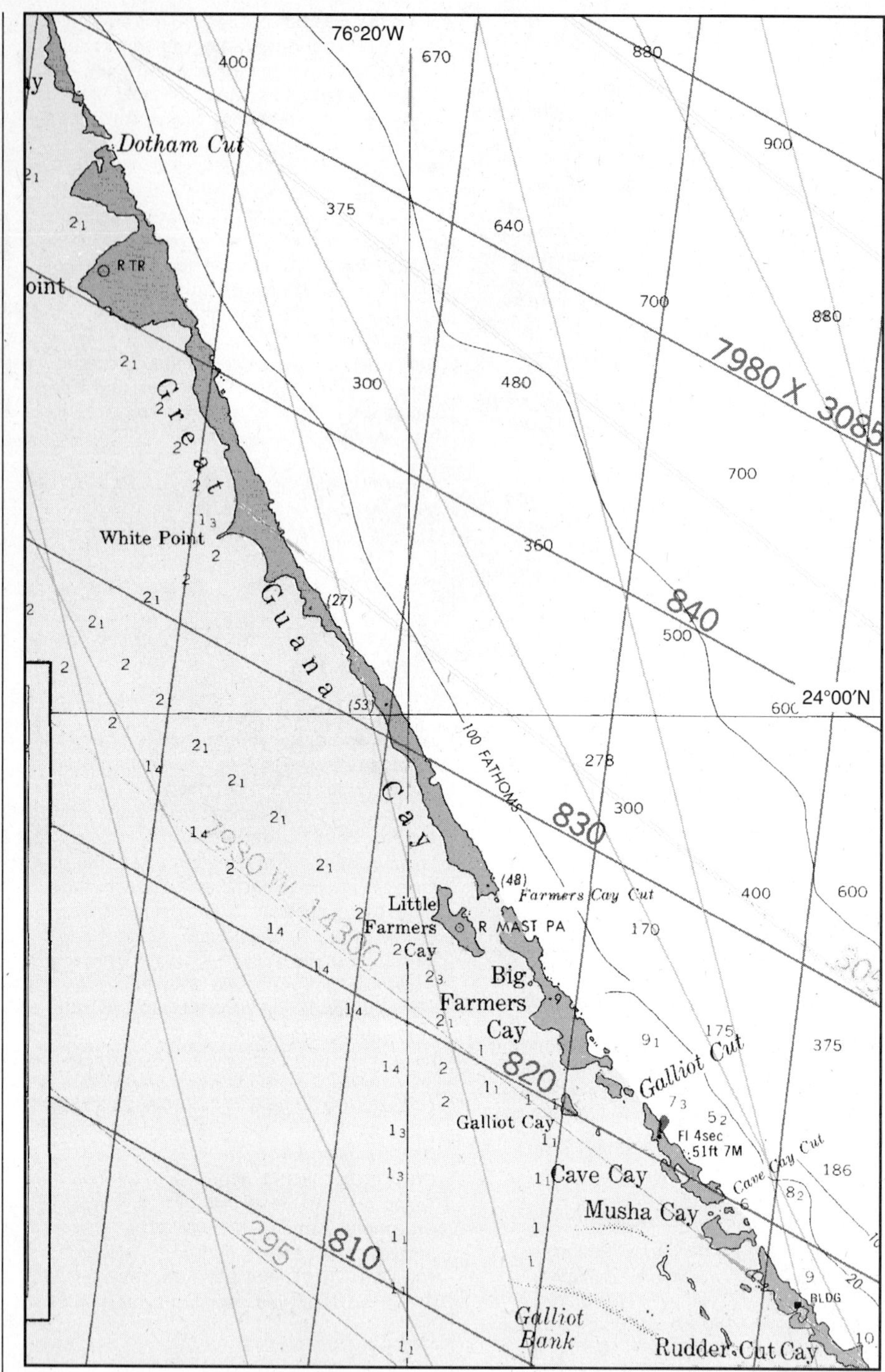

Little Farmer's Cay, soundings in fathoms *(from DMA 26305, 4th ed, 3/80)*

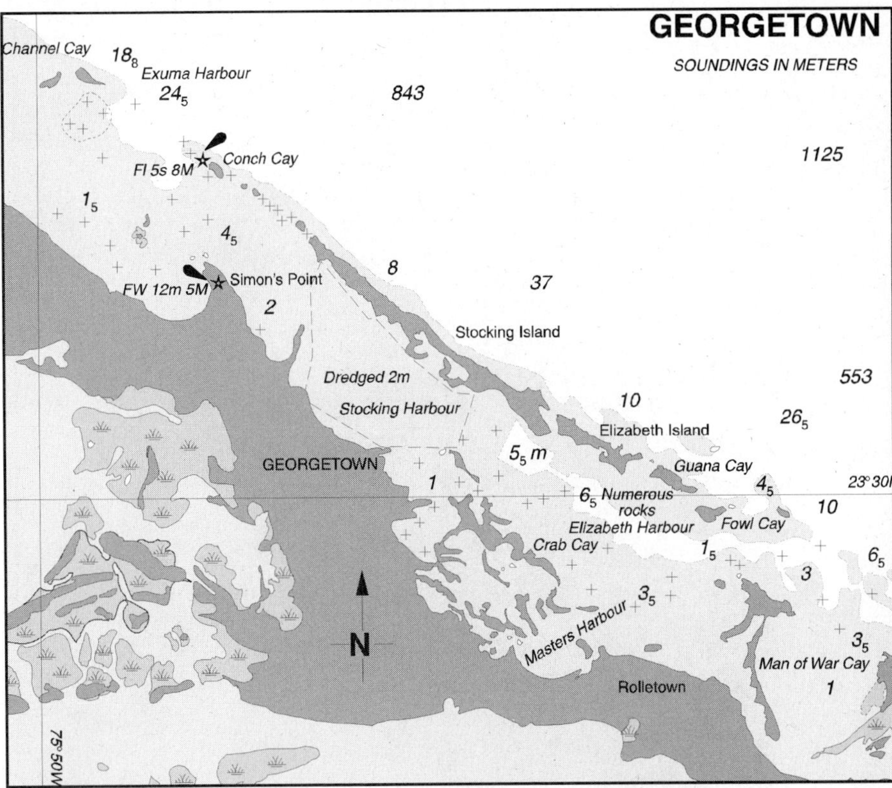

George Town approaches, soundings in meters

GEORGE TOWN

23 31N, 75 47W

Port of entry: Customs is located in the government building near the government wharf. You may tie to the wharf temporarily.

Dockage: Available at Exuma Docking Services in Kidd Cove (VHF channel 16, call sign, Sugar One; tel., 242-336-2578). The approaches to the dock at low tide are less than 6 feet.

Anchorage: Hundreds of cruising boats may be found here during the winter. A popular spot is off the "holes" at Stocking Island or in one of the "holes" themselves. Boats with drafts of around 6 feet may enter the holes with local knowledge. It is wise to scout the channels with your dinghy. Many boats anchor off the well-known Peace and Plenty Hotel (242-336-2551; in the U.S., 800-525-2210; fax 242-336-2093; e-mail, ssbpeace@aol.com; Web, www.peaceandplenty.com), where there is a convenient dinghy dock. There is another popular dinghy dock in Lake Victoria behind the Exuma Market.

Services: Fuel, water, ice, propane, groceries, some marine supplies, a telephone, restaurants, and bars are available. Good air connections can be made via Nassau or southern Florida.

George Town is the southernmost port of call for many Bahamas cruisers. There is bound to be a large gathering of boats here, but there always seems to be room for more. The popularity of this port has promoted the development of a good selection of shops oriented towards cruisers. This is the best restocking port in the southern Bahamas.

HAWKSBILL ROCKS, 23 25.5N, 76 05.6W. Fl W 3.3s, 32ft, 6M. White conical metal tower, black bands, 23ft.

JEWFISH CUT, 23 27.3N, 75 57.4W. Fl W 2.5s, 38ft, 8M. Black metal conical tower, white top, 23ft.

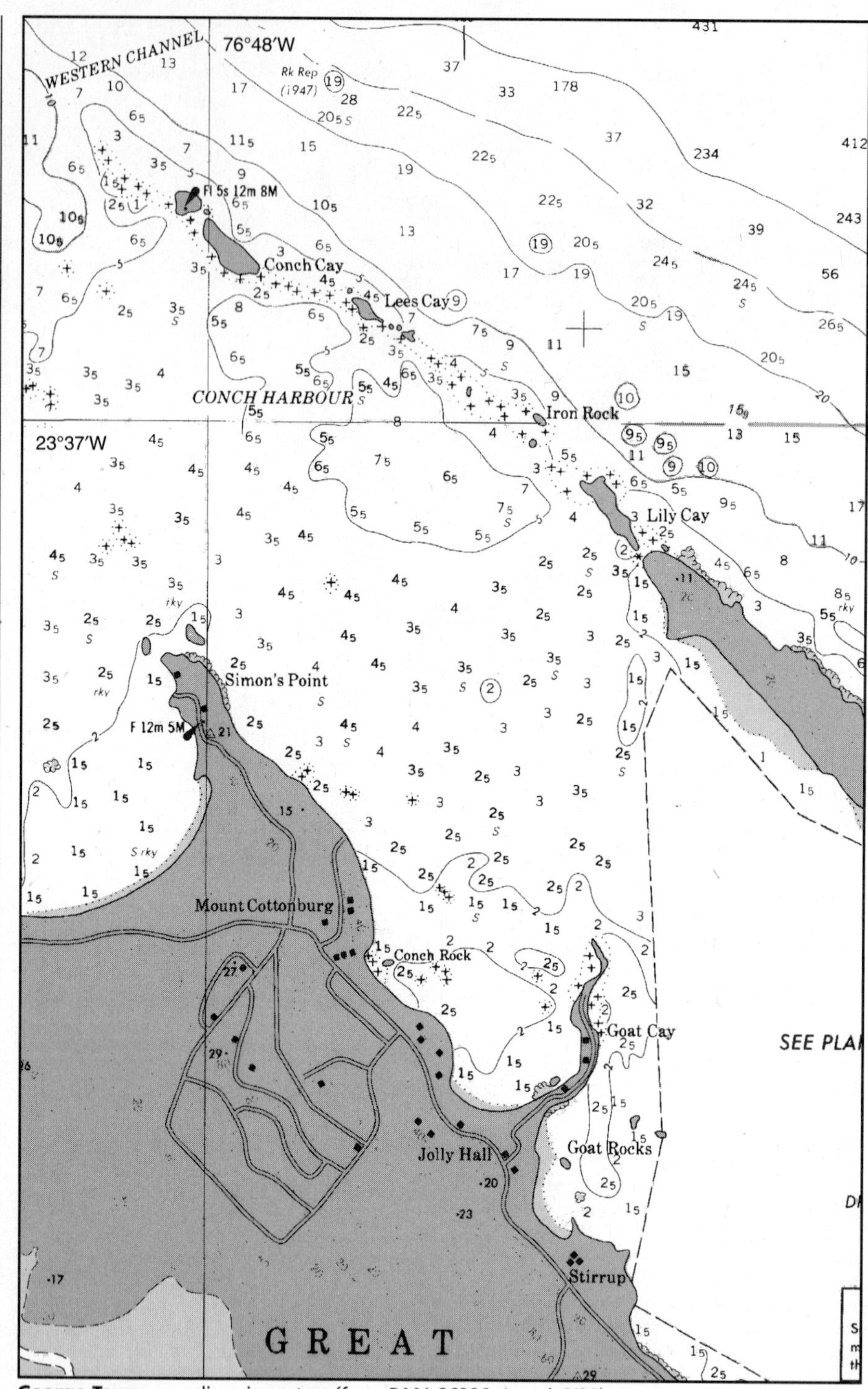

George Town, soundings in meters *(from DMA 26286, 1st ed, 8/86)*

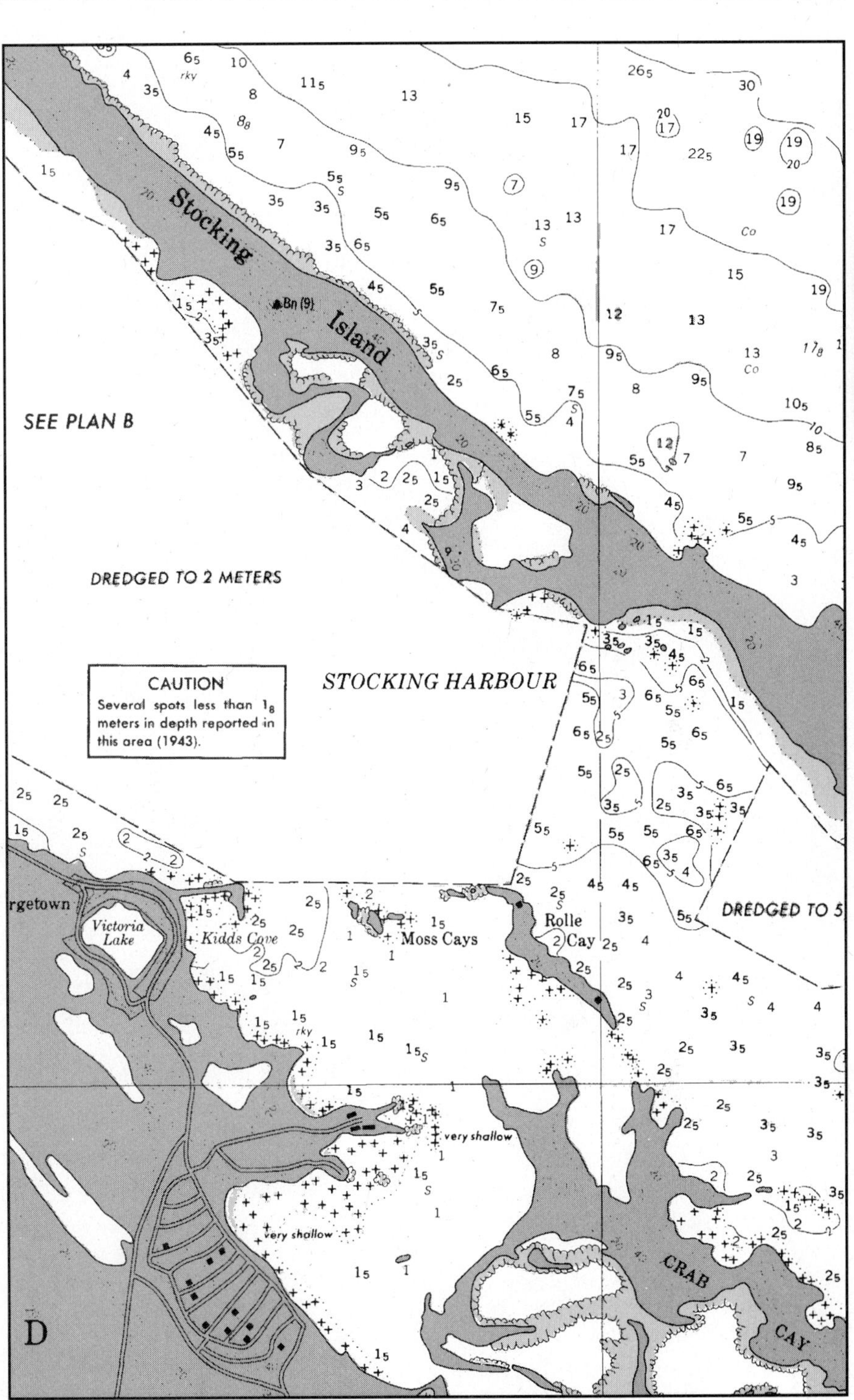

Stocking Island, soundings in meters *(from DMA 26286, 1st ed, 8/86)*

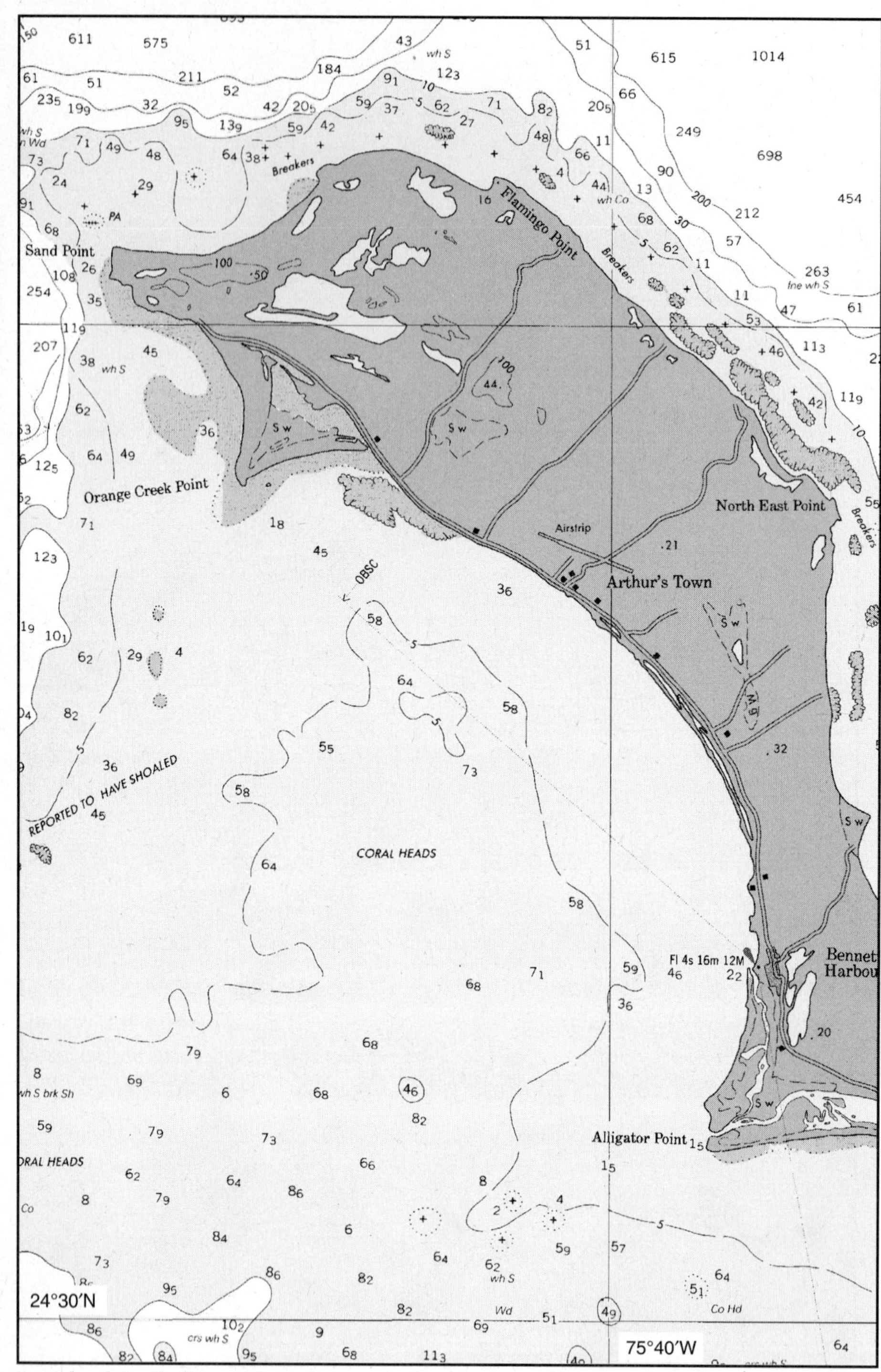

Cat Island North, Bennett's Harbour, soundings in meters *(from DMA 26284, 2nd ed, 9/93)*

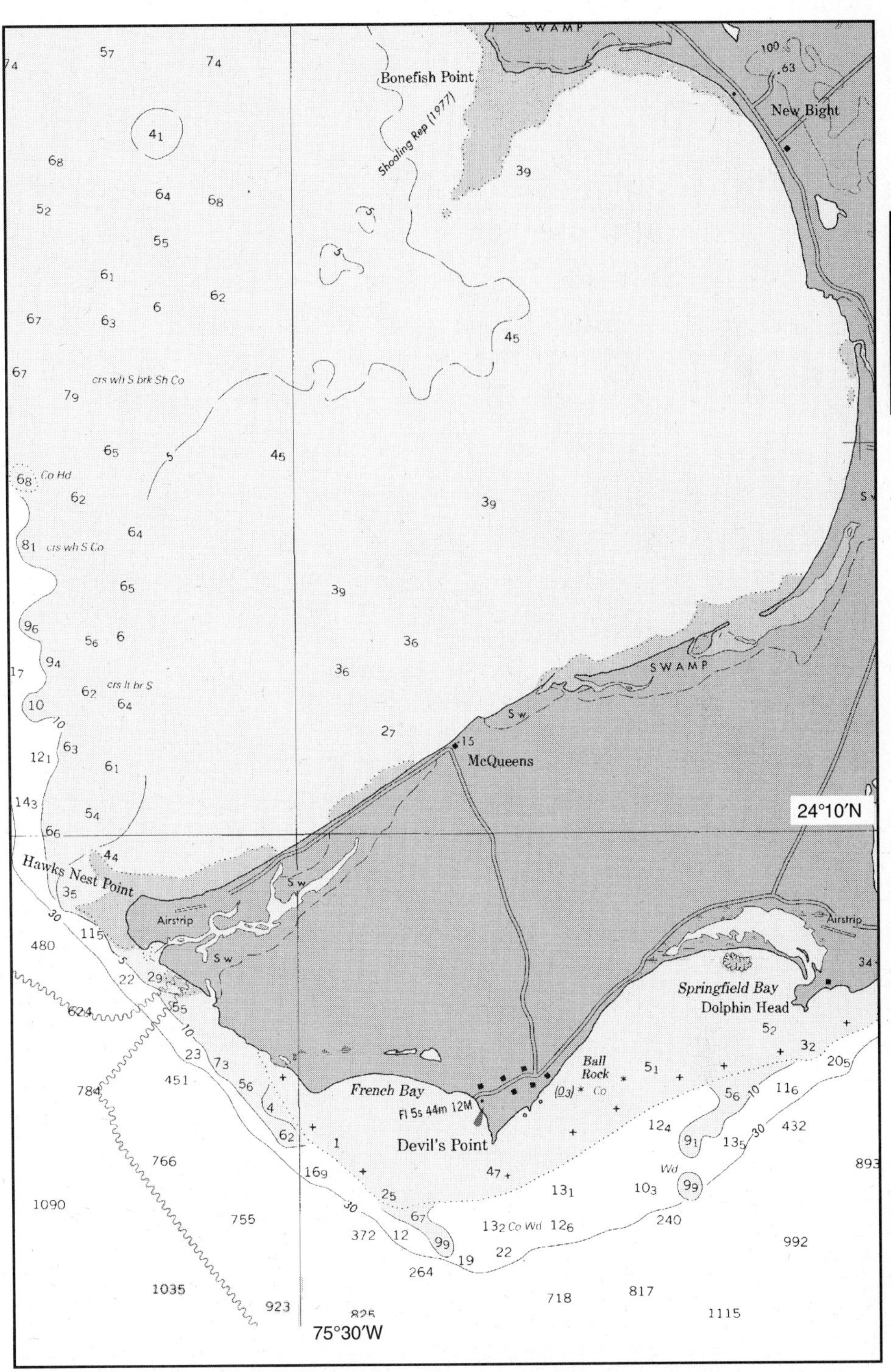

Cat Island South, Hawksnest Creek, soundings in meters *(from DMA 26284, 2nd ed, 9/93)*

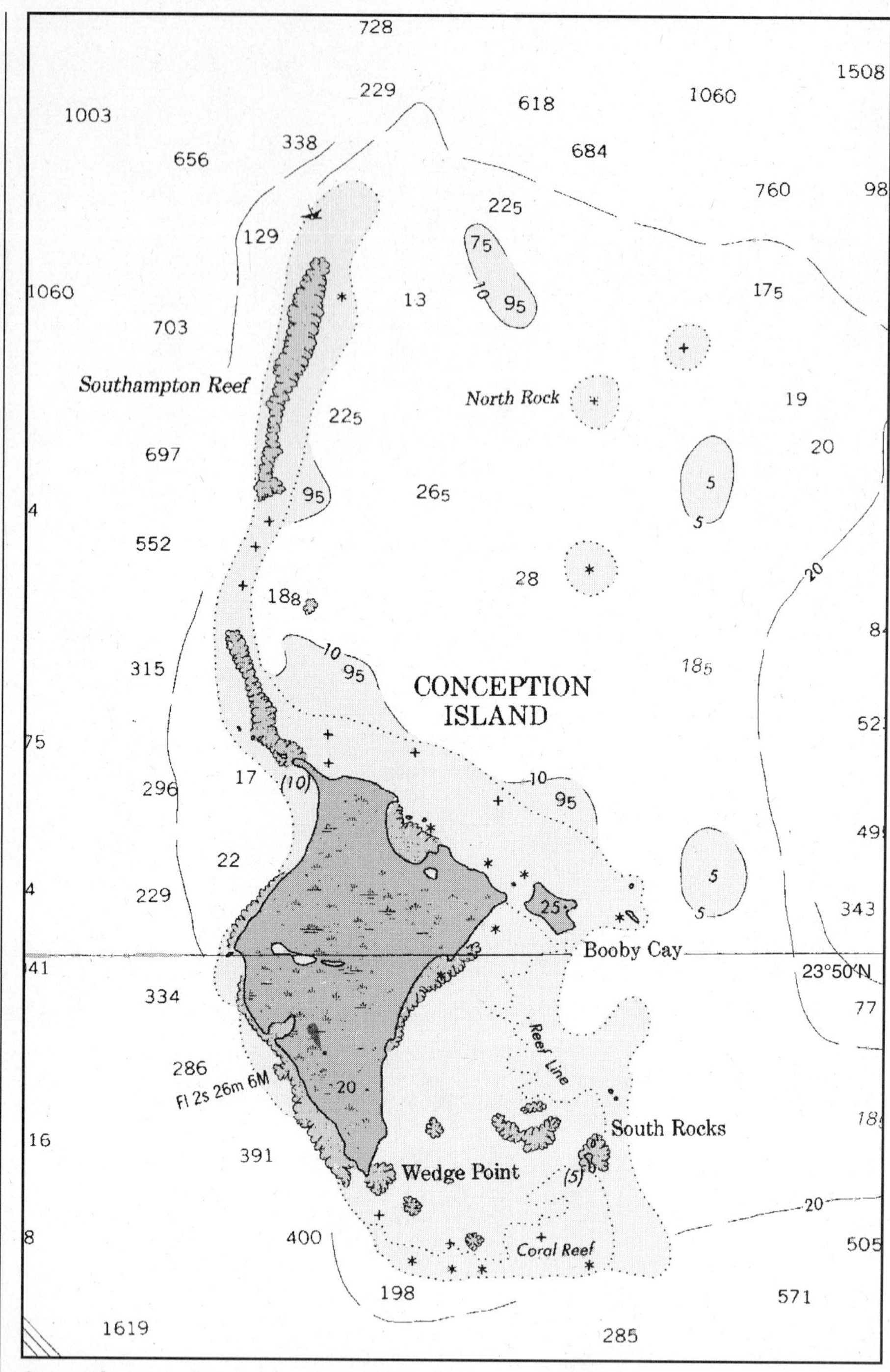

Conception, soundings in meters *(from DMA 26284, 2nd ed, 9/93)*

CAT ISLAND

BENNETT'S HARBOUR, entrance to Harbour Creek, 24 33.5N, 75 38.3W. Fl W 4s, 53ft, 12M. Light gray structure, 30ft. Visible 350°–130°.

BENNETT'S HARBOUR

24 33N, 75 38W
Dockage: You may be able to tie up temporarily to the government wharf.
Anchorage: There is not a lot of shelter off the west coast of Cat Island. Some of the creeks may offer shelter to shallow-draft boats.
Services: Services are very limited.

SMITH TOWN, 24 19.8N, 75 28.0W. Fl W 3.3s, 38ft, 7M. Gray metal framework tower.
DEVILS POINT LIGHT, on summit, 823 meters, NW of point, 24 07.5N, 75 28.0W. Fl W 5s, 143ft, 12M. White metal framework tower, 30ft.

HAWKSNEST CREEK

24 09N, 75 32W
Dockage: Contact the Hawks Nest Club Marina on VHF channel 16. Depths are reported to be about 6 feet.
Anchorage: You can anchor in the narrow channel between the shoals. Don't block the channel to the marina.
Services: Fuel, water, and ice are available.

CONCEPTION ISLAND

CONCEPTION ISLAND LIGHT, 23 50N, 75 08W. Fl W 2s, 84ft, 6M. Gray structure, 30ft.

CONCEPTION

23 50N, 75 07
Bahamas National Trust: This uninhabited island is a land and sea park overseen by The Bahamas National Trust, with regulations similar to those of the Exumas park. For more information contact: The Bahamas National Trust, P.O. Box N-4105, Nassau, Bahamas (tel., 242-393-1317; fax, 242-393-4978).
Anchorage only: Anchor in the bay at the northwest end of the island. Proceed in slowly as coral heads have been reported. The coral becomes considerably thicker within 0.2 mile of the beach.

RUM CAY

COTTON FIELD POINT LIGHT, 23 39.2N, 74 51.6W. L Fl W Y R 10s, 75ft, 10M. R 083.5°–006.5°, W 006.5°–014.5°, Y 014.5°–075.5°, W 075.5°–083.5°.
PORT NELSON LIGHT, near wharf, 23 38.4N, 74 40.0W. F W, 18ft, 5M. Mast, 13ft.

PORT NELSON

23 40N, 74 50W
Dockage: You may be able to tie up to the government dock temporarily. Depths are reported to be less than 6 feet. Dockage is also available at Sumner Point Marina (VHF channel 16; information in the U.S., 305-759-5563), just inside Sumner Point at the east end of St. George's Bay.
Anchorage: Anchor west of town, as depths permit. The charted approach takes you through a deep passage between several coral reefs. A bearing of 013°T on Cotton Field Point brings you safely across the reef. When the dock at Port Nelson bears 081°T, steer for the dock. Or anchor outside the entrance to Sumner Point marina or inside the marina's dredged basin (in both cases, call ahead for information).
Services: Fuel, electricity, water, and a new small restaurant and bar are available at the marina. Supplies are limited. Direct telephone service is under construction; Rum Cay is presently served by Batelco.

Port Nelson has fewer than 100 residents, so services are limited. The fishing is reported to be excellent. With the recent opening of Sumner Point Marina, Rum Cay is attracting cruisers en route to and from the Turks and Caicos who have already cleared in at Stella Maris or San Salvador.

SAN SALVADOR

DIXON HILL LIGHT, near NE point, 24 05.4N, 74 27.0W. Fl (2) W 10s, 163ft, 23M. White stone tower, dwelling each side, 72ft. Partially obscured 001°–008°, 010°–068°, and 076°–095°.
COCKBURN TOWN LIGHT, near landing, 24 03.0N, 74 32.0W. F W, 23ft, 1M. Mast, 12ft.

COCKBURN TOWN

24 03N, 74 32W
Port of entry: Boats anchored off Cockburn Town should call customs, whose office is at the airport. Boats at the marina may request

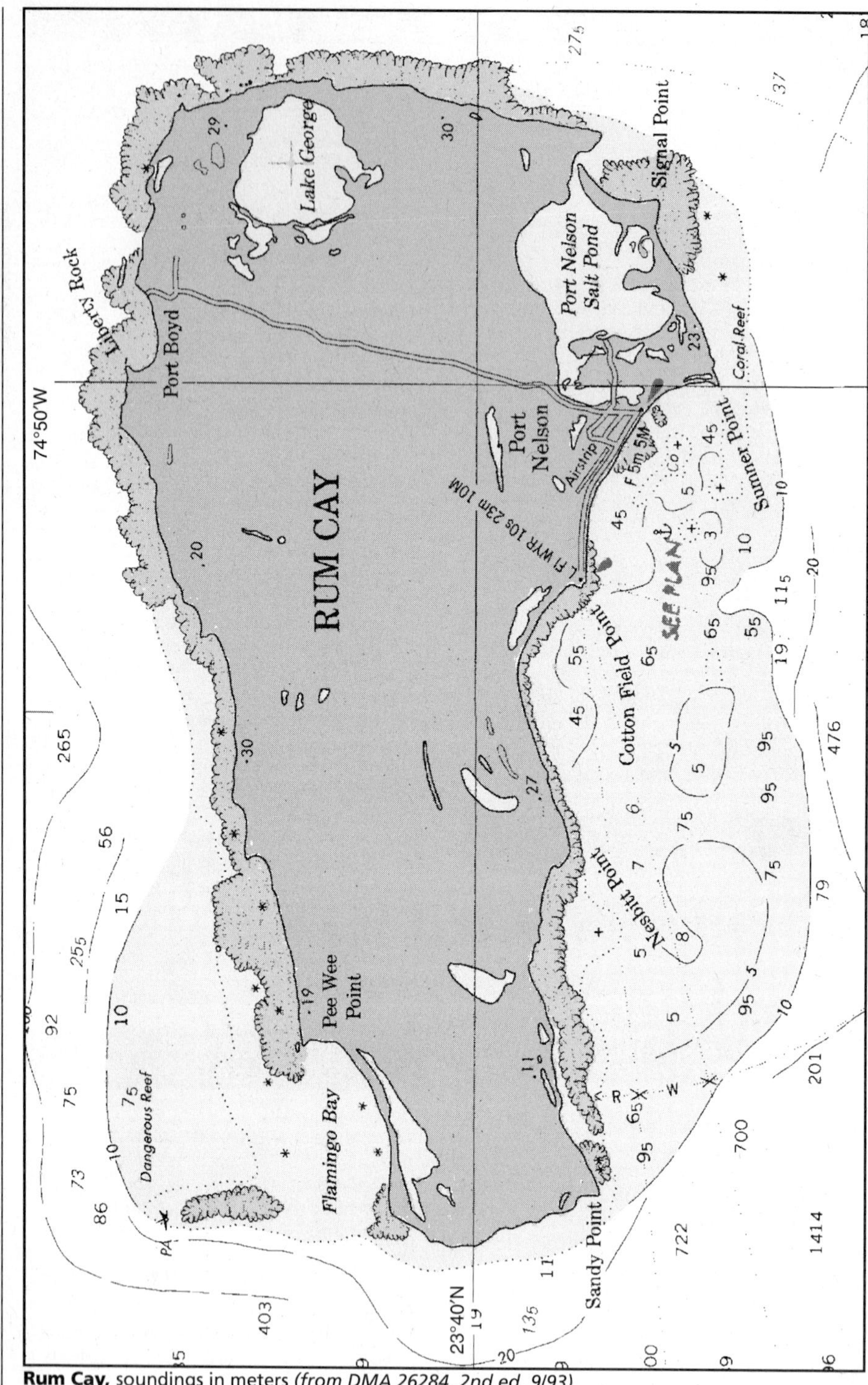

Rum Cay, soundings in meters *(from DMA 26284, 2nd ed, 9/93)*

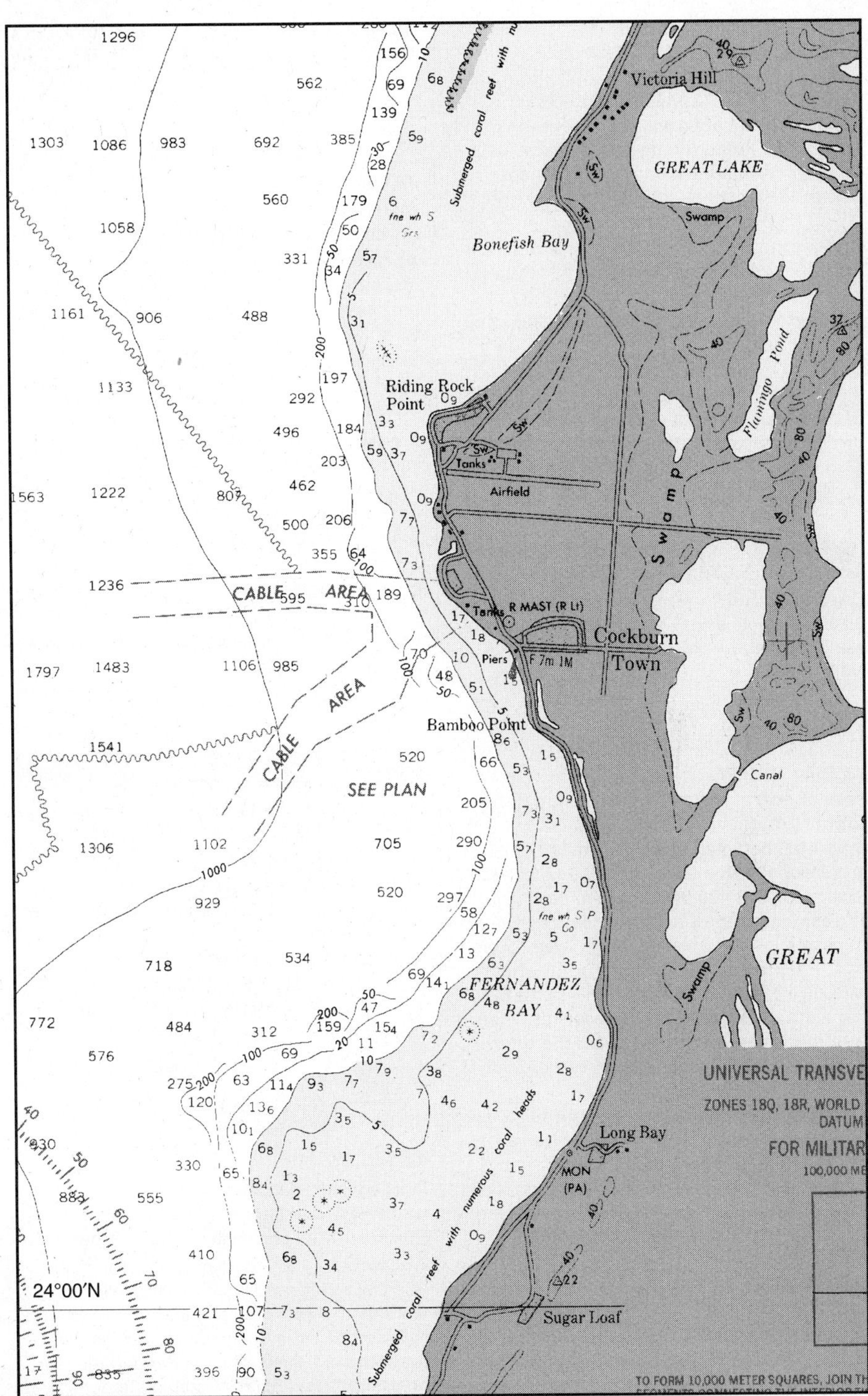

Cockburn Town, soundings in meters *(from DMA 26281, 3rd ed, 7/86)*

the marina to alert customs.
Dockage: Available at Riding Rock Marina (VHF channel 06; tel., 242-331-2631, information in the U.S., 800-272-1492, 954-359-8353) in a basin located about halfway between town and Riding Rock Point. A range (075°) leads boats safely in; the marina is undergoing expansion, scheduled to continue into 1998. Entrance depths are about 7 feet at low tide.
Anchorage: In good weather boats can anchor in the bay off of town. Be prepared to leave if bad weather threatens.
Services: Fuel, water, a telephone, and some supplies available.

This island is considered the first landing place of Columbus in the New World. A monument has been erected on the eastern side of the island and others well south of the government dock at Cockburn Town on the west side.

LONG ISLAND

CAPE ST. MARIA, near N extremity of island, 23 40.5N, 75 20.0W. Fl W 3.3s, 99ft, 14M. Gray metal framework tower, 30ft. Obscured 240°–340°.

STELLA MARIS

23 33N, 75 16W
Port of entry: The marina will alert customs to your arrival.
Dockage: Between them, the Stella Maris Inn and Marina (VHF channel 16; tel., 242-338-2051; in the U.S., 800-426-0466) have all facilities. The approach depths to the marina are reported to be about 4 feet at low tide and 7 feet at high tide. Tides run approximately 2 hours after Nassau. Head south about 1.5 miles past Dove Cay, until the sandbar to the east ends. Then turn to the east and head toward the bright yellow sign at the marina entrance (coordinates, 23°33.4′N, 75°16.5′W).
Services: A marine railway makes Stella Maris the only full-repair facility south of Nassau in the Bahamas. It also has fuel, water, ice, and supplies and offers engine and fiberglass repair services. Elsewhere on the island are banking services, a post office, and restaurants.

SIMMS LIGHT, 23 29N, 75 14W. F W, 23ft, 4M. Mast.
BOOBY ROCK LIGHT, approach to Clarence Town, 23 07.0N, 74 58.3W. Fl W 2s, 39ft, 8M. White tower, red bands, 26ft.
HARBOUR POINT LIGHT, approach to Clarence Town, 23 06.2N, 74 57.5W. F W, 25ft, 3M. White mast and hut, 18ft.

CLARENCE TOWN

23 06N, 74 58W
Dockage: You may be able to tie up to the government wharf temporarily.
Anchorage: Anchor off town, avoiding the charted cables. This area is wide open to north and west winds and may be untenable in bad weather.
Services: Fuel, water, groceries, a telephone, ice, and a restaurant are available.

The best approach is from the north, passing west of Booby Rocks. Approach in good light to avoid the many reefs and shoals in the area. As the administrative center of Long Island, it offers a variety of services.

SOUTH POINT LIGHT, Turbot Hill, 22 51.3N, 74 51.2W. Fl W 2.5s, 61ft, 12M. Gray metal structure, 30ft. Partially obscured 140°–245°.
GALLOWAY LANDING LIGHT, 23 04.3N, 75 59.2W. F W, 14ft, 2M. White mast, 3ft.

JUMENTOS CAYS

NUEVITAS ROCKS LIGHT, on eastern Jumentos Cays, 23 09.6N, 75 22.4W. Fl W 4s, 38ft, 10M. Gray metal tower, 30ft.
FLAMINGO CAY LIGHT, on hill, 22 53N, 75 52W. Fl W 6s, 138ft, 8M. Aluminum colored steel tower, 12ft.
RAGGED ISLAND LIGHT, Duncan Town, Man-O-War Hill, 22 11N, 75 44W. Fl W 3s, 118ft, 12M. Black pipe, platform. F W light shown from settlement wharf. Unreliable.

RAGGED ISLAND HARBOUR

22 14N, 75 44W
Dockage: The narrow, shallow channel from the northern end of Ragged Island to Duncan Town should be attempted with local knowledge only.
Anchorage: There is an anchorage between Hog Cay and Ragged Island, south of Pig Point. The sailing directions recommend an anchorage west of Little Ragged in a position where the south extremity of Little Ragged Island bears 097°T and Point Wilson, on Ragged Island, bears 004°T.

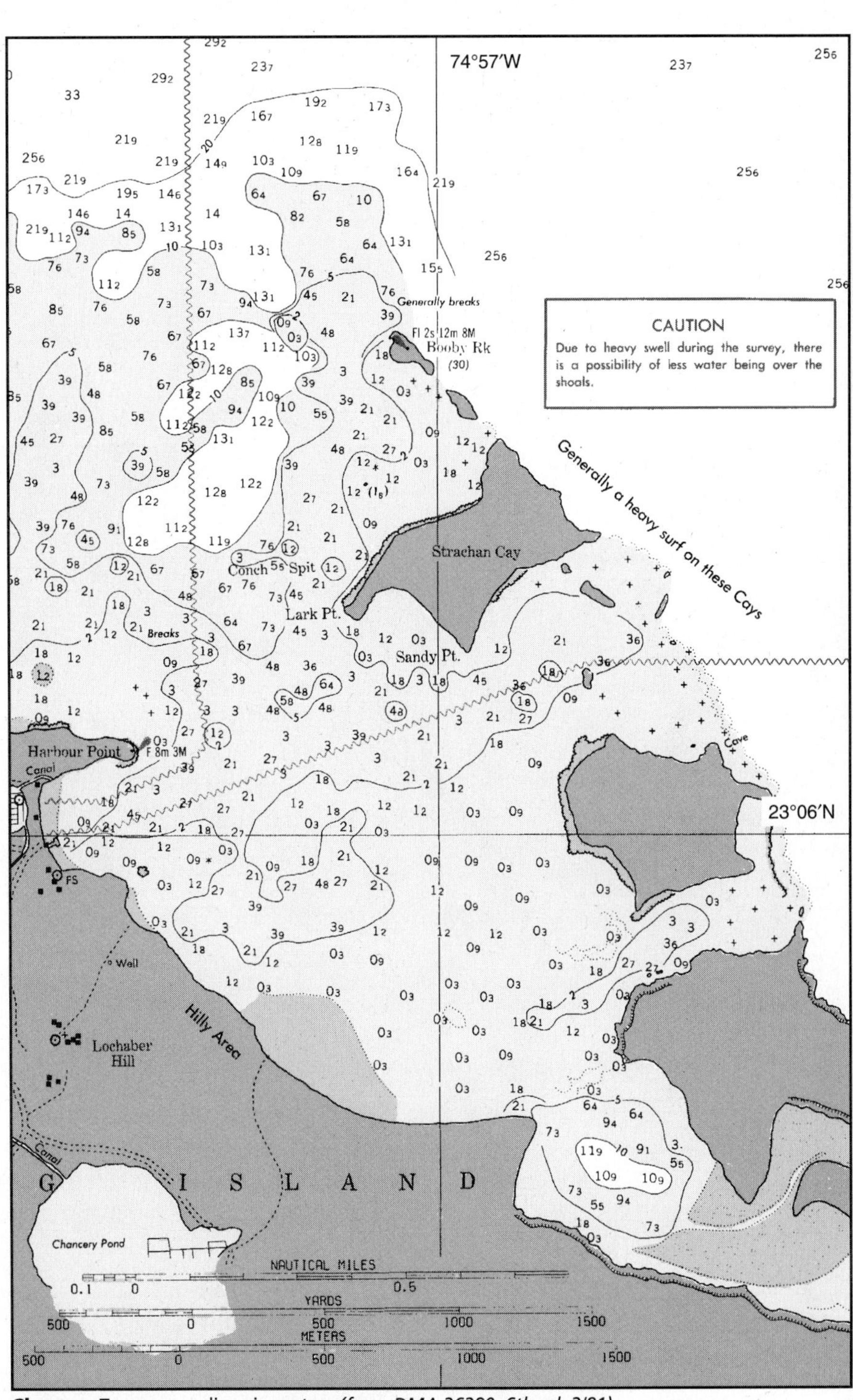

Clarence Town, soundings in meters *(from DMA 26280, 6th ed, 2/91)*

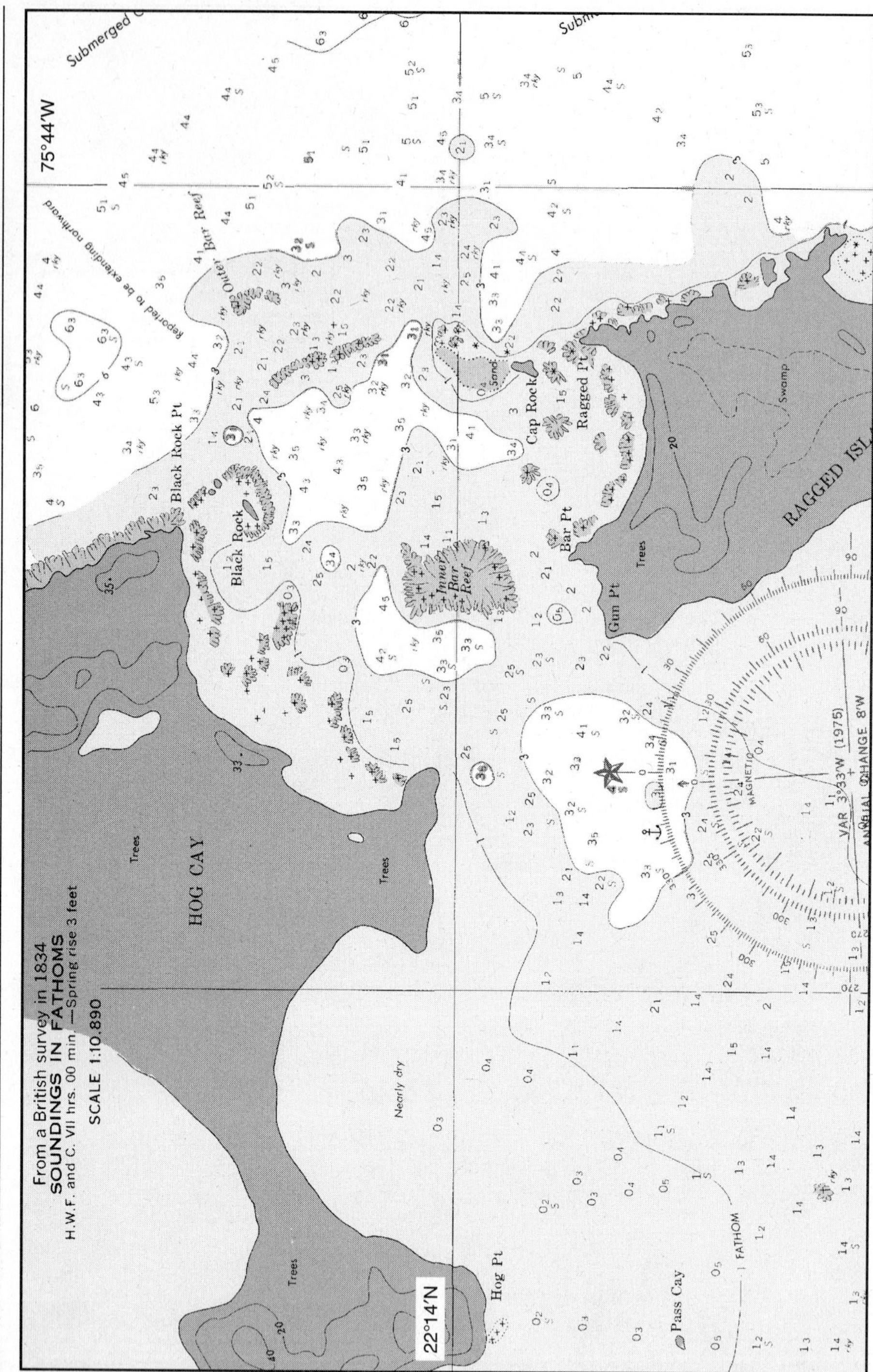

Ragged Island Harbour, soundings in fathoms and feet *(from DMA 26257, 14th ed, 1/79)*

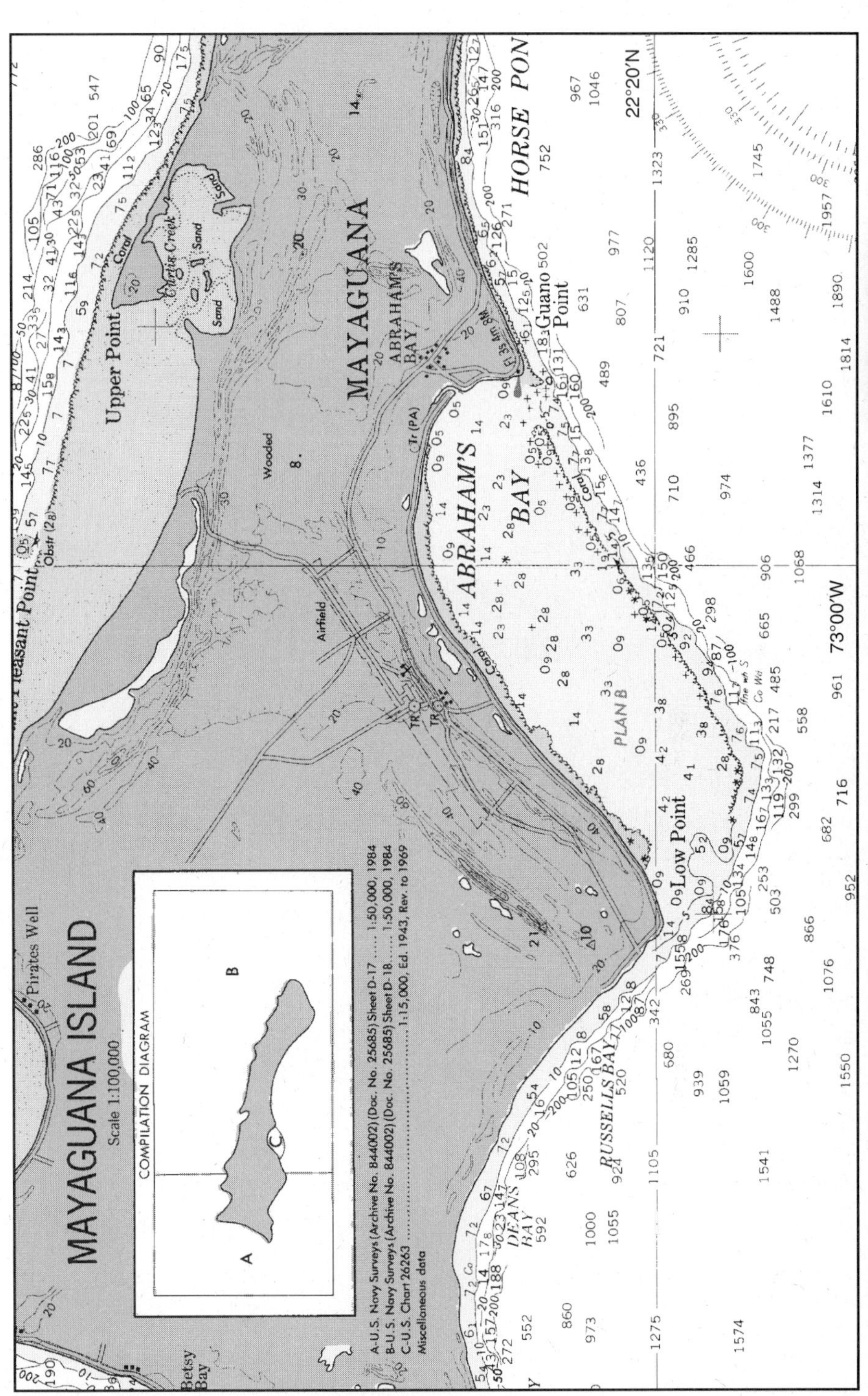

Abraham's Bay, soundings in meters *(from DMA 26263, 2nd ed, 8/86)*

P

Pilot

BAHAMAS

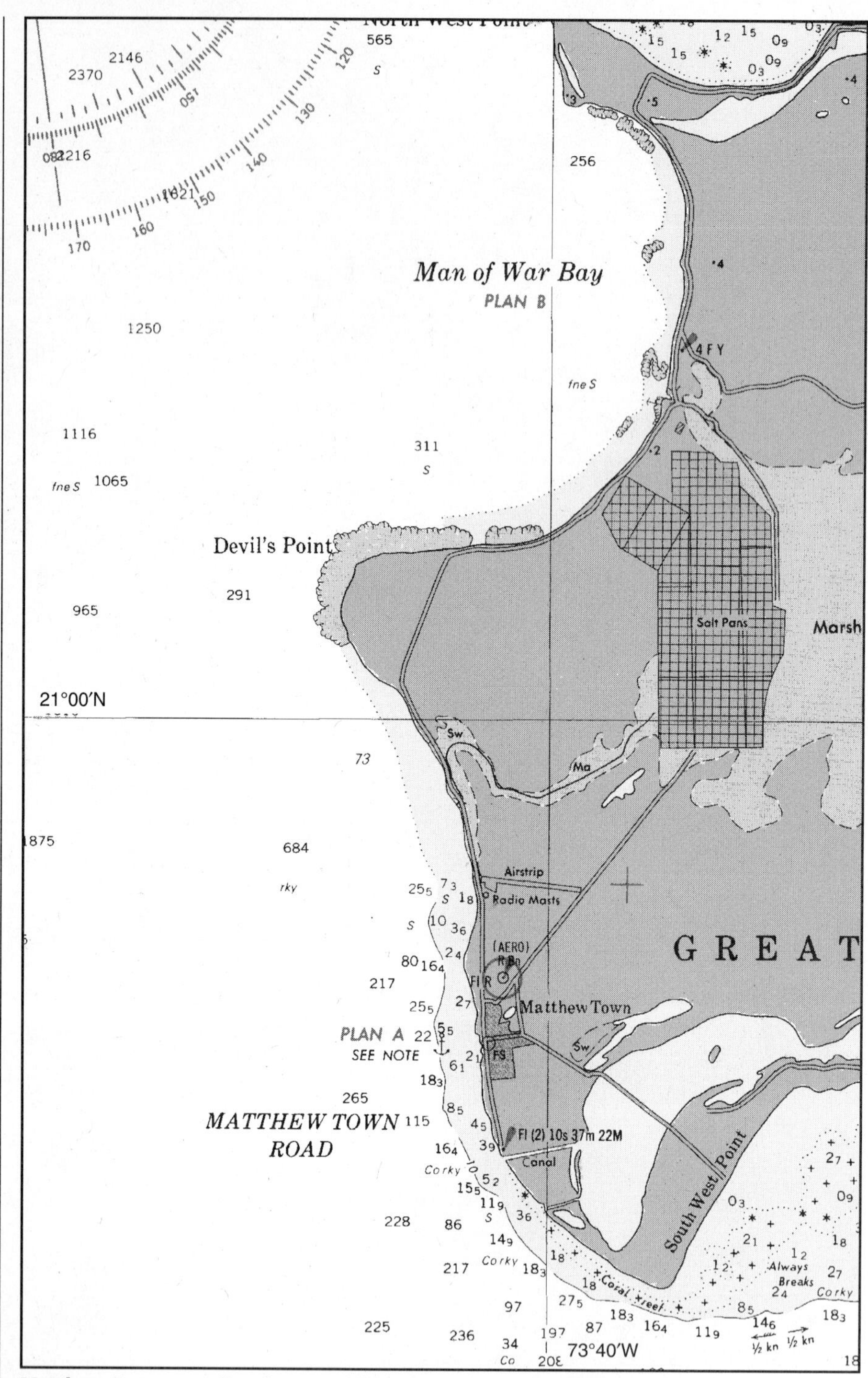

Matthew Town, soundings in meters *(from DMA 26267, 1st ed, 8/86)*

Services: There are some limited supplies, a bar, and a restaurant. Ragged Island is at the southern end of the Jumentos Cays, in one of the most remote areas of the Bahamas.

CAY SANTO DOMINGO LIGHT, 21 42N, 75 44W. Fl W 5s, 29ft, 7M. Aluminum and red steel structure. Reported extinguished (1990).

OLD BAHAMA CHANNEL

CAY LOBOS, near W end, 22 22.9N, 77 35.9W. Fl (2) W 20s, 145ft, 22M. White round metal tower, 148ft. Reported extinguished (1991).

CROOKED AND ACKLINS ISLANDS

WINDSOR POINT LIGHT, 22 34.0N, 74 22.5W. Fl W 3s, 36ft, 8M. Black structure, 36ft.

BIRD ROCK LIGHT, on summit, 22 50.9N, 74 21.5W. Fl W 15s, 112ft, 23M. White conical stone tower, 109ft.

MAJORS CAY LIGHT, 22 45N, 74 09W. F W, 30ft, 4M. Mast, 20ft. Unreliable.

ATWOOD HARBOUR LIGHT, near NE point of Acklins Island, 22 43.5N, 73 53.0W. Fl W 4.5s, 20ft, 5M. Metal tower.

ACKLINS ISLAND LIGHT, NE point, Hell Gate, 22 43.8N, 73 50.8W. Fl W 6s, 56ft, 10M. Gray steel tower, 32ft.

CASTLE ISLAND LIGHT, near SW point, 22 07.3N, 74 19.6W. Fl (2) W 20s, 131ft, 22M. White cylindrical tower, 111ft.

SPRING POINT LIGHT, 22 28.5N, 73 57.7W. F W, 19ft, 7M. White pole, 17ft.

LONG CAY LIGHT, E side of cay, 22 37.0N, 74 19.3W. F W, 60ft, 4M. White post.

PITTSTOWN POINT

22 50N, 74 21W

Anchorage: Many yachts anchor off the north-northwest side of the island, behind Bird Rock.

Services: Ashore here is the small resort of Pittstown Point Landing (VHF channel 16; information in the U.S., 800-752-2322). Cruisers dinghy to the beach to the small restaurant and bar; reservations are required.

MAYAGUANA

NORTHWEST POINT LIGHT, 22 27.8N, 73 07.6W. Fl W 5s, 70ft, 12M. White framework tower, red lantern.

GUANO POINT LIGHT, Abraham's Bay, 22 21.1N, 72 57.9W. Fl W 3s, 14ft, 8M. Gray structure, 6ft.

ABRAHAM'S BAY

22 21N, 73 00W

Anchorage: The best area to anchor is in Abraham's Bay, southwest of the settlement. There are two channels through the reef, one near Guano Point and the other southeast of Low Point. An approach in good light is recommended. Keep a good lookout for coral throughout Abraham's Bay.

Services: There is a telephone and limited supplies. You can jug fuel or water to your boat.

HOGSTY REEF

HOGSTY REEF LIGHT, on Northwest Cay, 21 41.4N, 73 50.8W. Fl W 4s, 29ft, 8M. Red mast, white bands.

GREAT INAGUA ISLAND

MAN OF WAR BAY RANGE (Front Light), 21 03.3N, 73 39.2W. F Y. Black and white square daymark. **(Rear Light),** 320 meters, 087° from front. F Y. Black and white square daymark.

MAN OF WAR BAY RANGE (Front Light), 21 03.3N, 73 39.3W. F Y. Black and white square daymark. **(Rear Light),** 120 meters, 130° from front. F Y. Black and white square daymark.

MATTHEW TOWN LIGHT, 20 56.5N, 73 40.2W. Fl (2) W 10s, 120ft, 22M. White cylindrical stone tower. Partially obscured 165°–183°.

MATTHEW TOWN

21 00N, 73 30W

Port of entry

Bahamas National Trust: Inagua National Park on the island is the home of a colony of some 60,000 flamingos. Entry is not authorized unless you are accompanied by a park warden. For more information contact: The Bahamas National Trust, P.O. Box N-4105, Nassau, Bahamas (tel., 242-393-1317; fax, 242-393-4978).

Anchorage: There is anchorage in Man of War Bay on the western side of the island. The wharf, extending 400 yards seaward, is owned by the Morton Salt Co. and is private. Vessels up to 35,000 tons call here. The terminal may be contacted on 2182 kHz or VHF channel 16 (call sign, Morton Salt). Matthew Town is the administrative center of the island. An open roadstead anchorage is available off the town.

Services: Some supplies and a telephone are available. There is an airstrip with flights available to Nassau.

Turks and Caicos Islands

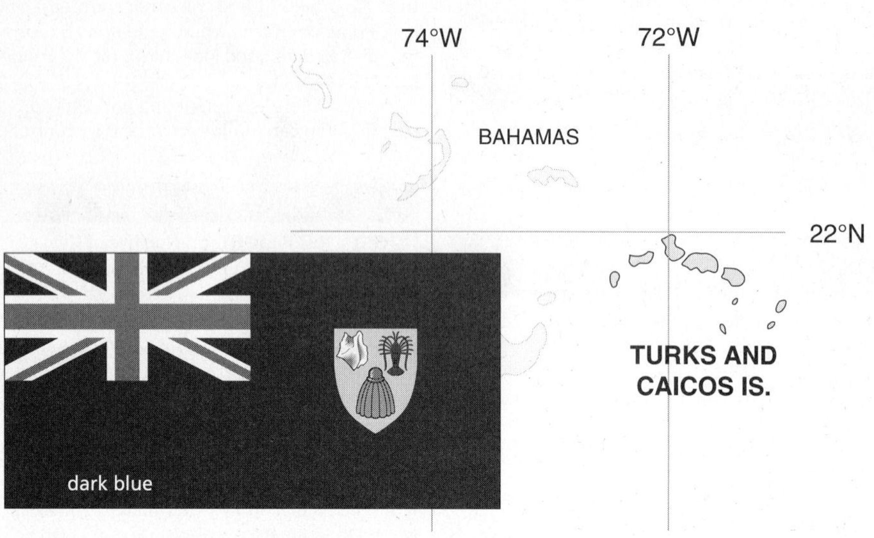

NOTE: Lights in the Turks and Caicos Islands are reported to be unreliable.

Please see cautions about chartlets and waypoints on page P 2.

ENTRY PROCEDURES

The Turks and Caicos are a British Crown Colony, with ports of entry on Providenciales, South Caicos, North Caicos (rarely a first port-of-call for cruising yachts), and Grand Turk, the islands' administrative center. Clearing in is a fairly straightforward process through the harbormaster, who has customs duties. Contact the harbormaster as you approach your port of entry, either directly on VHF channel 16 (call sign, Harbormaster) or through any of the marinas, who also monitor VHF channel 16.

Passports are required, except for citizens of the United States and Canada, who may substitute two valid forms of identification. Visas are not necessary for citizens of the United States, Canada, and Commonwealth countries but are required for all other nationals. Overtime charges apply to vessels clearing in during the lunch hour (1230–1400), after hours, Saturdays, Sundays, and holidays. Firearms should be declared, and those wishing to fish should check on the latest sportfishing regulations. There are restrictions regarding the use of all types of fishing spears.

For the latest information on clearing-in and departure procedures and entrance and boarding fees, on Providenciales contact customs at 809-946-4776 and immigration at 809-946-4233; on South Caicos, immigration at 809-946-3355; and on Grand Turk, immigration at 809-946-2939. There is no immigration office on North Caicos.

USEFUL INFORMATION

Language: English.
Currency: U.S. dollar ($).
Telephones: Area code, 809; local numbers, seven digits (first two are always 94). To call within the islands, dial only the last five digits. A new area code, 649, will become effective in the Turks and Caicos in May 1998; until then, callers can use both the old code (809) and the new code.

Public holidays (1998): New Year's Day, January 1; Commonwealth Day, March 10; Good Friday, April 10; Easter Monday, April 13; National Heroes Day, May 25; Queen's Official Birthday, June 13, observed June 15; Emancipation Day, August 1; National Youth Day, September 25; Columbus Day, October 12; International Human Rights Day, October 24; Christmas Day, December 25; Boxing Day, December 26.
Tourism offices: *On Providenciales:* Turks and Caicos Tourist Board, Turtle Cove Landing (tel., 809-946-4970; fax, 809-941-5494). *On Grand Turk:* Turks and Caicos Tourist Board, P.O. Box 128, Pond Street (809-946-2321; answering machine, 800-241-0824).

CAICOS ISLANDS

NORTHWEST POINT LIGHT, on NW point of Providenciales Island, 21 51.9N, 72 20.0W. Fl (3) W 15s, 14M. Reported Q (3) W 15s (August 1988).
WEST CAICOS LIGHT, S end, 21 37.9N, 72 27.9W. Q R, 52ft. Pillar.
PROVIDENCIALES ISLAND LIGHT, E end, 21 48.6N, 72 07.8W. Fl W 10s, 12M.
FRENCH CAY LIGHT, 21 30.3N, 72 12.1W. Fl R, 10ft. Pillar.
CAPE COMETE LIGHT, East Caicos Island, 21 43.4N, 71 28.3W. Fl (2) W 20s, 12M.
SOUTH CAICOS ISLAND LIGHT, Cockburn Harbor on Government Hill, 21 29.9N, 71 30.7W. F W, 50ft, 9M. White building, flat roof, 15ft. Visible 180°–090°.
LONG CAY LIGHT, E end, 21 29.0N, 71 32.1W. Fl R 2.5s, 5M.
DOVE CAY LIGHT, W end, 21 29.0N, 71 32.1W. Fl G 2.5s, 5M. White tower.
BUSH CAY LIGHT, 21 11.0N, 71 38.0W. Fl (2) W 10s, 14M.

PROVIDENCIALES

21 47N, 72 17W
Port of entry: Contact the harbormaster directly or via Turtle Cove Marina, Leeward Marina, or Caicos Marina. The harbormaster and marinas stand by on VHF channel 16.
Pilotage: The approach to Turtle Cove through the reef is made via Sellers Cut. New visitors are advised to call ahead for information concerning marks or to request a free pilot to help negotiate the channel. The approach to Leeward Marina is also through a cut in the reef marked by privately maintained buoys; yachts are advised to call before entering the channel. Both marinas report approach depths of more than 6 feet, although a sandbar is encroaching on the cut to Leeward Marina. Approaches to the south side of Providenciales are discussed under "Anchorages", below.
Dockage: Located on the north side of Providenciales are Turtle Cove Marina, which has recently built 36 new slips (dockmaster, 809-941-3781; office tel., 809-946-4303; fax, 809-946-4326), and Leeward Marina (dockmaster, 809-946-5553; office fax, 809-946-5674). On the south side dockage is available from the Caicos Marina and Shipyard (tel., 809-946-5600; fax, 809-946-5390).
Anchorage: A popular anchorage is off Sand Bore Channel, which leads into the harbors on the south side of Providenciales. Enter Sand Bore Channel through West Reef via a break south of a visible wreck. The channel then runs about 103°T toward the southern point of Providenciales. If in doubt, contact Caicos Marina on the VHF. A second approach is via Clear Sand Road. This channel is reported to run about 068° from the southern end of West Caicos. You can anchor in Clear Sand Road on the edge of the banks or west of Caicos Marina in Sapodilla Bay, which has depths of up to 8 feet. The government dock is nearby.
Services: All services are available, including haulouts and major repairs. There is a good selection of parts and supplies. Regular air service to the U.S. can bring in special items. There is good telephone service.

Providenciales has become a tourist, dive, and boating destination in recent years. The expansion of all facilities has made the Turks and Caicos Islands a good stop for those traveling between eastern North America and the Caribbean.

COCKBURN HARBOUR, SOUTH CAICOS

21 29N, 71 32W
Port of entry: Contact the harbormaster directly on VHF channel 16 or through Sea View Marina (tel./fax, 809-946-3219; monitors VHF channel 16 and 4417.0 kHz). The South Caicos immigration office can be reached at 809-946-3355.
Dockage: The Sea View Marina has 6.5 feet of water off its dock, with space for about 5 boats. Vessels with greater draft may be able to tie up at the government wharf.
Anchorage: Anchor off the town dock or north of Long Cay. Be wary of shallow spots and the charted reef north of Long Cay. The protection is good in most conditions.

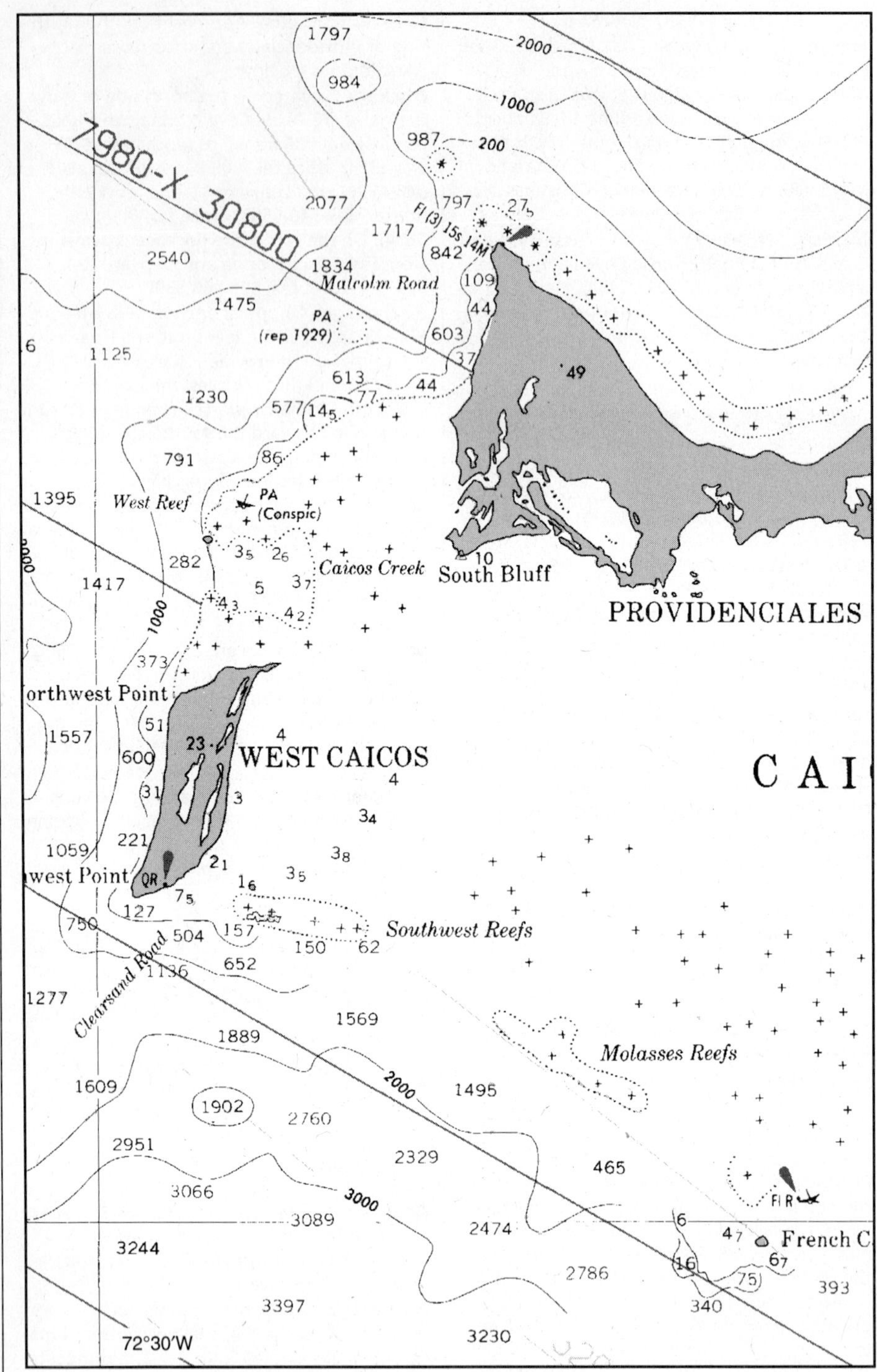

Providenciales, soundings in meters *(from DMA 26260, 5th ed, 2/88)*

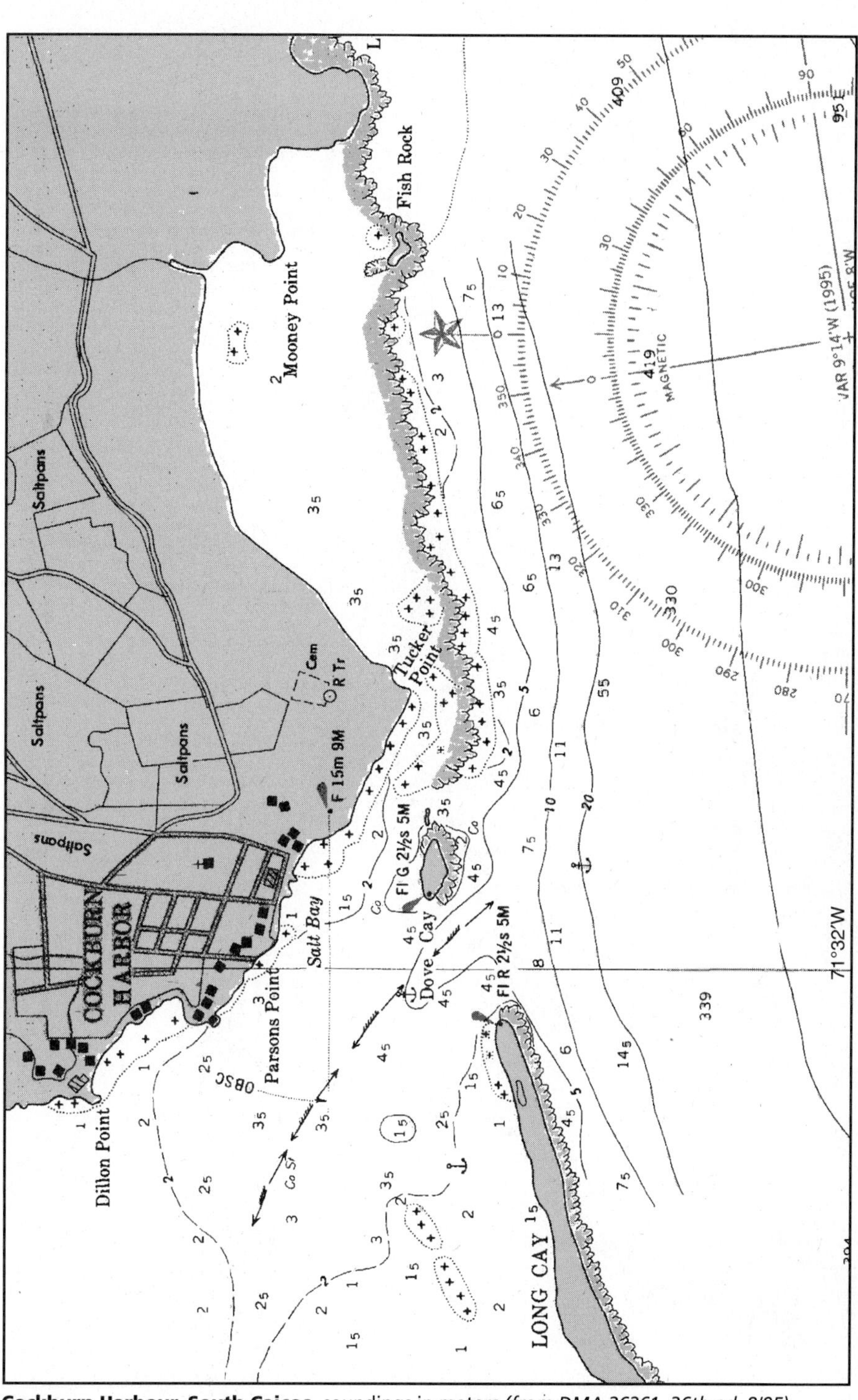

Cockburn Harbour, South Caicos, soundings in meters *(from DMA 26261, 26th ed, 8/95)*

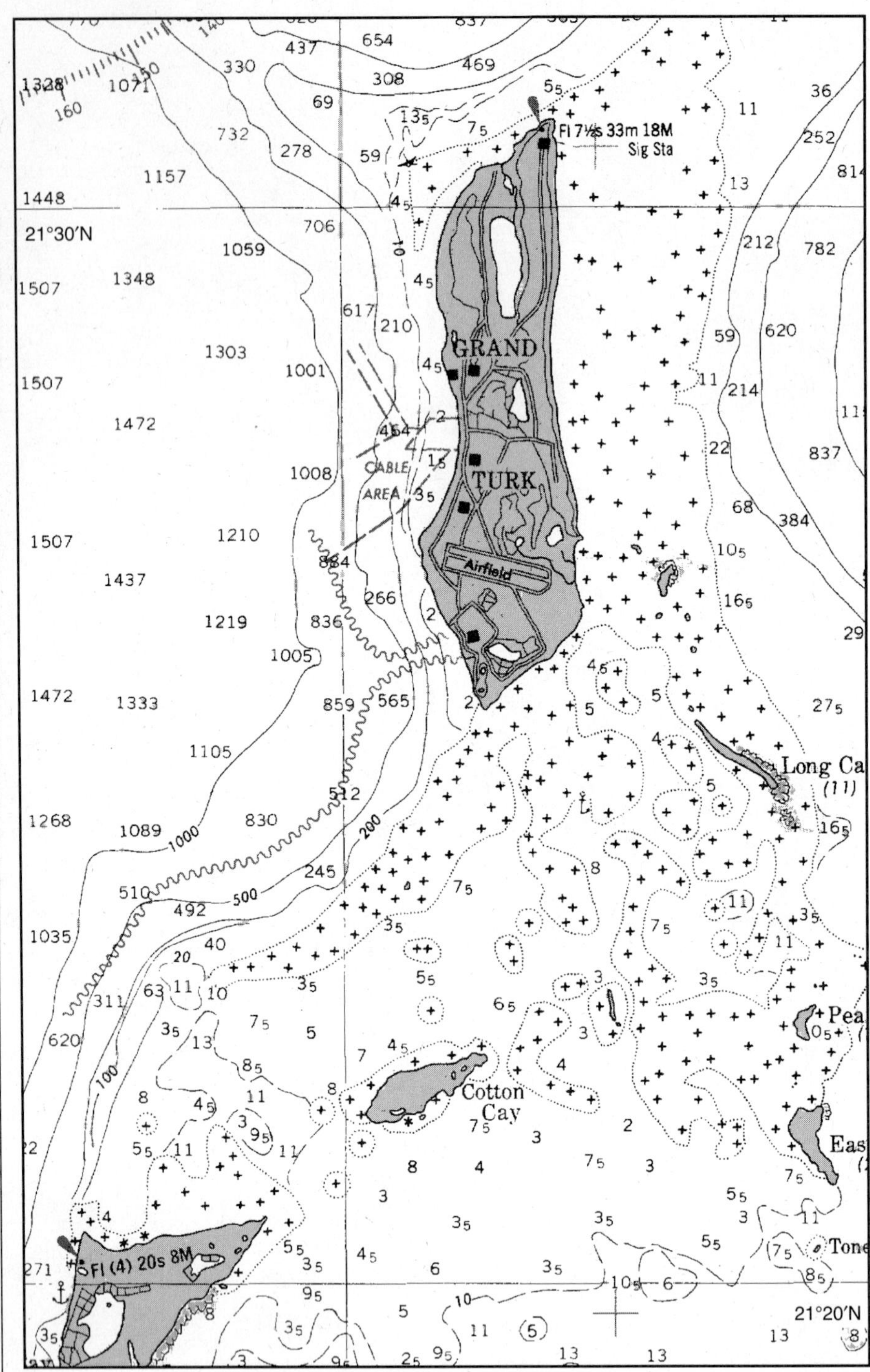

Grand Turk, soundings in meters *(from DMA 26261, 26th ed, 8/84)*

Services: Sea View Marina can provide fuel at its docks or the government wharf. It can also make arrangements for fresh water delivery, laundry, and small repairs. Some groceries are available at the market next door. Customers may use the marina's telephone, and restaurants are nearby.

The only harbor on South Caicos and the largest in the Caicos group, Cockburn Harbour offers decent protection in most conditions.

TURKS ISLANDS

GRAND TURK LIGHT, 21 30.9N, 71 07.9W. Fl W 7.5s, 108ft, 18M. White cylindrical iron tower, 60ft. Signal station. F W on tower 1.1M S; F R on tower 3M S.
SALT CAY LIGHT, NW point, 21 20.2N, 71 12.7W. Fl (4) W 20s, 8M.
SAND CAY LIGHT, 21 11.9N, 71 14.8W. Fl W 2s, 85ft, 10M. Red steel framework structure, 50ft. Reported extinguished (1985).

COCKBURN TOWN, GRAND TURK

21 28N, 71 09W
Port of entry: Contact the harbormaster on VHF channel 16 for instructions on clearing in (phone pager, 809-946-1008). Customs and immigration duties are generally performed in the south dock or town areas.
Pilotage: Reef extends along the west coast of Grand Turk, from half a mile to several hundred yards offshore. Mariners are urged to make their approach in good light.
Anchorage: The best anchorage is reported to be north of the wharf at the southern end of the island. A cut has been made through the reef at the island's north end, which allows access to sheltered North Creek. The entrance is marked by a pile of black stones on the east. Local knowledge is recommended for those attempting this pass. The Hawks Nest anchorage at the south end of the island also requires local knowledge.
Services: Some supplies and services, including engine repair, are available, and there is good telephone service. There are small shops, restaurants, and hotels.

Grand Turk is the administrative center of the Turks and Caicos, although the anchorages are not as sheltered as some in the Caicos area.

REED'S *NEEDS YOU!*

It is difficult to obtain accurate and up-to-date information on many areas in the Caribbean. To provide the mariner with the best possible product we use many sources. *Reed's* welcomes your contributions, suggestions, and updates. Please be as specific as possible, and include your address and phone number.
Send your comments to:
Editor, Thomas Reed Publications
13A Lewis Street, Boston, MA 02113, U.S.A.
Thank you!

CUBA

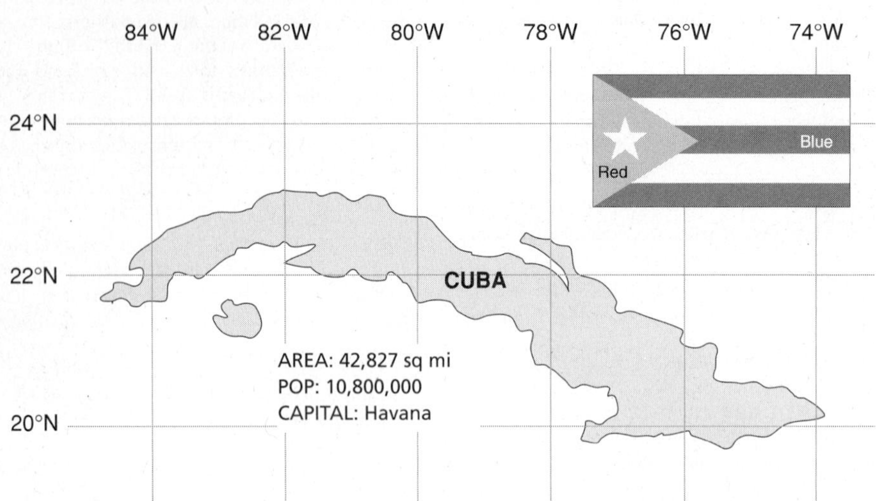

WARNING: Cuba claims a 12-mile territorial sea extending from straight baselines drawn from Cuban coastal points. The effect is that Cuba's claimed territorial sea extends in many areas well beyond the 12 miles from Cuba's physical coastline. These claims are not in conformance with international law and are not recognized by the United States. Nonetheless, in some cases, U.S. vessels have been stopped and boarded by Cuban authorities more than 20 miles from the Cuban coast. Within the limits of prudence and good judgment, mariners are advised to protest (but not physically resist) any improper attempt to stop, board, or seize U.S. vessels in international waters. Vessels in this situation should notify the U.S. Coast Guard of their status so that information about the situation can be relied to the Department of State for diplomatic attention.

NOTE: Many of the lights on the coast of Cuba have been reported to be irregular, unreliable, or extinguished.

Please see cautions about chartlets and waypoints on page P 2.

TRAFFIC SEPARATION SCHEMES: On September 1, 1990, seven Traffic Separation Schemes, approved by the International Maritime Organization, were implemented off the eastern, western, and northern coasts of Cuba. The government of Cuba has unilaterally established a mandatory ship-reporting system within the Old Bahama Channel to govern vessel movement within the area. While it is the U.S. position that the mandatory provisions of the ship-reporting system are inconsistent with international law, vessels should exercise caution while transiting the Old Bahama Channel.

The Traffic Separation Schemes are in the following areas:

Off Cabo San Antonio
Off La Tabla
Off Costa de Matanzas
In the Old Bahama Channel
Off Punta Maternillos
Off Punta Lucrecia
Off Cabo Maisi

CUBAN ASSETS CONTROL REGULATIONS

The Cuban Assets Control Regulations were issued by the U.S. government on July 8, 1963, under the Trading With the Enemy Act in response to certain hostile actions by the Cuban government. They are still in force today and affect all U.S. citizens and permanent residents wherever they are located, all people and organizations physically in the United States, and all branches and subsidiaries of U.S. organizations throughout the world. On August 20, 1994, President Clinton announced additional measures designed to limit Cuba's ability to accumulate foreign exchange. The regulations are administered by the U.S. Treasury Department's Office of Foreign Assets Control. The basic goal of the sanctions is to isolate Cuba economically and deprive it of U.S. dollars. Penalties for violating the sanctions range up to 10 years in prison and to $1,000,000 in corporate and $250,000 in individual fines. The regulations require those dealing with Cuba to maintain records and, upon request from the U.S. Treasury Department, to furnish information regarding such dealings.

Spending money in connection with tourist, business or recreational trips is prohibited, whether travelers go directly to Cuba or via a third country. Vessels are prohibited entry into any U.S. port if they are carrying passengers or goods to or from Cuba, unless specifically licensed by the Treasury Department. There are exceptions to these rules for family visitors, official government travelers, journalists, professional researchers, and sponsored travelers. Certain types of humanitarian assistance may be carried to Cuba. For the latest information on these regulations contact: Licensing Division, Office of Foreign Assets Control, U.S. Department of the Treasury, 1500 Pennsylvania Ave. NW, Washington, DC 20220 (202-622-2480).

When in Cuba U.S. citizens should register with the United States Interests Section, Switzerland Embassy, Calzada between L&M, Vedado, Havana (tel., 53-7-33-3550 through 33-3559).

ENTRY PROCEDURES

Mariners who leave for Cuba from Key West are urged to contact the U.S. Coast Guard there to obtain clearance. Recent reports indicate that vessels obtain clearance papers within 24 hours. Mariners are advised to enter Cuba at a marina where officials are familiar with the procedures for pleasure boats. Listed ports of entry (circling the island in a clockwise direction) are: Santa Lucía; Marina Hemingway in Barlovento (near Havana); the marinas in Veradero; Plaza Santa Lucía in Bahía de Nuevitas; Marina Bahía de Naranjo in the province of Holguin; Baracoa; Marina Punta Gorda in Bahía de Santiago de Cuba; Manzanillo; Ancon Peninsula, Casilda (near Trinidad); Jagua, Cienfuegos; Cayo Largo; and the Hotel Colony, Darsena de la Siguanea, Isla de la Juventud (Isla de Pinos). You must clear in and out whenever you move the boat from one point to another, and you should always check with authorities ashore, whether or not in a port of entry. Havana is a commercial port only and has no facilities for yachts.

You should have passports, a clearance from your last port, crew lists, and ship's papers. Make sure the crew list includes each crew's birthdate and nationality. Multiple copies of each are strongly recommended. Once you are inside the 12-mile limit of Cuban waters (see caution note, previous page), raise the Q flag, and notify port authorities; recommended procedures vary. If you are headed for a port of entry marina, try raising it on the VHF. The major marinas will usually also assist with the clearing-in process. If you are not headed for a marina, try raising the port authority of your destination on VHF channels 68, 72, or 16. The Ministry of Tourism also suggests calling tourism officials on VHF channel 19, Red Coast on 2760 kHz, or Red Tourist on 2790 kHz. In all cases, inform authorities of the name, color, type, and flag of your yacht, its port of registration, your last stop, intended entry port in Cuba, estimated time of arrival, and number of crew. A patrol boat may come out to meet you. You should expect to be met ashore by many officials, usually including an armed member of the national guard (Guarda Frontera).

Tourist cards or visas are usually granted at the port of entry for periods of time compatible with your intended length of stay and are usually renewable for a fee. For more information on traveling to Cuba contact the Cuban Interests Section in the Swiss embassy in Washington, DC (tel., 202-797-8510; fax, 202-797-8521).

To visit more than one port, you should apply for a cruising permit (fee) with stipulations on where and when you can go—be careful to avoid unplanned intermediate stops.

USEFUL INFORMATION

Language: Spanish.
Currency: Cuban peso. U.S. dollars are legal currency accepted by tourist-oriented businesses and preferred by many (officially other European currencies are also legal tender). There are special stores for visitors that take only U.S. dollars. The official exchange rate is 1 peso to US$1. U.S. citizens and permanent residents should check the restrictions on spending money discussed previously. U.S.-based credit cards or traveler's checks cannot be used, and all Cuban

currency should be changed before leaving the country. Non-U.S. citizens/residents may be able to use credit cards and traveler's checks that are not cleared through U.S. banks.

Telephones: Country code, 53; local numbers, number of digits varies. Direct service to Cuba from the U.S. and Canada is now available for most areas except Cayo Largo. To call out of the country from Cuba dial 8 plus the country's area code.

Public holidays (1998): Liberation Day, January 1; International Worker's Day, May 1; National Rebellion Days, July 25–27; Beginning of Independence War Day, October 10.

Tourism offices: *In Canada,* Cuban Tourist Office, 55 Queen Street East, Suite 705, Toronto, Ont. M5C 1R6 (tel., 416-362-0700; fax, 416-362-6799). Office also in Montréal.

Special note: Official Cuban charts, pilots, and other navigational information are published by the Instituto Cubano de Hydrografía and available in ports of entry. Marina Hemingway, where an array of yacht services can also be arranged, is a good source.

YUCATÁN CHANNEL

21 30N, 86 00W

Pilotage: This channel between Cabo San Antonio on Cuba and Isla Contoy off the Mexican coast is about 108 miles wide. It serves as the main route for shipping between the Gulf of Mexico and the Panama Canal. The east side of the channel is deep, shoaling gradually toward the Mexican coast.

There is a tremendous flow of water from the Caribbean Sea through the Yucatán Channel toward the Gulf of Mexico. Currents reach velocities of up to 4 knots on the stream's axis. The eastern border of the flow is about 20 miles off Cabo San Antonio, which is just beyond Cuba's 12-mile territorial limit. Twenty to 35 miles from Cabo San Antonio the rate is 1 knot; at 50 miles, 2 knots; at 65 miles, 3 knots; at 78 miles, 4 knots, and at 90 miles, or about 25 miles from Yucatán, 1 knot. The current axis is located about 35 miles off the Yucatán coast, or about 6 miles beyond the 100-fathom curve. The mean rate during April, May, and June along the axis is about 4 knots. The current increases in summer and decreases in winter. When the current is stronger, the width of the flow is wider, and when the flow is weaker, the width shrinks. There is also a noticeable daily variation in flow, particularly on the west side. The flow has been known to increase or decrease by as much as 3 knots in 5 hours.

Within 20 miles of Cabo San Antonio the flow is either northeast toward the Straits of Florida or southeast along the south coast of Cuba. At times the east-setting current can reach velocities of 4 knots, especially during southerly winds.

CAUTION: During periods of southerly winds it is advisable to avoid the coast of Cuba from Cabo San Antonio to Cabo Corrientes, due to dangerous currents.

CABO SAN ANTONIO TO BAHIA HONDA

CABO SAN ANTONIO LIGHT, W extremity of Cuba, 21 52.0N, 84 57.2W. Fl (2) W 10s, 103ft, 18M. Yellow round tower and dwelling, 75ft. Radar reflector. Aeromarine light.

CABO SAN ANTONIO

21 52N, 84 57W

Pilotage: Cabo San Antonio is the western extremity of Cuba. It is low and covered by trees that will be visible before the land. The curve of the coast is so gradual that the position of the cape can only be determined by the lighthouse. The area is reported to be radar conspicuous at a distance of 15 miles. Currents near the outer edge of the bank are confused, while near shore they flow north on a rising tide and south on a falling tide. Tide rips are also present.

Anchorage: Temporary anchorage is available south of the lighthouse, with the southeast extremity of the land bearing 135° and the west extremity, 023°.

BANCO SANCHO PARDO, 22 09.7N, 84 44.9W. Fl W 8s, 36ft, 10M. Red framework tower, 26ft.

LA TABLA, 22 18.3N, 84 40.1W. Fl W 5s, 12M. White framework tower on platform, 26ft. **RACON M** (– –), 30s, 12M.

ZORRITA, 22 22.3N, 84 34.9W. Fl W 7S, 36ft, 10M. Green framework tower on platform, 26ft.

EL PINTO, 22 25.0N, 84 31.2W. Fl W 15s, 36ft, 10M. Red framework tower, yellow bands on platform, 26ft.

CAYO BUENAVISTA, 22 24.1N, 84 26.7W. Fl W 5s, 108ft, 6M. Aluminum framework tower, 98ft.

QUEBRADO DE BUENAVISTA, 22 28.1N, 84 28.0W. Fl W 7s, 36ft, 10M. Metal framework tower on platform, 26ft.

CABEZO SECA, 22 32.0N, 84 20.0W. Fl W 12s, 36ft, 10M. White framework tower, red bands on platform, 26ft.

PUERTO DE LOS ARROYOS, Baja la Paila, 22 21.4N, 84 22.8W. Fl G 5s, 16ft, 3M. Green port-hand beacon with topmark.

GOLFO DE GUANAHACABIBES

22 08N, 84 35W

Pilotage: This area is located just to the northeast of Cabo San Antonio. The area is much encumbered with shoals and cays, and the intricate channels indicate local knowledge may be necessary.

LA FÉ

22 03N, 84 16W

Pilotage: This is a small port situated on the east bank 2 miles from the Río Guadiana in the southeast part of Golfo de Guanahacabibes. Vessels drawing up to 9 feet can anchor here. The area is marked with navigational aids, but they are considered unreliable. There is a wharf with fresh water available.

FONDEADERO LOS ARROYOS

22 22N, 84 26W

Pilotage: This anchorage is located in the extreme northeast part of Golfo de Guanahacabibes. The holding ground is reported to be good, and there is shelter from the prevailing winds. It is a loading place for the inland community of Mantua.

PUNTA TABACO, 22 34.7N, 84 15.3W. Fl W 8s, 36ft, 10M. Yellow framework tower on platform, 26ft.

RONCADORA, 22 38.3N, 84 11.6W. Fl W 10s, 36ft, 10M. Red framework tower on platform, 26ft.

ARCHIPIÉLAGO LOS COLORADOS

Pilotage: This group of sunken dangers fronts about 100 miles of the northeastern coast of Cuba from Golfo de Guanahacabibes to Bahía Honda. It consists of an almost uninterrupted series of reefs rising steeply from seaward. The shallow area between the reefs and the coast is also full of hazards and low-lying islets. The shore beyond the dangers is low and the dangers rarely marked by breakers, except during heavy weather. Boats should approach this coast with extreme caution. Those with local knowledge can use some shallow channels to proceed to towns along the coast. The coast becomes generally steep-to when east of the archipelago.

CAYO JUTÍAS LIGHT, NE point, 22 42.9N, 84 01.4W. L Fl W 15s, 138ft, 22M. Black metal octagonal framework tower, yellow bands, 135ft.

BAHÍA SANTA LUCÍA

22 42N, 83 58W

Pilotage: This shallow-water bay is located about 73 miles northeast of Cabo San Antonio. It is fronted by fields of mangroves and a low-lying terrain that rises gradually to hills in the interior. Seaward of the bay are the many dangers of the Archipiélago los Colorados. The island of Cayos Jutías is marked by a light, and the intricate entrance channel, Pasa Honda, begins 2.5 miles to its northeast. Many sunken dangers are scattered in this area, and a radar-conspicuous wreck is reported to lie on the reef about 2.5 miles northeast of Pasa Honda.

Port of entry

The port at Santa Lucía is under the jurisdiction of Puerto del Mariel, about 72 miles to the east-northeast. Ore is lightered out to larger vessels from the town.

PUNTA BANO, offshore on W side of entrance to channel, 22 41.2N, 83 58.2W. Fl R 4s, 16ft, 3M. Red starboard-hand beacon with topmark.

CAYO ARENAS LIGHT, 22 50.2N, 83 39.3W. Fl W 10s, 46ft, 10M. Aluminum framework tower, 39ft.

PUNTA GOBERNADORA LIGHT, 3 miles W of Bahía Honda entrance, 22 59.6N, 83 12.9W. Fl W 5s, 108ft, 27M. White conical iron tower, red bands, 105ft. Radar reflector. Aeromarine light. Reserve light range: 7M.

BAHÍA HONDA

BAHÍA HONDA, N of old fort on Punta del Morillo, 22 58.9N, 83 09.2W. Fl ~~(2)~~ W 10s, 89ft, 8M. White framework tower, 26ft.

BAHÍA HONDA RANGE, W end of Cayo del Muerto **(Front Light),** 22 57.4N, 83 09.8W. Q W, 16ft, 7M. Yellow diamond with black border on framework tower on piles, 13ft. **(Rear Light),** head of bay, 3.9 km, 183° from front, 22 55.3N, 83 10.0W. Fl W 6s, 46ft, 11M. White metal framework tower on piles, white diamond daymark with black border, 30ft. Visible 6° each side of rangeline.

BAHÍA HONDA

22 58N, 83 10W

Pilotage: This pocket bay is located about 120 miles northeast of Cabo San Antonio and 45 miles west of Havana. The saddle-shaped 2,270-foot summit of El Pan de Guajiaibon (22°47'N, 83°10'W) is a conspicuous landmark when approaching the bay from offshore.

The deep entrance channel leads through a coral reef into the bay. The port is part of the jurisdiction of Puerto del Mariel, 23 miles to the east. There are several deepwater docks and anchorages within the bay.

BAHÍA CABAÑAS LIGHT, Cerro Frias W side of entrance, 22 59.7N, 82 58.9W. Fl W 8s, 174ft, 10M. White framework tower, 26ft.
PUNTA ARENAS, edge of shoal S of Punta Arena (No.2A), 22 59.2N, 82 59.0W. Fl R 6s, 16ft, 3M. Red starboard-hand beacon with topmark.
PUNTA AFRICANA, NE point of Cayo Juan Tomas (No. 6), 22 59.2N, 82 58.4W. Fl R 4s, 16ft, 3M. Red starboard-hand beacon with topmark.
PIEDRA GLORIA BEACON NO. 8, 22 59.5N, 82 56.7W. Fl R 6s, 16ft, 3M. Red starboard-hand beacon with topmark.

CABAÑAS

23 00N, 82 58W
Pilotage: This is another sub-port of Mariel, which is located about 12 miles to the east. Puerto de Cabañas is divided into east and west parts by the Península Juan Tomas. The Canal Orozco leads into the western part, and the Canal Cabañas into the eastern part. Fuerte Reina Amalia stands in ruins on the north extremity of the peninsula and serves as a conspicuous landmark when one is approaching these channels from the sea.

Tidal currents in the entrance have a flow of about 1.5 knots during the ebb, with a somewhat stronger flow during the rainy season. There is a bulk sugar pier within the bay.

BAHÍA DEL MARIEL

PUERTO DEL MARIEL LIGHT, W side of entrance, 23 01.3N, 82 45.6W. Fl W 12s, 134ft, 11M. White framework tower, 102ft.
PUNTA REGULA ENTRANCE RANGE (Front Light), 23 00.9N, 82.45.5W. Q W, 16ft, 17M. Black diamond with white border on tower, 13ft.
(Rear Light), 612 meters, 182.9° from front. Q Y, 29ft, 16M. Black diamond with white border on tower, 23ft. Visible 6° each side of rangeline.

MARIEL

22 59N, 82 45W
Pilotage: This is one of the major pocket bays found along the north coast of Cuba. The terrain on either side of the narrow entrance channel is low-lying, except for the east side of the bay, where hills slope steeply upward directly from the water's edge. The west face of these hills has been quarried into a conspicuous white cliff.

Mesa del Mariel is a conspicuous plateau located about 8 miles east-northeast of the entrance to Bahía del Mariel. There is a distinctive terrace at the eastern end of the high elongated plateau and a remarkable steep slope at its west extremity. It has been reported as radar conspicuous.

There are several tall chimneys in a cement works on the east side of the bay; they may be seen up to 12 miles offshore. The light-colored smoke forms a dense distinctive cloud, which can be sighted from up to 25 miles. This is the best landmark on the north coast of Cuba.
The Cuban Naval Academy, a large group of white buildings, is located on the southeast side of the bay on the slopes behind the community of Mariel. This serves as a good landmark for the approach to the bay's entrance channel. Line up the center of the academy with a large white tower on the bay's entrance point. Range lights also mark the channel.

Mariel is an important port, particularly for cement and sugar cargoes. The harbor pilots may be contacted by VHF radio (call sign, Mariel Practice).

BOCA DEL RÍO BANES

23 02N, 82 38W
Pilotage: This small inlet is about 6.5 miles east of Bahía del Mariel. The channel is short and unencumbered. On its eastern shore there is a small sugar-loading facility.

RÍO SANTA ANA, 23 03.3N, 82 32.4W. Fl W 10s, 184ft, 15M. White water tank on blue columns, 26ft. Reserve light range, 8M.
RÍO SANTA ANA, BEACON NO. 5, 23 03.5N, 82 31.9W. Fl G 3s, 9ft, 3M. Green port-hand beacon with topmark.
RÍO SANTA ANA, BEACON NO. 6, 23 03.5N, 82 32.0W. Fl R 4s, 9ft, 4M. Red starboard-hand beacon with topmark.

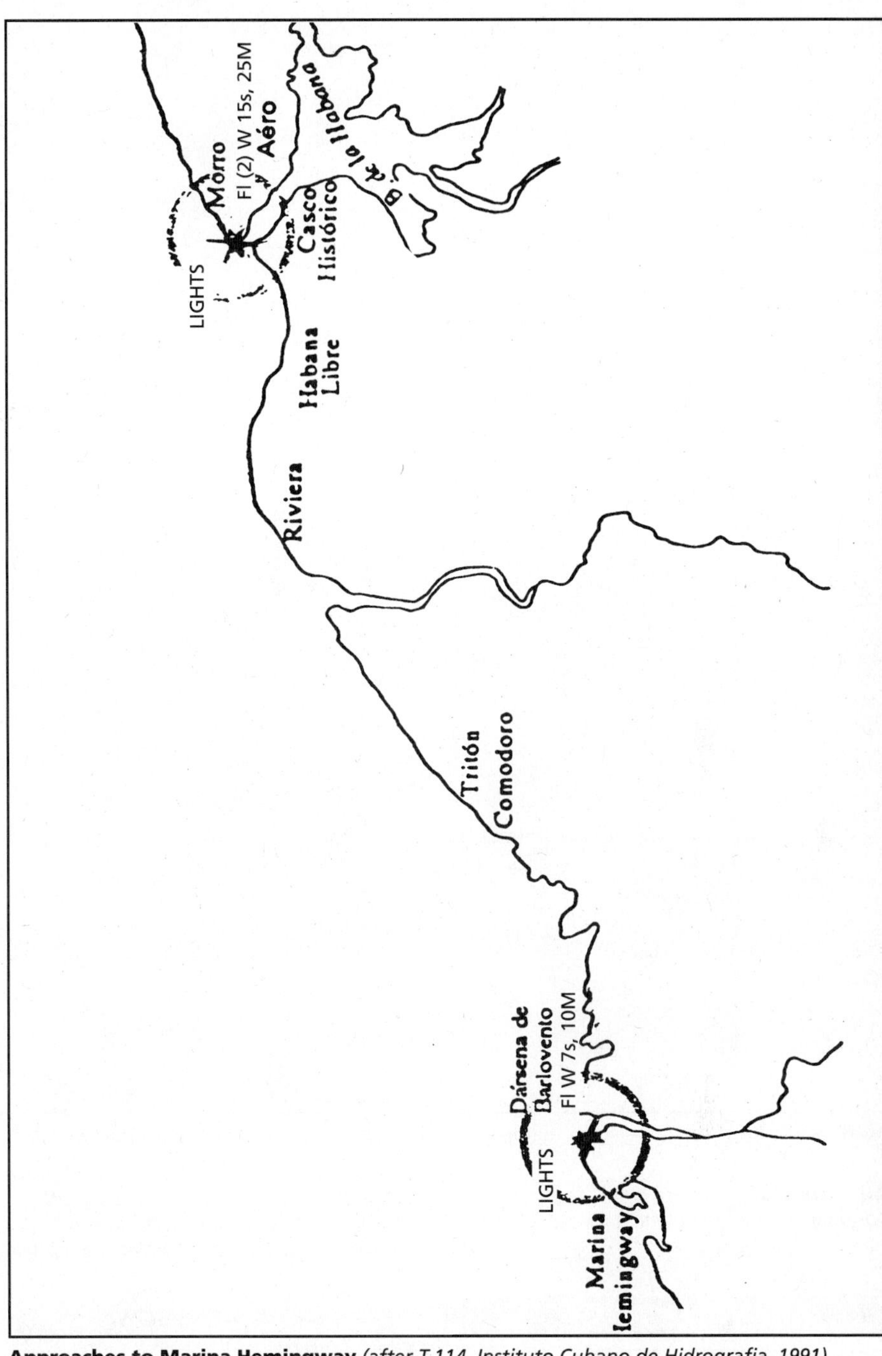

Approaches to Marina Hemingway *(after T-114, Instituto Cubano de Hidrografia, 1991)*

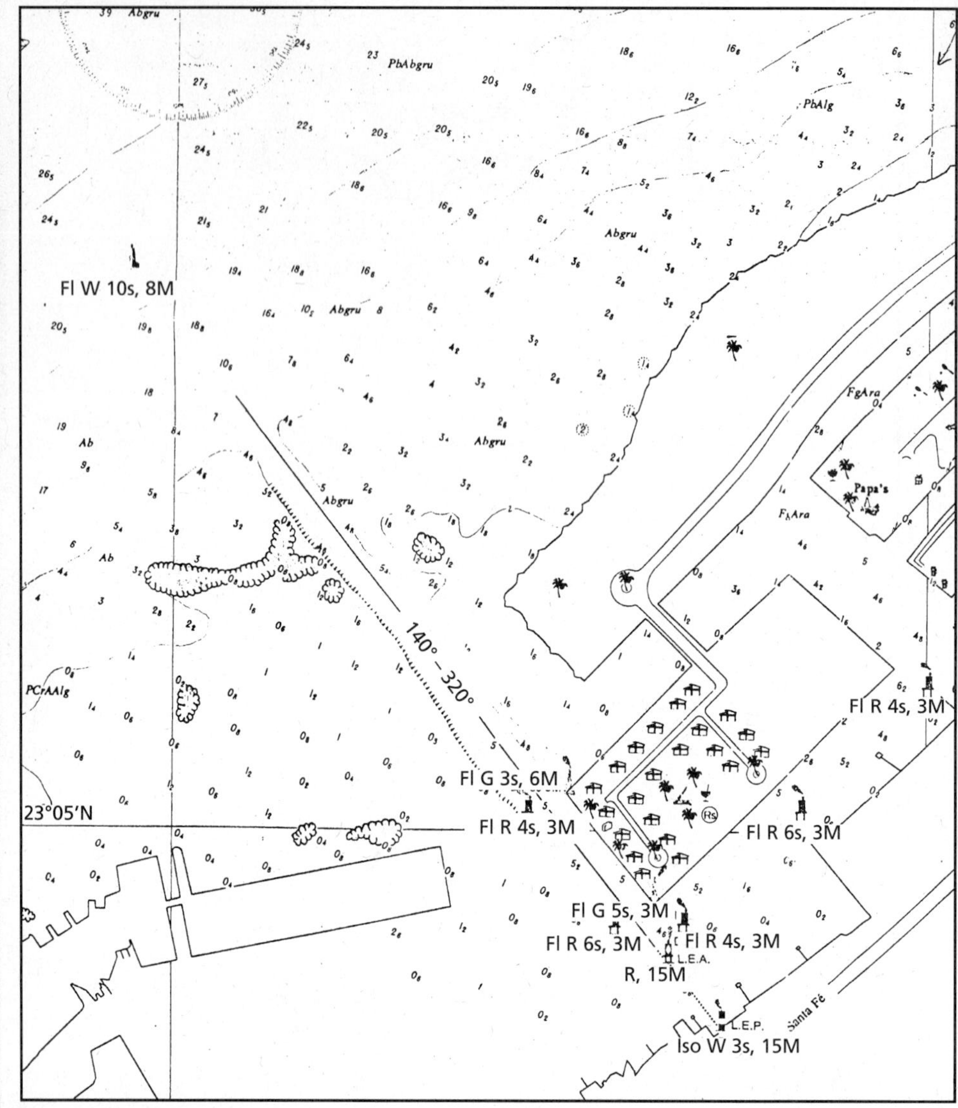

Marina Hemingway, soundings in meters *(after T-114, Instituto Cubano de Hidrografia, 1991)*

RÍO SANTA ANA, BEACON NO. 8, 23 03.5N, 82 32.0W. Fl R 6s, 9ft, 3M. Red starboard-hand beacon with topmark.

HAVANA AREA

DARSENA DE BARLOVENTO, 23 05.5N, 82 29.6W. Fl W 7s, 118ft, 10M. White metal framework tower on building, 26ft. Reserve light range, 10M.

SANTA FE DIRECTIONAL LIGHT, 23 04.9N, 82 30.2W. Q W, 16ft, 15M. White square day-mark with orange diamond on black framework tower.

MARINA HEMINGWAY

23 04N, 82 30W

Pilotage: The entrance to the marina is located about 8 miles west of the entrance to Havana and about 90 miles south-southwest of Key

West, Florida. A conspicuous beige hotel well west of Havana, and just west of a beach area, is the key for those who are hoping to find Marina Hemingway. The four-story hotel, with a government radio tower on its roof, is less than a mile east of the harbor entrance for Barlovento Bay, and is actually located within the gated complex of Marina Hemingway. The entrance to Barlovento is not easy to spot the first time; it tends to blend into the shoreline. A red-and-white vertically striped sea buoy will eventually become apparent. From the sea buoy a heading of 140°T will guide you in. Although the channel appears to be narrow, there's 17 to 18 feet of water throughout the entrance and up to the customs dock just inside.

Port of entry: The officials are located right in the marina. This is reported to be one of the best places for pleasure vessels to enter the country.

Dockage: There are several "canals" with more than 100 slips. Contact the government-run Marina Hemingway (VHF channel 16 or 72; tel. 537-24-1150; fax, 537-24-6653); the marina is an official government clearing agency. Dockage rates (summer 1997) were approximately 45 cents/foot/day, including electricity and water when available.

Services: There is a special market for visitors with U.S. dollars, and there are supermarkets nearby and in Havana. Repair facilities for yachts may be available, but services tend to be geared toward fishing and commercial vessels. Cuban charts, publications, and guides are available at the marina.

Marina Hemingway is the first stop for many pleasure vessels in Cuba and is a good place to clear in and collect information. The city of Havana is a short bus or taxi ride away. The marina is reported to be safe and well managed.

BAHÍA CHORRERA, W side of entrance (No. 2), 23 08.0N, 82 24.6W. Fl R 6s, 33ft, 3M. Red concrete tower, 26ft.

BAHÍA CHORRERA, BEACON NO. 3, on mole, 23 07.9N, 82 24.7W. Fl G 3s, 10ft, 4M. Green port-hand beacon with topmark.

BAHÍA DE LA CHORRERA

23 08N, 82 25W

Pilotage: This inlet is about 3 miles west of the entrance to Bahía de la Habana. The Río Almenderes flows into the harbor. The coast in the vicinity is low and ragged, with blackened coral honeycombed by the sea. Depths in the bay average 20 to 30 feet over coral sand, in a position just within the entrance. With winds from the northeast and northwest, through north, the anchorage is subject to surge.

CASTILLO DEL MORRO, E side of entrance, 23 09.0N, 82 21.4W. Fl (2) W 15s, 144ft, 25M. Yellow truncated conical tower, 82ft. Storm signals. Signal station. Aeromarine light. Reserve light range, 11M.

NEPTUNO, 23 08.5N, 82 20.9W. Fl R 4s, 26ft, 4M. Red starboard-hand beacon with topmark.

RANGE (Front Light), 23 07.9N, 82 19.8W. Q Y, 42ft, 10M. White pyramidal framework tower; yellow diamond daymark, 39ft. **(Rear Light),** 300 meters, 124° from front, 23 07.8N, 82 19.7W. Iso Y 6s, 62ft, 10M. White pyramidal framework tower; yellow diamond daymark, 59ft.

LA HABANA (HAVANA)

23 08N, 82 20W

Pilotage: Havana is one of the largest sugar shipping ports in the world and is the commercial and shipping center for all of Cuba. Castillo del Morro is east of the entrance and La Punta Castle is located to the west. A powerful light is shown from Castillo del Morro. The land east of Castillo del Morro is about 200 feet high and flat, with a prominent ridge, Sierras de Jaruco, located about 19 miles farther east. About the same distance west of the entrance is a distinctive notch in the east end of Mesa del Mariel. At night, the loom of the metropolis is reported visible up to 25 miles offshore.

CAUTION: Occasionally, powerful lights are displayed from the dome of the capitol in La Habana; these lights may be confused with the navigation light on Castillo del Morro.

The pilot station may be contacted on VHF channels 3 and 16 (call sign, Habana Practicos). The port signal station is located in Castillo del Morro (VHF channels 13, 16, and 68; call sign, Morro Habana). The port authorities may be reached on VHF channels 16 and 68 (call sign, Habana Capitonia). The harbor is not receptive to visiting yachts. There are no yachting facilities, no anchorages, and the water is often foul.

NOTE: Pleasure vessels should enter at Marina Hemingway, where facilities are geared to their needs.

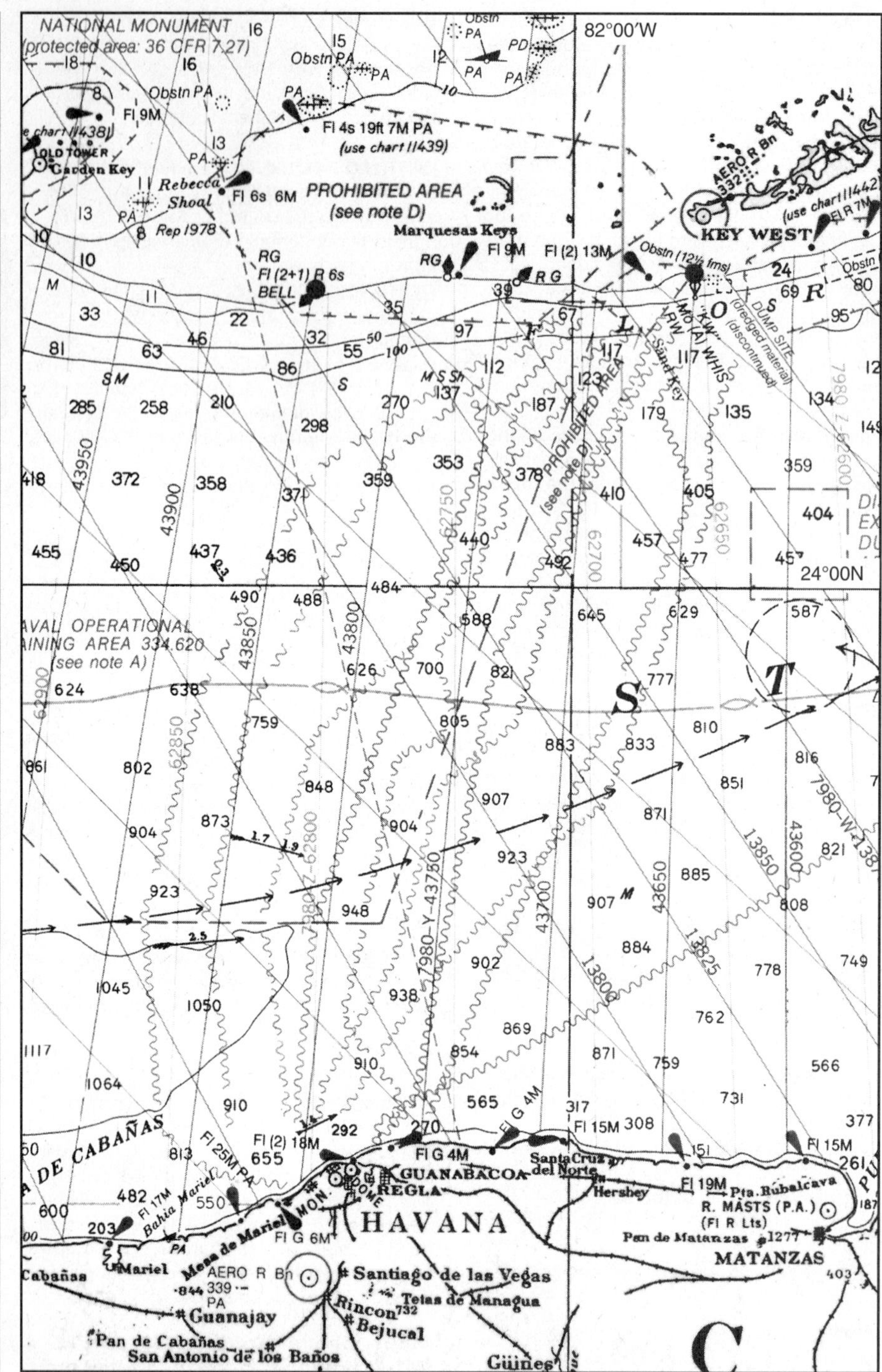

Approaches to Havana, soundings in fathoms *(from NOAA 11013, 40th ed, 10/93)*

RÍO GUANABO BEACON, 23 10.3N, 82 07.5W. Fl R 4s, 26ft, 3M. Red starboard-hand beacon with topmark.
RÍO JARUCO, W side of entrance, 23 10.8N, 82 00.6W. Fl W 10s, 56ft, 11M. White fiberglass tower, red bands, 33ft.

SANTA CRUZ DE NORTE

23 09N, 81 55W
Pilotage: This small community is located about 24 miles east of Bahía de La Habana. A distillery has a tall chimney and two storage tanks, which can be clearly seen from sea. Three tall chimneys of a sugar mill located 2 miles southwest of Santa Cruz del Norte standing on an elevation of 387 feet make a second excellent landmark. Between January and June the sugar mill may be brightly lit for nighttime operations.

CANASI LIGHT, 23 08.8N, 81 48.0W. Fl W 7s, 411ft, 15M. White fiberglass tower, red bands, 46ft. Reserve light range, 7M.
PUNTA SEBORUCO LIGHT, 23 09.1N, 81 36.4W. Fl W 15s, 115ft, 10M. Round concrete tower, red and white bands, 108ft.
PUNTA MAYA LIGHT, E side of entrance to Bahía de Matanzas, 23 05.6N, 81 28.5W. Fl W 8s, 112ft, 17M. White cylindrical tower, 105ft. F W light on wharf on N side of port.

MATANZAS

23 03N, 81 35W
Pilotage: This is an important port and city located about 45 miles east of La Habana. It is situated near the head of one of the north coast's largest bays. Two aeronautical lights shown from the high ground on the west side of the bay are visible for up to 35 miles.

Bajo Nuevo, a shoal with a least depth of 10 feet, and Bajo la Laja, with a least depth of 6 feet, lie offshore near the head of Bahía de Matanzas. Several deepwater passages lead in from the sea around and between these rocky dangers.

The pilots may be contacted on VHF channels 13, 16, and 68 (call sign, Matanzas Practicos).

VARADERO (CANALIZO PASO MALO)

KAWAMA WEST JETTY, 23 08.8N, 81 18.8W. Fl R 4s, 23ft, 3M. Red starboard-hand beacon with topmark, 20ft.
EAST JETTY, 23 08.0N, 81 18.7W. Fl G 3s, 23ft, 3M. Green port-hand beacon with topmark, 20ft.
RANGE (Front Light), 23 07.8N, 81 18.6W, Q W, 9ft, 10M. White concrete tower, square base, 7ft. **(Rear Light),** 108 meters, 152.6° from front. Iso W 6s, 23ft, 10M. White concrete tower, square base, 20ft.
LAGUNA DE PASO MALO, NO. 3, 23 07.8N, 81 18.3W. Fl G 3s, 13ft, 3M. Green port-hand beacon with topmark.
LAGUNA DE PASO MALO, NO. 6, 23 07.8N, 81 18.1W. Fl R 6s, 13ft, 3M. Red starboard-hand beacon with topmark.
LAGUNA DE PASO MALO, NO. 9, 23 07.8N, 81 17.7W. Fl G 5s, 13ft, 3M. Green port-hand beacon with topmark.
LAGUNA DE PASO MALO, NO. 11, 23 07.8N, 81 17.6W. Fl G 3s, 13ft, 3M. Green port-hand beacon with topmark.
LAGUNA DE PASO MALO, NO. 15, 23 08.0N, 81 17.0W. Fl G 5s, 13ft, 4M. Blue framework tower, 7ft.
LAGUNA DE PASO MALO, NO. 16, 35 meters S of No. 15. Fl R 6s, 13ft, 4M. Blue framework tower, 7ft.
LAGUNA DE PASO MALO, NO. 21, 23 07.8N, 81 16.7W. Fl G 3s, 13ft, 3M. Green port-hand beacon with topmark.
LAGUNA DE PASO MALO, NO. 22, 23 07.8N, 81 16.7W. Fl R 4s, 13ft, 3M. Red starboard-hand beacon with topmark.
CANAL CUEVA DE MUERTO, NO. 26, 23 7.9N, 81 16.5W. Fl R 6s. 20 ft, 3M. STARBOARD Red starboard-hand beacon with topmark.
CANAL CUEVA DE MUERTO, NO. 27, 23 8.2N, 81 16.1W. Fl G 5s. 20ft, 3M.Green port-hand beacon with topmark.

VARADERO

23 08N, 81 15W (approx.)
Pilotage: This resort community is located on a low part of the Península de Hicacos, near where it connects with the mainland. It is about 7.5 miles southwest of Punta Hicacos. This is one of the premier boating areas on the north coast of Cuba. From seaward this community is identified by tall trees, houses, windmills, and hotels. At night, the lights of the settlement are conspicuous.
Port of entry: This is a recommended place for pleasure vessels to clear in.
Dockage: Marina Acua is located just inside the canal. Marina Chapelín is in a mangrove-protected basin on the inside of the Península de Hicacos. Marina Gaviota and Marina Paradiso in this area are also listed as ports of entry.
Services: Fuel and other marine services are available. Because it is a tourist area, there are

nightclubs, restaurants, and shops. This is considered to be Cuba's premier beach.

PUNTA HICACOS

23 12N, 81 09W

Pilotage: This is the northern extremity of Península de Hicacos, which guards the northern shore of Bahía de Cárdenas. It is low and sandy but can be identified by the buildings of a salt works near it. Punta de Molas is the low-lying eastern extremity of the peninsula. The seaward side of the peninsula is a long fine beach, broken only in a few places by low cliffs. The highest cliff is situated 4.5 miles southwest of Punta Hicacos and is named Bernardino.

BAHÍA DE CÁRDENAS

CAYO PIEDRAS LIGHT, entrance to bay, 23 14.5N, 81 07.2W. Fl W 10s, 79ft, 10M. White round tower, 62ft.

CAYO MONITO, 23 13.8N, 81 08.6W. Fl (2) W 10s, 20ft, 5M. Isolated danger mark, black-red-black beacon with topmark.

CAYO DIANA, s side, 23 09.9N, 81 06.2W. Fl W 8s, 49ft, 10M. White framework tower, 39ft.

HICACOS, 23 9.0N, 81 7.0W. Fl (2)W 10s. 52 ft, 4M. Isolated danger mark, black-red-black beacon with topmark.

CAUTION: The lights and other navigational aids in Bahía de Cárdenas have been reported as unreliable.

CAYO DIANA, BEACON NO. 2, 23 10.8N, 81 07.7W. Fl R 6s, 13ft, 3M. Red starboard-hand beacon with topmark.

CAYO DIANA, BEACON NO. 3, 23 11.5N, 81 07.8W. Fl G 3s, 13ft, 3M. Green port-hand beacon with topmark.

CAYO DIANA, BEACON NO. 4, 23 11.2N, 81 07.7W. Fl R 4s, 13ft, 3M. Red starboard-hand beacon with topmark.

CHANNEL BEACON 4A, Punta Gorda, 23 10.0N, 81 10.3W. Fl R 4s, 13ft, 3M. Red starboard-hand beacon with topmark.

CHANNEL BEACON NO. 6A, 23 09.9N, 81 11.2W. Fl R 6s, 13ft, 3M. Red starboard-hand beacon with topmark.

CUPEY, NO. 1, 23 07.8N, 83 11.7W. Fl Y 7s, 10ft, 3M. Special mark, yellow beacon with topmark.

CANAL DE LA MANUY, NO. 1, 23 08.6N, 81 01.2W. Fl G 3s, 16ft, 3M. Green port-hand beacon with topmark.

CANAL DE LA MANUY, NO. 2, 23 08.3N, 81 01.1W. Fl R 4s, 16ft, 3M. Red starboard-hand beacon with topmark.

CANAL DE LA MANUY, NO. 3, 23 08.2N, 81 00.8W. Fl G 5s, 16ft, 3M. Green port-hand beacon with topmark.

CANAL DE LA MANUY, NO. 4, 23 07.8N, 81 00.6W. Fl R 6s, 16ft, 3M. Red starboard-hand beacon with topmark.

CANAL DE LA MANUY, NO. 7, 23 07.1N, 80 59.8W. Fl G 3s, 16ft, 3M. Green port-hand beacon with topmark.

CANAL DE LA MANUY, NO. 9, 23 05.2N, 80 57.8W. Fl G 5s, 16ft, 3M. Green port-hand beacon with topmark.

CANAL DE LOS BARCOS, NO. 2, 23 11.8N, 80 42.2W. Fl R 4s, 16ft, 3M. Red starboard-hand beacon with topmark.

CANAL DE LOS BARCOS, NO. 3, 23 10.1N, 80 42.1W. Fl G 5s, 16ft, 3M. Green port-hand beacon with topmark.

CÁRDENAS

23 02N, 81 12W

Pilotage: This is the second largest sugar exporting port in Cuba and also the location of a significant fishing industry. The port is located on the western side of Bahía de Cárdenas, under the shelter of Península de Hicacos to the northwest. The bay is scattered with low-lying mangrove-fringed islands, but there are several deepwater passages among them. The tall buildings in the resort of Varadero are prominent when approaching the bay from the north. The lighthouse on the low reef-fringed islet Cayo Piedras de Norte marks the entrance to the bay.

CAYO CRUZ DEL PADRE LIGHT, 23 16.9N, 80 53.9W. Fl W 7s, 82ft, 12M. White concrete tower on square base, 59ft. **RACON C** (– · – ·), 16M.

CAYO CRUZ DEL PADRE

23 16N, 80 55W

Pilotage: This small mangrove-fringed piece of land is located about 13 miles east-northeast of Punta Hicacos, making it the northernmost islet along the north coast of Cuba. To seaward of it is a dangerous, partially drying reef, which can be distinguished in calm weather by discoloration in the surrounding water. In heavy weather, waves break over it.

CAYO BAHÍA DE CADIZ LIGHT, N side of cay, 23 12.3N, 80 28.9W. Fl (3) W 15s, 177ft, 24M. White truncated pyramidal tower, black stripes, 161ft.

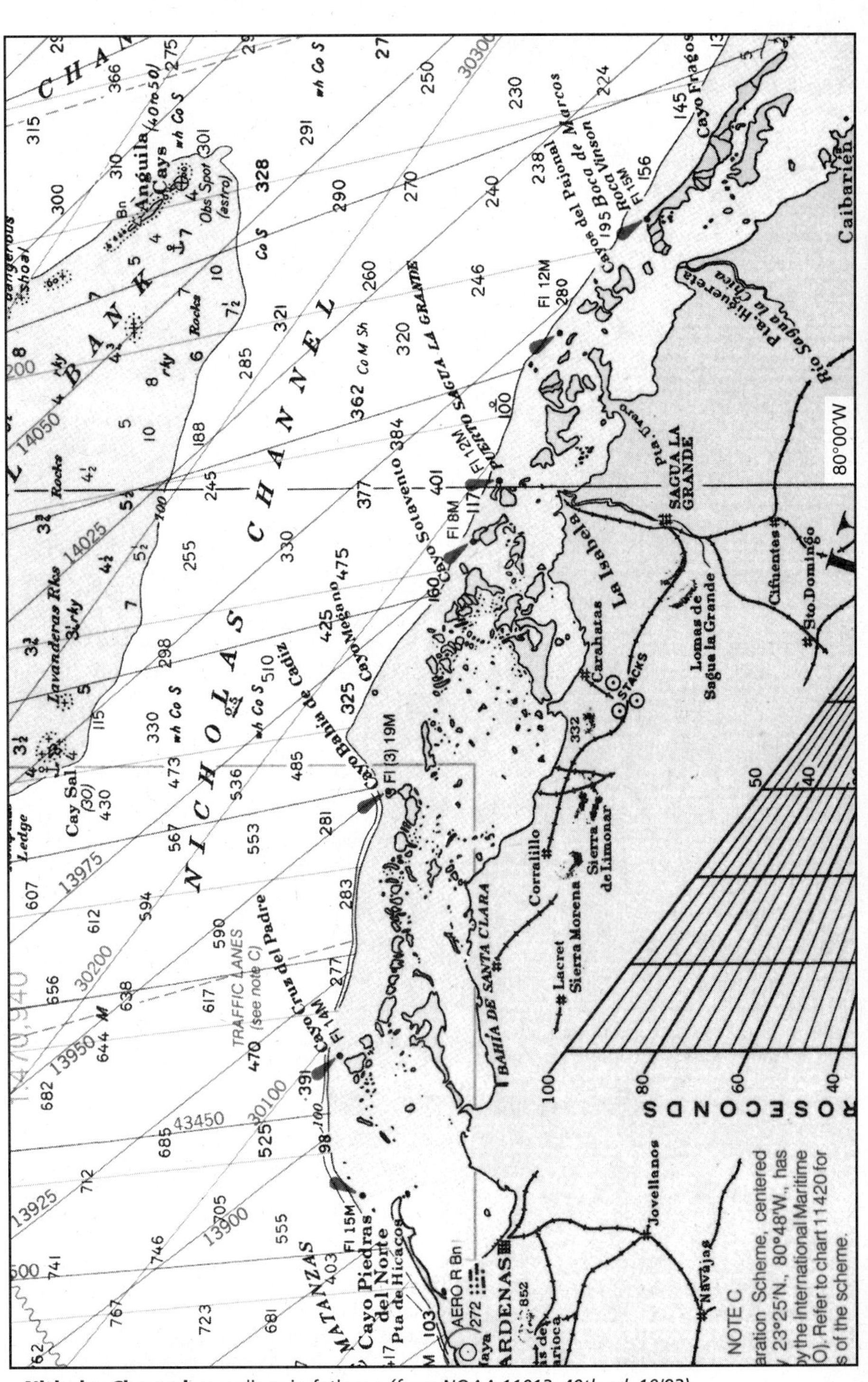

Nicholas Channel, soundings in fathoms *(from NOAA 11013, 40th ed, 10/93)*

CAYO BAHÍA DE CADIZ

23 12N, 80 29W
Pilotage: This flat islet is located about 24 miles east-southeast of Cayo Cruz del Padre. It is rocky on its north side and somewhat higher than other islets in the vicinity.
Anchorage: Bahía de Cadiz is a shoal bay located close southwest of the islet. Vessels with local knowledge can anchor here with some protection from the prevailing northeast winds. The anchorage is open to winds from the north.

SAGUA LA GRANDE

CAYO HICACAL, on Punta de la Rancheria, 23 04.3N, 80 05.2W. Fl W 8s, 36ft, 7M. White framework tower, 26ft.
CAYO HICACAL, NO. 4, 23 03.5N, 80 07.4W. Fl R 6s, 13ft, 3M. Red starboard-hand beacon with topmark.
CAYO DEL CRISTO, on Punta de los Practicos, 23 02.0N, 79 59.4W. Fl W 10s, 50ft, 10M. White framework tower, 39ft.
RÍO SAGUA ENTRANCE, E side, 22 56.8N, 80 00.0W. Fl G 3s, 13ft, 3M. Green port-hand beacon with topmark.
RÍO SAGUA, NO. 21, 22 56.9N. 80 00.6W. Fl G 5s, 13ft, 3M. Green port-hand beacon with topmark.
RÍO SAGUA, NO. 22, 22 56.9N. 80 00.6W. Fl R 6s, 13ft, 3M. Red starboard-hand beacon with topmark.
RÍO SAGUA, NO. 26, 22 56.8N. 80 01.0W. Fl R 6s, 13ft, 3M. Red starboard-hand beacon with topmark.

PUERTO SAGUA LA GRANDE

22 58N, 80 03W
Pilotage: Cayo del Cristo light, located about 6 miles north of Río Sagua la Grande, is the major identifying mark along this coast. The area off the mainland coast consists of small mangrove-covered islands. The main entrance to the port, which fronts the Río Sagua la Grande, is through Canal Boca de Maravillas. The dredged channel passes between Cayo de la Cruz and Cayo Maravillas. Being open to the northeast, there is frequently a very heavy sea in the entrance channel.

CAUTION: Dredged material has been deposited on each side of the channel where it lies uncharted and built up in the form of partially drying banks.

LA ISABELLA

22 57N, 80 00W
Pilotage: This small community located at the entrance to Río Sagua la Grande is a sugar-shipping center for Sagua la Grande, a well-populated community located about 12 miles upstream. The pilot may be contacted on VHF channel 16.

CANAL DEL SERON BEACON, E side, 22 58.0N, 79 55.6W. Fl G 5s, 13ft, 3M. Green port-hand mark with topmark.
BOCA DE JUTIAS BEACON, 22 58.0N, 79 51.9W. Fl G 5s, 13ft, 3M. Green port-hand mark with topmark.
CANAL DE CILINDRIN BEACON, 22 56.3N 79 49.0W. Fl G 5s, 13ft, 3M. Green port-hand mark with topmark.
CANAL DE LAS BARZAS BEACON, 22 54.0N, 79 45.6W. Fl G 3s, 13ft, 3M. Green port-hand mark with topmark.
CAYO ALTO, 22 51.9N, 79 46.5W. Fl G 5s, 16ft, 3M. Green port-hand beacon with topmark.
CAYO LA VELA, 22 56.6N, 79 45.4W. Fl W 12s, 39ft, 10M. White round metal tower, 30ft.
CAYO FRAGOSO, NW end, 22 48.5N, 79 34.7W. Fl W 15s, 68ft, 10M. White framework tower, 59ft.
CAYO FRAGOSO, BEACON NO. 2, 22 44.6N, 79 36.9W. Fl R 6s, 13ft, 3M. Red starboard-hand mark with topmark.
CAYO FRAGOSO, BEACON NO. 3, 22 42.6N, 79 36.2W. Fl G 5s, 13ft, 3M. Green port-hand mark with topmark.
CAYO FRAGOSO, BEACON NO. 5, 22 41.0N, 79 35.0W. Fl G 3s, 13ft, 3M. Green port-hand mark with topmark.
CAYO FRAGOSO, BEACON NO. 8, 22 40.1N, 79 34.2W. Fl R 4s, 13ft, 3M. Red starboard-hand mark with topmark.
CAYO FRANCES, W end at entrance to Caibarién, 22 38.5N, 79 13.8W. Fl W 10s, 32ft, 9M. White square concrete tower, 30ft.
CAYO FRANCES ANCHORAGE, 22 37.8N, 79 13.2W. Fl G 5s, 13ft, 3M. Green port-hand mark with topmark.
CAYO FRANCES, NO. 4, 22 35.6N, 79 17.4W. Fl R 4s, 16ft, 3M. Red starboard-hand beacon with topmark.
CAYO FRANCES, BEACON NO. 6, 22 34.4N, 79 18.5W. Fl R 6s, 16ft, 4M. Red starboard-hand mark with topmark.
CAYO FRANCES, BEACON NO. 8, 22 33.4N, 79 22.8W. Fl R 4s, 16ft, 4M. Red starboard-hand mark with topmark.

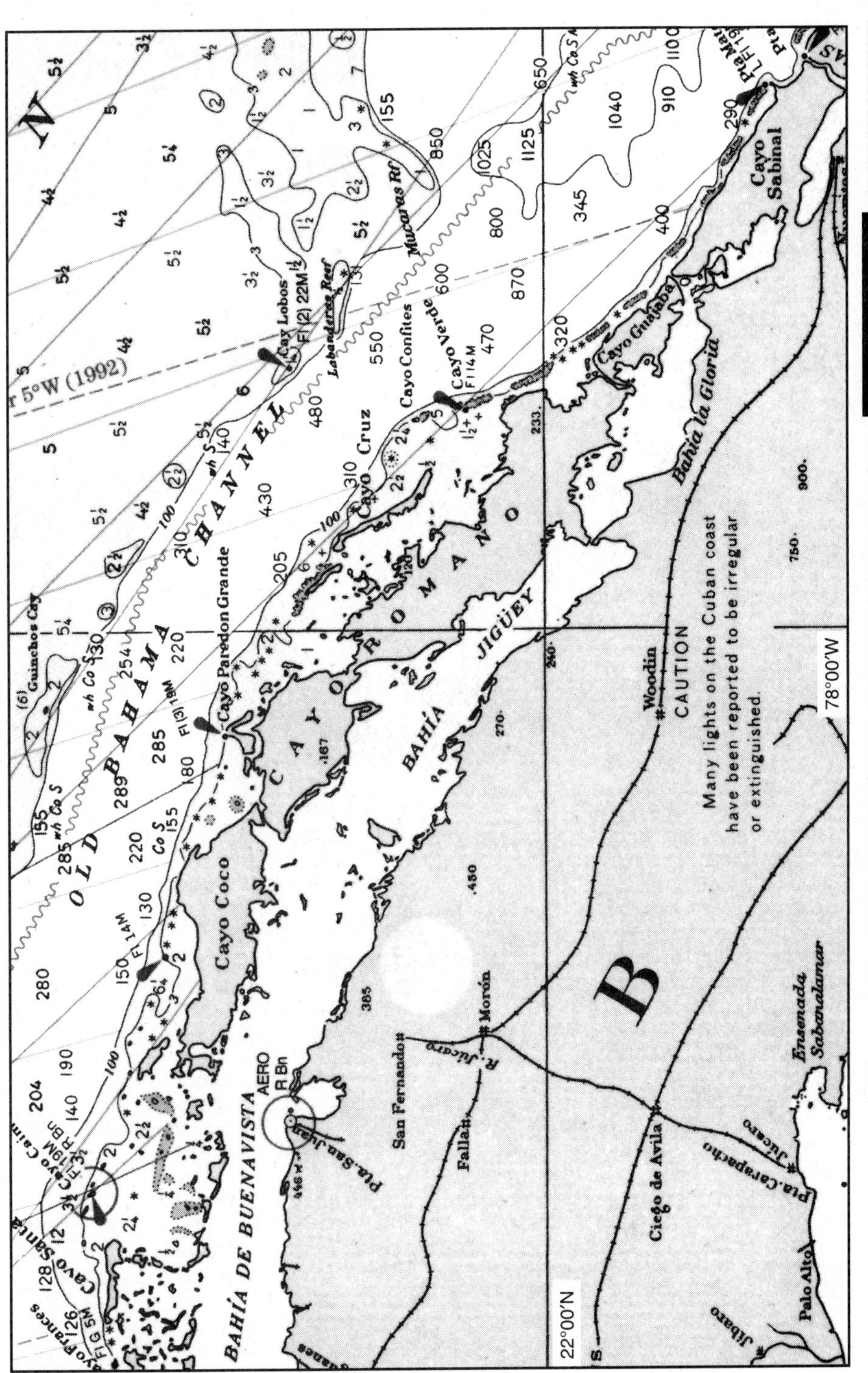

Old Bahama Channel, soundings in fathoms *(from NOAA 11013, 40th ed, 10/93)*

PUERTO DE CAIBARIÉN

22 37N, 79 15W

Pilotage: Located about 46 miles east-southeast of Puerto Sagua La Grande, with its seaward entrance near Cayo Frances (22° 38′N, 79° 13′W), this shoal-water area is the anchorage for Caibarién, a major sugar transshipment center located 16 miles to the west-southwest.

PUNTA BRAVA, BEACON NO. 9, 22 31.9N, 79 26.7W. Fl G 5s, 39ft, 4M. Green port-hand mark with topmark.

PUNTA BRAVA, BEACON NO. 11, 22 31.9N, 79 28.5W. Fl G 5s, 16ft, 4M. Green port-hand mark with topmark.

CAYO FIFA BEACON, SE, 22 35.5N, 79 28.5W. Fl R 6s, 16ft, 3M. Red starboard-hand mark with topmark.

CAYO FIFA BEACON, NE, 22 36.1N, 79 27.9W. Fl G 3s, 13ft, 3M. Green square daymark on concrete tower.

CANAL DEL REFUGIO, EAST BEACON, NO. 1, 22 31.8N, 79 27.5W. Fl G 5s, 13ft, 4M. Green port-hand beacon with topmark.

CANAL DEL REFUGIO, WEST BEACON, NO. 2, 22 31.5N, 79 27.3W. Fl R 6s, 13ft, 3M. Red starboard-hand beacon with topmark.

CANAL DE LAS PIRAGUAS BEACON, 22 37.0N, 79 13.2W. Fl R 4s, 16ft, 4M. Red starboard-hand beacon with topmark.

CANALIZO DE LOS BARCOS, BEACON NO. 2, 22 31.7N, 79 18.8W. Fl R 4s, 13ft, 4M. Red starboard-hand beacon with topmark.

CANALIZO DE LOS BARCOS, BEACON NO. 3, 22 31.0N, 79 18.9W. Fl G 3s, 16ft, 4M. Green port-hand beacon with topmark.

BAJO DEL MEDIO, BEACON NO. 5, off SW end, 22 29.8N, 79 17.4W. Fl G 5s, 16ft, 4M. Green port-hand beacon with topmark.

BAHÍA DE BUENAVISTA, BEACON NO. 6, 22 27.4N, 78 56.7W. Fl R 4s, 16ft, 3M. Red starboard-hand beacon with topmark.

CAYO BORRACHO BEACON, W end, 22 39.0N, 79 09.4W. Fl G 5s, 28ft, 3M. Green port-hand beacon with topmark.

CAYO CAIMAN GRANDE DE SANTA MARIA LIGHT, 22 41.1N, 78 53.0W. Fl W 5s, 158ft, 27M. Red conical tower, white bands, 105ft. ~~Radio-beacon~~. Aeromarine light. Reserve light range, 7M.

OLD BAHAMA CHANNEL

Pilotage: This channel separates the Great Bahama Bank from the north coast of Cuba and allows passage from the Atlantic via Crooked Island Passage to the Straits of Florida or the Gulf of Mexico. An IMO Traffic Separation Scheme has been established in the channel, which can best be seen on the appropriate chart.

The government of Cuba has unilaterally established a mandatory ship-reporting system to govern vessel movement in the area. See the introduction to the Cuba section and also "Special Notices" in chapter R for more information.

Two hours before entering the Traffic Separation Scheme vessels should contact the appropriate shore station in Cuba. The message should be prefixed "OLDBACHA" and should include information on the type of vessel, nationality, position, course, speed, and so on. For the western approach call "CLG-50 CAIMAN." For the eastern approach call "CLG-60 CONFITES." Communication with the stations is conducted on VHF channel 13, which should be monitored continuously while within the Traffic Separation Scheme. Vessels may contact the station in English or Spanish.

The southwest side of the channel is composed of low-lying islands and other dangers. There are a number of small lagoons suitable for exploration by small craft only. The southwest side of the channel from Cayo Paredon Grande Light (22°30′N, 78°10′W) to Cayo Confites (22°11′N, 77°40′W) is considered quite dangerous. For this 34-mile stretch vessels are recommended to stay in the middle of the channel and proceed with caution.

PASO MANUY WEST BEACON, 22 24.9N, 78 41.4W. Fl R 6s, 13ft, 4M. Red starboard-hand beacon with topmark.

CAYO JAULA, 22 34.2N, 78 30.9W. Fl W 10s, 68ft, 10M. White framework tower, 59ft.

BOCO DE MANATI BEACON, E side Manati Channel entrance, 22 15.7N, 78 29.8W. Fl G 5s, 13ft, 3M. Green port-hand beacon with topmark.

CAYO PAREDÓN GRANDE LIGHT, N side, 22 29.0N, 78 09.9W. Fl (3) W 15s, 157ft, 26M. Black and white checkered truncated pyramidal tower, 135ft. Aeromarine light. Reserve light range, 7M.

CAYO CONFITES, 22 11.4N, 77 39.7W. Fl W 7.5s, 75ft, 18M. White metal framework tower, 66ft.

CAYO CONFITES, NO. 2, 22 10.3N, 77 39.2W. Fl R 6s, 13ft, 3M. Red starboard-hand mark with topmark.

CAYO CONFITES, NO. 3, 22 10.2N, 77 38.6W. Fl G 5s, 13ft, 3M. Green port-hand mark with topmark.

CAYO CONFITES BEACON, 22 08.7N, 77 41.6W. Fl Y 7s, 13ft, 3M. Special mark, yellow with topmark.

CAYO CONFITES

22 10N, 77 40W

Pilotage: This low islet lies close within the outer edge of the bank along the south side of the Old Bahama Channel. A drying reef extends 1 mile south-southeast from the cay, and a channel 300 yards wide separates the cay from a reef that dries, which extends from it. A light is shown from a 66-foot-high tower standing on the north side of the cay.

Anchorage: Proceed to a position on the coastal edge of the bank where the light of Cayo Verde bears 191° and Cayo Confites's extremity bears 314°. From this point steers 270° until the south extremity of Cayo Confites bears 344°. Then haul to starboard and take a heading of about 323° until Cayo Confites's extremity bears 050°, distant 0.5 mile. Anchor in a charted depth of 22 feet.

BAHÍA DE NUEVITAS

PUNTA MATERNILLOS LIGHT, 21 39.8N, 77 08.5W. Fl W 15s, 174ft, 23M. White conical masonry tower, 171ft. Reserve light range, 7M.

PUNTA PRACTICOS, E side of entrance, 21 36.3N, 77 05.9W. Fl W 10s, 33ft, 6M. White pyramidal concrete tower, 30ft.

ENTRANCE RANGE (Front Light), 21 35.6N, 77 06.3W. Q W, 6ft, 8M. White diamond with black border on triangular structure, 6ft. **(Rear Light),** 170 meters, 185.6° from front. Iso W 3s, 23ft, 8M. White diamond with black border on triangular structure, 16ft.

PUNTA SALTEADORES LIGHTED BEACON NO. 3, 21 35.8N, 77 06.2W. Fl G 3s, 13ft, 3M. Green port-hand column with topmark.

PENA REDONDA BEACON NO. 4, 21 35.7N, 77 06.4W. Fl R 4s, 16ft, 3M. Red starboard-hand column with topmark.

PLAYA CHUCHU RANGE (Front Light), 21 35.6N, 77 06.9W. Q W, 10ft, 8M. White diamond with black border on triangular structure, 20ft. **(Rear Light),** 195 meters, 000.9° from front. Iso W 3s, 33ft, 8M. White diamond with black border on square concrete structure, 33ft.

LAS CALABAZAS RANGE (Front Light), 21 33.9N, 77 06.9W. Q W, 23ft, 8M. White diamond with black border on triangular structure, 20ft. **(Rear Light),** 190 meters, 180.9° from front. Iso W 3s, 33ft, 8M. White diamond with black border on square structure, 30ft.

BAJO DEL MEDIO RANGE (Front Light), 21 34.2N, 77 08.3W. Q W, 20ft, 8M. White diamond with black border on triangular structure, 13ft. **(Rear Light),** 265 meters, 269.5° from front. Iso W 3s, 33ft, 8M. White diamond with black border on square column, 30ft.

CAYO CAYITA RANGE (Common Front Light), 21 32.8N, 77 08.1W. Q W, 16ft, 8M. White diamond with black border on triangular structure, 13ft. **(Rear Light),** 290 meters, 179.5° from common front. Iso W 3s, 23ft, 8M. White diamond with black border on square column on pilings, 23ft. **(Rear Light),** 225 meters, 057.8° from common front. Iso W 3s, 33ft, 8M. White diamond with black border on square column, 30ft.

NUEVITAS

21 33N, 77 16W

Pilotage: The coast from Cayo Confites to Cayo Sabinal consists of sandy beaches, punctuated by numerous lagoons and swamps, with broken reefs adding to the dangers. Punta Maternillos light is about 4 miles northwest of the entrance to Puerto de Nuevitas and is the principal landmark in the area.

Bahía de Nuevitas is entered through a narrow deepwater channel 7 miles long. The bay is divided by a somewhat hilly and heavily scrub-covered peninsula extending 3 miles east from the southwest side of the bay. Bahía Nuevitas lies southeast of the peninsula and Bahía de Mayanabo lies northwest of it. The towns of Pastelillo and Puerto Tarafa, sub-ports of Nuevitas, lie respectively on the southeast and northwest sides of the peninsula. The terminal of Bufadero lies on the northeast side.

The surrounding coastal terrain is low-lying, flat, and without distinguishing features, except for the hills on the dividing peninsula and the nearby conical islets Cayo Ballenatos, which, rising above the lowland, are visible from the sea.

The entrance channel to Nuevitas has four right-angle turns in it and has a tidal current reaching 3 to 4.5 knots about 2 to 3 hours after high or low water. There is about a 20-minute slack.

The ports in Bahía de Nuevitas are the major sugar-shipping centers of Cuba. There is an IMO Traffic Separation Scheme in the waters off Bahía de Nuevitas.

BAHÍA DE MANATÍ TO BAHÍA DE BANES

PUNTA ROMA, 21 23.4N, 76 48.8W. Fl W 12s, 43ft, 10M. White conical truncated tower, 33ft.

PUERTO MANATÍ

21 22N, 76 50W
Pilotage: This small community is located on the west side of the pocket bay called Bahía de Manatí, located about 21 miles southeast of Puerto de Nuevitas and 165 miles west-northwest of Cabo Maisi. The light structure is reported to be visible by day up to 8 miles, and there is a gray brick chimney in the port. The port is a major sugar-shipping center.

PUERTO PADRE, Punta Masterlero, W side of entrance, 21 16.5N, 76 32.2W. Fl W 8s, 50ft, 10M. White conical truncated tower, 33ft.
PUERTO PADRE NO. 20, 21 13.0N, 76 34.4W. Fl R 6s, 16ft, 3M. Red starboard-hand beacon with topmark.

PUERTO PADRE

21 12N, 76 36W
Pilotage: This community is at the head of Bahía de Puerto Padre and is another major sugar-shipping point. The Punta Mastelero light structure stands at the seaward entrance to the bay and is reported visible during the day at a distance of 6 miles. The entrance itself is reported radar conspicuous at a distance of 5 miles.

PUNTA PIEDRA DEL MANGLE, 21 15.0N, 76 18.7W. Fl W 10s, 76ft, 10M. White metal framework tower, 59ft.
PUNTA RASA LIGHT, 21 09.0N, 76 07.8W. Fl W 15s, 112ft, 12M. White cylindrical concrete tower, red bands, 98ft.
PUERTO GIBARA, on Punta Peregrina, E side of entrance, 21 06.7N, 76 06.7W. Fl G 5s, 26ft, 5M. Green square on round green tower.

GIBARA

21 07N, 76 08W
Pilotage: This is the port for Holguin, the third largest city in Cuba. Silla de Gibara (21°02′N, 76°05′W) lies about 6 miles south-southeast of the entrance to Puerto Gibara. This saddle-shaped hill has a gray rocky summit rising to over 1,000 feet. Cerro Colorado (834 feet) and Cerro Yabazon (808 feet) are two conspicuous hills lying within miles west-southwest of Silla de Gibara. Lomas de Cupeicillo are a series of forested hills and conspicuous ridges rising to heights of 492 to 805 feet. They extend up to 10 miles west of Puerto Gibara.

Tidal currents are negligible in the entrance, but the flow from rivers in the rainy season can sometimes create a 0.25-knot current setting north.

BAHÍA DE BARIAY AND BAHÍA DE JURURU

21 05N, 76 01W
Pilotage: Bahía de Bariay is entered between Punta La Mula (Desiree) and a point 0.75 mile southwest. It is a shoal-water bay providing temporary anchorage for vessels with local knowledge in about 30 feet of water over white sand and coral.
Anchorage: Anchor close inshore off the second sandy beach south of Punta La Mulla. This anchorage is fully open to the north and is considered dangerous in the winter. Bahía de Jururu is located west of Bahía de Bariay and is completely sheltered. It is entered through a narrow channel blocked by a bar and is only suitable for small craft.

PUERTO DE VITA, Punta Barlovento, E side of entrance, 21 05.8N, 75 57.7W. Fl W 10s, 115ft, 10M. White cylindrical tower, 101ft.

PUERTO VITA

21 05N, 75 57W
Pilotage: The pocket bay of Bahía Vita is located about 9 miles east of Puerto Gibara and 115 miles west-northwest of the eastern extremity of Cuba. A narrow and intricate channel leads to a deepwater berthing facility at Puerto Vita. Because of the flat terrain in the area, vessels at Puerto Vita are discernible from sea.

The port is a sub-port of Gibara and another in the string of sugar-shipping centers. An excellent landmark on the approach is the tall white chimney of a sugar mill located about 4 miles south-southwest of the entrance. An isolated 397-foot-high hill, with a conspicuous outcropping of white rock on its top, is another good mark. The white rock looks like a vertical white stripe. The summit is reported to be visible 10 to 12 miles at sea. At night the Punta Barlovento light, on the east side of the port entrance, can be seen up to 10 miles.

BAHÍA NARANJO, E side of entrance, 21 06.8N, 75 52.6W. Fl W 6s, 59ft, 6M. White framework tower, 26ft.

BAHÍA DE NARANJO

21 06N, 75 53W

Pilotage: This is another pocket bay with no facilities except for a well-sheltered anchorage. On the west side of the bay is a flattened wooded hill rising to 344 feet. On the east side is an isolated sugarloaf hill. About 3 miles southeast of the entrance is a high flat-topped ridge having a white, precipitous west slope. A conspicuous red scarp at Punta Barlovento marks the entrance. The drying coastal reef north-northeast of Punta Barlovento extends almost 0.75 mile offshore.

Port of entry

BAHÍA SAMA, on Punta Bota Fuerte, E side of entrance, 21 07.5N, 75 46.1W. Fl W 8s, 98ft, 7M. White metal framework tower, 30ft.

SAMA

21 07N, 75 46W

Pilotage: This small community is located on the west side of Bahía Sama, which is about 11 miles east of Bahía Vita. The inlet is shoal, allowing vessels drawing less than 14.5 feet to enter. The mean tidal range here is about 2 feet. Pan de Sama, a rounded hill about 4 miles south-southwest of Bahía Sama, stands out well against a terrain of wooded flats and small undulations. This anchorage is open to winds from the north.

CABO LUCRECIA LIGHT, near extremity, 21 04.3N, 75 37.2W. Fl W 5s, 132ft, 25M. White stone tower on octagonal base, 121ft. Aeromarine light. Reserve light range, 7M. **RACON G** (– – ·), 14M.

CABO (PUNTA) LUCRECIA

21 04N, 75 37W

Pilotage: This is one of the principal landfalls for vessels proceeding along the north coast of Cuba. For a mile or two either side of the point the coast has a low profile and consists of a low white scarp partially interrupted by sandy beaches. Trees and mangroves are inland. A light is shown from a prominent stone tower standing 121 feet high. A stone dwelling stands behind the light tower. There is an IMO Traffic Separation Scheme off Cabo Lucrecia.

BAHÍA DE BANES, on Caracolillo Beach, S side of entrance, 20 52.6N, 75 39.7W. Fl W 8s, 43ft, 7M. White truncated conical tower, 33ft.

BAHÍA DE BANES, BEACON NO. 2, 20 52.9N, 75 39.8W. Fl R 4s, 13ft, 3M. Red starboard-hand column with topmark.

BAHÍA DE BANES, BEACON NO. 3, 20 52.8N, 75 39.8W. Fl G 5s, 13ft, 3M. Green port-hand column with topmark.

BAHÍA DE BANES, BEACON NO. 5, 20 52.8N, 75 39.9W. Fl G 3s, 13ft, 3M. Green port-hand column with topmark.

BAHÍA DE BANES, BEACON NO. 7, 20 52.7N, 75 40.1W. Fl G 5s, 13ft, 3M. Green port-hand column with topmark.

BAHÍA DE BANES, BEACON NO. 9, 20 52.4N, 75 40.3W. Fl G 3s, 10ft, 3M. Green port-hand column with topmark.

BAHÍA DE BANES, BEACON NO. 10, 20 52.5N, 75 40.3W. Fl R 6s, 13ft, 3M. Red starboard-hand column with topmark.

BAHÍA DE BANES, BEACON NO. 11, 20 52.9N, 75 40.6W. Fl G 3s, 13ft, 3M. Green port-hand column with topmark.

BAHÍA DE BANES, BEACON NO. 12, 20 52.8N, 75 40.8W. Fl R 4s, 15ft, 3M. Red starboard-hand column with topmark.

BAHÍA DE BANES, BEACON NO. 15, 20 53.0N, 75 41.1W. Fl G 5s, 15ft, 3M. Green port-hand column with topmark.

BAHÍA DE BANES, BEACON NO. 16, 20 53.4N, 75 41.4W. Fl R 4s, 20ft, 3M. Red starboard-hand column with topmark.

BAHÍA DE BANES, BEACON NO. 23, 20 54.6N, 75 42.5W. Fl G 3s, 15ft, 4M. Green port-hand column with topmark.

PUERTO BANES

20 55N, 75 42W

Pilotage: The deepwater pocket bay called Bahía de Banes is well sheltered and almost totally landlocked. The entrance channel is one of the most intricate on the Cuban coast. The bay is difficult to recognize from the sea, but from a position about 12 miles to the west three grouped hills, equal in elevation, can be seen. They are serrated in appearance and steep-to on the northeast side, but sloping on the southwest. Close northeast of the hills is a conspicuous rounded, or somewhat saddle-shaped, hill.

The entrance channel is marked by a light, but it should not be confused with the nearby light marking Bahía de Nipe. The sharp hairpin turns of the channel are crucial, and the seaward end is open to the northeast. Tidal currents run up to 6 knots, so vessels are advised to enter only at slack water during the day. The currents continue to run along the sides of the channel 40 to 45 minutes after high and low water.

Traffic signals are flown from the seaward entrance on the south side. A white flag means the channel is clear; a red flag means anchor and wait; a gray flag means wait, a vessel is outbound.

BAHÍA DE NIPE

PUNTA MAYARI LIGHT, E side of entrance, 20 47.5N, 75 31.5W. Fl W 6s, 115ft, 10M. White metal framework tower, 102ft.

PUNTA MAYARI ENTRANCE RANGE (Front Light), 20 46.3N, 75 32.8W. Q W, 25ft, 5M. Orange diamond on rectangular structure, 7ft. **(Rear Light),** 230 meters, 201.6° from front. Fl W 4s, 90ft, 8M. Orange diamond on rectangular structure, 7ft.

BAHÍA DE NIPE

20 47N, 75 42W

Pilotage: This is one of the largest pocket bays on the entire Cuban coast. The bay is extensive, well sheltered, and almost landlocked. The entrance channel is deep and easy, although strong tidal currents continue to run up to 45 minutes after high or low water. From the east the entrance appears as a steep-sided notch, while from the north it cannot be distinguished at any great distance. The Río Mayari empties into the bay and has cut a notch into Sierra de Cristal. This notch is visible well out at sea.

BAJO LA ESTRELLA, NO. 9, 20 47.5N, 75 36.7W. Fl G 3s, 15ft, 4M. Green port-hand mark with topmark.

BAJO SALINA GRANDE, NO. 10, S end, 20 48.5N, 75 41.8W. Fl R 6s, 13ft, 3M. Red starboard-hand mark with topmark.

ANTILLA CHANNEL

LENGUA TIERRA BEACON, NO. 11, 20 48.7N, 75 42.6W. Fl G 5s, 13ft, 4M. Green port-hand beacon with topmark.

BAJO MANATI, NO. 12, S end, 20 49.0N, 75 43.2W. Fl R 6s, 13ft, 4M. Red starboard-hand mark with topmark.

BAJO LENGUA DE TIERRA, NO. 13, N end, 20 48.9N, 75 43.8W. Fl G 5s, 13ft, 3M. Green port-hand mark with topmark.

BAJO MARABELLA, NO. 14, S end, 20 49.1N, 75 43.8W. Fl R 4s, 13ft, 4M. Red starboard-hand mark with topmark.

BAJO MARABELLA, NO. 15, 20 48.9N, 75 44.8W. Fl G 5s, 13ft, 4M. Green port-hand mark with topmark.

BAJO MARABELLA, NO. 17, 20 49.1N, 75 45.2W. Fl G 3s, 13ft, 4M. Green port-hand mark with topmark.

CANAL A, BEACON NO. 1A, 20 46.0N, 75 34.7W. Fl G 7s, 13ft, 4M. Green port-hand beacon with topmark.

BAJO PLANACA BEACON, NO. 3A, 20 44.7N, 75 34.6W. Fl G 7s, 13ft, 3M. Green port-hand beacon with topmark.

PUNTA LIBERAL, 20 44.8N, 75 28.8W. Fl W 10s, 56ft, 7M. White fiberglass tower, red bands, 33ft.

ANTILLA

20 50N, 75 44W

Pilotage: This is the principal shipping center for Bahía de Nipe and is located on the north side of the bay.

PRESTON

20 46N, 75 39W

Pilotage: This is another important sugar port located about 7.5 miles west-southwest of Punta Mayari. It is a sub-port of Antilla. The Río Mayari discharges near the berthing facilities, causing a great deal of silting.

FELTON

20 45N, 75 36W

Pilotage: This small community is on the west side of Bahía de Cajinaya and about 1 mile east of the mouth of the Río Mayari. Another sub-port of Antilla, it specializes in iron ore.

SAETIA

20 47N, 75 34W

Pilotage: This is the site of a small banana plantation and shipping facility.

BAHÍA DE LEVISA

BEACON NO. 17, 20 43.2N, 75 29.3W. Fl G 5s, 13ft, 3M. Green port-hand beacon with topmark.

BEACON NO. 19, 20 43.4N, 75 29.8W. Fl G 5s, 13ft, 3M. Green port-hand beacon with topmark.

BEACON NO. 22, 20 43.6N, 75 30.4W. Fl R 4s, 13ft, 3M. Red starboard-hand beacon with topmark.

BEACON NO. 23, 20 43.4N, 75 30.7W. Fl G 5s, 13ft, 3M. Green port-hand beacon with topmark.

BEACON NO. 27, 20 43.2N, 75 31.8W. Fl G 5s, 13ft, 3M. Green port-hand beacon with topmark.

BEACON NO. 28, 20 43.4N, 75 31.8W. Fl R 4s, 13ft, 3M. Red starboard-hand beacon with topmark.

BEACON NO. 29, 20 43.1N, 75 32.2W. Fl G 3s, 13ft, 3M. Green port-hand beacon with topmark.

BEACON NO. 32, 20 43.2N, 75 32.9W. Fl R 4s, 13ft, 3M. Red starboard-hand beacon with topmark.
CAYO GRANDE, 20 43.1N, 75 29.4W. Fl R 4s, 13ft, 3M. Red starboard-hand mark with topmark.

BAHÍA DE LEVISA

20 43N, 75 31W
Pilotage: This pocket bay is entered about 5 miles southwest of Bahía de Nipe. It is well sheltered and almost totally landlocked. It is entered by a very narrow and intricate channel. Nicaro (20°43′N, 75°33′W) is the main shipping port in the bay. It is another sub-port of Antilla in Bahía de Nipe. Sugar and nickel ore are handled here.

BAHÍA DE SAGUA DE TANAMO

PUNTA BARLOVENTO, E point of entrance, 20 43.2N, 75 19.2W. Fl W 8s, 43ft, 10M. White framework tower, 26ft.
BAHÍA DE SAGUA DE TANAMO ENTRANCE RANGE (Front Light), 20 42.6N, 75 19.5W. Q W, 22ft, 5M. Rectangular daymark, orange diamond, 7ft. **(Rear Light),** 210 meters, 180° from front. Iso W 6s, 43ft, 10M. Rectangular daymark, orange diamond, 7ft.
BAHÍA DE SAGUA DE TANAMO, NO. 3, W point, 20 42.6N, 75 19.8W. Fl G 3s, 13ft, 3M. Green port-hand beacon with topmark.
BAHÍA DE SAGUA DE TANAMO, NO. 5, Cayo Juanillo, W side, 20 42.3N, 75 20.2W. Fl G 5s, 13ft, 3M. Green port-hand beacon with topmark.
BAHÍA DE SAGUA DE TANAMO, NO. 8, Cayo Alto, E side, 20 41.7N, 75 20.2W. Fl R 6s, 13ft, 3M. Red starboard-hand beacon with topmark.
BAHÍA DE SAGUA DE TANAMO, NO. 11, Cayo Medio, W side, 20 41.1N, 75 19.4W. Fl G 3s, 13ft, 3M. Green port-hand beacon with topmark.

DE SAGUA DE TANAMO

20 42N, 75 19W
Pilotage: Bahía de Sagua de Tanamo is located about 9.5 miles east-southeast of Bahía de Levisa. It is entered via a deep but intricate channel. The surrounding terrain is hilly and rises inland in a succession of uneven hills to Sierra de Cristal, a conspicuous mountain range some 13 miles to the south. The entrance is concealed and difficult to identify from offshore. The light at the east entrance point is difficult to see by day at a distance greater than 3 miles. The tidal current runs over 3 knots in the channel. The port is a shipping center for sugar and molasses.

CAYO MOA TO CABO MAISI

CAYO MOA GRANDE, 20 41.6N, 74 54.4W. Fl W 10s, 76ft, 12M. Aluminum framework tower, 66ft.
BAJO GRANDE ENTRANCE RANGE (Front Light), 20 40.1N, 74 52.9W. Q W, 13ft, 14M. Red and yellow checkered rectangular daymark on white framework tower on concrete piles, 6ft. **(Rear Light),** 615 meters, 211° from front. Iso W 6s, 26ft, 14M. Red and yellow checkered redctangular daymark on white framework tower on concrete piles, 19ft.

PUERTO CAYO MOA

20 41N, 74 52W
Pilotage: Between Bahía de Sagua de Tanamo and Puerto Cayo Moa are several shallow inlets suitable for small craft with local knowledge. An unbroken barrier of drying reefs and sand flats extends about 2 miles offshore and is marked by breakers. There is a shoal-water lagoon between the reefs and the coastline. Behind the coastline the terrain rises rapidly to lofty interior mountains known as Cuchillas de Toa.

This port lies inshore of the outer barrier and is somewhat sheltered by it. An extremely deep gut passes east of the low-lying mangrove-covered islet Cayo Moa Grande and leads to the deepwater port facilities at Darsena de Yaguasey and Punta Gorda. Tidal currents seaward of the entrance set southwest on a rising tide at a velocity of about 1 knot and north on a falling tide. In the anchorage area south of Cayo Moa the currents set basically east and west.

PUNTA GORDA

20 38N, 74 51W
Pilotage: This small mining community is located about 2.75 miles southeast of Puerto Cayo Moa. Ships load chromium ore from lighters while on moorings. Punta Gorda is a sub-port of Bahía de Barocoa.

BAHÍA DE YAMANIQUEY

20 34N, 74 43W
Pilotage: This bay is located about 3 miles south of Punta Guarico. It is approached via a break in the offshore reefs and is suitable for small craft only. The entrance is dangerous, except in very calm weather. This is a sub-port of Baracoa.

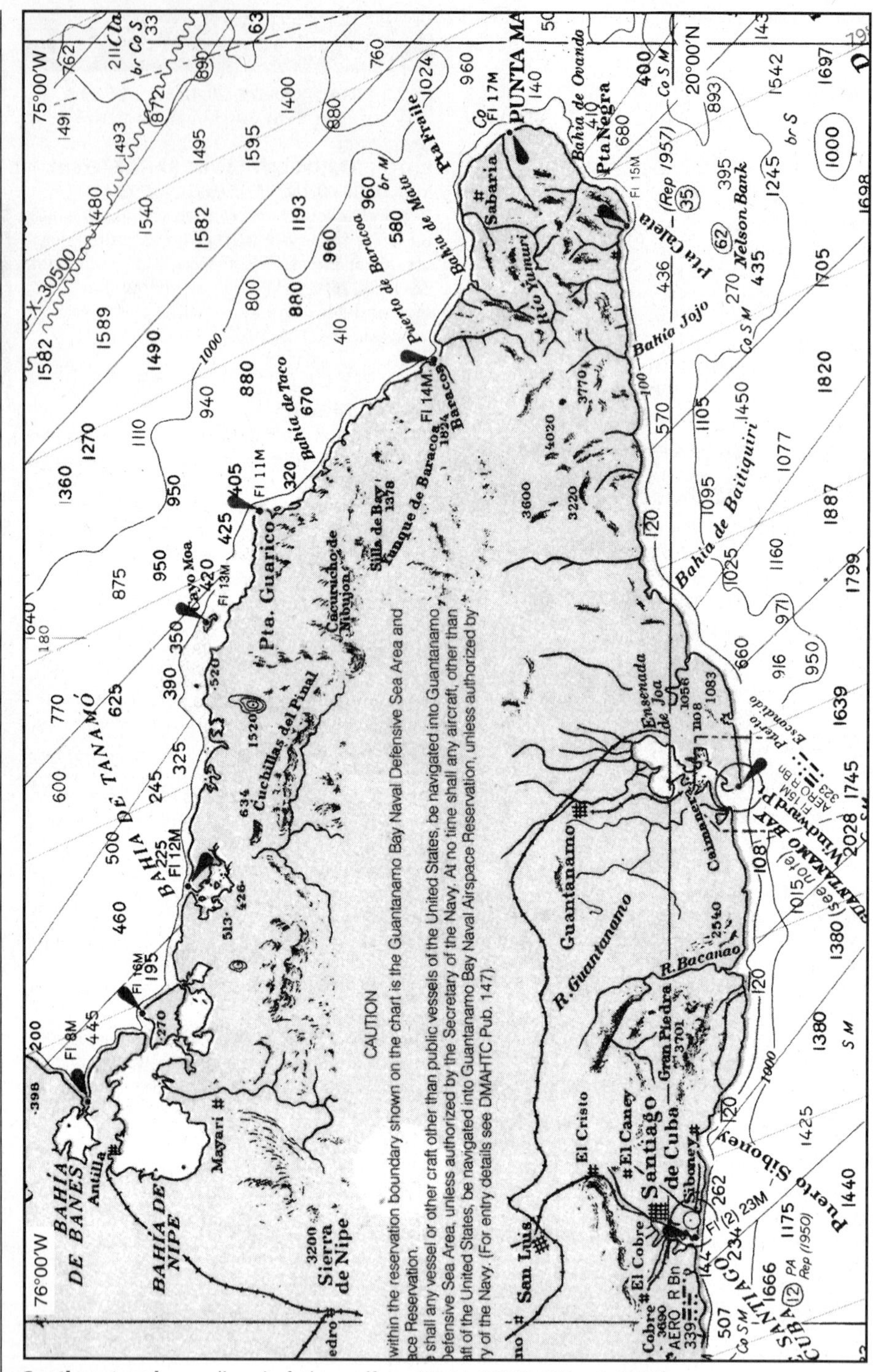

Southeast end, soundings in fathoms *(from NOAA 11013, 40th ed, 10/93)*

BAHÍA DE TACO

20 31N, 74 40W
Pilotage: This another miniature pocket bay, providing anchorage for small vessels with local knowledge. A short dog-legged channel has depths of about 30 feet. The sea breaks with considerable force on the rocky coast to the west of the entrance and tends to obscure the channel. Approaching vessels steer for the conspicuous south extremity of Punta Sotavento (20°32′N, 74°40′W) on a heading of 240°, then proceed in mid-channel to the anchorage.

PUNTA GUARICO, 20 37.1N, 74 43.9W. Fl W 6s, 37ft, 7M. White conical truncated tower, 33ft.

BAHÍA (PUERTO) NAVAS

Pilotage: This pocket bay is located about 10.5 miles southeast of Punta Guarico. It has an easy deepwater entrance and provides anchorage protected from the prevailing winds. The bay is open to the north.

BAHÍA DE (PUERTO) MARAVI

Pilotage: This deepwater estuary is located about 15 miles southeast of Punta Guarico. Anchorage is available in calm conditions, but the bay is open to the prevailing winds.

BAHÍA DE BARACOA, S entrance point, 20 21.1N, 74 29.9W. Fl W 6s, 63ft, 10M. Concrete tower on building, 59ft. Reserve light range, 7M.

BARACOA

20 21N, 74 30W
Pilotage: Bahía de Baracoa is located about 21 miles west-northwest of Cabo Maisi. It is a very small but deep pocket bay, open to the east and directly accessible from the open sea. The surrounding terrain is hilly and heavily scrub covered. Baracoa, on the east side of the bay, is one of the oldest communities in Cuba. Bahía de la Miel is a deepwater cove east of Baracoa, backed inland by a broad flat river valley leading to high interior hills.

Loma el Yunque, located 4 miles to the west, is the best approach landmark to the bay. It is a conspicuous steep-sided, flat-topped hill, rising to 1,932 feet. It is remarkable in its profile and can be seen for distances of 40 miles in clear weather. It is particularly visible from the northeast. Testas de Santa Teresa, two hills about 3.75 miles south-southeast of the bay, and Loma Majayara, close southeast of Bahía de la Miel, are three conspicuous hills, remarkable at a distance of 24 miles, with clear visibility.

CAUTION: This anchorage is subject to a heavy swell sent in by the prevailing winds, particularly during strong north and northeast winds in the winter.

Bahía de la Miel has anchorage somewhat sheltered from east winds, where vessels can anchor if they do not want to enter Baracoa.
Port of entry

BAHÍA DE MATA

20 18N, 74 23W
Pilotage: This bay will accommodate vessels up to about 300 feet, with drafts less than 15 feet. It has a deepwater entrance and is the first anchorage west of Cabo Maisi. Small vessels commonly drop anchor when abeam Punta Cuartel, tying their sterns to moorings farther in the bay. This is not a well-sheltered anchorage, particularly during the winter, when a heavy sea can set in.

~~**ROCA BUREN,** 20 21.3N, 74 30.1W. Fl G 5s, 39ft, 4M. Green port-hand mark~~. Removed 9/96.
QUEBRADO DEL MANGLE, 20 15.2N, 74 08.6W. Fl G 5s, 13ft, 4M. Green beacon, square daymark.
PUNTA MAISI LIGHT, on Punta de la Hembra, 20 14.6N, 74 08.6W. Fl W 5s, 122ft, 26M. White conical masonry tower and dwelling, 102ft.
RACON K (– • –), 16M.

CABO MAISI

20 13N, 74 08W
Pilotage: This is the eastern extremity of Cuba. It has a low-lying shore of white sand and is rounded. The land within the cape begins to rise 0.75 mile from the coast and when viewed from the north appears to form three steps, making useful landmarks. The terrain southwest of Punta Pintado becomes progressively steeper and more abrupt.

CAUTION: Cabo Maisi is a lee shore open to the effects of the sea and the prevailing east wind. Vessels navigating in the vicinity are cautioned to stand well offshore and are reminded that, if proceeding at night from the south, Cabo Maisi light is obscured to the west of a line bearing 359°.

An IMO Traffic Separation Scheme is in effect off Cabo Maisi.

WINDWARD PASSAGE

Pilotage: This passage lies between the eastern end of Cuba and the western part of Hispaniola, 45 miles to the east-southeast. An IMO Traffic Separation Scheme lies off Cabo Maisi, and the appropriate chart should be consulted when in the area. Vessels not using the traffic separation scheme should avoid it by as wide a margin as possible.

The current set through the middle of the channel is to the southwest at a rate usually less than 0.75 knot but may attain speeds of up to 2 knots. Near the coasts on either side of the passage tidal currents are strong and irregular.

On the east side of the passage (near Haiti) the current sets north at about 0.75 knot around Pearl Point (19°40′N, 73°25′W), but 6 miles offshore the current will be found to set west or west-southwest. North of Cap du Mole, on Haiti, there are ripples where the northerly flow meets the current running west along the north shore of the island.

On the west side of the passage (near Cuba) a north current sets around Cabo Maisi. This flow is affected by the tides and the wind. During the summer months and with southerly winds an east set is experienced. With northerly winds a southerly set is found. Frequently, especially during the winter, a westerly current of considerable strength will be experienced.

CABO MAISI TO BAHÍA DE GUANTÁNAMO

PUNTA CALETA LIGHT, 20 04.0N, 74 17.8W. Fl W 10s, 149ft, 10M. White skeleton tower.

BAHÍA DE BAITIQUIRI, N, 20 01.5N, 74 51.2W. Fl W 6s, 30ft, 8M. Red framework tower, 26ft.

PUERTO BAITIQUERI

20 01N, 74 51W

Pilotage: This very well sheltered pocket bay is located about 42 miles west of Cabo Maisi. It is clearly indicated by the opening between the hills on either side of the entrance. The entrance channel has a least depth of about 10 feet and is only 49 feet wide between the reefs on either side. The sea breaks heavily over these reefs and can be seen from 0.25 mile away.

PUERTO ESCONDIDO

19 55N, 75 03W

Pilotage: This port is located about 6 miles east of Bahía de Guantanamo. The entrance is quite deep but narrow, and the bay is landlocked. The coast on either side of the entrance appears as a continuous jagged bluff, and the entrance itself cannot be distinguished until very close in. A tower is situated on the summit of Mogate Peak, a hill that rises to 520 feet about 1.75 miles northwest of the entrance. A hill to the east of the entrance has a well-defined saddle-shaped summit. Vessels entering the port steer 336° and head for the extremity of a rocky scarp that lines up with the middle of the entrance channel.

Inside, the bay branches into numerous mangrove-fringed deepwater inlets that lead off into surrounding fields of drying tidal flats.

GUANTANAMO BAY

PUNTA BARLOVENTO (WINDWARD POINT LIGHT), on 110 meters peak, E side of entrance to bay, 19 53.7N, 75 09.6W. Fl W 5s, 378ft, 9M. On skeleton tower.

HICACAL BEACH RANGE, N side of entrance to bay **(Front Light),** 19 56.5N, 75 09.9W. Q W, 57ft, 17M. Red rectangular daymark, white stripe on skeleton tower. Higher intensity on rangeline. **(Rear Light),** 370 meters, 022° from front. Iso W 6s, 80ft, 15M. Red rectangular daymark, white stripe on skeleton tower, 37ft. Higher intensity on rangeline.

LEEWARD POINT AVIATION LIGHT, 19 54.2N, 75 12.3W. Al Fl W.

FISHERMAN POINT LIGHT, E side of entrance to bay, 19 55.2N, 75 09.7W. Fl W 4s, 28ft, 5M. Red skeleton tower on pyramidal base, black and white diamond-shaped daymark. Visible all around. Marks Hicacal Range exit. Various lights mark the channel in the cove south of Deer Light. A F R light is shown at each end of fuel berth N of Deer Point.

CORINASO POINT LIGHT, 19 55.1N, 75 09.3W. Fl G 6s, 42ft, 3M. Framework tower, black and white checkered diamond-shaped daymark.

McCALLA HILL, 19 54.9N, 75 09.6W. F R, 175ft, 4M. ~~Maintained by the US Navy~~. Private light.

RADIO POINT, LIGHT NO. 1, 19 55.3N, 75 09.0W. Fl G 4s, 15ft, 3M. Green square on black piling. Obscured 240°–310°.

RADIO POINT, LIGHT NO. 2, 19 55.4N, 75 08.9W. Q R, 14ft, 3M. Red triangular daymark on piles. Obscured 310°–050°.

RADIO POINT, LIGHT NO. 4, 19 55.3N, 75 08.8W. Fl R 4s, 11ft, 3M. Red triangular daymark on dolphin. Obscured 334°–139°.

DEER POINT, LIGHT NO. 1, 19 55.4N, 75 08.8W. Fl G 4s, 10ft, 3M. Green square daymark on dolphin.

DEER POINT, LIGHT NO. 2, Q R, 14ft, 3M. Red triangle on dolphin. Obscured 310°–050°.

DEER POINT, LIGHT NO. 3, 19 55.4N, 75 08.8W. Q G, 12ft, 2M. Green square daymark on dolphin.

DEER POINT, LIGHT NO. 4, Fl R 4s, 11ft, 3M. Red trangle on dolphin. Obscured 334°–139°.

DEER POINT, LIGHT NO. 5, 19 55.4N, 75 08.7W. Fl G 4s, 14ft, 3M. Green square daymark on dolphin.

JUNCTION LIGHT DP, 19 55.3N, 75 08.7W. Fl (2+1) R 6s, 13ft, 4M. Preferred channel mark, red-green-red on dolphin with topmark.

EVANS POINT, LIGHT NO. 8, 19 55.3N, 75 08.6W. Fl R 4s, 10ft, 3M. Red starboard-hand mark with topmark on dolphin.

MUD ISLAND, LIGHT NO. 1, 19 55.9N, 75 08.1W. Fl G 6s, 7ft, 3M. Green port-hand beacon with topmark.

MARINE SITE TWO, CHANNEL LIGHT NO. 1, 1.9 kilometers SE of Hospital Cay South Light, 19 55.9N, 75 08.1W. Fl G 4s, 10ft, 3M. Green port-hand mark with topmark on dolphin.

MARINE BOAT CHANNEL LIGHT NO. 2, 457 meters S of Caravela Point, 19 56.2N, 75 07.8W. Fl R 4s, 13ft, 3M. Triangular-shaped red daymark on dolphin.

GRANADILLO POINT WEST, LIGHT NO. 2, 19 56.8N, 75 07.5W. Fl R 4s, 13ft, 3M. Red starboard-hand mark with topmark on dolphin.

GRANADILLO BAY, EAGLE CHANNEL (CAYO TOMATE JUNCTION LIGHT CT), 19 56.9N, 75 07.9W, Fl R (2+1) 6s, 14ft, 3M. Preferred channel mark, red-green-red on dolphin with topmark.

Eagle Channel Lighted Buoy 1, Fl G 4s, 4M. Green.

HOSPITAL CAY, WATERGATE CHANNEL, NO. 2, 19 56.5N, 75 08.7W. Q R, 13ft, 3M. Red starboard-hand mark with topmark.

WATERGATE CHANNEL, NO. 4, 19 57.1N, 75 08.7W. Fl R 4s, 11ft, 4M. Red starboard-hand mark with topmark.

PALMA POINT SHOAL, NO. 13, 19 57.8N, 75 08.9W. Fl G 4s, 14ft, 4M. Green port-hand mark with topmark on dolphin.

DESEO, BEACON NO. 1, 19 58.9N, 75 08.5W, 16ft, 3M. Green port-hand beacon.

DESEO, BEACON NO. 2, 19 58.9N, 75 08.4W. Fl R 2s, 23ft, 3M. Red starboard-hand beacon.

CAYO RAMON, 19 59.2N, 75 07.8W. Fl (2+1), 14ft, 4M. Preferred channel beacon, green-red-green, with topmark.

BAHÍA DE GUANTÁNAMO

19 56N, 75 10W

Pilotage: This bay is located about 62 miles west-southwest of Cabo Maisi. It is spacious, well sheltered, easily entered, and largely landlocked. The bay is about 11 miles long and divides into outer and inner harbors. It is deep enough for any vessel.

REGULATIONS: Most of the bay is leased by the United States government and is part of the Guantanamo Bay Naval Defensive Sea Area. This area is closed to the public. Full details concerning entry control and application requirements will be found in Title 32, U.S. Code of Federal Regulations, Part 761.

Harbor Control Post: The U.S. Navy maintains a Harbor Entrance Control Post that challenges and identifies all vessels approaching Guantanamo Bay. Permission to enter the U.S. Naval Reservation boundaries should be secured in advance of arrival. Vessels sailing to Boquerón should forward their ETA and request permission to transit the waters of the naval reservation. Such passage is permitted only in daylight.

The Harbor Entrance Control Post operates from the signal station atop a building on McCalla Hill (19° 55′N, 75° 09′ 5W); the post monitors VHF channel 12 (call sign, Port Control). The pilots stand by on VHF channel 74.

BAHÍA DE SANTIAGO DE CUBA

MORRO DE CUBA LIGHT, on El Morro, E side of entrance to Santiago de Cuba, 19 58.0N, 75 52.1W. Fl (2) W 10s, 269ft, 26M. White round concrete tower, 44ft. Aeromarine light. Reserve light range, 7M.

SANTIAGO DE CUBA

20 01N, 75 50W

Pilotage: This is one of the oldest and most important port cities in Cuba. The bay is well sheltered and landlocked. A number of coves and inlets indent this natural harbor, which is entered by way of a 263-foot-wide channel. The entrance is marked by the fortifications of El Morro on the east side. Lower-lying fortifications at Punta Estrella are below the fort.

When entering, head for Punta Estrella on a course of 043° and proceed in mid-channel through most of the entrance fairway. The pilots monitor VHF channels 13 and 16 (call sign, Santiago Practicos).

A signal mast at the fortifications of El Morro indicates traffic conditions in the channel. International Code Flag P (see *Reed's Nautical Companion*) indicates a power-driven vessel is outbound; Flag P under a red pennant indicates a sailing vessel is outbound; and Flag P over a red pennant indicates the outbound vessel has anchored. *Caution: When a signal is displayed indicating a vessel is outbound, inbound vessels must not attempt to enter.* Contact the harbor authorities on VHF channel 16 before entering the harbor.
Port of entry
Dockage: Available at Marina Punta Gorda.

SANTIAGO DE CUBA TO CABO CRUZ

PUNTA ROMPE CANILLAS, 19 59.2N, 75 53.1W. Fl Y 7s, 13ft, 3M. Special yellow beacon with topmark.

PUERTO NIMA NIMA

19 57N, 75 59W
Pilotage: This port is located about 1.3 miles west-northwest of Punta Cabrera. There are numerous mooring buoys and a pier, which connects to the mines inland. A good mark on this part of the coast is a red hill excavated in terraces located about 1.5 miles west of Puerto Nima Nima.

ASERRADERO, 19 59.0N, 76 10.3W. Fl W 19s, 197ft, 10M. White framework tower, 66ft.

ASSERADO

19 57N, 76 09W
Bahía Aserradero indents the coast 9 miles west-northwest of Puerto Nima Nima. The Río Aserradero flows into this bay. A relic of the Spanish-American War, the wreck of the Spanish cruiser *Vizcaya,* is located on the west side of the bay.

PUERTO DE CHIRIVICO RANGE (Front Light), 19 58.2N, 76 23.8W. Q W, 13ft, 10M. White metal framework tower, 10ft. **(Rear Light),** on shore 225 meters, 337° from front. Fl W 4s, 33ft, 10M. White metal framework tower, 20ft. Visible 6° each side of rangeline.

PUERTO DE CHIVIRICO

19 58N, 76 24W
Pilotage: This port is located about 30 miles west of Bahía de Santiago de Cuba. It is west of Punta Tabacal, a 425-foot-high wooded conical hill with a grassy summit, which is easily identified from the east. The coast in this area is dominated by the rugged Sierra Maestra, which rise steeply. Pico Turquino (20°00'N, 76°50'W), the culminating summit of the mountain range, is the highest peak in Cuba at 6,560 feet. In clear weather the summit can be seen in Jamaica, 93 miles away.

The entrance to Puerto de Chivirico is between Cayo de Damas to the east and a peninsula that extends 600 yards from the coast to the west. The entrance channel is narrow. An abandoned ore-loading facility is in the harbor.

ENSENADA MAREA DEL PORTILLO, NO. 4, 19 54.7N, 77 11.2W. Fl R 6s, 10ft, 3M. Red starboard-hand beacon with topmark.
BAJO PUNTA DEL MEDIO, 19 54.6N, 77 11.4W. Fl (2) W 6s, 10ft, 4M. Isolated danger mark, black-red-black beacon with topmark.
PUNTA RASA, 19 54.8N, 77 11.1W. Fl Y 7s, 10ft, 3M. Special yellow beacon with topmark.

PUERTO PORTILLO

19 55N, 77 11W
Pilotage: The harbor is 45 miles west of Puerto de Chivirico. It is entered between Punta de Piedras and Punta de los Farallones, 0.5 mile to the west-southwest. Puerto Portillo can be identified by low, swampy mangrove-covered land on its east side and by the three perpendicular white cliffs on its west point. The bay is small and much encumbered.

ENSENADA DE MORA

19 54N, 77 18W
Pilotage: This break in the coast is located about 5 miles west of Puerto Portillo. Its entrance is between Cayo Blanco on the east, and Punta Icacos, 2 miles to the west-southwest. The bight consists of deep water indenting the low-lying coastal plain, with the westernmost part of the Sierra Maestra backing it. Close to the north-east is 1,248-foot-high Loma Aguada, which provides an excellent landmark. There are prominent cane fields west of the peak and a sugar mill on the northwest shore of the bay. The white spire of a church and a water tower with a red tank are north and northwest of the sugar mill.

The entrance is rather intricate, with a fairway partially blocked by rocky heads. The bay is encumbered with dangers. Head for Loma Aguada on a heading of 005°; then, when abeam of Punta Icacos, the low mangrove-covered west entrance of the bight, ease to port and steer for the light charted northeast of Cayo Pajaro. Pass north and east of Cayo Pajaro; then proceed to your destination.

PUERTO PILÓN RANGE (Front Light), 19 54.0N, 77 17.2W. Q W, 16ft, 5M. Black framework tower on piles, white diamond, yellow border, 6ft. **(Rear Light),** on shore 1.5 kilometers, 355.5° from front. Fl W 10s, 26ft, 10M. White triangular framework tower, white diamond, yellow border, 23ft.

BAJO LA JAVLA, 19 53.9N, 77 18.6W. Fl R 4s, 16ft, 3M. Red starboard-hand beacon with topmark.

PUERTO PILÓN BEACON NO. 11, 19 54.0N, 77 19.0W. Fl G 5s, 13ft, 3M. Green port-hand mark with topmark.

PILÓN

19 54N, 77 19W

Pilotage: This is the small town on the west side of Ensenada de Mora where the sugar mill is located. The maximum draft for this sub-port of Manzanillo is 20 feet.

CABO CRUZ LIGHT, 914 meters E of SW extremity of cape, 19 50.4N, 77 43.6W. Fl W 5s, 112ft, 22M. Yellow conical masonry tower, rectangular base,105ft. Aeromarine light. Reserve light range, 7M.

RESTINGA CABO CRUZ BEACON, 19 50.1N, 77 44.9W. Fl R 4s, 20ft, 5M. Red starboard-hand beacon with topmark.

RESTINGA CABO CRUZ BEACON NO. 2, 19 50.3N, 77 44.7W. Fl R 6s, 10ft, 2M. Red starboard-hand beacon with topmark.

RESTINGA CABO CRUZ BEACON NO. 3, 19 50.4N, 77 44.5W. Fl G 3s, 10ft., 3M. Green port-hand beacon with topmark.

CABO CRUZ

19 51N, 77 44W

Pilotage: Río Toro cuts a remarkable gorge through the coastal terraces 7 miles west of Ensenada de Mora. Just west of the river Ojo del Toro rises to a height of 1,750 feet. This is the westernmost peak of the Sierra Maestra and is very prominent. When viewed from the southwest, the summit of this mountain appears as two or three hummocks.

Cabo Cruz is low and sandy, backed by a relatively level, somewhat forested, plain. The plain continues inland as flat tableland, eventually rising into the foothills of the Sierra Maestra to the east. There is a pilot station at the village located on the sandy spit of Cabo Cruz, which consists of a few huts and a flagstaff. The light is about 0.5 mile east of the cape's extremity, on the rear of a large rectangular building. East of Punta del Ingles the light is obscured by high land when bearing less than 285°. It is reported the light is also beamed for use by aircraft. The sea breaks heavily on a reef awash located 1.5 miles west of the lighthouse. A light marks the outer end of the reef.

Pilots: The pilot station monitors VHF channels 13 and 16.

Anchorage: Anchorage is available in a depth of 24 feet over sand, northwest of Cabo Cruz.

GOLFO DE GUACANAYABO

20 28N, 77 30W

Pilotage: This large gulf on the south coast of Cuba lies between Cabo Cruz and Punta de las Angosturas, about 66 miles to the north-northwest. The gulf is interrupted by many shoals, reefs, and cays, including Bajo de Buena Esperanza in the center. Numerous channels lead through the groupings of above- and below-water dangers. These channels may be affected by silting caused by the rivers flowing into the gulf.

NIQUERO

20 03N, 77 35W

Pilotage: Bahía de Niquero is entered about 15 miles northeast of Cabo Cruz. Niquero is a sugar port, and the tall chimney of the sugar mill is an outstanding landmark. The pier is reported to be 525 feet long, taking a maximum draft of 23 feet alongside.

MEDIA LUNA

20 09N, 77 26W

Pilotage: This port lies about 25 miles northeast of Cabo Cruz. It is composed of a community and a sugar mill with several prominent chimneys. The sugar-loading pier is connected by rail to the town 1.25 miles inland. There is an anchorage northwest of the pier, with poor holding in soft mud reported. This bay is wide open to the northwest.

SAN RAMON

20 13N, 77 22W
Pilotage: This port is located 31 miles northeast of Cabo Cruz. The sugar mill is marked by a prominent chimney. There is a 660-foot wooden pier with a minimum depth of 21 feet alongside.

BANCO DE BUENA, CAYO MEDANO BEACON, 20 20.4N, 77 48.7W. Fl Y 7s, 16ft, 6M. Special mark, yellow beacon with topmark.

CANAL DE PALOMINO

BEACON NO. 9, 20 10.0N, 77 44.9W. Fl G 5s, 13ft, 3M. Green port-hand mark with topmark.
BEACON NO. 10, 20 09.2N, 77 45.1W. Fl R 6s, 13ft, 4M. Red starboard-hand mark with topmark.
BEACON NO. 12, 20 09.5N, 77 40.9W. Fl R 4s, 13ft, 3M. Red starboard-hand mark with topmark.
BEACON NO. 13, 20 09.8N, 77 39.8W. Fl G 5s, 13ft, 4M. Green port-hand mark with topmark.

BANCO FUSTETE BEACON, 20 11.3N, 77 35.6W. Fl G 3s, 13ft, 4M. Green port-hand mark with topmark.
BAJO OREJON GRANDE BEACON, 20 06.3N, 77 40.7W. Fl Y 7s, 16ft, 6M. Special mark, yellow beacon with topmark.

CIEBA HUECA

20 13N, 77 19W
Pilotage: This sugar-loading port is located 33 miles northeast of Cabo Cruz. A conspicuous chimney marks the location of a large sugar mill. Small tankers also use a pier here. Vessels anchor northwest of the pier.

CAMPECHUELA

20 14N, 77 17W
Pilotage: This is another sugar-loading town, located about 35 miles northeast of Cabo Cruz. There are berthing facilities for small craft only.

PUNTA GUA, 20 17.3N, 77 15.5W. Fl R 6s, 13ft, 4M. Red starboard-hand beacon with topmark.
ENSENADA GUA, 20 18.4N, 77 10.2W. Fl Y 5s, 13ft, 7M. Special mark, yellow beacon with topmark.

CAYOS MANZANILLO

CAYO PERLA, on S point, approach to Manzanillo, 20 21.4N, 77 14.6W. Fl W 5s, 36ft, 8M. White metal framework tower on piles, 29ft.
PUNTA SOCORRO, 20 20.8N, 77 12.9W. Fl G 5s, 12ft, 3M. Green port-hand beacon with topmark.
PUNTA CAIMANERA, 20 19.8N, 77 09.4W. Fl R 4s, 12ft, 5M. Red starboard-hand beacon with topmark.
CAYITA, 20 22.2N, 77 08.7W. Fl G 3s, 12ft, 4M. Green port-hand beacon with topmark.

MANZANILLO

20 21N, 77 07W
Pilotage: This small metropolis and sugar-shipping center is located in Bahía de Caimanera, about 40 miles northeast of Cabo Cruz. It is the port for Bayamo, one of the oldest cities in Cuba. The bay is fronted offshore by several low-lying, reef-fringed, mangrove-covered islets known as Cayos Manzanillo. The light-colored buildings of Manzanillo, at the head of the bight, may be seen over the cays at distances up to 20 miles under favorable conditions.
Port of entry

BAJO CUCHARILLAS, SW of Medano de Cauto, 20 31.6N, 77 17.3W. Fl R 4s, 13ft, 3M. Red starboard-hand beacon with topmark.
BAJO PATICOMBITO BEACON, 20 34.2N, 77 28.9W. Fl G 5s, 13ft, 3M. Green port-hand beacon with topmark.
CHINCHORRO BANK, 20 32.7N, 77 22.5W. Fl R 6s, 13ft, 4M. Red starboard-hand beacon with topmark.

PASO DE CHINCHORRO

BAJO DE SANTA CLARA, NE side of shoal, 20 31.6N, 77 24.0W. Fl G 3s, 13ft, 4M. Green port-hand beacon with topmark.
BANCO VIBORA BEACON, 20 33.8N, 77 41.5W. Fl R 6s, 13ft, 4M. Red starboard-hand beacon with topmark.

SANTA CRUZ, 20 42.2N, 77 59.0W. Fl W 7s, 95ft, 9M. Red concrete hut, white base, 10ft.
SANTA CRUZ BEACON NO. 2, 20 41.7N, 77 58.1W. Fl R 4s, 16ft, 4M. Red starboard-hand beacon with topmark.
PUNTA BONITA, NO. 3, 20 41.8N, 77 58.6W. Fl G 5s, 16ft, 4M. Green port-hand beacon with topmark.
PUNTA BONITA, NO. 4, 20 41.9N, 77 58.5W. Fl R 6s, 16ft, 4M. Red starboard-hand beacon with topmark.

MEDIA LUNA CHANNEL

CAYO PATRICIO BEACON, 20 35.8N, 77 49.6W. Fl G 5s, 13ft, 4M. Green port-hand beacon with topmark.

CAYO CULEBRA BEACON, 20 33.6N, 77 50.0W. Fl R 4s, 16ft, 3M. Red starboard-hand beacon with topmark.
CAYO MEDIO LUNA BEACON, 20 33.0N, 77 53.5W. Fl G 5s, 13ft, 4M. Green port-hand beacon with topmark.
CAYO ALACRAN BEACON, 20 39.4N, 77 44.8W. Q W, 10ft, 5M. North cardinal mark, black and yellow beacon with topmark.
CAYO BAJO BAYAMESES BEACON, 20 40.6N, 77 49.5W. Fl G 3s, 10ft, 3M, Green port-hand beacon with topmark.
CAYO LOS TRES LACIOS, 20 40.6N, 77 52.2W. Fl G 5s, 10ft, 3M. Green port-hand beacon with topmark.
CAYO JUAN SUAREZ BEACON NO. 2, 20 32.5N, 78 01.8W. Fl R 6s, 13ft, 3M. Red starboard-hand beacon with topmark.
CAYO JUAN SUAREZ BEACON NO. 3, 20 33.5N, 78 04.1W. Fl G 5s, 13ft, 3M. Green port-hand beacon with topmark.
CAYO JUAN SUAREZ BEACON NO. 4, 20 36.3N, 78 06.2W. Fl R 4s, 16ft, 3M. Red starboard-hand beacon with topmark.
CAYO JUAN SUAREZ BEACON NO. 5, 20 35.9N, 78 06.8W. Fl G 5s, 13ft, 3M. Green port-hand beacon with topmark.

CANAL DE CUATRO REALES

CAYO CARAPACHO, 20 26.9N, 78 02.5W. Fl W 10s, 47ft, 10M. White metal framework tower, 33ft.
BEACON NO. 5, 20 28.5N, 77 59.7W. Fl G 5s, 13ft, 5M. Green port-hand beacon with topmark.
BEACON NO. 8, 20 28.6N, 77 58.9W. Fl R 4s, 13ft, 4M. Red starboard-hand beacon with topmark.
BEACON NO. 9, 20 29.2N, 77 58.8W. Fl G 5s, 13ft, 3M. Green port-hand beacon with topmark.

SANTA CRUZ DEL SUR

20 42N, 77 59W

Pilotage: This community is located on the northwest shore of Golfo de Guacanayabo. The entrance channel is between La Ceiba bank and Carapacho Cay Lighthouse.

MUELLE MANOPLA

20 43N, 77 52W

Pilotage: This sub-port of Santa Cruz is located about 6.5 miles to its east. It is approached via Canal Media Luna and Bayameses Passage.

CANAL DE CABEZA DEL ESTE, W side of entrance, on Cabeza del Este Cay, 20 31.0N, 78 19.8W. Fl W 12s, 47ft, 10M. White framework tower, 39ft.
MEDANO DE MANUEL GOMEZ, 21 01.5N, 78 51.8W. Fl R 6s, 16ft, 3M. Red starboard-hand beacon with topmark.
CAYO MANUEL GOMEZ, NW extremity of reef, 21 04.5N, 78 50.8W. Fl R 4s, 16ft, 3M. Red starboard-hand beacon with topmark.
CAYO SANTA MARIA, SW part of the bank, 21 11.0N, 78 39.2W. Fl W 5s, 43ft, 7M. White framework tower, 39ft.
PUNTA VERTIENTES, 21 24.5N, 78 34.5W. Fl G 3s, 13ft, 3M. Green port-hand beacon with topmark.

CANAL BALANDRAS

BEACON NO. 1, 21 25.6N, 78 45.0W. Fl G 5s, 13ft, 3M. Green port-hand beacon with topmark.
BEACON NO. 3, 21 26.4N, 78 46.1W. Fl G 3s, 13ft, 3M. Green port-hand beacon with topmark.
MEDANO DE BALANDRAS, 21 26.0N, 78 49.0W. Fl G 5s, 13ft, 3M. Green port-hand beacon with topmark.
LIGHT BEACON NO. 6, 21 27.8N, 78 47.0W. Fl R 6s, 16ft, 4M. Red starboard-hand beacon with topmark.
CAYUELO SABICU NO. 9, 21 30.1N, 78 50.1W. Fl G 5s, 13ft, 3M. Green port-hand beacon with topmark.

CABEZO DEL FLAMENCO, 21 24.8N, 78 53.2W. Fl R 6s, 10ft, 3M. Red starboard-hand beacon with topmark.
PASA ANA MARIA, S side, 21 30.7N, 78 46.4W. Fl G 5s, 13ft, 3M. Green port-hand beacon with topmark.

GOLFO DE ANA MARIA

21 25N, 78 40W

Pilotage: Golfo de Ana Maria is part of a broad coastal indentation lying between Punta de Las Angostura and Punta Maria Aguilar about 112 miles to the west-northwest. The coast is similar to the Golfo de Guacanayabo in that it is low-lying and consists of a muddy shore overgrown by mangroves. Behind the shore is a coastal plain largely cultivated for sugar cane. In the far west-northwest the coast gradually rises into the foothills of the mountainous Sierra de Sancti Spiritus and Sierra de Trinidad. Conspicuous on this coast is Loma de Banao, one of the highest peaks in the chain of the Sierra de Sancti Spiritus. Also conspicuous is Pico Porterillo, the summit peak of Sierra de Trinidad.

The 70-mile islet chains Jardines de la Reina and Laberinto de Las Doce Leguas form the south side of the area. These islets are steep-to, reef fringed, and mangrove covered. A line of sunken dangers continues to Punta Maria Aguilar, and more islets string along to Punta de las Angosturas. Inside these barriers the area is somewhat obstructed by a considerable scattering of above- and below-water dangers, particularly in the east-southeast and west-northwest portions. The muddy bottom often discolors the water, making it difficult to see the dangers.

The gulf is entered by way of a number of passages leading through the offshore barriers to ports on the mainland.

FUERA CHANNEL

BOCA GRANDE INLET, E side of entrance, 21 33.0N, 78 40.6W. Fl G 5s, 16ft, 3M. Green port-hand beacon with topmark.

CANAL DE PALO ALTO

BEACON NO. 13, 21 34.8N, 78 58.3W. Fl G 3s, 10ft, 3M. Green port-hand mark with topmark.
BEACON NO. 14, 21 34.8N, 78 58.3W. Fl R 4s, 10ft, 3M. Red starboard-hand mark with topmark.
BEACON NO. 15, 21 34.9N, 78 58.3W. Fl G 5s, 10ft, 3M. Green port-hand mark with topmark.
BEACON NO. 16, 21 35.0N, 78 58.1W. Fl R, 6s, 10ft, 3M. Red starboard-hand mark with topmark.

CAYO ENCANTADO, 21 33.6N, 78 49.7W. Fl G 3s, 12ft, 3M. Green port-hand beacon with topmark.
JUCARO, W side of channel, 21 36.6N, 78 51.2W. Fl G 5s, 16ft, 3M. Green port-hand beacon with topmark.
JUCARO, MUELLE DE MAMBISAS, 21 36.9N, 78 51.2W. Fl R 4s, 13ft, 3M. Red starboard-hand mark with topmark.

CANAL DE BRETON

Pilotage: This is the primary deepwater access to the Golfo de Ana Maria area. It is entered about 127 miles northwest of Cabo Cruz. The east side of the entrance is marked by Cayo Breton, which is low and mangrove covered. There is a barely awash reef to seaward that breaks in heavy seas but is hard to see in calm weather. The conspicuous remains of a white concrete tower stand on the west of the reef. Vessels bound for Jucaro steer for the summit of Sierra de Sancti Spiritus on a heading of 354° and proceed so as to avoid the dangerous reef fronting Cayo Breton. When the light bears 098°, distant 4 miles, haul to starboard and proceed through Canal de Breton.

CAYO BRETON, on W side, 21 07.3N, 79 26.9W. Fl W 10s, 108ft, 12M. White metal framework tower, 98ft. **RACON K** (– • –), 12M.

CANAL DE BRETON

LA VELA, 21 13.3N, 79 33.3W. Fl R 4s, 16ft, 3M. Red starboard-hand beacon with topmark.
BEACON NO. 6, 21 13.8N, 79 26.3W. Fl R 4s, 13ft, 3M. Red starboard-hand beacon with topmark.
BEACON NO. 8, 21 24.1N, 79 22.1W. Fl R 4s, 10ft, 3M. Red starboard-hand beacon with topmark.

JUCARO

21 37N, 78 51W

Pilotage: This sugar-shipping port located about 123 miles northwest of Cabo Cruz is the harbor for the larger community of Ciego de Avila. The shipping anchorage (21°31′N, 78°53′W) lies west of the low-lying, mangrove-covered Cayo Guinea. This anchorage is about 40 miles northeast of the entrance to Canal de Breton.

ENSENADA DE SANTA MARIA

21 16N, 78 31W

Pilotage: This is a shoal coastal indentation located about 28 miles southeast of Jucaro. Sugar is lightered out to vessels from the harbor.

PALO ALTO

21 36N, 78 58W

Pilotage: This is a sub-port of Jucaro. It is marked by a conspicuous chimney that stands near the commercial pier. There are also four prominent gray molasses tanks.

Arrecife Palo Alto lies about 2.25 miles south of Palo Alto. It is awash near its south end, and a light marks its southeast side.

TOMEGUIN, N end of shoal, 21 19.1N, 79 13.0W. Fl R 6s, 13ft, 3M. Red starboard-hand mark with topmark.
CAYO CACHIBOCA LIGHT, 20 40.7N, 78 45.0W. Fl W 15s, 111ft, 10M. White framework tower on piles, 98ft.

CANAL TUNAS

Pilotage: This access to Golfo de Ana Maria is located about 148 miles northwest of Cabo Cruz. It is the most direct route to the port of Tunas de Zaza. Cayo Zaza de Fuera, the first above-water landmass southeast of Canal Tunas, is low-lying, heavily wooded, and sandy. Cayos Machos de Fuera lie northwest of the passage and are equally low and wooded. A sunken danger with visible boulders is reported to lie about 6.75 miles west-southwest of Cayo Zaza de Fuera.

Vessels proceed to a position about 9.5 miles west-northwest of Cayo Zaza de Fuera, then steer east-northeast until Cayo Blanco de Zaza light bears 351°, distant about 3 miles. If heading for Tunas de Zaza, steer for Central Siete de Noviembre chimney and proceed until Cayo Blanco de Zaza light bears 305° before heading toward town.

BOCA ESTERO DE TUNAS, NO. 8, 21 37.9N, 79 33.6W. Fl R 6s, 13ft, 3M. Red starboard-hand beacon with topmark.
CAYO BLANCO DE ZAZA, 21 35.9N, 79 35.9W. Fl W 12s, 48ft, 8M. White framework tower.

CANAL MULATAS

ESTERO TUNAS DE ZAZA, NO. 6, 21 36.4N. 79 33.0W. Fl R 4s, 13ft, 3M. Red starboard-hand beacon with topmark.
MUELLE DE TUNAS, 21 37.8N, 79 33.0W. Fl Y 7s, 13ft, 5M. Special mark, yellow beacon with topmark.

TUNAS DE ZAZA

21 38N, 79 33W
Pilotage: This sugar-shipping port is located about 149 miles northwest of Cabo Cruz.

CANAL DE JOBABO

Pilotage: This is the principal passage to Casilda. The entry is easy and the passage deep, but intricate. Cayo Blanco de Casilda on the west side of the entrance is a reef-fringed wooded islet of white rock and sand that has the appearance of a wedge when seen from the southwest. A largely uninterrupted line of awash and sunken dangers extends from the islet northwest to Punta Maria Aguilar. Banco Cascajal, a shoal-water sandbank, continues the islet northeast to the mainland.

CAYO BLANCO DE CASILDA, E end, 21 38.3N, 79 53.0W. Fl W 7s, 46ft, 10M. White metal framework tower, 39ft.

BAJO LOS GUAIROS, BEACON NO. 3, 21 40.3N, 79 53.4W. Fl G 5s, 13ft, 3M. Green port-hand column with topmark.
BAJO JOBABOS, BEACON NO. 4, 21 40.5N, 79 53.2W. Fl R 4s, 13ft, 3M. Red starboard-hand column with topmark.
CANAL DE LOS GUAIROS, BEACON NO. 5, 21 40.7N, 79 53.8W. Fl G 3s, 16ft, 3M. Green port-hand column with topmark.
BEACON NO. 9, 21 40.7N, 79 54.0W. Fl G 3s, 21ft, 3M. Green port-hand column with topmark.
BEACON NO. 11, 21 40.6N, 79 54.1W. Fl G 5s, 20ft, 3M. Green port-hand column with topmark.
BEACON NO. 12, 21 40.7N, 79 54.2W. Fl R 6s, 13ft, 3M. Red starboard-hand beacon with topmark.
BANCO DERRIBADA, 21 42.0N, 79 57.9W. Fl G 5s, 13ft, 3M. Green port-hand beacon with topmark.
BANO DEL GUEYO, BEACON NO. 14, 21 42.3N, 79 56.6W. Fl R 6s, 16ft, 3M. Red starboard-hand beacon with topmark.
BANO DEL MEDIO, BEACON NO. 17, 21 43.6N, 79 57.7W. Fl G 3s, 13ft, 3M. Green port-hand column with topmark.
PUNTA CASILDA, BEACON NO. 21, 21 43.8N, 79 58.2W. Fl G 3s, 13ft, 3M. Green port-hand column with topmark.
PUNTA LASTRE, BEACON NO. 25, 21 44.4N, 79 59.2W. Fl G 3s, 13ft, 3M. Green port-hand column with topmark.
BASE NAUTICA ANCON, NO. 2, 21 44.4N, 79 59.7W. Fl R 4s, 13ft, 3M. Starboard-hand pile with topmark.
BASE NAUTICA ANCON, NO. 3, 21 44.3N, 79 59.6W. Fl G 3s, 13ft, 3M. Port-hand pile with topmark.
BASE NAUTICA ANCON, NO. 5, 21 44.3N, 79 59.6W. Fl G 5s, 13ft, 3M. Port-hand pile with topmark.
CAYO RATON, BEACON NO. 26, 21 44.4N, 79 59.1W. Fl R 4s, 16ft, 3M. Red starboard-hand column with topmark.
BEACON NO. 28, 21 44.5N, 79 59.2W. Fl R 6S, 16ft, 3M. Red starboard-hand column with topmark.
BEACON NO. 31, 21 45.5N, 79 59.5W. Fl G 3s, 13ft, 3M. Green port-hand column with topmark.
BEACON NO. 32, 21 44.9N, 79 59.4W. Fl R 4s, 13ft, 3M. Red starboard-hand column with topmark.
BEACON NO. 37, 21 45.1N, 79 59.6W. Fl G 5s, 13ft, 3M. Green port-hand column with topmark.

CASILDA

21 45N, 79 59W
Pilotage: This port located just east of Punta Maria Aguilar is entered by the Canal de Jobabo. This is the port for the inland city of Trinidad. Sugar is the main product being shipped here.
Port of entry

PUNTA MARIA ANGUILAR TO CAYO LARGO

ANCON, Punta Maria Anguilar, 21 44.6N, 80 01.3W. Fl W 5s, 56ft, 15M. Pedestal on white water tank, 7ft.
RÍO YAGUANABO LIGHT, E entrance point, 21 51.4N, 80 12.4W. Fl W 10s, 190ft, 15M. White round concrete tower with hut at base, 111ft.
CIENFUEGOS, PUNTA DE LOS COLORADOS LIGHT, E side of entrance, 22 02.0N, 80 26.6W. Fl W 5s, 83ft, 23M. White conical masonry tower, rectangular base, 66ft. Reserve light range, 7M.
CIENFUEGOS RANGE NO. 1 (Front Light), 22 03.5N, 80 27.5W. F R, 5M. Orange diamond on white triangular concrete structure, 16ft. **(Rear Light),** 220 meters, 350.2° from front. F R, 5M. Orange diamond on white triangular concrete structure, 33ft.
CIENFUEGOS RANGE NO. 2 (Front Light), 22 03.8N, 80 27.9W. Fl W 1.5s, 13ft, 7M. Orange diamond on white triangular concrete structure, 16ft. **(Rear Light),** 60 meters, 322.8° from front. Fl W 1.5s, 13ft, 7M. Orange diamond on white triangular concrete structure, 16ft.
JURAGUA LIGHT, NO. 5, 22 03.6N, 80 27.9W. Fl G 3s, 22ft, 3M. Green square concrete tower on hut, square base.
PASA CABALLOS, 22 03.7N, 80 27.7W. Fl R 4s, 33ft, 4M. Red starboard-hand beacon with topmark.
CIENFUEGOS RANGE NO. 3 (Front Light), 22 03.6N, 80 27.8W. F W, 5M. Orange diamond on white triangular concrete structure, 10ft. **(Rear Light),** 50 meters, 204.9° from front. F W, 5M. Orange diamond on white triangular concrete structure, 13ft.
CAYO ALCATRAZ, SSW side, 22 04.3N, 80 26.6W. Fl G 5s, 13ft, 3M. Green port-hand beacon with topmark.
CIENFUEGOS RANGE NO. 4, on Cayo Carenas **(Front Light),** 22 05.1 N, 80 27.5W. Fl W 1.5s, 13ft, 7M. Orange diamond on white triangular concrete structure, 10ft. **(Rear Light),** 60 meters, 358.5° from front. Fl W 1.5s, 13ft, 7M. Orange diamond on white triangular concrete structure, 13ft.

PASA BAJO DE LA CUEVA

NO. 1, 22 05.8N, 80 27.3W. Fl G 3s, 13ft, 3M. Green port-hand pile with topmark.
NO. 2, 22 05.7N, 80 27.3W. Fl R 4s, 13ft, 3M. Red starboard-hand pile with topmark.
NO. 3, 22 05.7N, 80 27.2W. Fl G 5s, 13ft, 3M. Green port-hand pile with topmark.
NO. 4, 22 05.6N, 80 27.2W. Fl R 6s, 13ft, 3M. Red starboard-hand pile with topmark.

JUNCO SUR, 22 06.6N, 80 26.3W. Fl G 5s, 13ft, 3M. Green port-hand beacon with topmark.

ENSENADA DE COTICA

NO. 5B, 22 09.3N, 80 27.9W. Fl G 5s, 13ft, 4M. Green port-hand column with topmark.
NO. 6B, 22 09.4N, 80 27.8W. Fl R 4s, 14ft, 3M. Red starboard-hand column with topmark.
NO. 7B, 22 09.3N, 80 27.5W. Fl G 3s, 14ft, 3M. Green port-hand column with topmark.
NO. 8B, 22 09.1N, 80 27.4W. Fl R 4s, 14ft, 3M. Red starboard-hand column with topmark.

CIENFUEGOS

22 09N, 80 27W
Pilotage: This harbor is located about midway along the eastern side of Bahía de Cienfuegos. The bay is surrounded by level or undulating land, which is heavily cultivated with sugar cane, especially to the east. This is considered one of the most important cities on Cuba. Punta de los Colorados light marks the east side of the entrance to the bay. Pico Cuevita, 15.5 miles east-southeast, is an excellent landmark, with its sharp conspicuous crest, which seen from the west appears as the highest peak of Sierra de San Juan. Loma Guamo, about 6 miles north-northwest, is an irregular peak and useful mark for determining position offshore when plotted together with Pico La Cuevita and the light at Punta de los Colorados.

Tidal currents average 1 to 2 knots during the dry season, but can increase to 4 knots during the wet season. The ebb can be particularly strong when runoff adds to the flow. Entry on an ebb tide is recommended. Range lights mark the channel.
Port of entry

BANCO DE JAGUA

21 35N, 80 40W
Pilotage: This bank lies well offshore and to the east of Banco de los Jardines. It is an isolated large shoal-water patch of coral that rises steep-to. During the day, it can be seen at a distance of about 1 mile, but at night it is hard to spot and must be considered very dangerous.

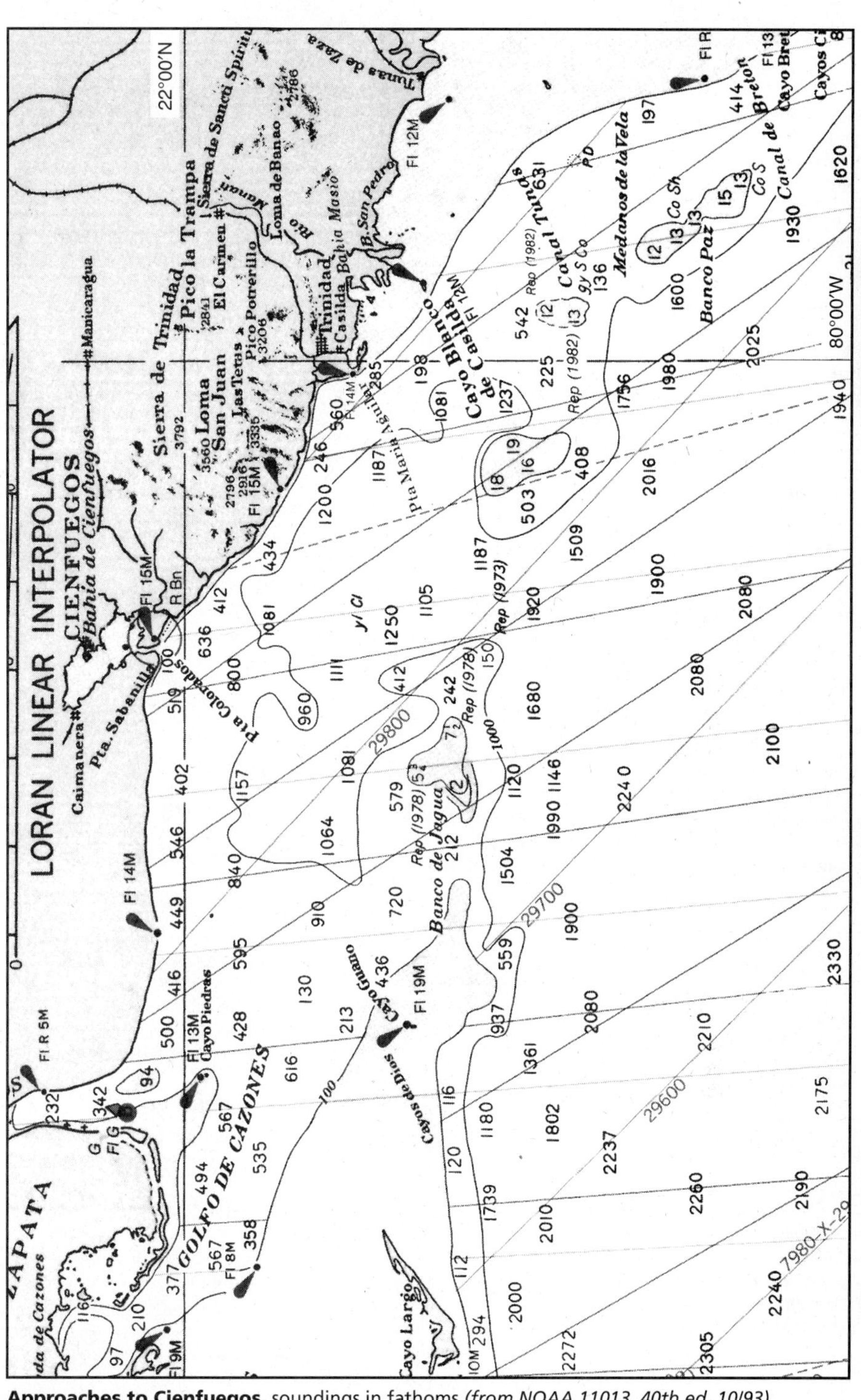

Approaches to Cienfuegos, soundings in fathoms *(from NOAA 11013, 40th ed, 10/93)*

CAYO PIEDRAS, W side of entrance to Bahía de Cochinos, 21 58.2N, 81 07.4W. Fl W 10s, 43ft, 8M. White cylindrical fiberglass tower, 33ft.
CAYO PIEDRAS, 21 58.1N, 81 07.6W, Fl G 5s, 13ft, 3M. Green port-hand beacon with topmark.
MUELLE, 21 58.0N, 81 07.5W. Fl G 3s, 13ft, 3M. Green port-hand beacon with topmark.
REEF, SW side, 21 57.7N, 81 09.4W. Fl R 4s, 13ft, 3M. Red starboard-hand beacon with topmark.

BAHÍA DE COCHINOS

22 07N, 81 10W
Pilotage: This is the deepest and most extensive of all the sleeve-like inlets indenting the Cuban coastline. Depths of over 100 fathoms are found over most of its area. Its west entrance point is Punta Palmillas, which is extended seaward to Cayo Piedras by a shoal-water spit that forms the only known danger in the immediate approaches. Playa Buenaventura lies at the head of the bay and is the only community of significance here. Anchorage near the inlet is not considered safe.

CAUTION: The western end of the Gulfo de Cazones, near Bahía de Cochinos (Bay of Pigs) is a prohibited zone. Coordinates are available on Cuban charts.

CAYO GUANO DEL ESTE LIGHT, near E extremity of Banco los Jardines, 21 39.7N, 81 02.5W. Fl (2) W 15s, 177ft, 25M. White round concrete tower, red bands on building, 148ft. Aeromarine light. Reserve light range, 7M.
PLAYA GIRON, 22 03.4N, 81 02.3W. Fl W 15s, 82ft, 8M. Pedestal on white water tank, 7ft.
BEACON NO. 1, 22 13.5N, 81 08.8W. Fl G 3s, 23ft, 3M. Green port-hand mark with topmark.
NO. 2, 22 13.6N, 81 08.7W. Fl R 6s, 13ft, 3M. Red starboard-hand beacon with topmark.
BEACON NO. 2, 22 12.6N, 81 08.4W. Fl R 4s, 26ft, 6M. Red starboard-hand beacon with topmark.
CABETA DE BUENAVENTURA, NO. 1, 22 12.6N, 81 08.4W. Fl G 3s, 13ft, 3M. Green port-hand beacon with topmark.
CABETA DE BUENAVENTURA, NO. 2, 22 16.7N, 81 12.5W. Fl R 4s, 13ft, 3M. Red starboard-hand beacon with topmark.
CABETA DE BUENAVENTURA, NO. 5, 22 16.9N, 81 12.6W. Fl G 5s, 13ft, 3M. Green port-hand beacon with topmark.
CABETA DE BUENAVENTURA, NO. 6, 22 16.9N, 81 12.6W. Fl R 4s, 13ft, 3M. Red starboard-hand beacon with topmark.
CAYO SIGUA, 21 53.4N, 81 25.1W. Fl W 7s, 30ft, 7M. White framework tower, 23ft.
PASA DE DIEGO PEREZ, on reef at entrance, 22 01.4N. 81 30.9W. Fl W 5s, 39ft, 7M. White framework tower on piles, 33ft.
CABEZO DEL VAPOR, 22 01.5N, 81 36.5W. Fl G 5s, 20ft, 3M. Green port-hand beacon with topmark.
CABEZO DEL CARBONERO, 22 05.9N, 81 46.0W. Fl G 3s, 20ft, 3M. Green port-hand beacon with topmark.
MEDANO DON CRISTOBAL, 22 07.3N, 81 49.2W. Fl R 6s, 13ft, 3M. Red starboard-hand beacon with topmark.
RESTINGA PRIETA, 22 15.8N, 81 58.6W. Fl G 5s, 13ft, 3M. Green port-hand beacon with topmark.
CAYOS BALLENATOS (LOS BALLENATOS), 21 34.7N, 81 38.3W. Fl W 10s, 30ft, 7M. White framework tower, 27ft.
CAYOS BALLENATOS BEACON NO. 25, 22 03.8N, 81 59.6W. Fl G 5s, 13ft, 3M. Green port-hand beacon with topmark.
CAYOS BALLENATOS BEACON NO. 28, 22 08.5N, 82 05.4W. Fl R 6s, 13ft, 3M. Red starboard-hand beacon with topmark.
BAJO LAS GORDAS, off SW side, 22 13.0N, 82 08.7W. Fl R 6s, 13ft, 3M. Red starboard-hand beacon with topmark.
CAYO AMBER, off SE side, 22 19.1N, 82 09.6W. Fl G 5s, 13ft, 3M. Green port-hand beacon with topmark.
PUNTA GORDA, 22 23.6N, 82 09.5W. Fl R 4s, 13ft, 3M. Red starboard-hand beacon with topmark.
CAYO LARGO, 21 38.4N, 81 33.9W. Fl W 6s, 43ft, 7M. White octagonal concrete tower, 23ft.
CAYO LARGO, NO. 6, 21 36.2N, 81 34.4W. Fl R 4s, 13ft, 4M. Red starboard-hand beacon with topmark.
CAYO LARGO, NO. 11, 21 37.5N, 81 34.2W. Fl G 3s, 13ft, 3M. Green port-hand beacon with topmark.
PLAYA SIRENA, 21 36.2N, 81 34.1W. Fl (2) W 10s, 6M. Isolated danger mark, black-red-black beacon with topmark.

CAYO LARGO

21 37N, 82 34W
Pilotage: This wooded cay in the Archipiélago de los Canarreos lies about 67 miles east of Isla de la Juventud. At the western end of the cay a marked channel leads to a tourist resort with a full-service marina. The marina is reported to monitor VHF channel 06.
Port of entry

CAYO AVALOS, S end, 21 32.3N, 82 09.9W. Fl W 8s, 30ft, 7M. White metal framework tower, 26ft.

CALETA DE CARAPACHIBEY LIGHT, S side of Isla de la Juventud, 21 26.9N, 82 55.5W. Fl W 7.5s, 184ft, 17M. White cylindrical concrete tower, yellow band, 170ft.

GOLFO DE BATABANO

PETATILLOS DEL NORTE, 22 28.0N, 82 39.8W. Q W, 20ft, 6M. North cardinal mark, black and yellow beacon with topmark.

BAJO LA PIPA, 22 09.5N, 82 58.0W. Fl R 5s, 13ft, 3M. Red starboard-hand beacon with topmark.

CAYO HAMBRE, 22 10.2N, 82 51.6W. Fl G 5s, 13ft, 3M. Green port-hand beacon with topmark.

PUNTA DE LOS BARCOS, 21 56.3N, 82 59.7W. Fl R 5s, 13ft, 3M. Red starboard-hand beacon with topmark.

PUNTA BUENAVISTA, 21 46.9N, 83 05.7W. Fl R 6s, 13ft, 3M. Red starboard-hand beacon with topmark.

DARSENA DE SIGUANEA NO. 1, W entrance, 21 37.0N, 82 59.1W. Fl G 5s, 13ft, 3M. Green port-hand beacon with topmark.

DARSENA DE SIGUANEA NO. 2, 21 37.0N, 82 59.1W. Fl R 6s, 16ft, 3M. Red starboard-hand beacon with topmark.

CAYOS LOS INDIOS, 21 43.1N, 83 10.0W. Fl G 5s, 13ft, 3M. Green port-hand beacon with topmark.

RÍO LAS CASAS, BEACON NO. 1, 21 55.2N, 82 47.7W. Fl G 3s, 16ft, 3M. Green port-hand beacon with topmark.

RÍO LAS CASAS, BEACON NO. 2, 21 55.2N, 82 47.8W. Fl R 5s, 16ft, 3M. Red starboard-hand beacon with topmark.

LOS COYUELOS, 21 38.4N, 83 11.2W. Q W, 13ft, 3M. North cardinal mark, black and yellow beacon with topmark.

GOLFO DE BATABANO LIGHT NO. 1, 21 59.5N, 82 43.4W. Fl G 3s, 13ft, 3M. Green port-hand beacon.

PASA LA MANTECA

BEACON NO. 2, 21 59.7N, 82 43.2W. Fl R 4s, 3M. Red starboard-hand beacon with topmark.

LIGHT NO. 13, 22 00.3N, 82 42.8W. Fl G 5s, 13ft, 3M. Green port-hand beacon with topmark.

LIGHT NO. 14, 22 00.2N, 82 42.7W. Fl R 4s, 13ft, 3M. Red starboard-hand beacon with topmark.

LIGHT NO. 21, 22 01.2N, 82 42.4W. Fl G 3s, 13ft, 3M. Green port-hand beacon.

LIGHT NO. 22, 22 01.3N, 82 42.3W. Fl R 4s, 13ft, 3M. Red starboard-hand beacon with topmark. Temporarily replaced by buoy (1994).

NO. 2, 21 55.7N, 82 39.4W, Fl R 4s, 13ft, 3M. Red starboard-hand beacon with topmark.

BEACON NO. 29, 21 56.1N, 82 37.5W. FL G 3S, 13ft, 3M. Green port-hand beacon with topmark.

NO. 30, 21 56.0N, 82 37.3W. Fl R 4s, 13ft, 3M. Red starboard-hand beacon with topmark. Pasa Quitasol marked by lights and beacons between nos. 2 and 30.

CANAL DEL INGLES BEACON, NE of Pasa de Quitasol, 21 57.2N, 82 36.5W. Fl G 5s, 13ft, 3M. Green port-hand beacon with topmark.

SURGIDERO DE BATABANO LIGHT, 22 41.1N, 82 17.8W. Fl W 10s, 101ft, 10M. White water tank, 6ft.

COMETA, 22 40.3N, 82 17.5W. Fl R 4s, 13ft, 3M. Red starboard-hand beacon with topmark.

REFUGE CANAL, E breakwater, 22 40.6N, 82 17.9W. Fl R 6s, 19ft, 3M. Red beacon with topmark.

REFUGE CANAL, W side, 22 40.8N, 82 17.9W. Fl G 5s, 13ft, 3M. Green port-hand column with topmark.

SURGIDERO DE BATABANO

22 41N, 82 18W

Pilotage: This city on the north side of the Golfo de Batabano has a partially sheltered roadstead anchorage. The area is exposed to southeast winds, which are common between July and October. Relatively shallow, the Golfo de Batabano lies between the mangrove-fringed island Cabo Diego Perez and Cabo Frances, 142 miles to the west. The numerous islets east and west of Isla de la Juventud form the gulf's southern boundary. Tidal action here is slight, but currents and water levels are strongly influenced by wind direction. A northeast wind lowers water levels, while a southeast wind raises them. Extreme lows occur with a northwest wind and extreme highs with a southwest one. Currents outside the gulf can be a concern as they tend to set vessels to the northwest, which is toward the dangers between Banco de Jardinillos and Isla de la Juventud. This is particularly so with southeast winds.

ISLA DE LA JUVENTUD (ISLA DE PINOS)

21 40W, 82 50W

Pilotage: This is the largest of the islands lying off the Cuban coast. It is quite flat, with the south part being very low, swampy, and densely wooded. The north two-thirds of the island has a wide, very flat coastal plain that merges with a scattering of high, often heavily forested interior hills and mountains. Loma la Canada (1,017 feet) is the highest and the first

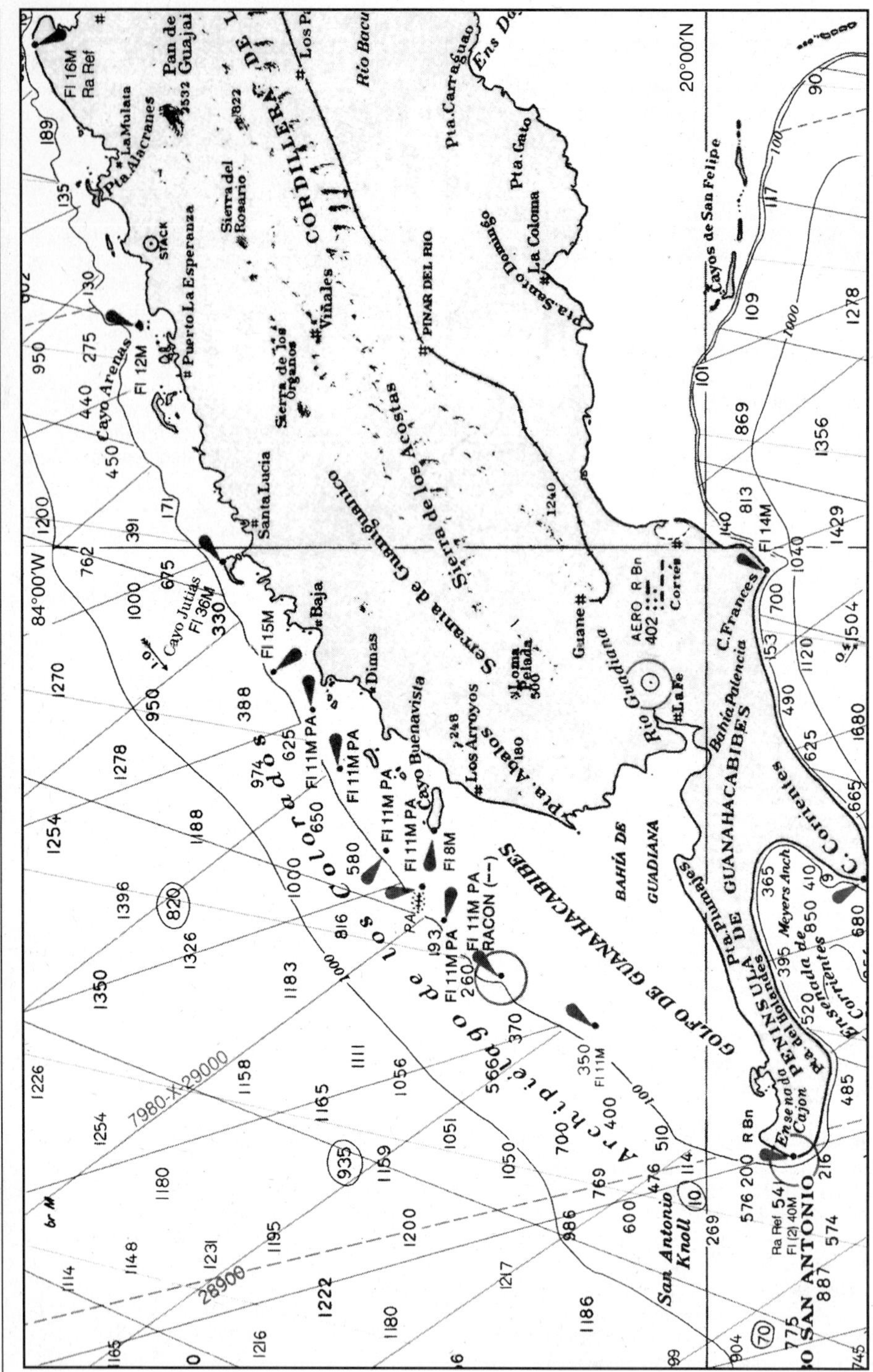

Western end, soundings in fathoms *(from NOAA 11013, 40th ed, 10/93)*

sighted when coming from the south. From the west it appears as a domed summit flanked by two sharp peaks. Loma Daguilla is the highest peak (612 feet) on the east side of the island; from the southeast it appears as a steep-sided isolated hill.

The Archipiélago de los Canarreos are the numerous islets lying scattered east of Isla de la Juventud for a distance of about 67 miles to the heavily wooded islet Cayo Largo. They continue on to Cayo Guano del Este, a group of high, closely spaced barren rocks, which form the easternmost above-water dangers fronting this section of the coast.

Ensenada de la Siguanea (21°38′N, 83°05′W), a spacious deepwater inlet on the west side of Isla de la Juventud, is entered between Punta Frances, the low-lying, mangrove-covered west extremity of the island, and Cayos los Indios, a group of low-lying heavily wooded islets that give a measure of shelter from the west. If approaching from the west, steer for Loma la Canada on a heading of 084°. When approaching from the south or southwest, pass no less than 2.25 miles northwest of Punta Frances before turning into the entrance.
Port of entry: Marina Colony, Darsena de la Siguanea.

Nueva Gerona (21°53′N, 82°48′W), a small riverine community lying somewhat inland on the north coast of Isla de la Juventud, is the principal community on the island. Vessels can approach this area by passaging around the west side of the island, via Ensenada de la Siguanea.

SUR BAJO BOQUERON, 22 21.1N, 82 25.7W. Fl G 5s, 16ft, 3M. Green port-hand beacon with topmark.
MONTERREY, 22 20.2N, 82 20.5W. Fl R 6s, 36ft, 5M. Red starboard-hand beacon with topmark, 26ft.
CAYO CRUZ, 22 28.1N, 82 16.8W. Fl G 3s, 36ft, 4M. Green port-hand beacon with topmark, 26ft.
BUENAVISTA, 22 30.0N, 82 21.2W. Fl R 6s, 13ft, 3M. Red starboard-hand beacon with topmark.
CAYO CULEBRA, 22 24.1N, 82 33.8W. Fl R 6s, 26ft, 3M. Red starboard-hand beacon with topmark.
CAYO CARABELA, 22 29.2N, 82 28.7W, Fl G 3s, 20ft, 3M. Green port-hand beacon with topmark.
BOQUERON DEL HACHA, 22 29.3N, 82 27.8W. Fl R 4s, 20ft, 3M. Red starboard-hand beacon with topmark.

ENSENADA DE COLOMA

LA COLOMA, 22 14.3N, 83 34.3W. Fl W 5s, 25M. Tower on green metal tank, 13ft.
SANTO DOMINGO APPROACH BEACON, 22 09.5N, 83 36.5W. L Fl W 10s, 19ft, 4M. Safe-water mark, red and white, topmark.
ENSENADA DE COLOMA, NO. 1, W side, 22 11.9N, 83 35.6W. Fl G 3s, 13ft, 3M. Green port-hand beacon with topmark.
ENSENADA DE COLOMA, NO. 4, E side, 22 12.4N, 83 35.3W. Fl R 4s, 13ft, 3M. Red starboard-hand beacon with topmark.
ENSENADA DE COLOMA, NO. 5, W side, 22 13.0N, 83 34.9W. Fl G 5s, 13ft, 3M. Green port-hand beacon with topmark.
ENSENADA DE COLOMA, NO. 8, E side, 22 13.5N, 83 34.6W. Fl R 4s, 13ft, 3M. Red starboard-hand beacon with topmark.
ENSENADA DE COLOMA, NO. 10, E side, 22 14.1N, 83 34.2W. Fl R 6s, 13ft, 3M. Red starboard-hand beacon with topmark.

ENSENADA DE CORTES

Pilotage: This coastal bight is located in the far western part of the Golfo de Batabano. The approaches are obstructed by a string of dangers extending west from Isla de la Juventud to the mainland. It is entered by means of a narrow passage west of a partially emerged sunken wreck charted about 12.5 miles north-northeast of Cabo Frances.

CABO FRANCES, 21 54.4N, 84 02.1W. Fl W 10s, 30ft, 10M. White framework tower, 26ft.
CABO CORRIENTES, 21 45.7N, 84 31.0W. Fl W 5s, 88ft, 10M. White framework tower, 72ft.

Cayman Islands

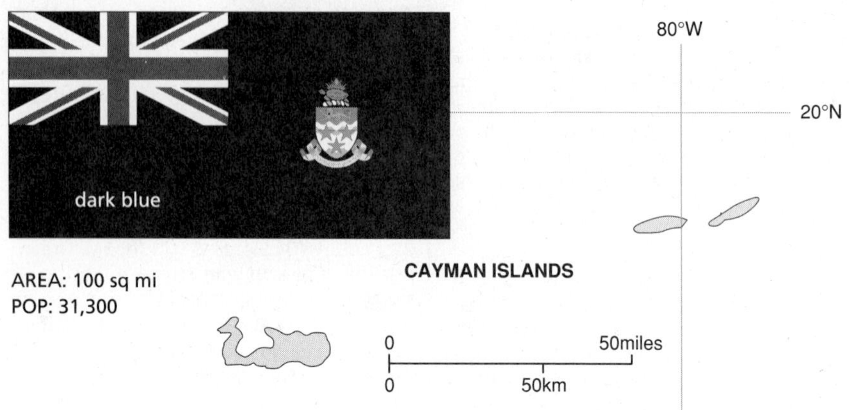

AREA: 100 sq mi
POP: 31,300

Please see cautions about chartlets and waypoints on page P 2.

ENTRY PROCEDURES

All vessels must display a Q flag upon arrival—there are stiff fines for not doing so—as well as the Cayman flag. All vessels must clear customs and immigration on Grand Cayman at George Town or on Cayman Brac at The Creek before proceeding to a marina or anchorage. At George Town, call Port Security on VHF channel 16, who will alert customs (809-949-2227) to your arrival, and anchor off the town or tie up to the government wharf to await further instructions; do not go ashore until instructed. If bad weather makes anchoring off George Town untenable, call port officials for information on an alternative port. The harbor is wide open to northwest winds.

The Caymans are a Dependent Territory of the United Kingdom. To clear in, vessels and crew are required to have clearance from the last port of call, ship's documents, proof of ownership, and passports and to fill out bill of health forms and ship's stores and personal effects lists. Officials from the public health department also spray entering vessels. There is a US$50 fee for vessels arriving after 1600 weekdays, after 1230 Saturdays, and on Sundays. The fee rises to US$60 on holidays. All guns, spear guns, pole spears, and Hawaiian slings must be declared and are held during your stay. The items are returned to the vessel upon departure from the islands. Marine conservation regulations are strict; they include no discharge of effluents anywhere in the Caymans and no anchoring within the marine park zone. For more information concerning marine park regulations, contact the Protection and Conservation Unit (809-949-8469).

USEFUL INFORMATION

Language: English.
Currency: Cayman dollar (CI$); exchange rate (summer 1997) US$1 = CI$.83. U.S currency is widely accepted, as are credit cards. The Caymans are one of the world's largest offshore financial centers.
Telephones: Area code, 345; local numbers, seven digits.
Public holidays (1998): New Year's Day, January 1; Ash Wednesday, February 25; Good Friday, April 10; Easter Monday, April 13; Discovery Day, May 18; Queen's Official Birthday (Monday in June, observance date set in late 1997); Constitution Day, July 6; Remembrance Day, November 9; Christmas Day, December 25; Boxing Day, December 26.

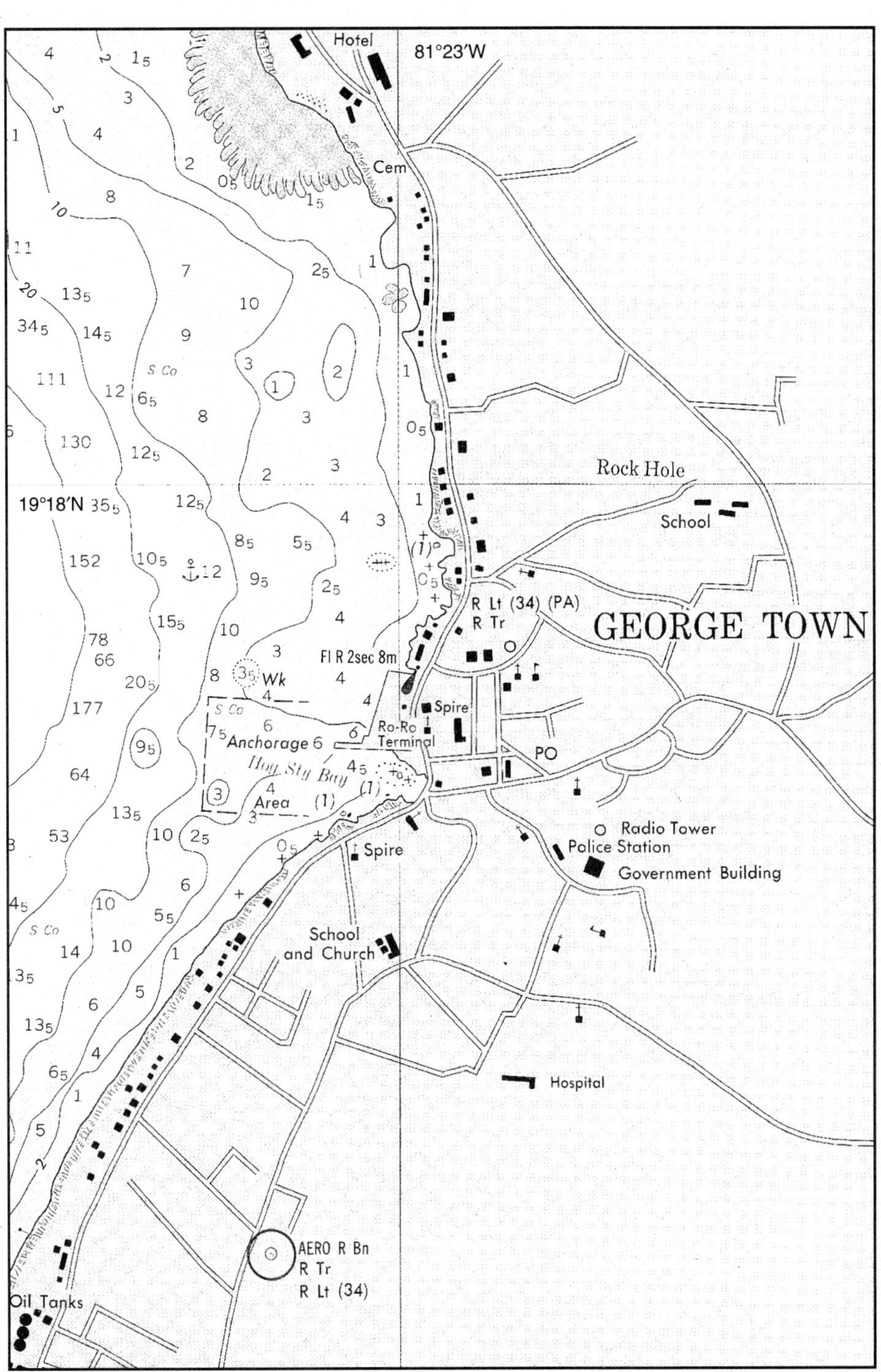

George Town, soundings in meters *(from DMA 27243, 1st ed, 3/81)*

Tourism offices: *On Grand Cayman:* Cayman Islands Department of Tourism, Government Building, P.O. Box 67, George Town (345-949-0623). *In the U.S.:* Cayman Islands Department of Tourism, 6100 Blue Lagoon Drive, Suite 150, Miami, FL 33126 (305-266-2300). Offices also in New York, Chicago, Houston, and Los Angeles. *In Canada:* Cayman Islands Department of Tourism, 234 Eglinton Avenue East, Suite 306, Toronto, Ont. M4P 1K5 (416-485-1550).

GRAND CAYMAN

GRAND CAYMAN LIGHT, George Town, near church, 19 17.8N, 81 23.0W. Q R, 41ft, 7M. Black steel tower, white base, 20ft.

BOATSWAIN POINT LIGHT, NW end of island, 19 23.1N, 81 24.6W. Fl W 15s, 90ft, 15M. White steel tower, black base, 20ft. Partially obscured 241°–257°, obscured 257°–shore.

GORLING BLUFF LIGHT, at SE end of island, 19 18.0N, 81 06.3W. Fl (2) W 20s, 72ft, 12M. White steel tower, black base, 27ft.

GEORGE TOWN AVIATION LIGHT, 19 17.5N, 81 21.5W. Al Fl W G, 39ft, 20M.

SOUTH WEST POINT LIGHT, 19 15.7N, 81 23.2W. Fl (2) W 10s, 30ft, 15M. White metal tower and base. Visible 253°–150°.

HOG STY BAY RANGE (Front Light), 19 17.7N, 81 23.1W. F G, 13ft. White metal mast, 7ft. **(Rear Light),** 100 meters, 90° from front. F G, 33ft, 3M. White metal mast, 30ft.

GEORGE TOWN

19 18N, 81 23W

Port of entry: Contact Port Security on VHF channel 16 upon arrival (see "Entry Procedures").

Dockage: You may be able to tie up to the government wharf in George Town temporarily while clearing customs. The Cayman Islands Yacht Club on North Sound (tel./fax, 345-945-4322) has 150 slips, many for visitors. Harbour House Marina (VHF channel 16; tel., 345-947-1307; fax, 345-947-6093), a boatyard at the south end of North Sound, can offer emergency dockage for vessels with a draft of 6 feet or less and a beam of 24 feet or less. Morgan's Harbour (VHF channel 81; tel., 345-945-1953) in West Bay off North Sound has 15 slips for boats drawing up to 10 feet; the channel to Morgan's Harbour is privately marked with red buoys.

Anchorage: There are sheltered anchorages in North Sound. In good weather you can anchor west of West Bay. When a norther threatens, boats anchored off George Town move around to the south end of the island.

Services: The Cayman Islands Yacht Club offers water, electricity, fuel, and ice. Harbour House Marina offers fuel, water, ice, a fully stocked chandlery, repair and some rigging services, and haulouts for vessels to 70 tons. Fuel, water, and ice are available at the small dock at Morgan's Harbour. There are good chandleries and grocery stores on the island. The offshore banking industry has promoted the development of modern facilities of all sorts.

Being situated well off the normal trade wind routes of the Caribbean, the Cayman Islands are less visited than other islands. The islands have world-famous diving opportunities and good shoreside facilities. The Lesser Caymans (Little Cayman and Cayman Brac) are located more than 60 miles northeast of Grand Cayman. They hold little attraction for private yachts, as there are few possibilities for shelter without local knowledge. Commercial transfer operations are carried out off both islands.

LITTLE CAYMAN

SOUTH WEST POINT LIGHT, 19 39.5N, 80 06.8W. Fl W 5s, 30ft, 10M. White steel tower, black base, 20ft.

EAST POINT LIGHT, 19 42.4N, 79 58.2W, Fl W 10s, 36ft, 10M. White tower and base.

CAYMAN BRAC

Port of entry: At The Creek. Call Port Security (VHF channel 16) ahead of your arrival.

CAYMAN BRAC LIGHT, SW point, 19 41.0N, 79 53.5W, Fl (2) W 15s, 41ft, 15M. White mast. Reported extinguished (1993).

NORTH EAST POINT LIGHT, 19 45N, 79 44W. L Fl W 20s, 150ft, 12M. White steel tower, black base, 20ft.

Jamaica

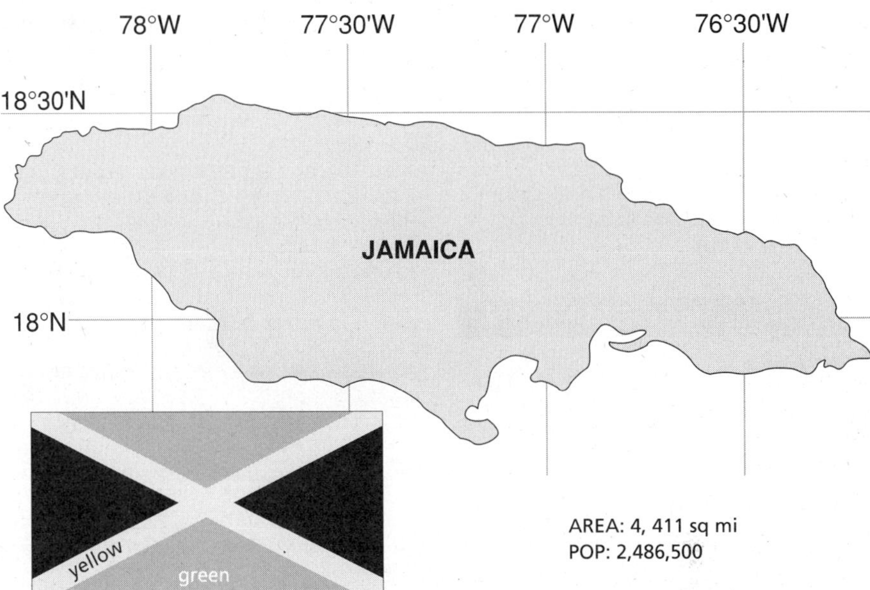

AREA: 4, 411 sq mi
POP: 2,486,500

P
Pilot
JAMAICA

Caution: The Port Authority of Jamaica is upgrading its navigation aids; buoys and lights may not flash with the frequencies that are noted on charts or in* Reed's. *The Port Authority "strongly recommends caution in this respect."

Please see cautions about chartlets and waypoints on page P 2.

ENTRY PROCEDURES

Most boaters make their first port of call Montego Bay, Port Antonio, or Kingston (Port Royal), but other entry ports are Ocho Rios, Discovery Bay, Bowden, Port Kaiser, and Port Esquivel—all of which have customs and immigrations facilities.

At *Reed's* press time (summer 1997), the government of Jamaica has legislation pending that will make customs and immigration a much simpler, more "cruiser friendly" process. Boaters will be issued a cruising permit (for approximately US$30) that will enable a visiting boat to cruise anywhere within Jamaica's waters. Until then, boaters must state their itinerary to receive a coastwise clearance and must clear in every time they enter one of the above ports of entry.

Upon arriving in Jamaica, boats should fly the Q flag and have the following documents: clearance from the last port of call, passports or identification for the crew, four copies of the crew's list, a stores list, and documentation and proof of ownership for the boat. Jamaica has recently prohibited the importation of any animals or birds. Absolutely no shipboard pets are allowed ashore under any circumstances. You may apply for a three-month stay in the country; a three-month extension is possible. Additional extensions may be requested. Normal working hours are generally 0800 to 1600 or 1700, Monday through Friday, with a midday lunchbreak. Overtime charges apply for after-hours clearances, with double time charged on Saturdays, Sundays, and public holidays. Actual fees are based upon the number of officials involved and the amount of time needed. If boat repairs necessitate importing parts from abroad, inquire at customs about transhipment forms so the parts will enter duty-free; there is a processing fee, but it is generally less than the duty would otherwise be.

Jamaica's Port Authority monitors VHF channels 11, 12, and 13; channel 68 is used primarily by pleasure craft.

When leaving the country you can wait for up to 24 hours after receiving your clearance if you do not have any firearms being held. Customs holds all firearms and ammunition for the duration of your stay.

USEFUL INFORMATION

Language: English.
Currency: Jamaican dollar (J$); exchange rate (summer 1997), J$35 = US$1. Credit cards are widely accepted in tourist areas.
Telephones: A new area code, 876, became effective November 1, 1997. Jamaica International Telecommunications Limited (JAMINTEL) provides telex and fax services. Its main office is at 15 North Street in Kingston (876-922-6031). Link calls can be made by calling "Kingston Radio" on VHF channel 16; you will then be asked to change to channel 26 or 27. Calls are charged on a collect basis. Boatphone (Cable and Wireless Co.) operates the most extensive cellular communications network in the eastern Caribbean, including Jamaica. Their number is 876-968-4000.
Public holidays (1998): New Year's Day, January 1; Ash Wednesday, February 25; Good Friday, April 10; Easter Monday, April 13; Labour Day, May 23; Independence Day, August 3; National Heroes Day, October 19; Christmas Day, December 25; Boxing Day, December 26.
Tourism offices: *In Jamaica:* Jamaica Tourist Board, 2 St. Lucia Avenue, Kingston (tel., 876-929-9200; fax, 876-929-9375). *In the U.S.:* Jamaica Tourist Board, 801 2nd Avenue, 20th Floor, New York, NY 10017 (800-233-4582, 212-856-9727). Offices also in Atlanta, Chicago, Dallas, Detroit, Los Angeles, Miami, Philadelphia, Shrewsbury (MA). *In Canada:* Jamaica Tourist Board, 1 Eglinton Avenue East, Suite 616, Toronto, Ont. M4P 3A1 (416-482-7850). The Jamaica Tourist Board in Florida offers boaters an excellent pamphlet, "Cruising in Jamaica," through their office in Miami at 305-665-0557; fax 305-665-7239.

LUCEA HARBOUR TO OCHO RIOS BAY

LUCEA HARBOUR LIGHT, on Flagstaff Reef, 18 27.1N, 78 09.8W. Fl R 4s, 12ft, 5M. Triangular steel skeleton structure.
MONTEGO BAY LOWER RANGE (Front Light), 18 28.2N, 77 55.5W. F R, 22ft, 7M. Red triangle on mast. Visible 108°–128°. ~~Reported extinguished.~~ **(Rear Light),** 465 meters, 118.5° from front. F R, 57ft, 5M. White triangle on mast, 52ft. Visible 108°–128°. Stopping lights. Reported destroyed (1997).
MONTEGO BAY UPPER RANGE (Front Light), 18 28.7N, 77 55.6W. F R, 44ft, 5M. Orange cylindrical metal tower on white fort, ball daymark. Visible 026°–046°. Stopping lights. ~~Reported destroyed (1991)~~. **(Rear Light),** 104 meters, 035.2° from front. F R, 113ft, 5M. Red round iron tower, ball daymark, 16ft. Visible 026°–046°. Stopping lights.
MONTEGO BAY ENTRANCE CHANNEL RANGE (Front Light), 18 27.6N, 77 56.3W. Oc W 2s, 26ft, 8M. Red triangle daymark on post. **(Rear Light),** 300 meters, 201° from front. Oc W 3s, 43ft, 13M. White triangular daymark on post. Reported extinguished (1997).
AVIATION LIGHT, 18 29.9N, 77 55.1W. Fl W 4s, 59ft, 10M. Control tower. ~~Reported extinguished (May 1986)~~.
MONTEGO BAY ENTRANCE CHANNEL LIGHT NO. 2, W side, 18 27.8N, 77 56.2W. Fl R 5s, 16ft. Red pile. Reported extinguished (February 1985).
MONTEGO BAY LIGHT NO. 3, W side, 18 27.8N, 77 56.3W. Fl R 3s, 16ft. Red pile. Reported extinguished (February 1985).
MONTEGO BAY LIGHT NO. 4, 18 27.7N, 77 56.5W. Fl R 5s, 16ft. Red steel column on pile.
MONTEGO BAY LIGHT NO. 5, 18 27.6N, 77 56.5W. Fl R 3s, 16ft. Red steel column on pile. Reported extinguished (1997).
MONTEGO BAY LIGHT NO. 6, 18 27.7N, 77 56.2W. F G, 20ft. Green steel column, 20ft. ~~Reported extinguished (May 1986)~~.

MONTEGO BAY

18 28N, 77 56W
Port of entry: If you anchor off or dock at the Montego Bay Yacht Club, club staff will call officials for you. Customs officials (876-979-8126) are located in the port handlers building by the main dock.
Dockage: The Montego Bay Yacht Club (VHF channel 16; tel., 876-979-8038; fax, 876-979-8262; e-mail, MBYC@infochan.com; Web site, www.mbyc.com.jm) is located in the bay west of town that forms the commercial port.

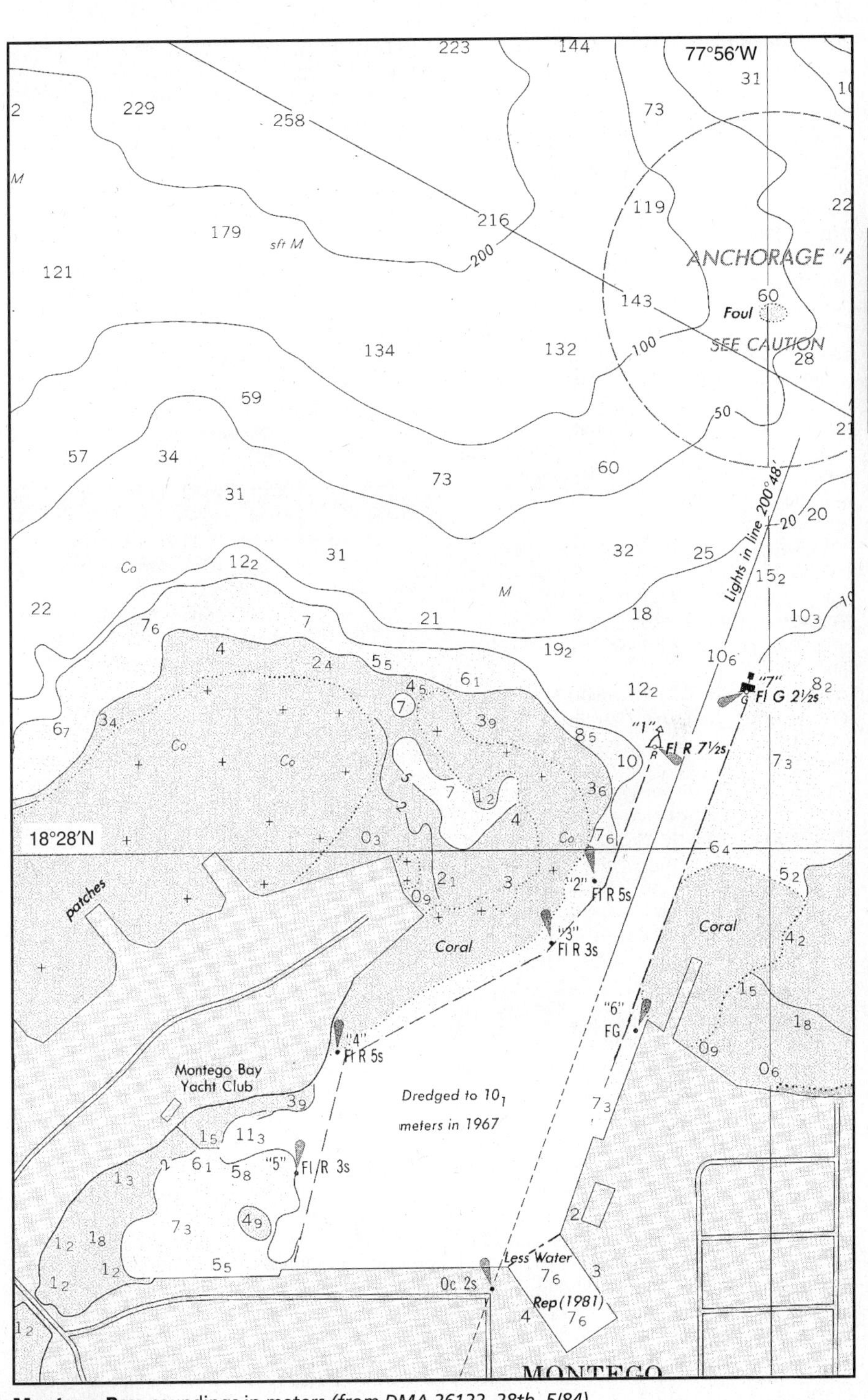

Montego Bay, soundings in meters *(from DMA 26122, 28th, 5/84)*

Depths along the club wharf vary, with a 15-foot maximum. There is a fee for anchoring off the club and using its facilities.
Services: The yacht club offers electricity, good drinking water, diesel fuel, laundry, and ice. (Boats requiring a lot of fuel or ice need to give one day's notice.) The first day's docking is free to members of reciprocal yacht clubs. Club facilities include a bar, restaurant, showers, and swimming. Minor repairs may be carried out at the docks; for emergencies, the club may be able to arrange a mobile crane for haulouts. There are no other haulout facilities here. All sorts of tourist diversions are available at the club, in town, and nearby.

Although the yacht club is situated a few minutes from downtown, its good facilities make it a popular stop for cruisers. Montego Bay is a well-sheltered harbor.

ROSE HALL LIGHT, 18 32.1N, 77 49.2W. Fl (5) W 30s, 106ft, 22M. Metal framework tower. Reported extinguished (April 1991).
FALMOUTH HARBOUR LIGHT, 18 29.3N, 77 39.1W. F R, 37ft. White structure, circular shaped.
DISCOVERY BAY, DRY HARBOUR RANGE (Front Light), 18 27.7N, 77 24.6W. F R, 25ft. White triangular daymark, point up, on post. **(Rear Light),** 220 meters, 193°52′ from front. F R, 40ft. Framework tower, white triangular daymark, point down.
ST. ANNS BAY RANGE (Front Light), 18 26.2N, 77 12.1W. F R, 59ft, 11M. White ball daymark on roof of Custom House, 23ft. **(Rear Light),** 580 meters, 193.5° from front. F R, 275ft, 11M. White iron column, ball daymark, 13ft.

OCHO RIOS BAY

Buoy, 18 25N, 77 07W. Fl G, 3s. Red buoy with triangular topmark. Radar reflector.
RANGE (Front Light), 18 24.5N, 77 06.9W. Oc R 5s, 42ft, 10M. Red and white triangular daymark. Shown when vessels are expected. **(Rear Light),** 384 meters, 169° from front. Oc R 5s, 150ft, 10M. Steel column, red and white triangular daymark. Synchronized with front. Shown when vessels are expected.
BEACON, 18 24.7N, 77 06.7W. Fl G 5s, 16ft. Square topmark on three-pile beacon.
BEACON, 18 24.6N, 77 06.6W. Fl G 1.5s, 16ft. Square topmark on three-pile beacon.

ORACABESSA BAY

RANGE (Front Light), E side of bay, 18 24.3N, 76 56.8W. Fl W 2.5s, 25ft, 7M. White circular iron column, white circular daymark. **(Rear Light),** 46 meters, 093°15′ from front. Q W, 41ft, 7M. White circular iron column, white circular daymark.
RANGE (Front Light), S part of bay, 18 24.0N, 76 57.1W. Fl W 5s, 12ft, 7M. White circular iron column, white diamond daymark. **(Rear Light),** 55 meters, 175°44′ from front. Q W, 18ft, 7M. White circular iron column, white diamond daymark.
GALINA POINT LIGHT, 18 25.2N, 76 55.1W. Fl W 12s, 62ft, 22M. White round concrete tower and hut, 28ft. F R on radio masts close W.

PORT ANTONIO

FOLLY POINT LIGHT, 18 11.3N, 76 26.6W. L Fl W 10s, 54ft, 23M. White tower, red bands, 49ft. Obscured by Wood Island.
FOLLY POINT RANGE (Front Light), 18 11.2N, 76 26.6W. F R. Beacon. When required. **(Rear Light),** 146 meters, 068°47′ from front. F R. Beacon. When required.
WEST HARBOUR RANGE (Front Light), on shore, 18 10.9N, 76 27.6W. F R, 25ft. White structure, 25ft. **(Rear Light),** 1,271 meters, 248°47′ from front. F R, 277ft. White structure, 20ft.
TITCHFIELD LIGHT, 18 11.0N, 76 27.1W. Q G. Green port-hand beacon.

PORT ANTONIO

18 11N, 76 27W
Port of entry: Port Antonio Marina (VHF channels 16 and 68; tel., 876-993-3209) and Huntress Marina (VHF channel 16; tel., 876-993-3053) will contact officials for yachts arriving at their docks.
Dockage: Navy Island Marina and Resort (VHF channel 72; tel., 876-993-2667) on Navy Island and Port Antonio Marina and Huntress Marina near the Boundbrook Wharf area all have dock space for visitors. A ferry offers transportation between Navy Island and the town's market. There is a free ferry between Navy Island and the mainland.
Services: All three marinas have water and electricity. Huntress and Port Antonio can arrange fuel delivery by truck; Navy Island Marina and Resort has a diesel fuel dock. All three marinas have a bar and restaurant; the Navy Island Marina also has tennis courts and shoreside accommodations. Huntress and Port Antonio also can arrange for minor repair services on

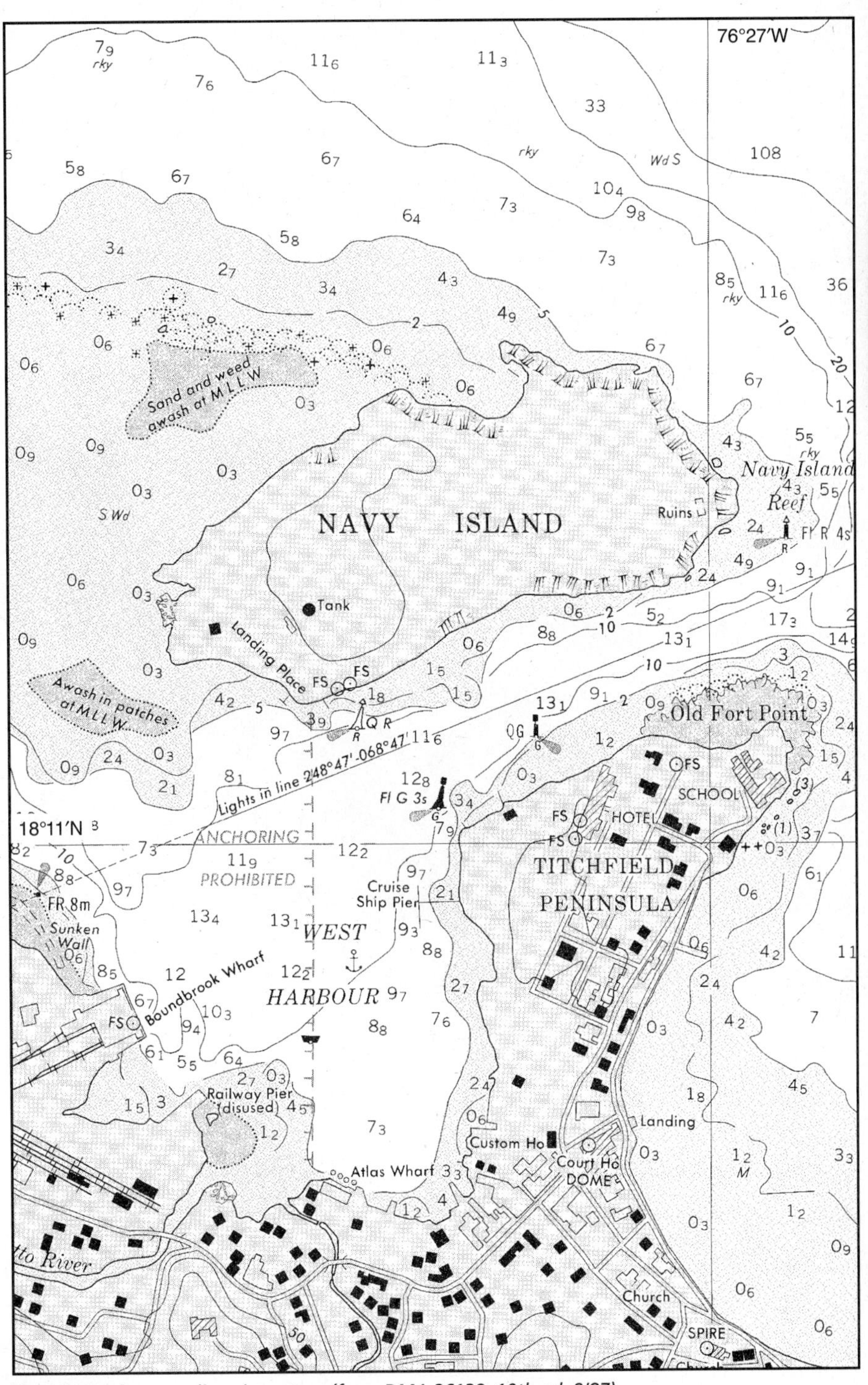

Port Antonio, soundings in meters *(from DMA 26129, 10th ed, 9/87)*

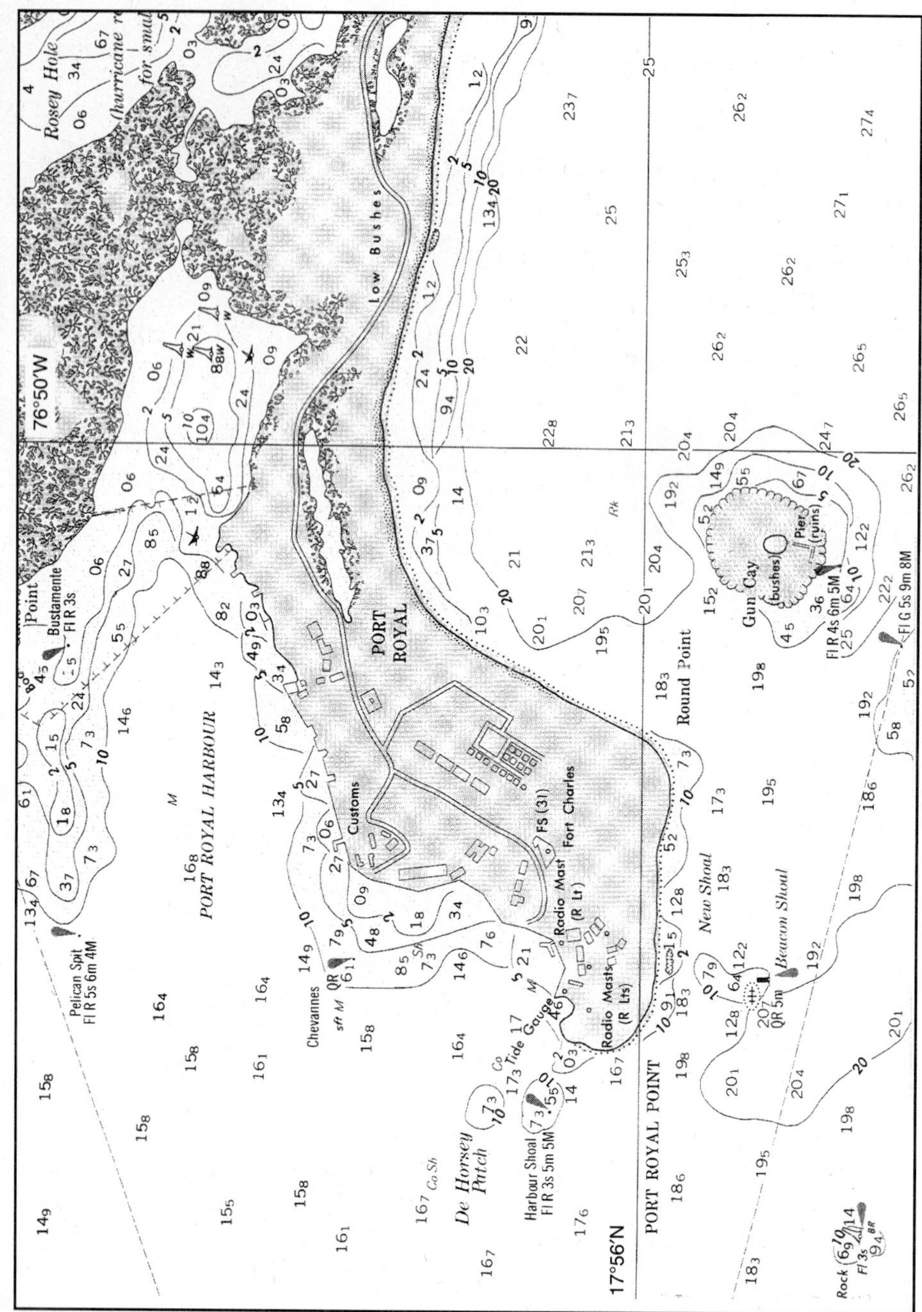

Port Royal, soundings in meters *(from DMA 26128, 49th ed, 10/94)*

site. Grocery stores and general supplies are available in town.

Port Antonio is one of the most beautiful towns on the island, and the marinas are a popular stop for cruisers. The mountain scenery is spectacular and makes a visit inland worthwhile. It is also a good place for yachts from the north to clear customs after a voyage down the Windward Passage.

MORANT POINT LIGHT, SE extremity of Jamaica, 17 54.9N, 76 11.1W. Fl (3) W 20s, 115ft, 22M. White metal tower, red bands, 95ft. Obscured when bearing more than 067°. Reported Fl (2) W 20s (April 1993).

MORANT POINT TO KINGSTON

MORANT CAYS LIGHT, Northeast Cay, 17 25.0N, 75 59.2W. L Fl W 10s, 75ft, 12M. Aluminum framework tower, 50ft. Radar reflector.

PORT MORANT RANGE (Front Light), 17 53.3N, 76 19.4W. F R, 34ft, 5M. White beacon. **(Rear Light),** 418 meters, 005°46′ from front. F R, 95ft, 5M. White beacon.

HARBOUR SHOAL SOUTH LIGHT, 17 52.4N, 76 19.6W. Fl G 5s.

HARBOUR SHOAL NORTH, Fl G 1.5s.

COTTON TREE SPIT, Fl R 1.5s.

LEITH HALL SPIT, Fl G 3s.

WATSON SPIT, Fl G 5s.

MORANT BAY LIGHT, 17 53N, 76 24W. Fl W 5s, 244ft, 14M. Red framework tower, 44ft. Visible 283°–078°; only top of tower is visible above trees. Reported extinguished (1988).

KINGSTON AND APPROACHES

PLUMB POINT LIGHT, 17 55.6N, 76 46.7W. Fl W R 9s, 70ft, 19M. White tower. W 297°–010°, R 010°–136°, W (uninstensified) 136°–181°, obscured elsewhere.

NORMAN MANLEY AVIATION LIGHT, 17 56.1N, 76 46.8W. Al Fl W G 8s, 55ft. Tower. Occasional.

Kingston Entrance Buoy, 17 54.5N, 76 45.5W. Mo (K) W, 10s. Safe water mark. Red and white pillar. **RACON K** (– • • –).

PORT ROYAL

GUN CAY BEACON, S end of shoal, 17 55.6N, 76 50.2W. Fl R 5s, 20ft, 5M. Beacon.

RACKHAMS CAY RANGE (Front Light), N extremity of shoal, 17 55.5N, 76 50.3W. Fl G 5s, 28ft, 8M. Green beacon. **(LAZARETTO, Rear Light),** 3.9 km, 284° from front, 17 56.0N, 76 52.5W. Fl W 3s, 92ft. White cairn. Lights mark the boat channel off Gallows Point when required.

BEACON SHOAL BEACON, S extremity, 17 55.7N, 76 50.8W. Fl R 1.5s, 18ft. Beacon.

HARBOUR SHOAL BEACON, SW edge of shoal, 17 55.9N, 76 51.0W. Fl R 3s, 16ft, 5M. Red starboard-hand beacon.

CHEVANNES BEACON, 17 56.2N, 76 50.8W. Q R. Red starboard-hand beacon. F R lights on four masts 600 meters S.

PORT ROYAL

17 56N, 76 50W

Port of entry: Tie up to the customs wharf at the Coast Guard station or anchor out with the Q flag flying. Customs officials will come out to you. After customs, you will be visited by immigration and quarantine officials. Morgan's Harbour (876-928-7223) will also alert customs, immigration, and quarantine officials for their customers. (The dive shop at Morgan's Harbour generally monitors VHF channel 16 from 0600 to 1700.) For more information contact customs at Port Royal (876-967-8069).

Dockage: Morgan's Harbour Hotel and Yacht Marina offers safe dockage and moorings in Port Royal at the entrance to Kingston harbor near the Coast Guard station. Or after clearing in, proceed to the Royal Jamaica Yacht Club (VHF channel 68; tel., 876-924-8685) located east of the airport on the north side of the Palisadoes Peninsula. The yacht club has a 160-foot dock for visiting boats (up to 10 feet draft), but you must be a member of a recognized yacht club to make use of its facilities (for which you will be charged).

Services: Morgan's Harbour and the yacht club both offer diesel, water, electricity, ice, and a restaurant and bar. The yacht club also sells unleaded gas, offers fax services, and has a crane and 25-ton marine railway. Morgan's Harbour has shoreside accommodations and a marine railway for boats with shallow drafts only. Visiting boats tie up stern-to on the main dock (12 feet depth). All other services are available in Port Royal or Kingston. From Port Royal a ferry offers service to Pier 2 in Kingston. It is about 12 miles to town by taxi. *Note: Because of crime threats, the U.S. embassy cautions visitors against travel after dark to and from Morgan's Harbour/Port Royal.*

Kingston is a busy commercial port and vessels should be aware of encountering large ships in the channels. It is a good idea to monitor VHF channel 16 when under way here. Port Royal is the former hangout of Captain Morgan, the pirate, and other nefarious characters. The area was subsequently converted to a naval base and features a maritime museum and a castle. The vibrant streets of Kingston are but a ferry ride away.

PELICAN SPIT, N side of Port Royal Harbour, 17 56.6N, 76 50.7W. Fl R 5s, 21ft, 4M. Red starboard-hand beacon.

BUSTAMENTE LIGHT, SW of Gallows Point, 17 56.6N, 76 50.3W. Fl R 3s. Red starboard-hand beacon.

KINGSTON HARBOUR

CURREYS GATE LIGHT, 17 57.0N, 76 50.9W. Fl W 1.5s. Red starboard-hand beacon.

DELBERT SICARD, 17 56.7N, 76 51.5W. Fl G 5s. Green port-hand beacon.

MORTON, Fl G 1.5s, 22ft, 6M. Green port-hand beacon.

BLOOMFIELD, 17 56.8N, 76 50.0W. Q R. Red starboard-hand beacon. In line 249.5° with Lazaretto, rear light.

ANGEL BEACON, 17 57.0N, 76 49.6W. Q R. Red starboard-hand beacon. In line 249.5° with Lazaretto, rear light.

TWO SISTERS LIGHT, near W extremity of Middle Ground, E side of ship channel, 17 57.5N, 76 50.7W. Q R, 18ft. Red starboard-hand beacon.

BURIAL GROUND LIGHT, W side of channel, 17 57.5N, 76 50.9W. Fl G 3s, 22ft. Green port-hand beacon.

SPHINX LIGHT, 17 57.6N, 76 50.6W. Fl R 5s, 18ft. Red starboard-hand beacon.

AUGUSTA, Fl G 5s. Green port-hand beacon. Reported destroyed (1988).

MAMMEE, E side of Ship Channel, Fl R 1.5s, 18ft. Red starboard-hand beacon.

ST. ALBANS, W side of Ship Channel. Fl G 1.5s, 22ft. Green port-hand beacon; 20ft. Reported destroyed, marked by green buoy 38 meters NNE (1988).

EAST HORSESHOE, N extremity of Middle Ground. Fl R 3s, 26 ft. Red starboard-hand beacon, 20ft.

HUNTS BAY BEACON, 17 57.9N, 76 49.9W. Fl G 3s. Green port-hand beacon.

GREENWICH, 17 58.3N, 76 49.6W. Fl R 3s, 17ft. Red starboard-hand beacon.

NEWPORT B, Q R. Red starboard-hand beacon.

MIDDLE GROUND, 17 57.6N, 76 49.5W. Fl R 5s, 17ft. Red starboard-hand beacon.

POND MOUTH, Fl G 1.5s, 22ft. Green port-hand beacon, 20ft.

PICKERING, 17 57.0N, 76 48.4W. Fl R 1.5s. Red starboard-hand beacon.

TUPPER, 17 56.9N, 76 47.9W. Fl R 3s. Red starboard-hand beacon.

ROYAL JAMAICAN YACHT CLUB MARINA LIGHT, E side of entrance, 17 56.6N, 76 46.5W. Fl G 3s. Pile.

SHELL PIER LIGHTS, on each end of pier, 17 57.8N, 76 44.7W. 2 Q R. Q R shown from eastern dolphin.

WRECK REEF, 17 49.7N, 76 55.3W. Fl R 5s, 25ft, 6M. Red and white banded iron column, 22ft. Radar reflector.

PORTLAND BIGHT

BARE BUSH CAY LIGHT, 17 45.2N, 77 02.0W. Fl R 10s, 28ft, 6M. White metal column, red bands, 13ft. Radar reflector.

PIGEON ISLAND RANGE (Common Front Light), 17 47.5N, 77 04.4W. Fl W 3s, 25ft, 9M. White metal column, round daymark on east-southeast side, 26ft. **(Rear Light),** on pumping station 6.07 mi 294°43′ from common front, 17 50.1N, 77 10.2W. Oc W 5s, 131ft, 16M. White metal structure, 39ft. **(Rear Light),** 0.5 mi 343°20′ from common front. L Fl W 6s, 80ft, 10M. White triangle daymark on white tower, 75ft. Visible 004°15′ either side of rangeline.

ROCKY POINT PIER LIGHT, 17 49.1N, 77 08.4W. 2 F R. One light shows from each end.

RANGE (Front Light), 17 53.4N, 77 08.4W. Fl R 3s, 26ft. White conical beacon. Q W and Q R are shown in harbor. **(Rear Light),** 677 meters, 300° from front, 17 53.5N, 77 08.8W. Q R, 41ft. White conical beacon.

SALT ISLAND LIGHT, N end, 17 50.0N, 77 08.2W. Fl W 5s, 21ft, 7M. White iron column, white disk topmark.

SALT RIVER LIGHT, near mouth, 17 50.0N, 77 09.7W. Q W, 31ft, 7M. White iron column, 28ft.

PORTLAND RIDGE LIGHT, summit, 17 44.4N, 77 09.5W. Fl (2) W 15s, 650ft, 20M. Steel framework tower, 115ft.

PORTLAND BIGHT TO SOUTH NEGRIL POINT

KAISER PIER RANGE (Front Light), outer end of pier, 17 51.5N, 77 36.2W. F R, 34ft. **(Rear Light),** 690 meters, 347.5° from front. F R, 138ft. Framework tower, 26ft.
LOVERS LEAP LIGHT, 17 52.0N, 77 39.7W Fl W 10s, 1739ft, 40M. White round tower, red bands, 49ft. Reported Fl W 18.5s (October 1985).
SAVANNA LA MAR LIGHT, N extremity, at Boat Stag Reef, 18 11.6N, 78 07.8W. Fl R 5s, 23ft, 9M. Iron column on concrete base, 17ft.
SAVANNA LA MAR RANGE (Front Light), on S side of ruined fort, 18 12.4N, 78 08.1W. F R, 13ft. Occasional. **(Rear Light),** 1006 meters, 032°10′ from front. F R, 40ft. Occasional.
SOUTH NEGRIL POINT LIGHT, W end of Jamaica, 18 14.7N, 78 21.7W. Fl W R 2s, 100ft, 15M. White tower, 89ft. R 297°–305°, W 305°–161°, R thence to the coast to the N (partially obscured by trees in this sector).
PEDRO BANK LIGHT, Northeast Cay, NW extremity of cay, 17 03.1N, 77 45.8W. Fl W 5s, 35ft, 11M. Beacon with square topmark, red and white bands.

NAVASSA ISLAND, U.S.A.

~~**NAVASSA ISLAND LIGHT,** summit, 18 23.8N, 75 00.7W. Fl W 10s, 395ft, 9M. Light gray cylindrical concrete tower, 162ft~~. Removed 1/97.

NAVASSA ISLAND, U.S.A.

18 24N, 75 01W
Navassa Island lies about 30 miles west of the western extremity of Haiti and is reported to be radar conspicuous at a distance of 20 miles. It is about 1.9 miles long and 1.1 miles wide. The island's shores are white cliffs that rise as high as 50 feet directly from the sea. The lighthouse is on the southeast side of the island.

The U.S. claimed possession of Navassa Island in 1857, and it was formally annexed in January 1916 by presidential proclamation. It is uninhabited, except for a few goats, and there is no water.

Lulu Bay, a small indentation on the island's southwest side, is the safest place to land. It was the site of a phosphate mining operation. Small craft can anchor here, but should be careful because of the frequent surge. Vessels can also anchor about 0.4 mile west-southwest of Lulu Bay with the light bearing about 080°T. The bottom is sand and coral in depths of around 80 feet. Mariners are cautioned that charted depths are based on old surveys.

A current with a rate of 1 to 2 knots sets along the southwest side of the island, in a northwesterly direction, changing to west at the last of the east-going tidal current.

The island is a reservation administered by the U.S. Coast Guard. Entry and landing are prohibited, except by permit. Contact the Commander, Seventh U.S. Coast Guard District, Brickell Plaza Building, 909 Southeast First Avenue, Miami, FL, 33131-3050.

HAITI

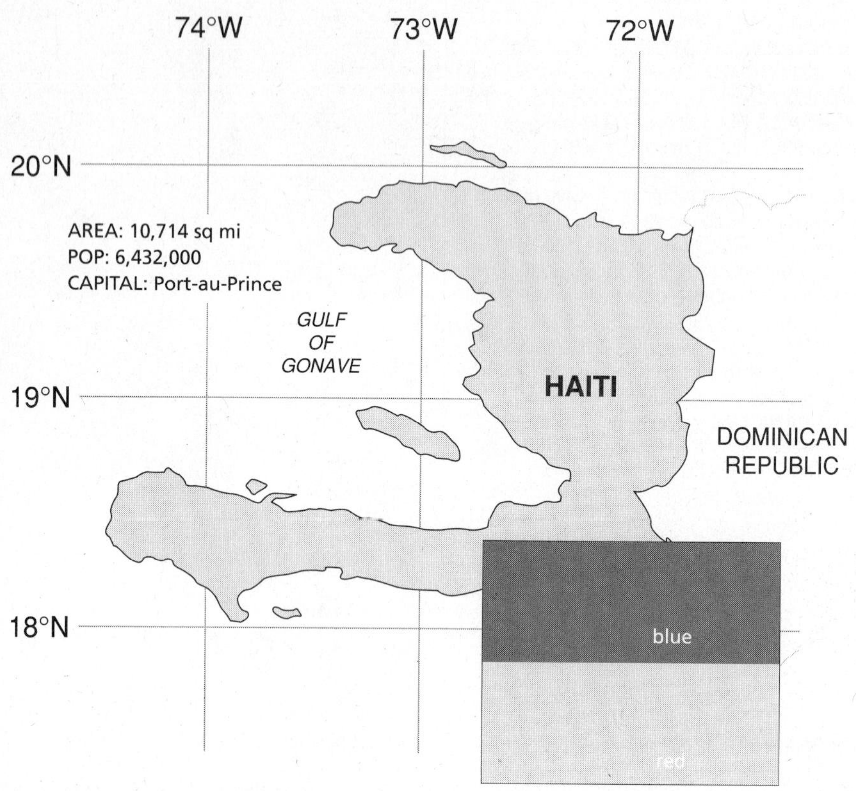

Caution: Lights along the coast of Haiti have been reported to be unreliable.

Please see cautions about chartlets and waypoints on page P 2.

ENTRY PROCEDURES

NOTE: Specific information concerning clearing-in procedures could not be confirmed at Reed's press time (summer 1997). Cruisers are advised that there are no fixed procedures in place at present to handle pleasure yachts. Those planning visits to Haiti are advised to contact the nearest Haiti consulate, the Haiti National Port Authority (main tel., 509-22-0299; other tel., 509-22-5300, 22-9052, 23-0473), or the Port-au-Prince customs administration office (509-46-4603) for up-to-date information.

Haiti is one of the least developed countries in the Western Hemisphere. There are shortages of goods and services throughout the country. Traditionally, the principal ports of entry have been Port-au-Prince and Cap-Haïtien. You should have a clearance from the last port of call, ship's papers, and crew lists; multiple copies may be helpful. Passports valid for at least six months are required (officially, U.S. and Canadian citizens need only proof of citizenship; in practice, however, immigration officials have been requiring passports of all visitors). Entry visas, which are valid for six months, must be obtained in advance of arrival, except that visas are not required for citizens of the U.S., Canada, Caricom member states, and a few other nations and that citizens of France, Italy, and Spain may request visas upon entry. All firearms and ammunition must be registered with both customs and immigration.

Clearing in at other ports of entry has been reported as difficult. Ports on Haiti's south coast are reported to be particularly poor, and mariners are advised not to cruise the area. Other former ports of entry include Fort Liberté, St. Marc, Petit Goâve, and Miragoâne.

For up-to-the minute-information on entry requirements and customs regulations, contact the Embassy of Haiti, 2311 Massachusetts Avenue NW, Washington, DC 20008 (tel., 202-332-4090; fax, 202-745-7215). In the U.S., Haiti also maintains consulates in Miami, Chicago, New York, Boston, and San Juan, Puerto Rico.

For assistance in Haiti, U.S. citizens may contact the Consular Section Annex of the U.S. Embassy (tel., 509-23-7011, 509-23-8971; fax, 509-23-9665) on Rue Oswald Durand, Port-au-Prince. The embassy is located on Harry Truman Boulevard, Port-au-Prince (P.O. Box 1761; tel., 509-22-0200; fax, 509-23-1641).

USEFUL INFORMATION

Language: Créole and French.
Currency: Gourde (G); exchange rate (summer 1997), G16.5 = $1. U.S. dollars are also officially accepted.
Telephones: Country code, 509; local numbers, six digits.
Public holidays (1998): Independence Day, January 1; Forefather's Day, January 2; Carnival (Mardi Gras), February 23–24*; Good Friday, April 10*; Pan-American Day, April 14*; Agriculture and Labor Day, May 1; Ascension Day, May 22; Flag and University Day, May 18; Corpus Christi, June 4*; Assumption Day, August 15*; Anniversary of the Death of Dessalines, October 17; United Nations Day, October 24*; All Saints Day, November 1*; All Souls' Day, November 2*; Battle of Vertières, November 18; Christmas Day, December 25*. **Religious and traditional holidays are not always observed throughout the country.*
Tourism office: *In Haiti:* Secretairerie d'Etat au Tourisme, 26 Rue L'Egitime, Port-au-Prince (509-23-5631).
Special remarks: Although crime remains a problem, there have been few reported incidents against U.S. citizens; however, visitors may be targeted by criminals, especially in urban areas. Mariners should expect continued periodic shortages of goods and services, including fuel, propane, and electricity. As the level of community sanitation is low, visitors are advised to drink only bottled water and bottled drinks. Medical facilities are few and generally below U.S. standards, and those seeking medical treatment should be prepared to pay in cash. Visitors to Haiti are also at risk for malaria and typhoid fever. For current health information, travelers can contact the Centers for Disease Control International Traveler's Hotline (voice recording, 404-332-4559; fax information service, 404-332-4565; Web, www.cdc.gov).

NORTH COAST

FORT LIBERTÉ

19 40N, 71 50W
Pilotage: The coast between Baie de Fort Liberté entrance and Pointe Jacquezy, about 8 miles west, is low with a sandy beach, fringed with reefs, and backed by mangroves. A long reef then extends another 8 miles in a west-northwest direction. The deep but narrow entrance to the bay, a landlocked basin, is in position 19°43.0'N, 71°50.8'W. The entrance can be distinguished from the north by a break in an otherwise white sandy beach.
Dockage: Depths are reported as shallow off the town docks.
Anchorage: This area is a good hurricane hole, with excellent holding ground reported.
Services: Fuel is generally available, but periodically is impossible to obtain. Be prepared to dinghy supplies out to your anchored boat.

To the west, somewhere between Passe Caracol and Cap-Haïtien, Columbus put the *Santa Maria* on the reefs. This area has many unmarked shoals and reefs and should be transited in good light if possible.

PORT-DE-CAP-HAÏTIEN, POINTE PICOLET, 19 47.5N, 72 11.2W. Fl W 4s, 30M.
PORT-DE-CAP-HAÏTIEN NO. 1, 19 47.8N, 72 10.7W. Fl G 5s. Green port-hand beacon with topmark. Channel marked by light beacons.
PORT-DE-CAP-HAÏTIEN, passenger berth, 19 46.3N, 72 11.5W. Fl R 5s. Red starboard-hand beacon with topmark.

CAP-HAÏTIEN

19 46N, 72 12W
Pilotage: The marked entrance channel to the harbor, which is somewhat protected from the east by reef and sandbank and from the

north by shoals, runs along the shore southward from Point Picolet. South of the main pier is an area where small craft and yachts can tie up. Pilots monitor VHF channel 16.

Dockage: Dockage may be available at the south side of the main docking basin.

Anchorage: Although the anchorage off town is unsheltered, Baie de l'Acul, 10 miles to the west of Pointe Picolet, is reported to be a good hurricane hole. Also, see Fort Liberté.

Services: There are groceries and general stores, but no marine supplies. Fuel for pleasure yachts may be limited.

Inland to the south the Palais San Souci and the Citadelle are historically interesting. The Citadelle is an enormous fort built between 1804 and 1817 by King Henri-Christophe at the cost of many lives to provide protection from an imagined attack by Napoléon's legions.

ÎLE DE LA TORTUE LIGHT, E point, 20 01N, 72 38W. Fl (2) W 6s, 77ft, 14M. White metal framework tower, triangular base, 46ft.

ÎLE DE LA TORTUE LIGHT, W point, 20 04.4N, 72 58.2W. V Q W.

PORT-DE-PAIX, W side of pier, 19 57.1N, 72 50.1W. F R, 7M. On second story of building. A F R light is shown from a post on NW corner of pier.

CAP DU MOLE ST. NICOLAS LIGHT, 19 49.4N, 73 25.0W. Oc W 3s, 15M. White tower. Reported extinguished (October 1993).

WEST COAST

POINTE LAPIERRE LIGHT, N side of entrance to Gonaives Bay, 19 27N, 72 46W. Fl W 6s, 318ft, 11M. White square stone tower.

POINTE DE ST. MARC LIGHT, 19 02.7N, 72 49.0W. Q (9) W 15s, 96ft, 9M. White framework tower.

LES ARCADINS LIGHT, NW point, 18 48.4N, 72 38.9W. Fl (2) W 5s, 41ft, 9M. Circular white iron tower, 31ft. Obscured 358°–012°.

PORT-AU-PRINCE AND APPROACHES

LAFITEAU RANGE (Front Light), on head of pier, 18 41.7N, 72 21.2W. F R. Black and white beacon on structure. **(Rear Light),** 304 meters, 044.5° from front. F R. Black and white beacon on building.

RANGE (Front Light), head of navy yard dock, 18 32.3N, 72 22.7W. F R. **(Rear Light),** inner end of dock, 120 meters, 183° from front, F G.

RANGE (Front Light), 18 32.9N, 72 20.3W. Oc W 3s, 12M. North tower of cathedral chimney. F R on port captain's office 0.5 mile WNW. Reported extinguished (1992). **(Rear Light),** near SW corner of Fort Alexander, 823 meters, 104° from front. Oc W 3s, 12M. Reported extinguished (1992).

POINTE DU LAMENTIN LIGHT, near shore 1280 meters E of point, 18 33.4N, 72 24.5W. Fl W 3s, 106ft, 16M. White circular iron tower, 93ft. Obscured by trees 109°–126°.

PORT-AU-PRINCE

18 33N, 72 21W

Pilotage: Mariners are warned that numerous fishing vessels may be encountered between Île de la Gonâve and the southeast head of Baie de Port-au-Prince. The forenoon haze may obscure the inner entrance range at distances greater than 4 miles. The most favorable time to enter port is in the afternoon or at dawn. Aids to navigation have been reported extinguished, missing, or showing incorrect characteristics.

Port of entry: Tie up at the commercial wharf or anchor out as directed by port control.

Port control: Contact port control on 2182 kHz or VHF channel 16. The pilot boats can also be reached on these channels. They are reported to respond to three whistle blasts.

Dockage: A small marina near the main wharf in Port-au-Prince may be able to furnish dockage. About 10 miles north of Port-au-Prince, a marina is located at Ibo Beach on the north side of Île à Cabrit, also known as Carenage Island. The island is surrounded by reefs and must be approached with care.

Anchorage: Anchoring is possible but not recommended off Port-au-Prince. There are good spots to anchor at Ibo Beach.

Supplies: Fuel and water should be available (visitors are encouraged to use bottled water for drinking). There are no marine supplies or repair facilities.

Port-au-Prince is the capital and principal city of Haiti. Although it has many political and economic problems, it has traditionally been a fascinating city to visit. Visitors should be prepared for much haggling and many requests for handouts—you may be carrying more cash than an average Haitian's yearly wages.

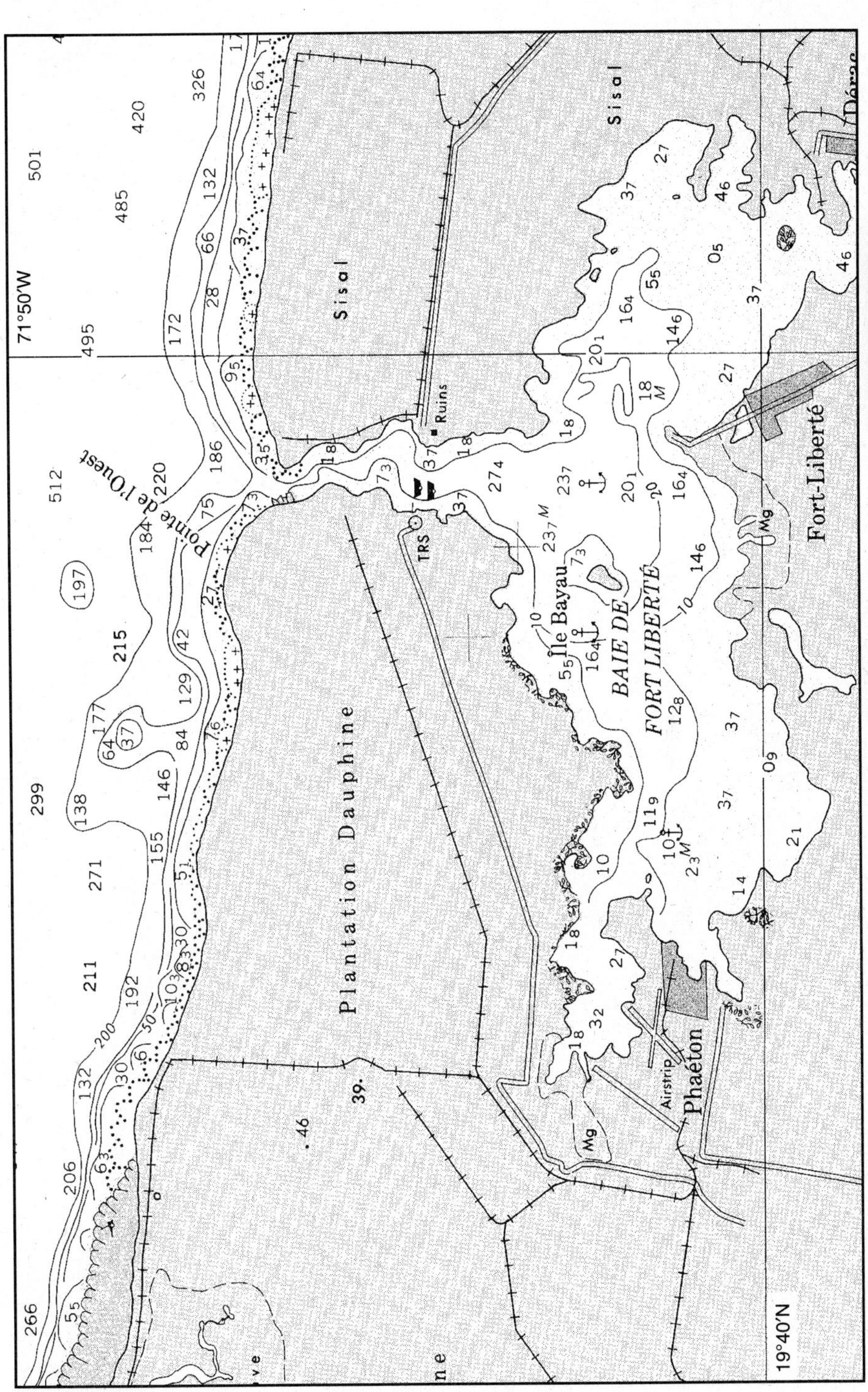

Fort Liberté, soundings in meters *(from DMA 26142, 9th ed, 7/90)*

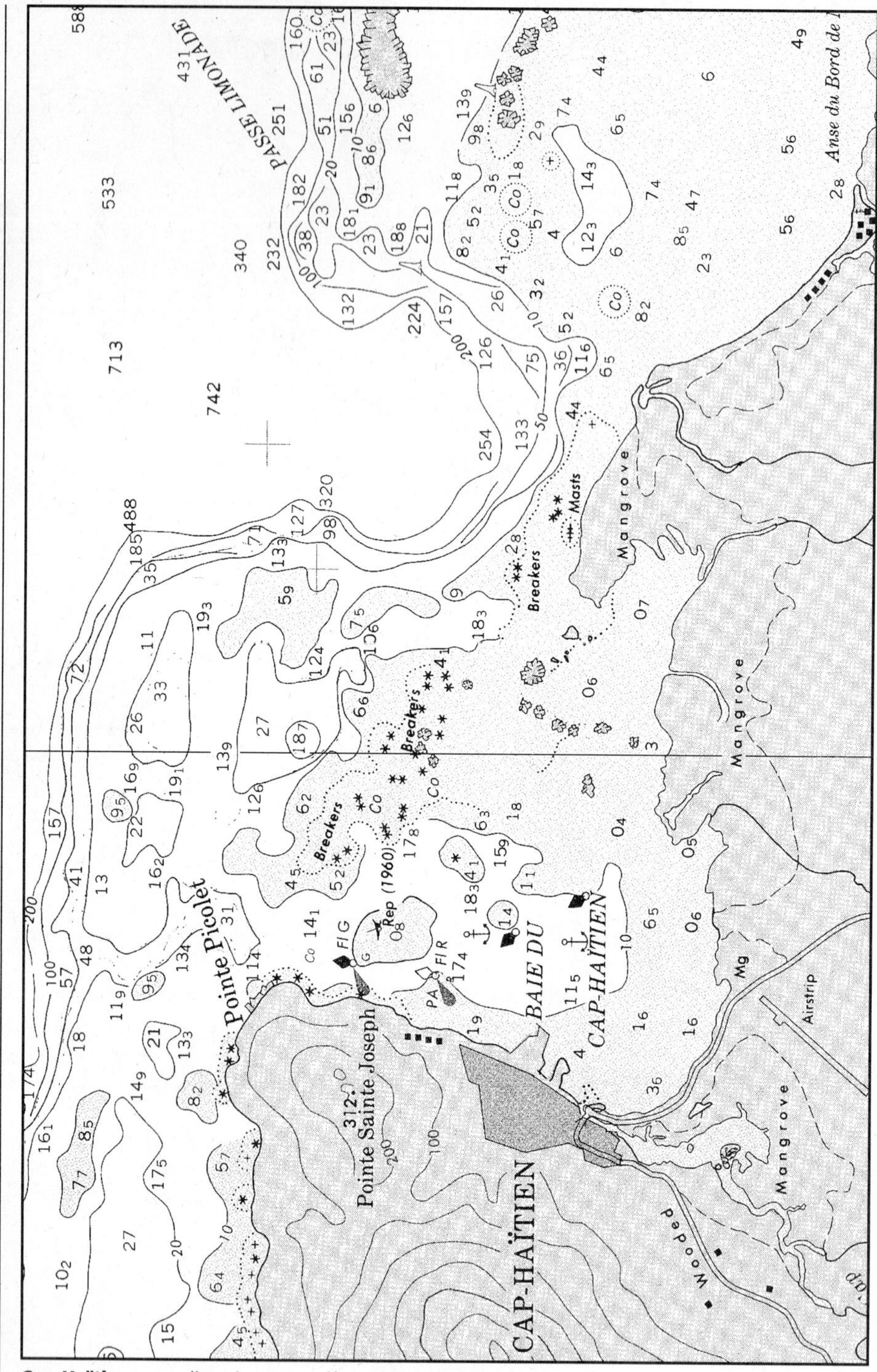

Cap-Haïtien, soundings in meters *(from DMA 26142, 9th ed, 7/90)*

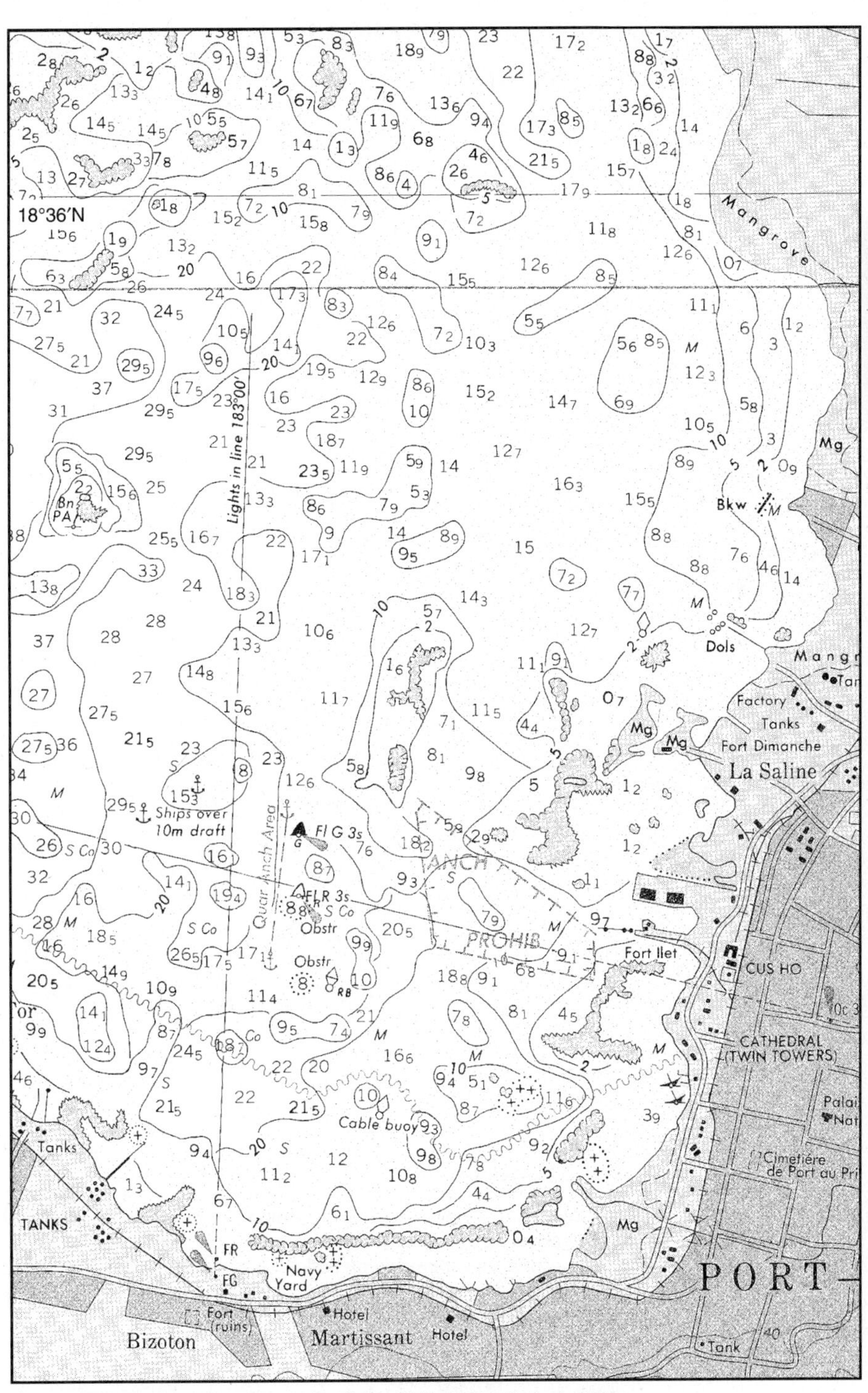

Port-au-Prince, soundings in meters *(from DMA 26184, 3rd ed, 7/89)*

PORT-AU-PRINCE TO CAP DAME MARIE

POINTE FANTASQUE LIGHT, SE end of Île de la Gonâve, 18 41.8N, 72 49.2W. Q (6) + L Fl W 15s, 50ft, 9M. Skeleton tower, black and white bands.

MIRAGOÂNE LIGHT, 18 27.2N, 73 06.4W. F R, 20ft. Corner of loading chute.

BANC DE ROCHELOIS LIGHT, in Canal de Sud, on Les Pirogues, 18 38.9N, 73 12.0W. Mo (A) W 10s, 30ft, 9M. White framework tower, black band, red lantern, hut on S side.

POINTE OUEST (WEST POINT) LIGHT, W end of Île de la Gonâve, 18 55.6N, 73 18.0W. Fl (4) W 15s, 279ft, 20M. White iron framework tower. Reported extinguished (February 1985).

GRANDE CAYEMITE LIGHT, N. point, 18 38.6N, 73 45.5W. V Q W, 54ft, 12M. White iron tower, triangular base, 44ft.

CAP DAME MARIE LIGHT, 18 36.3N, 74 25.7W. Iso W 5s, 123ft, 9M. White square concrete structure surmounted by skeleton framework, 15ft. Reported extinguished (October 1985).

SOUTH COAST

ÎLE VACHE LIGHT, 18 03.5N, 73 34.0W. Q (6) + L Fl W 15s, White square tower.

CAP JACMEL LIGHT, 18 10.0N, 72 32.0W. Fl W 6s, 127ft, 9M. White skeleton tower, red lantern, 20ft.

REED'S *NEEDS YOU!*

It is difficult to obtain accurate and up-to-date information on many areas in the Caribbean. To provide mariners with the best possible almanac, *Reed's* uses many sources. *Reed's* welcomes your contributions, suggestions, and updates. Please be as specific as possible and include your address and phone number.

Send this information to:
Editor, Thomas Reed Publications
13A Lewis Street, Boston, MA 02113, U.S.A.
e-mail: editor@treed.com
Thank you!

Dominican Republic

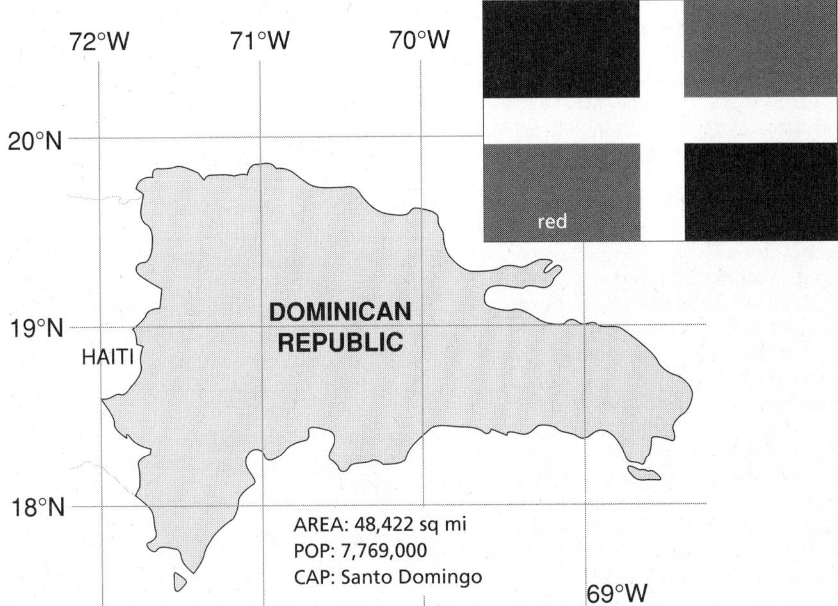

Caution: Lights along the coast of Haiti have been reported to be unreliable.

Please see cautions about chartlets and waypoints on page P 2.

ENTRY PROCEDURES

NOTE: Specific information regarding clearance procedures for pleasure yachts is difficult to obtain. Yachts planning to visit the Dominican Republic should be prepared to be patient and flexible.

Arriving yachts must enter the Dominican Republic at an official port of entry. Official ports of entry on the north coast are reported to be Pepillo Salcedo, in Manzanillo Bay; Luperon, in Puerto Blanco; Puerto Plata; and Samaná; on the south coast they are La Romana, San Pedro de Macoris, Santo Domingo, and Haina. Yachts have also been reported to clear in at Boca Chica without difficulty. Boats should fly the Q flag and await the arrival of officials. It is illegal to go ashore before you have cleared into the country.

You should have a clearance from the last port of call, passports for all crewmembers, and ship's documents. Firearms should be declared and will most likely be held until your departure. Everyone aboard who does not already have a visa must purchase a tourist card, which is good for 60 days and is renewable. You should obtain clearance to your next port of call, whether or not it is within the country.

In the U.S., current visa information may be obtained from the Embassy of the Dominican Republic, 1715 22nd St. NW, Washington, DC

20008 (202-332-6280). In the U.S. and Puerto Rico, there are also consulates in Boston, Houston, Jacksonville, Miami, Mobile, New Orleans, New York, Philadelphia, San Francisco, Mayaguez, and San Juan (PR).

USEFUL INFORMATION

Language: Spanish. At least a rudimentary knowledge of the language is advised.
Currency: Dominican Republic peso; exchange rate (summer 1997), between 12.85 pesos = US$1 (official) and approximately 13.15 pesos = US$1 (banks). Currency may be changed only at banks and hotel exchange booths.
Telephones: Country area code, 809; local numbers, seven digits.
Public holidays (1998): New Year's Day, January 1; Feast of the Three Kings, January 6; Altagracia, January 21; Duarte's Day, January 26; Independence Day, February 27; Good Friday, April 10; Labor Day, May 1; Corpus Christi, June 4; Restoration Day, August 16; Our Lady of Mercedes, September 24; Christmas Day, December 25.
Tourism offices: *In Santo Domingo:* from the U.S. only, 800-752-1151; also Tourism Promotion Council, Desiderio Arias No. 24, Bella Vista, Santo Domingo (809-535-3276). *In the U.S.:* Dominican Republic Tourist Office, 1501 Broadway, Suite 410, New York, NY 10036 (tel., 212-575-4966; fax, 212-575-5448). Office also in Miami (Coral Gables). *In Canada:* Dominican Republic Tourist Office, 2080 Crescent Street, Montréal, PQ H3G 2B8 (tel., 800-563-1611, 514-499-1918; fax, 514-499-1393).
Additional notes: Travelers to rural Dominican Republic, especially in the provinces bordering Haiti, are at risk for malaria. For current medical information, contact the Centers for Disease Control International Traveler's Hotline (voice recording, 404-332-4559; fax information service, 404-332-4565; Web, www.cdc.gov). U.S. citizens in need of assistance may contact the consular section of the U.S. Embassy in Santo Domingo (the embassy is located on the corner of Calle César Nicolas Penson and Calle Leopoldo Navarro, 809-221-2171; the consular section is located a half-mile away at the corner of Calle César Nicolas Penson and Maximo Gomez, 809-221-5030, 7:30–noon and 1–2 pm, Monday through Friday).

> ***Note***: Within an aid listing, underlined text indicates data that has changed or is new since September, 1996; ~~Strike-out~~ text indicates deleted data.

CAYO ARENAS TO PUERTO PLATA

CAYO ARENAS LIGHT, on Bahía de Monte Cristi, 19 52.9N, 71 52.0W. L Fl W 6s, 65ft, 13M. Red tower, black lantern.
PUERTO LIBERATADOR LIGHT, on head of steel pier, 19 43.2N, 71 44.7W. Iso R 20s, 50ft, 10M. White tower, red lantern. Reported removed (March 1992).

PEPILLO SALCEDO

19 43N, 71 45W
Port of entry
Pilotage: A light is shown from the end of the pier. A water tank 0.25 mile southwest of the pier and two oil tanks 600 yards southeast of the pier are good landmarks.
Dockage: If the surge is not too bad, you may be able to obtain permission to tie to the pier. It is about 740 feet long, with depths of 34 feet reported alongside. The pier is reported to be a good radar target for up to 15 miles.
Anchorage: With local knowledge you can bring boats into Estero Balsa, which is reported to be a good hurricane hole.
Services: Fuel should be available by truck. Water and groceries are available in limited quantities.

CABRA ISLAND LIGHT, NW side of Bahía de Monte Cristi, 19 54.1N, 71 40.3W. L Fl W 12s, 110ft, 13M. White pyramidal steel tower, 50ft.
EL MORRO DE MONTE CRISTI LIGHT, 19 54.2N, 71 39.1W. L Fl W 8s, 860ft, 25M. White metal tower, square base, 35ft. Reported destroyed (March 1992).
PUNTA RUCIA AVIATION LIGHT, 19 52.3N, 71 12.7W. Al Fl W G 10s.

LUPERON

19 54N, 70 57W
Port of entry
Pilotage: The small town of Luperon is situated in the southwestern arm of Puerto Blanco. The entrance to Puerto Blanco is about 4 miles east-southeast of Cabo Isabela, the northernmost point of Hispaniola. A coastal reef extends 1,200 yards seaward from the northwest side of the cape. Punta Patilla, about 5 miles east of Puerto Blanco, has a light and a reef extending 1 mile west of the point.
Anchorage: This anchorage offers excellent protection, and the holding ground is reported to be excellent.

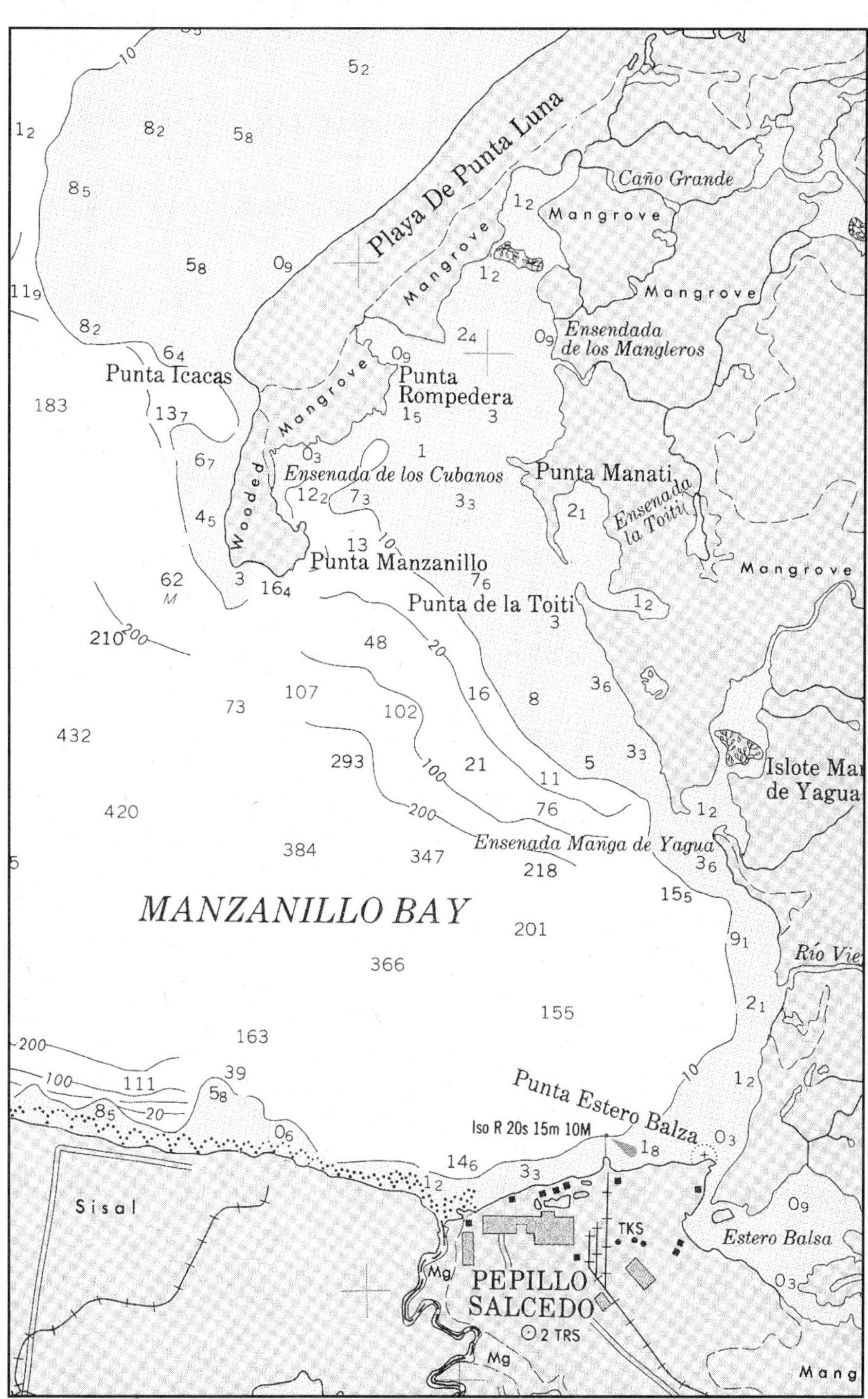

Pepillo Salcedo, soundings in meters *(from DMA 26142, 9th ed, 7/90)*

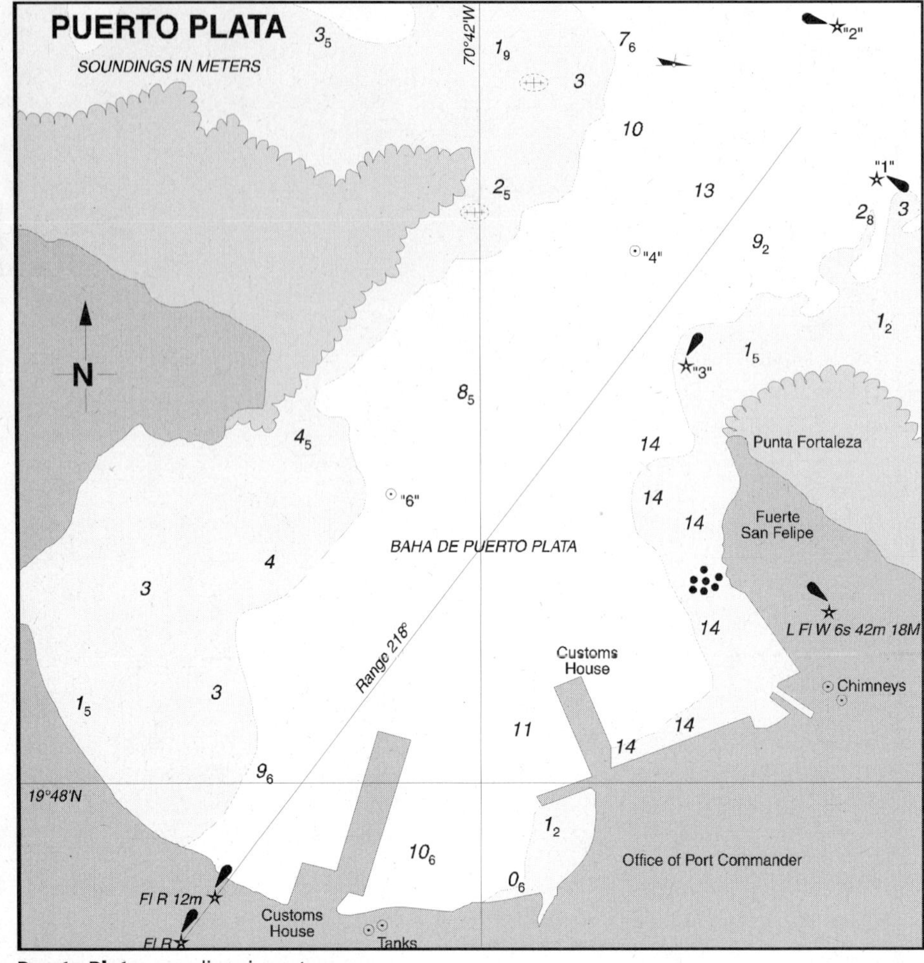

Puerto Plata, soundings in meters

Services: Fuel and water will have to be jugged to the boat. Other supplies are limited.

PUNTA PATILLA LIGHT, 19 54.8N, 70 49.9W. Mo (A) W 10s, 80ft, 12M. Yellow and black concrete tower, 30ft. Reported extinguished (March 1992).

PUERTO PLATA LIGHT, on hill near signal station, 19 48.8N, 70 41.5W. L Fl W 6s, 137ft, 18M. Yellow steel skeleton tower, black lantern, 59ft. Signal station. Occasional. Reported reduced intensity (January 1986).

PUERTO PLATA RANGE (Front Light), 19 48.6N, 70 42.0W. Fl R. White concrete tower, red lantern. (Reported destroyed 1995). **(Rear Light),** 230 meters, 218° from front. Fl R. White concrete tower, red lantern.

PUERTO PLATA AVIATION LIGHT, 19 46.2N, 70 33.5W. Al Fl W G 10s, 10M.

PUERTO PLATA

19 48N, 70 42W

Port of entry

Pilotage: A set of range lights stand at the head of the bay and when in line bearing 218°T lead through the entrance channel. The pilot boat may be contacted on VHF channel 16 (call sign, HIW8).

Dockage: The yacht dock is on the western side of the wharf to the east. Boats usually use a Mediterranean moor here to avoid damage from the surge.

Anchorage: There is some room to anchor near the yacht dock.

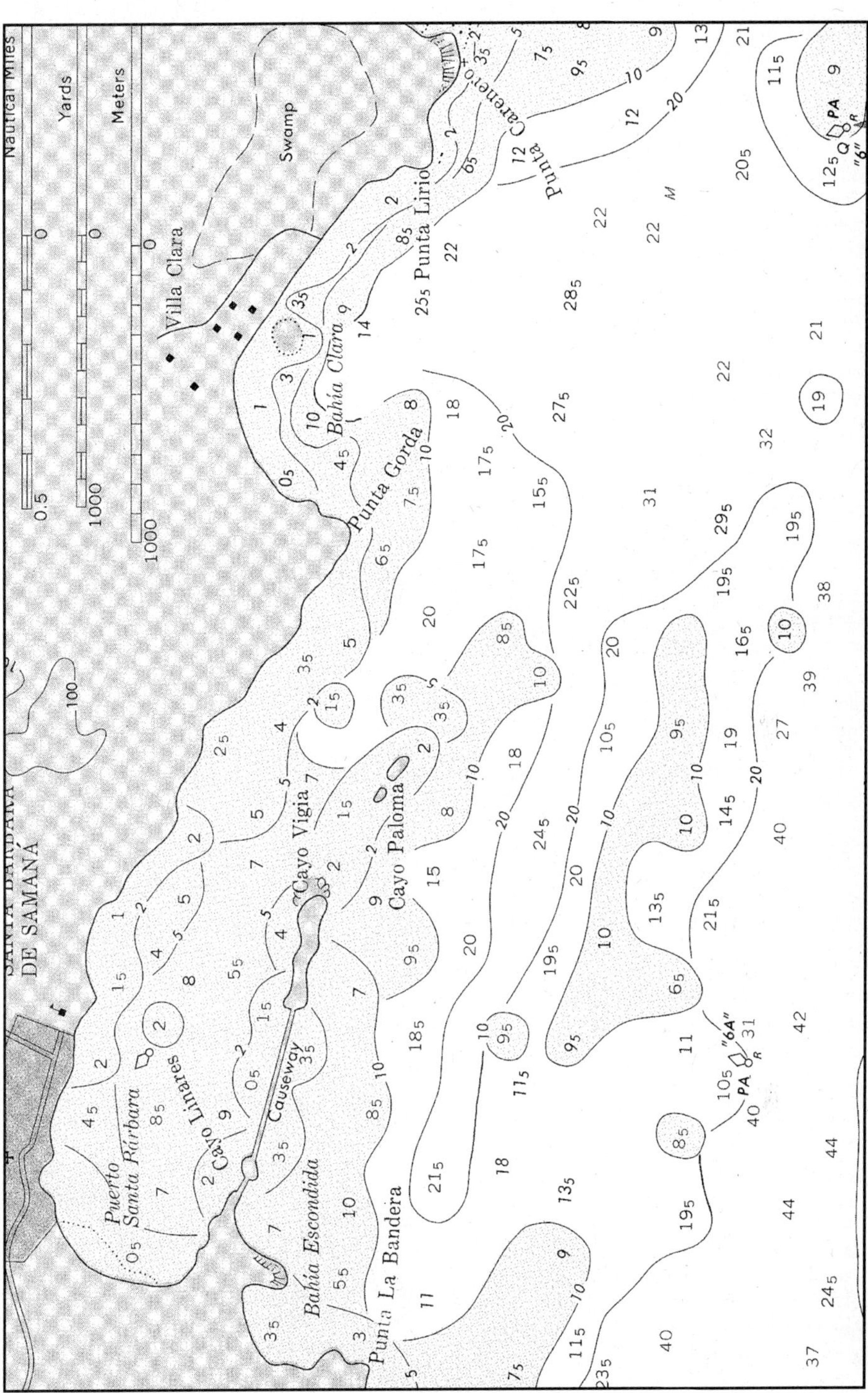

Samaná, soundings in meters *(from DMA 25723, 39th ed, 6/85)*

Services: Puerto Plata is the best place to restock along the north coast of the Dominican Republic. Fuel, water, and all groceries should be available. You may have to jug fuel. General repairs may be possible, but marine supplies are not available. There is a U.S. consular agent here (809-586-4204).

Puerto Plata is a popular stop for those trying to beat into the headwinds down the "Thorny Path" to the islands. The coastline is high (with elevations to 3,000 feet), and the water is deep close to shore. There are few official ports of entry on this coast, so you should plan your passage carefully. A good selection of charts is necessary to avoid the many reefs—it is wise to keep a good offing until sure of your position. The docking situation is rather rough and requires careful placement of anchors and heavy gear. Once you get the boat settled, you'll find an interesting city ashore, with further exploring possibilities inland.

PUERTO PLATA TO CABO ENGANO

CABO VIEJO FRANCES LIGHT, near edge of cliff, 19 40.5N, 69 54.6W. L Fl W 10s, 163ft, 18M. White pyramidal steel tower, 85ft. Visible 132°–304°.

PORT SANCHEZ LIGHT, at end of pier, 19 13.6N, 69 36.6W. Fl R 6s, 30ft, 6M. White pyramidal tower, 20ft. Reported removed (March 1992).

CABO SAMANÁ LIGHT, 19 18.5N, 69 08.6W. Fl W 5s, 463ft, 10M. White pyramidal skeleton tower, 62ft.

PUNTA BALANDRA LIGHT, on the bluff, 19 11.2N, 69 13.4W. Fl W 4s, 155ft, 10M. White skeleton tower, 25ft.

CAYO VIGIA LIGHT, 19 11.7N, 69 19.6W. Fl R, 23ft, 8M. White pyramidal metal tower.

SAMANÁ

19 12N, 69 20W

Charted name: Santa Bárbara de Samaná.

Port of entry: Tie up to the wharf and await the arrival of the officials. Approach carefully, as there are unconfirmed reports of depths less than 6 feet in the area.

Dockage: You may be able to tie up to the wharf for refueling, water, and loading supplies.

Anchorage: There is reported to be good holding in depths of 12 to 25 feet. Stay clear of the channel to the main wharf.

Services: Fuel, water, ice, water, and some groceries are available. A farmer's market offers fresh fruits and vegetables. There are no yacht repair facilities as such. There is a ferry from Samaná to Sabana de la Mar.

This is a good spot to plan your passage across the notorious Mona Passage or to clear in to the Dominican Republic from Puerto Rico. The scenery is reported to be spectacular, and a trip outside of town is recommended. Los Haitises National Park, on the other side of the bay, is reported to be quite interesting.

PUNTA NISIBON LIGHT, 18 58.5N, 68 46.2W. Fl (2) W 10s, 50ft, 12M. White pyramidal tower, red lantern, 35ft. Reported extinguished (March 1992).

CABO ENGANO LIGHT, E point of island, 18 36.8N, 68 19.5W. Fl W 5s, 141ft, 11M. Red and white steel tower, 66ft.

CABO ENGANO TO PUERTO DE ANDRÉS

PUNTA BARRACHANA LIGHT, 18 32.7N, 68 21.3W. Iso R 2s. Tower. Reported F R (1989).

BOCA DE YUMA LIGHT, 18 23.1N, 68 35.5W. Fl R 11s, 30ft, 10M. Red pyramidal steel tower, 30ft.

ISLA SAONA LIGHT, E end on Punta Cana, 18 06.6N, 68 34.5W. Fl W 10s, 105ft, 16M. White concrete tower. Reported destroyed (March 1992).

PUNTA LAGUNA LIGHT, 18 08.3N, 68 44.8W. Fl W 4s, 45ft, 10M. White concrete tower. Reported at reduced intensity (1989).

LA ROMANA LIGHT, E point at entrance to river, 18 24.8N, 68 57.1W. Fl W 6s, 90ft, 15M. Yellow framework tower, 69ft.

LA ROMANA

18 25N, 68 57W

Pilotage: The sugar mill on the west side of the mouth of the Río Dulce is conspicuous day or night. Its lights have been reported to be visible at up to 20 miles, and the smokestacks will be visible during the day. A navigation light is shown on the east side of the entrance to the river. The first large wharf to the west is for the sugar plant. The next wharf north is the government dock. Pilot boats monitor VHF channel 16, call sign, HIW 9.

Bridge height: Reported as 28 feet.

Port of entry: The government wharf is on the west side of the river before the bridge. Depths are reported as 5 to 6 feet alongside.

Dockage: There is a shipyard and a marina near the bridge on the eastern side of the river. There is a yacht club north of the bridge, on the eastern shore.
Anchorage: There is a possible anchorage in the small cove on the eastern side of the river about 200 yards inside the mouth.
Services: The shipyard can haul your boat and has a fuel dock. Most supplies are available.

SAN PEDRO DE MACORIS LIGHT, E point, 18 26.1N, 69 17.7W. L Fl W 10s, 49ft, 12M. White and red tower.

SAN PEDRO DE MACORIS

18 27N, 69 19W
Port of entry
Pilotage: Entry is recommended before 1100, as there is decreased swell in the channel entrance. The pilot boat monitors VHF radio, call sign, HIW 19.
Anchorage: This is primarily a commercial harbor, serving the needs of the sugar industry. Anchoring is possible, but no dockage is reported. Visiting yachtsmen may be better off visiting Club Nautico, in Boca Chica, about 17 miles to the west.

ISLA CATALINA LIGHT, Punta Berroa, 18 20.5N, 68 59.3W. L Fl W 8s, 40ft, 10M. Yellow and black concrete tower, 30ft. Reported destroyed (March 1992).

PUERTO DE ANDRÉS

LA CALETA AVIATION LIGHT, 18 25.9N, 69 40.2W, Al Fl W G 5s, 14M. On aero radiobeacon tower. On request.
RANGE (Front Light), 18 26.1N, 69 38.0W. Fl R, 6M. Yellow metal tower, 23ft. Reported extinguished (March 1992). **(Rear Light),** 118 meters, 300° from front. Fl R, 6M. Yellow metal tower 36ft. Reported extinguished (March 1992).

BOCA CHICA

18 26N, 69 38W
Port of entry (reported)
Pilotage: Boca Chica is also called Andrés on charts. The harbor is about 18 miles east of Santo Domingo at the head of a small bay. The port consists of a basin 500 yards long and 200 yards wide with wharves on its northwest and northeast sides and protected on its southwest side by a reef that has been reinforced to form a breakwater. The village of Andrés has a sugar factory with a conspicuous chimney, marked by a red obstruction light, with the name BOCA CHICA painted prominently on it. Range lights located on the south side of the port lead to the entrance channel when in line bearing 300°T. Vessels may enter only during the day. The port is under the jurisdiction of the Commander of the Port, Santo Domingo.
Dockage: Club Nautico yacht club has dockage, repairs, and luxurious facilities.
Services: Haulouts, fuel, water, repairs, and most supplies available.

PUNTA TORRECILLA LIGHT, 18 27.8N, 69 52.6W. L Fl (2) W 10s, 135ft, 13M. White tower, black diagonal bands.

PUERTO DE SANTO DOMINGO

ENTRANCE RANGE, E side of harbor **(Front Light),** 18 28.3N, 69 52.6W. F R. White concrete tower. **(Rear Light),** 155 meters, 047° from front. F R. White concrete tower.

SANTO DOMINGO

18 28N, 69 53W
Pilotage: The port of Santo Domingo lies at the mouth of the Río Ozama. West of the river the coast is low and rocky, with foothills rising 4 or 5 miles inland. East of the river, a coastal plain extends for 15 or 20 miles. Strong onshore winds can cause a rise in sea level considerably above normal in the bay. Currents in the river average 1.5 knots; however, during the rainy season (May to September) the current is reported to reach rates of 8 knots. The port may be entered from 0600 to 2100. Harbor authorities may be reached on the VHF radio.
Port of entry
Dockage and anchorage: This is primarily a commercial harbor. Boaters may be better off at Boca Chica, which is 18 miles by highway to the east.

Santo Domingo is one of the oldest cities in the Caribbean and one of the largest, with a population of around 2 million. The son of Christopher Columbus, Don Diego, began building a palace and fort here in 1510, as well as the first church in the New World, which is the supposed final resting place of the Navigator himself.

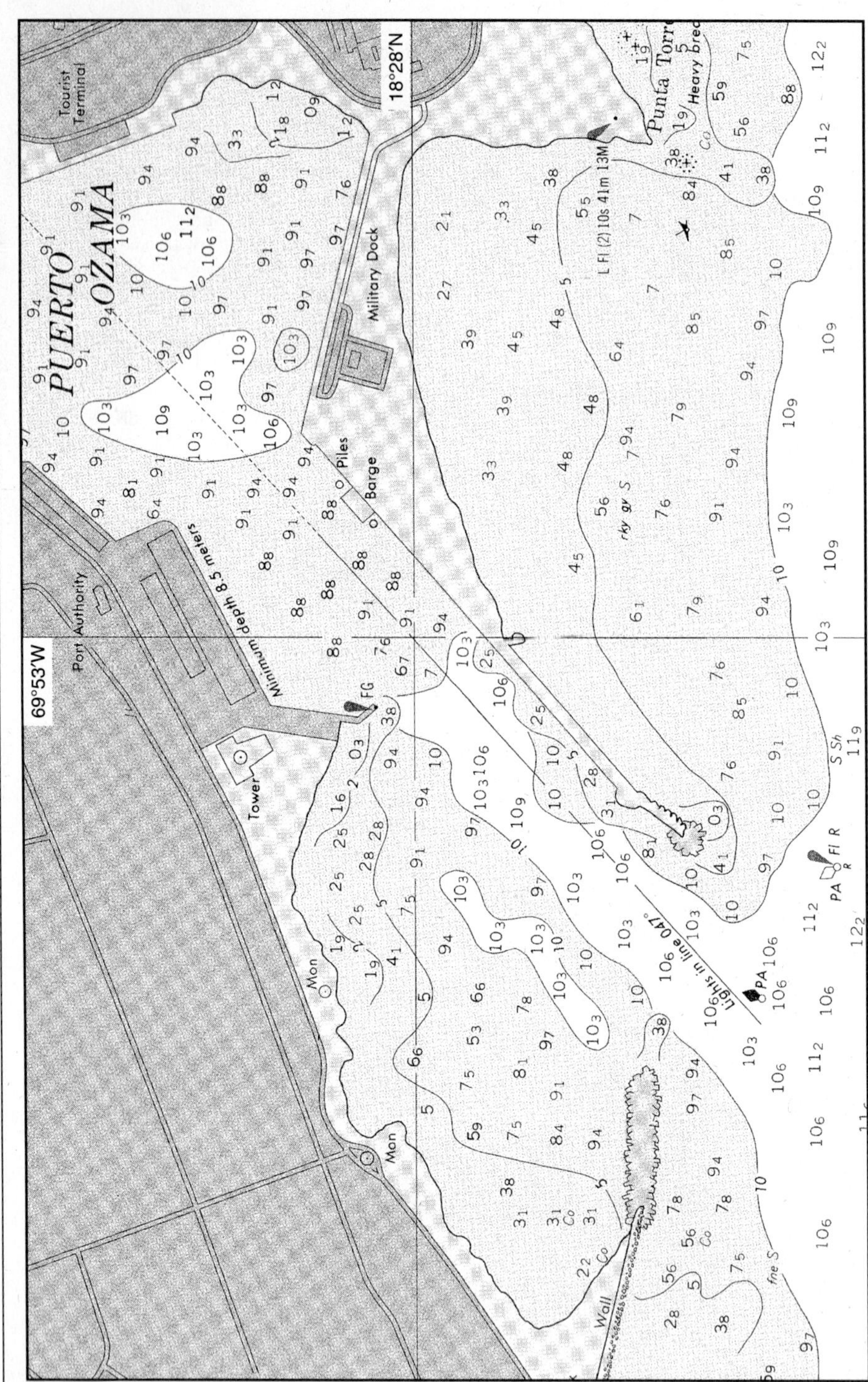

Santo Domingo, soundings in meters *(from DMA 25848, 24th ed, 11/86)*

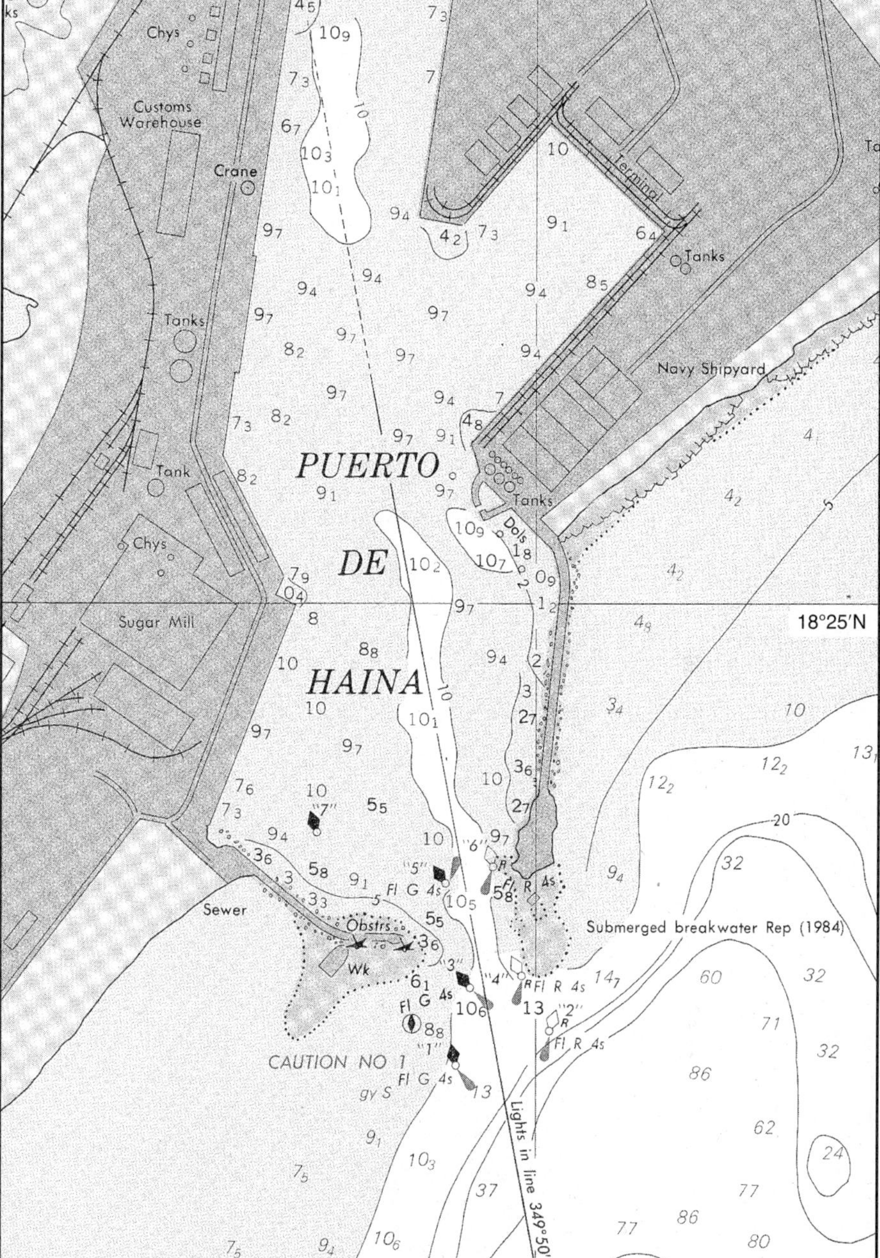

Puerto de Haina, soundings in meters *(from DMA 25848, 24th ed, 11/86)*

PUERTO DE HAINA TO CABO ROJO

PUERTO DE HAINA ENTRANCE RANGE (Front Light), 18 25.4N, 70 01.2W. Oc.R 4s. Yellow framework tower. R. lights mark two chimneys 800 meters SSW. Bearing is not reliable. **(Rear Light),** 150 meters, 351° from front. F R. Yellow framework tower.

PUERTO DE HAINA

18 25N, 70 01W

Pilotage: Two conspicuous chimneys, which show red lights, stand 450 yards north-northwest of the root of the west breakwater. The lights on the chimneys are visible for about 20 miles. There is an obstruction about 0.8 mile southeast of the harbor entrance. The river currents increase to 2.5 knots during the rainy season (May to September). The pilot boat may be contacted on VHF channel 16 or on 2182 kHz (call sign, HIW20).

Port of entry

Dockage and anchorage: Haina is primarily a commercial harbor.

PUNTA PALENQUE LIGHT, 18 13.5N, 70 09.0W. Fl W 7s, 45ft, 12M. White concrete tower, 30ft.

BOCA CANASTA LIGHT, 18 14.4N, 70 19.6W. F R; Fl R (vert), 22M. Mast.

PUNTA SALINAS LIGHT, Bahía de las Calderas, 18 12.3N, 70 31.8W. Fl W 3s, 98ft, 10M. Red pyramidal steel tower, 40ft.

PUERTO VIEJO DE AZUA LIGHT, 18 19.8N, 70 50.5W. L Fl (2) W 10s, 75ft, 15M. Black metal framework tower on concrete base, 29ft.

BARAHONA HARBOR RANGE (Front Light), 18 12.1N, 71 04.5W. F R, 6M. White pyramidal tower, 45ft. Occasional. **(Rear Light),** 85 meters, 243° from front. F R. White pyramidal tower. Occasional.

ISLA ALTO VELO, summit, 17 28.3N, 71 38.5W. Fl (2) W 10s, 535ft, 13M. Yellow truncated concrete tower, black triangular projections on each side. Reported extinguished (February 1986).

PUNTA BEATA LIGHT, 17 36.1N, 71 25.2W. Fl W 9s, 80ft, 14M. Concrete tower. Extinguished.

LOS FRAILES LIGHT, 17 38.0N, 71 40.9W. Q R, 32ft, 5M. Reported removed (March 1992).

ANSE JOSEPH LIGHT, head of jetty, 17 55.2N, 71 39.0W. F R, 8M.

PEDERNALES LIGHT, 18 02.2N, 71 44.9W. L Fl W 12s, 40ft, 11M. White pyramidal steel tower, red lantern, 30ft.

Puerto Rico

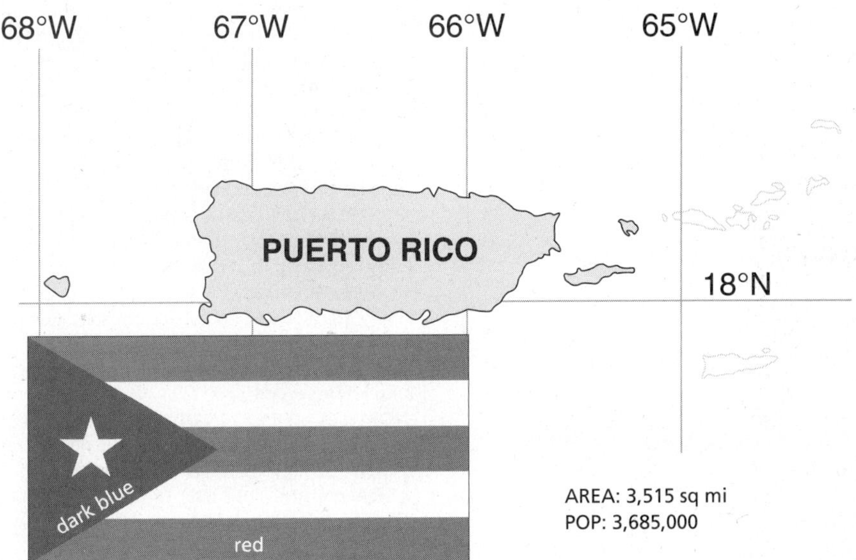

AREA: 3,515 sq mi
POP: 3,685,000

Please see cautions about chartlets and waypoints on page P 2.

ENTRY PROCEDURES

The Commonwealth of Puerto Rico is voluntarily associated with the United States. Many of the laws, rules, and regulations, including clearing in, are similar to those in effect in the mainland U.S. U.S. vessels returning from foreign waters, including the U.S. Virgins, should purchase the annual $25 customs user's decal in advance of arrival in Puerto Rico, if possible (see U.S. Customs Procedures, page P 7).

Ports of entry for recreational vessels are Mayagüez (787-831-3342 or -3343), Ponce (787-841-3130, -3131, or -3132), Fajardo (787-863-0950, -0102, -4075, or -0811), and San Juan (787-253-4533, -4534, -4535, or -4536). Customs inspectors on Vieques (787-741-8366) and Culebra (787-742-3531) also clear private vessels. Other ports of entry for commercial vessels only are Aguadilla, Guánica, Puerto de Jobos, and Humacao.

Yachts arriving on weekdays outside normal working hours (0800 to 1700) or on Saturdays, Sundays, and holidays must report to the San Juan station via telephone and await further instructions. U.S. vessels with U.S. crew may be able to clear in entirely by telephone. All vessels arriving from the U.S. Virgin Islands must report to customs; however, vessels traveling to the U.S. Virgin Islands from Puerto Rico do not need to clear out.

USEFUL INFORMATION

Languages: Spanish and English.
Currency: U.S. dollar (US$).
Telephones: Area code, 787; local numbers, seven digits.
Public holidays (1998): New Year's Day, January 1; Three King's Day, January 6; Eugenio Maria de Hostos Day, January 13; Martin Luther King, Jr.'s Birthday, January 19; President's Day, February 16; Emancipation Day, March 22; Good Friday, April 10; José de Diego Day, April 14; Memorial Day, May 25; Independence Day (US), July 4; Luis Muñoz Rivera Day, July 21; Constitution Day & José Celso Barbosa Day,

July 28; Labor Day, September 7; Columbus Day, October 12; Veteran's Day, November 11; Discovery Day, November 19; Thanksgiving, November 26; Christmas Day, December 25; New Year's Eve, December 31.

Tourist offices: *In Puerto Rico:* Department of Tourism, PO Box 902-3960, San Juan 00902-3960 (tel., 800-866-7827, 787-721-2400; fax, 787-725-4417). *In the U.S.:* Puerto Rico Tourism Co., 575 Fifth Avenue, New York, NY 10017 (voice mail, 800-223-6530 or 212-599-6262; fax, 212-818-1866). Offices also in Los Angeles and Miami (Coral Gables). *In Canada:* Puerto Rico Tourism Co., 41-43 Colbourne Street, Suite 301, Toronto, Ont. M5E 1E3 (tel., 800-667-0394, 416-368-2680; fax, 416-368-5350).

Emergencies, U.S. Coast Guard: San Juan station (VHF channel 16; 2182 kHz; tel., 787-729-6800).

WEST COAST

PUNTA BORINQUEN LIGHT, 18 29.8N, 67 08.9W. Fl (2) W 15s, 292ft, 24M. Gray cylindrical tower.

AGUADILLA JETTY LIGHT 1, 18 26.0N, 67 09.4W. Q G, 27ft, 4M. Square green daymark.

AGUADILLA

18 26N, 67 09W

Pilotage: A 1,208-foot-high naval communication tower south of town is quite prominent. Large vessels load raw sugar and molasses at the pier 1.1 miles north of town, and an Air Force pier is 1.8 miles north of town. The bay is exposed to north and west winds and experiences a frequent surge.

Dockage: None for pleasure craft.

Anchorage: Anchor off town, but be prepared for rough conditions in north winds.

Services: There are no yacht facilities. General supplies are available ashore.

PUNTA HIGUERO LIGHT, 18 21.7N, 67 16.2W. Oc W 4s, 90ft, 9M. Gray cylindrical tower.

RINCON SEWER EXTENSION LIGHT, on seaward side of sewer outfall, 18 20.1N, 67 15.3W. Q R, 12ft. On pile. Private aid.

BAHÍA DE MAYAGÜEZ

Manchas Exteriores Buoy 1, 18 14.6N, 67 12.7W. Green can.

Entrance Lighted Buoy 3, Fl G 4s, 4M. Green.

Entrance Lighted Buoy 4, Fl R 4s, 4M. Red.

RANGE (Front Light), 18 13.3N, 67 09.8W. Q G, 54ft. Rectangular red daymark, white stripe, on tower on roof of warehouse. Visible 4° each side of rangeline. **(Rear Light),** 310 yards, 092° from front. Oc G 4s, 87ft. Rectangular red daymark, white stripe, on tower on building. Visible 4° each side of rangeline.

Lighted Buoy 5, Fl G 4s, 4M. Green.

NORTH BREASTING DOLPHIN ONE OBSTRUCTION LIGHT, 18 13.3N, 67 10.2W. Fl W 2.5s, 13M. On dolphin. Private aid.

SOUTH BREASTING DOLPHIN SIX OBSTRUCTION LIGHT, Fl W 2.5s, 13M. On dolphin. Private aid.

Lighted Buoy 6, Fl R 4s, 3M. Red.

Buoy 8, red nun.

Buoy 10, marks SE end of 18-foot spot. Red nun.

Mayagüez Harbor Daybeacon, 18 12.5N, 67 09.5W. Diamond-shaped daymark, green and white, on dolphin.

NORTH SEWER EXTENSION LIGHT, marks edge of sewer outfall. Q W, 12ft. On pile. Private aid.

SOUTH SEWER EXTENSION LIGHT, marks end of sewer outfall, 18 12.1N, 67 09.4W. Q W, 12M. On pile. Private aid.

MAYAGÜEZ

18 12N, 67 09W

Pilotage: The open roadstead at Mayagüez is easy to enter day or night. The shipping terminal is in the north part of the bay. Prominent features are Punta Guanajibo, a 165-foot-high, flat-topped ridge 2 miles south of the city, and Cerro Anterior, a 433-foot-high saddle-shaped hill 1.5 miles inshore of the city. The city hall clock tower and a church are conspicuous above other buildings. Several red and white radio towers are visible along the south shore of the bay.

Port of entry: Boats should tie to the commercial wharf north of town and call customs.

Dockage: None for pleasure craft in town. Note that the Club Deportivo marina, about 6.5 miles south of the city, is for members only.

Anchorage: Pleasure boats anchor along the shore 1.2 miles south of the terminal.

Services: There are no pleasure-boat facilities in Mayagüez. The large city has all general supplies, good transportation, banking, and communications.

Mayagüez is primarily a commercial port, but makes a convenient first port of call for boats arriving from the west. The city is one of Puerto Rico's major ports and commercial centers.

CANAL DE GUANAJIBO

Punta Arenas Buoy 6, at NW edge of shoal point. Red nun.
Punta Ostiones Buoy 4, at SW edge of shoal spot off point. Red nun.
Punta Melones Shoal Buoy 1, green can.

PUERTO REAL

18 04N, 67 12W
Pilotage: The harbor is located about 2 miles north of the Bahía de Boquerón. It is a circular basin 0.7 mile in diameter used by local fishing vessels and small pleasure craft. Depths in the basin are 6 to 15 feet with shoal water toward the eastern end. The town of Puerto Real is on the north shore of the basin.
Dockage: There are a couple of small marinas with dockage and most services.
Anchorage: This is a well-protected spot.
Services: Water, fuel, some marine supplies, and groceries are available. There are haulout facilities for fishing boats that pleasure boats may be able to use. Take a *publico* (small bus) to Cabo Rojo for more extensive shopping.

BAHÍA DE BOQUERÓN

Bajo Enmedio South Lighted Buoy 1, at S end of shoal, 18 00.8N, 67 12.5W. Fl G 4s, 4M. Green.

BOQUERÓN

18 01N, 67 11W
Pilotage: For boats coming from the west the primary obstruction is Bajo Enmedio, stretching for about a mile across the entrance to the bay. It is marked by a buoy at its south end.
Dockage: The Club Nautico de Boquerón (VHF channel 68; tel., 787-851-1336) may have space at its docks near the north end of the waterfront. Low-tide depths are 4.5–5 feet.
Anchorage: Anchor south of the club docks. Hurricane protection among the mangroves is available in the Cano Boquerón a mile south. Depths of 10 feet are reported in the channel.
Services: Fuel, marine hardware, charts, and outboard parts are available from Boquerón Marine Center (VHF channel 68, call sign Blue Marlin; tel., 787-851-1674; fax, 787-254-2255). Water, electricity, and general supplies are also available. Boquerón is a beach hangout for university students, as well as a major gathering spot for cruisers.

MONA PASSAGE

Arrecife Tourmaline Lighted Buoy 8, 18 09.7N, 67 20.7W. Fl R 6s, 4M. Red.
ISLA DE MONA LIGHT, 18 06.6N, 67 54.5W. Fl W 5s, 323ft, 14M. On steel tower. Light may be obscured by land masses when viewed from approximate bearings 140° and 270°.
Canal de la Mona East Shoal Lighted Buoy 6, 18 05.3N, 67 25.4W. Fl R 6s, 4M. Red.
Canal de la Mona East Shoal Lighted Buoy 4, 18 00.4N, 67 23.0W. Fl R 6s, 4M. Red.
Canal de la Mona East Shoal Lighted Buoy 2, Fl R 6s, 4M. Red.
CABO ROJO LIGHT, 17 56.0N, 67 11.5W. Fl W 20s, 121ft, 20M. Gray hexagonal tower attached to flat-roofed dwelling. Emergency light when main light is extinguished shows Fl W 6s, 9M.

ISLA MAGUEYES

Buoy 1, black can. Private aid.
Buoy 2, red nun. Private aid.
Buoy 4, red nun. Private aid.
Buoy 6, red nun. Private aid.
Buoy 7, black can. Private aid.

BAHÍA DE GUÁNICA

Entrance Lighted Buoy 2, 17 55.2N, 66 54.6W. Fl R 4s, 4M. Red.
Buoy 3, green can.
GUÁNICA LIGHT, 17 57.1N, 66 54.3W. Fl W 6s, 132ft, 7M. White skeleton tower. Obscured by Punta Brea to seacoast traffic westward to 044°.
Lighted Buoy 4, on W end of shoal, 17 56.2N, 66 54.4W. Fl R 4s, 3M. Red.
Buoy 5, marks 27-foot shoal. Green can.
RANGE (Front Light), 17 57.9N, 66 54.6W. Q R, 27ft, 4M. Rectangular red daymark, white stripe, on skeleton tower. Visible 4° each side of rangeline. **(Rear Light),** 560 yards, 354.5° from front. Iso R 6s, 48ft, 5M. Rectangular red daymark, white stripe, on skeleton tower. Visible 4° each side of rangeline.
Lighted Buoy 6, Fl R 6s, 3M. Red.
Buoy 7, green can.
Buoy 8, red nun.
Buoy 9, 17 57.6N, 66 54.6W. Green can.
Buoy 10, at S end of shoal. Red nun.
Buoy 11, green can.
Buoy 12, red nun.
SUGAR PIER LIGHTS (2), 17 57.9N, 66 55.5W. F R, 16ft. On pier. Private aids.

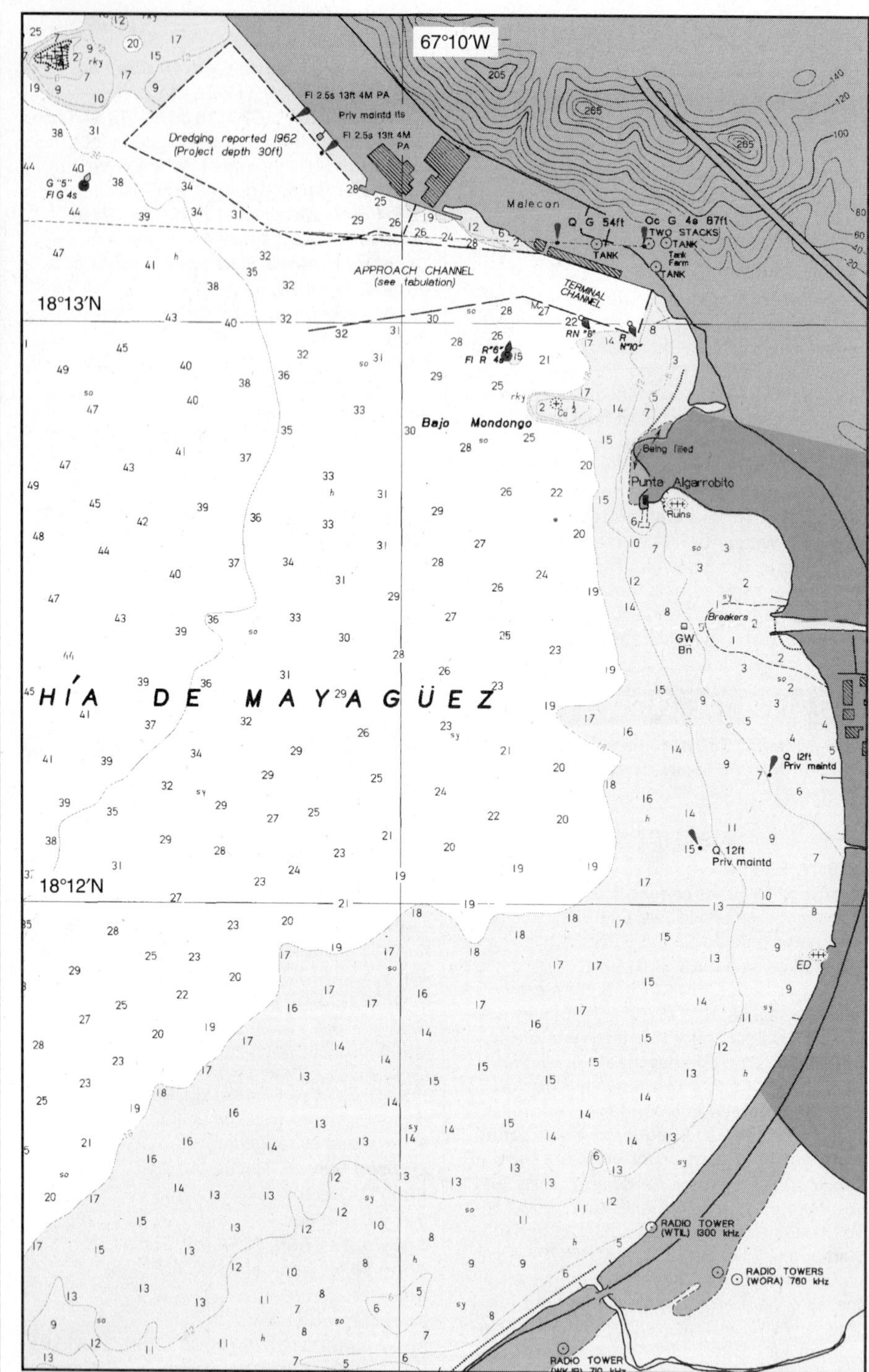

Mayagüez, soundings in feet *(from NOAA 25673, 14th ed, 10/90)*

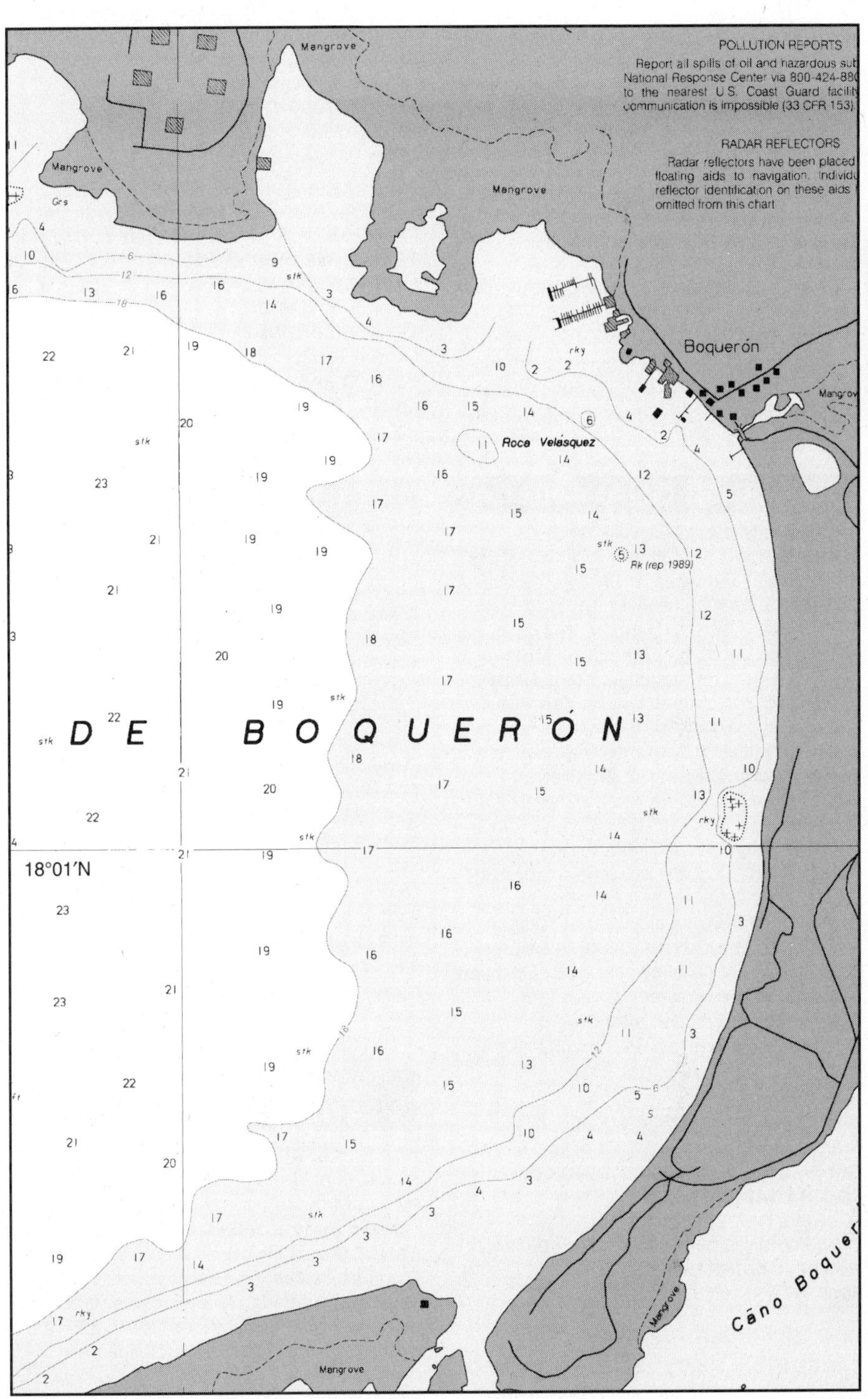

Boquerón, soundings in feet *(from NOAA 25675, 8th ed, 5/90)*

GUÁNICA

17 58N, 66 56W
Pilotage: A range of 354.5° leads up the main dredged channel to the town.
Anchorage: Anchor outside the dredged ship channels in well-protected waters. Depths become quite shallow in the north end of the bay off Playa de Guánica.
Services: General supplies are available.

Small Bahía de Guánica, 16 miles east of Cabo Rojo Light, is one of the better hurricane harbors in Puerto Rico. This is a quiet harbor with two completely sheltered bays. Prominent entrance features include Guánica Light, powerlines, and a water tower. The port serves large vessels loading fertilizer, molasses, and sugar.

BAHÍA DE GUAYANILLA

CAYO MARIA LANGA LIGHT, 17 58.0N, 66 45.2W. Fl W 2.5s, 42ft, 8M. Diamond-shaped daymark, red and white, on skeleton tower.
ENTRANCE RANGE (Front Light), 17 58.6N, 66 45.8W. Q Y, 16ft. Rectangular red daymark, white stripe, on dolphin. Visible all around, higher intensity on rangeline. **(Rear Light),** 1,000 yards, 358° from front. Iso Y 6s, 36ft. Rectangular red daymark, white stripe, on tower on piles. Visible 15° each side of rangeline.
Lighted Buoy 1, Fl G 2.5s, 3M. Green.
Lighted Buoy 2, 17 58.0N, 66 45.8W. Fl R 4s, 4M. Red.
Lighted Buoy 3, Fl G 4s, 4M. Green.
PUNTA GOTAY SOUTH MOORING DOLPHIN LIGHT, 17 58.8N, 66 45.8W. F R, 15ft. On square concrete deck. Private aid.
PUNTA GOTAY NORTH MOORING DOLPHIN LIGHT, 17 58.9N, 66 45.8W. F R, 15ft. On square concrete deck. Private aid.
Lighted Buoy 5, Fl G 6s, 4M. Green.
Lighted Buoy 6, marks end of shoal. Fl R 4s, 4M. Red.
Buoy 7, green can.
Buoy 8, marks end of shoal. Red nun.
Buoy 9, green can.
COMMONWEALTH OIL REFINING COMPANY SOUTH LOADING DOCK LIGHT, 17 59.6N, 66 46.0W. F R, 12ft. Private aid.
COMMONWEALTH OIL REFINING COMPANY NORTH LOADING DOCK LIGHTS (3), 17 59.8N, 66 45.9W. F R, 19ft. On pile. Private aids.

PPG INDUSTRIES

RANGE (Front Light), 18 00.3N, 66 46.0W. Q W, 20ft. Visible 4° each side of rangeline. Private aid. **(Rear Light),** 158 yards, 014° from front. Oc W 4s, 40ft. Visible 2° each side of rangeline. Private aid.
Lighted Buoy 1, 17 57.8N, 66 46.2W. Fl G 2.5s, Black. Private aid.
Lighted Buoy 2, Fl R 2.5s. Red. Private aid.
Lighted Buoy 3, Fl G 2.5s. Black. Private aid.
Lighted Buoy 4, Fl R 2.5s. Red. Private aid.
PIER LIGHT, F R, 13ft. On dolphin. Private aid.
TEXACO WEST MOORING DOLPHIN ONE LIGHT, 18 00.0N, 66 45.9W. Fl R 6s, 15ft. On dolphin. Private aid.
TEXACO LOADING DOCK LIGHT, F R, 30ft. On pile. Private aid.
TEXACO EAST MOORING DOLPHIN FOUR LIGHT, 18 00.0N, 66 45.8W. Fl R 6s, 15ft. On dolphin. Private aid.

BAHÍA DE TALLABOA

Lighted Buoy 1, 17 57.7N, 66 44.3W. Fl G 4s, 4M. Green.
Buoy 2, red nun.
Lighted Buoy 4, Fl R 4s, 3M. Red.
Lighted Buoy 5, 17 58.2N, 66 44.3W. Fl G 2.5s, 4M. Green.
Lighted Buoy 6, Q R, 2M. Red.
LIGHT 7A, 17 58.8N, 66 44.2W. Fl G 2.5s, 15ft. Square green daymark on pile. Private aid.
Buoy 7, green can.
LOADING PLATFORM LIGHTS (4), 17 59.0N, 66 44.5W. F W. Platform. Private aids.
Buoy 8, 17 58.7N, 66 43.7W. Red nun. Private aid.
Lighted Buoy 10, 17 59.0N, 66 43.8W. Fl R 2.5s. Red. Private aid.
PIER LIGHT, 17 59.2N, 66 43.9W. F R, 20ft. On pier. Private aid.
Buoy 12, 17 59.3N, 66 43.8W. Red nun. Private aid.

BAHÍA DE PONCE AND APPROACHES

ISLA DE CARDONA LIGHT, 17 57.4N, 66 38.1W. Fl W 4s, 46ft, 8M. White cylindrical tower on center of front of flat-roofed dwelling.
Lighted Buoy 1, Fl G 4s, 4M. Green.
RANGE (Front Light), 17 58.8N, 66 37.2W. Q G, 49ft. Rectangular red daymark, white stripe, on pyramidal skeleton tower. Visible 15° each side of rangeline. **(Rear Light),** 305 yards, 015° from front. Iso G 6s, 75ft. Rectangular red daymark, white stripe, on roof of building. Visible 4° each side of rangeline.

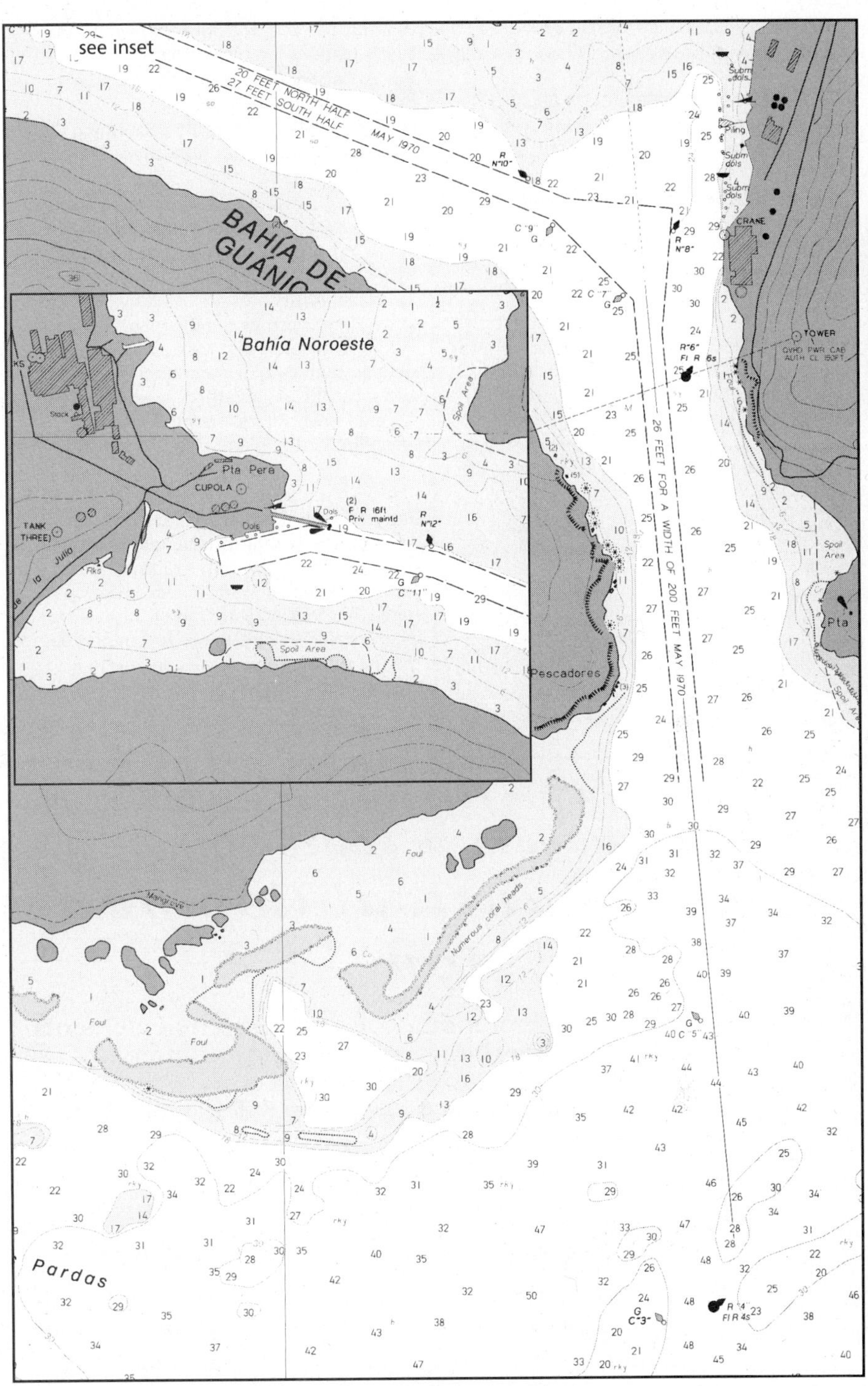

Guánica, soundings in feet *(from NOAA 25679, 9th ed, 6/90)*

Lighted Buoy 3, Fl G 4s, 4M. Green.
Lighted Buoy 4, on SW side of shoal. Fl R 2.5s, 3M. Red.
MUELLE MUNICIPAL PIER LIGHTS (2), 17 58.0N, 66 37.3W. F W, 10ft. On pier. Private aids.
PONCE HARBOR PIER 4 OBSTRUCTION LIGHT, 17 58.1N, 66 37.3W. F W, 5ft. Square white case. Light not visible 360°; may be obscured by vessels moored at pier. Private aid.
TRAILER WHARF OBSTRUCTION LIGHT, 17 58.4N, 66 37.1W. F W, 10ft. On steel pile. Private aid.
Lighted Buoy 5, Fl G 4s, 4M. Green.
SEWER EXTENSION LIGHT, marks end of sewer outfall, 17 58.9N, 66 37.8W. F W, 12ft. White rectangular framework on piles. Private aid.
RÍO BUCANA EAST JETTY LIGHT, 17 58.0N, 66 36.0W. Fl W 5s, 19ft. On tripod. Private aid.
RÍO BUCANA JETTY LIGHT, 17 58.1N, 66 36.0W. Fl W 5s, 19ft. On tripod. Private aid.

PONCE

17 59N, 66 37W

Pilotage: Prominent from offshore are the summit light of Isla Caja de Muertos, the cement factory stacks west of Ponce, the large microwave tower in Ponce, and the hotel on the hill back of Ponce. The charted radio towers can be seen from well offshore. The eastern entrance channel, marked by a lighted range running 015°T, leads to the waterfront.

Port of entry: Tie up at the yacht club on Isla de Gata and call customs.

Dockage: The Ponce Yacht and Fishing Club located on Isla de Gata (VHF channel 16; tel., 787-842-9003) has all facilities.

Anchorage: A designated anchorage is shown on the chart. Many cruisers anchor off the Ponce Yacht and Fishing Club, whose facilities they can use for a fee of $5/day.

Services: All services are available. Haulouts up to 50 tons can be handled, and repairs are readily available. Fuel, water, ice, electricity, showers, and a public telephone are available at the yacht club, and there are good grocery stores in town. There is good transportation to any point on the island.

Ponce is the second largest city in Puerto Rico and one of its most important ports. There are art museums, historic buildings, shops and restaurants, and excellent transportation facilities. You may want to leave your boat here while visiting San Juan or use the port as a crew transfer point. Ponce is your best bet for marine supplies and repairs on Puerto Rico's south coast.

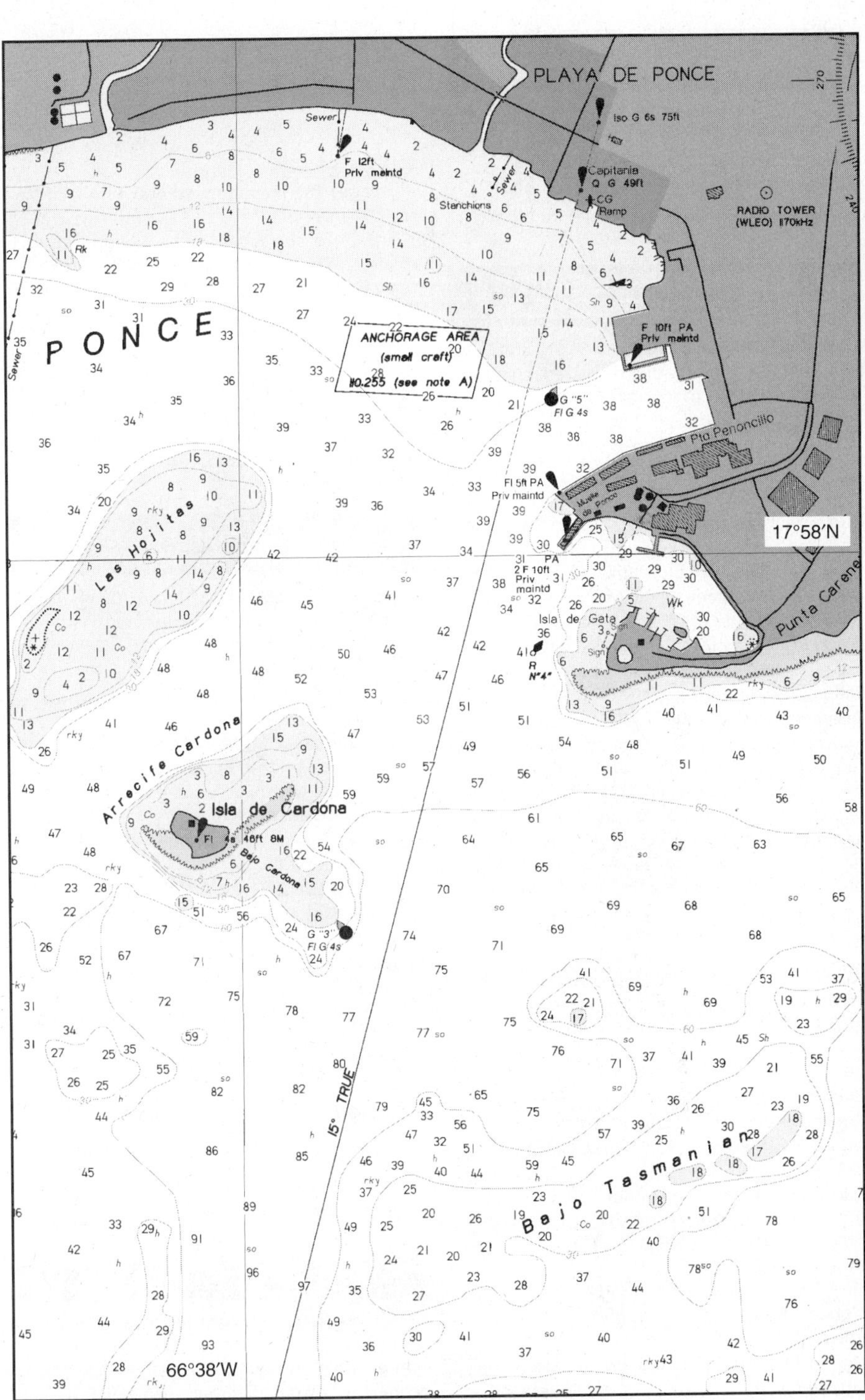

Playa de Ponce, soundings in feet *(from NOAA 25683, 14th ed, 12/91)*

ISLA CAJA DE MUERTOS

ISLA CAJA DE MUERTOS LIGHT, 17 53.6N, 66 31.3W. Fl W 30s, 297ft, 12M. Gray cylindrical tower on center of flat-roofed dwelling.

BARCAZA PIER LIGHT, 17 53.1N, 66 31.7W. Fl R 6s. On pier. Private aid.

BAHÍA DE JOBOS AND APPROACHES

CAYOS DE RATONES LIGHT, 17 56.0N, 66 17.0W. Fl W 2.5s, 48ft, 5M. Diamond-shaped daymark, red and white, on triangular skeleton tower.

Buoy 2, red nun.

SALINAS

17 58N, 66 18W

Pilotage: Enter the small bay east of Cayo Mata from the south.

Dockage: There are more than 100 slips, with depths of 3 to 8 feet, at the Marina de Salinas (VHF channel 16; tel., 787-824-3185).

Anchorage: Good protection and good holding are reported. Anchor anywhere off Playa de Salinas and dinghy in to the marina.

Services: Fuel, water, ice, electricity, laundry, and showers are available at the marina. The full facilities of the associated resort may be used by marina guests for a fee. Playa Marine, adjacent to the marina, has an array of repair and maintenance products and is a good contact for a wide range of services, from filling propane tanks to sail repair. Take the *publico* (local bus) to the town of Salinas a mile away or to Ponce for major shopping.

Salinas is a favorite anchorage on the south coast. Cruisers stop here for rest, repairs, and protection from the weather.

Cayo Puerca Buoy 3, green can.

Punta Colchones Cut Buoy 5, green can.

Punta Colchones Cut Buoy 6, red nun.

DISCHARGE PIPE EAST OBSTRUCTION LIGHT, on rock mound, 17 56.0N, 66 13.7W. Q R, 12ft. On piles. Private aid.

DISCHARGE PIPE WEST OBSTRUCTION LIGHT, on rock mound. Q R, 12ft. On piles. Private aid.

AGUIRRE POWER PLANT

Lighted Buoy 1, 17 56.6N, 66 13.5W. Q G, Green. Private aid.

Lighted Buoy 2, Q R. Red. Private aid.

Lighted Buoy 3, Fl G 4s. Green. Private aid.

Lighted Buoy 4, Fl R 4s. Red. Private aid.

Lighted Buoy 5, Fl G 4s. Green. Private aid.

Lighted Buoy 6, Fl R 4s. Red. Private aid.

Lighted Buoy 8, Fl R 4s. Red. Private aid.

PUERTO DE JOBOS

17 57N, 66 12W

Pilotage: The main buoyed entrance leads to the commercial wharves at Central Aguirre. The stacks of the sugar plant show up well from offshore. The southern small-boat channel through the reefs is known as Boca del Inferno. There is 11 feet of water over the bar, which breaks in heavy weather.

Dockage: There may be some dockage at a small yacht club in the indentation about 0.7 mile east of Punta Rodeo.

Anchorage: Good throughout the bay. Avoid the commercial area of Central Aguirre.

LAS MAREAS TO PUNTA TUNA

Lighted Buoy LM, 17 54.1N, 66 10.6W. Mo (A) W. Red and white stripes with red spherical topmark. Private aid.

RANGE (Front Light), 17 56.4N, 66 09.4W. Q W, 86ft. Rectangular red daymark, white stripe, on tower. Visible on rangeline only. Private aid. **(Rear Light),** 635 yards, 017.7° from front. Iso W 6s, 124ft. Rectangular red daymark, white stripe, on tower. Visible on rangeline only. Private aid.

Lighted Buoy 1, 17 55.6N, 66 09.9W. Fl G 4s, Black. Private aid.

Lighted Buoy 2, Fl R 2.5s. Red. Private aid.

LIGHT 3, Fl G 4s, 25ft. Square green daymark on pile. Private aid.

LIGHT 4, Fl R 2.5s, 50ft. Rectangular red daymark on tower. Private aid.

ARTICULATED LIGHT 5, Q G. On mast. Private aid.

LIGHT 6, Q R, 25ft. Rectangular red daymark on pile. Private aid.

PIER LIGHTS (3), F R, 10ft. On dolphin. Private aids.

PUNTA TUNA LIGHT, 17 59.4N, 65 53.1W. Fl (2) W 30s, 111ft, 16M. White octagonal tower on square flat-roofed dwelling.

PUERTO YABUCOA

CHANNEL RANGE (Front Light), 18 03.4N, 65 50.2W. Q W, 75ft. Rectangular red daymark, white stripe, on tower. Visible 1° each side of rangeline. Private aid. **(Rear Light),** 910 yards, 296.8° from front. Iso W 6s, 140ft. Rectangular

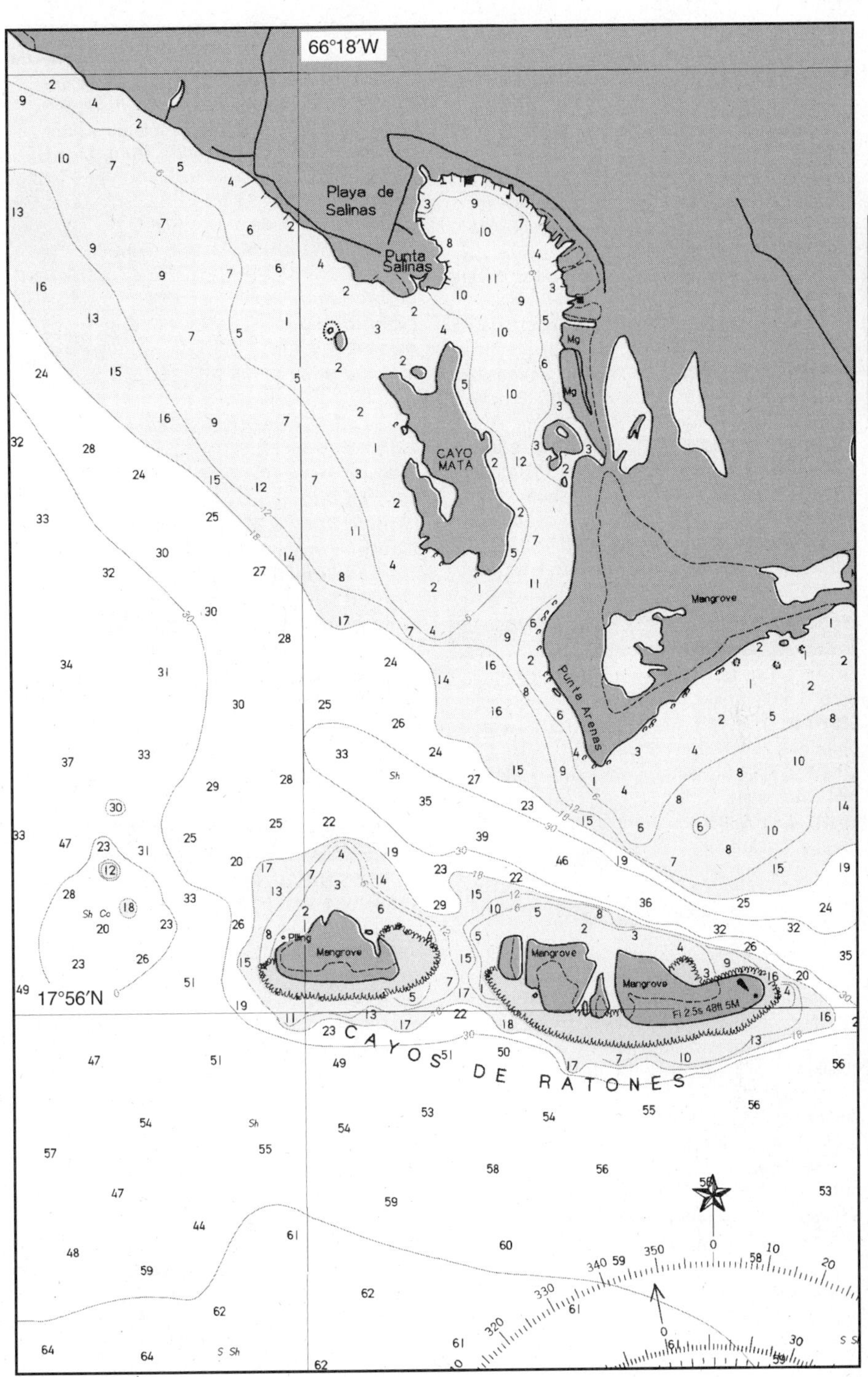

Salinas, soundings in feet *(from NOAA 25687, 11th ed, 1/91)*

red daymark, white stripe, on tower. Visible 1° each side of rangeline. Private aid.

Channel Lighted Buoy 2, 18 02.7N, 65 48.6W. Fl R 4s. Red. Private aid.

Channel Lighted Buoy 3, Q G. Black. Private aid.

CHANNEL LIGHT 5, Fl G 4s, 15ft. Square green daymark on dolphin. Radar reflector. Private aid.

CHANNEL LIGHT 6, Fl R 4s, 15ft. Rectangular red daymark on dolphin. Radar reflector. Private aid.

CHANNEL LIGHT 7, Fl G 4s, 15ft. Square green daymark on pile. Radar reflector. Private aid.

CHANNEL LIGHT 8, Fl R 4s, 15ft. Rectangular red daymark on pile. Radar reflector. Private aid.

CHANNEL BREAKWATER LIGHT, 18 03.3N, 65 49.7W. F R, 51ft. Radar reflector. Private aid.

CHANNEL LIGHT 9, Fl G 4s, 15ft. Square green daymark on pile. Radar reflector. Private aid.

CHANNEL LIGHT 10, Fl R 4s, 15ft. Rectangular red daymark on pile. Radar reflector. Private aid.

PALMAS DEL MAR

LIGHT 1, Fl G 6s, 12ft. Square green daymark on pile. Private aid.

LIGHT 2, Q W, 16ft. Rectangular red daymark on pile. R 203.5°–295°. Private aid.

LIGHT 3, Q G, 6ft. Square green daymark on pile. Private aid.

LIGHT 4, Q W, 6ft. Rectangular red daymark on pile. Private aid.

PALMAS DEL MAR

18 05N, 65 48W

Pilotage: The harbor entrance is marked by private lights. The Palmas del Mar Shipyard and Marine Store is on the south side of the harbor. It is reported that strong easterly winds cause breaking seas in the harbor entrance and surge inside the harbor.

Dockage: Available at the shipyard (VHF channel 16) with all facilities.

Anchorage: Anchor in the small basin north of the marina, inside the breakwall.

Services: All services are available, including haulouts and a marine store (787-850-2065).

This small artificial harbor is adjacent to a luxurious resort complex with restaurants and shopping.

PASAJE DE VIEQUES

RANGE (Front Light), 18 12.7N, 65 36.0W. Q W, 70ft. Rectangular white daymark, red stripe, on skeleton tower. Visible all around, Higher intensity on rangeline. **(Rear Light),** 1,715 yards, 025.4° from front. Iso W 6s, 142ft. Rectangular white daymark, red stripe, on tower. Visible 4° each side of rangeline.

NAVY PIER LIGHTS (2), W of Punta Puerca, 18 13.6N, 65 36.1W. F R. Maintained by U.S. Navy.

Lighted Buoy 2, Fl R 2.5s, 4M. Red.

Lighted Buoy 3, Fl G 2.5s, 4M. Green.

Buoy 4, red nun.

Lighted Buoy 6, Fl R 2.5s, 3M. Red.

RADAS ROOSEVELT PASSAGE

Lighted Buoy 11, Fl G 4s, 4M. Green.

Lighted Buoy 10, Fl R 4s, 3M. Red.

Lighted Buoy 9, Fl G 4s, 4M. Green.

Lighted Buoy 8, Fl R 2.5s, 3M. Red.

Lighted Buoy 7, Fl G 6s, 4M. Green.

Lighted Buoy 6, Q R, 3M. Red.

Lighted Buoy 5, Q G, 3M. Green.

Lighted Buoy 3, Fl G 2.5s, 3M. Green.

Lighted Buoy 2, Fl R 4s, 3M. Red.

Lighted Buoy 1, on W end of shoal, 18 13.4N, 65 28.1W. Fl G 6s, 4M. Green.

Sonda de Vieques Obstruction Lighted Buoy LAE, marks end of sewer outfall. IQ W. Black and red bands. Private aid.

ISLA DE VIEQUES

VIEQUES NAVAL PIER LIGHTS (2), 18 08.9N, 65 30.9W. F R, 15ft. On pier. Maintained by U.S. Navy.

VIEQUES NAVAL BREAKWATER LIGHT, 18 09.1N, 65 30.9W. Q R, 10ft. On pile. Maintained by U.S. Navy.

PUNTA MULAS LIGHT, 18 09.3N, 65 26.6W. Oc R 4s, 68ft, 7M. White octagonal tower on top of flat-roofed dwelling.

PUNTA ESTE LIGHT, 18 08.2N, 65 16.1W. Fl W 6s, 43ft, 7M. Diamond-shaped daymark, red and white, on tower.

PUNTA CONEJO LIGHT, 18 06.6N, 65 22.6W. Fl W 6s, 58ft, 7M. Diamond-shaped daymark, red and white, on tower.

PUERTO FERRO LIGHT, 18 05.9N, 65 25.4W. Fl W 4s, 56ft, 7M. Diamond-shaped daymark, red and white, on skeleton tower.

Puerto Real Shoal Buoy 1, on E end of shoal, 18 04.9N, 65 29.0W. Green can.

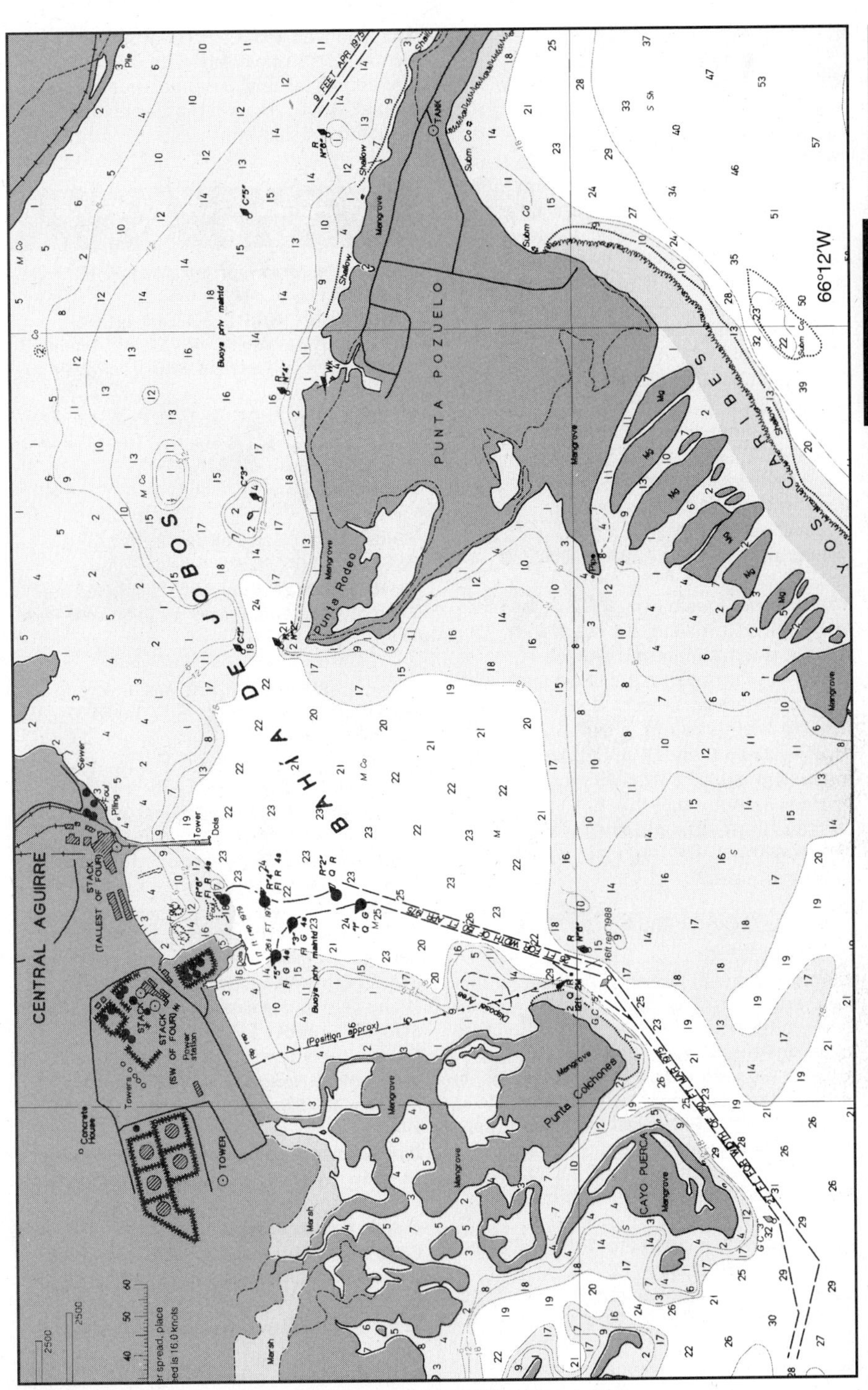

Puerto de Jobos, soundings in feet *(from NOAA 25687, 11th ed, 1/91)*

ROOSEVELT ROADS HARBOR

Channel Lighted Buoy 1, 18 11.8N, 65 36.6W. Fl G 6s, 4M. Green.

Channel Lighted Buoy 2, Fl R 6s, 4M. Red.

ENTRANCE CHANNEL RANGE (Front Light), 18 13.8N, 65 38.1W. Q W, 37ft. Rectangular red daymark, white stripe, on skeleton tower on piles. Visible 2° each side of rangeline. **(Rear Light),** 800 yards, 315° from front. Iso W 6s, 66ft. Rectangular red daymark, white stripe, on tower. Visible 2° each side of rangeline.

Channel Lighted Buoy 3, Fl G 4s, 4M. Green.

Channel Buoy 4, red nun.

Channel Buoy 5, green can.

Channel Lighted Buoy 6, Fl R 2.5s, 3M. Red.

CHANNEL LIGHT 7, Fl G 4s, 16ft, 4M. Square green daymark on dolphin.

Turning Basin Buoy 8, red nun.

Anchorage Buoy A, yellow nun.

Anchorage Buoy B, 18 13.1N, 65 37.0W. Yellow can.

Turning Basin Lighted Buoy 9, Fl G 4s, 4M. Green.

ROOSEVELT ROADS PIER 3 LIGHTS (2), F R, 10ft. On pier. Maintained by U.S. Navy.

ROOSEVELT ROADS PIER 2 LIGHTS (2), F R, 10ft. On pier. Maintained by U.S. Navy.

Turning Basin Buoy 11, green can.

Turning Basin Buoy 13, green can.

Anchorage Buoy C, yellow nun.

Anchorage Buoy D, yellow nun.

ENSENADA HONDA AVIATION LIGHT, 18 14N, 65 38W. Fl W, 60s, 119ft. Radiobeacon about 1 mile eastward.

ROOSEVELT ROADS

18 13N, 65 37W

Pilotage: No vessel shall enter or remain within the charted restricted areas extending about 1 mile to 1.5 miles offshore from the naval base unless on official business. Fishing is not permitted in the restricted areas. Military personnel (active, retired, or reservist) with IDs and federal employees with GS IDs may obtain permission to use the small-craft facilities of the base by calling the Roosevelt Roads Marina (787-865-3297). Make reservations before approaching the area.

Port control: Port control for the naval base stands by on VHF channel 12.

Dockage: The marina monitors VHF channel 16. The Roosevelt Roads Yacht Club is for members only.

Services: Fuel, water, and general marine supplies are available.

The Roosevelt Roads U.S. Naval Station (787-865-2000) is a convenient, safe port for active and retired military personnel, reservists, personnel with GS IDs, and those able to obtain a sponsor on the base.

ROOSEVELT ROADS TO FAJARDO

Cabeza de Perro Lighted Buoy 7, 18 13.6N, 65 33.7W. Fl G 6s, 4M. Green.

BAJOS CHINCHORRO DEL SUR LIGHT, 18 14.1N, 65 31.2W. Fl W 4s, 25ft, 8M. Diamond-shaped daymark, red and white, on tower on piles.

CABEZA DE PERRO LIGHT, 18 15.0N, 65 34.6W. Fl W 6s, 80ft, W 8M, R 6M. R 021°–031° and 066°–161°, obscured 031°–066°.

Roca Lavandera del Oeste Buoy 5, green can.

Bajos Largo Lighted Buoy 3, 18 17.7N, 65 34.8W. Fl G 2.5s, 3M. Green.

Punta Figueras Buoy 4, on shoal. Red nun. Buoy on edge of red sector of Cabeza de Perro Light.

PUERTO DEL RAY MARINA

North Channel Buoy 2, 18 17.5N, 65 37.7W. Red nun. Private aid.

NORTH CHANNEL LIGHT 4, Fl R 5s, 10ft. Triangular red daymark on pile. Private aid.

NORTH CHANNEL LIGHT 5, Fl G 5s, 28ft. Square green daymark on pile. Private aid.

South Channel Daybeacon 1, 18 17.0N, 65 37.5W. Square green daymark on pile. Private aid.

South Channel Daybeacon 3, square green daymark on pile. Private aid.

SOUTH CHANNEL LIGHT 4, Fl R 4s, 28ft. Triangular red daymark on pile. Private aid.

SOUTH CHANNEL LIGHT 5, Fl G 4s, 6ft. Square green daymark on pile. Private aid.

Punta Mata Redonda Obstruction Daybeacon, 18 18.3N, 65 37.2W. Diamond-shaped white daymark on pile, worded DANGER REEF.

Punta Barrancas Obstruction Daybeacon, 18 18.5N, 65 37.2W. Diamond-shaped white daymark on pile, worded DANGER REEF.

Isla de Ramos Buoy 2, red nun.

Cayo Largo Lighted Buoy 1A, midway of westerly shoal, 18 18.9N, 65 35.3W. Fl G 4s, 4M. Green.

Cayo Largo Buoy 1, N side of shoals. Green can.

Isla Palominos Lighted Buoy 2, 18 21.1N, 65 33.6W. Fl R 4s, 3M. Red.

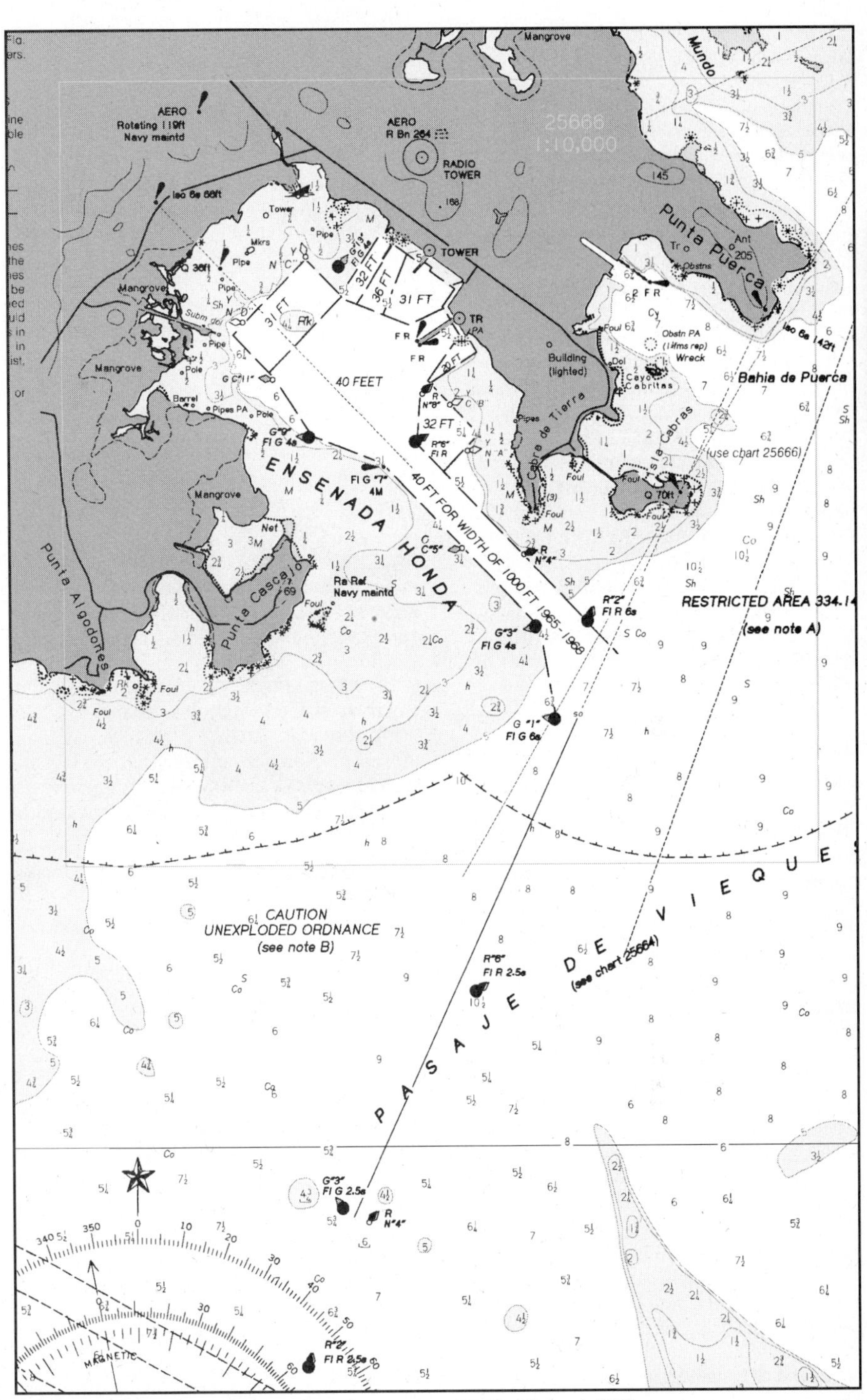

Roosevelt Roads, soundings in feet *(from NOAA 25663, 26th ed, 1/92)*

P

Pilot

PUERTO RICO

ISLA PALOMINOS MARINA

Buoy 1, 18 20.8N, 65 34.4W. Green can. Private aid.

NOTE: Five more alternating red nuns and green cans, numbered 2 through 6, mark this channel. All are private aids.

FAJARDO

Fifteen-foot Spot Lighted Buoy 1, Fl G 2.5s, 4M. Green.

Bajo Onaway Buoy 3, green can.

PUERTO CHICO MARINA BREAKWATER LIGHT, 18 20.8N, 65 38.0W. F R, 13ft. On pile. Private aid.

SEA LOVERS PIER A EAST LIGHT, 18 20.8N, 65 38.1W. F R, 10ft. On pier pile. Private aid.

SEA LOVERS PIER A WEST LIGHT, F R, 10ft. On pier pile. Private aid.

SEA LOVERS PIER B EAST LIGHT, F R, 10ft. On pier pile. Private aid.

SEA LOVERS PIER B WEST LIGHT, F R, 10ft. On pier pile. Private aid.

FAJARDO HARBOR PIER OBSTRUCTION LIGHTS (2), 18 20.1N, 65 37.8W. F R, 10ft. On pile. Private aids.

ISLETA MARINA, FAJARDO

PIER A LIGHT, 18 20.4N, 65 37.3W. Q W, 12ft. On pile. Private aid.

PIER B LIGHT, F G, 12ft. On pile. Private aid.

PIER C LIGHT, F G, 12ft. On pile. Private aid.

PIER D LIGHT, F G, 12ft. On pile. Private aid.

Daybeacon 2, rectangular red daymark on pile. Private aid.

FAJARDO

18 20N, 65 38W

Pilotage: The easiest entrance is from the north, via the unmarked ferry channel west of Bajo Laja. There are fixed red lights, privately maintained, on the ferry dock. The controlling depth is 11 feet to the 300-foot-long public wharf. Depths at the dock are 12 feet at the outer end and 8 feet alongside. The southern approach has more hazards and fewer marks, but should have about 9 to 11 feet.

Port of entry: Customs is on the waterfront near the ferry dock, which is west of the public pier.

Dockage: This area has a profusion of repair yards, marinas, and boating services. At the north end of Playa Sardinera is Puerto Chico Marina (787-863-0834). Villa Marina (VHF channels 16 and 68; tel., 787-863-5131) is located in a small dredged basin behind a breakwater just north of the public wharf. Isleta Marina (VHF channel 16; tel., 787-863-0370), with dockside depths to 14 feet, is on the island across from town. Puerto del Rey Marina (VHF channels 16 and 71; tel., 787-860-1000), behind a breakwater in Bahía Demajagua, about 3 miles south of the main waterfront, has depths for vessels drawing 18 feet or less.

Anchorage: There is good anchorage south of Isla Marina. Note the charted cable and pipeline areas between the island and Fajardo.

Services: All services are available. Isleta Marina offers free ferry service to downtown from Isla Marina.

Fajardo is one of the best places in Puerto Rico to obtain supplies or repair your boat. There is regular ferry service to Culebra and Vieques.

PUNTA GORDA

EAST JETTY LIGHT, 18 21.4N, 65 37.5W. F R, 18ft. Rectangular red daymark on pile on jetty. Private aid.

WEST JETTY LIGHT, F G, 10ft. Square green daymark on pile on jetty. Private aid.

EAST SEAWALL LIGHT, F R, 18ft. Rectangular red daymark on pile on seawall. Private aid.

WEST SEAWALL LIGHT, F G, 10ft. Square green daymark on pile on seawall. Private aid.

LAS CROABAS

Daybeacon 1, 18 21.8N, 65 37.3W. Square green daymark on pile.

Daybeacon 2, rectangular red daymark on pile.

Daybeacon 3, square green daymark on pile.

FAJARDO TO CULEBRA

Isla Palominos Lighted Buoy 2, 18 21.2N, 65 33.5W. Fl R, 4s, 3M. Red.

Bajo Blake South Buoy 3, 18 20.4N, 65 31.9W. Green can.

CAYO LOBITO LIGHT, 18 20.1N, 65 23.5W. Fl W 6s, 110ft, 8M. Diamond-shaped daymark, red and white, on skeleton tower.

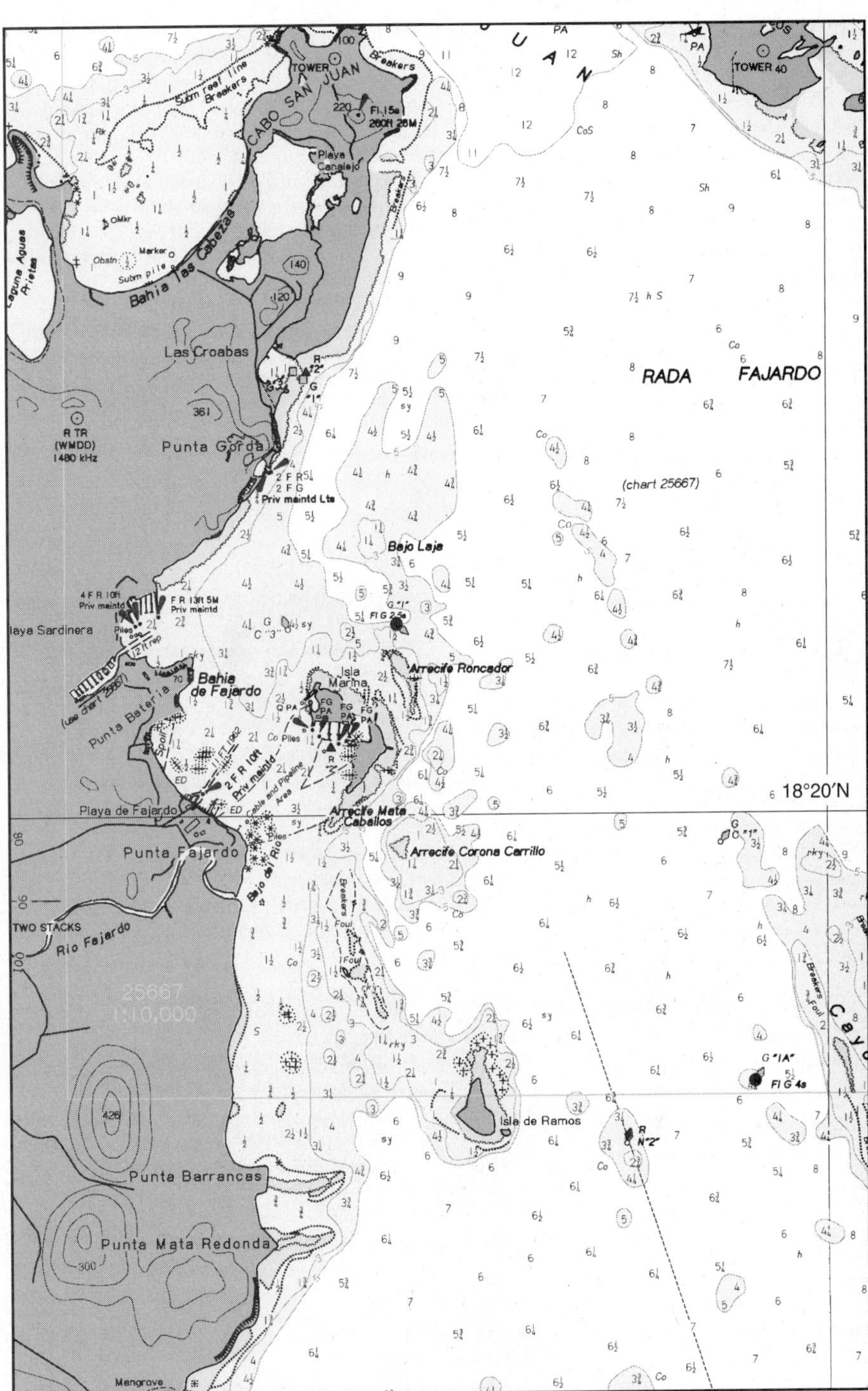

Fajardo, soundings in fathoms *(from NOAA 25663, 26th ed, 1/92)*

CULEBRA

PUNTA MELONES LIGHT, extremity of point, W side of island, 18 18.1N, 65 18.7W. Fl W 6s, 45ft, 6M. Diamond-shaped daymark, red and white, on tower, 18ft.

PUNTA DEL SOLDADO LIGHT, near S point of island, 18 16.7N, 65 17.2W. Fl W 2.5s, 65ft, 5M. Diamond-shaped daymark, red and white, on tower, 18ft.

Bajo Amarillo Lighted Buoy 2, 18 16.7N, 65 16.5W. Fl R 4s, 3M. Red.

Canal del Este Buoy 2, 18 16.6N, 65 15.1W. Red nun.

Cabezas Crespas Lighted Buoy 3, 18 16.8N, 65 15.4W. Fl G 4s, 4M. Green.

Cabezas Puercas Buoy 4, red nun.

ISLA DE CULEBRA OUTER RANGE (Front Daybeacon), 18 17.7N, 65 17.0W. Rectangular red daymark, white stripe, on skeleton tower. **(Rear Daybeacon),** 750 yards, 296° from front. Rectangular red daymark, white stripe, on skeleton tower.

Bajo Grouper Buoy 5, green can.

Bajo Camaron Buoy 6, red nun.

Bajo Snapper Lighted Buoy 8, 18 17.4N, 65 16.3W. Q R, 3M. Red.

ISLA DE CULEBRA INNER RANGE, on Punta Cemeterio **(Front Daybeacon),** 18 18.6N, 65 17.3W. Rectangular red daymark, white stripe, on skeleton tower. **(Rear Daybeacon),** 1,200 yards, 323° from front. Rectangular red daymark, white stripe, on skeleton tower.

Punta Colorada Lighted Buoy 9, Fl G 4s, 4M. Green.

Punta Caranero Buoy 10, red nun.

Punta Caranero Buoy 12, red nun.

Punta Cabras Buoy 14, red nun.

CULEBRA

18 18N, 65 18W

Pilotage: The area charted as Bahía de Sardinas is the location of the town's waterfront and ferry dock. You can anchor in the bay and dinghy in to the ferry wharf or the small creek just south of the wharf. This area is prone to a surge, or roll, under certain conditions. Most mariners prefer to enter Ensenada Honda and anchor west or southwest of Cayo Pirata. A large bay on the eastern side of Ensenada Honda has good hurricane anchorage among the mangroves at its head.

Port of entry: Mariners should call the airport customs station (787-742-3531) during office hours (0800 to 1200 and 1300 to 1700, Monday through Saturday) and San Juan customs after hours.

Dockage: You may be able to tie temporarily to the ferry wharf for fuel and supplies. There is no dockage for large boats in Ensenada Honda.

Anchorage: Excellent anchorages abound in Culebra. The most popular spot is east of town in Ensenada Honda. Dinghy ashore near the head of the small creek that passes through town. Anchorage can also be found in any of the bays along the shores of Ensenada Honda. A cool spot is west of the reef blocking the entrance to Ensenada Honda.

Services: There is fuel and water (via jugs), a hardware store, a canvas shop, a dive shop, several small restaurants, groceries, a post office, a bank, and telephones. Supplies can be limited. A ferry runs daily to Fajardo.

Culebra is only 20 miles west of St. Thomas, yet it is in another cruising world. The island is only quietly interested in tourists, but offers marvelous gunkholing and peaceful anchorages.

CULEBRA TO ST. THOMAS

ISLA CULEBRITA LIGHT, 18 18.8N, 65 13.7W. Fl W 10s, 305ft, 13M. Stone-colored cylindrical tower with red trim on flat-roofed dwelling. Obscured by cay 125°–142°.

Bajos Grampus South Lighted Buoy 2, S side of shoals, 18 14.3N, 65 12.5W. Fl R 4s, 3M. Red.

Sail Rock Lighted Buoy 1, 18 17.0N, 65 06.5W. Fl G 6s, 4M. Green.

SAVANA ISLAND LIGHT, 18 20.4N, 65 05.0W. Fl W 4s, 300ft, 7M. White tower.

CULEBRITA

18 19N, 65 14W

Pilotage: A powerful light marks the summit of Isla Culebrita. Approaching from the northeast, large seas may become steep breakers on the bar. Be prepared to leave if a norther threatens.

Anchorage: This is a lovely anchorage, off a crescent-shaped sandy beach. The island is all park land preserved in its natural state. There are no facilities here.

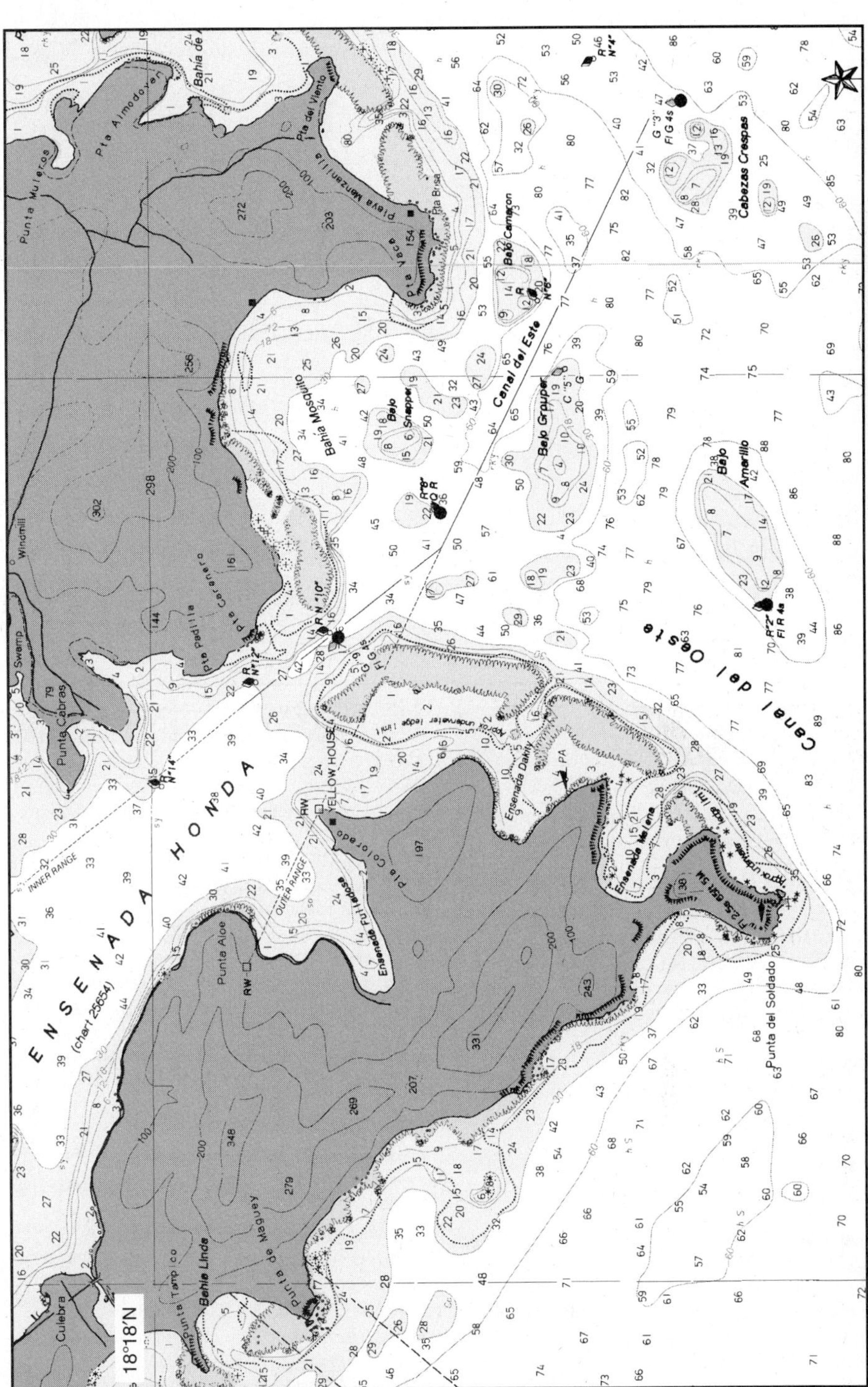

Culebra, soundings in feet *(from NOAA 25653, 11th ed, 12/92)*

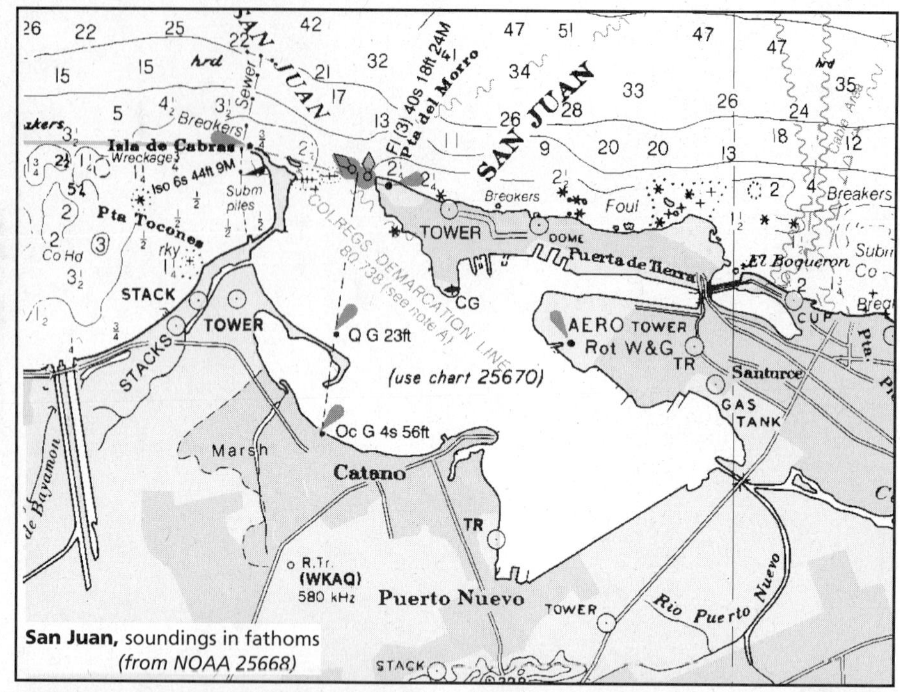

San Juan, soundings in fathoms (from NOAA 25668)

CABO SAN JUAN TO SAN JUAN HARBOR

CABO SAN JUAN LIGHT, 18 22.9N, 65 37.1W. Fl W 15s, 260ft, 26M. Cylindrical tower on front of white rectangular dwelling, black band around base.

LAS CUCARACHAS LIGHT, 18 24.0N, 65 36.7W. Fl W 6s, 38ft, 7M. Diamond-shaped daymark, green and white, on skeleton tower.

SAN JUAN AVIATION LIGHT, 18 26.5N, 66 00.3W. Al Fl W G 10s, 185ft.

Punta Picua Lighted Buoy WR2, NE of 2.5-fathom shoal, 18 26.6N, 65 45.6W. Fl R 4s, 4M. Red.

La Llave Buoy 1, 18 27.7N, 65 45.6W. Green can. Private aid.

La Llave Buoy 2, red nun. Private aid.

Lighted Buoy BC, 18 28.1N, 66 00.6W. Mo (A) W, 5M. Red and white stripes with red spherical topmark. Use only with local knowledge.

CABALLO CHANNEL RANGE (Front Lights, 2), 18 26.9N, 65 59.9W. F R, 20ft. Red and orange diamond daymark. Private aids. **(Rear Lights, 2),** 29 yards, 146.5° from front. F R, 25ft. Red and orange round daymark. Private aids.

PLATFORM LIGHT, 18 27.1N, 65 59.2W. F W, 12ft. On pile. Maintained by the FAA.

SAN JUAN HARBOR

PUERTO SAN JUAN LIGHT, 18 28.4N, 66 07.4W. Fl (3) W 40s, 181ft, 24M. Buff-colored square tower, octagonal base, on Morro Castle. Obscured 281°–061°.

CABRAS LIGHT, 18 28.5N, 66 08.4W. Iso W 6s, 44ft, 9M. Diamond-shaped daymark, red and white, on tower.

RANGE, W side of harbor **(Front Light),** 18 27.4N, 66 07.8W. Q G, 23ft. Rectangular red daymark, white stripe, on skeleton tower on piles. Visible 2° each side of rangeline. **(Rear Light),** 1,335 yards, 188° from front. Oc G 4s, 56ft. Rectangular red daymark, white stripe, on skeleton tower. Visible 2° each side of rangeline.

Lighted Buoy 1, 18 28.3N, 66 07.6W. Fl G 4s, 4M. Green.

Lighted Buoy 2, Fl R 6s, 3M. Red.

Lighted Buoy 3, 18 28.2N, 66 07.6W. Q G, 3M. Green.

Lighted Buoy 4, Fl R 4s, 3M. Red.

Lighted Buoy 5, 18 28.1N, 66 07.6W. Fl G 2.5s, 3M. Green.

Lighted Buoy 6, Fl R 2.5s, 2M. Red.

Buoy 6A, red nun.

Lighted Buoy 7, Fl G 4s, 4M. Green.

Lighted Buoy 8, Fl R 2.5s, 4M. Red.

Lighted Buoy 10, Fl R 2.5s, 4M. Red.

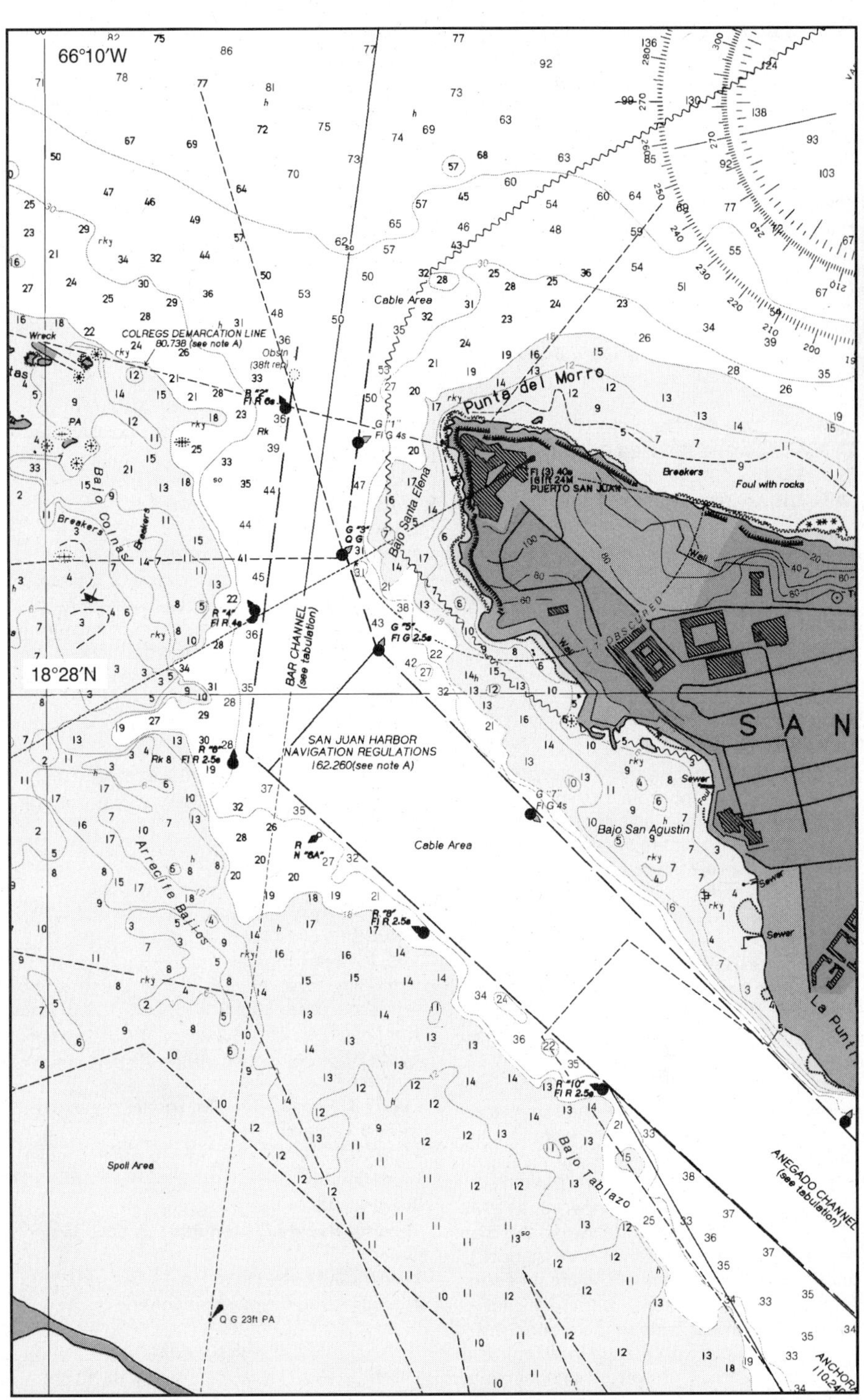

San Juan Channel, soundings in feet *(from NOAA 25670, 37th ed, 1/94)*

Lighted Buoy 11, Q G, 3M. Green.
Lighted Buoy 13, Fl G 4s, 3M. Green.
San Juan Harbor Anchorage F Buoy A, 18 27.0N, 66 07.0W. Yellow can.
Lighted Buoy 14, Q R, 2M. Red.
TMT OUTER DOLPHIN OBSTRUCTION LIGHT, 18 27.1N, 66 06.2W. Q W, 12ft. On mooring dolphin. Private aid.
TMT MIDDLE DOLPHIN OBSTRUCTION LIGHT, Q W, 12ft. On mooring dolphin. Private aid.
San Juan Harbor Anchorage E Buoy A, yellow can.
San Juan Harbor Anchorage E Buoy B, 18 27.0N, 66 06.1W. Yellow nun.

SAN ANTONIO CHANNEL

LA PUNTILLA FINGER PIER B LIGHT, 18 27.6N, 66 06.9W. F G, 10ft. On pier.
LA PUNTILLA FINGER PIER A LIGHT, F G, 10ft. On pier.
PIER 1 LIGHTS (3), 18 27.6N, 66 06.8W. F G, 10ft. On dolphins. Private aids.
PIER 3 LIGHTS (3), F G, 13ft. On dolphins. Private aids.
PIER 4 LIGHTS (3), 18 27.6N, 66 06.6W. F G. On pier. Private aids.

SAN JUAN

18 28N, 66 07W

Pilotage: The harbor entrance is deep and well marked. Approaching from offshore, be careful to stay well off the land until you are sure of your position, as there are many hazardous reefs along the coast. It can be hard to spot the charted landmarks from offshore, but you may be lucky enough to see a cruise ship entering or leaving. Steep seas can pile up on the bar at the entrance to the harbor.

Puerto Rico Ports Authority: The signal station at Fort San Cristobal is manned 24 hours a day for commercial traffic. The authorities monitor VHF channel 16.

Port of entry: Anchor off or dock at one of the marinas at the eastern end of the San Antonio Channel and immediately call customs.

Dockage: There are two marinas at the eastern end of the San Antonio Channel. The Club Náutico de San Juan (VHF channel 16; tel., 787-722-0177) is on the north side. Across the channel is the San Juan Bay Marina (VHF channel 16; tel., 787-721-8062). Reservations are suggested.

Anchorage: There is some room to anchor west of the marinas, among the moorings. Small white buoys with orange markings limit the anchoring area. The bottom is poor holding, and the area is swept by strong currents. You can dinghy in to the marina to the south.

Services: All services are available. There are facilities for haulouts and major repairs of engines, electronics, and hulls. There are several good chandleries and nearby grocery stores. San Juan is a good place for a major restocking of the ship's larder.

San Juan was founded in 1421 and is a delight for historically minded boaters. The approach to the harbor takes you dramatically below the ramparts of El Morro castle and past the walls of Old San Juan. Although limited in its cruising amenities, it is a fascinating port.

ARMY TERMINAL CHANNEL

Lighted Buoy A, Fl (2+1) G 6s, 3M. Green and red bands.
Lighted Buoy 2, Fl R 4s, 3M. Red.
Lighted Buoy 3, Fl R 4s, 3M. Red.
Buoy 4, red nun.
RANGE (Front Light), 18 25.7N, 66 06.5W. Q R, 28ft. Rectangular red daymark, white stripe, on tower. Visible 2° each side of rangeline. **(Rear Light),** 380 yards, 175.6° from front. Oc R 4s, 43ft. Rectangular red daymark, white stripe, on tower. Visible 2° each side of rangeline.
Lighted Buoy 5, Fl G 4s, 3M. Green.
Lighted Buoy 6, Fl R 4s, 3M. Red.
Lighted Buoy 7, Fl G 4s, 3M. Green.
Turning Basin Lighted Buoy 9, Fl G 2.5s, 3M. Green.

PUERTO NUEVO CHANNEL

Lighted Buoy 1, Fl G 4s, 4M. Green.
RANGE (Front Light), 18 26.5N, 66 05.2W. Q G, 45ft. Rectangular red daymark, white stripe, on tower on piles. Visible 2° each side of rangeline. **(Rear Light),** 610 yards, 058.4° from front. Oc G 4s, 79ft. Rectangular red daymark, white stripe, on tower. Visible 2° each side of rangeline.
LIGHT 3, Fl G 4s, 16ft, 4M. Square green daymark on dolphin.
LIGHT 5, Q G, 16ft, 2M. Square green daymark on dolphin.
Turning Basin Buoy 7, green can.
Turning Basin Lighted Buoy 9, Q G, 3M. Green.
TURNING BASIN LIGHT 10, Fl R 4s, 19ft, 2M. Triangular red daymark on dolphin.

MARTIN PENA CHANNEL

LIGHT 2, 18 26.5N, 66 05.1W. Fl R 4s, 10ft. Triangular red daymark on pile. Private aid.

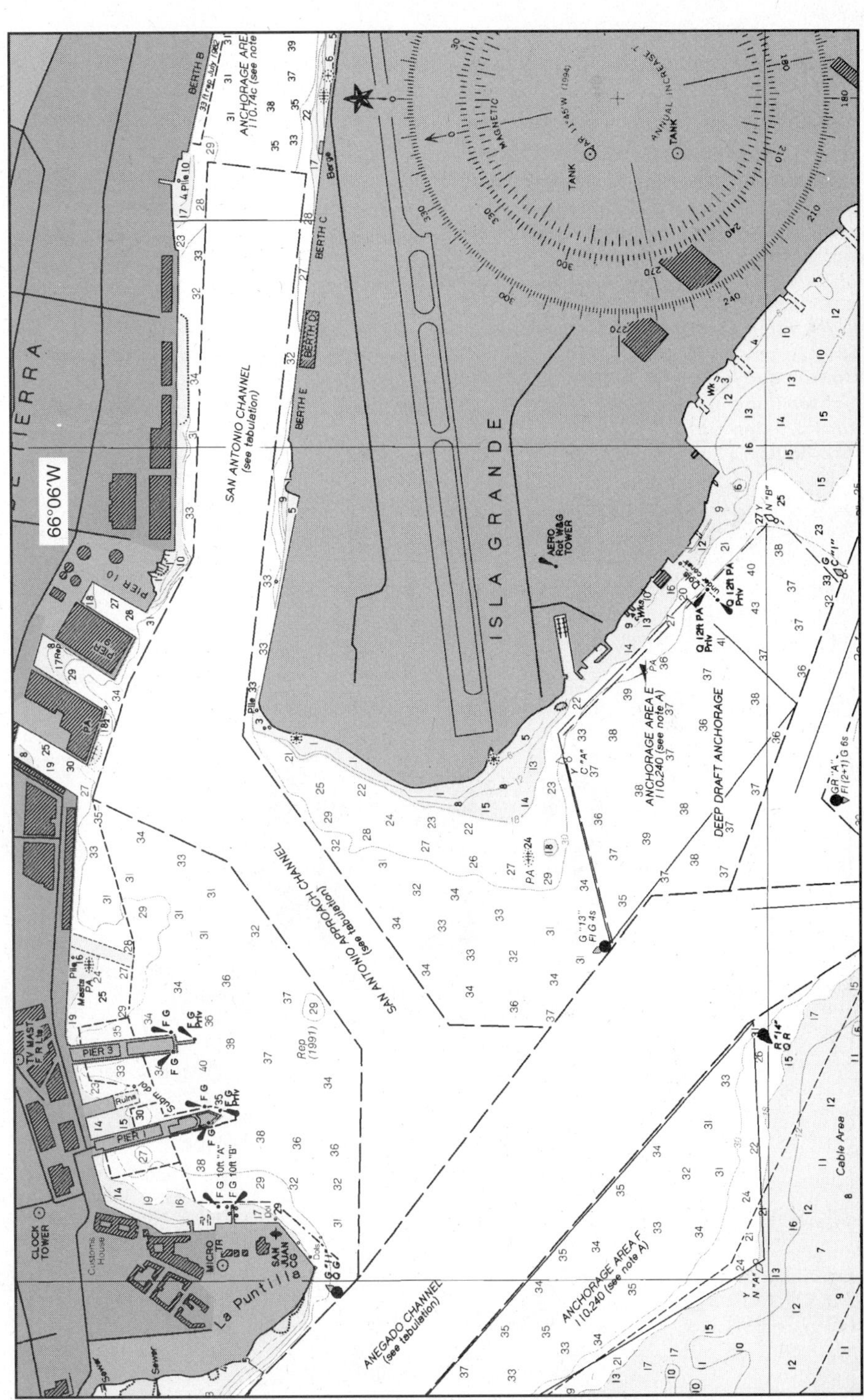

San Juan Harbor, soundings in feet *(from NOAA 25670, 37th ed, 1/94)*

NOTE: A series of 24 more green and red lights on piles mark Martin Pena Channel. All are private aids.

GRAVING DOCK CHANNEL

RANGE (Front Light), 18 26.5N, 66 05.2W. Q R, 41ft. Rectangular red daymark, white stripe, on skeleton tower. Visible all around, higher intensity on rangeline. **(Rear Light),** 535 yards 111.1° from front. Oc R 4s, 83ft. Rectangular red daymark, white stripe, on tower. Visible 2° each side of rangeline.
Buoy 1, green can.
Lighted Buoy 2, Fl R 4s, 3M. Red.
Lighted Buoy 3, Fl G 4s, 4M. Green.
Buoy 5, green can.
Turning Basin Buoy 6, red nun.

SAN JUAN TO PUNTA BORINQUEN

ARECIBO LIGHT, 18 29.0N, 66 41.9W. Fl W 5s, 120ft, 14M. White hexagonal tower attached to square flat-roofed dwelling.
Puerto Arecibo Buoy 1, green can.
Puerto Arecibo Buoy 2, red nun.
ARECIBO AVIATION LIGHT, 3.7km W of city, 18 27.1N, 66 40.5W. Oc R 1.5s, 225ft. Radio tower. Storm warnings.

ARECIBO

18 29N, 66 42W
Pilotage: Arecibo, located about 33 miles west of San Juan and marked by a powerful light, is one of the few places to take shelter on Puerto Rico's north coast. Buoys mark a dredged channel to a 600-foot-bulkhead wharf on the south side of the breakwater, which provides some protection. Depths range from 5 to 20 feet.
Anchorage: Anchor south of the breakwater.

RAMEY AIR FORCE BASE AVIATION LIGHT, 18 30.0N, 67 08.3W. Al W G 10s, 297ft.
PUNTA BORINQUEN LIGHT, 18 29.8N, 67 08.9W, Fl (2) W 15s, 292ft, 24M. Gray cylindrical tower.

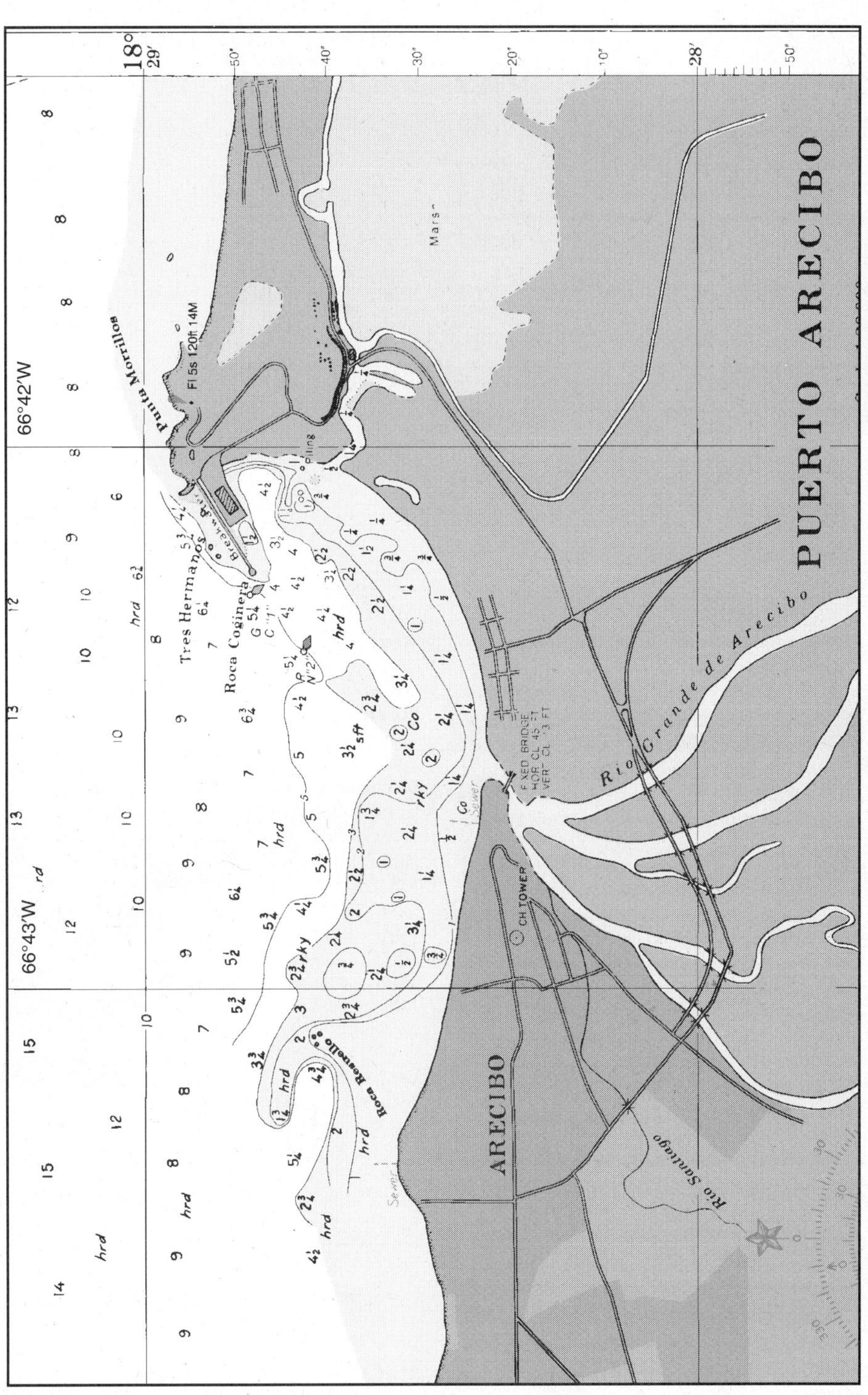

Arecibo, soundings in fathoms *(from NOAA 25668, 14th ed, 6/93)*

U.S. Virgin Islands

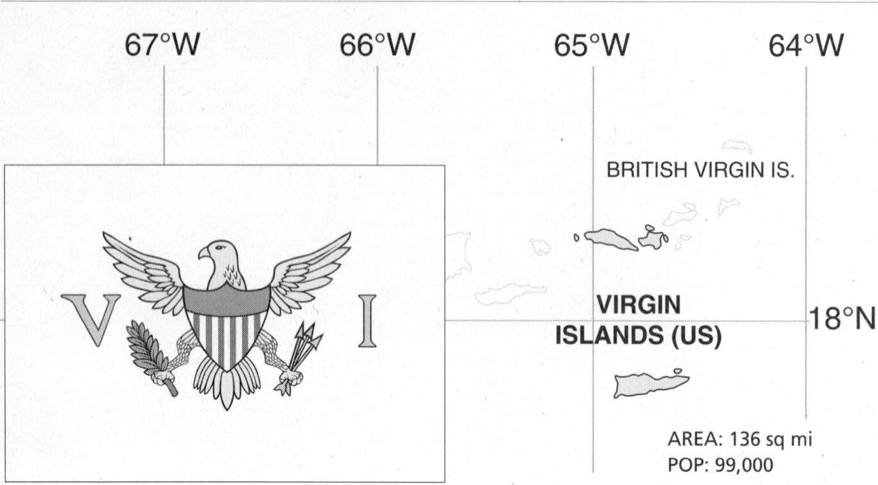

Please see cautions about chartlets and waypoints on page P 2.

ENTRY PROCEDURES

The U.S. Virgin Islands are a territory of the United States, but constitute a separate customs district. The three principal islands are St. Thomas, St. John, and St. Croix. U.S. vessels arriving from a foreign port (including the British Virgin Islands) and all foreign vessels must clear customs with authorities at Charlotte Amalie on St. Thomas, Cruz Bay on St. John, or Christiansted on St. Croix. Note that overtime is charged for Sunday and holiday arrivals. U.S. vessels arriving directly from the mainland U.S. and Puerto Rico do not need to clear in. Most foreign yachts have to obtain a cruising permit. See page P 7 for more information on U.S. customs procedures.

U.S.-registered vessels do not need to obtain a clearance when leaving the U.S. Virgin Islands but may need one when checking in with authorities in another country.

USEFUL INFORMATION

Languages: English; some Spanish, particularly on St. Croix.
Currency: U.S. dollar (US$).
Telephones: A new area code, 340, for the Virgin Islands went into effect on July 1, 1997. Local numbers are seven digits.
Public holidays (1998): New Year's Day, January 1; Three King's Day, January 6; Martin Luther King, Jr.'s Birthday, January 19; Presidents' Day, February 16; Holy Thursday, March 27; Good Friday, April 10; Easter Monday and Transfer Day, April 13; Children's Carnival Parade, April 25; Memorial Day, May 25; Organic Act Day, June 15; VI Emancipation Day, July 3; Independence Day (US), July 4; Supplication Day, July 27; Labor Day, September 7; Puerto Rico Friendship Day (Columbus Day, US), October 12; Local Thanksgiving Day, October 19; D. Hamilton Jackson Day, November 1 (Official holiday November 2); Thanksgiving (US), November 26; Christmas Day, December 25; Boxing Day, December 26.
Tourist offices: *In the U.S. Virgin Islands:* Department of Tourism, P.O. Box 6400, Charlotte Amalie, St. Thomas, VI 00804 (340-774-8784). Offices also in Cruz Bay, Christiansted, Frederiksted. *In the U.S.:* Department of Tourism, 1270 Avenue of the Americas, Suite 2108, New York, NY 10020 (800-372-8784, 212-332-2222). Offices also in Atlanta, Chicago, Los Angeles, Miami (Coral Gables), Washington DC, San Juan PR. *In Canada:* Department of Tourism, 3300 Bloor Street West, The Mutual Group Centre, Suite 3120–Centre Tower; Toronto, Ont. M8X 2X3 (800-465-8784, 416-233-1414).

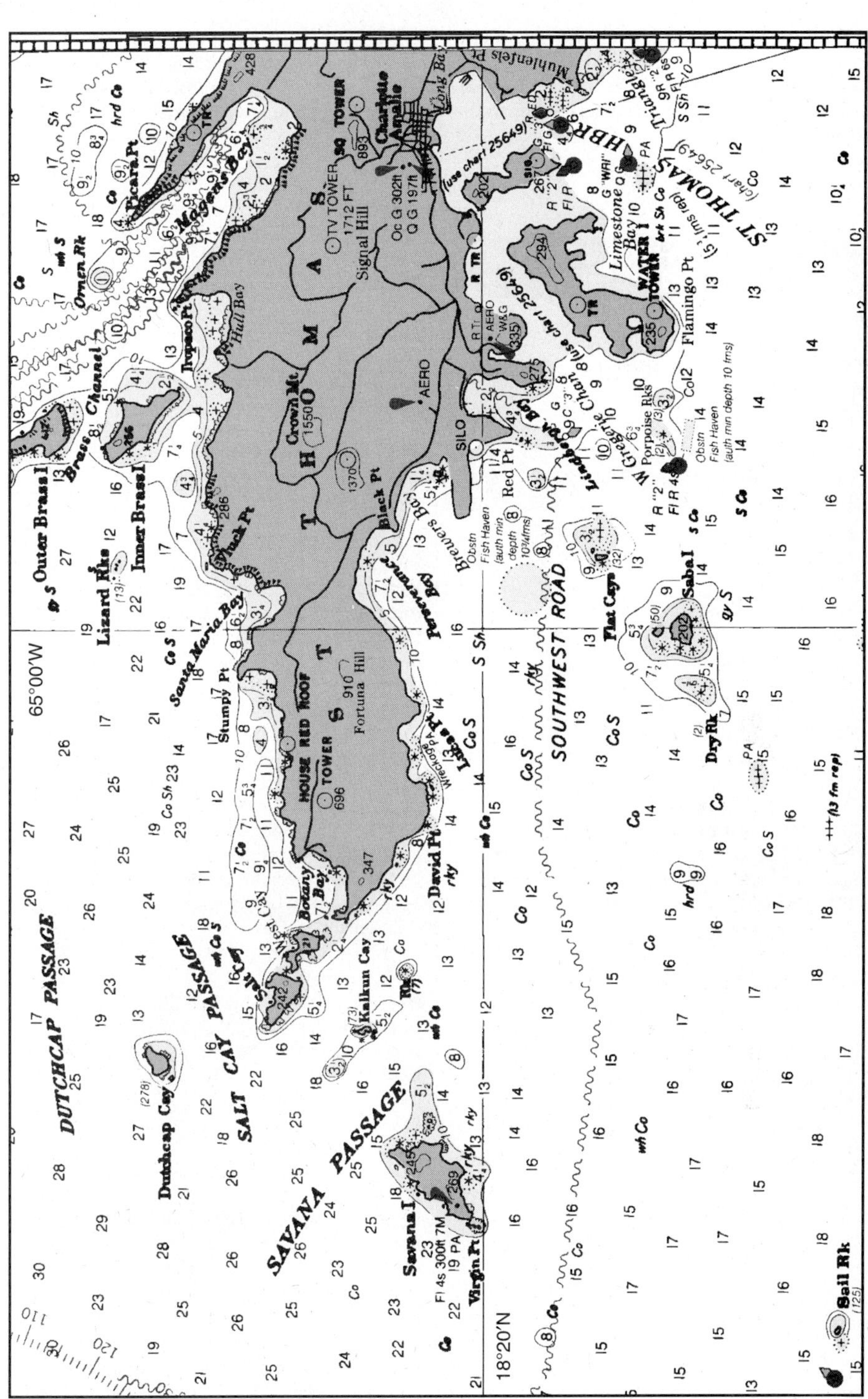

Savana Passage, St. Thomas, soundings in fathoms *(from NOAA 25650, 30th ed, 7/93)*

P

Pilot

U.S. VIRGIN ISLANDS

Emergencies, U.S. Coast Guard: Call on VHF channel 16 or 2182 kHz. Virgin Islands Radio (VHF channel 16; 2182 kHz; tel., 340-776-8282) can also help communicate emergencies to the proper authorities.
VHF: Working channels throughout the islands are VHF 25, 85, and 87

ST. THOMAS APPROACHES

Sail Rock Lighted Buoy 1, 18 17.0N, 65 06.5W. Fl G 6s, 4M. Green.
SAVANA ISLAND LIGHT, 18 20.4N, 65 05.0W. Fl W 4s, 300ft, 7M. White tower.

ST. THOMAS HARBOR AND APPROACHES

WEST GREGERIE CHANNEL

Lighted Buoy 2, W side, 18 18.4N, 65 33.5W. Fl R 4s, 3M. Red.
Buoy 3, green can.
Lighted Buoy 4, Fl R 2.5s, 3M. Red.
Lighted Buoy 5, Fl G 4s, 4M. Green.
LIGHT 6, 18 19.9N, 64 56.9W. Fl R 4s, 16ft, 4M. Red triangular daymark on dolphin.
CROWN BAY MOORING CELL LIGHT, 18 19.9N, 64 57.1W. Q W. On dolphin. Private aid.

KRUM BAY

SOUTH DOLPHIN OBSTRUCTION LIGHT, 18 19.6N, 64 57.7W. Q W. On dolphin. Sychronized with Pier Obstruction Light. Private aid.
PIER OBSTRUCTION LIGHT, 18 19.6N, 64 57.7W. Q W. On end of pier. Synchronized with North Mooring Dolphin Obstruction Light. Private aid.
NORTH MOORING DOLPHIN OBSTRUCTION LIGHT, 18 19.6N, 64 57.7W. Q W. On dolphin. Synchronized with Pipeway Obstruction Light. Private aid.
PIPEWAY OBSTRUCTION LIGHT, 18 19.6N, 64 57.7W. Q W. On end of pipeway. Synchronized with Pier Obstruction Light. Private aid.

EAST GREGERIE CHANNEL

Lighted Buoy WR1, 18 18.6N, 64 56.1W. Q G, 3M. Green.
Lighted Buoy 2, westerly side of shoal. Fl R 2.5s, 3M. Red.
Lighted Buoy 3, Fl G 4s, 4M. Green.

ST. THOMAS HARBOR

Entrance Lighted Buoy 2, S of rocks. Fl R 6s, 4M. Red.
BERG HILL RANGE (Front Light), 18 20.6N, 64 56.0W. Q G, 197ft. Rectangular red daymark, white stripe, on skeleton tower. Visible 2° each side of rangeline. **(Rear Light),** 125 yards, 344° from front. Oc G 4s, 302ft. Rectangular red daymark, white stripe, on skeleton tower. Visible 2° each side of rangeline.
Lighted Buoy 3, on shoal about center of entrance to harbor. Fl G 4s, 4M. Green.
Buoy 4, W of bank. Red nun.
Rupert Rock Daybeacon, white diamond-shaped daymark on pile, worded DANGER.
Lighted Buoy 6, off E side of entrance to harbor. Fl R 4s, 3M. Red.
WEST INDIAN DOCK LIGHT, 18 19.8N, 64 55.6W. F R, 12ft. On dock. Private aid.

CHARLOTTE AMALIE, ST. THOMAS

18 20N, 64 56W
Emergencies: Contact the U.S. Coast Guard on VHF channel 16 or 2182 kHz. Virgin Islands Radio can assist in emergencies.
Pilotage: Approaching from the north, St. Thomas's prominent, 1,500-plus-foot-high mountains make the island highly visible. At night, its loom is visible for many miles. The western approach to the main harbor of Charlotte Amalie, on the island's south side, passes through Savana Passage, marked by a light on Savana Island. At night, it is safer to pass west of the light to avoid unlit rocks in the passage. Several other unlit rocks clutter the approach from the west, but a safe bearing can be taken on the lights of the airport. Boats often anchor in West Gregerie Channel, while the channel east of Hassel Island is kept clear for the many cruise ships calling at the West Indian Docks. The entrance range of 344° will bring you safely into the main harbor.
Port of entry: See Entry Procedures, previous page, and the introduction to the Coast Pilot, page P 7. Vessels may tie up in a marina and call customs at 340-774-6755. The customs office is on the downtown waterfront in the big pink Edwin Wilmot Blyden Ferry Terminal building. Foreign vessels should fly the Q flag. Officials are available for private vessel clearances from 0800 to 1700, Monday through Friday. Officials are available at all other times (after hours, Saturday, Sunday, and holidays), but an overtime fee will be applied. Vessels arriving at night should report the next morning.
Virgin Islands Radio (WAH): This private station provides an important communications link in the islands. It broadcasts weather reports, relays telephone calls, and provides emergency

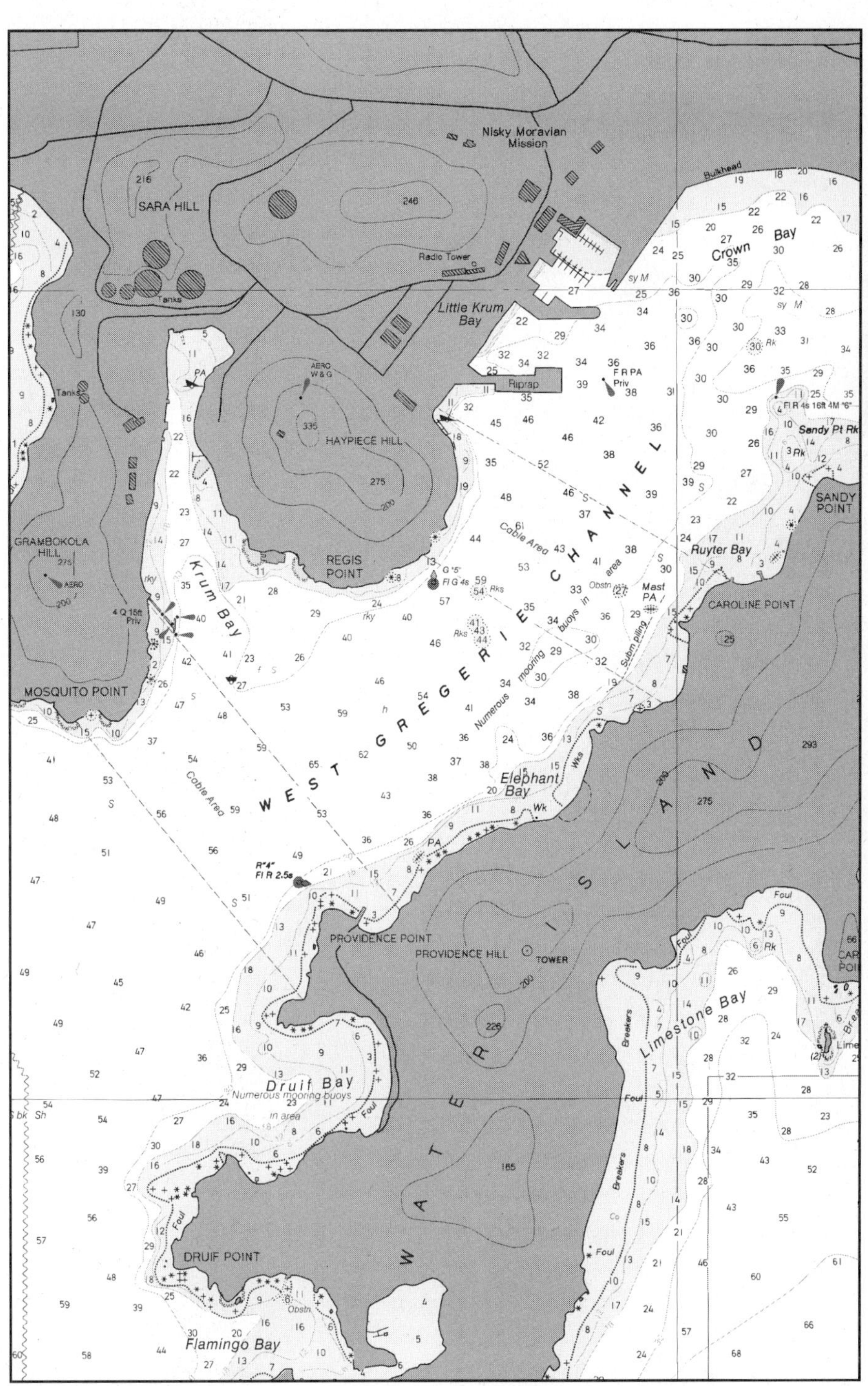

Charlotte Amalie (western part), soundings in feet *(from NOAA 25649, 16th ed, 5/93)*

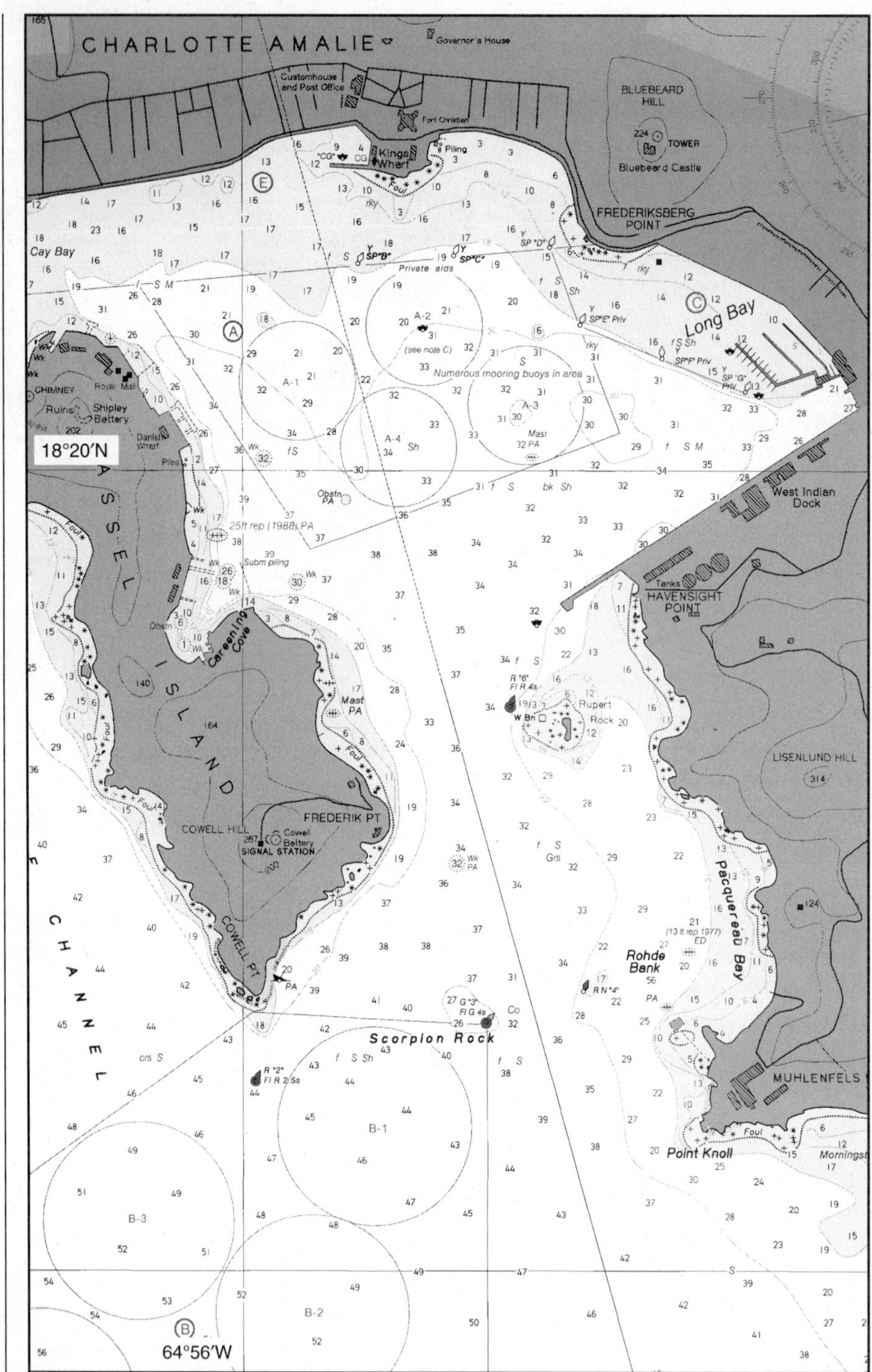

Charlotte Amalie (eastern part), soundings in feet *(from NOAA 25649, 16th ed, 5/93)*

assistance. The station can be contacted by radio (standby channels, VHF channel 16 and 2182 kHz) or by phone (340-776-8282/774-0444; toll-free at 888-732-8255; fax, 340-774-0996). For more information refer to the Communications section of Chapter R.

Dockage: Available on the west side of Crown Bay at Crown Bay Marina (tel., 340-774-2255; fax, 340-776-2760) and in the main harbor, near the cruise ship docks, at Yacht Haven Marina (340-774-6050). The marinas may also be contacted via VHF channel 16 or Virgin Islands Radio.

Moorings: The numerous moorings scattered around the harbor are privately owned.

Anchorages: There is usually room to anchor among the moored boats in Long Bay off Yacht Haven Marina, but take care to stay clear of the cruise ship docks and approaches to the marina and turning basin. The holding is good, but the harbor can be rough from wakes and surge from the south. Boats anchor all along West Gregerie Channel, so many at times they practically block passage. Other anchoring spots include Hassel Island and the cove on the west side of Water Island known locally as Honeymoon Harbor. A crowded dinghy dock is available at Yacht Haven. Locking your tender is recommended.

Repairs: Haulover Marine (VHF channel 16; tel., 340-776-2078; fax, 779-8426), located in West Gregerie Channel, is a full-service repair facility.

Services: All services are available. Large supermarkets, banks, a post office, hardware stores, marine stores, drug stores, and liquor stores are all within easy walking distance of the harbor front. A major airport has good connections to all points. If the part you need is not here, U.S. mail or commercial delivery services can get it to you quickly.

Charlotte Amalie is the capital of commerce and largest city in the Virgins. It is the place to take care of major repairs, stock up, and conduct business. The harbors are crowded with cruisers and charterboats from around the world, bringing fascinating variety to the waterfront. Ashore are all the pleasures of a major tourist destination, including historical sights and good restaurants. Charlotte Amalie is the center of action, but visitors should also exercise common sense during evening hours. Crime remains a problem.

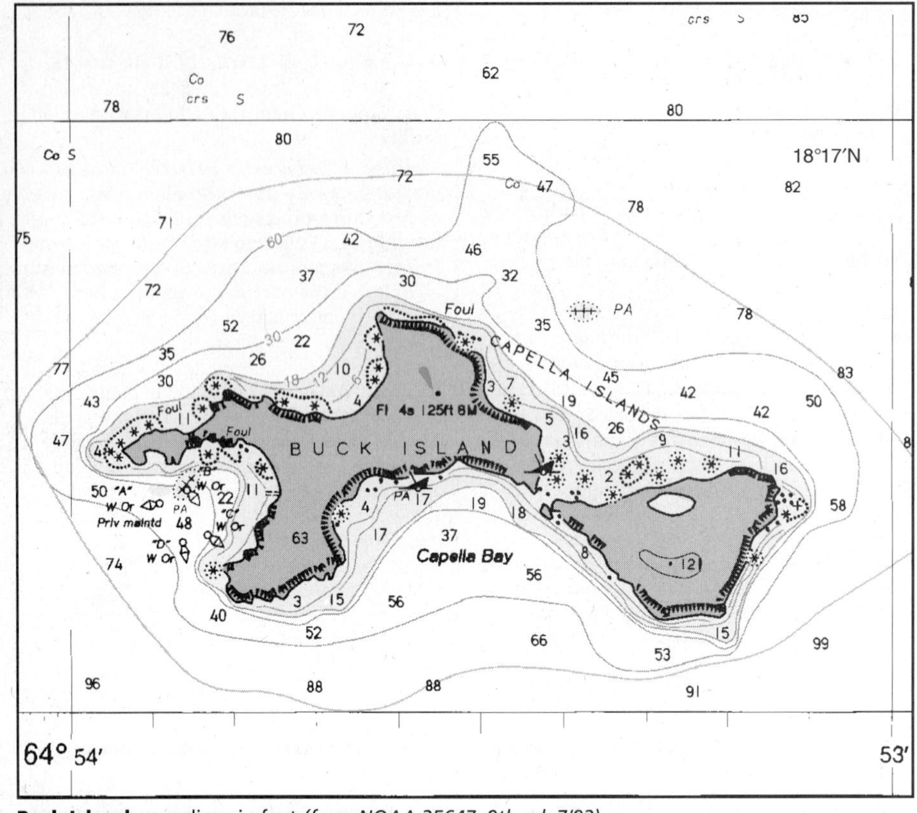

Buck Island, soundings in feet *(from NOAA 25647, 9th ed, 7/92)*

SOUTHEASTERN ST. THOMAS

BUCK ISLAND LIGHT, summit, 18 16.7N, 64 53.6W. Fl W 4s, 125ft, 8M. White square tower.

Packet Rock Buoy 2, on S side of rock, 18 17.8N, 64 53.4W. Red nun.

Red Point Buoy 1, 18 18.4N, 64 51.6W. Green can.

BENNER BAY CHANNEL

Lighted Buoy 1, 18 19.0N, 64 52.1W. Fl G 2s. Green. Private aid.

Lighted Buoy 2, Fl R 2s. Red. Private aid.

Buoy 3, green can. Private aid.

Buoy 4, red nun. Private aid.

Buoy 5, green can. Private aid.

Buoy 6, red nun. Private aid.

Buoy 7, green can. Private aid.

Buoy 8, red nun. Private aid.

Buoy 9, green can. Private aid.

Buoy 10, red nun. Private aid.

Buoy 12, red nun. Private aid.

Buoy 13, green can. Private aid.

Buoy 14, red nun. Private aid.

Buoy 15, green can. Private aid.

Buoy 16, red nun. Private aid.

Buoy 18, red nun. Private aid.

BENNER BAY, ST. THOMAS

18 19N, 64 52W

Pilotage: Do not pass between Cas Cay and Patricia Cay. The green can off the east end of Cas Cay is the main entrance mark. The channel into Benner Bay (known locally as the Lagoon) is privately marked by red and green buoys. The channel is reported to be shoal in places, with possible broken-off pilings outside the deeper water. You may want to call ahead for the latest depth information if you draw more than 6 feet.

Dockage: This area features several marinas and repair yards. Contact Compass Point Marina (340-775-6144; fax 340-779-2457), Tropical Marina/Ruan's (primarily powerboats, 6-foot draft maximum, 340-775-6595), or LaVita Marine Center/Independent Boat Yard (340-776-0466; fax 340-775-6576. They do not monitor VHF.) Compass Point Marina has 85 slips (although few are designated for transients). LaVita Marina Center has 80 slips.
Anchorage: There may be room to anchor along the channel leading into the bay. The water shoals rapidly outside the dredged area, and broken-off pilings have been reported.
Repairs: For repairs, contact Independent Boat Yard (340-766-0466).
Services: All services are available here or nearby on the island.

More sheltered than Charlotte Amalie or Red Hook, Benner Bay is a good place for major repairs or boat storage.

CURRENT CUT TO RED HOOK

CURRENT ROCK LIGHT, 18 18.9N, 64 50.1W. Fl W 6s, 20ft, 6M. Diamond-shaped daymark on skeleton tower. Higher intensity beams toward Buck Island and Two Brothers.
Cabrita Point Buoy 1, green can.
Redhook Bay Buoy 1, 18 19.7N, 64 50.6W. Green can. Private aid.
Redhook Bay Buoy 2, red nun. Private aid.
Redhook Bay Buoy 3, green can. Private aid.
Redhook Bay Buoy 4, red nun. Private aid.
Redhook Bay Junction Buoy P, 18 19.5N, 64 51.0W. Green and red bands, can. Private aid.
Redhook Bay South Channel Buoy 1, 18 19.5N, 64 51.0W. Green can. Private aid.
Redhook Bay South Channel Buoy 2, red nun. Private aid.
Redhook Bay Buoy 5, green can. Private aid.
Redhook Bay Buoy 6, red nun. Private aid.

RED HOOK, ST. THOMAS

18 20N, 64 51W
Dockage: Contact American Yacht Harbor (340-775-6454; fax 340-776-5970) in Red Hook Bay (106 slips, all rebuilt since 1995), or Sapphire Beach Resort and Marina (340-775-6100) in the small bay just north of Red Hook Point . Vessup Point Marina (340-779-2495), opposite American Yacht Harbor, has rebuilt one of its two docks that were destroyed in the fall 1995 hurricane. They now have space for 20 boats. All three establishments monitor VHF channel 16.
Anchorage: Many private moorings limit the available anchoring area, but some boats find room in Muller Bay, known locally as Vessup Bay. Boat traffic is heavy, and the ferries running to St. John kick up wakes.
Services: Most services are available here, except for haulouts. Supplies can be found near the docks or elsewhere on the island.

PILLSBURY SOUND AND ST. JOHN APPROACHES

Thatch Cay Fish Haven East Buoy, 18 21.1N, 64 51.1W. Orange and white banded can. Private aid.
Thatch Cay Fish Haven West Buoy, 18 21.1N, 64 51.5W. Orange and white banded can. Private aid.
TWO BROTHERS LIGHT, 18 20.6N, 64 49.0W. Fl W 6s, 23ft, 5M. Diamond-shaped daymark on skeleton tower.
STEVENS CAY LIGHT, 18 19.9N, 64 48.5W. Fl W 4s, 14ft, 5M. Diamond-shaped daymark on skeleton tower.
Mingo Rock Lighted Buoy 2, 18 19.4N, 64 48.2W. Fl R 4s, 4M. Red.

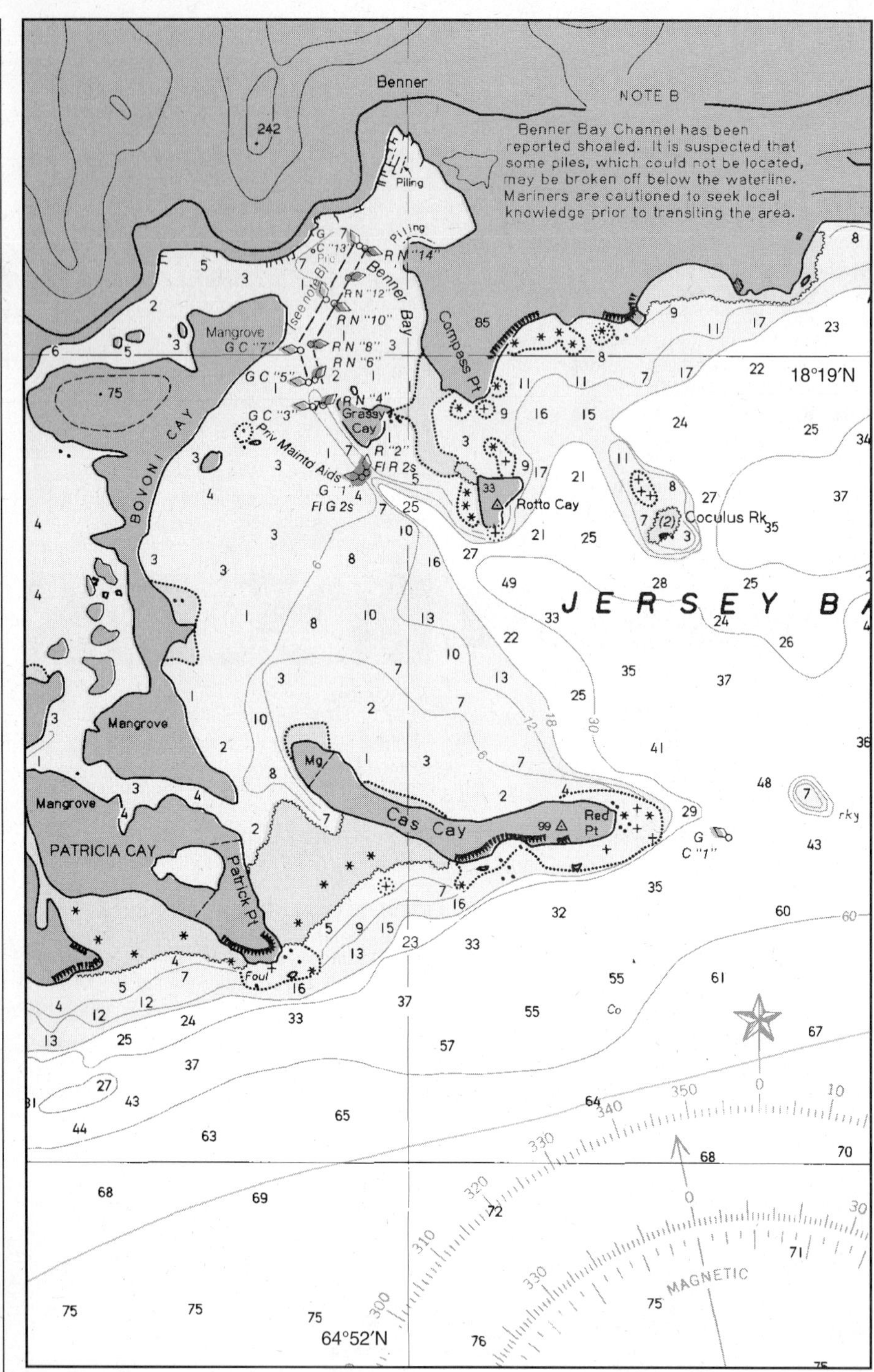

Benner Bay, soundings in feet *(from NOAA 25647, 9th ed, 7/92)*

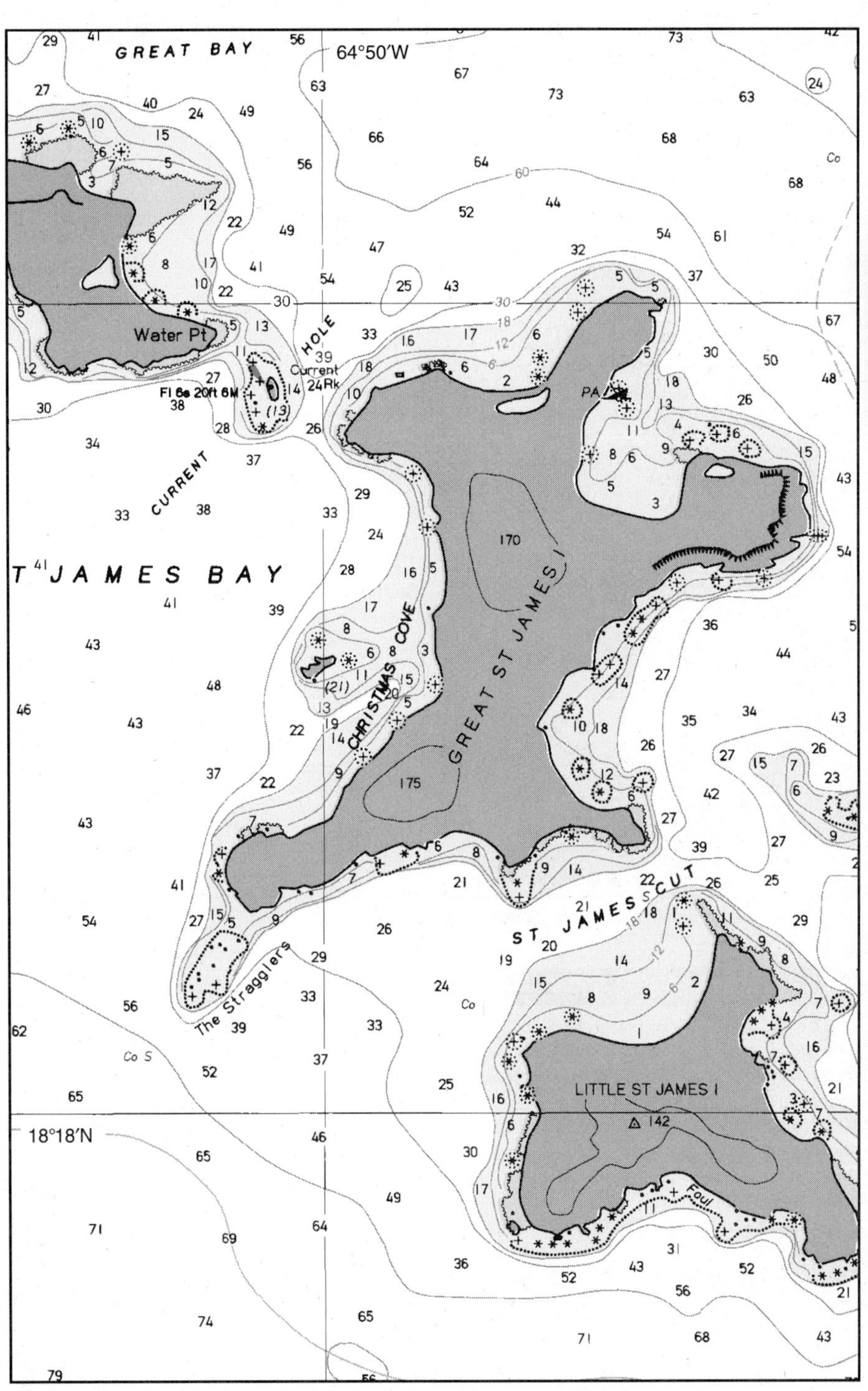

Current Cut and St. James Islands, soundings in feet *(from NOAA 25647, 9th ed, 7/92)*

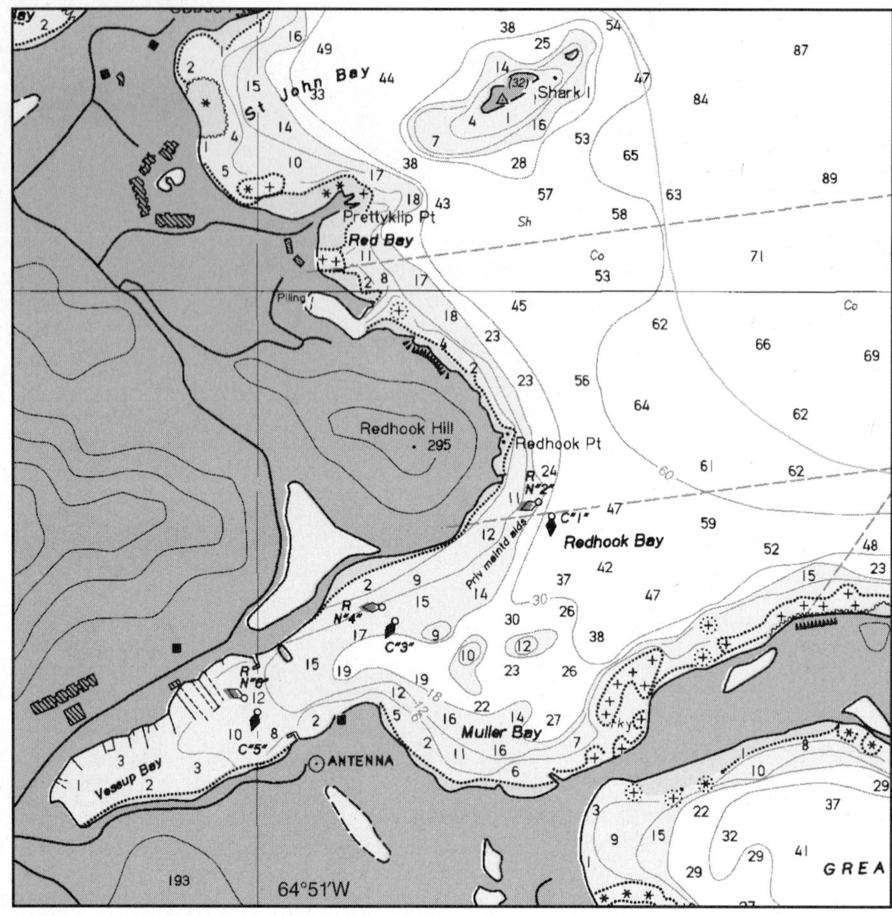

Red Hook, soundings in feet *(from NOAA 25647, 9th ed, 7/92)*

CRUZ BAY

CRUZ BAY LIGHT, 18 20.0N, 64 47.9W. Fl W 4s, 12ft, 5M. Diamond-shaped daymark on pile.
Buoy 1, green can. Private aid.
Buoy 2, red nun. Private aid.
Junction Buoy, 18 20.0N, 64 47.8W. Green and red bands. Private aid.
Buoy 3, green can. Private aid.
Buoy 4, red nun. Private aid.
Cruz Creek Buoy 1C, 18 20.0N, 64 47.7W. Green can. Private aid.
Cruz Creek Buoy 2C, red nun. Private aid.

CRUZ BAY, ST. JOHN

18 20N, 64 48W

Port of entry: Dinghy in to the customs dock (by the white building with latticework) in the northern arm of the harbor. Vessels arriving from the British Virgin Islands must check in with customs (340-776-6741). Hours are 0700 to 1800, seven days a week.

National Park Service: Much of St. John is part of the Virgin Islands National Park. A park service dock and visitors center are located in the northern arm of Cruz Bay, across the harbor from the customs office. Contact the park service office (340-776-6201) for the park's latest "Mooring and Anchoring Guide." The current limit on overnight stays within park waters is fourteen days a year. For boats to 55 feet overall, park moorings are located off Lind Point, in Little Lameshur Bay, Great Lameshur Bay, Leinster Bay, Hawksnest Bay, Reef Bay, Salt Pond Bay, Ram's Head Cove, Maho Bay, north Francis Bay just off Mary Point, off the southwest corner

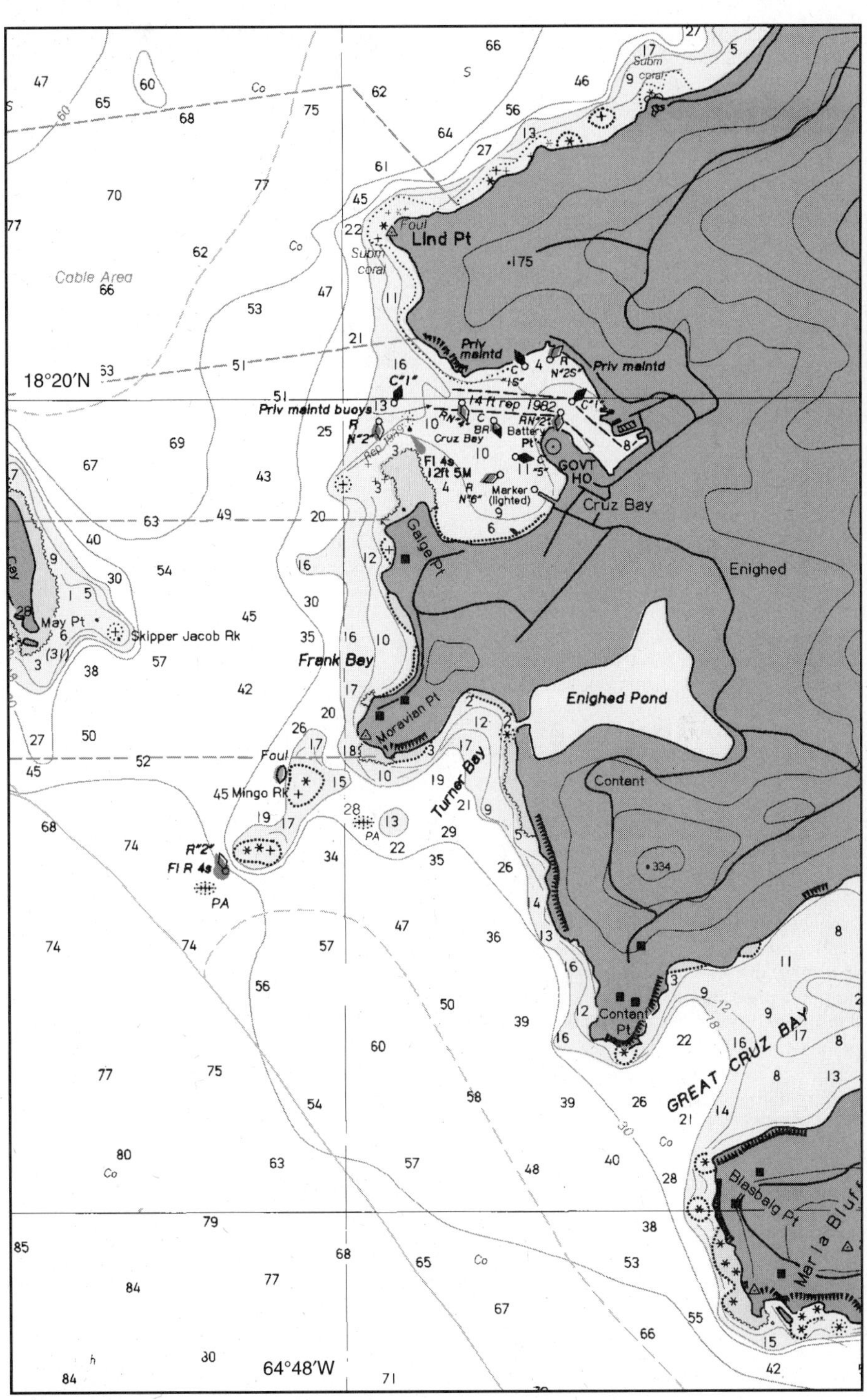

Cruz Bay, soundings in feet *(from NOAA 25647, 9th ed, 7/92)*

of Whistling Cay, and in Jumbie Bay. The 36 moorings (white with a blue stripe) are available free of charge on a first-come, first-served basis. Use of the moorings is limited to no more than 14 nights per year. There is one orange mooring that is limited to daytime use only. Visitors in park waters are required to use the dinghy channels, which are marked by red and green sausage-shaped buoys, when coming ashore and to remain outside the swimming areas, which are marked by white buoys. To protect critical habitat areas, anchoring is now prohibited in Reef, Little Lameshur, Great Lameshur, and Salt Pond bays. Boaters are requested to anchor in sand only—never in seagrass or coral. Personal watercraft and waterskiiing are prohibited in park waters, as is spearfishing.

Dockage: You can tie up for 15 minutes to the National Park Service dock at the finger pier in the cove off the northern channel in the harbor.

Anchorage: The anchorage areas in Cruz Bay are very small, crowded, and swept by wakes from the frequent ferries. Boats anchor both north and south of the main ferry channel in the southern arm of the harbor. Stay clear of the marked channels and the reef extending north from Galge Point. Temporary anchorage is available south of Lind Point. There may not be enough room to anchor safely in any of these spots. Dinghies may be landed north of the ferry wharf in the harbor's southern arm.

Services: Caneel Bay Shipyard (340-693-8771) has fuel, water, gas, and ice. They no longer do repairs of any kind , nor do they sell hardware. Limited groceries and general stores are available from several small shops in town. Connections (tel., 340-776-6922; fax, 340-776-6902) specializes in providing mariners with a wide array of communication services, including phone, fax, mail, Western Union, and reservation services. There are many small specialty shops, boutiques, and informal eateries.

Many people prefer the calmer pace of St. John to the big-city rush of Charlotte Amalie. Unfortunately, Cruz Bay is tiny, crowded, and much stirred up by ferries. With much of St. John being national park, the island's natural beauty is spectacular. It is well worth climbing one of the 1,000-foot-high peaks for a glimpse of the Virgin Islands spread before you.

WINDWARD PASSAGE

Johnson Reef Lighted Buoy 1JR, 18 21.9N, 64 46.4W. Fl G 4s, 4M. Green.

Johnson Reef South Buoy, 18 21.5N, 64 46.4W. Orange and white can worded DANGER SHOAL.

Lighted Buoy B, 18 20.1N, 64 48.2W, Fl Y 4s. Yellow.Private aid.

Lighted Buoy D, 18 22.2N, 64 46.6W, Fl Y 4s. Yellow. Private aid.

NORTHERN BAYS OF ST. JOHN

CANEEL BAY

Buoy 1, 18 20.6N, 64 47.3W. Black can. Private aid.

Buoy 2, red nun. Private aid.

Buoy 3, black can. Private aid.

Buoy 4, red nun. Private aid.

Buoy 5, black can. Private aid.

Buoy 6, red nun. Private aid.

HAWKSNEST BAY

Buoy 1, 18 20.8N, 64 46.7W. Black can. Private aid.

Buoy 2, red nun. Private aid.

TRUNK BAY

Buoy 1, 18 21.1N, 64 46.2W. Black can. Private aid.

Buoy 2, red nun. Private aid.

FRANCIS BAY

Buoy 1, 18 21.9N, 64 44.7W. Black can. Private aid.

Buoy 2, red nun. Private aid.

FRANCIS BAY, ST. JOHN

18 22N, 64 45W

National Park Service: Refer to the section on Cruz Bay for information on park services; NPS boats patrol park waters regularly.

Anchorage: This is an excellent anchorage area, with wonderful swimming, snorkeling, and beaching. Several private buoys mark an area where dinghies may proceed in to the beach. From here you can hike the road to the Annaberg Sugar Mill ruins near Leinster Bay.

Services: You may be able to purchase a few basic supplies or make a phone call at the resort in nearby Maho Bay.

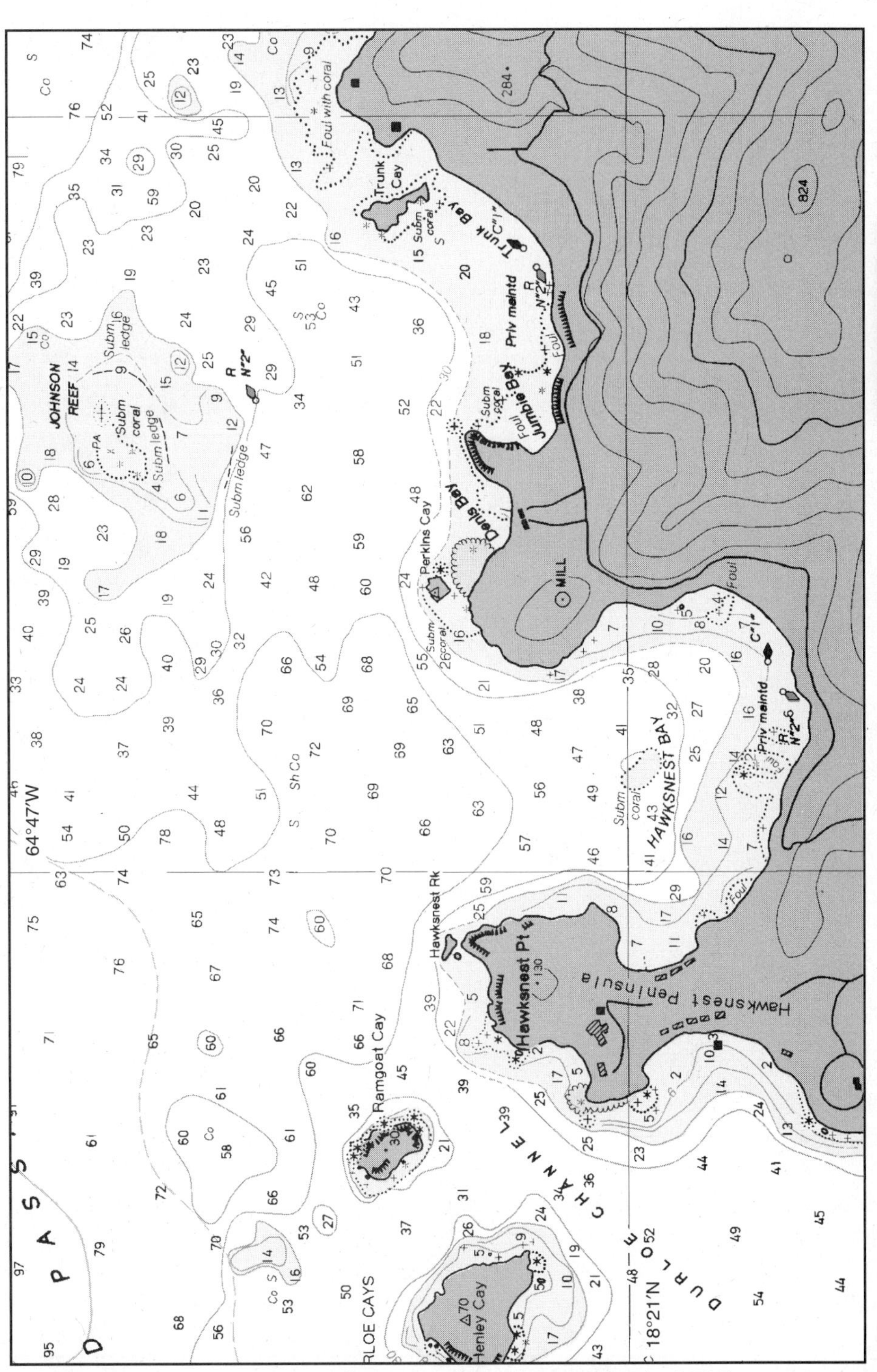

Hawksnest and Trunk Bays, soundings in feet *(from NOAA 25647, 9th ed, 7/92)*

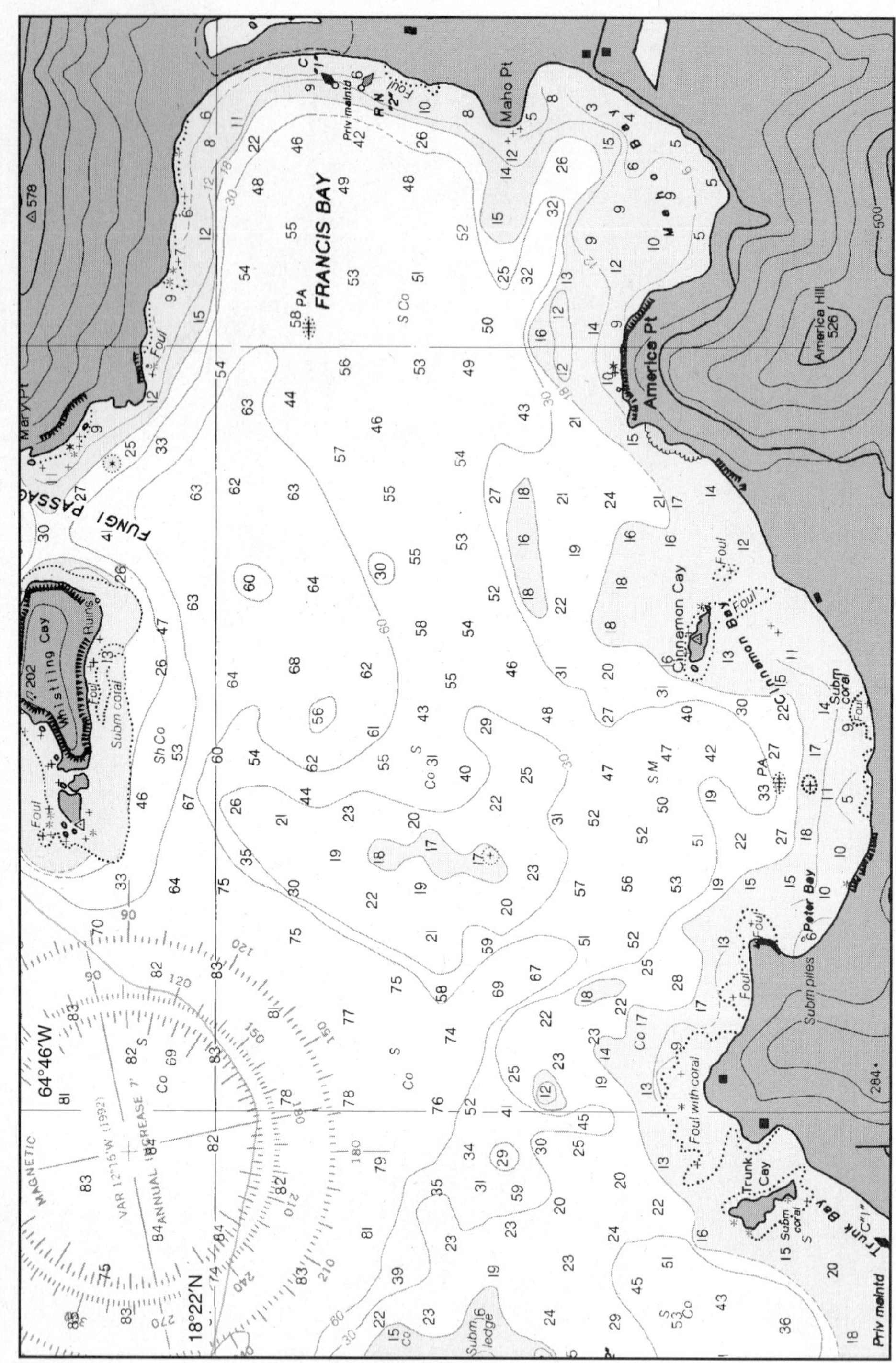

Francis Bay, soundings in feet *(from NOAA 25647, 9th ed, 7/92)*

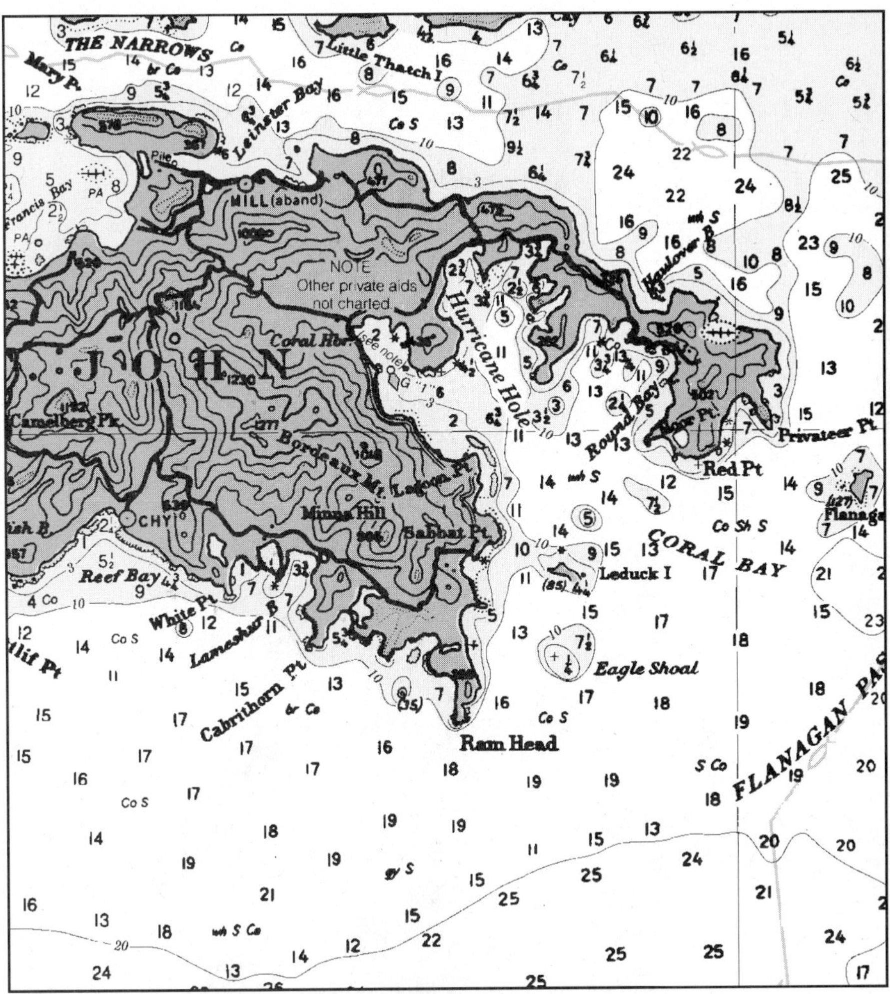

Coral Bay, soundings in fathoms *(from NOAA 25641)*

CORAL HARBOR CHANNEL

Buoy 1, 18 20.6N, 64 42.7W. Green can. Private aid.
Buoy 2, red nun. Private aid.
Buoy 3, green can. Private aid.
Buoy 4, red nun. Private aid.
Buoy 5, green can. Private aid.
Buoy 6, red nun. Private aid.

CORAL BAY AND HURRICANE HOLE, ST. JOHN

18 20N, 64 42W

Anchorage: There are many excellent possibilities for anchoring in any of the arms of the harbor. The area to the north is known as Hurricane Hole, because of the excellent protection afforded by the mangroves. The small boat wharf at the head of the bay is reported to have 3 feet of depth alongside.

Services: Coral Bay Marine (VHF channel 16; tel., 340-776-6859/6665) can offer some repairs, although there is no haulout facility. Coral Bay Marine sells ice and water (no fuel) and some basic marine supplies. Connections East (340-779-4994; fax, 340-776-6136) offers the same services here as the main Connections office in Cruz Bay. There are a few small shops, a grocery, and casual eateries, plus buses and taxis that run to Cruz Bay.

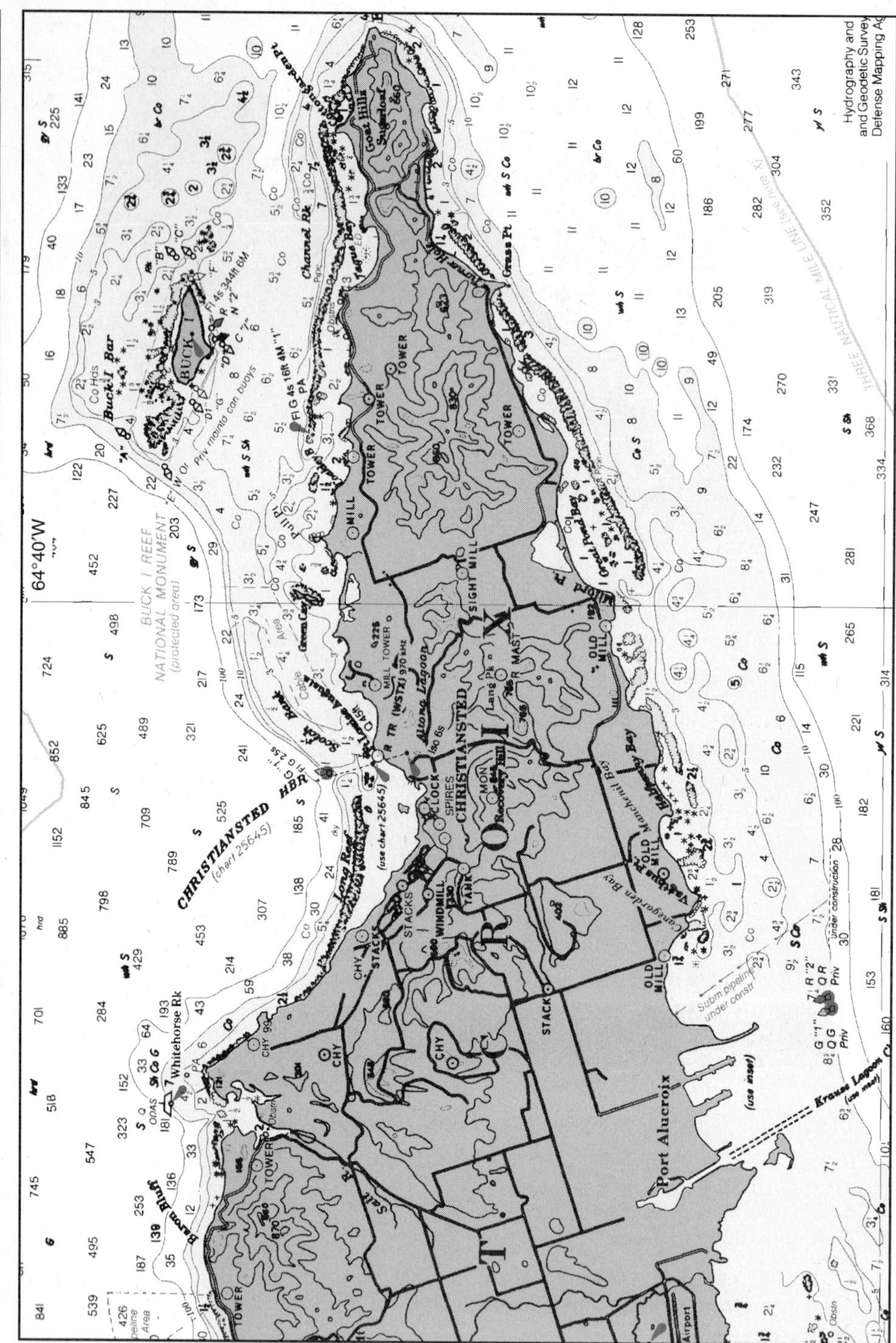

Christiansted approaches, soundings in fathoms *(from NOAA 25641, 22nd ed, 8/93)*

ST. CROIX, NORTH COAST

HAMS BLUFF LIGHT, NW point of island, 17 46.3N, 64 52.3W. Fl (2) W 30s, 394ft, 16M. White cylindrical tower. Visible 053°–265°, partially obscured 053°–062°.

St. Croix FAD Lighted Buoy A, 17 46.4N, 64 53.3W, Fl Y 4s. Yellow sphere. Private aid.

St. Croix FAD Lighted Buoy B, 17 48.5N, 64 35.4W, Fl Y 4s. Yellow sphere. Private aid.

National Undersea Research Habitat Lighted Buoy, off entrance to Salt River Bay, 17 47.5N, 64 45.3W. Q W. Large white hull lettered NOAA. Maintained by NOAA.

CHRISTIANSTED APPROACHES

Lighted Buoy 1, 17 45.8N, 64 41.8W. Fl G 2.5s, 4M. Green.

ENTRANCE CHANNEL RANGE, Fort Louisa Agusta **(Front Light),** 17 45.4N, 64 41.7W. Q W, 45ft. Red rectangular daymark on skeleton tower. Visible all around, higher intensity on rangeline. **(Rear Light),** 735 yards, 164° from front. Iso W, 6s, 93ft. Red rectangular daymark on skeleton tower. Visible all around, higher intensity on rangeline.

Lighted Buoy 3, at W point of bank. Fl G 4s. Green.

Buoy 4, at NE point of shoal. Red nun.

Light Buoy 6, Q R, 3M. Red.

LIGHT 7, Fl G 4s, 16ft, 4M. Green square daymark on dolphin.

Buoy 8, red nun.

Round Reef Northeast Junction Lighted Buoy RR, at NE point of reef, 17 45.4N, 64 41.8W. Fl (2+1) G 6s, 3M. Green and red bands, can.

LIGHT 9, Fl G 2.5s, 16ft, 5M. Green square daymark on dolphin.

LIGHT 10, Fl R 2.5s, 17ft, 3M. Red triangular daymark on pile.

LIGHT 11, Q G, 10ft, 2M. Green square daymark on dolphin.

Lighted Buoy 12, Fl R 4s, 3M. Red nun.

LIGHT 13, Fl G 4s, 16ft, 4M. Green square daymark on pile.

Lighted Buoy 14, Fl R 6s, 4M. Red.

Buoy 15, green can.

Daybeacon 16, red triangular daymark on pile.

SCHOONER CHANNEL

NOTE: This privately marked channel runs from east of Round Reef to the main harbor area.

Lighted Buoy 5, 17 45.3N, 64 41.8W. Q G, Green. Private aid.

Lighted Buoy 6, Q R. Red. Private aid.

Round Reef Southwest Daybeacon 2, 17 45.3N, 64 41.8W. Triangular red daymark on dolphin.

Lighted Buoy 7, Q G. Green. Private aid.

Lighted Buoy 8, Q R. Red. Private aid.

Lighted Buoy 9, Fl G 4s. Green. Private aid.

Lighted Buoy 10, Fl R 4s. Red. Private aid.

Lighted Buoy 11, Fl G 4s. Green. Private aid.

CHRISTIANSTED HARBOR WHARF LIGHTS (2), F W, 12ft. Private aids.

CHRISTIANSTED, ST. CROIX

17 45N, 64 42W

Pilotage: St. Croix is located about 37 miles south of St. Thomas. Because of extensive off-lying reefs, vessels should stay a mile or two offshore until sure of their position. When crossing from St. Thomas, allow for a westerly set to the current of about 0.5 knot—it is often prudent to lay a course for the east end of the island to avoid a final slog to windward. The initial entrance range is 164° and gets you through the reef. The channel then zigzags to the west and south to the waterfront.

Port of entry: The customs office is located near St. Croix Marine. During office hours (0800 to 1700, Monday through Friday), marina customers can request marina assistance in alerting customs (340-773-1011) to their arrival. On Saturdays, contact customs at 340-778-2257 or 773-1011; on Sundays, at 340-778-0216).

Dockage: Northeast of the main wharf, St. Croix Marine (340-773-0289) is a full-service yard with three new docks (built since Hurricane Luis) for 40 boats, fuel, water, electricity, laundry, restaurant, 300-ton railway, and 60-ton lift. About 3 miles east of Christiansted, 160 slips, electricity, water, ice, fuel, and hotel rooms are available at Green Cay Marina (340-773-1453). Enter via the channel west of Green Cay. In Salt River Bay about 4 miles west of Christiansted, Salt River Marina (340-778-9650) has two to three transient slips and one or two moorings (ask the harbormaster for mooring availability). All three marinas moniter VHF channel 16.

Anchorage: In Christiansted anchor southwest of Protestant Cay, avoiding the charted cable area, or off St. Croix Marine, which has a dinghy dock. Salt River Bay is a hurricane hole, with about 6 feet of water in the S-shaped reef entrance and good holding in mud and sand. Anchor wherever there is space. Call ahead to Salt River Marina (VHF channel 16)

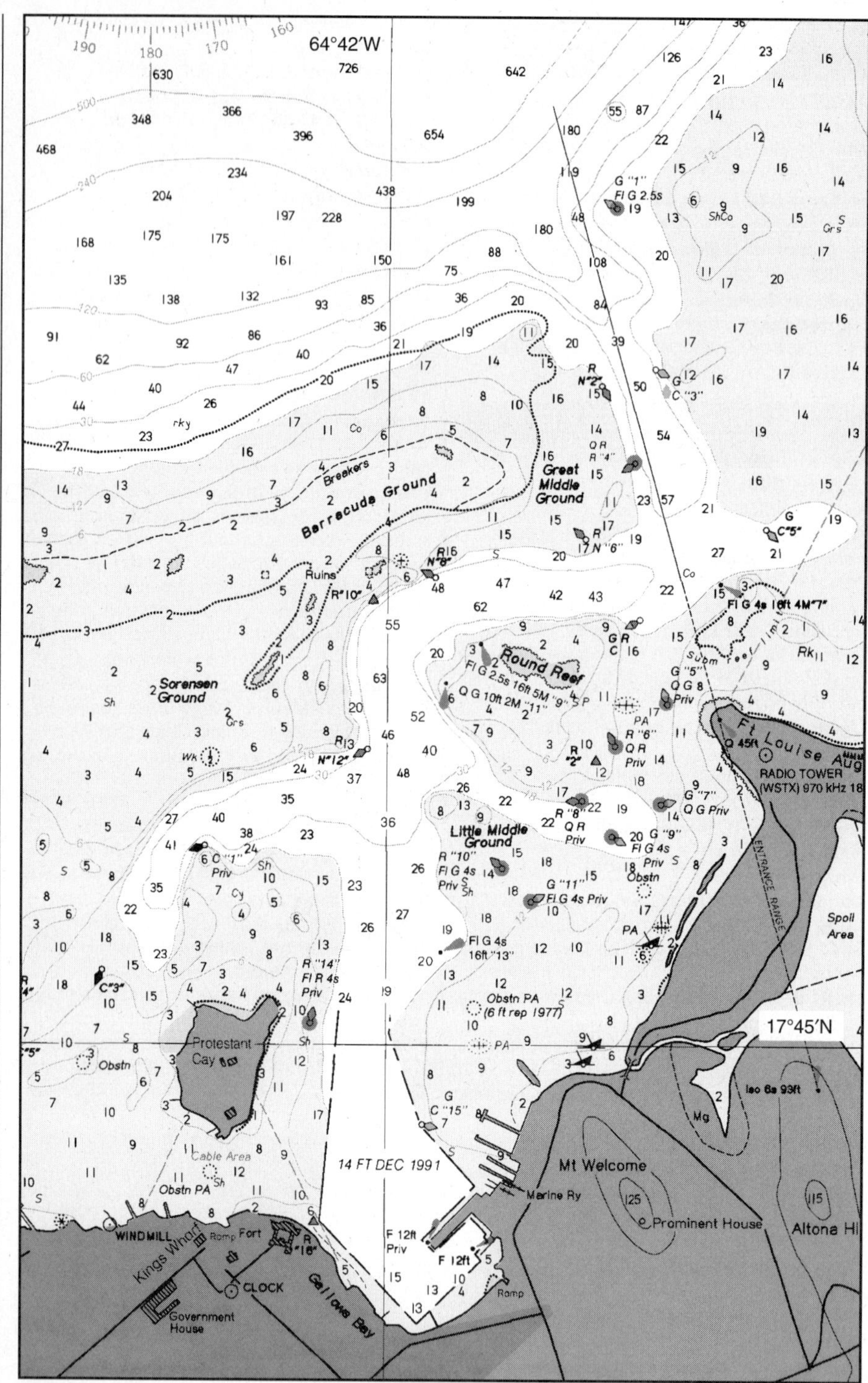

Christiansted, soundings in feet *(from NOAA 25645, 13th ed, 8/93)*

for the latest directions before entering. Private marks help define the channel.
Moorings: Both St. Croix Marine and Salt River Marina rent moorings.
Services: All services are available, including haulouts and repairs. A chandlery is located at St. Croix Marine. Within walking distance from the marina is Chandler's Wharf, with eateries, a hardware store, a travel agent, a post office, a bank, and other services. Public bus service is available on the island.

Because it is 37 miles south of the beaten track, Christiansted is less visited by cruisers. There are many historic buildings worth seeing, and the countryside is more open outside the towns. The combination of Danish, British, and West Indian heritage is more visible here than elsewhere in the U.S. Virgins.

DEVCON INTERNATIONAL CHANNEL

Note: This channel begins north of Protestant Cay and runs west to the industrial plant.

Buoy 1, 17 45.3N, 64 42.2W. Green can. Private aid.
Buoy 2, red nun. Private aid.
Buoy 3, green can. Private aid.
Buoy 4, red nun. Private aid.
Buoy 5, green can. Private aid.
Buoy 6, red nun. Private aid.
Buoy 7, green can. Private aid.
Buoy 8, red nun. Private aid.
Buoy 9, green can. Private aid.
Buoy 10, red nun. Private aid.
CHRISTIANSTED WATER INTAKE LIGHTS (4), 17 45.2N, 64 42.8W. F W, 12ft. On pile. Private aids.
CHRISTIANSTED WATER INTAKE SOUTH LIGHTS (4), 17 45.1N, 64 42.8W. F W. On piles. Private aids.

ST. CROIX, NORTHEAST COAST

COAKLEY BAY LIGHT 1, 17 46.1N, 64 38.2W. Fl G 4s, 16ft, 3M. Green square daymark on pile.
BUCK ISLAND LIGHT, 17 47.3N, 64 37.1W. Fl W 4s, 344ft, 6M. Red pyramidal skeleton tower.
Buck Island Channel Buoy 1, black can. Private aid.
Buck Island Channel Buoy 2, red nun. Private aid.

BUCK ISLAND REEF NATIONAL MONUMENT

Buoy A, 17 48.0N, 64 38.3W. Orange and white bands, can. Private aid.
Buoy B, Orange and white bands, can. Private aid.
Buoy C, Orange and white bands, can. Private aid.
Buoy D, Orange and white bands, can. Private aid.
Buoy D1, Orange and white bands, can. Private aid.
Buoy E, Orange and white bands, can. Private aid.
Buoy F, Orange and white bands, can. Private aid.
Buoy G, 17 47.3N, 64 37.7W. Orange and white bands, can. Private aid.

ST. CROIX, SOUTH COAST

LIME TREE BAY

CHANNEL RANGE (Front Light), 17 42.2N, 64 45.0W. F G, 165ft. Red rectangular daymark, white stripe, on skeleton tower. Visible 4° each side of rangeline. Private aid. **(Rear Light),** 500 yards, 334° from front. F G, 195ft. Red rectangular daymark, white stripe, on skeleton tower. Visible 4° each side of rangeline. Private aid.
CHANNEL EAST AUXILIARY RANGE (Front Light), 17 42.2N, 64 45.0W. F R, 55ft. Black and yellow bands, pole. Visible 4° each side of rangeline. Private aid. **(Rear Light),** 303 yards, 334° from front. F R, 70ft. Black and yellow bands, pole. Visible 4° each side of rangeline. Private aid.
Channel Lighted Buoy 1, 17 40.7N, 64 44.3W. Q G. Green. Spar station buoy. Private aid.
Channel Lighted Buoy 2, Q R. Red. Spar station buoy. Private aid.
Channel Lighted Buoy 3, Fl G 5s. Green. Private aid.
Channel Lighted Buoy 4, Fl R 5s. Red. Private aid.

NOTE: The following lights are located 35 feet outside channel limits.

CHANNEL LIGHT 5, Fl G 4s, 14ft. Green square daymark on pile. Private aid.

Channel Lighted Buoy 6, Fl R 4s, Red. Private aid.
CHANNEL LIGHT 7, Fl G 2.5s, 14ft. Green square daymark on pile. Private aid.
CHANNEL LIGHT 8, Fl R 2.5s, 14ft. Red triangular daymark on pile. Private aid.
CHANNEL LIGHT 9, Fl G 2.5s, 14ft. Green square daymark on pile. Private aid.
CHANNEL LIGHT 11, Fl G 2.5s, 14ft. Green square daymark on pile. Private aid.
CHANNEL JUNCTION LIGHT LK, Fl (2+1) G 6s, 14ft. Daymark with green and red bands on pile. Obscured 296°–026°. Private aid.
CONTAINER PORT BASIN LIGHT 13, Fl G 5s, 14ft. Green square daymark on pile. Private aid.
CONTAINER PORT BASIN LIGHT 15, Fl G 4s, 14 ft. Green square daymark on pile. Private aid.
CONTAINER PORT BASIN LIGHT 17, Fl G 2.5s, 14 ft. Green square daymark on pile. Private aid.

KRAUSE LAGOON CHANNEL

Entrance Lighted Buoy 1, 17 40.6N, 64 45.2W. Fl G 4s. Black. Private aid.
Entrance Lighted Buoy 2, Fl R 6s. Red. Private aid.
Lighted Buoy 3, Fl G 4s. Green. Private aid.
LIGHT 4, Fl R 4s, 18ft. Red triangular daymark on pile. Private aid.
Lighted Buoy 4A, Q R. Red. Private aid.
LIGHT 5, Q G, 14ft. Green square daymark on pile; on same structure as Krause Lagoon Cross Channel Range Front Light. Private aid.
LIGHT 6, Q R, 14ft. Red triangular daymark on pile. Private aid.
LIGHT 7, Fl G 4s, 15ft. Green square daymark on pile. Private aid.
LIGHT 8, Fl R 2.5s, 15ft. Red triangular daymark on pile. Private aid.
LIGHT 9, Fl G 4s, 16ft. Green square daymark on pile. Private aid.
LIGHT 10, Fl R 2.5s, 16ft. Red triangular daymark on pile. Private aid.
LIGHT 11, Fl G 4s, 16ft. Green square daymark on pile. Private aid.
LIGHT 12, Fl R 2.5s, 16ft. Red triangular daymark on pile. Private aid.
LIGHT 13, Fl G 4s, 16ft. Green square daymark on pile. Private aid.
LIGHT 14, Fl R 2.5s, 16ft. Red triangular daymark on pile. Private aid.
LIGHT 16, Fl R 2.5s, 16ft. Red triangular daymark on pile. Private aid.
LIGHT 17, Fl G 4s, 16ft. Green square daymark on pile. Private aid.
EAST MOORING DOLPHIN SIX LIGHT, Fl R 5s, 12ft. On dolphin. Private aid.
WEST TURNING DOLPHIN LIGHT, Fl W 5s, 12ft. On dolphin. Private aid.

KRAUSE LAGOON CROSS CHANNEL

RANGE (Front Light), 17 41.4N, 64 45.6W. Q G, 14ft. White rectangular daymark, black stripe, on pile; on same structure as Krause Lagoon Channel Light 5. Private aid. **(Rear Light),** 153 yards, 244° from front. Oc G 4s, 28ft. White rectangular daymark, black stripe, on pile. Private aid.
Lighted Buoy 1, Fl G 4s. Green. Private aid.
LIGHT 2, 17 41.3N, 64 45.5W. Fl R 4s, 14ft. Red triangular daymark on pile. Obscured 260°–350°. Private aid.
LIGHT 3, Fl G 2.5s, 14ft. Green square daymark on pile. Private aid.
LIGHT 4, Fl R 2.5s, 14ft. Red triangular daymark on pile. Obscured 288°–018°. Private aid.
Lighted Buoy 5, Q G. Green. Private aid.

FREDERIKSTED ROAD

AVIATION LIGHT, 17 42.5N, 64 47.8W. Al Fl W G 10s, 249ft.
SOUTHWEST CAPE LIGHT, 17 40.8N, 64 54.0W. Fl W 6s, 45ft, 7M. Gray skeleton tower.
Southwest Cape Shoal Buoy 2, marks shoal off cape, 17 39.0N, 64 54.6W. Red nun.
U.S. Navy Radar Reflector Buoy, 17 41.3N, 64 54.2W. Platform of three yellow spherical buoys, with orange radar reflector. Maintained by U.S. Navy.
FREDERIKSTED MOORING LIGHT, 17 42.8N, 64 53.4W. Fl W 4s, 18ft. On dolphin. Private aid.
FREDERIKSTED HARBOR LIGHT, S side of root of wharf, 17 43.0N, 64 53.1W. Fl W 4s (2 R sectors), 42ft, W 8M, R 6M. Diamond-shaped daymark on skeleton tower. R 000°–044.5° and 137°–180°.
FREDERIKSTED PIER LIGHTS, F R, 12ft. Private aids.

BRITISH VIRGIN ISLANDS

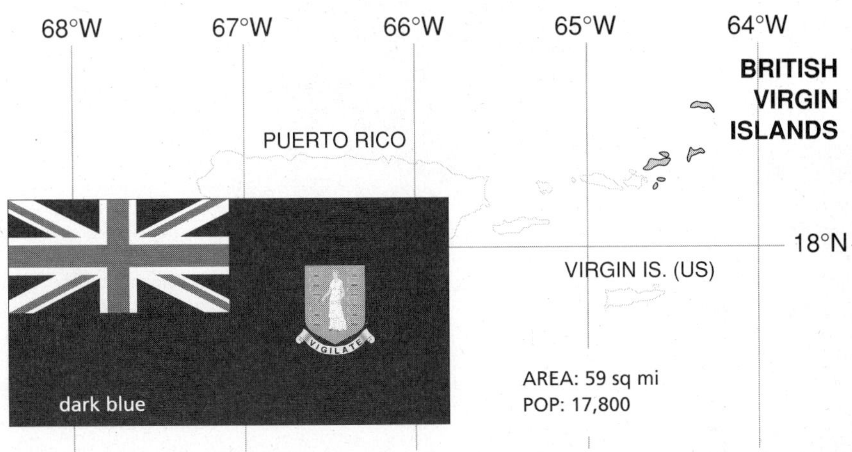

Please see cautions about chartlets and waypoints on page P 2.

ENTRY PROCEDURES

The British Virgin Islands (B.V.I.) include the islands of Jost Van Dyke, Tortola, Norman, Peter, Salt, Cooper, Ginger, Virgin Gorda, Guana, Great Camanoe, Anegada, and numerous small cays. The islands are a British dependency.

Ports of entry are located on Jost Van Dyke (Great Harbour), Tortola (Road Town and West End/Soper's Hole), and Virgin Gorda (Spanish Town/Virgin Gorda Yacht Harbour). If you clear in outside the hours of 0830 to 1530, Monday through Friday, and 0830 to 1230 on Saturday or on a public holiday, a small overtime fee may be assessed (actual working hours at individual offices may differ from non-overtime hours). For more information, call Road Town customs (284-494-3475). Vessels traveling between the U.S. Virgin Islands and the B.V.I. must clear customs and immigration.

You should have a clearance from the last port of call. You should also obtain a clearance before departing the B.V.I.s.

Immigration clearance will be given for the length of your visit or if requested, for 30 days. Extensions are possible for $10 per person. You can generally get permission for visits of up to six months, provided you can prove adequate finances. Passports are required, although citizens of the U.S. and Canada who are not asking for an extension of stay may substitute a birth certificate with a raised seal or a citizenship certificate and a photo identification. *(Note: Driver's licenses alone are not considered valid forms of identification).* Visitors from some countries may need a visa. For more information contact Road Town immigration (284-494-3471).

There are small fees for clearing in and out of the B.V.I. Private cruising boats are not charged for a cruising permit; there is a per diem charge for all charterboats. From December 1 through April 30 B.V.I.-registered charterboats pay $2 per person, per day. Boats registered elsewhere pay $4 per person, per day. From May 1 through November 30 the rates are, respectively, $0.75 and $4 per person, per day. Dive boats, sportfishing boats, and day charterboats should contact the B.V.I. customs department regarding cruising permits. All non-B.V.I. residents planning to fish must have a recreational fishing permit. For information, contact the Conservation and Fisheries Department in Road Town (284-494-5681).

Approximately 250 National Parks Trust moorings for day stops only have been placed in reef-sensitive areas and near many dive sites. The moorings are color coded by intended use. To pick up any of these moorings, by law you *must* have a permit, which is available from customs, charter and dive companies, and the National Parks Trust office in Road Town. Fees vary according to the type of yacht. For more information contact the Trust (tel., 284-494-3904/2069; fax, 284-494-6383).

USEFUL INFORMATION

Language: English.

Currency: U.S. dollar (US$); major credit cards widely accepted.

Telephones: A new area code, 284, went into effect on November 1, 1997. Local numbers are seven digits.

Public holidays (1998): New Year's Day, January 1; Commonwealth Day, March 9; Good Friday, April 10; Easter Monday, April 13; Whit Monday, May 19; Sovereign's Birthday, June (date not announced by *Reed's* press time); Territory Day, July 1; Festival Monday, Tuesday, Wednesday, August 3–5; St. Ursula's Day, October 21; Birthday of Heir to the Throne, November 14; Christmas Day, December 25; Boxing Day, December 26.

Tourism offices: *In the B.V.I.:* B.V.I. Tourist Board, P.O. Box 134, Road Town, Tortola (284-494-3134). *In the U.S.:* B.V.I. Tourist Board, 370 Lexington Ave., New York, NY 10017 (tel., 800-835-8530, 212-696-0400; fax, 212-949-8254). Office also in San Francisco.

Emergencies: Call Tortola Radio on VHF channel 16. Tortola Radio relays information to B.V.I. Fire and Rescue, who then pages the duty coordinator for Virgin Islands Search and Rescue (VISAR), a volunteer association. Stay on VHF 16 until contacted by VISAR. By telephone, dial 911 or 999 to reach Fire and Rescue directly. The VISAR office is in Road Town (284-494-4357).

JOST VAN DYKE

GREAT HARBOUR

18 27N, 64 45W

Pilotage: This bay is easy to enter with depths of 15 to 20 feet. A reef extends across most of the head of the bay, so don't venture in too far. A gap in the middle of the reef creates a safe route for dinghies traveling to the wharf. When anchoring, check your set well, as the holding is not all good and the wind tends to swing boats in strange ways.

Port of entry: Dinghy in to the main dock. Customs and immigration are near the wharf.

Dockage: None.

Anchorage: Little Harbour is a quiet, easy to enter lagoon and all three restaurants provide local food and atmosphere at the shore's edge. Great Harbour is the site of Foxy's Tamarind Bar and several other good West Indian restaurants. The anchorage is fairly well protected and the holding is good. White Bay lies west of Great Harbour and features a small hotel and restaurant; a channel through the center of the reef allows entrance to the anchorage, which is subject to winter swells. Sandy Cay, just to the east of Little Jost Van Dyke, is uninhabited and has a long stretch of white sandy beach. The water is deep almost until the shore; the area is prone to swells and is not a good year-round anchorage—but it's a great day stop.

Services: There are several well-known restaurant/bars, a few small stores, some groceries, and telephones, but no yacht services.

Jost Van Dyke is famous for its watering holes and relaxed atmosphere. The island is probably the easiest place to clear customs in the B.V.I.

TORTOLA

SOPERS HOLE LIGHT, passenger terminal, SW end, 18 23.4N, 64 42.2W. F R, 16ft.

WEST END

18 23N, 64 43W

Pilotage: This popular harbor is charted as Soper's Hole, and both names are used interchangeably. The entrance is deep and free of obstructions, except close to shore. Be prepared for currents up to 3 or 4 knots in the mouth of Soper's Hole, in Thatch Island Cut, and in the pass between Little Thatch and Frenchman's Cay. Stay clear of the docking ferries at the wharf on the north side of the harbor.

Port of entry: Officials are located in the building on the ferry wharf. Because of the constant flow of traffic at the wharf, it is best to dinghy over from your moored boat.

Dockage: Available from Soper's Hole Marina (VHF channels 12 and 16; tel., 284-495-4740, answers Sunsail).

Moorings: Sunsail and Frenchman's Cay Shipyard (VHF channel 16; tel., 284-495-4353) rent moorings on a first-come, first-served basis. Fees are generally collected in the early morning or evenings between 1700 and 1800. If you are picking up a mooring only for the time it

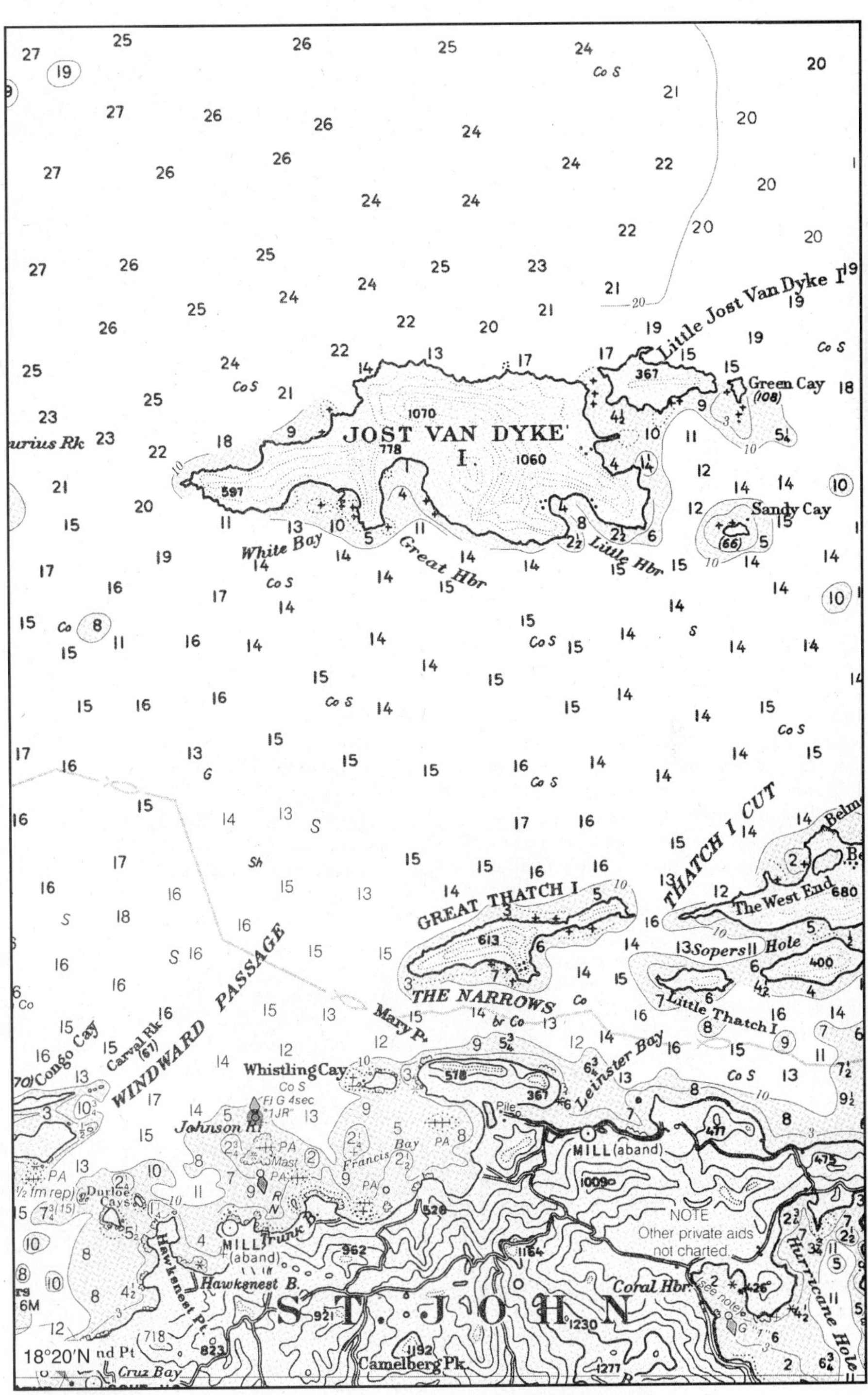

Jost Van Dyke, soundings in fathoms *(from NOAA 25611, 22nd ed, 8/93)*

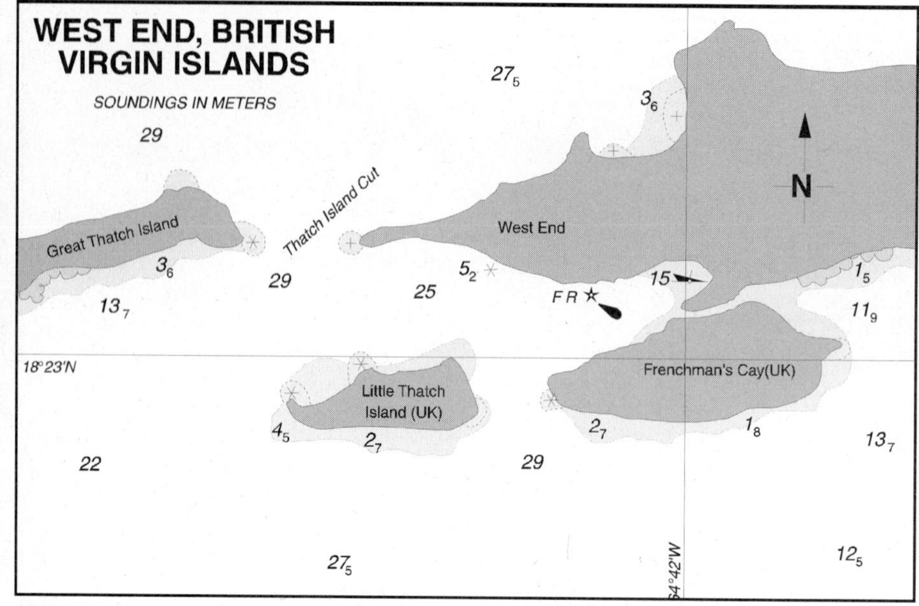

West End, soundings in meters *(after DMA 25611, 21st ed, 9/86)*

takes you to clear with officials, you may be charged just a nominal short-term rate or nothing at all.

Anchorage: With depths of 25 to 65 feet, lots of moorings, and frequent crowds of boats, anchoring is problematic at best. Renting a mooring is a viable alternative.

Services: Frenchman's Cay Shipyard has a 200-ton marine railway and complete repair facilities for boats of all types, including catamarans and wooden boats. Parts can be messengered from Nanny Cay and Road Town. Pusser's Rum and the Jolly Roger are popular bars and restaurants, and the Ample Hamper can supply most food and liquor needs. There are taxis to Road Town, telephones, and in the complex near the Sunsail docks, several shops.

West End is a popular and busy spot. The yacht facilities are good, and the entertainment ashore is geared to the many charterboats visiting here. The presence of first-class wooden boat specialists and repair facilities of all sorts makes Soper's Hole a unique harbor.

ROAD TOWN RANGE, Road Harbour **(Front Light),** 18 25.3N, 64 37.1W. F R, 37ft, 3M. Administration building. F R lights on radio mast 1.5 mi WNW; aero F R, F W (occasional) 3 mi E. **(Rear Light),** about 40 meters, 290° from front. F R, 52ft, 3M. Administration building.

BREAKWATER, 18 25.5N, 64 36.9W. Fl R 3s.
CRUISE SHIP DOCK, 18 25.3N, 64 36.7W. Fl W 3s. Dolphin.

ROAD TOWN

18 26N, 64 37W

Pilotage: The range of 290°T will take you in to the government wharf safely. There are several shoal areas within the harbor, some of which may not be marked. There is shallow water both north and south of the main wharf, off Baugher's Bay, and all along the shore north of the Fort Burt Marina area. Note that anchoring is prohibited in the central harbor area.

Port of entry: You may tie up temporarily to the main passenger wharf while clearing customs and immigration, but note that the wharf is subject to surge. The office is located just across the street from the wharf.

Dockage: Slips are available at the Nanny Cay Resort and Marina (VHF channel 16; tel., 284-494-2512) about 2 miles southwest of Road Harbour on the south coast of Tortola. Prospect Reef Resort and Marina just southwest of Fort Burt Point has limited dockage only for hotel guests. Slips are also available at Fort Burt Marina (284-494-4200), Tortola Marine Management at Road Reef Marina (VHF channel 12; tel., 284-494-2751) in the small cove south of Fort Burt Marina, at Village Cay Marina (VHF channel 16; tel., 284-494-2771) and B.V.I. Yacht

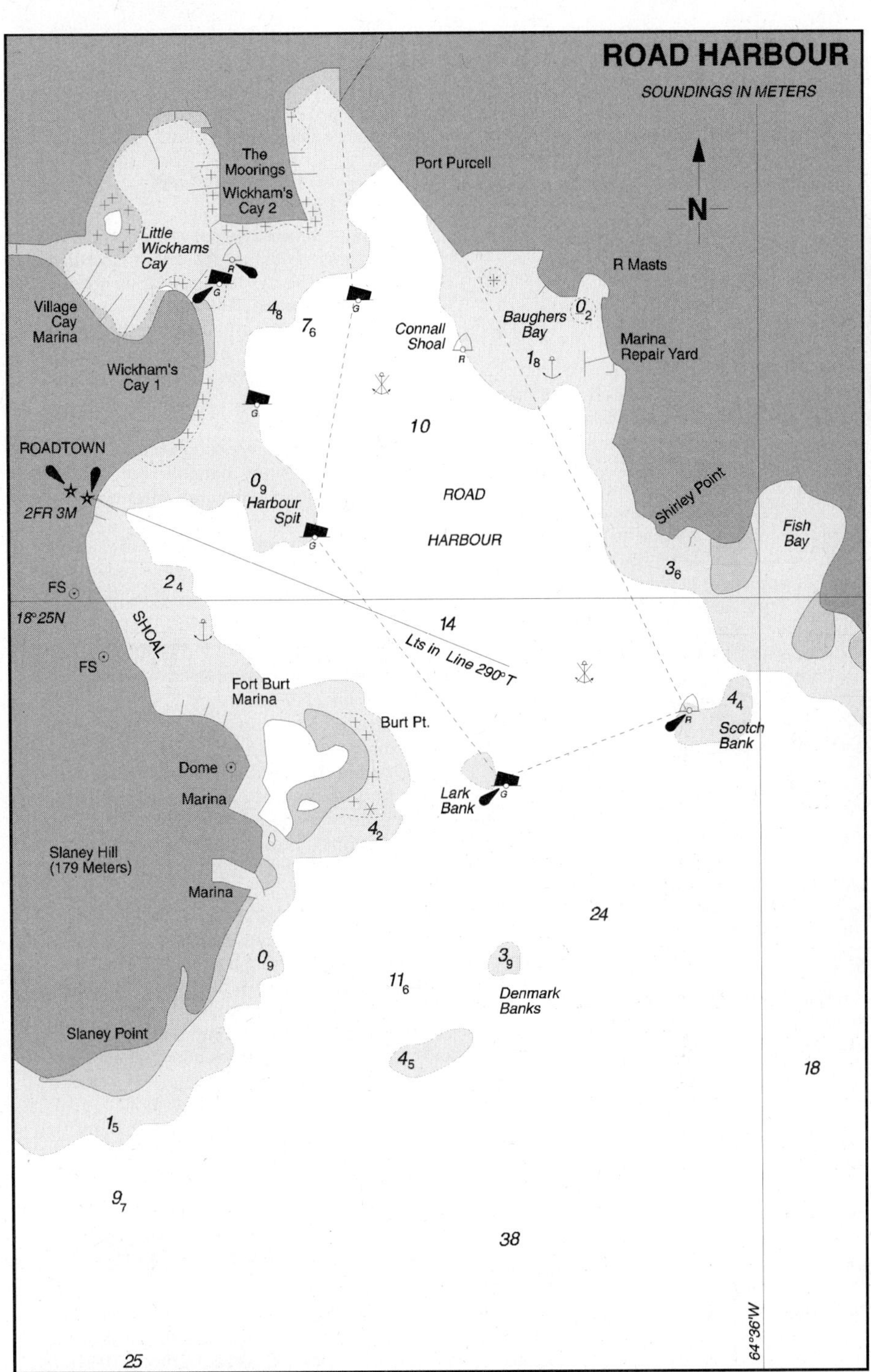

Road Harbour, soundings in meters *(after DMA 25611, 21st ed, 9/86)*

Charters (VHF channel 10; tel., 284-494-4289) in the Wickhams Cay basin, and through The Moorings at Wickham's Cay 2 (VHF channel 12; tel., 284-494-2331 x2202).
Anchorage: Most boats find room north of the Fort Burt Marina, where there is a dinghy dock. Check depths carefully, as the bottom shoals rapidly toward shore. There may be some room among the many moored boats in the Wickham's Cay basin. There is a dinghy dock near Village Cay Marina. It is also possible to anchor south of the charterboat marina in Baugher's Bay. Most of the harbor is subject to surge and swells, especially when the wind comes in south of east.
Services: Tortola Yacht Services (VHF channel 16; tel., 284-494-2124) on Wickham's Cay 2 is a full-service boatyard and chandlery. Nanny Cay also offers marine supplies and hauling and repair services. Fuel, water, and ice are available at Nanny Cay, Fort Burt Marina, The Moorings, and Inner Harbour Marina at Wickham's Cay basin. Fuel may be available at Village Cay. There is a propane fill-up within walking distance of Wickham's Cay 2. Road Town is the capital and commercial center of the British Virgin Islands. There are good telephone connections, banks, a post office, several grocery stores, many eateries, and good shopping. Everything is available in a compact area, easily accessible to those on foot.

Road Town is a delightful Caribbean city of small winding streets, with colorful wooden buildings leaning over the narrow sidewalks. Without the hectic pace of Charlotte Amalie, it is a pleasant place to stock up on supplies. The small bars and restaurants are often full of cruisers from around the world. Try a visit to the botanical garden or a trip into the hills via taxi or rental car.

FAT HOGS BAY LIGHT, 18 26.1N, 64 33.6W. Fl (2) W 5s, 25ft, 5M.
BELLAMY CAY LIGHT, off Beef Island, 18 27.0N, 64 31.9W. F W. White mast.

TRELLIS BAY

18 27N, 64 32W
Pilotage: There are rocks off Sprat Point on the east side of Trellis Bay, and a 3-foot-deep reef extends north from Conch Shell Point on the west side of the bay. A cable runs between the shore and the west side of Bellamy Cay, and there is a small shallow patch south of the cay.
Moorings: Available off Bellamy Cay from the Last Resort (VHF channel 16; tel., 284-495-2520).
Anchorage: Trellis Bay is a well-protected anchorage fringed by a semicircular beach.
Services: This is the location of Boardsailing BVI, the Conch Shell Point Restaurant, and the Loose Mongoose Beach Bar. At the bay's center is The Last Resort, an English-style restaurant that's a favorite of cruisers.

MARINA CAY

18 28N, 64 31W
Pilotage: A large conspicuous rock on the cay's east side marks the reef that extends south and then west of the cay. Approach the anchorage from the east between Marina Cay and Scrub Island or from the west along the shore of Great Camanoe. The reef-lined route from the north between Great Camanoe and Scrub should be negotiated in good light and sea conditions only.
Moorings: Available through Marina Cay.
Anchorage: This tiny island lies north of Trellis Bay. It's fringed with coral; enter from the north.
Services: Fuel, water, and ice are available at the small Marina Cay resort, which also has a restaurant, bar, and small Pusser's gift store.

NORMAN ISLAND

18 19N, 64 37W
Pilotage: The tall rocks west of Pelican Island are known as the Indians. This is a favorite dive site, where the National Parks Trust has placed moorings to prevent further damage to the coral. You must have a permit to use the moorings. South of Treasure Point is another reef and a series of caves, which are also a dive site. Use the moorings if possible.
Anchorage: Anchor as far in the Bight as you can to avoid the more-than-30-foot depths farther out. You can dinghy in to the small beach at the head of the cove. This anchorage is often crowded with charterboats in the winter. In settled weather, Benures Bay, on the north coast of the island, is a pleasant anchorage. Pass west of the reefs before looping up inside the hook of land.
Services: The anchored boat the *William Thornton* (VHF channel 16) serves as a bar and restaurant.

PETER ISLAND

18 21N, 64 35W
Pilotage: The only yacht facility on the island is located east of the spit of land forming the eastern boundary of Great Harbour. This body of water is called Sprat Bay.

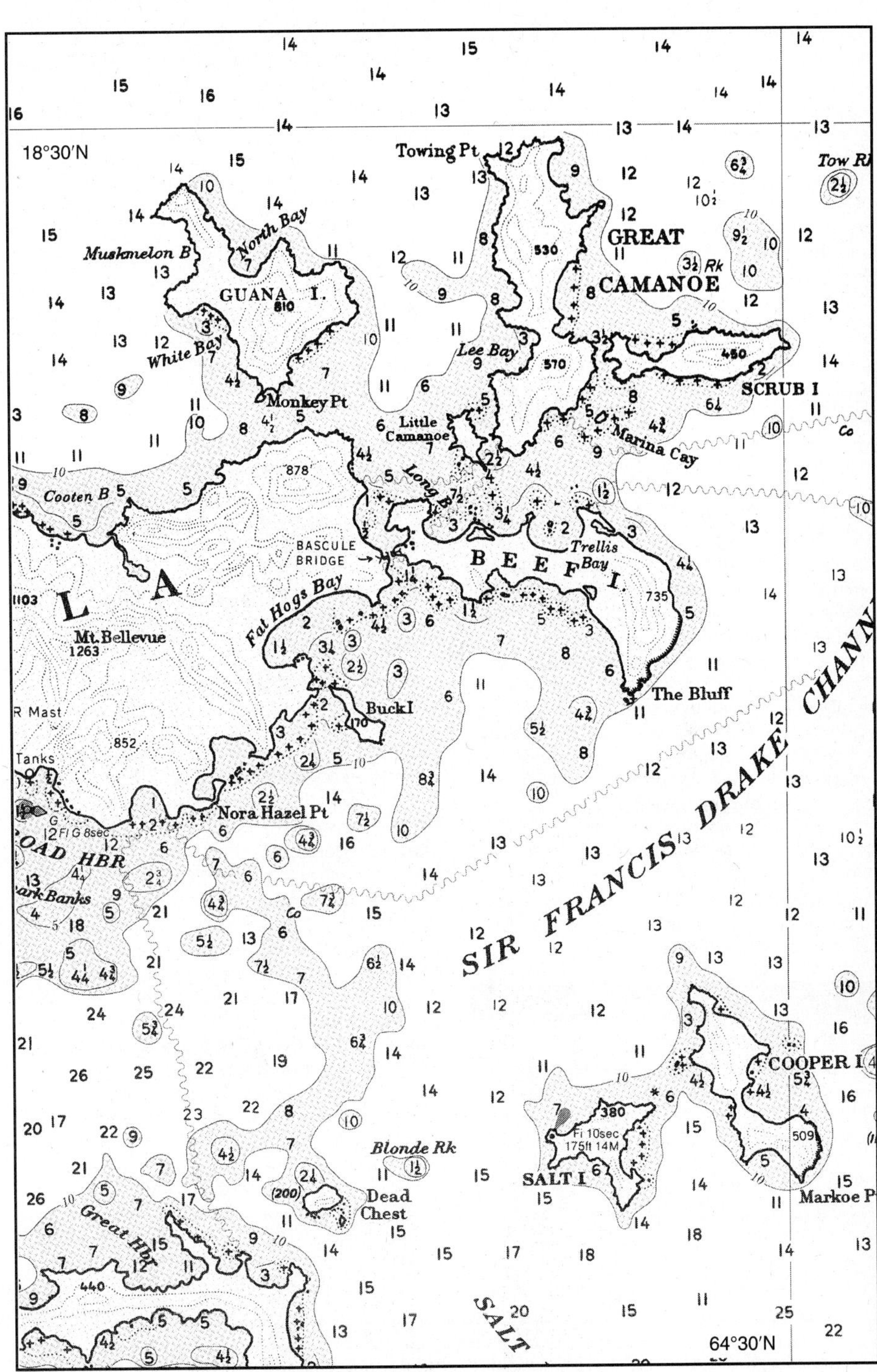

Trellis Bay and Marina Cay, soundings in fathoms *(from NOAA 25641, 22nd ed, 8/93)*

Dockage: Peter Island Resort and Yacht Harbour (VHF channel 16; tel., 284-495-2000) offers dockage for yachts as large as 170 feet or drawing as much as 12 feet.
Moorings: Available in Sprat Bay through Peter Island Resort.
Anchorage: Anchorages include Sprat Bay, Great Harbour, and Little Harbour. In settled weather you can anchor in White Bay on the south shore or west of Key Cay.
Services: Diesel, water, ice, and electricity are available at the Peter Island Resort docks. The resort's restaurant serves breakfast, lunch, and dinner.

SALT ISLAND

SALT ISLAND LIGHT, NW corner of island, 18 22.5N, 64 32.1W. Fl W 10s, 175ft, 14M.

SALT ISLAND

18 22N, 64 32W
Pilotage: Salt Island rises to a height of 380 feet. There is a rock awash off the northeast point of the island. There is a small settlement in Salt Island Bay on the north shore where workers once mined the island's salt ponds.This island was once a regular stopping off point for ships requiring salt for food preservation on the trade routes. It's also the location of the B.V.I.'s famed Wreck of the *Rhone* which sank during a hurricane in 1867. At Lee Bay, just north of the Rhone, moorings have been provided for those diving the wreck in order to minimize anchor damage. No anchoring at the site is allowed.
Anchorage: Both Lee Bay and Salt Pond Bay of the settlement can be rough anchorages and are recommended for day use only.

COOPER ISLAND

18 23N, 64 31W
Pilotage: Cooper Island is 1.7 miles long and rises to 530 feet at its south end. The passage between Salt and Cooper is constricted by a rock awash off the northeast point of Salt. A small islet off the west coast of Cooper forms the southern boundary of Manchioneel Bay.
Moorings: Available in Manchioneel Bay from Cooper Island Beach Club (VHF channel 16; answering machine only in Road Town, 284-494-3721).
Anchorage: The swirling winds, deep waters, and crowded moorings make this anchorage uncomfortable at times. You can dinghy in to the beach club dock.
Services: This is a good lunch stop for those sailing upwind to Virgin Gorda. The Cooper Island Beach Club offers lunch, dinner, and drinks; it's often the site of lively charterboat parties. Ice and scuba services are available, as are overnight accommodations.

GINGER ISLAND

GINGER ISLAND LIGHT, NE end of island, 18 23.6N, 64 28.3W. Fl W 5s, 498ft, 14M. Yellow tower.

VIRGIN GORDA

PAJAROS POINT LIGHT, 18 30.2N, 64 19.5W. Fl (3) W 15s, 200ft, 16M.

THE BATHS AND VIRGIN GORDA YACHT HARBOUR

18 26N, 64 27W
Pilotage: The Baths are located about 0.5 mile north of the southern tip of Virgin Gorda. Virgin Gorda Yacht Harbour in Spanish Town is located about 1.5 miles farther north. The marina is reached through a privately marked channel that takes you safely through the reef.
Port of entry: Contact customs at the marina.
Dockage: Contact Virgin Gorda Yacht Harbour (VHF channel 16; tel., 284-495-5500), which has all new docks for 100 boats and a special dock for megayachts.
Anchorage: The anchorage by the Baths can be very rolly and is only a day stop. Anchor off the huge boulders, and be very careful to avoid the many snorkelers if you motor in to the beach in your dinghy. There is no anchoring inside the Yacht Harbour.
Services: Virgin Gorda Yacht Harbour is a full-service marina with fuel, water, ice, laundromat, haulouts, repairs, and a chandlery. A shopping center, a grocery store, and several restaurants are nearby. You can buy stamps and mail letters in the Yacht Harbour complex. For additional postal services, the post office is a short taxi ride away. Taxi service to the Baths and the rest of the island is also available.

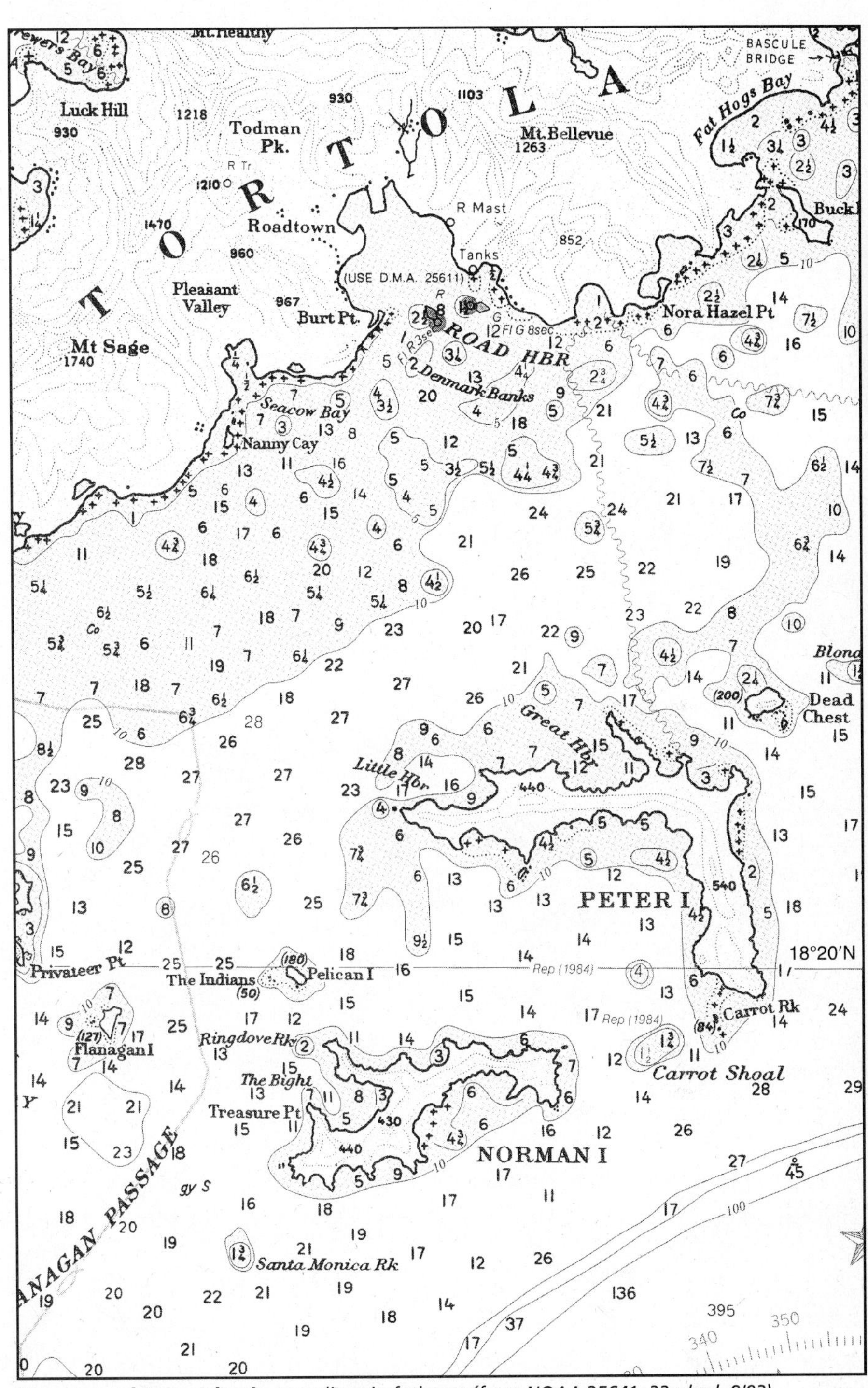

Norman and Peter Islands, soundings in fathoms *(from NOAA 25641, 22nd ed, 8/93)*

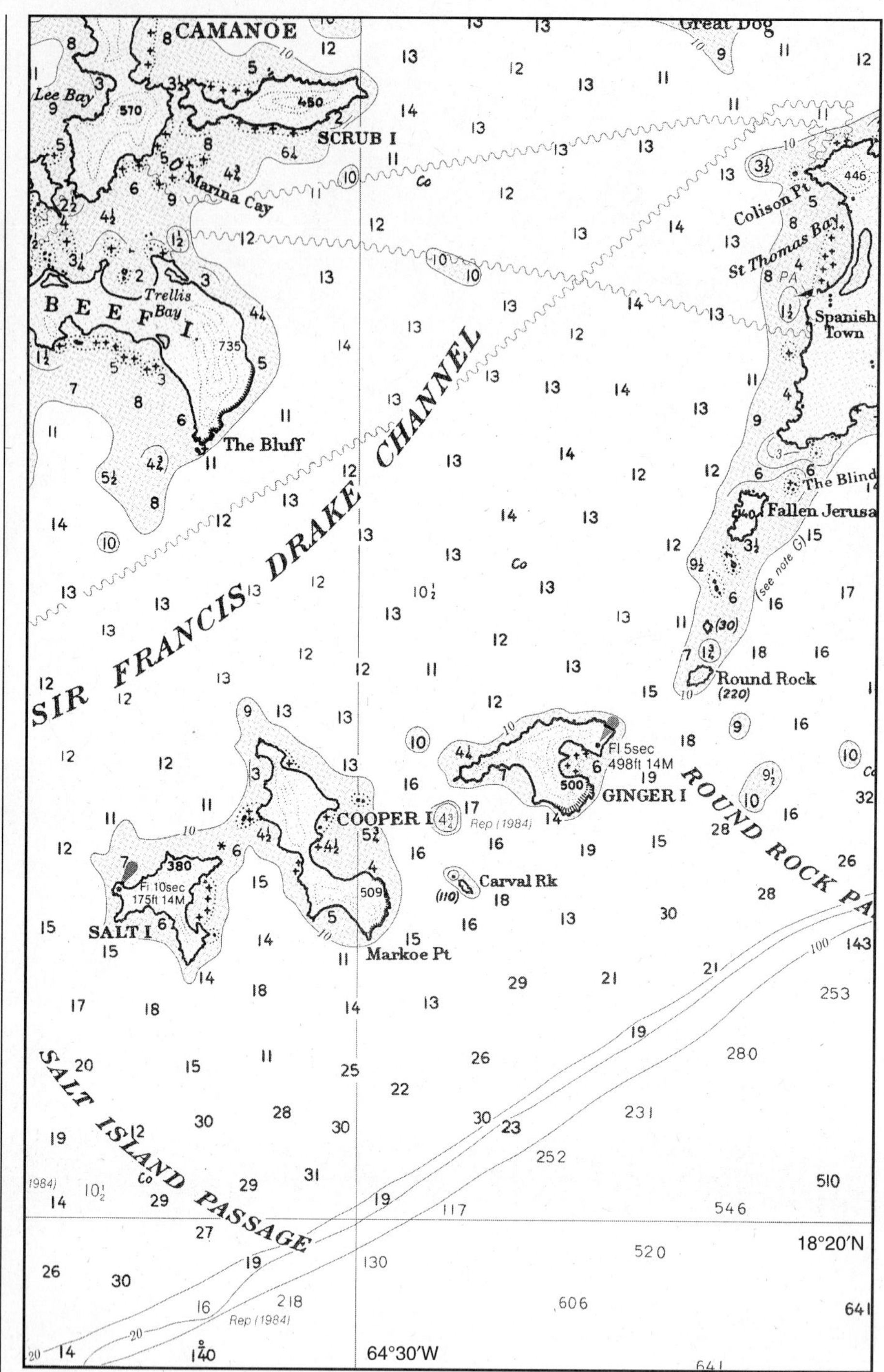

Salt, Cooper, and Ginger Islands, soundings in fathoms *(from NOAA 25641, 22nd ed, 8/93)*

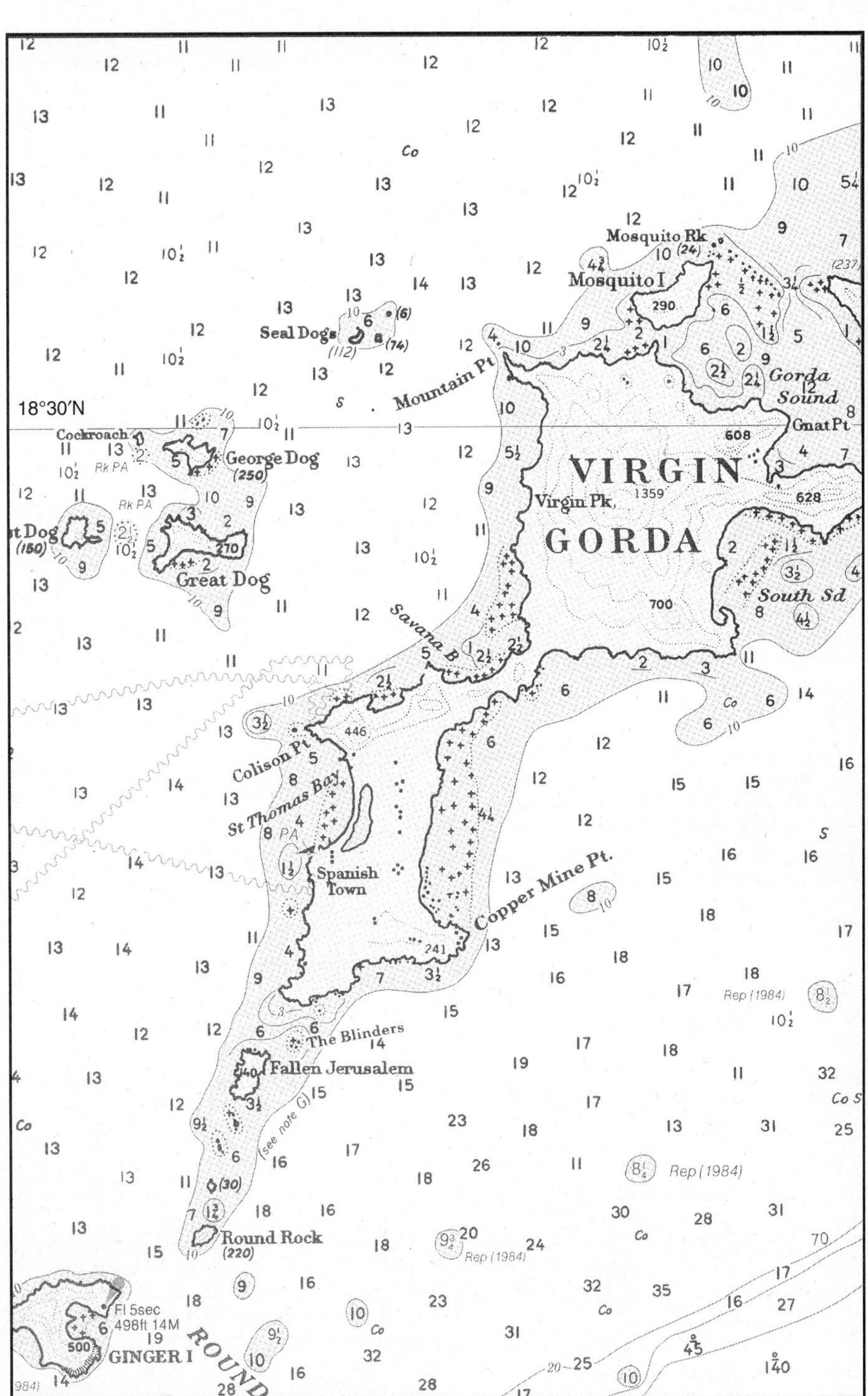

The Baths and Spanish Town, soundings in fathoms *(from NOAA 25641, 22nd ed, 8/93)*

GORDA SOUND

18 30N, 64 22W

Pilotage: The west entrance to the sound passes through a narrow channel south of Mosquito Island. Just east of the narrowest part (175 yards wide) is a shallow bar to the north. Controlling depth is about 5 or 6 feet. Attempt this passage only in good light. The northern entrance passes east of Colquhoun Reef and is marked by a red and green buoy at its narrowest point. Least depth is about 17 feet on this route.

Dockage: Available at Pusser's Leverick Bay (VHF channel 16; tel., 284-495-7369) in Leverick Bay along the south shore of Gorda Sound, at the Bitter End Yacht Club (VHF channel 16; tel., 284-494-2746) on the eastern side of the sound, and at Biras Creek Marina (VHF channel 16; tel., 284-494-3555), south of the Bitter End. Biras Creek facilities are reported to have been completely refurbished.

Moorings: Rental moorings are available at Pusser's, the Bitter End, Biras Creek, and Drake's Anchorage (VHF channel 16; tel., 284-494-2254) on Mosquito Island.

Anchorage: There are many excellent places to anchor in Gorda Sound. In addition to the anchorages off the establishments mentioned above, favorite spots include west of Prickly Pear Island, in Gun Creek, and near Saba Rock. Choose your spot according to the wind direction, and watch for coral heads in the shallows.

Services: All of the marinas and resorts, including Drake's Anchorage, have excellent bars and restaurants. Fuel, water, and ice are available at Pusser's, the Bitter End, and Biras Creek. The Bitter End offers mechanical repair services. Some provisions are available at the Bitter End and within walking distance of Pusser's. There are interesting shops at Pusser's and the Bitter End. Pirates Pub, a favorite cruiser's hangout on Saba Rock, can be contacted via VHF channel 16.

Gorda Sound is a miniature cruising ground all by itself. There are interesting anchorages, good restaurants, and beautiful views in all directions. Despite its popularity, a good anchoring spot always seems to be available.

ANEGADA

ANEGADA LIGHT, W end, 18 44.8N, 64 24.7W. Fl W 10s, 62ft, 10M.

ANEGADA

18 44N, 64 23W

Pilotage: Anegada is only 30 feet high and very difficult to spot from a distance. It is surrounded by coral reefs and shoals. Never approach the island after dark or in poor visibility. The direct course from Gorda Sound to the west end of Anegada is about 352°T. With the west end of Anegada lined up on your bow and Gorda Sound on your stern, you should clear all the coral heads. Currents are notoriously irregular, and you should judge leeway by watching Virgin Gorda behind you. Chances are you will be set well to the west. Once you can see the island, you may be able to spot the buoys marking the channel into the anchorage on the south side of the island. Stay in deep water until you are sure of your course. Several buoys should lead you on a course of about 60° or 70°T toward the wharf and hotels on shore. Jost Van Dyke should be on your stern when you are headed toward the hotel. If in doubt, call the Anegada Reef Hotel on VHF channel 16 or one of the boats in the anchorage. The controlling depth is about 7 feet to the hotel and then 6 feet in the anchorage area west of the hotel.

Anchorage: Anchor in the cove west of the hotel, watching for coral heads. This area is surprisingly sheltered, even in high winds. Boats sometimes anchor west of the restaurant on Pomato Point. Enter when the restaurant bears approximately 90°. Call Pomato Point on VHF channel 16 for directions.

Services: Neptune's Treasure (VHF channel 16; tel., 284-495-9439), rents tents and rooms and has a restaurant. The Anegada Reef Hotel (VHF channel 16; tel., 284-495-8002) offers rooms and meals as well as water and ice. Pomato Point Restaurant (VHF channel 16; tel., 284-495-9466) has an interesting museum of local artifacts. The Anegada Beach Club (VHF channel 16) also offers lunch and dinner dining. Call ahead for reservations and main entrée selections at all these establishments.

Anegada's fringing reefs have always been a treacherous trap for passing ships, creating a diving and treasure hunter's paradise. Most bareboat charters aren't permitted to visit here, so the anchorage tends to be less crowded than others in the B.V.I. Note that anchoring and fishing on Horseshoe Reef have been prohibited while fish stocks are restored.

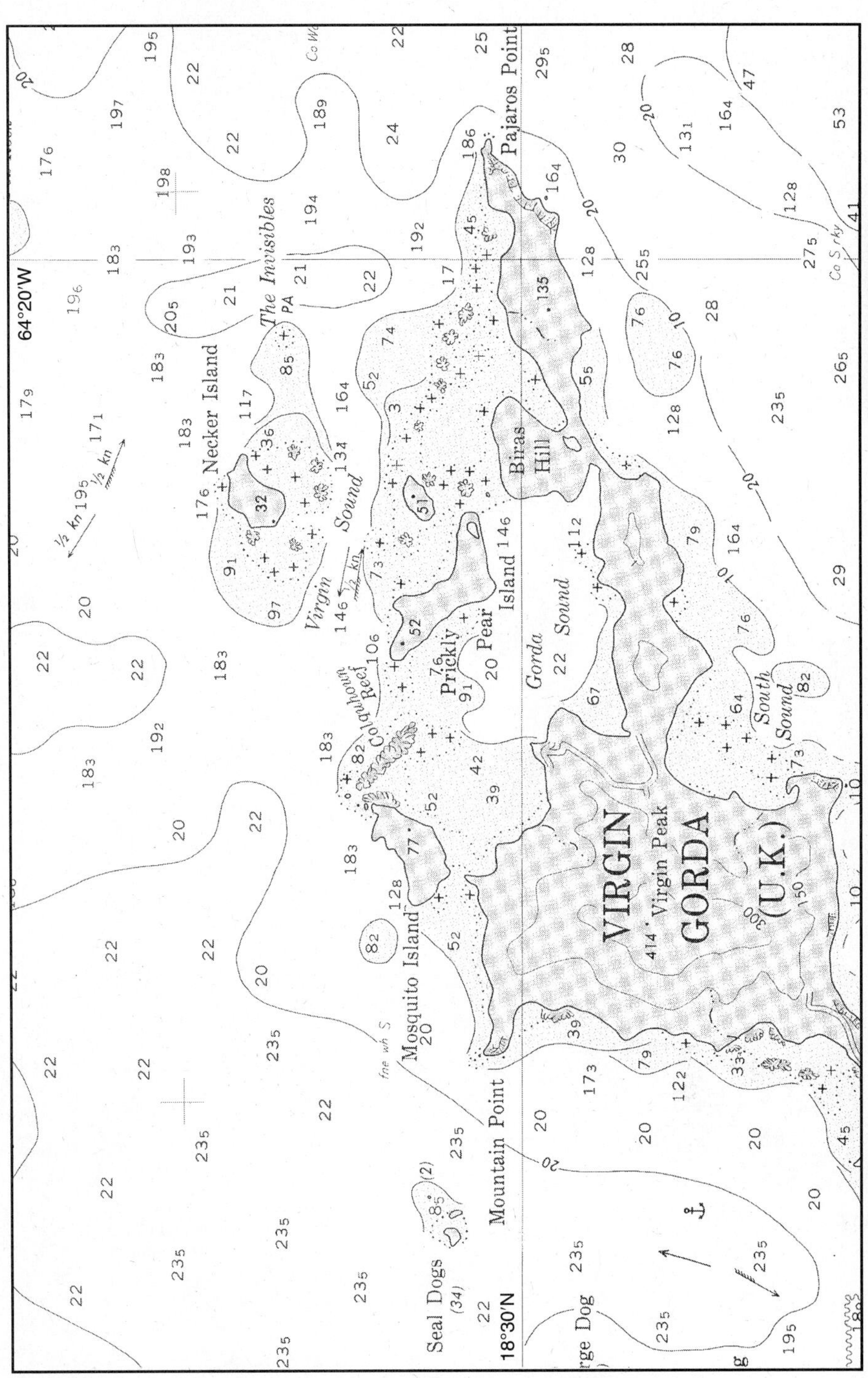

Gorda Sound, soundings in meters *(from DMA 25609, 4th ed, 10/86)*

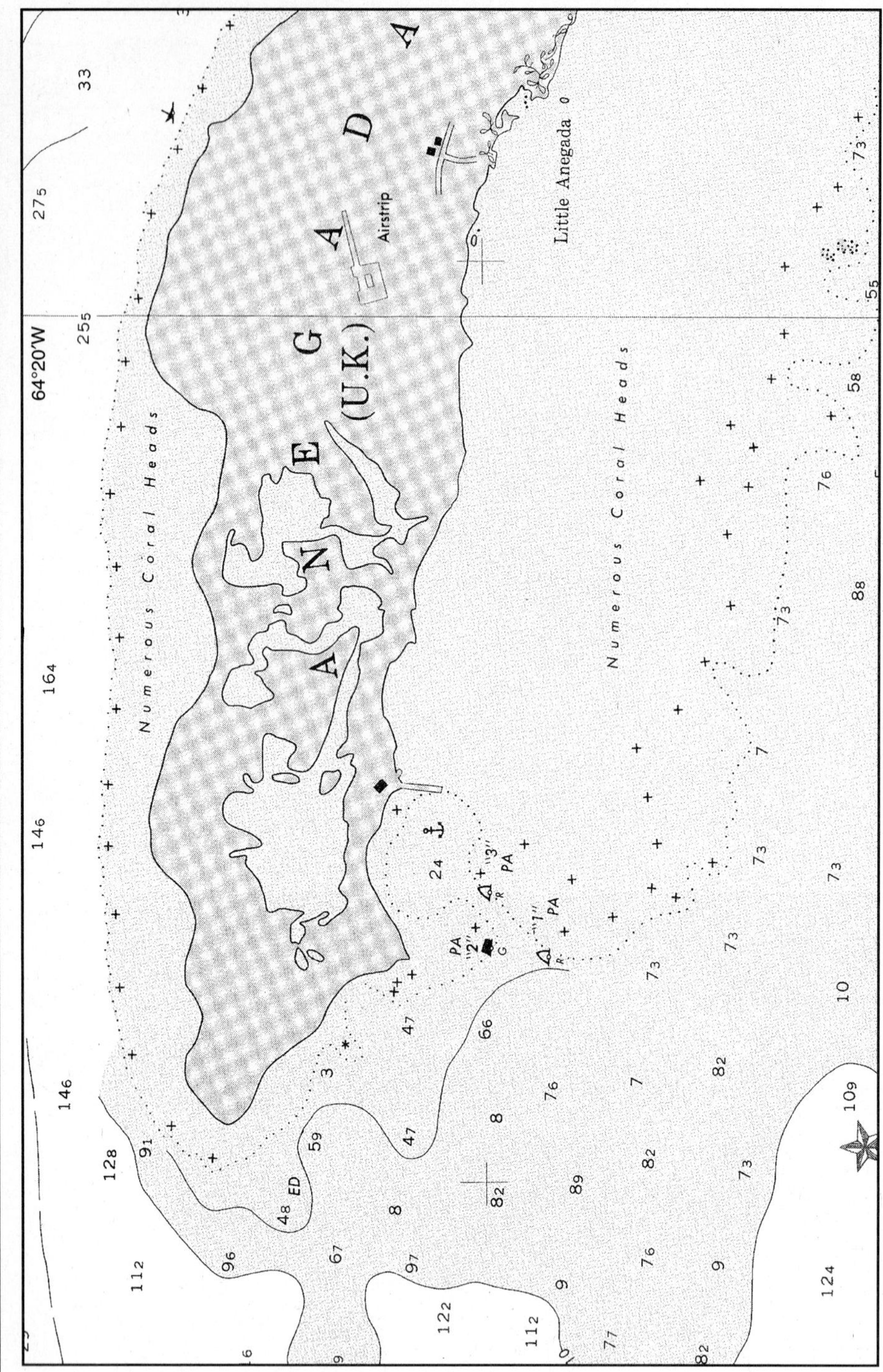

Anegada, soundings in meters *(from DMA 25609, 4th ed, 10/86)*

SOMBRERO, UNITED KINGDOM

18 36N, 63 25W

SOMBRERO LIGHT, SE side, 18 35.8N, 63 25.6W. Fl W 10s, 157ft, 20M. Red iron framework tower, 126ft. Irregular (October 1995).

This small island about 30 miles northwest of Anguilla lies on a small circular bank with depths of 10 to 30 fathoms. Its lighthouse is the only mark in the Anegada Passage. The island has precipitous sides and rises in sharp, jagged points. Vegetation is sparse, but the island supports abundant bird life.

A light is exhibited from a square metal framework tower on a white concrete base situated near the center of the island. The keeper's dwellings and a flag staff stand north of the light tower. A 20-foot-high circular concrete base of a former light tower stands east of the light. A ruined chimney about 33 feet high is about 175 yards southwest of the light.

There is an anchorage off the west side of the island in depths of about 12 fathoms.

ANGUILLA

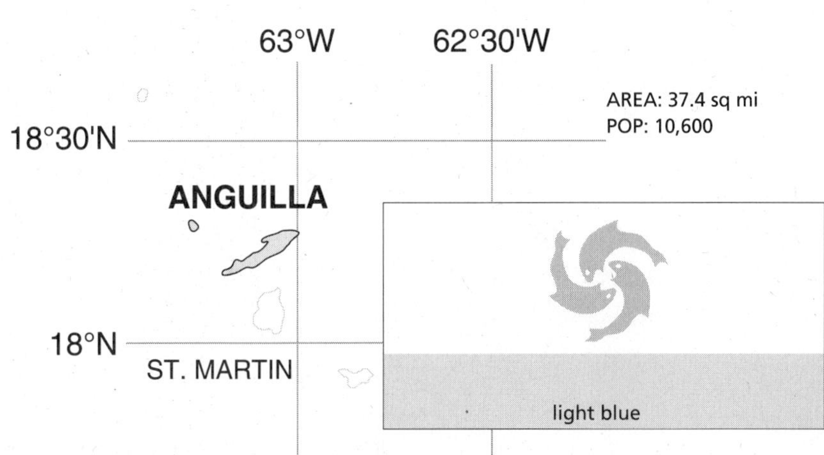

Please see cautions about chartlets and waypoints on page P 2.

ENTRY PROCEDURES

Anguilla is a British dependency, although it is basically self-governing.

Official ports of entry are Road Bay, on the north coast, and Blowing Point, on the south coast; however, Road Bay is the port of entry for all pleasure vessels. All boats must clear in before anchoring elsewhere off Anguilla or its cays. Customs and immigration offices are open from 0830 to 1200 and from 1300 to 1600 seven days a week (Road Bay customs, 264-497-5461).

All boats except those registered in Anguilla must purchase a cruising permit for the duration of their stay. For vessels to 5 tons the charge is $25 per day or $150 per week. For vessels from 5 to 20 tons the charge is $100 per day or $500 per week. For vessels over 20 tons the charge is $150 per day or $900 per week. Other fees are quoted for longer periods. Vessels over 20 tons must also pay port and lighthouse dues. The charge is $30 for vessels from 20 to 50 tons, $60 for vessels from 50 to 100 tons, and $120 for vessels from 100 to 250 tons. All charges are based on registered tonnage and are quoted in Eastern Caribbean dollars (EC$).

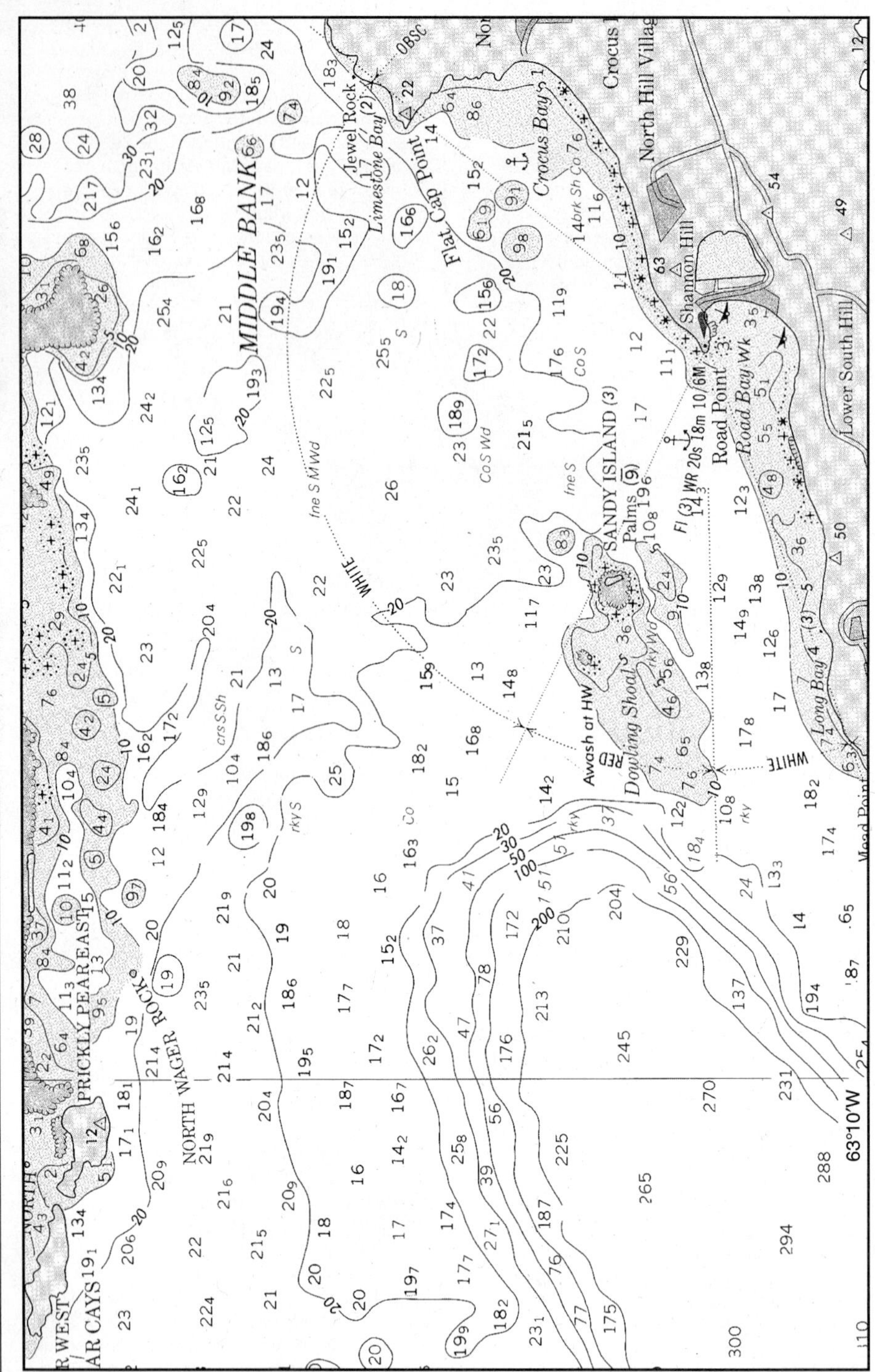

Road Bay, soundings in meters *(from DMA 25613, 1st ed, 9/84)*

Spearfishing and commercial fishing are prohibited aboard foreign boats. Check with customs for the areas where anchoring is prohibited.

The Department of Fisheries and Marine Resources has dropped white moorings at three marine park locations frequented by yachts: Sandy Island, Prickly Pear Cays, and Little Bay. The moorings are available for a fee of $15 from 0600 to 1900; no boats may anchor at the cays off Anguilla overnight. Mooring permits are issued at the port authority building.

USEFUL INFORMATION

Language: English.
Currency: Eastern Caribbean dollar (EC$); exchange rate (summer 1997), EC$2.70 = US$1. U.S. dollars are widely accepted.
Telephones: A new area code, 264, became effective on March 31, 1997; local numbers, seven digits. First three digits, 497; within the island, dial only the last four digits.
Public holidays (1998): New Year's Day, January 1; Good Friday, April 10; Easter Monday, April 13; Labour Day, May 4; Whit Monday, May 19; Anguilla Day, May 30; Queen's Official Birthday, June 16; August Monday, August 3; August Thursday, August 6; Constitution Day, August 7; Separation Day, December 19; Christmas Day, December 25; Boxing Day, December 26.
Tourism office: *In Anguilla:* Department of Tourism, Factory Plaza, The Valley (tel., 264-497-2759; fax, 264-497-2710; from the U.S., 800-553-4939).
Emergencies: Call Curaçao Radio, call sign PJC on VHF channel 16.

ANGUILLA APPROACHES

ANGUILLITA ISLAND LIGHT, 18 09.4N, 63 10.6W. Fl (2) W 15s, 48ft, 5M. Aluminum skeleton tower, red top, 26ft.
ANGUILLA ISLAND LIGHT, Road Point, 18 12.2N, 63 05.7W. Fl (2) W R 14s, 59ft, W 10M, R 6M. White concrete triangular structure,16ft. W 070°–089°, R 089°–116°, W 116°–218°.
WINDWARD POINT LIGHT, 18 16.8N, 62 58.1W. Fl (3) W, 14.5s, 82ft.

ROAD BAY

18 12N, 63 06W
Pilotage: Anguilla is quite low, with no conspicuous hills. The approach should be made in good light. Hazards include Dog Island, located about 10 miles northwest of Road Bay, and Prickly Pear Cays, located 5 miles north of the harbor. Seal Island Reefs extend about 5 miles in an easterly direction from Prickly Pear Cays. The water is quite deep right up to the edge of the reefs. Dowling Shoals and Sandy Island lie about 2 miles northwest of the harbor. The white-and-red sectored light shown from Road Point helps direct you around the hazards. Boats are warned to watch for anchored vessels not showing anchor lights in the bay.
Port of entry: Customs and immigration are located in the Marine Branch office of the Royal Anguilla Police, opposite the small (noncommercial) pier in Road Bay.
Anchorage: There is good anchorage in depths of 7 to 20 feet. Keep the channel to the commercial wharf clear. In the area immediately inside the hook of Road Point a coral reef and shallows extend toward town. You can dinghy in to the dock in the north part of the bay. There is a wide variety of uncrowded anchorages on other parts of the island and on nearby cays.
Services: There are few services oriented directly to yachts, but there are sources for fuel, water, ice, and groceries ashore. You'll have to ferry supplies to your boat via dinghy. There is a good variety of restaurants, bars, and shops.

Because Anguilla is somewhat off the beaten track and has a charterboat fee, it is visited by fewer boats than most islands in this area. There are few yacht-oriented services, but there are interesting anchorages, abundant dive sites, and pleasant eateries.

St. Martin

Please see cautions about chartlets and waypoints on page P 2.

ENTRY PROCEDURES

Situated at the top of the Guadeloupe Archipelago, the island, called St. Martin on the French side and Sint Maarten on the Dutch side, is 140 miles northwest of Guadeloupe, 144 miles east of Puerto Rico, and 1,460 miles south of New York. The capital is Marigot. St. Martin is half French and half Dutch, but there are no customs formalities between the two halves. The French courtesy flag should be flown here. The Dutch part of the island, Sint Maarten, is discussed in the next section of this chapter. The procedures for arriving yachts are handled informally, but you should check in with officials when arriving and leaving—this also means clearing in and out when passing between French St. Martin and Dutch Sint Maarten.

Yacht clearances are handled in Marigot. Officials are located in the port authority office near the ferry wharf. St. Martin has no duty on imports, so it is a good place to shop or have parts shipped to your boat.

USEFUL INFORMATION

Language: French is the official language, but English is spoken nearly everywhere.

Currency: French franc (F); exchange rate (summer 1997), 5.5F = US$1. U.S. dollars are widely accepted. Banks on the French side are open from 8:30 am to 1:30 pm weekdays and are closed on weekends, holidays, and afternoons preceding holidays.

Telephones: Country code, 590; local numbers, six digits. When calling within French St. Martin or to St. Barts (also part of the department of Guadeloupe), dial only the six-digit local number. To call Sint Maarten from St. Martin, dial 00-599-5, then the Dutch Sint Maarten number you want. When calling from the Dutch side to the French side, dial 06, then the six-digit French number. Calls are routed through Guadeloupe. For all calls from public phones on the French side, "Télécartes" are required. They look like credit cards and can be purchased at the Marigot post office, in book stores, and in some shops. There are no coin phones.

Public holidays (1998): New Year's Day, January 1; Carnival, February 21–25*; Good Friday, April 10*; Easter Monday, April 13*; Queen's Birthday (honoring the Queen of Holland), April 30; Labor Day, May 1; Ascension Day, May 22; Pentecost Monday, June 1;

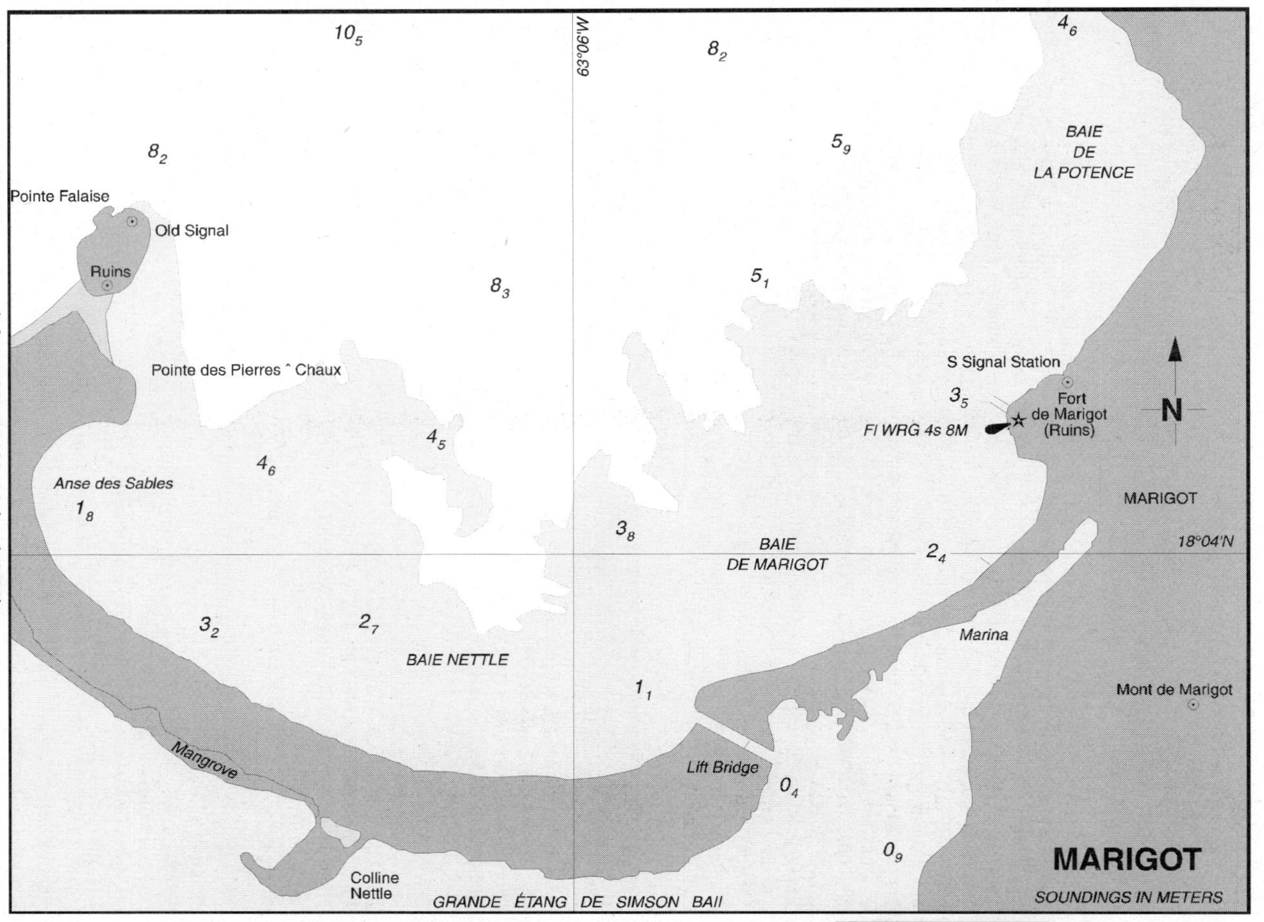

Marigot, soundings in meters *(after DMA 25608, 20th ed, 9/84)*

Bastille Day, July 14; Schoelcher Day, July 21; Assumption Day, August 15; Concordia Day, November 11; Christmas Day, December 25. *Many businesses closed. St. Martin hosts the Heineken Regatta in early March.

Pet regulations: Cats and dogs above the age of three months are admitted temporarily upon presentation of a certificate of origin and good health, or a certificate of antirabies inoculation, issued by a licensed veterinarian.

Tourism offices: *On St. Martin:* Office du Tourisme, Boulevard de France 97150, Marigot (590-87-57-21)—Monday through Friday, 8:30 am to 1 pm and 2:30 to 5:30 pm; Saturday 8 am to noon. *In the U.S.:* French Government Tourist Office, 444 Madison Avenue, New York, NY 10022 (900-990-0040, 9am–7pm; charge, $.95/minute).

Emergencies: Contact Curaçao Radio, call sign PJC, on VHF channel 16. Away from Marigot and Simpson Bay Lagoon you may not be able to call very far on VHF because of the island's high hills.

ST. MARTIN APPROACHES

BREAKWATER LIGHT, N end, head of breakwater, 18 06.9N, 63 03.0W. Fl R 2.5s, 10ft, 5M. White framework tower, red top, 6ft.

BREAKWATER SPUR LIGHT, 18 06.9N, 63 03.0W. Fl G 2.5s, 10ft, 5M. White framework tower, green top.

GAILSBAY JETTY, 18 5.0N, 63 5.4W. Fl (3) WRG 12s, 33 ft. W 6M, R 3M, G 3M.

BAIE DU MARIGOT LIGHT, 18 04.2N, 63 05.2W. Fl W R G 4s, 66ft, W 11M, R 8M, G 8M. White tower with red top, 33ft. R 104°–126°, W 126°–132°, G 132°–185°, obscured 185°–104°.

MARIGOT

18 04N, 63 05W

Pilotage: A flashing white, red, and green light is shown from the 200-foot-high hill topped by the ruins of Fort de Marigot. You can approach the commercial pier, below the light, on a bearing of 143°T.

Port of entry: The port authority office is near the ferry and commercial wharf.

Bridge to Simpson Bay Lagoon: The bridge that allows access into Simpson Bay Lagoon from the French side opens every day at 0900, 1330 or 1400, and 1730. For opening times of the bridge at the Dutch end of the lagoon, see the Sint Maarten section of this chapter.

Dockage: Most of the marinas are on the Dutch side of Simpson Bay Lagoon. In the east end of the lagoon, off Marigot, the Marina Port La Royal (tel., 590-87-20-43; fax, 590-87-55-95) and Caraibes Sport Boats (tel./fax, 590-87-89-38) offer all kinds of services, shops, and restaurants. Their facilities have been rebuilt since Hurricane Luis.

Anchorage: Marigot Bay is a good anchorage, although it can be subject to swells from the north. Anchor away from the approaches to the pier. In Simpson Bay Lagoon watch depths carefully. As a hurricane hole, it is sheltered from the ocean waves, but open to wind action.

Services: There is an extensive variety of yacht services here, and there are more in nearby Sint Maarten. There are excellent grocery, liquor, and gourmet shops, as befits a French island. UPS and Federal Express also maintain offices here if you need something in a hurry from the United States.

Because it is only 74 miles from Virgin Gorda, St. Martin is often a first port of call in the Leeward Islands. Marigot is the capital of the French half of the island and has a good harbor where you may clear in quite easily. Ashore are many fine restaurants and shops, which stock a wide variety of items at duty-free prices. With the relatively sheltered waters of Simpson Bay Lagoon nearby and a continental atmosphere, Marigot is a favorite long-term stop for many cruisers.

GRANDE CASE

18 06N, 63 03W

Pilotage: A shoal area north of the point forming the south end of Baie Grande Case is reported to be as shallow as 5 feet.

Anchorage: Anchor well in toward shore in the southeast part of the bay to avoid any swells rolling around Roche Crole off Bell Point. Depths are 15 to 20 feet, shoaling to 6 to 8 feet close to the beach.

Services: Marine services are limited, but dozens of fine restaurants line the main street. Provisions can be purchased, and there are general shops for browsing.

ANSE MARCEL

18 07N, 63 02W

Pilotage: This small bay is located west of the northernmost part of the island. It is the only major indentation on the northern coast. Do not pass between Marcel Rock located in the mouth of the bay and the mainland to the southwest, as the waters are quite shoal.

Dockage: Port Lonvilliers is entered by way of a narrow channel that begins in the southeast corner of the bay. Following Hurricane Luis, Port Lonvilliers Marina (VHF channels 11 and 16; tel., 590-87-31-94; fax, 590-87-33-96) rebuilt its docks. It now has space for 100 boats and offers amenities such as boutiques, a grocery, cafe, and La Capitainerie, a ship chandlery, open daily from 8 am to 7 pm Under most circumstances, this is a very protected spot.
Anchorage: There is sheltered anchorage in the bay, north of town. Depths run from 8 to 15 feet.
Services: Port Lonvilliers Marina has fuel, water, and ice. For repairs and marine supplies contact Yachting Maintenance (590-87-43-48). There is a small grocery store and several shops. Several restaurants are nearby.

OYSTER POND

18 03N, 63 01W
Pilotage: This small pond is located near the middle of the east coast of the island, where the French and Dutch sides meet. With an east-facing entrance surrounded by reefs, it is no place to enter after dark or in rough conditions. The entrance channel is privately marked by Captain Oliver's Marina. The controlling depth is reported to be about 10 feet as far as the fuel dock. If you are in doubt as to your navigation or the conditions, call the marina on VHF channels 16 or 67.
Dockage: Call Captain Oliver's (VHF channels 16 and 67; tel./fax, 590-87-33-47). The Moorings (590-87-32-55) charter fleet is also located here, but has no space for noncharterers.
Anchorage: The harbor is quite shoal, except for the area near the docks. There is a shoal in the middle of the harbor, with depths of 5 to 6 feet around most of the perimeter. Captain Oliver's dinghy dock is north of the marina.
Services: Fuel, water, ice, showers, small grocery, and ship chandlery are available at Captain Oliver's Marina, as is a dive shop, gourmet restaurant, and hotel. The Moorings has a ship's store and grocery, open to all. Other restaurants and shops are nearby.

Sint Maarten

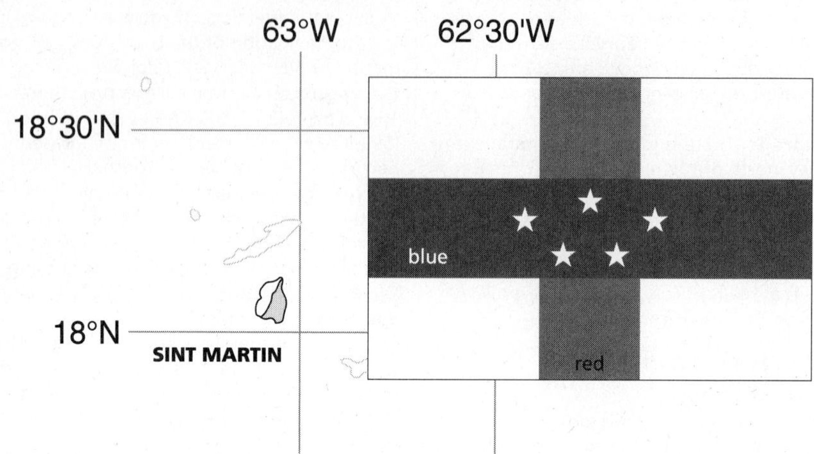

Please see cautions about chartlets and waypoints on page P 2.

ENTRY PROCEDURES

Dutch Sint Maarten, which is the southern half of the same island as French St. Martin, is part of the Netherlands Antilles. The Dutch courtesy flag should be flown while in these waters. Sint Maarten is a duty-free port and has no customs as such, but you should clear with immigration officials (599-5-22740) in the police station near the ferry dock in Philipsburg. You can reach the Philipsburg police at 599-5-22222. You should check in when traveling back and forth to the French side of the island. Formalities are swift and simple.

USEFUL INFORMATION

Language: Dutch. English is widely spoken.
Currency: Antillean guilder; exchange rate (summer 1997), 1.77 guilders = US$1. U.S. currency is widely accepted.
Telephones: Country code for northern Netherlands Antilles, 599; code for Sint Maarten, 5; local numbers, five digits. When calling the French side of the island from the Dutch side, use the code 06 before the six-digit French number. When calling the Dutch side from the French side, begin with the international access code of 19. The calls are routed through Guadeloupe.
Public holidays (1998): New Year's Day, January 1; Good Friday, April 10; Easter Monday, April 13; Queen's Birthday, April 30; Labor Day, May 1; Ascension Day, May 8; St. Maarten Day, November 11; Christmas Day, December 25; Boxing Day, December 26.
Tourism offices: *On Sint Maarten:* St. Maarten Tourist Bureau, Imperial Building, #23 W.J. Nisbeth Road, Philipsburg (599-5-22337). *In the U.S.:* St. Maarten Tourist Office, 675 Third Avenue, 18th Floor, New York, NY 10017 (800-786-2278 [800-STMAARTEN], 212-953-2084). *In Canada:* Sitmar Consultants (representative), 243 Ellerslie Avenue, Willowdale, Ont. M2N 1Y3 (tel., 416-223-3501; fax, 416-223-6887).
Emergencies: Contact Curaçao Radio, call sign PJC, on VHF channel 16.

SINT MAARTEN APPROACHES

SIMSON BAAI AVIATION LIGHT, 18 02.4N, 63 06.7W. Al Fl W G 6s, 52ft, 13M. On station building.

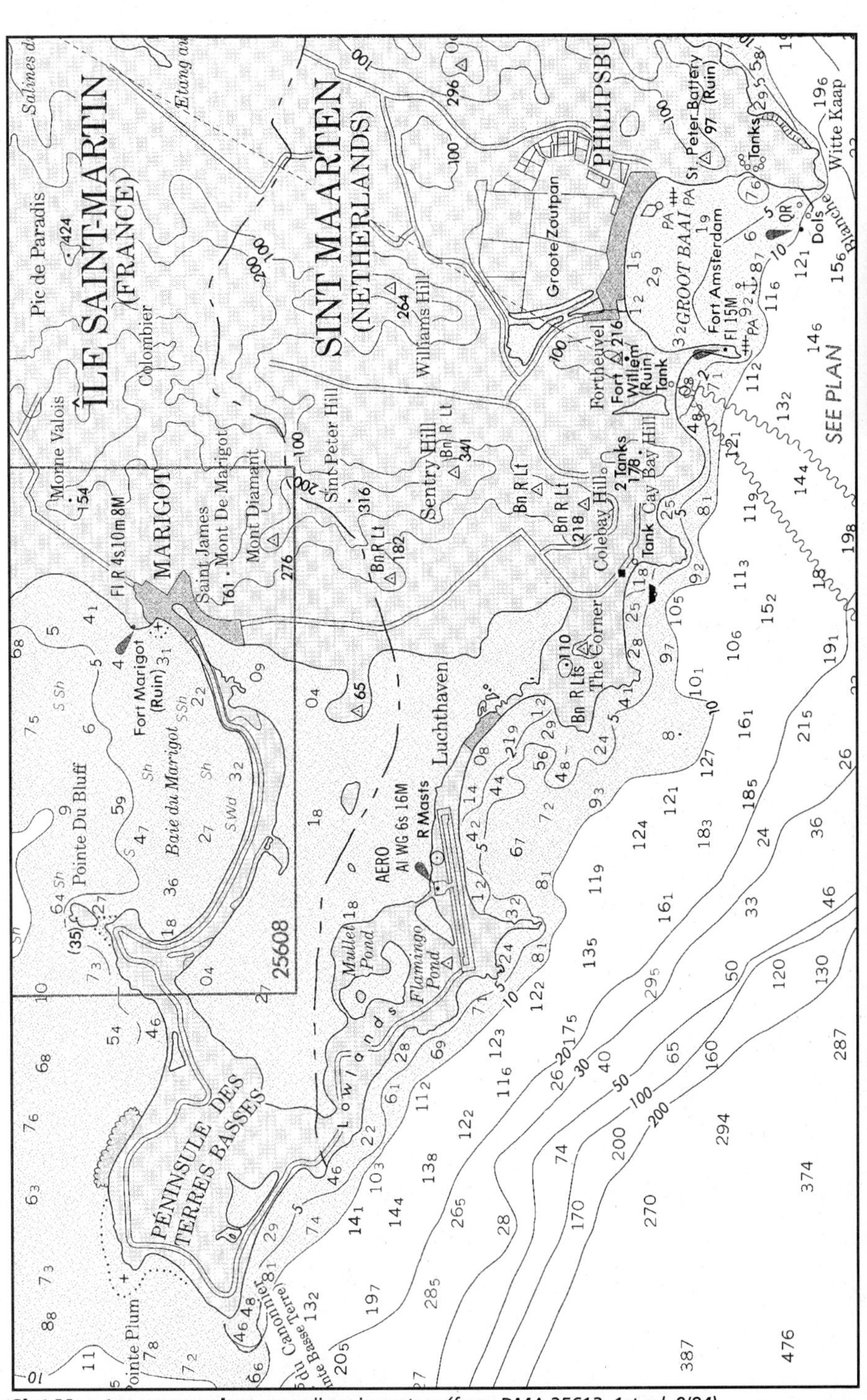

Sint Maarten approaches, soundings in meters *(from DMA 25613, 1st ed, 9/84)*

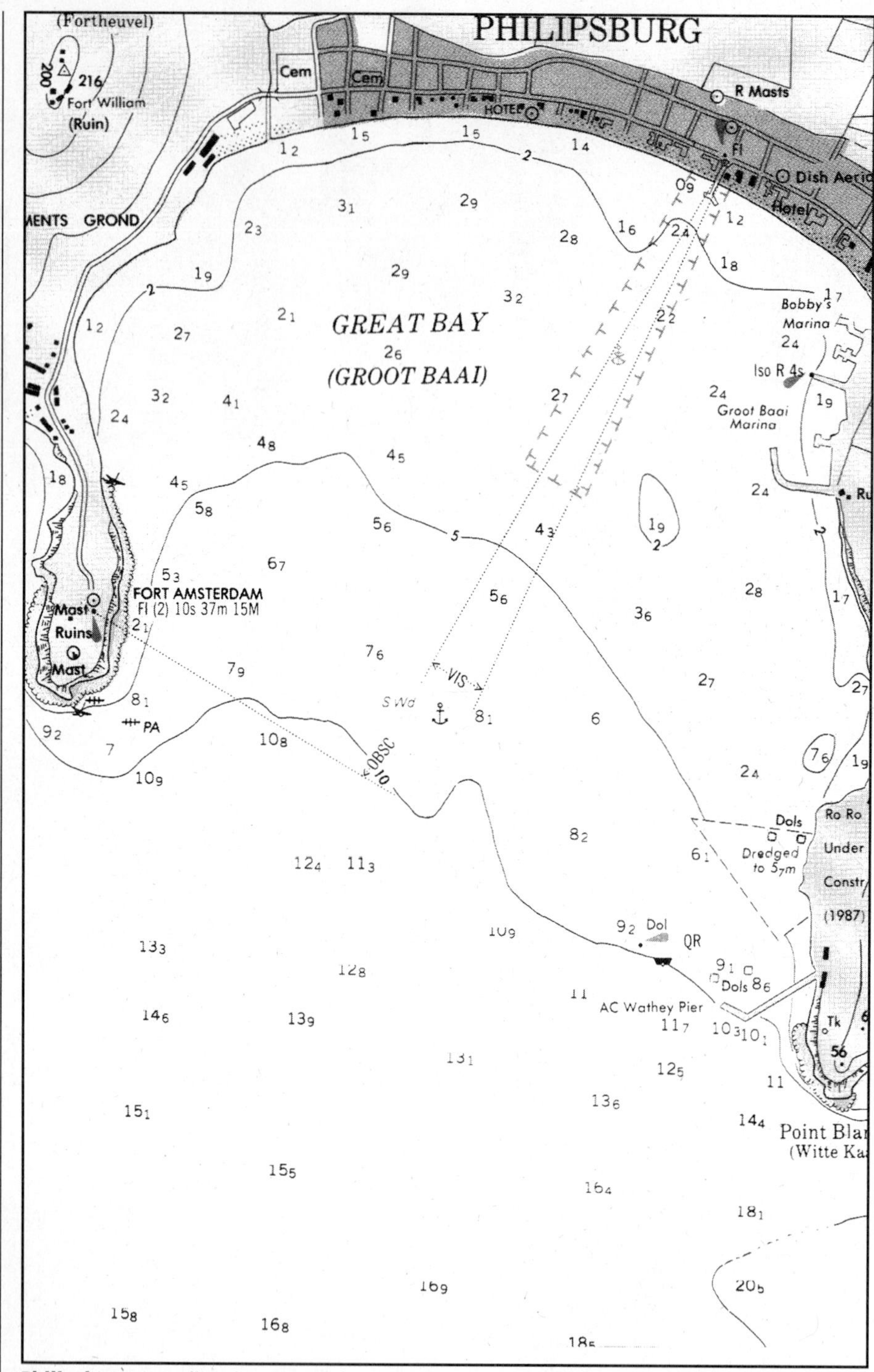

Philipsburg, soundings in meters *(after DMA 25613, inset, 1st ed, 9/84)*

GROOT BAAI, on side of old Fort Amsterdam, 18 00.8N, 63 03.6W. Fl (2) W 10s, 120ft, 15M. Wooden post, 6ft. Visible 300°–096°.
A.C. WATHEY PIER, 18 00.5N, 63 02.9W. Q R. Mooring dolphin. Two F R mark Ro Ro berth 380 meters NE. Fl R 10s on dolphin 120 meters SE.

PHILIPSBURG

18 01N, 63 03W
Pilotage: Groot Baai (Grand Bay) is the main port and the location of Sint Maarten's capital, Philipsburg. A light is shown from Fort Amsterdam on the western point of land at the bay's mouth. There is a stranded wreck about 200 yards south of the fort. At press time (summer 1997), the rebuilding of the cruise ship pier at the south end of the eastern bayfront was due to begin in August (following its devastation in the fall 1995 hurricane). A new town pier has been completed, with space for eight dinghies to tie up. Groot Baai is generally free of obstructions, except for shoaling in the northwest portion.
Port of entry
Dockage: Bobby's Marina (VHF channel 16; tel., 599-5-22366) and Great Bay Marina (VHF channel 16; tel., 599-5-22167) are on the eastern side of Groot Baai.
Anchorage: There is good anchorage in the bay, with depths from about 6 to 20 feet, although the fall 1995 hurricane Luis reportedly deepened the harbor somewhat.
Services: Almost any yacht service is available here and at very good prices for the Caribbean. Fuel, water, and ice are available at the marinas, and Bobby's Marina can haul boats to 90 tons. Among the many marine-oriented businesses, Dockside Management (VHF channel 16; 599-5-24096; fax, 599-5-22858) is a telephone, fax, information, message, and shipping center. The staff are very helpful and speak English. Dockside will also hold mail for cruisers (c/o P.O. Box 999, Philipsburg, Sint Maarten, Netherlands Antilles). Budget Marine (tel., 599-5-22068; fax, 599-5-23804), opposite Bobby's Marina, is among the best-stocked chandleries in the Caribbean and can order any gear or parts not in the store's current inventory. Tropical Sails (599-5-22842) specializes in navigation equipment, books, charts, and general marine gear. There are Federal Express and UPS offices on the island, so it is easy to get items quickly from the United States. There are several good grocery stores.

Philipsburg is a favorite stop for cruising boaters and one of the best places in the Caribbean for repairs or stocking up. The harbor is sometimes rolly, but is generally a good anchorage. Nearby Simpson Bay Lagoon provides better shelter and more marine services.

SIMSON BAAI

18 02N, 63 06W
Pilotage: Simson Baai, or Simpson Bay, is about 3 miles to the northwest of Groot Baai and Philipsburg. Princess Juliana Airport lies along the north side of the bay, and there is an aviation light shown from just north of the runway. The entrance to Simpson Bay Lagoon is on the eastern side of the bay.
Bridge to Simpson Bay Lagoon: The bridge at the Simpson Bay entrance to the lagoon currently opens at 1100 and 1800 seven days a week. The Simpson Bay bridge operator monitors VHF channel 12 (for information on the Marigot bridge entrance, see the St. Martin section of this chapter).
Dockage: There are several marinas in the Lagoon, among them, Simpson Bay Yacht Club Marina (VHF channel 16; tel., 599-5-43378), with luxurious services as well as one of Budget Marine's branch stores, and Island Water World (VHF channel 74; tel., 599-5-45310). Both the the Port de Plaisance Marina (tel., 599-5-45222; fax, 599-5-42315) and Lagoon Marina (599-5-45210) suffered significant damage in the hurricane. At *Reed's* press time, the Port de Plaisance resort was still closed, with no opening date available, and repairs for the Lagoon Marina docks were still in the planning stages. Lagoon Marina's other services, including laundry, showers, and a restaurant and bar, are fully operational, however.
Anchorage: The various anchorages in the lagoon are some of the most protected around. It is considered a hurricane hole, although there was significant damage here when it was directly hit by Hurricane Luis in fall 1995
Services: Extensive marine services are available here, in Philipsburg, and in Marigot on the French side. Fuel, water, and ice are available at the marinas. There are good grocery stores and restaurants nearby.

Simpson Bay Lagoon is one of the best places in the Caribbean for restocking and repairs. Most businesses and services have rebounded from the damage inflicted by Hurricane Luis in fall 1995. If you tire of being in Dutch waters, simply cross the lagoon to the French side.

St. Barthélemy

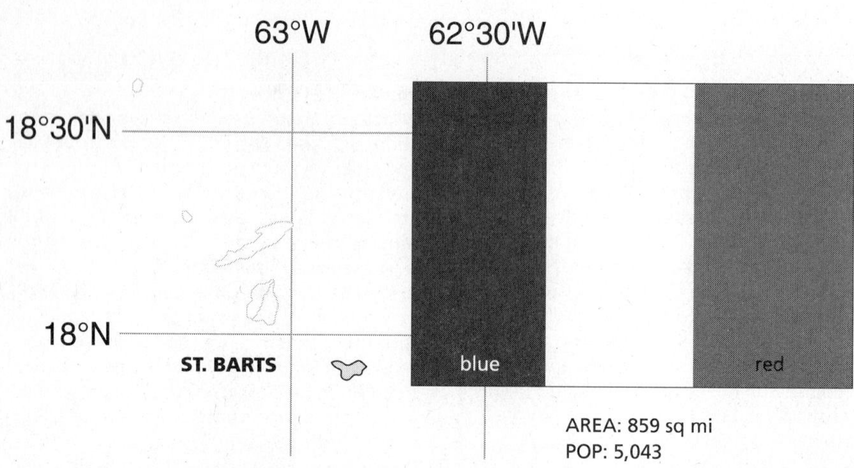

AREA: 859 sq mi
POP: 5,043

Please see cautions about chartlets and waypoints on page P 2.

ENTRY PROCEDURES

The island of St. Barthélemy (St. Barts) is a dependency of Guadeloupe, which in turn is an overseas Department and Region of France. St. Barts lies 125 miles northwest of Guadeloupe and 15 miles southeast of St. Martin. Its capital is Gustavia. The French courtesy flag should be flown here. Procedures for yachts clearing in and out are handled informally through the port authority. The office (VHF channel 16; tel., 590-27-66-97) is located on the main dock on the east side of Gustavia harbor.

USEFUL INFORMATION

Language: French is the official language, but English is widely spoken.

Currency: French franc (F); exchange rate (summer 1997), 5.5F = US$1. U.S. dollars are widely accepted. Banks are closed on weekends, holidays, and afternoons preceding holidays.

Telephones: The area code for St. Barts is 590. To call St. Barts from Dutch St. Maarten, dial 6 plus the six-digit St. Barts number. To call St. Barts from other French West Indies islands (Martinique, Guadeloupe, and French St. Martin), dial direct.

To call the U.S., dial 19, wait for second tone, then dial 1, area code, and number. (A one-minute call to the U.S. costs 12.85F at press time.) To call Sin. Maarten from St. Barts: 3 plus 5-digit St. Maarten number. To call Martinique, Guadeloupe, and French St. Martin, dial direct.

For local and international calls from public phones, use "Télécartes," which can be purchased at the Gustavia, St. Jean, and Lorient post offices and at the gas station near the airport. There are no coin booths on the island. The telex area code is 340.

Mail: St. Barts has one post office with two branches. The main office is on the corner of rue du Centenaire and rue Jeanne d'Arc in Gustavia and is open daily 8 am to noon and 2 pm to 4 pm; it closes on Wednesday and Saturday afternoons. The branch in Lorient is open from 7 to 11 am weekdays and from 8 to 10 am on Saturday. The branch at the St. Jean Commercial Center across from the airport opens daily from 9 am to noon and 3 to 5:30 pm, as well as Wednesday and Saturday mornings. All are closed on Sunday. Airmail stamps for postcards to the U.S. cost 3.10 F; airmail stamps for letters to the U.S. (less than 20 grams) cost 3.90F.

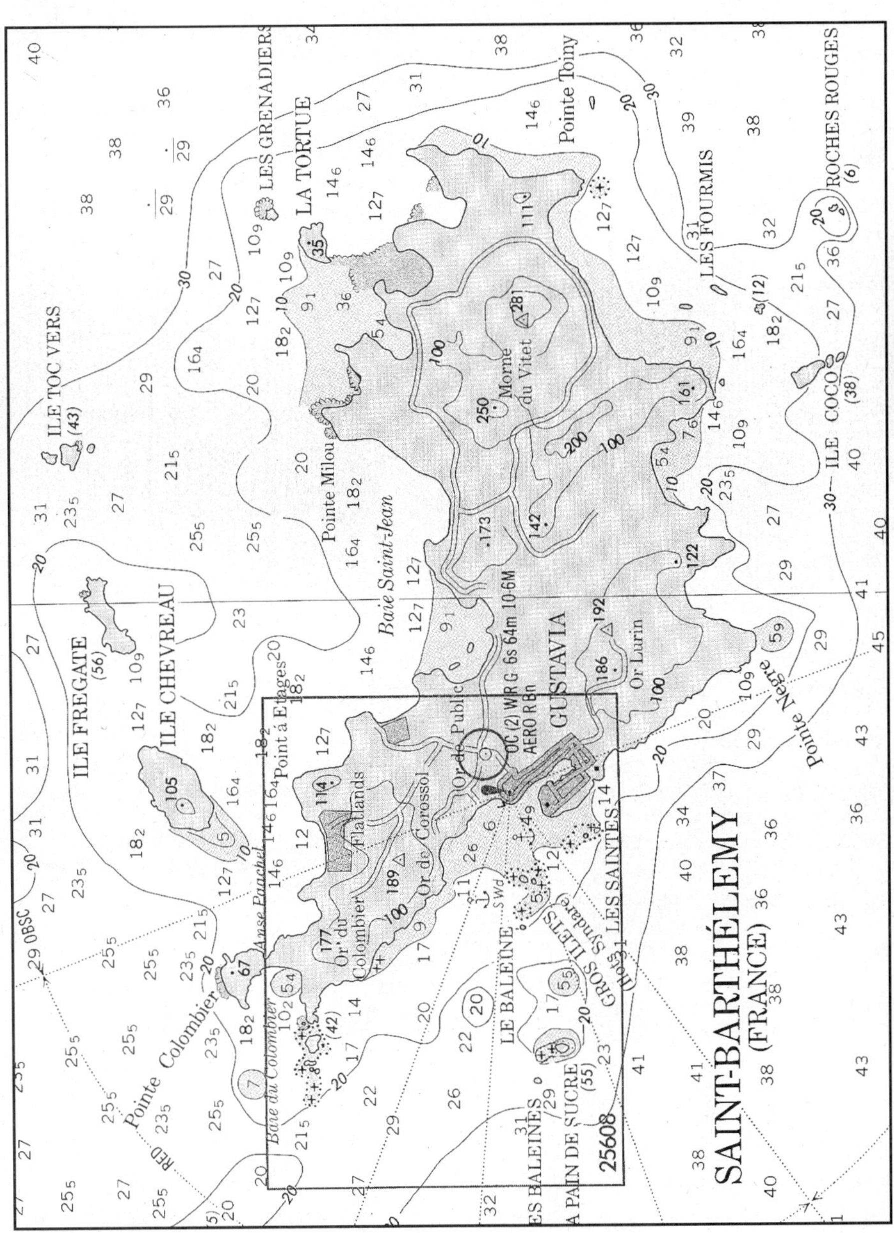

St. Barthélemy, soundings in meters *(from DMA 25613, 2nd ed, 1/95)*

Firearms: Yachts are permitted to have firearms aboard, but they must be declared.

Pet regulations: Dogs and cats over three months old are allowed in temporarily with certificates of origin and good health (or antirabies inoculation) issued by a licensed veterinarian.

Public holidays (1998): New Year's Day, January 1; Carnival, February 20–25*; Easter Monday, April 13; Labor Day, May 1; Armistice Day (World War II) and Ascension Thursday, May 21; Pentecost Monday, June 1; Slavery Abolition Day, May 27; Bastille Day, July 14; Assumption Day and St. Barts/Pitea Day, August

15; All Saints Day, November 1; All Souls Day, November 2; Armistice Day (World War I), November 11; Christmas Day, December 25. *Most businesses close, although not an official holiday.

Regattas: Annual regattas in St. Barts include the St. Barts Regatta, held just before Lent; the St. Barts Cup, held in April; and the three-day International Regatta, held in May, with races for gaffers, cruising, racing, etc.

Tourist offices: *On St. Barts:* Office du Tourisme, Quai du Général de Gaulle, BP 113, Gustavia 97098 (590-27-87-27). The office is located across from the Capitainerie; hours in peak season are 8:30 am to 6 pm, Monday through Friday, and 8:30 am to noon on Saturdays. From June through late September, the hours are a bit shorter. *In the U.S.:* French Government Tourist Office, 444 Madison Avenue, New York, NY 10022 (900-990-0040, 9am–7pm; charge, $.95/min.); in Beverly Hills, CA, 310-271-6665; and in Chicago, 312-751-7800. *In Canada:* French Government Tourist Office, Montreal, 514-288-4262/fax, 514-844-8901; in Toronto, 416-593-6427/fax, 416-979-7587.

Emergencies: Call Curaçao Radio, call sign PJC, on VHF channel 16.

Medical Facilities: Gustavia has a hospital (Tel, 27-60-35), five resident doctors, six dentists, and many specialists. There are pharmacies at La Savane Commercial Center in St. Jean (Tel, 27-66-61) and in Gustavia (Tel, 27-61-82).

ST. BARTS

FORT GUSTAVIA LIGHT, 17 54.3N, 62 51.2W. Fl (3) W R G, 12s, 210ft, W 11M, R 8M, G 8M. White truncated tower, red top, 29ft. R 340°–095°, W 095°–111°, G 111°–160°, obscured 160°–340°.

GUSTAVIA

17 54N, 62 51W

Pilotage: The port lies on the west coast of the island. The light is shown from Fort Gustave, which is north of the inner harbor. The safest approaches are made by coming in on the white sectors of the light. The port captain monitors VHF channel 16.

Port of entry

Dockage: It may be possible to tie up stern-to one of the quays in the inner harbor. Check with the port captain for availability and the daily charge.

Anchorage: Gustavia's harbor, 13 to 16 feet deep, has mooring and docking facilities for about 40 boats. You can anchor in the inner harbor if there is room or outside the marked channel into the inner harbor. Daily charges will be collected by a patrol boat or at the port captain's office. There are also fine anchorages at nearby Public, Corossol, and Colombier.

Services: St. Barts has all the conveniences boaters need: fuel, water, ice, a well-stocked chandlery, supermarkets, duty-free liquor, and fine French dining. For anything even remotely connected to boating, Loulou's Marine (590-27-62-74) is the first choice of many. The staff speaks English and the bulletin board is a treasure trove of yachting news. Fuel is available at the commercial wharf north of Fort Gustave by contacting St. Barth Marine (590-27-60-38, 590-27-50-50). St. Barth Marine also connects to an electronics sales and repair service. Le Ship (VHF channel 16; tel., 590-27-86-29; fax, 590-27-85-73) is a chandlery that represents several major hardware and maintenance manufacturers; parts not in stock can be ordered. Other services at Le Ship include sending faxes and renting cars. The AMC supermarket (tel., 590-27-60-09; fax, 590-27-85-71) is a source for groceries; call in advance and your provisions will be gathered for your arrival. There's a good selection of French restaurants. Being a duty-free port, Gustavia offers a wide variety of stores of all types. Prices are very good for the Caribbean.

Gustavia is a popular harbor and can be very crowded in season. You may have some difficulty finding a secure spot for your boat. Once ashore, you'll find a good variety of services and supplies, although not as extensive as in St. Martin and Sint Maarten. Because St. Barts is a French island, the dining can be particularly enjoyable.

SABA

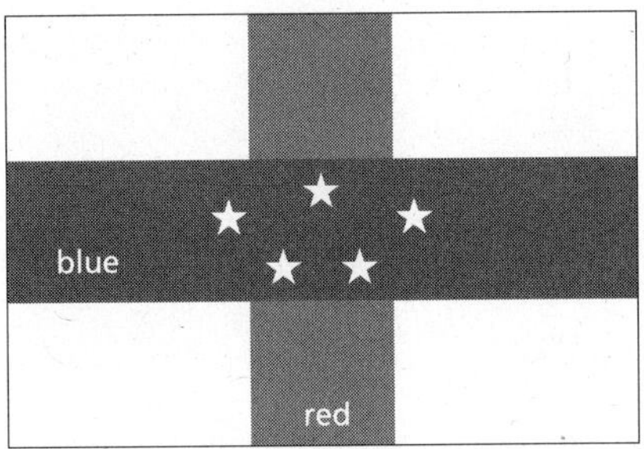

POP: 1,200

Please see cautions about chartlets and waypoints on page P 2.

ENTRY PROCEDURES

Saba lies about 26 miles southwest of St. Barthélemy and is part of the Netherlands Antilles. The Dutch courtesy flag should be flown. Saba is a free port, so there are no customs. However, vessels must clear in with the harbor office in Fort Baai (VHF channel 16, switch to VHF channel 11; tel., 599-4-63294). The harbormaster has immigration duties and collects harbor and clearance fees, which vary according to a vessel's length (40-foot yacht, US$8–12). By law, vessels must also check in with Saba Marine Park (VHF channel 16; tel., 599-4-63295) before proceeding to the Ladder Baai/Wells Baai mooring area. Formalities are swift and simple.

USEFUL INFORMATION

Language: Dutch (official). English is widely spoken.

Currency: Antillean guilder; exchange rate (summer 1997), 1.80 guilders = US$1. U.S. currency is widely accepted.

Telephones: Country code for the northern Netherlands Antilles, 599; Saba code (0)4; local numbers, five digits. Within Saba, dial five digits. (To call Saba from other northern Antillean islands, dial the prefix 04 and then the five-digit Saba number.)

Public holidays (1998): New Year's Day, January 1; Good Friday, April 10; Easter Monday, April 13; Coronation Day, April 30; Labor Day, May 1; Ascension Thursday, May 21; Carnival Monday, August 3 (tentative, official date set after *Reed's* press time); Saba Days, December 4–6; Christmas Day, December 25, Boxing Day, December 26.

Tourism offices: *On Saba:* Saba Tourist Bureau, P.O. Box 527, Windwardside (599-4-62231/62322; fax, 599-4-62350). *North American representative:* Classic Communications Intl., P.O. Box 6322, Boca Raton, FL 33472-6322 (800-722-2394, 407-394-8580; Web, www.turq.com/saba).

Emergencies: Saba Radio is no longer on the air. Curaçao Radio (599-9-63435), call sign PJC, has replaced Saba Radio as the the main communications service in this part of the Caribbean. For search and rescue emergencies Curaçao Radio is in contact with the U.S. Coast Guard, the Antillean Sea Rescue Foundation, and other island rescue agencies. The station also links medical emergency calls to appropriate doctors or facilities. The station can be contacted on VHF channels 16 and 26 seven days a week from 0600 to 2400. The daily weather is broadcast at the same time as traffic lists or on request. Traffic lists are broadcast at 0800 and 1600 and in between as calls come in. During hurricane season Curaçao Radio broadcasts hurricane advisories on the hour, as received, and on request.

SABA

AVIATION LIGHT, 17 38.1N, 63 14.3W. Fl W 1.5s, 3084ft. Radio mast.

ST. JOHN'S LIGHT, 17 37.1N, 63 14.6W. Fl (2) W 10s, 15M.

FORT BAAI BREAKWATER LIGHT, 17 37.0N, 63 15.1W. Fl (3) G.

FORT BAAI PIER LIGHT, 17 36.9N, 63 15.0W. Fl (3) R.

FORT BAAI

17 37N, 63 13W

Pilotage: The coast of Saba rises nearly perpendicular from the sea, with the 2,853-foot summit of the island usually enveloped in clouds. The island is a mass of rugged mountains with

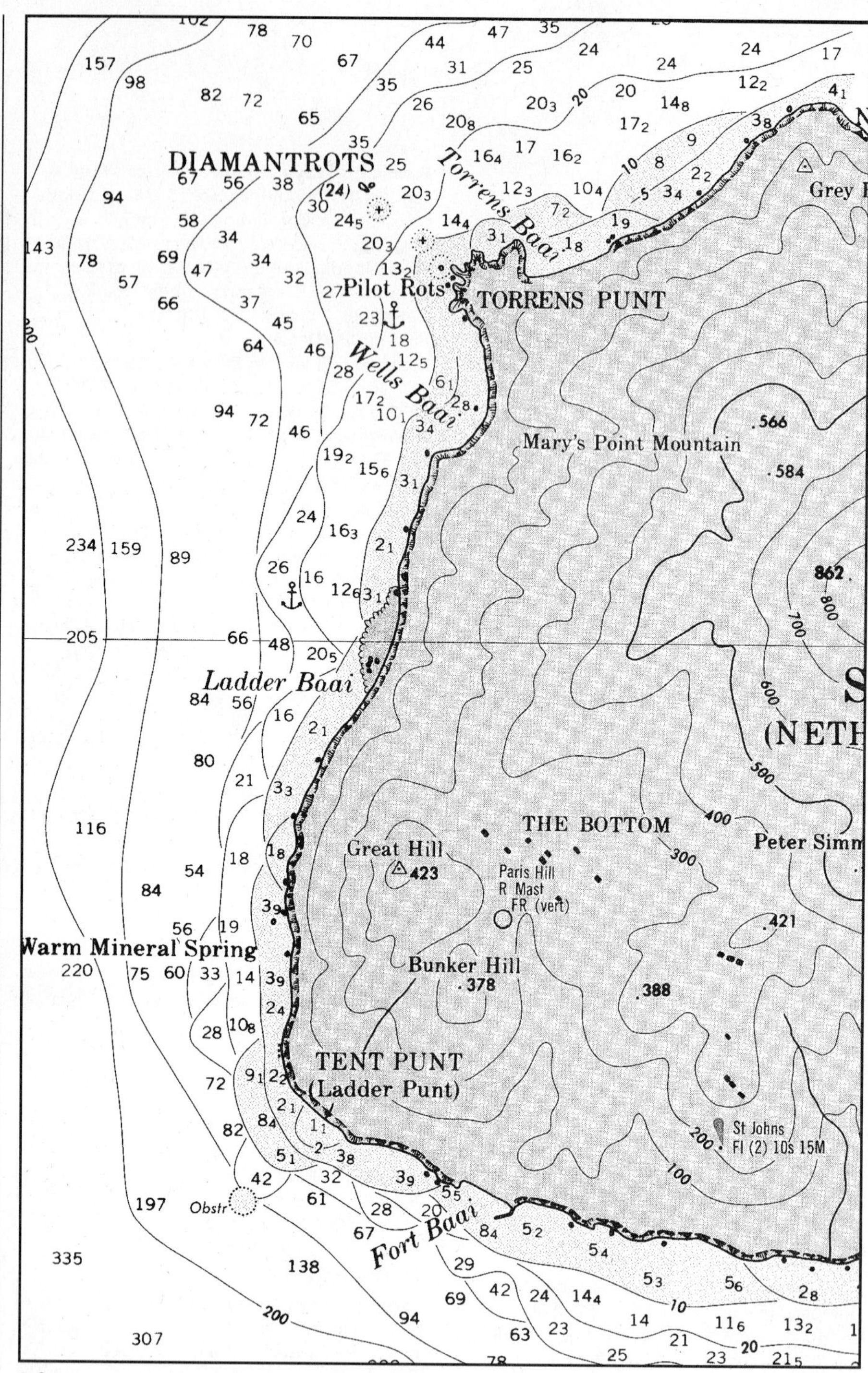

Saba, west coast, soundings in meters *(from DMA 25607, 3rd ed, 7/84)*

deep, precipitous ravines. Two conspicuous radio masts stand on the island. St. John's Light is exhibited about 0.5 mile east-northeast of Fort Baai. The Saba Bank, located south and west of the island, is deep enough to sail over, but should be avoided in heavy weather.

Port of entry

Dockage: You may be able to tie up inside the 100-yard-long jetty at Fort Baai while you clear. Check first with the harbormaster. Because the pier is small, it is for loading and unloading goods only, unless the harbormaster gives permission.

Anchorage: Arriving vessels usually anchor off Fort Baai and dinghy in to the dock to clear in. Saba's entire coast is part of the Saba Marine Park, and anchoring is restricted to two areas only, one off Fort Baai and the other off Ladder Baai and extending to Wells Baai. The Fort Baai anchorage is not recommended for overnight stays. Vessels must check in with the Saba Marine Park office in Fort Baai (VHF 16) before proceeding to the LadderBaai/Wells Baai anchorage area. The center of Wells Baai offers excellent holding in clear sand with plenty of swinging room. Ladder Baai offers good holding in depths of about 20 feet. There are rocks close in to shore.

Moorings: Mooring buoys around Saba are color-coded according to their intended use (dive site or overnight anchorage) and size. Specifics are available from the Saba Marine Park office (VHF 16). The two large orange metal buoys off Fort Baai may be used while vessels clear in, but call ahead to the harbormaster on VHF channel 16 for permission and availability. *Vessels are cautioned not to pick up the white or orange dive buoys anywhere off Saba.* There are five yellow mooring buoys designated specifically for yachts to 60 feet or 50 tons in the Ladder Baai/Wells Baai anchorage. The nominal yacht visitor and dive fees help support the Saba Marine Park. No fishing or collecting of any marine life, dead or alive, is allowed in the marine park. When diving or snorkeling you should display the dive flag. Be sure to stay 150 yards seaward of any dive operations when under way.

Services: You may be able to get some fuel via jug. Provisions are available at The Bottom, the first village you come to as you travel up the island road. There are several interesting restaurants and shops. Take a taxi tour of the island via the famous road that winds among the mountains. There are three dive shops in Fort Baai: Saba Deep (599-4-63347), Wilson's Dive Shop (599-4-63410), and Saba Reef Divers (599-4-62541). There is also one in Windwardside, Sea Saba (599-4-62246). Wilson's Dive Shop and Saba Reef Divers each also have an office and boutique in Windwardside. Information on the island's great diving opportunities is available from the Saba Tourist Bureau.

Sint Eustatius

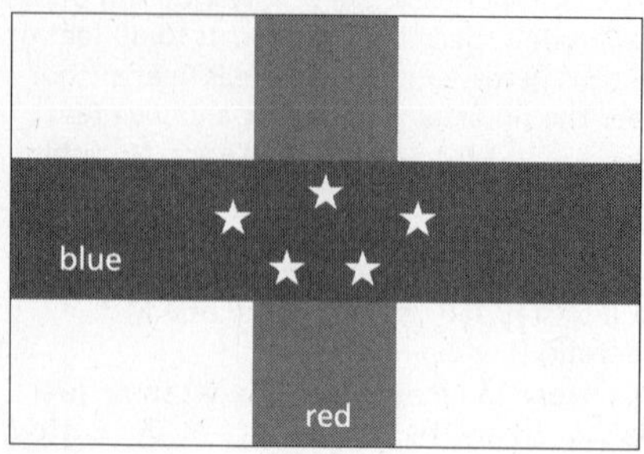

Please see cautions about chartlets and waypoints on page P 2.

ENTRY PROCEDURES

Sint Eustatius, known as Statia, is located about 38 miles south of Sint Maarten and is part of the Netherlands Antilles. The Dutch courtesy flag should be flown here. Statia is a duty-free port and has no customs as such; however, you should clear immigration with the harbor office (VHF channel 16; tel., 599-3-82205) in Oranjested near the foot of the large dock, where you register your vessel and number of crew. A valid passport is required, except for U.S. and Canadian citizens and residents, who may present a birth certificate, naturalization certificate, or re-entry visa, plus photo identification. Antilleans should present an ID card. Formalities are swift and simple. The harbormaster can be reached on VHF channel 16 or by phone at 599-3-82888.

USEFUL INFORMATION

Language: Dutch (official). English is widely spoken.
Currency: Antillean guilder; exchange rate (summer 1997), 1.77 guilders = US$1. U.S. currency and major credit cards are also widely accepted.
Telephones: Country code for the northern Netherlands Antilles, 599; island code, 3; local numbers, five digits. To call Statia from other northern Antillean islands, dial the prefix 03 and then the five-digit Statia number.
Public holidays (1998): New Year's Day, January 1; Good Friday, April 10; Easter Monday, April 13; Coronation Day, April 30; Labor Day, May 1; Ascension Day, May 8; Carnival Monday, July (date not yet set at press time); Statia America Day, November 16; Christmas Day, December 25; Boxing Day, December 26.
Tourism offices: *On St. Eustatius:* St. Eustatius Tourist Board, Fort Oranjestraat, Oranjestad (tel./fax, 599-3-82433). *North American representative:* Classic Communications International, P.O. Box 6322, Boca Raton, FL 33427 (tel., 800-722-2394, 407-394-8580; fax, 407-394-8588).
Emergencies: Contact Curaçao Radio, VHF channel 16.

WEST COAST

TUMBLEDOWN DICK BAY LIGHT, head of oil pier, 17 29.4N, 63 00.1W. Fl W 5s, 10M.
ORANJESTAD LIGHT, 17 28.8N, 62 59.2W. Fl (3) W 15s, 131ft, 17M.

ORANJESTAD

17 29N, 62 59W
Pilotage: The island is dominated by a perfectly formed extinct volcano 2,000 feet high and is one of the main attractions for visitors. A group of rugged hills is located on the northwest portion of the island. Fish pots may be encountered up to 3 miles off the west coast. About 1.2 miles west-southwest of the island's northern point, a submerged oil pipeline extends east-southeast toward shore from a moored buoy. Tumbledown Dick Bay Light is on the end of a 0.5-mile-long jetty located about 1 mile northwest of Oranjestad. The oil terminal at the jetty stands by on the VHF radio.
Port of entry
Anchorage: The anchorage off town, Oranje Baai, can be rolly. You may want to try a stern anchor to hold your boat into the swells.
Services: You can jug water and fuel to your boat, and there are good markets ashore. For a trip to Statia's delightful reefs, contact Dive Statia (VHF channel 16; tel., 599-3-82435) or Gold Rock Dive Center (599-3-82964). There are many historic buildings and several good restaurants, although the tourist industry is quite small.

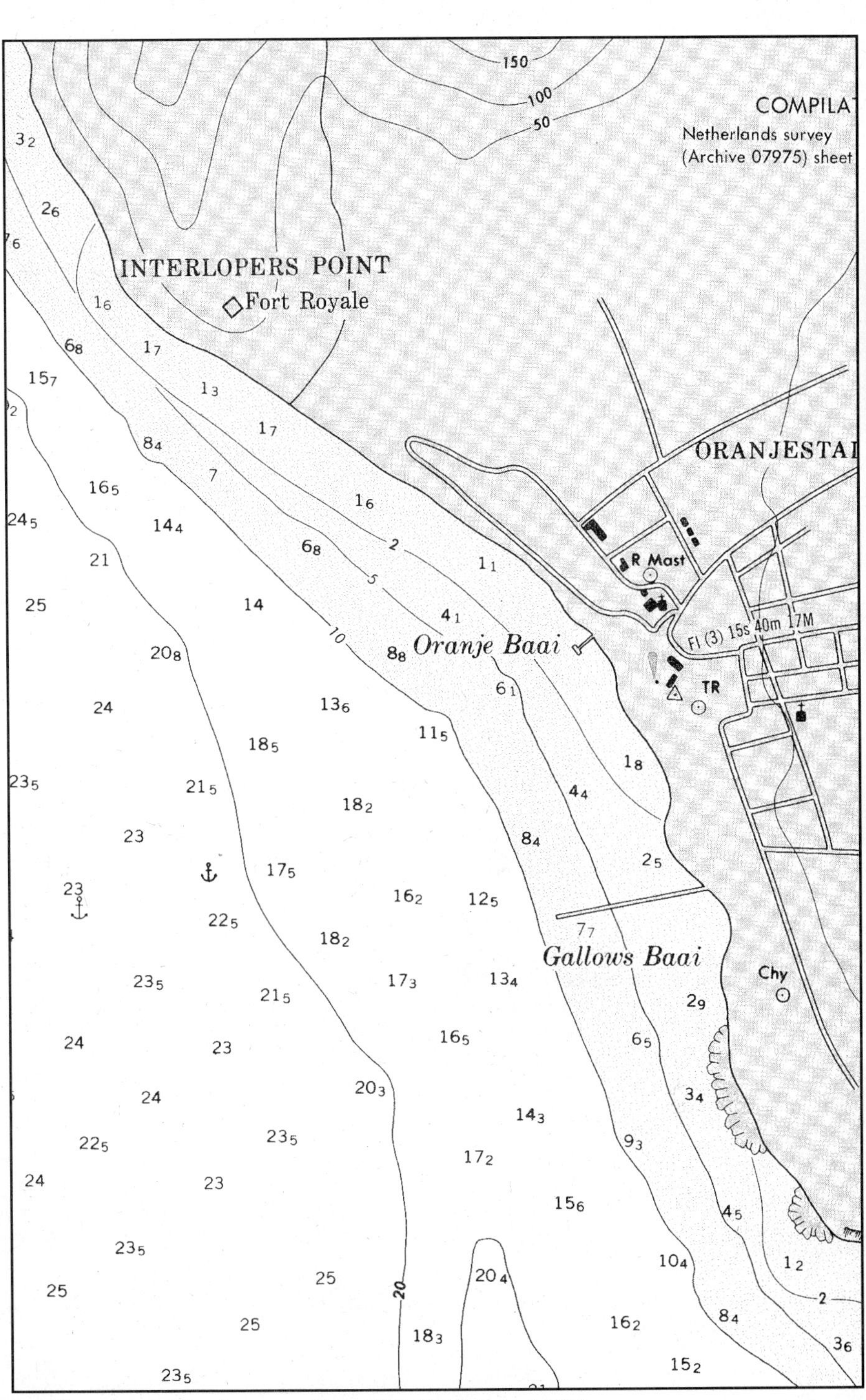

Oranjestad, soundings in meters *(from DMA 25607, 3rd ed, 7/84)*

St. Kitts and Nevis

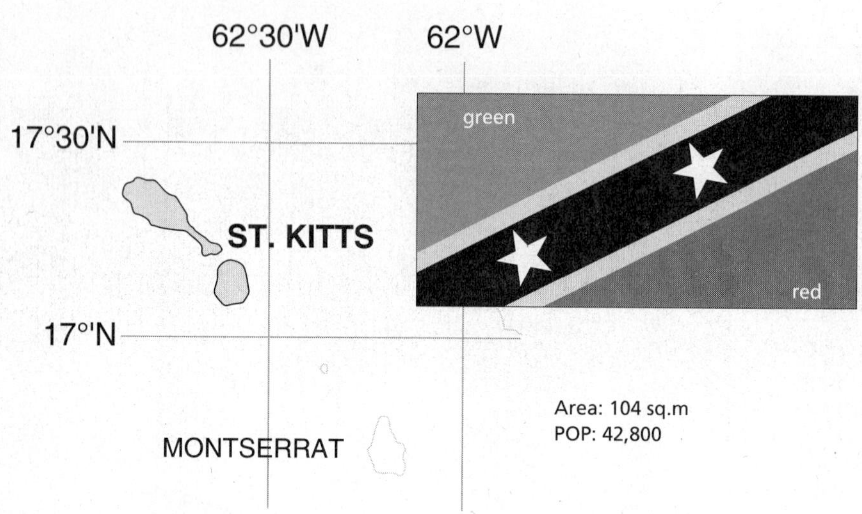

Area: 104 sq.m
POP: 42,800

SPECIAL NOTE: Caution is advised as the hydrography around St. Kitts and Nevis is incomplete.

Please see cautions about chartlets and waypoints on page P 2.

ENTRY PROCEDURES

St. Kitts (less commonly called St. Christopher) and Nevis are located about 40 miles west of Antigua. They are separate islands, but one country, sharing a British heritage.

The port of entry on St. Kitts is Basseterre; on Nevis it is Charlestown. All yachts are charged an EC$20 clearance fee, which is valid for both islands; additional customs fees are based on registered tonnage. Separate port authority charges apply to vessels over 10 tons. A cruising permit must be obtained to visit anchorages other than the ports of entry. Charterboats must pay an additional US$3 per passenger. When traveling between the islands you must obtain a boat pass and check in with the authorities when you arrive.

In Basseterre the customs office (869-465-8121, x100) is in the port authority building, and the immigration office is at the police station. In Charlestown customs is located at the foot of the pier (869-469-5521), and the immigration office is in the police station. Normal hours in both ports are 0800 to 1630 on Monday and Tuesday and 0800 to 1600 on Wednesday, Thursday, and Friday. Overtime fees are charged for clearing in outside business hours. If you arrive after hours in Basseterre, a port authority guard or one of the taxi drivers at the town dock can usually locate the on-duty customs officer. In Charlestown, it is best to wait until morning to locate officials.

You should have your passports, clearance from the last port of call, and ship's papers. Customs provides the necessary forms for crew lists. Firearms must be secured under lock and key while in port. Pets must remain aboard.

USEFUL INFORMATION

Language: English.
Currency: Eastern Caribbean dollar (EC$); exchange rate (summer 1997), EC$2.70 = US$1. U.S. dollars are widely accepted.

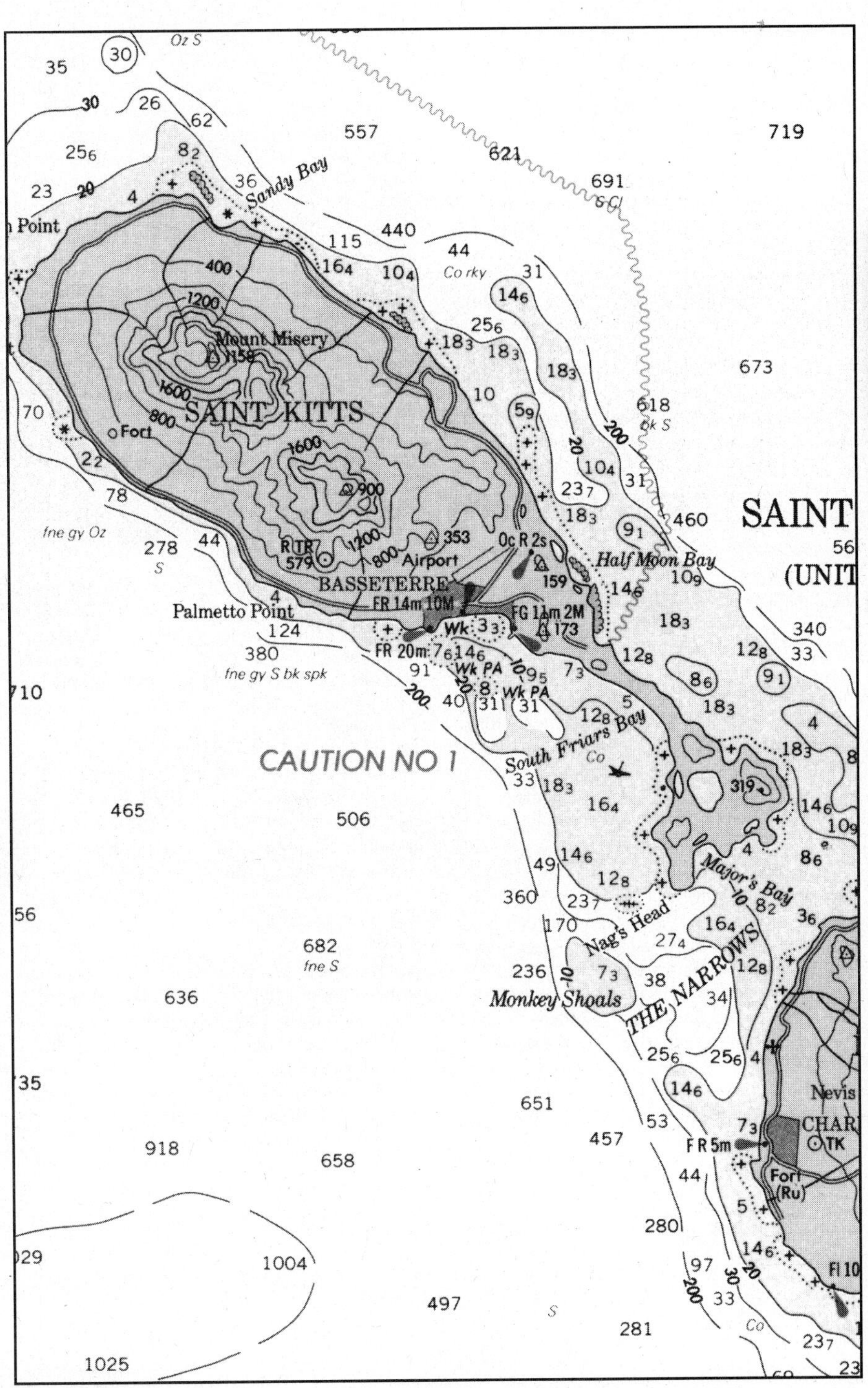

St. Kitts (St. Christopher), soundings in meters *(from DMA 25550, 2nd ed, 11/93)*

Telephones: Area code, 869; local numbers, seven digits. On St. Kitts, the first three numbers are 465 or 466; on Nevis, 469.
Public holidays (1998): New Year's Day, January 1; Good Friday, April 10; Easter Monday, April 13; Labour Day, May 4; Whit Monday, May 19; August Monday/Emancipation Day, August 3; Independence Day, September 19; Christmas Day, December 25; Boxing Day, December 26.
Tourism offices: *In St. Kitts:* Department of Tourism, Pelican Mall, Basseterre (869-465-4040). *In the U.S.:* St. Kitts & Nevis Department of Tourism, 414 East 75th St., New York, NY 10021 (212-535-1234). *In Canada:* St. Kitts & Nevis Department of Tourism, 11 Yorkville Avenue, Suite 508, Toronto, Ont. M4W 1L3 (416-921-7717). *Toll-free U.S. and Canada:* 800-582-6208.

ST. KITTS

SANDY POINT TOWN, on jetty, 17 22.0N, 62 51.4W. F R.
HALF MOON POINT LIGHT, 17 19.1N, 62 42.2W. Oc R 2s, F R. Tower.
FORT THOMAS LIGHT, W side of Basseterre harbor, 17 17.3N, 62 44.2W. F R, 67ft. Metal mast, 17ft. Aero Al Fl W G (occasional), 1.6M NE.
TREASURY BUILDINGS LIGHT, F R, 46ft, 10M.
FORT SMITH LIGHT, E side of Basseterre harbor, 17 17.3N, 62 42.6W. F G, 35ft, 2M. Concrete block. Fl W 2s, 2M, on dolphin 300 meters NW.

BASSETERRE

17 18N, 62 43W
Pilotage: Mount Misery at the north end of the island rises to 3,798 feet. The summit is usually covered with clouds. Only a narrow neck of sand connects the southeast end of the island to the main island. The port of Basseterre is situated on the southwest side of the main part of the island. A 283-foot-high white chimney is located about 0.5 mile north-northeast of the east end of town. The port authority can be reached on VHF channels 6 and 16.
Port of entry
Anchorage: The anchorage area is subject to swells when the wind shifts to the south. You might want to try the area off the Salt Ponds about a mile southeast of Basseterre.
Services: You can jug water and fuel to your boat. Caribe Yachts Ltd. (VHF channel 16; tel., 869-465-8411) offers repair services. This yard can haul catamarans up to 14 tons and sells marine supplies. There is a good variety of restaurants, grocery stores, and general shops. The tourist office is located in the center of Basseterre, about five minutes by taxi from the port authority.

NEVIS

CHARLESTOWN LIGHT, root of pier, 17 08.2N, 62 36.9W. F R, 15ft. Post. Unofficially reported unlit (1996). Iso R light on radio mast, 0.7M SSE.
DOGWOOD POINT LIGHT, S end of island, 17 05.7N, 62 36.4W. Fl W 10s, 29ft. Temporarily extinguished (1985). Unofficially reported missing (1996).

CHARLESTOWN

17 08N, 62 28W
Pilotage: Charlestown, located on the west coast of Nevis, is the island's capital. Nevis Peak rises to 3,230 feet in the center of the island and is usually hidden in the clouds. A concrete pier with depths from 11 to 15 feet alongside projects from the town waterfront. A prominent radio mast stands near the root of this pier, which is used by freighters and ferries. South of the commercial pier, a newer, smaller pier used by tourboat launches has a dinghy dock. Vessels approaching from Montserrat at night should take care not to confuse the red lights of the new radio station tower south of town, off which lies reef, with the (unlit) town pier light and anchorage area.
Port of entry
Anchorage: Anchor right off the town waterfront, in front of Pinney's Beach between the Four Seasons Hotel docks and the town piers. Use the dinghy dock at the pier south of the commercial pier.
Services: Fuel and water are available by jug or at the wharf via truck if you are getting a large quantity. There is a good variety of shops, restaurants, and grocery stores on the island. The Nevis tourist office is within walking distance of the pier.

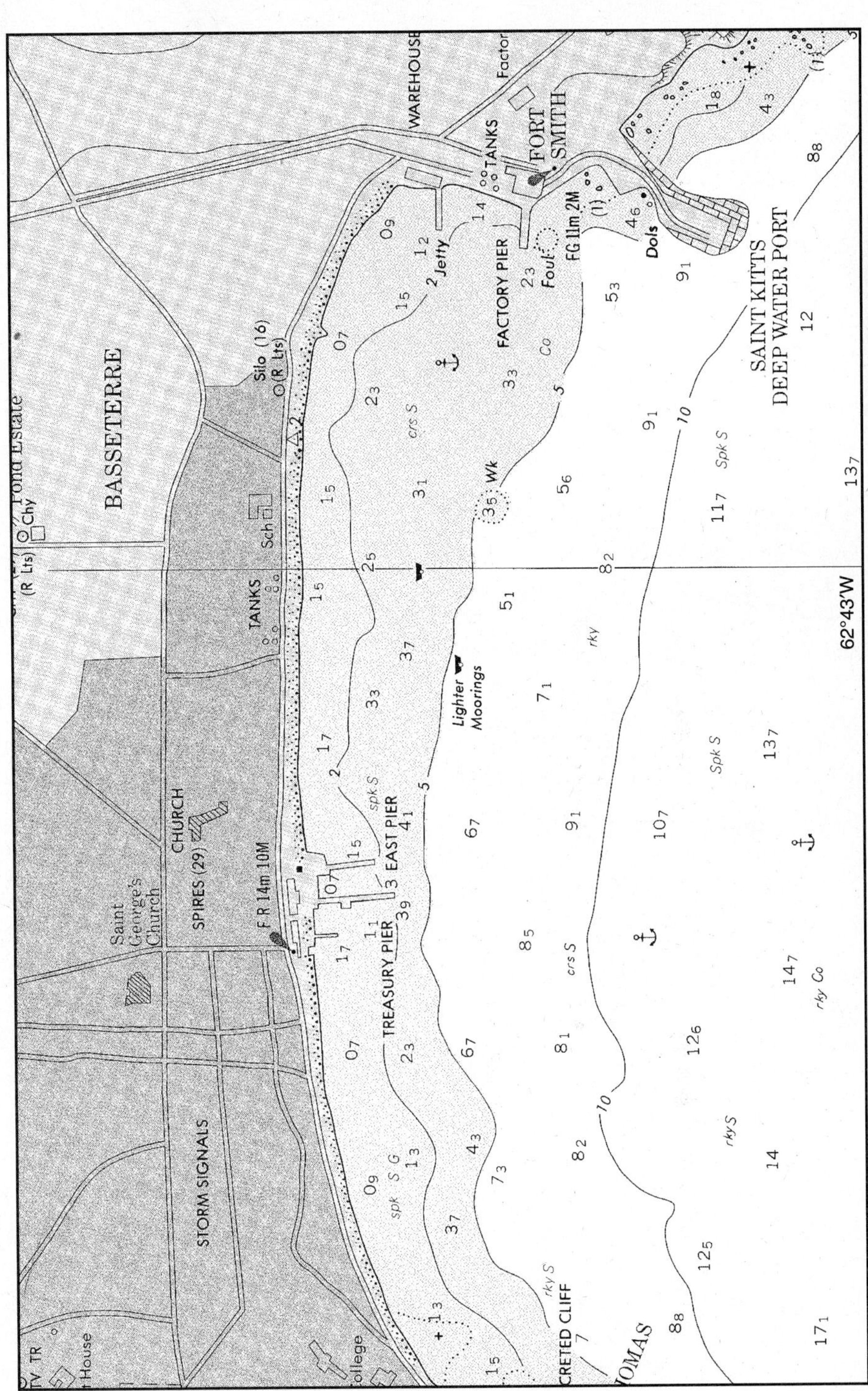

Basseterre, soundings in meters *(from DMA 25608, 20th ed, 9/84)*

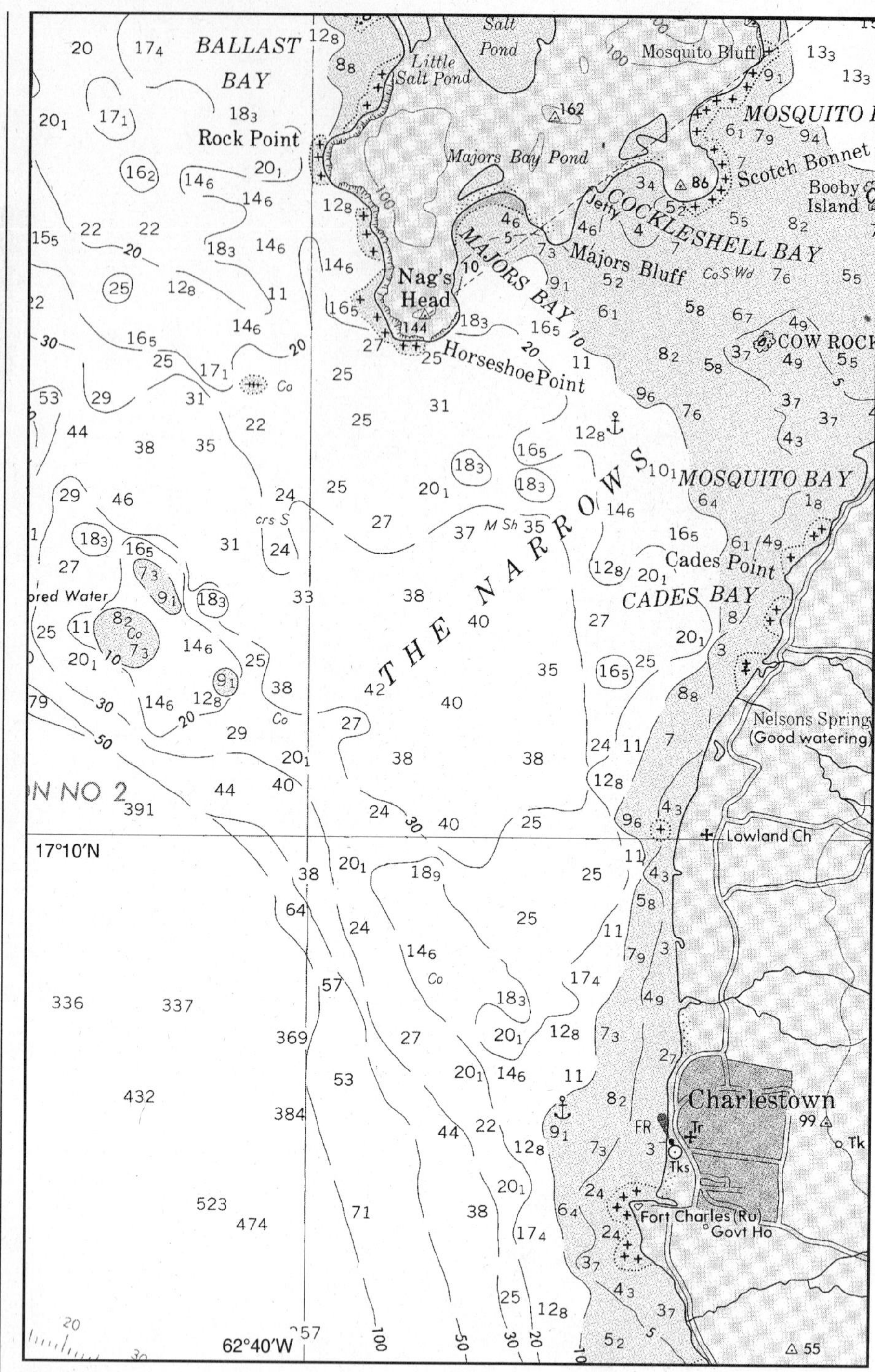

Charlestown, soundings in meters *(from DMA 25601, 33rd ed, 8/87)*

MONTSERRAT

62°30'W

ST. KITTS

ANTIGUA

17°'N

MONTSERRAT

AREA:39.5 sq mi
POP: 10,400

SPECIAL NOTE: The Soufriere Hills volcano at the southern end of Montserrat came to life in 1995, eventually forcing two-thirds of the British colony's population to move off the island. On August 4 and 5, 1997, the volcano erupted again, forcing hundreds to leave their homes in parts of the island that were previously considered beyond the volcano's reach. The capital city of Plymouth is reported to have been devastated in the August eruption; its landmark buildings are now "kindling," according to an eyewitness report. Cruisers are advised to call the Port Authority office (664-491-3293) for current information about visiting Montserrat. As Reed's goes to press, many services for cruisers, including customs and immigration offices, have been curtailed. Due to the possibility of a new, major eruption, the entire island may be abandoned.

Please see cautions about chartlets and waypoints on page P 2.

ENTRY PROCEDURES

Montserrat lies about 27 miles southwest of Antigua and is a British Crown Colony.

Because of the eruption of the Soufriere Hills volcano in August 1997, cruisers are advised to call the Port Authority (664-491-3293; fax, 664-491-8063) before traveling to Montserrat. Customs and immigration offices, which used to be located at Plymouth, were relocated to Old Towne prior to the August eruption. As Reed's goes to press, the Old Towne office was unable to speak with us because of the crisis.

Under normal conditions, firearms and ammunition must be locked on board if your stay in Montserrat is brief; if you are visiting longer than a couple of days, they are held at the police station. Again, under normal conditions, there are port clearance fees of EC$35.

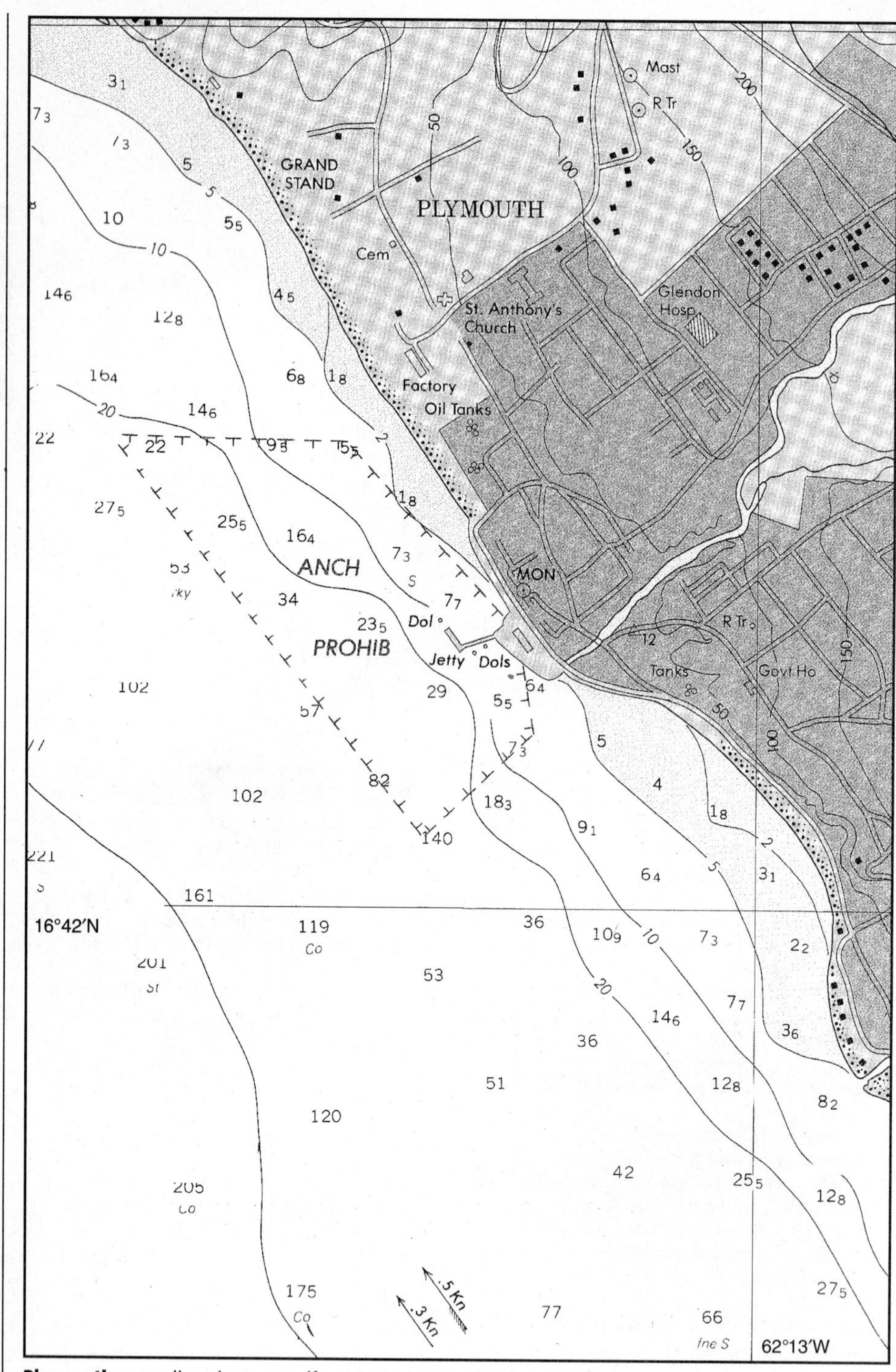

Plymouth, soundings in meters *(from DMA 25608, 20th ed, 9/84)*

Note: All information below may be inaccurate due the volcanic activity noted on the opening page of this section.

USEFUL INFORMATION

Language: English.
Currency: Eastern Caribbean dollar (EC$); exchange rate (summer 1997), EC$2.70 = US$1. U.S. dollars are widely accepted.
Telephones: Area code, 664; local numbers, seven digits. Dial all seven for on-island calls. Prior to the eruption of the volcano, excellent telecommunications were available on Montserrat, including Boatphone, the Caribbean cellular phone service. It is not known how the current crisis has affected the telephone service on the island.
Public holidays (1998): New Year's Day, January 1; St. Patrick's Day, March 17; Good Friday, April 10; Easter Monday, April 13; Labour Day, May 5; Whit Monday, May 19; Queen's Birthday, June 13; August Monday, August 3; Christmas Day, December 25; Boxing Day, December 26; Festival Day, December 31.
Tourism offices: *On Montserrat:* Montserrat Tourist Board, Church Road, P.O. Box 7; Plymouth (tel., 664-491-2230; fax, 664-491-7430). *In the U.S.:* Montserrat Tourist Information Office, 1208 Washington Drive, Centerport, NY 11721 (tel., 800-646-2002; fax, 516-425-0903). *In Canada:* New Concepts–Canada, 2455 Cawthra Road, Suite 70, Mississauga, Ont. L5A 3PL (tel., 905-803-0131; fax, 905-803-0132).
Emergencies: Call Curaçao Radio, call sign PJC, on VHF channel 16.
Radio Antilles: Called the "Big RA" by many cruisers, this station broadcasts from Montserrat on 930 kHz. Marine weather for the eastern Caribbean is broadcast at 1230 and 2225 UTC. General weather is given throughout the day.

MONTSERRAT APPROACHES

AVIATION LIGHT, 16 45.9N, 62 09.9W. Oc R 1.5s, 197ft. Obstruction light. Occasional. Reported visible 8 miles.
BLACKBURNE AIRPORT LIGHT, 16 45.5N, 62 09.5W. Aero Al Fl W G 6s, 33ft. Obstruction light. Occasional. Reported visible 8 miles.
CASTLE PEAK LIGHT, 16 42.5N, 62 10.9W. Aero Fl R 1.5s, 3,179 ft, 6M. Mast, 180ft. Obstruction light.
AVIATION LIGHT, 16 40.8N, 62 11.5W. Oc R 2s, 341ft, 22M. Radio tower. Obstruction light. R. lights (occas.) on 2 radio masts 500 meters ESE. Temporarily extinguished (1992).
PLYMOUTH LIGHT, jetty root, 16 42.3N, 62 13.3W. F R. Terminal building, 33ft. Difficult to distinguish.

PLYMOUTH

16 43N, 62 103W
Pilotage: The island presents a rugged and uneven appearance from seaward. The wooded Soufriere Hills at the south end rise to just over 3,000 feet. The coasts tend to be bold and steep-to. An L-shaped jetty projects 100 yards seaward at Plymouth. The ruins of an old dolphin near the end of the wharf have been reported to be a hazard. Another obstruction is marked by a buoy located about 115 feet northwest of the jetty head. The yacht club is situated south of the jetty. Anchoring is prohibited in much of the area off the town.
Port of entry
Dockage: You may be able to get permission from the port authority to tie alongside the wharf while taking on fuel or supplies. Be careful if a swell is running.
Anchorage: Anchorage is prohibited in much of the area directly off Plymouth. You might want to try directly off the yacht club, south of the main wharf. This area is reported to be very rolly. Old Road Bay, north of town, is reported to be a better anchorage, but you still may need a stern anchor to hold your boat into the swells. You can dinghy in to the Vue Pointe Hotel's pier in Old Road Bay.
Services: Large quantities of fuel and water can be obtained at the wharf by truck, but you'll have to jug smaller amounts. The tourist office, which is within walking distance of the port authority building, can be very helpful. There is a good variety of supermarkets, restaurants, and general shops. The Vue Pointe Hotel (tel., 664-491-5210; fax, 664-491-4813), located near the north end of Old Road Bay, is reported to be very friendly to yachtsmen. It has a landing, water, a telephone, garbage facilities, ice, and a bus into town. The hotel may be able to hold mail for you (c/o the Vue Pointe Hotel, P.O. Box 65, Montserrat, WI).

Antigua and Barbuda

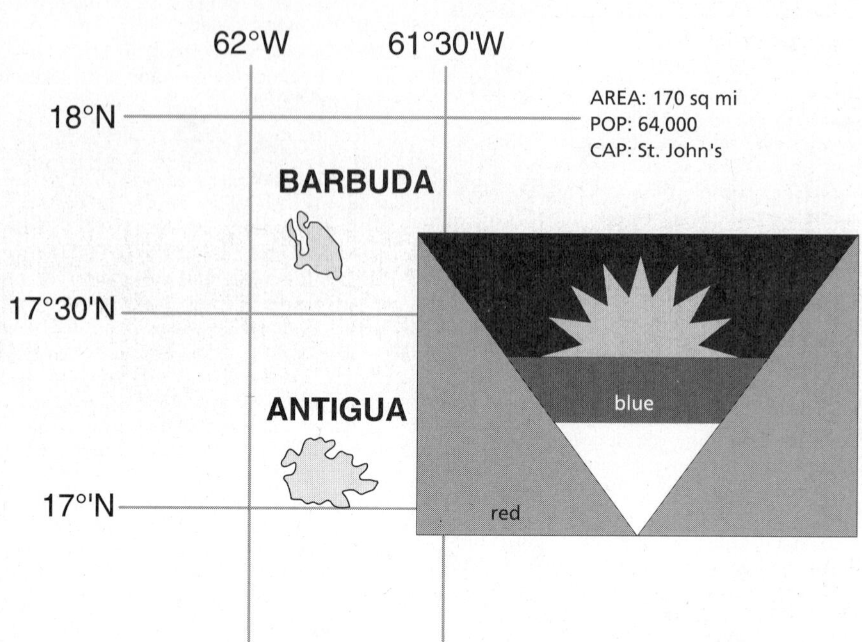

Please see cautions about chartlets and waypoints on page P 2.

ENTRY PROCEDURES

Antigua and Barbuda form an independent nation; until 1981, they were British colonies. Antigua, one of the first Caribbean islands to promote tourism (in the early 1960s) is the wealthiest. The small uninhabited island of Redonda, which lies 30 miles southwest of Antigua, also belongs to the country. There is little shelter from the ocean on the leeward west side of Redonda. The southwest coast provides the only decent anchorage, just north of the ruins of a wooden post office 100 yards offshore. The bottom is rocky so two anchors are recommended—one off the stern and the other off the bow with a tripping line.

On Barbuda, customs and immigration are located in Codrington; they can be reached on VHF channel 16. Although it is possible to clear into Barbuda, you must have a valid cruising permit that is obtainable only in Antigua. On Antigua, the ports of entry are Crabb's Slipway and Marina, St. John's, Falmouth Harbour, Jolly Harbour, and English Harbour.

Entering vessels must fill out an Antiguan Port Authority Clearance Form; required documents are clearance from the last port of call, ship's papers, and passports. Immigration requires that all persons on board have a valid passport, except U.S. citizens who may show a driver's license or birth certificate. The port entry fee and the required cruising permit fee are based on the length of the boat. For boats in the 40- to 80-foot range the two fees together are about US$16. Harbor fees are paid when vessels obtain outbound clearance.

USEFUL INFORMATION

Language: English.
Currency: Eastern Caribbean dollar (EC$); exchange rate (summer 1997), EC$2.70 = US$1.
Telephones: Area code, 268; local numbers, seven digits. Many phone numbers are restricted to local calls only, which means you may not be able to reach every number from overseas (local numbers are so designated in this section). The Boatphone cellular communications network is available in Antigua and Barbuda. Call 1-800-BOATFON to register.
Public holidays (1998): New Year's Day, January 1; Good Friday, April 10; Easter Monday, April 13; Labour Day, May 1; Whit Monday, May 19; Caricom Day, July 1; August Monday & Tuesday (last two days of Carnival), August 3 & 4*; Independence Day, November 1; Christmas Day, December 25; Boxing Day, December 26. *Tentative; dates not set as of *Reed's* press time (summer 1997).
Tourism offices: *In Antigua:* Antigua & Barbuda Department of Tourism, P.O. Box 363, St. John's (tel., 268-462-0480; fax, 268-462-2483). *In the U.S.:* Antigua & Barbuda Department of Tourism, 610 Fifth Avenue, Suite 311, New York, NY 10020 (tel., 212-541-4117, 888-268-4227; fax, 212-757-1607). Other office: Miami (Consulate General). *In Canada:* Antigua & Barbuda Department of Tourism, 60 St. Claire Avenue East, Suite 304, Toronto, Ont. M4T 1N5 (tel., 416-961-3085; fax, 416-961-7218).
Emergencies: Call the Antigua Coast Guard (VHF channels 16 and 68) or Curaçao Radio (VHF channel 16). VHF channel 68 is used in Antigua for calling other boats or shore-based stations. At peak times (for example, Antigua Sailing Week,) good manners suggest using low power whenever possible, keeping calls short and to the point,

BARBUDA

17 36N, 61 50W
Pilotage: This low island surrounded by extensive and treacherous reefs is not visible except in clear weather. *Caution: Charts are generally not reliable because of changing undersea growth.* The small-boat landing on Barbuda lies about 28 miles north of St. John's, Antigua. The best approach is from the west. A hotel is prominent on Coco Point, the southern point of the island. A Martello tower, partly ruined, is visible about 2 miles east of Palmetto Point, the island's southwest point. The boat harbor is located about 0.8 mile east of the Martello tower. Sand piles from a sand mining operation indicate the harbor's location. There are shoals and reef along the shore near the boat harbor. The channel to the dock is locally marked.
Port of entry: The customs station is located near the boat harbor. Officials monitor VHF channel 16. Customs reports that you can now obtain both inbound and outbound clearance here.
Anchorages: To clear customs or to visit the town of Codrington almost 3 miles north of the harbor by road, you can anchor in the small basin at the boat harbor in depths reported to be 5 to 9 feet. Most yachts anchor north of Palmetto Point or among the reefs along the south coast. In this area "eyeball navigation" is critical.
Services: Codrington Village is a sleepy small town with a post office and telephone service. The police station is near the dock. Provisioning in Barbuda is limited except for staples such as ice, beer, meat, and cheese. It is usually easy to locate someone willing to sell lobster and fish. Barbuda is a place for getting away from it all.

Palaster Reef, off the south coast of Barbuda, is a national park, where no fishing is permitted. This a wonderful place for snorkeling and diving on the coral reefs and coral heads. These add both to the beauty and to the difficulties. Be sure of your navigation, your charts, and your guides before heading to Barbuda.

NORTH ANTIGUA

PRICKLY PEAR ISLAND LIGHT, 17 10.5N, 61 48.9W. Q W, 26ft. Black round metal structure.
SANDY ISLAND LIGHT, near center of island, 17 08.1N, 61 55.4W. Fl W 15s, 53ft, 13M. White iron framework structure. Temporarily extinguished (1996).

PARHAM

17 08N, 61 45W
Pilotage: Approach the harbors on the north coast of Antigua via Boon Channel, north of St. John's. The major hazards are the reefs surrounding Prickly Pear Island, which is marked by a light. Parham Harbour, located on the north coast, has a channel leading into the harbour and to Crabbs Slipway and Marina. At publication time the channel on each side was

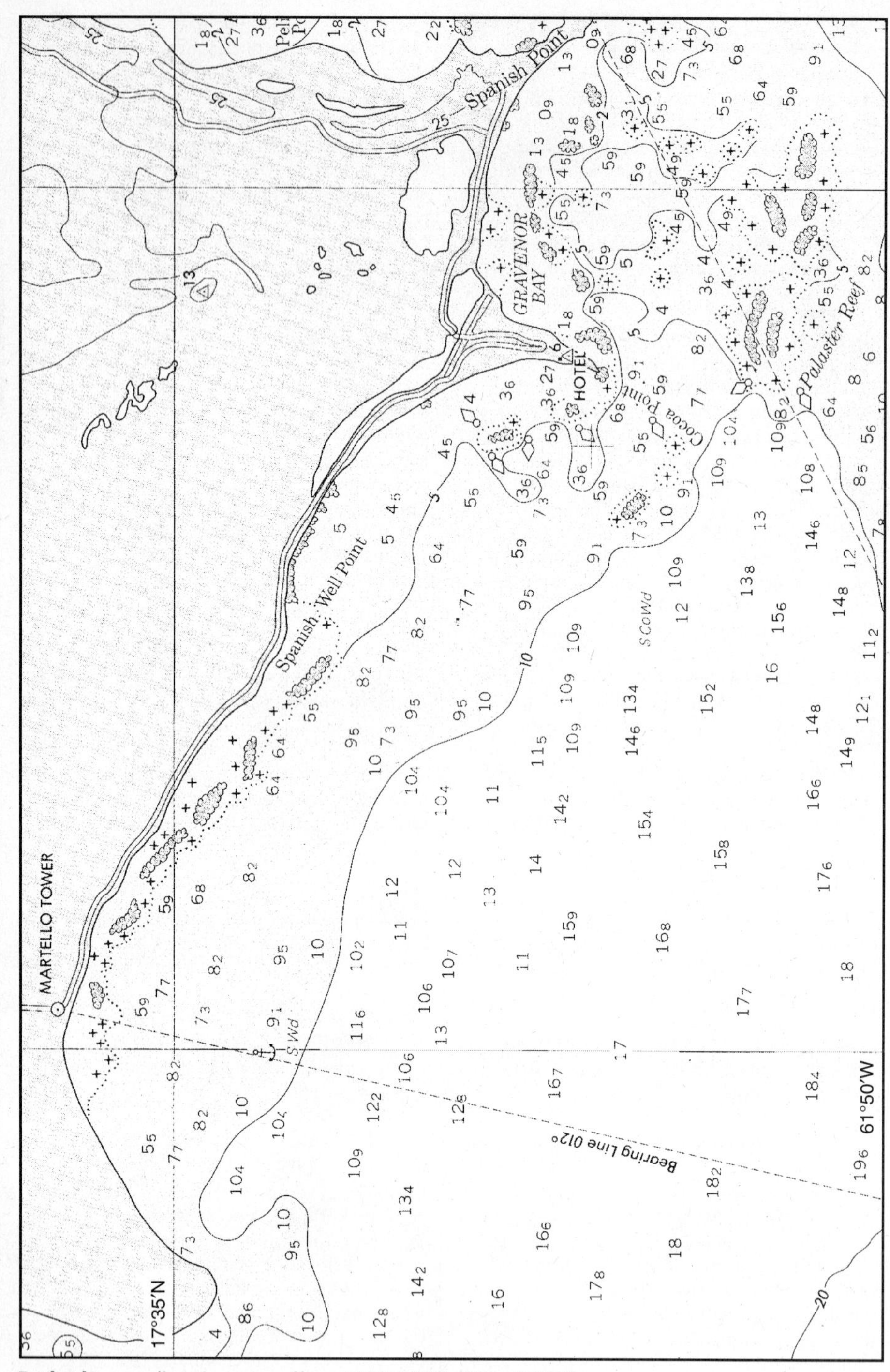

Barbuda, soundings in meters *(from DMA 25608, 20th ed, 9/84)*

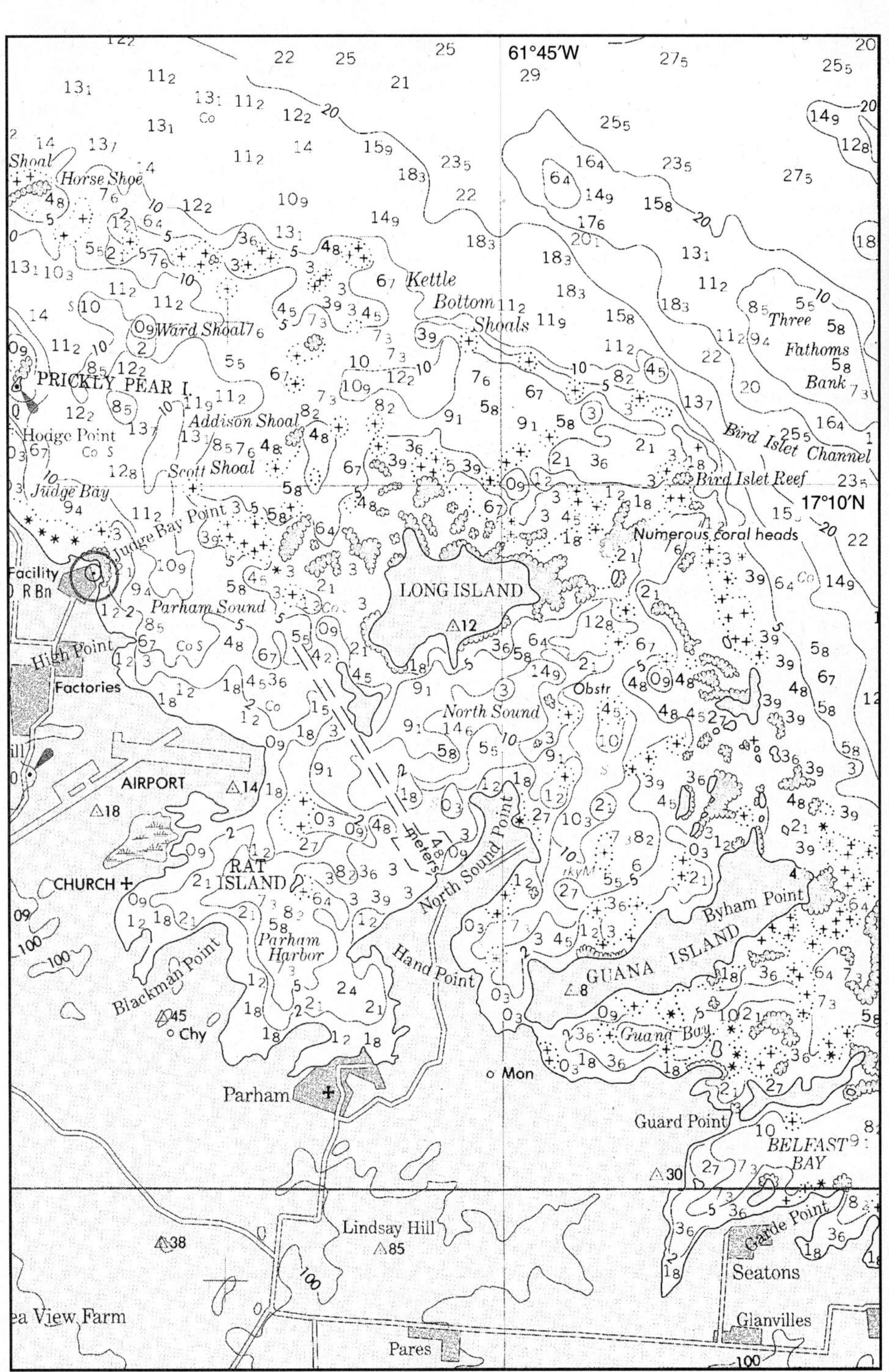

Parham, soundings in meters *(from DMA 25570, 2nd ed, 8.84)*

marked by two red balls just off Maiden Island. There are plans to place red buoys to starboard and black to port. The channel is easy to spot, but don't attempt to enter or leave except in daylight. Parham harbor is south of the west side of the point. Because of the many shoals and reefs, Antigua's north shore should be approached only in good light.

Port of entry: Clear in at Crabb's Slipway and Marina. Customs and immigration officials may be reached at 463-2372 (local number). The office generally opens at 0830 and closes at 1600 daily.

Dockage: Crabb's Slipway and Marina (VHF channels 16 and 68; tel., 268-463-2113) offers dockage and is a favorite gathering spot for cruisers.

Anchorage: Anchor off Crabb's or farther south near town. There are no anchorage fees. This area is well protected in most conditions.

Services: Crabb's Slipway and Marina has fuel and water and can do repairs and haul boats to 50 tons with a maximum beam of 15 feet and a draft of 8 feet. There is a supermarket ashore.

PILLAR ROCK LIGHT, S side of entrance to harbor, 17 07.8N, 61 52.6W. Fl G 4s, 106ft, 5M. White house, words PILLAR ROCK in black. Visible 067°–093°, obscured 093°–108°, visible thence to shore southeast of light.

RANGE (Front Light), 17 07.1N, 61 50.5W. Iso R 6s, 62ft, 6M. Metal framework tower, square daymark, 56ft. **(Rear Light),** 440 meters, 113° from front. Iso R 6s, 92ft, 6M. Metal framework tower, square daymark, 29ft.

FORT JAMES LIGHT, N side of entrance to harbor, 17 07.9N, 61 51.7W. Fl R 4s, 48ft, 5M. White pillar. Visible 344°–shore.

RANGE (Front Light), 17 01.4N, 61 46.3W. Q G, 35ft. Wooden pile, 33ft. **(Rear Light),** 290 meters, 029° from front. Iso G 2s, 75ft. Wooden pile, 33ft.

ST. JOHN'S HARBOUR

17 08N, 61 52W

Pilotage: Approaching from the north, be careful to stay west of the reefs and Diamond Bank. Sandy Island, marked by a light, lies in the western approaches. From the northwest, Fort James Light in alignment with the cathedral spires bearing 110° should keep you clear of dangers. When you get closer there is a charted range on a bearing of 113°. Warrington Bank, north of the channel, is less than 3 feet deep. The harbor pilot and harbor authorities monitor VHF channel 16 and the latter 2182 kHz. The oil terminal monitors 2182 kHz and VHF channel 16 (call sign Marine Center). The deepwater facilities for commercial vessels are on the north side of the harbor; the small-craft anchorage is in the inner harbor.

Port of entry: St. John's is primarily a commercial port, but private vessels can also clear with officials at Heritage Quay near the cruise ship pier in the inner harbor. Hours of operation are 0800 to 1600 most days. The local phone number for the Heritage Quay customs office is 462-0476.

Dockage: There are no yacht berthing facilities in St. John's Harbour, although construction has started on a dinghy dock and marina at the end of Redcliffe Quay. Dinghies can also be left in safety at the Treasury Pier.

Anchorage: Anchor outside the main ship channel. At times, you may encounter a swell from the north. The inner harbor has good holding for vessels drawing 9 feet or less. Make sure you are outside the dredged cruise ship turning basin—if you are in water deeper than 17 feet, you are in the turning basin. There are no harbor fees.

Just west of St. John's Harbour, Deep Harbour offers one of the most secure anchorages on the west coast. Certain weather conditions can make it rolly. There is 8 feet of water almost to the beach.

Services: Pleasure vessels can get fuel at the commercial dock in the harbor basin. There is good general shopping in St. John's, and provisions are more extensive and often less expensive than elsewhere on the island. The Redcliffe Quay area is oriented toward cruise ship passengers. Govee's Marine Supplies (268-462-2975) is the local chandlery, and Island Provision (tel., 268-480-5152/5153/5154, or voice recording, 268-480-5151; fax, 268-480-5165) offers a range of food goods to yachting clientele. There are many bars, restaurants, and hotels here and nearby.

St. John's is the capital and the major port of Antigua. The dinghy dock at the marina under development near the cruise ship dock allows yacht crews access to the town's excellent variety of shops and restaurants. You may want to visit here by land transport even if you are based in English Harbour.

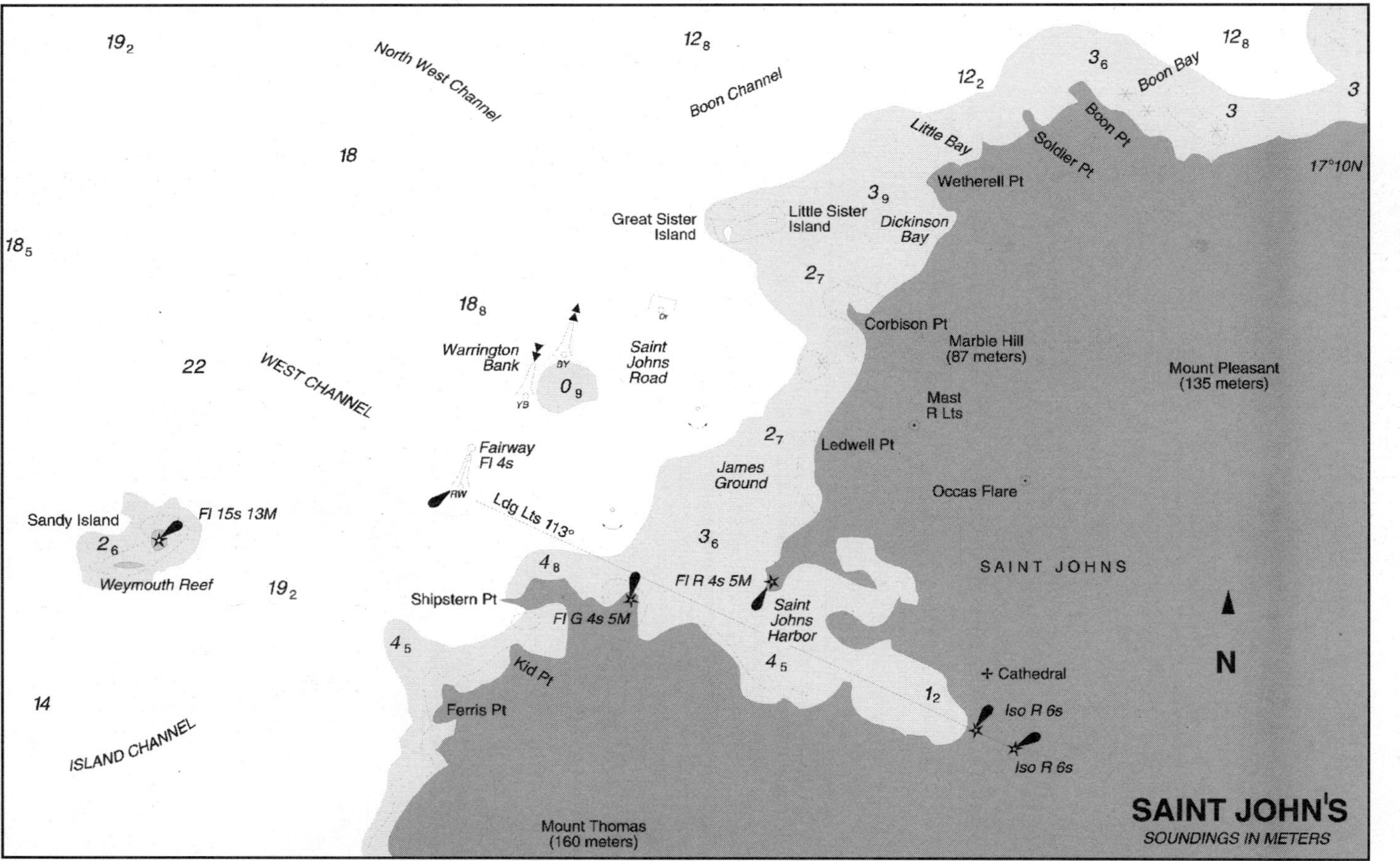

St. John's, soundings in meters *(after DMA 25570, 2nd ed, 8/84)*

WEST ANTIGUA

JOLLY HARBOUR

17 05N, 61 54W

Pilotage: The entrance to Jolly Harbour is just north of Reed Point, at the northern end of Morris Bay. Follow the marked, dredged channel into the inner lagoon, taking care to avoid the shoal patch off Reed Point.

Port of entry: Call ahead to the Jolly Harbour Port Authority (VHF channel 16; local number, 462-7929) before proceeding to the clearing station in the boatyard. Officials are available daily between 0800 and 1600.

Dockage: Available at Jolly Harbour Marina, the island's newest and most modern facility.

Anchorage: Boats can anchor in Morris Bay, but as is typical on the west coast, shoaling prevents boats lying close to shore. A direct line between Reed and Fryes Point has 5 feet of water. A quiet anchorage can be had between Pearns and Reed point, at the entrance marker to Jolly Harbour Marina and you can dinghy into the marina.

Services: Jolly Harbour (VHF channel 68; tel., 268-462-6042) offers a full range of marine services, including fuel, water, laundry, repair services, and haulouts. Budget Marine, a chandlery headquartered in Sint Maarten, has a branch store here (268-462-8726).

Jolly Harbour is a recently developed, sizable marina and condominium complex at Mosquito Cove on the west side of the island between Johnson Point and St. John's.

SOUTH ANTIGUA

FALMOUTH HARBOUR

17 01N, 61 47W

Pilotage: Falmouth Harbour is located 1 mile west of English Harbour. The approach to the anchorage off Antigua Yacht Club and Marina is marked by red and green channel markers. Coming into the harbor from the south, be wary of Bishop Shoal which is sometimes marked by a large buoy as shown on most charts. If the buoy isn't there, the shoal is easily spotted because it's always breaking, except in the calmest weather. Leave the shoal to starboard and then head to the eastern end of Falmouth Harbour and anchor off the yacht club dock, leaving the first buoy to port. If you want to go to the Catamaran Club, leave the first buoy to starboard and continue on the range lights.

Dockage: The Catamaran Club (VHF channels 16 and 68; tel., 268-460-1503) at the head of the bay has dockage for vessels to 200 feet LOA. Follow the range lights/orange daymarks in. The Antigua Yacht Club Marina (VHF channel 68; tel./fax, 268-460-1444; tel., 268-460-1543) in the eastern portion of the bay can also accommodate megayachts. Its docks have recently been extended. The approach is fairly straightforward. Vessels at both clubs are subject to the Falmouth/English Harbours' mooring and anchoring fee, which is based on vessel length. Vessels docked at Antigua Yacht Club are also subject to the National Parks Authority harbor dues.

Anchorage: There is good anchorage in the southeast portion of the bay. Be careful to observe the many shoal areas when searching for an anchoring spot. Nominal fees that are based on the boat's length are charged to anchor in Falmouth Harbour. Low-season rates are in effect from June 1 to mid-November, and high-season rates from mid-November to May 31.

Services: Fuel is available at the Catamaran Club. All other marina services are available at both facilities. Yacht services for most needs are available in the area (see English Harbour listings below). Fuel and water are also available at the south end of Falmouth Harbour near the Port Authority offices.

Falmouth and English harbors share many of the same shops and facilities. Falmouth may offer more room for those wanting to anchor out. This harbor is the center of many activities during the famous Antigua Race Week, held the last week in April.

RANGE (Front Light), 17 00.4N, 61 45.4W. Q R, 35ft. Wooden pile, 33ft. Reported extinguished (1985). **(Rear Light),** 100 meters, 055° from front. Iso R 2s, 85ft. Wooden pile, 33ft. Reported extinguished (1985).

CAPE SHIRLEY LIGHT, 17 00.1N, 61 44.7W. Fl (4) W 20s, 494ft, 20M. Metal mast, 69ft. Visible 264°–084°. Reported extinguished (February 1987).

ENGLISH HARBOUR

17 00N, 61 45W

Pilotage: Entering English Harbour during daylight is straightforward. The only danger is the reef off Charlotte Point. Be certain to favor Berkeley Point (also known as Barclay Point) on your approach. A good landmark is

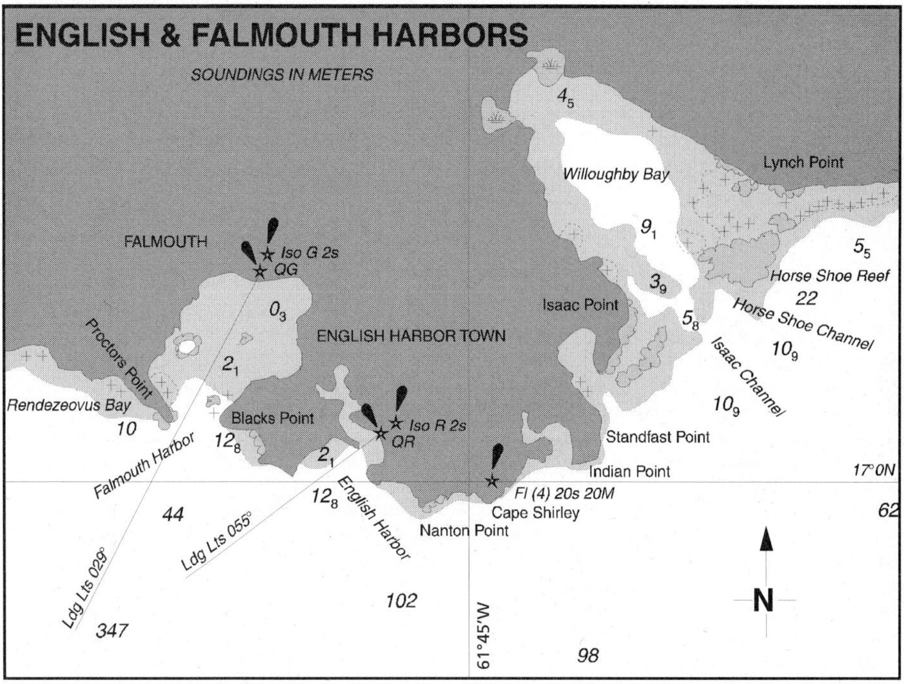

Falmouth and English Harbours, soundings in meters *(after DMA 25570, 2nd ed, 8/84)*

the Stone Hotel high on the hill behind Freeman's Point. Line it up with the largest and westernmost beach house on Freeman Bay (039° MAG). This will lead you past all dangers. Once in the harbor, there is a 4-knot speed limit for all craft.

Communications: Many businesses and services stand by on VHF channel 68 in the Falmouth and English harbors area.

English Harbour Radio V2MA: Nicholson Yacht Charters (268-460-1530) provides the local communications link for many boaters. Contact the company on VHF channel 68 or on 8294 kHz Monday through Friday from 0830 to 1630 and on Saturdays from 0830 until 1200 (no Saturday service between June 1 and October 31).

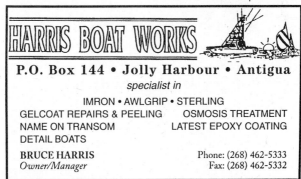

Port of entry: Flying the Q flag, vessels may anchor in Freeman Bay or farther in, taking care to stay clear of the channel, or tie up stern-to at the Dockyard. The boat's master should proceed immediately to the customs and immigration office (268-460-1397) in Nelson's Dockyard, on the western side of the channel leading into the harbor. Although the office is officially open from 0600 to 1800 every day, not all the officials are present at the beginning and end of the day; recommended clearing-in times are 0800 to 1600. A Dockyard fee of EC$5 per person is charged to all vessels entering English Harbour. Mooring/anchoring and National Parks Authority fees based on vessel length are charged here for all vessels, whether dockside or anchoring out.

Dockage: Contact Antigua Slipway (VHF channels 12 and 68; tel., 268-460-1056) on the eastern side of the inner harbor mouth for stern-to or side-to dockage. You can also tie up stern-to in the Dockyard area; rates are lower in the offseason (June 1–mid-November). Longer term rates are available. The Dockyard has public toilets and showers.

Anchoring: There is some room to anchor in Freeman Bay, outside the channel, or in the inner harbor. This can be a good hurricane hole, if you can get in before the crowd. As in Falmouth Harbour, anchorage fees are assessed here based on boat length. Rates are lower in the offseason.

Services: All services are available here or in Falmouth. Antigua Slipway can haul boats to 120 feet in length and 125 tons in displacement or drawing less than 13 feet and has full repair facilities, a chandlery, fuel, a laundromat, a restaurant, and shopping. Bailey's Supermarket is opposite the Catamaran Club in Falmouth. Sail repair facilities are also available. Antigua Yacht Services (VHF channel 68; tel., 268-460-2711) provides message and mail services of all types. Mail can be addressed c/o Antigua Yacht Services, P.O. Box 2242, St. Johns, Antigua, West Indies. Carib Marine (tel., 268-460-1521; fax, 268-460-1532) is a combination grocery and hardware store that can handle major provisioning. Seagull Services (268-460-3050), a complete engine and general repair facility, also services dinghies, rigging, and liferafts. This listing is just a sample of some of the many services in this area oriented to private yachts. There are many bars and restaurants here as well.

English Harbour is the location of the famous shipyard that once served the great British sailing fleets and now services some of the premier luxury charter yachts in the Caribbean. Consequently, there is limited anchoring and docking room. Many formal and informal regattas originate here, including the well-known Antigua Sailing Week, which begins the last Sunday in April. This area is a favorite watering hole for Caribbean yachties.

Guadeloupe

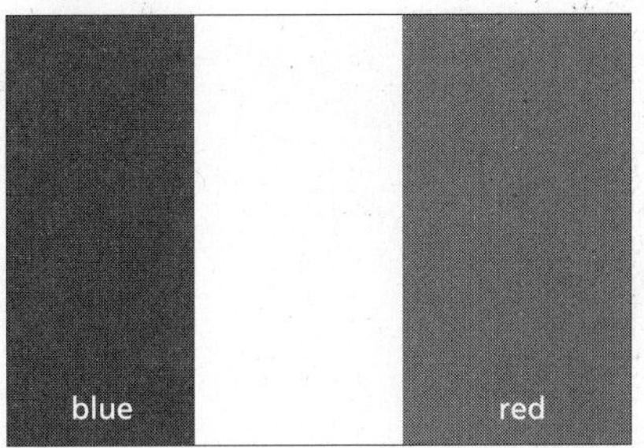

AREA: 530 sq mi
POP: 408,000

Please see cautions about chartlets and waypoints on page P 2.

ENTRY PROCEDURES

Guadeloupe is 1,845 miles from New York, 4,360 miles from Paris, and 2,138 miles from Montreal. It's also about 40 miles south of English Harbour, Antigua. An overseas department of France, it consists of two large islands, Basse-Terre and Grande-Terre, and the off-lying islands of Les Saintes, Marie-Galante, and La Désirade. St. Martin and St. Barts, described earlier in this chapter, are also part of this department. The French courtesy flag should be flown here.

On the two big islands, the ports of entry are Deshaies, Basse-Terre, and Pointe-à-Pitre. Grand-Bourg on Marie-Galante is also a port of entry. There is no charge for entering the country. U.S. boats that have only a state registration may or may not be permitted by officials to enter, so U.S. vessels should have federal documentation. No chartering is allowed by foreign yachts, although charterboats may pass through with non-French passengers.

If you visit Îles des Saintes before you have cleared in at an official port of entry, you should check in with the local police *(les gendarmes).*

USEFUL INFORMATION

Language: French and Creole. English is spoken in most hotels, restaurants, and tourist facilities, but it is useful to have some proficiency in French.

Currency: French franc (F); exchange rate (summer 1997), 5.5F = US$1. Legal tender is the French franc, but dollars are accepted everywhere. Major banks are located in Pointe-à-Pitre and are open 8am to noon and 2–4pm Monday through Friday. Some are also open on Saturday morning.

Telephones: Country code, 590; local numbers, six digits. On the island dial only the six-digit local number. You can call St. Barts, St. Martin, La Désirade, Marie-Galante, and Iles des Saintes directly using only the six-digit local numbers. Télécartes (phone cards) are sold at the post office and in many shops. Operator-assisted calls are higher in cost. The telex area code for Guadeloupe is 340.

Public holidays (1998): New Year's Day, January 1; Lenten Carnival, February 21–25*; Easter Monday, April 13; Labor Day, May 1; Armistice Day of 1945, May 8: Ascension Thursday, May 21; End of Slavery Day, May 27; Pentecost Monday, June 1; Bastille Day, July 14; Schoelcher Day, July 21; Assumption Day, August 15; All Saints' Day, November 1; Armistice Day of 1918, November 11; Christmas Day, December 25. *Most businesses close, although not an official holiday.

Tourism offices: *On Guadeloupe:* Office du Tourisme, 5 Square de la Banque, BP 1099, 97110 Pointe-à-Pitre (tel., 590-82-09-30; fax, 590-83-89-22; telex, 340 919715). *In the U.S.:* Guadeloupe Tourist Office, 161 Washington Valley Rd., Suite 205, Warren, NJ (888-4-GUADELOUPE; fax, 908-302-0809). *In Canada:* French Government Tourist Office, 1981 Ave. McGille College, Montreal, Quebec H31 2W9 (tel., 514-288-4264; fax, 514-844-8901).

Ferries: Ferries leave Pointe-à-Pitre harbor daily for round-trip service to Marie-Galante and Les Saintes for about US$32. For service to La Désirade, ferries connect at least twice daily (except Tuesday) from St. François (on Guadeloupe's eastern end). The cost is about US$20.

Pets: Cats and dogs over three months old are admitted temporarily with certificates of

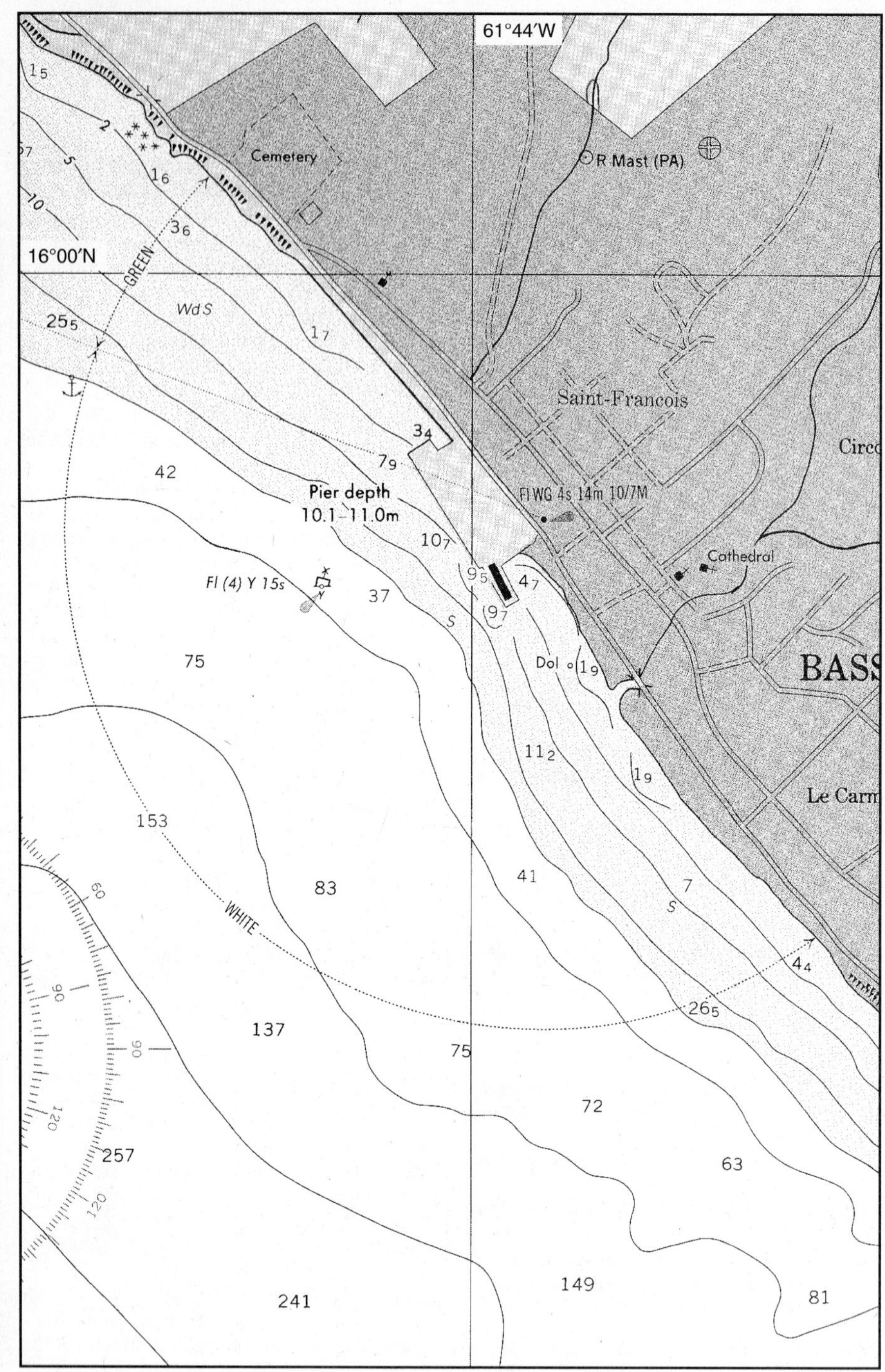

Basse-Terre, soundings in meters *(from DMA 25567, 15th ed, 10/90)*

origin and good health issued by a licensed veterinarian from the country of origin.
Firearms: Yachts are permitted to have firearms on board, but they must be declared.

BASSE-TERRE

ANSE A LA BARQUE, head of cove, 16 05.4N, 61 46.1W. Fl (2) W R G 6s, 36ft, W 8M, R 5M, G 5M. White tower, green top. R 050°–064°, W 064°–081°, G 081°–115°.
NORTH SIDE OF ENTRANCE LIGHT, 16 05.4N, 61 46.4W. Q (9) W 15s, 91ft, 9M. West cardinal mark, yellow-black-yellow tower with topmark, 23ft.
BASSE-TERRE LIGHT, on quay near customs house, 15 59.8N, 61 43.8W. Fl W G 4s, 46ft, W 10M, G 7M. White metal pylon, red lantern, 39ft. W 325°–110°, G 110°–135°.

BASSE-TERRE

15 59N, 61 44W
Pilotage: Between Guadeloupe and Antigua the current is basically west-going and off the southwest point of the island of Basse-Terre reaches speeds of 2 knots. The southwest part of the island is dark rock and quite steep-to. The three peaks of Soufrière may be visible if they are not obscured by clouds. A tower on a church and the steeple of the cathedral are prominent. A statue and a cross are on the church tower. The port authority monitors VHF channel 16.
Port of entry: There are customs offices in the Marina de Rivière Sens and in town.
Dockage: The Marina de Rivière Sens (tel., 590-81-77-61), located in a protected basin about 1.2 miles southeast of the commercial wharf, has slips.
Anchorage: Anchor north or south of the town wharf or off the entrance to the marina. There is no shelter in south or west winds.
Services: There are good yacht services in the area around the marina. The town has good general shopping and groceries.

Basse-Terre, the capital of Guadeloupe, was founded in 1643. The historic city rests in a dramatic setting below towering volcanic peaks. There are good markets and supplies for those who can find room in the marina.

POINTE DU VIEUX FORT, 15 56.9N, 61 42.6W. Fl (2+1) W 15s, 85ft, 22M. White tower, gray top. Obscured 297°–331° by Les Saintes. Reserve light F W.
TROIS RIVIERES LIGHT, on the point, 15 58.1N, 61 38.8W. Iso W R G 4s, 82ft, W 10M, R 7M, G 7M. White tower, red top, 29ft. R 275°–349°, W 349°–054°,G 054°–068°.
SOUTH BREAKWATER, 15 58.9N, 61 43.0W. Fl R 2.5s.
NORTH BREAKWATER, 15 58.9N, 61 43.0W. Fl(2) G 6s.
SAINTE MARIE LIGHT, root of E pier, 16 06.1N, 61 33.9W. Fl G 4s, 36ft, 6M. White square tower, green top, 33ft.
GOYAVE LIGHT, 16 08.4N, 61 34.2W. Fl(3) G 12s, 16ft, 5M. White tower, green top.
PETIT BOURG LIGHT, on jetty, 16 11.5N, 61 35.1W. Fl (2) G 6s, 4M. White tripod, green top.

REED'S *NEEDS YOU!*

It is difficult to obtain accurate and up-to-date information on many areas in the Caribbean. To provide mariners with the best possible almanac, *Reed's* uses many sources. *Reed's* welcomes your contributions, suggestions, and updates. Please be as specific as possible and include your address and phone number.
Send this information to:
Editor, Thomas Reed Publications
13A Lewis Street, Boston, MA 02113, U.S.A.
e-mail: editor@treed.com
Thank you!

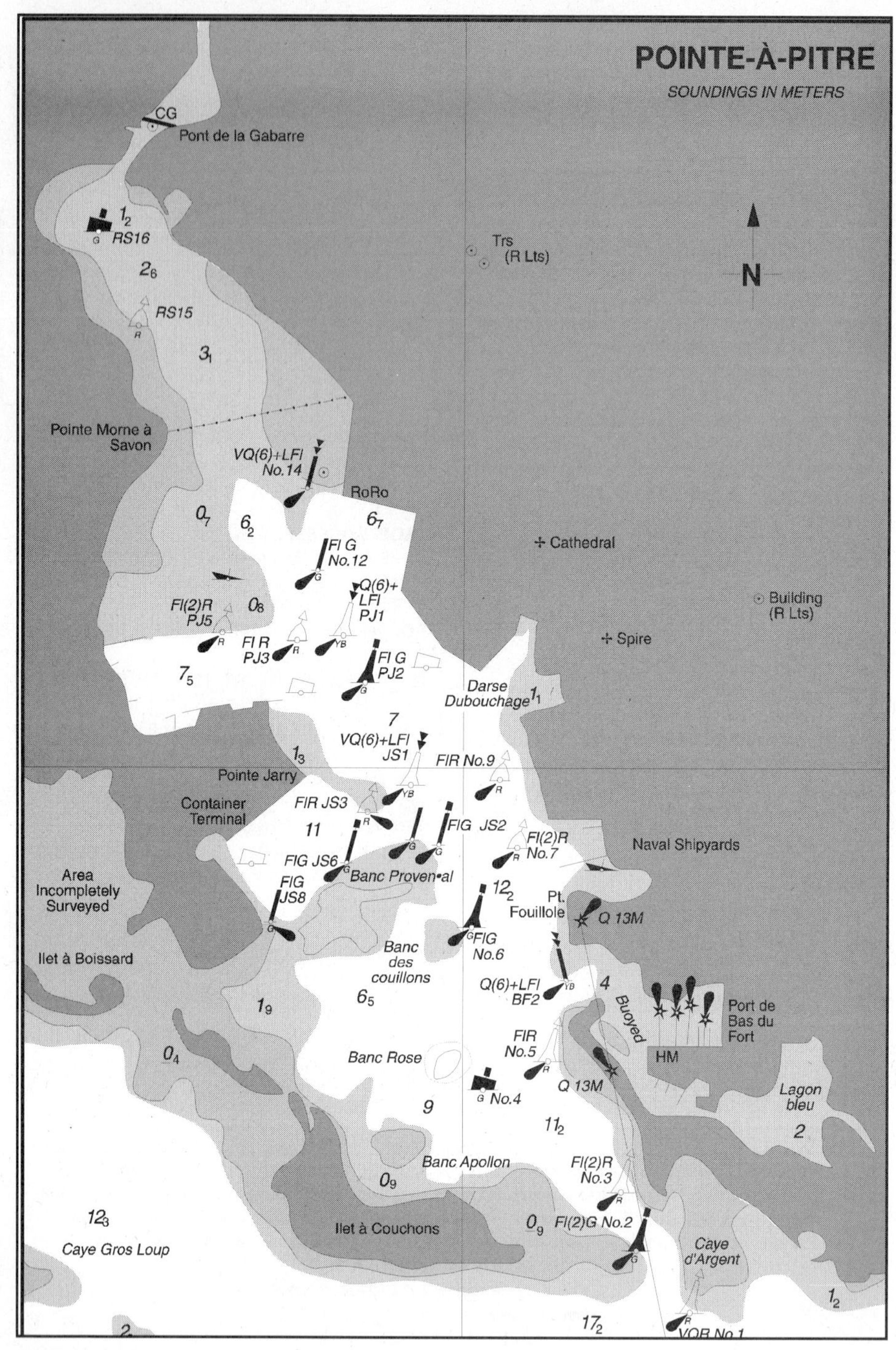

Pointe-à-Pitre, soundings in meters

POINTE-À-PITRE APPROACHES

ENTRANCE RANGE (Front Light), S end Monroux peninsula, 16 13.3N, 61 31.6W. Dir Q W, 49ft, 13M. White pylon, white rectangular topmark, 46ft. Intensified 3° either side of range. **(Rear Light),** Pointe Fouillole, 640 meters, 348° from front. Q W, 69ft, 13M. White mast, topmark white slats, 65ft.
AVIATION LIGHT, 16 15.8N, 61 31.7W. Mo (P) W 30s. Aero radiobeacon. Occasional.

POINTE-À-PITRE

16 14N, 61 32W
Pilotage: The approaches to the commercial port of Pointe-à-Pitre are well marked by buoys, lights, and ranges. The buoyage is laid out on the red-right-returning scheme. Beware of the coral reefs extending from the shore of Basse-Terre in the approaches. The west side of the inner harbor is commercial in nature, while the pleasure boat facilities are concentrated to the east. This area, which is known as Port du Plaisance, is the largest of the island's three boating basins. The Capitainerie (harbormaster) opens Monday through Friday, 8am to 1pm, and 3–5pm, and Saturday 8–11:30am (590-82-54-85).
Port of entry: Customs offices *(bureaux de douane)* are in the marina area and in town.
Dockage: Marina Bas du Fort (VHF channel 09; tel., 590-90-84-85), located in the first bay to the east as you enter the harbor, has 1,000 slips, 100 of which are available for visiting boats.
Anchorage: There are various places to anchor tucked just outside the marked channels near shore. Most boats try to be near the shipyard and marina complexes on the east side of the harbor. The marina rents moorings.

Services: Extensive services are available here, including several places for haulouts. There are many marine businesses in the Port du Plaisance area, rivaling the facilities available in Antigua. Among them are Karukera Marine (590-90-90-96) and Electro Nautique (590-82-18-35), both chandleries; a sail loft; diesel engine, electrics, and general engineering service; and an electronics repair service. There are supermarkets, restaurants, and shops in the area around the marina and in town.

Pointe-à-Pitre is a well-sheltered harbor, offering some of the best facilities and services for yachts available in the Caribbean. It is also a busy commercial port and the center of commerce for Guadeloupe.

GRANDE-TERRE

PORT LOUIS LIGHT, on the beach, 16 25.1N, 61 32.2W. Q (9) W 15s, 33ft, 9M. West cardinal mark, yellow-black-yellow beacon, 33ft. Visible 252°–162°.
ANSE BERTRAND LIGHT, 16 28.4N, 61 30.7W. Fl (2) W R G 6s, 46ft, W 9M, R 6M, G 6M. White concrete tower, 26ft. R 120°–163°, W 163°–170°, G 170°–200°, obscured 200°–120°.
PORT DU MOULE LIGHT, W side of entrance, 16 20.0N, 61 20.8W. Fl W R 4s, 39ft, W 9M, R 6M. White pylon, green top on hut, 36ft. R 110°–202°, W 202°–312°, R 312°–340°.
ENTRANCE LIGHT, E side. Dir Fl (2) W R G 6s, 23ft, W 7M, R 5M, G 5M. White pillar, red top, 23ft. R 353°–133°, W 133°–138°, G 138°–165° (W on Hastings Pass).
PORT DE SAINT FRANÇOIS LIGHT, directional light, 16 15.1N, 61 16.7W. Dir Q W R G, 33ft, W 9M, R 7M, G 7M. White tower. R 345°–358°, W 358°–002°, G 002°–015°.
PORT DE SAINT FRANÇOIS, on wharf, 16 15.1N, 61 16.6W. Fl G 4s, 30ft, 6M. White metal post, green top, 23ft.
ÎLET A GOZIER, 16 12.0N, 61 29.2W. Fl (2) R 10s, 79ft, 26M. White tower, red top, 69ft. Visible 259°–115°, obscured by trees on certain bearings toward Pointe Caraibe.

GRANDE-TERRE

Dockage: Near the southeastern tip of Grande-Terre is the Marina de la Grande Saline (590- 84-47-28), a major attraction of the St. François resort area. It has 250 berths, water, electricity, fuel, ice, and a slipway.

LA DÉSIRADE AND ÎLES DE LA PETITE TERRE

LA DESIRADE LIGHT, near SE point of island, 16 19.6N, 61 00.7W. Fl (2) W 10s, 165ft, 20M. White framework tower, red top, upper part enclosed, dwelling, 66ft.
BAIE MAHAULT RANGE (Front Light), 16 19.7N, 61 01.1W. Fl R 2s, 16ft, 4M. White pylon, red top. **(Rear Light),** 35 meters, 327° from front. Fl R 2s, 23ft, 4M. White pylon, red top. Synchronized with front.
GRANDE ANSE LIGHT, head of pier, 16 18.2N, 61 04.8W. Fl G 2.5s, 23ft, 1M. White mast.

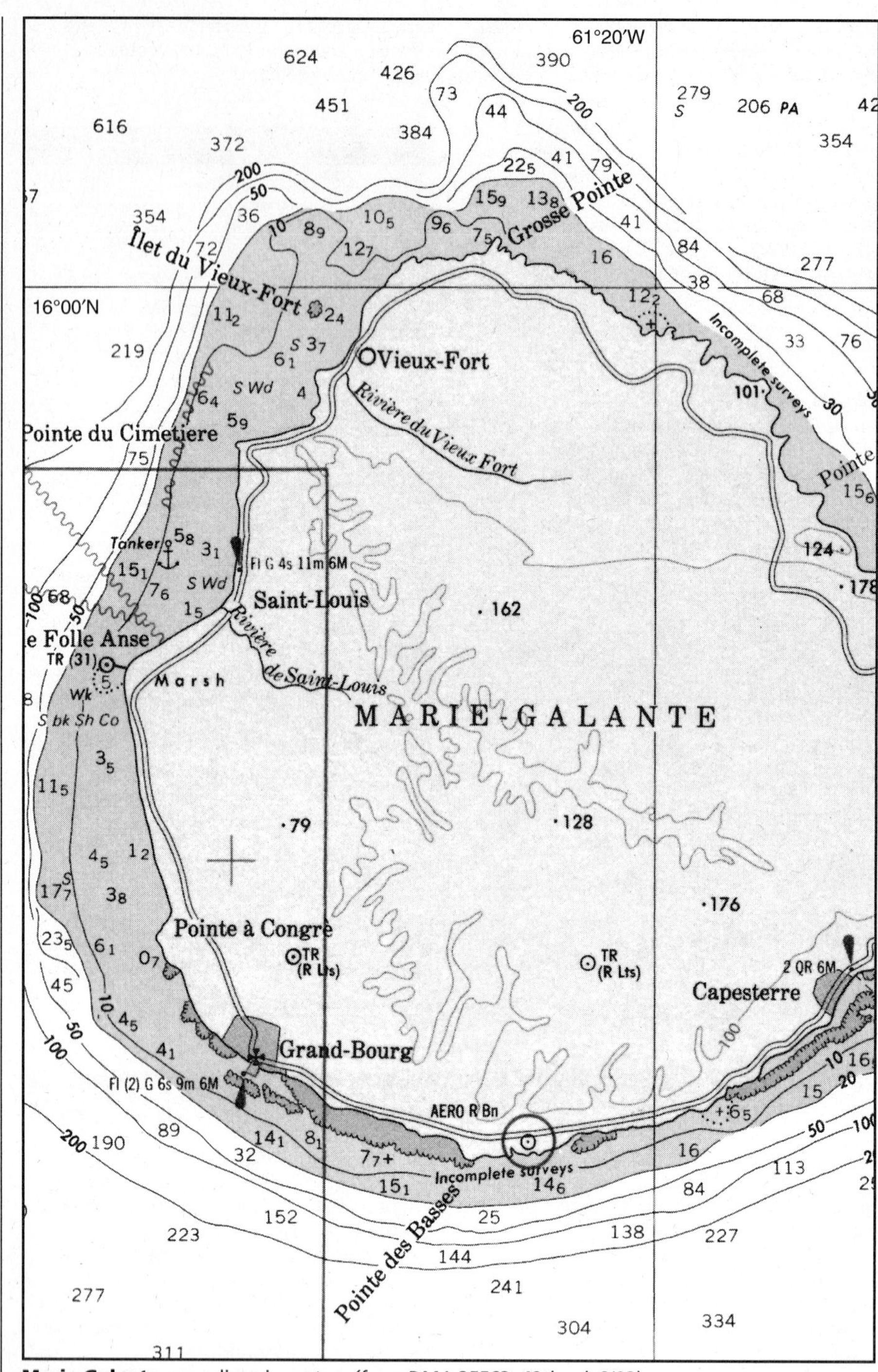

Marie-Galante, soundings in meters *(from DMA 25563, 49th ed, 8/93)*

LEADING LIGHT, Dir Oc (2) W R G 6s, 23ft, W 8M, R 6M, G 6M. White stone structure, red top, 23ft. R 250°–335°, W 335°–339°, G 339°–056°.
ÎLES DE PETITE TERRE, E extremity of Îlot Terre d'en Bas, 16 10.3N, 61 06.8W. Fl (3) W 12s, 108ft, 15M. Gray cylindrical tower. Square stone base, 85ft. Obscured by Île la Désirade 185°–213°.

LES SAINTES

BOURG DES SAINTES LIGHT, Terre d'en Haut, root of wharf, 15 52.1N, 61 35.2W. Fl W R G 4s, 30ft, W 10M, R 7M, G 7M. White metal framework tower, 26ft. R 075°–142°, W 142°–154°, G 154°–160°. Obstruction light on aerial.

BOURG DES SAINTES

15 52N, 61 35W
Pilotage: The town is located on the western side of Terre d'en Haut, near the middle of the island. A large conspicuous cross is illuminated at night and stands south of the church. Note the shoal patch, halfway to Îlet à Cabrit, marked by a buoy. A 100-foot-long pier projects from the waterfront, where a light is shown from near the root of the pier.
Clearance: Check with the gendarmes if arriving here from another country before having cleared in at an official port of entry.
Anchorage: Anchor away from the ferry dock and ferry channels.
Services: There is little in the way of marine services, but there is a good selection of boutiques, restaurants, and general stores.

MARIE-GALANTE

GRAND-BOURG LIGHT, on wharf, 15 52.9N, 61 19.2W. Fl (2) G 6s, 30ft, 6M. White metal post, green top, 29ft.

GRAND BOURG

15 53N, 61 19W
Pilotage: This town lies on the southwest side of Marie-Galante. The fort and the hospital on the northwest side of the town and the church with a belfry on the town's northeast side are all prominent landmarks. A conspicuous lighted TV tower stands about 1 mile north of town. Lighted buoys mark the edge of the reefs in the approach to the pier from the west.
Port of entry: You can clear in here, but you must get your outward clearance on Basse-Terre or Grande-Terre.
Anchorage: Anchor in or near the small basin near the ferry dock.
Services: General supplies are available.

ST. LOUIS LIGHT, W of church, 15 57.4N, 61 19.3W. Fl G 4s, 36ft, 6M. White square tower, green top, 30ft.
CAPESTERRE RANGE (Front Light), 15 53.9N, 61 13.2W. Q R, 39ft, 6M. Square tower, red and white bands, 23ft. Visible 246.5°–051.5°. **(Rear Light),** 100 meters, 313° from front. Q R, 52ft, 6M. Square tower, red and white bands, 19ft. Visible 246.5°–051.5°.

Dominica

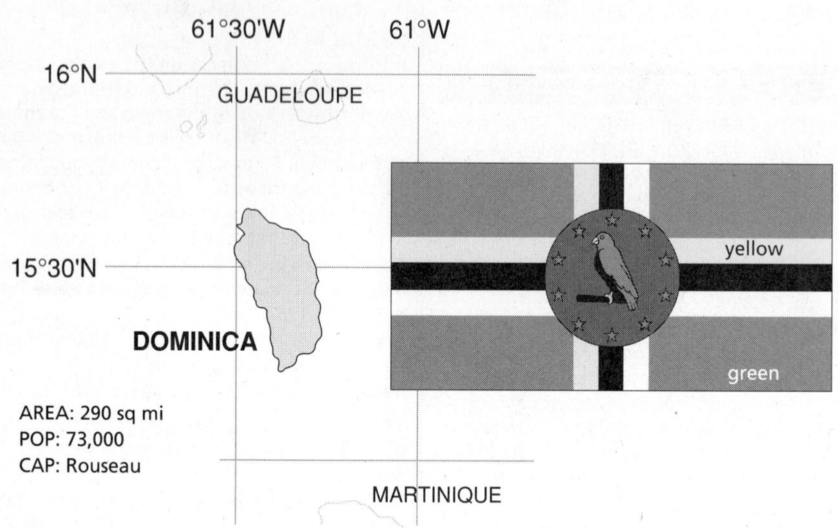

Please see cautions about chartlets and waypoints on page P 2.

ENTRY PROCEDURES

The northern tip of Dominica is about 38 miles south of Pointe-à-Pitre, Guadeloupe. Although the island has most recently been a British possession, a strong French influence seeps in from neighboring islands. The Commonwealth of Dominica is now an independent country.

Ports of entry are Portsmouth in Prince Rupert Bay and Roseau, the capital. Both are on the west coast of the island. Contact customs in Portsmouth at 809-445-5340 and in Roseau at 809-448-4462. The head immigration office in Roseau can be reached at 809-448-2222 x5159.

In Portsmouth try calling the Portsmouth Beach Hotel (VHF channel 16; tel., 809-445-5142) to see if you can contact officials from the facility's anchorage. In Roseau you might want to call the Anchorage Hotel (VHF channel 16; tel., 809-448-2638) for the same reason. These facilities are owned by the same family and specialize in making sailors feel welcome. They can assist with most of your needs while on the island.

In Portsmouth the officials are located near the commercial wharf south of town. In Roseau it is possible to clear in either at the customs office by the commercial wharves in Woodbridge Bay or at the ferry terminal on the bay front of Roseau Roads. Most private yachts clear in with customs and immigration at the ferry terminal. The area north of the mouth of the Roseau River, including the commercial wharves, is a Restricted Anchorage. To clear in at the commercial office, call the port manager on VHF channel 16 to obtain permission to anchor off the wharves. You can then dinghy in to clear with officials. Normal working hours for officials are 0800 to 1300 and 1400 to 1700 on Monday and 0800 to 1300 and 1400 to 1600 Tuesday through Friday. Overtime charges apply outside of these hours. A cruising permit should be obtained for your itinerary on the island.

The U.S. does not maintain an embassy in Dominica. U.S. citizens needing assistance should contact their embassy in Bridgetown, Barbados (246-436-4950).

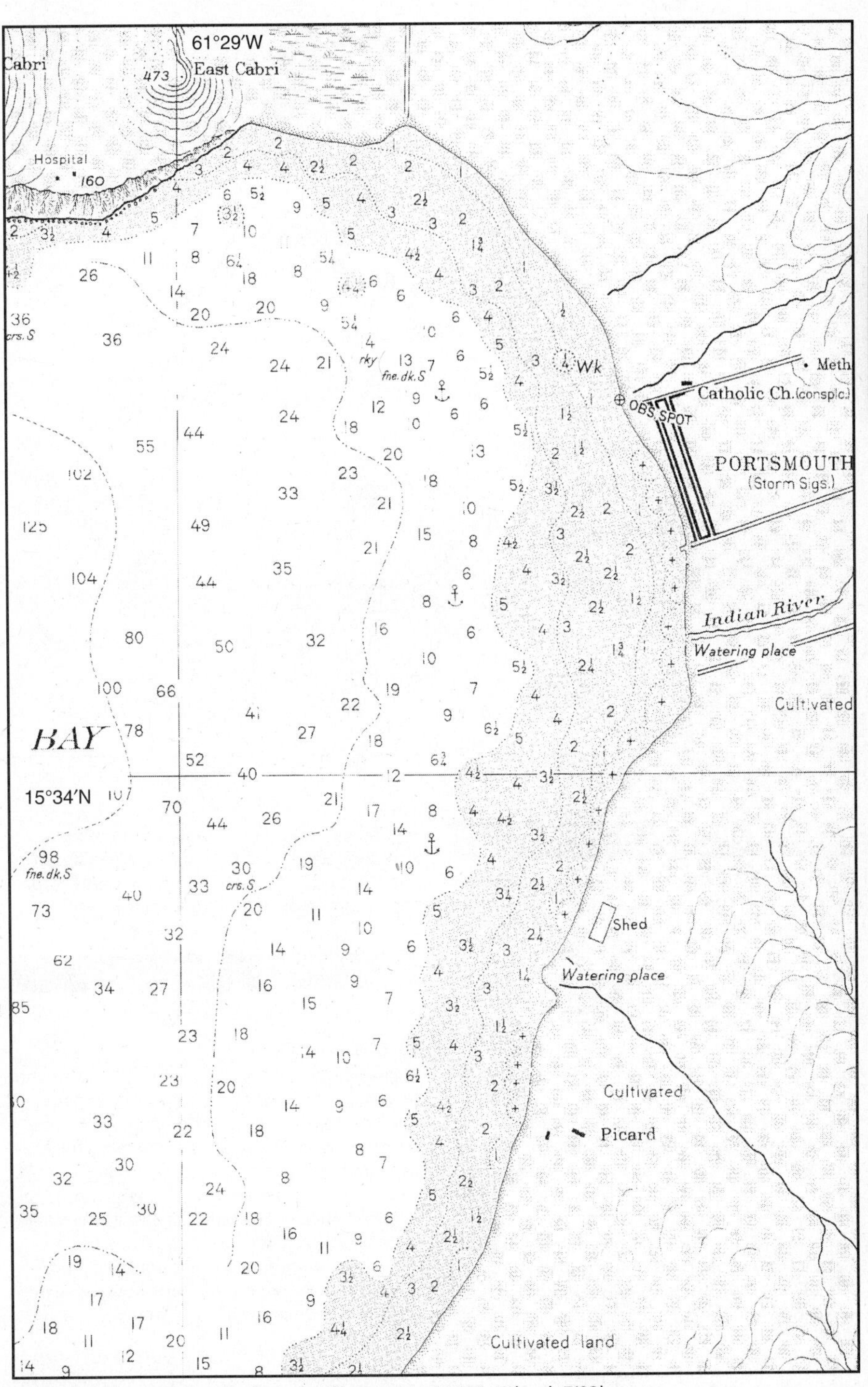

Portsmouth, soundings in fathoms *(from DMA 25562, 7th ed, 7/82)*

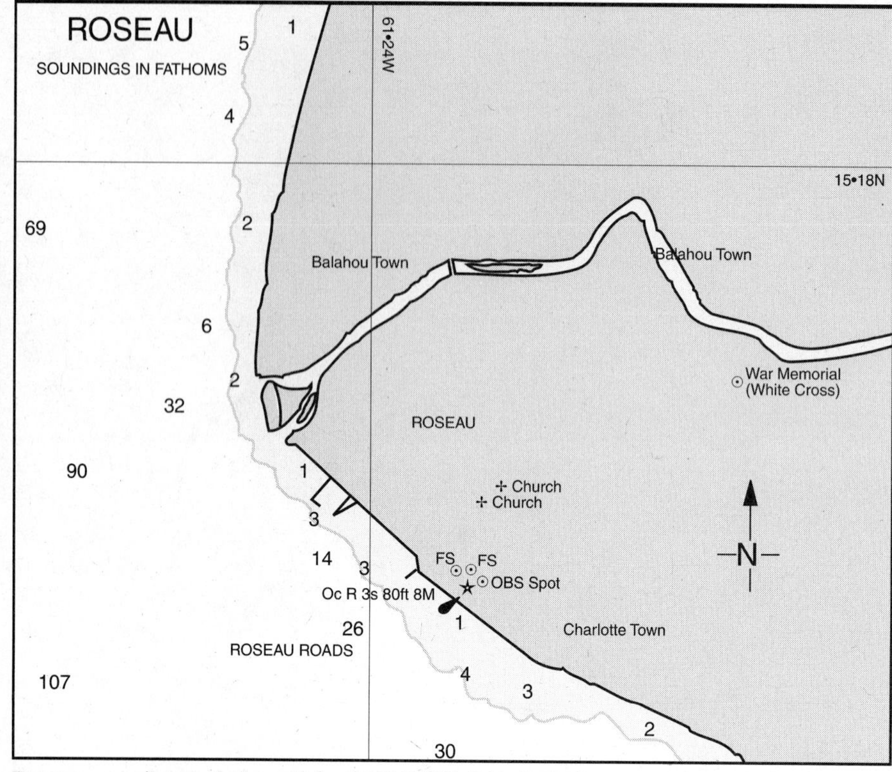

Roseau, soundings in fathoms *(after DMA 25562, 7th ed, 7/82)*

USEFUL INFORMATION

Languages: English (official), also French Creole because of the proximity of French islands.
Currency: Eastern Caribbean dollar (EC$); exchange rate (summer 1997), EC$2.70 = US$1. U.S. dollars may be accepted in tourist areas.
Telephones: A new area code, 767, replaced the old code, 809, in October 1997.
Public holidays (1998): New Year's Day, January 1; Carnival, February 23–24; Good Friday, April 10; Easter Monday, April 13; May Day (Labour Day), May 1; Whit Monday, June 1; August Monday, August 3; Independence Day, November 3; Community Service Day, November 4; Christmas Day, December 25; Boxing Day, December 26.
Tourism office: *In Dominica:* National Development Corporation, Division of Tourism, P.O. Box 293, Roseau (tel., 767-448-2045/2351; fax, 767-448-5840).
Medical: Medical care is limited in Dominica, and hospitals and doctors expect immediate cash payment. For health information about traveling to Dominca, contact the Centers for Disease Control and Prevention (tel., 404-332-4559; Web, www.cdc.gov).

WEST COAST

PORTSMOUTH

15 34N, 61 29W
Pilotage: Prince Rupert Bay, near the north end of the island's west coast, offers good shelter. There are no dangers in the approach to this steep-to coast. At the head of the bay, the gray spire and red roof of the Catholic church in the small town of Portsmouth are conspicuous. The white Methodist chapel to its east is prominent from offshore. There are two jetties and a prominent warehouse 1 mile south of Portsmouth (not shown on the Portsmouth chartlet) and a cruise ship dock to the north.
Port of entry

Dockage: The Portsmouth Beach Hotel may have room, either bow- or stern-to, at its jetty about 1 mile south of town. Moorings are also available.

Anchorage: Private yachts anchor near the Portsmouth Beach Hotel and dinghy in to the hotel jetty or else north of town near the Purple Turtle or Mamie's on the Beach restaurants.

Services: Services at the Portsmouth Beach Hotel were curtailed following the hurricanes of 1995. There is still a restaurant, and staff may assist with most projects on the island. Ask for recommendations on guides or taxis. Basic supplies and groceries are available in town. "Boat boys" offer their services here. Theft is sometimes an issue.

BARROUI LIGHT, 15 25.8N, 61 26.8W. 2 F R (vertical).

ROSEAU

15 18N, 61 24W

Pilotage: The port of Roseau comprises Woodbridge Bay and Roseau Roads. A deepwater wharf is situated 1,600 yards north of the mouth of the Roseau River. This area is a Restricted Anchorage, and boaters should contact the port manager on VHF channel 16 for permission to drop the hook. Morne Bruce, a tableland, rises to a height of 475 feet east of town. There are several old military buildings on its summit. There is a prominent flagstaff on the west corner of Fort Young at the south end of town. The Catholic church and Wesleyan church spires are also prominent. The red bridge crossing the mouth of the Roseau River is a good landmark.

Port of entry

Anchorage and moorings: Anchor near the Anchorage Hotel (VHF channel 16; tel., 767-448-2638) located about 1 mile south of town. The hotel also has moorings available.

Services: There are few yachting services here, but you can obtain most general supplies and groceries. The Anchorage Hotel has a dinghy dock, showers, water, pay telephones, dive facilities, and a mail drop; for use of these facilities, the hotel charges a yacht registration fee of US$10, good for three days. The staff can also assist you in obtaining fuel or other services. "Boat boys" may offer their services here.

ISLA AVES, VENEZUELA

15 40N, 63 37W
ISLA AVES LIGHT, 15 40.1N, 63 37.0W. FL W 5s, 49ft., 13M. Metal Structure. **RACON A** (· –), range 14M. Existence confirmed 4/97.

Venezuelan Coast Guard: Call sign, Simon Bolivar Coast Guard Station.

Isla Aves is a Venezuelan possession with a latitude slightly north of the northern tip of Dominica, but 128 miles to the west. It is nearly 600 yards long in a north–south direction and 100 yards wide. The maximum elevation of about 10 feet is near its northern end, and the sea breaks across the center in anything more than a moderate swell. The island is formed of coral, overlaid with sand, which supports some vegetation. Birds abound, and the island is a protected nature sanctuary. An oil-rig-style Texas tower with a radar reflector stands on the island. The structure on top of the tower is the size of an average house. The platform is approximately 60 feet high. This is a manned Venezuelan Coast Guard Station (call sign, Simon Bolivar Coast Guard Station). The island has been reported as a good radar target up to 30 miles distant. There is also a transmitting radar beacon (RACON).

About 100 yards from the south end of the island, a jetty with a depth of about 9 feet alongside its head projects from land about 88 yards. Depths less than 5 feet are found on several shoal patches in the western approaches to the jetty. In moderate weather you may be able to land near the center of the western side of the island. Here, a narrow sandy beach extends down to the low waterline here.

Vessels without local knowledge are advised to give the islands a berth of at least 1.5 miles. If you wish to anchor, the best approach is from the south-southwest. Keep a good lookout for coral heads. The recommended anchorage is about 600 yards offshore. Depths in this area should be about 17 feet, although there is a 9-foot patch reported about 100 yards east-northeast of this position.

Martinique

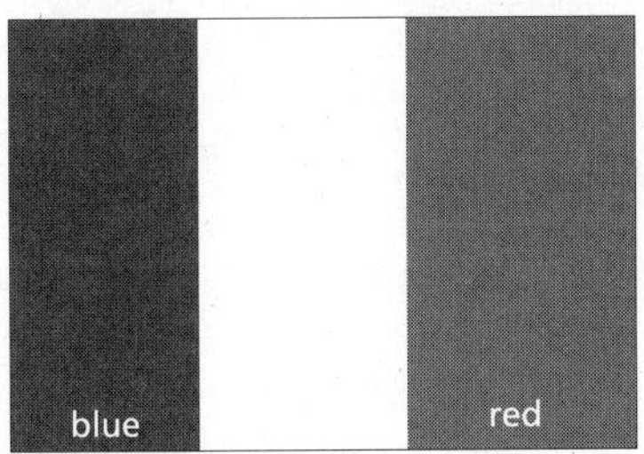

AREA: 425 sq mi
POP: 359,572
CAPITAL: Fort-de-France

Please see cautions about chartlets and waypoints on page P 2.

ENTRY PROCEDURES

This island country is a department of France. With a population of more than 350,000, it is one of the commercial centers of the Antilles. The French courtesy flag should be flown here.

The main port of entry is Fort-de-France. Clear in first with customs officials, who contact other officials. Hours are about 0830 to 1100 and 1500 to 1700 Monday through Saturday. You can also check in at St. Pierre, Trois-Ilets (Anse Mitan) or Marin, which maintain morning office hours only. The customs office by the cruise ship dock in Fort-de-France may be reached at 596-60-7172. You will need boat documents and passports. Firearms must be declared. Pets may come ashore if they are documented and vaccinated against rabies. Check on the latest regulations if you plan on staying in the country or leaving your boat for more than six months. You may be liable for duty after this period.

USEFUL INFORMATION

Language: French. Some proficiency is recommended, although some English is spoken by many yacht-oriented businesses.
Currency: Legal tender is the French franc, but U.S. and Canadian dollars are accepted almost everywhere. The exchange rate (summer 1997), 5.5F = US$1.
Telephones: Country code, 596; local numbers, six digits. Using a télécarte makes local and international calls easier and less expensive (a one-minute call to the U.S. costs about 11.5F, or $2.10). Télécartes are sold at all post offices and other outlets marked "Télécartes en Vente Ici"; they are used in special booths marked "Télécom," and there are about 90 such booths in and around Fort-de-France alone. Operator-assisted calls cost significantly more. The telex area code for Martinique is 300.
Public holidays (1998): New Year's Day, January 1; Carnival, February 8–12*; Easter Monday, April 13; Labor Day, May 1; Ascension Thursday, May 22; Pentecost Monday, June 1; Abolition of Slavery Day, May 22; Bastille Day, July 14; Assumption Day, August 15; All Saints' Day, November 1; Armistice Day, November 11; Christmas Day, December 25. *Most businesses closed, but not an official holiday.
Tourism offices: *In Martinique:* Office Départemental du Tourisme de la Martinique, BP 520, 97206 Fort-de-France (596-63-79-60). *In the U.S.:* Martinique Promotion Bureau, 444 Madison Avenue, 16th Floor, New York, NY 10022 (800-391-4909; also e-mail, Martinique @nyo.com; Web, www.nyo.com/martinique). *In Canada:* Martinique Promotion Bureau, 1981 Avenue McGill College, Suite 480, Montréal, PQ, H3A 2W9 (514-844-8566).

WEST COAST

PRECHEUR POINT LIGHT, 14 48.0N, 61 13.8W. Fl R 5s, 72ft, 19M. Iron tower, stone base, 9ft. Visible 338°–162°.
POINTE DES NEGRES LIGHT, 14 35.9N, 61 05.6W. Fl W 5s, 118ft, 25M. White metal framework tower, gray lantern, 92ft. Visible 276°–126°. Aero radiobeacon.

FORT-DE-FRANCE

FORT ST. LOUIS LIGHT, in SW part of fortress, 14 35.9N, 61 04.4W. Fl W R G, 102ft, W 14M, R 11M, G 11M. White iron framework structure, red and white daymarks, 20ft. R 320°–057°, W 057°–087°, G 087°–140°, obscured 140°–320°. 2 F R (horiz.) and 3 F R (vert.) on each of three radio masts 1.2–1.4 miles E.

BAIE DU CARENAGE RANGE (Front Light), 494 meters, 074° from signal staff, 14 36.3N, 61 03.7W. Iso G 4s, 138ft, 14M. Yellow rectangle, white metal framework tower, black bands, 39ft. Intensified 001°–007°. (Front range unofficially reported not intensified, June 1995). **(Rear Light),** 145 meters, 004° from front. Iso G 4s, 164ft, 14M. Yellow rectangle, white metal framework tower, black bands, 46ft. Synchronized with front. Intensified 001°–007°.

QUAI OUEST, SW corner, 14 35.9N, 61 04.1W. F R, 3ft, 6M. Red pedestal. Visible 304°–049°.

QUAI OUEST, SE corner. F G, 3ft, 6M. Green pedestal. Visible 284°–029°.

QUAI AUX HUILES, SW corner. F G, 3ft, 6M. White pedestal. Visible 306°–051°.

QUAI DES ANNEXES, Oc R 4s, 3ft, 6M. Green pedestal. Visible 315°–060°.

QUAI DES TOURELLES, Oc G 4s, 3ft, 6M. Green pedestal. Visible 315°–060°.

CONTAINER WHARF, Oc (2) R 6s, 3ft, 6M. Red pedestal.

LIGHT 7, 14 35.6N, 61 4.0W. Fl(2) R 6s. Red starboard-hand beacon with topmark.

LIGHT 3, 14 35.1N, 61 4.0W. Fl R 2.5s. Red starboard-hand beacon with topmark.

LE LAMENTIN AVIATION LIGHT, 14 35.7N, 61 00.0W. Fl (3+1) W 12s, 105ft, 20M. Occasional.

FORT-DE-FRANCE

14 36N, 61 04W

Pilotage: The island is very mountainous and easily identified by three outstanding peaks towering above the main mountain chain that traverses it in a northwest to southeast direction. Shoals, banks, and other dangers are marked by buoys in the approaches to the city. Note the 004° range leading into the main harbor. Local port authorities and pilots may be contacted on VHF channels 13 and 16. This city is a very busy port.

Port of entry: The officials are located behind the new cruise ship dock in the western portion of the waterfront.

Dockage: There may be some stern-to dockage on the wharf east of Fort Saint-Louis, but most boats anchor out. There should be some dockage available across the way on Pointe du Bout at Anse Mitan.

Anchorage: Most boats anchor in the western part of Baie des Flamands. For a more peaceful anchorage try one of the nearby bays or the area around Anse Mitan.

Ferries: Fort-de-France is linked round-trip with Pointe de Bout daily from early morning until after midnight, and with Anse Mitan, Anse-à-l'Ane, and Grande Anse d'Arlet from early morning until late afternoon. The arrival and departure point in Fort-de-France for all ferries is the Quai d'Esnambuc.

Services: Every type of marine service is here or nearby. Fuel and water are also available across the bay at Anse Mitan. The repair yards are Martinique Drydock (tel., 596-72-69-40; fax, 596-63-17-69), which is geared to large vessels; Ship Shop (tel., 596-73-73-99; fax, 596-71-43-40), which has a Travelift; and Multicap Caraibes (tel., 596-71-41-81; fax, 596-71-41-83), a boat builder with repair facilities for handling large multihulls. Ship Shop and Multicap Caraibes are located at Quai Ouest, but Ship Shop expects to move to new quarters in nearby Baie de Tourelles soon. There are several well-stocked chandleries in Fort-de-France, among them Sea Services (596-70-26-69), Littoral (596-70-28-70), and the Ship Shop. Other repair services cover sails and canvaswork, engines, metalwork, woodwork, and fiberglass. There are large first-class supermarkets and general stores of all types.

Fort-de-France is the busy and very European administrative center of Martinique. With more than 100,000 inhabitants, it is also the largest city in the Antilles and a great place to find the items and services a modern city can provide. Nearby are peaceful anchorages when you tire of the hustle-bustle.

POINTE DU BOUT

POINTE DU BOUT LIGHT, marina E jetty head, 14 33.4N, 61 03.5W. Fl G 2.5s, 13ft, 2M. Green post.

POINTE DU BOUT LIGHT, marina W jetty head. Fl R 2.5s, 13ft, 2M. Red post.

ANSE MITAN

14 34N, 61 03W

Pilotage: There is a buoyed reef, Caye de l'Anse Mitan, west of Pointe du Bout, the head of the Anse Mitan peninsula. There are also reefs close to shore around Pointe du Bout.

Dockage: The Ponton du Bakoua (VHF channel 16; tel., 596-66-05-45), about halfway down the peninsula on the western side, has some dockage. The Marina Pointe du Bout (VHF channel 09; tel., 596-66-07-74; fax, 596-66-00-50), in the square dredged basin just east of the point, also has slips.

Anchorage: Anchor along the western shore of the peninsula, taking care to avoid the small reef marked by a buoy. The reef is located west of the Ponton du Bakoua dock.

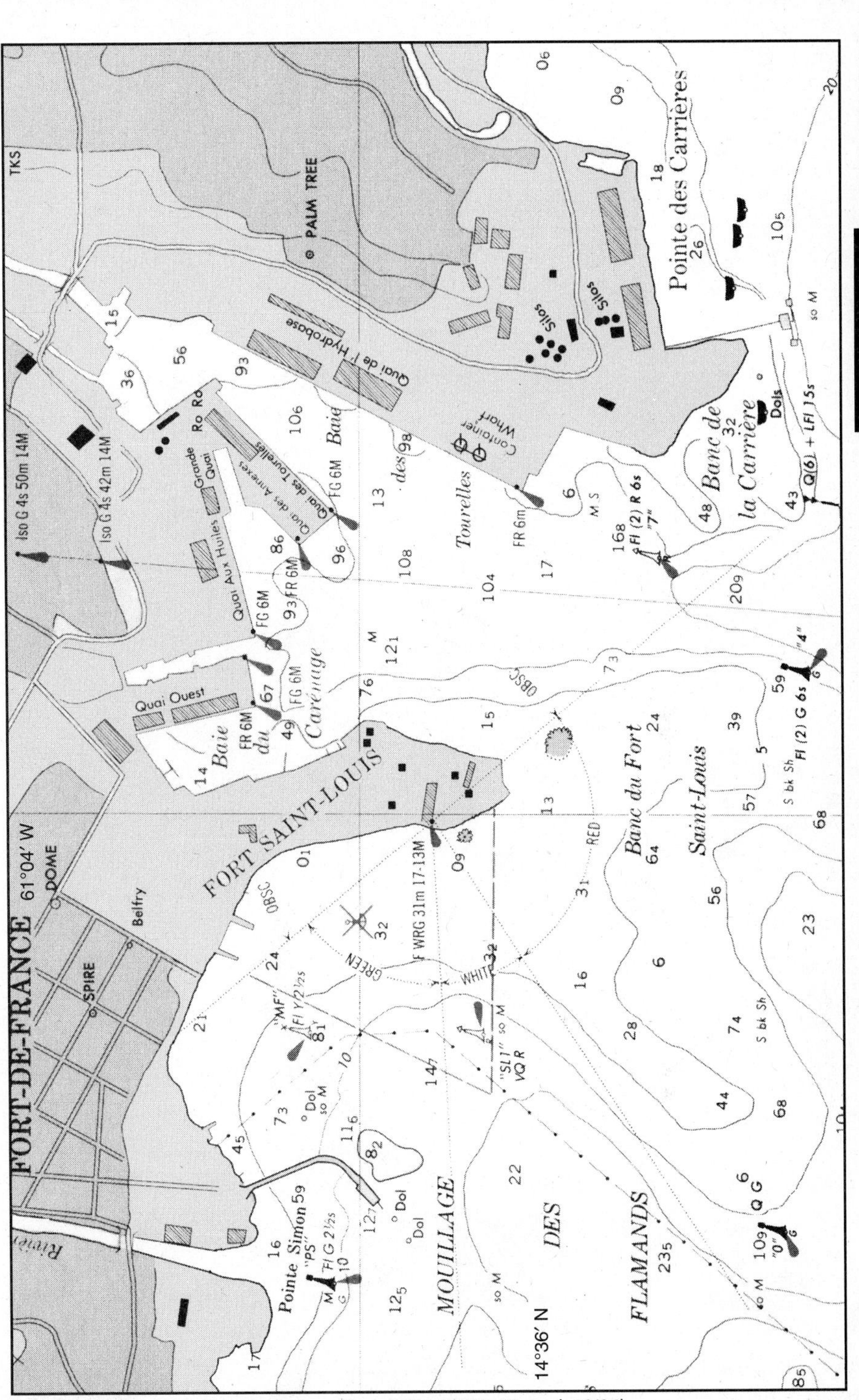

Fort-de-France, soundings in meters *(from DMA 25527, 31st ed, 12/94)*

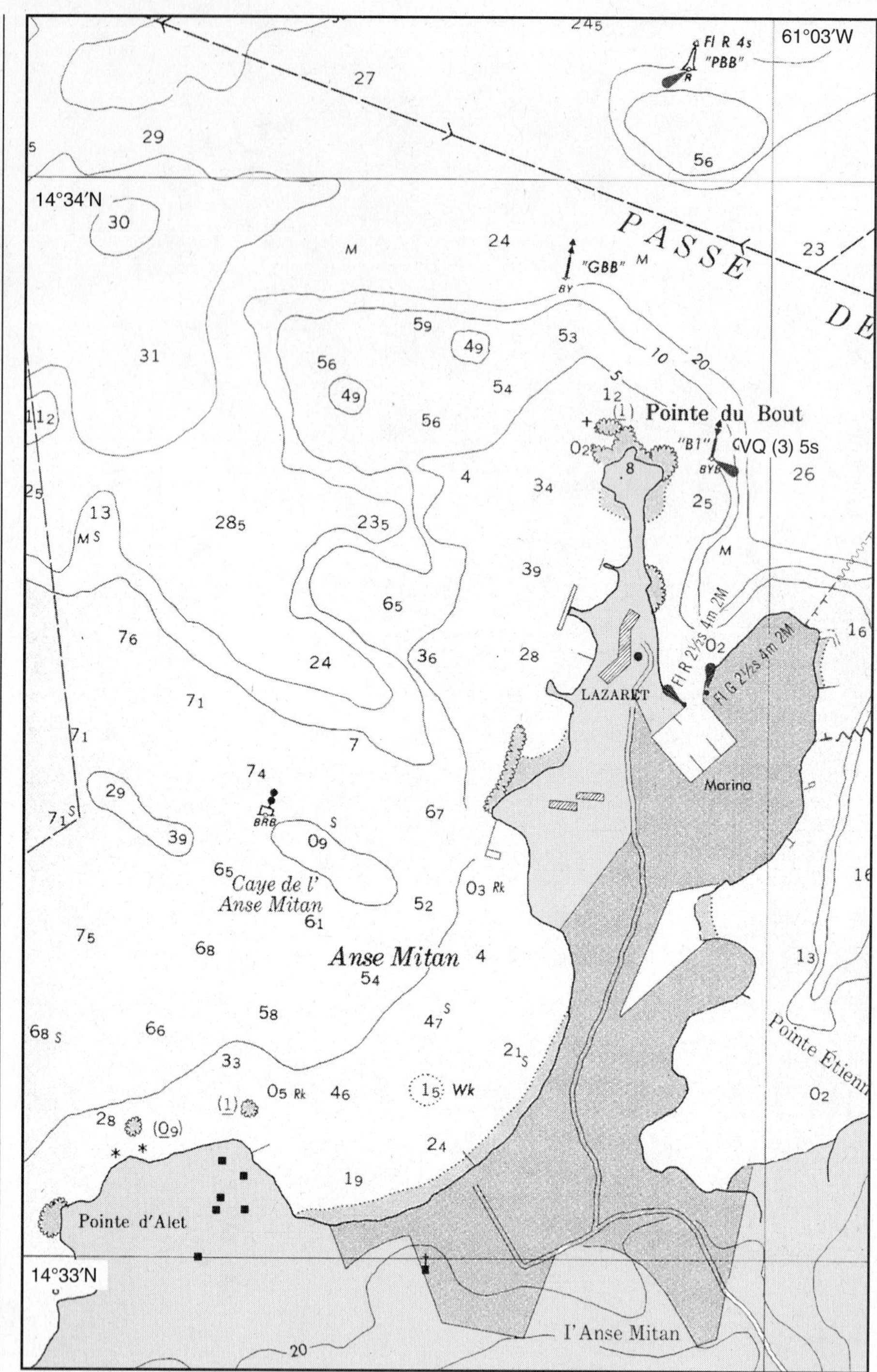

Anse Mitan, soundings in meters *(from DMA 25527, 31st ed, 12/94)*

Ferries: Ferries to Fort-de-France run frequently from the marina basin on Pointe du Bout and from the Anse Mitan dock in the big bay on the western side of the peninsula.
Services: Fuel and water are available at the Ponton du Bakoua. Mecanique Plaisance (596-66-05-40), located at the marina, services inboard engines and refrigeration systems. The local sail loft is Voilerie Caraibe Martinique (VHF channel 16; tel., 596-66-07-24). Nearby are supermarkets and general shops. For anything you can't find here, take the ferry across to Fort-de-France.

Anse Mitan is a convenient anchorage and is less commercial and more tourist oriented than Fort-de-France. The facilities are good, and the atmosphere less hectic than in the big city across the bay.

SOUTH COAST

POINTE DU MARIN, on W side of point, 14 26.9N, 60 53.2W. Dir Q W R G, 23ft, W 10M, R 7M, G 7M. White structure, green top, 23ft. R 015°–071°, W 071°–075°, G 075°–080°.
NORTH HEAD LIGHT, 14 27.1N, 60 53.1W. Fl R 2.5s, 16ft, 2M. Red starboard-hand beacon with topmark.
BANC MAJOR, LIGHT MA, 14 27.7N, 60 52.6W. Fl (2) G 6s, 7ft, 2M. Green port-hand beacon with topmark.
BANC DU MILIEU, LIGHT MI, 14 27.7N, 60 52.5W. Fl (2) R 6s, 7ft, 2M. Beacon.
BANC DE LA DOUANE, LIGHT MA 8, 14 28.0N, 60 52.2W. Fl G 2.5s, 7ft, 2M. Green port-hand beacon with topmark.
BANC DE LA DOUANE, LIGHT MA 10, 14 28.1N, 60 52.2W. Fl (2) G 6s, 7ft, 2M. Green port-hand beacon with topmark.

LE MARIN

14 28N, 60 52W
Pilotage: There are quite a few reefs in Cul-de-Sac Marin, but the clear water and good aids to navigation should make the approach easy. There are three conspicuous chimneys west of town, and a light is shown from Pointe du Marin. Because of the shoals, an approach in good light is recommended. There are channels to Carene Antilles and the SAEPP marina.
Port of entry: Officials maintain office hours from 0700 to 1200.
Dockage: The SAEPP marina (VHF channel 09; tel., 596-74-83-83; fax, 596-7492-20) is located in the lagoon at the eastern end of the waterfront. Space is apt to be tight due to the many charter operations in this area.
Anchorage: Anchor directly south of the waterfront, keeping the designated channels clear.
Services: Marin is the main base of the Martinique charter fleet, and all manner of repair facilities can be found here. Water, ice, electricity (220-volt), and fuel (duty free for vessels that have already received outbound clearance) are available at the SAEPP marina. West of the main docks is Carene Antilles (VHF channels 16 and 73; tel., 596-74-77-70; fax, 596-74-78-22), a large repair yard with a substantial chandlery on site. Captain's Ship Chandlery (tel., 596-74-87-55; fax, 596-74-96-71) is another source of marine hardware. Among other marine-oriented businesses are sail lofts, paint stores, and fiberglass, rigging, and mechanical repair facilities. There is a good selection of supermarkets and general shops.

The harbor of this small town offers good shelter and is a good alternative to busy Fort-de-France.

EAST COAST

BAIE TRINITE RANGE (Front Light), 14 45.2N, 60 58.2W. Iso W 4s, 20ft, 8M. White structure, 13ft. **(Rear Light),** 284.2° from front. Iso W 4s, 23ft, 8M. White structure, 20ft.
LA CARAVELLE LIGHT, 14 46.2N, 60 53.1W. Fl (3) W 15s, 423ft, 20M. Yellow square tower, white lantern, 46ft. Visible 113°–345°.
BAIE DU FRANÇOIS DIRECTIONAL LIGHT, 14 38.0N, 60 53.7W. Dir Q W R G, 108ft, W 8M, R 6M, G 6M. White tower, green top, 30ft. R 200°–245°, W 245°–248°, G 248°–280°, white sector indicates preferred channel.
PORT VAUCLIN LIGHT, N point, 14 33.0N, 60 50.4W. Q W R G, 46ft, W 11M, R 9M, G 9M. R 220°–230°, W 230°–232°, G 232°–250°.
DIQUE EST LIGHT, head, 14 32.7N, 60 50.1W. Fl G 2.5s, 13ft, 2M. White tower, green top, 10ft.
EPI OUEST LIGHT, head, 14 32.8N, 60 50.1W. Fl R 2.5s, 10ft, 2M. White tower, red top, 10ft.
PORT VAUCLIN DIRECTIONAL LIGHT, 305°, 14 26.0N, 60 49.8W. Dir Q W R G, W 10M, R 7M, G 7M. White pole. R 291°–304°, W 304°–306°, G 306°–319°.
ÎLET CABRIT LIGHT, N part, 14 23.4N, 60 52.3W. Fl R 5s, 138ft, 16M. Red pylon, hexagonal base. Visible 235°–106° and 107°–108°.

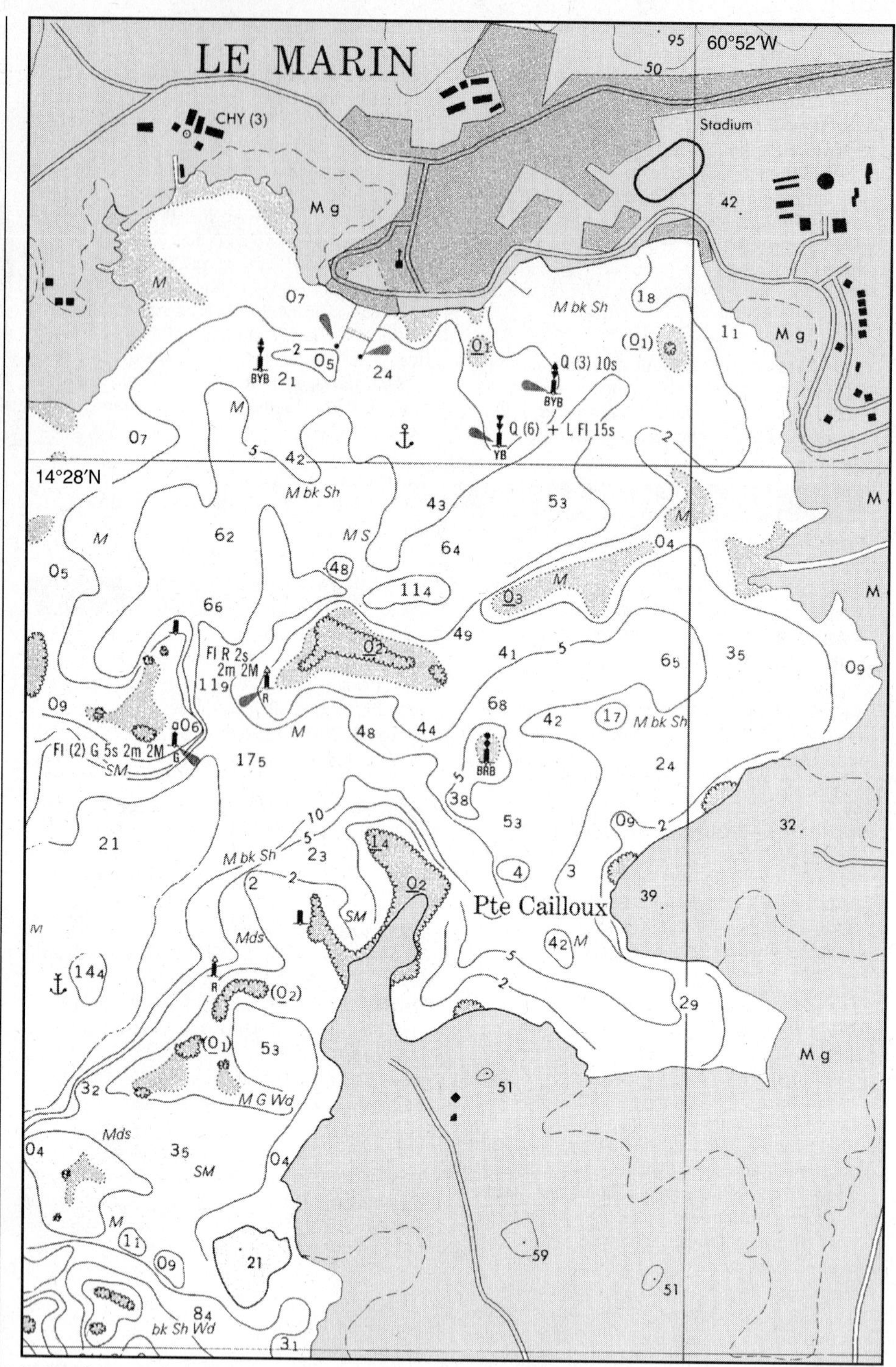

Le Marin, soundings in meters *(from DMA 25524, 42nd ed, 2/89)*

St. Lucia

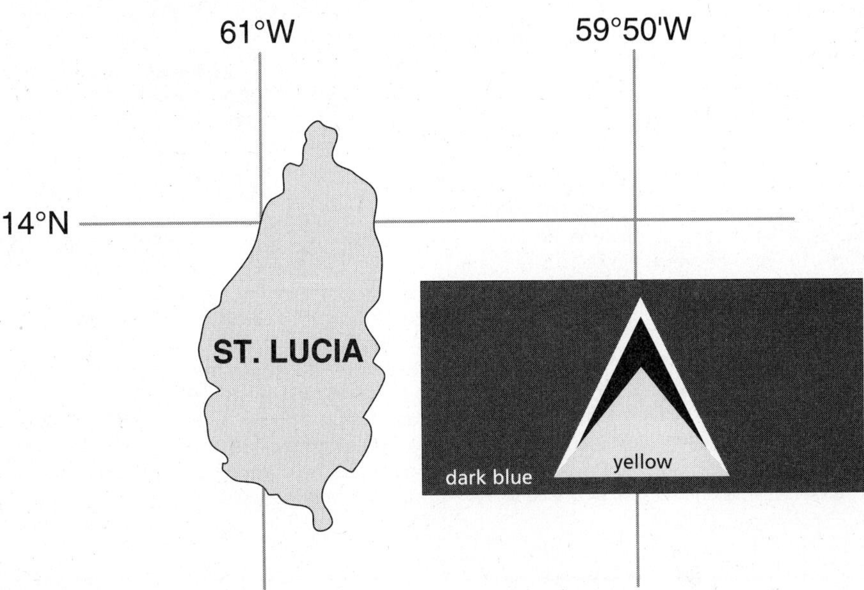

Please see cautions about chartlets and waypoints on page P 2.

ENTRY PROCEDURES

With France, England, and the Caribs battling for control of St. Lucia for three hundred years, the island has a dramatic history. St. Lucia was an English colony until it gained independence in 1979. It remains a member nation of the British Commonwealth.

Ports of entry are Rodney Bay, Castries, Marigot, and Vieux Fort. Navigational aids, practique, port, and clearance fees are charged at all ports of entry, with amounts depending on tonnage and length overall. The total for pleasure yachts under 40 feet is EC$30; over 40 feet, EC$40. Charter yachts are charged an additional EC$20–30.

In Rodney Bay customs, immigration, and harbour officials (758-452-0235) are located at Rodney Bay Marina. The office is normally open from 0800 to 1800 every day; overtime charges begin at 1630 and are charged for after-hours, weekend, and holiday clearances.

In Castries officials (758-452-3487) are located on North Wharf. In Marigot customs (758-451-4257) is on the south shore, west of the Moorings docks. In Vieux Fort customs (758-454-6526) is near the commercial wharf; calls to Vieux Fort customs on weekends are routed directly to customs at the airport for clearing-in instructions.

Necessary documents include clearance from the last port of call, ship's papers, three copies of crew's lists, and passports. Vessels planning to visit ports other than official ports of entry must request permission. Stopovers at Soufriere no longer require a special permit. Firearms and ammunition must be declared and may be held until you clear out for another island. Pets must remain on board.

USEFUL INFORMATION

Language: English. A French patois is also spoken.
Currency: Eastern Caribbean dollar (EC$); exchange rate (summer 1997), EC$2.70 = US$1. U.S. dollars are often accepted, especially in tourist areas.

Telephones: Area code, 758; local numbers, seven digits.
Public holidays (1998): New Year's Celebrations, January 1–2; Carnival Celebrations*, February 24–25; Independence Day, February 22; Good Friday, April 10; Easter Monday, April 13; Labour Day, May 1; Whit Monday, June 1; Corpus Christi, May 29; Emancipation Day, August 4; Thanksgiving Day, October 7; National Day, December 13; Christmas Day, December 25; Boxing Day, December 26. *Commercial holiday, not officially a government holiday.
Tourism offices: *In St. Lucia:* St. Lucia Tourist Board, Pointe Seraphine, P.O. Box 221, Castries (758-452-4094). *In the U.S.:* St. Lucia Tourist Board, 820 Second Avenue, 9th Floor, New York, NY 10017 (from U.S. and Canada, 800-456-3984; 212-867-2950).

APPROACHES TO RODNEY BAY

FOUREUR ROCK LIGHT, 14 04.2N, 60 58.7W. Fl (2) W 5s, 23ft, 2M.
MARINA LIGHT, N entrance, 14 04.6N, 60 57.3W. Q G, 3ft, 2M.
MARINA LIGHT, S entrance, 14 04.6N, 60 57.2W. Q R, 3ft, 2M.
MARINA RANGE (Front Light), 14 04.6N, 60 57.0W. Fl W 15ft, 2M. **(Rear Light),** 251 meters, 098.6° from front. Fl W 31ft, 2M.

RODNEY BAY

14 05N, 60 57W
Pilotage: Rodney Bay is a dredged basin off Gros Islet Bay on the northwest coast of the island. A light is shown from Foureur Islet in the approaches. There are lights at the entrance, and a range of 098.6° leads into Rodney Bay Marina. Approach depths are reported to be about 10 feet, with 6 to 12 feet in the lagoon. The tidal range averages about 0.8 foot to 1.2 feet. Call Rodney Bay Marina if you need assistance getting in.
Port of entry: The customs office is located in Rodney Bay Marina.
Dockage: Rodney Bay Marina (VHF channel 16; tel., 758-452-0324) has 250 berths.
Anchorage: You can anchor in Rodney Bay or outside along the beach south of the channel.
Services: The charter industry has spawned a variety of marine services here. Rodney Bay Marina offers fuel, water, electricity, showers,

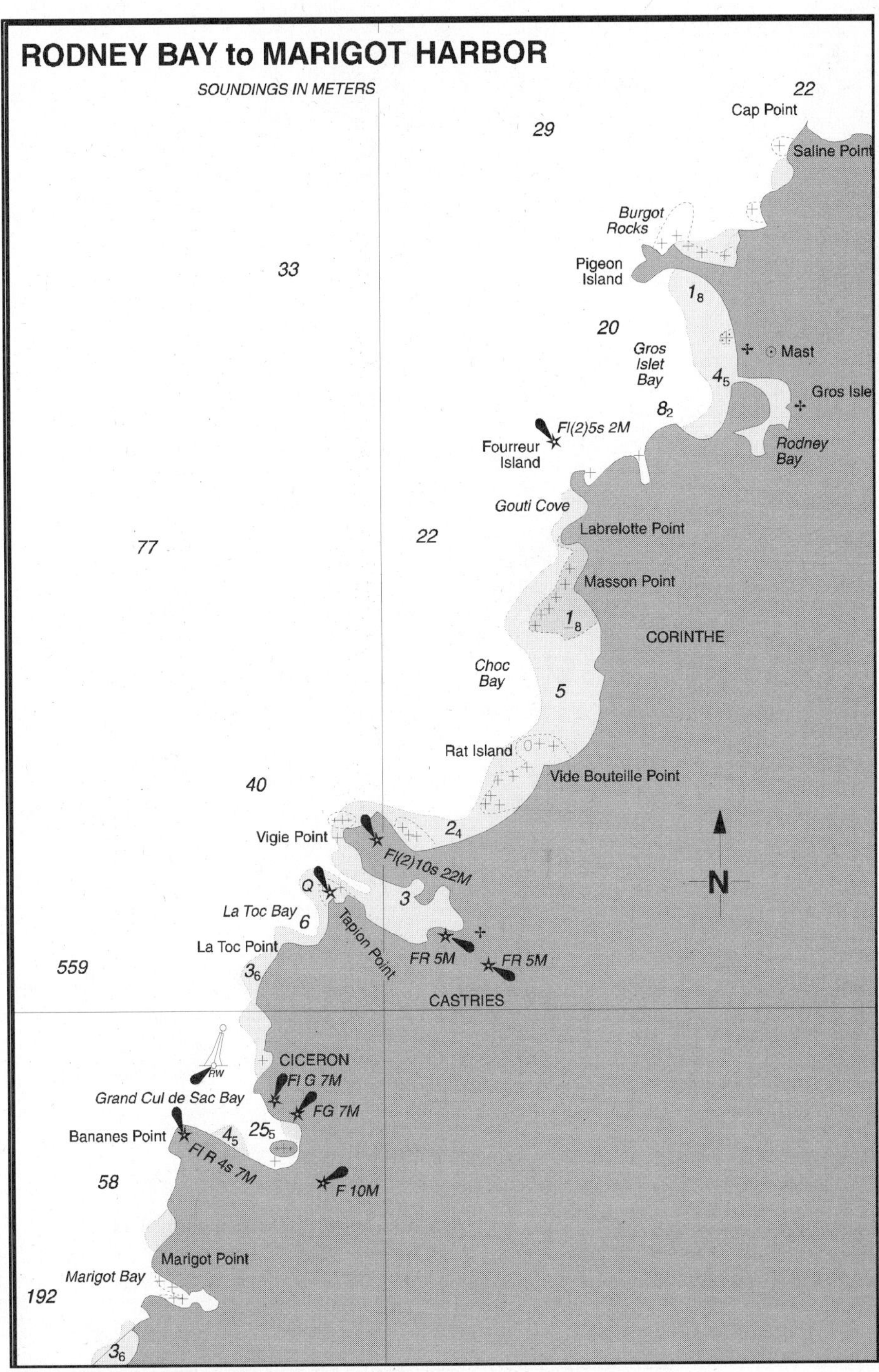
RODNEY BAY to MARIGOT HARBOR
SOUNDINGS IN METERS
Cap Point
Saline Point
Burgot Rocks
Pigeon Island
Mast
Gros Islet Bay
Gros Islet
Rodney Bay
Fl(2)5s 2M
Fourreur Island
Gouti Cove
Labrelotte Point
Masson Point
CORINTHE
Choc Bay
Rat Island
Vide Bouteille Point
Vigie Point
Fl(2)10s 22M
Q
La Toc Bay
La Toc Point
Tapion Point
FR 5M
FR 5M
CASTRIES
CICERON
Fl G 7M
FG 7M
Grand Cul de Sac Bay
Bananes Point
Fl R 4s 7M
F 10M
Marigot Point
Marigot Bay
N

Rodney Bay to Marigot Bay, soundings in meters

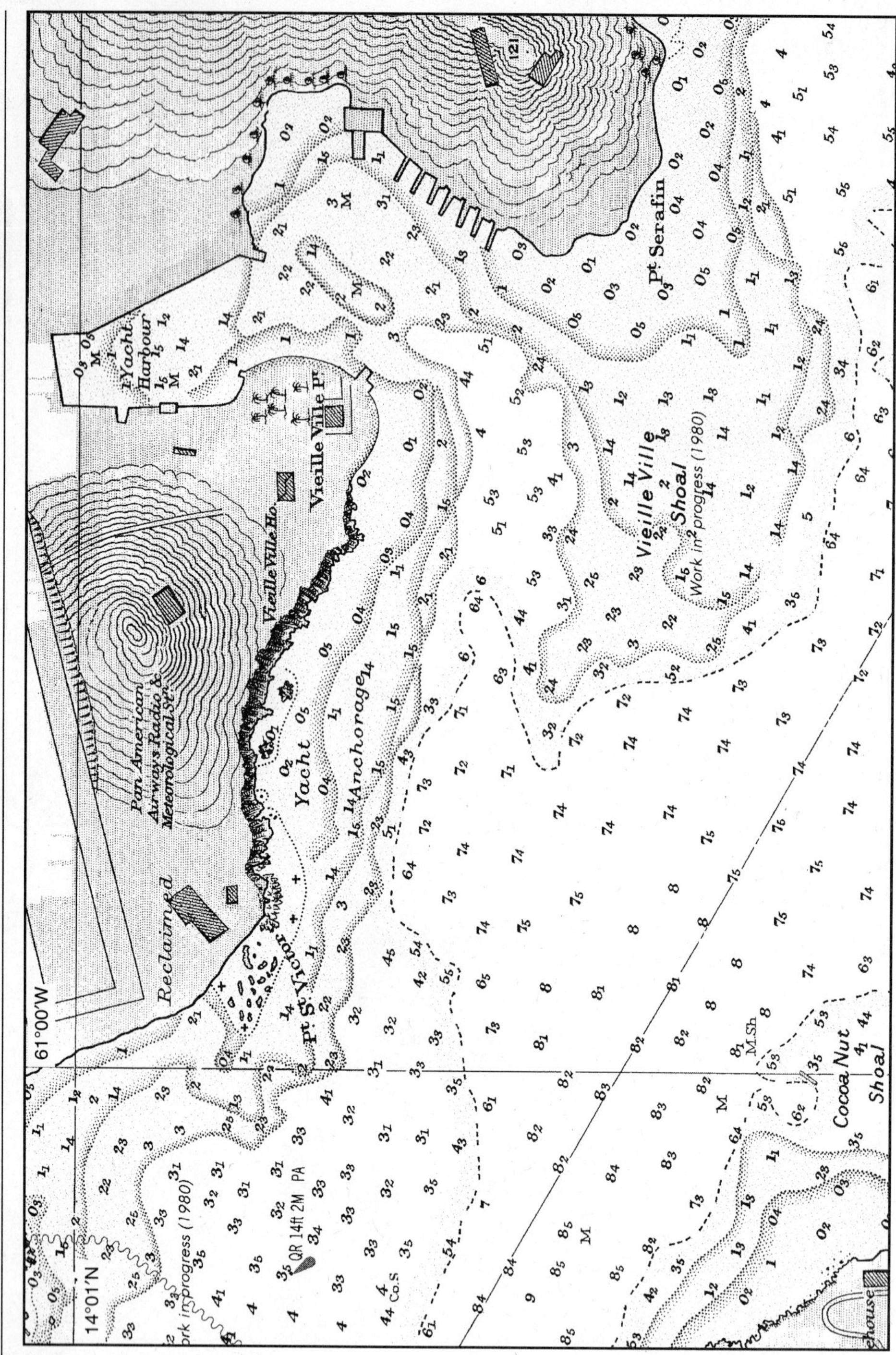

Castries, Vigie Yacht Harbour, soundings in fathoms *(from DMA 25528, 4th ed, 7/84)*

dry storage, haulouts, repairs, banking, and provisioning, all on site. Rodney Bay Ship Services (VHF channel 16, call sign, RBSS; tel., 758-452-9973) is a major chandlery and also services OMC outboards. Cay Electronics (VHF channel 09; tel., 758-452-9922) services and sells electrical and electronic devices, refrigeration units, and watermakers. Sunsail (VHF channel 16; tel., 758-452-8648) primarily services charter yachts, but will help arrange a variety of other repair work for noncharterers. Sunsail also has a sail loft with sail and rigging repair facilities and assists in sourcing, expediting, and brokering parts. It is also reported to run an excellent and inexpensive laundry service. There are many other marine-related services on the island that travel to Rodney Bay. Modestly stocked supermarkets and general shops are nearby.

CASTRIES TO MARIGOT

VIGIE LIGHT, summit N side of entrance to Castries, 14 01.2N, 61 00.0W. Fl (2) W 10s, 320ft, 22M. White circular masonry tower, red roof, 36ft. Visible 039°–212°, partially obscured by Pigeon Island 206°–209°. Signal station.

TAPION ROCK LIGHT, S side of entrance to Castries, 14 00.9N, 61 00.4W. Q W, 50ft, 8M. Old battery, 7M. Visible outside harbor 046°–192°, inside harbor 192°–287°. Temporarily extinguished (1980).

CASTRIES AIRFIELD EXTENSION LIGHT, 14 00.9N, 61 00.1W. Q G, 14ft, 2M. On sunken barge.

CASTRIES WEST WHARF (Front Light), 14 00.5N, 60 59.5W. F R, 43ft, 5M. White triangle, orange stripe, on metal skeleton tower. Emergency light: Q R. **(Rear Light),** 740 meters, 121° from front. F R, 110ft, 5M. White triangle, orange stripe, on metal skeleton tower. Emergency light: Q R.

CASTRIES NORTH WHARF LIGHT, corner of Berth No. 4 and 5, 14 00.6N, 60 59.6W. F Y, 1M.

CASTRIES

14 01N, 61 00W

Pilotage: Castries is the capital and principal commercial port on St. Lucia. On the north side of the harbor entrance Vigie Point, with Vigie Light on its summit, rises to a height of about 295 feet. To the south, Morne Fortune rises to 852 feet, with Fort Charlotte on its summit. The range of 121° should guide you in safely to the main wharf area. The pilots and port authorities (call sign, Castries Lighthouse) stand by on VHF channel 16 and 2182 kHz.

Port of entry: Proceed to the custom's wharf on North Wharf or anchor off and dinghy in.

Dockage: There is no dockage for transients.

Anchorage: The yacht anchorage is in Vigie Yacht Harbour or to its west. Many of Castries' boating services are concentrated in the Vigie area. (Note: The shoal area off Point Serafin by the yacht harbor entrance shown on the Vigie Harbour chartlet in this almanac is now reclaimed land.)

Services: St. Lucia Yacht Services (758-452-5057) in Vigie Yacht Harbour usually has fuel and water. Several yacht services are nearby, including engine and outboard repair concerns, woodworkers, and a fiberglass repair specialist. Castries Yacht Center (VHF channel 16; tel., 758-452-6234), just west of the yacht harbor, has fuel, water, electricity, showers, a small chandlery, and a 35-ton lift for haulouts and repairs. In Castries, there are good supermarkets and general shops, and the Castries airport is right next to the yacht harbor.

GRAND CUL DE SAC BAY

CUL DE SAC BAY RANGE (Front Light), 13 59.3N, 61 00.8W. Fl G 6s, 39ft, 7M. Green square on tower, 26ft. **(Rear Light),** 582 meters, 105.3° from front. F G, 66ft, 7M. Black round tower, yellow bands, green square daymark.

RANGE (Front Light), 13 58.7N, 61 005W. F G, 167ft, 7M. Black round tower, yellow bands, 89ft. Visible 128.1°–134.1°. **(Rear Light),** 582 meters, 131.1° from front. F G, 203ft, 7M. Red and white daymark on black round tower, yellow bands, 26ft. Visible 128.1°–134.1°.

BANANES POINT LIGHT, 13 59.0N, 61 02.0W. Fl R 4s, 85ft, 7M. Red triangle on tower, 16ft.

MARIGOT

13 58N, 61 02W

Pilotage: The entrance to this small bay is located about a mile south of the Hess oil terminal in Grande Cul de Sac Bay.

Port of entry: Customs is located on the south shore of the harbor, west of the docks. Guests at The Moorings complex may tie up at the marina docks and walk to the customs office.

Dockage: The Moorings Marina (VHF channels 16 and 25; tel., 758-451-4357) is in the inner harbor.

Anchorage: Anchor anywhere, avoiding the cable crossing at the narrow point in the harbor. The shelter is excellent, and some consider Marigot a hurricane hole.

Services: The Moorings complex features all the amenities of a modern resort, including hotels, swimming pools, restaurants, and a dive

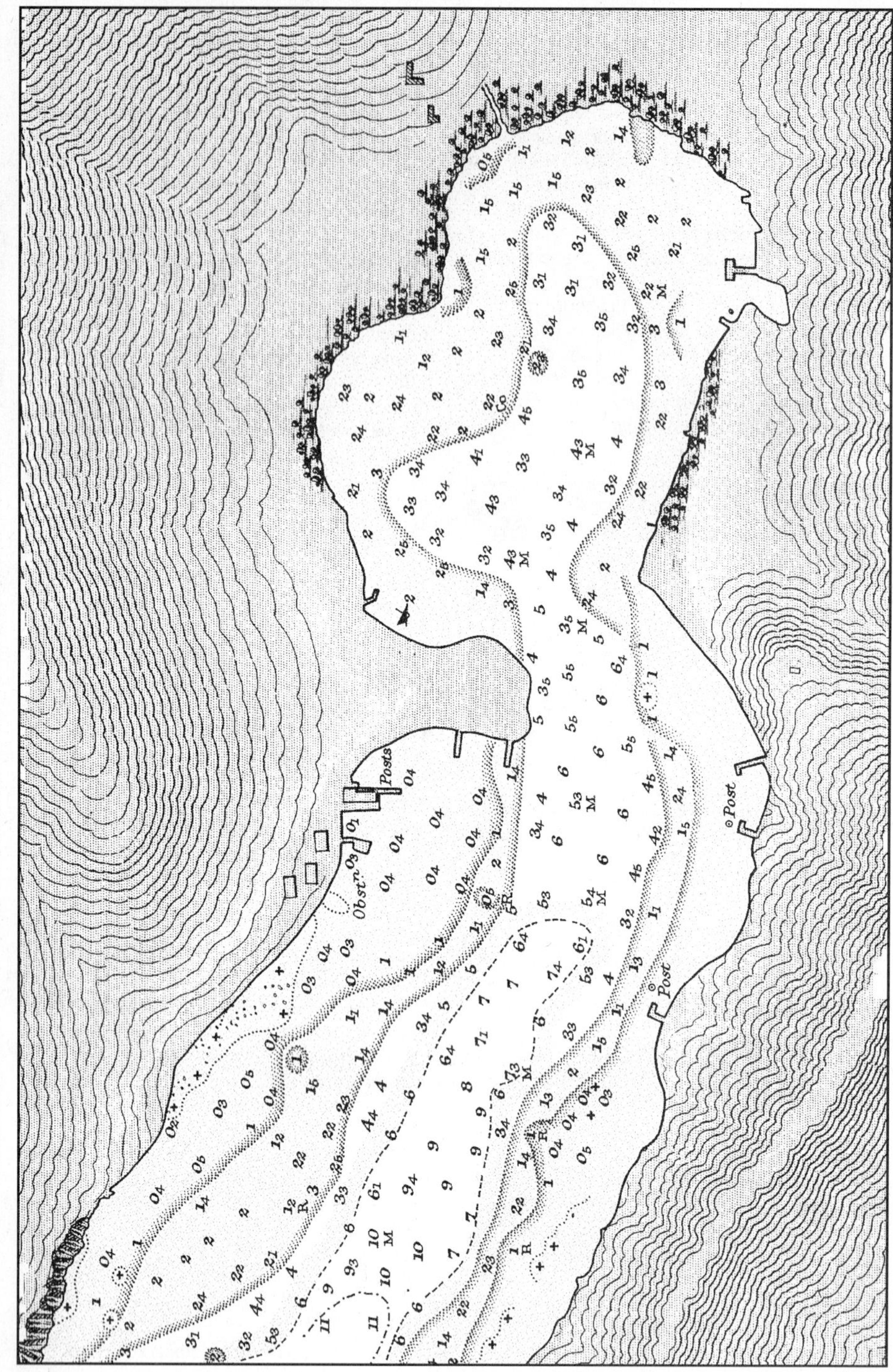

Marigot, soundings in meters *(from DMA 25528, 4th ed, 7/84)*

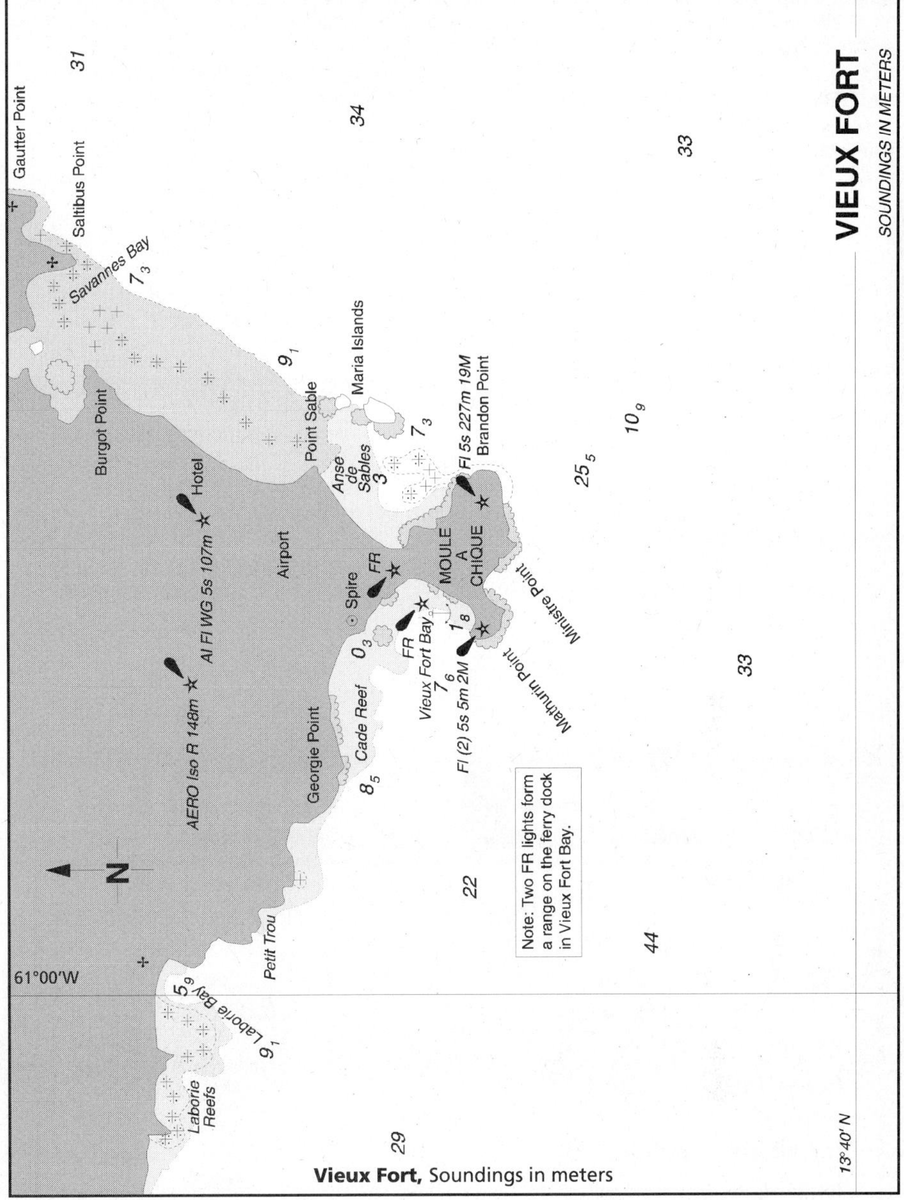

Vieux Fort, Soundings in meters

center. There is fuel, gasoline, propane, and a commissary at the docks and the hotel laundry is up the hill.

VIEUX FORT BAY

MATHURIN POINT LIGHT, 13 42.6N, 60 57.5W. V Q (2) W 5s, 16ft, 2M. Red skeleton tower, concrete base, 10ft.

VIEUX FORT BAY RANGE (Front Light), head of wharf, 13 43.1N, 60 57.2W. Q R, 26ft, 8M. Orange rectangle on building. **(Rear Light),** 460 meters, 060° from front. Iso R 6s, 65ft, 9M. Red metal framework tower, white bands, 33ft.

NORTH ENTRANCE CHANNEL, 13 43.3N, 60 57.5W. Q G, 13ft, 5M. Green pile.

SOUTH ENTRANCE CHANNEL, 13 43.1N, 60 57.2W. Q R, 13ft, 5M. Red pile.
BATTERY POINT, 13 43.3N, 60 57.3W. Fl Y 2.5, 13ft, 5M. Orange pillar.
BATTERY POINT, S, 13 43.1N, 60 57.2W. Fl Y 2.5s, 13ft, 5M. Orange pillar.
CAPE MOULE A CHIQUE LIGHT, Brandon Point, 13 42.6N, 60 56.5W. Fl W 5s, 745ft, 22M. Masonry tower, 29ft. Visible 197°–123°.

VIEUX FORT

13 43N, 60 57W
Pilotage: The town is located on the south tip of St. Lucia. Several obstructions in the bay are marked by buoys, and a range of 059.9° leads you in safely. The charted dome of the Custom House is prominent. The pilots and port officials (call sign, Vieux Fort Lighthouse) monitor VHF channel 16 and 2182 kHz.
Port of entry: Contact customs near the commercial wharf. Normal office hours are from 0800 to 1230 and 1330 to 1630, Monday through Friday. The customs office at the airport handles clearances after hours, on weekends, and on public holidays; calls to the Vieux Fort number (758-454-6526) are transferred automatically during those times to the airport office.
Anchorage: The recommended area is in the southeast portion of the bay, under the lee of Moule à Chique. This is south of the commercial wharf. Anchoring is prohibited in an area north of the wharf, because of pipelines and commercial moorings.
Services: There are few yacht services in Vieux Fort, but there are several restaurants, and most general supplies are available. The Il Pirata Restaurant (758-454-6610) has a dock northwest of town and showers for boaters.

EAST COAST

MOUNT TOURNEY AVIATION LIGHT, 13 44.4N, 60 57.9W. Iso R, 485ft. Red skeleton tower, white bands, 33ft.
MOUNT BELLEVUE AVIATION LIGHT, 13 44.4N, 60 56.7W. Al Fl W G 5s, 351ft. Beacon.
CAPE MARQUIS LIGHT, 14 03.3N, 60 53.6W. Fl (2) W 20s, 197ft. White square support.

St. Vincent and the Grenadines

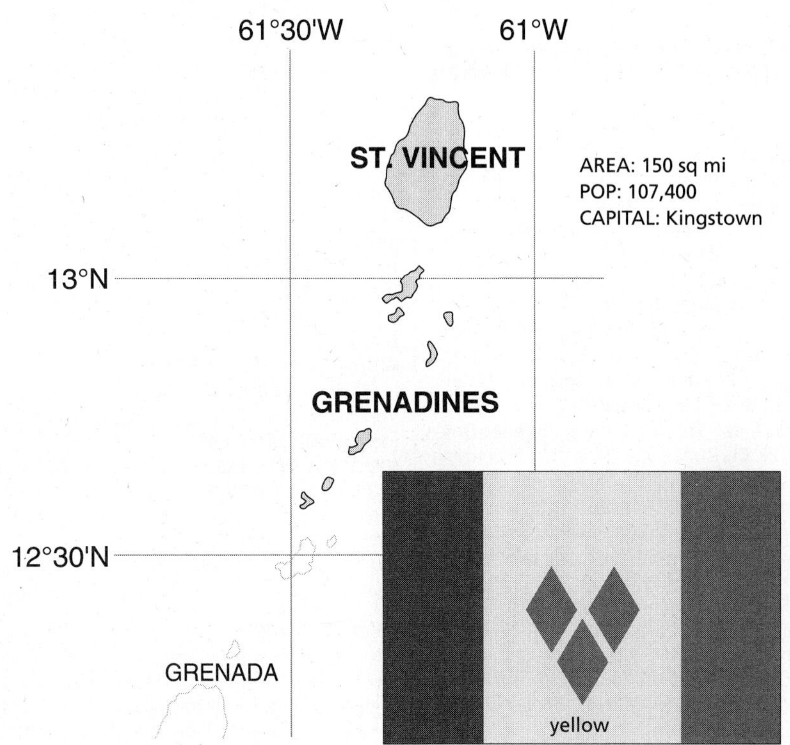

Please see cautions about chartlets and waypoints on page P 2.

ENTRY PROCEDURES

This nation includes the islands of St. Vincent, Bequia, Mustique, Canouan, the Tobago Cays, Union Island, and Petit St. Vincent (PSV). St. Vincent was a British colony until gaining independence in 1979. The islands of Carriacou and Petite Martinique, although geographically part of the Grenadines, belong to the nation of Grenada. All told, more than 100 islands, islets, and rocks extend for a distance of 52 miles between St. Vincent and Grenada.

Ports of entry on St. Vincent are Wallilabou Bay and Kingstown; on Bequia, Port Elizabeth; and on Union, Clifton. Vessels may also obtain clearances on Mustique and Canouan. Cruisers recommend Wallilabou Bay, Port Elizabeth, and Clifton as particularly convenient; in Kingstown, Ottley Hall Marina customers may find staff assistance helpful in contacting officials.

Customs offices handling yacht clearances can be reached at the following numbers: Kingstown, 809-456-1083; Port Elizabeth, 809-458-3211; Clifton, 809-458-8360; Canouan, 809-458-8050; and Mustique, 809-458-4621 (ask for the customs section). Generally, hours are about 0800 to 1600 or 1800, with a break for lunch from 1200 to 1300. For Wallilabou Bay, see the description under that heading. There is no charge for clearances in or out during normal office hours. After hours, the boarding fee is EC$20 (Sundays and holidays, EC$25). Stays longer than 24 hours require a separate clearing-out procedure. You should have clearance from your last port of call, ship's papers, and passports. A cruising permit is required to visit additional ports; the document fee is an exceedingly modest 15 cents (EC). Firearms and ammunition must be declared and secured aboard. Pets must remain aboard.

USEFUL INFORMATION

Language: English.
Currency: Eastern Caribbean dollar (EC$); exchange rate (summer 1997), EC$2.70 = US$1. U.S. dollars often accepted in tourist areas.
Telephones: Islands area code, 809; local numbers, seven digits. New area code, 784; effective date not announced as of *Reed's* press time (summer 1997).
Public holidays (1998): New Year's Day, January 1; National Heroes Day, January 22; Good Friday, April 10, Easter Monday, April 13; Labour Day, May 4; Whit Monday, June 1; Caricom Day, July 6; Carnival Tuesday, July 7; August Monday, August 3; Independence Day, October 27; Christmas Day, December 25; Boxing Day, December 26.
Tourism offices: *In St. Vincent:* St. Vincent and the Grenadines Tourism Office, P.O. Box 834, Kingstown (tel., 809-457-1502; fax, 809-456-2610). *In the U.S.:* St. Vincent and the Grenadines Tourist Office, 801 Second Avenue, 21st Floor, New York, NY 10017 (tel., 800-729-1726, 212-687-4981; fax, 212-949-5946). *In Canada:* St. Vincent and the Grenadines Tourist Office, 32 Park Road, Toronto, Ont. N4W 2N4 (416-924-5796).

ST. VINCENT

OWIA LIGHT (Cow and Calves), 13 22.3N, 61 09.0W. Fl W 10s, 118ft, 8M. White metal framework tower, 20ft. Visible 101°–307°.

DARK HEAD LIGHT, 13 16.8N, 61 16.0W. Fl W 5s, 338ft, 12M. Metal framework tower, 10ft. Visible 020°–211°. Oc R and 2 F R (vert.) 2.5 miles NE.

WALLILABOU BAY

13 15N, 61 17W
Pilotage: Wallilabou Bay, on the west coast of St. Vincent north of Kingstown, can be distinguished by a radio/television tower near the point of land that forms its southern entrance. Entering is straightforward.
Port of entry: Customs officials are geared to clearing in pleasure boats here; they stop by their office, opposite the Wallilabou Anchorage Restaurant, from approximately 1600 to 1800 every day except during the low season, when they check in more infrequently. If you are having difficulty reaching customs, ask the friendly staff at the restaurant for help.
Moorings: The Wallilabou Anchorage Restaurant (VHF channel 68; tel., 809-458-7270) has set out several free moorings for guests and nonguests alike.
Anchorage: Because there are depths of 35 to 50 feet right off the beach, you are advised to take a stern line ashore (boat boys can usually be hired to assist). Contact the restaurant if you are in doubt about a good location.
Services: In addition to breakfast, lunch, dinner, and bar service, the Wallilabou Anchorage Restaurant has water available from a hose at the dock and free showers and can assist with phone calls and tours of the island. The restaurant also sells ice and takes credit cards and foreign currency.

FORT CHARLOTTE LIGHT, 13 09.4N, 61 15.0W. Fl (3) W 20s, 640ft, 16M. Octagonal structure, 9ft. Visible shore–143°. Storm signals. Aero Oc R (occasional) 1.5M SE of light.

KINGSTOWN WHARF LIGHT, NW end of wharf, 13 09.1N, 61 14.1W. F R. Column.

KINGSTOWN WHARF LIGHT, SE end of wharf, 13 08.9N, 61 14.0W. F R. Column.

KINGSTOWN

13 09N, 61 14W
Pilotage: Kingstown is the capital of St. Vincent and its major port. The cathedral's square white tower in the northwestern part of town is conspicuous. The clock on the tower is lighted at night. The red cupola on the police station located north of the main wharf is also conspicuous. The pilots and port authorities use call sign ZQS and monitor VHF channel 16 and 2182 kHz.
Port of entry: Contact the officials (809-456-1083) in their office near the main wharf. A second customs post has been built at the entrance to Ottley Hall Marina, but is not yet approved for use (summer 1997). Until it does open, Ottley Hall Marina will alert customs officials on behalf of customers.
Dockage: Proceed either south to the Blue Lagoon (see the following heading) or west to Ottley Hall Marina (VHF channels 08 and 16; tel., 809-457-2178), a major facility in a small bay in the shadow of Fort Charlotte.
Anchorage: Anchor north or south of the main commercial wharf. During the day strong gusts of wind may blow down St. Vincent's mountain valleys with great violence. At night the breeze will usually be light.

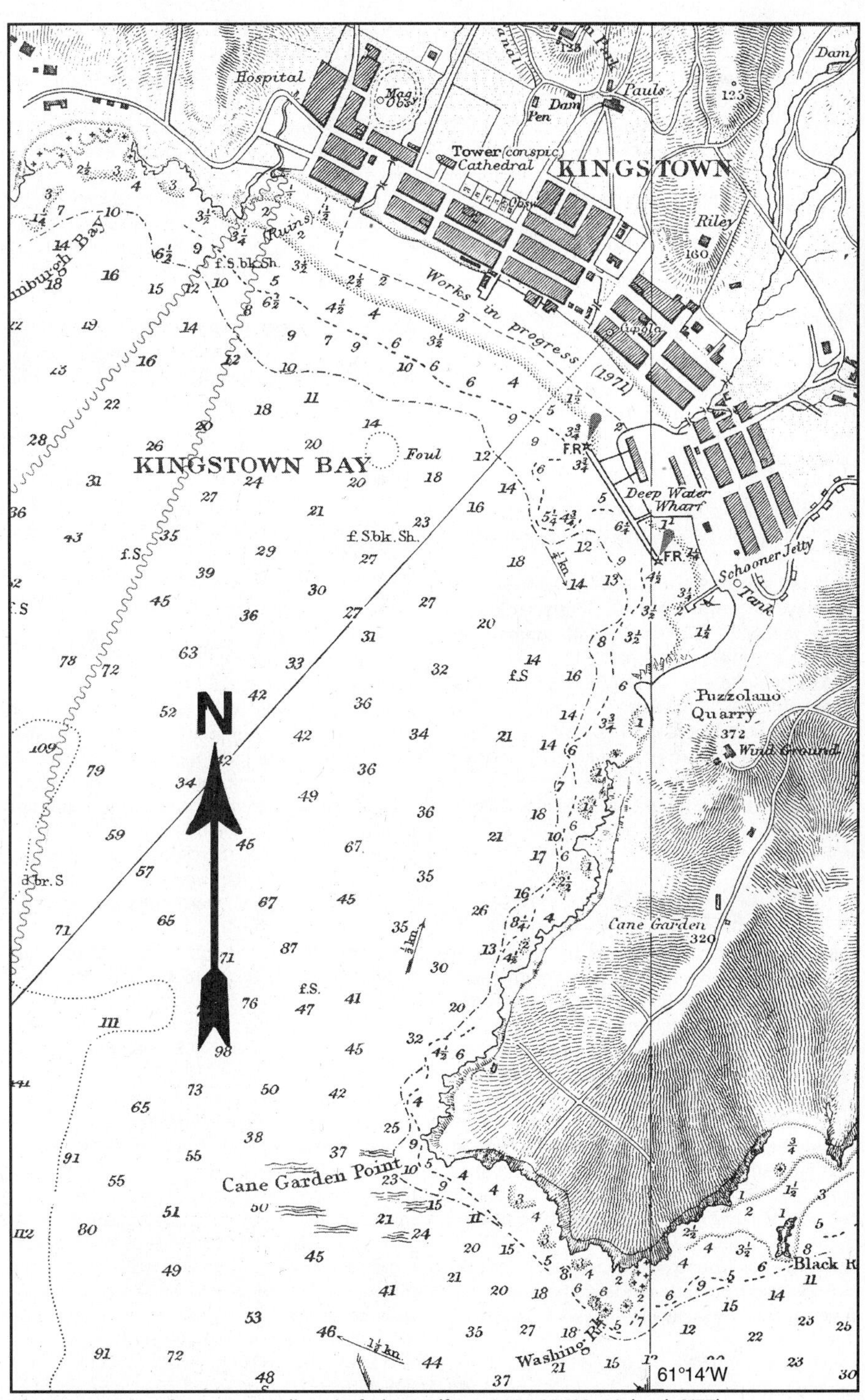

Kingstown, St. Vincent, soundings in fathoms *(from DMA 25483, 19th ed, 8/77)*

Services: In Kingstown fuel is available at the fish boat dock. However, it is easier to obtain at Ottley Hall or in Blue Lagoon. Ottley Hall is a new ultramodern marina, offering all repair and construction services, including haulout lifts and an under-cover drydock capable of handling superyachts. Elsewhere in Kingstown are metal fabricators, engine repair services, and a chandlery. There is a good selection of supermarkets and general provisioning stores in Kingstown.

Although the harbor of Kingstown offers few amenities for boaters, the nearby areas feature a good variety. Many boaters visit Kingstown by road or ferry, as it is more pleasant to clear customs elsewhere. The wild interior of the island is well worth a sightseeing tour or hiking trip.

YOUNG ISLAND CARENAGE LIGHT, 13 07.7N, 61 12.6W. Fl G 4s. Green port-hand pillar with topmark.
ROOKES POINT SHOAL LIGHT, 13 07.7N, 61 12.5W. V Q (6) + L Fl W 10s. South cardinal mark, yellow and black pillar with topmark.
DUVERNETTE ISLAND LIGHT, 13 07.5N, 61 12.8W. V Q (2) W 2s, 229ft, 6M. White metal framework tower, 29ft.
CALLIAQUA BAY LIGHT, 13 07.5N, 61 12.3W. Fl R 4s. Red starboard-hand mark, pillar with topmark.
BRIGHTON LIGHT, 13 07.3N, 61 10.6W. Fl W 4s, 118ft, 8M. White metal framework tower, 20ft. Visible 217°–077°.

YOUNG ISLAND CUT AND BLUE LAGOON

Pilotage: South of Kingstown, the small bight between Young Island and St. Vincent is easily entered from the west. The channel south to Calliaqua Bay, although marked, is narrow and demands care. Calliaqua Bay is open; entry to adjacent Blue Lagoon is by a narrow, marked, dredged channel whose depth varies between 5.6 and 7.6 feet. Do not use the south entrance.
Dockage: Slip space may be available at Lagoon Marina (VHF channel 68; tel., 809-458-4308) or from Barefoot Yacht Charters (VHF channel 68; tel., 809-456-9526) if charter yachts are out and space is free.
Moorings: In Young Island Cut, moorings may be available from independent operators. In Blue Lagoon, check availability with Barefoot Yacht Charters.
Anchorage: Both Young Island Cut and the lagoon require plenty of scope. In Young Island Cut, anchor with care and use a stern anchor because of the currents.
Services: There are numerous restaurants and bars in the Young Island Cut area. Lagoon Marina, on Blue Lagoon's north side, has fuel, laundry, showers, electricity, water, and a restaurant, bar, and accommodations. Barefoot Yacht Charters, in the southeast corner of the lagoon, has diesel, water, and ice and operates a restaurant. Barefoot is also a LIAT agent and can arrange for air ambulance service.

ADMIRALTY BAY, BEQUIA

DEVIL'S TABLE LIGHT, 13 00.7N, 61 15.6W. V Q (9) W 10s. West cardinal mark, yellow-black-yellow pillar with topmark.
ADMIRALTY BAY LIGHT, root of jetty, 13 00.5N, 61 14.7W. Fl W R G 4s, 19ft, 5M. White metal framework tower, 19ft. R shore–048°, W 048°–058°, G 058°–shore.
WEST CAY LIGHT, 12 59.3N, 61 18.0W. Fl W 10s, 42ft, 8M. White metal framework tower, 19ft.

PORT ELIZABETH, BEQUIA

13 00N, 61 15W
Pilotage: The town of Port Elizabeth is situated near the head of Admiralty Bay, on the west coast of Bequia. A radio mast south-southeast of the town is conspicuous. A reef, charted as Belmont Shoal, extends from the eastern shore of the bay south of the main wharf. A light is shown from the commercial wharf.
Port of entry: The customs office is located near the commercial wharf; entry procedures are reported as efficient.
Dockage: You may be able to tie up temporarily at Bequia Marina (809-458-3272), north of the main wharf.
Moorings: Daffodil's Marine Service (VHF channels 67 and 68; tel., 809-458-3942) can arrange moorings.
Anchorage: You can anchor west of the marina or south of the commercial wharf. Check the charts carefully for the shoal areas extending out from shore.
Services: You can get diesel and water at the marina or by calling Daffodil's Marine Service (VHF channels 67 and 68; tel., 809-458-3942), which delivers alongside. Daffodil's also offers mechanical and dinghy repairs and laundry services. Sail repairs and canvaswork are available from several sources. Grenadines Yachts and Equipment (VHF channels 16 and 68; tel., 809-458-3347; fax, 809-458-3696; e-mail,

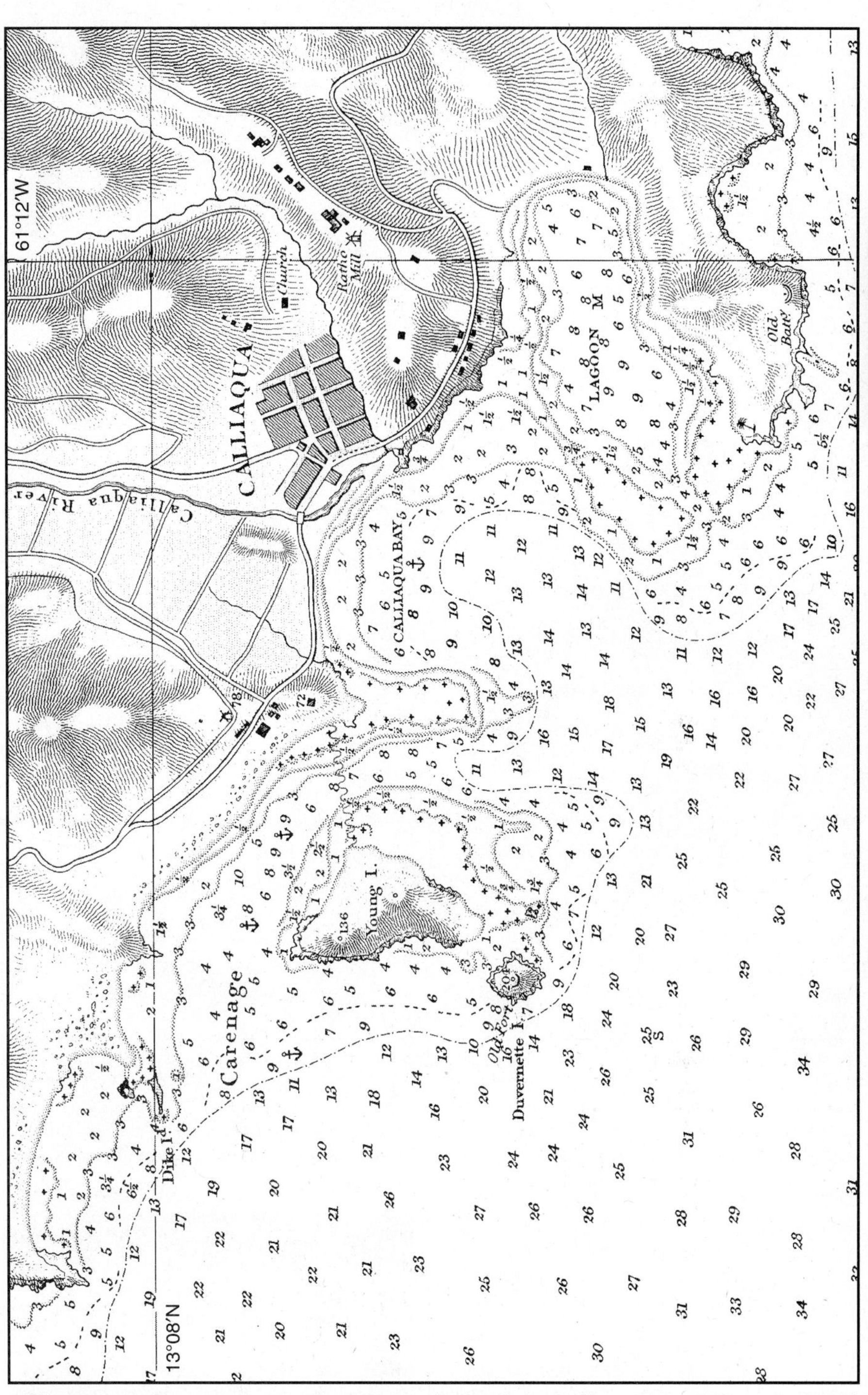

Calliaqua and Blue Lagoon, St. Vincent, soundings in fathoms *(from DMA 25483, 19th ed, 8/77)*

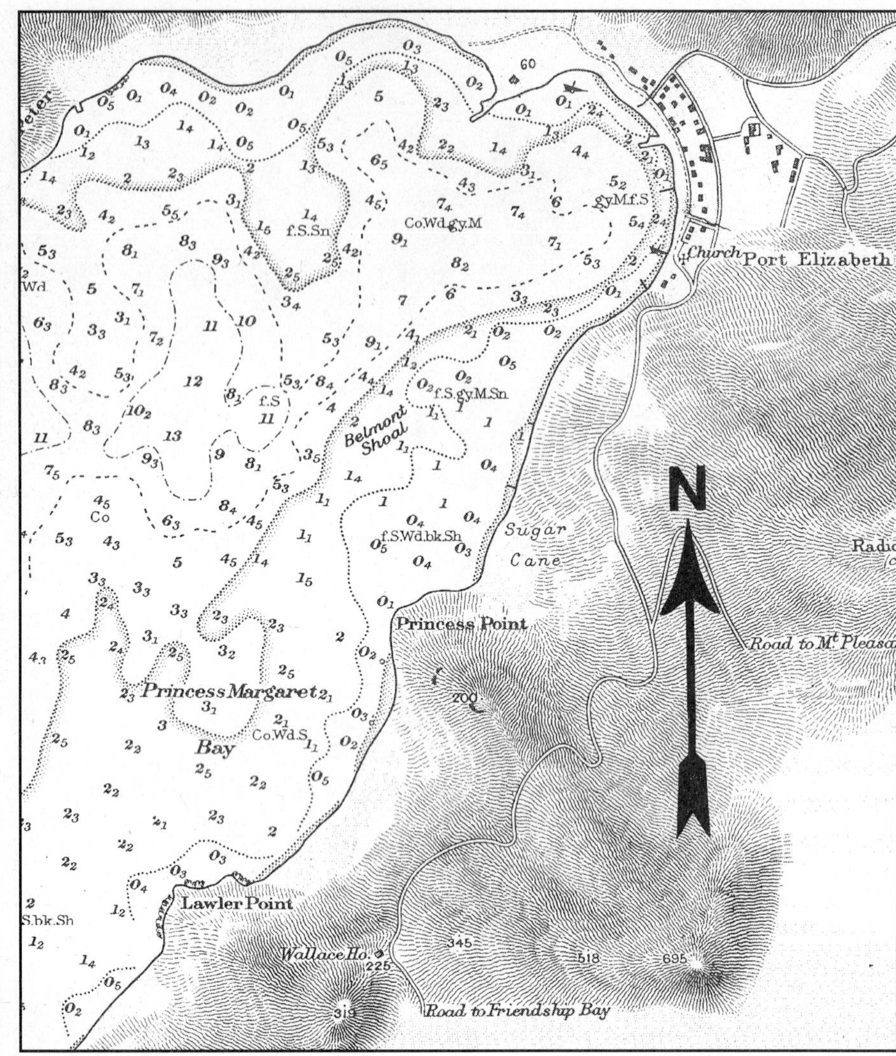

Port Elizabeth, Bequia, soundings in fathoms *(from DMA 25483, 19th ed, 8/77)*

danielfo@caribsurf.com) does all types of mechanical repairs and sells a wide variety of maintenance and repair items. Bo'sun's Locker (tel., 809-458-3246, or after hours, -3634; fax, 809-458-3925) is a complete chandlery. The Bequia Bookshop (809-458-3905) stocks charts, courtesy flags, and cruising guides. Other yacht services are available. There is a good variety of boutiques, restaurants, and supermarkets.

Bequia is a favorite port of call with sailors. The waters are often windy and boisterous, with gusts over 30 knots a matter of course. The gaps between some of the islands tend to funnel the wind into pockets of higher intensity. On Bequia you may be able find traditional West Indian boats being built under the trees on the beach.

BATTOWIA TO PETIT CANOUAN

BATTOWIA ISLAND LIGHT, 12 57.7N, 61 08.3W. Fl (2) W 20s, 708ft, 8M. White metal framework tower, 10ft.

MUSTIQUE LIGHT, Montezuma Shoal, 12 52.7N, 61 12.5W. Fl (2) W. Isolated danger mark, black-red-black pillar with topmark.

PETIT CANOUAN ISLAND LIGHT, 12 47.5N, 61 17.0W. Fl (4) W 40s, 252ft, 8M. White metal framework tower, 30ft.

CANOUAN

CHARLESTOWN BAY RANGE (Front Light), Canouan Island, 12 42.1N, 61 19.9W. F W, 18ft. Beacon. **(Rear Light),** 450 meters, 158°32′ from front light. F W, 165ft. White concrete tower.
RANGE (Front Light), marks the approach to the pier in Charlestown Bay, 12 42.7N, 61 19.5W. Iso W 4s, 46ft, 5M. Black and white square on white framework tower, 30ft. Visible on rangeline only. **(Rear Light),** 85 meters, 060° from front light. Fl W 5s, 91ft, 5M. Black and white square on white framework tower, 10ft. Visible on rangeline only.
CHARLESTOWN BAY PIER HEAD LIGHT, 12 42.2N, 61 19.8W, F W.
GRAND BAY, 12 42.6N, 61 20.0W. Fl G 4s. Green port-hand beacon with topmark.
GRAND BAY SOUTH, 12 42.4N, 61 20.0W. Fl R 4s. Red starboard-hand beacon with topmark.

CATHOLIC ISLAND TO UNION ISLAND

CATHOLIC ISLAND, 12 39.6N, 61 24.1W. Fl (2) W 20s, 144ft., 8M, White metal framework tower.
MAYREAU, Grand Col Point, 12 38.3N, 61 24.1W. V Q (9) W, 10s. West cardinal mark, yellow-black-yellow beacon with topmark.
JONDELL, 12 39.5N, 61 23.7W. V Q (3) 5s. East cardinal mark, black-yellow-black beacon with topmark.
CLIFTON HARBOR RANGE, Union Island **(Front Light),** 12 35.8N, 61 25.0W. F W, 13ft. White concrete tower. Lighted beacons mark reef and shoal. **(Rear Light),** 555 meters, 327.5° from front. F W, 125ft. White concrete tower.
THOMPSON REEF, SE edge, 12 35.6N, 61 24.7W. Fl R. Red starboard-hand beacon with topmark.
THOMPSON REEF, SW edge, 12 34.5N, 61 24.7W. FL R. Red starboard-hand beacon with topmark.
ROUNDABOUT REEF, SW edge, 12 35.6N, 61 24.8W. Fl R. Red starboard-hand beacon with topmark.
WESTWARD, 12 35.6N, 61 25.0W. Fl G. Green port-hand beacon with topmark.
GRAND DE COI LIGHT, 12 35.1N, 61 24.9W. V Q (9) W, 10s. West cardinal mark, yellow-black-yellow pillar with topmark.
MISS IRENE POINT LIGHT, 12 35.5N, 61 27.8W. Fl (2) W, 20s, 410ft, 8M. Metal framework tower, 20ft.

CLIFTON, UNION

12 36N, 61 25W
Pilotage: Clifton lies in a bay on the southeast end of Union Island. It is protected to the east by Thompson Reef. The new buoyage leading in to the harbor is the IALA B system, indicating you should follow the "red-right-returning" rule. This may vary with what charts show. *(Note: The list of navigational aids above is more accurate than shown on the Union Island chartlet; also an airport now extends from the island toward Red Island.)* Copper Reef, with a ruined concrete structure on its southwest end, lies in the middle of the harbor. There may be a range of 327.5° leading safely into the harbor.
Port of entry: Union Island is the southern check-in point for boats headed north through the Grenadines toward St. Vincent. Report to the customs (809-458-8360) and immigration officials located at the airport.
Dockage: Contact Anchorage Yacht Club (VHF channels 16 and 68; tel., 809-458-8221).
Anchorage: There is good anchorage in most of the harbor, but take care to avoid the various reefs. Depths are from 16 to 40 feet, with a sand bottom.
Services: You can jug fuel, but water and ice are available at Anchorage Yacht Club. The marina also has a mechanic, a sailmaker, and a marine railway. There is a good selection of small shops, restaurants, and supermarkets.

PETIT ST. VINCENT (PSV)

12 32N, 61 23W
Pilotage: Extensive reefs lie east, north, and northwest of the island. The anchorage is near the east end of the south coast of the island. Although PSV is a private resort island, yachtsmen are invited to visit. You can contact the Petit St. Vincent Resort at 809-458-8801 or on VHF channel 16.
Services: You may be able to purchase fuel from the resort in an emergency. Ice, bread, and phones are available during office hours. You can visit the bar, restaurant, and boutique, or ask about joining a hotel barbecue on the beach. Cottage rentals are available.

Note: Within an aid listing, underlined text indicates data that has changed or is new since September, 1996; ~~Strike out~~ text indicates deleted data. This should highlight recent changes for mariners, many of which may not be included in even the most recently published charts.

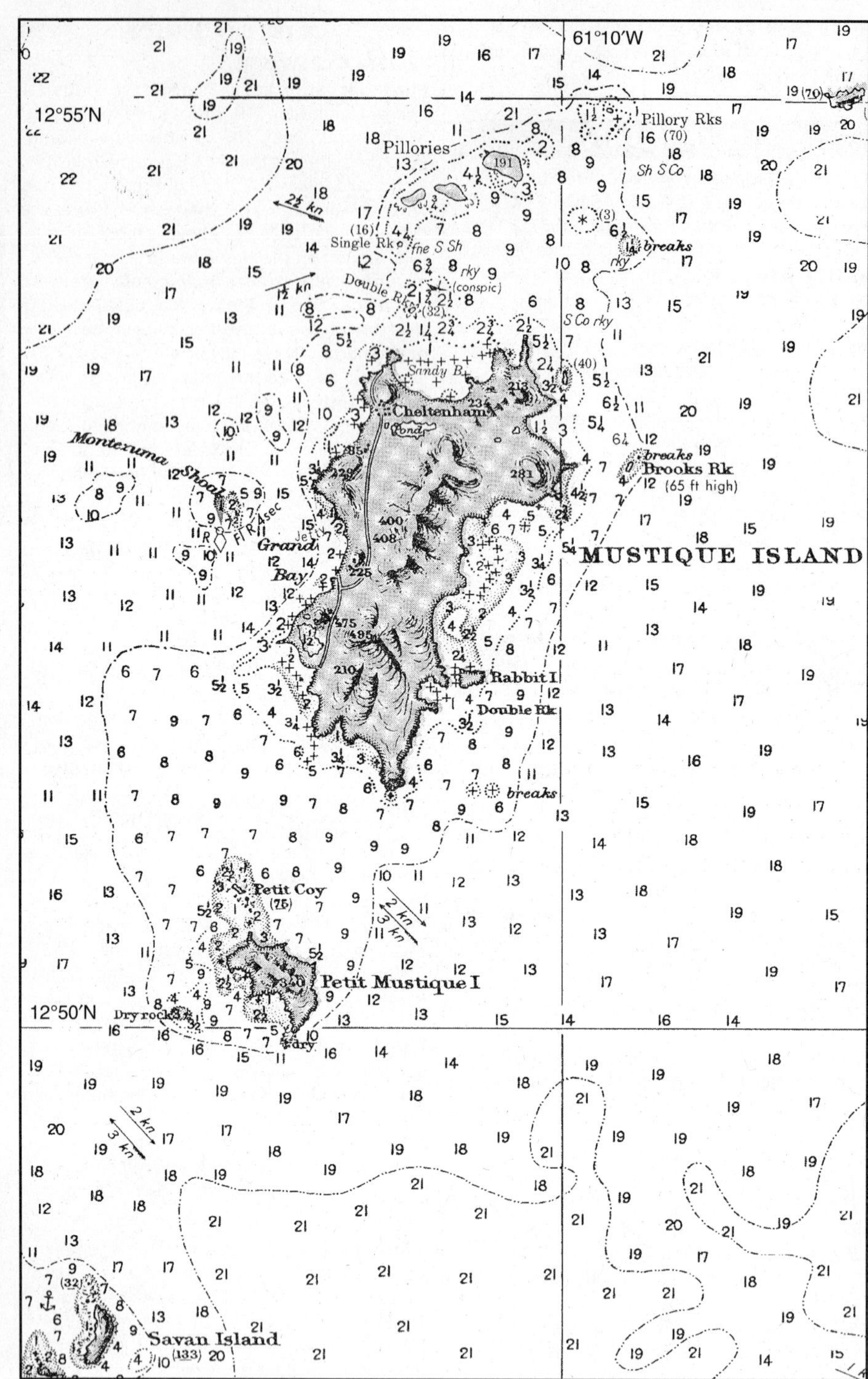

Mustique, soundings in fathoms *(from DMA 25482, 15th ed, 5/79)*

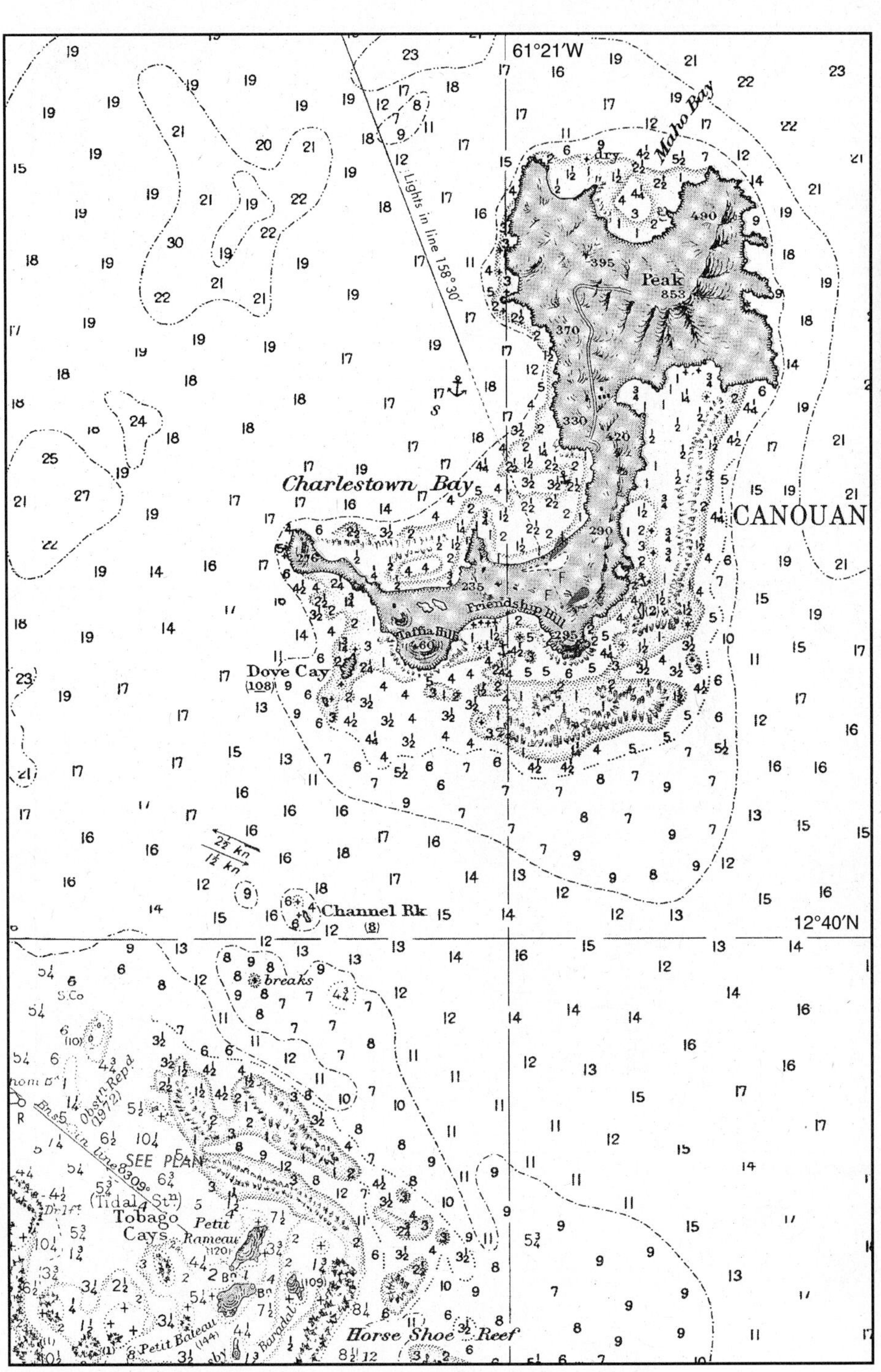

Canouan, soundings in fathoms *(from DMA 25482, 15th ed, 5/79)*

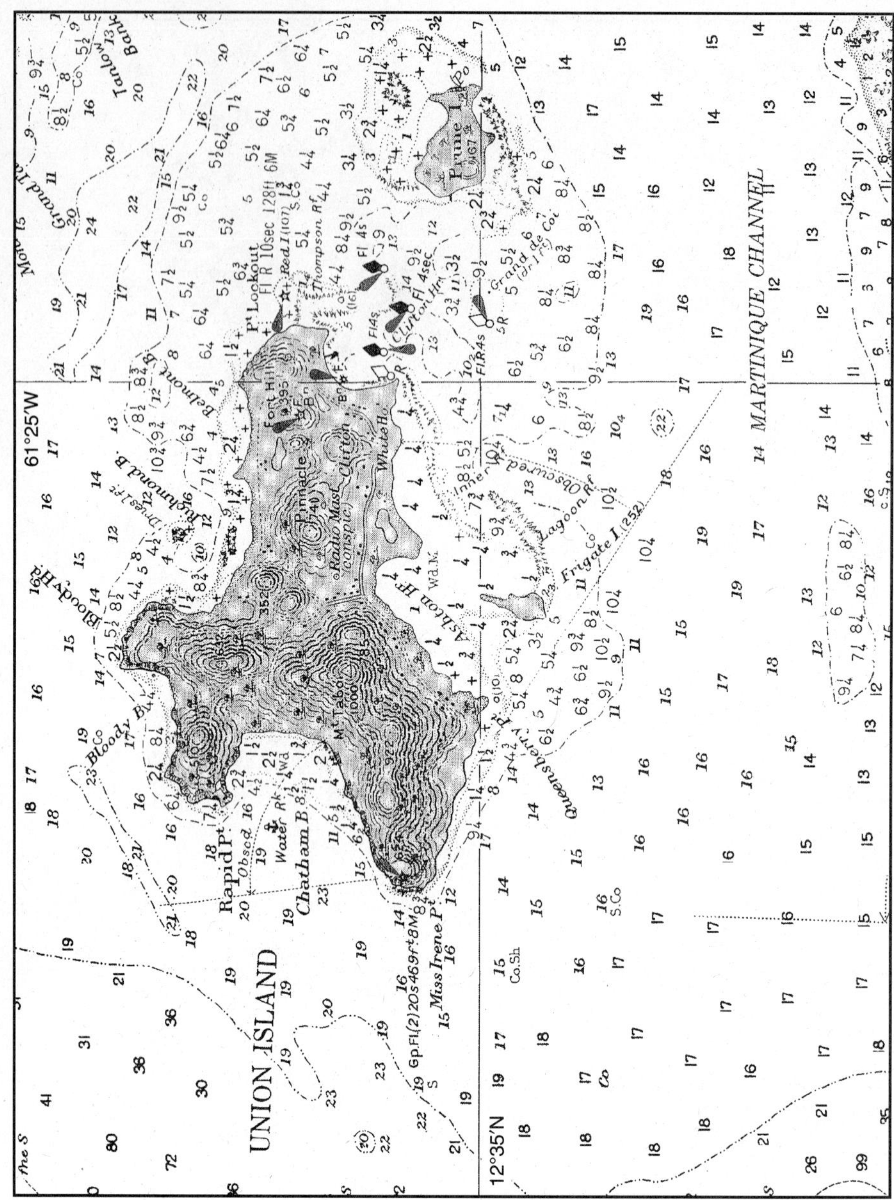

Union Island, soundings in fathoms *(from DMA 25482, 15th ed, 5/79)*

Grenada

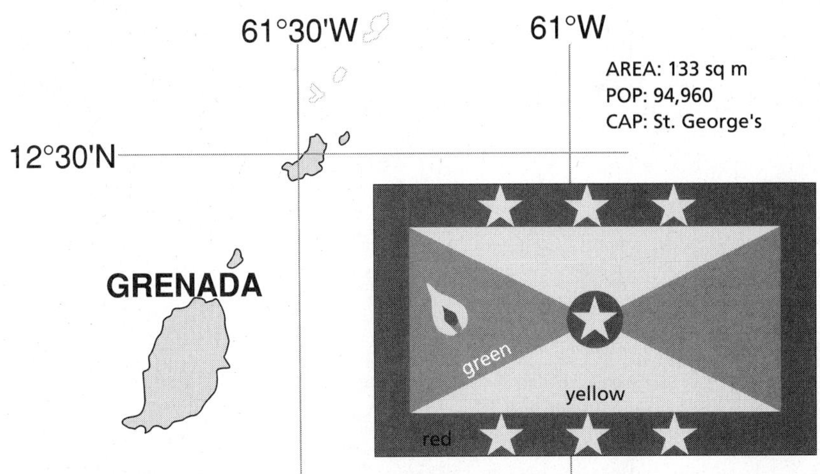

Please see cautions about chartlets and waypoints on page P 2.

ENTRY PROCEDURES

Grenada consists of three islands—Grenada, Petite Martinique, and Carriacou—which are located at the southern extremity of the windward islands, only 100 miles north of Venezuala. Grenada was a British colony until gaining independence in 1974. After a difficult period in the early 1980s, the government today is
stable, and the island is an independent nation member of the British Commonwealth.

The port of entry on Carriacou is Hillsborough. Unfortunately, you cannot clear customs on nearby Petite Martinique. On Grenada, the ports of entry are St. George's, Prickly Bay, and Grenville. Customs office hours are generally 0800 to 1600, Monday through Friday, with a break for lunch from 1200 to 1300. Overtime is charged for clearances conducted outside normal office hours. Customs in St. George's (473-440-2240) is located near the pier; in Prickly Bay (473-444-4549), at Spice Island Marine Services.

You should have a clearance from the last port of call, ship's papers, crew lists (four for clearing in and four for clearing out), and passports. As of September, 1997, port fees for boats up to 50 feet will be EC$30; 51–100 feet will be EC$45; over 100 feet will be EC$55. Additional fees are charged to charterboats; charterboats may contact the port authority in St. George's (473-440-3013) for the latest information. All vessels are charged EC$15 to clear out. Animals are prohibited without an import permit; they need proper health documents and a government officer must be notified at the port of entry if you bring an animal ashore.

USEFUL INFORMATION

Language: English.
Currency: Eastern Caribbean dollar (EC$), exchange rate (summer 1997), EC$2.67 = US$1. Many people accept U.S. dollars, especially in tourist areas.
Telephones: A new area code, 473, went into effect on October 1, 1997. Local numbers, seven digits. Boatphone, a mobile cellular phone service, is available for land and marine use. Coin and card phone services are available for both local and overseas calls.
Public holidays (1998): New Year's Day, January 1; Independence Day, February 7; Good Friday, April 10; Easter Monday, April 13;

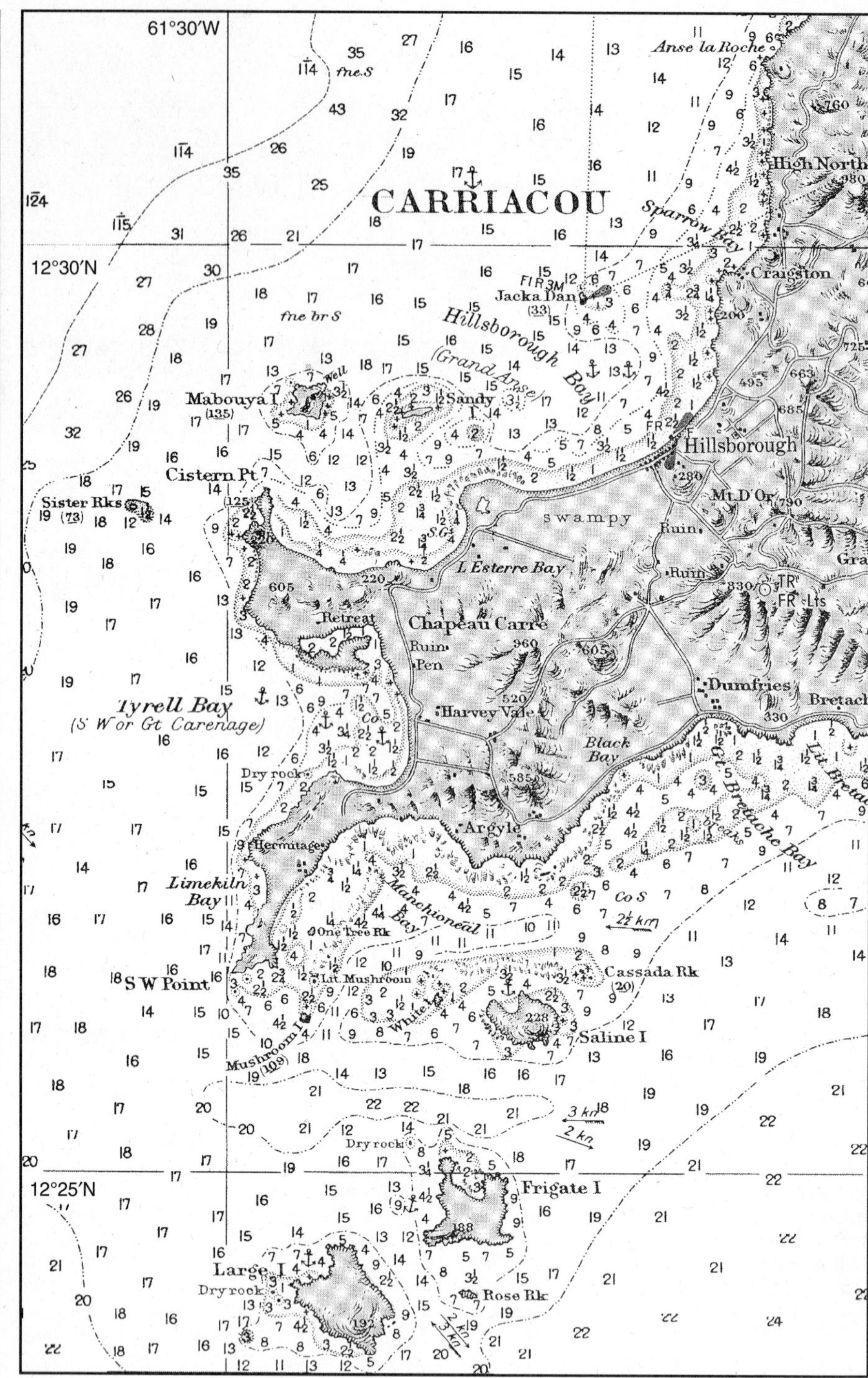

Carriacou, soundings in fathoms *(from DMA 25482, 15th ed, 5/79)*

Labour Day, May 1; Whit Monday, June 1; Corpus Christi, June 4; Emancipation Day, August 3; Carnival Monday & Tuesday, August 10–11; Thanksgiving Day, October 25; Christmas Day, December 25; Boxing Day, December 26. *Date not set as of *Reed's* press time, summer 1997)
Tourism offices: *In Grenada:* Grenada Board of Tourism, Burns Point, P.O. Box 293, St. George's (473-440-2279/1346; fax, 473-440-6637; e-mail, gbt@caribsurf.com). *In the U.S.:* Grenada Tourism Office, 820 Second Avenue, Suite 900D, New York, NY 10017 (tel., 212-687-9554; fax, 212-573-9731). *In Canada:* Grenada Consulate, 439 University Avenue, Suite 920, Toronto, Ont. M5G 1Y8 (tel., 416-595-1339; fax, 416-595-8278).

CARRIACOU

JACK A DAN LIGHT, 12 29.7N, 61 28.3W. Fl G 5s, 14ft, 3M. Visible 243°–182°.
SANDY ISLAND LIGHT, 12 29.2N, 61 28.7W. Q R. Red triangle on piles.

HILLSBOROUGH, CARRIACOU

12 29N, 61 28W
Pilotage: Hillsborough Bay is entered between Jack a Dan and Sandy Island, both of which have lights on them. The town has a 210-foot-long jetty with a depth of about 8 feet alongside its head. A church and tower stand at the southwest end of the village.
Port of entry: Clear here if you are headed south to Grenada or if you want to visit Petite Martinique. Customs is near the main dock.
Anchorage: The anchorage along the waterfront beach in depths of 10 to 20 feet is good. Dinghy in to the dock at the Silver Beach Resort to walk to customs or obtain other services.
Services: The Silver Beach Resort (VHF channel 16; tel., 473-443-7337) has a dinghy dock, showers, and a restaurant and bar. In summer 1997, the resort was rebuoying moorings and extending the dinghy dock. They hope to add water, ice, fuel, and a small vegetable and fish commissary to their services for yachts, but none of those services were in place at *Reed's* press time.

Ashore in Carriacou is a liquor store and a good selection of small markets and restaurants.

PETITE MARTINIQUE

12 33N, 61 23W
Petite Martinique, a small island 2.25 miles east of the north end of Carriacou and adjacent to Petit St. Vincent, is a good place to take on fuel and water. B & C Fuels (VHF channel 16, call sign, Golf Sierra; tel., 473-443-9110) on the northwest corner of the island has a convenient dock with about 14 feet of water alongside that is easily approached into the wind.

CARRIACOU TO GRENADA

RONDE ISLAND

12 18N, 61 35W
This small islet lies about 5 miles northeast of the north coast of Grenada. Reefs fringe the shores in places and extend across the bays on its north and east sides. Yachts may be able to anchor off the northwest side of the island, north of a shoal that extends along the western coast.

ISLE DE CAILLE

12 17N, 61 35W
This small islet lies south of Ronde Island. The narrow channel between the two islands has depths of 12 to 30 feet. During strong winds the sea breaks in this channel. You may be able to anchor under the lee of the island in an emergency.

THE SISTERS

12 18N, 61 36W
The Sisters are two groups of small islets just west of Ronde Island. *Caution: Volcanic activity has been reported as recently as 1989 in an area about 1.75 miles west of The Sisters.*

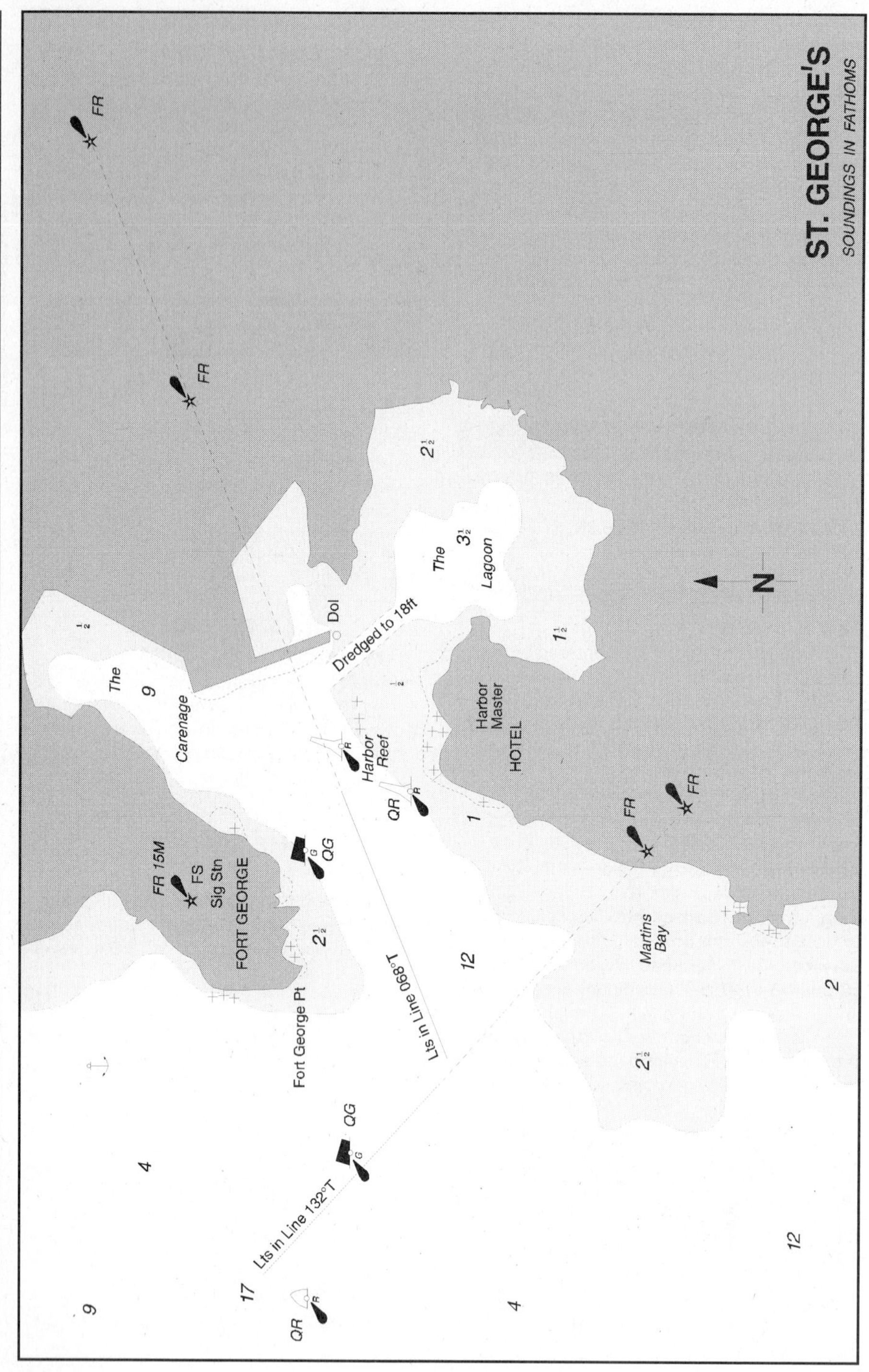

St. George's, soundings in fathoms

ST. GEORGE'S APPROACHES

West Buoy, 750 meters, 251.5° from St. George's Harbour Light, 12 03N, 61 46W. Q R. Red starboard-hand mark.

East Buoy, 530 meters, 238° from St. George's Harbor Light, 12 03N, 61 46W. Q G. Green port-hand mark.

ST. GEORGE'S HARBOUR LIGHT, extremity of north bastion of Fort George Point, 12 03.0N, 61 45.3W. F R, 188ft, 15M. Brick building. Visible 056.5°–151°. Red light on tower 1 mile E.

ST. GEORGE'S RANGE (Front Light), 12 02.5N, 61 45.2W, F R, 46ft. Red square daymark, white stripe on framework tower. Red lights on radio masts 1.7 miles SW. **(Rear Light),** 89 meters, 132° from front. F R, 92ft. Red triangular daymark, white stripe on framework tower.

RANGE (Front Light), 12 03.0N, 61 44.7W. F R, 102ft. Red triangular daymark, white stripe on framework tower. **(Rear Light),** 518 meters, 068.5° from front. F R, 299ft. Red square daymark, white stripe on framework tower.

ST. GEORGE'S, GRENADA

12 03N, 61 45W

Pilotage: St. George's is the capital of Grenada and the country's most important harbor. The inner part of the harbor has two basins; the small-craft facilities are located in the southern one known as the Lagoon. Two pairs of range lights are shown. The first pair, bearing 132°, lead from the north into Martins Bay. The second pair, bearing 068°, lead from Martins Bay into the inner harbor. Be careful not to cut the corner too close when turning into the channel leading to the Lagoon or you may discover Harbour Reef firsthand. You can contact port control (call sign J3YA) on VHF channel 16.

Port of entry: Yachts should proceed to the Grenada Yacht Services docks in the Lagoon, where a customs office is located.

Dockage: As of July, 1997, Grenada Yacht Services (VHF channel 16; tel., 473-440-2508) has been sold and gave notice to "all users of its premises" that it will not be responsible for any transactions after August 15, 1997.

Anchorage: The best shelter is in the Lagoon, but you can also try in the Carenage, which is very busy with commercial traffic.

Services: There are unconfirmed reports that a new $100 million yacht complex is in the works here. In the meantime, existing haulout facilities are not operational, although there are other marine services here and in nearby Prickly Bay.

St. George's is well sheltered and south of most hurricane tracks. Ashore there is a good variety of general shops, supermarkets, and restaurants.

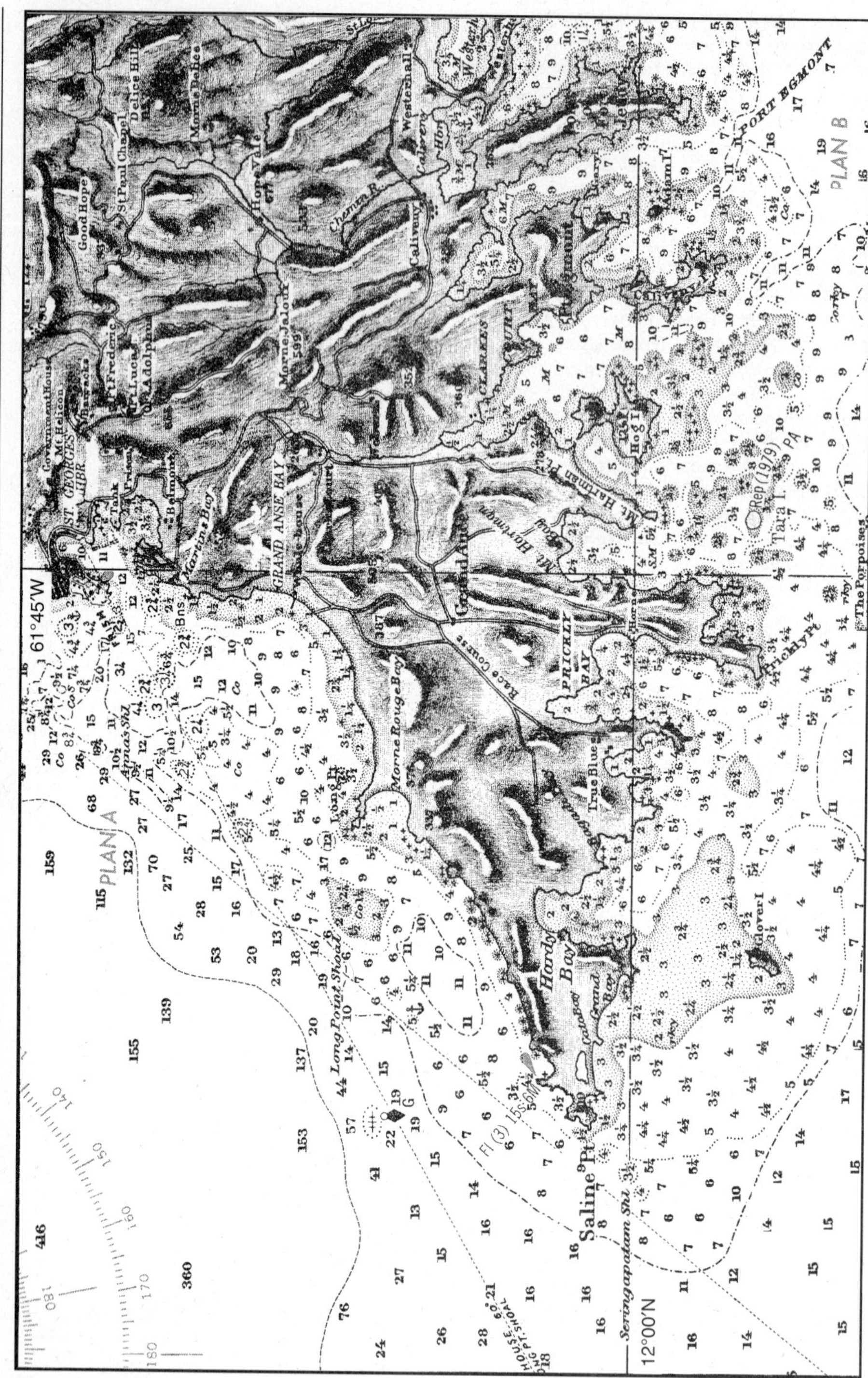

Grenada, south coast, soundings in fathoms *(from DMA 25481, 24th ed, 11/84)*

ST. GEORGE'S TO GRENVILLE BAY

PETIT CABRITS LIGHT, 12 01.0N, 61 46.4W. Fl (2+1) W 20s, 354ft, 18M. Red framework tower.
POINT SALINE LIGHT, 12 00.2N, 61 47.8W. Q (9) W 15s, 7M. West cardinal mark.
GLOVER ISLAND LIGHT, 11 59.2N, 61 47.2W. Q (6) + L Fl W 15s, 7M. South cardinal mark.

PRICKLY BAY, GRENADA

12 00N, 61 46W
Pilotage: Prickly Bay is located on the south coast of Grenada, about 2.5 miles east of Saline Point. When traveling from St. George's to Prickly Bay be sure to stay well outside Long Point Shoal. A course of 240°T, with Fort George Point in line with Government House (in St. George's), is reported to keep you clear of the shoal. There is a shoal spot in Prickly Bay, due west of Spice Island Marine Services.
Port of entry: Customs and immigration officials are located at Spice Island Marine Services and at the Moorings Secret Harbour Marina.
Dockage: Contact Spice Island Marine Services (VHF channel 16; tel., 473-444-4342; fax, 473-444-2816) on the eastern side of the bay. The Moorings Secret Harbour Marina (VHF channels 16 and 71; tel., 473-444-4449; fax, 473-444-2090) is located in the next cove to the east, known as Mount Hartmann Bay. This area is reported to be very safe and crime-free. The Rum Squall Bar at Secret Harbour is a popular gathering spot every evening for cruisers.
Anchorage: There is good anchorage in True Blue Bay, a small cove just west of Prickly Bay, and Prickly Bay.
Services: Fuel, water, ice, showers, haulouts, repairs, a chandlery, sail repairs, a canvas shop, a public telephone, minimarket, fax service, and a restaurant are all available at Spice Island Marine Services. The Moorings operation also has fuel, water, ice, laundry, showers, and a minimarket. There are markets and restaurants nearby. Other marine services in the area include sail repair, mechanical repairs, and woodworking.

The south coast of Grenada has several good anchorages, with convenient yachting services nearby. You are a short taxi or bus ride from St. George's, but in a much more natural setting. These harbors are also convenient to the airport near Saline Point.

GRENVILLE BAY, GRENADA

12 07N, 61 37W
Pilotage: This harbor, about midway up the east coast of the island, is protected by coral reefs to the east. The entrance channel is difficult to negotiate and should be attempted only in good weather, good light, and with local knowledge. During strong winds the sea breaks right across the entrance. There are reported to be two white beacons, on a bearing of 291°, that lead through the outer reef. The channel through the inner reef, called Luffing Channel, leads in a northerly direction. There may be some buoys marking these channels. The town pier is reported to have depths of 10 feet alongside its head.
Port of entry: This is the only port of entry on the east coast of Grenada. The officials are located near the pier.
Anchorage: Anchor in the area clear of reefs off the town pier.
Services: This is not a yachting area, but there are good markets and general shops.

This harbor is not visited by many boaters, because of its location on the east coast and its tricky reef entrance.

BARBADOS

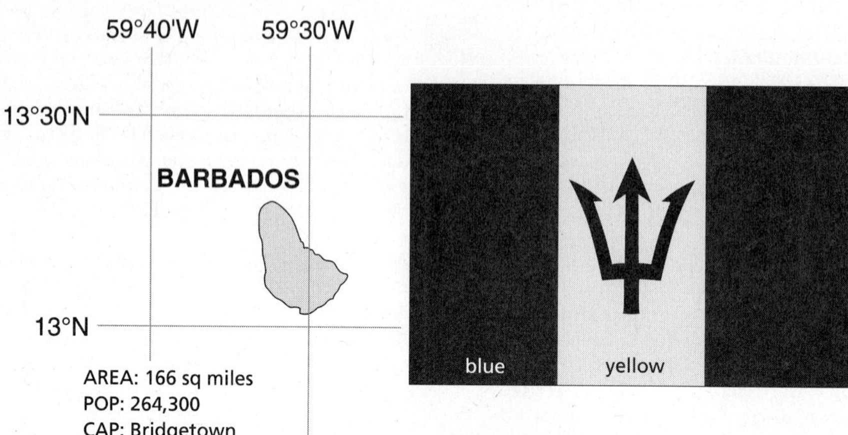

AREA: 166 sq miles
POP: 264,300
CAP: Bridgetown

Please see cautions about chartlets and waypoints on page P 2.

ENTRY PROCEDURES

Barbados is the easternmost of the Windward Islands, lying about 89 miles to windward of St. Vincent. A former British colony, it gained independence in 1966.

The port of entry is Bridgetown, located on the southwest side of the island. Vessels should contact the port authority signal station (VHF channels 12 and 16 and 2182 kHz; call sign, 8PA) for permission to enter Deepwater Harbour, east of the breakwater. Yachts should fly the Q flag. Customs officials (246-426-2604) are on duty from 0600 to 2200, seven days a week; the customs area is well sign-posted. The main customs office in town (246-427-5940) is open weekdays only from 0815 to 1630.

Entering vessels are visited by officers from customs, immigration, and public health, usually together. You should have clearance from the last port of call, ship's papers, crew lists, and passports. Request a coastwise clearance if you wish to anchor elsewhere on the island. Clearance and departure fees are based on the length of the boat. Overtime charges may apply after 1630, on weekends, and on public and bank holidays.

USEFUL INFORMATION

Language: English.
Currency: Barbados dollar (Bd$); exchange rate (summer 1997), Bd$1.98 = US$1.
Telephones: Area code, 246; local numbers, seven digits.
Public holidays (1998): New Year's Day, January 1; Errol Barrow Day, January 21; Good Friday, April 10; Easter Monday, April 13; May Day, May 5; Whit Monday, June 1; Kadooment Day, August 3; United Nations Day, October 5; Independence Day, November 30; Christmas Day, December 25; Boxing Day, December 26.
Tourism offices: *In Barbados:* Barbados Tourism Authority, Harbour Road., P.O. Box 242, Bridgetown (246-427-2623). *In the U.S.:* Barbados Tourism Authority, 800 Second Avenue, 2nd Floor, New York, NY 10017 (800-221-9831). Offices also in Coral Gables (FL) and Los Angeles (CA). *In Canada:* Barbados Tourism Authority, 5160 Yonge Street, Suite 1800, North York, Ont. M2N 6L9 (416-512-6569).

WEST COAST

HARRISON POINT LIGHT, NW side of island, 13 18.3N, 59 39.0W. Fl (2) W 15s, 193ft, 22M. White stone tower, 85ft.
MAYCOCKS BAY LIGHT, jetty, N end, 13 16.8N, 59 39.3W. Q R, 3M.

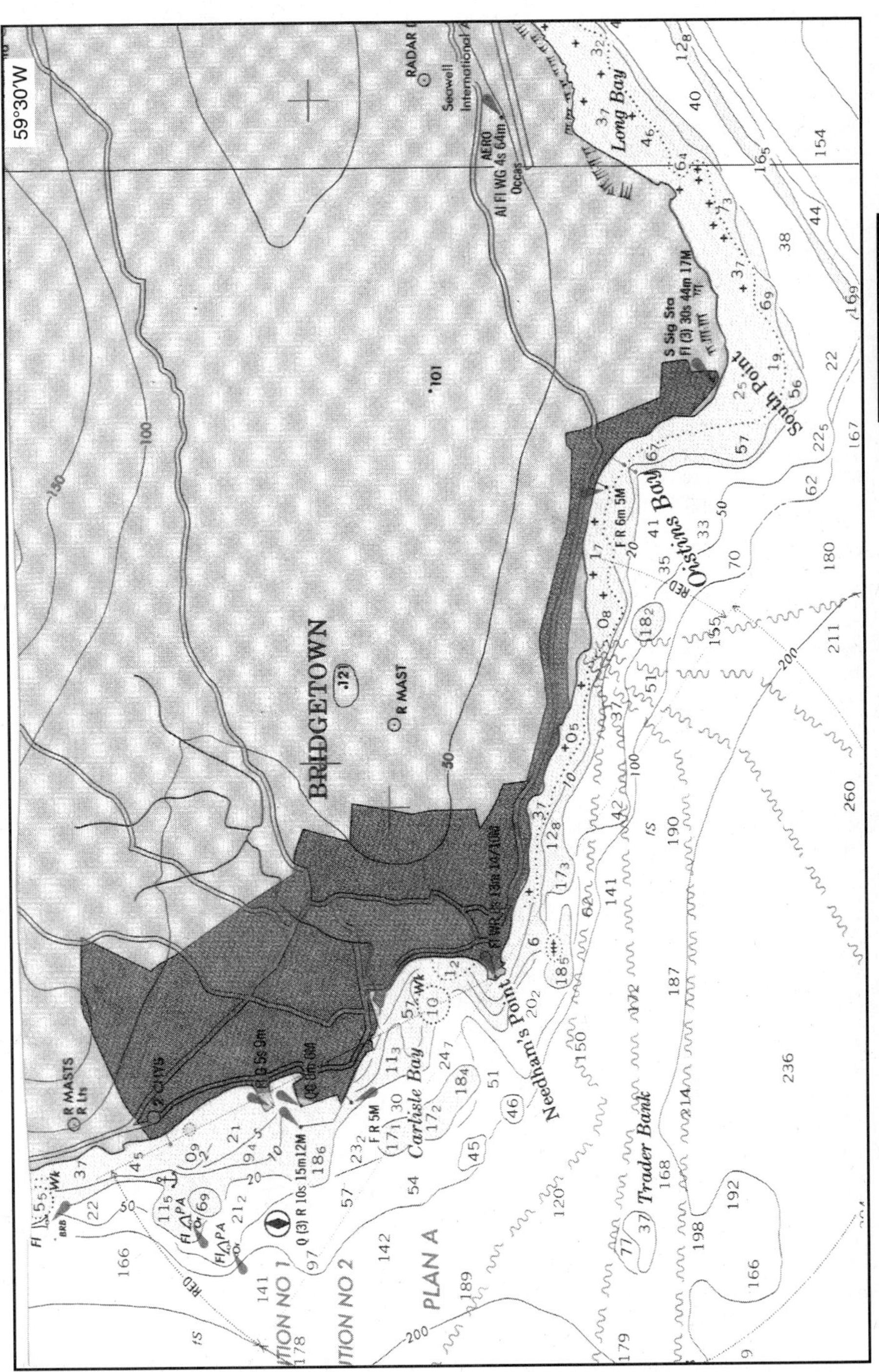

Approaches to Bridgetown, soundings in meters *(from DMA 25485, 45th ed, 12/94)*

MAYCOCKS BAY LIGHT, jetty, S end, 13 16.9N, 59 39.3W. Q G, 3M.

APPROACHES TO BRIDGETOWN

BULK FACILITY LIGHT, 13 06.5N, 59 37.8W. Fl G 5s, 29ft. Red metal mast, 13ft.

SHALLOW DRAFT JETTY, N, 13 06.4N, 59 37.5W. F R, 18M. Red mast.

SHALLOW DRAFT JETTY, S, 13 06.4N, 59 37.5W. F R, 18M. Red mast.

CONTAINER BERTH LIGHT, 13 06.2N, 59 37.9W. Q G, 26ft, 6M. Yellow rectangular daymark on red beacon, 36ft.

BRIDGETOWN BREAKWATER LIGHT, 13 06.3N, 59 38.0W. Q (3) R 10s, 49ft, 12M. Yellow triangular daymark, point up, on red beacon, 29ft. F W shown when required on flagstaff at shore end of pipeline at Spring Garden anchorage, 1.15 mile N.

OIL PIER LIGHT, SE end, 13 05.9N, 59 37.8W. Q R, 16ft, 5M. Silver metal mast, 13ft.

OIL PIER LIGHT, center. F R, 39ft, 5M. Silver metal mast, 33ft.

OIL PIER LIGHT, NW end. Q R, 16ft, 5M. Silver metal mast, 13ft.

FISHING HARBOR ENTRANCE LIGHT, S side, 13 05.7N, 59 37.2W. F G.

FISHING HARBOR ENTRANCE LIGHT, N side, 13 05.7N, 59 37.2W. F R.

BRIDGETOWN CAREENAGE LIGHT, S side of entrance, 13 05.6N, 59 37.1W. Fl (2) G 10s, 26ft, 2M. Silver metal framework structure, 16ft.

BRIDGETOWN

13 06N, 59 38W

Pilotage: The approaches to Bridgetown are generally free of hazards. The area known as The Shallows and the reef along the east coast of the island can generate very heavy seas when rounding South Point. Overfalls have been reported near the point.

Port of entry

Public correspondence: Call Barbados Radio, call sign 8PO, on 2182 kHz or VHF channel 16; working VHF channels are 26, 27, and 28. Shore-to-ship calls are placed via BARTEL, who then routes calls through Barbados Radio.

Dockage: It may be possible to tie up at an empty berth in the Careenage, which has a flashing green light at the entrance. Call the port signal station (246-436-6883) for availability and permission.

Anchorage: After clearing in, most yachts proceed to the anchorage in Carlisle Bay, north of Needham Point. You can land your dinghy by the Boatyard, the Barbados Yacht Club, the Barbados Cruising Club, or in the Careenage.

Services: The Boatyard (246-436-2622) in the northern part of Carlisle Bay can assist boaters with many services, including water, ice, laundry, communications, and making arrangements for repair. The whole property, including a restaurant and bar, has recently been rebuilt. At press time, The Boatyard had just received permission to build a 290 foot jetty for boats to berth while undergoing minor repairs. There is also a new dinghy dock under construction. You might also want to contact the Barbados Yacht Club (246-427-1125) or the small Barbados Cruising Club for assistance. They are located on either side of a hotel pier at the south end of Carlisle Bay, and both have bars overlooking the anchorage. Diesel is available in the Careenage and also in the fishing harbor just north of the Careenage, as is fresh water. Diesel and water can also be jugged from the Boatyard. Shallow-draft boats can be hauled in the Careenage. Bridgetown has good supermarkets and general stores.

Few sailors will make the rough trip against the trade winds to reach Barbados from the rest of the Antilles, but this island is often the first stop for those crossing from the Canaries. The approaches are easy.

NEEDHAM POINT LIGHT, 13 04.5N, 59 36.9W. Fl W R 8s, 43ft, W 14M, R 10M. Red metal mast, 30ft. R 274°–304°, W 304°–124°, R 124°–154°.

OISTINS FISHING JETTY LIGHT, 13 03.6N, 59 32.8W. F R, 20ft, 5M.

EAST COAST

RAGGED POINT LIGHT, 13 09.6N, 59 26.1W. Fl W 15s, 213ft, 21M. White circular coral stone tower, 97ft. Obscured when bearing less than 135°.

SEAWALL AVIATION LIGHT, 13 04.5N, 59 29.7W. Al Fl W G 4s, 210ft. Occasional.

SOUTH POINT LIGHT, 13 02.8N, 59 31.8W. Fl (3) W 30s, 145ft, 17M. Tower, alternate red and white bands, 90ft. Storm signal station.

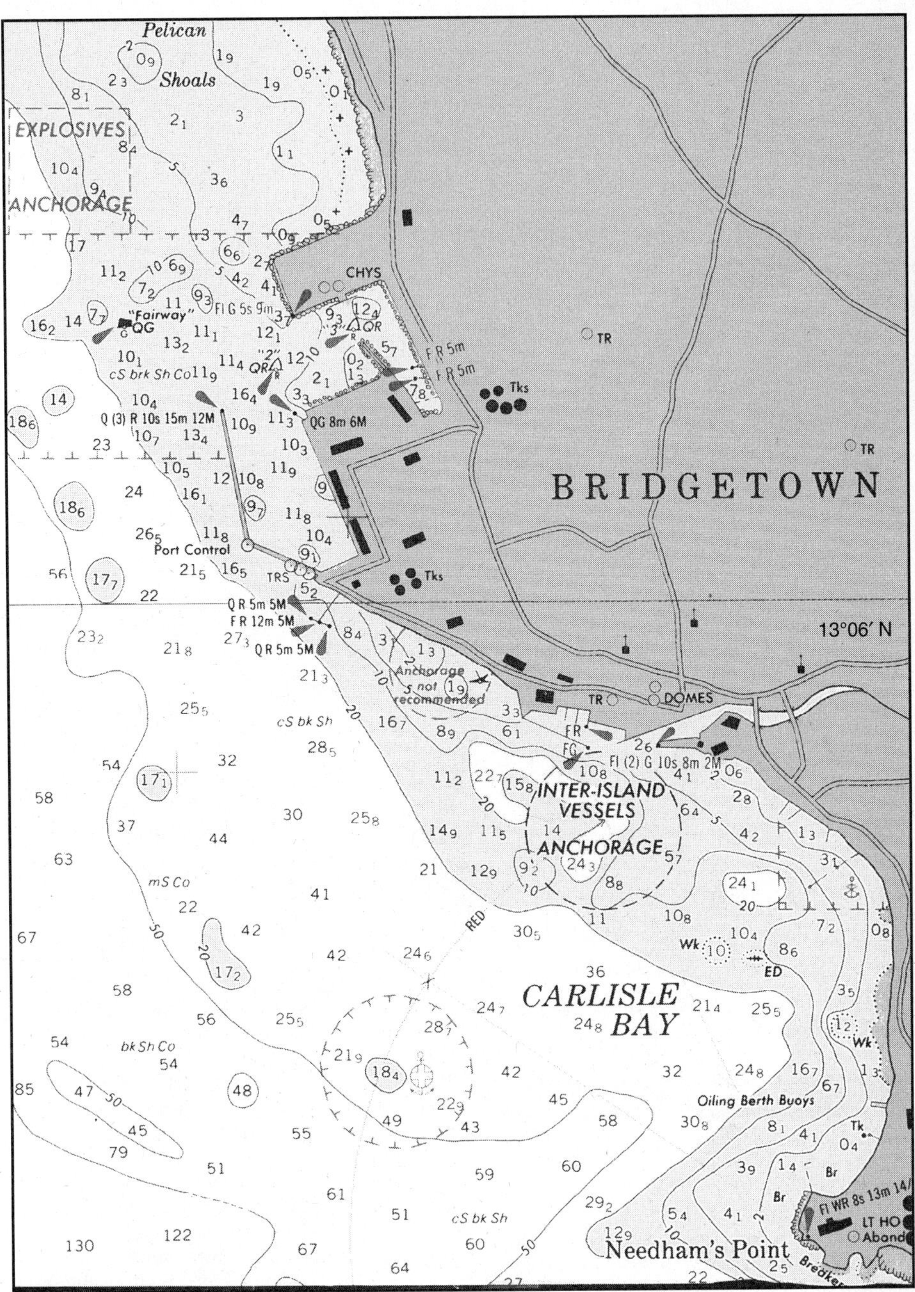

Bridgetown, soundings in meters *(from DMA 25485, 45th ed, 12/94)*

P
Pilot
BARBADOS

Trinidad and Tobago

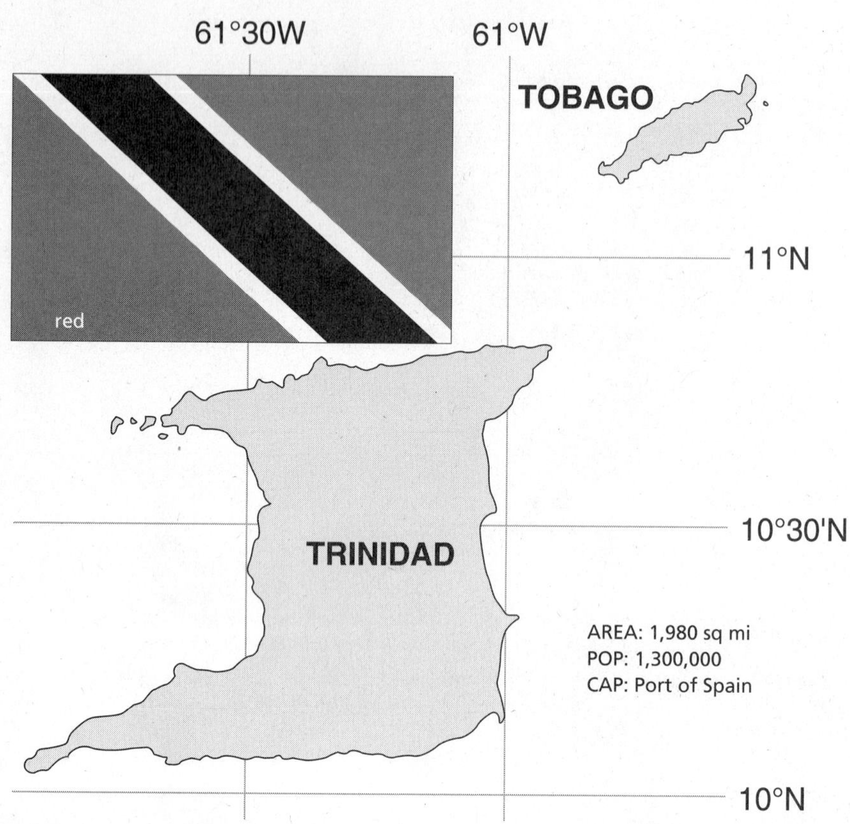

Please see cautions about chartlets and waypoints on page P 2.

ENTRY PROCEDURES

The islands of Trinidad and Tobago lie east of the Venezuelan coast. They form an independent country, with ties to the British Commonwealth.

The only port of entry on Tobago is Scarborough, located on the southwest end of the island. In Trinidad yachts may clear at the customs stations in Chaguaramas or Port of Spain; commercial vessels may clear only at Port of Spain. Customs recommends that yachts entering on Trinidad use Chaguaramas. Normal office hours for all stations are 0800 to 1600, Monday through Friday, except holidays. There are no entry fees, but there is an overtime fee of about US$45 for clearances outside office hours.

To clear in, you should have a clearance from the last port of call, ship's papers, crew and stores lists, and passports. Immigration normally grants a stay of three months, which may be extended another three months (extension fee). Guns and ammunition (including flare guns) must be declared. Once vessels have cleared in, they are free to move between anchorages on the same island without permission from authorities. Check with officials for the latest regulations governing movements between the two islands. If you plan to clear out from a different island than the one you entered, notify immigration on entering so that paperwork can be transferred to the exit port.

USEFUL INFORMATION

Language: English (official). Other languages are common, reflecting the varied origins of the islands' inhabitants.
Currency: Trinidad and Tobago dollar (TT$); exchange rate (summer 1997) TT$6.28 = US$1. U.S. currency is widely accepted.
Telephones: A new area code, 868, went into effect for both Trinidad and Tobago on June 1, 1997. For local numbers, dial seven digits.
Public holidays (1998): New Year's Day, January 1; Carnival, February 23–24*; Good Friday, April 10; Spiritual Baptist Liberation Shouter Day, Sunday, March 30; Easter Monday, April 13; Corpus Christi, June 4; Labour Day, June 19; Emancipation Day, August 1; Independence Day, August 31; Republic Day, September 24; Christmas Day, December 25; Boxing Day, December 26. The Hindu festival of Divali and the Muslim festival of Eid-Ul-Fitr are also holidays; their dates had not been officially set as of *Reed's* press time (summer 1997). *Not an official holiday, but most businesses are closed.
Tourism offices: *In Trinidad:* Tourism and Industrial Development Co. (TIDCO), 10–14 Philipps Street, Port of Spain (tel., 868-623-1932/4; fax, 868-623-4514; e-mail, tourism-info@tidco.co.tt; http://www.tidco.co.tt/~marine). *In Tobago:* TIDCO, Unit 12 ICD Mall, Sangster's Hill, Scarborough (tel., 868-639-4333; fax, 868-639-4514). *In the U.S.:* Sales, Marketing and Reservations Tourism Services, 7000 Boulevard East, Guttenberg, NJ 07093 (tel., 800-748-4224, 201-662-3403/3408; fax, 201-869-7628). *In Canada (representative):* RMR Group, 512 Duplex Avenue, Toronto, Ont. M4R 2E3 (tel., 416-484-4864; fax, 416-485-8256).
Additional notes: The government has published a boater's directory of marine services; ask at customs when you clear in. The government also has taken several steps to curb the growing crime rate. While crime is significantly lower on Tobago, travelers may wish to exercise normal precautions. Cases of yellow fever have occurred in Trinidad and Tobago, and a yellow fever vaccination is recommended. Medical care at some private facilities is better than at most public facilities. For the latest medical advisory, contact the Centers for Disease Control International Travelers Hotline (voice recording, 404-332-4559; fax information service, 404-332-4565; Web, www.cdc.gov).

Buccoo reef, off Tobago, was declared a national park in 1973; new, expanded boundaries are currently being proposed. The reef offers miles of good snorkeling. Within the park all marine life is protected. No fishing is allowed, nor is removing or harming coral, shells or other sea life. Anchoring is prohibited and park officials remind boaters not to drop dinghy anchors on the reef—use sandy spots only. The discharge of heads within 250 meters of the beach is also prohibited.

TOBAGO, NORTH COAST

ST. GILES ISLAND LIGHT, E end, 11 21.2N, 60 31.1W. Fl W 7.5s, 16M.
MAN OF WAR BAY LIGHT, 11 19.1N, 60 32.8W. Q W R G, 82ft, W 5M, R 4M, G 4M. R 098°–108°, W 108°–131°, G 131°–141°.
THE SISTERS LIGHT, 11 19.7N, 60 38.7W. Fl (2) W 10s, 8M.
COURTLAND POINT LIGHT, 11 13.2N, 60 46.8W. L Fl W 10s, 8M.
BOOBY POINT LIGHT, 11 10.9N, 60 48.8W. Fl Y 3s, 4M.
MILFORD BAY LIGHT, 11 09.4N, 60 50.4W. Q W R G, 23ft, W 5M, R 4M, G 4M. R 073°–083°, W 083°–128°, G 128°–138°.
CROWN POINT LIGHT, SW extremity of island, 11 08.7N, 60 50.7W. Fl (4) W 20s, 115ft, 11M. Steel lattice tower painted aluminum, 85ft.

TOBAGO, SOUTH COAST

LITTLE TOBAGO LIGHT, 11 17.5N, 60 29.5W. Fl (3) W 10s, 59ft, 5M.
SMITHS ISLAND LIGHT, 11 11.0N, 60 39.1W. Fl W R 5s, 59ft, W 7M, R 5M. R 068°–276°, W 276°–068°.
SCARBOROUGH LIGHT, on Fort George Point, 11 10.4N, 60 43.7W. Fl (2) W 20s, 462ft, 30M. White concrete building, yellow roof. Visible 258°–090°.
~~**SCARBOROUGH HARBOUR LIGHT,** E beacon, 11 10.6N, 60 44.2W. Q R, 18ft, 2M. White steel structure on piles.~~
RANGE (Front Light), 11 11.0N, 60 44.5W. V Q G, 13ft, 5M. Pile. **(Rear Light),** 115 meters, 329.5° from front. Iso W R G 2s, 49ft, W 7M, R 5M, G 5M. R 313.5°–323.5°, W 323.5°–335.5°, G 335.5°–345.5°.
~~**LOWER TOWN LIGHT,** 11 11.1N, 60 44.5W. Oc W 5s, 33ft, 11M~~. Removed 7/97.
BULLDOG SHOAL, 11 08.9N, 60 44.5W. VQ (6)+L Fl W 10s, 16 ft. South cardinal mark, yellow and black pillar with topmark.

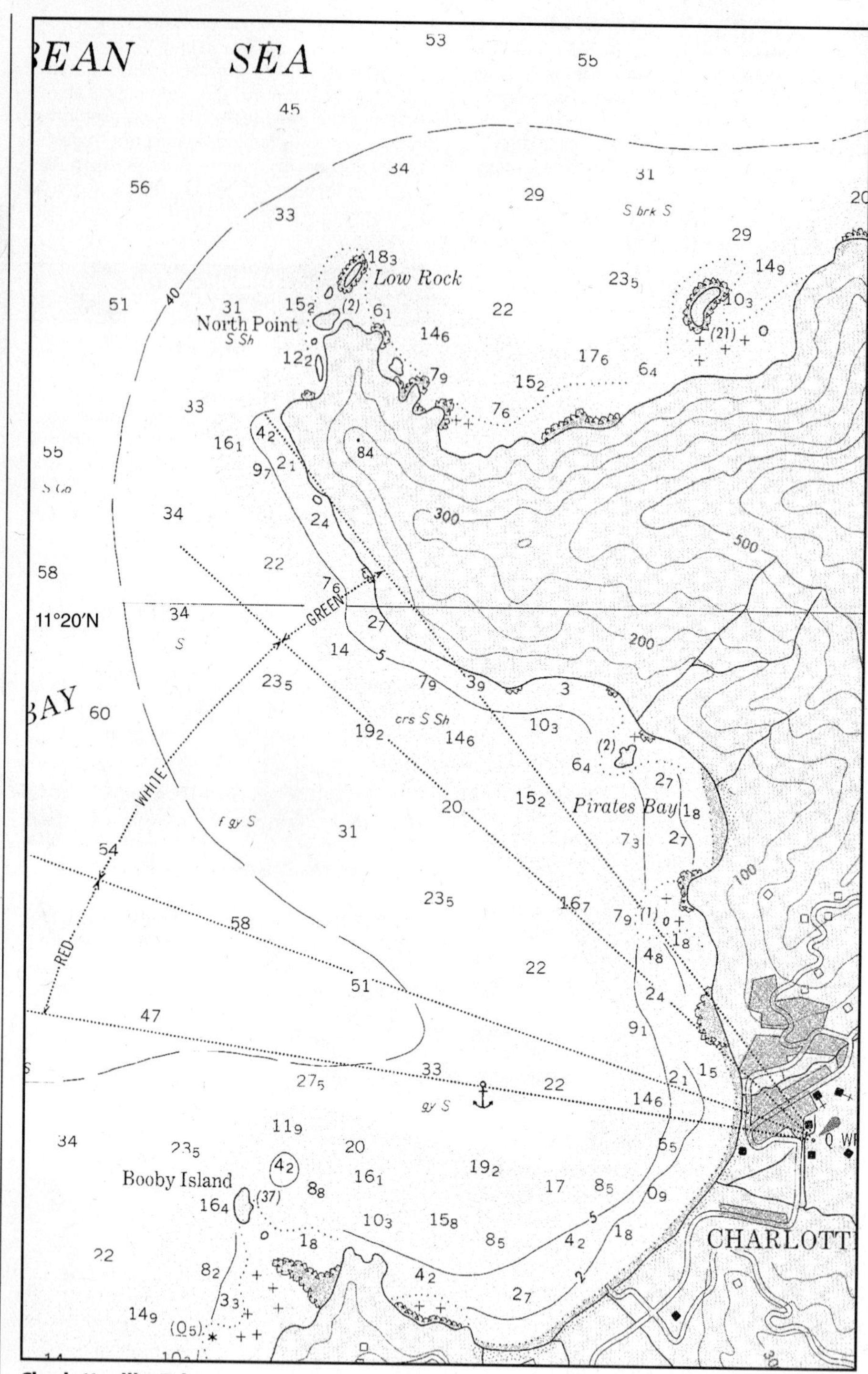

Charlotteville, Tobago, soundings in meters *(from DMA 24402, 16th ed, 9/87)*

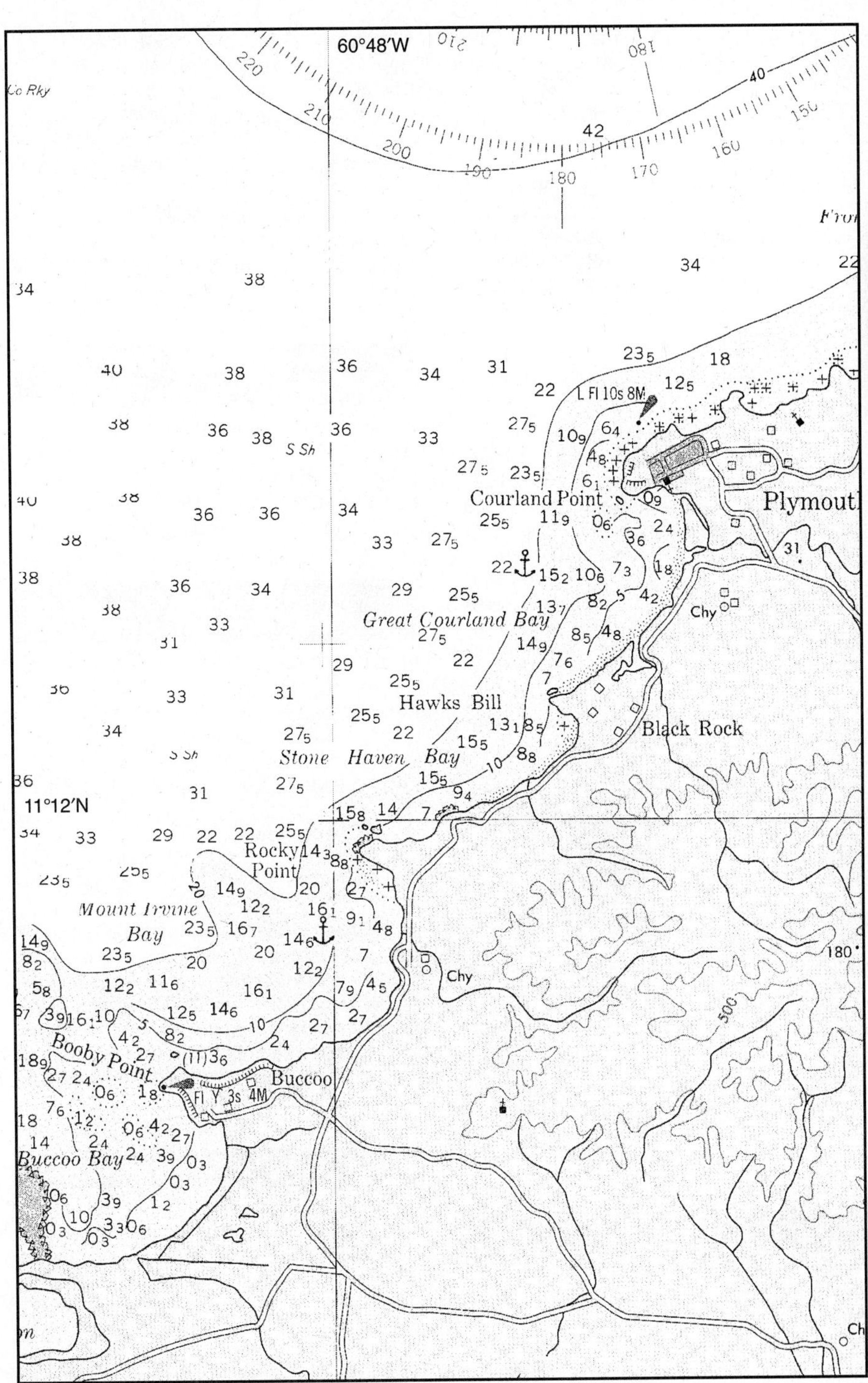

Plymouth, Tobago, soundings in meters *(from DMA 24402, 16th ed, 9/87)*

SCARBOROUGH HARBOUR NO. 1, 11 09.8N, 60 44.1W. Fl G 3s, 13ft, 4M. Green port-hand beacon.

SCARBOROUGH HARBOUR NO. 2, 11 09.9N, 60 43.2W. Fl R 3s, 16ft, 5M. Red starboard-hand beacon.

SCARBOROUGH HARBOUR CHANNEL EAST, 11 10.4N, 60 44.1W. Q.R. 13ft, 4M. Beacon.

SCARBOROUGH HARBOUR CHANNEL WEST, 11 10.3N, 60 44.3W. Fl G 3s, 16ft, 5M. Beacon.

SCARBOROUGH HARBOUR BREAKWATER HEAD, 11 10.6N, 60 44.3W. Q.R. 13ft, 5M. Beacon.

SCARBOROUGH, TOBAGO

11 11N, 60 44W

Pilotage: Tobago lies about 19 miles north-northeast of Trinidad and 78 miles southeast of Grenada. The equatorial current runs in a northwesterly direction between Tobago and Grenada and is very variable and unpredictable, running strong at some places and barely existent at others. Take careful note of the charted and buoyed shoals to the east and west of the entrance to Rockly Bay. Fort George and several nearby radio towers are conspicuous. A lighted range of 329.5° should lead you in safely. Watch for strong easterly currents entering Scarborough. When sailing to Trinidad avoid Wasp Shoal and Drew Bank, located about 3 miles off the southwestern tip of the island.

Port of entry: The customs station (868-639-2415) is located near the docks. Immigration can now be reached only at 868-639-0006.

Dockage: You may be able to tie up Mediterranean style (stern-to) behind the breakwater (not shown on the Scarborough chartlet on the following page).

Anchorage: Anchor outside the jetty in depths of around 10–12 feet. This area is reported to suffer from surge. Increasingly, boaters are choosing to anchor in Stove Bay and Mount Irvine, although there are no facilities there.

Services: You can get water and jug your fuel. While there are few yacht-oriented services, there are good general stores, markets, dive shops, and restaurants on the island.

TRINIDAD, SOUTH COAST

SOLDADO ROCK, 10 04.2N, 62 00.9W. Fl W 10s, 8M.

WOLF ROCK, 10 03.2N, 61 56.3W. Q (6) + L Fl W 15s, 13ft, 5M. Beacon. Temporarily extinguished (1988).

PUNTA DEL ARENAL, 10 02.8N, 61 55.7W. Fl W 7.5s, 72ft, 16M. White steel structure. Obscured when bearing less than 301°.

CHATHAM JETTY, head, 10 04.8N, 61 44.1W. F R, 3M.

TAPARO POINT, 10 03.4N, 61 37.7W. Fl (3) W 15s, 226ft, 14M.

LA LUNE POINT, 10 04.5N, 61 18.9W. Fl (4) W 20s, 148ft, 14M.

GALEOTA POINT, 10 08.5N, 60 59.7W. Fl W 5s, 285ft, 16M. White steel framework tower, 30ft.

TRINIDAD, EAST COAST

BRIGAND HILL, 10 29.4N, 61 04.2W. Fl (2+1) W 30s, 712ft, 20M. White steel framework tower, 36ft.

GALERA POINT, 10 50.0N, 60 54.3W. Oc W 10s, 141ft, 16M. White concrete tower, 75ft.

TRINIDAD, NORTH COAST

PETITE MATELOT POINT, 10 49.2N, 61 07.7W. Fl (3) W 15s, 7M.

CHUPARA POINT, 10 48.2N, 61 22.0W. Fl (2) W 10s, 325ft, 12M. White metal framework tower and hut.

SAUT D'EAU ISLAND, 10 46.2N, 61 30.7W. Q W, 7M.

NORTH POST, POINT A DIABLE, 10 44.7N, 61 33.8W. Fl W 5s, 747ft, 14M. Beacon. Visible 087°–252°.

NOTE: The current between Trinidad and Tobago sets to the northwest, usually with sufficient strength to prevent a sailing vessel from working through against it. The strength of the current is somewhat lessened by the ebb tidal stream. Current runs to the westward along the north shore of Trinidad.

NORTHWEST TRINIDAD

CHACACHACARE, 10 41.7N, 61 45.2W. Fl W 10s, 825ft, 26M. White concrete tower, 49ft.

CHACACHACARE BEACON, about 640 meters, 216° from main light, 10 41.4N, 61 45.4W. Fl W 2s, 503ft, 11M. White square on white metal framework tower, 26ft. In line 036° with Chacachacare.

POINT DE CABRAS, S end of Huevos, 10 40.9N, 61 43.2W. V Q (6) + L Fl W 10s, 40ft, 5M.

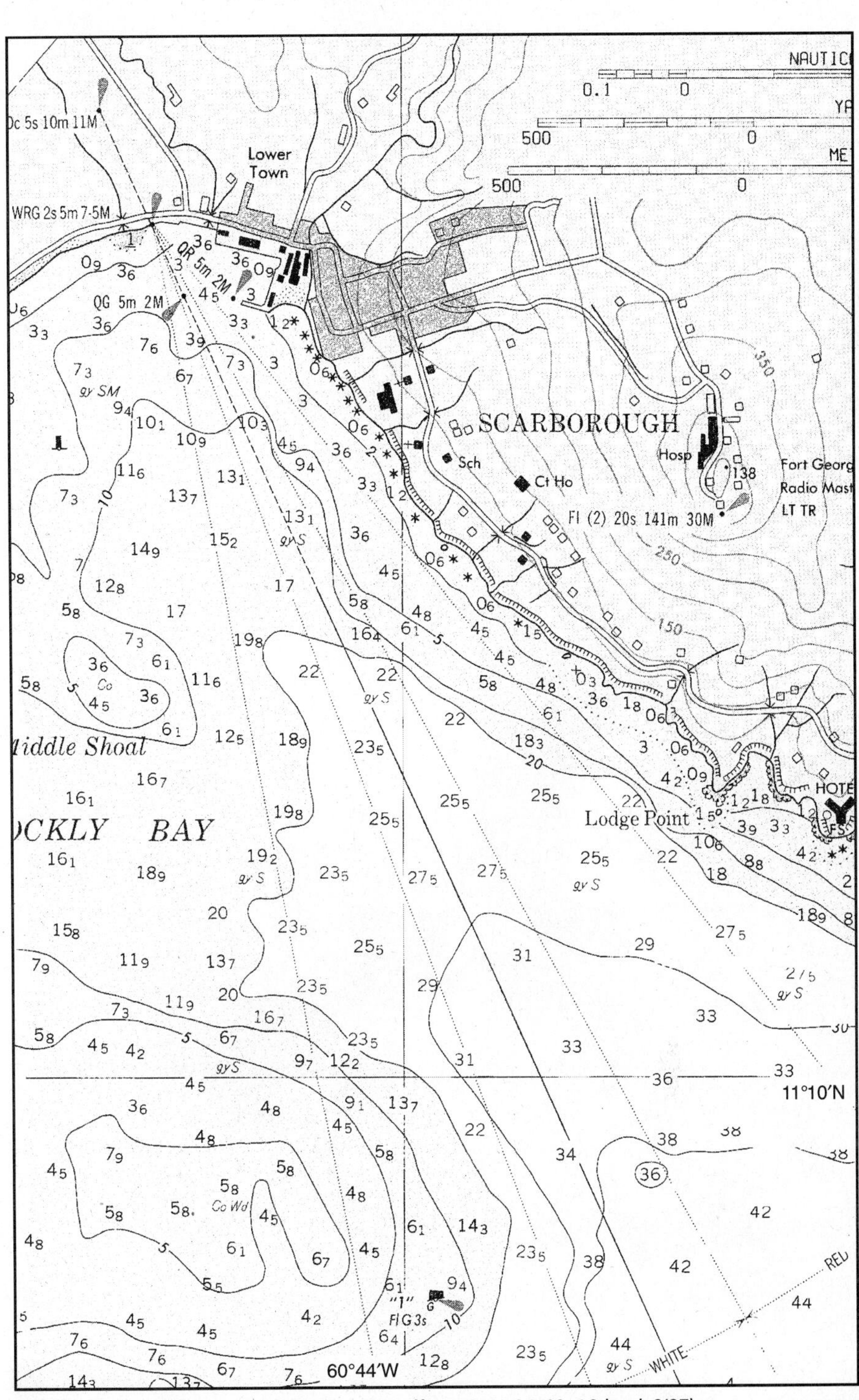

Scarborough, Tobago, soundings in meters *(from DMA 24402, 16th ed, 9/87)*

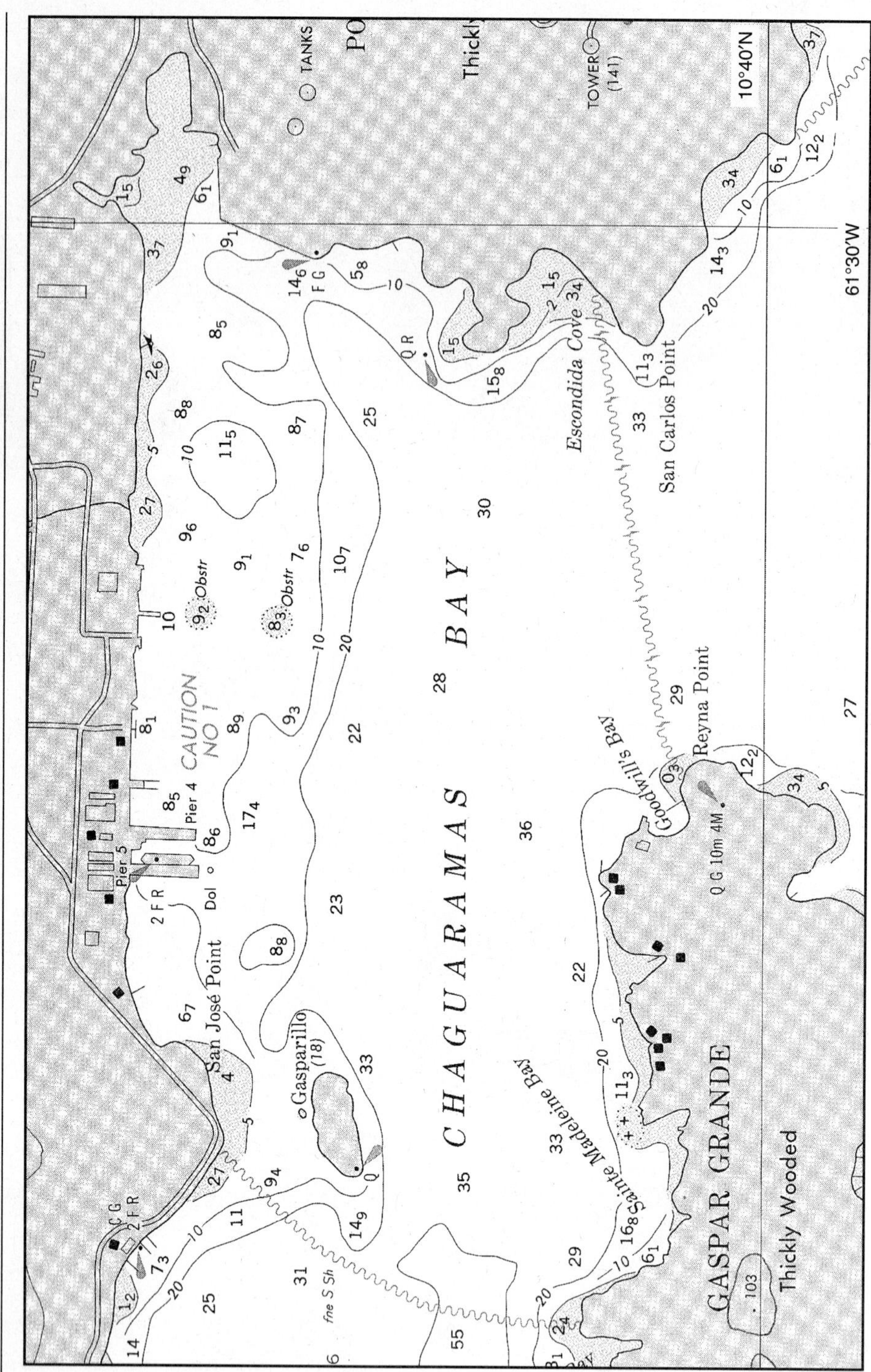

Chaguaramas, Trinidad, soundings in meters *(from DMA 24406, 29th ed, 7/91)*

LE CHAPEAU ROCK, 10 42.3N, 61 40.6W. Fl (3) W 10s.
TETERON ROCK, 10 40.8N, 61 40.1W. Fl G 4s, 24ft, 4M. White metal structure on concrete base. Reported extinguished.
LA RETRAITE COAST GUARD STATION, 10 40.7N, 61 39.5W. 2 F R.
GASPARILLO ISLAND, W end, 10 40.4N, 61 39.4W. Q W, 36ft. White mast.
ESPOLON POINT, SW point of Gaspar Grande Island, 10 39.8N, 61 40.1W. Fl W 4s, 42ft, 12M. White framework tower. Reported extinguished (December 1981).
CRONSTADT ISLAND, W end, 10 39.4N, 61 37.9W. Q R, 4M.
REYNA POINT, 10 40.0N, 61 38.8W. Q G, 33ft, 4M.
ESCONDIDA COVE, off N point, 10 40.3N, 61 38.3W. Q R, 17ft. White tripodal beacon. Unreliable.
FURNESS SMITH FLOATING DOCK, 10 40.7N, 61 38.9W. 2 F R.
~~**SOUTH END OF ALUMINUM COMPANY PIER,** W side of Point Gourde, Chaguaramas Bay, 10 40.5N, 61 38.1W. F G, 13ft. Unreliable.~~

CHAGUARAMAS, TRINIDAD

10 41N, 61 39W
Pilotage: Off the north coast of Trinidad the north- and northwest-going tidal streams attain velocities of 2 to 3 knots. Between July and October the influence may be felt 15 to 20 miles offshore. The Dragons Mouth has several channels leading south into the Gulf of Paria. In Boca de Monos (the easternmost channel) there is no south current during the flood, but the north-running ebb attains speeds of 2 to 3 knots. The tidal currents create eddies around piers 4 and 5 in Chaguaramas. A tidal surge, known locally as the Re-Mou, is strongest in July and August. It may appear as a clear line moving through the channel north of Gaspar Grande, achieving speeds of 1 to 5 knots. The customs office and the yacht facilities are located in the northeast portion of the bay.
Port of entry: Call Chaguaramas customs on VHF channel 16 to announce your arrival (tel., 868-634-4341). The customs dock is immediately below the customs office, near the aluminum plant. This station is now a full-service facility, with an immigration office in the same building as customs. Boaters can also now clear in at Crews Inn Marina. The staff there will expedite customs and immigration and the payment of harbormaster fees, all of which had to be done previously in Port of Spain.
Dockage: Twenty-two slips are available at Power Boats (VHF channel 72; tel., 868-634-4303; fax, 868-634-4327), which is primarily a repair yard, and Peake Yacht Services (VHF channel 68; tel., 868-634-4423; fax, 868-634-4387), which can accommodate large yachts. Fifty-five slips are also available at the new Crews Inn Marina (VHF channel 68; tel., 868-634-4384; fax, 868-634-4542).
Anchorage: Yachts anchor in front of the boatyards, but there is reportedly a slight roll. Visting yachts are asked to observe the proper anchoring areas as described in the chart of Chaguaramas Bay.
Repairs and services: Power Boats (which is not limited to power vessels) is a significant boat-hauling operation with with repair services that include a sail loft, electrical repair, woodworking, and upholstery. Also available on site are fuel, ice, water, showers, and a grocery. Peake Yacht Service, which specializes in fiberglass repair, has a new 150-ton Travelift, a new 120-foot paint shed, and a major chandlery (with a second store in Port of Spain). Water and electricity are free here. Both facilities have dinghy tie-ups. Industrial Marine Services (868-634-4337) has a 70-ton marine hoist, multilingual personnel, a new chandlery, restaurant, and laundry; they specialize in osmosis repair. The on-site repair contractor at IMS is Atlantic Yacht Services (VHF channel 68; tel., 868-634-4337), which handles wood, fiberglass, and metal work. Restaurants, bars, laundries, and communication facilities for yachts are all available in the area. A significant number of other marine-related services exist in Chaguaramas, including marine paint supplies, hardware, propeller repair, sailmakers, and upholsterers.

Chaguaramas has become a popular place to clear customs and immigration. The repair and storage facilities here are impressive and competitively priced.

NELSON, E end, 10 39.4N, 61 36.0W. Fl W 2.5s, 61ft, 5M. White mast.
POINT SINET RANGE (Front Light), 10 40.9N, 61 36.0W. Oc W 2.5s, 98ft, 14M. White square daymark, black stripe. Visible 038°–046°. Reduced visibility (5M) due to dust (occasional). Unreliable. **(Rear Light),** 180 meters, 042°14′ from front. Oc W 5s, 102ft, 14M. White square daymark, black stripe. Visible 038°–046°. Unreliable.

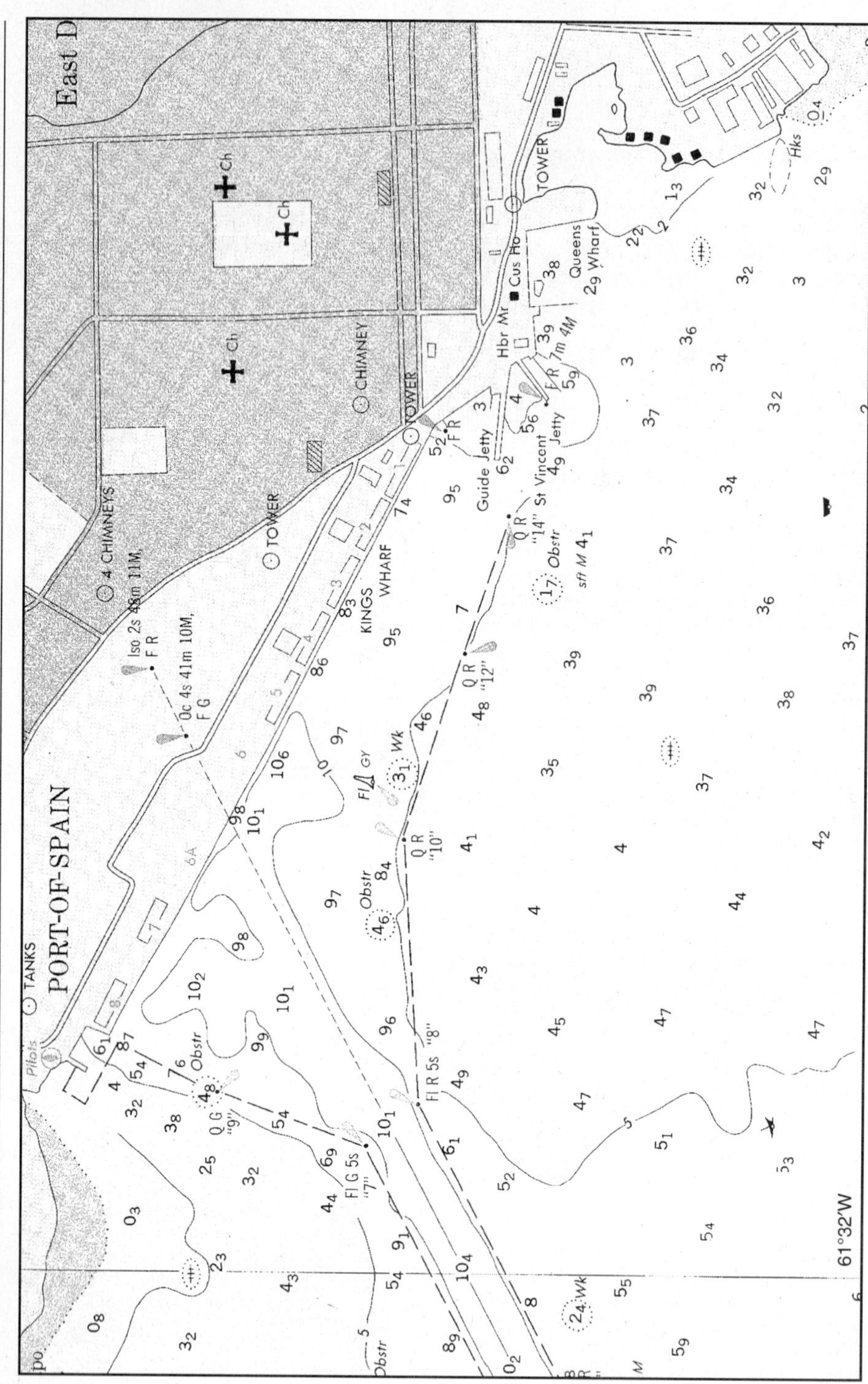

Port of Spain, Trinidad, soundings in meters *(from DMA 24406, 29th ed, 7/91)*

PORT OF SPAIN

GRIER CHANNEL RANGE (Front Light), 10 39.2N, 61 31.4W. Oc W 4s, 135ft, 10M; F G, 6M. White and orange checkered rectangular daymark on white framework tower. **(Rear Light),** 181 meters, 061°32′ from front. Iso W 2s, 157ft, 11M; F R, 6M. Orange rectangular daymark on white framework tower. Visible 058.4°–064.4°.
NO. 7, 10 38.9N, 61 31.9W. Fl G 5s. Concrete platform on pilings.
NO. 8, 10 38.8N, 61 31.9W. Fl R 5s. Concrete platform on pilings.
NO. 9, 10 39.1N, 61 31.9W. Q G. Concrete platform on pilings.
NO. 10, 10 38.9N, 61 31.5W. Q R. Concrete platform on pilings.
NO. 12, 10 38.8N, 61 31.3W. Q R. Concrete platform on pilings.
NO. 14, 10 38.7N, 61 31.1W. Q R. Concrete platform on pilings.
HEAD OF ST. VINCENT JETTY, 10 38.6N, 61 30.9W. F R, 23ft, 4M. Wooden post.
SEA LOTS CHANNEL NO. 1, entrance, N side, 10 37.6N, 61 32.0W. Fl (2) G 7.5s. Pile.
NO. 2, entrance, S side, 10 37.6N, 61 31.9W. V Q (9) W 10s. Pile.
NO. 3, 10 37.7N, 61 31.6W. Q G. Pile.
NO. 4, 10 37.7N, 61 31.6W. Q R. Pile.
NO. 5, 10 37.9N, 61 31.1W. V Q G. Pile.
NO. 6, 10 37.9N, 61 31.1W. V Q R. Pile.
NO. 7, 10 38.1N, 61 30.6W. Q G. Pile.
NO. 8, 10 38.1N, 61 30.6W. Q R. Pile.
NO. 9, 10 38.3N, 61 30.3W. L Fl G 5s. Pile.
NO. 10, 10 38.2N, 61 30.3W. L Fl R 5s. Pile.
NO. 12, 10 38.2N, 61 30.1W. Q R. Pile.
NO. 14, 10 38.1N, 61 30.0W. Q R. Pile.
SEA LOTS CHANNEL RANGE (Front Light), 10 38.4N, 61 29.8W. Oc W 10s, 85ft, 6M; F R. Orange rectangular daymark on white framework tower. **(Rear Light),** 640 meters, 069°10′ from front. Fl W 2s, 128ft, 6M; F R. Orange rectangular daymark on white framework tower.

PORT OF SPAIN, TRINIDAD

10 39N, 61 31W

Pilotage: Port of Spain is the principal port in Trinidad and also one of the most important ports in the West Indies. The Grier Channel range lights are often obscured by smoke, and there are several submerged obstructions outside the marked channels. Tidal currents set southeast at about 0.5 knot on the flood and 1.5 knots on the ebb. The harbormaster, Maritime Services Division of the Ministry of Works and Transport, monitors VHF channels 10 and 16 (call sign, Maritime Services).

Port of entry: The customs dock is located inshore of the dredged area at the southeast end of the Grier Channel. The main customs office in Port of Spain can be reached at 868-625-3311, -3312, -3314, or -3315. Customs at the harbor is through the tide surveyor's office (868-625-2501). Immigration can reached at 868-623-8147.

Dockage: You may be able to tie up right in Port of Spain. For information contact the Maritime Services Division (VHF channels 10 and 16; tel., 868-625-3858). Many vessels prefer to dock at the Trinidad and Tobago Yacht Club (VHF channel 68; tel./fax, 868-637-4260) in Cumana Bay, 4 miles northwest of town. This marina is the closest to Port of Spain and has 60 berths. During Carnival, the docks are likely to be very crowded, if not full. If you are headed for the yacht club, be sure to clear in with officials first.

Moorings: A small number of moorings may be available from the Trinidad and Tobago Yachting Association (VHF channel 68; tel./fax, 868-634-4376, -4210, -4519) in the northwest corner of Carnage Bay.

Anchorage: There are anchorages off both the yachting association and the yacht club. Both clubs have dinghy docks and offer the use of their facilities to visitors who purchase temporary memberships.

Services: Most services are available. Fuel, water, laundry, telephones, fax, restaurant, and bar are available at the Trinidad and Tobago Yacht Club. The yachting association offers haulouts, water, and showers but no fuel. Either facility would be a good place to start if searching for a particular marine service.

Venezuelan Embassy: If you are planning to visit Venezuela, you must have a visa before entering. The Venezuelan Embassy is located at 16 Victoria Avenue, Port of Spain (868-627-9821, -9823).

Port of Spain, with a population of more than 300,000, is a bustling commercial city. Its pre-Lenten Carnival festivities are world famous. Because it is outside the active hurricane zone, Trinidad has been attracting yachts in increasing numbers. Marine services are competitive in every sense. Yachts heading to Venezuela find Trinidad a convenient place to obtain visas.

POINT LISAS

RANGE, NO. 13 (Front Light), 10 22.7N, 61 29.0W. Q W, 30ft, 8M; F R, 30ft, 4M. Visible 083.8°–098.8°. **(No. 14, Rear Light),** 530 meters, 091°20′ from front. Oc W 5s, 52ft, 8M; F R, 56ft, 4M. Visible 087.3°–095.3°.

MARINE TERMINAL RANGE (Front Light), 10 20.9N, 61 27.9W. Q Y, 69ft, 10M. Building. **(Rear Light),** 260 meters, 081°25′ from front. Oc Y 3s, 82ft, 10M. Silo.

CHANNEL ENTRANCE NO. 1, N side, 10 22.7N, 61 31.0W. Fl (2) G 5s, 26ft, 8M. White beacon.

CHANNEL ENTRANCE NO. 2, S side, 10 22.6N, 61 31.0W. Fl W 3s, 26ft, 8M. White beacon.

SAVONNETA RANGE (Front Light), 10 24.2N, 61 29.5W. Fl W 2s, 98ft, 8M. Orange rectangle stripe. Visible 049.3°–055.3°. Identification light Fl R 4s. **(Rear Light),** 700 meters, 052°16′ from front. Fl W 2s, 135ft, 8M. Orange rectangle, white stripe. Synchronized with front. Identification light, Fl R 4s.

POINTE-A-PIERRE

LA CARRIERE, 10 19.3N, 61 27.6W. Fl W 2.5s, 233ft, 23M. White water tower. Emergency light range 18M.

HEAD OF PIPELINE VIADUCT, 10 18.8N, 61 28.8W. Fl (4) W 10s, 98ft, 14M. White metal tower.F R at 300-meter intervals along jetty.

TURNING BASIN, 10 18.8N, 61 28.7W. Fl R. Wooden pile.

OROPUCHE BANK BEACON, 10 16.8N, 61 33.9W. V Q W, 4M. Steel caisson.

BRIGHTON, head of pier, **(Front Light),** 10 14.9N, 61 38.1W. Fl (3) W 10s, 52ft, 15M. White metal tower, 41ft. Visible 134.2°–144.2°. Berthing signals. **(Rear Light),** 852 meters, 139°15′ from front. Fl W 5s, 100ft, 8M. Visible 134.2°–144.2°.

LA BREA, head of Pitch Point pier, 10 15.2N, 61 37.1W. Iso W 2s, 26ft, 10M. White framework tower.

POINT FORTIN

HEAD OF PIPELINE PIER, 10 12.6N, 61 42.2W. Fl (2) W 10s, 98ft, 14M. Aluminum framework tower, 79ft.

NORTH BREAKWATER, 10 11.3N, 61 41.6W. Fl G 3s.

SOUTH BREAKWATER, head, 10 11.2N, 61 41.6W. Fl R 3s.

REED'S NEEDS YOU!

It is difficult to obtain accurate and up-to-date information on many areas in the Caribbean. To provide the mariner with the best possible product we use many sources. *Reed's* welcomes your contributions, suggestions, and updates. Please be as specific as possible, and include your address and phone number. Send your comments to:

Editor, Thomas Reed Publications
13A Lewis Street, Boston, MA 02113, U.S.A.
e-mail: editor@treed.com

Thank you!

VENEZUELA

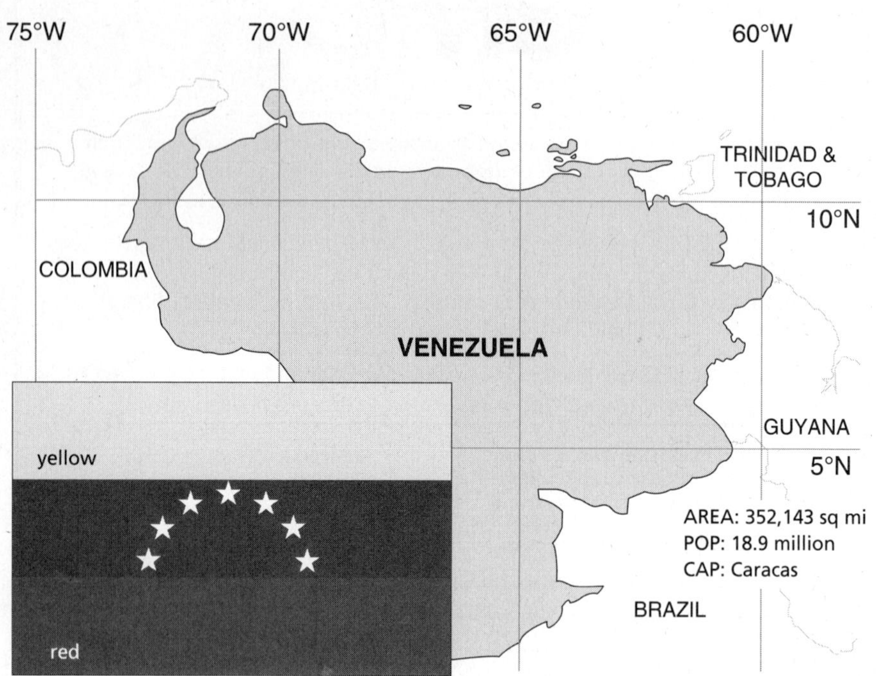

Please see cautions about chartlets and waypoints on page P 2.

ENTRY PROCEDURES

As the crow flies, it is almost 600 nautical miles from the Gulf of Paria to the Colombian border. The entire Caribbean coast of Venezuela measures about 1,700 miles. This area has become a popular cruising ground, especially during the hurricane season.

Before sailing to Venezuela you must obtain a visa. Application should be made at a Venezuelan embassy or consulate, such as the ones in Port of Spain, Trinidad; on Bonaire; or on Grenada. Passports and ship's papers are required; passports should have at least six months remaining before expiration. You will need passport photos and financial references to submit with your application. You also must provide details about your visit and your vessel in advance by fax or telex to the port captain of your intended port of entry and proof of having sent this information with your visa application. There are processing fees. Generally, you are issued a multiple-entry visa for one year, which is good for a 60-day stay initially, with two 60-day extensions available. It is wise to request clearance for a stay longer than you intend, as the best laid plans tend to change. Your stay is timed from the day you enter the country. At present, yachts that have reached the end of their allotted time must leave the country for three months before returning.

For more detailed information concerning visa applications, contact the nearest Venezuelan consulate. In the United States consulates are located at the Embassy of Venezuela, 1099 30th Street NW, Washington, DC 20007 (202-342-2214) and in Boston, Chicago, Houston, Miami, New Orleans, New York, San Francisco, and San Juan, Puerto Rico.

Official ports of entry include Güiria, Carúpano, Pampatar, Cumaná, Puerto La Cruz , La Guaira, Puerto Cabello, La Vela , Las Piedras, and Maracaibo. Maracaibo is primarily a commercial harbor and is not recommended for cruising boats. If you do enter there, you must use a pilot when entering. The Islas los Testigos are not an official port of entry; you should clear in at a mainland port of entry before cruising these islands. You should present your clearances in every harbor that has a port captain or customs office. Business hours are general 0830 to 1600, with a break for lunch.

Upon arrival in port you must proceed to customs *(aduana)*, the national guard, immigration, and then the port captain. Any weapons must be declared to the national guard and will be sealed on board. You must repeat this process in each new port, showing your clearance, or *zarpe*, from the last port. When requesting clearance to your next port, ask for permission to stop at any intermediate harbor along the way. To speed and simplify this process you may want to hire one of the commercial agents found in many ports.

USEFUL INFORMATION

Language: Spanish.
Currency: Bolívar (Bs); exchange rate (summer 1997), 494 bolívars = US$1.
Telephones: Country code, 58; a one- or two-digit city code follows, then a six- or seven-digit local number.
Public holidays (1998): New Year's Day, January 1; Day of Three Kings, January 6*; Carnival, February 19–20; Holy Thursday, April 9; Good Friday, April 10; Proclamation of Independence Day, April 19; Labor Day, May 1; Battle of Carabobo, June 24; Signing of Independence Day, July 5; Bolivar's Birthday, July 24; Columbus Day, October 12; All Saints Day, December 1; Christmas Day, December 25. *Not an official holiday, but observed by many businesses. *Note: Banks observe additional holidays.*
Tourism offices: *In Venezuela:* Corporación de Turismo de Venezuela, Av. Lecuna, Parque Central, Torre Oeste, Piso 37, Caracas (58-2-507-8815, -8816, -8827). *In the U.S.:* Oficina de Promoción del Turismo, Consulado General de Venezuela, 7 East 51st St., New York, NY 10022 (212-826-1678) and at other consulates. *In Canada:* At the embassy in Ottawa and consulates in Toronto and Montréal.

Additional notes: Malaria risk exists in rural areas of Venezuela, cholera and dengue fever occur throughout the country, and there is a risk of contracting yellow fever in the forest around Lake Maracaibo. Mariners should also be aware that some countries require yellow fever vaccinations of travelers coming from Venezuela. For the latest medical advisory, contact the Centers for Disease Control International Travelers Hotline (voice recording, 404-332-4559; fax information service, 404-332-4565; Web, www.cdc.gov). U.S. citizens needing assistance should contact the consular section of the U.S. Embassy, Calle Suspure and Calle F, Colinas de Valle Arriba (Caracas, 58-2-977-2011).

RÍO ORINOCO

Río Orinoco Approach Lighted Buoy "0.1," 8 56.1N, 60 11.0W. Fl W 10s, 4M. Safe water mark, red-white pillar. Radar reflector. **RACON O** (– – –), 14M.

GULF OF PARIA

PUNTA GORDA, 10 09.8N, 62 37.8W. Fl R 5s, 19ft, 3M. Aluminum and white column.

RÍO SAN JUAN

Buoy "E-1," 10 19N, 62 28W. Fl W 5s, 11ft. Black buoy.
LIGHT 1, 10 18.5N, 62 29.2W. Q W, 13ft, 6M.
LIGHT 1-B, 10 18.4N, 62 31.0W. Q W, 14ft, 6M.

~~***Güiria Approach Lighted Buoy,*** 10 34N, 62 15W. Fl W 3s~~. White buoy. Removed 4/97.
GÜIRIA EAST BREAKWATER, S extremity, 10 33.7N, 62 17.3W. Fl R 4.3s, 33ft, 10M. White round concrete tower with red bands.
F R lights on radio tower 0.9 mile N.
SOUTH BREAKWATER, E extremity, 10 33.8N, 62 17.5W. Fl G 9s, 16ft, 4M. Gray round concrete tower, 13ft.

GÜIRIA

10 34N, 62 18W
Pilotage: This small harbor located in the northwestern portion of the Gulf of Paria is about 30 miles west of the tip of the Península de Paria. The port may be identified by the red-roofed towers of a church and the radio towers and white tank on the high cliffs north of town. Several lights mark the harbor entrance. The pilot station for the Río San Juan and Orinoco area is here.

Port of entry: This is the closest port to Trinidad for those headed west. The customs house is inshore of the western docks.
Dockage: You may be able to tie up along the breakwater in the western part of the harbor.
Anchorage: The best shelter is inside the breakwaters, if there is room. Depths are reported to be better than 15 feet.
Services: Supplies, including fuel, are limited.

PUERTO DE HIERRO, 10 38.1N, 62 5.7W. Fl W 6s, 213ft, 10M.White iron framework tower with orange bands. Fl R on dolphin 780 meters, ESE. ~~Reported extinguished (1989)~~.
ISLA DE PATOS, 10 38.3N, 61 52.1W. Fl W 6s, 374ft, 11M. White metal framework tower-with orange bands. Q R marks platform 15 miles S. **RACON P** (• — — •), 15M.
ENSENADA MACURO, E head of pier, 10 39.0N, 61 56.6W. Fl W 2s, 52ft, 9M. Metal tower.
CABO TRES PUNTAS, 10 45.8N, 62 42.9W. Fl W 10s, 49ft, 19M. White fiberglass tower with orange bands.

LOS TESTIGOS

ISLA TESTIGO GRANDE, 11 22.8N, 63 07.2W. Fl W 4.5s, 1460ft, 10M. White iron framework tower with orange bands.

LOS TESTIGOS

11 23N, 63 07W
Pilotage: This group of several islets and above-water rocks is located about 48 miles northeast of Isla de Margarita, on the direct route from Grenada. Isla Testigo Grande, the largest island, rises to a height of more than 800 feet on its northwest part. A light stands near the summit, but it has been reported as not visible from a distance of 5 miles. The island has been reported to be a good radar target at distances up to 22 miles.

The channels between some of the islands are deep and clear. The north islands are mostly unlighted and surrounded by deep water. They should be given a wide berth at night, keeping in mind the constant northwestward-going current. This current varies in intensity and set. In August the rate may be as much as 1.5 knots. In February the set is more north-northwest, at a rate of 2 knots. In July, it may attain speeds to 3 knots.

There are several potential anchorages on the lee side of Isla Testigo Grande during periods of settled weather. Vessels should report to authorities ashore, but it is not a port of entry.

LOS FRAILES

11 12N, 63 44W
Pilotage: This group of rocky islets is located about 7 to 9 miles east-northeast of Cabo de la Isla on Isla de Margarita. The islets have sparse vegetation and are steep-to. The south islet is the largest, rising to a height of 300 feet. The group has been reported to be a good radar target at distances up to 24 miles.

Roca del Norte (11°16′N, 63°45′W) is a small rock about 10 feet high located about 3 miles north of Los Frailes. La Sola is a small rock about 25 feet high located about 10.8 miles east-northeast of Roca del Norte.

ISLA DE MARGARITA

CABO DE LA ISLA (CABO NEGRO), on summit of hill at point, 11 10.5N, 63 53.1W. Fl W 10s, 233ft, 15M. White fiberglass tower with orange bands.
AVIATION LIGHT, 10 55.0N, 63 58.2W. Al Fl W G 10s. On roof of control tower.
PUNTA MOSQUITO, 10 54.0N, 63 53.7W. Fl W 5.5s, 98ft, 15M. White fiberglass tower with orange bands, 33ft. R lights mark mast 7 miles NNW.

PAMPATAR

10 59N, 63 48W
Pilotage: The harbor is located on the eastern end of Isla de Margarita and is convenient for those voyaging from the Lesser Antilles via the Islas los Testigos. The bay is generally clear of dangers. A fort with two towers stands on the shore and is a good landmark. The bay is very exposed to the prevailing easterly winds, but a new breakwater provides some protection. You can land at the public pier.
Port of entry: Customs and immigration officials are located in this port. The procedure is complicated by the lack of banks in Pampatar for paying the requisite fees. Using an agent saves considerable time. Vessels may proceed directly to the next bay and Porlamar and initiate clearance procedures from there.
Anchorage: The anchorage area has some protection from the east but is wide open to the southeast.
Services: Limited supplies are available in Pampatar. You can take a bus to the larger Porlamar for the supermarkets and shopping.

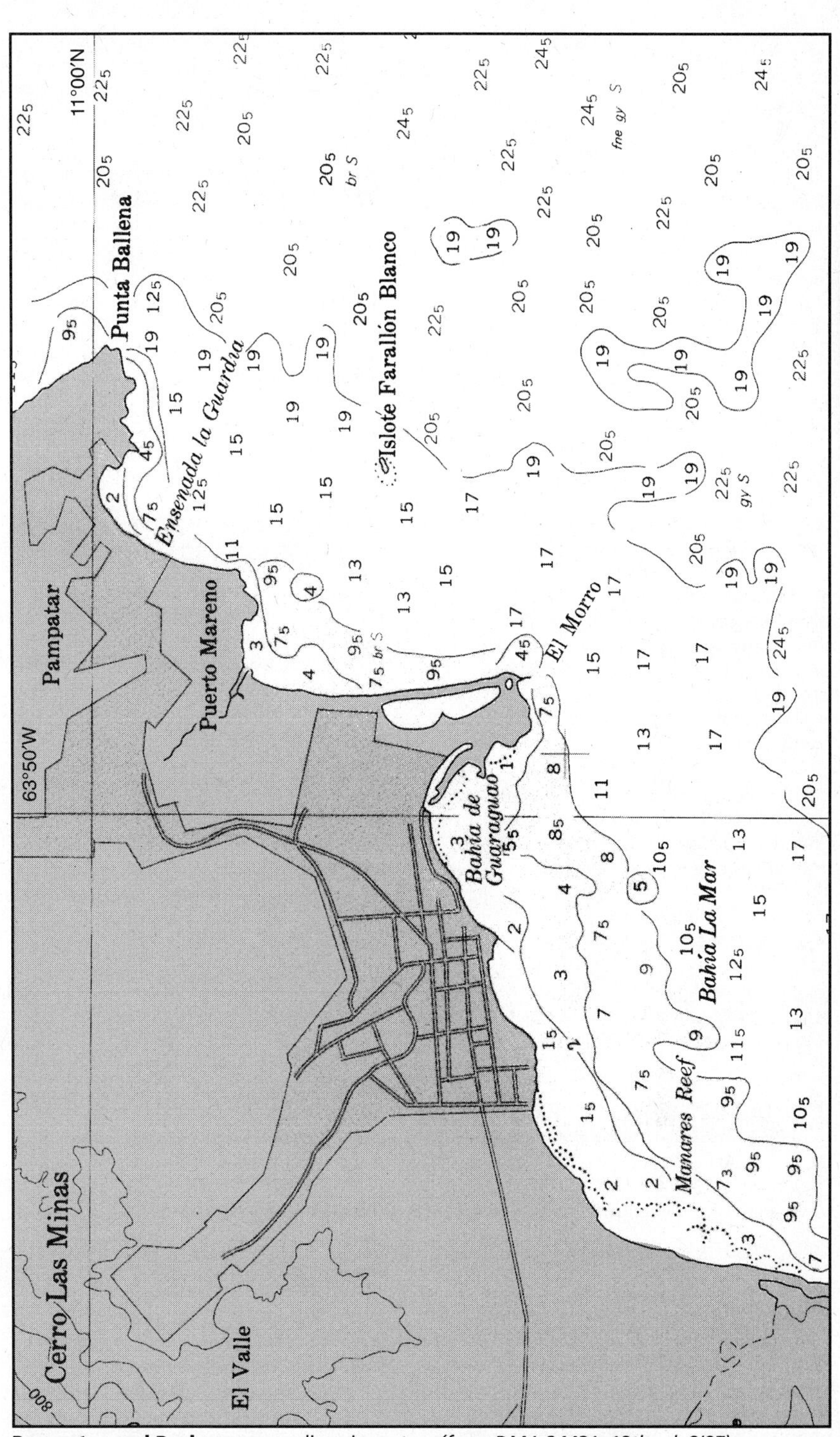

Pampatar and Porlamar, soundings in meters *(from DMA 24431, 13th ed, 8/87)*

PORLAMAR

10 57N, 63 51W
Pilotage: About 4 miles beyond Pampatar, on the north side of Bahía la Mar, is the substantial town of Porlamar, the largest on Isla de Margarita. The entrance is straightforward.
Dockage: The marinas are located at the eastern end of the bay.
Anchorage: Anchor off the marinas, taking care to avoid the shoal between the marinas and the main town.
Services: For a wide range of marine goods, contact Venezuelan Marine Supply (VHF channel 72 Monday through Saturday 0830–1830, call sign, Venezuelan Marine Supply or Venasca; tel., 58-95-64-16-46; fax, 58-95-64-25-29). Fuel is available at the commercial pier off the main town (6 feet minimum depths) and through other marine suppliers. Vessels that have not cleared in previously may do so from Porlamar by traveling on land to Pampatar. However, use of an agent is recommended here also. Venezuelan Marine Supply provides agent services at no additional charge beyond the government-assessed clearance fees. Customs officials prefer doing business with agents, and cruisers report that it can be a logistical nightmare to clear in without an agent's help.

Isla de Margarita is the largest of Venezuela's offshore islands and an important tourist destination, especially for Venezuelans. Some international services are available.

ISLA CUBAGUA

PUNTA CHARAGATO, NE extremity, 10 50.5N, 64 09.4W. Fl W 3s, 39ft, 13M. White fiberglass tower with orange bands, 10ft.
PUNTA BRASIL, NW side , 10 49.8N, 64 12.6W. Fl W 8s, 42ft, 10M. Fiberglass tower, 33ft.

LA BLANQUILLA

ISLA LA BLANQUILLA, S end, 11 49.7N, 64 36.1W. Fl W 5s, 66ft, 12M. White round fiberglass tower with orange bands. Visible 090°–260°.

LA BLANQUILLA

11 52N, 64 36W
Pilotage: This roundish island is about 5 miles in diameter, rising to a height of about 60 feet. It is located about 50 miles north-northwest of the western end of Isla de Margarita. The island is reported to be a good radar target at distances of up to 13 miles. The radar image is better when approaching from the west. There is a light on the south side. There are several possible anchorages on the southeastern side of the island.

LOS HERMANOS

11 47N, 64 25W
Pilotage: This chain of seven barren, rocky islets lie about 45 miles north of the west end of Isla de Margarita. All of the islets are steep-to and have clear deep passages between them. Isla Grueso, the largest, rises to a height of 650 feet and is reported to be a good radar target at distances up to 25 miles. Isla Chiquito at the southern end of the group is the smallest islet.

ISLA LA TORTUGA

PUNTA ORIENTAL, 10 54.0N, 65 12.3W. Fl W 7s, 56ft, 11M. Orange round fiberglass tower with white band. F R on radio towers 3.7 miles W, 8 miles WNW, and 4.1 miles NW.
CAYO HERRADURA, 10 59.8N, 65 23.1W. Fl W 10s, 59ft, 9M. Orange round fiberglass tower with white band.

ISLA LA TORTUGA

10 56N, 65 18W
Pilotage: This barren island is about 12 miles long (east to west) and 6 miles wide. It rises to a height of over 130 feet. The island is located about 46 miles west of Isla de Margarita. The south coast of the island is bold and steep-to, but the west and northwest sides are fringed by a sand and coral shoal that extends up to 2 miles offshore. Las Tortuguillas (two small islets) and Cayo Herradura lie near the outer edge of this shoal area. Several lighted radio towers stand in various locations on the island, and there are lights on Punta Oriental and Cayo Herradura. The island has been reported to be a good radar target at distances up to 24 miles. There are several possible anchorages along the west and north coasts.

ISLA ORCHILA

CERRO WALKER, 11 48.9N, 66 11.0W. Fl W 11.5s, 449ft, 15M. White hexagonal fiberglass tower with orange bands, 20ft.
LA ORCHILA, head of pier, 11 48.6N, 66 11.9W. Fl W 4s, 19ft, 5M. Concrete pyramid tower, orange and white bands, 10ft.

ISLA ORCHILA

11 48N, 66 08W

Pilotage: This small (6 miles long) island is generally low and flat, but has seven distinct hills on its north side. These hills are visible from 15 miles and when first sighted from the north or south appear as separate islands. A 262-foot-high hill stands on the west end of the island, and Cerro Walker, the summit of the island, rises to 457 feet about 1 mile to its east. The island is reported to be a good radar target at distances up to 16 miles.

Cayo Nordeste (11°52'N, 66°06'W) is the largest of a series of small cays lying on a bank extending north-northeast from the northeast end of Isla Orchila. Some ruined buildings can be seen on the north end of Cayo Nordeste.

CAUTION: Due to the presence of a naval base, the area around Isla Orchila is restricted and closed to general navigation.

ISLAS LOS ROQUES

CAYO GRANDE, Sebastopol, 11 46.8N, 66 34.9W. Fl W 6s, 52ft, 12M. White fiberglass tower with orange bands.

CAYO DE AGUA, 11 50.3N, 66 57.2W. Fl W 15s, 56ft, 10M. Cylindrical fiberglass tower with white and orange bands.

EL GRAN ROQUE, 11 57.5N, 66 41.1W. Fl W 10s, 374ft, 15M. White hexagonal fiberglass tower with orange bands, 20ft.

ISLAS LOS ROQUES

11 50N, 66 43W

Pilotage: This group of many cays and an extensive and dangerous coral reef lie 22 miles west of Isla Orchlla and about 70 miles north of La Guaira. The cays are all lower than 25 feet except for El Roque, which rises to 380 feet. El Roque lies on the northern extremity of the reef. The north extremity of the cay is in position 11° 58'N, 66° 41'W. A light is shown from this cay, and a disused lighthouse is nearby. El Roque has been reported as a good radar target at distances up to 23 miles.

A light is also shown from the southeast point of Cayo Grande, at the southeast end of the island group. There is a light at the western end of the group at 11°50'N, 66°57'W. The south side of the group is steep-to. Be careful approaching the eastern side as the strong west-northwest-going current and easterly winds can make it a dangerous lee shore.

Puerto El Roque (11°57'N, 66°39'W) is an anchorage area near the north part of Los Roques. Vessels should obtain local knowledge before entering. Los Roques makes an excellent cruising area in good conditions once you are behind the outer reefs.

ISLAS DE AVES

AVES DE BARLOVENTO, on S island of group, 11 56.6N, 67 26.6W. Fl W 10s, 52ft, 12M. White iron framework tower with orange bands.

AVES DE SOTAVENTO, on N island of group, 12 03.6N, 67 41.0W. Fl W 7s, 52ft, 12M. White iron framework tower with orange bands.

ISLAS DE AVES

12 00N, 67 24W

Pilotage: This group of cays about 28.5 miles west of Islas los Roques consists of two groups of low cays, about 8.5 miles apart lying on dangerous coral reefs. *Note: Another island with this name, also a Venezuelan possession, lies in position 15° 40'N, 63° 37'W (128 miles west of Dominica).*

The eastern group (11°58'N, 67°26'W), called Aves de Barlovento, is an area of small cays and shoals forming a circle about 5 miles in diameter. The east and north sides of the group form an almost continuous drying reef along which heavy breakers occur. The group is reported to be a good radar target up to 15 miles. There are good local-knowledge-only anchorages on the north sides of the cays.

Aves de Sotavento (12° 00'N, 67° 40'W) is the western group and is only about 33 miles east of Bonaire. A large cay on its southern side is mostly covered with mangroves. An unbroken drying reef extends from this cay along the east and north sides of the group. Heavy surf breaks along the entire reef, and there is foul ground up to 0.5 mile from it. The cay is reported to be a good radar target at distances up to 14 miles. There are possible anchorages within the reefs for those with local knowledge.

> ***Note***: Within an aid listing, underlined text indicates data that has changed or is new since September, 1996; ~~Strike out~~ text indicates deleted data.

CARÚPANO TO CUMANÁ

CARÚPANO, Cerro Miranda SW extremity of Hernan Vasquez Bay, 10 40.5N, 63 14.6W. Fl W 9s, 187ft, 24M. White metal framework tower with orange bands, 43ft.
HERNAN VASQUEZ, head of breakwater, 10 40.7N, 63 14.8W. Fl G 4s, 20ft. White concrete tower with green bands, 10ft.

CARÚPANO

10 40N, 63 15W
Pilotage: This harbor lies near the western end of the Península de Paria about 40 miles southeast of Isla de Margarita. Depths of 4 to 36 feet can be found up to a mile offshore in this area. To avoid the shoals and foul areas on either side of the harbor boats should stay in depths of more than 36 feet until directly off the port. Some radio masts 0.8 mile south of the breakwater, a cathedral, and a white church with twin spires are prominent landmarks. The port has a pier protected by a breakwater on its northeast side.
Port of entry: This is the first port of entry west of Trinidad on the north side of the Península de Paria.
Dockage: You may be able to tie up stern-to on one of the wharves.
Anchorage: Anchor off the docks. Reports of swells and rough conditions are frequent.
Services: Most basic supplies are available.

Vessels should stay at least 1.5 miles off the coast from Carúpano to Morro de Chacopata, 33 miles to the west. A strong westward-setting current will be found all along this coast.

MORRO DE CHACOPATA, 10 42.5N, 63 48.7W. Fl W 15s, 66ft. White iron framework tower with orange bands. ~~Reported extinguished (1990)~~.
MORRO DEL ROBLEDAR, 11 02.6N, 64 22.9W. Fl W 6s, 262ft, 13M. White fiberglass tower with orange bands.
CUMANÁ, Puerto Sucre, SE corner of pier, 10 27.6N, 64 11.7W. Fl W 4.4s, 20ft, 6M. White concrete tower with orange bands, 10ft.

CUMANÁ

10 28N, 64 11W
Pilotage: The harbor is at the entrance to the Golfo de Cariaco on Punta Caranero by the mouth of the Rio Manzanares. The airport light reported on the chart to be Al Fl W G is located south and inland of the commercial port, which is known as Puerto Sucre. The port's main pier projects southwest from the coast for 400 yards. The customs office is near the root of the pier. The yacht marina is located on the north side of the point, a bit northeast of the fishing harbor. There are reported to be red and green private marks indicating the dredged channel into the marina.
Port of entry
Anchorage: You can anchor off town, near the commercial pier.
Services: Fuel and water are available in the marina. There are haulout facilities, workboat-oriented repair shops, and several good chandleries.

BAHÍA GUANTA

ISLA REDONDA, 10 15.4N, 64 35.6W. Fl G 6s, 27ft, 5M. Concrete hexagonal pyramid tower with red and white bands, 10ft.
PUNTA QUEQUE, 10 15.1N, 64 35.5W. Fl G 4s, 23ft, 5M. Green fiberglass tower, white band, 10ft.
ISLA DE PLATA, 10 15.0N, 64 34.0W. Fl R 4s, 26ft, 9M. Red hexagonal pyramid, white band, 10ft.
ISLA PICUDA CHICA, 10 18.5N, 64 33.7W. Fl W 7s, 131ft, 11M. Orange hexagonal pyramid, white band, 20ft.
MORRO DE PITAHAYA, 10 15.4N, 64 35.7W. Fl R 3.5s, 65ft, 6M. Concrete hexagonal pyramid tower with red and white bands, 10ft.
ISLA CHIMANA SEGUNDA, W end, 10 17.4N, 64 36.4W. Fl W 10s, 164ft, 11M. White fiberglass tower with orange bands.
MORRO PELOTAS, summit, 10 18.3N, 64 41.1W. Fl (2) W 12s, 250ft, 15M. White framework tower, 19ft. Visible 320°–056°, 101°–180°. Reported irregular (1979).
CAYO BARRACHO RACON, 10 18.2N, 64 44.1W.
RACON B (– · · ·), 30M.

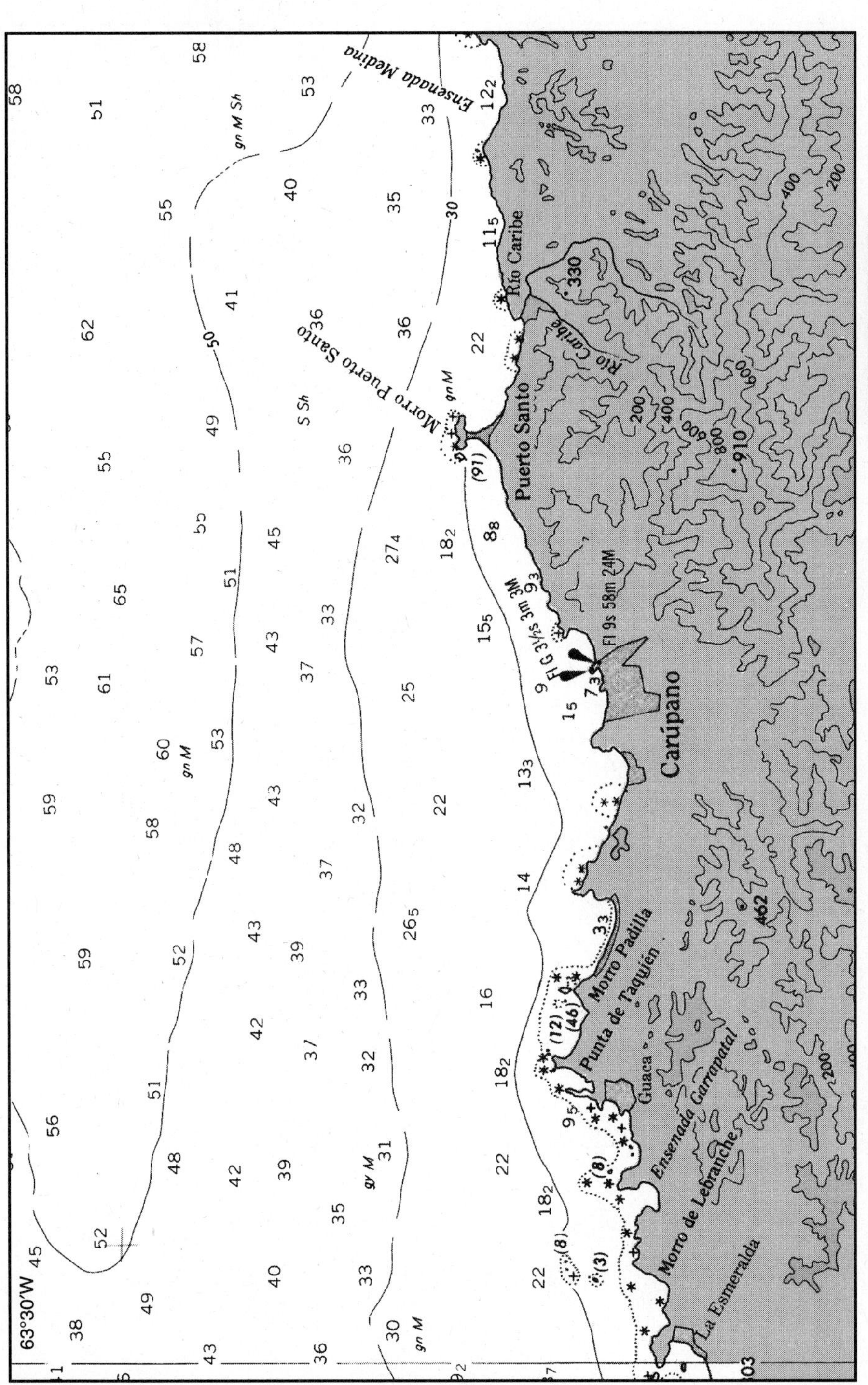

Carúpano, soundings in meters *(from DMA 24430, 5th ed, 8/93)*

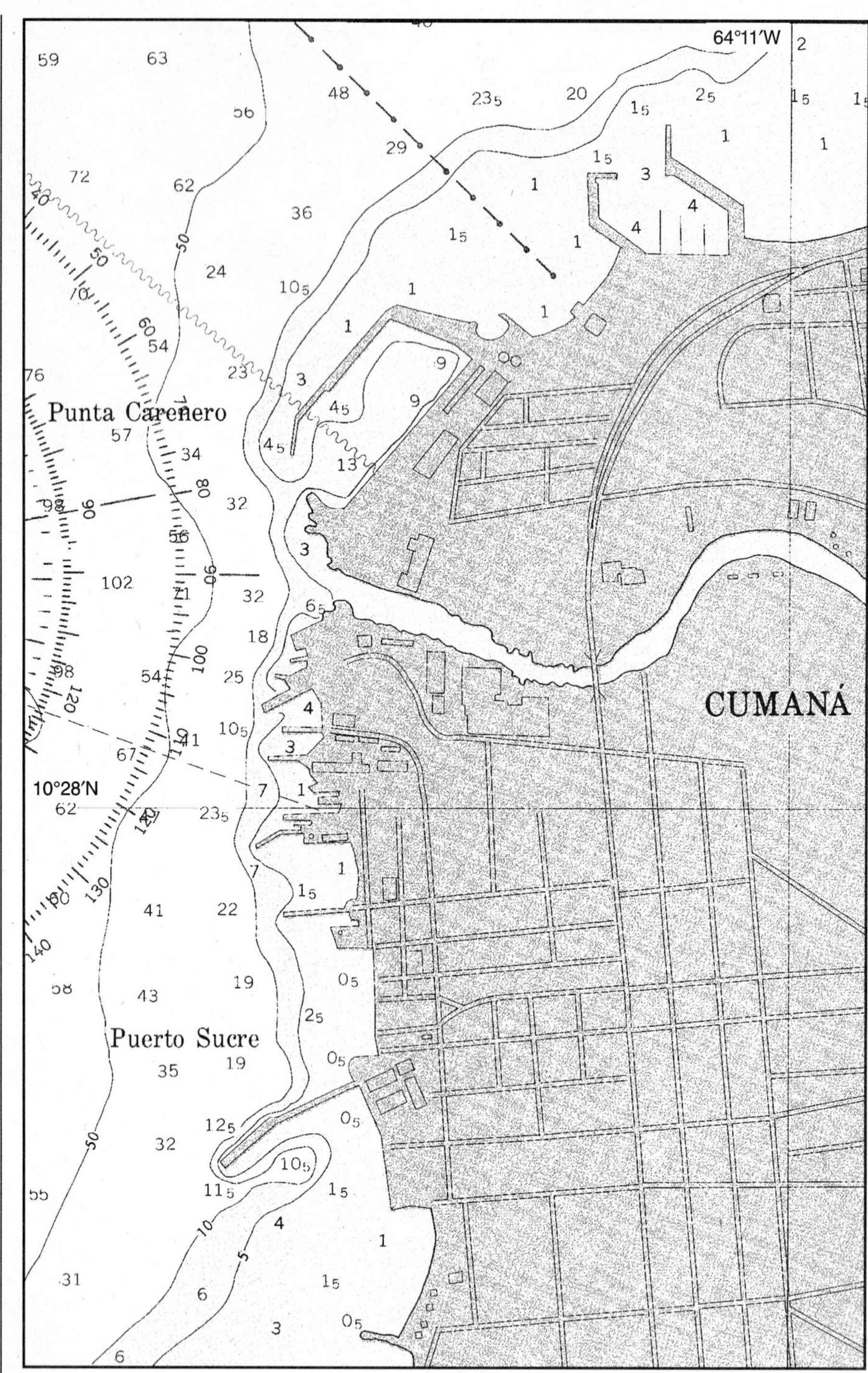

Cumaná, soundings in meters *(from DMA 24431, 13th ed, 8/87)*

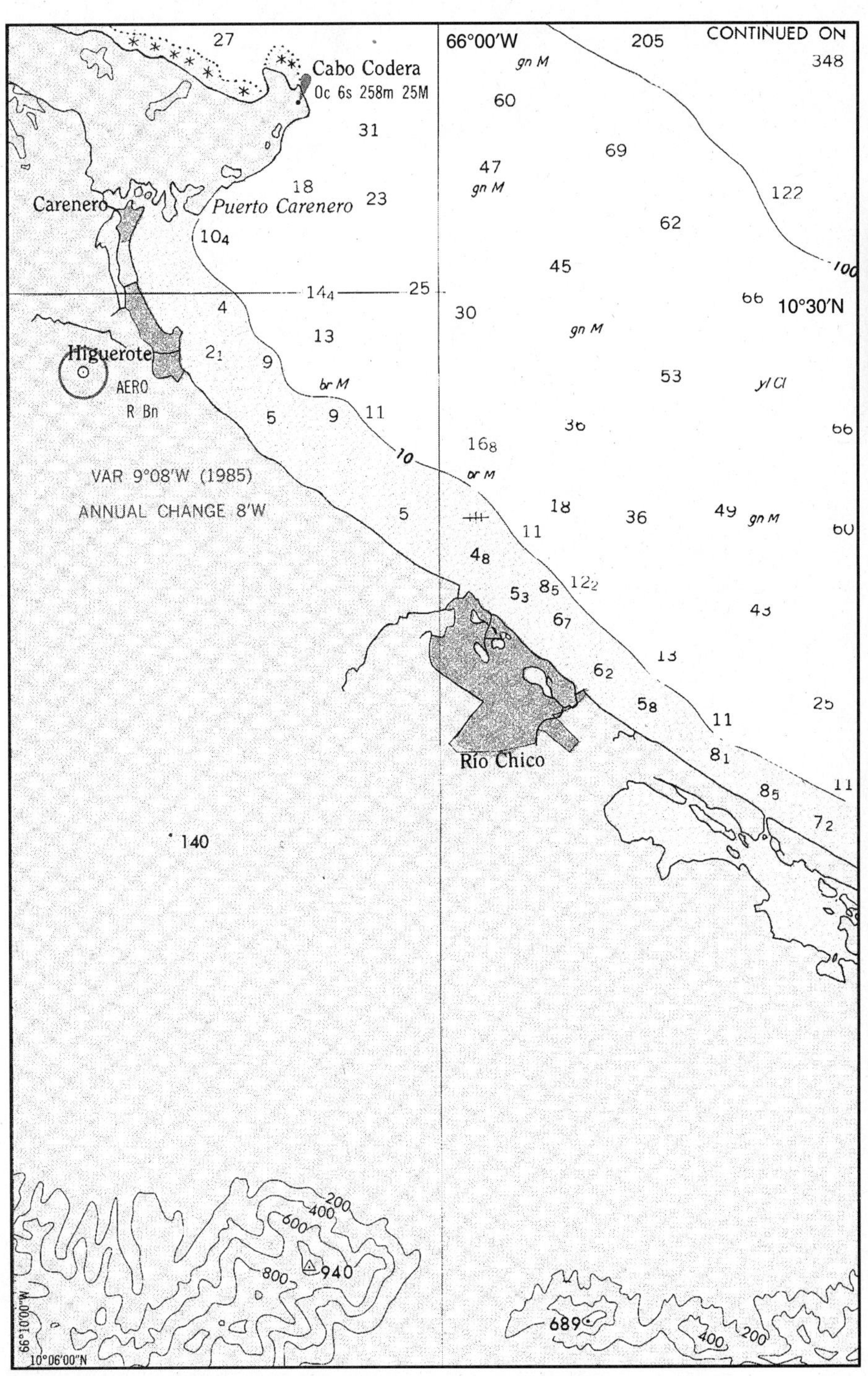

Higuerote and Carenero, soundings in meters *(from DMA 24430, 4th ed, 7/88)*

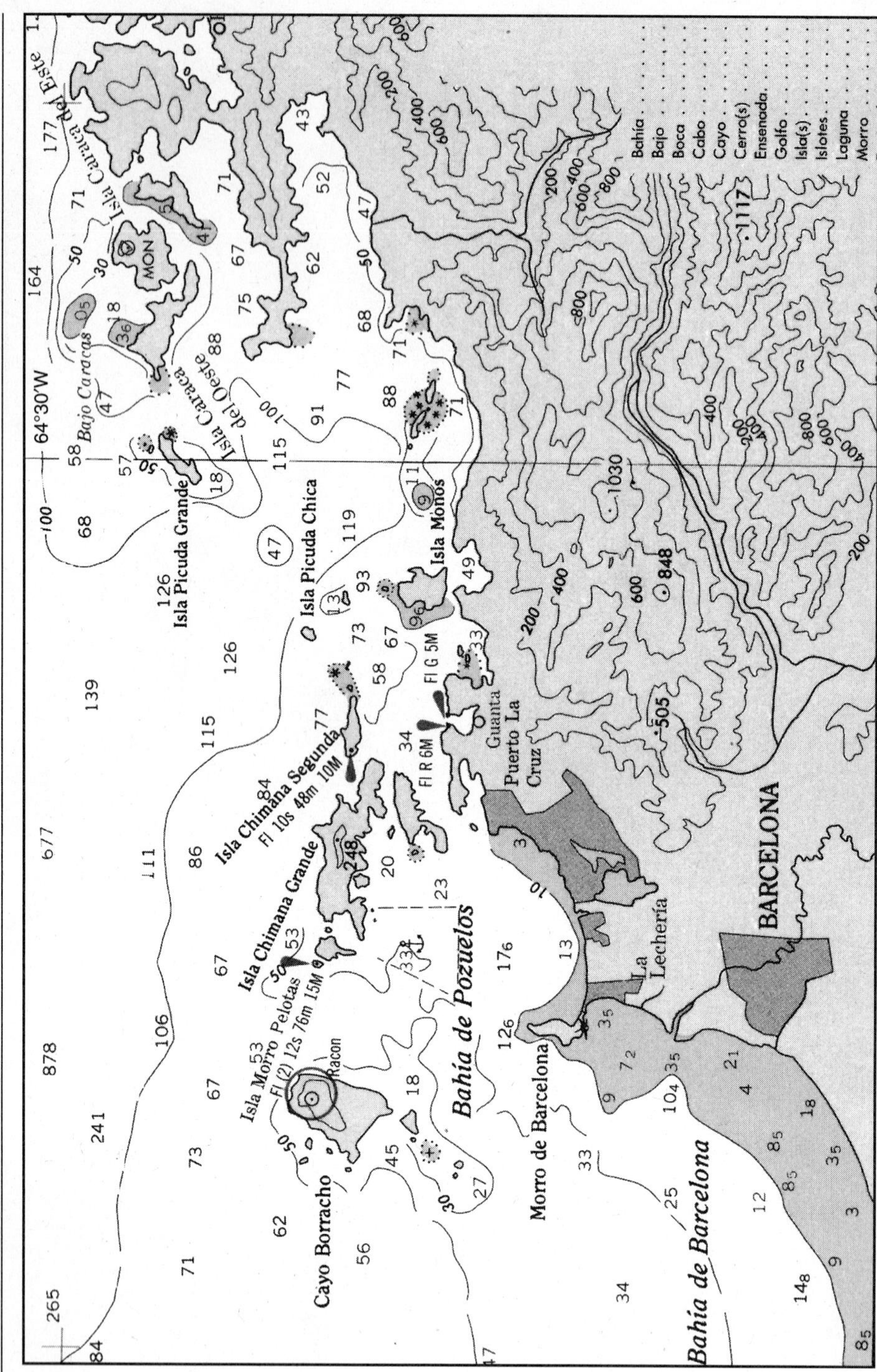

Puerto La Cruz area, soundings in meters *(from DMA 24430, 5th ed, 8/93)*

PUERTO LA CRUZ

10 14N, 64 38W
Pilotage: This is a major commercial and oil port, as well as a growing recreational boating center. The adjacent islands are relatively high and rocky with sparse vegetation. Radio towers south of the city, commercial cranes, and high-rises are prominent. There is a major light on the summit of Morro Pelotas 2 miles east of town.
Port of entry: Procedures are straightforward, if time-consuming. Agents are helpful and may be contacted through the marinas or on the VHF radio (several monitor VHF channel 77). An agency at Bahia Redonda Marina will handle customs and immigration for boaters who stay at the marina.
Dockage: There are several marinas representing a range of services in this area. Marina Puerto La Cruz (Paseo Colón) is in the center of town; several facilities, including Centro Marino de Oriente (CMO) and the new Bahía Redonda Marina with 200 berths, are located in the El Morro development; and Marina El Morro is located on the south side of the headland El Morro. The marinas variously monitor VHF channels 69, 71, 77, and 80.
Anchorage: Many boats anchor off the beach. This is a favorite area for those waiting out the hurricane season.
Services: Fuel is available in the marinas. Most marine services are available, including haul-outs and chandleries. The marinas here have tightened security and coordinated efforts with the Guardia National to reduce crime. There are good supermarkets and general stores of all types.

Puerto La Cruz is a popular stopover with cruisers exploring nearby Mochima National Park or planning visits to interior Venezuela. Marine and international services are available.

BAHÍA DE BARCELONA TO LA GUAIRA

BAHÍA DE BARCELONA

ISLA BURRO LIGHT 3, 10 14.7N, 64 38.1W. Fl G 3s, 13ft, 3M. Green port hand beacon.
LIGHT 4, 10 14.6N, 64 37.9W. Fl G 3s, 13ft, 3M. Green port hand beacon.
LIGHT 5, 10 14.8N, 64 37.8W. Fl R 3s, 13ft, 3M. Red starboard hand beacon. ~~F R on radio mast 1 mile ESE.~~
LIGHT 6, 10 14.9N, 64 37.7W. Fl G 3s, 13ft, 4M. Green port hand beacon.
LIGHT 7, 10 14.8N, 64 37.9W. Fl R 3s, 13ft, 3M. Red starboard hand beacon.
~~**CARGO DOCK,** W end, 10 14.4N, 64 38.1W. Q R, 3M.~~
BARCELONA AVIATION LIGHT, 10 07.2N, 64 41.0W. Al Fl W G 20s. Fl R on radio tower 4 miles N, 5 F R on radio tower 1 mile NW, F R on radio towers 5 miles NE, F W and F R on radio towers 9 miles NE.

ISLAS PIRITU, W end of W island, 10 09.5N, 64 58.2W. Fl W 10s, 66ft, 15M. White round fiberglass tower with orange bands.
ISLA BORRACHITOS DEL ESTE, 10 15.1N, 64 45.7W. Fl W 6s, 82ft, 11M. Orange hexagonal pyramid, white band, 20ft.
HIGUEROTE, E breakwater, head, 10 29.7N, 66 05.7W. Fl R.
PUERTO CARENERO, 10 31.6N, 66 07.0W. Fl W 9s, 138ft, 10M. White iron framework tower with orange bands.
CABO CODERA, 10 34.2N, 66 03.2W. Fl W 6s, 548 ft, 11M. White fiberglass tower, red bands.
ISLA FARALLIN, "El Centinela," 10 48.9N, 66 05.5W. Fl W 13s, 126ft, 15M. White hexagonal fiberglass tower with orange bands.

FARALLON CENTINELA

10 49N, 66 05W
Pilotage: This prominent, light-colored rock, from which a light is shown, rises to an elevation of about 90 feet. It lies about 14 miles north of Cabo Codera on the Venezuelan coast. The northeast side is steep, but the southwest side slopes gradually to the sea. A rock, which breaks, lies 0.3 mile to the northwest. Farallon Centinella has been reported to be a poor radar target at a distance of 7 miles.

PUNTA CAMURI–GRANDE, 10 37.7N, 66 43.0W. Fl W R 3s, 43ft, 7M. Yellow masonry tower, black bands, 33ft. W 029°–063°, R 063°–029°.
NAIGUATA, 10 36.9N, 66 45.0W. Iso W R G 2s, 52ft, 8M. White concrete tower, black bands.
~~**PUERTO DEL GUAICAMACUTO HOTEL,** W head, main breakwater, 10 37.6N, 66 50.7W.~~
CARBALLEDA, W. side of ent , 10 37.6N, 66 50.7W. Fl G 8s, 23ft, 4M. White hexagonal fiberglass tower with green bands.
CARBALLEDA, E. side of ent , 10 37.5N, 66 50.6W. Fl R 8s, 15ft, 4M. White hexagonal fiberglass tower with red bands.
CLUB, 10 37.5N, 66 44.5W. Fl W 11s, 180ft, 25M. White pipe on blue-and-white building.

LA GUAIRA

AVIATION LIGHT, 10 36.5N, 67 00.3W. Al Fl W G 10s, 465ft, 15M. Metal framework tower, 26ft.

LA GUAIRA, 10 34.9N, 66 56.8W. L Fl W 15.5s, 1,394ft, 25M. Orange tank, black letter **I** in center. Visible 291°–028°.

NORTH BREAKWATER, 10 36.5N, 66 57.1W. Fl G 3.7s, 46ft, 8M. White fiberglass tower with green bands, 33ft. ~~Extinguished (1984)~~.

SOUTH BREAKWATER, 10 36.1N, 66 57.2W. Fl R 3s, 43 ft, 8M. White fiberglass tower with red bands, 33ft.

BREAKWATER, head, 10 36.1N, 66 56.7W. Fl R 2.5s, 43ft, 8M. White fiberglass tower with red bands, 33ft.

LA GUAIRA

10 37N, 66 56W

Pilotage: This is an artificial harbor created by breakwaters to serve the city of Caracas. The north breakwater is 1,500 meters long. From October to March especially, the harbor experiences heavy ground swell with considerable scend. This harbor is very commercial and is generally avoided by pleasure boats except those wishing to clear in to the country. The nearby marinas offer better shelter to yachts.

Port of entry

Dockage: There are marinas at Caraballeda, 6 miles to the east, and in Catia La Mar, 4 miles to the west. There is another marina at Punta Naiguata, 6 miles east of Caraballeda. Most yachts will want to berth at one of these locations if visiting Caracas or La Guaira. You may be able to tie up at the main docks in La Guaira while clearing with officials.

Anchorage: This would be an emergency anchorage only. Better anchorages are off the marinas mentioned above.

Services: There are no yacht-oriented facilities in La Guaira, but most services are available at the marinas menitoned above.

LA GUAIRA TO LA VELA

PLAYA GRANDE YACHT BASIN WESTERN LIGHT, 10 37.2N, 67 00.9W. Fl G 6s, 30ft, 11M. Truncated tower, red bands.

PLAYA GRANDE YACHT BASIN EASTERN LIGHT, Q R, 16ft, 4M. Metal tower.

CATIA LA MAR, outer end of pier, 10 36.4N, 67 02.4W. Fl R 2s, 20ft, 9M. Triangular daymark, point up.

TACOA, 10 35.5N, 67 04.9W. Fl R.

MORRO CHORONI, 10 30.7N, 67 36.1W. Fl W 5s, 246ft, 15M. White iron framework tower with orange bands.

ISLA TURIAMO, 10 29.0N, 67 50.4W. Fl W 9s, 13ft, 9M. White framework tower, orange bands, 10ft.

BAHÍA DE TURIAMO, pier head, 10 27.5N, 67 50.7W. Fl W 3.3s, 23ft, 6M. White concrete tower, orange bands, 10ft.

BAHÍA DE TURIAMO, S end, 10 27.2N, 67 50.6W. Fl Y 1.5s, 13ft, 4M. Yellow framework tower.

NOTE: Due to the presence of a military naval base, the area around Isla Turiamo is restricted and closed to general navigation.

PUERTO CABELLO

ISLA ALCATRAZ, 10 30.5N, 67 58.5W. Fl W 6s, 62ft, 11M. White iron framework tower with orange bands. **RACON M (– –).**

PUNTA BRAVA, 10 29.4N, 68 00.6W. Fl W 6s, 121ft, 15M. White concrete tower, orange bands, 114ft. Visible 105°–255°. F R lights on radio mast 0.6 mile S.

FORT LIBERTADOR, on pier, 10 29.0N, 68 00.7W. Fl G 7s, 13ft, 5M. White concrete tower, green bands, 10ft.

PIER P.I.,10 29.0N, 68 00.5W. Oc R. Red mast.

NAVAL QUAY, SW end, 10 29.1N, 68 00.2W. Fl G, 20ft, 4M. Black framework tower, 10ft.

FORTIN SOLANO, 10 27.8N, 68 01.2W. Fl (3) W 15.5s, 551ft, 30M. Metal framework tower, 33ft.

PIERHEAD, 10 29.4N, 68 02.1W. Fl R.

ISLA GOAIGOAZA, 10 29.6N, 68 02.5W. Fl W 7.5s, 33ft, 8M. White pyramidal iron tower with orange bands.

PUERTO CABELLO AVIATION LIGHT, 10 28.6N, 68 04.5W. Al Fl W G 4s, 73ft, 14M. On airport control tower.

SILOS CARIBE, pier, 10 29.4N, 68 05.7W. Fl W 7s, 33ft, 4M. Black framework tower, 13ft.

PUNTA CHAVEZ, NW head of pier, 10 29.6N, 68 07.4W. Oc R 3s, 80ft, 10M. Aluminum tower, 79ft.

PUNTA CHAVEZ, SE head of pier. Fl G 3s, 80ft, 10M. Aluminum tower, 79ft.

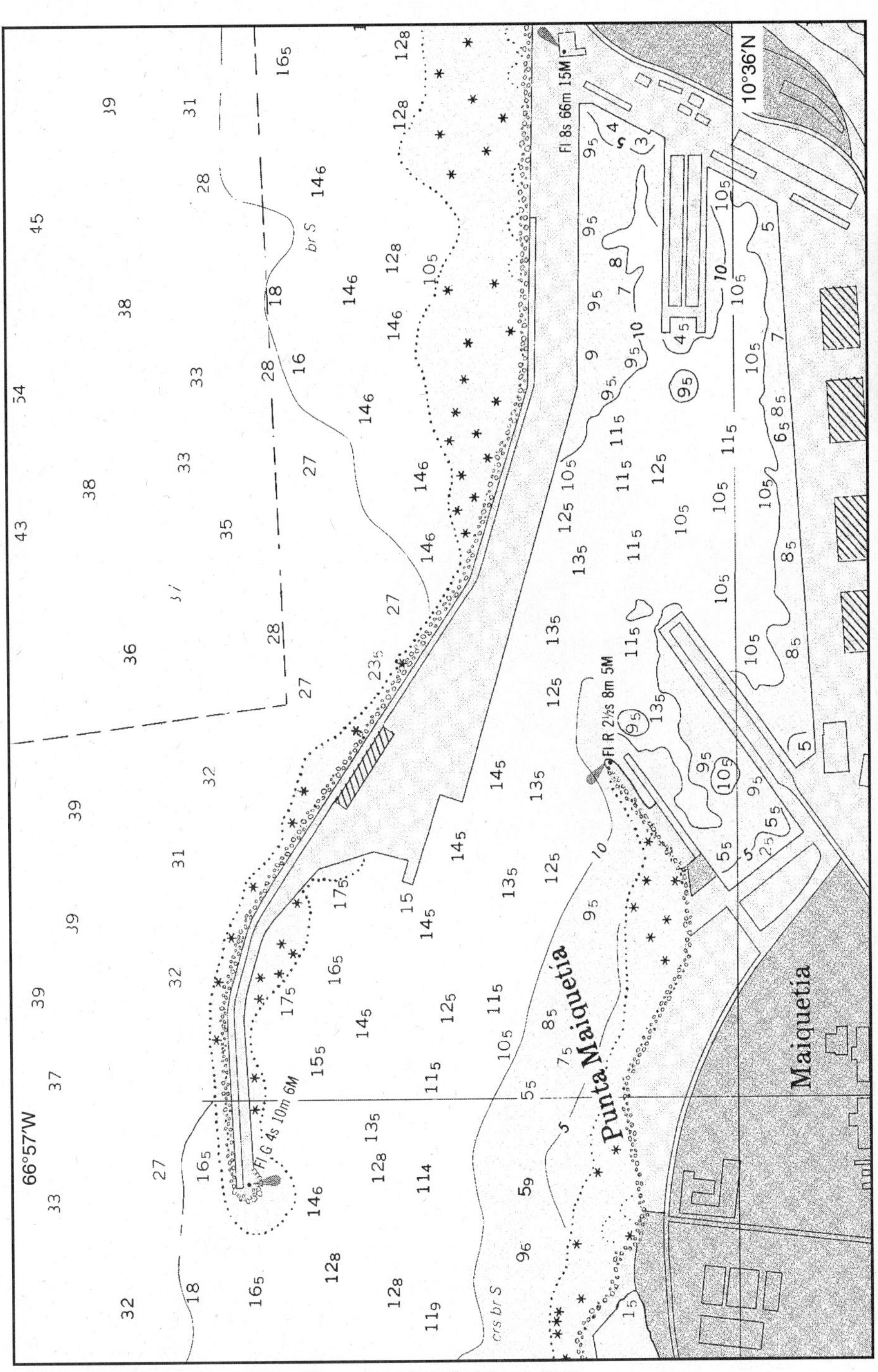

La Guaira, soundings in meters *(from DMA 24452, 6th ed, 8/86)*

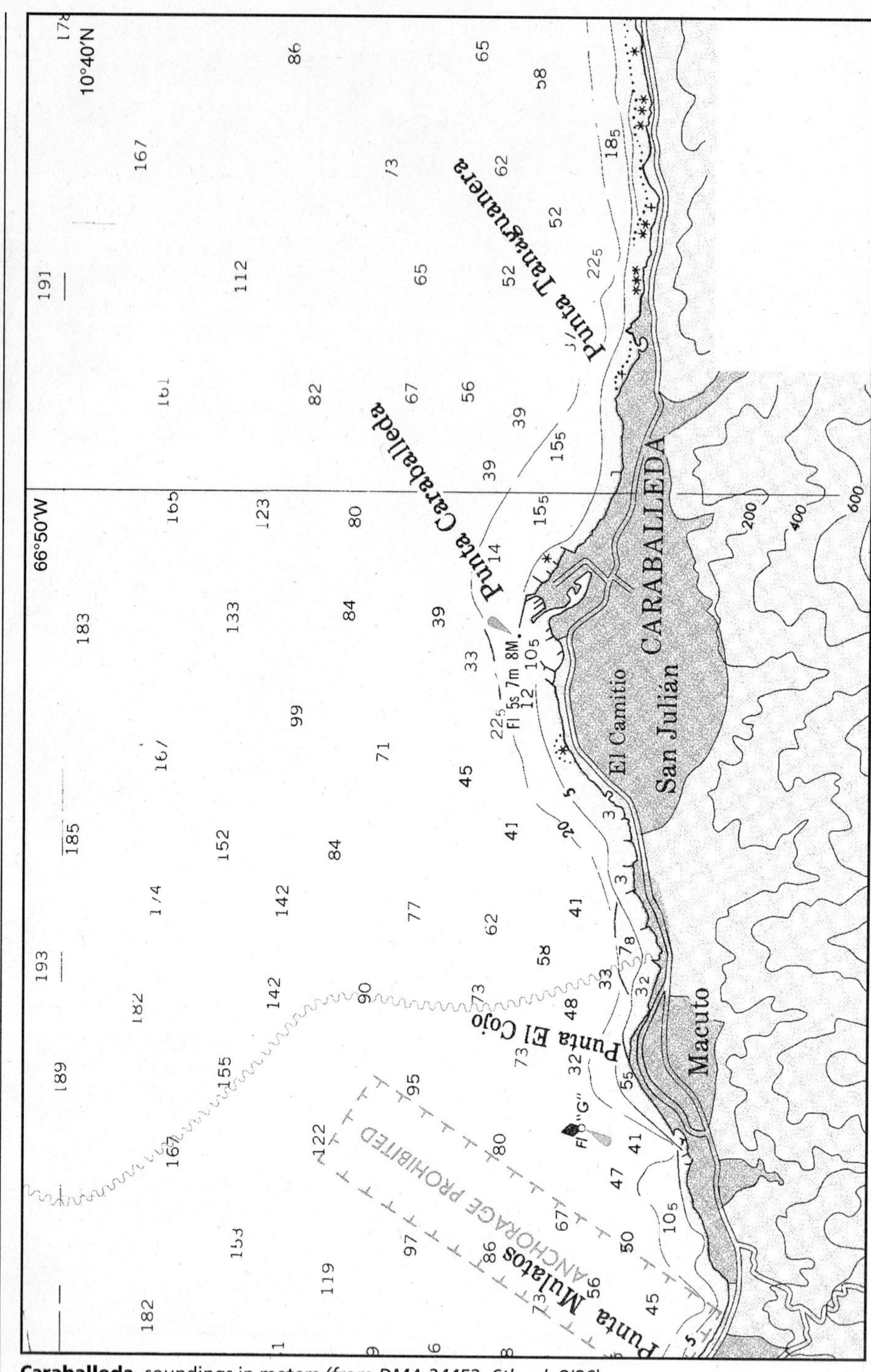

Caraballeda, soundings in meters *(from DMA 24452, 6th ed, 8/86)*

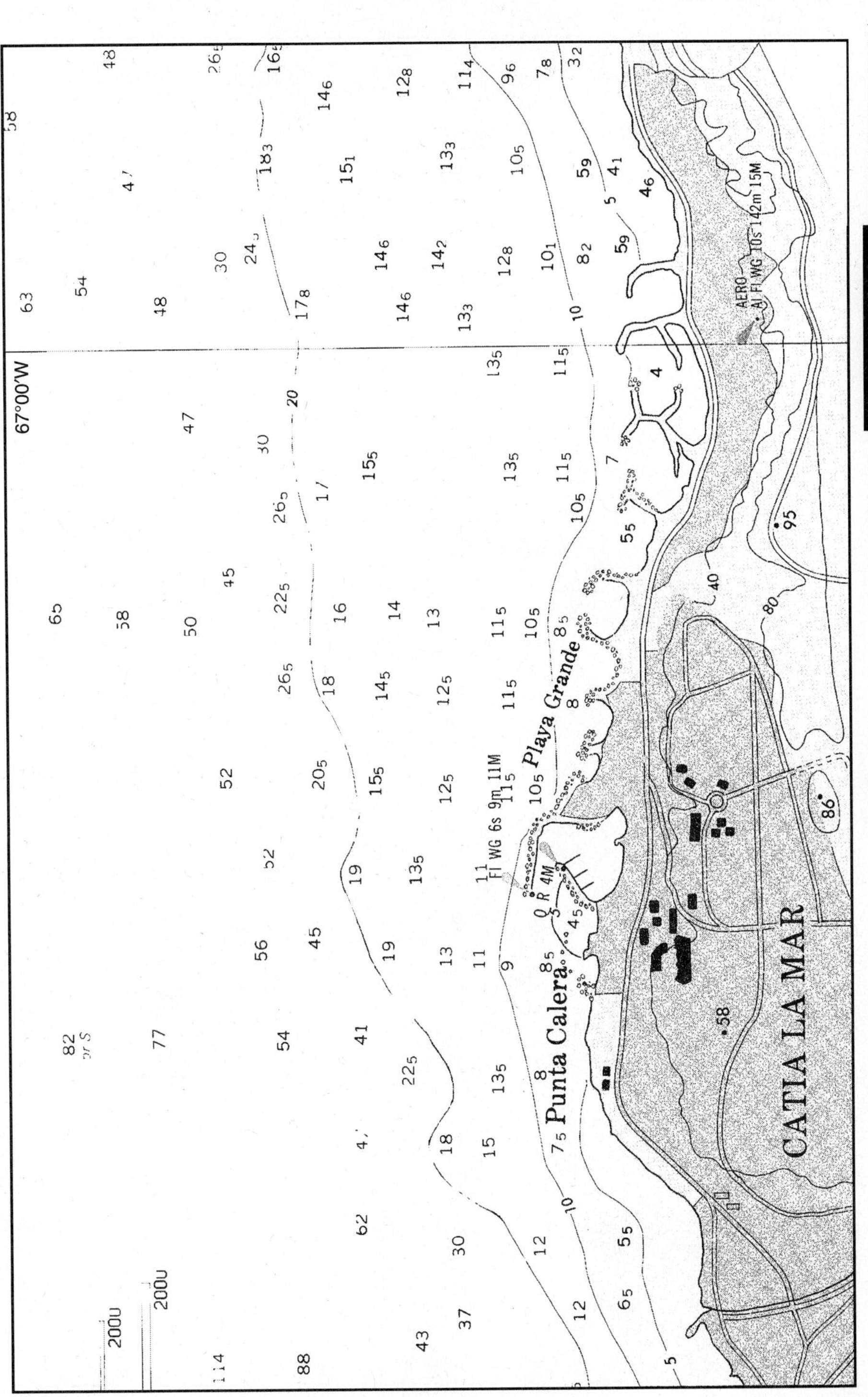

Catia La Mar, soundings in meters *(from DMA 24452, 6th ed, 8/86)*

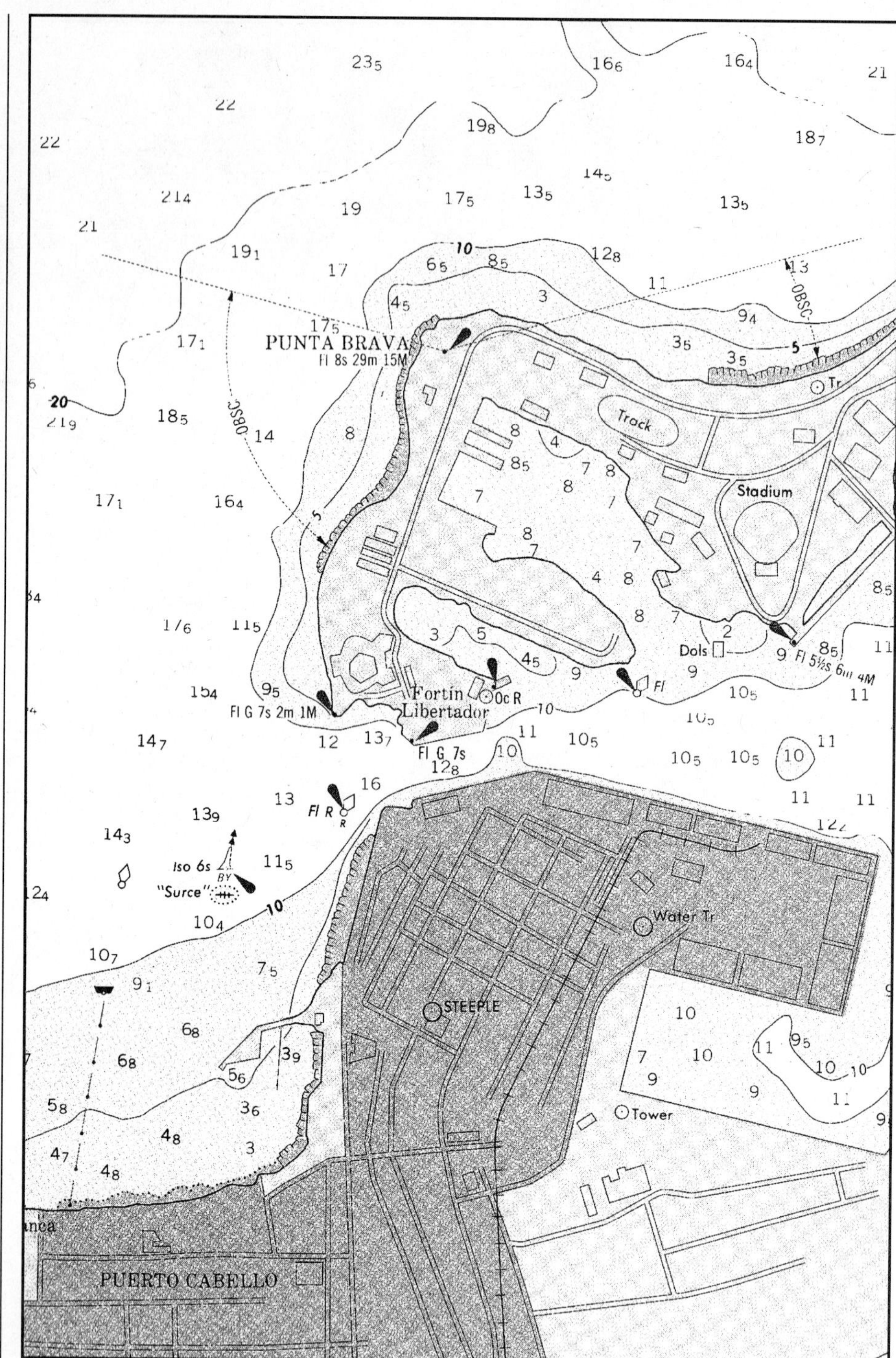

Puerto Cabello, soundings in meters *(from DMA 24454, 4th ed, 9/86)*

PUERTO CABELLO

10 29N, 68 00W

Pilotage: Puerto Cabello is the third ranking port in Venezuela. Bajo Larne, a 19- to 20-foot-deep shoal located about 0.3 mile northeast of Isla Goaigoaza, breaks when any sea is raised. There is a prominent power plant about 2 miles west of town. The marina is located in the outer harbor, south of the main ship channel. The northern part of the inner harbor is a naval area, and general traffic is prohibited. The port authority and customs house is located on the wharves to the south, near the entrance.

Port of entry

Dockage: Tie up in the Marina Puerto Cabello, located west of the charted steeple.

Anchorage: Anchor near the marina.

Services: Most services are available, and there are some marine supplies. In town are good supermarkets and general shops.

MORON OIL TERMINAL, 10 31.1N, 68 11.3W. Fl W 3s, 10M. Red post, white bands.

CAYO BORRACHO, 10 58.9N, 68 14.8W. Fl W 8s, 52ft, 12M. White pyramidal iron tower with orange bands.

CAYO NOROESTE, N side, 11 13.2N, 68 26.9W. Fl W 6s, 49ft, 12M. White pyramidal iron tower with orange bands.

PUNTA AGUIDE, 11 20.7N, 68 40.6W. L Fl W R 15s, 164ft, W 12M, R 7M. White metal framework tower with orange bands, 39ft. W shore–198°, R 198°–218°, W 218°–shore.

PUNTA ZAMURO, 11 26.6N, 68 50.0W. Fl W 10s, 49ft, 14M. Cylindrical fiberglass tower, white and orange bands.

PUNTA TOMORO ~~**MANZANILLO**~~**,** 11 32.0N, 69 16.2W. Fl W 6s, 148ft, 12M. White pyramidal iron tower with orange bands.

PUEBLO CUMAREBO, 11 29.0N, 69 21.2W. F R, 57ft, 9M. White masonry tower, 36ft.

PUNTA TAIMA TAIMA, 11 30.2N, 69 30.0W. Fl W 7s, 191ft, 12M. White pyramidal iron tower with orange bands.

LA VELA, 11 27.7N, 69 34.1W. Fl W, 65ft, 6M.

LA VELA

11 28N, 69 35W

Pilotage: Located in the south-east portion of Bahía Vela de Coro, La Vela, the port for Coro, is the last harbor before rounding the Península de Paraguaná. A pier is located about 1.8 miles northeast of the harbor. The customs house is a conspicuous white building with a red roof and the word "Aduana" painted on top.

CAUTION: An abnormal magnetic disturbance has been observed between Puerto Cumarebo and La Vela.

Port of entry

Anchorage: Punta Taimataima to the east is reported to provide little shelter. La Vela can be rough to untenable in strong winds. This is the last shelter for many miles if headed west along the Venezuela coast.

Services: General supplies are available here and in the nearby town of Coro.

PENÍNSULA DE PARAGUANÁ

PUNTA ADÍCORA, 11 56.7N, 69 48.0W. Fl W 16s, 56ft, 14M. White round fiberglass tower with orange bands.

CABO SAN ROMAN, N side of peninsula, 12 11.7N, 70 0.3W. Fl W 6s, 112ft, 19M. White octagonal framework tower with orange band, 79ft.

MACOLLA, NW side of peninsula, 12 05.7N, 70 12.5W. Fl W 10s, 157ft, 18M. Octagonal iron tower with concrete base, 135ft.

BAHÍA DE AMUAY (Front Light), 11 45.1N, 70 12.4W. F G, 122ft, 5M. Orange diamond daymark on aluminum tank. Oc R 2s and 2 F R on each of two towers, 0.93 and 1.2 miles ESE. **(Rear Light),** 365 meters, 074°45′ from front. F G, 156ft, 5M. Orange diamond daymark on aluminum tank.

LAS PIEDRAS, head of naval pier, 11 42.2N, 70 13.2W. Fl G 3s, 24ft, 8M. Concrete pyramid tower, white and green bands, 10ft.

LAS PIEDRAS

11 43N, 70 13W

Pilotage: This village is located on the west side of the Península de Paraguaná. There is a dangerous wreck located about 1 mile west-southwest of Punta Piedras. Conspicuous water tanks stand on the eastern shore of Bahía Boca de las Piedras. There are several commercial piers projecting from the town.

Port of entry: Customs is about 2 miles south of Boca de las Piedras in Caleta Guaranao.

Anchorage: Strong northeast trade winds have been reported here, but you should have good shelter with the land to the east.

PUNTA GORDA, 11 38.1N, 70 13.9W. Fl W R 10s, 204ft, 21M. Radio tower. R 340°–009°, W 009°–202°. Reported extinguished (1992).

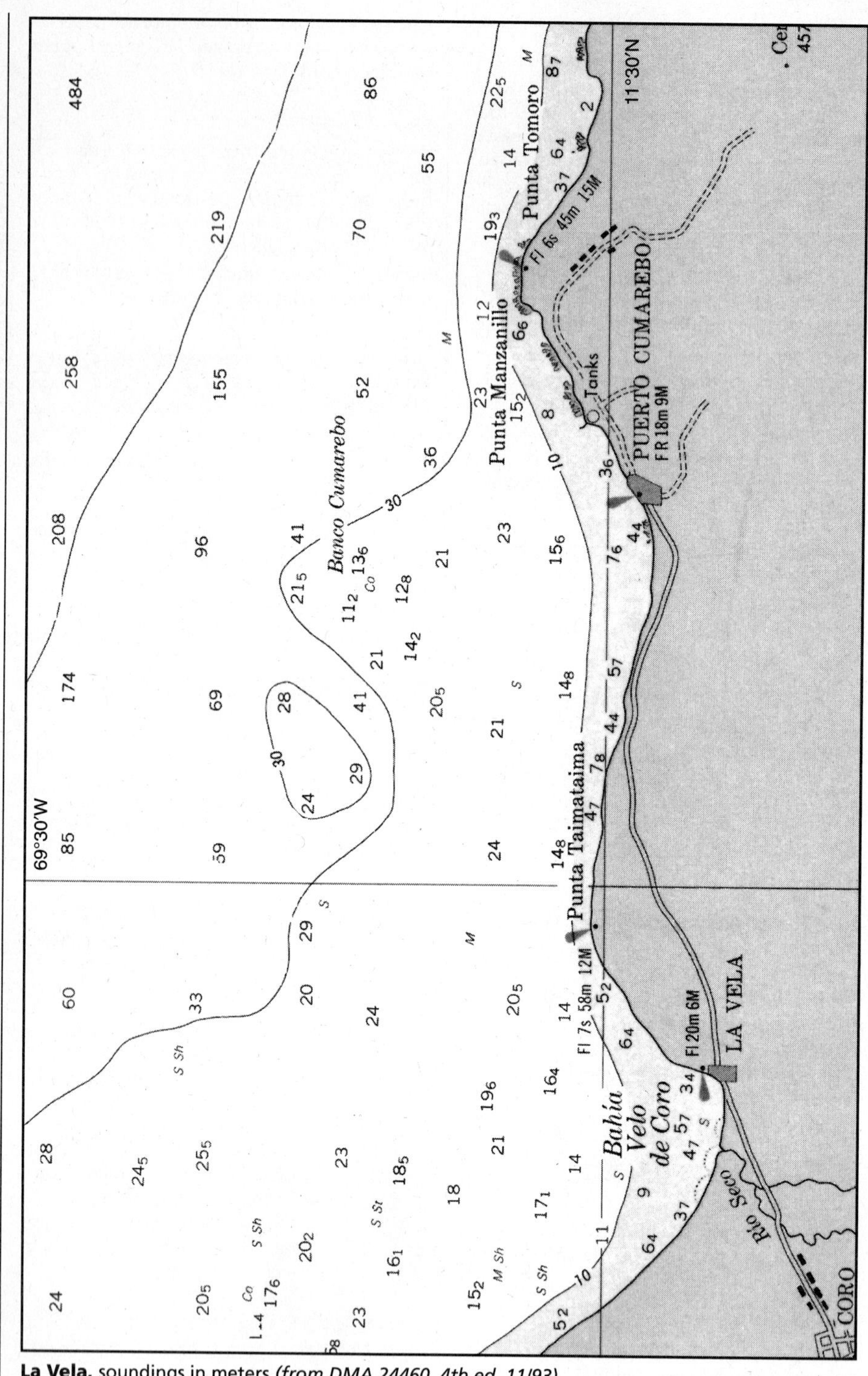

La Vela, soundings in meters *(from DMA 24460, 4th ed, 11/93)*

PUERTO CARDÓN

PIER NO. 3, 11 37.1N, 70 14.0W. F W, 7ft, 5M. Concrete structure. Chimneys marked by red lights 0.6 mile and 1.5 miles NE.

PIER NO. 1 (Front Light), 11 37.3N, 70 14.0W. F G, 13ft, 5M. Concrete structure. **(Rear Light),** 160 meters, 061° from front. F G, 43ft, 5M. Metal structure.

PIER NO. 2 (Front Light), 11 37.5N, 70 14.2W. F R, 13ft, 5M. Concrete structure. **(Rear Light),** 160 meters, 061° from front. F R, 43ft, 5M. Metal structure.

PIER NO. 4, 11 37.7N, 70 14.3W. F Blue, 13ft, 5M. Concrete structure.

GUARANAO

11 40N, 70 13W

GUARANAO PIER, S side, 11 40.3N, 70 13.0W. F R, 16ft, 3M.

GUARANAO PIER, N side, 11 40.3N, 70 13.0W. F G, 16ft, 3M.

GUARANAO, summit., 11 40.0N, 70 12.8W. Fl (2)W 12s, 117 ft, 15M White fiberglass tower with orange bands, 33ft.

MUELLE FLOTANTE, 11 40.3N, 70 12.9W. Fl R 4s, 26ft, 3M. Red fiberglass tower with white band, 10ft.

FIJO, 11 40.4N, 70 12.9W. Fl G 4s, 20ft, 3M Green fiberglass tower with white band, 10ft.

LAGO DE MARACAIBO ENTRANCE

MALECÓN DEL ESTE, E breakwater head, 11 01.4N, 71 34.9W. Fl Y 4s, 51ft, 10M. Fiberglass tower, white and orange bands, 33ft.

INNER CHANNEL

E31, 10 58.5N, 71 35.9W. Fl G 1.5s, 39ft.

E32, 10 58.5N, 71 36.1W. Fl R 1.5s, 39ft, 10M. Metal column on concrete piles.

E33, 10 57.5N, 71 36.1W. Fl G 2.5s, 118ft, 12M. Disused oil well derrick, concrete piles.

E34, 10 57.6N, 71 36.2W. Fl R 2.5s, 128ft, 12M. Truncated metal framework pyramid on concrete piles, 128ft. Tide gauge.

T35, 10 56.8N, 71 36.2W. Fl G 4s, 20ft, 5M. Green metal hut on piles, 20ft.

T36, 10 56.8N, 71 36.3W. Fl R 5s, 20ft, 3M. Aluminum and red hut on platform, on pile.

T37, 10 55.8N, 71 36.3W. Fl G 4s, 20ft, 3M. Aluminum and red hut on platform, on pile.

T38, 10 55.8N, 71 36.5W. Fl R 4s, 20ft, 5M. Red metal hut on piles, 20ft.

T39, 10 54.9N, 71 36.5W. Fl G 4s, 20ft, 4M. aluminum and green hut on platform, on pile.

T40, 10 54.9N, 71 36.6W. Fl R 4s, 20ft, 5M. Red metal hut on piles, 20ft.

T41, 10 53.9N, 71 36.6W. Fl G 4s, 20ft, 5M. Green metal hut on piles, 20ft.

T42, 10 53.9N, 71 36.8W. Fl R 4s, 20ft, 4M. Aluminum and red hut on platform, on pile.

T43, 10 52.9N, 71 36.8W. Fl G 4s, 20ft, 4M. Aluminum and green hut on platform, on pile. Tide gauge.

T44, 10 53.0N, 71 36.9W. Fl R 4s, 20ft, 5M. Red metal hut on piles, 20ft.

T45, 10 52.0N, 71 36.9W. Fl G 4s, 20ft, 5M. Green metal hut on piles, 20ft.

T46, 10 52.0N, 71 37.0W. Fl R 4s, 20ft, 4M. Aluminum and red hut on platform, on pile.

T47, 10 51.0N, 71 37.0W. Fl G 4s, 20ft, 5M. Green metal hut on piles, 20ft.

T48, 10 51.0N, 71 37.2W. Fl R 4s, 20ft, 5M. Red metal hut on piles, 20ft.

T49, 10 50.1N, 71 37.2W. Fl G 4s, 20ft, 5M. Green metal hut on piles, 20ft.

T50, 10 50.1N, 71 37.3W. Fl R 4s, 20ft, 4M. Aluminum and red hut on platform, on pile.

INNER CHANNEL RANGE Q (Front Light), 10 47.8N, 71 37.6W. Fl G 1.5s, 89ft, 7M. Metal framework tower on concrete piles. **(Rear Light),** 1,646 meters, 188.8° from front. Tide gauge.

RANGE M (Front Light), 10 46.6N, 71 37.0W. Fl W 1.5s, 49ft, 5M. Metal framework on concrete piles, 49ft. **(Rear Light),** 1,500 meters, 294°27′ from front.

COMMON REAR RANGE LIGHT P, Oc G 4.5s, 128ft, 8M. Metal framework tower on concrete piles.

MARACAIBO CHANNEL RANGE R (Front Light), 10 50.3N, 71 38.1W. Fl G 1.5s, 89ft, 7M. Metal framework tower on concrete piles. **(Rear Light S),** 1,829 meters, 330°49′ from front. Fl G 4s, 128ft, 8M. Truncated metal framework pyramid on concrete piles, 128ft.

T51, 10 49.4N, 71 37.3W. Oc G 2.5s, 20ft, 5M. Aluminum and green hut on platform, on pile.

T51A, 10 48.9N, 71 37.2W. Oc G 2.5s, 20ft, 4M. Green metal hut on piles, 20ft.

T52, 10 49.1N, 71 37.5W. Oc R 2.5s, 20ft, 5M. Red metal hut on piles, 20ft.

T53, 10 48.3N, 71 36.9W. Fl G 4s, 20ft, 4M. Aluminum and green hut on platform, on pile.

T54, 10 48.2N, 71 37.0W. Fl R 4s, 20ft, 5M. Red metal hut on piles, 20ft.

T55, 10 47.5N, 71 36.4W. Fl G 4s, 20ft, 5M. Green metal hut on piles, 20ft.

T56, 10 47.4N, 71 36.5W. Fl R 4s, 20ft, 5M. Red metal hut on piles, 20ft.

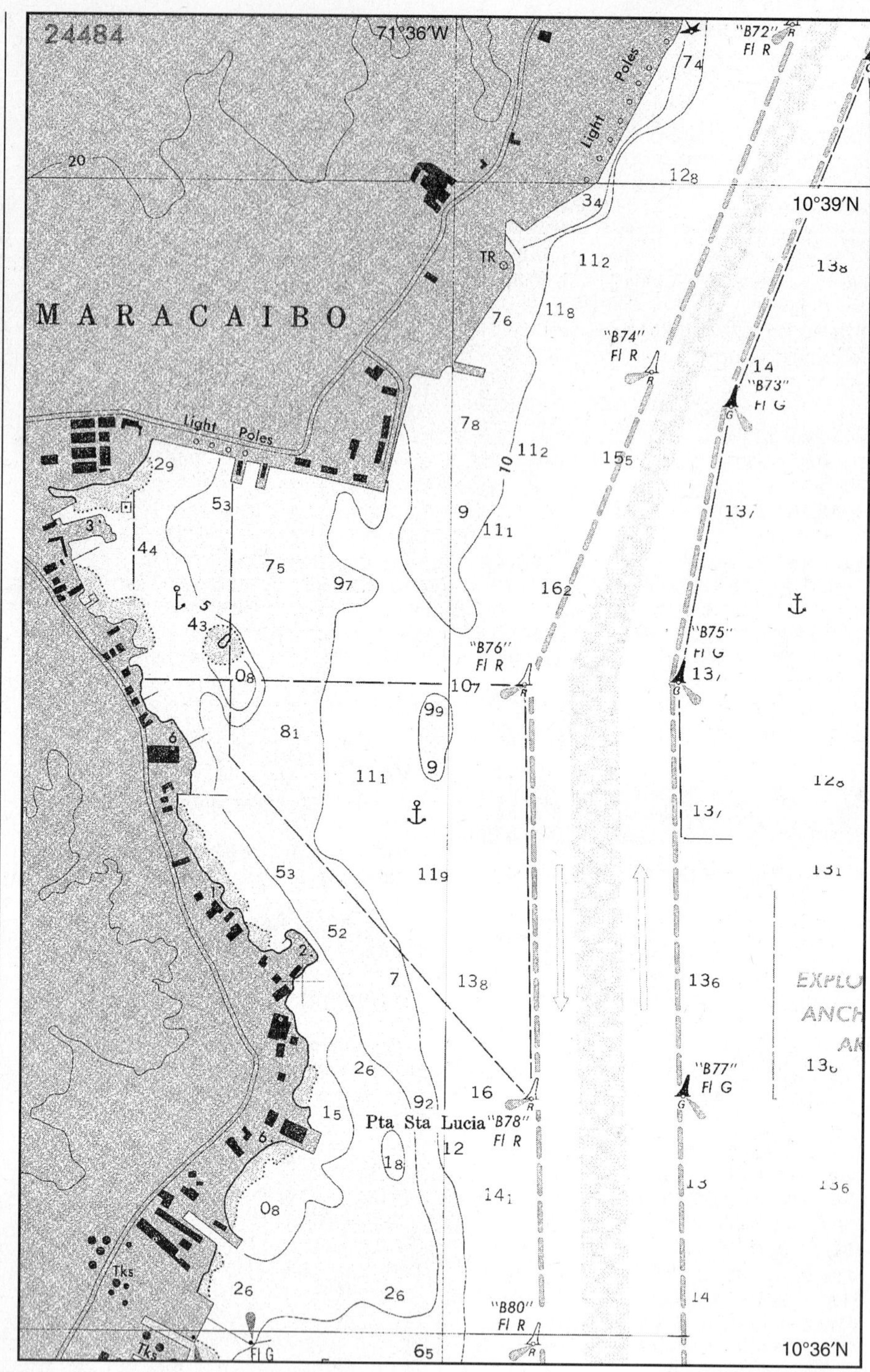

Maracaibo, soundings in meters *(from DMA 24482, 1st ed, 7/88)*

EP3, 10 46.2N, 71 37.7W. Fl G 5s, 13ft, 5M. Aluminum and green hut on platform, on pile.
SANTA CRUZ DE MARA JETTY, 10 47.9N, 71 39.9W. Fl W 5s, 20ft, 2M. Metal column.
LA ARREAGA PIER, 10 36.0N, 71 36.5W. Fl G.
LA ARREAGA ELBOW, 10 36.0N, 71 36.7W. 3 Fl G.

LAGO DE MARACAIBO

Note: There are nine more platform, pier, and oil terminal lights in the southern end of Lago de Maracaibo.

MARACAIBO

10 38N, 71 36W
Pilotage: This very commercial city is one of the busiest ports in Venezuela and a major oil industry center. Be wary of large commercial ships in the marked channels. Within 10 miles of the entrance to the outer channel, you should establish contact with the pilot station on Isla San Carlos (VHF channel 16, call sign San Carlos Pilot). You may be required to carry a pilot to Maracaibo. In any case, it is wise to monitor the VHF while in this area. Squalls, known locally as *chubascos,* occur frequently from May to August, usually between 1400 and 1900 hours. Winds are generally south to southeast with speeds to 50 knots. They may last for 30 minutes to 1 hour. A heavy rain usually follows. The Golfo de Venezuela is reported to be very rough with frequent high winds.
Port of entry
Dockage: Contact the Club Nautico de Maracaibo or the Los Andes Yacht Club.
Anchorage: You may be able to anchor near the clubs, but the depths are reported to be less than 6 feet.
Services: There is fuel, water, and a haulout facility at Club Nautico. The city has good supermarkets, general stores, and restaurants.

WEST AND NORTH GOLFO DE VENEZUELA

PUNTA PERRET, 11 47.5N, 71 20.5W. Fl W 15s, 33ft, 12M. White hexagonal fiberglass tower with orange band, 26ft.
MONJES DEL SUR, 12 21.3N, 70 54.1W. Fl W 10.6s, 253ft, 15M. Metal framework tower with masonry base, 5ft. **RACON M** (– –), 10M.

Bonaire

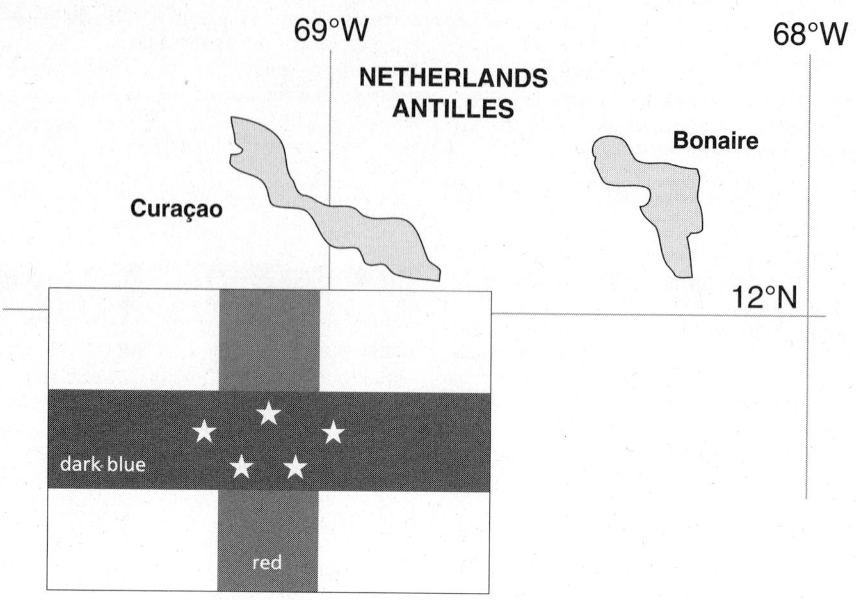

Please see cautions about chartlets and waypoints on page P 2.

ENTRY PROCEDURES

Bonaire, which lies off the coast of Venezuela, is part of the Netherlands Antilles, an autonomous territory of the Netherlands. It is also the eastern-most of what are familiarly known as the ABC islands, making it a good first stop for vessels headed from the Windwards or Venezuela to Panama.

The port of entry is the capital of Kralendijk, located midway along the island's western side, opposite the low rocky islet of Klein Bonaire. Customs and immigration are open 24 hours a day, seven days a week. Yachts may enter at the Harbour Village Marina (VHF channel 17; tel., 599-7-7419; fax, 599-7-5028) 1 mile north of the commercial pier and contact customs from there for instructions or tie up to the customs wharf at the commercial pier, an area that can be quite rough. Mariners have reported no problems traveling to town to visit the offices in person. A valid passport or other identification is required. Formalities are reported to be simple and swift once you reach officials.

Firearms and spearguns are held during your stay. There are strict rules preventing the taking of marine life, as most of the island's reefs are part of Bonaire Marine Park. Anchoring within the park is forbidden by law.

USEFUL INFORMATION

Language: Dutch. Papiamento is the local language, but English and Spanish are widely spoken.
Currency: Antillean guilder; exchange rate (summer 1997), 1.77 guilders = US$1. U.S. currency is widely accepted.
Telephones: Country code for the southern Netherlands Antilles, 599; Bonaire code, 7; local numbers, four digits.
Public holidays (1998): New Year's Day, January 1; Carnival's Rest Day, February 23; Good Friday, April 10; Easter Monday, April 13; Queen's Birthday, April 30; Labor Day, May 1;

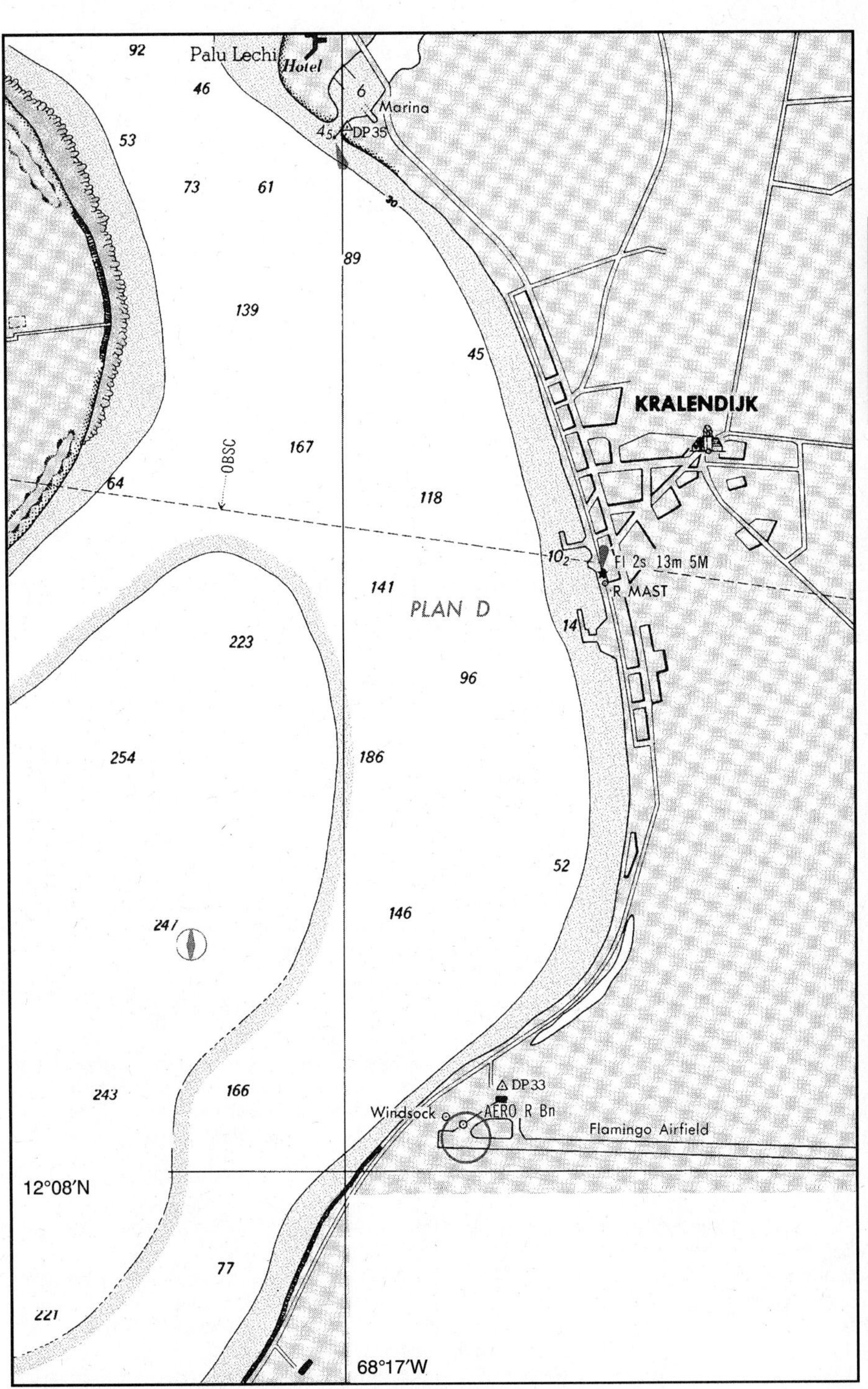

Kralendijk, soundings in meters *(from DMA 24461, 3rd ed, 7/80)*

Ascension Day, May 22; Bonaire Day, September 6; Christmas Day, December 25, Christmas Second Day, December 26.

Regattas: The Plaza Resort Marina (see Kralendijk dockage section) is hosting a new racing event two weeks after Antugua race week. This will include one of the few overnight races in the Caribbean. Contact the Plaza or Seaport Marina in Aruba.

Tourism offices: *In Bonaire:* Tourism Corporation Bonaire, Kaya Libertador Simón Bolivar, No. 12, Kralendijk (tel., 599-7-8322; fax, 599-7-8408). *In the U.S.:* Tourism Corporation Bonaire, 10 Rockefeller Plaza, Suite 900; New York, NY 10020 (tel., 800-826-6247, 212-956-5911; fax, 212-956-5913).

Marine park: Bonaire is world renowned for its coral reefs and clear waters. To protect these fragile reefs, the waters around the island were declared a marine park in 1979. The Bonaire Marine Park (VHF channel 77; tel., 599-7-8444) surrounds the island of Bonaire and Klein Bonaire, extending from the high-water mark to the 200-foot depth contour. Visiting Bonaire by boat means that you are sailing in the Bonaire Marine Park with strict rules and regulations for the sole purpose of preserving the corals and underwater life. Do not pump or drop anything overboard. Anchoring is forbidden everywhere except between the North Pier and the entrance to Harbour Village Marina. An exception is made for small boats less than 12 feet long (they may use a stone anchor).

Spearfishing is completely prohibited. It is illegal to have a speargun in your possession. There is a $10 annual fee for diving in Bonaire Marine Park. You must have an authorized dive tag with you whenever you dive; the park very much appreciates snorkelers also buying a tag as a donation.

Additional note: Venezuela has a consulate in Kralendijk (599-7-8275), where visa applications to visit that country may be submitted.

EAST, SOUTH, AND WEST COASTS

BOCA SPELONK, NE side of island, 12 13.0N, 68 11.7W. Fl W 5s, 100ft, 15M. White circular stone structure, 69ft. Visible 127°–002°, visibility variable on account of salt deposits that constantly form on the lenses.

LACRE PUNT, S point of island, 12 01.9N, 68 14.1W. Fl W 9s, 75ft, 13M. White round stone tower, red bands, 69ft. Visible 203°–147°, visibility variable on account of salt deposits that constantly form on the lenses.

PUNT VIERKANT, 12 07.0N, 68 17.6W. Fl (3) W 22s, 30ft, 5M. White square concrete tower, red lantern, 23ft.

KRALENDIJK, W side of island opposite Klein Bonaire, 12 09.1N, 68 16.5W. Fl W 2s, 44ft, 5M. White square stone structure, 23ft. Visible 027°–150°.

KLEIN BONAIRE, SW point, 12 09.4N, 68 19.8W. Fl (2) W 20s, 19ft, 9M. White round tower, 16ft. Visible 278°–143°, obscured elsewhere.

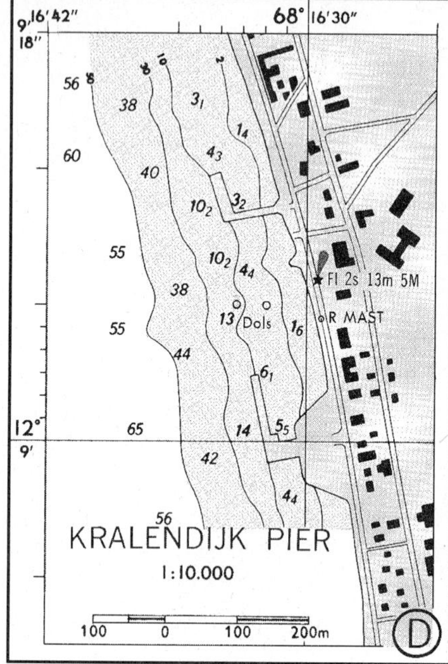

Kralendijk pier area, soundings in meters *(from DMA 24461, 3rd ed, 7/80)*

KRALENDIJK

12 09N, 68 17W

Pilotage: Bonaire is high at the north end, but low and sandy to the south. The highest peak is Brandaris, about 787 feet above sea level, and located on the northwest part of the island. The coast is all steep-to with no off-lying hazards. However, the east coast of Bonaire was where longtime voyager Peter Tangvald put his boat on the reefs and lost his life. There have been many reports of confusion in identifying the island's lights when approaching from offshore. The lights have been reported to be unreliable. Contact the harbormaster on VHF channel 16.

Port of entry

Communications: Yachts use VHF channel 77 for calling, switching to channel 71 to talk.

Dockage: Available at Harbour Village Marina, with completely rebuilt docks and new facilities in 1997 to accommodate vessels to 140 feet or with 12-foot draft (VHF channel 17; tel., 599-7-7419; fax, 599-7-5028). The marina is located in a small basin about 1 mile north of the commercial docks. It is reported to be safe to leave your boat here for extended periods. There are also facilities at Club Nautico (tel., 599-7-5800; fax, 599-7-5850). They have 12 slips, half a mile north of the North Pier. They can take boats up to 80 feet long; max. draft is 9 feet. This is a dock for settled weather only, but it is within walking distance of the supermarket and most restaurants. A dinghy dock is located inside the T-dock. The Plaza Resort Marina (VHF channel 18; tel., 599-7-2500; fax, 599-7-7133) is a new hotel and marina complex less than 1 mile south of the south pier. The marina has slips for 30 boats up to 70 feet in length; maximum draft is 7fi feet at high tide. The entrance is a narrow buoyed channel. Plans are to expand the marina to 70 slips and be accessible for boats with drafts of up to 13 feet. A 70-ton Travelift is planned for 1998.

Moorings: The Bonaire Marine Park has placed 40 moorings between the customs pier and Harbour Village Marina, exclusively for visiting boats. As of summer 1997, the moorings are free and there is no time restriction on their use; there is talk, though, that a small fee might be charged in the near future. You should be prepared to rapidly leave the moorings anchorage during wind reversals (they are most common between August and December).

Anchorage: Most boats anchor on the narrow shelf of shallow water between the marina and the commercial docks. Strong southwest winds can make the anchorage unsafe. These are usually experienced from September through early November. If you anchor out, dinghy in to the main wharf area or the dinghy dock at the Zee Zicht restaurant, located just north of the commercial wharves.

Some cruisers suggest not using more than one anchor here so you won't be delayed if you have to make a hasty departure. The holding ground in the anchorage is marginal—only a thin layer of sand covers a bank of hard coral. Local residents ask that you do not set an anchor over the dropoff because it will damage the coral.

In the event of a wind reversal, take shelter in one of the marinas or in the lee of Klein Bonaire.

Services: Harbour Village Marina offers fuel and water (desalinized and expensive), laundry, cable, showers, chandlery, and a new (1997) restaurant. There is no lift. The Plaza Resort Marina offers water and electricity; fax, mail & courier services (including FEDEX); ice, laundry, and showers; room/boat service, car rental and shore excursions; general repairs and maintenance. The staff speak Dutch, English, German, and Spanish. Blue Band Watersports and Toucan Diving are both located at the Plaza Resort Marina.

A bank, some restaurants, markets, and most general supplies can be found in Kralendkijk. Bonaire is a world-famous diving center, so those services are excellent. Contact the tourist board for information on parks and recreation opportunities. The large flocks of pink flamingoes are well worth seeing.

GOTO OIL TERMINAL

RANGE (Front Light), 12 13.5N, 68 23.8W. Q W, 6ft, 5M. **(Rear Light),** 420 meters, 023° from front and 400 meters, 338° from common front. 12 13.7N, 68 23.7W. F Y, 115ft, 5M.

COMMON RANGE (Front Light), 12 13.5N, 68 23.6W. F R 16ft, 4M. **(Rear Light),** 400 meters, 123° from common front, 450 meters, 338° from front. 12 13.7N, 68 23.5W. F Y, 138ft, 5M.

FRONT LIGHT, 12 13.4N, 68 23.4W. F G, 16ft, 4M.

RANGE NO. 1 (Front Light), 12 13.5N, 68 22.7W. F G, 46ft, 9M. **(Rear Light),** 197 meters, 067°47′ from front. F R, 82ft, 9M.

RANGE NO. 2 (Front Light), 12 13.4N, 68 21.4W. Q R, 26ft, 8M. **(Rear Light),** 126 meters, 081°04′ from front. Q G, 33ft, 6M.

NORTH COAST

PUNT WEKOEWA, 12 13.7N, 68 24.8W. Fl (3) W 20s, 49ft, 12M. White tower, 36ft. Visible 285°–155°, obscured elsewhere.

CERU BENTANA, 12 18.2N, 68 22.8W. Fl (4) W 22s, 144ft, 17M. Gray square stone building, 34ft. Visible 069.5°–073° and 074°–303°, obscured elsewhere.

Curaçao

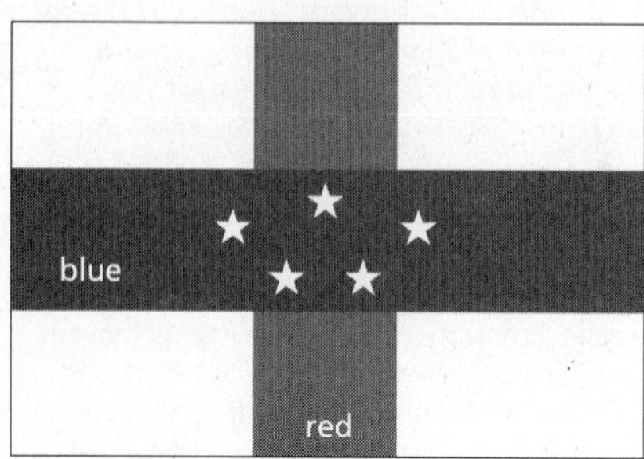

Please see cautions about chartlets and waypoints on page P 2.

ENTRY PROCEDURES

Located about 35 miles north of the Venezuelan coast and 30 miles west of Bonaire, Curaçao is the administrative center of the Netherlands Antilles. It is the middle and largest island of the so-called ABC group.

Ports of entry are Spaansche (Spanish) Water and Willemstad. Private vessels generally prefer to clear customs in Spaansche Water because of the commercial nature of Willemstad harbor. You still must travel overland to Willemstad to clear with immigration. All government offices are located at Willemstad (599-961-2000).

The entrance channel at Willemstad, called Sint Anna Baai, is crossed by two bridges. The port officials (VHF channels 12 and 16; call sign, Fort Nassau) are near the fuel dock, after the high-level bridge on the starboard side of the channel; call them for information and berthing instructions. In Spaansche Water, officials come to Sarifundy's Marina on a regular basis.

Immigration generally grants a three-month stay in the country. All firearms are held by officials. Formalities are reported to be simple and swift, once officials are contacted.

USEFUL INFORMATION

Language: Dutch. Papiamento is the local language. English and Spanish are widely spoken.
Currency: Antillean guilder (Naf); exchange rate (summer 1997), 1.77 guilders = US$1.
Telephones: Country code for the southern Netherlands Antilles, 599; island code, 9; local numbers, six digits.
Public holidays (1998): New Year's Day, January 1; Carnival Monday, February 23; Good Friday, April 10; Easter Monday, April 13; Queen's Birthday, April 30; Labor Day, May 1; Ascension Day, May 22; Curaçao Flag Day, July 2; Christmas Day, December 25; Christmas Second Day, December 26.
Tourism offices: *In Curaçao:* Curaçao Tourism Development Bureau, Pietermaai 19, P.O. Box 3266 (tel., 599-9-616000; fax, 599-9-612305). *In the U.S.:* Curaçao Tourist Board, 330 Biscayne Boulevard, Suite 808, Miami, FL 33132 (tel., 800-445-8266, 305-374-5811; fax, 305-374-6741). Office also in New York.

SOUTHWEST COAST

KLEIN CURAÇAO, near center of island, 11 59.5N, 68 39.0W. Fl (2) W 15s, 82ft, 15M. White tower, red top, 66ft.
PUNT KANON, SE point of island, 12 02.5N, 68 44.5W. Fl W 4s, 39ft, 8M. Round stone tower, red and white bands.
FUIKBAAI RANGE (Front Light), 12 03.2N, 68 49.7W. F G, 32ft, 2M. Red and white beacon with cross topmark. **(Rear Light),** 50 meters, 027.5° from front. F R, 42ft, 4M. Red and white beacon with triangular topmark.

SPAANSCHE WATER

12 04N, 68 51W
Pilotage: Spaansche Haven, the narrow channel leading into Spaansche (Spanish) Water, is reported to be hard to spot from offshore. The obscure entrance is identified by a cluster of oil storage tanks atop and fringing the side of the hill on the left bank of the entrance. The channel is 1.5 miles northwest of the marked entrance to Fuikbaai and 1.25 miles southeast of Lijhoek Light in Caracasbaai. The channel is reported to be privately marked and flanked by several shallow spots and reefs. An approach in good light is recommended.
Port of entry: Contact customs upon arrival.
Anchorage: Anchor near Sarifundy's, located in the northwest arm of Spaansche Water.

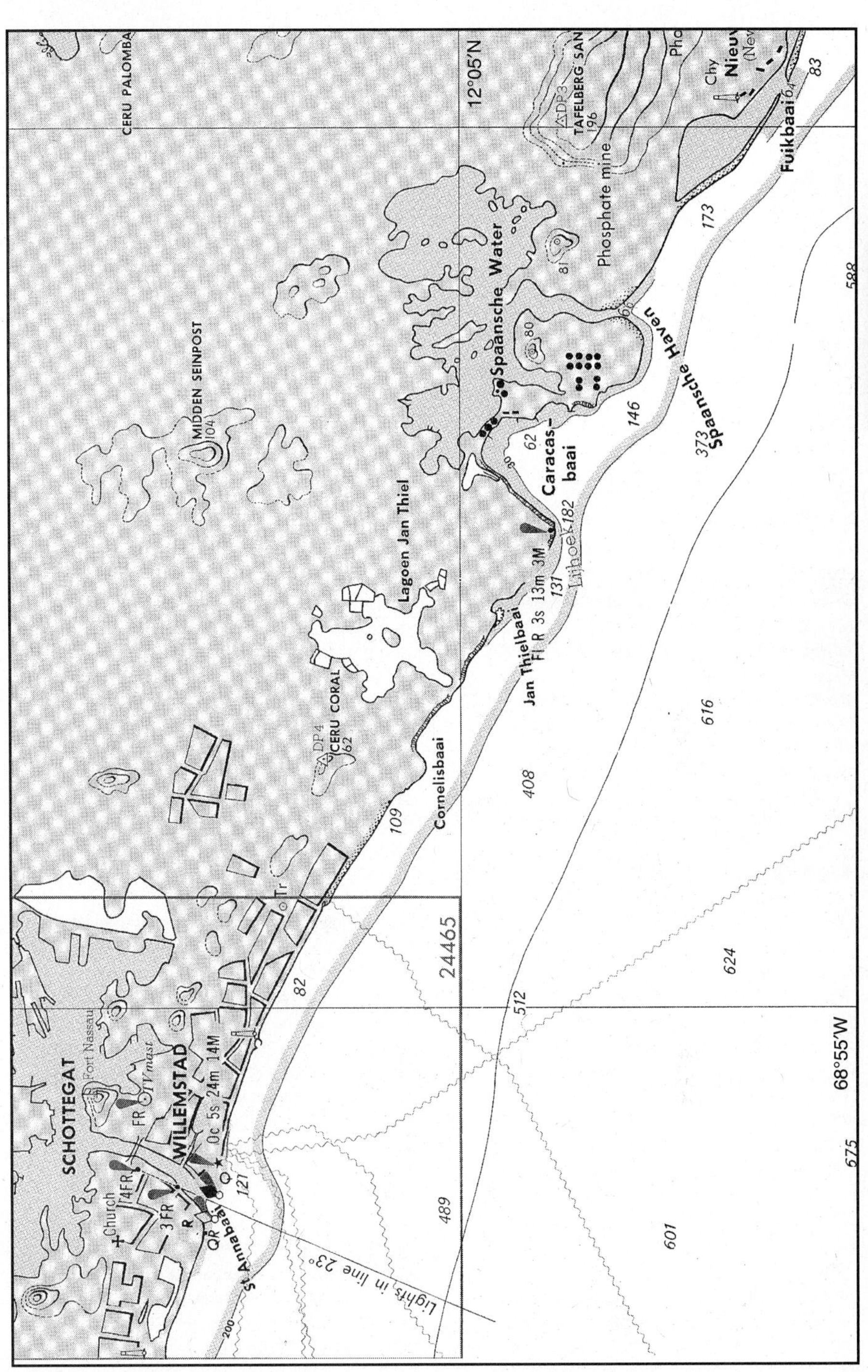

Spaansche Water, Curaçao, soundings in meters *(from DMA 24462, 4th ed, 4/82)*

Services: Sarifundy's Marina (tel., 599-9-677643; fax, -674672) has water, ice, laundry, showers, a dinghy dock, telephones, message services, rental cars, a restaurant, electronics repair services, and a small selection of groceries on site. Fuel is available next door at the Curaçao Yacht Club. Bus service is available from Sarifundy's four times a week to the local supermarket. There are several chandleries in nearby Willemstad. Buses also run several times a week into Willemstad from Spaansche Water. All general supplies are available.

CARACASBAAI ON LIJHOEK, W point of bay, 12 04.4N, 68 52.3W. Fl R 3s, 42ft, 4M. Red and white metal framework tower, red lantern, 19ft.

CARACASBAAI RANGE (Front Light), 12 04.8N, 68 51.8W. F R. Beacon with cross topmark. **(Rear Light),** 90 meters, 043.5° from front. F G. Beacon with triangular topmark, point down.

SINT ANNA BAAI

WILLEMSTAD LIGHT, about 247 meters, 263° from Riffort Light, 12 06.2N, 68 56.3W. Fl G 4s, 6ft, 1M. Black pole, 6ft.

RIFFORT LIGHT, SINT ANNA (WILLEMSTAD), E side of entrance, 12 06.4N, 68 55.9W. Oc W 5s, 83ft, 14M. Gray structure on battery wall, 56ft.

RANGE (Front Light), 12 06.8N, 68 55.7W. F R. Orange diamond daymark. **(Rear Light),** 580 meters, 042° from front. F R. Orange diamond daymark.

WEST SIDE OF ENTRANCE, 12 06.5N, 68 56.1W. F R, 23ft. Aluminum structure, 25ft.

EAST SIDE OF ENTRANCE, 12 06.5N, 68 56.0W. F G, 26ft. Aluminum bracket on fort wall.

RANGE (Front Light), 12 06.6N, 68 56.0W. 3 F R (vert), 54ft. Black metal mast, yellow bands, orange square shape. **(Rear Light),** 480 meters, 023.6° from front. 4 F R, 108ft. Steel mast, orange circular shape. Lights arranged in diamond shape.

WEST SIDE, near W wharf, Sint Anna Bay, 12 06.9N, 68 55.8W. F R, 23ft.

E SIDE OF ENTRANCE TO SCHOTTEGAT, 12 07.1N, 68 55.7W. F G, 67ft. Orange column, black bands. Visible only when entering or leaving Schottegat.

NORTH SIDE, 12 07.5N, 68 55.1W. F G.

BAAI MACOLA, 12 07.1N, 68 55.1W. Q W.

DR. ALBERT PLESMAN FIELD AVIATION LIGHT, 12 11.0N, 68 57.1W. Al Fl W G, 10s, 148ft, W 26M, G 21M. Concrete pillar with gallery. Occasional.

WILLEMSTAD

12 07N, 68 56W

Pilotage: The well-marked entrance to the harbor is called Sint Anna Baai. Approaching vessels should contact harbor control on VHF channels 12 or 16 (call sign, Fort Nassau). The first bridge, a floating pontoon structure, can be opened only by calling Fort Nassau. A single siren blast alerts land-based traffic 1 minute prior to the bridge's opening; no siren signals are used for boat traffic. The second bridge has a vertical clearance of 180 feet. At night Fort Nassau shows a green and white signal when vessel traffic can enter; a red and white signal means outgoing traffic is en route and incoming traffic must wait. By decree, no small boats or yachts are allowed to enter or leave the port between 1800 and 0600; the bridge will not open for smaller vessels during this period.

Port of entry: Tie up to the fueling wharf along the starboard side of Sint Anna Baai north of the high-level bridge. The harbor control office is nearby.

Dockage: Contact the harbor officials about the possibility of mooring to the bulkhead between the bridges, on the starboard side of the channel. The wharves are named Kleine Wharf, Groote Wharf, and Salazar Wharf, proceeding from seaward. The wharves on the west side of the channel are for larger ships.

Anchorage: Anchor in Spaansche Water.

Services: Willemstad is the commercial center of Curaçao and a major port. Most services are available, including haulouts and repairs. There are several chandleries.

BULLENBAAI

BULLENBAAI, 12 10.9N, 69 00.9W. Q W.

FIRST RANGE (Front Light), 12 12.1N, 69 02.1W. F R. Red daymark. **(Rear Light),** 251 meters, 043° from front, 12 12.2N, 69 01.9W. F G. Red daymark.

SECOND RANGE (Front Light), 12 11.7N, 69 01.2W. Q R. On tank. **(Rear Light),** 640 meters, 095° from front. Q G. On tank.

THIRD RANGE (Front Light), 12 11.2N, 69 00.9W. F G. **(Rear Light),** 200 meters, 123° from front. F R.

NORTHWEST COAST

KAAP ST. MARIE, 12 11.4N, 69 03.4W. Fl (2) W 9s, 41ft, 9M. Red rectangular iron tower, 21ft.

NOORDPUNT, 12 23.2N, 69 09.5W. Fl (3) W 15s, 138ft, 12M. White masonry structure, 20ft. Visible 006°–271°, also certain bearings N of 271°.

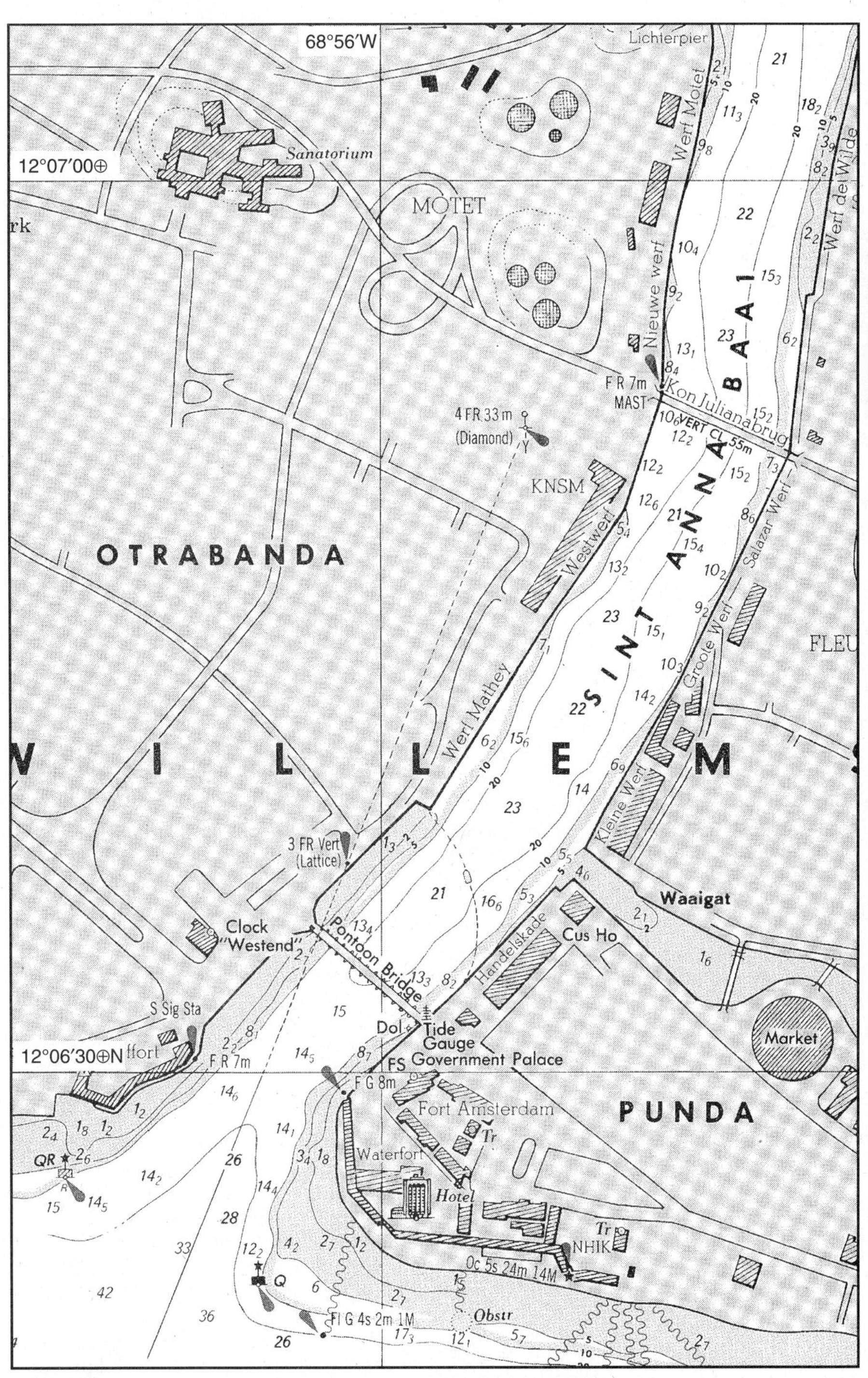

Willemstad, soundings in meters *(from DMA 24465, 5th ed, 8/83)*

Aruba

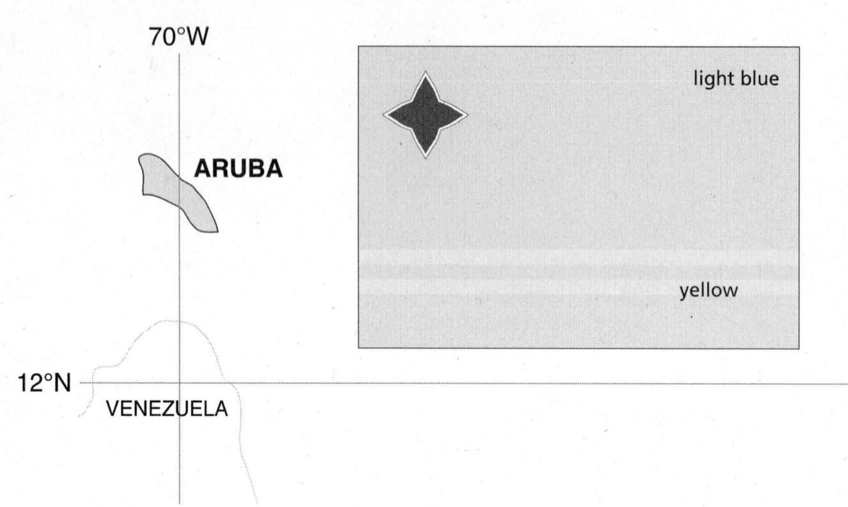

Please see cautions about chartlets and waypoints on page P 2.

ENTRY PROCEDURES

Aruba, the westernmost island of the ABC group, is located about 15 miles north of the Península de Paraguaná on the Venezuelan coast and about 60 miles west-northwest of the Willemstad area on Curaçao. The island is frequently the last stop for cruisers headed for Panama. Aruba is an independent nation, having separated from the Netherlands Antilles.

The port of entry is Oranjestad, about 12 miles northwest of the southern tip of the island. Contact the Aruba Ports on VHF channel 16 or 11. You must first dock in the commercial harbor to clear customs and immigration. Coming from the east, the southern harbor entrance is more difficult to identify. Keep the entrance buoy (white Fl 3s) to starboard, heading to the next channel marker [Fl(2) R 4s]. Be careful of the strong current at the entrance. At night, you can't see the reef on the port side, so stay on the starboard side of the channel. Firearms will be held until your departure.

USEFUL INFORMATION

Language: Dutch. Papiamento is the local language. English is widely spoken.
Currency: Aruban florin; exchange rate (summer 1997), 1.77 florins = US$1. U.S. currency is widely accepted.
Telephones: Country code, 297; Aruba code, 8; local numbers, five digits.
Public holidays (1998): New Year's Day, January 1; G.F. Croes Day, January 25; Carnival Monday occurs in February, but the exact date is not known at press time.; National Anthem & Flag Day, March 18; Good Friday, April 10; Easter Monday, April 13; Queen's Day, April 30; Labor Day, May 1; Ascension Day, May 22; Christmas Day, December 25; Boxing Day, December 26.
Tourism offices: *In Aruba:* Aruba Tourism Authority, L.G. Smith Boulevard 172, Oranjestad (297-8-21019). *In the U.S.:* Aruba Tourism Authority, 1000 Harbor Boulevard, Weehawken, NJ 07087 (tel., 800-862-7822, 201-330-0800; fax, 201-330-8757). Offices also in Atlanta, Houston, Miami. *In Canada:* Aruba Tourism Authority, 86 Bloor Street West, Suite 204, Toronto, Ont. M5S 1M5 (800-268-3042, 416-975-1950).

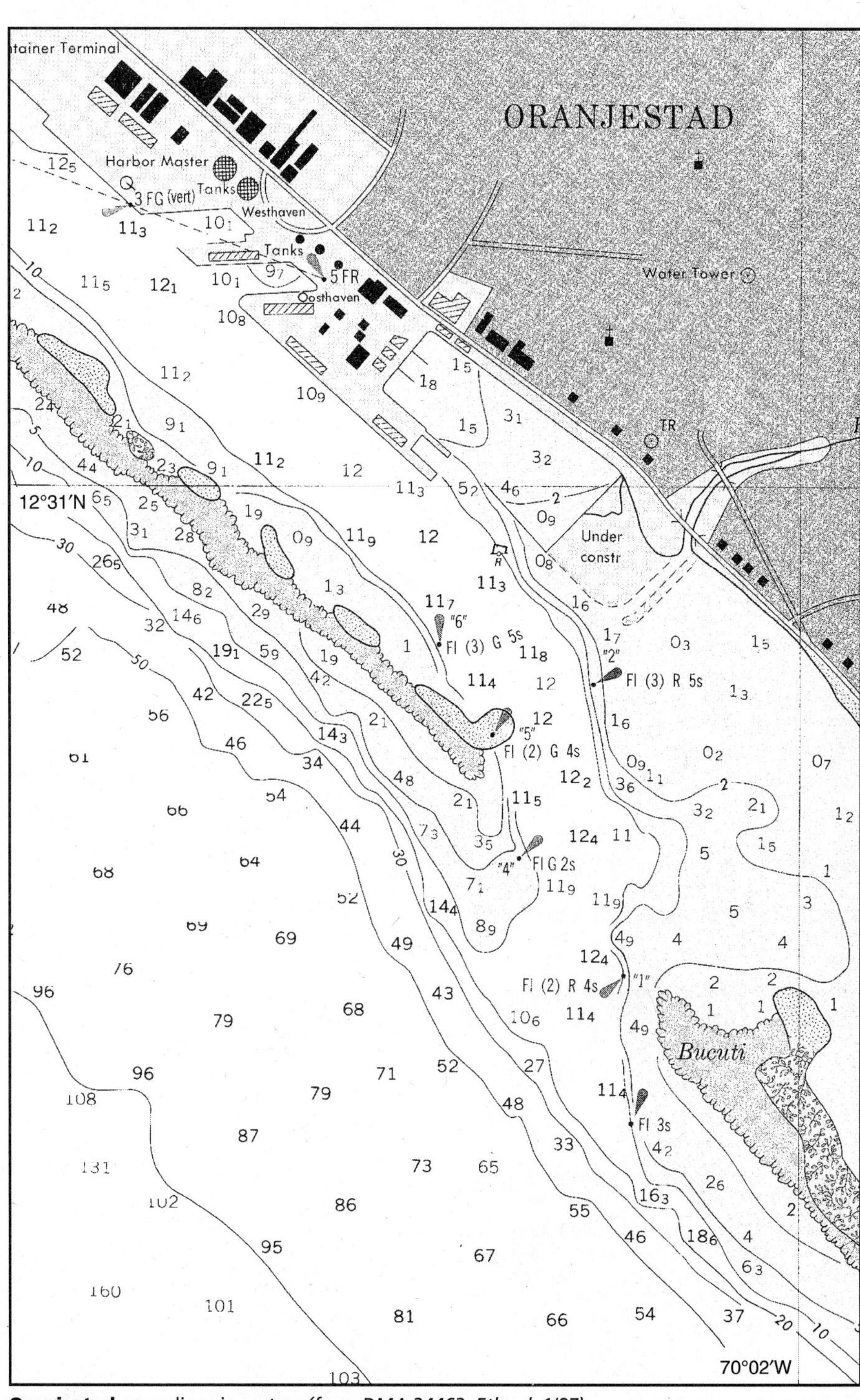

Oranjestad, soundings in meters *(from DMA 24463, 5th ed, 1/87)*

SOUTHEASTERN END

PUNT BASORA (COLORADO POINT), SE extremity of island, 12 25.3N, 69 52.1W. Fl W 6s, 167ft, 21M. Gray square stone tower, 33ft. Visible 160°–109°. Reported extinguished (1992).
INDIAANSKOP, NO. 10, 12 24.9N, 69 53.8W. Q G, 10ft. Steel trellis mast.
RODGERS LAGOON RANGE (Front Light), 12 25.2N, 69 53.1W. 2 F G. **(Rear Light),** 120 meters, 033° from front, 12 25.2N, 69 53.1W. 2 F R.

ST. NICHOLAS BAAI

RANGE (Front Light), 12 25.3N, 69 53.7W. Fl R 2s. Beacon. Occasional. **(Rear Light),** 140 meters, 087° from front. 12 25.3N, 69 53.6W. Fl R 2s. Beacon. Occasional.
BEACON NO. 8, 12 25.3N, 69 54.0W. Fl R 4s.
BEACON NO. 6, W side of entrance, E channel, 12 25.5N, 69 54.2W. Fl R 4s, 11ft. Pile.
BEACON NO. 5, W side of entrance, E channel, 12 25.6N, 69 54.2W. Q G, 11ft. Dolphin. F R and F G lights shown from heads of piers.
WEST ENTRANCE RANGE, near marine railway **(Front Light),** 12 25.9N, 69 54.4W. 4 F R, 102ft. Occasional. Lights arranged in diamond shape. **(Rear Light),** 240 meters, 083°14′ from front. 4 F R, 134ft. Occasional. Lights arranged in diamond shape. N side of channel marked by Fl R light on dolphin and two F R lights on dock. F G and F R lights are shown from outer corners of finger piers.
ENTRANCE NO. 2, N side, 12 25.9N, 69 55.1W. Fl R 3s, 23ft, 7M. Metal column.
MOORING POST, 12 25.8N, 69 54.8W. Fl R 3s. Dolphin.

COMMANDEURS BAAI

RANGE (Front Light), 12 27.1N, 69 56.8W. 2 F G (vert), 19ft, 1M. Pole. **(Rear Light),** 34 meters, 059° from front. 2 F R (vert), 23ft, 1M. Pole.

PAARDEN BAAI

ORANJESTAD, PRINSES BEATRIX AVIATION LIGHT, 12 31N, 70 00W. F R, 669ft. Tower, 62ft.
SOUTH ENTRANCE LIGHT, 12 30.3N, 70 02.2W. Q W. Concrete column.
SOUTH ENTRANCE LIGHT 1, 12 30.5N, 70 02.2W. Fl (2) R 4s. Concrete column.
SOUTH ENTRANCE LIGHT 4, 12 30.6N, 70 02.3W. Q G. Concrete column.
SOUTH ENTRANCE LIGHT 5, 12 30.7N, 70 02.4W. Fl (2) G 4s, 10ft. Concrete column.
SOUTH ENTRANCE LIGHT 2, 12 30.8N, 70 02.2W. Fl (3) R 5s, 10ft. Concrete column.
SOUTH ENTRANCE LIGHT 6, 12 30.8N, 70 02.4W. Fl (3) G 5s. Concrete column.
WEST ENTRANCE LIGHT, N side of channel, 12 31.6N, 70 03.4W. Fl R 2s. Concrete column.
WEST ENTRANCE LIGHT, 12 31.7N, 70 03.1W. Fl (2) R 4s.
WEST ENTRANCE LIGHT 7, S side of entrance, 12 31.3N, 70 03.0W. Fl (2) G 4s. Concrete column.
WESTERN ENTRANCE RANGE (Front Light), 12 31.3N, 70 02.8W. 3 F G (vert). Triangle on black mast with orange bands on N side of building, 75ft. **(Rear Light),** 430 meters, 110° from front, 5 F G. Orange diamond on black mast, orange bands, 95ft. F R lights are shown from NW and SW corners of Oosthaven.

ORANJESTAD

12 31N, 70 02W
Pilotage: The harbor is located in Paarden Baai and is protected by a barrier reef on its west side. Oranjestad is the capital of Aruba. There is a westward-flowing current in this area of 2 to 3 knots. Prominent landmarks are the white circular water tower and close by and to the northwest, the Roman Catholic church. Contact Oranjestad port radio on VHF channel 16 for information on entering the harbor, which is heavily commercial.
Port of entry: Tie up at the southern end of the main piers. The main customs office can be reached at 297-8-31414.
Dockage: Seaport Marina, opposite the commercial wharves in the Sonesta complex, has 40 slips. The marina office is located in the Seaport Marketplace #204 (0830 to 1600, Monday through Saturday, except holidays). The Aruba Nautical Club (297-8-53022) has limited facilities as *Reed's* goes to press (summer 1997); they are planning on building new docking facilities in 1997. Bucuti Yacht Club (297-8-23793), located to the southeast inside the reefs, may have space. Inquire locally for directions to these yacht clubs.
Anchorage: You can anchor southeast of the harbor area in the patch of deeper water extending east from marker 1. There are reports that boats anchor close in to the beach, avoiding the shallow spots on the way in. The depths in this area may be poorly charted.
Services: There are good supermarkets and general supplies. Air connections are excellent. Marine supplies are rather limited. Fuel and water are available at Seaport Marina and

Aruba Nautical Club. Seaport Marina, part of the Sonesta Suites complex, also has electricity, cable TV, telephone, and 24-hour security guards; guests may make use of all resort facilities, including showers, laundry, restaurants, nightclub, swimming pools, beaches, and tennis courts.

NORTHWEST POINT

NOORDWESTPUNT, 12 37N, 70 03W.
Fl (2) W R 10s, 180ft, 19M. Gray stone tower, 98ft. W 354°–005°, R 005°–013°, W 013°–295°.

REED'S NEEDS YOU!

It is difficult to obtain accurate and up-to-date information on many areas in the Caribbean. To provide mariners with the best possible almanac, *Reed's* uses many sources. *Reed's* welcomes your contributions, suggestions, and updates. Please be as specific as possible and include your address and phone number.

Send this information to:
Editor, Thomas Reed Publications
13A Lewis Street, Boston, MA 02113, U.S.A.
e-mail: editor@treed.com
Thank you!

Colombia

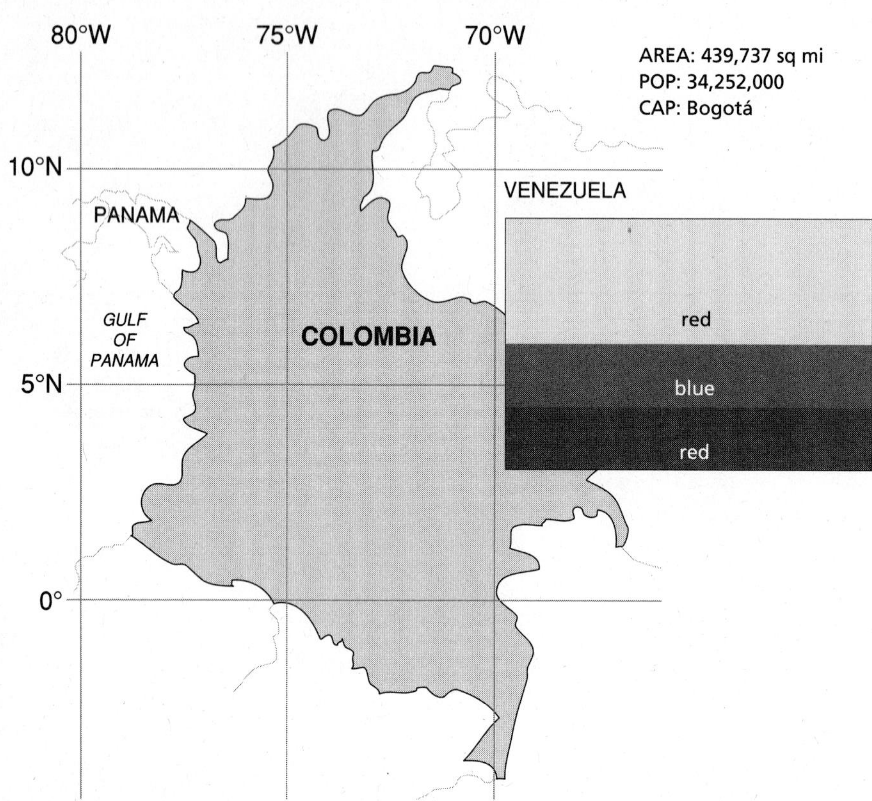

CAUTIONS: Colombian authorities advise vessels to exercise caution in coastal waters with depths of less than 30 meters, as the hydrographic information shown on the charts is derived from old survey data.

MARINE ADVISORY: The Colombian government has announced its intention to enforce its marine regulations more vigorously in an effort to control the flow of contraband. Mariners visiting Colombia's ports or transiting its claimed territorial waters are urged to comply with current regulations and are advised to heed orders of maritime authorities.

TRAVEL WARNING: The U.S. Department of State warns U.S. citizens of the dangers of travel to Colombia. With the exception of several popular tourist areas, violence continues to affect significant portions of the country. Recent attacks have targeted U.S. citizens and institutions. Colombia's north coast is among the areas considered unstable, except the tourist centers of Santa Marta, Barranquilla, Cartegena, and San Andrés. For the current status of this travel warning, call the U.S. Department of State Office of Overseas Citizens Services hotline (voice recording, 202-647-5225; fax information service, 202-647-3000). Some mariners advise staying well offshore when underway, and all vessels should avoid anchoring in remote areas.

Please see cautions about chartlets and waypoints on page P 2.

ENTRY PROCEDURES

From Punta Gallinas to the Panama Canal is about 530 miles. From Aruba to the canal is about 630 miles. Some cruisers break up this long run with a voyage along the Colombian coast or at least a stop in Cartagena and the Islas del Rosario. It is about 324 miles from Cartagena to the canal. The country's long Spanish heritage and beautiful scenery provide an interesting contrast to some of the Caribbean countries to the east.

Ports of entry on Colombia's Caribbean coast include Riohacha, Santa Marta, Barranquilla, Cartagena, Turbo, Zapzurro, and the island of San Andrés, which is discussed later in this chapter in the section "Western Caribbean: Offshore Islands."

A valid passport is required for entering Colombia. Stiff fines are imposed if passports are not stamped on arrival and if stays exceeding 30 days are not authorized by the Colombian Immigration Agency. Citizens of the U.S., Canada, Germany, the Netherlands, the United Kingdom, and many other countries are typically granted 90-day visas when they clear in; citizens of a few countries are required to have a visa on arrival. For more information contact the Colombian Embassy at 2118 LeRoy Place NW, Washington, DC 20008 (tel., 202-387-8338; fax, 202-232-8641) or a Colombian consulate. In the United States and Puerto Rico, Colombia has consulates in Atlanta, Boston, Chicago, Houston, Los Angeles, Miami, New Orleans, New York, San Francisco, Washington DC, and San Juan PR.

Yachts are generally required to use the services of an agent to handle clearing-in paperwork. A cruising permit, or *zarpe,* will be issued and should be presented in every port visited. The importation of firearms is strictly forbidden, and all weapons must be declared.

USEFUL INFORMATION

Language: Spanish.
Currency: Peso; exchange rate (summer 1997), approximately 1,064 pesos = US$1.
Telephones: Country code, 57; a city code follows, then the local number. The city code for Cartagena is 5 and Bogotá, 1.
Public holidays (1998): New Year's Day, January 1; Epiphany, January 6; St. Joseph's Day, March 19; Holy Thursday, April 9; Good Friday, April 10; Labor Day, May 1; Ascension Day, May 22; Corpus Christi, June 4; Feast of the Sacred Heart, June (varies); Saints Peter & Paul Day, late June; Independence Day, July 20; Battle of Boyacá, August 7; Assumption Day, August 15; Columbus Day, October 12 (observed October 13); All Saints' Day, November 1; Independence of Cartagena, November 17; Feast of the Immaculate Conception, December 8; Christmas Day, December 25.
Tourism offices: *In Colombia,* Corporacción Nationale de Turismo, Calle 28 No. 13A–15, Bogatá (57-1-283-9466, -284-3818). Also Corporacción Nationale de Turismo, Casa del Marques de Valvehoyos, Carrera Tres no. 36–57 Centro, Cartegena (57-5-664-7015, -7019).
Special note: Travelers to the Colombian coast may be exposed to tropical diseases, including malaria and dengue fever. Cholera has been reported in Colombia, generally outside the major cities and tourist areas. There is risk of typhoid fever and in the Uraba area, yellow fever. Travelers to Colombia should note that some Central and South American countries require yellow fever vaccinations for travelers coming from Colombia. For the latest medical advisory, contact the Centers for Disease Control International Travelers Hotline (voice recording, 404-332-4559; fax information service, 404-332-4565; Web, www.cdc.gov). In Colombia, U.S. citizens can seek advice and assistance at the consular section of the U.S. Embassy, Calle 22-D Bis, No. 47-51 (at the intersection of Avenida El Dorado and Carrera 50), Bogotá (57-1-315-0811; fax, 57-1-315-2197).

PENÍNSULA DE LA GUAJIRA

CASTILLETES, 11 51.4N, 71 19.5W. Fl W 9s, 72ft, 15M. White tower.
PUNTA ESPADA, 12 05.4N, 71 06.8W. Fl W 8s, 105ft, 20M. White metal tower, orange bands.
CHICHIBACOA, 12 17.7N, 71 13.1W. Fl W 10s, 89ft, 20M. Black and white tower, 72ft.
PUERTO ESTRELLA, 12 21.4N, 71 18.6W. Fl W 15s, 85ft, 18M. White tower, 72ft.
CHIMARE, 12 23.2N, 71 26.5W. Fl W 12s, 82ft, 20M. Orange tower, 72ft.
PUNTA GALLINAS, 12 28N, 71 40W. Fl W 10s, 110ft, 25M. Aluminum square tower, 112ft.
~~***Puerto Bolivar Bell Buoy No. 1,*** 12 17.5N, 71 58.4W. Fl G 2s. Green buoy. Bell. **RACON M (— —).**~~
PUERTO BOLIVAR RANGE (Front Light), 12 15.3N, 71 57.5W. Q W, 16ft, 5M. **(Rear Light),** 600 meters, 148° from front. Oc W 3s, 3M.

PUERTO BOLIVAR, coal loading pier, N, 12 15.6N, 71 57.9W. Oc R 3s, 23ft, 4M. Tower.
PUERTO BOLIVAR, coal loading pier, S, 12 15.3N, 71 57.6W. Oc R 3s, 23ft, 4M. Tower.
PUERTO BOLIVAR, S jetty, 12 15.4N, 71 57.8W. Oc R 3s, 26ft, 4M. Tower.
PUNTA LATATA, 12 15.5N, 71 58.5W. Fl W 3s, 13ft, 6M. Tower.
CABO DE LA VELA, 12 13.1N, 72 10.5W. Fl W 10s, 302ft, 16M. White tower. ~~Visible 358°–268°~~.

CABO DE LA VELA

12 13N, 72 11W
Pilotage: The cape is reported to give a good radar return up to a distance of 21 miles. There is a light on the point. There is possible anchorage partially sheltered from the wind and seas south of the outer extremity of the cape.

PUNTA MANAURE TO POZOS COLORADO

PUNTA MANAURE, 11 44.9N, 72 33.1W. Fl W 15s, 74ft, 14M. White tower.
PLATFORM CHUCHPA, 11 47.3N, 72 46.2W. L Fl W 20s, 98ft, 5M. Gray tower. Horn: 1 blast every 20s.
PIPELINE BEACON A ~~1~~, 11 44.4N, 72 44.1W. L Fl Y 10s, 16ft, 3M. Yellow tower.
PIPELINE BEACON B ~~2~~, L Fl Y 10s, 16ft, 3M. Yellow tower.
RIOHACHA, 11 32.7N, 72 56.0W. Fl W 10s, 115ft, 20M. White metal tower, 98ft. Aero radiobeacon about 2 miles ENE.

RIOHACHA

11 34N, 72 55W
Pilotage: The town lies about 60 miles southwest of Cabo de la Vela. The river is reported to have a navigable depth of about 12 feet, but cargo is lightered out to the anchorage to be loaded. Only small coasters frequent this port. There is a small pier off town.
Port of entry
Anchorage: Basically an open roadstead, the harbor can be rolly.
Services: Some supplies are available.

MORRO GRANDE, summit N side of bay, 11 15.0N, 74 13.8W. Fl (3) W 15s, 262ft, 18M. White tower. Obscured E of 203° by the high land of Aguja Island; visible over a small arc between the island and the mainland at 212°. **RACON M** (– –).

SANTA MARTA

11 15N, 74 13W
Pilotage: The harbor is located about 40 miles east-northeast of the entrance to Barranquilla. It is a small but commercially important port. Isla el Morro is a steep-sided islet about 0.5 mile west of the harbor. A light marks the island, and there are the ruins of a fort on it. A small rock, Morro Chico, lies just west of Punta Morrito at the harbor entrance. The northeast trades are reported to be gusty here. Between the months of March and December a local wind blows from the southwest between the hours of 1000 and 1300. A cathedral with two domes and a radio tower are conspicuous. A white stone monument stands on the summit of a hill about 0.8 mile north-northwest of the cathedral. There are also several tanks in the vicinity of the monument.
Port of entry: The customs house is near the wharves on the eastern side of the harbor.
Anchorage: Drop the hook south of the wharf area.
Services: Basic supplies are available.

POZOS COLORADOS (Front Light), 11 09.3N, 74 12.8W. Q R, 5M. White, orange, and red rectangular daymark on metal tower, 34ft.
(Rear Light), 395 meters, 085.5° from front. Iso R 6s, 5M. White, orange, and red rectangular daymark on metal tower, 34ft.
SIMON BOLIVAR AVIATION LIGHT, 11 07N, 74 14W. Al W G 10s. Airport control tower.

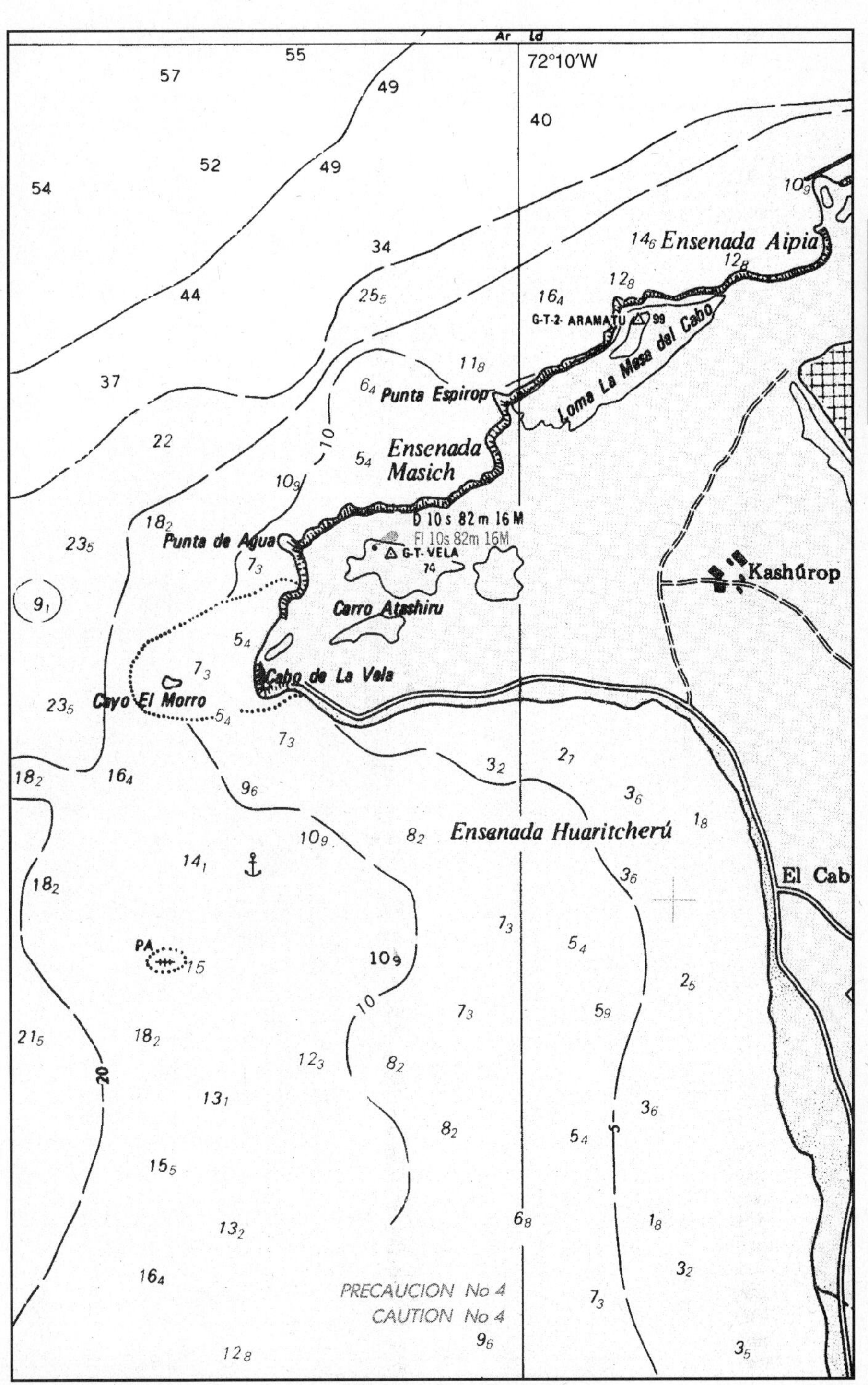

Cabo de la Vela, soundings in meters *(from DMA 24469, 1st ed, 8/87)*

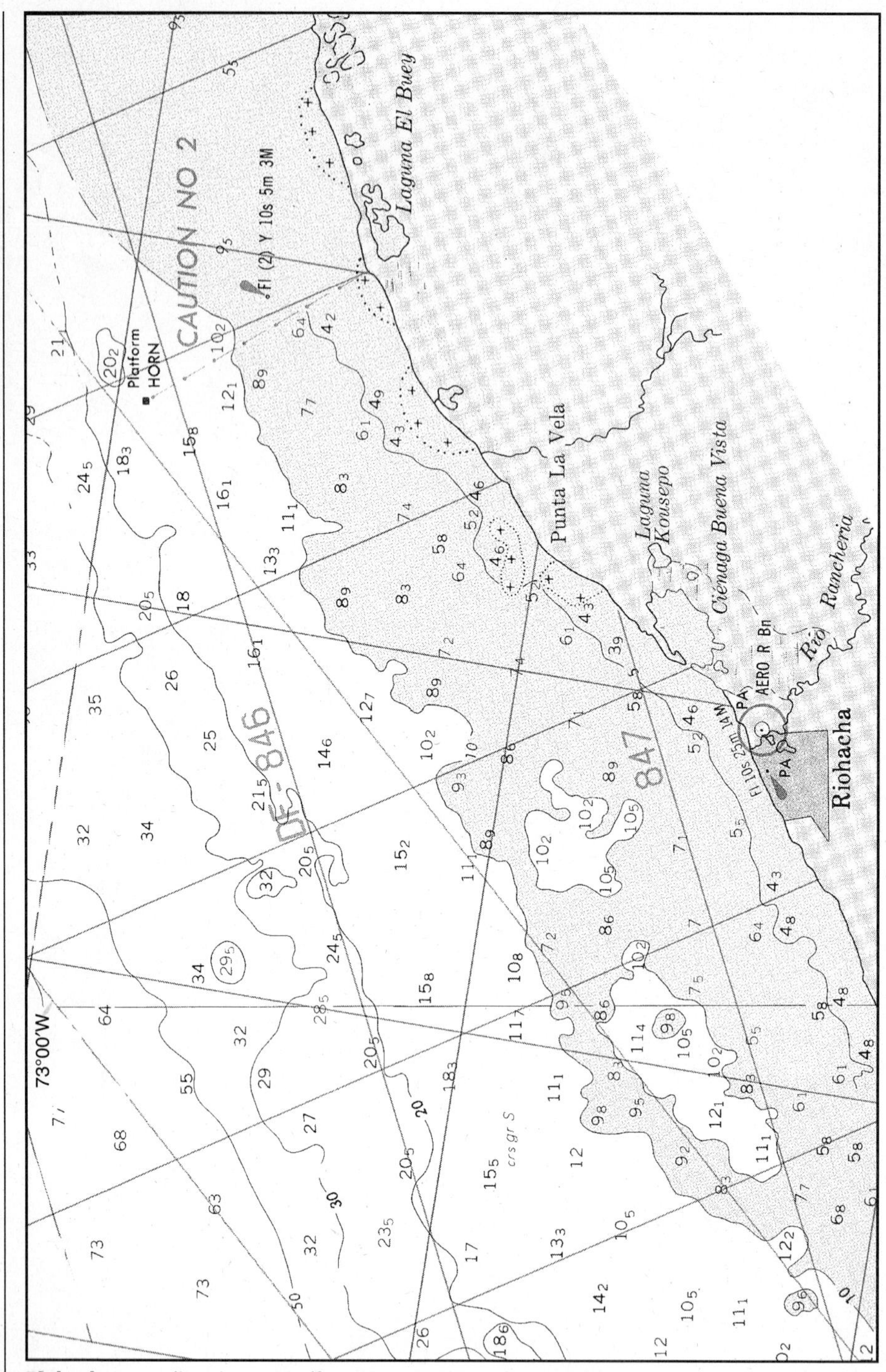

Riohacha, soundings in meters *(from DMA 24490, 2nd ed, 9/86)*

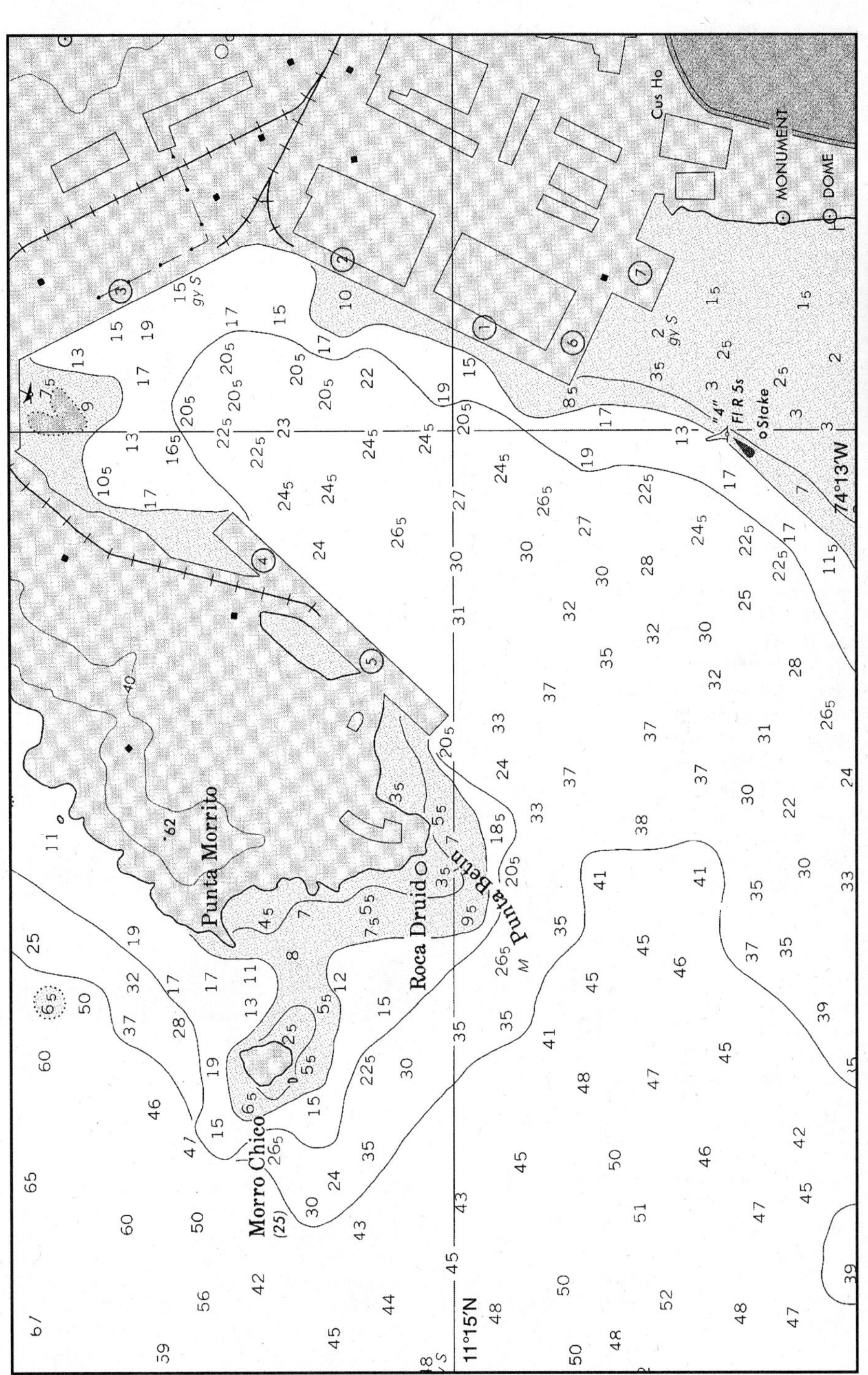

Santa Marta, soundings in meters *(from DMA 24492, 19th ed, 5/92)*

RÍO MAGDALENA

F-2, head of W breakwater, 11 06.4N, 74 51.3W. Fl R 5s, 75ft, 15M. Metal tower.
RACON B (– · · · ·), 23M.
F-1, head of E breakwater, 11 06.4N, 74 51.3W. Fl W 5s, 75ft, 15M. Metal structure.
E-1 ~~Z-3~~ RANGE (Front Light), on E breakwater, 11 06.2N, 74 50.9W. Iso.G 2s, 26ft, 8M. Metal tower with red daymark. **(E-3 ~~Z-5~~, Rear Light),** 260 meters, 136° from front. Iso.G 6s, 49ft, 10M. Metal tower with red daymark.
X-2, W side of river, 11 05.6N, 74 51.1W. Fl R 5s, 26ft, 5M. White round metal structure, red bands.
X-4, W side of river, 11 04.9N, 74 51.0W. Fl R 4s, 26ft, 4M. White round metal structure, red bands.
E-4 ~~Z-4~~ RANGE (Front Light), ~~downriver, on W breakwater,~~ 11 04.2N, 74 50.8W. Fl(2) WR 5s, 30ft, 4M.Metal tower. W. 313°-333°, R. 333°-313°. **(Rear Light),** 393 meters, 323° from front. Iso G 6s, 33ft, 7M. Red and white square on square metal tower, 29ft.
X-6, W side of river, 11 04.1N, 74 50.6W. Fl R 5s, 23ft, 4M. Red tower.
E-6 ~~Z-6~~ RANGE (Front Light), spur of W jetty, 11 03.9N, 74 50.7W. Iso W 2s, 33ft, 12M. Metal tower with red daymark. **(E-8 ~~Z-8~~ Rear Light),** 697 meters, 168° from front. Iso.W 4s, 75ft, 12M. Metal tower with red daymark.
E-12 ~~Z-10~~ RANGE (Front Light), W side of river, 11 03.2N, 74 50.2W. Iso W 2s, 33ft, 10M. Metal tower with red daymark. **(E-10 Rear Light),** 530 meters, 302° from front. Iso.W 4s, 75ft, 10M. Metal tower with red daymark.
X-8, W side of river, 11 03.5N, 74 50.4W. Fl (3)R 10s, 26ft, 5M. ~~Red tower.~~
X-10, W side of river below Las Flores, 11 03.0N, 74 50.0W. Fl (2)R 5s, 33ft, 4M. Metal structure, daymark.
~~**LAS FLORES RANGE (Front Light),** up river, W side of river, 11 02.6N, 74 49.6W. Fl G, 18ft, 7M....... **(Rear Light),** 500 meters, 142.9° from front. Iso G 6s, 36ft, 7M........~~
X-1, 11 01.4N, 74 47.2W. Fl (2)Y 5s, 23ft, 5M. White metal tower, red bands.
~~**LAS FLORES,** on pier, 11 02.5N, 74 49.2W. 2 F R (vert), 30ft, 3M. Tower.~~
E-5 ~~A-1~~ RANGE (Front Light), 11 00.9N, 74 46.4W. Iso W 2s, 33ft, 10M. Metal tower with red daymark. **(E-7 ~~A-2~~, Rear Light),** 460 meters, 122° from front. Iso.W 4s, 69ft, 10M. Metal tower with red daymark.
X-3, 11 01.1N, 74 46.7W. Fl (2)Y 5s, 23ft, 5M. White metal tower, red bands.
X-5, E side of river below new canal to Cienaga, 10 59.9N, 74 45.8W. Fl G 5s, 33ft, 4M. Green tower.
X-7, E side of river below Isla de Trupillo, 10 58.5N, 74 45.3W. Fl G 5s, 30ft, 4M. Green tower. Marks eastern limit turning basin off Isla de Trupillo.

BARRANQUILLA

11 00N, 74 48W
Pilotage: The port comprises the outer 10 miles of the Río Magdalena, the main artery of commercial traffic in Colombia. The river is navigable for 800 miles inland. Being in an area of strong northeast trades, it is sometimes not possible to enter the river safely. The waves at the entrance can be very short and steep. Generally, the winds and seas moderate at night.

When the water level is low in the river from February to April and August to October, the current does not exceed 2 knots. When the water level is high in July and November the current can reach 6 knots, and floating debris can be a problem.

The coast east of the river entrance is backed by low flat sand dunes. West of the entrance the coast consists of flat partly wooded islands. The breakwater heads are reported to give good radar returns up to 10 miles off.
Port of entry: Proceed 10 miles upstream to the city of Barranquilla.
Dockage: There may be some room at the yacht clubs.
Services: This is a large city and has good stocks of general supplies. Repairs and parts should be more available here than in the ports farther east.

PUNTA HERMOSA TO CARTAGENA

PUNTA HERMOSA, 10 57.8N, 75 01.0W. Fl (3) W 20s, 440ft, 22M. White square concrete tower, black stripe.

ISLA VERDE

11 02N, 75 00W
Pilotage: Isla Verde is actually a sandspit covered at all stages of the tide. This coast has many shoal areas within a distance of 3 miles from shore. Vessels should plot courses carefully until past the shoals along the coast for the 13 miles between Río Magdalena and Punta Hermosa. Generally, sailing beyond the 10-fathom line is safe. There is a light on Punta Hermosa and one on Punta de La Garita, as well as one on Punta Canoas.

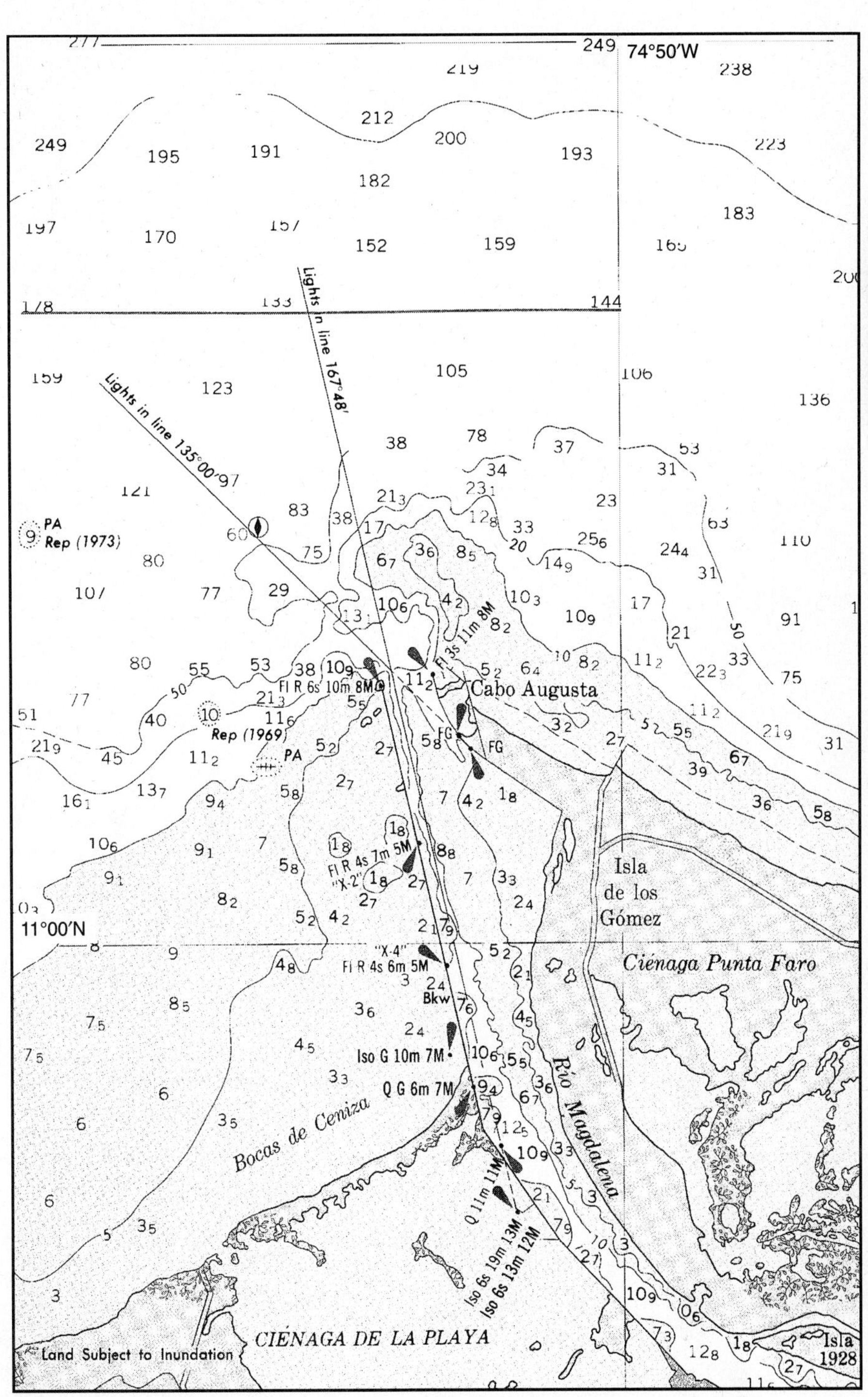

Approaches to Barranquilla, soundings in meters *(from DMA 24490, 2nd ed, 9/86)*

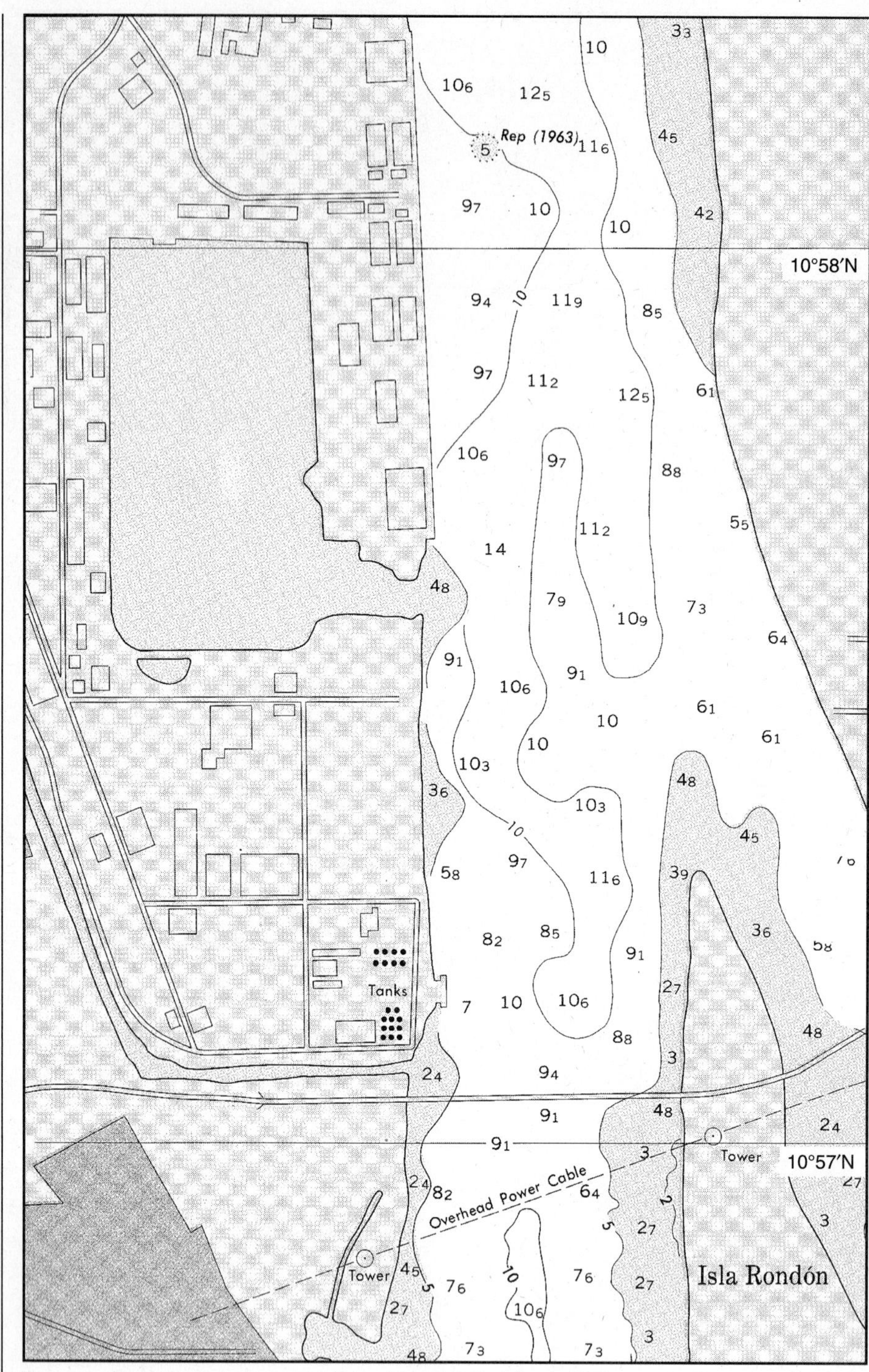

Barranquilla, soundings in meters *(from DMA 24502, 4th ed, 7/87)*

PUNTA GALERA, 10 48.1N, 75 15.6W. Fl W 6s, 82ft, 18M. White tower.
PUNTA CANOAS, N ide of point, 10 34.5N, 75 30.0W. Fl (2) W 20s, 302ft, 24M. White tower, 59ft.
CRESPO AVIATION LIGHT, 10 27.2N, 75 31.0W. Al W G 10s, 20M.

BANCOS DE SALMEDINA

10 23N, 75 40W
Pilotage: Two detached shoal areas of sand and coral lie about 5 miles northwest of the Boca Chica entrance buoy to Cartagena. Breakers have been reported on the western shoals, which have a least depth of about 15 feet. There is reported to be a light on the eastern bank.

CARTAGENA

ISLA TIERRA BOMBA, 10 20.6N, 75 35.1W. Fl (2) W 20s, 259ft, 23M. White metal framework tower, red bands, 39ft. ~~Radiobeacon~~.
ISLA BRUJAS, NO. 2, 10 20.0N, 75 31.1W. Fl W 3s, 65ft, 10M. White metal tower. F W on water tower 0.48 mile ENE.
CANAL DEL DIQUE, E side of canal, 10 17.8N, 75 31.6W. Fl W 10s, 52ft, 10M. White tower.
B2, 10 18.9N, 75 33.9W. Fl R, 10ft, 5M. Red starboard-hand beacon.
B1, 10 23.6N, 75 32.7W. Fl G 3s, 10ft, 5M. Green port-hand beacon.
CASTILLO GRANDE, 10 23.6N, 75 32.9W. Fl W 4s, 79ft, 14M. Stone tower, 72ft.
BANCO SALMEDINA, 10 22.5N, 75 39.5W. Fl W 10s, 72ft, 15M. Red, 72ft.

CARTAGENA

10 25N, 75 32W
Pilotage: This is the largest and most secure port on the north coast of Colombia. It is located about 324 miles east-northeast of the Panama Canal. The main entrance to the Bahía de Cartagena is Boca Chica, lying between two islands about 6.8 miles south-southwest of the city proper. The shipping channel is clearly marked. The northern entrance, known as Boca Grande, has a submerged rock barrier across its entire width. The small-boat passages through the barrier should be attempted only with local knowledge.

In the season of strong winds, January to June, the sea breeze sets in from the west and then veers northwest. At about noon it blows parallel to the coast. During the wet season, April to October, the climate is hot, and early-morning land breezes are usually preceded by rainsqualls of short duration. During the dry season from November to March, the winds are stronger.

The maximum tide range is about 2.6 feet and the mean range about 1 foot. The tidal current off the entrance to the inner harbor sets east-southeast on the flood at a velocity of 0.5 knot and west at the same speed during the maximum ebb.

There are several Restricted Areas in the harbor, and vessels should not travel or anchor in these areas. One fills most of the western harbor off Boca Grande and Castillo Grande, extending almost all the way east to Manga Island.

Port of entry: Many cruisers dock or anchor at the Club Nautico de Veleros on Manga Island, where you can contact an agent. An agent can also be secured through Club de Pesca.
Dockage: Cruisers recommend Club Nautico (VHF channel 16; call signs, Club Nautico Manga or Silva; tel., 575-665-0494), a marina located on the west coast of Manga Island. Depths are reported to be less than 7 feet at the docks. Cartagena's yacht club, Club de Pesca, is nearby on the north end of the island. Be prepared if strong winds come in from the south. There is reported to be dockage at the new Marina Manzanillo, located in the inlet behind the Navy School docks.
Anchorage: Yachts anchor off Club Nautico or north of the northwest corner of Manga island. The anchorages are protected and safe, but be prepared if strong winds come in from the south.
Services: Fuel is available at the Club de Pesca, and water, ice, and showers at Club Nautico. Repair facilities or assistance in finding services are all found at Club Nautico, whose English-speaking owner is reported to be very amiable. Haulouts can be arranged at several shipyards, including the new Marina Manzanillo. There are good supermarkets and general supplies of all types.

This city has a five-hundred-year history and has done a good job of preserving memories of it. The old walled city is a short walk from the marinas and is reported to be fascinating. The strategic location that so attracted the Spanish galleons still brings in the cruisers. Visiting yachts often participate in the local regattas that the yacht clubs sponsor every month or so.

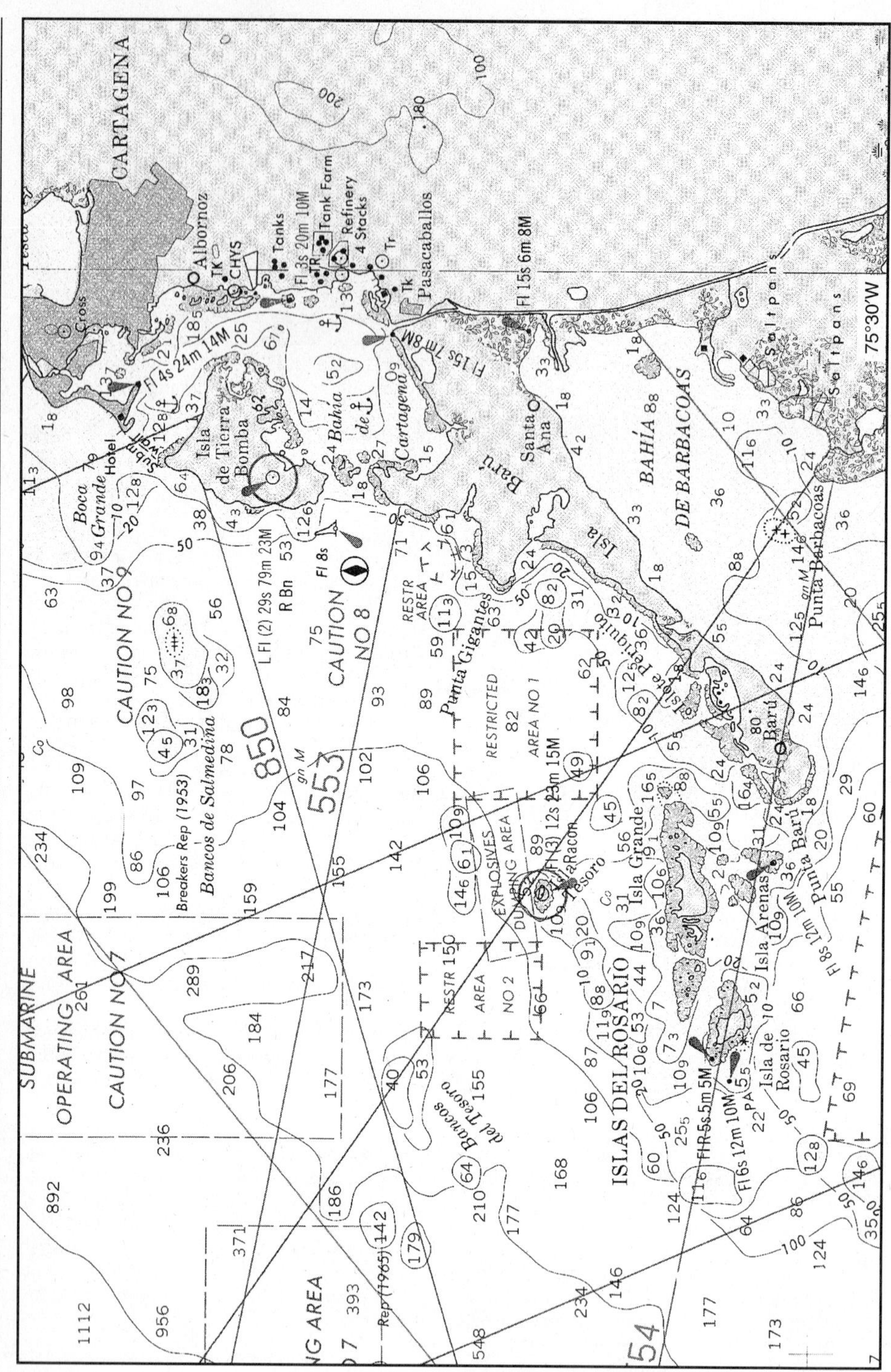

Approaches to Cartagena and Islas del Rosario, soundings in meters *(fr. DMA 24480, 1st ed, 6/90)*

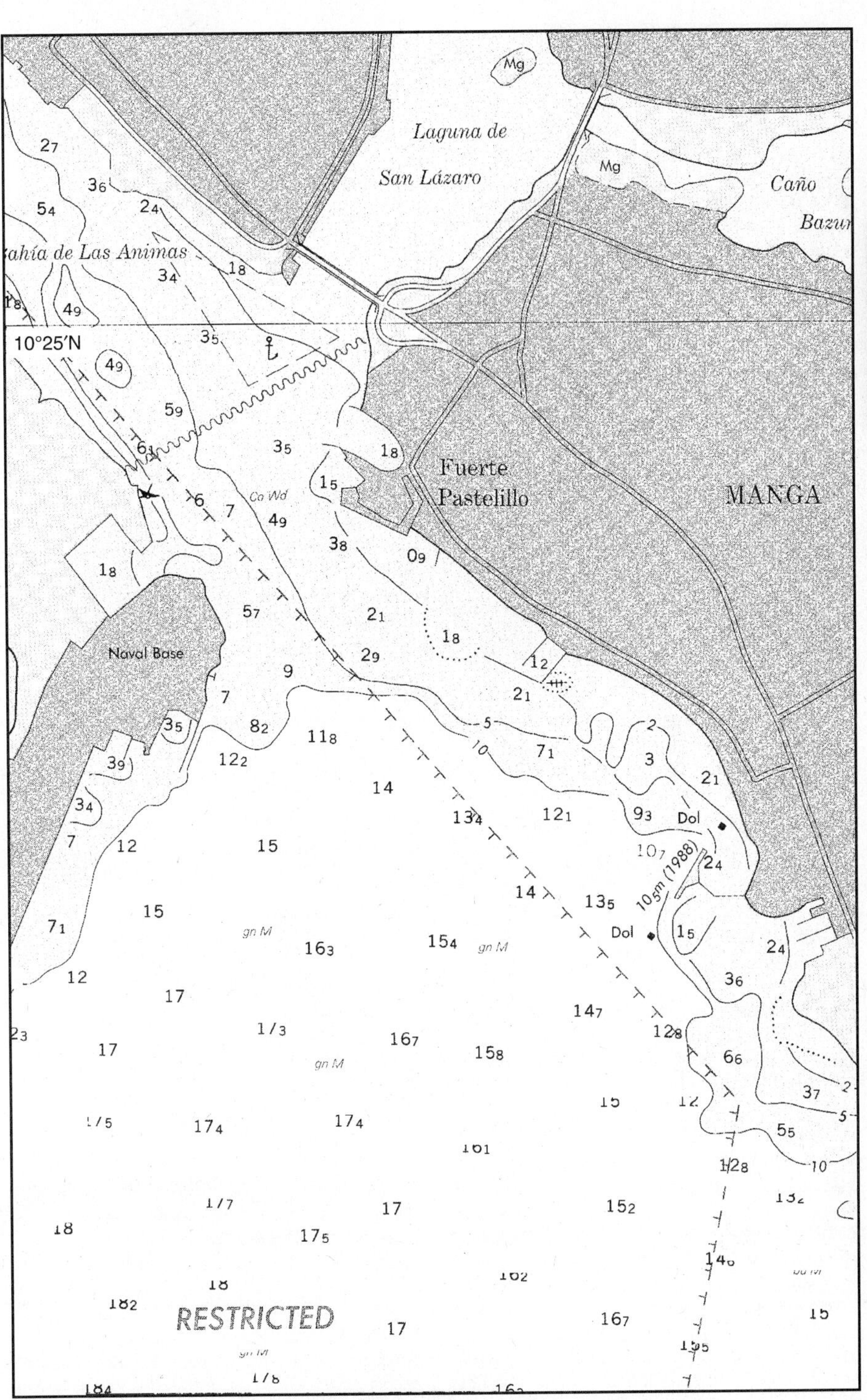

Cartagena Inner Harbor, soundings in meters *(from DMA 24509, 1st ed, 1/92)*

ISLAS DEL ROSARIO

ISLA DEL TESORO, 10 14.2N, 75 44.4W. Fl (3) W 12s, 75ft, 15M. White metal framework tower, 69ft. **RACON C** (– • – •).
ROSARIO LIGHTHOUSE, 10 09.7N, 75 48.6W. Fl W 6s, 39ft, 10M. White tower, 33ft.
ISLA DEL ROSARIO, 10 10.1N, 75 48.2W. Fl R 5s, 16ft, 5M. Tower.
ISLA ARENAS, 10 08.8N, 75 43.7W. Fl W 8s, 72ft, 10M. White metal framework tower, 20ft.

ISLAS DEL ROSARIO

10 11N, 75 45W
Pilotage: This group of small privately owned islands is located about 12 miles southwest of the Boca Chica entrance to Cartagena. They are surrounded by banks and shoals, making the area a danger to those approaching Cartagena from the southwest. This area is a popular cruising ground for local cruisers; pleasant anchorages are well protected.

The north island of the group, Isla del Tesoro, is marked by a light. There are two lights on the west end of the group and one on Isla Arenas. Note the charted Restricted Areas north and south of the islands.

Isla del Rosario, the south island of the group, lies about 5 miles west of Punta Baru and is covered by palm trees. Isla Grande, the largest of the group, lies about 2.3 miles east-northeast of Isla del Rosario. There are several good anchorages within the group, but bear in mind the depths between the islands are very irregular, and there are many rocks and reefs.

BAHÍA DE BARBACOAS, head of the bay, 10 14.5N, 75 31.4W. Fl W 15s, 19ft, 8M. Beacon, 16ft.
ISLA CEYCEN, 9 41.8N, 75 51.3W. Fl W 3s, 78ft, 15M. White tower.
ISLA MUCURA, 9 47.1N, 75 52.3W, Fl (2) W 8s, 79ft, 10M. White square metal tower.

ISLAS SAN BERNARDO

9 45N, 75 50W
Pilotage: This group of low rocks, wooded cays, and shoals lies about 20 miles south of Islas del Rosario. They extend for about 13.5 miles west of Punta San Bernardo. Note the Target and Restricted Areas in the vicinity of the group. There are lights on Isla Mucura and Isla Ceycen.

There is a small village, Islote, on a tiny island south of Isla Tintipan. There are several possible anchorages in this vicinity, but the many shoals and reefs must be avoided. An approach in good light is recommended. There are several deepwater channels between the group and the mainland. These islands are also a cruising destination for local boats from Cartagena.

COVENAS, 9 24.5N, 75 41.3W. Fl W 5s, 100ft, 16M. Top of cooling plant. F R on water tank 0.1 mile SW.
~~**COVENAS PIER HEAD,** 9 25.0N, 75 41.2W. 2 F R (vert), 50ft, 2M.~~

ROCA MORROSQUILLO

9 36N, 76 00W
Pilotage: This coral shoal consists of two major heads, one at a depth of about 30 feet and the other at about 69 feet. The group is reported to be marked by a lighted buoy, which would make a good position indicator if approaching the Islas San Bernardo from the south.

ISLA FUERTE, 9 23.6N, 76 10.6W. Fl W 10s, 135ft, 18M. White tower. **RACON Y** (– • – –).

ISLA FUERTE

9 23N, 76 11W
Pilotage: This low, wooded islet lies about 6.5 miles west-northwest of Punta Piedras and is surrounded by foul ground that extends up to 2.5 miles from the south side and nearly 1.5 miles from its west side. The islet is difficult to distinguish when approaching from the west, but there is a light on its eastern side. There is a small village on the south side of the islet and an area suitable for anchoring on the east side.

ISLA TORTUGUILLA, 9 01.9N, 76 20.6W. Fl W 10s, 59ft, 15M. White metal tower, 52ft.

ISLA TORTUGUILLA

9 02N, 76 20W
Pilotage: This tiny wooded islet lies about 5 miles west of Colina Tortugon. A navigation light shows from its northwest extremity. Depths of less than 30 feet lie within 0.8 mile of its shores. It may be possible to anchor here.

GOLFO DE URABÁ

PUNTA CARIBANA, 8 37.5N, 76 53.1W. Fl W 12s, 275ft, 20M. White square concrete tower, 13ft.

PUNTA CARIBANA

8 37N, 76 53W
Pilotage: This is the eastern point of the entrance to the Golfo de Urabá. It is low, wooded, and marked by a light. Cerro Aguilla, at over

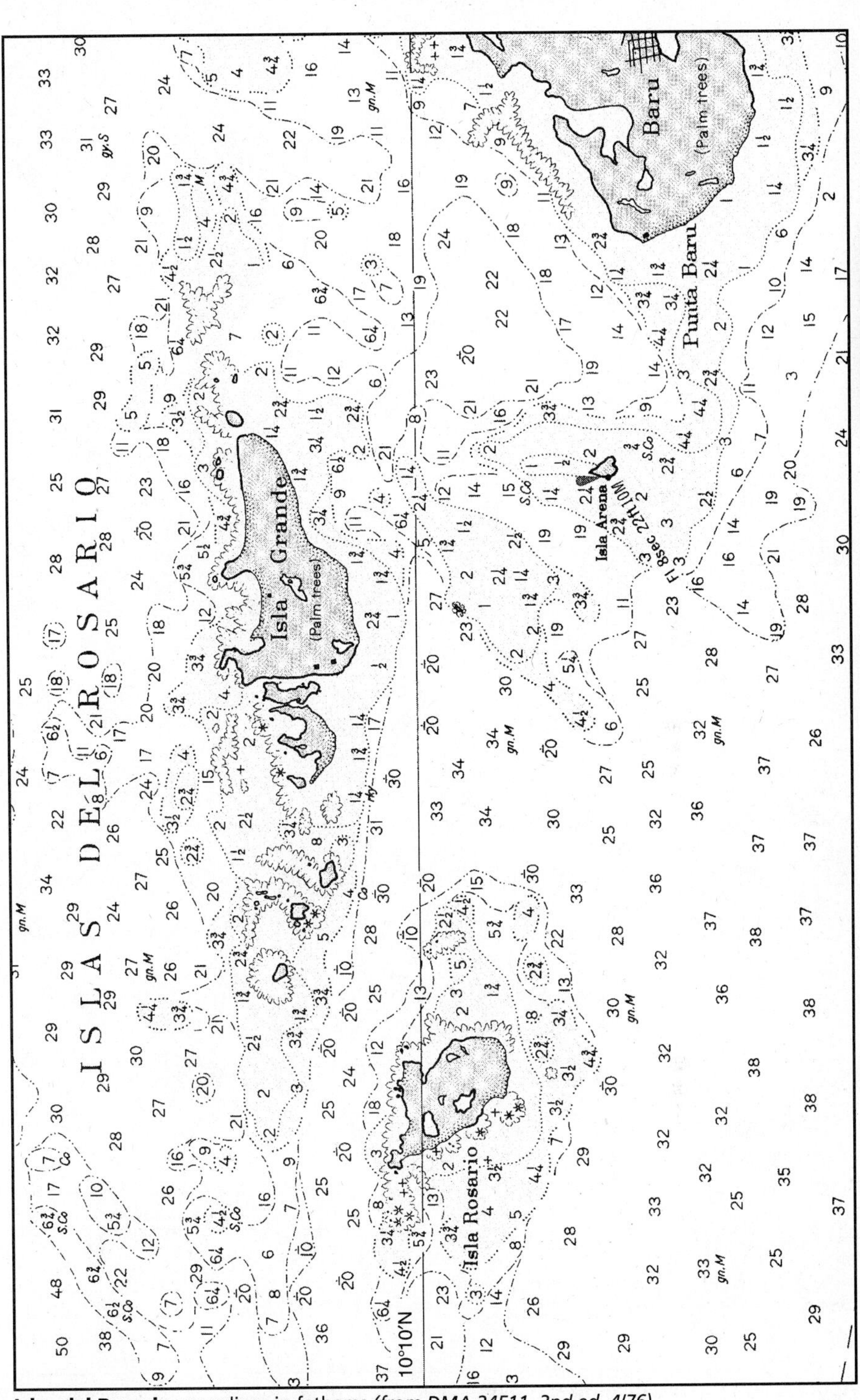

Islas del Rosario, soundings in fathoms *(from DMA 24511, 2nd ed, 4/76)*

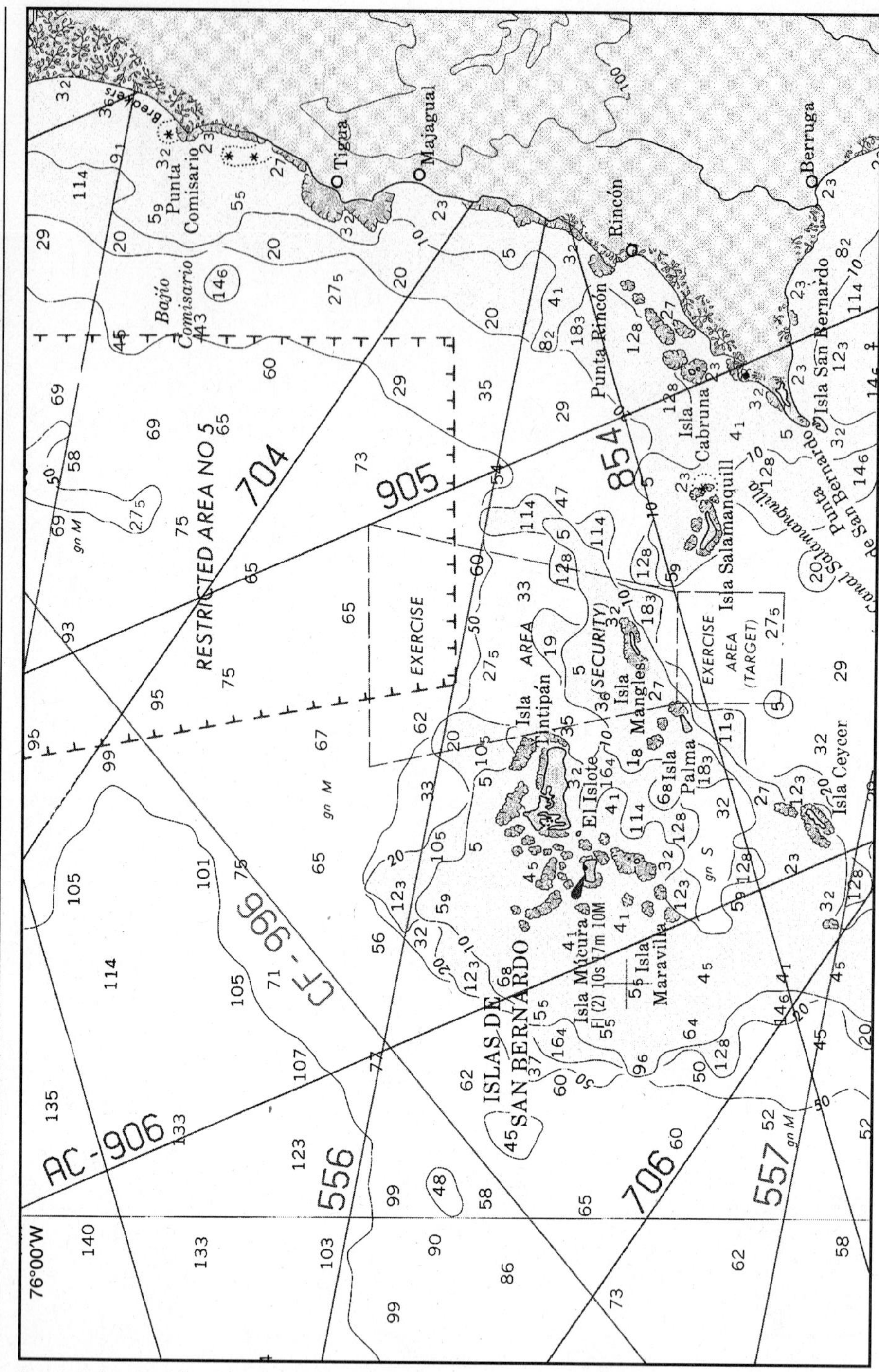

Approaches to Islas San Bernardo, soundings in meters *(from DMA 24480, 1st ed, 6/90)*

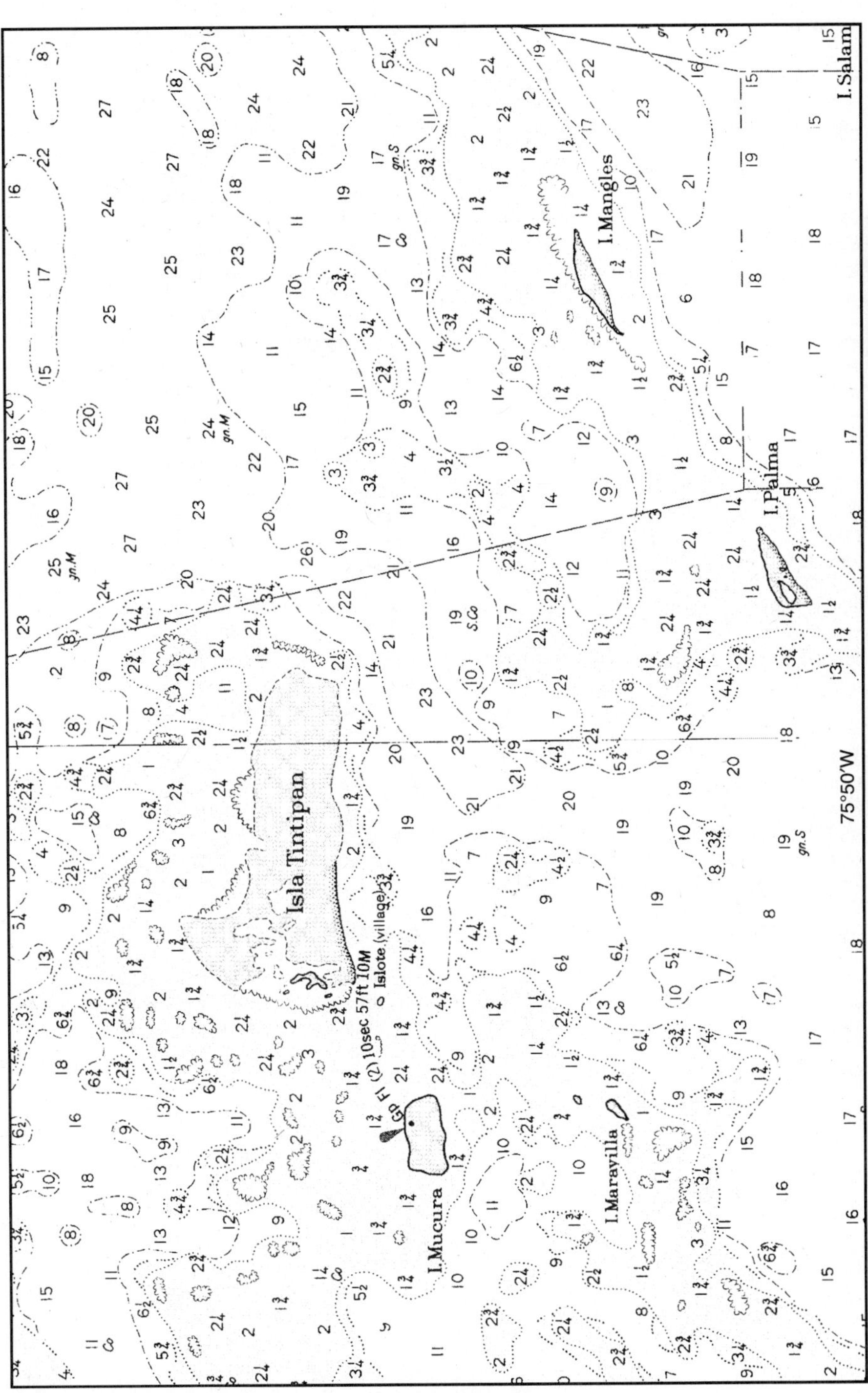

Islas San Bernardo, soundings in fathoms *(from DMA 24511, 2nd ed, 4/76)*

1,100 feet high, makes an excellent landmark. Foul ground, with some rocks awash, extends about 3.8 miles north-northwest from the point. This reef can be dangerous for coasting vessels, and a good offing is recommended.

The Golfo de Urubá experiences a northward-flowing current of up to 2 knots at times. This area experiences a severe local storm in the months from June to October. It is called Chocosanas and occurs most frequently between 2200 and 2400. It is preceded by light north winds and general lightning around the horizon. The wind gradually shifts from the north to the south and increases to near hurricane force in some instances. The storm lasts about fi hour and is accompanied by heavy rain and lightning. As much as 4 inches of rain may fall in one storm.

PUNTA ARENAS DEL NORTE, 8 33.3N, 76 56.2W. Fl W 9s, 44ft, 11M. White tower.
PUNTA CAIMAN, 8 16.5N, 76 46.4W. Fl W 7.5s, 59ft, 15M. White tower.
PUNTA DE LAS VACAS, 8 04.2N, 76 44.6W. Fl W 6s, 59ft, 12M. White metal framework tower, 33ft.
TURBO AIRFIELD TOWER, 8 04.5N, 76 44.1W. F R, 40M. Tower. Obstruction light. Aero Radiobeacon.

TURBO

8 06N, 76 43W
Pilotage: The harbor is located about 30 miles south of Punta Arenas del Norte. This coast is reported to be gradually creeping to the west. Vessels should keep a good offing to avoid shoal areas. There is a light shown from Punta Las Vacas at the entrance to Bahía Turbo. The bay is quite shallow. Check with the port officials for the latest depths in the channel. Officials board commercial vessels west of Punta Las Vacas, so you could anchor in this area to wait for clearance.
Port of entry
Services: Basic supplies and fuel are reported to be available.

BOCAS RÍO LEON, 7 55.9N, 76 45.0W. Fl W 8s, 69ft, 12M. White metal framework tower, 36ft.
PUNTA YARUMAL, 8 07.0N, 76 45.1W. Fl W 5s, 65ft, 15M. White metal tower, black bands, 49ft.
MATUTUNGO, 8 07.6N, 76 50.6W. Fl W 10s, 52ft, 12M. White tower.
ISLA DE LOS MUERTOS, E end, 8 07.9N, 76 49.3W. Fl W 6s, 23ft, 12M. White tower, 36ft.

ISLA NAPU

8 25N, 77 07W
Pilotage: This is a steep rocky island lying about 2.8 miles east-southeast of Punta de La Goleta. It is covered with brush and small trees.

ACANDI

8 31N, 77 16W
Pilotage: This town lies at the mouth of the Río Acandi about 11 miles southeast of Cabo Tiburón. An excellent landmark between Acandi and Cabo Tiburón is Terron de Azucar, a precipitous dark rock lying about 1.3 miles offshore midway between the two positions. A rocky ridge, upon which the sea breaks, connects the rock with the coast. South of Acandi is a sandy beach broken by hills 2 miles southeast of town.

CABO TIBURÓN, 8 40.5N, 77 21.6W. Fl (2) W 15s, 275ft, 15M. White square concrete tower, 13ft. **RACON G** (– – •).

CABO TIBURÓN

8 41N, 77 22W
Pilotage: This bold promontory marks the west entrance point of Golfo de Urabá and the border between Colombia and Panama. It rises to a height of about 405 feet, and there are two concrete beacons on its northwest extremity marking the border. The seaward beacon stands on a pinnacle rock.

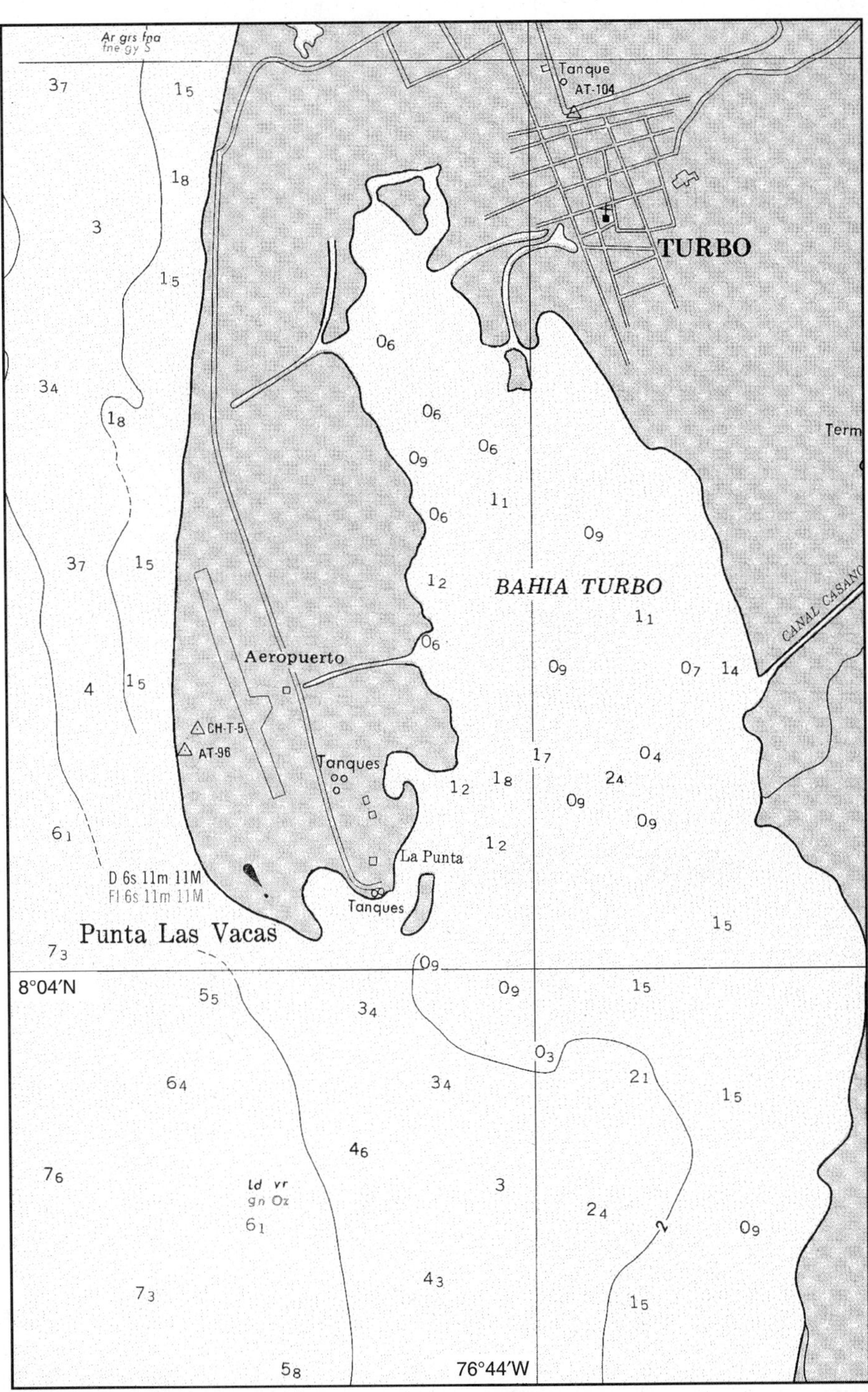

Turbo, soundings in meters *(from DMA 24517, 1st ed, 6/84)*

PANAMA

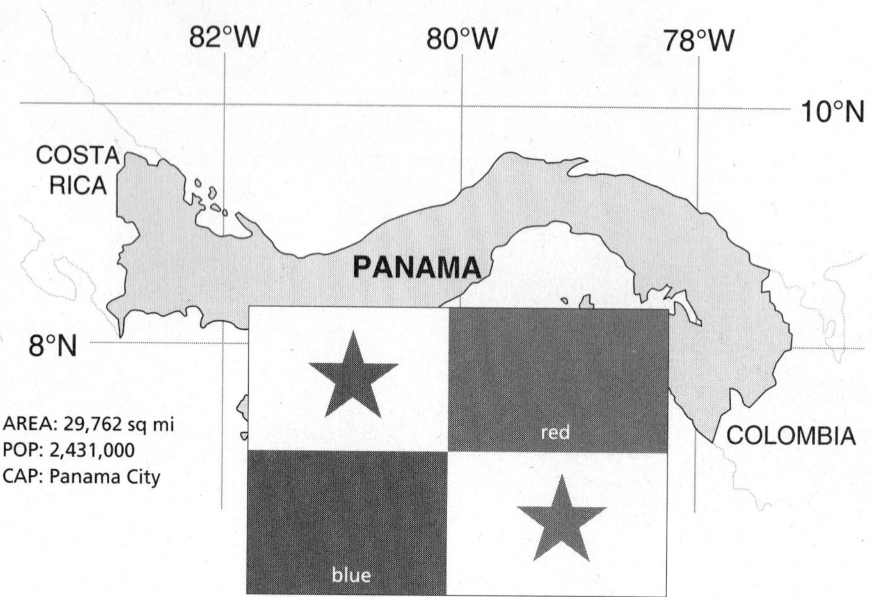

Please see cautions about chartlets and waypoints on page P 2.

ENTRY PROCEDURES

The Panama Canal in Panama is a natural crossroads for many Caribbean cruisers. Some are headed across the Pacific or around the world. Others will attempt the hard slog north along the west coast of North America. Some are headed to Hawaii, and others are just arriving in the Caribbean from the Pacific side. The canal is about 1,072 miles from St. Thomas, 1,202 from Martinique, 1,202 from Trinidad, 661 from Aruba, 324 from Cartagena, and 594 from Jamaica.

On Panama's Caribbean coast, there are ports of entry at Colón (in Puerto Cristóbal near the entrance to the Panama Canal) and at Almirante and Bocas del Toro near the Costa Rican border. The port of entry for the San Blas Islands (now also known as Kuna Yala) is Porvenir, but entering here does not clear you into Panama itself. After clearing in, you must purchase a three-month cruising permit from the office of the Consular y Naves, even if you are only transiting the canal; on the Caribbean coast, these offices are located in Porvenir, Colón, and Bocas del Toro. The cost of the permit varies according to boatlength.

Although U.S. citizens may enter Panama with only a certified birth certificate or a naturalization certificate, it is highly recommended that you do have a valid passport that is good for at least six months. All travelers to Panama must obtain a visa from a Panamanian embassy or consulate before traveling to Panama. (There are reports that immigration in Colón may, in special circumstances, issue a visa if you arrive without one.) The standard visa fee is US$10 per person; for citizens of countries with reciprocal fee arrangements with Panama, which includes the U.S., the fee is waived. Only U.S. citizens are issued multiple-entry visas, but regardless of nationality, each entry is limited to 30 days. Application for a 60-day extension may be made in Panama. In the U.S., for more information and visa applications contact the consular section of the Panama Embassy at 2862 McGill Terrace NW, Washington, DC 20008 (tel., 202-483-1407; fax, 202-483-8413). Panama also maintains consulates in the Caribbean at the following locations: Colombia (Barranquilla, Buenaventura, Cartagena, San Andres, and Santa Marta), Cayman Islands (Grand Cayman Island), Costa Rica (Punta Arenas), Cuba (Havana), Dominican Republic (Santo Domingo), Haiti (Port-au-Prince), Jamaica (Havana), Trinidad (Port of Spain), and Venezuela (Maracaibo).

In Colón the immigration office is next to the Panama Canal Yacht Club and should be your first stop in the clearance procedure. You will also pay a fee for quarantine inspection. You will be given a map showing you where to proceed for customs and the port authority to purchase the required cruising permit. Firearms will be held until your departure.

USEFUL INFORMATION

Language: Spanish. English is widely spoken. Canal operations are conducted in English.
Currency: U.S. dollar; Panamanian coins.
Telephones: Country code, 507; local numbers, seven digits.
Public holidays (1998): New Year's Day, January 1; National Martyrs' Day, January 9; Carnival Tuesday, February 11; Good Friday, April 10; Labor Day, May 1; Colombus Day, October 12; Independence Day, November 3; Flag Day, November 4; Emancipation from Spain Day, November 28; Mother's Day (Immaculate Conception), December 8; National Mourning Day, December 20; Christmas Day, December 25.
Additional notes: Cholera has been reported in Panama, and malaria is a risk in rural areas of the eastern provinces, including the San Blas, in the northwestern provinces of Veraguas and Bocas del Toro, and around Gatún Lake. There is no risk in the Canal Zone, Panama city, or surrounding area. For the latest health information contact the Centers for Disease Control International Travelers Hotline (404-332-4559; fax information service, 404-332-4565; Web, www.cdc.gov). The U.S. Southern Command Network (SCN) broadcasts current travel advisories for Panama on both television and radio. The U.S. Department of State notes that there is a medium incidence of crime in the Panama City and Colón areas. For assistance, U.S. citizens may contact the consular section of the U.S. Embassy at Balboa Avenue and 40th Street in Panama City (consular section, 507-227-1777; international mailing address, Apartado 6959, Panama 5, Republic of Panama; U.S. mailing address, Unit 0945, APO AA 34002). Office hours are 0900 to 1200 and 1300 to 1500.

THE PANAMA CANAL

It's recommended that you allow three full days for paperwork and two for transit through the canal. If you are planning to transit the canal, after you have cleared into the country, proceed to the Panama Canal Commission Admeasurers Office (507-443-2290). You make an appointment to have your boat measured, which is usually a quick process. These measurements determine the fee you pay to transit the canal. (The Admeasurer's Office publishes a guide called "The Panama Canal Guide for Small Yachts and Small Craft," which is useful to request in advance.)

Next, you proceed to Canal Operations Transit Scheduling office (507-443-2201). You will be informed of the equipment requirements for passage, given a schedule of passage, and notified of other safety information. Yachts are currently being allowed to transit every day, but the number of pleasure boats allowed to transit per day may be limited. You will usually be able to make your transit within two days of making your application.

You should have four lines (the Operations Office recommends 7/8-inch nylon) over 100 feet long and four line handlers. The captain of the boat is not counted as a line handler. Line handlers are often crews from other boats who wish to get a preview of the trip. Paid line handlers are available for about US$50 per day. They must also be housed, fed, and returned (via public transport) to Colón.

On the day before (preferably 24 hours prior to) your transit you must contact the Marine Traffic Control Office (507-252-4220 or the Cristóbal Signal Station on VHF channels 12 and 16). This office will confirm your transit and schedule the arrival of your pilot. You should be ready to go on time, as there are stiff fines for delays or cancellation.

There are basically three ways to negotiate the locks: sidewall tie-up, center tie-up, or rafted to a tug. Yachts report a preference for one of the first two options, as there seems to be less chance of damage. However, some people report being alongside a tug as the easiest method, as you don't constantly need to be handling your lines. Don't let the tug pull away from the lock wall with your boat tied alongside! Sometimes a group of yachts will be rafted together in the center of the lock. You will be given a choice of these methods at the Operations Office, but be prepared to change to one of the other methods along the way.

You are required to maintain a speed of at least 4 knots. Vessels that cannot sustain that speed are denied transit; however, they may be towed through the canal by a Panama Canal Commission launch for a fee (approximately $300). Slow boats will probably have to anchor overnight in Gatún Lake at Gamboa. If you are planning on stopping at the Pedro Miguel Yacht Club (tel., 507-232-4509; fax, 507-272-8105), you must make arrangements in advance and be able to show written confirmation of your reservations to your pilot. The club is located in Miraflores Lake near the Pedro Miguel Locks. Call the Canal Operations office when you are ready to resume your passage through the Canal.

There are three up-locks on the Caribbean side, and three down-locks on the Pacific side. They lift you 85 feet from sea level to Gatún Lake, then lower you into the Pacific Ocean on the other side. The lock chambers measure 1,000 feet long by 110 feet wide. The entire trip is about 50 miles in length.

The system of lighting and buoyage in the Canal utilizes range lights, generally green, in the longest reaches and light buoys and beacons along the sides. In general, there are red lights on one side and green lights on the other. The IALA (Region B) Maritime Buoyage System is used. The direction of buoyage changes at Pedro Miguel Locks (9°01′N, 79°37′W).

The Panama Canal Commission rigidly controls all radio communications in this area. All radio communications between vessels in the Panama Canal operating area or calls to stations outside the area must be forwarded through the Canal Radio Station (call sign, HPN) located in Balboa. Except for emergency traffic and routine bridge-to-bridge VHF communications, no vessel in transit through the canal shall communicate with any other station, local or distant.

The canal is currently controlled jointly by the United States and Panama, but will become the sole responsibility of Panama on January 1, 2000.

NORTHEAST COAST

PUERTO OBALDÍA, 8 40.0N, 77 25.0W.
L Fl W 10s, 39ft, 8M. White framework tower, 38ft.

PUERTO OBALDÍA

8 40N, 77 26W
Pilotage: This is a small cove located about 4 miles west of Cabo Tiburon and the Colombian border. There is a small village in the cove. The coast between here and the cape is steep-to and rocky, with heavily wooded hills rising inland. A prominent rock lies about 1 mile north of the east entrance point to the cove.

PUERTO PERME

8 45N, 77 32W
Pilotage: This is a narrow cove about 8 miles northwest of Puerto Obaldía. A village is located about 1 mile south-southwest of Puerto Perme. The port is abandoned but anchorage can be taken.

PUERTO CARRETO

Pilotage: This small cove is just west of Punta Carreto. Two steep-to rocky patches lie about 2.5 miles north of Punta Carreto, and seas break here in fresh breezes. The cove has depths of 18 to 54 feet. There is a small village near the mouth of a river on the west side of the cove.

Between Punta Carreto and Punta Brava, 39.5 miles to the northwest, the coast is fronted by many dangers extending up to 5 miles offshore. A prominent headland in this area is Punta Escoces, rising to a height of 580 feet, about 5.5 miles northwest of Puerto Carreto.

There are several roadstead anchorages along this coast. Isla de Pinos is prominent at 400 feet high. It stands about 0.3 mile offshore, about 2.5 miles northwest of Punta Sasardi. Isla Pajaros, a low islet surrounded by reefs, lies about 2.5 miles farther northwest. This island has been reported as a good landmark, as it is densely wooded.

PUNTA BRAVA

9 15N, 78 03W
Pilotage: Between here and Punta Mandinga, 56.5 miles to the west-northwest, the coast is fronted by the Archipiélago de las Mulatas, a group of small cays, reefs, and banks lying within 10 miles of the coast. There are a number of navigable channels in the group. Many vessels with local knowledge navigate this area.

RÍO DIABLO

9 27N, 78 35W
Pilotage: This village stands on two small islands near shore. They are located about 34 miles west-northwest of Punta Brava. Depths in this area are reported to be less than charted.

GOLFO DE SAN BLAS

9 30N, 79 00W
Pilotage: This area lies between Punta San Blas and Punta Mandinga. It can be entered from the north via Canal de San Blas, located about 2.5 miles east of Punta San Blas. There are numerous detached dangers in the south and west parts of the gulf. Numerous creeks and rivers discharge into the gulf, but their entrances are obstructed by bars. The north shore of the gulf is swampy and fringed with mangroves. The south shore is low, but east of Punta Mandinga the high land approaches the coast.

PUNTA SAN BLAS

9 34N, 78 58W
Pilotage: This point has a low extremity, but a 150-foot-high hill stands about 0.5 mile to the northwest. A 200-foot-high hill, the highest of a group of four, stands about 1.5 miles west-southwest of the end of the point. A conspicuous tower stands about 0.75 mile southwest of the end.

Between Punta San Blas and Punta Macolla, 28 miles to the west, the coast is low and wooded.

PORVENIR

9 33.5N, 78 57W
Pilotage: Isla Porvenir lies about 1 mile east of Punta San Blas and 0.3 mile north-northwest of Sail Rock. A government compound with two conspicuous buff-colored buildings and several lesser structures stands on the west side of the islet. All vessels in the Golfo de San Blas must clear in and out of this station.

CAUTION: Charts in this area are reported to be inaccurate. Passages should be made in good light.
Port of entry: You should check in at this station, whether or not you have obtained a cruising permit in Colón. This is the administrative center for the San Blas (Kuna Yala) area.

PUNTA MACOLLA

9 36N, 79 26W
Pilotage: This point is bold, high, and easily identified. Near the point the mountains, two of which have conspicuous high peaks, begin to approach the coast. From the west it appears as a dark bluff.

PUNTA PESCADOR

9 36N, 79 28W
Pilotage: This point is located a bit west of Punta Macolla. It is fringed by reefs.

PUNTA MANZANILLA

9 38N, 79 33W
Pilotage: This is the northernmost point in Panama. It is a high, precipitous projection, with two conical hillocks resembling a saddle. This is the termination of a mountain ridge that extends along the coast to the mouth of the Río Piedras.

Several off-lying rocks and islets are northeast and northwest of the point, including Islas los Magotes and Isla Tambor.

ISLA GRANDE, off Punta Manzanillo, 9 38.0N, 79 34.0W. Fl W 5s, 305ft, 12M. White metal tower, stone base, 85ft.

ISLA GRANDE

9 38N, 79 34W
Pilotage: This high, palm-covered island lies about 1 mile west of Punta Manzanilla and is reported to give a good radar return up to a distance of 20 miles. There is a light on the northeast part of the island. A confused sea and tide rips are found in the vicinity.

Isla Tambor is connected to the island by reefs. The coast of the island is foul ground except for the northeast part, but a deep channel runs between the island and Punta Manzanilla.

PUERTO GAROTE

9 36N, 79 35W
Pilotage: This small, protected harbor about 27 miles east of Colón is entered about 0.5 mile west of Isla Grande. It is formed by the coast and several off-lying islands. The narrow entrance has depths of 36 to 72 feet, while the inlet leading into the harbor has a depth of 34 feet. The center of the harbor is about 20 feet deep. There is a small pier and anchoring room.
Services: Fresh produce, modest groceries, gasoline, and propane are available in the village on the mainland.

FARALLON SUCIO ROCK, summit, 9 39N, 79 38W. Fl R, 5s, 107ft, 12M. White concrete pyramidal tower. Reported at reduced intensity (1984).

LOS FARALLONES

9 39N, 79 38W
Pilotage: This group of rocks lies about 2 miles offshore to the north-northwest of Punta Cacique. They are about 4 miles west of Isla Tambor. There is a light on the westernmost rock.

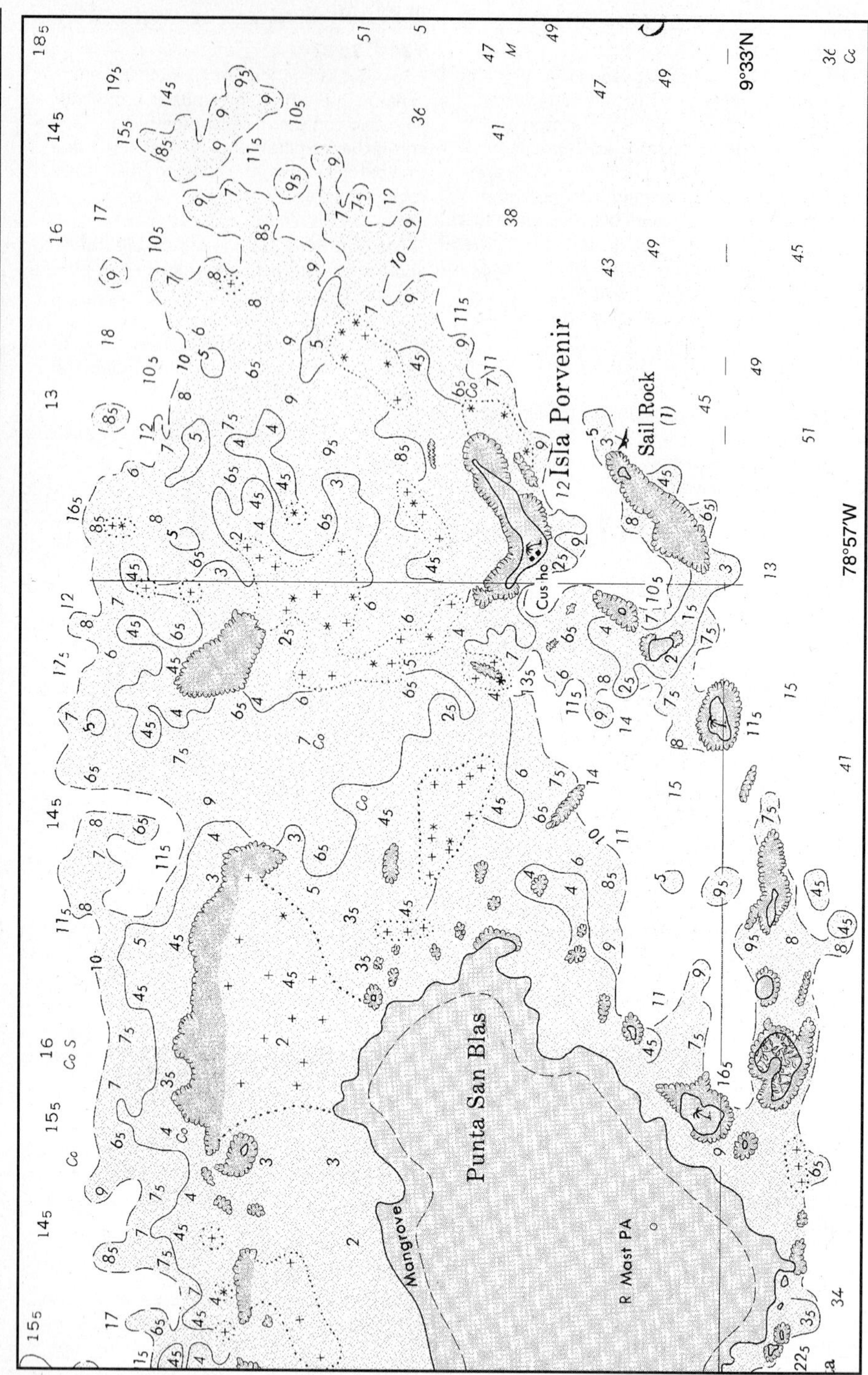

Porvenir, soundings in meters *(from DMA26065, 3rd ed, 6/84)*

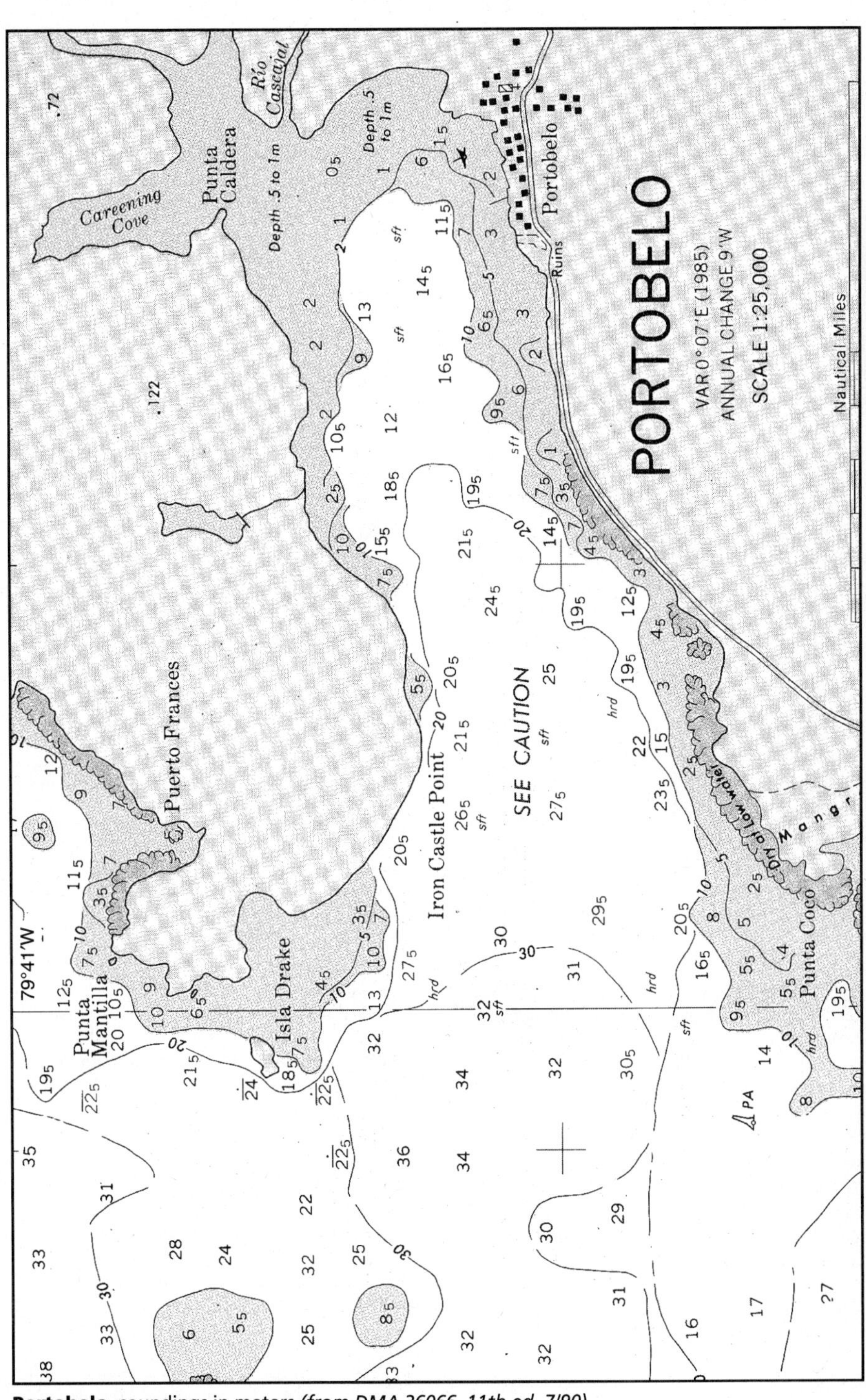

Portobelo, soundings in meters *(from DMA 26066, 11th ed, 7/90)*

PORTOBELO

9 33N, 79 40W

Pilotage: This good harbor of refuge lies about 18 miles northeast of the entrance to Puerto Cristóbal and 11 miles southwest of Isla Grande. Bajo Salmedina, a coral reef upon which the sea breaks, is located about fl mile west of Punta Mantilla. Depths in the harbor may be less than charted due to silting. A church with a large red roof and small white tower is conspicuous. This is reported to be a good anchorage, with some supplies available in town. Without local knowledge, vessels should stay at least 3 miles offshore from here to the entrance to Puerto Cristóbal.

BAHÍA LAS MINAS

RANGE (Front Light), 9 23.9N, 79 49.5W. F G. Orange diamond daymark on beacon. **(Rear Light),** 80 meters, 171° from front. F G. Orange diamond daymark on beacon.

RANGE (Front Light), 9 23.4N, 79 48.8W. F G. Orange diamond daymark on beacon. **(Rear Light),** 160 meters, 148°45′ from front. F G. Orange diamond daymark on beacon.

NORTHWEST PIER, N end, 9 23.8N, 79 49.1W. F R.

TANKER JETTY, N end. 9 23.6N, 79 49.1W. F R.

PUERTO DE BAHÍA DE LAS MINAS

9 24N, 79 49W

Pilotage: This commercial port is entered via a well-marked channel located about 6 miles northeast of the entrance to Puerto Cristóbal. There is a large refinery located in the bay. Without local knowledge, vessels should stay at least 3 miles offshore from here to the entrance to Puerto Cristóbal.

COLÓN — CRISTÓBAL

TORO POINT, W. side of entrance to Limón Bay, 9 22.4N, 79 57.1W. L Fl W 30s, 108ft, 16M. White metal tower, stone base.

WEST BREAKWATER, inside head, 9 23.4N, 79 55.5W. V Q (2) R 2s, 100ft, 16M. Red square on metal tower. Radar reflector.

EAST BREAKWATER, head, 9 23.3N, 79 54.9W. Mo (U) G 20s, 100ft, 16M. Green triangle on green metal tower. F R on chimney 5.9M E; F R on signal station 1.7M E; Q R and three F R (vert) on mast 1.9M SSE; four F R on each of two water tanks 3.5M SSE; F R 2.1M SSE. **RACON U (• • –)**.

EXPLOSIVE ANCHORAGE, near W limit, 9 22.4N, 79 56.7W. Fl Y 2s, 15ft. Special mark, yellow.

CRISTÓBAL HARBOR, BEACON NO. 2, 9 20.9N, 79 54.4W. Q R. Red starboard-hand mark.

BEACON NO. 3, 9 20.8N, 79 54.4W. Oc G 5s. Green port-hand mark.

BEACON NO. 4, 9 20.8N, 79 54.3W. Oc R 5s. Red starboard-hand mark.

BEACON NO. 5, 9 20.8N, 79 54.2W. Oc G 5s. Green port-hand mark.

BEACON NO. 6, 9 20.8N, 79 54.3W. Oc R 5s. Red starboard-hand mark.

BEACON NO. 8, 9 20.7N, 79 54.3W. Oc R 5s. Red starboard-hand mark.

BEACON NO. 7, 9 20.6N, 79 54.2W. Oc G 5s. Green port-hand mark.

BEACON NO. 9, Q G. Green port-hand mark.

BEACON NO. 10, 9 20.6N, 79 54.3W. Oc R 5s. Red starboard-hand mark.

PANAMA CANAL, ATLANTIC ENTRANCE RANGE (Front Light), 9 17.7N, 79 55.4W. F G. Red lights shown on W side and green lights shown on E side of dredged channel and in Limón Bay. The canal is marked by lights, and leading lights mark the center of the channel. **(Middle Light),** 1,037 meters, 180° from front. F G, 98ft, 15M. Concrete conical tower, 75ft. Visible on rangeline only. **(Rear Light),** 2,278 meters, 180° from front. Oc G, 158ft, 15M. Concrete conical tower, 46ft. Visible on rangeline only. F R lights shown on each of two radio towers 1.1 miles NE.

PUERTO CRISTÓBAL

9 21N, 79 55W

Pilotage: This is the entrance to the Panama Canal. Entrance to the harbor is made between two breakwaters about 0.3 mile apart. There are lights at the end of both breakwaters and an offshore approach buoy. All vessels must contact the Cristóbal Signal Station (VHF channels 12 or 16) before entering the harbor. The station is located on the western end of the Muelle Cristóbal, which extends westward into the harbor from Colón. Vessels should maintain a continuous watch on channel 12 while in the harbor.

Dockage: The Panama Canal Yacht Club (fax, 507-417-7752) is located on the east side of Canal French, northeast of Anchorage Area F. This is where you can begin your trek to the various port offices. Check with the club for advice on street crime before heading out. If you are planning to stop at the Pedro Miguel Boat Club, you must obtain written permission in advance, or your canal pilot will not let you stop. The boat club reserves a certain number of slips for transients. Club Nautico is located on the eastern shore of Colón.

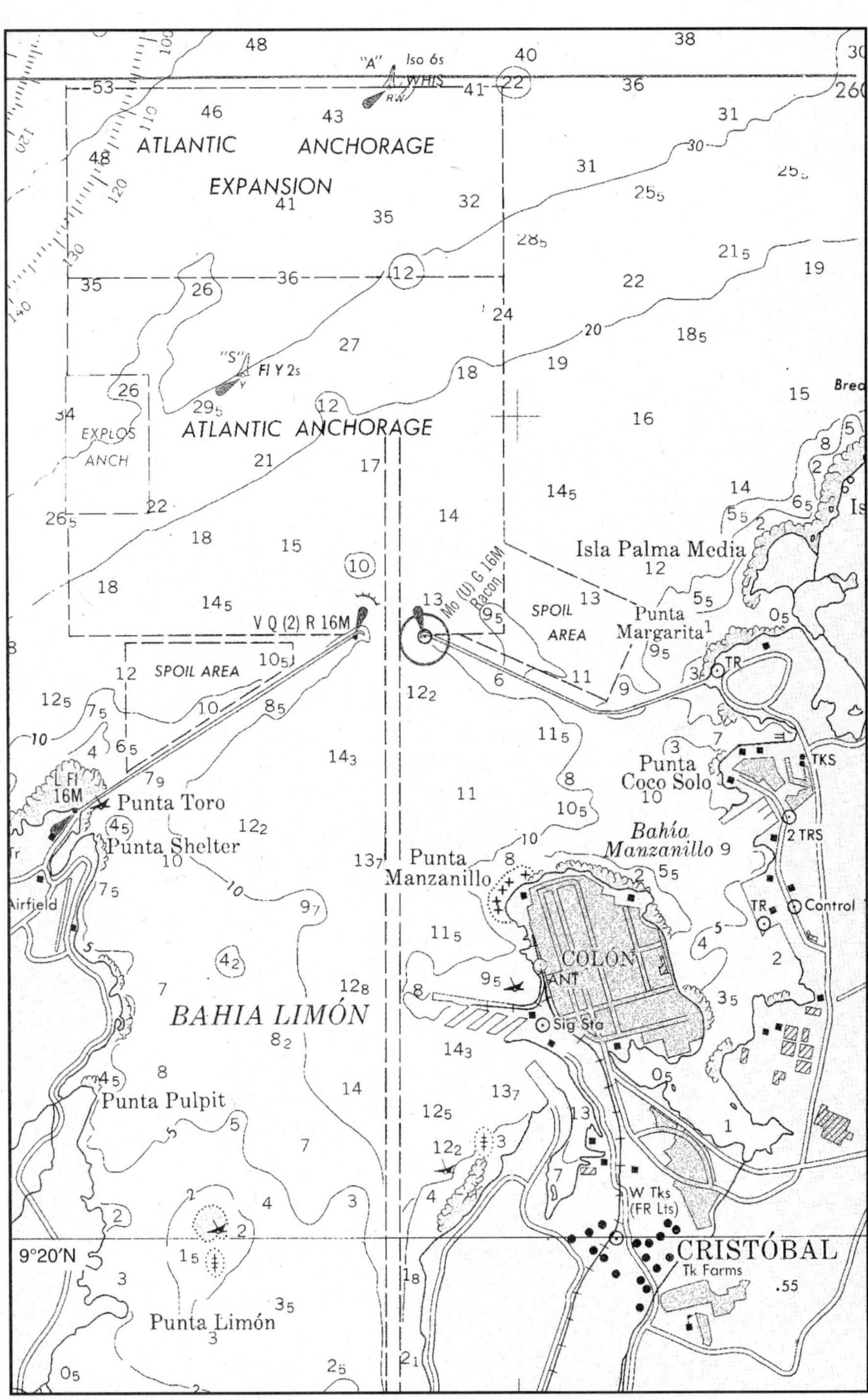

Approaches to Cristóbal, soundings in meters *(from DMA 26066, 11th ed, 7/90)*

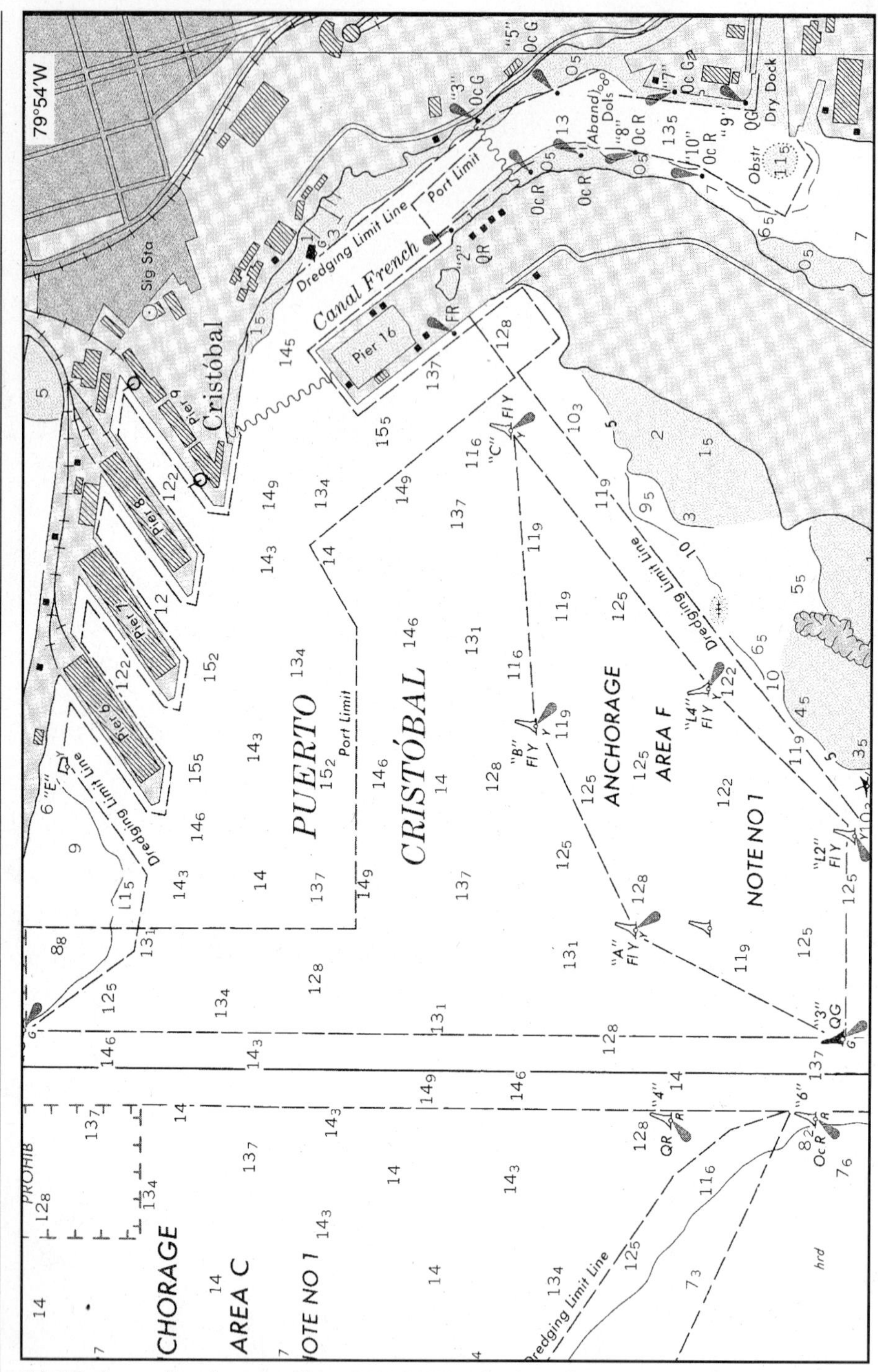

Cristóbal, soundings in meters *(from DMA 26068, 8th ed, 2/90)*

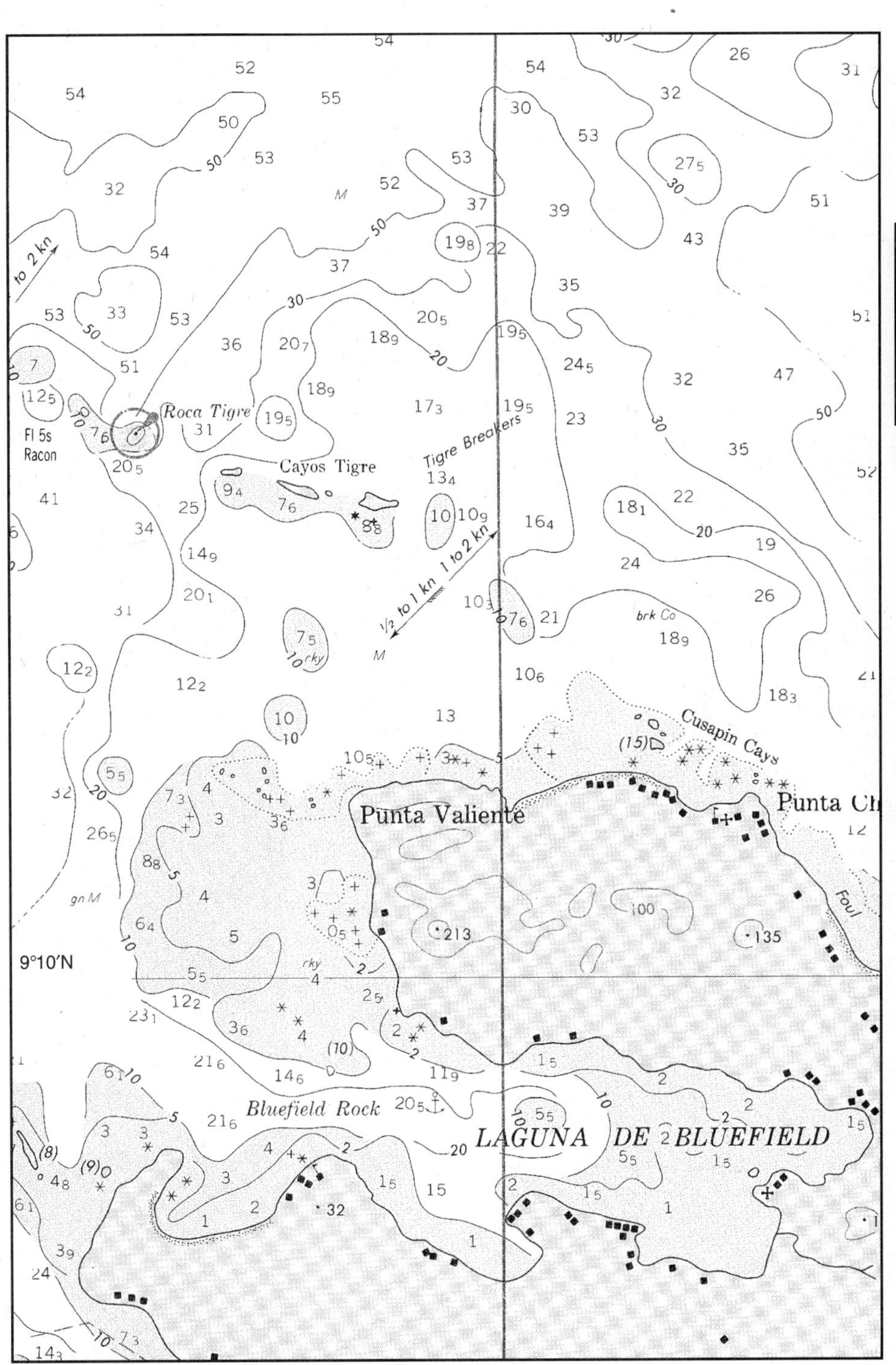

Laguna de Bluefield, soundings in meters *(from DMA 28041, 3rd ed, 7/85)*

Anchorage: Yachts should anchor in Anchorage Area F, located south of the Muelle Cristóbal. This area is marked by flashing yellow buoys and is known as "The Flats."

Services: Fuel, water, showers, laundry, haul-out facilities, and a bar and restaurant can all be found at the yacht club. This is also the place to get advice on transiting the canal or locating something you might need. The Pedro Miguel Boat Club inland along the canal is set up for those who want "do-it-yourself" type repairs, including haulouts. It is reported to be a good place to store your boat while traveling ashore.

Panama Canal: See description on preceding pages.

LAGUNA DE CHIRIQUÍ

ESCUDO DE VERAGUAS, 9 05.3N, 81 32.2W. Fl W 7s, 120ft, 6M. White square framework tower, 19ft.

CHIRIQUÍ RANGE (Front Light), 8 56.5N, 82 06.8W. Q W. White cylindrical tower with black stripe. F R obstruction lights on tower 1.4M W; F Y on oil tanks 1.5 miles W. Visible 202°–222°. **(Rear Light),** 60 meters, 212° from front. Fl W 4s. White cylindrical tower with black stripe. Visible 202°–222°.

LAGUNA DE BLUEFIELD

9 09N, 81 54W

Pilotage: Located south of Punta Valiente, in the entrance to Laguna de Chiriquí, this is a possible anchorage. The ground is foul west of Punta Valiente and along the shores in Laguna de Chiriquí. The bottom is reported to be mud.

LAGUNA DE CHIRIQUÍ

8 56N, 82 07W

Pilotage: There are many anchoring possibilities in the Laguna de Chiriquí. The entrance, through Canal del Tigre, is well marked because of the oil terminal at Chiriquí Grande. Currents in the inlet are generally weaker than 1 knot. The Chiriquí Grande pilots stand by on VHF channel 14, and the terminal can be contacted by calling "Rambala Control" on VHF channel 16. The terminal is located on the south side of the bay. There are several lit moorings and a range of 212° for the terminal.

PUNTA VALIENTE

9 11N, 81 55W

Pilotage: This is the eastern point at the entrance to Laguna de Chiriquí. It is about 120 miles west of the entrance to the Canal. There is foul ground west of the point, extending south to the entrance to Laguna de Bluefield. Valiente Peak, a conspicuous 758-foot-high peak, stands about 1 mile south of Punta Valiente.

ROCA TIGRE, 9 13.0N, 81 56.5W. Fl W 5s, 50ft. Metal tower. **RACON P** (• – – •).

BAHÍA DE ALMIRANTE

HOSPITAL POINT, 9 19.9N, 82 13.2W. Fl (2) W 10s, 36ft, 13M. Metal pedestal, 6ft. F R lights on radio masts 1.2M, 1.9M, and 2.2M WNW.

BEACON 8, 9 20.4N, 82 13.6W. Q R, 16ft. Red.

BOCAS DEL TORO, pier NE corner, 9 20.0N, 82 14.4W. Fl G, 27ft. White pedestal, 20ft.

PIER, SW corner, 9 20.0N, 82 14.4W. Fl R.

BOCAS DEL TORO

9 21N, 82 14W

Port of entry

Pilotage: The harbor is entered via Canal de Bocas Del Toro, which begins west of Cabo Toro (9°22′N, 82°12′W). There is a breaking reef around the shores of Cabo Toro, and Toro lies awash about 350 yards northeast of the cape. Heavy seas break on the shoals extending up to a mile off Long Bay Point to the north. The channel to town is marked with lighted aids to navigation.

Anchorage: Anchor off town as depths permit. Some general supplies may be available. This is the last good anchorage before reaching the border of Costa Rica.

JUAN POINT, 9 18.2N, 82 17.8W. Fl G 4s, 15ft, 5M. Green tower.

PONDSOCK REEF, 9 17.3N, 82 19.8W. Fl W 6s, 15ft, 5M. Black tower, red bands.

ALMIRANTE, pier, 9 17.3N, 82 23.5W. Fl R. Post. Occasional.

ISLA PASTORES, NW end, 9 14.7N, 82 21.2W Fl W 10s, 60ft, 9M. Metal windmill tower, 49ft.

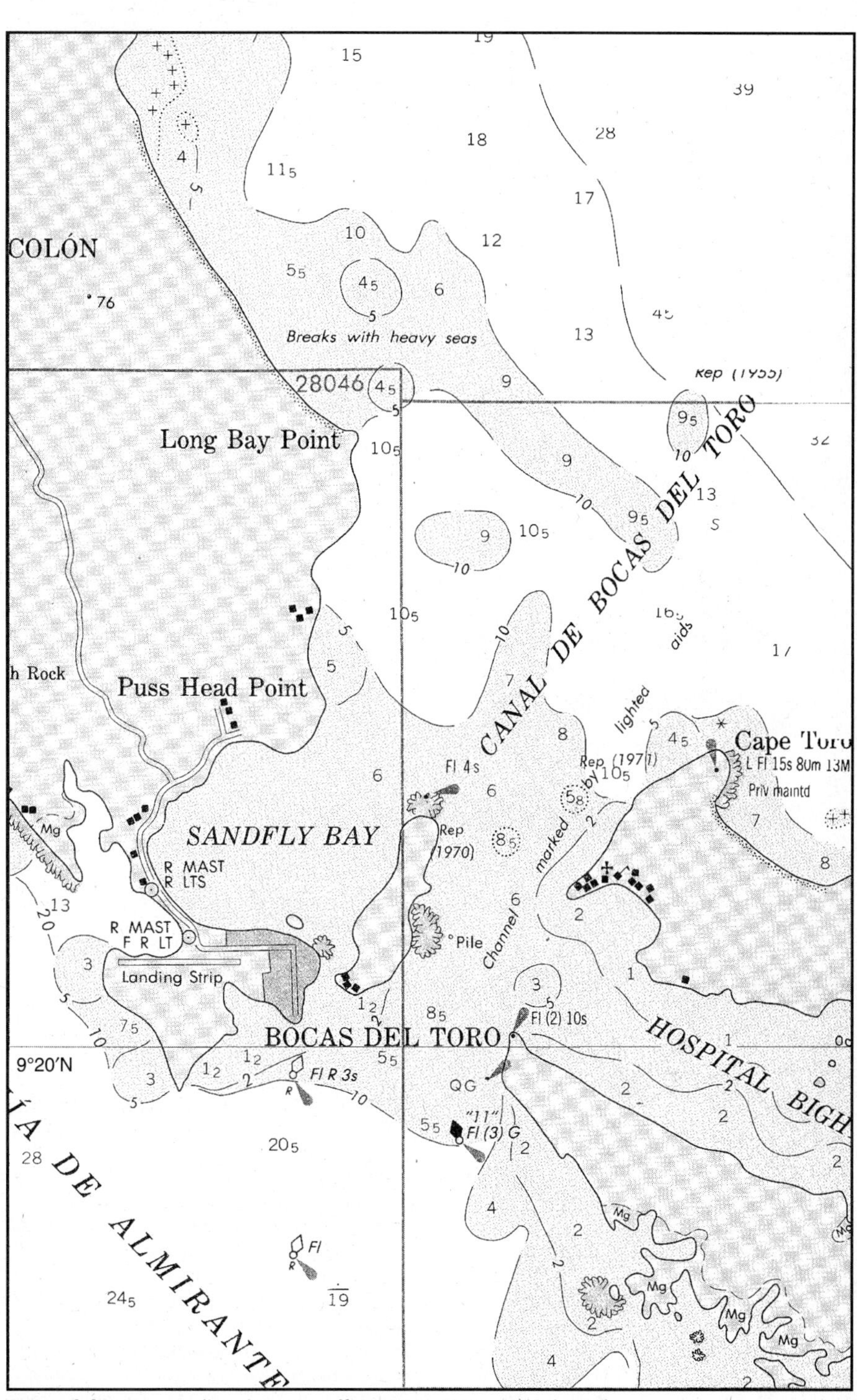

Bocas del Toro, soundings in meters *(from DMA 28041, 3rd ed, 7/85)*

Costa Rica

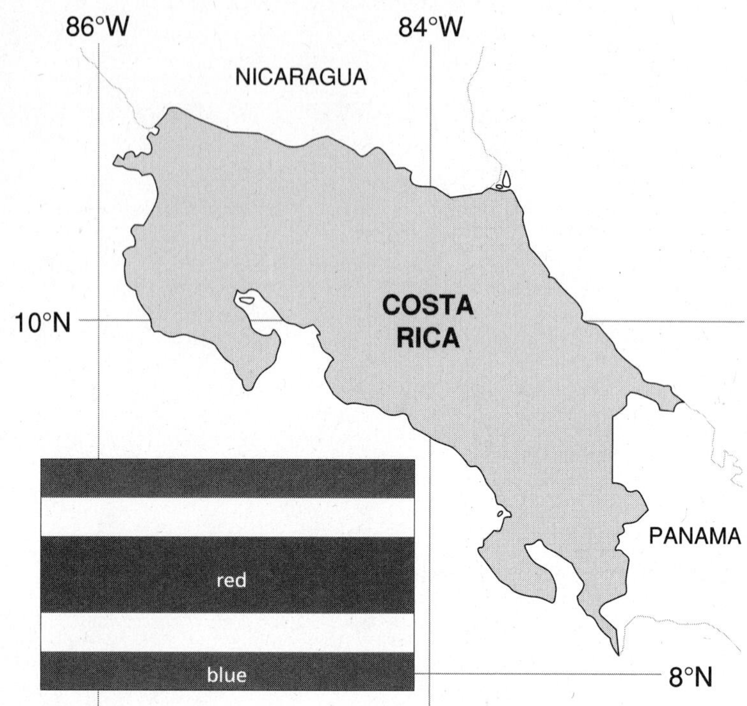

Please see cautions about chartlets and waypoints on page P 2.

ENTRY PROCEDURES

The only harbor of interest to private vessels on the Caribbean coast of Costa Rica is Puerto Limón. It is a port of entry and a major shipping center for the country. Bananas and cocoa are the principal exports, but considerable general cargo is handled. Nearby commercial Puerto Moín is also a port of entry.

Visitors should have the usual papers—ship's documents, clearance from the last port of call, crew lists, and passports. U.S., Canadian, Central and South American, and some European and Caribbean citizens are issued tourist cards upon arrival; citizens of other countries should check with Costa Rican officials before heading for this port. For more information on entry requirements contact the Consulate of Costa Rica, 2112 S Street NW, Washington, DC 20008 (202-328-6628). In the U.S. and Puerto Rico, there are additional Costa Rican consulates in Atlanta, Chicago, Houston, Los Angeles, Miami, New Orleans, New York, San Antonio (TX), San Diego, San Francisco, Tampa (FL), and San Juan, Puerto Rico.

USEFUL INFORMATION

Language: Spanish. English is spoken frequently, especially in tourist areas and cities.
Currency: Colón; exchange rate (summer 1997), 210 colónes = US$1.
Telephones: Country code, 506; local numbers, seven digits.
Public holidays (1998): New Year's Day, January 1; Feast of St. Joseph, March 19; Holy Thursday, April 9; Good Friday, April 10; Santa Rosa Battle Anniversary, April 11; Labor Day, May 1; Annexation Day, July 25; Virgin of Los

Angeles, August 2; Mother's Day, August 15; Independence Day, September 15; Columbus Day, October 12; Immaculate Conception Day, December 8; Christmas Day, December 25. Government offices are officially open but most businesses close the week between Christmas and New Year's.
Tourism office: *In Costa Rica:* Costa Rica Tourist Board, ICT PO Box 777-1000, San Jose (from the U.S. and Canada, tel., 800-343-6332; fax, 506-255-4997).
Special notes: The U.S. Department of State reports that crime is rising in Costa Rica, with tourists as well as residents the victims. Incidents have occurred at tourist areas. There is limited risk of malaria in rural areas, and cases of cholera have been reported. For the latest medical advisory, contact the Centers for Disease Control International Travelers Hotline (voice recording, 404-332-4559; fax information service, 404-332-4565; Web, www.cdc.gov). For assistance within Costa Rica, U.S. citizens may contact the Consular Section of the U.S. Embassy in Pavas, San José (506-220-3050; emergencies outside business hours, 506-220-3127).

CARIBBEAN COAST

SIXAOLA, 9 34.1N, 82 33.6W. Fl W 10s, 46ft, 11M. White skeleton steel tower, 39ft. Visible 114°–315°. **RACON K** (– · –), 20M.
LIMÓN, ISLA UVITA, summit, 9 59.6N, 83 00.7W. Fl W 10s, 134ft, 5M. White metal framework tower, 82ft. Fl R 2s and F R on radio tower 0.9M WNW; Aero Al Fl W G 10s 2M SW; R lights on radio tower 1.5M SW.
LIMÓN CONTAINER PIER, S end. 9 59N, 83 01W. F R, 49ft, 10M.

PUERTO LIMÓN

10 00N, 83 01W
Pilotage: This open coastal harbor is about 234 miles west-northwest of the entrance to the Panama Canal. The approaches are safe and easy to access. At night the loom of the bright pier lights is seen well out to sea. Two radio masts on Punta Blanca are reported to be prominent. A south-setting current is generally experienced between Isla Uvita and Puerto Limón. During the dry season it attains a velocity of less than 1 knot, but during the rainy season may run at 0.75 to 1.25 knots.

In general, the current sets southeast along the coast from Limón north to the Nicaraguan border. There is a strong discharge from the rivers during the rainy season. A continual swell from the northeast breaks along the beaches of this entire coast.

Volcan Turrialba (10°01'N, 83°45'W), a 11,349-foot-high inactive volcano, is located about 43 miles west of Puerto Limón. Seen from the east, the peak and crater appear clearly defined, with the hollow of the crater being on the north side. The peak is generally hidden by clouds. Close south of the volcano stands 11,250-foot Monte Irazú. On a clear day, these conspicuous peaks can be seen for a considerable distance.

Port of entry: The customs office is across the street from the gates of the port area.
Radio communications: Call for the port captain (capitania del puerto) or Puerto Limón port authorities on VHF channel 16 or 2182 kHz (call sign, TIM).
Anchorage: The anchorage is rather exposed and open to swells. There is some shelter from Isla Uvita. There is reported to be a strong surge along the docks here. Be prepared to leave quickly if bad weather threatens.
Services: There are good markets and general stores. Fuel and water can be jugged. Public transportation connects to San José and other parts of the country. Many residents here speak English.

BAHÍA MOÍN, ISLA PAJAROS, 10 01.0N, 83 04.6W. Fl W 2.5s. Tower.
PUERTO MOÍN RANGE (Front Light), 10 00.2N, 83 04.6W. F W. Tower. **(Rear Light),** 130 meters, 151° from front. F W. Tower.

BAHÍA MOÍN

10 00N, 83 05W
Pilotage: This bay is the location of the commercial port facility Puerto Moín, which is on the west side of the same peninsula as Puerto Limón. It is primarily a banana-exporting and oil-importing terminal. There is a light on Isla Pájaros and a rock close to the northwest. The tanker moorings in the bay are subject to frequent changes of positions, colors, and characteristics. There is a continual surge in this harbor. Depths have been reported (1993) to be less than charted due to earthquake activity.
Port of entry: Cruising yachts should clear in Puerto Limón.
Anchorage: Private yachts need special permission from the port authorities to anchor in this harbor. There are bus connections to Limón.

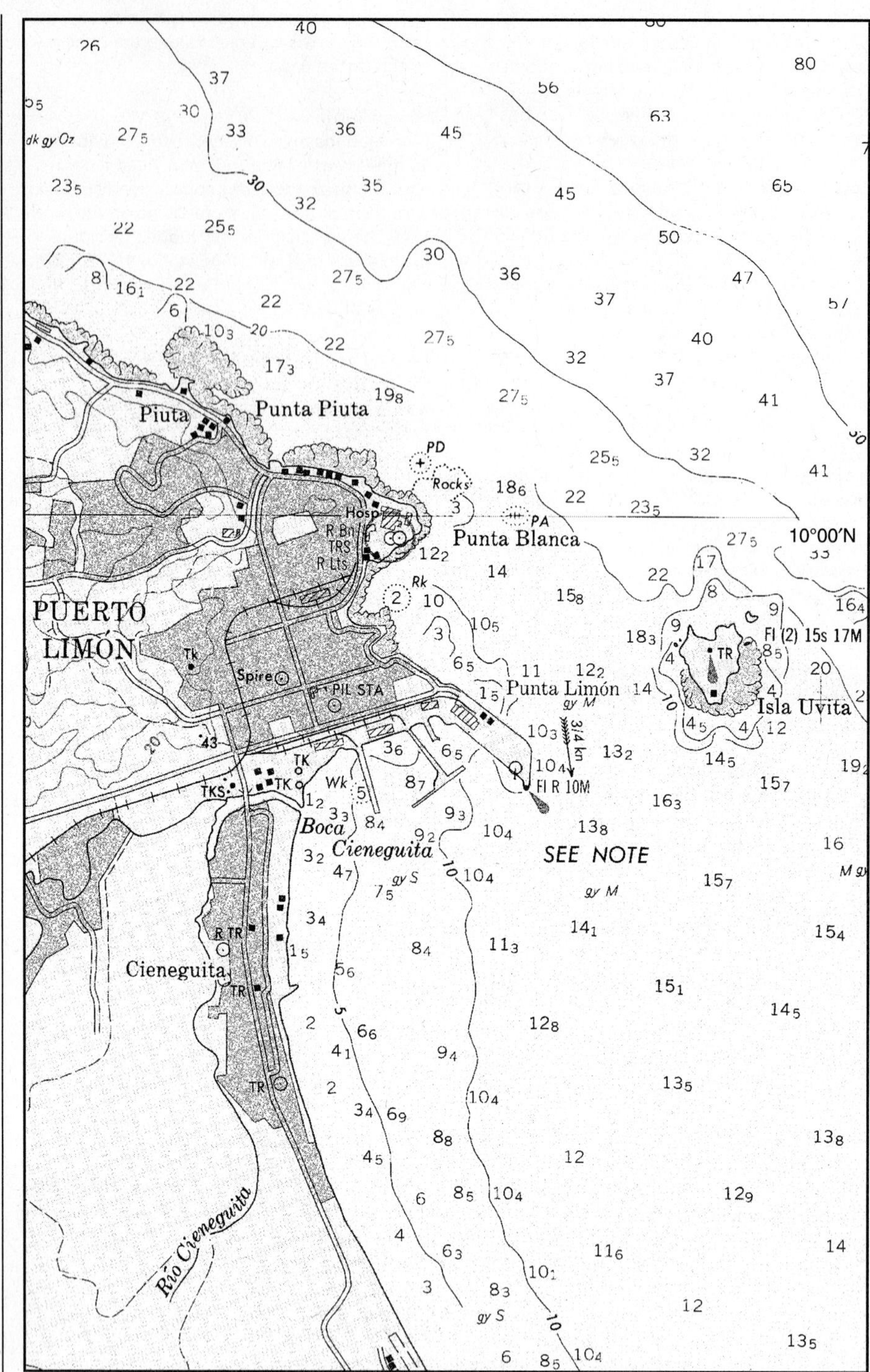

Puerto Limón, soundings in meters *(from DMA 28049, 12th ed, 10/89)*

Nicaragua

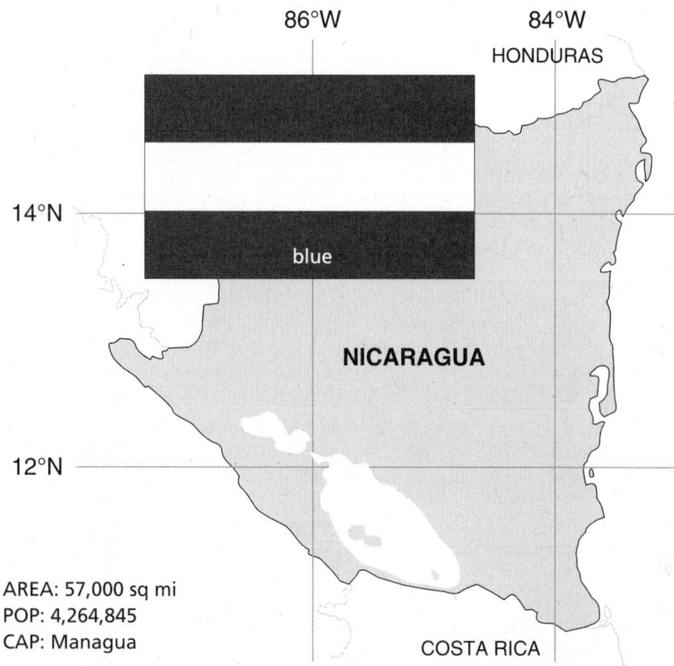

AREA: 57,000 sq mi
POP: 4,264,845
CAP: Managua

CAUTION: Mariners operating small vessels such as yachts and fishing boats are advised to avoid the Caribbean ports and waters of Nicaragua until further notice. There have been several cases of foreign flag vessels being seized off the Nicaraguan coast by Nicaraguan authorities. While in all cases passengers and crews have been released within a period of weeks, in some cases the ships have been searched, personal gear and navigational equipment have been stolen, and there have been excessive delays in releasing vessels. Prompt U.S. embassy consular access to detained U.S. citizens may not be possible because of non-notification of the embassy by the Nicaraguan government. It should also be noted that there have been incidents of piracy in the Caribbean waters off the coast of Nicaragua.

Please see cautions about chartlets and waypoints on page P 2.

ENTRY PROCEDURES

The port of entry on the Caribbean coast is Bluefields. This harbor is located about 275 miles northwest of the Panama Canal and 120 miles north of Puerto Limón, Costa Rica.

Passports should be valid at least six months beyond your intended departure date, and all persons should be prepared to demonstrate financial resources of at least about US$200. You should also have the usual papers: clearance from the last port, ship's documents, and crew lists. Visas may be required, depending on your nationality and your intended length of stay. Many nationals, including U.S. citizens, can get a tourist card valid for 30 days (extendable) upon entering the country. Citizens of Canada and most Caribbean countries should apply for a visa in advance of their visit. For visa information contact the Nicaraguan Consulate, 1627 New Hampshire Avenue NW, Washington, DC 20009 (202-939-6570; fax, 202-939-6545). The consulate is part of the Nicaraguan Embassy.

For latest advisories about traveling to Nicaragua, contact the U.S. State Department Citizens' Emergency Center (202-647-5225) or the Canadian Travel Advisory Line (1-800-267-6788).

USEFUL INFORMATION

Language: Spanish. English is spoken in some Caribbean coastal and island communities.
Currency: Córdoba; exchange rate (summer 1997), 8.5 córdobas = US$1. Credit cards are accepted by more hotels and restaurants than in the past, but use is still not widespread. Traveler's checks are accepted at a few major tourist hotels but otherwise may be changed at banks and official exchange stations *(casas de cambio).*
Telephones: Country code, 505; a one-, two-, three-, or four-digit city code follows, and then the local number. The city code for Leone is 311; Managua, 2; San Juan del Sur, 4682; San Marcos, 432; and Bluefields (Autonomous Atlantic Region), 822.
Public holidays (1998): New Year's Day, January 1; Anniversary of the Elections, February 25; Labor Day, May 1; Anniversary of the Revolution, July 19; Independence Day, September 14–15; Columbus Day, October 12; Immaculate Conception Day, December 8; Christmas Day, December 25.
Tourism office: *In the U.S.*, National Travel Information Center, P.O. Box 140357, Miami, FL 33314 (tel., 305-860-0747 or 1-800-660-7253).
Special notes: Travelers to Nicaragua may be exposed to tropical diseases. Malaria is present in rural areas. Cases of cholera and dengue fever have been reported. For the latest medical advisories, contact the Centers for Disease Control International Travelers Hotline (voice recording, 404-332-4559; fax information service, 404-332-4565; Web, www.cdc.gov). For information or assistance while in the country, U.S. citizens may contact the United States Embassy, Kilometer 4fi Carretera Sur, Managua (505-2-66-6010).

Inland travelers should be aware of occasional flare-ups of armed violence, especially in the northern part of the country. The roads connecting Nicaragua and Honduras may be dangerous. All roads are considered dangerous at night. Vessels should be aware that boundary disputes with Honduras and Costa Rica concerning Caribbean waters persist, as does a dispute with Colombia concerning Isla de San Andrés and surrounding waters.

CURRENTS

Between Punta del Mono (Punta Mico) and Punta Gorda (near Cayos Miskitos) the currents on the edge of the 200-meter depth curve are affected by the offshore equatorial current, which sets west-northwest. Generally, part of the current recurves to the southwest in the vicinity of Isla de Providencia, while the main flow continues northwest across Miskito Bank. Subsequently, the currents along the coast are variable and subject to great and sudden change, being influenced to a large extent by the wind. However, they tend to run south at a velocity varying from 0.5 to 3 knots. This southerly set at times reverses its direction for several days before resuming its normal flow.

The current is stronger in the vicinity of Punta Mico, where it sometimes sets east, than between Cayos de Perlas and Puerto Cabezas. Between Puerto Cabezas and Punta Gorda the current is variable, but tends to set north. However, this current, too, may be completely reversed by a norther. A countercurrent setting south may be experienced close inshore.

PUNTA CASTILLO TO EL BLUFF (BLUEFIELDS)

PUNTA CASTILLO

10 56N, 83 40W
Pilotage: This low point marked by breakers is about 10 miles north-northwest of the Río Colorado. The entrance to San Juan del Norte is located about 2.8 miles west of the point. Morris Shoal, about 4 miles long and 1 mile wide, lies about 10 miles east of the point. Depths over the shoal range from 60 feet to about 72 feet.

SAN JUAN DEL NORTE

10 56N, 83 42W
Pilotage: This harbor was once the proposed terminus of the Nicaraguan ship canal. The project was abandoned in 1893, and the harbor is now closed to ocean-shipping because of silting, but it is still possible for small craft to enter. The village and abandoned equipment are reported to be obscured by trees.

North of the harbor the coast is low and sandy for about 29 miles to a bold rocky point. A small wooded island lies near the point. The surf breaks constantly along these sandy beaches. The twin peaks of Round Hill rise to about 617 feet about 20 miles northwest of Punta Castillo. The coast then becomes higher and extends irregularly northeast to Punta Gorda (11°26′N, 83°48′W), a prominent rocky point. Several steep-to islets lie close offshore. Río de Punta Gorda discharges into the sea about 4.5 miles northeast of the point. There are several settlements at the river mouth. From the point the coast extends irregularly for 12.5 miles northeast to Punta del Mono (Punta Mico).

ISLA DEL PAJARO BOBO

11 30N, 83 43W

Pilotage: This 154-foot-high wooded islet lies about 5.8 miles southwest of Punta Mico and 3 miles offshore. When seen from the east it appears as a small green conical hill, but from the south it appears wedge-shaped, with the higher end to the west.

PUNTA DEL MONO (PUNTA MICO)

11 36N, 83 40W

Pilotage: This is the south extremity of a bold rocky peninsula that extends about 1.5 miles southeast and is about 2 miles wide. In the middle of the promontory stands Red Hill, about 100 feet high and red in color. Red cliffs extend about 1 mile northwest from Black Bluff on the northeast end of the promontory. Shoal water and several cays lie within 1 mile of the point. From here a low coast extends about 45 miles north to Punta Mosquito at the entrance to Laguna de Perlas. El Bluff, a bold promontory, stands about midway along this stretch of coast.

FRENCH CAY

11 44N, 83 37W

Pilotage: This small, flat, wooded islet rises to a height of 90 feet. It stands about 7.5 miles north-northeast of Black Bluff, on the north part of a narrow reef about 1.25 miles long. Sister Cays, three small islets, stand 8.5 miles north of Black Bluff. Two breaking reefs lie about 3 miles southeast and 2.5 miles east of Green Point. Cayo de la Paloma is a 108-foot- high cay with a saddle-shaped summit lying about 1.5 miles east-northeast of Green Point. It is the largest cay in this area. Other hazards in the area include White Rock and Guano Cay. Caution should be used in this area because of the many reefs and dangers.

EL BLUEFIELDS BLUFF SUMMIT LIGHT, 12 00N, 83 41W. Fl W 3.8s, 163ft, 14M. Red metal framework tower, 26ft.

EL BLUFF

12 00N, 83 41W

Pilotage: This bold promontory is more than 137 feet high and wooded. It stands on the east side of the entrance to Laguna de Bluefields, but looks like an island from seaward. Red cliffs stand on the east side of the promontory.

BLUEFIELDS

12 01N, 83 45W

Pilotage: Cayo Casaba, a low wooded islet with a smaller islet to its west, lies in the middle of the entrance to Laguna de Bluefields. The navigable channel between this islet and El Bluff is about 200 to 400 yards wide. The town of Bluefields is located on the west side of the lagoon about 3.5 miles inside the entrance. The port for the town is on the west and northwest sides of El Bluff. This is where all cargo is handled. The entrance bar has a least depth of about 9 feet, and the harbor has depths of less than 18 feet. Currents off the entrance generally set south at about 1.5 to 2 knots, but can reverse for a day for no apparent reason. A conspicuous white house on the northwest side of El Bluff and the radio towers in town are landmarks. The pilots monitor VHF channel 16.

Port of entry: Check in at the office on the El Bluff docks.

Anchorage: About 1.5 miles S of El Bluff, the bottom is soft mud in about 24 feet of water. Inside the bar is a good, although confined, anchorage with depths of about 26 feet west of the wharf on the northwest side of El Bluff.

MISKITO BANK

Pilotage: From a position about 25 miles south of Punta del Mono (Punta Mico) the 200-meter curve extends north-northeast to a position 28 miles east of the point and then extends north-northeast to a position about 72 miles east of Punta Gorda near the Honduran border. Miskito Bank lies within this 200-meter-depth curve.

The depths within the curve are irregular, and numerous cays, islands, and other dangers exist in the area. There are numerous detached shallow patches—some with depths of less than 6 feet. Most of the bank has depths of around 78 to 108 feet. Coral reefs grow about 0.2 foot annually, and depths may be less than charted in some areas.

The turtle fishermen in this area are reported to know where most of the rocks are and are adept at estimating depths. This is an area for eyeball navigation during the day. Travel when the sun is high and preferably behind you.

CAUTION: Great care is required in navigating this area because of reef growth and the unreliability and/or age of chart surveys. New shoals have been reported.

BLOWING ROCK

12 02N, 83 02W

Pilotage: This rock is about 4 feet high and has a hole through the center from which water is occasionally forced, which looks like the spouting of a whale. It lies about 38 miles east of El Bluff.

ISLA DEL MAIZ GRANDE (GREAT CORN ISLAND)

12 10N, 83 03W

Pilotage: This island is about 2.5 miles long by 2 miles wide and lies about 38 miles east-northeast of El Bluff. Mount Pleasant rises to a 371-foot wooded peak in the middle of the north part of the island. A 98-foot-high rocky bluff stands at the south end of the island. The island is fringed by foul ground extending about 0.25 to 1.5 miles offshore. A stranded wreck about 1.25 miles north-northeast of Mount Pleasant is reported to give a good radar return.

Anchorage: There is an anchorage in Southwest Bay on the southwest side of the island. Depths run about 27 to 30 feet. There are also anchorages in Brig Bay on the island's west side and in Long Bay on the southeast side. Care must be taken to avoid the reefs and shoal patches in or near these anchorages.

A 321-foot-long pier extends from the shore near a shrimp processing plant in the head of Southwest Bay. Depths range from 6 to 13 feet alongside. There is a conspicuous building that is lighted at night. A similar building, also lit at night, stands at the head of Brig Bay.

ISLA DEL MAIZ PEQUEÑA (LITTLE CORN ISLAND)

12 18N, 82 59W

Pilotage: This 125-foot-high island is about 0.5 mile long and 0.5 mile wide. It lies about 7.5 miles north-northeast of Isla del Maiz Grande. Its north and northeast sides are fringed by reefs extending about 0.5 to 1 mile offshore. The west side of the island is fairly steep-to seaward of the 10-meter curve, which lies between 0.25 and 0.5 mile offshore.

Anchorage: There is an anchorage in Pelican Bay on the southwest side of the island. Depths are reported to be about 36 feet with the west tangent of the island bearing 342° and the south tangent of the island bearing 106°.

CAYOS DE PERLAS

12 29N, 83 19W

Pilotage: These cays and reefs lie between Punta de Perlas and a position about 130.5 miles east. They extend about 12 miles to the north within the 20-meter-depth curve, which runs about 8 to 17 miles offshore.

Although there are depths of 36 to 60 feet in this area, navigation is very hazardous because of the numerous charted and uncharted reefs and shoals. Some of these may not be visible because of the turbid waters. A mud bottom may be only a thin covering over coral, and you should anchor with care.

SEAL CAY

12 25N, 83 17W

Pilotage: The southeasternmost of the Cayos de Perlas, Seal Cay is a small coral ridge about 3 feet high, lying about 12.5 miles east of Punta de Perlas. Foul ground extends about 0.25 mile northwest and about 1 mile south-southwest from the cay, but the southeast side is steep-to.

The south limits of Cayos de Perlas are marked by Columbilla Cay, a steep-to reef-fringed islet, 110 feet high to the tops of the trees, located about 6.5 miles southwest of Seal Cay. Maroon Cay, a similar islet on the edge of a reef extending about 2 miles east from Punta de Perlas, marks the southwest extremity of the area.

BODEN REEF AND NORTHEAST CAYS

12 30N, 83 19W

Pilotage: Boden Reef is about 0.8 mile long, with a least depth of 14 feet. It lies along the eastern edge of the Cayos de Perlas about midway between Seal Cay and the Northeast Cays located 8 miles northwest of Seal Cay. The Northeast Cays are a small group of reef-fringed islets with several rocks awash about 1 mile to the northwest. They mark the northeastern extremity of the Cayos de Perlas. Numerous rocks awash and 9- to 18-foot patches lie up to about 5 miles west-southwest of these cays. Within this area lie the Crawl Cays and the Tungawarra Cays, as well as numerous small cays and shoals.

Anchorage: Good anchorage can be taken about 0.3 mile southwest of a mooring buoy lying 0.3 mile west of Little Tungawarra Cay. This anchorage should be approached from the south. Other anchorages may be found in the lee of the larger cays.

GREAT KING CAY

12 45N, 83 21W

Pilotage: This 70-foot-high cay is the largest of a group of islets lying about 12 miles north of the Cayos de Perlas and about 9 to 13 miles off the mainland. Little King Cay is about 32 feet high and stands about 0.8 of a mile east of Great King. The two Rocky Cays lie about 1 mile to the northwest, and 8-foot-high Little Tyra Cay lies about 2 to 3 miles west-southwest of Great King.

GREAT TYRA CAY

12 52N, 83 23W

Pilotage: This cay is the largest of a group of islets and shoals lying about 8 miles north of Great King Cay. Foul ground extends to its north about 0.5 mile. Seal Cay, two barren rocks about 10 feet high, lies about 0.8 mile to its south, and a dangerous shoal patch lies about 1 mile east of Seal Cay. Several detached reefs lie between 1 and 3 miles southwest of Great Tyra Cay. Shoals are also located about 1 mile and 5 miles northwest of Great Tyra.

Tyra Rock is about 8 feet high and located about 4.5 miles northeast of Great Tyra Cay. Numerous detached reefs and shoals, over which the sea breaks, lie up to 3.5 miles west and 1 mile north of Tyra Rock.

CAYOS MAN O WAR

13 01N, 83 23W

Pilotage: This cluster of islets lies about 11 miles offshore and 6.8 miles northwest of Tyra Rock. The largest rises to a height of about 50 feet. An oil barge, formerly used as a storage tank, stands on the west cay. In a sheltered bight on the west side of this cay are pilings where vessels formerly moored to load lightered cargo. Depths alongside range from 13 to 22 feet.

Numerous detached reefs and shoals lie up to about 3 miles south-southwest and west-northwest of Cayos Man O War.

CAUTION: Many vessels have reported striking coral heads inside the 20-meter curve in the vicinity of Cayos Man O War and between the cay and Puerto Cabezas.

EGG ROCK

13 02N, 83 22W

Pilotage: This 6-foot-high steep-to rock is located about 1.3 miles northeast of Cayos Man O War. A dangerous shoal lies about 13 miles north-northwest of Cayos Man O War.

PUNTA PERLAS TO CABO GRACIAS A DIOS

RÍO GRANDE DE MATAGALPA

12 54N, 83 32W

Pilotage: This river is navigable by barges for 106 miles upstream. The bar entrance has a depth of about 5 feet. Río Grande village is on the north bank of the river near the entrance. A fruit station and a wharf are on the south bank, and two radio towers to the south are conspicuous. Sandy Bay village is 3.5 miles north of the river.

PUERTO ISABEL RANGE (Front Light), 13 21.0N, 83 33.0W. F R. Roof of shed. **(Rear Light),** 90 meters, 276° from front. F W.

PUERTO ISABEL

13 22N, 83 34W

Pilotage: This open roadstead is a privately owned port. There is a 0.5-mile-long pier extending east from shore. From March to April currents run north and from May to February south at a velocity of 1 to 2 knots.

RÍO PRINZAPOLCA

13 25N, 83 34W

Pilotage: There is a small village on the south bank of the river. Shoal water, with depths of less than 6 feet, extends 1 mile east from the river.

RÍO HUALPASIXA

13 29N, 83 33W

Pilotage: The sea breaks heavily on the bar at the entrance to this river, which is located about 4 miles north of the Río Prinzapolca. There is a village at the river mouth and another called Wounta on the south side of a small lagoon about 4 miles north of the river. Foul ground extends about 3 miles east from just north of the lagoon entrance.

Several prominent mounds, each about 80 feet high, stand near the coast about 14 miles north of Wounta. From east of these mounds south to a position 11 miles north of the Río Grande, the area within the 20-meter curve is restricted by numerous dangers. From the mounds north to Puerto Cabezas, there are few dangers within the 10-meter curve, which lies within 2 to 4 miles of the coast.

RÍO HUAHUA

13 53N, 83 27W

Pilotage: This river is about 21 miles north of Wounta. After heavy rains, muddy river water discolors the sea for some distance offshore. From the river the coast extends about 10 miles north to Bragman Bluff, a bold headland about 98 feet high with red cliffs extending about 0.5 mile along its east edge.

BRAGMAN BLUFF LIGHT, 14 01.5N, 83 23.3W. Mo (B) W 10.6s, 168ft, 5M. Southern square water tower. Visible 201.5°–021.5°. Private light.

PUERTO CABEZAS

14 01N, 83 23W

Pilotage: This harbor is an open roadstead with a government wharf extending about 0.5 mile southeast from Bragman Bluff. It is a banana and lumber export port. Some water towers, the radio towers on the east side of town, and several chimneys in its southeast part are prominent landmarks. At night, the loom of the town's lights may be seen long before the navigational light can be spotted. The beach shows poorly on radar, although railroad cars on the siding are reported to make an excellent return. The pier is reported to have a good radar return up to 14 miles. The current in the vicinity of the pier sets south or south-southwest at a velocity of about 1.3 knots. This is a possible port of entry, although the harbor is not sheltered and makes a poor anchorage.

CAYOS MISKITOS AND SURROUNDING CAYS

14 23N, 82 46W

Pilotage: Cayos Miskitos is the center of a group of cays and reefs with a diameter of about 19 miles. Approaches to this area are dangerous because of the many detached reefs and shoals and the lack of any navigational aids. Cayos Miskitos is the largest of the group, being about 2.5 miles in diameter. Several smaller islands and islets lie adjacent to it. This area is part of a reef that extends 19 miles north-northeast from a position about 21 miles east of Punta Gorda. Currents in this vicinity generally run north to northwest at speeds of 0.25 to 1 knot. Inside the 20-meter curve close to the coast there may be a southerly countercurrent. These currents are variable and may even reverse, especially with a norther. Porgee Channel (14°26'N, 82°41'W) is a narrow and intricate route bisecting the reef from east to west about 0.5 mile northeast of the northeasternmost island.

TSIANKUALAIA ROCK

14 20N, 83 04W

Pilotage: This rock, with a depth of about 9 feet over it, lies on the southwest side of the Cayos Miskito group, about 7.8 miles east of Punta Gorda. Several rocks within 6 feet of the surface lie within 0.5 mile south and southeast of this rock. Waham Cay, 3 feet high, stands about 3.5 miles north-northeast of the rock. Toro Cay and Kisura Cay, two similar islets, stand about 2 miles north-northwest and 2 miles north, respectively, of Waham Cay.

Alice Agnes Rocks consist of several awash rocks that lie on the southwest limits of the Cayos Miskitos group. They are about 13 miles west-southwest of Cayos Miskitos and about the same distance southeast of Punta Gorda.

NED THOMAS CAY

14 10N, 82 48W

Pilotage: This cay and The Witties are about in the center of a group of reefs. The group is about 8 miles in diameter and lies about 11.5 miles south of Cayos Miskitos. Sea Devil Reef and Franklin Reef lie on the south limits of the Cayos Miskitos group. They have depths of at least 15 feet over them.

SOUTHEAST ROCK

14 10N, 82 29W

Pilotage: This rock lying within 9 feet of the surface is on the southeast extremity of the Cayos Miskitos group.

HANNIBAL SHOALS

14 26N, 82 31W

Pilotage: This 25-foot-deep shoal marks the eastern side of the Cayos Miskitos group.

BLUE CHANNEL

14 25N, 82 50W

Pilotage: This channel is about 1.5 to 3 miles wide and runs parallel to the west side of Miskito Reef. Depths range from 30 to 84 feet. The Morrison Dennis Cays and the Valpatara Reefs (14°27'N, 82°58'W) extend about 12 miles north.

AUIAPUNI REEF

14 31N, 83 05W

Pilotage: This group of shoal patches about 1.5 miles in extent lies about 7.5 miles southwest of Outer Mohegan. They are at the west extremity of the Cayos Miskitos group.

HAMKERA

14 34N, 82 58W

Pilotage: This small group of islets and reefs lies about 15 miles north-northwest of Cayos Miskitos. From Outer Mohegan, the largest of the Hamkera group, numerous detached reefs extend about 3.5 miles north to several rocks awash. This group is the most northwestern portion of the Cayos Miskitos.

EDINBURGH REEF

14 50N, 82 39W

Pilotage: This 4-mile-long reef lies awash about 8.25 miles north of the northern limits of the Cayos Miskitos group. Edinburgh Cay is about 1.5 miles west-northwest of the southwest end of the reef. Edinburgh Channel is a clear passage between this reef and the ones to the south.

Cock Rocks are a series of drying rocks about 0.5 mile in extent that lie about 4.5 miles north of Edinburgh Reef.

PUNTA GORDA

14 21N, 83 12W
Pilotage: This mainland point is wooded and low and has almost no identifying features. The Cayos Miskitos group lies about 7.5 miles to the east.

CABO GRACIAS A DIOS

15 00N, 83 10W
Pilotage: This is a small town on the south side of the cape of the same name, at the mouth of the Río Coco. The river is obstructed by several cays at its mouth. The depth over the bar is about 6 feet, but is subject to change because of silting. This town, which is the seat of the governor, is a timber- and banana-exporting port. Except for some radio masts, the town cannot be seen from the south because of heavy foliage.

WESTERN CARIBBEAN: OFFSHORE ISLANDS AND BANKS

SPECIAL NOTE: Vessels should be aware that boundary disputes persist between Nicaragua and Colombia concerning Isla de San Andrés and surrounding waters.

Please see cautions about chartlets and waypoints on page P 2.

ENTRY PROCEDURES

For information on entry procedures to the various countries that govern these islands refer to the separate section on that country. The Cayos de Albuquerque, Cayos del Este Sudeste, Isla de San Andrés, and Isla de Providencia are all under the jurisdiction of Colombia. Only San Andrés and Providencia have settlements with customs stations where formalities may be complicated, but you should be prepared to check in with any officials found on the smaller cays. The Islas Santanilla (Swan Islands) are under the jurisdiction of Honduras.

CAYOS DE ALBUQUERQUE (COLOMBIA)

12 10N, 81 51W
CAYOS DE ALBUQUERQUE LIGHT, on Cayo del Norte, 12 10.2N, 81 50.0W. Fl (2) W 20s, 72ft, 14M. Black and white banded tower.

Pilotage: These small cays are about 111 miles east-northeast of Punta del Mono (Punta Mico), Nicaragua, and about 204 miles northwest of the Panama Canal. They lie on a bank about 5 miles in extent, with steep-to sides. Numerous rocky heads and drying reefs exist, particularly near the bank's east and south sides.

Cayo del Norte and Cayo del Sur, near the bank's east side, stand about 6 feet high and 4 feet high respectively and are reported to give a good radar return at distances up to 12 miles. Both cays are reported heavily wooded with palms. A light is shown from Cayo del Norte.

CAYOS DEL ESTE SUDESTE (COLOMBIA)

12 24N, 81 28W
CAYOS DEL ESTE SUDESTE LIGHT, Courtown Cays, Cayo Bolivar, 12 24.0N, 81 27.9W. Fl W 15s, 72ft, 17M. Metal tower, 95ft. Radar reflector. **RACON B** (– · · ·).

Pilotage: These cays lie on a 7-mile-long (north–south), 2-mile-wide coral bank about 24 miles northeast of Cayos de Albuquerque. A reef extends across the northern portion of the bank, and a broken reef extends along the bank's eastern side. The middle and western parts have numerous shoal patches. There is a stranded wreck about 3.3 miles northwest of the light on Cayo Bolivar.

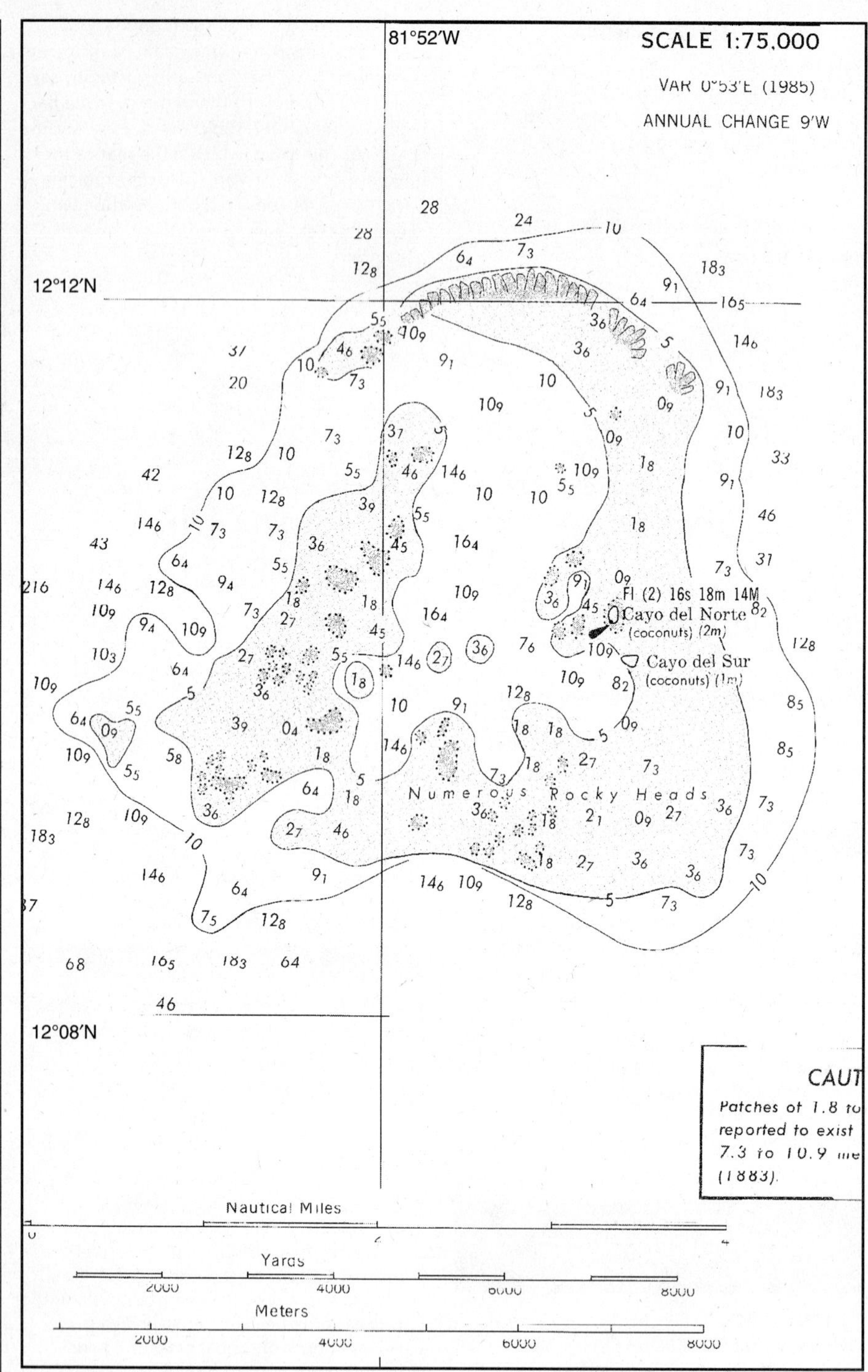

Cayos de Albuquerque, soundings in meters *(from DMA 26081, 8th ed, 5/85)*

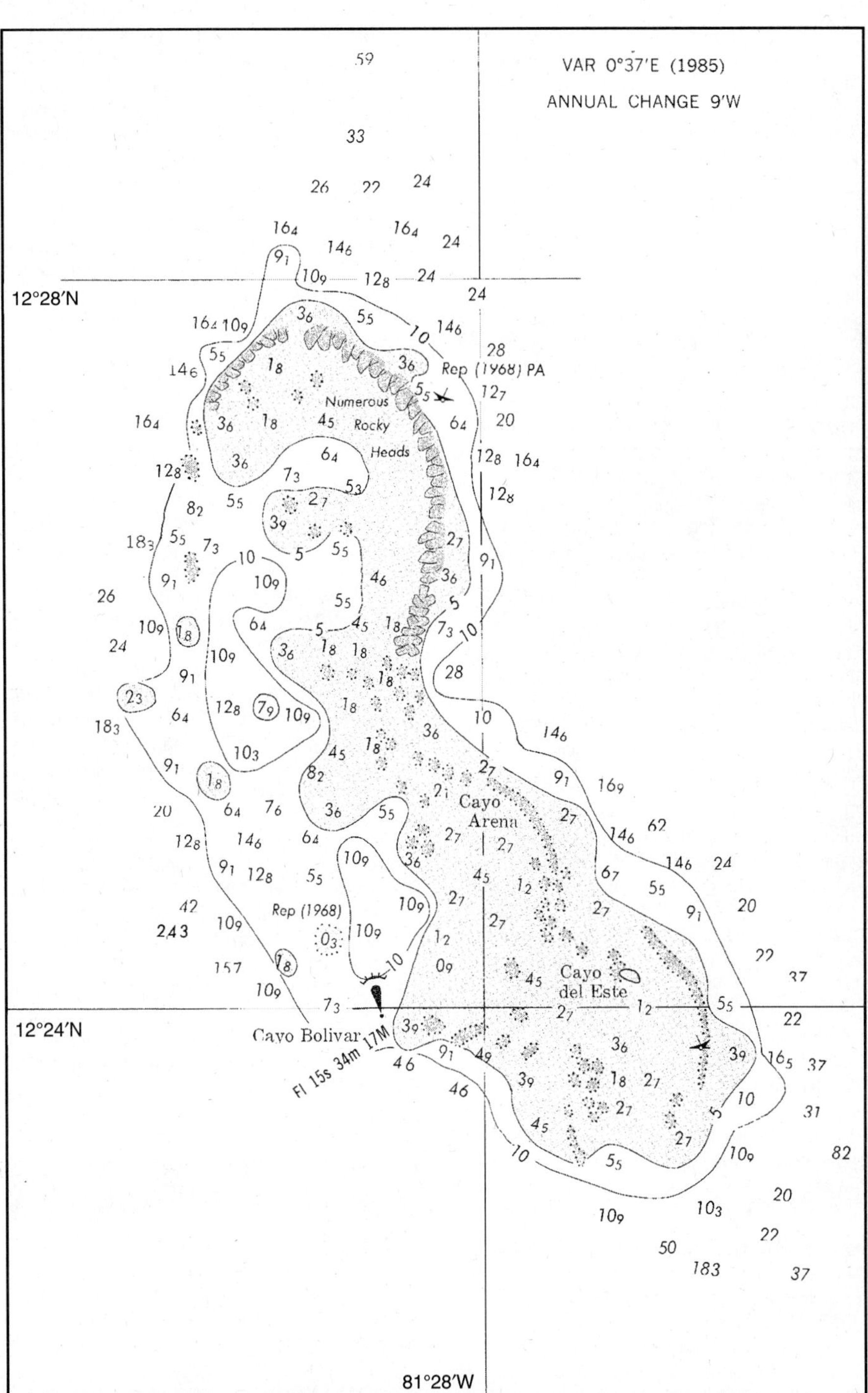

Cayos del Este Sudeste, soundings in meters *(from DMA 26081, 8th ed, 5/85)*

Cayo del Este is on the southeast portion of the bank and is thickly wooded with palm trees. Cayo Bolivar, about 1.3 miles west of Cayo del Este, is about 6 feet high with several palms on it. West Cay is a small low cay about 800 yards west-northwest of Cayo del Este. Cayo Arena, a small dry sandbank about 1.3 miles northwest of Cayo del Este, is reported to give a good radar return at 14 miles. Fishermen's huts may be found on these cays. Anchorage, with care, may be found on the bank's western edge.

ISLA DE SAN ANDRÉS (COLOMBIA)

12 33N, 81 43W

PUNTA SUR LIGHT (SOPLADOR), 12 28.9N, 81 43.8W. Fl W 9s, 82ft, 14M. White tower, 75ft.

COVE LIGHT, 12 31.0N, 81 43.8W. Fl R 3s, 65ft, 8M. Red tower, 59ft.

PUNTA EVANS LIGHT, 12 31.9N, 81 44.1W. Fl G 5s, 65ft, 8M. Green tower, white bands, 60ft.

AVIATION LIGHT, 12 35.1N, 81 41.7W. Al Fl W G 10s, 20M. Radio mast.

AERO RADIOBEACON, 12 35.1N, 81 42.3W. **SPP** (· · · · – – – · – – ·), 387kHz NON, A2A, 192M.

CAYO CORDOBA LIGHT, SE end, 12 33.1N, 81 41.3W. Fl W 15s, 131ft, 24M. White tower, 62ft.

Pilotage: The island lies about 16 miles west-northwest of Cayos del Este Sudeste, or about 218 miles northwest of the Panama Canal. It is about 7 miles long (north–south) and 1.5 miles wide. A ridge of hills extends most of its length. Three flat-topped summits rise above the ridge, but they appear as only two hills when viewed from north or south. A cliff at the north end of the ridge is distinctive. The island is reported to give a good radar return at distances to 23 miles.

PUNTA NORTE

12 36N, 81 42W

Pilotage: Punta Norte, the northern extremity of the island, is fringed by foul ground up to 1 mile north of the point. Blowing Rocks, where the sea breaks heavily during north winds, are located about 0.8 mile north of the point.

Cayo Johnny (Cayo Sucre), low and covered by palm trees, stands on the northwest end of a detached reef about 1 mile east-northeast of Punta Norte. The coastal reef extends about 0.5 mile north and east of Cayo Johnny, then about 4 miles south to a position about 1.8 miles east-northeast of Punta Sterthemberg. The south side of this reef is indented by a narrow, irregular channel that leads north about 2.3 miles to Bahía de San Andrés. South of Punta Sterthemberg the reef lies within 0.3 mile of the eastern coastline of the island. The west side of the island is steep-to and free of dangers.

PUNTA SUR

12 29N, 81 44W

Pilotage: This south extremity of the island is wooded. The western side of the island is mainly rocky cliffs. It is indented about 2.5 miles north of Punta Sur by Rada El Cove. Temporary anchorage may be found off the mouth of the cove or for small craft, inside.

BAHÍA DE SAN ANDRÉS

12 35N, 81 42W

Pilotage: This bay is formed by the reef to the east of the island and the east side of Isla de San Andrés. A 300-meter-long wharf lies parallel to the shore in the southwest part of the bay. Proceed with caution when entering the channel and when maneuvering within the harbor—charted depths and shoals may be inaccurate.

Port of entry: An agent is required to clear in here.

Dockage: Several marinas are located here.

Anchorage: Yachts can anchor on either side of Cayo Santander, but should pass around the south side of the cay, as a shallow bar connects to the mainland on the north side. Yachts report anchoring off Club Nautico in the deep water east of the cay. Dinghies can be landed at the marina east of Club Nautico. Proceed cautiously, as depths and shoals may not be charted accurately. The anchorage is reported to be cool and comfortable, with clean water.

Services: This is a duty-free port and a tourist center for Colombians. There are good supermarkets, fine restaurants, and very good shops.

With its good anchorage and fine shopping, Isla de San Andrés makes a good stopping point for vessels making the run between Honduras and Panama.

ISLA DE PROVIDENCIA (COLOMBIA)

13 21N, 81 22W

PROVIDENCE ISLAND LIGHT, south end, 13 19.3N, 81 23.5W. Fl (2) W 14s, 220ft, 18M. White metal tower, 59ft. Visible 318°–157°.

PALMA CAY LIGHT, 13 24.0N, 81 22.1W. Fl W 10s, 82ft, 15M. White tower, 59ft.

LOW CAY, 13 31.6N, 81 20.6W. Fl W 10s, 66ft, 14M. White tower, 59ft.

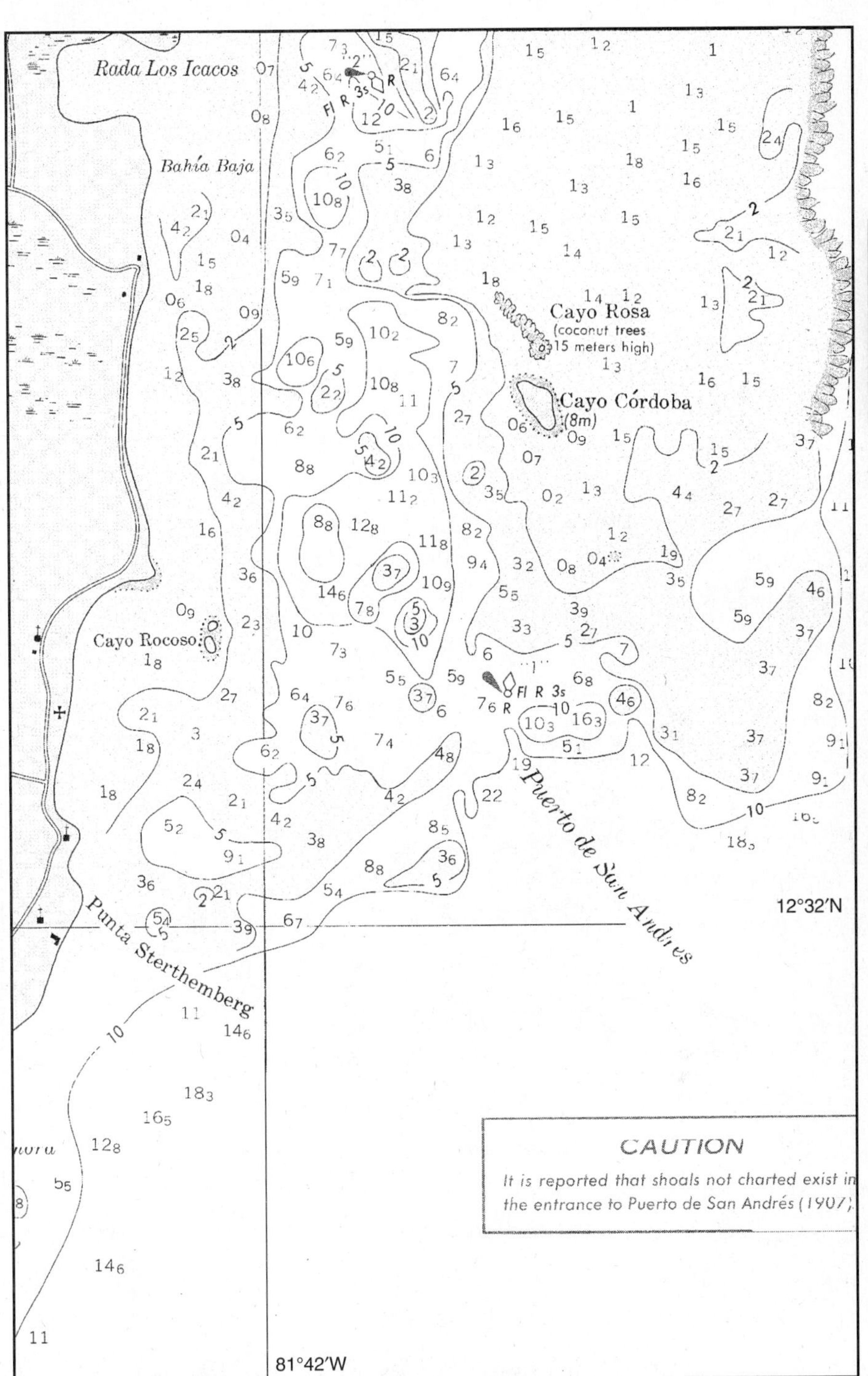

Approaches to San Andrés, soundings in meters *(from DMA 26081, 8th ed, 5/85)*

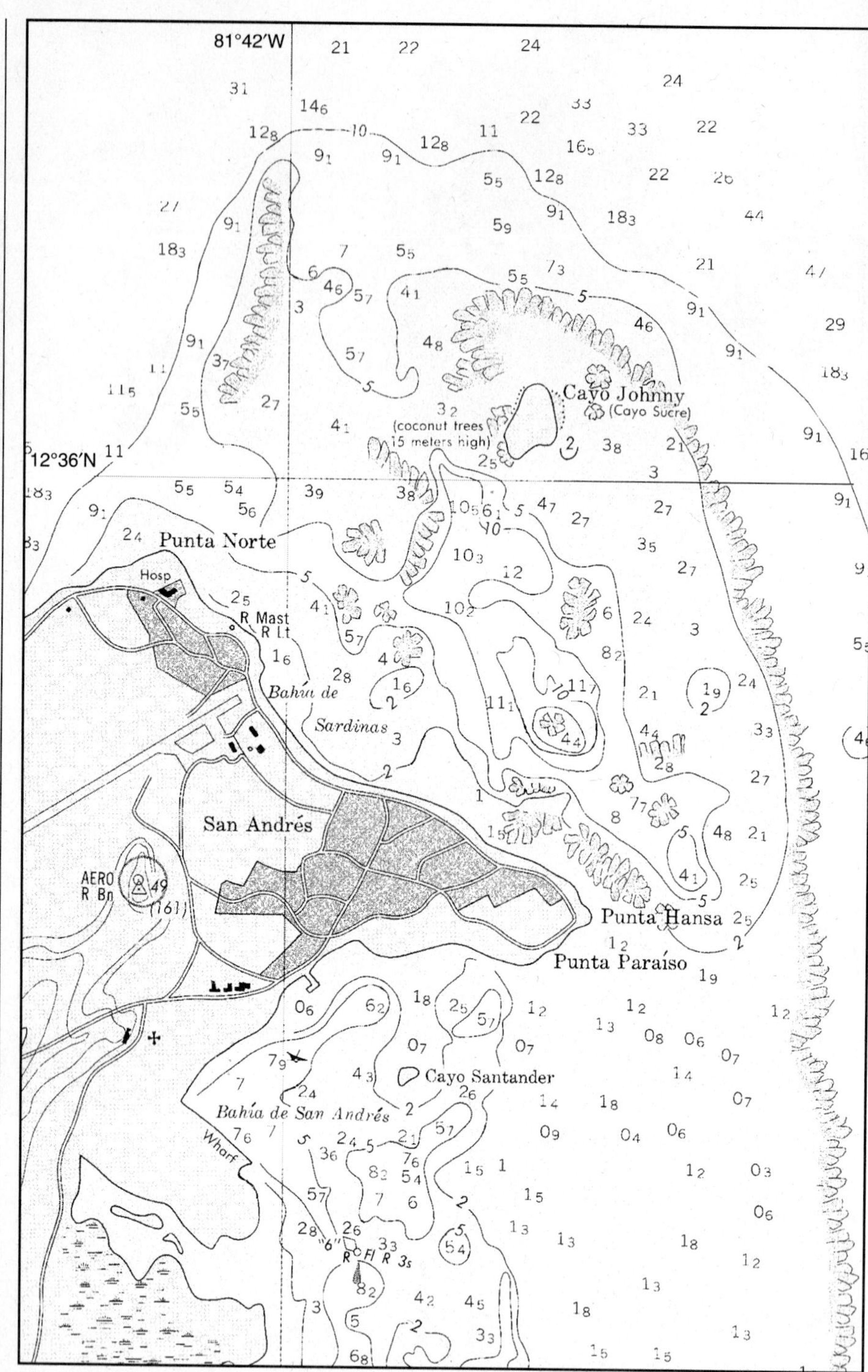

San Andrés, soundings in meters *(from DMA 26081, 8th ed, 5/85)*

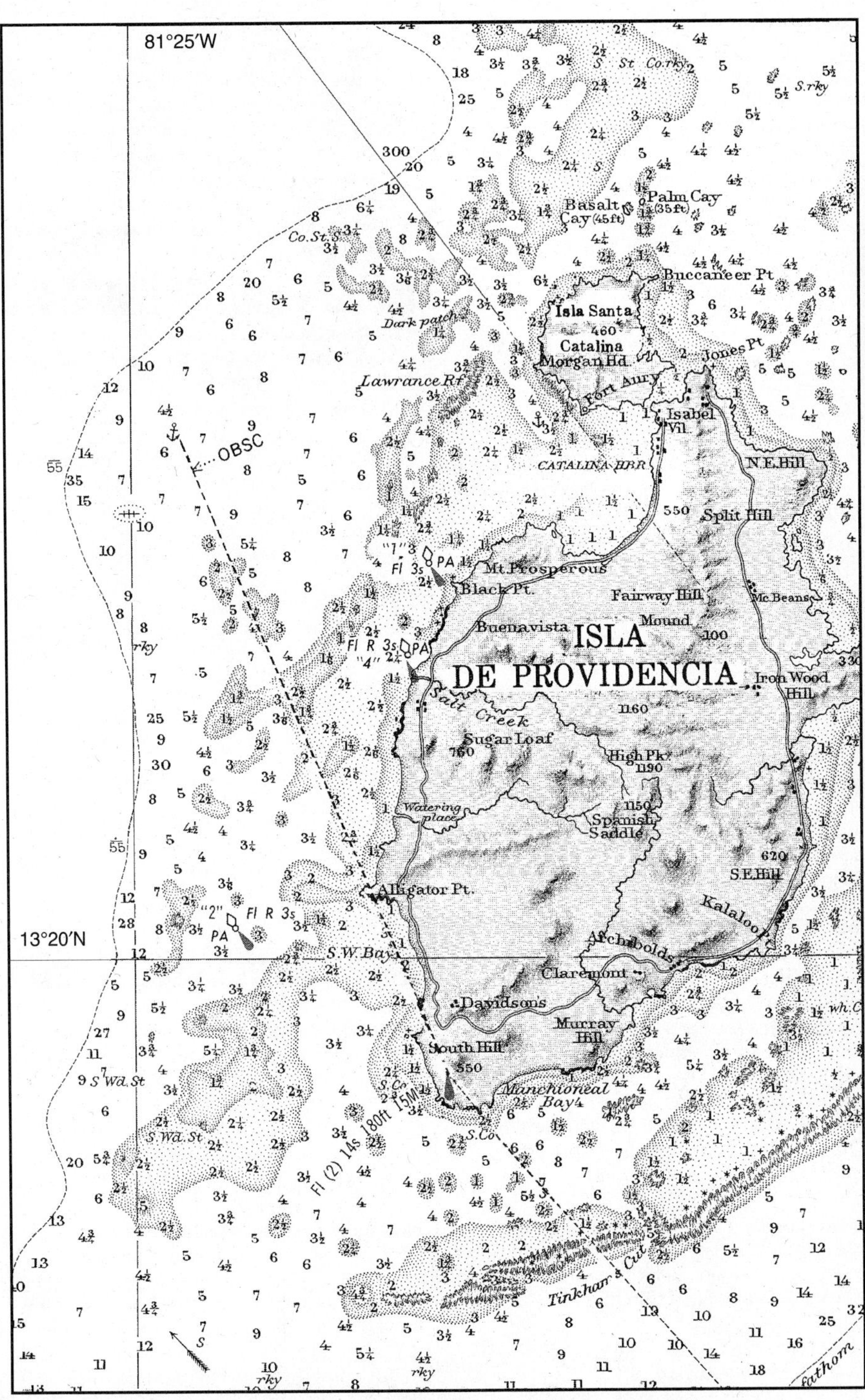

Isla de Providencia, soundings in fathoms *(from DMA 26083, 14th ed, 11/83)*

Pilotage: Isla de Providencia lies about 50 miles north-northeast of Isla de San Andrés. The island and Isla Santa Catalina, nearby to the north, together extend about 4.5 miles in a north–south direction. The mountainous center rises to three peaks of about the same elevation, with the highest being over 1,190 feet. The north extremity of Isla de Providencia is Jones Point. On a spur extending south from this point rises 551-foot-high Spit Hill. From the northwest or southeast a rocky chasm is prominent on Spit Hill.

Isla de Providencia is reported to give a good radar return up to 37 miles. A reef lies within about 2 miles of the east and south sides of the island and up to 8 miles north from its north end. The west side of the island is foul except for an area about 1.5 miles wide west of Catalina Harbor. Strong and irregular currents exist in the vicinity of Isla de Providencia.

CATALINA HARBOR

13 23N, 81 23W

Pilotage: This harbor is located on the west side of the northern part of Isla de Providencia. Morgan Head, the west extremity of the island, is a prominent 40-foot-high rock. Isla Santa Catalina forms the northern side of the harbor area. Basalt Cay and Palm Cay are located about 0.25 mile north of Isla Santa Catalina. The suggested approach from the northwest is with Morgan Head in line with Fairway Hill bearing 143°. The approach from the southwest may be marked by some buoys, and runs very close to shore. Any approach should be made in good light, with the sun behind you, as there are many shoal patches and reefs.

Port of entry: Check with the government officials in Isabel Village, located in the northeastern part of Catalina Harbor.

Anchorage: The harbor is reported to be well sheltered, with good holding in sand. Depths off of Isabel Village range from 6 to 9 feet across most of the anchorage area.

Services: This is a quiet little island settlement. There are telephones, general stores, a post office, and a government office.

Isla de Providencia has a largely English-speaking population of about 4,500 and direct flights to San Andrés, which has air connections to several Caribbean destinations.

LOW CAY LIGHT, 13 31.6N, 81 20.6W. Fl W 10s, 66ft, 14M. White tower, 59ft.

LOW CAY

13 32N, 81 21W

Pilotage: This cay is located on the northwest extremity of the reef, about 8.5 miles north of Jones Point on Isla de Providencia. A reef that dries in places extends about 0.5 mile south. There is a light on the cay.

Anchorage: The area 1 mile south of the cay is reported to be a good anchorage. There are shoal patches in this area.

RONCADOR BANK

13 34N, 80 04W

RONCADOR BANK LIGHT, on Roncador Cay, N end of bank, 13 34.9N, 80 05.3W. Fl W 11s, 79ft, 15M. Red metal tower.

Pilotage: The bank lies about 75 miles east-northeast of Isla de Providencia. It is very steep-to and is about 7 miles long in a north-south direction with a maximum width of about 3.5 miles. Roncador Cay, composed of sand and blocks of coral, lies on the north part of the bank and is 13 feet high. A light is shown from the cay. The bank is mostly covered by reefs, drying sandbanks, and coral heads. A stranded wreck on the south end of the bank is reported to be radar conspicuous. A strong northwest current usually sets over the bank. *Caution: The wreck and the south end of the bank were reported (1981) to be 1.5 miles southeast of their charted positions.*

Anchorage: Good anchorage can be taken on the bank's west edge, but take care to avoid the coral heads, which can be easily seen.

SERRANA BANK

14 24N, 80 16W

SERRANA BANK LIGHT, Southwest Cay, 14 16.3N, 80 23.5W. Fl W 10s, 82ft, 15M. White metal tower, black band.

Pilotage: This extensive dangerous shoal area lies with its southwest end about 44 miles north-northwest of Roncador Cay. The bank is steep-to and about 20 miles long (northeast–southwest) and about 6 miles wide. All sides, except the west and southwest edges, are fringed by a nearly unbroken reef. The sea breaking over the reef on the east side of the bank is visible for several miles farther than the cays that stand on it. Mariners are advised to use extreme caution in the vicinity of Serrana Bank because of strong currents. On the west and southwest side of the bank are numerous live coral heads with less than a meter of water over them. *Caution: It has been reported (1989) that Serrana Bank might lie about 5 miles east of its charted position.*

SOUTHWEST CAY

14 16N, 80 24W

Pilotage: This small cay, composed of sand covered with grass and stunted brushwood, is about 32 feet high and the largest of the few cays on Serrana Bank. It is reported to be a good radar target at distances up to 10 miles. A steep-to reef extends about 9 miles northeast from the cay, but is not always visible. A ledge on the edge of the reef 6 miles northeast of Southwest Cay is about 2 feet high. A drying sandbank stands about 1.5 miles farther northeast.

Anchorage: Temporary anchorage for small craft may be found in depths of 42 to 54 feet about 0.6 mile north-northwest of the cay.

SOUTH CAY CHANNEL

14 21N, 80 15W

Pilotage: This 0.3-mile-wide channel is located about midway along the southeast side of Serrana Bank. Depths in the fairway range from 24 feet to 42 feet. The currents in the channel run from 1.5 to 2 knots.

Anchorage: Temporary anchorage for small craft can be found with a depth of about 27 feet in a position 1 mile northeast of the channel entrance or in depths of about 42 feet midway between the cays at the channel entrance.

EAST CAY CHANNEL

14 21N, 80 11W

Pilotage: This channel is located about 4 miles east of South Cay Channel. It is about 0.5 mile wide, with depths of 60 to 84 feet in the fairway. East Cay lies on the west side of the entrance and a spur of the reef extends about 2 miles north from it. The current sets in and out of this channel at a rate of 1.5 to 2 knots.

Anchorage: Small craft may anchor in depth of 42 to 54 feet about 0.8 mile northeast of East Cay.

NORTH CAY

14 28N, 80 17W

Pilotage: This cay is about 13.5 miles northeast of Southwest Cay on the north end of the reef. It is small and low, with a reef extending about 3 miles from the cay to the southwest. An object resembling a light pylon was reported to lie on or near this cay; it should not be mistaken for the light on Southwest Cay. A stranded wreck was reported (1971) to lie about 1 mile southwest of North Cay. Northwest Rocks and the border of the bank are visible by radar from distances of less than 10 miles. Turtle fishermen visit this area from March to August. On occasion, the masts of their vessels and their temporary huts may be sighted before the reefs themselves. Currents run at 1.5 to 2 knots.

QUITA SUENO BANK

14 15N, 81 15W

QUITA SUENO BANK LIGHT, N end, 14 29.2N, 81 08.2W. Fl W 9s, 75ft, 15M. White metal tower.

Pilotage: The bank's south end is about 39 miles north-northeast of Isla de Providencia. It is very steep-to and dangerous. As defined by the 200-meter curve, the bank extends about 34 miles north and has a maximum width of 13.5 miles. A 22-mile-long reef lies along its east side. *Caution: Take great care when passing east of the bank as the current here sets strongly to the west. Furthermore, the description of this area is based on an 1833 survey.* Two wrecks stranded on the reef are reported to give a good radar return. There are reported to be other wrecks on the reef. A detached shoal with depths of 17 to 22 feet lies about 14 miles west-northwest of the north edge of the reef. A vessel struck a coral head about 14 miles southwest, and a dangerous sunken rock is charted about 19 miles south-southwest of the north edge of the reef. It seems probable that another unsurveyed reef exists west of Quita Sueno Bank. A depth of 60 feet was reported (1964) to lie 38 miles west-northwest of Quita Sueno Light.

Anchorage: Good anchorage can be taken in about 60 feet, clear sand, and coral west of the rocky ground that lies near the middle of the reef.

SERRANILLA BANK

15 55N, 79 54W

SERRANILLA BANK LIGHT, Beacon Cay, 15 47.7N, 79 50.7W. Fl (2) W 20s, 108ft, 25M. Orange structure, concrete base.

RACON Z (– – · ·).

Pilotage: This bank lies about 78 miles north-northeast of Serrana Bank. As defined by the 200-meter curve, the bank is about 24 miles long and 20 miles wide and very steep-to, with depths generally ranging between 30 feet and 120 feet. There are shoal areas in the vicinity of the cays on the east and south parts of the bank. There is no perceptible current on the bank, but the current runs west-northwest at 0.3 to 1 knot in the vicinity.

BEACON CAY

15 47N, 79 50W
Pilotage: This is the largest of the three cays on Serranilla Bank. It lies about 7.5 miles southwest of East Cay. It is about 8 feet high, covered by grass, and marked with a coral stone beacon on its west end. A light is shown from this cay. A 7-foot-deep shoal is located about 5 miles south of the light. There are numerous obstructions north of Beacon and East cays for a distance of about 4.5 miles.
Anchorage: Good anchorage is reported in depths of 36 feet about 1 mile northwest of Beacon Cay. Take care to avoid the coral heads on the bank.

WEST BREAKER

15 48N, 79 59W
Pilotage: This dangerous breaking ledge about 2 feet high lies almost 8 miles west of Beacon Cay and is the westernmost danger on the bank.

EAST CAY

15 52N, 79 44W
Pilotage: This is the easternmost above-water feature on Serranilla Bank. It is small, covered by bushes, and about 7 feet high. It lies about 3 miles west of the east edge of the bank. Foul ground extends about 2.5 miles north and northeast from the cay. Three miles northeast of East Cay lies Northeast Breaker, a coral ledge with a rock awash on its south side.

BAJO NUEVO (HONDURAS)

15 53N, 78 33W (east end)
BAJO NUEVO LIGHT, located on Low Cay, 15 51.2N, 78 38.0W. Fl (2) W 15s, 69ft, 15M. White tower, black stripes.

Pilotage: This bank, which is about 5 miles wide and 14 miles long on a northeast to southwest axis, has not been well surveyed. It has been reported (1967) to extend west to about 15°48'N, 78°55'W. Its southwest end is not well defined, but the northwest side of the bank is reported to be clear of known dangers. Depths of 12 feet are reported to extend up to 10 miles west of Bajo Nuevo. Seals gather on the reefs here and are hunted in March and April.

East Reef and West Reef, consisting of numerous rocky heads and separated by a 0.5-mile-wide opening, extend along the southeast side of the bank. They are over 2.5 miles wide and steep-to on the southeast and north sides. Sand accumulates on the reefs, forming low ridges, sometimes barely awash. A stranded wreck at the northeast end of East Reef is reported to be visible on radar. Another stranded wreck lies on the west end of West Reef. The current in the vicinity of the bank sets west and southwest at speeds of up to 2 knots.

LOW CAY

15 52N, 78 39W
Pilotage: This cay at the north end of West Reef is about 5 feet high and barren. It is composed of broken coral, driftwood, and sand.
Anchorage: In moderate weather exposed anchorage can be taken in depths of 42 to 48 feet about 1.5 miles west of Low Cay. The bottom is sand and coral. Approach the anchorage from the west, taking care to avoid coral heads.

ROSALIND BANK

16 26N, 80 31W
Pilotage: The south extremity of the bank is located about 167 miles east-northeast of Cabo Gracias a Dios. As defined by the 200-meter-depth curve, the bank is about 63 miles long and 35 miles wide. General depths range from 60 feet to 121 feet over coarse sand and coral. The shallowest spots are about 24 feet deep.

The current generally sets northwest at a velocity of 1.5 knots over the bank. On striking the ledge near the southeast edge, the current causes a race that has the appearance of breakers.

Caution: An extensive bank, about 41 miles long and 10 miles wide, lies about 11 miles west of Rosalind Bank. Depths range from 24 feet to 216 feet. The shallowest detached patches are found along the east edge of the bank. A detached 36-foot-patch lies on the north part of the bank.

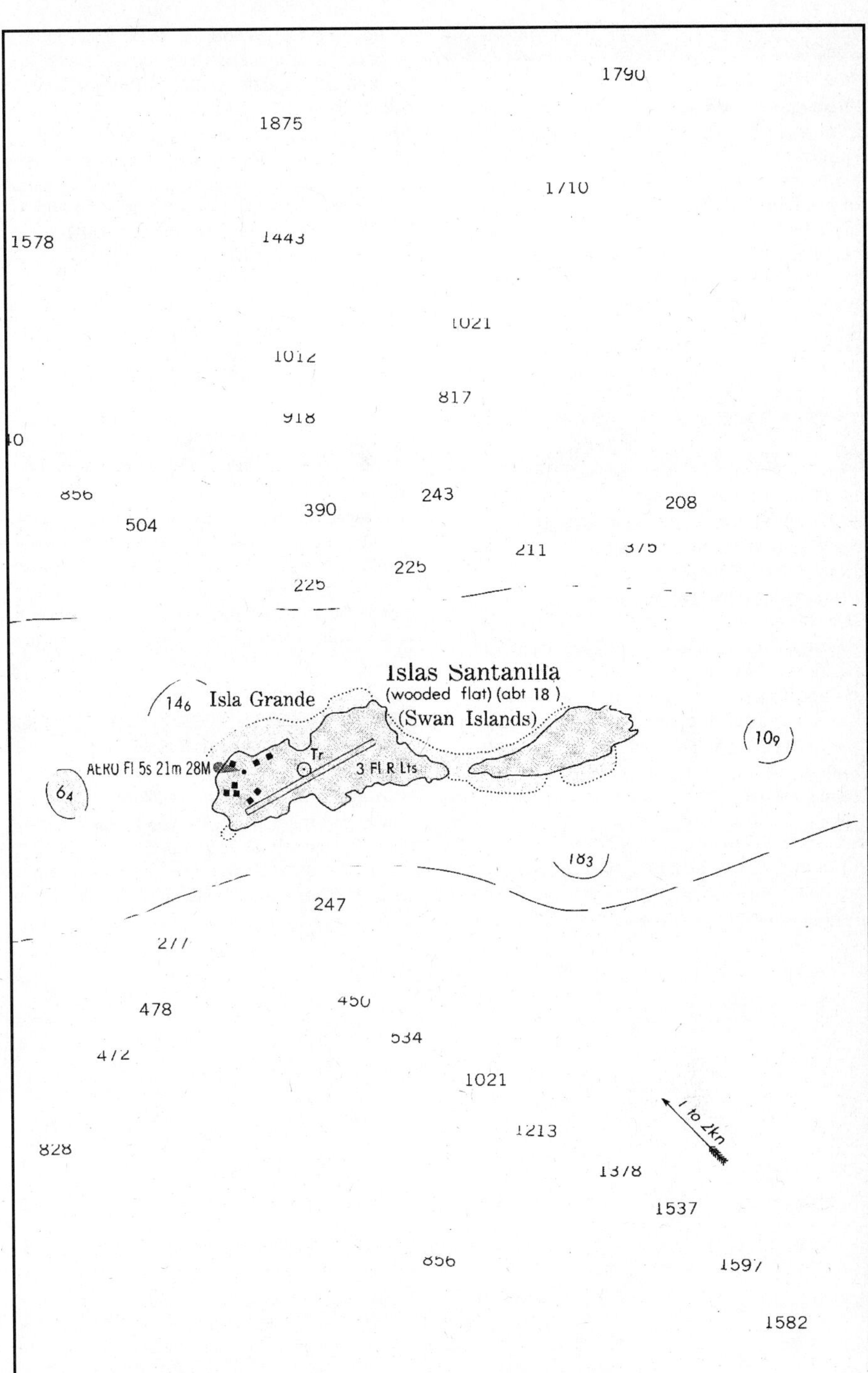

Swan Islands, soundings in meters *(from DMA 28150, 2nd ed, 4/89)*

THUNDER KNOLL

16 27N, 81 20W

Pilotage: This bank is about 11 miles in extent and composed of coral sand. It lies about 4 miles west of the northern part of the extensive bank mentioned above. Depths range from 79 feet to 230 feet. Two detached shoals with depths of 42 feet to 51 feet and 115 feet to 121 feet lie between 4.5 and 8.5 miles west of Thunder Knoll. A detached 36-foot patch was reported to lie about 4 miles southwest of the southwest part of Thunder Knoll.

ISLAS SANTANILLA (HONDURAS)

17 25N, 83 56W

ISLAS DE EL CISNE AVIATION LIGHT, near the W end of the largest island, 17 24.5N, 83 56.5W. Fl W 5s, 70ft, 28M. 3 Fl R 1.5s (vert) on tower 1 mile E. Occasional.

Pilotage: The Islas Santanilla (Swan Islands) are located about 150 miles north-northwest of Cabo Gracias a Dios. There are two small islands lying close together near the west part of a narrow bank about 18 miles long. The east island is 1.5 miles long and 60 feet high and has a bold rocky shore. The island is densely covered by trees and bushes. The west island is flat, about 1.8 miles long, and also tree-covered. A Honduran navy post and a cattle farm exist on the southwest end of the west island. The passage separating the islands is foul. The islands have been reported to be good radar targets at distances up to 25 miles. The light with an aerobeacon stands on the northwest side of the west island.

Anchorage: When the wind is in the northeast or east, anchor off the sandy bay at the west end of the west island. During north winds anchor close inshore south of the west end of the island. During south and southwest gales anchorage can be taken north of the islands, nearer the east than the west end.

These islands were ceded by the U.S. to Honduras in 1972; the former U.S. weather station here was handed over to Honduras in 1980.

ROSARIO BANK

18 30N, 84 04W

Pilotage: This bank is about 10 miles in extent and lies about 62 miles north of the Islas Santanilla. Depths range from 60 feet to 210 feet. There are other shallow banks in the vicinity. A detached 24-foot patch was reported to lie about 6 miles east of the east side of Rosario Bank.

MISTERIOSA BANK

18 51N, 83 50W

Pilotage: This bank is about 24 miles long and from 2 to 7 miles wide. It is centered about 87 miles north of the Islas Santanilla. Depths over the reddish coral bank range from 42 feet to 161 feet. There are other shallow banks in the vicinity.

HONDURAS

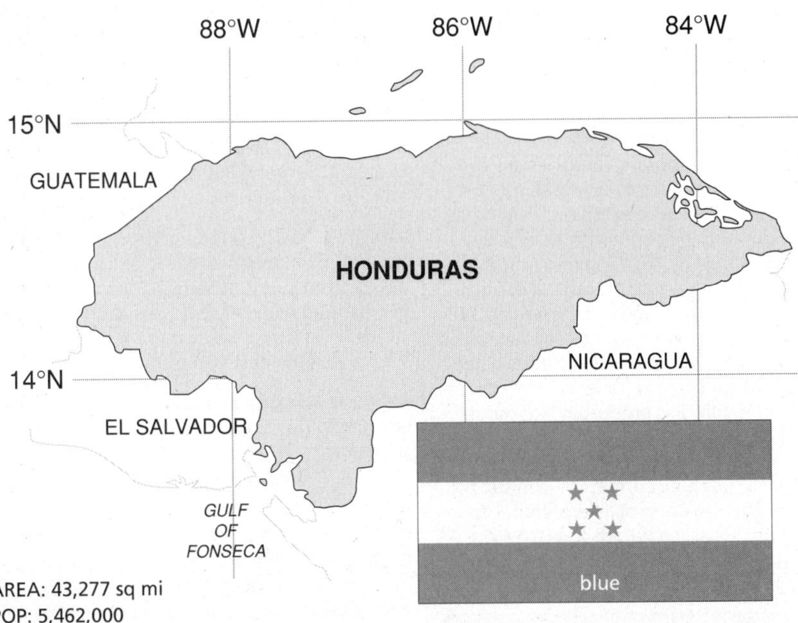

AREA: 43,277 sq mi
POP: 5,462,000
CAP: Tegucigalpa

CAUTION: Many lights in Honduras have been reported as irregular or unreliable.

Please see cautions about chartlets and waypoints on page P 2.

ENTRY PROCEDURES

Ports of entry in the Bay Islands (Islas de la Bahía) include Guanaja settlement on Isla de Guanaja and Coxen Hole on Roatán. On the mainland, you may clear at Puerto Castilla, La Ceiba, and Puerto Cortés.

You should have passports, clearance from the last port of call, ship's papers, and the proper courtesy flag. Fly the yellow Q flag upon arrival. Obtain a *zarpe,* or cruising permit, from officials, and be sure to list any place you are planning to visit. You will have to clear in and out of each port as you cruise through Honduras.

Visas are usually issued for stays of 90 days, with renewals possible. Visas are not required for U.S. citizens and citizens of several other countries. For specific information on entry requirements contact the Embassy of Honduras, 3007 Tilden Street NW, Washington, DC 20008 (tel., 202-966-7702; fax, 202-966-9751). Consulates in the U.S. are located in Chicago, Houston, Los Angeles, Miami, New Orleans, New York, San Francisco, and Tampa (FL).

USEFUL INFORMATION

Language: Spanish. English is widely spoken in the Bay Islands.
Currency: Lempira (L); exchange rate (summer 1997), L13 = US$1. It may be difficult to arrange for transfer of funds from abroad to Honduras, particularly to the Bay Islands, where banking facilities are limited.
Telephones: Country code, 504; local numbers, six digits.
Public holidays (1998): New Year's Day, January 1; Good Friday, April 10 (Most businesses are also closed on the Saturday before Easter); Easter, April 12; Easter Monday, April 13;

Day of the Americas, April 14; Labor Day, May 1; Independence Day, September 15; Soldier's Day, October 3; Columbus Day, October 12; Armed Forces Day, October 21; Christmas Day, December 25.

Tourist offices: *In the U.S.:* Tourism Section, Embassy of Honduras, 3007 Tilden Street NW, Washington, DC 20008 (202-966-7702). Also, Information Office of the Honduras Institute of Tourism, P.O. Box 140458, Coral Gables, FL 33114 (tel., 800-410-9608; fax, 305-461-0602).

Special notes: The border areas between Honduras and neighboring countries are considered potentially hazardous and should be avoided. The presence of extensive minefields on both sides of the border with Nicaragua, particularly along the Río Coco and in the Atlantic coast region, also presents a danger. Both rural and urban crime rates are rising. U.S. citizens may check with the Consular Section of the U.S. Embassy, Avenida la Paz, Tegucigalpa (504-36-9320) for current travel and security information. Cholera is present in Honduras, and there is risk in rural areas of malaria. There have been recent epidemics of dengue fever. For additional health information, contact the Centers for Disease Control International Travelers Hotline (voice recording, 404-332-4559; fax information service, 404-332-4565; Web, www.cdc.gov).

CABO GRACIAS A DIOS

15 00N, 83 09W

Pilotage: This low, swampy, tree-covered point marks the border between Nicaragua and Honduras. The 151-foot-high radio masts standing close west of the disused lighthouse are the best landmarks during the day.

CAUTION: The cape has been reported to lie about 2 miles east of its charted position. Less depths than charted have been reported to lie 4 miles southeast and from 4 to 10 miles east of the cape. Vessels should give the cape a berth of at least 10 miles and in thick weather should keep in depths of over 60 feet.

ARRECIFE DE LA MEDIA LUNA (HALF MOON REEF)

15 13N, 82 38W

Pilotage: The north extremity of this group of reefs and cays is about 34 miles east-northeast of Cabo Gracias a Dios. This area extends about 20 miles on the north–south axis and about 11 miles on the east–west, as defined by the 20-meter curve. The southern extremity is marked by Cock Rocks. Logwood Cay (Cayo Modera) stands on the west side of the bank.

Cayo Media Luna (Half Moon Cay) lies about 2.5 miles south of Logwood Cay, and a crescent-shaped reef extends about 0.5 mile east, then 0.8 mile north from it.

Bobel Cay lies about 4 miles south-southeast of Cayo Media Luna. Two cays close together lie between 6.3 and 7 miles east-southeast of Cayo Media Luna. Several rocky heads lie in the vicinity of these cays.

Savanna Cut, a narrow passage with depths of 36 to 60 feet, lies between Arrecife de Media Luna and Savanna Reefs (15°10′N, 82°25′W), which are located about 5 miles to the east.

Arrecife Alagardo (Alargate Reef) is the easternmost visible danger on the Miskito Bank. It lies about 5 miles east of the Savanna Reefs, where it is marked by heavy breakers. During periods of fresh northeast winds there is often a strong set toward the east side of the reef.

NOTE: It has been reported that Arrecife Alagardo lies 2 miles east of its charted position.

BANCO DEL CABO (MAIN CAPE SHOAL)

15 16N, 82 57W

Pilotage: This nearly awash shoal lies about 17 miles northeast of Cabo Gracias a Dios. It is about 4 miles long. The sea seldom breaks over it, but it may be spotted by the discoloration of the water in the area.

Main Cape Channel, the passage between Banco del Cabo and Arrecife de la Media Luna, is clear of known dangers and has depths ranging from 60 feet to 95 feet.

GORDA BANK

15 36N, 82 13W

Pilotage: This bank, with depths of less than 66 feet, extends for about 52 miles northwest of a position about 18 miles east-northeast of the northeast extremity of Arrecife Alagardo. Shallower depths than those charted may exist on the bank. The bottom is clearly visible, and on the north side there are a number of patches of flat coral covered with dark weed.

CAYO GORDA

15 52N, 82 24W

Pilotage: This barren cay stands on the north edge of Gorda Bank. It is about 12 feet high and is composed of sand, broken coral, and large stones. A reef extends about 1.8 miles northwest from it, but its east and south sides are steep-to.

CAUTION: An obstruction was reported to lie about 4.5 miles southwest of Cayo Gorda.

Farral Rock, located 5 miles east of Cayo Gorda, breaks in heavy weather and can be identified by its dark appearance, in contrast to the white sandy bank on which it lies. The two parts of the stranded wreck near the rock were reported to be good radar targets up to 10 miles distant.

BANCOS DEL CABO FALSO (FALSE CAPE BANK)

15 32N, 83 03W
Pilotage: This is a dangerous, steep-to, breaking bank, lying about 32 miles north-northeast of Cabo Gracias a Dios.

CAYOS COCOROCUMA

15 43N, 83 00W
Pilotage: This reef lies about 44 miles north-northeast of Cabo Gracias a Dios. It is about 5 miles long, but a detached coral patch, on which the sea breaks, lies about 0.8 mile west of its north end. A group of seven small cays, not over 2 feet high and 1 mile in extent, lie on the south end of the reef. The southernmost, and largest, cay, is covered with bushes and some coconut trees on its east end. Another cay, about 1 mile to the north, has a square clump of brushwood that resembles an isolated rock when seen from a distance.

CAYOS PICHONES (PIGEON CAYS)

15 45N, 82 56W
Pilotage: These two cays lie 3 miles east of Cayos Cocorocuma. The westernmost is a small islet at the south end of a dangerous, steep-to, half-moon-shaped reef. The reef is about 0.8 mile in extent, and the sea breaks heavily here in stormy weather. A steep-to reef lies about 2.8 miles southeast of this islet.

BANCO VIVORILLO (VIVARIO BANK)

15 54N, 83 22W
Pilotage: This 10-mile-long coral bank lies with its northwest extremity about 59 miles north of Cabo Gracias a Dios. Depths over the bank range from about 9 feet to 33 feet.

CAYOS VIVORILLO LIGHT, 15 50.0N, 83 17.7W. Fl W 10s, 13M.

CAYOS VIVORILLO (VIVARIO CAYS)

15 50N, 83 18W
Pilotage: These bush- and tree-covered cays lie on a coral reef at the southeast end of Banco Vivorillo. A continuous line of breakers front the steep-to east side of the reef. Practically all of this reef is usually dry or awash.

CAYOS BECERRO

15 55N, 83 16W
Pilotage: This group of eight small cays lie about 5.5 miles northeast of Cayos Vivorillo. They lie on a coral ledge about 3.5 miles long by 1 mile wide. The sea always breaks along the east and north sides of the reef.

Grand Becerro, the largest cay, consists of two parts. It stands near the south part of the ledge and is marked by mangroves.

El Becerro, a rock over which the sea usually breaks, stands 1.25 miles southeast of Grand Becerro. It is surrounded by a coral reef with depths of 12 to 30 feet. Rocky pinnacles, with depths of about 30 feet, lie about 5.5 miles north-northeast of El Becerro.

Hannibal Banks consists of two small shoals about 0.8 mile apart, lying in the southeast part of the passage between Cayos Vivorillo and Cayos Becerro. The north shoal has a least depth of 34 feet and the south shoal a least depth of 42 feet.

CAYOS CARATASCA

16 02N, 83 20W
Pilotage: This group of seven small cays lies about in the middle of a shoal bank located about 9 miles northwest of Cayos Becerro. The southernmost cay has some vegetation, but the others are barren.

CAYOS CAJONES (HOBBIES)

16 06N, 83 13W
Pilotage: This steep-to reef, about 13 miles long in an east–west direction, lies centered about 11 miles north of El Becerro. The narrow west part nearly always dries, but it does not always break. A small cay with bushes and coconut trees stands about 3 miles west of the east end of the reef. A 23-foot patch is reported (1970) to lie about 5.5 miles northwest of the middle part of Cayos Cajones. Another 30-foot patch lies about 6 miles south of the middle of the reef.

CAUTION: Exercise great caution when approaching this reef. A vessel approaching from the north at night or in hazy weather should not come within depths of less than 180 feet, which lie up to 11 miles north of the reef. Depths of 121 feet will be found about 4 miles off. Within that depth the bottom is coral and sand, and outside it is mud.

During periods of strong northeast winds there is often a strong current set toward the north side of Cayos Cajones, which adds considerably to their danger.

CABO FALSO

15 12N, 83 20W
CABO FALSO LIGHT, 15 15.2N, 83 23.7W.
Fl W 5s, 75ft, 19M.

Pilotage: This low point is about 21.5 miles northwest of Cabo Gracias a Dios. It is backed by several isolated trees and brushwood. It should be approached with caution, as a hard sandbank, with depths of less than 18 feet, lies about 3 miles to its northeast. The sea usually breaks over this bank, and the inner part dries in places. The 36-foot-line lies about 6.5 miles northeast of the cape.

CAUTION: Cabo Falso has been reported to lie about 3 miles west of its charted position. The area in the vicinity of the cape has not been thoroughly examined, and passing vessels should give it a wide berth.

The Río Cruta lies about 3.5 miles northwest of Cabo Falso. Its shallow mouth is marked by high trees, which have the appearance of a bluff. When viewed from the west, they may be mistaken for Cabo Falso. In 1973 the mouth of the river was easily identified by radar. A light is shown from the mouth of the river.

PUNTA PATUCA

15 49N, 84 17W
PUNTA PATUCA LIGHT, 15 49.0N, 84 18.2W.
Fl W 10s, 73ft, 19M.

Pilotage: This low, but prominent, point lies about 64 miles northwest of Río Cruta. The coast presents the same generally low aspect as the coast southeast of Río Cruta.

The southern entrance to the Laguna Caratasca lies about 27 miles west-northwest of the Río Cruta and can be identified by a large group of trees on either side. This large freshwater lagoon parallels the coast for about 35 miles and is separated from the sea by a low, narrow, thinly wooded ridge of sand. In 1973, the prominent point on the east side of the entrance was reported to be a good radar target.

Estero Tabacunta, the west outlet for the Laguna Caratasca, lies about 28 miles northwest of the more southerly entrance. Both entrances are fronted by shallow bars, but the latter entrance has a channel with a depth of about 6 feet. Low white sand cliffs serve to identify the coast in the vicinity of Estero Tabacunta.

Río Patuca lies about 9 miles north-northwest of Estero Tabacunta. Punta Patuca marks the west entrance to the 150-mile-long river, one of the longest in Honduras. The mouth is about 225 yards wide, but is difficult to make out unless a vessel is close-in. A light stands at the mouth. The controlling depth over the bar is about 6 feet during the dry season and from 8 to 10 feet in the wet season. The outgoing current, even in the dry season, attains a rate of 1.5 knots. The only landmarks are a series of light-colored bluffs, which stand southeast of the river mouth, and a low rounded hill in which the land to the east seems to end. The east entrance point of the river is low and sandy.

LAGUNA DE BRUS

Pilotage: The entrance to this lagoon is at its western end, which is about 22 miles west of Punta Patuca. The entrance is marked on its west side by a clump of trees, higher than those elsewhere in the vicinity. This entrance is hard to spot from seaward. The entrance bar has depths of 6 to 7 feet in the dry season and is usually fronted by heavy breakers. There are depths of 10 to 11 feet within the lagoon, but there are many shoals and shallow areas.

RÍO SICO

Pilotage: The river entrance is about 16 miles west-northwest of the entrance to the Laguna de Brus, or about 38 miles from Punta Patuca. This is about where 3,700-foot-high Cerro Payas (15°45'N, 84°56'W) rises abruptly from the low land along the coast. This mountain forms the eastern end of the Sierras La Cruz, an irregular mountain chain. The peak is frequently obscured by clouds, but 2,050-foot-high Pico Panoche, about 5 miles to the north, is usually visible. A vessel proceeding west will sight these peaks soon after passing Punta Patuca.

The bar at the river entrance has a least depth of about 5 feet in the dry season and as much as 9 feet in the rainy season. It is passable by boats only in moderate weather.

The land rises abruptly on both sides of the river, and the mountains approach fairly close to the coast. The flat, swampy part of Honduras ends here, and the land to the west is traversed by numerous ridges that reach the coast in places.

CABO CAMARÓN

16 00N, 85 00W
CABO CAMARÓN LIGHT, 15 59.2N, 85 01.9W. Fl W 5s, 73ft, 19M.

Pilotage: This point is located about 5 miles west-northwest of the Río Sico. It is low, rounded, and topped by trees. The land is flat for some distance inland. A light is shown from the cape.

The coast between Cabo Camarón and the Río Aquan, 42 miles to the west, is indented by a bight extending about 6 miles to the south. The east part of the bight is low and sandy, and the west part is a low, thinly wooded beach topped by some sand hills 40 to 60 feet high.

From Cabo Camarón the coast trends about 15 miles west-southwest to Piedracito, a small distinctive rocky bluff. Cabeza Piedra Grande (15°54'N, 85°29'W), a rocky bluff about 400 feet high, is located 11 miles farther west. Cerro Sangrelaya (15°52'N, 85°09'W) rises to 6,150 feet in a position about 5 miles southeast of Piedracito. Seven miles south of the same point, a conical peak rises to 3,149 feet. A saddle-shaped summit stands about 5 miles southwest of the latter peak.

Iriona, a small settlement located 8 miles southwest of Cabo Camarón, is the seat of government for the territory east to the Nicaraguan border.

A lower ridge of mountains rises south of Iriona and extends west to a position close south of Cabeza Piedra Grande. From a position about midway between Cabo Camarón and Piedracito, as far west as Cabeza Piedra Grande, the lower slopes of the mountains nearly reach the coast.

RÍO AQUAN

15 58N, 85 44W
Pilotage: This river enters the sea via two mouths located about 2.5 miles apart. The river extends 120 miles inland. The east mouth is located 16 miles west-northwest of Cabeza Piedra Grande and is shallow. The east side of this entrance forms a distinctive point; about 3 miles to its southeast is a 79-foot-high hill close to the coast. The hill appears round when seen from the east or west, but from the north its west end appears as a flattened summit and its east end as a sugarloaf hill. The two ends are separated by a chasm.

The settlement of Santa Rosa de Aquan stands on the east bank of the river, about 1 mile within the east entrance. Vessels bound for the Río Aquan should call at Trujillo for clearance in entering and departing.

PUNTA CAXINAS

16 02N, 86 01W
PUNTA CAXINAS LIGHT, 16 01.5N, 86 00.6W. Fl W 7s, 75ft, 19M.

Pilotage: This point (referred to in some sources as Punta Castilla) is located at the west extremity of Cabo de Honduras, a narrow, 5-mile-long neck of low land some 24 miles south of Isla de Guanaja in the Bay Islands. The cape is bordered by a scantily wooded beach with some scattered 40- to 60-foot sand hills. A light is shown from here.

Depths are about 60 feet 3 miles offshore and less than 240 feet up to 17 miles out. Within 17 miles of Cabo de Honduras the depths become irregular, with shoals of between 30 and 60 feet to the northeast, 57 feet to the north, and 30 to 56 feet to the northwest.

The mountain range south of Cabeza Piedra Grande extends west for about 14 miles and then appears to terminate rather abruptly in a saddle-shaped summit 2,499 feet high. This summit is about 8 miles south of the mouth of the Río Aquan. A much smaller sugarloaf peak stands west of the summit. A wide valley lies between this peak and Montanas de Trujillo.

ISLAS DE LA BAHÍA (BAY ISLANDS)

Pilotage: This group of islands fronts the coast for a distance of about 75 miles in a west-

southwest direction from a position about 30 miles north-northeast of Punta Caxinas. The group consists of Isla de Guanaja, Isla de Roatán, Isla de Utila, and three smaller islands.
Currents: The currents around the islands are extremely uncertain, particularly during the summer. The equatorial current north of the islands sets west, but when the northers have ceased, its surface influence is felt on the islands. The countercurrent generally sets in the opposite direction south of the islands. Currents may be greatly altered, or even reversed, by winds and tides. The range of the tropic tide at Isla de Roatán is greater than anywhere else in the area. The current sets west and north with a rising tide and south and east with a falling tide. A counterclockwise eddy is observed north of Isla de Utila.
Winds: The prevailing winds on the sheltered south sides of the islands are from the southeast and at times attain a maximum velocity of 45 knots. During the winter months the winds may come from any direction.

ISLA DE GUANAJA (BONACCA ISLAND)

16 28N, 85 54W
BLACK ROCK POINT LIGHT, 16 29.9N, 85 49.0W. Fl W 10s, 200ft, 25M.
POND CAY LIGHT, Isla de Guanaja, 16 26.3N, 85 52.8W. Q G, 2M. Gray concrete column, 29ft.

Pilotage: This easternmost island of the Bay Islands group is about 8.3 miles long and 2.5 miles wide at its widest part. The island is composed of densely wooded hills that rise to a height of 1,200 feet near its center. The northeast extremity is a bold peninsula terminating in 102-foot-high East Cliff, which is ochre in color. Ochre Bluff at the southwest end of the island is of the same height and color. Several anchorages are available within the coastal reefs, especially on the southeast side of the island. Numerous reefs, shoals, and small cays fringe the island, especially on its southeast side.

GUANAJA SETTLEMENT

16 26N, 85 54W
Pilotage: This small town is located on Sheen Cays, 0.3 mile off the southeast side of the island. Some of the buildings stand on piles around the cays. The approach depths are reported to be around 30 feet, with depths of 9 to 18 feet alongside the piers. You should be able to contact the port captain by VHF radio.
Port of entry
Anchorage: Anchor west of the main cay, in depths of 8 to 30 feet.
Services: Fuel, water, groceries, and basic supplies should be available. There are restaurants, a post office, and communication facilities.

ISLA DE ROATÁN

16 25N, 86 23W
PUNTA OUESTE LIGHT, 16 16.1N, 86 36.1W. Fl W 5s, 74ft, 19M.

Pilotage: The largest island of the Bay Islands group, Roatán lies about 15 miles west of Isla de Guanaja and is about 28 miles long and 2 miles wide. Isla Santo Elena, Isla Morat, and Isla Barbareta stand close off its east end.

The island is densely wooded and hilly, with general elevations rising 298 to 499 feet. A 735-foot-high peak rises about 7 miles from the east end, and an 800-foot-high peak rises about 6 miles from the west end.

Punta Oueste is the southwest extremity of the island. A light is shown from the point. A conspicuous white church with a red roof and a square bell tower stands about 2 miles east-northeast of Punta Oueste.

The south shore of the island is indented with bays and coves suitable for small craft. The west and southwest parts of the island are steep-to, but elsewhere the island is fringed by a steep-to reef that extends up to about 1 mile offshore. Isla Barbareta, off the east end, is fronted by a reef that lies up to about 2 miles off its east and south sides.

By keeping the west extremity of Isla de Roatán bearing 272°, a vessel will pass south of all off-lying dangers when approaching from the east.

PUERTO REAL

16 25N, 86 19W
Pilotage: This harbor is located on the south side of the island, near the eastern end. Shelter is provided from the south by George Reef, George Cay, and Long Reef. George Reef extends about 1 mile west from the east side of the harbor entrance. George Cay, wooded and low with the ruins of a fort at its west end, stands about 250 yards from the west end of the reef. Long Reef is separated from George Reef by a channel 200 yards wide. Long Reef is a nearly dry ledge, 0.8 mile long, that protects

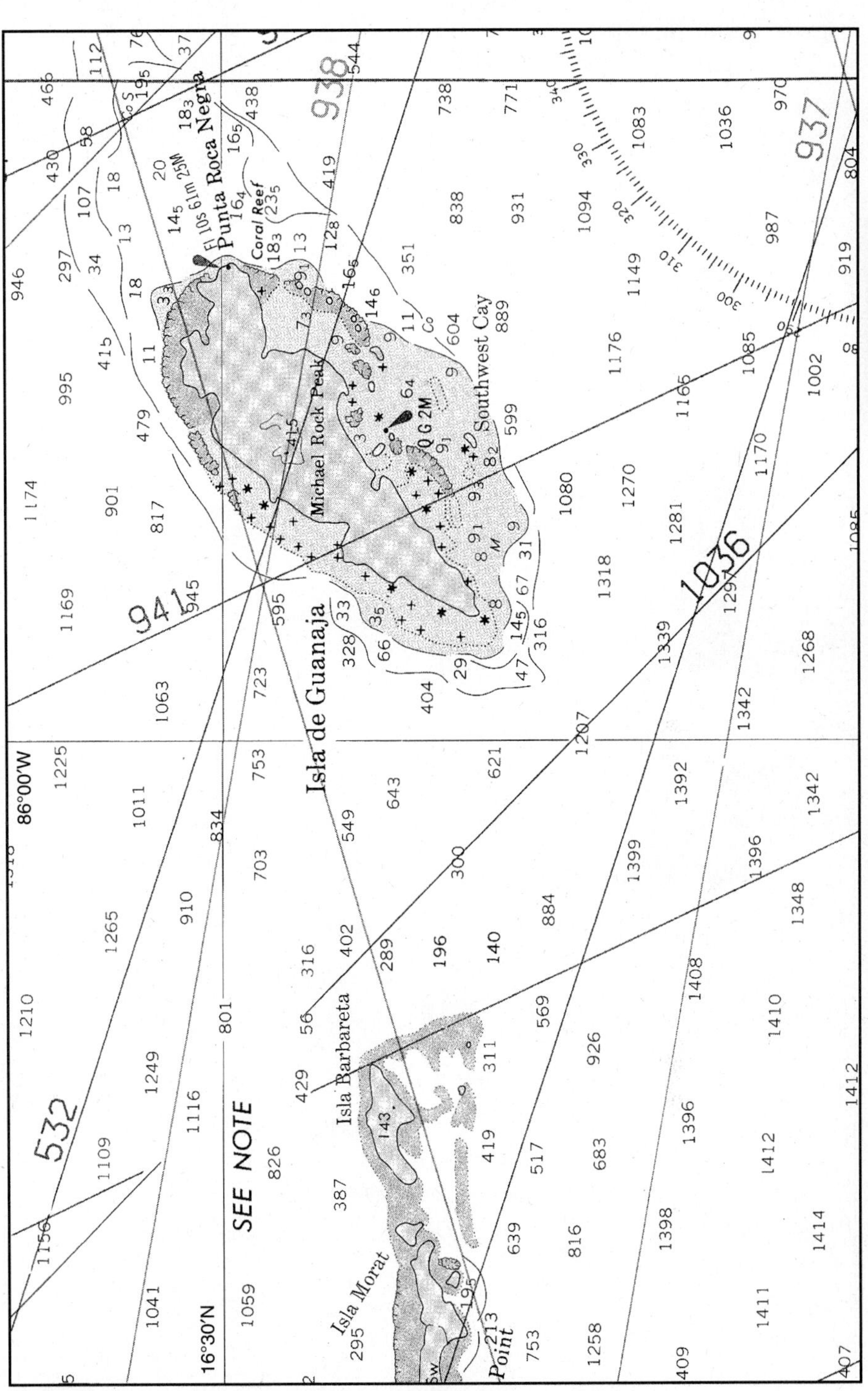

Isla de Guanaja, soundings in meters *(from DMA 28150, 2nd ed, 4/89)*

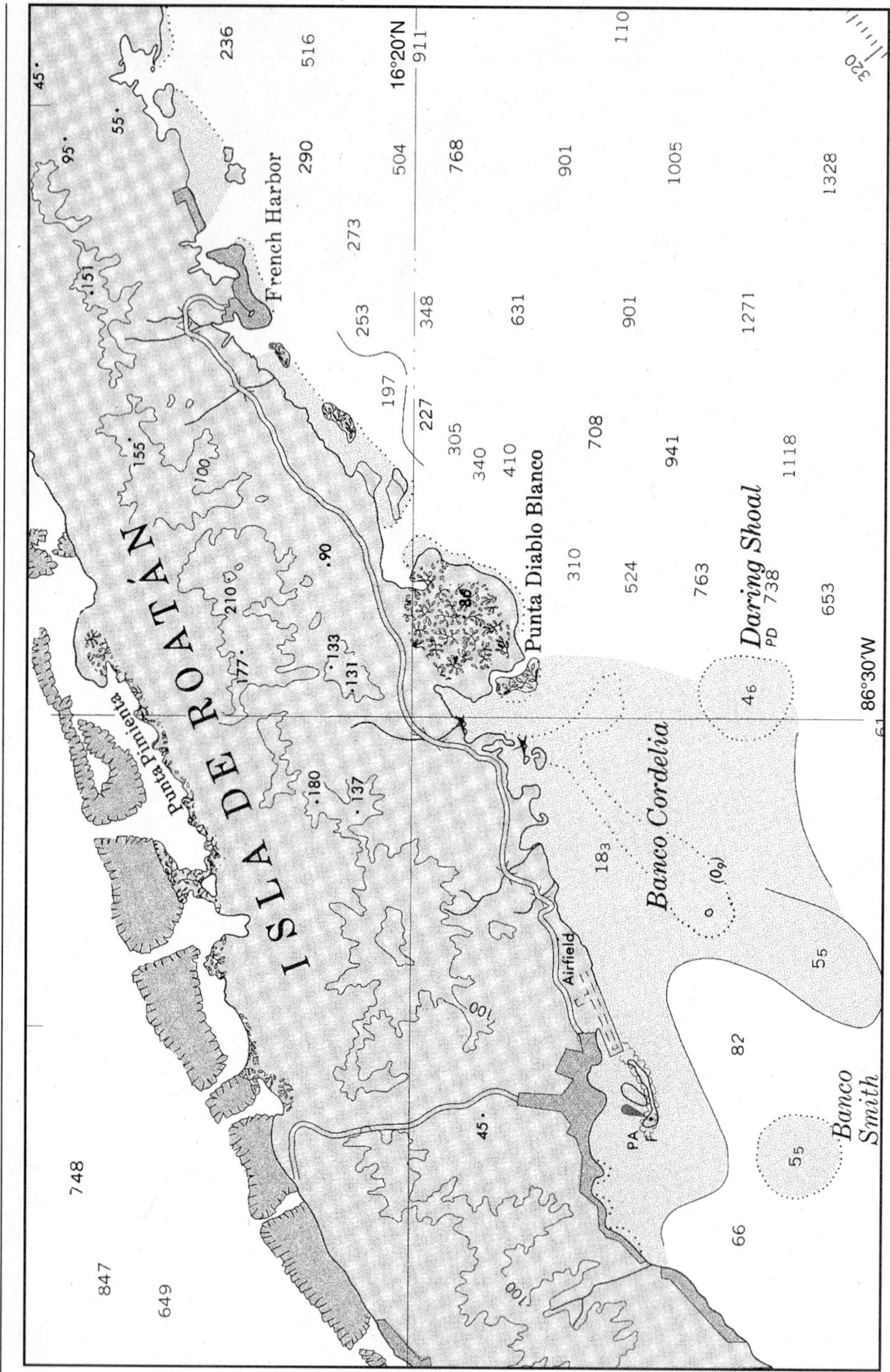

French Harbour and Coxen Hole, soundings in meters *(from DMA 28154, 1st ed, 3/88)*

the west side of the harbor. Enter the harbor via the channel between the reefs, which has reported depths of 18 to 27 feet. The reefs are steep-to and easily seen. A good landmark is the 735-foot-peak that stands about 0.8 mile west of the harbor.
Anchorage: Anchor in the eastern part of the protected water north of the reefs.

OAK RIDGE HARBOUR

Pilotage: This anchorage is located about 4 miles west of Puerto Real. It is entered through a narrow channel with a least depth of 20 feet, which leads north to the town. Depths off town are about 20 feet. A conspicuous stranded wreck lies close west of the entrance to the channel and is a good mark in the approach. A pier extends from the shore of the harbor.
Services: Fuel and water should be available at the fish docks.

~~**FRENCH HARBOUR LIGHT,** 16 23.5N, 86 23.0W. Q W, 15M. Mast, 29ft.~~

FRENCH HARBOUR

Pilotage: This harbor is about 10 miles west of Puerto Real. The town is almost surrounded by water, as extensive lagoons back it.
Dockage: Contact the French Harbour Yacht Club.
Anchorage: Anchor off the French Harbour Yacht Club in the lagoon east of town.
Services: Fuel, water, ice, groceries, and general supplies are available.

~~**COXEN HOLE REEF LIGHT,** SW extremity of reef at entrance, 16 18.6N, 86 32.3W. F W, 25ft. Reported extinguished (1989).~~

COXEN HOLE

16 18N, 86 35W
Pilotage: This is the principal town on the island and the seat of government for the Bay Islands. It occupies the east part of a bight on the south shore of the island, about 3.8 miles east-northeast of Punta Oueste. Its west side is bordered by dark 20-foot-high cliffs and its east and south sides by the reef upon which Coxen Cay stands. Identifiable landmarks are Carib Point, about in the middle of the bight, Hendricks Hill rising to 298 feet about 0.5 mile north of the point, and an 899-foot-peak rising about 0.8 mile northeast of Hendricks Hill. When Hendricks Hill bears 020° a vessel will be in the fairway of the west channel and may steer in on this bearing until abeam of Coxen Cay. Care should be taken to clear the reef off the west side of the cay. The settlement is on the north shore of the bight, north of the cay.

Banco Becerro (Seal Bank) lies nearly awash about 0.3 mile southwest of Coxen Cay. The channel between this reef and the reef southwest of Coxen Cay is about 300 yards wide.

Banco Smith (16°17'N, 86°35'W) lies about 1 mile south-southwest of Coxen Cay. It has a least depth of about 21 feet.

Banco Cordelia, which dries, lies about 1 mile east-southeast of Banco Smith. A bank about 1 mile in extent with depths of 18 to 72 feet lies 1.5 miles southeast of Coxen Cay.

Daring Shoal is a detached 15-foot patch about 2.5 miles east of Coxen Cay. It is reported joined to Banco Cordelia by a narrow ridge, with depths of less than 60 feet.

Port of entry: Coxen Hole is the administrative center for the Bay Islands.
Dockage: The town wharf is reported to have depths of 6 to 10 feet alongside.
Anchorage: Anchor either northwest or northeast of Coxen Cay. The airport runway partially extends along the reef toward Coxen Cay in the eastern part of the harbor. The bottom is reported to be sand and coral heads.
Services: Fuel, water, ice, groceries, and general supplies are available, along with banks, a post office, and communication facilities.

ISLA DE UTILA

16 06N, 86 56W
UTILA PEAK LIGHT, NE end of island on summit, 16 07N, 86 53W. Fl W 10s, 325ft, 20M. Red metal framework tower, 36ft. Temporarily destroyed (1996).
Pilotage: This is the westernmost of the Bay Islands. It lies about 18 miles southwest of Isla de Roatán. The island is about 7.5 miles long and from 1.3 to 2.8 miles wide. It is generally low, swampy, and thickly wooded. A range of 60- to 70-foot-high hills is located near its east end. Pumpkin Hill, 289 feet high and conical, stands near the island's northeast extremity. A disused, black framework light structure, 102 feet high, stands near the island's west extremity.

NOTE: In 1962, Isla de Utila and Isla de Roatán were reported to lie 2 to 4 miles farther apart than charted.

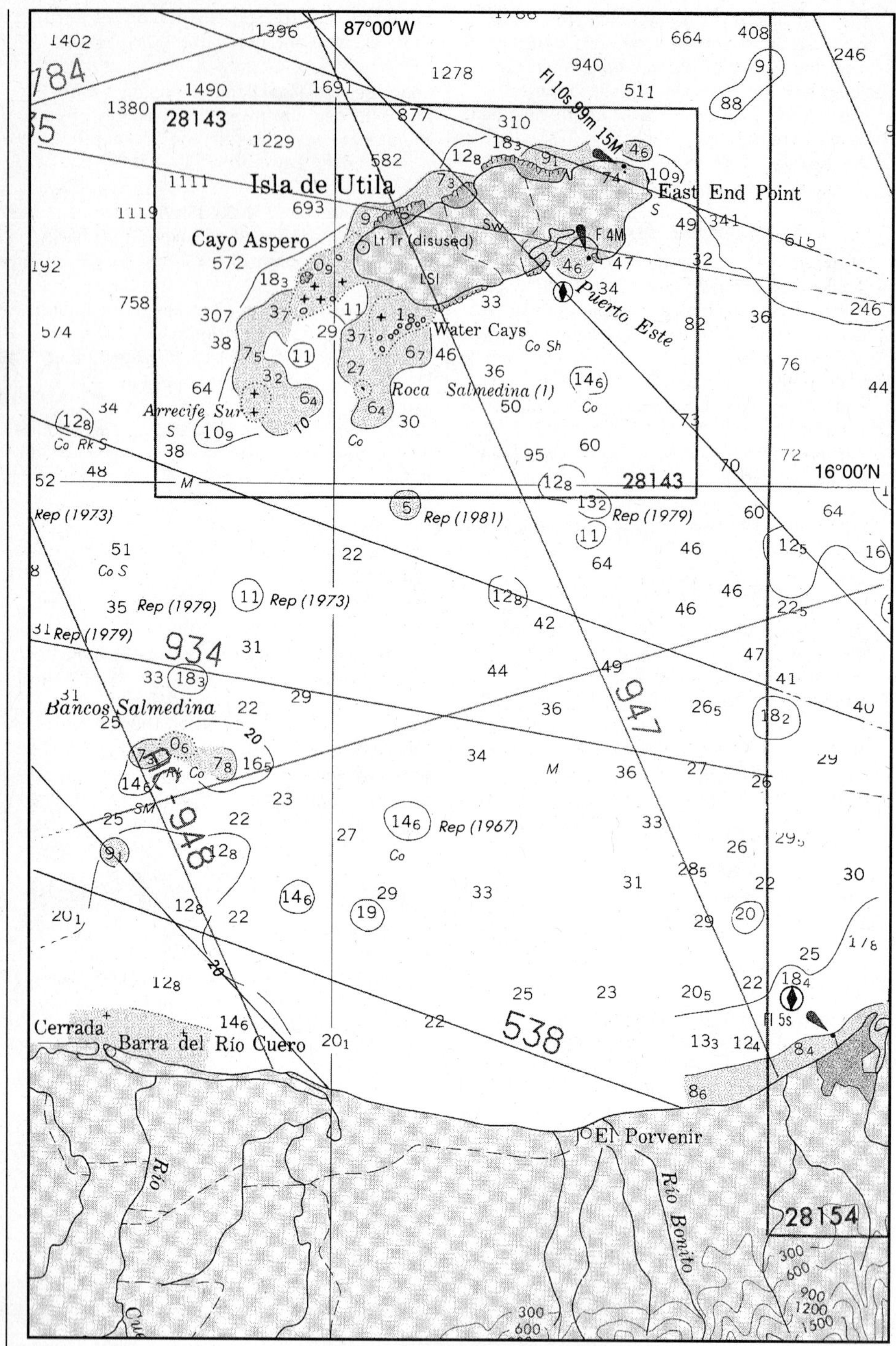

Isla de Utila, soundings in meters *(from DMA 28150, 2nd ed, 4/89)*

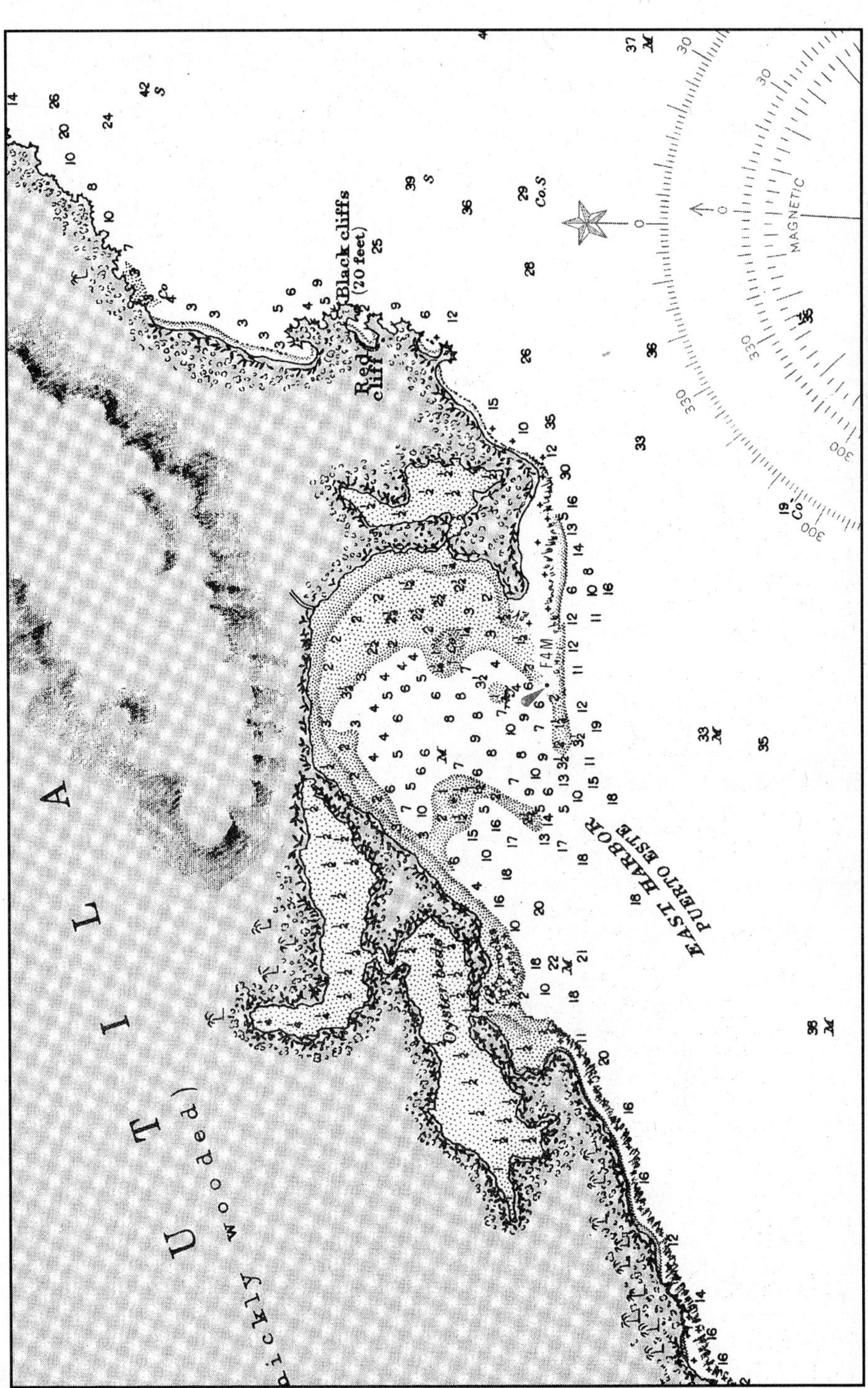

Puerto Este, soundings in fathoms *(from DMA 28143, 6th ed, 3/85)*

The island is steep-to, except for its southwest side. The north side of the island is indented by several shallow bays. An area of foul ground extends about 48 miles southwest from the southwest side of the island and is about 3 to 4 miles wide. Several good anchorages lie within this area, for those with local knowledge.

PUERTO ESTE LIGHT, East Reef, 16 05N, 86 55W. F W, 4M. Wooden building.

PUERTO ESTE

16 06N, 86 54W

Pilotage: This harbor lies about 1 mile west of the 20-foot black and red cliffs that form the southeast extremity of Isla de Utila. The harbor is about 0.8 mile in extent and provides anchorage in depths of 12 to 40 feet, with a clay bottom over coral.

Reefs narrow the entrance to about 300 yards, but depths run 30 to 36 feet. Two small coral heads lie about 0.3 mile from the east shore of the harbor. A church steeple in range 020° with a prominent tree is the leading mark used by local pilots. It has been reported that entering vessels should steer 040°, passing about 200 yards off a stake on the coral reef to the east.

BAHÍA DE TRUJILLO

15 58N, 86 00W

Pilotage: This bay is about 7 miles wide between Punta Caxinas at the end of Cabo de Honduras and the coast to the south. From the south entrance point of the bay the coast extends west for about 115 miles to Punta Caballos. The shore is generally low, wooded, and bordered by sandy beaches. Mountain ranges reach the shore in places, but are 10 miles inland in others. The mountains extend west to Tela, about 85 miles away. South of Tela the ranges curve inland and extend to the southwest. An extensive low, densely wooded plain lies between the base of the ranges and Montanas de Omoa, about 30 miles to the west.

The mountain ranges rise to high prominent peaks south and southwest of Bahía de Trujillo and south and southeast of Punta Congrejal. These peaks are good landmarks from offshore, particularly Cerro Congrejal, an 8,049-foot-high mountain that is located about 10 miles southwest of Punta Congrejal.

There are few navigational aids along this coast. Puerto Castilla is on the north side of the bay, and Trujillo on the southeast side. A shallow channel at the eastern end of the bay leads into a spacious lagoon. Depths in the central part of the bay average 39 to 184 feet. When approaching from the north give Punta Caxinas a berth of 1 to 1.5 miles. The tank in Puerto Castilla is higher than the surrounding vegetation and can be seen from about 12 miles to the north.

The north and east shores of the bay are low, swampy, and wooded, with no prominent landmarks. The south shore is backed by a high mountain chain that extends almost to the shore at Trujillo. Pico Colentura, a 3,198-foot peak, rises 3 miles south of the town. A *vigia,* or lookout hill, 2,499 feet high, stands at the northeast end of the mountain chain. These mountains are sometimes referred to as Montanas de Trujillo.

Callo Blanco is a dangerous reef located about 5 miles south of Punta Caxinas.

Currents: The usual set of the current off this coast is east within the 100-fathom curve. This current is uncertain due to the influence of the tides and winds. Within Bahía de Trujillo there is very little current during calms or east winds. With west winds the current sets east and counterclockwise around the bay at rates up to 2 knots.

Winds: The winds along the coast of Honduras are easterly for most of the year, with a pronounced diurnal variation. Calms and light offshore winds are frequent during the late night and early morning. Strong winds seldom blow in the early morning, except during November and December. During these months there are several days with northerly winds that attain gale force. During periods of strong northwest to northeast winds a heavy swell enters Bahía de Trujillo.

PUERTO CASTILLA

16 00N, 85 58W

Pilotage: This port is located on the south side of Cabo de Honduras, about 2 miles southeast of Punta Caxinas, on the north side of Bahía de Trujillo.

Port of entry

Anchorage: Anchor east of the commercial port area. Check locally for the latest depths over the bar at the entrance to the lagoon in the eastern part of the bay.

Services: Fuel, water, and general supplies should be available.

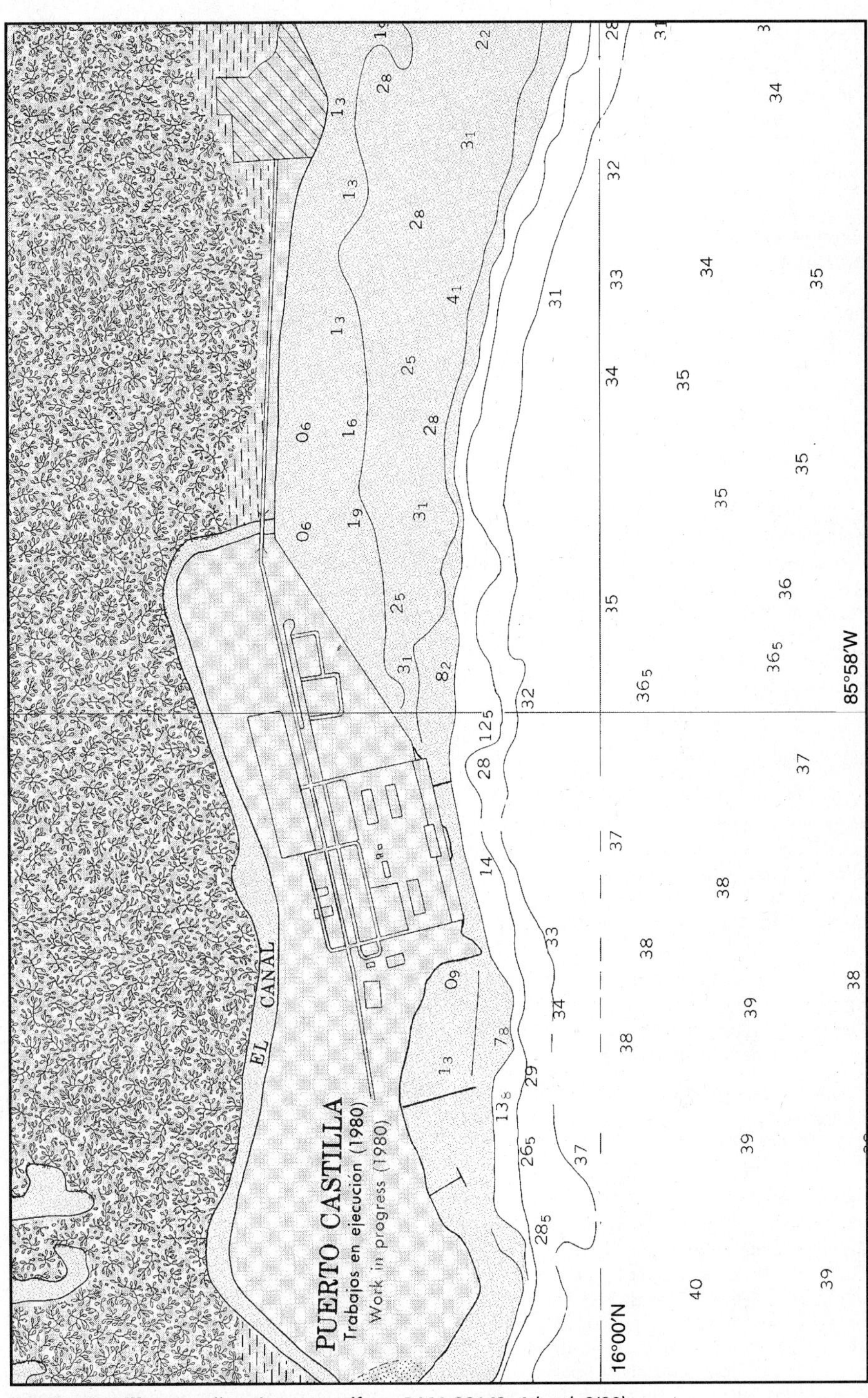

Puerto Castilla, soundings in meters *(from DMA 28142, 4th ed, 9/80)*

TRUJILLO

15 55N, 85 57W

Pilotage: The port is located on the southeast shore of Bahía de Trujillo. The twin spires of the church in town are conspicuous when making your approach. A small fort looks out over the bay. The town pier is about 200 feet long and has a least depth of 16 feet along its south side.

Port of entry

Anchorage: In calm weather you can anchor here, but you should move north in the bay when swells invade.

Services: Fuel, water, and general supplies should be available.

COCHINO GRANDE LIGHT, 15 58.6N, 86 28.5W. Fl W 7s, 516ft, 40M.

CAYOS COCHINOS

15 58N, 86 34W

Pilotage: This group of cays lies about 27 miles west of Punta Caxinas and 9 miles north of Punta Catchabutan. The island farthest to the east is densely wooded and rises to a height of 430 feet. The north side of the island is steep-to, but a coral spit with depths of 24 to 36 feet extends 1.3 miles from the east side. A group of cays and rocks lie a short distance off the south side. A light stands on the east side of the island.

A wooded island about 1 mile farther southwest rises to a height of 499 feet. A steep-to coral ledge with numerous cays and sandbanks extends about 3 miles southwest from the islands. The channel between the islands has depths of 85 to 95 feet. Depths of 25 to 37 feet lie off the northwest side of these two islands.

BANCO PROVIDENCIA

15 55N, 86 38W

Pilotage: This dangerous bank lies 9 miles northwest of Punta Catchabutan.

CAUTION: Dangerous uncharted shoals are likely to be encountered anywhere within the 200-meter contour in this area.

BANCO SALMEDINA

15 55N, 87 05W

Pilotage: This bank lies 25 miles west of Banco Providencia and about 10 miles offshore. It is a dangerous steep-to patch of coral, with a least depth of 2 feet near its east end. This reef breaks when there is any swell. A detached 24-foot patch lies 1 mile south-southeast of the bank and a 15-foot patch lies 6 miles northeast of it. Banco Salmedina should be given a berth of at least 2 miles.

PUNTA CONGREJAL

15 47N, 86 51W

Pilotage: This low sandy point is marked by the trunks of trees and the discolored water from the Río Congrejal, which extends some distance out to sea. Depths of less than 36 feet extend up to 1.5 miles off the point.

The mountain chain that backs this part of the coast rises to Cerro Nana Cruz, a 6,098-foot peak located 9 miles southeast of the point. Cerro Congrejal (Bonito Peak) is 8,049 feet high and lies about 10 miles southwest of the point. This peak appears as a well-defined, sharp cone when seen from the northeast, but when seen from the northwest has a small, flat shoulder projecting east from just below the summit.

LA CEIBA LIGHT, head of pier, 15 47.5N, 86 47.8W. Fl W 5s, 16ft. Wooden tower.

PUERTO LA CEIBA

15 46N, 86 48W

Pilotage: This open roadstead harbor, one of the principal ports of Honduras, is about 1 mile southwest of Punta Congrejal. A 1,423-foot-long wooden pier projects from the town.

Currents: The current in the area has been reported to be westward, attaining a velocity of 2 knots at times. During northers the current sets south.

Winds: The prevailing winds during the day are northeast and at night southwest. Normally the weather is calm with gentle breezes, except during the season of the northers (November and December) when winds of gale force occur.

Port of entry

Anchorage: Anchorage is prohibited east of the pier. On the approach of a norther vessels are advised to proceed to sea. Swells are likely to be felt in this anchorage.

PUNTA OBISPO

15 51N, 87 23W

PUNTA OBISPO LIGHT, 15 50.9N, 87 22.5W. Fl W 5s, 132ft, 20M. White framework tower, 49ft.

Pilotage: Punta Obispo is about 34 miles west of Punta Congrejal. The flat coastal plain along this coast gradually widens as the mountains become more sloping and recede inland. There are swamps and marshes a short distance inland, and numerous streams discharge into the sea. The shore is covered with trees and thick vegetation that almost reach the water's edge.

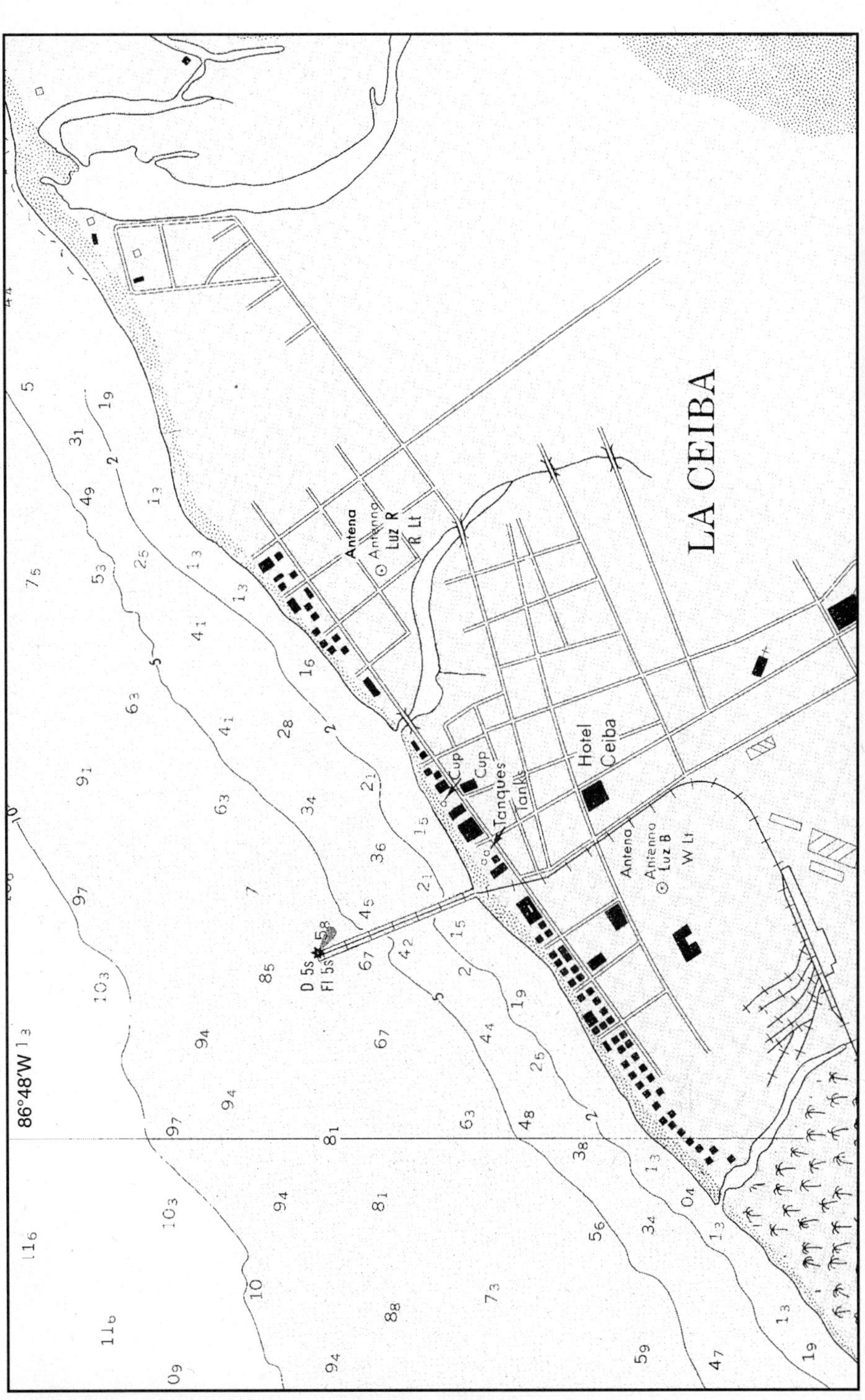

La Ceiba, soundings in meters *(from DMA 28144, 3rd ed, 10/85)*

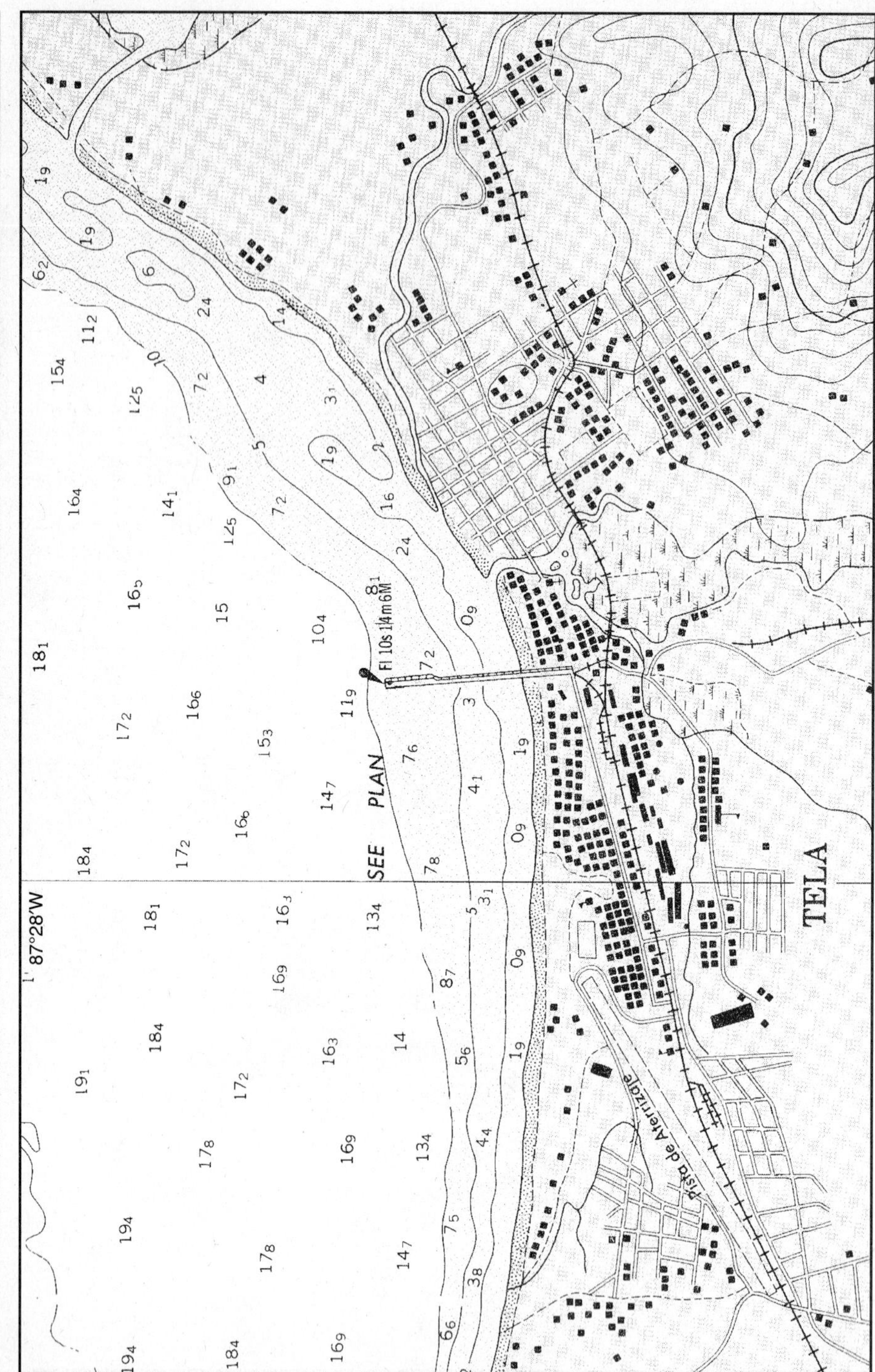

Puerto de Tela, soundings in meters *(from DMA 28161, 3rd ed, 12/82, corr. to 1/86)*

The point is a bluff, rocky, tree-covered headland that is the termination of a conspicuous, conical, grassy hill. The Clerks, a group of 20-foot-high rocks, lie close off the point. Depths of 121 feet lie within 0.5 mile of the rocks. The entire area northeast to northwest of the point has depths of as little as 21 feet in places.

Cabo Triunfo, a bold rocky projection, stands about 4 miles southwest of Punta Obispo.

BAHÍA DE TELA

15 47N, 87 27W
Pilotage: This bay is entered between Cabo Triunfo and Punta Sal, about 12.5 miles to the northwest. It is bordered by a low sandy coast, but is backed by a high mountain ridge about 8 miles south of Punta Obispo. A range of mountains 2,000 to 3,000 feet high backs the coast in the vicinity of Tela, but just west of the town it veers inland and extends south. Laguna de los Micos, a large, shallow, body of water, backs the bay and is entered about 6 miles northwest of Tela.

TELA LIGHT, head of pier, 15 47.2N, 87 27.6W. Fl W 10s, 45ft, 6M. Red metal tower, 60ft.

TELA

15 47N, 87 27W
Pilotage: This open roadstead harbor is located in the southeast part of Bahía de Tela, near the eastern entrance point. It is the second ranking port in Honduras. The prevailing winds are east and northeast, but strong northers may occur in the winter months. The average tidal range is about 1 foot, but wind also raises and lowers the water level. The current off the pier has been reported to set west in the morning and east in the afternoon. A pier about 2,000 feet long extends north from shore in front of the town. During north winds it is impossible to remain at the dock.
Anchorage: The anchorage is west of the pier, but a swell is likely here. Be prepared to put to sea if a norther threatens.

PUNTA SAL

15 55N, 87 36W
PUNTA SAL LIGHT, 15 55.5N, 87 36.1W. Fl (4) W 30s, 275ft, 15M. White framework tower, 49ft.

Pilotage: This bold, rocky promontory is the western boundary of Bahía de Tela. It projects 2 miles northeast from the coast, rising to wooded, irregular hills. It appears as an island when viewed against the low land to the south. A light is shown from the north end of the point. Three or four rocks, similar to The Clerks but much higher, lie about 0.5 mile off the east extremity of the point.

The coast between Punta Sal and Punta Caballos, about 22 miles to the west-southwest, is low, sandy, and densely wooded. The Montanas de Omoa back the western part of this coast.

Puerto Escondido lies 2 miles southwest of Punta Sal. This small cove is a good anchorage for small craft.

The small bay of Laguna Tinto lies 1 mile southwest of Puerto Escondido and should be approached with local knowledge. There is reported to be good anchorage for small craft.

Río Ulua is entered about 4 miles west of the entrance to Laguna Tinto. Punta Ulua, at the river entrance, is low and well defined. This large river is navigable by small river steamers for about 139 miles. The muddy discharge from the river discolors the sea for some distance offshore. A dangerous rock lies about 8 miles south-southwest of Punta Ulua. It lies within the 10-meter curve, about 1 mile offshore.

Río Chamelecon is located about 6 miles west-southwest of the Río Ulua. An isolated 636-foot conical hill is nearby. This is about 9 miles east-northeast of Punta Caballos.

PUNTA CABALLOS

15 50N, 87 58W
PUNTA CABALLOS LIGHT, 15 51.0N, 87 57.7W. Fl W 5s, 190ft, 20M. Tower. F R lights 0.22, 0.25, and 3 miles ESE; Fl R on tank 2.34 miles SE. Reported to lie 0.25 mile NNE (1993).
NO. 1 LIGHT, 15 51.2N, 87 57.8W. Q (3) G. Green articulated light. Reported missing (March 1986).
NO. 2 LIGHT, 15 51.0N, 87 58.2W. Q (3) R. Red articulated light. Reported Q R (1993).
NO. 3 LIGHT, 15 51.2N, 87 57.7W. Q (3) G. Articulated light. Reported destroyed (1993).
NO. 4 LIGHT, 15 50.5N, 87 58.3W. Q (3) R. Articulated light. Reported Q R (1993).
PIER LIGHT, on shoal off head, 15 50.3N, 87 57.3W. Fl R 2s. Red cylindrical daymark on pile. Destroyed.

P
Pilot
HONDURAS

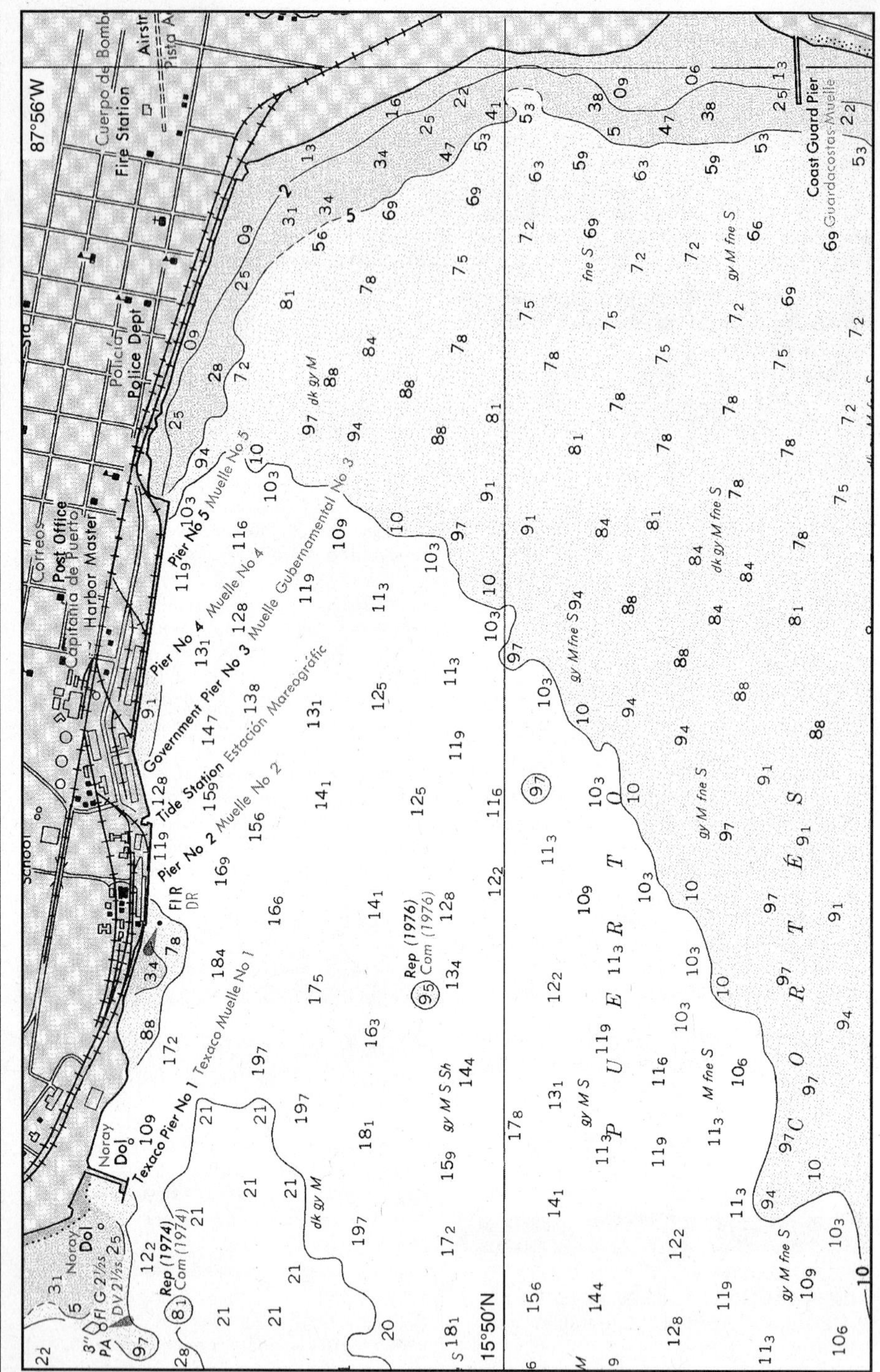

Puerto Cortés, soundings in meters *(from DMA 28163, 22nd ed, 3/84)*

Pilotage: This is the west extremity of a low, wooded peninsula that forms the north side of the harbor of Puerto Cortés. The radio masts on the point are good landmarks, as are the towers of a refinery located about 0.3 mile east of the light. Caution is necessary when navigating in this area as it has not been completely examined. A dangerous below-water rock lies 1.5 miles northwest of Cerro Cardona (15°53'N, 87°51'W), an isolated conical 518-foot-high hill situated 7.5 miles east-northeast of Punta Caballos.

The coast from Punta Caballos to Cabo Tres Puntas, about 39 miles west-northwest, forms a bight that indents the coast about 11 miles to the south. Puerto Cortés occupies the east part of this bight and Ensenada de Omoa occupies the south part. The Río Montagua enters the sea near the west part of Ensenada de Omoa and forms the boundary between Honduras and Guatemala. The east side of the bight is bounded by the base of the Montanas de Omoa, which rise to several prominent peaks. This range extends inland to the southwest from the head of the bight, and the land to the west becomes low and swampy.

PUERTO CORTÉS

15 50N, 87 57W

Pilotage: This is the major port of Honduras. It is situated on the north side of Bahía de Cortés. The red roof of a hotel and a water tank 2.25 miles southeast of Punta Caballos are good landmarks. The pilots monitor VHF channels 6 and 16.

Port of entry

Anchorage: The best anchorage is reported to be off the naval base southeast of town. This area is near the charted "Coast Guard Pier."

Services: Fuel, water, provisions, and some repairs are reported to be available. There are banks, a post office, restaurants, and communication facilities. The naval base may be able to haul your vessel and assist with repairs.

PUNTA DE OMOA

15 47N, 88 03W

Pilotage: This low but prominent point is about 6.3 miles southwest of Punta Caballos. Red cliffs stand on the coast about 3 miles east of the point. The coast is low, sandy, and backed by high, wooded ground and mountains. A disused lighthouse stands on Punta de Omoa. A submarine pipeline extends from shore to a group of mooring buoys about 0.5 mile north of the point.

Omoa is a small sheltered port that stands close south of Punta de Omoa. A small wharf extends from the town.

Ensenada de Omoa is a bight between Punta de Omoa and the mouth of the Río Montagua, about 10.5 miles to the west-southwest. A heavy swell rolls in during northers.

The Montanas de Omoa back this section of coast. Pico de Montagua, a 7,308-foot peak, lies about 9 miles southwest of the head of Ensenada de Omoa. This is the termination of the mountain chain. The other prominent peaks in this range are usually obscured.

RÍO MONTAGUA

15 44N, 88 13W

Pilotage: This river marks the boundary between Honduras and Guatemala. It is a fairly large, shallow river, navigable by riverboats to about 35 miles above its mouth.

GUATEMALA

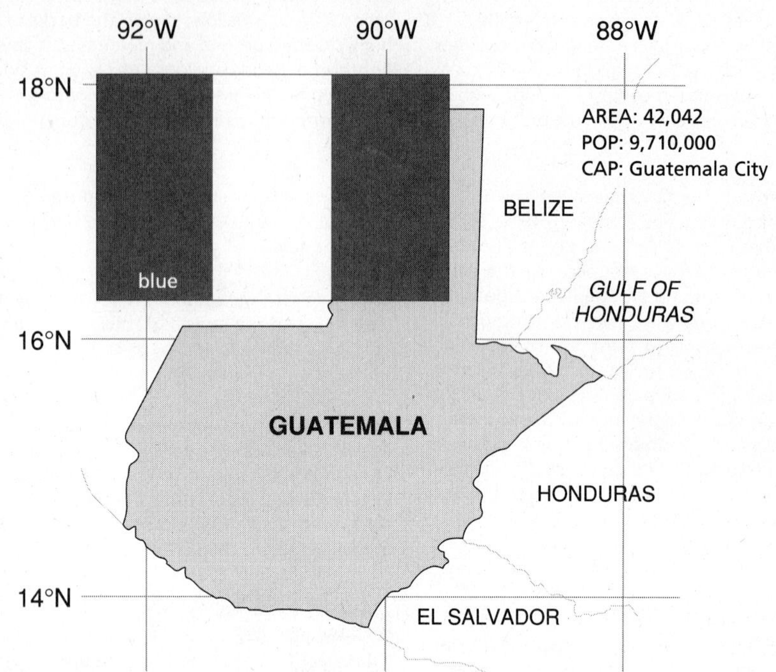

CAUTION: Many lights in Guatemala have been reported as irregular or unreliable.

Please see cautions about chartlets and waypoints on page P 2.

ENTRY PROCEDURES

Guatemala has a democratic government, which was temporarily suspended in May 1993 but was quickly restored. Cruising visitors seem to be able to avoid some of the problems inland; many report pleasant visits to Guatemala and the Río Dulce. Ports of entry on the Caribbean coast are Puerto Barrios, Santo Tomas de Castilla, and Livingston. Port officials are reported to be strict about showing the proper flags.

You must have passports, clearance from the last port of call, and ship's papers. You must declare all firearms, which will be held by the port captain. Citizens of most countries should have visas in advance of arrival; U.S. citizens no longer need a visa or tourist card for stays up to three months. The three-month stay can be extended upon application. Citizens of most other countries should have visas in advance of arrival. You should request a cruising permit, or *zarpe,* stating where you intend to visit. Permits are usually good for 90 days. You should carry some form of identification with you, other than your passport, in case of robbery or theft.

For the latest regulations contact the Guatemalan Embassy, Consular Section, 2220 R Street NW, Washington, DC 20008 (202-745-4952) or any of the Guatemalan consulates in Chicago, Houston, Los Angeles, Miami, New Orleans, New York, and San Francisco.

USEFUL INFORMATION

Language: Spanish.

Currency: Quetzal (Q); exchange rate (summer 1997), 5.86 quetzals = US$1.

Telephones: Country code, 502; a one-digit city code follows (Guatemala City is 2; all other locations use 9). Local numbers have six digits.

Public holidays (1998): New Year's Day, January 1; Holy Week, April 8–12; Labor Day, May 1; Army Day, June 30; Assumption of the Virgin Mary (Guatemala City only), August 15; Independence Day, September 15; Revolution Day, October 20; All Saints Day, November 1; Christmas Day, December 25; New Year's Eve, December 31. Other religious holidays are celebrated but banks and government offices are officially open.

Tourism offices: *In Guatemala:* Guatemala Tourist Commission, 7 Avenue 1-17, Zona 4, Guatemala City (tel., 502-2-31-1333; fax, 502-2-31-8893). *In the U.S.:* Guatemala Tourist Commission, 299 Alhambra Circle, Suite 510, Coral Gables, FL 33134 (tel., 800-742-4529; fax, 305-442-1013).

Special notes: Clashes between guerrilla and government forces occur periodically, and much of the countryside is currently considered unsafe for travel.

WARNING: The government of Guatemala and URNG guerrillas signed a final peace accord on December 29, 1996, to end the country's 36-year internal conflict. There have been no armed encounters between the Guatemalan army and guerilla forces since March 1996. Periodically, unfounded rumors that foreigners are involved in the theft of children for the purposes of using their organs for transplant has led to threats and incidents of mob violence in several areas of the country. The last such incidents occurred in 1994. Although the threat of further incidents is not currently considered immediate, travelers should be aware that in rural areas there exists greater likelihood, albeit small, of such an incident. Travelers should avoid contact with Guatemalan children. For the current status of travel in Guatemala, call the U.S. Department of State Consular Affairs hotline (voice recording, 202-647-5225; fax information service, 202-647-3000) or visit them on the Web at www.state.gov.

Unfortunately, violent crime is a very serious and growing problem throughout Guatemala; no area can be definitively characterized as "safe." Reports of car-jackings, hold-ups, rapes, and murders are frequent in both tourist areas and the countryside. Intercity travel at night is particularly dangerous. For assistance or updated information while in Guatemala, U.S. citizens should contact the Consular Section of the U.S. Embassy, Avenida la Reforma 7-01, Zone 10, Guatemala City (505-2-31-15-41); consular office hours for American citizens' services are 0800 to 1200 and 1300 to 1500.

Cholera and dengue fever are present in Guatemala, and there is risk of malaria in rural areas outside the central highlands. For updated health information contact the Centers for Disease Control International Travelers Hotline (voice recording, 404-332-4559; fax information service, 404-332-4565; Web, www.cdc.gov).

RÍO MONTAGUA

15 44N, 88 13W

Pilotage: This river marks the boundary between Honduras and Guatemala. It is fairly large and shallow and is navigable by river boats to about 35 miles above its mouth.

Between the Río Montagua and Cabo Tres Puntas, about 27 miles to the northwest, the low, swampy coast is bordered by a dark sandy beach, backed by trees.

Río San Francisco del Mar discharges about 14 miles northwest of the Río Montagua, and a branch of the river leads to Bahía la Graciosa. A conspicuous 600-foot-high tableland stands about 8 miles southwest of the mouth of this river.

CABO TRES PUNTAS (CAPE THREE POINTS)

15 58N, 88 37W

CABO TRES PUNTAS LIGHT, 15 57.4N, 88 36.2W. Fl W 10s, 132ft, 17M. White framework tower.

Pilotage: This is a prominent, well-wooded point, which is the northwest extremity of a low, wooded peninsula about 11 miles long that borders the northeast side of Bahía de Amatique. A conspicuous tower stands on the cape. Steep-to foul ground, which usually breaks, extends about 0.5 mile west from the cape.

BAHÍA DE AMATIQUE (HONDURAS BAY)

15 56N, 88 44W

Pilotage: The bay is entered between Cabo Tres Puntas and Punta Gorda, about 13.5 miles to the northwest in Belize. It has general depths of 30 feet to 102 feet over its central portion, with gradual shoaling toward the eastern shore. The east side of the bay recedes about 13 miles south to the narrow entrance of Bahía de Santo Tomas de Castilla, which recedes an additional 2.5 miles south to its head. The land on the east side of the bay is generally flat, swampy, and densely wooded. From the entrance to Bahía de Santo Tomas de Castilla the west side of the bay extends about 23 miles northwest and then about 10 miles northeast to Punta Gorda. The land on the west side of the bay is higher, densely wooded, and backed by mountain ranges.

From Cabo Tres Puntas the east side of the bay extends about 1.5 miles south-southeast to Punta Manabique and then about 5.7 miles southeast to Firewood Point.

Bahía la Graciosa (Hospital Bight) is a shallow bight that recedes about 4 miles to the south-east and is entered between Firewood Point and Punta Manglar. A sandbar with a depth of 14 feet extends across the entrance. From Punta Manglar, the coast, which is low and swampy, extends about 7.5 miles south-southwest to the east entrance point of Bahía de Santo Tomas de Castilla.

OX TONGUE SHOAL LIGHT, 15 53.9N, 88 41.1W. Fl W 3s, 13ft, 12M. White and orange framework structure. Radar reflector.
HEREDIA SHOAL LIGHT, 15 50.8N, 88 40.4W. Fl R 6s, 20ft. White and orange framework structure. Reported extinguished (1989).

OX TONGUE SHOAL

15 53N, 88 38W
Pilotage: This narrow shoal has depths of 18 feet and less. It extends about 7.5 miles west-northwest from Punta Manglar. A light marks the west extremity of the shoal. It has been reported (1984) that because of reef buildup, this light should be kept at least 1.3 miles to the east when entering or leaving port.

Heredia Shoal, with a least depth of 18 feet, lies about 3 miles south-southeast of the light on Ox Tongue Shoal. A buoy marks the north side of the shoal.
A shoal with depths of 28 feet lies about 3.3 miles southeast of Heredia Shoal, and another with depths of 42 feet lies a further 1.5 miles southeast.

BAJO VILLEDO LIGHT, 15 44.7N, 88 36.9W. Fl W 2s, 17ft. Aluminum framework tower. Radar reflector.

BAJO VILLEDO

15 45N, 88 37W
Pilotage: This reef has a least charted depth of about 15 feet. It is marked by a light and lies west of the range lights leading to Puerto Barrios.

BAHÍA SANTO TOMAS DE CASTILLA

Pilotage: The bay is entered between Punta Manglar and Punta Palma about 1.5 miles west-northwest. Its densely wooded shores are bordered by a mud flat, leaving a navigable basin about 2 miles in extent within the 5-meter-depth curve.

PUERTO BARRIOS LIGHT, 15 43.8N, 88 35.8W. Fl W 8s, 25M.
HEAD OF PIER LIGHT, 15 44.5N, 88 36.6W. Oc W 4s, 5M. Three F R mark jetty ruins (1982).

PUERTO BARRIOS

15 44N, 88 36W
Pilotage: This port, formerly the busiest in Guatemala, is located on the east side of Bahía de Santo Tomas de Castilla. A dredged channel leads south from a position 2 miles north-northeast of Bajo Villedo. Lights at Santo Tomas de Castilla lead through this channel, bearing 189.5°. A 1,000-foot-long pier extends west from the shore. Ruins, marked by lights, extend 1,100 feet farther to the west-northwest. There is a minimum depth of about 24 feet at the pier.
Winds: Land and sea breezes are the predominating winds at Puerto Barrios. The sea breeze blows from the north during the day, gradually diminishing toward evening. The land breeze blows from the south from about midnight to sunrise. This regular cycle is altered by northers during the winter months.
Currents: The current off the pier is reported to be diurnal in nature. During the morning the current sets northwest at a rate of 0.4 knot and occasionally at 2.3 knots. The velocity increases after a strong norther has abated and during periods of heavy rain. In the afternoon the current reverses and sets southeast at a rate of 0.2 to 0.6 knot. The mean rise and fall of the tide is less than 1 foot.
Port of entry
Anchorage: Anchorage is prohibited north of the pier, due to incoming traffic. Anchor south of the pier.
Services: Fuel, water, groceries, and general supplies should be available.

RANGE (Front Light), 15 41.7N, 88 37.4W. F R. **(Rear Light),** about 290 meters, 237° from front. F R.
MATIAS DE GALVEZ RANGE (Front Light), 15 41.6N, 88 37.2W. Q Y, 35ft. Metal framework tower, red triangular-shaped daymark, point up. Visible on rangeline only. **(Rear Light),** S,

about 347 meters, 189.5° from front, 15 41.4N, 88 37.3W. Oc Y 4s, 80ft. Metal framework tower, red triangular-shaped daymark, point up. Visible on rangeline only.

PUERTO SANTO TOMAS DE CASTILLA

15 42N, 88 37W

Pilotage: This is the main commercial port facility for Guatemala. It is located on the south shore of Bahía de Santo Tomas de Castilla. There is a single 3,000-foot-long wharf with a depth of about 33 feet alongside. Commercial vessels often moor stern-to this wharf. The wharf may be extended beyond this length.

Winds: Sea breezes predominate between 1100 and sunset, reaching force 4, or occasionally force 6, at about 1500. At other times it is calm, or there are light southerly breezes. This pattern is disturbed by the passage of a depression to the north, when squalls and southwest winds may be expected.

Currents: A weak current, seldom exceeding 0.5 knot, sets southwest across the turning basin.

Port of entry: This is the primary commercial port of entry for Guatemala, but cruisers report that pleasure boats are discouraged from entering here. You can enter at nearby Puerto Barrios.

PUNTA HERRERIA

15 49N, 88 44W

Pilotage: This point is located about 7 miles northwest of Punta Palma, which is the western point at the entrance to Bahía de Santo Tomas de Castilla. The south shore of Bahía de Amatique is backed by high ground that reaches an elevation of 1,118 feet about 3 miles south of Livingston Bay.

LIVINGSTON BAY

Pilotage: The bay is entered between Punta Herreria and a point 1.3 miles northwest. The Río Dulce enters the head of the bay. The 6-foot-depth contour extends across the entrance to the bay and about 0.5 mile farther seaward. Depths reach about 18 feet a further 0.5 mile offshore.

LIVINGSTON

15 50N, 88 45W

Pilotage: This is an important city for the transit of goods on the Río Dulce and a port of entry. A marked 6-foot-deep channel leads across the entrance bar and to the customs house wharf in Livingston. A refinery with four conspicuous chimneys is situated 1 mile northwest of Livingston.

Winds: The winds in this area are mostly northeast, with the strongest breezes being from May to September, with frequent heavy thunder squalls at night. The air is moist in all seasons, but May to October is the wettest.

Port of entry

Anchorage: Cruisers generally anchor near the Texaco fuel dock, which is located a bit southwest of the town dock.

Services: Fuel, water, groceries, and general supplies are available.

RÍO QUEHUECHE

15 51N, 88 46W

Pilotage: This river discharges into the sea about 1.3 miles northwest of Livingston. It may be identified by a waterfall located 1 mile to its northwest and by the refinery with four conspicuous chimneys located about 0.25 mile to the southeast.

PUNTA COCOLI

15 53N, 88 49W

Pilotage: This prominent round bluff is located about 3.8 miles northwest of the entrance to Río Quehueche. Río Cocoli flows into the sea about 0.8 mile southwest of the bluff.

From Punta Cocoli the coast trends 2.5 miles west-southwest to Punta San Martin and then about 3 miles north-northwest to the mouth of the Sarstoon River. The coast is low, but about 2 miles inland it is backed by heavily wooded mountains, which rise to a height of 1,394 feet. These mountains are about 2.8 miles south-southwest of the mouth of the Sarstoon River. They extend about 6 miles to the west at the same height.

SARSTOON RIVER

Pilotage: This river forms the boundary between Guatemala and Belize. The river's banks are low, swampy, and covered with mangroves. The entrance is about 0.5 mile wide, with a bar across the mouth. The bar has a depth of about 6 feet and generally breaks heavily. Depths of less than 18 feet lie about 1 mile east-northeast of the river's mouth.

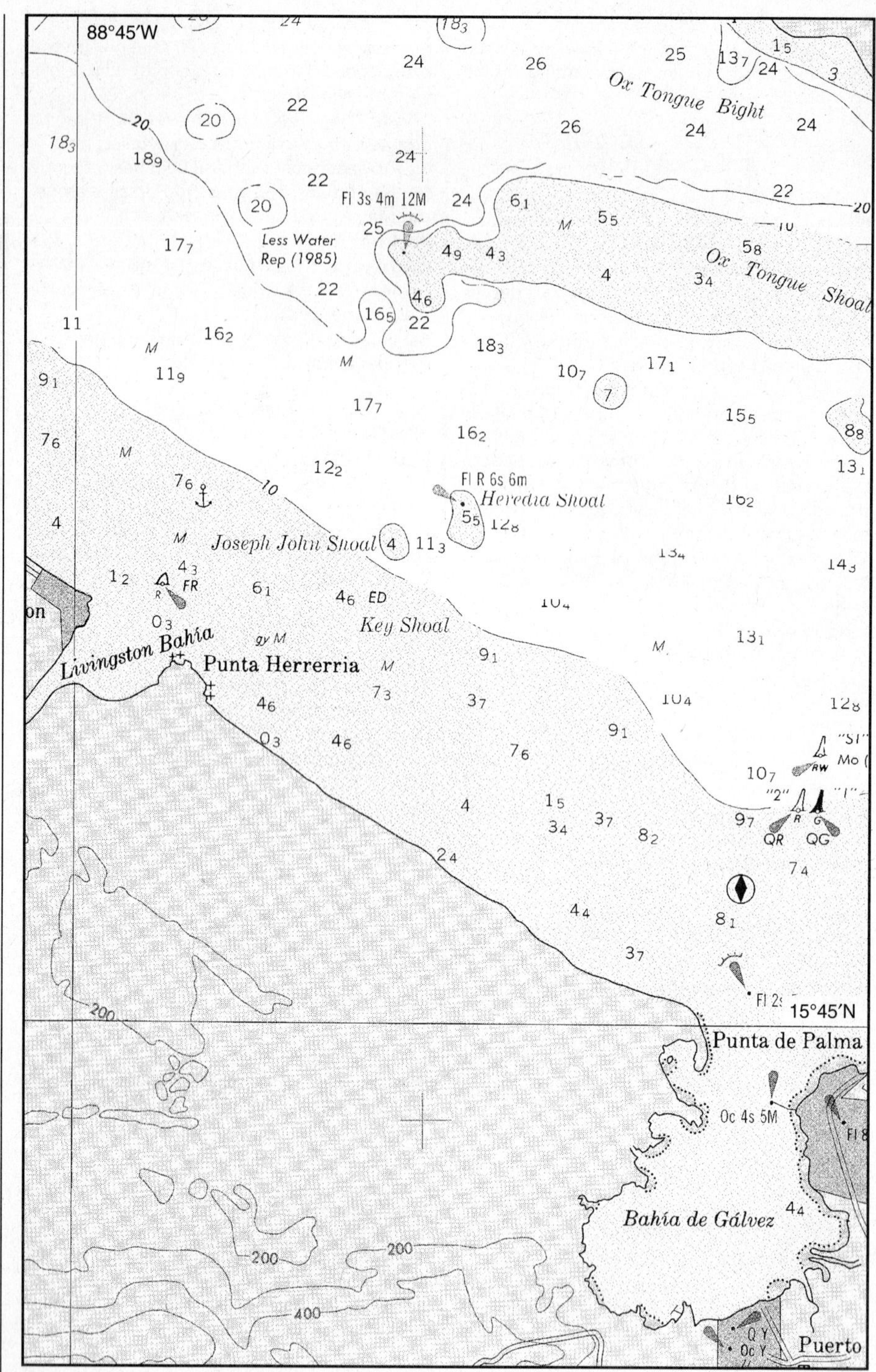

Approaches to Livingston and Puerto Barrios, *soundings in meters (fr. DMA 28162, 29th ed, 2/85)*

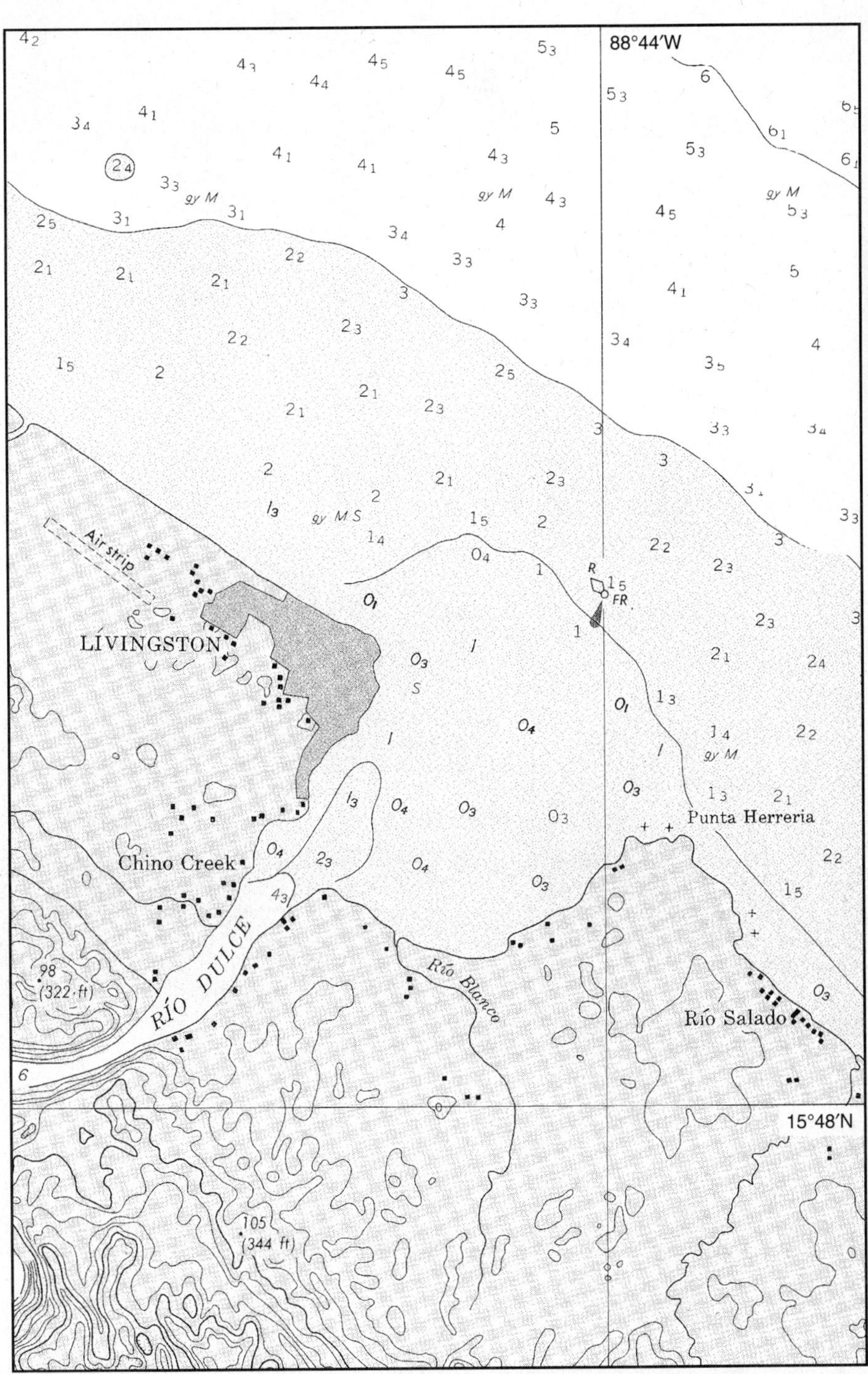

Livingston and Río Dulce, soundings in fathoms and feet *(from DMA 28164, 16th ed, 2/85)*

Belize

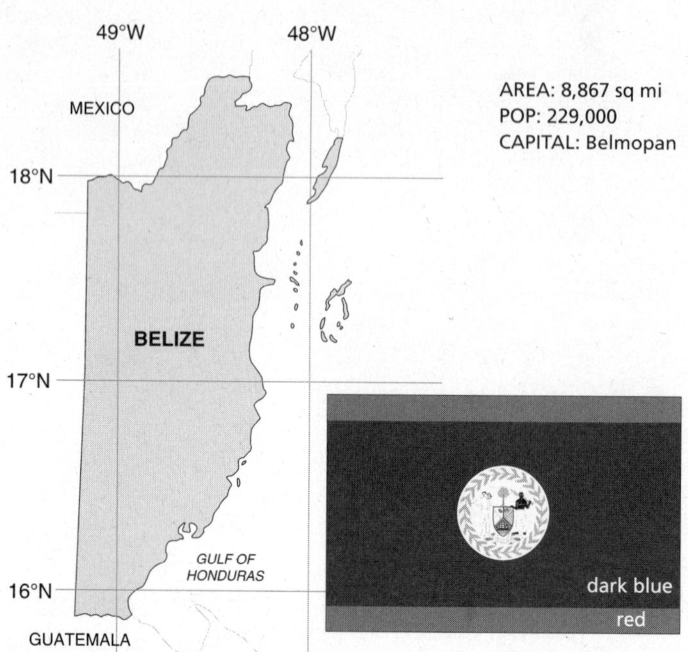

AREA: 8,867 sq mi
POP: 229,000
CAPITAL: Belmopan

CAUTION: Many lights in Belize have been reported as irregular or unreliable.

Please see cautions about chartlets and waypoints on page P 2.

ENTRY PROCEDURES

Belize was formerly a British Crown Colony, but gained independence in 1981. It is an independent country with a democratically elected government.

Ports of entry include Punta Gorda, Dangriga (formerly Stann Creek), Big Creek, Belize City, and San Pedro. Although there are no customs officials at San Pedro and you must pay for the official's airfare from Belize City, it is an immense advantage if you are coming from the north because it saves you a trip to Belize City. This process is reported to cost about US$50 to $60. Vessels arriving from the south clear in at Punta Gorda or Dangriga, where formalities are easier than in Belize City, which is geared to commercial traffic. Contact the harbormaster when you enter port.

You should have passports, clearance from the last port of call, crew lists, stores lists, and the ship's papers. Citizens of most countries, including the U.S., are granted a 30-day stay, but a permit is needed for longer visits. Apply through the immigration department or the local police department. You may be requested to prove you have sufficient funds for your stay. U.S., Canadian, and U.K. citizens do not need a visa; citizens of other countries should obtain a visa before entering Belize. Visitors are allowed a one-month stay without an extension from the Belize Immigration Department. It's recommended that you contact your local embassy or consulate for details. For the latest entry information, contact the Embassy of Belize, 2535 Massachusetts Avenue NW, Washington, DC 20008 (202-332-9636).

USEFUL INFORMATION

Language: English (official). Spanish is widely spoken; other languages are Creole, Garifuna, and Mayan.
Currency: Belize dollar (Bz$); exchange rate (summer 1997), about Bz$2 = US$1.
Telephones: Country code, 501; a district code follows and then the local number. The district code for Punta Gorda is 07; Dangriga, 05; Placencia, 06; Belize City, 02; San Pedro, 026; Corozal, 04; and San Ignacio, 092. International calls do not dial the 0 in the district number. 800 numbers are not accessible from Belize.
Public holidays (1998): New Year's Day, January 1; Baron Bliss Day, March 9; Good Friday, April 10; Holy Saturday, April 11; Easter Monday, April 13; Labour Day, May 1; Commonwealth Day, May 25*; St. Georges Cay Day, September 10; Independence Day, September 21; Columbus Day, October 12; Garifuna Day, November 19; Christmas Day, December 25; Boxing Day, December 26. *Date not officially set until December 1997.
Tourism offices: *In Belize:* Belize Tourist Board, 83 North Front Street, P.O. Box 325, Belize City (501-[0]2-77213). *In the U.S.:* Belize Tourist Board, 421 Seventh Avenue, Suite 1110, New York, NY 10001 (tel., 800-624-0686; 212-563-6011; fax, 212-563-6033).
Additional notes: Tropical diseases, including dengue fever, may be encountered in Belize. Malaria is a risk in rural areas, including the offshore islands, but not in the central coastal district. Cholera has been reported. For current information, contact the U.S. Centers for Disease Control International Travelers Hotline (voice recording, 404-332-4559; fax information service, 404-332-4565; Web, www.cdc.gov). For assistance in Belize, U.S. citizens may contact the Consular Section of the U.S. Embassy, 29 Gabourel Lane, Belize City (tel., 501-[0]2-77161; fax, 501-[0]2-30802).

SARSTOON RIVER TO PUNTA GORDA

Pilotage: The Sarstoon River forms the boundary between Guatemala and Belize. The river's banks are low, swampy, and covered with mangroves. The entrance is about 0.5 mile wide, with a bar across the mouth. The bar has a depth of about 6 feet and generally breaks heavily. Depths of less than 18 feet lie about 1 mile east-northeast of the river's mouth.

From the mouth of the river the coast trends about 5 miles north-northwest to the mouth of the Temash River. The coast then trends about 5 miles northeast to Mother Point, a prominent high bluff. About 9 miles north-northwest of the Sarstoon River, a small isolated hill rises to a height of 400 feet. There are red cliffs between the Temash River and Mother Point. About 1.5 miles north-northeast of Mother Point and about 2.8 miles southwest of Orange Point the River Moho enters the sea.

The sea breaks heavily on the bars at the mouths of the Temash and the Moho rivers. Their banks are swampy and fringed with impenetrable mangroves for about 40 or 50 miles inland, where they become firm and covered with mahogany trees. The current in these rivers runs about 1 knot.

ORANGE POINT

16 05N, 88 49W
Pilotage: This point is located about 12 miles north-northeast of the Sarstoon River and the Guatemala border, just south of Punta Gorda. The point is readily identified, as the land is about 30 feet high, dropping abruptly to the coast. The coast and the land are flat and densely wooded for a considerable distance inland.

PUNTA GORDA TO SITTEE POINT

PUNTA GORDA LIGHT, at Carib Settlement, 16 06.3N, 88 47.9W. F W, 56ft, 9M. White mast, 56ft. R lights on radio tower 0.6 mile SW.

PUNTA GORDA

16 06N, 88 48W
Pilotage: The town is located just north of Orange Point. Gorda Hill is a conspicuous saddle-shaped hill located about 2.8 miles north-northwest of Punta Gorda. A mountain range rises to an elevation of 1,000 feet about 6 miles west of Gorda Hill. This range forms part of a chain of mountains that parallels the coast, from 10 to 20 miles inland, to within 13 miles southwest of Belize City. A light is shown from Punta Gorda.
Port of entry: Formalities are reported to be easier here than in Belize City. This is a good place to clear in if you are arriving from Guatemala or Honduras. Contact the harbormaster.
Anchorage: Anchor off the town dock, where you can dinghy in. The dock is reported to have less than 6 feet of water alongside. The roadstead is open and subject to swells. Be prepared to get under way if conditions get rough.

Services: Groceries, communications, and general supplies are available.

PORK AND DOUGHBOY POINT

16 11N, 88 44W

Pilotage: This point lies about 6 miles northeast of Punta Gorda on the mainland. The Río Grande flows into the sea about 2.3 miles south-southwest of the point.

Between Pork and Doughboy Point and Punta Ycacos (Icacos Point), 9.5 miles northeast, is Port Honduras. This extensive bay is fouled with many dangers.

Deep River, or Río Hondo, flows into the north part of Port Honduras, about 4.5 miles northwest of Punta Ycacos. Depths over the bar are about 2 feet, and inside depths run about 12 to 18 feet.

PUNTA YCACOS

16 15N, 88 35W

Pilotage: This is the southern extremity of a small cay situated close south of a tongue of land that extends 1 mile south. The cay is covered with pine trees.

Wilson Cay is located about 0.8 mile south of Punta Ycacos, and about 1.5 miles west of the point are the Bedford Cays. There are numerous other cays in Port Honduras.

EAST SNAKE CAY LIGHT, 16 12.5N, 88 30.4W. Fl W 3s, 65ft, 13M. Concrete framework tower, 65ft.

SNAKE CAYS

16 12N, 88 32W

Pilotage: These cays lie just outside the 20-meter-depth curve, about 30 miles northeast of Punta Gorda and 4 miles southeast of Punta Ycacos. They cover an area about 8 miles long from southwest to northeast and about 4 miles from southeast to northwest. There are four densely wooded cays and numerous shoal patches with depths of 6 to 30 feet. East Snake Cay, the farthest northeast, is marked by a light.

PUNTA NEGRA

16 16N, 88 33W

Pilotage: This conspicuous bluff stands about 2.8 miles northeast of Punta Ycacos. Between Punta Negra and Monkey River, 6.5 miles to the north-northeast, the coast is more elevated and bordered by a sandy beach.

MONKEY RIVER LIGHT, N side of entrance, 16 22.0N, 88 29.1W. F W, 52ft, 8M. White mast, 52ft. Reported extinguished (1985).

MONKEY RIVER

16 22N, 88 29W

Pilotage: This area may be identified by some of the prominent houses in the town, centered principally on the south bank of the river. A light is also on the south bank at the river's entrance.

The 10-meter-depth curve is well defined and lies 1 mile east of the river's mouth. Numerous dangers lie within the 10-meter curve along this coast.

From Monkey River the coast extends north-northeast about 11 miles to Placencia Point and then continues north-northeast another 19 miles to Sittee Point. This coast is low, swampy, and wooded. The shore is fringed by low, wooded cays and intersected by numerous small creeks and streams.

MONKEY SHOAL

16 23N, 88 25W

Pilotage: About 4 miles east-northeast of the light at Monkey River is Monkey Shoal; other isolated patches lie in the vicinity. A light buoy is moored 1 mile east of Monkey Shoal, which has general depths over 13 feet.

Penguin Shoals has a least depth of about 10 feet and lies about 3 miles north-northeast of Monkey Shoal.

Middle Shoal is about 0.5 mile north-northeast of Penguin Shoal, with a least depth of about 13 feet. A light buoy is moored 1 mile northeast of Middle Shoal.

Potts Shoal, with a least depth of 12 feet, lies about 1.5 miles north of Middle Shoal.

BUGLE CAYS LIGHT, SW cay, E side of the Narrows, main channel leading to Belize City from the S, 16 29.4N, 88 19.2W. Fl (2) W 10s, 61ft, 10M. White framework tower, 62ft.

BUGLE CAYS

16 29N, 88 19W

Pilotage: These two cays lie about 3 miles southeast of Placencia Point, on the east side of the channel. Without local knowledge, these cays will not be distinguished from a distance of more than 4 miles.

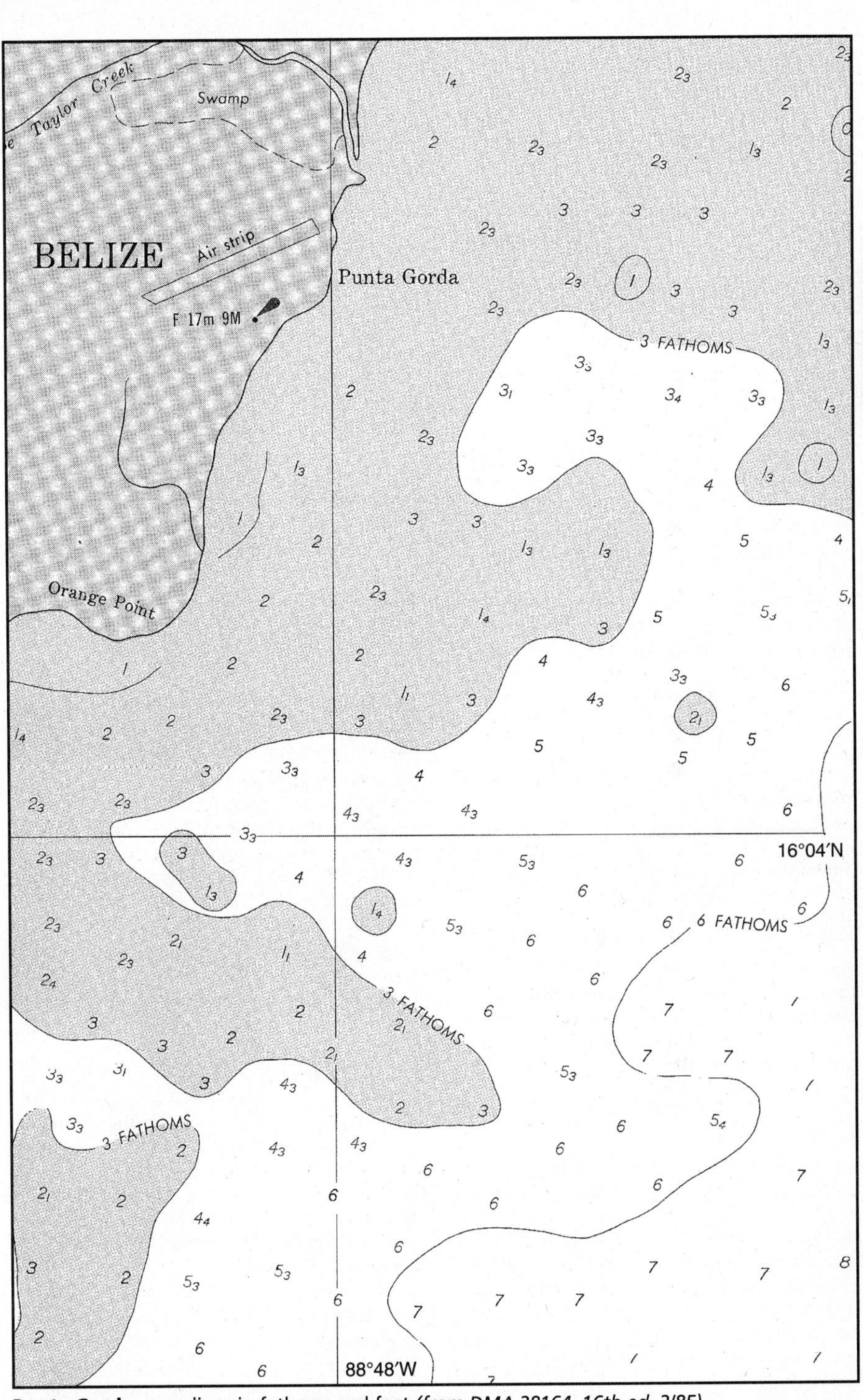

Punta Gorda, soundings in fathoms and feet *(from DMA 28164, 16th ed, 2/85)*

Shoals with depths of 6 to 18 feet extend about 4.8 miles south of Bugle Cays. The Narrows lie between these dangers, Potts Shoal, and the other dangers on the west side of the passage. The Narrows are about 1.5 miles wide, with general depths of 33 to 85 feet.

HARVEST CAY

Pilotage: This cay is about 2 miles south of the entrance to Placencia Lagoon and has a conspicuous 79-foot-high table hill on it. This is just south of the buoyed channel leading into Big Creek.

BIG CREEK

16 30N, 88 25W

Pilotage: This narrow river lies about 2.5 miles west of Placencia Point. It is approached by a buoyed channel, dredged to a depth of 22 feet, entered east of Harvest Cay. A small 34-foot-long pier with depths of 18 feet alongside extends from the right bank of the creek. Vessels up to 230 feet long berth port-side-to, heading downstream. Big Creek is primarily a commercial port, and pilots from Belize City (VHF channel 16 or 2182 kHz) operate here.

PLACENCIA LAGOON AVIATION LIGHT, 16 32.0N, 88 25.2W. F R, 315ft, 35M. Mast.

PLACENCIA POINT

16 31N, 88 22W

Pilotage: This is the southern point on the narrow ridge of low land that protects the east side of Placencia Lagoon. This shallow lagoon parallels the coast for about 10 miles to the north.

Placencia Cay stands close east of the point, and the bight thus formed is a good anchorage for small craft. There is a small settlement on Placencia Point with some groceries, a post office, a hotel, and some restaurants.

FALSE CAY

16 36N, 88 20W

Pilotage: This cay is about 0.8 mile offshore and about 5 miles north-northeast of Placencia Point. The cay is low, narrow, and covered with bushes and palms. Foul ground extends about 0.3 mile from it to the northeast through east, to its southwest side. The west and northwest sides are steep-to, with depths up to 34 feet between the cay and shore.

Anchorage: There is reported to be good anchorage between the cay and shore in depths of about 34 feet.

From Placencia Point the coast runs about 8.8 miles north-northeast to Jonathan Point. A series of small bays are formed on this coast by Rum Point, False Point, and Rocky Point. They lie 1.8 miles, 4.3 miles, and 5.3 miles north of Placencia Point, respectively. A pier extends about 500 feet from shore at the village of Riverdale, about 2.8 miles north of Jonathan Point.

From Jonathan Point the coast trends about 4 miles north to the mouth of South Stann Creek, then about 3 miles north to the entrance to Sapodilla Lagoon. Between Placencia Point and the mouth of South Stann Creek the coast is backed by an extensive plain of ridges from 50 to 100 feet high. About 10 miles inland is a ridge of mountains, which rises to a height of 3,680 feet about 18 miles west of Sapodilla Lagoon. Two peaks of 2,034 feet and 1,679 feet rise, respectively, about 11.3 miles northwest and 9.5 miles west-southwest of the entrance to Sapodilla Lagoon.

South Stann Creek flows into the sea across a shallow bar about 2 miles north-northeast of Riverdale.

SITTEE POINT TO MULLINS RIVER

16 48N, 88 15W

SITTEE POINT LIGHT, 16 48.4N, 88 14.8W. Fl W 5s, 30ft, 8M. White metal framework tower. F W shown at all times in Sapodilla Lagoon 3 miles WSW and on pier head at Riverdale 7.8 miles SSW.

Pilotage: This is a well-defined, wooded point that extends about 0.8 mile from the coast. It forms the south entrance point to the Sittee River.

The river has a depth of about 3 feet over the bar at its entrance. Dead tree stumps rise from the bottom. Depths of less than 18 feet extend about 0.5 mile east from the river. Some ledges situated 3.8 miles southeast and 4 miles northeast of the light on Sittee Point have depths over them of 13 and 15 feet respectively. The ledges restrict the fairway of the Inner Channel to 3 miles in width.

False Sittee Point is a narrow projection of land located about 2 miles north-northwest of Sittee Point. Shoal ground extends about 0.8 mile east and northeast from the point.

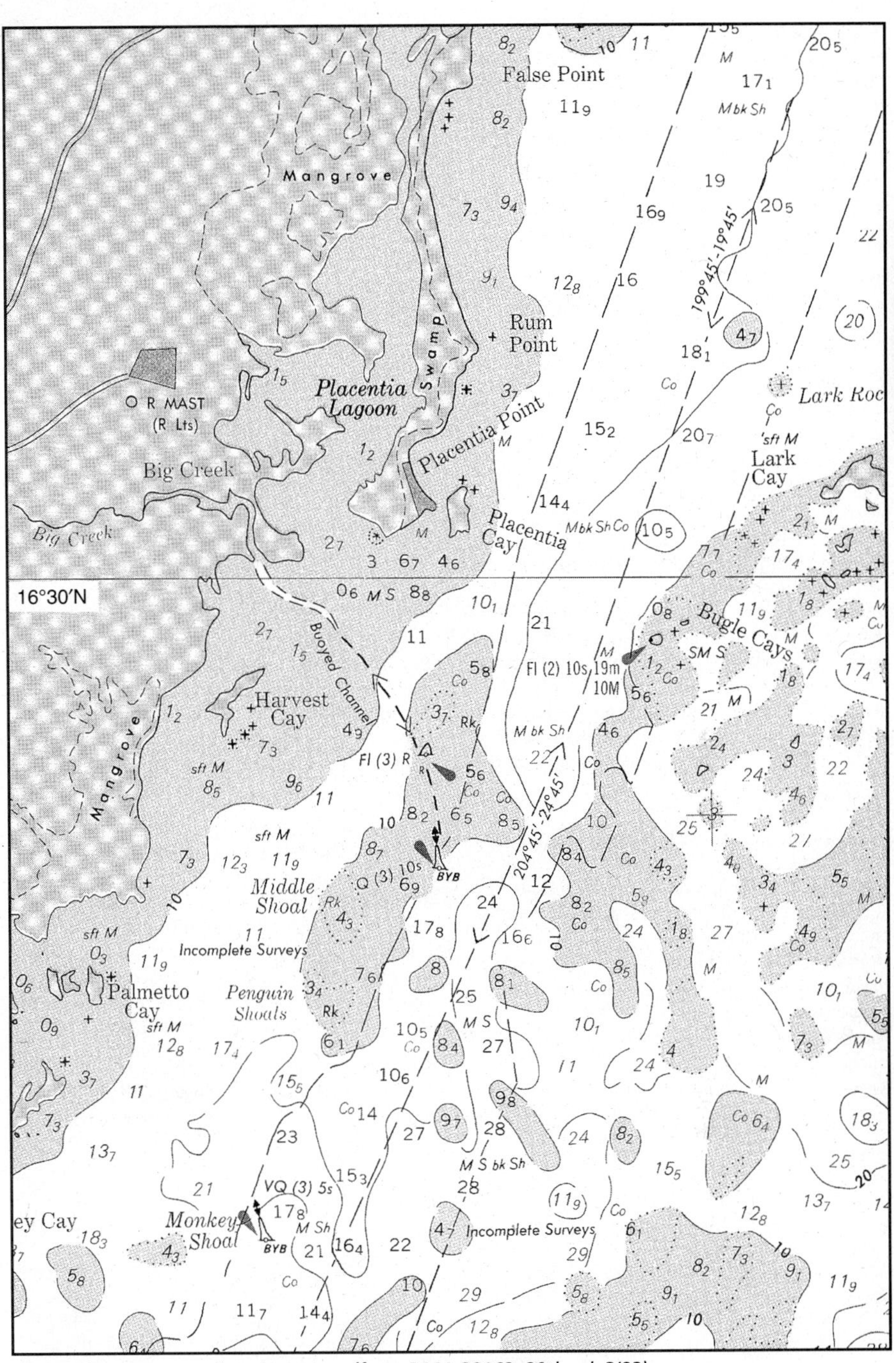

Placentia area, soundings in meters *(from DMA 28162, 29th ed, 3/92)*

Commerce Bight is entered north of False Sittee Point and extends about 3 miles to the west. The depths within the bight decrease gradually from 36 feet in the entrance to 18 feet about 0.3 mile to 0.5 mile offshore. *Caution: Less water than charted has been reported (1994).* A conspicuous white building stands close to the coast, about 2 miles north-northwest of False Sittee Point.

A pier with depths of 5 feet alongside the west and south sides lies 0.5 mile east of the mouth of Yemeri Creek. This is about 5.3 miles north-northeast of False Sittee Point. A road connects the pier with Stann Creek 2.3 miles to the northeast. A light stands on the pier.

DANGRIGA (STANN CREEK)

16 58N, 88 13W

Pilotage: The town, known as Dangriga, is located on North Stann Creek, just north of Commerce Bight. The village is on both shores of the creek and fronts the coast for about 1 mile. This is the seat of government for the district of Stann Creek. There is a 400-foot pier, with shallow depths alongside, located about 0.3 mile north-northwest of the mouth of the creek. There are many prominent buildings between the jetty and the creek. A private 600-foot jetty is located about 0.5 mile south of the creek.

In 1972 it was reported that a church with a white tower and a conspicuous square tank on a metal framework tower stood in the town. A conspicuous radio mast stands close west of the tank.

Port of entry: Formalities are reported to be easier here than in Belize City. Contact the harbormaster.

Anchorage: Anchor off the town docks north of the river mouth.

Services: Groceries, fuel (by jug), a post office, telephones, and general supplies are available.

COLSON POINT LIGHT, 17 04.4N, 88 14.2W. Fl W 10s, 39ft, 9M. White metal framework tower.

COLSON POINT

17 04N, 88 15W

Pilotage: The point is located about 6.3 miles north-northwest of North Stann Creek. The 10-meter depth curve lies about 2.3 miles off the point. A light marks the point. A rock with a depth of 6.5 feet over it lies 0.3 mile north-northwest of Colson Point light structure. From Colson Point the coast trends about 1.3 miles west, then turns north-northwest. Mullins River flows into the sea 2.8 miles northwest of Colson Point. The town of Mullins River is located on the north shore of the river near the mouth. The bar at the mouth of this river shifts frequently and is considered dangerous.

MANATEE RIVER LIGHT, 17 13.8N, 88 18.2W. F W, 33ft, 2M. Occasional.

MANATEE RIVER

17 13N, 88 18W

Pilotage: This river discharges about 7.8 miles north of Mullins River. It leads to a shallow lagoon, and the shifting bar at the mouth is considered dangerous. A light is occasionally exhibited from a post on the south side of the river mouth.

Dolphin Head, a 400-foot-high hill, is located about 5 miles northwest of the Manatee River light structure. The Paps, 351 feet high, and 298-foot Saddle Hill are two hummocks situated about 2 and 5.3 miles north-northwest of Dolphin Head. These hills are the terminus of the mountain range that backs the coast of Belize. The coast to the north and the terrain well into the interior is low and swampy in places.

There are several shoals with depths of around 20 feet located west of the Manatee River. The Río Sibun flows into the sea about 12 miles north-northeast of Manatee River.

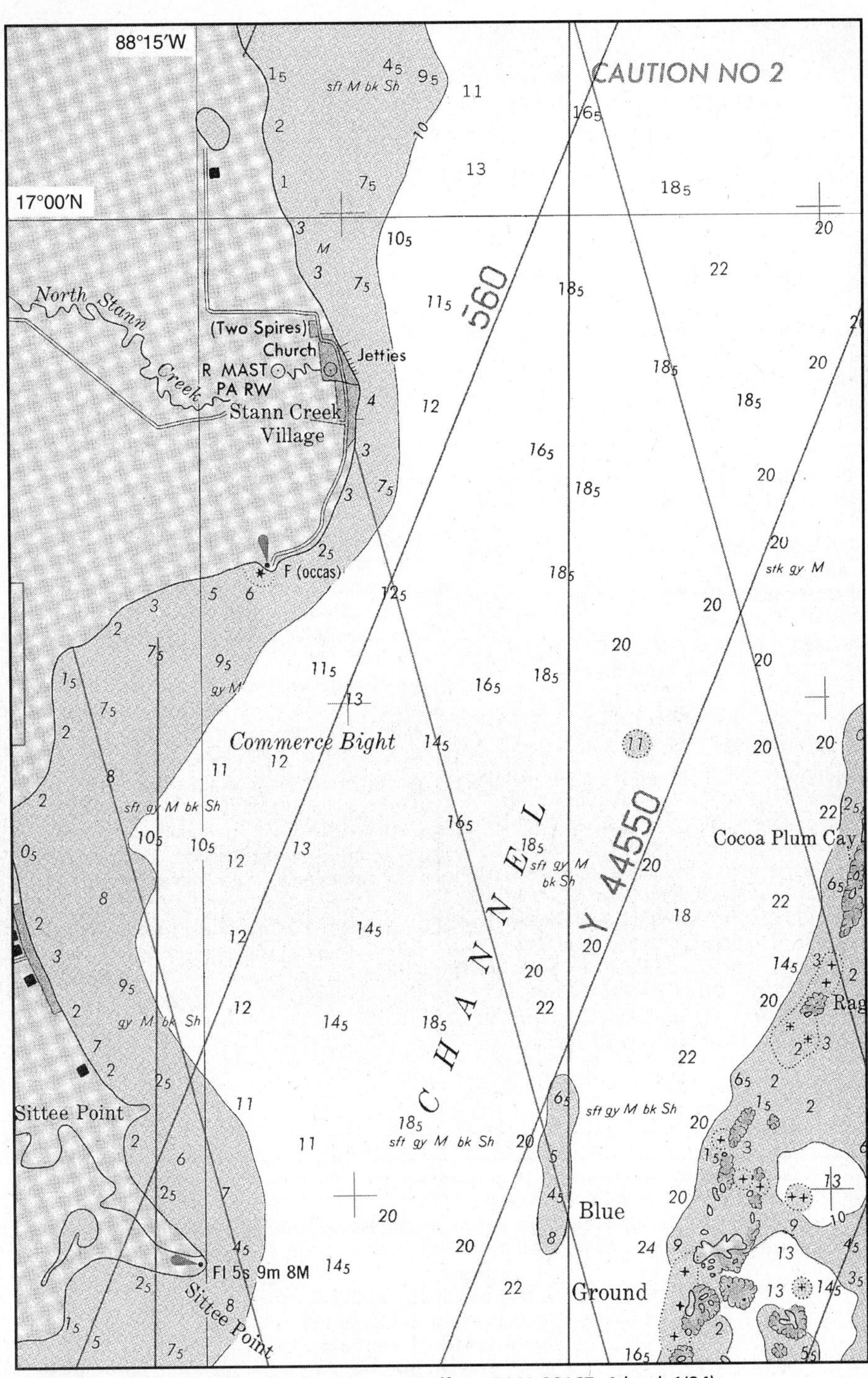

Dangriga (Stann Creek), soundings in meters *(from DMA 28167, 4th ed, 1/84)*

BELIZE CITY AND APPROACHES

TRIANGLES LIGHT, 17 21.5N, 88 12.3W. Fl G 5s, 16ft, 5M. Black concrete pile, 16ft.

ROBINSON POINT LIGHT, W point of island, 17 22.0N, 88 11.7W. Q W, 38ft, 8M. White framework tower, 39ft.

FRANK KNOLL LIGHT, 17 23.8N, 88 11.7W. Q W, 16ft, 5M. Concrete pile, 16ft.

SUGAR BERTH B LIGHT, 17 23.5N, 88 10.6W. Fl R 2.5s, 16ft, 5M. Black concrete pile, 16ft.

SUGAR BERTH A LIGHT, 17 25.5N, 88 08.9W. Fl G 2.5s, 16ft, 5M. Black concrete pile, 16ft.

WESTWARD PATCH LIGHT, 17 25.5N, 88 11.3W. Q R, 16ft, 5M. Red concrete pile, 16ft.

MIDDLE GROUND LIGHT, 17 28.0N, 88 10.5W. Fl R 2.5s, 16ft, 5M. Red concrete pile, 16ft.

FORT GEORGE LIGHT, 17 29.6N, 88 10.7W. Fl R 5s, 52ft, 8M. White concrete pillar, red band on base, 52ft. F R lights on radio mast 670 meters WNW; F R and Fl R lights on radio mast 1.3 miles NW.

NORTH DROWNED CAY, 17 29.8N, 88 08.8W. Q W. Black beacon.

BELIZE CITY AVIATION LIGHT, 17 32.5N, 88 18.5W. Q W, 30M. Radio mast.

BELIZE CITY

17 30N, 88 11W

Pilotage: Belize City is located on the south branch of the Belize River, about 17 miles north-northeast of Manatee River. The harbor is an open roadstead with limited docking facilities for large vessels. It is approached from inside the reef via Inner or Main Channel and from outside the reef via Eastern Channel (see the description later in this section). The city is situated on both sides of Haulover Creek, which is the southern part of the Belize River.

Winds: Easterly and southeast winds prevail from the middle of February to the end of September, with the greatest period of calm occurring at the end of February. Northeast through northwest winds prevail the rest of the year. The average wind velocity is about 10 knots. Northers are most likely to occur in November and December, but they rarely exceed force 4 or 5.

Currents: The current generally sets south through the harbor at a rate of 1.5 knots, but during the season of the northers this rate may increase to 3 knots. North currents, which attain a velocity of 1.5 to 2 knots, may be experienced during the rainy season. The mean range of the tide is negligible; however, east winds raise the water level and north winds lower it. During northers the fall may be as much as 2 to 2.5 feet.

Port of entry: This harbor is geared to the needs of commercial ships, and the formalities for yachts may be less convenient than at some other ports. Customs is near the pier at Fort George Point.

Dockage: The 1,500-foot wharf at Fort George Point is reported to have depths of 1 to 6 feet alongside. Cruisers dinghy in here, but report there is a problem with petty theft. The Ramada Royal Reef Marina has slips for vessels drawing up to 5.5 feet.

Anchorage: Cruisers usually anchor off the pier at Fort George Point. This area is open to the wind and swells.

Services: Fuel and water are available at the pier on Fort George Point. Fuel, water, electricity, and ice are available to shoal-draft vessels at the Ramada Royal Reef Marina. The marina also can arrange for repair services and is adjacent to an informal dockside restaurant and bar.

In Belize City there are good grocery stores, general supplies, communication links, banks, post office, and hardware stores, and there is a marine railway in Haulover Creek.

REED'S *NEEDS YOU!*

It is difficult to obtain accurate and up-to-date information on many areas in the Caribbean. To provide mariners with the best possible almanac, *Reed's* uses many sources. *Reed's* welcomes your contributions, suggestions, and updates. Please be as specific as possible and include your address and phone number.

Send this information to:
Editor, Thomas Reed Publications
13A Lewis Street, Boston, MA 02113, U.S.A.
e-mail: editor@treed.com
Thank you!

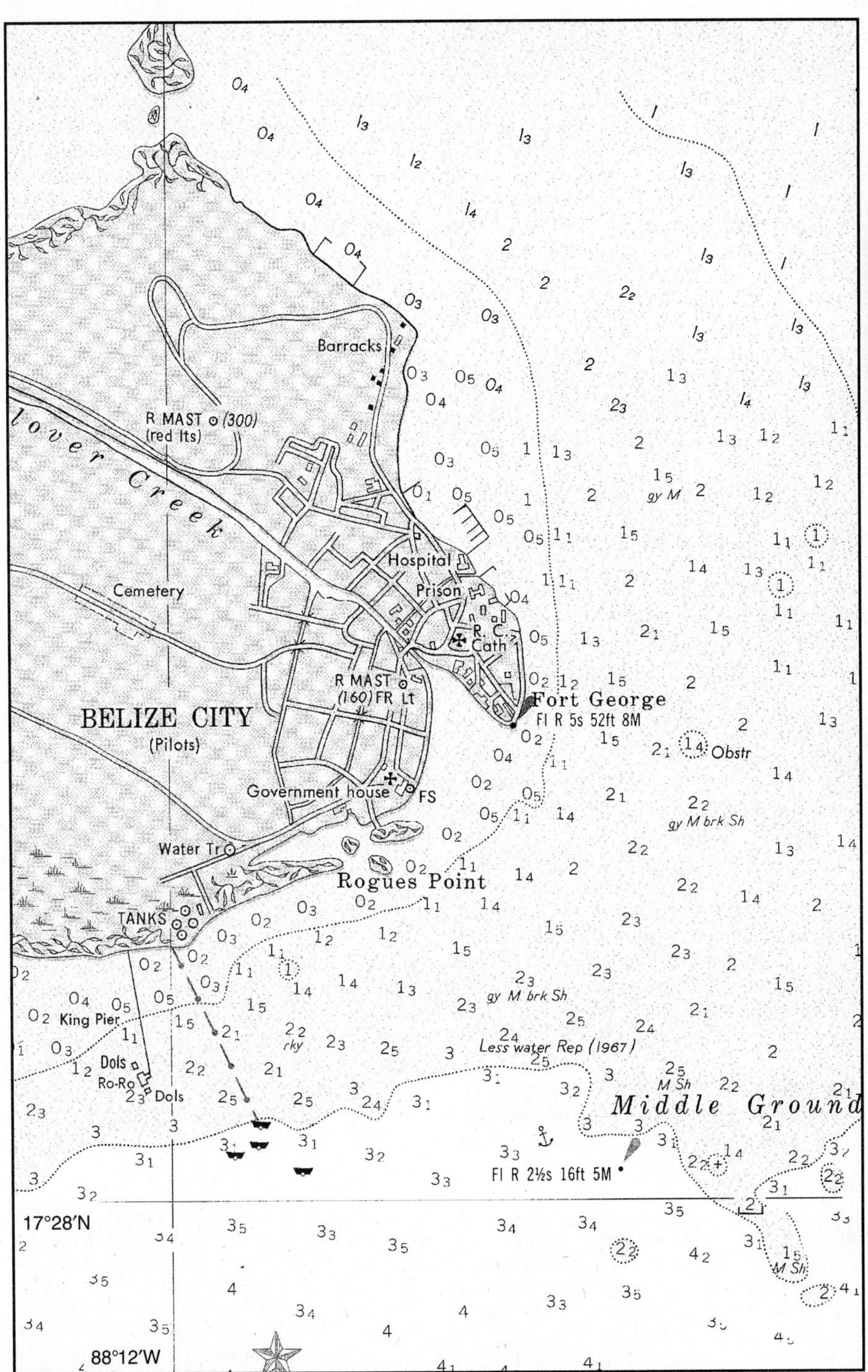

Belize City, soundings in fathoms and feet *(from DMA 28168, 21st ed, 1/86)*

BELIZE CITY TO AMBERGRIS CAY

Pilotage: From St. George's Cay the reef trends about 19 miles north to abreast the south end of Ambergris Cay. The reef's eastern edge is steep-to. The area west of the barrier reef is shoal, with general depths of 4 to 15 feet. The route between Belize City and Ambergris Cay inside the reef is shoal and should be attempted only by boats drawing less than 6 or 7 feet. Numerous cays are located in this shoal area.

ROCK POINT

17 40N, 88 15W

Pilotage: This point is about 12 miles north-northwest of Fort George in Belize City. Between Belize City and here lie Peter's Bluff, a low cay covered with trees about 2.5 miles north of Fort George, and Riders Cay, a small islet north of the former cay. There are depths of about 6 to 7 feet in the area east of the cays.

Hicks Cays lie east of Rock Point and about midway between the coast and the barrier reef. Depths in the area run 6 to 10 feet.

CAY CORKER (CAY CAULKER)

17 47N, 88 01W

Pilotage: This cay is about 5 miles south of the south end of Ambergris Cay. There is a small town with tourist facilities and an anchorage in the bight on the southwest part of the island.

CANGREJO CAY

17 52N, 88 02W

Pilotage: This cay is located about 2.5 miles out on the reef extending southwest from the southern end of Ambergris Cay. The description of Ambergris Cay is under the heading "The Barrier Reef."

CHETUMAL BAY

BULKHEAD LIGHT, 17 56N, 88 08W. Fl W 1.5s, 30ft, 10M. Red metal framework tower, concrete base, 19ft.

Pilotage: This area is entered via an intricate channel located between Cangrejo Cay and Northern River, 10 miles away on the mainland to the west. It is about 22 miles from Belize to the Northern River, as the crow flies. The coast is low and covered with mangroves, which extend to the shore. The water is shoal.

The channel leading north into Chetumal Bay has a least charted depth of about 6 feet. This channel takes you to the mouth of the Hondo River, which forms part of the boundary with Mexico. Parts of the channel are marked with stake beacons, but its use is recommended only to those with local knowledge.

COROZAL LIGHT, 18 22N, 88 27W. F R, 5M.

COROZAL

18 22N, 88 24W

Pilotage: This small town stands near the west head of Chetumal Bay and is fronted by a small pier with a depth of 4 feet alongside. Anchorage can be taken about 300 yards to the southeast of the pier.

CONSEJO POINT LIGHT, 18 27N, 88 20W. F W, 50ft, 5M. Mast.

HONDO RIVER (RÍO HONDO)

Pilotage: Consejo Point, about 8 miles northeast of Corozal, is the south entrance point to the Hondo River, which forms part of the border with Mexico. The Mexican settlement of Payo Obispo is located about 2.5 miles north of Consejo Point. The river is navigable by small craft drawing less than 4 feet for about 70 miles. A bar with a depth of about 5 feet fronts the river mouth.

THE BARRIER REEF

Pilotage: This reef fronts the coast of Belize and lies from 10 to 22 miles offshore. It extends from the Sapodilla Cays, which are about 31 miles east of Punta Gorda, north for 118 miles to Ambergris Cay. Abreast of Belize City the reef is about 8 miles offshore. There are general depths of 6 to 18 feet on the reef, but numerous small cays, rocks, and coral banks are interspersed along its entire length.
The seaward side is extremely steep-to, with depths over 656 feet close off the outer edge. A heavy sea usually breaks along the entire seaward extremity.

In the area west of the reef between Sapodilla Cays and Blue Ground Range, about 44 miles to the north-northeast, foul ground with numerous cays and shoals extends west to the fairway of the Inner or Main Channel. In the vicinity of Blue Ground Range the west side of the reef becomes regular and fairly steep-to as it extends north to the dangers on the south side of Belize City.

NOTE: Navigation through the various intricate openings in the barrier reef should not be attempted without local knowledge or the assistance of a pilot.

Many of the cays have been planted with coconut palms, and some have houses on them. Hurricanes periodically devastate the area, or some part of it, and the descriptions of the cays that follow may be out of date. Although the vegetation usually recovers quickly, the shape of individual cays may be permanently altered and some may disappear completely.

HUNTING CAY LIGHT, 16 06.6N, 88 15.8W. L Fl W 10s, 57ft, 13M. White metal framework tower, 56ft.

SAPODILLA CAYS

16 07N, 88 16W

Pilotage: This group of small cays is located about 20 miles southeast of Punta Ycacos. This group is the farthest south on the great barrier reef. They extend about 3.8 miles northeast from Sapodilla Cay to Grassy Cay. The reef extends about 32 miles north-northeast from here to Gladden Spit. Between Sapodilla Cays and Gladden Spit the barrier reef is generally broken, and there are numerous cays, rocks, and openings along its entire length.

Sapodilla Cay (16°05′N, 88°17′W) has a conspicuous clump of coconut trees on it. Hunting and Nicolas cays, the middle cays of the group, are densely wooded. They are inhabited. There are numerous dangerous heads and irregular depths in the area.

SEAL CAY

16 10N, 88 20W

Pilotage: This islet is situated on the northeast part of a small circular reef enclosing a lagoon. The former Seal Cays, 2.5 miles south, are reported to have been destroyed by a hurricane, and in 1972 only a sand bore remained. Shoal patches are in the area.

LAWRENCE ROCK

16 10N, 88 20W

Pilotage: This rock lies about 3.3 miles west of Seal Cay and has a least depth of under 4 feet. This rock is dangerous because the water over it is not sufficiently discolored to indicate its position. The ground is foul between the rock and Seal Cay.

RANGUANA CAY

16 20N, 88 10W

Pilotage: This tree-covered cay lies about midway along reef between Sapodilla Cay and Gladden Spit and about 1.3 miles from the seaward side. Ranguana Entrance, an opening through the reef about 0.3 mile wide, lies about 1.5 miles southeast of the cays. There are depths of 20 to 24 feet in this passage.

POMPION CAY

16 24N, 88 06W

Pilotage: This group of wooded islands that lie about 4.8 miles northeast of Ranguana Cay is reported to be inhabited.

Little Water Cay and Hatchet Cay are two wooded cays lying about 3 miles north and 4.3 miles north-northeast of Pompion Cay, respectively. The two cays have been reported to be good radar targets.

GLADDEN SPIT

16 31N, 87 59W

Pilotage: This is the easternmost projection of the barrier reef.

CAUTION: This spit has been reported to lie about 2 miles east of its charted position.

Gladden and Queen Cay entrances lie about 1.5 and 4.5 miles southwest of Gladden Spit. They have least charted depths of about 8 feet and 13 feet, respectively. These openings lead to the Inner or Main Channel through narrow intricate passages. They should not be attempted without local knowledge or a pilot.

Victoria Channel is the navigable passage that lies between the dangers extending east from Bugle Cays and those extending west from Gladden Spit. Victoria Channel may be used as an alternate route to Inner or Main Channel. Laughing Bird Cay (16°26′N, 88°12′W), Moho Cay (16°30′N, 88°10′W), Bakers Rendezvous, and Crawl Cay are all on the western side of Victoria Channel. Quamino Cay (16°39′N, 88°13′W) is located on the north side of the northwest end of Victoria Channel.

For about 14 miles to South Cut, the barrier reef extends north-northwest in a solid coral barrier, with no cays. Then for about 6 miles farther north-northwest to Water Cay, it is broken with numerous drying sandbanks and some above-water rocky heads. There are several cuts along this part of the barrier reef, which are usable by small vessels that have local knowledge. The west side of the reef is irregular and the area west to Inner or Main Channel is interspersed with numerous cays and shoals.

WATER CAY

16 49N, 88 05W
Pilotage: This is a fairly large, wooded cay.

BLUE GROUND RANGE

16 48N, 88 09W
Pilotage: This group of cays is located on the west side of the reef opposite South Water Cay and about 5.5 miles east of Sittee Point. They are about 2.5 miles in extent in a north–south direction.

TOBACCO CAY

16 54N, 88 09W
Pilotage: This small wooded cay lies on the north side of Tobacco Cay Entrance, 5 miles north of Water Cay.

Tobacco Cay Entrance is a 13-foot-deep passage that leads west toward Tobacco Range, 1 mile west of the entrance. The passage leads to the Inner or Main Channel, where it joins about 2.5 miles north of Cocoa Plum Cay (16°53′N, 88°07′W).

Tobacco Reef, which in many places is nearly dry, lies between Water Cay and Tobacco Cay entrance.

From Tobacco Cay to Glory Cay, about 12.5 miles to the north, the reef is almost continuous and practically dry. This section of the reef is known as Columbus Reef.

Cross Cay and Columbus Cay, both with trees, lie close west of the barrier reef. They lie 5 miles and 6.5 miles north-northeast of Tobacco Cay, respectively. Several cays lie west of Cross and Columbus Cays.

GLORY CAY

17 06N, 88 01W
Pilotage: This small, sandy cay is located 0.75 mile northwest of Columbus Reef. Southern Long Cay is a wooded cay located inside the barrier reef 1 mile southwest of Glory Cay.

From Glory Cay the barrier reef, which is broken in many places, trends north-northwest about 28 miles to St. George's Cay, which is north of Eastern Channel. The cays west of the barrier reef and those northwest to north-northwest of Glory Cay are wooded.

SKIFF SAND

17 13N, 88 03W
Pilotage: Skiff Sand, about 1 meter high, is located 7.3 miles north-northwest of Glory Cay.

Rendezvous Cay stands about 1.5 miles north of Skiff Sand in the middle of a break in the barrier reef. It is covered with vegetation.

Bluefield Range, a group of mangrove-covered cays, stand near the west side of the barrier reef west-northwest of Skiff Sand.

The barrier reef between Rendezvous Cay and English Cay, about 5 miles to the north, is broken and marked by numerous banks, cays, and coral reefs. Immediately north of English Cay is Eastern Channel. Goffs Cay is located about 13 miles north of English Cay and north of the entrance to Eastern Channel. This cay is small and sandy, and a drying coral head lies about 0.5 mile to its southeast.

EASTERN CHANNEL RANGE (Front Light), 17 19.7N, 88 02.6W. Q W, 23ft, 9M. Concrete pillar, 23ft. **(Rear Light),** on English Cay, 300 meters, 300° from front. Fl W 2.5s, 62ft, 11M. White tower, 62ft.
GOFFS CAY, SANDBORE LIGHT, 17 20.4N, 88 02.0W. Fl R 5s, 16ft, 3M. Red concrete pile, 16ft.
WATER CAY SPIT LIGHT, 17 21.4N, 88 04.4W. Q R, 16ft, 5M. Red concrete pile, 16ft.
NORTHEAST SPIT LIGHT, 17 22.8N, 88 05.3W. Q G, 16ft. Green concrete pile, 16ft.
WHITE GROUNDS SPIT LIGHT, 17 22.7N, 88 07.0W. Fl W 2.5s, 16ft, 5M. Yellow concrete pile, 16ft.
SPANISH CAY SPIT LIGHT, 17 22.5N, 88 08.5W. Q G, 16ft, 5M. Green concrete pile, 16ft.
HALFWAY LIGHT, 17 22.2N, 88 09.5W. Q R, 16ft, 5M. Red concrete pile, 16ft.
SOUTHWEST SIDE LIGHT, 17 22.0N, 88 09.7W. Q G, 16ft, 5M. Black concrete pile, 16ft.

EASTERN CHANNEL

17 20N, 88 02W (ENGLISH CAY)
Pilotage: Approaching Eastern Channel from the north, steer for a position about 7.5 miles west of Mauger Cay (17°36′N, 87°46′W), which is the northernmost of the Turneffe Islands. Steer a south-southwest course to pass near mid-channel between Turneffe Islands and the barrier reef, until the lights on English Cay are in line on a bearing of 300°.

Approaching Eastern Channel from the south, steer for a position 2.8 miles south of Cay Bokel (17°10′N, 87°54′W), at the southern end of the Turneffe Islands. Then steer a mid-channel course between the Turneffe Islands Reef and Rendezvous Cay until the lights on English Cay are in line on a bearing of 300°.

To enter Eastern Channel steer for the lights on English Cay bearing 300°, until the east side of Water Cay (17°23′N, 88°04′W) bears 340°. Alter course to 340° and maintain it until Eastern Channel opens to the west-northwest. Then proceed through the sinuous passage, maintaining a mid-channel course until One Man Cay Channel is reached.

The east entrance to One Man Cay Channel is about 400 yards wide between the reefs, each of which is marked by a light. Steer a northwest course through the channel until Robinson Point (17°22′N, 88°12′W), marked by a light, bears 233°. Then steer as required to the anchorage off Belize City.

PAUNCH CAY

17 24N, 88 02W
Pilotage: This barren cay lies close to the edge of the reef about 3 miles north of Goffs Cay. From Paunch Cay the steep-to reef extends about 9.5 miles north to St. George's Cay. This low sandy cay is easily identified by the houses and coconut palms on it. From St. George's Cay the reef trends about 19 miles north to abreast the south end of Ambergris Cay.

Drowned Cays are located about 1 to 2 miles west of the barrier reef.

AMBERGRIS CAY

18 02N, 87 55W
Pilotage: This cay's northern end is separated from Mexico, and the mainland, by Boca Bacalar Chico, a narrow boat channel with very shallow depths. The island appears to be part of the coast from offshore. The east extremity of the island, known as Reef Point, is about 0.5 to 1.5 miles inside the reef line. The reefs gradually approach the coast of the mainland about 9 miles north of Reef Point.

CAUTION: The current generally sets strongly west toward the barrier reef, accompanied by a heavy swell, so that every possible caution should be observed here.

San Pedro: This settlement is on the east side of Ambergris Cay, toward the south end. It is inside the barrier reef, and the only approaches from the sea are via reef channels. The entrance is reported to be dangerous when a sea is running and should not be attempted without local knowledge. A shallow channel leads south from here, inside the reefs to Belize City. You can fly government officials to here from Belize City to clear in.
Dockage: The Belize Yacht Club is "not a yacht club as our name suggests, but we do have a dock which can house boats with drafts of 6 feet or less." The club has water, electricity, and fuel.

OFFSHORE REEFS

Pilotage: The dangers offshore consist of Glover Reef, Lighthouse Reef, and the Turneffe Islands. They are all located east of the barrier reef described previously. There are deep passages between all three reefs and the barrier reef.
Currents: In the vicinity of the reefs, currents in November, December, and January depend on the winds. A north-setting current occurs during west winds, and a south-setting current with north winds. During February and March the current usually sets north at a rate of about 1.5 knots. In April and May it usually sets south at a rate of 1.5 knots. In June, July, and August the current usually sets north at a rate of 1.5 knots. In September and October the north-setting current increases to 2 knots.

SOUTHWEST CAYS LIGHT, 16 42.9N, 87 50.6W. Fl (2) W 5s, 9M. White metal framework tower, 36ft.
NORTHEAST SIDE OF REEF LIGHT, 16 54.6N, 87 42.0W. Fl W 5s, 9M. White metal framework tower, 36ft. Radar reflector.

GLOVER REEF

16 50N, 87 47W
Pilotage: The south extremity of this reef is located about 13 miles north-northeast of Gladden Spit, on the barrier reef. It is about 14 miles long and 6 miles wide. There is a barrier reef around the perimeter, which is impassable except for a small opening at its southern extremity, west of Southwest Cays, and another north of Long Cay on the southeastern side. The latter entrance is dangerous because of its location on the windward side of the reef.

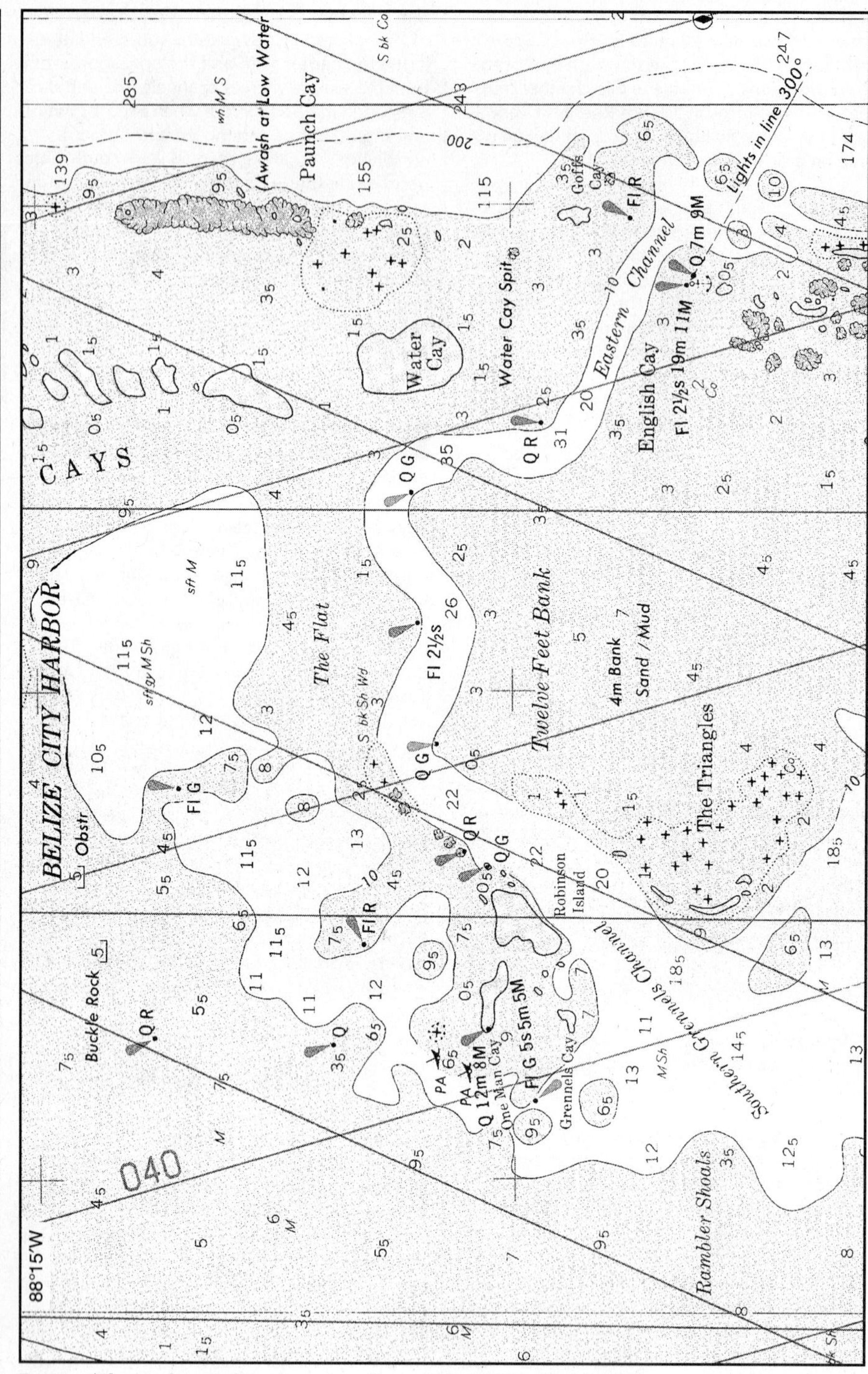

Eastern Channel, soundings in meters *(from DMA 28167, 4th ed, 1/84)*

CAUTION: Glover Reef must be approached with great care, especially from the north, as the reef is low and not always visible from a distance. The reef is steep-to, and soundings give little warning. A strong west-going current has been experienced on several occasions between Glover Reef and Lighthouse Reef.

The north point of Glover Reef is known as Amounme Point. A light is exhibited on the northeast edge of the reef, 4 miles southeast of Amounme Point.

Southwest Cays (16°43'N, 87°51'W) are the farthest south of the five cays situated on the south extremity of Glover Reef. The opening in the reef west of Southwest Cays accommodates vessels drawing up to 12 feet, with local knowledge.

HALF MOON CAY LIGHT, E end, 17 12.3N, 87 31.6W. Fl (4) W 15s, 80ft, 14M. White metal framework tower, 79ft.
SANDBORE CAY LIGHT, N end of reef, 17 28.1N, 87 29.2W. Fl W 10s, 83ft, 17M. Red metal framework tower, 82ft.

LIGHTHOUSE REEF

17 16N, 87 32W
Pilotage: This reef is the danger farthest to the east of Belize's barrier reef. It lies about 13 .5 miles north-northeast of the northeast extremity of Glover Reef. The reef is steep-to and unbroken except in the vicinity of Half Moon Cay, where there is a passage. There is reported to be a pass at the northwest tip of the reef.

CAUTION: Lighthouse Reef should be approached with great care as there is considerable doubt about the exact position of the edges of the reef. The western edge and the southern part of the reef are reported to lie up to 1 mile west of their charted positions.

Half Moon Cay (17°12'N, 87°32'W) is marked by a light. It lies within the southeast part of the reef. Numerous coral heads surround the islet, but there is a shallow opening about 0.5 mile to the west. The islet has been reported to give a good radar return at distances up to 20 miles. Sandbore Cay (17°28'N, 87°30'W) lies near the north extremity of the reef. It is a small, tree-covered islet. Four small white buildings and a wooden jetty stand close to the lighthouse on the north side of the cay. Northern Cay, about 0.75 mile southwest of Sandbore Cay, is wooded.

CAY BOKEL LIGHT, S extremity of Turneffe Islands, 17 09.8N, 87 54.4W. Fl (3) W 15s, 33ft, 8M. White metal framework tower, 33ft.
MAUGER CAY LIGHT, 17 36.5N, 87 46.2W. Fl (2) W 10s, 61ft, 13M. White metal framework tower, 62ft.

TURNEFFE ISLANDS

17 22N, 87 51W
Pilotage: This extensive group of mangrove-covered islands and cays on a coral and sand reef lies about 12 miles west of Lighthouse Reef. The barrier reef surrounding these islands is about 30 miles long and up to 10 miles wide. These islands are so closely grouped that from a distance they appear as one large flat island. It was reported in 1964 that radar returns from Grand Point, the south end of the islands, may be easily mistaken for Cay Bokel.

CAUTION: The east side of the Turneffe Islands is charted from old surveys and should be approached with caution.

The reef that fringes this group lies at an average distance of 0.5 mile off the east side and 1 mile off the west side. The reef on the west side is awash in many places, but the seaward edge on all sides is steep-to. There are several shallow openings and numerous lagoons between the various islands, which are used by small craft with local knowledge.

Cay Bokel (17°10'N, 87°54'W) is a small patch of sand that lies at the south extremity of the reef extending south from Grand Point. There are several detached cays near the south extremity of the reef, but from a distance, they appear as part of the main group of islands. On one of these, Big Cay Bokel, there are several fishing lodges.

Several detached cays stand on the reef north of the main group of islands. Mauger Cay (17°36'N, 87°46'W) is the northernmost cay in the Turneffe group. The lighthouse on the cay is prominent and is a good landmark if heading toward Eastern Channel from the north.

Mexico

Hydrographic and cartographic information in this section has been supplied, in part, courtesy of the Secretaría de Marina, Dirección de Oceanografía Naval of Mexico.

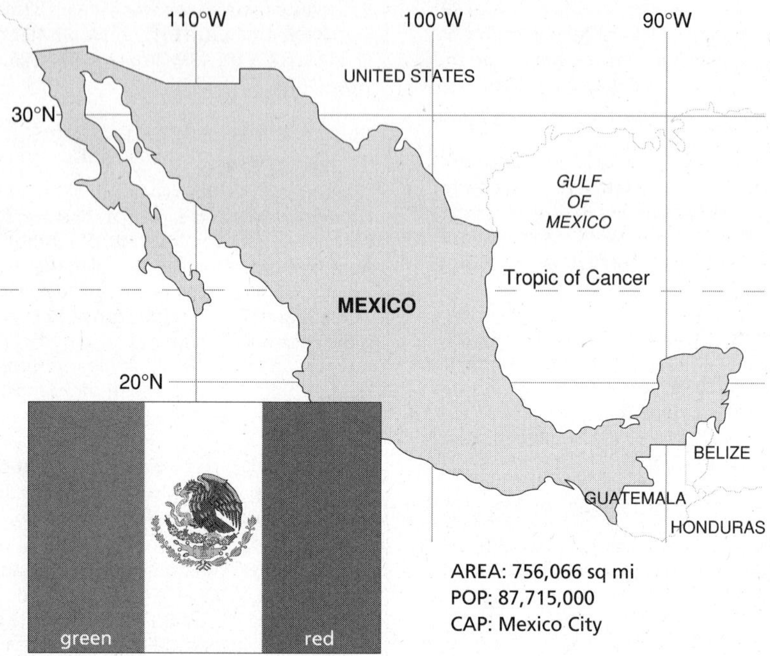

AREA: 756,066 sq mi
POP: 87,715,000
CAP: Mexico City

CAUTION: In the past, some aids have been reported as unreliable, missing, off position, or showing incorrect characteristics. According to the NIMA Notice to Mariners, many aids have been changed or reidentified in these waters over the last year, as indicated in the following listing. This may mean that the aids will be more reliable than in the past, but it also means that charts are out of date.

Please see cautions about chartlets and waypoints on page P 2.

ENTRY PROCEDURES

The Yucatán coast of Mexico is becoming a more popular cruising destination, as it is located only about 340 miles southwest of Key West, Florida. Some cruisers slog out against the current from Florida, while others visit the Yucatán as part of a clockwise circuit of the Caribbean basin. The area west of the Yucatán peninsula is still seldom visited by pleasure boaters.

Ports of entry on the Yucatán coast include Xcalak, San Miguel on Cozumel, Puerto Morelos, and Puerto Mujeres on Isla Mujeres. On the north side of the Yucatán and in the Gulf of Mexico, you can clear at Progreso, Campeche, Frontera, Coatzacoalcos, Veracruz, and Tampico. When you check in you will be issued a coastwise clearance, or *zarpe,* on which you should list all the ports you are planning to visit while in Mexico. You should have the officials grant permission to visit intermediate ports, if possible. You will have to check in with the port officials in any port you stop at along the way.

Mexican paperwork is reported to be time-consuming and exacting. It may help to have multiple copies of your crew list typed in *Spanish*, which you will probably have to give to officials in each port. For each crewmember, you should have either a tourist card or visa (which depends on each crew's nationality). If you are arriving from Belize, these items are available from the Mexican Consulate in Belize City. Citizens of the U.S., the United Kingdom, Canada, and most west European countries do not need a visa for entry and should be able to obtain tourist cards on arrival. You must turn in your tourist card when leaving the country. Generally a tourist card is good for 30 to 90 days.

Tourist cards are available in advance at these offices in the United States and Canada: Chicago (312-565-2786), Houston (713-880-5153), Los Angeles (310-203-8191), Miami (305-443-9160), New York (212-755-7261, -7262, -7263), Washington (202-728-1750, -1754), Montréal (514-871-1052), and Toronto, (416-364-2255).

Upon your arrival at your first port of entry in Mexico, you must go to the immigration office and obtain the necessary Tourist Entry Form for each passenger aboard, if the forms have not been previously obtained. You should have passports. U.S. citizens should have at least proof of citizenship and a photo identification, but a passport is preferable. You should also have departure clearance from your last port of call. Some vessels will also need a "health permit," depending on the departure point; check with the nearest Mexican consulate if you think this regulation may apply. A $10 (payable in U.S. currency or credit card) fee will be assessed each boat for paperwork.

You then proceed to customs to obtain a Temporary Import Permit by presenting the following information: vessel ownership title; document verifying that the vessel's owner resides in a foreign country (Tourist Entry Form); and a credit card, bond, or deposit mechanism.

All firearms must be declared and have proper permits in advance from a Mexican consulate. Your guns will be held for the duration of your stay, unless you have a Mexican hunting license for them, which you should apply for in advance of your arrival. For more information on hunting licenses, inquire at the nearest consulate or write to: Dirección General de Caza, Serdan 27, Mexico, D.F. Penalties for violating the firearms importation regulations are stiff.

Fishing licenses must be purchased for each crewmember and the boat if you have any fishing gear on board. Again, this license is best applied for before arriving in Mexico. For fishing information write to: Dirección General de Pesca, Av. Alvaro Obregón 269, Mexico 7, D.F.

To obtain information on the latest entry requirements contact the Mexican Consulate at 2827 16th Street NW, Washington, DC 20009 (202-736-1000) or the consulate nearest you. In the U.S. there are consular offices in Alabama, Arizona, California, Colorado, Florida, Georgia, Hawaii, Illinois, Louisiana, Massachusetts, Michigan, Missouri, New Mexico, New York, and North Carolina. For assistance in Mexico, U.S. citizens may contact the United States Embassy in Mexico City (52-5-211-0042). The U.S. also maintains a number of consulates and consular agents in Mexico, including consular agents in Cancún, Veracruz, and Tampico. The Consulado General de México in Austin, TX (tel. 512-478-1866; fax 512-478-8008) publishes an Internet guide to taking your boat to Mexico. The Web site is: www.conr.com.

USEFUL INFORMATION

Language: Spanish. In tourist areas many people speak some English.
Currency: Peso (new); exchange rate (summer 1997), about 7.5 pesos = US$1.
Telephones: Country code, 52; a city code follows, and then the local number. The city code for Cancún is 988; Veracruz, 28; Tampico, 131; and Mexico City, 5.
Public holidays (1998): New Year's Day, January 1; Three Kings Day, January 6; Constitution Day, February 5; Flag Day, February 24; Birth of Benito Juarez, March 21; Maundy Thursday, April 9; Good Friday, April 10; Labor Day, May 1; Cinco de Mayo, May 5; Independence Day, September 16; Columbus Day, October 12; Day of the Dead, November 1–2; Anniversary of the Revolution, November 20; Day of Our Lady of Guadelupe, December 12; Christmas, December 24–25. Other religious and local holidays are celebrated, but banks and government offices may be officially open.
Tourism offices: *In Mexico:* Ministry of Tourism, 726 Av. Mariano Escobedo, Mexico City (52-5-254-8967; toll-free international, 800-482-9832). *In the U.S.:* Mexican Government Tourism Office, 405 Park Avenue, Suite 1401, New York, NY 10022 (tel., 212-755-7261; tollfree U.S. and Canada, 800-446-3942; fax, 212-753-2874). Office also in Chicago. *In Canada:* Mexican Government Tourist Office, 2 Bloor Street West,

Suite 1801, Toronto, ON M4W 3E2 (tel., 416-925-2753; fax, 416-925-6061). Office also in Montréal, PQ.
Additional notes: The brochure "Official Guide, Taking Your Boat to Mexico" is available free from the Mexican Government Tourism Office. Visitors are also reminded that tropical diseases may be encountered in Mexico. There is some risk of malaria in rural areas of the Yucatán peninsula. Dengue fever has been reported, as has cholera. Typhoid fever is also more common in Mexico. Care should be taken to drink only sterile water, well-cooked vegetables, and well-cooked shellfish. For up-to-date health notices, contact the Centers for Disease Control International Travelers Hotline (voice recording, 404-332-4559; fax information service, 404-332-4565; Web, www.cdc.gov).

BAHÍA DE CHETUMAL

CIUDAD CHETUMAL (Payo Obispo), 18 31.2N, 88 16.5W. Fl W 6s, 59ft, 17M. Round concrete tower, 49ft. Visible 115°–064°.
HEAD OF FISCAL MOLE, Q W, 23ft. Wooden post, 13ft.
AVIATION LIGHT, 18 30.7N, 88 19.5W. Al Fl W G 10s, 57ft. Tower, 38ft. ~~Radiobeacon~~.

PAYO OBISPO (CIUDAD CHETUMAL)

18 30N, 88 17W
Pilotage: This settlement is in Chetumal on the north side of the Río Hondo, which forms the border with Belize. It is the site of a Mexican naval base, which maintains the radio station in the town. A T-head pier, about 130 feet long and 80 feet across the face, with a depth of 6 feet alongside the face and 3 feet alongside its west side, extends from the shore in front of town. Lighters and tugs are used to unload cargo from freighters anchored off the west side of Cay Corker, in Belize.

Chetumal Bay extends about 26 miles north-northeast from the Río Hondo. It narrows in its northern part where the Río Kik flows into it.

LA AGUADA, 18 14.0N, 87.55.0W. Fl (2) W 10s, 40ft, 11M. Concrete cylindrical tower, 33ft.
XCALAK, 18 15.8N, 87 50.2W. Fl (3) W 12s, 43ft, 17M. Round concrete tower, 39ft. Visible 214°–360°. ~~A Fl W light shows close by~~.

BOCA BACALAR CHICO TO COZUMAL

BOCA BACALAR CHICO

18 11N, 87 52W
Pilotage: This narrow boat passage separates the north end of Ambergris Cay, in Belize, from the Mexican mainland to the north. Xcalak is a small coastal port of entry located about 6 miles north of Boca Bacalar Chico. Vessels drawing less than 15 feet can enter and anchor inside the reef; local knowledge is recommended. The bottom is reported to be mud and rock, with patches of white sand, providing good holding.

From Xcalak the coast trends about 65 miles north-northeast to Punta Herrero. It is low, flat, and wooded. Nearly all of this stretch of coast is fronted by a reef that lies from 1 to 1.5 miles offshore. There are several channels through the reef, but they should not be attempted without local knowledge.

CAYO LOBOS, center of cay, 18 23.5N, 87 23.2W. Fl (3) W 12s, 46ft, 11M. Truncated pyramidal metal tower, 42ft.
RACON C (– • – •), 20M.
PUNTA GAVILAN, 18 25.2N, 87 46.1W. Fl W 6s, 36ft, 11M. White truncated pyramidal aluminum tower, 39ft.
CAYO CENTRO, 18 35.5N, 87 19.9W. Fl (2) W 10s, 50ft, 11M. Truncated pyramidal aluminum tower, 49ft.
EL MAJAHUAL, 18 43.8N, 87 41.8W. Fl (4) W 16s, 33ft, 11M. Truncated pyramidal metal tower, 30ft.
BANCO CHINCHORRO, N of the two islets that form Cayo Norte, 18 45.8N, 87 18.9W. Fl W 6s, 52ft, 11M. White round concrete tower, 33ft.

BANCO CHINCHORRO

18 35N, 87 27W
Pilotage: This dangerous steep-to shoal lies about 28 miles northeast of Ambergris Cay in Belize. It is about 14 to 16 miles offshore. The greater part of the shoal has depths ranging from 6 to 24 feet, but there are numerous rocky heads and sandbanks. The stranded wrecks that lie along the east side of the shoal have been reported to be conspicuous visually and on radar. It has been reported that Banco Chinchorro is a good radar target for southbound vessels.

Cayo Lobos is a small cay near the south extremity of the bank. There are several openings through the reef to the west and northwest of the cay; local knowledge is recommended.

Cayo Centro (18°36′N, 87°20′W) lies in the middle of the bank, about 1.5 miles from its east side. This low cay is about 2.5 miles long. It is composed of sand with a covering of bushes and coconut trees. A 1-mile-long salt water lagoon lies in the middle of the cay. A dangerous strong current sets into Firefly Bight, about 2 miles southeast of Cayo Centro.

Cayo Norte (18°45′N, 87°19′W) lies about 1.5 miles south of the northern extremity of the bank. It actually consists of two cays, covered with dense vegetation and trees. A disused lighthouse and building stand close south of the current lighthouse.

CAUTION: In the vicinity of Banco Chinchorro there is usually a very strong current setting toward its east side.

The passage between the bank and the coast of the mainland to the west is clear of dangers and deep.

EL UBERO, 19 04.5N, 87 33.5W. Fl (3) W 12s, 33ft, 11M. Truncated pyramidal aluminum tower, 33ft.
PUNTA HERRERO, S side of entrance to Bahía del Espiritu Santo, 19 18.7N, 87 26.8W. Fl (2) W 10s, 75ft, 15M. White truncated pyramidal metal tower, 72ft. ~~Visible 090°–022°~~.

PUNTA HERRERO

19 18N, 87 26W
Pilotage: This is the southern entrance point to Bahía del Espiritu Santo. It is fronted by foul ground extending about 1 mile offshore. The entrance to the bay to the north is also generally foul.

The coast extends about 53 miles north from here to Salta Iman. The southern part of this coast is indented by Bahía del Espiritu Santo, about 10 miles wide, and by Bahía de la Ascension, about 10 miles farther north.

The coast between Bahía de la Ascension and Salta Iman is quite regular, low, and densely wooded. Punta Yaan is a group of conspicuous cliffs that stand on an otherwise flat section of the coast.

The coast from Punta Yaan extends northeast for about 50 miles to Puerto Morelos. Moderately elevated land extends about 6 miles north from Punta Yaan, then becomes low and flat. Isla Cozumel lies about 17 miles south of Puerto Morelos and 9 miles offshore. The coast about 5 miles southwest of Puerto Morelos is fringed by a steep-to reef which extends up to 1.3 miles offshore.

The coast from Puerto Morelos extends north-northeast about 14 miles and then gradually curves north-northwest for about 36 miles to Cabo Catoche. Numerous islands, cays, and other dangers lie within 1 to 15 miles offshore in this area.

BAHÍA DEL ESPIRITU SANTO

19 22N, 87 28W
Pilotage: The bay is entered between Punta Herrero and Punta Fupar, about 11 miles to the north. It recedes about 16 miles to the southwest and is about 7 to 10 miles wide. The depths in the entrance and for a short distance within the bay range from 20 to 30 feet, but in 1943, it was reported the actual depths in this area were 45 to 48 feet. The general depths within the bay range from 8 to 15 feet, about 2.8 miles northwest of Punta Herrero.

From Punta Fupar the coast, which is bordered by a rocky ledge, trends about 7.8 miles north-northeast to Punta Pájaros (19°34′N, 87°25′W). A drying reef extends about 0.5 mile east from Punta Pájaros.

PUNTA OWEN, 19 19.7N, 87 26.6W. Fl (3) W 12s, 39ft, 11M. White truncated pyramidal aluminum tower, 33ft.
PUNTA PÁJAROS, 19 35.8N, 87 24.8W. Fl W 6s, 36ft, 10M. Truncated pyramidal aluminum tower, 30ft.
PUNTA NOHKU, 19 38.7N, 87 27.3W. Fl (2) W 10s, 39ft, 11M. White truncated pyramidal aluminum tower, 33ft.
CAYO CULEBRAS, 19 42N, 87 28W. Fl (3) W 12s, 46ft, 11M. White cylindrical concrete tower, 39ft.
PUNTA VIGIA CHICO, 19 46.3N, 87 35.1W. Fl (4) W 16s, 39ft, 11M. White cylindrical concrete tower, 33ft.
PUNTA ALLEN, Bahía de la Ascension, 19 46.9N, 87 28.0W. Fl (4) W 16s, 72ft, 16M. White round concrete tower, 66ft. ~~Visible 200°–070°~~.

BAHÍA DE LA ASCENSION

19 41N, 87 30W
Pilotage: This bay is entered between Punta Nohku (19°37′N, 87°28′W), about 3.25 miles northwest of Punta Pájaros, and Punta Allen, about 8 miles north. The bay recedes about 16 miles southwest and is from 5 to 11 miles wide.

P
Pilot
MEXICO

The opening through the reef stretching across the entrance is about 2 miles wide. Depths are reported to be 18 to 22 feet through this opening and as far west as the seaward side of the bar blocking the entrance to the bay itself. The opening north of Cayos Culebra is about 3.75 miles wide, but the entrance to the bay is obstructed by a bar with a depth of under 8 feet. Cayos Culebra are a group of mangrove cays that stand in the middle of the entrance just within the bar. Two yellow range beacons stand on these cays. There are depths of 10 to 18 feet within the bar, but the inner reaches of the bay have not been examined.

The opening south of Cayos Culebra, between them and Punta Nohku, is shallow and should not be attempted.

The coast between Punta Allen and Punta Yaan, about 23 miles to the north, is low, flat, and densely wooded.

TULUM, Salta Iman, 20 12.0N, 87 26.8W. Fl W 6s, 75ft, 15M. White truncated pyramidal concrete tower, 36ft.

PUNTA YAAN

20 11N, 87 27W

Pilotage: Punta Yaan is conspicuous, with the only cliffs along this coast. They are about 79 feet high and front the coast for about 3 miles. The ruins of a large, square watchtower stand at their north end.

Tulum is a small settlement fronted by a white beach located about 4 miles north-northeast of Punta Yaan. A small pier extends from the shore abreast of the settlement. A conspicuous, small stone temple on a truncated pyramid stands about 0.5 mile inland. It is overgrown with vegetation. There is a lighthouse at Tulum, and some cruisers report anchoring inside the reef here.

The coast between Tulum and Puerto Morelos extends about 45 miles northeast and again becomes low and flat.

The only known off-lying dangers are Isla de Cozumel and Cozumel Bank, which together front the coast for almost 30 miles and lie from 9 to 13 miles offshore. The passage between the island and the mainland is deep and clear of any known dangers.

Puerto Aventura is a marina located on the mainland about 11.5 miles southwest of Playa del Carmen (20°37'N, 87°04'W). Cruisers report anchoring off the latter town, inside the barrier reef. Ferries run from here to San Miguel, on Isla de Cozumel, so there are good navigational aids in the approach to the pier.

ISLA DE COZUMEL

20 26N, 86 53W

Pilotage: The south extremity of the island lies about 25 miles east-northeast of Punta Yaan on the mainland. The island is generally low and densely wooded and extends for 24 miles along the coast. The average width is about 9 miles. The passage between the island and the mainland to the west is deep and clear of dangers.

Punta Celerain (South Point) is low but well defined. It is fringed by a steep-to reef, which extends about 0.5 mile offshore.

The east side of the island is composed of sandy beaches separated by rocky points. The coast extends about 24 miles northeast to Punta Molas.

The west side of the island extends about 8 miles north-northwest, and then about 13 miles north-northeast to its northwest extremity. The shore is bordered by a narrow coral beach, with vegetation extending down almost to the water's edge.

Caleta Bay is a small body of water, lying about 8 miles south-southwest of the northwest extremity of the island. The entrance channel is narrow and shoal. A wharf, some buildings, and some fuel tanks stand on the shores of the bay, and a conspicuous hotel lies close southwest of it. It may be possible to anchor inside the small bay, but check locally for the latest information on the rocks and depths inside. The controlling depth is reported to be less than 6 feet.

The town of San Miguel lies about 3 miles north-northeast of Caleta Bay.

Punta Norte, the northwest extremity of Isla de Cozumel, lies about 3.25 miles northeast of the entrance to **Banco Playa** yacht harbor. The shore between is lined with numerous hotels, one of which, a white building, is conspicuous and can be seen from a considerable distance seaward. A conspicuous clump of trees stands about 2.5 miles east of the northwest extremity of the island. A reef extends up to 1.75 miles offshore. Some coral heads lie within the 10-meter-depth curve, which lies up to 2.75 miles offshore.

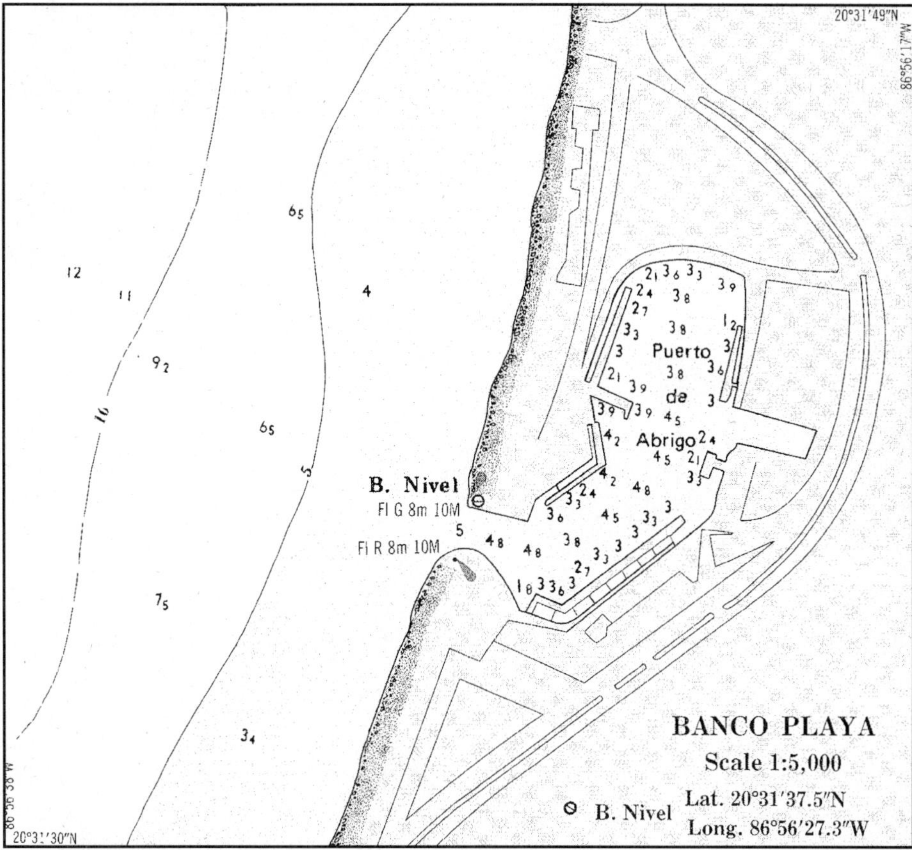

Banco Playa, soundings in meters *(after DMA 28197, 2nd ed, 6/88)*

La Laguna is an almost landlocked body of water lying across the northern part of the island.

The north side of the island extends about 10 miles east-northeast from its northwest extremity to Punta Molas (20°35'N, 86°44'W), the north-eastern extremity of the island. The point is marked by a light.

Cozumel Bank extends north from the north side of the island and lies up to 6 miles north of Punta Molas. The bank has been reported to extend up to 14 miles north-northeast and 4 miles east from Punta Molas. General depths on the bank range from 30 to 132 feet. Ripples, which at times have the appearance of breakers, mark the east edge of this bank.

PUNTA CELARAIN, S point of island, 20 16.0N, 86 59.5W. Fl W 5s, 85ft, 20M. White round concrete tower with dwelling, 82ft. ~~Visible 240°–155°~~.

BANCO PLAYA, entrance, N side, 20 31.6N, 86 56.4W. Fl G 5s, 26ft, 6M. Green fiberglass post, 18ft.

BANCO PLAYA, S side, 20 31.6N, 86 56.4W. Fl R 5s, 30ft, 6M. Red fiberglass post, 18ft.

GOVERNMENT WHARF, 20 30.7N, 86 57.1W. Iso R 2s, 20ft, 6M.

SAN MIGUEL DE COZUMEL, 20 30.4N, 86 57.3W. Fl (2) W 5s, 56ft, 15M. White round concrete tower, 49ft. ~~Visible 031°–182°~~.

AVIATION LIGHT, 20 30.6N, 86 56.6W. Al Fl W G 10s, 82ft, 14M. Skeleton tower. ~~Aero radiobeacon~~.

PUNTA MOLAS, N point of island, 20 35.2N, 86 43.9W. Fl (3) W 12s, 69ft, 20M. White round concrete tower with red bands, 66ft. ~~Visible 070°–337°~~.

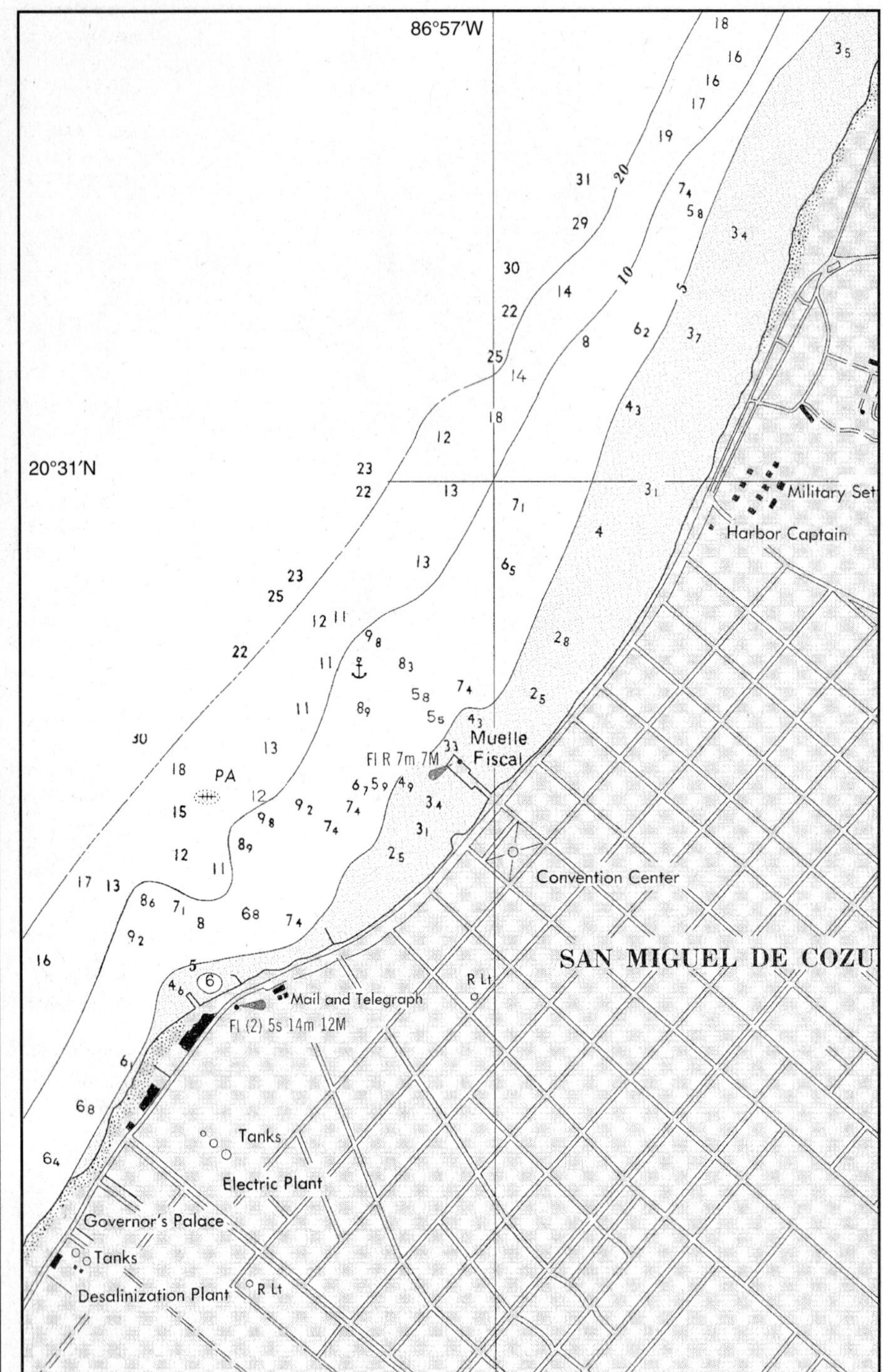

San Miguel de Cozumel, soundings in meters *(from DMA 28197, 2nd ed, 6/88)*

SAN MIGUEL DE COZUMEL

20 30N, 86 58W

Pilotage: The town lies about 3 miles north-northeast of Caleta Bay. It is the principal settlement on the island and a popular tourist destination. The International Pier, formerly known as Mulle del Transbordador, is located about 1 mile northeast of Caleta Bay. This is the primary cruise ship dock. Strong currents of up to 3 knots have been reported (1995) in the vicinity of the pier. A radiobeacon is located close southeast of the International Pier. A light is shown from the 43-foot-high lighthouse located 2 miles northeast of the International Pier. There is a dolphin berth for small tankers near the lighthouse. An aeronautical lightbeacon is shown from a tower about 1.5 miles east of the lighthouse. An aeronautical radiobeacon is near the aeronautical lightbeacon. A pier extends northwest from the shore adjacent to town about 1 mile northeast of the lighthouse. It is 180 feet long, with a depth of about 11 feet at its head. Ferries run from here to Playa del Carmen and Puerto Morelos.

Landmarks: A conspicuous hotel stands close east of the International Pier. A conspicuous clock tower, painted white and dimly illuminated at night, stands near the root of the town pier at San Miguel. A blue-domed water tower and a framework radio tower, both conspicuous, stand close northeast and 0.5 mile southwest respectively, of the clock tower. A prominent airport control tower stands about 1.5 miles northeast of the town.

Port of entry: Cruisers report that formalities can be difficult here. Some people report good success using the services of an agent available to those docked in the marina. Other agents are available to those anchored out.

Dockage: The Club Nautico marina is entered about 1 mile northeast of the town dock in San Miguel. This basin is charted as the Banco Playa Yacht Harbor. The marina is reported to respond to "Puerto Abbrigo" on the VHF radio. The docks are likely to be crowded with sportfishing boats.

Anchorage: Cruisers usually anchor north of the town dock. Be prepared to leave quickly if the wind is shifting to the north or northwest.

Services: Fuel and water are available in the marina. There are boutiques, restaurants, and tourist shops of all sorts ashore. Most types of groceries and general supplies are available.

YUCATÁN COAST: COZUMEL TO CABO CATOCHE

CALETA DE XEL-HA, 20 28.2N, 87 16.0W. Fl W 6s, 39ft, 11M. Truncated pyramidal aluminum tower, 26ft.

XCARET RANGE (Front Light), 20 35N, 87 06W. Fl W 3s. (Rear Light), 20 35N, 87 06W. Iso W 2s.

XCARET JETTY, N side, 20 35 N, 87 06W. Fl G 5s.

XCARET JETTY, S side, 20 35 N, 87 06W. Fl R 5s.

XCARET WHARF, W side, 20 20.0 N, 86 59.0W. Fl G 3s.

XCARET WHARF, E side, 20 20.0 N, 86 59.0W. Fl R 3s.

CALETA DE CHACHALET, 20 29.5N, 87 14.0W. Fl (2) W 10s, 33ft, 11M. Truncated pyramidal aluminum tower, 26ft.

PLAYA DEL CARMEN, 20 37.0N, 87 4.7W. Fl (3) W 12s, 39ft, 11M. Truncated pyramidal aluminum tower, 30ft.

PUNTA MAROMA, 20 43.5N, 86 58.0W. Fl (2) W 10s, 36ft, 11M. Truncated pyramidal aluminum tower, 30ft.

PUNTA BRAVA, 20 48.6N, 86 54.8W. Fl (4) W 16s, 33ft, 11M. Truncated pyramidal aluminum tower, 30ft.

PUERTO MORELOS, near wharf, 20 50.7N, 86 53.0W. Fl W 6s, 52ft, 18M. White round concrete tower, 49ft. Visible 205°–035°.

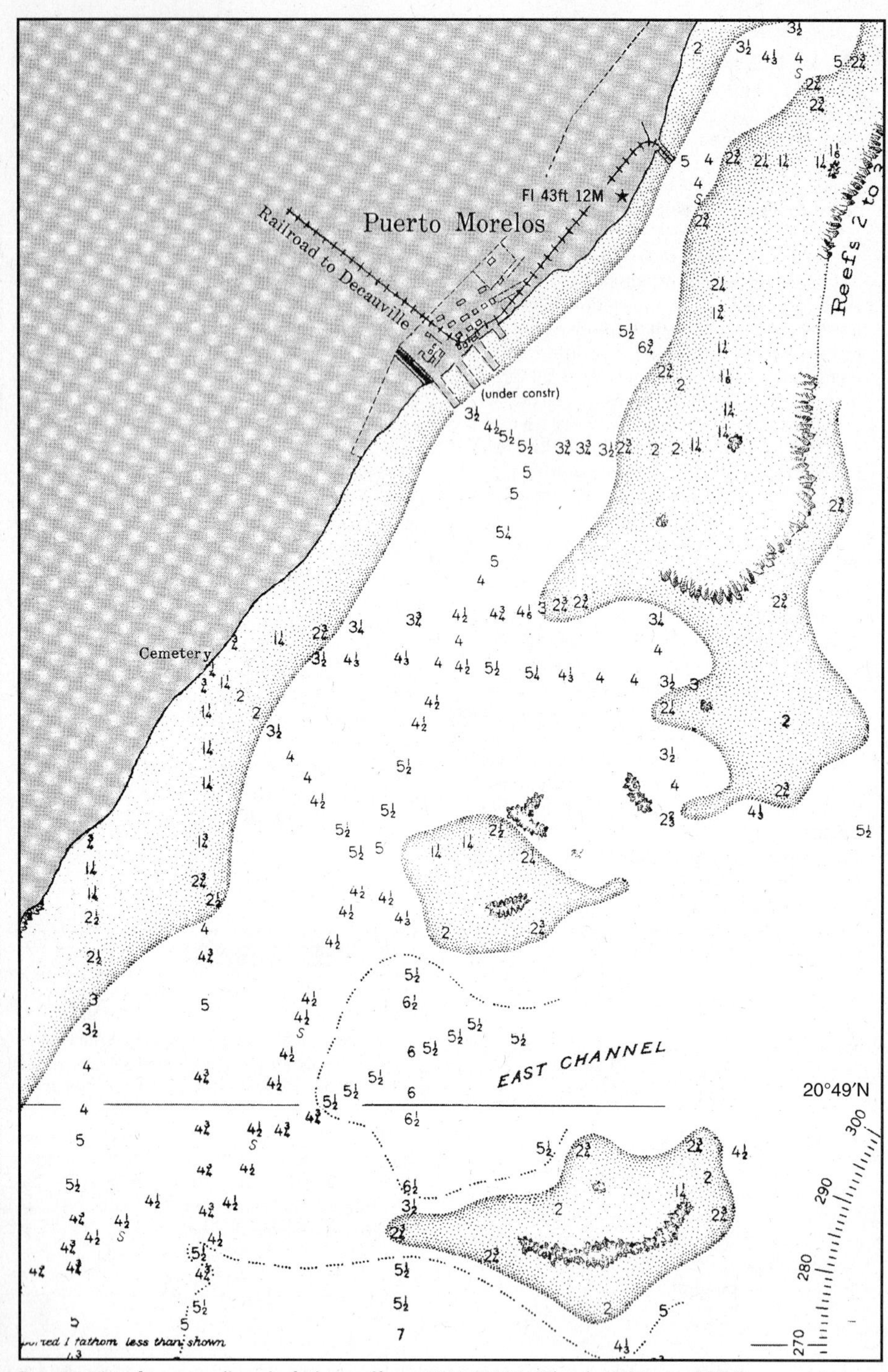

Puerto Morelos, soundings in fathoms *(from DMA 28201, 7th ed, 11/20, rev. 12/74)*

PUERTO MORELOS

20 51N, 86 54W

Pilotage: This is a small fishing and ferry port, with depths of 22 to 33 feet. Ferries run from here out to San Miguel de Cozumel. The village is on the mainland about 1.5 miles north of the reef entrance. The reefs lie about 0.3 mile offshore from the village and up to 1 mile offshore abreast the south part of the harbor. Some of these reefs are 2 to 3 feet high in places. A small pier with depths of 16 feet at its head extends from the shore north of the village. The coast should be approached from the east to a position about 1.8 miles south of the light, which stands north of the village. When this light bears 024° a vessel should steer for it on that bearing.

Port of entry

Anchorage: Anchor north of the town pier, where the ferries come in from Isla de Cozumel. The off-lying reef provides adequate shelter.

Services: Water, fuel, and some basic supplies of groceries and general stores are available.

The coast between Puerto Morelos and Cabo Catoche extends about 14 miles northeast and then curves north-northwest for 36 miles. The terrain is generally low and wooded. Sand hills stand along some parts of the coast. The terrain in the north part of this area becomes more elevated.

DUQUE DE ALBA, 20 50.4N, 86 52.7W. Iso.W 2s, 16ft, 11M. Metal framework tower, 10ft.
PUNTA NIZUK, 21 2.0N, 86 47.0W. Fl (4) W 16s, 33ft, 11M. Truncated pyramidal aluminum tower, 30ft.
CANCÚN, 21 08.3N, 86 44.5W. Fl (3) W 12s, 49ft, 11M. White cylindrical concrete tower, 39ft.

ISLA CANCÚN

21 05N, 86 47W

Pilotage: This island lies about 14 miles north-northeast of Puerto Morelos. It is composed of 40- to 60-foot-high sand hills. The north and south extremities of the island curve to the west and almost reach the mainland. There are two bridges connecting the island with the mainland. A 1970 report stated that the south end of the island appeared to be connected to the mainland when viewed by radar.
A conspicuous tower, marked by a fixed red obstruction light and a structure in the shape of a truncated pyramid, stand near Punta Cancún. Another tower, marked by a fixed red obstruction light, stands 2.8 miles south-southwest of the point.
Isla Cancún encloses a series of interconnected lagoons having shallow channels to the sea. A natural harbor for small craft is located south of Punta Nizuk, the south extremity of the island, and is protected by a reef. Local knowledge is required to enter this harbor. Club Med has a facility located on Punta Nizuk. The harbors on Isla Cancún are suitable only for shallow-draft craft that can clear the low bridges to the mainland. This is a major international tourist destination.

ARROWSMITH BANK

21 05N, 86 25W

Pilotage: This bank lies about 16 miles east of Isla Cancún. The bank is about 20 miles long and from 1.5 to 5 miles wide. There are depths of 54 to 150 feet over it, with the shoalest part lying on the center of its west part.

CAUTION: The east edge of this bank lies about 2 miles farther east than charted.

A north-northeast-setting current crosses Arrowsmith Bank, attaining a rate of from 2 to 3 knots. The south end of the bank is marked by strong rips.
In 1980 an area of heavily breaking seas was reported centered in position 20°50'N, 86°35'W. This is south of the bank, and there was a moderate southeast breeze at the time of the observation.
A shoal with a depth of 27 feet over it was reported to lie near the northeast extremity of Arrowsmith Bank.

EL MECO, 7.4 kilometers N of ruins, 21 12.7N, 86 48.4W. Fl W 6s, 26ft, 11M. White truncated pyramidal concrete tower, 17ft.
ROCA DE LA BANDERA (Becket Rock), 21 9.9N, 86 44.0W. Fl (2) W 10s, 13ft, 11M. Red round concrete tower with black bands, 10ft. Visible 306°–077°.

BAHÍA MUJERES

21 13N, 86 46W

Pilotage: This bay lies between Isla Mujeres and the mainland. It is entered between the northeast extremity of Isla Cancún and the south extremity of Isla Mujeres. A sandbank extends about 2 miles west from the south extremity of Isla Mujeres, then trends north about 2 miles toward the island. The 5-meter depth curve marks this sandbank. There are many coral heads on this bank, and some of them near the south end nearly dry. The southwest edge of the bank is marked by a light buoy. The general depths in this bay are from 6.5 feet to 26 feet.

P
Pilot
MEXICO

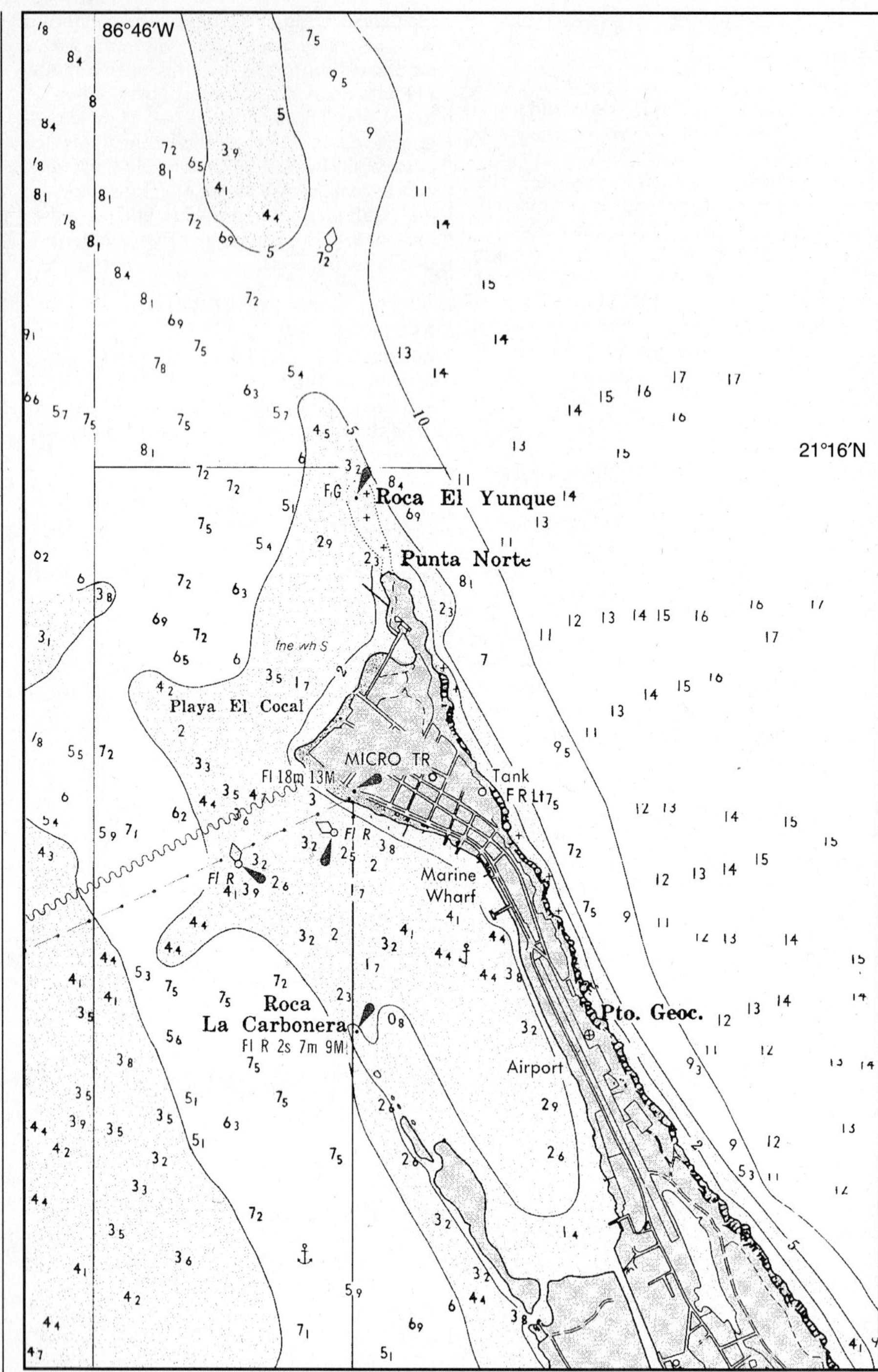

Isla Mujeres, soundings in meters *(from DMA 28202, 20th ed, 1/84)*

A passage between this bank and the shoals to the west is about 0.3 mile wide and leads north past Puerto Mujeres where it joins Canal las Pailas. This passage rounds the north end of Isla Mujeres and turns back toward the sea about 0.5 mile north of El Yunque.

Puerto Juarez is located in the southwest part of Bahía Mujeres, on the mainland about 5.5 miles west-southwest of the lighthouse on the south end of Isla Mujeres. The town serves as a terminal for ferries servicing Isla Mujeres. An 800-foot-long pier has depths of about 6 feet alongside. Punta Sam is located 3 miles north of Puerto Juarez and is also a ferry terminal.

ROCA YUNKE, N point of island, 21 15.9N, 86 45.4W. Fl G 5s, 16ft, 6M. Square concrete and aluminum tower, 13ft.
PIEDRA LA CARBONERA RANGE (Front Light), 21 14.8N, 86 45.4W. Fl W 3s, 20ft, 6M. White square concrete tower, 16ft. **(Rear Light),** 980 meters, 000° from front, 21 15.3N, 86 45.4W. Fl (2) W 10s, 62ft, 20M. White round concrete tower, red bands, 52ft.
ISLA MUJERES, S extremity, 21 12.1N, 86 43.1W. Fl (4) W 16s, 75ft, 14M. White octagonal concrete tower, 34ft. Visible 150°–131°.

ISLA MUJERES

21 14N, 86 45W
Pilotage: This island lies about 4.5 miles north-northeast of Isla Cancún. Its location is strategic for cruisers arriving from or departing to the United States via the Yucatán Channel. See the section on Cuba for more information on the Yucatán Channel.

The island is about 4 miles long, low, narrow, and wooded. The south part is slightly elevated. The ruins of a square watch tower stand near the south part of the island. The east side of the island is composed of fairly steep-to rocky shelves. They terminate at the north end in El Yunque (Anvil Rock), which is square, black, and about 6 feet high. Some white cliffs stand about midway along the east side of the island. A large, square, white hotel stands on the north extremity of the island and has been reported to be a good radar target. The channel leads along the western shore of the island into the harbor. There is a shoal area north of the light on Roca la Carbonera.
Port of entry: Enter at Puerto Mujeres (21°16′N, 86°45′W), located at the north end of the island, on its west side.
Dockage: Marina Paraiso is located in the southern part of the harbor, with reported dockside depths of about 9 feet. It has showers and a laundry.
Anchorage: Anchor south of the main commercial piers.
Services: The fuel dock is located just southeast of the light in the harbor. There are reported to be good supplies of groceries and general stores. You can take a ferry across to the tourist pleasures of Cancún.

ISLA BLANCA

21 22N, 86 49W
Pilotage: This is a narrow ridge of sand, about 7 feet high, which is a continuation of the ridge extending south into Bahía Mujeres. From the south end of the sand ridge a reef fringes the coast extending past Isla Blanca to Isla Contoy, about 4 miles farther north. This reef begins about 3 miles north-northwest of the northern extremity of Isla Mujeres. There are one or two openings for boats in this reef.

ISLA CONTOY, near N point, 21 31.6N, 86 48.3W. Fl W 7s, 105ft, 21M. White round concrete tower with dwelling, 103ft. Visible 010°–350°. **RACON T** (–).

ISLA CONTOY

21 30N, 86 49W
Pilotage: This island lies about 4 miles north-northeast of Isla Blanca and 6.5 miles offshore. Its east side is composed of a narrow ridge of sand hills covered with bushes and trees to within 1.5 miles of its northwest extremity. A narrow ridge of coral with depths of 18 feet extends about 3 miles north-northwest from two rocks close off the northwest extremity of the island. The greater part of the west side of the island is intersected by numerous small lagoons. This island is a bird and wildlife sanctuary.
Anchorage: Cruisers take temporary anchorage off the west side of the island. You may be checked by the Mexican navy or other government officials if anchored here.

CAYO SUCIO

21 25N, 86 53W
Pilotage: This cay is close to the coast, about 3 miles northwest of Isla Blanca. The coast trends about 8 miles to the northwest from here and is more elevated with dense woods ashore. Some trees stand up to 131 feet above sea level. The conspicuous ruins of a church stand at the north end of this coast.

From the ruins of the church the coast extends about 6 miles west-northwest to abreast of Cabo Catoche, which is on the north side of Isla Holbox.

CABO CATOCHE

21 36N, 87 04W

CABO CATOCHE, 21 36.3N, 87 6.1W. Fl (4) W 20s, 49ft, 25M. White square concrete tower with red band, dwelling, 46ft. Visible 080°–285°.

Pilotage: This is a sand projection on the northern side of Isla Holbox. Depths of 18 to 30 feet extend 10 miles to the north and east.

The coast extends from here about 85 miles west to Punta Yalkubul and then about 105 miles west-southwest to Punta Boxcohuo. This north coast of the Península de Yucatán is low and arid, with few conspicuous landmarks. The coast is fringed with many detached shoals and coral patches. Lagartos Lagoon lies parallel to almost all of this section of coast. A shallow coastal bank, as defined by the 5-meter-depth curve, lies between 1 and 5 miles from this shore.

Tides and currents: Between Cabo Catoche and Punta Boxcohuo the current sets west at a rate of 0.5 to 1 knot, following the trend of the coast about 30 miles offshore. The mean rise at springs ranges from about 1.3 to 2.3 feet. The water on the off-lying banks is influenced by the winds as well as the tides. The wind may offset the normal tides for several days.

~~**CHIQUILA,** 21 26N, 87 15W. Fl W, 29ft, 9M. Red iron column, hut. Marks outer limit of anchorage~~. Removed 12/97.

CHIQUILA RANGE (Front Light), 21 26N, 87 17W. Fl W 3s, 15ft, 11M. Truncated pyramidal aluminum tower, 10ft. **(Rear Light),** Iso W 2s, 46ft, 11M. Truncated pyramidal aluminum tower, 23ft.

ISLA HOLBOX, W end 21 32.1N, 87 18.1W. Fl W 6s, 26ft, 14M. Aluminum tower, 30ft.

HOLBOX RANGE (Front Light), 21 33N, 87 18W. Fl W 3s, 43ft, 12M. Truncated pyramidal iron tower, 34ft. **(Rear Light),** Iso W 2s, 61ft, 12M. Truncated pyramidal iron tower, 51ft.

PUNTA FRANCISCA (Mosquito), 21 32.1N, 87 18.1W. Fl (2) W 10s, 36ft, 14M. Truncated pyramidal aluminum tower, 30ft.

ISLA HOLBOX

21 33N, 87 14W

Pilotage: This is one of a chain of low, narrow islands that front the coast between Cabo Catoche and Boca de Conil to the west. Punta Francisca is the northwest extremity of the island.

A chain of cays, fronted by numerous sand ridges, with depths of 8 to 30 feet, extends 8 miles northeast from Cabo Catoche and 9 miles northwest from Punta Francisca. Bajo Corsario, with a least depth of 15 feet, lies 8 miles north of Punta Francisca.

CAMPECHE BANK

Pilotage: This bank extends about 155 miles north from the north coast of the Yucatán peninsula and about 120 miles west from the peninsula's west side. The limits of the bank can best be seen on the appropriate chart, such as NOAA 411.

This steep-to bank has very irregular depths and is marked on its east and north edges by heavy ripples and a confused sea. Some of the dangers on the banks are marked by discolored water. Many cays, shoals, and reefs are found within its limits. Small vessels have reported encountering very heavy seas on parts of the bank during periods of stormy weather.

CAUTION: Banco de Campeche has not been surveyed for many years, and as reports of new shoals are constantly received, it is reasonable to assume that many more dangers exist than are shown on the charts. Many isolated shallow patches, with depths of as little as 18 feet, have been reported north of the Yucatán Peninsula.

BOCA DE CONIL

21 29N, 87 35W

Pilotage: This is the 3-mile-wide west entrance to Laguna de Yalahua, which is a shallow body of water lying between Isla Holbox and the mainland. Foul ground extends about 7 miles northwest from Boca de Conil.

The low, sandy coast between Boca de Conil and Cuyo, about 16 miles to the west, is marked by conspicuous groves of trees near its east end.

MONTE DE CUYO, summit, 21 30.9N, 87 41.1W. Fl (3) W 12s, 85ft, 14M. Red round concrete tower. Visible 100°–273°.

~~**PIER**~~ **BREAKWATER,** W side, 21 30N, 87 12W. Fl R 5s, 39ft, 7M. Truncated pyramidal metal tower, 24ft.

~~PIER~~ BREAKWATER, E side, 21 30N, 87 12W. Fl G 5s, 39ft, 7M. Truncated pyramidal metal tower, 24ft.

EL CUYO RANGE (Front Light), 21 30.9N, 87 41.9W. Fl W 3s, 51ft, 11M. Truncated pyramidal metal tower, 39ft. **(Rear Light),** Iso W 2s, 62ft, 11M. Truncated pyramidal metal tower, 52ft.

CUYO

21 32N, 87 41W

Pilotage: This town has a wharf over 350 feet long, with a depth of almost 7 feet at its outer end. The town stands on the narrow strip of land between the sea and the lagoon that parallels this coast. El Cuyo, a 40-foot-high hill, stands close to the town. A 12-foot shoal lies about 5 miles north of town.

COLORADOS, 21 35.5N, 87 59.5W. Fl (2) W 10s, 43ft, 11M. Truncated pyramidal metal tower, 39ft.

RÍO LAGARTOS, 21 35.9N, 88 12.2W. Fl (2) W 10s, 69ft, 11M. White square concrete tower, 59ft.

BREAKWATER, W side, 21 40N, 88 12W. Fl R 5s, 33ft, 6M. Concrete structure, 15ft.

BREAKWATER, E side, 21 40N, 88 12W. Fl G 5s, 33ft, 6M. Concrete structure, 15ft.

PIER, W side, 21 36N, 88 12W. Fl R 5s, 40ft, 7M. Aluminum tower, 24ft.

PIER, E side, 21 36N, 88 12W. Fl W 5s, 40ft, 7M. Aluminum tower, 24ft.

~~**RÍO LAGARTOS SOUTH,** 21 35.1N, 88 12.1W. Fl (2) W 10s, 69ft. Square concrete tower, 59ft.~~

LAS COLORADOS

21 37N, 88 01W

Pilotage: This open roadstead harbor has two small piers. The principal export is salt, which is barged out to the anchorage. From the north, depths in the approach range from 24 to 30 feet, and are quite regular. The two piers form good radar targets from a distance of about 17 miles.

The coast between Cuyo and the Río Lagartos, 30 miles to the west, is fronted by a low, sandy beach. The Río Lagartos is the outlet of the narrow lagoon that parallels this coast.

Bajo Antonieta is a shoal that uncovers to a height of 2 feet. It is located about 4 miles west-northwest of the Río Lagartos.

Between the Río Lagartos and Punta Yalkubul, about 22 miles to the west, the coast is more elevated. It is 170 feet higher about 6 miles east of the point.

PUNTA YALKUBUL TO PROGRESO

YALKUBUL, 21 31.2N, 88 36.7W. Fl (4) W 16s, 69ft, 14M. White round concrete tower, red bands, dwelling, 66ft. Visible 057°–251°.

Punta Yalkubul is low and marked by trees. Bajo Carmelita, a shoal with depths of less than 15 feet, lies 11 miles northwest of the point. Bajo Pawashick, a shoal with a least depth of 9 feet, lies 9 miles west of the point.

The coast between Punta Yalkubul and Punta Arenas, 13 miles to the southwest, remains low. A shallow lagoon entrance lies close east of the latter point.

DZILAM DE BRAVO, 21 21.8N, 88 54.8W. Fl W 6s, 59ft, 14M. White round concrete tower.

BREAKWATER, W side, 21 22.2N, 88 53.6W. Fl R 5s, 39ft, 7M. Truncated pyramidal aluminum tower.

BREAKWATER, E side, 21 22.2N, 88 53.3W. Fl G 5s, 40ft, 7M. Truncated pyramidal aluminum tower.

DZILAM DE BRAVO RANGE (Front Light), 21 21.9N, 88 53.6W. Fl W 3s, 46ft, 11M. Truncated pyramidal aluminum tower, 39ft. **(Rear Light),** Iso W 2s, 59ft, 11M. Truncated pyramidal aluminum tower, 52ft.

Dzilam (Silan) is a small town fronted by a pier located 6 miles west of Punta Arenas.

Between Punta Arenas and Progreso, about 48 miles to the west, the coast consists of low sandy beach, slightly wooded, and backed by swampy ground. Several villages lie along this section of coast.

Chicxulub is a small village fronted by a pier located about 2 miles east of Progreso.

TELCHAC, 21 19.8N, 89 15.1W. Fl (3) W 8s, 72ft, 14M. White round tower.

BREAKWATER, W side, 21 20N, 89 16W. Fl R 5s, 40ft, 7M. Truncated pyramidal aluminum tower, 24ft.

BREAKWATER, E side, 21 20N, 89 16W. Fl G 5s, 40ft, 7M. Truncated pyramidal metal tower, 24ft.

TELCHAC RANGE (Front Light), 21 20N, 89 16W. Fl W 3s, 46ft, 11M. Truncated pyramidal concrete tower, 43ft. **(Rear Light),** Iso W 2s, 59ft, 11M. Truncated pyramidal concrete tower, 52ft.

PROGRESO

PROGRESO, near pier, 21 17.1N, 89 40.1W. Fl W 6s, 108ft, 16M. White truncated conical concrete tower, 108ft. Visible 066°–250°.
PIER LIGHT, 21 18.9N, 89 40.4W. Iso.R 2s, 30ft, 7M. Tower, 22ft.
STATE WHARF, W side, 21 18.2N, 89 40.3W. Iso R 2s, 33ft, 7M. Tower, 28ft.
STATE WHARF, E side, 21 18.2N, 89 40.3W. Iso.R 2s, 33ft, 7M. Tower, 28ft.
TERMINAL REMOTA, NW corner, 21 20.7N, 89 40.8W. Fl (2) G 10s, 33ft, 7M. Tower, 28ft.
TERMINAL REMOTA, NE corner, 21 20.7N, 89 40.6W. Fl (2) W 10s, 36ft, 8M. Tower, 28ft.
RACON Y (– • – –), 16M.

PROGRESO

21 17N, 89 40W
Pilotage: Progreso is the most important port on the Yucatán peninsula. The harbor is an open roadstead, with a long pier fronting the town. Sisal is the principal export. The major danger to small craft is a wreck lying 1.5 miles north of New Pier. When approaching the port, the lighthouse is usually the first object sighted. A square gray tower, about 60 feet high, stands close west of the lighthouse. East of the town, and somewhat detached from it, is a large square building close to the coast and partly surrounded by trees. The town itself appears as a group of low gray or white buildings. The piers, together with the warehouses on them, are prominent from a distance of 4 miles.

Terminal Remota has two berths (for commercial vessels) protected by an artificial island. It is situated at the north end of a long causeway and pier, about 0.3 mile in length, with its root near Progreso Light. Muelle Fiscal also has two berths located on the same causeway about 5,970 feet from shore. Pino Suárez Pier is located close to the west of the other two piers. It is about 200 yards long with depths of 8 to 9 feet alongside.
Winds: The prevailing winds are from the northeast to southeast. Northers occur between October and March. Storm signals are displayed from a flagstaff near the lighthouse.
Currents: The current along this coast sets to the west.
Port of entry

PUERTO DE YUCALPETÉN RANGE (Front Light), 21 16.2N, 89 42.4W. Fl W 3s, 46ft, 5M. Red truncated pyramidal metal tower. **(Rear Light),** 195 meters 169° from front. Iso.W 2s, 59ft, 5M. Red truncated pyramidal aluminum tower, 52ft.
HEAD OF JETTY, E side, 21 17.0N, 89 42.5W. Fl G 5s, 39ft, 7M. Red truncated pyramidal metal tower, 24ft.
HEAD OF JETTY, W side, 21 16.8N, 89 42.5W. Fl R 5s, 38ft, 7M. Red truncated pyramidal metal tower, 24ft.

PUERTO DE YUKALPETÉN

21 17N, 89 43W
Pilotage: This new "free port" is located about 3 miles west of Progreso. The east breakwater extends approximately 600 feet from the shore, while the west breakwater is less prominent. The least depth between the breakwaters is just under 8 feet. Inside the harbor, alongside the two piers, there is a least depth of under 7 feet.

~~**CHUBERNA,** 21 14.0N, 89 50.0W. Fl G 5s, 23ft, 7M~~. Removed 7/97.
SISAL, on old white fort, 21 09.8N, 90 02.1W. Fl (3) W 12s, 72ft, 12M. White round concrete tower with red bands. ~~Visible 055°–260°~~.

The coast between Puerto de Yukalpetén and Sisal, about 20 miles to the west-southwest, is marked by several villages and is more wooded than elsewhere in the vicinity. The village of Sisal is nearly abandoned.

ARRECIFE SISAL

21 21N, 90 09W
Pilotage: This is a coral reef with an obstruction on it located about 12 miles north-northwest of Sisal. It may be marked by discolored water under certain weather conditions. A 21-foot-deep coral shoal lies about 6.5 miles west-northwest of Sisal.

The coast between Sisal and Punta Boxcohuo, about 17 miles southwest, is low and sandy. A building in ruins, about 5 miles southwest of Sisal, is the only conspicuous landmark. The coast between Sisal and a position about 5 miles south of Celestun is radar conspicuous.

ISLA PEREZ, on S part of Alacrán Reef, 22 23.5N, 89 41.8W. Fl (2) W 8s, 69ft, 13M. Red round concrete tower with dwelling, 59ft.
RACON Z (– – • •), 25M.
ISLA DESTERREDA, on N part of Alacrán Reef, 22 31.7N, 89 47.4W. Fl (4) W 16s, 46ft, 10M. Tubular metal tower, 24ft.

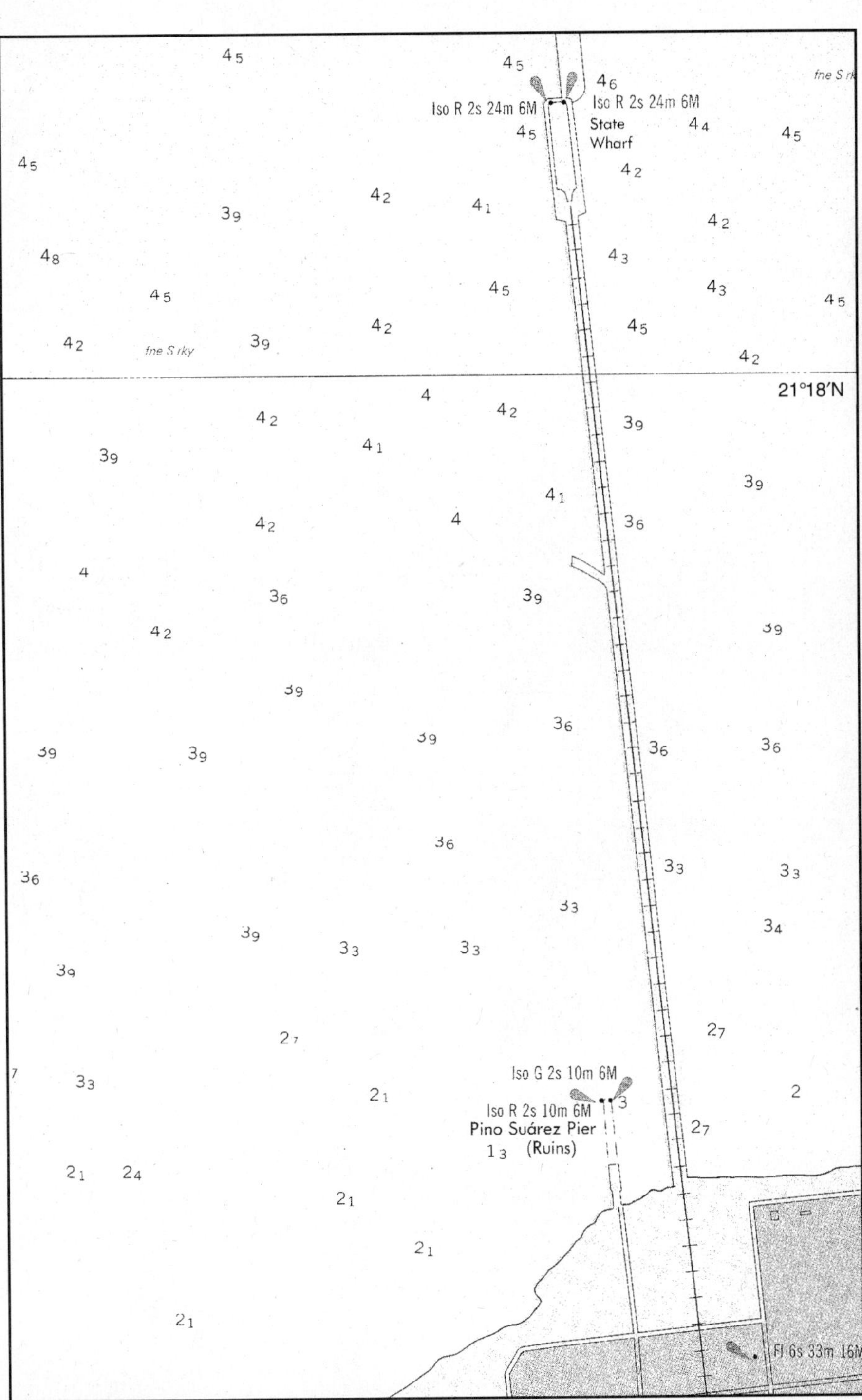

Progreso, soundings in meters *(from DMA 28223, 3rd ed, 6/91)*

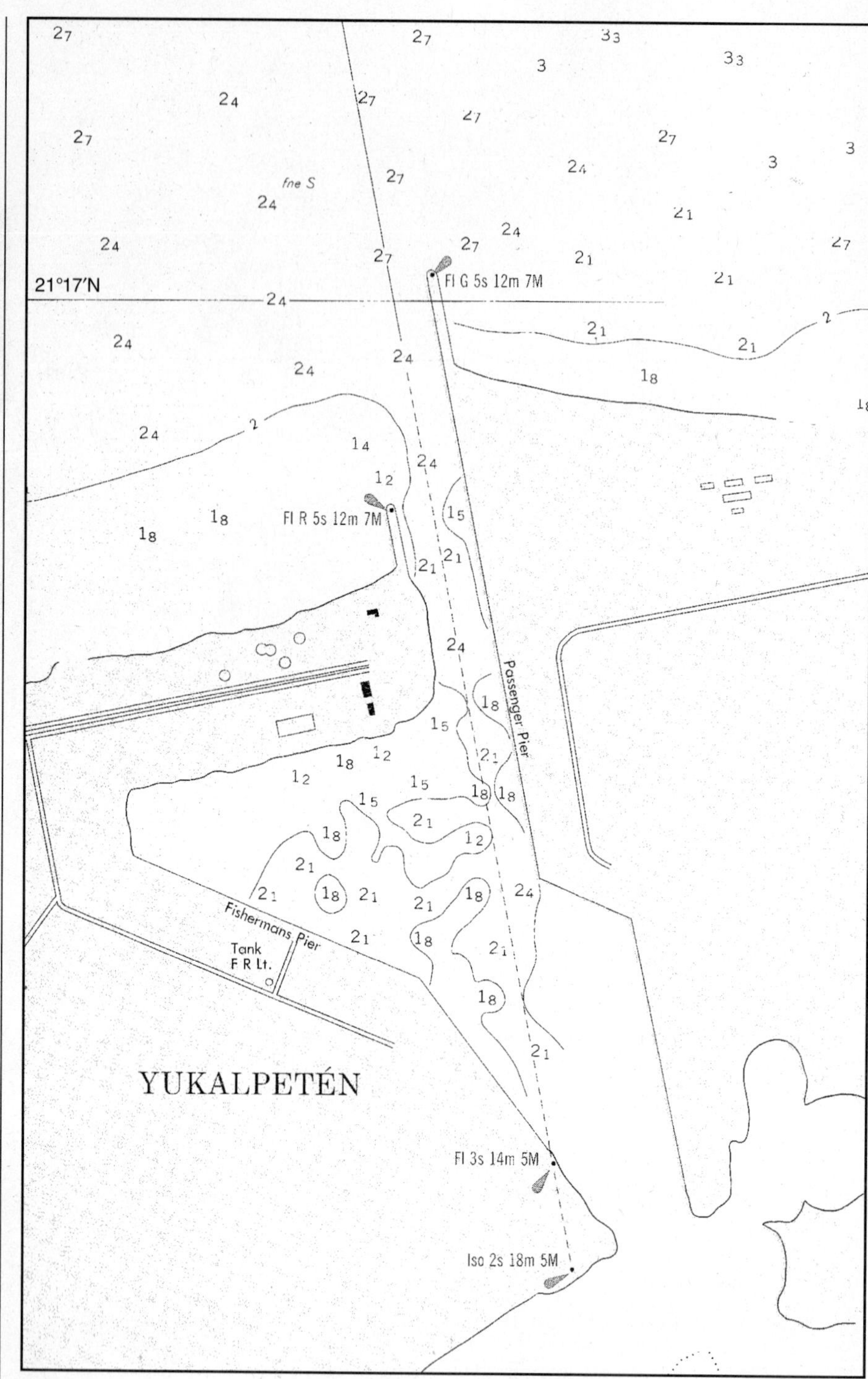

Yukalpetén, soundings in meters *(from DMA 28223, 3rd ed, 6/91)*

ARRECIFE ALACRÁN

22 29N, 89 42W

Pilotage: This steep-to, half-moon-shaped reef lies about 34 miles northwest of Granville Shoal. This is about 68 miles north-northwest of Progreso, on the Campeche Bank. The main part of the reef consists of a compact mass of coral heads about 14 miles long and 8 miles wide. The entire northeast side of the reef is awash. Isla Chica and Isla Pájaros, two small, low cays about 0.3 mile apart, stand near the south end of the reef. Isla Desterrada is a small 10-foot-high cay located on the northwest end of Arrecife Alacrán.

Isla Perez is a narrow cay about 14 feet high. It lies near the south end of the reef. The light structure on this cay has been reported to be a good radar target at distances up to 18 miles. The light is shown from a round red masonry tower with a parapet. A gray masonry tower stands adjacent to the light. The lighthouse and tower are fully visible when approaching from the west. A stranded wreck close east of the light structure was reported to be a good radar target at distances up to 13 miles.

Anchorage: Small vessels anchor in depths of 36 to 60 feet over fine sand, mud, and coral, about 0.3 mile east of Isla Perez. The anchorage is approached from the south through an unmarked channel with a depth of about 24 feet. This channel should not be attempted without local knowledge.

Isla Desertora (not to be confused with Isla Desterrada farther to the north-northwest) is a small 12-foot-high cay located about 3 miles northwest of Isla Perez.

Currents: The current usually sets west at about 1 knot in the vicinity of these reefs, but may set north or south, depending upon the wind.

ARRECIFE MADAGASCAR

21 26N, 90 18W

Pilotage: This narrow coral ledge, with a least depth of under 9 feet, lies about 69 miles south-southwest of Arrecife Alacrán. The sea does not break on this ledge, which is covered with weed that appears the same color as the water.

Breakers were reported about 6 miles north-northeast of the ledge in 1909.

Arrecife de La Serpiente, with a least depth of 27 feet, lies about 11.5 miles west of Arrecife Madagascar.

PUNTA PALMAS, 21 03.8N, 90 15.2W. Fl (2) W 10s, 138ft, 12M. White round concrete tower, dwelling, 128ft. Visible 048°–236°.
CELESTUN, center of town, near shore, 20 51N, 90 24W. Fl (3) W 12s, 69ft, 10M. White round concrete tower. Visible 002°–188°.
NORTH BREAKWATER, 20 51N, 90 53W. Fl G 5s, 33ft., 7M. White tubular tower, 26ft.
SOUTH BREAKWATER, 20 51N, 90 53W, Fl R 5s, 33ft, 7M. White tubular tower, 26ft.
CELESTUN RANGE (Front Light), 20 51N, 90 23W. Fl W 3s, 46ft, 9M. Truncated pyramidal aluminum tower, 43ft. **(Rear Light),** Iso W 2s, 59ft, 9M. Aluminum tower, 56ft.
ISLA ARENA, 20 36N, 90 28W. Fl W 6s, 46ft, 10M. White round concrete tower, 33ft.
CAYO ARENAS, 22 08N, 91 24W. L Fl W 12s, 79ft, 12M. White concrete tower, 65ft. **RACON X** (– · · · –).

CAYO ARENAS

22 07N, 91 24W

Pilotage: This guano-covered cay is about 20 feet high and is located on the southeast edge of a detached reef 0.8 mile long. The cay lies about 93 miles west-southwest of Arrecife Alacrán. The horns of the reef extend 0.5 mile northwest and 0.3 mile west from the cay. A small wharf stands on the northwest side of the cay. A detached reef, 1.25 miles long with a rocky patch 7 feet high on its south end and a small coral patch 2 feet high off its northwest end, lies 1 mile east of Cayo Arenas. The intervening channel is about 0.5 mile wide at its north entrance. At its south end are three coral patches, with depths of 14 to 19 feet. These patches are about 0.8 mile southeast of Cayo Arenas.

Anchorage: Vessels can anchor in this channel in depths of 60 to 72 feet, about 0.5 mile east of the cay. Depths of 24 to 36 feet can be found between the horns of the reef 0.3 mile northwest of the cay.

BAJO NUEVO

21 50N, 92 05W

Pilotage: This reef has depths of less than 6 feet. It lies about 40 miles west-southwest of Cayo Arenas and is marked by breakers.

BANCOS INGLESES

21 49N, 91 56W

Pilotage: These two banks lie with their least depths of 30 to 133 feet about 8 miles southeast of Bajo Nuevo.

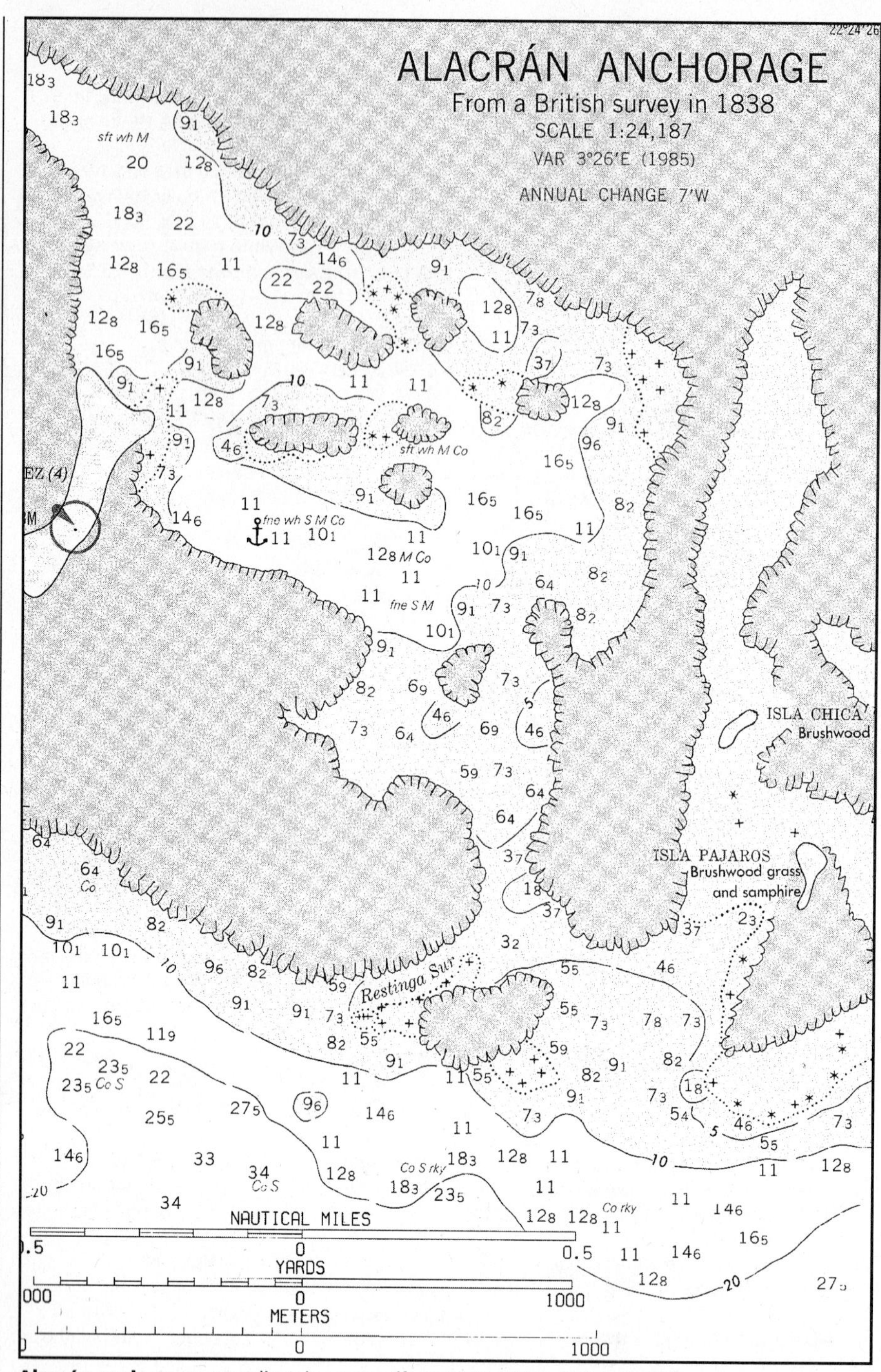

Alacrán anchorage, soundings in meters *(from DMA 28221, 16th ed, 8/90)*

ISLA TRIANGULOS OEST, 20 59N, 92 18W. Fl (3) W 20s, 79ft, 12M. Red rectangular concrete tower, 72ft.

ARRECIFES TRIÁNGULOS

20 57N, 92 14W
Pilotage: This group of two reefs lie about 82 miles southwest of Cayos Arenas. The two groups are about 6 miles apart.

Triángulo Oest is a 0.8-mile-long reef with an 11-foot-high cay on its southwest end. A disused lighthouse stands near the lighthouse on the cay near the southwest end.

CAUTION: A dangerous wreck lies about 22 miles northeast of the lighthouse on Triángulo Oest. An awash rock was reported to lie about 18 miles north-northeast of Triángulo Oest.

Triángulo Este (20°55′N, 92°13′W) and Triángulo Sur (20°54′N, 92°14′W), which nearly dry, are separated by a channel 0.3 mile wide with depths of 42 to 60 feet. A 24-foot-high cay stands on the south end of Triángulo Este. A reef extends about 1 mile northeast from the cay, and a coral ledge extends about 1 mile farther northeast. Triángulo Sur has a ledge extending about 1 mile southwest from the southwesternmost cay.
Anchorage: Vessels can anchor off the southwestern end of Triángulo Sur.

OBISPO NORTE AND OBISPO SUR

20 29N, 92 12W
Pilotage: These shoals lie about 25 miles south of Triángulo Oest. They have depths of 24 to 60 feet. They are marked by discolored water. Obispo Norte has a depth of 15 feet near its north end. A dangerous wreck lies between the two shoals.

BANCO PERA

20 42N, 91 56W
Pilotage: This shoal bank has depths of 54 feet and is located about 28 miles southeast of Triángulo Oest. Banco Nuevo has a depth of 48 feet in a position about 38 miles southeast of Triángulo Oest.

CAYOS ARCAS RANGE (Front Light), W side of Cay del Centro, 20 12.7N, 91 58.2W. Fl W 3s, 42ft, 10M. Metal tower. **(Rear Light),** 80 meters, 107° from front. Fl (2) W 15s, 72ft, 13M. White round concrete tower and house. Radar reflector.

CAYOS ARCAS

20 13N, 91 58W
Pilotage: This is the southernmost group of dangers on the Banco de Campeche. It consists of a group of three islets, with surrounding reefs. They lie about 44 miles south-southeast of Triángulo Este.

Cayo del Centro is the northernmost and largest. It is composed of sand and rises at its south end to a height of 21 feet. It lies on the southeast end of a reef that extends about 1 mile northwest and about 0.5 mile west from it. This reef is always visible. The islet is scantily covered with grass, some bushes, and several clumps of palm trees. A pair of lighted beacons in range 107° stand on the cay. A stranded wreck stands on the reef about 800 yards northwest of the north extremity of Cayo del Centro.
Anchorage: Vessels use the range to work into the anchorage west-northwest of the cay.

Cayo del Este is a small 10-foot-high cay on a detached reef about 0.3 mile southeast of Cayo del Centro. The intervening channel has depths of 36 to 84 feet. A stranded wreck stands on the edge of the reef about 300 yards northeast of the islet.

Cayo del Oeste is a small 6-foot-high cay lying on a small detached reef about 0.8 mile west of the south end of Cayo del Centro. The narrow intervening passage has depths of 24 to 27 feet.

PLATFORM PR-1 REBOMBEO, 18 56.7N, 92 37.2W. Q W, 56ft, 10M. Four lights, one installed on each corner of platform.
RACON O (– – –).
PLATFORM POOL-C, 19 13.4N, 92 15.8W. Q W, 56ft, 10M. Four lights, one installed on each corner of platform.
PLATFORM CAAN-A, 19 13.3N, 92 05.4W. Q W, 56ft, 10M. Four lights, one installed on each corner of platform.
PLATFORM ECO-1, 19 01.8N, 92 01.1W. Q W, 56ft, 10M. Four lights, one installed on each corner of platform.
PLATFORM CHUC-1, 19 10.3N, 92 17.1W. Q W, 56ft, 10M. Four lights, one installed on each corner of platform.
PLATFORM POOL-D, 19 14.5N, 92 17.3W. Q W, 56ft, 10M. Four lights, one installed on each corner of platform.
PLATFORM POOL-B, 19 14.2N, 92 16.6W. Q W, 56ft, 10M. Four lights, one installed on each corner of platform.

PLATFORM ABKATUM-D, 19 17.9N, 92 12.1W. Q W, 56ft, 10M. Four lights, one installed on each corner of platform.

PLATFORM ABKATUM-C, 19 19.7N, 92 11.3W. Q W, 56ft, 10M. Four lights, one installed on each corner of platform.

PLATFORM ABKATUM-F, 19 16.7N, 92 09.5W. Q W, 56ft, 10M. Four lights, one installed on each corner of platform.

PLATFORM ABKATUM-G, 19 18.0N, 92 08.4W. Q W, 56ft, 10M. Four lights, one installed on each corner of platform.

PLATFORM ABKATUM-J, 19 16.4N, 92 07.9W. Q W, 56ft, 10M. Four lights, one installed on each corner of platform.

PLATFORM AKAL-F, 19 23.9N, 92 03.7W. Q W, 56ft, 10M. Four lights, one installed on each corner of platform.

PLATFORM ABKATUM-H, 19 20.5N, 92 13.2W. Q W, 56ft, 10M. Four lights, one installed on each corner of platform.

PLATFORM ABKATUM-93, 19 18.0N, 92 10.9W. Q W, 56ft, 10M. Four lights, one installed on each corner of platform.

PLATFORM ABKATUM-E, 19 16.4N, 92 11.1W. Q W, 56ft, 10M. Four lights, one installed on each corner of platform.

PLATFORM ABKATUM-A, 19 17.7N, 92 10.2W. Q W, 56ft, 10M. Four lights, one installed on each corner of platform.

PLATFORM NOHOCH-A, 19 22.0N, 92 00.3W. Q W, 56ft, 10M. Four lights, one installed on each corner of platform.

PLATFORM ABKATUM-I, 19 15.3N, 92 10.0W. Q W, 56ft, 13M. Four lights, one installed on each corner of platform.

PLATFORM AKAL-J, 19 25.5N, 92 04.6W. Q W, 56ft, 13M. Four lights, one installed on each corner of platform.

PLATFORM AKAL-G, 19 22.8N, 92 03.0W. Q W, 56ft, 13M. Four lights, one installed on each corner of platform.

PLATFORM KU-1, 19 29.9N, 92 08.3W. Q W, 56ft, 13M. Four lights, one installed on each corner of platform.

PLATFORM AKAL-I, 19 23.9N, 92 00.9W. Q W, 56ft, 13M. Four lights, one installed on each corner of platform.

PLATFORM AKAL-M, 19 27.7N, 92 03.7W. Q W, 56ft, 13M. Four lights, one installed on each corner of platform.

PLATFORM IXTOC-1, 19 24.4N, 92 12.7W. Q W, 56ft, 13M. Four lights, one installed on each corner of platform.

PLATFORM AKAL-E, 19 25.1N, 92 03.0W. Q W, 56ft, 13M. Four lights, one installed on each corner of platform.

PLATFORM AKAL-N, 19 26.2N, 92 03.7W. Q W, 56ft, 13M. Four lights, one installed on each corner of platform.

PLATFORM AKAL-O, 19 24.4N, 92 04.9W. Q W, 56ft, 10M. Four lights, one installed on each corner of platform.

PLATFORM AKAL-D, 19 25.0N, 92 01.6W. Q W, 56ft, 10M. Four lights, one installed on each corner of platform.

PLATFORM POOL-A, 19 14.3N, 92 15.2W. Q W, 56ft, 10M. Four lights, one installed on each corner of platform.

PLATFORM AKAL-R, 19 20.9N, 92 02.9W. Q W, 56ft, 10M. Four lights, one installed on each corner of platform.

PLATFORM NOHO CH-B, 19 20.6N, 92 00.3W. Q W, 56ft, 10M. Four lights, one installed on each corner of platform.

PLATFORM KU-A, 19 31.2N, 92 11.3W. Q W, 56ft, 10M. Four lights, one installed on each corner of platform.

PLATFORM AKAL-C, 19 23.9N, 92 02.3W. Q W, 56ft, 13M. Four lights, one installed on each corner of platform. **RACON Y** (– · – –).

PLATFORM KU-H, 19 35.3N, 92 12.0W. Q W, 56ft, 10M. Four lights, one installed on each corner of platform.

PLATFORM KU-F, 19 29.7N, 92 10.4W. Q W, 56ft, 10M. Four lights, one installed on each corner of platform.

PLATFORM KU-G, 19 30.8N, 92 09.3W. Q W, 56ft, 10M. Four lights, one installed on each corner of platform.

PLATFORM KU-M, 19 33.8N, 92 11.0W. Q W, 56ft, 10M. Four lights, one installed on each corner of platform.

PLATFORM ECO-1, 19 01.8N, 92 01.1W. Q W, 56ft, 10M. Four lights, one installed on each corner of platform. **RACON Q** (– – · –).

PLATFORM AKAL-P, 19 22.8N, 92 04.4W. Q W, 56ft, 10M. Four lights, one installed on each corner of platform.

PLATFORM CHUC-A, 19 10.8N, 92 17.2W. Q W, 56ft, 10M. Four lights, one installed on each corner of platform.

PLATFORM CHUC-B, 19 08.7N, 92 18.3W. Q W, 56ft, 10M. Four lights, one installed on each corner of platform.

CAYO ARCAS OIL TERMINAL, 20 11.5N, 91 59.9W. Q W, 56ft, 10M. Four lights, one installed on each corner of platform. **RACON C** (– · – ·).

CAYO ARCAS TERMINAL

20 10N, 91 59W
Pilotage: This oil production and export facility lies about 1 to 3 miles south of Cayos Arcas. The distribution platform is the north end of a pipeline carrying oil from a marine oil field 45 miles to the south. Pipelines are laid from this platform to two tanker moorings, known as SBMs. The terminal monitors VHF channel 9.

CAUTION: A major offshore oilfield is located between 40 and 58 miles south-southwest of Cayos Arcas. A number of production platforms interconnected by submarine pipelines are located throughout this area. Mobile drilling rigs, platforms, and associated structures, sometimes unlit, may be encountered anywhere in the area. Numerous pipelines, many uncharted, exist within the oilfields and between them and shore.

An oil platform is located at 19°24′N, 92°02′W.

Akbatum A Oil/Gas Platform is located in position 19°17.9′N, 92°10.3′W. Contact "Marine Control" on VHF channels 9 and 16.

PLATFORM ABKATUM-O, 19 16.9N, 92 13.7W. Q W, 56ft, 10M. Four lights, one installed on each corner of platform.
PLATFORM AKAL-H, 19 22.4N, 92 01.9W. Q W, 56ft, 10M. Four lights, one installed on each corner of platform.
PLATFORM AKAL-L, 19 26.9N, 92 05.0W. Q W, 56ft, 10M. Four lights, one installed on each corner of platform.
PLATFORM AKAL-S, 19 21.7N, 92 04.2W. Q W, 56ft, 10M. Four lights, one installed on each corner of platform.
PLATFORM BATAB-1A, 19 17.6N, 92 18.6W. Q W, 56ft, 10M. Four lights, one installed on each corner of platform.
PLATFORM BATAB-A, 19 17.8N, 92 19.0W. Q W, 56ft, 10M. Four lights, one installed on each corner of platform.
PLATFORM CAAN-1, 19 12.2N, 92 05.1W. Q W, 56ft, 10M. Four lights, one installed on each corner of platform.
PLATFORM VECH-A, 19 05.8N, 92 31.9W. Q W, 56ft, 10M. Four lights, one installed on each corner of platform.
PLATFORM YUM-1, 18 46.9N, 92 34.1W. Q W, 56ft, 10M. Four lights, one installed on each corner of platform.
BAY OF CAMPECHE RADAR STATION PR-1, 18 56.6N, 92 37.3W. Q W, 56ft, 10M. Four lights, one installed on each corner of platform.
RACON C (– • – •).

YUCATÁN, WEST COAST

21 02N, 90 18W
Pilotage: Punta Boxcohuo is the northwest extremity of the Yucatán Peninsula. It is a low and sandy projection. Several shoals are reported to lie near the 10-meter-depth curve, about 9 miles seaward of the point. A 9-foot shoal lies 3 miles north of the point. A shoal, with depth unknown, is reported to lie 20 miles west-northwest of the lighthouse on the point. The section of coast forming the west side of the Yucatán peninsula and the east side of the Bay of Campeche extends 13 miles south-southwest from Punta Boxcohua to Celestún, then 60 miles farther south to the town of Campeche. A wide and shallow bank fronts this portion of the coast.

Between Campeche and Punta Morro, about 13 miles to the southwest, and Champoton, 20 miles farther south, the coast is bolder and backed by prominent ridges of hills.

Between Champoton and Punta Xicalango, about 80 miles to the southwest, the coast is low, wooded, and fronted by a continuous sandy beach. Laguna de Terminos indents the south part of the coast. For a distance of 30 miles southwest of Champoton numerous sand and shell patches, with depths of 14 to 30 feet, extend as far as 12 miles offshore.

From Punta Xicalango the coast extends to the west for about 45 miles to Punta Buey and is low and free of dangers. There are no ports of importance along this section of coast.

Winds: During the rainy season, from June to September, squalls blow at times with considerable force. The prevailing winds are from northeast to southeast.
Currents: There is seldom any current off the west coast of the Yucatán peninsula.

REAL DE LAS SALINAS

20 45N, 90 26W
Pilotage: This is the west entrance to the lagoon that parallels the north coast of the Yucatán peninsula. It lies about 6 miles south of Celestún.

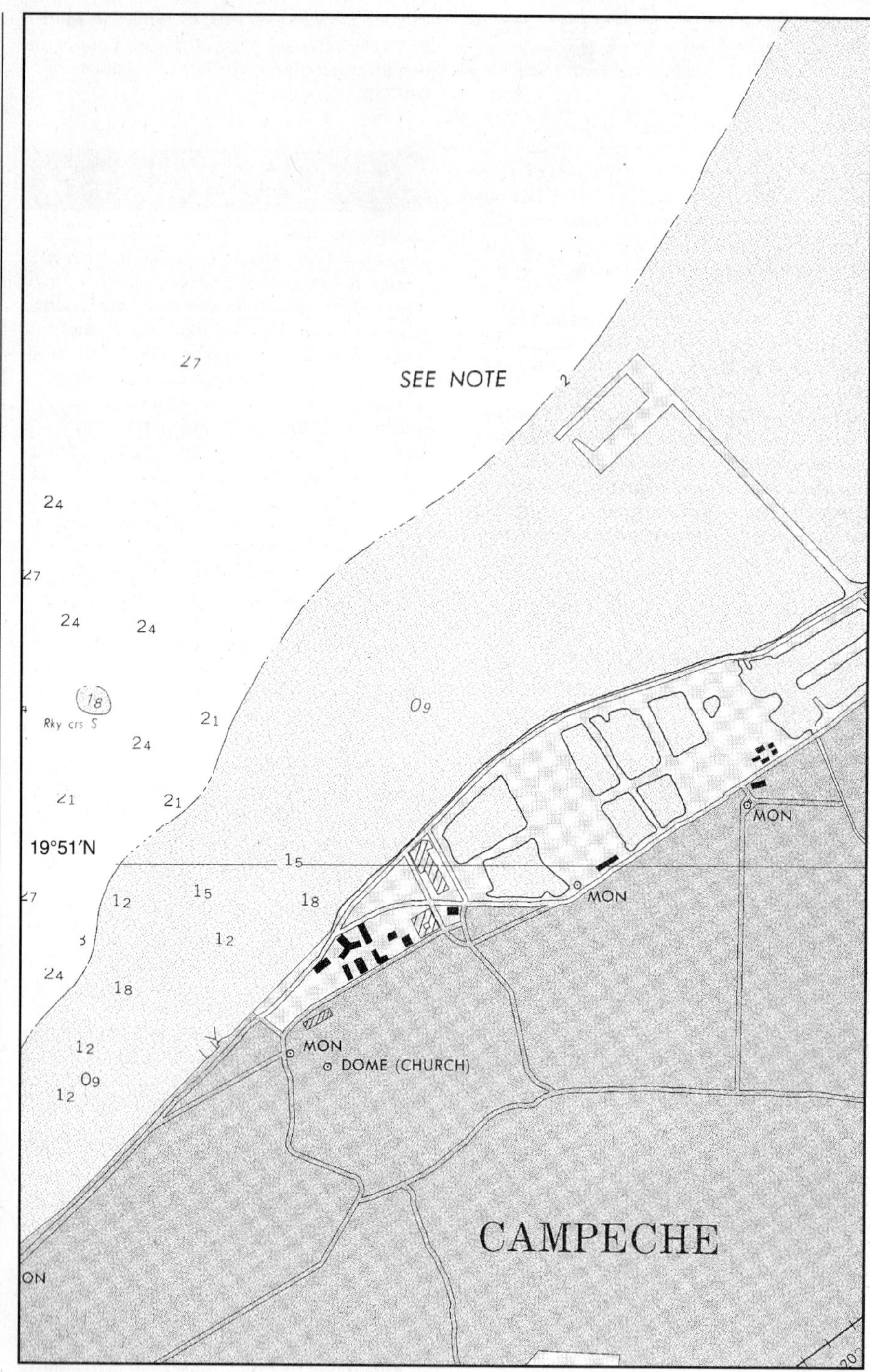

Campeche, soundings in meters *(from DMA 28265, 1st ed, 9/88)*

Between Real Salinas and Campeche, about 55 miles to the south, the low coast is bordered by swampy ground for about 25 miles and then becomes more elevated, rising to heights of 100 to 280 feet north of Campeche.

Isla de Piedras is about 30 feet high and located close offshore, about 23 miles north of Campeche.

SAN BARTOLO HILL, Campeche, 19 49.0N, 90 35.0W. Fl (2) W 12s, 256ft, 26M. Square white concrete tower, 39ft. ~~A F W light is shown from the mole~~.
LERMA RANGE (Front Light), 19 48.4N, 90 36.0W. Fl W 3s, 85ft, 8M. Truncated pyramidal metal tower, 48ft. **(Rear Light),** 190 meters, 139° from front. Iso W 2s, 245ft, 9M. Truncated pyramidal metal tower, 14.ft.
WHARF, at head, 19 49.0N, 90 35.6W. Iso W 2s, 30ft, 9M. White round tower, 20ft.
LERMA PIER NO. 2, 19 48.8N, 90 35.9W. Iso G 2s, 21ft, 6M. Round concrete tower, 13ft.
LERMA PIER NO. 1, 19 48.8N, 90 35.9W. Iso R 2s, 21ft, 6M. Round concrete tower, 13ft.
BRETON PIER NO. 3, 19 49.1N, 90 35.6W. Iso G 2s, 30ft, 6M. Green pipe, 21ft.
BRETON PIER NO. 2, 19 49.1N, 90 35.7W. Iso R 2s, 30ft, 6M. Red pipe, 21ft.

CAMPECHE

19 51N, 90 33W
Pilotage: This is the capital of the state of Campeche and a major sisal-exporting port. It stands on a plain bordered on three sides by a small amphitheater of hills. Two forts stand 1.3 miles northeast, and two similar forts stand 1.8 miles southwest of the city. A signal station is located atop an old fort near the waterfront. A blue flag indicates bad weather and a red flag indicates that the port is closed.

A pier extending from the shore abreast of the city was reported to have a dredged channel with a depth of 18 feet leading up to it.
Port of entry

Muelle Castillo Breton, a new 550-yard-long pier with a depth of 12 feet alongside stands at the village of Lerma, about 3.5 miles southwest of Campeche.

Puerto de Abrigo is situated about 0.3 mile southeast of Muelle Castillo Breton, and there is a small pier 0.3 mile northeast of the same jetty.

Between Campeche and Punta Mastun Grande, a wooded 423-foot-high headland about 9 miles southwest, and Punta Morro, about 5 miles farther south-southwest, the coast is bold and backed by ridges of hills.

CAUTION: Dangerous wrecks lie 10 miles northwest and 18 miles west of Punta Mastun.

Between Punta Morro and the Río Champoton, about 20 miles to the south, the coast is backed by a ridge of hills that are more broken than those to the north. The ridge terminates in a prominent 360-foot-high hill about 8 miles south of the river.

PUNTA MORRO, 19 40.6N, 90 42.2W. Fl (3) W 6s, 177ft, 27M. White octagonal concrete tower, 52ft. Visible 000°–197°.
RÍO CHAMPOTON, S side of entrance, 19 21.6N, 90 43.3W. Fl W 5s, 82ft, 25M. White square concrete tower, 72ft. Visible 045°–180°.

CHAMPOTON

19 22N, 90 43W
Pilotage: This is a small village on the south side of the entrance to the Río Champoton. An old fort and two churches stand in the town. Observation Cay, 5 feet high, stands close off the river entrance.

Between the Río Champoton and Barra de Puerto Real, about 55 miles to the southwest, the low, wooded coast is fronted by a continuous stretch of sandy beach. Numerous sand and shell patches, with depths of 8 to 18 feet, lie up to 5 miles off this section of coast.

BANCOS CHAMPOTON

19 23N, 90 50W
Pilotage: This is a group of detached patches with depths of 14 to 27 feet extending 10 to 15 miles west from Champoton.

BANCOS DE SABUNCUY

19 10N, 91 16W
Pilotage: This is a group of patches with depths of 14 to 30 feet lying about 32 miles southwest of Champoton. They extend up to 12 miles offshore.

SABUNCUY, 19 00N, 91 11W. Fl (3) W 12s, 59ft, 11M. White round concrete tower, 33ft.

LAGUNA DE TERMINOS

ISLA AGUADA RANGE (Front Light), on Punta del Tigre entrance, 18 47.2N, 91 29.7W. Fl W 3s, 39ft, 11M. White round concrete tower, 39ft. **(Rear Light),** 400 meters, 151° from front, Fl (4) W 12s, 75ft, 18M. White concrete tower, 66ft.

BOCA DE LOS PARGOS, 18 37.5N, 91 17.2W. Fl W 6s, 26ft, 10M. White concrete tower, 23ft.

LAGUNA DE TERMINOS, Tio Campo, S end, 18 30.7N, 91 47.3W. Fl W 6s, 33ft, 10M. Truncated conical tower, 30ft.

PUNTA DEL ZACATAL, 18 36.7N, 91 51.4W. Fl W 5s, 49ft, 17M. Truncated pyramidal aluminum tower, 46ft.

PUNTA ATALAYA, W end of Isla del Carmen, 18 38.7N, 91 50.7W. Fl (3) W R 12s, 69ft, 21M. White truncated conical masonry tower, red cupola, 66ft. R 086°–113°, W 113°–086°.

BARRA DE PUERTO REAL

18 47N, 91 30W

Pilotage: This is the east entrance to Laguna de Terminos. It is 2 miles wide and has depths ranging from 6 to 12 feet. A pair of lighted beacons in range 152° stand on Punta del Tigre on the north side of the entrance. A shallow, breaking spit extends 2.5 miles west-northwest from the point. A channel for vessels drawing less than 10 feet leads to an anchorage off Punta del Tigre, but should not be attempted without local knowledge.

Isla del Carmen, low and sparsely wooded, fronts Laguna de Terminos between Barra de Puerto Real and Barra Principal, the west entrance of the lagoon, which is located about 20 miles west-southwest.

CARMEN

18 39N, 91 50W

Pilotage: Barra Principal is about 6 miles wide between Punta Atalaya, 0.5 mile southeast of the southwest end of Isla del Carmen, and Punta Xicalango to the northwest. A breaking spit extends 4.5 miles northwest from Punta Atalaya, and depths of 6 to 12 feet extend 3 miles north from Punta Xicalango. A buoyed channel, with depths of 12 to 14 feet, leads through these shoals to the town of Carmen on the southeast side of Punta Atalaya. Two lighted beacons in range 180° stand on the west side of the entrance and indicate the fairway of this channel.

CAUTION: Considerable changes to depths, navigation aids, and the coastline have taken place in the approaches to Carmen. The most recent editions of appropriate charts should be consulted.

The wharves that front the town have a depth of 11 feet alongside. A pier, with a depth of 18 feet alongside, extends from the shore at the naval base near the southeast end of the town. In 1979 an artificial harbor was under construction in the area north of Punta Atalaya.

Between Barra Principal and the Río Grijalva, about 46 miles to the west, the coast is low and has no prominent features.

RÍO SAN PEDRO, 18 39.1N, 92 27.9. Fl G 5s, 30ft, 8M. Green round concrete tower, 30ft.

~~***Entrance Buoy,*** 18 41N, 92 42W. Fl W 2s. Black buoy.~~

DARSENA DE ASTATA, 18 42.0N, 92 05.3W. Fl W6s, 30 ft, 9M. White round concrete tower, 26ft.

CAMPO PESQUERO, 18 41.0N, 92 13.2W. Fl W 6s, 30 ft, 9M. White round concrete tower, 26ft.

XICALANGO RANGE (Front Light), 18 37.8N, 91 54.0W. Fl W 3s, 43ft, 11M. Truncated pyramidal metal tower, 39ft. **(Rear Light),** 185 meters, 180° from front. Fl W 6s, 105ft, 20M. Red round concrete tower, 92ft.

LAGUNA AZUL, E breakwater, 18 37.8N, 91 50.0W. Fl R 5s, 20ft, 4M. Red round metal tower, 13ft.

LAGUNA AZUL, W breakwater, 18 37.4N, 91 50.0N. Fl G 5s, 20 ft, 4M. Green round metal tower, 13ft.

GOVERNMENT PIER, at head, N side, 18 38.2N, 91 50.2W. Iso G 2s, 26ft, 4M. Round concrete tower, 16ft.

GOVERNMENT PIER, S side, 18 38.2N, 91 50.2W. Iso G 2s, 26ft, 4M. Round concrete tower, 16ft.

FRONTERA, 18 36.8N, 92 41.5W. Fl W 6s, 98ft, 14M. White skeleton iron tower with eight columns, 95ft. Visible 045°–245°. Two F W range lights mark entrance to Grijalva Canal. Radar reflector.

EAST BREAKWATER, 18 37.1N, 92 41.2W. Fl G 5s, 23ft, 7M. White truncated pyramidal concrete tower, 16ft.

PUNTA BUEY

18 39N, 92 43W
Pilotage: This low point has been reported to be extending north and northwest. Depths of 18 to 24 feet extend about 1.5 miles north and northwest from the point.

The coast recedes about 8 miles southwest from Punta Buey, then extends in a general west-southwest direction for about 95 miles to Coatzacoalcos, then west for 10 miles, then north-northwest for an additional 25 miles to Punta Zapotitlan. This low, marshy coast is indented by several lagoons and is covered with very heavy vegetation except in the tidal marshes. Several rivers discharge into the gulf, and hills stand 3 to 8 miles inland on the west part of this coast.

Winds: The winds tend to blow from the north. Northers occur at about eight-day intervals between October and March.
Current: From October to March the current near the shore sets east at a rate of 1 to 1.5 knots.

FRONTERA (ALVARO OBREGÓN)

18 35N, 92 39W
Pilotage: Puerto Frontera stands on the east bank of the Río Grijalva about 5 miles south of the entrance. The port is used mainly by small coast vessels. Depths of less than 18 feet extend about 0.8 mile north and 1 mile northwest from the east entrance point, and similar depths extend about 1.5 miles north and 0.5 mile west from Punta Buey. The bar lying north of the entrance points has been reported to be dredged to a depth of 20 feet. The buoyed channel within the river has a navigable width of about 250 yards with depths ranging from 32 to 40 feet. The preferred channel passes east of Isla Buey about 3 miles south of the entrance.

Grijalva Canal, which crosses the peninsula about 1.5 miles south of Punta Buey, was formerly used to enter the river. In 1974 this canal was reported no longer in use.

When the river is at its highest level the bar has the least depth over it, and when the river is at its lowest the channel is scoured out, so the depths are better. At the end of the rainy season in December, the least depths may be expected, but the first norther will increase the depth by about 3 feet

There is only one tidal rise every 24 hours, and the rise at springs is only about 2 feet. An overhead cable north of Frontera has a clearance of about 118 feet. Fiscal Wharf parallels the shore abreast town, with depths of 18 to 20 feet alongside. Anchorage can be taken in the river off the town, according to the direction of the port authority.
Port of entry

Between Punta Buey and Coatzacoalcos, about 100 miles to the west-southwest, the coast is low, fairly steep-to, and bordered by mangroves and palm trees.

CHILTEPEC, at river entrance, 18 26.3N, 93 05.7W. Fl (3) W 12s, 49ft, 14M. White round concrete tower, 43ft.
~~**DOS BOCAS,** 18 26.1N, 93 11.5W.~~
~~**RACON B** (-----)~~. Removed 4/97

RÍO GONZALEZ

18 26N, 93 04W
Pilotage: This river is located about 23 miles southwest of Punta Buey. It is shallow, but is navigated by small-draft craft up to 90 miles inland. Barra de Chiltepec is the east entrance of this river and has the village of Chiltepec on its west entrance point. Small coastal vessels frequent this port.

Barra de Dos Bocas is the entrance to a lagoon about 3.5 miles west of the Río Gonzalez. A dangerous wreck lies about 9 miles northeast of Barra de Dos Bocas.

Dos Bocas Terminal, 5 miles west of the Chiltepec lighthouse, lies at the end of the pipeline from the marine oil field to the north. See the previous section on the Cayo Arcas Terminal. The terminal here has a 3,600-foot-long berth and three moorings for large tankers located 11 miles to the north. Pipelines connect the tanker moorings to shore. Anchoring is prohibited within an area shown on the chart enclosing the moorings and pipelines.
TUPILCO, 18 25.0N, 93 26.2W. Fl (4) W 16s, 79ft, 14M. White round concrete tower, 75ft.

BARRA DE TUPILCO

18 26N, 93 25W
Pilotage: This is a shallow entrance to a lagoon. It is located about 16 miles west of Barra de Dos Bocas. A tower stands on the west side of the entrance.
Laguna del Carmen is located about 26 miles west of Barra de Tupilco. In 1976 an awash rock was reported to lie 17 miles north-northwest of Laguna del Carmen.

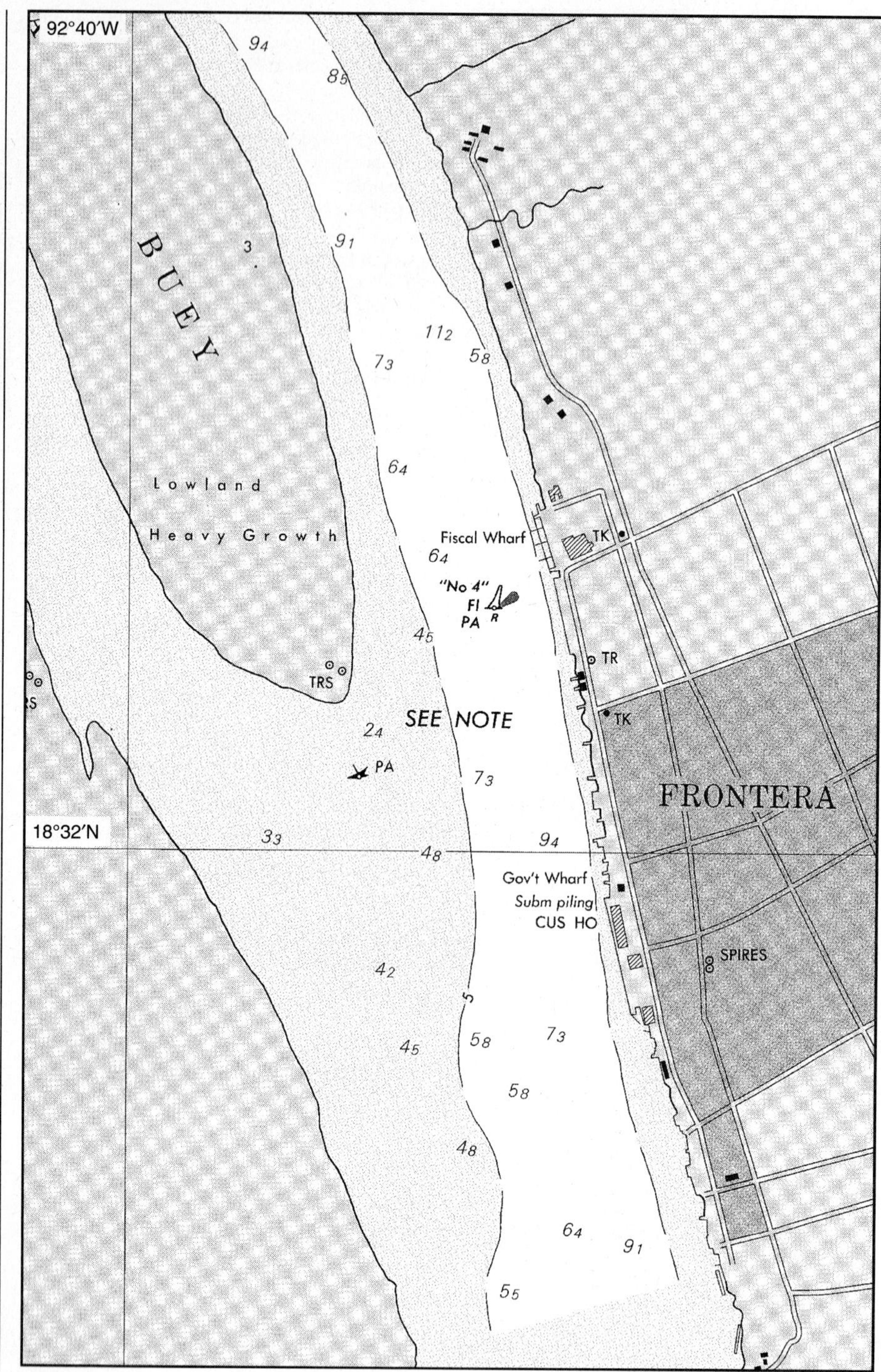

Frontera, soundings in meters *(from DMA 28261, 8th ed, 7/88)*

TONALA, W side of entrance to Río Tonala, 18 12.8N, 94 07.9W. Fl (3) W 12s, 75ft, 16M. Red square masonry tower with white bands, 56ft. ~~Visible 081°–246°.~~ ~~Two F W range lights mark entrance to Tonala River.~~

RÍO TONALA

18 13N, 94 08W

Pilotage: This river has a bar with depths of under 8 feet at its entrance. It lies about 16 miles west of Laguna del Carmen. Each entrance point is marked by a conspicuous sand hill. Local knowledge is needed to enter the river. Vessels anchor here to load mahogany.

COATZACOALCOS

CERRO DEL GAVILAN, 18 09.0N, 94 24.0W. Fl (2)W 18s, 177ft, 20M. White prismatic octagonal masonry tower with dwelling, 46ft. Radar reflector. ~~Iso R 2s marks top of mast~~.

RANGE, E side of river **(Front Light),** 18 08.2N, 94 24.3W. Fl W 3s, 144ft, 9M. White truncated pyramidal metal tower with orange bands. **(Rear Light),** 290 meters, 162° from front. Iso.W 2s, 167ft, 9M. White truncated pyramidal metal tower with orange bands.

PIEDRA NANCHITAL, 18 06.0N, 94 25.0W. Fl (2) W 10s, 33ft, 10M. White framework tower, 29ft.

EAST BREAKWATER HEAD, 18 10.0N, 94 4.8W. Fl G 5s, 56ft, 7M. Red truncated pyramidal metal beacon, white band, 49ft.

WEST BREAKWATER HEAD, 18 10.0N, 94 24.9W. Fl R 5s, 46ft, 7M. Red truncated pyramidal metal beacon, white band, 46ft.

DARSENA DE PAJARITOS RANGE (Front Light), 18 07.3N, 94 24.3W. Fl Blue 3s, 52ft, 10M. White round metal tower with red bands, 49ft. **(Rear Light),** 200 meters, 180° from front. Iso W 2s, 82ft, 10M. White truncated pyramidal metal tower with red bands, 74ft.

BERTH 6 RANGE (Front Light), 18 07.5N, 94 24.4W. Q Y. White metal column, red bands. **(Rear Light),** 100 meters, 270° from front. Fl (3) Y 10s. White metal column, red bands.

BASIN LIGHT, E, 18 07.9N, 94 24.2W. Fl G 3s. Metal column, 16ft.

BASIN LIGHT, w, 18 07.9N, 94 24.3W. Fl R 3s. Metal column.

Note: Within an aid listing, underlined text indicates data that has changed or is new since September, 1996; ~~Strike out~~ text indicates deleted data.

COATZACOALCOS

18 08N, 94 25W

Pilotage: This port was formerly known as Puerto Mexico. It lies on the west bank of the Río Coatzacoalcos, close within the entrance. There is a free port on the same side of the river, south of town. The town, with its buildings and radio towers, is prominent from seaward and can be identified during the day by smoke and at night by the loom of its lights. Approaching from seaward several high dark green hills are observed east of the entrance. On the east side of the entrance is a round hill. Westward of the entrance appear some low sand hills, the tops of which are covered with green vegetation. Low land fronts the hills on either side. The water is muddy for about 1.5 miles outside the heads of the breakwater. The breakwaters are conspicuous, and the east side of the river mouth is a radar target at 40 miles. The range leads up the main channel bearing 162°.

This port is occasionally closed for up to several days during strong northers, which generally occur from November to March. These storms can raise the water level considerably in the port. Maximum draft crossing the bar is 37 feet, but it has been reported that the depths over the bar are liable to decrease considerably because of unusually strong currents setting to the west. When the river is flooding from June to October, depths are liable to decrease even more.

The wharf along the west side of the river is over 6,200 feet long, with depths of over 28 feet alongside. There are strong currents in the vicinity of the wharves in the river. The inner harbor basin on the east side of the river is called Darsena de Pajaritos. Several oil loading facilities lie upriver.

Between Coatzacoalcos and the Río Barilla, about 12 miles to the west, the coast is low and unvaried. The Río Barilla is the entrance to a lagoon, which is connected to Coatzacoalcos by a river south of the city.

Port of entry

Tides and currents: Offshore the current sets northwest, but near the breakwaters it sets east. The current in the river varies with the stage of the tide, attaining its maximum rate of 5 to 5.5 knots about 2 hours after high water. During the first 3 hours of the flood the rate ranges from 2.5 to 3 knots. The rise of the tide is about 2 feet.

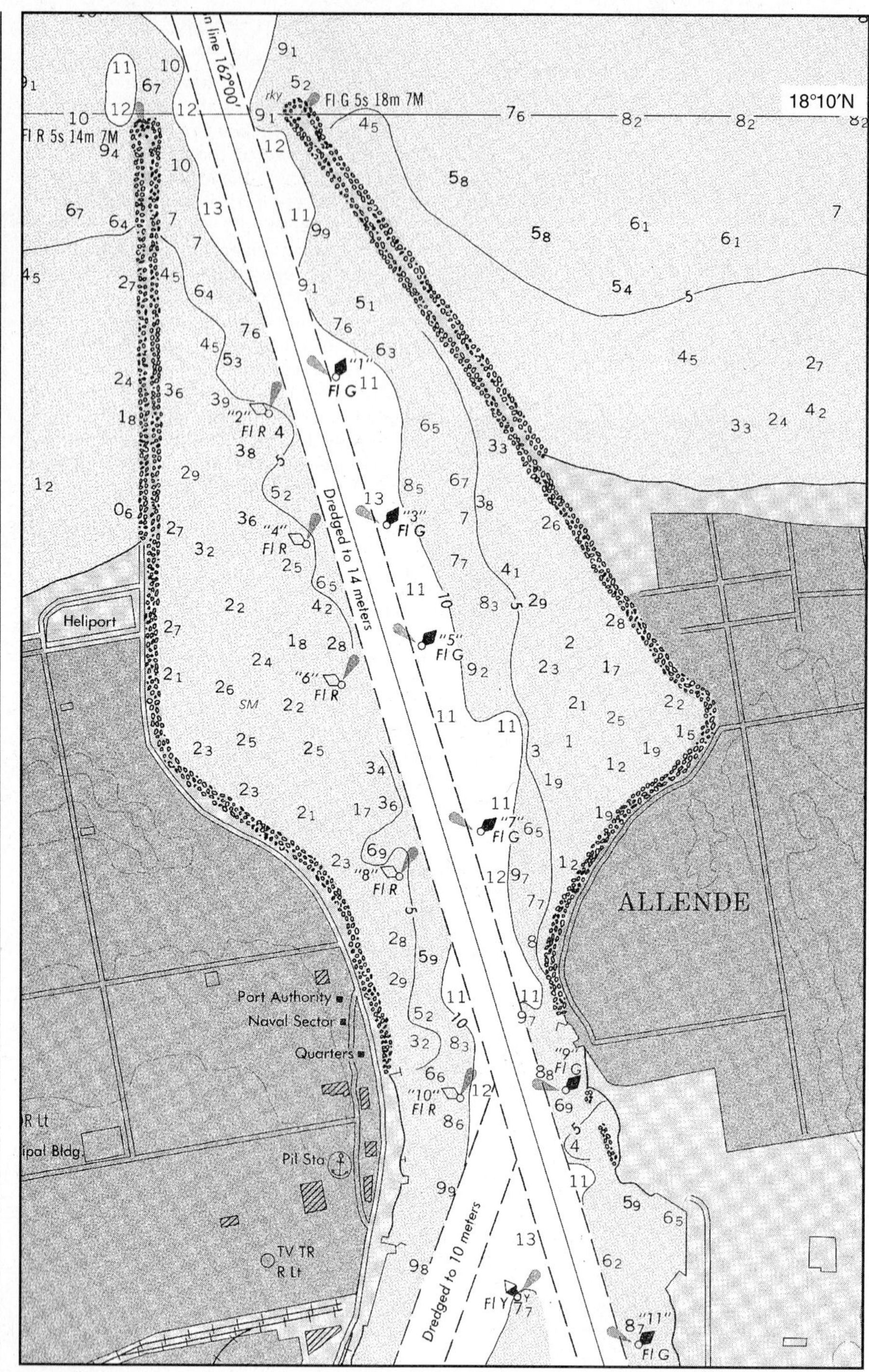

Coatzacoalcos, soundings in meters *(from DMA 28281, 31st ed, 4/92)*

PUNTA SAN JUAN

18 17N, 94 37W

Pilotage: This point is located about 5 miles north of the Río Barilla and has a small islet lying close off it.

Between Punta San Juan and Punta Zapotitlan, about 20 miles north-northwest, the coast is backed by mountain ranges about 3 to 8 miles inland. Cerro San Martin, conspicuous from seaward, stands about 8 miles inland west of Punta San Juan.

PUNTA ZAPOTITLAN, 18 31.4N, 94 48.0W. Fl W 6s, 98ft, 20M. Truncated conical concrete tower, 92ft. ~~Visible 116°–324°~~.

PUNTA ZAPOTITLAN

18 33N, 94 48W

Pilotage: This prominent point is bordered by a reef that extends 0.5 mile offshore. Depths of greater than 60 feet lie about 1.5 miles off the point. An old disused lighthouse stands near the lighthouse on the point.

The coast between Punta Zapotitlan and Cabo Rojo extends in a general northwest direction for about 230 miles. Mountain ranges with some conspicuous peaks back the coastal plain. The more prominent coastal features and dangers are lighted. Numerous islets, shoals, and other dangers lie off this section of the coast.

Between Punta Zapotitlan and Alvarado, about 56 miles west-northwest, the first 33 miles of coast are backed by ranges of hills about 3 to 8 miles inland. The remaining part of the coast is lower, being composed of sand hills 50 to 200 feet high.

Barra Zontecomapan obstructs the entrance to the Laguna Coxcoapan, about 10 miles west of Punta Zapotitlan. This bar has a depth of about 6 feet over it. The entrance may be identified by a bluff a little to its west.

Punta Morrillo is a bold, rounded bluff standing about 15 miles west-northwest of Punta Zapotitlan.

Punta Roca Partida (18°42′N, 95°11′W) is a group of perpendicular cliffs that stand about 8 miles west-northwest of Punta Morrillo. A rocky islet stands close off the point.

Volcan San Martin Tuxtla (18°33′N, 95°12′W) is a 4,600-foot-high volcano about 10 miles south of Punta Roca Partida. This can be seen from a great distance in clear weather. When active, the column of smoke by day and the flames at night make this volcano an excellent landmark.

Currents: The currents off this coast are variable and uncertain. They are usually dependent on the force and direction of the wind. The coastal current usually sets south in the winter and north in the summer.

Winds: The trade winds blow from northeast to east-northeast. After the season of the northers, from October to March, there are light north breezes, calms, squally rains, and thick weather up to the middle of August, when the trades resume again.

ROCA PARTIDA, 18 42.2N, 94 11.2W. Fl (4) W 18s, 318ft, 22M. Square concrete tower, 43ft. ~~Visible 112°–294°~~.

ALVARADO, GASODUCTO DE PEMEX, 18 47.0N, 94 45.0W. Fl Y 2s, 29ft, 11M. Yellow round tower, 13ft.

EAST FORTIN (Punta Alvarado), on point SE of town, 18 46.3N, 95 45.9W. Fl W 6s, 30ft, 13M. Square concrete tower, 22ft. ~~Visible 005.5°–021.5°~~.

ALVARADO, on E entrance point of Río Papaloapan, 18 46.9N, 95 44.7W. Fl (3) W 12s, 125ft, 21M. Square concrete tower and dwelling, 26ft. Visible 136°–274°.

ALVARADO, E breakwater, 18 47.4N, 95 44.5W. Fl G 5s, 29ft, 11M. Green aluminum tower, 24ft.

ALVARADO, W breakwater, 18 47.7N, 95 45.0W. Fl R 5s, 29ft, 11M. Red aluminum framework tower, 24ft.

ALVARADO

18 47N, 95 46W

Pilotage: This is primarily a fishing port. It lies within the entrance to a lagoon of the same name located about 32 miles west-northwest of Punta Roca Partida. A 14-foot-deep shoal lies 1.8 miles north of the east entrance point. Two patches with depths of 8 and 7 feet lie in mid-channel about 0.3 mile northwest and 0.8 mile south-southwest of the east entrance point. Depths of less than 36 feet extend 1 mile north of the east entrance point.

The bar is situated about 0.5 mile outside the entrance points. It is constantly shifting and the sea nearly always breaks on it. In 1985 a depth of 26 feet was reported over the bar.

A conspicuous sandy bluff and a beacon stand on the east entrance point. A beacon stands on the west entrance point.

The coast between Alvarado and Punta Coyal, about 20 miles northwest, is bordered by two large lagoons along its southern half and by low land along its northwest half.

ANEGADA DE AFUERA, on NW point of reef, 19 10.3N, 95 52.2W. Fl (4) W 16s, 33ft, 10M. Red truncated pyramidal metal tower with black bands, 39ft.
ARRECIFE SANTIAGUILLO, center of island, 19 08.6N, 95 48.4W. Fl (2) W R 10s, 118ft, 22M. Red concrete cylindrical tower. R 002°–056°, W 056°–084°, R 084°–126°, W 126°–002°.
RACON O (– – –), 25M.
ARRECIFE DE CABEZO, N end, 19 05.6N, 95 51.9W. Fl (2) W 10s, 30ft, 11M. Square concrete tower, 23ft. ~~Reported extinguished (1989)~~.
ARRECIFE DE CABEZO, S end, 19 03.1N, 95 49.7W. Fl R 6s, 26ft, 8M. Square concrete tower, 23ft.
ARRECIFE EL RIZO, S end, 19 03.3N, 95 55.2W. Fl (3) W 12s, 39ft, 11M. Rectangular concrete tower, 33ft.

PUNTA COYAL

19 03N, 95 58W

Pilotage: This blunt point is composed of low sand hills. A 262-foot-high sand hill stands about 3 miles west of the point. The village of Anton Lizardo stands on the north side of this point. Arrecife El Giote extends 0.8 mile from the coast close west of Anton Lizardo.

Anton Lizardo Anchorage (19°04′N, 95°59′W) lies between Arrecife Chopas and Punta Coyal. This anchorage is used by commercial vessels sheltering from northers.

Between Punta Coyal and Punta Mocambo, about 9 miles to the northwest, the coast recedes about 2 miles to form a sandy bay bordered by low sand hills. A reef lies about 0.8 mile southeast of Punta Mocambo. The Río Jamapa, with depths of 3 to 6 feet over the bar, discharges into the gulf about 2.5 miles south of Punta Mocambo.

The coast between Punta Mocambo and Puerto Veracruz, 4 miles to the northwest, is low and sandy.

ARRECIFE SANTIAGUILLO

19 09N, 95 48W

Pilotage: This reef stands at the northeast end of a group of reefs located 11 miles east-northeast of Punta Coyal. It stands about 8 feet high.

Arrecife Anegadilla is the outermost reef of the group off Punta Coyal. It lies 0.5 mile east-southeast of Arrecife Santiaguillo.

ARRECIFE ANEGADA DE AFUERA

19 09N, 95 51W

Pilotage: This reef is about 2.5 miles long, and it lies about 1.5 miles west-northwest of Arrecife Santiaguillo. There is a cay on a small reef close off its south end.

Arrecife Cabeza is about 3.5 miles long and 1.5 miles wide. It lies about 4 miles south-southwest of Arrecife Santiaguillo.

Arrecife de Enmedio, with shoal patches close off its west side and a small reef and another patch close off its north end, lies about 3.5 miles northeast of Punta Coyal. A small cay stands on the south end of the reef.

Arrecife Rizo, about 1.5 miles long, lies 1.5 miles south of Arrecife de Enmedio. A spit with a depth of 10 feet over its outer end extends a little more than 0.5 mile north from the north extremity of Arrecife Rizo.

ANTON LIZARDO ANCHORAGE

BLANCA REEF, 19 05.0N, 96 00.0W. Fl (2) W 10s, 35ft, 11M. Concrete pyramidal tower.
EL GIOTE REEF, 19 04.0N, 96 00.0W. Fl R 6s, 30ft, 8M. Square concrete tower, 24ft.
~~**ISLA SALMEDINA,** S part, 19 04.1N, 95 57.2W. Fl W 1.5s, 23ft, 3M. Metal mast, black and red bands.~~
ISLA DE ENMEDIO, 19 06.0N, 95 56.3W. Fl (3) W R G, 46ft, W 13M, R 8M, G 7M. Truncated conical concrete tower, with dwelling, 46ft. W 147°–220°, R 220°–260°, W 260°–268°, G 268°–299°, W 299°–327°, R 327°–032°, W 032°–057°, R 057°–147°.

ARRECIFE CHOPAS

19 05N, 95 58W

Pilotage: This reef is about 3.3 miles long and lies about 2 miles north of Punta Coyal. A small reef lies close south of a grass-covered cay on the south end of the reef. Several small reefs lie close off the north and northwest ends.

Arrecife Blanca is about 0.3 mile in extent with a small cay on it. It lies about 1 mile west of Arrecife Chopas.

NOTE: Vessels without local knowledge should not attempt the passages between Arrecife Rizo, Arrecife de Enmedio, and Arrecife Chopas. Foul ground lies in the passage between Arrecife Chopas and Arrecife Blanca; this passage should not be used.

ARRECIFE ANEGADA DE LA ADENTRO

19 14N, 96 04W

Pilotage: This is the outermost reef in the approaches to Veracruz. These reefs all break and are easily identified in clear weather. This one is located 4 miles east-northeast of the harbor entrance.

Isla Verde is a low white cay standing on the south end of a reef about 1.5 miles south-southwest of Arrecife Anegada de la Adentro. The intervening channel is about 1 mile wide and deep.

Bajo Paducah (19°12′N, 96°05′W) is a shoal with a least depth of almost 15 feet. It lies about 0.3 mile west of the north end of the reef on which Isla Verde stands.

Arrecife de Pájaros is about 1 mile long and lies about 1.3 mile west-southwest of Isla Verde.

Isla de Sacrificios is a small cay on the south end of a reef close south of Arrecife de Pájaros. The passage between here and Arrecife de Pájaros is narrow and foul.

Bajo Mersey (19°11′N, 96°06′W) has a least depth of 15 feet.

Arrecife Blanquilla lies about 1.8 miles west of Arrecife Anegada de la Adentro. The channel between the two reefs is 1.5 miles wide and deep. The channel west of Arrecife Blanquilla has depths of 60 feet and is 0.8 mile wide.

Arrecife de la Gallega extends 1.3 miles north from the north side of the harbor entrance, and Arreciffe Galleguilla lies close northeast of its north end.

Arrecife Hornos extends about 0.3 mile east from the root of the southeast breakwater.

RÍO JAMAPA, N breakwater, 19 06.2N, 96 05.8W. Fl R 5s, 29ft, 7M. Red metal tower, 21ft.
RÍO JAMAPA, S breakwater, 19 06.0N, 96 05.9W. Fl G 5s, 23ft, 7M. Green metal tower, 21ft.

VERACRUZ

LA GALLEGUILLA, on E side of reef, 19 13.8N, 96 07.3W. Fl (2) 10s, 36ft, 15M. White cylindrical concrete tower, 30ft.
ARRECIFE BLANQUILLA, 19 13.6N, 96 06.1W. Fl (4) G 16s, 49ft, 9M. Red truncated conical concrete tower with white bands, 46ft. Notice to Mariners indicates change from Red to Green light. We are dubious!
ARRECIFE LA BLANQUILLA, S side of reef, 19 13.3N, 96 05.8W. Fl (2) R 10s, 46ft, 9M. Red cylindrical concrete tower, 46ft. ~~Occasional~~. ~~**RACON B (– · · ·)**~~.
ANEGADA DE ADENTRO, NW extremity, 19 13.7N, 96 03.7W. Fl (3) G 12s, 36ft, 9M. Green round concrete tower, 31ft.
ISLA VERDE, on bands S of island, 19 11.9N, 96 04.9W. Fl (4) W 16s, 26ft, 15M. Red truncated conical concrete tower, 21ft.
ARRECIFE PAJAROS, NW end, 19 11.6N, 96 05.7W. Fl W 6s, 23ft, 15M. White truncated conical concrete tower, 21ft.
ISLA DE SACRIFICIOS, 19 10.4N, 96 05.5W. Fl W 15s, 127ft, 25M; Fl (2) W R G 15s, 46ft, W 22M, R 11M, G 9M. White round concrete tower, black bands, 143ft. Lower light: R 134°–157°, W 157°–163°, G 163°–187°, W 187°–195°, R 195°–238°, W 238°– 334°, obscured 334°–134°. **RACON Z** (– – · ·), 25M.
ARRECIFE HORNOS, 19 11.5N, 96 07.3W. Fl (2) W 10s, 10ft, 7M. Black framework tower with two red bands, 10ft.
EL MAREGRAPO, 19 11.4N, 96 07.4W. Fl G 2s, 19ft, 6M. Green square concrete tower.
HEAD OF NORTHEAST BREAKWATER, 19 12.2N, 96 07.2W. Fl R 5s, 39ft, 10M. Red truncated pyramidal concrete tower, 23ft.
HEAD OF SOUTHEAST BREAKWATER, 19 12.0N, 96 07.3W. Fl G 5s, 33ft, 10M. Green truncated pyramidal concrete tower, 23ft.
HEAD OF INNER BREAKWATER NO. 3, 19 12.1N, 96 07.6W. Iso W 2s, 49ft, 11M. White concrete tower with small house.
PILOT PIER, N side, 19 12.0N, 96 07.9W. Iso G 2s, 16ft, 6M. White concrete tower.
PILOT PIER, S side. Iso G 2s, 16ft, 6M. White concrete tower, 8ft.
MUELLE DE PEMEX, E head, 19 12.2N, 96 07.4W. Iso R 2s, 13ft, 6M. Red tower.
MUELLE DE PEMEX, W side, 19 12.2N, 96 07.4W. Iso R 2s, 13ft, 6M. Red tower.
CATHEDRAL, dome, 19 11.9N, 96 08.2W. Iso W 2s, 102ft, 25M. On cathedral.

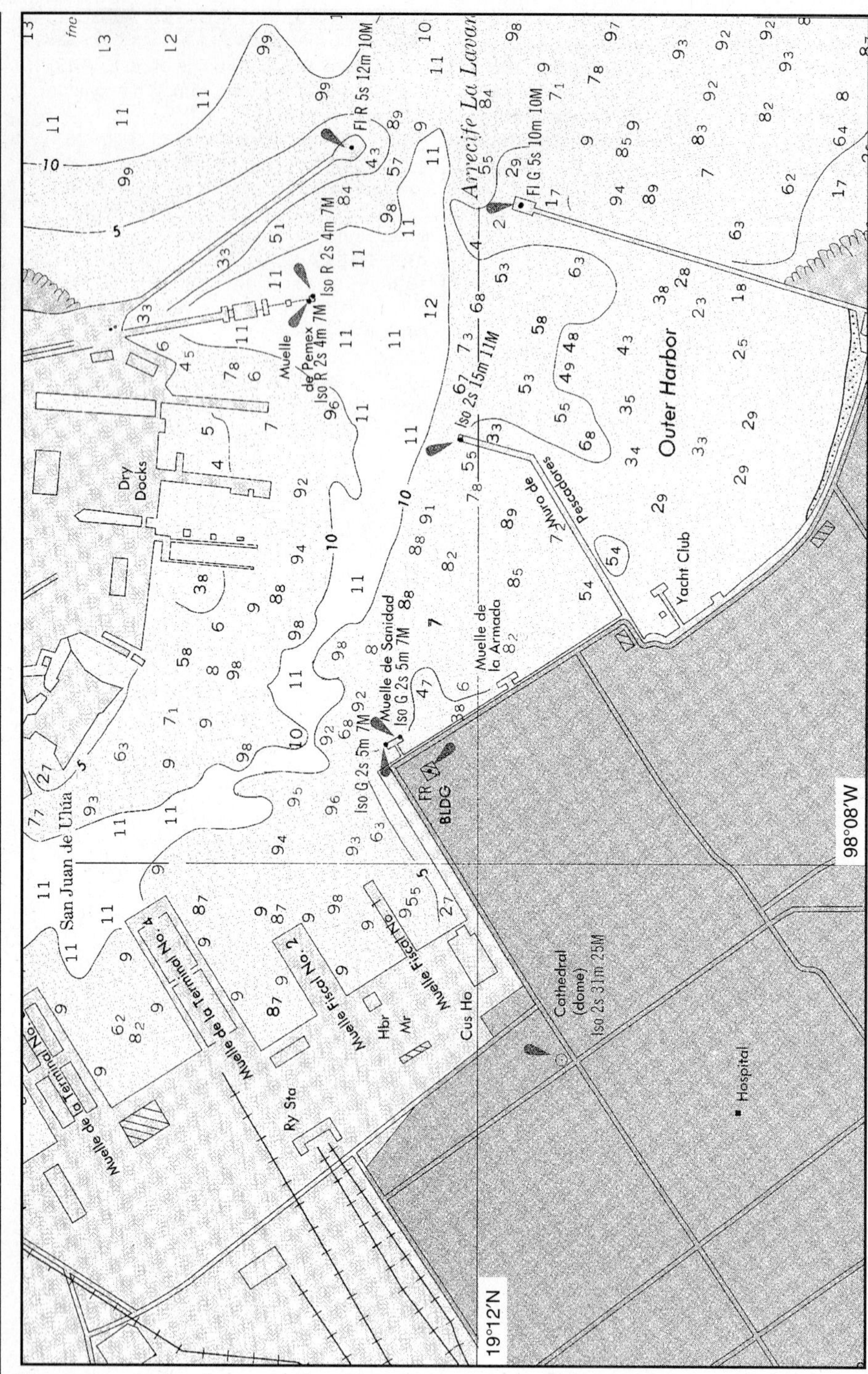

Veracruz, soundings in meters *(from DMA 28302, 14th ed, 8/91)*

PUERTO VERACRUZ

19 12N, 96 08W

Pilotage: This is one of the principal ports of Mexico. The harbor is an artificial basin protected by breakwaters on three sides.

Volcan Citaletepet (Pico de Orizaba) is a 17,392-foot-high volcano located 63 miles west of the city. The volcano is inactive, but its 3.5-mile-diameter crater can easily be distinguished from a considerable distance on a clear day. Cerro Nauhcampatepetl (19°29′N, 97°08′W) is a 14,040-foot-high mountain with a peculiar shape. It stands about 25 miles north-northeast of the above volcano. Snow falls on isolated spots at higher elevations.

A prominent high radio tower stands about 0.8 mile south of the root of the southeast breakwater. To enter the harbor steer for the cathedral dome about 0.3 mile west-southwest of Benito Juarez tower on a heading of 261°. At night steer for the light on Muro de Pescadores on a heading of 270° until the light structure on the east corner of Muelle de la Terminal bears 290°, which will lead through the entrance in mid-channel between the breakwater heads.

Storm signals are displayed from a red and white banded flagstaff on Benito Juarez tower.

Port of entry

Winds: With the exception of land breezes at night, the winds usually blow from seaward and are heavily saturated with moisture. March and April are the least humid months. Offshore the tradewinds blow from northeast to east-southeast. During the season of the northers, from October to March, winds of up to 50 knots occur, making it impossible for vessels to enter the harbor. After April, there are light north winds, calms, squally rains, and unsettled weather, until about the middle of August when the tradewinds resume. Land and sea breezes alternate regularly, even in the intervals between northers. The former begin shortly after sunset and the latter at about 0900.

Tides and currents: The diurnal range of the tide is about 1.6 feet, but the water level is influenced by the force and direction of the winds. The tidal currents are weak and are overcome by the coastal currents, which are also affected by the winds. During the winter, the current usually sets south, and in the summer it sets north.

VERACRUZ TO RÍO TUXPAN

Pilotage: Between Arrecife de la Gallega, which extends north from Puerto Veracruz, and Punta Gorda about 3 miles northwest, the coast recedes about 2 miles, forming Bahía de Veragua.

From Punta Gorda the coast extends about 6.8 miles west-northwest to Punta Antigua, on the south side of the entrance to the Río de la Antigua. It then extends about 9.5 miles to Punta Zempoala. Punta Antigua has been reported to be a good radar target at distances up to 29 miles and is reported to be identifiable with charted features by radar at distances up to 19 miles.

Bajo Zempoala has depths of 16 to 26 feet about 4 miles north of Punta Zempoala and 2 miles offshore.

Between Punta Zempoala and Punta del Morro, about 25 miles north-northwest, the coast has several reef-fringed points and is marked by several conspicuous peaks. Pico Zempoala (19°33′N, 96°27′W) is the southernmost peak. It is 2,299-feet high and stands about 10 miles northwest of Punta Zempoala. Los Dos Antriscos, a peak 2,620 feet high, stands 5 miles southwest of Punta del Morro. A bare chimney rock, 875 feet high and prominent, stands 2 miles west-northwest of Punta Penon, which lies about 9 miles north-northwest of Punta Zempoala.

Isla Bernal Chico (19°40′N, 96°23′W) is a 144-foot-high island lying close offshore about 12 miles north-northwest of Punta Zempoala. A prominent bare rocky hill, over 300 feet high, stands on the coast 1.8 miles west-southwest of this island.

The coast between Punta Gorda and the Río Nautla, about 30 miles to the northwest, and the Río Tecolutla 18 miles farther northwest, is fronted by a narrow, thickly wooded strip of land paralleled by a narrow lagoon. A range of hills stands about 5 miles inland.

Dos Hermanos (20°06′N, 96°52′W) is a conspicuous peak rising to an elevation of 1,169 feet about 9 miles south-southwest of the Río Nautla. Cerro Burras stands about 9 miles southwest of the Río Tecolutla. A dangerous wreck lies 4 miles east of the north side of the mouth of Río Tecolutla.

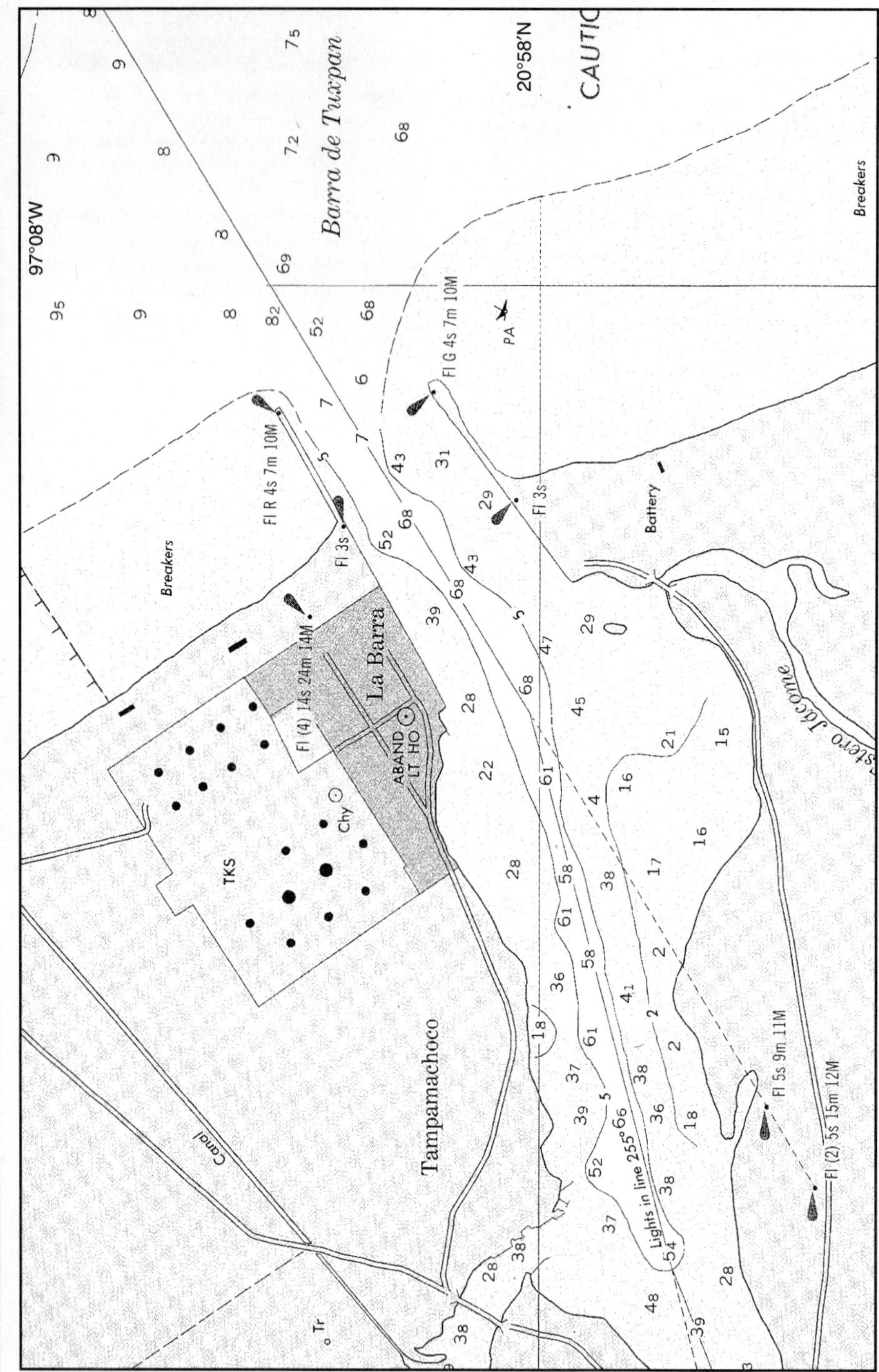

Río Tuxpan entrance, soundings in meters *(from DMA 28321, 35th ed, 9/88)*

Between the Río Tecolutla and Punta de Piedras, a reef-fringed point about 25 miles northwest, several rivers discharge into the gulf from the lagoon that backs the coast.

Between Punta Piedras and the Río de Tuxpan, about 9 miles northwest, the coast is about 100 feet high.

CHACHALACAS, 19 24.8N, 96 19.3W. Fl W 6s, 30ft, 5M. White fiberglass tower, 16ft.
PUNTA DELGADA, located on Punta del Morro, 19 51.5N, 96 27.5W. Fl (3) W 25s, 151ft, 19M. White cylindrical concrete tower, 72ft.
RÍO NAUTLA, S bank, 20 14N, 96 47W. Fl W 6s, 98ft, 25M. White concrete cylindrical tower, 72ft. Visible 150°–300°. ~~Radiobeacon (about 5 miles SSE)~~.
TECOLUTLA, N side of Río Tecolutla, 20 28.6N, 97 00.1W. Fl (2) W 10s, 82ft, 25M. White truncated conical concrete tower, 74ft. Visible 149°–301°.

BAJO BLAKE

20 45N, 96 58W
Pilotage: This shoal has a depth of 26 feet in a location about 13 miles southeast of Punta de Piedras.

SOUTH BREAKWATER, Fl G 5s, 27ft, 8M. Silver metal tower, 26ft.
NORTH BREAKWATER, Fl R 5s, 27ft, 8M. Silver metal tower, 26ft.
ESCUALO, 20 41.5N, 97 56.5W. Q W, 84ft, 10M. Yellow rectangular metal tower, 81ft.
MORSA, 20 46.2N, 97 58.2W. Q W, 84ft, 10M. Yellow rectangular metal platform, 81ft.
ATUM C, 20 53.6N, 97 01.0W. Q W, 84ft, 10M. Yellow rectangular metal tower. 81ft.
ATUM B, 20 52.6N, 97 00.9W. Q W, 84ft, 10M. Yellow rectangular metal platform, 81ft.
ATUM A, 20 51.6N, 97 00.1W. Q W, 84ft, 10M. Yellow rectangular metal platform, 81ft.
BARGE B, 20 57.9N, 97 02.3W. Q W, 84ft, 10M. Yellow rectangular metal platform, 81ft.
BARGE A, 20 56.6N, 97 02.3W. Q W, 84ft, 10M. Yellow rectangular metal platform, 81ft.
MARSOPA, 21 03.5N, 97 02.5W. Q W, 84ft, 10M. Yellow rectangular metal platform, 81ft.
TIBURON, 21 24.1N, 97 08.0W. Q W, 84ft, 10M. Yellow rectangular metal platform, 81ft.
ARENQUE B, 22 16.7N, 97 31.0W. Q W, 84ft, 10M. Yellow rectangular metal platform, 81ft.
ARENQUE A, 22 14.0N, 97 30.1W. Q W, 84ft, 10M. Yellow rectangular metal platform, 81ft.

BARRA DE CAZONES, 20 43.3N, 97 12.0W. Fl W 5s, 85ft, 15M. White concrete tower, 20ft.
PUNTA CAZONES, 20 46N, 97 12W. Fl (2) W 10s, 49ft, 9M. White truncated pyramidal metal tower with orange daymark, 36ft.

TUXPAN

NORTH BREAKWATER HEAD, 20 58.4N, 97 18.2W. Fl R 5s, 27ft, 8M. White round concrete tower, rectangular base, 25ft.
NORTH BREAKWATER ROOT, 20 58.3N, 97 18.4W. Fl W 3s, 23 ft, 5M. White round fiberglass tower, orange bands, 11ft.
SOUTH BREAKWATER HEAD, 20 58.2N, 97 18.2W. Fl G 5s, 26ft, 8M. White round concrete towe with rectangular base, 25ft.
SOUTH BREAKWATER ROOT, 20 58.0N, 97 18.3W. Fl W 3s, 23 ft, 5M. White round fiberglass tower with orange bands, 11ft.
LA BARRA, north side, mouth of river, 20 58.3N, 97 18.5W. Fl (4) W 7s, 79ft, 15M. White truncated conical masonry tower and hut, 72ft. Visible 180°–325°.
RANGE (Front Light), 20 57.4N, 97 20.6W. Fl W 3s, 38ft, 16M. White truncated pyramidal metal tower with orange daymark, 36ft. **(Rear Light),** 95 meters, 248° from front. Iso W 2s, 55ft, 16M. White truncated pyramidal metal tower with orange daymark, 43ft.
SOUTH SIDE RANGE (Front Light), 20 57.7N, 97 19.3W. Fl W 3s, 56ft, 20M. White truncated pyramidal metal tower, orange daymark, 54ft. **(Rear Light),** 270 meters, 238° from front. Iso W 2s, 75ft, 20M. White truncated pyramidal metal tower with orange daymark, 74ft.
NORTH SIDE RANGE 1 (Front Light), 20 57.6N, 97 20.2W. Fl W 3s, 39ft, 20M. White truncated pyramidal metal tower with orange daymark, 36ft. **(Rear Light),** 250 meters, 255° from front. Iso W 2s, 56ft, 16M. White truncated pyramidal metal tower with orange daymark, 52ft.
COBOS, 20 58.0N, 97 21.5W. Fl G 3s, 36ft, 3M. Truncated pyramidal tower, 25ft.
SEMINARIO, 20 58.0N, 97 22.1W. Fl W 3s, 49ft, 3M. Truncated pyramidal tower, 41ft.

RÍO TUXPAN

20 58N, 97 19W
Pilotage: The entrance has a maximum depth of 16 feet over the bar and is fronted by breakwaters that extend 0.3 mile offshore from the entrance points. It was reported in 1991 the breakwaters were being lengthened and the harbor was to be dredged to 36 feet to accommodate a new container terminal. Within the bar depths increase considerably. The least depth,

as far as the town of Tuxpan, is under 9 feet. Vessels drawing under 7 feet can proceed up to 36 miles above the entrance.

From seaward several tall stacks and oil tanks on the banks of the river are conspicuous. The navigation lights at the river entrance are hard to identify and are often confused with the numerous fishing vessels found in the vicinity.

Tuxpan is located about 5 miles above the entrance to the river and is a port of entry. Between the Río Tuxpan and Cabo Rojo the low coast extends northwest for about 19 miles, then north-northeast an additional 18 miles forming a bight called Puerto Lobos. This bight provides some protection from the northers that blow with considerable strength during the winter.

ARRECIFE TUXPAN, Bajo Centro, 21 01.7N, 97 11.7W. Fl (2) W 10s, 36ft, 11M. Truncated pyramidal concrete tower, 33ft. Orange truncated pyramidal concrete tower, 34ft. **RACON X** (– · · –), Range 20.
ARRECIFE TANGUIJO, 21 08.6N, 97 16.4W. Fl (3) W 12s, 36ft, 11M. Orange truncated pyramidal concrete tower, 34ft.

BAJO DE TUXPAN

21 01N, 97 12W
Pilotage: This small steep-to reef has a low cay on it, in a position about 6 miles east-northeast of the Río de Tuxpan. Bajo de Enmedio is a small reef lying about 3 miles northwest of Bajo de Tuxpan.

Bajo de Tanguijo lies about 7 miles northwest of Bajo de Tuxpan. It is a small, steep-to drying reef.

ARRECIFE MEDIO, 21 30.7N, 97 15.2W. Fl R 3s, 33ft, 8M. White truncated pyramidal metal tower, 24ft.
ARRECIFE LA BLANQUILLA, 21 32.8N, 97 16.9W. Fl W 6s, 33ft, 11M. Orange truncated pyramidal concrete tower, 31ft.
ISLA DE LOBOS, 21 27.2N, 97 13.7W. Fl W 5s, 105ft, 20M. White truncated conical concrete tower, 98ft. **RACON O** (– – –), 20M.

ISLA LOBOS

21 28N, 97 13W
Pilotage: This small islet is about 30 feet high to the tops of the trees. It lies 9 miles southeast of Cabo Rojo. A reef extends about 1 mile north from the islet.

Arrecife Medio is a small, steep-to, breaking reef, lying about 2.5 miles north-northwest of Isla Lobos.

BAJO DE LA BLANQUILLA

21 32N, 97 16W
Pilotage: This is a breaking reef with a drying sandbank on it. It lies 6 miles northwest of Isla Lobos. A disused light structure stands on the reef. The passage between here and Cabo Rojo is of doubtful safety as the soundings are irregular.

CABO ROJO

21 33N, 97 20W
Pilotage: This blunt headland is composed of sand hills about 35 feet high. It is bordered by a reef that extends about 1.8 miles east from it.

The coast trends northwest for 49 miles to the Río Panuco, then about 160 miles north and an additional 67 miles north-northeast to the Río Grande. In general, the coast is composed of sand dunes and wooded hummocks. Mountain ranges with conspicuous peaks lie farther inland along the south part of this coast. Several extensive lagoons, separated from the coast by narrow strips of land, lie along this section. Several rivers discharge into the gulf.

Between Cabo Rojo and the Río Panuco the coast is bordered by a narrow strip of land, a few hundred feet to 5 miles wide, fronting the Laguna de Tamiahua. Canal del Chijol crosses this lagoon and is frequented by small craft going from Tuxpan to Tampico. The sand hills backing the coast about 4 miles north of Cabo Rojo are 70 feet high and rise to heights of 348 feet about 18 miles farther north. The coast then becomes very low as far north as the Río Panuco.

An 8-foot shoal lies 5 miles south-southeast of the entrance to Tampico.

Winds: The winds are east-southeast from August to April and east from April to June. During the summer the land breezes blow from midnight to 0900 and then yield to the sea breezes as far north as 26°, where the mountain ranges terminate.
Currents: During the winter the current sets north and in the opposite direction during the summer. About 15 to 20 miles offshore the current generally sets north at a rate of about 1 knot.

BAJOS DE BURROS, 21 42.5N, 97 36.2W. Fl (2) W 10s, 19ft, 5M. White round fiberglass tower, orange bands, 18ft.

LA LAJA, 21 42.1N, 97 41.4W. Fl R 6s, 19ft, 8M. White truncated pyramidal metal tower, 18ft.

~~**MOCTEZUMA,** 21 44.2N, 97 45.5W. Fl R 5s, 19ft, 11M. White concrete tower, 16ft.~~

PUNTA MORALES, 21 46.5N, 97 37.2W. Fl W 5s, 19ft, 11M. White truncated pyramidal concrete tower, 18ft.

PUNTA MANGLES, 21 54.6N, 97 41.7W. Fl W 5s, 19ft, 11M. White truncated pyramidal concrete tower, 18ft.

TAMPICO

TAMPICO, N point of river entrance, 21 15.8N, 97 47.7W. Fl (3) W 6s, 141ft, 24M. Hexagonal aluminum tower. Radar reflector. Fl R and Fl G lights mark the bridge at Tamos, above Tampico.

ENTRANCE RANGE, N bank of river **(Front Light),** 22 15.5N, 97 48.0W. Fl W 3s, 69ft, 24M. Truncated pyramidal metal tower, 52ft. **(Rear Light),** 480 yards 257° from front. Iso W 2s, 118ft, 24M. Truncated pyramidal metal tower, 102ft.

HEAD OF NORTH BREAKWATER, ~~ESCOLLERA NORTE~~, 22 16.0N, 97 46.4W. Fl R 5s, 39ft, 9M. White square concrete tower, 19ft. Visible 188°–322°. **RACON Q** (– – • –), 20M.

HEAD OF SOUTH BREAKWATER, ~~ESCOLLERA SUR~~, 22 15.8N, 97 46.5W. Fl G 5s, 33ft, 9M. White square concrete tower, 19ft. Visible 188°–322°.

SOUTH BREAKWATER DIRECTIONAL LIGHT, root, 22 15.5N, 97 47.5W. Dir Fl W 3s, 33ft, 11M. Truncated pyramidal aluminum tower, 24ft.

CANAL DE CHIJOL, ~~on W bank~~, 22 14.7N, 97 49.2W. Fl W 3s, 29ft, 10M. Square concrete tower, 20ft.

ARBOL GRANDE, on E bank, 22 14.0N, 97 50.0W. Fl W 3s, 23ft, 10M. Square concrete tower, 15ft.

COLONIA DEL GOLFO, on E bank, 22 13.8N, 97 50.2W. Fl W 3s, 33ft, 10M. Concrete tower, 19ft.

COLONIA DE LA ISLETA, on E bank, 22 12.9N, 97 50.0W. Fl W 3s, 23ft, 10M. White square concrete tower and hut, 15ft.

ISLA PEREZ, on E bank, 22 12.5N, 97 50.2W. Fl W 3s, 30ft, 10M. White pyramidal concrete tower, 18ft.

CANAL DE PUEBLO VIEJO RANGE (Front Light), 22 12.5N, 07 50.7W. Fl W 3s, 22ft, 11M. White concrete tower, 15ft. **(Rear Light),** 710 yards, 110° from front. Iso.W 2s, 69ft, 13M. White concrete tower, 30ft.

LA RIVERA DE TAMPICO ALTO, 22 06.0N, 97 46.9W. Fl W 3s, 30ft, 5M. Truncated pyramidal metal tower, 25ft.

PUNTA DE ANIMAS, 22 03.9N, 97 45.8W. Fl W 3s, 26ft, 5M. White truncated pyramidal tower, orange daymark, 26ft.

PUNTA BUSTOS, 21 59.3N, 97 43.5W. Fl W 5s, 20ft, 11M. White truncated pyramidal concrete tower, 18ft.

RIHL FIELD AVIATION LIGHT, 22 12N, 97 49W. Al Fl W G 5s, 78ft. Beacon.

ARENQUE TO TIERRA, pipeline **(Front Light),** 22 17.1N, 97 48.1W. Fl W 3s. **(Rear Light),** 184 meters from front. Fl W 6s.

TAMPICO

22 16N, 97 50W

Pilotage: Tampico is on the lower reaches of the Río Panuco. This is an important petroleum port. A 100-yard-wide channel leads across the bar and midway between the breakwaters. The bar has a least depth of about 30 feet. During the rainy season, depths decrease due to heavy silting. The harbor can be identified from offshore by the numerous chimneys and tanks of the oil refineries. The hills to the south of the river are grass covered and higher than those to the north, which are composed of whitish gray sand. On closer approach Tampico lighthouse, on the north side of the river entrance, and the light structures on the breakwater heads, will be sighted.

CAUTION: The smoke from the refineries may make Tampico light appear red when the wind is from the southwest and the humidity is high.

The entrance range is on a bearing of 257°. The current sets north across the entrance at times, and vessels usually enter at a good speed to offset its effect.

Several offshore oil platforms lie about 18 miles in a generally easterly direction from the harbor. Submarine pipelines run from these platforms to the shore. Anchorage in the vicinity of these platforms and pipelines is prohibited.

Winds: Hot southwest winds blow in the months of March and April from about 1100, sometimes lasting until 1500 or 1600, before switching back to the east-southeast. Northers are frequent during the winter season and usually last from 8 to 24 hours. At such times the port is closed.

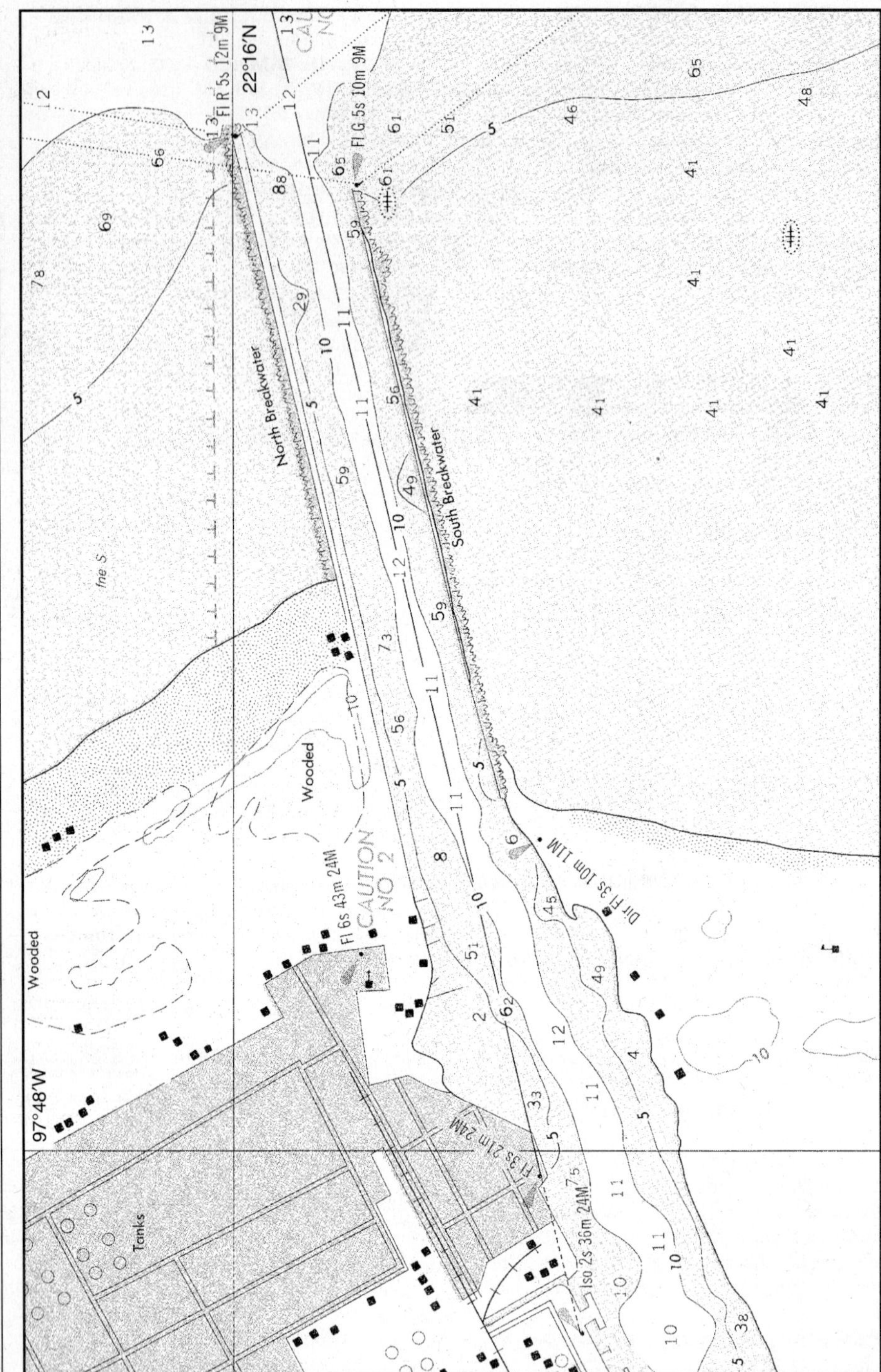

Tampico entrance channel, soundings in meters *(from DMA 28325, 1st ed, 3/91)*

Tides and currents: The tides are irregular and in the vicinity of the bar are greatly affected by the prevailing winds and the rate of discharge of the Río Panuco. The maximum rise above mean low water is about 2 feet. When the river is in flood, currents attain a velocity of 8.5 knots between the sea walls, and a velocity of 6 knots in the upper reaches. Normally the current in the river flows at a rate of 3 knots.

ALTAMIRA

ALTAMIRA, 22 29.5N, 97 51.9W. Fl W 6s, 138ft, 14M. Concrete octagonal tower, 120ft.

~~**ESCOLLERA SUR**~~ **SOUTH BREAKWATER,** 22 28.9N, 97 51.0W. Fl G 5s, 36ft, 9M. Round concrete tower, 25ft.

~~**ESCOLLERA NORTE**~~ **NORTH BREAKWATER,** 22 29.5N, 97 50.9W. Fl R 5s, 42ft, 9M. Truncated pyramidal tower, 23ft.

~~**ESCOLLERA NORTE,** W, 22 29.5N, 97 51.2W. Fl R 5s, 237ft, 7M. Metal tower, 10ft.~~

RANGE (Front Light), 22 29.2N, 97 53.3W. Fl W 3s, 36ft, 11M. Truncated pyramidal metal tower, 24ft. **(Rear Light),** 400 meters, 270° from front. Iso W 2s, 56ft, 11M. Truncated pyramidal tower, 24ft. ~~Reported extinguished (1993).~~

~~**ESPIGON SUR**~~ **SOUTH ENTRANCE,** 22 29.0N, 97 51.5W. Fl G 5s, 36ft, 9M. Round concrete tower, 23ft.

~~**ESPIGON NORTE**~~ **NORTH ENTRANCE,** 22 29.4N, 97 51.6W. Fl R 5s, 36ft, 9M. Round concrete tower, 25ft.

CANAL DE NAVIGATION, NO. 3, 22 28.7N, 97 53.1W. Fl G 3s, 42ft, 7M. Truncated pyramidal aluminum tower, 24ft.

NO. 1, 22 29.7N, 97 52.7W. Fl G 3s, 33ft, 7M. Aluminum tower, 22ft.

~~**NO. 6,** 22 28.7N, 97 53.4W. Fl R 3s, 41ft, 7M. Aluminum tower, 22ft.~~

NO. 4, 22 29.4N, 97 52.8W. Fl R 3s, 28ft, 7M. Truncated pyramidal tower, 24ft.

NO. 2, 22 29.5N, 97 52.1W. Fl R 3s, 33ft, 7M. Truncated pyramidal metal tower, 24ft.

ALTAMIRA

22 25N, 97 55W

Pilotage: This port is reported to be a general cargo terminal.

> ***Note***: . Within an aid listing, underlined text indicates data that has changed or is new since September, 1996; ~~Strike out~~ text indicates deleted data.

TAMPICO TO RÍO SAN FERNANDO

Pilotage: Between Tampico and Barra de Chavarria and Barra de la Trinidad, two shallow lagoon entrances about 23 and 30 miles north, the coast is backed by wooded hills about 200 feet high. Shallow water lies up to 1.5 miles offshore between the entrances to the lagoons.

An 18-foot patch was reported to lie about 9 miles north of Tampico entrance in 1912. An obstruction was reported in 1928 about 17 miles north of Tampico entrance.

Cerro Metate (22°47′N, 97°58′W) is a flat-topped hill 866 feet high, located about 32 miles north-northwest of Tampico.

Between Barra de la Trinidad and Barra del Torda, a shallow lagoon entrance about 18 miles north, the coast is low and sandy. Some rocks lie up to 2 miles offshore in places.

Between Barra del Torda and Río Indios Morales (23°24′N, 97°46′W), about 28 miles to the north, the coast is backed by wooded hummocks 70 feet high. From abreast Punta Jerez a range of hills, Sierra de San Jose de Las Rusias, extends about 52 miles north at a distance of from 7 to 12 miles inland. A conspicuous sugarloaf peak stands about 18 miles southwest of the Río Indios Morales.

The coast between the Río Indios Morales and the Río Soto de la Marina, about 21 miles to the north, is bordered by a narrow strip of land that fronts a lagoon.

Between the Río Soto de la Marina and Boquillas Ceradas (25°02′N, 97°30′W), about 77 miles to the north, and the Río Fernando, about 25 miles farther north-northeast, the coast is bordered by a strip of land 1 to 5 miles wide fronting Laguna de La Madre. The high hills in the interior terminate about 24 miles north of the Río Soto de la Marina.

A shoal, about 3 miles in extent with a least depth of 15 feet, lies about 25 miles north of the Río Soto de la Marina and 2 miles offshore. Boquillas Ceradas are four nearly closed entrances to Laguna de la Madre.

The Río Fernando (25°58′N, 97°09′W), with a depth of 3 feet over the bar, drains a lagoon in the interior.

A shoal, 3 miles long with a least depth of 10 feet, lies with its south end 5 miles north of the Río San Fernando and about 3 miles offshore.

PUNTA JEREZ, on low sandy beach, 22 53.5N, 97 46.1W. Fl W 6s, 78ft, 24M. White round concrete tower, 66ft. ~~Visible 176°–014°~~.
BARRA CARRIZO, 23 22.1N, 97 46.0W. Fl (2) W 10s, 30ft, 11M. Truncated pyramidal tower, 24ft.
LA PESA RANGE (Front Light), 23 46.7N, 97 45.1W. Fl W 3s, 28ft, 11M. Truncated pyramidal aluminum tower, 24ft. **(Rear Light),** Iso W 2s, 56ft, 11M. Truncated pyramidal aluminum tower, 52ft.
NORTH BREAKWATER, 23 46.2N, 97 43.9W. Fl G 5s, 22ft, 9M. Round concrete tower, 25ft.
SOUTH BREAKWATER, 23 46.0N, 97 44.1W. Fl G 5s, 25ft, 9M. Truncated pyramidal metal tower, 24ft.
LA PESCA, 23 46.5N, 97 44.2W. Fl (4) W 16s, 60ft, 24M. White round concrete tower and hut, 52ft.
LA CARBONERA, 24 37.8N, 97 43.0W. Fl W 6s. 39 ft, 14M. White round concrete tower, red bands, 33ft.
SOTO LA MARINA, 23 46.5N, 97 47.5W. Fl W 2s, 52ft, 13M. Metal tower, 39ft. Radar reflector.
CANAL DE CHAVEZ, 25 52.0N, 97 10.1W. Fl W 6s, 38ft, 14M. White round concrete tower, 33ft.
EL MEZQUITAL, 25 15.0N, 97 26.5W. Fl (3) W 10s, 82ft, 18M. White round concrete tower and hut, 85ft.
EL MEZQUITAL, S breakwater, 25 14.3N, 97 25.3W. Fl G 5s, 22ft, 7M. Round concrete tower.
RÍO BRAVO, 25 56.8N, 97 08.7W. Fl W 6s, 59ft, 14M. Round concrete tower, 52ft.

RÍO GRANDE

25 58N, 97 09W
Pilotage: This river is located about 36 miles north of the Río San Fernando. It forms the boundary with the United States. By international agreement the river is closed to navigation.

TIDES

T

TIDE TABLE INTRODUCTION

Accurate tide predictions are the result of good science and careful data collection by various national hydrographic offices. Reed's *gratefully acknowledges the work of these agencies, and is proud to present their data in the most complete set of tide tables for the Caribbean basin in print.*

The first section of chapter T contains tide prediction tables for 14 primary reference stations organized geographically, starting with Bermuda, then southeastern Florida, then clockwise through the islands, Caribbean South America and Central America.

The second section, the tide differences table, contains time and height differences for more than 500 other locations (often called subordinate stations). Instructions on the use of the tide differences tables are found on page T 47.

All these tables are based on data collected and maintained by the U.S. National Ocean Service (NOS) and the U.K. Hydrographic Office. The tables are presented in *Reed's* custom format, using software developed by Nautical Software, Inc. *Reed's* tables are compact, and include sunrise and sunset times for the middle and end of each month (new this year).

Caution: Tide heights are subject to weather influences that can not be predicted in the long term.

Heights: High water is the maximum height reached by each rising tide, and low water is the minimum height reached by each falling tide. High and low waters can be selected from the predictions by the comparison of consecutive heights. Because of diurnal inequality at certain places, however, there may be a difference of only a few tenths of a foot between one high water and low water on a given day, but a marked difference in height between the other high water and low water. Therefore, in using the tide tables it is essential to note carefully the heights as well as the times of the tides.

Time: The kind of time used for the predictions at each reference station is indicated by its name and the time meridian at the top of each page. Unlike *Reed's* North American East and West Coast Almanacs, these tables are not adjusted for Daylight Saving Time (2 am April 5 – 2 am October 25). We made this decision because many Caribbean nations do not use Daylight Saving Time. We have noted those places that do use DST in both the tide and tide differences tables.

Tide curves: All three tide curve types occur in the Caribbean. In a few locations, there are classical *semidiurnal* tide curves, where there are two tides a day of similar height. Throughout much of the are, the tidal curve is either *mixed* (where the height difference between one high and low is significantly less than the difference between the tidal day's other high and low, or during part of the tidal month there is only one high and low per tidal day) or *diurnal* (only one high and low per tidal day).

Datum: Tidal datums are very much like chart datums; they define the height of water that the chart or table maker has chosen as "0." There are a remarkable number of datum names and definitions in use throughout the world. Theoretically, the tide datum should not concern the mariner, as it should be matched with the chart datum in use for that location. For instance, if you are using a tide table marked *U.S. Datum,* 0 feet represents Mean Lower Low Water (MLLW)—the average of the lower of the two low waters of each day. NOS charts use the same datum. Therefore, you may simply apply the tide heights in the table to the charted depths; at the time of a 6-foot tide over a charted 8-foot bottom there should be 14 feet of water.

The tide tables marked *chart datum* are less definitive. NOS says on the subject: "For foreign coasts a datum approximating to mean low water springs (MLWS), Indian spring low water, or the lowest possible low water is generally used." The tide datums are supposed to match the local charts for these waters, but often there are redundant charts originating in several different countries, and they may use different chart datums. For maximum accuracy, you should check the datum used on your chart and then scan the low-tide heights in the table you are using to approximate its datum. Tide tables (like U.S.) with numerous tides less than 0 feet per month use a MLLW-type datum. Tables with almost no minus tides per month use a datum like MLWS; and tables with no minus tides probably use Lowest Astronomical Tide (LAT).

Variation in sea level: Changes in winds and barometric conditions cause variations in sea level from day to day. In general, with onshore winds or a low barometer the heights of both the high and low waters will be higher than predicted, while with offshore winds or a high barometer they will be lower. There are also seasonal variations in sea level, but these variations have been included in the predictions for each station. At ocean stations the seasonal variation in sea level is usually less than half a foot. At stations on tidal rivers the average seasonal variation in river level due to freshets and droughts may be considerably more than a foot. The predictions for these stations include an allowance for this seasonal variation, representing average freshet and drought conditions. Unusual freshets or droughts, however, will cause the tides to be higher or lower, respectively, than predicted.

Number of tides: At locations with *semidiurnal* or *mixed* tide curves, there are usually but not always two high and two low waters in a day. Tides follow the moon more closely than they do the sun, and the lunar or tidal day is about 50 minutes longer than the solar day. This causes the tide to occur later each day, and a tide that has occurred near the end of one calendar day will be followed by a corresponding tide that may skip the next day and occur in the early morning of the third day. Thus on certain days of each month only a single high or a single low water occurs. At some stations, during portions of each month, the tide becomes diurnal—that is, only one high and one low water will occur during the period of a lunar day.

Relation of tide to current: In using these tide prediction tables, bear in mind that they give the times and heights of high and low waters and not the times of turning of the current or slack water. For stations on the outer coast there is usually a small difference between the time of high or low water and the beginning of ebb or flood current, but for places in narrow channels, landlocked harbors, or on tidal rivers, the time of slack water may differ by several hours from the time of high or low water stand. The relation of the times of high and low water to the turning of the current depends upon a number of factors, so no simple or general rule can be given. See Currents Chapter for more information.

Sunrise and Sunset: This year *Reed's* is pleased to include sunrise and sunset times in our tide tables. The small table at the bottom left-hand corner of each monthly column gives the times for the 15th and the last day of the month. Intermediate times may be easily interpolated.

	Sun↑	Sun↓
15	0721	1737
31	0714	1752

For more information on the physics of tides, see *Reed's Nautical Companion*.

Tide Watchers Wanted

Reed's seeks local information about tides and currents. There are many locations, especially in the Caribbean, where official government tide and current data is scarce. Often, local boaters know how local tides and currents work. *Reed's* wants to help make that information available to all boaters.

For instance, you might know a place that usually has a high tide about 50 minutes after a regular reference station. Or you might spend a lot of time in such a place and be willing to devote a bit of time and effort to measuring the tide or current.

We can give you some tips on how to evaluate tides and currents, and we can integrate your local knowledge into our tide differences table. Of course, we will carefully mark such data as local knowledge, and not nearly as precise as the differences developed by hydrographic offices. Approximate information, properly handled, is better than no information.

If you are interested in being a Tide Watcher, please contact *Reed's* for further information. The reward will be helping your fellow mariners to better know and use the power of tides.

ST. GEORGE'S, BERMUDA

HIGH & LOW WATER 1998 — Chart Datum — 32°23'N 64°42'W

Atlantic Standard Time (60°W) — Add 1H Daylight Saving Time: April 5 – October 24

JANUARY

Day	Time	ft	Day	Time	ft
1 Th	0346	0.7	16 F	0432	0.9
	1012	4.0		1034	3.6
	1640	0.7		1707	0.9
	2235	3.3		2302	3.2
2 F	0438	0.7	17 Sa	0515	1.1
	1057	3.8		1114	3.4
	1729	0.7		1746	1.0
	2325	3.3		2346	3.1
3 Sa	0536	0.8	18 Su	0556	1.2
	1146	3.7		1157	3.2
	1821	0.7		1820	1.0
4 Su	0020	3.4	19 M	0034	3.0
	0641	1.0		0636	1.3
	1239	3.5		1242	3.0
	1917	0.8		1855	1.1
5 M	0120	3.4	20 Tu	0124	2.9
	0749	1.1		0737	1.4
	1337	3.3		1331	2.8
	☾ 2015	0.8		☽ 1940	1.1
6 Tu	0226	3.4	21 W	0219	2.9
	0857	1.1		0856	1.5
	1441	3.1		1426	2.6
	2114	0.8		2033	1.1
7 W	0334	3.5	22 Th	0318	3.0
	1005	1.1		0958	1.4
	1549	3.0		1526	2.6
	2212	0.8		2130	1.1
8 Th	0440	3.6	23 F	0418	3.1
	1111	1.0		1055	1.3
	1654	3.0		1628	2.6
	2308	0.8		2227	1.0
9 F	0539	3.7	24 Sa	0515	3.3
	1213	0.9		1146	1.2
	1754	3.0		1728	2.7
				2321	0.9
10 Sa	0001	0.7	25 Su	0609	3.5
	0630	3.8		1233	1.0
	1308	0.8		1824	2.9
	1846	3.1			
11 Su	0051	0.7	26 M	0013	0.7
	0717	3.9		0658	3.7
	1356	0.8		1317	0.8
	1934	3.2		1915	3.0
12 M	0137	0.7	27 Tu	0104	0.6
	0800	3.9		0744	3.9
	1437	0.8		1401	0.7
	○ 2018	3.2		2002	3.2
13 Tu	0221	0.7	28 W	0153	0.5
	0841	3.9		0829	4.0
	1514	0.8		1446	0.5
	2100	3.3		● 2048	3.4
14 W	0305	0.7	29 Th	0242	0.4
	0919	3.8		0912	4.0
	1550	0.8		1530	0.5
	2140	3.3		2133	3.5
15 Th	0349	0.8	30 F	0332	0.4
	0956	3.7		0955	4.0
	1628	0.8		1615	0.4
	2220	3.2		2219	3.6
			31 Sa	0424	0.5
				1039	3.8
				1701	0.4
				2307	3.6

	Sun↑	Sun↓
15	0721	1737
31	0714	1752

FEBRUARY

Day	Time	ft	Day	Time	ft
1 Su	0520	0.6	16 M	0507	1.1
	1126	3.6		1124	3.1
	1748	0.5		1724	1.0
	2359	3.6		2357	3.2
2 M	0623	0.7	17 Tu	0545	1.2
	1216	3.4		1202	2.9
	1841	0.6		1803	1.0
3 Tu	0054	3.5	18 W	0043	3.1
	0730	0.9		0634	1.3
	1311	3.1		1236	2.8
	☾ 1940	0.7		1849	1.1
4 W	0155	3.4	19 Th	0135	3.0
	0839	1.0		0731	1.4
	1412	2.9		1318	2.6
	2044	0.8		☽ 1943	1.1
5 Th	0303	3.4	20 F	0234	3.0
	0950	1.1		0839	1.4
	1521	2.7		1436	2.5
	2147	0.9		2043	1.1
6 F	0414	3.4	21 Sa	0337	3.1
	1100	1.0		0954	1.3
	1632	2.7		1550	2.6
	2248	0.9		2147	1.1
7 Sa	0520	3.4	22 Su	0438	3.3
	1205	1.0		1101	1.2
	1736	2.8		1656	2.7
	2344	0.8		2249	0.9
8 Su	0615	3.5	23 M	0536	3.5
	1258	0.9		1156	1.0
	1831	2.9		1755	2.9
				2349	0.8
9 M	0035	0.8	24 Tu	0629	3.7
	0703	3.6		1247	0.8
	1341	0.9		1849	3.2
	1919	3.1			
10 Tu	0121	0.8	25 W	0045	0.6
	0746	3.7		0718	3.9
	1412	0.8		1335	0.6
	2001	3.2		1939	3.5
11 W	0204	0.7	26 Th	0138	0.4
	0824	3.7		0805	4.0
	1442	0.8		1421	0.4
	○ 2040	3.3		● 2026	3.7
12 Th	0247	0.8	27 F	0230	0.3
	0857	3.7		0850	4.0
	1516	0.8		1505	0.3
	2116	3.4		2113	3.8
13 F	0328	0.8	28 Sa	0321	0.3
	0930	3.6		0935	3.9
	1553	0.8		1548	0.3
	2153	3.4		2200	3.9
14 Sa	0408	0.9			
	1005	3.5			
	1627	0.9			
	2232	3.3			
15 Su	0441	1.0			
	1044	3.3			
	1654	0.9			
	2314	3.3			

	Sun↑	Sun↓
15	0702	1805
28	0648	1816

MARCH

Day	Time	ft	Day	Time	ft
1 Su	0413	0.3	16 M	0414	1.0
	1021	3.8		1017	3.2
	1630	0.3		1616	0.9
	2248	3.9		2244	3.4
2 M	0508	0.5	17 Tu	0442	1.0
	1108	3.5		1053	3.1
	1713	0.4		1650	1.0
	2338	3.8		2323	3.3
3 Tu	0608	0.6	18 W	0521	1.1
	1158	3.3		1121	2.9
	1802	0.6		1729	1.0
				2358	3.2
4 W	0031	3.6	19 Th	0608	1.2
	0714	0.8		1152	2.8
	1252	3.0		1815	1.1
	1903	0.7			
5 Th	0129	3.4	20 F	0036	3.2
	0822	1.0		0704	1.3
	1351	2.8		1240	2.7
	☾ 2014	0.9		1910	1.1
6 F	0235	3.3	21 Sa	0145	3.1
	0932	1.1		0806	1.3
	1459	2.6		1345	2.6
	2124	1.0		☽ 2013	1.2
7 Sa	0347	3.2	22 Su	0259	3.2
	1042	1.1		0914	1.3
	1612	2.7		1518	2.7
	2230	1.0		2120	1.1
8 Su	0456	3.2	23 M	0404	3.3
	1146	1.0		1020	1.2
	1718	2.8		1628	2.9
	2329	1.0		2226	1.0
9 M	0553	3.3	24 Tu	0504	3.4
	1237	1.0		1121	1.0
	1812	2.9		1728	3.1
				2330	0.8
10 Tu	0021	0.9	25 W	0600	3.6
	0642	3.4		1215	0.8
	1315	1.0		1823	3.4
	1859	3.1			
11 W	0105	0.9	26 Th	0030	0.6
	0726	3.5		0651	3.8
	1338	0.9		1306	0.6
	1940	3.3		1914	3.7
12 Th	0146	0.9	27 F	0126	0.4
	0804	3.5		0740	3.9
	1407	0.9		1353	0.4
	2018	3.4		● 2003	4.0
13 F	0226	0.8	28 Sa	0220	0.3
	0836	3.5		0827	3.9
	1443	0.9		1438	0.3
	○ 2052	3.5		2051	4.1
14 Sa	0306	0.8	29 Su	0313	0.2
	0906	3.5		0914	3.8
	1518	0.8		1519	0.3
	2128	3.5		2139	4.1
15 Su	0343	0.9	30 M	0406	0.3
	0940	3.4		1002	3.6
	1548	0.9		1600	0.3
	2205	3.5		2227	4.1
			31 Tu	0501	0.4
				1050	3.4
				1644	0.4
				2318	3.9

	Sun↑	Sun↓
15	0630	1827
31	0609	1838

APRIL

Day	Time	ft	Day	Time	ft
1 W	0558	0.6	16 Th	0507	1.0
	1141	3.2		1056	2.9
	1735	0.6		1704	1.0
				2327	3.4
2 Th	0010	3.7	17 F	0555	1.1
	0700	0.8		1136	2.8
	1235	2.9		1750	1.1
	1836	0.8			
3 F	0107	3.4	18 Sa	0011	3.3
	0803	0.9		0648	1.2
	1334	2.7		1228	2.7
	☾ 1949	1.0		1848	1.1
4 Sa	0210	3.2	19 Su	0111	3.3
	0908	1.0		0748	1.2
	1441	2.7		1335	2.7
	2101	1.1		☽ 1954	1.2
5 Su	0320	3.1	20 M	0224	3.2
	1012	1.1		0850	1.1
	1552	2.7		1453	2.8
	2208	1.1		2103	1.1
6 M	0428	3.1	21 Tu	0332	3.3
	1110	1.1		0952	1.0
	1655	2.8		1602	3.0
	2309	1.1		2209	1.0
7 Tu	0526	3.1	22 W	0434	3.4
	1158	1.1		1050	0.9
	1748	3.0		1703	3.3
				2314	0.8
8 W	0001	1.1	23 Th	0531	3.5
	0615	3.2		1145	0.7
	1228	1.0		1758	3.6
	1833	3.2			
9 Th	0045	1.0	24 F	0014	0.6
	0659	3.2		0624	3.6
	1256	1.0		1237	0.5
	1915	3.4		1850	3.9
10 F	0125	0.9	25 Sa	0112	0.4
	0737	3.3		0715	3.7
	1332	0.9		1325	0.4
	1952	3.5		1940	4.1
11 Sa	0204	0.9	26 Su	0208	0.3
	0810	3.3		0804	3.7
	1408	0.9		1410	0.3
	○ 2027	3.6		● 2028	4.2
12 Su	0243	0.9	27 M	0302	0.3
	0841	3.3		0853	3.6
	1443	0.9		1453	0.3
	2103	3.6		2117	4.2
13 M	0319	0.9	28 Tu	0356	0.3
	0916	3.2		0942	3.4
	1515	0.9		1537	0.3
	2140	3.6		2206	4.1
14 Tu	0352	0.9	29 W	0450	0.4
	0951	3.1		1031	3.3
	1547	0.9		1624	0.5
	2217	3.5		2257	3.9
15 W	0427	1.0	30 Th	0544	0.6
	1024	3.0		1123	3.1
	1624	0.9		1717	0.6
	2252	3.5		2348	3.6

	Sun↑	Sun↓
15	0550	1849
30	0534	1900

ST. GEORGE'S, BERMUDA

HIGH & LOW WATER 1998 — Chart Datum — 32°23'N 64°42'W

Atlantic Standard Time (60°W) — Add 1H Daylight Saving Time: April 5 – October 24

MAY

Day		Time	ft	Time	ft	Time	ft	Time	ft
1	F	0640	0.7	1217	2.9	1817	0.8		
2	Sa	0043	3.4	0736	0.9	1315	2.8	1926	1.0
3	Su	0143	3.2	0833	1.0	1420	2.7	☾ 2033	1.1
4	M	0250	3.0	0927	1.0	1528	2.7	2137	1.2
5	Tu	0356	2.9	1016	1.1	1628	2.9	2236	1.2
6	W	0453	2.9	1058	1.0	1719	3.0	2330	1.1
7	Th	0542	2.9	1137	1.0	1804	3.2		
8	F	0017	1.0	0626	3.0	1216	0.9	1845	3.3
9	Sa	0059	1.0	0704	3.0	1255	0.9	1923	3.5
10	Su	0140	0.9	0739	3.0	1333	0.8	2000	3.5
11	M	0218	0.9	0816	3.0	1409	0.8	○ 2039	3.6
12	Tu	0255	0.9	0853	3.0	1446	0.8	2117	3.6
13	W	0333	0.9	0929	3.0	1523	0.8	2155	3.6
14	Th	0413	0.9	1003	2.9	1603	0.9	2232	3.6
15	F	0457	0.9	1042	2.8	1646	0.9	2312	3.5
16	Sa	0545	1.0	1128	2.8	1736	1.0	2359	3.4
17	Su	0637	1.0	1222	2.8	1836	1.1		
18	M	0054	3.3	0733	1.0	1324	2.9	1943	1.1
19	Tu	0156	3.3	0830	0.9	1432	3.0	☽ 2049	1.0
20	W	0301	3.3	0927	0.8	1538	3.2	2154	0.9
21	Th	0404	3.3	1023	0.7	1639	3.5	2257	0.8
22	F	0503	3.3	1118	0.6	1735	3.7	2358	0.6
23	Sa	0558	3.4	1210	0.4	1828	3.9		
24	Su	0057	0.5	0651	3.4	1300	0.3	1918	4.1
25	M	0154	0.4	0742	3.4	1347	0.3	● 2008	4.2
26	Tu	0248	0.4	0832	3.3	1434	0.3	2057	4.1
27	W	0341	0.4	0922	3.3	1520	0.4	2146	4.0
28	Th	0432	0.5	1012	3.1	1609	0.5	2235	3.8
29	F	0522	0.6	1102	3.0	1702	0.7	2324	3.6
30	Sa	0611	0.7	1154	2.9	1759	0.8		
31	Su	0015	3.4	0658	0.8	1249	2.8	1900	1.0

	Sun↑	Sun↓
15	0521	1910
31	0513	1921

JUNE

Day		Time	ft	Time	ft	Time	ft	Time	ft
1	M	0108	3.1	0746	0.9	1348	2.8	☾ 2000	1.1
2	Tu	0205	2.9	0834	1.0	1452	2.8	2059	1.2
3	W	0307	2.8	0922	1.0	1552	2.8	2156	1.2
4	Th	0407	2.7	1009	1.0	1644	3.0	2251	1.2
5	F	0458	2.7	1055	0.9	1729	3.1	2342	1.1
6	Sa	0542	2.8	1139	0.9	1810	3.2		
7	Su	0029	1.0	0625	2.8	1220	0.8	1851	3.4
8	M	0112	1.0	0707	2.8	1300	0.8	1933	3.5
9	Tu	0152	0.9	0749	2.9	1339	0.8	2015	3.6
10	W	0232	0.8	0831	2.9	1420	0.7	○ 2056	3.7
11	Th	0312	0.8	0912	2.9	1502	0.7	2137	3.7
12	F	0356	0.8	0951	2.9	1546	0.8	2216	3.7
13	Sa	0442	0.8	1032	2.9	1633	0.8	2258	3.6
14	Su	0530	0.8	1118	3.0	1727	0.9	2345	3.5
15	M	0621	0.8	1210	3.0	1828	0.9		
16	Tu	0036	3.4	0714	0.8	1307	3.1	1931	1.0
17	W	0132	3.3	0809	0.8	1409	3.2	☽ 2035	1.0
18	Th	0232	3.2	0904	0.7	1513	3.3	2138	0.9
19	F	0335	3.2	1000	0.7	1615	3.5	2241	0.8
20	Sa	0436	3.2	1054	0.6	1714	3.7	2342	0.7
21	Su	0535	3.2	1148	0.5	1809	3.9		
22	M	0041	0.6	0631	3.2	1239	0.4	1901	4.0
23	Tu	0138	0.6	0723	3.2	1330	0.4	● 1951	4.0
24	W	0231	0.5	0814	3.2	1419	0.4	2039	4.0
25	Th	0320	0.5	0903	3.2	1507	0.5	2126	3.9
26	F	0406	0.6	0951	3.2	1555	0.6	2212	3.8
27	Sa	0450	0.7	1039	3.1	1645	0.7	2257	3.6
28	Su	0533	0.8	1126	3.0	1737	0.9	2341	3.4
29	M	0617	0.9	1213	3.0	1830	1.0		
30	Tu	0025	3.2	0703	0.9	1301	2.9	1925	1.2

	Sun↑	Sun↓
15	0512	1928
30	0515	1930

JULY

Day		Time	ft	Time	ft	Time	ft	Time	ft
1	W	0112	3.0	0750	1.0	1352	2.9	☾ 2021	1.2
2	Th	0201	2.8	0838	1.0	1447	2.9	2117	1.3
3	F	0254	2.7	0927	1.0	1543	3.0	2212	1.3
4	Sa	0350	2.7	1015	1.0	1638	3.1	2305	1.2
5	Su	0448	2.7	1100	1.0	1730	3.2	2355	1.1
6	M	0543	2.7	1144	0.9	1818	3.4		
7	Tu	0041	1.1	0635	2.8	1227	0.9	1905	3.6
8	W	0125	1.0	0724	2.9	1311	0.8	1951	3.7
9	Th	0207	0.9	0811	3.0	1356	0.7	○ 2036	3.8
10	F	0251	0.8	0855	3.1	1442	0.7	2118	3.9
11	Sa	0335	0.7	0938	3.2	1531	0.7	2200	3.9
12	Su	0422	0.7	1020	3.3	1621	0.7	2243	3.8
13	M	0510	0.7	1106	3.3	1716	0.8	2328	3.7
14	Tu	0600	0.7	1155	3.4	1815	0.8		
15	W	0017	3.5	0652	0.7	1249	3.4	1917	0.9
16	Th	0110	3.4	0746	0.8	1346	3.4	☽ 2019	1.0
17	F	0208	3.2	0842	0.8	1449	3.5	2122	1.0
18	Sa	0309	3.1	0938	0.8	1554	3.6	2225	1.0
19	Su	0414	3.1	1035	0.7	1658	3.7	2327	0.9
20	M	0518	3.1	1130	0.7	1756	3.8		
21	Tu	0027	0.9	0617	3.1	1223	0.7	1849	3.9
22	W	0122	0.8	0710	3.2	1314	0.7	1939	4.0
23	Th	0211	0.8	0800	3.3	1403	0.7	● 2025	4.0
24	F	0254	0.8	0847	3.4	1450	0.7	2109	3.9
25	Sa	0334	0.8	0931	3.4	1537	0.8	2150	3.8
26	Su	0414	0.9	1013	3.4	1623	0.9	2228	3.6
27	M	0456	0.9	1054	3.3	1710	1.0	2308	3.5
28	Tu	0538	1.0	1136	3.2	1758	1.2	2349	3.3
29	W	0621	1.1	1220	3.2	1848	1.3		
30	Th	0034	3.1	0705	1.2	1308	3.1	1940	1.4
31	F	0122	2.9	0749	1.2	1359	3.1	☾ 2035	1.4

	Sun↑	Sun↓
15	0523	1927
31	0533	1918

AUGUST

Day		Time	ft	Time	ft	Time	ft	Time	ft
1	Sa	0214	2.8	0834	1.2	1455	3.1	2131	1.5
2	Su	0310	2.7	0924	1.2	1553	3.2	2226	1.4
3	M	0409	2.7	1015	1.2	1651	3.3	2319	1.3
4	Tu	0509	2.8	1107	1.1	1746	3.5		
5	W	0009	1.2	0605	2.9	1157	1.0	1837	3.7
6	Th	0055	1.1	0658	3.1	1247	0.9	1926	3.9
7	F	0141	1.0	0748	3.3	1336	0.8	○ 2013	4.0
8	Sa	0226	0.9	0834	3.5	1426	0.7	2057	4.1
9	Su	0312	0.8	0919	3.6	1516	0.7	2140	4.1
10	M	0359	0.7	1003	3.7	1608	0.7	2224	4.0
11	Tu	0446	0.7	1049	3.8	1703	0.7	2310	3.8
12	W	0536	0.7	1137	3.8	1801	0.8	2358	3.7
13	Th	0628	0.8	1230	3.7	1902	1.0		
14	F	0051	3.4	0723	0.9	1326	3.7	☽ 2004	1.1
15	Sa	0148	3.3	0821	0.9	1429	3.6	2108	1.2
16	Su	0251	3.1	0920	1.0	1538	3.6	2212	1.2
17	M	0400	3.1	1018	1.0	1646	3.7	2315	1.2
18	Tu	0507	3.1	1115	1.0	1746	3.8		
19	W	0015	1.1	0606	3.2	1209	1.0	1839	3.9
20	Th	0106	1.1	0658	3.4	1300	1.0	1927	3.9
21	F	0146	1.1	0745	3.5	1346	1.0	● 2011	3.9
22	Sa	0223	1.1	0829	3.6	1431	1.0	2052	3.9
23	Su	0300	1.1	0910	3.7	1515	1.0	2127	3.8
24	M	0340	1.1	0947	3.7	1559	1.1	2200	3.7
25	Tu	0420	1.1	1024	3.6	1642	1.2	2237	3.6
26	W	0459	1.2	1104	3.5	1723	1.3	2318	3.4
27	Th	0532	1.3	1147	3.5	1800	1.4		
28	F	0003	3.3	0600	1.4	1235	3.4	1834	1.5
29	Sa	0051	3.1	0639	1.4	1326	3.3	1928	1.6
30	Su	0143	3.0	0729	1.5	1421	3.3	☾ 2039	1.6
31	M	0240	2.9	0827	1.5	1519	3.3	2145	1.6

	Sun↑	Sun↓
15	0543	1904
31	0554	1845

ST. GEORGE'S, BERMUDA

HIGH & LOW WATER 1998 — Chart Datum — 32°23'N 64°42'W

Atlantic Standard Time (60°W) — Add 1H Daylight Saving Time: April 5 – October 24

SEPTEMBER

Day	Time	ft	Day	Time	ft
1 Tu	0340 0929 1618 2242	2.9 1.4 3.5 1.5	16 W	0455 1102 1730 2359	3.3 1.3 3.7 1.3
2 W	0440 1033 1715 2336	3.0 1.3 3.6 1.4	17 Th	0551 1157 1822	3.4 1.3 3.8
3 Th	0538 1132 1809	3.2 1.2 3.8	18 F	0043 0641 1245 1909	1.3 3.6 1.2 3.9
4 F	0026 0632 1227 1859	1.2 3.4 1.0 4.0	19 Sa	0115 0726 1328 1952	1.3 3.7 1.2 3.9
5 Sa	0113 0722 1319 1947	1.1 3.7 0.9 4.2	20 Su	0149 0808 1410 ● 2031	1.3 3.8 1.2 3.9
6 Su	0200 0810 1411 ○ 2032	0.9 3.9 0.8 4.2	21 M	0226 0846 1452 2102	1.3 3.9 1.2 3.8
7 M	0247 0856 1503 2117	0.8 4.1 0.7 4.2	22 Tu	0306 0919 1533 2132	1.3 3.9 1.3 3.7
8 Tu	0333 0941 1556 2203	0.7 4.2 0.7 4.1	23 W	0343 0955 1613 2209	1.3 3.9 1.3 3.6
9 W	0420 1029 1651 2250	0.7 4.2 0.8 3.9	24 Th	0416 1035 1647 2250	1.3 3.8 1.4 3.5
10 Th	0509 1118 1749 2340	0.8 4.1 0.9 3.7	25 F	0443 1118 1715 2334	1.4 3.7 1.5 3.3
11 F	0602 1211 1849	0.9 4.0 1.1	26 Sa	0519 1205 1757	1.5 3.6 1.6
12 Sa	0034 0701 1308 ☽ 1951	3.5 1.0 3.9 1.2	27 Su	0022 0604 1255 1850	3.2 1.5 3.5 1.7
13 Su	0132 0802 1412 2054	3.3 1.1 3.7 1.3	28 M	0115 0656 1351 ☾ 1952	3.1 1.6 3.5 1.7
14 M	0239 0903 1524 2159	3.2 1.2 3.7 1.4	29 Tu	0213 0755 1449 2103	3.0 1.6 3.5 1.7
15 Tu	0350 1004 1632 2303	3.2 1.3 3.7 1.4	30 W	0314 0900 1548 2206	3.1 1.6 3.6 1.6

	Sun↑	Sun↓
15	0603	1826
30	0613	1806

OCTOBER

Day	Time	ft	Day	Time	ft
1 Th	0415 1008 1646 2303	3.2 1.5 3.7 1.4	16 F	0532 1141 1759	3.5 1.4 3.6
2 F	0512 1112 1740 2355	3.5 1.3 3.9 1.3	17 Sa	0006 0620 1228 1845	1.4 3.7 1.4 3.7
3 Sa	0606 1210 1830	3.7 1.1 4.1	18 Su	0039 0704 1309 1927	1.3 3.8 1.3 3.7
4 Su	0045 0656 1305 1919	1.1 4.0 0.9 4.2	19 M	0115 0744 1349 2004	1.3 3.9 1.3 3.7
5 M	0133 0743 1358 ○ 2006	0.9 4.3 0.8 4.2	20 Tu	0154 0820 1429 ● 2033	1.3 4.0 1.3 3.6
6 Tu	0219 0830 1451 2052	0.8 4.4 0.7 4.2	21 W	0232 0853 1509 2105	1.3 4.0 1.3 3.6
7 W	0305 0918 1545 2140	0.7 4.5 0.7 4.0	22 Th	0307 0928 1547 2142	1.3 3.9 1.4 3.5
8 Th	0352 1007 1640 2229	0.7 4.4 0.8 3.9	23 F	0338 1008 1621 2223	1.3 3.9 1.4 3.4
9 F	0440 1057 1737 2321	0.8 4.3 0.9 3.7	24 Sa	0410 1050 1655 2306	1.4 3.8 1.5 3.3
10 Sa	0537 1151 1836	1.0 4.1 1.1	25 Su	2306 0450 1135 1738	3.3 1.5 3.7 1.6
11 Su	0016 0639 1249 1936	3.5 1.1 3.9 1.2	26 M	0536 1223 1829	1.5 3.6 1.6
12 M	0117 0742 1354 ☽ 2037	3.3 1.3 3.7 1.3	27 Tu	0044 0629 1318 1929	3.1 1.6 3.6 1.6
13 Tu	0226 0845 1506 2138	3.2 1.4 3.6 1.4	28 W	0145 0730 1416 ☾ 2032	3.1 1.6 3.5 1.6
14 W	0338 0947 1612 2238	3.3 1.4 3.6 1.4	29 Th	0247 0839 1516 2134	3.2 1.6 3.6 1.5
15 Th	0439 1047 1709 2328	3.4 1.4 3.6 1.4	30 F	0348 0949 1613 2231	3.4 1.5 3.7 1.3
			31 Sa	0445 1054 1708 2324	3.6 1.3 3.8 1.1

	Sun↑	Sun↓
15	0623	1747
31	0636	1730

NOVEMBER

Day	Time	ft	Day	Time	ft
1 Su	0539 1153 1800	3.9 1.1 3.9	16 M	0003 0639 1248 1858	1.2 3.7 1.3 3.4
2 M	0015 0629 1250 1850	0.9 4.2 0.9 4.0	17 Tu	0043 0717 1328 1932	1.2 3.8 1.2 3.4
3 Tu	0104 0718 1344 1939	0.8 4.4 0.8 4.0	18 W	0122 0752 1408 2004	1.1 3.9 1.2 3.4
4 W	0151 0806 1438 ○ 2027	0.7 4.5 0.7 3.9	19 Th	0200 0826 1447 ● 2039	1.1 3.9 1.2 3.3
5 Th	0238 0855 1532 2117	0.6 4.6 0.7 3.8	20 F	0234 0904 1524 2118	1.1 3.9 1.2 3.3
6 F	0324 0944 1626 2207	0.7 4.5 0.7 3.7	21 Sa	0308 0943 1600 2158	1.2 3.9 1.3 3.2
7 Sa	0413 1035 1721 2300	0.8 4.3 0.9 3.5	22 Su	0344 1024 1638 2238	1.2 3.8 1.3 3.1
8 Su	0510 1128 1817 2356	0.9 4.1 1.0 3.4	23 M	0424 1104 1721 2320	1.3 3.7 1.3 3.1
9 M	0614 1224 1913	1.1 3.8 1.1	24 Tu	0509 1145 1810	1.4 3.6 1.4
10 Tu	0057 0718 1327 ☽ 2009	3.2 1.3 3.6 1.2	25 W	0008 0603 1234 1905	3.1 1.5 3.6 1.4
11 W	0206 0821 1438 2105	3.2 1.4 3.4 1.3	26 Th	0108 0708 1332 ☾ 2003	3.1 1.5 3.5 1.3
12 Th	0318 0922 1545 2158	3.2 1.4 3.4 1.3	27 F	0212 0819 1434 2102	3.2 1.5 3.5 1.2
13 F	0418 1022 1642 2244	3.3 1.4 3.3 1.3	28 Sa	0316 0930 1535 2159	3.4 1.4 3.5 1.1
14 Sa	0510 1118 1732 2324	3.5 1.4 3.3 1.3	29 Su	0415 1035 1633 2254	3.6 1.2 3.5 0.9
15 Su	0556 1206 1817	3.6 1.4 3.3	30 M	0511 1136 1729 2347	3.9 1.0 3.6 0.8

	Sun↑	Sun↓
15	0649	1719
30	0702	1714

DECEMBER

Day	Time	ft	Day	Time	ft
1 Tu	0604 1234 1822	4.1 0.8 3.6	16 W	0011 0647 1307 1857	1.0 3.6 1.2 3.0
2 W	0038 0655 1329 1914	0.6 4.3 0.7 3.7	17 Th	0051 0723 1348 1935	1.0 3.7 1.1 3.1
3 Th	0127 0744 1423 ○ 2005	0.5 4.4 0.6 3.6	18 F	0129 0801 1426 ● 2016	1.0 3.8 1.1 3.1
4 F	0215 0833 1516 2055	0.5 4.5 0.6 3.6	19 Sa	0205 0841 1502 2057	0.9 3.9 1.0 3.1
5 Sa	0303 0922 1607 2146	0.5 4.4 0.6 3.5	20 Su	0242 0921 1538 2138	1.0 3.9 1.0 3.1
6 Su	0353 1011 1658 2237	0.6 4.2 0.7 3.4	21 M	0321 0959 1617 2216	1.0 3.8 1.0 3.1
7 M	0446 1102 1749 2330	0.8 4.0 0.8 3.3	22 Tu	0402 1036 1700 2256	1.0 3.8 1.0 3.1
8 Tu	0545 1153 1839	1.0 3.7 1.0	23 W	0447 1115 1745 2341	1.1 3.7 1.0 3.1
9 W	0027 0645 1247 1929	3.2 1.1 3.4 1.1	24 Th	0539 1159 1835	1.2 3.5 1.0
10 Th	0129 0746 1348 ☽ 2018	3.1 1.3 3.2 1.2	25 F	0034 0645 1251 1930	3.2 1.2 3.4 1.0
11 F	0242 0847 1459 2107	3.1 1.4 3.0 1.2	26 Sa	0133 0758 1350 ☾ 2029	3.2 1.3 3.3 1.0
12 Sa	0347 0946 1604 2155	3.1 1.4 2.9 1.2	27 Su	0237 0909 1453 2128	3.4 1.2 3.2 0.9
13 Su	0442 1044 1657 2242	3.2 1.4 2.9 1.1	28 M	0342 1015 1557 2226	3.5 1.1 3.2 0.8
14 M	0529 1137 1743 2327	3.3 1.3 2.9 1.1	29 Tu	0444 1118 1700 2322	3.7 1.0 3.2 0.7
15 Tu	0610 1224 1821	3.5 1.2 3.0	30 W	0542 1218 1759	3.9 0.8 3.3
			31 Th	0016 0636 1314 1854	0.5 4.1 0.7 3.3

	Sun↑	Sun↓
15	0713	1716
31	0721	1724

MIAMI, FLORIDA (Harbor Entrance)

HIGH & LOW WATER 1998 — U.S. Datum — 25°46.2'N 80°07.9'W

Eastern Standard Time (75°W) — Add 1H Daylight Saving Time: April 5 – October 24

JANUARY

Day	Time	ft	Time	ft	Time	ft	Time	ft
1 Th	0337	-0.4	0958	3.0	1609	-0.1	2213	2.8
2 F	0427	-0.3	1047	2.9	1700	-0.1	2307	2.7
3 Sa	0520	-0.2	1138	2.8	1754	-0.1		
4 Su	0006	2.6	0618	-0.0	1233	2.7	1853	-0.1
5 M	0110	2.6	0721	0.1	1333	2.6	☾ 1955	-0.1
6 Tu	0217	2.5	0827	0.2	1436	2.5	2059	-0.2
7 W	0325	2.6	0934	0.2	1538	2.5	2202	-0.2
8 Th	0429	2.6	1037	0.2	1639	2.5	2301	-0.3
9 F	0528	2.7	1136	0.1	1735	2.6	2355	-0.4
10 Sa	0621	2.8	1229	0.0	1827	2.6		
11 Su	0045	-0.4	0710	2.8	1318	-0.0	1916	2.6
12 M	0132	-0.5	0755	2.8	1403	-0.1	○ 2001	2.6
13 Tu	0215	-0.4	0837	2.8	1446	-0.1	2045	2.5
14 W	0257	-0.3	0917	2.7	1527	-0.1	2127	2.5
15 Th	0338	-0.2	0956	2.6	1607	-0.0	2209	2.4
16 F	0418	-0.1	1035	2.5	1648	0.0	2252	2.3
17 Sa	0459	0.0	1114	2.4	1729	0.1	2336	2.2
18 Su	0542	0.2	1155	2.3	1812	0.1		
19 M	0024	2.1	0628	0.3	1238	2.2	1859	0.2
20 Tu	0116	2.0	0719	0.4	1326	2.1	☽ 1950	0.2
21 W	0212	2.0	0815	0.5	1418	2.0	2043	0.1
22 Th	0310	2.0	0914	0.5	1513	2.0	2138	0.1
23 F	0407	2.1	1011	0.4	1608	2.1	2232	-0.1
24 Sa	0501	2.3	1106	0.3	1702	2.2	2323	-0.2
25 Su	0550	2.4	1156	0.1	1753	2.3		
26 M	0012	-0.4	0637	2.6	1244	-0.0	1843	2.5
27 Tu	0100	-0.5	0723	2.7	1330	-0.2	1931	2.6
28 W	0147	-0.7	0808	2.8	1416	-0.4	● 2020	2.7
29 Th	0235	-0.7	0853	2.9	1503	-0.5	2109	2.8
30 F	0323	-0.7	0940	2.9	1551	-0.5	2200	2.8
31 Sa	0413	-0.6	1028	2.8	1641	-0.5	2254	2.7

	Sun↑	Sun↓
15	0709	1752
31	0705	1804

FEBRUARY

Day	Time	ft	Time	ft	Time	ft	Time	ft
1 Su	0505	-0.5	1118	2.7	1734	-0.5	2350	2.6
2 M	0601	-0.3	1212	2.6	1831	-0.4		
3 Tu	0051	2.5	0701	-0.1	1310	2.4	☾ 1932	-0.3
4 W	0157	2.4	0806	0.0	1413	2.3	2037	-0.3
5 Th	0305	2.4	0914	0.1	1518	2.2	2143	-0.3
6 F	0412	2.4	1021	0.1	1622	2.2	2245	-0.3
7 Sa	0513	2.4	1122	0.1	1721	2.3	2341	-0.3
8 Su	0607	2.5	1215	0.0	1814	2.3		
9 M	0031	-0.4	0654	2.5	1302	-0.1	1902	2.4
10 Tu	0116	-0.4	0736	2.5	1344	-0.1	1945	2.4
11 W	0157	-0.4	0815	2.6	1423	-0.2	○ 2026	2.4
12 Th	0236	-0.3	0851	2.5	1501	-0.2	2104	2.4
13 F	0313	-0.3	0927	2.5	1537	-0.2	2142	2.4
14 Sa	0350	-0.2	1001	2.4	1613	-0.2	2221	2.3
15 Su	0427	-0.1	1036	2.3	1649	-0.1	2300	2.2
16 M	0505	0.1	1113	2.2	1728	-0.0	2343	2.1
17 Tu	0546	0.2	1152	2.1	1810	0.0		
18 W	0029	2.1	0633	0.3	1236	2.0	1858	0.1
19 Th	0122	2.0	0726	0.4	1327	2.0	☽ 1952	0.1
20 F	0221	2.0	0826	0.4	1426	2.0	2052	0.0
21 Sa	0323	2.1	0929	0.4	1528	2.0	2153	-0.1
22 Su	0423	2.2	1030	0.3	1630	2.2	2252	-0.2
23 M	0518	2.4	1125	0.1	1728	2.4	2347	-0.4
24 Tu	0609	2.6	1217	-0.2	1822	2.6		
25 W	0039	-0.6	0658	2.7	1306	-0.4	1913	2.8
26 Th	0129	-0.7	0745	2.9	1354	-0.6	● 2004	2.9
27 F	0218	-0.8	0831	3.0	1442	-0.7	2054	3.0
28 Sa	0307	-0.7	0919	3.0	1531	-0.8	2145	3.0

	Sun↑	Sun↓
15	0656	1814
28	0645	1822

MARCH

Day	Time	ft	Time	ft	Time	ft	Time	ft
1 Su	0357	-0.7	1007	2.9	1621	-0.7	2237	2.9
2 M	0448	-0.5	1057	2.8	1713	-0.6	2332	2.8
3 Tu	0542	-0.3	1150	2.6	1808	-0.5		
4 W	0031	2.6	0641	-0.1	1247	2.4	1909	-0.3
5 Th	0135	2.5	0746	0.1	1351	2.3	☾ 2014	-0.2
6 F	0243	2.3	0855	0.2	1459	2.2	2122	-0.1
7 Sa	0351	2.3	1003	0.2	1606	2.2	2226	-0.1
8 Su	0453	2.3	1104	0.2	1707	2.2	2323	-0.1
9 M	0546	2.4	1156	0.1	1759	2.3		
10 Tu	0013	-0.1	0632	2.4	1241	0.0	1845	2.4
11 W	0056	-0.2	0711	2.5	1320	-0.1	1926	2.4
12 Th	0136	-0.2	0748	2.5	1357	-0.1	○ 2004	2.5
13 F	0212	-0.2	0822	2.5	1431	-0.2	2040	2.5
14 Sa	0248	-0.1	0855	2.5	1505	-0.2	2116	2.5
15 Su	0323	-0.1	0928	2.5	1539	-0.1	2152	2.5
16 M	0358	0.0	1001	2.4	1613	-0.1	2229	2.4
17 Tu	0434	0.2	1036	2.3	1649	-0.0	2308	2.3
18 W	0513	0.3	1113	2.2	1729	0.0	2352	2.3
19 Th	0556	0.4	1156	2.1	1816	0.1		
20 F	0042	2.2	0647	0.5	1247	2.1	1910	0.2
21 Sa	0139	2.2	0747	0.5	1349	2.1	☽ 2013	0.2
22 Su	0243	2.2	0853	0.4	1456	2.2	2119	0.1
23 M	0346	2.3	0957	0.3	1603	2.3	2223	-0.0
24 Tu	0446	2.5	1056	0.1	1705	2.5	2322	-0.2
25 W	0540	2.7	1151	-0.2	1802	2.8		
26 Th	0017	-0.4	0631	2.9	1242	-0.4	1855	3.0
27 F	0109	-0.5	0720	3.0	1332	-0.6	● 1946	3.2
28 Sa	0159	-0.6	0808	3.1	1421	-0.7	2037	3.3
29 Su	0249	-0.6	0856	3.1	1510	-0.8	2128	3.3
30 M	0339	-0.5	0945	3.1	1559	-0.7	2219	3.2
31 Tu	0430	-0.3	1035	2.9	1651	-0.6	2313	3.0

	Sun↑	Sun↓
15	0630	1830
31	0613	1837

APRIL

Day	Time	ft	Time	ft	Time	ft	Time	ft
1 W	0524	-0.1	1128	2.7	1746	-0.4		
2 Th	0010	2.8	0621	0.1	1226	2.5	1845	-0.1
3 F	0112	2.6	0725	0.3	1329	2.4	☾ 1949	0.0
4 Sa	0217	2.4	0832	0.4	1437	2.2	2056	0.2
5 Su	0323	2.4	0939	0.4	1545	2.2	2201	0.2
6 M	0424	2.4	1038	0.3	1646	2.3	2258	0.2
7 Tu	0516	2.4	1129	0.2	1737	2.4	2348	0.2
8 W	0600	2.5	1212	0.2	1822	2.5		
9 Th	0030	0.1	0639	2.5	1250	0.1	1901	2.6
10 F	0109	0.1	0715	2.6	1325	-0.0	1938	2.6
11 Sa	0145	0.1	0749	2.6	1359	-0.1	○ 2014	2.7
12 Su	0221	0.1	0822	2.6	1433	-0.1	2049	2.7
13 M	0256	0.1	0855	2.5	1506	-0.1	2125	2.7
14 Tu	0331	0.2	0929	2.5	1541	-0.1	2201	2.6
15 W	0407	0.3	1004	2.4	1618	0.0	2240	2.6
16 Th	0446	0.4	1043	2.4	1658	0.1	2323	2.5
17 F	0530	0.5	1127	2.3	1745	0.2		
18 Sa	0012	2.4	0620	0.5	1220	2.2	1840	0.2
19 Su	0108	2.4	0719	0.5	1323	2.3	☽ 1943	0.2
20 M	0210	2.4	0824	0.4	1433	2.3	2051	0.2
21 Tu	0314	2.5	0929	0.3	1541	2.5	2157	0.1
22 W	0414	2.6	1030	0.0	1645	2.7	2259	-0.0
23 Th	0511	2.8	1126	-0.2	1743	3.0	2355	-0.2
24 F	0604	3.0	1218	-0.4	1837	3.2		
25 Sa	0048	-0.3	0655	3.1	1309	-0.6	1929	3.3
26 Su	0140	-0.4	0744	3.2	1359	-0.7	● 2019	3.4
27 M	0230	-0.4	0834	3.2	1448	-0.7	2110	3.3
28 Tu	0320	-0.3	0923	3.1	1538	-0.6	2201	3.2
29 W	0411	-0.2	1014	2.9	1629	-0.5	2253	3.1
30 Th	0504	-0.0	1107	2.7	1723	-0.2	2348	2.9

	Sun↑	Sun↓
15	0558	1844
30	0545	1852

MIAMI, FLORIDA (Harbor Entrance)

HIGH & LOW WATER 1998 — U.S. Datum — 25°46.2'N 80°07.9'W

Eastern Standard Time (75°W) — Add 1H Daylight Saving Time: April 5 – October 24

MAY

Day	Time	ft	Day	Time	ft
1 F	0600	0.2	16 Sa	0510	0.4
	1203	2.5		1108	2.4
	1819	-0.0		1724	0.1
				2349	2.5
2 Sa	0045	2.7	17 Su	0601	0.4
	0700	0.3		1203	2.4
	1305	2.4		1819	0.2
	1920	0.2			
3 Su	0146	2.5	18 M	0043	2.5
	0803	0.4		0658	0.3
	1410	2.3		1306	2.4
	☾ 2024	0.3		☽ 1921	0.2
4 M	0247	2.4	19 Tu	0143	2.5
	0906	0.4		0801	0.3
	1515	2.2		1414	2.4
	2126	0.4		2028	0.2
5 Tu	0344	2.4	20 W	0245	2.5
	1002	0.4		0904	0.1
	1615	2.3		1521	2.6
	2223	0.4		2134	0.1
6 W	0435	2.4	21 Th	0346	2.6
	1052	0.3		1005	-0.1
	1706	2.4		1625	2.7
	2313	0.4		2236	0.0
7 Th	0520	2.4	22 F	0444	2.8
	1135	0.2		1102	-0.3
	1751	2.5		1724	2.9
	2357	0.3		2334	-0.1
8 F	0600	2.5	23 Sa	0539	2.9
	1214	0.1		1157	-0.5
	1831	2.6		1819	3.1
9 Sa	0037	0.3	24 Su	0029	-0.2
	0638	2.5		0632	3.0
	1251	0.0		1249	-0.6
	1910	2.7		1912	3.2
10 Su	0116	0.3	25 M	0121	-0.2
	0714	2.5		0723	3.0
	1326	-0.1		1339	-0.7
	1946	2.7		● 2003	3.3
11 M	0152	0.2	26 Tu	0212	-0.3
	0749	2.5		0813	3.0
	1401	-0.1		1429	-0.7
	○ 2023	2.8		2053	3.2
12 Tu	0229	0.2	27 W	0302	-0.2
	0824	2.5		0903	2.9
	1437	-0.1		1518	-0.6
	2100	2.8		2142	3.1
13 W	0306	0.3	28 Th	0353	-0.1
	0901	2.5		0953	2.8
	1514	-0.1		1608	-0.4
	2138	2.7		2232	3.0
14 Th	0344	0.3	29 F	0444	0.0
	0939	2.5		1045	2.6
	1553	-0.1		1659	-0.2
	2217	2.7		2323	2.8
15 F	0425	0.3	30 Sa	0536	0.1
	1021	2.4		1139	2.5
	1636	0.0		1751	0.0
	2301	2.6			
			31 Su	0015	2.6
				0631	0.2
				1236	2.3
				1847	0.2

	Sun↑	Sun↓
15	0535	1900
31	0530	1908

JUNE

Day	Time	ft	Day	Time	ft
1 M	0108	2.5	16 Tu	0022	2.6
	0727	0.3		0639	0.1
	1336	2.2		1251	2.4
	☾ 1944	0.3		1903	0.1
2 Tu	0203	2.3	17 W	0119	2.5
	0824	0.3		0739	0.0
	1436	2.2		1356	2.5
	2043	0.4		☽ 2007	0.1
3 W	0256	2.3	18 Th	0220	2.5
	0918	0.3		0841	-0.1
	1534	2.2		1503	2.5
	2139	0.5		2112	0.1
4 Th	0347	2.3	19 F	0321	2.6
	1008	0.2		0943	-0.2
	1627	2.3		1607	2.7
	2231	0.5		2216	0.1
5 F	0434	2.3	20 Sa	0421	2.6
	1053	0.1		1042	-0.4
	1714	2.4		1707	2.8
	2319	0.4		2316	-0.0
6 Sa	0518	2.3	21 Su	0518	2.7
	1135	0.0		1138	-0.5
	1758	2.5		1804	2.9
7 Su	0002	0.4	22 M	0012	-0.1
	0559	2.4		0612	2.8
	1215	-0.1		1231	-0.6
	1838	2.6		1857	3.0
8 M	0044	0.3	23 Tu	0105	-0.1
	0638	2.4		0705	2.8
	1254	-0.1		1322	-0.6
	1918	2.7		● 1947	3.0
9 Tu	0123	0.2	24 W	0156	-0.2
	0717	2.4		0755	2.8
	1332	-0.2		1411	-0.6
	○ 1957	2.7		2035	3.0
10 W	0203	0.2	25 Th	0245	-0.2
	0757	2.5		0845	2.7
	1411	-0.2		1459	-0.5
	2036	2.7		2122	2.9
11 Th	0242	0.2	26 F	0333	-0.1
	0837	2.5		0933	2.7
	1451	-0.3		1546	-0.4
	2116	2.7		2208	2.8
12 F	0323	0.2	27 Sa	0420	-0.0
	0919	2.5		1022	2.5
	1533	-0.2		1633	-0.2
	2158	2.7		2254	2.7
13 Sa	0406	0.1	28 Su	0508	0.0
	1005	2.5		1111	2.4
	1619	-0.2		1721	-0.0
	2242	2.7		2340	2.5
14 Su	0453	0.1	29 M	0556	0.1
	1055	2.5		1202	2.3
	1708	-0.1		1809	0.2
	2330	2.6			
15 M	0544	0.1	30 Tu	0026	2.4
	1150	2.4		0645	0.2
	1803	0.0		1255	2.2
				1900	0.3

	Sun↑	Sun↓
15	0529	1914
30	0533	1916

JULY

Day	Time	ft	Day	Time	ft
1 W	0115	2.3	16 Th	0057	2.6
	0736	0.2		0718	-0.2
	1350	2.1		1339	2.6
	☾ 1954	0.4		☽ 1948	0.1
2 Th	0204	2.2	17 F	0157	2.6
	0827	0.2		0820	-0.2
	1447	2.1		1445	2.6
	2049	0.5		2054	0.2
3 F	0255	2.2	18 Sa	0300	2.5
	0918	0.2		0924	-0.2
	1542	2.2		1551	2.6
	2144	0.5		2159	0.2
4 Sa	0345	2.2	19 Su	0402	2.5
	1007	0.1		1025	-0.3
	1633	2.2		1653	2.7
	2236	0.5		2301	0.1
5 Su	0433	2.2	20 M	0502	2.6
	1054	0.0		1123	-0.4
	1721	2.4		1750	2.8
	2325	0.4		2358	0.1
6 M	0520	2.3	21 Tu	0558	2.7
	1139	-0.1		1217	-0.4
	1806	2.5		1843	2.9
7 Tu	0011	0.3	22 W	0051	0.0
	0605	2.3		0651	2.7
	1222	-0.2		1308	-0.4
	1849	2.6		1931	2.9
8 W	0054	0.2	23 Th	0140	-0.0
	0649	2.4		0740	2.7
	1305	-0.3		1355	-0.4
	1930	2.7		● 2017	2.9
9 Th	0136	0.1	24 F	0226	-0.1
	0732	2.5		0827	2.7
	1348	-0.3		1440	-0.3
	○ 2012	2.8		2100	2.9
10 F	0219	0.0	25 Sa	0310	-0.1
	0816	2.6		0912	2.7
	1431	-0.4		1523	-0.2
	2054	2.8		2141	2.8
11 Sa	0302	-0.0	26 Su	0352	-0.0
	0902	2.6		0956	2.6
	1516	-0.4		1606	-0.1
	2137	2.8		2222	2.7
12 Su	0347	-0.1	27 M	0434	0.1
	0950	2.6		1040	2.5
	1603	-0.3		1648	0.1
	2222	2.8		2302	2.6
13 M	0434	-0.1	28 Tu	0516	0.1
	1041	2.6		1125	2.4
	1653	-0.2		1731	0.2
	2310	2.7		2343	2.5
14 Tu	0525	-0.1	29 W	0600	0.2
	1136	2.6		1212	2.3
	1747	-0.1		1816	0.4
15 W	0002	2.7	30 Th	0026	2.3
	0619	-0.1		0645	0.3
	1235	2.6		1302	2.2
	1845	0.0		1905	0.5
			31 F	0112	2.2
				0734	0.3
				1356	2.2
				☾ 1959	0.6

	Sun↑	Sun↓
15	0539	1915
31	0547	1908

AUGUST

Day	Time	ft	Day	Time	ft
1 Sa	0202	2.2	16 Su	0242	2.6
	0827	0.3		0906	0.0
	1453	2.2		1536	2.7
	2056	0.7		2145	0.4
2 Su	0256	2.2	17 M	0348	2.6
	0920	0.3		1011	0.0
	1550	2.3		1639	2.8
	2153	0.6		2249	0.4
3 M	0350	2.2	18 Tu	0451	2.7
	1013	0.2		1111	-0.0
	1643	2.4		1737	2.8
	2247	0.6		2346	0.3
4 Tu	0444	2.3	19 W	0548	2.7
	1104	0.1		1205	-0.0
	1732	2.5		1828	2.9
	2337	0.4			
5 W	0534	2.4	20 Th	0036	0.3
	1153	-0.0		0639	2.8
	1818	2.7		1254	-0.1
				1914	3.0
6 Th	0024	0.3	21 F	0122	0.2
	0623	2.6		0725	2.9
	1240	-0.2		1338	-0.0
	1903	2.8		● 1955	3.0
7 F	0110	0.1	22 Sa	0204	0.2
	0710	2.8		0809	2.9
	1326	-0.3		1419	0.0
	○ 1946	3.0		2034	3.0
8 Sa	0154	-0.0	23 Su	0243	0.1
	0758	2.9		0849	2.9
	1412	-0.3		1459	0.1
	2030	3.0		2111	2.9
9 Su	0239	-0.1	24 M	0321	0.2
	0845	3.0		0929	2.8
	1459	-0.3		1537	0.2
	2114	3.1		2147	2.9
10 M	0325	-0.2	25 Tu	0358	0.2
	0934	3.0		1008	2.8
	1547	-0.3		1615	0.3
	2201	3.1		2223	2.8
11 Tu	0413	-0.2	26 W	0436	0.3
	1025	3.0		1048	2.7
	1637	-0.2		1654	0.5
	2249	3.0		2300	2.7
12 W	0504	-0.2	27 Th	0515	0.4
	1120	3.0		1131	2.6
	1730	-0.0		1735	0.6
	2340	2.9		2339	2.6
13 Th	0558	-0.2	28 F	0557	0.5
	1218	2.9		1217	2.5
	1828	0.1		1821	0.8
14 F	0036	2.8	29 Sa	0023	2.5
	0657	-0.1		0643	0.5
	1321	2.8		1308	2.5
	☽ 1930	0.3		☾ 1912	0.9
15 Sa	0137	2.7	30 Su	0113	2.4
	0801	-0.0		0736	0.6
	1428	2.7		1405	2.4
	2037	0.4		2010	0.9
			31 M	0210	2.4
				0834	0.6
				1505	2.5
				2112	0.9

	Sun↑	Sun↓
15	0554	1857
31	0601	1841

MIAMI, FLORIDA (Harbor Entrance)

HIGH & LOW WATER 1998 — U.S. Datum — 25°46.2'N 80°07.9'W

Eastern Standard Time (75°W) — Add 1H Daylight Saving Time: April 5 – October 24

SEPTEMBER

		Time	ft
1	Tu	0311	2.4
		0934	0.5
		1603	2.6
		2211	0.8
2	W	0411	2.6
		1031	0.4
		1657	2.8
		2305	0.6
3	Th	0507	2.7
		1125	0.2
		1747	2.9
		2355	0.4
4	F	0600	3.0
		1216	0.1
		1834	3.1
5	Sa	0042	0.2
		0650	3.2
		1304	-0.0
		1919	3.3
6	Su	0129	0.0
		0739	3.4
		1352	-0.1
		○ 2005	3.4
7	M	0215	-0.1
		0827	3.5
		1440	-0.1
		2050	3.4
8	Tu	0302	-0.2
		0917	3.5
		1528	-0.1
		2138	3.4
9	W	0351	-0.2
		1008	3.5
		1619	0.0
		2227	3.3
10	Th	0442	-0.1
		1102	3.4
		1712	0.2
		2319	3.2
11	F	0537	0.0
		1159	3.2
		1810	0.4
12	Sa	0016	3.1
		0636	0.2
		1302	3.1
		☽ 1913	0.6
13	Su	0119	2.9
		0741	0.3
		1410	3.0
		2022	0.7
14	M	0227	2.8
		0849	0.4
		1518	2.9
		2131	0.7
15	Tu	0336	2.8
		0956	0.4
		1623	3.0
		2235	0.7
16	W	0440	2.8
		1057	0.4
		1719	3.0
		2330	0.6
17	Th	0536	2.9
		1150	0.4
		1808	3.1
18	F	0018	0.5
		0625	3.0
		1236	0.4
		1850	3.1
19	Sa	0100	0.5
		0708	3.1
		1318	0.4
		1929	3.1
20	Su	0138	0.4
		0747	3.1
		1356	0.4
		● 2004	3.1
21	M	0214	0.4
		0825	3.2
		1432	0.5
		2038	3.1
22	Tu	0249	0.4
		0901	3.1
		1508	0.5
		2112	3.1
23	W	0323	0.4
		0937	3.1
		1544	0.6
		2146	3.0
24	Th	0358	0.5
		1015	3.0
		1620	0.8
		2221	2.9
25	F	0434	0.6
		1054	3.0
		1659	0.9
		2258	2.8
26	Sa	0514	0.7
		1137	2.9
		1743	1.0
		2341	2.7
27	Su	0559	0.8
		1226	2.8
		1832	1.1
28	M	0031	2.6
		0652	0.8
		1321	2.7
		☾ 1930	1.1
29	Tu	0130	2.6
		0752	0.8
		1422	2.8
		2033	1.1
30	W	0236	2.7
		0857	0.8
		1524	2.9
		2136	1.0

	Sun↑	Sun↓
15	0607	1825
30	0613	1809

OCTOBER

		Time	ft
1	Th	0342	2.8
		1000	0.7
		1622	3.0
		2233	0.8
2	F	0442	3.0
		1058	0.5
		1714	3.2
		2326	0.5
3	Sa	0537	3.3
		1151	0.4
		1804	3.4
4	Su	0016	0.3
		0629	3.5
		1242	0.2
		1852	3.6
5	M	0104	0.1
		0719	3.7
		1331	0.1
		○ 1939	3.7
6	Tu	0152	-0.1
		0809	3.9
		1420	0.1
		2027	3.7
7	W	0240	-0.2
		0859	3.9
		1510	0.1
		2115	3.7
8	Th	0330	-0.1
		0950	3.8
		1601	0.2
		2205	3.6
9	F	0421	-0.0
		1043	3.7
		1654	0.4
		2258	3.4
10	Sa	0516	0.2
		1140	3.5
		1752	0.6
		2356	3.2
11	Su	0615	0.4
		1242	3.3
		1855	0.8
12	M	0100	3.0
		0719	0.6
		1348	3.2
		☽ 2003	0.9
13	Tu	0210	2.9
		0828	0.7
		1455	3.1
		2111	0.9
14	W	0319	2.9
		0936	0.8
		1558	3.0
		2214	0.9
15	Th	0423	3.0
		1036	0.8
		1653	3.1
		2307	0.8
16	F	0518	3.0
		1128	0.7
		1740	3.1
		2353	0.7
17	Sa	0605	3.1
		1213	0.7
		1821	3.1
18	Su	0033	0.6
		0646	3.2
		1253	0.7
		1858	3.2
19	M	0109	0.5
		0723	3.3
		1330	0.7
		1932	3.2
20	Tu	0143	0.5
		0759	3.3
		1405	0.7
		● 2006	3.2
21	W	0217	0.5
		0834	3.3
		1440	0.7
		2039	3.1
22	Th	0250	0.5
		0910	3.3
		1516	0.8
		2113	3.1
23	F	0325	0.5
		0946	3.2
		1552	0.9
		2148	3.0
24	Sa	0401	0.6
		1024	3.1
		1630	1.0
		2225	2.9
25	Su	0440	0.7
		1105	3.0
		1712	1.1
		2308	2.8
26	M	0524	0.8
		1152	3.0
		1801	1.1
		2359	2.8
27	Tu	0617	0.9
		1245	2.9
		1857	1.1
28	W	0059	2.7
		0717	0.9
		1344	2.9
		☾ 1959	1.1
29	Th	0207	2.8
		0824	0.9
		1446	3.0
		2103	0.9
30	F	0315	2.9
		0930	0.8
		1546	3.1
		2203	0.7
31	Sa	0418	3.2
		1031	0.6
		1642	3.2
		2258	0.4

	Sun↑	Sun↓
15	0620	1754
31	0629	1740

NOVEMBER

		Time	ft
1	Su	0515	3.4
		1127	0.5
		1735	3.4
		2350	0.2
2	M	0609	3.6
		1220	0.3
		1826	3.6
3	Tu	0041	-0.0
		0701	3.8
		1311	0.2
		1915	3.7
4	W	0130	-0.2
		0751	3.9
		1401	0.1
		○ 2004	3.7
5	Th	0220	-0.2
		0842	3.9
		1452	0.2
		2054	3.7
6	F	0310	-0.2
		0933	3.8
		1543	0.3
		2145	3.5
7	Sa	0401	-0.1
		1025	3.7
		1636	0.4
		2239	3.4
8	Su	0455	0.1
		1120	3.5
		1733	0.6
		2336	3.2
9	M	0552	0.4
		1218	3.3
		1834	0.7
10	Tu	0039	3.0
		0654	0.6
		1319	3.1
		☽ 1938	0.8
11	W	0146	2.8
		0800	0.7
		1422	3.0
		2042	0.8
12	Th	0253	2.8
		0905	0.8
		1523	2.9
		2142	0.8
13	F	0356	2.8
		1005	0.9
		1617	2.9
		2235	0.7
14	Sa	0451	2.9
		1058	0.8
		1704	2.9
		2320	0.6
15	Su	0537	3.0
		1143	0.8
		1745	2.9
16	M	0000	0.5
		0618	3.0
		1224	0.8
		1823	3.0
17	Tu	0037	0.4
		0657	3.1
		1302	0.7
		1859	3.0
18	W	0112	0.4
		0733	3.2
		1339	0.7
		● 1934	3.0
19	Th	0147	0.3
		0809	3.2
		1415	0.7
		2009	3.0
20	F	0222	0.3
		0845	3.2
		1451	0.7
		2045	2.9
21	Sa	0258	0.3
		0922	3.1
		1528	0.7
		2122	2.9
22	Su	0335	0.4
		1000	3.1
		1607	0.8
		2201	2.8
23	M	0415	0.4
		1040	3.0
		1649	0.8
		2245	2.7
24	Tu	0500	0.5
		1125	2.9
		1736	0.8
		2336	2.7
25	W	0551	0.6
		1215	2.8
		1830	0.8
26	Th	0036	2.7
		0650	0.7
		1311	2.8
		☾ 1930	0.7
27	F	0142	2.7
		0755	0.7
		1412	2.8
		2032	0.6
28	Sa	0249	2.8
		0901	0.6
		1513	2.9
		2134	0.4
29	Su	0354	3.0
		1004	0.5
		1612	3.0
		2232	0.1
30	M	0454	3.2
		1104	0.4
		1709	3.2
		2327	-0.1

	Sun↑	Sun↓
15	0639	1732
30	0650	1730

DECEMBER

		Time	ft
1	Tu	0551	3.4
		1200	0.2
		1802	3.3
2	W	0020	-0.3
		0644	3.5
		1253	0.1
		1854	3.4
3	Th	0112	-0.4
		0735	3.6
		1344	0.0
		○ 1945	3.4
4	F	0202	-0.4
		0826	3.6
		1435	0.0
		2036	3.3
5	Sa	0252	-0.4
		0916	3.5
		1526	0.1
		2127	3.2
6	Su	0343	-0.3
		1006	3.4
		1618	0.2
		2220	3.1
7	M	0434	-0.1
		1058	3.2
		1711	0.3
		2314	2.9
8	Tu	0528	0.1
		1150	3.0
		1806	0.4
9	W	0012	2.7
		0624	0.4
		1245	2.8
		1904	0.5
10	Th	0113	2.6
		0723	0.5
		1341	2.7
		☽ 2002	0.5
11	F	0216	2.5
		0823	0.7
		1437	2.6
		2059	0.5
12	Sa	0317	2.5
		0923	0.7
		1530	2.5
		2152	0.5
13	Su	0413	2.5
		1017	0.7
		1620	2.5
		2240	0.4
14	M	0503	2.6
		1107	0.7
		1705	2.5
		2323	0.3
15	Tu	0547	2.7
		1151	0.6
		1747	2.6
16	W	0004	0.2
		0628	2.8
		1233	0.5
		1827	2.6
17	Th	0042	0.1
		0707	2.8
		1312	0.5
		1905	2.6
18	F	0120	0.0
		0745	2.9
		1350	0.4
		● 1943	2.6
19	Sa	0158	-0.0
		0822	2.9
		1428	0.4
		2022	2.6
20	Su	0236	-0.0
		0900	2.9
		1507	0.4
		2101	2.6
21	M	0315	-0.0
		0938	2.9
		1546	0.3
		2143	2.6
22	Tu	0357	0.0
		1019	2.8
		1629	0.3
		2229	2.6
23	W	0442	0.1
		1102	2.7
		1715	0.3
		2320	2.5
24	Th	0532	0.2
		1150	2.7
		1806	0.3
25	F	0016	2.5
		0627	0.3
		1244	2.6
		1903	0.2
26	Sa	0119	2.5
		0729	0.3
		1342	2.6
		☾ 2005	0.1
27	Su	0226	2.6
		0835	0.3
		1444	2.6
		2108	-0.0
28	M	0332	2.7
		0941	0.3
		1546	2.7
		2209	-0.2
29	Tu	0436	2.8
		1044	0.2
		1646	2.8
		2308	-0.4
30	W	0534	3.0
		1143	0.0
		1744	2.8
31	Th	0004	-0.5
		0629	3.1
		1238	-0.1
		1838	2.9

	Sun↑	Sun↓
15	0700	1732
31	0707	1741

VACA KEY, FLORIDA

HIGH & LOW WATER 1998 U.S. Datum 24°42.6'N 81°06.4'W

Eastern Standard Time (75°W) Add 1H Daylight Saving Time: April 5 – October 24

JANUARY

Day	Time	ft	Day	Time	ft
1	0526	-0.4	16	0559	-0.2
	1210	1.2		1221	1.2
Th	1658	0.3	F	1744	0.2
	2337	1.9			
2	0610	-0.3	17	0006	1.5
	1253	1.3		0634	-0.0
F	1754	0.2	Sa	1255	1.2
				1834	0.2
3	0029	1.8	18	0048	1.3
	0655	-0.2		0710	0.1
Sa	1339	1.3	Su	1332	1.2
	1859	0.2		1930	0.3
4	0127	1.5	19	0136	1.2
	0744	-0.0		0747	0.2
Su	1429	1.4	M	1413	1.2
	2014	0.2		2035	0.3
5	0237	1.3	20	0233	1.0
	0835	0.1		0828	0.3
M	1523	1.5	Tu	1500	1.3
☾	2135	0.1	☽	2149	0.2
6	0402	1.1	21	0345	0.9
	0930	0.2		0916	0.4
Tu	1622	1.6	W	1553	1.3
	2255	0.0		2300	0.1
7	0535	1.0	22	0513	0.8
	1028	0.3		1010	0.4
W	1723	1.7	Th	1651	1.4
8	0005	-0.1	23	0002	0.0
	0656	1.0		0635	0.8
Th	1126	0.4	F	1106	0.4
	1821	1.8		1749	1.5
9	0106	-0.3	24	0054	-0.1
	0759	1.0		0736	0.9
F	1221	0.3	Sa	1159	0.4
	1915	1.9		1844	1.6
10	0158	-0.3	25	0140	-0.3
	0849	1.1		0824	0.9
Sa	1313	0.3	Su	1249	0.3
	2005	1.9		1935	1.7
11	0245	-0.4	26	0222	-0.4
	0933	1.1		0906	1.0
Su	1401	0.3	M	1337	0.2
	2050	1.9		2023	1.8
12	0328	-0.4	27	0303	-0.5
	1011	1.1		0945	1.1
M	1447	0.2	Tu	1424	0.1
○	2132	1.9		2110	1.9
13	0408	-0.4	28	0343	-0.5
	1045	1.1		1023	1.1
Tu	1532	0.2	W	1512	0.0
	2211	1.9	●	2157	1.9
14	0446	-0.3	29	0423	-0.5
	1118	1.1		1101	1.2
W	1615	0.2	Th	1602	-0.1
	2249	1.8		2244	1.9
15	0523	-0.3	30	0503	-0.5
	1149	1.1		1139	1.3
Th	1659	0.2	F	1654	-0.1
	2327	1.6		2333	1.7
			31	0543	-0.4
				1219	1.4
			Sa	1751	-0.2

	Sun↑	Sun↓
15	0711	1758
31	0708	1810

FEBRUARY

Day	Time	ft	Day	Time	ft
1	0026	1.5	16	0026	1.2
	0625	-0.2		0619	0.1
Su	1301	1.4	M	1242	1.3
	1853	-0.1		1852	0.0
2	0123	1.3	17	0110	1.1
	0709	-0.1		0649	0.2
M	1347	1.5	Tu	1318	1.3
	2002	-0.1		1946	0.1
3	0230	1.1	18	0201	0.9
	0757	0.1		0723	0.3
Tu	1440	1.5	W	1400	1.3
☾	2119	-0.1		2051	0.1
4	0352	0.9	19	0305	0.8
	0852	0.2		0806	0.4
W	1542	1.5	Th	1450	1.3
	2239	-0.2	☽	2205	0.0
5	0528	0.8	20	0431	0.7
	0955	0.3		0904	0.4
Th	1652	1.5	F	1552	1.3
	2352	-0.2		2317	-0.0
6	0650	0.8	21	0604	0.7
	1101	0.3		1015	0.5
F	1803	1.6	Sa	1703	1.4
7	0055	-0.3	22	0019	-0.1
	0751	0.9		0710	0.8
Sa	1205	0.3	Su	1124	0.4
	1905	1.6		1813	1.5
8	0147	-0.3	23	0110	-0.3
	0837	0.9		0758	0.9
Su	1302	0.2	M	1225	0.3
	1958	1.7		1914	1.6
9	0232	-0.3	24	0155	-0.4
	0915	1.0		0838	1.0
M	1353	0.2	Tu	1321	0.1
	2043	1.7		2008	1.8
10	0310	-0.3	25	0236	-0.4
	0948	1.0		0915	1.1
Tu	1439	0.1	W	1413	-0.0
	2123	1.7		2100	1.8
11	0346	-0.3	26	0315	-0.5
	1017	1.1		0951	1.3
W	1522	0.0	Th	1504	-0.2
○	2200	1.6	●	2149	1.8
12	0419	-0.3	27	0354	-0.4
	1044	1.1		1027	1.4
Th	1603	-0.0	F	1556	-0.3
	2235	1.6		2239	1.8
13	0451	-0.2	28	0433	-0.4
	1111	1.2		1104	1.5
F	1643	-0.0	Sa	1648	-0.4
	2311	1.5		2329	1.6
14	0521	-0.1			
	1139	1.2			
Sa	1724	-0.0			
	2347	1.4			
15	0550	-0.0			
	1209	1.3			
Su	1806	-0.0			

	Sun↑	Sun↓
15	0659	1819
28	0649	1827

MARCH

Day	Time	ft	Day	Time	ft
1	0512	-0.3	16	0507	0.1
	1143	1.6		1127	1.5
Su	1743	-0.4	M	1742	-0.1
2	0021	1.4	17	0010	1.2
	0552	-0.1		0534	0.2
M	1224	1.7	Tu	1158	1.5
	1842	-0.4		1823	-0.1
3	0117	1.2	18	0052	1.1
	0635	0.0		0603	0.3
Tu	1310	1.7	W	1232	1.5
	1947	-0.3		1910	-0.1
4	0222	1.0	19	0141	1.0
	0723	0.2		0636	0.4
W	1402	1.6	Th	1311	1.4
	2059	-0.2		2008	-0.0
5	0341	0.8	20	0243	0.9
	0819	0.3		0718	0.5
Th	1506	1.5	F	1400	1.4
☾	2217	-0.2		2117	0.0
6	0515	0.8	21	0403	0.8
	0930	0.4		0818	0.5
F	1624	1.5	Sa	1503	1.4
	2332	-0.1	☽	2232	-0.0
7	0635	0.8	22	0531	0.8
	1046	0.4		0939	0.5
Sa	1745	1.5	Su	1623	1.4
				2338	-0.1
8	0035	-0.1	23	0636	0.9
	0732	0.9		1101	0.5
Su	1157	0.3	M	1744	1.5
	1854	1.5			
9	0126	-0.2	24	0033	-0.1
	0814	1.0		0723	1.0
M	1256	0.2	Tu	1210	0.3
	1948	1.5		1854	1.6
10	0208	-0.2	25	0119	-0.2
	0848	1.1		0802	1.2
Tu	1346	0.1	W	1309	0.1
	2033	1.6		1955	1.7
11	0243	-0.2	26	0201	-0.2
	0916	1.2		0839	1.4
W	1431	0.0	Th	1404	-0.1
	2112	1.6		2050	1.7
12	0315	-0.1	27	0241	-0.2
	0941	1.2		0915	1.6
Th	1511	-0.0	F	1457	-0.3
○	2147	1.5	●	2142	1.7
13	0345	-0.1	28	0320	-0.2
	1006	1.3		0952	1.7
F	1550	-0.1	Sa	1548	-0.5
	2221	1.5		2233	1.6
14	0413	-0.0	29	0400	-0.1
	1031	1.4		1030	1.9
Sa	1627	-0.1	Su	1640	-0.5
	2256	1.4		2323	1.5
15	0441	0.0	30	0439	-0.0
	1058	1.4		1110	1.9
Su	1704	-0.2	M	1733	-0.5
	2332	1.3			
			31	0015	1.3
				0520	0.1
			Tu	1152	1.9
				1829	-0.5

	Sun↑	Sun↓
15	0634	1834
31	0618	1841

APRIL

Day	Time	ft	Day	Time	ft
1	0110	1.2	16	0042	1.1
	0604	0.2		0527	0.4
W	1238	1.8	Th	1157	1.7
	1929	-0.3		1846	-0.1
2	0211	1.0	17	0131	1.0
	0654	0.4		0603	0.5
Th	1331	1.7	F	1236	1.6
	2036	-0.2		1940	-0.1
3	0325	0.9	18	0231	1.0
	0755	0.5		0649	0.6
F	1434	1.6	Sa	1326	1.6
☾	2149	-0.1		2043	-0.0
4	0450	0.9	19	0342	0.9
	0912	0.5		0755	0.6
Sa	1553	1.5	Su	1430	1.5
	2300	0.0	☽	2151	0.0
5	0605	1.0	20	0454	1.0
	1035	0.5		0921	0.6
Su	1721	1.4	M	1552	1.5
				2255	0.0
6	0001	0.0	21	0554	1.1
	0659	1.1		1047	0.5
M	1149	0.4	Tu	1720	1.5
	1834	1.4		2350	0.0
7	0050	0.1	22	0641	1.3
	0738	1.2		1159	0.3
Tu	1248	0.3	W	1838	1.5
	1930	1.4			
8	0130	0.1	23	0038	-0.0
	0809	1.3		0722	1.5
W	1337	0.2	Th	1300	0.0
	2016	1.4		1943	1.6
9	0205	0.1	24	0122	0.0
	0836	1.4		0801	1.7
Th	1419	0.0	F	1356	-0.2
	2055	1.4		2041	1.6
10	0236	0.1	25	0204	0.0
	0901	1.5		0840	1.9
F	1458	-0.1	Sa	1448	-0.4
	2131	1.4		2135	1.6
11	0305	0.1	26	0245	0.1
	0926	1.6		0919	2.0
Sa	1534	-0.1	Su	1539	-0.5
○	2206	1.4	●	2226	1.5
12	0333	0.2	27	0326	0.1
	0953	1.6		1000	2.1
Su	1609	-0.2	M	1630	-0.6
	2241	1.4		2316	1.4
13	0400	0.2	28	0408	0.2
	1021	1.7		1042	2.1
M	1644	-0.2	Tu	1721	-0.5
	2319	1.3			
14	0427	0.3	29	0006	1.3
	1051	1.7		0452	0.3
Tu	1721	-0.2	W	1126	2.1
	2358	1.2		1814	-0.4
15	0456	0.4	30	0059	1.2
	1122	1.7		0538	0.4
W	1801	-0.2	Th	1213	2.0
				1910	-0.3

	Sun↑	Sun↓
15	0603	1847
30	0550	1854

VACA KEY, FLORIDA

HIGH & LOW WATER 1998 | U.S. Datum | 24°42.6'N 81°06.4'W

Eastern Standard Time (75°W) | Add 1H Daylight Saving Time: April 5 – October 24

MAY

Day	Time	ft	Day	Time	ft
1 F	0156	1.1	16 Sa	0122	1.1
	0631	0.5		0543	0.6
	1304	1.8		1216	1.8
	2011	-0.1		1919	-0.1
2 Sa	0300	1.0	17 Su	0216	1.1
	0735	0.6		0636	0.6
	1403	1.6		1308	1.7
	2115	0.0		2014	-0.1
3 Su	0411	1.1	18 M	0315	1.1
	0854	0.6		0747	0.6
	1516	1.4		1412	1.6
	☾ 2218	0.1		☽ 2113	0.0
4 M	0518	1.1	19 Tu	0414	1.2
	1019	0.6		0913	0.6
	1640	1.3		1532	1.5
	2314	0.2		2212	0.1
5 Tu	0610	1.2	20 W	0509	1.3
	1132	0.5		1037	0.4
	1759	1.3		1701	1.4
				2306	0.1
6 W	0002	0.2	21 Th	0558	1.5
	0649	1.3		1150	0.2
	1232	0.3		1824	1.4
	1902	1.3		2356	0.2
7 Th	0043	0.3	22 F	0643	1.7
	0721	1.4		1252	-0.1
	1320	0.2		1934	1.4
	1951	1.3			
8 F	0119	0.3	23 Sa	0043	0.2
	0750	1.6		0727	1.9
	1402	0.1		1348	-0.3
	2034	1.3		2034	1.4
9 Sa	0151	0.3	24 Su	0128	0.2
	0819	1.7		0810	2.1
	1440	-0.1		1440	-0.5
	2112	1.3		2128	1.4
10 Su	0222	0.3	25 M	0213	0.2
	0848	1.7		0853	2.2
	1516	-0.1		1530	-0.5
	2150	1.3		● 2218	1.3
11 M	0251	0.4	26 Tu	0257	0.3
	0918	1.8		0937	2.2
	1551	-0.2		1619	-0.6
	○ 2228	1.3		2306	1.3
12 Tu	0321	0.4	27 W	0342	0.3
	0949	1.8		1021	2.2
	1627	-0.2		1708	-0.5
	2307	1.3		2354	1.2
13 W	0351	0.4	28 Th	0428	0.4
	1022	1.9		1106	2.1
	1704	-0.2		1757	-0.4
	2348	1.2			
14 Th	0424	0.5	29 F	0041	1.2
	1056	1.9		0517	0.4
	1744	-0.2		1152	2.0
				1847	-0.2
15 F	0033	1.2	30 Sa	0131	1.1
	0500	0.5		0611	0.5
	1134	1.8		1240	1.8
	1829	-0.2		1940	-0.1
			31 Su	0223	1.1
				0714	0.6
				1333	1.6
				2033	0.1

	Sun↑	Sun↓
15	0541	1902
31	0536	1910

JUNE

Day	Time	ft	Day	Time	ft
1 M	0318	1.2	16 Tu	0240	1.3
	0829	0.6		0744	0.5
	1434	1.4		1403	1.5
	☾ 2127	0.2		2037	0.0
2 Tu	0414	1.2	17 W	0331	1.4
	0950	0.6		0906	0.4
	1547	1.3		1520	1.4
	2219	0.3		☽ 2130	0.1
3 W	0505	1.3	18 Th	0424	1.5
	1104	0.5		1027	0.3
	1707	1.2		1649	1.2
	2306	0.3		2224	0.2
4 Th	0548	1.4	19 F	0517	1.7
	1206	0.3		1140	0.1
	1821	1.1		1816	1.2
	2349	0.4		2317	0.3
5 F	0627	1.5	20 Sa	0609	1.9
	1257	0.2		1244	-0.1
	1920	1.1		1929	1.2
6 Sa	0028	0.4	21 Su	0008	0.3
	0703	1.6		0700	2.0
	1340	0.1		1340	-0.3
	2009	1.2		2029	1.2
7 Su	0104	0.4	22 M	0058	0.3
	0738	1.7		0749	2.1
	1420	-0.1		1432	-0.4
	2052	1.2		2121	1.2
8 M	0139	0.4	23 Tu	0147	0.3
	0813	1.8		0836	2.2
	1457	-0.2		1520	-0.5
	2133	1.2		● 2208	1.2
9 Tu	0213	0.4	24 W	0235	0.3
	0849	1.9		0922	2.2
	1533	-0.2		1607	-0.4
	○ 2213	1.2		2252	1.2
10 W	0247	0.5	25 Th	0322	0.3
	0925	2.0		1007	2.2
	1610	-0.3		1652	-0.4
	2253	1.2		2334	1.2
11 Th	0323	0.5	26 F	0410	0.4
	1002	2.0		1050	2.1
	1648	-0.3		1736	-0.3
	2335	1.2			
12 F	0401	0.5	27 Sa	0015	1.2
	1040	2.0		0459	0.4
	1728	-0.3		1133	1.9
				1820	-0.2
13 Sa	0018	1.2	28 Su	0055	1.2
	0444	0.5		0551	0.4
	1122	1.9		1217	1.8
	1811	-0.2		1903	-0.0
14 Su	0103	1.2	29 M	0137	1.2
	0533	0.5		0649	0.5
	1207	1.8		1304	1.6
	1857	-0.2		1947	0.1
15 M	0150	1.2	30 Tu	0219	1.3
	0632	0.5		0756	0.5
	1300	1.7		1355	1.4
	1945	-0.1		2032	0.2

	Sun↑	Sun↓
15	0536	1915
30	0539	1918

JULY

Day	Time	ft	Day	Time	ft
1 W	0304	1.3	16 Th	0249	1.6
	0910	0.5		0855	0.3
	1455	1.2		1513	1.3
	☾ 2118	0.4		☽ 2051	0.3
2 Th	0352	1.4	17 F	0343	1.7
	1024	0.5		1015	0.2
	1608	1.1		1643	1.1
	2205	0.4		2147	0.4
3 F	0441	1.5	18 Sa	0443	1.9
	1130	0.4		1130	0.1
	1731	1.0		1812	1.1
	2251	0.5		2245	0.5
4 Sa	0530	1.6	19 Su	0544	2.0
	1226	0.2		1236	-0.1
	1845	1.0		1925	1.1
	2336	0.5		2343	0.5
5 Su	0617	1.7	20 M	0643	2.1
	1314	0.1		1333	-0.2
	1942	1.1		2023	1.1
6 M	0019	0.5	21 Tu	0039	0.5
	0701	1.8		0738	2.1
	1356	-0.0		1423	-0.2
	2030	1.1		2110	1.2
7 Tu	0100	0.5	22 W	0132	0.4
	0744	1.9		0827	2.2
	1436	-0.1		1509	-0.2
	2113	1.1		2152	1.2
8 W	0140	0.5	23 Th	0222	0.4
	0825	2.0		0913	2.2
	1513	-0.2		1551	-0.2
	2153	1.2		● 2230	1.3
9 Th	0221	0.5	24 F	0310	0.4
	0906	2.0		0956	2.1
	1551	-0.3		1630	-0.2
	○ 2233	1.2		2305	1.3
10 F	0303	0.4	25 Sa	0357	0.4
	0948	2.1		1036	2.0
	1629	-0.3		1708	-0.1
	2312	1.3		2339	1.4
11 Sa	0347	0.4	26 Su	0443	0.4
	1030	2.1		1116	1.9
	1708	-0.3		1745	0.0
	2352	1.3			
12 Su	0435	0.4	27 M	0012	1.4
	1115	2.0		0531	0.4
	1748	-0.2		1155	1.8
				1821	0.1
13 M	0033	1.4	28 Tu	0045	1.4
	0528	0.4		0621	0.4
	1203	1.9		1236	1.6
	1830	-0.1		1857	0.3
14 Tu	0115	1.4	29 W	0121	1.5
	0628	0.4		0717	0.5
	1257	1.7		1321	1.4
	1914	0.0		1934	0.4
15 W	0200	1.5	30 Th	0201	1.5
	0738	0.3		0821	0.5
	1358	1.5		1414	1.3
	2000	0.2		2013	0.5
			31 F	0245	1.5
				0932	0.5
				1519	1.1
				☾ 2057	0.6

	Sun↑	Sun↓
15	0545	1917
31	0552	1910

AUGUST

Day	Time	ft	Day	Time	ft
1 Sa	0337	1.6	16 Su	0415	2.0
	1044	0.5		1118	0.2
	1641	1.1		1809	1.2
	2149	0.7		2223	0.7
2 Su	0434	1.7	17 M	0527	2.0
	1149	0.4		1225	0.1
	1807	1.1		1918	1.2
	2245	0.7		2330	0.7
3 M	0532	1.7	18 Tu	0635	2.1
	1243	0.3		1322	0.1
	1914	1.1		2010	1.3
	2339	0.7			
4 Tu	0627	1.9	19 W	0031	0.6
	1329	0.1		0732	2.2
	2005	1.2		1409	0.1
				2052	1.4
5 W	0029	0.6	20 Th	0126	0.6
	0718	2.0		0822	2.2
	1410	0.0		1450	0.1
	2047	1.2		2128	1.4
6 Th	0116	0.6	21 F	0215	0.5
	0806	2.1		0906	2.2
	1448	-0.1		1527	0.1
	2126	1.3		● 2200	1.5
7 F	0203	0.5	22 Sa	0301	0.4
	0851	2.2		0945	2.1
	1526	-0.1		1601	0.1
	○ 2203	1.4		2229	1.6
8 Sa	0249	0.4	23 Su	0345	0.4
	0936	2.2		1022	2.1
	1603	-0.1		1634	0.2
	2240	1.5		2257	1.6
9 Su	0338	0.3	24 M	0427	0.4
	1022	2.2		1058	2.0
	1640	-0.1		1706	0.3
	2317	1.6		2325	1.7
10 M	0428	0.3	25 Tu	0509	0.4
	1109	2.1		1134	1.8
	1718	-0.0		1736	0.4
	2355	1.7		2355	1.7
11 Tu	0522	0.2	26 W	0553	0.4
	1159	1.9		1212	1.7
	1758	0.1		1807	0.5
12 W	0035	1.8	27 Th	0028	1.8
	0621	0.2		0639	0.5
	1253	1.7		1254	1.6
	1840	0.3		1838	0.6
13 Th	0119	1.9	28 F	0104	1.8
	0727	0.2		0733	0.5
	1355	1.5		1343	1.4
	1925	0.4		1911	0.7
14 F	0209	1.9	29 Sa	0147	1.8
	0842	0.2		0837	0.6
	1509	1.3		1443	1.3
	☽ 2016	0.5		☾ 1952	0.8
15 Sa	0307	2.0	30 Su	0238	1.8
	1001	0.2		0951	0.6
	1639	1.2		1602	1.2
	2116	0.6		2047	0.9
			31 M	0340	1.8
				1104	0.5
				1733	1.2
				2157	0.9

	Sun↑	Sun↓
15	0559	1859
31	0606	1845

T Tides

VACA KEY, FLORIDA

HIGH & LOW WATER 1998 — U.S. Datum — 24°42.6'N 81°06.4'W

Eastern Standard Time (75°W) — Add 1H Daylight Saving Time: April 5 – October 24

SEPTEMBER

Day	Time	ft	Day	Time	ft
1	0449	1.9	16	0627	2.1
	1206	0.4		1301	0.4
Tu	1844	1.3	W	1947	1.5
	2305	0.9			
2	0556	2.0	17	0029	0.8
	1256	0.3		0726	2.1
W	1934	1.4	Th	1345	0.4
				2024	1.6
3	0005	0.8	18	0123	0.7
	0655	2.1		0814	2.1
Th	1338	0.2	F	1422	0.4
	2015	1.5		2056	1.7
4	0059	0.7	19	0209	0.6
	0748	2.2		0855	2.1
F	1416	0.2	Sa	1455	0.4
	2051	1.6		2123	1.8
5	0150	0.5	20	0252	0.5
	0838	2.3		0932	2.1
Sa	1453	0.1	Su	1526	0.5
	2127	1.7	●	2149	1.9
6	0240	0.4	21	0332	0.4
	0926	2.3		1006	2.0
Su	1530	0.1	M	1556	0.5
○	2202	1.9		2214	2.0
7	0329	0.2	22	0410	0.4
	1014	2.3		1041	2.0
M	1607	0.2	Tu	1624	0.6
	2238	2.0		2241	2.0
8	0421	0.1	23	0448	0.4
	1103	2.1		1116	1.9
Tu	1645	0.3	W	1652	0.7
	2316	2.1		2310	2.0
9	0514	0.1	24	0526	0.4
	1154	2.0		1153	1.8
W	1724	0.4	Th	1720	0.8
	2357	2.2		2342	2.0
10	0611	0.1	25	0608	0.5
	1248	1.8		1234	1.7
Th	1806	0.5	F	1749	0.8
11	0042	2.2	26	0017	2.0
	0714	0.2		0654	0.5
F	1350	1.6	Sa	1321	1.5
	1852	0.7		1821	0.9
12	0134	2.2	27	0058	2.0
	0825	0.3		0750	0.6
Sa	1504	1.4	Su	1419	1.4
☽	1947	0.8		1901	1.0
13	0237	2.1	28	0147	1.9
	0943	0.4		0900	0.6
Su	1632	1.3	M	1534	1.4
	2055	0.9	☾	1958	1.1
14	0352	2.1	29	0250	1.9
	1100	0.4		1015	0.6
M	1757	1.4	Tu	1700	1.4
	2212	0.9		2118	1.1
15	0514	2.1	30	0406	1.9
	1208	0.4		1121	0.6
Tu	1900	1.4	W	1808	1.4
	2326	0.9		2240	1.0

	Sun↑	Sun↓
15	0611	1829
30	0617	1813

OCTOBER

Day	Time	ft	Day	Time	ft
1	0524	2.0	16	0025	0.8
	1214	0.5		0712	2.0
Th	1857	1.6	F	1310	0.6
	2348	0.9		1948	1.8
2	0632	2.1	17	0116	0.7
	1258	0.4		0800	2.0
F	1936	1.7	Sa	1345	0.6
				2018	1.9
3	0046	0.7	18	0201	0.5
	0731	2.2		0840	2.0
Sa	1338	0.4	Su	1418	0.6
	2012	1.9		2044	2.0
4	0139	0.5	19	0240	0.4
	0825	2.2		0917	2.0
Su	1416	0.4	M	1448	0.7
	2048	2.1		2109	2.1
5	0231	0.2	20	0317	0.4
	0916	2.2		0951	1.9
M	1454	0.4	Tu	1516	0.7
○	2124	2.2	●	2135	2.1
6	0321	0.1	21	0353	0.3
	1006	2.2		1025	1.9
Tu	1533	0.5	W	1545	0.7
	2202	2.4		2203	2.2
7	0412	-0.0	22	0428	0.3
	1056	2.1		1101	1.8
W	1612	0.5	Th	1612	0.8
	2242	2.5		2234	2.2
8	0504	-0.0	23	0505	0.3
	1147	1.9		1138	1.7
Th	1653	0.6	F	1641	0.9
	2325	2.5		2306	2.2
9	0559	0.1	24	0543	0.4
	1242	1.8		1220	1.6
F	1736	0.7	Sa	1711	0.9
				2341	2.1
10	0012	2.4	25	2341	2.1
	0659	0.2		0627	0.4
Sa	1342	1.6	Su	1307	1.5
	1825	0.8		1745	1.0
11	0106	2.3	26	0021	2.1
	0806	0.3		0718	0.5
Su	1452	1.5	M	1403	1.5
	1925	1.0		1829	1.0
12	0210	2.2	27	0109	2.0
	0919	0.4		0819	0.5
M	1613	1.5	Tu	1510	1.4
☽	2040	1.0		1929	1.1
13	0328	2.1	28	0211	1.9
	1032	0.5		0927	0.6
Tu	1730	1.5	W	1622	1.4
	2205	1.0	☾	2053	1.1
14	0455	2.0	29	0329	1.9
	1136	0.6		1031	0.6
W	1829	1.6	Th	1723	1.5
	2322	0.9		2221	1.0
15	0612	2.0	30	0454	1.9
	1228	0.6		1126	0.5
Th	1914	1.7	F	1812	1.7
				2334	0.8
			31	0611	1.9
				1214	0.5
			Sa	1854	1.9

	Sun↑	Sun↓
15	0623	1758
31	0631	1745

NOVEMBER

Day	Time	ft	Day	Time	ft
1	0035	0.5	16	0146	0.4
	0717	2.0		0822	1.7
Su	1257	0.5	M	1337	0.7
	1933	2.1		2004	2.0
2	0130	0.3	17	0225	0.3
	0815	2.0		0900	1.7
M	1338	0.5	Tu	1409	0.7
	2012	2.3		2033	2.1
3	0222	0.0	18	0301	0.2
	0908	2.0		0936	1.6
Tu	1419	0.5	W	1439	0.7
	2051	2.4	●	2103	2.1
4	0312	-0.1	19	0337	0.1
	0959	1.9		1011	1.6
W	1500	0.5	Th	1509	0.7
○	2132	2.6		2134	2.1
5	0402	-0.2	20	0411	0.1
	1049	1.8		1048	1.6
Th	1542	0.6	F	1539	0.7
	2216	2.6		2207	2.1
6	0453	-0.2	21	0447	0.1
	1139	1.7		1126	1.5
F	1626	0.6	Sa	1611	0.7
	2301	2.5		2242	2.1
7	0546	-0.1	22	0525	0.1
	1231	1.6		1208	1.5
Sa	1712	0.7	Su	1645	0.8
	2350	2.4		2318	2.1
8	0642	0.1	23	0606	0.2
	1326	1.5		1253	1.4
Su	1804	0.8	M	1724	0.8
				2359	2.0
9	0043	2.3	24	0653	0.2
	0742	0.2		1343	1.4
M	1428	1.5	Tu	1812	0.8
	1907	0.9			
10	0144	2.1	25	0047	1.9
	0847	0.4		0745	0.3
Tu	1537	1.5	W	1438	1.4
☽	2024	0.9		1916	0.9
11	0257	1.9	26	0146	1.8
	0952	0.5		0842	0.4
W	1646	1.5	Th	1536	1.4
	2151	0.9	☾	2037	0.8
12	0422	1.8	27	0301	1.7
	1052	0.6		0940	0.4
Th	1744	1.6	F	1632	1.5
	2309	0.8		2204	0.7
13	0543	1.7	28	0428	1.6
	1142	0.6		1036	0.4
F	1829	1.7	Sa	1724	1.7
				2319	0.5
14	0012	0.7	29	0554	1.6
	0648	1.7		1128	0.5
Sa	1225	0.6	Su	1811	1.9
	1905	1.8			
15	0103	0.5	30	0024	0.2
	0740	1.7		0706	1.6
Su	1303	0.7	M	1216	0.5
	1935	1.9		1857	2.1

	Sun↑	Sun↓
15	0641	1738
30	0652	1735

DECEMBER

Day	Time	ft	Day	Time	ft
1	0120	-0.0	16	0207	0.0
	0808	1.6		0842	1.3
Tu	1303	0.5	W	1331	0.5
	1941	2.2		2002	1.9
2	0213	-0.2	17	0245	-0.0
	0902	1.6		0920	1.3
W	1348	0.5	Th	1406	0.5
	2027	2.4		2038	1.9
3	0303	-0.4	18	0320	-0.1
	0952	1.5		0956	1.3
Th	1433	0.5	F	1439	0.5
○	2112	2.4	●	2113	2.0
4	0353	-0.4	19	0355	-0.2
	1040	1.5		1034	1.3
F	1519	0.4	Sa	1514	0.5
	2159	2.4		2150	2.0
5	0442	-0.4	20	0431	-0.2
	1127	1.5		1112	1.3
Sa	1606	0.5	Su	1550	0.5
	2245	2.4		2227	2.0
6	0531	-0.3	21	0508	-0.2
	1213	1.4		1151	1.3
Su	1655	0.5	M	1629	0.5
	2333	2.2		2306	1.9
7	0621	-0.1	22	0547	-0.1
	1302	1.4		1232	1.2
M	1749	0.6	Tu	1713	0.5
				2348	1.8
8	0024	2.0	23	0628	-0.1
	0713	0.0		1315	1.3
Tu	1352	1.3	W	1805	0.5
	1850	0.6			
9	0118	1.8	24	0036	1.7
	0806	0.2		0713	0.0
W	1447	1.3	Th	1400	1.3
	2002	0.7		1909	0.5
10	0219	1.6	25	0133	1.5
	0901	0.3		0801	0.1
Th	1543	1.4	F	1448	1.4
☽	2123	0.6		2024	0.4
11	0332	1.4	26	0244	1.4
	0956	0.4		0853	0.2
F	1639	1.4	Sa	1541	1.5
	2240	0.6	☾	2147	0.3
12	0456	1.3	27	0410	1.2
	1047	0.5		0948	0.3
Sa	1728	1.5	Su	1636	1.6
	2346	0.4		2304	0.1
13	0613	1.3	28	0542	1.1
	1134	0.5		1044	0.4
Su	1811	1.6	M	1732	1.8
14	0041	0.3	29	0012	-0.1
	0714	1.2		0700	1.1
M	1216	0.6	Tu	1140	0.4
	1849	1.7		1827	1.9
15	0127	0.2	30	0112	-0.3
	0802	1.3		0804	1.2
Tu	1255	0.6	W	1233	0.4
	1926	1.8		1921	2.1
			31	0205	-0.4
				0857	1.2
			Th	1325	0.3
				2012	2.1

	Sun↑	Sun↓
15	0702	1739
31	0709	1747

KEY WEST, FLORIDA

HIGH & LOW WATER 1998 | U.S. Datum | 24°33.2'N 81°48.5'W

Eastern Standard Time (75°W) | Add 1H Daylight Saving Time: April 5 – October 24

JANUARY

Day	Time	ft	Day	Time	ft
1	0526	-0.4	16	0559	-0.2
	1210	1.2		1221	1.2
	Th 1658	0.3		F 1744	0.2
	2337	1.9			
2	0610	-0.3	17	0006	1.5
	1253	1.3		0634	-0.0
	F 1754	0.2		Sa 1255	1.2
				1834	0.2
3	0029	1.8	18	0048	1.3
	0655	-0.2		0710	0.1
	Sa 1339	1.3		Su 1332	1.2
	1859	0.2		1930	0.3
4	0127	1.5	19	0136	1.2
	0744	-0.0		0747	0.2
	Su 1429	1.4		M 1413	1.2
	2014	0.2		2035	0.3
5	0237	1.3	20	0233	1.0
	0835	0.1		0828	0.3
	M 1523	1.5		Tu 1500	1.3
	☾ 2135	0.1		☽ 2149	0.2
6	0402	1.1	21	0345	0.9
	0930	0.2		0916	0.4
	Tu 1622	1.6		W 1553	1.3
	2255	0.0		2300	0.1
7	0535	1.0	22	0513	0.8
	1028	0.3		1010	0.4
	W 1723	1.7		Th 1651	1.4
8	0005	-0.1	23	0002	0.0
	0656	1.0		0635	0.8
	Th 1126	0.4		F 1106	0.4
	1821	1.8		1749	1.5
9	0106	-0.3	24	0054	-0.1
	0759	1.0		0736	0.9
	F 1221	0.3		Sa 1159	0.4
	1915	1.9		1844	1.6
10	0158	-0.3	25	0140	-0.3
	0849	1.1		0824	0.9
	Sa 1313	0.3		Su 1249	0.3
	2005	1.9		1935	1.7
11	0245	-0.4	26	0222	-0.4
	0933	1.1		0906	1.0
	Su 1401	0.3		M 1337	0.2
	2050	1.9		2023	1.8
12	0328	-0.4	27	0303	-0.5
	1011	1.1		0945	1.1
	M 1447	0.2		Tu 1424	0.1
	○ 2132	1.9		2110	1.9
13	0408	-0.4	28	0343	-0.5
	1045	1.1		1023	1.1
	Tu 1532	0.2		W 1512	0.0
	2211	1.9		● 2157	1.9
14	0446	-0.3	29	0423	-0.5
	1118	1.1		1101	1.2
	W 1615	0.2		Th 1602	-0.1
	2249	1.8		2244	1.9
15	0523	-0.3	30	0503	-0.5
	1149	1.1		1139	1.3
	Th 1659	0.2		F 1654	-0.1
	2327	1.6		2333	1.7
			31	0543	-0.4
				1219	1.4
				Sa 1751	-0.2

	Sun↑	Sun↓
15	0714	1801
31	0710	1813

FEBRUARY

Day	Time	ft	Day	Time	ft
1	0026	1.5	16	0026	1.2
	0625	-0.2		0619	0.1
	Su 1301	1.4		M 1242	1.3
	1853	-0.1		1852	0.0
2	0123	1.3	17	0110	1.1
	0709	-0.1		0649	0.2
	M 1347	1.5		Tu 1318	1.3
	2002	-0.1		1946	0.1
3	0230	1.1	18	0201	0.9
	0757	0.1		0723	0.3
	Tu 1440	1.5		W 1400	1.3
	☾ 2119	-0.1		2051	0.1
4	0352	0.9	19	0305	0.8
	0852	0.2		0806	0.4
	W 1542	1.5		Th 1450	1.3
	2239	-0.2		☽ 2205	0.0
5	0528	0.8	20	0431	0.7
	0955	0.3		0904	0.4
	Th 1652	1.5		F 1552	1.3
	2352	-0.2		2317	-0.0
6	0650	0.8	21	0604	0.7
	1101	0.3		1015	0.5
	F 1803	1.6		Sa 1703	1.4
7	0055	-0.3	22	0019	-0.1
	0751	0.9		0710	0.8
	Sa 1205	0.3		Su 1124	0.4
	1905	1.6		1813	1.5
8	0147	-0.3	23	0110	-0.3
	0837	0.9		0758	0.9
	Su 1302	0.2		M 1225	0.3
	1958	1.7		1914	1.6
9	0232	-0.3	24	0155	-0.4
	0915	1.0		0838	1.0
	M 1353	0.2		Tu 1321	0.1
	2043	1.7		2008	1.8
10	0310	-0.3	25	0236	-0.4
	0948	1.0		0915	1.1
	Tu 1439	0.1		W 1413	-0.0
	2123	1.7		2100	1.8
11	0346	-0.3	26	0315	-0.5
	1017	1.1		0951	1.3
	W 1522	0.0		Th 1504	-0.2
	○ 2200	1.6		● 2149	1.8
12	0419	-0.3	27	0354	-0.4
	1044	1.1		1027	1.4
	Th 1603	-0.0		F 1556	-0.3
	2235	1.6		2239	1.8
13	0451	-0.2	28	0433	-0.4
	1111	1.2		1104	1.5
	F 1643	-0.0		Sa 1648	-0.4
	2311	1.5		2329	1.6
14	0521	-0.1			
	1139	1.2			
	Sa 1724	-0.0			
	2347	1.4			
15	0550	-0.0			
	1209	1.3			
	Su 1806	-0.0			

	Sun↑	Sun↓
15	0702	1822
28	0651	1830

MARCH

Day	Time	ft	Day	Time	ft
1	0512	-0.3	16	0507	0.1
	1143	1.6		1127	1.5
	Su 1743	-0.4		M 1742	-0.1
2	0021	1.4	17	0010	1.2
	0552	-0.1		0534	0.2
	M 1224	1.7		Tu 1158	1.5
	1842	-0.4		1823	-0.1
3	0117	1.2	18	0052	1.1
	0635	0.0		0603	0.3
	Tu 1310	1.7		W 1232	1.5
	1947	-0.3		1910	-0.1
4	0222	1.0	19	0141	1.0
	0723	0.2		0636	0.4
	W 1402	1.6		Th 1311	1.4
	2059	-0.2		2008	-0.0
5	0341	0.8	20	0243	0.9
	0819	0.3		0718	0.5
	Th 1506	1.5		F 1400	1.4
	☾ 2217	-0.2		2117	0.0
6	0515	0.8	21	0403	0.8
	0930	0.4		0818	0.5
	F 1624	1.5		Sa 1503	1.4
	2332	-0.1		☽ 2232	-0.0
7	0635	0.8	22	0531	0.8
	1046	0.4		0939	0.5
	Sa 1745	1.5		Su 1623	1.4
				2338	-0.1
8	0035	-0.1	23	0636	0.9
	0732	0.9		1101	0.5
	Su 1157	0.3		M 1744	1.5
	1854	1.5			
9	0126	-0.2	24	0033	-0.1
	0814	1.0		0723	1.0
	M 1256	0.2		Tu 1210	0.3
	1948	1.5		1854	1.6
10	0208	-0.2	25	0119	-0.2
	0848	1.1		0802	1.2
	Tu 1346	0.1		W 1309	0.1
	2033	1.6		1955	1.7
11	0243	-0.2	26	0201	-0.2
	0916	1.2		0839	1.4
	W 1431	0.0		Th 1404	-0.1
	2112	1.6		2050	1.7
12	0315	-0.1	27	0241	-0.2
	0941	1.2		0915	1.6
	Th 1511	-0.0		F 1457	-0.3
	○ 2147	1.5		● 2142	1.7
13	0345	-0.1	28	0320	-0.2
	1006	1.3		0952	1.7
	F 1550	-0.1		Sa 1548	-0.5
	2221	1.5		2233	1.6
14	0413	-0.0	29	0400	-0.1
	1031	1.4		1030	1.9
	Sa 1627	-0.1		Su 1640	-0.5
	2256	1.4		2323	1.5
15	0441	0.0	30	0439	-0.0
	1058	1.4		1110	1.9
	Su 1704	-0.2		M 1733	-0.5
	2332	1.3			
			31	0015	1.3
				0520	0.1
				Tu 1152	1.9
				1829	-0.5

	Sun↑	Sun↓
15	0637	1837
31	0621	1844

APRIL

Day	Time	ft	Day	Time	ft
1	0110	1.2	16	0042	1.1
	0604	0.2		0527	0.4
	W 1238	1.8		Th 1157	1.7
	1929	-0.3		1846	-0.1
2	0211	1.0	17	0131	1.0
	0654	0.4		0603	0.5
	Th 1331	1.7		F 1236	1.6
	2036	-0.2		1940	-0.1
3	0325	0.9	18	0231	1.0
	0755	0.5		0649	0.6
	F 1434	1.6		Sa 1326	1.6
	☾ 2149	-0.1		2043	-0.0
4	0450	0.9	19	0342	0.9
	0912	0.5		0755	0.6
	Sa 1553	1.5		Su 1430	1.5
	2300	0.0		☽ 2151	0.0
5	0605	1.0	20	0454	1.0
	1035	0.5		0921	0.6
	Su 1721	1.4		M 1552	1.5
				2255	0.0
6	0001	0.0	21	0554	1.1
	0659	1.1		1047	0.5
	M 1149	0.4		Tu 1720	1.5
	1834	1.4		2350	0.0
7	0050	0.1	22	0641	1.3
	0738	1.2		1159	0.3
	Tu 1248	0.3		W 1838	1.5
	1930	1.4			
8	0130	0.1	23	0038	-0.0
	0809	1.3		0722	1.5
	W 1337	0.2		Th 1300	0.0
	2016	1.4		1943	1.6
9	0205	0.1	24	0122	0.0
	0836	1.4		0801	1.7
	Th 1419	0.0		F 1356	-0.2
	2055	1.4		2041	1.6
10	0236	0.1	25	0204	0.0
	0901	1.5		0840	1.9
	F 1458	-0.1		Sa 1448	-0.4
	2131	1.4		2135	1.6
11	0305	0.1	26	0245	0.1
	0926	1.6		0919	2.0
	Sa 1534	-0.1		Su 1539	-0.5
	○ 2206	1.4		● 2226	1.5
12	0333	0.2	27	0326	0.1
	0953	1.6		1000	2.1
	Su 1609	-0.2		M 1630	-0.6
	2241	1.4		2316	1.4
13	0400	0.2	28	0408	0.2
	1021	1.7		1042	2.1
	M 1644	-0.2		Tu 1721	-0.5
	2319	1.3			
14	0427	0.3	29	0006	1.3
	1051	1.7		0452	0.3
	Tu 1721	-0.2		W 1126	2.1
	2358	1.2		1814	-0.4
15	0456	0.4	30	0059	1.2
	1122	1.7		0538	0.4
	W 1801	-0.2		Th 1213	2.0
				1910	-0.3

	Sun↑	Sun↓
15	0606	1850
30	0553	1857

KEY WEST, FLORIDA

HIGH & LOW WATER 1998 U.S. Datum 24°33.2'N 81°48.5'W

Eastern Standard Time (75°W) Add 1H Daylight Saving Time: April 5 – October 24

MAY

Day	Time	ft	Day	Time	ft
1	0156	1.1	16	0122	1.1
	0631	0.5		0543	0.6
F	1304	1.8	Sa	1216	1.8
	2011	-0.1		1919	-0.1
2	0300	1.0	17	0216	1.1
	0735	0.6		0636	0.6
Sa	1403	1.6	Su	1308	1.7
	2115	0.0		2014	-0.1
3	0411	1.1	18	0315	1.1
	0854	0.6		0747	0.6
Su	1516	1.4	M	1412	1.6
☾	2218	0.1	☽	2113	0.0
4	0518	1.1	19	0414	1.2
	1019	0.6		0913	0.6
M	1640	1.3	Tu	1532	1.5
	2314	0.2		2212	0.1
5	0610	1.2	20	0509	1.3
	1132	0.5		1037	0.4
Tu	1759	1.3	W	1701	1.4
				2306	0.1
6	0002	0.2	21	0558	1.5
	0649	1.3		1150	0.2
W	1232	0.3	Th	1824	1.4
	1902	1.3		2356	0.2
7	0043	0.3	22	0643	1.7
	0721	1.4		1252	-0.1
Th	1320	0.2	F	1934	1.4
	1951	1.3			
8	0119	0.3	23	0043	0.2
	0750	1.6		0727	1.9
F	1402	0.1	Sa	1348	-0.3
	2034	1.3		2034	1.4
9	0151	0.3	24	0128	0.2
	0819	1.7		0810	2.1
Sa	1440	-0.1	Su	1440	-0.5
	2112	1.3		2128	1.4
10	0222	0.3	25	0213	0.2
	0848	1.7		0853	2.2
Su	1516	-0.1	M	1530	-0.5
	2150	1.3	●	2218	1.3
11	0251	0.4	26	0257	0.3
	0918	1.8		0937	2.2
M	1551	-0.2	Tu	1619	-0.6
○	2228	1.3		2306	1.3
12	0321	0.4	27	0342	0.3
	0949	1.8		1021	2.2
Tu	1627	-0.2	W	1708	-0.5
	2307	1.3		2354	1.2
13	0351	0.4	28	0428	0.4
	1022	1.9		1106	2.1
W	1704	-0.2	Th	1757	-0.4
	2348	1.2			
14	0424	0.5	29	0041	1.2
	1056	1.9		0517	0.4
Th	1744	-0.2	F	1152	2.0
				1847	-0.2
15	0033	1.2	30	0131	1.1
	0500	0.5		0611	0.5
F	1134	1.8	Sa	1240	1.8
	1829	-0.2		1940	-0.1
			31	0223	1.1
				0714	0.6
			Su	1333	1.6
				2033	0.1

	Sun↑	Sun↓
15	0544	1904
31	0539	1912

JUNE

Day	Time	ft	Day	Time	ft
1	0318	1.2	16	0240	1.3
	0829	0.6		0744	0.5
M	1434	1.4	Tu	1403	1.5
☾	2127	0.2		2037	0.0
2	0414	1.2	17	0331	1.4
	0950	0.6		0906	0.4
Tu	1547	1.3	W	1520	1.4
	2219	0.3	☽	2130	0.1
3	0505	1.3	18	0424	1.5
	1104	0.5		1027	0.3
W	1707	1.2	Th	1649	1.2
	2306	0.3		2224	0.2
4	0548	1.4	19	0517	1.7
	1206	0.3		1140	0.1
Th	1821	1.1	F	1816	1.2
	2349	0.4		2317	0.3
5	0627	1.5	20	0609	1.9
	1257	0.2		1244	-0.1
F	1920	1.1	Sa	1929	1.2
6	0028	0.4	21	0008	0.3
	0703	1.6		0700	2.0
Sa	1340	0.1	Su	1340	-0.3
	2009	1.2		2029	1.2
7	0104	0.4	22	0058	0.3
	0738	1.7		0749	2.1
Su	1420	-0.1	M	1432	-0.4
	2052	1.2		2121	1.2
8	0139	0.4	23	0147	0.3
	0813	1.8		0836	2.2
M	1457	-0.2	Tu	1520	-0.5
	2133	1.2	●	2208	1.2
9	0213	0.4	24	0235	0.3
	0849	1.9		0922	2.2
Tu	1533	-0.2	W	1607	-0.4
○	2213	1.2		2252	1.2
10	0247	0.5	25	0322	0.3
	0925	2.0		1007	2.2
W	1610	-0.3	Th	1652	-0.4
	2253	1.2		2334	1.2
11	0323	0.5	26	0410	0.4
	1002	2.0		1050	2.1
Th	1648	-0.3	F	1736	-0.3
	2335	1.2			
12	0401	0.5	27	0015	1.2
	1040	2.0		0459	0.4
F	1728	-0.3	Sa	1133	1.9
				1820	-0.2
13	0018	1.2	28	0055	1.2
	0444	0.5		0551	0.4
Sa	1122	1.9	Su	1217	1.8
	1811	-0.2		1903	-0.0
14	0103	1.2	29	0137	1.2
	0533	0.5		0649	0.5
Su	1207	1.8	M	1304	1.6
	1857	-0.2		1947	0.1
15	0150	1.2	30	0219	1.3
	0632	0.5		0756	0.5
M	1300	1.7	Tu	1355	1.4
	1945	-0.1		2032	0.2

	Sun↑	Sun↓
15	0539	1918
30	0542	1921

JULY

Day	Time	ft	Day	Time	ft
1	0304	1.3	16	0249	1.6
	0910	0.5		0855	0.3
W	1455	1.2	Th	1513	1.3
☾	2118	0.4	☽	2051	0.3
2	0352	1.4	17	0343	1.7
	1024	0.5		1015	0.2
Th	1608	1.1	F	1643	1.1
	2205	0.4		2147	0.4
3	0441	1.5	18	0443	1.9
	1130	0.4		1130	0.1
F	1731	1.0	Sa	1812	1.1
	2251	0.5		2245	0.5
4	0530	1.6	19	0544	2.0
	1226	0.2		1236	-0.1
Sa	1845	1.0	Su	1925	1.1
	2336	0.5		2343	0.5
5	0617	1.7	20	0643	2.1
	1314	0.1		1333	-0.2
Su	1942	1.1	M	2023	1.1
6	0019	0.5	21	0039	0.5
	0701	1.8		0738	2.1
M	1356	-0.0	Tu	1423	-0.2
	2030	1.1		2110	1.2
7	0100	0.5	22	0132	0.4
	0744	1.9		0827	2.2
Tu	1436	-0.1	W	1509	-0.2
	2113	1.1		2152	1.2
8	0140	0.5	23	0222	0.4
	0825	2.0		0913	2.2
W	1513	-0.2	Th	1551	-0.2
	2153	1.2	●	2230	1.3
9	0221	0.5	24	0310	0.4
	0906	2.0		0956	2.1
Th	1551	-0.3	F	1630	-0.2
○	2233	1.2		2305	1.3
10	0303	0.4	25	0357	0.4
	0948	2.1		1036	2.0
F	1629	-0.3	Sa	1708	-0.1
	2312	1.3		2339	1.4
11	0347	0.4	26	0443	0.4
	1030	2.1		1116	1.9
Sa	1708	-0.3	Su	1745	0.0
	2352	1.3			
12	0435	0.4	27	0012	1.4
	1115	2.0		0531	0.4
Su	1748	-0.2	M	1155	1.8
				1821	0.1
13	0033	1.4	28	0045	1.4
	0528	0.4		0621	0.4
M	1203	1.9	Tu	1236	1.6
	1830	-0.1		1857	0.3
14	0115	1.4	29	0121	1.5
	0628	0.4		0717	0.5
Tu	1257	1.7	W	1321	1.4
	1914	0.0		1934	0.4
15	0200	1.5	30	0201	1.5
	0738	0.3		0821	0.5
W	1358	1.5	Th	1414	1.3
	2000	0.2		2013	0.5
			31	0245	1.5
				0932	0.5
			F	1519	1.1
			☾	2057	0.6

	Sun↑	Sun↓
15	0548	1919
31	0556	1912

AUGUST

Day	Time	ft	Day	Time	ft
1	0337	1.6	16	0415	2.0
	1044	0.5		1118	0.2
Sa	1641	1.1	Su	1809	1.2
	2149	0.7		2223	0.7
2	0434	1.7	17	0527	2.0
	1149	0.4		1225	0.1
Su	1807	1.1	M	1918	1.2
	2245	0.7		2330	0.7
3	0532	1.7	18	0635	2.1
	1243	0.3		1322	0.1
M	1914	1.1	Tu	2010	1.3
	2339	0.7			
4	0627	1.9	19	0031	0.6
	1329	0.1		0732	2.2
Tu	2005	1.2	W	1409	0.1
				2052	1.4
5	0029	0.6	20	0126	0.6
	0718	2.0		0822	2.2
W	1410	0.0	Th	1450	0.1
	2047	1.2		2128	1.4
6	0116	0.6	21	0215	0.5
	0806	2.1		0906	2.2
Th	1448	-0.1	F	1527	0.1
	2126	1.3	●	2200	1.5
7	0203	0.5	22	0301	0.4
	0851	2.2		0945	2.1
F	1526	-0.1	Sa	1601	0.1
○	2203	1.4		2229	1.6
8	0249	0.4	23	0345	0.4
	0936	2.2		1022	2.1
Sa	1603	-0.1	Su	1634	0.2
	2240	1.5		2257	1.6
9	0338	0.3	24	0427	0.4
	1022	2.2		1058	2.0
Su	1640	-0.1	M	1706	0.3
	2317	1.6		2325	1.7
10	0428	0.3	25	0509	0.4
	1109	2.1		1134	1.8
M	1718	-0.0	Tu	1736	0.4
	2355	1.7		2355	1.7
11	0522	0.2	26	0553	0.4
	1159	1.9		1212	1.7
Tu	1758	0.1	W	1807	0.5
12	0035	1.8	27	0028	1.8
	0621	0.2		0639	0.5
W	1253	1.7	Th	1254	1.6
	1840	0.3		1838	0.6
13	0119	1.9	28	0104	1.8
	0727	0.2		0733	0.5
Th	1355	1.5	F	1343	1.4
	1925	0.4		1911	0.7
14	0209	1.9	29	0147	1.8
	0842	0.2		0837	0.6
F	1509	1.3	Sa	1443	1.3
☽	2016	0.5	☾	1952	0.8
15	0307	2.0	30	0238	1.8
	1001	0.2		0951	0.6
Sa	1639	1.2	Su	1602	1.2
	2116	0.6		2047	0.9
			31	0340	1.8
				1104	0.5
			M	1733	1.2
				2157	0.9

	Sun↑	Sun↓
15	0602	1902
31	0609	1847

KEY WEST, FLORIDA

HIGH & LOW WATER 1998 — U.S. Datum — 24°33.2'N 81°48.5'W

Eastern Standard Time (75°W) — Add 1H Daylight Saving Time: April 5 – October 24

T

Tides

SEPTEMBER

Day	Time	ft	Time	ft	Time	ft	Time	ft
1 Tu	0449	1.9	1206	0.4	1844	1.3	2305	0.9
2 W	0556	2.0	1256	0.3	1934	1.4		
3 Th	0005	0.8	0655	2.1	1338	0.2	2015	1.5
4 F	0059	0.7	0748	2.2	1416	0.2	2051	1.6
5 Sa	0150	0.5	0838	2.3	1453	0.1	2127	1.7
6 Su	0240	0.4	0926	2.3	1530	0.1	○ 2202	1.9
7 M	0329	0.2	1014	2.3	1607	0.2	2238	2.0
8 Tu	0421	0.1	1103	2.1	1645	0.3	2316	2.1
9 W	0514	0.1	1154	2.0	1724	0.4	2357	2.2
10 Th	0611	0.1	1248	1.8	1806	0.5		
11 F	0042	2.2	0714	0.2	1350	1.6	1852	0.7
12 Sa	0134	2.2	0825	0.3	1504	1.4	☽ 1947	0.8
13 Su	0237	2.1	0943	0.4	1632	1.3	2055	0.9
14 M	0352	2.1	1100	0.4	1757	1.4	2212	0.9
15 Tu	0514	2.1	1208	0.4	1900	1.4	2326	0.9
16 W	0627	2.1	1301	0.4	1947	1.5		
17 Th	0029	0.8	0726	2.1	1345	0.4	2024	1.6
18 F	0123	0.7	0814	2.1	1422	0.4	2056	1.7
19 Sa	0209	0.6	0855	2.1	1455	0.4	2123	1.8
20 Su	0252	0.5	0932	2.1	1526	0.5	● 2149	1.9
21 M	0332	0.4	1006	2.0	1556	0.5	2214	2.0
22 Tu	0410	0.4	1041	2.0	1624	0.6	2241	2.0
23 W	0448	0.4	1116	1.9	1652	0.7	2310	2.0
24 Th	0526	0.4	1153	1.8	1720	0.8	2342	2.0
25 F	0608	0.5	1234	1.7	1749	0.8		
26 Sa	0017	2.0	0654	0.5	1321	1.5	1821	0.9
27 Su	0058	2.0	0750	0.6	1419	1.4	1901	1.0
28 M	0147	1.9	0900	0.6	1534	1.4	☾ 1958	1.1
29 Tu	0250	1.9	1015	0.6	1700	1.4	2118	1.1
30 W	0406	1.9	1121	0.6	1808	1.4	2240	1.0

	Sun↑	Sun↓
15	0614	1832
30	0619	1816

OCTOBER

Day	Time	ft	Time	ft	Time	ft	Time	ft
1 Th	0524	2.0	1214	0.5	1857	1.6	2348	0.9
2 F	0632	2.1	1258	0.4	1936	1.7		
3 Sa	0046	0.7	0731	2.2	1338	0.4	2012	1.9
4 Su	0139	0.5	0825	2.2	1416	0.4	2048	2.1
5 M	0231	0.2	0916	2.2	1454	0.4	○ 2124	2.2
6 Tu	0321	0.1	1006	2.2	1533	0.5	2202	2.4
7 W	0412	-0.0	1056	2.1	1612	0.5	2242	2.5
8 Th	0504	-0.0	1147	1.9	1653	0.6	2325	2.5
9 F	0559	0.1	1242	1.8	1736	0.7		
10 Sa	0012	2.4	0659	0.2	1342	1.6	1825	0.8
11 Su	0106	2.3	0806	0.3	1452	1.5	1925	1.0
12 M	0210	2.2	0919	0.4	1613	1.5	☽ 2040	1.0
13 Tu	0328	2.1	1032	0.5	1730	1.5	2205	1.0
14 W	0455	2.0	1136	0.6	1829	1.6	2322	0.9
15 Th	0612	2.0	1228	0.6	1914	1.7		
16 F	0025	0.8	0712	2.0	1310	0.6	1948	1.8
17 Sa	0116	0.7	0800	2.0	1345	0.6	2018	1.9
18 Su	0201	0.5	0840	2.0	1418	0.6	2044	2.0
19 M	0240	0.4	0917	2.0	1448	0.7	2109	2.1
20 Tu	0317	0.4	0951	1.9	1516	0.7	● 2135	2.1
21 W	0353	0.3	1025	1.9	1545	0.7	2203	2.2
22 Th	0428	0.3	1101	1.8	1612	0.8	2234	2.2
23 F	0505	0.3	1138	1.7	1641	0.9	2306	2.2
24 Sa	0543	0.4	1220	1.6	1711	0.9	2341	2.1
25 Su	2341	2.1	0627	0.4	1307	1.5	1745	1.0
26 M	0021	2.1	0718	0.5	1403	1.5	1829	1.0
27 Tu	0109	2.0	0819	0.5	1510	1.4	1929	1.1
28 W	0211	1.9	0927	0.6	1622	1.4	☾ 2053	1.1
29 Th	0329	1.9	1031	0.6	1723	1.5	2221	1.0
30 F	0454	1.9	1126	0.5	1812	1.7	2334	0.8
31 Sa	0611	1.9	1214	0.5	1854	1.9		

	Sun↑	Sun↓
15	0626	1801
31	0634	1748

NOVEMBER

Day	Time	ft	Time	ft	Time	ft	Time	ft
1 Su	0035	0.5	0717	2.0	1257	0.5	1933	2.1
2 M	0130	0.3	0815	2.0	1338	0.5	2012	2.3
3 Tu	0222	0.0	0908	2.0	1419	0.5	2051	2.4
4 W	0312	-0.1	0959	1.9	1500	0.5	○ 2132	2.6
5 Th	0402	-0.2	1049	1.8	1542	0.6	2216	2.6
6 F	0453	-0.2	1139	1.7	1626	0.6	2301	2.5
7 Sa	0546	-0.1	1231	1.6	1712	0.7	2350	2.4
8 Su	0642	0.1	1326	1.5	1804	0.8		
9 M	0043	2.3	0742	0.2	1428	1.5	1907	0.9
10 Tu	0144	2.1	0847	0.4	1537	1.5	☽ 2024	0.9
11 W	0257	1.9	0952	0.5	1646	1.5	2151	0.9
12 Th	0422	1.8	1052	0.6	1744	1.6	2309	0.8
13 F	0543	1.7	1142	0.6	1829	1.7		
14 Sa	0012	0.7	0648	1.7	1225	0.6	1905	1.8
15 Su	0103	0.5	0740	1.7	1303	0.7	1935	1.9
16 M	0146	0.4	0822	1.7	1337	0.7	2004	2.0
17 Tu	0225	0.3	0900	1.7	1409	0.7	2033	2.1
18 W	0301	0.2	0936	1.6	1439	0.7	● 2103	2.1
19 Th	0337	0.1	1011	1.6	1509	0.7	2134	2.1
20 F	0411	0.1	1048	1.6	1539	0.7	2207	2.1
21 Sa	0447	0.1	1126	1.5	1611	0.7	2242	2.1
22 Su	0525	0.1	1208	1.5	1645	0.8	2318	2.1
23 M	0606	0.2	1253	1.4	1724	0.8	2359	2.0
24 Tu	0653	0.2	1343	1.4	1812	0.8		
25 W	0047	1.9	0745	0.3	1438	1.4	1916	0.9
26 Th	0146	1.8	0842	0.4	1536	1.4	☾ 2037	0.8
27 F	0301	1.7	0940	0.4	1632	1.5	2204	0.7
28 Sa	0428	1.6	1036	0.4	1724	1.7	2319	0.5
29 Su	0554	1.6	1128	0.5	1811	1.9		
30 M	0024	0.2	0706	1.6	1216	0.5	1857	2.1

	Sun↑	Sun↓
15	0644	1741
30	0654	1739

DECEMBER

Day	Time	ft	Time	ft	Time	ft	Time	ft
1 Tu	0120	-0.0	0808	1.6	1303	0.5	1941	2.2
2 W	0213	-0.2	0902	1.6	1348	0.5	2027	2.4
3 Th	0303	-0.4	0952	1.5	1433	0.5	○ 2112	2.4
4 F	0353	-0.4	1040	1.5	1519	0.4	2159	2.4
5 Sa	0442	-0.4	1127	1.5	1606	0.5	2245	2.4
6 Su	0531	-0.3	1213	1.4	1655	0.5	2333	2.2
7 M	0621	-0.1	1302	1.4	1749	0.6		
8 Tu	0024	2.0	0713	0.0	1352	1.3	1850	0.6
9 W	0118	1.8	0806	0.2	1447	1.3	2002	0.7
10 Th	0219	1.6	0901	0.3	1543	1.4	☽ 2123	0.6
11 F	0332	1.4	0956	0.4	1639	1.4	2240	0.6
12 Sa	0456	1.3	1047	0.5	1728	1.5	2346	0.4
13 Su	0613	1.3	1134	0.5	1811	1.6		
14 M	0041	0.3	0714	1.2	1216	0.6	1849	1.7
15 Tu	0127	0.2	0802	1.3	1255	0.6	1926	1.8
16 W	0207	0.0	0842	1.3	1331	0.5	2002	1.9
17 Th	0245	-0.0	0920	1.3	1406	0.5	2038	1.9
18 F	0320	-0.1	0956	1.3	1439	0.5	● 2113	2.0
19 Sa	0355	-0.2	1034	1.3	1514	0.5	2150	2.0
20 Su	0431	-0.2	1112	1.3	1550	0.5	2227	2.0
21 M	0508	-0.2	1151	1.3	1629	0.5	2306	1.9
22 Tu	0547	-0.1	1232	1.2	1713	0.5	2348	1.8
23 W	0628	-0.1	1315	1.3	1805	0.5		
24 Th	0036	1.7	0713	0.0	1400	1.3	1909	0.5
25 F	0133	1.5	0801	0.1	1448	1.4	2024	0.4
26 Sa	0244	1.4	0853	0.2	1541	1.5	☾ 2147	0.3
27 Su	0410	1.2	0948	0.3	1636	1.6	2304	0.1
28 M	0542	1.1	1044	0.4	1732	1.8		
29 Tu	0012	-0.1	0700	1.1	1140	0.4	1827	1.9
30 W	0112	-0.3	0804	1.2	1233	0.4	1921	2.1
31 Th	0205	-0.4	0857	1.2	1325	0.3	2012	2.1

	Sun↑	Sun↓
15	0704	1742
31	0712	1750

NASSAU, BAHAMAS

HIGH & LOW WATER 1998 — Chart Datum — 25°05'N 77°21'W

Eastern Standard Time (75°W) — Add 1H Daylight Saving Time: April 5 – October 24

JANUARY

Day	Time	ft	Time	ft	Time	ft	Time	ft
1 Th	0304	0.9	0934	4.5	1601	0.9	2159	3.6
2 F	0358	0.9	1023	4.3	1650	0.9	2254	3.7
3 Sa	0454	1.0	1115	4.2	1741	0.9	2353	3.8
4 Su	0556	1.1	1210	4.0	1834	0.9		
5 M	0054	3.8	0701	1.2	1308	3.8	☾ 1929	0.9
6 Tu	0157	3.9	0809	1.3	1408	3.6	2026	0.9
7 W	0300	4.0	0917	1.3	1509	3.5	2122	0.9
8 Th	0400	4.1	1022	1.3	1610	3.4	2218	0.9
9 F	0457	4.2	1123	1.2	1708	3.4	2311	0.9
10 Sa	0551	4.3	1218	1.2	1802	3.4		
11 Su	0002	0.9	0640	4.4	1309	1.1	1853	3.4
12 M	0050	0.9	0727	4.3	1355	1.1	○ 1942	3.3
13 Tu	0136	1.0	0810	4.3	1438	1.1	2027	3.3
14 W	0221	1.1	0851	4.2	1519	1.1	2111	3.3
15 Th	0304	1.2	0930	4.0	1557	1.2	2154	3.3
16 F	0346	1.3	1007	3.9	1635	1.2	2238	3.3
17 Sa	0430	1.4	1045	3.7	1712	1.2	2322	3.3
18 Su	0515	1.5	1124	3.6	1750	1.2		
19 M	0008	3.4	0604	1.6	1205	3.4	1830	1.3
20 Tu	0056	3.4	0657	1.7	1251	3.3	☽ 1914	1.2
21 W	0147	3.5	0755	1.7	1341	3.2	2000	1.2
22 Th	0240	3.6	0854	1.7	1436	3.2	2050	1.1
23 F	0332	3.8	0952	1.6	1532	3.2	2141	1.1
24 Sa	0424	4.0	1047	1.4	1628	3.2	2233	1.0
25 Su	0514	4.1	1139	1.3	1722	3.3	2325	0.8
26 M	0603	4.3	1228	1.1	1815	3.5		
27 Tu	0017	0.7	0651	4.4	1316	1.0	1907	3.6
28 W	0108	0.7	0739	4.5	1403	0.8	● 1958	3.7
29 Th	0200	0.6	0827	4.5	1450	0.7	2049	3.9
30 F	0253	0.7	0916	4.4	1537	0.7	2142	3.9
31 Sa	0347	0.7	1005	4.3	1625	0.7	2236	4.0

	Sun↑	Sun↓
15	0657	1742
31	0653	1754

FEBRUARY

Day	Time	ft	Time	ft	Time	ft	Time	ft
1 Su	0443	0.9	1057	4.1	1716	0.7	2333	4.0
2 M	0543	1.0	1151	3.9	1809	0.8		
3 Tu	0033	4.0	0646	1.2	1248	3.7	☾ 1904	0.9
4 W	0136	4.0	0754	1.3	1349	3.5	2003	1.0
5 Th	0240	4.0	0903	1.4	1453	3.4	2102	1.0
6 F	0343	4.1	1010	1.4	1556	3.3	2201	1.0
7 Sa	0442	4.1	1110	1.3	1656	3.3	2257	1.0
8 Su	0536	4.1	1204	1.3	1750	3.3	2349	1.1
9 M	0624	4.1	1251	1.2	1839	3.4		
10 Tu	0037	1.1	0708	4.1	1333	1.2	1924	3.4
11 W	0121	1.1	0748	4.1	1410	1.2	○ 2005	3.5
12 Th	0202	1.2	0825	4.0	1446	1.2	2044	3.5
13 F	0242	1.2	0900	3.9	1519	1.2	2122	3.5
14 Sa	0321	1.3	0934	3.8	1552	1.2	2201	3.5
15 Su	0401	1.4	1009	3.7	1626	1.2	2240	3.6
16 M	0442	1.5	1045	3.5	1701	1.3	2322	3.6
17 Tu	0527	1.6	1124	3.4	1740	1.3		
18 W	0008	3.6	0617	1.7	1208	3.3	1823	1.3
19 Th	0058	3.7	0713	1.7	1259	3.2	☽ 1913	1.3
20 F	0153	3.7	0813	1.7	1357	3.2	2008	1.3
21 Sa	0250	3.9	0914	1.6	1458	3.3	2106	1.2
22 Su	0348	4.0	1012	1.5	1559	3.4	2206	1.1
23 M	0443	4.2	1107	1.3	1658	3.5	2303	0.9
24 Tu	0536	4.3	1158	1.1	1753	3.7	2359	0.8
25 W	0627	4.5	1247	0.9	1846	4.0		
26 Th	0053	0.7	0717	4.5	1334	0.8	● 1938	4.2
27 F	0147	0.6	0806	4.5	1422	0.7	2030	4.3
28 Sa	0240	0.6	0855	4.5	1509	0.7	2122	4.4

	Sun↑	Sun↓
15	0644	1804
28	0634	1812

MARCH

Day	Time	ft	Time	ft	Time	ft	Time	ft
1 Su	0334	0.7	0945	4.3	1558	0.7	2215	4.4
2 M	0429	0.9	1037	4.1	1649	0.8	2311	4.3
3 Tu	0528	1.0	1131	3.9	1742	0.9		
4 W	0010	4.2	0631	1.2	1229	3.7	1839	1.1
5 Th	0113	4.2	0737	1.4	1332	3.5	☾ 1940	1.2
6 F	0218	4.1	0846	1.5	1438	3.4	2043	1.3
7 Sa	0322	4.0	0952	1.5	1543	3.3	2146	1.3
8 Su	0422	4.0	1050	1.5	1643	3.4	2244	1.3
9 M	0515	4.0	1140	1.4	1735	3.5	2335	1.3
10 Tu	0602	4.0	1223	1.4	1821	3.6		
11 W	0022	1.3	0643	4.0	1301	1.4	1902	3.7
12 Th	0104	1.3	0720	4.0	1335	1.3	○ 1939	3.7
13 F	0143	1.3	0755	3.9	1407	1.3	2015	3.8
14 Sa	0221	1.4	0828	3.8	1438	1.3	2050	3.9
15 Su	0258	1.4	0901	3.8	1510	1.3	2126	3.9
16 M	0336	1.5	0935	3.7	1542	1.3	2203	3.9
17 Tu	0415	1.6	1011	3.6	1617	1.4	2243	3.9
18 W	0458	1.6	1050	3.5	1656	1.4	2327	3.9
19 Th	0546	1.7	1135	3.4	1741	1.5		
20 F	0016	3.9	0640	1.7	1228	3.3	1833	1.5
21 Sa	0112	3.9	0739	1.7	1328	3.3	☽ 1933	1.5
22 Su	0213	4.0	0840	1.6	1433	3.4	2038	1.4
23 M	0314	4.1	0939	1.5	1537	3.6	2143	1.3
24 Tu	0413	4.2	1035	1.3	1637	3.9	2245	1.1
25 W	0508	4.4	1127	1.1	1733	4.1	2343	0.9
26 Th	0602	4.5	1217	0.9	1826	4.4		
27 F	0038	0.8	0653	4.5	1305	0.8	● 1918	4.6
28 Sa	0132	0.7	0743	4.5	1353	0.7	2009	4.7
29 Su	0226	0.7	0833	4.4	1441	0.7	2101	4.8
30 M	0320	0.8	0924	4.3	1530	0.8	2154	4.7
31 Tu	0415	1.0	1016	4.1	1621	0.9	2248	4.6

	Sun↑	Sun↓
15	0619	1819
31	0603	1826

APRIL

Day	Time	ft	Time	ft	Time	ft	Time	ft
1 W	0512	1.2	1111	3.9	1715	1.1	2346	4.5
2 Th	0613	1.3	1211	3.7	1813	1.3		
3 F	0047	4.3	0718	1.5	1315	3.5	☾ 1915	1.5
4 Sa	0150	4.1	0823	1.6	1422	3.5	2021	1.6
5 Su	0253	4.0	0925	1.6	1526	3.5	2126	1.7
6 M	0352	4.0	1019	1.6	1624	3.6	2224	1.7
7 Tu	0444	3.9	1106	1.6	1713	3.7	2316	1.6
8 W	0529	3.9	1146	1.5	1756	3.8		
9 Th	0001	1.6	0609	3.9	1221	1.5	1834	3.9
10 F	0042	1.5	0646	3.9	1254	1.4	1910	4.1
11 Sa	0121	1.5	0721	3.8	1326	1.4	○ 1945	4.1
12 Su	0159	1.5	0755	3.8	1358	1.4	2020	4.2
13 M	0236	1.5	0830	3.7	1430	1.4	2055	4.2
14 Tu	0314	1.6	0905	3.7	1504	1.4	2132	4.2
15 W	0354	1.6	0943	3.6	1540	1.5	2211	4.2
16 Th	0437	1.7	1024	3.5	1621	1.5	2255	4.2
17 F	0523	1.7	1111	3.5	1708	1.6	2345	4.2
18 Sa	0615	1.7	1206	3.5	1804	1.6		
19 Su	0040	4.1	0712	1.7	1308	3.5	☽ 1908	1.6
20 M	0140	4.1	0810	1.6	1413	3.7	2017	1.6
21 Tu	0243	4.2	0908	1.5	1517	3.9	2124	1.5
22 W	0343	4.2	1004	1.3	1617	4.2	2228	1.3
23 Th	0441	4.3	1056	1.1	1713	4.5	2327	1.1
24 F	0536	4.4	1147	1.0	1807	4.7		
25 Sa	0024	1.0	0628	4.4	1236	0.8	1859	4.9
26 Su	0118	0.9	0720	4.4	1325	0.8	● 1950	5.0
27 M	0212	0.9	0811	4.3	1414	0.8	2041	5.0
28 Tu	0305	1.0	0903	4.2	1504	0.9	2133	4.9
29 W	0400	1.1	0956	4.0	1555	1.1	2226	4.8
30 Th	0455	1.3	1051	3.8	1649	1.3	2321	4.6

	Sun↑	Sun↓
15	0548	1833
30	0535	1840

T Tides

NASSAU, BAHAMAS

HIGH & LOW WATER 1998 — Chart Datum — 25°05'N 77°21'W

Eastern Standard Time (75°W) — Add 1H Daylight Saving Time: April 5 – October 24

MAY

Day	Time	ft	Day	Time	ft
1 F	0553	1.4	16 Sa	0505	1.6
	1151	3.7		1055	3.6
	1746	1.5		1648	1.6
				2321	4.4
2 Sa	0018	4.4	17 Su	0555	1.6
	0653	1.6		1151	3.6
	1254	3.6		1746	1.7
	1848	1.7			
3 Su	0116	4.2	18 M	0015	4.3
	0752	1.6		0648	1.6
	1358	3.6		1252	3.7
☾	1953	1.9	☽	1851	1.7
4 M	0215	4.0	19 Tu	0114	4.2
	0847	1.7		0744	1.5
	1459	3.6		1356	3.9
	2056	1.9		1959	1.7
5 Tu	0311	3.9	20 W	0215	4.2
	0938	1.7		0840	1.4
	1554	3.7		1459	4.1
	2155	1.9		2107	1.6
6 W	0401	3.8	21 Th	0316	4.2
	1021	1.6		0935	1.3
	1641	3.9		1559	4.4
	2247	1.9		2212	1.5
7 Th	0447	3.8	22 F	0415	4.2
	1101	1.6		1028	1.1
	1723	4.0		1655	4.6
	2333	1.8		2313	1.3
8 F	0528	3.8	23 Sa	0511	4.2
	1137	1.5		1120	1.0
	1802	4.2		1749	4.9
9 Sa	0016	1.7	24 Su	0010	1.2
	0607	3.8		0606	4.2
	1212	1.5		1211	0.9
	1839	4.3		1842	5.0
10 Su	0056	1.7	25 M	0105	1.2
	0645	3.8		0659	4.2
	1246	1.4		1301	0.9
	1915	4.4	●	1933	5.1
11 M	0135	1.6	26 Tu	0159	1.1
	0723	3.7		0751	4.1
	1321	1.4		1351	1.0
○	1951	4.5		2023	5.1
12 Tu	0215	1.6	27 W	0251	1.2
	0800	3.7		0843	4.0
	1356	1.4		1441	1.1
	2028	4.5		2113	5.0
13 W	0254	1.6	28 Th	0343	1.3
	0839	3.6		0936	3.9
	1433	1.4		1531	1.3
	2107	4.5		2204	4.8
14 Th	0335	1.6	29 F	0436	1.4
	0920	3.6		1030	3.8
	1513	1.5		1623	1.5
	2148	4.5		2254	4.6
15 F	0418	1.6	30 Sa	0528	1.5
	1005	3.6		1127	3.7
	1557	1.6		1718	1.7
	2232	4.4		2345	4.4
			31 Su	0620	1.6
				1225	3.7
				1815	1.9

	Sun↑	Sun↓
15	0525	1847
31	0520	1855

JUNE

Day	Time	ft	Day	Time	ft
1 M	0037	4.1	16 Tu	0624	1.5
	0712	1.6		1236	4.0
	1324	3.7		1837	1.7
☾	1916	2.0			
2 Tu	0129	4.0	17 W	0052	4.3
	0801	1.7		0718	1.4
	1421	3.7		1338	4.1
	2017	2.1	☽	1944	1.7
3 W	0220	3.8	18 Th	0151	4.2
	0847	1.7		0813	1.3
	1513	3.8		1440	4.3
	2115	2.1		2052	1.7
4 Th	0310	3.8	19 F	0252	4.1
	0930	1.6		0909	1.3
	1601	4.0		1541	4.5
	2209	2.0		2158	1.6
5 F	0357	3.7	20 Sa	0352	4.1
	1011	1.6		1004	1.2
	1645	4.1		1639	4.7
	2258	2.0		2300	1.5
6 Sa	0443	3.7	21 Su	0451	4.0
	1051	1.5		1058	1.1
	1726	4.3		1734	4.9
	2344	1.9		2359	1.5
7 Su	0527	3.7	22 M	0547	4.0
	1130	1.5		1150	1.1
	1806	4.4		1827	5.0
8 M	0028	1.8	23 Tu	0054	1.4
	0609	3.7		0641	4.0
	1209	1.4		1242	1.1
	1845	4.6	●	1918	5.0
9 Tu	0110	1.7	24 W	0146	1.4
	0652	3.7		0734	4.0
	1248	1.4		1332	1.2
○	1925	4.7		2007	5.0
10 W	0152	1.6	25 Th	0236	1.4
	0734	3.7		0826	3.9
	1328	1.4		1421	1.3
	2005	4.7		2054	4.9
11 Th	0234	1.6	26 F	0324	1.4
	0817	3.7		0916	3.9
	1410	1.4		1510	1.4
	2046	4.7		2140	4.7
12 F	0316	1.6	27 Sa	0411	1.5
	0902	3.7		1007	3.8
	1455	1.4		1559	1.6
	2129	4.7		2225	4.5
13 Sa	0400	1.5	28 Su	0456	1.6
	0950	3.7		1057	3.8
	1543	1.5		1648	1.8
	2214	4.6		2310	4.3
14 Su	0445	1.5	29 M	0541	1.6
	1041	3.8		1148	3.8
	1635	1.6		1740	2.0
	2303	4.5		2354	4.1
15 M	0533	1.5	30 Tu	0625	1.7
	1137	3.9		1240	3.8
	1733	1.7		1833	2.1
	2355	4.4			

	Sun↑	Sun↓
15	0520	1901
30	0523	1904

JULY

Day	Time	ft	Day	Time	ft
1 W	0039	4.0	16 Th	0032	4.3
	0708	1.7		0653	1.4
	1332	3.8		1319	4.4
☾	1930	2.2	☽	1930	1.8
2 Th	0126	3.8	17 F	0130	4.2
	0752	1.7		0749	1.4
	1424	3.9		1422	4.5
	2027	2.2		2038	1.8
3 F	0215	3.7	18 Sa	0232	4.0
	0836	1.7		0846	1.4
	1514	4.1		1524	4.6
	2124	2.2		2146	1.8
4 Sa	0306	3.7	19 Su	0334	4.0
	0921	1.7		0944	1.4
	1601	4.2		1625	4.7
	2218	2.1		2249	1.8
5 Su	0357	3.6	20 M	0436	3.9
	1006	1.6		1041	1.3
	1647	4.4		1721	4.8
	2309	2.0		2348	1.7
6 M	0447	3.7	21 Tu	0534	3.9
	1051	1.5		1136	1.3
	1732	4.5		1814	4.9
	2356	1.9			
7 Tu	0535	3.7	22 W	0042	1.6
	1136	1.5		0628	4.0
	1815	4.7		1228	1.4
				1904	4.9
8 W	0042	1.8	23 Th	0131	1.6
	0623	3.8		0720	4.0
	1221	1.4		1318	1.4
	1859	4.8	●	1950	4.9
9 Th	0126	1.7	24 F	0217	1.6
	0710	3.8		0808	4.0
	1306	1.4		1405	1.5
○	1942	4.9		2034	4.8
10 F	0209	1.6	25 Sa	0300	1.6
	0756	3.9		0854	4.0
	1353	1.3		1450	1.6
	2026	4.9		2115	4.7
11 Sa	0253	1.5	26 Su	0340	1.6
	0844	4.0		0939	4.0
	1441	1.4		1535	1.7
	2110	4.8		2155	4.5
12 Su	0337	1.4	27 M	0419	1.7
	0933	4.1		1023	4.0
	1531	1.4		1619	1.9
	2157	4.8		2233	4.3
13 M	0422	1.4	28 Tu	0457	1.7
	1025	4.2		1107	4.0
	1625	1.5		1704	2.0
	2245	4.6		2312	4.2
14 Tu	0510	1.4	29 W	0535	1.8
	1120	4.2		1153	4.0
	1722	1.6		1751	2.2
	2337	4.5		2352	4.0
15 W	0600	1.4	30 Th	0614	1.8
	1218	4.3		1240	4.0
	1824	1.7		1843	2.3
			31 F	0036	3.9
				0656	1.8
				1330	4.1
			☾	1938	2.3

	Sun↑	Sun↓
15	0529	1902
31	0537	1855

AUGUST

Day	Time	ft	Day	Time	ft
1 Sa	0124	3.8	16 Su	0217	4.0
	0742	1.8		0827	1.6
	1422	4.2		1507	4.7
	2037	2.3		2134	2.0
2 Su	0218	3.7	17 M	0323	3.9
	0832	1.8		0930	1.6
	1515	4.3		1610	4.7
	2135	2.2		2238	1.9
3 M	0314	3.7	18 Tu	0426	3.9
	0923	1.8		1030	1.6
	1607	4.4		1708	4.8
	2231	2.1		2335	1.9
4 Tu	0410	3.7	19 W	0524	4.0
	1015	1.7		1126	1.6
	1657	4.6		1800	4.8
	2322	2.0			
5 W	0505	3.8	20 Th	0026	1.8
	1107	1.6		0617	4.1
	1745	4.7		1218	1.6
				1847	4.8
6 Th	0010	1.8	21 F	0110	1.8
	0556	4.0		0705	4.1
	1158	1.5		1305	1.6
	1832	4.9	●	1930	4.7
7 F	0056	1.7	22 Sa	0151	1.7
	0646	4.1		0749	4.2
	1248	1.4		1349	1.7
○	1918	5.0		2010	4.6
8 Sa	0141	1.5	23 Su	0228	1.7
	0735	4.3		0830	4.2
	1338	1.3		1431	1.8
	2004	5.0		2047	4.5
9 Su	0225	1.4	24 M	0303	1.8
	0824	4.4		0909	4.3
	1428	1.3		1511	1.9
	2050	5.0		2122	4.4
10 M	0310	1.4	25 Tu	0337	1.8
	0914	4.5		0947	4.3
	1520	1.3		1551	2.0
	2138	4.9		2157	4.3
11 Tu	0356	1.3	26 W	0411	1.8
	1006	4.6		1026	4.3
	1614	1.4		1631	2.1
	2227	4.7		2232	4.1
12 W	0444	1.4	27 Th	0445	1.9
	1100	4.6		1107	4.2
	1710	1.6		1715	2.2
	2318	4.5		2310	4.0
13 Th	0534	1.4	28 F	0523	1.9
	1157	4.7		1151	4.2
	1811	1.7		1802	2.3
				2352	3.9
14 F	0013	4.3	29 Sa	0605	2.0
	0628	1.5		1239	4.3
	1258	4.6		1856	2.3
☽	1916	1.9	☾		
15 Sa	0113	4.2	30 Su	0041	3.8
	0726	1.5		0653	2.0
	1402	4.7		1333	4.3
	2025	1.9		1954	2.3
			31 M	0138	3.8
				0747	2.0
				1429	4.4
				2054	2.3

	Sun↑	Sun↓
15	0544	1845
31	0550	1830

NASSAU, BAHAMAS

HIGH & LOW WATER 1998 Chart Datum 25°05'N 77°21'W

Eastern Standard Time (75°W) Add 1H Daylight Saving Time: April 5 – October 24

T Tides

SEPTEMBER

Day		Time	ft	Time	ft	Time	ft	Time	ft
1	Tu	0238	3.8	0845	1.9	1526	4.5	2152	2.2
2	W	0339	3.9	0945	1.8	1621	4.6	2246	2.0
3	Th	0437	4.0	1042	1.7	1714	4.8	2336	1.8
4	F	0532	4.3	1137	1.5	1804	4.9		
5	Sa	0023	1.6	0623	4.5	1231	1.4	1853	5.0
6	Su	0109	1.5	0713	4.7	1323	1.3	○ 1940	5.0
7	M	0155	1.3	0803	4.9	1414	1.2	2028	5.0
8	Tu	0241	1.3	0853	5.0	1507	1.3	2117	4.9
9	W	0328	1.3	0945	5.0	1601	1.4	2207	4.7
10	Th	0417	1.3	1039	5.0	1657	1.5	2300	4.5
11	F	0509	1.4	1136	4.9	1758	1.7	2357	4.3
12	Sa ☽	0605	1.6	1237	4.8	1902	1.9		
13	Su	0059	4.1	0706	1.7	1342	4.7	2011	2.0
14	M	0205	4.0	0811	1.8	1448	4.6	2118	2.0
15	Tu	0313	4.0	0916	1.9	1552	4.6	2220	2.0
16	W	0417	4.0	1019	1.9	1649	4.6	2314	1.9
17	Th	0513	4.1	1115	1.8	1740	4.6		
18	F	0001	1.9	0602	4.2	1205	1.8	1824	4.5
19	Sa	0041	1.8	0646	4.3	1250	1.8	1904	4.5
20	Su	0118	1.8	0725	4.4	1332	1.8	● 1941	4.4
21	M	0151	1.8	0802	4.4	1410	1.9	2015	4.3
22	Tu	0223	1.8	0837	4.5	1448	1.9	2049	4.2
23	W	0255	1.8	0913	4.5	1525	2.0	2122	4.1
24	Th	0327	1.8	0949	4.5	1604	2.0	2157	4.0
25	F	0401	1.9	1027	4.4	1645	2.1	2235	3.9
26	Sa	0438	1.9	1109	4.4	1731	2.2	2318	3.8
27	Su	0521	2.0	1156	4.4	1822	2.2		
28	M	0009	3.8	0611	2.0	1249	4.4	☾ 1918	2.2
29	Tu	0107	3.7	0709	2.0	1347	4.4	2017	2.2
30	W	0210	3.8	0813	2.0	1447	4.4	2114	2.0

	Sun↑	Sun↓
15	0556	1814
30	0602	1758

OCTOBER

Day		Time	ft	Time	ft	Time	ft	Time	ft
1	Th	0313	4.0	0918	1.9	1546	4.5	2209	1.8
2	F	0413	4.2	1020	1.7	1641	4.6	2300	1.6
3	Sa	0508	4.5	1118	1.5	1734	4.8	2349	1.4
4	Su	0601	4.7	1213	1.3	1825	4.8		
5	M	0037	1.3	0652	5.0	1307	1.2	○ 1915	4.8
6	Tu	0124	1.2	0742	5.2	1359	1.2	2005	4.8
7	W	0212	1.1	0833	5.2	1453	1.2	2055	4.7
8	Th	0301	1.1	0925	5.2	1547	1.3	2147	4.5
9	F	0351	1.2	1019	5.1	1643	1.4	2242	4.3
10	Sa	0445	1.4	1115	5.0	1743	1.6	2341	4.1
11	Su	0542	1.6	1215	4.8	1846	1.7		
12	M	0045	4.0	0645	1.8	1318	4.6	☽ 1952	1.8
13	Tu	0152	3.9	0752	1.9	1423	4.5	2055	1.9
14	W	0300	3.9	0900	2.0	1525	4.4	2153	1.9
15	Th	0401	4.0	1003	2.0	1621	4.3	2243	1.8
16	F	0454	4.1	1058	1.9	1710	4.2	2326	1.8
17	Sa	0540	4.2	1147	1.9	1753	4.2		
18	Su	0004	1.7	0620	4.3	1231	1.9	1832	4.1
19	M	0039	1.7	0657	4.4	1311	1.8	1907	4.1
20	Tu	0111	1.7	0733	4.5	1349	1.8	● 1942	4.0
21	W	0143	1.7	0807	4.5	1426	1.8	2016	3.9
22	Th	0215	1.7	0842	4.5	1503	1.8	2051	3.9
23	F	0248	1.7	0918	4.5	1542	1.9	2128	3.8
24	Sa	0323	1.8	0956	4.5	1622	1.9	2208	3.7
25	Su	0402	1.8	1037	4.4	1707	2.0	2253	3.6
26	M	0447	1.9	1123	4.4	1755	2.0	2345	3.6
27	Tu	0539	1.9	1214	4.3	1848	1.9		
28	W	0044	3.7	0640	1.9	1311	4.3	☾ 1943	1.9
29	Th	0147	3.8	0747	1.9	1411	4.3	2039	1.7
30	F	0249	4.0	0854	1.8	1511	4.3	2134	1.5
31	Sa	0349	4.2	0958	1.6	1609	4.4	2226	1.3

	Sun↑	Sun↓
15	0608	1743
31	0617	1730

NOVEMBER

Day		Time	ft	Time	ft	Time	ft	Time	ft
1	Su	0445	4.5	1059	1.4	1705	4.4	2317	1.2
2	M	0539	4.8	1156	1.2	1758	4.5		
3	Tu	0007	1.0	0631	5.0	1251	1.1	1851	4.5
4	W	0056	0.9	0722	5.2	1345	1.1	○ 1942	4.4
5	Th	0146	0.9	0814	5.2	1438	1.1	2035	4.3
6	F	0236	1.0	0906	5.2	1533	1.1	2129	4.2
7	Sa	0328	1.1	0959	5.0	1629	1.2	2225	4.0
8	Su	0422	1.3	1054	4.8	1726	1.4	2324	3.9
9	M	0520	1.5	1151	4.6	1825	1.5		
10	Tu	0027	3.8	0622	1.7	1250	4.4	☽ 1925	1.6
11	W	0132	3.7	0728	1.8	1350	4.2	2022	1.6
12	Th	0236	3.8	0834	1.9	1448	4.0	2115	1.6
13	F	0334	3.8	0936	1.9	1541	3.9	2202	1.6
14	Sa	0425	4.0	1032	1.9	1630	3.8	2244	1.6
15	Su	0509	4.1	1121	1.8	1713	3.8	2322	1.5
16	M	0549	4.2	1205	1.8	1753	3.7	2357	1.5
17	Tu	0627	4.3	1246	1.7	1832	3.7		
18	W	0031	1.4	0703	4.4	1325	1.7	● 1909	3.6
19	Th	0106	1.4	0739	4.4	1404	1.6	1947	3.6
20	F	0141	1.4	0815	4.4	1442	1.6	2025	3.5
21	Sa	0217	1.4	0852	4.4	1522	1.6	2105	3.5
22	Su	0255	1.5	0931	4.4	1603	1.6	2147	3.5
23	M	0337	1.5	1012	4.3	1646	1.6	2234	3.5
24	Tu	0425	1.6	1058	4.3	1732	1.6	2326	3.5
25	W	0518	1.6	1148	4.2	1821	1.5		
26	Th	0024	3.6	0619	1.7	1242	4.1	☾ 1913	1.4
27	F	0125	3.7	0725	1.6	1341	4.0	2007	1.3
28	Sa	0227	3.9	0833	1.6	1440	4.0	2102	1.2
29	Su	0327	4.2	0939	1.4	1540	4.0	2155	1.0
30	M	0424	4.4	1041	1.3	1638	4.0	2248	0.9

	Sun↑	Sun↓
15	0627	1722
30	0637	1720

DECEMBER

Day		Time	ft	Time	ft	Time	ft	Time	ft
1	Tu	0519	4.7	1140	1.1	1734	4.0	2341	0.8
2	W	0613	4.9	1236	1.0	1829	4.0		
3	Th	0032	0.7	0705	5.0	1331	1.0	○ 1923	4.0
4	F	0124	0.7	0757	5.0	1425	0.9	2017	3.9
5	Sa	0215	0.8	0848	4.9	1518	1.0	2111	3.8
6	Su	0308	0.9	0940	4.8	1611	1.0	2206	3.7
7	M	0401	1.1	1032	4.5	1704	1.1	2303	3.6
8	Tu	0457	1.3	1124	4.3	1757	1.2		
9	W	0002	3.6	0555	1.5	1216	4.1	1849	1.3
10	Th	0102	3.5	0656	1.7	1309	3.8	☽ 1940	1.4
11	F	0200	3.6	0758	1.8	1402	3.6	2028	1.4
12	Sa	0255	3.6	0858	1.8	1453	3.5	2113	1.4
13	Su	0346	3.7	0955	1.8	1541	3.4	2156	1.3
14	M	0431	3.9	1046	1.7	1628	3.4	2236	1.3
15	Tu	0514	4.0	1133	1.7	1712	3.3	2316	1.2
16	W	0554	4.1	1217	1.6	1755	3.3	2355	1.2
17	Th	0633	4.2	1259	1.5	1838	3.3		
18	F	0033	1.2	0711	4.3	1340	1.4	● 1919	3.3
19	Sa	0113	1.1	0750	4.3	1420	1.3	2002	3.3
20	Su	0154	1.2	0830	4.3	1500	1.3	2045	3.3
21	M	0236	1.2	0910	4.3	1541	1.3	2130	3.4
22	Tu	0321	1.2	0953	4.2	1623	1.2	2217	3.4
23	W	0410	1.3	1037	4.1	1707	1.2	2309	3.5
24	Th	0504	1.3	1126	4.0	1754	1.1		
25	F	0004	3.6	0603	1.4	1218	3.9	1845	1.1
26	Sa	0103	3.7	0708	1.4	1315	3.8	☾ 1938	1.0
27	Su	0205	3.9	0815	1.4	1414	3.7	2033	0.9
28	M	0306	4.1	0921	1.3	1515	3.6	2129	0.8
29	Tu	0405	4.3	1026	1.2	1616	3.6	2225	0.7
30	W	0503	4.5	1127	1.1	1715	3.6	2320	0.7
31	Th	0558	4.6	1224	1.0	1812	3.6		

	Sun↑	Sun↓
15	0647	1723
31	0655	1731

SAN JUAN, PUERTO RICO

HIGH & LOW WATER 1998 — Chart Datum — 18°28'N 66°01'W

Atlantic Standard Time (60°W) — Daylight Saving Time not observed

JANUARY

Day	Time	ft	Day	Time	ft
1 Th	0357 1108 1744 2248	-0.3 1.8 0.2 0.9	16 F	0453 1147 1821 2348	-0.1 1.4 0.2 0.8
2 F	0450 1151 1827 2352	-0.2 1.7 0.1 1.0	17 Sa	0537 1219 1855	0.0 1.3 0.1
3 Sa	0548 1236 1912	-0.1 1.6 0.0	18 Su	0041 0625 1253 1930	0.8 0.1 1.1 0.1
4 Su	0102 0653 1322 2000	1.0 0.0 1.5 -0.1	19 M	0138 0718 1328 2007	0.9 0.2 1.0 0.0
5 M	0216 0805 1411 ☾ 2050	1.2 0.2 1.3 -0.2	20 Tu	0238 0821 1405 ☽ 2046	0.9 0.3 0.9 -0.0
6 Tu	0330 0925 1503 2142	1.3 0.3 1.2 -0.3	21 W	0339 0931 1447 2128	1.0 0.4 0.9 -0.1
7 W	0442 1047 1559 2235	1.4 0.4 1.0 -0.4	22 Th	0438 1045 1532 2212	1.1 0.4 0.8 -0.2
8 Th	0548 1205 1657 2329	1.5 0.4 0.9 -0.4	23 F	0534 1154 1621 2258	1.2 0.4 0.7 -0.3
9 F	0647 1314 1754	1.6 0.4 0.9	24 Sa	0624 1253 1712 2344	1.3 0.4 0.7 -0.3
10 Sa	0021 0742 1414 1850	-0.4 1.7 0.3 0.8	25 Su	0711 1343 1805	1.5 0.3 0.7
11 Su	0111 0832 1506 1944	-0.4 1.7 0.3 0.8	26 M	0032 0756 1427 1858	-0.4 1.5 0.3 0.7
12 M	0159 0917 1552 ○ 2034	-0.4 1.7 0.3 0.8	27 Tu	0120 0838 1507 1952	-0.5 1.6 0.2 0.8
13 Tu	0245 0959 1633 2122	-0.4 1.6 0.2 0.8	28 W	0209 0920 1546 ● 2047	-0.5 1.6 0.1 0.8
14 W	0329 1037 1711 2210	-0.3 1.6 0.2 0.8	29 Th	0300 1001 1626 2145	-0.5 1.6 0.0 0.9
15 Th	0411 1113 1746 2258	-0.2 1.5 0.2 0.8	30 F	0353 1043 1707 2244	-0.4 1.5 -0.1 1.0
			31 Sa	0448 1125 1750 2347	-0.3 1.5 -0.2 1.1

	Sun↑	Sun↓
15	0700	1808
31	0658	1818

FEBRUARY

Day	Time	ft	Day	Time	ft
1 Su	0548 1209 1836	-0.2 1.3 -0.3	16 M	0005 0603 1207 1828	0.9 0.1 0.9 -0.0
2 M	0054 0653 1255 1925	1.2 -0.0 1.2 -0.3	17 Tu	0054 0653 1239 1904	1.0 0.2 0.9 -0.1
3 Tu	0203 0805 1345 ☾ 2018	1.2 0.1 1.0 -0.4	18 W	0149 0751 1314 1944	1.0 0.3 0.8 -0.1
4 W	0315 0923 1440 2114	1.3 0.2 0.9 -0.4	19 Th	0248 0857 1355 ☽ 2030	1.0 0.3 0.7 -0.2
5 Th	0426 1043 1539 2213	1.4 0.3 0.8 -0.4	20 F	0349 1007 1444 2122	1.1 0.4 0.7 -0.2
6 F	0532 1159 1642 2311	1.4 0.3 0.7 -0.5	21 Sa	0449 1115 1541 2217	1.2 0.4 0.7 -0.3
7 Sa	0633 1304 1744	1.5 0.3 0.7	22 Su	0544 1212 1643 2314	1.3 0.3 0.7 -0.4
8 Su	0008 0727 1358 1843	-0.5 1.5 0.2 0.7	23 M	0635 1300 1745	1.4 0.3 0.7
9 M	0101 0815 1444 1936	-0.4 1.5 0.2 0.7	24 Tu	0010 0722 1343 1846	-0.4 1.4 0.2 0.8
10 Tu	0150 0858 1523 2024	-0.4 1.4 0.2 0.8	25 W	0105 0806 1424 1944	-0.5 1.5 0.1 0.9
11 W	0235 0935 1558 ○ 2110	-0.3 1.4 0.1 0.8	26 Th	0200 0849 1504 ● 2042	-0.5 1.5 -0.0 1.1
12 Th	0317 1009 1629 2153	-0.3 1.3 0.1 0.8	27 F	0255 0931 1546 2140	-0.4 1.4 -0.2 1.2
13 F	0357 1039 1658 2235	-0.2 1.2 0.1 0.9	28 Sa	0351 1014 1628 2238	-0.4 1.4 -0.3 1.3
14 Sa	0437 1108 1727 2319	-0.1 1.1 0.0 0.9			
15 Su	0518 1137 1757	0.0 1.0 0.0			

	Sun↑	Sun↓
15	0652	1825
28	0644	1830

MARCH

Day	Time	ft	Day	Time	ft
1 Su	0448 1057 1713 2338	-0.3 1.3 -0.4 1.4	16 M	0508 1055 1659 2334	0.1 0.9 -0.0 1.2
2 M	0549 1142 1800	-0.1 1.1 -0.4	17 Tu	0551 1123 1730	0.2 0.8 -0.0
3 Tu	0040 0653 1231 1851	1.4 0.0 1.0 -0.4	18 W	0018 0638 1154 1806	1.2 0.3 0.8 -0.1
4 W	0145 0803 1324 1947	1.4 0.1 0.9 -0.4	19 Th	0108 0732 1229 1848	1.2 0.3 0.7 -0.1
5 Th	0253 0917 1423 ☾ 2047	1.4 0.2 0.8 -0.4	20 F	0203 0831 1314 1939	1.2 0.4 0.7 -0.1
6 F	0402 1032 1528 2151	1.4 0.2 0.8 -0.3	21 Sa	0302 0935 1410 ☽ 2038	1.3 0.4 0.7 -0.1
7 Sa	0509 1141 1636 2255	1.4 0.2 0.7 -0.3	22 Su	0403 1035 1518 2142	1.3 0.4 0.7 -0.2
8 Su	0609 1240 1741 2356	1.4 0.2 0.8 -0.3	23 M	0500 1128 1629 2248	1.3 0.3 0.8 -0.2
9 M	0702 1328 1839	1.4 0.2 0.8	24 Tu	0553 1216 1737 2353	1.4 0.2 0.9 -0.2
10 Tu	0051 0748 1409 1930	-0.2 1.3 0.2 0.9	25 W	0642 1259 1840	1.4 0.1 1.1
11 W	0141 0828 1443 2016	-0.2 1.3 0.1 0.9	26 Th	0054 0729 1342 1940	-0.3 1.4 -0.0 1.2
12 Th	0226 0902 1513 2057	-0.2 1.2 0.1 1.0	27 F	0154 0813 1424 ● 2037	-0.3 1.4 -0.2 1.4
13 F	0308 0933 1540 ○ 2136	-0.1 1.1 0.1 1.0	28 Sa	0252 0858 1507 2132	-0.2 1.3 -0.3 1.5
14 Sa	0348 1001 1606 2215	-0.0 1.1 0.0 1.1	29 Su	0350 0942 1551 2228	-0.2 1.2 -0.4 1.6
15 Su	0427 1028 1632 2253	0.0 1.0 0.0 1.1	30 M	0448 1028 1638 2325	-0.1 1.1 -0.4 1.7
			31 Tu	0548 1116 1727	0.0 1.0 -0.4

	Sun↑	Sun↓
15	0633	1834
31	0620	1838

APRIL

Day	Time	ft	Day	Time	ft
1 W	0023 0650 1208 1819	1.7 0.1 0.9 -0.4	16 Th	0629 1114 1720	0.4 0.8 -0.1
2 Th	0124 0756 1305 1917	1.6 0.2 0.9 -0.3	17 F	0037 0718 1154 1805	1.4 0.4 0.7 -0.1
3 F	0227 0904 1409 ☾ 2019	1.5 0.2 0.8 -0.2	18 Sa	0128 0811 1247 1859	1.4 0.4 0.7 -0.1
4 Sa	0332 1011 1519 2127	1.5 0.3 0.8 -0.1	19 Su	0223 0905 1353 ☽ 2004	1.4 0.4 0.8 -0.0
5 Su	0435 1112 1630 2235	1.4 0.2 0.8 -0.0	20 M	0319 0958 1509 2115	1.4 0.3 0.8 0.0
6 M	0533 1204 1736 2339	1.3 0.2 0.9 0.0	21 Tu	0415 1047 1624 2229	1.4 0.2 1.0 0.0
7 Tu	0624 1247 1832	1.3 0.2 1.0	22 W	0509 1134 1733 2340	1.4 0.1 1.2 0.0
8 W	0037 0708 1324 1921	0.0 1.2 0.1 1.1	23 Th	0600 1219 1836	1.4 -0.0 1.4
9 Th	0129 0746 1355 2004	0.1 1.2 0.1 1.2	24 F	0047 0649 1304 1935	0.0 1.3 -0.2 1.6
10 F	0216 0820 1423 2042	0.1 1.1 0.1 1.2	25 Sa	0150 0737 1348 2030	0.0 1.3 -0.3 1.7
11 Sa	0259 0850 1449 ○ 2119	0.1 1.0 0.1 1.3	26 Su	0250 0824 1434 ● 2124	0.0 1.2 -0.4 1.8
12 Su	0340 0918 1515 2155	0.2 1.0 0.0 1.4	27 M	0349 0912 1520 2217	0.1 1.1 -0.4 1.9
13 M	0420 0946 1541 2231	0.2 0.9 0.0 1.4	28 Tu	0446 1000 1607 2310	0.1 1.0 -0.4 1.9
14 Tu	0501 1013 1609 2310	0.3 0.8 -0.0 1.4	29 W	0544 1051 1657	0.2 1.0 -0.4
15 W	0544 1041 1642 2351	0.3 0.8 -0.0 1.4	30 Th	0005 0643 1146 1750	1.8 0.2 0.9 -0.3

	Sun↑	Sun↓
15	0608	1842
30	0558	1846

SAN JUAN, PUERTO RICO

HIGH & LOW WATER 1998 — Chart Datum — 18°28'N 66°01'W

Atlantic Standard Time (60°W) — Daylight Saving Time not observed

MAY

Day	Time	ft	Day	Time	ft
1 F	0100	1.7	16 Sa	0013	1.6
	0743	0.2		0702	0.4
	1246	0.9		1133	0.8
	1847	-0.1		1738	-0.1
2 Sa	0157	1.6	17 Su	0100	1.6
	0843	0.3		0747	0.4
	1353	0.8		1235	0.8
	1949	0.0		1835	-0.0
3 Su	0254	1.5	18 M	0148	1.5
	0940	0.3		0834	0.3
	1505	0.9		1347	0.9
	☾ 2056	0.1		1941	0.1
4 M	0351	1.4	19 Tu	0240	1.5
	1033	0.2		0922	0.2
	1617	0.9		1506	1.0
	2206	0.2		☽ 2056	0.2
5 Tu	0444	1.3	20 W	0333	1.4
	1119	0.2		1010	0.1
	1721	1.0		1621	1.2
	2314	0.3		2215	0.2
6 W	0532	1.2	21 Th	0426	1.3
	1159	0.1		1057	-0.1
	1816	1.1		1729	1.4
				2332	0.2
7 Th	0016	0.3	22 F	0518	1.3
	0615	1.1		1145	-0.2
	1234	0.1		1832	1.6
	1903	1.2			
8 F	0111	0.3	23 Sa	0043	0.3
	0654	1.1		0611	1.2
	1305	0.1		1232	-0.3
	1945	1.3		1929	1.8
9 Sa	0202	0.3	24 Su	0148	0.2
	0730	1.0		0702	1.1
	1335	0.0		1319	-0.4
	2023	1.4		2023	1.9
10 Su	0248	0.3	25 M	0249	0.2
	0802	0.9		0753	1.1
	1402	0.0		1407	-0.5
	2059	1.5		● 2115	2.0
11 M	0331	0.3	26 Tu	0346	0.2
	0833	0.9		0844	1.0
	1430	-0.0		1455	-0.5
	○ 2135	1.6		2206	2.0
12 Tu	0413	0.3	27 W	0441	0.2
	0902	0.8		0935	1.0
	1459	-0.1		1543	-0.4
	2212	1.6		2256	1.9
13 W	0455	0.4	28 Th	0535	0.2
	0932	0.8		1029	0.9
	1531	-0.1		1633	-0.3
	2250	1.6		2345	1.9
14 Th	0536	0.4	29 F	0627	0.3
	1005	0.8		1125	0.9
	1607	-0.1		1724	-0.2
	2330	1.6			
15 F	0618	0.4	30 Sa	0034	1.7
	1044	0.8		0719	0.3
	1649	-0.1		1225	0.9
				1818	-0.0
			31 Su	0123	1.6
				0810	0.2
				1331	0.9
				1916	0.1

	Sun↑	Sun↓
15	0551	1851
31	0547	1857

JUNE

Day	Time	ft	Day	Time	ft
1 M	0211	1.5	16 Tu	0118	1.6
	0859	0.2		0800	0.2
	1441	0.9		1344	1.0
	☾ 2020	0.3		1930	0.2
2 Tu	0259	1.4	17 W	0205	1.5
	0945	0.2		0847	0.1
	1551	1.0		1501	1.1
	2128	0.4		☽ 2047	0.3
3 W	0346	1.2	18 Th	0255	1.4
	1028	0.1		0935	-0.1
	1654	1.1		1614	1.3
	2239	0.4		2208	0.4
4 Th	0431	1.1	19 F	0348	1.2
	1107	0.1		1026	-0.2
	1749	1.2		1723	1.5
	2346	0.5		2328	0.4
5 F	0515	1.1	20 Sa	0443	1.1
	1143	0.0		1116	-0.3
	1837	1.3		1825	1.7
6 Sa	0047	0.5	21 Su	0041	0.4
	0557	1.0		0539	1.1
	1217	0.0		1207	-0.4
	1920	1.4		1922	1.8
7 Su	0142	0.4	22 M	0147	0.4
	0636	0.9		0635	1.0
	1249	-0.0		1258	-0.4
	2000	1.5		2015	1.9
8 M	0232	0.4	23 Tu	0246	0.4
	0713	0.9		0730	1.0
	1322	-0.1		1348	-0.4
	2038	1.6		● 2105	2.0
9 Tu	0317	0.4	24 W	0340	0.3
	0748	0.8		0823	1.0
	1354	-0.1		1437	-0.4
	2115	1.7		2153	1.9
10 W	0359	0.4	25 Th	0430	0.3
	0822	0.8		0917	0.9
	1429	-0.2		1526	-0.3
	○ 2153	1.7		2239	1.9
11 Th	0439	0.4	26 F	0517	0.3
	0859	0.8		1010	0.9
	1506	-0.2		1614	-0.2
	2231	1.7		2323	1.8
12 F	0517	0.4	27 Sa	0602	0.3
	0940	0.8		1104	0.9
	1547	-0.2		1702	-0.1
	2310	1.7			
13 Sa	0556	0.4	28 Su	0005	1.7
	1028	0.8		0645	0.3
	1632	-0.1		1201	0.9
	2351	1.7		1751	0.1
14 Su	0635	0.3	29 M	0045	1.5
	1125	0.9		0727	0.2
	1724	-0.1		1302	1.0
				1843	0.2
15 M	0033	1.6	30 Tu	0125	1.4
	0716	0.3		0809	0.2
	1231	0.9		1405	1.0
	1823	0.0		1941	0.4

	Sun↑	Sun↓
15	0548	1902
30	0551	1905

JULY

Day	Time	ft	Day	Time	ft
1 W	0205	1.3	16 Th	0135	1.4
	0850	0.2		0812	-0.0
	1510	1.1		1451	1.4
	☾ 2046	0.5		☽ 2044	0.4
2 Th	0246	1.2	17 F	0225	1.3
	0931	0.1		0904	-0.1
	1613	1.1		1603	1.5
	2157	0.5		2205	0.5
3 F	0330	1.1	18 Sa	0319	1.2
	1011	0.1		0958	-0.2
	1710	1.3		1711	1.7
	2309	0.6		2326	0.6
4 Sa	0414	1.0	19 Su	0418	1.1
	1050	0.1		1054	-0.3
	1802	1.4		1814	1.8
5 Su	0016	0.6	20 M	0039	0.5
	0500	0.9		0519	1.0
	1129	-0.0		1150	-0.3
	1848	1.5		1912	1.9
6 M	0115	0.6	21 Tu	0142	0.5
	0544	0.9		0619	1.0
	1208	-0.1		1244	-0.3
	1931	1.6		2005	1.9
7 Tu	0207	0.5	22 W	0236	0.5
	0627	0.9		0717	1.0
	1247	-0.1		1336	-0.3
	2012	1.7		2053	1.9
8 W	0251	0.5	23 Th	0324	0.4
	0710	0.9		0812	1.0
	1327	-0.2		1426	-0.2
	2052	1.8		● 2137	1.9
9 Th	0331	0.5	24 F	0408	0.4
	0753	0.9		0904	1.1
	1407	-0.2		1513	-0.1
	○ 2130	1.8		2218	1.8
10 F	0408	0.5	25 Sa	0447	0.4
	0838	0.9		0955	1.1
	1450	-0.2		1559	-0.0
	2209	1.8		2256	1.7
11 Sa	0444	0.4	26 Su	0525	0.4
	0928	0.9		1045	1.1
	1536	-0.2		1644	0.1
	2247	1.8		2332	1.6
12 Su	0520	0.4	27 M	0600	0.3
	1022	1.0		1135	1.1
	1626	-0.1		1729	0.2
	2326	1.7			
13 M	0558	0.3	28 Tu	0005	1.5
	1122	1.1		0635	0.3
	1720	0.0		1227	1.1
				1816	0.4
14 Tu	0007	1.7	29 W	0039	1.4
	0640	0.2		0711	0.3
	1227	1.2		1322	1.2
	1820	0.1		1908	0.5
15 W	0049	1.5	30 Th	0114	1.3
	0724	0.1		0749	0.3
	1337	1.3		1421	1.2
	1928	0.3		2008	0.6
			31 F	0151	1.2
				0829	0.2
				1521	1.3
				☾ 2116	0.7

	Sun↑	Sun↓
15	0556	1905
31	0602	1900

AUGUST

Day	Time	ft	Day	Time	ft
1 Sa	0232	1.1	16 Su	0302	1.2
	0912	0.2		0935	-0.1
	1621	1.4		1655	1.8
	2228	0.7		2320	0.7
2 Su	0319	1.1	17 M	0406	1.2
	0957	0.2		1037	-0.1
	1718	1.5		1759	1.9
	2338	0.7			
3 M	0409	1.0	18 Tu	0029	0.7
	1043	0.1		0513	1.2
	1810	1.6		1138	-0.0
				1857	1.9
4 Tu	0039	0.7	19 W	0126	0.6
	0500	1.0		0616	1.2
	1130	0.0		1236	-0.0
	1857	1.7		1948	1.9
5 W	0129	0.7	20 Th	0215	0.6
	0552	1.0		0714	1.2
	1217	-0.0		1330	0.0
	1941	1.8		2034	1.9
6 Th	0212	0.6	21 F	0257	0.6
	0644	1.0		0807	1.3
	1304	-0.1		1419	0.1
	2022	1.8		● 2115	1.8
7 F	0249	0.6	22 Sa	0334	0.5
	0735	1.1		0856	1.3
	1352	-0.1		1505	0.2
	○ 2101	1.9		2151	1.7
8 Sa	0325	0.5	23 Su	0407	0.5
	0827	1.2		0941	1.4
	1440	-0.1		1548	0.2
	2140	1.9		2224	1.7
9 Su	0401	0.4	24 M	0438	0.5
	0921	1.3		1025	1.4
	1531	-0.0		1630	0.4
	2218	1.8		2254	1.6
10 M	0439	0.3	25 Tu	0508	0.5
	1018	1.4		1108	1.4
	1624	0.1		1712	0.5
	2258	1.8		2324	1.5
11 Tu	0518	0.2	26 W	0538	0.4
	1117	1.5		1153	1.4
	1720	0.2		1756	0.6
	2339	1.7		2354	1.4
12 W	0602	0.1	27 Th	0611	0.4
	1219	1.6		1240	1.5
	1822	0.3		1844	0.7
13 Th	0023	1.5	28 F	0026	1.3
	0648	0.0		0646	0.4
	1325	1.6		1332	1.5
	1929	0.5		1939	0.7
14 F	0110	1.4	29 Sa	0101	1.3
	0740	-0.0		0726	0.4
	1435	1.7		1428	1.5
	☽ 2044	0.6		2042	0.8
15 Sa	0203	1.3	30 Su	0143	1.2
	0835	-0.0		0812	0.4
	1546	1.8		1528	1.6
	2203	0.7		☾ 2150	0.8
			31 M	0231	1.2
				0903	0.3
				1628	1.6
				2257	0.9

	Sun↑	Sun↓
15	0606	1852
31	0610	1840

T Tides

SAN JUAN, PUERTO RICO

HIGH & LOW WATER 1998 — Chart Datum — 18°28'N 66°01'W

Atlantic Standard Time (60°W) — Daylight Saving Time not observed

SEPTEMBER

Date	Time	ft	Date	Time	ft
1 Tu	0328 0959 1725 2355	1.1 0.3 1.7 0.8	16 W	0007 0514 1128 1832	0.7 1.3 0.3 1.9
2 W	0429 1055 1815	1.2 0.2 1.8	17 Th	0059 0618 1229 1922	0.7 1.4 0.3 1.9
3 Th	0042 0530 1152 1901	0.8 1.2 0.2 1.9	18 F	0142 0713 1324 2005	0.7 1.5 0.4 1.8
4 F	0123 0629 1246 1944	0.7 1.3 0.2 1.9	19 Sa	0219 0802 1413 2042	0.6 1.5 0.4 1.7
5 Sa	0201 0725 1340 2025	0.6 1.4 0.1 1.9	20 Su	0251 0846 1458 ● 2116	0.6 1.6 0.5 1.7
6 Su	0239 0820 1433 ○ 2106	0.5 1.6 0.2 1.9	21 M	0321 0927 1540 2145	0.6 1.7 0.5 1.6
7 M	0317 0915 1528 2146	0.4 1.7 0.2 1.8	22 Tu	0348 1005 1621 2214	0.6 1.7 0.6 1.5
8 Tu	0357 1010 1623 2228	0.3 1.8 0.3 1.7	23 W	0415 1043 1701 2241	0.5 1.7 0.7 1.4
9 W	0440 1107 1721 2311	0.2 1.9 0.4 1.6	24 Th	0443 1123 1744 2310	0.5 1.8 0.7 1.4
10 Th	0526 1207 1823 2358	0.1 2.0 0.5 1.5	25 F	0513 1205 1829 2341	0.5 1.8 0.8 1.3
11 F	0615 1310 1930	0.1 2.0 0.6	26 Sa	0548 1251 1920	0.5 1.8 0.9
12 Sa	0050 0710 1416 ☽ 2041	1.4 0.1 2.0 0.7	27 Su	0017 0629 1343 2017	1.3 0.5 1.8 0.9
13 Su	0148 0810 1524 2156	1.4 0.1 2.0 0.8	28 M	0100 0717 1439 ☾ 2117	1.2 0.5 1.8 0.9
14 M	0253 0915 1632 2306	1.3 0.2 2.0 0.8	29 Tu	0155 0814 1537 2215	1.2 0.5 1.8 0.9
15 Tu	0404 1022 1735	1.3 0.2 2.0	30 W	0300 0918 1634 2307	1.3 0.5 1.8 0.9

	Sun↑	Sun↓
15	0612	1827
30	0615	1814

OCTOBER

Date	Time	ft	Date	Time	ft
1 Th	0410 1024 1727 2352	1.3 0.5 1.9 0.8	16 F	0021 0614 1217 1843	0.7 1.5 0.6 1.7
2 F	0517 1129 1815	1.4 0.4 1.9	17 Sa	0101 0706 1314 1924	0.6 1.6 0.6 1.7
3 Sa	0034 0619 1231 1901	0.7 1.6 0.4 1.9	18 Su	0135 0752 1405 2000	0.6 1.7 0.6 1.6
4 Su	0114 0717 1330 1945	0.5 1.8 0.4 1.8	19 M	0205 0833 1451 2033	0.5 1.8 0.7 1.5
5 M	0155 0813 1428 ○ 2028	0.4 1.9 0.4 1.8	20 Tu	0233 0910 1533 ● 2103	0.5 1.8 0.7 1.4
6 Tu	0237 0907 1525 2112	0.3 2.1 0.4 1.7	21 W	0300 0946 1614 2131	0.5 1.9 0.7 1.4
7 W	0320 1001 1623 2157	0.2 2.2 0.5 1.6	22 Th	0326 1021 1655 2159	0.5 1.9 0.8 1.3
8 Th	0406 1056 1721 2245	0.1 2.3 0.6 1.6	23 F	0355 1058 1736 2228	0.5 1.9 0.8 1.2
9 F	0454 1153 1822 2336	0.1 2.3 0.6 1.5	24 Sa	0426 1138 1819 2300	0.4 1.9 0.8 1.2
10 Sa	0546 1252 1926	0.1 2.2 0.7	25 Su	0502 1220 1906 2339	0.4 1.9 0.9 1.2
11 Su	0033 0643 1353 2033	1.4 0.2 2.2 0.7	26 M	0544 1307 1955	0.4 1.9 0.9
12 M	0137 0745 1457 ☽ 2139	1.4 0.3 2.1 0.7	27 Tu	0029 0635 1357 2045	1.2 0.5 1.9 0.9
13 Tu	0248 0853 1601 2241	1.3 0.4 2.0 0.7	28 W	0131 0735 1450 ☾ 2134	1.2 0.5 1.8 0.8
14 W	0403 1005 1701 2335	1.4 0.5 1.9 0.7	29 Th	0244 0844 1544 2221	1.3 0.5 1.8 0.7
15 Th	0512 1114 1755	1.4 0.5 1.8	30 F	0359 0957 1636 2306	1.4 0.6 1.8 0.6
			31 Sa	0508 1110 1727 2350	1.6 0.6 1.7 0.4

	Sun↑	Sun↓
15	0618	1802
31	0624	1752

NOVEMBER

Date	Time	ft	Date	Time	ft
1 Su	0611 1219 1815	1.8 0.5 1.7	16 M	0049 0735 1352 1911	0.4 1.7 0.7 1.3
2 M	0033 0709 1323 1903	0.3 2.0 0.5 1.6	17 Tu	0120 0814 1440 1946	0.3 1.8 0.7 1.2
3 Tu	0118 0804 1424 1951	0.1 2.1 0.5 1.6	18 W	0149 0851 1524 2018	0.3 1.8 0.7 1.2
4 W	0203 0858 1523 ○ 2039	0.0 2.3 0.5 1.5	19 Th	0218 0926 1606 ● 2049	0.3 1.9 0.7 1.1
5 Th	0250 0951 1620 2129	-0.1 2.4 0.5 1.4	20 F	0247 1002 1647 2119	0.2 1.9 0.7 1.1
6 F	0338 1044 1718 2221	-0.1 2.4 0.6 1.4	21 Sa	0318 1038 1727 2151	0.2 1.9 0.7 1.0
7 Sa	0428 1137 1816 2316	-0.0 2.3 0.6 1.3	22 Su	0352 1116 1806 2228	0.2 1.9 0.7 1.0
8 Su	0521 1232 1915	0.1 2.2 0.6	23 M	0431 1156 1846 2314	0.2 1.9 0.7 1.0
9 M	0017 0618 1328 2014	1.3 0.2 2.1 0.6	24 Tu	0515 1238 1928	0.2 1.8 0.7
10 Tu	0125 0720 1425 ☽ 2112	1.3 0.3 2.0 0.6	25 W	0011 0608 1322 2010	1.0 0.3 1.8 0.6
11 W	0238 0828 1522 2206	1.3 0.4 1.8 0.5	26 Th	0119 0710 1409 ☾ 2054	1.1 0.4 1.7 0.5
12 Th	0352 0940 1616 2255	1.3 0.6 1.7 0.5	27 F	0234 0821 1458 2139	1.2 0.5 1.6 0.4
13 F	0500 1052 1706 2337	1.4 0.6 1.6 0.5	28 Sa	0349 0939 1549 2225	1.3 0.5 1.5 0.2
14 Sa	0600 1158 1752	1.5 0.7 1.5	29 Su	0459 1057 1641 2312	1.5 0.5 1.5 0.0
15 Su	0015 0650 1259 1834	0.4 1.6 0.7 1.4	30 M	0602 1210 1733	1.7 0.5 1.4

	Sun↑	Sun↓
15	0631	1747
30	0640	1747

DECEMBER

Date	Time	ft	Date	Time	ft
1 Tu	0000 0700 1318 1826	-0.1 1.9 0.5 1.3	16 W	0037 0750 1423 1859	0.0 1.6 0.5 0.9
2 W	0048 0755 1421 1919	-0.2 2.1 0.5 1.3	17 Th	0111 0829 1509 1935	0.0 1.7 0.5 0.9
3 Th	0137 0848 1519 ○ 2012	-0.3 2.2 0.5 1.2	18 F	0144 0906 1550 ● 2011	-0.0 1.7 0.5 0.8
4 F	0227 0940 1614 2106	-0.3 2.2 0.4 1.2	19 Sa	0218 0942 1629 2046	-0.1 1.7 0.5 0.8
5 Sa	0317 1030 1708 2201	-0.3 2.2 0.4 1.1	20 Su	0253 1018 1705 2125	-0.1 1.8 0.5 0.8
6 Su	0408 1120 1801 2259	-0.2 2.1 0.4 1.1	21 M	0331 1054 1740 2210	-0.1 1.7 0.5 0.8
7 M	0501 1210 1853	-0.1 2.0 0.4	22 Tu	0414 1131 1815 2302	-0.1 1.7 0.4 0.9
8 Tu	0000 0556 1259 1944	1.1 0.0 1.9 0.4	23 W	0501 1210 1851	-0.0 1.6 0.3
9 W	0107 0655 1348 2034	1.1 0.2 1.7 0.3	24 Th	0003 0555 1250 1931	0.9 0.1 1.6 0.2
10 Th	0217 0758 1436 ☽ 2122	1.1 0.4 1.5 0.3	25 F	0111 0657 1333 2014	1.0 0.2 1.5 0.1
11 F	0328 0907 1523 2206	1.2 0.5 1.4 0.2	26 Sa	0224 0809 1419 ☾ 2100	1.1 0.3 1.3 -0.0
12 Sa	0435 1020 1610 2248	1.2 0.6 1.3 0.2	27 Su	0338 0928 1509 2150	1.3 0.4 1.2 -0.2
13 Su	0533 1130 1655 2326	1.3 0.6 1.2 0.1	28 M	0447 1049 1603 2241	1.5 0.4 1.1 -0.3
14 M	0624 1235 1739	1.4 0.6 1.1	29 Tu	0552 1206 1700 2334	1.6 0.4 1.0 -0.4
15 Tu	0003 0709 1332 1820	0.1 1.5 0.6 1.0	30 W	0651 1314 1759	1.8 0.4 1.0
			31 Th	0027 0746 1415 1857	-0.5 1.9 0.3 1.0

	Sun↑	Sun↓
15	0649	1751
31	0656	1759

CHARLOTTE AMALIE, U.S.V.I.

HIGH & LOW WATER 1998 | Chart Datum | 18°20'N 64°56'W

Atlantic Standard Time (60°W) | Daylight Saving Time not observed

T Tides

JANUARY

Day		Time	ft	Time	ft	Time	ft	Time	ft
1	Th	0304	-0.3	1138	0.7				
2	F	0356	-0.2	1221	0.6	2037	0.0	2315	0.0
3	Sa	0458	-0.1	1303	0.5	2046	-0.0		
4	Su	0132	0.1	0619	-0.0	1344	0.4	2107	-0.1
5	M ☾	0327	0.2	0816	0.1	1421	0.3	2133	-0.1
6	Tu	0450	0.3	1034	0.1	1453	0.2	2203	-0.2
7	W	0554	0.4	1253	0.1	1513	0.1	2236	-0.3
8	Th	0648	0.5	2312	-0.3				
9	F	0737	0.6	2348	-0.4				
10	Sa	0822	0.7						
11	Su	0025	-0.4	0906	0.7				
12	M ○	0102	-0.4	0946	0.7				
13	Tu	0138	-0.4	1024	0.6				
14	W	0212	-0.3	1058	0.6				
15	Th	0243	-0.3	1128	0.5				
16	F	0312	-0.2	1153	0.4	2107	-0.1	2240	-0.0
17	Sa	0334	-0.1	1212	0.4	2041	-0.1		
18	Su	0214	-0.0	0331	-0.0	1224	0.3	2041	-0.1
19	M	1228	0.2	2051	-0.2				
20	Tu ☽	0554	0.2	0821	0.2	1209	0.2	2109	-0.2
21	W	0600	0.3	2133	-0.3				
22	Th	0623	0.4	2203	-0.3				
23	F	0652	0.4	2238	-0.4				
24	Sa	0724	0.5	2316	-0.4				
25	Su	0759	0.6	2358	-0.4				
26	M	0835	0.6						
27	Tu	0042	-0.5	0913	0.6				
28	W ●	0128	-0.4	0952	0.6				
29	Th	0218	-0.4	1031	0.5	1815	-0.0	2037	-0.0
30	F	0312	-0.3	1109	0.5	1829	-0.1	2213	0.0
31	Sa	0412	-0.2	1147	0.4	1851	-0.1	2352	0.1

	Sun↑	Sun↓
15	0655	1804
31	0654	1814

FEBRUARY

Day		Time	ft	Time	ft	Time	ft	Time	ft
1	Su	0524	-0.1	1223	0.3	1919	-0.1		
2	M	0135	0.1	0659	-0.0	1254	0.2	1952	-0.2
3	Tu ☾	0311	0.2	0910	0.0	1318	0.1	2029	-0.3
4	W	0431	0.3	1206	0.0	1312	0.0	2109	-0.3
5	Th	0535	0.4	2153	-0.4				
6	F	0629	0.5	2237	-0.4				
7	Sa	0718	0.6	2322	-0.4				
8	Su	0802	0.6						
9	M	0005	-0.4	0842	0.6				
10	Tu	0047	-0.4	0918	0.5				
11	W ○	0128	-0.4	0950	0.5				
12	Th	0206	-0.3	1017	0.4	1810	-0.0	2038	-0.0
13	F	0244	-0.2	1039	0.3	1809	-0.1	2202	-0.0
14	Sa	0322	-0.1	1055	0.3	1815	-0.1	2330	0.0
15	Su	0405	-0.1	1105	0.2	1827	-0.1		
16	M	0108	0.1	0505	0.0	1108	0.2	1846	-0.2
17	Tu	0243	0.1	0658	0.1	1055	0.1	1911	-0.2
18	W	0352	0.2	1944	-0.3				
19	Th ☽	0442	0.3	2024	-0.3				
20	F	0525	0.4	2109	-0.3				
21	Sa	0605	0.5	2159	-0.4				
22	Su	0645	0.5	2250	-0.4				
23	M	0724	0.5	2344	-0.4				
24	Tu	0804	0.5	1603	-0.0	1731	-0.0		
25	W	0038	-0.4	0843	0.5	1601	-0.0	1859	0.0
26	Th ●	0134	-0.4	0922	0.5	1615	-0.0	2013	0.1
27	F	0232	-0.3	0959	0.4	1634	-0.1	2126	0.1
28	Sa	0335	-0.2	1034	0.3	1658	-0.1	2241	0.2

	Sun↑	Sun↓
15	0648	1821
28	0640	1826

MARCH

Day		Time	ft	Time	ft	Time	ft	Time	ft
1	Su	0445	-0.1	1105	0.2	1726	-0.1		
2	M	0001	0.2	0611	-0.0	1132	0.1	1800	-0.2
3	Tu	0124	0.3	0803	0.0	1150	0.1	1839	-0.2
4	W	0245	0.4	1924	-0.3				
5	Th ☾	0359	0.4	2014	-0.3				
6	F	0503	0.5	2108	-0.3				
7	Sa	0558	0.5	2204	-0.3				
8	Su	0646	0.5	2259	-0.3				
9	M	0729	0.5	2352	-0.3				
10	Tu	0805	0.5	1551	-0.0	1823	0.0		
11	W	0042	-0.2	0837	0.4	1551	-0.0	1930	0.1
12	Th	0130	-0.2	0902	0.4	1555	-0.0	2030	0.1
13	F ○	0217	-0.1	0922	0.3	1601	-0.0	2127	0.1
14	Sa	0305	-0.1	0936	0.3	1611	-0.1	2224	0.2
15	Su	0359	0.0	0945	0.2	1625	-0.1	2323	0.2
16	M	0505	0.1	0947	0.2	1644	-0.1		
17	Tu	0025	0.3	0639	0.1	0936	0.1	1710	-0.2
18	W	0129	0.3	1743	-0.2				
19	Th	0231	0.4	1826	-0.2				
20	F	0330	0.4	1917	-0.2				
21	Sa ☽	0423	0.5	2018	-0.3				
22	Su	0513	0.5	2124	-0.3				
23	M	0600	0.5	1433	0.0	1534	0.0	2233	-0.2
24	Tu	0644	0.5	1411	0.0	1724	0.1	2340	-0.2
25	W	0726	0.5	1420	0.0	1838	0.1		
26	Th	0047	-0.2	0805	0.4	1435	-0.0	1943	0.2
27	F ●	0154	-0.2	0842	0.4	1455	-0.0	2045	0.3
28	Sa	0301	-0.1	0916	0.3	1518	-0.1	2147	0.4
29	Su	0414	-0.1	0945	0.2	1545	-0.1	2251	0.4
30	M	0536	0.0	1008	0.1	1616	-0.2	2357	0.5
31	Tu	0720	0.1	1018	0.1	1651	-0.2		

	Sun↑	Sun↓
15	0629	1830
31	0615	1834

APRIL

Day		Time	ft	Time	ft	Time	ft	Time	ft
1	W	0106	0.5	1731	-0.2				
2	Th	0214	0.6	1818	-0.2				
3	F ☾	0321	0.6	1914	-0.2				
4	Sa	0421	0.6	2018	-0.2				
5	Su	0515	0.6	2129	-0.1				
6	M	0600	0.5	1410	0.0	1708	0.0	2239	-0.1
7	Tu	0639	0.5	1407	0.0	1823	0.1	2346	-0.1
8	W	0711	0.4	1409	0.0	1920	0.2		
9	Th	0049	-0.0	0736	0.4	1414	-0.0	2009	0.2
10	F	0149	0.0	0755	0.3	1422	-0.0	2054	0.3
11	Sa ○	0248	0.1	0807	0.2	1432	-0.1	2138	0.4
12	Su	0350	0.1	0813	0.2	1446	-0.1	2221	0.4
13	M	0502	0.1	0812	0.2	1506	-0.1	2306	0.5
14	Tu	1530	-0.2	2353	0.5				
15	W	1601	-0.2						
16	Th	0043	0.5	1637	-0.2				
17	F	0137	0.6	1722	-0.2				
18	Sa	0232	0.6	1820	-0.2				
19	Su ☽	0327	0.6	1933	-0.1				
20	M	0420	0.6	1252	0.0	1508	0.1	2059	-0.1
21	Tu	0510	0.5	1242	0.0	1655	0.1	2228	-0.1
22	W	0556	0.5	1251	0.0	1808	0.2	2352	-0.0
23	Th	0638	0.4	1308	-0.0	1909	0.3		
24	F	0111	-0.0	0716	0.3	1329	-0.1	2005	0.4
25	Sa	0228	0.0	0749	0.3	1353	-0.1	2100	0.5
26	Su ●	0345	0.0	0816	0.2	1419	-0.2	2154	0.6
27	M	0509	0.1	0835	0.1	1449	-0.2	2250	0.7
28	Tu	0659	0.1	0831	0.1	1521	-0.2	2346	0.7
29	W	1556	-0.2						
30	Th	0043	0.7	1634	-0.2				

	Sun↑	Sun↓
15	0604	1837
30	0554	1841

CHARLOTTE AMALIE, U.S.V.I.

HIGH & LOW WATER 1998 — Chart Datum — 18°20'N 64°56'W

Atlantic Standard Time (60°W) — Daylight Saving Time not observed

MAY

Date	Time	ft	Date	Time	ft
1 F	0141	0.7	16 Sa	0057	0.7
	1717	-0.2		1641	-0.2
2 Sa	0237	0.6	17 Su	0147	0.7
	1806	-0.1		1741	-0.1
3 Su ☾	0329	0.6	18 M	0238	0.6
	1911	-0.0		1133	0.0
				1354	0.1
				1903	-0.0
4 M	0417	0.5	19 Tu ☽	0328	0.6
	1248	0.0		1120	0.0
	1702	0.1		1607	0.1
	2043	0.0		2050	0.0
5 Tu	0457	0.5	20 W	0417	0.5
	1242	0.0		1131	0.0
	1817	0.2		1727	0.3
	2225	0.1		2240	0.1
6 W	0529	0.4	21 Th	0501	0.4
	1244	0.0		1148	-0.0
	1906	0.2		1828	0.4
	2357	0.1			
7 Th	0553	0.3	22 F	0020	0.1
	1250	-0.0		0540	0.3
	1946	0.3		1211	-0.1
				1922	0.5
8 F	0118	0.2	23 Sa	0151	0.1
	0609	0.3		0613	0.2
	1258	-0.1		1237	-0.2
	2022	0.4		2013	0.6
9 Sa	0235	0.2	24 Su	0319	0.1
	0618	0.2		0638	0.2
	1310	-0.1		1305	-0.2
	2056	0.5		2103	0.7
10 Su	0353	0.2	25 M ●	0454	0.1
	0616	0.2		0648	0.1
	1326	-0.1		1336	-0.3
	2131	0.5		2152	0.8
11 M ○	1347	-0.2	26 Tu	1409	-0.3
	2206	0.6		2241	0.8
12 Tu	1412	-0.2	27 W	1443	-0.3
	2244	0.6		2330	0.8
13 W	1441	-0.2	28 Th	1517	-0.2
	2325	0.7			
14 Th	1515	-0.2	29 F	0018	0.7
				1552	-0.2
15 F	0009	0.7	30 Sa	0105	0.7
	1554	-0.2		1625	-0.1
			31 Su	0149	0.6
				1653	-0.0

	Sun↑	Sun↓
15	0547	1847
31	0543	1853

JUNE

Date	Time	ft	Date	Time	ft
1 M ☾	0230	0.6	16 Tu	0154	0.6
	1138	0.0		0953	0.1
				1441	0.2
				1902	0.1
2 Tu	0304	0.5	17 W ☽	0239	0.5
	1125	0.0		1007	0.0
				1624	0.3
				2106	0.2
3 W	0331	0.4	18 Th	0321	0.4
	1126	-0.0		1028	-0.0
	1857	0.3		1734	0.4
	2229	0.2		2315	0.2
4 Th	0349	0.3	19 F	0358	0.3
	1132	-0.0		1054	-0.1
	1921	0.4		1831	0.5
5 F	0045	0.3	20 Sa	0109	0.2
	0355	0.3		0429	0.3
	1142	-0.1		1124	-0.2
	1947	0.5		1922	0.7
6 Sa	1156	-0.1	21 Su	0258	0.2
	2014	0.5		0446	0.2
				1156	-0.2
				2011	0.8
7 Su	1215	-0.2	22 M	1231	-0.3
	2043	0.6		2057	0.8
8 M	1238	-0.2	23 Tu ●	1306	-0.3
	2113	0.7		2143	0.9
9 Tu	1305	-0.3	24 W	1342	-0.3
	2146	0.7		2226	0.9
10 W ○	1336	-0.3	25 Th	1418	-0.2
	2222	0.8		2309	0.8
11 Th	1411	-0.3	26 F	1453	-0.2
	2300	0.8		2348	0.8
12 F	1450	-0.3	27 Sa	1525	-0.1
	2341	0.8			
13 Sa	1533	-0.2	28 Su	0025	0.7
				1552	-0.0
14 Su	0024	0.7	29 M	0057	0.6
	1624	-0.1		1018	0.1
				1328	0.1
				1559	0.1
15 M	0109	0.7	30 Tu	0123	0.5
	1002	0.1		0959	0.1
	1201	0.1			
	1729	-0.0			

	Sun↑	Sun↓
15	0544	1858
30	0547	1900

JULY

Date	Time	ft	Date	Time	ft
1 W ☾	0140	0.5	16 Th ☽	0147	0.5
	1000	0.0		0858	0.1
				1615	0.5
				2152	0.3
2 Th	0143	0.4	17 F	0219	0.4
	1009	0.0		0930	-0.0
	1848	0.4		1723	0.6
3 F	0007	0.4	18 Sa	0017	0.3
	0051	0.4		0239	0.3
	1022	-0.0		1006	-0.1
	1903	0.5		1820	0.7
4 Sa	1041	-0.1	19 Su	1044	-0.1
	1924	0.6		1911	0.8
5 Su	1104	-0.1	20 M	1125	-0.2
	1949	0.7		1958	0.9
6 M	1131	-0.2	21 Tu	1205	-0.2
	2017	0.7		2043	0.9
7 Tu	1202	-0.2	22 W	1246	-0.2
	2048	0.8		2125	0.9
8 W	1237	-0.2	23 Th ●	1326	-0.1
	2122	0.8		2204	0.9
9 Th ○	1315	-0.2	24 F	1406	-0.1
	2157	0.9		2240	0.8
10 F	1357	-0.2	25 Sa	1443	-0.0
	2234	0.8		2312	0.8
11 Sa	1442	-0.2	26 Su	1518	0.1
	2313	0.8		2339	0.7
12 Su	1532	-0.1	27 M	0753	0.2
	2353	0.7		1104	0.3
				1551	0.2
				2359	0.6
13 M	0756	0.2	28 Tu	0757	0.2
	1040	0.2		1331	0.3
	1632	0.0		1620	0.3
14 Tu	0032	0.7	29 W	0010	0.6
	0808	0.2		0807	0.2
	1247	0.3			
	1747	0.1			
15 W	0111	0.6	30 Th	0006	0.5
	0830	0.1		0823	0.2
	1444	0.4		1722	0.5
	1933	0.2		2108	0.5
			31 F ☾	0843	0.1
				1747	0.6

	Sun↑	Sun↓
15	0552	1900
31	0558	1855

AUGUST

Date	Time	ft	Date	Time	ft
1 Sa	0909	0.1	16 Su	0921	0.1
	1814	0.7		1758	0.9
2 Su	0940	0.0	17 M	1010	0.0
	1842	0.8		1849	1.0
3 M	1015	-0.0	18 Tu	1059	0.0
	1912	0.8		1936	1.0
4 Tu	1054	-0.0	19 W	1148	0.0
	1943	0.9		2018	1.0
5 W	1136	-0.1	20 Th	1235	0.1
	2017	0.9		2057	1.0
6 Th	1221	-0.1	21 F ●	0521	0.4
	2052	0.9		0615	0.4
				1321	0.1
				2131	0.9
7 F ○	1308	-0.0	22 Sa	0512	0.4
	2129	0.9		0742	0.4
				1406	0.2
				2201	0.9
8 Sa	0538	0.4	23 Su	0518	0.4
	0645	0.4		0855	0.5
	1358	-0.0		1450	0.3
	2206	0.9		2225	0.8
9 Su	0533	0.4	24 M	0528	0.4
	0823	0.4		1009	0.5
	1451	0.1		1534	0.4
	2243	0.8		2241	0.7
10 M	0548	0.3	25 Tu	0539	0.4
	0949	0.3		1126	0.6
	1551	0.2		1624	0.4
	2319	0.8		2249	0.7
11 Tu	0611	0.3	26 W	0554	0.4
	1119	0.5		1251	0.6
	1701	0.3		1730	0.5
	2353	0.7		2245	0.6
12 W	0638	0.3	27 Th	0614	0.3
	1254	0.6		1418	0.7
	1831	0.4		1935	0.6
				2209	0.6
13 Th	0024	0.6	28 F	0639	0.3
	0712	0.2		1528	0.7
	1427	0.7			
	2031	0.4			
14 F ☽	0048	0.5	29 Sa	0712	0.2
	0750	0.2		1622	0.8
	1550	0.8			
	2314	0.5			
15 Sa	0051	0.5	30 Su ☾	0751	0.2
	0834	0.1		1706	0.9
	1659	0.9			
			31 M	0838	0.2
				1745	0.9

	Sun↑	Sun↓
15	0602	1847
31	0606	1835

CHARLOTTE AMALIE, U.S.V.I.

HIGH & LOW WATER 1998 | Chart Datum | 18°20'N 64°56'W

Atlantic Standard Time (60°W) | Daylight Saving Time not observed

SEPTEMBER

Date	Day	Time	ft
1	Tu	0929 1824	0.2 1.0
2	W	1024 1902	0.2 1.0
3	Th	1119 1940	0.2 1.0
4	F	0330 0524 1216 2017	0.5 0.5 0.2 1.0
5	Sa	0328 0648 1314 2054	0.5 0.6 0.2 1.0
6	Su ○	0340 0759 1414 2130	0.5 0.6 0.3 0.9
7	M	0358 0907 1517 2204	0.5 0.7 0.3 0.8
8	Tu	0421 1016 1627 2236	0.4 0.7 0.4 0.7
9	W	0449 1129 1749 2303	0.4 0.8 0.5 0.7
10	Th	0521 1246 1934 2322	0.3 0.9 0.5 0.6
11	F	0600 1404	0.3 0.9
12	Sa ☽	0645 1518	0.3 1.0
13	Su	0737 1625	0.2 1.0
14	M	0835 1724	0.2 1.1
15	Tu	0936 1816	0.2 1.1
16	W	1038 1901	0.3 1.1
17	Th	0307 0513 1138 1941	0.5 0.5 0.3 1.0
18	F	0307 0631 1234 2016	0.5 0.6 0.3 1.0
19	Sa	0314 0735 1329 2044	0.5 0.6 0.4 0.9
20	Su ●	0323 0833 1422 2105	0.5 0.7 0.4 0.8
21	M	0332 0928 1517 2119	0.5 0.7 0.5 0.8
22	Tu	0344 1022 1616 2124	0.5 0.8 0.6 0.7
23	W	0358 1117 1728 2118	0.4 0.8 0.6 0.7
24	Th	0416 1213 1930 2030	0.4 0.9 0.7 0.7
25	F	0440 1311	0.4 0.9
26	Sa	0510 1408	0.3 0.9
27	Su	0548 1503	0.3 1.0
28	M ☾	0637 1555	0.3 1.0
29	Tu	0737 1644	0.3 1.0
30	W	0846 1730	0.3 1.0

	Sun↑	Sun↓
15	0608	1823
30	0611	1810

OCTOBER

Date	Day	Time	ft
1	Th	0218 0313 1001 1813	0.5 0.5 0.3 1.0
2	F	0145 0513 1114 1855	0.5 0.6 0.4 1.0
3	Sa	0150 0626 1225 1934	0.5 0.7 0.4 0.9
4	Su	0204 0728 1334 2010	0.5 0.8 0.4 0.9
5	M ○	0223 0827 1444 2044	0.5 0.8 0.4 0.8
6	Tu	0247 0926 1557 2113	0.4 0.9 0.5 0.7
7	W	0313 1026 1718 2137	0.4 1.0 0.5 0.6
8	Th	0344 1128 1857 2150	0.3 1.0 0.6 0.6
9	F	0419 1232	0.3 1.1
10	Sa	0458 1337	0.3 1.1
11	Su	0544 1442	0.3 1.1
12	M ☽	0639 1544	0.3 1.1
13	Tu	0745 1640	0.3 1.1
14	W	0159 0256 0901 1729	0.5 0.5 0.4 1.0
15	Th	0131 0505 1021 1812	0.5 0.5 0.4 1.0
16	F	0134 0619 1137 1847	0.5 0.6 0.5 0.9
17	Sa	0140 0717 1248 1914	0.5 0.7 0.5 0.8
18	Su	0149 0807 1355 1934	0.5 0.8 0.5 0.7
19	M	0159 0853 1501 1944	0.4 0.8 0.6 0.7
20	Tu ●	0211 0935 1610 1944	0.4 0.9 0.6 0.6
21	W	0226 1016 1740 1923	0.3 0.9 0.6 0.6
22	Th	0244 1057	0.3 1.0
23	F	0306 1140	0.3 1.0
24	Sa	0334 1225	0.2 1.0
25	Su	0407 1312	0.2 1.0
26	M	0447 1402	0.2 1.0
27	Tu	0537 1453	0.3 1.0
28	W ☾	0645 1545	0.3 1.0
29	Th	0042 0239 0813 1634	0.5 0.5 0.4 0.9
30	F	0019 0441 0951 1720	0.4 0.5 0.4 0.9
31	Sa	0025 0554 1124 1802	0.4 0.6 0.4 0.8

	Sun↑	Sun↓
15	0614	1758
31	0619	1748

NOVEMBER

Date	Day	Time	ft
1	Su	0040 0653 1249 1841	0.4 0.7 0.4 0.7
2	M	0101 0747 1409 1914	0.3 0.8 0.4 0.6
3	Tu	0125 0840 1529 1942	0.3 0.9 0.4 0.6
4	W ○	0152 0932 1653 2002	0.2 1.0 0.5 0.5
5	Th	0222 1025 1841 1959	0.1 1.1 0.4 0.5
6	F	0256 1119	0.1 1.1
7	Sa	0332 1214	0.1 1.1
8	Su	0411 1309	0.1 1.1
9	M	0453 1404	0.2 1.0
10	Tu ☽	0543 1457	0.2 0.9
11	W	0647 1545	0.3 0.9
12	Th	0015 0442 0821 1628	0.3 0.4 0.4 0.8
13	F	0013 0604 1012 1702	0.3 0.5 0.4 0.7
14	Sa	0019 0657 1155 1727	0.3 0.6 0.5 0.6
15	Su	0028 0740 1327 1743	0.3 0.7 0.5 0.6
16	M	0040 0817 1455 1744	0.2 0.7 0.5 0.5
17	Tu	0053 0853	0.2 0.8
18	W	0109 0926	0.1 0.9
19	Th ●	0129 0959	0.1 0.9
20	F	0152 1034	0.0 0.9
21	Sa	0220 1110	0.0 0.9
22	Su	0251 1150	0.0 0.9
23	M	0327 1232	0.0 0.9
24	Tu	0409 1317	0.1 0.9
25	W	0501 1403 2329	0.1 0.8 0.3
26	Th ☾	0057 0614 1451 2257	0.3 0.2 0.8 0.2
27	F	0342 0801 1538 2304	0.3 0.3 0.7 0.2
28	Sa	0508 1002 1622 2321	0.4 0.3 0.6 0.1
29	Su	0610 1153 1701 2344	0.6 0.3 0.5 0.1
30	M	0703 1331 1735	0.7 0.3 0.4

	Sun↑	Sun↓
15	0627	1743
30	0635	1742

DECEMBER

Date	Day	Time	ft
1	Tu	0011 0753 1501 1800	0.0 0.8 0.3 0.3
2	W	0040 0841 1640 1809	-0.1 0.9 0.3 0.3
3	Th ○	0113 0930	-0.1 0.9
4	F	0147 1018	-0.1 1.0
5	Sa	0223 1105	-0.2 1.0
6	Su	0300 1153	-0.1 0.9
7	M	0338 1239	-0.1 0.9
8	Tu	0416 1323	-0.0 0.8
9	W	0455 1404 2301	0.1 0.7 0.1
10	Th ☽	0338 0534 1440 2255	0.2 0.2 0.6 0.1
11	F	0557 0733 1508 2301	0.3 0.3 0.5 0.1
12	Sa	0635 1030 1524 2312	0.4 0.3 0.4 0.0
13	Su	0708 1304 1518 2325	0.5 0.3 0.4 -0.0
14	M	0739 2340	0.6 -0.1
15	Tu	0808	0.6
16	W	0000 0837	-0.1 0.7
17	Th	0022 0906	-0.2 0.7
18	F ●	0049 0936	-0.2 0.8
19	Sa	0118 1008	-0.2 0.8
20	Su	0151 1042	-0.3 0.8
21	M	0228 1119	-0.2 0.8
22	Tu	0309 1158	-0.2 0.7
23	W	0357 1238 2143 2306	-0.1 0.7 0.1 0.1
24	Th	0455 1320 2124	-0.0 0.6 0.0
25	F	0157 0619 1402 2137	0.1 0.1 0.5 -0.0
26	Sa ☾	0353 0821 1442 2158	0.2 0.1 0.4 -0.1
27	Su	0508 1038 1518 2226	0.4 0.2 0.3 -0.1
28	M	0607 1243 1548 2257	0.5 0.2 0.2 -0.2
29	Tu	0659 1442 1602 2331	0.6 0.1 0.1 -0.3
30	W	0747	0.7
31	Th	0008 0834	-0.3 0.8

	Sun↑	Sun↓
15	0644	1747
31	0652	1755

T

Tides

FORT-DE-FRANCE, MARTINIQUE

HIGH & LOW WATER 1998 | Chart Datum | 14°36'N 61°05'W

Atlantic Standard Time (60°W) | Daylight Saving Time not observed

JANUARY

Day	Time	ft	Day	Time	ft
1 Th	0000 1604	1.0 2.1	16 F	0010 0755 1045 1645	1.1 1.7 1.6 1.9
2 F	0018 0807 1019 1655	1.0 1.7 1.6 2.0	17 Sa	0039 0815 1211 1725	1.2 1.7 1.6 1.8
3 Sa	0057 0824 1155 1752	1.1 1.7 1.6 1.9	18 Su	0105 0837 1347 1809	1.2 1.7 1.6 1.7
4 Su	0135 0848 1334 1900	1.2 1.7 1.6 1.8	19 M	0125 0859 1532 1908	1.3 1.8 1.5 1.6
5 M	0212 0917 1516 ☾ 2026	1.2 1.8 1.5 1.7	20 Tu	0141 0924 1707 ☽ 2050	1.4 1.8 1.5 1.5
6 Tu	0248 0951 1649 2215	1.3 1.9 1.4 1.6	21 W	0148 0950 1811 2341	1.4 1.9 1.4 1.5
7 W	0321 1028 1806	1.4 1.9 1.3	22 Th	0128 1020 1855	1.5 1.9 1.3
8 Th	0017 0350 1108 1910	1.5 1.5 2.0 1.2	23 F	1052 1933	2.0 1.2
9 F	0221 0411 1150 2005	1.6 1.5 2.1 1.1	24 Sa	1128 2009	2.0 1.1
10 Sa	1234 2055	2.1 1.0	25 Su	1208 2044	2.1 1.1
11 Su	1317 2140	2.1 1.0	26 M	1251 2121	2.1 1.0
12 M ○	1401 2222	2.1 1.0	27 Tu	1336 2157	2.1 1.0
13 Tu	1443 2301	2.1 1.0	28 W ●	1423 2234	2.1 1.0
14 W	1524 2337	2.0 1.0	29 Th	0608 0818 1513 2311	1.6 1.6 2.1 1.1
15 Th	0738 0924 1605	1.7 1.6 2.0	30 F	0620 0938 1606 2347	1.7 1.6 2.0 1.1
			31 Sa	0641 1056 1703	1.7 1.6 1.9

	Sun↑	Sun↓
15	0634	1755
31	0633	1803

FEBRUARY

Day	Time	ft	Day	Time	ft
1 Su	0022 0706 1217 1807	1.2 1.7 1.5 1.8	16 M	0010 0708 1331 1831	1.4 1.8 1.5 1.7
2 M	0056 0738 1342 1921	1.3 1.8 1.5 1.7	17 Tu	0024 0733 1445 1947	1.5 1.9 1.5 1.6
3 Tu	0127 0814 1509 ☾ 2054	1.4 1.9 1.4 1.6	18 W	0032 0801 1558 2143	1.5 1.9 1.4 1.6
4 W	0155 0854 1633 2250	1.5 1.9 1.3 1.6	19 Th ☽	0024 0833 1701	1.5 2.0 1.4
5 Th	0219 0939 1747	1.5 2.0 1.2	20 F	0912 1755	2.0 1.3
6 F	0113 0226 1027 1850	1.6 1.6 2.1 1.1	21 Sa	0955 1841	2.0 1.2
7 Sa	1118 1944	2.1 1.1	22 Su	1044 1924	2.1 1.2
8 Su	1208 2032	2.1 1.1	23 M	1136 2005	2.1 1.2
9 M	1258 2116	2.1 1.1	24 Tu	1231 2045	2.1 1.2
10 Tu	1346 2155	2.1 1.1	25 W	0417 0626 1327 2123	1.7 1.7 2.1 1.2
11 W	0529 0746 1433 ○ 2230	1.7 1.7 2.0 1.1	26 Th	0423 0749 1423 ● 2201	1.7 1.6 2.1 1.2
12 Th	0545 0857 1518 2301	1.7 1.6 2.0 1.2	27 F	0439 0900 1521 2236	1.7 1.6 2.0 1.3
13 F	0605 1004 1602 2328	1.7 1.6 1.9 1.3	28 Sa	0501 1008 1621 2310	1.8 1.5 2.0 1.4
14 Sa	0624 1110 1647 2351	1.8 1.6 1.8 1.3			
15 Su	0645 1218 1734	1.8 1.6 1.8			

	Sun↑	Sun↓
15	0629	1809
28	0622	1813

MARCH

Day	Time	ft	Day	Time	ft
1 Su	0528 1115 1724 2342	1.8 1.5 1.9 1.4	16 M	0523 1211 1804 2306	1.9 1.5 1.8 1.6
2 M	0559 1225 1834	1.9 1.4 1.8	17 Tu	0546 1305 1913 2315	2.0 1.5 1.7 1.6
3 Tu	0011 0634 1338 1955	1.5 2.0 1.4 1.7	18 W	0613 1401 2045 2313	2.0 1.4 1.7 1.7
4 W	0036 0715 1453 2134	1.6 2.0 1.3 1.7	19 Th	0644 1459	2.0 1.4
5 Th	0059 0801 1606 ☾ 2343	1.6 2.1 1.3 1.7	20 F	0722 1556	2.1 1.3
6 F	0105 0852 1715	1.7 2.1 1.2	21 Sa ☽	0807 1651	2.1 1.3
7 Sa	0947 1816	2.1 1.2	22 Su	0902 1744	2.1 1.3
8 Su	1046 1910	2.1 1.2	23 M	1004 1833	2.1 1.3
9 M	1145 1957	2.1 1.2	24 Tu	0246 0401 1112 1918	1.8 1.8 2.1 1.3
10 Tu	0334 0559 1242 2038	1.7 1.7 2.1 1.3	25 W	0235 0555 1220 2001	1.8 1.7 2.1 1.3
11 W	0348 0720 1338 2115	1.8 1.7 2.0 1.3	26 Th	0247 0714 1328 2041	1.8 1.7 2.0 1.4
12 Th	0405 0827 1431 2146	1.8 1.7 2.0 1.4	27 F	0306 0821 1434 ● 2118	1.8 1.6 2.0 1.4
13 F	0423 0927 1522 ○ 2212	1.8 1.6 1.9 1.4	28 Sa	0329 0923 1540 2151	1.9 1.5 2.0 1.5
14 Sa	0442 1023 1613 2235	1.9 1.6 1.9 1.5	29 Su	0357 1024 1646 2223	2.0 1.4 1.9 1.6
15 Su	0502 1117 1706 2253	1.9 1.5 1.8 1.5	30 M	0427 1124 1756 2251	2.0 1.4 1.9 1.6
			31 Tu	0502 1224 1911 2317	2.1 1.3 1.8 1.7

	Sun↑	Sun↓
15	0613	1815
31	0601	1817

APRIL

Day	Time	ft	Day	Time	ft
1 W	0541 1327 2038 2338	2.1 1.3 1.8 1.7	16 Th	0508 1325	2.1 1.3
2 Th	0623 1430 2229 2347	2.2 1.3 1.8 1.8	17 F	0543 1413	2.1 1.3
3 F ☾	0711 1534	2.2 1.3	18 Sa	0624 1503	2.1 1.3
4 Sa	0805 1635	2.1 1.3	19 Su ☽	0714 1555	2.1 1.3
5 Su	0905 1732	2.1 1.3	20 M	0817 1647	2.1 1.3
6 M	0152 0328 1012 1823	1.8 1.8 2.1 1.3	21 Tu	0100 0314 0932 1737	1.8 1.8 2.0 1.4
7 Tu	0159 0523 1122 1907	1.8 1.8 2.0 1.4	22 W	0101 0509 1057 1825	1.9 1.8 2.0 1.4
8 W	0215 0649 1232 1946	1.9 1.7 2.0 1.4	23 Th	0117 0632 1221 1908	1.9 1.7 2.0 1.5
9 Th	0233 0756 1338 2018	1.9 1.7 1.9 1.5	24 F	0139 0740 1342 1948	1.9 1.6 1.9 1.5
10 F	0252 0853 1441 2045	1.9 1.6 1.9 1.6	25 Sa	0204 0840 1457 2023	2.0 1.5 1.9 1.6
11 Sa	0311 0943 1541 ○ 2108	2.0 1.5 1.8 1.6	26 Su	0233 0936 1610 ● 2055	2.1 1.4 1.9 1.7
12 Su	0330 1028 1641 2125	2.0 1.5 1.8 1.7	27 M	0305 1030 1721 2124	2.1 1.3 1.9 1.7
13 M	0351 1112 1743 2139	2.0 1.4 1.8 1.7	28 Tu	0340 1123 1836 2149	2.2 1.2 1.8 1.8
14 Tu	0413 1155 1851 2147	2.1 1.4 1.8 1.7	29 W	0418 1217 1957 2209	2.2 1.2 1.8 1.8
15 W	0439 1239 2021 2140	2.1 1.4 1.8 1.8	30 Th	0458 1310	2.2 1.2

	Sun↑	Sun↓
15	0551	1819
30	0543	1822

FORT-DE-FRANCE, MARTINIQUE

HIGH & LOW WATER 1998 — Chart Datum — 14°36'N 61°05'W

Atlantic Standard Time (60°W) — Daylight Saving Time not observed

MAY

		Time	ft
1	F	0541 1404	2.2 1.2
2	Sa	0627 1457	2.2 1.3
3	**Su** ☾	0719 1548	2.1 1.3
4	M	0022 0217 0820 1637	1.9 1.9 2.0 1.4
5	Tu	0032 0429 0934 1721	1.9 1.8 2.0 1.4
6	W	0049 0610 1101 1800	1.9 1.8 1.9 1.5
7	Th	0109 0723 1230 1834	2.0 1.7 1.8 1.6
8	F	0128 0818 1353 1901	2.0 1.6 1.8 1.6
9	Sa	0148 0903 1509 1923	2.0 1.5 1.8 1.7
10	**Su**	0208 0944 1620 1939	2.1 1.4 1.8 1.7
11	M ○	0230 1021 1732 1949	2.1 1.4 1.8 1.8
12	Tu	0254 1058	2.2 1.3
13	W	0319 1135	2.2 1.3
14	Th	0349 1213	2.2 1.3
15	F	0422 1253	2.2 1.2
16	Sa	0501 1336	2.2 1.3
17	**Su**	0546 1421	2.2 1.3
18	M	0640 1507 2322	2.1 1.3 1.9
19	Tu ☽	0156 0748 1554 2331	1.9 2.1 1.4 1.9
20	W	0401 0914 1641 2351	1.8 2.0 1.5 2.0
21	Th	0537 1055 1725	1.7 1.9 1.5
22	F	0016 0651 1235 1806	2.0 1.6 1.9 1.6
23	Sa	0044 0752 1407 1843	2.1 1.5 1.9 1.7
24	**Su**	0116 0847 1531 1916	2.2 1.4 1.9 1.8
25	M ●	0150 0938 1649 1945	2.2 1.3 1.9 1.8
26	Tu	0227 1028 1806 2010	2.3 1.2 1.9 1.8
27	W	0306 1116	2.3 1.2
28	Th	0345 1203	2.3 1.2
29	F	0427 1249	2.3 1.2
30	Sa	0509 1334	2.3 1.2
31	**Su**	0553 1417 2249	2.2 1.3 1.9

	Sun↑	Sun↓
15	0537	1825
31	0535	1830

JUNE

		Time	ft
1	M ☾	0047 0641 1458 2304	1.9 2.1 1.4 2.0
2	Tu	0303 0737 1535 2324	1.9 2.0 1.5 2.0
3	W	0509 0856 1609 2345	1.8 1.9 1.6 2.0
4	Th	0639 1046 1637	1.7 1.8 1.6
5	F	0007 0737 1245 1659	2.1 1.6 1.8 1.7
6	Sa	0029 0820 1436 1711	2.1 1.6 1.8 1.8
7	**Su**	0051 0857	2.2 1.5
8	M	0116 0931	2.2 1.4
9	Tu	0141 1004	2.3 1.3
10	W ○	0210 1037	2.3 1.3
11	Th	0241 1111	2.3 1.3
12	F	0316 1147	2.3 1.3
13	Sa	0355 1224	2.3 1.3
14	**Su**	0438 1303	2.3 1.3
15	M	0528 1344 2141	2.2 1.4 2.0
16	Tu	0034 0626 1424 2157	1.9 2.2 1.4 2.0
17	W ☽	0230 0740 1505 2222	1.9 2.1 1.5 2.1
18	Th	0417 0916 1544 2250	1.8 2.0 1.6 2.1
19	F	0544 1110 1622 2323	1.7 1.9 1.7 2.2
20	Sa	0653 1304 1657	1.6 1.9 1.8
21	**Su**	0000 0751 1447 1728	2.3 1.4 1.9 1.9
22	M	0038 0843 1621 1753	2.4 1.3 1.9 1.9
23	Tu ●	0118 0931	2.4 1.3
24	W	0200 1018	2.5 1.2
25	Th	0242 1102	2.5 1.2
26	F	0324 1144	2.4 1.3
27	Sa	0406 1223 2048 2157	2.4 1.3 2.0 2.0
28	**Su**	0448 1300 2104 2334	2.3 1.4 2.0 2.0
29	M	0531 1335 2125	2.2 1.5 2.1
30	Tu	0120 0616 1405 2149	2.0 2.1 1.6 2.1

	Sun↑	Sun↓
15	0536	1835
30	0539	1838

JULY

		Time	ft
1	W ☾	0320 0711 1430 2213	2.0 2.0 1.7 2.2
2	Th	0515 0835 1449 2239	1.9 1.9 1.7 2.2
3	F	0634 1053 1458 2305	1.8 1.8 1.8 2.3
4	Sa	0723 2331	1.7 2.3
5	**Su**	0800 2359	1.6 2.4
6	M	0833	1.5
7	Tu	0029 0905	2.4 1.5
8	W	0102 0937	2.4 1.4
9	Th ○	0138 1010	2.5 1.4
10	F	0216 1044	2.5 1.4
11	Sa	0258 1119	2.5 1.4
12	**Su**	0344 1155 1940 2157	2.5 1.4 2.1 2.0
13	M	0433 1231 1954 2327	2.4 1.5 2.1 2.0
14	Tu	0529 1308 2016	2.3 1.6 2.1
15	W	0101 0634 1343 2044	2.0 2.2 1.7 2.2
16	Th ☽	0239 0754 1417 2118	1.9 2.1 1.8 2.3
17	F	0412 0937 1448 2156	1.8 2.0 1.8 2.4
18	Sa	0533 1140 1516 2237	1.7 2.0 1.9 2.4
19	**Su**	0640 1350 1536 2322	1.6 2.0 2.0 2.5
20	M	0738	1.5
21	Tu	0008 0829	2.6 1.5
22	W	0054 0916	2.6 1.4
23	Th ●	0141 0959	2.6 1.4
24	F	0226 1040 1834 2008	2.6 1.5 2.1 2.1
25	Sa	0312 1117 1852 2125	2.5 1.5 2.1 2.1
26	**Su**	0356 1151 1913 2241	2.5 1.6 2.2 2.1
27	M	0441 1221 1937	2.4 1.7 2.2
28	Tu	0000 0526 1248 2002	2.1 2.3 1.7 2.2
29	W	0127 0615 1309 2028	2.1 2.2 1.8 2.3
30	Th	0302 0718 1324 2056	2.0 2.1 1.9 2.3
31	F ☾	0435 0856 1328 2125	1.9 2.0 1.9 2.4

	Sun↑	Sun↓
15	0543	1838
31	0548	1834

AUGUST

		Time	ft
1	Sa	0547 2157	1.9 2.4
2	**Su**	0638 2231	1.8 2.5
3	M	0718 2308	1.7 2.5
4	Tu	0754 2348	1.7 2.6
5	W	0829	1.6
6	Th	0031 0904	2.6 1.6
7	F ○	0116 0938	2.6 1.6
8	Sa	0204 1013 1741 2003	2.6 1.6 2.2 2.1
9	**Su**	0254 1048 1751 2121	2.6 1.6 2.2 2.1
10	M	0347 1123 1809 2235	2.5 1.7 2.2 2.1
11	Tu	0443 1156 1834 2351	2.5 1.8 2.3 2.0
12	W	0546 1229 1903	2.4 1.8 2.3
13	Th	0110 0658 1259 1938	2.0 2.3 1.9 2.4
14	F ☽	0232 0825 1327 2019	1.9 2.2 2.0 2.5
15	Sa	0354 1014 1351 2104	1.8 2.1 2.1 2.5
16	**Su**	0509 1228 1403 2154	1.8 2.1 2.1 2.6
17	M	0615 2247	1.7 2.6
18	Tu	0713 2341	1.6 2.7
19	W	0804	1.6
20	Th	0035 0850 1632 1822	2.7 1.6 2.2 2.2
21	F ●	0128 0931 1648 1941	2.6 1.7 2.2 2.2
22	Sa	0219 1008 1707 2052	2.6 1.7 2.2 2.2
23	**Su**	0309 1041 1729 2158	2.5 1.8 2.3 2.1
24	M	0358 1110 1751 2302	2.5 1.8 2.3 2.1
25	Tu	0448 1134 1814	2.4 1.9 2.3
26	W	0008 0539 1153 1839	2.1 2.3 2.0 2.4
27	Th	0115 0639 1207 1905	2.0 2.2 2.0 2.4
28	F	0225 0756 1212 1934	2.0 2.2 2.1 2.5
29	Sa	0334 0956 1153 2007	1.9 2.1 2.1 2.5
30	**Su** ☾	0438 2045	1.9 2.5
31	M	0532 2128	1.8 2.6

	Sun↑	Sun↓
15	0551	1828
31	0553	1817

FORT-DE-FRANCE, MARTINIQUE

HIGH & LOW WATER 1998 | Chart Datum | 14°36'N 61°05'W

Atlantic Standard Time (60°W) | Daylight Saving Time not observed

SEPTEMBER

		Time	ft
1	Tu	0620 2217	1.8 2.6
2	W	0703 2311	1.8 2.6
3	Th	0743	1.7
4	F	0007 0822 1553 1814	2.6 1.7 2.2 2.2
5	Sa	0105 0859 1556 1936	2.6 1.7 2.2 2.2
6	Su ○	0203 0935 1610 2045	2.6 1.8 2.3 2.1
7	M	0302 1010 1631 2150	2.5 1.8 2.3 2.0
8	Tu	0402 1043 1656 2254	2.5 1.9 2.4 2.0
9	W	0506 1114 1726	2.4 2.0 2.4
10	Th	0000 0615 1143 1800	1.9 2.3 2.0 2.5
11	F	0107 0733 1209 1840	1.8 2.3 2.1 2.5
12	Sa ☽	0217 0906 1232 1925	1.8 2.2 2.2 2.6
13	Su	0328 1101 1247 2016	1.8 2.2 2.2 2.6
14	M	0437 2112	1.7 2.6
15	Tu	0540 2214	1.7 2.6
16	W	0637 1443 1606 2318	1.7 2.2 2.2 2.6
17	Th	0727 1453 1750	1.7 2.2 2.2
18	F	0021 0811 1511 1911	2.5 1.8 2.3 2.2
19	Sa	0122 0850 1531 2019	2.5 1.8 2.3 2.1
20	Su ●	0221 0923 1552 2120	2.4 1.9 2.3 2.0
21	M	0318 0952 1614 2216	2.4 1.9 2.3 2.0
22	Tu	0414 1016 1635 2310	2.3 2.0 2.4 1.9
23	W	0510 1034 1658	2.3 2.0 2.4
24	Th	0002 0611 1047 1722	1.9 2.2 2.1 2.4
25	F	0053 0723 1052 1748	1.9 2.2 2.1 2.5
26	Sa	0146 0903 1037 1818	1.8 2.1 2.1 2.5
27	Su	0239 1853	1.8 2.5
28	M ☾	0332 1935	1.8 2.5
29	Tu	0424 2027	1.7 2.5
30	W	0516 2129	1.7 2.5

	Sun↑	Sun↓
15	0554	1806
30	0554	1755

OCTOBER

		Time	ft
1	Th	0604 1430 1542 2240	1.7 2.2 2.2 2.4
2	F	0649 1411 1743 2353	1.7 2.2 2.1 2.4
3	Sa	0731 1421 1901	1.8 2.2 2.1
4	Su	0104 0811 1438 2007	2.4 1.8 2.3 2.0
5	M ○	0214 0848 1500 2107	2.4 1.9 2.3 1.9
6	Tu	0321 0921 1527 2205	2.3 1.9 2.4 1.8
7	W	0429 0953 1557 2302	2.3 2.0 2.4 1.7
8	Th	0539 1021 1631	2.2 2.0 2.5
9	F	0000 0653 1047 1709	1.6 2.2 2.1 2.5
10	Sa	0058 0815 1110 1751	1.6 2.1 2.1 2.5
11	Su	0158 0954 1127 1838	1.6 2.1 2.1 2.5
12	M ☽	0259 1931	1.6 2.5
13	Tu	0359 2032	1.6 2.4
14	W	0457 1306 1514 2140	1.6 2.1 2.1 2.4
15	Th	0549 1322 1707 2256	1.6 2.2 2.1 2.3
16	F	0637 1342 1835	1.7 2.2 2.0
17	Sa	0012 0717 1403 1946	2.2 1.7 2.2 1.9
18	Su	0126 0752 1424 2044	2.2 1.8 2.2 1.9
19	M	0235 0821 1446 2135	2.1 1.9 2.3 1.8
20	Tu ●	0341 0845 1508 2222	2.1 1.9 2.3 1.7
21	W	0445 0903 1530 2305	2.1 2.0 2.3 1.6
22	Th	0551 0913 1553 2346	2.0 2.0 2.4 1.6
23	F	0707 0916 1618	2.0 2.0 2.4
24	Sa	0028 1646	1.6 2.4
25	Su	0110 1718	1.5 2.4
26	M	0153 1755	1.5 2.4
27	Tu	0239 1841	1.5 2.3
28	W ☾	0327 1938	1.5 2.3
29	Th	0415 1244 1447 2051	1.5 2.1 2.0 2.2
30	F	0503 1238 1652 2218	1.6 2.1 2.0 2.1
31	Sa	0549 1252 1818 2350	1.6 2.1 1.9 2.1

	Sun↑	Sun↓
15	0556	1745
31	0600	1737

NOVEMBER

		Time	ft
1	Su	0633 1312 1925	1.7 2.1 1.8
2	M	0117 0713 1337 2023	2.1 1.7 2.2 1.6
3	Tu	0237 0749 1405 2118	2.0 1.8 2.2 1.5
4	W ○	0352 0822 1437 2210	2.0 1.8 2.3 1.4
5	Th	0505 0851 1512 2301	2.0 1.9 2.3 1.3
6	F	0619 0917 1550 2353	2.0 1.9 2.4 1.3
7	Sa	0736 0940 1631	2.0 1.9 2.4
8	Su	0044 1714	1.3 2.4
9	M	0136 1801	1.3 2.3
10	Tu ☽	0227 1853	1.3 2.2
11	W	0318 1137 1356 1953	1.4 2.0 2.0 2.1
12	Th	0406 1155 1604 2106	1.4 2.0 1.9 2.0
13	F	0451 1217 1749 2236	1.5 2.0 1.8 1.9
14	Sa	0531 1239 1907	1.6 2.0 1.7
15	Su	0013 0606 1302 2006	1.9 1.6 2.1 1.6
16	M	0145 0634 1325 2054	1.8 1.7 2.1 1.5
17	Tu	0309 0655 1349 2136	1.8 1.7 2.2 1.4
18	W	0428 0709 1412 2214	1.8 1.8 2.2 1.4
19	Th ●	0555 0707 1436 2250	1.8 1.8 2.2 1.3
20	F	1502 2325	2.2 1.3
21	Sa	1531	2.2
22	Su	0001 1602	1.2 2.2
23	M	0037 1638	1.2 2.2
24	Tu	0116 1718	1.2 2.2
25	W	0156 1808	1.3 2.1
26	Th ☾	0238 1103 1323 1909	1.3 1.9 1.8 2.0
27	F	0322 1107 1535 2030	1.3 1.9 1.8 1.9
28	Sa	0405 1124 1716 2214	1.4 1.9 1.7 1.8
29	Su	0448 1148 1832	1.5 2.0 1.6
30	M	0002 0528 1217 1933	1.8 1.6 2.0 1.4

	Sun↑	Sun↓
15	0606	1733
30	0613	1734

DECEMBER

		Time	ft
1	Tu	0142 0605 1249 2027	1.7 1.6 2.1 1.3
2	W	0311 0639 1324 2118	1.7 1.7 2.2 1.2
3	Th ○	0432 0709 1401 2206	1.7 1.7 2.2 1.1
4	F	0550 0734 1441 2254	1.8 1.7 2.2 1.0
5	Sa	1522 2340	2.2 1.0
6	Su	1605	2.2
7	M	0026 1649	1.0 2.2
8	Tu	0110 0947 1037 1735	1.1 1.8 1.8 2.1
9	W	0152 1003 1231 1824	1.2 1.8 1.8 2.0
10	Th ☽	0233 1025 1432 1922	1.2 1.8 1.7 1.9
11	F	0310 1050 1634 2039	1.3 1.8 1.7 1.7
12	Sa	0344 1115 1812 2227	1.4 1.9 1.6 1.6
13	Su	0411 1141 1918	1.5 1.9 1.5
14	M	0032 0432 1207 2007	1.6 1.5 2.0 1.4
15	Tu	0239 0437 1233 2047	1.6 1.6 2.0 1.3
16	W	1259 2122	2.1 1.2
17	Th	1326 2155	2.1 1.1
18	F ●	1354 2227	2.1 1.1
19	Sa	1425 2259	2.1 1.1
20	Su	1459 2332	2.1 1.0
21	M	1535	2.1
22	Tu	0006 1617	1.0 2.1
23	W	0042 1703	1.1 2.0
24	Th	0119 0914 1206 1757	1.1 1.7 1.7 1.9
25	F	0156 0929 1358 1905	1.2 1.7 1.6 1.8
26	Sa ☾	0233 0952 1545 2035	1.3 1.8 1.6 1.7
27	Su	0310 1020 1716 2229	1.3 1.9 1.4 1.6
28	M	0345 1054 1827	1.4 1.9 1.3
29	Tu	0031 0418 1131 1927	1.6 1.5 2.0 1.2
30	W	0223 0448 1211 2019	1.6 1.6 2.1 1.1
31	Th	0406 0510 1253 2108	1.6 1.6 2.1 1.0

	Sun↑	Sun↓
15	0622	1738
31	0630	1746

PORT OF SPAIN, TRINIDAD

HIGH & LOW WATER 1998 | Chart Datum | 10°39'N 61°31'W

Atlantic Standard Time (60°W) | Daylight Saving Time not observed

JANUARY

Day	Time	ft	Day	Time	ft
1	0000	0.5	16	0021	0.7
	0622	3.4		0642	3.3
Th	1157	1.4	F	1229	1.3
	1802	3.9		1819	3.4
2	0036	0.6	17	0057	0.9
	0705	3.4		0720	3.2
F	1246	1.4	Sa	1313	1.4
	1850	3.7		1859	3.3
3	0120	0.7	18	0131	1.1
	0749	3.4		0758	3.2
Sa	1340	1.4	Su	1400	1.5
	1943	3.6		1945	3.0
4	0207	0.9	19	0209	1.3
	0837	3.4		0839	3.1
Su	1442	1.5	M	1453	1.6
	2042	3.4		2039	2.8
5	0258	1.1	20	0253	1.5
	0929	3.4		0926	3.0
M	1551	1.4	Tu	1553	1.6
	☾ 2148	3.2		☽ 2142	2.6
6	0355	1.3	21	0345	1.6
	1028	3.5		1019	3.0
Tu	1707	1.4	W	1700	1.6
	2306	3.0		2255	2.5
7	0457	1.5	22	0446	1.8
	1131	3.5		1116	3.1
W	1819	1.2	Th	1810	1.4
8	0029	3.0	23	0014	2.5
	0600	1.6		0549	1.8
Th	1233	3.6	F	1213	3.1
	1923	1.0		1914	1.2
9	0139	3.0	24	0128	2.6
	0701	1.6		0649	1.8
F	1329	3.7	Sa	1308	3.3
	2018	0.8		2005	1.0
10	0236	3.0	25	0225	2.7
	0757	1.5		0744	1.7
Sa	1420	3.7	Su	1359	3.5
	2106	0.7		2049	0.7
11	0324	3.1	26	0311	2.9
	0847	1.5		0834	1.5
Su	1506	3.8	M	1447	3.7
	2149	0.6		2130	0.5
12	0406	3.2	27	0353	3.1
	0934	1.4		0921	1.3
M	1549	3.8	Tu	1533	3.8
	○ 2228	0.5		2210	0.4
13	0446	3.2	28	0434	3.3
	1019	1.3		1007	1.2
Tu	1630	3.8	W	1618	3.9
	2307	0.6		● 2250	0.3
14	0525	3.3	29	0514	3.4
	1103	1.3		1053	1.0
W	1708	3.7	Th	1703	4.0
	2345	0.6		2330	0.3
15	0603	3.3	30	0553	3.6
	1147	1.3		1140	0.9
Th	1744	3.6	F	1749	3.9
			31	0010	0.3
				0633	3.6
			Sa	1230	0.9
				1837	3.7

	Sun↑	Sun↓
15	0629	1803
31	0630	1810

FEBRUARY

Day	Time	ft	Day	Time	ft
1	0053	0.5	16	0050	1.0
	0715	3.6		0705	3.2
Su	1323	0.9	M	1322	1.1
	1929	3.5		1915	3.0
2	0138	0.7	17	0122	1.2
	0800	3.6		0741	3.1
M	1421	0.9	Tu	1406	1.2
	2026	3.2		2003	2.8
3	0228	1.0	18	0157	1.4
	0852	3.5		0823	3.0
Tu	1528	1.0	W	1500	1.3
	☾ 2131	3.0		2101	2.5
4	0324	1.2	19	0238	1.6
	0951	3.4		0912	3.0
W	1643	1.0	Th	1604	1.3
	2250	2.8		☽ 2211	2.4
5	0428	1.4	20	0337	1.8
	1058	3.3		1011	2.9
Th	1759	1.0	F	1714	1.3
				2331	2.4
6	0018	2.7	21	0457	1.8
	0539	1.5		1119	3.0
F	1211	3.3	Sa	1822	1.1
	1909	0.9			
7	0132	2.8	22	0048	2.5
	0649	1.5		0610	1.8
Sa	1316	3.4	Su	1227	3.1
	2008	0.7		1922	0.9
8	0228	2.9	23	0150	2.7
	0750	1.5		0714	1.6
Su	1412	3.5	M	1329	3.4
	2056	0.6		2013	0.7
9	0313	3.0	24	0238	3.0
	0842	1.3		0810	1.4
M	1500	3.6	Tu	1424	3.6
	2137	0.5		2058	0.5
10	0352	3.1	25	0322	3.3
	0927	1.2		0900	1.1
Tu	1543	3.6	W	1515	3.8
	2214	0.5		2141	0.3
11	0428	3.2	26	0402	3.5
	1010	1.1		0949	0.8
W	1621	3.6	Th	1603	4.0
	○ 2250	0.5		● 2222	0.2
12	0503	3.3	27	0442	3.7
	1051	1.0		1036	0.6
Th	1656	3.6	F	1650	4.0
	2324	0.6		2302	0.2
13	0537	3.3	28	0520	3.8
	1129	1.0		1123	0.4
F	1727	3.5	Sa	1737	3.9
	2355	0.7		2343	0.3
14	0607	3.3			
	1206	1.0			
Sa	1758	3.4			
15	0022	0.9			
	0634	3.3			
Su	1242	1.1			
	1834	3.2			

	Sun↑	Sun↓
15	0627	1815
28	0622	1817

MARCH

Day	Time	ft	Day	Time	ft
1	0600	3.9	16	0551	3.4
	1212	0.4		1211	0.8
Su	1825	3.8	M	1812	3.2
2	0027	0.5	17	0013	1.1
	0643	3.8		0622	3.3
M	1304	0.4	Tu	1247	0.9
	1917	3.5		1852	3.0
3	0113	0.7	18	0044	1.3
	0730	3.7		0657	3.2
Tu	1400	0.5	W	1327	1.0
	2013	3.2		1938	2.8
4	0203	1.0	19	0117	1.4
	0822	3.5		0735	3.1
W	1503	0.7	Th	1413	1.1
	2117	2.9		2032	2.6
5	0300	1.3	20	0155	1.6
	0922	3.3		0821	3.0
Th	1615	0.8	F	1511	1.1
	☾ 2235	2.7		2137	2.5
6	0407	1.5	21	0252	1.8
	1033	3.2		0921	3.0
F	1734	0.9	Sa	1622	1.2
				☽ 2253	2.5
7	0005	2.7	22	0417	1.8
	0524	1.6		1034	3.0
Sa	1152	3.1	Su	1734	1.1
	1849	0.9			
8	0118	2.8	23	0007	2.7
	0641	1.5		0540	1.7
Su	1305	3.2	M	1153	3.1
	1950	0.8		1839	1.0
9	0211	2.9	24	0110	2.9
	0744	1.4		0649	1.5
M	1404	3.3	Tu	1304	3.4
	2038	0.8		1935	0.8
10	0252	3.1	25	0202	3.2
	0834	1.2		0748	1.2
Tu	1451	3.4	W	1406	3.6
	2117	0.7		2025	0.6
11	0328	3.2	26	0247	3.5
	0916	1.1		0841	0.9
W	1533	3.4	Th	1500	3.8
	2153	0.7		2110	0.5
12	0402	3.3	27	0329	3.8
	0956	0.9		0931	0.5
Th	1610	3.5	F	1550	4.0
	2227	0.7		● 2153	0.4
13	0434	3.4	28	0409	4.0
	1033	0.8		1019	0.3
F	1642	3.5	Sa	1638	4.0
	○ 2257	0.8		2235	0.4
14	0503	3.4	29	0449	4.1
	1107	0.8		1106	0.1
Sa	1709	3.4	Su	1726	3.9
	2323	0.9		2318	0.5
15	0526	3.4	30	0531	4.1
	1139	0.8		1154	0.1
Su	1737	3.3	M	1814	3.7
	2345	1.0			
			31	0003	0.7
				0615	4.0
			Tu	1244	0.2
				1905	3.5

	Sun↑	Sun↓
15	0614	1817
31	0604	1817

APRIL

Day	Time	ft	Day	Time	ft
1	0050	0.9	16	0017	1.4
	0703	3.8		0626	3.4
W	1338	0.4	Th	1259	0.8
	2000	3.2		1921	2.9
2	0141	1.2	17	0053	1.5
	0756	3.5		0706	3.3
Th	1436	0.6	F	1344	0.9
	2103	2.9		2013	2.8
3	0239	1.4	18	0139	1.6
	0857	3.2		0754	3.2
F	1543	0.8	Sa	1437	1.0
	☾ 2216	2.8		2114	2.8
4	0347	1.6	19	0239	1.8
	1008	3.0		0856	3.1
Sa	1659	1.0	Su	1543	1.1
	2340	2.8		☽ 2222	2.8
5	0505	1.7	20	0358	1.8
	1130	3.0		1009	3.1
Su	1816	1.1	M	1654	1.1
				2331	3.0
6	0050	2.8	21	0519	1.7
	0626	1.6		1130	3.2
M	1247	3.0	Tu	1801	1.1
	1919	1.1			
7	0139	3.0	22	0032	3.2
	0729	1.4		0630	1.4
Tu	1347	3.1	W	1245	3.4
	2007	1.1		1900	1.0
8	0219	3.1	23	0125	3.5
	0816	1.2		0730	1.1
W	1435	3.2	Th	1349	3.6
	2047	1.1		1953	0.9
9	0254	3.3	24	0212	3.8
	0857	1.0		0824	0.7
Th	1515	3.3	F	1446	3.7
	2123	1.1		2040	0.8
10	0328	3.4	25	0256	4.0
	0934	0.9		0914	0.4
F	1551	3.3	Sa	1537	3.8
	2156	1.1		2126	0.8
11	0358	3.5	26	0338	4.1
	1009	0.8		1002	0.2
Sa	1622	3.3	Su	1626	3.9
	○ 2224	1.1		● 2210	0.8
12	0422	3.5	27	0421	4.2
	1040	0.7		1049	0.1
Su	1649	3.3	M	1714	3.8
	2246	1.1		2254	0.9
13	0446	3.5	28	0504	4.1
	1110	0.6		1136	0.1
M	1719	3.3	Tu	1802	3.6
	2312	1.2		2340	1.0
14	0516	3.5	29	0550	4.0
	1143	0.6		1224	0.2
Tu	1755	3.2	W	1851	3.4
	2343	1.3			
15	0550	3.5	30	0028	1.2
	1220	0.7		0638	3.7
W	1836	3.0	Th	1314	0.4
				1944	3.2

	Sun↑	Sun↓
15	0556	1818
30	0549	1819

PORT OF SPAIN, TRINIDAD

HIGH & LOW WATER 1998 — Chart Datum — 10°39'N 61°31'W

Atlantic Standard Time (60°W) — Daylight Saving Time not observed

MAY

Date	Day	Time	ft	Time	ft	Time	ft	Time	ft
1	F	0119	1.4	0730	3.5	1408	0.7	2041	3.0
2	Sa	0215	1.6	0829	3.2	1506	0.9	2144	2.9
3	Su	0320	1.7	0937	3.0	1612	1.2	☾ 2254	2.8
4	M	0435	1.8	1053	2.8	1723	1.3	2358	2.9
5	Tu	0552	1.7	1210	2.8	1828	1.4		
6	W	0050	3.0	0657	1.5	1315	2.9	1921	1.4
7	Th	0133	3.2	0746	1.3	1405	3.0	2006	1.4
8	F	0212	3.3	0828	1.1	1448	3.1	2044	1.4
9	Sa	0246	3.4	0906	0.9	1525	3.2	2116	1.4
10	Su	0314	3.5	0941	0.8	1558	3.2	2142	1.4
11	M	0342	3.6	1013	0.6	1628	3.2	○ 2210	1.3
12	Tu	0413	3.6	1046	0.6	1702	3.2	2243	1.4
13	W	0448	3.6	1121	0.5	1741	3.2	2319	1.4
14	Th	0525	3.6	1200	0.6	1823	3.1	2356	1.5
15	F	0604	3.5	1240	0.6	1908	3.1		
16	Sa	0038	1.6	0648	3.4	1325	0.8	1958	3.0
17	Su	0128	1.7	0739	3.3	1416	0.9	2053	3.0
18	M	0229	1.7	0840	3.2	1514	1.1	2153	3.1
19	Tu	0343	1.7	0951	3.2	1620	1.2	☽ 2256	3.2
20	W	0502	1.6	1110	3.2	1726	1.2	2355	3.4
21	Th	0613	1.4	1227	3.3	1828	1.2		
22	F	0050	3.7	0715	1.0	1334	3.4	1923	1.2
23	Sa	0139	3.9	0809	0.7	1432	3.6	2012	1.2
24	Su	0226	4.0	0859	0.5	1525	3.6	2100	1.2
25	M	0311	4.1	0946	0.3	1613	3.6	● 2145	1.2
26	Tu	0355	4.1	1032	0.2	1700	3.6	2230	1.2
27	W	0440	4.1	1117	0.2	1745	3.5	2316	1.3
28	Th	0525	3.9	1202	0.3	1831	3.3		
29	F	0003	1.4	0611	3.7	1248	0.6	1918	3.2
30	Sa	0052	1.6	0700	3.5	1335	0.8	2008	3.1
31	Su	0145	1.7	0753	3.2	1425	1.1	2102	3.0

	Sun↑	Sun↓
15	0545	1821
31	0543	1825

JUNE

Date	Day	Time	ft	Time	ft	Time	ft	Time	ft
1	M	0244	1.8	0852	3.0	1520	1.3	☾ 2158	3.0
2	Tu	0352	1.8	1000	2.8	1619	1.5	2256	3.0
3	W	0504	1.8	1113	2.7	1721	1.7	2350	3.1
4	Th	0612	1.6	1224	2.8	1820	1.7		
5	F	0039	3.2	0709	1.4	1324	2.8	1911	1.7
6	Sa	0121	3.3	0756	1.2	1414	2.9	1953	1.7
7	Su	0158	3.5	0836	1.0	1455	3.0	2028	1.6
8	M	0232	3.6	0913	0.8	1533	3.1	2103	1.6
9	Tu	0309	3.7	0948	0.7	1609	3.1	2140	1.5
10	W	0347	3.8	1025	0.6	1647	3.2	○ 2218	1.5
11	Th	0426	3.8	1103	0.5	1727	3.2	2257	1.5
12	F	0506	3.8	1143	0.5	1809	3.2	2339	1.5
13	Sa	0548	3.7	1223	0.6	1852	3.3		
14	Su	0024	1.6	0634	3.7	1307	0.7	1938	3.3
15	M	0114	1.6	0726	3.5	1354	0.9	2027	3.3
16	Tu	0214	1.6	0824	3.4	1447	1.1	2120	3.4
17	W	0325	1.6	0932	3.3	1546	1.3	☽ 2218	3.4
18	Th	0444	1.5	1049	3.2	1651	1.4	2317	3.6
19	F	0558	1.3	1210	3.2	1755	1.5		
20	Sa	0015	3.7	0702	1.1	1321	3.3	1854	1.5
21	Su	0110	3.8	0757	0.8	1421	3.3	1947	1.5
22	M	0201	4.0	0847	0.6	1514	3.4	2036	1.5
23	Tu	0248	4.0	0933	0.5	1601	3.4	● 2123	1.5
24	W	0334	4.1	1015	0.4	1644	3.4	2208	1.5
25	Th	0418	4.0	1057	0.5	1724	3.4	2253	1.5
26	F	0501	3.9	1138	0.6	1805	3.3	2337	1.6
27	Sa	0543	3.7	1219	0.8	1846	3.3		
28	Su	0022	1.6	0624	3.5	1259	1.0	1928	3.2
29	M	0109	1.7	0707	3.3	1339	1.2	2011	3.2
30	Tu	0200	1.8	0755	3.1	1420	1.5	2057	3.1

	Sun↑	Sun↓
15	0545	1829
30	0548	1832

JULY

Date	Day	Time	ft	Time	ft	Time	ft	Time	ft
1	W	0259	1.8	0853	2.9	1505	1.7	☾ 2146	3.1
2	Th	0406	1.8	1000	2.8	1559	1.8	2239	3.2
3	F	0517	1.8	1115	2.7	1701	1.9	2332	3.2
4	Sa	0623	1.6	1231	2.7	1802	2.0		
5	Su	0023	3.4	0719	1.4	1336	2.8	1856	1.9
6	M	0111	3.5	0805	1.2	1427	2.9	1945	1.9
7	Tu	0156	3.6	0845	1.0	1510	3.0	2030	1.8
8	W	0240	3.8	0924	0.8	1550	3.2	2113	1.7
9	Th	0324	3.9	1003	0.6	1629	3.3	○ 2155	1.6
10	F	0407	4.0	1043	0.5	1709	3.4	2238	1.5
11	Sa	0450	4.0	1123	0.5	1749	3.5	2322	1.5
12	Su	0534	4.0	1203	0.6	1828	3.6		
13	M	0008	1.4	0620	3.9	1245	0.8	1910	3.6
14	Tu	0059	1.5	0711	3.7	1329	1.0	1954	3.6
15	W	0156	1.5	0807	3.5	1418	1.2	2043	3.6
16	Th	0304	1.5	0911	3.3	1512	1.4	☽ 2138	3.6
17	F	0424	1.5	1027	3.2	1614	1.6	2239	3.7
18	Sa	0542	1.4	1155	3.1	1723	1.8	2344	3.7
19	Su	0650	1.2	1312	3.1	1829	1.8		
20	M	0047	3.8	0747	1.0	1414	3.2	1928	1.8
21	Tu	0143	3.9	0837	0.9	1505	3.3	2019	1.8
22	W	0234	3.9	0920	0.8	1548	3.4	2106	1.7
23	Th	0319	4.0	0959	0.8	1625	3.4	● 2150	1.6
24	F	0401	4.0	1037	0.8	1700	3.5	2232	1.6
25	Sa	0440	3.9	1113	0.9	1735	3.5	2313	1.6
26	Su	0517	3.8	1148	1.0	1809	3.5	2352	1.6
27	M	0551	3.7	1221	1.2	1840	3.5		
28	Tu	0032	1.6	0628	3.5	1251	1.4	1912	3.4
29	W	0114	1.7	0711	3.3	1323	1.5	1948	3.4
30	Th	0202	1.8	0800	3.1	1400	1.7	2033	3.4
31	F	0302	1.8	0900	2.9	1445	1.9	☾ 2125	3.3

	Sun↑	Sun↓
15	0552	1833
31	0555	1831

AUGUST

Date	Day	Time	ft	Time	ft	Time	ft	Time	ft
1	Sa	0412	1.8	1013	2.8	1545	2.1	2224	3.3
2	Su	0525	1.7	1135	2.7	1700	2.2	2327	3.4
3	M	0632	1.6	1256	2.8	1809	2.2		
4	Tu	0028	3.5	0728	1.4	1358	3.0	1909	2.1
5	W	0124	3.7	0814	1.1	1445	3.2	2002	1.9
6	Th	0215	3.9	0857	0.9	1526	3.4	2050	1.8
7	F	0304	4.1	0939	0.8	1606	3.6	○ 2135	1.6
8	Sa	0350	4.2	1019	0.7	1644	3.7	2220	1.4
9	Su	0435	4.3	1059	0.7	1722	3.9	2306	1.3
10	M	0520	4.2	1139	0.7	1800	4.0	2353	1.3
11	Tu	0607	4.1	1220	0.9	1839	4.0		
12	W	0043	1.3	0657	3.9	1304	1.1	1922	4.0
13	Th	0138	1.3	0752	3.7	1351	1.4	2010	3.9
14	F	0242	1.4	0854	3.4	1444	1.7	☽ 2105	3.8
15	Sa	0359	1.5	1010	3.2	1545	1.9	2207	3.7
16	Su	0522	1.5	1146	3.1	1656	2.1	2318	3.7
17	M	0635	1.4	1307	3.1	1811	2.1		
18	Tu	0032	3.7	0735	1.3	1407	3.2	1916	2.1
19	W	0134	3.8	0824	1.2	1453	3.4	2009	2.0
20	Th	0226	3.9	0904	1.2	1530	3.5	2054	1.8
21	F	0310	3.9	0940	1.1	1602	3.6	● 2136	1.7
22	Sa	0349	4.0	1014	1.2	1634	3.7	2215	1.6
23	Su	0424	4.0	1047	1.2	1704	3.7	2251	1.6
24	M	0455	3.9	1117	1.3	1730	3.8	2326	1.5
25	Tu	0526	3.8	1144	1.4	1754	3.8		
26	W	0000	1.6	0603	3.7	1212	1.5	1824	3.7
27	Th	0039	1.6	0644	3.5	1244	1.7	1901	3.7
28	F	0122	1.7	0731	3.3	1319	1.9	1942	3.6
29	Sa	0213	1.8	0826	3.1	1357	2.1	2029	3.5
30	Su	0317	1.8	0933	2.9	1447	2.2	☾ 2127	3.5
31	M	0431	1.8	1054	2.9	1608	2.4	2236	3.5

	Sun↑	Sun↓
15	0557	1825
31	0557	1817

PORT OF SPAIN, TRINIDAD

HIGH & LOW WATER 1998 — Chart Datum — 10°39'N 61°31'W

Atlantic Standard Time (60°W) — Daylight Saving Time not observed

SEPTEMBER

Day	Time	ft	Day	Time	ft
1 Tu	0543	1.7	16 W	0021	3.6
	1217	3.0		0715	1.6
	1731	2.4		1350	3.4
	2348	3.6		1906	2.2
2 W	0646	1.6	17 Th	0126	3.7
	1325	3.2		0802	1.5
	1839	2.2		1430	3.5
				1958	2.0
3 Th	0054	3.8	18 F	0217	3.8
	0739	1.4		0840	1.5
	1414	3.4		1503	3.6
	1937	2.0		2042	1.9
4 F	0153	4.0	19 Sa	0259	3.9
	0827	1.2		0915	1.5
	1457	3.7		1534	3.8
	2029	1.8		2121	1.7
5 Sa	0246	4.2	20 Su	0336	3.9
	0910	1.0		0949	1.5
	1537	3.9		1605	3.9
	2117	1.5		● 2158	1.6
6 Su	0335	4.4	21 M	0409	3.9
	0952	0.9		1020	1.5
	1615	4.1		1632	3.9
	○ 2204	1.3		2231	1.5
7 M	0422	4.4	22 Tu	0438	3.9
	1033	0.9		1047	1.6
	1653	4.3		1654	4.0
	2250	1.2		2304	1.5
8 Tu	0509	4.4	23 W	0509	3.8
	1114	1.0		1114	1.6
	1730	4.3		1719	4.0
	2338	1.1		2337	1.4
9 W	0557	4.2	24 Th	0546	3.7
	1156	1.2		1144	1.7
	1810	4.3		1752	3.9
10 Th	0028	1.1	25 F	0015	1.5
	0647	4.0		0628	3.6
	1241	1.4		1218	1.9
	1854	4.2		1829	3.8
11 F	0122	1.2	26 Sa	0056	1.6
	0741	3.7		0714	3.4
	1329	1.6		1254	2.0
	1943	4.1		1908	3.7
12 Sa	0222	1.3	27 Su	0143	1.7
	0842	3.4		0808	3.2
	1422	1.9		1334	2.2
	☽ 2039	3.9		1953	3.6
13 Su	0333	1.5	28 M	0240	1.8
	0959	3.2		0911	3.1
	1524	2.1		1427	2.4
	2143	3.7		☾ 2049	3.6
14 M	0455	1.6	29 Tu	0349	1.8
	1137	3.2		1025	3.1
	1637	2.3		1543	2.5
	2259	3.6		2158	3.6
15 Tu	0613	1.6	30 W	0501	1.8
	1255	3.2		1142	3.2
	1757	2.3		1705	2.4
				2315	3.6

	Sun↑	Sun↓
15	0556	1807
30	0555	1758

OCTOBER

Day	Time	ft	Day	Time	ft
1 Th	0607	1.7	16 F	0108	3.5
	1248	3.4		0727	1.8
	1815	2.3		1355	3.6
				1940	2.0
2 F	0029	3.8	17 Sa	0200	3.6
	0704	1.6		0808	1.8
	1340	3.7		1429	3.7
	1916	2.0		2024	1.8
3 Sa	0133	4.0	18 Su	0242	3.7
	0754	1.4		0845	1.8
	1424	3.9		1503	3.9
	2010	1.7		2103	1.6
4 Su	0229	4.2	19 M	0320	3.7
	0840	1.3		0921	1.7
	1505	4.2		1535	3.9
	2100	1.4		2141	1.5
5 M	0321	4.3	20 Tu	0354	3.8
	0924	1.2		0953	1.7
	1545	4.4		1602	4.0
	○ 2148	1.2		● 2215	1.4
6 Tu	0410	4.4	21 W	0425	3.8
	1006	1.2		1021	1.7
	1623	4.5		1626	4.0
	2235	1.0		2248	1.3
7 W	0458	4.3	22 Th	0458	3.7
	1049	1.3		1051	1.8
	1703	4.5		1655	4.0
	2323	0.9		2322	1.3
8 Th	0547	4.2	23 F	0536	3.6
	1133	1.4		1124	1.9
	1745	4.4		1729	3.9
9 F	0013	0.9	24 Sa	0000	1.3
	0637	4.0		0619	3.5
	1220	1.6		1201	2.0
	1831	4.3		1806	3.9
10 Sa	0106	1.1	25 Su	0042	1.4
	0731	3.7		0706	3.4
	1309	1.8		1240	2.1
	1921	4.1		1846	3.8
11 Su	0203	1.3	26 M	0128	1.5
	0832	3.5		0758	3.3
	1404	2.1		1325	2.3
	2018	3.8		1932	3.7
12 M	0306	1.5	27 Tu	0220	1.6
	0943	3.3		0857	3.2
	1506	2.3		1420	2.4
	☽ 2123	3.6		2027	3.6
13 Tu	0419	1.6	28 W	0321	1.7
	1112	3.3		1003	3.2
	1618	2.3		1530	2.4
	2239	3.5		☾ 2134	3.5
14 W	0534	1.8	29 Th	0427	1.7
	1226	3.3		1110	3.3
	1737	2.3		1645	2.4
				2249	3.6
15 Th	0001	3.5	30 F	0531	1.7
	0638	1.8		1211	3.5
	1316	3.5		1755	2.2
	1847	2.2			
			31 Sa	0005	3.7
				0629	1.7
				1303	3.8
				1857	1.9

	Sun↑	Sun↓
15	0555	1749
31	0558	1743

NOVEMBER

Day	Time	ft	Day	Time	ft
1 Su	0114	3.8	16 M	0220	3.4
	0721	1.6		0813	1.8
	1350	4.0		1431	3.8
	1953	1.6		2046	1.5
2 M	0213	4.0	17 Tu	0302	3.5
	0809	1.5		0852	1.8
	1433	4.2		1506	3.9
	2045	1.3		2126	1.3
3 Tu	0307	4.1	18 W	0340	3.5
	0855	1.4		0928	1.8
	1515	4.4		1536	3.9
	2134	1.0		2203	1.2
4 W	0358	4.1	19 Th	0417	3.5
	0940	1.4		1000	1.8
	1556	4.5		1605	3.9
	○ 2221	0.8		● 2238	1.1
5 Th	0447	4.0	20 F	0453	3.5
	1025	1.5		1033	1.8
	1638	4.5		1638	3.9
	2309	0.8		2313	1.1
6 F	0536	3.9	21 Sa	0532	3.4
	1112	1.6		1110	1.8
	1723	4.4		1714	3.9
	2358	0.8		2352	1.1
7 Sa	0625	3.8	22 Su	0614	3.3
	1200	1.7		1149	1.9
	1811	4.2		1752	3.8
8 Su	0049	0.9	23 M	0033	1.2
	0718	3.6		0700	3.3
	1251	1.9		1230	2.0
	1902	4.0		1832	3.7
9 M	0141	1.1	24 Tu	0116	1.2
	0814	3.4		0748	3.2
	1346	2.0		1316	2.1
	1958	3.7		1917	3.6
10 Tu	0238	1.4	25 W	0203	1.4
	0915	3.3		0840	3.2
	1447	2.2		1410	2.2
	☽ 2059	3.5		2010	3.5
11 W	0338	1.6	26 Th	0256	1.5
	1023	3.3		0935	3.3
	1554	2.2		1515	2.2
	2209	3.3		☾ 2112	3.4
12 Th	0443	1.7	27 F	0353	1.6
	1130	3.3		1033	3.4
	1707	2.2		1626	2.1
	2324	3.3		2223	3.4
13 F	0545	1.8	28 Sa	0453	1.6
	1226	3.4		1131	3.5
	1816	2.0		1737	1.9
				2340	3.4
14 Sa	0035	3.3	29 Su	0552	1.6
	0641	1.9		1225	3.7
	1311	3.6		1841	1.6
	1914	1.9			
15 Su	0132	3.3	30 M	0054	3.5
	0730	1.9		0647	1.6
	1352	3.7		1315	3.9
	2003	1.7		1939	1.3

	Sun↑	Sun↓
15	0602	1740
30	0609	1742

DECEMBER

Day	Time	ft	Day	Time	ft
1 Tu	0158	3.6	16 W	0246	3.1
	0739	1.6		0826	1.8
	1402	4.1		1438	3.7
	2033	1.1		2115	1.1
2 W	0254	3.7	17 Th	0330	3.1
	0828	1.5		0905	1.7
	1448	4.2		1513	3.7
	2122	0.8		2154	1.0
3 Th	0346	3.7	18 F	0410	3.2
	0916	1.5		0941	1.7
	1533	4.3		1547	3.8
	○ 2210	0.7		● 2229	0.9
4 F	0435	3.7	19 Sa	0447	3.2
	1004	1.5		1018	1.7
	1618	4.3		1623	3.8
	2256	0.6		2305	0.8
5 Sa	0522	3.6	20 Su	0525	3.2
	1052	1.5		1056	1.7
	1705	4.2		1700	3.8
	2343	0.6		2341	0.8
6 Su	0609	3.5	21 M	0604	3.2
	1141	1.6		1136	1.7
	1752	4.0		1738	3.7
7 M	0030	0.7	22 Tu	0020	0.8
	0657	3.4		0645	3.2
	1232	1.7		1218	1.7
	1841	3.8		1818	3.6
8 Tu	0118	0.9	23 W	0059	0.9
	0747	3.4		0727	3.2
	1326	1.8		1303	1.8
	1933	3.6		1902	3.5
9 W	0207	1.2	24 Th	0140	1.0
	0840	3.3		0811	3.2
	1423	1.9		1354	1.8
	2029	3.4		1952	3.4
10 Th	0300	1.4	25 F	0225	1.2
	0935	3.3		0858	3.3
	1525	1.9		1454	1.8
	☽ 2131	3.2		2050	3.3
11 F	0355	1.6	26 Sa	0315	1.3
	1033	3.3		0948	3.4
	1632	1.9		1603	1.7
	2238	3.0		☾ 2156	3.2
12 Sa	0454	1.7	27 Su	0412	1.5
	1132	3.3		1044	3.5
	1741	1.8		1717	1.6
	2351	3.0		2314	3.1
13 Su	0553	1.8	28 M	0513	1.5
	1226	3.4		1143	3.6
	1846	1.6		1827	1.3
14 M	0059	3.0	29 Tu	0036	3.1
	0650	1.8		0614	1.6
	1315	3.5		1242	3.7
	1943	1.4		1929	1.1
15 Tu	0157	3.0	30 W	0146	3.2
	0741	1.8		0713	1.5
	1359	3.6		1337	3.9
	2031	1.3		2024	0.8
			31 Th	0245	3.3
				0808	1.5
				1429	4.0
				2114	0.6

	Sun↑	Sun↓
15	0616	1747
31	0624	1755

Tides

PUNTA GORDA, VENEZUELA

HIGH & LOW WATER 1998 | Chart Datum | 10°10'N 62°38'W

Atlantic Standard Time (60°W) | Daylight Saving Time not observed

JANUARY

Date	Time	ft	Date	Time	ft
1 Th	0111	-0.9	16 F	0145	-0.5
	0657	6.3		0731	5.8
	1324	0.4		1359	0.3
	1903	6.9		1930	6.1
2 F	0155	-0.8	17 Sa	0222	-0.2
	0742	6.3		0805	5.7
	1412	0.4		1440	0.5
	1950	6.6		2008	5.8
3 Sa	0242	-0.6	18 Su	0301	0.1
	0829	6.3		0841	5.5
	1506	0.5		1523	0.6
	2043	6.3		2049	5.5
4 Su	0335	-0.3	19 M	0343	0.5
	0922	6.2		0921	5.4
	1608	0.5		1613	0.8
	2142	5.9		2135	5.2
5 M	0435	0.1	20 Tu	0431	0.8
	1021	6.0		1008	5.3
	1715	0.5		1710	0.9
☾	2251	5.5	☽	2229	4.9
6 Tu	0539	0.3	21 W	0528	1.1
	1125	5.9		1102	5.1
	1825	0.4		1814	0.9
				2332	4.6
7 W	0008	5.3	22 Th	0631	1.2
	0645	0.5		1203	5.1
	1234	5.9		1917	0.7
	1931	0.1			
8 Th	0127	5.2	23 F	0043	4.5
	0750	0.5		0734	1.2
	1342	6.0		1307	5.2
	2033	-0.3		2017	0.3
9 F	0239	5.4	24 Sa	0153	4.6
	0849	0.4		0833	1.1
	1444	6.2		1409	5.5
	2129	-0.7		2111	-0.1
10 Sa	0340	5.6	25 Su	0255	4.9
	0944	0.2		0925	0.8
	1538	6.4		1504	5.8
	2220	-0.9		2200	-0.5
11 Su	0431	5.8	26 M	0348	5.3
	1033	0.1		1013	0.5
	1625	6.5		1553	6.1
	2306	-1.1		2245	-0.9
12 M	0514	5.9	27 Tu	0434	5.6
	1119	0.0		1059	0.2
	1707	6.6		1640	6.5
○	2349	-1.1		2328	-1.2
13 Tu	0553	5.9	28 W	0517	6.0
	1201	0.0		1142	-0.1
	1745	6.5		1724	6.7
			●		
14 W	0029	-1.0	29 Th	0011	-1.4
	0627	5.9		0559	6.2
	1242	0.1		1226	-0.3
	1820	6.4		1808	6.8
15 Th	0108	-0.8	30 F	0053	-1.4
	0700	5.8		0641	6.4
	1321	0.2		1311	-0.5
	1855	6.3		1853	6.8
			31 Sa	0137	-1.3
				0724	6.5
				1358	-0.5
				1940	6.5

	Sun↑	Sun↓
15	0633	1808
31	0634	1815

FEBRUARY

Date	Time	ft	Date	Time	ft
1 Su	0223	-1.0	16 M	0222	-0.0
	0810	6.4		0800	5.7
	1449	-0.4		1443	0.1
	2031	6.2		2016	5.6
2 M	0314	-0.6	17 Tu	0256	0.3
	0859	6.2		0836	5.5
	1546	-0.2		1525	0.3
	2127	5.7		2057	5.3
3 Tu	0410	-0.1	18 W	0336	0.7
	0953	5.9		0918	5.4
	1650	-0.1		1617	0.5
☾	2231	5.2		2146	4.9
4 W	0513	0.3	19 Th	0428	1.0
	1055	5.6		1008	5.2
	1759	0.0		1721	0.6
	2347	4.9	☽	2245	4.6
5 Th	0621	0.5	20 F	0537	1.3
	1205	5.4		1108	5.0
	1908	-0.1		1832	0.6
				2356	4.4
6 F	0110	4.8	21 Sa	0652	1.3
	0729	0.6		1219	5.0
	1319	5.4		1939	0.4
	2013	-0.3			
7 Sa	0227	4.9	22 Su	0113	4.5
	0831	0.5		0801	1.1
	1428	5.6		1332	5.2
	2111	-0.6		2039	-0.1
8 Su	0329	5.2	23 M	0224	4.8
	0927	0.2		0900	0.7
	1525	5.8		1438	5.5
	2202	-0.9		2133	-0.5
9 M	0418	5.4	24 Tu	0322	5.2
	1017	-0.0		0952	0.2
	1613	6.0		1534	6.0
	2247	-1.0		2221	-1.0
10 Tu	0458	5.6	25 W	0412	5.7
	1101	-0.2		1039	-0.3
	1653	6.1		1624	6.4
	2329	-1.1		2306	-1.3
11 W	0532	5.7	26 Th	0456	6.2
	1141	-0.3		1124	-0.7
	1729	6.2		1711	6.7
○			●	2349	-1.5
12 Th	0006	-1.0	27 F	0539	6.5
	0602	5.8		1209	-1.0
	1219	-0.4		1757	6.8
	1802	6.2			
13 F	0042	-0.8	28 Sa	0033	-1.5
	0631	5.8		0621	6.7
	1255	-0.3		1254	-1.2
	1833	6.1		1842	6.8
14 Sa	0116	-0.6			
	0659	5.8			
	1330	-0.3			
	1905	6.0			
15 Su	0149	-0.3			
	0728	5.7			
	1405	-0.1			
	1939	5.8			

	Sun↑	Sun↓
15	0631	1820
28	0626	1821

MARCH

Date	Time	ft	Date	Time	ft
1 Su	0117	-1.3	16 M	0117	-0.1
	0703	6.7		0653	6.0
	1340	-1.1		1335	-0.4
	1928	6.6		1912	5.9
2 M	0202	-1.0	17 Tu	0147	0.2
	0746	6.6		0724	5.9
	1429	-0.9		1408	-0.2
	2017	6.2		1947	5.7
3 Tu	0252	-0.5	18 W	0217	0.5
	0833	6.3		0758	5.8
	1524	-0.6		1446	0.0
	2110	5.7		2027	5.4
4 W	0347	0.0	19 Th	0251	0.8
	0924	5.9		0838	5.6
	1625	-0.3		1533	0.3
	2212	5.1		2113	5.1
5 Th	0449	0.5	20 F	0337	1.2
	1024	5.5		0927	5.4
	1732	-0.1		1634	0.5
☾	2325	4.8		2210	4.8
6 F	0558	0.8	21 Sa	0450	1.4
	1135	5.1		1026	5.1
	1842	0.0		1749	0.6
			☽	2319	4.6
7 Sa	0050	4.6	22 Su	0616	1.5
	0708	0.8		1140	5.0
	1255	5.0		1902	0.4
	1948	-0.1			
8 Su	0209	4.8	23 M	0038	4.7
	0812	0.6		0732	1.2
	1410	5.2		1300	5.1
	2048	-0.3		2007	0.1
9 M	0310	5.1	24 Tu	0152	5.0
	0908	0.3		0835	0.7
	1510	5.4		1413	5.5
	2139	-0.5		2104	-0.4
10 Tu	0356	5.4	25 W	0254	5.6
	0956	-0.0		0929	0.0
	1557	5.7		1515	6.0
	2223	-0.7		2154	-0.8
11 W	0433	5.6	26 Th	0346	6.1
	1039	-0.3		1019	-0.6
	1636	5.9		1608	6.4
	2304	-0.8		2241	-1.1
12 Th	0504	5.8	27 F	0432	6.6
	1118	-0.5		1105	-1.1
	1710	6.0		1657	6.8
	2340	-0.7	●	2327	-1.3
13 F	0532	5.9	28 Sa	0515	6.9
	1154	-0.6		1150	-1.4
	1741	6.1		1743	6.9
○					
14 Sa	0014	-0.6	29 Su	0011	-1.2
	0559	5.9		0558	7.1
	1229	-0.6		1235	-1.5
	1810	6.1		1828	6.9
15 Su	0046	-0.4	30 M	0055	-1.0
	0625	6.0		0640	7.0
	1302	-0.6		1321	-1.5
	1840	6.0		1914	6.6
			31 Tu	0141	-0.6
				0722	6.8
				1409	-1.2
				2002	6.2

	Sun↑	Sun↓
15	0618	1822
31	0609	1822

APRIL

Date	Time	ft	Date	Time	ft
1 W	0230	-0.1	16 Th	0147	0.8
	0807	6.4		0727	6.1
	1502	-0.8		1418	-0.1
	2053	5.7		2004	5.6
2 Th	0324	0.4	17 F	0221	1.1
	0857	5.9		0808	6.0
	1559	-0.3		1503	0.1
	2152	5.3		2050	5.4
3 F	0426	0.8	18 Sa	0309	1.3
	0954	5.4		0857	5.7
	1704	0.1		1601	0.4
☾	2301	4.9		2146	5.2
4 Sa	0534	1.1	19 Su	0422	1.6
	1104	5.0		0956	5.4
	1812	0.3		1713	0.6
			☽	2252	5.1
5 Su	0021	4.8	20 M	0548	1.5
	0643	1.1		1110	5.2
	1225	4.9		1827	0.5
	1918	0.3			
6 M	0138	4.9	21 Tu	0008	5.2
	0747	0.8		0705	1.2
	1343	5.0		1232	5.2
	2017	0.1		1934	0.3
7 Tu	0237	5.2	22 W	0121	5.5
	0843	0.4		0810	0.6
	1445	5.3		1350	5.5
	2109	-0.0		2034	-0.1
8 W	0322	5.5	23 Th	0224	6.0
	0931	0.1		0907	-0.1
	1533	5.5		1455	6.0
	2154	-0.2		2127	-0.4
9 Th	0359	5.7	24 F	0318	6.5
	1013	-0.3		0958	-0.7
	1612	5.8		1551	6.4
	2234	-0.2		2217	-0.7
10 F	0430	5.9	25 Sa	0407	6.9
	1052	-0.5		1046	-1.2
	1646	6.0		1642	6.7
	2311	-0.2		2304	-0.8
11 Sa	0459	6.1	26 Su	0452	7.2
	1128	-0.7		1132	-1.5
	1717	6.1		1729	6.8
○	2345	-0.1	●	2350	-0.7
12 Su	0526	6.2	27 M	0535	7.3
	1203	-0.7		1218	-1.6
	1747	6.1		1815	6.8
13 M	0018	0.1	28 Tu	0035	-0.5
	0553	6.3		0617	7.2
	1236	-0.6		1303	-1.5
	1817	6.1		1901	6.6
14 Tu	0048	0.3	29 W	0121	-0.2
	0622	6.3		0700	6.9
	1308	-0.5		1350	-1.2
	1849	6.0		1947	6.3
15 W	0117	0.5	30 Th	0210	0.3
	0652	6.3		0744	6.5
	1342	-0.4		1440	-0.8
	1924	5.8		2036	5.9

	Sun↑	Sun↓
15	0601	1822
30	0554	1823

PUNTA GORDA, VENEZUELA

HIGH & LOW WATER 1998 | Chart Datum | 10°10'N 62°38'W

Atlantic Standard Time (60°W) | Daylight Saving Time not observed

MAY

Day	Time	ft	Day	Time	ft
1	0303	0.7	16	0207	1.2
	0831	6.0		0747	6.2
F	1534	-0.3	Sa	1445	-0.0
	2130	5.5		2033	5.7
2	0401	1.1	17	0259	1.4
	0924	5.5		0837	6.0
Sa	1633	0.2	**Su**	1539	0.2
	2231	5.2		2127	5.6
3	0505	1.2	18	0407	1.5
	1028	5.1		0936	5.7
Su	1736	0.5	M	1644	0.4
☾	2340	5.0		2230	5.6
4	0611	1.2	19	0524	1.4
	1143	4.9		1047	5.4
M	1839	0.6	Tu	1754	0.5
			☽	2339	5.7
5	0049	5.1	20	0639	1.0
	0714	1.0		1207	5.3
Tu	1300	4.9	W	1902	0.4
	1938	0.6			
6	0149	5.3	21	0050	5.9
	0810	0.6		0745	0.5
W	1406	5.1	Th	1326	5.5
	2031	0.5		2004	0.2
7	0237	5.6	22	0154	6.3
	0859	0.2		0844	-0.2
Th	1458	5.3	F	1435	5.9
	2118	0.4		2101	0.0
8	0317	5.8	23	0251	6.7
	0944	-0.1		0938	-0.7
F	1542	5.6	Sa	1535	6.2
	2201	0.3		2153	-0.2
9	0352	6.0	24	0343	7.0
	1024	-0.4		1028	-1.2
Sa	1619	5.8	**Su**	1628	6.5
	2240	0.3		2243	-0.3
10	0424	6.2	25	0430	7.2
	1102	-0.6		1115	-1.5
Su	1653	5.9	M	1717	6.6
	2316	0.4	●	2330	-0.2
11	0455	6.4	26	0515	7.3
	1138	-0.7		1201	-1.5
M	1725	6.0	Tu	1803	6.6
○	2351	0.5			
12	0524	6.4	27	0017	-0.1
	1213	-0.7		0558	7.2
Tu	1757	6.0	W	1246	-1.4
				1848	6.5
13	0023	0.7	28	0103	0.2
	0555	6.5		0640	6.9
W	1247	-0.6	Th	1332	-1.1
	1830	6.0		1931	6.3
14	0055	0.8	29	0151	0.5
	0628	6.5		0723	6.6
Th	1322	-0.5	F	1418	-0.7
	1907	6.0		2016	6.0
15	0128	1.0	30	0240	0.8
	0705	6.4		0807	6.1
F	1400	-0.3	Sa	1507	-0.2
	1947	5.9		2102	5.7
			31	0333	1.1
				0855	5.7
			Su	1558	0.2
				2153	5.5

	Sun↑	Sun↓
15	0550	1825
31	0548	1829

JUNE

Day	Time	ft	Day	Time	ft
1	0431	1.2	16	0352	1.1
	0949	5.3		0921	5.9
M	1654	0.6	Tu	1618	0.3
☾	2248	5.3		2207	6.1
2	0531	1.3	17	0501	1.0
	1052	5.0		1028	5.6
Tu	1753	0.9	W	1724	0.5
	2348	5.3	☽	2311	6.1
3	0632	1.1	18	0612	0.8
	1202	4.8		1144	5.4
W	1852	1.0	Th	1831	0.6
4	0048	5.4	19	0019	6.2
	0730	0.8		0720	0.4
Th	1312	4.9	F	1304	5.4
	1948	1.0		1936	0.6
5	0142	5.6	20	0125	6.4
	0823	0.5		0822	-0.1
F	1413	5.1	Sa	1418	5.6
	2039	0.9		2037	0.5
6	0230	5.8	21	0227	6.7
	0911	0.2		0919	-0.6
Sa	1505	5.3	**Su**	1522	5.9
	2126	0.9		2133	0.3
7	0312	6.0	22	0323	6.9
	0955	-0.2		1011	-1.0
Su	1549	5.5	M	1618	6.2
	2209	0.8		2225	0.2
8	0350	6.2	23	0413	7.1
	1036	-0.4		1100	-1.3
M	1628	5.7	Tu	1707	6.4
	2249	0.8	●	2314	0.2
9	0425	6.4	24	0459	7.1
	1115	-0.6		1146	-1.3
Tu	1704	5.8	W	1752	6.5
	2326	0.8			
10	0459	6.5	25	0001	0.2
	1153	-0.7		0542	7.1
W	1739	6.0	Th	1230	-1.2
○				1834	6.4
11	0002	0.9	26	0046	0.3
	0534	6.6		0623	6.9
Th	1229	-0.7	F	1313	-1.0
	1815	6.1		1913	6.3
12	0038	0.9	27	0130	0.5
	0610	6.7		0703	6.6
F	1307	-0.6	Sa	1355	-0.6
	1853	6.1		1951	6.2
13	0117	1.0	28	0215	0.7
	0650	6.6		0743	6.3
Sa	1346	-0.5	**Su**	1437	-0.2
	1934	6.2		2030	6.0
14	0159	1.0	29	0301	0.9
	0734	6.5		0825	5.9
Su	1430	-0.3	M	1521	0.2
	2019	6.1		2111	5.8
15	0250	1.1	30	0351	1.1
	0823	6.2		0911	5.6
M	1520	-0.0	Tu	1609	0.7
	2110	6.1		2155	5.7

	Sun↑	Sun↓
15	0550	1833
30	0553	1836

JULY

Day	Time	ft	Day	Time	ft
1	0445	1.2	16	0438	0.7
	1003	5.2		1009	5.8
W	1702	1.0	Th	1656	0.6
☾	2246	5.5	☽	2243	6.4
2	0544	1.2	17	0547	0.6
	1102	5.0		1123	5.4
Th	1800	1.3	F	1804	0.9
	2341	5.5		2349	6.3
3	0644	1.1	18	0656	0.4
	1210	4.8		1244	5.3
F	1859	1.4	Sa	1913	1.0
4	0040	5.5	19	0059	6.3
	0742	0.8		0802	0.1
Sa	1319	4.8	**Su**	1404	5.5
	1956	1.4		2017	0.9
5	0137	5.7	20	0207	6.5
	0836	0.5		0902	-0.3
Su	1422	5.0	M	1513	5.8
	2049	1.4		2117	0.8
6	0229	5.9	21	0308	6.7
	0925	0.2		0956	-0.7
M	1516	5.3	Tu	1609	6.1
	2138	1.3		2210	0.6
7	0316	6.1	22	0401	6.9
	1010	-0.2		1044	-0.9
Tu	1602	5.5	W	1657	6.3
	2222	1.1		2259	0.4
8	0358	6.4	23	0447	7.0
	1052	-0.4		1129	-1.0
W	1643	5.8	Th	1739	6.5
	2304	1.0	●	2344	0.3
9	0438	6.6	24	0529	7.0
	1132	-0.6		1211	-0.9
Th	1722	6.1	F	1815	6.5
○	2344	0.9			
10	0518	6.8	25	0027	0.3
	1211	-0.7		0607	6.9
F	1800	6.3	Sa	1251	-0.7
				1849	6.5
11	0024	0.8	26	0107	0.4
	0557	6.9		0643	6.7
Sa	1250	-0.7	**Su**	1329	-0.4
	1838	6.5		1922	6.4
12	0104	0.7	27	0147	0.5
	0639	6.9		0719	6.5
Su	1330	-0.7	M	1406	-0.0
	1919	6.6		1954	6.4
13	0148	0.7	28	0227	0.7
	0724	6.8		0755	6.3
M	1413	-0.4	Tu	1444	0.4
	2002	6.7		2028	6.2
14	0237	0.7	29	0310	0.9
	0812	6.5		0834	6.0
Tu	1501	-0.1	W	1524	0.8
	2050	6.6		2106	6.1
15	0334	0.7	30	0358	1.1
	0907	6.2		0919	5.6
W	1555	0.2	Th	1609	1.2
	2143	6.5		2149	5.9
			31	0452	1.2
				1010	5.3
			F	1703	1.6
			☾	2239	5.8

	Sun↑	Sun↓
15	0557	1837
31	0600	1834

AUGUST

Day	Time	ft	Day	Time	ft
1	0553	1.3	16	0633	0.6
	1111	5.0		1228	5.4
Sa	1806	1.9	**Su**	1853	1.5
	2337	5.7			
2	0657	1.2	17	0036	6.2
	1222	4.9		0742	0.4
Su	1911	1.9	M	1353	5.5
				2001	1.4
3	0042	5.7	18	0151	6.3
	0758	0.9		0844	0.1
M	1336	4.9	Tu	1504	5.8
	2013	1.9		2102	1.2
4	0146	5.9	19	0256	6.5
	0853	0.6		0938	-0.2
Tu	1441	5.2	W	1558	6.2
	2108	1.6		2155	0.9
5	0244	6.1	20	0350	6.8
	0943	0.2		1027	-0.4
W	1535	5.6	Th	1642	6.5
	2157	1.4		2242	0.6
6	0334	6.5	21	0435	6.9
	1028	-0.2		1110	-0.5
Th	1620	6.0	F	1719	6.7
	2242	1.0	●	2325	0.4
7	0420	6.8	22	0514	7.0
	1110	-0.5		1149	-0.4
F	1701	6.4	Sa	1751	6.8
○	2325	0.7			
8	0503	7.1	23	0005	0.3
	1151	-0.7		0549	7.0
Sa	1740	6.8	**Su**	1226	-0.2
				1821	6.8
9	0007	0.5	24	0042	0.3
	0545	7.3		0622	6.9
Su	1231	-0.7	M	1301	0.0
	1820	7.0		1849	6.8
10	0049	0.3	25	0118	0.4
	0628	7.3		0653	6.8
M	1311	-0.6	Tu	1334	0.3
	1900	7.2		1917	6.8
11	0133	0.2	26	0154	0.6
	0713	7.2		0726	6.6
Tu	1354	-0.4	W	1407	0.7
	1942	7.2		1948	6.7
12	0221	0.2	27	0232	0.8
	0800	6.9		0801	6.4
W	1440	-0.0	Th	1440	1.1
	2028	7.1		2022	6.6
13	0314	0.4	28	0313	1.0
	0853	6.5		0841	6.1
Th	1532	0.5	F	1518	1.5
	2118	6.9		2101	6.4
14	0415	0.5	29	0402	1.3
	0953	6.0		0927	5.7
F	1633	0.9	Sa	1604	1.9
☽	2215	6.6		2147	6.2
15	0522	0.7	30	0502	1.5
	1104	5.6		1023	5.4
Sa	1741	1.3	**Su**	1709	2.3
	2322	6.4	☾	2243	6.0
			31	0611	1.5
				1131	5.1
			M	1826	2.4
				2350	5.8

	Sun↑	Sun↓
15	0602	1829
31	0602	1821

T Tides

PUNTA GORDA, VENEZUELA

HIGH & LOW WATER 1998 — Chart Datum — 10°10'N 62°38'W

Atlantic Standard Time (60°W) — Daylight Saving Time not observed

SEPTEMBER

Date	Day	Time	ft	Time	ft	Time	ft	Time	ft
1	Tu	0718	1.3	1250	5.1	1938	2.3		
2	W	0104	5.9	0819	1.0	1404	5.4	2039	1.9
3	Th	0212	6.2	0913	0.6	1504	5.9	2132	1.5
4	F	0310	6.6	1001	0.1	1553	6.4	2220	1.0
5	Sa	0401	7.0	1045	-0.2	1636	6.9	2304	0.5
6	Su	0447	7.4	1127	-0.4	1717	7.4	○ 2347	0.2
7	M	0532	7.6	1208	-0.5	1757	7.7		
8	Tu	0031	-0.1	0616	7.6	1250	-0.4	1838	7.8
9	W	0115	-0.1	0701	7.5	1334	-0.1	1920	7.8
10	Th	0203	-0.1	0748	7.2	1420	0.3	2005	7.6
11	F	0255	0.2	0839	6.7	1512	0.8	2054	7.2
12	Sa	0353	0.5	0937	6.2	1612	1.4	☽ 2149	6.8
13	Su	0459	0.8	1047	5.8	1721	1.8	2256	6.4
14	M	0610	0.9	1211	5.6	1834	1.9		
15	Tu	0015	6.2	0719	0.8	1337	5.7	1943	1.7
16	W	0135	6.2	0822	0.6	1446	6.0	2043	1.4
17	Th	0243	6.4	0916	0.4	1538	6.4	2136	1.1
18	F	0336	6.7	1003	0.2	1618	6.7	2221	0.7
19	Sa	0419	6.9	1045	0.1	1652	6.9	2302	0.5
20	Su	0456	7.0	1123	0.1	1722	7.0	● 2340	0.4
21	M	0528	7.1	1159	0.3	1749	7.1		
22	Tu	0016	0.3	0559	7.1	1232	0.5	1815	7.1
23	W	0050	0.4	0628	7.0	1303	0.8	1842	7.1
24	Th	0124	0.5	0659	6.8	1333	1.1	1912	7.1
25	F	0158	0.7	0733	6.6	1403	1.5	1944	7.0
26	Sa	0236	1.0	0811	6.4	1435	1.8	2022	6.8
27	Su	0320	1.3	0854	6.1	1515	2.2	2106	6.5
28	M	0416	1.5	0947	5.7	1618	2.5	☾ 2200	6.3
29	Tu	0525	1.6	1052	5.5	1744	2.6	2308	6.1
30	W	0637	1.6	1209	5.5	1903	2.5		

	Sun↑	Sun↓
15	0601	1812
30	0600	1802

OCTOBER

Date	Day	Time	ft	Time	ft	Time	ft	Time	ft
1	Th	0026	6.1	0742	1.3	1325	5.8	2009	2.0
2	F	0141	6.3	0840	0.9	1429	6.3	2106	1.4
3	Sa	0246	6.7	0930	0.4	1522	6.9	2156	0.8
4	Su	0341	7.2	1017	0.1	1608	7.4	2242	0.3
5	M	0430	7.5	1102	-0.1	1651	7.8	○ 2327	-0.2
6	Tu	0517	7.7	1145	-0.2	1733	8.1		
7	W	0012	-0.4	0602	7.8	1229	-0.1	1815	8.2
8	Th	0057	-0.4	0648	7.6	1314	0.2	1858	8.1
9	F	0145	-0.3	0735	7.3	1402	0.7	1943	7.8
10	Sa	0236	0.0	0826	6.9	1454	1.2	2031	7.3
11	Su	0332	0.4	0923	6.4	1554	1.6	2126	6.8
12	M	0435	0.8	1030	6.0	1701	2.0	☽ 2231	6.4
13	Tu	0543	1.0	1148	5.8	1812	2.0	2349	6.1
14	W	0651	1.1	1309	5.9	1919	1.9		
15	Th	0110	6.1	0752	1.0	1415	6.1	2019	1.5
16	F	0219	6.2	0847	0.8	1505	6.5	2110	1.1
17	Sa	0313	6.5	0934	0.7	1545	6.7	2156	0.8
18	Su	0356	6.7	1016	0.6	1618	6.9	2237	0.5
19	M	0432	6.8	1055	0.6	1648	7.1	2314	0.3
20	Tu	0505	6.9	1130	0.7	1716	7.2	● 2350	0.3
21	W	0535	6.9	1203	0.9	1743	7.2		
22	Th	0025	0.3	0605	6.9	1235	1.1	1811	7.3
23	F	0058	0.4	0636	6.8	1305	1.4	1841	7.2
24	Sa	0132	0.6	0710	6.6	1334	1.7	1914	7.1
25	Su	0208	0.8	0747	6.4	1406	1.9	1951	7.0
26	M	0249	1.0	0830	6.2	1446	2.2	2036	6.7
27	Tu	0341	1.3	0921	6.0	1547	2.4	2130	6.4
28	W	0446	1.4	1022	5.8	1710	2.5	☾ 2235	6.2
29	Th	0557	1.4	1133	5.9	1830	2.3	2352	6.1
30	F	0704	1.3	1247	6.1	1939	1.8		
31	Sa	0111	6.2	0805	0.9	1353	6.6	2038	1.2

	Sun↑	Sun↓
15	0600	1754
31	0602	1748

NOVEMBER

Date	Day	Time	ft	Time	ft	Time	ft	Time	ft
1	Su	0220	6.6	0859	0.6	1450	7.1	2131	0.5
2	M	0319	7.0	0950	0.3	1540	7.6	2221	-0.1
3	Tu	0412	7.3	1037	0.1	1626	7.9	2308	-0.5
4	W	0502	7.5	1124	0.1	1711	8.1	○ 2354	-0.7
5	Th	0549	7.5	1210	0.2	1755	8.2		
6	F	0041	-0.7	0636	7.4	1256	0.4	1838	8.0
7	Sa	0128	-0.6	0723	7.1	1345	0.8	1923	7.6
8	Su	0218	-0.2	0813	6.8	1437	1.2	2011	7.2
9	M	0311	0.2	0906	6.4	1534	1.5	2103	6.7
10	Tu	0409	0.6	1006	6.0	1636	1.8	☽ 2204	6.2
11	W	0511	1.0	1113	5.8	1742	1.9	2314	5.8
12	Th	0615	1.1	1223	5.8	1847	1.7		
13	F	0030	5.7	0715	1.2	1327	6.0	1946	1.4
14	Sa	0140	5.8	0810	1.1	1420	6.2	2039	1.0
15	Su	0238	6.0	0859	1.0	1503	6.4	2126	0.7
16	M	0325	6.2	0944	0.9	1541	6.6	2208	0.4
17	Tu	0404	6.3	1024	0.9	1614	6.8	2248	0.2
18	W	0440	6.4	1102	0.9	1645	6.9	2326	0.1
19	Th	0513	6.5	1137	1.0	1715	7.0	●	
20	F	0002	0.0	0545	6.5	1211	1.2	1745	7.1
21	Sa	0037	0.1	0617	6.4	1243	1.4	1816	7.1
22	Su	0112	0.2	0651	6.4	1315	1.5	1851	7.0
23	M	0148	0.3	0729	6.3	1350	1.7	1930	6.9
24	Tu	0229	0.5	0811	6.2	1433	1.8	2015	6.6
25	W	0316	0.7	0900	6.1	1530	1.9	2108	6.4
26	Th	0414	0.9	0957	6.0	1642	1.9	☾ 2211	6.1
27	F	0519	1.0	1102	6.0	1758	1.7	2324	5.9
28	Sa	0627	1.0	1211	6.2	1909	1.3		
29	Su	0042	5.9	0731	0.8	1318	6.5	2012	0.7
30	M	0155	6.1	0830	0.6	1419	6.9	2109	0.1

	Sun↑	Sun↓
15	0606	1745
30	0612	1747

DECEMBER

Date	Day	Time	ft	Time	ft	Time	ft	Time	ft
1	Tu	0300	6.4	0925	0.4	1514	7.3	2201	-0.4
2	W	0357	6.7	1016	0.2	1605	7.6	2251	-0.8
3	Th	0449	6.9	1105	0.1	1652	7.7	○ 2339	-1.1
4	F	0538	7.0	1153	0.1	1738	7.7		
5	Sa	0026	-1.1	0625	6.9	1240	0.3	1822	7.6
6	Su	0112	-0.9	0710	6.7	1328	0.5	1906	7.3
7	M	0159	-0.6	0756	6.5	1418	0.8	1952	6.8
8	Tu	0248	-0.2	0843	6.2	1510	1.1	2040	6.4
9	W	0339	0.2	0933	5.9	1605	1.3	2132	5.9
10	Th	0434	0.6	1027	5.7	1705	1.4	☽ 2231	5.5
11	F	0532	0.9	1126	5.6	1806	1.3	2338	5.3
12	Sa	0630	1.1	1226	5.6	1905	1.2		
13	Su	0047	5.2	0727	1.2	1324	5.7	2001	0.9
14	M	0151	5.3	0820	1.1	1415	5.9	2052	0.5
15	Tu	0247	5.4	0909	1.0	1500	6.1	2139	0.2
16	W	0333	5.6	0953	1.0	1540	6.3	2222	-0.0
17	Th	0414	5.7	1035	0.9	1616	6.4	2302	-0.2
18	F	0451	5.8	1113	0.9	1650	6.6	● 2341	-0.3
19	Sa	0526	5.9	1150	0.9	1724	6.7		
20	Su	0018	-0.4	0600	6.0	1225	1.0	1758	6.7
21	M	0054	-0.4	0635	6.1	1300	1.0	1835	6.7
22	Tu	0131	-0.3	0713	6.1	1338	1.0	1915	6.6
23	W	0211	-0.1	0754	6.1	1422	1.1	2000	6.5
24	Th	0255	0.0	0841	6.1	1514	1.1	2051	6.2
25	F	0347	0.3	0933	6.0	1618	1.1	2150	5.9
26	Sa	0447	0.5	1032	6.0	1729	1.0	☾ 2259	5.6
27	Su	0553	0.6	1137	6.0	1840	0.7		
28	M	0016	5.4	0701	0.6	1246	6.2	1947	0.2
29	Tu	0133	5.5	0805	0.5	1352	6.4	2048	-0.3
30	W	0244	5.7	0904	0.3	1453	6.7	2144	-0.8
31	Th	0346	6.0	0959	0.1	1548	6.9	2236	-1.1

	Sun↑	Sun↓
15	0620	1752
31	0628	1800

AMUAY, VENEZUELA

HIGH & LOW WATER 1998 — Chart Datum — 11°45'N 70°13'W

Atlantic Standard Time (60°W) — Daylight Saving Time not observed

JANUARY

Date	Day	Time	ft	Time	ft	Time	ft	Time	ft	Moon
1	Th	0007	-0.3	0805	0.8	1112	0.6	1708	1.0	
2	F	0053	-0.2	0852	0.8	1236	0.6	1836	0.9	
3	Sa	0142	-0.2	0938	0.9	1403	0.5	2032	0.8	
4	Su	0232	-0.1	1022	1.0	1527	0.3	2209	0.7	
5	M	0322	-0.0	1105	1.1	1642	0.2	2330	0.7	☾
6	Tu	0412	0.1	1148	1.1	1746	-0.0			
7	W	0042	0.7	0502	0.1	1229	1.2	1842	-0.2	
8	Th	0149	0.7	0551	0.2	1309	1.2	1933	-0.3	
9	F	0252	0.7	0639	0.3	1347	1.2	2021	-0.4	
10	Sa	0354	0.7	0728	0.3	1421	1.2	2109	-0.5	
11	Su	0455	0.7	0817	0.4	1448	1.2	2156	-0.5	
12	M	0556	0.7	0909	0.5	1457	1.1	2243	-0.5	○
13	Tu	0657	0.7	1004	0.5	1506	1.0	2330	-0.4	
14	W	0757	0.7	1105	0.5	1529	0.9			
15	Th	0017	-0.3	0854	0.8	1213	0.5	1602	0.8	
16	F	0105	-0.2	0945	0.8	1330	0.5	1641	0.6	
17	Sa	0152	-0.1	1030	0.8	1453	0.5	1730	0.5	
18	Su	0239	-0.0	1105	0.8	1613	0.3	2242	0.4	
19	M	0324	0.1	1127	0.8	1717	0.2			
20	Tu	0005	0.4	0405	0.2	1126	0.8	1806	0.1	☽
21	W	0109	0.4	0440	0.2	1116	0.9	1846	-0.1	
22	Th	0159	0.4	0506	0.3	1132	0.9	1923	-0.2	
23	F	0233	0.4	0524	0.3	1202	1.0	1957	-0.3	
24	Sa	0259	0.4	0546	0.4	1238	1.1	2030	-0.4	
25	Su	0329	0.4	0622	0.4	1317	1.1	2102	-0.4	
26	M	0406	0.5	0708	0.4	1358	1.1	2136	-0.4	
27	Tu	0448	0.5	0801	0.3	1443	1.1	2212	-0.4	
28	W	0533	0.5	0900	0.3	1532	1.0	2251	-0.4	●
29	Th	0622	0.6	1004	0.3	1629	0.9	2334	-0.3	
30	F	0712	0.6	1115	0.3	1744	0.8			
31	Sa	0021	-0.3	0802	0.7	1230	0.2	1919	0.7	

	Sun↑	Sun↓
15	0706	1836
31	0706	1844

FEBRUARY

Date	Day	Time	ft	Time	ft	Time	ft	Time	ft	Moon
1	Su	0110	-0.2	0853	0.8	1349	0.1	2054	0.6	
2	M	0201	-0.1	0942	0.8	1508	0.0	2219	0.6	
3	Tu	0255	0.0	1031	0.9	1621	-0.1	2334	0.6	☾
4	W	0348	0.1	1118	1.0	1724	-0.3			
5	Th	0041	0.6	0441	0.1	1203	1.0	1820	-0.4	
6	F	0142	0.6	0534	0.2	1247	1.1	1911	-0.5	
7	Sa	0240	0.6	0625	0.2	1329	1.1	1959	-0.5	
8	Su	0336	0.6	0716	0.2	1408	1.0	2045	-0.5	
9	M	0431	0.6	0808	0.3	1440	1.0	2130	-0.5	
10	Tu	0525	0.6	0900	0.3	1458	0.9	2215	-0.4	
11	W	0618	0.6	0955	0.3	1505	0.8	2259	-0.3	○
12	Th	0712	0.6	1053	0.3	1529	0.7	2344	-0.2	
13	F	0803	0.6	1156	0.3	1605	0.6			
14	Sa	0029	-0.1	0850	0.6	1305	0.3	1653	0.5	
15	Su	0114	0.0	0926	0.6	1417	0.2	1804	0.4	
16	M	0159	0.1	0941	0.6	1527	0.1	2258	0.3	
17	Tu	0243	0.2	0915	0.6	1628	0.0			
18	W	0010	0.4	0323	0.3	0927	0.7	1718	-0.1	
19	Th	0100	0.4	0356	0.3	1004	0.8	1801	-0.2	☽
20	F	0126	0.4	0425	0.4	1046	0.9	1839	-0.3	
21	Sa	0144	0.4	0458	0.4	1131	0.9	1914	-0.3	
22	Su	0209	0.5	0540	0.3	1219	1.0	1947	-0.4	
23	M	0242	0.5	0627	0.3	1307	1.0	2021	-0.4	
24	Tu	0319	0.5	0718	0.3	1358	1.0	2056	-0.4	
25	W	0400	0.6	0811	0.2	1452	1.0	2135	-0.3	
26	Th	0443	0.6	0908	0.2	1552	0.9	2216	-0.3	●
27	F	0529	0.6	1009	0.1	1701	0.8	2300	-0.2	
28	Sa	0618	0.7	1114	0.0	1821	0.7	2348	-0.1	

	Sun↑	Sun↓
15	0703	1848
28	0657	1851

MARCH

Date	Day	Time	ft	Time	ft	Time	ft	Time	ft	Moon
1	Su	0709	0.7	1222	-0.0	1946	0.7			
2	M	0039	0.0	0803	0.8	1334	-0.1	2109	0.7	
3	Tu	0134	0.1	0858	0.8	1445	-0.2	2224	0.7	
4	W	0231	0.2	0953	0.9	1554	-0.3	2332	0.7	
5	Th	0330	0.3	1047	0.9	1656	-0.3			☾
6	F	0033	0.7	0428	0.3	1139	1.0	1751	-0.4	
7	Sa	0129	0.7	0525	0.3	1230	1.0	1842	-0.4	
8	Su	0221	0.8	0620	0.3	1318	1.0	1929	-0.4	
9	M	0310	0.7	0713	0.3	1404	0.9	2014	-0.3	
10	Tu	0357	0.7	0805	0.3	1446	0.9	2058	-0.3	
11	W	0442	0.7	0857	0.2	1525	0.8	2141	-0.1	
12	Th	0524	0.7	0949	0.2	1554	0.7	2223	-0.0	
13	F	0600	0.6	1043	0.2	1614	0.6	2305	0.1	○
14	Sa	0618	0.6	1138	0.2	1702	0.5	2346	0.2	
15	Su	0516	0.6	1236	0.2	2013	0.5			
16	M	0027	0.3	0531	0.7	1335	0.1	2157	0.5	
17	Tu	0106	0.4	0609	0.7	1435	0.1	2321	0.5	
18	W	0145	0.5	0658	0.8	1531	-0.0			
19	Th	0015	0.6	0224	0.5	0757	0.8	1621	-0.1	
20	F	0029	0.6	0311	0.6	0901	0.9	1705	-0.1	
21	Sa	0035	0.6	0402	0.5	1006	0.9	1745	-0.2	☽
22	Su	0057	0.7	0453	0.5	1108	1.0	1822	-0.2	
23	M	0127	0.7	0544	0.4	1208	1.0	1859	-0.2	
24	Tu	0201	0.8	0634	0.3	1307	1.0	1936	-0.2	
25	W	0238	0.8	0726	0.2	1407	1.0	2015	-0.1	
26	Th	0317	0.8	0818	0.1	1509	1.0	2057	-0.0	
27	F	0357	0.9	0913	0.1	1616	0.9	2140	0.1	●
28	Sa	0439	0.9	1010	-0.0	1728	0.9	2226	0.2	
29	Su	0522	0.9	1110	-0.1	1844	0.8	2315	0.3	
30	M	0610	0.9	1212	-0.1	2002	0.8			
31	Tu	0009	0.4	0704	0.9	1316	-0.2	2116	0.8	

	Sun↑	Sun↓
15	0649	1852
31	0639	1853

APRIL

Date	Day	Time	ft	Time	ft	Time	ft	Time	ft	Moon
1	W	0108	0.5	0806	1.0	1421	-0.2	2224	0.9	
2	Th	0211	0.5	0911	1.0	1524	-0.2	2325	0.9	
3	F	0317	0.5	1016	1.0	1623	-0.2			☾
4	Sa	0020	1.0	0422	0.5	1118	1.0	1718	-0.2	
5	Su	0110	1.0	0524	0.5	1216	0.9	1808	-0.2	
6	M	0156	1.0	0621	0.4	1312	0.9	1855	-0.1	
7	Tu	0238	1.0	0715	0.4	1406	0.9	1939	-0.0	
8	W	0316	0.9	0806	0.3	1459	0.8	2021	0.1	
9	Th	0347	0.9	0855	0.2	1552	0.8	2101	0.2	
10	F	0400	0.9	0943	0.2	1650	0.7	2140	0.3	
11	Sa	0325	0.9	1030	0.2	1758	0.7	2216	0.5	○
12	Su	0324	0.9	1117	0.1	1923	0.7	2247	0.6	
13	M	0347	0.9	1205	0.1	2107	0.7	2307	0.7	
14	Tu	0419	1.0	1254	0.1					
15	W	0458	1.0	1342	0.0					
16	Th	0544	1.0	1431	-0.0					
17	F	0645	1.0	1519	-0.0	2333	0.8			
18	Sa	0234	0.8	0804	1.0	1604	-0.0	2349	0.9	
19	Su	0351	0.7	0932	1.0	1648	-0.0			☽
20	M	0016	1.0	0453	0.6	1053	1.0	1730	-0.0	
21	Tu	0049	1.0	0548	0.5	1206	1.0	1811	0.0	
22	W	0125	1.1	0639	0.4	1313	1.0	1853	0.1	
23	Th	0201	1.1	0730	0.2	1419	1.0	1936	0.2	
24	F	0238	1.2	0821	0.1	1525	1.0	2019	0.3	
25	Sa	0315	1.2	0913	-0.0	1633	1.0	2105	0.4	
26	Su	0351	1.2	1007	-0.1	1743	1.0	2152	0.5	●
27	M	0427	1.2	1102	-0.2	1856	1.0	2244	0.6	
28	Tu	0459	1.2	1159	-0.2	2007	1.0	2340	0.7	
29	W	0522	1.1	1257	-0.2	2115	1.0			
30	Th	0044	0.7	0537	1.1	1356	-0.2	2217	1.1	

	Sun↑	Sun↓
15	0630	1853
30	0623	1855

T Tides

AMUAY, VENEZUELA

HIGH & LOW WATER 1998 — Chart Datum — 11°45'N 70°13'W

Atlantic Standard Time (60°W) — Daylight Saving Time not observed

MAY

Day	Time	ft
1 F	0154	0.8
	0617	1.0
	1454	-0.1
	2312	1.1
2 Sa	0308	0.8
	0943	1.0
	1550	-0.1
3 Su ☾	0001	1.2
	0421	0.7
	1100	0.9
	1642	-0.0
4 M	0046	1.2
	0527	0.6
	1208	0.9
	1731	0.1
5 Tu	0126	1.2
	0626	0.5
	1312	0.9
	1817	0.2
6 W	0201	1.2
	0717	0.4
	1412	0.8
	1900	0.3
7 Th	0227	1.1
	0805	0.3
	1511	0.8
	1939	0.4
8 F	0233	1.1
	0849	0.2
	1611	0.8
	2016	0.5
9 Sa	0209	1.1
	0932	0.1
	1716	0.8
	2047	0.6
10 Su	0211	1.2
	1013	0.1
	1833	0.8
	2108	0.7
11 M ○	0233	1.2
	1054	0.0
12 Tu	0302	1.2
	1134	0.0
13 W	0335	1.2
	1214	-0.0
14 Th	0411	1.2
	1255	-0.0
15 F	0452	1.2
	1338	-0.0
	2229	1.0
16 Sa	0016	0.9
	0547	1.1
	1424	-0.0
	2238	1.0
17 Su	0211	0.9
	0717	1.0
	1510	0.0
	2306	1.1
18 M	0341	0.8
	0915	1.0
	1557	0.1
	2340	1.1
19 Tu ☽	0450	0.6
	1050	1.0
	1644	0.1
20 W	0015	1.2
	0548	0.5
	1210	1.0
	1729	0.2
21 Th	0051	1.3
	0639	0.3
	1321	1.0
	1815	0.3
22 F	0128	1.3
	0729	0.1
	1429	1.0
	1900	0.4
23 Sa	0204	1.4
	0819	-0.0
	1535	1.0
	1946	0.5
24 Su	0239	1.4
	0908	-0.2
	1641	1.0
	2033	0.6
25 M ●	0311	1.4
	0959	-0.2
	1748	1.0
	2123	0.7
26 Tu	0336	1.4
	1050	-0.3
	1856	1.0
	2216	0.8
27 W	0346	1.3
	1142	-0.3
	2002	1.1
	2316	0.8
28 Th	0400	1.2
	1236	-0.2
	2104	1.1
29 F	0024	0.9
	0427	1.2
	1329	-0.2
	2201	1.2
30 Sa	0140	0.9
	0504	1.0
	1423	-0.1
	2252	1.2
31 Su	0301	0.8
	0552	0.9
	1515	0.0
	2338	1.2

	Sun↑	Sun↓
15	0618	1858
31	0616	1902

JUNE

Day	Time	ft
1 M ☾	0422	0.7
	1049	0.8
	1606	0.1
2 Tu	0018	1.3
	0532	0.6
	1206	0.8
	1654	0.2
3 W	0053	1.3
	0628	0.5
	1315	0.8
	1738	0.3
4 Th	0119	1.2
	0714	0.3
	1419	0.8
	1819	0.5
5 F	0127	1.2
	0757	0.2
	1521	0.8
	1855	0.6
6 Sa	0109	1.2
	0837	0.1
	1625	0.8
	1924	0.7
7 Su	0111	1.3
	0915	0.0
	1735	0.8
	1940	0.8
8 M	0132	1.3
	0951	-0.0
9 Tu	0201	1.4
	1027	-0.1
10 W ○	0233	1.4
	1101	-0.1
11 Th	0307	1.4
	1136	-0.1
12 F	0344	1.3
	1214	-0.1
	2054	0.9
	2227	0.9
13 Sa	0426	1.2
	1255	-0.1
	2115	1.0
14 Su	0015	0.9
	0522	1.1
	1340	-0.0
	2149	1.1
15 M	0155	0.8
	0709	1.0
	1428	0.1
	2225	1.2
16 Tu	0326	0.7
	0921	0.9
	1517	0.1
	2303	1.2
17 W ☽	0440	0.5
	1057	0.9
	1606	0.2
	2342	1.3
18 Th	0540	0.3
	1216	0.9
	1654	0.3
19 F	0021	1.4
	0633	0.1
	1326	0.9
	1742	0.4
20 Sa	0059	1.4
	0722	-0.0
	1432	1.0
	1830	0.5
21 Su	0136	1.5
	0810	-0.2
	1536	1.0
	1919	0.6
22 M	0211	1.5
	0857	-0.3
	1639	1.0
	2008	0.7
23 Tu ●	0241	1.5
	0945	-0.3
	1741	1.0
	2100	0.7
24 W	0301	1.4
	1033	-0.3
	1844	1.1
	2156	0.8
25 Th	0310	1.3
	1122	-0.3
	1945	1.1
	2258	0.9
26 F	0329	1.2
	1212	-0.2
	2043	1.1
27 Sa	0007	0.9
	0358	1.1
	1301	-0.1
	2137	1.2
28 Su	0125	0.9
	0433	1.0
	1352	0.0
	2225	1.2
29 M	0250	0.8
	0516	0.8
	1442	0.1
	2308	1.2
30 Tu	0416	0.7
	1047	0.7
	1530	0.3
	2345	1.2

	Sun↑	Sun↓
15	0618	1906
30	0621	1909

JULY

Day	Time	ft
1 W ☾	0526	0.5
	1208	0.7
	1617	0.4
2 Th	0013	1.2
	0617	0.4
	1319	0.7
	1700	0.5
3 F	0024	1.2
	0659	0.2
	1424	0.7
	1738	0.6
4 Sa	0011	1.2
	0738	0.1
	1526	0.8
	1809	0.7
5 Su	0012	1.3
	0814	0.0
	1628	0.8
	1827	0.8
6 M	0034	1.4
	0849	-0.0
7 Tu	0104	1.4
	0922	-0.1
8 W	0138	1.4
	0954	-0.1
9 Th ○	0215	1.4
	1026	-0.1
	1826	0.9
	2012	0.8
10 F	0255	1.4
	1100	-0.1
	1856	0.9
	2127	0.9
11 Sa	0338	1.3
	1137	-0.1
	1934	1.0
	2247	0.8
12 Su	0430	1.2
	1218	-0.0
	2016	1.0
13 M	0012	0.8
	0546	1.1
	1303	0.1
	2100	1.1
14 Tu	0140	0.7
	0749	0.9
	1352	0.2
	2143	1.2
15 W	0305	0.6
	0937	0.9
	1443	0.3
	2227	1.3
16 Th ☽	0420	0.4
	1104	0.9
	1534	0.4
	2309	1.4
17 F	0522	0.2
	1219	0.9
	1626	0.5
	2352	1.4
18 Sa	0617	0.0
	1326	1.0
	1718	0.5
19 Su	0033	1.5
	0706	-0.1
	1428	1.0
	1809	0.6
20 M	0113	1.5
	0754	-0.2
	1527	1.0
	1901	0.7
21 Tu	0151	1.5
	0840	-0.3
	1625	1.0
	1953	0.7
22 W	0224	1.5
	0926	-0.3
	1722	1.1
	2047	0.8
23 Th ●	0249	1.4
	1012	-0.2
	1818	1.1
	2144	0.8
24 F	0258	1.3
	1058	-0.1
	1915	1.1
	2246	0.8
25 Sa	0316	1.2
	1144	-0.0
	2009	1.1
	2354	0.8
26 Su	0346	1.1
	1231	0.1
	2101	1.1
27 M	0108	0.8
	0423	0.9
	1319	0.2
	2147	1.2
28 Tu	0229	0.7
	0919	0.8
	1407	0.3
	2227	1.2
29 W	0350	0.6
	1055	0.7
	1454	0.5
	2258	1.2
30 Th	0456	0.5
	1214	0.7
	1540	0.6
	2310	1.2
31 F ☾	0546	0.3
	1324	0.8
	1622	0.7
	2256	1.2

	Sun↑	Sun↓
15	0625	1910
31	0628	1907

AUGUST

Day	Time	ft
1 Sa	0629	0.2
	1425	0.8
	1658	0.8
	2302	1.3
2 Su	0706	0.1
	1518	0.8
	1724	0.8
	2329	1.4
3 M	0742	0.1
	1556	0.9
	1740	0.8
4 Tu	0004	1.4
	0815	0.0
	1605	0.9
	1805	0.9
5 W	0044	1.5
	0846	-0.0
	1612	0.9
	1848	0.8
6 Th	0125	1.5
	0916	-0.0
	1635	0.9
	1942	0.8
7 F ○	0210	1.4
	0948	-0.0
	1709	1.0
	2040	0.8
8 Sa	0258	1.4
	1022	0.0
	1749	1.0
	2144	0.8
9 Su	0354	1.3
	1101	0.1
	1833	1.1
	2253	0.7
10 M	0504	1.2
	1143	0.2
	1920	1.1
11 Tu	0007	0.6
	0639	1.1
	1230	0.3
	2009	1.2
12 W	0124	0.5
	0821	1.0
	1320	0.4
	2058	1.3
13 Th	0241	0.4
	0951	1.0
	1414	0.5
	2148	1.4
14 F ☽	0353	0.3
	1109	1.0
	1509	0.6
	2236	1.4
15 Sa	0457	0.1
	1217	1.1
	1606	0.7
	2324	1.5
16 Su	0552	0.0
	1318	1.1
	1702	0.7
17 M	0011	1.5
	0643	-0.1
	1415	1.1
	1757	0.7
18 Tu	0056	1.5
	0730	-0.1
	1509	1.2
	1852	0.8
19 W	0140	1.5
	0816	-0.1
	1601	1.2
	1946	0.8
20 Th	0221	1.4
	0900	-0.0
	1651	1.2
	2041	0.8
21 F ●	0258	1.4
	0944	0.0
	1742	1.2
	2138	0.8
22 Sa	0323	1.2
	1028	0.1
	1831	1.2
	2238	0.8
23 Su	0331	1.1
	1112	0.3
	1920	1.2
	2341	0.8
24 M	0359	1.0
	1157	0.4
	2006	1.2
25 Tu	0049	0.7
	0446	0.9
	0754	0.9
	1242	0.5
26 W	0159	0.7
	0940	0.9
	1328	0.7
	2110	1.2
27 Th	0309	0.6
	1109	0.9
	1414	0.8
	2042	1.2
28 F	0410	0.5
	1226	0.9
	1500	0.9
	2042	1.2
29 Sa	0501	0.4
	1329	1.0
	1543	0.9
	2119	1.3
30 Su ☾	0545	0.3
	1416	1.0
	1620	1.0
	2204	1.4
31 M	0624	0.2
	1436	1.0
	1654	1.0
	2252	1.4

	Sun↑	Sun↓
15	0630	1901
31	0631	1852

AMUAY, VENEZUELA

HIGH & LOW WATER 1998 — Chart Datum — 11°45'N 70°13'W

Atlantic Standard Time (60°W) — Daylight Saving Time not observed

SEPTEMBER

Day	Time	ft	Day	Time	ft
1 Tu	0659	0.2	16 W	0045	1.5
	1435	1.0		0700	0.1
	1731	1.0		1444	1.4
	2341	1.5		1850	0.9
2 W	0731	0.2	17 Th	0137	1.5
	1443	1.1		0745	0.2
	1815	0.9		1529	1.4
				1946	0.8
3 Th	0031	1.5	18 F	0228	1.4
	0802	0.2		0829	0.3
	1507	1.1		1613	1.4
	1904	0.9		2040	0.8
4 F	0123	1.5	19 Sa	0320	1.3
	0833	0.2		0911	0.4
	1538	1.2		1655	1.3
	1956	0.8		2135	0.7
5 Sa	0218	1.5	20 Su	0416	1.2
	0907	0.2		0953	0.5
	1615	1.2		1732	1.3
	2052	0.8		● 2230	0.7
6 Su	0318	1.4	21 M	0525	1.1
	0944	0.3		1035	0.6
	1655	1.3		1759	1.3
	○ 2150	0.7		2327	0.7
7 M	0426	1.3	22 Tu	0652	1.1
	1024	0.4		1116	0.8
	1738	1.3		1731	1.3
	2253	0.6			
8 Tu	0546	1.2	23 W	0026	0.6
	1109	0.5		0829	1.0
	1825	1.3		1156	0.9
	2359	0.5		1709	1.3
9 W	0713	1.2	24 Th	0125	0.6
	1157	0.6		1006	1.1
	1915	1.4		1236	1.0
				1737	1.3
10 Th	0108	0.4	25 F	0224	0.5
	0839	1.2		1138	1.1
	1250	0.7		1315	1.1
	2009	1.4		1818	1.4
11 F	0218	0.3	26 Sa	0319	0.5
	0958	1.2		1256	1.2
	1348	0.8		1356	1.2
	2105	1.5		1908	1.4
12 Sa	0325	0.2	27 Su	0409	0.4
	1108	1.2		1338	1.2
	1449	0.9		1450	1.2
	☽ 2202	1.5		2008	1.4
13 Su	0427	0.2	28 M	0453	0.4
	1209	1.3		1337	1.2
	1552	0.9		1549	1.2
	2258	1.5		☾ 2114	1.4
14 M	0522	0.1	29 Tu	0532	0.3
	1305	1.3		1322	1.2
	1654	0.9		1643	1.1
	2353	1.5		2221	1.5
15 Tu	0613	0.1	30 W	0607	0.3
	1356	1.4		1327	1.3
	1753	0.9		1734	1.1
				2325	1.5

	Sun↑	Sun↓
15	0631	1842
30	0630	1832

OCTOBER

Day	Time	ft	Day	Time	ft
1 Th	0640	0.3	16 F	0140	1.4
	1349	1.3		0711	0.4
	1823	1.0		1454	1.5
				1948	0.8
2 F	0028	1.5	17 Sa	0238	1.3
	0713	0.4		0754	0.5
	1419	1.4		1529	1.5
	1913	0.9		2040	0.7
3 Sa	0131	1.5	18 Su	0339	1.2
	0748	0.4		0834	0.6
	1453	1.4		1555	1.4
	2004	0.7		2130	0.6
4 Su	0234	1.4	19 M	0444	1.2
	0825	0.5		0913	0.7
	1529	1.5		1556	1.4
	2056	0.6		2220	0.6
5 M	0342	1.4	20 Tu	0557	1.1
	0905	0.6		0950	0.9
	1607	1.5		1524	1.4
	○ 2151	0.5		● 2310	0.5
6 Tu	0454	1.3	21 W	0724	1.1
	0948	0.7		1023	1.0
	1647	1.5		1535	1.5
	2249	0.4			
7 W	0611	1.3	22 Th	0000	0.5
	1035	0.8		0908	1.1
	1729	1.5		1045	1.1
	2349	0.4		1601	1.5
8 Th	0730	1.3	23 F	0049	0.4
	1126	0.9		1635	1.5
	1816	1.6			
9 F	0051	0.3	24 Sa	0139	0.4
	0847	1.3		1715	1.5
	1222	1.0			
	1910	1.5			
10 Sa	0154	0.2	25 Su	0227	0.4
	0958	1.3		1803	1.5
	1326	1.1			
	2016	1.5			
11 Su	0256	0.2	26 M	0313	0.4
	1100	1.4		1906	1.4
	1434	1.1			
	2126	1.5			
12 M	0355	0.2	27 Tu	0355	0.3
	1156	1.4		1218	1.3
	1544	1.1		1526	1.2
	☽ 2234	1.5		2030	1.4
13 Tu	0449	0.2	28 W	0435	0.3
	1246	1.5		1217	1.3
	1652	1.0		1639	1.1
	2339	1.5		☾ 2159	1.3
14 W	0540	0.2	29 Th	0512	0.4
	1332	1.5		1238	1.4
	1755	0.9		1736	1.0
				2320	1.3
15 Th	0040	1.4	30 F	0549	0.4
	0627	0.3		1307	1.4
	1414	1.5		1827	0.8
	1853	0.9			
			31 Sa	0034	1.3
				0627	0.4
				1340	1.5
				1916	0.7

	Sun↑	Sun↓
15	0631	1823
31	0634	1816

NOVEMBER

Day	Time	ft	Day	Time	ft
1 Su	0143	1.3	16 M	0352	1.0
	0706	0.5		0754	0.7
	1414	1.6		1436	1.4
	2005	0.5		2121	0.4
2 M	0250	1.3	17 Tu	0500	1.0
	0747	0.6		0829	0.8
	1450	1.6		1415	1.4
	2056	0.4		2205	0.3
3 Tu	0359	1.3	18 W	0618	1.0
	0830	0.7		0856	0.9
	1526	1.6		1426	1.5
	2148	0.2		2249	0.2
4 W	0510	1.2	19 Th ●	1451	1.5
	0915	0.8		2331	0.2
	1601	1.6			
	○ 2241	0.2			
5 Th	0622	1.2	20 F	1522	1.5
	1004	0.9			
	1635	1.6			
	2337	0.1			
6 F	0735	1.2	21 Sa	0013	0.2
	1057	1.0		1556	1.5
	1706	1.6			
7 Sa	0033	0.1	22 Su	0055	0.2
	0845	1.3		1634	1.4
	1158	1.1			
	1734	1.5			
8 Su	0131	0.1	23 M	0136	0.1
	0949	1.3		1719	1.4
	1307	1.1			
	1809	1.4			
9 M	0228	0.1	24 Tu	0218	0.1
	1046	1.4		1118	1.1
	1422	1.1		1256	1.1
	2045	1.3		1822	1.3
10 Tu	0324	0.1	25 W	0259	0.2
	1137	1.4		1105	1.2
	1539	1.0		1504	1.0
	☽ 2215	1.3		2005	1.2
11 W	0416	0.2	26 Th	0341	0.2
	1223	1.5		1127	1.2
	1653	0.9		1628	0.9
	2331	1.2		☾ 2156	1.1
12 Th	0505	0.3	27 F	0422	0.2
	1305	1.5		1157	1.3
	1759	0.8		1731	0.7
				2326	1.1
13 F	0040	1.1	28 Sa	0505	0.3
	0552	0.3		1231	1.4
	1343	1.5		1823	0.5
	1856	0.7			
14 Sa	0144	1.1	29 Su	0042	1.0
	0635	0.4		0547	0.4
	1415	1.5		1306	1.5
	1948	0.6		1912	0.3
15 Su	0248	1.1	30 M	0152	1.0
	0716	0.6		0630	0.4
	1438	1.4		1342	1.5
	2035	0.4		2001	0.1

	Sun↑	Sun↓
15	0638	1814
30	0645	1815

DECEMBER

Day	Time	ft	Day	Time	ft
1 Tu	0259	1.0	16 W	0508	0.7
	0714	0.5		0740	0.7
	1418	1.5		1328	1.3
	2049	-0.0		2144	-0.1
2 W	0406	1.0	17 Th	0636	0.7
	0800	0.6		0743	0.7
	1453	1.6		1352	1.3
	2139	-0.1		2222	-0.1
3 Th	0513	1.0	18 F ●	1423	1.3
	0847	0.7		2259	-0.1
	1526	1.5			
	○ 2229	-0.2			
4 F	0621	1.0	19 Sa	1456	1.3
	0939	0.8		2336	-0.1
	1554	1.5			
	2321	-0.2			
5 Sa	0728	1.1	20 Su	1533	1.3
	1035	0.8			
	1613	1.4			
6 Su	0013	-0.2	21 M	0012	-0.1
	0832	1.1		1614	1.2
	1139	0.9			
	1633	1.3			
7 M	0106	-0.2	22 Tu	0049	-0.1
	0931	1.1		0930	0.8
	1251	0.9		1052	0.8
	1704	1.2		1704	1.1
8 Tu	0200	-0.1	23 W	0129	-0.1
	1024	1.2		0938	0.9
	1411	0.9		1258	0.8
	1746	1.0		1818	1.0
9 W	0252	-0.0	24 Th	0211	-0.0
	1112	1.2		1007	0.9
	1534	0.8		1441	0.7
	1850	0.9		2019	0.8
10 Th	0343	0.1	25 F	0256	0.0
	1156	1.3		1042	1.0
	1653	0.6		1607	0.5
	☽ 2329	0.8		2208	0.8
11 F	0432	0.2	26 Sa	0342	0.1
	1235	1.3		1119	1.1
	1759	0.5		1714	0.3
				☾ 2334	0.7
12 Sa	0042	0.8	27 Su	0429	0.1
	0517	0.3		1158	1.2
	1308	1.3		1810	0.1
	1853	0.4			
13 Su	0149	0.8	28 M	0048	0.7
	0600	0.4		0515	0.2
	1332	1.3		1237	1.3
	1940	0.2		1900	-0.1
14 M	0254	0.7	29 Tu	0155	0.8
	0639	0.5		0602	0.3
	1336	1.2		1315	1.3
	2023	0.1		1948	-0.2
15 Tu	0359	0.7	30 W	0259	0.8
	0714	0.6		0649	0.3
	1321	1.2		1353	1.4
	2104	0.0		2036	-0.4
			31 Th	0401	0.8
				0738	0.4
				1431	1.3
				2124	-0.4

	Sun↑	Sun↓
15	0653	1820
31	0701	1828

MALECÓN-ZAPARA, VENEZUELA

HIGH & LOW WATER 1998 — Chart Datum — 11°00'N 71°35'W

Atlantic Standard Time (60°W) — Daylight Saving Time not observed

JANUARY

Date	Time	ft	Date	Time	ft
1 Th	0045	0.7	16 F	0222	1.2
	0722	4.0		0851	3.8
	1301	1.9		1501	2.0
	1908	4.5		2030	3.9
2 F	0138	0.7	17 Sa	0309	1.4
	0818	4.1		0936	3.8
	1406	1.8		1552	2.0
	2009	4.4		2121	3.8
3 Sa	0234	0.8	18 Su	0353	1.6
	0915	4.3		1017	3.8
	1513	1.6		1638	1.9
	2113	4.4		2210	3.7
4 Su	0332	0.9	19 M	0432	1.7
	1011	4.4		1055	3.8
	1618	1.4		1719	1.8
	2218	4.3		2258	3.7
5 M	0431	1.0	20 Tu	0507	1.7
	1105	4.5		1131	3.9
	1721	1.1		1755	1.6
	☾ 2324	4.3		☽ 2344	3.7
6 Tu	0530	1.1	21 W	0540	1.7
	1159	4.6		1207	4.0
	1821	0.9		1829	1.4
7 W	0028	4.3	22 Th	0029	3.7
	0629	1.2		0614	1.7
	1252	4.7		1243	4.1
	1918	0.6		1903	1.2
8 Th	0130	4.2	23 F	0113	3.7
	0727	1.3		0650	1.7
	1344	4.7		1319	4.2
	2014	0.5		1939	1.0
9 F	0230	4.2	24 Sa	0156	3.8
	0824	1.4		0729	1.6
	1435	4.7		1357	4.3
	2109	0.4		2018	0.8
10 Sa	0328	4.1	25 Su	0239	3.8
	0921	1.5		0811	1.6
	1526	4.7		1436	4.4
	2203	0.4		2100	0.6
11 Su	0425	4.1	26 M	0323	3.9
	1018	1.6		0857	1.5
	1616	4.6		1519	4.5
	2256	0.5		2145	0.5
12 M	0521	4.0	27 Tu	0409	3.9
	1115	1.7		0946	1.5
	1707	4.4		1605	4.6
	○ 2349	0.6		2233	0.4
13 Tu	0617	3.9	28 W	0458	4.0
	1212	1.9		1040	1.5
	1757	4.3		1655	4.6
				● 2323	0.5
14 W	0041	0.8	29 Th	0550	4.0
	0711	3.8		1139	1.4
	1309	2.0		1750	4.6
	1848	4.1			
15 Th	0132	1.0	30 F	0017	0.5
	0803	3.8		0645	4.1
	1406	2.0		1242	1.4
	1939	4.0		1850	4.5
			31 Sa	0114	0.6
				0744	4.2
				1348	1.3
				1956	4.4

	Sun↑	Sun↓
15	0710	1843
31	0711	1850

FEBRUARY

Date	Time	ft	Date	Time	ft
1 Su	0214	0.8	16 M	0310	1.7
	0845	4.2		0925	3.5
	1456	1.1		1546	1.6
	2104	4.3		2138	3.5
2 M	0318	0.9	17 Tu	0350	1.8
	0947	4.3		1006	3.5
	1603	0.9		1625	1.5
	2214	4.2		2227	3.5
3 Tu	0422	1.0	18 W	0427	1.8
	1047	4.4		1047	3.5
	1708	0.6		1702	1.3
	☾ 2322	4.2		2314	3.5
4 W	0526	1.1	19 Th	0503	1.8
	1146	4.5		1126	3.6
	1809	0.5		1739	1.2
				☽ 2358	3.6
5 Th	0027	4.2	20 F	0540	1.7
	0628	1.1		1204	3.8
	1242	4.5		1817	0.9
	1907	0.3			
6 F	0128	4.2	21 Sa	0040	3.6
	0727	1.2		0619	1.6
	1335	4.5		1242	4.0
	2003	0.3		1857	0.7
7 Sa	0225	4.1	22 Su	0122	3.7
	0823	1.2		0701	1.5
	1426	4.5		1323	4.1
	2056	0.3		1940	0.6
8 Su	0318	4.0	23 M	0203	3.8
	0916	1.3		0746	1.3
	1515	4.4		1405	4.3
	2147	0.4		2025	0.4
9 M	0409	3.9	24 Tu	0246	3.9
	1008	1.4		0834	1.2
	1603	4.3		1451	4.5
	2237	0.6		2112	0.4
10 Tu	0458	3.8	25 W	0332	4.0
	1059	1.5		0926	1.1
	1649	4.2		1541	4.6
	2325	0.8		2202	0.4
11 W	0544	3.7	26 Th	0421	4.1
	1149	1.6		1022	1.0
	1735	4.0		1635	4.6
	○			● 2255	0.4
12 Th	0012	1.1	27 F	0513	4.1
	0629	3.6		1121	0.9
	1239	1.7		1734	4.5
	1821	3.9		2351	0.6
13 F	0058	1.3	28 Sa	0610	4.2
	0714	3.5		1225	0.8
	1328	1.8		1838	4.4
	1909	3.7			
14 Sa	0144	1.5			
	0758	3.5			
	1417	1.8			
	1958	3.6			
15 Su	0228	1.6			
	0842	3.5			
	1503	1.7			
	2048	3.5			

	Sun↑	Sun↓
15	0708	1855
28	0702	1857

MARCH

Date	Time	ft	Date	Time	ft
1 Su	0053	0.7	16 M	0142	1.8
	0711	4.1		0745	3.3
	1331	0.6		1409	1.4
	1947	4.3		2022	3.4
2 M	0158	0.9	17 Tu	0226	1.9
	0816	4.1		0830	3.2
	1439	0.5		1449	1.3
	2058	4.2		2112	3.4
3 Tu	0307	1.0	18 W	0309	1.9
	0923	4.2		0916	3.2
	1546	0.4		1530	1.2
	2210	4.2		2201	3.4
4 W	0416	1.1	19 Th	0351	1.9
	1028	4.2		1000	3.3
	1651	0.2		1610	1.1
	2317	4.2		2246	3.5
5 Th	0523	1.1	20 F	0431	1.8
	1130	4.2		1043	3.4
	1752	0.2		1652	0.9
	☾			2329	3.6
6 F	0020	4.2	21 Sa	0512	1.7
	0625	1.0		1124	3.6
	1228	4.3		1734	0.8
	1850	0.2		☽	
7 Sa	0118	4.1	22 Su	0009	3.7
	0722	1.0		0555	1.5
	1322	4.3		1207	3.8
	1944	0.2		1818	0.6
8 Su	0210	4.1	23 M	0049	3.8
	0815	1.0		0640	1.3
	1413	4.2		1252	4.0
	2036	0.4		1904	0.5
9 M	0258	4.0	24 Tu	0130	4.0
	0905	1.1		0727	1.1
	1500	4.2		1339	4.2
	2124	0.6		1951	0.4
10 Tu	0342	3.9	25 W	0213	4.1
	0952	1.2		0817	0.9
	1545	4.1		1429	4.4
	2210	0.8		2041	0.4
11 W	0422	3.7	26 Th	0258	4.2
	1038	1.3		0911	0.7
	1628	3.9		1523	4.4
	2254	1.1		2133	0.5
12 Th	0501	3.6	27 F	0347	4.3
	1122	1.3		1007	0.5
	1712	3.8		1621	4.4
	2336	1.3		● 2229	0.6
13 F	0539	3.5	28 Sa	0441	4.3
	1204	1.4		1107	0.3
	1756	3.7		1724	4.4
	○			2329	0.8
14 Sa	0018	1.6	29 Su	0538	4.3
	0619	3.4		1209	0.2
	1246	1.4		1830	4.3
	1842	3.5			
15 Su	0059	1.7	30 M	0035	1.0
	0700	3.3		0641	4.2
	1327	1.4		1315	0.1
	1931	3.5		1941	4.2
			31 Tu	0145	1.1
				0749	4.1
				1421	0.1
				2053	4.2

	Sun↑	Sun↓
15	0654	1858
31	0644	1858

APRIL

Date	Time	ft	Date	Time	ft
1 W	0258	1.2	16 Th	0229	2.1
	0858	4.0		0824	3.2
	1527	0.1		1442	0.9
	2203	4.2		2136	3.5
2 Th	0410	1.2	17 F	0317	2.1
	1006	4.0		0911	3.3
	1631	0.1		1526	0.9
	2308	4.3		2220	3.6
3 F	0516	1.1	18 Sa	0403	2.0
	1110	4.1		0959	3.4
	1731	0.1		1611	0.8
	☾			2301	3.8
4 Sa	0006	4.3	19 Su	0449	1.8
	0616	1.0		1047	3.6
	1209	4.1		1657	0.7
	1827	0.2		☽ 2340	3.9
5 Su	0059	4.2	20 M	0536	1.5
	0711	1.0		1135	3.8
	1303	4.1		1744	0.6
	1920	0.4			
6 M	0146	4.2	21 Tu	0020	4.1
	0801	0.9		0624	1.3
	1353	4.0		1226	3.9
	2008	0.7		1832	0.6
7 Tu	0228	4.1	22 W	0101	4.3
	0848	1.0		0714	1.0
	1439	3.9		1318	4.1
	2053	0.9		1922	0.6
8 W	0305	4.0	23 Th	0144	4.4
	0931	1.0		0806	0.7
	1522	3.8		1414	4.2
	2135	1.2		2014	0.7
9 Th	0340	3.8	24 F	0231	4.5
	1012	1.0		0900	0.4
	1605	3.7		1512	4.3
	2214	1.4		2109	0.8
10 F	0414	3.7	25 Sa	0321	4.5
	1050	1.1		0956	0.2
	1647	3.6		1613	4.3
	2251	1.7		2208	1.0
11 Sa	0449	3.6	26 Su	0414	4.5
	1127	1.1		1055	-0.0
	1732	3.6		1717	4.3
	○ 2329	1.8		● 2311	1.2
12 Su	0526	3.5	27 M	0513	4.4
	1203	1.1		1156	-0.1
	1819	3.5		1825	4.3
13 M	0008	2.0	28 Tu	0020	1.4
	0607	3.4		0616	4.3
	1240	1.1		1259	-0.1
	1908	3.5		1935	4.3
14 Tu	0052	2.1	29 W	0134	1.5
	0650	3.3		0724	4.1
	1319	1.1		1402	-0.1
	1958	3.5		2045	4.3
15 W	0139	2.1	30 Th	0248	1.5
	0736	3.2		0833	4.0
	1359	1.0		1506	0.0
	2048	3.5		2151	4.3

	Sun↑	Sun↓
15	0636	1858
30	0629	1859

MALECÓN–ZAPARA, VENEZUELA

HIGH & LOW WATER 1998 — Chart Datum — 11°00'N 71°35'W

Atlantic Standard Time (60°W) — Daylight Saving Time not observed

MAY

Date	Day	Time	ft	Time	ft	Time	ft	Time	ft
1	F	0359	1.4	0941	4.0	1607	0.2	2252	4.4
2	Sa	0503	1.3	1045	3.9	1705	0.3	2346	4.4
3	Su ☾	0601	1.2	1143	3.9	1759	0.6		
4	M	0033	4.4	0654	1.1	1237	3.9	1849	0.8
5	Tu	0115	4.3	0741	1.0	1326	3.8	1934	1.1
6	W	0151	4.2	0825	1.0	1412	3.8	2015	1.3
7	Th	0225	4.1	0904	1.0	1455	3.7	2052	1.6
8	F	0257	4.1	0941	0.9	1538	3.6	2127	1.8
9	Sa	0330	4.0	1015	0.9	1621	3.6	2200	1.9
10	Su	0404	3.9	1049	0.9	1706	3.6	2236	2.1
11	M	0441	3.8	1123	0.9	1753	3.6	○ 2317	2.2
12	Tu	0520	3.6	1159	0.8	1842	3.6		
13	W	0003	2.3	0602	3.6	1238	0.8	1932	3.6
14	Th	0054	2.3	0647	3.5	1319	0.8	2021	3.7
15	F	0148	2.3	0735	3.5	1404	0.8	2108	3.8
16	Sa	0243	2.2	0827	3.5	1449	0.7	2152	3.9
17	Su	0337	2.1	0920	3.6	1537	0.7	2234	4.0
18	M	0429	1.9	1015	3.7	1625	0.7	2314	4.2
19	Tu	0521	1.6	1111	3.8	1715	0.7	☽ 2355	4.4
20	W	0612	1.2	1208	3.9	1806	0.8		
21	Th	0038	4.6	0704	0.9	1306	4.0	1859	0.9
22	F	0123	4.7	0757	0.5	1405	4.1	1954	1.0
23	Sa	0211	4.8	0852	0.2	1506	4.2	2052	1.2
24	Su	0302	4.8	0947	-0.0	1609	4.2	2153	1.4
25	M	0356	4.7	1044	-0.1	1713	4.2	● 2259	1.6
26	Tu	0454	4.5	1143	-0.2	1820	4.2		
27	W	0009	1.7	0555	4.4	1243	-0.1	1927	4.3
28	Th	0122	1.8	0700	4.2	1343	0.0	2033	4.3
29	F	0234	1.7	0806	4.0	1442	0.2	2134	4.4
30	Sa	0342	1.7	0912	3.9	1540	0.5	2229	4.4
31	Su	0444	1.5	1014	3.8	1635	0.7	2318	4.4

	Sun↑	Sun↓
15	0624	1902
31	0623	1906

JUNE

Date	Day	Time	ft	Time	ft	Time	ft	Time	ft
1	M ☾	0540	1.4	1112	3.8	1726	1.0		
2	Tu	0001	4.4	0630	1.3	1205	3.7	1812	1.2
3	W	0038	4.4	0715	1.2	1254	3.7	1853	1.5
4	Th	0112	4.4	0756	1.1	1340	3.6	1930	1.7
5	F	0144	4.3	0832	1.0	1424	3.6	2004	1.9
6	Sa	0216	4.3	0905	0.9	1507	3.6	2036	2.0
7	Su	0250	4.2	0937	0.8	1551	3.6	2110	2.1
8	M	0325	4.2	1010	0.7	1636	3.7	2149	2.2
9	Tu	0402	4.1	1044	0.7	1723	3.7	2232	2.3
10	W	0441	4.0	1121	0.6	1811	3.7	○ 2321	2.3
11	Th	0522	3.9	1202	0.6	1859	3.8		
12	F	0014	2.4	0607	3.9	1245	0.6	1948	3.9
13	Sa	0111	2.4	0657	3.8	1331	0.6	2035	4.0
14	Su	0212	2.3	0752	3.8	1419	0.7	2121	4.2
15	M	0312	2.1	0852	3.8	1509	0.8	2206	4.3
16	Tu	0411	1.8	0953	3.8	1601	0.9	2250	4.5
17	W	0508	1.5	1056	3.9	1654	1.0	☽ 2335	4.7
18	Th	0603	1.1	1159	3.9	1749	1.1		
19	F	0021	4.8	0657	0.7	1301	4.0	1845	1.2
20	Sa	0109	4.9	0751	0.4	1403	4.1	1943	1.4
21	Su	0159	4.9	0845	0.1	1505	4.1	2043	1.6
22	M	0250	4.9	0939	-0.0	1607	4.2	2146	1.7
23	Tu	0344	4.8	1034	-0.1	1709	4.2	● 2250	1.8
24	W	0440	4.6	1130	-0.0	1812	4.2	2358	1.9
25	Th	0537	4.4	1226	0.1	1914	4.3		
26	F	0106	2.0	0637	4.2	1322	0.3	2013	4.3
27	Sa	0214	2.0	0738	4.0	1417	0.6	2109	4.3
28	Su	0319	1.9	0840	3.9	1511	0.9	2200	4.4
29	M	0419	1.8	0940	3.7	1602	1.1	2244	4.4
30	Tu	0513	1.6	1037	3.6	1649	1.4	2323	4.4

	Sun↑	Sun↓
15	0624	1910
30	0628	1913

JULY

Date	Day	Time	ft	Time	ft	Time	ft	Time	ft
1	W	0601	1.5	1130	3.6	1732	1.6	☾ 2358	4.4
2	Th	0643	1.4	1219	3.5	1809	1.8		
3	F	0031	4.4	0720	1.2	1306	3.5	1844	1.9
4	Sa	0104	4.4	0754	1.1	1350	3.6	1916	2.0
5	Su	0138	4.4	0825	0.9	1433	3.6	1950	2.1
6	M	0213	4.4	0856	0.8	1517	3.7	2028	2.1
7	Tu	0249	4.4	0930	0.6	1601	3.8	2110	2.2
8	W	0327	4.4	1006	0.6	1646	3.8	2155	2.2
9	Th	0407	4.4	1046	0.5	1732	3.9	○ 2246	2.2
10	F	0450	4.3	1128	0.5	1820	4.0	2341	2.2
11	Sa	0537	4.3	1214	0.5	1908	4.1		
12	Su	0041	2.2	0630	4.2	1302	0.6	1958	4.2
13	M	0144	2.1	0729	4.1	1353	0.7	2047	4.4
14	Tu	0249	1.8	0834	4.0	1447	0.9	2138	4.5
15	W	0353	1.5	0941	3.9	1543	1.0	2228	4.7
16	Th	0454	1.2	1050	3.9	1642	1.2	☽ 2319	4.8
17	F	0553	0.8	1157	4.0	1741	1.3		
18	Sa	0010	4.9	0649	0.5	1302	4.0	1842	1.5
19	Su	0101	4.9	0744	0.3	1404	4.1	1942	1.6
20	M	0152	4.9	0837	0.1	1504	4.1	2042	1.7
21	Tu	0244	4.9	0930	0.0	1603	4.2	2142	1.8
22	W	0335	4.8	1022	0.1	1700	4.2	2243	1.9
23	Th	0428	4.6	1115	0.3	1756	4.2	● 2344	2.0
24	F	0521	4.4	1206	0.5	1852	4.2		
25	Sa	0047	2.0	0615	4.2	1258	0.8	1945	4.2
26	Su	0149	2.0	0710	4.0	1349	1.0	2035	4.2
27	M	0250	2.0	0807	3.8	1438	1.3	2121	4.2
28	Tu	0347	1.9	0904	3.6	1525	1.5	2202	4.2
29	W	0439	1.8	1000	3.5	1609	1.7	2240	4.2
30	Th	0524	1.6	1054	3.5	1649	1.9	2316	4.2
31	F	0603	1.5	1144	3.5	1726	2.0	☾ 2351	4.3

	Sun↑	Sun↓
15	0632	1914
31	0635	1911

AUGUST

Date	Day	Time	ft	Time	ft	Time	ft	Time	ft
1	Sa	0638	1.3	1230	3.5	1800	2.0		
2	Su	0026	4.3	0709	1.1	1315	3.6	1835	2.1
3	M	0102	4.4	0740	0.9	1358	3.7	1912	2.1
4	Tu	0138	4.5	0814	0.7	1440	3.8	1953	2.0
5	W	0215	4.5	0850	0.6	1522	3.9	2037	2.0
6	Th	0254	4.6	0929	0.5	1605	4.0	2125	2.0
7	F	0337	4.6	1011	0.4	1650	4.1	○ 2218	2.0
8	Sa	0423	4.6	1055	0.5	1737	4.2	2314	1.9
9	Su	0514	4.5	1144	0.5	1826	4.3		
10	M	0016	1.8	0611	4.4	1235	0.7	1918	4.4
11	Tu	0121	1.6	0714	4.2	1330	0.9	2014	4.5
12	W	0228	1.4	0823	4.1	1430	1.0	2111	4.6
13	Th	0335	1.1	0935	4.0	1532	1.2	2208	4.7
14	F	0439	0.8	1047	4.0	1637	1.4	☽ 2305	4.8
15	Sa	0540	0.6	1156	4.1	1741	1.5		
16	Su	0000	4.9	0638	0.3	1300	4.1	1843	1.5
17	M	0054	4.9	0733	0.2	1401	4.2	1943	1.6
18	Tu	0146	4.9	0826	0.2	1457	4.2	2041	1.7
19	W	0237	4.8	0917	0.2	1551	4.2	2137	1.7
20	Th	0326	4.7	1007	0.4	1642	4.2	2232	1.8
21	F	0415	4.5	1056	0.6	1731	4.2	● 2328	1.9
22	Sa	0504	4.3	1143	0.9	1819	4.1		
23	Su	0024	2.0	0553	4.1	1230	1.2	1905	4.1
24	M	0120	2.0	0644	3.8	1317	1.5	1949	4.0
25	Tu	0214	2.0	0738	3.7	1402	1.7	2032	4.0
26	W	0307	1.9	0832	3.5	1446	1.9	2114	4.0
27	Th	0355	1.8	0927	3.5	1528	2.0	2154	4.0
28	F	0437	1.6	1020	3.4	1608	2.1	2233	4.0
29	Sa	0514	1.5	1110	3.5	1645	2.1	2311	4.1
30	Su	0548	1.3	1156	3.5	1721	2.1	☾ 2348	4.2
31	M	0620	1.1	1240	3.6	1800	2.1		

	Sun↑	Sun↓
15	0637	1906
31	0637	1857

MALECÓN-ZAPARA, VENEZUELA

HIGH & LOW WATER 1998 — Chart Datum — 11°00'N 71°35'W

Atlantic Standard Time (60°W) — Daylight Saving Time not observed

SEPTEMBER

Day	Time	ft	Day	Time	ft
1	0025	4.3	16	0137	4.8
	0654	0.9		0809	0.4
Tu	1321	3.8	W	1441	4.3
	1840	2.0		2034	1.6
2	0103	4.5	17	0226	4.7
	0731	0.7		0859	0.5
W	1401	3.9	Th	1529	4.3
	1924	1.9		2126	1.6
3	0142	4.6	18	0314	4.5
	0810	0.6		0946	0.8
Th	1441	4.0	F	1614	4.2
	2011	1.8		2217	1.7
4	0225	4.7	19	0400	4.3
	0852	0.5		1032	1.1
F	1523	4.2	Sa	1656	4.2
	2101	1.7		2307	1.7
5	0310	4.7	20	0446	4.1
	0936	0.5		1115	1.3
Sa	1607	4.3	Su	1737	4.1
	2155	1.6		● 2356	1.8
6	0401	4.7	21	0532	3.9
	1024	0.6		1158	1.6
Su	1655	4.4	M	1817	4.0
	○ 2253	1.4			
7	0456	4.6	22	0045	1.8
	1116	0.7		0620	3.8
M	1746	4.5	Tu	1239	1.8
	2356	1.3		1857	3.9
8	0557	4.5	23	0132	1.8
	1211	0.9		0711	3.6
Tu	1842	4.6	W	1321	2.0
				1939	3.9
9	0101	1.1	24	0218	1.8
	0704	4.3		0804	3.5
W	1312	1.1	Th	1403	2.2
	1942	4.6		2022	3.8
10	0209	0.9	25	0302	1.7
	0816	4.2		0858	3.5
Th	1417	1.3	F	1445	2.3
	2045	4.7		2105	3.8
11	0316	0.7	26	0342	1.6
	0931	4.2		0951	3.5
F	1526	1.4	Sa	1527	2.3
	2148	4.7		2148	3.9
12	0421	0.5	27	0419	1.4
	1043	4.2		1039	3.6
Sa	1634	1.5	Su	1607	2.3
	☽ 2250	4.8		2228	4.0
13	0523	0.4	28	0455	1.3
	1151	4.2		1124	3.7
Su	1740	1.5	M	1648	2.2
	2349	4.8		☾ 2308	4.1
14	0622	0.3	29	0532	1.1
	1253	4.3		1205	3.8
M	1842	1.5	Tu	1730	2.1
				2348	4.3
15	0044	4.8	30	0610	0.9
	0717	0.3		1244	3.9
Tu	1349	4.3	W	1814	2.0
	1940	1.5			

	Sun↑	Sun↓
15	0636	1848
30	0636	1838

OCTOBER

Day	Time	ft	Day	Time	ft
1	0028	4.4	16	0210	4.5
	0650	0.8		0835	0.9
Th	1323	4.1	F	1500	4.4
	1901	1.8		2110	1.5
2	0112	4.6	17	0256	4.4
	0732	0.7		0920	1.2
F	1402	4.3	Sa	1538	4.3
	1950	1.6		2156	1.5
3	0158	4.7	18	0341	4.2
	0817	0.6		1002	1.5
Sa	1444	4.5	Su	1615	4.2
	2042	1.4		2241	1.6
4	0249	4.7	19	0425	4.0
	0905	0.7		1041	1.7
Su	1530	4.6	M	1651	4.1
	2138	1.2		2324	1.6
5	0344	4.7	20	0510	3.8
	0956	0.8		1119	2.0
M	1619	4.7	Tu	1728	4.0
	○ 2236	1.0		●	
6	0443	4.6	21	0005	1.6
	1051	0.9		0557	3.7
Tu	1713	4.7	W	1156	2.1
	2338	0.8		1807	3.9
7	0547	4.5	22	0046	1.6
	1151	1.1		0646	3.6
W	1812	4.7	Th	1235	2.3
				1849	3.9
8	0043	0.7	23	0127	1.6
	0657	4.4		0738	3.6
Th	1256	1.3	F	1316	2.4
	1915	4.7		1933	3.8
9	0150	0.6	24	0207	1.5
	0810	4.3		0831	3.6
F	1407	1.5	Sa	1400	2.4
	2021	4.7		2017	3.8
10	0257	0.4	25	0247	1.4
	0924	4.3		0922	3.6
Sa	1519	1.6	Su	1446	2.5
	2128	4.7		2101	3.9
11	0402	0.4	26	0327	1.3
	1034	4.4		1009	3.7
Su	1629	1.6	M	1532	2.4
	2232	4.7		2145	4.0
12	0504	0.3	27	0407	1.2
	1139	4.4		1052	3.8
M	1734	1.5	Tu	1618	2.3
	☽ 2332	4.7		2228	4.1
13	0602	0.4	28	0448	1.1
	1237	4.5		1131	4.0
Tu	1834	1.5	W	1704	2.2
				☾ 2312	4.3
14	0029	4.7	29	0530	1.0
	0657	0.5		1209	4.1
W	1330	4.5	Th	1752	1.9
	1929	1.5		2358	4.4
15	0121	4.6	30	0614	0.9
	0748	0.7		1248	4.3
Th	1417	4.5	F	1842	1.7
	2021	1.5			
			31	0046	4.5
				0659	0.9
			Sa	1328	4.5
				1934	1.4

	Sun↑	Sun↓
15	0636	1829
31	0638	1823

NOVEMBER

Day	Time	ft	Day	Time	ft
1	0138	4.6	16	0317	4.0
	0747	0.9		0925	1.8
Su	1412	4.7	M	1533	4.3
	2028	1.1		2209	1.4
2	0233	4.6	17	0400	3.9
	0838	0.9		1001	2.0
M	1459	4.8	Tu	1608	4.2
	2124	0.8		2247	1.4
3	0331	4.6	18	0444	3.8
	0933	1.1		1034	2.1
Tu	1551	4.9	W	1644	4.1
	2222	0.6		2324	1.4
4	0433	4.5	19	0530	3.7
	1031	1.2		1109	2.3
W	1647	4.9	Th	1723	4.1
	○ 2324	0.5		●	
5	0539	4.5	20	0001	1.4
	1135	1.4		0618	3.7
Th	1747	4.9	F	1147	2.4
				1804	4.0
6	0027	0.4	21	0039	1.4
	0649	4.4		0709	3.7
F	1244	1.6	Sa	1229	2.5
	1852	4.8		1847	4.0
7	0132	0.3	22	0119	1.3
	0801	4.4		0759	3.7
Sa	1356	1.7	Su	1317	2.5
	1959	4.7		1931	4.0
8	0237	0.4	23	0200	1.3
	0912	4.4		0849	3.7
Su	1508	1.7	M	1407	2.5
	2106	4.7		2017	4.0
9	0340	0.4	24	0242	1.2
	1019	4.5		0935	3.8
M	1617	1.7	Tu	1459	2.4
	2210	4.7		2104	4.0
10	0441	0.5	25	0326	1.2
	1120	4.5		1018	4.0
Tu	1720	1.6	W	1551	2.3
	☽ 2311	4.6		2152	4.1
11	0539	0.6	26	0411	1.1
	1215	4.6		1058	4.1
W	1818	1.5	Th	1643	2.1
				☾ 2242	4.2
12	0007	4.5	27	0457	1.1
	0632	0.8		1138	4.3
Th	1303	4.6	F	1735	1.8
	1912	1.5		2335	4.3
13	0100	4.4	28	0544	1.1
	0721	1.1		1219	4.5
F	1346	4.5	Sa	1828	1.5
	2001	1.4			
14	0148	4.3	29	0029	4.4
	0806	1.3		0634	1.1
Sa	1424	4.5	Su	1302	4.7
	2047	1.4		1922	1.1
15	0234	4.2	30	0125	4.4
	0848	1.6		0725	1.1
Su	1459	4.4	M	1348	4.9
	2130	1.4		2016	0.8

	Sun↑	Sun↓
15	0643	1820
30	0650	1822

DECEMBER

Day	Time	ft	Day	Time	ft
1	0223	4.5	16	0330	3.8
	0820	1.2		0917	2.0
Tu	1438	5.0	W	1529	4.3
	2113	0.6		2206	1.2
2	0323	4.5	17	0413	3.8
	0917	1.3		0950	2.1
W	1532	5.0	Th	1606	4.2
	2211	0.4		2240	1.2
3	0426	4.4	18	0457	3.7
	1018	1.5		1025	2.2
Th	1629	5.0	F	1645	4.2
	○ 2310	0.3		● 2317	1.1
4	0532	4.4	19	0544	3.7
	1123	1.6		1104	2.3
F	1729	4.9	Sa	1725	4.2
				2355	1.1
5	0012	0.3	20	0632	3.7
	0639	4.4		1149	2.3
Sa	1231	1.7	Su	1807	4.1
	1832	4.8			
6	0114	0.3	21	0036	1.1
	0748	4.4		0721	3.7
Su	1342	1.8	M	1239	2.3
	1937	4.7		1852	4.1
7	0216	0.4	22	0120	1.1
	0855	4.4		0810	3.8
M	1452	1.8	Tu	1333	2.3
	2042	4.6		1940	4.1
8	0318	0.6	23	0205	1.1
	0958	4.4		0857	3.9
Tu	1559	1.7	W	1429	2.2
	2145	4.5		2031	4.1
9	0417	0.8	24	0252	1.1
	1055	4.5		0942	4.0
W	1700	1.6	Th	1527	2.1
	2246	4.4		2126	4.1
10	0513	1.0	25	0341	1.1
	1146	4.5		1026	4.2
Th	1757	1.5	F	1625	1.8
	☽ 2342	4.3		2223	4.1
11	0604	1.2	26	0432	1.1
	1231	4.5		1111	4.3
F	1848	1.5	Sa	1721	1.5
				☾ 2321	4.2
12	0033	4.2	27	0524	1.2
	0651	1.4		1156	4.5
Sa	1310	4.5	Su	1817	1.1
	1935	1.4			
13	0121	4.1	28	0020	4.2
	0734	1.6		0618	1.2
Su	1345	4.4	M	1244	4.7
	2017	1.3		1912	0.8
14	0205	4.0	29	0119	4.3
	0811	1.8		0713	1.2
M	1419	4.4	Tu	1334	4.8
	2056	1.3		2007	0.5
15	0247	3.9	30	0218	4.3
	0846	1.9		0810	1.3
Tu	1453	4.3	W	1426	4.9
	2132	1.2		2103	0.3
			31	0318	4.3
				0909	1.4
			Th	1520	4.9
				2200	0.2

	Sun↑	Sun↓
15	0657	1827
31	0705	1835

CRISTÓBAL–COLÓN, PANAMA

HIGH & LOW WATER 1998 — Chart Datum — 9°21'N 79°55'W

Eastern Standard Time (75°W) — Daylight Saving Time not observed

JANUARY

Day	Time	ft	Time	ft	Time	ft	Time	ft
1 Th	0000	-0.1	0500	0.3	1416	1.2	2216	-0.3
2 F	0509	0.5	0828	0.4	1509	1.1	2247	-0.3
3 Sa	0538	0.6	1009	0.4	1605	0.9	2318	-0.3
4 Su	0613	0.8	1137	0.3	1701	0.8	2350	-0.3
5 M ☾	0651	1.0	1257	0.2	1800	0.6		
6 Tu	0023	-0.3	0731	1.1	1411	0.1	1902	0.5
7 W	0057	-0.2	0814	1.2	1522	0.0	2006	0.4
8 Th	0131	-0.2	0859	1.3	1629	-0.1	2116	0.2
9 F	0206	-0.1	0945	1.4	1735	-0.2	2233	0.2
10 Sa	0241	-0.0	1032	1.4	1838	-0.2		
11 Su	0001	0.1	0315	0.1	1120	1.3	1936	-0.2
12 M	0142	0.2	0348	0.1	1209	1.2	○ 2029	-0.2
13 Tu	1257	1.1	2115	-0.2				
14 W	0450	0.3	1345	1.0	2155	-0.2		
15 Th	0516	0.4	0724	0.4	1431	0.9	2228	-0.2
16 F	0540	0.5	0905	0.4	1517	0.8	2255	-0.2
17 Sa	0603	0.6	1031	0.4	1601	0.7	2318	-0.1
18 Su	0625	0.7	1144	0.4	1644	0.6	2339	-0.1
19 M	0648	0.8	1249	0.3	1729	0.5	2358	-0.1
20 Tu ☽	0711	0.8	1347	0.2	1814	0.4		
21 W	0016	-0.1	0738	0.9	1442	0.1	1900	0.3
22 Th	0034	-0.1	0806	1.0	1536	0.1	1948	0.2
23 F	0054	-0.0	0838	1.1	1628	0.0	2040	0.2
24 Sa	0115	-0.0	0913	1.1	1720	-0.1	2137	0.1
25 Su	0140	-0.0	0952	1.2	1810	-0.1	2245	0.1
26 M	0212	-0.0	1034	1.2	1857	-0.2		
27 Tu	0003	0.1	0254	0.0	1121	1.2	1939	-0.2
28 W	0123	0.1	0358	0.1	1211	1.1	● 2018	-0.2
29 Th	0229	0.2	0531	0.2	1306	1.1	2055	-0.2
30 F	0320	0.4	0718	0.2	1404	1.0	2131	-0.3
31 Sa	0406	0.5	0900	0.2	1505	0.8	2207	-0.3

	Sun↑	Sun↓
15	0641	1819
31	0642	1826

FEBRUARY

Day	Time	ft	Time	ft	Time	ft	Time	ft
1 Su	0451	0.7	1031	0.1	1607	0.7	2243	-0.3
2 M	0535	0.9	1150	0.1	1710	0.6	2321	-0.3
3 Tu	0619	1.1	1300	-0.0	1813	0.5	☾ 2359	-0.2
4 W	0704	1.2	1404	-0.1	1914	0.4		
5 Th	0038	-0.2	0750	1.2	1506	-0.2	2015	0.3
6 F	0117	-0.1	0836	1.3	1606	-0.2	2117	0.2
7 Sa	0156	-0.1	0923	1.2	1705	-0.2	2222	0.2
8 Su	0235	-0.0	1011	1.2	1804	-0.2	2334	0.2
9 M	0314	0.0	1059	1.1	1900	-0.2		
10 Tu	0054	0.2	0357	0.1	1147	1.0	1953	-0.2
11 W	0217	0.2	0451	0.2	1236	0.9	○ 2039	-0.1
12 Th	0323	0.3	0612	0.3	1326	0.8	2117	-0.1
13 F	0406	0.4	0750	0.3	1417	0.7	2147	-0.1
14 Sa	0436	0.5	0922	0.3	1509	0.6	2212	-0.0
15 Su	0501	0.5	1037	0.2	1559	0.5	2232	-0.0
16 M	0524	0.6	1138	0.2	1647	0.4	2251	-0.0
17 Tu	0548	0.7	1230	0.1	1733	0.4	2311	-0.0
18 W	0615	0.8	1317	0.0	1816	0.3	2331	-0.0
19 Th	0645	0.9	1401	-0.0	1857	0.3	☽ 2355	-0.0
20 F	0717	1.0	1445	-0.1	1938	0.2		
21 Sa	0023	-0.1	0753	1.0	1529	-0.1	2020	0.2
22 Su	0056	-0.1	0832	1.1	1612	-0.1	2107	0.2
23 M	0137	-0.1	0914	1.1	1657	-0.1	2200	0.2
24 Tu	0226	-0.0	1001	1.1	1741	-0.2	2302	0.2
25 W	0328	-0.0	1053	1.0	1825	-0.2		
26 Th	0008	0.3	0446	0.0	1150	0.9	● 1909	-0.2
27 F	0114	0.4	0619	0.1	1253	0.8	1953	-0.2
28 Sa	0215	0.6	0755	0.1	1402	0.7	2037	-0.2

	Sun↑	Sun↓
15	0639	1829
28	0635	1831

MARCH

Day	Time	ft	Time	ft	Time	ft	Time	ft
1 Su	0311	0.7	0924	-0.0	1513	0.6	2122	-0.2
2 M	0404	0.9	1040	-0.1	1621	0.5	2207	-0.2
3 Tu	0455	1.0	1147	-0.2	1726	0.4	2253	-0.1
4 W	0544	1.1	1247	-0.3	1826	0.4	2338	-0.1
5 Th ☾	0633	1.2	1343	-0.3	1923	0.4		
6 F	0023	-0.1	0721	1.2	1437	-0.3	2017	0.3
7 Sa	0109	-0.1	0809	1.1	1529	-0.3	2111	0.3
8 Su	0154	-0.0	0856	1.0	1621	-0.2	2205	0.3
9 M	0240	0.0	0944	1.0	1712	-0.2	2301	0.3
10 Tu	0331	0.1	1033	0.8	1802	-0.1		
11 W	0000	0.3	0428	0.1	1124	0.7	1849	-0.0
12 Th	0059	0.3	0540	0.2	1220	0.6	○ 1932	0.0
13 F	0152	0.4	0707	0.2	1321	0.5	2009	0.1
14 Sa	0235	0.4	0832	0.2	1428	0.4	2039	0.1
15 Su	0310	0.5	0944	0.1	1532	0.4	2104	0.1
16 M	0341	0.6	1040	0.0	1629	0.3	2126	0.1
17 Tu	0411	0.7	1126	-0.0	1716	0.3	2148	0.1
18 W	0441	0.8	1206	-0.1	1755	0.3	2213	0.1
19 Th	0514	0.8	1243	-0.1	1828	0.3	2243	0.1
20 F	0549	0.9	1319	-0.2	1900	0.2	2318	0.0
21 Sa	0627	1.0	1355	-0.2	1932	0.2	☽ 2359	0.0
22 Su	0707	1.0	1432	-0.2	2008	0.3		
23 M	0046	-0.0	0750	1.0	1509	-0.2	2049	0.3
24 Tu	0140	-0.0	0837	0.9	1547	-0.2	2137	0.4
25 W	0245	-0.0	0928	0.9	1628	-0.2	2230	0.5
26 Th	0359	0.0	1027	0.8	1711	-0.1	2328	0.6
27 F ●	0525	-0.0	1133	0.6	1757	-0.1		
28 Sa	0028	0.7	0655	-0.1	1250	0.5	1846	-0.1
29 Su	0128	0.8	0819	-0.1	1410	0.4	1938	-0.1
30 M	0226	0.9	0933	-0.2	1528	0.4	2033	-0.0
31 Tu	0322	1.0	1037	-0.3	1638	0.4	2129	-0.0

	Sun↑	Sun↓
15	0627	1831
31	0618	1831

APRIL

Day	Time	ft	Time	ft	Time	ft	Time	ft
1 W	0416	1.1	1134	-0.4	1739	0.4	2224	0.0
2 Th	0508	1.1	1226	-0.4	1834	0.4	2318	0.0
3 F ☾	0558	1.1	1315	-0.4	1925	0.4		
4 Sa	0011	0.0	0647	1.1	1401	-0.4	2013	0.4
5 Su	0103	0.0	0735	1.0	1446	-0.3	2059	0.4
6 M	0157	0.1	0823	0.9	1528	-0.2	2145	0.5
7 Tu	0253	0.1	0911	0.7	1607	-0.1	2229	0.5
8 W	0355	0.1	1001	0.6	1644	-0.0	2313	0.5
9 Th	0507	0.1	1058	0.5	1717	0.0	2356	0.5
10 F	0628	0.1	1208	0.4	1746	0.1		
11 Sa	0036	0.6	0749	0.1	1338	0.3	○ 1810	0.2
12 Su	0113	0.6	0857	-0.0	1517	0.3	1830	0.2
13 M	0149	0.7	0950	-0.1	1645	0.3	1851	0.3
14 Tu	0225	0.7	1032	-0.1	1745	0.3	1918	0.3
15 W	0301	0.8	1108	-0.2	1814	0.3	2003	0.3
16 Th	0339	0.9	1141	-0.2	1829	0.3	2055	0.2
17 F	0418	0.9	1211	-0.3	1843	0.3	2150	0.2
18 Sa	0458	1.0	1241	-0.3	1902	0.3	2245	0.2
19 Su	0540	1.0	1311	-0.3	1927	0.4	☽ 2344	0.1
20 M	0624	0.9	1342	-0.3	1959	0.4		
21 Tu	0047	0.1	0711	0.9	1414	-0.3	2037	0.5
22 W	0156	0.1	0802	0.8	1448	-0.2	2121	0.7
23 Th	0311	0.1	0859	0.6	1525	-0.2	2209	0.8
24 F	0433	0.0	1005	0.5	1605	-0.1	2302	0.9
25 Sa	0557	-0.1	1124	0.4	1649	-0.1	2356	1.0
26 Su ●	0718	-0.2	1253	0.3	1739	-0.0		
27 M	0053	1.1	0829	-0.3	1424	0.3	1837	0.0
28 Tu	0149	1.2	0931	-0.4	1544	0.3	1941	0.1
29 W	0245	1.2	1026	-0.5	1650	0.3	2047	0.1
30 Th	0339	1.2	1116	-0.5	1745	0.4	2153	0.2

	Sun↑	Sun↓
15	0610	1830
30	0604	1831

CRISTÓBAL–COLÓN, PANAMA

HIGH & LOW WATER 1998 | Chart Datum | 9°21'N 79°55'W

Eastern Standard Time (75°W) | Daylight Saving Time not observed

MAY

Date	Day	Time	ft	Time	ft	Time	ft	Time	ft	Moon
1	F	0432	1.1	1202	-0.5	1834	0.5	2256	0.2	
2	Sa	0522	1.0	1244	-0.5	1920	0.5	2358	0.2	
3	Su	0610	0.9	1324	-0.4	2002	0.6			☾
4	M	0059	0.2	0657	0.8	1400	-0.3	2042	0.6	
5	Tu	0201	0.2	0742	0.7	1432	-0.2	2121	0.7	
6	W	0307	0.2	0829	0.5	1500	-0.1	2157	0.7	
7	Th	0419	0.2	0920	0.4	1523	-0.0	2232	0.7	
8	F	0536	0.1	1024	0.3	1539	0.1	2306	0.8	
9	Sa	0653	0.0	1155	0.2	1543	0.1	2340	0.8	
10	Su	0801	-0.0							
11	M	0014	0.9	0855	-0.1					○
12	Tu	0050	0.9	0937	-0.2					
13	W	0127	0.9	1013	-0.3					
14	Th	0207	1.0	1045	-0.3					
15	F	0249	1.0	1113	-0.3					
16	Sa	0332	1.0	1140	-0.4	1848	0.3	2106	0.3	
17	Su	0416	1.0	1206	-0.4	1854	0.4	2231	0.3	
18	M	0501	0.9	1232	-0.4	1916	0.5	2348	0.2	☽
19	Tu	0549	0.9	1300	-0.3	1947	0.6			
20	W	0103	0.2	0640	0.7	1330	-0.3	2024	0.8	
21	Th	0221	0.1	0735	0.6	1401	-0.3	2105	0.9	
22	F	0341	0.0	0838	0.4	1435	-0.2	2151	1.1	
23	Sa	0500	-0.1	0952	0.3	1512	-0.2	2240	1.2	
24	Su	0616	-0.2	1121	0.2	1553	-0.1	2332	1.3	
25	M	0726	-0.3	1259	0.1	1640	0.0			●
26	Tu	0025	1.3	0827	-0.4	1435	0.2	1737	0.1	
27	W	0119	1.3	0921	-0.5	1555	0.2	1846	0.2	
28	Th	0213	1.2	1011	-0.5	1656	0.3	2004	0.2	
29	F	0306	1.2	1055	-0.5	1745	0.4	2121	0.3	
30	Sa	0357	1.1	1136	-0.5	1828	0.5	2236	0.3	
31	Su	0446	1.0	1213	-0.4	1907	0.6	2346	0.3	

	Sun↑	Sun↓
15	0600	1833
31	0559	1837

JUNE

Date	Day	Time	ft	Time	ft	Time	ft	Time	ft	Moon
1	M	0532	0.8	1246	-0.4	1944	0.7			☾
2	Tu	0055	0.3	0616	0.7	1315	-0.3	2018	0.7	
3	W	0204	0.2	0701	0.6	1340	-0.2	2051	0.8	
4	Th	0314	0.2	0746	0.4	1400	-0.1	2122	0.9	
5	F	0426	0.1	0838	0.3	1415	-0.0	2152	0.9	
6	Sa	0537	0.1	0943	0.2	1423	0.0	2224	1.0	
7	Su	0645	-0.0	1117	0.1	1417	0.1	2256	1.0	
8	M	0743	-0.1	2330	1.0					
9	Tu	0832	-0.2							○
10	W	0007	1.1	0911	-0.3					
11	Th	0047	1.1	0945	-0.3					
12	F	0129	1.1	1015	-0.3					
13	Sa	0213	1.1	1041	-0.4					
14	Su	0300	1.0	1106	-0.4	1816	0.4	2105	0.3	
15	M	0348	1.0	1130	-0.4	1828	0.5	2242	0.3	
16	Tu	0437	0.9	1157	-0.4	1854	0.7			
17	W	0008	0.3	0529	0.7	1224	-0.3	1928	0.9	☽
18	Th	0127	0.2	0625	0.6	1254	-0.3	2006	1.0	
19	F	0244	0.1	0725	0.4	1326	-0.3	2048	1.2	
20	Sa	0358	-0.0	0833	0.3	1400	-0.2	2133	1.3	
21	Su	0509	-0.1	0949	0.2	1437	-0.2	2221	1.3	
22	M	0616	-0.2	1117	0.1	1515	-0.1	2311	1.4	
23	Tu	0718	-0.3	1255	0.1	1558	0.0			●
24	W	0002	1.3	0815	-0.4	1433	0.1	1650	0.1	
25	Th	0055	1.3	0906	-0.4	1554	0.2	1800	0.2	
26	F	0147	1.2	0952	-0.4	1651	0.3	1928	0.3	
27	Sa	0238	1.1	1032	-0.4	1734	0.4	2058	0.3	
28	Su	0327	1.0	1108	-0.4	1811	0.5	2222	0.4	
29	M	0415	0.8	1140	-0.3	1844	0.6	2339	0.3	
30	Tu	0500	0.7	1207	-0.3	1914	0.7			

	Sun↑	Sun↓
15	0601	1841
30	0604	1844

JULY

Date	Day	Time	ft	Time	ft	Time	ft	Time	ft	Moon
1	W	0050	0.3	0545	0.6	1231	-0.2	1943	0.8	☾
2	Th	0157	0.2	0629	0.5	1252	-0.1	2012	0.9	
3	F	0301	0.2	0716	0.4	1310	-0.1	2041	1.0	
4	Sa	0403	0.1	0806	0.2	1326	-0.1	2111	1.0	
5	Su	0504	0.0	0904	0.2	1338	-0.0	2142	1.1	
6	M	0603	-0.0	1016	0.1	1345	0.0	2216	1.1	
7	Tu	0658	-0.1	1157	0.0	1335	0.0	2253	1.1	
8	W	0746	-0.2	2332	1.2					
9	Th	0827	-0.2							○
10	F	0015	1.1	0901	-0.2					
11	Sa	0101	1.1	0931	-0.3					
12	Su	0150	1.1	0958	-0.3	1659	0.4	1944	0.3	
13	M	0242	1.0	1025	-0.3	1718	0.5	2135	0.3	
14	Tu	0336	0.9	1053	-0.3	1748	0.7	2306	0.3	
15	W	0432	0.7	1123	-0.3	1823	0.9			
16	Th	0026	0.2	0531	0.6	1154	-0.3	1902	1.1	☽
17	F	0139	0.1	0631	0.5	1228	-0.2	1944	1.2	
18	Sa	0247	-0.0	0733	0.3	1303	-0.2	2028	1.3	
19	Su	0352	-0.1	0839	0.2	1340	-0.2	2114	1.3	
20	M	0456	-0.2	0949	0.2	1419	-0.1	2202	1.4	
21	Tu	0557	-0.2	1107	0.1	1459	-0.0	2252	1.3	
22	W	0656	-0.2	1233	0.1	1543	0.1	2342	1.3	
23	Th	0752	-0.3	1403	0.2	1636	0.2			●
24	F	0034	1.2	0842	-0.3	1522	0.3	1748	0.2	
25	Sa	0126	1.1	0926	-0.2	1619	0.4	1921	0.3	
26	Su	0217	0.9	1003	-0.2	1659	0.5	2056	0.4	
27	M	0307	0.8	1035	-0.2	1732	0.6	2222	0.3	
28	Tu	0357	0.7	1103	-0.1	1802	0.7	2335	0.3	
29	W	0444	0.6	1126	-0.1	1829	0.8			
30	Th	0038	0.2	0531	0.5	1148	-0.1	1856	0.9	
31	F	0135	0.2	0616	0.4	1208	-0.0	1924	0.9	☾

	Sun↑	Sun↓
15	0608	1845
31	0611	1842

AUGUST

Date	Day	Time	ft	Time	ft	Time	ft	Time	ft	Moon
1	Sa	0227	0.1	0702	0.3	1228	-0.0	1954	1.0	
2	Su	0317	0.1	0748	0.3	1248	0.0	2025	1.1	
3	M	0408	0.0	0835	0.2	1309	0.0	2059	1.1	
4	Tu	0459	-0.0	0927	0.2	1332	0.0	2136	1.2	
5	W	0548	-0.0	1026	0.1	1400	0.0	2216	1.2	
6	Th	0635	-0.1	1135	0.1	1436	0.1	2300	1.1	
7	F	0717	-0.1	1250	0.2	1532	0.1	2347	1.1	○
8	Sa	0755	-0.1	1357	0.3	1657	0.2			
9	Su	0040	1.0	0830	-0.1	1449	0.4	1844	0.3	
10	M	0137	0.9	0903	-0.1	1535	0.6	2029	0.3	
11	Tu	0238	0.8	0938	-0.1	1619	0.7	2202	0.2	
12	W	0341	0.7	1013	-0.1	1702	0.9	2320	0.1	
13	Th	0445	0.6	1050	-0.1	1746	1.1			
14	F	0029	0.0	0546	0.5	1129	-0.1	1831	1.2	☽
15	Sa	0132	-0.1	0647	0.4	1209	-0.1	1917	1.3	
16	Su	0231	-0.1	0746	0.4	1250	-0.1	2005	1.3	
17	M	0328	-0.2	0845	0.3	1332	-0.0	2053	1.3	
18	Tu	0425	-0.2	0946	0.3	1417	0.0	2142	1.3	
19	W	0522	-0.1	1050	0.3	1503	0.1	2232	1.2	
20	Th	0618	-0.1	1200	0.3	1555	0.2	2323	1.1	
21	F	0711	-0.1	1313	0.3	1658	0.2			●
22	Sa	0017	1.0	0801	-0.0	1421	0.4	1818	0.3	
23	Su	0112	0.9	0844	0.0	1515	0.5	1950	0.3	
24	M	0209	0.8	0920	0.0	1557	0.6	2117	0.3	
25	Tu	0307	0.7	0950	0.1	1631	0.7	2229	0.3	
26	W	0402	0.6	1016	0.1	1701	0.8	2328	0.2	
27	Th	0452	0.5	1039	0.1	1729	0.8			
28	F	0018	0.2	0539	0.5	1101	0.1	1758	0.9	
29	Sa	0102	0.1	0621	0.4	1123	0.1	1829	1.0	☾
30	Su	0144	0.1	0700	0.4	1148	0.1	1901	1.0	
31	M	0225	0.0	0737	0.4	1216	0.1	1936	1.1	

	Sun↑	Sun↓
15	0612	1837
31	0611	1829

CRISTÓBAL–COLÓN, PANAMA

HIGH & LOW WATER 1998 — Chart Datum — 9°21'N 79°55'W

Eastern Standard Time (75°W) — Daylight Saving Time not observed

SEPTEMBER

Day	Time	ft
1 Tu	0307	0.0
	0815	0.3
	1247	0.1
	2013	1.1
2 W	0348	0.0
	0856	0.3
	1325	0.1
	2054	1.1
3 Th	0430	0.0
	0942	0.3
	1411	0.1
	2138	1.1
4 F	0511	0.0
	1036	0.4
	1510	0.1
	2227	1.0
5 Sa	0552	0.0
	1135	0.4
	1624	0.2
	2322	1.0
6 Su ○	0634	0.0
	1237	0.6
	1753	0.2
7 M	0025	0.9
	0716	0.0
	1337	0.7
	1929	0.2
8 Tu	0134	0.8
	0759	0.0
	1434	0.8
	2057	0.1
9 W	0247	0.7
	0844	0.0
	1527	1.0
	2213	0.0
10 Th	0357	0.6
	0931	0.0
	1619	1.1
	2319	-0.1
11 F	0503	0.6
	1018	0.0
	1709	1.2
12 Sa	0017	-0.2
	0603	0.5
	1106	0.1
	☽ 1759	1.3
13 Su	0112	-0.2
	0659	0.5
	1155	0.1
	1848	1.3
14 M	0204	-0.2
	0752	0.5
	1243	0.1
	1938	1.3
15 Tu	0254	-0.2
	0844	0.5
	1333	0.1
	2027	1.2
16 W	0344	-0.1
	0937	0.5
	1425	0.2
	2117	1.1
17 Th	0433	-0.0
	1030	0.5
	1521	0.2
	2208	1.0
18 F	0521	0.0
	1125	0.5
	1625	0.3
	2303	0.9
19 Sa	0608	0.1
	1221	0.6
	1741	0.3
20 Su	0003	0.8
	0652	0.2
	1313	0.6
	● 1906	0.3
21 M	0112	0.6
	0732	0.2
	1400	0.7
	2029	0.3
22 Tu	0226	0.6
	0807	0.3
	1440	0.7
	2138	0.2
23 W	0337	0.5
	0837	0.3
	1516	0.8
	2231	0.1
24 Th	0437	0.5
	0904	0.4
	1549	0.9
	2315	0.1
25 F	0525	0.5
	0931	0.4
	1622	0.9
	2354	0.0
26 Sa	0603	0.5
	0959	0.3
	1655	1.0
27 Su	0029	0.0
	0634	0.5
	1030	0.3
	1730	1.0
28 M	0103	-0.0
	0702	0.5
	1106	0.3
	☾ 1806	1.1
29 Tu	0137	-0.0
	0731	0.5
	1146	0.3
	1845	1.1
30 W	0210	-0.0
	0801	0.5
	1232	0.2
	1926	1.1

	Sun↑	Sun↓
15	0610	1821
30	0609	1811

OCTOBER

Day	Time	ft
1 Th	0244	-0.0
	0837	0.5
	1325	0.2
	2010	1.1
2 F	0318	0.0
	0918	0.6
	1427	0.2
	2059	1.0
3 Sa	0354	0.0
	1005	0.7
	1540	0.2
	2155	0.9
4 Su	0432	0.1
	1057	0.8
	1702	0.2
	2300	0.8
5 M ○	0514	0.1
	1153	0.9
	1830	0.1
6 Tu	0016	0.6
	0600	0.1
	1251	1.0
	1954	0.0
7 W	0140	0.6
	0651	0.2
	1348	1.2
	2107	-0.1
8 Th	0301	0.5
	0748	0.2
	1445	1.3
	2210	-0.2
9 F	0413	0.5
	0846	0.2
	1540	1.3
	2306	-0.2
10 Sa	0516	0.6
	0946	0.2
	1633	1.4
	2358	-0.3
11 Su	0611	0.6
	1045	0.2
	1725	1.3
12 M	0046	-0.3
	0702	0.6
	1142	0.2
	☽ 1816	1.3
13 Tu	0132	-0.2
	0750	0.7
	1240	0.3
	1907	1.2
14 W	0216	-0.2
	0837	0.7
	1339	0.3
	1956	1.1
15 Th	0258	-0.1
	0923	0.7
	1440	0.3
	2047	0.9
16 F	0337	0.0
	1008	0.8
	1547	0.3
	2140	0.8
17 Sa	0413	0.1
	1052	0.8
	1703	0.3
	2240	0.7
18 Su	0445	0.2
	1134	0.8
	1824	0.3
	2354	0.5
19 M	0513	0.3
	1215	0.9
	1943	0.2
20 Tu	0131	0.5
	0533	0.4
	1253	0.9
	● 2050	0.1
21 W	0327	0.4
	0544	0.4
	1329	0.9
	2141	0.1
22 Th	1405	1.0
	2222	0.0
23 F	1441	1.0
	2258	-0.0
24 Sa	1518	1.1
	2329	-0.1
25 Su	2329	-0.1
	0655	0.5
	1556	1.1
	2358	-0.1
26 M	0656	0.5
	0924	0.5
	1635	1.1
27 Tu	0026	-0.1
	0706	0.5
	1024	0.4
	1715	1.1
28 W	0053	-0.1
	0724	0.6
	1125	0.4
	☾ 1757	1.1
29 Th	0120	-0.1
	0749	0.6
	1228	0.4
	1842	1.0
30 F	0149	-0.1
	0821	0.8
	1336	0.3
	1930	0.9
31 Sa	0219	-0.0
	0900	0.9
	1450	0.3
	2025	0.8

	Sun↑	Sun↓
15	0608	1804
31	0610	1758

NOVEMBER

Day	Time	ft
1 Su	0251	-0.0
	0943	1.0
	1610	0.2
	2128	0.7
2 M	0327	0.0
	1032	1.1
	1733	0.1
	2244	0.5
3 Tu	0406	0.1
	1124	1.3
	1852	0.0
4 W	0013	0.4
	0453	0.1
	1218	1.3
	○ 2002	-0.1
5 Th	0147	0.4
	0547	0.2
	1314	1.4
	2104	-0.2
6 F	0313	0.4
	0651	0.3
	1410	1.4
	2159	-0.3
7 Sa	0423	0.5
	0802	0.3
	1506	1.4
	2249	-0.3
8 Su	0521	0.6
	0914	0.4
	1600	1.4
	2336	-0.4
9 M	0611	0.6
	1024	0.4
	1653	1.3
10 Tu	0019	-0.3
	0657	0.7
	1132	0.4
	☽ 1743	1.2
11 W	0100	-0.3
	0741	0.8
	1238	0.4
	1833	1.0
12 Th	0137	-0.2
	0823	0.8
	1346	0.4
	1922	0.9
13 F	0211	-0.1
	0903	0.9
	1456	0.3
	2011	0.7
14 Sa	0240	0.0
	0942	1.0
	1610	0.3
	2105	0.6
15 Su	0304	0.1
	1018	1.0
	1728	0.3
	2210	0.5
16 M	0321	0.2
	1053	1.0
	1845	0.2
	2341	0.4
17 Tu	0325	0.3
	1126	1.1
	1951	0.1
18 W ●	1200	1.1
	2046	0.0
19 Th	1235	1.1
	2128	-0.0
20 F	1311	1.1
	2205	-0.1
21 Sa	1348	1.1
	2236	-0.1
22 Su	1428	1.1
	2303	-0.2
23 M	1509	1.1
	2328	-0.2
24 Tu	1551	1.1
	2351	-0.2
25 W	0657	0.6
	1000	0.5
	1634	1.1
26 Th	0014	-0.2
	0708	0.7
	1122	0.5
	☾ 1720	1.0
27 F	0038	-0.2
	0731	0.8
	1240	0.4
	1808	0.9
28 Sa	0103	-0.1
	0803	0.9
	1357	0.3
	1901	0.7
29 Su	0132	-0.1
	0841	1.1
	1515	0.2
	2001	0.6
30 M	0203	-0.1
	0924	1.2
	1632	0.1
	2111	0.4

	Sun↑	Sun↓
15	0614	1756
30	0620	1758

DECEMBER

Day	Time	ft
1 Tu	0238	-0.0
	1010	1.4
	1747	0.0
	2234	0.3
2 W	0317	0.0
	1100	1.4
	1855	-0.1
3 Th	0009	0.3
	0401	0.1
	1153	1.5
	○ 1957	-0.2
4 F	0147	0.3
	0455	0.2
	1247	1.5
	2052	-0.3
5 Sa	0314	0.3
	0603	0.3
	1342	1.4
	2143	-0.4
6 Su	0421	0.4
	0723	0.3
	1437	1.4
	2229	-0.4
7 M	0515	0.5
	0846	0.4
	1531	1.3
	2311	-0.4
8 Tu	0601	0.6
	1007	0.4
	1622	1.2
	2350	-0.3
9 W	0644	0.7
	1124	0.4
	1712	1.0
10 Th	0026	-0.3
	0723	0.8
	1238	0.4
	☽ 1801	0.9
11 F	0057	-0.2
	0800	0.9
	1350	0.4
	1849	0.7
12 Sa	0125	-0.1
	0835	1.0
	1501	0.3
	1938	0.6
13 Su	0148	-0.0
	0909	1.0
	1613	0.2
	2030	0.4
14 M	0206	0.1
	0941	1.1
	1723	0.2
	2134	0.3
15 Tu	0216	0.1
	1013	1.1
	1830	0.1
	2302	0.2
16 W	0212	0.2
	1045	1.1
	1930	0.0
17 Th	1118	1.2
	2020	-0.1
18 F ●	1153	1.2
	2102	-0.1
19 Sa	1231	1.2
	2137	-0.2
20 Su	1310	1.2
	2206	-0.2
21 M	1352	1.1
	2230	-0.2
22 Tu	1436	1.1
	2252	-0.2
23 W	0626	0.5
	0819	0.5
	1521	1.0
	2313	-0.2
24 Th	0619	0.6
	1012	0.5
	1609	0.9
	2335	-0.2
25 F	0637	0.7
	1141	0.4
	1700	0.8
26 Sa	0000	-0.2
	0705	0.9
	1300	0.3
	☾ 1754	0.7
27 Su	0027	-0.2
	0740	1.1
	1415	0.2
	1852	0.5
28 M	0057	-0.2
	0820	1.2
	1527	0.1
	1956	0.4
29 Tu	0131	-0.2
	0904	1.4
	1635	-0.0
	2107	0.3
30 W	0208	-0.1
	0950	1.4
	1740	-0.1
	2227	0.2
31 Th	0248	-0.1
	1040	1.5
	1843	-0.2
	2357	0.2

	Sun↑	Sun↓
15	0628	1803
31	0636	1811

T — Tides

TAMPICO, MEXICO

HIGH & LOW WATER 1998 — Chart Datum — 22°13'N 97°51'W

Central Standard Time (90°W) — Daylight Saving Time not observed

JANUARY

Day	Time	ft	Day	Time	ft
1 Th	1003 1842	-0.6 1.2	16 F	0228 1041 1840	0.7 -0.2 0.8
2 F	1049 1909	-0.4 1.1	17 Sa	0011 0338 1112 1847	0.6 0.6 -0.0 0.8
3 Sa	0127 0334 1137 1930	0.7 0.7 -0.2 1.0	18 Su	0055 0507 1140 1850	0.4 0.5 0.2 0.7
4 Su	0201 0557 1229 1944	0.5 0.6 0.1 0.9	19 M	0142 0713 1206 1851	0.3 0.5 0.4 0.7
5 M ☾	0243 0840 1330 1955	0.2 0.6 0.4 0.9	20 Tu ☽	0230 1020 1221 1852	0.2 0.5 0.5 0.7
6 Tu	0329 1117 1454 2001	-0.1 0.7 0.6 0.9	21 W	0317 1852	0.0 0.8
7 W	0416 1303 1657 2002	-0.3 0.9 0.8 0.9	22 Th	0403 1351 1658 1835	-0.1 0.8 0.8 0.8
8 Th	0505 1410	-0.5 1.1	23 F	0447 1415	-0.3 1.0
9 F	0553 1502	-0.6 1.2	24 Sa	0531 1445	-0.4 1.1
10 Sa	0640 1547	-0.7 1.2	25 Su	0615 1518	-0.6 1.1
11 Su	0726 1628	-0.7 1.2	26 M	0659 1551	-0.7 1.2
12 M ○	0810 1705	-0.7 1.2	27 Tu	0743 1623 2139 2332	-0.7 1.2 0.9 0.9
13 Tu	0851 1738 2240	-0.6 1.1 0.9	28 W ●	0828 1653 2158	-0.7 1.1 0.8
14 W	0021 0931 1805 2302	0.9 -0.5 1.0 0.8	29 Th	0051 0913 1718 2227	0.9 -0.6 1.1 0.7
15 Th	0125 1007 1826 2333	0.8 -0.4 0.9 0.7	30 F	0210 0958 1739 2304	0.9 -0.4 1.0 0.6
			31 Sa	0334 1044 1754 2349	0.8 -0.2 0.9 0.3

	Sun↑	Sun↓
15	0714	1809
31	0711	1820

FEBRUARY

Day	Time	ft	Day	Time	ft
1 Su	0511 1131 1805	0.7 0.1 0.8	16 M	0537 1114 1707	0.7 0.4 0.8
2 M	0041 0708 1222 1813	0.1 0.7 0.4 0.8	17 Tu	0006 0715 1140 1706	0.2 0.7 0.6 0.8
3 Tu ☾	0138 0933 1325 1816	-0.1 0.8 0.7 0.8	18 W	0058 0941 1208 1703	0.1 0.8 0.7 0.8
4 W	0239 1155 1520 1806	-0.3 0.9 0.9 0.9	19 Th ☽	0156 1227 1353 1639	0.0 0.9 0.9 0.9
5 Th	0341 1319	-0.4 1.1	20 F	0258 1302	-0.1 1.0
6 F	0442 1411	-0.5 1.1	21 Sa	0359 1334	-0.2 1.1
7 Sa	0539 1452	-0.6 1.2	22 Su	0457 1405	-0.3 1.2
8 Su	0631 1528 2055 2140	-0.6 1.2 1.0 1.0	23 M	0551 1436 2008 2136	-0.4 1.2 1.0 1.0
9 M	0719 1558 2045 2320	-0.5 1.1 0.9 0.9	24 Tu	0642 1504 2008 2319	-0.4 1.2 0.9 1.0
10 Tu	0803 1624 2059	-0.4 1.0 0.8	25 W	0731 1528 2026	-0.4 1.2 0.8
11 W ○	0029 0843 1644 2119	0.9 -0.3 0.9 0.7	26 Th ●	0039 0819 1549 2054	1.0 -0.3 1.1 0.7
12 Th	0128 0918 1657 2143	0.9 -0.2 0.9 0.6	27 F	0155 0907 1604 2128	1.0 -0.2 1.0 0.5
13 F	0225 0951 1705 2211	0.9 -0.1 0.8 0.5	28 Sa	0311 0955 1616 2209	1.1 0.1 0.9 0.3
14 Sa	0322 1020 1708 2244	0.8 0.1 0.8 0.4			
15 Su	0424 1048 1707 2322	0.8 0.3 0.8 0.3			

	Sun↑	Sun↓
15	0703	1829
28	0654	1835

MARCH

Day	Time	ft	Day	Time	ft
1 Su	0432 1045 1625 2256	1.0 0.3 0.9 0.1	16 M	0453 1038 1522 2222	1.0 0.6 0.9 0.2
2 M	0601 1137 1630 2349	1.0 0.6 0.9 -0.1	17 Tu	0557 1112 1520 2302	1.0 0.8 0.9 0.1
3 Tu	0747 1239 1630	1.0 0.8 0.9	18 W	0718 1158 1513 2350	1.0 0.9 0.9 0.1
4 W	0049 0951 1427 1604	-0.2 1.1 1.0 1.0	19 Th	0904	1.0
5 Th ☾	0156 1142	-0.2 1.2	20 F	0049 1047	0.0 1.1
6 F	0307 1249	-0.3 1.2	21 Sa ☽	0158 1151	-0.0 1.2
7 Sa	0418 1335	-0.3 1.2	22 Su	0310 1233	-0.1 1.3
8 Su	0523 1410 1942 2121	-0.2 1.2 1.0 1.0	23 M	0420 1306	-0.1 1.3
9 M	0620 1437 1936 2307	-0.2 1.1 0.9 1.0	24 Tu	0523 1333 1902 2240	-0.1 1.2 0.9 1.0
10 Tu	0709 1458 1949	-0.1 1.1 0.8	25 W	0622 1355 1915	-0.1 1.2 0.7
11 W	0019 0753 1513 2008	1.0 -0.0 1.0 0.7	26 Th	0011 0718 1412 1941	1.1 0.0 1.1 0.5
12 Th ○	0120 0832 1523 2029	1.0 0.1 0.9 0.6	27 F ●	0129 0812 1425 2013	1.2 0.2 1.0 0.3
13 F	0214 0906 1527 2053	1.0 0.2 0.9 0.5	28 Sa	0242 0905 1435 2051	1.2 0.4 0.9 0.1
14 Sa	0305 0938 1526 2119	1.0 0.4 0.8 0.4	29 Su	0355 1000 1442 2133	1.3 0.6 0.9 -0.1
15 Su	0357 1008 1524 2148	1.0 0.5 0.8 0.3	30 M	0511 1057 1445 2220	1.3 0.8 1.0 -0.2
			31 Tu	0632 1206 1441 2312	1.3 1.0 1.0 -0.3

	Sun↑	Sun↓
15	0641	1841
31	0626	1847

APRIL

Day	Time	ft	Day	Time	ft
1 W	0803	1.3	16 Th	0722 2309	1.2 -0.1
2 Th	0010 0938	-0.3 1.3	17 F	0839	1.2
3 F ☾	0116 1059	-0.2 1.3	18 Sa	0005 0951	-0.1 1.3
4 Sa	0230 1155	-0.1 1.3	19 Su ☽	0110 1045	-0.1 1.3
5 Su	0347 1234	-0.0 1.2	20 M	0224 1124	-0.0 1.3
6 M	0457 1302 1841 2231	0.1 1.1 0.8 0.9	21 Tu	0341 1153 1810 2141	0.0 1.2 0.7 0.8
7 Tu	0559 1322 1851 2358	0.2 1.0 0.7 1.0	22 W	0454 1215 1815 2333	0.1 1.1 0.6 0.9
8 W	0652 1335 1910	0.3 1.0 0.6	23 Th	0602 1231 1837	0.2 1.0 0.3
9 Th	0104 0738 1342 1930	1.0 0.4 0.9 0.4	24 F	0058 0707 1242 1908	1.1 0.4 1.0 0.1
10 F	0200 0819 1344 1953	1.1 0.5 0.9 0.3	25 Sa	0212 0810 1251 1944	1.2 0.6 0.9 -0.2
11 Sa ○	0250 0857 1342 2017	1.1 0.6 0.9 0.2	26 Su ●	0321 0913 1257 2025	1.4 0.8 0.9 -0.3
12 Su	0338 0933 1339 2043	1.1 0.7 0.9 0.1	27 M	0428 1019 1258 2108	1.4 0.9 1.0 -0.5
13 M	0425 1011 1335 2111	1.2 0.8 0.9 0.0	28 Tu	0536 1146 1242 2155	1.5 1.1 1.1 -0.5
14 Tu	0516 1054 1329 2144	1.2 0.9 0.9 -0.0	29 W	0646 2245	1.4 -0.5
15 W	0614 2223	1.2 -0.1	30 Th	0758 2340	1.4 -0.4

	Sun↑	Sun↓
15	0612	1852
30	0601	1858

TAMPICO, MEXICO

HIGH & LOW WATER 1998 | Chart Datum | 22°13'N 97°51'W

Central Standard Time (90°W) | Daylight Saving Time not observed

MAY

Day	Time	ft
1 F	0907	1.3
2 Sa	0040 1005	-0.2 1.2
3 Su ☾	0148 1048	-0.0 1.1
4 M	0301 1117 1753 2113	0.1 1.1 0.7 0.7
5 Tu	0415 1138 1756 2315	0.3 1.0 0.5 0.8
6 W	0524 1150 1813	0.4 0.9 0.4
7 Th	0037 0625 1157 1835	0.9 0.5 0.9 0.2
8 F	0139 0720 1158 1858	1.0 0.6 0.8 0.1
9 Sa	0232 0810 1155 1922	1.1 0.7 0.8 -0.0
10 Su	0319 0859 1151 1948	1.1 0.8 0.9 -0.1
11 M ○	0402 0950 1143 2016	1.2 0.9 0.9 -0.2
12 Tu	0446 2047	1.2 -0.3
13 W	0532 2122	1.2 -0.3
14 Th	0623 2201	1.3 -0.3
15 F	0718 2246	1.3 -0.3
16 Sa	0813 2337	1.3 -0.3
17 Su	0902	1.3
18 M ☽	0036 0941	-0.2 1.2
19 Tu	0144 1010 1719 2005	-0.0 1.1 0.6 0.6
20 W	0302 1032 1713 2244	0.2 1.0 0.4 0.7
21 Th	0424 1046 1735	0.4 1.0 0.1
22 F	0025 0547 1057 1807	0.9 0.6 0.9 -0.1
23 Sa	0144 0706 1105 1844	1.1 0.7 0.9 -0.4
24 Su	0251 0824 1109 1924	1.3 0.9 1.0 -0.6
25 M ●	0352 0948 1103 2008	1.4 1.0 1.0 -0.7
26 Tu	0451 2052	1.5 -0.7
27 W	0549 2138	1.4 -0.7
28 Th	0644 2226	1.4 -0.6
29 F	0737 2315	1.3 -0.4
30 Sa	0824	1.2
31 Su	0006 0903	-0.2 1.1

	Sun↑	Sun↓
15	0552	1904
31	0548	1912

JUNE

Day	Time	ft
1 M ☾	0100 0931 1717 1806	0.0 1.0 0.6 0.6
2 Tu	0201 0950 1651 2146	0.2 0.9 0.5 0.6
3 W	0310 1002 1709 2352	0.4 0.9 0.3 0.7
4 Th	0428 1007 1733	0.6 0.8 0.1
5 F	0113 0548 1007 1759	0.9 0.7 0.8 -0.0
6 Sa	0210 0703 1004 1827	1.0 0.8 0.9 -0.1
7 Su	0256 0817 0956 1855	1.1 0.9 0.9 -0.3
8 M	0336 1925	1.2 -0.3
9 Tu ○	0414 1957	1.2 -0.4
10 W	0453 2032	1.3 -0.5
11 Th	0533 2109	1.3 -0.5
12 F	0615 2150	1.3 -0.5
13 Sa	0656 2233	1.3 -0.4
14 Su	0734 2320	1.3 -0.3
15 M	0806	1.2
16 Tu	0012 0831 1542 1817	-0.1 1.1 0.6 0.6
17 W ☽	0112 0849 1553 2128	0.2 1.1 0.4 0.7
18 Th	0227 0902 1623 2344	0.5 1.0 0.1 0.9
19 F	0401 0911 1701	0.7 1.0 -0.1
20 Sa	0116 0547 0915 1743	1.1 0.9 1.0 -0.4
21 Su	0223 0738 0911 1827	1.3 1.1 1.1 -0.6
22 M	0320 1912	1.4 -0.7
23 Tu ●	0411 1957	1.5 -0.7
24 W	0459 2042	1.5 -0.6
25 Th	0543 2127	1.4 -0.5
26 F	0624 2210	1.4 -0.4
27 Sa	0659 2253	1.3 -0.2
28 Su	0728 2334	1.2 0.0
29 M	0750 1421 1625	1.1 0.8 0.8
30 Tu	0014 0804 1457 1900	0.2 1.0 0.6 0.7

	Sun↑	Sun↓
15	0548	1917
30	0551	1920

JULY

Day	Time	ft
1 W ☾	0054 0813 1534 2204	0.5 1.0 0.5 0.7
2 Th	0139 0816 1610	0.7 1.0 0.3
3 F	0039 0254 0814 1644	0.9 0.9 1.0 0.2
4 Sa	0151 0500 0808 1719	1.0 1.0 1.0 0.0
5 Su	0230 1754	1.1 -0.1
6 M	0303 1829	1.3 -0.2
7 Tu	0334 1905	1.3 -0.3
8 W	0407 1943	1.4 -0.3
9 Th ○	0439 2021	1.5 -0.4
10 F	0512 2101	1.5 -0.3
11 Sa	0544 2142	1.5 -0.3
12 Su	0613 1200 1346 2225	1.4 1.1 1.1 -0.1
13 M	0637 1232 1524 2310	1.4 1.0 1.0 0.1
14 Tu	0656 1312 1723 2358	1.3 0.8 0.9 0.4
15 W	0710 1358 1950	1.2 0.6 0.9
16 Th ☽	0053 0719 1448 2232	0.7 1.2 0.3 1.1
17 F	0206 0725 1539	1.0 1.2 0.1
18 Sa	0037 0408 0722 1632	1.3 1.2 1.3 -0.1
19 Su	0149 1724	1.5 -0.2
20 M	0242 1815	1.6 -0.3
21 Tu	0327 1904	1.6 -0.4
22 W	0407 1951	1.7 -0.3
23 Th ●	0442 0945 1118 2035	1.6 1.4 1.4 -0.2
24 F	0513 0959 1234 2117	1.5 1.3 1.4 -0.1
25 Sa	0539 1024 1340 2156	1.5 1.2 1.3 0.1
26 Su	0600 1056 1445 2232	1.4 1.1 1.2 0.3
27 M	0614 1135 1555 2306	1.3 1.0 1.2 0.5
28 Tu	0623 1218 1718 2337	1.3 0.9 1.1 0.7
29 W	0626 1307 1909	1.2 0.8 1.1
30 Th	0003 0626 1359 2201	0.9 1.2 0.7 1.1
31 F ☾	0016 0622 1452	1.1 1.3 0.5

	Sun↑	Sun↓
15	0557	1919
31	0604	1913

AUGUST

Day	Time	ft
1 Sa	0614 1543	1.3 0.4
2 Su	0206 1632	1.4 0.3
3 M	0217 1718	1.5 0.2
4 Tu	0240 1802	1.6 0.1
5 W	0306 1845	1.7 0.1
6 Th	0334 0858 1009 1928	1.7 1.5 1.5 0.0
7 F ○	0401 0904 1134 2010	1.7 1.5 1.5 0.0
8 Sa	0426 0926 1248 2053	1.7 1.4 1.5 0.1
9 Su	0448 0954 1401 2137	1.7 1.3 1.5 0.3
10 M	0507 1029 1519 2222	1.6 1.2 1.5 0.5
11 Tu	0521 1111 1647 2308	1.5 1.0 1.4 0.8
12 W	0531 1159 1832 2358	1.5 0.8 1.4 1.1
13 Th	0538 1254 2043	1.5 0.6 1.5
14 F ☽	0058 0539 1356 2307	1.3 1.5 0.4 1.6
15 Sa	0242 0525 1500	1.6 1.6 0.3
16 Su	0046 1606	1.8 0.2
17 M	0143 1708	1.9 0.2
18 Tu	0226 1805	1.9 0.2
19 W	0302 0831 0924 1857	1.9 1.7 1.7 0.2
20 Th	0331 0819 1114 1944	1.8 1.6 1.7 0.3
21 F ●	0355 0834 1228 2027	1.8 1.5 1.7 0.4
22 Sa	0415 0856 1332 2107	1.7 1.4 1.7 0.6
23 Su	0429 0922 1431 2143	1.6 1.3 1.6 0.7
24 M	0438 0951 1529 2216	1.5 1.2 1.6 0.9
25 Tu	0442 1023 1631 2247	1.5 1.1 1.6 1.1
26 W	0441 1059 1741 2315	1.5 1.0 1.5 1.3
27 Th	0438 1141 1910 2342	1.5 0.9 1.5 1.4
28 F	0434 1230 2121	1.6 0.9 1.6
29 Sa ☾	0008 0425 1327	1.5 1.6 0.8
30 Su	0016 0144 0337 1432	1.7 1.7 1.7 0.7
31 M	0052 1537	1.8 0.7

	Sun↑	Sun↓
15	0609	1903
31	0614	1850

TAMPICO, MEXICO

HIGH & LOW WATER 1998 — Chart Datum — 22°13'N 97°51'W

Central Standard Time (90°W) — Daylight Saving Time not observed

SEPTEMBER

Date	Day	Phase	Time	ft	Time	ft	Time	ft	Time	ft
1	Tu		0119	1.8	1636	0.6				
2	W		0146	1.9	1731	0.5				
3	Th		0212	1.9	0744	1.7	0933	1.7	1822	0.5
4	F		0236	1.9	0744	1.6	1113	1.7	1911	0.5
5	Sa		0258	1.9	0802	1.5	1231	1.8	1959	0.6
6	Su	○	0317	1.8	0829	1.4	1344	1.8	2047	0.7
7	M		0332	1.8	0901	1.2	1457	1.9	2135	0.9
8	Tu		0343	1.7	0939	1.0	1613	1.9	2225	1.2
9	W		0351	1.7	1023	0.8	1737	1.9	2318	1.4
10	Th		0355	1.7	1113	0.7	1913	1.9		
11	F		0022	1.7	0352	1.8	1209	0.6	2105	2.0
12	Sa	☽	0217	1.8	0313	1.8	1314	0.5	2255	2.0
13	Su		1425	0.5						
14	M		0012	2.1	1539	0.5				
15	Tu		0102	2.1	1649	0.6				
16	W		0138	2.0	0734	1.8	0846	1.8	1752	0.6
17	Th		0206	1.9	0717	1.7	1053	1.8	1846	0.7
18	F		0227	1.9	0730	1.6	1214	1.8	1935	0.8
19	Sa		0243	1.8	0751	1.4	1319	1.8	2018	1.0
20	Su	●	0253	1.7	0814	1.3	1416	1.9	2057	1.1
21	M		0258	1.7	0839	1.2	1510	1.9	2134	1.2
22	Tu		0259	1.6	0905	1.1	1602	1.9	2208	1.4
23	W		0255	1.6	0934	1.0	1657	1.9	2242	1.5
24	Th		0250	1.6	1005	0.9	1756	1.8	2317	1.6
25	F		0243	1.7	1042	0.9	1909	1.8		
26	Sa		0005	1.7	0230	1.7	1126	0.9	2042	1.8
27	Su		1220	0.8	2220	1.9				
28	M	☾	1325	0.8	2323	1.9				
29	Tu		1438	0.8						
30	W		0004	2.0	1549	0.8				

	Sun↑	Sun↓
15	0619	1835
30	0623	1820

OCTOBER

Date	Day	Phase	Time	ft	Time	ft	Time	ft	Time	ft
1	Th		0035	2.0	1655	0.8				
2	F		0101	2.0	0644	1.6	1028	1.7	1756	0.8
3	Sa		0122	1.9	0656	1.4	1159	1.8	1854	0.9
4	Su		0138	1.8	0719	1.2	1317	1.9	1949	1.0
5	M	○	0151	1.7	0750	1.0	1428	2.0	2045	1.2
6	Tu		0201	1.7	0826	0.8	1539	2.1	2141	1.4
7	W		0207	1.7	0907	0.6	1651	2.1	2242	1.6
8	Th		0210	1.7	0952	0.4	1808	2.1	2355	1.8
9	F		0202	1.8	1042	0.4	1932	2.1		
10	Sa		1137	0.4	2059	2.1				
11	Su		1240	0.4	2219	2.1				
12	M	☽	1351	0.5	2318	2.0				
13	Tu		1507	0.6	2359	1.9				
14	W		1622	0.8						
15	Th		0027	1.8	0623	1.5	1010	1.6	1729	0.9
16	F		0048	1.7	0635	1.3	1148	1.6	1829	1.0
17	Sa		0103	1.6	0654	1.2	1300	1.7	1922	1.1
18	Su		0112	1.6	0718	1.0	1400	1.8	2009	1.2
19	M		0115	1.5	0741	0.9	1453	1.8	2054	1.3
20	Tu	●	0113	1.5	0806	0.8	1542	1.8	2136	1.4
21	W		0108	1.5	0832	0.7	1629	1.8	2218	1.5
22	Th		0059	1.5	0900	0.6	1717	1.8	2309	1.6
23	F		0043	1.6	0930	0.6	1809	1.8		
24	Sa		1006	0.5	1908	1.8				
25	Su		1048	0.5	2015	1.8				
26	M		1137	0.5	2119	1.8				
27	Tu		1236	0.5	2210	1.8				
28	W	☾	1345	0.6	2248	1.8				
29	Th		1501	0.6	2317	1.7				
30	F		0555	1.3	0914	1.3	1617	0.7	2339	1.7
31	Sa		0555	1.1	1116	1.4	1730	0.9	2355	1.6

	Sun↑	Sun↓
15	0628	1807
31	0635	1755

NOVEMBER

Date	Day	Phase	Time	ft	Time	ft	Time	ft	Time	ft
1	Su		0616	0.8	1243	1.6	1840	1.0		
2	M		0008	1.5	0646	0.6	1357	1.7	1946	1.2
3	Tu	○	0016	1.5	0722	0.3	1506	1.9	2054	1.3
4	W		0021	1.5	0801	0.1	1611	2.0	2206	1.5
5	Th		0021	1.5	0845	-0.0	1717	2.0		
6	F		0931	-0.1	1823	2.0				
7	Sa		1021	-0.1	1931	1.9				
8	Su		1114	0.0	2035	1.8				
9	M		1211	0.1	2130	1.7				
10	Tu	☽	1315	0.3	2212	1.6				
11	W		1425	0.5	2243	1.5				
12	Th		0531	1.1	0838	1.1	1539	0.7	2304	1.4
13	F		0537	0.9	1059	1.2	1654	0.8	2319	1.3
14	Sa		0557	0.7	1230	1.3	1803	1.0	2327	1.3
15	Su		0622	0.6	1338	1.4	1907	1.1	2329	1.2
16	M		0647	0.4	1434	1.5	2007	1.1	2325	1.2
17	Tu		0713	0.3	1522	1.5	2106	1.2	2315	1.2
18	W	●	0740	0.2	1605	1.5				
19	Th		0807	0.1	1647	1.6				
20	F		0837	0.0	1729	1.6				
21	Sa		0910	0.0	1812	1.6				
22	Su		0945	-0.0	1859	1.6				
23	M		1026	-0.0	1946	1.5				
24	Tu		1111	0.0	2029	1.5				
25	W		1203	0.1	2105	1.5				
26	Th	☾	1304	0.3	2133	1.4				
27	F		0506	0.9	0708	0.9	1416	0.4	2155	1.3
28	Sa		0450	0.6	1013	0.9	1540	0.6	2210	1.2
29	Su		0511	0.4	1205	1.1	1709	0.8	2221	1.2
30	M		0543	0.1	1326	1.3	1837	1.0	2229	1.1

	Sun↑	Sun↓
15	0644	1749
30	0654	1747

DECEMBER

Date	Day	Phase	Time	ft	Time	ft	Time	ft	Time	ft
1	Tu		0620	-0.2	1434	1.5	2004	1.1	2231	1.2
2	W		0701	-0.4	1535	1.6				
3	Th	○	0745	-0.5	1633	1.7				
4	F		0831	-0.6	1729	1.7				
5	Sa		0918	-0.6	1822	1.6				
6	Su		1006	-0.5	1912	1.5				
7	M		1055	-0.3	1955	1.4				
8	Tu		1145	-0.1	2031	1.3				
9	W		1237	0.1	2058	1.2				
10	Th	☽	0412	0.7	0600	0.7	1334	0.3	2118	1.1
11	F		0421	0.5	0918	0.7	1440	0.5	2132	1.0
12	Sa		0447	0.4	1139	0.8	1600	0.7	2139	0.9
13	Su		0517	0.2	1310	0.9	1731	0.8	2139	0.9
14	M		0547	0.0	1411	1.0	1901	0.9	2132	0.9
15	Tu		0618	-0.1	1459	1.1				
16	W		0648	-0.2	1539	1.2				
17	Th		0718	-0.3	1615	1.2				
18	F	●	0750	-0.4	1650	1.2				
19	Sa		0822	-0.4	1724	1.3				
20	Su		0857	-0.4	1758	1.3				
21	M		0933	-0.4	1832	1.2				
22	Tu		1013	-0.4	1905	1.2				
23	W		1055	-0.3	1933	1.1				
24	Th		1142	-0.1	1956	1.1				
25	F		0259	0.6	0530	0.6	1235	0.1	2013	1.0
26	Sa	☾	0318	0.3	0838	0.6	1341	0.4	2026	0.9
27	Su		0352	0.1	1111	0.7	1511	0.6	2034	0.9
28	M		0432	-0.2	1254	1.0	1707	0.8	2038	0.9
29	Tu		0515	-0.5	1404	1.2	1919	1.0	2024	1.0
30	W		0601	-0.7	1501	1.3				
31	Th		0648	-0.8	1551	1.4				

	Sun↑	Sun↓
15	0704	1751
31	0711	1759

TIDE DIFFERENCES TABLE

The following pages provide information for calculating the tides at various locations around the Caribbean and in Bermuda, along Florida's east coast, and in the Bahamas. This information is keyed to the tide tables for the primary reference stations found earlier in this chapter.

This comprehensive table of tide difference information has been created exclusively for *Reed's*. While *Reed's* has made every effort to ensure the table's accuracy and completeness, hydrographic data for the Caribbean is unfortunately sparse and less accurate than information available for continental North America. Nonetheless, *Reed's* feels this table represents the best tidal information ever printed for the Caribbean.

Caution #1: Bear in mind that local phenomena can greatly alter water depths. In some places, strong winds can affect tidal depths—for example, by as much as 3 feet along parts of the Mexican coast. Hurricanes can push surges ahead of them more than 10 feet high. Harbors at river mouths often experience seasonal fluctuations in depths, according to the amount of rainfall in the river's watershed.

Caution #2: The time and height values in the table are average differences derived from comparisons of simultaneous tide observations at the subordinate location and its reference station. Because these values are constant, they may not always provide for the daily variations of the actual tide, especially if the subordinate station is some distance from the primary reference station. Therefore, although the application of the time and height differences will generally provide fairly accurate approximations, they cannot result in predictions as accurate as those at the reference station.

HOW TO USE THE TABLE

The table of tide differences is organized geographically by *tide difference stations.* It begins with stations in Bermuda, moves south to Florida and the Bahamas, and then circles the entire Caribbean basin in a clockwise direction, from Cuba to Mexico.

In the **Place** column, locate the tide difference station in which you are interested. Then note the name of the *primary reference tide station* listed in bold type under the **Differences** heading above the station you have chosen (e.g., **on MIAMI, p. T 7).** The first reference above your chosen station is the table to which you should apply these differences to obtain the prediction. For your convenience, the page number for that table is also given.

The two columns under the **Position** heading list the tide difference station's approximate latitude and longitude.

Under the **Differences** heading, the two **Time** columns give the *time differences* to be added or subtracted to the times given in the primary reference table for the day in which you are interested. Note that all time differences are given in *hours* and *minutes.* The first column gives the time difference for *high tide*, and the second column gives the difference for *low tide*. A plus sign (+) indicates that the time should be added to the time listed in the primary reference tide table. A minus sign (–) indicates that the time should be subtracted from the primary reference table time.

The two **Height** columns give the *height differences* for high and low tide that should be applied to the *tide heights* listed in the primary reference table. You compute the difference one of two ways, depending on how the data is presented in the tide differences table. If there are no parentheses around the entries, simply multiply when you see an asterisk (*), add when there is a plus sign (+), and subtract when there is a minus sign (–). The first difference applies to the high tide height from the reference table; and the second, to the low tide value from the reference table.

If parentheses enclose the numbers in the tide differences table, the tide height from the primary reference table must be multiplied by the first number inside the parentheses, which is preceded by an asterisk (*). The second value in the parentheses is then added to or subtracted from the value derived from the multiplication.

The heights derived using these tables are based on the same tidal datum as the reference station. Be careful that this datum matches the datum of the chart you are using. See page T 2 for more about datums.

The **Range** column gives either the average *spring range* or the average *diurnal range* in feet for the tide difference station. The heading in this column closest above the station specifies which range applies. In most instances in the Caribbean basin, the range shown is diurnal, which is the average range between the lowest and highest daily tides. Spring range is the

T

Tides

average range during a months larger tides. Keep in mind that these values are averages and that the actual range may be strongly influenced by local winds, rainfall, or other factors.

Note: In the table, dashes (– –) are substituted for data that is unknown, indeterminate, unreliable, or not applicable.

EXAMPLE

Suppose you want tidal information for the Allans–Pensacola Cays area of the Abacos in the Bahamas on a certain day. Find the Allans–Pensacola Cays listing in the tide differences table (the first entry under the Abacos region subheading in the Bahamas on page T 54). Note that differences for this station are keyed to the tide table for Nassau, Bahamas (found on page T 17). The latitude and longitude of the Allans–Pensacola Cays station is 26° 59′ north latitude, 77° 40′ west longitude.

The **time differences** at the Allans–Pensacola Cays area are +35 minutes for high tide and +45 minutes for low tide.

On March 1 of our example year, the tide in Nassau looks like this (note that this is example data, not 1998 data):

1	0035	2.6
	0639	0.8
Sa	1247	2.6
	1900	0.6

To find the corresponding times at Allans–Pensacola, add 35 minutes to the Nassau highs and 45 minutes to the lows:

0035 + 35 = 0110 (high)
0639 + 45 = 0724 (low)
1247 + 35 = 0122 (high)
1900 + 45 = 1945 (low).

In the tide differences table, the **height differences** for Allans–Pensacola Cays are enclosed by parentheses (*0.95 – 0.64). Therefore, to determine the heights of the lows and highs at Allans–Pensacola, the tide heights from the Nassau tide table for March 1 must first be multiplied by the first number, the ratio 0.95. Because the second number in parentheses, 0.64, is preceded by a minus sign (–), it is then subtracted from the result of the multiplication.

Reading from the Nassau table, on March 1, the height of the first high is 2.6 feet above the chart datum of soundings; the first low, 0.8 feet; the second high, 2.6 feet; and the second low, 0.6 feet. To find the corresponding approximate heights at Allans–Pensacola Cays, multiply the heights by 0.95 and then subtract 0.64:

2.6 feet (0.95) = 2.5; 2.5 – 0.64 = 1.8 feet (high)
0.8 feet (0.95) = 0.8; 0.8 – 0.64 = 0.2 feet (low)
2.6 feet (0.95) = 2.5; 2.5 – 0.64 = 1.8 feet (high)
0.6 feet (0.95) = 0.6; 0.6 – 0.64 = 0.0 feet (low).

Reading directly from the table, the average spring tide **range** for Allans–Pensacola Cays is 2.9 feet.

Tide Watchers Wanted

Reed's seeks local information about tides and currents. There are many locations, especially in the Caribbean, where official government tide and current data is scarce. Often, local boaters know how local tides and currents work. *Reed's* wants to help make that information available to all boaters.

For instance, you might know a place that usually has a high tide about 50 minutes after a regular reference station. Or you might spend a lot of time in such a place and be willing to devote a bit of time and effort to measuring the tide or current.

We can give you some tips on how to evaluate tides and currents, and we can integrate your local knowledge into our tide differences table. Of course, we will carefully mark such data as local knowledge, and not nearly as precise as the differences developed by hydrographic offices. Approximate information, properly handled, is better than no information.

If you are interested in being a Tide Watcher, please contact *Reed's* for further information. The reward will be helping your fellow mariners to better know and use the power of tides.

PLACE	POSITION		DIFFERENCES				RANGE
			Time		Height		
	north latitude	west longitude	high h m	low h m	high ft	low ft	spring ft
FLORIDA							
Eastern Standard Time (75°)			*Add 1 hour Daylight Saving April 5 – October 24*				
East coast			**on MIAMI, p. T 7**				
Ponce de Leon Inlet	29°04′	80°55′	+0:04	+0:21	*0.92	*0.94	2.7
Cape Canaveral	28°26′	80°34′	–0:43	–0:40	*1.39	*1.38	4.1
Port Canaveral entrance	28°24.5′	80°36.0′	–0:36	–0:31	*1.42	*1.19	4.3
Patrick Air Force Base	28°14.7′	80°36.0′	–0:41	–0:34	*1.39	*1.25	4.2
Canova Beach	28°08.3′	80°34.7′	–0:30	–0:22	*1.37	*1.31	4.1
Indian River							
Micco	27°52.4′	80°29.8′	+1:37	+2:23	*0.13	*0.50	0.3
Sebastian	27°48.7′	80°27.8′	+1:55	+2:35	*0.14	*0.44	0.4
Wabasso	27°45.3′	80°25.6′	+2:43	+3:28	*0.16	*0.38	0.5
Vero Beach	27°38.0′	80°22.5′	+3:19	+3:45	*0.34	*0.69	1.0
Oslo	27°35.6′	80°21.4′	+3:23	+4:03	*0.31	*0.44	0.9
St. Lucie	27°28.7′	80°20.0′	+1:04	+1:50	*0.45	*0.88	1.3
Sebastian Inlet bridge	27°51.6′	80°26.9′	–0:25	–0:20	*0.86	*0.88	2.6
Vero Beach, ocean	27°40.2′	80°21.6′	–0:33	–0:33	*1.37	*1.31	4.1
Fort Pierce Inlet, south jetty	27°28.2′	80°17.3′	–0:08	–0:14	*1.06	*1.31	3.1
Fort Pierce Inlet	27°28.1′	80°17.8′	+0:09	+0:03	*0.76	*1.12	2.2
Indian River (cont.)							
Fort Pierce	27°27.4′	80°19.4′	+1:12	+1:05	*0.52	*1.00	1.5
Ankona	27°21.3′	80°16.5′	+2:39	+3:07	*0.46	*0.75	1.3
Eden, Nettles Island	27°17.2′	80°13.6′	+2:58	+3:35	*0.42	*0.81	1.2
Jensen Beach	27°14.1′	80°12.6′	+2:40	+3:08	*0.44	*0.81	1.3
St. Lucie River							
North Fork	27°14.6′	80°18.8′	+2:51	+3:32	*0.42	*0.81	1.2
Stuart	27°12.0′	80°15.5′	+2:36	+3:34	*0.37	*0.75	1.1
South Fork	27°09.9′	80°15.3′	+2:55	+3:36	*0.40	*0.81	1.1
Sewall Point	27°10.5′	80°11.3′	+1:36	+2:13	*0.40	*0.81	1.1
Port Salerno, Manatee Pocket	27°09.1′	80°11.7′	+1:14	+1:50	*0.39	*0.81	1.1
Seminole Shores	27°11.0′	80°09.5′	–0:36	–0:31	*1.19	*1.12	3.6
Great Pocket	27°09.1′	80°10.3′	+1:18	+1:46	*0.46	*0.88	1.3
Peck Lake, ICW	27°06.8′	80°08.7′	+1:36	+2:14	*0.53	*0.88	1.5
Gomez, South Jupiter Narrows	27°05.7′	80°08.2′	+1:56	+2:41	*0.55	*0.94	1.6
Hobe Sound bridge	27°03.8′	80°07.4′	+1:51	+2:29	*0.63	*0.88	1.8
Hobe Sound, Jupiter Island	27°02.2′	80°06.4′	+1:39	+2:16	*0.70	*0.88	2.1
Conch Bar, Jupiter Sound	26°59.3′	80°05.6′	+1:19	+1:38	*0.69	*0.94	2.0
Jupiter Sound, south end	26°57.1′	80°04.7′	+0:45	+0:49	*0.81	*1.19	2.4
Jupiter Inlet, south jetty	26°56.6′	80°04.4′	+0:13	–0:05	*1.00	*1.25	3.0
Jupiter Inlet, US Highway 1 bridge	26°56.9′	80°05.1′	+0:51	+1:09	*0.79	*1.00	2.4
Loxahatchee River							
A1A highway bridge	26°56.8′	80°05.4′	+0:57	+0:58	*0.81	*1.00	2.4
Tequesta	26°57.0′	80°06.1′	+1:22	+2:02	*0.75	*1.00	2.2
Tequesta, North Fork entrance	26°57.1′	80°06.1′	+1:14	+1:46	*0.72	*0.81	2.2
Tequesta, North Fork	26°57.6′	80°06.3′	+1:37	+2:17	*0.70	*0.88	2.1
North Fork, 2 mi above entrance	26°58.6′	80°06.9′	+1:27	+1:59	*0.79	*1.00	2.3
3 mi above A1A highway bridge	26°58.2′	80°07.5′	+1:19	+1:53	*0.80	*1.00	2.4
Boy Scout dock	26°59.2′	80°08.5′	+1:24	+2:01	*0.85	*1.19	2.5
Southwest Fork, 0.5 mi abv entr	26°56.6′	80°07.2′	+1:04	+1:39	*0.82	*1.25	2.4
Southwest Fork, spillway	26°56.1′	80°08.6′	+1:15	+1:49	*0.79	*1.12	2.3

*Heights for this station are found by multiplying heights in the appropriate tide tables by the *ratio listed here. When parentheses are used, multiply heights by the *ratio and then add or subtract the second number as indicated.

T Tides

PLACE	POSITION		DIFFERENCES				RANGE
			Time		Height		
	north latitude	west longitude	high h m	low h m	high ft	low ft	spring ft
FLORIDA (cont.)							
Eastern Standard Time (75°)			*Add 1 hour Daylight Saving April 5 – October 24*				
			on MIAMI, p. T 7				
Jupiter, Lake Worth Creek, ICW	26°56.1′	80°05.1′	+0:57	+1:16	*0.84	*1.12	2.5
Lake Worth Crk, daymark 19, ICW	26°54.7′	80°04.8′	+0:52	+1:12	*0.85	*1.06	2.5
Donald Ross Bridge, ICW	26°52.9′	80°04.2′	+0:43	+0:54	*0.93	*1.06	2.8
PGA Boulevard bridge, ICW	26°50.6′	80°04.0′	+0:21	+0:35	*1.07	*1.19	3.2
Lake Worth							
North Palm Beach	26°49.6′	80°03.3′	+0:05	+0:17	*1.15	*1.19	3.4
Port of Palm Beach	26°46.2′	80°03.1′	+0:02	+0:08	*1.09	*1.19	3.3
Palm Beach	26°44.0′	80°02.5′	+0:10	+0:19	*1.10	*1.19	3.3
Palm Beach, Hwy 704 bridge	26°42.3′	80°02.7′	+0:42	+0:45	*1.02	*1.00	3.1
West Palm Beach Canal	26°38.7′	80°02.7′	+1:10	+1:37	*1.02	*1.12	3.0
Boynton Beach	26°32.9′	80°03.2′	+1:26	+2:09	*1.01	*1.00	3.0
Lake Worth pier, ocean	26°36.7′	80°02.0′	–0:17	–0:13	*1.12	*1.00	3.4
Ocean Ridge, ICW	26°31.6′	80°03.2′	+1:39	–2:14	*1.01	*1.12	3.1
Delray Beach, ICW	26°28.4′	80°03.7′	+1:42	+2:01	*1.00	*1.12	3.0
South Delray Beach, ICW	26°26.8′	80°03.9′	+1:42	+2:01	*1.00	*1.12	3.0
Yamato, ICW	26°24.2′	80°04.2′	+1:43	+1:59	*0.97	*1.12	2.9
Lake Wyman, ICW	26°22.2′	80°04.2′	+1:38	+1:51	*0.91	*1.12	2.7
Boca Raton, Lake Boca Raton	26°20.6′	80°04.6′	+0:46	+1:11	*0.90	*1.00	2.7
Deerfield Beach, Hillsboro River	26°18.8′	80°04.9′	+0:51	+1:07	*0.94	*0.94	2.8
Hillsboro Beach, ICW	26°16.5′	80°04.8′	+0:25	+0:38	*0.98	*0.94	3.0
Hillsboro Inlet, CG light station	26°15.5′	80°04.9′	+0:09	+0:07	*1.00	*1.12	3.0
Hillsboro Inlet Marina	25°15.6′	80°05.1′	+0:17	+0:28	*0.98	*1.00	2.9
Hillsboro Inlet, ocean	25°15.4′	80°04.8′	0:00	+0:04	*1.04	*1.06	3.1
Lauderdale-by-the-Sea, fish pier	26°11.3′	80°05.6′	–0:11	–0:09	*1.06	*1.12	3.2
Fort Lauderdale							
Bahia Mar Yacht Club	26°06.8′	80°06.5′	+0:18	+0:37	*0.97	*1.06	2.9
Andrews Ave. bridge, New River	26°07.1′	80°08.7′	+0:38	+0:55	*0.85	*0.94	2.6
Mayan Lake	26°06.0′	80°06.5′	+0:43	+1:06	*0.84	*0.88	2.5
Port Everglades, Turning Basin	26°05.5′	80°07.4′	–0:08	–0:09	*1.03	*1.06	3.1
South Port Everglades, ICW	26°04.9′	80°07.0′	0:00	+0:01	*1.02	*1.25	3.0
Whiskey Creek, north end	26°04.8′	80°06.7′	0:00	–0:02	*1.01	*1.12	3.0
Port Laudania, Dania cutoff canal	26°03.6′	80°07.8′	+0:24	+0:15	*0.93	*1.06	2.8
Whiskey Creek, south end	26°03.3′	80°06.8′	+0:26	+0:33	*0.91	*1.06	2.7
Hollywood Beach, W Lake, N end	26°02.6′	80°07.6′	+1:31	+1:46	*0.78	*0.94	2.3
Hollywood Beach, W Lake, S end	26°02.0′	80°07.4′	+1:25	+1:49	*0.82	*1.00	2.4
Hollywood Beach	26°02.4′	80°06.9′	+1:00	+1:08	*0.84	*1.00	2.5
Golden Beach, ICW	25°58.0′	80°07.4′	+1:36	+2:01	*0.84	*0.94	2.5
Dumfoundling Bay	26°56.5′	80°07.5′	+1:39	+2:09	*0.83	*0.94	2.5
Sunny Isles, Biscayne Creek	25°56′	80°08′	+2:21	+2:28	*0.71	*0.69	2.2
North Miami Beach, fishing pier	25°55.8′	80°07.2′	0:00	+0:02	*1.02	*1.12	3.0
North Miami Beach, Haulover pier	26°54.2′	80°07.2′	–0:04	–0:02	*1.00	*0.81	3.0
Bakers Haulover Inlet, inside	25°54.2′	80°07.6′	+1:16	+1:34	*0.82	*0.89	2.4
Indian Creek Golf Club, ICW	25°52.5′	80°09′	+1:36	+1:50	*0.85	*0.81	2.6
MIAMI HARBOR ENTRANCE	25°46.2′	80°08′	daily predictions				3.0
Government Cut, Miami Hbr entr	25°45.8′	80°08′	+0:30	+0:04	*0.93	*1.00	2.8
Biscayne Bay							
Miami, 79th St. Causeway	25°51′	80°10′	+1:43	+2:14	*0.78	*0.75	2.4

*Heights for this station are found by multiplying heights in the appropriate tide tables by the *ratio listed here. When parentheses are used, multiply heights by the *ratio and then add or subtract the second number as indicated.

PLACE	POSITION		DIFFERENCES				RANGE
			Time		Height		
	north latitude	west longitude	high h m	low h m	high ft	low ft	spring ft
FLORIDA (cont.)							
Eastern Standard Time (75°)			*Add 1 hour Daylight Saving April 5 – October 24*				
			on MIAMI, p. T 7				
San Marino Island	25°47.6′	80°09.8′	+1:00	+1:02	*0.85	*0.88	2.6
Miami, marina	25°46.7′	80°11.1′	+0:54	+0:56	*0.88	*0.88	2.6
Miami, causeway, east end	25°46′	80°09′	+1:17	+1:11	*0.78	*0.81	2.4
Dodge Island, Fishermans Chan	25°46.2′	80°10.1′	+0:57	+1:14	*0.84	*0.88	2.5
Dinner Key Marina	25°43.6′	80°14.2′	+1:17	+1:52	*0.78	*0.81	2.3
Florida Keys							
Key Biscayne Yacht Club	25°41.9′	80°10.2′	+1:07	+1:35	*0.80	*0.81	2.4
Cutler, Biscayne Bay	25°37.0′	80°18.3′	+1:23	+2:00	*0.79	*0.88	2.4
Soldier Key	25°35′	80°10′	+0:53	+1:20	*0.74	*0.75	2.3
Fowey Rocks	25°35′	80°06′	+0:01	+0:03	*0.97	*0.94	2.9
Ragged Keys, Biscayne Bay	25°32.0′	80°10.3′	+1:07	+1:25	*0.66	*0.66	1.7
Elliott Key, outside	25°29′	80°11′	–0:04	0:00	*0.93	*0.94	2.8
Elliott Key Harbor	25°27.2′	80°11.8′	+2:19	+3:04	*0.59	*0.56	1.5
Adams Key, Biscayne Bay	25°23.8′	80°14.0′	+1:24	+1:12	*0.61	*0.61	1.5
Christmas Point, Elliot Key	25°23.5′	80°13.8′	+0:36	+0:41	*0.73	*0.73	1.8
Turkey Point, Biscayne Bay	25°26.2′	80°19.7′	+2:33	+3:25	*0.65	*0.65	1.6
Totten Key	25°22.7′	80°15.4′	+2:42	+3:25	*0.50	*0.50	1.3
Ocean Reef Harbor, Key Largo	25°18.6′	80°16.8′	+0:13	+0:18	*0.93	*0.93	2.3
Carysfort Reef	25°13.3′	88°12.7′	+0.42	+0.43	*0.93	*0.93	2.3
Largo Sound, Key Largo	25°08.3′	80°23.7′	+2.36	+3.07	*0.32	*0.32	0.8
Key Largo, South Sound, Key Largo	25°06.8′	80°25′	+0.46	+1.53	*0.61	*0.56	1.6
Garden Cove, Key Largo	25°10.3′	80°22′	+0:22	+0:29	*0.86	*0.86	2.2
Mosquito Bank	25°04′	80°24′	+0:22	+0:31	*0.85	*0.88	2.6
Molasses Reef	25°01′	80°23′	+0:14	+0:12	*0.88	*0.88	2.6
Pumpkin Key, Card Sound	25°19.5′	80°17.6′	+2:58	+2:56	*0.25	*0.25	0.6
Tavernier, Hawk Channel	25°00.2′	80°31.0′	+0:31	+0:29	*0.83	*0.83	2.1
Plantation Key, Hawk Channel	24°58.4′	88°33′	+0.28	+0.16	*0.88	*0.88	2.2
Alligator Reef Light	24°51.0′	80°37.1′	+0:31	+0:28	*0.77	*0.77	1.9
Channel Five, east side, Hawk Chan	24°50.2′	80°46.0′	–0:54	–0:42	*0.90	*0.58	1.3
Channel Five, west side, Hawk Chan	24°50.4′	80°46.8′	–0:58	–0:41	*1.00	*0.67	1.4
Florida Bay			**on KEY WEST, p. T 13**				
Long Key Channel, east	24°48′	80°51′	–1:10	–1:07	*0.84	*0.42	1.5
Long Key Channel, west end	24°48′	80°53′	+5:58	+5:40	*0.79	*0.38	1.4
Toms Harbor Cut	24°47.0′	80°54.4′	–1.19	–0.30	*0.37	*0.38	0.5
Duck Key, Hawk Channel	24°46.0′	80°54.8′	–1:11	–0:40	*0.96	*0.50	1.4
Grassy Key, north side	24°46.4′	80°56.5′	+5:41	+6:49	*0.66	*0.66	1.0
Grassy Key, south side	24°45.3′	80°57.5′	–0:52	–0:26	*0.71	*0.72	2.1
Flamingo	25°08.5′	80°55.4′	+5:28	+7:20	*1.47	*1.08	2.0
Fat Deer Key	24°44.0′	81°01.1′	+5:09	+6:26	*0.87	*0.87	1.4
Boot Key Harbor bridge	24°42.2′	81°06.3′	–1:03	–0:37	*1.13	*0.75	1.6
Key Colony Beach	24°43.1′	81°81.1′	–1.17	–0.53	*1.22	*0.83	1.7
Vaca Key, USCG station, Florida Bay	24°42.6′	81°06.4′	daily predictions				0.9
Sombrero Key, Hawk Channel	24°37.6′	81°06.7′	–1:03	–0:39	*1.18	*0.79	1.6
Knight Key Channel	24°42.4′	81°07.5′	–2:00	–0:18	*0.54	*0.50	0.7
Pigeon Key, south side	24°42.2′	81°09.3′	–0:55	–0:26	*0.81	*0.50	1.1
Pigeon Key, north side	24°42.3′	81°09.4′	–0:10	+0:45	*0.46	*0.46	0.6
Molasses Key	24°41.0′	81°11.5′	–0:56	–0:16	*0.79	*0.50	1.1
Money Key	24°41.0′	81°12.9′	+0:03	+1:17	*0.58	*0.58	0.8

*Heights for this station are found by multiplying heights in the appropriate tide tables by the *ratio listed here. When parentheses are used, multiply heights by the *ratio and then add or subtract the second number as indicated.

T

Tides

PLACE	POSITION		DIFFERENCES				RANGE
			Time		Height		
	north latitude	west longitude	high h m	low h m	high ft	low ft	spring ft
FLORIDA (cont.)							
Eastern Standard Time (75°)			*Add 1 hour Daylight Saving April 5 – October 24*				
			on KEY WEST, p. T 13				
Little Duck Key, east end, Hawk Ch	24°40.9′	81°13.7′	–0.49	+0.05	*0.67	*0.67	0.9
West Bahia Honda Key	24°46.8′	81°16.3′	+3:59	+4:01	*0.97	*1.00	1.6
Horseshoe Keys, south end	24°46.0′	81°17.0′	+3:54	+3:09	*0.86	*1.00	1.4
Johnson Keys, south end	24°44.6′	81°18.0′	+3:36	+2:33	*0.72	*0.96	1.1
Johnson Keys, north end	24°46.0′	81°19.4′	+3:35	+4:22	*1.31	*1.38	2.1
Bahia Honda Key, Bahia Honda Ch	24°39.3′	81°16.9′	–0:45	–0:27	*0.86	*0.62	1.5
Big Pine Key							
Spanish Harbor	24°38.9′	81°19.8′	–0:44	–0:03	*0.75	*0.42	1.3
Doctors Arm, Bogie Channel	24°41.4′	81°21.4′	+0:41	+1:47	*0.63	*0.71	1.0
Bogie Channel bridge	24°41.9′	81°20.9′	+2:10	+2:11	*0.65	*0.83	1.0
No Name Key, east side	24°41.9′	81°19.1′	+1:35	+1:33	*0.58	*0.83	0.9
Little Pine Key, south end	24°42.8′	81°18.2′	+1:07	+1:07	*0.56	*0.79	0.9
Big Spanish Channel							
Porpoise Key	24°43.1′	81°21.1′	+3:23	+2:29	*0.72	*1.00	1.1
Water Key, west end	24°44.4′	81°20.5′	+3:23	+2:37	*0.81	*1.04	1.3
Mayo Key	24°44.0′	81°21.7′	+3:35	+3:01	*0.92	*1.08	1.5
Crawl Key	24°45.4′	81°21.5′	+3:34	+4:13	*1.33	*1.33	2.2
Annette Key, north end	24°45.5′	81°23.4′	+3:30	+4:33	*1.44	*1.29	2.4
Little Pine Key, north end	24°45.0′	81°19.7′	+3:38	+3:28	*1.05	*1.21	1.7
Big Pine Key, northeast shore	24°43.7′	81°23.2′	+3:19	+2:30	*0.86	*1.08	1.4
Big Pine Key, north end	24°44.7′	81°23.7′	+4:24	+5:56	*0.96	*0.83	1.6
Little Spanish Key, Spanish Banks	24°46.5′	81°22.2′	+3:25	+4:30	*1.74	*1.62	2.9
Big Spanish Key	24°47.3′	81°24.7′	+3:19	+4:29	*1.97	*1.50	3.4
Munson Is, Newfound Hbr Chan	24°37.4′	81°24.2′	–0:40	–0:12	*0.98	*0.67	1.7
Ramrod Key, Newfound Harbor	24°39.0′	81°24.2′	–0:41	+0:05	*0.90	*0.50	1.6
Middle Torch Ky, Torch Ramrod Ch	24°39.7′	81°24.1′	–0:16	+1:29	*0.69	*0.38	1.2
Little Torch Key, Torch Channel	24°39.9′	81°23.7′	+0:11	+1:45	*0.57	*0.33	1.0
Big Pine Key, Newfound Hbr Ch	24°39.1′	81°22.5′	–0:09	+0:44	*0.82	*0.46	1.5
Big Pine Key, Coupon Bight	24°39.1′	81°21.0′	–0:20	+0:49	*0.87	*0.54	1.5
Pine Channel Bridge							
Little Torch Key, south side	24°39.9′	81°23.3′	–0:15	+0:57	*0.68	*0.33	1.0
Little Torch Key, north side	24°39.9′	81°23.2′	–0:13	+0:54	*0.69	*0.38	1.0
Big Pine Key, south side	24°40.1′	81°22.3′	–0:13	+1:03	*0.67	*0.33	1.2
Big Pine Key, north side	24°40.2′	81°22.1′	+0:03	+1:44	*0.57	*0.33	1.0
Big Pine Key, west side, Pine Chan	24°41.4′	81°23.0′	+0:21	+1:52	*0.52	*0.42	0.9
Howe Key, S end, Harbor Channel	24°43.5′	81°24.4′	+4:43	+4:49	*0.72	*0.62	1.2
Big Torch Key, Harbor Channel	24°44.3′	81°26.6′	+3:47	+5:51	*1.58	*1.29	2.7
Water Keys, S end, Harbor Chan	24°44.8′	81°27.0′	+3:42	+5:41	*1.52	*1.00	2.6
Howe Key, northwest end	24°45.5′	81°25.7′	+3:29	+5:22	*1.68	*1.33	2.9
Summerland Key, Niles Channel S	24°39.1′	81°26.1′	–0:36	+0:11	*0.85	*0.71	1.4
Summerland Key, Niles Chan brdg	24°39.6′	81°26.2′	–0:10	+0:56	*0.67	*0.58	1.1
Ramrod Key, Niles Channel bridge	24°39.6′	81°25.4′	–0:13	+1:12	*0.67	*0.46	1.2
Big Torch Key, Niles Channel	24°42.3′	81°26.0′	+3:15	+2:05	*0.61	*0.71	1.0
Knockemdown Key, north end	24°42.9′	81°28.7′	+3:30	+4:54	*1.35	*1.21	2.3
Raccoon Key, east side	24°44.5′	81°29.0′	+3:20	+5:09	*1.50	*1.21	2.6
Content Key, Content Passage	24°47.4′	81°29.0′	+2:47	+3:50	*2.13	*1.83	3.6
Key Lois, southeast end	24°36.4′	81°28.2′	–1:15	–0:45	*1.06	*0.75	1.8

*Heights for this station are found by multiplying heights in the appropriate tide tables by the *ratio listed here. When parentheses are used, multiply heights by the *ratio and then add or subtract the second number as indicated.

PLACE	POSITION		DIFFERENCES				RANGE
			Time		Height		
	north latitude	west longitude	high h m	low h m	high ft	low ft	spring ft
FLORIDA (cont.)							
Eastern Standard Time (75°)			*Add 1 hour Daylight Saving April 5 – October 24*				
			on KEY WEST, p. T 13				
Sugarloaf Key, E side, Tarpon Crk	24°37.7′	81°30.6′	–0:41	+0:15	*0.89	*0.58	1.6
Gopher Key, Cudjoe Bay	24°38.5′	81°29.1′	–0:46	+0:17	*0.90	*0.71	1.5
Sugarloaf Key, Pirates Cove	24°39.2′	81°30.9′	–0:48	+1:41	*0.59	*0.75	0.9
Cudjoe Key, Cudjoe Bay	24°39.6′	81°29.5′	–0:38	+0:41	*0.87	*0.71	1.5
Summerland Key, SW side, Kemp C	24°39.0′	81°26.8′	–0:26	+0:50	*0.81	*0.54	1.4
Cudjoe Key, Kemp Channel bridge	24°39.7′	81°28.1′	---	---	*0.59	*0.59	0.9
Cudjoe Key, NE side, Kemp Chan	24°41.2′	81°29.0′	+3:45	---	---	---	--
Cudjoe Key, N end, Kemp Chan	24°42.0′	81°30.6′	+3:32	+4:40	*1.63	*1.46	2.7
Sugarloaf Key, NE side, Bow Chan	24°40.3′	81°32.0′	+3:47	+3:24	*1.01	*0.71	1.8
Cudjoe Key, Pirates Cove	24°39.8′	81°30.8′	+3:50	+2:55	*0.77	*0.79	1.3
Sugarloaf Key, north end, Bow Ch	24°41.6′	81°33.3′	+3:37	+5:20	*1.29	*0.75	2.3
Pumpkin Key, Bow Channel	24°43.0′	81°33.7′	+3:17	+4:39	*1.56	*1.17	2.7
Sawyer Key, outside, Cudjoe Chan	24°45.5′	81°33.7′	+2:45	+5:24	*1.57	*0.50	2.9
Sawyer Key, inside, Cudjoe Chan	24°45.5′	81°33.7′	+2:37	+5:19	*1.43	*0.50	2.6
Johnston Key, SW end, Turkey Bsn	24°42.6′	81°35.6′	+3:26	+5:38	*1.10	*0.50	2.0
Upper Sugarloaf Sound							
Perky	24°38.9′	81°34.2′	+5:37	+8:25	*0.28	*0.08	0.5
Park Channel bridge	24°39.3′	81°32.4′	+5:47	+8:33	*0.26	*0.29	0.4
North Harris Channel	24°39.0′	81°33.2′	+5:32	+8:04	*0.25	*0.25	0.4
Tarpon Creek	24°37.8′	81°31.0′	–0:29	+0:17	*0.35	*0.38	0.6
Snipe Keys							
Inner Narrows, southeast end	24°39.5′	81°36.5′	+3:25	+5:39	*1.28	*0.83	2.2
Middle Narrows	24°40.0′	81°37.8′	+3:44	+5:54	*1.02	*0.67	1.8
Snipe Point	24°41.5′	81°40.4′	+2:15	+3:33	*1.69	*1.29	2.9
Waltz Key Basin							
Waltz Key	24°38.8′	81°39.2′	+3:53	+5:33	*1.03	*0.96	1.7
Duck Key Point, Duck Key	24°37.4′	81°41.1′	+3:27	+4:57	*1.19	*0.96	2.0
O'Hara Key, north end	24°37.0′	81°38.7′	+3:53	+5:39	*1.03	*0.83	1.8
Big Coppitt Key, northeast side	24°36.1′	81°39.3′	+4:21	+6:54	*0.84	*0.33	1.5
Saddlebunch Keys							
Channel No. 5	24°36.7′	81°37.5′	+4:32	+6:58	*0.66	*1.12	1.0
Channel No. 4	24°36.9′	81°37.0′	+4:35	+5:36	*0.54	*0.29	1.0
Channel No. 3	24°37.4′	81°36.2′	+1:44	–0:10	*0.43	*0.21	0.8
Similar Sound							
Bird Key	24°35.3′	81°38.3′	–0:21	+1:03	*0.59	*0.42	1.0
Shark Key, southeast end	24°36.2′	81°38.7′	+0:18	+1:51	*0.52	*0.46	0.9
Saddlebunch Keys	24°36.0′	81°37.3′	+0:39	+2:41	*0.37	*0.21	0.7
Rockland Key, Rockland Chan brdg	24°35.5′	81°40.1′	+5:02	+6:06	*0.76	*0.88	1.2
Boca Chica Key, Long Point	24°36.2′	81°41.9′	+3:54	+5:22	*0.94	*0.71	1.6
Channel Key, west side	24°36.2′	81°43.5′	+3:09	+3:07	*0.70	*0.71	1.1
Boca Chica Channel bridge	24°34.6′	81°43.2′	+1:23	+1:29	*0.57	*0.67	0.9
Key Haven, Stock Island Channel	24°34.8′	81°44.3′	+2:25	+2:57	*0.73	*0.79	1.2
Sigsbee Park, Garrison Bight Chan	24°35.1′	81°46.5′	+1:59	+2:06	*0.81	*0.88	1.0
Key West, south side, Hawk Chan	24°32.7′	81°47.0′	–0:52	–0:30	*1.07	*0.92	1.8
KEY WEST	24°33.2′	81°48.5′	daily predictions				1.6
Sand Key Lighthouse, Sand Key Ch	24°27.2′	81°52.6′	–1:03	–0:39	*0.94	*0.79	1.6
Garden Key, Dry Tortugas	24°37.6′	82°52.3′	+0:29	+0:33	*0.94	*1.33	1.4

*Heights for this station are found by multiplying heights in the appropriate tide tables by the *ratio listed here. When parentheses are used, multiply heights by the *ratio and then add or subtract the second number as indicated.

PLACE	POSITION		DIFFERENCES				RANGE
			Time		Height		
	north latitude	west longitude	high h m	low h m	high ft	low ft	spring ft
BERMUDA							
Atlantic Standard Time (60°)			*Add 1 hour Daylight Saving April 5 – October 24*				
			on ST. GEORGE'S, p. T 4				
ST. GEORGE'S ISLAND	32°23′	64°42′	daily predictions				3.0
St. David's Island	32°22′	64°39′	–0:07	–0:07	(*0.96	+0.23)	2.6
Great Sound	32°19′	64°50′	+0:15	+0:15	(*1.08	–0.14)	3.2
BAHAMA ISLANDS							
Eastern Standard Time (75°)			*Add 1 hour Daylight Saving April 5 – October 24*				
Grand Bahama			**on NASSAU, p. T 17**				
Freeport Harbour	26°31′	78°46′	0:00	0:00	(*1.13	–1.15)	3.3
Little Bahama Bank							
Memory Rock	26°57′	79°07′	+0:24	+0:29	*0.88	*0.88	2.7
Walkers Cay	27°16′	78°24′	+1:25	+1:25	(*0.95	–0.64)	2.9
The Abacos							
Allans–Pensacola Cays	26°59′	77°40′	+0:35	+0:45	(*0.95	–0.64)	2.9
Green Turtle Cay	26°46′	77°18′	+0:05	+0:05	(*1.00	–0.95)	3.0
Pelican Harbour	26°23′	76°58′	+0:25	+0:25	(*0.95	–0.64)	2.9
Great Bahama Bank							
North Bimini	25°44′	79°18′	+0:13	+0:25	*0.92	*0.92	2.9
Cat Cay	25°33′	79°17′	+0:23	+0:23	(*0.88	–0.47)	2.6
South Riding Rock	25°14′	79°10′	+0:40	+0:40	(*0.88	–0.48)	2.6
Guinchos Cay	22°45′	78°07′	+0:14	+0:19	*0.81	*0.81	2.6
Cay Sal Bank							
Elbow Cay	23°57′	80°28′	+1:26	+1:31	*0.81	*0.81	2.6
Berry Islands							
Great Stirrup Cay	24°49′	77°55′	+0:25	+0:25	(*0.88	–0.48)	2.6
Andros Island							
Mastic Point	25°03′	77°58′	+0:05	+0:05	(*0.97	–0.73)	2.6
Fresh Creek	24°42′	77°46′	+0:05	+0:05	(*1.08	–0.77)	3.2
New Providence							
NASSAU	25°05′	77°21′	daily predictions				3.0
Eleuthera							
Royal Island Harbour	25°31′	76°51′	+0:05	+0:05	(*0.88	–0.48)	2.6
Wide Opening	25°25′	76°41′	+0:25	+0:25	(*0.95	–0.64)	2.9
Tracking Station	25°15′	76°19′	+2:17	+2:36	*0.92	*0.92	2.9
Eleuthera Island, east coast	24°56′	76°09′	+0:15	+0:18	(*0.88	–0.47)	2.6
The Exumas							
Ship Channel	24°52′	76°48′	–0:15	–0:15	(*0.95	–0.64)	2.9
Steventon	23°40′	75°58′	–0:05	–0:05	(*0.88	–0.48)	2.6
George Town	23°32′	75°49′	–0:20	–0:20	(*0.83	–0.30)	2.6
Long Island							
Clarence Harbour	23°06′	74°59′	+0:49	+0:54	*1.00	*1.00	3.1
Hard Bargain	23°00′	74°57′	+0:40	+0:40	(*0.76	–0.15)	2.3
Jumentos Cays							
Nurse Channel	22°31′	75°51′	+0:15	+0:10	(*0.92	–0.46)	2.9
Cat Island							
The Bight	24°19′	75°26′	–0:35	–0:35	(*0.92	–0.46)	2.9
San Salvador	24°06′	74°26′	–0:35	–0:35	(*0.92	–0.46)	2.9

*Heights for this station are found by multiplying heights in the appropriate tide tables by the *ratio listed here. When parentheses are used, multiply heights by the *ratio and then add or subtract the second number as indicated.

PLACE	POSITION		DIFFERENCES				RANGE
			Time		Height		
	north latitude	west longitude	high h m	low h m	high ft	low ft	spring ft
BAHAMA ISLANDS (cont.)							
Eastern Standard Time (75°)			*Add 1 hour Daylight Saving April 5 – October 24*				
Acklins Island			**on NASSAU, p. T 17**				
Datum Bay	22°10′	74°18′	–0:15	–0:15	(*0.80	–0.17)	2.6
Mayaguana							
Abraham's Bay	22°22′	73°00′	+0:10	+0:13	*0.77	*0.77	2.5
Start Point	22°20′	73°03′	+0:25	+0:25	(*0.51	+0.32)	1.6
Little Inagua	21°27′	73°01′	+0:10	+0:10	(*0.88	–0.46)	2.6
Great Inagua							
Matthew Town	20°57′	73°41′	+0:15	+0:15	(*0.52	+0.38)	1.6
TURKS & CAICOS							
Eastern Standard Time (75°)			*Add 1 hour Daylight Saving April 5 – October 24*				
Turks Islands			**on NASSAU, p. T 17**				
Grand Turk	21°26′	71°09′	–0:15	–0:15	(*0.64	–0.14)	1.9
Hawks Nest Anchorage	21°26′	71°07′	–0:19	–0:14	*0.81	*0.81	2.6
North Caicos Island							
Sandy Point	21°56′	72°03′	+0:38	+0:38	(*0.56	–0.00)	1.6
CUBA							
Eastern Standard Time (75°)			*Add 1 hour Daylight Saving April 5 – October 24*				
North coast			**on PORT OF SPAIN, p. T 29**				
Los Arroyos	22°22′	84°23′	+4:55	+6:05	(*0.31	+0.18)	1.0
			on KEY WEST, p. T 13				
Bahía Honda	22°58′	83°13′	–1:04	– 0:23	*0.76	*0.76	1.4
Havana	23°09′	82°20′	–0:48	– 0:40	*0.76	*0.76	1.5
Matanzas	23°04′	81°32′	–0:59	– 0:59	*0.92	*0.92	1.5
Cardenas	23°04′	81°12′	–0:11	+ 0:34	*1.08	*1.08	1.8
			on NASSAU, p. T 17				
La Isabela	22°56′	80°00′	+1:41	+1:42	*0.62	*0.62	2.0
Cayo Paredón Grande	22°29′	78°09′	+0:15	+0:05	(*0.79	–0.99)	2.6
Bahía Nuevitas							
Entrance	21°38′	77°07′	+1:15	+1:05	(*0.44	–0.51)	1.3
Nuevitas	21°35′	77°15′	+2:55	+2:55	(*0.51	–0.67)	1.6
Puerto Padre	21°12′	76°36′	+1:25	+1:15	(*0.76	–1.02)	2.3
Puerto Gibara	21°06′	76°08′	+0:25	+0:20	(*0.64	–0.80)	2.0
Bahía Nipe							
Punta Caranero	20°47′	75°34′	+0:30	+0:25	(*0.76	–1.02)	2.3
Antilla	20°50′	75°44′	+0:50	+0:50	(*0.76	–1.02)	2.3
Bahía de Levisa entrance	20°45′	75°28′	+0:18	+0:19	*0.74	*0.74	2.2
Puerto Tánamo	20°43′	75°19′	+0:20	+0:20	(*0.64	–0.80)	2.0
Baracoa	20°21′	74°30′	+0:10	+0:10	(*0.60	–0.91)	1.9
Puerto Maisí	20°15′	74°08′	+0:05	+0:05	(*0.79	–0.99)	2.6
South coast			**on SAN JUAN, p. T 20**				**diurnal**
Guantánamo Bay	19°54′	75°09′	–0:17	–0:23	*0.89	*0.89	1.4
Puerto de Santiago de Cuba	19°59′	75°52′	+0:30	+0:17	*0.89	*0.89	1.4
Puerto de Pilón	19°54′	77°19′	+0:11	+0:13	*0.72	*0.72	1.2
Manzanillo	20°21′	77°07′	+1:41	+1:38	*1.39	*1.39	2.2

*Heights for this station are found by multiplying heights in the appropriate tide tables by the *ratio listed here. When parentheses are used, multiply heights by the *ratio and then add or subtract the second number as indicated.

PLACE	POSITION		DIFFERENCES				RANGE
			Time		Height		
	north latitude	west longitude	high h m	low h m	high ft	low ft	diurnal ft
CUBA (cont.)							
Eastern Standard Time (75°)			*Add 1 hour Daylight Saving April 5 – October 24*				
			on SAN JUAN, p. T 20				
Casilda	21°45′	79°59′	+1:04	+0:52	*0.65	*0.65	1.0
Bahía de Cienfuegos							
Punta Pasacaballos	22°04′	80°27′	+0:49	+0:58	*0.80	*0.80	1.3
Cienfuegos	22°08′	80°27′	+0:51	+0:58	*0.81	*0.81	1.3
Carapachibey, Isla de Pinos	21°27′	82°55′	+0:43	+0:52	*0.54	*0.54	0.9
La Coloma	22°14′	83°34′	+2:04	+ 2:23	*0.54	*0.54	0.9
Cabo San Antonio	21°52′	84°58′	–0:50	–0:07	*0.92	*0.92	1.5
CAYMAN ISLANDS							
Eastern Standard Time (75°)			*Daylight Saving Time not observed*				
			on CHARLOTTE AMALIE, p. T 23				
Grand Cayman†	19°20′	81°20′	+0:11	+0:15	*1.63	*1.63	1.3
JAMAICA							
Eastern Standard Time (75°)			*Daylight Saving Time not observed*				
North coast			**on CHARLOTTE AMALIE, p. T 23**				
South Negril Point†	18°18′	78°24′	+5:25	+5:29	*3.88	*2.12	1.7
Montego Bay	18°28′	77°55′	+1:28	+1:36	*1.25	*1.25	1.0
St. Anns Bay	18°25′	77°14′	+0:55	+0:59	*1.00	*1.00	0.8
			on CRISTÓBAL, p. T 41				
Port Antonio	18°11′	76°27°	–2:45	–3:00	(*0.89	+0.26)	0.9
South coast							
Port Morant†	17°53′	76°19°	–1:55	–2:15	(*0.90	+0.63)	0.9
Port Royal†	17°57′	76°50°	–1:25	–1:40	(*0.60	+0.87)	0.6
Port Esquivel	17°53′	77°08′	–1:22	–1:45	(*0.70	+0.24)	0.6
Black River	18°01′	77°51′	–1:38	–2:15	(*0.58	+0.69)	0.6
Savanna la Mar	18°12′	78°08′	–1:53	–2:30	(*1.00	+0.67)	0.9
HAITI							
Eastern Standard Time (75°)			*Add 1 hour Daylight Saving April 5 – October 24*				
			on SAN JUAN, p. T 20				
Massacre, Rivière du entrance	19°43′	71°46′	–1:01	–1:01	*1.45	*1.45	2.3
Port-au-Prince	18°33′	72°21′	–0:32	–0:32	+0.1	+0.1	0.6
			on CHARLOTTE AMALIE, p. T 23				
Jacmel†	18°13′	72°34′	–1:48	–1:44	(*2.68	–0.31)	1.3

†The tide at this location is chiefly diurnal.

*Heights for this station are found by multiplying heights in the appropriate tide tables by the *ratio listed here. When parentheses are used, multiply heights by the *ratio and then add or subtract the second number as indicated.

PLACE	POSITION		DIFFERENCES				RANGE
			Time		Height		
	north latitude	west longitude	high h m	low h m	high ft	low ft	diurnal ft
DOMINICAN REPUBLIC							
Eastern Standard Time (75°)			*Daylight Saving Time not observed*				
			on SAN JUAN, p. T 20				
Puerto Plata	19°49′	70°42′	–1:09	–1:14	*1.45	*1.45	2.3
Sanchez	19°13′	69°36′	–0:37	–0:37	*2.05	*2.05	3.3
Santa Barbara de Samaná	19°12′	69°20′	–0:51	–0:47	*1.25	*1.25	2.0
			on CHARLOTTE AMALIE, p. T 23				
Santo Domingo†	18°27′	69°53′	+1:44	–2:45	*1.00	*1.00	0.8
PUERTO RICO							
Atlantic Standard Time (60°)			*Daylight Saving Time not observed*				
West coast			**on SAN JUAN, p. T 20**				
Puerto Real	18°05′	67°11′	–0:33	–0:26	*0.72	*0.72	1.2
Mayagüez	18°13′	67°09′	–0:30	–0:21	*0.99	*0.99	1.6
North coast							
SAN JUAN	18°28′	66°01′	daily predictions				1.6
East coast							
Playa de Fajardo	18°20′	65°38′	–0:10	–0:13	*0.99	*0.99	1.6
Radas Roosevelt (Roosevelt Roads)	18°14′	65°37′	+0:02	+0:20	*0.63	*0.63	1.0
Culebra							
Ensenada Honda	18°18′	65°17′	–0:34	–0:15	*0.99	*0.99	1.0
			on CHARLOTTE AMALIE, p. T 23				
Isla Culebrita†	18°19′	65°14′	–0:57	+0:24	*1.39	*1.39	1.1
Isla de Vieques			**on SAN JUAN, p. T 20**				
Punta Mulas	18°09′	65°26′	–0:14	–0:17	*0.72	*0.72	1.2
			on CHARLOTTE AMALIE, p. T 23				
Puerto Ferro†	18°06′	65°26′	–0:49	+0:45	*1.00	*1.00	0.8
South coast							
Puerto Maunabo†	18°00′	65°53′	+0:41	–103	*0.88	*0.88	0.7
Arroyo†	17°58′	66°04′	+2:29	–2:03	*1.00	*1.00	0.8
Playa Cortada†	17°59′	66°27′	+1:15	–2:53	*1.00	*1.00	0.8
Playa de Ponce†	17°58′	66°37′	+0:58	–2:29	*1.00	*1.00	0.8
Guánica†	17°58′	66°55′	+0:15	–1:58	*0.88	*0.88	0.7
Isla Magueyes†	17°58′	67°03′	+1:35	–2:08	*0.88	*0.88	0.7
U.S. VIRGIN ISLANDS							
Atlantic Standard Time (60°)			*Daylight Saving Time not observed*				
			on SAN JUAN, p. T 20				
Magens Bay, St. Thomas	18°22′	64°55′	–0:16	–0:07	*0.89	*0.89	1.4
			on CHARLOTTE AMALIE, p. T 23				
CHARLOTTE AMALIE, St. Thomas†	18°20′	64°56′	daily predictions				0.8
St. John	18°20′	64°48′	+1:12	+1:16	(*0.82	+0.31)	1.3
Christiansted, St. Croix†	17°45′	64°42′	–1:05	–1:07	*1.00	*1.00	0.8

†The tide at this location is chiefly diurnal.

*Heights for this station are found by multiplying heights in the appropriate tide tables by the *ratio listed here. When parentheses are used, multiply heights by the *ratio and then add or subtract the second number as indicated.

PLACE	POSITION		DIFFERENCES				RANGE
			Time		Height		
	north latitude	west longitude	high h m	low h m	high ft	low ft	diurnal ft
BRITISH VIRGIN ISLANDS							
Atlantic Standard Time (60°)			*Daylight Saving Time not observed*				
			on CHARLOTTE AMALIE, p. T 23				
Tortola	18°26′	64°37′	+0:07	+0:11	(*0.82	+1.01)	1.0
Anegada	18°44′	64°23′	–0:58	–0:04	(*2.33	–0.48)	1.9
ANGUILLA							
Atlantic Standard Time (60°)			*Daylight Saving Time not observed*				
			on CHARLOTTE AMALIE, p. T 23				
Road Bay	18°12′	63°06′	–0:18	–0:14	(*0.84	+0.66)	1.3
ST. BARTHELEMY							
Atlantic Standard Time (60°)			*Daylight Saving Time not observed*				
			on CHARLOTTE AMALIE, p. T 23				
St. Barthélemy†	17°54′	62°51′	–1:48	–1:02	*1.75	*1.75	1.4
ANTIGUA							
Atlantic Standard Time (60°)			*Daylight Saving Time not observed*				
			on CHARLOTTE AMALIE, p. T 23				
Parham	17°06′	61°46′	---	---	---	---	0.9
St. John's	17°08′	61°52′	+0:12	+0:12	(*0.81	+0.62)	1.3
GUADELOUPE							
Atlantic Standard Time (60°)			*Daylight Saving Time not observed*				
			on CHARLOTTE AMALIE, p. T 23				
Sainte Rose	16°20′	61°42′	–2:28	–2:24	(*1.28	+0.67)	0.9
Pointe-à-Pitre	16°14′	61°32′	–1:58	–1:54	(*1.05	+1.59)	0.9
Iles des Saintes	15°52′	61°36′	–3:33	–3:34	(*0.84	+1.12)	0.6
DOMINICA							
Atlantic Standard Time (60°)			*Daylight Saving Time not observed*				
			on CHARLOTTE AMALIE, p. T 23				
Portsmouth	15°34′	61°28′	+0:08	+0:06	(*1.09	+1.33)	1.0
Woodbridge Bay	15°19′	61°24′	+0:12	+0:16	(*1.47	+0.69)	1.3
			on FORT-DE-FRANCE, p. T 26				
Roseau	15°18′	61°24′	+0:26	+0:13	*1.71	*1.71	1.2

†The tide at this location is chiefly diurnal.

*Heights for this station are found by multiplying heights in the appropriate tide tables by the *ratio listed here. When parentheses are used, multiply heights by the *ratio and then add or subtract the second number as indicated.

PLACE	POSITION		DIFFERENCES				RANGE
			Time		Height		
	north latitude	west longitude	high h m	low h m	high ft	low ft	diurnal ft
MARTINIQUE							
Atlantic Standard Time (60°)			*Daylight Saving Time not observed*				
			on FORT-DE-FRANCE, p. T 26				
FORT-DE-FRANCE	14°35′	61°03′	daily predictions				0.5
ST. LUCIA							
Atlantic Standard Time (60°)			*Daylight Saving Time not observed*				
			on FORT-DE-FRANCE, p. T 26				
Castries	14°01′	61°00′	–0:10	–0:52	(*1.14	–0.65)	1.0
Vieux Fort Bay	13°44′	60°58′	+0:53	+0:40	*1.82	*1.82	1.0
ST. VINCENT & THE GRENADINES							
Atlantic Standard Time (60°)			*Daylight Saving Time not observed*				
			on FORT-DE-FRANCE, p. T 26				
Kingstown, St. Vincent	13°10′	61°13′	–0:10	–1:12	(*0.35	+1.12)	1.0
Mustique	12°51′	61°11′	+0:55	+0:23	(*1.50	–1.88)	1.2
Charlestown Bay, Canouan	12°42′	61°20′	+0:17	–0:32	(*1.50	–1.23)	1.3
Tobago Cays	12°38′	61°21′	+0:30	–0:22	(*1.49	–1.37)	1.3
Clifton Harbour, Union Island	12°36′	61°25′	+0:40	–0:44	(*1.14	+0.01)	0.9
GRENADA							
Atlantic Standard Time (60°)			*Daylight Saving Time not observed*				
			on FORT-DE-FRANCE, p. T 26				
Hillsborough Bay, Carriacou	12°29′	61°27′	+0:10	–0:15	(*1.14	–0.99)	0.9
St. George's Harbour	12°03′	61°45′	–0:10	–0:22	(*0.66	+0.41)	0.7
Prickly Bay	12°00′	61°45′	+0:35	–0:34	(*1.60	–1.36)	1.3
BARBADOS							
Atlantic Standard Time (60°)			*Daylight Saving Time not observed*				
			on PORT OF SPAIN, p. T 29				
Carlisle Bay, Bridgetown	13°06′	59°37′	–0:35	–0:36	(*0.89	–0.48)	1.9
TRINIDAD & TOBAGO							
Atlantic Standard Time (60°)			*Daylight Saving Time not observed*				
Tobago			**on PORT OF SPAIN, p. T 29**				
Plymouth	11°13′	60°47′	–1:05	–1:05	(*0.87	+0.06)	2.0
Man of War Bay	11°19′	60°32′	–0:35	–0:35	(*0.98	–1.12)	2.3
Scarborough	11°11′	60°44′	–0:14	–0:13	(*0.98	–0.12)	2.3
Trinidad, north coast							**spring**
Las Cuevas Bay	10°47′	61°24′	–1:01	–0:58	(*0.82	–0.20)	1.9
Toco	10°50′	60°56′	–1:10	–1:10	(*1.32	–0.61)	2.9

*Heights for this station are found by multiplying heights in the appropriate tide tables by the *ratio listed here. When parentheses are used, multiply heights by the *ratio and then add or subtract the second number as indicated.

PLACE	POSITION		DIFFERENCES				RANGE
			Time		Height		
	north latitude	west longitude	high h m	low h m	high ft	low ft	spring ft
TRINIDAD & TOBAGO (cont.)							
Atlantic Standard Time (60°)			*Daylight Saving Time not observed*				
Trinidad, east coast			**on PORT OF SPAIN, p. T 30**				
Guayamare Point	10°45′	60°58′	–1:00	–1:00	(*1.39	–0.52)	3.0
			on PUNTA GORDA, p. T 32				
Nariva River	10°24′	61°02′	–1:06	–2:16	(*0.41	+1.3)	3.1
Trinidad, south coast							
Guayaguayare Bay	10°09′	61°01′	–1:32	–2:09	(*0.53	+1.3)	3.8
Erin Bay	10°04′	61°39′	–0:50	–1:41	–0.3	+1.2	5.6
Trinidad, west coast							
Bonasse pier	10°05′	61°52′	–0:43	–1:15	–1.0	+1.4	4.4
Point Fortin	10°11′	61°42′	–0:54	–1:27	(*0.47	1.26)	3.6
			on PORT OF SPAIN, p. T 30				
Lisas Point	10°23′	61°29′	–0:09	–0:08	(*1.42	–0.87)	3.2
PORT OF SPAIN	10°39′	61°31′	daily predictions				2.3
Carenage Bay	10°41′	61°36′	0:00	0:00	*1.00	*1.00	2.6
			on PUNTA GORDA, p. T 32				
Staubles Bay	10°41′	61°39′	–1:07	–2:02	(*0.33	+1.7)	2.5
VENEZUELA							
Atlantic Standard Time (60°)			*Daylight Saving Time not observed*				
			on PUNTA GORDA, p. T 32				
Río Orinoco entr, Isla Ramon Isidro	8°39′	60°35′	+0:07	–0:12	+0.2	+1.0	6.7
Gulf of Paria							
PUNTA GORDA, Río San Juan	10°10′	62°38′	daily predictions				7.1
Barra de Maturin, channel entr.	10°18′	62°31′	–0:22	–0:45	–1.0	+0.2	5.7
Boca Pedernales entrance	10°01′	62°12′	–0:03	–0:34	–1.3	+0.2	5.4
Puerto de Hierro	10°37′	62°05′	–0:46	–1:19	*0.59	*0.59	4.2
Macuro	10°39′	61°56′	–1:15	–2:05	*0.38	*0.38	2.7
			on AMUAY, p. T 35				**diurnal**
Carúpano†	10°40′	63°15′	–1:17	–0:42	+0.2	0.0	1.4
Porlamar, Isla de Margarita†	10°57′	63°51′	–1:9	–0:59	+0.6	0.0	1.8
Cumaná†	10°28′	64°11′	-2:37	–1:02	–0.1	0.0	1.1
Carenero†	10°32′	66°07′	–1:51	–1:59	+0.8	+1.0	1.0
La Guaira†	10°36′	66°56′	-2:29	–1:59	+0.8	+1.0	1.0
AMUAY	11°45′	70°13′	daily predictions				1.2
			on MALECÓN, p. T 38				**spring**
MALECÓN–ISLA ZAPARA	11°00′	71°35′	daily predictions				3.6
Bahía de Tablazos	10°53′	71°35′	+0:30	+0:11	*0.61	*0.31	2.3
Punta de Palmas	10°48′	71°37′	+0:35	+0:16	*0.49	*0.31	1.8
BONAIRE							
Atlantic Standard Time (60°)			*Daylight Saving Time not observed*				
			on CRISTÓBAL, p. T 41				**diurnal**
Kralendijk	12°09′	68°17′	+0:35	+1:42	(*0.65	+0.98)	1.0

†The tide at this location is chiefly diurnal.

*Heights for this station are found by multiplying heights in the appropriate tide tables by the *ratio listed here. When parentheses are used, multiply heights by the *ratio and then add or subtract the second number as indicated.

PLACE	POSITION		DIFFERENCES				RANGE
			Time		Height		
	north latitude	west longitude	high h m	low h m	high ft	low ft	diurnal ft
CURAÇAO							
Atlantic Standard Time (60°)			*Daylight Saving Time not observed*				
			on CRISTÓBAL, p. T 41				
Schottegat†	12°07′	68°56′	+0:25	+1:09	*0.82	*0.82	0.9
Willemstad	12°06′	68°56′	+0:42	+1:38	(*0.65	+0.98)	1.0
ARUBA							
Atlantic Standard Time (60°)			*Daylight Saving Time not observed*				
			on CRISTÓBAL, p. T 41				
St. Nicolaas Bay	12°26′	69°54′	---	---	---	---	0.8
Oranjestad	12°31′	70°03′	+0:21	+1:16	(*0.65	+0.98)	1.0
Malmok Bay	12°36′	70°03′	+0:33	+1:36	(*0.65	+0.98)	1.0
COLOMBIA							
Eastern Standard Time (75°)			*Daylight Saving Time not observed*				
			on CRISTÓBAL, p. T 41				
Ríohacha	11°33′	72°55′	–0:35	–0:35	(*1.01	–0.18)	1.0
Santa Marta	11°15′	74°13′	–1:19	–1:08	(*0.99	–0.06)	1.0
Puerto Colombia	11°00′	74°58′	–0:52	–1:08	(*1.29	–0.16)	1.3
Cartagena	10°26′	75°34′	–1:16	–0:48	(*0.90	–0.05)	1.0
Puerto Coveñas	9°20′	75°40′	–1:06	–0:46	(*0.99	–0.06)	1.0
Turbo	8°10′	76°45′	–0:49	–0:30	*1.43	*1.43	1.4
Punta Yarumal	8°07′	76°45′	–0:49	–0:30	(*1.62	–0.16)	1.6
PANAMA							
Eastern Standard Time (75°)			*Daylight Saving Time not observed*				
			on CRISTÓBAL, p. T 41				
Puerto Caledonia	8°54′	77°41′	+0:12	0:00	(*1.01	+0.04)	1.0
Puerto Mandinga	9°30′	79°03′	+0:08	–0:05	(*1.01	+0.04)	1.0
CRISTÓBAL–COLÓN	9°21′	79°55′	daily predictions				1.1
Escudo de Veragua	9°07′	81°34′	+0:30	+0:15	(*0.89	–0.07)	1.0
Bocas del Toro	9°21′	82°15′	+0:20	+0:25	(*0.99	–0.06)	1.0
COSTA RICA							
Central Standard Time (90°)			*Daylight Saving Time not observed*				
			on CRISTÓBAL, p. T 41				
Puerto Limón	10°00′	83°02′	–0:39	–0:37	(*0.90	–0.02)	1.0
NICARAGUA							
Central Standard Time (90°)			*Daylight Saving Time not observed*				
			on PORT OF SPAIN, p. T 29				
Greytown (San Juan del Norte)	10°55′	83°42′	–2:30	–1:50	(*0.50	–0.58)	1.3
Bluefields Bluff	12°00′	83°40′	–2:35	–1:55	(*0.33	–0.40)	1.0
Isla del Maíz Grande (Great Corn Is)	12°10′	83°03′	–1:55	–1:40	(*0.49	–0.36)	1.3
Cayos del Perlas (Pearl Cays)	12°25′	83°25′	–1:40	–1:20	(*0.49	–0.36)	1.3

†The tide at this location is chiefly diurnal.

*Heights for this station are found by multiplying heights in the appropriate tide tables by the *ratio listed here. When parentheses are used, multiply heights by the *ratio and then add or subtract the second number as indicated.

PLACE	POSITION		DIFFERENCES				RANGE
			Time		Height		
	north	west	high	low	high	low	diurnal
	latitude	longitude	h m	h m	ft	ft	ft
NICARAGUA (cont.)							
Central Standard Time (90°)			*Daylight Saving Time not observed*				
			on NASSAU, p. T 17				
Puerto Cabezas (Bragman's Bluff)	14°01′	83°23′	+4:26	+4:37	*0.54	*0.54	1.9
Cabo Gracias á Dios	15°00′	83°10′	+1:44	+0:54	*0.54	*0.54	1.6
HONDURAS							
Central Standard Time (90°)			*Daylight Saving Time not observed*				
			on KEY WEST, p. T 13				
Islas Santanilla (Swan Is), Hbr Bay	17°24′	83°42′	–1:18	–0:33	*0.51	*0.51	0.9
Isla de Guanaja (Bonacca)	16°29′	85°54′	–1:26	–1:42	*0.72	*0.72	1.3
Isla de Roatán, Port Royal	16°24′	86°20′	–2:41	–2:35	*0.92	*0.92	1.4
Puerto Castilla	16°00′	86°02′	–0:48	–0:13	*0.46	*0.46	0.8
Puerto Cortés	15°50′	87°57′	–0:43	–0:02	*0.38	*0.38	0.6
GUATEMALA							
Central Standard Time (90°)			*Daylight Saving Time not observed*				
			on KEY WEST, p. T 13				
Río Dulce entrance	15°50′	88°49′	–1:25	–1:35	*0.92	*0.92	1.5
			on KEY WEST, p. T 13				
BELIZE							
Central Standard Time (90°)			*Daylight Saving Time not observed*				
Punta Gorda	16°06′	88°49′	–0:27	+0:30	*0.46	*0.46	0.8
Belize City	17°30′	88°11′	+0:14	+0:47	*0.46	*0.46	0.7
			on TAMPICO, p. T 44				
MEXICO							
Central Standard Time (90°)			*Daylight Saving Time not observed*				
Progreso†	21°18′	89°40′	+1:19	+0:23	*1.29	*1.29	1.8
Campeche	19°51′	90°32′	–0:15	+0:50	(*2.14	–0.60)	2.0
Cuidad del Carmen	18°40′	91°51′	–0:40	+0:30	(*1.00	+0.70)	1.4
Frontera (Alvaro Obregón)†	18°32′	92°39′	–0:18	–0:27	*1.14	*1.14	1.6
Coatzacoalcos (Puerto México)†	18°09′	94°25′	–0:40	+0:05	*1.07	*1.07	1.5
Alvarado†	18°46′	95°46′	+0:51	+0:27	*0.93	*0.93	1.3
Veracruz†	19°12′	96°08′	–0:19	–0:12	*1.21	*1.21	1.7
Tuxpan†	21°00′	97°20′	+0:02	+0:04	*1.21	*1.21	1.7
TAMPICO HARBOR (Madero)†	22°13′	97°51′	daily predictions				1.4
Matamoros†	25°53′	97°31′	+0:55	+0:40	*1.00	*1.00	1.4

†The tide at this location is chiefly diurnal.

Heights for this station are found by multiplying heights in the appropriate tide tables by the *ratio listed here. When parentheses are used, multiply heights by the *ratio and then add or subtract the second number as indicated.

HEIGHT OF TIDE AT ANY TIME

There are several methods to estimate the height of tide for a given time between predicted highs and lows. On these pages, we present the Rule of Twelfths, NOS's Table 3, and a graphic solution.

Caution: All of the methods presented are based on the assumption that the rise and fall conform to simple cosine curves. The heights obtained will be approximate. The roughness of approximation will vary as the tide curve differs from a cosine curve. Semi-diurnal tide curves tend to be closer to cosine curves than mixed or especially diurnal curves. See the beginning of this chapter and/or *Reed's Companion* for a fuller explanation of tide types.

THE RULE OF TWELFTHS

A tide rises or falls (approximately)

$^1/_{12}$	of its range during the	**1st hour**
$^2/_{12}$	of its range during the	**2nd hour**
$^3/_{12}$	of its range during the	**3rd hour**
$^3/_{12}$	of its range during the	**4th hour**
$^2/_{12}$	of its range during the	**5th hour**
$^1/_{12}$	of its range during the	**6th hour**

This is very much an approximation, but it does give you a sense of how the tides accelerate and decelerate; and it gives you some tools with which to approximate intermediate heights.

For example, if the tide will rise 10 feet during its 6-hour cycle, the law of twelfths suggests that after 2 hours it will have risen 2.5 feet:

$(^1/_{12} + ^2/_{12}) = ^3/_{12} = ^1/_4 \times 10 = 2.5$

If a tide will fall 7 feet in its 6-hour cycle, after 4 hours it will have fallen 5.25 feet:

$^9/_{12} = ^3/_4 \times 7 = 5.25$

TABLE 3

Read the footnote to Table 3 on the next page first. If confused, you may wish to try this example.

We'll find the height of tide at 0755 on a day when the predicted tides before and after 0755 are given as

morning low: 0522, 0.1 feet.
morning high: 1114, 4.2 feet.

Therefore, the duration of rise is $11^h 14^m – 5^h 22^m = 5^h 52^m$. The time from the nearest high or low water is then $7^h 55^m$ - $5^h 22^m = 2^h 33^m$ (from low). The range of tide is given as 4.2 - 0.1 = 4.1 feet.

We then enter the left-hand boldfaced column of the table and find the value nearest our value for the duration of rise and fall ($5^h 52^m$), in this case $6^h 00^m$. Following across that row, we look for the tabular time that is closest to $2^h 33^m$, the time from the nearest tide—in this case $2^h 36^m$. Staying in the $2^h 36^m$ column, we move into the bottom section of the table (Correction to height) and look left across to rows to find the tabular value in the left-hand boldfaced column that is closest to our range of tide value (4.1 feet)--in this case 4.0 feet. Matching the 2^h 36^m column and the 4.0 row, we arrive at a correction of 1.6 feet. Because the nearest tide was low, we add the correction to the low:

0.1 + 1.6 = 1.7-foot tide height at 0755.

GRAPHIC METHOD

You can graph a typical tide using the *one-quarter, one-tenth rule*:

Plot the high- and low-water points in the order of their occurrence, measuring time horizontally and height vertically. Draw a light straight line connecting the points. Divide this line into four equal parts. At the quarter point adjacent to high water draw a vertical line above the point, and at the quarter point adjacent to low water draw a vertical line below the point, making the length of these lines equal to one-tenth of the range between the high and low waters used. Finally, draw a smooth curve through the points of high and low waters and the intermediate points, making the curve well rounded near high and low waters. This curve will approximate the tide curve, and heights for any time of the day may be scaled from it.

An example of the graphic method is illustrated below. Using the same predicted tides as in the above example, the approximate height at $1^h 00^m$ is 1 foot.

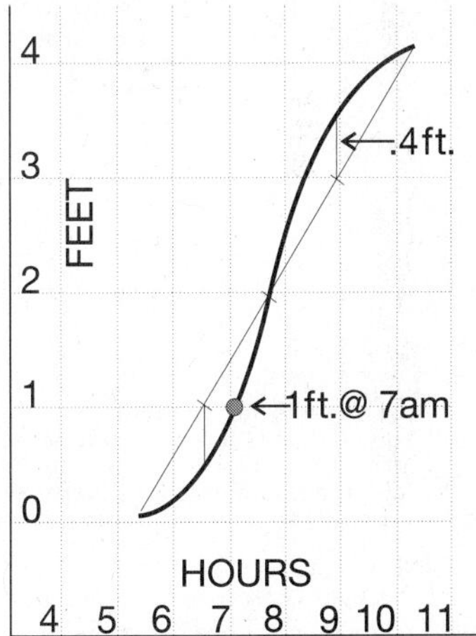

T
Tides

TABLE 3 – HEIGHT OF TIDE AT ANY TIME

Duration of Rise or Fall	Time from the nearest high water or low water														
h.m.	h.m.	h.m.	h.m.	h.m.	h.m.	h.m.	h.m.	h.m.	h.m.	h.m.	h.m.	h.m.	h.m.	h.m.	h.m.
4 00	0 08	0 16	0 24	0 32	0 40	0 48	0 56	1 04	1 12	1 20	1 28	1 36	1 44	1 52	2 00
4 20	0 09	0 17	0 26	0 35	0 43	0 52	1 01	1 09	1 18	1 27	1 35	1 44	1 53	2 01	2 10
4 40	0 09	0 19	0 28	0 37	0 47	0 56	1 05	1 15	1 24	1 33	1 43	1 52	2 01	2 11	2 20
5 00	0 10	0 20	0 30	0 40	050	1 00	1 10	1 20	1 30	1 40	1 50	2 00	2 10	2 20	2 30
5 20	0 11	0 21	0 32	0 43	0 53	1 04	1 15	1 25	1 36	1 47	1 57	2 08	2 19	2 29	2 40
5 40	0 11	0 23	0 34	0 45	0 57	1 08	1 19	1 31	1 42	1 53	2 05	2 16	2 27	2 39	2 50
6 00	0 12	0 24	0 36	0 48	1 00	1 12	1 24	1 36	1 48	2 00	2 12	2 24	2 36	2 48	3 00
6 20	0 13	0 25	0 38	0 51	1 03	1 16	1 29	1 41	1 54	2 07	2 19	2 32	2 45	2 57	3 10
6 40	0 13	0 27	0 40	0 53	1 07	1 20	1 33	1 47	2 00	2 13	2 27	2 40	2 53	3 07	3 20
7 00	0 14	0 28	0 42	0 56	1 10	1 24	1 38	1 52	2 06	2 20	2 34	2 48	3 02	3 16	3 30
7 20	0 15	0 29	0 44	0 59	1 13	1 28	1 43	1 57	2 12	2 27	2 41	2 56	3 11	3 25	3 40
7 40	0 15	0 31	0 46	1 01	1 17	1 32	1 47	2 03	2 18	2 33	2 49	3 04	3 19	3 35	3 50
8 00	0 16	0 32	0 48	1 04	1 20	1 36	1 52	2 08	2 24	2 40	2 56	3 12	3 28	3 44	4 00
8 20	0 17	0 33	0 50	1 07	1 23	1 40	1 57	2 13	2 30	2 47	3 03	3 20	3 37	3 53	4 10
8 40	0 17	0 35	0 52	1 09	1 27	1 44	2 01	2 19	2 36	2 53	3 11	3 28	3 45	4 03	4 20
9 00	0 18	0 36	0 54	1 12	1 30	1 48	2 06	2 24	2 42	3 00	3 18	3 36	3 54	4 12	4 30
9 20	0 19	0 37	0 56	1 15	1 33	1 52	2 11	2 29	2 48	3 07	3 25	3 44	4 03	4 21	4 40
9 40	0 19	0 39	0 58	1 17	1 37	1 56	2 15	2 35	2 54	3 13	3 33	3 52	4 11	4 31	4 50
10 00	0 20	0 40	1 00	1 20	1 40	2 00	2 20	2 40	3 00	3 20	3 40	4 00	4 20	4 40	5 00
10 20	0 21	0 41	1 02	1 23	1 43	2 04	2 25	2 245	3 06	3 27	3 47	4 08	4 29	4 49	5 10
10 40	0 21	0 43	1 04	1 25	1 47	2 08	2 29	2 51	3 12	3 33	3 55	4 16	4 37	4 59	5 20

Range of Tide	Correction to height														
Ft.	Ft.	Ft.	Ft.	Ft.	Ft.	Ft.	Ft.	Ft.	Ft.	Ft.	Ft.	Ft.	Ft.	Ft.	Ft.
0.5	0.0	0.0	0.0	0.0	0.0	0.0	0.1	0.1	0.1	0.1	0.1	0.2	0.2	0.2	0.2
1.0	0.0	0.0	0.0	0.0	0.1	0.1	0.1	0.2	0.2	0.2	0.3	0.3	0.4	0.4	0.5
1.5	0.0	0.0	0.0	0.1	0.1	0.1	0.2	0.2	0.3	0.4	0.4	0.5	0.6	0.7	0.8
2.0	0.0	0.0	0.0	0.1	0.1	0.2	0.3	0.3	0.4	0.5	0.6	0.7	0.8	0.9	1.0
2.5	0.0	0.0	0.1	0.1	0.2	0.2	0.3	0.4	0.5	0.6	0.7	0.9	1.0	1.1	1.2
3.0	0.0	0.0	0.1	0.1	0.2	0.3	0.4	0.5	0.6	0.8	0.9	1.0	1.2	1.3	1.5
3.5	0.0	0.0	0.1	0.2	0.2	0.3	0.4	0.6	0.7	0.9	1.0	1.2	1.4	1.6	1.8
4.0	0.0	0.0	0.1	0.2	0.3	0.4	0.5	0.7	0.8	1.0	1.2	1.4	1.6	1.8	2.0
4.5	0.0	0.0	0.1	0.2	0.3	0.4	0.6	0.7	0.9	1.1	1.3	1.6	1.8	2.0	2.2
5.0	0.0	0.1	0.1	0.2	0.3	0.5	0.6	0.8	1.0	1.2	1.5	1.7	2.0	2.2	2.5
5.5	0.0	0.1	0.1	0.2	0.4	0.5	0.7	0.9	1.1	1.4	1.6	1.9	2.2	2.5	2.8
6.0	0.0	0.1	0.1	0.3	0.4	0.6	0.8	1.0	1.2	1.5	1.8	2.1	2.4	2.7	3.0
6.5	0.0	0.1	0.2	0.3	0.4	0.6	0.8	1.1	1.3	1.6	1.9	2.2	2.6	2.9	3.2
7.0	0.0	0.1	0.2	0.3	0.5	0.7	0.9	1.2	1.4	1.8	2.1	2.4	2.8	3.1	3.5
7.5	0.0	0.1	0.2	0.3	0.5	0.7	1.0	1.2	1.5	1.9	2.2	2.6	3.0	3.4	3.8
8.0	0.0	0.1	0.2	0.3	0.5	0.8	1.0	1.3	1.6	2.0	2.4	2.8	3.2	3.6	4.0
8.5	0.0	0.1	0.2	0.4	0.6	0.8	1.1	1.4	1.8	2.1	2.5	2.9	3.4	3.8	4.2
9.0	0.0	0.1	0.2	0.4	0.6	0.9	1.2	1.5	1.9	2.2	2.7	3.1	3.6	4.0	4.5
9.5	0.0	0.1	0.2	0.4	0.6	0.9	1.2	1.6	2.0	2.4	2.8	3.3	3.8	4.3	4.8
10.0	0.0	0.1	0.2	0.4	0.7	1.0	1.3	1.7	2.1	2.5	3.0	3.5	4.0	4.5	5.0
10.5	0.0	0.1	0.3	0.5	0.7	1.0	1.3	1.7	2.2	2.6	3.1	3.6	4.2	4.7	5.2
11.0	0.0	0.1	0.3	0.5	0.7	1.1	1.4	1.8	2.3	2.8	3.3	3.8	4.4	4.9	5.5
11.5	0.0	0.1	0.3	0.5	0.8	1.1	1.5	1.9	2.4	2.9	3.4	4.0	4.6	5.1	5.8
12.0	0.0	0.1	0.3	0.5	0.8	1.1	1.5	2.0	2.5	3.0	3.6	4.1	4.8	5.4	6.0
12.5	0.0	0.1	0.3	0.5	0.8	1.2	1.6	2.1	2.6	3.1	3.7	4.3	5.0	5.6	6.2
13.0	0.0	0.1	0.3	0.6	0.9	1.2	1.7	2.2	2.7	3.2	3.9	4.5	5.1	5.8	6.5
13.5	0.0	0.1	0.3	0.6	0.9	1.3	1.7	2.2	2.8	3.4	4.0	4.7	5.3	6.0	6.8
14.0	0.0	0.2	0.3	0.6	0.9	1.3	1.8	2.3	2.9	3.5	4.2	4.8	5.5	6.3	7.0
14.5	0.0	0.2	0.4	0.6	1.0	1.4	1.9	2.4	3.0	3.6	4.3	5.0	5.7	6.5	7.2
15.0	0.0	0.2	0.4	0.6	1.0	1.4	1.9	2.5	3.1	3.8	4.4	5.2	5.9	6.7	7.5
15.5	0.0	0.2	0.4	0.7	1.0	1.5	2.0	2.6	3.2	3.9	4.6	5.4	6.1	6.9	7.8
16.0	0.0	0.2	0.4	0.7	1.1	1.5	2.1	2.6	3.3	4.0	4.7	5.5	6.3	7.2	8.0
16.5	0.0	0.2	0.4	0.7	1.1	1.6	2.1	2.7	3.4	4.1	4.9	5.7	6.5	7.4	8.2
17.0	0.0	0.2	0.4	0.7	1.1	1.6	2.2	2.8	3.5	4.2	5.0	5.9	6.7	7.6	8.5
17.5	0.0	0.2	0.4	0.8	1.2	1.7	2.2	2.9	3.6	4.4	5.2	6.0	6.9	7.8	8.8
18.0	0.0	0.2	0.4	0.8	1.2	1.7	2.3	3.0	3.7	4.5	5.3	6.2	7.1	8.1	9.0
18.5	0.1	0.2	0.5	0.8	1.2	1.8	2.4	3.1	3.8	4.6	5.5	6.4	7.3	8.3	9.2
19.0	0.1	0.2	0.5	0.8	1.3	1.8	2.4	3.1	3.9	4.8	5.6	6.6	7.5	8.5	9.5
19.5	0.1	0.2	0.5	0.8	1.3	1.9	2.5	3.2	4.0	4.9	5.8	6.7	7.7	8.7	9.8
20.0	0.1	0.2	0.5	0.9	1.3	1.9	2.6	3.3	4.1	5.0	5.9	6.9	7.9	9.0	10.0

Before using this table, get the times and heights of the two tide events that your desired time of height falls between. Calculate the duration of time and the range of heights between the two events, and the time difference between your desired time and the nearest high or low.

Enter the table with the Duration of Rise or Fall boldfaced in the upper left column that is closest to your calculation. Scan that row to find the Time from Nearest High or Low closest to your differance. Use that column to enter the bottom section of the table. Find the row whose boldfaced value Range of Tide is closest to yours, and get your correction value where the row and column cross.

When the nearest tide is high water, subtract the correction; and vice versa.

CURRENTS

TIDAL CURRENT TABLES

Unfortunately, there has been very little data collected on tidal currents around the Caribbean rim. On the following pages are all the tidal current predictions available for this area from the U.S. National Ocean Service. The area covered is only southeast Florida and Puerto Rico. Explanatory notes are below. Starting on page C 27 is information about the Gulf Stream and Caribbean ocean currents.

The three primary reference tidal current tables on the following pages indicate for each day of 1998 the times of all slacks (minimum current flow), times of maximum flood for both ebbs and both floods, and the maximum current speed, in knots, for ebb and flood currents. The approximate direction in degrees true of the ebb and flood currents is also indicated in the top right-hand corner of each tidal current table. As in the tide tables, times in the current tables are local, on a 24-hour clock. They have not been adjusted for Daylight Saving Time (2 am, April 5–2 am, October 25) to stay in keeping with our Caribbean tide tables (*Reed's* East and West Coast of North America Almanacs have all tables corrected for DST).

There are usually four slacks and four maximums each day. If one is missing on a given day, it may occur after midnight as the first slack or maximum of the following day. At some stations where the diurnal inequality is large, there may be on certain days a continuous flood or ebb current with varying speed throughout half the day, giving only two slacks and two maximums on that particular day.

The outstanding feature of the currents in the area covered is the diurnal inequality—that is, the differences in speed of two consecutive flood or ebb maximums. This inequality varies mainly with the moon's declination; consequently it tends to disappear when the moon is near the equator. In certain places, the inequality is chiefly in the flood currents. At other places the inequality is chiefly in the ebb currents, while at still other places, there is a marked inequality in both flood and ebb currents. The effect of the inequality at some places is such that at times the current may be erratic or one flood or ebb current of the day may be quite weak. Therefore, in using the predictions it is essential to note carefully the speeds as well as the times.

To alert mariners to the moon's effect, the current tables carry symbols indicating the days of the new moon, quarter moons, and full moon. The monthly pages of the ephemeris chapter contain information on the Moon's declination, as well as the times of its apogee and perigee.

Current predictions for additional stations in southeast Florida and Puerto Rico can be calculated from information provided in the current differences table that begins on page C 24. Instructions on how to use the differences table are found on pages C 22–3.

It is important to notice that the predicted slacks and strengths given in this table refer to the horizontal motion of the water and not to the vertical rise and fall of the tide. The relation of current to tide is not constant, but varies from place to place, and the time of slack water does not generally coincide with the time of high or low water, nor does the time of maximum speed of the current usually coincide with the time of most rapid change in the vertical height of the tide. At stations located on a tidal river or bay the time of slack water may differ from 1 to 3 hours from the time of high or low water.

In using this table, bear in mind that actual times of slack or maximum occasionally differ from the predicted times by as much as half an hour and in rare instances the difference may be as much as an hour. Comparisons of predicted with observed times of slack water indicate that more than 90 percent of the slack waters occurred within half an hour of the predicted times. To make sure, therefore, of getting the full advantage of a favorable current or slack water, the navigator should reach the entrance or strait at least half an hour before the predicted time of the desired condition of current. Currents are frequently disturbed by wind or variations in river discharge. On days when the current is affected by such disturbing influences the times and speeds will differ from those given in the table, but local knowledge may enable one to make proper allowance for these effects.

For more information on tidal currents in general, see *Reed's Nautical Companion.*

SPECIAL NOTICES

BISCAYNE BAY/PORT OF MIAMI, FL – #1

The Biscayne Bay Pilots report variances between predicted and actual currents. Cross-channel current variations in Government Cut are particularly difficult to negotiate. Caution should be exercised when entering Government Cut from the sea during flood tide with northeasterly winds; a strong turning torque occurs when the bow is just inside the north jetty. A similar but less serious situation occurs when leaving the port during ebb tide. Horizontal current gradients occur in the turning basin immediately west of Government Cut which may make maneuvering difficult. The Coast Guard reports that ships may encounter current anomalies at the mouth of the Miami River which have caused occasional groundings. (Issued September 29, 1983)

BISCAYNE BAY/PORT OF MIAMI, FL – #2

The Biscayne Bay Pilots report that recent dredging and construction by the US Corps of Engineers (COE) supporting Miami port expansion has significantly effected the currents in Miami Harbor. Both flood and ebb currents should be expected to be stronger than indicated in official published predictions. The actual times for maximum and slack currents should be expected to deviate from the published predictions. Funding to support a survey to obtain new data for more accurate tidal current predictions is not available at this time. Installation of a Physical Oceanographic Real Time System (PORTS), like the one in operation in Tampa Bay, would be the best solution for long term marine safety. (Issued July 17, 1997)

Tide Watchers Wanted

Reed's seeks local information about tides and currents. There are many locations, especially in the Caribbean, where official government tide and current data is scarce. Often, local boaters know how local tides and currents work. *Reed's* wants to help make that information available to all boaters.

For instance, you might know a place that usually has a high tide about 50 minutes after a regular reference station. Or you might spend a lot of time in such a place and be willing to devote a bit of time and effort to measuring the tide or current.

We can give you some tips on how to evaluate tides and currents, and we can integrate your local knowledge into our tide differences table. Of course, we will carefully mark such data as local knowledge, and not nearly as precise as the differences developed by hydrographic offices. Approximate information, properly handled, is better than no information.

If you are interested in being a Tide Watcher, please contact *Reed's* for further information. The reward will be helping your fellow mariners to better know and use the power of tides.

MIAMI, FLORIDA (Harbor Entrance)

CURRENT TABLE 1998 — 25°45.9'N 80°08.2W — Flood 293° Ebb 112°

Eastern Time (75°W) — Add 1 hour for Daylight Saving Time: April 5 – October 24

JANUARY

Day	Slack time	Max time	Fld knots	Ebb knots
1 Th		0138		1.9
	0525	0753	2.3	
	1133	1416		1.9
	1757	2017	2.0	
	2347			
2 F		0226		1.9
	0614	0845	2.2	
	1219	1504		1.9
	1847	2115	2.0	
3 Sa	0040	0320		1.8
	0707	0940	2.1	
	1307	1554		1.9
	1942	2211	2.0	
4 Su	0136	0416		1.7
	0804	1031	2.0	
	1359	1647		1.8
	2039	2304	1.9	
5 M	0236	0522		1.5
	0903	1123	1.8	
	1455	1817		1.7
☾	2136			
6 Tu		0011	1.8	
	0339	0746		1.5
	1002	1228	1.6	
	1552	2006		1.7
	2233			
7 W		0208	1.9	
	0444	0848		1.5
	1103	1430	1.6	
	1653	2103		1.8
	2332			
8 Th		0310	1.9	
	0550	0946		1.6
	1204	1533	1.6	
	1754	2202		1.8
9 F	0030	0409	2.0	
	0650	1047		1.7
	1303	1632	1.7	
	1851	2304		1.9
10 Sa	0125	0504	2.1	
	0743	1143		1.8
	1356	1726	1.8	
	1943	2358		2.0
11 Su	0216	0553	2.2	
	0830	1232		1.9
	1446	1812	1.8	
	2031			
12 M		0044		2.0
	0304	0636	2.2	
	0916	1315		1.9
○	1533	1854	1.8	
	2118			
13 Tu		0126		1.9
	0349	0717	2.1	
	0959	1357		1.8
	1618	1934	1.8	
	2204			
14 W		0205		1.8
	0432	0756	2.0	
	1040	1437		1.7
	1700	2013	1.7	
	2248			
15 Th		0242		1.7
	0514	0833	1.9	
	1119	1514		1.6
	1741			
	2031			

Day	Slack time	Max time	Fld knots	Ebb knots
16 F		0308		1.5
	0556	0829	1.8	
	1157	1536		1.5
	1823	2048	1.6	
17 Sa	0013	0305		1.4
	0639	0906	1.7	
	1236	1525		1.4
	1908	2130	1.6	
18 Su	0058	0327		1.3
	0726	0947	1.6	
	1317	1547		1.4
	1956	2212	1.5	
19 M	0146	0402		1.2
	0816	1028	1.5	
	1402	1621		1.3
	2045	2254	1.5	
20 Tu	0237	0443		1.1
	0908	1111	1.4	
	1450	1703		1.3
☽	2135	2341	1.4	
21 W	0332	0534		1.0
	1000	1159	1.3	
	1541	1757		1.2
	2226			
22 Th		0035	1.4	
	0429	0849		0.9
	1055	1256	1.2	
	1635	1917		1.2
	2320			
23 F		0139	1.4	
	0529	0935		1.0
	1153	1357	1.2	
	1732	2024		1.3
24 Sa	0015	0245	1.5	
	0626	1034		1.1
	1249	1456	1.3	
	1827	2112		1.4
25 Su	0108	0439	1.7	
	0717	1129		1.3
	1340	1600	1.4	
	1919	2224		1.5
26 M	0158	0524	1.9	
	0805	1209		1.5
	1429	1712	1.6	
	2009	2350		1.7
27 Tu	0246	0552	2.1	
	0852	1239		1.7
	1516	1752	1.9	
	2059			
28 W		0025		1.9
	0334	0619	2.3	
	0939	1305		1.9
●	1603	1832	2.1	
	2151			
29 Th		0101		2.0
	0421	0656	2.4	
	1026	1335		2.0
	1649	1915	2.2	
	2242			
30 F		0142		2.1
	0508	0739	2.4	
	1112	1413		2.1
	1736	2005	2.2	
	2333			
31 Sa		0230		2.1
	0557	0830	2.3	
	1158	1459		2.1
	1825	2104	2.2	

FEBRUARY

Day	Slack time	Max time	Fld knots	Ebb knots
1 Su	0025	0323		1.9
	0648	0926	2.2	
	1245	1547		2.0
	1918	2202	2.2	
2 M	0119	0417		1.8
	0743	1017	2.0	
	1336	1637		1.9
	2014	2254	2.1	
3 Tu	0216	0524		1.6
	0841	1106	1.8	
	1430	1752		1.7
☾	2112			
4 W		0004	1.9	
	0317	0726		1.5
	0940	1207	1.6	
	1528	1948		1.7
	2209			
5 Th		0146	1.8	
	0420	0831		1.5
	1039	1411	1.5	
	1628	2049		1.7
	2308			
6 F		0250	1.8	
	0527	0929		1.5
	1141	1514	1.5	
	1732	2147		1.7
7 Sa	0009	0349	1.9	
	0631	1028		1.6
	1242	1613	1.6	
	1833	2248		1.8
8 Su	0106	0445	1.9	
	0725	1125		1.7
	1337	1708	1.7	
	1927	2343		1.8
9 M	0157	0535	2.0	
	0812	1214		1.8
	1426	1756	1.7	
	2014			
10 Tu		0030		1.9
	0244	0619	2.0	
	0854	1257		1.8
	1511	1838	1.8	
	2100			
11 W		0111		1.9
	0329	0658	2.0	
	0934	1337		1.8
○	1554	1915	1.8	
	2143			
12 Th		0149		1.8
	0410	0734	1.9	
	1013	1413		1.7
	1634	1948	1.7	
	2225			
13 F		0222		1.7
	0451	0757	1.8	
	1050	1442		1.6
	1713	1942	1.7	
	2306			
14 Sa		0239		1.6
	0530	0751	1.8	
	1126	1436		1.6
	1752	2010	1.7	
	2346			
15 Su		0228		1.5
	0610	0828	1.7	
	1203	1438		1.5
	1832	2053	1.7	

Day	Slack time	Max time	Fld knots	Ebb knots
16 M	0027	0253		1.4
	0652	0911	1.7	
	1242	1507		1.5
	1916	2137	1.7	
17 Tu	0111	0328		1.4
	0739	0955	1.6	
	1323	1545		1.5
	2003	2221	1.7	
18 W	0159	0409		1.3
	0830	1039	1.5	
	1409	1626		1.4
	2054	2306	1.6	
19 Th	0252	0453		1.1
	0922	1125	1.4	
	1500	1713		1.3
☽	2146	2356	1.5	
20 F	0348	0548		1.0
	1017	1217	1.3	
	1555	1813		1.2
	2241			
21 Sa		0055	1.5	
	0448	0715		1.0
	1115	1318	1.2	
	1655	1933		1.3
	2340			
22 Su		0201	1.5	
	0549	0859		1.1
	1215	1422	1.3	
	1757	2042		1.4
23 M	0038	0309	1.6	
	0646	1043		1.3
	1310	1529	1.5	
	1855	2155		1.5
24 Tu	0132	0455	1.9	
	0737	1137		1.5
	1402	1707	1.7	
	1949	2338		1.8
25 W	0223	0540	2.1	
	0825	1216		1.8
	1451	1751	2.0	
	2041			
26 Th		0023		2.0
	0313	0613	2.3	
	0913	1249		2.0
●	1539	1828	2.3	
	2134			
27 F		0102		2.2
	0402	0647	2.4	
	1001	1323		2.2
	1627	1907	2.4	
	2226			
28 Sa		0143		2.2
	0450	0727	2.4	
	1048	1402		2.3
	1714	1954	2.5	
	2317			

MIAMI, FLORIDA (Harbor Entrance)

CURRENT TABLE 1998 — 25°45.9'N 80°08.2W — Flood 293° Ebb 112°

Eastern Time (75°W) — Add 1 hour for Daylight Saving Time: April 5 – October 24

MARCH

Day	Slack time	Max time	Fld knots	Ebb knots
1		0230		2.2
	0538	0815	2.3	
Su	1135	1447		2.3
	1803	2054	2.4	
2	0008	0322		2.1
	0628	0912	2.2	
M	1223	1536		2.2
	1854	2155	2.3	
3	0059	0415		1.9
	0722	1006	2.0	
Tu	1312	1625		2.0
	1949	2248	2.1	
4	0154	0519		1.6
	0819	1053	1.8	
W	1406	1735		1.7
	2047	2354	1.9	
5	0252	0703		1.5
	0917	1207	1.5	
Th	1504	1929		1.6
☾	2145			
6		0121	1.7	
	0354	0812		1.5
F	1016	1350	1.4	
	1605	2033		1.6
	2244			
7		0227	1.7	
	0459	0908		1.5
Sa	1117	1454	1.4	
	1710	2129		1.6
	2345			
8		0326	1.7	
	0606	1005		1.5
Su	1218	1552	1.5	
	1815	2227		1.6
9	0043	0422	1.7	
	0701	1101		1.6
M	1313	1646	1.6	
	1910	2323		1.7
10	0135	0513	1.8	
	0746	1151		1.7
Tu	1402	1735	1.7	
	1957			
11		0010		1.8
	0222	0557	1.9	
W	0826	1234		1.8
	1445	1817	1.8	
	2039			
12		0052		1.8
	0306	0636	1.9	
Th	0904	1312		1.8
○	1527	1853	1.8	
	2121			
13		0129		1.8
	0347	0709	1.8	
F	0941	1345		1.7
	1606	1923	1.8	
	2201			
14		0200		1.7
	0426	0723	1.7	
Sa	1018	1406		1.6
	1644	1914	1.8	
	2241			
15		0213		1.6
	0505	0721	1.7	
Su	1055	1347		1.6
	1721	1939	1.8	
	2320			
16		0159		1.5
	0543	0755	1.7	
M	1132	1401		1.6
	1759	2019	1.8	
	2359			
17		0222		1.5
	0623	0838	1.7	
Tu	1209	1433		1.6
	1839	2104	1.8	
18	0041	0258		1.5
	0707	0924	1.6	
W	1249	1512		1.6
	1925	2151	1.8	
19	0126	0339		1.4
	0756	1010	1.5	
Th	1334	1555		1.5
	2016	2237	1.7	
20	0217	0424		1.3
	0849	1056	1.5	
F	1425	1642		1.4
	2110	2325	1.6	
21	0312	0515		1.2
	0944	1147	1.4	
Sa	1523	1738		1.3
☽	2207			
22		0021	1.6	
	0412	0623		1.1
Su	1042	1247	1.3	
	1625	1855		1.3
	2308			
23		0126	1.5	
	0514	0814		1.2
M	1142	1355	1.4	
	1731	2025		1.4
24	0009	0237	1.6	
	0613	0928		1.4
Tu	1241	1507	1.6	
	1833	2145		1.6
25	0107	0417	1.8	
	0707	1054		1.6
W	1334	1650	1.9	
	1930	2320		1.8
26	0200	0518	2.1	
	0757	1147		1.9
Th	1425	1741	2.2	
	2023			
27		0012		2.0
	0252	0601	2.3	
F	0846	1228		2.2
●	1514	1822	2.5	
	2116			
28		0054		2.2
	0342	0637	2.4	
Sa	0935	1306		2.3
	1603	1900	2.6	
	2208			
29		0136		2.3
	0431	0714	2.4	
Su	1024	1346		2.4
	1651	1944	2.6	
	2259			
30		0223		2.2
	0519	0759	2.3	
M	1112	1432		2.3
	1740	2042	2.5	
	2349			
31		0315		2.1
	0608	0859	2.1	
Tu	1200	1523		2.2
	1830	2146	2.3	

APRIL

Day	Slack time	Max time	Fld knots	Ebb knots
1	0039	0409		1.9
	0700	1001	2.0	
W	1250	1614		1.9
	1924	2239	2.1	
2	0131	0510		1.7
	0756	1052	1.7	
Th	1343	1721		1.7
	2021	2337	1.9	
3	0226	0637		1.5
	0854	1206	1.5	
F	1440	1906		1.5
☾	2119			
4		0053	1.7	
	0325	0749		1.4
Sa	0952	1326	1.4	
	1541	2012		1.5
	2217			
5		0200	1.6	
	0426	0845		1.5
Su	1050	1429	1.4	
	1646	2107		1.5
	2316			
6		0258	1.5	
	0529	0937		1.5
M	1149	1525	1.4	
	1751	2202		1.5
7	0015	0353	1.6	
	0626	1031		1.5
Tu	1244	1619	1.6	
	1848	2256		1.6
8	0108	0444	1.6	
	0712	1122		1.6
W	1332	1709	1.7	
	1934	2346		1.6
9	0155	0531	1.7	
	0752	1206		1.7
Th	1416	1752	1.8	
	2016			
10		0029		1.7
	0239	0610	1.7	
F	0830	1244		1.7
	1456	1828	1.9	
	2056			
11		0106		1.7
	0320	0644	1.7	
Sa	0907	1316		1.7
○	1536	1857	1.9	
	2136			
12		0137		1.6
	0401	0654	1.7	
Su	0945	1330		1.6
	1614	1849	1.9	
	2215			
13		0151		1.5
	0440	0653	1.7	
M	1024	1313		1.6
	1652	1912	1.9	
	2255			
14		0136		1.5
	0518	0726	1.7	
Tu	1102	1330		1.6
	1730	1949	1.9	
	2334			
15		0156		1.5
	0557	0808	1.6	
W	1140	1403		1.7
	1809	2034	1.9	
16	0015	0232		1.5
	0640	0855	1.6	
Th	1221	1444		1.7
	1853	2123	1.9	
17	0059	0315		1.5
	0728	0944	1.6	
F	1306	1530		1.6
	1944	2211	1.8	
18	0149	0401		1.4
	0822	1033	1.5	
Sa	1359	1618		1.5
	2040	2300	1.7	
19	0243	0451		1.3
	0918	1124	1.5	
Su	1458	1713		1.4
☽	2139	2353	1.7	
20	0340	0554		1.2
	1015	1222	1.5	
M	1602	1828		1.3
	2239			
21		0056	1.6	
	0440	0747		1.3
Tu	1114	1332	1.5	
	1708	2022		1.4
	2341			
22		0209	1.7	
	0541	0900		1.5
W	1212	1454	1.7	
	1813	2139		1.6
23	0041	0335	1.8	
	0637	1007		1.8
Th	1308	1629	2.0	
	1911	2300		1.8
24	0137	0454	2.0	
	0729	1117		2.0
F	1400	1725	2.3	
	2005	2357		2.0
25	0230	0544	2.2	
	0819	1207		2.2
Sa	1450	1810	2.6	
	2057			
26		0043		2.2
	0321	0625	2.3	
Su	0909	1250		2.3
●	1540	1851	2.6	
	2149			
27		0127		2.2
	0410	0704	2.3	
M	0959	1332		2.3
	1629	1934	2.6	
	2240			
28		0214		2.2
	0459	0747	2.2	
Tu	1049	1418		2.2
	1717	2028	2.5	
	2329			
29		0306		2.1
	0548	0849	2.0	
W	1138	1510		2.1
	1807	2131	2.3	
30	0017	0400		1.9
	0639	0954	1.9	
Th	1227	1605		1.8
	1859	2225	2.1	

MIAMI, FLORIDA (Harbor Entrance)

CURRENT TABLE 1998 | 25°45.9'N 80°08.2W | Flood 293° Ebb 112°

Eastern Time (75°W) | Add 1 hour for Daylight Saving Time: April 5 – October 24

MAY

Day	Slack time	Max time	Fld knots	Ebb knots
1 F	0107	0455		1.7
	0733	1047	1.7	
	1319	1706		1.6
	1954	2316	1.8	
2 Sa	0159	0607		1.5
	0829	1145	1.5	
	1415	1836		1.4
	2051			
3 Su ☾		0019	1.6	
	0253	0720		1.4
	0925	1256	1.4	
	1514	1946		1.4
	2147			
4 M		0127	1.5	
	0349	0817		1.4
	1020	1400	1.4	
	1615	2041		1.4
	2243			
5 Tu		0226	1.4	
	0445	0906		1.5
	1114	1455	1.4	
	1718	2132		1.4
	2340			
6 W		0319	1.4	
	0541	0955		1.5
	1208	1547	1.5	
	1816	2224		1.4
7 Th	0034	0410	1.4	
	0630	1046		1.5
	1257	1637	1.6	
	1905	2316		1.5
8 F	0124	0459	1.5	
	0714	1134		1.6
	1342	1723	1.8	
	1948			
9 Sa		0002		1.5
	0209	0541	1.5	
	0753	1214		1.6
	1424	1802	1.8	
	2028			
10 Su		0041		1.6
	0252	0616	1.6	
	0832	1246		1.6
	1505	1832	1.9	
	2109			
11 M ○		0114		1.5
	0334	0628	1.6	
	0912	1259		1.6
	1545	1827	1.9	
	2150			
12 Tu		0134		1.5
	0414	0628	1.6	
	0953	1246		1.6
	1625	1848	2.0	
	2231			
13 W		0120		1.5
	0454	0701	1.6	
	1034	1306		1.7
	1704	1925	2.0	
	2311			
14 Th		0137		1.5
	0535	0742	1.6	
	1115	1340		1.7
	1744	2009	2.0	
	2353			
15 F		0212		1.5
	0617	0830	1.6	
	1159	1422		1.7
	1828	2058	1.9	
16 Sa	0037	0256		1.5
	0705	0922	1.6	
	1246	1510		1.6
	1919	2149	1.9	
17 Su	0125	0344		1.5
	0759	1013	1.6	
	1340	1601		1.6
	2016	2239	1.9	
18 M ☽	0217	0435		1.5
	0855	1105	1.6	
	1440	1656		1.4
	2115	2331	1.8	
19 Tu	0313	0534		1.4
	0951	1201	1.6	
	1543	1810		1.3
	2215			
20 W		0030	1.7	
	0411	0726		1.5
	1048	1312	1.7	
	1648	2024		1.4
	2315			
21 Th		0143	1.7	
	0510	0842		1.7
	1146	1454	1.9	
	1753	2130		1.6
22 F	0016	0311	1.8	
	0608	0940		1.9
	1243	1608	2.1	
	1853	2241		1.8
23 Sa	0114	0431	1.9	
	0703	1051		2.0
	1336	1707	2.4	
	1947	2342		2.0
24 Su	0208	0527	2.0	
	0754	1150		2.2
	1428	1757	2.5	
	2039			
25 M ●		0032		2.1
	0300	0614	2.1	
	0845	1238		2.3
	1518	1841	2.6	
	2130			
26 Tu		0118		2.1
	0350	0656	2.1	
	0936	1322		2.2
	1608	1924	2.5	
	2221			
27 W		0205		2.1
	0439	0740	2.0	
	1027	1409		2.1
	1656	2014	2.4	
	2309			
28 Th		0255		2.0
	0527	0838	1.9	
	1116	1500		2.0
	1744	2113	2.2	
	2355			
29 F		0346		1.8
	0616	0939	1.8	
	1205	1553		1.8
	1833	2205	2.0	
30 Sa	0041	0435		1.7
	0707	1030	1.7	
	1254	1644		1.5
	1925	2251	1.8	
31 Su	0128	0531		1.5
	0800	1118	1.5	
	1347	1754		1.3
	2019	2338	1.6	

JUNE

Day	Slack time	Max time	Fld knots	Ebb knots
1 M ☾	0218	0641		1.4
	0854	1216	1.4	
	1442	1912		1.2
	2113			
2 Tu		0042	1.4	
	0308	0742		1.4
	0945	1322	1.3	
	1539	2010		1.2
	2207			
3 W		0146	1.3	
	0400	0832		1.4
	1036	1419	1.4	
	1638	2059		1.3
	2301			
4 Th		0240	1.3	
	0452	0918		1.4
	1128	1510	1.5	
	1736	2148		1.3
	2355			
5 F		0331	1.3	
	0544	1005		1.4
	1219	1601	1.6	
	1830	2241		1.3
6 Sa	0048	0422	1.3	
	0633	1055		1.4
	1307	1650	1.7	
	1916	2332		1.4
7 Su	0136	0509	1.4	
	0717	1141		1.5
	1352	1733	1.8	
	1959			
8 M		0015		1.4
	0222	0547	1.4	
	0759	1217		1.5
	1435	1807	1.8	
	2041			
9 Tu ○		0051		1.5
	0306	0604	1.5	
	0841	1233		1.5
	1517	1812	1.9	
	2124			
10 W		0117		1.5
	0349	0605	1.6	
	0924	1227		1.6
	1559	1828	2.0	
	2207			
11 Th		0114		1.5
	0431	0640	1.6	
	1009	1249		1.7
	1641	1904	2.0	
	2250			
12 F		0126		1.6
	0513	0721	1.7	
	1054	1324		1.7
	1724	1947	2.1	
	2332			
13 Sa		0200		1.6
	0557	0809	1.7	
	1141	1407		1.7
	1809	2037	2.0	
14 Su	0016	0244		1.6
	0644	0902	1.7	
	1230	1457		1.7
	1859	2130	2.0	
15 M	0103	0332		1.7
	0736	0956	1.8	
	1324	1550		1.6
	1955	2220	1.9	
16 Tu	0153	0422		1.7
	0832	1049	1.8	
	1423	1645		1.5
	2054	2311	1.9	
17 W ☽	0247	0519		1.6
	0928	1144	1.8	
	1525	1800		1.4
	2152			
18 Th		0007	1.7	
	0344	0707		1.6
	1024	1255	1.8	
	1628	2021		1.5
	2252			
19 F		0118	1.7	
	0442	0831		1.8
	1121	1446	1.9	
	1733	2121		1.6
	2353			
20 Sa		0302	1.7	
	0542	0928		1.9
	1219	1551	2.1	
	1834	2225		1.7
21 Su	0052	0413	1.8	
	0639	1036		2.0
	1315	1651	2.3	
	1930	2328		1.8
22 M	0147	0512	1.9	
	0733	1139		2.1
	1408	1743	2.4	
	2021			
23 Tu ●		0020		2.0
	0240	0602	2.0	
	0824	1230		2.1
	1459	1829	2.4	
	2112			
24 W		0107		2.0
	0330	0647	2.0	
	0915	1315		2.1
	1548	1913	2.4	
	2201			
25 Th		0153		2.0
	0419	0731	1.9	
	1006	1400		2.0
	1635	1958	2.2	
	2247			
26 F		0240		1.9
	0506	0821	1.8	
	1054	1447		1.9
	1721	2050	2.1	
	2331			
27 Sa		0326		1.8
	0551	0916	1.7	
	1141	1534		1.7
	1807	2140	1.9	
28 Su	0013	0410		1.7
	0638	1005	1.6	
	1227	1616		1.5
	1855	2221	1.8	
29 M	0056	0450		1.5
	0727	1044	1.5	
	1316	1656		1.3
	1945	2232	1.6	
30 Tu	0140	0537		1.3
	0818	1055	1.5	
	1407	1816		1.2
	2037	2254	1.5	

MIAMI, FLORIDA (Harbor Entrance)

CURRENT TABLE 1998 — 25°45.9'N 80°08.2W — Flood 293° Ebb 112°

Eastern Time (75°W) — Add 1 hour for Daylight Saving Time: April 5 – October 24

JULY

Day	Slack time	Max time	Fld knots	Ebb knots
1 W ☾	0227	0654		1.3
	0908	1123	1.4	
	1500	1931		1.1
	2129	2334	1.3	
2 Th	0315	0752		1.2
	0957	1332	1.3	
	1245	2025		1.1
	1555			
	2221			
3 F		0024	1.2	
	0406	0838		1.3
	1048	1429	1.4	
	1652	2112		1.1
	2315			
4 Sa	0153	0244	1.1	
	0458	0920		1.3
	1140	1521	1.4	
	1749	2202		1.2
5 Su	0010	0337	1.2	
	0551	1006		1.3
	1231	1613	1.5	
	1842	2257		1.2
6 M	0102	0430	1.2	
	0641	1100		1.3
	1320	1701	1.7	
	1929	2347		1.3
7 Tu	0151	0516	1.3	
	0728	1144		1.4
	1406	1741	1.8	
	2013			
8 W		0026		1.4
	0237	0541	1.4	
	0813	1210		1.5
	1451	1802	1.9	
	2057			
9 Th ○		0056		1.5
	0322	0547	1.6	
	0859	1218		1.6
	1536	1813	2.0	
	2142			
10 F		0109		1.6
	0407	0623	1.7	
	0947	1242		1.7
	1620	1847	2.1	
	2227			
11 Sa		0122		1.7
	0451	0704	1.8	
	1036	1317		1.8
	1705	1929	2.2	
	2311			
12 Su		0152		1.8
	0535	0751	1.9	
	1125	1400		1.8
	1751	2017	2.1	
	2355			
13 M		0235		1.8
	0622	0845	1.9	
	1215	1450		1.8
	1840	2111	2.1	
14 Tu	0041	0323		1.9
	0713	0942	2.0	
	1308	1544		1.7
	1935	2203	2.0	
15 W	0130	0412		1.8
	0808	1034	2.0	
	1405	1638		1.6
	2033	2253	1.9	
16 Th ☽	0223	0505		1.8
	0905	1128	1.9	
	1506	1759		1.4
	2131	2346	1.7	
17 F	0319	0656		1.7
	1001	1243	1.9	
	1608	2011		1.5
	2230			
18 Sa		0058	1.6	
	0418	0824		1.8
	1059	1433	1.9	
	1713	2110		1.6
	2330			
19 Su		0254	1.6	
	0519	0922		1.8
	1158	1535	2.0	
	1816	2211		1.6
20 M	0031	0358	1.7	
	0620	1027		1.9
	1256	1635	2.1	
	1913	2313		1.8
21 Tu	0128	0458	1.8	
	0716	1129		1.9
	1350	1728	2.2	
	2005			
22 W		0006		1.9
	0221	0549	1.9	
	0807	1220		2.0
	1440	1815	2.3	
	2053			
23 Th ●		0053		1.9
	0310	0634	1.9	
	0857	1305		2.0
	1528	1858	2.2	
	2139			
24 F		0136		1.9
	0357	0716	1.9	
	0946	1348		1.9
	1614	1939	2.1	
	2222			
25 Sa		0219		1.9
	0442	0759	1.8	
	1032	1430		1.8
	1658	2022	2.0	
	2303			
26 Su		0300		1.8
	0524	0845	1.7	
	1116	1510		1.7
	1741	2104	1.8	
	2342			
27 M		0336		1.6
	0607	0928	1.7	
	1159	1540		1.5
	1824	2108	1.7	
28 Tu	0021	0356		1.5
	0651	0932	1.6	
	1243	1542		1.4
	1910	2137	1.6	
29 W	0102	0350		1.4
	0738	1002	1.6	
	1330	1600		1.2
	1959	2215	1.5	
30 Th	0146	0413		1.3
	0827	1041	1.5	
	1420	1634		1.1
	2050	2255	1.4	
31 F ☾	0233	0450		1.3
	0917	1124	1.5	
	1513	1718		1.0
	1842	1944		0.9
	2142	2341	1.3	

AUGUST

Day	Slack time	Max time	Fld knots	Ebb knots
1 Sa	0322	0538		1.2
	1008	1215	1.4	
	1609	2036		1.0
	2235			
2 Su		0034	1.2	
	0415	0831		1.1
	1100	1317	1.4	
	1707	2123		1.0
	2331			
3 M		0135	1.1	
	0511	0847		1.2
	1155	1527	1.4	
	1805	2217		1.1
4 Tu	0028	0235	1.2	
	0607	0906		1.3
	1248	1625	1.6	
	1857	2312		1.2
5 W	0120	0336	1.3	
	0700	1101		1.4
	1338	1712	1.7	
	1944	2357		1.4
6 Th	0209	0516	1.5	
	0749	1146		1.5
	1426	1745	1.9	
	2030			
7 F ○		0029		1.5
	0255	0536	1.7	
	0837	1214		1.7
	1513	1802	2.1	
	2115			
8 Sa		0051		1.7
	0341	0610	1.9	
	0927	1242		1.9
	1559	1833	2.2	
	2201			
9 Su		0112		1.9
	0426	0650	2.1	
	1018	1316		2.0
	1645	1913	2.3	
	2247			
10 M		0143		2.0
	0512	0735	2.2	
	1108	1358		2.0
	1732	1959	2.2	
	2332			
11 Tu		0223		2.0
	0559	0828	2.2	
	1158	1447		1.9
	1821	2052	2.2	
12 W	0018	0311		2.0
	0649	0927	2.2	
	1250	1540		1.8
	1914	2146	2.1	
13 Th	0107	0400		2.0
	0744	1021	2.1	
	1346	1635		1.7
	2012	2236	1.9	
14 F ☽	0200	0452		1.8
	0841	1114	2.0	
	1445	1820		1.5
	2110	2328	1.7	
15 Sa	0257	0653		1.7
	0939	1256	1.8	
	1547	1959		1.5
	2209			
16 Su		0126	1.5	
	0357	0816		1.7
	1038	1417	1.8	
	1652	2058		1.5
	2310			
17 M		0242	1.5	
	0500	0915		1.7
	1138	1518	1.9	
	1758	2156		1.6
18 Tu	0012	0343	1.6	
	0604	1016		1.8
	1238	1617	2.0	
	1857	2255		1.7
19 W	0110	0441	1.7	
	0702	1115		1.9
	1332	1710	2.0	
	1947	2349		1.8
20 Th	0202	0533	1.8	
	0753	1207		1.9
	1422	1758	2.1	
	2032			
21 F ●		0035		1.9
	0249	0618	1.9	
	0840	1251		1.9
	1508	1839	2.1	
	2114			
22 Sa		0116		1.9
	0334	0658	1.9	
	0925	1331		1.9
	1552	1917	2.0	
	2154			
23 Su		0154		1.8
	0415	0734	1.8	
	1009	1408		1.8
	1634	1950	1.9	
	2233			
24 M		0228		1.7
	0455	0803	1.8	
	1051	1440		1.6
	1714	1946	1.8	
	2310			
25 Tu		0250		1.6
	0535	0759	1.7	
	1131	1448		1.5
	1754	2013	1.7	
	2347			
26 W		0235		1.5
	0615	0836	1.7	
	1212	1447		1.4
	1837	2054	1.6	
27 Th	0026	0255		1.5
	0658	0921	1.7	
	1255	1517		1.3
	1923	2138	1.5	
28 F	0107	0330		1.5
	0746	1005	1.6	
	1342	1554		1.2
	2013	2222	1.5	
29 Sa ☾	0152	0410		1.4
	0836	1049	1.6	
	1433	1637		1.1
	2105	2307	1.3	
30 Su	0242	0455		1.3
	0929	1136	1.5	
	1528	1727		1.0
	2159	2357	1.2	
31 M	0337	0549		1.2
	1023	1232	1.4	
	1627	1846		0.9
	1926	2045		0.9
	2255			

MIAMI, FLORIDA (Harbor Entrance)

CURRENT TABLE 1998 — 25°45.9'N 80°08.2W — Flood 293° Ebb 112°

Eastern Time (75°W) — Add 1 hour for Daylight Saving Time: April 5 – October 24

SEPTEMBER

Day	Slack time	Max time	Fld knots	Ebb knots
1 Tu		0055	1.2	
	0435	0710		1.2
	1119	1337	1.4	
	1727	2129		1.0
	2353			
2 W		0200	1.2	
	0536	0831		1.3
	1217	1445	1.5	
	1824	2224		1.2
3 Th	0049	0304	1.4	
	0634	0931		1.4
	1311	1634	1.7	
	1914	2317		1.4
4 F	0139	0443	1.6	
	0727	1117		1.6
	1401	1719	1.9	
	2001	2356		1.6
5 Sa	0227	0529	1.9	
	0817	1202		1.8
	1450	1750	2.1	
	2047			
6 Su ○		0025		1.9
	0314	0602	2.2	
	0908	1237		2.0
	1538	1820	2.3	
	2134			
7 M		0054		2.1
	0401	0639	2.4	
	1000	1313		2.1
	1625	1857	2.3	
	2221			
8 Tu		0128		2.2
	0448	0721	2.4	
	1050	1354		2.1
	1713	1940	2.3	
	2308			
9 W		0208		2.2
	0535	0811	2.4	
	1141	1443		2.1
	1801	2032	2.2	
	2355			
10 Th		0256		2.2
	0625	0911	2.3	
	1232	1537		1.9
	1854	2129	2.0	
11 F	0044	0347		2.0
	0719	1010	2.2	
	1326	1634		1.7
	1950	2223	1.9	
12 Sa ☽	0138	0443		1.8
	0818	1106	2.0	
	1423	1813		1.5
	2050	2316	1.7	
13 Su	0236	0644		1.6
	0917	1242	1.8	
	1525	1942		1.5
	2149			
14 M		0116	1.5	
	0337	0805		1.6
	1017	1358	1.7	
	1629	2042		1.5
	2250			
15 Tu		0226	1.5	
	0443	0903		1.7
	1117	1459	1.7	
	1735	2137		1.6
	2351			
16 W		0325	1.6	
	0549	0959		1.7
	1218	1555	1.8	
	1835	2233		1.7
17 Th	0049	0421	1.7	
	0649	1056		1.8
	1313	1649	1.9	
	1924	2326		1.8
18 F	0139	0512	1.8	
	0738	1148		1.8
	1401	1736	1.9	
	2006			
19 Sa		0012		1.8
	0225	0557	1.9	
	0822	1232		1.9
	1446	1817	1.9	
	2044			
20 Su ●		0052		1.8
	0307	0636	1.9	
	0904	1311		1.8
	1528	1853	1.9	
	2122			
21 M		0127		1.8
	0347	0709	1.9	
	0944	1345		1.7
	1609	1920	1.8	
	2200			
22 Tu		0154		1.7
	0425	0721	1.8	
	1024	1412		1.6
	1648	1910	1.7	
	2237			
23 W		0151		1.6
	0503	0724	1.8	
	1103	1401		1.5
	1726	1938	1.7	
	2314			
24 Th		0148		1.6
	0542	0800	1.8	
	1143	1409		1.5
	1806	2018	1.6	
	2352			
25 F		0215		1.6
	0622	0844	1.8	
	1224	1441		1.4
	1849	2104	1.5	
26 Sa	0032	0253		1.5
	0706	0931	1.7	
	1308	1521		1.3
	1938	2151	1.5	
27 Su	0116	0336		1.5
	0757	1017	1.7	
	1357	1605		1.2
	2031	2237	1.4	
28 M ☾	0206	0422		1.4
	0851	1105	1.6	
	1452	1653		1.1
	2126	2326	1.3	
29 Tu	0303	0513		1.2
	0948	1157	1.5	
	1549	1753		1.0
	2222			
30 W		0023	1.3	
	0404	0622		1.2
	1046	1259	1.4	
	1649	2026		1.1
	2320			

OCTOBER

Day	Slack time	Max time	Fld knots	Ebb knots
1 Th		0129	1.3	
	0508	0807		1.3
	1145	1408	1.5	
	1748	2109		1.3
2 F	0017	0239	1.5	
	0610	0917		1.5
	1243	1520	1.7	
	1842	2209		1.5
3 Sa	0110	0406	1.8	
	0706	1044		1.7
	1336	1645	1.9	
	1931	2313		1.8
4 Su	0200	0512	2.1	
	0758	1144		1.9
	1426	1731	2.1	
	2018	2357		2.0
5 M ○	0248	0553	2.4	
	0849	1226		2.1
	1516	1806	2.3	
	2106			
6 Tu		0034		2.2
	0336	0630	2.6	
	0941	1306		2.2
	1604	1842	2.4	
	2155			
7 W		0112		2.3
	0424	0709	2.6	
	1032	1348		2.2
	1653	1923	2.3	
	2244			
8 Th		0153		2.3
	0512	0755	2.6	
	1122	1437		2.1
	1742	2013	2.2	
	2333			
9 F		0242		2.2
	0602	0857	2.4	
	1212	1533		2.0
	1833	2116	2.0	
10 Sa	0023	0337		2.0
	0656	1005	2.2	
	1304	1633		1.7
	1929	2220	1.8	
11 Su	0116	0438		1.8
	0754	1105	2.0	
	1400	1756		1.5
	2028	2329	1.6	
12 M ☽	0215	0626		1.5
	0853	1220	1.8	
	1459	1920		1.5
	2128			
13 Tu		0056	1.5	
	0317	0746		1.5
	0953	1333	1.6	
	1601	2021		1.5
	2227			
14 W		0205	1.5	
	0422	0844		1.6
	1052	1434	1.6	
	1704	2113		1.6
	2326			
15 Th		0302	1.5	
	0530	0938		1.6
	1152	1529	1.6	
	1804	2206		1.7
16 F	0022	0356	1.6	
	0630	1032		1.7
	1247	1621	1.7	
	1853	2258		1.7
17 Sa	0112	0447	1.8	
	0719	1124		1.7
	1336	1710	1.7	
	1934	2345		1.7
18 Su	0156	0533	1.9	
	0801	1209		1.7
	1421	1752	1.8	
	2012			
19 M		0026		1.8
	0238	0612	1.9	
	0840	1249		1.7
	1502	1828	1.7	
	2049			
20 Tu ●		0100		1.7
	0317	0645	1.9	
	0919	1323		1.7
	1543	1854	1.7	
	2126			
21 W		0124		1.6
	0356	0656	1.9	
	0958	1348		1.6
	1622	1842	1.6	
	2205			
22 Th		0113		1.6
	0434	0657	1.9	
	1037	1336		1.5
	1701	1909	1.6	
	2243			
23 F		0117		1.6
	0512	0730	1.9	
	1117	1343		1.5
	1740	1947	1.6	
	2322			
24 Sa		0145		1.6
	0551	0812	1.9	
	1157	1413		1.5
	1822	2033	1.5	
25 Su	0003	0223		1.6
	0633	0900	1.8	
	1240	1454		1.4
	1908	2122	1.5	
26 M	0046	0307		1.5
	0722	0949	1.7	
	1327	1539		1.3
	2001	2211	1.5	
27 Tu	0137	0355		1.4
	0818	1038	1.7	
	1419	1627		1.3
	2056	2301	1.4	
28 W ☾	0235	0447		1.3
	0916	1129	1.6	
	1515	1723		1.2
	2152	2356	1.4	
29 Th	0337	0551		1.2
	1014	1227	1.5	
	1613	1847		1.2
	2249			
30 F		0101	1.5	
	0442	0743		1.3
	1115	1335	1.6	
	1712	2030		1.4
	2346			
31 Sa		0215	1.6	
	0546	0905		1.5
	1214	1445	1.7	
	1809	2123		1.7

MIAMI, FLORIDA (Harbor Entrance)

CURRENT TABLE 1998 — 25°45.9'N 80°08.2W — Flood 293° Ebb 112°

Eastern Time (75°W) — Add 1 hour for Daylight Saving Time: April 5 – October 24

NOVEMBER

Day	Slack time	Max time	Fld knots	Ebb knots
1 Su	0041	0342	1.9	
	0645	1017		1.7
	1311	1605	1.9	
	1902	2228		1.9
2 M	0133	0453	2.2	
	0739	1125		1.9
	1403	1710	2.1	
	1951	2331		2.1
3 Tu	0223	0541	2.5	
	0830	1215		2.1
	1454	1754	2.2	
	2040			
4 W ○		0018		2.3
	0313	0622	2.7	
	0922	1258		2.2
	1544	1832	2.3	
	2130			
5 Th		0059		2.4
	0402	0701	2.7	
	1013	1342		2.2
	1633	1911	2.2	
	2221			
6 F		0143		2.3
	0451	0746	2.6	
	1103	1432		2.1
	1722	1959	2.1	
	2312			
7 Sa		0233		2.2
	0541	0849	2.4	
	1152	1528		2.0
	1813	2115	1.9	
8 Su	0002	0331		2.0
	0633	0957	2.2	
	1242	1625		1.8
	1907	2222	1.8	
9 M	0055	0434		1.7
	0728	1052	2.0	
	1335	1732		1.6
	2005	2320	1.6	
10 Tu ☽	0152	0558		1.5
	0827	1152	1.7	
	1430	1850		1.5
	2102			
11 W		0029	1.5	
	0252	0720		1.4
	0925	1301	1.6	
	1527	1954		1.5
	2159			
12 Th		0136	1.4	
	0355	0820		1.4
	1022	1403	1.5	
	1624	2046		1.6
	2254			
13 F		0234	1.5	
	0459	0911		1.5
	1119	1458	1.5	
	1721	2134		1.6
	2348			
14 Sa		0326	1.6	
	0601	1003		1.5
	1215	1549	1.5	
	1813	2224		1.6
15 Su	0039	0417	1.7	
	0652	1055		1.5
	1306	1639	1.5	
	1857	2313		1.6
16 M	0125	0505	1.8	
	0734	1143		1.6
	1352	1724	1.6	
	1937	2357		1.6
17 Tu	0207	0546	1.9	
	0814	1225		1.6
	1435	1803	1.6	
	2016			
18 W ●		0034		1.6
	0248	0621	1.9	
	0853	1302		1.6
	1516	1831	1.6	
	2055			
19 Th		0059		1.6
	0328	0641	1.9	
	0933	1331		1.5
	1557	1820	1.6	
	2135			
20 F		0051		1.6
	0408	0636	1.9	
	1013	1330		1.5
	1637	1845	1.6	
	2216			
21 Sa		0054		1.6
	0447	0707	2.0	
	1053	1325		1.5
	1717	1922	1.6	
	2257			
22 Su		0122		1.6
	0526	0747	1.9	
	1134	1353		1.5
	1758	2007	1.6	
	2339			
23 M		0200		1.6
	0607	0834	1.9	
	1216	1433		1.5
	1843	2057	1.6	
24 Tu	0024	0245		1.6
	0654	0924	1.8	
	1301	1519		1.5
	1934	2149	1.6	
25 W	0114	0335		1.5
	0749	1014	1.8	
	1350	1607		1.5
	2029	2240	1.6	
26 Th ☾	0212	0427		1.4
	0847	1104	1.7	
	1444	1700		1.4
	2124	2333	1.6	
27 F	0314	0527		1.3
	0946	1159	1.6	
	1540	1809		1.4
	2219			
28 Sa		0034	1.6	
	0418	0716		1.3
	1046	1303	1.6	
	1638	1958		1.6
	2316			
29 Su		0151	1.8	
	0522	0853		1.5
	1147	1415	1.7	
	1737	2058		1.8
30 M	0014	0325	2.0	
	0624	0959		1.7
	1246	1536	1.8	
	1834	2200		1.9

DECEMBER

Day	Slack time	Max time	Fld knots	Ebb knots
1 Tu	0109	0435	2.3	
	0720	1109		1.8
	1341	1653	1.9	
	1926	2313		2.1
2 W	0201	0529	2.5	
	0812	1204		2.0
	1433	1746	2.1	
	2017			
3 Th ○		0008		2.2
	0252	0615	2.6	
	0904	1251		2.1
	1524	1829	2.2	
	2109			
4 F		0054		2.3
	0342	0658	2.6	
	0955	1336		2.2
	1614	1910	2.1	
	2201			
5 Sa		0139		2.3
	0432	0743	2.5	
	1045	1425		2.1
	1703	2000	2.0	
	2252			
6 Su		0229		2.1
	0521	0840	2.3	
	1132	1518		2.0
	1752	2108	1.9	
	2342			
7 M		0325		1.9
	0610	0940	2.1	
	1219	1610		1.8
	1843	2207	1.8	
8 Tu	0033	0420		1.7
	0702	1031	1.9	
	1307	1703		1.6
	1937	2258	1.7	
9 W	0126	0522		1.5
	0756	1118	1.7	
	1356	1809		1.5
	2031	2353	1.5	
10 Th ☽	0222	0643		1.3
	0852	1217	1.5	
	1447	1918		1.4
	2125			
11 F		0058	1.4	
	0319	0748		1.3
	0946	1323	1.4	
	1539	2013		1.4
	2217			
12 Sa		0158	1.4	
	0419	0841		1.3
	1040	1421	1.3	
	1632	2100		1.4
	2309			
13 Su		0252	1.5	
	0519	0930		1.3
	1136	1513	1.3	
	1726	2147		1.5
14 M	0001	0343	1.6	
	0615	1022		1.3
	1230	1604	1.3	
	1816	2238		1.5
15 Tu	0051	0433	1.7	
	0703	1114		1.4
	1320	1653	1.4	
	1902	2327		1.5
16 W	0136	0519	1.8	
	0746	1201		1.5
	1406	1737	1.4	
	1944			
17 Th		0008		1.5
	0220	0558	1.8	
	0827	1241		1.5
	1450	1811	1.5	
	2026			
18 F ●		0039		1.5
	0302	0627	1.9	
	0908	1314		1.5
	1533	1807	1.5	
	2108			
19 Sa		0045		1.5
	0344	0622	1.9	
	0950	1331		1.5
	1614	1826	1.6	
	2151			
20 Su		0042		1.6
	0424	0648	2.0	
	1031	1320		1.5
	1655	1903	1.6	
	2235			
21 M		0108		1.7
	0505	0726	2.0	
	1113	1341		1.6
	1737	1946	1.7	
	2320			
22 Tu		0145		1.7
	0547	0812	2.0	
	1154	1418		1.6
	1820	2036	1.7	
23 W	0006	0230		1.7
	0633	0902	2.0	
	1237	1503		1.7
	1909	2129	1.8	
24 Th	0056	0320		1.6
	0725	0953	1.9	
	1324	1551		1.7
	2002	2220	1.8	
25 F	0152	0412		1.5
	0822	1042	1.9	
	1416	1641		1.6
	2057	2312	1.8	
26 Sa ☾	0252	0509		1.4
	0921	1134	1.8	
	1511	1741		1.6
	2152			
27 Su		0010	1.8	
	0354	0646		1.3
	1020	1233	1.6	
	1608	1928		1.6
	2250			
28 M		0128	1.8	
	0459	0844		1.4
	1121	1347	1.6	
	1708	2045		1.8
	2349			
29 Tu		0313	2.0	
	0603	0947		1.6
	1222	1528	1.7	
	1809	2150		1.9
30 W	0047	0420	2.2	
	0702	1056		1.7
	1320	1641	1.8	
	1905	2305		2.0
31 Th	0142	0517	2.4	
	0756	1154		1.9
	1414	1737	2.0	
	1959	2359		2.1

KEY WEST, FLORIDA (0.3 nm west of Fort Taylor)

CURRENT TABLE 1998 — 24°32.9'N 81°49.0'W — Flood 20° Ebb 195°

Eastern Time (75°W) — Add 1 hour for Daylight Saving Time: April 5 – October 24

JANUARY

Day	Slack time	Max time	Fld knots	Ebb knots
1 Th		0238		2.4
	0617	0901	1.6	
	1212	1516		2.0
	1852	2113	1.1	
	2356			
2 F		0327		2.2
	0707	0947	1.4	
	1256	1604		1.9
	1945	2206	1.0	
3 Sa	0054	0421		2.0
	0803	1038	1.2	
	1342	1657		1.9
	2045	2308	0.9	
4 Su	0200	0520		1.8
	0906	1136	1.0	
	1433	1755		1.8
	2150			
5 M		0022	0.9	
	0315	0627		1.6
	1017	1243	0.8	
☾	1529	1859		1.8
	2259			
6 Tu		0152	0.9	
	0433	0743		1.4
	1132	1405	0.7	
	1630	2007		1.8
7 W	0006	0315	1.0	
	0548	0904		1.4
	1245	1524	0.7	
	1732	2115		1.9
8 Th	0107	0421	1.2	
	0656	1015		1.5
	1349	1628	0.7	
	1831	2216		2.0
9 F	0202	0516	1.4	
	0754	1113		1.6
	1442	1721	0.8	
	1925	2308		2.1
10 Sa	0251	0604	1.5	
	0844	1200		1.7
	1528	1807	0.9	
	2015	2354		2.2
11 Su	0335	0647	1.5	
	0928	1241		1.7
	1609	1847	0.9	
	2101			
12 M		0035		2.2
	0417	0725	1.5	
	1008	1317		1.8
○	1647	1923	1.0	
	2143			
13 Tu		0113		2.2
	0456	0758	1.4	
	1045	1350		1.8
	1724	1955	0.9	
	2223			
14 W		0150		2.1
	0534	0828	1.3	
	1119	1423		1.8
	1800	2026	0.9	
	2302			
15 Th		0227		2.0
	0612	0856	1.2	
	1152	1457		1.7
	1839	2059	0.9	
	2341			
16 F		0305		1.9
	0651	0925	1.1	
	1225	1533		1.7
	1920	2136	0.8	
17 Sa	0021	0346		1.8
	0733	0959	0.9	
	1258	1613		1.6
	2005	2217	0.7	
18 Su	0105	0430		1.6
	0821	1038	0.8	
	1333	1658		1.5
	2057	2305	0.6	
19 M	0156	0520		1.4
	0915	1123	0.6	
	1411	1747		1.5
	2154			
20 Tu		0003	0.6	
	0300	0617		1.2
	1019	1217	0.4	
☽	1456	1843		1.4
	2256			
21 W		0115	0.5	
	0413	0722		1.1
	1130	1321	0.4	
	1552	1943		1.5
	2357			
22 Th		0242	0.6	
	0527	0833		1.1
	1238	1435	0.3	
	1656	2045		1.6
23 F	0052	0354	0.8	
	0633	0940		1.2
	1335	1544	0.4	
	1758	2143		1.7
24 Sa	0141	0447	1.1	
	0730	1037		1.4
	1421	1638	0.6	
	1856	2236		2.0
25 Su	0225	0530	1.3	
	0821	1125		1.6
	1502	1724	0.8	
	1948	2324		2.2
26 M	0309	0609	1.5	
	0906	1208		1.8
	1541	1807	1.0	
	2038			
27 Tu		0010		2.4
	0351	0648	1.6	
	0948	1250		2.0
	1621	1849	1.2	
	2127			
28 W		0055		2.5
	0434	0726	1.7	
	1029	1331		2.1
●	1701	1932	1.3	
	2215			
29 Th		0140		2.6
	0517	0805	1.7	
	1109	1413		2.2
	1744	2016	1.4	
	2304			
30 F		0226		2.5
	0603	0846	1.6	
	1149	1456		2.2
	1830	2103	1.4	
	2355			
31 Sa		0314		2.3
	0651	0929	1.5	
	1230	1542		2.2
	1920	2153	1.3	

FEBRUARY

Day	Slack time	Max time	Fld knots	Ebb knots
1 Su	0050	0404		2.1
	0743	1016	1.2	
	1312	1631		2.1
	2016	2250	1.2	
2 M	0150	0459		1.8
	0843	1108	1.0	
	1359	1725		1.9
	2119	2357	1.0	
3 Tu	0257	0602		1.5
	0952	1210	0.7	
	1451	1826		1.8
☾	2230			
4 W		0126	0.9	
	0412	0718		1.3
	1111	1333	0.5	
	1553	1937		1.7
	2344			
5 Th		0257	1.0	
	0528	0847		1.2
	1234	1508	0.5	
	1701	2054		1.7
6 F	0053	0408	1.1	
	0638	1005		1.3
	1344	1619	0.6	
	1808	2204		1.8
7 Sa	0153	0505	1.2	
	0737	1104		1.4
	1436	1714	0.7	
	1909	2300		1.9
8 Su	0243	0553	1.3	
	0826	1150		1.6
	1518	1800	0.8	
	2002	2346		2.0
9 M	0326	0635	1.4	
	0908	1228		1.7
	1554	1839	0.9	
	2049			
10 Tu		0025		2.1
	0405	0710	1.4	
	0944	1259		1.7
	1627	1912	1.0	
	2131			
11 W		0100		2.1
	0441	0740	1.3	
	1018	1328		1.8
○	1659	1940	1.0	
	2209			
12 Th		0132		2.1
	0515	0805	1.3	
	1049	1356		1.8
	1731	2006	1.1	
	2246			
13 F		0205		2.1
	0548	0827	1.2	
	1119	1426		1.9
	1805	2033	1.1	
	2322			
14 Sa		0240		2.0
	0623	0852	1.1	
	1147	1500		1.8
	1840	2105	1.0	
	2358			
15 Su		0317		1.8
	0700	0922	1.0	
	1215	1536		1.8
	1919	2141	0.9	
16 M	0036	0357		1.7
	0741	0957	0.8	
	1242	1616		1.7
	2004	2223	0.8	
17 Tu	0120	0442		1.5
	0829	1037	0.7	
	1311	1702		1.6
	2056	2313	0.7	
18 W	0214	0534		1.3
	0928	1124	0.5	
	1345	1754		1.5
	2158			
19 Th		0014	0.6	
	0324	0635		1.1
	1040	1223	0.3	
☽	1435	1854		1.5
	2307			
20 F		0131	0.6	
	0443	0746		1.1
	1157	1336	0.3	
	1552	2002		1.5
21 Sa	0013	0305	0.7	
	0557	0901		1.2
	1302	1500	0.4	
	1718	2109		1.7
22 Su	0111	0415	1.0	
	0700	1007		1.4
	1353	1612	0.6	
	1831	2211		1.9
23 M	0202	0506	1.2	
	0753	1100		1.6
	1436	1706	0.9	
	1933	2305		2.2
24 Tu	0249	0549	1.5	
	0839	1146		1.9
	1516	1753	1.1	
	2028	2354		2.4
25 W	0334	0629	1.6	
	0922	1228		2.1
	1557	1837	1.4	
	2119			
26 Th		0041		2.6
	0418	0708	1.7	
	1003	1310		2.3
●	1638	1920	1.6	
	2209			
27 F		0126		2.6
	0502	0747	1.7	
	1042	1351		2.4
	1721	2004	1.7	
	2258			
28 Sa		0212		2.5
	0547	0827	1.6	
	1122	1434		2.5
	1806	2049	1.7	
	2348			

KEY WEST, FLORIDA (0.3 nm west of Fort Taylor)

CURRENT TABLE 1998 — 24°32.9'N 81°49.0'W — Flood 20° Ebb 195°

Eastern Time (75°W) — Add 1 hour for Daylight Saving Time: April 5 – October 24

MARCH

Day	Slack time	Max time	Fld knots	Ebb knots
1 Su		0258		2.4
	0634	0909	1.5	
	1201	1518		2.4
	1855	2137	1.5	
2 M	0040	0347		2.1
	0724	0953	1.2	
	1242	1605		2.2
	1949	2230	1.3	
3 Tu	0136	0439		1.8
	0822	1042	0.9	
	1326	1657		2.0
	2051	2334	1.1	
4 W	0238	0539		1.5
	0930	1140	0.6	
	1416	1756		1.8
	2202			
5 Th ☾		0100	0.9	
	0348	0653		1.2
	1053	1306	0.4	
	1520	1907		1.6
	2321			
6 F		0236	0.8	
	0502	0826		1.1
	1223	1454	0.4	
	1635	2033		1.5
7 Sa	0037	0349	0.9	
	0611	0948		1.2
	1334	1606	0.5	
	1750	2150		1.6
8 Su	0141	0447	1.0	
	0710	1046		1.4
	1423	1700	0.7	
	1854	2248		1.7
9 M	0231	0534	1.1	
	0758	1129		1.5
	1500	1745	0.9	
	1948	2333		1.9
10 Tu	0312	0614	1.2	
	0838	1204		1.7
	1532	1823	1.0	
	2035			
11 W		0010		2.0
	0348	0648	1.2	
	0913	1233		1.8
	1602	1854	1.1	
	2115			
12 Th ○		0042		2.0
	0421	0715	1.2	
	0945	1259		1.9
	1631	1919	1.2	
	2153			
13 F		0112		2.0
	0452	0737	1.2	
	1015	1326		1.9
	1701	1942	1.2	
	2228			
14 Sa		0142		2.0
	0524	0757	1.1	
	1043	1355		2.0
	1732	2007	1.2	
	2302			
15 Su		0215		2.0
	0557	0820	1.0	
	1109	1427		2.0
	1805	2037	1.2	
	2337			
16 M		0250		1.9
	0631	0849	1.0	
	1133	1502		1.9
	1841	2111	1.1	
17 Tu	0013	0328		1.7
	0710	0922	0.8	
	1157	1541		1.8
	1922	2151	1.0	
18 W	0054	0412		1.5
	0754	1001	0.7	
	1224	1625		1.7
	2011	2237	0.9	
19 Th	0143	0501		1.4
	0850	1047	0.5	
	1257	1716		1.6
	2111	2334	0.8	
20 F	0247	0600		1.2
	1000	1144	0.4	
	1347	1816		1.5
	2222			
21 Sa ☽		0045	0.7	
	0403	0709		1.1
	1118	1258	0.3	
	1510	1926		1.5
	2335			
22 Su		0215	0.7	
	0519	0824		1.2
	1227	1427	0.4	
	1652	2039		1.7
23 M	0041	0338	0.9	
	0624	0934		1.4
	1321	1548	0.7	
	1813	2147		1.9
24 Tu	0137	0436	1.2	
	0719	1030		1.7
	1407	1647	1.0	
	1919	2246		2.1
25 W	0228	0523	1.4	
	0807	1119		2.0
	1450	1737	1.3	
	2016	2337		2.4
26 Th	0314	0605	1.5	
	0851	1203		2.3
	1532	1822	1.6	
	2109			
27 F ●		0025		2.5
	0359	0646	1.6	
	0932	1245		2.5
	1615	1906	1.8	
	2159			
28 Sa		0111		2.5
	0444	0726	1.6	
	1013	1327		2.6
	1658	1950	1.9	
	2248			
29 Su		0156		2.5
	0529	0806	1.5	
	1052	1410		2.6
	1744	2034	1.8	
	2337			
30 M		0242		2.3
	0616	0847	1.3	
	1132	1454		2.5
	1832	2121	1.6	
31 Tu	0027	0329		2.0
	0707	0930	1.1	
	1213	1540		2.3
	1925	2212	1.4	

APRIL

Day	Slack time	Max time	Fld knots	Ebb knots
1 W	0119	0420		1.7
	0803	1018	0.8	
	1257	1630		2.0
	2025	2312	1.1	
2 Th	0217	0517		1.4
	0911	1117	0.5	
	1348	1728		1.7
	2134			
3 F ☾		0033	0.9	
	0320	0627		1.2
	1033	1245	0.4	
	1453	1838		1.5
	2253			
4 Sa		0207	0.8	
	0429	0756		1.1
	1201	1432	0.4	
	1612	2005		1.4
5 Su	0012	0320	0.8	
	0534	0916		1.2
	1309	1543	0.5	
	1729	2126		1.5
6 M	0118	0418	0.9	
	0632	1014		1.4
	1355	1637	0.7	
	1834	2225		1.6
7 Tu	0208	0505	1.0	
	0719	1057		1.5
	1430	1721	0.9	
	1929	2310		1.7
8 W	0249	0545	1.0	
	0800	1131		1.7
	1502	1758	1.0	
	2015	2347		1.8
9 Th	0324	0618	1.1	
	0836	1200		1.8
	1532	1829	1.1	
	2056			
10 F		0018		1.9
	0357	0645	1.0	
	0908	1226		1.9
	1601	1855	1.2	
	2133			
11 Sa ○		0048		1.9
	0429	0705	1.0	
	0938	1254		2.0
	1631	1918	1.3	
	2209			
12 Su		0119		1.9
	0500	0725	1.0	
	1005	1324		2.0
	1702	1943	1.3	
	2244			
13 M		0151		1.9
	0533	0750	0.9	
	1031	1357		2.0
	1734	2012	1.3	
	2319			
14 Tu		0226		1.8
	0607	0819	0.9	
	1055	1432		2.0
	1810	2047	1.3	
	2356			
15 W		0305		1.7
	0645	0854	0.8	
	1121	1512		1.9
	1851	2126	1.2	
16 Th	0037	0348		1.6
	0730	0933	0.7	
	1151	1556		1.8
	1939	2212	1.0	
17 F	0124	0437		1.4
	0824	1021	0.5	
	1231	1648		1.7
	2037	2307	0.9	
18 Sa	0222	0534		1.3
	0930	1120	0.4	
	1329	1748		1.6
	2146			
19 Su ☽		0014	0.8	
	0331	0640		1.3
	1044	1233	0.4	
	1457	1858		1.6
	2300			
20 M		0135	0.8	
	0441	0751		1.4
	1151	1402	0.5	
	1636	2012		1.7
21 Tu	0010	0257	0.9	
	0545	0900		1.6
	1248	1525	0.8	
	1757	2123		1.8
22 W	0111	0402	1.1	
	0641	0958		1.9
	1338	1627	1.2	
	1905	2225		2.0
23 Th	0204	0453	1.3	
	0731	1049		2.1
	1424	1719	1.5	
	2003	2319		2.2
24 F	0254	0539	1.4	
	0817	1136		2.4
	1508	1806	1.7	
	2057			
25 Sa		0008		2.3
	0341	0622	1.4	
	0900	1220		2.6
	1553	1851	1.9	
	2147			
26 Su ●		0055		2.4
	0426	0704	1.4	
	0943	1304		2.6
	1638	1936	1.9	
	2236			
27 M		0141		2.3
	0513	0745	1.3	
	1024	1347		2.6
	1724	2020	1.8	
	2324			
28 Tu		0226		2.1
	0600	0827	1.2	
	1105	1431		2.5
	1812	2106	1.7	
29 W	0012	0312		1.9
	0650	0911	1.0	
	1147	1517		2.2
	1904	2155	1.4	
30 Th	0101	0401		1.7
	0746	0959	0.7	
	1233	1607		2.0
	2000	2250	1.1	

KEY WEST, FLORIDA (0.3 nm west of Fort Taylor)

CURRENT TABLE 1998 — 24°32.9'N 81°49.0'W — Flood 20° Ebb 195°

Eastern Time (75°W) — Add 1 hour for Daylight Saving Time: April 5 – October 24

MAY

Day	Slack time	Max time	Fld knots	Ebb knots
1 F	0153	0455		1.5
	0850	1056	0.5	
	1325	1702		1.7
	2105	2359	0.9	
2 Sa	0249	0557		1.3
	1004	1218	0.4	
	1429	1807		1.5
	2217			
3 Su ☾		0124	0.7	
	0348	0711		1.2
	1121	1357	0.4	
	1544	1925		1.3
	2332			
4 M		0238	0.7	
	0448	0826		1.3
	1226	1509	0.5	
	1659	2045		1.3
5 Tu	0039	0338	0.7	
	0543	0926		1.4
	1314	1604	0.7	
	1805	2148		1.4
6 W	0133	0426	0.8	
	0632	1011		1.5
	1353	1650	0.9	
	1901	2236		1.5
7 Th	0217	0507	0.8	
	0715	1048		1.7
	1427	1728	1.0	
	1949	2316		1.6
8 F	0255	0541	0.9	
	0753	1121		1.8
	1459	1801	1.1	
	2032	2350		1.7
9 Sa	0330	0609	0.9	
	0828	1151		1.9
	1531	1829	1.2	
	2112			
10 Su		0023		1.8
	0404	0632	0.9	
	0859	1222		2.0
	1603	1854	1.3	
	2150			
11 M ○		0055		1.8
	0437	0655	0.9	
	0929	1255		2.1
	1635	1922	1.4	
	2227			
12 Tu		0129		1.8
	0511	0723	0.8	
	0957	1330		2.1
	1710	1953	1.4	
	2304			
13 W		0206		1.8
	0547	0755	0.8	
	1025	1408		2.1
	1747	2028	1.3	
	2343			
14 Th		0246		1.7
	0626	0832	0.8	
	1057	1449		2.0
	1829	2109	1.3	
15 F	0024	0329		1.6
	0711	0915	0.7	
	1135	1535		2.0
	1917	2155	1.2	

Day	Slack time	Max time	Fld knots	Ebb knots
16 Sa	0110	0418		1.5
	0804	1005	0.6	
	1224	1627		1.8
	2014	2248	1.0	
17 Su	0203	0513		1.5
	0905	1104	0.6	
	1329	1727		1.7
	2119	2350	0.9	
18 M ☽	0301	0614		1.5
	1012	1216	0.6	
	1454	1834		1.6
	2229			
19 Tu		0101	0.9	
	0404	0720		1.6
	1118	1339	0.7	
	1623	1947		1.6
	2339			
20 W		0217	0.9	
	0505	0826		1.7
	1217	1502	0.9	
	1742	2059		1.7
21 Th	0044	0326	1.0	
	0602	0926		2.0
	1311	1608	1.2	
	1850	2204		1.9
22 F	0142	0424	1.1	
	0655	1021		2.2
	1400	1703	1.5	
	1950	2301		2.0
23 Sa	0234	0514	1.2	
	0744	1111		2.4
	1448	1753	1.7	
	2044	2353		2.1
24 Su	0323	0601	1.2	
	0830	1158		2.5
	1535	1840	1.8	
	2135			
25 M ●		0041		2.1
	0411	0645	1.2	
	0915	1243		2.6
	1621	1925	1.8	
	2223			
26 Tu		0126		2.1
	0457	0728	1.1	
	0959	1327		2.5
	1707	2009	1.8	
	2309			
27 W		0211		2.0
	0544	0810	1.0	
	1042	1411		2.4
	1754	2053	1.6	
	2354			
28 Th		0255		1.8
	0633	0854	0.9	
	1126	1457		2.2
	1843	2138	1.4	
29 F	0039	0341		1.7
	0725	0941	0.7	
	1213	1544		1.9
	1936	2225	1.1	
30 Sa	0125	0429		1.5
	0822	1033	0.6	
	1304	1635		1.7
	2032	2319	0.9	
31 Su	0212	0521		1.4
	0924	1138	0.5	
	1402	1732		1.5
	2135			

JUNE

Day	Slack time	Max time	Fld knots	Ebb knots
1 M ☾		0023	0.7	
	0302	0619		1.3
	1029	1301	0.4	
	1510	1836		1.3
	2242			
2 Tu		0136	0.6	
	0355	0722		1.3
	1132	1419	0.5	
	1620	1947		1.3
	2348			
3 W		0241	0.6	
	0447	0822		1.4
	1225	1521	0.6	
	1726	2055		1.3
4 Th	0047	0336	0.6	
	0538	0915		1.5
	1310	1611	0.8	
	1826	2152		1.4
5 F	0138	0422	0.6	
	0625	1000		1.6
	1350	1654	1.0	
	1918	2239		1.5
6 Sa	0222	0500	0.7	
	0708	1040		1.8
	1427	1731	1.1	
	2006	2320		1.6
7 Su	0302	0532	0.7	
	0746	1116		1.9
	1502	1804	1.2	
	2049	2357		1.6
8 M	0339	0600	0.7	
	0822	1153		2.0
	1537	1834	1.3	
	2130			
9 Tu ○		0033		1.7
	0415	0628	0.8	
	0857	1230		2.1
	1613	1904	1.4	
	2210			
10 W		0110		1.7
	0451	0701	0.8	
	0931	1308		2.2
	1650	1937	1.4	
	2249			
11 Th		0148		1.8
	0528	0737	0.8	
	1007	1348		2.2
	1730	2014	1.4	
	2329			
12 F		0229		1.8
	0608	0817	0.8	
	1047	1432		2.2
	1813	2055	1.4	
13 Sa	0010	0312		1.7
	0652	0902	0.8	
	1133	1519		2.1
	1901	2140	1.3	
14 Su	0053	0400		1.7
	0742	0953	0.8	
	1227	1611		2.0
	1954	2230	1.2	
15 M	0140	0452		1.7
	0839	1051	0.8	
	1333	1708		1.8
	2055	2326	1.0	

Day	Slack time	Max time	Fld knots	Ebb knots
16 Tu	0232	0549		1.7
	0942	1159	0.8	
	1449	1812		1.7
	2201			
17 W ☽		0030	0.9	
	0327	0650		1.7
	1046	1318	0.9	
	1609	1923		1.6
	2312			
18 Th		0141	0.8	
	0426	0754		1.8
	1149	1441	1.0	
	1726	2036		1.6
19 F	0020	0254	0.8	
	0525	0857		2.0
	1248	1552	1.2	
	1835	2146		1.7
20 Sa	0123	0359	0.9	
	0621	0956		2.1
	1342	1651	1.5	
	1936	2247		1.7
21 Su	0219	0456	0.9	
	0715	1050		2.3
	1433	1743	1.6	
	2032	2341		1.8
22 M	0310	0546	1.0	
	0806	1140		2.4
	1521	1831	1.7	
	2122			
23 Tu ●		0029		1.9
	0358	0632	1.0	
	0854	1227		2.4
	1608	1916	1.7	
	2208			
24 W		0114		1.9
	0443	0716	1.0	
	0940	1311		2.4
	1653	1958	1.6	
	2251			
25 Th		0155		1.9
	0528	0757	1.0	
	1025	1354		2.3
	1738	2038	1.5	
	2332			
26 F		0236		1.8
	0612	0838	0.9	
	1109	1437		2.1
	1823	2116	1.3	
27 Sa	0012	0316		1.7
	0658	0919	0.8	
	1153	1520		1.9
	1909	2154	1.1	
28 Su	0051	0357		1.6
	0746	1003	0.7	
	1240	1605		1.7
	1957	2234	0.9	
29 M	0131	0441		1.5
	0838	1052	0.6	
	1331	1654		1.5
	2050	2319	0.8	
30 Tu	0213	0529		1.5
	0934	1150	0.5	
	1429	1748		1.4
	2149			

KEY WEST, FLORIDA (0.3 nm west of Fort Taylor)

CURRENT TABLE 1998 — 24°32.9'N 81°49.0'W — Flood 20° Ebb 195°

Eastern Time (75°W) — Add 1 hour for Daylight Saving Time: April 5 – October 24

JULY

Day	Slack time	Max time	Fld knots	Ebb knots
1 W ☾		0011	0.6	
	0258	0621		1.4
	1033	1301	0.5	
	1533	1848		1.3
	2252			
2 Th		0113	0.5	
	0347	0718		1.4
	1131	1419	0.6	
	1641	1954		1.2
	2357			
3 F		0222	0.4	
	0439	0815		1.5
	1224	1525	0.7	
	1746	2100		1.2
4 Sa	0057	0324	0.5	
	0531	0910		1.6
	1312	1617	0.8	
	1845	2158		1.3
5 Su	0149	0414	0.5	
	0621	0959		1.7
	1355	1701	1.0	
	1938	2248		1.4
6 M	0233	0455	0.6	
	0708	1044		1.9
	1435	1740	1.2	
	2025	2332		1.5
7 Tu	0313	0531	0.7	
	0751	1127		2.0
	1514	1814	1.3	
	2109			
8 W		0012		1.6
	0351	0607	0.8	
	0833	1208		2.1
	1553	1848	1.4	
	2151			
9 Th ○		0051		1.7
	0428	0644	0.9	
	0916	1250		2.3
	1633	1923	1.5	
	2231			
10 F		0130		1.8
	0506	0723	1.0	
	0959	1333		2.3
	1714	2001	1.5	
	2310			
11 Sa		0211		1.9
	0546	0805	1.0	
	1045	1417		2.3
	1757	2040	1.5	
	2349			
12 Su		0253		1.9
	0629	0851	1.1	
	1134	1504		2.2
	1844	2123	1.4	
13 M	0030	0339		2.0
	0718	0940	1.1	
	1228	1555		2.1
	1935	2210	1.3	
14 Tu	0113	0428		1.9
	0812	1036	1.0	
	1329	1650		1.9
	2032	2302	1.1	
15 W	0200	0522		1.9
	0912	1140	1.0	
	1438	1751		1.7
	2137			

Day	Slack time	Max time	Fld knots	Ebb knots
16 Th ☽		0001	0.9	
	0252	0621		1.9
	1017	1256	1.0	
	1554	1900		1.5
	2248			
17 F		0110	0.7	
	0350	0725		1.9
	1125	1424	1.0	
	1709	2016		1.4
18 Sa	0002	0229	0.7	
	0452	0832		1.9
	1230	1541	1.2	
	1820	2133		1.4
19 Su	0110	0345	0.7	
	0555	0938		2.0
	1329	1643	1.3	
	1923	2238		1.5
20 M	0210	0447	0.8	
	0654	1037		2.1
	1423	1737	1.5	
	2018	2333		1.6
21 Tu	0301	0540	0.9	
	0749	1129		2.2
	1512	1824	1.5	
	2106			
22 W		0019		1.7
	0346	0626	0.9	
	0839	1216		2.3
	1557	1906	1.6	
	2149			
23 Th ●		0100		1.8
	0428	0707	1.0	
	0926	1258		2.2
	1640	1944	1.5	
	2228			
24 F		0136		1.8
	0507	0744	1.0	
	1010	1337		2.2
	1720	2018	1.4	
	2305			
25 Sa		0211		1.8
	0546	0819	1.0	
	1052	1415		2.1
	1800	2049	1.3	
	2339			
26 Su		0245		1.8
	0625	0853	0.9	
	1133	1454		2.0
	1840	2118	1.1	
27 M	0013	0321		1.7
	0706	0928	0.9	
	1214	1534		1.8
	1922	2150	1.0	
28 Tu	0047	0400		1.7
	0750	1008	0.8	
	1258	1617		1.6
	2007	2227	0.8	
29 W	0121	0442		1.6
	0839	1054	0.7	
	1347	1705		1.4
	2059	2309	0.6	
30 Th	0158	0530		1.5
	0935	1148	0.6	
	1444	1758		1.3
	2200	2359	0.5	
31 F ☾	0241	0623		1.5
	1035	1255	0.6	
	1552	1900		1.2
	2308			

AUGUST

Day	Slack time	Max time	Fld knots	Ebb knots
1 Sa		0100	0.4	
	0332	0721		1.5
	1137	1420	0.6	
	1703	2009		1.1
2 Su	0017	0212	0.3	
	0434	0823		1.5
	1234	1536	0.7	
	1809	2118		1.2
3 M	0117	0325	0.4	
	0536	0922		1.6
	1324	1631	0.9	
	1908	2217		1.3
4 Tu	0205	0422	0.5	
	0635	1016		1.8
	1410	1715	1.1	
	1959	2306		1.5
5 W	0246	0508	0.7	
	0728	1104		2.0
	1452	1753	1.3	
	2045	2348		1.7
6 Th	0324	0549	0.9	
	0818	1150		2.2
	1533	1829	1.5	
	2126			
7 F ○		0029		1.9
	0401	0629	1.1	
	0906	1234		2.4
	1615	1905	1.6	
	2206			
8 Sa		0108		2.0
	0440	0710	1.2	
	0953	1318		2.4
	1657	1943	1.6	
	2245			
9 Su		0149		2.1
	0520	0752	1.3	
	1041	1403		2.4
	1740	2022	1.6	
	2323			
10 M		0231		2.2
	0604	0837	1.4	
	1130	1449		2.3
	1826	2103	1.5	
11 Tu	0002	0315		2.2
	0651	0925	1.4	
	1223	1538		2.2
	1916	2148	1.3	
12 W	0043	0403		2.1
	0744	1019	1.3	
	1320	1631		1.9
	2012	2237	1.0	
13 Th	0128	0454		2.0
	0844	1120	1.1	
	1425	1730		1.6
	2116	2333	0.8	
14 F ☽	0218	0552		1.9
	0951	1236	1.0	
	1537	1839		1.4
	2230			
15 Sa		0044	0.6	
	0317	0658		1.8
	1104	1411	1.0	
	1652	2001		1.3
	2351			

Day	Slack time	Max time	Fld knots	Ebb knots
16 Su		0217	0.5	
	0426	0813		1.8
	1216	1531	1.1	
	1804	2125		1.3
17 M	0105	0341	0.6	
	0536	0927		1.8
	1320	1634	1.2	
	1907	2231		1.4
18 Tu	0205	0444	0.7	
	0641	1031		1.9
	1415	1726	1.3	
	2000	2323		1.6
19 W	0252	0535	0.9	
	0738	1122		2.0
	1503	1811	1.4	
	2045			
20 Th		0005		1.7
	0331	0618	1.0	
	0829	1206		2.1
	1545	1850	1.4	
	2124			
21 F ●		0040		1.8
	0407	0655	1.1	
	0914	1244		2.1
	1624	1924	1.4	
	2159			
22 Sa		0112		1.8
	0441	0727	1.1	
	0955	1319		2.1
	1700	1952	1.3	
	2232			
23 Su		0141		1.9
	0515	0755	1.1	
	1034	1352		2.1
	1735	2016	1.2	
	2303			
24 M		0211		1.9
	0549	0823	1.1	
	1111	1426		2.0
	1810	2041	1.1	
	2332			
25 Tu		0244		1.9
	0625	0854	1.0	
	1148	1503		1.8
	1847	2109	1.0	
26 W	0001	0320		1.8
	0705	0929	1.0	
	1226	1542		1.7
	1928	2143	0.8	
27 Th	0029	0400		1.7
	0749	1009	0.9	
	1309	1626		1.5
	2016	2222	0.6	
28 F	0058	0444		1.6
	0840	1057	0.7	
	1400	1717		1.3
	2113	2308	0.5	
29 Sa ☾	0132	0535		1.5
	0941	1155	0.6	
	1505	1816		1.1
	2224			
30 Su		0005	0.3	
	0219	0634		1.4
	1048	1311	0.6	
	1620	1924		1.1
	2340			
31 M		0116	0.3	
	0332	0740		1.5
	1154	1446	0.7	
	1732	2039		1.1

KEY WEST, FLORIDA (0.3 nm west of Fort Taylor)

CURRENT TABLE 1998 | 24°32.9'N 81°49.0'W | Flood 20° Ebb 195°

Eastern Time (75°W) | Add 1 hour for Daylight Saving Time: April 5 – October 24

SEPTEMBER

Day	Slack time	Max time	Fld knots	Ebb knots
1 Tu	0045	0241	0.3	
	0457	0848		1.6
	1253	1557	0.9	
	1835	2145		1.3
2 W	0135	0354	0.5	
	0610	0949		1.8
	1343	1646	1.1	
	1928	2237		1.5
3 Th	0216	0446	0.8	
	0711	1043		2.0
	1429	1728	1.3	
	2014	2322		1.8
4 F	0255	0531	1.1	
	0805	1131		2.3
	1512	1806	1.5	
	2056			
5 Sa		0003		2.0
	0334	0613	1.3	
	0856	1217		2.4
	1555	1843	1.6	
	2136			
6 Su		0044		2.2
	0413	0655	1.5	
	0944	1302		2.5
○	1638	1921	1.6	
	2215			
7 M		0124		2.4
	0454	0737	1.6	
	1033	1346		2.5
	1721	2000	1.6	
	2253			
8 Tu		0206		2.5
	0538	0822	1.7	
	1122	1432		2.4
	1807	2040	1.4	
	2332			
9 W		0250		2.4
	0626	0909	1.6	
	1213	1520		2.2
	1857	2124	1.2	
10 Th	0013	0337		2.3
	0718	1000	1.4	
	1308	1612		1.9
	1952	2212	1.0	
11 F	0056	0428		2.1
	0818	1100	1.2	
	1409	1710		1.6
	2058	2309	0.7	
12 Sa	0147	0526		1.9
	0926	1218	1.0	
	1518	1820		1.3
☽	2217			
13 Su		0025	0.5	
	0249	0634		1.7
	1044	1356	0.9	
	1631	1948		1.2
	2344			
14 M		0214	0.4	
	0406	0757		1.6
	1202	1516	1.0	
	1742	2114		1.3
15 Tu	0059	0335	0.5	
	0523	0919		1.7
	1310	1618	1.1	
	1843	2217		1.4
16 W	0153	0434	0.7	
	0631	1023		1.8
	1405	1709	1.2	
	1934	2305		1.6
17 Th	0235	0522	0.9	
	0728	1112		1.9
	1450	1751	1.2	
	2016	2343		1.7
18 F	0310	0603	1.1	
	0817	1152		2.0
	1529	1828	1.2	
	2053			
19 Sa		0014		1.8
	0342	0637	1.2	
	0859	1226		2.0
	1604	1858	1.2	
	2126			
20 Su		0042		1.9
	0413	0706	1.2	
	0938	1257		2.0
●	1637	1922	1.2	
	2156			
21 M		0109		1.9
	0444	0730	1.2	
	1014	1328		2.0
	1709	1943	1.1	
	2225			
22 Tu		0138		2.0
	0516	0754	1.2	
	1049	1359		1.9
	1742	2005	1.0	
	2251			
23 W		0210		2.0
	0549	0823	1.2	
	1124	1434		1.8
	1817	2033	0.9	
	2316			
24 Th		0244		1.9
	0625	0856	1.1	
	1200	1512		1.7
	1855	2105	0.8	
	2340			
25 F		0323		1.8
	0705	0934	1.0	
	1239	1554		1.5
	1940	2143	0.6	
26 Sa	0006	0406		1.7
	0754	1019	0.9	
	1326	1642		1.3
	2035	2228	0.5	
27 Su	0037	0456		1.6
	0852	1113	0.7	
	1425	1739		1.2
	2145	2324	0.3	
28 M	0123	0555		1.5
	1002	1222	0.6	
	1539	1846		1.1
☾	2302			
29 Tu		0036	0.3	
	0243	0703		1.4
	1114	1350	0.7	
	1653	2000		1.2
30 W	0009	0204	0.3	
	0427	0815		1.6
	1220	1514	0.8	
	1757	2109		1.4

OCTOBER

Day	Slack time	Max time	Fld knots	Ebb knots
1 Th	0101	0326	0.6	
	0549	0923		1.8
	1315	1612	1.0	
	1852	2205		1.7
2 F	0145	0424	0.9	
	0655	1021		2.0
	1404	1657	1.3	
	1939	2252		2.0
3 Sa	0226	0511	1.3	
	0752	1112		2.2
	1450	1738	1.4	
	2023	2335		2.2
4 Su	0306	0555	1.6	
	0844	1159		2.4
	1534	1818	1.5	
	2104			
5 M		0018		2.5
	0348	0638	1.8	
	0934	1245		2.5
○	1618	1857	1.6	
	2144			
6 Tu		0100		2.6
	0431	0722	1.9	
	1022	1330		2.5
	1702	1937	1.5	
	2223			
7 W		0142		2.6
	0516	0806	1.9	
	1111	1416		2.3
	1749	2018	1.4	
	2303			
8 Th		0226		2.5
	0603	0853	1.7	
	1201	1503		2.1
	1839	2102	1.1	
	2345			
9 F		0313		2.4
	0656	0943	1.5	
	1254	1554		1.8
	1935	2151	0.9	
10 Sa	0029	0404		2.1
	0754	1042	1.2	
	1351	1651		1.5
	2041	2248	0.6	
11 Su	0121	0501		1.8
	0902	1157	1.0	
	1455	1759		1.3
	2201			
12 M		0010	0.4	
	0227	0610		1.6
	1020	1333	0.8	
☽	1603	1925		1.2
	2328			
13 Tu		0201	0.4	
	0347	0736		1.5
	1140	1452	0.8	
	1709	2049		1.3
14 W	0040	0317	0.5	
	0506	0901		1.5
	1250	1553	0.9	
	1808	2150		1.4
15 Th	0131	0415	0.8	
	0615	1004		1.6
	1345	1642	1.0	
	1858	2236		1.6
16 F	0210	0502	0.9	
	0711	1053		1.7
	1430	1724	1.0	
	1939	2313		1.7
17 Sa	0243	0541	1.1	
	0759	1131		1.8
	1507	1800	1.1	
	2016	2343		1.8
18 Su	0314	0615	1.2	
	0841	1204		1.9
	1541	1829	1.1	
	2049			
19 M		0010		1.9
	0345	0642	1.3	
	0919	1234		1.9
	1613	1851	1.0	
	2119			
20 Tu		0038		2.0
	0415	0706	1.3	
	0955	1304		1.9
●	1644	1911	1.0	
	2147			
21 W		0107		2.0
	0446	0730	1.3	
	1029	1335		1.9
	1717	1934	0.9	
	2213			
22 Th		0139		2.0
	0518	0757	1.3	
	1104	1409		1.8
	1751	2002	0.8	
	2238			
23 F		0214		2.0
	0553	0830	1.2	
	1140	1447		1.7
	1829	2035	0.7	
	2302			
24 Sa		0252		1.9
	0633	0907	1.1	
	1218	1528		1.5
	1912	2113	0.6	
	2330			
25 Su	2331	0335		1.8
	0719	0951	1.0	
	1303	1615		1.4
	2005	2159	0.5	
26 M	0005	0425		1.7
	0814	1042	0.9	
	1356	1710		1.3
	2110	2255	0.4	
27 Tu	0057	0523		1.5
	0921	1145	0.7	
	1501	1813		1.2
	2222			
28 W		0005	0.4	
	0221	0630		1.5
	1034	1302	0.7	
☾	1611	1923		1.3
	2329			
29 Th		0131	0.5	
	0405	0744		1.5
	1144	1425	0.8	
	1715	2031		1.5
30 F	0025	0256	0.7	
	0530	0855		1.7
	1245	1531	1.0	
	1811	2130		1.8
31 Sa	0113	0400	1.1	
	0638	0957		1.9
	1338	1624	1.1	
	1902	2221		2.1

KEY WEST, FLORIDA (0.3 nm west of Fort Taylor)

CURRENT TABLE 1998 — 24°32.9'N 81°49.0'W — Flood 20° Ebb 195°

Eastern Time (75°W) — Add 1 hour for Daylight Saving Time: April 5 – October 24

NOVEMBER

Day	Slack time	Max time	Fld knots	Ebb knots
1 Su	0158	0452	1.4	
	0738	1052		2.1
	1427	1710	1.3	
	1948	2308		2.4
2 M	0242	0539	1.7	
	0832	1142		2.3
	1514	1753	1.4	
	2032	2353		2.6
3 Tu	0326	0624	1.9	
	0922	1229		2.3
	1600	1835	1.4	
	2114			
4 W ○		0037		2.7
	0411	0709	2.0	
	1011	1315		2.3
	1646	1917	1.4	
	2156			
5 Th		0121		2.7
	0457	0754	1.9	
	1059	1401		2.2
	1733	2000	1.2	
	2239			
6 F		0206		2.6
	0545	0840	1.8	
	1148	1448		2.0
	1823	2045	1.1	
	2322			
7 Sa		0252		2.4
	0636	0929	1.5	
	1238	1537		1.8
	1918	2133	0.8	
8 Su	0008	0342		2.1
	0732	1023	1.3	
	1330	1631		1.6
	2021	2230	0.6	
9 M	0101	0437		1.8
	0836	1130	1.0	
	1425	1732		1.4
	2134	2348	0.5	
10 Tu ☽	0206	0542		1.6
	0948	1253	0.8	
	1525	1846		1.3
	2253			
11 W		0129	0.4	
	0322	0700		1.4
	1105	1412	0.7	
	1625	2003		1.3
12 Th	0003	0247	0.5	
	0439	0824		1.4
	1216	1516	0.7	
	1722	2107		1.4
13 F	0056	0345	0.7	
	0548	0931		1.4
	1315	1607	0.8	
	1812	2156		1.5
14 Sa	0137	0434	0.9	
	0646	1023		1.5
	1402	1651	0.8	
	1856	2235		1.7
15 Su	0213	0515	1.1	
	0735	1104		1.6
	1441	1727	0.9	
	1936	2308		1.8
16 M	0246	0550	1.2	
	0819	1139		1.7
	1516	1758	0.9	
	2011	2338		1.9
17 Tu	0318	0620	1.3	
	0858	1211		1.7
	1550	1822	0.9	
	2044			
18 W ●		0008		2.0
	0349	0646	1.3	
	0936	1242		1.8
	1623	1844	0.8	
	2114			
19 Th		0040		2.1
	0422	0711	1.3	
	1012	1315		1.8
	1656	1909	0.8	
	2142			
20 F		0114		2.1
	0455	0739	1.3	
	1048	1349		1.7
	1731	1939	0.8	
	2210			
21 Sa		0150		2.1
	0531	0811	1.3	
	1125	1427		1.7
	1808	2013	0.8	
	2239			
22 Su		0229		2.0
	0610	0848	1.2	
	1203	1508		1.6
	1850	2053	0.7	
	2313			
23 M		0312		1.9
	0654	0931	1.1	
	1245	1553		1.5
	1939	2139	0.6	
	2356			
24 Tu		0401		1.8
	0746	1019	1.0	
	1332	1645		1.5
	2036	2234	0.6	
25 W	0054	0458		1.7
	0847	1116	0.9	
	1426	1743		1.4
	2141	2340	0.5	
26 Th ☾	0213	0602		1.6
	0956	1222	0.8	
	1527	1846		1.5
	2247			
27 F		0059	0.6	
	0345	0713		1.5
	1107	1336	0.8	
	1629	1952		1.6
	2348			
28 Sa		0224	0.8	
	0509	0826		1.6
	1214	1449	0.9	
	1729	2055		1.9
29 Su	0043	0337	1.1	
	0621	0933		1.8
	1314	1551	1.0	
	1824	2151		2.1
30 M	0134	0435	1.4	
	0723	1033		1.9
	1407	1644	1.1	
	1915	2243		2.4

DECEMBER

Day	Slack time	Max time	Fld knots	Ebb knots
1 Tu	0222	0526	1.7	
	0819	1126		2.0
	1457	1733	1.2	
	2003	2332		2.5
2 W	0309	0614	1.9	
	0911	1216		2.1
	1545	1818	1.2	
	2050			
3 Th ○		0018		2.6
	0356	0700	1.9	
	1000	1302		2.1
	1632	1903	1.2	
	2135			
4 F		0104		2.6
	0442	0745	1.9	
	1047	1348		2.1
	1719	1947	1.1	
	2220			
5 Sa		0149		2.5
	0530	0829	1.8	
	1132	1433		2.0
	1807	2031	1.0	
	2306			
6 Su		0235		2.4
	0619	0914	1.5	
	1218	1519		1.8
	1858	2118	0.9	
	2353			
7 M		0322		2.1
	0710	1002	1.3	
	1303	1606		1.7
	1954	2210	0.7	
8 Tu	0043	0413		1.8
	0806	1054	1.0	
	1350	1658		1.5
	2055	2312	0.6	
9 W	0141	0508		1.6
	0908	1155	0.8	
	1439	1755		1.4
	2202			
10 Th ☽		0032	0.5	
	0247	0612		1.4
	1016	1309	0.6	
	1531	1858		1.4
	2309			
11 F		0157	0.5	
	0359	0725		1.3
	1127	1421	0.6	
	1625	2003		1.4
12 Sa	0008	0305	0.6	
	0508	0839		1.2
	1232	1521	0.6	
	1718	2101		1.5
13 Su	0058	0359	0.8	
	0611	0941		1.3
	1327	1611	0.6	
	1808	2149		1.6
14 M	0140	0445	0.9	
	0705	1031		1.4
	1412	1653	0.6	
	1853	2230		1.7
15 Tu	0217	0525	1.1	
	0754	1113		1.5
	1452	1727	0.7	
	1934	2307		1.8
16 W	0253	0600	1.2	
	0837	1149		1.6
	1529	1756	0.7	
	2011	2342		2.0
17 Th	0327	0629	1.3	
	0918	1223		1.7
	1603	1822	0.8	
	2046			
18 F ●		0018		2.1
	0402	0657	1.4	
	0956	1257		1.7
	1638	1850	0.8	
	2120			
19 Sa		0054		2.1
	0437	0726	1.4	
	1033	1333		1.7
	1712	1922	0.8	
	2154			
20 Su		0132		2.2
	0514	0758	1.4	
	1110	1410		1.7
	1749	1958	0.8	
	2230			
21 M		0212		2.2
	0553	0834	1.4	
	1147	1450		1.7
	1828	2039	0.9	
	2311			
22 Tu		0256		2.1
	0636	0914	1.3	
	1226	1533		1.7
	1913	2125	0.8	
	2358			
23 W		0343		2.0
	0724	0959	1.2	
	1307	1621		1.7
	2005	2217	0.8	
24 Th	0056	0436		1.8
	0820	1050	1.0	
	1353	1714		1.7
	2104	2318	0.8	
25 F	0206	0536		1.7
	0923	1147	0.9	
	1446	1812		1.7
	2208			
26 Sa ☾		0030	0.8	
	0327	0643		1.5
	1033	1254	0.8	
	1544	1916		1.8
	2314			
27 Su		0154	0.9	
	0448	0757		1.5
	1145	1408	0.7	
	1647	2021		1.9
28 M	0017	0316	1.1	
	0603	0911		1.6
	1252	1522	0.8	
	1748	2125		2.1
29 Tu	0114	0423	1.4	
	0709	1018		1.7
	1352	1626	0.9	
	1846	2223		2.3
30 W	0208	0518	1.6	
	0807	1116		1.8
	1446	1721	1.0	
	1941	2316		2.4
31 Th	0258	0608	1.7	
	0859	1207		1.9
	1534	1810	1.1	
	2032	2359		2.5

VIEQUES PASSAGE, PUERTO RICO

CURRENT TABLE 1998 — 18°11.3'N 65°37.1'W — Flood 250° Ebb 55°

Eastern Time (60°W) — Daylight Saving Time not observed

JANUARY

Day	Slack time	Max time	Fld knots	Ebb knots
1 Th		0106	0.6	
	0405	0649		0.5
	0927	1249	0.8	
	1553	1924		1.0
	2259			
2 F		0151	0.7	
	0455	0744		0.6
	1029	1342	0.7	
	1641	2010		0.9
	2338			
3 Sa		0238	0.7	
	0548	0841		0.6
	1136	1438	0.6	
	1731	2058		0.9
4 Su	0019	0327	0.8	
	0642	0941		0.7
	1247	1538	0.6	
	1824	2148		0.8
5 M ☾	0102	0418	0.8	
	0737	1043		0.7
	1401	1641	0.5	
	1920	2241		0.7
6 Tu	0147	0511	0.9	
	0834	1146		0.8
	1516	1747	0.4	
	2021	2336		0.6
7 W	0234	0606	0.9	
	0930	1249		0.8
	1629	1855	0.4	
	2125			
8 Th		0034		0.6
	0324	0701	0.9	
	1025	1350		0.9
	1738	2001	0.4	
	2232			
9 F		0133		0.5
	0415	0756	0.9	
	1119	1448		0.9
	1839	2103	0.4	
	2339			
10 Sa		0232		0.5
	0508	0849	0.9	
	1210	1541		0.9
	1933	2201	0.4	
11 Su	0042	0329		0.5
	0602	0941	0.8	
	1258	1631		1.0
	2020	2253	0.5	
12 M ○	0141	0424		0.4
	0655	1030	0.8	
	1343	1716		1.0
	2103	2342	0.5	
13 Tu	0235	0516		0.4
	0748	1117	0.7	
	1425	1759		0.9
	2142			
14 W		0026	0.5	
	0326	0606		0.4
	0840	1202	0.7	
	1505	1839		0.9
	2218			
15 Th		0108	0.6	
	0413	0654		0.4
	0933	1247	0.6	
	1544	1918		0.8
	2252			
16 F		0148	0.6	
	0458	0742		0.5
	1027	1331	0.5	
	1622	1956		0.8
	2325			
17 Sa		0227	0.6	
	0542	0829		0.5
	1122	1417	0.5	
	1701	2034		0.7
	2357			
18 Su		0306	0.6	
	0625	0918		0.5
	1220	1505	0.4	
	1742	2113		0.7
19 M	0031	0346	0.7	
	0709	1008		0.5
	1322	1556	0.4	
	1826	2155		0.6
20 Tu ☽	0105	0428	0.7	
	0753	1059		0.6
	1427	1650	0.3	
	1914	2239		0.5
21 W	0141	0511	0.7	
	0837	1152		0.6
	1532	1748	0.3	
	2008	2326		0.5
22 Th	0220	0556	0.7	
	0923	1245		0.7
	1634	1847	0.3	
	2107			
23 F		0016		0.4
	0301	0643	0.7	
	1008	1336		0.7
	1732	1946	0.3	
	2208			
24 Sa		0109		0.4
	0346	0732	0.7	
	1055	1426		0.8
	1823	2041	0.3	
	2310			
25 Su		0203		0.4
	0436	0821	0.8	
	1142	1514		0.9
	1908	2132	0.4	
26 M	0009	0257		0.4
	0529	0911	0.8	
	1228	1601		0.9
	1950	2221	0.4	
27 Tu	0103	0352		0.5
	0626	1002	0.8	
	1315	1646		0.9
	2029	2307	0.5	
28 W ●	0156	0445		0.5
	0724	1053	0.8	
	1403	1731		1.0
	2107	2352	0.6	
29 Th	0247	0539		0.6
	0824	1144	0.8	
	1450	1816		1.0
	2146			
30 F		0038	0.7	
	0338	0633		0.6
	0925	1237	0.8	
	1538	1902		0.9
	2225			
31 Sa		0124	0.8	
	0429	0729		0.7
	1028	1331	0.7	
	1627	1949		0.9
	2306			

FEBRUARY

Day	Slack time	Max time	Fld knots	Ebb knots
1 Su		0212	0.8	
	0522	0825		0.7
	1133	1427	0.7	
	1718	2037		0.8
	2348			
2 M		0301	0.9	
	0616	0924		0.8
	1241	1526	0.6	
	1811	2127		0.8
3 Tu ☾	0033	0353	0.9	
	0712	1024		0.8
	1351	1628	0.5	
	1907	2220		0.7
4 W	0120	0447	0.9	
	0809	1126		0.8
	1502	1732	0.4	
	2007	2317		0.6
5 Th	0210	0543	0.9	
	0906	1228		0.8
	1612	1838	0.4	
	2111			
6 F		0016		0.5
	0303	0640	0.8	
	1003	1329		0.8
	1717	1943	0.4	
	2218			
7 Sa		0116		0.5
	0358	0737	0.8	
	1057	1427		0.8
	1816	2044	0.4	
	2324			
8 Su		0217		0.4
	0455	0832	0.8	
	1149	1520		0.9
	1907	2140	0.4	
9 M	0025	0315		0.4
	0552	0925	0.7	
	1237	1608		0.9
	1951	2230	0.5	
10 Tu	0121	0409		0.4
	0648	1014	0.7	
	1322	1652		0.8
	2031	2315	0.5	
11 W ○	0211	0459		0.5
	0742	1100	0.7	
	1403	1733		0.8
	2106	2355	0.5	
12 Th	0256	0546		0.5
	0833	1144	0.6	
	1443	1811		0.8
	2139			
13 F		0034	0.6	
	0338	0631		0.5
	0924	1227	0.6	
	1521	1847		0.7
	2210			
14 Sa		0110	0.6	
	0418	0714		0.5
	1014	1310	0.5	
	1559	1923		0.7
	2241			
15 Su		0147	0.6	
	0458	0758		0.6
	1105	1353	0.5	
	1637	1959		0.7
	2312			
16 M		0223	0.6	
	0538	0842		0.6
	1157	1438	0.4	
	1718	2037		0.6
	2344			
17 Tu		0302	0.7	
	0619	0928		0.6
	1253	1526	0.4	
	1801	2117		0.5
18 W	0018	0342	0.7	
	0702	1017		0.6
	1351	1617	0.3	
	1848	2201		0.5
19 Th ☽	0056	0426	0.7	
	0747	1108		0.7
	1450	1712	0.3	
	1940	2248		0.4
20 F	0137	0513	0.7	
	0835	1200		0.7
	1549	1809	0.3	
	2037	2341		0.4
21 Sa	0225	0603	0.7	
	0925	1254		0.7
	1644	1906	0.3	
	2137			
22 Su		0037		0.4
	0318	0657	0.7	
	1017	1346		0.8
	1735	2002	0.4	
	2238			
23 M		0136		0.4
	0417	0753	0.7	
	1109	1438		0.8
	1820	2055	0.4	
	2337			
24 Tu		0235		0.5
	0519	0848	0.8	
	1202	1528		0.8
	1903	2144	0.5	
25 W	0033	0332		0.6
	0621	0944	0.8	
	1253	1616		0.9
	1944	2233	0.6	
26 Th ●	0126	0428		0.7
	0723	1038	0.8	
	1344	1704		0.9
	2025	2320	0.7	
27 F	0218	0523		0.7
	0824	1132	0.8	
	1434	1751		0.9
	2106			
28 Sa		0007	0.8	
	0310	0617		0.8
	0925	1226	0.8	
	1525	1838		0.9
	2148			

VIEQUES PASSAGE, PUERTO RICO

CURRENT TABLE 1998 — 18°11.3'N 65°37.1'W — Flood 250° Ebb 55°

Eastern Time (60°W) — Daylight Saving Time not observed

MARCH

Day	Slack time	Max time	Fld knots	Ebb knots
1 Su		0055	0.9	
	0401	0712		0.9
	1026	1321	0.7	
	1615	1926		0.8
	2231			
2 M		0144	0.9	
	0454	0807		0.9
	1127	1416	0.7	
	1707	2016		0.8
	2317			
3 Tu		0235	0.9	
	0548	0904		0.9
	1230	1513	0.6	
	1801	2108		0.7
4 W	0004	0327	0.9	
	0643	1002		0.9
	1335	1613	0.5	
	1857	2202		0.6
5 Th ☾	0055	0422	0.8	
	0740	1102		0.8
	1440	1715	0.5	
	1958	2300		0.5
6 F	0149	0519	0.8	
	0837	1202		0.8
	1544	1818	0.4	
	2102			
7 Sa		0001		0.5
	0248	0618	0.7	
	0934	1302		0.8
	1644	1921	0.4	
	2207			
8 Su		0103		0.5
	0349	0717	0.7	
	1030	1358		0.8
	1739	2019	0.4	
	2309			
9 M		0205		0.5
	0451	0814	0.6	
	1123	1451		0.7
	1826	2111	0.5	
10 Tu	0006	0302		0.5
	0551	0908	0.6	
	1212	1538		0.7
	1908	2158	0.5	
11 W	0056	0354		0.5
	0648	0958	0.6	
	1258	1621		0.7
	1945	2240	0.5	
12 Th	0140	0441		0.5
	0741	1044	0.6	
	1341	1701		0.7
	2019	2319	0.6	
13 F ○	0221	0525		0.6
	0830	1128	0.5	
	1421	1738		0.7
	2051	2355	0.6	
14 Sa	0259	0607		0.6
	0917	1210	0.5	
	1501	1814		0.6
	2122			
15 Su		0029	0.6	
	0336	0647		0.6
	1004	1251	0.5	
	1540	1850		0.6
	2152			
16 M		0104	0.6	
	0413	0727		0.7
	1050	1333	0.5	
	1619	1926		0.5
	2224			
17 Tu		0140	0.7	
	0450	0808		0.7
	1138	1416	0.4	
	1700	2004		0.5
	2257			
18 W		0218	0.7	
	0530	0852		0.7
	1227	1502	0.4	
	1744	2044		0.5
	2333			
19 Th		0259	0.7	
	0613	0937		0.7
	1317	1550	0.4	
	1830	2129		0.4
20 F	0014	0344	0.7	
	0659	1026		0.7
	1410	1641	0.4	
	1921	2219		0.4
21 Sa ☽	0102	0433	0.6	
	0749	1117		0.7
	1502	1735	0.4	
	2017	2314		0.4
22 Su	0158	0529	0.6	
	0843	1211		0.7
	1553	1830	0.4	
	2116			
23 M		0014		0.4
	0302	0628	0.6	
	0939	1306		0.7
	1641	1925	0.5	
	2214			
24 Tu		0116		0.5
	0409	0729	0.7	
	1037	1400		0.8
	1728	2018	0.5	
	2312			
25 W		0217		0.6
	0516	0829	0.7	
	1134	1453		0.8
	1812	2109	0.6	
26 Th	0007	0316		0.7
	0621	0928	0.7	
	1230	1545		0.8
	1857	2159	0.7	
27 F ●	0100	0413		0.8
	0724	1025	0.7	
	1325	1635		0.8
	1941	2248	0.8	
28 Sa	0152	0507		0.9
	0824	1121	0.7	
	1418	1724		0.8
	2025	2337	0.9	
29 Su	0243	0601		1.0
	0923	1215	0.7	
	1511	1814		0.8
	2110			
30 M		0026	0.9	
	0335	0654		1.0
	1021	1309	0.7	
	1603	1904		0.7
	2157			
31 Tu		0116	0.9	
	0426	0748		1.0
	1118	1404	0.7	
	1657	1955		0.7
	2246			

APRIL

Day	Slack time	Max time	Fld knots	Ebb knots
1 W		0207	0.9	
	0519	0842		1.0
	1216	1459	0.6	
	1752	2048		0.6
	2337			
2 Th		0300	0.8	
	0612	0937		0.9
	1314	1557	0.6	
	1850	2145		0.6
3 F ☾	0032	0355	0.8	
	0706	1033		0.9
	1411	1655	0.5	
	1951	2244		0.5
4 Sa	0132	0453	0.7	
	0802	1130		0.8
	1508	1754	0.5	
	2053	2347		0.5
5 Su	0236	0553	0.6	
	0858	1227		0.7
	1601	1852	0.5	
	2153			
6 M		0049		0.5
	0343	0654	0.5	
	0954	1322		0.7
	1650	1946	0.5	
	2250			
7 Tu		0150		0.5
	0449	0753	0.5	
	1049	1413		0.7
	1734	2035	0.5	
	2341			
8 W		0245		0.5
	0551	0848	0.5	
	1140	1500		0.6
	1814	2120	0.6	
9 Th	0026	0335		0.6
	0647	0939	0.5	
	1229	1544		0.6
	1851	2200	0.6	
10 F	0107	0420		0.6
	0738	1026	0.5	
	1315	1624		0.6
	1925	2238	0.6	
11 Sa ○	0144	0501		0.7
	0825	1110	0.5	
	1359	1703		0.5
	1958	2313	0.7	
12 Su	0220	0541		0.7
	0910	1153	0.5	
	1441	1740		0.5
	2029	2348	0.7	
13 M	0256	0620		0.8
	0954	1234	0.5	
	1523	1817		0.5
	2101			
14 Tu		0023	0.7	
	0331	0659		0.8
	1037	1315	0.5	
	1605	1854		0.4
	2135			
15 W		0100	0.7	
	0408	0738		0.8
	1120	1357	0.4	
	1647	1934		0.4
	2211			
16 Th		0138	0.7	
	0448	0819		0.8
	1203	1440	0.4	
	1731	2016		0.4
	2252			
17 F		0221	0.7	
	0530	0903		0.8
	1247	1526	0.4	
	1818	2104		0.4
	2341			
18 Sa		0308	0.6	
	0616	0949		0.8
	1332	1614	0.4	
	1908	2157		0.4
19 Su ☽	0038	0401	0.6	
	0708	1040		0.8
	1418	1705	0.5	
	2002	2255		0.4
20 M	0143	0500	0.6	
	0803	1133		0.7
	1504	1758	0.5	
	2059	2358		0.5
21 Tu	0254	0603	0.6	
	0903	1228		0.7
	1550	1851	0.6	
	2155			
22 W		0101		0.6
	0407	0708	0.6	
	1005	1324		0.7
	1637	1944	0.7	
	2251			
23 Th		0203		0.7
	0517	0812	0.6	
	1106	1419		0.7
	1724	2037	0.8	
	2345			
24 F		0302		0.8
	0623	0914	0.6	
	1207	1513		0.7
	1811	2128	0.9	
25 Sa	0037	0358		0.9
	0724	1012	0.7	
	1305	1606		0.7
	1858	2219	0.9	
26 Su ●	0129	0452		1.0
	0823	1108	0.7	
	1401	1658		0.7
	1946	2309	1.0	
27 M	0219	0545		1.1
	0919	1203	0.7	
	1456	1750		0.7
	2034	2359	1.0	
28 Tu	0310	0636		1.1
	1013	1257	0.7	
	1551	1842		0.6
	2124			
29 W		0049	0.9	
	0359	0728		1.1
	1106	1350	0.6	
	1647	1935		0.6
	2216			
30 Th		0140	0.9	
	0449	0819		1.0
	1158	1443	0.6	
	1743	2030		0.5
	2311			

VIEQUES PASSAGE, PUERTO RICO

CURRENT TABLE 1998 — 18°11.3'N 65°37.1'W — Flood 250° Ebb 55°

Eastern Time (60°W) — Daylight Saving Time not observed

MAY

Day	Slack time	Max time	Fld knots	Ebb knots
1 F		0233	0.8	
	0539	0910		0.9
	1248	1537	0.6	
	1841	2127		0.5
2 Sa	0010	0327	0.7	
	0630	1002		0.9
	1338	1632	0.6	
	1939	2227		0.5
3 Su ☾	0114	0424	0.6	
	0722	1054		0.8
	1426	1726	0.6	
	2037	2328		0.5
4 M	0222	0524	0.5	
	0815	1146		0.7
	1512	1818	0.6	
	2132			
5 Tu		0029		0.5
	0333	0624	0.4	
	0909	1238		0.6
	1556	1908	0.6	
	2223			
6 W		0127		0.5
	0441	0724	0.4	
	1005	1328		0.6
	1637	1954	0.6	
	2310			
7 Th		0221		0.6
	0543	0821	0.4	
	1100	1415		0.5
	1716	2037	0.6	
	2352			
8 F		0309		0.6
	0639	0914	0.4	
	1153	1500		0.5
	1753	2118	0.7	
9 Sa	0031	0354		0.7
	0730	1004	0.4	
	1245	1543		0.5
	1829	2156	0.7	
10 Su	0108	0435		0.8
	0816	1050	0.4	
	1334	1625		0.4
	1903	2233	0.7	
11 M ○	0144	0515		0.8
	0859	1133	0.4	
	1420	1705		0.4
	1938	2310	0.7	
12 Tu	0220	0553		0.9
	0941	1215	0.4	
	1505	1744		0.4
	2013	2347	0.7	
13 W	0256	0632		0.9
	1021	1257	0.4	
	1549	1825		0.4
	2051			
14 Th		0025	0.7	
	0334	0711		0.9
	1100	1338	0.5	
	1633	1908		0.4
	2134			
15 F		0106	0.7	
	0414	0751		0.9
	1139	1419	0.5	
	1717	1954		0.4
	2223			
16 Sa		0151	0.7	
	0457	0834		0.9
	1218	1503	0.5	
	1804	2044		0.4
	2319			
17 Su		0241	0.6	
	0544	0919		0.8
	1258	1549	0.5	
	1854	2140		0.4
18 M	0024	0337	0.6	
	0635	1008		0.8
	1339	1638	0.6	
	1947	2241		0.5
19 Tu ☽	0135	0438	0.5	
	0730	1059		0.7
	1422	1729	0.7	
	2042	2344		0.6
20 W	0251	0543	0.5	
	0831	1154		0.7
	1507	1822	0.7	
	2137			
21 Th		0047		0.7
	0406	0650	0.5	
	0934	1250		0.7
	1553	1915	0.8	
	2231			
22 F		0148		0.8
	0517	0756	0.5	
	1039	1347		0.6
	1642	2009	0.9	
	2325			
23 Sa		0247		0.9
	0622	0859	0.5	
	1142	1444		0.6
	1731	2101	0.9	
24 Su	0017	0343		1.0
	0723	0959	0.6	
	1244	1539		0.6
	1821	2153	1.0	
25 M ●	0108	0437		1.1
	0819	1056	0.6	
	1344	1634		0.6
	1912	2244	1.0	
26 Tu	0158	0528		1.1
	0911	1150	0.6	
	1442	1727		0.6
	2004	2335	0.9	
27 W	0247	0618		1.1
	1001	1243	0.6	
	1538	1821		0.5
	2056			
28 Th		0025	0.9	
	0334	0707		1.1
	1049	1334	0.6	
	1634	1915		0.5
	2151			
29 F		0115	0.8	
	0421	0755		1.0
	1135	1424	0.6	
	1729	2010		0.5
	2248			
30 Sa		0206	0.7	
	0507	0842		0.9
	1219	1514	0.6	
	1824	2106		0.5
	2349			
31 Su		0258	0.6	
	0553	0928		0.8
	1302	1603	0.6	
	1919	2204		0.5

JUNE

Day	Slack time	Max time	Fld knots	Ebb knots
1 M ☾	0054	0353	0.5	
	0640	1015		0.8
	1343	1651	0.6	
	2011	2302		0.5
2 Tu	0203	0450	0.4	
	0728	1102		0.7
	1423	1739	0.6	
	2102			
3 W		0000		0.5
	0314	0549	0.3	
	0820	1150		0.6
	1502	1825	0.7	
	2149			
4 Th		0056		0.6
	0422	0649	0.3	
	0915	1239		0.5
	1541	1910	0.7	
	2233			
5 F		0149		0.6
	0526	0748	0.3	
	1013	1327		0.5
	1620	1953	0.7	
	2315			
6 Sa		0238		0.7
	0623	0844	0.3	
	1112	1415		0.4
	1658	2035	0.7	
	2355			
7 Su		0323		0.8
	0713	0936	0.3	
	1209	1501		0.4
	1736	2116	0.7	
8 M	0033	0406		0.8
	0759	1024	0.4	
	1303	1546		0.4
	1814	2155	0.7	
9 Tu	0111	0447		0.9
	0842	1110	0.4	
	1354	1630		0.4
	1854	2235	0.7	
10 W ○	0149	0527		0.9
	0921	1153	0.4	
	1442	1714		0.3
	1936	2316	0.7	
11 Th	0228	0606		0.9
	0959	1234	0.5	
	1527	1759		0.4
	2021	2358	0.7	
12 F	0307	0646		0.9
	1035	1314	0.5	
	1612	1846		0.4
	2112			
13 Sa		0043	0.7	
	0349	0726		0.9
	1111	1356	0.5	
	1657	1935		0.4
	2208			
14 Su		0131	0.7	
	0433	0809		0.9
	1147	1439	0.6	
	1745	2029		0.5
	2310			
15 M		0223	0.6	
	0520	0853		0.9
	1224	1524	0.6	
	1835	2126		0.5
16 Tu	0018	0320	0.6	
	0611	0941		0.8
	1304	1612	0.7	
	1928	2226		0.6
17 W ☽	0131	0421	0.5	
	0705	1032		0.7
	1346	1703	0.8	
	2022	2329		0.7
18 Th	0247	0526	0.5	
	0804	1126		0.7
	1431	1756	0.8	
	2117			
19 F		0031		0.8
	0402	0634	0.4	
	0908	1222		0.6
	1518	1850	0.9	
	2213			
20 Sa		0133		0.8
	0513	0740	0.4	
	1014	1320		0.6
	1608	1945	0.9	
	2307			
21 Su		0232		0.9
	0617	0845	0.4	
	1120	1419		0.6
	1700	2039	0.9	
	2359			
22 M		0328		1.0
	0716	0945	0.5	
	1225	1517		0.5
	1753	2132	0.9	
23 Tu ●	0050	0421		1.0
	0810	1042	0.5	
	1327	1613		0.5
	1847	2224	0.9	
24 W	0139	0512		1.1
	0859	1135	0.6	
	1426	1709		0.5
	1941	2314	0.9	
25 Th	0226	0559		1.0
	0944	1225	0.6	
	1522	1803		0.5
	2036			
26 F		0004	0.8	
	0311	0645		1.0
	1027	1313	0.6	
	1616	1856		0.5
	2132			
27 Sa		0052	0.7	
	0355	0729		1.0
	1107	1359	0.6	
	1708	1949		0.5
	2229			
28 Su		0141	0.6	
	0437	0812		0.9
	1145	1444	0.6	
	1758	2042		0.5
	2328			
29 M		0230	0.5	
	0519	0854		0.8
	1222	1528	0.7	
	1847	2135		0.5
30 Tu	0031	0321	0.5	
	0601	0936		0.7
	1258	1612	0.7	
	1935	2229		0.5

VIEQUES PASSAGE, PUERTO RICO

CURRENT TABLE 1998 18°11.3'N 65°37.1'W Flood 250° Ebb 55°

Eastern Time (60°W) Daylight Saving Time not observed

JULY

Day	Slack time	Max time	Fld knots	Ebb knots
1	0136	0414	0.4	
	0646	1019		0.7
W	1335	1656	0.7	
☾	2021	2323		0.5
2	0244	0511	0.3	
	0735	1104		0.6
Th	1412	1740	0.7	
	2107			
3		0017		0.6
	0352	0610	0.3	
F	0829	1152		0.5
	1450	1824	0.7	
	2151			
4		0110		0.6
	0456	0709	0.3	
Sa	0928	1241		0.4
	1529	1909	0.7	
	2234			
5		0200		0.7
	0554	0808	0.3	
Su	1029	1331		0.4
	1610	1954	0.7	
	2317			
6		0248		0.8
	0646	0902	0.3	
M	1130	1421		0.4
	1652	2038	0.7	
	2359			
7		0333		0.8
	0732	0953	0.3	
Tu	1228	1511		0.4
	1736	2122	0.8	
8	0040	0417		0.9
	0813	1039	0.4	
W	1321	1559		0.4
	1824	2207	0.8	
9	0122	0458		0.9
	0851	1123	0.4	
Th	1410	1648		0.4
○	1914	2252	0.8	
10	0204	0539		0.9
	0927	1205	0.5	
F	1457	1737		0.4
	2007	2338	0.8	
11	0247	0620		0.9
	1002	1246	0.6	
Sa	1543	1827		0.5
	2104			
12		0026	0.7	
	0331	0702		0.9
Su	1037	1328	0.6	
	1631	1919		0.5
	2204			
13		0117	0.7	
	0416	0745		0.9
M	1113	1412	0.7	
	1719	2013		0.6
	2307			
14		0210	0.6	
	0504	0830		0.9
Tu	1151	1458	0.7	
	1811	2110		0.6
15	0015	0307	0.6	
	0554	0918		0.8
W	1232	1547	0.8	
	1904	2209		0.7
16	0125	0407	0.5	
	0648	1008		0.7
Th	1315	1638	0.8	
☽	2000	2311		0.8
17	0239	0512	0.4	
	0746	1102		0.7
F	1402	1732	0.9	
	2056			
18		0014		0.8
	0352	0618	0.4	
Sa	0849	1200		0.6
	1452	1828	0.9	
	2153			
19		0116		0.9
	0501	0725	0.4	
Su	0955	1300		0.5
	1544	1924	0.9	
	2248			
20		0215		0.9
	0605	0829	0.4	
M	1103	1400		0.5
	1640	2020	0.9	
	2342			
21		0312		0.9
	0702	0930	0.4	
Tu	1209	1500		0.5
	1736	2115	0.9	
22	0033	0404		1.0
	0752	1025	0.5	
W	1311	1558		0.5
	1833	2208	0.8	
23	0121	0453		1.0
	0837	1116	0.5	
Th	1408	1653		0.5
●	1930	2258	0.8	
24	0207	0538		0.9
	0918	1203	0.6	
F	1501	1746		0.5
	2025	2346	0.7	
25	0250	0621		0.9
	0956	1247	0.6	
Sa	1550	1836		0.5
	2120			
26		0032	0.7	
	0331	0701		0.9
Su	1032	1328	0.6	
	1637	1925		0.5
	2214			
27		0118	0.6	
	0410	0740		0.8
M	1106	1408	0.7	
	1721	2013		0.5
	2310			
28		0204	0.5	
	0450	0819		0.7
Tu	1139	1448	0.7	
	1805	2101		0.5
29	0007	0251	0.4	
	0530	0858		0.7
W	1213	1529	0.7	
	1849	2151		0.6
30	0106	0341	0.4	
	0613	0939		0.6
Th	1248	1610	0.7	
	1933	2241		0.6
31	0208	0434	0.3	
	0700	1022		0.5
F	1325	1653	0.7	
☾	2018	2333		0.6

AUGUST

Day	Slack time	Max time	Fld knots	Ebb knots
1	0312	0531	0.3	
	0752	1109		0.5
Sa	1403	1739	0.7	
	2104			
2		0026		0.7
	0415	0629	0.3	
Su	0850	1159		0.4
	1445	1826	0.7	
	2150			
3		0118		0.7
	0513	0728	0.3	
M	0952	1252		0.4
	1530	1914	0.7	
	2237			
4		0208		0.7
	0605	0824	0.3	
Tu	1053	1346		0.4
	1620	2004	0.7	
	2323			
5		0257		0.8
	0651	0915	0.3	
W	1151	1440		0.4
	1713	2053	0.7	
6	0010	0343		0.8
	0732	1003	0.4	
Th	1245	1533		0.4
	1808	2143	0.8	
7	0056	0427		0.9
	0810	1047	0.5	
F	1335	1626		0.5
○	1905	2233	0.8	
8	0142	0511		0.9
	0846	1130	0.6	
Sa	1424	1717		0.6
	2003	2323	0.8	
9	0228	0554		0.9
	0922	1213	0.6	
Su	1512	1809		0.6
	2102			
10		0014	0.7	
	0315	0637		0.9
M	0959	1257	0.7	
	1601	1902		0.7
	2203			
11		0106	0.7	
	0403	0722		0.9
Tu	1038	1343	0.8	
	1651	1956		0.8
	2305			
12		0200	0.7	
	0451	0808		0.8
W	1118	1431	0.8	
	1743	2052		0.8
13	0009	0256	0.6	
	0542	0857		0.8
Th	1202	1521	0.9	
	1838	2150		0.8
14	0116	0355	0.5	
	0637	0949		0.7
F	1248	1614	0.9	
☽	1934	2251		0.8
15	0225	0457	0.5	
	0735	1044		0.6
Sa	1338	1709	0.9	
	2032	2353		0.8
16	0334	0603	0.4	
	0838	1144		0.6
Su	1433	1807	0.8	
	2130			
17		0055		0.8
	0440	0708	0.4	
M	0945	1246		0.5
	1530	1907	0.8	
	2227			
18		0155		0.9
	0541	0812	0.4	
Tu	1052	1348		0.5
	1631	2005	0.8	
	2322			
19		0251		0.9
	0635	0910	0.5	
W	1156	1449		0.5
	1732	2101	0.8	
20	0014	0343		0.9
	0722	1003	0.5	
Th	1254	1547		0.5
	1831	2154	0.7	
21	0103	0430		0.8
	0804	1050	0.5	
F	1346	1640		0.5
●	1928	2244	0.7	
22	0147	0513		0.8
	0842	1134	0.6	
Sa	1434	1729		0.5
	2022	2330	0.6	
23	0229	0553		0.8
	0917	1214	0.6	
Su	1517	1815		0.6
	2113			
24		0015	0.6	
	0309	0631		0.7
M	0950	1252	0.6	
	1558	1859		0.6
	2203			
25		0058	0.5	
	0348	0708		0.7
Tu	1022	1329	0.7	
	1639	1943		0.6
	2253			
26		0141	0.5	
	0426	0744		0.6
W	1054	1406	0.7	
	1718	2026		0.6
	2344			
27		0226	0.4	
	0507	0822		0.6
Th	1127	1444	0.7	
	1759	2111		0.6
28	0037	0312	0.4	
	0550	0902		0.5
F	1201	1524	0.7	
	1842	2158		0.6
29	0133	0402	0.3	
	0636	0946		0.5
Sa	1239	1607	0.7	
	1926	2248		0.7
30	0230	0455	0.3	
	0727	1033		0.4
Su	1321	1654	0.7	
☾	2014	2340		0.7
31	0327	0551	0.3	
	0823	1124		0.4
M	1408	1744	0.6	
	2103			

VIEQUES PASSAGE, PUERTO RICO

CURRENT TABLE 1998 — 18°11.3'N 65°37.1'W — Flood 250° Ebb 55°

Eastern Time (60°W) — Daylight Saving Time not observed

SEPTEMBER

Day	Slack time	Max time	Fld knots	Ebb knots
1		0033		0.7
	0422	0647	0.3	
Tu	0922	1220		0.4
	1501	1837	0.7	
	2155			
2		0125		0.7
	0512	0742	0.4	
W	1022	1318		0.4
	1600	1933	0.7	
	2247			
3		0216		0.8
	0558	0834	0.4	
Th	1118	1416		0.5
	1701	2028	0.7	
	2339			
4		0305		0.8
	0640	0922	0.5	
F	1212	1512		0.5
	1802	2123	0.7	
5	0030	0353		0.8
	0719	1009	0.6	
Sa	1302	1606		0.6
	1903	2216	0.7	
6	0121	0439		0.8
	0758	1054	0.7	
Su	1352	1659		0.7
○	2003	2309	0.7	
7	0211	0525		0.8
	0838	1140	0.8	
M	1442	1752		0.8
	2102			
8		0002	0.7	
	0300	0611		0.8
Tu	0919	1226	0.8	
	1532	1844		0.9
	2200			
9		0055	0.7	
	0350	0659		0.8
W	1001	1314	0.9	
	1623	1938		0.9
	2300			
10		0149	0.7	
	0441	0747		0.8
Th	1046	1403	0.9	
	1715	2033		0.9
11	0000	0244	0.6	
	0534	0838		0.7
F	1133	1455	0.9	
	1809	2130		0.9
12	0102	0342	0.6	
	0630	0932		0.6
Sa	1224	1550	0.9	
☽	1906	2228		0.9
13	0205	0443	0.5	
	0729	1030		0.6
Su	1320	1647	0.8	
	2004	2328		0.8
14	0308	0546	0.5	
	0833	1132		0.5
M	1420	1747	0.7	
	2102			
15		0029		0.8
	0409	0649	0.5	
Tu	0938	1236		0.5
	1524	1849	0.7	
	2201			
16		0128		0.8
	0505	0749	0.5	
W	1042	1339		0.5
	1630	1950	0.6	
	2257			
17		0223		0.8
	0555	0844	0.5	
Th	1140	1440		0.5
	1734	2047	0.6	
	2351			
18		0314		0.7
	0640	0933	0.5	
F	1233	1535		0.5
	1834	2141	0.6	
19	0040	0400		0.7
	0720	1018	0.6	
Sa	1319	1624		0.6
	1929	2230	0.6	
20	0125	0442		0.7
	0756	1058	0.6	
Su	1402	1710		0.6
●	2020	2315	0.5	
21	0208	0521		0.6
	0829	1136	0.6	
M	1441	1752		0.7
	2108	2358	0.5	
22	0249	0558		0.6
	0901	1211	0.7	
Tu	1518	1833		0.7
	2154			
23		0040	0.5	
	0328	0635		0.6
W	0933	1247	0.7	
	1555	1913		0.7
	2239			
24		0121	0.5	
	0409	0711		0.5
Th	1005	1323	0.7	
	1632	1953		0.7
	2325			
25		0204	0.4	
	0450	0749		0.5
F	1039	1400	0.7	
	1711	2035		0.7
26	0012	0248	0.4	
	0534	0830		0.4
Sa	1115	1441	0.6	
	1752	2119		0.7
27	0100	0335	0.4	
	0620	0914		0.4
Su	1157	1524	0.6	
	1836	2206		0.7
28	0150	0424	0.4	
	0710	1003		0.4
M	1244	1613	0.6	
☾	1924	2255		0.7
29	0239	0516	0.4	
	0804	1057		0.4
Tu	1339	1706	0.6	
	2017	2347		0.7
30	0328	0609	0.4	
	0900	1155		0.4
W	1442	1804	0.6	
	2112			

OCTOBER

Day	Slack time	Max time	Fld knots	Ebb knots
1		0041		0.7
	0415	0702	0.5	
Th	0956	1256		0.5
	1548	1905	0.6	
	2209			
2		0134		0.7
	0500	0753	0.5	
F	1050	1355		0.6
	1656	2005	0.6	
	2307			
3		0226		0.7
	0544	0843	0.6	
Sa	1143	1453		0.7
	1801	2104	0.6	
4	0003	0317		0.7
	0627	0932	0.7	
Su	1234	1548		0.8
	1902	2200	0.7	
5	0058	0407		0.7
	0710	1020	0.8	
M	1324	1642		0.9
○	2002	2255	0.7	
6	0152	0456		0.7
	0754	1108	0.9	
Tu	1415	1735		1.0
	2059	2349	0.7	
7	0244	0545		0.7
	0839	1157	0.9	
W	1505	1827		1.0
	2155			
8		0043	0.7	
	0337	0635		0.7
Th	0926	1246	0.9	
	1556	1919		1.0
	2252			
9		0137	0.7	
	0431	0727		0.7
F	1015	1337	0.9	
	1648	2013		1.0
	2348			
10		0232	0.6	
	0526	0820		0.6
Sa	1107	1430	0.9	
	1741	2107		1.0
11	0044	0329	0.6	
	0624	0917		0.6
Su	1203	1526	0.8	
	1835	2203		0.9
12	0141	0427	0.6	
	0725	1017		0.5
M	1304	1624	0.7	
☽	1932	2300		0.8
13	0236	0526	0.6	
	0827	1120		0.5
Tu	1410	1726	0.6	
	2029	2358		0.8
14	0330	0624	0.5	
	0928	1225		0.5
W	1520	1828	0.6	
	2127			
15		0054		0.7
	0420	0720	0.6	
Th	1026	1327		0.5
	1630	1930	0.5	
	2224			
16		0148		0.7
	0506	0811	0.6	
F	1119	1425		0.6
	1735	2029	0.5	
	2319			
17		0237		0.6
	0548	0858	0.6	
Sa	1206	1517		0.6
	1835	2123	0.5	
18	0011	0323		0.6
	0627	0940	0.6	
Su	1249	1604		0.7
	1928	2212	0.5	
19	0059	0405		0.5
	0702	1019	0.7	
M	1328	1647		0.7
	2016	2258	0.5	
20	0145	0445		0.5
	0736	1056	0.7	
Tu	1404	1728		0.8
●	2101	2341	0.5	
21	0228	0524		0.5
	0809	1131	0.7	
W	1440	1806		0.8
	2144			
22		0022	0.5	
	0311	0601		0.4
Th	0842	1207	0.7	
	1515	1845		0.8
	2226			
23		0103	0.4	
	0354	0639		0.4
F	0916	1243	0.7	
	1551	1923		0.8
	2307			
24		0145	0.4	
	0437	0719		0.4
Sa	0953	1321	0.7	
	1629	2003		0.8
	2349			
25	2350	0227	0.4	
	0521	0801		0.4
Su	1034	1402	0.6	
	1709	2045		0.8
26	0030	0311	0.4	
	0607	0847		0.4
M	1121	1447	0.6	
	1753	2129		0.8
27	0113	0356	0.4	
	0655	0939		0.4
Tu	1216	1538	0.6	
	1841	2216		0.7
28	0155	0444	0.5	
	0746	1035		0.4
W	1320	1634	0.5	
☾	1934	2306		0.7
29	0238	0534	0.5	
	0839	1135		0.5
Th	1430	1736	0.5	
	2031	2359		0.7
30	0322	0625	0.6	
	0932	1236		0.6
F	1542	1840	0.5	
	2132			
31		0053		0.7
	0406	0716	0.7	
Sa	1026	1337		0.7
	1652	1944	0.5	
	2233			

VIEQUES PASSAGE, PUERTO RICO

CURRENT TABLE 1998 18°11.3'N 65°37.1'W Flood 250° Ebb 55°

Eastern Time (60°W) Daylight Saving Time not observed

NOVEMBER

Day	Slack time	Max time	Fld knots	Ebb knots
1 Su		0148		0.7
	0452	0808	0.8	
	1118	1435		0.8
	1759	2045	0.6	
	2335			
2 M		0242		0.7
	0538	0859	0.8	
	1210	1531		0.9
	1900	2144	0.6	
3 Tu	0034	0335		0.7
	0625	0949	0.9	
	1301	1625		1.0
	1958	2241	0.6	
4 W	0132	0428		0.7
	0713	1039	1.0	
	1351	1718		1.1
○	2054	2336	0.7	
5 Th	0228	0520		0.6
	0803	1130	1.0	
	1441	1809		1.1
	2148			
6 F		0030	0.7	
	0324	0613		0.6
	0854	1221	1.0	
	1532	1901		1.1
	2240			
7 Sa		0124	0.7	
	0420	0707		0.6
	0947	1313	0.9	
	1622	1952		1.1
	2332			
8 Su		0218	0.7	
	0517	0803		0.6
	1044	1406	0.8	
	1713	2044		1.0
9 M	0022	0312	0.6	
	0616	0901		0.5
	1144	1501	0.7	
	1804	2136		0.9
10 Tu	0112	0406	0.6	
	0715	1002		0.5
	1249	1559	0.6	
☽	1857	2229		0.8
11 W	0200	0501	0.6	
	0813	1105		0.5
	1359	1700	0.5	
	1951	2321		0.7
12 Th	0247	0554	0.6	
	0910	1207		0.5
	1512	1802	0.4	
	2046			
13 F		0014		0.7
	0331	0645	0.6	
	1003	1307		0.6
	1623	1904	0.4	
	2142			
14 Sa		0105		0.6
	0413	0733	0.7	
	1051	1403		0.6
	1729	2003	0.4	
	2239			
15 Su		0154		0.5
	0453	0817	0.7	
	1135	1453		0.7
	1828	2058	0.4	
	2334			
16 M		0240		0.5
	0531	0859	0.7	
	1215	1539		0.7
	1920	2149	0.4	
17 Tu	0027	0325		0.4
	0607	0939	0.7	
	1253	1622		0.8
	2006	2237	0.4	
18 W	0118	0407		0.4
	0643	1016	0.7	
	1330	1701		0.8
	2050	2321	0.4	
19 Th	0206	0448		0.4
	0719	1054	0.7	
	1405	1740		0.9
●	2130			
20 F		0003	0.4	
	0253	0529		0.4
	0755	1131	0.7	
	1441	1818		0.9
	2209			
21 Sa		0044	0.4	
	0338	0610		0.3
	0833	1209	0.7	
	1517	1856		0.9
	2247			
22 Su		0125	0.4	
	0421	0652		0.3
	0915	1249	0.7	
	1556	1935		0.9
	2324			
23 M		0205	0.5	
	0505	0737		0.3
	1003	1332	0.6	
	1636	2015		0.9
24 Tu	0000	0246	0.5	
	0550	0826		0.4
	1057	1419	0.6	
	1720	2057		0.8
25 W	0037	0329	0.5	
	0637	0919		0.4
	1159	1511	0.6	
	1807	2142		0.8
26 Th	0115	0414	0.6	
	0726	1016		0.5
	1308	1609	0.5	
☾	1859	2231		0.7
27 F	0155	0502	0.6	
	0817	1116		0.5
	1422	1712	0.5	
	1956	2323		0.7
28 Sa	0237	0553	0.7	
	0910	1218		0.6
	1536	1818	0.5	
	2057			
29 Su		0017		0.7
	0321	0645	0.8	
	1003	1319		0.8
	1648	1924	0.5	
	2202			
30 M		0113		0.6
	0408	0737	0.9	
	1056	1418		0.9
	1754	2027	0.5	
	2306			

DECEMBER

Day	Slack time	Max time	Fld knots	Ebb knots
1 Tu		0210		0.6
	0457	0830	0.9	
	1149	1514		1.0
	1855	2129	0.5	
2 W	0010	0306		0.6
	0548	0923	1.0	
	1240	1609		1.1
	1952	2227	0.6	
3 Th	0112	0403		0.6
	0640	1015	1.0	
	1331	1701		1.1
○	2045	2322	0.6	
4 F	0212	0458		0.6
	0734	1107	1.0	
	1421	1752		1.1
	2136			
5 Sa		0016	0.6	
	0310	0554		0.6
	0829	1159	0.9	
	1510	1842		1.1
	2224			
6 Su		0108	0.6	
	0407	0650		0.5
	0926	1251	0.9	
	1558	1931		1.1
	2311			
7 M		0159	0.7	
	0504	0746		0.5
	1025	1344	0.8	
	1646	2019		1.0
	2356			
8 Tu		0250	0.7	
	0600	0843		0.5
	1127	1437	0.7	
	1733	2107		0.9
9 W	0039	0340	0.7	
	0655	0942		0.5
	1233	1532	0.5	
	1821	2154		0.8
10 Th	0121	0429	0.7	
	0749	1041		0.5
	1342	1630	0.4	
☽	1910	2242		0.7
11 F	0202	0518	0.7	
	0841	1140		0.5
	1454	1729	0.4	
	2001	2330		0.6
12 Sa	0242	0605	0.7	
	0930	1237		0.6
	1605	1830	0.3	
	2055			
13 Su		0019		0.5
	0321	0651	0.7	
	1016	1331		0.6
	1710	1930	0.3	
	2153			
14 M		0107		0.5
	0400	0735	0.7	
	1059	1422		0.7
	1809	2027	0.3	
	2252			
15 Tu		0156		0.4
	0439	0818	0.7	
	1140	1508		0.8
	1901	2121	0.3	
	2351			
16 W		0243		0.4
	0518	0900	0.7	
	1219	1552		0.8
	1948	2210	0.3	
17 Th	0047	0330		0.4
	0557	0940	0.7	
	1257	1633		0.9
	2030	2257	0.4	
18 F	0140	0415		0.3
	0637	1020	0.7	
	1335	1713		0.9
●	2109	2340	0.4	
19 Sa	0228	0459		0.3
	0719	1101	0.7	
	1412	1752		0.9
	2146			
20 Su		0020	0.4	
	0314	0544		0.3
	0804	1142	0.7	
	1451	1830		0.9
	2221			
21 M		0100	0.5	
	0358	0629		0.4
	0853	1225	0.7	
	1530	1909		0.9
	2254			
22 Tu		0139	0.5	
	0441	0716		0.4
	0947	1310	0.6	
	1612	1948		0.9
	2328			
23 W		0219	0.6	
	0525	0806		0.4
	1046	1359	0.6	
	1656	2030		0.8
24 Th	0002	0301	0.6	
	0612	0900		0.5
	1151	1453	0.6	
	1743	2114		0.8
25 F	0038	0345	0.7	
	0701	0957		0.6
	1300	1551	0.5	
	1834	2202		0.7
26 Sa	0117	0433	0.7	
	0753	1057		0.6
	1414	1653	0.4	
☾	1930	2253		0.7
27 Su	0200	0524	0.8	
	0846	1159		0.7
	1528	1758	0.4	
	2030	2347		0.6
28 M	0245	0617	0.9	
	0941	1300		0.8
	1639	1905	0.4	
	2135			
29 Tu		0045		0.6
	0335	0712	0.9	
	1036	1400		0.9
	1745	2010	0.4	
	2243			
30 W		0144		0.6
	0427	0807	0.9	
	1130	1458		1.0
	1846	2113	0.5	
	2349			
31 Th		0244		0.5
	0522	0902	1.0	
	1222	1552		1.0
	1941	2211	0.5	

C Currents

CURRENT DIFFERENCES TABLE

The current differences table that follows provides information for calculating tidal current information for various locations in southeastern Florida and in Puerto Rico. This information must be used in conjunction with one of the primary reference current tables found earlier in this chapter.

Unfortunately, tidal current data is extremely limited for most of the Caribbean. There is more information about general currents, however, and the introduction to the Coast Pilot (Chapter 2) includes two current charts illustrating the trend of the average currents to be found in the Caribbean. Local currents are discussed in the appropriate country section of the Coast Pilot.

Mariners should bear in mind the effects of local winds and weather conditions. For instance, tidal currents off the Yucatán coast of Mexico can be reversed by a strong norther. Rivers often create local currents near their mouths that may reinforce coastal currents or may flow in opposition to local currents.

For more information on tides and currents in the Caribbean see Reed's Nautical Companion.

HOW TO USE THE TABLE

The current differences table on the following pages is organized geographically beginning with current difference stations in Florida and finishing with those in Puerto Rico.

In the **Place** column, find the location of the station in which you are interested. The name of the primary reference current table to which that station is keyed is listed above the chosen location in bold type under the Differences column, along with the page number in this chapter where the table may be found (e.g., **on Miami, p. 3.9).**

The two **Position** columns to the right of the station's name give its approximate latitude and longitude. Because the positions are listed according to the exactness recorded in the original survey records, the preciseness of the stations' locations varies.

The **Time Differences** columns express the differences that must be applied to the times derived from the primary reference table. Differences are expressed in *hours* and *minutes.* The term *slack* in the headings refers to the time when the current is at a minimum (usually, but not always, 0), before turning to flood or ebb.

The time differences for *slack before flood* and *slack before ebb* should be added or subtracted as indicated by the plus (+) or minus (–) sign to the times from the column labeled *slack time* on the primary reference table. When determining *slack before flood,* use the entry in the primary table under *slack time* that shows a current speed (in knots) under the adjacent *flood* column, to the right of the *max. time* column. When determining *slack before ebb,* use the entry under *slack time* that shows a current speed (in knots) under the *ebb* column.

The **Speed Ratio** columns give ratios for determining approximate current speeds at the current difference station for a specific day at the time of maximum flow. The appropriate maximum flood or maximum ebb speed from the primary reference table is multiplied by the relevant flood or ebb *speed ratio* given for the current difference station.

The **Current Direction and Max Speed** columns give the average directions of current flow at the times of the maximum flood and ebb (in degrees true), with corresponding average speeds of maximum flood and ebb (which may differ from the speeds calculated for a specific day using the speed ratios). Figures for these columns are read directly from the table, making it easy to make quick assessments of average local maximum currents.

Note: In the table, dashes (– –) are substituted for data that is unknown, indeterminate, unreliable, or not applicable.

EXAMPLE

Suppose you are interested in tidal currents in the New River, Fort Lauderdale, Florida, on March 25. This location is listed in the current differences table in the **Place** column, under the Florida, East coast, subheading. This location is keyed to the Miami, Florida (Harbor Entrance) current table found on page C 5. The current differences are calculated for an approximate **position** in the New River of 26° 06.73′ north latitude, 80° 07.18′ west longitude.

Now you would turn to the time and speed entries in the Miami current table for March 25. For the example's sake, let's say we find predictions that look like:

	Slack time	Max time	Fld knots	Ebb knots
		0133		1.6
25	0431	0647	1.8	
Tu	1018	1308		1.6
	1646	1905	2.0	
	2246			

The current is at maximum ebb (1.6 knots) at 0133, slack at 0431, at maximum flood at 0647 (1.8 knots), slack at 1018, at maximum ebb at 1308 (1.6 knots), slack at 1646, at maximum flood at 1905 (2.0 knots), and slack again at 2246.

To find the **time of slack before ebb** in the New River, the time difference of –0:06 minutes under the *Slack before ebb* heading in the current differences table should be subtracted from the slack times before the ebbs (1018 and 2246) in the Miami table. Minimum currents before ebbs will occur at 1012 and 2240 in the New River.

Similarly, the **time of maximum ebb** in the New River can be computed by subtracting the time difference of –0.16 minutes shown under *max. ebb* in the current differences table from the times of maximum ebb in Miami (0133 and 1308). Maximum ebbs in the New River will occur at 0117 and 1252.

To find the **time of slack before flood** in the New River, the time difference of –0:43 minutes under the *slack before flood* heading should be subtracted from the slack times before the floods noted above for Miami (0431 and 1646). In the New River minimum currents before floods will thus occur at 0348 and 1603.

Similarly, the **time of maximum flood** in the New River can be computed by subtracting the time difference of –0:39 minutes shown under the *max. flood* heading from the times of maximum floods shown in the Miami table (0647 and 1905). Maximum floods in the New River will occur at 0608 and 1826.

The **Speed Ratio** figures allow you to determine current speeds at the current difference station at the time of maximum flow for a specific day (which may differ from the *average* maximum flow speeds presented in the final two columns of the table).

To figure the approximate maximum flood speeds in the New River on March 25, 1997, multiply the maximum current speed under the *Fld* heading for the relevant times in Miami (1.8 for the first flood and 2.0 for the second) by the speed ratio for the New River given under the *flood* heading (0.4) in the current differences table:

1.8 knots (0.4) = 0.7 knots
2.0 knots (0.4) = 0.8 knots.

To compute the approximate maximum ebb speeds in the New River on the same day, multiply the speeds under the *ebb* heading for Miami (1.6 for both ebbs) by the speed ratio under the *ebb* heading (0.3) for the New River:

1.6 knots (0.3) = 0.5 knots

Information in the **Current Direction and Max Speed** columns is read directly. At the time of maximum flood, the New River current flows 005° true; at the time of maximum ebb, 130° true. The average flood maximum speed is 0.8 knots; the average ebb maximum speed, 0.5 knots.

C

Currents

Tide Watchers Wanted

Reed's seeks local information about tides and currents to include in this almanac. Please see the notice on page C 3.

PLACE	POSITION		TIME DIFFERENCES				SPEED RATIOS		CURRENT DIRECTION & MAX SPEED			
	north latitude	west longitude	slack before flood h m	max flood h m	slack before ebb h m	max ebb h m	flood	ebb	flood dir	ebb dir	flood knots	ebb knots
FLORIDA												
Eastern Standard Time (GMT –5)							*Add 1 hour Daylight Saving April 5 – October 24*					
Atlantic coast			**on MIAMI, p. C 4**									
Fort Pierce Inlet	27°28.3′	80°17.5′	+1:19	+0:39	+0:48	+0:35	1.5	2.0	250°	072°	2.6	3.1
Lake Worth Inlet, between jetties	26°46.33′	80°02.13′	+0:13	–0:07	–0:01	0:00	1.3	2.3	273°	094°	2.4	3.6
Fort Lauderdale, New River	26°06.73′	80°07.18′	–0:43	–0:39	–0:06	–0:16	0.4	0.3	005°	130°	0.8	0.5
Port Everglades												
Entrance, between jetties	26°05.58′	80°06.32′	–0:08	–0:49	–0:43	–0:34	0.3	0.4	275°	095°	0.6	0.7
Entrance, from southward (canal)	26°05.2′	80°06.9′	+0:40	+0:07	+0:31	–0:09	0.7	1.1	167°	358°	1.3	1.7
Turning basin	26°05.70′	80°07.05′	–1:01	–1:07	–1:02	–1:11	0.1	0.3	320°	155°	0.2	0.5
Turning basin, 300 yards north of	26°05.8′	80°07.1′	–0:20	–1:09	–0:27	–0:14	0.5	1.1	349°	160°	0.9	1.8
17th Street bridge	26°06.02′	80°07.13′	–0:38	–0:53	–0:28	–0:55	1.1	1.2	350°	170°	1.9	1.9
Miami Harbor												
Bakers Haulover Cut	25°54.0′	80°07.4′	–0:01	+0:07	+0:14	–0:17	1.6	1.6	270°	090°	2.9	2.5
Government Cut												
East entrance, off north jetty	25°45.59′	80°07.35′	–0:02	–0:19	–0:08	–0:26	0.4	0.9	236°	092°	0.6	1.5
East entrance, inside south jetty	25°45.61′	80°07.66′	–0:07	–0:06	–0:04	0:00	1.2	1.1	343°	116°	2.1	1.8
Midway, north side	25°45.84′	80°07.96′	–0:12	–0:03	–0:07	–0:08	0.7	0.5	292°	108°	1.2	0.7
MIAMI HARBOR ENTRANCE	25°45.90′	80°08.17′	daily predictions				– –	– –	293°	112°	1.8	1.6
West entrance, south side	25°45.85′	80°08.25′	+0:09	+0:10	–0:04	+0:01	0.9	1.6	288°	100°	1.6	2.5
Main Channel												
Causeway Island, 0.2 mi SE of <1>	25°46.06′	80°08.58′	+0:01	+0:23	–0:01	–0:14	0.8	0.4	306°	131°	1.4	0.7
Lummus Island, NE corner <2>	25°46.02′	80°08.70′	–0:07	–0:02	+0:06	–0:04	0.1	0.4	265°	104°	0.2	0.7
Lummus Is, 0.18 mi off NW corner	25°46.32′	80°09.45′	–0:05	+0:50	+0:28	–0:08	0.3	0.3	285°	102°	0.5	0.5
Dodge Is, 0.1 mi off NW corner	25°46.89′	80°10.90′	+0:17	–0:14	+0:01	+0:04	0.2	0.3	277°	093°	0.4	0.4
Fishermans Channel												
Lummus Island, SE corner <3>	25°45.95′	80°08.71′	+0:08	+0:06	+0:19	–0:20	0.5	0.2	208°	086°	0.9	0.4
Fisher Island, 0.2 mi northwest of	25°45.87′	80°09.08′	+0:14	+0:38	+0:17	+0:39	0.6	0.7	280°	090°	1.0	1.1
Lummus Is, 0.15 mi off SW corner	25°45.89′	80°09.69′	+0:20	–0:20	+0:10	+0:22	0.3	0.6	271°	095°	0.6	0.9
West end, southwest of Dodge Is	25°46.36′	80°10.74′	–0:05	–0:32	–0:15	–0:21	0.1	0.2	277°	089°	0.2	0.3

FLORIDA (cont.)

Eastern Standard Time (GMT –5) — *Add 1 hour Daylight Saving April 5 – October 24*

Miami River entrance	25°46.21′	80°11.23′	+0:15	–0:02	–0:01	+0:46	0.1	0.4	261°	071°	0.2	0.6
Fowey Rocks Light, 1.5 mi SW of	25°35′	80°07′	current weak and variable				– –	– –	– – –	– – –	– –	– –
Florida Keys			**on KEY WEST, p. C 10**									
Caesar Creek, Biscayne Bay	25°23.2′	80°13.6′	+0:07	–0:08	–0:14	–0:05	1.2	1.0	316°	123°	1.2	1.8
Long Key, drawbridge east of	24°50.4′	80°46.2′	+0:58	+1:27	+2:21	+1:33	1.1	0.7	000°	202°	1.1	1.2
Long Key viaduct	24°48.1′	80°51.9′	+1:34	+1:28	+2:02	+1:57	0.9	0.7	349°	170°	0.9	1.2
Moser Channel, swing bridge	24°42.0′	81°10.2′	+1:07	+1:30	+1:50	+1:47	1.4	1.0	339°	166°	1.4	1.8
Bahía Honda Harbor, bridge	24°39.4′	81°17.3′	+1:01	+0:39	+1:53	+1:05	1.4	1.2	004°	182°	1.4	2.1
No Name Key, northeast of	24°42.3′	81°18.8′	+0:55	+1:24	+1:20	+0:53	0.7	0.5	312°	142°	0.7	0.9
Key West												
Main Ship Channel entrance	24°28.4′	81°48.1′	–0:44	–0:12	+0:10	+0:10	0.2	0.3	040°	178°	0.2	0.4
Main Ship Channel <4>	24°30.5′	81°48.3′	– – –	– – –	– – –	+0:30	– –	0.2	064°	133°	– –	0.4
KEY WEST, 0.3 mi W of Ft. Taylor	24°32.9′	81°49.0′	daily predictions				– –	– –	022°	194°	1.0	1.7
Fort Taylor, 0.6 mi north of	24°33.5′	81°48.6′	+0:20	+0:13	–0:11	+0:13	0.6	0.7	042°	202°	0.6	1.2
Turning basin	24°34.0′	81°48.25′	+0:43	+0:44	+0:29	+1:06	0.8	0.6	048°	216°	0.8	1.1
Northwest Channel	24°35.0′	81°50.9′	–0:08	–0:03	–0:09	–0:07	1.2	0.8	353°	162°	1.2	1.4
Northwest Channel	24°37.3′	81°52.8′	–0:28	–0:19	–0:20	–0:20	0.6	0.4	346°	168°	0.6	0.6
Boca Grande Channel	24°34′	82°04′	–0:40	–0:45	–0:01	–0:06	1.1	0.8	353°	194°	1.1	1.2
New Ground <5>	24°39.0′	82°25.0′	+1:36	+1:55	+1:28	+1:18	0.7	0.4	068°	244°	0.7	0.7
Isaac Shoal	24°33.5′	82°32.2′	+1:00	+0:54	+1:52	+1:55	1.0	0.5	002°	181°	1.0	0.8
Southeast Channel	24°37.62′	82°51.07′	–0:27	–0:06	+0:37	+0:36	0.6	0.4	004°	172°	0.6	0.6
Southwest Channel	24°36.92′	82°54.70′	+0:45	+0:59	+1:25	+2:04	0.4	0.4	001°	209°	0.4	0.6

PUERTO RICO

Atlantic Standard Time (GMT – 4) — *Daylight Saving Time not observed*

			on VIEQUES PASSAGE, p. C 16									
Punta Ostiones, 1.5 mi west of	18°05.20′	67°13.6′	–0:26	–0:52	–0:04	–0:35	1.7	1.3	187°	001°	1.0	0.9
VIEQUES PASSAGE	18°11.3′	65°37.1′	daily predictions				– –	– –	250°	057°	0.6	0.7
Vieques Sound	18°15.87′	65°34.20′	–0:44	–1:16	–1:28	–1:05	0.7	0.9	180°	355°	0.4	0.6
Largo Shoals, west of	18°19′	65°35′	–0:52	–1:28	–1:33	–1:08	0.7	1.0	186°	330°	0.4	0.7
Ramos Cay, 0.3 mi southeast of	18°18.6′	65°36.4′	– – –	–0:42	– – –	–0:44	0.3	0.1	120°	284°	0.2	0.1
Palominos Is, 0.9 mi SW of <6>	18°20.1′	65°34.8′	– – –	– – –	– – –	–0:48	– –	0.7	– – –	307°	– –	0.5

Currents C

PLACE	POSITION		TIME DIFFERENCES				SPEED RATIOS		CURRENT DIRECTION & MAX SPEED			
	north latitude	west longitude	slack before flood h m	max flood h m	slack before ebb h m	max ebb h m	flood	ebb	flood dir	ebb dir	flood knots	ebb knots
PUERTO RICO (cont.)												
Atlantic Standard Time (GMT – 4)									*Daylight Saving Time not observed*			
			on VIEQUES PASSAGE, p. C 16									
Fajardo Harbor, channel	18°20′	65°37′	–1:13	–1:52	–2:27	–1:45	0.5	1.6	162°	339°	0.3	1.1
Isla Marina, 0.2 mi west of <6>	18°20.50′	65°37.38′	– – –	– – –	– – –	–2:06	– –	1.0	– – –	335°	– –	0.7
Coronala Laja, 0.4 mi NW of <6>	18°21.6′	65°37.3′	– – –	– – –	– – –	–1:33	– –	0.4	– – –	000°	– –	0.3
Pasaje de San Juan <6>	18°23.9′	65°36.6′	– – –	– – –	– – –	–1:15	– –	1.7	– – –	310°	– –	1.2
Bahía de San Juan	18°27.23′	66°06.6′	current weak and variable				– –	– –	– – –	– – –	– –	– –

<1> Maximum ebb time difference is for the middle of the phase. The speed is near 0.7 knot throughout most of ebb phase. Speeds a short distance away may vary significantly.
<2> Maximum flood time difference is for the middle of the phase. The speed is very low throughout most of the flood phase.
<3> Values may vary significantly a short distance away.
<4> The flood is weak and variable.
<5> The current tends to rotate clockwise. At the times for minimum before flood there may be a weak current flowing northward, while at times for minimum before ebb there may be a weak current flowing southeastward.
<6> The current seldom floods.

GULF STREAM INFORMATION

The region where the Gulf of Mexico narrows to form the channel between the Florida Keys and Cuba may be regarded as the head of the Gulf Stream. From this region the stream sets eastward and northward through the Straits of Florida and after passing Little Bahama Bank continues northward and then northeastward, following the general direction of the 100-fathom curve as far as Cape Hatteras, North Carolina. The flow in the Straits is frequently referred to as the Florida Current.

Shortly after emerging from the Straits of Florida, the stream is joined by the Antilles Current, which flows northwesterly along the open ocean side of the West Indies before uniting with the water that has passed through the Straits. Beyond Cape Hatteras the combined current turns more and more eastward under the combined effects of the deflecting force of the Earth's rotation and the eastward-trending coastline, until it reaches the region of the Grand Banks of Newfoundland.

Eastward of the Grand Banks the whole surface is slowly driven eastward and northeastward by the prevailing westerly winds to the coastal waters of northwestern Europe. For distinction, this broad and variable wind-driven surface movement is sometimes referred to as the North Atlantic Drift or Gulf Stream Drift. In general, the Gulf Stream as it issues into the sea through the Straits of Florida may be characterized as a swift, highly saline current of blue water whose upper stratum is composed of warm water.

On its western or inner side, the Gulf Stream is separated from the coastal waters by a zone of rapidly falling temperature, to which the term "cold wall" has been applied. It is most clearly marked north of Cape Hatteras, but extends more or less well defined from the Straits of Florida to the Grand Banks.

Throughout the whole stretch of 400 miles in the Straits of Florida, the stream flows with considerable speed. Abreast of Havana, the average surface speed in the axis of the stream is about 2.5 knots. As the cross-sectional area of the stream decreases, the speed increases gradually, until abreast of Cape Florida it becomes about 3.5 knots. From this point within the narrows of the Straits, the speed along the axis gradually decreases to about 2.5 knots off Cape Hatteras. These values are for the axis of the stream where the current is a maximum; the speed of the stream decreases gradually from the axis as the edges of the stream are approached. The speed of the stream, furthermore, is subject to fluctuations brought about by variations in winds and barometric pressure.

The Gulf Stream Mean Surface Speed table (next page) gives the mean surface speed of the Gulf Stream in two cross-sections in the Straits of Florida. Vessels crossing the Gulf Stream at Jupiter or Fowey Rocks should make an average allowance of 2.5 knots in a northerly direction for the current. Vessels crossing the stream from Havana should make a fair allowance for the average current between 100-fathom curves of 1.1 knots in an east-northeasterly direction.

From within the Straits, the axis of the Gulf Stream runs approximately parallel with the 100-fathom curve as far as Cape Hatteras. Since this stretch of coastline sweeps northward in a sharper curve than does the 100-fathom line, the stream lies at varying distances from the shore. The lateral boundaries of the current within the Straits are fairly well fixed, but when the stream flows into the sea the eastern boundary becomes somewhat vague. On the western side, the limits can be defined approximately since the waters of the stream differ in color, temperature, salinity, and flow from the inshore coastal waters. On the east, however, the Antilles Current combines with the Gulf Stream, so that its waters here merge gradually with the waters of the open Atlantic. Observations of the National Ocean Service indicate that, in general, the average position of the inner edge of the Gulf Stream as far as Cape Hatteras lies inside the 50-fathom curve. The Gulf Stream, however, shifts somewhat with the seasons and is considerably influenced by the winds, which cause fluctuations in its position, direction, and speed; consequently any limits that are assigned refer to mean or average positions. For the approximate mean positions of the inner edge and axis (point where greatest speed may be found) see the Approximate Mean Position of the Gulf Stream table on the next page.

At the western edge of the Straits of Florida the limits of the Gulf Stream are not well defined, and for this reason the location of the inner edge has been omitted for Havana, Cuba, and Key West, Florida, in this table. Between Fowey Rocks and Jupiter Inlet the inner edge is deflected westward and lies very close to the shoreline.

Along the Florida Reefs, between Alligator Reef and Dry Tortugas, the distance of the northerly edge of the Gulf Stream from the edge of the reefs gradually increases towards the west. Off Alligator Reef it is quite close inshore, while off Rebecca Shoal and Dry Tortugas it is possibly 15 to 20 miles south of the 100-fathom curve. Between the reefs and the northern edge of the Gulf Stream, the currents are ordinarily tidal and are subject at all times to considerable modification by local winds and barometric conditions. This neutral zone varies in both length and breadth; it may extend along the reefs a greater or lesser distance than stated, and its width varies as the northern edge of the Gulf Stream approaches or recedes from the reefs.

The approximate position of the axis of the Gulf Stream for various regions is shown on the following National Ocean Service charts: No. 11013, Straits of Florida and Approaches; No. 411, Gulf of Mexico; No. 11460, Cape Canaveral to Key West; No. 11420, Havana to Tampa Bay. Chart No. 11009 shows the axis and position of the inner edge of the Gulf Stream from Cape Hatteras to the Straits of Florida.

GULF STREAM MEAN SURFACE SPEED

Between Rebecca Shoal and Cuba		Between Fowey Rocks and Gun Cay	
Distance south of Rebecca Shoal	Mean surface speed observed	Distance east of Fowey Rocks	Mean surface speed observed
Nautical miles	Knots	Nautical Miles	Knots
20	0.3	8	2.7
35	0.7	11.5	3.5
50	2.2	15	3.2
68	2.2	22	2.7
86	0.8	29	2.1
		36	1.7

APPROXIMATE MEAN POSITION OF THE GULF STREAM

Locality	Inner Edge Nautical miles	Axis Nautical miles
North of Havana, Cuba		25
Southeast of Key West, FL		45
East of Fowey Rocks, FL		10
East of Miami Beach, FL		15
East of Palm Beach, FL		15
East of Jupiter Inlet, FL		20
East of Cape Canaveral, FL	10	45
East of Daytona Beach, FL	25	75
East of Ormond Beach, FL	25	75
East of St. Augustine, FL (coast line)	40	85
East of Jacksonville, FL (coast line)	55	90
Southeast of Savannah, GA (coast line)	65	95
Southeast of Charleston, SC (coast line)	55	90
Southeast of Myrtle Beach, SC	60	100
Southeast of Cape Fear, NC (light)	35	75
Southeast of Cape Lookout, NC (light)	20	50
Southeast of Cape Hatteras, NC	10	35
Southeast of Virginia Beach, VA	85	115
Southeast of Atlantic City, NJ	120	
Southeast of Sandy Hook, NJ	150	

CARIBBEAN SUMMER CURRENTS

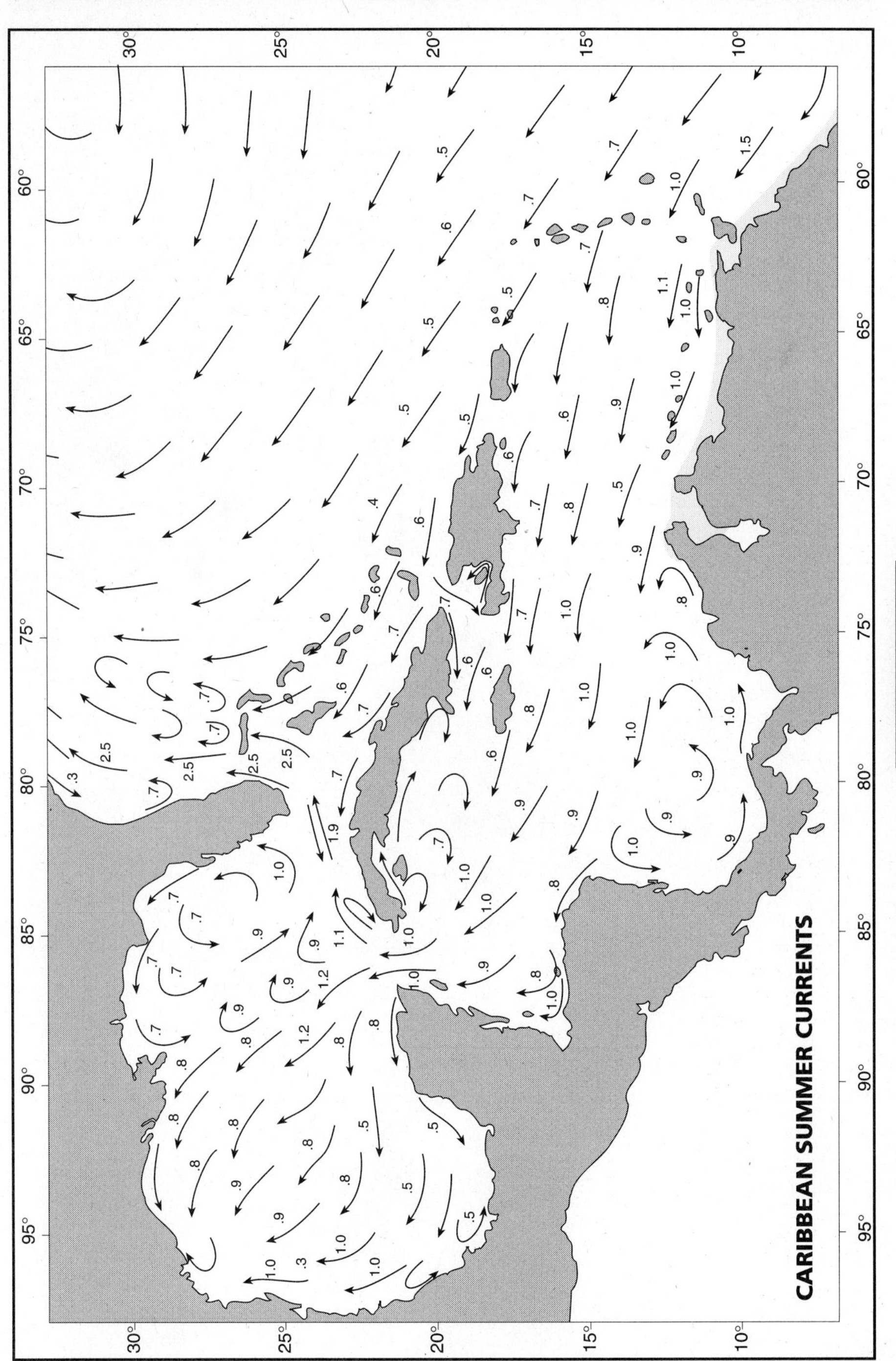

CARIBBEAN WINTER CURRENTS

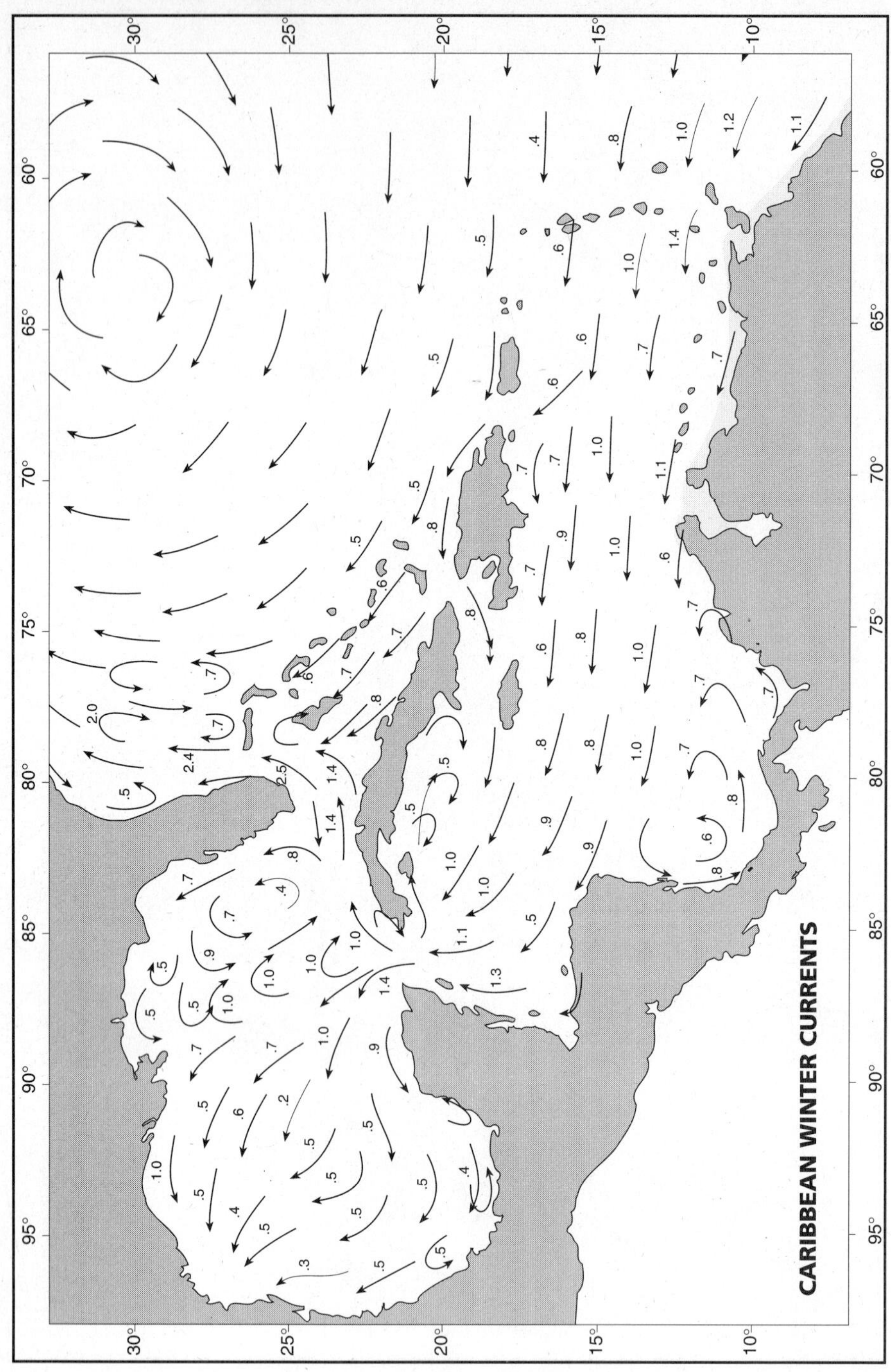

Resources

COMMUNICATIONS & WEATHER

ELECTRONIC NAVIGATION

NOTICES TO MARINERS

INTRODUCTION

Chapter R includes everything that used to be in *Reed's* Communications & Weather Services, Electronic Aids to Navigation, and Notice to Mariners chapters—and more! We encourage you to skim this chapter quickly just to get a sense of what's here and how it's organized.

In recent years, advances in communications, information, and geopositioning technology have brought many benefits to mariners. Along with the benefits, however, have come a high degree of complexity and a rapid rate of change. *Reed's* advises mariners to stay abreast of these fields by reading Notices to Mariners (see page R 36) and nautical periodicals. Another excellent source is the Coast Guard's Web site at: www.navcen.uscg.mil (see page R 37).

CONTACTING THE U.S. COAST GUARD

VHF

In coastal waters of the United States, vessels may contact the Coast Guard on VHF channel 16. Channel 16 may be used only for distress and calling. Routine radio checks on this frequency are prohibited. After initial contact on channel 16 you will usually be asked to switch to channel 22A or channel 12.

CELLULAR PHONE

The Coast Guard does not advocate cellular phones as a substitute for regular maritime radio distress and safety systems, particularly VHF maritime radio. The reasons are that cellular phones generally cannot provide ship to ship safety communications or communications with rescue vessels; cellular coverage offshore is limited, and may change without notice; and the Coast Guard can not home into a cellular signal as they can a VHF signal.

However, the Coast Guard recognizes cellular telephones as an additional way for boaters to contact them in times of emergency. If you make such a call, the Coast Guard recommends that you have information concerning your cellular service immediately available (cellular phone number, service provider, complete roaming number, etc.); and that you know your location and carry a complete set of visual distress signals.

Note: Reed's provides telephone numbers for Coast Guard's search and rescue stations at the beginning of each *Coast Pilot* section. Note also that some cellular services have established *CG as a quick dial access to the Coast Guard.

MEDIUM-FREQUENCY SSB

The international hailing and distress frequency is 2182 kHz in the medium-frequency band. This frequency is monitored by all Coast Guard stations, any vessel equipped with the proper equipment while under way, and many coast stations. After initial contact on 2182 kHz you will usually be asked to switch to 2670 kHz, which is the U.S. Coast Guard working frequency for weather broadcasts, notices to mariners, and marine safety broadcasts.

U.S. Coast Guard communication stations and cutters have discontinued watchkeeping on distress frequency 500 kHz and ceased all Morse code services in the medium-frequency and high-frequency radiotelegraphy bands.

HIGH-FREQUENCY-SSB

The U.S. Coast Guard monitors the following carrier frequencies for distress, emergency, and safety traffic. Allow at least one minute for a response before switching channels. These frequencies are also used for voice weather broadcasts, navigation warnings, and medical communications. The Coast Guard monitors many channels at once and may not be able to respond immediately.

Location, call sign, and hours of watch:

Portsmouth, VA (NMN)

424 ....2300-1100 UTC
601 ....24 Hrs
816 ....24 Hrs
1205 ....1100-2300 UTC
1625 ....(on request only)

Boston, MA (NMF) (remote from NMN)

424 ....2230-1030 UTC
601 ....24 Hours
816 ....24 Hours
1205 ....1030-2230 UTC
1625 ....(0n request only)

Miami, FL (NMA) (remote from NMN)

601 ....24 Hours
1205 ....24 Hours
1625 ....24 Hours

New Orleans, LA (NMG)

424 . . . . .24 Hours
601 . . . . .24 Hours
816 . . . . .24 Hours
1205 . . . . .24 Hours
1625 . . . .(on request only)

Channel	Transmit	Receive
424	4134 kHz	4426 kHz
601	6200 kHz	6501 kHz
816	8240 kHz	8764 kHz
1205	12242 kHz	13089 kHz
1625	16432 kHz	17314 kHz

HF-RADIOTELEX/SITOR

The U.S. Coast Guard guards the following frequencies for medium and long-range radio-teletype communications concerning distress, emergency, and safety traffic. SITOR is a protocol of narrow band direct printing teletype for error free data communications from ship to ship or ship to shore. The Coast Guard sends SITOR navigational warnings broadcasts using the FEC mode. Note that there are techniques available to monitor radiotelex and SITOR communications with a SSB or shortwave radio and a computer. Also note that frequencies listed are assigned. Carrier frequency is located 1700Hz below the assigned frequency.

Location, call sign, and hours of watch:

Portsmouth, VA (NMN): SITOR ID 1097.

404 . . . . on request
604 . . . . 2300–1100 UTC
824 . . . . .24 hours
1227 . . . . .24 hours
1627 . . . . .24 hours
2227 . . . . .1100–2300 UTC

ITU SITOR Channel	Ship Frequency	Coast Frequency
404	4174.0 kHz	4212.0 kHz
604	6264.5 kHz	6316.0 kHz
824	8380.0 kHz	8428.0 kHz
1227	12490.0 kHz	12592.5 kHz
1627	16696.5 kHz	16819.5 kHz
2227	22297.5 kHz	22389.5 kHz

SENDING A DISTRESS CALL

The Coast Guard suggests that a distress call be transmitted in the following manner:

1. If you have an HF radiotelephone tuned to 2182 kHz, send the radiotelephone alarm signal if one is available. If you have a VHF marine radio, tune it to channel 16. Unless you know you are outside VHF range of shore and ships, call on channel 16 first.
2. Distress signal "MAYDAY", spoken three times.
3. The words "THIS IS", spoken once.
4. Name of vessel in distress (spoken three times) and call sign or boat registration number, spoken once.
5. Repeat "MAYDAY" and name of vessel, spoken once.
6. Give position of vessel by latitude or longitude or by bearing (true or magnetic, state which) and distance to a well-know landmark such as a navigational aid or small island, or in any terms which will assist a responding station in locating the vessel in distress. Include any information on vessel movement such as course, speed and destination.
7. Nature of distress (sinking, fire etc.).
8. Kind of assistance desired.
9. Number of persons on board.
10. Any other information which might facilitate rescue, such as length or tonnage of vessel, number of persons needing medical attention, color hull, cabin, masks, etc.
11. The word "OVER"
12. Stay by the radio if possible. Even after the message has been received, the Coast Guard can find you more quickly if you can transmit a signal on which a rescue boat or aircraft can home.

***Note:* Many Caribbean nations have their own search and rescue operations. You will find these resources listed in the *Useful Information* sections of the *Pilot* chapter, and their radio components listed in the *Coastal Radio Resources* section in this chapter.**

VHF CHANNELS

#	Transmit	Receive	Use
01A	156.050	156.050	Port Operations and Commercial. VTS in selected areas.
05A	156.250	156.250	Port Operations. VTS in Seattle
06	156.300	156.300	Intership Safety
07A	156.350	156.350	Commercial
08	156.400	156.400	Commercial (Intership only)
09	156.450	156.450	Boater Calling. Commercial and Non-Commercial.
10	156.500	156.500	Commercial
11	156.550	156.550	Commercial. VTS in selected areas.
12	156.600	156.600	Port Operations. VTS in selected areas.
13	156.650	156.650	**Intership Navigation Safety.** (Ships over 20 meter length must maintain a listening watch on this channel in US waters.)
14	156.700	156.700	Port Operations. VTS in selected areas.
15	--	156.750	Environmental (Receive only). Used by Class C EPIRBs.
16	156.800	156.800	**International Distress, Safety and Calling. (**Ships required to carry a radio, the U.S.C.G., and most coast stations maintain a listening watch on this channel.)
17	156.850	156.850	State Control
18A	156.900	156.900	Commercial
19A	156.950	156.950	Commercial
20	157.000	161.600	Port Operations (duplex)
20A	157.000	157.000	Port Operations
21A	157.050	157.050	U.S. Government only
22A	157.100	157.100	Coast Guard
23A	157.150	157.150	U.S. Government only
24	157.200	161.800	Public Correspondence (Marine Operator)
25	157.250	161.850	Public Correspondence (Marine Operator)
26	157.300	161.900	Public Correspondence (Marine Operator)
27	157.350	161.950	Public Correspondence (Marine Operator)
28	157.400	162.000	Public Correspondence (Marine Operator)
63A	156.175	156.175	Port Operations and Commercial. VTS in selected areas.
65A	156.275	156.275	Port Operations
66A	156.325	156.325	Port Operations
67	156.375	156.375	Commercial (Intership only)
68	156.425	156.425	Non-Commercial
69	156.475	156.475	Non-Commercial
70	156.525	156.525	Digital Selective Calling (voice communications not allowed)
71	156.575	156.575	Non-Commercial
72	156.625	156.625	Non-Commercial (Intership only)
73	156.675	156.675	Port Operations
74	156.725	156.725	Port Operations
77	156.875	156.875	Port Operations (Intership only)
78A	156.925	156.925	Non-Commercial
79A	156.975	156.975	Commercial
80A	157.025	157.025	Commercial
81A	157.075	157.075	U.S. Government only - Environmental protection operations.
82A	157.125	157.125	U.S. Government only
83A	157.175	157.175	U.S. Government only
84	157.225	161.825	Public Correspondence (Marine Operator)
85	157.275	161.875	Public Correspondence (Marine Operator)
86	157.325	161.925	Public Correspondence (Marine Operator)
87	157.375	161.975	Public Correspondence (Marine Operator)
88	157.425	162.025	Public Correspondence in selected areas only.
88A	157.425	157.425	Commercial (Intership only)

SSB CHANNELS

These simplex single sideband radiotelephone frequencies are provided for worldwide common use by ships of all categories, for communications with coast stations or other ships. These frequencies are shared and are not available for the exclusive use of any station.

Note: All duplex and simplex frequencies are upper sideband (USB), with assigned frequency 1.4 kHz above the listed carrier frequency.

HF SIMPLEX CHANNELS

4146	6224	8294	12,353
4149	6227	8297	12,356
	6230		12,359
			12,362
			12,365
16,528	18,825	22,159	25,100
16,531	18,828	22,162	25,103
16,534	18,831	22,165	25,106
16,537	18,834	22,168	25,109
16,540	18,837	22,171	25,112
16,543	18,840	22,174	25,115
16,546	18,843	22,177	25,118

GMDSS RADIOTELEPHONE CHANNELS

These simplex frequencies are designated under the Global Maritime Distress & Safety System for distress and safety communications, and (except for 2182 kHz) are not normally guarded.

2182 kHz	8291 kHz
4125	12,290
6215	16,420

This HF radiotelephone channel and frequency information was derived from Appendix 16 of the International Telecommunications Union (ITU) Radio Regulations.

EMERGENCY MEDICAL ADVICE

In U.S. waters, contact the U.S. Coast Guard on 2182 kHz or on one of the high-frequency single-sideband channels listed above. In addition, high-seas radiotelephone and coastal marine operators can transfer emergency calls to the nearest Coast Guard station or search and rescue organization. There are no charges when the ship states the call is an emergency involving the safety of life or property at sea.

A partial list of stations and organizations in the Caribbean that assist with emergency medical advice include:

Bahamas: BASRA (VHF channel 16, 2182 kHz).
Barbados: Coast station Barbados (8PO).
Bermuda: Bermuda Radio (VRT, ZBM).
British Virgin Islands: Tortola Radio.
Costa Rica: Coast station Limón (TIM).
Cuba: Coast station Habana (CLA, CBT); coast station Santiago de Cuba (CLM).
Dominican Republic: Coast station Santo Domingo Piloto (HIA).
Guadeloupe: Coast station Pointe-à-Pitre (FFQ). Message must be in French.
Jamaica: Coast station Kingston (6YI).
Martinique: Coast station Fort-de-France (FFP). Message must be in French. This service assumes no liability.
Netherlands Antilles: Coast station Curaçao (PJC).
Panama: Coast station Canal (HPN60). Canal guards 500 kHz.
Saba: Saba Radio (PJS) *See entry on page R 10.*
United States: Coast Guard stations Boston, MA; Camslant Chesapeake (Portsmouth), VA; Miami, FL; and San Juan, Puerto Rico (NMR); coast station Virgin Islands Radio (WAH).

RADIO WATCHKEEPING

In the waters of the United States, every power-driven vessel of 20m length or greater, every vessel of 100 tons and upward carrying one or more passengers for hire, every towing vessel of 26 ft length or greater, and every dredge and floating plant near a channel or fairway, must maintain a watch on both VHF channels 13 (156.650 MHZ) and 16 (156.800 MHZ) while the vessel is underway. These rules apply to all foreign vessels, recreational vessels, federal government and military vessels as well as commercial vessels, operating within U.S. territorial waters.

Ships participating in a vessel traffic service area must also keep watch on the VTS radio channel designated for that area. Such ships may discontinue their watch on VHF channel 16 while in the VTS area. Vessels not required to carry a marine radio (e.g. recreational vessels less than 20m length), but which voluntarily carry a radio, must maintain a watch on channel 16 whenever the radio is operating and not being used to communicate. Such vessels may alternatively maintain a watch on VHF channel 09, the boater calling channel.

International Radio Watchkeeping Regulations require that ships 300 tons and over and passenger ships on international voyages must maintain, where practicable, a continuous listening watch on VHF channel 16.

COASTAL RADIO RESOURCES

Vessels operating along the North American coast have excellent safety, weather, and communications resources available via VHF and medium-frequency SSB radio, as well as from AM and FM stations. Around the Caribbean rim, services vary widely. In the following pages, *Reed's* explains the services of several common types of stations, and lists selected specfic stations from north to south.

NOTE: Most times in this section are in UTC (Universal Coordinated Time), which for all practical purposes is the same as GMT (Greenwich Mean Time).

NOAA Weather Radio operates VHF-FM radio stations on frequencies 162.55, 162.40, and 162.475 MHz. These stations provide continuous weather broadcasts, including both regional and local weather. Broadcast tapes are generally updated every three hours during the day and at least every six hours.
Broadcasts vary, but generally include the following information:

1) Marine forecasts and warnings for coastal waters out to 60 miles
2) Offshore waters forecast for waters 60 to 250 miles from land
3) State forecasts and local forecasts
4) Selected weather observations from the Coast Guard, buoys, and other stations.

Whenever severe weather warnings are necessary, the tape is updated and the transmission devoted to "up-to-the-minute" information on storm dangers.

Outside the continental United States NOAA weather broadcasts are transmitted from Puerto Rico and the U.S. Virgin Islands. Bermuda and the Bahamas have local VHF broadcasts as well.

WX-1: 162.55 MHz
WX-2: 162.40 MHz
WX-3: 162.475 MHz

The United States Coast Guard also transmits navigation information on medium frequency (usually 2670.0 kHz). The broadcast is announced on 2182 kHz and VHF channel 16 before commencing. Coast Guard broadcasts usually include notices to mariners, marine warnings, hurricane information, and weather. The Coast Guard's high-seas weather information is transmitted via station NMN (Portsmouth, Virgina). See Offshore section.

Coast stations should respond to calls on the international hailing and distress frequencies: VHF channel 16 or 2182 kHz. These stations include pilots, port authorities, and public correspondence stations. The latter can be used to make telephone calls. Most stations monitoring 2182 kHz also respond to calls on VHF channel 16. A partial listing of coast stations follows.

Commercial broadcast stations are another source of information of use to mariners. In most cases, these stations are transmit-only. *Reed's* has included in its list some of the more prominent English-language stations around the Caribbean basin. Schedules and frequencies are subject to frequent change. Stations of particular interest to mariners are noted by an asterix (*).

BERMUDA

WEATHER

ZBM*, Bermuda Harbour Radio, St. Georges. VHF channel 27 and 2582 kHz at 1235 and 2035 UTC.
WX-2: 162.4 MHz, continuous.

COAST STATIONS

ZBM, Bermuda Harbour Radio, St. Georges. Stands by on VHF channel 16 and 2182 kHz.
VRT, Bermuda Radio, public correspondence. VHF 16, switch to VHF channels 26 and 28.

GENERAL BROADCAST STATIONS

ZBM, 1340 kHz 1kW, FM 89.1MHz.
ZFB, 1230 kHz 1kW, FM 94.9MHz.
VSB, 1160 kHz, 1280 kHz, 1450 kHz 1kW, FM 106.2MHz.

FLORIDA, ATLANTIC COAST

NOAA WEATHER

KHB-39, Jacksonville. WX-1.
KIH-26, Daytona Beach. WX-2.
WXJ-70, Melbourne. WX-1.
KEC-50, West Palm Beach. WX-3.
KHB-34, Miami. WX-1.
WXJ-95, Key West. WX-2.

U.S. COAST GUARD STATIONS

NMA-10, Mayport. 1215 and 2215 UTC on VHF channel 22A. 0620 and 1820 UTC on 2670.0 kHz.

NCF, Miami Beach. 1230 and 2230 UTC on VHF channel 22A. 0350 and 1550 UTC on 2670.0 kHz.

NOK, Key West. 1200 and 2200 UTC on VHF channel 22A.

COAST STATIONS

WOM*. AT&T Coast Station. Weather at 1300 and 2300 UTC on 4363.0, 8722.0, 13092.0, 17242.0, and 22738.0 kHz (channels 403, 802, 1206, 1601, 2215). Traffic on the same channels on the hour, every odd hour.

BAHAMAS

WEATHER

C6N, Nassau. VHF channel 27 every even hour.

Radio Bahamas*. 810 kHz 1kW, 1240 kHz 1kW, 1540 kHz 20kW, FM 107.1 MHz, FM 107.9 MHz. Weather at 1230 and 2330 UTC and with the news on the hour.

Radio Abaco FM 93.5 broadcasts weather for cruisers in the Abacos at 0900, 1300, and toward the end of the 6 p.m. evening news (at approximately 1825). **Radio Abaco** also broadcasts local weather (Green Turtle Cay to Cherokee Sound) over VHF channel 68 at 0815 daily. Also there is a 24-hour recorded weather at 242-367-4531.

COAST STATIONS

Numerous coast stations monitor VHF channel 16 in the Bahamas. Most marinas stand by, and there is usually a vessel within radio range willing to assist. A partial listing only of some widely used stations follows.

Call sign: BASRA*. Bahamas Air Sea Rescue Association. Contact on VHF channel 16 or 2182 kHz.

Call sign: Jack Tar. West End Marina. Contact on VHF channel 16.

Call sign: Walker's Cay. Walker's Cay Marina. Contact on VHF channel 16.

C6X2, Marsh Harbour. Public correspondence. Stands by on 2182 kHz.

Call sign: Treasure Cay. Treasure Cay Marina. Contact on VHF channel 16.

Call sign: ZFP-81. Freeport Harbour Control. VHF channel 16 or 2182 kHz.

Call sign: Bortow Pilots. Freeport pilots. Contact on VHF channels 6, 10, 13, 14, and 16.

Call sign: Borco Marine. Bahamas Oil Refining Company, Freeport. Contact on VHF channel 16.

Call sign: Cat Cay Club. Cat Cay, Bahamas. Contact on VHF channel 16.

Call sign: Chub Cay. Chub Cay Marina. Contact on VHF channel 16.

Call sign: Great Harbour Cay Marina. Great Harbour, Berry Islands. Contact on VHF channel 16.

Call sign: Nassau Harbour Control. Contact on VHF channels 6, 10, 13, 14 and 16. Stands by on 2182 kHz.

C6N3, Nassau. Public correspondence. VHF channel 27.

C6N2, Nassau. Public correspondence. Stands by on 2182 kHz.

Call sign: Wee Watin. Highborne Cay, Exumas. Contact on VHF channel 16.

Call sign: Sampson. Sampson Cay, Exumas. Contact on VHF channel 16.

Call sign: Stella Maris. Stella Maris Marina, Long Island. Contact on VHF channel 16.

Call sign: Morton Salt. commercial wharf on Great Inagua. Contact on VHF channel 16 or 2182 kHz.

TURKS AND CAICOS

Caicos Pilots. Contact on VHF channel 16.

VSI, Grand Turk. Public correspondence. VHF channel 16 or 2182 kHz.

Cockburn Town Pilots. Grand Turk. Contact on VHF channel 16.

Radio Turks and Caicos. 1460 kHz 2.5kW.

Atlantic Beacon. 1570 kHz 50kW.

Coral Radio. 89.3 MHz, 89.9 MHz, 90.5 MHz, FM 92.5 MHz.

WPRT Radio. FM 88.7 MHz.

CUBA

WEATHER

CLT, CLA, Habana. 0105, 1305, 2005, and 2205 UTC on 2760 kHz. Broadcasts in Spanish.

CLX, Casa Blanca. 1700, 1945, and 2200 on 6995 kHz. Broadcasts in Spanish.

COAST STATIONS

Call sign: Mariel Practice. Mariel pilots. Contact on VHF channel 16.

CLT, CLA, Habana. Local navigational warnings in Spanish on 2760 kHz at 2310 UTC. Public correspondence. Stands by on 2182 kHz.

Call sign: Habana Practicos. Habana pilots. Contact on VHF channels 13 and 16.

Call sign: Morro Habana. Habana port signal station. Contact on VHF channels 13, 16, and 68.
Call sign: Habana Capitonia. Habana Port Captain. Contact on channels 16 and 68.
Call sign: Terminales Contenedores. Container terminal. Contact on VHF channels 16 and 74.
CLX, Casa Blanca. Stands by on 2182 kHz.
CLW, Cardenas. Public correspondence. Stands by on 2182 kHz.
CLC2, Isabela de Sagua. Public correspondence. Stands by on 2182 kHz.
Isabela de Sagua Pilots. Contact on VHF channel 16.
Call sign: CLG-50 Caiman. Old Bahama Channel Traffic Control. Contact on VHF channel 13.
Call sign: CLG-60 Confites. Old Bahama Channel Traffic Control. Contact on VHF channel 13.
CLC3, Caibarien. Public correspondence. Stands by on 2182 kHz.
CLK, Nuevitas. Local navigational warnings in Spanish on 2760 kHz at 2320 UTC. Public correspondence. Stands by on 2182 kHz.
CLM4, Guardalabarca. Public correspondence. Stands by on 2182 kHz.
CLM3, Baracoa. Public correspondence. Stands by on 2182 kHz.
Call sign: Guantanamo Port Control. Contact on VHF channel 12.
Guantanamo Bay Pilots. Contact on VHF channel 74.
CLM, Santiago de Cuba. Local navigational warnings in Spanish on 2760 kHz at 2340 UTC. Stands by on 2182 kHz.
Call sign: Santiago Practicos. Santiago de Cuba pilots. Contact on VHF channels 13 and 16.
Call sign: Castilda Practicos. Golfo de Guancanayabo pilots. Contact on VHF channels 13 and 16.
CLK2, Santa Cruz del Sur. Public correspondence. Stands by on 2182 kHz.
CLC, Cienfuegos. Local navigational warnings in Spanish on 2760 kHz at 2300 UTC. Public correspondence. Stands by on 2182 kHz.
CLT2, Batabano. Public correspondence. Stands by on 2182 kHz.
CLT3, Nueva Gerona. Public correspondence. Stands by on 2182 kHz.
CLF3, El Morrillo. Public correspondence. Stands by on 2182 kHz.
CLF-2, Arroyos de Mantua. Local navigational warnings in Spanish on 2760 kHz at 2350 UTC. Stands by on 2182 kHz.

GENERAL BROADCASTS

There are dozens of Spanish language broadcast stations, on all segments of the dial.
Radio Taíno. 830 kHz 300kW, 1160 kHz 300kW. English 0100–0200, 0300–0430, and 2000–2200 UTC.
Radio Habana. Shortwave. English 0000–0400 UTC on 6000 kHz, 0100–0500 on 9830 USB, 0400–0500 UTC on 6180 kHz, and 0500–0700 UTC on 9820 kHz, 2100–2200 UTC on 11705 kHz (summer) and 11720 (winter), and 2200–2300 on 6180 kHz and 11960 USB.
AFRTS. Armed Forces Radio, Guantánamo Bay, 1340 kHz 0.25kW, FM 102 MHz, FM 103 MHz.

CAYMAN ISLANDS

Call sign: Cayman Harbor. Cayman Brac Agent. Contact on VHF channel 16.
Call sign: Cayman Energy. Cayman Brac operations control. Contact on VHF channel 16.
Call sign: Bonito. Cayman Brac pilots. Contact on VHF channel 16.
Radio Cayman. 1205 kHz 1kW, 1555 kHz 10kW, FM 89.9 MHz, FM 91.9 MHz, FM 105.3 MHz.

JAMAICA

WEATHER

6YX*, Jamaica Coast Guard. 0130, 1430, and 1900 UTC on VHF channel 13. 1330 and 1830 UTC on 2738 kHz.

COAST STATIONS

6YX, Kingston, Jamaica Coast Guard. Stands by on 2182 kHz.
6YI, Kingston. Public correspondence. Stands by on 2182 kHz.
Kingston Pilots. Contact on VHF channels 11 and 16.
Call sign: Silver Spray. Port Esquivel pilots. Contact on VHF channel 16.
Port Antonio Pilots. Contact on VHF channel 16.
Discovery Bay Pilots. Contact on VHF channel 16.
Montego Bay Pilots. Contact on VHF channel 16.

GENERAL BROADCASTS

Jamaica Broadcasting Corporation. 560 kHz 5kW, 620 kHz 5kW, 700 kHz 10kW, 750 kHz 10kW, 850 kHz 10kW, 1090 kHz 1kW, FM 91.1 MHz, FM 92.1 MHz, FM 93.3 MHz, FM 97.1 MHz, FM 97.3 MHz, FM 98.7 MHz, FM 99.7 MHz, FM 100.3 MHz, FM 103.9 MHz,

FM 105.7 MHz. Weather and fishermen's forecast weekdays at 2248 UTC.
Radio Jamaica. 550 kHz 5kW, 580 kHz 10kW, 720 kHz 10kW, 770 kHz 5kW, FM 90.5 MHz, FM 91.5 MHz, FM 92.7 MHz, FM 92.9 MHz, FM 94.5 MHz, FM 95.7 MHz, FM 98.1 MHz, FM 101.3 MHz, FM 104.5 MHz.
Island Broadcasting Services. KLAS, FM 89.3 MHz.

HAITI

COAST STATIONS

MF, Port-au-Prince. Public correspondence. Stands by on VHF channel 16 and 2182 kHz.
Port-au-Prince Pilots. Contact on VHF channel 16.
Cap-Haïtien Pilots. Stands by on VHF channel 16.

GENERAL BROADCASTS

Radio 4VEH, 1030 kHz 10kW. Cap-Haïtien. News and weather at 1200 UTC. English 1100–1400 and 2300–2400 UTC.

DOMINICAN REPUBLIC

HIA, Santo Domingo Piloto. Public correspondence. Stands by on VHF channel 16 or on 2182 kHz.
Call sign: HIW 19. San Pedro de Macoris pilots. Stands by on VHF channel 16 and 2182 kHz.
Call sign: HIW 20. Puerto de Haina pilots. Stands by on VHF channel 16 or on 2182 kHz.
Call sign: HIW 9. Romano pilots. Stands by on VHF channel 16.
Call sign: HIW 8. Puerto Plata pilots. Stands by on VHF channel 16.

There are dozens of Spanish-language broadcast stations on all segments of the dial.

PUERTO RICO

NOAA WEATHER

WXJ-69, Maricao. WX-1.
WXJ-68*, San Juan. WX-2.

U.S. COAST GUARD STATION

NMR-1*, Greater Antilles Section, San Juan. 1210 and 2210 UTC on VHF channel 22A. 0030 and 1430 UTC on 2670.0 kHz.

COAST STATIONS

KRV, Ponce Playa. Public correspondence. VHF channel 28.
WHU 645, Luquillo. Public correspondence. VHF channel 86.
WCU 243, Maricao. Public correspondence. VHF channel 27.
WCT, Saturce. Public correspondence. VHF channel 26.
KMD 214, San Juan. Public correspondence. VHF channel 28.

GENERAL BROADCASTS

There are dozens of Spanish-language broadcast stations on all segments of the dial. This is a partial list of English broadcasts:
WOSO*, San Juan. 1030 kHz 10kW, All-English news and talk.
WBMJ, San Juan. 1190 kHz, 10kW, 0930–0300 UTC, English and Spanish religious.
WIVV, Vieques. 1370 kHz, 5kW, 0925–0230 UTC, English and Spanish religious.
AFRTS, Armed Forces Radio, U.S. Air Force. 780 kHz 0.05kW Aguadilla, 1040 kHz 0.05kW San Juan, 1200 kHz 0.05kW Roosevelt Roads, 1460 kHz.

U.S. VIRGIN ISLANDS

WEATHER

WXM-96*, St. Thomas. WX-3.
WAH*, Virgin Islands Radio, St. Thomas. Detailed offshore and Caribbean weather is announced on 2182 kHz and VHF channel 16 and broadcast on ITU channel 401 and VHF channel 25, 85, or 87 at 0600, 1400, and 2200 local time. Local VI and Puerto Rico marine weather is announced on VHF channel 16 and broadcast on VHF channel 25, 85, or 87 at 0800 and 2000 local time.

COAST STATION

WAH*. Virgin Islands Radio, Public correspondence, St. Thomas. Stands by on 2182 kHz and VHF channel 16; switches to VHF channels 25, 85, or 87; also ITU channel 401. See also "High Seas Radiotelephone Service" section later in this chapter.

GENERAL BROADCASTS

WSTX. 970 kHz 5kW, FM 100.3 MHz.
WVWI. 1000 kHz 5kW.
WGOD. 1090 kHz 0.25kW, FM 97.9 MHz.
WRRA. 1290 kHz 0.5kW.
WSTA. 1340 kHz 1kW.
WIUJ. FM 88.9 MHz.
WDCM. FM 92.3 MHz.
WAVI. FM 93.5 MHz.
WIVI. FM 99.5 MHz.

BRITISH VIRGIN ISLANDS

WEATHER

Tortola Radio. VHF channel 27 at 1300 and 1600 UTC.

COAST STATIONS

Tortola Radio. Contact on VHF channel 16.
Tortola Pilots. Contact on VHF channel 16.

GENERAL BROADCASTS

ZBVI. 780 kHz 10kW. Tortola.
Caribbean Broadcasting System. FM 91.7 MHz (Z Gold), FM 94.3 MHz (The Heat), FM 103.7 (ZROD).

ANGUILLA

Radio Anguilla. 1505 kHz 1kW.
ZJF-FM. 105 MHz.
The Caribbean Beacon. 690 kHz 15kW, 1610 kHz 50kW, FM 100.1 MHz.

ST. MARTIN

Radiodiffusion Française D'Outre-Mer (RFO). FM 99.2 MHz. Broadcasts in French.
Radio Caraibes International. FM 104.7 MHz. Broadcasts in French.
Radio St. Martin. FM 95.3 MHz. Broadcasts in French and English.
Radio Voix Chretiennes de St. Martin. FM 106 MHz. Broadcasts in French and English.

ST. MAARTEN

WEATHER

WX-1. Philipsburg.

COAST STATION

Philipsburg Harbor Master. Contact on VHF channel 16.

SABA

WEATHER

PJS*, Saba Radio. Currently in state of transition. Please read entry below.

COAST STATION

PJS*, Saba Radio. Public correspondence. This used to be the main communications service in the Leeward Islands, with a 200-mile range on VHF (its antenna is 3,000 feet above sea level). At *Reed's* press time (summer, 1997) the Saba Radio office was shut down but calls were being handled by Curaçao Radio via remote hookup. When vessels call Saba Radio, someone in Curaçao will answer. Only VHF channel 16 is being monitored, but it is watched 24 hours a day. Traffic lists are no longer broadcast. The only transmission from Curaçao Radio is weather broadcasts on VHF channel 26 at 1505 UTC (Saba, Sint Maarten, & Statia) and at 1135 UTC (Aruba, Bonaire, & Curaçao). The telephone for Curaçao Radio is 599-736-4069; fax, 599-736-4871
Reportedly, the Antillean Sea Rescue Foundation plans to establish a new station on Sint Maarten, which will replace many of the functions of Saba Radio.

ST. EUSTATIUS

Oranjestad Pilots. Call on VHF channel 16.

ST. KITTS AND NEVIS

Basseterre Pilots and Port Authorities. Contact on VHF channel 16.
Radio ZIZ. 555 kHz 20kW, FM 90 MHz.
Radio Paradise. 825 kHz 50kW.
Voice of Nevis. 895 kHz 10kW.

MONTSERRAT

Radio Antill0es*. 930 kHz 20kw. Marine weather forecasts at 1230 and 2225 UTC. Radio Antilles is a good source of weather information for the Eastern Caribbean.

ANTIGUA AND BARBUDA

COAST STATIONS

English Harbour Radio 2VMA*, Nicholson Yacht Charters (268-460-1530). Contact on VHF channel 68 or on 8294.0 kHz Monday through Friday 0830 to 1630 local time, Saturday 0830 until 1200 (no Saturday broadcast June 1–October 31).
English Harbour Port Authority. Contact on VHF channel 16.
Saint John's Harbour Authorities. Contact on VHF channel 16 or 2182 kHz.
Call sign: Marine Center. Saint John's oil terminal. Contact on VHF channel 16 or 2182 kHz.

GENERAL BROADCASTS

Antigua and Barbuda Broadcast Service. 620 kHz 10Kw.

Radio ZDK. 1100 kHz 10kW, FM 99.0 MHz.
Caribbean Radio Lighthouse. 1165 kHz 10kW, FM 90.0 MHz.
BBC Relay Station. 5975 kHz, 6110 kHz, 6195 kHz, 9640 kHz, 15205 kHz, 15220 kHz.
VOA Relay Station. 1580 kHz 50kW

GUADELOUPE

Pointe-à-Pitre Pilots. Contact on VHF channel 16.
Basse-Terre Port Authority. Contact on VHF channel 16.
Radiodiffusion Française D'Outre-Mer (RFO). 640 kHz 40kW, 1420 kHz 5kW, FM 97.0 MHz. Broadcasts in French.

DOMINICA

Roseau Pilots. Contact on VHF channel 16.
Dominica Broadcasting Corporation. 595 kHz 10kW, FM 88.1 MHz.
Gospel Broadcasting Corporation. ZGBC, 740 kHz 10kW, FM 102.1 MHz.

MARTINIQUE

WEATHER

FFP, Fort-de-France. Every hour +30 min. 1100–2300 (July 1–Oct 31), 1130, and 2230 on VHF channel 26. Every hour +30 min. 2300–1100 (July 1–Oct 31), 0030, and 1500 on VHF channel 11. Every hour +30 min. 1100–2300 (July 1–Oct 31), 1333, and 2215 UTC on 2545 kHz.

COAST STATIONS

FFP, Fort-de-France. Location navigational warnings in French on 2545 kHz at 1133 and 1733. Public correspondence. Stands by on 2182 kHz.
Fort-de-France Port Authorities. Contact on VHF channels 13 and 16.

GENERAL BROADCASTS

Société Nationale de Radio-Télévision Française d'Outre Mer. 1310 kHz 20kW, FM 92 MHz, FM 95 MHz. Broadcasts in French.
Radio Caraibes International. 1090 kHz 20kW, FM 89.9 MHz, FM 92.5 MHz, FM 98.7 MHz, FM 106.1 MHz. Broadcasts in French.

ST. LUCIA

COAST STATIONS

Call sign: Vieux Fort Lighthouse. Vieux Fort port officials. Contact on VHF channel 16 or 2182 kHz.
Call sign: Hess St. Lucia. Oil terminal in Grande Cul de Sac Bay. Contact on VHF channel 16.
Call sign: Castries Lighthouse. Port Castries port authorities. Contact on VHF channel 16 or 2182 kHz.

GENERAL BROADCASTS

Radio St. Lucia, 660 kHz 10kW, FM 97.3 MHz, 99.5 MHz, 107.3 MHz. In English and Creole.
Radio Caribbean International. 840 kHz 20kW, FM 95.5 MHz, FM 99.1 MHz, FM 101.1 MHz. In English and Creole.

ST. VINCENT AND THE GRENADINES

Call sign: ZQS. Kingstown port authorities. Contact on VHF channel 16 and 2182 kHz.
St. Vincent and the Grenadines National Broadcasting Corporation. 705 kHz 10kW.

GRENADA

Call sign: J3YA. St. George Port Control. Contact on VHF channel 16.
Call sign: J3YB, St. George pilots. Contact on VHF channel 16.
Radio Grenada. 535 kHz 20kW.

BARBADOS

COAST STATIONS

8PA. Bridgetown. Port Signal Station. Stands by on VHF channels 12 and 16. Monitors 2182 kHz.
8PO, Bridgetown. Public correspondence. Stands by on 2182 kHz.

GENERAL BROADCASTS

Caribbean Broadcasting. 900 kHz 10kW, FM 98.1 MHz.
Radio Liberty. FM 98.1 MHz.
Barbados Rediffusion. 790 kHz 20kW, FM 104.1 MHz.
Voice of Barbados. 790 kHz.
Yess Ten Four. FM 104.1 MHz.
Barbados Broadcasting. FM 90.7 MHz, FM 102.1 MHz.

TRINIDAD AND TOBAGO

WEATHER

9YL, North Post. 1340 and 2040 UTC on VHF channels 24, 25, 26, and 27. 1250 and 1850 UTC on 2735 kHz and 3165 kHz.

R
Resources

COAST STATION

9YL, North Post. Public correspondence. Stands by on 2182 kHz.

GENERAL BROADCASTS

National Broadcasting Service. 610 kHz 50kW, FM 98.9 MHz, 100.0 MHz.

Trinidad Broadcasting Company. 730 kHz 20kW, FM 91.1 MHz, FM 95.1 MHz, FM 105.1 MHz.

VENEZUELA

YVG, La Guaira. Public correspondence. Stands by on 2182 kHz.

Puerto Cabello Pilots. Monitor VHF channel 16.

Puerto El Guamache, Isla de Margarita. Contact on VHF channel 16.

Call sign: Meneven Puerto La Cruz. Puerto La Cruz pilots. Contact on VHF channel 16.

Call sign: San Carlos Pilot. Pilots for Lago de Maracaibo. Contact on VHF channel 16.

Puerto Cabello Pilots. Contact on VHF channel 16.

There are hundreds of Spanish-language general broadcast stations on all segments of the dial.

BONAIRE

WEATHER

Trans World Radio, Bonaire*. 800 kHz 500 kW. News and weather in English 1100 UTC, and 0145 UTC.

COAST STATIONS

Pilots. Contact Curaçao Coast Radio PJC.

Goto Oil Terminal. Contact on VHF channel 16.

GENERAL BROADCASTS

Trans World Radio. See above.

CURAÇAO

WEATHER

PJC, Curaçao Coast Radio. Weather broadcasts on VHF channel 26 at 1505 UTC (Saba, Sint Maarten, & Statia) and at 1135 UTC (Aruba, Bonaire, & Curaçao).

COAST STATIONS

Call sign: Fort Nassau. Port Operations, VHF channels 12 and 16.

PJC, Curaçao Coast Radio. Public correspondence. Stands by on 2182 kHz and VHF channel 16 24 hours a day. The telephone for Curaçao Radio is 599-736-4069; fax, 599-736-4871.

GENERAL BROADCASTS

PJC-9. 1500 kHz 3kW, FM 105.1 MHz. Dutch and some English.

PJL-3. 1100 kHz 0.25kW. Voice of America News at 1800 UTC. Sunday religious broadcast in English 1200–1400 and 2200–2300 UTC

Radio Korsou. FM 93.9 MHz, FM 101.1 MHz. English Tuesdays at 0000 UTC.

ARUBA

Sint Nicolaas Baai Pilots. Contact on VHF channels 14 and 16.

Oranjestad Port Radio. Contact on VHF channel 16.

Radio Victoria. 960 kHz 10kW, FM 93.1 MHz. Dutch.

Radio 1270. 1270 kHz 1.5kW. Dutch.

Voice of Aruba. 1320 kHz 1kW, FM 89.9 MHz. Dutch.

Radio Kelkboom. 1440 kHz 1kW, FM 106.7 MHz. Dutch.

Radio Caruso Booy. FM 97.9 MHz. Dutch.

Radio Carina. FM 103.5 MHz. Dutch.

COLOMBIA

HKB, Barranquilla. Public correspondence. Stands by on 2182 kHz.

Call sign: Puerto Bolivar. Port authorities. Contact on VHF channels 6 and 16.

Cartagena Pilots. Contact on VHF channels 11 and 16.

There are hundreds of Spanish-language general broadcast stations on all segments of the dial.

PANAMA

Puerto de Bahia de las Minas Pilots. Contact on VHF channel 16.

Puerto Cristobal Signal Station. Contact on VHF channel 12.

HPN, Canal Radio Station, Balboa. All vessels must communicate with the Port Captain through HPN. Other than emergency traffic and routine bridge-to-bridge VHF communications, no vessels in transit through the Canal shall communicate with any other station, local or distant. Contact on VHF channel 16 or 2182 kHz.

HPP. Public correspondence. Stands by on 2182 kHz.

Call sign: Rambala Control. Oil terminal in Laguna de Chiriqui. Contact on VHF channel 16.
Call sign: Chiriqui Grande Pilots. Laguna de Chiriqui. Contact on VHF channel 16.
ACA20. 790 kHz 10kW. AFRTS Armed Forces Radio, Southern Command Network, Ft. Clayton, Canal Zone.

There are dozens of Spanish-language general broadcast stations on all segments of the dial.

COSTA RICA

TIM, Limón. Public correspondence. Stands by on 2182 kHz.
Limón Pilots. Contact on VHF channel 16.

GENERAL BROADCASTS

There are dozens of Spanish-language general broadcast stations on all segments of the dial.
AWR. Shortwave 5030 kHz, 5970 kHz, 6150 kHz, 9725 kHz, 11870 kHz, 13750 kHz and 15460 kHz 50kW. English 1100–1300 UTC and 2300–0100 UTC. Saturday at 1230 UTC

NICARAGUA

Bluefields Pilots. Contact on VHF channel 16.

There are dozens of Spanish-language general broadcast stations on all segments of the dial.

HONDURAS

Puerto Cortes Pilots. Contact on VHF channels 6 and 16

GENERAL BROADCASTS

There are dozens of Spanish-language general broadcast stations on all segments of the dial.
La Voz Evangélica. 810 kHz 1kW, 1310 kHz 1kW, 1390 kHz 10kW. Shortwave 4820.2 kHz. English from 0300 to 0500 Monday. Religious.

GUATEMALA

GENERAL BROADCASTS

There are dozens of Spanish-language general broadcast stations on all segments of the dial.
Radio Cultural. 730 kHz 10kW. English 0300–0430 UTC and Sundays 2345–0430 UTC.
Unión Radio. 1330 kHz 10 kW. English 0200–0400 UTC.

BELIZE

WEATHER

Radio Belize*. 830 kHz 10kW, 910 kHz 1kW, 930 kHz 1kW, 940 kHz 1kW, FM 88.9 MHz, FM 91.1 MHz. News and weather in English at 0100, 0300, 1300, 1500, 1700, 1830, 2100, and 2300 UTC.

COAST STATIONS

Belize Pilot Station. Stands by on 2182 kHz.
Belize Customs Control. Stands by on 2750 kHz.
Belize City Pilots. Contact on VHF channel 16.

GENERAL BROADCASTS

Radio Belize. See above.
British Forces Broadcast Service. FM 93.1 MHz, FM 99.1 MHz.
VOA Relay Station. 1530 kHz 50kW, 1580 kHz 50kW.

MEXICO

COAST STATIONS

XFP, Chetumal. Public correspondence. Stands by on 2182 kHz.
San Miguel de Cozumel Pilot. Contact on VHF channel 16.
XFC, Cozumel. Public correspondence. Stands by on 2182 kHz.
Cayo Arcas Terminal. Contact on VHF channel 09.
XFN, Progreso. Public correspondence. Stands by on 2182 kHz.
XFF, Coatzacoalcos. Public correspondence. Stands by on 2182 kHz.
XFU, Veracruz. Public correspondence. Stands by on 2182 kHz.
XFS,Tampico. Public correspondence. Stands by on 2182 kHz.

There are hundreds of Spanish-language general broadcast stations on all segments of the dial.

OFFSHORE RADIO RESOURCES

Vessels venturing more than about 100 miles offshore usually use SSB and/or shortwave radio for weather and safety information, and for communications. Reed's lists a number of important SSB/shortwave resources in this section. See the following sections for information about SSB phone and data services, as well as satellite communications.

U.S. COAST GUARD, PORTSMOUTH, VA (NMN)

The U.S. Coast Guard, Camslant Chesapeake (Portsmouth), VA (call sign NMN), provides the most complete offshore weather information available in this region. Broadcasts include forecasts and a synopsis for New England, the west central North Atlantic, the southwest North Atlantic (including the Bahamas), the Gulf of Mexico, and the Caribbean. Note that some of the earlier broadcast times have changed in 1997, and NMN unfortunately no longer broadcasts Gulf Stream information.

0330, 0500, 0930 UTC on 4426.0, 6501.0, and 8764.0 kHz.

1130, 1600, 2200, 2330 UTC on 6501.0, 8764.0, and 13089.0 kHz.

1730 UTC on 8764.0, 13089.0, and 17314.0 kHz.

0203 and 1333 UTC on 2679 kHz.

0230 and 1120 UTC on VHF channel 22A.

These forecasts are delivered by a droning synthesized voice known as "Perfect Paul". The times listed above are the beginning of the broadcast cycle. Note that the various area forecasts are also repeated on a regular basis. For instance, New England Offshore weather is given at 1000, 1600, and 2200 UTC. By noting the times of the particular section you want, you can avoid waiting the next time. A very useful tool for getting the most out of these forecasts is a plastic weather area map used with dry markers.

HAM RADIO NETS

Amateur or "Ham" radio can be an important communications asset for offshore boaters. There are a number of marine-oriented "Nets" providing weather and safety information to all listeners, and communications to licensed users.

Of particular note is the Waterway Net which is reported to have excellent weather reports for the waters south of Norfolk,VA; accurate hurricane reporting; and an extensive group of boaters keeping in touch and sharing information. The Maritime Mobile Service Net provides licensed operators with inexpensive phone connections. These calls will not be private, and business communications are not permitted.

If you are interested in becoming a licensed Ham operator, it is reported that Radio Shack offers good introductory materials.

Below is a list of nets that have been active at one time. Those in boldface type are known to be active at press time. *Reed's* welcomes current information on nets for this section.

UTC	Frequency MHz	Net Name, Area, Type of Information
0100	3.935	Gulf Coast Hurricane (Gulf, WX/TFC)
0200	14.334	Brazil, East Coast (East Coast, Atlantic, WX/TFC)
0630	14.313	International Maritime Mobile (Atlantic, Mediterranean, Caribbean, TFC)
1030	3.815	Caribbean Weather (Caribbean, WX/TFC)
1100*	**7.240**	**Caribbean Maritime Mobile** (Caribbean, TFC)
1100	14.283	Caribus Traffic (East Coast, Caribbean, TFC)
1110	3.930	Puerto Rico Weather (Puerto Rico, Virgin Islands, WX)
1215	4.003	Caribbean SSB Weather (Caribbean WX)
1220*	**3.696**	**Bahamas Weather Net** * One hour earlier during Daylight Saving Time.
1230	7.185	Barbados Info (Caribbean, TFC)
1230	8.104	Caribbean SSB Weather (Caribbean, WX)
1245*	**7.268**	**East Coast Waterway** (East Coast, Caribbean, TFC, WX).
1300	21.400	TransAtlantic (Atlantic, Mediterranean, Caribbean, TFC)
1400	7.292	Florida Coast (Florida, TFC)
1700	**14.300**	**Maritime Mobile Service (Worldwide)**
1800	14.303	United Kingdom Maritime (Atlantic, Med., TFC)

2215	8.107	Caribbean SSB Weather (Caribbean, hurricane season only, WX)
2230	3.815	Caribbean Weather (Caribbean, WX/TFC)
2310	3.930	Puerto Rico Weather (Puerto Rico, Virgin Islands, WX)
2400	14.313	Maritime Mobile Service (Worldwide, TFC)
as nec.	**14.325**	**Hurricane** (Atlantic, Caribbean, Pacific, emergency, WX) Also uses 14.175 and 14.275 as needed

* One hour earlier during Daylight Saving Time
WX = Weather information
TFC = Traffic

HERB HILGENBERG SOUTHBOUND TWO

Herb Hilgenberg is an amateur weather forcaster who operates a very popular SSB weather net for offshore sailors. For many years Herb was based in Bermuda, but he now broadcasts from Lake Ontario, Canada. Herb's station is VAX-498 and he begins his broadcasts at 4pm Eastern Daylight Time (2000 UTC) daily on 12.359MHz upper sideband. Herb moves in a geographic pattern around the Atlantic, receiving weather condition reports from vessels and dispensing individual forecasts to them. If you have a ship's license to transmit, you may join the net; but listen first to learn Herb's protocol. Before your voyage, you may wish to contact Herb by fax at 905-681-7114.

WWV

In addition to broadcasting time signals, radio station WWV, Fort Collins, Colorado, broadcasts marine storm warnings for the North Atlantic Ocean and GPS status and outage reports. These voice announcements follow the regular time announcements; storm warnings are made at 8 through 10 minutes after the hour (an additional segment, at 11 minutes, may be used when there are unusually widespread storm conditions) and GPS reports at 14 to 15 minutes after the hour. Frequencies are 2500, 5000, 10000, 15000, and 20000 kHz. See the section "Time Signals" later in this chapter for more information on station WWV.

The National Weather Service issues warning updates at 0500, 1100, 1700, and 2300 UTC. The updates are announced at the next broadcast following time of issue. Topics include locations of storm centers, low-pressure systems, and hurricanes and their predicted positions for the next 12 to 24 hours.

VOICE OF AMERICA (VOA)

Voice of America
330 Independence Avenue SW
Washington, DC 20547
202-619-2538

Reception varies by the season of the year and the time of day; try all frequencies to find the best one. VOA broadcasts to Africa are also often heard in the Caribbean. Key frequencies for English-language broadcasts are listed below.

Directed to the Caribbean:
930 kHz (via Radio Antilles, Montserrat); 5995 kHz; 6130 kHz; 6165 kHz; 7405 kHz; 9455 kHz; 9590 kHz; 9775 kHz; 11695 kHz; 13740 kHz.

BBC WORLD SERVICE

British Broadcasting Corporation
World Service
International Audience Correspondence
Bush House, Strand
London, WC2B 4PH, UK
tel., 44-171-257-2875; subscriptions tel., 44-171-257-2211; fax, 44-171-257-8254; e-mail, worldservice.letters@bbc.co.uk

Reception varies by the season of the year and the time of day; try all frequencies to find the best one. Lower frequencies are generally best at night. Key frequencies for English-language broadcasts are listed below.

Directed to east coast of Canada and U.S.:
5965 kHz; 5975 kHz; 6175 kHz; 6195 kHz; 7325 kHz; 9515 kHz; 9590 kHz; 11775 kHz; 11865 kHz.

Directed to the Caribbean basin and Central America:
930 kHz; 5975 kHz; 6175 kHz; 6195 kHz; 9590 kHz; 9895 kHz; 15220 kHz; 15390 kHz; 17715 kHz; 17840 kHz.

NAVTEX

NAVTEX is an internationally coordinated method of broadcasting distress, urgent, and safety messages and weather forecasts and warnings using small, low-cost printing receivers designed to be installed in the pilothouse of a vessel. A series of coast stations transmit NAVTEX radio teletype safety messages on medium frequency 518 kHz during preset time slots. Routine messages are normally broadcast four to six times daily. Urgent messages are broadcast on receipt, provided that an adjacent station is not transmitting. The coverage of NAVTEX is reasonably continuous out to 200 nautical miles from the transmitting station. Interference from or receipt of stations farther away occasionally occurs at night. In the U.S., the Coast Guard is the responsible agency for NAVTEX operation.

To receive these broadcasts of notices and marine weather mariners need a special NAVTEX receiver. Mariners who do not have NAVTEX receivers but have SITOR radio equipment can also receive these broadcasts by operating it in the FEC mode and tuning to 518 kHz. There are techniques to receive NAVTEX with a SSB or shortwave radio and a computer.

Sydney, NS, VCO, Identifier Q (English), 0240, 0640, 1040, 1440, 1840, and 2240 UTC; identifier S (French), 0255, 0655, 1055, 1455, 1855, and 2255 UTC.

Yarmouth, NS, VAU, Identifier U (English), 0320, 0720, 1120, 1520, 1920 and 2320 UTC; identifier V (French), 0335, 0735, 1555, 1955, and 2355 UTC.

Portsmouth, VA, NMN, Identifier N. 0130, 0530, 0930, 1330, 1730, and 2130 UTC.

Miami, FL, NMA, Identifier A. 0000, 0400, 0800, 1200, 1600, and 2000 UTC.

New Orleans, LA, NMG, Identifier G. 0300, 0700, 1100, 1500, 1900, and 2300 UTC.

San Juan, PR, NMR, Identifier R. 0200, 0600, 1000, 1400, 1800, and 2200 UTC.

St. George's, Bermuda (Bermuda Harbour Radio ZBM), Identifier B. 0100, 0500, 0900, 1300, 1700, and 2100 UTC.

WEATHER FACSIMILE BROADCASTS

Notes:

Select a frequency 1.9 kHz below those listed above when using a single sideband radio to receive these broadcasts.

Contrary to rumors circulating over the last few years, neither the U.S. Coast Guard nor the U.S. Navy have plans to discontinue weather facsimile broadcasts in the near future. Despite budget constraints, the U.S. Coast Guard plans to continue its radiofacsimile broadcasts to meet both the needs of its cutters and public safety needs. The U.S. Navy has delayed suspending its broadcasts of marine weather information until at least 1998 because of delays in installing equipment in its ships. Navy radio weather facsimile broadcasts from Cutler, ME, will cease once Navy ships have been fitted with new satellite equipment.

U.S. NAVY CUTLER, ME (NAM)

Broadcast times and frequencies: 0000–1200 UTC on 3357 kHz; and 1200–0000 UTC on 10865 kHz.

Time UTC		Subject
0015		Fax schedule 1, (0000Z–1200Z)
0345		Fax schedule 2, (1200Z–0000Z)
	1215	Fax schedule 1, (0000Z–1200Z)
	1545	Fax schedule 2, (1200Z–0000Z)

Maximum area covered: 0N–80N; 110W–30E

U.S. COAST GUARD MARSHFIELD, MA (NMF)

Broadcast frequencies: 6340.5, 9110.0, and 12750.0 kHz.

Times, all frequencies: 0230–0438, 0800–1028, 1430–1551, 1720–1759, and 1900–2228 UTC. Continuously on 6340.5 and 12750.0 kHz

Time UTC		Subject
0230	1420	Test pattern (start of broadcast)
0233/1433		Preliminary surface analysis, Area 1
0243		Fax schedule, Part 1
0254		Fax schedule, Part 2
0305		Request for comments
0315		Product notice bulletin
	1443	96-hour 500 mb forecast, Area 4
	1453	96-hour surface forecast, Area 4
	1503	Satellite image, Area 5
	1515	Sea state analysis (meters), Area 4
0325	1525	Surface analysis (E. Atlantic), Area 2

Time	UTC	Subject
0338	1538	Surface analysis (W. Atlantic), Area 3
0351		Satellite image, Area 5
	1551	End of broadcast
	1720	Test pattern (start of broadcast)
0402	1723	Surface analysis (E. Atlantic), Area 2
0415	1736	Surface analysis (W. Atlantic), Area 3
0428	1749	500 mb analysis, Area 4
0438	1759	End of broadcast
	1900	Test pattern (start of broadcast)
	1903	Fax schedule, Part 1
	1914	Fax schedule, Part 2
	1925	Request for comments/ Product notice bulletin
	1935	Gulf Stream analysis, Areas 7, 8
	1945	Gulf Stream analysis (Part A), Areas 9,10
	1955	Gulf Stream analysis (Part B), Areas 7, 11, 12
0800		Test pattern (start of broadcast)
0805	2005	Preliminary surface analysis, Area 1
0815	2015	24-hour surface forecast, Area 1
0825	2025	24-hour wind/wave forecast, Area 1
0835	2035	36-hour surface forecast, Area 1
0845	2045	36 hour wind/wave forecast, Area 1
0855	2055	48-hour surface forecast, Area 4
0905	2105	48-hour sea state forecast, Area 4
0915	2115	48-hour 500 mb forecast, Area 4
0925	2125	Surface analysis (E. Atlantic), Area 2
0938	2138	Surface analysis (W. Atlantic), Area 3
0951	2151	Satellite imagery, Area 6
1002	2202	Surface analysis (E. Atlantic), Area 2
1015	2215	Surface analysis (W. Atlantic), Area 3
1028	2228	End of broadcast

Area 1: 28N–52N; 45W–85W
Area 2: 15N–65N; 10E–45W
Area 3: 15N–65N; 40W–95W
Area 4: 15N–65N; 10E–95W
Area 5: 20N–55N, 55W–95W
Area 6: 00N–60N; 30W–100W
Area 7: 32N–50N; 43W–80W
Area 8: 18N–38N; 62W–98W
Area 9: 32N–50N; 43W–80W
Area 10: 18N–38N; 65W–82W
Area 11: 30N–40N; 65W–82W
Area 12: 33N–47N; 65W–77W

U.S. COAST GUARD BELLE CHASE, LA (NMG)

Broadcast times and frequencies: continuous on 8503.9; 1100–2300 on 17148.3; and 2300–1100 on 12789.9 kHz.

Note: Although the frequencies of current broadcasts (summer 1996) from New Orleans provide good long-range coverage, they do not provide adequate short- to medium-range coverage. To correct this deficiency, the Coast Guard is considering transmitting broadcasts on a frequency near 4 MHz, to replace those on 12789.9 and 17148.3 kHz. Broadcasts will begin on this new frequency once the National Weather Service completes a review of facsimile products. The Coast Guard has decided to begin transmitting marine radiofax broadcasts from New Orleans on a third frequency once equipment is installed. Broadcasts on the new frequencies is expected to begin in fall 1996.

Time	UTC	Subject
0005		18/36 hr Gulf surface prognosis
0230		Gulf surface analysis
0735		U.S. surface analysis
0920		Gulf surface analysis
1100		High seas forecast
1125		Tropical surface analysis
	1205	18/36 hr Gulf surface prognosis
	1430	Gulf surface analysis
	1500	Fax schedule
	1910	U.S. surface analysis
	2020	Gulf surface analysis
	2100	Oceanographic products when available. Surface temp analysis Mon, Wed, Fri. Gulf stream analysis Tue, Thu, Sat.
	2300	High seas forecast
	2325	Tropical surface analysis

Maximum area covered: 5S–50N; 125W–0W

R Resources

CELLULAR SERVICE

BATELCO

Bahamas Telecommunications Corporation (Batelco)
The Mall at Marathon
P.O. Box N3048
Nassau, NP, Bahamas
242-394-4000.

Cellular service is available throughout most of the Bahamas, except the southernmost islands of Mayaguana and Great Inagua. Batelco has roaming agreements with a few U.S. companies, rents equipment, and offers permanent Batelco numbers. Batelco service is reported to be expensive, and a multi-month contract may be required.

BOATPHONE

Boatphone Marketing Ltd. (Group Headquarters)
P.O. Box 1516
Lower Newgate Street
St. Johns, Antigua
For information or advance roamer, rental, or charter registration:
800-BOATFON (800-262-8366), from U.S.
800-567-8366, from Canada
268-462-5051, from elsewhere
268-462-5052, fax

Cellular phone network throughout the eastern Caribbean, from the British Virgin Islands to Grenada, plus Jamaica and the Cayman Islands. Users not registered in advance dial "0" and "SND" (*0SND in Jamaica) from their cellular phones to register locally or call a local office (listed below). Other services include: operator assistance, roaming, voice and fax messaging, and call waiting, forwarding, and conferencing.

Cayman Islands, B System, 809-949-7800
Jamaica, B System, 809-968-4000
British Virgin Islands, A System; 809-494-3825
St. Martin, A System, 590-273000
St. Maarten, B System, 599-5-22100
St. Barts, A System, 590-273000
St. Kitts & Nevis, A System, 809-465-3003
Montserrat, A System, 664-491-2112
Antigua & Barbuda, A System; 268-462-5051
Guadeloupe, B System, 590-93-79-39
Martinique, B System, 596-504888
St. Lucia, A System, 758-452-0361
St. Vincent/Grenadines, A System, 809-457-4600
Grenada, A System, 809-440-1111 (local); 809-452-0361 (overseas).

VITEL CELLULAR

Vitel Cellular
Fort Mylner Commercial Center, Suite 2
St. Thomas, USVI 00801
809-777-8899
or
Sunny Isle Shopping Center
St. Croix, USVI
809-773-9991

Providing cellular telephone service throughout the USVI and surrounding waters, Vitel Cellular is a B carrier on the AMPS system. Credit card roaming is available by setting the phone system select setting to B. Users who have not preregistered can activate service by dialing "211" on their cellular phones.

BARBADOS COMMUNICATION SERVICES (CELLCOM)

Barbados Communication Services (Cellcom)
Enfield House, Upper Collymore Rock
St. Michael, Barbados
246-230-4BCS (-4227); 246-431-4000
Cellular telephone service on Barbados is offered through BCS. Users can receive temporary registration by phone using a MasterCard, Visa, or American Express credit card. There are registration and per diem fees, as well as per call charges.

HIGH SEAS RADIO-TELEPHONE SERVICE

AT&T COAST STATION WOO

P.O. Box 550
End of Beach Avenue
Manahawkin, NJ 08050

Technical information: 800-538-5936 or 954-587-0910 (collect)

Digital Selective Calling ID: 00-366-0002
Shore-to-ship calls: 800-SEA-CALL
Ship registration set-up: 800-752-0279 or 407-850-4895 (collect)

Cost: $14.93 for the first 3 minutes, $4.98 each additional minute or fraction thereof (3-minute minimum charge per call). You are not charged until the person you request to speak to has come on the line. High-Seas Direct service cost: $3.50/minute (1-minute minimum).

ITU Channel	Coast Station Transmit	Ship Station Transmit
410	4384.0 kHz	4092.0 kHz
411	4387.0	4095.0
416	4402.0	4110.0
422*	4420.0	4128.0
808	8740.0	8216.0
811	8749.0	8225.0
815	8761.0	8237.0

TU Channel	Coast Station Transmit	Ship Station Transmit
826*	8794.0	8270.0
1203	13083.0	12236.0
1210	13104.0	12257.0
1211	13107.0	12260.0
1228*	13158.0	12311.0
1605	17254.0	16372.0
1620	17299.0	16417.0
1626	17317.0	16435.0
1631*	17332.0	16450.0
2201	22696.0	22000.0
2205	22708.0	22012.0
2210	22723.0	22027.0
2236	22801.0	22105.0

*High-Seas Direct service

Voice broadcasts
Traffic lists on the hour every even hour UTC on channels 411 and 811.

National Weather Service broadcasts
Weather at 1200 and 2200 UTC on channels 411 and 811.

SITOR/Digital Selective Calling data broadcasts
Frequencies (center): 4212.5, 6328.0, 8433.0, 12632.0 kHz.
Weather at 20 minutes past even UTC hours.

AT&T COAST STATION WOM

1340 N.W. 40th Avenue
Fort Lauderdale, FL 33313

Technical information: 800-538-5936 or 954-587-0910 (collect)

Digital Selective Calling ID: 00-366-0001

Shore-to-ship calls: 800-SEA-CALL
Ship registration set-up: 800-752-0279 or 407-850-4895 (collect)

Cost: $14.93 for the first 3 minutes, $4.98 each additional minute or fraction thereof (3-minute minimum charge per call). You are not charged until the person you request to speak to has come on the line. High-Seas Direct service cost: $3.50/minute (1-minute minimum).

ITU Channel	Coast Station Transmit	Ship Station Transmit
209	2490.0 kHz	2031.5 kHz
221	2514.0	2118.0
245	2566.0	2390.0
247	2442.0	2406.0
403	4363.0	4071.0
412	4390.0	4098.0
417	4405.0	4113.0
423*	4423.0	4131.0
802	8722.0	8198.0
805*	8731.0	8207.0
810	8746.0	8222.0
814	8758.0	8234.0
825	8791.0	8267.0
831	8809.0	8285.0
1206	13092.0	12245.0
1208	13098.0	12251.0
1209	13101.0	12254.0
1215	13119.0	12272.0
1223	13143.0	12296.0
1230*	13164.0	12317.0
1601	17242.0 kHz	16360.0 kHz
1609*	17266.0	16384.0
1610	17269.0	16387.0
1611	17272.0	16390.0
1616	17287.0	16405.0
2215	22738.0	22042.0
2216	22741.0	22045.0
2222	22759.0	22063.0

*High-Seas Direct service

Voice broadcasts
Traffic lists on the hour every odd hour UTC on channels 403, 802, 1206, 1601, and 2215.

National Weather Service broadcasts
Weather 1300 and 2300 UTC and on request at no charge on channels 403, 802, 1206, 1601, and 2215. Also, hurricane warnings broadcast with every traffic list when they are in effect.

SITOR/Digital Selective Calling data broadcasts (transmit only)
Frequencies (center): 4215.5, 6327.5, 8432.5, 12631.0, and 22425.5 kHz.
Weather at 40 minutes past even UTC hours.

R
Resources

COAST STATION WLO

Mobile Marine Radio, Inc.
7700 Rinla Avenue
Mobile, AL 36619

Customer service department: tel., 334-660-9804; fax, 334-660-9850; e-mail, MMRCUSVC@aol.com
DSC ID: 00-366-0003

Mobile Marine Radio offers the following services: SITOR (marine telex), marine Morse code telegram (CW), and voice communications via high-seas and VHF radiotelephone. Specialized services include: e-mail addressed deliveries, marine weather, automatic radiotelex access, and other specialized information delivery.

Shore-to-ship calls: dial "O" and ask for the Mobile Alabama Marine Operator, or call WLO direct: 800-633-1634 or 334-666-3555.

ITU Channel	Coast Station Transmit	Ship Station Transmit
205	2572.0 kHz	2430.0 kHz
405#	4369.0	4077.0
414#	4396.0	4104.0
419#	4411.0	4119.0
428	4351.0	4060.0
607#	6519.0	6218.0
824#	8788.0	8264.0
829#	8803.0	8279.0
830#&	8806.0	8282.0
836	8713.0	8113.0
1212#	13110.0	12263.0
1225#	13149.0	12302.0
1226#&	13152.0	12305.0
1233	13173.0	12326.0
1235	13179.0	12332.0
1607#	17260.0	16378.0
1641#	17362.0	16480.0
1643	17368.0	16486.0
1647	17380.0	16498.0
1807	19773.0	18798.0
2237	22804.0	22108.0
2242	22819.0	22123.0
2246	22831.0	22135.0
2503	26151.0	25076.0

#Guarded channels
&DSC channels

Voice broadcasts
Traffic lists every hour on the hour on channels 414, 830, 1226, 1641, and 2237.

SITOR/Digital Selective Calling data/CW broadcasts
Frequencies: Call WLO for station lists.

COAST STATION WAH

Virgin Islands Radio
Avalon Communication Corp.
7843 Mountain Top
St. Thomas, VI 00802

Shore-to-ship calls and information:
tel., 809-776-8282 or toll-free 888-732-8255 (former number, 800-LEEWARD, temporarily out of service); fax, 809-776-3116

Cost $1.25/minute for radio time (3-minute minimum) plus the usual telephone rates. Forms of payment include: major bank and telephone credit cards, collect calls, third-party billing, and marine identification numbers. VI Radio also offers repeater, fax, answering, and message services.

ITU Channel	Coast Station Transmit	Ship Station Transmit
	2182.0 kHz	2182.0 kHz
401	4357.0	4065.0

Note: Coastal channel 2506.0 kHz and ITU channels 409, 604, 605, 804, and 1201 were discontinued following the hurricanes of fall 1995. At Reed's *press time (summer 1996) it was uncertain whether the station would resume use of these channels.*

Voice broadcasts
Notice of traffic list announcements and the appropriate VHF channel are given hourly on VHF channel 16 and 2182 kHz. Traffic is broadcast on VHF channels 25, 85, or 87 and on ITU channel 401.

Weather broadcasts
Detailed offshore and Caribbean weather is announced on 2182 kHz and VHF channel 16 and broadcast on ITU channel 401 and VHF channel 25, 85, or 87 at 0600, 1400, and 2200 local time. Local VI and Puerto Rico marine weather is announced on VHF channel 16 and broadcast on VHF channel 25, 85, or 87 at 0800 and 2000 local time.

DATA COMMUNICATION SERVICE

Globe Wireless
One Meyn Road
Half Moon Bay, CA 94019
tel., 415-726-6588; fax, 415-726-8604
e-mail: dgilbert@globewireless.com

Globe Wireless offers 24-hour HF radio data communication services worldwide. Services include e-mail, facsimile, telex, SITOR, marine telegram (CW), and Inmarsat radio communications through twelve networked HF coast stations and a centralized traffic center, which will also transcribe incoming voice messages from anywhere ashore for redelivery to subscriber vessels. Different service packages address different vessel needs.

Globe Wireless broadcasts combined traffic lists from each station once an hour. Weather is available 24 hours a day, seven days a week on selected channels.

Globe Wireless stations are now located at the following sites around the world:
VCT, Tors Cove, Newfoundland, Canada
WCC, Chatham, MA
WNU, New Orleans (Slidell Radio), LA
KFS, Palo Alto, CA
KPH, San Francisco, CA
KEJ, Maui, HI
ZLA, Awanui, New Zealand
VIP, Perth, Australia
AM9, Manama, Bahrain
ZSC, Cape Town, South Africa
LSD, Buenos Aires, Argentina
8PO, Bridgetown, Barbados
SAB, Göteborg, Sweden

Global Radio Network
7700 Rinla Avenue
Mobile, AL 36619
(800-633-1312).

WLO Global Radio Network is equipped to handle e-mail and facsimile transmissions as well as regular SSB voice traffic:

PinOak Digital Inc.
PO Box 360
Gladstone, New Jersey 07934
Voice 908-234-2020
(800-PIN-OAK1)

PinOak Digital maintains a private high-frequiency station for economical e-mail and facsimile transmissions:

SATELLITE COMMUNICATIONS

1998 may be the year that satellite communications systems become small enough and affordable enough for small and medium-size vessels.

ORBCOMM has activated a system of low level satellites that provide "short message" data communications with hand-held transceivers:

ORBCOMM
21700 Atlantic Boulevard
Dulles, VA 20166-6801
(800-ORBCOMM)

COMSAT's enhanced Inmarsat-C service provides data communications (data, telex, fax and e-mail) to and from anywhere on earth:

COMSAT Mobile Communications
22300 COMSAT Drive
Clarksburg, MD 20871
(301-428-2493)

COMSAT is also introducing Planet1, a portable voice system using the new Inmarsat 3 satellite constellation.

American Mobile Satellite Corporation has introduced the Skycell network, with coverage up to 200 miles offshore and compatability with regular cellular service (520-792-9429).

Meanwhile, both Motorola and Inmarsat are working on new satellite systems promising quality data and voice communications with low-powered, economical transmitters.

TIME SIGNALS

The U.S. Naval Observatory (USNO) provides recorded time announcements by telephone (202-762-1401 or 900-410-8463; identical recordings both numbers, but fee charged for 900 number call).

Station WWV from Fort Collins, Colorado, and station WWVH from Keka'a, Kauai, Hawaii, broadcast continuous time announcements. WWV transmits on 2500, 5000, 10000, 15000, and 20000 kHz. WWVH transmits on 2500, 5000, 10000, and 15000 kHz. These broadcasts are prepared by the National Institute of Standards and Technology, Time and Frequency Division, Boulder, CO 80303.

Time announcements are made every minute, commencing at 15 seconds before the minute on WWVH and 7 1/2 seconds before the minute on WWV. The time given is Coordinated Universal Time (UTC), which is essentially the same as Greenwich Mean Time (the time at the Prime Meridian, 0 longitude, Greenwich, England).

Reed's uses UTC whenever possible in listing radio broadcast times. See Chapter 6, Nautical Ephemeris, for a listing of local, or zone times, for most maritime nations of the world.

WWV and WWVH also broadcast storm warnings, which are explained in the section "Voice Radio Broadcasts" earlier in this chapter, and GPS updates, which are discussed in Chapter 4.

Station CHU broadcasts Canada's official time, which is based on UTC, from a station near Ottawa, Ontario, Canada (45 17.8N, 75 45.4W). This station can be heard in the North Atlantic. The signal is transmitted continuously on 3330, 7335, and 14670 kHz, and the time is announced in alternate minutes in French and English. This broadcast is the responsibility of the National Research Council's Institute for National Measurement Standards, Ottawa, Ontario, Canada.

BEAUFORT WEATHER SCALE

Beaufort Force Scale	Wind knots	Wind mph	Seas feet	Common name	Description
0	under 1	under 1	-	Calm	Sea like mirror
1	1 - 3	1 - 3	0.25	Light air	Ripples with appearance of scales: no foam crests
2	4 - 6	4 - 7	0.5 - 1	Light breeze	Small wavelets; crests of glassy appearance, not breaking
3	7 - 10	8 - 12	2 - 3	Gentle breeze	Large wavelets; crests begin to break; scattered whitecaps
4	11 - 16	13 - 18	31/2 - 5	Moderate breeze	Small waves, becoming longer numerous whitecaps
5	17 - 21	19 - 24	6 - 8	Fresh breeze	Moderate waves, taking longer form; many whitecaps; some spray
6	22 - 27	25 - 31	91/2 - 13	Strong breeze	Larger waves forming; whitecaps everywhere; more spray
7	28 - 33	32 - 38	131/2 - 19	Near gale	Sea heaps up; white foam from breaking waves begins to be blown in streaks
8	34 - 40	39 - 46	18 - 25	Gale	Moderately high waves of greater length; edges of crests begin to break into spindrift; foam is blown in well- marked streaks
9	41 - 47	47 - 54	23 - 32	Strong gale	High waves; sea begins to roll; dense streaks of foam; spray may reduce visibility
10	48 - 55	55 - 63	29 - 41	Storm	Very high waves with overhanging crests; sea takes white appearance as foam is blown in very dense streaks; rolling is heavy and visibility is reduced
11	56 - 63	64 - 72	37 - 52	Violent storm	Exceptionally high waves; sea covered with white foam patches; visibility still more reduced
12	64 and over	73 and over	45 and over	Hurricane	Air filled with foam; sea completely white with driving spray; visibility greatly reduced

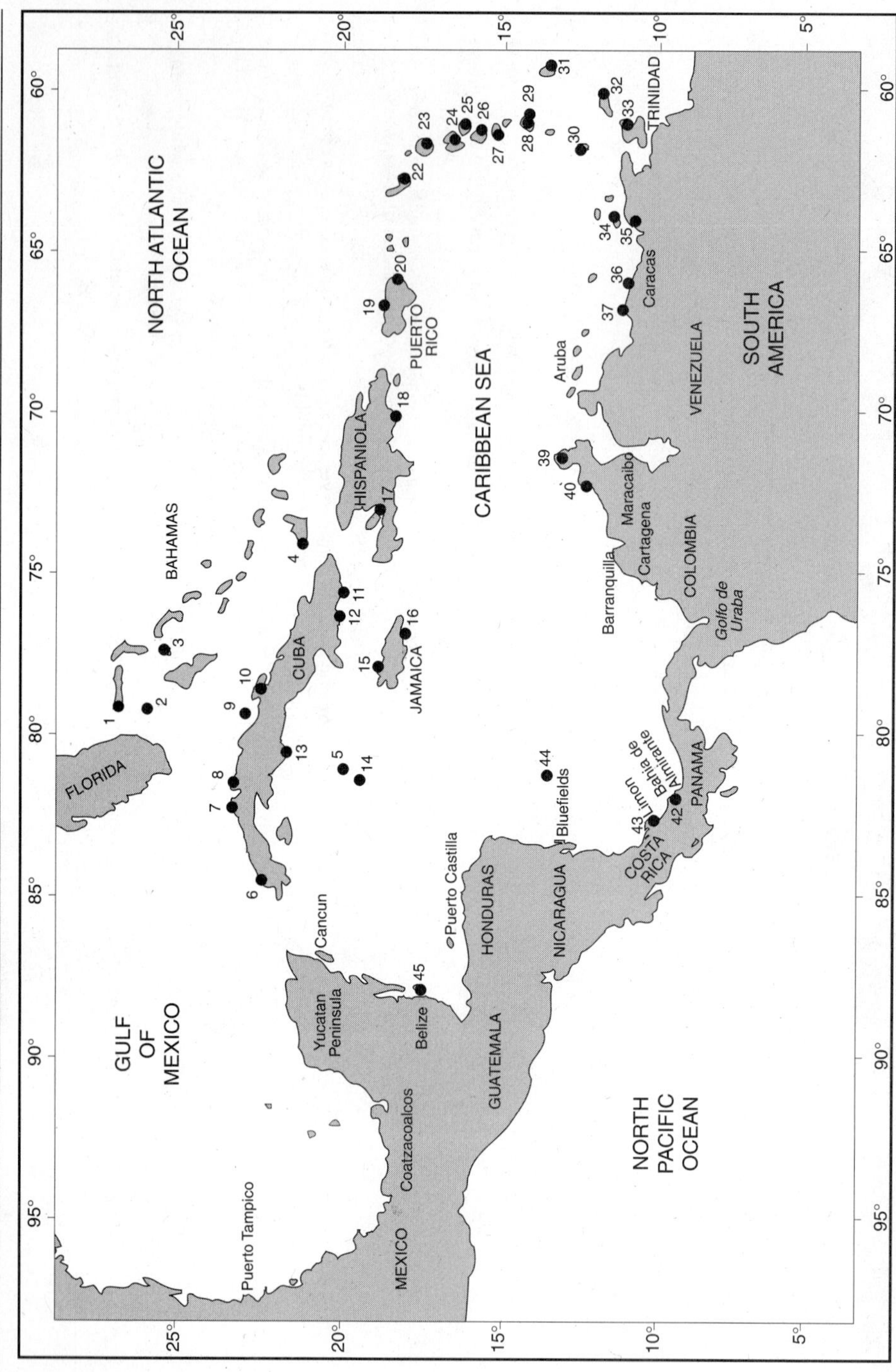

Radiobeacon Reference Map

MORSE CODE FOR IDENTIFICATION OF RADIOBEACONS

A ·–	I ··	Q ––·–	Y –·––	1 ·––––	9 ––––·
B –···	J ·–––	R ·–·	Z ––··	2 ··–––	0 –––––
C –·–·	K –·–	S ···		3 ···––	
D –··	L ·–··	T –		4 ····–	
E ·	M ––	U ··–		5 ·····	
F ··–·	N –·	V ···–		6 –····	
G ––·	O –––	W ·––		7 ––···	
H ····	P ·––·	X –··–		8 –––··	

RADIOBEACONS

The radiobeacon numbers in the following list refer to the map on the preceding page (missing numbers represent discontinued beacons).

Most radiobeacons on Caribbean coasts or islands are aero beacons, which are located to best serve the needs of aircraft and should be used with caution. All radiobeacons are subject to interference and refraction errors. As with all aids to navigation in the Caribbean, users are cautioned that radiobeacons have been reported as unreliable. Users are also reminded that when the distance to a radiobeacon is greater than 50 miles, a correction is usually applied to the bearing before plotting on a mercator chart. Refer to *Reed's Nautical Companion* for more radiobeacon information.

BERMUDA

NOTE: This radiobeacon is not shown on the radiobeacon reference map.

St. Davids Head, 32 22.0N, 64 38.9W. BSD (–··· ··· –··), 323 kHz, A2A, 150M.

FLORIDA

NOTE: These radiobeacons are not shown on the radiobeacon reference map.

Palm Beach Aero, 26 41.0N, 80 13.0W. PB (·––· –···), 356 kHz, A2A.

Miami, 25 44.0N, 80 09.6W. U (··–), 322 kHz, A2A, 100M. Discontinued.

Marathon Aero, 24 42.7N, 81 05.7W. MTH (–– – ····), 260 kHz, NON, A2A, 50M.

BAHAMA ISLANDS

1) West End International Airport Aero, 26 41.3N, 78 58.7W. ZWE (––·· ·–– ·), 317 kHz, A2A, 100M. Reported discontinued 1997.

2) Bimini Islands Aero, 25 42.5N, 79 16.3W. ZBB (––·· –··· –···), 396 kHz, A2A, 300M. Reported destroyed 1997.

3) Nassau International Airport Aero, 25 02.0N, 77 28.0W. ZQA (––·· ––·– ·–), 251 kHz, A1A, 240M.

4) Great Inagua Aero, 20 57.6N, 73 40.5W. ZIN (––·· ·· –·), 376 kHz, A2A, 50M.

CUBA

6) San Julian Aero, 22 05.0N, 84 13.0W. USJ (··– ··· ·–––), 402 kHz, 155M.

7) Havana Aero, 22 58.0N, 82 26.0W. A (·–), 339 kHz, 256M.

8) Varder Aero, 23 05.0N, 81 22.0W. UVR (··– ···– ·–·), 272 kHz, A2A, 216M.

9) Cayo Caiman Grande, 22 41.1N, 78 53.0W. CCG (–·–· –·–· ––·), 290 kHz, NON, A2A, 155M.

10) Punta Alegre, 22 22.5N, 78 47.3W. UPA (··– ·––· ·–), 382kHz, A2A, 70M.

11) Guantanamo Bay Aero, 19 54.0N, 75 10.0W. NBW (–· –··· ·––), 276.2 kHz.

12) Santiago de Cuba Aero, 19 58.0N, 75 50.0W. UCU (··– –·–· ··–), 339 kHz A2A, 70M.

R Resources

CAYMAN ISLANDS

14) Grand Cayman Aero, 19 17.0N, 81 23.0W. ZIY (––·· ·· –·––), 344 kHz, NON, A2A, 240M.

5) Cayman Brac Aero, 19 41.4N, 79 51.4W. CBC (–·–· –··· –·–·), 415 kHz, 240M.

JAMAICA

15) Montego Bay Aero, 18 30.0N, 77 55.0W. MBJ (–– –··· ·–––) 248 kHz, NON, A2A, 150M.

16) Kingston Aero, 17 58.0N, 76 53.0W. KIN (–·– ·· –·), 360 kHz, NON, A2A, 250M.

HAITI

17) Port-au-Prince Aero, 18 35.0N, 72 17.0W. HHP (···· ···· ·––·), 270 kHz, A2A, 240M.

DOMINICAN REPUBLIC

18) Punta Caucedo Aero, 18 27.0N, 69 40.0W. HIV (···· ·· ···–), 400 kHz, A2A, 240M.

PUERTO RICO

19) Dorado Aero, 18 28.0N, 66 25.0W. DDP (–·· –·· ·––·), 391 kHz, NON, A2A, 365M.

20) Roosevelt Roads Aero, 18 14.0N, 65 37.0W. NRR (–· ·–· ·–·), 264 kHz, NON, A2A, 115M.

LEEWARD ISLANDS

22) St. Barthélemy Aero, 17 54.0N, 62 51.0W. BY (–··· –·––), 338 kHz, A1A, 50M.

23) Antigua (Coolidge) Aero, 17 07.5N, 61 47.6W. ZDX (––·· –·· –··–) 369 kHz, NON, A2A, 256M.

24) Guadeloupe (Point-à-Pitre) Aero, 16 16.0N, 61 32.0W. FXG (··–· –··– ––·), 300 kHz, NON, A2A, 250M.

25) Guadeloupe (Marie–Galante/Grand Bourg) Aero, 15 52.0N, 61 16.0W. MG (–– ––·), 376 kHz, A2A, 50M.

26) Dominica Aero, 15 32.8N, 61 18.4W. DOM (–·· ––– ––), 273 kHz, 100M.

WINDWARD ISLANDS

27) Martinique (Fort de France) Aero, 14 36.0N, 61 06.0W, FXF (··–· –··– ··–·) 314 kHz, A2A, 100M.

28) Saint Lucia (Hewanorra) Aero, 13 44.0N, 60 59.0W. BNE (–··· –· ·), 305 kHz, 100M.

29) Saint Vincent (Arnos Vale) Aero, 13 09.0N, 61 13.0W. SV (··· ···–) 403 kHz, 143M.

30) Grenada Aero, 12 00.5N, 61 46.8W. GND (––· –· –··), 362 kHz, NON, A2A, 115M.

31) Barbados (Seawell Aerodome) Aero, 13 04.0N, 59 30.0W. BGI (–··· ––· ··), 345 kHz, NON, A2A, 256M.

TRINIDAD & TOBAGO

32) Crown Point, Tobago, Aero, 11 09.0N, 60 51.0W. TAB (– ·– –···), 323 kHz, A2A, 100M.

33) Piarco, Trinidad, Aero, 10 36.0N, 61 26.0W. POS (·––· ––– ···), 382 kHz, NON, A2A, 256M.

VENEZUELA

34) Margarita Aero, 10 55.3N, 63 57.4W. MTA (–– – ·–), 206 kHz, A2A, 100M.

35) Higuerote Aero, 10 28.0N, 66 05.7W. HOT (···· ––– –), 353 kHz, NON A2A, 240M.

36) La Orchila, 11 48.0N, 66 07.0W. ORC (––– ·–· –·–·) 320 kHz, A2A, 180M.

37) Maiquetia Aero, 10 37.0N, 66 59.0W. MIQ (–– ·· ––·–), 292 kHz, A2A, 130M.

NETHERLANDS ANTILLES

38) Curaçao Aero, 12 13.0N, 69 04.0W. Discontinued

COLOMBIA — ATLANTIC COAST

39) Portete Aero, 12 14.2N, 71 59.7W. PTE (·––· – ·), 420 kHz.

40) Riohacha Aero, 11 33.0N, 72 54.0W. RHC (·–· ···· –·–·), 295 kHz, NON, A2A, 75M.

41) Isla Tierra Bomba Light Station, Discontinued.

PANAMA — ATLANTIC COAST

42) Almirante, 9 17.4N, 82 23.6W. B (–···), 290 kHz, A2A. Transmits continuously 0500–1300.

COSTA RICA

43) Limón (TIM), 10 00.1N, 83 01.6W. M (– –), period 600s, 290 kHz, A2A, 50M. Transmits upon request through Limon (TIM) at 00 and 30 minutes past the hour requested.

SAN ANDRÉS ISLAND — COLOMBIA

44) Isla San Andrés, 12 35.1N, 81 42.3W. SPP (• • • • – – • • – – •), 387 kHz, N0N, A2A, 40M.

BELIZE

45) Belize Aero, 17 32.0N, 88 18.0W. BZE (– • • • – – • • •), 392 kHz, N0N, A2A, 192M.

MEXICO

All Mexican radiobeacons were discontinued in 1997.

RADAR BEACONS (RACONS)

Radar beacons (RACONs) are radar responder devices designed to produce a distinctive image on a vessel's radar set, thus enabling mariners to determine position with greater certainty than would be possible using a normal radar display alone.
RACON marks displayed on a radar screen are Morse characters typically of 1 to 2 miles in length, always start with a dash, and always extend radially outward from the radar target marked by the beacon. The display represents the approximate range and bearing to the RACON. RACONs are used to mark aids that may or may not have lateral significance; the RACON signal itself is for identification purposes only. RACON locations and identifications are included in most marine navigational charts and in *Reed's* are noted in individual aid-to-navigation descriptions.

RACONs in U.S. waters are used to mark and identify points on shore; channel separation, LNBs and other buoys; channel entrances under bridges; and uncharted hazards to navigation (the Morse letter D—dash, dot, dot [– • •]—has been reserved for this purpose).

LORAN-C

LORAN-C is of some use in the northern regions of the Caribbean. Groundwave coverage extends from the northern Bahamas to Cuba, to the Gulf of Honduras, and into the Gulf of Mexico along the Mexican coast.

Many users have reported good success in returning to positions where a LORAN reading was previously recorded. Even though the latitude/longitude readouts may be inaccurate, the readings received at a particular location have proven to be repeatable. The best results will be obtained by recording (or storing in the unit's memory) on-site Time Difference (TD) readings.

Keep in mind the possible differences between the NAD-83 Coordinate System and the NAD-27 System used on some Caribbean charts. Differences of more than a mile have been reported when comparing latitude/longitude positions in the two systems. Government charts list the coordinate system used.

Skywave coverage is much less accurate and is subject to false readings. Skywave fixes must be used with great caution and should be confirmed with other navigational means. Skywaves may be usable as far southeast as the Virgin Islands; as far south as Maracaibo, Venezuela; and into Columbia, Panama, and Costa Rica. *Reed's* has received reports of usable signals being picked up throughout the Caribbean.
The Southeast U.S. 7980 Chain is in use in this area. It consists of the following broadcast sites:

Malone, FL, 31 00N, 85 10W. 7980–Master.
Grangeville, LA, 30 44N, 90 50W. 7980–W.
Raymondville, TX, 26 32N, 97 50W. 7980–X.
Jupiter, FL, 27 02N, 80 07W. 7980–Y.
Carolina Beach, NC, 34 04N, 77 55W. 7980–Z.

Current LORAN-C status reports are available from a number of sources: the Local Notices to Mariners, District 7; the U.S. Coast Guard's NIS at NAVCEN (see below); and NAVTEX broadcasts and U.S. Coast Guard VHF and HF-SSB radio broadcasts (see chapter 5, Weather and Communications Services). Current information on these resources can be heard in a voice recording announcement (904-569-5241). LORAN-C status reports for chain 7980 are also available from the Chain Operations watchstander in Malone, FL (904-569-1108).

For best use of LORAN-C in the Caribbean mariners should obtain charts with a LORAN grid overlay. Increased position accuracy is possible by plotting LORAN time differences on these charts. The latitude/longitude conversion programs on many LORAN sets have been found to be much less accurate in these waters than off the continental U.S.

GPS AND DGPS

The Global Positioning System (GPS) has reached initial operational capability. The GPS constellation consists of 24 satellites. Selective availability (SA) is on, offering approximately 100-meter accuracy.

Differential GPS (DGPS) is the regular Global Positioning System (GPS) with an added correction (differential) signal. This correction signal improves the absolute accuracy of the indicated position to 1-5 meters (compared with 100 meters for standard GPS) and can be broadcast over any authorized communication channel.

Differential signals are broadcast commercially by several methods, including satellites and FM radio stations. Most mariners, however, will use the signals transmitted by the U.S. Coat Guard from former radiobeacon sites. These signals in the MF/LF band (285-325kHz) must be received on equipment designed for this purpose, and the GPS receiver being used must be able to accept the corrections. Coverage is limited to the range of each transmitter, see below. The Initial Operational Capability phase of DGPS was entered in January 1996, but work continues to improve it's reliability and coverage.

During DGPS system implementation, Coast Guard DPGS broadcasts are unmonitored. Users may experience service interruptions without advance notice. Signals should not be used in any circumstance where a sudden system failure or accuracy degradation could constitute a safety hazard. Furthermore, DGPS signals are dependent upon the availability of GPS.

As with LORAN, mariners should keep in mind possible differences between the NAD-83 Coordinate System and the NAD-27 System used on some Caribbean charts. Differences of over a mile have been reported when comparing latitude/longitude positions in the two systems. Government charts list the coordinate system used, and your GPS unit should be adjusted accordingly.

DGPS stations located in the area covered by this edition of *Reed's* are:
Miami, FL (Virginia Key), 25 44N, 80 09.6W. 322 kHz, 100 bps, 75M.
Isabella, PR, 18 27.7N, 67 04.0W. 295 kHz, 100 bps, 125M.

Current information concerning the status of GPS and DGPS is available through a variety of communications media (see NAVCEN Services, below; and in chapter 5, the following sections: Voice Radio Broadcasts, U.S. Coast Guard; NAVTEX; and Time Signals).

REED'S NEEDS YOU!

It is difficult to obtain accurate and up-to-date information on many areas in the Caribbean. To provide mariners with the best possible almanac, *Reed's* uses many sources. *Reed's* welcomes your contributions, suggestions, and updates. Please be as specific as possible and include your address and phone number.

Send this information to:
Editor, Thomas Reed Publications
13A Lewis Street, Boston, MA 02113, U.S.A.
e-mail: editor@treed.com
Thank you!

NOTICES TO MARINERS

The following are various special Warnings and special Notices to Mariners that we feel are of importance to our readers. Most are published exactly as they are stated in NIMA NTM #1/1997. We highly recommend that readers also use the U.S. Coast Guard's *local* Notice to Mariners; and we have provided pertinent information about obtaining them at the end of this section.

SPECIAL WARNINGS & NOTICES

SPECIAL WARNINGS FOR CUBA

SOUTH FLORIDA SECURITY ZONE: Permit to Depart for Cuban Waters. In March 1996, measures implemented under Presidential Proclamation No. 6867 established a temporary security zone in all U.S. territorial waters (extending to the 3-nautical-mile limit) adjacent to the state of Florida south of 26°19′N. Noncommercial vessels less than 50 meters (approximately 164 feet) in length may not depart from this security zone with the intent of entering Cuban territorial waters without express authorization from one of the following U.S. Coast Guard officials or designees: Commander, Seventh Coast Guard District; Captain of the Port, Miami; Captain of the Port, Tampa. Vessels and persons violating this regulation may be subject to seizure and forfeiture of the vessel, a fine, and/or imprisonment, plus suspension or revocation of Coast Guard licenses. The authorization permit is free; a one-page application form and a brief information packet are available from the Coast Guard's Marine Safety Office, Miami (305-536-5693), or Seventh District Command, Operations Center, Miami (305-536-5611).

CUBA: Territorial Seas. Mariners are advised to use extreme caution in transiting the waters surrounding Cuba. Cuba vigorously enforces a 12-nautical-mile territorial sea extending from straight baselines drawn from Cuban coastal points and offshore islands. The effect is that Cuba's claimed territorial sea extends in many areas well beyond 12 miles from Cuba's physical coastline. These claims are not in conformance with international law and are not recognized by the United States. Nonetheless, within distances extending in some cases upwards of 20 miles from the Cuban coast, U.S. vessels have been stopped and boarded by Cuban authorities. In a public statement issued in July 1995, the Cuban government asserted its "firm determination" to take actions necessary to defend Cuban territorial sovereignty and to prevent unauthorized incursions into Cuban territorial waters and airspace. The Cuban government warns that any boat violating Cuba's territorial sea can be sunk. The U.S. government takes this statement seriously. Within the limits of prudence and good judgment, mariners are advised to protest (but not physically resist) any improper attempt to stop, board, or seize U.S. vessels in international waters. They should also notify the U.S. Coast Guard of their status so the information can be relayed to the U.S. Department of State for diplomatic attention.

TRADE WITH CUBA: An embargo on all trade with Cuba proclaimed on 7 February 1962 remains in effect. Among other provisions, it is prohibited to engage in the transportation of goods or merchandise from anywhere to Cuba unless specifically licensed by the U.S. government, and such licenses are seldom granted. No person subject to U.S. jurisdiction may engage in any trans-action with regards to Cuban goods or services; thus it is un-lawful for U.S. cruising boats to put into Cuban ports for fuel, supplies, dockage, etc. Persons who violate provisions of the embargo may be subject to civil or criminal sanctions, or both, and vessels may be subject to seizure and forfeiture.

OLD BAHAMA CHANNEL: Traffic Separation Scheme. On September 1, 1990, seven Traffic Separation Schemes, approved by the International Maritime Organization, were implemented off the eastern, western, and northern coasts of Cuba. The government of Cuba has unilaterally established a mandatory ship-reporting system within the Old Bahama Channel to govern vessel movement within the area. While it is the U.S. position that the mandatory provisions of the ship-reporting system are inconsistent with international law, vessels should exercise caution while transiting the Old Bahama Channel.

The Traffic Separation Schemes are located:

Off Cabo San Antonio
Off La Tabla
Off Costa de Matanzas
In the Old Bahama Channel

Off Punta Maternillos
Off Punta Lucrecia
Off Cabo Maisi

STRAITS OF FLORIDA: EEZ Advisory. In view of the recent (winter 1996) U.S. and Cuban maritime surveillance activities near the territorial seas of Cuba, the U.S. government is seeking to ensure that U.S.-registered vessels do not conduct commercial fishing activities in the Cuban Exclusive Economic Zone (EEZ). The maritime boundary between the U.S. and Cuban EEZ off the north coast of Cuba, agreed to in 1977, is essentially an equidistant line starting west of Cay Sal Bank and proceeding west and then northwest. Consistent with the agreement, the U.S. exercises sovereign rights and jurisdiction with respect to all fishing activity north of the line and Cuba south of the line. The beginning and ending coordinates of the line are: 23°55′30⊕N, 81°12′55⊕W and 25°12′25⊕N, 86°33′12⊕W.

SPECIAL WARNING FOR NICARAGUA

Mariners operating small vessels such as yachts and fishing boats are advised to avoid both the Caribbean and Pacific ports and waters of Nicaragua until further notice. Nicaraguan law mandates the payment of a fine equal to 200 percent of the value of any boat caught fishing illegally within Nicaragua's Exclusive Economic Zone (EEZ). There have been several cases of foreign-flagged vessels being seized by Nicaraguan authorities. While in all cases passengers and crew have been released within a period of several weeks, in some cases the ships have been searched, personal gear and navigational equipment have been stolen, and there have been excessive delays in releasing vessels. Prompt U.S. embassy consular access to U.S. citizens who are detained may not be possible due to non-notification of the embassy by the Nicaraguan government.

Mariners should also note that there have been recent incidents of piracy in Caribbean and Pacific waters off the coasts of Nicaragua.

SPECIAL NAVIGATIONAL NOTICES

FLORIDA DRAWBRIDGES: VHF-FM Channel to be Used. Because of heavy congestion being experienced on VHF channel 13 for vessel communications, the Tampa office of the Federal Communications Commission encourages all commercial and noncommercial vessels to use VHF channel 9 as a *calling* and working frequency when hailing drawbridges in the state of Florida. Drawtenders may no longer monitor VHF channel 13, but will continue to monitor VHF channel 16 as a calling channel. When contacted, bridgetenders in Florida will normally use VHF channel 9 as a working frequency.

ST. LUCIE CROSSROADS, FL: Shoaling. Because of constant shifting shoaling at the intersection of the St. Lucie Inlet, St. Lucie River, and the Intracoastal Waterway, mariners are requested to maintain slow speed and use extreme caution throughout the area. The aids to navigation have been relocated to reflect the best water, but areas may become shoaled over into the marked channel.

FORT LAUDERDALE–PORT EVERGLADES, FL: Hazards to Navigation. The U.S. Coast Guard Captain of the Port, Miami, advises that large vessels arriving or departing from Port Everglades are to exercise extreme caution due to heavy small-craft traffic. This advisory includes personal watercraft (i.e., jet skis, wave runners). Any vessel operator who observes personal watercraft engaging in any reckless operation such as jumping a wake too close to another vessel or swerving at the last moment to avoid a collision should immediately report such activities to the U.S. Coast Guard or Florida Marine Patrol.

HAWK CHANNEL, FL: Shoaling. Severe shoaling exists at the east side entrance to Snake Creek Channel and extends easterly for approximately 300 yards. Vessels with drafts greater than 3 feet are advised not to transit this area during periods of minus tides.

BAHAMAS: Aid to Navigation. Castle Island Light (22 07.3N, 74 19.6W) has been reported to no longer flash all round the horizon and is not visible on a southerly approach.

GRENADA: Aid to Navigation. Petit Cabrits Light (12 01.0N, 61 46.5W) has been reported as extinguished.

VENEZUELA: Aid to Navigation. Monjes del Sur Light (12 21.3N, 70 54.1W) has been reported as extinguished.

NAVIGATION RELATED NOTICES

CAUTION CONCERNING RELIANCE UPON AIDS TO NAVIGATION

The aids to navigation depicted on charts comprise a system consisting of fixed and floating aids with varying degrees of reliability. Therefore, prudent mariners will not rely solely on any single aid to navigation, particularly a floating aid.

The buoy symbol is used to indicate the approximate position of the buoy body and the sinker that secures the buoy to the seabed. The approximate position is used because of practical limitations in positioning and maintaining buoys and their sinkers in precise geographical locations. These limitations include, but are not limited to: inherent imprecisions in position-fixing methods, prevailing atmospheric and sea conditions, the slope and the material of the seabed, the fact that buoys are moored to sinkers by varying lengths of chain, and the fact that buoy and sinker positions are not under continuous surveillance but are checked only during periodic maintenance visits, which often occur more than a year apart. The position of the buoy body can be expected to shift inside and outside the charting symbol because of the forces of nature. The mariner is also cautioned that buoys are liable to be carried away, shifted, capsized, sunk, etc. Lighted buoys may be extinguished or sound signals may not function as the result of ice or other natural causes, collisions, or other accidents.

For the foregoing reasons, a prudent mariner must not rely completely upon the position or operation of floating aids to navigation, but will utilize bearings from fixed objects and aids to navigation on shore. Further, a vessel attempting to pass close aboard always risks collision with a yawing buoy or with the obstruction that the buoy marks.

CHART NOTES REGARDING DATUMS

Particular attention should be exercised during a passage when transferring the navigational plot to an adjacent chart upon a different geodetic datum or when transferring positions from one chart to another chart of the same area that is based upon a different datum. The transfer of positions should be done by bearings and distances from common features.

Notes on charts should be read with care, as they give important information not graphically presented. Notes in connection with chart title include the horizontal geodetic datum that serves as a reference for the values of the latitude and longitude of any point or object on the chart. The latitude and longitudes of the same points or objects on a second chart of the same area that is based upon a different datum will differ from those of the first chart. The difference may be navigationally significant.

It is particularly necessary to note the chart datum when using the Global Positioning System (GPS). Some features, such as isolated islands, may be as much as several miles from their charted location, especially if the chart has been produced from older surveys.

WARNING ON USE OF FLOATING AIDS TO NAVIGATION.

The aids to navigation depicted on charts comprise a system consisting of fixed and floating aids with varying degrees of reliability. Therefore, prudent mariners will not rely solely on any single aid to navigation, particularly a floating aid.

The buoy symbol is used to indicate the approximate position of the buoy body and the sinker which secures the buoy to the seabed. The approximate position is used because of practical limitations in positioning and maintaining buoys and their sinkers in precise geographical locations. These limitations include, but are not limited to, inherent imprecisions in position fixing methods, prevailing atmospheric and sea conditions, the slope of and the material making up the seabed, the fact that buoys are moored to sinkers by varying lengths of chain, and the fact that buoy and/or sinker positions are not under continuous surveillance but are normally checked only during periodic maintenance visits which often occur more than a year apart. The position of the buoy body can be expected to shift inside and outside the charting symbol due to the forces of nature. The mariner is also cautioned that buoys are liable to be carried away, shifted, capsized, sunk, etc. Lighted buoys may be extinguished or sound signals may not function as the result of ice of other natural causes, collisions, or other accidents.

For the foregoing reasons, a prudent mariner must not rely completely upon the position or operation of floating aids to navigation, but will also utilize bearings from fixed objects and aids to navigation on shore. Further, a vessel attempting to pass close aboard always risks collision with a yawing buoy or with the obstruction the buoy marks.

CAUTION IN CLOSE APPROACH TO MOORED OFFSHORE AIDS

Courses should invariably be set to pass these aids with sufficient clearance to avoid the possibility of collision. Errors of observation, current and wind effects, other vessels in the vicinity, and defects in steering gear may be, and have been, the cause of collisions. Experience shows that buoys cannot be safely used as leading marks to be passed close aboard, and should always be left broad off the course, whenever searoom permits.

It should be borne in mind that most large buoys are anchored to a very long scope of chain and, as a result, the radius of their swinging circle is considerable. The charted position is the approximate location. Furthermore, under certain conditions of wind and current, they are subject to sudden and unexpected sheers which are certain to hazard a vessel attempting to pass close aboard.

Watch (station) buoys are sometimes moored near large buoys to mark the approximate station of the large buoy. Since these buoys are always unlighted and, in some cases, moored as much as a mile from the large buoy, the danger of a closely passing vessel colliding with them is always present, particularly so during darkness or periods of reduced visibility.

REPORTING OF DANGERS TO NAVIGATION.

Mariners will occasionally discover uncharted shoals, malfunctions of important navigational aids, or other dangerous situations which should be made known to other navigators. In addition to broadcasting a safety message addressed CQ (to all ships), those items which can be classified as urgent should be reported via radio to the closest responsible charting authority. NIMA additionally requests the reporting of such urgent navigational data directly to its Bethesda, MD. office via any U.S. Government radio station. General criteria for important data is "that information, without which, a mariner might expose his vessel to unnecessary danger." Follow-up letter reports are encouraged to provide NIMA additional evaluation background. The initial report should be brief, but must contain:

What - description of danger
When - GMT and date
Where - Latitude and Longitude (Reference chart in use)
Who - Reporting vessel and observer

Additional guidance in preparing a report of navigational information is contained in Pub. 117 Radio Navigational Aids.

CAUTION IN USING FOREIGN CHARTS.

In the interest of safe navigation, caution should be exercised in the use of foreign charts not maintained through U.S. Notice to Mariners.

Foreign produced charts are occasionally mentioned in NIMA Sailing Directions when such charts may be of a better scale than U.S. produced charts. The mariner is advised that if or when such foreign charts are used for navigation it is his responsibility to maintain those charts from the Notice to Mariners of the foreign country producing the charts.
The mariner is warned that the buoyage systems, shapes, colors, and light rhythms used by other countries often have a different significance than the U.S. system.

IALA MARITIME BUOYAGE SYSTEM

The IALA Maritime (combined Cardinal/Lateral) Buoyage System has been, is being, or will be implemented by nearly every maritime buoyage jurisdiction worldwide as either REGION A buoyage (red to port) or REGION B buoyage (red to starboard). The actual conversion in REGION A began in 1977 and is continuing. Conversion in REGION B has begun, and is near completion.

The terms "REGION A" and "REGION B" are used to determine which type of buoyage region is in effect or undergoing conversion in a particular area. The major difference between the two buoyage regions concerns the lateral marks. When viewed from sea, the lateral marks in REGION A will be red to port; in REGION B they will be red to starboard. Shapes of lateral marks will be the same in both regions: can to port; cone (nun) to starboard. Cardinal and other marks will continue to follow current guidelines and may be found in both regions.

A modified lateral mark, indicating the preferred channel where a channel divides has been introduced for use in both regions. It is intended that after a reasonable period, each chart will make reflection to REGION A or REGION B to indicate which type of lateral buoyage is in use in that particular area.

NOTE: All North American, Central American, South American, and Caribbean coasts fall inside IALA REGION B; and Bermuda has now completed its changeover to IALA REGION B.

COMMUNICATIONS NOTICES

FCC RADIO LICENSES

The Telecommunications Act of 1996 gave the Federal Communications Commission discretion to eliminate the radio station licensing requirement for vessels operating domestically that are not required by law to carry a radio. This has been effected by changes in the FCC Rules and Regulations.

The licensing requirement was eliminated only for recreational vessels operating solely in domestic waters and carrying aboard only the following equipment: VHF radio, any type of emergency-position indicating beacon (EPIRB), or any type of radar. Mariners aboard these vessels must continue to operate their equipment in accordance with FCC rules. For the status of procedures to obtain refunds on the unused user portion of a license obtained after July 1994, call the FCC tollfree information line (888-225-5322). *(Note: CB radios, cellular telephones, and receive-only equipment are also exempted from licensing requirements).*

For FCC licensing purposes, the following vessels are not considered recreational: power-driven vessels over 20 meters (approximately 65 feet), any Coast Guard-inspected vessel carrying more than six passengers for hire, and any vessel documented for commercial use (including commercial fishing vessels).

Recreational vessels that use MF/HF single-sideband radio, satellite communications, or telegraphy are required to have a Ship Station License. To operate this equipment or to communicate with a foreign station by VHF radio, you must also have a Restricted Radiotelephone Operator Permit. The Ship Station License must be on board the vessel; it may be posted at the location of the set or kept in a more secure, dry location.

To apply for a Ship Station License request FCC Form 506, fee-type code PASR. New and renewal licenses are valid for 10 years; the filing fee is currently (summer 1997) $95. If you later install additional equipment not covered by the original ship's license, you must request a modification, also on Form 506, fee type code PASM; filing fee, $45. To obtain a Restricted Radiotelephone Operators Permit, request FCC Form 753. This license is valid for your lifetime and the cost is currently $45. Applications may be obtained from the FCC Forms Distribution Center, 2803 52nd Avenue, Hyattsville, MD 20781 (800-418-3676). Additional information is available from the Public Inquiries Branch of the FCC in Washington, DC (202-418-0220), the FCC tollfree information line (888-225-5322), and the FCC office in Gettysburg, PA (717-337-1212).

Mariners should be aware that the U.S. Coast Guard is authorized to check an FCC Ship Station License during boarding inspections on all vessels carrying equipment covered by the regulations or required to carry a VHF radio. Mariners who fail to obtain required FCC authorizations are liable for criminal misdemeanor penalties.

VHF CHANNEL 09 NONCOMMERCIAL CALLING CHANNEL

In an effort to eliminate traffic congestion on channel 16 VHF-FM, the Federal Communications Commission (FCC) has designated channel 9 VHF-FM as the nationwide *Recreational Calling Channel* for use by noncommercial boaters within the United States. Use of this channel by recreational boaters is strictly voluntary but is strongly encouraged by the Coast Guard and the FCC.

It is important to note, however, that the Coast Guard does not monitor channel 9 VHF-FM for calls requiring assistance. Vessels in distress should make their calls on channel 16 VHF-FM.

DIGITAL SELECTIVE CALLING DISTRESS ALERT RELAYS

Digital selective calling (DSC) is a new capability being offered with some VHF and HF maritime radios, intended to initiate voice calls and provide distress alert information to the U.S. Coast Guard and other rescue coordination centers. DSC is a major element of the Global Maritime Distress and Safety System (MDSS), an International Maritime Organization-mandated telecommunications system required to be carried on cargo and passenger vessels on international voyages; many ships are now so equipped and all must be by 1999.

The Coast Guard has MF/HF DSC capabilities at the Communications Area Master Station Atlantic, Portsmouth, VA (call sign NMN), which also remotely operates transmitters at Boston (NMF) and Miami (NMA); and at Communications Station, New Orleans (NMG). Channel frequencies for all stations are: 2187.5, 4207.5, 6312, 8414.5, 12577, and 16804.5; carrier frequencies are 1700 Hz below the assigned frequencies above. Plans exist to install a coast-wide DSC capability beginning in the year 2000. Until then the Coast Guard cannot receive a VHF DSC distress alert unless a mariner with a DSC compatible radio receives an alert and relays it to the Coast Guard. Mariners receiving a VHF distress alert should attempt to contact the vessel sending the distress alert and obtain information concerning the distress and then contact the Coast Guard to pass on this information. The Coast Guard will treat these alerts as legitimate distress calls. Continue listening on the working channel to ensure communications between the Coast Guard and ship in distress is established. Finally, be ready to provide further assistance if asked by the Coast Guard.

HURRICANE SEASON: JUNE 1–NOVEMBER 30

Hurricanes often develop with little warning. Extensive damage to small craft attended by loss of life often results. All mariners, particularly operators of small fishing vessels, are reminded that advanced planning to prevent loss of vessel and crew should include: (a) instruction of crew and passengers in location of emergency equipment and procedures, (b) presailing check of vessel, equipment, and systems for seaworthiness, (c) installation of strong ground tackle, (d) review of storm center evasion procedures, (e) knowledge of nearest hurricane shelter, and (f) constant radio watch on VHF channel 16 and frequent monitoring of VHF-FM weather channels and any other available source of storm information. On-line information can be found at www.nhc.noaa.gov/products.htm.

A Tropical Cyclone Advisory message is issued every six hours, with intermediate bulletins as needed. Two levels of alerts are used. **A HURRICANE WATCH** is a preliminary alert that a hurricane may threaten a specified portion of the coast. **A HURRICANE WARNING** indicates that hurricane conditions *are expected* within 24 hours for the specified area. To provide additional information, the National Hurricane Center also issues **Marine Advisories** that provide forecast positions, movements, and intensities for up to 72 hours ahead; these include probabilities of hurricane strikes for coastal locations and offshore coordinates.

Storm Surges. A considerable rise or fall in the level of the sea along a particular coast may result from strong winds and sharp changes in barometric pressure. In cases where the water level is raised, higher waves can form with greater depth; this condition can be destructive to low-lying regions, particularly at high stages of tide. This type of wave occurs especially in coastal regions bordering on shallow waters that are subject to tropical storms.

Maneuvering to Avoid a Storm Center. If you find yourself caught at sea in a storm area, the proper action to take depends in part on your position relative to the storm center and its direction of travel. The circular area of the storm can be di-vided into two parts. The part to the right of the storm track (facing in the direction in which the storm is moving) is the dangerous semicircle. This is because the forward motion of the storm center adds to the wind velocity, and this in turn creates higher seas; also the wind directions tend to push your vessel closer to the storm track. The part of the storm track to the left of the storm track is called the navigable semicircle where conditions will be somewhat less severe, and the winds will push your vessel away from the storm. In any event, set a course that will take you as far as possible from the storm track, but still prepare for the worst.

During the hurricane season, *drawbridges* may deviate from normal operating procedures. Some may be unable to open because of high winds. Others may be authorized to close for extended periods to facilitate evacuation of land traffic. Mariners should anticipate bridge closures by listening to weather broadcasts on hurricane conditions and seek passage through drawbridges well in advance of the arrival of gale-force winds. In the U.S., drawbridges are authorized to close upon the approach of gale-force winds of 34 knots or greater.

Locks and navigation structures are operated

until water levels and wind conditions warrant operations unsafe, when they are closed in accordance with hurricane preparation procedures. Structures may be closed on short notice well in advance of an actual storm; mariners are advised to seek safe harbor as soon as possible.

Aids to navigation, particularly buoys, may be moved from their charted position, damaged, destroyed, or otherwise made inoperative as a result of storms or other natural causes. Mariners should not rely completely upon the position or operation of an aid to navigation.

MISCELLANEOUS NOTICES

SUBMARINE TELECOMMUNICATIONS CABLES

Vessels intending to anchor and fishing vessels working in cable areas should maintain a buffer zone around cables to prevent unnecessary entanglement. AT&T advises that fishing gear should be kept a distance of 1 nautical mile from the charted location of all submarine cables. Most modern submarine telecommunications cables carry up to 8,000 volts. Attempts to cut the cable or untangle gear could be lethal. Mariners who become entangled in a cable should call the AT&T 24-hour operations center collect at 908-234-6771.

REPORTING OIL AND CHEMICAL SPILLS, DERELICT DRUMS

With many exotic chemicals being transported in drums as deck cargo, increasingly more reports are received of loss overboard of these potentially dangerous containers. Even empty drums may contain residues that are extremely hazardous to touch or smell, and a few vapors may be explosive.

When coming upon derelict drums, whether afloat or from the sea bottom, this danger should be considered. Identifying labels give adequate warning, but containers are more likely to be found with caution labels washed off. Avoid direct contact. For sightings in U.S. waters, notify the U.S. Coast Guard's National Response Center (24-hour toll-free reporting number, 800-424-8802; fax, 202-479-7165; telex, 892427; 24-hour access to language interpretation is available for reports not received in English).

Oil and chemical discharges in U.S. waters should also be called into this center. Derelict drums and oil and chemical discharges near any foreign shore should be reported to the government authorities of the nearest port.

COMPLIANCE WITH THE ACT TO PREVENT POLLUTION FROM SHIPS

The Act to Prevent Pollution from ships (33 U.S.C. 1901) implements into U.S. law the International Convention for the Prevention of Pollution from Ships, as modified by the Protocol of 1978 (commonly referred to as MARIPOL 73/78). Annex V of these documents deals with the marine prevention of pollution by plastics and other garbage produced during vessel operations.

Annex V is applicable to all recreational, fishing, uninspected and inspected vessels, and foreign flag vessels on the navigable waters and all other waters subject to the jurisdiction of the United States, out to and including the Exclusive Economic Zone (200 miles).

Annex V prohibits the disposal of any and all plastic material from any vessel anywhere in the marine environment. Dunnage, lining, and packing material that float may be disposed of beyond 25 miles from the nearest land. Other garbage that will not float may be disposed of beyond 12 miles from land, except that garbage that can pass through a 25mm mesh screen (approximately 1 square inch) may be disposed of beyond 3 miles. Dishwater is not considered is not considered garbage within the meaning of Annex V when it is the liquid residue from the manual or automatic washing of dishes or cooking utensils. More restrictive disposal regimes apply in waters designated as "Special Areas." This Annex requires terminals to provide reception facilities at ports and terminals to receive plastics and other garbage from visiting vessels,

The civil penalty for each violation of MARIPOL 73/78 is not more than #25,000. The criminal penalty for a person who knowingly violates the MARIPOL Protocol, or its implementing regulations, consists of a fine of not more than $250,000 and/or imprisonment for not more than 5 years, or both. U.S. law also provides criminal penalties of up to #500,000 against organizations that violate MARIPOL.

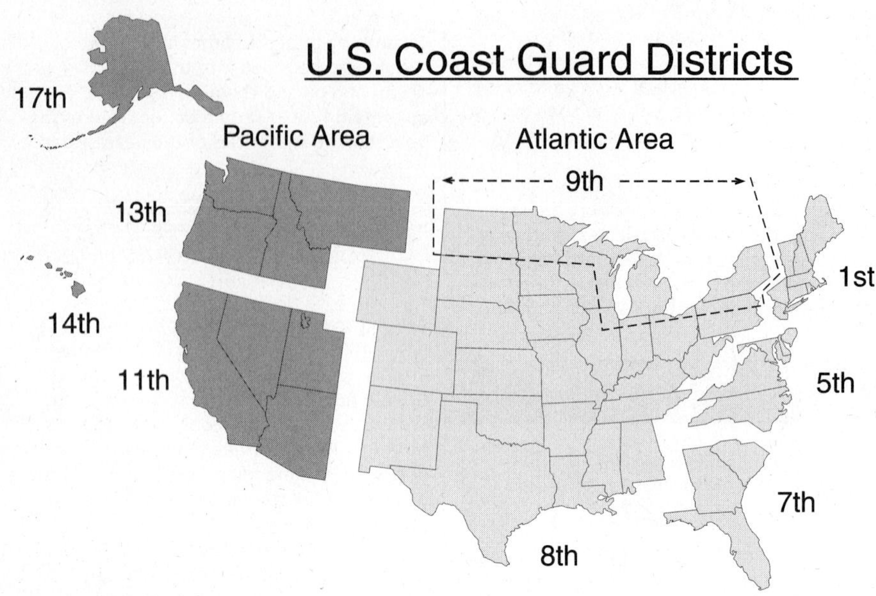

Local Notice to Mariners are published weekly by each Coast Guard District, and contain the latest information about special Warnings, changes to aids to navigation, and more.
If you would like to receive the LNM for your District, contact the appropriate office listed below.
LNMs are also available on the internet at www.navcen.uscg.mil

First District
408 Atlantic Avenue
Boston, MA 02110-3350
Day: 617-223-8338
Night: 617-223-8558

Fifth District
Federal Building
431 Crawford Street
Portsmouth, VA 23704-5004
Day: 804-398-6489
Night: 804-398-6231

Seventh District
Brickell Plaza Federal Building
909 SE 1st Avenue, Rm. 406
Miami, FL 33131-3050
Day: 305-536-5621
Night: 305-536-5611

Eighth District
Hale Boggs Federal Building
501 Magazine Street
New Orleans, LA 70130-3396
Day: 504-589-6234
Night: 504-589-6225

Ninth District
1240 East 9th Street
Cleveland, OH 44199-2060
Day: 216-522-3991
Night: 216-522-3984

Eleventh District
Building 50-6
Coast Guard Island
Alameda, CA 94501-5100
Day: 510-437-2940
Night: 510-437-3700

Thirteenth District
915 Second Avenue
Seattle, WA 98174-1067
Day: 206-220-7280
Night: 206-220-7004

Fourteenth District
Prince Kalanianaole Fed. Bldg.
9th Floor, Room 9139
300 Ala Moana Blvd.
Honolulu, HI 96850-4982
Day: 808-541-2317
Night: 808-541-2500

Seventeenth District
P.O. Box 25517
Juneau, AK 99802-5517
Day: 907-463-2262
Night: 907-463-2000

SEND FOR YOUR FREE 1998 SUPPLEMENT

Just return the post-paid reply card and questionnaire to receive your free list of navigation updates, changes, and special notices. Publication date—Spring 1998.

OTHER RESOURCES

WEATHER INFORMATION

National Hurricane Center
Coral Gables, Florida
305-229-4483, advisory updates, voice recording
305-229-4470, main office; 305-229-9901, fax;
internet: http://www.nhc.noaa.gov

The voice recording number provides current information on developing tropical storms and hurricanes 24 hours a day. The voice menu options on the main office number also provide access to local Florida weather recordings on a 24-hour basis. Use the fax number to receive current Gulf Stream information from south of Cape Hatteras to the Straits of Florida and the Gulf of Mexico, including current speeds, water temperature, and eddy locations.

NOAA Weather Service
Internet: http://www.nws.noaa.gov

Information available includes forecasts and warnings, up-to-date marine weather charts, and the NOAA Weather Radio Guide.

National Climatic Data Center
151 Patton Avenue, Room 120
Asheville, NC 28801-5001
tel., 704-271-4800; fax, 704-271-4876;
nternet: http://www.ncdc.noaa.gov

Detailed climatological charts and information for all areas of the world is available for a fee. Among the products of interest to mariners are Gulf Stream charts, climatological studies of Atlantic, Gulf of Mexico, and Caribbean waters and coastal zones, and tropical storm data. Ask for Marine Publications and Microforms. You can receive data by mail or fax.

Bermuda Weather Service
P.O. Box 123
St. George's, Bermuda GE BX
441-293-6659, general information
441-293-6658, fax

Located at the airport, this operation provides a weather chart packet for yachts, general forecasts, marine forecasts, and tropical storm warnings. It also operates a dedicated automatic telephone weather system. From overseas, call 441- 297-7977. In Bermuda, dial: 9770 for general forecasts; 9771 for current weather; 9772 for marine forecasts; and 9773 for weather warnings. Other information may be available on request.

Nassau Meteorological Service
Nassau, Bahamas

For up-to-date weather forecasts for Bahamian waters, call 242-377-7040. The number is manned 24 hours a day. For a recorded announcement, in Nassau, dial only 915.

Eastern Caribbean Meteorological Information

On most islands in the Leewards and Windwards, current weather forecasts may be available by calling meteorological offices or overseas airlines offices located at airports served by overseas airlines.

NAVCEN SERVICES

Mariners now have direct access by a number of communication means to the U.S. Coast Guard's Navigation Information Center (NIS), a branch of the Coast Guard's centralized navigation center in Alexandria, Virginia. NIS provides Loran-C, OMEGA, GPS, DGPS, marine radiobeacon, and general radionavigation user information and status, as well as access to weekly editions of the various local notices to mariners issued by individual Coast Guard districts (U.S. territory covered by this edition of *Reed's Nautical Almanac* is under the jurisdiction of U.S. Coast Guard District 7 with headquarters at: Brickell Plaza, Federal Building, 909 SE 1st Avenue, Miami, FL 33131-3050; tel., 305-536-5621).

NIS watchstanders are available 24 hours a day to handle telephone, fax, and mail inquiries (tel., 703-313-5900; fax, -5920; mail, NAVCEN, 7323 Telegraph Road, Alexandria, VA 22310-3998). Specifically for GPS users, a 24-hour voice recording (703-313-5907) provides access to a 90-second message of current GPS status. Forecast outages, historical outages, and other changes in GPS are included as time permits.

Additionally, NIS information is available through an automatic fax-back information service. Users access the system by dialing 703-313-5931 or -5932 and requesting Document 1, an index listing information available by fax. Topics include GPS, DGPS, Omega, LORAN, and general radionavigation information, plus Local Notices to Mariners issued within the past 30 days.

NIS information is also available through a computer bulletin board system (BBS). Connection to the bulletin board can be made via phone (703-313-5910; modem parameters, 300 bd–28.8kb; 8 data bits, no parity, 1 stop bit [8N1], asynchronous comms, full duplex).

NIS also disseminates radionavigational information through the Internet at the following addresses: HTTP://www.navcen.uscg.mil; and FTP://ftp.navcen.uscg.mil.

NIS safety and radionavigational messages may also be heard through U.S. Coast Guard broadcast stations on VHF-FM channel 22A and HF-SSB 2670 kHz (see chapter 5, Voice Radio Broadcasts, for a listing of station locations).

Finally, NIS also updates radionavigation information on the National Imagery and Mapping Agency (formerly DMA) bulletin board service NAVINFONET. Users must register off-line before they can use NAVINFONET. To register, contact Marine Navigation Department, NAVINFONET D-44, NIMA Hydrographic/Topographic Center, 4600 Sangamore Road, Bethesda, MD 20816-5003. Information by telephone may be gotten by calling 800-455-0899 or 314-260-1236, but registration must be by mail. NAVINFONET also includes chart and catalogue correction databases, broadcast warnings, light lists, antishipping activity messages, and mobile offshore drilling unit information.

CRUISING GUIDES

Reed's believes that the best complement to proper charts and a copy of our Almanac is a good cruising guide. We are building a database of all the guides available for North America. It is by no means complete—some of these listings are missing information or are somewhat out-dated. If you would like to contribute information or an opinion on a guide you have used, please contact the editor. Note that this database is available at our web site, with a form for submitting comments.

Many thanks to the Nautical Mind bookstore in Toronto for helping us to get this list started.

CARIBBEAN, GENERAL

Landfall Legalese, Vol. 2: The Caribbean
Spears, Alan
A compendium of the legal requirements and protocols for entering and clearing the popular cruising ports throughout the Caribbean. Appendix contains documentation forms and procedures, a crew contract, and other valuable information. 1994, sc, 120p., $26.00

Courtesy Flags Made Easy: Caribbean
Conger, Mary
Includes separate instructions as well as full-sized patterns and designs for the flags of 36 countries located in and around the Caribbean basin or en route to it.
1989, spiral, 131p., illus., $22.00

Cruising Ports: Florida to California via Panama - *Rains, Capt. John*
Invaluable guide for those making this trip. Author is a professinal delivery skipper, and text is based on 100,000 miles of experience.
1993, 3rd ed., sc, 260p., illus., $25.00

The Gentleman's Guide to Passages South
Van Sant, Bruce
5th ed. A handy guide aimed at the live-aboard, long-term cruiser, describing 17 windward passages between Florida and Venezuela. Each passage is divided into an average of four safe and easy sails to a total of 70 anchorages. Includes information on hauling out and reprovisioning, on communications, customs and immigration, and tips of special interest. New edition contains GPS coordinates and Spanish Virgin Islands.
1996, sc, 306p., illus., $25.00

A Cruising Guide to the Caribbean - *Stone, William & Anne Hays*
4th ed. Includes the North coast of South America, Central America, and the Yucatan. Most comprehensive guide available for anyone planning to cruise these waters. Includes a checklist of planning details, description of weather and sea conditions, how to get there, plus sailing directions.
1993, hc, 640p., photos & illus., $35.00

Tales of the Caribbean - *Seyfarth, Fritz*
1991, $11.00

EASTERN CARIBBEAN

A Cruising Guide to the Caribbean
Marshall, Michael W.
Information needed to cruise the waters from Antigua to Venezuela, including Barbados, Trinidad, and Tobago. A brief history of each island is followed by details of the best anchorages, approaches, beaches, diving locations, shore facillties, formalities, places of interest, inter-island routes, landfalls, local radio station frequencies, and hurricane holes.
1992, sc, 160p., photos, $25.00

Cruising Guide to Trinidad & Tobago (1997/98) - *Doyle, Chris*
Features include information about facilities, weather, navigation, stars, customs and immigration, sightseeing, diving, and more. Contains many colour photographs and sketch charts, as well as munerous GPS coordinates.
1996, spiral, 144p., illus., $19.00

The Exuma Guide - *Pavlidis, Stephen J.*
A comprehensive guide loaded with information, detailed sketch charts, and local knowl-

edge. Covers GPS waypoints, marine facilities, customs regulations, beacons and navigational aids, distances, ham and weather radio broadcasts and stations, anchoring tips, tide and current information, and phone numbers.
1995, sc, 214p., photos & illus., $27.00

Following in the Wake of the Buccaneers: Your First Bahamas Voyage - *Grabowski, Steve*
A reassuring, fact-filled guide for novices to Bahamas cruising. Covers history, essential gear and spares, etiquette, documents, navigation, anchoring, and more. Good aerial photos. 1995, sc, 178p., photos & illus., $18.00

Cruising Guide to Trinidad and Tobago, Venezuela and Bonaire - *Doyle, Chris*
1994, $25.00

The Cruising Guide to the Northwest Caribbean - *Calder, Nigel*
Contains 100 chartlets, plus sketches and photographs of difficult anchorage entrances and coral reef passages of the Yucatan coast of Mexico, Belize, Guatemala, Honduras, and the Bay Islands. Covers history, politics, languages, culture, currencies, and boating facilities of the various regions. Detailed passage-making and pilotage instructions.
1991, hc, 288p., photos & illus., $30.00

Cruising Guide to the Leeward Islands
Doyle, Chris 4th ed. A guide to sailing, diving and exploring the Leeward Islands from Anguilla through Dominica. Sketch charts, street maps, colour aerial photos, some local advertising plus lots of piloting information and history. New edition contains GPS co-ordinates and extensive Mariner's Directory.
1996, sc, 448p., photos & illus., $21.00

Cruising Guide to Cuba - *Charles, Simon*
Simon Charles has visited Cuba several times and circumnavigated the island aboard a 34-foot trawler. This guide, based on his detailed logs and observations, covers physical features of the island, weather, population, industry and trade, modern history, travel restrictions, yacht preparation, travel permits and dealing with officials, strategy, routes and destinations, navigation (inc. GPS coordinates), and anchorages.1994, sc, 250p., photos & illus., $26.00

Bermuda Yachting Guide
Bermuda Maritime Museum
Includes information for navigating the island, GPS waypoints, sketch charts, customs, and shopping tips. 1994, sc, $17.00

BVI Bermuda Bound - *Tighe, Capt. T. R.*
1986, $54.00

Abaco: The History of an Out Island and its Cays - *Dodge, Steve*
1983, $14.00

Bahamas-Out Island Odyssey - *Jeffrey, Nan*
A visitor's view of Out Island life. Covers all the related island groups: Bahama Out Islands, Exuma, Eleuthera, Great Abaco, and others.
1995, sc, 200p., photos, $15.00

Bahamas GPS & Loran Handbook
Better Boating Association 1990, $39.00

The Cruising Guide to Abaco 1997
Dodge, Steve Contains diagrams, tide tables, snorkelling and diving maps and tips, advice on crossing from 46lorida to Abaco, tropical medicine and fishing, and a brief history of Abaco. Seventeen charts of the island are reproduced in black and white.
annual, sc, 92p., illus., $11.00

Cove Hopping South to the Virgin Islands
Rogers, J.A.
Covers more than 40 anchorages with illustrations and candid information not found in other cruising guides for the route from the Keys to the Virgins via the Bahamas, the Turks and Caicos, Hispanola, and Puerto Rico.
1996, sc, 256p., photos & illus., $16.00

Virgin Anchorages - *The Moorings*
8th ed.A beautifully illustrated guide featuring colour aerial photographs of 41 of the most exquisite Virgin Island anchorages. Includes sailing directions. A valuable guide and a book to keep as a memento of your voyage. 1995, sc, 86p., photos, $23.00

The Lesser Antilles: Barbados/Grenada to the Virgin Islands
R.C.C. Pilotage Foundation
A comprehensive guide to the Eastern Caribbean covering Barbados, the Grenadines, St. Vincent, St. Lucia, Martinique, Dominica, Guadeloupe, Antigua, Monserrat, St. Barts, and the Virgin Islands. Includes geography, approaches, formalities, and anchorages for each area. Many sketch charts, photos of approaches.
1991, hc, 248p., photos & illus., $57.00

Yachtsman's Guide To The Virgin Islands (1996 ed)(mar/97) - *Fields, Meredith, ed.*
1995, $17.00

Yachtsman's Guide to the Bahamas - *Fields, Meredith, ed.*
annual, $38.00

Yachtsman's Guide to Jamaica
Lethbridge, John
A complete cruising guide to Jamaica, including making landfall, putting together a cruise, weather patterns, sea conditions, and detailed instructions on how to use some 50 ports, harbours and anchorages.
1996, sc, 176p., photos & illus., $31.00

Cruising Guide to the Virgin Islands
Scott, Simon & Nancy Scott, eds.
8th ed. The most popular guide for anyone sailing, diving or fishing the British or U.S. Virgin Islands. Information on navigation, customs, services and facilities, and more. Many sketch charts, photos, some advertising, piloting information blended with local history. Spectacular colour aerial photos.
1996, sc, 288p., photos, $21.00

Virgin Islands Skipper's Handbook
Better Boating Association
A handbook containing charts and aerial views of each harbour, plus information on scuba diving, snorkelling, night life, customs, provisions, and more.
1990, spiral, 96p., illus., $21.00

Yachting Guide to Bermuda
Harris, J. & E. Harris, Eds.
Contains information on planning your trip, navigation, anchorages, and more. 20 Aerial photographs and sketch charts. Contains advertising. First produced in 1977 by Michael Voegeli. 1994, spiral, 144p., illus., $20.00

The Turks and Caicos Islands
Smithers, Amelia 1995, $10.00

Sailor's Guide to the Windward Islands
Doyle, Chris
8th. ed. Covers Martinique, St. Lucia, St. Vincent, the Grenadines, and Grenada. Complete cruising guide, with sketch charts and details about anchorages, as well as a guide for the traveller and tourist to shops, restaurant, and local recipes. 1996, sc, 320p., illus., $21.00

Street's Cruising Guide to the Caribbean: Puerto Rico & Virgin Islands
Street, Donald M., Jr.
Puerto Rico, Passage Islands, U.S. & British Virgin Islands "This newly revised edition by the acknowledged master of Caribbean waters includes coves, harbors, lagoons, straits, beaches, and rocks around and among these islands. It is complete with many new charts of previously uncharted areas as well as those more frequently visited." W. W. Norton & Company, Inc.— ISBN: 0-393-03896-3, 2nd ed 1995, sc, spiral, 256p., illus., $38.00

Street's Cruising Guide to the Caribbean: Martinique to Trinidad
Street, Donald M.
Including Martinique, St. Lucia, St. Vincent, Barbados, Northern Grenadines, Southern Grenadines, Grenada, and Trinidad & Tobago. W. W. Norton & Company, Inc.-- ISBN: 0-393-03523-9, 2nd ed. 1995, sc, 190p., illus., $35.00

Street's Cruising Guide to the Caribbean: Anguilla to Dominica
Street, Donald M., Jr.
Including Anguilla, St. Martin, St. Barts, Saba, Statia, St. Kitts, Nevis, Antigua, Barbuda, Montserrat, Rodonda, Guadeloupe, and Dominica. W. W. Norton & Company, Inc. ISBN: 0-393-03525-5, 1993, sc, 175p., illus., $35.00

St. Maarten Area Cruising Guide
Eiman, William, ed. 1989, $20.00

Sailing Haiti Single Handed - *Rogers, J.A.*
From the Turks & Caicos Islands to Cap Haitien and along Haiti's northern, western and southern coasts.
1996, sc, 92p., photos & illus., $9.00

Street's Cruising Guide to the Caribbean: Venezuela - *Street, Donald M., Jr.*
Including the Golfo de Paria, Isla Margarita, the Peninsula de Araya, El Morro de Barcelona, Puerto Tucacas, the Eastern and Western Offshore Islands, Aruba, Bonaire, and Curacao. W. W. Norton & Company, Inc. ISBN: 0-393-03345-7, sc, 162p., illus., $28.00

CARIBBEAN CENTRAL AMERICA

Panama by Water - *Lambert, Trish*
A comprehensive cruising guide to Panama, its Atlantic and Pacific coasts, and the Canal.
1994, sc, 85p., photos & illus., $21.00

Destination Mexico
Mehaffy, Carolyn & Bob Mehaffy
Everything you need to know about getting yourself and your boat ready for a cruise to Mexico. Includes provisioning, personal gear, medicines, and equipment necessary for a safe voyage. 1995, sc, 183p., illus., $21.00

Cruising Guide to Belize & Mexico's Caribbean Coast - *Rauscher*
2nd ed. Covers chartering, fishing, snorkelling and diving, how to get there, customs, transportation, suggested itineraries, weather, navigation, communications, history, and more. Complete descriptions of the greatest Mayan ruins. Two large fold-out charts of the entire area, plus many photos-aerial and other-and more than 100 sketch charts.
1996, sc, 304p., photos & illus., $38.00

NOTES

NOTES

NOTES

R

Resources

NOTES

NOTES

R

Resources

NOTES

NOTES

R

Resources

NOTES

EPHEMERIS

INTRODUCTION TO REED'S NAUTICAL EPHEMERIS

The ephemeris section of *Reed's Nautical Almanac* continues to carry all the information necessary to calculate the position of a boat by celestial navigation. It has several other uses as well (see page E 19).

This ephemeris is the functional equivalent of the *Nautical Almanac* published annually by the United States Naval Observatory, but it is organized in quite a different and more compact fashion. If you are familiar with the U.S. almanac, please take a moment to read the following section to reorient yourself.

In the U.S. almanac, you are used to finding all the critical data presented three days at a time on a two-page table known as the Daily Pages. *Reed's* presents the same information by *month*, in six pages of tables. In the U.S. book, most data is given hourly; in *Reed's*, the increment varies according to necessity. Some bodies are tabulated for every 2 hours, some for 6 hours, and some for 24 hours. Here is an overview of the **monthly tables:**

First page: Sun and moon general information such as rises and sets, transit (meridian passage), and time of civil twilight. Several phenomena are calculated for 30°N latitude with corrections supplied for other latitudes.

Second page: Sixty navigation stars with not only the customary declination and sidereal hour angle (SHA) but also Greenwich hour angle (GHA) and the astronomer's right ascension (RA). Note that a transit time is also supplied for each star (with daily corrections), a very useful aid for star identification.

Third and fourth pages: GHA and declination of the sun and Aries for every 2 hours.

Fifth page: GHA, declination, and meridian passage of the four navigation planets (Venus, Jupiter, Mars, and Saturn) for each day of the month.

Sixth page: GHA and declination of the moon for every 6 hours.

Reed's correction tables are also different. In the U.S. almanac you find "increments" pages where corrections for all bodies are presented in one continuous table, 2 minutes per page. *Reed's* has different correction tables for different bodies, and most present the correction in several parts. Also, moon and planet GHA and declination corrections are all based on a rate of change given in the monthly tables, a concept somewhat like *v* and *d* corrections.

The various altitude correction tables are similar to the U.S. versions, except in many cases they nicely include the dip correction. Also, both the moon and Polaris tables are much easier to use.

In sum, users of the U.S. *Nautical Almanac* will find *Reed's* ephemeris foreign at first. It takes a little study to comprehend *Reed's* layout, and usually you will have to do a little more addition to compute your corrections. However, *Reed's* is quite compact, and it works!

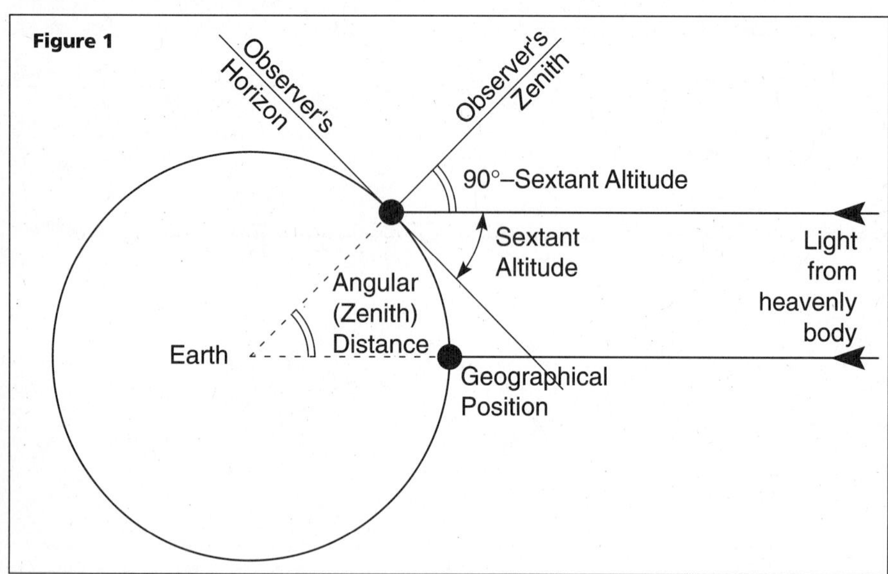

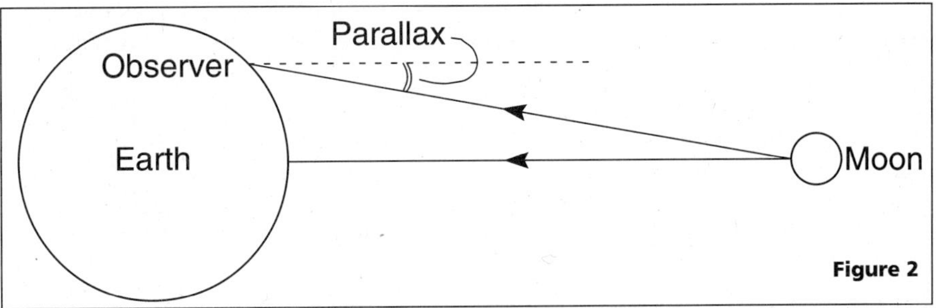

Figure 2

It is worth noting that *Reed's* habit of listing transit times of all bodies makes its ephemeris particularly good for general star-gazing.

CELESTIAL NAVIGATION: AN OVERVIEW

This section is intended as a brief overview of the principles and terminology of celestial navigation.

Figure 1 suggests that by measuring the angle between a heavenly body and the horizon, it is possible to find the angular distance, or *zenith distance* (ZD), between your position and the spot on the earth's surface where the body is directly overhead (its *geographical position,* or GP). By definition, an arc from the center of the earth along its surface can be converted directly to nautical miles. Hence, the ZD and GP provide a position line, just as the range of a charted terrestrial landmark can be used to derive a position line in coastal navigation. For instance, if the corrected sextant altitude in the figure was 42°, then the ZD would be 48° or a 2,880-nautical-mile range between your vessel and the body's GP (60 miles/degree x 48). For this information to be of any practical value, you must:

(a) be able to see a suitable heavenly body;
(b) know the body's geographical position;
(c) be able to correct the sextant altitude for errors such as atmospheric refraction; and
(d) be able to plot the position line on a chart.

FINDING THE POSITION OF A HEAVENLY BODY

By convention, celestial navigation assumes that the sun, moon, planets, and stars are located on the inner surface of a celestial sphere whose center is at the center of the earth and whose poles, equator, and prime meridian are precisely aligned with their terrestrial counterparts. Latitude and longitude, however, are not used.

Declination (Dec) is the celestial equivalent of latitude. It is measured in degrees north or south of the celestial equator. *Greenwich hour angle* (GHA) is equivalent to longitude, except that it is always measured in a westerly direction. Therefore, a star whose declination is N39° and whose GHA is 40° would appear overhead to an observer at 39°N 40°W, while one whose declination is S17° and whose GHA is 219° would be overhead when seen from 17°S 141°E.

The ephemeris predicts the declination and GHA of the sun, the moon, four planets, and sixty stars for every second of one year.

CORRECTING THE SEXTANT ALTITUDE

The *sextant altitude* (Hs) is subject to a number of errors.

Index error (IE) is caused by imperfect adjustment of the sextant. A quick check of index error can be achieved by moving the index arm until the horizon—seen through the sextant—appears as an unbroken line. The sextant should then read 0, but there is often a small discrepancy. Note the size of this discrepancy and whether it is "on the arc" (the index arm is to the left of 0) or "off the arc" (the index arm is to the right of 0).

To apply index error to the sextant altitude, the simple rule is "If it's on the arc, take it off; if it's off the arc, add it on"—that is, an index error "on the arc" should be subtracted from the sextant altitude.

Dip correction arises from the fact that a perfect 90° horizon is always slightly below an observer whose *height of eye* (HE) is above sea level.

The result of applying dip and index error to the sextant altitude is called *apparent altitude* (Ha), the altitude shot with any sextant from any height deck. Apparent altitude is the common value needed to enter many tables in order to determine the final three corrections.

Parallax is caused by the size of the earth and the fact that the geometry of celestial navigation assumes that sights are taken from the center of the earth (Figures 1 and 2). Because the earth is relatively tiny, parallax is really significant only in the case of the moon.

Refraction is caused by the atmosphere distorting the light from heavenly bodies. It is greatest for bodies near the horizon, and as it can vary in extreme conditions, sights with altitudes of less than 10° should be regarded with skepticism.

Semi-diameter arises from the physical size of the body being observed and the fact that tabulated values refer to the center of the body concerned. In the case of the sun and moon, this is about 15′ higher than the lower edge of the visible disc.

PLOTTING A POSITION LINE

Since an angular measurement through the earth's surface from the center can be directly converted to nautical miles, the zenith distance gives the diameter in miles of a circular position line centered on the body's geographical position. In other words, 90° less the corrected altitude is the actual distance from the observer to the body. In practice, this distance is usually far too large to be plotted on a chart, so the results of sights have to be "reduced" to more manageable dimensions (see page E 6.) However, zenith distance is used in both noon sights of the sun and Pole Star sights to calculate and plot your latitude.

USING THE TABLES

The following subsections describe in more detail how to use the ephemeris tables to calculate the necessary data for various types of celestial problems.

SUN SIGHTS

Finding the sun's position

The declination and GHA of the sun are given on the third and fourth pages of each monthly section, tabulated at 2-hour intervals for each day.

The sun's declination changes so slowly that the necessary accuracy can be achieved by mental interpolation. The correction may be positive or negative according to the season. Its sign can be found by looking at the entry that follows to see whether the declination is increasing (+) or decreasing (–).

For instance, if you take a sight at 0925 on January 1, you will observe that the declination is S23°00.4′ at 0800 and S23°00.0′ at 1000. Therefore, the rate of change is –0.4′ every 2 hours. For 0925, you can interpolate a correction of –0.3′ to apply to the 0800 value.

The sun's GHA increases at a rate of 15° per hour, but to simplify the job of interpolating accurately, a GHA correction table for the sun appears on page E 106. Note that the GHA correction must always be added because GHA is always measured to the west of Greenwich.

Sun altitude corrections

Refraction, dip, and parallax tables are on page E 104, and the sun's semi-diameter is given on the first page of each monthly section, but for convenience, all four corrections have been combined in the Sun Altitude Total Correction Table on page E 105. For greater accuracy, another small correction will be found at the foot of this table to account for the sun's apparently changing diameter.

Note that the Sun Altitude Total Correction Table relates only to sights of the sun's lower limb. Sights of the upper limb must be corrected either by using the separate tables or by subtracting twice the semi-diameter. Note also that the separate tables must be used for sun sights of less than 9°.

STAR SIGHTS

Finding the time of twilight

Star sights can be taken only when the star and the horizon are both visible—that is, at twilight. Strictly speaking, *civil twilight* occurs when the sun is 6° below the horizon and *nautical twilight* when the sun is 12° below the horizon. The best time to take star sights is around civil twilight, so this is the twilight used in *Reed's*. It is found on the first of the monthly pages.

The times given for twilight, sunrise, and sunset are correct for 30°N, 0°W. They must therefore be corrected to take account of your own latitude and longitude.

The *latitude correction* for twilight (and sunrise and sunset) is given in the subsidiary table on the right-hand side of the same page. The *longitude correction* is found by adding 4 minutes for every degree of westerly longitude or subtracting 4 minutes for every degree of easterly longitude. The Arc-to-Time Conversion Table on page E 103 helps with this calculation.

Finding the position of a star

The essential information for sixty stars that are bright enough to be visible at twilight is given in the star list on the second page of each monthly section. This list notes the GHA of each star at 0000 on the first day of the month. The tables on page E 106 can be used to correct this value for date and time. An alternative method is to use the star's *sidereal hour angle* (SHA), which gives its position measured westward relative to a meridian on the celestial equator called the *First Point of Aries.*

To find the GHA of a star, first find the GHA of Aries. This value is tabulated at 2-hour intervals on the third and fourth pages of each monthly section. The correction tables on page E 106 simplify the job of interpolation. To the GHA of Aries add the SHA listed for your chosen star. If the result is more than 360°, subtract 360°.

The *declinations* of stars change so slowly that for navigational purposes they can be regarded as being fixed throughout a month and can be taken directly from the monthly star list.

Right ascension (RA) refers to the star's position measured eastward from the First Point of Aries and is usually given in terms of time (1 hour=15°). This value is useful in astronomy, where it is the common measurement of star positions.

Star altitude corrections

For stars, errors due to semi-diameter and parallax are negligible, so sextant altitudes need only be corrected for dip and refraction. It is possible to allow for each error individually, using the tables on page E 104. Both should be subtracted from the observed altitude (Ho). For convenience, refraction and dip have been combined into the Star or Planet Altitude Total Correction Table on page E 108.

PLANET SIGHTS

Of the planets, only Venus, Jupiter, Mars, and Saturn are of navigational significance. The section on planets, pages E 19–21, gives you complete information on the visibility of these planets during the year.

Finding the position of a planet

The *declination* and *GHA* of the navigational planets are presented on the fifth page of each monthly section. They are tabulated at midnight (00h 00m) GMT for each day.

The GHAs of planets increase at about 15° per hour. The rate of change on any particular day is given more accurately in the column headed *Mean var/hr.* To save space, only the minutes are shown: a Mean var/hr value of 3.2 means an hourly increase of 15°03.2′, while a value of 59.4 means an hourly increase of 14°59.4′.

Once you know the rate of change, the Planets GHA Correction Table on pages E 113–14 allows you to interpolate to the nearest hour, while the table on page E 115 permits you to interpolate to the nearest minute and second.

The planets' declination is also shown for midnight GMT on each day, with the mean hourly variation. The Planet Declination Correction Table is on pages E 116–17. As with the sun, it is important to note whether the declination is increasing or decreasing.

Planet altitude corrections

Corrections for planet sights are exactly the same as those for stars, page E 108.

MOON SIGHTS

Finding the position of the moon

The moon is so much closer to earth than any other heavenly body and moves so much faster that finding its position and correcting its observed altitude are much more complicated and error-prone. The layout of the tables and the processes involved, however, are similar to those for the planets.

The moon's *declination* and *GHA* are given on the sixth page of each monthly section, tabulated at 6-hour intervals for each day.

The moon's GHA increases at about 14°30' per hour. The precise rate of increase is given in the column headed Mean var/hr, but to save space only the minutes are shown. Therefore, a Mean var/hr of 27.7 means an hourly increase of 14°27.7′.

Knowing the rate of change, you can use the Moon GHA Correction Table on page E 108 to interpolate to the nearest hour, while the table on pages E 109–10 allows you to interpolate to the nearest minute and second.

Similarly, the declination is shown along with its mean hourly variation. Hourly corrections are calculated by simply multiplying the number of hours times the variation, while the table on page E 111 interpolates for minutes.

Moon altitude corrections

The corrections to be applied to moon sights can be found from the Moon Altitude Total Correction Table on page E 107.

First, however, you need to know the *horizontal parallax* (the parallax error that would occur if the observed altitude were 0°), which can be obtained from the first page of the monthly sections. Taking care to choose the correct part of the table—depending on whether the sight is of the upper or lower limb—find the column in the table corresponding to the value of horizontal parallax and the row corresponding to the sextant altitude. The correction obtained must be added, unless it is for an observation of the upper limb in excess of 64°, in which case it must be subtracted.

Finally, add the correction for *height of eye*, found from the second table on page E 107.

Note: Do not use the standard dip tables in conjunction with the Moon Altitude Total Correction Table.

THE NOON SIGHT

The so-called noon sight—taken at approximately noon local time—gives the simplest of all astro position lines. It is more correctly called a *meridian passage* sight, because it is taken at the moment when the sun transits the observer's meridian—that is, when it is at its highest and directly south or north of the observer.

The first step in taking a noon sight is to determine when meridian passage will occur. This time can be found from the column headed Transit in the top half of the first page in the appropriate monthly section. Here you will find the time (GMT) when the sun crosses the Greenwich meridian (0°).

Use the Arc to Time Conversion Table on page E 103, or calculate using 1°=4 minutes, to correct the time of transit for your *dead-reckoning* (DR) longitude. An approximation is all that is required, because in practice it is usual to measure the sun's altitude a few minutes

Figure 3

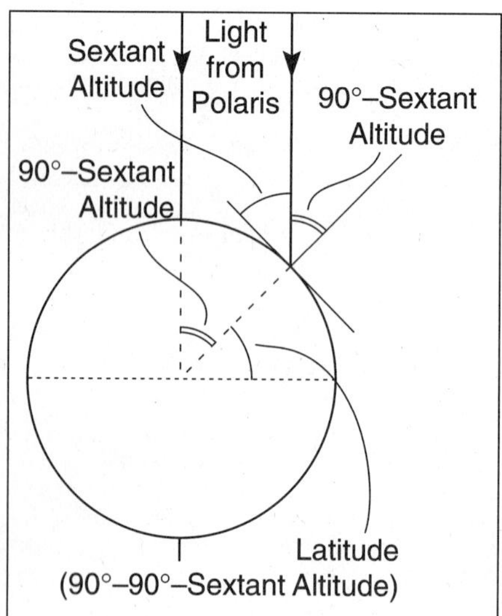

before the expected time and to repeat the observation at intervals until the sun stops rising in the sky and starts to descend.

Having corrected the sextant altitude in the usual way, subtract the observed altitude from 90° to determine the *zenith distance* (ZD).

Latitude is then equal to either the sum or the difference of the declination and the zenith distance. It is usually obvious whether declination should be added or subtracted, but if there is any ambiguity, note whether the sun is to your north or south, and give the zenith distance the opposite name. If the zenith distance and declination have the same name, add them together. If they have different names, take the smaller from the larger, and name the latitude after the larger.

LATITUDE BY POLARIS

If Polaris were perfectly aligned with the pole, as its alternative name of Pole Star suggests, its altitude (after correction for the usual errors) would be exactly the same as the observer's latitude. Figure 3 illustrates the principle. It is not, in fact, in perfect alignment but is sufficiently close that only minor arithmetical adjustments are required. Polaris is not a particularly bright star, but it can usually be found at twilight by setting the sextant index arm to your approximate latitude and scanning the northern sky with the sextant.

A table giving the correction to be applied to the observed altitude is on page E 102. It is first necessary to find the *local hour angle* (LHA) of Aries. This is done by finding the GHA of Aries and subtracting from it your longitude if you are west of Greenwich or adding to it your longitude if you are east of Greenwich.

SIGHT REDUCTION

THE INTERCEPT METHOD

Astro position lines, aside from noon and Polaris latitude lines, are most easily plotted using the *Marc St. Hilaire, or Intercept, method.* This method involves calculating what the altitude of a heavenly body would be as seen from some convenient nearby position and comparing this value with its *observed altitude* (Ho). If the observed altitude (Ho) is greater than the *calculated altitude* (Hc), the true position must be closer to the heavenly body from the *assumed position* (AP), and vice versa. The size of the discrepancy shows how much closer (or farther away) and is called the *intercept.*

Figure 4 illustrates the whole celestial plot. The beauty of the intercept method is that only the assumed postion (AP), intercept, and position line need to be drawn, all of which can be done on a manageable scale. You do need to know the direction as well as the length of the intercept. This bearing will be the same as the actual bearing of the heavenly body, but it is easier and more accurate to calculate it than to measure it.

Figure 4 also illustrates the concept that although a range plotted directly from the *geographical position* (GP) would form a circle, this position circle is so big that a section up to about 60 miles in length can be regarded as a straight line that runs at right angles to the intercept.

The calculated altitude and bearing can both be found from precomputed tables such as the *Sight Reduction Tables for Air Navigation (H.O. 249);* by the more compact versine and log cosine tables and the ABC tables, found in *Reed's* on pages E 124–43 and E 118–23 respectively; or with an electronic calculator.

Whichever method you choose, the calculations require the declination of the body; the latitude of an assumed position (it is usually necessary to round out your DR to the nearest whole degree); and the local hour angle (LHA).

The LHA is the difference in longitude between your assumed position and the geographical position of the heavenly body. It is found by subtracting westerly longitude from the GHA (+360° if necessary) or by adding easterly longitude (–360° if necessary). You will find it most convenient to make the assumed position (AP) such that LHA becomes a whole number of degrees. If the DR is 34°50′N, 36°40′W, for instance, and the GHA of the body is 110°35.4′, a suitable AP would be 35°N, 36°35.4′W.

The various sight reduction methods eventually give you a calculated altitude (Hc) and an azimuth. The azimuth given by the calculator method or *Reed's* tables is always less than 90° and is the bearing relative to the North or South Pole, measuring east or west. For instance, an azimuth of 70° could correspond to a bearing of 070° (N70°E), 110° (S70°E), 250° (S70°W), or 290° (N70°W). The easiest way to decide which is correct is to note the approximate bearing of the heavenly body at the time of the sight. Alternatively, it can be found from the rules at the end of the C tables (see pages E 122–23).

ELECTRONIC SIGHT REDUCTION

For electronic sight reduction a calculator capable of handling trigonometric functions is essential, and at least one memory is useful, but be prepared to store data on a separate notepad as well.

The first task is to convert the declination into degrees and decimals and to select a suitable assumed position that gives a latitude and an LHA in whole degrees.

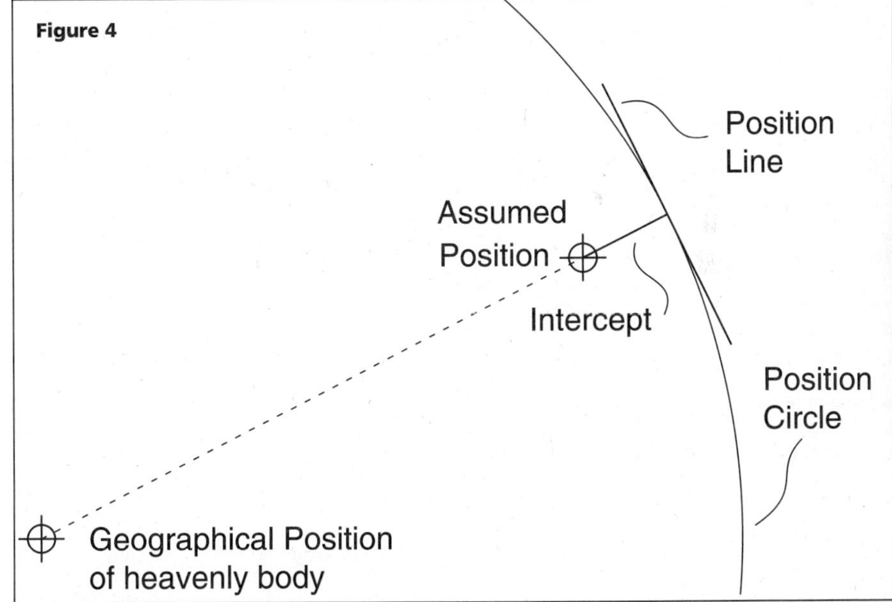

Then the calculated altitude (Hc) is given by the formula:

Sin Hc = Cos LHA x Cos Lat x Cos Dec ± Sin Lat x Sin Dec

Note that if Lat and Dec have the same name, they should be added, and if they have opposite names, the smaller value should be subtracted.

The *azimuth angle* is given by the formula:

$$\text{Sin Az} = \frac{\text{Sin LHA} \times \text{Cos Dec}}{\text{Cos Hc}}$$

The exact process required to carry out these calculations depends on the notations and defaults of the calculator being used. The sun sight example following should get you started.

Figure 5

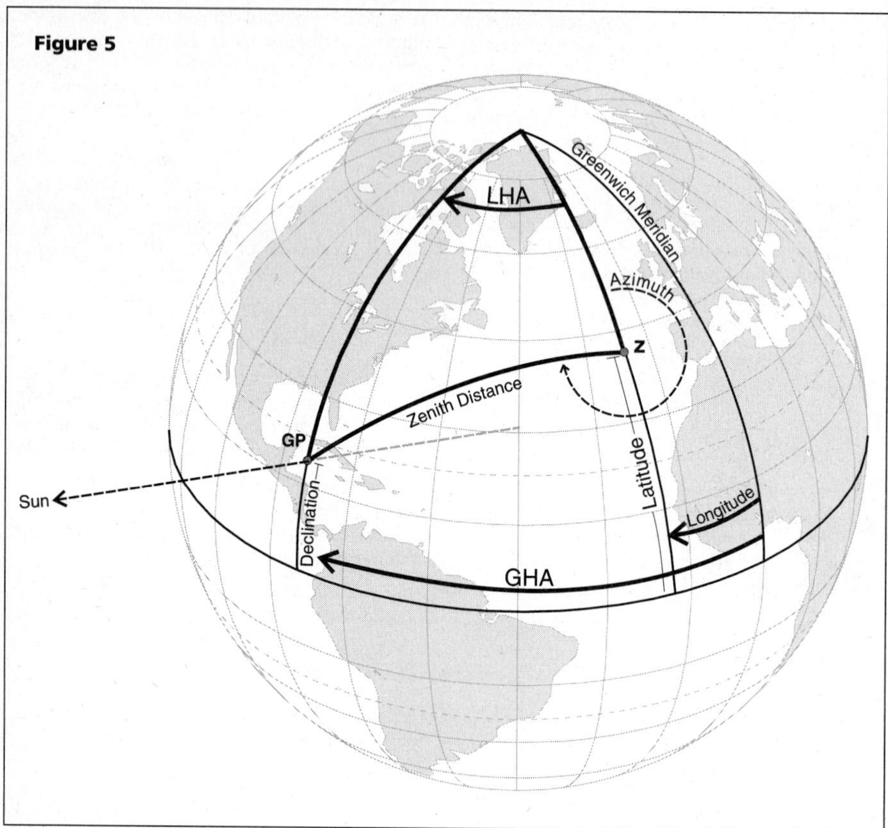

The Celestial Triangle Illustrated

In Figure 5, the navigator is on passage from Bermuda to Europe. His sun sight gives him his zenith distance in angular form (**Ho**) from the **G**eographical **P**osition of the sun, which he can determine exactly by taking the exact time and date of his sight to the ephemeris and looking up the sun's **Dec**lination and **G**reenwich **H**our **A**ngle. To make use of this information, the navigator then solves the celestial triangle with any of several means of sight reduction. He estimates an **A**ssumed **P**osition close to his real position so that he can derive two sides and one angle of the triangle. 90°– assumed latitude = one side. 90°– **Dec**. sun = second side. GHA sun – assumed longitude = the angle **LHA**. Sight reduction yields Hc (90° – the Zenith Distance), which is compared to Ho (90° – measured Zenith Distance). The resulting intercept is plotted using the AP and the Azimuth of the sun (also derived from the sight reduction). The final result is a line of position (see Figure 4).

Note that from the sun's GP we can deduce that the season is either late spring or early summer and that this sight was taken in the late afternoon.

SIGHT REDUCTION BY TABLES

The same formulae could be used in conjunction with a set of mathematical tables instead of a pocket calculator, but the calculations can be made simpler by applying the *versine formula* (the versine of an angle is equal to 1 minus its cosine):

Vers Zenith Distance = Vers LHA x Cos Lat x Cos Dec + Vers (Lat ± Dec)

Note that if Lat and Dec have the same name, they should be *subtracted,* and if they have opposite names, they should be *added.*

Once you find the computed versine value in the table, you read the column and row headings as the zenith distance. Hc is then found by subtracting the zenith distance from 90°.

To simplify the multiplication process, the versine tables (pages E 124–37) include logarithms of versines in bold type, and the tables on pages E 138–43 are log cosines. Adding logs is the equivalent of multiplying. You may subtract 10 as needed (as you would subtract 360° from an angle greater than 360°). Hence, the formula in terms of *Reed's* tables is:

Natural Versine Zenith Distance = Nat Vers (Log Vers LHA + Log Cos Lat + Log Cos Dec) + Nat Vers (Lat ± Dec)

The azimuth is calculated with the ABC Tables on pages E 118–23.

A is found from the A table, using LHA to find the correct column and latitude to find the row. If the LHA is between 90° and 270°, *A* is negative.

B is found from the B table, again with LHA determining the correct column, but this time with declination determining the row. B is negative if the latitude and declination have the same name (i.e., are both south or both north).

The C table is used to convert the sum of *A* and *B* into the azimuth. If C is positive, the azimuth is the opposite name (north or south) as your latitude, and if the LHA is less than 180°, the azimuth is westerly.

WORKED EXAMPLES

Noon sight
Mer Pass November 4; DR: 35°00′N, 35°40′W
Sext Alt of sun lower limb (LL): 39°15.4′
Index Error (IE), 0; Height of Eye (HE), 10ft

1. Find the time of meridian passage on Nov. 4.

Transit (from page E 90)	1144 GMT
Correct for DR longitude	
35°40′ to time (from page E 103)	0223
	1407 GMT

2. Calculate declination of sun from the monthly ephemeris.

Dec at 1400 GMT (page E 92)	S 15°25.1′
Correct for 7 min by interpolation	+ 0.1′
(7m/2h = .058 x +1.5 (2h diff)= +0.1′)	
Dec at 1407 GMT	S 15°25.2′

3. Calculate observed altitude (Ho).

Sextant altitude	39°15.4′
Total correction (from page E 105)	+ 11.9′
Monthly correction (from page E 105)	+ 0.2′
Observed altitude (Ho)	39°27.5′

4. Calculate latitude.

Zenith	90°00.0′
Subtract Ho	– 39°27.5′
Zenith distance	(N) 50°32.5′
Declination	–S 15°25.2′
Latitude	35°07.3′N

Morning/afternoon sun sight
15h 22m 42s GMT Nov 4; DR: 34°54′N, 35°50′W
Sext Alt of sun LL: 36°16.3′
IE, 0; HE, 10ft

1. Find the sun's position at 15h 22m 42s GMT on November 4 from the monthly ephemeris.

GHA at 1400 (from page E 92) GHA	34°06.3′
Correct for 1h 22m (from page E 106)	20°30.0′
Correct for 42s (from page E 106)	10.5′
GHA at 15h 22m 42s	54°46.8′

Dec at 1400 (from page E 92)	S 15°25.1′
Correct for 1h 22m by interpolation	+ 1.0′
Dec at 15h 22m 42s	S 15°26.1′

2. Correct the sextant altitude.

Sextant altitude	36°16.3′
Sun total correction (from page E 105)	+ 11.7′
Monthly correction (from page E 105)	+ 0.2′
Observed altitude	36°28.2′

3. Determine the assumed position and calculate the LHA:

GHA	54°46.8′
Assumed longitude	35°46.8′W
LHA	19°00.0′
Assumed latitude	35°N

4. Reduce the sight. This example uses the calculator method. See the next example to use *Reed's* tables.

Convert declination to degrees and decimals:

15 degrees	15.000°
26.1 minutes ÷ 60	0.435°
	15.435°

Find the intercept. Depending on the particular calculator the key sequence looks something like this (remember to use degree notation).

LHA [COS] [×] Lat [COS] [×] Dec [COS] [=]

19 [COS] [×] 35 [COS] [×] 15.435 [COS] [=]

In this case the answer is 0.7466. Write down this interim answer (or use the store function of your calculator) for use in the next calculation (the – is because Lat is N and Dec is S):

Lat [SIN] [×] Dec [SIN] [–] 0.7466 [=] Hc [aSIN]

35 [SIN] [×] 15.435 [SIN] [–] 0.7466 [=] Hc [aSIN]

In this case the answer is 36.437 (ignore sign), which is the calculated altitude. Write it down (for safety), and then, with the answer still on the display, convert the decimal degrees to minutes:

0.437 [×] 60 [=] 26.2

Calculated altitude (Hc) =	36°26.2′
Observed altitude (Ho) =	36°28.2′
Intercept	toward 2.0′

Find the azimuth/bearing:

LHA [SIN] [×] Dec [COS] [/] Hc [COS] [=]
Azimuth [ASIN]

19 [SIN] [×] 15.435 [COS] [/] 36.437 [COS] [=]
Azimuth [ASIN]

In this case the answer is 22.960, so the azimuth angle is 23° — actually S23°W, or 203°.

Evening stars/planet sight

2000 GMT Nov 4; DR: 34°41′N, 36°35′W
Sext Alt Jupiter: 32°21.4′ at 19h 56m 26s GMT
Sext Alt Altair: 62°41.5′ at 20h 01m 19s GMT
IE, 0; HE, 10ft

1. Find the time of evening twilight at 35°N, 36°35′W on November 4.

Twilight (from page E 90)	1736 GMT
Latitude corr (from page E 90)	– 0008
	1728 GMT
Longitude corr (from page E 103)	+ 0226
Corrected time of twilight	1954 GMT

2. Find the position of Jupiter at 19h 56m 26s GMT on November 4.

GHA at 0000 (from page E 94)	53°12.2′
Mean Var per hour: 15°02.5′	
Correction for 19h (from page E 114)	285°47.5′
Correction for 56m (from page E 115)	14°02.3′
Correction for 26s (from page E 115)	6.5′
GHA at 19h 56m 26s	353°08.5′

Dec at 0000 (from page E 94)	S 5°58.2′
Mean Var per hour: 0.0′	
Corr for 19h 56m*	+ 0.4′
Dec	S 5°58.6′

*When Mean Var < 0.5, it is often more accurate to interpolate to next day

3. Correct the sextant altitude.

Sextant altitude	32°21.4′
Planet total corr (from page E 108)	– 4.6′
Observed altitude (Ho)	32°16.8′

4. Determine the assumed position and calculate the LHA.

GHA	353°08.5′
Assumed longitude	36°08.5′W
LHA	317°00.0′
Assumed latitude	35°N

5. Reduce the sight with *Reed's* tables. Using the versine tables, find the intercept.

Log versine LHA (from page E 130)	9.4292
Log cos Lat (from page E 140)	9.9134
Log cos Dec (from page E 139)	9.9976
Total	29.3402
Delete tens digit	9.3402
Convert Log to Nat (from page E 129)	0.2188

Add Lat and Dec (because opposite names):

Dec	S 5°58.6′
Lat	N 35°00.0′
	40°58.6′

Nat versine of 40°58.6′ (page E 129)	0.2450
Add previous answer #	0.2188
Nat versine ZD (page E 132)	0.4638
Zenith Distance (ZD)	57°34.5′
Subtract from 90° (= 89° 60)	90°00.0′
Hc	32°25.5′
Ho	32°16.8′
Intercept	away 8.7′

6. Find the bearing.

A (from page E 120)	+ 0.75
B (from page E 120)	+ 0.15
A + B	+ 0.90
C (from page E 122)	S 54°E
Bearing (180° – 54°)	126°

7. Find the position of Altair at 20h 01m 19s GMT on November 4.

GHA Aries at 2000 (from page E 92)	343°51.4′
Corr for 0h 1m (from page E 106)	0°15.0′
Correction for 19s (from page E 106)	4.8′
GHA Aries at 20h 01m 19s	344°11.2′
SHA Altair (from page E 91)	62°19.4′
GHA Altair	406°30.6′
Less 360°	46°30.6′
Assumed longitude	W 36°30.6′
LHA	10°00.0′
Assumed latitude	N 35°
Dec Altair (from page E 91)	N 8°52.1′

Calculate #7 by electronic or tabular methods: Intercept 19.2′ toward, bearing 202°.

Latitude by Polaris

20h 02m GMT Nov 4; DR: 34°41′N, 36°35′W

Sext Alt Polaris: 34°45.2′
IE, 0; HE, 10ft

GHA Aries at 2000 (from page E 92)	343°51.4′
Corr for 0h 2m (from page E 106)	0°30.1′
GHA Aries at 2002	344°21.5′
DR longitude	36°35.0′
LHA Aries	307°46.5′
Sextant altitude	34°45.2′
Star altitude corr (from page E 108)	– 4.4′
Observed altitude (Ho)	34°40.8′
Polaris correction (from page E 102)	+ 0.0′
Latitude	34°40.8′

Moon sight

21h 43m 07s GMT Nov 4; DR: 34°34′N, 36°52′W
Sext Alt Moon LL: 16°03.3′
IE, 0; HE, 10ft

GHA moon at 1800 (from page E 95)	265°08.7′
Mean Var per hour (p. E 95): 14°24.4′	
Correction for 3h (from page E 108)	43°13.2′
Correction for 43m (from page E 109)	10°19.5′
Correction for 07s (from page E 109)	1.7′
GHA at 21h 43m 07s	318°43.1′
Dec at 1800 (from page E 95)	N 12°52.5′
Mean Var per hour: +9.3′	
Correction for 3h (3 ↔ 9.3)	27.9′
Correction for 43m (from page E 111)	6.6′
Dec at 21h 43m 27s	N 13°27.0′
Sextant altitude (Hs)	16°03.3′
Horizontal par (from page E 90): 61.4′	
Moon total corr (from page E 107)	62.2′
Moon dip corr (from page E 107)	6.7′
Observed altitude (Ho)	17°12.2′

Calculate intercept by electronic or tabular methods:

AP: N 35° W 36°43.1′
Intercept 11.8′ away, bearing 086°.

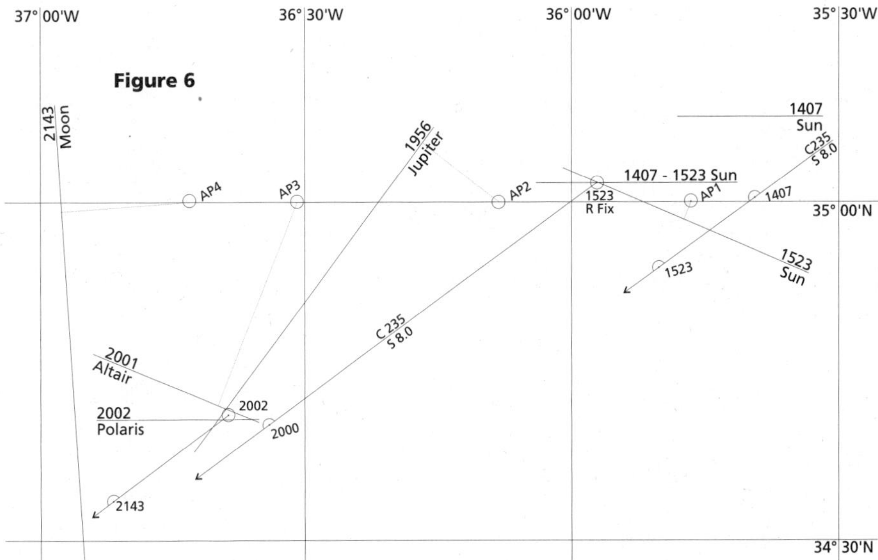

Figure 6 is a plot of the various example sights. The noon sun line was run with the DR to create a 15:35 running fix. This vessel seems to be experiencing a westerly current set.

USING A CALCULATOR FOR SIGHT REDUCTION

GHA AND DECLINATION OF THE SUN

Greenwich hour angle (GHA) and declination (Dec) of the sun for 1998 can be calculated using an electronic calculator and the following sets of monthly polynomial coefficients.

The date and time in GMT are used to form the interpolation factor *p*. This can be expressed as $p = d/32$ where *d* is the sum of the month and decimal of a day. Then GHA–GMT in hours and Dec in degrees are calculated from polynomial expressions of the form:

$$a_0 + a_1 p + a_2 p^2 + a_3 p^3 + a_4 p^4$$

where *p* is the interpolating factor. The simplest way of evaluating this is to use the nested form:

$$(((a_4 p + a_3) p + a_2) p + a_1) p + a_0$$

Example. Calculate GHA and Dec of sun on July 23, 1998, at 17h 02m 15s GMT.

GMT = 17.0375h and the interpolation factor is

$p = (23 + 17.0375/24)/32 = 0.740934$

$\text{GHA} - \text{GMT} = 11.94133\text{h} - 0.10694\text{h}p + 0.03860\text{h}p^2 + 0.02382\text{h}p^3 - 0.00283\text{h}p^4$

$= 11.89212\text{h}$

Hence, GHA = 11.89212h + GMT = 11.89212h + 17.0375h

Remove multiples of 24h from GHA and multiply by 15 to convert from hours to degrees.

Then, GHA = 73.9443° = 73°56.7′

$\text{Dec} = 23.1939° - 1.8427°p - 3.4716°p^2 + 0.1562°p^3 + 0.0835°p^4$

= 20.0114° = N20°00.7′

Semi-diameter of the sun

Jan 1–Feb 4	16.3′	Apr 20–May 14	15.9′	Oct 12–Nov 3	16.1′
Feb 5–Mar 5	16.2′	May 15–Aug 25	15.8′	Nov 4–Dec 2	16.2′
Mar 6–Mar 28	16.1′	Aug 26–Sep 20	15.9′	Dec 3–Dec 31	16.3′
Mar 29–Apr 19	16.0′	Sep 21–Oct 11	16.0′		

To correct for the effect of parallax of the sun it is normally sufficient to add 0.1′ to all observed altitudes less than 70°. If greater accuracy is required, the correction is 0.15′ x cos altitude.

Monthly polynomial coefficients for the sun, 1998

		JANUARY		FEBRUARY		MARCH		APRIL	
		GHA–GMT h	Dec °	GHA–GMT h	Dec °	GHA–GMT h	Dec °	GHA–GMT h	Dec °
Sun	a_0	11.95322	−23.1082	11.77735	−17.4958	11.78897	−8.1231	11.92698	3.9908
	a_1	−0.25705	2.2863	−0.08379	8.8751	0.09493	12.0792	0.15857	12.4148
	a_2	0.04534	3.9577	0.12537	2.6332	0.08344	0.9807	−0.00343	−0.6624
	a_3	0.04292	−0.2405	−0.01398	−0.6032	−0.03789	−0.5667	−0.04208	−0.4861
	a_4	−0.00958	−0.1113	−0.00288	−0.0083	0.00248	0.0081	0.00897	−0.0064
check sum		11.77485	−17.2160	11.80207	−6.5990	11.93193	4.3782	12.04901	15.2507

		MAY		JUNE		JULY		AUGUST	
		GHA-GMT h	Dec °	GHA-GMT h	Dec °	GHA-GMT h	Dec °	GHA-GMT h	Dec °
Sun	a_0	12.04488	14.6420	12.04047	21.8526	11.94133	23.1939	11.89313	18.3664
	a_1	0.07022	9.8704	−0.07650	4.6355	−0.10694	−1.8427	0.02414	−7.8273
	a_2	−0.06673	−2.0777	−0.05420	−3.2476	0.03860	−3.4716	0.09279	−2.5518
	a_3	−0.02476	−0.4783	0.01655	−0.2860	0.02382	0.1562	−0.01266	0.4122
	a_4	0.01441	0.0378	0.00848	0.1106	−0.00283	0.0835	−0.00087	0.0315
check sum		12.03802	21.9942	11.93480	23.0651	11.89398	18.1193	11.99653	8.4310

		SEPTEMBER		OCTOBER		NOVEMBER		DECEMBER	
		GHA-GMT h	Dec °	GHA-GMT h	Dec °	GHA-GMT h	Dec °	GHA-GMT h	Dec °
Sun	a_0	11.99132	8.7914	12.16313	−2.6340	12.27228	−13.9668	12.19282	−21.5701
	a_1	0.16493	−11.4976	0.17606	−12.4319	0.02582	−10.4156	−0.18847	−5.2878
	a_2	0.04960	−1.1778	−0.03327	0.2370	−0.11008	1.8820	−0.09969	3.4855
	a_3	−0.03512	0.4386	−0.04128	0.4924	−0.02362	0.6448	0.03441	0.4873
	a_4	0.00327	0.0355	0.00835	0.0461	0.01626	−0.0312	0.00834	−0.1674
check sum		12.17400	−3.4099	12.27299	−14.2904	12.18066	−21.8868	11.94741	−23.0525

GHA ARIES

Calculate the apparent Greenwich hour angle of Aries, in degrees, for any date and time (UT) during 1998, from the following formula:

$$\text{GHA Aries} = 99.4596° + 0.9856474\,(D + \text{UT}/24) + 15\,\text{UT} + \text{NUT}$$

where

$\text{NUT} = -0.0044° \sin(163.8° - 0.053D) - 0.0003° \sin(198.9° + 1.971D)$

D = day of year (D = 1 on January 1)

UT = Universal time in hours

If necessary put GHA Aries in the range 0 to 360 degrees.

The day of the year (D) within any given year may be calculated from the following expressions,

$$D = [\,275\,\boldsymbol{m}/9\,] - (\boldsymbol{g}+1)\,[\,(\boldsymbol{m}+9)/12\,] - 30 + \boldsymbol{d}$$

where

m is the month number (January = 1, February = 2, etc.),

d is the day of the month,

[x] means take the integer part of x, and

g = 1 for a non-leap year

g = 0 for a leap year.

For 1998, g =1

During 1998, this formula is accurate to 0.02′. For other dates the error will be larger.

Example:

What is GHA Aries at 20h 01m 19s GMT on November 4th?

UT = 20 + (1/60) + (19/3600) = 20.02194

D = [275 * 11 / 9] – (1 + 1) * [(11 + 9) / 12] – 30 + 4

D = 336 – 2 – 30 + 4 = 308 (verify with data on page E 90)

NUT = –0.0044 * sin(163.8 – (0.053 * 308)) – 0.0003 * sin(198.9 + (1.971 * 308))

NUT = –0.0044 * sin(147.476) – 0.0003 * sin(805.968) = –0.0026649

GHA Aries = 99.4596 + 0.9856474 * (308 + 20.02194 / 24) + 15 * 20.02194 – 0.0026649

GHA Aries = 99.4596 + 304.40167 + 300.3291 – 0.0026649 = 704.18771 or 344.18771°

GHA Aries = 344°11.3′ (compare this with 344°11.2′ in Altair example page E 11)

Both these pages were prepared by HM Nautical Almanac Office, Royal Greenwich Observatory, reproduced with permission from data supplied by the Science and Engineering Research Council.

SIGHT REDUCTION FORM FOR USE WITH REED'S TABLES

Date			Sight No.	
WT	– –		Body	
WE –F +S	– –		HE	ft / m
ZT	– –		Hs	° . '
ZD	– –		IC	° . '
UT	– –		Dip	. '
G Date			Ha	° . '
DR Lat	° . '	N S	Main Corr.	° . '
DR Lo	° . '	W E	Addl Corr.	° . '
Asm Lat	° . '	N S	Ho	° . '

GHA _____at		GMT	° . '
Mean Var/Hr	° . '		
Corr. For		hr	° . '
Corr. For		min	° . '
Corr. For		sec	° . '
Total GHA		Less 360°	° . '

SHA_______		° . '
Total GHA	Less 360°	° . '

Dec	GMT	° . '	N S
Mean Var/Hr	° . '		
Corr. For	hr	° . '	
Corr. For	min	° . '	
Total Dec		° . '	N S

West Longitude		East Longitude	
GHA	° '	Asm LHA	° 00'
Asm LHA (–)	° 00 '	GHA (–)	° '
Asm Lo	° ' W	Asm Lo	° ' E

Log Versine Asm LHA	
Log Cos Asm Lat (+)	
Log Cos Dec (+)	
Total	
Delete Tens Digit	
Convert Log to Nat Versine	

Larger Of Asm Lat Or Dec	° . '
Smaller Of Asm Lat Or Dec	° . '
If Same Name Subtract If Opposite Name Add	° . '
Nat Versine Of Sum/Diff	
Nat Versine From Above (+)	
Natural Versine Of ZD	
	89 ° 60 . 0 '
Zenith Distance (–)	° . '
Hc	° . '
Ho (from above)	° . '
Intercept Ho > Hc **T** oward Hc > Ho **A** way (–)	° . '

A	B	A+B	C	Zn

TIME

Zone Time is a system that rigidly divides the world into 25 time zones, all relating to the Prime Meridian that passes through Greenwich, England. Zone Time 0 is also known as both Greenwich Mean Time (GMT) and Universal Coordinated Time (UTC).

Standard Time zones are created by individual countries according to their needs. While Standard Times often equal Zone Times, they do not necessarily do so. The table on the opposite page lists countries where Standard Time varies from Zone Time.

All the tables in *Reed's* ephemeris section are based on GMT. Some countries keep **Daylight Saving Time** during the summer months. When Daylight Saving Time is in effect, clocks are advanced 1 hour over Standard Time.

MEASURING TIME: DEFINITIONS

Mean time. The earth does not orbit uniformly round the sun, but for the purposes of time-keeping it is assumed that it does.

Mean solar day (1) = 24 mean solar hours. The average time taken for the sun to make two successive transits of the same meridian. This represents slightly more than a 360° rotation of the earth, because the earth is constantly moving on in its orbit and will therefore have to revolve a little farther in order to present the same face to the sun once more.

Sidereal day (1) = 23h 56m 04.1s of mean solar time. The time taken for the earth to complete a 360° rotation. Because the stars are so distant, earth's progress along its orbit is relatively insignificant. The sidereal day can therefore be treated as the time taken for a star to make two transits of the same meridian.

Lunar day (1) = approx 24h 50m of mean solar time (average).

Calendar month (1) = 28, 29, 30, or 31 days, depending on the month.

Lunar month (or **lunation**, or **synodical month**). The time interval between successive new moons—about 29½ mean solar days. As with the mean solar day, this represents slightly more than a 360° rotation.

Common and leap years. Under the Gregorian calendar, a common year consists of 365 calendar days, a leap year of 366 calendar days. This would suggest that with leap years every 4 years, the average year works out at 365.25 days. In fact, this figure is close, but not close enough, to the length of the mean solar year, so leap years are defined as those years that are divisible by 4 (as 1992, 1996, etc.), excluding century years unless they are divisible by 400; thus the year 2000 will be a leap year, but 2100 a common year. One complete cycle of the Gregorian calendar takes 400 years, in which there are:

97 (leap) years of 366 days each	=	35,502
303 years of 365 days each	=	110,595
Therefore, the total number of days	=	146,097

The mean length of year over the whole cycle is thus 146,097/400 days, or 365.2425 days. This is known as the **civil year.**

TIME ZONES

Zone	Boundary
	0° Greenwich meridian
ZONE **0**	
	7°30′W
ZONE **+1**	
	22°30′W
ZONE **+2**	
	37°30′W
ZONE **+3**	
	52°30′W
ZONE **+4**	
	67°30′W
ZONE **+5**	
	82°30′W
ZONE **+6**	
	97°30′W
ZONE **+7**	
	112°30′W
ZONE **+8**	
	127°30′W
ZONE **+9**	
	142°30′W
ZONE **+10**	
	157°30′W
ZONE **+11**	
	172°30′W
ZONE **+12**	
	180°00′ Intl Date Line
ZONE **–12**	
	172°30′E
ZONE **–11**	
	157°30′E
ZONE **–10**	
	142°30′E
ZONE **–9**	
	127°30′E
ZONE **–8**	
	112°30′E
ZONE **–7**	
	97°30′E
ZONE **–6**	
	82°30′E
ZONE **–5**	
	67°30′E
ZONE **–4**	
	52°30′E
ZONE **–3**	
	37°30′E
ZONE **–2**	
	22°30′E
ZONE **–1**	
	7°30′E
ZONE **0**	
	0°Greenwich meridian

Note: The International Date Line does not follow the 180° meridian throughout its length, but deviates in order to avoid populated areas.

STANDARD TIMES

Places normally keeping GMT

Ascension Island
Burkina-Faso
† Canary Islands
† Channel Islands
† Faeroes
Gambia
Ghana
Guinea-Bissau
Guinea
Iceland
† Ireland (Rep. of)
Ivory Coast
Liberia
† Madeira
Mali
Mauritania
Morocco
St Helena
São Tomé/Principe
Senegal
Sierra Leone
Togo Republic
Tristan da Cunha
† United Kingdom

† May keep British Summer Time, one hour in advance of GMT; in 1998 effective period March 29 0100h to Oct 25 0100h GMT.

Areas keeping times fast on GMT

Add to GMT to find Standard Time; subtract from Standard Time to find GMT.

* Albania 1h
Algeria 1h
Andaman Is 5fih
* Angola 1h
Australia:
* Aus Cap Ter 10h
* Lord Howe I 10fih
* NSW 10h
Queensland 10h
* Tasmania 10h
* Victoria 10h
* South Aust 9fih
N Territory 9fih
West Aust 8h
* Austria 1h
Bahrain 3h
* Balearic Is 1h
Bangladesh 6h
* Belgium 1h
* Bulgaria 2h
Cameroon 1h
China 8h
Congo 1h
* Corsica 1h
* Crete 2h
* Cyprus 2h
* Denmark 1h
* Egypt 2h
* Estonia 2h
Ethiopia 3h
Fiji 12h
* Finland 2h
* France 1h
* Germany 1h
* Gibraltar 1h
* Greece 2h
Guam 10h
Hong Kong 8h
* Hungary 1h
India 5fih
Indonesia 7–9h
* Iran 3fih
* Iraq 3h
* Israel 2h
* Italy 1h
Japan 9h
* Jordan 2h
Kenya 3h
Korea (N & S) 9h
Laos 7h
* Latvia 2h
* Lebanon 2h
* Libya 1h
* Lithuania 2h
Madagascar 3h
Malawi 2h
Malaysia 8h
* Malta 1h
Marshall Is 12h
Mauritius 4h
* Monaco 1h
Mozambique 2h
Myanmar 6fih
* New Zealand 12h
Nigeria 1h
* Netherlands 1h
* Norway 1h
Pakistan 5h
Papua Nw Guin 10h
Philippines 8h
* Poland 1h
* Portugal 1h
* Romania 2h
Russia 3fih
Saudi Arabia 3h
Seychelles 4h
Singapore 8h
Solomon Is 11h
Somalia 3h
South Africa 2h
* Spain 1h
Sri Lanka 5fih
* Sweden 1h
* Switzerland 1h
* Syria 2h
Taiwan 8h
Tanzania 3h
Thailand 7h
Tunisia 1h
* Turkey 2h
Vanuatu 11h
Vietnam 7h
* Yugoslavia (former) 1h
Zambia 2h
Zimbabwe 2h

* May keep Daylight Saving Time, effective dates vary.

Areas keeping times slow on GMT

Subtract from GMT to find Standard Time; add to Standard Time to find GMT.

Argentina 3h
Aruba 4h
* Azores 1h
* Bahamas 5h
Barbados 4h
Belize 6h
* Bermuda 4h
* Brazil E 3h
* Brazil W 4h
Canada:
* Newfndland 3fih
* Atlantic Zone 4h
* Eastern Zone 5h
* Central Zone 6h
* Mountain Zn 7h
* Pacific Zone 8h
Cape Verde Is 1h
Cayman Islands 5h
* Chile 4h
Cocos(Keeling) 6fih
Colombia 5h
* Cook Is 10h
Costa Rica 6h
* Cuba 5h
Dominican Rep 4h
Ecuador 5h
* Easter Island 6h
El Salvador 6h
* Falkland Is 4h
French Guiana 3h
Fr Polynesia 9–10h
Galapagos Is 6h
* Greenland 1–4h
Grenada 4h
Guadeloupe 4h
Guatemala 6h
Guyana 4h
* Haiti 5h
Honduras 6h
Jamaica 5h
Leeward Is 4h
Martinique 4h
Mexico 6–8h
Midway Is 11h
Netherlands Antilles 4h
Nicaragua 6h
Panama 5h
* Paraguay 4h
Peru 5h
Puerto Rico 4h
* St Pierre and Miquelon 3h
Samoa 11h
South Georgia 2h
Trinidad/Tobago 4h
* Turks/Caicos Is 5h
United States:
* Eastern Zone 5h
* Central Zone 6h
* Mountain Zn 7h
* Pacific Zone 8h
* Alaska Zone 9hr
Hawaiian Is 10h
Uruguay 3h
Venezuela 4h
Virgin Is (UK/US) 4h
Windward Is 4h

* May keep Daylight Saving Time, effective dates vary.

WORLD MAP OF TIME ZONES

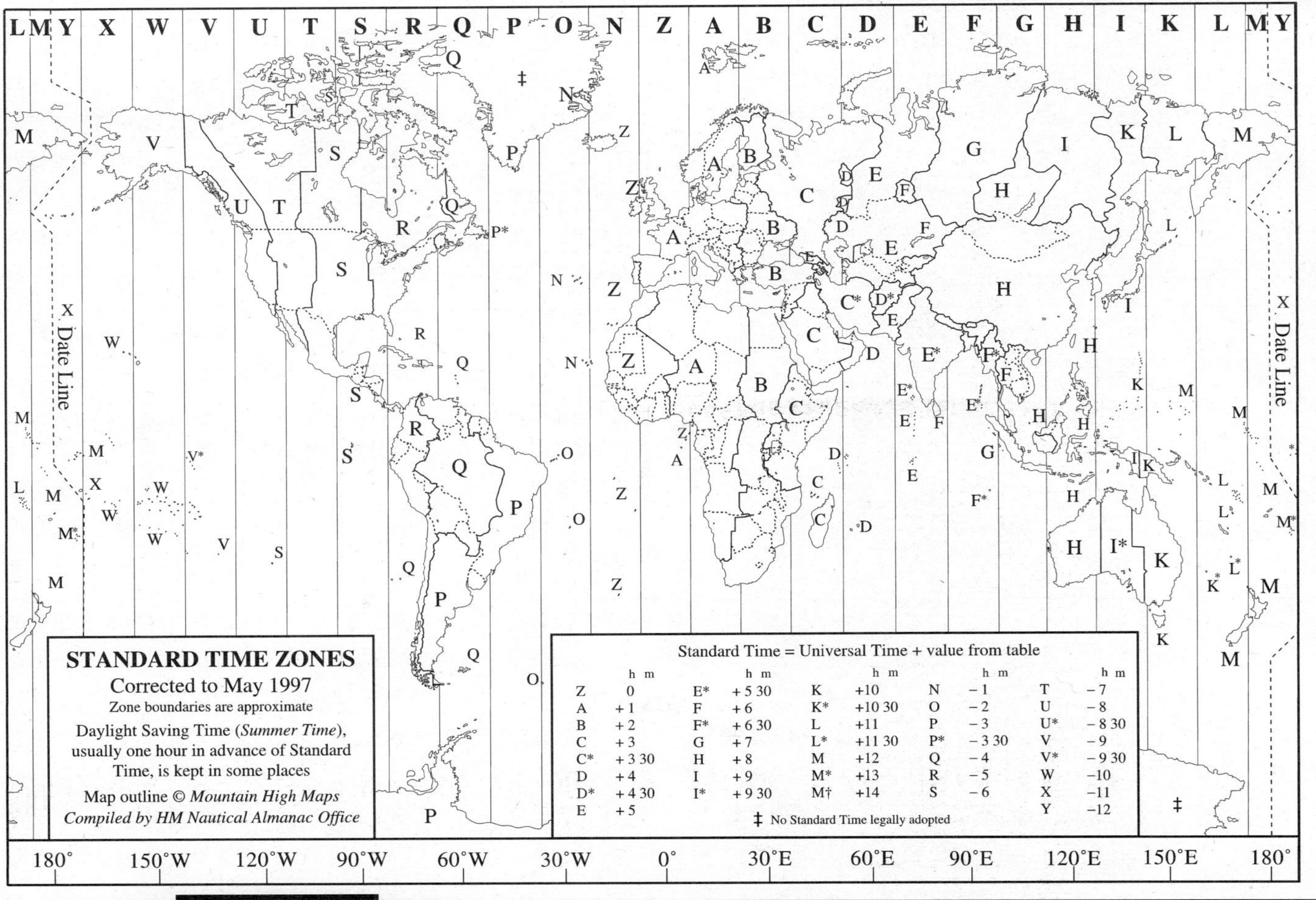
Date Line
STANDARD TIME ZONES
Corrected to May 1997
Zone boundaries are approximate
Daylight Saving Time (*Summer Time*), usually one hour in advance of Standard Time, is kept in some places
Map outline © *Mountain High Maps*
Compiled by HM Nautical Almanac Office
Standard Time = Universal Time + value from table
Z 0; A +1; B +2; C +3; C* +3 30; D +4; D* +4 30; E +5
E* +5 30; F +6; F* +6 30; G +7; H +8; I +9; I* +9 30
K +10; K* +10 30; L +11; L* +11 30; M +12; M* +13; M† +14
N −1; O −2; P −3; P* −3 30; Q −4; R −5; S −6
T −7; U −8; U* −8 30; V −9; V* −9 30; W −10; X −11; Y −12
‡ No Standard Time legally adopted
180° 150°W 120°W 90°W 60°W 30°W 0° 30°E 60°E 90°E 120°E 150°E 180°

Ephemeris

1998

January

S	M	T	W	T	F	S
				1	2	3
4	5	6	7	8	9	10
11	12	13	14	15	16	17
18	19	20	21	22	23	24
25	26	27	28	29	30	31

February

S	M	T	W	T	F	S
1	2	3	4	5	6	7
8	9	10	11	12	13	14
15	16	17	18	19	20	21
22	23	24	25	26	27	28

March

S	M	T	W	T	F	S
1	2	3	4	5	6	7
8	9	10	11	12	13	14
15	16	17	18	19	20	21
22	23	24	25	26	27	28
29	30	31				

April

S	M	T	W	T	F	S
			1	2	3	4
5	6	7	8	9	10	11
12	13	14	15	16	17	18
19	20	21	22	23	24	25
26	27	28	29	30		

May

S	M	T	W	T	F	S
					1	2
3	4	5	6	7	8	9
10	11	12	13	14	15	16
17	18	19	20	21	22	23
24	25	26	27	28	29	30
31						

June

S	M	T	W	T	F	S
	1	2	3	4	5	6
7	8	9	10	11	12	13
14	15	16	17	18	19	20
21	22	23	24	25	26	27
28	29	30				

July

S	M	T	W	T	F	S
			1	2	3	4
5	6	7	8	9	10	11
12	13	14	15	16	17	18
19	20	21	22	23	24	25
26	27	28	29	30	31	

August

S	M	T	W	T	F	S
						1
2	3	4	5	6	7	8
9	10	11	12	13	14	15
16	17	18	19	20	21	22
23	24	25	26	27	28	29
30	31					

September

S	M	T	W	T	F	S
		1	2	3	4	5
6	7	8	9	10	11	12
13	14	15	16	17	18	19
20	21	22	23	24	25	26
27	28	29	30			

October

S	M	T	W	T	F	S
				1	2	3
4	5	6	7	8	9	10
11	12	13	14	15	16	17
18	19	20	21	22	23	24
25	26	27	28	29	30	31

November

S	M	T	W	T	F	S
1	2	3	4	5	6	7
8	9	10	11	12	13	14
15	16	17	18	19	20	21
22	23	24	25	26	27	28
29	30					

December

S	M	T	W	T	F	S
		1	2	3	4	5
6	7	8	9	10	11	12
13	14	15	16	17	18	19
20	21	22	23	24	25	26
27	28	29	30	31		

1999

January

S	M	T	W	T	F	S
					1	2
3	4	5	6	7	8	9
10	11	12	13	14	15	16
17	18	19	20	21	22	23
24	25	26	27	28	29	30
31						

February

S	M	T	W	T	F	S
	1	2	3	4	5	6
7	8	9	10	11	12	13
14	15	16	17	18	19	20
21	22	23	24	25	26	27
28						

March

S	M	T	W	T	F	S
	1	2	3	4	5	6
7	8	9	10	11	12	13
14	15	16	17	18	19	20
21	22	23	24	25	26	27
28	29	30	31			

April

S	M	T	W	T	F	S
				1	2	3
4	5	6	7	8	9	10
11	12	13	14	15	16	17
18	19	20	21	22	23	24
25	26	27	28	29	30	

May

S	M	T	W	T	F	S
						1
2	3	4	5	6	7	8
9	10	11	12	13	14	15
16	17	18	19	20	21	22
23	24	25	26	27	28	29
30	31					

June

S	M	T	W	T	F	S
		1	2	3	4	5
6	7	8	9	10	11	12
13	14	15	16	17	18	19
20	21	22	23	24	25	26
27	28	29	30			

July

S	M	T	W	T	F	S
				1	2	3
4	5	6	7	8	9	10
11	12	13	14	15	16	17
18	19	20	21	22	23	24
25	26	27	28	29	30	31

August

S	M	T	W	T	F	S
1	2	3	4	5	6	7
8	9	10	11	12	13	14
15	16	17	18	19	20	21
22	23	24	25	26	27	28
29	30	31				

September

S	M	T	W	T	F	S
			1	2	3	4
5	6	7	8	9	10	11
12	13	14	15	16	17	18
19	20	21	22	23	24	25
26	27	28	29	30		

October

S	M	T	W	T	F	S
					1	2
3	4	5	6	7	8	9
10	11	12	13	14	15	16
17	18	19	20	21	22	23
24	25	26	27	28	29	30
31						

November

S	M	T	W	T	F	S
	1	2	3	4	5	6
7	8	9	10	11	12	13
14	15	16	17	18	19	20
21	22	23	24	25	26	27
28	29	30				

December

S	M	T	W	T	F	S
			1	2	3	4
5	6	7	8	9	10	11
12	13	14	15	16	17	18
19	20	21	22	23	24	25
26	27	28	29	30	31	

OTHER USES FOR REED'S EPHEMERIS

This ephemeris is a critical tool for celestial navigation; it also has a variety of other uses.

Moon phenomena: Tidal forces (and some say 'other' forces) are greatly influenced by the gyrations of the Moon through our sky. The first page of each monthly section gives the moon's rising and setting, it's declination, it's apparent phases, and the extremes of its orbit (perigee and apogee). Tracking these phenomena against the tide curves in a particular place can lead to a better understanding of what drives the tides.

Compass adjustment: With the amplitudes section on p. E 25, you can get the sun's bearing at sunrise or sunset to check your compass. You can also reference true south by using the sun's meridian passage, listed on the first page of each monthly section, along with the correction for your longitude from page E 103 to get the exact time of your local apparent noon. (Some readers have used this to properly align solar homes.)

Star-gazing: One of the joys of offshore passages is the occasional clear, calm night with unobstructed views of the heavens. The ephemeris is a great resource for understanding celestial mechanics and identifying planets and stars. The next few pages contain excellent information on the locations and habits of the more visible planets. Following this are some basic star charts. The second page of each monthly section, the stars table, gives the transit times, GHAs, and declinations of 60 major stars. You can use these to relate the bodies to eachother and to the sun, moon, and planets.

Try projecting the celestial sphere onto the night sky. Identify stars or planets with near zero declinations to define the celestial equator. Match GHAs and declinations with known places on earth to further define the celestial coordinates. Enjoy.

VISIBILITY OF PLANETS 1998

VENUS is a brilliant object in the evening sky until midway through the second week of January when it becomes too close to the sun for observation. It reappears at the beginning of the third week of January in the morning sky, where it stays until the end of the third week of September, when it again becomes too close to the sun; from mid-December it is visible in the evening sky. Venus is in conjunction with Mercury on Jan. 26, Aug. 25 and Sept. 11, with Jupiter on Apr. 23, with Saturn on May 29 and with Mars on Aug. 5.

MARS can be seen in the evening sky in Capricornus, then Aquarius from the last week of January. It passes into Pisces at the end of February and remains visible there until just after the first week of March, when it becomes too close to the sun for observation. It reappears in the morning sky during the second week of July in Gemini (passing 6° S of Pollux on Aug. 11), moving into Cancer in mid-August, Leo in mid-September (passing 1° N of Regulus on October 6) and Virgo beginning at the second half of November. Mars is in conjunction with Jupiter on Jan. 21, with Mercury on Mar. 11 and Mar. 30, and with Venus on Aug. 5.

JUPITER can be seen in the evening sky in Capricornus until it passes into Aquarius at the end of January. It becomes too close to the sun for observation after the first week of February and reappears in the morning sky during the second week of March. Its westward elongation increases until after mid-June it can be seen for more than half the night, passing into Pisces in early June and back to Aquarius in late August. It is at opposition on Sept. 16 when it is visible all night. Its eastward elongation then decreases, and after mid-December it can only be seen in the evening sky. Jupiter is in conjunction with Mars on Jan. 21, and with Venus on Apr. 23.

SATURN is visible in the evening sky in Pisces until late March when it becomes too close to the sun for observation. From the beginning of May it can be seen in the morning sky, passing into Cetus in the second half of July and returning to Pisces from mid-September. It is at opposition on Oct. 23 when it can be seen throughout the night. For the remainder of the year its eastward elongation decreases being visible for most of the night. Saturn is in conjunction with Mercury on May 12 and with Venus on May 29.

MERCURY can be seen only low in the east before sunrise, or low in the west after sunset (about the time of the beginning or end of civil twilight). It is visible in the mornings between the following approximate dates:

January 1 (–0.1) to February 10 (–0.8)
April 15 (+2.8) to June 3 (–1.4)
August 22 (+2.2) to September 16 (–1.4)
December 7 (+1.8) to December 31 (–0.4)

The planet is brighter at the end of each period.

Mercury (cont.) It is visible in the evenings between the following approximate dates:

March 4 (–1.4) to March 30 (+2.3)
June 18 (–1.3) to August 6 (+2.7)
October 9 (–0.7) to November 26 (+1.8)

The planet is brighter at the start of each period. The figures in parentheses are the magnitudes.

PLANET DIAGRAMS

The graph below shows the changing declinations of the major planets during the year. The diagram opposite shows the local mean time of meridian passage (transit) of the sun, of the five major planets, and of each 30° of SHA during the year. Together, these graphs provide a general picture of the relative positions and availability of these bodies for observation.

The trick to the meridian passage graph is to select your day and then to read the meridian passages from bottom to top for 24 hours and to understand them as defining the relationships of the bodies throughout the day. For instance, observe the events of November 20:

Mars transits the meridian at about 0800, 4 hours before the sun, so it should be visible early in the morning and be about halfway up the eastern sky at dawn. The **sun** transits before 1200, indicating a negative equation of time also known as a "fast" sun. **Venus** transits right after the sun and therefore (as indicated by the hatched band) will probably not be visible at any time of the day. **Mercury** transits at about 1300, an hour behind the sun; so it should be very low in the western sky at dusk. **Jupiter** transits 7 hours after the sun, and thus should be very high in the eastern sky at dusk with **Saturn** about 45° (or 3 hours) below it. Both will be visible much of the night.

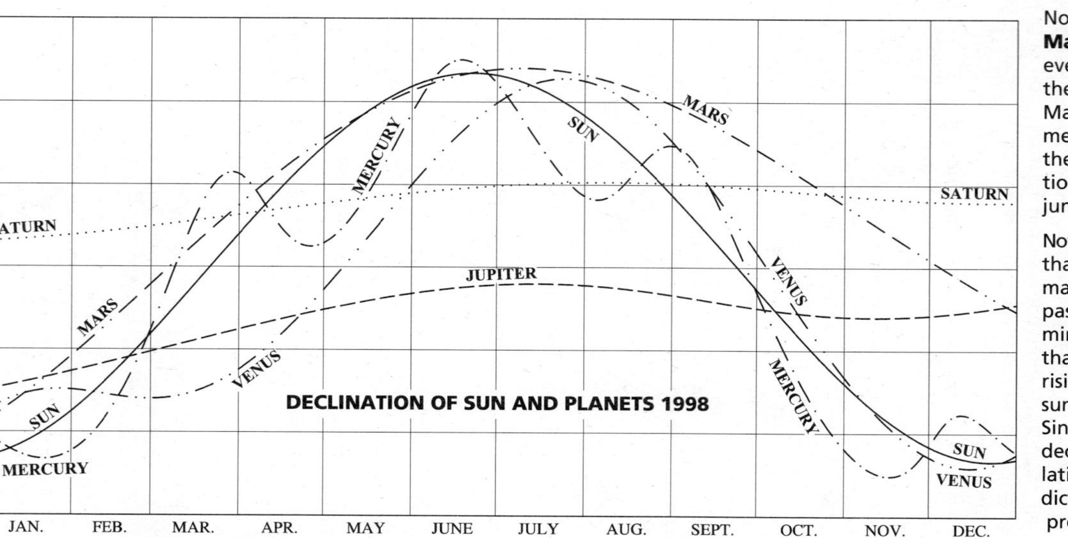

Note how **Saturn, Jupiter, Mars,** and **Venus** gather in evening sky at the beginning of the year. Observe that when Mars and Jupiter make similar meridian passage in late January, they also have similar declinations. They will be in close conjunction.

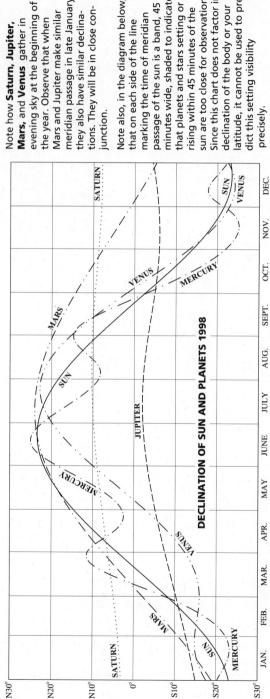

Note also, in the diagram below, that on each side of the line marking the time of meridian passage of the sun is a band, 45 minutes wide, shaded to indicate that planets and stars setting or rising within 45 minutes of the sun are too close for observation. Since this chart does not factor in declination of the body or your latitude, it cannot be used to predict this setting visibility precisely.

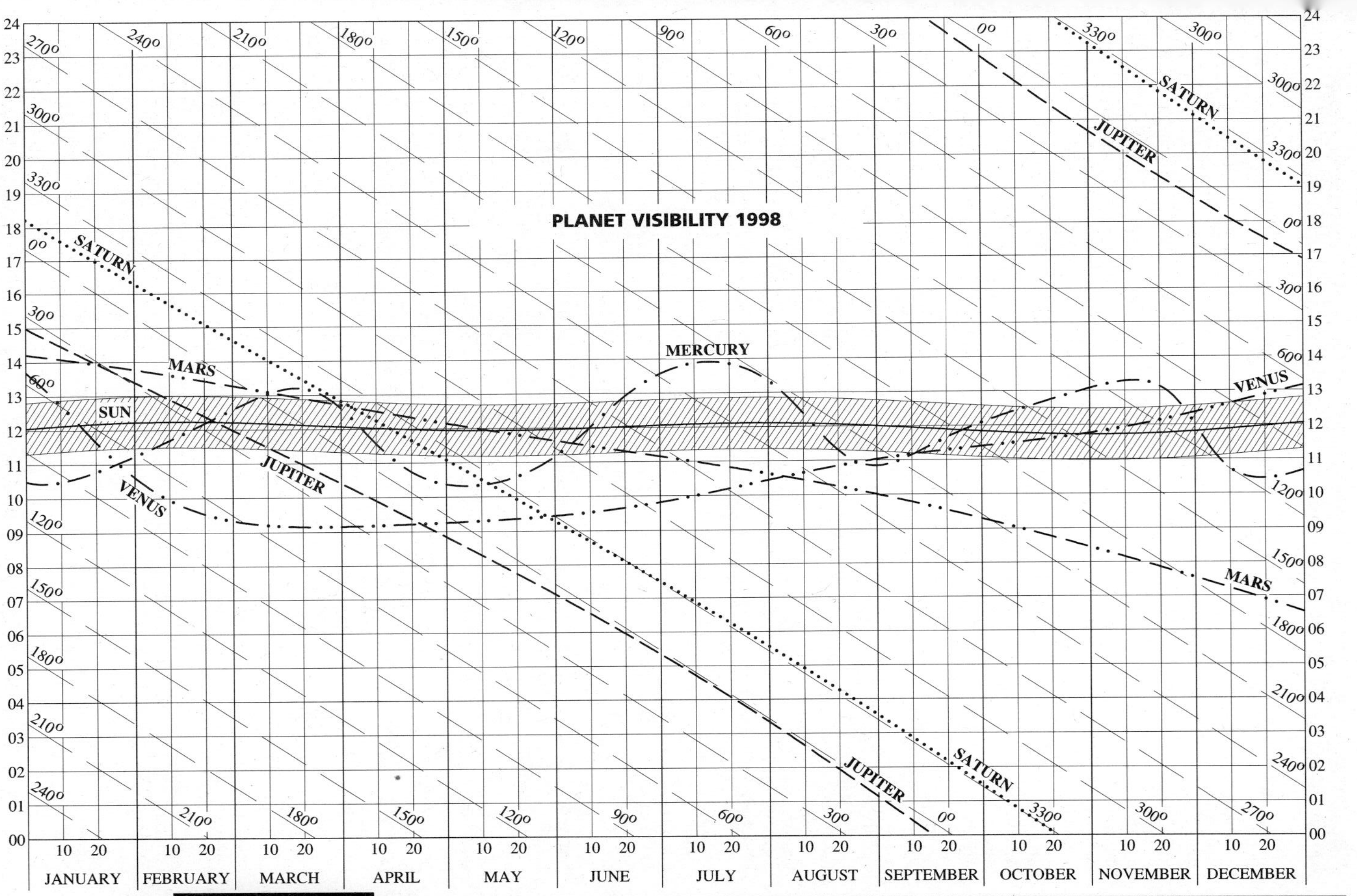
PLANET VISIBILITY 1998
SUN
MERCURY
VENUS
MARS
JUPITER
SATURN
JANUARY
FEBRUARY
MARCH
APRIL
MAY
JUNE
JULY
AUGUST
SEPTEMBER
OCTOBER
NOVEMBER
DECEMBER

Ephemeris

ALPHABETICAL INDEX OF PRINCIPAL STARS

With their approximate places, 1998

Proper Name	Bayer's Name (Constellation)	Mag	RA (h, m)	Dec (°)	SHA (°)	No
Acamar	θ Eridani	3.1	2 58	S 40	315	8
Achernar	α Eridani	0.6	1 38	S 57	336	5
Acrux	α Crucis	1.1	12 27	S 63	173	32
Adhara	ε Canis Majoris	1.6	6 59	S 29	255	20
Aldebaran	α Tauri	1.1	4 36	N 17	291	11
Alioth	ε Ursae Majoris	1.7	12 54	N 56	167	35
Alkaid	η Ursae Majoris	1.9	13 47	N 49	153	37
Al Na'ir	α Gruis	2.2	22 08	S 47	28	58
Alnilam	ε Orionis	1.8	5 36	S 1	276	16
Alphard	α Hydrae	2.2	9 28	S 9	218	27
Alphecca	α Coronae Borealis	2.3	15 35	N 27	126	44
Alpheratz	α Andromedae	2.2	0 08	N 29	358	1
Altair	α Aquilae	0.9	19 51	N 9	62	54
Ankaa	α Phoenicis	2.4	0 26	S 42	353	2
Antares	α Scorpii	1.2	16 29	S 26	113	45
Arcturus	α Bootis	0.2	14 16	N 19	146	40
Atria	α Triang Australis	1.9	16 48	S 69	108	46
Avior	ε Carinae	1.7	8 22	S 60	234	24
Bellatrix	γ Orionis	1.7	5 25	N 6	279	14
Betelgeuse	α Orionis	0.1–1.2	5 55	N 7	271	17
Canopus	α Carinae	−0.9	6 24	S 53	264	18
Capella	α Aurigae	0.2	5 17	N 46	281	13
Castor	α Geminorum	1.6	7 34	N 32	246	21
Deneb	α Cygni	1.3	20 41	N 45	50	56
Denebola	β Leonis	2.2	11 49	N 15	183	30
Diphda	β Ceti	2.2	0 44	S 18	349	4
Dubhe	α Ursae Majoris	2.0	11 04	N 62	194	29
Elnath	β Tauri	1.8	5 26	N 29	278	15
Eltanin	γ Draconis	2.4	17 56	N 51	91	50
Enif	ε Pegasi	2.5	21 44	N 10	34	57
Fomalhaut	α Piscis Austrini	1.3	22 58	S 30	16	59
Gacrux	γ Crucis	1.6	12 31	S 57	172	33
Gienah	γ Corvi	2.8	12 16	S 18	176	31
Hadar	β Centauri	0.9	14 04	S 60	149	38
Hamal	α Arietis	2.2	2 07	N 23	328	7
Kaus Australis	ε Sagittarii	2.0	18 24	S 34	84	51
Kochab	β Ursae Minoris	2.2	14 51	N 74	137	43
Markab	α Pegasi	2.6	23 05	N 15	14	60
Menkar	α Ceti	2.8	3 02	N 4	314	9
Menkent	θ Centauri	2.3	14 07	S 36	148	39
Miaplacidus	β Carinae	1.8	9 13	S 70	222	26
Mimosa	β Crucis	1.5	12 48	S 60	168	34
Mirfak	α Persei	1.9	3 24	N 50	309	10
Nunki	σ Sagittarii	2.1	18 55	S 26	76	53
Peacock	α Pavonis	2.1	20 26	S 57	54	55
POLARIS	α Ursae Minoris	2.1	2 31	N 89	323	6
Pollux	β Geminorum	1.2	7 45	N 28	244	23
Procyon	α Canis Minoris	0.5	7 39	N 5	245	22
Rasalhague	α Ophiuchi	2.1	17 35	N 13	96	49
Regulus	α Leonis	1.3	10 08	N 12	208	28
Rigel	β Orionis	0.3	5 14	S 8	281	12
Rigil Kent	α Centauri	0.1	14 40	S 61	140	41
Sabik	η Ophiuchi	2.6	17 10	S 16	102	47
Schedar	α Cassiopeiae	2.5	0 40	N 57	350	3
Shaula	γ Scorpii	1.7	17 33	S 37	97	48
Sirius	α Canis Majoris	−1.6	6 45	S 17	259	19
Spica	α Virginis	1.2	13 25	S 11	159	36
Suhail	γ Velorum	2.2	9 08	S 43	223	25
Vega	α Lyrae	0.1	18 37	N 39	81	52
Zuben'ubi	α Librae	2.9	14 51	S 16	137	42

The last column refers to the number given to the star in this almanac. The star's exact position may be found according to this number on the monthly pages.

STAR CHART OF NORTHERN HEMISPHERE

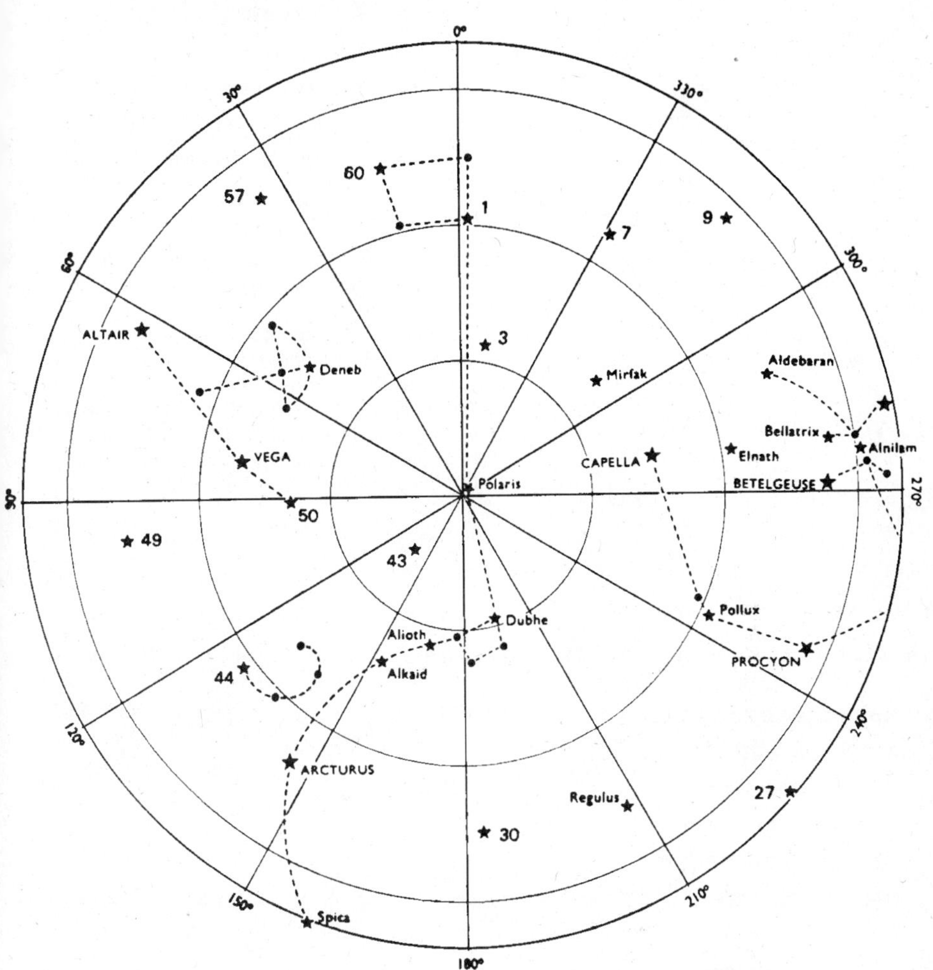

*** Stars of the first magnitude**
(capital letters)

● Stars of magnitude 2.0 to 1.0
(small letters)

Key to numbered stars

1. Alpheratz	27. Alphard	49. Rasalhague
3. Schedar	30. Denebola	50. Eltanin
7. Hamal	43. Kochab	57. Enif
9. Menkar	44. Alphecca	60. Markab

AUXILIARY STAR CHARTS

POLE-DIPPER-DUBHE-BENETNASCH-KOCHAB

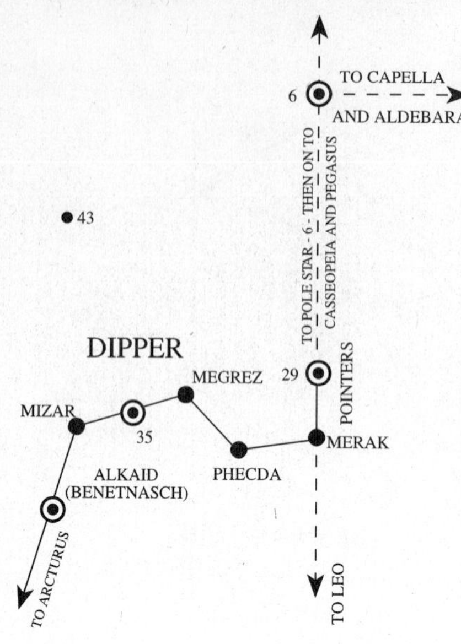

No. 6. POLARIS.

No. 29. DUBHE.

No. 35. ALIOTH.

No. 43. KOCHAB.

No. 11. ALDERBARAN.

No. 13. CAPELLA.

Polaris – the Pole Star – familiar in the Northern Hemishpere is always seen in the same part of the heavens, over the Pole of the Earth. It is the brightest star in the Little Bear (Ursa Minor). The position of Polaris in the Little Bear corresponds to the position of Alkaid (Benetnasch) 37, in the Great Bear. The Dipper or Great Bear (Ursa Major) is the easiest recognizable constellation in the northern heavens, a straight line through Merak and Dubhe – the Pointers – leads to the Pole Star.

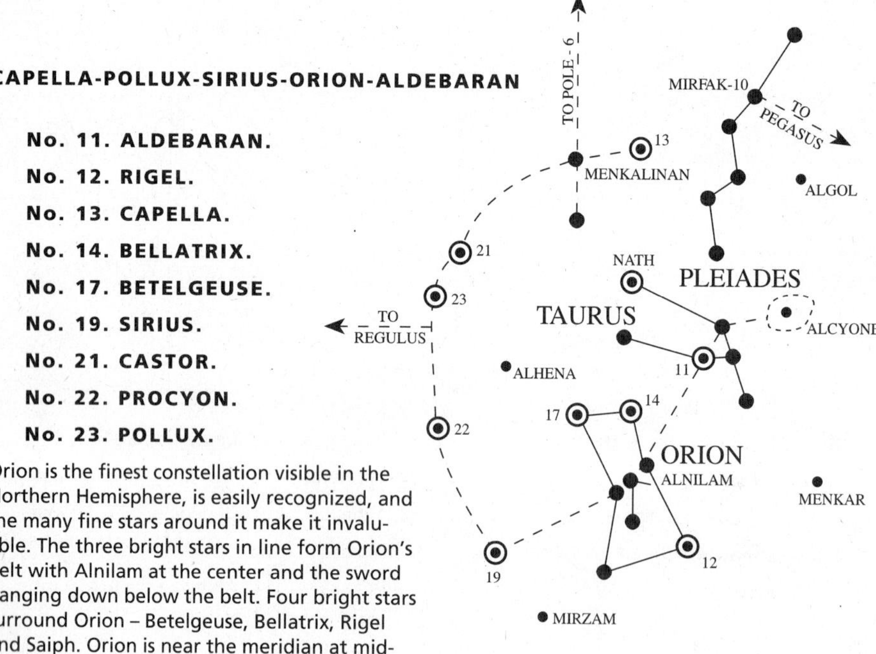

CAPELLA-POLLUX-SIRIUS-ORION-ALDEBARAN

No. 11. ALDEBARAN.

No. 12. RIGEL.

No. 13. CAPELLA.

No. 14. BELLATRIX.

No. 17. BETELGEUSE.

No. 19. SIRIUS.

No. 21. CASTOR.

No. 22. PROCYON.

No. 23. POLLUX.

Orion is the finest constellation visible in the Northern Hemisphere, is easily recognized, and the many fine stars around it make it invaluable. The three bright stars in line form Orion's belt with Alnilam at the center and the sword hanging down below the belt. Four bright stars surround Orion – Betelgeuse, Bellatrix, Rigel and Saiph. Orion is near the meridian at midnight late in the year, and therefore is only visible in northern latitudes in winter and early spring.

AMPLITUDES

The bearing of the sun when rising or setting is known as its *amplitude*.

It is useful for calculating compass error, as accurate time or sight reduction is not required. All that is required is to know the approximate latitude from the chart and the sun's approximate declination from this ephemeris.

With these values, the table on following pages will give you the true bearing of the sun at sunrise or sunset in any part of the world, up to Latitude 66°. This bearing can be compared with your compass bearing at sunrise or sunset to deduce any deviation it might have.

Notes:

The "theoretical sunrise" is considered to take place at the moment when the sun's center is on the edge of the horizon to the eastward.

Due to refraction (i.e., the bending of rays of light when passing through the atmosphere), the sun appears higher than it actually is. It appears to rise before it is actually above the horizon, and it has actually set when you can still see a small portion of the limb. The rule of thumb is to take amplitude bearings, both at rising and setting, when the sun's lower limb is about half the sun's diameter above the horizon, as it is then that the center of the sun is actually on the horizon.

This table can also be used to find the true bearing at rising and setting of any celestial body other than the moon, within these declinations.

Example:

November 18th, on passage to Bermuda, DR 28°N, 64°30′ W, the sun rose bearing by compass 130°. The Variation from the chart was 14°W. Find the deviation.

On page E 93, you will find that the sun's declination on this day is about S 19°. Going to the table on pages E 26–27 with our declination and latitude, we get a bearing of 68°. The important note at the bottom of the table states:

Name the Bearing the same as the Declination NORTH or SOUTH and EAST if rising, WEST if setting.

Therefore, our answer is S 68°E—in other words 68° east from 180°, or 112° on the compass rose. Now apply this to a typical compass table:

+ W	**T**rue	112°	from table
– E	**V**ariation	14°W	from chart
	Magnetic	126°	
	Deviation	4°W	deduce
↓	**C**ompass	130°	actual reading

You may also wish to draw the problem as shown below.

Note that this procedure will give you only the compass deviation for the heading that you are on when you took the bearing. To generate a deviation card, you can get separate bearings while steering the vessel on eight or more cardinal headings.

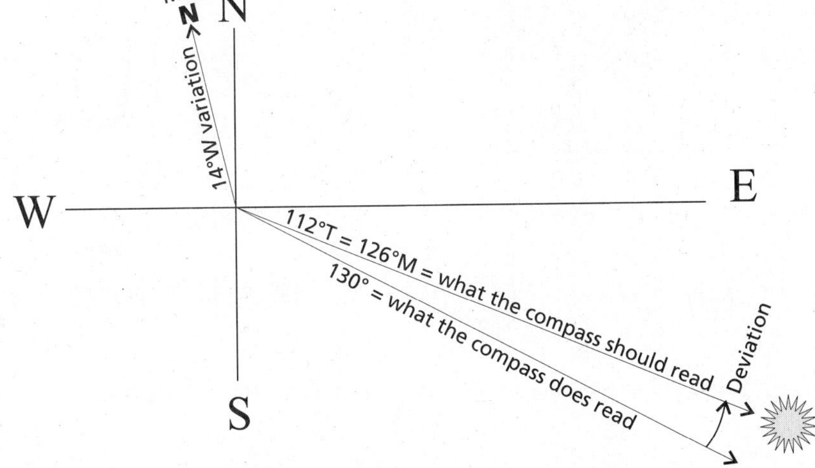

SUN'S TRUE BEARING AT SUNRISE AND SUNSET

LATITUDES 0° to 66° DECLINATIONS 0° to 11°

	DECLINATION											
LAT	0°	1°	2°	3°	4°	5°	6°	7°	8°	9°	10°	11°
	°	°	°	°	°	°	°	°	°	°	°	°
0° to 5°	90	89	88	87	86	85	84	83	82	81	80	79
6°	90	89	88	87	86	85	84	83	82	81	79.9	78.9
7°	90	89	88	87	86	85	84	83	81.9	80.9	79.9	78.9
8°	90	89	88	87	86	85	84	82.9	81.9	80.9	79.9	78.9
9°	90	89	88	87	86	85	83.9	82.9	81.9	80.9	79.8	78.9
10°	90	89	88	87	86	84.9	83.9	82.9	81.9	80.9	79.8	78.8
11°	90	89	88	87	86	84.9	83.9	82.9	81.9	80.8	79.8	78.8
12°	90	89	88	87	85.9	84.9	83.9	82.9	81.8	80.8	79.8	78.8
13°	90	89	88	86.9	85.9	84.9	83.8	82.8	81.8	80.8	79.7	78.7
14°	90	89	88	86.9	85.9	84.8	83.8	82.8	81.8	80.7	79.7	78.7
15°	90	89	88	86.9	85.9	84.8	83.8	82.8	81.7	80.7	79.6	78.6
16°	90	89	87.9	86.9	85.8	84.8	83.8	82.7	81.7	80.6	79.6	78.6
17°	90	89	87.9	86.9	85.8	84.8	83.7	82.7	81.6	80.6	79.5	78.5
18°	90	89	87.9	86.9	85.8	84.8	83.7	82.6	81.6	80.5	79.5	78.4
19°	90	89	87.9	86.8	85.8	84.7	83.7	82.6	81.5	80.5	79.4	78.4
20°	90	88.9	87.9	86.8	85.8	84.7	83.6	82.6	81.5	80.4	79.4	78.3
21°	90	88.9	87.9	86.8	85.7	84.7	83.6	82.5	81.4	80.4	79.3	78.2
22°	90	88.9	87.9	86.8	85.7	84.6	83.5	82.5	81.4	80.3	79.2	78.1
23°	90	88.9	87.9	86.7	85.7	84.6	83.5	82.4	81.3	80.2	79.1	78.0
24°	90	88.9	87.8	86.7	85.6	84.5	83.4	82.3	81.2	80.1	79.0	78.0
25°	90	88.9	87.8	86.7	85.6	84.5	83.4	82.3	81.2	80.1	79.0	77.9
26°	90	88.9	87.8	86.7	85.6	84.4	83.3	82.2	81.1	80.0	78.9	77.8
27°	90	88.9	87.8	86.6	85.5	84.4	83.3	82.1	81.0	79.9	78.8	77.6
28°	90	88.9	87.8	86.6	85.5	84.4	83.2	82.1	80.9	79.8	78.7	77.5
29°	90	88.9	87.8	86.6	85.5	84.3	83.1	82.0	80.9	79.7	78.6	77.4
30°	90	88.9	87.7	86.5	85.4	84.2	83.1	81.9	80.8	79.6	78.5	77.3
31°	90	88.9	87.7	86.5	85.4	84.2	83.0	81.8	80.7	79.5	78.3	77.1
32°	90	88.9	87.7	86.5	85.3	84.1	82.9	81.7	80.6	79.4	78.2	77.0
33°	90	88.8	87.7	86.4	85.3	84.0	82.8	81.7	80.5	79.3	78.0	76.9
34°	90	88.8	87.6	86.4	85.2	84.0	82.7	81.5	80.3	79.1	77.9	76.7
35°	90	88.8	87.5	86.3	85.1	83.9	82.7	81.4	80.2	79.0	77.8	76.5
36°	90	88.8	87.5	86.3	85.0	83.8	82.6	81.3	80.1	78.8	77.6	76.3
37°	90	88.7	87.5	86.2	85.0	83.7	82.5	81.2	80.0	78.7	77.4	76.2
38°	90	88.7	87.5	86.2	84.9	83.6	82.4	81.1	79.8	78.5	77.3	76.0
39°	90	88.7	87.4	86.1	84.8	83.6	82.3	81.0	79.7	78.4	77.1	75.8
40°	90	88.7	87.4	86.1	84.8	83.5	82.1	80.8	79.5	78.2	76.9	75.6
41°	90	88.7	87.3	86.0	84.7	83.4	82.0	80.7	79.4	78.0	76.7	75.3
42°	90	88.6	87.3	86.0	84.6	83.3	81.9	80.6	79.2	77.8	76.5	75.1
43°	90	88.6	87.3	85.9	84.5	83.1	81.8	80.4	79.0	77.6	76.4	74.9
44°	90	88.6	87.2	85.8	84.4	83.0	81.6	80.2	78.8	77.4	76.0	74.6
45°	90	88.6	87.2	85.7	84.3	82.9	81.5	80.1	78.6	77.2	75.8	74.3
46°	90	88.6	87.1	85.7	84.2	82.8	81.3	79.9	78.4	77.0	75.5	74.0
47°	90	88.5	87.1	85.6	84.1	82.6	81.2	79.7	78.2	76.7	75.2	73.7
48°	90	88.5	87.0	85.5	84.0	82.5	81.0	79.5	78.0	76.5	75.0	73.4
49°	90	88.5	86.9	85.4	83.9	82.4	80.8	79.3	77.7	76.2	74.6	73.1
50°	90	88.4	86.9	85.3	83.8	82.2	80.7	79.1	77.5	75.9	74.3	72.7
51°	90	88.4	86.8	85.2	83.6	82.0	80.4	78.8	77.2	75.6	74.0	72.3
52°	90	88.4	86.7	85.1	83.5	81.9	80.2	78.6	76.9	75.3	73.6	71.9
53°	90	88.3	86.7	85.0	83.3	81.7	80.0	78.3	76.6	74.9	73.2	71.5
54°	90	88.3	86.6	84.9	83.2	81.5	79.7	78.0	76.3	74.6	72.8	71.0
55°	90	88.2	86.5	84.8	83.0	81.3	79.5	77.7	75.9	74.2	72.4	70.6
56°	90	88.2	86.4	84.6	82.8	81.0	79.2	77.4	75.6	73.7	71.9	70.0
57°	90	88.2	86.3	84.5	82.6	80.8	78.9	77.1	75.2	73.3	71.4	69.5
58°	90	88.1	86.2	84.3	82.4	80.5	78.6	76.7	74.8	72.8	70.9	68.9
59°	90	88.0	86.1	84.2	82.2	80.2	78.3	76.3	74.3	72.3	70.3	68.2
60°	90	88.0	86.0	84.0	82.0	80.0	77.9	75.9	73.8	71.8	69.7	67.6
61°	90	87.9	85.9	83.8	81.7	79.6	77.5	75.4	73.3	71.2	69.0	66.8
62°	90	87.9	85.7	83.6	81.4	79.3	77.1	74.9	72.7	70.5	68.3	66.0
63°	90	87.8	85.6	83.4	81.2	78.9	76.7	74.4	72.1	69.8	67.5	65.1
64°	90	87.7	85.4	83.1	80.8	78.5	76.2	73.9	71.5	69.1	66.7	64.2
65°	90	87.6	85.3	82.9	80.5	78.1	75.7	73.2	70.8	68.3	65.7	63.2
66°	90	87.5	85.1	82.6	80.1	77.6	75.1	72.6	70.0	67.4	64.7	62.0

Name the Bearing the same as the Declination NORTH or SOUTH and EAST if rising, WEST if setting.

For example of how to use this table, see previous page.

SUN'S TRUE BEARING AT SUNRISE AND SUNSET

LATITUDES 0° to 66° DECLINATIONS 12° to 23°

	DECLINATION											
LAT.	12°	13°	14°	15°	16°	17°	18°	19°	20°	21°	22°	23°
	°	°	°	°	°	°	°	°	°	°	°	°
0° to 5°	77.9	76.9	75.9	74.9	73.9	72.9	71.9	70.9	69.9	68.8	67.9	66.9
6°	77.9	76.9	75.9	74.9	73.9	72.9	71.9	70.9	69.9	68.8	67.9	66.9
7°	77.9	76.9	75.9	74.9	73.9	72.9	71.9	70.8	69.8	68.8	67.8	66.8
8°	77.9	76.9	75.9	74.8	73.8	72.8	71.8	70.8	69.8	68.8	67.8	66.8
9°	77.8	76.8	75.8	74.8	73.8	72.8	71.8	70.7	69.7	68.7	67.7	66.7
10°	77.8	76.8	75.8	74.8	73.7	72.7	71.7	70.7	69.7	68.7	67.6	66.6
11°	77.8	76.8	75.7	74.7	73.7	72.7	71.6	70.6	69.6	68.6	67.6	66.5
12°	77.7	76.7	75.7	74.6	73.6	72.6	71.6	70.6	69.5	68.5	67.5	66.4
13°	77.7	76.6	75.6	74.6	73.6	72.5	71.5	70.5	69.4	68.4	67.4	66.4
14°	77.6	76.6	75.6	74.5	73.5	72.5	71.4	70.4	69.4	68.3	67.3	66.2
15°	77.6	76.5	75.5	74.4	73.4	72.4	71.3	70.3	69.3	68.2	67.2	66.1
16°	77.5	76.5	75.4	74.4	73.3	72.3	71.2	70.2	69.1	68.1	67.1	66.0
17°	77.4	76.4	75.3	74.3	73.3	72.2	71.1	70.1	69.0	68.0	66.9	65.9
18°	77.4	76.3	75.3	74.2	73.2	72.1	71.0	70.0	68.9	67.9	66.8	65.7
19°	77.4	76.2	75.2	74.1	73.0	72.0	70.9	69.9	68.8	67.7	66.7	65.6
20°	77.2	76.1	75.1	74.0	72.9	71.9	70.8	69.7	68.6	67.6	66.5	65.4
21°	77.1	76.0	75.0	73.9	72.8	71.7	70.7	69.6	68.5	67.4	66.3	65.2
22°	77.0	76.0	74.9	73.8	72.7	71.6	70.5	69.4	68.3	67.3	66.2	65.1
23°	76.9	75.9	74.8	73.7	72.6	71.5	70.4	69.3	68.2	67.1	66.0	64.9
24°	76.8	75.7	74.6	73.5	72.5	71.3	70.2	69.1	68.0	66.9	65.8	64.7
25°	76.7	75.6	74.5	73.4	72.3	71.2	70.1	68.9	67.8	66.7	65.6	64.5
26°	76.6	75.5	74.4	73.3	72.1	71.0	69.9	68.8	67.6	66.5	65.4	64.2
27°	76.5	75.4	74.3	73.1	72.0	70.8	69.7	68.6	67.4	66.3	65.1	64.0
28°	76.4	75.2	74.1	73.0	71.8	70.7	69.5	68.4	67.2	66.1	64.9	63.8
29°	76.2	75.1	73.9	72.8	71.6	70.5	69.3	68.2	67.0	65.8	64.6	63.5
30°	76.1	75.0	73.8	72.6	71.4	70.3	69.1	67.9	66.7	65.5	64.4	63.2
31°	76.0	74.8	73.6	72.4	71.2	70.0	68.9	67.6	66.5	65.3	64.1	62.9
32°	75.8	74.6	73.4	72.2	71.0	69.8	68.6	67.4	66.2	65.0	63.8	62.6
33°	75.7	74.4	73.2	72.0	70.8	69.6	68.4	67.1	65.9	64.7	63.5	62.2
34°	75.5	74.2	73.0	71.8	70.6	69.3	68.1	66.9	65.6	64.4	63.1	61.9
35°	75.3	74.1	72.8	71.6	70.3	69.1	67.8	66.6	65.3	64.0	62.8	61.5
36°	75.1	73.8	72.6	71.3	70.1	68.8	67.5	66.3	65.0	63.7	62.4	61.1
37°	74.9	73.6	72.4	71.1	69.8	68.5	67.2	65.9	64.6	63.3	62.0	60.7
38°	74.7	73.4	72.1	70.8	69.5	68.2	66.9	65.6	64.3	62.9	61.6	60.3
39°	74.5	73.2	71.9	70.5	69.2	67.9	66.6	65.2	63.9	62.5	61.2	59.8
40°	74.2	72.9	71.6	70.2	68.9	67.6	66.2	64.8	63.5	62.1	60.7	59.3
41°	74.0	72.7	71.3	70.0	68.6	67.2	65.8	64.4	63.0	61.6	60.2	58.8
42°	73.7	72.4	71.0	69.6	68.2	66.8	65.4	64.0	62.6	61.2	59.7	58.3
43°	73.5	72.1	70.7	69.3	67.9	66.4	65.0	63.6	62.1	60.7	59.2	57.7
44°	73.2	71.8	70.3	68.9	67.5	66.0	64.6	63.1	61.6	60.1	58.6	57.1
45°	72.9	71.4	70.0	68.5	67.0	65.6	64.1	62.6	61.1	59.5	58.0	56.4
46°	72.6	71.1	69.6	68.1	66.6	65.1	63.6	62.1	60.5	58.9	57.4	55.8
47°	72.2	70.7	69.2	67.7	66.2	64.6	63.0	61.5	59.9	58.3	56.7	55.0
48°	71.9	70.3	68.8	67.2	65.7	64.1	62.5	60.9	59.3	57.6	55.9	54.3
49°	71.5	69.9	68.4	66.8	65.1	63.5	61.9	60.2	58.6	56.9	55.2	53.4
50°	71.1	69.5	67.9	66.2	64.6	63.0	61.3	59.6	57.8	56.1	54.3	52.6
51°	70.7	69.0	67.4	65.7	64.0	62.3	60.6	58.8	57.1	55.3	53.5	51.6
52°	70.3	68.6	66.9	65.1	63.4	61.6	59.9	58.1	56.2	54.4	52.5	50.6
53°	69.8	68.0	66.3	64.5	62.7	60.9	59.1	57.2	55.4	53.4	51.5	49.5
54°	69.3	67.5	65.7	63.9	62.0	60.2	58.3	56.4	54.4	52.4	50.4	48.3
55°	68.7	67.9	65.0	63.2	61.3	59.3	57.4	55.4	53.4	51.3	49.2	47.1
56°	68.2	66.3	64.4	62.4	60.5	58.5	56.4	54.4	52.3	50.1	47.9	45.7
57°	67.6	65.6	63.6	61.6	59.6	57.5	55.4	53.3	51.1	48.8	46.5	44.2
58°	66.9	64.9	62.8	60.8	58.6	56.5	54.3	52.1	49.8	47.4	45.0	42.5
59°	66.2	64.1	62.0	59.8	57.6	55.4	53.1	50.8	48.4	45.9	43.3	40.6
60°	65.4	63.3	61.1	58.8	56.5	54.2	51.8	49.4	46.8	44.2	41.5	38.6
61°	64.6	62.3	60.1	57.7	55.3	52.9	50.4	47.8	45.1	42.3	39.4	36.3
62°	63.7	61.4	59.0	56.5	54.0	51.5	48.8	46.1	43.2	40.2	37.1	33.7
63°	62.7	60.3	57.8	55.2	52.6	49.9	47.1	44.2	41.1	37.9	34.4	30.6
64°	61.7	59.1	56.5	53.8	51.0	48.2	45.2	42.0	38.7	35.2	31.3	27.0
65°	60.5	57.8	55.1	52.2	49.3	46.2	43.0	39.6	36.0	32.0	27.6	22.4
66°	59.2	56.4	53.5	50.5	47.3	44.0	40.5	36.8	32.8	28.2	22.9	16.1

Name the Bearing the same as the Declination NORTH or SOUTH and EAST if rising. WEST if setting.

FEBRUARY 26, 1998, ECLIPSE

Diagram shows paths and times of total and partial eclipse visibility. Carribean cruisers take note.

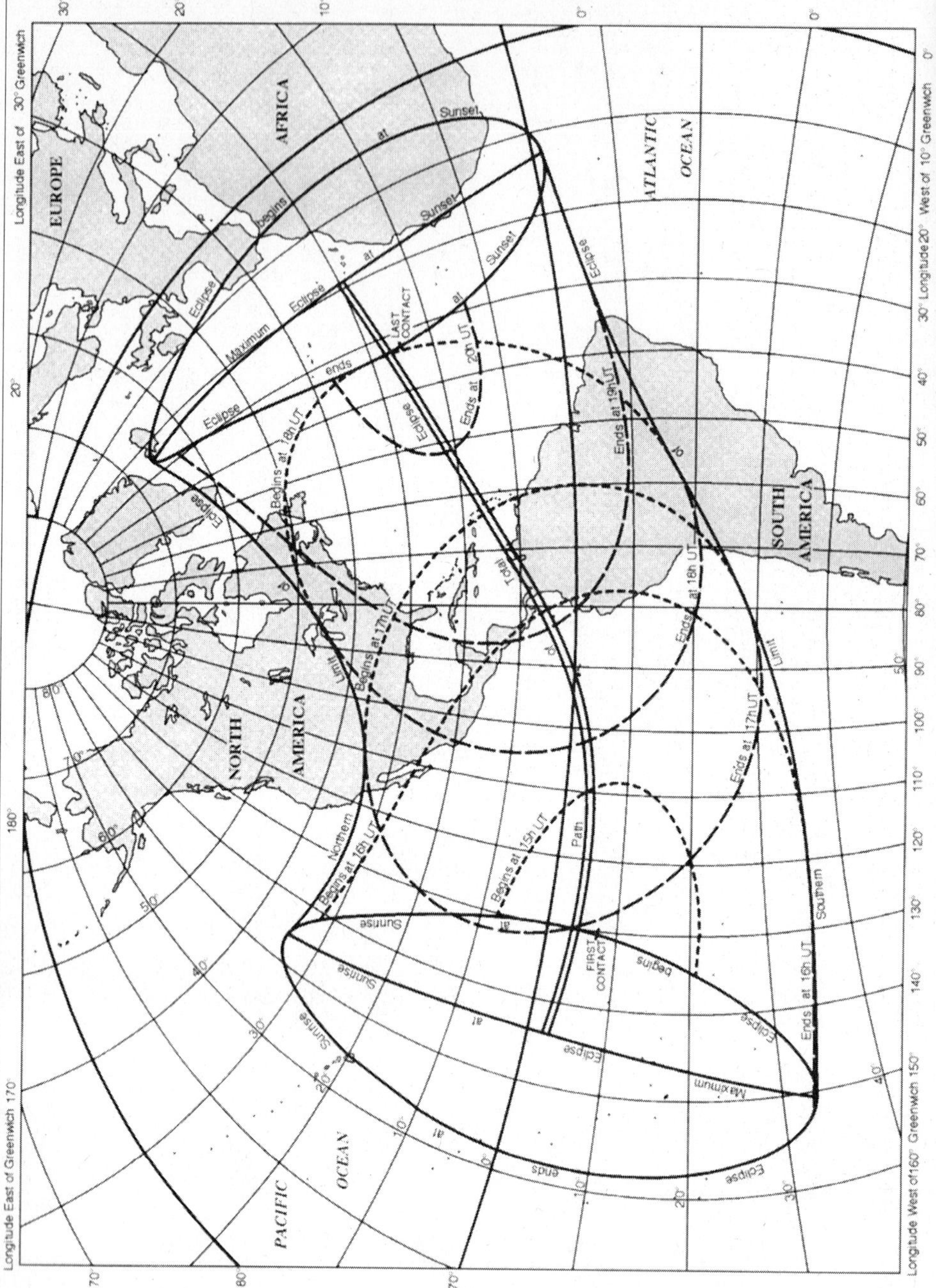

Printed with permission from the Astronomical Applications Dept. of the U.S. Naval Observatory

CELESTIAL PHENOMENA 1998

SEASONS 1998

Vernal equinox. Spring begins when the sun enters Aries, 19h 55m March 20.

Summer solstice. Summer begins when the sun enters Cancer, 14h 03m June 21.

Autumn equinox. Autumn begins when the sun enters Libra, 23h 05m September 23.

Winter solstice. Winter begins when the sun enters Capricorn, 01h 56m December 22.

Earth is at perihelion (closest to the sun) 21h January 4; at aphelion (farthest from the sun), 00h July 4.

ECLIPSES 1998

In 1998 there are two eclipses of the sun and three eclipses of the moon:

February 26. The year's major eclipse. See diagram on opposite page for details.

March 13. Partial eclipse of the moon begins 2h 14m, ends 06h 25m. Visible in all North, South and Central Americas except Alaska and NW Canada.

August 8. Partial eclipse of the moon begins 1h 32m, ends 03h 18m. Visible in South, Central, and eastern North America.

August 21–22. Annular eclipse of the of the sun begins 23h 10m, ends 05h 02m. Visible in Australia, New Zealand, and SW Pacific.

September 6. Partial eclipse of the moon begins 9h 14m, ends 13h 06m. Visible in South, Central, and North America except extreme east at beginning, visibility moving west.

Further celestial phenomena can be found in the diary section on the second of each set of monthly pages. The phases of the moon are on the first page of each section.

USABILITY IN 1999

If necessary, the 1998 data given in these tables may be used for 1999 sun and star sights. For the sun, take out the GHA and Dec for the same date, but for a time 5h 48m *earlier* than the UT of observation; *add* 87°00' to the GHA so obtained. For stars, calculate GHA and Dec for the same date and time, but *subtract* 15.1' from the GHA so found. The error in either case is unlikely to exceed 0.4'. The almanac cannot be used for 1999 moon or planet observations.

INDEX OF TABLES

JANUARY 1998 SUN & MOON GMT

SUN

Day of Yr	Day of Mth	Day of Week	Equation of Time 0h	Equation of Time 12h	Transit	Semi-diam	Lat 30°N Twi-light	Lat 30°N Sun-rise	Lat 30°N Sun-set	Lat 30°N Twi-light
			m s	m s	h m	′	h m	h m	h m	h m
1	**1**	**Th**	+03 17	+03 31	**12 04**	16.3	06 30	06 56	17 11	17 38
2	**2**	**Fri**	+03 46	+04 00	**12 04**	16.3	06 30	06 56	17 12	17 38
3	**3**	**Sat**	+04 14	+04 27	**12 04**	16.3	06 30	06 56	17 13	17 39
4	**4**	**Sun**	+04 41	+04 55	**12 05**	16.3	06 30	06 57	17 13	17 40
5	**5**	**Mon**	+05 08	+05 22	**12 05**	16.3	06 30	06 57	17 14	17 40
6	**6**	**Tu**	+05 35	+05 48	**12 06**	16.3	06 31	06 57	17 15	17 41
7	**7**	**Wed**	+06 01	+06 14	**12 06**	16.3	06 31	06 57	17 16	17 42
8	**8**	**Th**	+06 27	+06 40	**12 07**	16.3	06 31	06 57	17 17	17 43
9	**9**	**Fri**	+06 53	+07 05	**12 07**	16.3	06 31	06 57	17 17	17 43
10	**10**	**Sat**	+07 17	+07 29	**12 07**	16.3	06 31	06 57	17 18	17 44
11	**11**	**Sun**	+07 41	+07 53	**12 08**	16.3	06 31	06 57	17 19	17 45
12	**12**	**Mon**	+08 05	+08 17	**12 08**	16.3	06 31	06 57	17 20	17 46
13	**13**	**Tu**	+08 28	+08 39	**12 09**	16.3	06 31	06 57	17 21	17 47
14	**14**	**Wed**	+08 50	+09 01	**12 09**	16.3	06 31	06 57	17 21	17 47
15	**15**	**Th**	+09 12	+09 23	**12 09**	16.3	06 31	06 57	17 22	17 48
16	**16**	**Fri**	+09 33	+09 43	**12 10**	16.3	06 31	06 57	17 23	17 49
17	**17**	**Sat**	+09 54	+10 03	**12 10**	16.3	06 31	06 56	17 24	17 50
18	**18**	**Sun**	+10 13	+10 23	**12 10**	16.3	06 30	06 56	17 25	17 51
19	**19**	**Mon**	+10 32	+10 41	**12 11**	16.3	06 30	06 56	17 26	17 51
20	**20**	**Tu**	+10 51	+10 59	**12 11**	16.3	06 30	06 56	17 27	17 52
21	**21**	**Wed**	+11 08	+11 17	**12 11**	16.3	06 30	06 55	17 27	17 53
22	**22**	**Th**	+11 25	+11 33	**12 12**	16.3	06 30	06 55	17 28	17 54
23	**23**	**Fri**	+11 41	+11 49	**12 12**	16.3	06 29	06 55	17 29	17 55
24	**24**	**Sat**	+11 56	+12 04	**12 12**	16.3	06 29	06 54	17 30	17 56
25	**25**	**Sun**	+12 11	+12 18	**12 12**	16.3	06 29	06 54	17 31	17 56
26	**26**	**Mon**	+12 25	+12 31	**12 13**	16.3	06 28	06 54	17 32	17 57
27	**27**	**Tu**	+12 38	+12 44	**12 13**	16.3	06 28	06 53	17 33	17 58
28	**28**	**Wed**	+12 50	+12 56	**12 13**	16.3	06 27	06 53	17 34	17 59
29	**29**	**Th**	+13 01	+13 07	**12 13**	16.3	06 27	06 52	17 34	18 00
30	**30**	**Fri**	+13 12	+13 17	**12 13**	16.3	06 26	06 52	17 35	18 00
31	**31**	**Sat**	+13 22	+13 26	**12 13**	16.3	06 26	06 51	17 36	18 01

Lat Corr to Sunrise, Sunset etc

Lat	Twi-light	Sun-rise	Sun-set	Twi-light
°	h m	h m	h m	h m
N60	+1 25	+1 52	−1 52	−1 25
55	+1 02	+1 19	−1 19	−1 02
50	+0 45	+0 56	−0 56	−0 45
45	+0 31	+0 38	−0 38	−0 31
N40	+0 20	+0 23	−0 23	−0 20
35	+0 09	+0 11	−0 11	−0 09
30	0 00	0 00	0 00	0 00
25	−0 09	−0 10	+0 10	+0 09
N20	−0 17	−0 19	+0 19	+0 17
15	−0 24	−0 27	+0 27	+0 24
10	−0 32	−0 35	+0 35	+0 32
N 5	−0 40	−0 43	+0 43	+0 40
0	−0 47	−0 51	+0 51	+0 47
S 5	−0 55	−0 59	+0 59	+0 55
10	−1 04	−1 07	+1 07	+1 04
15	−1 13	−1 15	+1 15	+1 13
S20	−1 22	−1 24	+1 24	+1 22
25	−1 33	−1 33	+1 33	+1 33
30	−1 45	−1 44	+1 44	+1 45
35	−1 58	−1 55	+1 55	+1 58
S40	−2 14	−2 09	+2 09	+2 14
45	−2 34	−2 25	+2 25	+2 34
50	−3 00	−2 44	+2 44	+3 00

NOTES

Latitude corrections are for mid-month.

Equation of time = mean time (LHA mean sun) minus apparent time (LHA true sun).

MOON

Day of Yr	Day of Mth	Day of Week	Age	Transit (Upper)	Diff	Semi-diam	Hor Par	Lat 30°N Moon-rise	Lat 30°N Moon-set
			days	h m	m	′	′	h m	h m
1	**1**	**Th**	03	**14 37**	**55**	16.1	59.2	08 59	20 20
2	**2**	**Fri**	04	**15 32**	**52**	16.2	59.3	09 44	21 23
3	**3**	**Sat**	05	**16 24**	**52**	16.2	59.4	10 27	22 27
4	**4**	**Sun**	06	**17 16**	**52**	16.2	59.3	11 08	23 30
5	**5**	**Mon**	07	**18 08**	**51**	16.1	59.2	11 49	- -
6	**6**	**Tu**	08	**18 59**	**53**	16.1	58.9	12 29	00 32
7	**7**	**Wed**	09	**19 52**	**53**	16.0	58.7	13 12	01 35
8	**8**	**Th**	10	**20 45**	**55**	15.9	58.3	13 57	02 37
9	**9**	**Fri**	11	**21 40**	**54**	15.8	58.0	14 44	03 38
10	**10**	**Sat**	12	**22 34**	**54**	15.7	57.5	15 35	04 37
11	**11**	**Sun**	13	**23 28**	**52**	15.5	57.0	16 29	05 33
12	**12**	**Mon**	14	**24 20**	**-**	15.4	56.5	17 24	06 26
13	**13**	**Tu**	15	**00 20**	**50**	15.3	56.0	18 19	07 14
14	**14**	**Wed**	16	**01 10**	**48**	15.1	55.5	19 14	07 58
15	**15**	**Th**	17	**01 58**	**46**	15.0	55.0	20 09	08 38
16	**16**	**Fri**	18	**02 44**	**44**	14.9	54.6	21 02	09 14
17	**17**	**Sat**	19	**03 28**	**42**	14.8	54.3	21 54	09 49
18	**18**	**Sun**	20	**04 10**	**42**	14.8	54.2	22 45	10 22
19	**19**	**Mon**	21	**04 52**	**43**	14.8	54.2	23 36	10 55
20	**20**	**Tu**	22	**05 35**	**43**	14.8	54.4	- -	11 29
21	**21**	**Wed**	23	**06 18**	**45**	14.9	54.8	00 28	12 04
22	**22**	**Th**	24	**07 03**	**48**	15.1	55.3	01 22	12 42
23	**23**	**Fri**	25	**07 51**	**51**	15.3	56.0	02 16	13 24
24	**24**	**Sat**	26	**08 42**	**53**	15.5	56.8	03 11	14 10
25	**25**	**Sun**	27	**09 35**	**55**	15.7	57.6	04 08	15 01
26	**26**	**Mon**	28	**10 30**	**57**	15.9	58.4	05 03	15 58
27	**27**	**Tu**	29	**11 27**	**57**	16.1	59.2	05 58	16 59
28	**28**	**Wed**	00	**12 24**	**57**	16.3	59.8	06 49	18 03
29	**29**	**Th**	01	**13 21**	**55**	16.4	60.1	07 38	19 09
30	**30**	**Fri**	02	**14 16**	**55**	16.4	60.3	08 23	20 15
31	**31**	**Sat**	03	**15 11**	**53**	16.4	60.2	09 06	21 20

Phases of the Moon

	d	h	m
☾ First Quarter	5	14	18
○ Full Moon	12	17	24
☽ Last Quarter	20	19	40
● New Moon	28	06	01
Perigee	3	09	
Apogee	18	21	
Perigee	30	14	
On 0°	4	20	
Farthest N of 0°	11	09	
On 0°	18	16	
Farthest S of 0°	25	22	

NOTES

Diff equals daily change in transit time.

To correct moonrise, moonset, or transit times for latitude and/or longitude see page E 112. For further information about these tables see page E 4. For arc-to-time correction see page E 103.

JANUARY 1998 — STARS — 0h GMT January 1

No	Name	Mag	Transit h m	Dec ° ′	GHA ° ′	RA h m	SHA ° ′
♈	**ARIES**		**17 15**		**100 26.6**		
1	**Alpheratz**	2.2	**17 24**	N29 04.9	**98 22.4**	0 08	357 55.8
2	**Ankaa**	2.4	**17 41**	S 42 19.3	**93 54.1**	0 26	353 27.5
3	**Schedar**	2.5	**17 56**	N56 31.8	**90 20.6**	0 40	349 54.0
4	**Diphda**	2.2	**17 59**	S 18 00.1	**89 34.4**	0 43	349 07.8
5	**Achernar**	0.6	**18 53**	S 57 15.2	**76 02.2**	1 38	335 35.6
6	**POLARIS**	2.1	**19 46**	N89 15.5	**62 48.1**	2 31	322 21.5
7	**Hamal**	2.2	**19 22**	N23 27.2	**68 40.6**	2 07	328 14.0
8	**Acamar**	3.1	**20 13**	S 40 19.1	**55 53.8**	2 58	315 27.2
9	**Menkar**	2.8	**20 17**	N 4 04.8	**54 53.9**	3 02	314 27.3
10	**Mirfak**	1.9	**20 39**	N49 51.3	**49 23.6**	3 24	308 57.0
11	**Aldebaran**	1.1	**21 50**	N16 30.2	**31 29.3**	4 36	291 02.7
12	**Rigel**	0.3	**22 29**	S 8 12.5	**21 49.7**	5 14	281 23.1
13	**Capella**	0.2	**22 31**	N45 59.7	**21 18.1**	5 17	280 51.5
14	**Bellatrix**	1.7	**22 40**	N 6 20.7	**19 11.0**	5 25	278 44.4
15	**Elnath**	1.8	**22 41**	N28 36.2	**18 53.8**	5 26	278 27.2
16	**Alnilam**	1.8	**22 51**	S 1 12.4	**16 24.7**	5 36	275 58.1
17	**Betelgeuse**	0.1-1.2	**23 09**	N 7 24.2	**11 40.4**	5 55	271 13.8
18	**Canopus**	-0.9	**23 38**	S 52 41.9	**4 27.4**	6 24	264 00.8
19	**Sirius**	-1.6	**0 03**	S 16 43.0	**359 10.4**	6 45	258 43.8
20	**Adhara**	1.6	**0 17**	S 28 58.4	**355 48.0**	6 59	255 21.4
21	**Castor**	1.6	**0 53**	N31 53.4	**346 49.2**	7 34	246 22.6
22	**Procyon**	0.5	**0 57**	N 5 13.6	**345 38.3**	7 39	245 11.7
23	**Pollux**	1.2	**1 03**	N28 01.7	**344 08.4**	7 45	243 41.8
24	**Avior**	1.7	**1 40**	S 59 30.3	**334 48.7**	8 23	234 22.1
25	**Suhail**	2.2	**2 26**	S 43 25.5	**323 27.2**	9 08	223 00.6
26	**Miaplacidus**	1.8	**2 31**	S 69 42.5	**322 07.7**	9 13	221 41.1
27	**Alphard**	2.2	**2 45**	S 8 39.1	**318 34.0**	9 28	218 07.4
28	**Regulus**	1.3	**3 26**	N11 58.5	**308 22.4**	10 08	207 55.8
29	**Dubhe**	2.0	**4 21**	N61 45.4	**294 32.4**	11 04	194 05.8
30	**Denebola**	2.2	**5 06**	N14 34.9	**283 12.2**	11 49	182 45.6
31	**Gienah**	2.8	**5 33**	S 17 31.8	**276 30.9**	12 16	176 04.3
32	**Acrux**	1.1	**5 44**	S 63 05.0	**273 48.8**	12 27	173 22.2
33	**Gacrux**	1.6	**5 48**	S 57 05.8	**272 40.4**	12 31	172 13.8
34	**Mimosa**	1.5	**6 05**	S 59 40.4	**268 32.2**	12 48	168 05.6
35	**Alioth**	1.7	**6 11**	N55 58.0	**266 57.6**	12 54	166 31.0
36	**Spica**	1.2	**6 42**	S 11 09.0	**259 10.3**	13 25	158 43.7
37	**Alkaid**	1.9	**7 05**	N49 19.2	**253 34.9**	13 47	153 08.3
38	**Hadar**	0.9	**7 21**	S 60 21.4	**249 31.3**	14 04	149 04.7
39	**Menkent**	2.3	**7 24**	S 36 21.4	**248 48.1**	14 07	148 21.5
40	**Arcturus**	0.2	**7 33**	N19 11.5	**246 33.2**	14 16	146 06.6
41	**Rigil Kent.**	0.1	**7 56**	S 60 49.2	**240 34.6**	14 39	140 08.0
42	**Zuben'ubi**	2.9	**8 08**	S 16 01.9	**237 45.2**	14 51	137 18.6
43	**Kochab**	2.2	**8 08**	N74 09.6	**237 46.8**	14 51	137 20.2
44	**Alphecca**	2.3	**8 51**	N26 43.2	**226 47.8**	15 35	126 21.2
45	**Antares**	1.2	**9 46**	S 26 25.5	**213 07.6**	16 29	112 41.0
46	**Atria**	1.9	**10 05**	S 69 01.1	**208 20.5**	16 48	107 53.9
47	**Sabik**	2.6	**10 27**	S 15 43.2	**202 53.0**	17 10	102 26.4
48	**Shaula**	1.7	**10 50**	S 37 05.9	**197 04.9**	17 33	96 38.3
49	**Rasalhague**	2.1	**10 51**	N12 33.7	**196 44.3**	17 35	96 17.7
50	**Eltanin**	2.4	**11 13**	N51 29.4	**191 18.7**	17 57	90 52.1
51	**Kaus Aust.**	2.0	**11 40**	S 34 23.0	**184 26.5**	18 24	83 59.9
52	**Vega**	0.1	**11 53**	N38 47.0	**181 14.0**	18 37	80 47.4
53	**Nunki**	2.1	**12 11**	S 26 17.8	**176 39.9**	18 55	76 13.3
54	**Altair**	0.9	**13 07**	N 8 51.9	**162 46.7**	19 51	62 20.1
55	**Peacock**	2.1	**13 41**	S 56 44.5	**154 05.0**	20 25	53 38.4
56	**Deneb**	1.3	**13 57**	N45 16.5	**150 06.5**	20 41	49 39.9
57	**Enif**	2.5	**15 00**	N 9 52.0	**134 25.6**	21 44	33 59.0
58	**Al Na'ir**	2.2	**15 24**	S 46 58.4	**128 25.5**	22 08	27 58.9
59	**Fomalhaut**	1.3	**16 13**	S 29 38.1	**116 03.9**	22 58	15 37.3
60	**Markab**	2.6	**16 20**	N15 11.7	**114 16.9**	23 05	13 50.3

Star Transit Corr Table

Date	Corr h m	Date	Corr h m
1	0 00	17	–1 03
2	–0 04	18	–1 07
3	–0 08	19	–1 11
4	–0 12	20	–1 15
5	–0 16	21	–1 19
6	–0 20	22	–1 23
7	–0 24	23	–1 27
8	–0 28	24	–1 30
9	–0 31	25	–1 34
10	–0 35	26	–1 38
11	–0 39	27	–1 42
12	–0 43	28	–1 46
13	–0 47	29	–1 50
14	–0 51	30	–1 54
15	–0 55	31	–1 58
16	–0 59		

STAR TRANSIT

To find the approximate time of transit of a star for any day of the month use the table above. All corrections are subtractive.

If the value taken from the table is greater than the time of transit for the first of the month, add 23h 56min to the time of transit before subtracting the correction.

Example: What time will Adhara (No 20) transit the meridian on January 30?

	h min
Transit on Jan 1	00 17
	+23 56
	24 13
Corr for Jan 30	–01 54
Transit on Jan 30	22 19

JANUARY DIARY

d h	
1 03	Mars 4° S of Moon
1 23	Jupiter 3° S of Moon
4 21	Earth at perihelion
5 12	Saturn 0.2° N of Moon
6 15	Mercury greatest elong. W.(23°)
9 13	Aldebaran 0.4° S of Moon
9 17	Venus 4° N of Neptune
16 11	Venus in conjunction
19 23	Neptune in conjunction
21 01	Mars 0.2° S of Jupiter
26 17	Mercury 8° S of Venus
27 00	Venus 3° N of Moon
27 01	Mercury 5° S of Moon
28 20	Uranus in conjunction
29 17	Jupiter 2° S of Moon
30 01	Mars 1.7° S of Moon

NOTES

Star declinations may be used as is for the whole month. To correct GHA for day, hour, minute, and second see page E 106; for further explanation of this table see page E 4.

JANUARY 1998 — SUN & ARIES — GMT

Time	SUN GHA ° ′	Dec ° ′	ARIES GHA ° ′
Thursday, 1st January			
00	179 10.6	S23 02.0	100 26.6
02	209 10.0	23 01.6	130 31.6
04	239 09.4	23 01.2	160 36.5
06	269 08.8	23 00.8	190 41.4
08	299 08.2	23 00.4	220 46.4
10	329 07.6	23 00.0	250 51.3
12	359 07.0	22 59.5	280 56.2
14	29 06.4	22 59.1	311 01.1
16	59 05.8	22 58.7	341 06.1
18	89 05.2	22 58.3	11 11.0
20	119 04.6	22 57.9	41 15.9
22	149 04.1	S22 57.4	71 20.9
Friday, 2nd January			
00	179 03.5	S22 57.0	101 25.8
02	209 02.9	22 56.6	131 30.7
04	239 02.3	22 56.1	161 35.6
06	269 01.7	22 55.7	191 40.6
08	299 01.1	22 55.2	221 45.5
10	329 00.5	22 54.8	251 50.4
12	359 00.0	22 54.3	281 55.4
14	28 59.4	22 53.9	312 00.3
16	58 58.8	22 53.4	342 05.2
18	88 58.2	22 53.0	12 10.1
20	118 57.6	22 52.5	42 15.1
22	148 57.0	S22 52.0	72 20.0
Saturday, 3rd January			
00	178 56.5	S22 51.6	102 24.9
02	208 55.9	22 51.1	132 29.9
04	238 55.3	22 50.6	162 34.8
06	268 54.7	22 50.1	192 39.7
08	298 54.2	22 49.6	222 44.6
10	328 53.6	22 49.2	252 49.6
12	358 53.0	22 48.7	282 54.5
14	28 52.4	22 48.2	312 59.4
16	58 51.9	22 47.7	343 04.4
18	88 51.3	22 47.2	13 09.3
20	118 50.7	22 46.7	43 14.2
22	148 50.1	S22 46.2	73 19.1
Sunday, 4th January			
00	178 49.6	S22 45.7	103 24.1
02	208 49.0	22 45.2	133 29.0
04	238 48.4	22 44.6	163 33.9
06	268 47.8	22 44.1	193 38.8
08	298 47.3	22 43.6	223 43.8
10	328 46.7	22 43.1	253 48.7
12	358 46.1	22 42.5	283 53.6
14	28 45.6	22 42.0	313 58.6
16	58 45.0	22 41.5	344 03.5
18	88 44.4	22 40.9	14 08.4
20	118 43.9	22 40.4	44 13.3
22	148 43.3	S22 39.9	74 18.3
Monday, 5th January			
00	178 42.8	S22 39.3	104 23.2
02	208 42.2	22 38.8	134 28.1
04	238 41.6	22 38.2	164 33.1
06	268 41.1	22 37.7	194 38.0
08	298 40.5	22 37.1	224 42.9
10	328 40.0	22 36.5	254 47.8
12	358 39.4	22 36.0	284 52.8
14	28 38.8	22 35.4	314 57.7
16	58 38.3	22 34.8	345 02.6
18	88 37.7	22 34.3	15 07.6
20	118 37.2	22 33.7	45 12.5
22	148 36.6	S22 33.1	75 17.4

Time	SUN GHA ° ′	Dec ° ′	ARIES GHA ° ′
Tuesday, 6th January			
00	178 36.1	S22 32.5	105 22.3
02	208 35.5	22 31.9	135 27.3
04	238 35.0	22 31.3	165 32.2
06	268 34.4	22 30.8	195 37.1
08	298 33.9	22 30.2	225 42.1
10	328 33.3	22 29.6	255 47.0
12	358 32.8	22 29.0	285 51.9
14	28 32.2	22 28.4	315 56.8
16	58 31.7	22 27.7	346 01.8
18	88 31.1	22 27.1	16 06.7
20	118 30.6	22 26.5	46 11.6
22	148 30.0	S22 25.9	76 16.6
Wednesday, 7th January			
00	178 29.5	S22 25.3	106 21.5
02	208 28.9	22 24.7	136 26.4
04	238 28.4	22 24.0	166 31.3
06	268 27.9	22 23.4	196 36.3
08	298 27.3	22 22.8	226 41.2
10	328 26.8	22 22.1	256 46.1
12	358 26.3	22 21.5	286 51.0
14	28 25.7	22 20.9	316 56.0
16	58 25.2	22 20.2	347 00.9
18	88 24.6	22 19.6	17 05.8
20	118 24.1	22 18.9	47 10.8
22	148 23.6	S22 18.3	77 15.7
Thursday, 8th January			
00	178 23.0	S22 17.6	107 20.6
02	208 22.5	22 16.9	137 25.5
04	238 22.0	22 16.3	167 30.5
06	268 21.5	22 15.6	197 35.4
08	298 20.9	22 14.9	227 40.3
10	328 20.4	22 14.3	257 45.3
12	358 19.9	22 13.6	287 50.2
14	28 19.3	22 12.9	317 55.1
16	58 18.8	22 12.2	348 00.0
18	88 18.3	22 11.6	18 05.0
20	118 17.8	22 10.9	48 09.9
22	148 17.2	S22 10.2	78 14.8
Friday, 9th January			
00	178 16.7	S22 09.5	108 19.8
02	208 16.2	22 08.8	138 24.7
04	238 15.7	22 08.1	168 29.6
06	268 15.2	22 07.4	198 34.5
08	298 14.7	22 06.7	228 39.5
10	328 14.1	22 06.0	258 44.4
12	358 13.6	22 05.3	288 49.3
14	28 13.1	22 04.6	318 54.3
16	58 12.6	22 03.8	348 59.2
18	88 12.1	22 03.1	19 04.1
20	118 11.6	22 02.4	49 09.0
22	148 11.1	S22 01.7	79 14.0
Saturday, 10th January			
00	178 10.5	S22 00.9	109 18.9
02	208 10.0	22 00.2	139 23.8
04	238 09.5	21 59.5	169 28.8
06	268 09.0	21 58.7	199 33.7
08	298 08.5	21 58.0	229 38.6
10	328 08.0	21 57.2	259 43.5
12	358 07.5	21 56.5	289 48.5
14	28 07.0	21 55.8	319 53.4
16	58 06.5	21 55.0	349 58.3
18	88 06.0	21 54.2	20 03.3
20	118 05.5	21 53.5	50 08.2
22	148 05.0	S21 52.7	80 13.1

Time	SUN GHA ° ′	Dec ° ′	ARIES GHA ° ′	Time
Sunday, 11th January				
00	178 04.5	S21 52.0	110 18.0	00
02	208 04.0	21 51.2	140 23.0	02
04	238 03.5	21 50.4	170 27.9	04
06	268 03.0	21 49.6	200 32.8	06
08	298 02.5	21 48.9	230 37.8	08
10	328 02.0	21 48.1	260 42.7	10
12	358 01.5	21 47.3	290 47.6	12
14	28 01.1	21 46.5	320 52.5	14
16	58 00.6	21 45.7	350 57.5	16
18	88 00.1	21 44.9	21 02.4	18
20	117 59.6	21 44.1	51 07.3	20
22	147 59.1	S21 43.3	81 12.3	22
Monday, 12th January				
00	177 58.6	S21 42.5	111 17.2	00
02	207 58.1	21 41.7	141 22.1	02
04	237 57.7	21 40.9	171 27.0	04
06	267 57.2	21 40.1	201 32.0	06
08	297 56.7	21 39.3	231 36.9	08
10	327 56.2	21 38.5	261 41.8	10
12	357 55.7	21 37.7	291 46.7	12
14	27 55.3	21 36.9	321 51.7	14
16	57 54.8	21 36.0	351 56.6	16
18	87 54.3	21 35.2	22 01.5	18
20	117 53.8	21 34.4	52 06.5	20
22	147 53.4	S21 33.6	82 11.4	22
Tuesday, 13th January				
00	177 52.9	S21 32.7	112 16.3	00
02	207 52.4	21 31.9	142 21.2	02
04	237 51.9	21 31.0	172 26.2	04
06	267 51.5	21 30.2	202 31.1	06
08	297 51.0	21 29.4	232 36.0	08
10	327 50.5	21 28.5	262 41.0	10
12	357 50.1	21 27.7	292 45.9	12
14	27 49.6	21 26.8	322 50.8	14
16	57 49.1	21 25.9	352 55.7	16
18	87 48.7	21 25.1	23 00.7	18
20	117 48.2	21 24.2	53 05.6	20
22	147 47.8	S21 23.3	83 10.5	22
Wednesday, 14th January				
00	177 47.3	S21 22.5	113 15.5	00
02	207 46.8	21 21.6	143 20.4	02
04	237 46.4	21 20.7	173 25.3	04
06	267 45.9	21 19.9	203 30.2	06
08	297 45.5	21 19.0	233 35.2	08
10	327 45.0	21 18.1	263 40.1	10
12	357 44.6	21 17.2	293 45.0	12
14	27 44.1	21 16.3	323 50.0	14
16	57 43.7	21 15.4	353 54.9	16
18	87 43.2	21 14.5	23 59.8	18
20	117 42.8	21 13.6	54 04.7	20
22	147 42.3	S21 12.7	84 09.7	22
Thursday, 15th January				
00	177 41.9	S21 11.8	114 14.6	00
02	207 41.4	21 10.9	144 19.5	02
04	237 41.0	21 10.0	174 24.5	04
06	267 40.5	21 09.1	204 29.4	06
08	297 40.1	21 08.2	234 34.3	08
10	327 39.7	21 07.3	264 39.2	10
12	357 39.2	21 06.3	294 44.2	12
14	27 38.8	21 05.4	324 49.1	14
16	57 38.3	21 04.5	354 54.0	16
18	87 37.9	21 03.6	24 59.0	18
20	117 37.5	21 02.6	55 03.9	20
22	147 37.0	S21 01.7	85 08.8	22

NOTES

Sun and Aries GHA corrections for additional hour, minutes, and seconds are on page E 106.
Correct sun declination by interpolation. For an example see page E 4.

JANUARY 1998 — SUN & ARIES — GMT

Friday, 16th January

Time	SUN GHA ° ′	Dec ° ′	ARIES GHA ° ′
00	177 36.6	**S21 00.8**	115 13.7
02	207 36.2	**20 59.8**	145 18.7
04	237 35.8	**20 58.9**	175 23.6
06	267 35.3	**20 57.9**	205 28.5
08	297 34.9	**20 57.0**	235 33.4
10	327 34.5	**20 56.0**	265 38.4
12	357 34.0	**20 55.1**	295 43.3
14	27 33.6	**20 54.1**	325 48.2
16	57 33.2	**20 53.2**	355 53.2
18	87 32.8	**20 52.2**	25 58.1
20	117 32.4	**20 51.2**	56 03.0
22	147 31.9	**S20 50.3**	86 07.9

Saturday, 17th January

Time	SUN GHA ° ′	Dec ° ′	ARIES GHA ° ′
00	177 31.5	**S20 49.3**	116 12.9
02	207 31.1	**20 48.3**	146 17.8
04	237 30.7	**20 47.3**	176 22.7
06	267 30.3	**20 46.4**	206 27.7
08	297 29.9	**20 45.4**	236 32.6
10	327 29.4	**20 44.4**	266 37.5
12	357 29.0	**20 43.4**	296 42.4
14	27 28.6	**20 42.4**	326 47.4
16	57 28.2	**20 41.4**	356 52.3
18	87 27.8	**20 40.4**	26 57.2
20	117 27.4	**20 39.4**	57 02.2
22	147 27.0	**S20 38.4**	87 07.1

Sunday, 18th January

Time	SUN GHA ° ′	Dec ° ′	ARIES GHA ° ′
00	177 26.6	**S20 37.4**	117 12.0
02	207 26.2	**20 36.4**	147 16.9
04	237 25.8	**20 35.4**	177 21.9
06	267 25.4	**20 34.4**	207 26.8
08	297 25.0	**20 33.4**	237 31.7
10	327 24.6	**20 32.4**	267 36.7
12	357 24.2	**20 31.4**	297 41.6
14	27 23.8	**20 30.3**	327 46.5
16	57 23.4	**20 29.3**	357 51.4
18	87 23.0	**20 28.3**	27 56.4
20	117 22.6	**20 27.2**	58 01.3
22	147 22.2	**S20 26.2**	88 06.2

Monday, 19th January

Time	SUN GHA ° ′	Dec ° ′	ARIES GHA ° ′
00	177 21.8	**S20 25.2**	118 11.1
02	207 21.5	**20 24.1**	148 16.1
04	237 21.1	**20 23.1**	178 21.0
06	267 20.7	**20 22.1**	208 25.9
08	297 20.3	**20 21.0**	238 30.9
10	327 19.9	**20 20.0**	268 35.8
12	357 19.5	**20 18.9**	298 40.7
14	27 19.2	**20 17.9**	328 45.6
16	57 18.8	**20 16.8**	358 50.6
18	87 18.4	**20 15.7**	28 55.5
20	117 18.0	**20 14.7**	59 00.4
22	147 17.7	**S20 13.6**	89 05.4

Tuesday, 20th January

Time	SUN GHA ° ′	Dec ° ′	ARIES GHA ° ′
00	177 17.3	**S20 12.5**	119 10.3
02	207 16.9	**20 11.5**	149 15.2
04	237 16.5	**20 10.4**	179 20.1
06	267 16.2	**20 09.3**	209 25.1
08	297 15.8	**20 08.2**	239 30.0
10	327 15.4	**20 07.2**	269 34.9
12	357 15.1	**20 06.1**	299 39.9
14	27 14.7	**20 05.0**	329 44.8
16	57 14.3	**20 03.9**	359 49.7
18	87 14.0	**20 02.8**	29 54.6
20	117 13.6	**20 01.7**	59 59.6
22	147 13.2	**S20 00.6**	90 04.5

Wednesday, 21st January

Time	SUN GHA ° ′	Dec ° ′	ARIES GHA ° ′
00	177 12.9	**S19 59.5**	120 09.4
02	207 12.5	**19 58.4**	150 14.4
04	237 12.2	**19 57.3**	180 19.3
06	267 11.8	**19 56.2**	210 24.2
08	297 11.5	**19 55.1**	240 29.1
10	327 11.1	**19 54.0**	270 34.1
12	357 10.8	**19 52.9**	300 39.0
14	27 10.4	**19 51.7**	330 43.9
16	57 10.1	**19 50.6**	0 48.9
18	87 09.7	**19 49.5**	30 53.8
20	117 09.4	**19 48.4**	60 58.7
22	147 09.0	**S19 47.3**	91 03.6

Thursday, 22nd January

Time	SUN GHA ° ′	Dec ° ′	ARIES GHA ° ′
00	177 08.7	**S19 46.1**	121 08.6
02	207 08.3	**19 45.0**	151 13.5
04	237 08.0	**19 43.9**	181 18.4
06	267 07.7	**19 42.7**	211 23.3
08	297 07.3	**19 41.6**	241 28.3
10	327 07.0	**19 40.4**	271 33.2
12	357 06.7	**19 39.3**	301 38.1
14	27 06.3	**19 38.1**	331 43.1
16	57 06.0	**19 37.0**	1 48.0
18	87 05.7	**19 35.8**	31 52.9
20	117 05.3	**19 34.7**	61 57.8
22	147 05.0	**S19 33.5**	92 02.8

Friday, 23rd January

Time	SUN GHA ° ′	Dec ° ′	ARIES GHA ° ′
00	177 04.7	**S19 32.4**	122 07.7
02	207 04.3	**19 31.2**	152 12.6
04	237 04.0	**19 30.0**	182 17.6
06	267 03.7	**19 28.9**	212 22.5
08	297 03.4	**19 27.7**	242 27.4
10	327 03.0	**19 26.5**	272 32.3
12	357 02.7	**19 25.3**	302 37.3
14	27 02.4	**19 24.2**	332 42.2
16	57 02.1	**19 23.0**	2 47.1
18	87 01.8	**19 21.8**	32 52.1
20	117 01.5	**19 20.6**	62 57.0
22	147 01.1	**S19 19.4**	93 01.9

Saturday, 24th January

Time	SUN GHA ° ′	Dec ° ′	ARIES GHA ° ′
00	177 00.8	**S19 18.2**	123 06.8
02	207 00.5	**19 17.0**	153 11.8
04	237 00.2	**19 15.8**	183 16.7
06	266 59.9	**19 14.6**	213 21.6
08	296 59.6	**19 13.4**	243 26.6
10	326 59.3	**19 12.2**	273 31.5
12	356 59.0	**19 11.0**	303 36.4
14	26 58.7	**19 09.8**	333 41.3
16	56 58.4	**19 08.6**	3 46.3
18	86 58.1	**19 07.4**	33 51.2
20	116 57.8	**19 06.2**	63 56.1
22	146 57.5	**S19 05.0**	94 01.1

Sunday, 25th January

Time	SUN GHA ° ′	Dec ° ′	ARIES GHA ° ′
00	176 57.2	**S19 03.8**	124 06.0
02	206 56.9	**19 02.5**	154 10.9
04	236 56.6	**19 01.3**	184 15.8
06	266 56.3	**19 00.1**	214 20.8
08	296 56.0	**18 58.8**	244 25.7
10	326 55.7	**18 57.6**	274 30.6
12	356 55.5	**18 56.4**	304 35.6
14	26 55.2	**18 55.1**	334 40.5
16	56 54.9	**18 53.9**	4 45.4
18	86 54.6	**18 52.7**	34 50.3
20	116 54.3	**18 51.4**	64 55.3
22	146 54.0	**S18 50.2**	95 00.2

Monday, 26th January

Time	SUN GHA ° ′	Dec ° ′	ARIES GHA ° ′	Time
00	176 53.8	**S18 48.9**	125 05.1	00
02	206 53.5	**18 47.7**	155 10.1	02
04	236 53.2	**18 46.4**	185 15.0	04
06	266 52.9	**18 45.2**	215 19.9	06
08	296 52.7	**18 43.9**	245 24.8	08
10	326 52.4	**18 42.6**	275 29.8	10
12	356 52.1	**18 41.4**	305 34.7	12
14	26 51.8	**18 40.1**	335 39.6	14
16	56 51.6	**18 38.8**	5 44.5	16
18	86 51.3	**18 37.6**	35 49.5	18
20	116 51.0	**18 36.3**	65 54.4	20
22	146 50.8	**S18 35.0**	95 59.3	22

Tuesday, 27th January

Time	SUN GHA ° ′	Dec ° ′	ARIES GHA ° ′	Time
00	176 50.5	**S18 33.7**	126 04.3	00
02	206 50.2	**18 32.5**	156 09.2	02
04	236 50.0	**18 31.2**	186 14.1	04
06	266 49.7	**18 29.9**	216 19.0	06
08	296 49.5	**18 28.6**	246 24.0	08
10	326 49.2	**18 27.3**	276 28.9	10
12	356 49.0	**18 26.0**	306 33.8	12
14	26 48.7	**18 24.7**	336 38.8	14
16	56 48.5	**18 23.4**	6 43.7	16
18	86 48.2	**18 22.1**	36 48.6	18
20	116 48.0	**18 20.8**	66 53.5	20
22	146 47.7	**S18 19.5**	96 58.5	22

Wednesday, 28th January

Time	SUN GHA ° ′	Dec ° ′	ARIES GHA ° ′	Time
00	176 47.5	**S18 18.2**	127 03.4	00
02	206 47.2	**18 16.9**	157 08.3	02
04	236 47.0	**18 15.6**	187 13.3	04
06	266 46.7	**18 14.3**	217 18.2	06
08	296 46.5	**18 13.0**	247 23.1	08
10	326 46.3	**18 11.7**	277 28.0	10
12	356 46.0	**18 10.4**	307 33.0	12
14	26 45.8	**18 09.0**	337 37.9	14
16	56 45.5	**18 07.7**	7 42.8	16
18	86 45.3	**18 06.4**	37 47.8	18
20	116 45.1	**18 05.1**	67 52.7	20
22	146 44.9	**S18 03.7**	97 57.6	22

Thursday, 29th January

Time	SUN GHA ° ′	Dec ° ′	ARIES GHA ° ′	Time
00	176 44.6	**S18 02.4**	128 02.5	00
02	206 44.4	**18 01.1**	158 07.5	02
04	236 44.2	**17 59.7**	188 12.4	04
06	266 43.9	**17 58.4**	218 17.3	06
08	296 43.7	**17 57.0**	248 22.3	08
10	326 43.5	**17 55.7**	278 27.2	10
12	356 43.3	**17 54.4**	308 32.1	12
14	26 43.1	**17 53.0**	338 37.0	14
16	56 42.8	**17 51.7**	8 42.0	16
18	86 42.6	**17 50.3**	38 46.9	18
20	116 42.4	**17 48.9**	68 51.8	20
22	146 42.2	**S17 47.6**	98 56.8	22

Friday, 30th January

Time	SUN GHA ° ′	Dec ° ′	ARIES GHA ° ′	Time
00	176 42.0	**S17 46.2**	129 01.7	00
02	206 41.8	**17 44.9**	159 06.6	02
04	236 41.6	**17 43.5**	189 11.5	04
06	266 41.4	**17 42.1**	219 16.5	06
08	296 41.1	**17 40.8**	249 21.4	08
10	326 40.9	**17 39.4**	279 26.3	10
12	356 40.7	**17 38.0**	309 31.2	12
14	26 40.5	**17 36.7**	339 36.2	14
16	56 40.3	**17 35.3**	9 41.1	16
18	86 40.1	**17 33.9**	39 46.0	18
20	116 39.9	**17 32.5**	69 51.0	20
22	146 39.7	**S17 31.1**	99 55.9	22

Saturday, 31st January

Time	SUN GHA ° ′	Dec ° ′	ARIES GHA ° ′
00	176 39.6	**S17 29.7**	130 00.8
02	206 39.4	**17 28.4**	160 05.7
04	236 39.2	**17 27.0**	190 10.7
06	266 39.0	**S17 25.6**	220 15.6
08	296 38.8	**S17 24.2**	250 20.5
10	326 38.6	**17 22.8**	280 25.5
12	356 38.4	**17 21.4**	310 30.4
14	26 38.2	**S17 20.0**	340 35.3
16	56 38.0	**S17 18.6**	10 40.2
18	86 37.9	**17 17.2**	40 45.2
20	116 37.7	**17 15.8**	70 50.1
22	146 37.5	**S17 14.4**	100 55.0

JANUARY

Ephemeris

JANUARY 1998 PLANETS 0h GMT

VENUS Mer Pass h m	VENUS GHA ° ′	VENUS Mean Var/hr 14°+	VENUS Dec ° ′	VENUS Mean Var/hr ′	Day		JUPITER GHA ° ′	JUPITER Mean Var/hr 15°+	JUPITER Dec ° ′	JUPITER Mean Var/hr ′	JUPITER Mer Pass h m
13 36	155 14.8	3.2	S17 24.2	0.5	1	Th	135 32.7	2.0	S14 55.0	0.2	14 56
13 31	156 31.8	3.3	S17 12.6	0.5	2	Fri	136 19.5	2.0	S14 50.9	0.2	14 53
13 26	157 51.3	3.4	S17 01.4	0.5	3	Sat	137 06.4	1.9	S14 46.8	0.2	14 50
13 20	159 13.2	3.5	S16 50.5	0.4	4	SUN	137 53.1	1.9	S14 42.6	0.2	14 47
13 14	160 37.5	3.6	S16 40.1	0.4	5	Mon	138 39.8	1.9	S14 38.4	0.2	14 43
13 09	162 04.1	3.7	S16 30.0	0.4	6	Tu	139 26.5	1.9	S14 34.2	0.2	14 40
13 03	163 32.7	3.8	S16 20.4	0.4	7	Wed	140 13.1	1.9	S14 29.9	0.2	14 37
12 56	165 03.4	3.8	S16 11.2	0.4	8	Th	140 59.6	1.9	S14 25.6	0.2	14 34
12 50	166 35.8	3.9	S16 02.5	0.3	9	Fri	141 46.1	1.9	S14 21.3	0.2	14 31
12 44	168 09.9	4.0	S15 54.3	0.3	10	Sat	142 32.5	1.9	S14 16.9	0.2	14 28
12 38	169 45.4	4.0	S15 46.5	0.3	11	SUN	143 18.8	1.9	S14 12.5	0.2	14 25
12 31	171 22.1	4.1	S15 39.2	0.3	12	Mon	144 05.1	1.9	S14 08.1	0.2	14 22
12 25	172 59.7	4.1	S15 32.4	0.3	13	Tu	144 51.4	1.9	S14 03.6	0.2	14 19
12 18	174 38.0	4.1	S15 26.1	0.2	14	Wed	145 37.6	1.9	S13 59.2	0.2	14 16
12 12	176 16.6	4.1	S15 20.3	0.2	15	Th	146 23.8	1.9	S13 54.7	0.2	14 13
12 05	177 55.4	4.1	S15 15.0	0.2	16	Fri	147 09.9	1.9	S13 50.1	0.2	14 10
11 58	179 34.1	4.1	S15 10.1	0.2	17	Sat	147 56.0	1.9	S13 45.6	0.2	14 06
11 52	181 12.5	4.1	S15 05.8	0.2	18	SUN	148 42.0	1.9	S13 41.0	0.2	14 03
11 45	182 50.3	4.0	S15 01.9	0.1	19	Mon	149 28.0	1.9	S13 36.4	0.2	14 00
11 39	184 27.1	4.0	S14 58.5	0.1	20	Tu	150 14.0	1.9	S13 31.7	0.2	13 57
11 33	186 02.9	3.9	S14 55.6	0.1	21	Wed	150 59.9	1.9	S13 27.1	0.2	13 54
11 27	187 37.4	3.9	S14 53.3	0.1	22	Th	151 45.8	1.9	S13 22.4	0.2	13 51
11 20	189 10.2	3.8	S14 51.4	0.1	23	Fri	152 31.6	1.9	S13 17.7	0.2	13 48
11 14	190 41.3	3.7	S14 50.0	0.0	24	Sat	153 17.4	1.9	S13 12.9	0.2	13 45
11 09	192 10.5	3.6	S14 49.0	0.0	25	SUN	154 03.2	1.9	S13 08.1	0.2	13 42
11 03	193 37.7	3.5	S14 48.5	0.0	26	Mon	154 48.9	1.9	S13 03.4	0.2	13 39
10 57	195 02.6	3.4	S14 48.4	0.0	27	Tu	155 34.6	1.9	S12 58.6	0.2	13 36
10 52	196 25.2	3.3	S14 48.7	0.0	28	Wed	156 20.3	1.9	S12 53.7	0.2	13 33
10 47	197 45.4	3.2	S14 49.4	0.0	29	Th	157 06.0	1.9	S12 48.9	0.2	13 30
10 42	199 03.1	3.1	S14 50.5	0.1	30	Fri	157 51.6	1.9	S12 44.0	0.2	13 27
10 37	200 18.3	3.0	S14 51.9	0.1	31	Sat	158 37.2	1.9	S12 39.1	0.2	13 24

VENUS, Av. Mag. -4.3
S.H.A. January
5 56; 10 59; 15 62; 20 65; 25 68; 30 70.

JUPITER, Av. Mag.-2.0
S.H.A. January
5 34; 10 33; 15 32; 20 31; 25 30; 30 29.

MARS Mer Pass h m	MARS GHA ° ′	MARS Mean Var/hr 15°+	MARS Dec ° ′	MARS Mean Var/hr ′	Day		SATURN GHA ° ′	SATURN Mean Var/hr 15°+	SATURN Dec ° ′	SATURN Mean Var/hr ′	SATURN Mer Pass h m
14 12	146 53.4	0.5	S18 39.4	0.6	1	Th	86 48.6	2.4	N 3 06.8	0.0	18 10
14 11	147 04.7	0.5	S18 26.1	0.6	2	Fri	87 46.2	2.4	N 3 07.7	0.0	18 06
14 10	147 16.2	0.5	S18 12.5	0.6	3	Sat	88 43.7	2.4	N 3 08.7	0.0	18 02
14 10	147 27.8	0.5	S17 58.8	0.6	4	SUN	89 41.2	2.4	N 3 09.7	0.0	17 58
14 09	147 39.5	0.5	S17 45.0	0.6	5	Mon	90 38.5	2.4	N 3 10.8	0.0	17 55
14 08	147 51.2	0.5	S17 30.9	0.6	6	Tu	91 35.7	2.4	N 3 11.8	0.0	17 51
14 07	148 03.2	0.5	S17 16.6	0.6	7	Wed	92 32.8	2.4	N 3 13.0	0.0	17 47
14 07	148 15.2	0.5	S17 02.2	0.6	8	Th	93 29.9	2.4	N 3 14.1	0.1	17 43
14 06	148 27.3	0.5	S16 47.6	0.6	9	Fri	94 26.8	2.4	N 3 15.3	0.1	17 39
14 05	148 39.5	0.5	S16 32.8	0.6	10	Sat	95 23.7	2.4	N 3 16.6	0.1	17 36
14 04	148 51.9	0.5	S16 17.8	0.6	11	SUN	96 20.4	2.4	N 3 17.9	0.1	17 32
14 03	149 04.4	0.5	S16 02.7	0.6	12	Mon	97 17.1	2.4	N 3 19.2	0.1	17 28
14 02	149 16.9	0.5	S15 47.4	0.6	13	Tu	98 13.6	2.4	N 3 20.5	0.1	17 24
14 02	149 29.6	0.5	S15 31.9	0.7	14	Wed	99 10.1	2.3	N 3 21.9	0.1	17 21
14 01	149 42.4	0.5	S15 16.3	0.7	15	Th	100 06.5	2.3	N 3 23.4	0.1	17 17
14 00	149 55.3	0.5	S15 00.6	0.7	16	Fri	101 02.8	2.3	N 3 24.8	0.1	17 13
13 59	150 08.4	0.5	S14 44.7	0.7	17	Sat	101 58.9	2.3	N 3 26.3	0.1	17 09
13 58	150 21.5	0.6	S14 28.6	0.7	18	SUN	102 55.1	2.3	N 3 27.9	0.1	17 06
13 57	150 34.7	0.6	S14 12.4	0.7	19	Mon	103 51.1	2.3	N 3 29.4	0.1	17 02
13 56	150 48.1	0.6	S13 56.1	0.7	20	Tu	104 47.0	2.3	N 3 31.0	0.1	16 58
13 55	151 01.5	0.6	S13 39.6	0.7	21	Wed	105 42.8	2.3	N 3 32.7	0.1	16 55
13 54	151 15.1	0.6	S13 23.0	0.7	22	Th	106 38.6	2.3	N 3 34.3	0.1	16 51
13 54	151 28.7	0.6	S13 06.2	0.7	23	Fri	107 34.3	2.3	N 3 36.0	0.1	16 47
13 53	151 42.5	0.6	S12 49.4	0.7	24	Sat	108 29.8	2.3	N 3 37.8	0.1	16 43
13 52	151 56.3	0.6	S12 32.4	0.7	25	SUN	109 25.3	2.3	N 3 39.6	0.1	16 40
13 51	152 10.3	0.6	S12 15.3	0.7	26	Mon	110 20.8	2.3	N 3 41.4	0.1	16 36
13 50	152 24.4	0.6	S11 58.0	0.7	27	Tu	111 16.1	2.3	N 3 43.2	0.1	16 32
13 49	152 38.5	0.6	S11 40.7	0.7	28	Wed	112 11.3	2.3	N 3 45.1	0.1	16 29
13 48	152 52.8	0.6	S11 23.3	0.7	29	Th	113 06.5	2.3	N 3 47.0	0.1	16 25
13 47	153 07.1	0.6	S11 05.7	0.7	30	Fri	114 01.6	2.3	N 3 48.9	0.1	16 21
13 46	153 21.6	0.6	S10 48.1	0.7	31	Sat	114 56.6	2.3	N 3 50.8	0.1	16 18

MARS, Av. Mag. +1.2
S.H.A. January
5 43; 10 39; 15 35; 20 32; 25 28; 30 24.

NOTE: Planet corrections on pages E113–15.

SATURN, Av. Mag. +0.7
S.H.A. January
5 346; 10 346; 15 346; 20 346; 25 345; 30 345.

JANUARY 1998 — MOON

Day	GMT hr	GHA ° ′	Mean Var/hr 14°+	Dec ° ′	Mean Var/hr ′
1 Th	**0**	148 46.5	26.6	**S 15 07.2**	7.1
	6	235 26.2	26.8	**S 14 24.8**	7.7
	12	322 06.9	27.0	**S 13 39.1**	8.2
	18	48 48.7	27.1	**S 12 50.3**	8.6
2 Fri	**0**	135 31.7	27.4	**S 11 58.6**	9.1
	6	222 15.8	27.6	**S 11 04.3**	9.5
	12	309 01.1	27.8	**S 10 07.5**	9.9
	18	35 47.6	27.9	**S 9 08.4**	10.2
3 Sat	**0**	122 35.2	28.2	**S 8 07.5**	10.5
	6	209 23.9	28.3	**S 7 04.8**	10.7
	12	296 13.6	28.5	**S 6 00.6**	10.9
	18	23 04.2	28.6	**S 4 55.2**	11.1
4 Sun	**0**	109 55.6	28.7	**S 3 48.8**	11.2
	6	196 47.8	28.8	**S 2 41.6**	11.3
	12	283 40.5	28.9	**S 1 34.0**	11.3
	18	10 33.6	29.0	**S 0 26.2**	11.4
5 Mon	**0**	97 27.1	29.0	**N 0 41.7**	11.3
	6	184 20.8	29.0	**N 1 49.3**	11.2
	12	271 14.5	29.0	**N 2 56.4**	11.1
	18	358 08.1	28.9	**N 4 02.8**	10.9
6 Tu	**0**	85 01.6	28.8	**N 5 08.3**	10.7
	6	171 54.7	28.7	**N 6 12.6**	10.4
	12	258 47.3	28.6	**N 7 15.5**	10.2
	18	345 39.4	28.5	**N 8 16.9**	9.8
7 Wed	**0**	72 30.9	28.5	**N 9 16.4**	9.5
	6	159 21.7	28.3	**N 10 13.8**	9.2
	12	246 11.7	28.2	**N 11 09.0**	8.8
	18	333 00.9	28.1	**N 12 01.8**	8.4
8 Th	**0**	59 49.3	27.9	**N 12 52.0**	7.9
	6	146 36.8	27.7	**N 13 39.3**	7.3
	12	233 23.5	27.7	**N 14 23.7**	6.8
	18	320 09.4	27.5	**N 15 04.9**	6.3
9 Fri	**0**	46 54.6	27.4	**N 15 42.8**	5.7
	6	133 39.1	27.4	**N 16 17.3**	5.1
	12	220 23.1	27.2	**N 16 48.2**	4.5
	18	307 06.6	27.2	**N 17 15.5**	3.9
10 Sat	**0**	33 49.8	27.2	**N 17 39.0**	3.3
	6	120 33.0	27.2	**N 17 58.7**	2.6
	12	207 16.1	27.2	**N 18 14.5**	1.9
	18	293 59.4	27.3	**N 18 26.4**	1.3
11 Sun	**0**	20 43.1	27.4	**N 18 34.4**	0.7
	6	107 27.4	27.5	**N 18 38.5**	0.0
	12	194 12.5	27.6	**N 18 38.7**	0.6
	18	280 58.5	27.9	**N 18 35.0**	1.3
12 Mon	**0**	7 45.6	28.1	**N 18 27.6**	1.9
	6	94 34.0	28.4	**N 18 16.5**	2.5
	12	181 23.9	28.6	**N 18 01.8**	3.1
	18	268 15.3	28.9	**N 17 43.6**	3.7
13 Tu	**0**	355 08.4	29.2	**N 17 22.1**	4.1
	6	82 03.3	29.5	**N 16 57.4**	4.6
	12	169 00.2	29.8	**N 16 29.6**	5.1
	18	255 58.9	30.1	**N 15 58.8**	5.6
14 Wed	**0**	342 59.7	30.5	**N 15 25.3**	6.0
	6	70 02.5	30.8	**N 14 49.3**	6.5
	12	157 07.3	31.2	**N 14 10.7**	6.8
	18	244 14.2	31.5	**N 13 30.0**	7.2
15 Th	**0**	331 23.0	31.9	**N 12 47.1**	7.5
	6	58 33.8	32.2	**N 12 02.3**	7.8
	12	145 46.4	32.4	**N 11 15.7**	8.0
	18	233 00.8	32.7	**N 10 27.6**	8.3
16 Fri	**0**	320 16.9	33.0	**N 9 37.9**	8.5
	6	47 34.7	33.2	**N 8 47.0**	8.7
	12	134 53.9	33.5	**N 7 54.9**	8.9
	18	222 14.5	33.7	**N 7 01.8**	9.0
17 Sat	**0**	309 36.4	33.9	**N 6 07.8**	9.1
	6	36 59.3	34.0	**N 5 13.1**	9.2
	12	124 23.3	34.1	**N 4 17.8**	9.3
	18	211 48.1	34.3	**N 3 21.9**	9.4
18 Sun	**0**	299 13.5	34.4	**N 2 25.7**	9.5
	6	26 39.5	34.4	**N 1 29.3**	9.4
	12	114 05.8	34.4	**N 0 32.7**	9.4
	18	201 32.4	34.5	**S 0 23.8**	9.4
19 Mon	**0**	288 59.0	34.5	**S 1 20.3**	9.4
	6	16 25.6	34.4	**S 2 16.6**	9.3
	12	103 51.8	34.3	**S 3 12.6**	9.3
	18	191 17.6	34.2	**S 4 08.1**	9.2
20 Tu	**0**	278 42.9	34.1	**S 5 03.2**	9.1
	6	6 07.3	34.0	**S 5 57.5**	8.9
	12	93 30.9	33.7	**S 6 51.1**	8.8
	18	180 53.4	33.5	**S 7 43.8**	8.6
21 Wed	**0**	268 14.6	33.3	**S 8 35.5**	8.4
	6	355 34.5	33.0	**S 9 26.1**	8.2
	12	82 52.9	32.8	**S 10 15.5**	8.0
	18	170 09.5	32.5	**S 11 03.4**	7.7
22 Th	**0**	257 24.4	32.2	**S 11 49.8**	7.4
	6	344 37.4	31.8	**S 12 34.5**	7.1
	12	71 48.3	31.5	**S 13 17.4**	6.8
	18	158 57.1	31.1	**S 13 58.4**	6.5
23 Fri	**0**	246 03.6	30.6	**S 14 37.1**	6.0
	6	333 07.8	30.3	**S 15 13.6**	5.6
	12	60 09.6	29.8	**S 15 47.7**	5.2
	18	147 09.1	29.4	**S 16 19.1**	4.7
24 Sat	**0**	234 06.1	29.0	**S 16 47.6**	4.3
	6	321 00.6	28.7	**S 17 13.3**	3.7
	12	47 52.8	28.3	**S 17 35.7**	3.2
	18	134 42.6	27.8	**S 17 54.9**	2.6
25 Sun	**0**	221 30.1	27.5	**S 18 10.5**	2.0
	6	308 15.5	27.2	**S 18 22.6**	1.3
	12	34 58.8	26.8	**S 18 30.9**	0.6
	18	121 40.1	26.5	**S 18 35.3**	0.0
26 Mon	**0**	208 19.8	26.3	**S 18 35.6**	0.7
	6	294 57.8	26.1	**S 18 31.9**	1.4
	12	21 34.6	25.9	**S 18 24.0**	2.1
	18	108 10.1	25.8	**S 18 11.9**	2.7
27 Tu	**0**	194 44.8	25.7	**S 17 55.5**	3.5
	6	281 18.8	25.6	**S 17 34.9**	4.2
	12	7 52.4	25.5	**S 17 10.1**	4.9
	18	94 25.7	25.5	**S 16 41.0**	5.6
28 Wed	**0**	180 59.1	25.6	**S 16 08.0**	6.2
	6	267 32.6	25.7	**S 15 30.9**	6.8
	12	354 06.6	25.8	**S 14 50.0**	7.5
	18	80 41.1	25.9	**S 14 05.5**	8.1
29 Th	**0**	167 16.3	26.0	**S 13 17.5**	8.6
	6	253 52.4	26.2	**S 12 26.2**	9.1
	12	340 29.4	26.4	**S 11 31.9**	9.5
	18	67 07.5	26.6	**S 10 34.8**	10.0
30 Fri	**0**	153 46.6	26.7	**S 9 35.2**	10.4
	6	240 26.9	26.9	**S 8 33.3**	10.7
	12	327 08.2	27.1	**S 7 29.4**	11.0
	18	53 50.7	27.2	**S 6 23.9**	11.1
31 Sat	**0**	140 34.2	27.4	**S 5 17.0**	11.3
	6	227 18.7	27.6	**S 4 09.0**	11.5
	12	314 04.2	27.8	**S 3 00.2**	11.6
	18	40 50.5	27.8	**S 1 50.9**	11.6

NOTE: Moon GHA corrections for additional hours, minutes, and seconds are on page E 108–10. Moon declination corrections are on page E 111. The mean var/hr value is needed for both corrections.

FEBRUARY 1998 SUN & MOON GMT

SUN

Day of Yr	Day of Mth	Day of Week	Equation of Time 0h	Equation of Time 12h	Transit	Semi-diam	Lat 30°N Twi-light	Lat 30°N Sun-rise	Lat 30°N Sun-set	Lat 30°N Twi-light
			m s	m s	h m	′	h m	h m	h m	h m
32	**1**	**Sun**	+13 31	+13 35	**12 14**	16.3	06 25	06 50	17 37	18 02
33	**2**	**Mon**	+13 39	+13 42	**12 14**	16.3	06 25	06 50	17 38	18 03
34	**3**	**Tu**	+13 46	+13 49	**12 14**	16.3	06 24	06 49	17 39	18 04
35	**4**	**Wed**	+13 52	+13 55	**12 14**	16.3	06 24	06 49	17 40	18 04
36	**5**	**Th**	+13 58	+14 00	**12 14**	16.2	06 23	06 48	17 40	18 05
37	**6**	**Fri**	+14 03	+14 05	**12 14**	16.2	06 22	06 47	17 41	18 06
38	**7**	**Sat**	+14 06	+14 08	**12 14**	16.2	06 22	06 47	17 42	18 07
39	**8**	**Sun**	+14 10	+14 11	**12 14**	16.2	06 21	06 46	17 43	18 08
40	**9**	**Mon**	+14 12	+14 13	**12 14**	16.2	06 20	06 45	17 44	18 08
41	**10**	**Tu**	+14 13	+14 14	**12 14**	16.2	06 20	06 44	17 45	18 09
42	**11**	**Wed**	+14 14	+14 14	**12 14**	16.2	06 19	06 44	17 45	18 10
43	**12**	**Th**	+14 14	+14 14	**12 14**	16.2	06 18	06 43	17 46	18 11
44	**13**	**Fri**	+14 13	+14 13	**12 14**	16.2	06 17	06 42	17 47	18 11
45	**14**	**Sat**	+14 12	+14 11	**12 14**	16.2	06 17	06 41	17 48	18 12
46	**15**	**Sun**	+14 09	+14 08	**12 14**	16.2	06 16	06 40	17 49	18 13
47	**16**	**Mon**	+14 06	+14 05	**12 14**	16.2	06 15	06 39	17 49	18 14
48	**17**	**Tu**	+14 03	+14 01	**12 14**	16.2	06 14	06 38	17 50	18 14
49	**18**	**Wed**	+13 58	+13 56	**12 14**	16.2	06 13	06 37	17 51	18 15
50	**19**	**Th**	+13 53	+13 51	**12 14**	16.2	06 12	06 36	17 52	18 16
51	**20**	**Fri**	+13 48	+13 44	**12 14**	16.2	06 11	06 36	17 52	18 17
52	**21**	**Sat**	+13 41	+13 38	**12 14**	16.2	06 10	06 35	17 53	18 17
53	**22**	**Sun**	+13 34	+13 30	**12 14**	16.2	06 09	06 34	17 54	18 18
54	**23**	**Mon**	+13 27	+13 23	**12 13**	16.2	06 08	06 33	17 55	18 19
55	**24**	**Tu**	+13 18	+13 14	**12 13**	16.2	06 07	06 32	17 55	18 19
56	**25**	**Wed**	+13 10	+13 05	**12 13**	16.2	06 06	06 31	17 56	18 20
57	**26**	**Th**	+13 00	+12 55	**12 13**	16.2	06 05	06 29	17 57	18 21
58	**27**	**Fri**	+12 50	+12 45	**12 13**	16.2	06 04	06 28	17 57	18 22
59	**28**	**Sat**	+12 40	+12 34	**12 13**	16.2	06 03	06 27	17 58	18 22

Lat Corr to Sunrise, Sunset etc

Lat	Twi-light	Sun-rise	Sun-set	Twi-light
°	h m	h m	h m	h m
N60	+0 40	+0 59	−0 59	−0 40
55	+0 30	+0 43	−0 43	−0 30
50	+0 22	+0 31	−0 31	−0 22
45	+0 16	+0 21	−0 21	−0 16
N40	+0 10	+0 13	−0 13	−0 10
35	+0 05	+0 06	−0 06	−0 05
30	0 00	0 00	0 00	0 00
25	−0 05	−0 06	+0 06	+0 05
N20	−0 09	−0 11	+0 11	+0 09
15	−0 13	−0 16	+0 16	+0 13
10	−0 18	−0 21	+0 21	+0 18
N 5	−0 22	−0 25	+0 25	+0 22
0	−0 26	−0 30	+0 30	+0 26
S 5	−0 31	−0 34	+0 34	+0 31
10	−0 36	−0 39	+0 39	+0 36
15	−0 42	−0 44	+0 44	+0 42
S20	−0 47	−0 49	+0 49	+0 47
25	−0 54	−0 55	+0 55	+0 54
30	−1 01	−1 01	+1 01	+1 01
35	−1 09	−1 07	+1 07	+1 09
S40	−1 19	−1 15	+1 15	+1 19
45	−1 31	−1 24	+1 24	+1 31
50	−1 46	−1 35	+1 35	+1 46

NOTES

Latitude corrections are for mid-month.

Equation of time = mean time (LHA mean sun) minus apparent time (LHA true sun).

MOON

Day of Yr	Day of Mth	Day of Week	Age	Transit (Upper)	Diff	Semi-diam	Hor Par	Lat 30°N Moon-rise	Lat 30°N Moon-set
			days	h m	m	′	′	h m	h m
32	**1**	**Sun**	04	**16 04**	**52**	16.3	59.9	09 48	22 25
33	**2**	**Mon**	05	**16 56**	**53**	16.2	59.5	10 30	23 28
34	**3**	**Tu**	06	**17 49**	**53**	16.1	59.0	11 12	- -
35	**4**	**Wed**	07	**18 42**	**53**	15.9	58.4	11 56	00 31
36	**5**	**Th**	08	**19 35**	**54**	15.8	57.8	12 43	01 32
37	**6**	**Fri**	09	**20 29**	**53**	15.6	57.3	13 31	02 31
38	**7**	**Sat**	10	**21 22**	**52**	15.5	56.8	14 23	03 27
39	**8**	**Sun**	11	**22 14**	**50**	15.3	56.3	15 16	04 20
40	**9**	**Mon**	12	**23 04**	**48**	15.2	55.8	16 11	05 09
41	**10**	**Tu**	13	**23 52**	**46**	15.1	55.4	17 05	05 54
42	**11**	**Wed**	14	**24 38**	**-**	15.0	55.0	18 00	06 35
43	**12**	**Th**	15	**00 38**	**45**	14.9	54.6	18 53	07 13
44	**13**	**Fri**	16	**01 23**	**43**	14.8	54.3	19 46	07 48
45	**14**	**Sat**	17	**02 06**	**43**	14.8	54.2	20 37	08 22
46	**15**	**Sun**	18	**02 49**	**42**	14.7	54.1	21 29	08 55
47	**16**	**Mon**	19	**03 31**	**42**	14.7	54.1	22 20	09 28
48	**17**	**Tu**	20	**04 13**	**44**	14.8	54.3	23 12	10 03
49	**18**	**Wed**	21	**04 57**	**46**	14.9	54.7	- -	10 39
50	**19**	**Th**	22	**05 43**	**48**	15.0	55.2	00 05	11 18
51	**20**	**Fri**	23	**06 31**	**50**	15.2	55.9	00 59	12 01
52	**21**	**Sat**	24	**07 21**	**54**	15.5	56.7	01 53	12 49
53	**22**	**Sun**	25	**08 15**	**55**	15.7	57.6	02 48	13 41
54	**23**	**Mon**	26	**09 10**	**56**	16.0	58.6	03 42	14 39
55	**24**	**Tu**	27	**10 06**	**57**	16.2	59.5	04 34	15 41
56	**25**	**Wed**	28	**11 03**	**57**	16.4	60.3	05 24	16 46
57	**26**	**Th**	29	**12 00**	**56**	16.6	60.8	06 12	17 53
58	**27**	**Fri**	01	**12 56**	**56**	16.6	61.0	06 57	19 01
59	**28**	**Sat**	02	**13 52**	**55**	16.6	61.0	07 41	20 08

Phases of the Moon

	d	h	m
☾ First Quarter	3	22	53
○ Full Moon	11	10	23
☽ Last Quarter	19	15	27
● New Moon	26	17	26
Apogee	15	15	
Perigee	27	20	
On 0°	1	04	
Farthest N of 0°	7	17	
On 0°	15	00	
Farthest S of 0°	22	08	
On 0°	28	13	

NOTES

Diff equals daily change in transit time.

To correct moonrise, moonset, or transit times for latitude and/or longitude see page E 112. For further information about these tables see page E 4. For arc-to-time correction see page E 103.

FEBRUARY 1998 — STARS — 0h GMT February 1

No	Name	Mag	Transit h m	Dec ° ′	GHA ° ′	RA h m	SHA ° ′
♈	**ARIES**		**15 14**		**131 00.0**		
1	**Alpheratz**	2.2	**15 22**	N29 04.8	**128 55.9**	0 08	357 55.9
2	**Ankaa**	2.4	**15 40**	S 42 19.3	**124 27.6**	0 26	353 27.6
3	**Schedar**	2.5	**15 54**	N56 31.7	**120 54.3**	0 40	349 54.3
4	**Diphda**	2.2	**15 57**	S 18 00.1	**120 07.9**	0 43	349 07.9
5	**Achernar**	0.6	**16 51**	S 57 15.2	**106 35.9**	1 38	335 35.9
6	**POLARIS**	2.1	**17 43**	N89 15.5	**93 33.4**	2 30	322 33.4
7	**Hamal**	2.2	**17 20**	N23 27.1	**99 14.1**	2 07	328 14.1
8	**Acamar**	3.1	**18 11**	S 40 19.1	**86 27.3**	2 58	315 27.3
9	**Menkar**	2.8	**18 15**	N 4 04.7	**85 27.4**	3 02	314 27.4
10	**Mirfak**	1.9	**18 37**	N49 51.3	**79 57.2**	3 24	308 57.2
11	**Aldebaran**	1.1	**19 49**	N16 30.2	**62 02.8**	4 36	291 02.8
12	**Rigel**	0.3	**20 27**	S 8 12.5	**52 23.2**	5 14	281 23.2
13	**Capella**	0.2	**20 29**	N45 59.8	**51 51.6**	5 17	280 51.6
14	**Bellatrix**	1.7	**20 38**	N 6 20.7	**49 44.5**	5 25	278 44.5
15	**Elnath**	1.8	**20 39**	N28 36.3	**49 27.3**	5 26	278 27.3
16	**Alnilam**	1.8	**20 49**	S 1 12.4	**46 58.1**	5 36	275 58.1
17	**Betelgeuse**	0.1-1.2	**21 08**	N 7 24.2	**42 13.8**	5 55	271 13.8
18	**Canopus**	-0.9	**21 36**	S 52 42.1	**35 01.0**	6 24	264 01.0
19	**Sirius**	-1.6	**21 57**	S 16 43.1	**29 43.8**	6 45	258 43.8
20	**Adhara**	1.6	**22 11**	S 28 58.5	**26 21.5**	6 59	255 21.5
21	**Castor**	1.6	**22 47**	N31 53.4	**17 22.6**	7 34	246 22.6
22	**Procyon**	0.5	**22 51**	N 5 13.6	**16 11.7**	7 39	245 11.7
23	**Pollux**	1.2	**22 57**	N28 01.7	**14 41.8**	7 45	243 41.8
24	**Avior**	1.7	**23 35**	S 59 30.4	**5 22.1**	8 23	234 22.1
25	**Suhail**	2.2	**0 24**	S 43 25.7	**354 00.6**	9 08	223 00.6
26	**Miaplacidus**	1.8	**0 29**	S 69 42.7	**352 41.1**	9 13	221 41.1
27	**Alphard**	2.2	**0 43**	S 8 39.2	**349 07.3**	9 28	218 07.3
28	**Regulus**	1.3	**1 24**	N11 58.4	**338 55.7**	10 08	207 55.7
29	**Dubhe**	2.0	**2 19**	N61 45.5	**325 05.5**	11 04	194 05.5
30	**Denebola**	2.2	**3 04**	N14 34.8	**313 45.4**	11 49	182 45.4
31	**Gienah**	2.8	**3 31**	S 17 31.9	**307 04.1**	12 16	176 04.1
32	**Acrux**	1.1	**3 42**	S 63 05.1	**304 21.8**	12 27	173 21.8
33	**Gacrux**	1.6	**3 46**	S 57 06.0	**303 13.5**	12 31	172 13.5
34	**Mimosa**	1.5	**4 03**	S 59 40.5	**299 05.3**	12 48	168 05.3
35	**Alioth**	1.7	**4 09**	N55 58.0	**297 30.7**	12 54	166 30.7
36	**Spica**	1.2	**4 40**	S 11 09.1	**289 43.5**	13 25	158 43.5
37	**Alkaid**	1.9	**5 03**	N49 19.2	**284 08.0**	13 47	153 08.0
38	**Hadar**	0.9	**5 19**	S 60 21.5	**280 04.3**	14 04	149 04.3
39	**Menkent**	2.3	**5 22**	S 36 21.5	**279 21.3**	14 07	148 21.3
40	**Arcturus**	0.2	**5 31**	N19 11.4	**277 06.4**	14 16	146 06.4
41	**Rigil Kent.**	0.1	**5 55**	S 60 49.3	**271 07.6**	14 39	140 07.6
42	**Zuben'ubi**	2.9	**6 06**	S 16 01.9	**268 18.4**	14 51	137 18.4
43	**Kochab**	2.2	**6 06**	N74 09.6	**268 19.5**	14 51	137 19.5
44	**Alphecca**	2.3	**6 49**	N26 43.1	**257 21.0**	15 35	126 21.0
45	**Antares**	1.2	**7 44**	S 26 25.5	**243 40.8**	16 29	112 40.8
46	**Atria**	1.9	**8 03**	S 69 01.1	**238 53.4**	16 48	107 53.4
47	**Sabik**	2.6	**8 25**	S 15 43.2	**233 26.2**	17 10	102 26.2
48	**Shaula**	1.7	**8 48**	S 37 05.9	**227 38.1**	17 33	96 38.1
49	**Rasalhague**	2.1	**8 49**	N12 33.7	**227 17.5**	17 35	96 17.5
50	**Eltanin**	2.4	**9 11**	N51 29.2	**221 51.9**	17 57	90 51.9
51	**Kaus Aust.**	2.0	**9 38**	S 34 23.0	**214 59.7**	18 24	83 59.7
52	**Vega**	0.1	**9 51**	N38 46.8	**211 47.2**	18 37	80 47.2
53	**Nunki**	2.1	**10 09**	S 26 17.8	**207 13.1**	18 55	76 13.1
54	**Altair**	0.9	**11 05**	N 8 51.8	**193 20.0**	19 51	62 20.0
55	**Peacock**	2.1	**11 40**	S 56 44.4	**184 38.3**	20 25	53 38.3
56	**Deneb**	1.3	**11 55**	N45 16.4	**180 39.9**	20 41	49 39.9
57	**Enif**	2.5	**12 58**	N 9 51.9	**164 59.0**	21 44	33 59.0
58	**Al Na'ir**	2.2	**13 22**	S 46 58.3	**158 59.0**	22 08	27 59.0
59	**Fomalhaut**	1.3	**14 11**	S 29 38.1	**146 37.3**	22 58	15 37.3
60	**Markab**	2.6	**14 18**	N15 11.6	**144 50.3**	23 05	13 50.3

Star Transit Corr Table

Date	Corr h m	Date	Corr h m
1	0 00	17	−1 03
2	−0 04	18	−1 07
3	−0 08	19	−1 11
4	−0 12	20	−1 15
5	−0 16	21	−1 19
6	−0 20	22	−1 23
7	−0 24	23	−1 27
8	−0 28	24	−1 30
9	−0 31	25	−1 34
10	−0 35	26	−1 38
11	−0 39	27	−1 42
12	−0 43	28	−1 46
13	−0 47	29	−1 50
14	−0 51	30	−1 54
15	−0 55	31	−1 58
16	−0 59		

STAR TRANSIT

To find the approximate time of transit of a star for any day of the month use the table above. All corrections are subtractive.

If the value taken from the table is greater than the time of transit for the first of the month, add 23h 56min to the time of transit before subtracting the correction.

Example: What time will Suhail (No 25) transit the meridian on February 28?

	h min
Transit on Feb 1	00 24
	+23 56
	24 20
Corr for Feb 28	−01 46
Transit on Feb 28	22 34

FEBRUARY DIARY

d h	
1 21	Saturn 0.6° N of Moon
2 11	Mercury 2° S of Neptune
5 18	Aldebaran 0.2° S of Moon
5 18	Venus stationary
8 05	Mercury 1.4° S of Uranus
14 12	Pallas in conjunction
20 02	Venus greatest brilliancy
22 08	Mercury in conjunction
23 09	Jupiter in conjunction
23 17	Venus 1.6° N of Moon
24 06	Neptune 3° S of Moon
24 22	Uranus 3° S of Moon
27 22	Mars 0.7° N of Moon

NOTES

Star declinations may be used as is for the whole month. To correct GHA for day, hour, minute, and second see page E 106; for further explanation of this table see page E 4.

FEBRUARY 1998 — SUN & ARIES — GMT

Sunday, 1st February

Time	SUN GHA ° ′	Dec ° ′	ARIES GHA ° ′
00	176 37.3	S17 13.0	131 00.0
02	206 37.2	17 11.5	161 04.9
04	236 37.0	17 10.1	191 09.8
06	266 36.8	17 08.7	221 14.7
08	296 36.6	17 07.3	251 19.7
10	326 36.5	17 05.9	281 24.6
12	356 36.3	17 04.4	311 29.5
14	26 36.1	17 03.0	341 34.5
16	56 36.0	17 01.6	11 39.4
18	86 35.8	17 00.2	41 44.3
20	116 35.6	16 58.7	71 49.2
22	146 35.5	S16 57.3	101 54.2

Monday, 2nd February

Time	SUN GHA ° ′	Dec ° ′	ARIES GHA ° ′
00	176 35.3	S16 55.9	131 59.1
02	206 35.2	16 54.4	162 04.0
04	236 35.0	16 53.0	192 08.9
06	266 34.8	16 51.5	222 13.9
08	296 34.7	16 50.1	252 18.8
10	326 34.5	16 48.7	282 23.7
12	356 34.4	16 47.2	312 28.7
14	26 34.2	16 45.8	342 33.6
16	56 34.1	16 44.3	12 38.5
18	86 33.9	16 42.8	42 43.4
20	116 33.8	16 41.4	72 48.4
22	146 33.6	S16 39.9	102 53.3

Tuesday, 3rd February

Time	SUN GHA ° ′	Dec ° ′	ARIES GHA ° ′
00	176 33.5	S16 38.5	132 58.2
02	206 33.4	16 37.0	163 03.2
04	236 33.2	16 35.5	193 08.1
06	266 33.1	16 34.1	223 13.0
08	296 32.9	16 32.6	253 17.9
10	326 32.8	16 31.1	283 22.9
12	356 32.7	16 29.7	313 27.8
14	26 32.5	16 28.2	343 32.7
16	56 32.4	16 26.7	13 37.7
18	86 32.3	16 25.2	43 42.6
20	116 32.2	16 23.8	73 47.5
22	146 32.0	S16 22.3	103 52.4

Wednesday, 4th February

Time	SUN GHA ° ′	Dec ° ′	ARIES GHA ° ′
00	176 31.9	S16 20.8	133 57.4
02	206 31.8	16 19.3	164 02.3
04	236 31.7	16 17.8	194 07.2
06	266 31.5	16 16.3	224 12.2
08	296 31.4	16 14.8	254 17.1
10	326 31.3	16 13.3	284 22.0
12	356 31.2	16 11.8	314 26.9
14	26 31.1	16 10.3	344 31.9
16	56 31.0	16 08.8	14 36.8
18	86 30.8	16 07.3	44 41.7
20	116 30.7	16 05.8	74 46.7
22	146 30.6	S16 04.3	104 51.6

Thursday, 5th February

Time	SUN GHA ° ′	Dec ° ′	ARIES GHA ° ′
00	176 30.5	S16 02.8	134 56.5
02	206 30.4	16 01.3	165 01.4
04	236 30.3	15 59.8	195 06.4
06	266 30.2	15 58.3	225 11.3
08	296 30.1	15 56.8	255 16.2
10	326 30.0	15 55.3	285 21.2
12	356 29.9	15 53.7	315 26.1
14	26 29.8	15 52.2	345 31.0
16	56 29.7	15 50.7	15 35.9
18	86 29.6	15 49.2	45 40.9
20	116 29.5	15 47.6	75 45.8
22	146 29.4	S15 46.1	105 50.7

Friday, 6th February

Time	SUN GHA ° ′	Dec ° ′	ARIES GHA ° ′
00	176 29.3	S15 44.6	135 55.6
02	206 29.2	15 43.1	166 00.6
04	236 29.2	15 41.5	196 05.5
06	266 29.1	15 40.0	226 10.4
08	296 29.0	15 38.5	256 15.4
10	326 28.9	15 36.9	286 20.3
12	356 28.8	15 35.4	316 25.2
14	26 28.7	15 33.8	346 30.1
16	56 28.7	15 32.3	16 35.1
18	86 28.6	15 30.7	46 40.0
20	116 28.5	15 29.2	76 44.9
22	146 28.4	S15 27.6	106 49.9

Saturday, 7th February

Time	SUN GHA ° ′	Dec ° ′	ARIES GHA ° ′
00	176 28.4	S15 26.1	136 54.8
02	206 28.3	15 24.5	166 59.7
04	236 28.2	15 23.0	197 04.6
06	266 28.1	15 21.4	227 09.6
08	296 28.1	15 19.9	257 14.5
10	326 28.0	15 18.3	287 19.4
12	356 27.9	15 16.7	317 24.4
14	26 27.9	15 15.2	347 29.3
16	56 27.8	15 13.6	17 34.2
18	86 27.8	15 12.0	47 39.1
20	116 27.7	15 10.5	77 44.1
22	146 27.6	S15 08.9	107 49.0

Sunday, 8th February

Time	SUN GHA ° ′	Dec ° ′	ARIES GHA ° ′
00	176 27.6	S15 07.3	137 53.9
02	206 27.5	15 05.7	167 58.9
04	236 27.5	15 04.2	198 03.8
06	266 27.4	15 02.6	228 08.7
08	296 27.4	15 01.0	258 13.6
10	326 27.3	14 59.4	288 18.6
12	356 27.3	14 57.8	318 23.5
14	26 27.2	14 56.2	348 28.4
16	56 27.2	14 54.7	18 33.4
18	86 27.1	14 53.1	48 38.3
20	116 27.1	14 51.5	78 43.2
22	146 27.1	S14 49.9	108 48.1

Monday, 9th February

Time	SUN GHA ° ′	Dec ° ′	ARIES GHA ° ′
00	176 27.0	S14 48.3	138 53.1
02	206 27.0	14 46.7	168 58.0
04	236 26.9	14 45.1	199 02.9
06	266 26.9	14 43.5	229 07.9
08	296 26.9	14 41.9	259 12.8
10	326 26.8	14 40.3	289 17.7
12	356 26.8	14 38.7	319 22.6
14	26 26.8	14 37.1	349 27.6
16	56 26.7	14 35.5	19 32.5
18	86 26.7	14 33.9	49 37.4
20	116 26.7	14 32.2	79 42.4
22	146 26.7	S14 30.6	109 47.3

Tuesday, 10th February

Time	SUN GHA ° ′	Dec ° ′	ARIES GHA ° ′
00	176 26.6	S14 29.0	139 52.2
02	206 26.6	14 27.4	169 57.1
04	236 26.6	14 25.8	200 02.1
06	266 26.6	14 24.2	230 07.0
08	296 26.6	14 22.5	260 11.9
10	326 26.5	14 20.9	290 16.8
12	356 26.5	14 19.3	320 21.8
14	26 26.5	14 17.7	350 26.7
16	56 26.5	14 16.0	20 31.6
18	86 26.5	14 14.4	50 36.6
20	116 26.5	14 12.8	80 41.5
22	146 26.5	S14 11.1	110 46.4

Wednesday, 11th February

Time	SUN GHA ° ′	Dec ° ′	ARIES GHA ° ′	Time
00	176 26.5	S14 09.5	140 51.3	00
02	206 26.5	14 07.9	170 56.3	02
04	236 26.5	14 06.2	201 01.2	04
06	266 26.4	14 04.6	231 06.1	06
08	296 26.4	14 03.0	261 11.1	08
10	326 26.4	14 01.3	291 16.0	10
12	356 26.4	13 59.7	321 20.9	12
14	26 26.4	13 58.0	351 25.8	14
16	56 26.5	13 56.4	21 30.8	16
18	86 26.5	13 54.7	51 35.7	18
20	116 26.5	13 53.1	81 40.6	20
22	146 26.5	S13 51.4	111 45.6	22

Thursday, 12th February

Time	SUN GHA ° ′	Dec ° ′	ARIES GHA ° ′	Time
00	176 26.5	S13 49.8	141 50.5	00
02	206 26.5	13 48.1	171 55.4	02
04	236 26.5	13 46.4	202 00.3	04
06	266 26.5	13 44.8	232 05.3	06
08	296 26.5	13 43.1	262 10.2	08
10	326 26.5	13 41.5	292 15.1	10
12	356 26.6	13 39.8	322 20.1	12
14	26 26.6	13 38.1	352 25.0	14
16	56 26.6	13 36.5	22 29.9	16
18	86 26.6	13 34.8	52 34.8	18
20	116 26.6	13 33.1	82 39.8	20
22	146 26.7	S13 31.5	112 44.7	22

Friday, 13th February

Time	SUN GHA ° ′	Dec ° ′	ARIES GHA ° ′	Time
00	176 26.7	S13 29.8	142 49.6	00
02	206 26.7	13 28.1	172 54.5	02
04	236 26.7	13 26.4	202 59.5	04
06	266 26.8	13 24.8	233 04.4	06
08	296 26.8	13 23.1	263 09.3	08
10	326 26.8	13 21.4	293 14.3	10
12	356 26.9	13 19.7	323 19.2	12
14	26 26.9	13 18.0	353 24.1	14
16	56 26.9	13 16.3	23 29.0	16
18	86 27.0	13 14.7	53 34.0	18
20	116 27.0	13 13.0	83 38.9	20
22	146 27.0	S13 11.3	113 43.8	22

Saturday, 14th February

Time	SUN GHA ° ′	Dec ° ′	ARIES GHA ° ′	Time
00	176 27.1	S13 09.6	143 48.8	00
02	206 27.1	13 07.9	173 53.7	02
04	236 27.2	13 06.2	203 58.6	04
06	266 27.2	13 04.5	234 03.5	06
08	296 27.2	13 02.8	264 08.5	08
10	326 27.3	13 01.1	294 13.4	10
12	356 27.3	12 59.4	324 18.3	12
14	26 27.4	12 57.7	354 23.3	14
16	56 27.4	12 56.0	24 28.2	16
18	86 27.5	12 54.3	54 33.1	18
20	116 27.5	12 52.6	84 38.0	20
22	146 27.6	S12 50.9	114 43.0	22

Sunday, 15th February

Time	SUN GHA ° ′	Dec ° ′	ARIES GHA ° ′	Time
00	176 27.6	S12 49.2	144 47.9	00
02	206 27.7	12 47.5	174 52.8	02
04	236 27.8	12 45.8	204 57.8	04
06	266 27.8	12 44.0	235 02.7	06
08	296 27.9	12 42.3	265 07.6	08
10	326 27.9	12 40.6	295 12.5	10
12	356 28.0	12 38.9	325 17.5	12
14	26 28.1	12 37.2	355 22.4	14
16	56 28.1	12 35.5	25 27.3	16
18	86 28.2	12 33.7	55 32.2	18
20	116 28.3	12 32.0	85 37.2	20
22	146 28.3	S12 30.3	115 42.1	22

NOTES

Sun and Aries GHA corrections for additional hour, minutes, and seconds are on page E 106.
Correct sun declination by interpolation. For an example see page E 4.

FEBRUARY 1998 SUN & ARIES GMT

Monday, 16th February

Time	SUN GHA ° ′	Dec ° ′	ARIES GHA ° ′
00	176 28.4	S12 28.6	145 47.0
02	206 28.5	12 26.8	175 52.0
04	236 28.5	12 25.1	205 56.9
06	266 28.6	12 23.4	236 01.8
08	296 28.7	12 21.6	266 06.7
10	326 28.8	12 19.9	296 11.7
12	356 28.8	12 18.2	326 16.6
14	26 28.9	12 16.4	356 21.5
16	56 29.0	12 14.7	26 26.5
18	86 29.1	12 13.0	56 31.4
20	116 29.2	12 11.2	86 36.3
22	146 29.2	S12 09.5	116 41.2

Tuesday, 17th February

Time	SUN GHA ° ′	Dec ° ′	ARIES GHA ° ′
00	176 29.3	S12 07.7	146 46.2
02	206 29.4	12 06.0	176 51.1
04	236 29.5	12 04.2	206 56.0
06	266 29.6	12 02.5	237 01.0
08	296 29.7	12 00.8	267 05.9
10	326 29.8	11 59.0	297 10.8
12	356 29.9	11 57.3	327 15.7
14	26 30.0	11 55.5	357 20.7
16	56 30.0	11 53.7	27 25.6
18	86 30.1	11 52.0	57 30.5
20	116 30.2	11 50.2	87 35.5
22	146 30.3	S11 48.5	117 40.4

Wednesday, 18th February

Time	SUN GHA ° ′	Dec ° ′	ARIES GHA ° ′
00	176 30.4	S11 46.7	147 45.3
02	206 30.5	11 45.0	177 50.2
04	236 30.6	11 43.2	207 55.2
06	266 30.7	11 41.4	238 00.1
08	296 30.8	11 39.7	268 05.0
10	326 30.9	11 37.9	298 09.9
12	356 31.0	11 36.1	328 14.9
14	26 31.2	11 34.4	358 19.8
16	56 31.3	11 32.6	28 24.7
18	86 31.4	11 30.8	58 29.7
20	116 31.5	11 29.1	88 34.6
22	146 31.6	S11 27.3	118 39.5

Thursday, 19th February

Time	SUN GHA ° ′	Dec ° ′	ARIES GHA ° ′
00	176 31.7	S11 25.5	148 44.4
02	206 31.8	11 23.7	178 49.4
04	236 31.9	11 22.0	208 54.3
06	266 32.0	11 20.2	238 59.2
08	296 32.2	11 18.4	269 04.2
10	326 32.3	11 16.6	299 09.1
12	356 32.4	11 14.8	329 14.0
14	26 32.5	11 13.1	359 18.9
16	56 32.6	11 11.3	29 23.9
18	86 32.8	11 09.5	59 28.8
20	116 32.9	11 07.7	89 33.7
22	146 33.0	S11 05.9	119 38.7

Friday, 20th February

Time	SUN GHA ° ′	Dec ° ′	ARIES GHA ° ′
00	176 33.1	S11 04.1	149 43.6
02	206 33.3	11 02.3	179 48.5
04	236 33.4	11 00.6	209 53.4
06	266 33.5	10 58.8	239 58.4
08	296 33.6	10 57.0	270 03.3
10	326 33.8	10 55.2	300 08.2
12	356 33.9	10 53.4	330 13.2
14	26 34.0	10 51.6	0 18.1
16	56 34.2	10 49.8	30 23.0
18	86 34.3	10 48.0	60 27.9
20	116 34.5	10 46.2	90 32.9
22	146 34.6	S10 44.4	120 37.8

Saturday, 21st February

Time	SUN GHA ° ′	Dec ° ′	ARIES GHA ° ′
00	176 34.7	S10 42.6	150 42.7
02	206 34.9	10 40.8	180 47.7
04	236 35.0	10 39.0	210 52.6
06	266 35.2	10 37.2	240 57.5
08	296 35.3	10 35.4	271 02.4
10	326 35.4	10 33.5	301 07.4
12	356 35.6	10 31.7	331 12.3
14	26 35.7	10 29.9	1 17.2
16	56 35.9	10 28.1	31 22.1
18	86 36.0	10 26.3	61 27.1
20	116 36.2	10 24.5	91 32.0
22	146 36.3	S10 22.7	121 36.9

Sunday, 22nd February

Time	SUN GHA ° ′	Dec ° ′	ARIES GHA ° ′
00	176 36.5	S10 20.9	151 41.9
02	206 36.6	10 19.0	181 46.8
04	236 36.8	10 17.2	211 51.7
06	266 36.9	10 15.4	241 56.6
08	296 37.1	10 13.6	272 01.6
10	326 37.3	10 11.8	302 06.5
12	356 37.4	10 09.9	332 11.4
14	26 37.6	10 08.1	2 16.4
16	56 37.7	10 06.3	32 21.3
18	86 37.9	10 04.5	62 26.2
20	116 38.1	10 02.6	92 31.1
22	146 38.2	S10 00.8	122 36.1

Monday, 23rd February

Time	SUN GHA ° ′	Dec ° ′	ARIES GHA ° ′
00	176 38.4	S 9 59.0	152 41.0
02	206 38.6	9 57.1	182 45.9
04	236 38.7	9 55.3	212 50.9
06	266 38.9	9 53.5	242 55.8
08	296 39.1	9 51.6	273 00.7
10	326 39.2	9 49.8	303 05.6
12	356 39.4	9 48.0	333 10.6
14	26 39.6	9 46.1	3 15.5
16	56 39.7	9 44.3	33 20.4
18	86 39.9	9 42.5	63 25.4
20	116 40.1	9 40.6	93 30.3
22	146 40.3	S 9 38.8	123 35.2

Tuesday, 24th February

Time	SUN GHA ° ′	Dec ° ′	ARIES GHA ° ′
00	176 40.4	S 9 36.9	153 40.1
02	206 40.6	9 35.1	183 45.1
04	236 40.8	9 33.2	213 50.0
06	266 41.0	9 31.4	243 54.9
08	296 41.2	9 29.5	273 59.9
10	326 41.3	9 27.7	304 04.8
12	356 41.5	9 25.9	334 09.7
14	26 41.7	9 24.0	4 14.6
16	56 41.9	9 22.2	34 19.6
18	86 42.1	9 20.3	64 24.5
20	116 42.3	9 18.5	94 29.4
22	146 42.5	S 9 16.6	124 34.4

Wednesday, 25th February

Time	SUN GHA ° ′	Dec ° ′	ARIES GHA ° ′
00	176 42.6	S 9 14.7	154 39.3
02	206 42.8	9 12.9	184 44.2
04	236 43.0	9 11.0	214 49.1
06	266 43.2	9 09.2	244 54.1
08	296 43.4	9 07.3	274 59.0
10	326 43.6	9 05.5	305 03.9
12	356 43.8	9 03.6	335 08.8
14	26 44.0	9 01.7	5 13.8
16	56 44.2	8 59.9	35 18.7
18	86 44.4	8 58.0	65 23.6
20	116 44.6	8 56.1	95 28.6
22	146 44.8	S 8 54.3	125 33.5

Thursday, 26th February

Time	SUN GHA ° ′	Dec ° ′	ARIES GHA ° ′	Time
00	176 45.0	S 8 52.4	155 38.4	00
02	206 45.2	8 50.6	185 43.3	02
04	236 45.4	8 48.7	215 48.3	04
06	266 45.6	8 46.8	245 53.2	06
08	296 45.8	8 44.9	275 58.1	08
10	326 46.0	8 43.1	306 03.1	10
12	356 46.2	8 41.2	336 08.0	12
14	26 46.4	8 39.3	6 12.9	14
16	56 46.6	8 37.5	36 17.8	16
18	86 46.8	8 35.6	66 22.8	18
20	116 47.1	8 33.7	96 27.7	20
22	146 47.3	S 8 31.8	126 32.6	22

Friday, 27th February

Time	SUN GHA ° ′	Dec ° ′	ARIES GHA ° ′	Time
00	176 47.5	S 8 30.0	156 37.6	00
02	206 47.7	8 28.1	186 42.5	02
04	236 47.9	8 26.2	216 47.4	04
06	266 48.1	8 24.3	246 52.3	06
08	296 48.3	8 22.5	276 57.3	08
10	326 48.6	8 20.6	307 02.2	10
12	356 48.8	8 18.7	337 07.1	12
14	26 49.0	8 16.8	7 12.1	14
16	56 49.2	8 14.9	37 17.0	16
18	86 49.4	8 13.0	67 21.9	18
20	116 49.7	8 11.2	97 26.8	20
22	146 49.9	S 8 09.3	127 31.8	22

Saturday, 28th February

Time	SUN GHA ° ′	Dec ° ′	ARIES GHA ° ′	Time
00	176 50.1	S 8 07.4	157 36.7	00
02	206 50.3	8 05.5	187 41.6	02
04	236 50.6	8 03.6	217 46.5	04
06	266 50.8	8 01.7	247 51.5	06
08	296 51.0	7 59.8	277 56.4	08
10	326 51.2	7 57.9	308 01.3	10
12	356 51.5	7 56.0	338 06.3	12
14	26 51.7	7 54.2	8 11.2	14
16	56 51.9	7 52.3	38 16.1	16
18	86 52.2	7 50.4	68 21.0	18
20	116 52.4	7 48.5	98 26.0	20
22	146 52.6	S 7 46.6	128 30.9	22

FEBRUARY 1998 PLANETS 0h GMT

VENUS Mer Pass h m	VENUS GHA ° ′	VENUS Mean Var/hr 14°+	VENUS Dec ° ′	VENUS Mean Var/hr ′	Day		JUPITER GHA ° ′	JUPITER Mean Var/hr 15°+	JUPITER Dec ° ′	JUPITER Mean Var/hr ′	JUPITER Mer Pass h m
10 32	201 30.9	2.9	S14 53.6	0.1	1	SUN	159 22.8	1.9	S12 34.2	0.2	13 21
10 27	202 41.0	2.8	S14 55.6	0.1	2	Mon	160 08.4	1.9	S12 29.3	0.2	13 18
10 23	203 48.4	2.7	S14 57.8	0.1	3	Tu	160 53.9	1.9	S12 24.3	0.2	13 15
10 19	204 53.3	2.6	S15 00.3	0.1	4	Wed	161 39.4	1.9	S12 19.3	0.2	13 12
10 15	205 55.6	2.5	S15 02.9	0.1	5	Th	162 24.9	1.9	S12 14.3	0.2	13 09
10 11	206 55.4	2.4	S15 05.8	0.1	6	Fri	163 10.4	1.9	S12 09.3	0.2	13 06
10 07	207 52.6	2.3	S15 08.8	0.1	7	Sat	163 55.9	1.9	S12 04.3	0.2	13 03
10 03	208 47.3	2.2	S15 11.9	0.1	8	SUN	164 41.4	1.9	S11 59.3	0.2	13 00
10 00	209 39.6	2.1	S15 15.0	0.1	9	Mon	165 26.9	1.9	S11 54.2	0.2	12 57
09 57	210 29.6	2.0	S15 18.3	0.1	10	Tu	166 12.3	1.9	S11 49.2	0.2	12 54
09 54	211 17.1	1.9	S15 21.5	0.1	11	Wed	166 57.7	1.9	S11 44.1	0.2	12 51
09 51	212 02.4	1.8	S15 24.8	0.1	12	Th	167 43.2	1.9	S11 39.0	0.2	12 48
09 48	212 45.5	1.7	S15 28.0	0.1	13	Fri	168 28.6	1.9	S11 33.9	0.2	12 44
09 45	213 26.5	1.6	S15 31.2	0.1	14	Sat	169 14.0	1.9	S11 28.7	0.2	12 41
09 43	214 05.3	1.5	S15 34.3	0.1	15	SUN	169 59.4	1.9	S11 23.6	0.2	12 38
09 40	214 42.2	1.5	S15 37.3	0.1	16	Mon	170 44.8	1.9	S11 18.4	0.2	12 35
09 38	215 17.1	1.4	S15 40.1	0.1	17	Tu	171 30.3	1.9	S11 13.3	0.2	12 32
09 36	215 50.1	1.3	S15 42.8	0.1	18	Wed	172 15.7	1.9	S11 08.1	0.2	12 29
09 34	216 21.3	1.2	S15 45.3	0.1	19	Th	173 01.1	1.9	S11 02.9	0.2	12 26
09 32	216 50.7	1.2	S15 47.6	0.1	20	Fri	173 46.5	1.9	S10 57.7	0.2	12 23
09 30	217 18.5	1.1	S15 49.7	0.1	21	Sat	174 31.9	1.9	S10 52.5	0.2	12 20
09 28	217 44.7	1.0	S15 51.6	0.1	22	SUN	175 17.3	1.9	S10 47.3	0.2	12 17
09 27	218 09.3	1.0	S15 53.2	0.1	23	Mon	176 02.7	1.9	S10 42.1	0.2	12 14
09 25	218 32.5	0.9	S15 54.5	0.0	24	Tu	176 48.1	1.9	S10 36.9	0.2	12 11
09 24	218 54.2	0.8	S15 55.5	0.0	25	Wed	177 33.6	1.9	S10 31.6	0.2	12 08
09 23	219 14.6	0.8	S15 56.2	0.0	26	Th	178 19.0	1.9	S10 26.4	0.2	12 05
09 21	219 33.6	0.7	S15 56.6	0.0	27	Fri	179 04.5	1.9	S10 21.1	0.2	12 02
09 20	219 51.5	0.7	S15 56.6	0.0	28	Sat	179 49.9	1.9	S10 15.9	0.2	11 59

VENUS, Av. Mag. -4.6
S.H.A. February
5 71; 10 71; 15 69; 20 67; 25 64; 28 62.

JUPITER, Av. Mag.-2.0
S.H.A. February
5 27; 10 26; 15 25; 20 24; 25 23; 28 22.

MARS Mer Pass h m	MARS GHA ° ′	MARS Mean Var/hr 15°+	MARS Dec ° ′	MARS Mean Var/hr ′	Day		SATURN GHA ° ′	SATURN Mean Var/hr 15°+	SATURN Dec ° ′	SATURN Mean Var/hr ′	SATURN Mer Pass h m
13 45	153 36.1	0.6	S10 30.3	0.7	1	SUN	115 51.5	2.3	N 3 52.8	0.1	16 14
13 44	153 50.8	0.6	S10 12.5	0.7	2	Mon	116 46.4	2.3	N 3 54.8	0.1	16 10
13 43	154 05.5	0.6	S 9 54.6	0.8	3	Tu	117 41.2	2.3	N 3 56.9	0.1	16 07
13 42	154 20.4	0.6	S 9 36.5	0.8	4	Wed	118 35.9	2.3	N 3 59.0	0.1	16 03
13 41	154 35.3	0.6	S 9 18.4	0.8	5	Th	119 30.5	2.3	N 4 01.1	0.1	16 00
13 40	154 50.3	0.6	S 9 00.3	0.8	6	Fri	120 25.0	2.3	N 4 03.2	0.1	15 56
13 39	155 05.4	0.6	S 8 42.0	0.8	7	Sat	121 19.5	2.3	N 4 05.3	0.1	15 52
13 38	155 20.6	0.6	S 8 23.7	0.8	8	SUN	122 13.9	2.3	N 4 07.5	0.1	15 49
13 37	155 35.9	0.6	S 8 05.3	0.8	9	Mon	123 08.3	2.3	N 4 09.7	0.1	15 45
13 36	155 51.2	0.6	S 7 46.8	0.8	10	Tu	124 02.5	2.3	N 4 11.9	0.1	15 41
13 35	156 06.7	0.6	S 7 28.3	0.8	11	Wed	124 56.7	2.3	N 4 14.2	0.1	15 38
13 34	156 22.2	0.6	S 7 09.7	0.8	12	Th	125 50.9	2.3	N 4 16.5	0.1	15 34
13 33	156 37.8	0.7	S 6 51.1	0.8	13	Fri	126 45.0	2.3	N 4 18.8	0.1	15 31
13 32	156 53.4	0.7	S 6 32.4	0.8	14	Sat	127 39.0	2.2	N 4 21.1	0.1	15 27
13 31	157 09.2	0.7	S 6 13.7	0.8	15	SUN	128 32.9	2.2	N 4 23.4	0.1	15 24
13 30	157 25.0	0.7	S 5 54.9	0.8	16	Mon	129 26.8	2.2	N 4 25.8	0.1	15 20
13 29	157 40.8	0.7	S 5 36.1	0.8	17	Tu	130 20.6	2.2	N 4 28.2	0.1	15 16
13 28	157 56.8	0.7	S 5 17.2	0.8	18	Wed	131 14.4	2.2	N 4 30.6	0.1	15 13
13 27	158 12.8	0.7	S 4 58.3	0.8	19	Th	132 08.1	2.2	N 4 33.0	0.1	15 09
13 25	158 28.8	0.7	S 4 39.4	0.8	20	Fri	133 01.7	2.2	N 4 35.4	0.1	15 06
13 24	158 45.0	0.7	S 4 20.4	0.8	21	Sat	133 55.3	2.2	N 4 37.9	0.1	15 02
13 23	159 01.1	0.7	S 4 01.5	0.8	22	SUN	134 48.8	2.2	N 4 40.4	0.1	14 59
13 22	159 17.4	0.7	S 3 42.5	0.8	23	Mon	135 42.3	2.2	N 4 42.9	0.1	14 55
13 21	159 33.7	0.7	S 3 23.4	0.8	24	Tu	136 35.7	2.2	N 4 45.4	0.1	14 51
13 20	159 50.0	0.7	S 3 04.4	0.8	25	Wed	137 29.0	2.2	N 4 47.9	0.1	14 48
13 19	160 06.4	0.7	S 2 45.3	0.8	26	Th	138 22.3	2.2	N 4 50.5	0.1	14 44
13 18	160 22.8	0.7	S 2 26.3	0.8	27	Fri	139 15.6	2.2	N 4 53.1	0.1	14 41
13 17	160 39.3	0.7	S 2 07.2	0.8	28	Sat	140 08.8	2.2	N 4 55.6	0.1	14 37

MARS, Av. Mag. +1.2
S.H.A. February
5 20; 10 16; 15 12; 20 9; 25 5; 28 3.

SATURN, Av. Mag. +0.7
S.H.A. February
5 345; 10 344; 15 344; 20 343; 25 343; 28 343.

NOTE: Planet corrections on pages E113-15

FEBRUARY 1998 MOON

Day	GMT hr	GHA ° ′	Mean Var/hr 14°+	Dec ° ′	Mean Var/hr ′
1 Sun	**0**	127 37.6	28.0	**S 0 41.3**	11.5
	6	214 25.4	28.0	**N 0 28.2**	11.6
	12	301 13.8	28.1	**N 1 37.4**	11.4
	18	28 02.7	28.3	**N 2 46.0**	11.3
2 Mon	**0**	114 51.9	28.3	**N 3 53.7**	11.1
	6	201 41.4	28.3	**N 5 00.3**	10.8
	12	288 31.0	28.3	**N 6 05.6**	10.6
	18	15 20.8	28.3	**N 7 09.4**	10.3
3 Tu	**0**	102 10.4	28.3	**N 8 11.3**	10.0
	6	189 00.0	28.3	**N 9 11.2**	9.6
	12	275 49.4	28.2	**N 10 09.0**	9.2
	18	2 38.6	28.2	**N 11 04.3**	8.7
4 Wed	**0**	89 27.4	28.1	**N 11 57.0**	8.3
	6	176 16.0	28.0	**N 12 47.0**	7.8
	12	263 04.2	28.0	**N 13 34.1**	7.3
	18	349 52.0	28.0	**N 14 18.2**	6.8
5 Th	**0**	76 39.6	27.9	**N 14 59.0**	6.3
	6	163 26.8	27.8	**N 15 36.6**	5.6
	12	250 13.8	27.8	**N 16 10.7**	5.1
	18	337 00.6	27.8	**N 16 41.4**	4.4
6 Fri	**0**	63 47.3	27.8	**N 17 08.4**	3.8
	6	150 33.9	27.8	**N 17 31.8**	3.3
	12	237 20.7	27.8	**N 17 51.5**	2.6
	18	324 07.6	27.8	**N 18 07.4**	2.0
7 Sat	**0**	50 54.8	28.0	**N 18 19.6**	1.3
	6	137 42.5	28.1	**N 18 28.0**	0.7
	12	224 30.8	28.2	**N 18 32.7**	0.1
	18	311 19.8	28.4	**N 18 33.6**	0.6
8 Sun	**0**	38 09.7	28.4	**N 18 30.8**	1.2
	6	125 00.5	28.7	**N 18 24.4**	1.7
	12	211 52.5	28.9	**N 18 14.4**	2.3
	18	298 45.6	29.1	**N 18 00.9**	2.8
9 Mon	**0**	25 40.1	29.4	**N 17 44.1**	3.4
	6	112 36.1	29.6	**N 17 23.9**	4.0
	12	199 33.5	29.9	**N 17 00.6**	4.4
	18	286 32.6	30.2	**N 16 34.3**	4.9
10 Tu	**0**	13 33.3	30.4	**N 16 05.0**	5.4
	6	100 35.7	30.7	**N 15 32.9**	5.8
	12	187 39.8	31.0	**N 14 58.3**	6.2
	18	274 45.6	31.3	**N 14 21.1**	6.6
11 Wed	**0**	1 53.2	31.5	**N 13 41.6**	6.9
	6	89 02.5	31.9	**N 13 00.0**	7.4
	12	176 13.5	32.1	**N 12 16.3**	7.6
	18	263 26.2	32.4	**N 11 30.7**	7.9
12 Th	**0**	350 40.4	32.7	**N 10 43.5**	8.1
	6	77 56.2	32.9	**N 9 54.6**	8.4
	12	165 13.5	33.1	**N 9 04.4**	8.6
	18	252 32.1	33.3	**N 8 12.9**	8.8
13 Fri	**0**	339 52.0	33.5	**N 7 20.3**	9.0
	6	67 13.1	33.7	**N 6 26.8**	9.1
	12	154 35.2	33.9	**N 5 32.4**	9.2
	18	241 58.4	34.0	**N 4 37.4**	9.3
14 Sat	**0**	329 22.4	34.2	**N 3 41.7**	9.4
	6	56 47.1	34.2	**N 2 45.7**	9.4
	12	144 12.4	34.4	**N 1 49.4**	9.4
	18	231 38.2	34.4	**N 0 52.9**	9.4
15 Sun	**0**	319 04.3	34.4	**S 0 03.6**	9.4
	6	46 30.7	34.4	**S 1 00.0**	9.3
	12	133 57.1	34.4	**S 1 56.2**	9.4
	18	221 23.4	34.4	**S 2 52.2**	9.2
16 Mon	**0**	308 49.5	34.3	**S 3 47.6**	9.1
	6	36 15.3	34.2	**S 4 42.5**	9.0
	12	123 40.6	34.1	**S 5 36.8**	8.9
	18	211 05.3	34.0	**S 6 30.3**	8.7
17 Tu	**0**	298 29.1	33.8	**S 7 22.8**	8.6
	6	25 52.1	33.7	**S 8 14.4**	8.4
	12	113 14.1	33.4	**S 9 04.8**	8.2
	18	200 34.8	33.2	**S 9 53.9**	8.0
18 Wed	**0**	287 54.2	33.0	**S 10 41.7**	7.7
	6	15 12.2	32.7	**S 11 28.0**	7.4
	12	102 28.7	32.4	**S 12 12.6**	7.1
	18	189 43.4	32.1	**S 12 55.5**	6.8
19 Th	**0**	276 56.4	31.9	**S 13 36.5**	6.5
	6	4 07.5	31.5	**S 14 15.4**	6.1
	12	91 16.7	31.1	**S 14 52.2**	5.7
	18	178 23.8	30.8	**S 15 26.6**	5.3
20 Fri	**0**	265 28.7	30.5	**S 15 58.6**	4.9
	6	352 31.5	30.1	**S 16 28.0**	4.4
	12	79 32.1	29.7	**S 16 54.6**	3.9
	18	166 30.5	29.3	**S 17 18.3**	3.4
21 Sat	**0**	253 26.7	29.0	**S 17 38.9**	2.9
	6	340 20.6	28.6	**S 17 56.4**	2.3
	12	67 12.4	28.3	**S 18 10.4**	1.7
	18	154 02.0	27.9	**S 18 21.0**	1.1
22 Sun	**0**	240 49.6	27.6	**S 18 28.0**	0.4
	6	327 35.2	27.3	**S 18 31.2**	0.2
	12	54 19.0	27.0	**S 18 30.6**	0.8
	18	141 01.0	26.7	**S 18 26.0**	1.5
23 Mon	**0**	227 41.5	26.5	**S 18 17.4**	2.2
	6	314 20.6	26.3	**S 18 04.6**	2.8
	12	40 58.4	26.1	**S 17 47.8**	3.6
	18	127 35.1	26.0	**S 17 26.8**	4.3
24 Tu	**0**	214 10.9	25.8	**S 17 01.6**	4.9
	6	300 46.1	25.8	**S 16 32.3**	5.7
	12	27 20.7	25.7	**S 15 58.9**	6.3
	18	113 54.9	25.7	**S 15 21.5**	6.9
25 Wed	**0**	200 28.9	25.7	**S 14 40.2**	7.6
	6	287 02.9	25.6	**S 13 55.1**	8.2
	12	13 37.0	25.7	**S 13 06.4**	8.7
	18	100 11.4	25.7	**S 12 14.4**	9.2
26 Th	**0**	186 46.1	25.8	**S 11 19.1**	9.7
	6	273 21.2	25.9	**S 10 20.8**	10.2
	12	359 56.8	26.0	**S 9 19.9**	10.6
	18	Eclipse of the Sun occurs today			
27 Fri	**0**	173 09.7	26.3	**S 7 10.9**	11.3
	6	259 47.1	26.4	**S 6 03.5**	11.6
	12	346 25.1	26.5	**S 4 54.6**	11.7
	18	73 03.7	26.6	**S 3 44.5**	11.9
28 Sat	**0**	159 42.9	26.6	**S 2 33.5**	11.9
	6	246 22.6	26.7	**S 1 21.9**	11.9
	12	333 02.9	26.7	**S 0 10.1**	12.0
	18	59 43.6	26.8	**N 1 01.6**	11.8

NOTE: Moon GHA corrections for additional hours, minutes, and seconds are on page E 108–10. Moon declination corrections are on page E 111. The mean var/hr value is needed for both corrections.

MARCH 1998 SUN & MOON GMT

SUN

Yr	Day of Mth	Week	Equation of Time 0h	Equation of Time 12h	Transit	Semi-diam	Lat 30°N Twi-light	Lat 30°N Sun-rise	Lat 30°N Sun-set	Lat 30°N Twi-light
			m s	m s	h m	′	h m	h m	h m	h m
60	**1**	**Sun**	+12 29	+12 23	**12 12**	16.2	06 02	06 26	17 59	18 23
61	**2**	**Mon**	+12 17	+12 11	**12 12**	16.2	06 01	06 25	18 00	18 24
62	**3**	**Tu**	+12 05	+11 59	**12 12**	16.2	06 00	06 24	18 00	18 24
63	**4**	**Wed**	+11 53	+11 46	**12 12**	16.2	05 59	06 23	18 01	18 25
64	**5**	**Th**	+11 39	+11 33	**12 12**	16.2	05 58	06 22	18 02	18 26
65	**6**	**Fri**	+11 26	+11 19	**12 11**	16.1	05 57	06 21	18 02	18 26
66	**7**	**Sat**	+11 12	+11 05	**12 11**	16.1	05 56	06 20	18 03	18 27
67	**8**	**Sun**	+10 58	+10 50	**12 11**	16.1	05 55	06 18	18 04	18 28
68	**9**	**Mon**	+10 43	+10 35	**12 11**	16.1	05 53	06 17	18 04	18 28
69	**10**	**Tu**	+10 28	+10 20	**12 10**	16.1	05 52	06 16	18 05	18 29
70	**11**	**Wed**	+10 12	+10 04	**12 10**	16.1	05 51	06 15	18 06	18 29
71	**12**	**Th**	+09 56	+09 48	**12 10**	16.1	05 50	06 14	18 06	18 30
72	**13**	**Fri**	+09 40	+09 32	**12 10**	16.1	05 49	06 13	18 07	18 31
73	**14**	**Sat**	+09 24	+09 16	**12 09**	16.1	05 48	06 11	18 07	18 31
74	**15**	**Sun**	+09 07	+08 59	**12 09**	16.1	05 46	06 10	18 08	18 32
75	**16**	**Mon**	+08 50	+08 42	**12 09**	16.1	05 45	06 09	18 09	18 33
76	**17**	**Tu**	+08 33	+08 24	**12 08**	16.1	05 44	06 08	18 09	18 33
77	**18**	**Wed**	+08 16	+08 07	**12 08**	16.1	05 43	06 07	18 10	18 34
78	**19**	**Th**	+07 58	+07 50	**12 08**	16.1	05 42	06 05	18 11	18 35
79	**20**	**Fri**	+07 41	+07 32	**12 08**	16.1	05 40	06 04	18 11	18 35
80	**21**	**Sat**	+07 23	+07 14	**12 07**	16.1	05 39	06 03	18 12	18 36
81	**22**	**Sun**	+07 05	+06 56	**12 07**	16.1	05 38	06 02	18 12	18 36
82	**23**	**Mon**	+06 47	+06 38	**12 07**	16.1	05 37	06 01	18 13	18 37
83	**24**	**Tu**	+06 29	+06 20	**12 06**	16.1	05 36	05 59	18 14	18 38
84	**25**	**Wed**	+06 11	+06 02	**12 06**	16.1	05 34	05 58	18 14	18 38
85	**26**	**Th**	+05 53	+05 44	**12 06**	16.1	05 33	05 57	18 15	18 39
86	**27**	**Fri**	+05 35	+05 26	**12 05**	16.1	05 32	05 56	18 16	18 39
87	**28**	**Sat**	+05 17	+05 08	**12 05**	16.0	05 31	05 55	18 16	18 40
88	**29**	**Sun**	+04 59	+04 50	**12 05**	16.0	05 29	05 53	18 17	18 41
89	**30**	**Mon**	+04 41	+04 32	**12 05**	16.0	05 28	05 52	18 17	18 41
90	**31**	**Tu**	+04 23	+04 14	**12 04**	16.0	05 27	05 51	18 18	18 42

Lat Corr to Sunrise, Sunset etc

Lat	Twi-light	Sun-rise	Sun-set	Twi-light
°	h m	h m	h m	h m
N60	–0 10	+0 07	–0 07	+0 10
55	–0 06	+0 06	–0 06	+0 06
50	–0 04	+0 04	–0 04	+0 04
45	–0 02	+0 03	–0 03	+0 02
N40	–0 01	+0 02	–0 02	+0 01
35	0 00	+0 01	–0 01	0 00
30	0 00	0 00	0 00	0 00
25	0 00	–0 01	+0 01	0 00
N20	0 00	–0 02	+0 02	0 00
15	0 00	–0 02	+0 02	0 00
10	0 00	–0 03	+0 03	0 00
N 5	–0 01	–0 04	+0 04	+0 01
0	–0 02	–0 05	+0 05	+0 02
S 5	–0 02	–0 05	+0 05	+0 02
10	–0 03	–0 06	+0 06	+0 03
15	–0 05	–0 07	+0 07	+0 05
S20	–0 06	–0 08	+0 08	+0 06
25	–0 08	–0 09	+0 09	+0 08
30	–0 11	–0 10	+0 10	+0 11
35	–0 13	–0 12	+0 12	+0 13
S40	–0 17	–0 13	+0 13	+0 17
45	–0 21	–0 15	+0 15	+0 21
50	–0 26	–0 17	+0 17	+0 26

NOTES

Latitude corrections are for mid-month.

Equation of time = mean time (LHA mean sun) minus apparent time (LHA true sun).

MOON

Yr	Day of Mth	Week	Age	Transit (Upper)	Diff	Semi-diam	Hor Par	Lat 30°N Moon-rise	Lat 30°N Moon-set
			days	h m	m	′	′	h m	h m
60	**1**	**Sun**	03	**14 47**	**55**	16.5	60.7	08 25	21 15
61	**2**	**Mon**	04	**15 42**	**54**	16.4	60.1	09 08	22 20
62	**3**	**Tu**	05	**16 36**	**55**	16.2	59.4	09 53	23 24
63	**4**	**Wed**	06	**17 31**	**54**	16.0	58.6	10 40	- -
64	**5**	**Th**	07	**18 25**	**53**	15.7	57.8	11 29	00 25
65	**6**	**Fri**	08	**19 18**	**52**	15.5	57.0	12 20	01 23
66	**7**	**Sat**	09	**20 10**	**51**	15.4	56.4	13 13	02 17
67	**8**	**Sun**	10	**21 01**	**48**	15.2	55.8	14 06	03 07
68	**9**	**Mon**	11	**21 49**	**46**	15.1	55.3	15 00	03 53
69	**10**	**Tu**	12	**22 35**	**45**	14.9	54.8	15 54	04 34
70	**11**	**Wed**	13	**23 20**	**44**	14.9	54.5	16 47	05 13
71	**12**	**Th**	14	**24 04**	**-**	14.8	54.3	17 40	05 49
72	**13**	**Fri**	15	**00 04**	**42**	14.7	54.1	18 32	06 23
73	**14**	**Sat**	16	**00 46**	**42**	14.7	54.0	19 23	06 56
74	**15**	**Sun**	17	**01 28**	**43**	14.7	54.0	20 15	07 29
75	**16**	**Mon**	18	**02 11**	**43**	14.7	54.1	21 06	08 03
76	**17**	**Tu**	19	**02 54**	**45**	14.8	54.3	21 59	08 38
77	**18**	**Wed**	20	**03 39**	**46**	14.9	54.7	22 51	09 16
78	**19**	**Th**	21	**04 25**	**49**	15.0	55.1	23 45	09 57
79	**20**	**Fri**	22	**05 14**	**50**	15.2	55.8	- -	10 41
80	**21**	**Sat**	23	**06 04**	**53**	15.4	56.5	00 38	11 30
81	**22**	**Sun**	24	**06 57**	**54**	15.6	57.4	01 30	12 24
82	**23**	**Mon**	25	**07 51**	**55**	15.9	58.4	02 22	13 22
83	**24**	**Tu**	26	**08 46**	**56**	16.2	59.3	03 11	14 24
84	**25**	**Wed**	27	**09 42**	**56**	16.4	60.2	03 59	15 29
85	**26**	**Th**	28	**10 38**	**56**	16.6	60.9	04 45	16 36
86	**27**	**Fri**	29	**11 34**	**56**	16.7	61.3	05 29	17 44
87	**28**	**Sat**	00	**12 30**	**56**	16.7	61.4	06 14	18 52
88	**29**	**Sun**	01	**13 26**	**57**	16.7	61.2	06 58	20 00
89	**30**	**Mon**	02	**14 23**	**57**	16.5	60.6	07 44	21 08
90	**31**	**Tu**	03	**15 20**	**56**	16.3	59.9	08 31	22 12

Phases of the Moon

	d	h	m
☾ First Quarter	5	08	41
○ Full Moon	13	04	34
☽ Last Quarter	21	07	38
● New Moon	28	03	14
Apogee	15	01	
Perigee	28	07	
Farthest N of 0°	6	22	
On 0°	14	06	
Farthest S of 0°	21	17	
On 0°	28	00	

NOTES

Diff equals daily change in transit time.

To correct moonrise, moonset, or transit times for latitude and/or longitude see page E 112. For further information about these tables see page E 4. For arc-to-time correction see page E 103.

MARCH 1998 — STARS — 0h GMT March 1

No	Name	Mag	Transit h m	Dec ° ′	GHA ° ′	RA h m	SHA ° ′
♈	**ARIES**		**13 23**		**158 35.8**		
1	**Alpheratz**	2.2	**13 32**	N29 04.7	**156 31.8**	0 08	357 56.0
2	**Ankaa**	2.4	**13 49**	S 42 19.2	**152 03.5**	0 26	353 27.7
3	**Schedar**	2.5	**14 04**	N56 31.6	**148 30.2**	0 40	349 54.4
4	**Diphda**	2.2	**14 07**	S 18 00.0	**147 43.8**	0 43	349 08.0
5	**Achernar**	0.6	**15 01**	S 57 15.1	**134 11.8**	1 38	335 36.0
6	**POLARIS**	2.1	**15 52**	N89 15.5	**121 19.2**	2 29	322 43.4
7	**Hamal**	2.2	**15 30**	N23 27.1	**126 50.0**	2 07	328 14.2
8	**Acamar**	3.1	**16 21**	S 40 19.1	**114 03.3**	2 58	315 27.5
9	**Menkar**	2.8	**16 25**	N 4 04.7	**113 03.3**	3 02	314 27.5
10	**Mirfak**	1.9	**16 47**	N49 51.3	**107 33.2**	3 24	308 57.4
11	**Aldebaran**	1.1	**17 58**	N16 30.2	**89 38.7**	4 36	291 02.9
12	**Rigel**	0.3	**18 37**	S 8 12.5	**79 59.1**	5 14	281 23.3
13	**Capella**	0.2	**18 39**	N45 59.8	**79 27.6**	5 17	280 51.8
14	**Bellatrix**	1.7	**18 48**	N 6 20.6	**77 20.4**	5 25	278 44.6
15	**Elnath**	1.8	**18 49**	N28 36.3	**77 03.2**	5 26	278 27.4
16	**Alnilam**	1.8	**18 59**	S 1 12.5	**74 34.0**	5 36	275 58.2
17	**Betelgeuse**	0.1-1.2	**19 18**	N 7 24.2	**69 49.7**	5 55	271 13.9
18	**Canopus**	-0.9	**19 46**	S 52 42.1	**62 37.0**	6 24	264 01.2
19	**Sirius**	-1.6	**20 07**	S 16 43.1	**57 19.7**	6 45	258 43.9
20	**Adhara**	1.6	**20 21**	S 28 58.6	**53 57.4**	6 59	255 21.6
21	**Castor**	1.6	**20 57**	N31 53.5	**44 58.5**	7 34	246 22.7
22	**Procyon**	0.5	**21 01**	N 5 13.6	**43 47.6**	7 39	245 11.8
23	**Pollux**	1.2	**21 07**	N28 01.7	**42 17.7**	7 45	243 41.9
24	**Avior**	1.7	**21 45**	S 59 30.6	**32 58.1**	8 23	234 22.3
25	**Suhail**	2.2	**22 30**	S 43 25.8	**21 36.5**	9 08	223 00.7
26	**Miaplacidus**	1.8	**22 35**	S 69 42.9	**20 17.1**	9 13	221 41.3
27	**Alphard**	2.2	**22 49**	S 8 39.3	**16 43.1**	9 28	218 07.3
28	**Regulus**	1.3	**23 30**	N11 58.4	**6 31.5**	10 08	207 55.7
29	**Dubhe**	2.0	**0 29**	N61 45.6	**352 41.2**	11 04	194 05.4
30	**Denebola**	2.2	**1 14**	N14 34.8	**341 21.1**	11 49	182 45.3
31	**Gienah**	2.8	**1 41**	S 17 32.0	**334 39.8**	12 16	176 04.0
32	**Acrux**	1.1	**1 52**	S 63 05.3	**331 57.5**	12 27	173 21.7
33	**Gacrux**	1.6	**1 56**	S 57 06.1	**330 49.2**	12 31	172 13.4
34	**Mimosa**	1.5	**2 13**	S 59 40.6	**326 40.9**	12 48	168 05.1
35	**Alioth**	1.7	**2 19**	N55 58.1	**325 06.3**	12 54	166 30.5
36	**Spica**	1.2	**2 50**	S 11 09.1	**317 19.1**	13 25	158 43.3
37	**Alkaid**	1.9	**3 13**	N49 19.2	**311 43.6**	13 47	153 07.8
38	**Hadar**	0.9	**3 29**	S 60 21.7	**307 39.8**	14 04	149 04.0
39	**Menkent**	2.3	**3 32**	S 36 21.6	**306 56.9**	14 07	148 21.1
40	**Arcturus**	0.2	**3 41**	N19 11.4	**304 42.0**	14 16	146 06.2
41	**Rigil Kent.**	0.1	**4 04**	S 60 49.4	**298 43.1**	14 40	140 07.3
42	**Zuben'ubi**	2.9	**4 16**	S 16 02.0	**295 54.0**	14 51	137 18.2
43	**Kochab**	2.2	**4 16**	N74 09.6	**295 54.8**	14 51	137 19.0
44	**Alphecca**	2.3	**4 59**	N26 43.1	**284 56.6**	15 35	126 20.8
45	**Antares**	1.2	**5 54**	S 26 25.5	**271 16.3**	16 29	112 40.5
46	**Atria**	1.9	**6 13**	S 69 01.1	**266 28.6**	16 48	107 52.8
47	**Sabik**	2.6	**6 35**	S 15 43.3	**261 01.7**	17 10	102 25.9
48	**Shaula**	1.7	**6 58**	S 37 05.9	**255 13.6**	17 33	96 37.8
49	**Rasalhague**	2.1	**6 59**	N12 33.6	**254 53.1**	17 35	96 17.3
50	**Eltanin**	2.4	**7 21**	N51 29.2	**249 27.4**	17 57	90 51.6
51	**Kaus Aust.**	2.0	**7 48**	S 34 22.9	**242 35.2**	18 24	83 59.4
52	**Vega**	0.1	**8 01**	N38 46.7	**239 22.8**	18 37	80 47.0
53	**Nunki**	2.1	**8 19**	S 26 17.8	**234 48.7**	18 55	76 12.9
54	**Altair**	0.9	**9 15**	N 8 51.7	**220 55.6**	19 51	62 19.8
55	**Peacock**	2.1	**9 49**	S 56 44.3	**212 13.9**	20 25	53 38.1
56	**Deneb**	1.3	**10 05**	N45 16.3	**208 15.5**	20 41	49 39.7
57	**Enif**	2.5	**11 08**	N 9 51.9	**192 34.7**	21 44	33 58.9
58	**Al Na'ir**	2.2	**11 32**	S 46 58.2	**186 34.7**	22 08	27 58.9
59	**Fomalhaut**	1.3	**12 21**	S 29 38.0	**174 13.0**	22 58	15 37.2
60	**Markab**	2.6	**12 28**	N15 11.6	**172 26.1**	23 05	13 50.3

Star Transit Corr Table

Date	Corr h m	Date	Corr h m
1	0 00	17	−1 03
2	−0 04	18	−1 07
3	−0 08	19	−1 11
4	−0 12	20	−1 15
5	−0 16	21	−1 19
6	−0 20	22	−1 23
7	−0 24	23	−1 27
8	−0 28	24	−1 30
9	−0 31	25	−1 34
10	−0 35	26	−1 38
11	−0 39	27	−1 42
12	−0 43	28	−1 46
13	−0 47	29	−1 50
14	−0 51	30	−1 54
15	−0 55	31	−1 58
16	−0 59		

STAR TRANSIT

To find the approximate time of transit of a star for any day of the month use the table above. All corrections are subtractive.

If the value taken from the table is greater than the time of transit for the first of the month, add 23h 56min to the time of transit before subtracting the correction.

Example: What time will Dubhe (No 29) transit the meridian on March 30?

	h min
Transit on Mar 1	00 29
	+23 56
	24 25
Corr for Mar 30	−01 54
Transit on Mar 30	22 31

MARCH DIARY

d h	
1 09	Saturn 1.0° N of Moon
7 10	Venus 4° N of Neptune
11 15	Mercury 1.2° N of Mars
12 16	Pluto stationary
13 20	Juno 0.9° N of Moon
19 07	Venus 3° N of Uranus
20 04	Mercury greatest elong. E.(19°)
20 20	Equinox
23 17	Neptune 3° S of Moon
24 10	Uranus 3° S of Moon
24 19	Venus 0.1° S of Moon
26 12	Jupiter 0.8° S of Moon
27 15	Mercury stationary
27 19	Venus greatest elong. W.(47°)
30 05	Mercury 4° N of Mars

NOTES

Star declinations may be used as is for the whole month. To correct GHA for day, hour, minute, and second see page E 106; for further explanation of this table see page E 4.

MARCH 1998 SUN & ARIES GMT

Time	SUN GHA ° ′	Dec ° ′	ARIES GHA ° ′	Time	SUN GHA ° ′	Dec ° ′	ARIES GHA ° ′	Time	SUN GHA ° ′	Dec ° ′	ARIES GHA ° ′	Time
	Sunday, 1st March				**Friday, 6th March**				**Wednesday, 11th March**			
00	176 52.9	S 7 44.7	158 35.8	00	177 08.6	S 5 49.7	163 31.5	00	177 27.0	S 3 52.7	168 27.2	00
02	206 53.1	7 42.8	188 40.8	02	207 08.9	5 47.7	193 36.4	02	207 27.4	3 50.7	198 32.1	02
04	236 53.3	7 40.9	218 45.7	04	237 09.2	5 45.8	223 41.4	04	237 27.7	3 48.7	228 37.1	04
06	266 53.6	7 39.0	248 50.6	06	267 09.4	5 43.8	253 46.3	06	267 28.0	3 46.8	258 42.0	06
08	296 53.8	7 37.1	278 55.5	08	297 09.7	5 41.9	283 51.2	08	297 28.3	3 44.8	288 46.9	08
10	326 54.1	7 35.2	309 00.5	10	327 10.0	5 40.0	313 56.2	10	327 28.7	3 42.9	318 51.9	10
12	356 54.3	7 33.3	339 05.4	12	357 10.3	5 38.0	344 01.1	12	357 29.0	3 40.9	348 56.8	12
14	26 54.5	7 31.4	9 10.3	14	27 10.6	5 36.1	14 06.0	14	27 29.3	3 38.9	19 01.7	14
16	56 54.8	7 29.5	39 15.3	16	57 10.9	5 34.2	44 10.9	16	57 29.7	3 37.0	49 06.6	16
18	86 55.0	7 27.6	69 20.2	18	87 11.2	5 32.2	74 15.9	18	87 30.0	3 35.0	79 11.6	18
20	116 55.3	7 25.7	99 25.1	20	117 11.5	5 30.3	104 20.8	20	117 30.3	3 33.0	109 16.5	20
22	146 55.5	S 7 23.8	129 30.0	22	147 11.8	S 5 28.3	134 25.7	22	147 30.7	S 3 31.1	139 21.4	22
	Monday, 2nd March				**Saturday, 7th March**				**Thursday, 12th March**			
00	176 55.8	S 7 21.9	159 35.0	00	177 12.1	S 5 26.4	164 30.7	00	177 31.0	S 3 29.1	169 26.4	00
02	206 56.0	7 20.0	189 39.9	02	207 12.4	5 24.4	194 35.6	02	207 31.3	3 27.1	199 31.3	02
04	236 56.3	7 18.1	219 44.8	04	237 12.7	5 22.5	224 40.5	04	237 31.7	3 25.2	229 36.2	04
06	266 56.5	7 16.2	249 49.8	06	267 13.0	5 20.6	254 45.4	06	267 32.0	3 23.2	259 41.1	06
08	296 56.8	7 14.2	279 54.7	08	297 13.3	5 18.6	284 50.4	08	297 32.3	3 21.2	289 46.1	08
10	326 57.0	7 12.3	309 59.6	10	327 13.6	5 16.7	314 55.3	10	327 32.7	3 19.3	319 51.0	10
12	356 57.3	7 10.4	340 04.5	12	357 13.9	5 14.7	345 00.2	12	357 33.0	3 17.3	349 55.9	12
14	26 57.5	7 08.5	10 09.5	14	27 14.2	5 12.8	15 05.2	14	27 33.3	3 15.3	20 00.8	14
16	56 57.8	7 06.6	40 14.4	16	57 14.5	5 10.8	45 10.1	16	57 33.7	3 13.4	50 05.8	16
18	86 58.0	7 04.7	70 19.3	18	87 14.8	5 08.9	75 15.0	18	87 34.0	3 11.4	80 10.7	18
20	116 58.3	7 02.8	100 24.2	20	117 15.1	5 06.9	105 19.9	20	117 34.3	3 09.4	110 15.6	20
22	146 58.5	S 7 00.9	130 29.2	22	147 15.4	S 5 05.0	135 24.9	22	147 34.7	S 3 07.5	140 20.6	22
	Tuesday, 3rd March				**Sunday, 8th March**				**Friday, 13th March**			
00	176 58.8	S 6 59.0	160 34.1	00	177 15.7	S 5 03.0	165 29.8	00	177 35.0	S 3 05.5	170 25.5	00
02	206 59.0	6 57.0	190 39.0	02	207 16.0	5 01.1	195 34.7	02	207 35.4	3 03.5	200 30.4	02
04	236 59.3	6 55.1	220 44.0	04	237 16.3	4 59.2	225 39.7	04	237 35.7	3 01.6	230 35.3	04
06	266 59.6	6 53.2	250 48.9	06	267 16.6	4 57.2	255 44.6	06	267 36.0	2 59.6	260 40.3	06
08	296 59.8	6 51.3	280 53.8	08	297 16.9	4 55.3	285 49.5	08	297 36.4	2 57.6	290 45.2	08
10	327 00.1	6 49.4	310 58.7	10	327 17.2	4 53.3	315 54.4	10	327 36.7	2 55.7	320 50.1	10
12	357 00.3	6 47.5	341 03.7	12	357 17.5	4 51.4	345 59.4	12	357 37.1	2 53.7	350 55.1	12
14	27 00.6	6 45.5	11 08.6	14	27 17.8	4 49.4	16 04.3	14	27 37.4	2 51.7	21 00.0	14
16	57 00.9	6 43.6	41 13.5	16	57 18.1	4 47.5	46 09.2	16	57 37.8	2 49.8	51 04.9	16
18	87 01.1	6 41.7	71 18.5	18	87 18.4	4 45.5	76 14.2	18	87 38.1	2 47.8	81 09.8	18
20	117 01.4	6 39.8	101 23.4	20	117 18.7	4 43.6	106 19.1	20	117 38.4	2 45.8	111 14.8	20
22	147 01.7	S 6 37.9	131 28.3	22	147 19.1	S 4 41.6	136 24.0	22	147 38.8	S 2 43.8	141 19.7	22
	Wednesday, 4th March				**Monday, 9th March**				**Saturday, 14th March**			
00	177 01.9	S 6 35.9	161 33.2	00	177 19.4	S 4 39.6	166 28.9	00	177 39.1	S 2 41.9	171 24.6	00
02	207 02.2	6 34.0	191 38.2	02	207 19.7	4 37.7	196 33.9	02	207 39.5	2 39.9	201 29.6	02
04	237 02.5	6 32.1	221 43.1	04	237 20.0	4 35.7	226 38.8	04	237 39.8	2 37.9	231 34.5	04
06	267 02.7	6 30.2	251 48.0	06	267 20.3	4 33.8	256 43.7	06	267 40.2	2 36.0	261 39.4	06
08	297 03.0	6 28.3	281 53.0	08	297 20.6	4 31.8	286 48.7	08	297 40.5	2 34.0	291 44.3	08
10	327 03.3	6 26.3	311 57.9	10	327 20.9	4 29.9	316 53.6	10	327 40.9	2 32.0	321 49.3	10
12	357 03.6	6 24.4	342 02.8	12	357 21.2	4 27.9	346 58.5	12	357 41.2	2 30.0	351 54.2	12
14	27 03.8	6 22.5	12 07.7	14	27 21.6	4 26.0	17 03.4	14	27 41.6	2 28.1	21 59.1	14
16	57 04.1	6 20.6	42 12.7	16	57 21.9	4 24.0	47 08.4	16	57 41.9	2 26.1	52 04.0	16
18	87 04.4	6 18.6	72 17.6	18	87 22.2	4 22.1	77 13.3	18	87 42.2	2 24.1	82 09.0	18
20	117 04.6	6 16.7	102 22.5	20	117 22.5	4 20.1	107 18.2	20	117 42.6	2 22.2	112 13.9	20
22	147 04.9	S 6 14.8	132 27.5	22	147 22.8	S 4 18.1	137 23.1	22	147 42.9	S 2 20.2	142 18.8	22
	Thursday, 5th March				**Tuesday, 10th March**				**Sunday, 15th March**			
00	177 05.2	S 6 12.8	162 32.4	00	177 23.2	S 4 16.2	167 28.1	00	177 43.3	S 2 18.2	172 23.8	00
02	207 05.5	6 10.9	192 37.3	02	207 23.5	4 14.2	197 33.0	02	207 43.6	2 16.2	202 28.7	02
04	237 05.8	6 09.0	222 42.2	04	237 23.8	4 12.3	227 37.9	04	237 44.0	2 14.3	232 33.6	04
06	267 06.0	6 07.1	252 47.2	06	267 24.1	4 10.3	257 42.9	06	267 44.4	2 12.3	262 38.5	06
08	297 06.3	6 05.1	282 52.1	08	297 24.4	4 08.4	287 47.8	08	297 44.7	2 10.3	292 43.5	08
10	327 06.6	6 03.2	312 57.0	10	327 24.8	4 06.4	317 52.7	10	327 45.1	2 08.3	322 48.4	10
12	357 06.9	6 01.3	343 02.0	12	357 25.1	4 04.4	347 57.6	12	357 45.4	2 06.4	352 53.3	12
14	27 07.2	5 59.3	13 06.9	14	27 25.4	4 02.5	18 02.6	14	27 45.8	2 04.4	22 58.3	14
16	57 07.4	5 57.4	43 11.8	16	57 25.7	4 00.5	48 07.5	16	57 46.1	2 02.4	53 03.2	16
18	87 07.7	5 55.5	73 16.7	18	87 26.0	3 58.6	78 12.4	18	87 46.5	2 00.4	83 08.1	18
20	117 08.0	5 53.5	103 21.7	20	117 26.4	3 56.6	108 17.4	20	117 46.8	1 58.5	113 13.0	20
22	147 08.3	S 5 51.6	133 26.6	22	147 26.7	S 3 54.6	138 22.3	22	147 47.2	S 1 56.5	143 18.0	22

NOTES

Sun and Aries GHA corrections for additional hour, minutes, and seconds are on page E 106.
Correct sun declination by interpolation. For an example see page E 4.

MARCH 1998 — SUN & ARIES — GMT

Monday, 16th March

Time	SUN GHA ° ′	Dec ° ′	ARIES GHA ° ′
00	177 47.5	S 1 54.5	173 22.9
02	207 47.9	1 52.5	203 27.8
04	237 48.2	1 50.6	233 32.8
06	267 48.6	1 48.6	263 37.7
08	297 48.9	1 46.6	293 42.6
10	327 49.3	1 44.6	323 47.5
12	357 49.7	1 42.7	353 52.5
14	27 50.0	1 40.7	23 57.4
16	57 50.4	1 38.7	54 02.3
18	87 50.7	1 36.7	84 07.3
20	117 51.1	1 34.8	114 12.2
22	147 51.5	S 1 32.8	144 17.1

Tuesday, 17th March

Time	SUN GHA ° ′	Dec ° ′	ARIES GHA ° ′
00	177 51.8	S 1 30.8	174 22.0
02	207 52.2	1 28.8	204 27.0
04	237 52.5	1 26.9	234 31.9
06	267 52.9	1 24.9	264 36.8
08	297 53.2	1 22.9	294 41.7
10	327 53.6	1 20.9	324 46.7
12	357 54.0	1 19.0	354 51.6
14	27 54.3	1 17.0	24 56.5
16	57 54.7	1 15.0	55 01.5
18	87 55.1	1 13.0	85 06.4
20	117 55.4	1 11.1	115 11.3
22	147 55.8	S 1 09.1	145 16.2

Wednesday, 18th March

Time	SUN GHA ° ′	Dec ° ′	ARIES GHA ° ′
00	177 56.1	S 1 07.1	175 21.2
02	207 56.5	1 05.1	205 26.1
04	237 56.9	1 03.2	235 31.0
06	267 57.2	1 01.2	265 36.0
08	297 57.6	0 59.2	295 40.9
10	327 58.0	0 57.2	325 45.8
12	357 58.3	0 55.2	355 50.7
14	27 58.7	0 53.3	25 55.7
16	57 59.1	0 51.3	56 00.6
18	87 59.4	0 49.3	86 05.5
20	117 59.8	0 47.3	116 10.5
22	148 00.1	S 0 45.4	146 15.4

Thursday, 19th March

Time	SUN GHA ° ′	Dec ° ′	ARIES GHA ° ′
00	178 00.5	S 0 43.4	176 20.3
02	208 00.9	0 41.4	206 25.2
04	238 01.2	0 39.4	236 30.2
06	268 01.6	0 37.5	266 35.1
08	298 02.0	0 35.5	296 40.0
10	328 02.3	0 33.5	326 45.0
12	358 02.7	0 31.5	356 49.9
14	28 03.1	0 29.5	26 54.8
16	58 03.4	0 27.6	56 59.7
18	88 03.8	0 25.6	87 04.7
20	118 04.2	0 23.6	117 09.6
22	148 04.6	S 0 21.6	147 14.5

Friday, 20th March

Time	SUN GHA ° ′	Dec ° ′	ARIES GHA ° ′
00	178 04.9	S 0 19.7	177 19.4
02	208 05.3	0 17.7	207 24.4
04	238 05.7	0 15.7	237 29.3
06	268 06.0	0 13.7	267 34.2
08	298 06.4	0 11.8	297 39.2
10	328 06.8	0 09.8	327 44.1
12	358 07.1	0 07.8	357 49.0
14	28 07.5	0 05.8	27 53.9
16	58 07.9	0 03.9	57 58.9
18	88 08.2	0 01.9	88 03.8
20	118 08.6	N 0 00.1	118 08.7
22	148 09.0	N 0 02.1	148 13.7

Saturday, 21st March

Time	SUN GHA ° ′	Dec ° ′	ARIES GHA ° ′
00	178 09.4	N 0 04.0	178 18.6
02	208 09.7	0 06.0	208 23.5
04	238 10.1	0 08.0	238 28.4
06	268 10.5	0 10.0	268 33.4
08	298 10.8	0 11.9	298 38.3
10	328 11.2	0 13.9	328 43.2
12	358 11.6	0 15.9	358 48.2
14	28 12.0	0 17.9	28 53.1
16	58 12.3	0 19.8	58 58.0
18	88 12.7	0 21.8	89 02.9
20	118 13.1	0 23.8	119 07.9
22	148 13.5	N 0 25.8	149 12.8

Sunday, 22nd March

Time	SUN GHA ° ′	Dec ° ′	ARIES GHA ° ′
00	178 13.8	N 0 27.7	179 17.7
02	208 14.2	0 29.7	209 22.7
04	238 14.6	0 31.7	239 27.6
06	268 14.9	0 33.7	269 32.5
08	298 15.3	0 35.6	299 37.4
10	328 15.7	0 37.6	329 42.4
12	358 16.1	0 39.6	359 47.3
14	28 16.4	0 41.6	29 52.2
16	58 16.8	0 43.5	59 57.2
18	88 17.2	0 45.5	90 02.1
20	118 17.6	0 47.5	120 07.0
22	148 17.9	N 0 49.5	150 11.9

Monday, 23rd March

Time	SUN GHA ° ′	Dec ° ′	ARIES GHA ° ′
00	178 18.3	N 0 51.4	180 16.9
02	208 18.7	0 53.4	210 21.8
04	238 19.1	0 55.4	240 26.7
06	268 19.4	0 57.3	270 31.7
08	298 19.8	0 59.3	300 36.6
10	328 20.2	1 01.3	330 41.5
12	358 20.6	1 03.3	0 46.4
14	28 20.9	1 05.2	30 51.4
16	58 21.3	1 07.2	60 56.3
18	88 21.7	1 09.2	91 01.2
20	118 22.1	1 11.1	121 06.1
22	148 22.4	N 1 13.1	151 11.1

Tuesday, 24th March

Time	SUN GHA ° ′	Dec ° ′	ARIES GHA ° ′
00	178 22.8	N 1 15.1	181 16.0
02	208 23.2	1 17.1	211 20.9
04	238 23.6	1 19.0	241 25.9
06	268 23.9	1 21.0	271 30.8
08	298 24.3	1 23.0	301 35.7
10	328 24.7	1 24.9	331 40.6
12	358 25.1	1 26.9	1 45.6
14	28 25.4	1 28.9	31 50.5
16	58 25.8	1 30.8	61 55.4
18	88 26.2	1 32.8	92 00.4
20	118 26.6	1 34.8	122 05.3
22	148 26.9	N 1 36.7	152 10.2

Wednesday, 25th March

Time	SUN GHA ° ′	Dec ° ′	ARIES GHA ° ′
00	178 27.3	N 1 38.7	182 15.1
02	208 27.7	1 40.7	212 20.1
04	238 28.1	1 42.6	242 25.0
06	268 28.4	1 44.6	272 29.9
08	298 28.8	1 46.6	302 34.9
10	328 29.2	1 48.5	332 39.8
12	358 29.6	1 50.5	2 44.7
14	28 30.0	1 52.5	32 49.6
16	58 30.3	1 54.4	62 54.6
18	88 30.7	1 56.4	92 59.5
20	118 31.1	1 58.4	123 04.4
22	148 31.5	N 2 00.3	153 09.4

Thursday, 26th March

Time	SUN GHA ° ′	Dec ° ′	ARIES GHA ° ′
00	178 31.8	N 2 02.3	183 14.3
02	208 32.2	2 04.3	213 19.2
04	238 32.6	2 06.2	243 24.1
06	268 33.0	2 08.2	273 29.1
08	298 33.3	2 10.1	303 34.0
10	328 33.7	2 12.1	333 38.9
12	358 34.1	2 14.1	3 43.8
14	28 34.5	2 16.0	33 48.8
16	58 34.8	2 18.0	63 53.7
18	88 35.2	2 20.0	93 58.6
20	118 35.6	2 21.9	124 03.6
22	148 36.0	N 2 23.9	154 08.5

Friday, 27th March

Time	SUN GHA ° ′	Dec ° ′	ARIES GHA ° ′
00	178 36.4	N 2 25.8	184 13.4
02	208 36.7	2 27.8	214 18.3
04	238 37.1	2 29.8	244 23.3
06	268 37.5	2 31.7	274 28.2
08	298 37.9	2 33.7	304 33.1
10	328 38.2	2 35.6	334 38.1
12	358 38.6	2 37.6	4 43.0
14	28 39.0	2 39.5	34 47.9
16	58 39.4	2 41.5	64 52.8
18	88 39.7	2 43.5	94 57.8
20	118 40.1	2 45.4	125 02.7
22	148 40.5	N 2 47.4	155 07.6

Saturday, 28th March

Time	SUN GHA ° ′	Dec ° ′	ARIES GHA ° ′
00	178 40.9	N 2 49.3	185 12.6
02	208 41.2	2 51.3	215 17.5
04	238 41.6	2 53.2	245 22.4
06	268 42.0	2 55.2	275 27.3
08	298 42.4	2 57.1	305 32.3
10	328 42.7	2 59.1	335 37.2
12	358 43.1	3 01.1	5 42.1
14	28 43.5	3 03.0	35 47.1
16	58 43.9	3 05.0	65 52.0
18	88 44.2	3 06.9	95 56.9
20	118 44.6	3 08.9	126 01.8
22	148 45.0	N 3 10.8	156 06.8

Sunday, 29th March

Time	SUN GHA ° ′	Dec ° ′	ARIES GHA ° ′
00	178 45.4	N 3 12.8	186 11.7
02	208 45.8	3 14.7	216 16.6
04	238 46.1	3 16.7	246 21.5
06	268 46.5	3 18.6	276 26.5
08	298 46.9	3 20.6	306 31.4
10	328 47.3	3 22.5	336 36.3
12	358 47.6	3 24.5	6 41.3
14	28 48.0	3 26.4	36 46.2
16	58 48.4	3 28.4	66 51.1
18	88 48.8	3 30.3	96 56.0
20	118 49.1	3 32.3	127 01.0
22	148 49.5	N 3 34.2	157 05.9

Monday, 30th March

Time	SUN GHA ° ′	Dec ° ′	ARIES GHA ° ′
00	178 49.9	N 3 36.1	187 10.8
02	208 50.3	3 38.1	217 15.8
04	238 50.6	3 40.0	247 20.7
06	268 51.0	3 42.0	277 25.6
08	298 51.4	3 43.9	307 30.5
10	328 51.7	3 45.9	337 35.5
12	358 52.1	3 47.8	7 40.4
14	28 52.5	3 49.7	37 45.3
16	58 52.9	3 51.7	67 50.3
18	88 53.2	3 53.6	97 55.2
20	118 53.6	3 55.6	128 00.1
22	148 54.0	N 3 57.5	158 05.0

Tuesday, 31st March

Time	SUN GHA ° ′	Dec ° ′	ARIES GHA ° ′
00	178 54.4	N 3 59.5	188 10.0
02	208 54.7	4 01.4	218 14.9
04	238 55.1	4 03.3	248 19.8
06	268 55.5	N 4 05.3	278 24.8
08	298 55.9	N 4 07.2	308 29.7
10	328 56.2	4 09.1	338 34.6
12	358 56.6	4 11.1	8 39.5
14	28 57.0	N 4 13.0	38 44.5
16	58 57.3	N 4 14.9	68 49.4
18	88 57.7	4 16.9	98 54.3
20	118 58.1	4 18.8	128 59.2
22	148 58.5	N 4 20.8	159 04.2

MARCH 1998 PLANETS 0h GMT

VENUS						JUPITER				
Mer Pass h m	GHA ° ′	Mean Var/hr 14°+	Dec ° ′	Mean Var/hr ′	Day	GHA ° ′	Mean Var/hr 15°+	Dec ° ′	Mean Var/hr ′	Mer Pass h m
09 19	**220 08.1**	0.7	**S15 56.3**	0.0	**1 SUN**	**180 35.4**	1.9	**S10 10.6**	0.2	**11 56**
09 18	**220 23.7**	0.6	**S15 55.7**	0.0	**2 Mon**	**181 20.9**	1.9	**S10 05.3**	0.2	**11 53**
09 17	**220 38.1**	0.6	**S15 54.6**	0.1	**3 Tu**	**182 06.4**	1.9	**S10 00.1**	0.2	**11 50**
09 16	**220 51.5**	0.5	**S15 53.2**	0.1	**4 Wed**	**182 51.9**	1.9	**S 9 54.8**	0.2	**11 47**
09 15	**221 04.0**	0.5	**S15 51.4**	0.1	**5 Th**	**183 37.4**	1.9	**S 9 49.5**	0.2	**11 44**
09 15	**221 15.4**	0.4	**S15 49.1**	0.1	**6 Fri**	**184 23.0**	1.9	**S 9 44.3**	0.2	**11 41**
09 14	**221 26.0**	0.4	**S15 46.4**	0.1	**7 Sat**	**185 08.5**	1.9	**S 9 39.0**	0.2	**11 38**
09 13	**221 35.8**	0.4	**S15 43.4**	0.2	**8 SUN**	**185 54.1**	1.9	**S 9 33.7**	0.2	**11 35**
09 13	**221 44.7**	0.3	**S15 39.8**	0.2	**9 Mon**	**186 39.7**	1.9	**S 9 28.4**	0.2	**11 32**
09 12	**221 52.8**	0.3	**S15 35.9**	0.2	**10 Tu**	**187 25.4**	1.9	**S 9 23.1**	0.2	**11 29**
09 12	**222 00.2**	0.3	**S15 31.5**	0.2	**11 Wed**	**188 11.0**	1.9	**S 9 17.9**	0.2	**11 26**
09 11	**222 06.8**	0.3	**S15 26.6**	0.2	**12 Th**	**188 56.7**	1.9	**S 9 12.6**	0.2	**11 23**
09 11	**222 12.8**	0.2	**S15 21.3**	0.2	**13 Fri**	**189 42.4**	1.9	**S 9 07.3**	0.2	**11 20**
09 11	**222 18.2**	0.2	**S15 15.5**	0.3	**14 Sat**	**190 28.2**	1.9	**S 9 02.1**	0.2	**11 17**
09 10	**222 22.9**	0.2	**S15 09.3**	0.3	**15 SUN**	**191 13.9**	1.9	**S 8 56.8**	0.2	**11 14**
09 10	**222 27.1**	0.2	**S15 02.5**	0.3	**16 Mon**	**191 59.7**	1.9	**S 8 51.5**	0.2	**11 11**
09 10	**222 30.7**	0.1	**S14 55.4**	0.3	**17 Tu**	**192 45.6**	1.9	**S 8 46.3**	0.2	**11 08**
09 10	**222 33.8**	0.1	**S14 47.7**	0.3	**18 Wed**	**193 31.4**	1.9	**S 8 41.0**	0.2	**11 04**
09 10	**222 36.5**	0.1	**S14 39.6**	0.4	**19 Th**	**194 17.3**	1.9	**S 8 35.8**	0.2	**11 01**
09 09	**222 38.6**	0.1	**S14 31.0**	0.4	**20 Fri**	**195 03.2**	1.9	**S 8 30.5**	0.2	**10 58**
09 09	**222 40.3**	0.1	**S14 21.9**	0.4	**21 Sat**	**195 49.2**	1.9	**S 8 25.3**	0.2	**10 55**
09 09	**222 41.7**	0.0	**S14 12.3**	0.4	**22 SUN**	**196 35.1**	1.9	**S 8 20.1**	0.2	**10 52**
09 09	**222 42.6**	0.0	**S14 02.3**	0.4	**23 Mon**	**197 21.2**	1.9	**S 8 14.8**	0.2	**10 49**
09 09	**222 43.2**	0.0	**S13 51.8**	0.5	**24 Tu**	**198 07.2**	1.9	**S 8 09.6**	0.2	**10 46**
09 09	**222 43.5**	0.0	**S13 40.9**	0.5	**25 Wed**	**198 53.3**	1.9	**S 8 04.4**	0.2	**10 43**
09 09	**222 43.4**	0.0	**S13 29.5**	0.5	**26 Th**	**199 39.4**	1.9	**S 7 59.2**	0.2	**10 40**
09 09	**222 43.1**	0.0	**S13 17.6**	0.5	**27 Fri**	**200 25.6**	1.9	**S 7 54.1**	0.2	**10 37**
09 09	**222 42.5**	0.0	**S13 05.3**	0.5	**28 Sat**	**201 11.8**	1.9	**S 7 48.9**	0.2	**10 34**
09 09	**222 41.7**	0.0	**S12 52.5**	0.6	**29 SUN**	**201 58.1**	1.9	**S 7 43.7**	0.2	**10 31**
09 09	**222 40.6**	14	**S12 39.3**	0.6	**30 Mon**	**202 44.4**	1.9	**S 7 38.6**	0.2	**10 28**
09 09	**222 39.3**	59.9	**S12 25.6**	0.6	**31 Tu**	**203 30.8**	1.9	**S 7 33.5**	0.2	**10 25**

VENUS, Av. Mag. -4.5
S.H.A. March
5 59; 10 54; 15 50; 20 45; 25 40; 30 35.

JUPITER, Av. Mag.-2.0
S.H.A. March
5 21; 10 20; 15 19; 20 18; 25 17; 30 16.

MARS						SATURN				
Mer Pass h m	GHA ° ′	Mean Var/hr 15°+	Dec ° ′	Mean Var/hr ′	Day	GHA ° ′	Mean Var/hr 15°+	Dec ° ′	Mean Var/hr ′	Mer Pass h m
13 16	**160 55.8**	0.7	**S 1 48.1**	0.8	**1 SUN**	**141 01.9**	2.2	**N 4 58.2**	0.1	**14 34**
13 15	**161 12.4**	0.7	**S 1 29.1**	0.8	**2 Mon**	**141 55.0**	2.2	**N 5 00.9**	0.1	**14 30**
13 13	**161 29.0**	0.7	**S 1 10.0**	0.8	**3 Tu**	**142 48.1**	2.2	**N 5 03.5**	0.1	**14 27**
13 12	**161 45.6**	0.7	**S 0 50.9**	0.8	**4 Wed**	**143 41.1**	2.2	**N 5 06.1**	0.1	**14 23**
13 11	**162 02.3**	0.7	**S 0 31.9**	0.8	**5 Th**	**144 34.1**	2.2	**N 5 08.8**	0.1	**14 20**
13 10	**162 19.0**	0.7	**S 0 12.8**	0.8	**6 Fri**	**145 27.0**	2.2	**N 5 11.5**	0.1	**14 16**
13 09	**162 35.8**	0.7	**N 0 06.2**	0.8	**7 Sat**	**146 19.9**	2.2	**N 5 14.1**	0.1	**14 13**
13 08	**162 52.5**	0.7	**N 0 25.2**	0.8	**8 SUN**	**147 12.7**	2.2	**N 5 16.8**	0.1	**14 09**
13 07	**163 09.4**	0.7	**N 0 44.2**	0.8	**9 Mon**	**148 05.5**	2.2	**N 5 19.5**	0.1	**14 06**
13 06	**163 26.2**	0.7	**N 1 03.1**	0.8	**10 Tu**	**148 58.3**	2.2	**N 5 22.3**	0.1	**14 02**
13 05	**163 43.1**	0.7	**N 1 22.0**	0.8	**11 Wed**	**149 51.0**	2.2	**N 5 25.0**	0.1	**13 59**
13 03	**163 59.9**	0.7	**N 1 40.9**	0.8	**12 Th**	**150 43.7**	2.2	**N 5 27.7**	0.1	**13 55**
13 02	**164 16.8**	0.7	**N 1 59.8**	0.8	**13 Fri**	**151 36.3**	2.2	**N 5 30.5**	0.1	**13 52**
13 01	**164 33.8**	0.7	**N 2 18.6**	0.8	**14 Sat**	**152 28.9**	2.2	**N 5 33.2**	0.1	**13 48**
13 00	**164 50.7**	0.7	**N 2 37.4**	0.8	**15 SUN**	**153 21.5**	2.2	**N 5 36.0**	0.1	**13 45**
12 59	**165 07.7**	0.7	**N 2 56.1**	0.8	**16 Mon**	**154 14.0**	2.2	**N 5 38.7**	0.1	**13 41**
12 58	**165 24.6**	0.7	**N 3 14.8**	0.8	**17 Tu**	**155 06.6**	2.2	**N 5 41.5**	0.1	**13 38**
12 57	**165 41.6**	0.7	**N 3 33.4**	0.8	**18 Wed**	**155 59.0**	2.2	**N 5 44.3**	0.1	**13 34**
12 55	**165 58.6**	0.7	**N 3 52.0**	0.8	**19 Th**	**156 51.5**	2.2	**N 5 47.1**	0.1	**13 31**
12 54	**166 15.6**	0.7	**N 4 10.6**	0.8	**20 Fri**	**157 43.9**	2.2	**N 5 49.9**	0.1	**13 27**
12 53	**166 32.5**	0.7	**N 4 29.0**	0.8	**21 Sat**	**158 36.3**	2.2	**N 5 52.7**	0.1	**13 24**
12 52	**166 49.5**	0.7	**N 4 47.5**	0.8	**22 SUN**	**159 28.6**	2.2	**N 5 55.5**	0.1	**13 20**
12 51	**167 06.5**	0.7	**N 5 05.8**	0.8	**23 Mon**	**160 21.0**	2.2	**N 5 58.3**	0.1	**13 17**
12 50	**167 23.5**	0.7	**N 5 24.1**	0.8	**24 Tu**	**161 13.3**	2.2	**N 6 01.1**	0.1	**13 13**
12 49	**167 40.5**	0.7	**N 5 42.3**	0.8	**25 Wed**	**162 05.6**	2.2	**N 6 03.9**	0.1	**13 10**
12 48	**167 57.4**	0.7	**N 6 00.5**	0.8	**26 Th**	**162 57.8**	2.2	**N 6 06.8**	0.1	**13 06**
12 46	**168 14.4**	0.7	**N 6 18.6**	0.8	**27 Fri**	**163 50.1**	2.2	**N 6 09.6**	0.1	**13 03**
12 45	**168 31.3**	0.7	**N 6 36.6**	0.7	**28 Sat**	**164 42.3**	2.2	**N 6 12.4**	0.1	**12 59**
12 44	**168 48.3**	0.7	**N 6 54.5**	0.7	**29 SUN**	**165 34.5**	2.2	**N 6 15.2**	0.1	**12 56**
12 43	**169 05.2**	0.7	**N 7 12.3**	0.7	**30 Mon**	**166 26.7**	2.2	**N 6 18.1**	0.1	**12 52**
12 42	**169 22.1**	0.7	**N 7 30.1**	0.7	**31 Tu**	**167 18.8**	2.2	**N 6 20.9**	0.1	**12 49**

MARS, Av. Mag. +1.2
S.H.A. March
5 359; 10 356; 15 352; 20 349; 25 345; 30 342.

NOTE: Planet corrections on pages E113-15

SATURN, Av. Mag. +0.7
S.H.A. March
5 342; 10 342; 15 341; 20 340; 25 340; 30 339.

MARCH 1998 MOON

Day	GMT hr	GHA	Mean Var/hr	Dec	Mean Var/hr
		° ′	14°+	° ′	′
1 Sun	**0**	146 24.6	26.9	**N 2 12.8**	11.7
	6	233 06.1	26.9	**N 3 23.3**	11.6
	12	319 47.8	27.0	**N 4 32.8**	11.3
	18	46 29.7	27.0	**N 5 41.0**	11.1
2 Mon	**0**	133 11.8	27.0	**N 6 47.5**	10.7
	6	219 54.1	27.0	**N 7 52.1**	10.4
	12	306 36.4	27.1	**N 8 54.7**	10.0
	18	33 18.8	27.0	**N 9 54.8**	9.6
3 Tu	**0**	120 01.3	27.1	**N 10 52.4**	9.1
	6	206 43.8	27.1	**N 11 47.1**	8.6
	12	293 26.3	27.1	**N 12 38.9**	8.0
	18	20 08.9	27.1	**N 13 27.6**	7.5
4 Wed	**0**	106 51.5	27.1	**N 14 13.0**	6.9
	6	193 34.3	27.2	**N 14 54.9**	6.3
	12	280 17.2	27.2	**N 15 33.3**	5.7
	18	7 00.4	27.2	**N 16 08.1**	5.2
5 Th	**0**	93 43.9	27.3	**N 16 39.2**	4.5
	6	180 27.7	27.4	**N 17 06.6**	3.9
	12	267 12.1	27.5	**N 17 30.1**	3.3
	18	353 57.0	27.6	**N 17 49.8**	2.6
6 Fri	**0**	80 42.6	27.8	**N 18 05.7**	2.0
	6	167 29.0	27.9	**N 18 17.7**	1.3
	12	254 16.3	28.1	**N 18 25.9**	0.7
	18	341 04.6	28.2	**N 18 30.3**	0.0
7 Sat	**0**	67 54.0	28.4	**N 18 31.1**	0.5
	6	154 44.6	28.7	**N 18 28.1**	1.2
	12	241 36.6	28.9	**N 18 21.6**	1.8
	18	328 29.9	29.1	**N 18 11.5**	2.3
8 Sun	**0**	55 24.6	29.4	**N 17 58.0**	2.8
	6	142 20.9	29.6	**N 17 41.2**	3.4
	12	229 18.8	30.0	**N 17 21.2**	3.9
	18	316 18.3	30.2	**N 16 58.1**	4.4
9 Mon	**0**	43 19.5	30.5	**N 16 32.1**	4.9
	6	130 22.3	30.8	**N 16 03.2**	5.3
	12	217 26.9	31.1	**N 15 31.6**	5.7
	18	304 33.1	31.3	**N 14 57.4**	6.2
10 Tu	**0**	31 41.1	31.7	**N 14 20.8**	6.5
	6	118 50.6	31.9	**N 13 41.9**	6.9
	12	206 01.8	32.2	**N 13 00.8**	7.2
	18	293 14.6	32.4	**N 12 17.7**	7.5
11 Wed	**0**	20 28.9	32.6	**N 11 32.8**	7.8
	6	107 44.6	32.9	**N 10 46.1**	8.1
	12	195 01.7	33.0	**N 9 57.8**	8.3
	18	282 20.1	33.3	**N 9 08.1**	8.5
12 Th	**0**	9 39.8	33.5	**N 8 17.2**	8.7
	6	97 00.5	33.6	**N 7 25.0**	8.8
	12	184 22.3	33.8	**N 6 31.8**	9.0
	18	271 45.0	34.0	**N 5 37.8**	9.1
13 Fri	**0**	359 08.6	34.0	**N 4 43.0**	9.2
	6	86 32.9	34.1	**N 3 47.5**	9.3
	12	173 57.8	34.2	**N 2 51.6**	9.3
	18	261 23.2	34.3	**N 1 55.4**	9.4
14 Sat	**0**	348 48.9	34.4	**N 0 58.9**	9.5
	6	76 15.0	34.3	**N 0 02.4**	9.4
	12	163 41.2	34.4	**S 0 54.1**	9.4
	18	251 07.4	34.3	**S 1 50.5**	9.4
15 Sun	**0**	338 33.5	34.3	**S 2 46.5**	9.3
	6	65 59.5	34.2	**S 3 42.2**	9.2
	12	153 25.1	34.2	**S 4 37.3**	9.0
	18	240 50.3	34.1	**S 5 31.7**	8.9
16 Mon	**0**	328 14.9	34.0	**S 6 25.3**	8.8
	6	55 38.9	33.9	**S 7 18.0**	8.6
	12	143 02.1	33.8	**S 8 09.7**	8.4
	18	230 24.4	33.5	**S 9 00.3**	8.2
17 Tu	**0**	317 45.7	33.3	**S 9 49.5**	8.0
	6	45 05.8	33.1	**S 10 37.4**	7.7
	12	132 24.8	33.0	**S 11 23.7**	7.4
	18	219 42.5	32.7	**S 12 08.3**	7.1
18 Wed	**0**	306 58.8	32.4	**S 12 51.2**	6.8
	6	34 13.6	32.1	**S 13 32.2**	6.4
	12	121 26.9	31.9	**S 14 11.1**	6.1
	18	208 38.5	31.7	**S 14 47.8**	5.8
19 Th	**0**	295 48.5	31.4	**S 15 22.3**	5.3
	6	22 56.7	31.0	**S 15 54.3**	4.8
	12	110 03.2	30.7	**S 16 23.8**	4.4
	18	197 07.9	30.4	**S 16 50.5**	3.9
20 Fri	**0**	284 10.7	30.2	**S 17 14.5**	3.5
	6	11 11.7	29.9	**S 17 35.5**	3.0
	12	98 10.9	29.5	**S 17 53.5**	2.4
	18	185 08.3	29.2	**S 18 08.3**	1.8
21 Sat	**0**	272 03.9	29.0	**S 18 19.8**	1.3
	6	358 57.7	28.7	**S 18 27.9**	0.7
	12	85 49.9	28.4	**S 18 32.4**	0.1
	18	172 40.5	28.2	**S 18 33.4**	0.5
22 Sun	**0**	259 29.6	28.0	**S 18 30.6**	1.2
	6	346 17.2	27.7	**S 18 24.1**	1.8
	12	73 03.5	27.5	**S 18 13.8**	2.4
	18	159 48.5	27.3	**S 17 59.6**	3.1
23 Mon	**0**	246 32.4	27.1	**S 17 41.5**	3.7
	6	333 15.3	27.0	**S 17 19.4**	4.4
	12	59 57.3	26.9	**S 16 53.4**	5.0
	18	146 38.6	26.8	**S 16 23.5**	5.7
24 Tu	**0**	233 19.1	26.6	**S 15 49.7**	6.3
	6	319 59.1	26.6	**S 15 12.0**	6.9
	12	46 38.6	26.5	**S 14 30.7**	7.5
	18	133 17.7	26.5	**S 13 45.7**	8.1
25 Wed	**0**	219 56.6	26.5	**S 12 57.2**	8.7
	6	306 35.2	26.4	**S 12 05.3**	9.3
	12	33 13.7	26.3	**S 11 10.2**	9.7
	18	119 52.0	26.3	**S 10 12.2**	10.1
26 Th	**0**	206 30.3	26.4	**S 9 11.4**	10.6
	6	293 08.5	26.3	**S 8 08.0**	11.0
	12	19 46.6	26.4	**S 7 02.4**	11.3
	18	106 24.8	26.3	**S 5 54.8**	11.6
27 Fri	**0**	193 02.8	26.4	**S 4 45.5**	11.9
	6	279 40.8	26.3	**S 3 34.7**	12.0
	12	6 18.7	26.3	**S 2 22.9**	12.1
	18	92 56.5	26.3	**S 1 10.4**	12.1
28 Sat	**0**	179 34.1	26.2	**N 0 02.6**	12.1
	6	266 11.6	26.2	**N 1 15.6**	12.1
	12	352 48.8	26.1	**N 2 28.3**	12.0
	18	79 25.7	26.1	**N 3 40.4**	11.8
29 Sun	**0**	166 02.4	26.0	**N 4 51.5**	11.6
	6	252 38.8	26.0	**N 6 01.3**	11.3
	12	339 14.9	26.0	**N 7 09.4**	11.0
	18	65 50.7	25.9	**N 8 15.7**	10.7
30 Mon	**0**	152 26.2	25.9	**N 9 19.6**	10.2
	6	239 01.4	25.9	**N 10 21.1**	9.8
	12	325 36.4	25.8	**N 11 19.7**	9.2
	18	52 11.3	25.8	**N 12 15.3**	8.7
31 Tu	**0**	138 46.0	25.8	**N 13 07.7**	8.1
	6	225 20.7	25.8	**N 13 56.6**	7.4
	12	311 55.5	25.8	**N 14 41.8**	6.8
	18	38 30.5	25.9	**N 15 23.3**	6.3

NOTE: Moon GHA corrections for additional hours, minutes, and seconds are on page E 108–10. Moon declination corrections are on page E 111. The mean var/hr value is needed for both corrections.

APRIL 1998 SUN & MOON GMT

SUN

Day of Yr	Mth	Week	Equation of Time 0h	Equation of Time 12h	Transit	Semi-diam	Lat 30°N Twi-light	Lat 30°N Sun-rise	Lat 30°N Sun-set	Lat 30°N Twi-light
			m s	m s	h m	′	h m	h m	h m	h m
91	**1**	**Wed**	+04 05	+03 56	**12 04**	16.0	05 26	05 50	18 19	18 43
92	**2**	**Th**	+03 47	+03 38	**12 04**	16.0	05 25	05 49	18 19	18 43
93	**3**	**Fri**	+03 30	+03 21	**12 03**	16.0	05 23	05 47	18 20	18 44
94	**4**	**Sat**	+03 12	+03 03	**12 03**	16.0	05 22	05 46	18 20	18 45
95	**5**	**Sun**	+02 54	+02 46	**12 03**	16.0	05 21	05 45	18 21	18 45
96	**6**	**Mon**	+02 37	+02 29	**12 02**	16.0	05 20	05 44	18 22	18 46
97	**7**	**Tu**	+02 20	+02 12	**12 02**	16.0	05 18	05 43	18 22	18 46
98	**8**	**Wed**	+02 03	+01 55	**12 02**	16.0	05 17	05 42	18 23	18 47
99	**9**	**Th**	+01 47	+01 38	**12 02**	16.0	05 16	05 40	18 23	18 48
100	**10**	**Fri**	+01 30	+01 22	**12 01**	16.0	05 15	05 39	18 24	18 48
101	**11**	**Sat**	+01 14	+01 06	**12 01**	16.0	05 14	05 38	18 25	18 49
102	**12**	**Sun**	+00 58	+00 50	**12 01**	16.0	05 13	05 37	18 25	18 50
103	**13**	**Mon**	+00 42	+00 35	**12 01**	16.0	05 11	05 36	18 26	18 50
104	**14**	**Tu**	+00 27	+00 19	**12 00**	16.0	05 10	05 35	18 26	18 51
105	**15**	**Wed**	+00 12	+00 05	**12 00**	16.0	05 09	05 34	18 27	18 52
106	**16**	**Th**	-00 03	-00 10	**12 00**	16.0	05 08	05 32	18 28	18 52
107	**17**	**Fri**	-00 17	-00 24	**12 00**	16.0	05 07	05 31	18 28	18 53
108	**18**	**Sat**	-00 31	-00 37	**11 59**	16.0	05 06	05 30	18 29	18 54
109	**19**	**Sun**	-00 44	-00 51	**11 59**	15.9	05 05	05 29	18 30	18 54
110	**20**	**Mon**	-00 57	-01 03	**11 59**	15.9	05 03	05 28	18 30	18 55
111	**21**	**Tu**	-01 10	-01 16	**11 59**	15.9	05 02	05 27	18 31	18 56
112	**22**	**Wed**	-01 22	-01 28	**11 59**	15.9	05 01	05 26	18 31	18 56
113	**23**	**Th**	-01 33	-01 39	**11 58**	15.9	05 00	05 25	18 32	18 57
114	**24**	**Fri**	-01 45	-01 50	**11 58**	15.9	04 59	05 24	18 33	18 58
115	**25**	**Sat**	-01 55	-02 01	**11 58**	15.9	04 58	05 23	18 33	18 58
116	**26**	**Sun**	-02 06	-02 11	**11 58**	15.9	04 57	05 22	18 34	18 59
117	**27**	**Mon**	-02 15	-02 20	**11 58**	15.9	04 56	05 21	18 35	19 00
118	**28**	**Tu**	-02 25	-02 29	**11 58**	15.9	04 55	05 20	18 35	19 00
119	**29**	**Wed**	-02 33	-02 37	**11 57**	15.9	04 54	05 19	18 36	19 01
120	**30**	**Th**	-02 42	-02 45	**11 57**	15.9	04 53	05 18	18 37	19 02

Lat Corr to Sunrise, Sunset etc

Lat	Twi-light	Sun-rise	Sun-set	Twi-light
°	h m	h m	h m	h m
N60	–1 11	–0 49	+0 49	+1 11
55	–0 50	–0 36	+0 36	+0 50
50	–0 35	–0 26	+0 26	+0 35
45	–0 24	–0 18	+0 18	+0 24
N40	–0 14	–0 11	+0 11	+0 14
35	–0 07	–0 05	+0 05	+0 07
30	0 00	0 00	0 00	0 00
25	+0 06	+0 05	–0 05	–0 06
N20	+0 11	+0 09	–0 09	–0 11
15	+0 15	+0 13	–0 13	–0 15
10	+0 19	+0 16	–0 16	–0 19
N 5	+0 23	+0 20	–0 20	–0 23
0	+0 27	+0 23	–0 23	–0 27
S 5	+0 30	+0 27	–0 27	–0 30
10	+0 33	+0 30	–0 30	–0 33
15	+0 36	+0 33	–0 33	–0 36
S20	+0 39	+0 37	–0 37	–0 39
25	+0 42	+0 41	–0 41	–0 42
30	+0 45	+0 45	–0 45	–0 45
35	+0 49	+0 50	–0 50	–0 49
S40	+0 52	+0 55	–0 55	–0 52
45	+0 56	+1 01	–1 01	–0 56
50	+1 00	+1 08	–1 08	–1 00

NOTES

Latitude corrections are for mid-month.

Equation of time = mean time (LHA mean sun) minus apparent time (LHA true sun).

MOON

Day of Yr	Mth	Week	Age	Transit (Upper)	Diff	Semi-diam	Hor Par	Lat 30°N Moon-rise	Lat 30°N Moon-set
			days	h m	m	′	′	h m	h m
91	**1**	**Wed**	04	**16 16**	**56**	16.1	59.0	09 21	23 14
92	**2**	**Th**	05	**17 12**	**54**	15.8	58.0	10 13	- -
93	**3**	**Fri**	06	**18 06**	**52**	15.6	57.1	11 07	00 11
94	**4**	**Sat**	07	**18 58**	**49**	15.3	56.3	12 01	01 04
95	**5**	**Sun**	08	**19 47**	**47**	15.2	55.6	12 56	01 51
96	**6**	**Mon**	09	**20 34**	**45**	15.0	55.0	13 50	02 34
97	**7**	**Tu**	10	**21 19**	**43**	14.9	54.6	14 43	03 14
98	**8**	**Wed**	11	**22 02**	**43**	14.8	54.3	15 35	03 50
99	**9**	**Th**	12	**22 45**	**42**	14.7	54.1	16 27	04 24
100	**10**	**Fri**	13	**23 27**	**42**	14.7	54.0	17 19	04 58
101	**11**	**Sat**	14	**24 09**	**-**	14.7	54.0	18 10	05 31
102	**12**	**Sun**	15	**00 09**	**43**	14.7	54.1	19 02	06 04
103	**13**	**Mon**	16	**00 52**	**45**	14.8	54.2	19 54	06 39
104	**14**	**Tu**	17	**01 37**	**46**	14.8	54.5	20 47	07 16
105	**15**	**Wed**	18	**02 23**	**47**	14.9	54.8	21 40	07 55
106	**16**	**Th**	19	**03 10**	**50**	15.1	55.3	22 33	08 38
107	**17**	**Fri**	20	**04 00**	**51**	15.2	55.8	23 25	09 25
108	**18**	**Sat**	21	**04 51**	**52**	15.4	56.5	- -	10 17
109	**19**	**Sun**	22	**05 43**	**53**	15.6	57.3	00 16	11 12
110	**20**	**Mon**	23	**06 36**	**54**	15.8	58.1	01 05	12 10
111	**21**	**Tu**	24	**07 30**	**53**	16.1	59.0	01 51	13 12
112	**22**	**Wed**	25	**08 23**	**54**	16.3	59.8	02 36	14 16
113	**23**	**Th**	26	**09 17**	**55**	16.5	60.5	03 20	15 21
114	**24**	**Fri**	27	**10 12**	**56**	16.6	61.0	04 03	16 28
115	**25**	**Sat**	28	**11 08**	**56**	16.7	61.2	04 46	17 36
116	**26**	**Sun**	00	**12 04**	**58**	16.7	61.2	05 31	18 44
117	**27**	**Mon**	01	**13 02**	**59**	16.6	60.8	06 18	19 52
118	**28**	**Tu**	02	**14 01**	**58**	16.4	60.1	07 08	20 57
119	**29**	**Wed**	03	**14 59**	**56**	16.1	59.2	08 00	21 59
120	**30**	**Th**	04	**15 55**	**55**	15.9	58.3	08 55	22 55

Phases of the Moon

	d	h	m
☾ First Quarter	3	20	18
○ Full Moon	11	22	23
☽ Last Quarter	19	19	53
● New Moon	26	11	41
Apogee	11	02	
Perigee	25	18	
Farthest N of 0°	3	05	
On 0°	10	12	
Farthest S of 0°	17	23	
On 0°	24	11	
Farthest N of 0°	30	14	

NOTES

Diff equals daily change in transit time.

To correct moonrise, moonset, or transit times for latitude and/or longitude see page E 112. For further information about these tables see page E 4. For arc-to-time correction see page E 103.

APRIL 1998 — STARS — 0h GMT April 1

No	Name	Mag	Transit h m	Dec ° ′	GHA ° ′	RA h m	SHA ° ′
♈	**ARIES**		**11 22**		**189 09.1**		
1	**Alpheratz**	2.2	**11 30**	N 29 04.6	**187 05.0**	0 08	357 55.9
2	**Ankaa**	2.4	**11 48**	S 42 19.0	**182 36.7**	0 26	353 27.6
3	**Schedar**	2.5	**12 02**	N 56 31.5	**179 03.4**	0 40	349 54.3
4	**Diphda**	2.2	**12 05**	S 17 59.9	**178 17.0**	0 43	349 07.9
5	**Achernar**	0.6	**12 59**	S 57 14.9	**164 45.2**	1 38	335 36.1
6	**POLARIS**	2.1	**13 50**	N 89 15.3	**151 57.8**	2 29	322 48.7
7	**Hamal**	2.2	**13 28**	N 23 27.0	**157 23.4**	2 07	328 14.3
8	**Acamar**	3.1	**14 19**	S 40 19.0	**144 36.7**	2 58	315 27.6
9	**Menkar**	2.8	**14 23**	N 4 04.7	**143 36.7**	3 02	314 27.6
10	**Mirfak**	1.9	**14 45**	N 49 51.2	**138 06.6**	3 24	308 57.5
11	**Aldebaran**	1.1	**15 57**	N 16 30.2	**120 12.1**	4 36	291 03.0
12	**Rigel**	0.3	**16 35**	S 8 12.5	**110 32.6**	5 14	281 23.5
13	**Capella**	0.2	**16 37**	N 45 59.7	**110 01.1**	5 17	280 52.0
14	**Bellatrix**	1.7	**16 46**	N 6 20.7	**107 53.8**	5 25	278 44.7
15	**Elnath**	1.8	**16 47**	N 28 36.2	**107 36.7**	5 26	278 27.6
16	**Alnilam**	1.8	**16 57**	S 1 12.4	**105 07.5**	5 36	275 58.4
17	**Betelgeuse**	0.1-1.2	**17 16**	N 7 24.2	**100 23.2**	5 55	271 14.1
18	**Canopus**	-0.9	**17 44**	S 52 42.1	**93 10.6**	6 24	264 01.5
19	**Sirius**	-1.6	**18 05**	S 16 43.1	**87 53.2**	6 45	258 44.1
20	**Adhara**	1.6	**18 19**	S 28 58.6	**84 30.9**	6 59	255 21.8
21	**Castor**	1.6	**18 55**	N 31 53.5	**75 31.9**	7 34	246 22.8
22	**Procyon**	0.5	**18 59**	N 5 13.6	**74 21.1**	7 39	245 12.0
23	**Pollux**	1.2	**19 05**	N 28 01.8	**72 51.1**	7 45	243 42.0
24	**Avior**	1.7	**19 43**	S 59 30.7	**63 31.7**	8 22	234 22.6
25	**Suhail**	2.2	**20 28**	S 43 25.9	**52 09.9**	9 08	223 00.8
26	**Miaplacidus**	1.8	**20 33**	S 69 43.0	**50 50.8**	9 13	221 41.7
27	**Alphard**	2.2	**20 47**	S 8 39.3	**47 16.5**	9 28	218 07.4
28	**Regulus**	1.3	**21 28**	N 11 58.4	**37 04.8**	10 08	207 55.7
29	**Dubhe**	2.0	**22 23**	N 61 45.7	**23 14.6**	11 04	194 05.5
30	**Denebola**	2.2	**23 09**	N 14 34.9	**11 54.4**	11 49	182 45.3
31	**Gienah**	2.8	**23 35**	S 17 32.0	**5 13.1**	12 16	176 04.0
32	**Acrux**	1.1	**23 46**	S 63 05.5	**2 30.7**	12 27	173 21.6
33	**Gacrux**	1.6	**23 51**	S 57 06.3	**1 22.4**	12 31	172 13.3
34	**Mimosa**	1.5	**0 11**	S 59 40.8	**357 14.1**	12 48	168 05.0
35	**Alioth**	1.7	**0 17**	N 55 58.2	**355 39.5**	12 54	166 30.4
36	**Spica**	1.2	**0 48**	S 11 09.2	**347 52.3**	13 25	158 43.2
37	**Alkaid**	1.9	**1 11**	N 49 19.3	**342 16.7**	13 47	153 07.6
38	**Hadar**	0.9	**1 27**	S 60 21.8	**338 12.9**	14 04	149 03.8
39	**Menkent**	2.3	**1 30**	S 36 21.7	**337 30.0**	14 07	148 20.9
40	**Arcturus**	0.2	**1 39**	N 19 11.5	**335 15.2**	14 16	146 06.1
41	**Rigil Kent.**	0.1	**2 03**	S 60 49.5	**329 16.2**	14 40	140 07.1
42	**Zuben'ubi**	2.9	**2 14**	S 16 02.0	**326 27.1**	14 51	137 18.0
43	**Kochab**	2.2	**2 14**	N 74 09.8	**326 27.7**	14 51	137 18.6
44	**Alphecca**	2.3	**2 58**	N 26 43.2	**315 29.7**	15 35	126 20.6
45	**Antares**	1.2	**3 52**	S 26 25.6	**301 49.4**	16 29	112 40.3
46	**Atria**	1.9	**4 11**	S 69 01.2	**297 01.4**	16 49	107 52.3
47	**Sabik**	2.6	**4 33**	S 15 43.3	**291 34.8**	17 10	102 25.7
48	**Shaula**	1.7	**4 56**	S 37 06.0	**285 46.6**	17 34	96 37.5
49	**Rasalhague**	2.1	**4 57**	N 12 33.6	**285 26.2**	17 35	96 17.1
50	**Eltanin**	2.4	**5 19**	N 51 29.2	**280 00.4**	17 57	90 51.3
51	**Kaus Aust.**	2.0	**5 46**	S 34 22.9	**273 08.3**	18 24	83 59.2
52	**Vega**	0.1	**5 59**	N 38 46.8	**269 55.8**	18 37	80 46.7
53	**Nunki**	2.1	**6 18**	S 26 17.8	**265 21.8**	18 55	76 12.7
54	**Altair**	0.9	**7 13**	N 8 51.8	**251 28.7**	19 51	62 19.6
55	**Peacock**	2.1	**7 48**	S 56 44.2	**242 46.8**	20 25	53 37.7
56	**Deneb**	1.3	**8 03**	N 45 16.2	**238 48.6**	20 41	49 39.5
57	**Enif**	2.5	**9 06**	N 9 51.9	**223 07.8**	21 44	33 58.7
58	**Al Na'ir**	2.2	**9 30**	S 46 58.0	**217 07.8**	22 08	27 58.7
59	**Fomalhaut**	1.3	**10 19**	S 29 37.9	**204 46.2**	22 58	15 37.1
60	**Markab**	2.6	**10 26**	N 15 11.6	**202 59.3**	23 05	13 50.2

Star Transit Corr Table

Date	Corr h m	Date	Corr h m
1	0 00	17	–1 03
2	–0 04	18	–1 07
3	–0 08	19	–1 11
4	–0 12	20	–1 15
5	–0 16	21	–1 19
6	–0 20	22	–1 23
7	–0 24	23	–1 27
8	–0 28	24	–1 30
9	–0 31	25	–1 34
10	–0 35	26	–1 38
11	–0 39	27	–1 42
12	–0 43	28	–1 46
13	–0 47	29	–1 50
14	–0 51	30	–1 54
15	–0 55	31	–1 58
16	–0 59		

STAR TRANSIT

To find the approximate time of transit of a star for any day of the month use the table above. All corrections are subtractive.

If the value taken from the table is greater than the time of transit for the first of the month, add 23h 56min to the time of transit before subtracting the correction.

Example: What time will Spica (No 36) transit the meridian on April 30?

	h min
Transit on Apr 1	00 48
	+23 56
	24 44
Corr for Apr 30	–01 54
Transit on Apr 30	22 50

APRIL DIARY

d h	
1 08	Aldebaran 0.2° S of Moon
6 17	Mercury in conjunction
8 23	Ceres in conjunction
13 12	Saturn in conjunction
19 02	Mercury stationary
20 02	Neptune 3° S of Moon
20 20	Uranus 3° S of Moon
23 02	Venus 0.3° N of Jupiter
23 07	Jupiter 0.2° S of Moon
23 08	Venus 0.1° N of Moon
24 19	Mercury 0.9° N of Moon
28 18	Aldebaran 0.4° S of Moon

NOTES

Star declinations may be used as is for the whole month. To correct GHA for day, hour, minute, and second see page E 106; for further explanation of this table see page E 4.

APRIL 1998 SUN & ARIES GMT

Time	SUN GHA ° ′	Dec ° ′	ARIES GHA ° ′	Time	SUN GHA ° ′	Dec ° ′	ARIES GHA ° ′	Time	SUN GHA ° ′	Dec ° ′	ARIES GHA ° ′	Time
	Wednesday, 1st April				**Monday, 6th April**				**Saturday, 11th April**			
00	178 58.8	**N 4 22.7**	189 09.1	**00**	179 20.8	**N 6 17.5**	194 04.8	**00**	179 41.6	**N 8 09.6**	199 00.5	**00**
02	208 59.2	**4 24.6**	219 14.0	**02**	209 21.1	**6 19.4**	224 09.7	**02**	209 42.0	**8 11.5**	229 05.4	**02**
04	238 59.6	**4 26.5**	249 19.0	**04**	239 21.5	**6 21.3**	254 14.7	**04**	239 42.3	**8 13.3**	259 10.3	**04**
06	269 00.0	**4 28.5**	279 23.9	**06**	269 21.9	**6 23.2**	284 19.6	**06**	269 42.6	**8 15.1**	289 15.3	**06**
08	299 00.3	**4 30.4**	309 28.8	**08**	299 22.2	**6 25.1**	314 24.5	**08**	299 43.0	**8 17.0**	319 20.2	**08**
10	329 00.7	**4 32.3**	339 33.7	**10**	329 22.6	**6 27.0**	344 29.4	**10**	329 43.3	**8 18.8**	349 25.1	**10**
12	359 01.1	**4 34.3**	9 38.7	**12**	359 22.9	**6 28.9**	14 34.4	**12**	359 43.6	**8 20.7**	19 30.1	**12**
14	29 01.4	**4 36.2**	39 43.6	**14**	29 23.3	**6 30.8**	44 39.3	**14**	29 43.9	**8 22.5**	49 35.0	**14**
16	59 01.8	**4 38.1**	69 48.5	**16**	59 23.6	**6 32.6**	74 44.2	**16**	59 44.3	**8 24.3**	79 39.9	**16**
18	89 02.2	**4 40.1**	99 53.5	**18**	89 24.0	**6 34.5**	104 49.2	**18**	89 44.6	**8 26.2**	109 44.8	**18**
20	119 02.5	**4 42.0**	129 58.4	**20**	119 24.4	**6 36.4**	134 54.1	**20**	119 44.9	**8 28.0**	139 49.8	**20**
22	149 02.9	**N 4 43.9**	160 03.3	**22**	149 24.7	**N 6 38.3**	164 59.0	**22**	149 45.3	**N 8 29.8**	169 54.7	**22**
	Thursday, 2nd April				**Tuesday, 7th April**				**Sunday, 12th April**			
00	179 03.3	**N 4 45.8**	190 08.2	**00**	179 25.1	**N 6 40.2**	195 03.9	**00**	179 45.6	**N 8 31.6**	199 59.6	**00**
02	209 03.7	**4 47.8**	220 13.2	**02**	209 25.4	**6 42.1**	225 08.9	**02**	209 45.9	**8 33.5**	230 04.6	**02**
04	239 04.0	**4 49.7**	250 18.1	**04**	239 25.8	**6 44.0**	255 13.8	**04**	239 46.3	**8 35.3**	260 09.5	**04**
06	269 04.4	**4 51.6**	280 23.0	**06**	269 26.1	**6 45.8**	285 18.7	**06**	269 46.6	**8 37.1**	290 14.4	**06**
08	299 04.8	**4 53.5**	310 28.0	**08**	299 26.5	**6 47.7**	315 23.7	**08**	299 46.9	**8 39.0**	320 19.3	**08**
10	329 05.1	**4 55.5**	340 32.9	**10**	329 26.8	**6 49.6**	345 28.6	**10**	329 47.2	**8 40.8**	350 24.3	**10**
12	359 05.5	**4 57.4**	10 37.8	**12**	359 27.2	**6 51.5**	15 33.5	**12**	359 47.6	**8 42.6**	20 29.2	**12**
14	29 05.9	**4 59.3**	40 42.7	**14**	29 27.5	**6 53.4**	45 38.4	**14**	29 47.9	**8 44.4**	50 34.1	**14**
16	59 06.2	**5 01.2**	70 47.7	**16**	59 27.9	**6 55.2**	75 43.4	**16**	59 48.2	**8 46.3**	80 39.0	**16**
18	89 06.6	**5 03.1**	100 52.6	**18**	89 28.2	**6 57.1**	105 48.3	**18**	89 48.5	**8 48.1**	110 44.0	**18**
20	119 07.0	**5 05.1**	130 57.5	**20**	119 28.6	**6 59.0**	135 53.2	**20**	119 48.9	**8 49.9**	140 48.9	**20**
22	149 07.3	**N 5 07.0**	161 02.5	**22**	149 28.9	**N 7 00.9**	165 58.1	**22**	149 49.2	**N 8 51.7**	170 53.8	**22**
	Friday, 3rd April				**Wednesday, 8th April**				**Monday, 13th April**			
00	179 07.7	**N 5 08.9**	191 07.4	**00**	179 29.3	**N 7 02.7**	196 03.1	**00**	179 49.5	**N 8 53.5**	200 58.8	**00**
02	209 08.1	**5 10.8**	221 12.3	**02**	209 29.6	**7 04.6**	226 08.0	**02**	209 49.8	**8 55.4**	231 03.7	**02**
04	239 08.4	**5 12.7**	251 17.2	**04**	239 30.0	**7 06.5**	256 12.9	**04**	239 50.1	**8 57.2**	261 08.6	**04**
06	269 08.8	**5 14.7**	281 22.2	**06**	269 30.3	**7 08.3**	286 17.9	**06**	269 50.5	**8 59.0**	291 13.5	**06**
08	299 09.2	**5 16.6**	311 27.1	**08**	299 30.7	**7 10.2**	316 22.8	**08**	299 50.8	**9 00.8**	321 18.5	**08**
10	329 09.5	**5 18.5**	341 32.0	**10**	329 31.0	**7 12.1**	346 27.7	**10**	329 51.1	**9 02.6**	351 23.4	**10**
12	359 09.9	**5 20.4**	11 37.0	**12**	359 31.4	**7 14.0**	16 32.6	**12**	359 51.4	**9 04.4**	21 28.3	**12**
14	29 10.3	**5 22.3**	41 41.9	**14**	29 31.7	**7 15.8**	46 37.6	**14**	29 51.7	**9 06.2**	51 33.3	**14**
16	59 10.6	**5 24.2**	71 46.8	**16**	59 32.1	**7 17.7**	76 42.5	**16**	59 52.1	**9 08.0**	81 38.2	**16**
18	89 11.0	**5 26.1**	101 51.7	**18**	89 32.4	**7 19.6**	106 47.4	**18**	89 52.4	**9 09.9**	111 43.1	**18**
20	119 11.4	**5 28.1**	131 56.7	**20**	119 32.8	**7 21.4**	136 52.4	**20**	119 52.7	**9 11.7**	141 48.0	**20**
22	149 11.7	**N 5 30.0**	162 01.6	**22**	149 33.1	**N 7 23.3**	166 57.3	**22**	149 53.0	**N 9 13.5**	171 53.0	**22**
	Saturday, 4th April				**Thursday, 9th April**				**Tuesday, 14th April**			
00	179 12.1	**N 5 31.9**	192 06.5	**00**	179 33.5	**N 7 25.2**	197 02.2	**00**	179 53.3	**N 9 15.3**	201 57.9	**00**
02	209 12.5	**5 33.8**	222 11.5	**02**	209 33.8	**7 27.0**	227 07.1	**02**	209 53.6	**9 17.1**	232 02.8	**02**
04	239 12.8	**5 35.7**	252 16.4	**04**	239 34.2	**7 28.9**	257 12.1	**04**	239 54.0	**9 18.9**	262 07.8	**04**
06	269 13.2	**5 37.6**	282 21.3	**06**	269 34.5	**7 30.7**	287 17.0	**06**	269 54.3	**9 20.7**	292 12.7	**06**
08	299 13.6	**5 39.5**	312 26.2	**08**	299 34.8	**7 32.6**	317 21.9	**08**	299 54.6	**9 22.5**	322 17.6	**08**
10	329 13.9	**5 41.4**	342 31.2	**10**	329 35.2	**7 34.5**	347 26.9	**10**	329 54.9	**9 24.3**	352 22.5	**10**
12	359 14.3	**5 43.3**	12 36.1	**12**	359 35.5	**7 36.3**	17 31.8	**12**	359 55.2	**9 26.1**	22 27.5	**12**
14	29 14.7	**5 45.2**	42 41.0	**14**	29 35.9	**7 38.2**	47 36.7	**14**	29 55.5	**9 27.9**	52 32.4	**14**
16	59 15.0	**5 47.1**	72 45.9	**16**	59 36.2	**7 40.0**	77 41.6	**16**	59 55.8	**9 29.7**	82 37.3	**16**
18	89 15.4	**5 49.0**	102 50.9	**18**	89 36.6	**7 41.9**	107 46.6	**18**	89 56.1	**9 31.5**	112 42.3	**18**
20	119 15.7	**5 50.9**	132 55.8	**20**	119 36.9	**7 43.7**	137 51.5	**20**	119 56.4	**9 33.3**	142 47.2	**20**
22	149 16.1	**N 5 52.9**	163 00.7	**22**	149 37.2	**N 7 45.6**	167 56.4	**22**	149 56.8	**N 9 35.1**	172 52.1	**22**
	Sunday, 5th April				**Friday, 10th April**				**Wednesday, 15th April**			
00	179 16.5	**N 5 54.8**	193 05.7	**00**	179 37.6	**N 7 47.5**	198 01.3	**00**	179 57.1	**N 9 36.9**	202 57.0	**00**
02	209 16.8	**5 56.7**	223 10.6	**02**	209 37.9	**7 49.3**	228 06.3	**02**	209 57.4	**9 38.7**	233 02.0	**02**
04	239 17.2	**5 58.6**	253 15.5	**04**	239 38.3	**7 51.2**	258 11.2	**04**	239 57.7	**9 40.4**	263 06.9	**04**
06	269 17.6	**6 00.5**	283 20.4	**06**	269 38.6	**7 53.0**	288 16.1	**06**	269 58.0	**9 42.2**	293 11.8	**06**
08	299 17.9	**6 02.4**	313 25.4	**08**	299 38.9	**7 54.9**	318 21.1	**08**	299 58.3	**9 44.0**	323 16.7	**08**
10	329 18.3	**6 04.3**	343 30.3	**10**	329 39.3	**7 56.7**	348 26.0	**10**	329 58.6	**9 45.8**	353 21.7	**10**
12	359 18.6	**6 06.2**	13 35.2	**12**	359 39.6	**7 58.6**	18 30.9	**12**	359 58.9	**9 47.6**	23 26.6	**12**
14	29 19.0	**6 08.0**	43 40.2	**14**	29 39.9	**8 00.4**	48 35.8	**14**	29 59.2	**9 49.4**	53 31.5	**14**
16	59 19.4	**6 09.9**	73 45.1	**16**	59 40.3	**8 02.2**	78 40.8	**16**	59 59.5	**9 51.2**	83 36.5	**16**
18	89 19.7	**6 11.8**	103 50.0	**18**	89 40.6	**8 04.1**	108 45.7	**18**	89 59.8	**9 52.9**	113 41.4	**18**
20	119 20.1	**6 13.7**	133 54.9	**20**	119 41.0	**8 05.9**	138 50.6	**20**	120 00.1	**9 54.7**	143 46.3	**20**
22	149 20.4	**N 6 15.6**	163 59.9	**22**	149 41.3	**N 8 07.8**	168 55.6	**22**	150 00.4	**N 9 56.5**	173 51.2	**22**

NOTES

Sun and Aries GHA corrections for additional hour, minutes, and seconds are on page E 106.
Correct sun declination by interpolation. For an example see page E 4.

APRIL 1998 SUN & ARIES GMT

Time	SUN GHA ° ′	Dec ° ′	ARIES GHA ° ′	Time	SUN GHA ° ′	Dec ° ′	ARIES GHA ° ′	Time	SUN GHA ° ′	Dec ° ′	ARIES GHA ° ′	Time
	Thursday, 16th April				**Tuesday, 21st April**				**Sunday, 26th April**			
00	180 00.7	**N 9 58.3**	203 56.2	**00**	180 17.5	**N11 42.8**	208 51.9	**00**	180 31.4	**N13 22.6**	213 47.6	**00**
02	210 01.0	**10 00.1**	234 01.1	**02**	210 17.7	**11 44.6**	238 56.8	**02**	210 31.7	**13 24.2**	243 52.5	**02**
04	240 01.3	**10 01.8**	264 06.0	**04**	240 18.0	**11 46.3**	269 01.7	**04**	240 31.9	**13 25.9**	273 57.4	**04**
06	270 01.6	**10 03.6**	294 11.0	**06**	270 18.2	**11 48.0**	299 06.7	**06**	270 32.1	**13 27.5**	304 02.3	**06**
08	300 01.9	**10 05.4**	324 15.9	**08**	300 18.5	**11 49.7**	329 11.6	**08**	300 32.3	**13 29.1**	334 07.3	**08**
10	330 02.2	**10 07.2**	354 20.8	**10**	330 18.8	**11 51.4**	359 16.5	**10**	330 32.5	**13 30.7**	4 12.2	**10**
12	0 02.5	**10 08.9**	24 25.7	**12**	0 19.0	**11 53.1**	29 21.4	**12**	0 32.7	**13 32.3**	34 17.1	**12**
14	30 02.8	**10 10.7**	54 30.7	**14**	30 19.3	**11 54.7**	59 26.4	**14**	30 32.9	**13 33.9**	64 22.1	**14**
16	60 03.1	**10 12.5**	84 35.6	**16**	60 19.5	**11 56.4**	89 31.3	**16**	60 33.1	**13 35.5**	94 27.0	**16**
18	90 03.4	**10 14.2**	114 40.5	**18**	90 19.8	**11 58.1**	119 36.2	**18**	90 33.3	**13 37.1**	124 31.9	**18**
20	120 03.7	**10 16.0**	144 45.5	**20**	120 20.0	**11 59.8**	149 41.2	**20**	120 33.5	**13 38.7**	154 36.8	**20**
22	150 04.0	**N10 17.8**	174 50.4	**22**	150 20.3	**N12 01.5**	179 46.1	**22**	150 33.7	**N13 40.3**	184 41.8	**22**
	Friday, 17th April				**Wednesday, 22nd April**				**Monday, 27th April**			
00	180 04.3	**N10 19.5**	204 55.3	**00**	180 20.5	**N12 03.2**	209 51.0	**00**	180 33.9	**N13 41.9**	214 46.7	**00**
02	210 04.6	**10 21.3**	235 00.2	**02**	210 20.8	**12 04.9**	239 55.9	**02**	210 34.1	**13 43.5**	244 51.6	**02**
04	240 04.9	**10 23.1**	265 05.2	**04**	240 21.0	**12 06.6**	270 00.9	**04**	240 34.3	**13 45.1**	274 56.6	**04**
06	270 05.1	**10 24.8**	295 10.1	**06**	270 21.2	**12 08.3**	300 05.8	**06**	270 34.5	**13 46.7**	305 01.5	**06**
08	300 05.4	**10 26.6**	325 15.0	**08**	300 21.5	**12 10.0**	330 10.7	**08**	300 34.7	**13 48.3**	335 06.4	**08**
10	330 05.7	**10 28.4**	355 20.0	**10**	330 21.7	**12 11.6**	0 15.6	**10**	330 34.9	**13 49.9**	5 11.3	**10**
12	0 06.0	**10 30.1**	25 24.9	**12**	0 22.0	**12 13.3**	30 20.6	**12**	0 35.0	**13 51.5**	35 16.3	**12**
14	30 06.3	**10 31.9**	55 29.8	**14**	30 22.2	**12 15.0**	60 25.5	**14**	30 35.2	**13 53.1**	65 21.2	**14**
16	60 06.6	**10 33.6**	85 34.7	**16**	60 22.5	**12 16.7**	90 30.4	**16**	60 35.4	**13 54.7**	95 26.1	**16**
18	90 06.9	**10 35.4**	115 39.7	**18**	90 22.7	**12 18.3**	120 35.4	**18**	90 35.6	**13 56.3**	125 31.1	**18**
20	120 07.2	**10 37.1**	145 44.6	**20**	120 22.9	**12 20.0**	150 40.3	**20**	120 35.8	**13 57.9**	155 36.0	**20**
22	150 07.4	**N10 38.9**	175 49.5	**22**	150 23.2	**N12 21.7**	180 45.2	**22**	150 36.0	**N13 59.5**	185 40.9	**22**
	Saturday, 18th April				**Thursday, 23rd April**				**Tuesday, 28th April**			
00	180 07.7	**N10 40.6**	205 54.5	**00**	180 23.4	**N12 23.4**	210 50.1	**00**	180 36.2	**N14 01.0**	215 45.8	**00**
02	210 08.0	**10 42.4**	235 59.4	**02**	210 23.7	**12 25.0**	240 55.1	**02**	210 36.4	**14 02.6**	245 50.8	**02**
04	240 08.3	**10 44.1**	266 04.3	**04**	240 23.9	**12 26.7**	271 00.0	**04**	240 36.6	**14 04.2**	275 55.7	**04**
06	270 08.6	**10 45.9**	296 09.2	**06**	270 24.1	**12 28.4**	301 04.9	**06**	270 36.7	**14 05.8**	306 00.6	**06**
08	300 08.9	**10 47.6**	326 14.2	**08**	300 24.4	**12 30.0**	331 09.9	**08**	300 36.9	**14 07.3**	336 05.5	**08**
10	330 09.1	**10 49.4**	356 19.1	**10**	330 24.6	**12 31.7**	1 14.8	**10**	330 37.1	**14 08.9**	6 10.5	**10**
12	0 09.4	**10 51.1**	26 24.0	**12**	0 24.8	**12 33.4**	31 19.7	**12**	0 37.3	**14 10.5**	36 15.4	**12**
14	30 09.7	**10 52.9**	56 29.0	**14**	30 25.1	**12 35.0**	61 24.6	**14**	30 37.5	**14 12.1**	66 20.3	**14**
16	60 10.0	**10 54.6**	86 33.9	**16**	60 25.3	**12 36.7**	91 29.6	**16**	60 37.7	**14 13.6**	96 25.3	**16**
18	90 10.3	**10 56.3**	116 38.8	**18**	90 25.5	**12 38.4**	121 34.5	**18**	90 37.8	**14 15.2**	126 30.2	**18**
20	120 10.5	**10 58.1**	146 43.7	**20**	120 25.8	**12 40.0**	151 39.4	**20**	120 38.0	**14 16.8**	156 35.1	**20**
22	150 10.8	**N10 59.8**	176 48.7	**22**	150 26.0	**N12 41.7**	181 44.4	**22**	150 38.2	**N14 18.3**	186 40.0	**22**
	Sunday, 19th April				**Friday, 24th April**				**Wednesday, 29th April**			
00	180 11.1	**N11 01.6**	206 53.6	**00**	180 26.2	**N12 43.3**	211 49.3	**00**	180 38.4	**N14 19.9**	216 45.0	**00**
02	210 11.4	**11 03.3**	236 58.5	**02**	210 26.4	**12 45.0**	241 54.2	**02**	210 38.5	**14 21.5**	246 49.9	**02**
04	240 11.6	**11 05.0**	267 03.4	**04**	240 26.7	**12 46.6**	271 59.1	**04**	240 38.7	**14 23.0**	276 54.8	**04**
06	270 11.9	**11 06.8**	297 08.4	**06**	270 26.9	**12 48.3**	302 04.1	**06**	270 38.9	**14 24.6**	306 59.8	**06**
08	300 12.2	**11 08.5**	327 13.3	**08**	300 27.1	**12 49.9**	332 09.0	**08**	300 39.1	**14 26.1**	337 04.7	**08**
10	330 12.5	**11 10.2**	357 18.2	**10**	330 27.3	**12 51.6**	2 13.9	**10**	330 39.2	**14 27.7**	7 09.6	**10**
12	0 12.7	**11 12.0**	27 23.2	**12**	0 27.6	**12 53.2**	32 18.9	**12**	0 39.4	**14 29.2**	37 14.5	**12**
14	30 13.0	**11 13.7**	57 28.1	**14**	30 27.8	**12 54.9**	62 23.8	**14**	30 39.6	**14 30.8**	67 19.5	**14**
16	60 13.3	**11 15.4**	87 33.0	**16**	60 28.0	**12 56.5**	92 28.7	**16**	60 39.8	**14 32.3**	97 24.4	**16**
18	90 13.5	**11 17.1**	117 37.9	**18**	90 28.2	**12 58.2**	122 33.6	**18**	90 39.9	**14 33.9**	127 29.3	**18**
20	120 13.8	**11 18.9**	147 42.9	**20**	120 28.5	**12 59.8**	152 38.6	**20**	120 40.1	**14 35.4**	157 34.3	**20**
22	150 14.1	**N11 20.6**	177 47.8	**22**	150 28.7	**N13 01.4**	182 43.5	**22**	150 40.3	**N14 37.0**	187 39.2	**22**
	Monday, 20th April				**Saturday, 25th April**				**Thursday, 30th April**			
00	180 14.3	**N11 22.3**	207 52.7	**00**	180 28.9	**N13 03.1**	212 48.4	**00**	180 40.4	**N14 38.5**	217 44.1	**00**
02	210 14.6	**11 24.0**	237 57.7	**02**	210 29.1	**13 04.7**	242 53.3	**02**	210 40.6	**14 40.1**	247 49.0	**02**
04	240 14.9	**11 25.7**	268 02.6	**04**	240 29.3	**13 06.4**	272 58.3	**04**	240 40.8	**14 41.6**	277 54.0	**04**
06	270 15.1	**11 27.5**	298 07.5	**06**	270 29.5	**13 08.0**	303 03.2	**06**	270 40.9	**14 43.1**	307 58.9	**06**
08	300 15.4	**11 29.2**	328 12.4	**08**	300 29.8	**13 09.6**	333 08.1	**08**	300 41.1	**14 44.7**	338 03.8	**08**
10	330 15.7	**11 30.9**	358 17.4	**10**	330 30.0	**13 11.3**	3 13.1	**10**	330 41.2	**14 46.2**	8 08.8	**10**
12	0 15.9	**11 32.6**	28 22.3	**12**	0 30.2	**13 12.9**	33 18.0	**12**	0 41.4	**14 47.7**	38 13.7	**12**
14	30 16.2	**11 34.3**	58 27.2	**14**	30 30.4	**13 14.5**	63 22.9	**14**	30 41.6	**14 49.3**	68 18.6	**14**
16	60 16.4	**11 36.0**	88 32.2	**16**	60 30.6	**13 16.1**	93 27.8	**16**	60 41.7	**14 50.8**	98 23.5	**16**
18	90 16.7	**11 37.7**	118 37.1	**18**	90 30.8	**13 17.8**	123 32.8	**18**	90 41.9	**14 52.3**	128 28.5	**18**
20	120 17.0	**11 39.4**	148 42.0	**20**	120 31.0	**13 19.4**	153 37.7	**20**	120 42.0	**14 53.9**	158 33.4	**20**
22	150 17.2	**N11 41.1**	178 46.9	**22**	150 31.2	**N13 21.0**	183 42.6	**22**	150 42.2	**N14 55.4**	188 38.3	**22**

APRIL 1998 PLANETS 0h GMT

VENUS						JUPITER				
Mer Pass h m	GHA ° ′	Mean Var/hr 14°+	Dec ° ′	Mean Var/hr ′	Day	GHA ° ′	Mean Var/hr 15°+	Dec ° ′	Mean Var/hr ′	Mer Pass h m
09 10	222 37.9	59.9	S12 11.5	0.6	1 Wed	204 17.2	1.9	S 7 28.3	0.2	10 22
09 10	222 36.2	59.9	S11 57.0	0.6	2 Th	205 03.6	1.9	S 7 23.2	0.2	10 18
09 10	222 34.4	59.9	S11 42.0	0.6	3 Fri	205 50.1	1.9	S 7 18.2	0.2	10 15
09 10	222 32.5	59.9	S11 26.7	0.7	4 Sat	206 36.6	1.9	S 7 13.1	0.2	10 12
09 10	222 30.4	59.9	S11 10.9	0.7	5 SUN	207 23.2	1.9	S 7 08.0	0.2	10 09
09 10	222 28.1	59.9	S10 54.7	0.7	6 Mon	208 09.9	1.9	S 7 03.0	0.2	10 06
09 10	222 25.7	59.9	S10 38.1	0.7	7 Tu	208 56.6	1.9	S 6 58.0	0.2	10 03
09 11	222 23.3	59.9	S10 21.1	0.7	8 Wed	209 43.4	2.0	S 6 53.0	0.2	10 00
09 11	222 20.7	59.9	S10 03.7	0.7	9 Th	210 30.2	1.9	S 6 48.0	0.2	09 57
09 11	222 18.0	59.9	S 9 46.0	0.8	10 Fri	211 17.0	2.0	S 6 43.0	0.2	09 54
09 11	222 15.2	59.9	S 9 27.9	0.8	11 Sat	212 04.0	2.0	S 6 38.1	0.2	09 50
09 11	222 12.4	59.9	S 9 09.4	0.8	12 SUN	212 51.0	2.0	S 6 33.2	0.2	09 47
09 11	222 09.5	59.9	S 8 50.6	0.8	13 Mon	213 38.0	2.0	S 6 28.3	0.2	09 44
09 12	222 06.4	59.9	S 8 31.4	0.8	14 Tu	214 25.1	2.0	S 6 23.4	0.2	09 41
09 12	222 03.4	59.9	S 8 11.9	0.8	15 Wed	215 12.3	2.0	S 6 18.5	0.2	09 38
09 12	222 00.2	59.9	S 7 52.0	0.8	16 Th	215 59.5	2.0	S 6 13.7	0.2	09 35
09 12	221 57.0	59.9	S 7 31.9	0.9	17 Fri	216 46.8	2.0	S 6 08.9	0.2	09 32
09 12	221 53.8	59.9	S 7 11.4	0.9	18 Sat	217 34.2	2.0	S 6 04.1	0.2	09 28
09 13	221 50.5	59.9	S 6 50.6	0.9	19 SUN	218 21.6	2.0	S 5 59.4	0.2	09 25
09 13	221 47.1	59.9	S 6 29.5	0.9	20 Mon	219 09.1	2.0	S 5 54.6	0.2	09 22
09 13	221 43.7	59.9	S 6 08.2	0.9	21 Tu	219 56.6	2.0	S 5 49.9	0.2	09 19
09 13	221 40.3	59.9	S 5 46.6	0.9	22 Wed	220 44.3	2.0	S 5 45.3	0.2	09 16
09 14	221 36.8	59.9	S 5 24.7	0.9	23 Th	221 32.0	2.0	S 5 40.6	0.2	09 13
09 14	221 33.3	59.9	S 5 02.5	0.9	24 Fri	222 19.7	2.0	S 5 36.0	0.2	09 09
09 14	221 29.7	59.8	S 4 40.2	0.9	25 Sat	223 07.6	2.0	S 5 31.4	0.2	09 06
09 14	221 26.1	59.8	S 4 17.6	1.0	26 SUN	223 55.5	2.0	S 5 26.8	0.2	09 03
09 15	221 22.5	59.8	S 3 54.7	1.0	27 Mon	224 43.5	2.0	S 5 22.3	0.2	09 00
09 15	221 18.8	59.8	S 3 31.7	1.0	28 Tu	225 31.6	2.0	S 5 17.8	0.2	08 57
09 15	221 15.1	59.8	S 3 08.4	1.0	29 Wed	226 19.8	2.0	S 5 13.3	0.2	08 53
09 15	221 11.4	59.8	S 2 45.0	1.0	30 Th	227 08.0	2.0	S 5 08.9	0.2	08 50

VENUS, Av. Mag. -4.2
S.H.A. April
5 29; 10 24; 15 19; 20 14; 25 9; 30 3.

JUPITER, Av. Mag.-2.1
S.H.A. April
5 14; 10 13; 15 12; 20 11; 25 10; 30 9.

MARS						SATURN				
Mer Pass h m	GHA ° ′	Mean Var/hr 15°+	Dec ° ′	Mean Var/hr ′	Day	GHA ° ′	Mean Var/hr 15°+	Dec ° ′	Mean Var/hr ′	Mer Pass h m
12 41	169 39.0	0.7	N 7 47.8	0.7	1 Wed	168 11.0	2.2	N 6 23.7	0.1	12 45
12 40	169 55.9	0.7	N 8 05.3	0.7	2 Th	169 03.1	2.2	N 6 26.6	0.1	12 42
12 39	170 12.7	0.7	N 8 22.8	0.7	3 Fri	169 55.2	2.2	N 6 29.4	0.1	12 38
12 37	170 29.6	0.7	N 8 40.2	0.7	4 Sat	170 47.3	2.2	N 6 32.2	0.1	12 35
12 36	170 46.4	0.7	N 8 57.5	0.7	5 SUN	171 39.4	2.2	N 6 35.1	0.1	12 32
12 35	171 03.2	0.7	N 9 14.7	0.7	6 Mon	172 31.5	2.2	N 6 37.9	0.1	12 28
12 34	171 20.0	0.7	N 9 31.8	0.7	7 Tu	173 23.6	2.2	N 6 40.7	0.1	12 25
12 33	171 36.7	0.7	N 9 48.8	0.7	8 Wed	174 15.6	2.2	N 6 43.5	0.1	12 21
12 32	171 53.5	0.7	N10 05.6	0.7	9 Th	175 07.7	2.2	N 6 46.4	0.1	12 18
12 31	172 10.2	0.7	N10 22.4	0.7	10 Fri	175 59.7	2.2	N 6 49.2	0.1	12 14
12 30	172 26.8	0.7	N10 39.0	0.7	11 Sat	176 51.8	2.2	N 6 52.0	0.1	12 11
12 29	172 43.5	0.7	N10 55.6	0.7	12 SUN	177 43.8	2.2	N 6 54.8	0.1	12 07
12 27	173 00.1	0.7	N11 12.0	0.7	13 Mon	178 35.8	2.2	N 6 57.6	0.1	12 04
12 26	173 16.6	0.7	N11 28.3	0.7	14 Tu	179 27.8	2.2	N 7 00.4	0.1	12 00
12 25	173 33.2	0.7	N11 44.5	0.7	15 Wed	180 19.9	2.2	N 7 03.2	0.1	11 57
12 24	173 49.7	0.7	N12 00.5	0.7	16 Th	181 11.9	2.2	N 7 06.0	0.1	11 53
12 23	174 06.1	0.7	N12 16.4	0.7	17 Fri	182 03.9	2.2	N 7 08.7	0.1	11 50
12 22	174 22.5	0.7	N12 32.2	0.7	18 Sat	182 55.9	2.2	N 7 11.5	0.1	11 47
12 21	174 38.9	0.7	N12 47.9	0.6	19 SUN	183 47.9	2.2	N 7 14.3	0.1	11 43
12 20	174 55.3	0.7	N13 03.4	0.6	20 Mon	184 40.0	2.2	N 7 17.0	0.1	11 40
12 19	175 11.5	0.7	N13 18.8	0.6	21 Tu	185 32.0	2.2	N 7 19.8	0.1	11 36
12 18	175 27.8	0.7	N13 34.0	0.6	22 Wed	186 24.0	2.2	N 7 22.5	0.1	11 33
12 17	175 44.0	0.7	N13 49.2	0.6	23 Th	187 16.1	2.2	N 7 25.3	0.1	11 29
12 15	176 00.1	0.7	N14 04.1	0.6	24 Fri	188 08.1	2.2	N 7 28.0	0.1	11 26
12 14	176 16.3	0.7	N14 19.0	0.6	25 Sat	189 00.2	2.2	N 7 30.7	0.1	11 22
12 13	176 32.3	0.7	N14 33.6	0.6	26 SUN	189 52.2	2.2	N 7 33.4	0.1	11 19
12 12	176 48.3	0.7	N14 48.2	0.6	27 Mon	190 44.3	2.2	N 7 36.1	0.1	11 15
12 11	177 04.3	0.7	N15 02.6	0.6	28 Tu	191 36.4	2.2	N 7 38.8	0.1	11 12
12 10	177 20.2	0.7	N15 16.8	0.6	29 Wed	192 28.5	2.2	N 7 41.5	0.1	11 08
12 09	177 36.1	0.7	N15 30.8	0.6	30 Th	193 20.6	2.2	N 7 44.1	0.1	11 05

MARS, Av. Mag. +1.3
S.H.A. April
5 338; 10 334; 15 331; 20 327; 25 323; 30 320.

NOTE: Planet corrections on pages E113-15

SATURN, Av. Mag. +0.5
S.H.A. April
5 339; 10 338; 15 337; 20 337; 25 336; 30 336.

APRIL 1998 — MOON

Day	GMT hr	GHA ° ′	Mean Var/hr 14°+	Dec ° ′	Mean Var/hr ′
1 Wed	0	125 05.9	25.9	N 16 01.0	5.5
	6	211 41.6	26.1	N 16 34.6	4.8
	12	298 18.0	26.2	N 17 04.2	4.2
	18	24 55.0	26.3	N 17 29.6	3.5
2 Th	0	111 33.0	26.6	N 17 51.0	2.8
	6	198 12.0	26.7	N 18 08.2	2.2
	12	284 52.1	26.9	N 18 21.3	1.5
	18	11 33.6	27.1	N 18 30.3	0.8
3 Fri	0	98 16.5	27.4	N 18 35.3	0.1
	6	185 01.0	27.7	N 18 36.4	0.5
	12	271 47.2	28.0	N 18 33.7	1.2
	18	358 35.1	28.3	N 18 27.2	1.7
4 Sat	0	85 24.9	28.6	N 18 17.0	2.3
	6	172 16.7	29.0	N 18 03.3	2.9
	12	259 10.4	29.4	N 17 46.3	3.4
	18	346 06.2	29.7	N 17 25.9	4.0
5 Sun	0	73 04.0	30.0	N 17 02.5	4.4
	6	160 03.9	30.3	N 16 36.1	5.0
	12	247 05.8	30.7	N 16 06.8	5.3
	18	334 09.8	31.0	N 15 34.8	5.8
6 Mon	0	61 15.7	31.4	N 15 00.3	6.2
	6	148 23.6	31.6	N 14 23.4	6.5
	12	235 33.4	31.9	N 13 44.3	6.9
	18	322 45.0	32.2	N 13 03.0	7.2
7 Tu	0	49 58.4	32.5	N 12 19.8	7.6
	6	137 13.4	32.7	N 11 34.7	7.8
	12	224 29.9	33.0	N 10 47.9	8.1
	18	311 47.9	33.2	N 9 59.6	8.3
8 Wed	0	39 07.3	33.4	N 9 09.9	8.5
	6	126 28.0	33.6	N 8 18.9	8.7
	12	213 49.7	33.8	N 7 26.7	8.9
	18	301 12.5	34.0	N 6 33.5	9.0
9 Th	0	28 36.2	34.1	N 5 39.5	9.1
	6	116 00.6	34.2	N 4 44.6	9.3
	12	203 25.7	34.3	N 3 49.1	9.3
	18	290 51.4	34.4	N 2 53.2	9.4
10 Fri	0	18 17.4	34.4	N 1 56.8	9.4
	6	105 43.8	34.4	N 1 00.2	9.5
	12	193 10.3	34.5	N 0 03.4	9.5
	18	280 36.8	34.4	S 0 53.4	9.5
11 Sat	0	8 03.3	34.4	S 1 50.0	9.4
	6	95 29.6	34.3	S 2 46.4	9.3
	12	182 55.6	34.2	S 3 42.4	9.2
	18	270 21.1	34.2	S 4 38.0	9.1
12 Sun	0	357 46.1	34.1	S 5 32.9	9.0
	6	85 10.5	33.9	S 6 27.0	8.9
	12	172 34.1	33.8	S 7 20.2	8.7
	18	259 56.9	33.6	S 8 12.5	8.5
13 Mon	0	347 18.7	33.5	S 9 03.6	8.3
	6	74 39.5	33.2	S 9 53.3	8.1
	12	161 59.2	33.1	S 10 41.7	7.8
	18	249 17.6	32.8	S 11 28.5	7.5
14 Tu	0	336 34.8	32.6	S 12 13.7	7.2
	6	63 50.6	32.4	S 12 57.0	6.8
	12	151 05.1	32.2	S 13 38.4	6.5
	18	238 18.0	31.9	S 14 17.7	6.2
15 Wed	0	325 29.5	31.7	S 14 54.8	5.7
	6	52 39.5	31.4	S 15 29.5	5.3
	12	139 47.9	31.1	S 16 01.8	4.9
	18	226 54.8	30.8	S 16 31.5	4.5
16 Th	0	314 00.1	30.6	S 16 58.4	3.9
	6	41 03.9	30.4	S 17 22.6	3.5
	12	128 06.2	30.1	S 17 43.7	2.9
	18	215 06.9	29.8	S 18 01.8	2.4
17 Fri	0	302 06.2	29.6	S 18 16.7	1.9
	6	29 04.1	29.4	S 18 28.4	1.3
	12	116 00.7	29.2	S 18 36.7	0.8
	18	202 56.1	29.0	S 18 41.6	0.2
18 Sat	0	289 50.2	28.8	S 18 42.9	0.4
	6	16 43.2	28.7	S 18 40.7	1.1
	12	103 35.2	28.5	S 18 34.9	1.7
	18	190 26.3	28.4	S 18 25.4	2.2
19 Sun	0	277 16.5	28.3	S 18 12.1	2.8
	6	4 06.0	28.1	S 17 55.2	3.5
	12	90 54.8	28.0	S 17 34.5	4.1
	18	177 43.1	28.0	S 17 10.1	4.8
20 Mon	0	264 30.8	27.9	S 16 42.1	5.3
	6	351 18.1	27.8	S 16 10.3	5.9
	12	78 05.1	27.7	S 15 35.0	6.6
	18	164 51.7	27.7	S 14 56.1	7.1
21 Tu	0	251 38.1	27.7	S 14 13.8	7.6
	6	338 24.3	27.6	S 13 28.1	8.2
	12	65 10.2	27.6	S 12 39.2	8.7
	18	151 56.0	27.6	S 11 47.2	9.2
22 Wed	0	238 41.5	27.5	S 10 52.2	9.7
	6	325 26.8	27.5	S 9 54.5	10.1
	12	52 11.9	27.5	S 8 54.2	10.5
	18	138 56.7	27.4	S 7 51.4	10.8
23 Th	0	225 41.2	27.3	S 6 46.5	11.1
	6	312 25.3	27.2	S 5 39.7	11.4
	12	39 08.9	27.2	S 4 31.2	11.7
	18	125 52.1	27.1	S 3 21.3	11.9
24 Fri	0	212 34.7	27.0	S 2 10.2	12.0
	6	299 16.6	26.9	S 0 58.3	12.0
	12	25 57.8	26.7	N 0 14.1	12.1
	18	112 38.3	26.6	N 1 26.8	12.0
25 Sat	0	199 17.9	26.5	N 2 39.3	12.0
	6	285 56.7	26.3	N 3 51.3	11.9
	12	12 34.5	26.1	N 5 02.6	11.7
	18	99 11.3	26.0	N 6 12.7	11.4
26 Sun	0	185 47.1	25.8	N 7 21.4	11.1
	6	272 22.0	25.6	N 8 28.3	10.8
	12	358 55.9	25.5	N 9 33.0	10.3
	18	85 28.9	25.4	N 10 35.3	9.9
27 Mon	0	172 01.0	25.2	N 11 34.8	9.3
	6	258 32.3	25.1	N 12 31.3	8.8
	12	345 02.9	25.1	N 13 24.5	8.3
	18	71 33.0	25.0	N 14 14.2	7.6
28 Tu	0	158 02.7	25.0	N 15 00.2	6.9
	6	244 32.2	24.9	N 15 42.2	6.2
	12	331 01.5	24.9	N 16 20.1	5.6
	18	57 31.1	25.0	N 16 53.8	4.8
29 Wed	0	144 01.0	25.1	N 17 23.2	4.1
	6	230 31.5	25.3	N 17 48.3	3.4
	12	317 02.8	25.4	N 18 08.9	2.7
	18	43 35.1	25.6	N 18 25.1	1.9
30 Th	0	130 08.7	25.9	N 18 36.9	1.2
	6	216 43.8	26.1	N 18 44.3	0.5
	12	303 20.4	26.4	N 18 47.5	0.2
	18	29 59.0	26.7	N 18 46.5	0.9

NOTE: Moon GHA corrections for additional hours, minutes, and seconds are on page E 108–10. Moon declination corrections are on page E 111. The mean var/hr value is needed for both corrections.

APRIL Ephemeris E

MAY 1998 — SUN & MOON — GMT

SUN

Yr	Mth	Week	Equation of Time 0h	Equation of Time 12h	Transit	Semi-diam	Lat 30°N Twi-light	Lat 30°N Sun-rise	Lat 30°N Sun-set	Lat 30°N Twi-light
			m s	m s	h m	′	h m	h m	h m	h m
121	**1**	**Fri**	-02 49	-02 53	**11 57**	15.9	04 52	05 17	18 37	19 03
122	**2**	**Sat**	-02 56	-03 00	**11 57**	15.9	04 51	05 17	18 38	19 03
123	**3**	**Sun**	-03 03	-03 06	**11 57**	15.9	04 50	05 16	18 38	19 04
124	**4**	**Mon**	-03 09	-03 12	**11 57**	15.9	04 49	05 15	18 39	19 05
125	**5**	**Tu**	-03 15	-03 18	**11 57**	15.9	04 49	05 14	18 40	19 05
126	**6**	**Wed**	-03 20	-03 22	**11 57**	15.9	04 48	05 13	18 40	19 06
127	**7**	**Th**	-03 25	-03 27	**11 57**	15.9	04 47	05 12	18 41	19 07
128	**8**	**Fri**	-03 29	-03 30	**11 56**	15.9	04 46	05 12	18 42	19 07
129	**9**	**Sat**	-03 32	-03 33	**11 56**	15.9	04 45	05 11	18 42	19 08
130	**10**	**Sun**	-03 35	-03 36	**11 56**	15.9	04 44	05 10	18 43	19 09
131	**11**	**Mon**	-03 37	-03 38	**11 56**	15.9	04 44	05 09	18 44	19 10
132	**12**	**Tu**	-03 39	-03 40	**11 56**	15.9	04 43	05 09	18 44	19 10
133	**13**	**Wed**	-03 40	-03 40	**11 56**	15.9	04 42	05 08	18 45	19 11
134	**14**	**Th**	-03 41	-03 41	**11 56**	15.8	04 41	05 07	18 46	19 12
135	**15**	**Fri**	-03 41	-03 40	**11 56**	15.8	04 41	05 07	18 46	19 12
136	**16**	**Sat**	-03 40	-03 40	**11 56**	15.8	04 40	05 06	18 47	19 13
137	**17**	**Sun**	-03 39	-03 38	**11 56**	15.8	04 39	05 06	18 47	19 14
138	**18**	**Mon**	-03 37	-03 36	**11 56**	15.8	04 39	05 05	18 48	19 14
139	**19**	**Tu**	-03 35	-03 33	**11 56**	15.8	04 38	05 04	18 49	19 15
140	**20**	**Wed**	-03 32	-03 30	**11 56**	15.8	04 38	05 04	18 49	19 16
141	**21**	**Th**	-03 28	-03 27	**11 57**	15.8	04 37	05 03	18 50	19 17
142	**22**	**Fri**	-03 24	-03 22	**11 57**	15.8	04 36	05 03	18 51	19 17
143	**23**	**Sat**	-03 20	-03 17	**11 57**	15.8	04 36	05 03	18 51	19 18
144	**24**	**Sun**	-03 15	-03 12	**11 57**	15.8	04 35	05 02	18 52	19 18
145	**25**	**Mon**	-03 09	-03 06	**11 57**	15.8	04 35	05 02	18 52	19 19
146	**26**	**Tu**	-03 03	-03 00	**11 57**	15.8	04 35	05 01	18 53	19 20
147	**27**	**Wed**	-02 57	-02 53	**11 57**	15.8	04 34	05 01	18 54	19 20
148	**28**	**Th**	-02 49	-02 46	**11 57**	15.8	04 34	05 01	18 54	19 21
149	**29**	**Fri**	-02 42	-02 38	**11 57**	15.8	04 33	05 00	18 55	19 22
150	**30**	**Sat**	-02 34	-02 30	**11 57**	15.8	04 33	05 00	18 55	19 22
151	**31**	**Sun**	-02 26	-02 21	**11 58**	15.8	04 33	05 00	18 56	19 23

Lat Corr to Sunrise, Sunset etc

Lat	Twi-light	Sun-rise	Sun-set	Twi-light
°	h m	h m	h m	h m
N60	−2 21	−1 44	+1 44	+2 21
55	−1 35	−1 14	+1 14	+1 35
50	−1 05	−0 52	+0 52	+1 05
45	−0 43	−0 36	+0 36	+0 43
N40	−0 26	−0 22	+0 22	+0 26
35	−0 12	−0 10	+0 10	+0 12
30	0 00	0 00	0 00	0 00
25	+0 10	+0 09	−0 09	−0 10
N20	+0 20	+0 17	−0 17	−0 20
15	+0 28	+0 25	−0 25	−0 28
10	+0 36	+0 32	−0 32	−0 36
N 5	+0 43	+0 39	−0 39	−0 43
0	+0 50	+0 46	−0 46	−0 50
S 5	+0 57	+0 53	−0 53	−0 57
10	+1 04	+1 00	−1 00	−1 04
15	+1 10	+1 07	−1 07	−1 10
S20	+1 17	+1 14	−1 14	−1 17
25	+1 24	+1 22	−1 22	−1 24
30	+1 31	+1 31	−1 31	−1 31
35	+1 39	+1 40	−1 40	−1 39
S40	+1 48	+1 51	−1 51	−1 48
45	+1 58	+2 04	−2 04	−1 58
50	+2 10	+2 19	−2 19	−2 10

NOTES

Latitude corrections are for mid-month.

Equation of time = mean time (LHA mean sun) minus apparent time (LHA true sun).

MOON

Yr	Mth	Week	Age	Transit (Upper)	Diff	Semi-diam	Hor Par	Lat 30°N Moon-rise	Lat 30°N Moon-set
			days	h m	m	′	′	h m	h m
121	**1**	**Fri**	05	**16 50**	**52**	15.6	57.3	09 51	23 47
122	**2**	**Sat**	06	**17 42**	**48**	15.4	56.5	10 47	- -
123	**3**	**Sun**	07	**18 30**	**46**	15.2	55.7	11 43	00 32
124	**4**	**Mon**	08	**19 16**	**45**	15.0	55.1	12 37	01 14
125	**5**	**Tu**	09	**20 01**	**42**	14.9	54.6	13 30	01 51
126	**6**	**Wed**	10	**20 43**	**42**	14.8	54.3	14 22	02 26
127	**7**	**Th**	11	**21 25**	**43**	14.7	54.1	15 14	03 00
128	**8**	**Fri**	12	**22 08**	**42**	14.7	54.0	16 05	03 32
129	**9**	**Sat**	13	**22 50**	**44**	14.7	54.1	16 57	04 05
130	**10**	**Sun**	14	**23 34**	**46**	14.8	54.3	17 49	04 40
131	**11**	**Mon**	15	**24 20**	**-**	14.9	54.5	18 42	05 16
132	**12**	**Tu**	16	**00 20**	**47**	14.9	54.8	19 36	05 54
133	**13**	**Wed**	17	**01 07**	**50**	15.0	55.2	20 30	06 37
134	**14**	**Th**	18	**01 57**	**51**	15.2	55.7	21 22	07 23
135	**15**	**Fri**	19	**02 48**	**52**	15.3	56.2	22 14	08 13
136	**16**	**Sat**	20	**03 40**	**52**	15.5	56.7	23 03	09 06
137	**17**	**Sun**	21	**04 32**	**52**	15.6	57.4	23 49	10 03
138	**18**	**Mon**	22	**05 24**	**53**	15.8	58.0	- -	11 03
139	**19**	**Tu**	23	**06 17**	**52**	16.0	58.7	00 34	12 04
140	**20**	**Wed**	24	**07 09**	**52**	16.2	59.3	01 16	13 07
141	**21**	**Th**	25	**08 01**	**53**	16.3	59.9	01 57	14 11
142	**22**	**Fri**	26	**08 54**	**55**	16.4	60.4	02 39	15 16
143	**23**	**Sat**	27	**09 49**	**56**	16.5	60.6	03 21	16 22
144	**24**	**Sun**	28	**10 45**	**57**	16.5	60.6	04 06	17 29
145	**25**	**Mon**	29	**11 42**	**59**	16.4	60.3	04 53	18 36
146	**26**	**Tu**	01	**12 41**	**59**	16.3	59.8	05 45	19 40
147	**27**	**Wed**	02	**13 40**	**57**	16.1	59.1	06 39	20 41
148	**28**	**Th**	03	**14 37**	**54**	15.9	58.3	07 36	21 36
149	**29**	**Fri**	04	**15 31**	**52**	15.6	57.4	08 34	22 26
150	**30**	**Sat**	05	**16 23**	**48**	15.4	56.5	09 31	23 10
151	**31**	**Sun**	06	**17 11**	**46**	15.2	55.8	10 27	23 50

Phases of the Moon

Phase	d	h	m
☾ First Quarter	3	10	04
○ Full Moon	11	14	29
☽ Last Quarter	19	04	35
● New Moon	25	19	32
Apogee	8	09	
Perigee	24	00	
On 0°	7	19	
Farthest S of 0°	15	06	
On 0°	21	20	
Farthest N of 0°	28	00	

NOTES

Diff equals daily change in transit time.

To correct moonrise, moonset, or transit times for latitude and/or longitude see page E 112. For further information about these tables see page E 4. For arc-to-time correction see page E 103.

MAY 1998 — STARS — 0h GMT May 1

No	Name	Mag	Transit h m	Dec ° ′	GHA ° ′	RA h m	SHA ° ′
♈	**ARIES**		**9 24**		**218 43.3**		
1	**Alpheratz**	2.2	**9 32**	N29 04.6	**216 39.0**	0 08	357 55.7
2	**Ankaa**	2.4	**9 50**	S 42 18.9	**212 10.8**	0 26	353 27.5
3	**Schedar**	2.5	**10 04**	N56 31.4	**208 37.4**	0 40	349 54.1
4	**Diphda**	2.2	**10 07**	S 17 59.8	**207 51.1**	0 43	349 07.8
5	**Achernar**	0.6	**11 01**	S 57 14.7	**194 19.3**	1 38	335 36.0
6	**POLARIS**	2.1	**11 52**	N89 15.2	**181 29.9**	2 29	322 46.6
7	**Hamal**	2.2	**11 30**	N23 27.0	**186 57.5**	2 07	328 14.2
8	**Acamar**	3.1	**12 21**	S 40 18.8	**174 10.9**	2 58	315 27.6
9	**Menkar**	2.8	**12 25**	N 4 04.8	**173 10.8**	3 02	314 27.5
10	**Mirfak**	1.9	**12 47**	N49 51.1	**167 40.8**	3 24	308 57.5
11	**Aldebaran**	1.1	**13 59**	N16 30.2	**149 46.4**	4 36	291 03.1
12	**Rigel**	0.3	**14 37**	S 8 12.5	**140 06.8**	5 14	281 23.5
13	**Capella**	0.2	**14 39**	N45 59.7	**139 35.4**	5 17	280 52.1
14	**Bellatrix**	1.7	**14 48**	N 6 20.7	**137 28.1**	5 25	278 44.8
15	**Elnath**	1.8	**14 49**	N28 36.2	**137 11.0**	5 26	278 27.7
16	**Alnilam**	1.8	**14 59**	S 1 12.4	**134 41.8**	5 36	275 58.5
17	**Betelgeuse**	0.1-1.2	**15 18**	N 7 24.2	**129 57.5**	5 55	271 14.2
18	**Canopus**	-0.9	**15 46**	S 52 42.0	**122 45.0**	6 24	264 01.7
19	**Sirius**	-1.6	**16 08**	S 16 43.1	**117 27.5**	6 45	258 44.2
20	**Adhara**	1.6	**16 21**	S 28 58.5	**114 05.2**	6 59	255 21.9
21	**Castor**	1.6	**16 57**	N31 53.5	**105 06.3**	7 34	246 23.0
22	**Procyon**	0.5	**17 02**	N 5 13.6	**103 55.4**	7 39	245 12.1
23	**Pollux**	1.2	**17 07**	N28 01.8	**102 25.4**	7 45	243 42.1
24	**Avior**	1.7	**17 45**	S 59 30.7	**93 06.2**	8 22	234 22.9
25	**Suhail**	2.2	**18 30**	S 43 25.9	**81 44.3**	9 08	223 01.0
26	**Miaplacidus**	1.8	**18 35**	S 69 43.0	**80 25.4**	9 13	221 42.1
27	**Alphard**	2.2	**18 50**	S 8 39.3	**76 50.8**	9 28	218 07.5
28	**Regulus**	1.3	**19 30**	N11 58.5	**66 39.1**	10 08	207 55.8
29	**Dubhe**	2.0	**20 25**	N61 45.8	**52 49.1**	11 04	194 05.8
30	**Denebola**	2.2	**21 11**	N14 34.9	**41 28.6**	11 49	182 45.3
31	**Gienah**	2.8	**21 37**	S 17 32.1	**34 47.3**	12 16	176 04.0
32	**Acrux**	1.1	**21 48**	S 63 05.6	**32 05.0**	12 27	173 21.7
33	**Gacrux**	1.6	**21 53**	S 57 06.4	**30 56.7**	12 31	172 13.4
34	**Mimosa**	1.5	**22 09**	S 59 40.9	**26 48.4**	12 48	168 05.1
35	**Alioth**	1.7	**22 15**	N55 58.3	**25 13.8**	12 54	166 30.5
36	**Spica**	1.2	**22 46**	S 11 09.2	**17 26.5**	13 25	158 43.2
37	**Alkaid**	1.9	**23 09**	N49 19.5	**11 51.0**	13 47	153 07.7
38	**Hadar**	0.9	**23 25**	S 60 21.9	**7 47.0**	14 04	149 03.7
39	**Menkent**	2.3	**23 28**	S 36 21.7	**7 04.2**	14 07	148 20.9
40	**Arcturus**	0.2	**23 37**	N19 11.5	**4 49.3**	14 16	146 06.0
41	**Rigil Kent.**	0.1	**0 05**	S 60 49.7	**358 50.3**	14 40	140 07.0
42	**Zuben'ubi**	2.9	**0 16**	S 16 02.1	**356 01.3**	14 51	137 18.0
43	**Kochab**	2.2	**0 16**	N74 09.9	**356 01.9**	14 51	137 18.6
44	**Alphecca**	2.3	**1 00**	N26 43.3	**345 03.8**	15 35	126 20.5
45	**Antares**	1.2	**1 54**	S 26 25.6	**331 23.5**	16 29	112 40.2
46	**Atria**	1.9	**2 13**	S 69 01.3	**326 35.2**	16 49	107 51.9
47	**Sabik**	2.6	**2 35**	S 15 43.3	**321 08.8**	17 10	102 25.5
48	**Shaula**	1.7	**2 58**	S 37 06.0	**315 20.6**	17 34	96 37.3
49	**Rasalhague**	2.1	**2 59**	N12 33.7	**315 00.2**	17 35	96 16.9
50	**Eltanin**	2.4	**3 21**	N51 29.3	**309 34.4**	17 57	90 51.1
51	**Kaus Aust.**	2.0	**3 49**	S 34 22.9	**302 42.2**	18 24	83 58.9
52	**Vega**	0.1	**4 01**	N38 46.9	**299 29.8**	18 37	80 46.5
53	**Nunki**	2.1	**4 20**	S 26 17.8	**294 55.8**	18 55	76 12.5
54	**Altair**	0.9	**5 15**	N 8 51.8	**281 02.7**	19 51	62 19.4
55	**Peacock**	2.1	**5 50**	S 56 44.1	**272 20.6**	20 26	53 37.3
56	**Deneb**	1.3	**6 06**	N45 16.3	**268 22.5**	20 41	49 39.2
57	**Enif**	2.5	**7 08**	N 9 51.9	**252 41.8**	21 44	33 58.5
58	**Al Na'ir**	2.2	**7 32**	S 46 57.9	**246 41.7**	22 08	27 58.4
59	**Fomalhaut**	1.3	**8 21**	S 29 37.8	**234 20.2**	22 58	15 36.9
60	**Markab**	2.6	**8 28**	N15 11.6	**232 33.3**	23 05	13 50.0

Star Transit Corr Table

Date	Corr h m	Date	Corr h m
1	0 00	17	–1 03
2	–0 04	18	–1 07
3	–0 08	19	–1 11
4	–0 12	20	–1 15
5	–0 16	21	–1 19
6	–0 20	22	–1 23
7	–0 24	23	–1 27
8	–0 28	24	–1 30
9	–0 31	25	–1 34
10	–0 35	26	–1 38
11	–0 39	27	–1 42
12	–0 43	28	–1 46
13	–0 47	29	–1 50
14	–0 51	30	–1 54
15	–0 55	31	–1 58
16	–0 59		

STAR TRANSIT

To find the approximate time of transit of a star for any day of the month use the table above. All corrections are subtractive.

If the value taken from the table is greater than the time of transit for the first of the month, add 23h 56min to the time of transit before subtracting the correction.

Example: What time will Kochab (No 43) transit the meridian on May 30?

	h min
Transit on May 1	00 16
	+23 56
	24 12
Corr for May 30	–01 54
Transit on May 30	22 18

MAY DIARY

d h	
4 11	Neptune stationary
4 17	Mercury greatest elong. W.(27°)
8 18	Juno stationary
12 16	Mercury 0.8° S of Saturn
12 20	Mars in conjunction
17 07	Neptune 3° S of Moon
17 20	Uranus stationary
18 03	Uranus 3° S of Moon
20 23	Jupiter 0.4° N of Moon
22 22	Venus 1.7° N of Moon
23 08	Saturn 1.7° N of Moon
24 11	Mercury 3° N of Moon
28 05	Pluto at opposition
29 02	Venus 0.3° N of Saturn

NOTES

Star declinations may be used as is for the whole month. To correct GHA for day, hour, minute, and second see page E 106; for further explanation of this table see page E 4.

MAY 1998 SUN & ARIES GMT

Time	SUN GHA ° ′	Dec ° ′	ARIES GHA ° ′	Time	SUN GHA ° ′	Dec ° ′	ARIES GHA ° ′	Time	SUN GHA ° ′	Dec ° ′	ARIES GHA ° ′	Time
	Friday, 1st May				**Wednesday, 6th May**				**Monday, 11th May**			
00	180 42.4	N14 56.9	218 43.3	00	180 50.0	N16 25.0	223 38.9	00	180 54.3	N17 46.2	228 34.6	00
02	210 42.5	14 58.4	248 48.2	02	210 50.1	16 26.4	253 43.9	02	210 54.3	17 47.5	258 39.6	02
04	240 42.7	14 59.9	278 53.1	04	240 50.2	16 27.8	283 48.8	04	240 54.4	17 48.8	288 44.5	04
06	270 42.8	15 01.5	308 58.0	06	270 50.3	16 29.2	313 53.7	06	270 54.4	17 50.1	318 49.4	06
08	300 43.0	15 03.0	339 03.0	08	300 50.4	16 30.6	343 58.7	08	300 54.5	17 51.4	348 54.3	08
10	330 43.1	15 04.5	9 07.9	10	330 50.5	16 32.0	14 03.6	10	330 54.5	17 52.7	18 59.3	10
12	0 43.3	15 06.0	39 12.8	12	0 50.6	16 33.4	44 08.5	12	0 54.5	17 54.0	49 04.2	12
14	30 43.4	15 07.5	69 17.8	14	30 50.7	16 34.8	74 13.4	14	30 54.6	17 55.2	79 09.1	14
16	60 43.6	15 09.0	99 22.7	16	60 50.8	16 36.2	104 18.4	16	60 54.6	17 56.5	109 14.1	16
18	90 43.7	15 10.5	129 27.6	18	90 50.9	16 37.6	134 23.3	18	90 54.6	17 57.8	139 19.0	18
20	120 43.9	15 12.0	159 32.5	20	120 51.0	16 39.0	164 28.2	20	120 54.7	17 59.1	169 23.9	20
22	150 44.0	N15 13.5	189 37.5	22	150 51.1	N16 40.4	194 33.2	22	150 54.7	N18 00.3	199 28.8	22
	Saturday, 2nd May				**Thursday, 7th May**				**Tuesday, 12th May**			
00	180 44.2	N15 15.0	219 42.4	00	180 51.2	N16 41.8	224 38.1	00	180 54.7	N18 01.6	229 33.8	00
02	210 44.3	15 16.5	249 47.3	02	210 51.3	16 43.2	254 43.0	02	210 54.8	18 02.9	259 38.7	02
04	240 44.4	15 18.0	279 52.3	04	240 51.3	16 44.6	284 47.9	04	240 54.8	18 04.1	289 43.6	04
06	270 44.6	15 19.5	309 57.2	06	270 51.4	16 46.0	314 52.9	06	270 54.8	18 05.4	319 48.6	06
08	300 44.7	15 21.0	340 02.1	08	300 51.5	16 47.4	344 57.8	08	300 54.8	18 06.7	349 53.5	08
10	330 44.9	15 22.5	10 07.0	10	330 51.6	16 48.7	15 02.7	10	330 54.9	18 07.9	19 58.4	10
12	0 45.0	15 24.0	40 12.0	12	0 51.7	16 50.1	45 07.7	12	0 54.9	18 09.2	50 03.3	12
14	30 45.1	15 25.5	70 16.9	14	30 51.8	16 51.5	75 12.6	14	30 54.9	18 10.4	80 08.3	14
16	60 45.3	15 27.0	100 21.8	16	60 51.8	16 52.9	105 17.5	16	60 54.9	18 11.7	110 13.2	16
18	90 45.4	15 28.5	130 26.7	18	90 51.9	16 54.2	135 22.4	18	90 55.0	18 12.9	140 18.1	18
20	120 45.6	15 30.0	160 31.7	20	120 52.0	16 55.6	165 27.4	20	120 55.0	18 14.2	170 23.1	20
22	150 45.7	N15 31.4	190 36.6	22	150 52.1	N16 57.0	195 32.3	22	150 55.0	N18 15.4	200 28.0	22
	Sunday, 3rd May				**Friday, 8th May**				**Wednesday, 13th May**			
00	180 45.8	N15 32.9	220 41.5	00	180 52.2	N16 58.3	225 37.2	00	180 55.0	N18 16.7	230 32.9	00
02	210 46.0	15 34.4	250 46.5	02	210 52.2	16 59.7	255 42.2	02	210 55.0	18 17.9	260 37.8	02
04	240 46.1	15 35.9	280 51.4	04	240 52.3	17 01.1	285 47.1	04	240 55.0	18 19.2	290 42.8	04
06	270 46.2	15 37.4	310 56.3	06	270 52.4	17 02.4	315 52.0	06	270 55.1	18 20.4	320 47.7	06
08	300 46.3	15 38.8	341 01.2	08	300 52.5	17 03.8	345 56.9	08	300 55.1	18 21.6	350 52.6	08
10	330 46.5	15 40.3	11 06.2	10	330 52.5	17 05.2	16 01.9	10	330 55.1	18 22.9	20 57.6	10
12	0 46.6	15 41.8	41 11.1	12	0 52.6	17 06.5	46 06.8	12	0 55.1	18 24.1	51 02.5	12
14	30 46.7	15 43.2	71 16.0	14	30 52.7	17 07.9	76 11.7	14	30 55.1	18 25.3	81 07.4	14
16	60 46.9	15 44.7	101 21.0	16	60 52.7	17 09.2	106 16.6	16	60 55.1	18 26.6	111 12.3	16
18	90 47.0	15 46.2	131 25.9	18	90 52.8	17 10.6	136 21.6	18	90 55.1	18 27.8	141 17.3	18
20	120 47.1	15 47.6	161 30.8	20	120 52.9	17 11.9	166 26.5	20	120 55.1	18 29.0	171 22.2	20
22	150 47.2	N15 49.1	191 35.7	22	150 52.9	N17 13.3	196 31.4	22	150 55.1	N18 30.2	201 27.1	22
	Monday, 4th May				**Saturday, 9th May**				**Thursday, 14th May**			
00	180 47.4	N15 50.5	221 40.7	00	180 53.0	N17 14.6	226 36.4	00	180 55.2	N18 31.4	231 32.1	00
02	210 47.5	15 52.0	251 45.6	02	210 53.1	17 15.9	256 41.3	02	210 55.2	18 32.7	261 37.0	02
04	240 47.6	15 53.5	281 50.5	04	240 53.1	17 17.3	286 46.2	04	240 55.2	18 33.9	291 41.9	04
06	270 47.7	15 54.9	311 55.5	06	270 53.2	17 18.6	316 51.1	06	270 55.2	18 35.1	321 46.8	06
08	300 47.8	15 56.4	342 00.4	08	300 53.3	17 20.0	346 56.1	08	300 55.2	18 36.3	351 51.8	08
10	330 48.0	15 57.8	12 05.3	10	330 53.3	17 21.3	17 01.0	10	330 55.2	18 37.5	21 56.7	10
12	0 48.1	15 59.3	42 10.2	12	0 53.4	17 22.6	47 05.9	12	0 55.2	18 38.7	52 01.6	12
14	30 48.2	16 00.7	72 15.2	14	30 53.4	17 23.9	77 10.9	14	30 55.2	18 39.9	82 06.6	14
16	60 48.3	16 02.2	102 20.1	16	60 53.5	17 25.3	107 15.8	16	60 55.2	18 41.1	112 11.5	16
18	90 48.4	16 03.6	132 25.0	18	90 53.6	17 26.6	137 20.7	18	90 55.2	18 42.3	142 16.4	18
20	120 48.5	16 05.0	162 30.0	20	120 53.6	17 27.9	167 25.6	20	120 55.2	18 43.5	172 21.3	20
22	150 48.7	N16 06.5	192 34.9	22	150 53.7	N17 29.2	197 30.6	22	150 55.2	N18 44.7	202 26.3	22
	Tuesday, 5th May				**Sunday, 10th May**				**Friday, 15th May**			
00	180 48.8	N16 07.9	222 39.8	00	180 53.7	N17 30.6	227 35.5	00	180 55.2	N18 45.9	232 31.2	00
02	210 48.9	16 09.3	252 44.7	02	210 53.8	17 31.9	257 40.4	02	210 55.1	18 47.1	262 36.1	02
04	240 49.0	16 10.8	282 49.7	04	240 53.8	17 33.2	287 45.4	04	240 55.1	18 48.3	292 41.0	04
06	270 49.1	16 12.2	312 54.6	06	270 53.9	17 34.5	317 50.3	06	270 55.1	18 49.5	322 46.0	06
08	300 49.2	16 13.6	342 59.5	08	300 53.9	17 35.8	347 55.2	08	300 55.1	18 50.6	352 50.9	08
10	330 49.3	16 15.1	13 04.5	10	330 54.0	17 37.1	18 00.1	10	330 55.1	18 51.8	22 55.8	10
12	0 49.4	16 16.5	43 09.4	12	0 54.0	17 38.4	48 05.1	12	0 55.1	18 53.0	53 00.8	12
14	30 49.5	16 17.9	73 14.3	14	30 54.1	17 39.7	78 10.0	14	30 55.1	18 54.2	83 05.7	14
16	60 49.6	16 19.3	103 19.2	16	60 54.1	17 41.0	108 14.9	16	60 55.1	18 55.3	113 10.6	16
18	90 49.7	16 20.7	133 24.2	18	90 54.2	17 42.3	138 19.9	18	90 55.1	18 56.5	143 15.5	18
20	120 49.8	16 22.2	163 29.1	20	120 54.2	17 43.6	168 24.8	20	120 55.0	18 57.7	173 20.5	20
22	150 49.9	N16 23.6	193 34.0	22	150 54.3	N17 44.9	198 29.7	22	150 55.0	N18 58.9	203 25.4	22

NOTES

Sun and Aries GHA corrections for additional hour, minutes, and seconds are on page E 106.
Correct sun declination by interpolation. For an example see page E 4.

MAY 1998 — SUN & ARIES — GMT

Saturday, 16th May

Time	SUN GHA ° ′	Dec ° ′	ARIES GHA ° ′
00	180 55.0	N19 00.0	233 30.3
02	210 55.0	19 01.2	263 35.3
04	240 55.0	19 02.3	293 40.2
06	270 54.9	19 03.5	323 45.1
08	300 54.9	19 04.7	353 50.0
10	330 54.9	19 05.8	23 55.0
12	0 54.9	19 07.0	53 59.9
14	30 54.8	19 08.1	84 04.8
16	60 54.8	19 09.3	114 09.8
18	90 54.8	19 10.4	144 14.7
20	120 54.8	19 11.6	174 19.6
22	150 54.7	N19 12.7	204 24.5

Sunday, 17th May

Time	SUN GHA ° ′	Dec ° ′	ARIES GHA ° ′
00	180 54.7	N19 13.8	234 29.5
02	210 54.7	19 15.0	264 34.4
04	240 54.6	19 16.1	294 39.3
06	270 54.6	19 17.2	324 44.3
08	300 54.6	19 18.4	354 49.2
10	330 54.5	19 19.5	24 54.1
12	0 54.5	19 20.6	54 59.0
14	30 54.5	19 21.7	85 04.0
16	60 54.4	19 22.9	115 08.9
18	90 54.4	19 24.0	145 13.8
20	120 54.4	19 25.1	175 18.8
22	150 54.3	N19 26.2	205 23.7

Monday, 18th May

Time	SUN GHA ° ′	Dec ° ′	ARIES GHA ° ′
00	180 54.3	N19 27.3	235 28.6
02	210 54.2	19 28.4	265 33.5
04	240 54.2	19 29.5	295 38.5
06	270 54.1	19 30.6	325 43.4
08	300 54.1	19 31.7	355 48.3
10	330 54.0	19 32.8	25 53.3
12	0 54.0	19 33.9	55 58.2
14	30 54.0	19 35.0	86 03.1
16	60 53.9	19 36.1	116 08.0
18	90 53.9	19 37.2	146 13.0
20	120 53.8	19 38.3	176 17.9
22	150 53.7	N19 39.4	206 22.8

Tuesday, 19th May

Time	SUN GHA ° ′	Dec ° ′	ARIES GHA ° ′
00	180 53.7	N19 40.5	236 27.8
02	210 53.6	19 41.6	266 32.7
04	240 53.6	19 42.6	296 37.6
06	270 53.5	19 43.7	326 42.5
08	300 53.5	19 44.8	356 47.5
10	330 53.4	19 45.9	26 52.4
12	0 53.3	19 46.9	56 57.3
14	30 53.3	19 48.0	87 02.2
16	60 53.2	19 49.1	117 07.2
18	90 53.2	19 50.1	147 12.1
20	120 53.1	19 51.2	177 17.0
22	150 53.0	N19 52.3	207 22.0

Wednesday, 20th May

Time	SUN GHA ° ′	Dec ° ′	ARIES GHA ° ′
00	180 53.0	N19 53.3	237 26.9
02	210 52.9	19 54.4	267 31.8
04	240 52.8	19 55.4	297 36.7
06	270 52.8	19 56.5	327 41.7
08	300 52.7	19 57.5	357 46.6
10	330 52.6	19 58.6	27 51.5
12	0 52.6	19 59.6	57 56.5
14	30 52.5	20 00.6	88 01.4
16	60 52.4	20 01.7	118 06.3
18	90 52.3	20 02.7	148 11.2
20	120 52.3	20 03.7	178 16.2
22	150 52.2	N20 04.8	208 21.1

Thursday, 21st May

Time	SUN GHA ° ′	Dec ° ′	ARIES GHA ° ′
00	180 52.1	N20 05.8	238 26.0
02	210 52.0	20 06.8	268 31.0
04	240 51.9	20 07.8	298 35.9
06	270 51.9	20 08.9	328 40.8
08	300 51.8	20 09.9	358 45.7
10	330 51.7	20 10.9	28 50.7
12	0 51.6	20 11.9	58 55.6
14	30 51.5	20 12.9	89 00.5
16	60 51.5	20 13.9	119 05.5
18	90 51.4	20 14.9	149 10.4
20	120 51.3	20 15.9	179 15.3
22	150 51.2	N20 16.9	209 20.2

Friday, 22nd May

Time	SUN GHA ° ′	Dec ° ′	ARIES GHA ° ′
00	180 51.1	N20 17.9	239 25.2
02	210 51.0	20 18.9	269 30.1
04	240 50.9	20 19.9	299 35.0
06	270 50.8	20 20.9	329 39.9
08	300 50.7	20 21.9	359 44.9
10	330 50.6	20 22.9	29 49.8
12	0 50.5	20 23.9	59 54.7
14	30 50.5	20 24.9	89 59.7
16	60 50.4	20 25.8	120 04.6
18	90 50.3	20 26.8	150 09.5
20	120 50.2	20 27.8	180 14.4
22	150 50.1	N20 28.8	210 19.4

Saturday, 23rd May

Time	SUN GHA ° ′	Dec ° ′	ARIES GHA ° ′
00	180 50.0	N20 29.7	240 24.3
02	210 49.9	20 30.7	270 29.2
04	240 49.8	20 31.7	300 34.2
06	270 49.7	20 32.6	330 39.1
08	300 49.6	20 33.6	0 44.0
10	330 49.4	20 34.5	30 48.9
12	0 49.3	20 35.5	60 53.9
14	30 49.2	20 36.5	90 58.8
16	60 49.1	20 37.4	121 03.7
18	90 49.0	20 38.4	151 08.7
20	120 48.9	20 39.3	181 13.6
22	150 48.8	N20 40.2	211 18.5

Sunday, 24th May

Time	SUN GHA ° ′	Dec ° ′	ARIES GHA ° ′
00	180 48.7	N20 41.2	241 23.4
02	210 48.6	20 42.1	271 28.4
04	240 48.5	20 43.1	301 33.3
06	270 48.3	20 44.0	331 38.2
08	300 48.2	20 44.9	1 43.2
10	330 48.1	20 45.8	31 48.1
12	0 48.0	20 46.8	61 53.0
14	30 47.9	20 47.7	91 57.9
16	60 47.8	20 48.6	122 02.9
18	90 47.6	20 49.5	152 07.8
20	120 47.5	20 50.4	182 12.7
22	150 47.4	N20 51.4	212 17.7

Monday, 25th May

Time	SUN GHA ° ′	Dec ° ′	ARIES GHA ° ′
00	180 47.3	N20 52.3	242 22.6
02	210 47.2	20 53.2	272 27.5
04	240 47.0	20 54.1	302 32.4
06	270 46.9	20 55.0	332 37.4
08	300 46.8	20 55.9	2 42.3
10	330 46.7	20 56.8	32 47.2
12	0 46.5	20 57.7	62 52.1
14	30 46.4	20 58.6	92 57.1
16	60 46.3	20 59.5	123 02.0
18	90 46.1	21 00.4	153 06.9
20	120 46.0	21 01.2	183 11.9
22	150 45.9	N21 02.1	213 16.8

Tuesday, 26th May

Time	SUN GHA ° ′	Dec ° ′	ARIES GHA ° ′	Time
00	180 45.8	N21 03.0	243 21.7	00
02	210 45.6	21 03.9	273 26.6	02
04	240 45.5	21 04.8	303 31.6	04
06	270 45.4	21 05.6	333 36.5	06
08	300 45.2	21 06.5	3 41.4	08
10	330 45.1	21 07.4	33 46.4	10
12	0 44.9	21 08.2	63 51.3	12
14	30 44.8	21 09.1	93 56.2	14
16	60 44.7	21 10.0	124 01.1	16
18	90 44.5	21 10.8	154 06.1	18
20	120 44.4	21 11.7	184 11.0	20
22	150 44.2	N21 12.5	214 15.9	22

Wednesday, 27th May

Time	SUN GHA ° ′	Dec ° ′	ARIES GHA ° ′	Time
00	180 44.1	N21 13.4	244 20.9	00
02	210 44.0	21 14.2	274 25.8	02
04	240 43.8	21 15.1	304 30.7	04
06	270 43.7	21 15.9	334 35.6	06
08	300 43.5	21 16.7	4 40.6	08
10	330 43.4	21 17.6	34 45.5	10
12	0 43.2	21 18.4	64 50.4	12
14	30 43.1	21 19.3	94 55.4	14
16	60 42.9	21 20.1	125 00.3	16
18	90 42.8	21 20.9	155 05.2	18
20	120 42.6	21 21.7	185 10.1	20
22	150 42.5	N21 22.6	215 15.1	22

Thursday, 28th May

Time	SUN GHA ° ′	Dec ° ′	ARIES GHA ° ′	Time
00	180 42.3	N21 23.4	245 20.0	00
02	210 42.2	21 24.2	275 24.9	02
04	240 42.0	21 25.0	305 29.9	04
06	270 41.9	21 25.8	335 34.8	06
08	300 41.7	21 26.6	5 39.7	08
10	330 41.6	21 27.4	35 44.6	10
12	0 41.4	21 28.2	65 49.6	12
14	30 41.3	21 29.0	95 54.5	14
16	60 41.1	21 29.8	125 59.4	16
18	90 40.9	21 30.6	156 04.4	18
20	120 40.8	21 31.4	186 09.3	20
22	150 40.6	N21 32.2	216 14.2	22

Friday, 29th May

Time	SUN GHA ° ′	Dec ° ′	ARIES GHA ° ′	Time
00	180 40.5	N21 33.0	246 19.1	00
02	210 40.3	21 33.8	276 24.1	02
04	240 40.1	21 34.6	306 29.0	04
06	270 40.0	21 35.4	336 33.9	06
08	300 39.8	21 36.1	6 38.9	08
10	330 39.6	21 36.9	36 43.8	10
12	0 39.5	21 37.7	66 48.7	12
14	30 39.3	21 38.5	96 53.6	14
16	60 39.1	21 39.2	126 58.6	16
18	90 39.0	21 40.0	157 03.5	18
20	120 38.8	21 40.8	187 08.4	20
22	150 38.6	N21 41.5	217 13.4	22

Saturday, 30th May

Time	SUN GHA ° ′	Dec ° ′	ARIES GHA ° ′	Time
00	180 38.5	N21 42.3	247 18.3	00
02	210 38.3	21 43.0	277 23.2	02
04	240 38.1	21 43.8	307 28.1	04
06	270 38.0	21 44.5	337 33.1	06
08	300 37.8	21 45.3	7 38.0	08
10	330 37.6	21 46.0	37 42.9	10
12	0 37.4	21 46.8	67 47.9	12
14	30 37.3	21 47.5	97 52.8	14
16	60 37.1	21 48.2	127 57.7	16
18	90 36.9	21 49.0	158 02.6	18
20	120 36.7	21 49.7	188 07.6	20
22	150 36.6	N21 50.4	218 12.5	22

Sunday, 31st May

Time	SUN GHA ° ′	Dec ° ′	ARIES GHA ° ′
00	180 36.4	N21 51.2	248 17.4
02	210 36.2	21 51.9	278 22.3
04	240 36.0	21 52.6	308 27.3
06	270 35.8	N21 53.3	338 32.2
08	300 35.7	N21 54.0	8 37.1
10	330 35.5	21 54.7	38 42.1
12	0 35.3	21 55.5	68 47.0
14	30 35.1	N21 56.2	98 51.9
16	60 34.9	N21 56.9	128 56.8
18	90 34.7	21 57.6	159 01.8
20	120 34.6	21 58.3	189 06.7
22	150 34.4	N21 59.0	219 11.6

MAY 1998 PLANETS 0h GMT

VENUS / JUPITER

Venus Mer Pass h m	Venus GHA ° ′	Venus Mean Var/hr 14°+	Venus Dec ° ′	Venus Mean Var/hr ′	Day		Jupiter GHA ° ′	Jupiter Mean Var/hr 15°+	Jupiter Dec ° ′	Jupiter Mean Var/hr ′	Jupiter Mer Pass h m
09 16	**221 07.6**	59.8	**S 2 21.4**	1.0	**1**	**Fri**	**227 56.3**	2.0	**S 5 04.5**	0.2	**08 47**
09 16	**221 03.8**	59.8	**S 1 57.6**	1.0	**2**	**Sat**	**228 44.7**	2.0	**S 5 00.1**	0.2	**08 44**
09 16	**220 59.9**	59.8	**S 1 33.7**	1.0	**3**	**SUN**	**229 33.2**	2.0	**S 4 55.8**	0.2	**08 41**
09 16	**220 55.9**	59.8	**S 1 09.6**	1.0	**4**	**Mon**	**230 21.8**	2.0	**S 4 51.5**	0.2	**08 37**
09 17	**220 51.9**	59.8	**S 0 45.4**	1.0	**5**	**Tu**	**231 10.5**	2.0	**S 4 47.3**	0.2	**08 34**
09 17	**220 47.9**	59.8	**S 0 21.0**	1.0	**6**	**Wed**	**231 59.2**	2.0	**S 4 43.1**	0.2	**08 31**
09 17	**220 43.8**	59.8	**N 0 03.4**	1.0	**7**	**Th**	**232 48.1**	2.0	**S 4 38.9**	0.2	**08 28**
09 17	**220 39.6**	59.8	**N 0 28.0**	1.0	**8**	**Fri**	**233 37.0**	2.0	**S 4 34.7**	0.2	**08 24**
09 18	**220 35.3**	59.8	**N 0 52.7**	1.0	**9**	**Sat**	**234 26.0**	2.1	**S 4 30.6**	0.2	**08 21**
09 18	**220 31.0**	59.8	**N 1 17.4**	1.0	**10**	**SUN**	**235 15.2**	2.0	**S 4 26.6**	0.2	**08 18**
09 18	**220 26.6**	59.8	**N 1 42.3**	1.0	**11**	**Mon**	**236 04.4**	2.1	**S 4 22.5**	0.2	**08 15**
09 19	**220 22.1**	59.8	**N 2 07.2**	1.0	**12**	**Tu**	**236 53.7**	2.1	**S 4 18.6**	0.2	**08 11**
09 19	**220 17.4**	59.8	**N 2 32.1**	1.0	**13**	**Wed**	**237 43.1**	2.1	**S 4 14.6**	0.2	**08 08**
09 19	**220 12.7**	59.8	**N 2 57.1**	1.0	**14**	**Th**	**238 32.6**	2.1	**S 4 10.7**	0.2	**08 05**
09 20	**220 07.9**	59.8	**N 3 22.1**	1.0	**15**	**Fri**	**239 22.2**	2.1	**S 4 06.9**	0.2	**08 01**
09 20	**220 02.9**	59.8	**N 3 47.2**	1.0	**16**	**Sat**	**240 11.9**	2.1	**S 4 03.0**	0.2	**07 58**
09 20	**219 57.9**	59.8	**N 4 12.2**	1.0	**17**	**SUN**	**241 01.7**	2.1	**S 3 59.3**	0.2	**07 55**
09 21	**219 52.7**	59.8	**N 4 37.3**	1.0	**18**	**Mon**	**241 51.6**	2.1	**S 3 55.6**	0.2	**07 51**
09 21	**219 47.4**	59.8	**N 5 02.4**	1.0	**19**	**Tu**	**242 41.6**	2.1	**S 3 51.9**	0.2	**07 48**
09 21	**219 41.9**	59.8	**N 5 27.4**	1.0	**20**	**Wed**	**243 31.8**	2.1	**S 3 48.2**	0.1	**07 45**
09 22	**219 36.3**	59.8	**N 5 52.4**	1.0	**21**	**Th**	**244 22.0**	2.1	**S 3 44.7**	0.1	**07 41**
09 22	**219 30.6**	59.8	**N 6 17.3**	1.0	**22**	**Fri**	**245 12.3**	2.1	**S 3 41.1**	0.1	**07 38**
09 23	**219 24.7**	59.7	**N 6 42.2**	1.0	**23**	**Sat**	**246 02.8**	2.1	**S 3 37.6**	0.1	**07 35**
09 23	**219 18.6**	59.7	**N 7 07.1**	1.0	**24**	**SUN**	**246 53.3**	2.1	**S 3 34.2**	0.1	**07 31**
09 23	**219 12.4**	59.7	**N 7 31.8**	1.0	**25**	**Mon**	**247 44.0**	2.1	**S 3 30.8**	0.1	**07 28**
09 24	**219 06.0**	59.7	**N 7 56.5**	1.0	**26**	**Tu**	**248 34.8**	2.1	**S 3 27.5**	0.1	**07 25**
09 24	**218 59.4**	59.7	**N 8 21.0**	1.0	**27**	**Wed**	**249 25.7**	2.1	**S 3 24.2**	0.1	**07 21**
09 25	**218 52.7**	59.7	**N 8 45.4**	1.0	**28**	**Th**	**250 16.8**	2.1	**S 3 21.0**	0.1	**07 18**
09 25	**218 45.8**	59.7	**N 9 09.8**	1.0	**29**	**Fri**	**251 07.9**	2.1	**S 3 17.8**	0.1	**07 14**
09 26	**218 38.6**	59.7	**N 9 33.9**	1.0	**30**	**Sat**	**251 59.2**	2.1	**S 3 14.7**	0.1	**07 11**
09 26	**218 31.3**	59.7	**N 9 58.0**	1.0	**31**	**SUN**	**252 50.6**	2.1	**S 3 11.6**	0.1	**07 08**

VENUS, Av. Mag. -4.1
S.H.A. May
5 358; 10 353; 15 348; 20 342; 25 337; 30 331.

JUPITER, Av. Mag.-2.2
S.H.A. May
5 9; 10 8; 15 7; 20 6; 25 5; 30 5.

MARS / SATURN

Mars Mer Pass h m	Mars GHA ° ′	Mars Mean Var/hr 15°+	Mars Dec ° ′	Mars Mean Var/hr ′	Day		Saturn GHA ° ′	Saturn Mean Var/hr 15°+	Saturn Dec ° ′	Saturn Mean Var/hr ′	Saturn Mer Pass h m
12 08	**177 52.0**	0.7	**N15 44.8**	0.6	**1**	**Fri**	**194 12.7**	2.2	**N 7 46.8**	0.1	**11 02**
12 07	**178 07.8**	0.7	**N15 58.5**	0.6	**2**	**Sat**	**195 04.8**	2.2	**N 7 49.4**	0.1	**10 58**
12 06	**178 23.5**	0.7	**N16 12.1**	0.6	**3**	**SUN**	**195 57.0**	2.2	**N 7 52.1**	0.1	**10 55**
12 05	**178 39.2**	0.6	**N16 25.5**	0.6	**4**	**Mon**	**196 49.1**	2.2	**N 7 54.7**	0.1	**10 51**
12 04	**178 54.8**	0.7	**N16 38.8**	0.5	**5**	**Tu**	**197 41.3**	2.2	**N 7 57.3**	0.1	**10 48**
12 03	**179 10.5**	0.6	**N16 51.9**	0.5	**6**	**Wed**	**198 33.5**	2.2	**N 7 59.9**	0.1	**10 44**
12 02	**179 26.0**	0.6	**N17 04.8**	0.5	**7**	**Th**	**199 25.7**	2.2	**N 8 02.4**	0.1	**10 41**
12 01	**179 41.5**	0.6	**N17 17.6**	0.5	**8**	**Fri**	**200 18.0**	2.2	**N 8 05.0**	0.1	**10 37**
12 00	**179 57.0**	0.6	**N17 30.2**	0.5	**9**	**Sat**	**201 10.3**	2.2	**N 8 07.5**	0.1	**10 34**
11 59	**180 12.4**	0.6	**N17 42.6**	0.5	**10**	**SUN**	**202 02.6**	2.2	**N 8 10.1**	0.1	**10 30**
11 58	**180 27.8**	0.6	**N17 54.8**	0.5	**11**	**Mon**	**202 54.9**	2.2	**N 8 12.6**	0.1	**10 27**
11 57	**180 43.1**	0.6	**N18 06.9**	0.5	**12**	**Tu**	**203 47.2**	2.2	**N 8 15.1**	0.1	**10 23**
11 56	**180 58.5**	0.6	**N18 18.8**	0.5	**13**	**Wed**	**204 39.6**	2.2	**N 8 17.5**	0.1	**10 20**
11 55	**181 13.6**	0.6	**N18 30.4**	0.5	**14**	**Th**	**205 32.0**	2.2	**N 8 20.0**	0.1	**10 16**
11 54	**181 28.8**	0.6	**N18 41.9**	0.5	**15**	**Fri**	**206 24.4**	2.2	**N 8 22.4**	0.1	**10 13**
11 53	**181 43.9**	0.6	**N18 53.2**	0.5	**16**	**Sat**	**207 16.8**	2.2	**N 8 24.9**	0.1	**10 09**
11 52	**181 59.0**	0.6	**N19 04.4**	0.5	**17**	**SUN**	**208 09.3**	2.2	**N 8 27.3**	0.1	**10 06**
11 51	**182 14.1**	0.6	**N19 15.3**	0.4	**18**	**Mon**	**209 01.8**	2.2	**N 8 29.7**	0.1	**10 02**
11 50	**182 29.1**	0.6	**N19 26.1**	0.4	**19**	**Tu**	**209 54.4**	2.2	**N 8 32.1**	0.1	**09 59**
11 49	**182 44.0**	0.6	**N19 36.7**	0.4	**20**	**Wed**	**210 46.9**	2.2	**N 8 34.4**	0.1	**09 55**
11 48	**182 58.9**	0.6	**N19 47.1**	0.4	**21**	**Th**	**211 39.5**	2.2	**N 8 36.8**	0.1	**09 52**
11 47	**183 13.8**	0.6	**N19 57.3**	0.4	**22**	**Fri**	**212 32.2**	2.2	**N 8 39.1**	0.1	**09 48**
11 46	**183 28.6**	0.6	**N20 07.2**	0.4	**23**	**Sat**	**213 24.8**	2.2	**N 8 41.4**	0.1	**09 45**
11 45	**183 43.4**	0.6	**N20 17.0**	0.4	**24**	**SUN**	**214 17.6**	2.2	**N 8 43.7**	0.1	**09 41**
11 44	**183 58.1**	0.6	**N20 26.7**	0.4	**25**	**Mon**	**215 10.3**	2.2	**N 8 45.9**	0.1	**09 38**
11 43	**184 12.8**	0.6	**N20 36.1**	0.4	**26**	**Tu**	**216 03.1**	2.2	**N 8 48.2**	0.1	**09 34**
11 42	**184 27.5**	0.6	**N20 45.3**	0.4	**27**	**Wed**	**216 55.9**	2.2	**N 8 50.4**	0.1	**09 31**
11 41	**184 42.1**	0.6	**N20 54.3**	0.4	**28**	**Th**	**217 48.8**	2.2	**N 8 52.6**	0.1	**09 27**
11 40	**184 56.7**	0.6	**N21 03.1**	0.4	**29**	**Fri**	**218 41.7**	2.2	**N 8 54.8**	0.1	**09 24**
11 39	**185 11.2**	0.6	**N21 11.7**	0.3	**30**	**Sat**	**219 34.7**	2.2	**N 8 56.9**	0.1	**09 20**
11 38	**185 25.8**	0.6	**N21 20.1**	0.3	**31**	**SUN**	**220 27.7**	2.2	**N 8 59.1**	0.1	**09 17**

MARS, Av. Mag. +1.3
S.H.A. May
5 316; 10 313; 15 309; 20 305; 25 302; 30 298.

NOTE: Planet corrections on pages E113-15

SATURN, Av. Mag. +0.5
S.H.A. May
5 335; 10 334; 15 334; 20 333; 25 333; 30 332.

MAY 1998

MOON

Day	GMT hr	GHA ° ′	Mean Var/hr 14°+	Dec ° ′	Mean Var/hr ′
1 Fri	**0**	116 39.5	27.2	**N 18 41.4**	1.6
	6	203 22.1	27.5	**N 18 32.4**	2.2
	12	290 07.0	27.9	**N 18 19.5**	2.8
	18	16 54.3	28.3	**N 18 03.1**	3.4
2 Sat	**0**	103 43.9	28.7	**N 17 43.1**	3.9
	6	190 36.1	29.1	**N 17 19.8**	4.4
	12	277 30.7	29.6	**N 16 53.3**	5.0
	18	4 27.8	29.9	**N 16 23.9**	5.4
3 Sun	**0**	91 27.4	30.4	**N 15 51.6**	5.9
	6	178 29.4	30.8	**N 15 16.7**	6.2
	12	265 33.8	31.1	**N 14 39.3**	6.7
	18	352 40.5	31.5	**N 13 59.7**	7.0
4 Mon	**0**	79 49.4	31.8	**N 13 17.9**	7.3
	6	167 00.5	32.2	**N 12 34.0**	7.6
	12	254 13.5	32.5	**N 11 48.4**	8.0
	18	341 28.5	32.8	**N 11 01.1**	8.1
5 Tu	**0**	68 45.2	33.1	**N 10 12.2**	8.4
	6	156 03.6	33.4	**N 9 22.0**	8.6
	12	243 23.5	33.6	**N 8 30.5**	8.8
	18	330 44.7	33.8	**N 7 37.8**	8.9
6 Wed	**0**	58 07.2	34.0	**N 6 44.2**	9.1
	6	145 30.7	34.1	**N 5 49.7**	9.2
	12	232 55.2	34.3	**N 4 54.5**	9.3
	18	320 20.5	34.4	**N 3 58.6**	9.4
7 Th	**0**	47 46.3	34.4	**N 3 02.2**	9.5
	6	135 12.7	34.4	**N 2 05.5**	9.5
	12	222 39.5	34.5	**N 1 08.5**	9.5
	18	310 06.4	34.5	**N 0 11.3**	9.5
8 Fri	**0**	37 33.3	34.5	**S 0 45.9**	9.5
	6	125 00.2	34.4	**S 1 43.0**	9.5
	12	212 26.9	34.4	**S 2 39.9**	9.4
	18	299 53.2	34.3	**S 3 36.4**	9.4
9 Sat	**0**	27 19.0	34.2	**S 4 32.5**	9.2
	6	114 44.2	34.0	**S 5 28.0**	9.1
	12	202 08.7	33.9	**S 6 22.8**	9.0
	18	289 32.3	33.7	**S 7 16.8**	8.8
10 Sun	**0**	16 54.9	33.6	**S 8 09.7**	8.6
	6	104 16.5	33.4	**S 9 01.6**	8.4
	12	191 36.9	33.1	**S 9 52.2**	8.2
	18	278 56.1	33.0	**S 10 41.5**	8.0
11 Mon	**0**	6 13.9	32.7	**S 11 29.2**	7.7
	6	93 30.2	32.4	**S 12 15.3**	7.4
	12	180 45.1	32.2	**S 12 59.5**	7.0
	18	267 58.5	32.0	**S 13 41.8**	6.7
12 Tu	**0**	355 10.3	31.7	**S 14 22.0**	6.3
	6	82 20.5	31.4	**S 15 00.0**	5.9
	12	169 29.1	31.2	**S 15 35.6**	5.5
	18	256 36.1	30.9	**S 16 08.7**	5.0
13 Wed	**0**	343 41.5	30.6	**S 16 39.2**	4.5
	6	70 45.3	30.3	**S 17 06.8**	4.1
	12	157 47.5	30.1	**S 17 31.6**	3.6
	18	244 48.3	29.9	**S 17 53.3**	3.1
14 Th	**0**	331 47.6	29.6	**S 18 11.9**	2.5
	6	58 45.6	29.5	**S 18 27.3**	2.0
	12	145 42.3	29.3	**S 18 39.3**	1.4
	18	232 37.9	29.0	**S 18 47.9**	0.8
15 Fri	**0**	319 32.4	28.9	**S 18 53.0**	0.2
	6	46 26.0	28.8	**S 18 54.5**	0.4
	12	133 18.7	28.6	**S 18 52.4**	1.0
	18	220 10.7	28.5	**S 18 46.7**	1.6
16 Sat	**0**	307 02.1	28.4	**S 18 37.2**	2.2
	6	33 53.1	28.5	**S 18 24.1**	2.9
	12	120 43.7	28.4	**S 18 07.3**	3.5
	18	207 34.0	28.3	**S 17 46.9**	4.1

Day	GMT hr	GHA ° ′	Mean Var/hr 14°+	Dec ° ′	Mean Var/hr ′
17 Sun	**0**	294 24.2	28.4	**S 17 22.8**	4.7
	6	21 14.4	28.4	**S 16 55.1**	5.3
	12	108 04.5	28.4	**S 16 23.8**	5.8
	18	194 54.8	28.4	**S 15 49.1**	6.4
18 Mon	**0**	281 45.1	28.4	**S 15 11.1**	7.0
	6	8 35.6	28.4	**S 14 29.7**	7.4
	12	95 26.4	28.5	**S 13 45.2**	7.9
	18	182 17.3	28.5	**S 12 57.6**	8.5
19 Tu	**0**	269 08.3	28.6	**S 12 07.1**	8.9
	6	355 59.6	28.6	**S 11 13.8**	9.4
	12	82 50.9	28.6	**S 10 17.9**	9.8
	18	169 42.3	28.6	**S 9 19.6**	10.1
20 Wed	**0**	256 33.7	28.5	**S 8 19.0**	10.4
	6	343 25.0	28.5	**S 7 16.3**	10.8
	12	70 16.2	28.5	**S 6 11.8**	11.1
	18	157 07.0	28.4	**S 5 05.6**	11.3
21 Th	**0**	243 57.5	28.4	**S 3 58.0**	11.5
	6	330 47.5	28.3	**S 2 49.1**	11.7
	12	57 36.8	28.1	**S 1 39.4**	11.8
	18	144 25.5	28.0	**S 0 28.9**	11.8
22 Fri	**0**	231 13.3	27.8	**N 0 42.0**	11.9
	6	318 00.2	27.7	**N 1 53.1**	11.9
	12	44 46.0	27.4	**N 3 04.0**	11.7
	18	131 30.7	27.2	**N 4 14.4**	11.6
23 Sat	**0**	218 14.1	27.0	**N 5 24.1**	11.5
	6	304 56.2	26.7	**N 6 32.8**	11.2
	12	31 36.9	26.5	**N 7 40.1**	10.9
	18	118 16.2	26.3	**N 8 45.8**	10.6
24 Sun	**0**	204 54.1	26.1	**N 9 49.5**	10.2
	6	291 30.6	25.9	**N 10 50.9**	9.8
	12	18 05.6	25.6	**N 11 49.7**	9.3
	18	104 39.3	25.4	**N 12 45.6**	8.8
25 Mon	**0**	191 11.7	25.2	**N 13 38.4**	8.2
	6	277 42.9	25.0	**N 14 27.8**	7.6
	12	4 13.2	24.9	**N 15 13.6**	6.9
	18	90 42.6	24.8	**N 15 55.5**	6.2
26 Tu	**0**	177 11.4	24.7	**N 16 33.4**	5.5
	6	263 39.7	24.7	**N 17 07.1**	4.8
	12	350 07.9	24.7	**N 17 36.4**	4.1
	18	76 36.1	24.7	**N 18 01.3**	3.4
27 Wed	**0**	163 04.7	24.9	**N 18 21.7**	2.6
	6	249 33.9	25.0	**N 18 37.5**	1.8
	12	336 04.0	25.2	**N 18 48.9**	1.1
	18	62 35.2	25.5	**N 18 55.7**	0.4
28 Th	**0**	149 07.8	25.8	**N 18 58.1**	0.4
	6	235 42.1	26.1	**N 18 56.1**	1.1
	12	322 18.3	26.4	**N 18 49.8**	1.8
	18	48 56.5	26.8	**N 18 39.4**	2.4
29 Fri	**0**	135 37.0	27.2	**N 18 25.0**	3.1
	6	222 19.8	27.6	**N 18 06.9**	3.7
	12	309 05.2	28.0	**N 17 45.0**	4.2
	18	35 53.2	28.4	**N 17 19.8**	4.8
30 Sat	**0**	122 43.8	29.0	**N 16 51.2**	5.3
	6	209 37.1	29.4	**N 16 19.6**	5.8
	12	296 33.1	29.8	**N 15 45.1**	6.3
	18	23 31.8	30.2	**N 15 07.9**	6.6
31 Sun	**0**	110 33.1	30.6	**N 14 28.3**	7.0
	6	197 37.0	31.1	**N 13 46.3**	7.3
	12	284 43.4	31.5	**N 13 02.3**	7.7
	18	11 52.1	31.9	**N 12 16.2**	8.0

NOTE: Moon GHA corrections for additional hours, minutes, and seconds are on page E 108–10. Moon declination corrections are on page E 111. The mean var/hr value is needed for both corrections.

JUNE 1998 — SUN & MOON — GMT

SUN

Day of Yr	Day of Mth	Day of Week	Equation of Time 0h	Equation of Time 12h	Transit	Semi-diam	Lat 30°N Twi-light	Lat 30°N Sun-rise	Lat 30°N Sun-set	Lat 30°N Twi-light
			m s	m s	h m	′	h m	h m	h m	h m
152	**1**	**Mon**	-02 17	-02 12	**11 58**	15.8	04 32	04 59	18 56	19 23
153	**2**	**Tu**	-02 08	-02 03	**11 58**	15.8	04 32	04 59	18 57	19 24
154	**3**	**Wed**	-01 58	-01 53	**11 58**	15.8	04 32	04 59	18 57	19 25
155	**4**	**Th**	-01 48	-01 43	**11 58**	15.8	04 32	04 59	18 58	19 25
156	**5**	**Fri**	-01 38	-01 33	**11 58**	15.8	04 32	04 59	18 58	19 26
157	**6**	**Sat**	-01 28	-01 22	**11 59**	15.8	04 31	04 59	18 59	19 26
158	**7**	**Sun**	-01 17	-01 11	**11 59**	15.8	04 31	04 58	18 59	19 27
159	**8**	**Mon**	-01 06	-01 00	**11 59**	15.8	04 31	04 58	19 00	19 27
160	**9**	**Tu**	-00 54	-00 49	**11 59**	15.8	04 31	04 58	19 00	19 28
161	**10**	**Wed**	-00 43	-00 37	**11 59**	15.8	04 31	04 58	19 01	19 28
162	**11**	**Th**	-00 31	-00 25	**12 00**	15.8	04 31	04 58	19 01	19 28
163	**12**	**Fri**	-00 19	-00 13	**12 00**	15.8	04 31	04 58	19 01	19 29
164	**13**	**Sat**	-00 06	00 00	**12 00**	15.8	04 31	04 58	19 02	19 29
165	**14**	**Sun**	+00 06	+00 12	**12 00**	15.8	04 31	04 58	19 02	19 30
166	**15**	**Mon**	+00 19	+00 25	**12 00**	15.8	04 31	04 58	19 02	19 30
167	**16**	**Tu**	+00 31	+00 38	**12 01**	15.8	04 31	04 59	19 03	19 30
168	**17**	**Wed**	+00 44	+00 51	**12 01**	15.8	04 31	04 59	19 03	19 31
169	**18**	**Th**	+00 57	+01 04	**12 01**	15.8	04 31	04 59	19 03	19 31
170	**19**	**Fri**	+01 10	+01 17	**12 01**	15.8	04 31	04 59	19 04	19 31
171	**20**	**Sat**	+01 23	+01 30	**12 01**	15.8	04 32	04 59	19 04	19 31
172	**21**	**Sun**	+01 37	+01 43	**12 02**	15.8	04 32	04 59	19 04	19 32
173	**22**	**Mon**	+01 50	+01 56	**12 02**	15.8	04 32	05 00	19 04	19 32
174	**23**	**Tu**	+02 03	+02 09	**12 02**	15.8	04 32	05 00	19 04	19 32
175	**24**	**Wed**	+02 16	+02 22	**12 02**	15.8	04 33	05 00	19 05	19 32
176	**25**	**Th**	+02 29	+02 35	**12 03**	15.8	04 33	05 00	19 05	19 32
177	**26**	**Fri**	+02 42	+02 48	**12 03**	15.8	04 33	05 01	19 05	19 32
178	**27**	**Sat**	+02 54	+03 01	**12 03**	15.8	04 34	05 01	19 05	19 32
179	**28**	**Sun**	+03 07	+03 13	**12 03**	15.8	04 34	05 01	19 05	19 32
180	**29**	**Mon**	+03 19	+03 25	**12 03**	15.8	04 34	05 02	19 05	19 33
181	**30**	**Tu**	+03 31	+03 37	**12 04**	15.8	04 35	05 02	19 05	19 33

Lat Corr to Sunrise, Sunset etc

Lat	Twi-light	Sun-rise	Sun-set	Twi-light
°	h m	h m	h m	h m
N60	–3 37	–2 23	+2 23	+3 37
55	–2 08	–1 38	+1 38	+2 08
50	–1 25	–1 08	+1 08	+1 25
45	–0 56	–0 46	+0 46	+0 56
N40	–0 33	–0 28	+0 28	+0 33
35	–0 15	–0 13	+0 13	+0 15
30	0 00	0 00	0 00	0 00
25	+0 13	+0 12	–0 12	–0 13
N20	+0 25	+0 22	–0 22	–0 25
15	+0 35	+0 32	–0 32	–0 35
10	+0 45	+0 41	–0 41	–0 45
N 5	+0 54	+0 50	–0 50	–0 54
0	+1 03	+0 58	–0 58	–1 03
S 5	+1 12	+1 07	–1 07	–1 12
10	+1 20	+1 16	–1 16	–1 20
15	+1 29	+1 25	–1 25	–1 29
S20	+1 37	+1 34	–1 34	–1 37
25	+1 47	+1 44	–1 44	–1 47
30	+1 56	+1 55	–1 55	–1 56
35	+2 07	+2 07	–2 07	–2 07
S40	+2 18	+2 21	–2 21	–2 18
45	+2 32	+2 38	–2 38	–2 32
50	+2 48	+2 59	–2 59	–2 48

NOTES

Latitude corrections are for mid-month.

Equation of time = mean time (LHA mean sun) minus apparent time (LHA true sun).

MOON

Day of Yr	Day of Mth	Day of Week	Age	Transit (Upper)	Diff	Semi-diam	Hor Par	Lat 30°N Moon-rise	Lat 30°N Moon-set
			days	h m	m	′	′	h m	h m
152	**1**	**Mon**	07	**17 57**	**43**	15.0	55.1	11 22	- -
153	**2**	**Tu**	08	**18 40**	**43**	14.9	54.6	12 15	00 26
154	**3**	**Wed**	09	**19 23**	**42**	14.8	54.3	13 07	01 01
155	**4**	**Th**	10	**20 05**	**42**	14.8	54.2	13 59	01 34
156	**5**	**Fri**	11	**20 47**	**44**	14.8	54.2	14 50	02 06
157	**6**	**Sat**	12	**21 31**	**45**	14.8	54.3	15 43	02 40
158	**7**	**Sun**	13	**22 16**	**47**	14.9	54.6	16 36	03 15
159	**8**	**Mon**	14	**23 03**	**49**	15.0	54.9	17 29	03 53
160	**9**	**Tu**	15	**23 52**	**51**	15.1	55.4	18 24	04 34
161	**10**	**Wed**	16	**24 43**	**-**	15.2	55.8	19 18	05 19
162	**11**	**Th**	17	**00 43**	**53**	15.3	56.3	20 10	06 08
163	**12**	**Fri**	18	**01 36**	**53**	15.5	56.8	21 01	07 01
164	**13**	**Sat**	19	**02 29**	**53**	15.6	57.3	21 49	07 58
165	**14**	**Sun**	20	**03 22**	**52**	15.8	57.8	22 34	08 57
166	**15**	**Mon**	21	**04 14**	**52**	15.9	58.3	23 17	09 58
167	**16**	**Tu**	22	**05 06**	**51**	16.0	58.7	23 58	10 59
168	**17**	**Wed**	23	**05 57**	**51**	16.1	59.1	- -	12 02
169	**18**	**Th**	24	**06 48**	**52**	16.2	59.5	00 38	13 05
170	**19**	**Fri**	25	**07 40**	**54**	16.3	59.7	01 18	14 09
171	**20**	**Sat**	26	**08 34**	**55**	16.3	59.8	02 00	15 13
172	**21**	**Sun**	27	**09 29**	**57**	16.3	59.8	02 45	16 18
173	**22**	**Mon**	28	**10 26**	**58**	16.2	59.5	03 33	17 23
174	**23**	**Tu**	29	**11 24**	**58**	16.1	59.1	04 25	18 25
175	**24**	**Wed**	00	**12 22**	**56**	16.0	58.6	05 20	19 23
176	**25**	**Th**	01	**13 18**	**53**	15.8	57.9	06 18	20 16
177	**26**	**Fri**	02	**14 11**	**51**	15.6	57.1	07 16	21 03
178	**27**	**Sat**	03	**15 02**	**48**	15.4	56.4	08 14	21 46
179	**28**	**Sun**	04	**15 50**	**45**	15.2	55.7	09 11	22 24
180	**29**	**Mon**	05	**16 35**	**44**	15.0	55.1	10 05	23 00
181	**30**	**Tu**	06	**17 19**	**42**	14.9	54.7	10 59	23 34

Phases of the Moon

	d	h	m
☾ First Quarter	2	01	45
○ Full Moon	10	04	18
☽ Last Quarter	17	10	38
● New Moon	24	03	50
Apogee	5	00	
Perigee	20	17	
On 0°	4	03	
Farthest S of 0°	11	13	
On 0°	18	04	
Farthest N of 0°	24	11	

NOTES

Diff equals daily change in transit time.

To correct moonrise, moonset, or transit times for latitude and/or longitude see page E 112. For further information about these tables see page E 4. For arc-to-time correction see page E 103.

JUNE 1998 — STARS — 0h GMT June 1

No	Name	Mag	Transit h m	Dec ° ′	GHA ° ′	RA h m	SHA ° ′
	ARIES		**7 22**		**249 16.6**		
1	**Alpheratz**	2.2	**7 30**	N29 04.7	**247 12.1**	0 08	357 55.5
2	**Ankaa**	2.4	**7 48**	S 42 18.7	**242 43.8**	0 26	353 27.2
3	**Schedar**	2.5	**8 02**	N56 31.4	**239 10.4**	0 40	349 53.8
4	**Diphda**	2.2	**8 05**	S 17 59.7	**238 24.2**	0 43	349 07.6
5	**Achernar**	0.6	**8 59**	S 57 14.5	**224 52.4**	1 38	335 35.8
6	**POLARIS**	2.1	**9 51**	N89 15.1	**211 55.6**	2 29	322 39.0
7	**Hamal**	2.2	**9 28**	N23 27.1	**217 30.6**	2 07	328 14.0
8	**Acamar**	3.1	**10 19**	S 40 18.7	**204 44.1**	2 58	315 27.5
9	**Menkar**	2.8	**10 23**	N 4 04.9	**203 44.0**	3 02	314 27.4
10	**Mirfak**	1.9	**10 45**	N49 51.1	**198 13.9**	3 24	308 57.3
11	**Aldebaran**	1.1	**11 57**	N16 30.2	**180 19.6**	4 36	291 03.0
12	**Rigel**	0.3	**12 35**	S 8 12.4	**170 40.1**	5 14	281 23.5
13	**Capella**	0.2	**12 37**	N45 59.6	**170 08.6**	5 17	280 52.0
14	**Bellatrix**	1.7	**12 46**	N 6 20.7	**168 01.3**	5 25	278 44.7
15	**Elnath**	1.8	**12 47**	N28 36.2	**167 44.2**	5 26	278 27.6
16	**Alnilam**	1.8	**12 57**	S 1 12.3	**165 15.0**	5 36	275 58.4
17	**Betelgeuse**	0.1-1.2	**13 16**	N 7 24.2	**160 30.7**	5 55	271 14.1
18	**Canopus**	-0.9	**13 45**	S 52 41.9	**153 18.4**	6 24	264 01.8
19	**Sirius**	-1.6	**14 06**	S 16 43.0	**148 00.8**	6 45	258 44.2
20	**Adhara**	1.6	**14 19**	S 28 58.4	**144 38.6**	6 59	255 22.0
21	**Castor**	1.6	**14 55**	N31 53.5	**135 39.6**	7 34	246 23.0
22	**Procyon**	0.5	**15 00**	N 5 13.6	**134 28.7**	7 39	245 12.1
23	**Pollux**	1.2	**15 06**	N28 01.8	**132 58.8**	7 45	243 42.2
24	**Avior**	1.7	**15 43**	S 59 30.6	**123 39.8**	8 22	234 23.2
25	**Suhail**	2.2	**16 28**	S 43 25.8	**112 17.8**	9 08	223 01.2
26	**Miaplacidus**	1.8	**16 33**	S 69 43.0	**110 59.1**	9 13	221 42.5
27	**Alphard**	2.2	**16 48**	S 8 39.2	**107 24.2**	9 27	218 07.6
28	**Regulus**	1.3	**17 28**	N11 58.5	**97 12.5**	10 08	207 55.9
29	**Dubhe**	2.0	**18 23**	N61 45.9	**83 22.6**	11 04	194 06.0
30	**Denebola**	2.2	**19 09**	N14 35.0	**72 02.0**	11 49	182 45.4
31	**Gienah**	2.8	**19 35**	S 17 32.1	**65 20.7**	12 16	176 04.1
32	**Acrux**	1.1	**19 46**	S 63 05.7	**62 38.6**	12 27	173 22.0
33	**Gacrux**	1.6	**19 51**	S 57 06.5	**61 30.2**	12 31	172 13.6
34	**Mimosa**	1.5	**20 07**	S 59 41.0	**57 21.8**	12 48	168 05.2
35	**Alioth**	1.7	**20 14**	N55 58.4	**55 47.3**	12 54	166 30.7
36	**Spica**	1.2	**20 45**	S 11 09.2	**47 59.9**	13 25	158 43.3
37	**Alkaid**	1.9	**21 07**	N49 19.6	**42 24.4**	13 47	153 07.8
38	**Hadar**	0.9	**21 23**	S 60 22.1	**38 20.4**	14 04	149 03.8
39	**Menkent**	2.3	**21 26**	S 36 21.8	**37 37.5**	14 07	148 20.9
40	**Arcturus**	0.2	**21 35**	N19 11.6	**35 22.7**	14 16	146 06.1
41	**Rigil Kent.**	0.1	**21 59**	S 60 49.8	**29 23.6**	14 40	140 07.0
42	**Zuben'ubi**	2.9	**22 10**	S 16 02.1	**26 34.5**	14 51	137 17.9
43	**Kochab**	2.2	**22 10**	N74 10.1	**26 35.4**	14 51	137 18.8
44	**Alphecca**	2.3	**22 54**	N26 43.4	**15 37.1**	15 35	126 20.5
45	**Antares**	1.2	**23 48**	S 26 25.6	**1 56.7**	16 29	112 40.1
46	**Atria**	1.9	**0 11**	S 69 01.4	**357 08.3**	16 49	107 51.7
47	**Sabik**	2.6	**0 33**	S 15 43.3	**351 42.0**	17 10	102 25.4
48	**Shaula**	1.7	**0 56**	S 37 06.0	**345 53.8**	17 34	96 37.2
49	**Rasalhague**	2.1	**0 58**	N12 33.8	**345 33.4**	17 35	96 16.8
50	**Eltanin**	2.4	**1 19**	N51 29.5	**340 07.6**	17 57	90 51.0
51	**Kaus Aust.**	2.0	**1 47**	S 34 23.0	**333 15.3**	18 24	83 58.7
52	**Vega**	0.1	**1 59**	N38 47.0	**330 02.9**	18 37	80 46.3
53	**Nunki**	2.1	**2 18**	S 26 17.8	**325 28.9**	18 55	76 12.3
54	**Altair**	0.9	**3 13**	N 8 51.9	**311 35.8**	19 51	62 19.2
55	**Peacock**	2.1	**3 48**	S 56 44.1	**302 53.6**	20 26	53 37.0
56	**Deneb**	1.3	**4 04**	N45 16.4	**298 55.6**	20 41	49 39.0
57	**Enif**	2.5	**5 06**	N 9 52.0	**283 14.9**	21 44	33 58.3
58	**Al Na'ir**	2.2	**5 30**	S 46 57.9	**277 14.7**	22 08	27 58.1
59	**Fomalhaut**	1.3	**6 19**	S 29 37.7	**264 53.3**	22 58	15 36.7
60	**Markab**	2.6	**6 27**	N15 11.7	**263 06.4**	23 05	13 49.8

Star Transit Corr Table

Date	Corr h m	Date	Corr h m
1	0 00	17	−1 03
2	−0 04	18	−1 07
3	−0 08	19	−1 11
4	−0 12	20	−1 15
5	−0 16	21	−1 19
6	−0 20	22	−1 23
7	−0 24	23	−1 27
8	−0 28	24	−1 30
9	−0 31	25	−1 34
10	−0 35	26	−1 38
11	−0 39	27	−1 42
12	−0 43	28	−1 46
13	−0 47	29	−1 50
14	−0 51	30	−1 54
15	−0 55	31	−1 58
16	−0 59		

STAR TRANSIT

To find the approximate time of transit of a star for any day of the month use the table above. All corrections are subtractive.

If the value taken from the table is greater than the time of transit for the first of the month, add 23h 56min to the time of transit before subtracting the correction.

Example: What time will Atria (No 46) transit the meridian on June 30?

	h min
Transit on Jun 1	00 11
	+23 56
	24 07
Corr for Jun 30	−01 54
Transit on Jun 30	22 13

JUNE DIARY

d h	
1 04	Regulus 1.0° N of Moon
9 19	Vesta in conjunction
10 07	Mercury in conjunction
13 12	Neptune 2° S of Moon
14 08	Uranus 3° S of Moon
17 11	Jupiter 0.8° N of Moon
19 20	Saturn 2° N of Moon
21 14	Solstice
21 14	Venus 3° N of Moon
22 14	Aldebaran 0.4° S of Moon
25 13	Mercury 5° N of Moon
27 11	Mercury 5° S of Pollux
28 12	Regulus 0.8° N of Moon

NOTES

Star declinations may be used as is for the whole month. To correct GHA for day, hour, minute, and second see page E 106; for further explanation of this table see page E 4.

JUNE 1998 — SUN & ARIES — GMT

Monday, 1st June

Time	SUN GHA ° ′	Dec ° ′	ARIES GHA ° ′
00	180 34.2	N21 59.7	249 16.6
02	210 34.0	22 00.3	279 21.5
04	240 33.8	22 01.0	309 26.4
06	270 33.6	22 01.7	339 31.3
08	300 33.4	22 02.4	9 36.3
10	330 33.2	22 03.1	39 41.2
12	0 33.0	22 03.8	69 46.1
14	30 32.9	22 04.4	99 51.1
16	60 32.7	22 05.1	129 56.0
18	90 32.5	22 05.8	160 00.9
20	120 32.3	22 06.4	190 05.8
22	150 32.1	N22 07.1	220 10.8

Tuesday, 2nd June

Time	SUN GHA ° ′	Dec ° ′	ARIES GHA ° ′
00	180 31.9	N22 07.8	250 15.7
02	210 31.7	22 08.4	280 20.6
04	240 31.5	22 09.1	310 25.6
06	270 31.3	22 09.7	340 30.5
08	300 31.1	22 10.4	10 35.4
10	330 30.9	22 11.0	40 40.3
12	0 30.7	22 11.7	70 45.3
14	30 30.5	22 12.3	100 50.2
16	60 30.3	22 13.0	130 55.1
18	90 30.1	22 13.6	161 00.1
20	120 29.9	22 14.2	191 05.0
22	150 29.7	N22 14.9	221 09.9

Wednesday, 3rd June

Time	SUN GHA ° ′	Dec ° ′	ARIES GHA ° ′
00	180 29.5	N22 15.5	251 14.8
02	210 29.3	22 16.1	281 19.8
04	240 29.1	22 16.8	311 24.7
06	270 28.9	22 17.4	341 29.6
08	300 28.7	22 18.0	11 34.5
10	330 28.5	22 18.6	41 39.5
12	0 28.3	22 19.2	71 44.4
14	30 28.1	22 19.8	101 49.3
16	60 27.9	22 20.4	131 54.3
18	90 27.7	22 21.0	161 59.2
20	120 27.5	22 21.6	192 04.1
22	150 27.2	N22 22.2	222 09.0

Thursday, 4th June

Time	SUN GHA ° ′	Dec ° ′	ARIES GHA ° ′
00	180 27.0	N22 22.8	252 14.0
02	210 26.8	22 23.4	282 18.9
04	240 26.6	22 24.0	312 23.8
06	270 26.4	22 24.6	342 28.8
08	300 26.2	22 25.2	12 33.7
10	330 26.0	22 25.8	42 38.6
12	0 25.8	22 26.4	72 43.5
14	30 25.6	22 26.9	102 48.5
16	60 25.3	22 27.5	132 53.4
18	90 25.1	22 28.1	162 58.3
20	120 24.9	22 28.7	193 03.3
22	150 24.7	N22 29.2	223 08.2

Friday, 5th June

Time	SUN GHA ° ′	Dec ° ′	ARIES GHA ° ′
00	180 24.5	N22 29.8	253 13.1
02	210 24.3	22 30.4	283 18.0
04	240 24.0	22 30.9	313 23.0
06	270 23.8	22 31.5	343 27.9
08	300 23.6	22 32.0	13 32.8
10	330 23.4	22 32.6	43 37.8
12	0 23.2	22 33.1	73 42.7
14	30 23.0	22 33.7	103 47.6
16	60 22.7	22 34.2	133 52.5
18	90 22.5	22 34.7	163 57.5
20	120 22.3	22 35.3	194 02.4
22	150 22.1	N22 35.8	224 07.3

Saturday, 6th June

Time	SUN GHA ° ′	Dec ° ′	ARIES GHA ° ′
00	180 21.8	N22 36.3	254 12.2
02	210 21.6	22 36.9	284 17.2
04	240 21.4	22 37.4	314 22.1
06	270 21.2	22 37.9	344 27.0
08	300 21.0	22 38.4	14 32.0
10	330 20.7	22 39.0	44 36.9
12	0 20.5	22 39.5	74 41.8
14	30 20.3	22 40.0	104 46.7
16	60 20.1	22 40.5	134 51.7
18	90 19.8	22 41.0	164 56.6
20	120 19.6	22 41.5	195 01.5
22	150 19.4	N22 42.0	225 06.5

Sunday, 7th June

Time	SUN GHA ° ′	Dec ° ′	ARIES GHA ° ′
00	180 19.1	N22 42.5	255 11.4
02	210 18.9	22 43.0	285 16.3
04	240 18.7	22 43.5	315 21.2
06	270 18.5	22 44.0	345 26.2
08	300 18.2	22 44.5	15 31.1
10	330 18.0	22 45.0	45 36.0
12	0 17.8	22 45.4	75 41.0
14	30 17.5	22 45.9	105 45.9
16	60 17.3	22 46.4	135 50.8
18	90 17.1	22 46.9	165 55.7
20	120 16.8	22 47.3	196 00.7
22	150 16.6	N22 47.8	226 05.6

Monday, 8th June

Time	SUN GHA ° ′	Dec ° ′	ARIES GHA ° ′
00	180 16.4	N22 48.3	256 10.5
02	210 16.1	22 48.7	286 15.5
04	240 15.9	22 49.2	316 20.4
06	270 15.7	22 49.6	346 25.3
08	300 15.4	22 50.1	16 30.2
10	330 15.2	22 50.5	46 35.2
12	0 14.9	22 51.0	76 40.1
14	30 14.7	22 51.4	106 45.0
16	60 14.5	22 51.9	136 50.0
18	90 14.2	22 52.3	166 54.9
20	120 14.0	22 52.8	196 59.8
22	150 13.8	N22 53.2	227 04.7

Tuesday, 9th June

Time	SUN GHA ° ′	Dec ° ′	ARIES GHA ° ′
00	180 13.5	N22 53.6	257 09.7
02	210 13.3	22 54.1	287 14.6
04	240 13.0	22 54.5	317 19.5
06	270 12.8	22 54.9	347 24.5
08	300 12.6	22 55.3	17 29.4
10	330 12.3	22 55.7	47 34.3
12	0 12.1	22 56.2	77 39.2
14	30 11.8	22 56.6	107 44.2
16	60 11.6	22 57.0	137 49.1
18	90 11.3	22 57.4	167 54.0
20	120 11.1	22 57.8	197 58.9
22	150 10.8	N22 58.2	228 03.9

Wednesday, 10th June

Time	SUN GHA ° ′	Dec ° ′	ARIES GHA ° ′
00	180 10.6	N22 58.6	258 08.8
02	210 10.4	22 59.0	288 13.7
04	240 10.1	22 59.4	318 18.7
06	270 09.9	22 59.8	348 23.6
08	300 09.6	23 00.1	18 28.5
10	330 09.4	23 00.5	48 33.4
12	0 09.1	23 00.9	78 38.4
14	30 08.9	23 01.3	108 43.3
16	60 08.6	23 01.7	138 48.2
18	90 08.4	23 02.0	168 53.2
20	120 08.1	23 02.4	198 58.1
22	150 07.9	N23 02.8	229 03.0

Thursday, 11th June

Time	SUN GHA ° ′	Dec ° ′	ARIES GHA ° ′	Time
00	180 07.6	N23 03.1	259 07.9	00
02	210 07.4	23 03.5	289 12.9	02
04	240 07.1	23 03.9	319 17.8	04
06	270 06.9	23 04.2	349 22.7	06
08	300 06.6	23 04.6	19 27.7	08
10	330 06.4	23 04.9	49 32.6	10
12	0 06.1	23 05.3	79 37.5	12
14	30 05.9	23 05.6	109 42.4	14
16	60 05.6	23 05.9	139 47.4	16
18	90 05.4	23 06.3	169 52.3	18
20	120 05.1	23 06.6	199 57.2	20
22	150 04.9	N23 06.9	230 02.2	22

Friday, 12th June

Time	SUN GHA ° ′	Dec ° ′	ARIES GHA ° ′	Time
00	180 04.6	N23 07.3	260 07.1	00
02	210 04.4	23 07.6	290 12.0	02
04	240 04.1	23 07.9	320 16.9	04
06	270 03.9	23 08.3	350 21.9	06
08	300 03.6	23 08.6	20 26.8	08
10	330 03.3	23 08.9	50 31.7	10
12	0 03.1	23 09.2	80 36.7	12
14	30 02.8	23 09.5	110 41.6	14
16	60 02.6	23 09.8	140 46.5	16
18	90 02.3	23 10.1	170 51.4	18
20	120 02.1	23 10.4	200 56.4	20
22	150 01.8	N23 10.7	231 01.3	22

Saturday, 13th June

Time	SUN GHA ° ′	Dec ° ′	ARIES GHA ° ′	Time
00	180 01.5	N23 11.0	261 06.2	00
02	210 01.3	23 11.3	291 11.2	02
04	240 01.0	23 11.6	321 16.1	04
06	270 00.8	23 11.9	351 21.0	06
08	300 00.5	23 12.2	21 25.9	08
10	330 00.3	23 12.5	51 30.9	10
12	0 00.0	23 12.7	81 35.8	12
14	29 59.7	23 13.0	111 40.7	14
16	59 59.5	23 13.3	141 45.7	16
18	89 59.2	23 13.6	171 50.6	18
20	119 59.0	23 13.8	201 55.5	20
22	149 58.7	N23 14.1	232 00.4	22

Sunday, 14th June

Time	SUN GHA ° ′	Dec ° ′	ARIES GHA ° ′	Time
00	179 58.4	N23 14.3	262 05.4	00
02	209 58.2	23 14.6	292 10.3	02
04	239 57.9	23 14.9	322 15.2	04
06	269 57.6	23 15.1	352 20.2	06
08	299 57.4	23 15.4	22 25.1	08
10	329 57.1	23 15.6	52 30.0	10
12	359 56.9	23 15.9	82 34.9	12
14	29 56.6	23 16.1	112 39.9	14
16	59 56.3	23 16.3	142 44.8	16
18	89 56.1	23 16.6	172 49.7	18
20	119 55.8	23 16.8	202 54.7	20
22	149 55.5	N23 17.0	232 59.6	22

Monday, 15th June

Time	SUN GHA ° ′	Dec ° ′	ARIES GHA ° ′	Time
00	179 55.3	N23 17.3	263 04.5	00
02	209 55.0	23 17.5	293 09.4	02
04	239 54.7	23 17.7	323 14.4	04
06	269 54.5	23 17.9	353 19.3	06
08	299 54.2	23 18.2	23 24.2	08
10	329 53.9	23 18.4	53 29.1	10
12	359 53.7	23 18.6	83 34.1	12
14	29 53.4	23 18.8	113 39.0	14
16	59 53.2	23 19.0	143 43.9	16
18	89 52.9	23 19.2	173 48.9	18
20	119 52.6	23 19.4	203 53.8	20
22	149 52.4	N23 19.6	233 58.7	22

NOTES

Sun and Aries GHA corrections for additional hour, minutes, and seconds are on page E 106.
Correct sun declination by interpolation. For an example see page E 4.

JUNE 1998 — SUN & ARIES — GMT

Tuesday, 16th June

Time	SUN GHA ° ′	Dec ° ′	ARIES GHA ° ′
00	179 52.1	N23 19.8	264 03.6
02	209 51.8	23 20.0	294 08.6
04	239 51.5	23 20.2	324 13.5
06	269 51.3	23 20.3	354 18.4
08	299 51.0	23 20.5	24 23.4
10	329 50.7	23 20.7	54 28.3
12	359 50.5	23 20.9	84 33.2
14	29 50.2	23 21.1	114 38.1
16	59 49.9	23 21.2	144 43.1
18	89 49.7	23 21.4	174 48.0
20	119 49.4	23 21.6	204 52.9
22	149 49.1	N23 21.7	234 57.9

Wednesday, 17th June

Time	SUN GHA ° ′	Dec ° ′	ARIES GHA ° ′
00	179 48.9	N23 21.9	265 02.8
02	209 48.6	23 22.0	295 07.7
04	239 48.3	23 22.2	325 12.6
06	269 48.1	23 22.3	355 17.6
08	299 47.8	23 22.5	25 22.5
10	329 47.5	23 22.6	55 27.4
12	359 47.2	23 22.8	85 32.4
14	29 47.0	23 22.9	115 37.3
16	59 46.7	23 23.0	145 42.2
18	89 46.4	23 23.2	175 47.1
20	119 46.2	23 23.3	205 52.1
22	149 45.9	N23 23.4	235 57.0

Thursday, 18th June

Time	SUN GHA ° ′	Dec ° ′	ARIES GHA ° ′
00	179 45.6	N23 23.6	266 01.9
02	209 45.3	23 23.7	296 06.9
04	239 45.1	23 23.8	326 11.8
06	269 44.8	23 23.9	356 16.7
08	299 44.5	23 24.0	26 21.6
10	329 44.3	23 24.1	56 26.6
12	359 44.0	23 24.3	86 31.5
14	29 43.7	23 24.4	116 36.4
16	59 43.4	23 24.5	146 41.3
18	89 43.2	23 24.6	176 46.3
20	119 42.9	23 24.7	206 51.2
22	149 42.6	N23 24.7	236 56.1

Friday, 19th June

Time	SUN GHA ° ′	Dec ° ′	ARIES GHA ° ′
00	179 42.4	N23 24.8	267 01.1
02	209 42.1	23 24.9	297 06.0
04	239 41.8	23 25.0	327 10.9
06	269 41.5	23 25.1	357 15.8
08	299 41.3	23 25.2	27 20.8
10	329 41.0	23 25.2	57 25.7
12	359 40.7	23 25.3	87 30.6
14	29 40.4	23 25.4	117 35.6
16	59 40.2	23 25.5	147 40.5
18	89 39.9	23 25.5	177 45.4
20	119 39.6	23 25.6	207 50.3
22	149 39.3	N23 25.6	237 55.3

Saturday, 20th June

Time	SUN GHA ° ′	Dec ° ′	ARIES GHA ° ′
00	179 39.1	N23 25.7	268 00.2
02	209 38.8	23 25.8	298 05.1
04	239 38.5	23 25.8	328 10.1
06	269 38.3	23 25.8	358 15.0
08	299 38.0	23 25.9	28 19.9
10	329 37.7	23 25.9	58 24.8
12	359 37.4	23 26.0	88 29.8
14	29 37.2	23 26.0	118 34.7
16	59 36.9	23 26.0	148 39.6
18	89 36.6	23 26.1	178 44.6
20	119 36.3	23 26.1	208 49.5
22	149 36.1	N23 26.1	238 54.4

Sunday, 21st June

Time	SUN GHA ° ′	Dec ° ′	ARIES GHA ° ′
00	179 35.8	N23 26.1	268 59.3
02	209 35.5	23 26.2	299 04.3
04	239 35.2	23 26.2	329 09.2
06	269 35.0	23 26.2	359 14.1
08	299 34.7	23 26.2	29 19.1
10	329 34.4	23 26.2	59 24.0
12	359 34.2	23 26.2	89 28.9
14	29 33.9	23 26.2	119 33.8
16	59 33.6	23 26.2	149 38.8
18	89 33.3	23 26.2	179 43.7
20	119 33.1	23 26.2	209 48.6
22	149 32.8	N23 26.2	239 53.5

Monday, 22nd June

Time	SUN GHA ° ′	Dec ° ′	ARIES GHA ° ′
00	179 32.5	N23 26.2	269 58.5
02	209 32.2	23 26.2	300 03.4
04	239 32.0	23 26.1	330 08.3
06	269 31.7	23 26.1	0 13.3
08	299 31.4	23 26.1	30 18.2
10	329 31.1	23 26.1	60 23.1
12	359 30.9	23 26.0	90 28.0
14	29 30.6	23 26.0	120 33.0
16	59 30.3	23 26.0	150 37.9
18	89 30.1	23 25.9	180 42.8
20	119 29.8	23 25.9	210 47.8
22	149 29.5	N23 25.8	240 52.7

Tuesday, 23rd June

Time	SUN GHA ° ′	Dec ° ′	ARIES GHA ° ′
00	179 29.2	N23 25.8	270 57.6
02	209 29.0	23 25.8	301 02.5
04	239 28.7	23 25.7	331 07.5
06	269 28.4	23 25.6	1 12.4
08	299 28.1	23 25.6	31 17.3
10	329 27.9	23 25.5	61 22.3
12	359 27.6	23 25.5	91 27.2
14	29 27.3	23 25.4	121 32.1
16	59 27.1	23 25.3	151 37.0
18	89 26.8	23 25.2	181 42.0
20	119 26.5	23 25.2	211 46.9
22	149 26.2	N23 25.1	241 51.8

Wednesday, 24th June

Time	SUN GHA ° ′	Dec ° ′	ARIES GHA ° ′
00	179 26.0	N23 25.0	271 56.8
02	209 25.7	23 24.9	302 01.7
04	239 25.4	23 24.8	332 06.6
06	269 25.2	23 24.7	2 11.5
08	299 24.9	23 24.7	32 16.5
10	329 24.6	23 24.6	62 21.4
12	359 24.4	23 24.5	92 26.3
14	29 24.1	23 24.4	122 31.3
16	59 23.8	23 24.3	152 36.2
18	89 23.5	23 24.1	182 41.1
20	119 23.3	23 24.0	212 46.0
22	149 23.0	N23 23.9	242 51.0

Thursday, 25th June

Time	SUN GHA ° ′	Dec ° ′	ARIES GHA ° ′
00	179 22.7	N23 23.8	272 55.9
02	209 22.5	23 23.7	303 00.8
04	239 22.2	23 23.6	333 05.8
06	269 21.9	23 23.4	3 10.7
08	299 21.7	23 23.3	33 15.6
10	329 21.4	23 23.2	63 20.5
12	359 21.1	23 23.1	93 25.5
14	29 20.9	23 22.9	123 30.4
16	59 20.6	23 22.8	153 35.3
18	89 20.3	23 22.6	183 40.3
20	119 20.1	23 22.5	213 45.2
22	149 19.8	N23 22.3	243 50.1

Friday, 26th June

Time	SUN GHA ° ′	Dec ° ′	ARIES GHA ° ′	Time
00	179 19.5	N23 22.2	273 55.0	00
02	209 19.3	23 22.0	304 00.0	02
04	239 19.0	23 21.9	334 04.9	04
06	269 18.7	23 21.7	4 09.8	06
08	299 18.5	23 21.6	34 14.8	08
10	329 18.2	23 21.4	64 19.7	10
12	359 17.9	23 21.2	94 24.6	12
14	29 17.7	23 21.1	124 29.5	14
16	59 17.4	23 20.9	154 34.5	16
18	89 17.1	23 20.7	184 39.4	18
20	119 16.9	23 20.5	214 44.3	20
22	149 16.6	N23 20.4	244 49.3	22

Saturday, 27th June

Time	SUN GHA ° ′	Dec ° ′	ARIES GHA ° ′	Time
00	179 16.4	N23 20.2	274 54.2	00
02	209 16.1	23 20.0	304 59.1	02
04	239 15.8	23 19.8	335 04.0	04
06	269 15.6	23 19.6	5 09.0	06
08	299 15.3	23 19.4	35 13.9	08
10	329 15.0	23 19.2	65 18.8	10
12	359 14.8	23 19.0	95 23.7	12
14	29 14.5	23 18.8	125 28.7	14
16	59 14.3	23 18.6	155 33.6	16
18	89 14.0	23 18.4	185 38.5	18
20	119 13.7	23 18.2	215 43.5	20
22	149 13.5	N23 17.9	245 48.4	22

Sunday, 28th June

Time	SUN GHA ° ′	Dec ° ′	ARIES GHA ° ′	Time
00	179 13.2	N23 17.7	275 53.3	00
02	209 13.0	23 17.5	305 58.2	02
04	239 12.7	23 17.3	336 03.2	04
06	269 12.5	23 17.1	6 08.1	06
08	299 12.2	23 16.8	36 13.0	08
10	329 11.9	23 16.6	66 18.0	10
12	359 11.7	23 16.4	96 22.9	12
14	29 11.4	23 16.1	126 27.8	14
16	59 11.2	23 15.9	156 32.7	16
18	89 10.9	23 15.6	186 37.7	18
20	119 10.7	23 15.4	216 42.6	20
22	149 10.4	N23 15.1	246 47.5	22

Monday, 29th June

Time	SUN GHA ° ′	Dec ° ′	ARIES GHA ° ′	Time
00	179 10.1	N23 14.9	276 52.5	00
02	209 09.9	23 14.6	306 57.4	02
04	239 09.6	23 14.4	337 02.3	04
06	269 09.4	23 14.1	7 07.2	06
08	299 09.1	23 13.8	37 12.2	08
10	329 08.9	23 13.6	67 17.1	10
12	359 08.6	23 13.3	97 22.0	12
14	29 08.4	23 13.0	127 27.0	14
16	59 08.1	23 12.8	157 31.9	16
18	89 07.9	23 12.5	187 36.8	18
20	119 07.6	23 12.2	217 41.7	20
22	149 07.4	N23 11.9	247 46.7	22

Tuesday, 30th June

Time	SUN GHA ° ′	Dec ° ′	ARIES GHA ° ′	Time
00	179 07.1	N23 11.6	277 51.6	00
02	209 06.9	23 11.3	307 56.5	02
04	239 06.6	23 11.1	338 01.5	04
06	269 06.4	23 10.8	8 06.4	06
08	299 06.1	23 10.5	38 11.3	08
10	329 05.9	23 10.2	68 16.2	10
12	359 05.6	23 09.9	98 21.2	12
14	29 05.4	23 09.5	128 26.1	14
16	59 05.1	23 09.2	158 31.0	16
18	89 04.9	23 08.9	188 35.9	18
20	119 04.7	23 08.6	218 40.9	20
22	149 04.4	N23 08.3	248 45.8	22

JUNE 1998 PLANETS 0h GMT

VENUS						JUPITER				
Mer Pass h m	GHA ° ′	Mean Var/hr 14°+	Dec ° ′	Mean Var/hr ′	Day	GHA ° ′	Mean Var/hr 15°+	Dec ° ′	Mean Var/hr ′	Mer Pass h m
09 27	218 23.8	59.7	N10 21.8	1.0	1 Mon	253 42.1	2.2	S 3 08.6	0.1	07 04
09 27	218 16.1	59.7	N10 45.5	1.0	2 Tu	254 33.8	2.2	S 3 05.6	0.1	07 01
09 28	218 08.1	59.7	N11 09.1	1.0	3 Wed	255 25.6	2.2	S 3 02.7	0.1	06 57
09 28	218 00.0	59.7	N11 32.4	1.0	4 Th	256 17.5	2.2	S 2 59.9	0.1	06 54
09 29	217 51.6	59.6	N11 55.5	1.0	5 Fri	257 09.5	2.2	S 2 57.1	0.1	06 50
09 29	217 42.9	59.6	N12 18.4	0.9	6 Sat	258 01.7	2.2	S 2 54.4	0.1	06 47
09 30	217 34.1	59.6	N12 41.1	0.9	7 SUN	258 54.0	2.2	S 2 51.7	0.1	06 43
09 31	217 24.9	59.6	N13 03.5	0.9	8 Mon	259 46.5	2.2	S 2 49.1	0.1	06 40
09 31	217 15.6	59.6	N13 25.7	0.9	9 Tu	260 39.0	2.2	S 2 46.6	0.1	06 36
09 32	217 06.0	59.6	N13 47.6	0.9	10 Wed	261 31.8	2.2	S 2 44.1	0.1	06 33
09 33	216 56.1	59.6	N14 09.3	0.9	11 Th	262 24.6	2.2	S 2 41.7	0.1	06 29
09 33	216 46.0	59.6	N14 30.6	0.9	12 Fri	263 17.6	2.2	S 2 39.3	0.1	06 26
09 34	216 35.5	59.6	N14 51.7	0.9	13 Sat	264 10.7	2.2	S 2 37.0	0.1	06 22
09 35	216 24.9	59.5	N15 12.5	0.8	14 SUN	265 04.0	2.2	S 2 34.8	0.1	06 19
09 35	216 13.9	59.5	N15 32.9	0.8	15 Mon	265 57.4	2.2	S 2 32.6	0.1	06 15
09 36	216 02.7	59.5	N15 53.0	0.8	16 Tu	266 51.0	2.2	S 2 30.5	0.1	06 12
09 37	215 51.2	59.5	N16 12.8	0.8	17 Wed	267 44.7	2.2	S 2 28.4	0.1	06 08
09 38	215 39.4	59.5	N16 32.2	0.8	18 Th	268 38.6	2.3	S 2 26.5	0.1	06 05
09 39	215 27.3	59.5	N16 51.3	0.8	19 Fri	269 32.6	2.3	S 2 24.5	0.1	06 01
09 39	215 14.9	59.5	N17 09.9	0.8	20 Sat	270 26.8	2.3	S 2 22.7	0.1	05 57
09 40	215 02.3	59.5	N17 28.2	0.7	21 SUN	271 21.1	2.3	S 2 20.9	0.1	05 54
09 41	214 49.3	59.5	N17 46.1	0.7	22 Mon	272 15.6	2.3	S 2 19.2	0.1	05 50
09 42	214 36.1	59.4	N18 03.6	0.7	23 Tu	273 10.2	2.3	S 2 17.6	0.1	05 46
09 43	214 22.6	59.4	N18 20.6	0.7	24 Wed	274 05.0	2.3	S 2 16.0	0.1	05 43
09 44	214 08.9	59.4	N18 37.2	0.7	25 Th	275 00.0	2.3	S 2 14.5	0.1	05 39
09 45	213 54.8	59.4	N18 53.4	0.7	26 Fri	275 55.1	2.3	S 2 13.1	0.1	05 35
09 46	213 40.5	59.4	N19 09.1	0.6	27 Sat	276 50.4	2.3	S 2 11.8	0.1	05 32
09 47	213 25.9	59.4	N19 24.4	0.6	28 SUN	277 45.9	2.3	S 2 10.5	0.0	05 28
09 48	213 11.0	59.4	N19 39.2	0.6	29 Mon	278 41.5	2.3	S 2 09.3	0.0	05 24
09 49	212 55.9	59.4	N19 53.4	0.6	30 Tu	279 37.3	2.3	S 2 08.2	0.0	05 21

VENUS, Av. Mag. -4.0
S.H.A. June
5 325; 10 319; 15 313; 20 307; 25 301; 30 295.

JUPITER, Av. Mag.-2.4
S.H.A. June
5 4; 10 3; 15 3; 20 2; 25 2; 30 2.

MARS						SATURN				
Mer Pass h m	GHA ° ′	Mean Var/hr 15°+	Dec ° ′	Mean Var/hr ′	Day	GHA ° ′	Mean Var/hr 15°+	Dec ° ′	Mean Var/hr ′	Mer Pass h m
11 37	185 40.3	0.6	N21 28.3	0.3	1 Mon	221 20.7	2.2	N 9 01.2	0.1	09 13
11 36	185 54.8	0.6	N21 36.3	0.3	2 Tu	222 13.8	2.2	N 9 03.2	0.1	09 10
11 35	186 09.2	0.6	N21 44.0	0.3	3 Wed	223 06.9	2.2	N 9 05.3	0.1	09 06
11 34	186 23.6	0.6	N21 51.6	0.3	4 Th	224 00.1	2.2	N 9 07.4	0.1	09 03
11 33	186 38.1	0.6	N21 59.0	0.3	5 Fri	224 53.4	2.2	N 9 09.4	0.1	08 59
11 32	186 52.4	0.6	N22 06.1	0.3	6 Sat	225 46.7	2.2	N 9 11.4	0.1	08 56
11 31	187 06.8	0.6	N22 13.1	0.3	7 SUN	226 40.0	2.2	N 9 13.3	0.1	08 52
11 30	187 21.2	0.6	N22 19.8	0.3	8 Mon	227 33.4	2.2	N 9 15.3	0.1	08 48
11 29	187 35.5	0.6	N22 26.3	0.3	9 Tu	228 26.8	2.2	N 9 17.2	0.1	08 45
11 28	187 49.8	0.6	N22 32.6	0.3	10 Wed	229 20.3	2.2	N 9 19.1	0.1	08 41
11 27	188 04.1	0.6	N22 38.7	0.2	11 Th	230 13.9	2.2	N 9 21.0	0.1	08 38
11 26	188 18.4	0.6	N22 44.6	0.2	12 Fri	231 07.5	2.2	N 9 22.8	0.1	08 34
11 25	188 32.7	0.6	N22 50.3	0.2	13 Sat	232 01.1	2.2	N 9 24.6	0.1	08 31
11 24	188 47.0	0.6	N22 55.7	0.2	14 SUN	232 54.8	2.2	N 9 26.4	0.1	08 27
11 23	189 01.2	0.6	N23 01.0	0.2	15 Mon	233 48.6	2.2	N 9 28.2	0.1	08 24
11 23	189 15.5	0.6	N23 06.0	0.2	16 Tu	234 42.5	2.2	N 9 29.9	0.1	08 20
11 22	189 29.7	0.6	N23 10.8	0.2	17 Wed	235 36.3	2.3	N 9 31.7	0.1	08 16
11 21	189 44.0	0.6	N23 15.4	0.2	18 Th	236 30.3	2.3	N 9 33.3	0.1	08 13
11 20	189 58.2	0.6	N23 19.8	0.2	19 Fri	237 24.3	2.3	N 9 35.0	0.1	08 09
11 19	190 12.5	0.6	N23 24.0	0.2	20 Sat	238 18.4	2.3	N 9 36.6	0.1	08 06
11 18	190 26.8	0.6	N23 27.9	0.2	21 SUN	239 12.6	2.3	N 9 38.2	0.1	08 02
11 17	190 41.0	0.6	N23 31.7	0.1	22 Mon	240 06.8	2.3	N 9 39.8	0.1	07 58
11 16	190 55.3	0.6	N23 35.2	0.1	23 Tu	241 01.0	2.3	N 9 41.4	0.1	07 55
11 15	191 09.6	0.6	N23 38.5	0.1	24 Wed	241 55.4	2.3	N 9 42.9	0.1	07 51
11 14	191 23.9	0.6	N23 41.6	0.1	25 Th	242 49.8	2.3	N 9 44.4	0.1	07 48
11 13	191 38.2	0.6	N23 44.5	0.1	26 Fri	243 44.3	2.3	N 9 45.8	0.1	07 44
11 12	191 52.6	0.6	N23 47.1	0.1	27 Sat	244 38.8	2.3	N 9 47.3	0.1	07 40
11 11	192 07.0	0.6	N23 49.6	0.1	28 SUN	245 33.5	2.3	N 9 48.7	0.1	07 37
11 10	192 21.4	0.6	N23 51.8	0.1	29 Mon	246 28.2	2.3	N 9 50.1	0.1	07 33
11 09	192 35.8	0.6	N23 53.8	0.1	30 Tu	247 22.9	2.3	N 9 51.4	0.1	07 29

MARS, Av. Mag. +1.5
S.H.A. June
5 293; 10 290; 15 286; 20 282; 25 278; 30 275.

NOTE: Planet corrections on pages E113-15

SATURN, Av. Mag. +0.5
S.H.A. June
5 332; 10 331; 15 331; 20 330; 25 330; 30 330.

JUNE 1998

MOON

Day	GMT hr	GHA ° ′	Mean Var/hr 14°+	Dec ° ′	Mean Var/hr ′
1 Mon	**0**	99 03.1	32.2	**N 11 28.5**	8.3
	6	186 16.2	32.5	**N 10 39.1**	8.5
	12	273 31.4	32.8	**N 9 48.3**	8.7
	18	0 48.4	33.1	**N 8 56.2**	8.9
2 Tu	**0**	88 07.1	33.4	**N 8 03.0**	9.0
	6	175 27.4	33.6	**N 7 08.7**	9.2
	12	262 49.1	33.8	**N 6 13.6**	9.3
	18	350 12.0	34.0	**N 5 17.8**	9.4
3 Wed	**0**	77 36.0	34.2	**N 4 21.3**	9.5
	6	165 00.9	34.3	**N 3 24.4**	9.5
	12	252 26.5	34.4	**N 2 27.1**	9.6
	18	339 52.7	34.5	**N 1 29.6**	9.7
4 Th	**0**	67 19.3	34.4	**N 0 31.9**	9.7
	6	154 46.1	34.5	**S 0 25.8**	9.6
	12	242 13.0	34.5	**S 1 23.4**	9.6
	18	329 39.8	34.4	**S 2 20.8**	9.5
5 Fri	**0**	57 06.4	34.4	**S 3 17.8**	9.4
	6	144 32.5	34.3	**S 4 14.5**	9.3
	12	231 58.2	34.2	**S 5 10.5**	9.2
	18	319 23.1	34.0	**S 6 05.9**	9.1
6 Sat	**0**	46 47.2	33.8	**S 7 00.5**	8.9
	6	134 10.3	33.6	**S 7 54.2**	8.8
	12	221 32.3	33.4	**S 8 46.8**	8.5
	18	308 53.1	33.3	**S 9 38.2**	8.3
7 Sun	**0**	36 12.6	33.0	**S 10 28.4**	8.1
	6	123 30.6	32.7	**S 11 17.0**	7.8
	12	210 47.1	32.4	**S 12 04.1**	7.5
	18	298 02.0	32.2	**S 12 49.4**	7.2
8 Mon	**0**	25 15.2	31.9	**S 13 32.8**	6.9
	6	112 26.6	31.6	**S 14 14.2**	6.5
	12	199 36.2	31.3	**S 14 53.5**	6.1
	18	286 44.1	31.0	**S 15 30.3**	5.7
9 Tu	**0**	13 50.1	30.6	**S 16 04.7**	5.3
	6	100 54.2	30.4	**S 16 36.5**	4.8
	12	187 56.6	30.1	**S 17 05.4**	4.3
	18	274 57.2	29.8	**S 17 31.5**	3.8
10 Wed	**0**	1 56.1	29.5	**S 17 54.5**	3.2
	6	88 53.3	29.2	**S 18 14.3**	2.7
	12	175 49.0	29.0	**S 18 30.8**	2.1
	18	262 43.3	28.8	**S 18 43.9**	1.5
11 Th	**0**	349 36.3	28.6	**S 18 53.4**	0.9
	6	76 28.1	28.4	**S 18 59.4**	0.3
	12	163 18.8	28.3	**S 19 01.7**	0.3
	18	250 08.7	28.2	**S 19 00.2**	0.9
12 Fri	**0**	336 57.9	28.1	**S 18 55.0**	1.5
	6	63 46.5	28.0	**S 18 46.0**	2.2
	12	150 34.7	27.9	**S 18 33.2**	2.8
	18	237 22.6	28.0	**S 18 16.6**	3.5
13 Sat	**0**	324 10.5	28.0	**S 17 56.2**	4.0
	6	50 58.4	28.0	**S 17 32.1**	4.7
	12	137 46.4	28.0	**S 17 04.4**	5.3
	18	224 34.8	28.2	**S 16 33.0**	5.8
14 Sun	**0**	311 23.6	28.2	**S 15 58.2**	6.5
	6	38 12.8	28.3	**S 15 19.9**	7.0
	12	125 02.5	28.4	**S 14 38.4**	7.5
	18	211 52.9	28.5	**S 13 53.8**	8.0
15 Mon	**0**	298 43.8	28.6	**S 13 06.2**	8.4
	6	25 35.4	28.7	**S 12 15.7**	8.9
	12	112 27.5	28.8	**S 11 22.6**	9.3
	18	199 20.2	28.9	**S 10 27.0**	9.7
16 Tu	**0**	286 13.4	28.9	**S 9 29.1**	10.1
	6	13 07.1	29.1	**S 8 29.1**	10.4
	12	100 01.1	29.1	**S 7 27.1**	10.6
	18	186 55.4	29.1	**S 6 23.5**	10.9
17 Wed	**0**	273 49.9	29.1	**S 5 18.3**	11.1
	6	0 44.4	29.1	**S 4 11.8**	11.3
	12	87 38.8	29.0	**S 3 04.3**	11.4
	18	174 33.0	29.0	**S 1 56.0**	11.5
18 Th	**0**	261 26.9	28.9	**S 0 47.0**	11.5
	6	348 20.3	28.8	**N 0 22.4**	11.6
	12	75 13.1	28.6	**N 1 31.8**	11.6
	18	162 05.2	28.5	**N 2 41.1**	11.5
19 Fri	**0**	248 56.3	28.4	**N 3 50.1**	11.4
	6	335 46.5	28.2	**N 4 58.4**	11.2
	12	62 35.5	28.0	**N 6 05.8**	11.1
	18	149 23.2	27.8	**N 7 12.0**	10.7
20 Sat	**0**	236 09.7	27.5	**N 8 16.7**	10.5
	6	322 54.7	27.2	**N 9 19.8**	10.1
	12	49 38.3	27.0	**N 10 20.8**	9.8
	18	136 20.4	26.8	**N 11 19.6**	9.3
21 Sun	**0**	223 00.9	26.5	**N 12 15.9**	8.9
	6	309 40.0	26.2	**N 13 09.4**	8.4
	12	36 17.6	26.0	**N 13 59.9**	7.8
	18	122 53.9	25.8	**N 14 47.1**	7.2
22 Mon	**0**	209 28.8	25.6	**N 15 30.8**	6.6
	6	296 02.6	25.4	**N 16 10.8**	6.0
	12	22 35.5	25.3	**N 16 47.0**	5.3
	18	109 07.5	25.2	**N 17 19.1**	4.6
23 Tu	**0**	195 38.9	25.2	**N 17 47.0**	3.8
	6	282 10.0	25.2	**N 18 10.6**	3.1
	12	8 41.0	25.2	**N 18 29.9**	2.4
	18	95 12.1	25.2	**N 18 44.7**	1.7
24 Wed	**0**	181 43.7	25.4	**N 18 55.0**	1.0
	6	268 16.0	25.5	**N 19 00.9**	0.2
	12	354 49.2	25.8	**N 19 02.4**	0.5
	18	81 23.7	26.0	**N 18 59.5**	1.3
25 Th	**0**	167 59.7	26.3	**N 18 52.4**	1.9
	6	254 37.3	26.6	**N 18 41.1**	2.6
	12	341 16.9	27.0	**N 18 25.7**	3.3
	18	67 58.6	27.3	**N 18 06.5**	3.9
26 Fri	**0**	154 42.5	27.8	**N 17 43.6**	4.5
	6	241 28.9	28.1	**N 17 17.1**	5.0
	12	328 17.7	28.6	**N 16 47.3**	5.5
	18	55 09.0	29.0	**N 16 14.4**	6.0
27 Sat	**0**	142 03.0	29.5	**N 15 38.5**	6.5
	6	228 59.5	29.9	**N 14 59.9**	6.9
	12	315 58.7	30.3	**N 14 18.8**	7.2
	18	43 00.4	30.7	**N 13 35.3**	7.7
28 Sun	**0**	130 04.6	31.1	**N 12 49.7**	8.0
	6	217 11.2	31.5	**N 12 02.2**	8.2
	12	304 20.2	31.9	**N 11 12.9**	8.5
	18	31 31.3	32.3	**N 10 22.0**	8.7
29 Mon	**0**	118 44.6	32.6	**N 9 29.7**	8.9
	6	205 59.8	32.8	**N 8 36.3**	9.1
	12	293 16.9	33.1	**N 7 41.7**	9.2
	18	20 35.6	33.4	**N 6 46.2**	9.4
30 Tu	**0**	107 55.9	33.7	**N 5 49.9**	9.5
	6	195 17.5	33.8	**N 4 53.0**	9.6
	12	282 40.3	34.0	**N 3 55.7**	9.6
	18	10 04.1	34.1	**N 2 57.9**	9.7

NOTE: Moon GHA corrections for additional hours, minutes, and seconds are on page E 108–10. Moon declination corrections are on page E 111. The mean var/hr value is needed for both corrections.

JULY 1998 SUN & MOON GMT

SUN

Yr	Mth	Week	Equation of Time 0h	Equation of Time 12h	Transit	Semi-diam	Lat 30°N Twi-light	Lat 30°N Sun-rise	Lat 30°N Sun-set	Lat 30°N Twi-light
			m s	m s	h m	′	h m	h m	h m	h m
182	**1**	**Wed**	+03 43	+03 49	**12 04**	15.8	04 35	05 02	19 05	19 33
183	**2**	**Th**	+03 55	+04 00	**12 04**	15.8	04 35	05 03	19 05	19 32
184	**3**	**Fri**	+04 06	+04 12	**12 04**	15.8	04 36	05 03	19 05	19 32
185	**4**	**Sat**	+04 17	+04 22	**12 04**	15.8	04 36	05 04	19 05	19 32
186	**5**	**Sun**	+04 28	+04 33	**12 05**	15.8	04 37	05 04	19 05	19 32
187	**6**	**Mon**	+04 38	+04 43	**12 05**	15.8	04 37	05 04	19 05	19 32
188	**7**	**Tu**	+04 48	+04 53	**12 05**	15.8	04 38	05 05	19 05	19 32
189	**8**	**Wed**	+04 58	+05 02	**12 05**	15.8	04 38	05 05	19 05	19 32
190	**9**	**Th**	+05 07	+05 11	**12 05**	15.8	04 39	05 06	19 04	19 31
191	**10**	**Fri**	+05 15	+05 20	**12 05**	15.8	04 39	05 06	19 04	19 31
192	**11**	**Sat**	+05 24	+05 28	**12 05**	15.8	04 40	05 07	19 04	19 31
193	**12**	**Sun**	+05 32	+05 36	**12 06**	15.8	04 40	05 07	19 04	19 31
194	**13**	**Mon**	+05 39	+05 43	**12 06**	15.8	04 41	05 08	19 03	19 30
195	**14**	**Tu**	+05 46	+05 50	**12 06**	15.8	04 41	05 08	19 03	19 30
196	**15**	**Wed**	+05 53	+05 56	**12 06**	15.8	04 42	05 09	19 03	19 30
197	**16**	**Th**	+05 59	+06 02	**12 06**	15.8	04 43	05 09	19 02	19 29
198	**17**	**Fri**	+06 04	+06 07	**12 06**	15.8	04 43	05 10	19 02	19 29
199	**18**	**Sat**	+06 10	+06 12	**12 06**	15.8	04 44	05 11	19 02	19 28
200	**19**	**Sun**	+06 14	+06 16	**12 06**	15.8	04 44	05 11	19 01	19 28
201	**20**	**Mon**	+06 18	+06 20	**12 06**	15.8	04 45	05 12	19 01	19 27
202	**21**	**Tu**	+06 22	+06 23	**12 06**	15.8	04 46	05 12	19 00	19 27
203	**22**	**Wed**	+06 25	+06 26	**12 06**	15.8	04 46	05 13	19 00	19 26
204	**23**	**Th**	+06 27	+06 28	**12 06**	15.8	04 47	05 13	18 59	19 26
205	**24**	**Fri**	+06 29	+06 30	**12 06**	15.8	04 48	05 14	18 59	19 25
206	**25**	**Sat**	+06 30	+06 30	**12 07**	15.8	04 48	05 15	18 58	19 24
207	**26**	**Sun**	+06 31	+06 31	**12 07**	15.8	04 49	05 15	18 57	19 24
208	**27**	**Mon**	+06 31	+06 31	**12 07**	15.8	04 50	05 16	18 57	19 23
209	**28**	**Tu**	+06 30	+06 30	**12 06**	15.8	04 50	05 16	18 56	19 22
210	**29**	**Wed**	+06 29	+06 28	**12 06**	15.8	04 51	05 17	18 56	19 22
211	**30**	**Th**	+06 27	+06 26	**12 06**	15.8	04 52	05 18	18 55	19 21
212	**31**	**Fri**	+06 25	+06 23	**12 06**	15.8	04 52	05 18	18 54	19 20

Lat Corr to Sunrise, Sunset etc

Lat	Twi-light	Sun-rise	Sun-set	Twi-light
°	h m	h m	h m	h m
N60	−2 58	−2 06	+2 06	+2 58
55	−1 54	−1 28	+1 28	+1 54
50	−1 17	−1 02	+1 02	+1 17
45	−0 51	−0 42	+0 42	+0 51
N40	−0 30	−0 25	+0 25	+0 30
35	−0 14	−0 12	+0 12	+0 14
30	0 00	0 00	0 00	0 00
25	+0 12	+0 11	−0 11	−0 12
N20	+0 23	+0 20	−0 20	−0 23
15	+0 33	+0 29	−0 29	−0 33
10	+0 42	+0 37	−0 37	−0 42
N 5	+0 50	+0 46	−0 46	−0 50
0	+0 58	+0 54	−0 54	−0 58
S 5	+1 06	+1 01	−1 01	−1 06
10	+1 14	+1 09	−1 09	−1 14
15	+1 22	+1 18	−1 18	−1 22
S20	+1 29	+1 26	−1 26	−1 29
25	+1 38	+1 36	−1 36	−1 38
30	+1 47	+1 46	−1 46	−1 47
35	+1 56	+1 57	−1 57	−1 56
S40	+2 07	+2 10	−2 10	−2 07
45	+2 19	+2 25	−2 25	−2 19
50	+2 33	+2 43	−2 43	−2 33

NOTES

Latitude corrections are for mid-month.

Equation of time = mean time (LHA mean sun) minus apparent time (LHA true sun).

MOON

Yr	Mth	Week	Age	Transit (Upper)	Diff	Semi-diam	Hor Par	Lat 30°N Moon-rise	Lat 30°N Moon-set
			days	h m	m	′	′	h m	h m
182	**1**	**Wed**	07	**18 01**	**42**	14.8	54.4	11 51	- -
183	**2**	**Th**	08	**18 43**	**43**	14.8	54.3	12 42	00 07
184	**3**	**Fri**	09	**19 26**	**44**	14.8	54.3	13 34	00 40
185	**4**	**Sat**	10	**20 10**	**46**	14.9	54.5	14 26	01 14
186	**5**	**Sun**	11	**20 56**	**48**	14.9	54.9	15 20	01 50
187	**6**	**Mon**	12	**21 44**	**51**	15.1	55.3	16 14	02 29
188	**7**	**Tu**	13	**22 35**	**53**	15.2	55.9	17 08	03 13
189	**8**	**Wed**	14	**23 28**	**53**	15.4	56.5	18 02	04 00
190	**9**	**Th**	15	**24 21**	**-**	15.6	57.1	18 55	04 53
191	**10**	**Fri**	16	**00 21**	**55**	15.7	57.6	19 45	05 49
192	**11**	**Sat**	17	**01 16**	**53**	15.8	58.2	20 32	06 49
193	**12**	**Sun**	18	**02 09**	**53**	16.0	58.6	21 17	07 50
194	**13**	**Mon**	19	**03 02**	**52**	16.0	58.9	21 59	08 53
195	**14**	**Tu**	20	**03 54**	**52**	16.1	59.1	22 39	09 55
196	**15**	**Wed**	21	**04 46**	**51**	16.1	59.3	23 19	10 58
197	**16**	**Th**	22	**05 37**	**53**	16.2	59.3	- -	12 01
198	**17**	**Fri**	23	**06 30**	**53**	16.2	59.3	00 00	13 05
199	**18**	**Sat**	24	**07 23**	**55**	16.1	59.2	00 43	14 08
200	**19**	**Sun**	25	**08 18**	**56**	16.1	58.9	01 28	15 11
201	**20**	**Mon**	26	**09 14**	**56**	16.0	58.6	02 17	16 13
202	**21**	**Tu**	27	**10 10**	**56**	15.9	58.2	03 10	17 11
203	**22**	**Wed**	28	**11 06**	**54**	15.7	57.8	04 05	18 05
204	**23**	**Th**	29	**12 00**	**52**	15.6	57.2	05 03	18 55
205	**24**	**Fri**	01	**12 52**	**50**	15.4	56.6	06 01	19 40
206	**25**	**Sat**	02	**13 42**	**46**	15.3	56.0	06 58	20 20
207	**26**	**Sun**	03	**14 28**	**45**	15.1	55.4	07 54	20 58
208	**27**	**Mon**	04	**15 13**	**43**	15.0	54.9	08 48	21 33
209	**28**	**Tu**	05	**15 56**	**42**	14.9	54.6	09 41	22 06
210	**29**	**Wed**	06	**16 38**	**43**	14.8	54.3	10 33	22 39
211	**30**	**Th**	07	**17 21**	**43**	14.8	54.2	11 25	23 12
212	**31**	**Fri**	08	**18 04**	**45**	14.8	54.3	12 17	23 48

Phases of the Moon

	d	h	m
☾ First Quarter	1	18	43
○ Full Moon	9	16	01
☽ Last Quarter	16	15	13
● New Moon	23	13	44
☾ First Quarter	31	12	05
Apogee	2	17	
Perigee	16	14	
Apogee	30	12	
On 0°	1	12	
Farthest S of 0°	8	21	
On 0°	15	10	
Farthest N of 0°	21	20	
On 0°	28	21	

NOTES

Diff equals daily change in transit time.

To correct moonrise, moonset, or transit times for latitude and/or longitude see page E 112. For further information about these tables see page E 4. For arc-to-time correction see page E 103.

JULY 1998 STARS 0h GMT July 1

No	Name	Mag	Transit h m	Dec ° ′	GHA ° ′	RA h m	SHA ° ′
♈	**ARIES**		**5 24**		**278 50.7**		
1	**Alpheratz**	2.2	**5 32**	N29 04.8	**276 45.9**	0 08	357 55.2
2	**Ankaa**	2.4	**5 50**	S 42 18.6	**272 17.6**	0 26	353 26.9
3	**Schedar**	2.5	**6 04**	N56 31.4	**268 44.1**	0 40	349 53.4
4	**Diphda**	2.2	**6 07**	S 17 59.6	**267 58.0**	0 44	349 07.3
5	**Achernar**	0.6	**7 01**	S 57 14.4	**254 26.1**	1 38	335 35.4
6	**POLARIS**	2.1	**7 53**	N89 15.0	**241 18.1**	2 30	322 27.4
7	**Hamal**	2.2	**7 30**	N23 27.1	**247 04.5**	2 07	328 13.8
8	**Acamar**	3.1	**8 21**	S 40 18.5	**234 18.0**	2 58	315 27.3
9	**Menkar**	2.8	**8 25**	N 4 04.9	**233 17.9**	3 02	314 27.2
10	**Mirfak**	1.9	**8 47**	N49 51.0	**227 47.7**	3 24	308 57.0
11	**Aldebaran**	1.1	**9 59**	N16 30.2	**209 53.5**	4 36	291 02.8
12	**Rigel**	0.3	**10 37**	S 8 12.3	**200 14.1**	5 14	281 23.4
13	**Capella**	0.2	**10 39**	N45 59.6	**199 42.5**	5 17	280 51.8
14	**Bellatrix**	1.7	**10 48**	N 6 20.8	**197 35.3**	5 25	278 44.6
15	**Elnath**	1.8	**10 49**	N28 36.2	**197 18.2**	5 26	278 27.5
16	**Alnilam**	1.8	**10 59**	S 1 12.3	**194 49.0**	5 36	275 58.3
17	**Betelgeuse**	0.1-1.2	**11 18**	N 7 24.3	**190 04.7**	5 55	271 14.0
18	**Canopus**	-0.9	**11 47**	S 52 41.7	**182 52.5**	6 24	264 01.8
19	**Sirius**	-1.6	**12 08**	S 16 42.9	**177 34.9**	6 45	258 44.2
20	**Adhara**	1.6	**12 21**	S 28 58.3	**174 12.6**	6 59	255 21.9
21	**Castor**	1.6	**12 57**	N31 53.4	**165 13.7**	7 34	246 23.0
22	**Procyon**	0.5	**13 02**	N 5 13.7	**164 02.8**	7 39	245 12.1
23	**Pollux**	1.2	**13 08**	N28 01.7	**162 32.8**	7 45	243 42.1
24	**Avior**	1.7	**13 45**	S 59 30.4	**153 14.0**	8 22	234 23.3
25	**Suhail**	2.2	**14 30**	S 43 25.7	**141 52.0**	9 08	223 01.3
26	**Miaplacidus**	1.8	**14 35**	S 69 42.9	**140 33.5**	9 13	221 42.8
27	**Alphard**	2.2	**14 50**	S 8 39.2	**136 58.3**	9 27	218 07.6
28	**Regulus**	1.3	**15 30**	N11 58.5	**126 46.7**	10 08	207 56.0
29	**Dubhe**	2.0	**16 26**	N61 45.8	**112 56.9**	11 04	194 06.2
30	**Denebola**	2.2	**17 11**	N14 35.0	**101 36.2**	11 49	182 45.5
31	**Gienah**	2.8	**17 37**	S 17 32.0	**94 54.9**	12 16	176 04.2
32	**Acrux**	1.1	**17 48**	S 63 05.7	**92 12.9**	12 27	173 22.2
33	**Gacrux**	1.6	**17 53**	S 57 06.5	**91 04.5**	12 31	172 13.8
34	**Mimosa**	1.5	**18 09**	S 59 41.0	**86 56.2**	12 48	168 05.5
35	**Alioth**	1.7	**18 16**	N55 58.4	**85 21.6**	12 54	166 30.9
36	**Spica**	1.2	**18 47**	S 11 09.1	**77 34.0**	13 25	158 43.3
37	**Alkaid**	1.9	**19 09**	N49 19.6	**71 58.6**	13 47	153 07.9
38	**Hadar**	0.9	**19 25**	S 60 22.1	**67 54.7**	14 04	149 04.0
39	**Menkent**	2.3	**19 28**	S 36 21.8	**67 11.7**	14 07	148 21.0
40	**Arcturus**	0.2	**19 37**	N19 11.7	**64 56.8**	14 16	146 06.1
41	**Rigil Kent.**	0.1	**20 01**	S 60 49.8	**58 57.9**	14 40	140 07.2
42	**Zuben'ubi**	2.9	**20 12**	S 16 02.1	**56 08.7**	14 51	137 18.0
43	**Kochab**	2.2	**20 12**	N74 10.1	**56 10.0**	14 51	137 19.3
44	**Alphecca**	2.3	**20 56**	N26 43.5	**45 11.2**	15 35	126 20.5
45	**Antares**	1.2	**21 50**	S 26 25.6	**31 30.8**	16 29	112 40.1
46	**Atria**	1.9	**22 10**	S 69 01.5	**26 42.5**	16 49	107 51.8
47	**Sabik**	2.6	**22 31**	S 15 43.2	**21 16.1**	17 10	102 25.4
48	**Shaula**	1.7	**22 54**	S 37 06.1	**15 27.8**	17 34	96 37.1
49	**Rasalhague**	2.1	**22 56**	N12 33.9	**15 07.5**	17 35	96 16.8
50	**Eltanin**	2.4	**23 17**	N51 29.6	**9 41.7**	17 57	90 51.0
51	**Kaus Aust.**	2.0	**23 45**	S 34 23.0	**2 49.3**	18 24	83 58.6
52	**Vega**	0.1	**0 02**	N38 47.2	**359 37.0**	18 37	80 46.3
53	**Nunki**	2.1	**0 20**	S 26 17.8	**355 02.9**	18 55	76 12.2
54	**Altair**	0.9	**1 15**	N 8 52.0	**341 09.8**	19 51	62 19.1
55	**Peacock**	2.1	**1 50**	S 56 44.2	**332 27.4**	20 26	53 36.7
56	**Deneb**	1.3	**2 06**	N45 16.5	**328 29.5**	20 41	49 38.8
57	**Enif**	2.5	**3 08**	N 9 52.2	**312 48.8**	21 44	33 58.1
58	**Al Na'ir**	2.2	**3 32**	S 46 57.9	**306 48.5**	22 08	27 57.8
59	**Fomalhaut**	1.3	**4 21**	S 29 37.6	**294 27.1**	22 58	15 36.4
60	**Markab**	2.6	**4 29**	N15 11.8	**292 40.3**	23 05	13 49.6

Star Transit Corr Table

Date	Corr h m	Date	Corr h m
1	0 00	17	−1 03
2	−0 04	18	−1 07
3	−0 08	19	−1 11
4	−0 12	20	−1 15
5	−0 16	21	−1 19
6	−0 20	22	−1 23
7	−0 24	23	−1 27
8	−0 28	24	−1 30
9	−0 31	25	−1 34
10	−0 35	26	−1 38
11	−0 39	27	−1 42
12	−0 43	28	−1 46
13	−0 47	29	−1 50
14	−0 51	30	−1 54
15	−0 55	31	−1 58
16	−0 59		

STAR TRANSIT

To find the approximate time of transit of a star for any day of the month use the table above. All corrections are subtractive.

If the value taken from the table is greater than the time of transit for the first of the month, add 23h 56min to the time of transit before subtracting the correction.

Example: What time will Nunki (No 53) transit the Meridian on July 30?

	h min
Transit on Jul 1	00 20
	+23 56
	24 16
Corr for Jul 30	−01 54
Transit on Jul 30	22 22

JULY DIARY

d h	
3 05	Venus 4° N of Aldebaran
4 00	Earth at aphelion
10 18	Neptune 2° S of Moon
11 13	Uranus 3° S of Moon
14 19	Jupiter 1.0° N of Moon
17 03	Mercury greatest elong. E.(27°)
17 05	Saturn 2° N of Moon
18 18	Jupiter stationary
18 23	Ceres 1.1° S of Moon
19 21	Aldebaran 0.3° S of Moon
21 12	Venus 4° N of Moon
22 03	Mars 5° N of Moon
23 20	Neptune at opposition
25 14	Mercury 2° S of Moon
25 21	Regulus 0.7° N of Moon
30 05	Mercury stationary

NOTES

Star declinations may be used as is for the whole month. To correct GHA for day, hour, minute, and second see page E 106; for further explanation of this table see page E 4.

JULY 1998 SUN & ARIES GMT

Wednesday, 1st July

Time	SUN GHA ° ′	Dec ° ′	ARIES GHA ° ′
00	179 04.2	N23 08.0	278 50.7
02	209 03.9	23 07.7	308 55.7
04	239 03.7	23 07.3	339 00.6
06	269 03.4	23 07.0	9 05.5
08	299 03.2	23 06.7	39 10.4
10	329 02.9	23 06.3	69 15.4
12	359 02.7	23 06.0	99 20.3
14	29 02.5	23 05.7	129 25.2
16	59 02.2	23 05.3	159 30.2
18	89 02.0	23 05.0	189 35.1
20	119 01.7	23 04.6	219 40.0
22	149 01.5	N23 04.3	249 44.9

Thursday, 2nd July

Time	SUN GHA ° ′	Dec ° ′	ARIES GHA ° ′
00	179 01.3	N23 03.9	279 49.9
02	209 01.0	23 03.6	309 54.8
04	239 00.8	23 03.2	339 59.7
06	269 00.6	23 02.8	10 04.7
08	299 00.3	23 02.5	40 09.6
10	329 00.1	23 02.1	70 14.5
12	358 59.8	23 01.7	100 19.4
14	28 59.6	23 01.4	130 24.4
16	58 59.4	23 01.0	160 29.3
18	88 59.1	23 00.6	190 34.2
20	118 58.9	23 00.2	220 39.2
22	148 58.7	N22 59.8	250 44.1

Friday, 3rd July

Time	SUN GHA ° ′	Dec ° ′	ARIES GHA ° ′
00	178 58.4	N22 59.4	280 49.0
02	208 58.2	22 59.1	310 53.9
04	238 58.0	22 58.7	340 58.9
06	268 57.7	22 58.3	11 03.8
08	298 57.5	22 57.9	41 08.7
10	328 57.3	22 57.5	71 13.7
12	358 57.1	22 57.1	101 18.6
14	28 56.8	22 56.7	131 23.5
16	58 56.6	22 56.2	161 28.4
18	88 56.4	22 55.8	191 33.4
20	118 56.1	22 55.4	221 38.3
22	148 55.9	N22 55.0	251 43.2

Saturday, 4th July

Time	SUN GHA ° ′	Dec ° ′	ARIES GHA ° ′
00	178 55.7	N22 54.6	281 48.1
02	208 55.5	22 54.2	311 53.1
04	238 55.2	22 53.7	341 58.0
06	268 55.0	22 53.3	12 02.9
08	298 54.8	22 52.9	42 07.9
10	328 54.6	22 52.4	72 12.8
12	358 54.4	22 52.0	102 17.7
14	28 54.1	22 51.6	132 22.6
16	58 53.9	22 51.1	162 27.6
18	88 53.7	22 50.7	192 32.5
20	118 53.5	22 50.2	222 37.4
22	148 53.2	N22 49.8	252 42.4

Sunday, 5th July

Time	SUN GHA ° ′	Dec ° ′	ARIES GHA ° ′
00	178 53.0	N22 49.3	282 47.3
02	208 52.8	22 48.9	312 52.2
04	238 52.6	22 48.4	342 57.1
06	268 52.4	22 47.9	13 02.1
08	298 52.2	22 47.5	43 07.0
10	328 51.9	22 47.0	73 11.9
12	358 51.7	22 46.5	103 16.9
14	28 51.5	22 46.1	133 21.8
16	58 51.3	22 45.6	163 26.7
18	88 51.1	22 45.1	193 31.6
20	118 50.9	22 44.6	223 36.6
22	148 50.7	N22 44.1	253 41.5

Monday, 6th July

Time	SUN GHA ° ′	Dec ° ′	ARIES GHA ° ′
00	178 50.5	N22 43.6	283 46.4
02	208 50.2	22 43.2	313 51.4
04	238 50.0	22 42.7	343 56.3
06	268 49.8	22 42.2	14 01.2
08	298 49.6	22 41.7	44 06.1
10	328 49.4	22 41.2	74 11.1
12	358 49.2	22 40.7	104 16.0
14	28 49.0	22 40.2	134 20.9
16	58 48.8	22 39.7	164 25.9
18	88 48.6	22 39.1	194 30.8
20	118 48.4	22 38.6	224 35.7
22	148 48.2	N22 38.1	254 40.6

Tuesday, 7th July

Time	SUN GHA ° ′	Dec ° ′	ARIES GHA ° ′
00	178 48.0	N22 37.6	284 45.6
02	208 47.8	22 37.1	314 50.5
04	238 47.6	22 36.5	344 55.4
06	268 47.4	22 36.0	15 00.4
08	298 47.2	22 35.5	45 05.3
10	328 47.0	22 34.9	75 10.2
12	358 46.8	22 34.4	105 15.1
14	28 46.6	22 33.9	135 20.1
16	58 46.4	22 33.3	165 25.0
18	88 46.2	22 32.8	195 29.9
20	118 46.0	22 32.2	225 34.8
22	148 45.8	N22 31.7	255 39.8

Wednesday, 8th July

Time	SUN GHA ° ′	Dec ° ′	ARIES GHA ° ′
00	178 45.6	N22 31.1	285 44.7
02	208 45.4	22 30.6	315 49.6
04	238 45.2	22 30.0	345 54.6
06	268 45.0	22 29.5	15 59.5
08	298 44.8	22 28.9	46 04.4
10	328 44.6	22 28.3	76 09.3
12	358 44.4	22 27.8	106 14.3
14	28 44.2	22 27.2	136 19.2
16	58 44.0	22 26.6	166 24.1
18	88 43.8	22 26.0	196 29.1
20	118 43.7	22 25.5	226 34.0
22	148 43.5	N22 24.9	256 38.9

Thursday, 9th July

Time	SUN GHA ° ′	Dec ° ′	ARIES GHA ° ′
00	178 43.3	N22 24.3	286 43.8
02	208 43.1	22 23.7	316 48.8
04	238 42.9	22 23.1	346 53.7
06	268 42.7	22 22.5	16 58.6
08	298 42.5	22 21.9	47 03.6
10	328 42.4	22 21.3	77 08.5
12	358 42.2	22 20.7	107 13.4
14	28 42.0	22 20.1	137 18.3
16	58 41.8	22 19.5	167 23.3
18	88 41.6	22 18.9	197 28.2
20	118 41.4	22 18.3	227 33.1
22	148 41.3	N22 17.7	257 38.1

Friday, 10th July

Time	SUN GHA ° ′	Dec ° ′	ARIES GHA ° ′
00	178 41.1	N22 17.1	287 43.0
02	208 40.9	22 16.5	317 47.9
04	238 40.7	22 15.8	347 52.8
06	268 40.6	22 15.2	17 57.8
08	298 40.4	22 14.6	48 02.7
10	328 40.2	22 14.0	78 07.6
12	358 40.0	22 13.3	108 12.6
14	28 39.9	22 12.7	138 17.5
16	58 39.7	22 12.0	168 22.4
18	88 39.5	22 11.4	198 27.3
20	118 39.3	22 10.8	228 32.3
22	148 39.2	N22 10.1	258 37.2

Saturday, 11th July

Time	SUN GHA ° ′	Dec ° ′	ARIES GHA ° ′	Time
00	178 39.0	N22 09.5	288 42.1	00
02	208 38.8	22 08.8	318 47.1	02
04	238 38.7	22 08.2	348 52.0	04
06	268 38.5	22 07.5	18 56.9	06
08	298 38.3	22 06.8	49 01.8	08
10	328 38.2	22 06.2	79 06.8	10
12	358 38.0	22 05.5	109 11.7	12
14	28 37.8	22 04.9	139 16.6	14
16	58 37.7	22 04.2	169 21.6	16
18	88 37.5	22 03.5	199 26.5	18
20	118 37.3	22 02.8	229 31.4	20
22	148 37.2	N22 02.2	259 36.3	22

Sunday, 12th July

Time	SUN GHA ° ′	Dec ° ′	ARIES GHA ° ′	Time
00	178 37.0	N22 01.5	289 41.3	00
02	208 36.9	22 00.8	319 46.2	02
04	238 36.7	22 00.1	349 51.1	04
06	268 36.5	21 59.4	19 56.1	06
08	298 36.4	21 58.7	50 01.0	08
10	328 36.2	21 58.0	80 05.9	10
12	358 36.1	21 57.3	110 10.8	12
14	28 35.9	21 56.6	140 15.8	14
16	58 35.8	21 55.9	170 20.7	16
18	88 35.6	21 55.2	200 25.6	18
20	118 35.5	21 54.5	230 30.5	20
22	148 35.3	N21 53.8	260 35.5	22

Monday, 13th July

Time	SUN GHA ° ′	Dec ° ′	ARIES GHA ° ′	Time
00	178 35.2	N21 53.1	290 40.4	00
02	208 35.0	21 52.4	320 45.3	02
04	238 34.9	21 51.7	350 50.3	04
06	268 34.7	21 51.0	20 55.2	06
08	298 34.6	21 50.2	51 00.1	08
10	328 34.4	21 49.5	81 05.0	10
12	358 34.3	21 48.8	111 10.0	12
14	28 34.1	21 48.1	141 14.9	14
16	58 34.0	21 47.3	171 19.8	16
18	88 33.8	21 46.6	201 24.8	18
20	118 33.7	21 45.8	231 29.7	20
22	148 33.5	N21 45.1	261 34.6	22

Tuesday, 14th July

Time	SUN GHA ° ′	Dec ° ′	ARIES GHA ° ′	Time
00	178 33.4	N21 44.4	291 39.5	00
02	208 33.3	21 43.6	321 44.5	02
04	238 33.1	21 42.9	351 49.4	04
06	268 33.0	21 42.1	21 54.3	06
08	298 32.8	21 41.4	51 59.3	08
10	328 32.7	21 40.6	82 04.2	10
12	358 32.6	21 39.9	112 09.1	12
14	28 32.4	21 39.1	142 14.0	14
16	58 32.3	21 38.3	172 19.0	16
18	88 32.2	21 37.6	202 23.9	18
20	118 32.0	21 36.8	232 28.8	20
22	148 31.9	N21 36.0	262 33.8	22

Wednesday, 15th July

Time	SUN GHA ° ′	Dec ° ′	ARIES GHA ° ′	Time
00	178 31.8	N21 35.2	292 38.7	00
02	208 31.6	21 34.5	322 43.6	02
04	238 31.5	21 33.7	352 48.5	04
06	268 31.4	21 32.9	22 53.5	06
08	298 31.3	21 32.1	52 58.4	08
10	328 31.1	21 31.3	83 03.3	10
12	358 31.0	21 30.6	113 08.2	12
14	28 30.9	21 29.8	143 13.2	14
16	58 30.7	21 29.0	173 18.1	16
18	88 30.6	21 28.2	203 23.0	18
20	118 30.5	21 27.4	233 28.0	20
22	148 30.4	N21 26.6	263 32.9	22

NOTES

Sun and Aries GHA corrections for additional hour, minutes, and seconds are on page E 106.
Correct sun declination by interpolation. For an example see page E 4.

JULY 1998 — SUN & ARIES — GMT

Thursday, 16th July

Time	SUN GHA ° ′	Dec ° ′	ARIES GHA ° ′
00	178 30.3	N21 25.8	293 37.8
02	208 30.1	21 25.0	323 42.7
04	238 30.0	21 24.1	353 47.7
06	268 29.9	21 23.3	23 52.6
08	298 29.8	21 22.5	53 57.5
10	328 29.7	21 21.7	84 02.5
12	358 29.5	21 20.9	114 07.4
14	28 29.4	21 20.1	144 12.3
16	58 29.3	21 19.2	174 17.2
18	88 29.2	21 18.4	204 22.2
20	118 29.1	21 17.6	234 27.1
22	148 29.0	N21 16.8	264 32.0

Friday, 17th July

Time	SUN GHA ° ′	Dec ° ′	ARIES GHA ° ′
00	178 28.9	N21 15.9	294 37.0
02	208 28.7	21 15.1	324 41.9
04	238 28.6	21 14.2	354 46.8
06	268 28.5	21 13.4	24 51.7
08	298 28.4	21 12.6	54 56.7
10	328 28.3	21 11.7	85 01.6
12	358 28.2	21 10.9	115 06.5
14	28 28.1	21 10.0	145 11.5
16	58 28.0	21 09.1	175 16.4
18	88 27.9	21 08.3	205 21.3
20	118 27.8	21 07.4	235 26.2
22	148 27.7	N21 06.6	265 31.2

Saturday, 18th July

Time	SUN GHA ° ′	Dec ° ′	ARIES GHA ° ′
00	178 27.6	N21 05.7	295 36.1
02	208 27.5	21 04.8	325 41.0
04	238 27.4	21 04.0	355 46.0
06	268 27.3	21 03.1	25 50.9
08	298 27.2	21 02.2	55 55.8
10	328 27.1	21 01.3	86 00.7
12	358 27.0	21 00.5	116 05.7
14	28 26.9	20 59.6	146 10.6
16	58 26.8	20 58.7	176 15.5
18	88 26.7	20 57.8	206 20.4
20	118 26.6	20 56.9	236 25.4
22	148 26.5	N20 56.0	266 30.3

Sunday, 19th July

Time	SUN GHA ° ′	Dec ° ′	ARIES GHA ° ′
00	178 26.4	N20 55.1	296 35.2
02	208 26.4	20 54.2	326 40.2
04	238 26.3	20 53.3	356 45.1
06	268 26.2	20 52.4	26 50.0
08	298 26.1	20 51.5	56 54.9
10	328 26.0	20 50.6	86 59.9
12	358 25.9	20 49.7	117 04.8
14	28 25.8	20 48.8	147 09.7
16	58 25.8	20 47.9	177 14.7
18	88 25.7	20 47.0	207 19.6
20	118 25.6	20 46.1	237 24.5
22	148 25.5	N20 45.1	267 29.4

Monday, 20th July

Time	SUN GHA ° ′	Dec ° ′	ARIES GHA ° ′
00	178 25.4	N20 44.2	297 34.4
02	208 25.4	20 43.3	327 39.3
04	238 25.3	20 42.4	357 44.2
06	268 25.2	20 41.4	27 49.2
08	298 25.1	20 40.5	57 54.1
10	328 25.1	20 39.6	87 59.0
12	358 25.0	20 38.6	118 03.9
14	28 24.9	20 37.7	148 08.9
16	58 24.8	20 36.7	178 13.8
18	88 24.8	20 35.8	208 18.7
20	118 24.7	20 34.9	238 23.7
22	148 24.6	N20 33.9	268 28.6

Tuesday, 21st July

Time	SUN GHA ° ′	Dec ° ′	ARIES GHA ° ′
00	178 24.6	N20 32.9	298 33.5
02	208 24.5	20 32.0	328 38.4
04	238 24.4	20 31.0	358 43.4
06	268 24.4	20 30.1	28 48.3
08	298 24.3	20 29.1	58 53.2
10	328 24.2	20 28.1	88 58.2
12	358 24.2	20 27.2	119 03.1
14	28 24.1	20 26.2	149 08.0
16	58 24.1	20 25.2	179 12.9
18	88 24.0	20 24.3	209 17.9
20	118 23.9	20 23.3	239 22.8
22	148 23.9	N20 22.3	269 27.7

Wednesday, 22nd July

Time	SUN GHA ° ′	Dec ° ′	ARIES GHA ° ′
00	178 23.8	N20 21.3	299 32.7
02	208 23.8	20 20.3	329 37.6
04	238 23.7	20 19.4	359 42.5
06	268 23.7	20 18.4	29 47.4
08	298 23.6	20 17.4	59 52.4
10	328 23.6	20 16.4	89 57.3
12	358 23.5	20 15.4	120 02.2
14	28 23.5	20 14.4	150 07.2
16	58 23.4	20 13.4	180 12.1
18	88 23.4	20 12.4	210 17.0
20	118 23.3	20 11.4	240 21.9
22	148 23.3	N20 10.4	270 26.9

Thursday, 23rd July

Time	SUN GHA ° ′	Dec ° ′	ARIES GHA ° ′
00	178 23.2	N20 09.4	300 31.8
02	208 23.2	20 08.4	330 36.7
04	238 23.1	20 07.4	0 41.7
06	268 23.1	20 06.3	30 46.6
08	298 23.1	20 05.3	60 51.5
10	328 23.0	20 04.3	90 56.4
12	358 23.0	20 03.3	121 01.4
14	28 22.9	20 02.2	151 06.3
16	58 22.9	20 01.2	181 11.2
18	88 22.9	20 00.2	211 16.1
20	118 22.8	19 59.2	241 21.1
22	148 22.8	N19 58.1	271 26.0

Friday, 24th July

Time	SUN GHA ° ′	Dec ° ′	ARIES GHA ° ′
00	178 22.8	N19 57.1	301 30.9
02	208 22.7	19 56.0	331 35.9
04	238 22.7	19 55.0	1 40.8
06	268 22.7	19 54.0	31 45.7
08	298 22.7	19 52.9	61 50.6
10	328 22.6	19 51.9	91 55.6
12	358 22.6	19 50.8	122 00.5
14	28 22.6	19 49.8	152 05.4
16	58 22.6	19 48.7	182 10.4
18	88 22.5	19 47.6	212 15.3
20	118 22.5	19 46.6	242 20.2
22	148 22.5	N19 45.5	272 25.1

Saturday, 25th July

Time	SUN GHA ° ′	Dec ° ′	ARIES GHA ° ′
00	178 22.5	N19 44.5	302 30.1
02	208 22.5	19 43.4	332 35.0
04	238 22.4	19 42.3	2 39.9
06	268 22.4	19 41.2	32 44.9
08	298 22.4	19 40.2	62 49.8
10	328 22.4	19 39.1	92 54.7
12	358 22.4	19 38.0	122 59.6
14	28 22.4	19 36.9	153 04.6
16	58 22.3	19 35.9	183 09.5
18	88 22.3	19 34.8	213 14.4
20	118 22.3	19 33.7	243 19.4
22	148 22.3	N19 32.6	273 24.3

Sunday, 26th July

Time	SUN GHA ° ′	Dec ° ′	ARIES GHA ° ′	Time
00	178 22.3	N19 31.5	303 29.2	00
02	208 22.3	19 30.4	333 34.1	02
04	238 22.3	19 29.3	3 39.1	04
06	268 22.3	19 28.2	33 44.0	06
08	298 22.3	19 27.1	63 48.9	08
10	328 22.3	19 26.0	93 53.9	10
12	358 22.3	19 24.9	123 58.8	12
14	28 22.3	19 23.8	154 03.7	14
16	58 22.3	19 22.7	184 08.6	16
18	88 22.3	19 21.6	214 13.6	18
20	118 22.3	19 20.5	244 18.5	20
22	148 22.3	N19 19.3	274 23.4	22

Monday, 27th July

Time	SUN GHA ° ′	Dec ° ′	ARIES GHA ° ′	Time
00	178 22.3	N19 18.2	304 28.4	00
02	208 22.3	19 17.1	334 33.3	02
04	238 22.3	19 16.0	4 38.2	04
06	268 22.3	19 14.9	34 43.1	06
08	298 22.3	19 13.7	64 48.1	08
10	328 22.4	19 12.6	94 53.0	10
12	358 22.4	19 11.5	124 57.9	12
14	28 22.4	19 10.3	155 02.8	14
16	58 22.4	19 09.2	185 07.8	16
18	88 22.4	19 08.1	215 12.7	18
20	118 22.4	19 06.9	245 17.6	20
22	148 22.4	N19 05.8	275 22.6	22

Tuesday, 28th July

Time	SUN GHA ° ′	Dec ° ′	ARIES GHA ° ′	Time
00	178 22.5	N19 04.6	305 27.5	00
02	208 22.5	19 03.5	335 32.4	02
04	238 22.5	19 02.3	5 37.3	04
06	268 22.5	19 01.2	35 42.3	06
08	298 22.5	19 00.0	65 47.2	08
10	328 22.6	18 58.9	95 52.1	10
12	358 22.6	18 57.7	125 57.1	12
14	28 22.6	18 56.6	156 02.0	14
16	58 22.6	18 55.4	186 06.9	16
18	88 22.7	18 54.2	216 11.8	18
20	118 22.7	18 53.1	246 16.8	20
22	148 22.7	N18 51.9	276 21.7	22

Wednesday, 29th July

Time	SUN GHA ° ′	Dec ° ′	ARIES GHA ° ′	Time
00	178 22.8	N18 50.7	306 26.6	00
02	208 22.8	18 49.6	336 31.6	02
04	238 22.8	18 48.4	6 36.5	04
06	268 22.9	18 47.2	36 41.4	06
08	298 22.9	18 46.0	66 46.3	08
10	328 22.9	18 44.8	96 51.3	10
12	358 23.0	18 43.7	126 56.2	12
14	28 23.0	18 42.5	157 01.1	14
16	58 23.0	18 41.3	187 06.1	16
18	88 23.1	18 40.1	217 11.0	18
20	118 23.1	18 38.9	247 15.9	20
22	148 23.2	N18 37.7	277 20.8	22

Thursday, 30th July

Time	SUN GHA ° ′	Dec ° ′	ARIES GHA ° ′	Time
00	178 23.2	N18 36.5	307 25.8	00
02	208 23.3	18 35.3	337 30.7	02
04	238 23.3	18 34.1	7 35.6	04
06	268 23.4	18 32.9	37 40.5	06
08	298 23.4	18 31.7	67 45.5	08
10	328 23.5	18 30.5	97 50.4	10
12	358 23.5	18 29.3	127 55.3	12
14	28 23.6	18 28.1	158 00.3	14
16	58 23.6	18 26.9	188 05.2	16
18	88 23.7	18 25.6	218 10.1	18
20	118 23.7	18 24.4	248 15.0	20
22	148 23.8	N18 23.2	278 20.0	22

Friday, 31st July

Time	SUN GHA ° ′	Dec ° ′	ARIES GHA ° ′
00	178 23.8	N18 22.0	308 24.9
02	208 23.9	18 20.8	338 29.8
04	238 23.9	18 19.5	8 34.8
06	268 24.0	N18 18.3	38 39.7
08	298 24.1	N18 17.1	68 44.6
10	328 24.1	18 15.8	98 49.5
12	358 24.2	18 14.6	128 54.5
14	28 24.3	N18 13.4	158 59.4
16	58 24.3	N18 12.1	189 04.3
18	88 24.4	18 10.9	219 09.3
20	118 24.5	18 09.6	249 14.2
22	148 24.5	N18 08.4	279 19.1

JULY 1998 **PLANETS** **0h GMT**

VENUS Mer Pass h m	GHA ° ′	Mean Var/hr 14°+	Dec ° ′	Mean Var/hr ′	Day	JUPITER GHA ° ′	Mean Var/hr 15°+	Dec ° ′	Mean Var/hr ′	Mer Pass h m
09 50	**212 40.5**	59.3	**N20 07.2**	0.6	**1 Wed**	**280 33.2**	2.3	**S 2 07.1**	0.0	**05 17**
09 51	**212 24.8**	59.3	**N20 20.5**	0.5	**2 Th**	**281 29.3**	2.3	**S 2 06.1**	0.0	**05 13**
09 52	**212 08.9**	59.3	**N20 33.3**	0.5	**3 Fri**	**282 25.6**	2.4	**S 2 05.2**	0.0	**05 09**
09 53	**211 52.8**	59.3	**N20 45.5**	0.5	**4 Sat**	**283 22.1**	2.4	**S 2 04.4**	0.0	**05 06**
09 54	**211 36.4**	59.3	**N20 57.2**	0.5	**5 SUN**	**284 18.7**	2.4	**S 2 03.6**	0.0	**05 02**
09 55	**211 19.7**	59.3	**N21 08.4**	0.4	**6 Mon**	**285 15.5**	2.4	**S 2 02.9**	0.0	**04 58**
09 56	**211 02.8**	59.3	**N21 19.0**	0.4	**7 Tu**	**286 12.4**	2.4	**S 2 02.3**	0.0	**04 54**
09 57	**210 45.7**	59.3	**N21 29.0**	0.4	**8 Wed**	**287 09.6**	2.4	**S 2 01.8**	0.0	**04 51**
09 59	**210 28.4**	59.3	**N21 38.5**	0.4	**9 Th**	**288 06.9**	2.4	**S 2 01.3**	0.0	**04 47**
10 00	**210 10.9**	59.3	**N21 47.4**	0.3	**10 Fri**	**289 04.4**	2.4	**S 2 00.9**	0.0	**04 43**
10 01	**209 53.1**	59.3	**N21 55.7**	0.3	**11 Sat**	**290 02.1**	2.4	**S 2 00.6**	0.0	**04 39**
10 02	**209 35.2**	59.2	**N22 03.5**	0.3	**12 SUN**	**290 59.9**	2.4	**S 2 00.4**	0.0	**04 35**
10 03	**209 17.1**	59.2	**N22 10.6**	0.3	**13 Mon**	**291 57.9**	2.4	**S 2 00.2**	0.0	**04 31**
10 05	**208 58.8**	59.2	**N22 17.1**	0.2	**14 Tu**	**292 56.1**	2.4	**S 2 00.2**	0.0	**04 28**
10 06	**208 40.3**	59.2	**N22 23.0**	0.2	**15 Wed**	**293 54.5**	2.4	**S 2 00.2**	0.0	**04 24**
10 07	**208 21.7**	59.2	**N22 28.3**	0.2	**16 Th**	**294 53.0**	2.4	**S 2 00.2**	0.0	**04 20**
10 08	**208 03.0**	59.2	**N22 33.0**	0.2	**17 Fri**	**295 51.8**	2.5	**S 2 00.4**	0.0	**04 16**
10 10	**207 44.1**	59.2	**N22 37.0**	0.1	**18 Sat**	**296 50.7**	2.5	**S 2 00.6**	0.0	**04 12**
10 11	**207 25.0**	59.2	**N22 40.4**	0.1	**19 SUN**	**297 49.8**	2.5	**S 2 00.9**	0.0	**04 08**
10 12	**207 05.9**	59.2	**N22 43.2**	0.1	**20 Mon**	**298 49.1**	2.5	**S 2 01.3**	0.0	**04 04**
10 13	**206 46.7**	59.2	**N22 45.3**	0.1	**21 Tu**	**299 48.5**	2.5	**S 2 01.8**	0.0	**04 00**
10 15	**206 27.4**	59.2	**N22 46.8**	0.0	**22 Wed**	**300 48.2**	2.5	**S 2 02.3**	0.0	**03 56**
10 16	**206 08.0**	59.2	**N22 47.7**	0.0	**23 Th**	**301 48.0**	2.5	**S 2 03.0**	0.0	**03 52**
10 17	**205 48.6**	59.2	**N22 47.9**	0.0	**24 Fri**	**302 48.0**	2.5	**S 2 03.7**	0.0	**03 48**
10 19	**205 29.1**	59.2	**N22 47.4**	0.0	**25 Sat**	**303 48.2**	2.5	**S 2 04.4**	0.0	**03 44**
10 20	**205 09.6**	59.2	**N22 46.3**	0.1	**26 SUN**	**304 48.5**	2.5	**S 2 05.3**	0.0	**03 40**
10 21	**204 50.1**	59.2	**N22 44.5**	0.1	**27 Mon**	**305 49.1**	2.5	**S 2 06.2**	0.0	**03 36**
10 23	**204 30.6**	59.2	**N22 42.1**	0.1	**28 Tu**	**306 49.8**	2.5	**S 2 07.2**	0.0	**03 32**
10 24	**204 11.1**	59.2	**N22 39.0**	0.2	**29 Wed**	**307 50.7**	2.5	**S 2 08.3**	0.0	**03 28**
10 25	**203 51.6**	59.2	**N22 35.3**	0.2	**30 Th**	**308 51.8**	2.6	**S 2 09.5**	0.0	**03 24**
10 26	**203 32.2**	59.2	**N22 30.9**	0.2	**31 Fri**	**309 53.0**	2.6	**S 2 10.7**	0.1	**03 20**

VENUS, Av. Mag. -3.9
S.H.A. July
5 289; 10 282; 15 276; 20 270; 25 263; 30 256.

JUPITER, Av. Mag.-2.6
S.H.A. July
5 2; 10 1; 15 1; 20 1; 25 1; 30 1.

MARS Mer Pass h m	GHA ° ′	Mean Var/hr 15°+	Dec ° ′	Mean Var/hr ′	Day	SATURN GHA ° ′	Mean Var/hr 15°+	Dec ° ′	Mean Var/hr ′	Mer Pass h m
11 08	**192 50.2**	0.6	**N23 55.6**	0.1	**1 Wed**	**248 17.8**	2.3	**N 9 52.7**	0.1	**07 26**
11 07	**193 04.7**	0.6	**N23 57.2**	0.1	**2 Th**	**249 12.7**	2.3	**N 9 54.0**	0.1	**07 22**
11 06	**193 19.3**	0.6	**N23 58.6**	0.1	**3 Fri**	**250 07.7**	2.3	**N 9 55.2**	0.1	**07 18**
11 05	**193 33.8**	0.6	**N23 59.8**	0.0	**4 Sat**	**251 02.8**	2.3	**N 9 56.5**	0.0	**07 15**
11 04	**193 48.4**	0.6	**N24 00.7**	0.0	**5 SUN**	**251 57.9**	2.3	**N 9 57.6**	0.0	**07 11**
11 03	**194 03.1**	0.6	**N24 01.5**	0.0	**6 Mon**	**252 53.2**	2.3	**N 9 58.8**	0.0	**07 07**
11 02	**194 17.8**	0.6	**N24 02.0**	0.0	**7 Tu**	**253 48.5**	2.3	**N 9 59.9**	0.0	**07 04**
11 01	**194 32.5**	0.6	**N24 02.3**	0.0	**8 Wed**	**254 43.9**	2.3	**N10 01.0**	0.0	**07 00**
11 00	**194 47.3**	0.6	**N24 02.4**	0.0	**9 Th**	**255 39.3**	2.3	**N10 02.1**	0.0	**06 56**
10 59	**195 02.2**	0.6	**N24 02.3**	0.0	**10 Fri**	**256 34.9**	2.3	**N10 03.1**	0.0	**06 53**
10 58	**195 17.0**	0.6	**N24 02.0**	0.0	**11 Sat**	**257 30.5**	2.3	**N10 04.1**	0.0	**06 49**
10 57	**195 32.0**	0.6	**N24 01.5**	0.0	**12 SUN**	**258 26.2**	2.3	**N10 05.1**	0.0	**06 45**
10 56	**195 47.0**	0.6	**N24 00.8**	0.0	**13 Mon**	**259 22.0**	2.3	**N10 06.0**	0.0	**06 41**
10 55	**196 02.0**	0.6	**N23 59.8**	0.0	**14 Tu**	**260 17.9**	2.3	**N10 06.9**	0.0	**06 38**
10 54	**196 17.1**	0.6	**N23 58.7**	0.1	**15 Wed**	**261 13.9**	2.3	**N10 07.8**	0.0	**06 34**
10 53	**196 32.3**	0.6	**N23 57.4**	0.1	**16 Th**	**262 09.9**	2.3	**N10 08.6**	0.0	**06 30**
10 52	**196 47.5**	0.6	**N23 55.8**	0.1	**17 Fri**	**263 06.1**	2.3	**N10 09.4**	0.0	**06 27**
10 51	**197 02.8**	0.6	**N23 54.1**	0.1	**18 Sat**	**264 02.3**	2.3	**N10 10.2**	0.0	**06 23**
10 50	**197 18.2**	0.6	**N23 52.1**	0.1	**19 SUN**	**264 58.6**	2.3	**N10 10.9**	0.0	**06 19**
10 49	**197 33.6**	0.6	**N23 50.0**	0.1	**20 Mon**	**265 55.0**	2.4	**N10 11.6**	0.0	**06 15**
10 48	**197 49.1**	0.7	**N23 47.7**	0.1	**21 Tu**	**266 51.5**	2.4	**N10 12.3**	0.0	**06 12**
10 47	**198 04.7**	0.7	**N23 45.1**	0.1	**22 Wed**	**267 48.1**	2.4	**N10 12.9**	0.0	**06 08**
10 46	**198 20.4**	0.7	**N23 42.4**	0.1	**23 Th**	**268 44.8**	2.4	**N10 13.5**	0.0	**06 04**
10 45	**198 36.1**	0.7	**N23 39.5**	0.1	**24 Fri**	**269 41.6**	2.4	**N10 14.1**	0.0	**06 00**
10 44	**198 51.9**	0.7	**N23 36.4**	0.1	**25 Sat**	**270 38.5**	2.4	**N10 14.6**	0.0	**05 56**
10 43	**199 07.8**	0.7	**N23 33.1**	0.1	**26 SUN**	**271 35.4**	2.4	**N10 15.1**	0.0	**05 53**
10 42	**199 23.8**	0.7	**N23 29.6**	0.2	**27 Mon**	**272 32.5**	2.4	**N10 15.6**	0.0	**05 49**
10 41	**199 39.8**	0.7	**N23 25.9**	0.2	**28 Tu**	**273 29.6**	2.4	**N10 16.0**	0.0	**05 45**
10 40	**199 56.0**	0.7	**N23 22.0**	0.2	**29 Wed**	**274 26.9**	2.4	**N10 16.4**	0.0	**05 41**
10 39	**200 12.2**	0.7	**N23 17.9**	0.2	**30 Th**	**275 24.2**	2.4	**N10 16.8**	0.0	**05 37**
10 38	**200 28.6**	0.7	**N23 13.7**	0.2	**31 Fri**	**276 21.7**	2.4	**N10 17.1**	0.0	**05 34**

MARS, Av. Mag. +1.6
S.H.A. July
5 271; 10 267; 15 264; 20 260; 25 256; 30 253.

NOTE: Planet corrections on pages E113-15

SATURN, Av. Mag. +0.5
S.H.A. July
5 329; 10 329; 15 329; 20 328; 25 328; 30 328.

JULY 1998 — MOON

Day	GMT hr	GHA ° ′	Mean Var/hr 14°+	Dec ° ′	Mean Var/hr ′
1 Wed	0	97 28.8	34.3	N 1 59.9	9.7
	6	184 54.2	34.3	N 1 01.8	9.7
	12	272 20.1	34.4	N 0 03.6	9.7
	18	359 46.3	34.4	S 0 54.4	9.7
2 Th	0	87 12.7	34.4	S 1 52.2	9.6
	6	174 39.1	34.4	S 2 49.7	9.5
	12	262 05.3	34.3	S 3 46.7	9.4
	18	349 31.2	34.2	S 4 43.2	9.3
3 Fri	0	76 56.6	34.1	S 5 39.0	9.2
	6	164 21.4	34.0	S 6 34.0	9.0
	12	251 45.3	33.8	S 7 28.2	8.8
	18	339 08.3	33.7	S 8 21.3	8.7
4 Sat	0	66 30.2	33.5	S 9 13.3	8.4
	6	153 50.9	33.2	S 10 04.0	8.2
	12	241 10.2	32.9	S 10 53.3	7.9
	18	328 28.0	32.7	S 11 41.1	7.7
5 Sun	0	55 44.2	32.4	S 12 27.2	7.4
	6	142 58.7	32.1	S 13 11.6	7.1
	12	230 11.4	31.8	S 13 54.0	6.7
	18	317 22.3	31.4	S 14 34.3	6.3
6 Mon	0	44 31.2	31.1	S 15 12.4	6.0
	6	131 38.1	30.8	S 15 48.1	5.5
	12	218 43.0	30.5	S 16 21.2	5.1
	18	305 45.9	30.2	S 16 51.7	4.5
7 Tu	0	32 46.8	29.7	S 17 19.3	4.1
	6	119 45.8	29.5	S 17 44.0	3.6
	12	206 42.8	29.1	S 18 05.5	3.0
	18	293 37.9	28.9	S 18 23.7	2.4
8 Wed	0	20 31.3	28.6	S 18 38.5	1.9
	6	107 23.0	28.3	S 18 49.9	1.2
	12	194 13.2	28.2	S 18 57.6	0.6
	18	281 02.0	27.9	S 19 01.5	0.0
9 Th	0	7 49.5	27.7	S 19 01.7	0.6
	6	94 36.0	27.6	S 18 58.0	1.3
	12	181 21.7	27.5	S 18 50.4	2.0
	18	268 06.6	27.4	S 18 38.8	2.7
10 Fri	0	354 51.0	27.4	S 18 23.3	3.3
	6	81 35.1	27.4	S 18 03.9	3.9
	12	168 19.0	27.3	S 17 40.5	4.6
	18	255 02.9	27.3	S 17 13.4	5.3
11 Sat	0	341 47.1	27.4	S 16 42.4	5.9
	6	68 31.6	27.5	S 16 07.8	6.4
	12	155 16.5	27.6	S 15 29.6	7.0
	18	242 02.0	27.7	S 14 48.0	7.6
12 Sun	0	328 48.2	27.8	S 14 03.1	8.0
	6	55 35.1	28.0	S 13 15.1	8.5
	12	142 22.7	28.1	S 12 24.2	9.0
	18	229 11.2	28.2	S 11 30.5	9.4
13 Mon	0	316 00.5	28.3	S 10 34.3	9.8
	6	42 50.6	28.5	S 9 35.8	10.1
	12	129 41.4	28.6	S 8 35.1	10.5
	18	216 33.0	28.7	S 7 32.6	10.8
14 Tu	0	303 25.2	28.8	S 6 28.4	11.0
	6	30 18.0	28.9	S 5 22.7	11.1
	12	117 11.2	29.0	S 4 15.9	11.3
	18	204 04.7	29.0	S 3 08.1	11.5
15 Wed	0	290 58.5	29.0	S 1 59.6	11.5
	6	17 52.5	29.0	S 0 50.7	11.5
	12	104 46.3	29.0	N 0 18.5	11.6
	18	191 40.1	28.9	N 1 27.6	11.4
16 Th	0	278 33.5	28.8	N 2 36.5	11.4
	6	5 26.5	28.8	N 3 44.8	11.2
	12	92 19.0	28.6	N 4 52.4	11.1
	18	179 10.8	28.5	N 5 58.9	10.8

Day	GMT hr	GHA ° ′	Mean Var/hr 14°+	Dec ° ′	Mean Var/hr ′
17 Fri	0	266 01.8	28.4	N 7 04.2	10.6
	6	352 51.9	28.2	N 8 08.1	10.4
	12	79 40.9	28.0	N 9 10.1	9.9
	18	166 28.9	27.8	N 10 10.2	9.6
18 Sat	0	253 15.8	27.6	N 11 08.1	9.2
	6	340 01.5	27.4	N 12 03.6	8.7
	12	66 45.9	27.1	N 12 56.4	8.2
	18	153 29.1	26.9	N 13 46.3	7.7
19 Sun	0	240 11.1	26.8	N 14 33.2	7.2
	6	326 52.0	26.6	N 15 16.8	6.6
	12	53 31.7	26.5	N 15 56.9	6.0
	18	140 10.5	26.3	N 16 33.4	5.4
20 Mon	0	226 48.4	26.2	N 17 06.0	4.8
	6	313 25.6	26.1	N 17 34.8	4.1
	12	40 02.2	26.0	N 17 59.6	3.4
	18	126 38.5	26.0	N 18 20.2	2.7
21 Tu	0	213 14.5	26.0	N 18 36.6	2.0
	6	299 50.6	26.1	N 18 48.8	1.3
	12	26 27.0	26.2	N 18 56.7	0.5
	18	113 03.9	26.2	N 19 00.4	0.1
22 Wed	0	199 41.6	26.5	N 18 59.8	0.8
	6	286 20.2	26.7	N 18 55.0	1.5
	12	13 00.0	26.9	N 18 46.2	2.2
	18	99 41.1	27.2	N 18 33.3	2.9
23 Th	0	186 23.9	27.5	N 18 16.5	3.5
	6	273 08.4	27.8	N 17 56.0	4.1
	12	359 54.8	28.1	N 17 31.8	4.6
	18	86 43.3	28.5	N 17 04.2	5.2
24 Fri	0	173 33.9	28.9	N 16 33.3	5.7
	6	260 26.7	29.2	N 15 59.3	6.2
	12	347 21.8	29.6	N 15 22.4	6.7
	18	74 19.2	30.0	N 14 42.8	7.1
25 Sat	0	161 19.0	30.4	N 14 00.7	7.5
	6	248 21.0	30.7	N 13 16.3	7.8
	12	335 25.3	31.1	N 12 29.7	8.1
	18	62 31.8	31.5	N 11 41.2	8.4
26 Sun	0	149 40.5	31.9	N 10 51.0	8.6
	6	236 51.3	32.1	N 9 59.2	8.9
	12	324 04.0	32.4	N 9 06.1	9.0
	18	51 18.6	32.8	N 8 11.7	9.3
27 Mon	0	138 34.9	33.0	N 7 16.3	9.4
	6	225 52.8	33.2	N 6 20.1	9.5
	12	313 12.3	33.5	N 5 23.1	9.6
	18	40 33.0	33.6	N 4 25.6	9.6
28 Tu	0	127 55.0	33.9	N 3 27.6	9.7
	6	215 17.9	34.0	N 2 29.4	9.8
	12	302 41.8	34.1	N 1 31.0	9.8
	18	30 06.4	34.2	N 0 32.6	9.7
29 Wed	0	117 31.5	34.2	S 0 25.8	9.6
	6	204 57.1	34.3	S 1 23.9	9.6
	12	292 22.9	34.4	S 2 21.6	9.5
	18	19 48.8	34.3	S 3 18.9	9.4
30 Th	0	107 14.6	34.3	S 4 15.7	9.3
	6	194 40.2	34.2	S 5 11.8	9.2
	12	282 05.4	34.1	S 6 07.0	9.1
	18	9 30.0	33.9	S 7 01.4	8.9
31 Fri	0	96 53.9	33.8	S 7 54.8	8.7
	6	184 17.0	33.7	S 8 47.0	8.4
	12	271 39.1	33.4	S 9 38.0	8.2
	18	359 00.1	33.3	S 10 27.6	8.0

NOTE: Moon GHA corrections for additional hours, minutes, and seconds are on page E 108–110. Moon declination corrections are on page E 111. The mean var/hr value is needed for both corrections.

AUGUST 1998 — SUN & MOON — GMT

SUN

Yr	Day of Mth	Week	Equation of Time 0h	Equation of Time 12h	Transit	Semi-diam	Lat 30°N Twi-light	Lat 30°N Sun-rise	Lat 30°N Sun-set	Lat 30°N Twi-light
			m s	m s	h m	′	h m	h m	h m	h m
213	**1**	**Sat**	+06 22	+06 20	**12 06**	15.8	04 53	05 19	18 53	19 19
214	**2**	**Sun**	+06 18	+06 16	**12 06**	15.8	04 53	05 19	18 53	19 19
215	**3**	**Mon**	+06 14	+06 11	**12 06**	15.8	04 54	05 20	18 52	19 18
216	**4**	**Tu**	+06 09	+06 06	**12 06**	15.8	04 55	05 21	18 51	19 17
217	**5**	**Wed**	+06 03	+06 00	**12 06**	15.8	04 55	05 21	18 50	19 16
218	**6**	**Th**	+05 57	+05 54	**12 06**	15.8	04 56	05 22	18 50	19 15
219	**7**	**Fri**	+05 50	+05 47	**12 06**	15.8	04 57	05 22	18 49	19 14
220	**8**	**Sat**	+05 43	+05 39	**12 06**	15.8	04 57	05 23	18 48	19 13
221	**9**	**Sun**	+05 35	+05 31	**12 06**	15.8	04 58	05 24	18 47	19 12
222	**10**	**Mon**	+05 26	+05 22	**12 05**	15.8	04 59	05 24	18 46	19 12
223	**11**	**Tu**	+05 17	+05 13	**12 05**	15.8	04 59	05 25	18 45	19 11
224	**12**	**Wed**	+05 08	+05 03	**12 05**	15.8	05 00	05 25	18 44	19 10
225	**13**	**Th**	+04 57	+04 52	**12 05**	15.8	05 01	05 26	18 43	19 09
226	**14**	**Fri**	+04 47	+04 41	**12 05**	15.8	05 01	05 27	18 42	19 08
227	**15**	**Sat**	+04 35	+04 30	**12 04**	15.8	05 02	05 27	18 41	19 07
228	**16**	**Sun**	+04 24	+04 18	**12 04**	15.8	05 03	05 28	18 40	19 05
229	**17**	**Mon**	+04 11	+04 05	**12 04**	15.8	05 03	05 28	18 39	19 04
230	**18**	**Tu**	+03 59	+03 52	**12 04**	15.8	05 04	05 29	18 38	19 03
231	**19**	**Wed**	+03 45	+03 38	**12 04**	15.8	05 05	05 29	18 37	19 02
232	**20**	**Th**	+03 32	+03 24	**12 03**	15.8	05 05	05 30	18 36	19 01
233	**21**	**Fri**	+03 17	+03 10	**12 03**	15.8	05 06	05 31	18 35	19 00
234	**22**	**Sat**	+03 03	+02 55	**12 03**	15.8	05 06	05 31	18 34	18 59
235	**23**	**Sun**	+02 47	+02 40	**12 03**	15.8	05 07	05 32	18 33	18 58
236	**24**	**Mon**	+02 32	+02 24	**12 02**	15.8	05 08	05 32	18 32	18 57
237	**25**	**Tu**	+02 16	+02 08	**12 02**	15.8	05 08	05 33	18 31	18 56
238	**26**	**Wed**	+01 59	+01 51	**12 02**	15.9	05 09	05 33	18 30	18 54
239	**27**	**Th**	+01 43	+01 34	**12 02**	15.9	05 09	05 34	18 29	18 53
240	**28**	**Fri**	+01 25	+01 17	**12 01**	15.9	05 10	05 35	18 28	18 52
241	**29**	**Sat**	+01 08	+00 59	**12 01**	15.9	05 11	05 35	18 26	18 51
242	**30**	**Sun**	+00 50	+00 40	**12 01**	15.9	05 11	05 36	18 25	18 50
243	**31**	**Mon**	+00 31	+00 22	**12 00**	15.9	05 12	05 36	18 24	18 48

Lat Corr to Sunrise, Sunset etc

Lat	Twi-light	Sun-rise	Sun-set	Twi-light
°	h m	h m	h m	h m
N60	−1 40	−1 14	+1 14	+1 40
55	−1 10	−0 54	+0 54	+1 10
50	−0 49	−0 38	+0 38	+0 49
45	−0 33	−0 26	+0 26	+0 33
N40	−0 20	−0 16	+0 16	+0 20
35	−0 09	−0 07	+0 07	+0 09
30	0 00	0 00	0 00	0 00
25	+0 08	+0 07	−0 07	−0 08
N20	+0 15	+0 13	−0 13	−0 15
15	+0 21	+0 18	−0 18	−0 21
10	+0 27	+0 24	−0 24	−0 27
N 5	+0 33	+0 29	−0 29	−0 33
0	+0 38	+0 34	−0 34	−0 38
S 5	+0 43	+0 39	−0 39	−0 43
10	+0 48	+0 44	−0 44	−0 48
15	+0 53	+0 49	−0 49	−0 53
S20	+0 57	+0 55	−0 55	−0 57
25	+1 02	+1 01	−1 01	−1 02
30	+1 07	+1 07	−1 07	−1 07
35	+1 13	+1 14	−1 14	−1 13
S40	+1 19	+1 22	−1 22	−1 19
45	+1 26	+1 31	−1 31	−1 26
50	+1 33	+1 42	−1 42	−1 33

NOTES

Latitude corrections are for mid-month.
Equation of time = mean time (LHA mean sun) minus apparent time (LHA true sun).

MOON

Yr	Day of Mth	Week	Age	Transit (Upper)	Diff	Semi-diam	Hor Par	Lat 30°N Moon-rise	Lat 30°N Moon-set
			days	h m	m	′	′	h m	h m
213	**1**	**Sat**	09	**18 49**	47	14.9	54.6	13 09	- -
214	**2**	**Sun**	10	**19 36**	49	15.0	55.0	14 03	00 25
215	**3**	**Mon**	11	**20 25**	51	15.1	55.6	14 57	01 06
216	**4**	**Tu**	12	**21 16**	53	15.3	56.2	15 51	01 51
217	**5**	**Wed**	13	**22 09**	55	15.5	57.0	16 44	02 41
218	**6**	**Th**	14	**23 04**	55	15.7	57.7	17 36	03 36
219	**7**	**Fri**	15	**23 59**	55	15.9	58.4	18 25	04 35
220	**8**	**Sat**	16	**24 54**	-	16.1	59.0	19 12	05 37
221	**9**	**Sun**	17	**00 54**	54	16.2	59.5	19 56	06 40
222	**10**	**Mon**	18	**01 48**	53	16.3	59.8	20 38	07 45
223	**11**	**Tu**	19	**02 41**	53	16.3	59.8	21 19	08 49
224	**12**	**Wed**	20	**03 34**	52	16.3	59.8	22 00	09 54
225	**13**	**Th**	21	**04 26**	54	16.2	59.5	22 43	10 58
226	**14**	**Fri**	22	**05 20**	54	16.1	59.2	23 27	12 02
227	**15**	**Sat**	23	**06 14**	55	16.0	58.8	- -	13 05
228	**16**	**Sun**	24	**07 09**	55	15.9	58.4	00 15	14 06
229	**17**	**Mon**	25	**08 04**	55	15.8	57.9	01 05	15 04
230	**18**	**Tu**	26	**08 59**	54	15.6	57.4	01 59	15 59
231	**19**	**Wed**	27	**09 53**	52	15.5	56.9	02 54	16 49
232	**20**	**Th**	28	**10 45**	50	15.4	56.4	03 51	17 35
233	**21**	**Fri**	29	**11 35**	47	15.2	55.9	04 48	18 17
234	**22**	**Sat**	00	**12 22**	45	15.1	55.4	05 44	18 55
235	**23**	**Sun**	01	**13 07**	44	15.0	55.0	06 39	19 31
236	**24**	**Mon**	02	**13 51**	43	14.9	54.6	07 32	20 05
237	**25**	**Tu**	03	**14 34**	42	14.8	54.3	08 25	20 38
238	**26**	**Wed**	04	**15 16**	43	14.8	54.2	09 17	21 12
239	**27**	**Th**	05	**15 59**	44	14.7	54.1	10 09	21 46
240	**28**	**Fri**	06	**16 43**	45	14.8	54.2	11 00	22 22
241	**29**	**Sat**	07	**17 28**	47	14.9	54.5	11 53	23 01
242	**30**	**Sun**	08	**18 15**	50	15.0	55.0	12 46	23 43
243	**31**	**Mon**	09	**19 05**	51	15.1	55.6	13 39	- -

Phases of the Moon

		d	h	m
○	Full Moon	8	02	10
☽	Last Quarter	14	19	48
●	New Moon	22	02	03
☾	First Quarter	30	05	06
	Perigee	11	12	
	Apogee	27	06	
	Farthest S of 0°	5	06	
	On 0°	11	17	
	Farthest N of 0°	18	03	
	On 0°	25	05	

NOTES

Diff equals daily change in transit time.
To correct moonrise, moonset, or transit times for latitude and/or longitude see page E 112. For further information about these tables see page E 4. For arc-to-time correction see page E 103.

AUGUST 1998 — STARS — 0h GMT August 1

No	Name	Mag	Transit h m	Dec ° ′	GHA ° ′	RA h m	SHA ° ′
♈	**ARIES**		**3 22**		**309 24.0**		
1	**Alpheratz**	2.2	**3 30**	N29 04.9	**307 19.0**	0 08	357 55.0
2	**Ankaa**	2.4	**3 48**	S 42 18.6	**302 50.7**	0 26	353 26.7
3	**Schedar**	2.5	**4 02**	N56 31.6	**299 17.1**	0 40	349 53.1
4	**Diphda**	2.2	**4 05**	S 17 59.5	**298 31.1**	0 44	349 07.1
5	**Achernar**	0.6	**4 59**	S 57 14.4	**284 59.1**	1 38	335 35.1
6	**POLARIS**	2.1	**5 52**	N89 15.1	**271 38.3**	2 31	322 14.3
7	**Hamal**	2.2	**5 29**	N23 27.2	**277 37.5**	2 07	328 13.5
8	**Acamar**	3.1	**6 20**	S 40 18.5	**264 51.0**	2 58	315 27.0
9	**Menkar**	2.8	**6 24**	N 4 05.0	**263 51.0**	3 02	314 27.0
10	**Mirfak**	1.9	**6 46**	N49 51.1	**258 20.7**	3 24	308 56.7
11	**Aldebaran**	1.1	**7 57**	N16 30.3	**240 26.6**	4 36	291 02.6
12	**Rigel**	0.3	**8 35**	S 8 12.2	**230 47.2**	5 14	281 23.2
13	**Capella**	0.2	**8 38**	N45 59.5	**230 15.6**	5 17	280 51.6
14	**Bellatrix**	1.7	**8 46**	N 6 20.8	**228 08.4**	5 25	278 44.4
15	**Elnath**	1.8	**8 47**	N28 36.2	**227 51.3**	5 26	278 27.3
16	**Alnilam**	1.8	**8 57**	S 1 12.2	**225 22.1**	5 36	275 58.1
17	**Betelgeuse**	0.1-1.2	**9 16**	N 7 24.4	**220 37.9**	5 55	271 13.9
18	**Canopus**	-0.9	**9 45**	S 52 41.6	**213 25.6**	6 24	264 01.6
19	**Sirius**	-1.6	**10 06**	S 16 42.8	**208 08.0**	6 45	258 44.0
20	**Adhara**	1.6	**10 19**	S 28 58.1	**204 45.8**	6 59	255 21.8
21	**Castor**	1.6	**10 55**	N31 53.4	**195 46.8**	7 34	246 22.8
22	**Procyon**	0.5	**11 00**	N 5 13.7	**194 36.0**	7 39	245 12.0
23	**Pollux**	1.2	**11 06**	N28 01.7	**193 06.0**	7 45	243 42.0
24	**Avior**	1.7	**11 43**	S 59 30.3	**183 47.2**	8 22	234 23.2
25	**Suhail**	2.2	**12 28**	S 43 25.6	**172 25.2**	9 08	223 01.2
26	**Miaplacidus**	1.8	**12 33**	S 69 42.7	**171 06.9**	9 13	221 42.9
27	**Alphard**	2.2	**12 48**	S 8 39.1	**167 31.6**	9 27	218 07.6
28	**Regulus**	1.3	**13 28**	N11 58.5	**157 20.0**	10 08	207 56.0
29	**Dubhe**	2.0	**14 24**	N61 45.7	**143 30.3**	11 04	194 06.3
30	**Denebola**	2.2	**15 09**	N14 35.0	**132 09.6**	11 49	182 45.6
31	**Gienah**	2.8	**15 36**	S 17 32.0	**125 28.3**	12 16	176 04.3
32	**Acrux**	1.1	**15 46**	S 63 05.6	**122 46.5**	12 27	173 22.5
33	**Gacrux**	1.6	**15 51**	S 57 06.4	**121 38.0**	12 31	172 14.0
34	**Mimosa**	1.5	**16 07**	S 59 41.0	**117 29.7**	12 48	168 05.7
35	**Alioth**	1.7	**16 14**	N55 58.4	**115 55.1**	12 54	166 31.1
36	**Spica**	1.2	**16 45**	S 11 09.1	**108 07.4**	13 25	158 43.4
37	**Alkaid**	1.9	**17 07**	N49 19.6	**102 32.1**	13 47	153 08.1
38	**Hadar**	0.9	**17 23**	S 60 22.1	**98 28.3**	14 04	149 04.3
39	**Menkent**	2.3	**17 26**	S 36 21.8	**97 45.2**	14 07	148 21.2
40	**Arcturus**	0.2	**17 35**	N19 11.7	**95 30.3**	14 16	146 06.3
41	**Rigil Kent.**	0.1	**17 59**	S 60 49.8	**89 31.5**	14 40	140 07.5
42	**Zuben'ubi**	2.9	**18 10**	S 16 02.0	**86 42.1**	14 51	137 18.1
43	**Kochab**	2.2	**18 10**	N74 10.1	**86 43.9**	14 51	137 19.9
44	**Alphecca**	2.3	**18 54**	N26 43.5	**75 44.7**	15 35	126 20.7
45	**Antares**	1.2	**19 48**	S 26 25.6	**62 04.2**	16 29	112 40.2
46	**Atria**	1.9	**20 08**	S 69 01.6	**57 16.0**	16 49	107 52.0
47	**Sabik**	2.6	**20 29**	S 15 43.2	**51 49.5**	17 10	102 25.5
48	**Shaula**	1.7	**20 52**	S 37 06.1	**46 01.2**	17 34	96 37.2
49	**Rasalhague**	2.1	**20 54**	N12 34.0	**45 40.9**	17 35	96 16.9
50	**Eltanin**	2.4	**21 15**	N51 29.7	**40 15.1**	17 57	90 51.1
51	**Kaus Aust.**	2.0	**21 43**	S 34 23.0	**33 22.7**	18 24	83 58.7
52	**Vega**	0.1	**21 56**	N38 47.3	**30 10.4**	18 37	80 46.4
53	**Nunki**	2.1	**22 14**	S 26 17.8	**25 36.2**	18 55	76 12.2
54	**Altair**	0.9	**23 09**	N 8 52.1	**11 43.1**	19 51	62 19.1
55	**Peacock**	2.1	**23 44**	S 56 44.3	**3 00.6**	20 26	53 36.6
56	**Deneb**	1.3	**0 04**	N45 16.7	**359 02.8**	20 41	49 38.8
57	**Enif**	2.5	**1 06**	N 9 52.2	**343 22.0**	21 44	33 58.0
58	**Al Na'ir**	2.2	**1 30**	S 46 57.9	**337 21.6**	22 08	27 57.6
59	**Fomalhaut**	1.3	**2 20**	S 29 37.6	**325 00.2**	22 58	15 36.2
60	**Markab**	2.6	**2 27**	N15 11.9	**323 13.4**	23 05	13 49.4

Star Transit Corr Table

Date	Corr h m	Date	Corr h m
1	0 00	17	–1 03
2	–0 04	18	–1 07
3	–0 08	19	–1 11
4	–0 12	20	–1 15
5	–0 16	21	–1 19
6	–0 20	22	–1 23
7	–0 24	23	–1 27
8	–0 28	24	–1 30
9	–0 31	25	–1 34
10	–0 35	26	–1 38
11	–0 39	27	–1 42
12	–0 43	28	–1 46
13	–0 47	29	–1 50
14	–0 51	30	–1 54
15	–0 55	31	–1 58
16	–0 59		

STAR TRANSIT

To find the approximate time of transit of a star for any day of the month use the table above. All corrections are subtractive.

If the value taken from the table is greater than the time of transit for the first of the month, add 23h 56min to the time of transit before subtracting the correction.

Example: What time will Deneb (No 56) transit the Meridian on August 30?

	h min
Transit on Aug 1	00 04
	+23 56
	24 00
Corr for Aug 30	–01 54
Transit on Aug 30	22 06

AUGUST DIARY

d h	
3 07	Uranus at opposition
5 03	Venus 0.8° S of Mars
7 01	Neptune 2° S of Moon
7 19	Uranus 3° S of Moon
8 18	Venus 7° S of Pollux
11 00	Jupiter 0.9° N of Moon
11 21	Mars 6° S of Pollux
13 12	Saturn 2° N of Moon
14 00	Mercury in conjunction
16 03	Aldebaran 0.2° S of Moon
16 16	Saturn stationary
19 20	Mars 4° N of Moon
20 14	Venus 3° N of Moon
23 05	Mercury stationary
25 23	Mercury 3° S of Venus
31 09	Mercury greatest elong. W.(18°)

NOTES

Star declinations may be used as is for the whole month. To correct GHA for day, hour, minute, and second see page E 106; for further explanation of this table see page E 4.

AUGUST 1998 SUN & ARIES GMT

Time	SUN GHA ° ′	Dec ° ′	ARIES GHA ° ′	Time	SUN GHA ° ′	Dec ° ′	ARIES GHA ° ′	Time	SUN GHA ° ′	Dec ° ′	ARIES GHA ° ′	Time
	Saturday, 1st August				**Thursday, 6th August**				**Tuesday, 11th August**			
00	178 24.6	N18 07.2	309 24.0	00	178 30.8	N16 48.7	314 19.7	00	178 40.7	N15 23.5	319 15.4	00
02	208 24.7	18 05.9	339 29.0	02	208 30.9	16 47.3	344 24.7	02	208 40.9	15 22.0	349 20.4	02
04	238 24.7	18 04.7	9 33.9	04	238 31.1	16 46.0	14 29.6	04	238 41.1	15 20.5	19 25.3	04
06	268 24.8	18 03.4	39 38.8	06	268 31.2	16 44.6	44 34.5	06	268 41.3	15 19.1	49 30.2	06
08	298 24.9	18 02.1	69 43.8	08	298 31.3	16 43.2	74 39.4	08	298 41.5	15 17.6	79 35.1	08
10	328 25.0	18 00.9	99 48.7	10	328 31.5	16 41.9	104 44.4	10	328 41.7	15 16.1	109 40.1	10
12	358 25.0	17 59.6	129 53.6	12	358 31.6	16 40.5	134 49.3	12	358 41.9	15 14.6	139 45.0	12
14	28 25.1	17 58.4	159 58.5	14	28 31.7	16 39.1	164 54.2	14	28 42.1	15 13.1	169 49.9	14
16	58 25.2	17 57.1	190 03.5	16	58 31.9	16 37.7	194 59.2	16	58 42.3	15 11.6	199 54.9	16
18	88 25.3	17 55.8	220 08.4	18	88 32.0	16 36.3	225 04.1	18	88 42.5	15 10.2	229 59.8	18
20	118 25.4	17 54.6	250 13.3	20	118 32.2	16 35.0	255 09.0	20	118 42.7	15 08.7	260 04.7	20
22	148 25.4	N17 53.3	280 18.2	22	148 32.3	N16 33.6	285 13.9	22	148 42.9	N15 07.2	290 09.6	22
	Sunday, 2nd August				**Friday, 7th August**				**Wednesday, 12th August**			
00	178 25.5	N17 52.0	310 23.2	00	178 32.5	N16 32.2	315 18.9	00	178 43.1	N15 05.7	320 14.6	00
02	208 25.6	17 50.8	340 28.1	02	208 32.6	16 30.8	345 23.8	02	208 43.4	15 04.2	350 19.5	02
04	238 25.7	17 49.5	10 33.0	04	238 32.8	16 29.4	15 28.7	04	238 43.6	15 02.7	20 24.4	04
06	268 25.8	17 48.2	40 38.0	06	268 32.9	16 28.0	45 33.7	06	268 43.8	15 01.2	50 29.4	06
08	298 25.9	17 46.9	70 42.9	08	298 33.1	16 26.6	75 38.6	08	298 44.0	14 59.7	80 34.3	08
10	328 26.0	17 45.7	100 47.8	10	328 33.2	16 25.2	105 43.5	10	328 44.2	14 58.2	110 39.2	10
12	358 26.0	17 44.4	130 52.7	12	358 33.4	16 23.8	135 48.4	12	358 44.4	14 56.7	140 44.1	12
14	28 26.1	17 43.1	160 57.7	14	28 33.5	16 22.4	165 53.4	14	28 44.6	14 55.2	170 49.1	14
16	58 26.2	17 41.8	191 02.6	16	58 33.7	16 21.0	195 58.3	16	58 44.8	14 53.7	200 54.0	16
18	88 26.3	17 40.5	221 07.5	18	88 33.8	16 19.6	226 03.2	18	88 45.1	14 52.2	230 58.9	18
20	118 26.4	17 39.2	251 12.5	20	118 34.0	16 18.2	256 08.2	20	118 45.3	14 50.7	261 03.8	20
22	148 26.5	N17 37.9	281 17.4	22	148 34.2	N16 16.8	286 13.1	22	148 45.5	N14 49.1	291 08.8	22
	Monday, 3rd August				**Saturday, 8th August**				**Thursday, 13th August**			
00	178 26.6	N17 36.6	311 22.3	00	178 34.3	N16 15.4	316 18.0	00	178 45.7	N14 47.6	321 13.7	00
02	208 26.7	17 35.3	341 27.2	02	208 34.5	16 14.0	346 22.9	02	208 45.9	14 46.1	351 18.6	02
04	238 26.8	17 34.0	11 32.2	04	238 34.6	16 12.6	16 27.9	04	238 46.1	14 44.6	21 23.6	04
06	268 26.9	17 32.7	41 37.1	06	268 34.8	16 11.2	46 32.8	06	268 46.4	14 43.1	51 28.5	06
08	298 27.0	17 31.4	71 42.0	08	298 35.0	16 09.7	76 37.7	08	298 46.6	14 41.6	81 33.4	08
10	328 27.1	17 30.1	101 47.0	10	328 35.1	16 08.3	106 42.7	10	328 46.8	14 40.0	111 38.3	10
12	358 27.2	17 28.8	131 51.9	12	358 35.3	16 06.9	136 47.6	12	358 47.0	14 38.5	141 43.3	12
14	28 27.3	17 27.5	161 56.8	14	28 35.5	16 05.5	166 52.5	14	28 47.3	14 37.0	171 48.2	14
16	58 27.4	17 26.2	192 01.7	16	58 35.6	16 04.1	196 57.4	16	58 47.5	14 35.5	201 53.1	16
18	88 27.5	17 24.9	222 06.7	18	88 35.8	16 02.6	227 02.4	18	88 47.7	14 33.9	231 58.1	18
20	118 27.6	17 23.6	252 11.6	20	118 36.0	16 01.2	257 07.3	20	118 47.9	14 32.4	262 03.0	20
22	148 27.7	N17 22.3	282 16.5	22	148 36.1	N15 59.8	287 12.2	22	148 48.2	N14 30.9	292 07.9	22
	Tuesday, 4th August				**Sunday, 9th August**				**Friday, 14th August**			
00	178 27.8	N17 20.9	312 21.5	00	178 36.3	N15 58.3	317 17.2	00	178 48.4	N14 29.4	322 12.8	00
02	208 28.0	17 19.6	342 26.4	02	208 36.5	15 56.9	347 22.1	02	208 48.6	14 27.8	352 17.8	02
04	238 28.1	17 18.3	12 31.3	04	238 36.7	15 55.5	17 27.0	04	238 48.8	14 26.3	22 22.7	04
06	268 28.2	17 17.0	42 36.2	06	268 36.8	15 54.0	47 31.9	06	268 49.1	14 24.7	52 27.6	06
08	298 28.3	17 15.6	72 41.2	08	298 37.0	15 52.6	77 36.9	08	298 49.3	14 23.2	82 32.6	08
10	328 28.4	17 14.3	102 46.1	10	328 37.2	15 51.2	107 41.8	10	328 49.5	14 21.7	112 37.5	10
12	358 28.5	17 13.0	132 51.0	12	358 37.4	15 49.7	137 46.7	12	358 49.8	14 20.1	142 42.4	12
14	28 28.6	17 11.6	162 56.0	14	28 37.5	15 48.3	167 51.7	14	28 50.0	14 18.6	172 47.3	14
16	58 28.8	17 10.3	193 00.9	16	58 37.7	15 46.8	197 56.6	16	58 50.3	14 17.0	202 52.3	16
18	88 28.9	17 09.0	223 05.8	18	88 37.9	15 45.4	228 01.5	18	88 50.5	14 15.5	232 57.2	18
20	118 29.0	17 07.6	253 10.7	20	118 38.1	15 43.9	258 06.4	20	118 50.7	14 13.9	263 02.1	20
22	148 29.1	N17 06.3	283 15.7	22	148 38.3	N15 42.5	288 11.4	22	148 51.0	N14 12.4	293 07.1	22
	Wednesday, 5th August				**Monday, 10th August**				**Saturday, 15th August**			
00	178 29.2	N17 05.0	313 20.6	00	178 38.5	N15 41.0	318 16.3	00	178 51.2	N14 10.8	323 12.0	00
02	208 29.4	17 03.6	343 25.5	02	208 38.6	15 39.6	348 21.2	02	208 51.4	14 09.3	353 16.9	02
04	238 29.5	17 02.3	13 30.5	04	238 38.8	15 38.1	18 26.1	04	238 51.7	14 07.7	23 21.8	04
06	268 29.6	17 00.9	43 35.4	06	268 39.0	15 36.7	48 31.1	06	268 51.9	14 06.2	53 26.8	06
08	298 29.7	16 59.6	73 40.3	08	298 39.2	15 35.2	78 36.0	08	298 52.2	14 04.6	83 31.7	08
10	328 29.9	16 58.2	103 45.2	10	328 39.4	15 33.8	108 40.9	10	328 52.4	14 03.1	113 36.6	10
12	358 30.0	16 56.9	133 50.2	12	358 39.6	15 32.3	138 45.9	12	358 52.7	14 01.5	143 41.5	12
14	28 30.1	16 55.5	163 55.1	14	28 39.8	15 30.8	168 50.8	14	28 52.9	13 59.9	173 46.5	14
16	58 30.2	16 54.2	194 00.0	16	58 40.0	15 29.4	198 55.7	16	58 53.2	13 58.4	203 51.4	16
18	88 30.4	16 52.8	224 05.0	18	88 40.1	15 27.9	229 00.6	18	88 53.4	13 56.8	233 56.3	18
20	118 30.5	16 51.4	254 09.9	20	118 40.3	15 26.4	259 05.6	20	118 53.7	13 55.2	264 01.3	20
22	148 30.6	N16 50.1	284 14.8	22	148 40.5	N15 25.0	289 10.5	22	148 53.9	N13 53.7	294 06.2	22

NOTES

Sun and Aries GHA corrections for additional hour, minutes, and seconds are on page E 106.
Correct sun declination by interpolation. For an example see page E 4.

AUGUST 1998 SUN & ARIES GMT

Time	SUN GHA ° ′	Dec ° ′	ARIES GHA ° ′	Time	SUN GHA ° ′	Dec ° ′	ARIES GHA ° ′	Time	SUN GHA ° ′	Dec ° ′	ARIES GHA ° ′	Time
	Sunday, 16th August				**Friday, 21st August**				**Wednesday, 26th August**			
00	178 54.2	N13 52.1	324 11.1	00	179 10.7	N12 15.2	329 06.8	00	179 30.2	N10 33.4	334 02.5	00
02	208 54.4	13 50.5	354 16.0	02	209 11.0	12 13.5	359 11.7	02	209 30.6	10 31.7	4 07.4	02
04	238 54.7	13 49.0	24 21.0	04	239 11.3	12 11.9	29 16.7	04	239 30.9	10 29.9	34 12.4	04
06	268 54.9	13 47.4	54 25.9	06	269 11.7	12 10.2	59 21.6	06	269 31.3	10 28.2	64 17.3	06
08	298 55.2	13 45.8	84 30.8	08	299 12.0	12 08.5	89 26.5	08	299 31.6	10 26.5	94 22.2	08
10	328 55.4	13 44.2	114 35.8	10	329 12.3	12 06.9	119 31.5	10	329 32.0	10 24.7	124 27.1	10
12	358 55.7	13 42.6	144 40.7	12	359 12.6	12 05.2	149 36.4	12	359 32.3	10 23.0	154 32.1	12
14	28 55.9	13 41.1	174 45.6	14	29 12.9	12 03.5	179 41.3	14	29 32.7	10 21.3	184 37.0	14
16	58 56.2	13 39.5	204 50.5	16	59 13.2	12 01.9	209 46.2	16	59 33.0	10 19.5	214 41.9	16
18	88 56.4	13 37.9	234 55.5	18	89 13.5	12 00.2	239 51.2	18	89 33.4	10 17.8	244 46.9	18
20	118 56.7	13 36.3	265 00.4	20	119 13.8	11 58.5	269 56.1	20	119 33.7	10 16.0	274 51.8	20
22	148 57.0	N13 34.7	295 05.3	22	149 14.1	N11 56.9	300 01.0	22	149 34.1	N10 14.3	304 56.7	22
	Monday, 17th August				**Saturday, 22nd August**				**Thursday, 27th August**			
00	178 57.2	N13 33.1	325 10.3	00	179 14.4	N11 55.2	330 06.0	00	179 34.5	N10 12.5	335 01.6	00
02	208 57.5	13 31.5	355 15.2	02	209 14.7	11 53.5	0 10.9	02	209 34.8	10 10.8	5 06.6	02
04	238 57.8	13 30.0	25 20.1	04	239 15.0	11 51.8	30 15.8	04	239 35.2	10 09.0	35 11.5	04
06	268 58.0	13 28.4	55 25.0	06	269 15.4	11 50.2	60 20.7	06	269 35.5	10 07.3	65 16.4	06
08	298 58.3	13 26.8	85 30.0	08	299 15.7	11 48.5	90 25.7	08	299 35.9	10 05.5	95 21.4	08
10	328 58.5	13 25.2	115 34.9	10	329 16.0	11 46.8	120 30.6	10	329 36.2	10 03.8	125 26.3	10
12	358 58.8	13 23.6	145 39.8	12	359 16.3	11 45.1	150 35.5	12	359 36.6	10 02.0	155 31.2	12
14	28 59.1	13 22.0	175 44.8	14	29 16.6	11 43.4	180 40.5	14	29 37.0	10 00.3	185 36.1	14
16	58 59.3	13 20.4	205 49.7	16	59 16.9	11 41.8	210 45.4	16	59 37.3	9 58.5	215 41.1	16
18	88 59.6	13 18.8	235 54.6	18	89 17.2	11 40.1	240 50.3	18	89 37.7	9 56.8	245 46.0	18
20	118 59.9	13 17.2	265 59.5	20	119 17.6	11 38.4	270 55.2	20	119 38.0	9 55.0	275 50.9	20
22	149 00.2	N13 15.6	296 04.5	22	149 17.9	N11 36.7	301 00.2	22	149 38.4	N 9 53.3	305 55.8	22
	Tuesday, 18th August				**Sunday, 23rd August**				**Friday, 28th August**			
00	179 00.4	N13 14.0	326 09.4	00	179 18.2	N11 35.0	331 05.1	00	179 38.8	N 9 51.5	336 00.8	00
02	209 00.7	13 12.3	356 14.3	02	209 18.5	11 33.3	1 10.0	02	209 39.1	9 49.7	6 05.7	02
04	239 01.0	13 10.7	26 19.3	04	239 18.9	11 31.6	31 14.9	04	239 39.5	9 48.0	36 10.6	04
06	269 01.2	13 09.1	56 24.2	06	269 19.2	11 29.9	61 19.9	06	269 39.9	9 46.2	66 15.6	06
08	299 01.5	13 07.5	86 29.1	08	299 19.5	11 28.3	91 24.8	08	299 40.2	9 44.5	96 20.5	08
10	329 01.8	13 05.9	116 34.0	10	329 19.8	11 26.6	121 29.7	10	329 40.6	9 42.7	126 25.4	10
12	359 02.1	13 04.3	146 39.0	12	359 20.1	11 24.9	151 34.7	12	359 41.0	9 40.9	156 30.3	12
14	29 02.3	13 02.7	176 43.9	14	29 20.5	11 23.2	181 39.6	14	29 41.3	9 39.2	186 35.3	14
16	59 02.6	13 01.1	206 48.8	16	59 20.8	11 21.5	211 44.5	16	59 41.7	9 37.4	216 40.2	16
18	89 02.9	12 59.4	236 53.8	18	89 21.1	11 19.8	241 49.4	18	89 42.1	9 35.6	246 45.1	18
20	119 03.2	12 57.8	266 58.7	20	119 21.5	11 18.1	271 54.4	20	119 42.4	9 33.9	276 50.1	20
22	149 03.5	N12 56.2	297 03.6	22	149 21.8	N11 16.4	301 59.3	22	149 42.8	N 9 32.1	306 55.0	22
	Wednesday, 19th August				**Monday, 24th August**				**Saturday, 29th August**			
00	179 03.7	N12 54.6	327 08.5	00	179 22.1	N11 14.7	332 04.2	00	179 43.2	N 9 30.3	336 59.9	00
02	209 04.0	12 52.9	357 13.5	02	209 22.4	11 13.0	2 09.2	02	209 43.6	9 28.5	7 04.8	02
04	239 04.3	12 51.3	27 18.4	04	239 22.8	11 11.3	32 14.1	04	239 43.9	9 26.8	37 09.8	04
06	269 04.6	12 49.7	57 23.3	06	269 23.1	11 09.5	62 19.0	06	269 44.3	9 25.0	67 14.7	06
08	299 04.9	12 48.1	87 28.3	08	299 23.4	11 07.8	92 23.9	08	299 44.7	9 23.2	97 19.6	08
10	329 05.2	12 46.4	117 33.2	10	329 23.8	11 06.1	122 28.9	10	329 45.1	9 21.4	127 24.6	10
12	359 05.4	12 44.8	147 38.1	12	359 24.1	11 04.4	152 33.8	12	359 45.4	9 19.7	157 29.5	12
14	29 05.7	12 43.2	177 43.0	14	29 24.4	11 02.7	182 38.7	14	29 45.8	9 17.9	187 34.4	14
16	59 06.0	12 41.5	207 48.0	16	59 24.8	11 01.0	212 43.7	16	59 46.2	9 16.1	217 39.3	16
18	89 06.3	12 39.9	237 52.9	18	89 25.1	10 59.3	242 48.6	18	89 46.6	9 14.3	247 44.3	18
20	119 06.6	12 38.3	267 57.8	20	119 25.4	10 57.6	272 53.5	20	119 46.9	9 12.5	277 49.2	20
22	149 06.9	N12 36.6	298 02.7	22	149 25.8	N10 55.8	302 58.4	22	149 47.3	N 9 10.8	307 54.1	22
	Thursday, 20th August				**Tuesday, 25th August**				**Sunday, 30th August**			
00	179 07.2	N12 35.0	328 07.7	00	179 26.1	N10 54.1	333 03.4	00	179 47.7	N 9 09.0	337 59.1	00
02	209 07.5	12 33.3	358 12.6	02	209 26.5	10 52.4	3 08.3	02	209 48.1	9 07.2	8 04.0	02
04	239 07.8	12 31.7	28 17.5	04	239 26.8	10 50.7	33 13.2	04	239 48.5	9 05.4	38 08.9	04
06	269 08.1	12 30.0	58 22.5	06	269 27.1	10 49.0	63 18.1	06	269 48.8	9 03.6	68 13.8	06
08	299 08.4	12 28.4	88 27.4	08	299 27.5	10 47.2	93 23.1	08	299 49.2	9 01.8	98 18.8	08
10	329 08.7	12 26.8	118 32.3	10	329 27.8	10 45.5	123 28.0	10	329 49.6	9 00.0	128 23.7	10
12	359 08.9	12 25.1	148 37.2	12	359 28.2	10 43.8	153 32.9	12	359 50.0	8 58.2	158 28.6	12
14	29 09.2	12 23.5	178 42.2	14	29 28.5	10 42.1	183 37.9	14	29 50.4	8 56.5	188 33.6	14
16	59 09.5	12 21.8	208 47.1	16	59 28.8	10 40.3	213 42.8	16	59 50.7	8 54.7	218 38.5	16
18	89 09.8	12 20.1	238 52.0	18	89 29.2	10 38.6	243 47.7	18	89 51.1	8 52.9	248 43.4	18
20	119 10.1	12 18.5	268 57.0	20	119 29.5	10 36.9	273 52.6	20	119 51.5	8 51.1	278 48.3	20
22	149 10.4	N12 16.8	299 01.9	22	149 29.9	N10 35.1	303 57.6	22	149 51.9	N 8 49.3	308 53.3	22
					Monday, 31st August							
00	179 52.3	N 8 47.5	338 58.2	08	299 53.8	N 8 40.3	99 17.9	16	59 55.4	N 8 33.1	219 37.6	16
02	209 52.7	8 45.7	9 03.1	10	329 54.2	8 38.5	129 22.8	18	89 55.8	8 31.3	249 42.5	18
04	239 53.1	8 43.9	39 08.0	12	359 54.6	8 36.7	159 27.8	20	119 56.2	8 29.5	279 47.5	20
06	269 53.5	N 8 42.1	69 13.0	14	29 55.0	N 8 34.9	189 32.7	22	149 56.6	N 8 27.7	309 52.4	22

AUGUST 1998 PLANETS 0h GMT

VENUS						JUPITER				
Mer Pass h m	GHA ° ′	Mean Var/hr 14°+	Dec ° ′	Mean Var/hr ′	Day	GHA ° ′	Mean Var/hr 15°+	Dec ° ′	Mean Var/hr ′	Mer Pass h m
10 28	**203 12.8**	59.2	**N22 25.9**	0.2	**1 Sat**	**310 54.5**	2.6	**S 2 12.0**	0.1	**03 16**
10 29	**202 53.4**	59.2	**N22 20.2**	0.3	**2 SUN**	**311 56.1**	2.6	**S 2 13.4**	0.1	**03 12**
10 30	**202 34.2**	59.2	**N22 13.9**	0.3	**3 Mon**	**312 57.8**	2.6	**S 2 14.8**	0.1	**03 08**
10 32	**202 15.0**	59.2	**N22 06.9**	0.3	**4 Tu**	**313 59.8**	2.6	**S 2 16.3**	0.1	**03 03**
10 33	**201 55.9**	59.2	**N21 59.3**	0.3	**5 Wed**	**315 01.9**	2.6	**S 2 17.9**	0.1	**02 59**
10 34	**201 36.9**	59.2	**N21 51.1**	0.4	**6 Th**	**316 04.2**	2.6	**S 2 19.6**	0.1	**02 55**
10 35	**201 18.0**	59.2	**N21 42.2**	0.4	**7 Fri**	**317 06.6**	2.6	**S 2 21.3**	0.1	**02 51**
10 37	**200 59.3**	59.2	**N21 32.7**	0.4	**8 Sat**	**318 09.3**	2.6	**S 2 23.1**	0.1	**02 47**
10 38	**200 40.7**	59.2	**N21 22.5**	0.4	**9 SUN**	**319 12.0**	2.6	**S 2 25.0**	0.1	**02 43**
10 39	**200 22.2**	59.2	**N21 11.8**	0.5	**10 Mon**	**320 15.0**	2.6	**S 2 26.9**	0.1	**02 39**
10 40	**200 03.9**	59.2	**N21 00.4**	0.5	**11 Tu**	**321 18.1**	2.6	**S 2 28.9**	0.1	**02 34**
10 41	**199 45.7**	59.2	**N20 48.4**	0.5	**12 Wed**	**322 21.3**	2.6	**S 2 31.0**	0.1	**02 30**
10 43	**199 27.7**	59.3	**N20 35.8**	0.6	**13 Th**	**323 24.8**	2.6	**S 2 33.1**	0.1	**02 26**
10 44	**199 09.9**	59.3	**N20 22.6**	0.6	**14 Fri**	**324 28.3**	2.7	**S 2 35.3**	0.1	**02 22**
10 45	**198 52.3**	59.3	**N20 08.9**	0.6	**15 Sat**	**325 32.1**	2.7	**S 2 37.5**	0.1	**02 17**
10 46	**198 34.8**	59.3	**N19 54.5**	0.6	**16 SUN**	**326 35.9**	2.7	**S 2 39.8**	0.1	**02 13**
10 47	**198 17.6**	59.3	**N19 39.6**	0.6	**17 Mon**	**327 39.9**	2.7	**S 2 42.2**	0.1	**02 09**
10 48	**198 00.5**	59.3	**N19 24.1**	0.7	**18 Tu**	**328 44.1**	2.7	**S 2 44.6**	0.1	**02 05**
10 50	**197 43.7**	59.3	**N19 08.0**	0.7	**19 Wed**	**329 48.4**	2.7	**S 2 47.1**	0.1	**02 00**
10 51	**197 27.1**	59.3	**N18 51.4**	0.7	**20 Th**	**330 52.8**	2.7	**S 2 49.6**	0.1	**01 56**
10 52	**197 10.7**	59.3	**N18 34.3**	0.7	**21 Fri**	**331 57.4**	2.7	**S 2 52.2**	0.1	**01 52**
10 53	**196 54.5**	59.3	**N18 16.6**	0.8	**22 Sat**	**333 02.1**	2.7	**S 2 54.9**	0.1	**01 48**
10 54	**196 38.6**	59.3	**N17 58.5**	0.8	**23 SUN**	**334 06.9**	2.7	**S 2 57.5**	0.1	**01 43**
10 55	**196 22.9**	59.4	**N17 39.8**	0.8	**24 Mon**	**335 11.9**	2.7	**S 3 00.3**	0.1	**01 39**
10 56	**196 07.4**	59.4	**N17 20.6**	0.8	**25 Tu**	**336 17.0**	2.7	**S 3 03.0**	0.1	**01 35**
10 57	**195 52.2**	59.4	**N17 00.9**	0.8	**26 Wed**	**337 22.1**	2.7	**S 3 05.9**	0.1	**01 30**
10 58	**195 37.2**	59.4	**N16 40.7**	0.9	**27 Th**	**338 27.4**	2.7	**S 3 08.7**	0.1	**01 26**
10 59	**195 22.4**	59.4	**N16 20.1**	0.9	**28 Fri**	**339 32.8**	2.7	**S 3 11.6**	0.1	**01 22**
11 00	**195 07.9**	59.4	**N15 59.0**	0.9	**29 Sat**	**340 38.3**	2.7	**S 3 14.5**	0.1	**01 17**
11 01	**194 53.6**	59.4	**N15 37.4**	0.9	**30 SUN**	**341 43.9**	2.7	**S 3 17.5**	0.1	**01 13**
11 02	**194 39.5**	59.4	**N15 15.4**	0.9	**31 Mon**	**342 49.6**	2.7	**S 3 20.5**	0.1	**01 08**

VENUS, Av. Mag. -3.9
S.H.A. August
5 249; 10 242; 15 236; 20 229; 25 223; 30 217.

JUPITER, Av. Mag.-2.8
S.H.A. August
5 2; 10 2; 15 2; 20 3; 25 3; 30 4.

MARS						SATURN				
Mer Pass h m	GHA ° ′	Mean Var/hr 15°+	Dec ° ′	Mean Var/hr ′	Day	GHA ° ′	Mean Var/hr 15°+	Dec ° ′	Mean Var/hr ′	Mer Pass h m
10 37	**200 45.0**	0.7	**N23 09.3**	0.2	**1 Sat**	**277 19.2**	2.4	**N10 17.4**	0.0	**05 30**
10 35	**201 01.5**	0.7	**N23 04.7**	0.2	**2 SUN**	**278 16.8**	2.4	**N10 17.6**	0.0	**05 26**
10 34	**201 18.2**	0.7	**N22 59.9**	0.2	**3 Mon**	**279 14.5**	2.4	**N10 17.8**	0.0	**05 22**
10 33	**201 34.9**	0.7	**N22 54.9**	0.2	**4 Tu**	**280 12.4**	2.4	**N10 18.0**	0.0	**05 18**
10 32	**201 51.7**	0.7	**N22 49.8**	0.2	**5 Wed**	**281 10.3**	2.4	**N10 18.2**	0.0	**05 14**
10 31	**202 08.6**	0.7	**N22 44.4**	0.2	**6 Th**	**282 08.3**	2.4	**N10 18.3**	0.0	**05 11**
10 30	**202 25.6**	0.7	**N22 39.0**	0.2	**7 Fri**	**283 06.4**	2.4	**N10 18.4**	0.0	**05 07**
10 29	**202 42.7**	0.7	**N22 33.3**	0.2	**8 Sat**	**284 04.7**	2.4	**N10 18.4**	0.0	**05 03**
10 28	**202 59.9**	0.7	**N22 27.5**	0.3	**9 SUN**	**285 03.0**	2.4	**N10 18.4**	0.0	**04 59**
10 26	**203 17.3**	0.7	**N22 21.5**	0.3	**10 Mon**	**286 01.4**	2.4	**N10 18.4**	0.0	**04 55**
10 25	**203 34.7**	0.7	**N22 15.3**	0.3	**11 Tu**	**286 59.9**	2.4	**N10 18.3**	0.0	**04 51**
10 24	**203 52.2**	0.7	**N22 09.0**	0.3	**12 Wed**	**287 58.5**	2.4	**N10 18.2**	0.0	**04 47**
10 23	**204 09.8**	0.7	**N22 02.5**	0.3	**13 Th**	**288 57.3**	2.4	**N10 18.1**	0.0	**04 43**
10 22	**204 27.5**	0.7	**N21 55.8**	0.3	**14 Fri**	**289 56.1**	2.5	**N10 17.9**	0.0	**04 40**
10 20	**204 45.4**	0.7	**N21 49.0**	0.3	**15 Sat**	**290 55.0**	2.5	**N10 17.7**	0.0	**04 36**
10 19	**205 03.3**	0.8	**N21 42.0**	0.3	**16 SUN**	**291 54.0**	2.5	**N10 17.5**	0.0	**04 32**
10 18	**205 21.3**	0.8	**N21 34.9**	0.3	**17 Mon**	**292 53.1**	2.5	**N10 17.2**	0.0	**04 28**
10 17	**205 39.5**	0.8	**N21 27.6**	0.3	**18 Tu**	**293 52.4**	2.5	**N10 16.9**	0.0	**04 24**
10 16	**205 57.7**	0.8	**N21 20.1**	0.3	**19 Wed**	**294 51.7**	2.5	**N10 16.6**	0.0	**04 20**
10 14	**206 16.1**	0.8	**N21 12.6**	0.3	**20 Th**	**295 51.1**	2.5	**N10 16.2**	0.0	**04 16**
10 13	**206 34.6**	0.8	**N21 04.8**	0.3	**21 Fri**	**296 50.6**	2.5	**N10 15.8**	0.0	**04 12**
10 12	**206 53.2**	0.8	**N20 56.9**	0.3	**22 Sat**	**297 50.3**	2.5	**N10 15.4**	0.0	**04 08**
10 11	**207 11.9**	0.8	**N20 48.9**	0.3	**23 SUN**	**298 50.0**	2.5	**N10 14.9**	0.0	**04 04**
10 09	**207 30.7**	0.8	**N20 40.7**	0.3	**24 Mon**	**299 49.8**	2.5	**N10 14.4**	0.0	**04 00**
10 08	**207 49.7**	0.8	**N20 32.4**	0.4	**25 Tu**	**300 49.7**	2.5	**N10 13.8**	0.0	**03 56**
10 07	**208 08.7**	0.8	**N20 24.0**	0.4	**26 Wed**	**301 49.8**	2.5	**N10 13.3**	0.0	**03 52**
10 06	**208 27.9**	0.8	**N20 15.4**	0.4	**27 Th**	**302 49.9**	2.5	**N10 12.6**	0.0	**03 48**
10 04	**208 47.1**	0.8	**N20 06.7**	0.4	**28 Fri**	**303 50.1**	2.5	**N10 12.0**	0.0	**03 44**
10 03	**209 06.5**	0.8	**N19 57.8**	0.4	**29 Sat**	**304 50.4**	2.5	**N10 11.3**	0.0	**03 40**
10 02	**209 26.1**	0.8	**N19 48.8**	0.4	**30 SUN**	**305 50.9**	2.5	**N10 10.6**	0.0	**03 36**
10 00	**209 45.7**	0.8	**N19 39.7**	0.4	**31 Mon**	**306 51.4**	2.5	**N10 09.9**	0.0	**03 32**

MARS, Av. Mag. +1.7
S.H.A. August
5 249; 10 245; 15 242; 20 238; 25 235; 30 231.

NOTE: Planet corrections on pages E113-15

SATURN, Av. Mag. +0.3
S.H.A. August
5 328; 10 328; 15 328; 20 328; 25 328; 30 328.

AUGUST 1998 MOON

Day	GMT hr	GHA	Mean Var/hr	Dec	Mean Var/hr	Day	GMT hr	GHA	Mean Var/hr	Dec	Mean Var/hr
		° ′	14°+	° ′	′			° ′	14°+	° ′	′
1 Sat	0	86 19.8	33.0	S 11 15.8	7.7	17 Mon	0	243 28.2	26.7	N 18 15.9	2.7
	6	173 38.1	32.7	S 12 02.3	7.4		6	330 08.3	26.7	N 18 32.3	2.0
	12	260 54.9	32.5	S 12 47.1	7.1		12	56 48.5	26.7	N 18 44.6	1.3
	18	348 10.0	32.2	S 13 30.1	6.8		18	143 29.0	26.8	N 18 52.7	0.6
2 Sun	0	75 23.4	31.9	S 14 11.1	6.5	18 Tu	0	230 10.1	27.0	N 18 56.8	0.0
	6	162 35.1	31.5	S 14 49.9	6.0		6	316 51.9	27.2	N 18 56.8	0.7
	12	249 44.8	31.3	S 15 26.4	5.7		12	43 34.6	27.3	N 18 52.8	1.4
	18	336 52.5	31.0	S 16 00.6	5.3		18	130 18.4	27.5	N 18 44.9	2.0
3 Mon	0	63 58.3	30.6	S 16 32.2	4.8	19 Wed	0	217 03.4	27.7	N 18 33.0	2.7
	6	151 02.0	30.2	S 17 01.1	4.3		6	303 49.7	28.0	N 18 17.4	3.3
	12	238 03.6	29.9	S 17 27.2	3.8		12	30 37.6	28.2	N 17 58.1	3.9
	18	325 03.2	29.5	S 17 50.2	3.3		18	117 27.1	28.5	N 17 35.2	4.4
4 Tu	0	52 00.8	29.3	S 18 10.2	2.8	20 Th	0	204 18.4	28.9	N 17 09.0	4.9
	6	138 56.4	28.9	S 18 26.9	2.2		6	291 11.5	29.2	N 16 39.5	5.5
	12	225 50.1	28.6	S 18 40.2	1.6		12	18 06.6	29.5	N 16 06.9	5.9
	18	312 41.9	28.3	S 18 49.9	0.9		18	105 03.6	29.9	N 15 31.4	6.4
5 Wed	0	39 32.0	28.1	S 18 56.1	0.3	21 Fri	0	192 02.6	30.2	N 14 53.2	6.8
	6	126 20.5	27.8	S 18 58.5	0.3		6	279 03.6	30.6	N 14 12.4	7.3
	12	213 07.5	27.6	S 18 57.1	0.9		12	6 06.6	30.8	N 13 29.2	7.6
	18	299 53.2	27.4	S 18 51.7	1.6		18	93 11.6	31.2	N 12 43.7	8.0
6 Th	0	26 37.7	27.3	S 18 42.5	2.2	22 Sat	0	Eclipse of	the Su	noccurs t	oday
	6	113 21.2	27.1	S 18 29.2	2.9		6	267 27.5	31.8	N 11 07.0	8.5
	12	200 03.9	27.0	S 18 11.9	3.6		12	354 38.3	32.1	N 10 16.1	8.8
	18	286 45.9	26.9	S 17 50.6	4.2		18	81 50.8	32.4	N 9 23.6	9.0
7 Fri	0	13 27.5	26.9	S 17 25.3	4.9	23 Sun	0	169 05.0	32.7	N 8 29.8	9.1
	6	100 08.8	26.9	S 16 56.0	5.5		6	256 20.8	32.9	N 7 34.9	9.3
	12	186 50.0	26.9	S 16 22.9	6.2		12	343 38.2	33.1	N 6 39.1	9.5
	18	273 31.2	26.9	S 15 46.0	6.8		18	70 56.9	33.4	N 5 42.3	9.5
8 Sat	0	0 12.7	26.9	S 15 05.4	7.4	24 Mon	0	158 16.9	33.5	N 4 45.0	9.7
	6	86 54.4	27.1	S 14 21.2	8.0		6	245 38.0	33.8	N 3 47.1	9.7
	12	173 36.6	27.1	S 13 33.7	8.5		12	333 00.1	33.8	N 2 48.9	9.8
	18	260 19.3	27.3	S 12 43.0	9.0		18	60 23.2	34.0	N 1 50.4	9.7
9 Sun	0	347 02.6	27.3	S 11 49.3	9.5	25 Tu	0	147 47.0	34.1	N 0 51.9	9.7
	6	73 46.5	27.5	S 10 52.8	9.8		6	235 11.4	34.2	S 0 06.6	9.7
	12	160 31.1	27.6	S 9 53.7	10.3		12	322 36.3	34.2	S 1 04.9	9.7
	18	247 16.5	27.6	S 8 52.3	10.6		18	50 01.6	34.3	S 2 02.9	9.6
10 Mon	0	334 02.5	27.8	S 7 48.8	10.9	26 Wed	0	137 27.1	34.3	S 3 00.4	9.5
	6	60 49.1	27.9	S 6 43.5	11.2		6	224 52.6	34.3	S 3 57.4	9.3
	12	147 36.4	28.0	S 5 36.6	11.3		12	312 18.1	34.2	S 4 53.7	9.2
	18	234 24.3	28.1	S 4 28.4	11.5		18	39 43.3	34.1	S 5 49.2	9.1
11 Tu	0	321 12.7	28.2	S 3 19.2	11.7	27 Th	0	127 08.2	34.0	S 6 43.8	9.0
	6	48 01.5	28.2	S 2 09.3	11.7		6	214 32.6	34.0	S 7 37.4	8.7
	12	134 50.6	28.3	S 0 59.0	11.8		12	301 56.4	33.9	S 8 29.8	8.5
	18	221 39.9	28.2	N 0 11.6	11.8		18	29 19.5	33.6	S 9 21.0	8.3
12 Wed	0	308 29.4	28.2	N 1 22.0	11.7	28 Fri	0	116 41.6	33.5	S 10 10.8	8.0
	6	35 18.9	28.3	N 2 32.1	11.6		6	204 02.8	33.4	S 10 59.1	7.8
	12	122 08.3	28.2	N 3 41.5	11.4		12	291 22.8	33.1	S 11 45.9	7.5
	18	208 57.6	28.1	N 4 50.0	11.2		18	18 41.6	32.9	S 12 30.9	7.2
13 Th	0	295 46.5	28.1	N 5 57.4	10.9	29 Sat	0	105 59.1	32.6	S 13 14.1	6.9
	6	22 35.1	28.0	N 7 03.3	10.7		6	193 15.1	32.4	S 13 55.3	6.5
	12	109 23.2	27.9	N 8 07.6	10.4		12	280 29.6	32.1	S 14 34.4	6.1
	18	196 10.7	27.8	N 9 10.0	10.0		18	7 42.4	31.8	S 15 11.4	5.7
14 Fri	0	282 57.6	27.7	N 10 10.2	9.7	30 Sun	0	94 53.5	31.5	S 15 46.0	5.3
	6	9 43.8	27.6	N 11 08.1	9.2		6	182 02.9	31.3	S 16 18.2	4.9
	12	96 29.3	27.4	N 12 03.4	8.7		12	269 10.5	30.9	S 16 47.7	4.5
	18	183 14.0	27.4	N 12 55.9	8.2		18	356 16.2	30.7	S 17 14.6	4.0
15 Sat	0	269 58.0	27.2	N 13 45.4	7.7	31 Mon	0	83 20.0	30.3	S 17 38.6	3.4
	6	356 41.3	27.1	N 14 31.8	7.2		6	170 22.0	30.0	S 17 59.7	2.9
	12	83 23.8	27.0	N 15 14.8	6.5		12	257 22.0	29.7	S 18 17.7	2.4
	18	170 05.7	26.9	N 15 54.4	5.9		18	344 20.2	29.4	S 18 32.4	1.8
16 Sun	0	256 47.0	26.8	N 16 30.4	5.3						
	6	343 27.8	26.7	N 17 02.6	4.7						
	12	70 08.1	26.7	N 17 31.0	4.0						
	18	156 48.3	26.6	N 17 55.5	3.3						

NOTE: Moon GHA corrections for additional hours, minutes, and seconds are on page E 108–110. Moon declination corrections are on page E 111. The mean var/hr value is needed for both corrections.

SEPTEMBER 1998 SUN & MOON GMT

SUN

Yr	Day of Mth	Week	Equation of Time 0h	Equation of Time 12h	Transit	Semi-diam	Lat 30°N Twi-light	Lat 30°N Sun-rise	Lat 30°N Sun-set	Lat 30°N Twi-light
			m s	m s	h m	′	h m	h m	h m	h m
244	**1**	**Tu**	+00 13	+00 03	**12 00**	15.9	05 12	05 37	18 23	18 47
245	**2**	**Wed**	-00 07	-00 16	**12 00**	15.9	05 13	05 37	18 22	18 46
246	**3**	**Th**	-00 26	-00 36	**11 59**	15.9	05 14	05 38	18 20	18 45
247	**4**	**Fri**	-00 45	-00 55	**11 59**	15.9	05 14	05 38	18 19	18 43
248	**5**	**Sat**	-01 05	-01 15	**11 59**	15.9	05 15	05 39	18 18	18 42
249	**6**	**Sun**	-01 26	-01 36	**11 58**	15.9	05 15	05 40	18 17	18 41
250	**7**	**Mon**	-01 46	-01 56	**11 58**	15.9	05 16	05 40	18 16	18 40
251	**8**	**Tu**	-02 06	-02 17	**11 58**	15.9	05 16	05 41	18 14	18 39
252	**9**	**Wed**	-02 27	-02 38	**11 57**	15.9	05 17	05 41	18 13	18 37
253	**10**	**Th**	-02 48	-02 59	**11 57**	15.9	05 18	05 42	18 12	18 36
254	**11**	**Fri**	-03 09	-03 20	**11 57**	15.9	05 18	05 42	18 11	18 35
255	**12**	**Sat**	-03 30	-03 41	**11 56**	15.9	05 19	05 43	18 09	18 33
256	**13**	**Sun**	-03 51	-04 02	**11 56**	15.9	05 19	05 43	18 08	18 32
257	**14**	**Mon**	-04 13	-04 23	**11 56**	15.9	05 20	05 44	18 07	18 31
258	**15**	**Tu**	-04 34	-04 45	**11 55**	15.9	05 20	05 44	18 06	18 30
259	**16**	**Wed**	-04 55	-05 06	**11 55**	15.9	05 21	05 45	18 04	18 28
260	**17**	**Th**	-05 17	-05 27	**11 55**	15.9	05 21	05 45	18 03	18 27
261	**18**	**Fri**	-05 38	-05 49	**11 54**	15.9	05 22	05 46	18 02	18 26
262	**19**	**Sat**	-05 59	-06 10	**11 54**	15.9	05 23	05 46	18 01	18 25
263	**20**	**Sun**	-06 20	-06 31	**11 53**	16.0	05 23	05 47	17 59	18 23
264	**21**	**Mon**	-06 42	-06 52	**11 53**	16.0	05 24	05 48	17 58	18 22
265	**22**	**Tu**	-07 03	-07 13	**11 53**	16.0	05 24	05 48	17 57	18 21
266	**23**	**Wed**	-07 24	-07 34	**11 52**	16.0	05 25	05 49	17 56	18 20
267	**24**	**Th**	-07 45	-07 55	**11 52**	16.0	05 25	05 49	17 54	18 18
268	**25**	**Fri**	-08 06	-08 16	**11 52**	16.0	05 26	05 50	17 53	18 17
269	**26**	**Sat**	-08 26	-08 37	**11 51**	16.0	05 26	05 50	17 52	18 16
270	**27**	**Sun**	-08 47	-08 57	**11 51**	16.0	05 27	05 51	17 51	18 15
271	**28**	**Mon**	-09 07	-09 17	**11 51**	16.0	05 28	05 51	17 50	18 13
272	**29**	**Tu**	-09 27	-09 37	**11 50**	16.0	05 28	05 52	17 48	18 12
273	**30**	**Wed**	-09 47	-09 57	**11 50**	16.0	05 29	05 53	17 47	18 11

Lat Corr to Sunrise, Sunset etc

Lat	Twi-light	Sun-rise	Sun-set	Twi-light
°	h m	h m	h m	h m
N60	–0 36	–0 18	+0 18	+0 36
55	–0 25	–0 13	+0 13	+0 25
50	–0 18	–0 09	+0 09	+0 18
45	–0 12	–0 06	+0 06	+0 12
N40	–0 07	–0 04	+0 04	+0 07
35	–0 03	–0 02	+0 02	+0 03
30	0 00	0 00	0 00	0 00
25	+0 03	+0 02	–0 02	–0 03
N20	+0 05	+0 03	–0 03	–0 05
15	+0 07	+0 04	–0 04	–0 07
10	+0 08	+0 06	–0 06	–0 08
N 5	+0 10	+0 07	–0 07	–0 10
0	+0 11	+0 08	–0 08	–0 11
S 5	+0 12	+0 09	–0 09	–0 12
10	+0 13	+0 10	–0 10	–0 13
15	+0 14	+0 11	–0 11	–0 14
S20	+0 14	+0 12	–0 12	–0 14
25	+0 15	+0 13	–0 13	–0 15
30	+0 15	+0 15	–0 15	–0 15
35	+0 15	+0 16	–0 16	–0 15
S40	+0 14	+0 17	–0 17	–0 14
45	+0 14	+0 19	–0 19	–0 14
50	+0 13	+0 21	–0 21	–0 13

NOTES

Latitude corrections are for mid-month.

Equation of time = mean time (LHA mean sun) minus apparent time (LHA true sun).

MOON

Yr	Day of Mth	Week	Age	Transit (Upper)	Diff	Semi-diam	Hor Par	Lat 30°N Moon-rise	Lat 30°N Moon-set
			days	h m	m	′	′	h m	h m
244	**1**	**Tu**	10	**19 56**	**54**	15.3	56.3	14 31	00 30
245	**2**	**Wed**	11	**20 50**	**54**	15.6	57.2	15 23	01 22
246	**3**	**Th**	12	**21 44**	**55**	15.8	58.1	16 13	02 18
247	**4**	**Fri**	13	**22 39**	**55**	16.1	58.9	17 01	03 18
248	**5**	**Sat**	14	**23 34**	**55**	16.3	59.7	17 47	04 21
249	**6**	**Sun**	15	**24 29**	**-**	16.4	60.3	18 31	05 27
250	**7**	**Mon**	16	**00 29**	**54**	16.5	60.6	19 14	06 33
251	**8**	**Tu**	17	**01 23**	**55**	16.5	60.7	19 56	07 39
252	**9**	**Wed**	18	**02 18**	**55**	16.5	60.5	20 40	08 46
253	**10**	**Th**	19	**03 13**	**55**	16.4	60.1	21 25	09 52
254	**11**	**Fri**	20	**04 08**	**56**	16.2	59.5	22 12	10 57
255	**12**	**Sat**	21	**05 04**	**56**	16.0	58.9	23 02	12 00
256	**13**	**Sun**	22	**06 00**	**55**	15.9	58.2	23 55	13 00
257	**14**	**Mon**	23	**06 55**	**54**	15.7	57.5	- -	13 56
258	**15**	**Tu**	24	**07 49**	**52**	15.5	56.9	00 50	14 47
259	**16**	**Wed**	25	**08 41**	**50**	15.3	56.3	01 46	15 34
260	**17**	**Th**	26	**09 31**	**47**	15.2	55.8	02 42	16 16
261	**18**	**Fri**	27	**10 18**	**46**	15.1	55.3	03 37	16 55
262	**19**	**Sat**	28	**11 04**	**44**	15.0	54.9	04 32	17 31
263	**20**	**Sun**	29	**11 48**	**43**	14.9	54.6	05 26	18 05
264	**21**	**Mon**	01	**12 31**	**42**	14.8	54.3	06 18	18 39
265	**22**	**Tu**	02	**13 13**	**43**	14.7	54.1	07 10	19 12
266	**23**	**Wed**	03	**13 56**	**43**	14.7	54.0	08 02	19 45
267	**24**	**Th**	04	**14 39**	**45**	14.7	54.0	08 54	20 21
268	**25**	**Fri**	05	**15 24**	**45**	14.8	54.1	09 46	20 58
269	**26**	**Sat**	06	**16 09**	**48**	14.8	54.4	10 38	21 39
270	**27**	**Sun**	07	**16 57**	**50**	15.0	54.9	11 30	22 23
271	**28**	**Mon**	08	**17 47**	**51**	15.1	55.5	12 22	23 11
272	**29**	**Tu**	09	**18 38**	**52**	15.3	56.2	13 13	- -
273	**30**	**Wed**	10	**19 30**	**54**	15.6	57.1	14 02	00 04

Phases of the Moon

	d	h	m
○ Full Moon	6	11	21
☽ Last Quarter	13	01	58
● New Moon	20	17	01
☾ First Quarter	28	21	11
Perigee	8	06	
Apogee	23	22	
Farthest S of 0°	1	16	
On 0°	8	01	
Farthest N of 0°	14	09	
On 0°	21	12	
Farthest S of 0°	29	01	

NOTES

Diff equals daily change in transit time.

To correct moonrise, moonset, or transit times for latitude and/or longitude see page E 112. For further information about these tables see page E 4. For arc-to-time correction see page E 103.

SEPTEMBER 1998 — STARS — 0h GMT September 1

No	Name	Mag	Transit h m	Dec ° ′	GHA ° ′	RA h m	SHA ° ′
♈	**ARIES**		**1 20**		**339 57.3**		
1	**Alpheratz**	2.2	**1 28**	N29 05.0	**337 52.2**	0 08	357 54.9
2	**Ankaa**	2.4	**1 46**	S 42 18.7	**333 23.8**	0 26	353 26.5
3	**Schedar**	2.5	**2 00**	N56 31.7	**329 50.2**	0 40	349 52.9
4	**Diphda**	2.2	**2 03**	S 17 59.5	**329 04.3**	0 44	349 07.0
5	**Achernar**	0.6	**2 57**	S 57 14.5	**315 32.1**	1 38	335 34.8
6	**POLARIS**	2.1	**3 51**	N89 15.2	**302 00.2**	2 32	322 02.9
7	**Hamal**	2.2	**3 27**	N23 27.3	**308 10.6**	2 07	328 13.3
8	**Acamar**	3.1	**4 18**	S 40 18.5	**295 24.0**	2 58	315 26.7
9	**Menkar**	2.8	**4 22**	N 4 05.1	**294 24.1**	3 02	314 26.8
10	**Mirfak**	1.9	**4 44**	N49 51.2	**288 53.7**	3 24	308 56.4
11	**Aldebaran**	1.1	**5 55**	N16 30.3	**270 59.7**	4 36	291 02.4
12	**Rigel**	0.3	**6 34**	S 8 12.1	**261 20.3**	5 14	281 23.0
13	**Capella**	0.2	**6 36**	N45 59.5	**260 48.5**	5 17	280 51.2
14	**Bellatrix**	1.7	**6 44**	N 6 20.9	**258 41.5**	5 25	278 44.2
15	**Elnath**	1.8	**6 45**	N28 36.2	**258 24.3**	5 26	278 27.0
16	**Alnilam**	1.8	**6 55**	S 1 12.2	**255 55.2**	5 36	275 57.9
17	**Betelgeuse**	0.1-1.2	**7 14**	N 7 24.4	**251 10.9**	5 55	271 13.6
18	**Canopus**	-0.9	**7 43**	S 52 41.5	**243 58.6**	6 24	264 01.3
19	**Sirius**	-1.6	**8 04**	S 16 42.7	**238 41.1**	6 45	258 43.8
20	**Adhara**	1.6	**8 17**	S 28 58.1	**235 18.9**	6 59	255 21.6
21	**Castor**	1.6	**8 53**	N31 53.4	**226 19.9**	7 34	246 22.6
22	**Procyon**	0.5	**8 58**	N 5 13.7	**225 09.1**	7 39	245 11.8
23	**Pollux**	1.2	**9 04**	N28 01.7	**223 39.1**	7 45	243 41.8
24	**Avior**	1.7	**9 41**	S 59 30.1	**214 20.3**	8 22	234 23.0
25	**Suhail**	2.2	**10 26**	S 43 25.5	**202 58.4**	9 08	223 01.1
26	**Miaplacidus**	1.8	**10 32**	S 69 42.5	**201 40.0**	9 13	221 42.7
27	**Alphard**	2.2	**10 46**	S 8 39.1	**198 04.8**	9 28	218 07.5
28	**Regulus**	1.3	**11 27**	N11 58.5	**187 53.2**	10 08	207 55.9
29	**Dubhe**	2.0	**12 22**	N61 45.5	**174 03.6**	11 04	194 06.3
30	**Denebola**	2.2	**13 07**	N14 34.9	**162 42.9**	11 49	182 45.6
31	**Gienah**	2.8	**13 34**	S 17 31.9	**156 01.6**	12 16	176 04.3
32	**Acrux**	1.1	**13 44**	S 63 05.5	**153 19.9**	12 26	173 22.6
33	**Gacrux**	1.6	**13 49**	S 57 06.3	**152 11.4**	12 31	172 14.1
34	**Mimosa**	1.5	**14 05**	S 59 40.8	**148 03.2**	12 48	168 05.9
35	**Alioth**	1.7	**14 12**	N55 58.3	**146 28.5**	12 54	166 31.2
36	**Spica**	1.2	**14 43**	S 11 09.1	**138 40.8**	13 25	158 43.5
37	**Alkaid**	1.9	**15 05**	N49 19.5	**133 05.6**	13 47	153 08.3
38	**Hadar**	0.9	**15 21**	S 60 22.0	**129 01.8**	14 04	149 04.5
39	**Menkent**	2.3	**15 24**	S 36 21.7	**128 18.6**	14 07	148 21.3
40	**Arcturus**	0.2	**15 33**	N19 11.7	**126 03.7**	14 16	146 06.4
41	**Rigil Kent.**	0.1	**15 57**	S 60 49.8	**120 05.1**	14 39	140 07.8
42	**Zuben'ubi**	2.9	**16 08**	S 16 02.0	**117 15.5**	14 51	137 18.2
43	**Kochab**	2.2	**16 08**	N74 10.0	**117 17.7**	14 51	137 20.4
44	**Alphecca**	2.3	**16 52**	N26 43.5	**106 18.1**	15 35	126 20.8
45	**Antares**	1.2	**17 47**	S 26 25.6	**92 37.6**	16 29	112 40.3
46	**Atria**	1.9	**18 06**	S 69 01.6	**87 49.7**	16 49	107 52.4
47	**Sabik**	2.6	**18 27**	S 15 43.2	**82 22.9**	17 10	102 25.6
48	**Shaula**	1.7	**18 51**	S 37 06.1	**76 34.6**	17 34	96 37.3
49	**Rasalhague**	2.1	**18 52**	N12 34.0	**76 14.3**	17 35	96 17.0
50	**Eltanin**	2.4	**19 14**	N51 29.8	**70 48.7**	17 57	90 51.4
51	**Kaus Aust.**	2.0	**19 41**	S 34 23.1	**63 56.1**	18 24	83 58.8
52	**Vega**	0.1	**19 54**	N38 47.4	**60 43.8**	18 37	80 46.5
53	**Nunki**	2.1	**20 12**	S 26 17.8	**56 09.6**	18 55	76 12.3
54	**Altair**	0.9	**21 07**	N 8 52.1	**42 16.4**	19 51	62 19.1
55	**Peacock**	2.1	**21 42**	S 56 44.4	**33 34.0**	20 26	53 36.7
56	**Deneb**	1.3	**21 58**	N45 16.8	**29 36.2**	20 41	49 38.9
57	**Enif**	2.5	**23 01**	N 9 52.3	**13 55.3**	21 44	33 58.0
58	**Al Na'ir**	2.2	**23 24**	S 46 58.0	**7 54.9**	22 08	27 57.6
59	**Fomalhaut**	1.3	**0 18**	S 29 37.7	**355 33.5**	22 58	15 36.2
60	**Markab**	2.6	**0 25**	N15 12.0	**353 46.6**	23 05	13 49.3

Star Transit Corr Table

Date	Corr h m	Date	Corr h m
1	0 00	17	–1 03
2	–0 04	18	–1 07
3	–0 08	19	–1 11
4	–0 12	20	–1 15
5	–0 16	21	–1 19
6	–0 20	22	–1 23
7	–0 24	23	–1 27
8	–0 28	24	–1 30
9	–0 31	25	–1 34
10	–0 35	26	–1 38
11	–0 39	27	–1 42
12	–0 43	28	–1 46
13	–0 47	29	–1 50
14	–0 51	30	–1 54
15	–0 55	31	–1 58
16	–0 59		

STAR TRANSIT

To find the approximate time of transit of a star for any day of the month use the table above. All corrections are subtractive.

If the value taken from the table is greater than the time of transit for the first of the month, add 23h 56min to the time of transit before subtracting the correction.

Example: What time will Fomalhaut (No 59) transit the Meridian on Sept. 30?

	h min
Transit on Sep 1	00 18
	+23 56
	24 14
Corr for Sep 30	–01 54
Transit on Sep 30	22 20

SEPTEMBER DIARY

d h	
3 0	eptune 2° S of Moon
4 03	Uranus 3° S of Moon
6 10	Venus 0.8° N of Regulus
7 04	Jupiter 0.5° N of Moon
7 19	Mercury 0.8° N of Regulus
9 18	Saturn 2° N of Moon
11 00	Mercury 0.4° N of Venus
12 08	Aldebaran 0.3° S of Moon
12 11	Ceres 0.9° S of Moon
16 03	Jupiter at opposition
16 05	Pallas at opposition
17 12	Mars 2° N of Moon
18 11	Regulus 0.6° N of Moon
23 06	Equinox
25 20	Mercury in conjunction
30 19	Neptune 2° S of Moon

NOTES

Star declinations may be used as is for the whole month. To correct GHA for day, hour, minute, and second see page E 106; for further explanation of this table see page E 4.

SEPTEMBER 1998 SUN & ARIES GMT

Time	SUN GHA ° ′	Dec ° ′	ARIES GHA ° ′	Time	SUN GHA ° ′	Dec ° ′	ARIES GHA ° ′	Time	SUN GHA ° ′	Dec ° ′	ARIES GHA ° ′	Time
	Tuesday, 1st September				**Sunday, 6th September**				**Friday, 11th September**			
00	179 57.0	N 8 25.9	339 57.3	00	180 21.5	N 6 35.8	344 53.0	00	180 47.4	N 4 43.1	349 48.7	00
02	209 57.4	8 24.1	10 02.3	02	210 21.9	6 34.0	14 58.0	02	210 47.8	4 41.2	19 53.6	02
04	239 57.8	8 22.2	40 07.2	04	240 22.3	6 32.1	45 02.9	04	240 48.2	4 39.3	49 58.6	04
06	269 58.2	8 20.4	70 12.1	06	270 22.8	6 30.2	75 07.8	06	270 48.7	4 37.4	80 03.5	06
08	299 58.5	8 18.6	100 17.0	08	300 23.2	6 28.4	105 12.7	08	300 49.1	4 35.5	110 08.4	08
10	329 58.9	8 16.8	130 22.0	10	330 23.6	6 26.5	135 17.7	10	330 49.6	4 33.6	140 13.4	10
12	359 59.3	8 15.0	160 26.9	12	0 24.0	6 24.7	165 22.6	12	0 50.0	4 31.7	170 18.3	12
14	29 59.7	8 13.2	190 31.8	14	30 24.4	6 22.8	195 27.5	14	30 50.4	4 29.8	200 23.2	14
16	60 00.1	8 11.4	220 36.8	16	60 24.9	6 20.9	225 32.4	16	60 50.9	4 27.9	230 28.1	16
18	90 00.5	8 09.6	250 41.7	18	90 25.3	6 19.1	255 37.4	18	90 51.3	4 26.0	260 33.1	18
20	120 00.9	8 07.7	280 46.6	20	120 25.7	6 17.2	285 42.3	20	120 51.8	4 24.1	290 38.0	20
22	150 01.3	N 8 05.9	310 51.5	22	150 26.1	N 6 15.3	315 47.2	22	150 52.2	N 4 22.2	320 42.9	22
	Wednesday, 2nd September				**Monday, 7th September**				**Saturday, 12th September**			
00	180 01.7	N 8 04.1	340 56.5	00	180 26.6	N 6 13.5	345 52.2	00	180 52.6	N 4 20.3	350 47.8	00
02	210 02.1	8 02.3	11 01.4	02	210 27.0	6 11.6	15 57.1	02	210 53.1	4 18.4	20 52.8	02
04	240 02.5	8 00.5	41 06.3	04	240 27.4	6 09.7	46 02.0	04	240 53.5	4 16.5	50 57.7	04
06	270 02.9	7 58.6	71 11.3	06	270 27.9	6 07.9	76 06.9	06	270 54.0	4 14.6	81 02.6	06
08	300 03.3	7 56.8	101 16.2	08	300 28.3	6 06.0	106 11.9	08	300 54.4	4 12.7	111 07.6	08
10	330 03.7	7 55.0	131 21.1	10	330 28.7	6 04.1	136 16.8	10	330 54.8	4 10.8	141 12.5	10
12	0 04.1	7 53.2	161 26.0	12	0 29.1	6 02.3	166 21.7	12	0 55.3	4 08.8	171 17.4	12
14	30 04.5	7 51.3	191 31.0	14	30 29.6	6 00.4	196 26.7	14	30 55.7	4 06.9	201 22.3	14
16	60 04.9	7 49.5	221 35.9	16	60 30.0	5 58.5	226 31.6	16	60 56.2	4 05.0	231 27.3	16
18	90 05.4	7 47.7	251 40.8	18	90 30.4	5 56.6	256 36.5	18	90 56.6	4 03.1	261 32.2	18
20	120 05.8	7 45.9	281 45.8	20	120 30.9	5 54.8	286 41.4	20	120 57.1	4 01.2	291 37.1	20
22	150 06.2	N 7 44.0	311 50.7	22	150 31.3	N 5 52.9	316 46.4	22	150 57.5	N 3 59.3	321 42.1	22
	Thursday, 3rd September				**Tuesday, 8th September**				**Sunday, 13th September**			
00	180 06.6	N 7 42.2	341 55.6	00	180 31.7	N 5 51.0	346 51.3	00	180 57.9	N 3 57.4	351 47.0	00
02	210 07.0	7 40.4	12 00.5	02	210 32.1	5 49.1	16 56.2	02	210 58.4	3 55.5	21 51.9	02
04	240 07.4	7 38.5	42 05.5	04	240 32.6	5 47.3	47 01.2	04	240 58.8	3 53.6	51 56.8	04
06	270 07.8	7 36.7	72 10.4	06	270 33.0	5 45.4	77 06.1	06	270 59.3	3 51.7	82 01.8	06
08	300 08.2	7 34.9	102 15.3	08	300 33.4	5 43.5	107 11.0	08	300 59.7	3 49.7	112 06.7	08
10	330 08.6	7 33.1	132 20.3	10	330 33.9	5 41.6	137 15.9	10	331 00.2	3 47.8	142 11.6	10
12	0 09.0	7 31.2	162 25.2	12	0 34.3	5 39.8	167 20.9	12	1 00.6	3 45.9	172 16.6	12
14	30 09.4	7 29.4	192 30.1	14	30 34.7	5 37.9	197 25.8	14	31 01.0	3 44.0	202 21.5	14
16	60 09.8	7 27.5	222 35.0	16	60 35.2	5 36.0	227 30.7	16	61 01.5	3 42.1	232 26.4	16
18	90 10.2	7 25.7	252 40.0	18	90 35.6	5 34.1	257 35.7	18	91 01.9	3 40.2	262 31.3	18
20	120 10.7	7 23.9	282 44.9	20	120 36.0	5 32.2	287 40.6	20	121 02.4	3 38.3	292 36.3	20
22	150 11.1	N 7 22.0	312 49.8	22	150 36.5	N 5 30.4	317 45.5	22	151 02.8	N 3 36.3	322 41.2	22
	Friday, 4th September				**Wednesday, 9th September**				**Monday, 14th September**			
00	180 11.5	N 7 20.2	342 54.7	00	180 36.9	N 5 28.5	347 50.4	00	181 03.3	N 3 34.4	352 46.1	00
02	210 11.9	7 18.4	12 59.7	02	210 37.3	5 26.6	17 55.4	02	211 03.7	3 32.5	22 51.1	02
04	240 12.3	7 16.5	43 04.6	04	240 37.8	5 24.7	48 00.3	04	241 04.1	3 30.6	52 56.0	04
06	270 12.7	7 14.7	73 09.5	06	270 38.2	5 22.8	78 05.2	06	271 04.6	3 28.7	83 00.9	06
08	300 13.1	7 12.8	103 14.5	08	300 38.6	5 20.9	108 10.1	08	301 05.0	3 26.8	113 05.8	08
10	330 13.5	7 11.0	133 19.4	10	330 39.1	5 19.0	138 15.1	10	331 05.5	3 24.8	143 10.8	10
12	0 14.0	7 09.1	163 24.3	12	0 39.5	5 17.2	168 20.0	12	1 05.9	3 22.9	173 15.7	12
14	30 14.4	7 07.3	193 29.2	14	30 39.9	5 15.3	198 24.9	14	31 06.4	3 21.0	203 20.6	14
16	60 14.8	7 05.5	223 34.2	16	60 40.4	5 13.4	228 29.9	16	61 06.8	3 19.1	233 25.6	16
18	90 15.2	7 03.6	253 39.1	18	90 40.8	5 11.5	258 34.8	18	91 07.2	3 17.2	263 30.5	18
20	120 15.6	7 01.8	283 44.0	20	120 41.2	5 09.6	288 39.7	20	121 07.7	3 15.2	293 35.4	20
22	150 16.0	N 6 59.9	313 49.0	22	150 41.7	N 5 07.7	318 44.6	22	151 08.1	N 3 13.3	323 40.3	22
	Saturday, 5th September				**Thursday, 10th September**				**Tuesday, 15th September**			
00	180 16.5	N 6 58.1	343 53.9	00	180 42.1	N 5 05.8	348 49.6	00	181 08.6	N 3 11.4	353 45.3	00
02	210 16.9	6 56.2	13 58.8	02	210 42.6	5 03.9	18 54.5	02	211 09.0	3 09.5	23 50.2	02
04	240 17.3	6 54.4	44 03.7	04	240 43.0	5 02.0	48 59.4	04	241 09.5	3 07.6	53 55.1	04
06	270 17.7	6 52.5	74 08.7	06	270 43.4	5 00.2	79 04.4	06	271 09.9	3 05.6	84 00.1	06
08	300 18.1	6 50.7	104 13.6	08	300 43.9	4 58.3	109 09.3	08	301 10.4	3 03.7	114 05.0	08
10	330 18.5	6 48.8	134 18.5	10	330 44.3	4 56.4	139 14.2	10	331 10.8	3 01.8	144 09.9	10
12	0 19.0	6 47.0	164 23.5	12	0 44.7	4 54.5	169 19.1	12	1 11.2	2 59.9	174 14.8	12
14	30 19.4	6 45.1	194 28.4	14	30 45.2	4 52.6	199 24.1	14	31 11.7	2 57.9	204 19.8	14
16	60 19.8	6 43.3	224 33.3	16	60 45.6	4 50.7	229 29.0	16	61 12.1	2 56.0	234 24.7	16
18	90 20.2	6 41.4	254 38.2	18	90 46.1	4 48.8	259 33.9	18	91 12.6	2 54.1	264 29.6	18
20	120 20.6	6 39.5	284 43.2	20	120 46.5	4 46.9	289 38.9	20	121 13.0	2 52.2	294 34.5	20
22	150 21.1	N 6 37.7	314 48.1	22	150 46.9	N 4 45.0	319 43.8	22	151 13.5	N 2 50.2	324 39.5	22

NOTES

Sun and Aries GHA corrections for additional hour, minutes, and seconds are on page E 106.
Correct sun declination by interpolation. For an example see page E 4.

SEPTEMBER 1998 — SUN & ARIES — GMT

Time	SUN GHA ° ′	Dec ° ′	ARIES GHA ° ′	Time	SUN GHA ° ′	Dec ° ′	ARIES GHA ° ′	Time	SUN GHA ° ′	Dec ° ′	ARIES GHA ° ′	Time
	Wednesday, 16th September				**Monday, 21st September**				**Saturday, 26th September**			
00	181 13.9	**N 2 48.3**	354 44.4	**00**	181 40.5	**N 0 52.2**	359 40.1	**00**	182 06.7	**S 1 04.6**	4 35.8	**00**
02	211 14.4	**2 46.4**	24 49.3	**02**	211 41.0	**0 50.2**	29 45.0	**02**	212 07.1	**1 06.6**	34 40.7	**02**
04	241 14.8	**2 44.5**	54 54.3	**04**	241 41.4	**0 48.3**	59 49.9	**04**	242 07.6	**1 08.5**	64 45.6	**04**
06	271 15.2	**2 42.5**	84 59.2	**06**	271 41.9	**0 46.3**	89 54.9	**06**	272 08.0	**1 10.5**	94 50.6	**06**
08	301 15.7	**2 40.6**	115 04.1	**08**	301 42.3	**0 44.4**	119 59.8	**08**	302 08.4	**1 12.4**	124 55.5	**08**
10	331 16.1	**2 38.7**	145 09.0	**10**	331 42.7	**0 42.4**	150 04.7	**10**	332 08.8	**1 14.4**	155 00.4	**10**
12	1 16.6	**2 36.7**	175 14.0	**12**	1 43.2	**0 40.5**	180 09.7	**12**	2 09.3	**1 16.3**	185 05.3	**12**
14	31 17.0	**2 34.8**	205 18.9	**14**	31 43.6	**0 38.6**	210 14.6	**14**	32 09.7	**1 18.3**	215 10.3	**14**
16	61 17.5	**2 32.9**	235 23.8	**16**	61 44.1	**0 36.6**	240 19.5	**16**	62 10.1	**1 20.2**	245 15.2	**16**
18	91 17.9	**2 31.0**	265 28.8	**18**	91 44.5	**0 34.7**	270 24.4	**18**	92 10.6	**1 22.2**	275 20.1	**18**
20	121 18.4	**2 29.0**	295 33.7	**20**	121 44.9	**0 32.7**	300 29.4	**20**	122 11.0	**1 24.1**	305 25.1	**20**
22	151 18.8	**N 2 27.1**	325 38.6	**22**	151 45.4	**N 0 30.8**	330 34.3	**22**	152 11.4	**S 1 26.1**	335 30.0	**22**
	Thursday, 17th September				**Tuesday, 22nd September**				**Sunday, 27th September**			
00	181 19.3	**N 2 25.2**	355 43.5	**00**	181 45.8	**N 0 28.8**	0 39.2	**00**	182 11.8	**S 1 28.0**	5 34.9	**00**
02	211 19.7	**2 23.2**	25 48.5	**02**	211 46.3	**0 26.9**	30 44.2	**02**	212 12.3	**1 30.0**	35 39.8	**02**
04	241 20.1	**2 21.3**	55 53.4	**04**	241 46.7	**0 24.9**	60 49.1	**04**	242 12.7	**1 31.9**	65 44.8	**04**
06	271 20.6	**2 19.4**	85 58.3	**06**	271 47.1	**0 23.0**	90 54.0	**06**	272 13.1	**1 33.8**	95 49.7	**06**
08	301 21.0	**2 17.4**	116 03.3	**08**	301 47.6	**0 21.0**	120 58.9	**08**	302 13.5	**1 35.8**	125 54.6	**08**
10	331 21.5	**2 15.5**	146 08.2	**10**	331 48.0	**0 19.1**	151 03.9	**10**	332 14.0	**1 37.7**	155 59.6	**10**
12	1 21.9	**2 13.6**	176 13.1	**12**	1 48.5	**0 17.2**	181 08.8	**12**	2 14.4	**1 39.7**	186 04.5	**12**
14	31 22.4	**2 11.6**	206 18.0	**14**	31 48.9	**0 15.2**	211 13.7	**14**	32 14.8	**1 41.6**	216 09.4	**14**
16	61 22.8	**2 09.7**	236 23.0	**16**	61 49.3	**0 13.3**	241 18.7	**16**	62 15.2	**1 43.6**	246 14.3	**16**
18	91 23.3	**2 07.8**	266 27.9	**18**	91 49.8	**0 11.3**	271 23.6	**18**	92 15.6	**1 45.5**	276 19.3	**18**
20	121 23.7	**2 05.8**	296 32.8	**20**	121 50.2	**0 09.4**	301 28.5	**20**	122 16.1	**1 47.5**	306 24.2	**20**
22	151 24.1	**N 2 03.9**	326 37.8	**22**	151 50.7	**N 0 07.4**	331 33.4	**22**	152 16.5	**S 1 49.4**	336 29.1	**22**
	Friday, 18th September				**Wednesday, 23rd September**				**Monday, 28th September**			
00	181 24.6	**N 2 02.0**	356 42.7	**00**	181 51.1	**N 0 05.5**	1 38.4	**00**	182 16.9	**S 1 51.4**	6 34.1	**00**
02	211 25.0	**2 00.0**	26 47.6	**02**	211 51.5	**0 03.5**	31 43.3	**02**	212 17.3	**1 53.3**	36 39.0	**02**
04	241 25.5	**1 58.1**	56 52.5	**04**	241 52.0	**0 01.6**	61 48.2	**04**	242 17.8	**1 55.3**	66 43.9	**04**
06	271 25.9	**1 56.2**	86 57.5	**06**	271 52.4	**S 0 00.4**	91 53.2	**06**	272 18.2	**1 57.2**	96 48.8	**06**
08	301 26.4	**1 54.2**	117 02.4	**08**	301 52.8	**0 02.3**	121 58.1	**08**	302 18.6	**1 59.1**	126 53.8	**08**
10	331 26.8	**1 52.3**	147 07.3	**10**	331 53.3	**0 04.3**	152 03.0	**10**	332 19.0	**2 01.1**	156 58.7	**10**
12	1 27.3	**1 50.4**	177 12.3	**12**	1 53.7	**0 06.2**	182 07.9	**12**	2 19.4	**2 03.0**	187 03.6	**12**
14	31 27.7	**1 48.4**	207 17.2	**14**	31 54.1	**0 08.1**	212 12.9	**14**	32 19.9	**2 05.0**	217 08.6	**14**
16	61 28.1	**1 46.5**	237 22.1	**16**	61 54.6	**0 10.1**	242 17.8	**16**	62 20.3	**2 06.9**	247 13.5	**16**
18	91 28.6	**1 44.6**	267 27.0	**18**	91 55.0	**0 12.0**	272 22.7	**18**	92 20.7	**2 08.9**	277 18.4	**18**
20	121 29.0	**1 42.6**	297 32.0	**20**	121 55.5	**0 14.0**	302 27.6	**20**	122 21.1	**2 10.8**	307 23.3	**20**
22	151 29.5	**N 1 40.7**	327 36.9	**22**	151 55.9	**S 0 15.9**	332 32.6	**22**	152 21.5	**S 2 12.8**	337 28.3	**22**
	Saturday, 19th September				**Thursday, 24th September**				**Tuesday, 29th September**			
00	181 29.9	**N 1 38.7**	357 41.8	**00**	181 56.3	**S 0 17.9**	2 37.5	**00**	182 21.9	**S 2 14.7**	7 33.2	**00**
02	211 30.4	**1 36.8**	27 46.7	**02**	211 56.8	**0 19.8**	32 42.4	**02**	212 22.4	**2 16.7**	37 38.1	**02**
04	241 30.8	**1 34.9**	57 51.7	**04**	241 57.2	**0 21.8**	62 47.4	**04**	242 22.8	**2 18.6**	67 43.1	**04**
06	271 31.2	**1 32.9**	87 56.6	**06**	271 57.6	**0 23.7**	92 52.3	**06**	272 23.2	**2 20.5**	97 48.0	**06**
08	301 31.7	**1 31.0**	118 01.5	**08**	301 58.1	**0 25.7**	122 57.2	**08**	302 23.6	**2 22.5**	127 52.9	**08**
10	331 32.1	**1 29.0**	148 06.5	**10**	331 58.5	**0 27.6**	153 02.1	**10**	332 24.0	**2 24.4**	157 57.8	**10**
12	1 32.6	**1 27.1**	178 11.4	**12**	1 58.9	**0 29.6**	183 07.1	**12**	2 24.4	**2 26.4**	188 02.8	**12**
14	31 33.0	**1 25.2**	208 16.3	**14**	31 59.4	**0 31.5**	213 12.0	**14**	32 24.9	**2 28.3**	218 07.7	**14**
16	61 33.5	**1 23.2**	238 21.2	**16**	61 59.8	**0 33.5**	243 16.9	**16**	62 25.3	**2 30.3**	248 12.6	**16**
18	91 33.9	**1 21.3**	268 26.2	**18**	92 00.2	**0 35.4**	273 21.9	**18**	92 25.7	**2 32.2**	278 17.5	**18**
20	121 34.3	**1 19.3**	298 31.1	**20**	122 00.7	**0 37.4**	303 26.8	**20**	122 26.1	**2 34.2**	308 22.5	**20**
22	151 34.8	**N 1 17.4**	328 36.0	**22**	152 01.1	**S 0 39.3**	333 31.7	**22**	152 26.5	**S 2 36.1**	338 27.4	**22**
	Sunday, 20th September				**Friday, 25th September**				**Wednesday, 30th September**			
00	181 35.2	**N 1 15.5**	358 41.0	**00**	182 01.5	**S 0 41.3**	3 36.6	**00**	182 26.9	**S 2 38.0**	8 32.3	**00**
02	211 35.7	**1 13.5**	28 45.9	**02**	212 02.0	**0 43.2**	33 41.6	**02**	212 27.3	**2 40.0**	38 37.3	**02**
04	241 36.1	**1 11.6**	58 50.8	**04**	242 02.4	**0 45.2**	63 46.5	**04**	242 27.7	**2 41.9**	68 42.2	**04**
06	271 36.6	**1 09.6**	88 55.7	**06**	272 02.8	**0 47.1**	93 51.4	**06**	272 28.2	**2 43.9**	98 47.1	**06**
08	301 37.0	**1 07.7**	119 00.7	**08**	302 03.3	**0 49.0**	123 56.4	**08**	302 28.6	**2 45.8**	128 52.0	**08**
10	331 37.4	**1 05.8**	149 05.6	**10**	332 03.7	**0 51.0**	154 01.3	**10**	332 29.0	**2 47.7**	158 57.0	**10**
12	1 37.9	**1 03.8**	179 10.5	**12**	2 04.1	**0 52.9**	184 06.2	**12**	2 29.4	**2 49.7**	189 01.9	**12**
14	31 38.3	**1 01.9**	209 15.5	**14**	32 04.6	**0 54.9**	214 11.1	**14**	32 29.8	**2 51.6**	219 06.8	**14**
16	61 38.8	**0 59.9**	239 20.4	**16**	62 05.0	**0 56.8**	244 16.1	**16**	62 30.2	**2 53.6**	249 11.8	**16**
18	91 39.2	**0 58.0**	269 25.3	**18**	92 05.4	**0 58.8**	274 21.0	**18**	92 30.6	**2 55.5**	279 16.7	**18**
20	121 39.7	**0 56.0**	299 30.2	**20**	122 05.8	**1 00.7**	304 25.9	**20**	122 31.0	**2 57.5**	309 21.6	**20**
22	151 40.1	**N 0 54.1**	329 35.2	**22**	152 06.3	**S 1 02.7**	334 30.9	**22**	152 31.4	**S 2 59.4**	339 26.5	**22**

SEPTEMBER 1998 PLANETS 0h GMT

VENUS						JUPITER				
Mer Pass	GHA	Mean Var/hr	Dec	Mean Var/hr	Day	GHA	Mean Var/hr	Dec	Mean Var/hr	Mer Pass
h m	° ′	14°+	° ′	′		° ′	15°+	° ′	′	h m
11 03	**194 25.7**	59.4	**N14 53.0**	1.0	**1 Tu**	**343 55.3**	2.7	**S 3 23.5**	0.1	**01 04**
11 04	**194 12.1**	59.4	**N14 30.1**	1.0	**2 Wed**	**345 01.2**	2.7	**S 3 26.6**	0.1	**01 00**
11 04	**193 58.7**	59.5	**N14 06.9**	1.0	**3 Th**	**346 07.1**	2.8	**S 3 29.7**	0.1	**00 55**
11 05	**193 45.6**	59.5	**N13 43.3**	1.0	**4 Fri**	**347 13.1**	2.8	**S 3 32.8**	0.1	**00 51**
11 06	**193 32.7**	59.5	**N13 19.2**	1.0	**5 Sat**	**348 19.2**	2.8	**S 3 35.9**	0.1	**00 47**
11 07	**193 20.0**	59.5	**N12 54.8**	1.0	**6 SUN**	**349 25.3**	2.8	**S 3 39.0**	0.1	**00 42**
11 08	**193 07.5**	59.5	**N12 30.1**	1.0	**7 Mon**	**350 31.5**	2.8	**S 3 42.2**	0.1	**00 38**
11 09	**192 55.2**	59.5	**N12 05.0**	1.1	**8 Tu**	**351 37.7**	2.8	**S 3 45.4**	0.1	**00 33**
11 09	**192 43.2**	59.5	**N11 39.5**	1.1	**9 Wed**	**352 44.0**	2.8	**S 3 48.5**	0.1	**00 29**
11 10	**192 31.3**	59.5	**N11 13.8**	1.1	**10 Th**	**353 50.4**	2.8	**S 3 51.7**	0.1	**00 25**
11 11	**192 19.6**	59.5	**N10 47.7**	1.1	**11 Fri**	**354 56.7**	2.8	**S 3 55.0**	0.1	**00 20**
11 12	**192 08.1**	59.5	**N10 21.3**	1.1	**12 Sat**	**356 03.1**	2.8	**S 3 58.2**	0.1	**00 16**
11 13	**191 56.8**	59.5	**N 9 54.6**	1.1	**13 SUN**	**357 09.6**	2.8	**S 4 01.4**	0.1	**00 11**
11 13	**191 45.6**	59.5	**N 9 27.6**	1.1	**14 Mon**	**358 16.0**	2.8	**S 4 04.6**	0.1	**00 07**
11 14	**191 34.6**	59.5	**N 9 00.4**	1.1	**15 Tu**	**359 22.5**	2.8	**S 4 07.8**	0.1	**00 02**
11 15	**191 23.8**	59.6	**N 8 32.9**	1.2	**16 Wed**	**0 29.0**	2.8	**S 4 11.0**	0.1	**23 54**
11 15	**191 13.1**	59.6	**N 8 05.2**	1.2	**17 Th**	**1 35.5**	2.8	**S 4 14.2**	0.1	**23 49**
11 16	**191 02.6**	59.6	**N 7 37.2**	1.2	**18 Fri**	**2 41.9**	2.8	**S 4 17.4**	0.1	**23 45**
11 17	**190 52.2**	59.6	**N 7 09.0**	1.2	**19 Sat**	**3 48.4**	2.8	**S 4 20.6**	0.1	**23 40**
11 18	**190 41.9**	59.6	**N 6 40.6**	1.2	**20 SUN**	**4 54.9**	2.8	**S 4 23.7**	0.1	**23 36**
11 18	**190 31.8**	59.6	**N 6 12.0**	1.2	**21 Mon**	**6 01.3**	2.8	**S 4 26.9**	0.1	**23 32**
11 19	**190 21.7**	59.6	**N 5 43.3**	1.2	**22 Tu**	**7 07.7**	2.8	**S 4 30.0**	0.1	**23 27**
11 20	**190 11.8**	59.6	**N 5 14.3**	1.2	**23 Wed**	**8 14.1**	2.8	**S 4 33.1**	0.1	**23 23**
11 20	**190 02.0**	59.6	**N 4 45.2**	1.2	**24 Th**	**9 20.5**	2.8	**S 4 36.2**	0.1	**23 18**
11 21	**189 52.2**	59.6	**N 4 16.0**	1.2	**25 Fri**	**10 26.8**	2.8	**S 4 39.3**	0.1	**23 14**
11 21	**189 42.5**	59.6	**N 3 46.6**	1.2	**26 Sat**	**11 33.1**	2.8	**S 4 42.3**	0.1	**23 10**
11 22	**189 32.9**	59.6	**N 3 17.1**	1.2	**27 SUN**	**12 39.3**	2.8	**S 4 45.3**	0.1	**23 05**
11 23	**189 23.4**	59.6	**N 2 47.4**	1.2	**28 Mon**	**13 45.5**	2.8	**S 4 48.2**	0.1	**23 01**
11 23	**189 13.9**	59.6	**N 2 17.7**	1.2	**29 Tu**	**14 51.5**	2.8	**S 4 51.2**	0.1	**22 56**
11 24	**189 04.4**	59.6	**N 1 47.9**	1.2	**30 Wed**	**15 57.6**	2.7	**S 4 54.0**	0.1	**22 52**

VENUS, Av. Mag. -3.9
S.H.A. September
5 210; 10 204; 15 198; 20 192; 25 186; 30 181.

JUPITER, Av. Mag.-2.9
S.H.A. September
5 4; 10 5; 15 6; 20 6; 25 7; 30 7.

MARS						SATURN				
Mer Pass	GHA	Mean Var/hr	Dec	Mean Var/hr	Day	GHA	Mean Var/hr	Dec	Mean Var/hr	Mer Pass
h m	° ′	15°+	° ′	′		° ′	15°+	° ′	′	h m
09 59	**210 05.4**	0.8	**N19 30.5**	0.4	**1 Tu**	**307 52.0**	2.5	**N10 09.1**	0.0	**03 28**
09 58	**210 25.3**	0.8	**N19 21.1**	0.4	**2 Wed**	**308 52.7**	2.5	**N10 08.3**	0.0	**03 24**
09 56	**210 45.3**	0.8	**N19 11.6**	0.4	**3 Th**	**309 53.5**	2.5	**N10 07.5**	0.0	**03 20**
09 55	**211 05.4**	0.8	**N19 02.0**	0.4	**4 Fri**	**310 54.4**	2.5	**N10 06.6**	0.0	**03 16**
09 54	**211 25.6**	0.8	**N18 52.3**	0.4	**5 Sat**	**311 55.4**	2.5	**N10 05.7**	0.0	**03 12**
09 52	**211 45.9**	0.8	**N18 42.4**	0.4	**6 SUN**	**312 56.5**	2.5	**N10 04.8**	0.0	**03 08**
09 51	**212 06.3**	0.9	**N18 32.5**	0.4	**7 Mon**	**313 57.6**	2.6	**N10 03.8**	0.0	**03 04**
09 50	**212 26.9**	0.9	**N18 22.4**	0.4	**8 Tu**	**314 58.9**	2.6	**N10 02.8**	0.0	**03 00**
09 48	**212 47.5**	0.9	**N18 12.2**	0.4	**9 Wed**	**316 00.2**	2.6	**N10 01.8**	0.0	**02 55**
09 47	**213 08.3**	0.9	**N18 01.9**	0.4	**10 Th**	**317 01.7**	2.6	**N10 00.8**	0.0	**02 51**
09 45	**213 29.2**	0.9	**N17 51.5**	0.4	**11 Fri**	**318 03.2**	2.6	**N 9 59.7**	0.0	**02 47**
09 44	**213 50.2**	0.9	**N17 41.0**	0.4	**12 Sat**	**319 04.8**	2.6	**N 9 58.6**	0.0	**02 43**
09 43	**214 11.3**	0.9	**N17 30.4**	0.4	**13 SUN**	**320 06.5**	2.6	**N 9 57.5**	0.1	**02 39**
09 41	**214 32.5**	0.9	**N17 19.7**	0.5	**14 Mon**	**321 08.3**	2.6	**N 9 56.3**	0.0	**02 35**
09 40	**214 53.8**	0.9	**N17 08.8**	0.5	**15 Tu**	**322 10.2**	2.6	**N 9 55.2**	0.1	**02 31**
09 38	**215 15.2**	0.9	**N16 57.9**	0.5	**16 Wed**	**323 12.1**	2.6	**N 9 54.0**	0.1	**02 27**
09 37	**215 36.8**	0.9	**N16 46.9**	0.5	**17 Th**	**324 14.2**	2.6	**N 9 52.7**	0.1	**02 23**
09 36	**215 58.4**	0.9	**N16 35.8**	0.5	**18 Fri**	**325 16.3**	2.6	**N 9 51.5**	0.1	**02 19**
09 34	**216 20.2**	0.9	**N16 24.6**	0.5	**19 Sat**	**326 18.5**	2.6	**N 9 50.2**	0.1	**02 14**
09 33	**216 42.1**	0.9	**N16 13.3**	0.5	**20 SUN**	**327 20.7**	2.6	**N 9 48.9**	0.1	**02 10**
09 31	**217 04.1**	0.9	**N16 01.9**	0.5	**21 Mon**	**328 23.1**	2.6	**N 9 47.6**	0.1	**02 06**
09 30	**217 26.2**	0.9	**N15 50.5**	0.5	**22 Tu**	**329 25.5**	2.6	**N 9 46.2**	0.1	**02 02**
09 28	**217 48.4**	0.9	**N15 38.9**	0.5	**23 Wed**	**330 28.0**	2.6	**N 9 44.9**	0.1	**01 58**
09 27	**218 10.7**	0.9	**N15 27.3**	0.5	**24 Th**	**331 30.6**	2.6	**N 9 43.5**	0.1	**01 54**
09 25	**218 33.1**	0.9	**N15 15.6**	0.5	**25 Fri**	**332 33.2**	2.6	**N 9 42.1**	0.1	**01 49**
09 24	**218 55.7**	0.9	**N15 03.8**	0.5	**26 Sat**	**333 35.9**	2.6	**N 9 40.6**	0.1	**01 45**
09 22	**219 18.3**	1.0	**N14 52.0**	0.5	**27 SUN**	**334 38.6**	2.6	**N 9 39.2**	0.1	**01 41**
09 21	**219 41.1**	0.9	**N14 40.0**	0.5	**28 Mon**	**335 41.5**	2.6	**N 9 37.7**	0.1	**01 37**
09 19	**220 03.9**	1.0	**N14 28.0**	0.5	**29 Tu**	**336 44.3**	2.6	**N 9 36.2**	0.1	**01 33**
09 18	**220 26.9**	1.0	**N14 15.9**	0.5	**30 Wed**	**337 47.3**	2.6	**N 9 34.7**	0.1	**01 29**

MARS, Av. Mag. +1.7
S.H.A. September
5 228; 10 224; 15 221; 20 218; 25 215; 30 212.

NOTE: Planet corrections on pages E113-15

SATURN, Av. Mag. +0.2
S.H.A. September
5 328; 10 328; 15 328; 20 329; 25 329; 30 329.

SEPTEMBER 1998 MOON

Day	GMT hr	GHA ° ′	Mean Var/hr 14°+	Dec ° ′	Mean Var/hr ′	Day	GMT hr	GHA ° ′	Mean Var/hr 14°+	Dec ° ′	Mean Var/hr ′
1 Tu	**0**	71 16.6	29.0	**S 18 43.9**	1.3	17 Th	**0**	221 57.2	30.3	**N 15 32.6**	6.4
	6	158 11.1	28.8	**S 18 51.9**	0.7		**6**	308 58.8	30.6	**N 14 54.9**	6.7
	12	245 04.0	28.5	**S 18 56.3**	0.1		**12**	36 02.3	30.9	**N 14 14.6**	7.1
	18	331 55.2	28.3	**S 18 57.1**	0.6		**18**	123 07.7	31.3	**N 13 31.9**	7.5
2 Wed	**0**	58 44.9	28.0	**S 18 54.1**	1.2	18 Fri	**0**	210 15.0	31.6	**N 12 47.0**	7.9
	6	145 33.2	27.8	**S 18 47.3**	1.8		**6**	297 24.1	31.8	**N 12 00.2**	8.1
	12	232 20.1	27.6	**S 18 36.7**	2.5		**12**	24 35.0	32.1	**N 11 11.5**	8.5
	18	319 05.9	27.4	**S 18 22.1**	3.1		**18**	111 47.6	32.4	**N 10 21.1**	8.7
3 Th	**0**	45 50.6	27.3	**S 18 03.5**	3.8	19 Sat	**0**	199 01.8	32.6	**N 9 29.1**	8.9
	6	132 34.4	27.2	**S 17 41.0**	4.5		**6**	286 17.6	32.8	**N 8 35.9**	9.1
	12	219 17.4	27.1	**S 17 14.6**	5.1		**12**	13 34.7	33.1	**N 7 41.4**	9.2
	18	305 59.8	27.0	**S 16 44.2**	5.8		**18**	100 53.2	33.3	**N 6 45.9**	9.5
4 Fri	**0**	32 41.7	27.0	**S 16 09.9**	6.4	20 Sun	**0**	188 12.9	33.5	**N 5 49.5**	9.6
	6	119 23.3	26.9	**S 15 31.7**	7.0		**6**	275 33.8	33.7	**N 4 52.3**	9.6
	12	206 04.6	26.9	**S 14 49.9**	7.6		**12**	2 55.6	33.8	**N 3 54.6**	9.7
	18	292 45.7	26.8	**S 14 04.5**	8.2		**18**	90 18.3	34.0	**N 2 56.5**	9.8
5 Sat	**0**	19 26.8	26.9	**S 13 15.6**	8.8	21 Mon	**0**	177 41.8	34.0	**N 1 58.0**	9.7
	6	106 08.0	26.8	**S 12 23.4**	9.3		**6**	265 05.9	34.1	**N 0 59.4**	9.7
	12	192 49.3	27.0	**S 11 28.0**	9.8		**12**	352 30.5	34.2	**N 0 00.8**	9.8
	18	279 30.8	27.0	**S 10 29.8**	10.2		**18**	79 55.6	34.3	**S 0 57.6**	9.7
6 Sun	**0**	6 12.5	27.0	**S 9 28.9**	10.6	22 Tu	**0**	167 20.9	34.2	**S 1 55.8**	9.6
	6	92 54.4	27.0	**S 8 25.5**	11.0		**6**	254 46.4	34.2	**S 2 53.6**	9.5
	12	179 36.6	27.1	**S 7 19.9**	11.3		**12**	342 11.9	34.2	**S 3 50.9**	9.5
	18	266 19.1	27.1	**S 6 12.4**	11.5		**18**	69 37.4	34.2	**S 4 47.6**	9.3
7 Mon	**0**	353 01.7	27.2	**S 5 03.3**	11.8	23 Wed	**0**	157 02.6	34.2	**S 5 43.4**	9.2
	6	79 44.6	27.2	**S 3 52.8**	11.9		**6**	244 27.5	34.1	**S 6 38.4**	9.0
	12	166 27.6	27.2	**S 2 41.3**	12.1		**12**	331 52.0	34.0	**S 7 32.4**	8.8
	18	253 10.7	27.2	**S 1 29.1**	12.1		**18**	59 16.0	33.9	**S 8 25.2**	8.6
8 Tu	**0**	339 53.9	27.2	**S 0 16.5**	12.1	24 Th	**0**	146 39.3	33.8	**S 9 16.7**	8.3
	6	66 37.1	27.2	**N 0 56.2**	12.1		**6**	234 01.8	33.6	**S 10 06.9**	8.1
	12	153 20.1	27.2	**N 2 08.7**	12.0		**12**	321 23.5	33.4	**S 10 55.5**	7.9
	18	240 03.1	27.1	**N 3 20.6**	11.8		**18**	48 44.2	33.3	**S 11 42.6**	7.6
9 Wed	**0**	326 45.8	27.1	**N 4 31.7**	11.6	25 Fri	**0**	136 03.9	33.1	**S 12 27.9**	7.2
	6	53 28.3	27.0	**N 5 41.6**	11.4		**6**	223 22.4	32.9	**S 13 11.3**	6.9
	12	140 10.4	27.0	**N 6 50.0**	11.1		**12**	310 39.8	32.7	**S 13 52.8**	6.5
	18	226 52.2	26.9	**N 7 56.7**	10.7		**18**	37 55.8	32.4	**S 14 32.2**	6.2
10 Th	**0**	313 33.5	26.8	**N 9 01.4**	10.4	26 Sat	**0**	125 10.5	32.2	**S 15 09.4**	5.8
	6	40 14.4	26.7	**N 10 03.7**	9.9		**6**	212 23.8	31.9	**S 15 44.3**	5.4
	12	126 54.9	26.7	**N 11 03.5**	9.5		**12**	299 35.6	31.7	**S 16 16.7**	5.0
	18	213 34.8	26.5	**N 12 00.6**	8.9		**18**	26 45.9	31.4	**S 16 46.6**	4.5
11 Fri	**0**	300 14.4	26.5	**N 12 54.6**	8.5	27 Sun	**0**	113 54.7	31.2	**S 17 13.9**	4.0
	6	26 53.5	26.5	**N 13 45.5**	7.9		**6**	201 01.9	30.9	**S 17 38.3**	3.6
	12	113 32.2	26.4	**N 14 32.9**	7.3		**12**	288 07.5	30.6	**S 17 59.9**	3.1
	18	200 10.7	26.3	**N 15 16.8**	6.6		**18**	15 11.6	30.4	**S 18 18.5**	2.5
12 Sat	**0**	286 48.9	26.3	**N 15 57.0**	6.0	28 Mon	**0**	102 14.2	30.1	**S 18 34.0**	2.0
	6	13 26.9	26.3	**N 16 33.4**	5.4		**6**	189 15.2	29.9	**S 18 46.3**	1.5
	12	100 04.9	26.3	**N 17 05.8**	4.7		**12**	276 14.7	29.6	**S 18 55.3**	0.8
	18	186 43.0	26.4	**N 17 34.3**	4.0		**18**	3 12.7	29.4	**S 19 00.9**	0.3
13 Sun	**0**	273 21.3	26.5	**N 17 58.6**	3.4	29 Tu	**0**	90 09.3	29.2	**S 19 03.1**	0.3
	6	0 00.0	26.5	**N 18 18.8**	2.6		**6**	177 04.6	28.9	**S 19 01.7**	0.9
	12	86 39.2	26.6	**N 18 34.9**	1.9		**12**	263 58.6	28.8	**S 18 56.7**	1.5
	18	173 19.0	26.8	**N 18 46.8**	1.3		**18**	350 51.4	28.6	**S 18 48.0**	2.1
14 Mon	**0**	259 59.6	26.9	**N 18 54.6**	0.5	30 Wed	**0**	77 43.0	28.4	**S 18 35.6**	2.7
	6	346 41.2	27.1	**N 18 58.3**	0.1		**6**	164 33.7	28.3	**S 18 19.5**	3.4
	12	73 23.9	27.4	**N 18 57.9**	0.8		**12**	251 23.3	28.1	**S 17 59.5**	4.0
	18	160 07.9	27.6	**N 18 53.6**	1.4		**18**	338 12.1	28.0	**S 17 35.8**	4.7
15 Tu	**0**	246 53.2	27.8	**N 18 45.3**	2.1						
	6	333 40.0	28.1	**N 18 33.2**	2.7						
	12	60 28.4	28.4	**N 18 17.5**	3.2						
	18	147 18.6	28.7	**N 17 58.1**	3.9						
16 Wed	**0**	234 10.5	29.0	**N 17 35.4**	4.4						
	6	321 04.3	29.3	**N 17 09.3**	4.9						
	12	48 00.0	29.6	**N 16 40.0**	5.4						
	18	134 57.7	30.0	**N 16 07.8**	5.8						

NOTE: Moon GHA corrections for additional hours, minutes, and seconds are on page E 108–110. Moon declination corrections are on page E 111. The mean var/hr value is needed for both corrections.

OCTOBER 1998 SUN & MOON GMT

SUN

Yr	Mth	Week	Equation of Time 0h	Equation of Time 12h	Transit	Semi-diam	Lat 30°N Twi-light	Lat 30°N Sun-rise	Lat 30°N Sun-set	Lat 30°N Twi-light
			m s	m s	h m	′	h m	h m	h m	h m
274	**1**	**Th**	-10 07	-10 17	**11 50**	16.0	05 29	05 53	17 46	18 10
275	**2**	**Fri**	-10 26	-10 36	**11 49**	16.0	05 30	05 54	17 45	18 08
276	**3**	**Sat**	-10 45	-10 55	**11 49**	16.0	05 30	05 54	17 43	18 07
277	**4**	**Sun**	-11 04	-11 14	**11 49**	16.0	05 31	05 55	17 42	18 06
278	**5**	**Mon**	-11 23	-11 32	**11 48**	16.0	05 32	05 55	17 41	18 05
279	**6**	**Tu**	-11 41	-11 50	**11 48**	16.0	05 32	05 56	17 40	18 04
280	**7**	**Wed**	-11 59	-12 07	**11 48**	16.0	05 33	05 57	17 39	18 03
281	**8**	**Th**	-12 16	-12 25	**11 48**	16.0	05 33	05 57	17 37	18 01
282	**9**	**Fri**	-12 33	-12 41	**11 47**	16.0	05 34	05 58	17 36	18 00
283	**10**	**Sat**	-12 49	-12 57	**11 47**	16.0	05 34	05 58	17 35	17 59
284	**11**	**Sun**	-13 05	-13 13	**11 47**	16.0	05 35	05 59	17 34	17 58
285	**12**	**Mon**	-13 21	-13 28	**11 47**	16.1	05 36	06 00	17 33	17 57
286	**13**	**Tu**	-13 36	-13 43	**11 46**	16.1	05 36	06 00	17 32	17 56
287	**14**	**Wed**	-13 50	-13 57	**11 46**	16.1	05 37	06 01	17 31	17 55
288	**15**	**Th**	-14 04	-14 11	**11 46**	16.1	05 38	06 02	17 30	17 54
289	**16**	**Fri**	-14 18	-14 24	**11 46**	16.1	05 38	06 02	17 28	17 53
290	**17**	**Sat**	-14 30	-14 36	**11 45**	16.1	05 39	06 03	17 27	17 52
291	**18**	**Sun**	-14 42	-14 48	**11 45**	16.1	05 39	06 04	17 26	17 51
292	**19**	**Mon**	-14 54	-15 00	**11 45**	16.1	05 40	06 04	17 25	17 50
293	**20**	**Tu**	-15 05	-15 10	**11 45**	16.1	05 41	06 05	17 24	17 49
294	**21**	**Wed**	-15 15	-15 20	**11 45**	16.1	05 41	06 06	17 23	17 48
295	**22**	**Th**	-15 25	-15 29	**11 45**	16.1	05 42	06 06	17 22	17 47
296	**23**	**Fri**	-15 34	-15 38	**11 44**	16.1	05 43	06 07	17 21	17 46
297	**24**	**Sat**	-15 42	-15 46	**11 44**	16.1	05 43	06 08	17 20	17 45
298	**25**	**Sun**	-15 50	-15 53	**11 44**	16.1	05 44	06 08	17 19	17 44
299	**26**	**Mon**	-15 57	-16 00	**11 44**	16.1	05 45	06 09	17 18	17 43
300	**27**	**Tu**	-16 03	-16 06	**11 44**	16.1	05 45	06 10	17 18	17 42
301	**28**	**Wed**	-16 08	-16 11	**11 44**	16.1	05 46	06 11	17 17	17 41
302	**29**	**Th**	-16 13	-16 15	**11 44**	16.1	05 47	06 11	17 16	17 40
303	**30**	**Fri**	-16 17	-16 19	**11 44**	16.1	05 48	06 12	17 15	17 39
304	**31**	**Sat**	-16 20	-16 22	**11 44**	16.1	05 48	06 13	17 14	17 39

Lat Corr to Sunrise, Sunset etc

Lat	Twi-light	Sun-rise	Sun-set	Twi-light
°	h m	h m	h m	h m
N60	+0 19	+0 36	–0 36	–0 19
55	+0 15	+0 27	–0 27	–0 15
50	+0 11	+0 19	–0 19	–0 11
45	+0 08	+0 13	–0 13	–0 08
N40	+0 05	+0 08	–0 08	–0 05
35	+0 03	+0 04	–0 04	–0 03
30	0 00	0 00	0 00	0 00
25	–0 03	–0 04	+0 04	+0 03
N20	–0 05	–0 07	+0 07	+0 05
15	–0 08	–0 10	+0 10	+0 08
10	–0 10	–0 13	+0 13	+0 10
N 5	–0 13	–0 16	+0 16	+0 13
0	–0 16	–0 19	+0 19	+0 16
S 5	–0 19	–0 22	+0 22	+0 19
10	–0 22	–0 25	+0 25	+0 22
15	–0 26	–0 28	+0 28	+0 26
S20	–0 30	–0 32	+0 32	+0 30
25	–0 34	–0 35	+0 35	+0 34
30	–0 39	–0 39	+0 39	+0 39
35	–0 45	–0 44	+0 44	+0 45
S40	–0 52	–0 49	+0 49	+0 52
45	–1 01	–0 55	+0 55	+1 01
50	–1 11	–1 02	+1 02	+1 11

NOTES

Latitude corrections are for mid-month.

Equation of time = mean time (LHA mean sun) minus apparent time (LHA true sun).

MOON

Yr	Mth	Week	Age	Transit (Upper)	Diff	Semi-diam	Hor Par	Lat 30°N Moon-rise	Lat 30°N Moon-set
			days	h m	m	′	′	h m	h m
274	**1**	**Th**	11	**20 24**	**54**	15.8	58.0	14 50	01 01
275	**2**	**Fri**	12	**21 18**	**54**	16.1	59.0	15 36	02 01
276	**3**	**Sat**	13	**22 12**	**55**	16.3	59.9	16 20	03 04
277	**4**	**Sun**	14	**23 07**	**55**	16.5	60.6	17 03	04 10
278	**5**	**Mon**	15	**24 02**	**-**	16.7	61.1	17 46	05 17
279	**6**	**Tu**	16	**00 02**	**56**	16.7	61.3	18 30	06 25
280	**7**	**Wed**	17	**00 58**	**58**	16.7	61.2	19 16	07 33
281	**8**	**Th**	18	**01 56**	**58**	16.5	60.7	20 04	08 41
282	**9**	**Fri**	19	**02 54**	**58**	16.4	60.0	20 55	09 48
283	**10**	**Sat**	20	**03 52**	**57**	16.1	59.2	21 48	10 51
284	**11**	**Sun**	21	**04 49**	**56**	15.9	58.4	22 44	11 50
285	**12**	**Mon**	22	**05 45**	**53**	15.7	57.5	23 41	12 44
286	**13**	**Tu**	23	**06 38**	**51**	15.4	56.7	- -	13 33
287	**14**	**Wed**	24	**07 29**	**48**	15.3	56.0	00 37	14 17
288	**15**	**Th**	25	**08 17**	**46**	15.1	55.4	01 33	14 56
289	**16**	**Fri**	26	**09 03**	**44**	15.0	54.9	02 28	15 33
290	**17**	**Sat**	27	**09 47**	**42**	14.9	54.5	03 21	16 07
291	**18**	**Sun**	28	**10 29**	**43**	14.8	54.2	04 14	16 40
292	**19**	**Mon**	29	**11 12**	**42**	14.7	54.0	05 06	17 13
293	**20**	**Tu**	00	**11 54**	**43**	14.7	53.9	05 57	17 46
294	**21**	**Wed**	01	**12 37**	**44**	14.7	53.9	06 49	18 21
295	**22**	**Th**	02	**13 21**	**45**	14.7	54.0	07 41	18 57
296	**23**	**Fri**	03	**14 06**	**47**	14.8	54.2	08 33	19 37
297	**24**	**Sat**	04	**14 53**	**49**	14.8	54.5	09 26	20 19
298	**25**	**Sun**	05	**15 42**	**49**	14.9	54.9	10 17	21 06
299	**26**	**Mon**	06	**16 31**	**51**	15.1	55.4	11 08	21 56
300	**27**	**Tu**	07	**17 22**	**52**	15.3	56.1	11 56	22 50
301	**28**	**Wed**	08	**18 14**	**51**	15.5	56.9	12 43	23 47
302	**29**	**Th**	09	**19 05**	**53**	15.7	57.8	13 28	- -
303	**30**	**Fri**	10	**19 58**	**53**	16.0	58.7	14 11	00 47
304	**31**	**Sat**	11	**20 51**	**53**	16.2	59.6	14 53	01 49

Phases of the Moon

		d	h	m
○	Full Moon	5	20	12
☽	Last Quarter	12	11	11
●	New Moon	20	10	09
☾	First Quarter	28	11	46
	Perigee	6	13	
	Apogee	21	05	
	On 0°	5	12	
	Farthest N of 0°	11	15	
	On 0°	18	18	
	Farthest S of 0°	26	08	

NOTES

Diff equals daily change in transit time.

To correct moonrise, moonset, or transit times for latitude and/or longitude see page E 112. For further information about these tables see page E 4. For arc-to-time correction see page E 103.

OCTOBER 1998 — STARS — 0h GMT October 1

No	Name	Mag	Transit h m	Dec ° ′	GHA ° ′	RA h m	SHA ° ′
	ARIES		23 18		9 31.5		
1	Alpheratz	2.2	23 26	N 29 05.1	7 26.4	0 08	357 54.9
2	Ankaa	2.4	23 44	S 42 18.8	2 58.0	0 26	353 26.5
3	Schedar	2.5	0 02	N 56 31.9	359 24.4	0 40	349 52.9
4	Diphda	2.2	0 05	S 17 59.6	358 38.4	0 44	349 06.9
5	Achernar	0.6	0 59	S 57 14.6	345 06.2	1 38	335 34.7
6	POLARIS	2.1	1 54	N 89 15.3	331 27.0	2 32	321 55.5
7	Hamal	2.2	1 29	N 23 27.4	337 44.7	2 07	328 13.2
8	Acamar	3.1	2 20	S 40 18.5	324 58.1	2 58	315 26.6
9	Menkar	2.8	2 24	N 4 05.1	323 58.1	3 02	314 26.6
10	Mirfak	1.9	2 46	N 49 51.3	318 27.6	3 24	308 56.1
11	Aldebaran	1.1	3 57	N 16 30.3	300 33.7	4 36	291 02.2
12	Rigel	0.3	4 36	S 8 12.2	290 54.3	5 14	281 22.8
13	Capella	0.2	4 38	N 45 59.6	290 22.4	5 17	280 50.9
14	Bellatrix	1.7	4 46	N 6 20.9	288 15.5	5 25	278 44.0
15	Elnath	1.8	4 47	N 28 36.2	287 58.3	5 26	278 26.8
16	Alnilam	1.8	4 57	S 1 12.2	285 29.2	5 36	275 57.7
17	Betelgeuse	0.1-1.2	5 16	N 7 24.4	280 44.9	5 55	271 13.4
18	Canopus	-0.9	5 45	S 52 41.5	273 32.5	6 24	264 01.0
19	Sirius	-1.6	6 06	S 16 42.7	268 15.1	6 45	258 43.6
20	Adhara	1.6	6 19	S 28 58.1	264 52.9	6 59	255 21.4
21	Castor	1.6	6 55	N 31 53.3	255 53.9	7 35	246 22.4
22	Procyon	0.5	7 00	N 5 13.7	254 43.1	7 39	245 11.6
23	Pollux	1.2	7 06	N 28 01.6	253 13.1	7 45	243 41.6
24	Avior	1.7	7 43	S 59 30.1	243 54.2	8 22	234 22.7
25	Suhail	2.2	8 28	S 43 25.4	232 32.4	9 08	223 00.9
26	Miaplacidus	1.8	8 34	S 69 42.4	231 13.9	9 13	221 42.4
27	Alphard	2.2	8 48	S 8 39.1	227 38.9	9 28	218 07.4
28	Regulus	1.3	9 29	N 11 58.5	217 27.3	10 08	207 55.8
29	Dubhe	2.0	10 24	N 61 45.4	203 37.6	11 04	194 06.1
30	Denebola	2.2	11 09	N 14 34.9	192 17.0	11 49	182 45.5
31	Gienah	2.8	11 36	S 17 31.9	185 35.8	12 16	176 04.3
32	Acrux	1.1	11 46	S 63 05.3	182 54.1	12 26	173 22.6
33	Gacrux	1.6	11 51	S 57 06.2	181 45.6	12 31	172 14.1
34	Mimosa	1.5	12 08	S 59 40.7	177 37.3	12 48	168 05.8
35	Alioth	1.7	12 14	N 55 58.1	176 02.7	12 54	166 31.2
36	Spica	1.2	12 45	S 11 09.1	168 15.0	13 25	158 43.5
37	Alkaid	1.9	13 07	N 49 19.4	162 39.8	13 47	153 08.3
38	Hadar	0.9	13 23	S 60 21.9	158 36.1	14 04	149 04.6
39	Menkent	2.3	13 26	S 36 21.6	157 52.8	14 07	148 21.3
40	Arcturus	0.2	13 35	N 19 11.6	155 37.9	14 16	146 06.4
41	Rigil Kent.	0.1	13 59	S 60 49.7	149 39.4	14 39	140 07.9
42	Zuben'ubi	2.9	14 10	S 16 02.0	146 49.8	14 51	137 18.3
43	Kochab	2.2	14 10	N 74 09.9	146 52.3	14 51	137 20.8
44	Alphecca	2.3	14 54	N 26 43.4	135 52.4	15 35	126 20.9
45	Antares	1.2	15 49	S 26 25.6	122 11.9	16 29	112 40.4
46	Atria	1.9	16 08	S 69 01.6	117 24.3	16 48	107 52.8
47	Sabik	2.6	16 29	S 15 43.2	111 57.2	17 10	102 25.7
48	Shaula	1.7	16 53	S 37 06.1	106 09.0	17 34	96 37.5
49	Rasalhague	2.1	16 54	N 12 34.0	105 48.6	17 35	96 17.1
50	Eltanin	2.4	17 16	N 51 29.8	100 23.1	17 57	90 51.6
51	Kaus Aust.	2.0	17 43	S 34 23.1	93 30.5	18 24	83 59.0
52	Vega	0.1	17 56	N 38 47.4	90 18.2	18 37	80 46.7
53	Nunki	2.1	18 14	S 26 17.8	85 43.9	18 55	76 12.4
54	Altair	0.9	19 09	N 8 52.2	71 50.8	19 51	62 19.3
55	Peacock	2.1	19 44	S 56 44.5	63 08.4	20 26	53 36.9
56	Deneb	1.3	20 00	N 45 16.9	59 10.5	20 41	49 39.0
57	Enif	2.5	21 03	N 9 52.3	43 29.6	21 44	33 58.1
58	Al Na'ir	2.2	21 27	S 46 58.1	37 29.2	22 08	27 57.7
59	Fomalhaut	1.3	22 16	S 29 37.7	25 07.7	22 58	15 36.2
60	Markab	2.6	22 23	N 15 12.1	23 20.8	23 05	13 49.3

Star Transit Corr Table

Date	Corr h m	Date	Corr h m
1	0 00	17	−1 03
2	−0 04	18	−1 07
3	−0 08	19	−1 11
4	−0 12	20	−1 15
5	−0 16	21	−1 19
6	−0 20	22	−1 23
7	−0 24	23	−1 27
8	−0 28	24	−1 30
9	−0 31	25	−1 34
10	−0 35	26	−1 38
11	−0 39	27	−1 42
12	−0 43	28	−1 46
13	−0 47	29	−1 50
14	−0 51	30	−1 54
15	−0 55	31	−1 58
16	−0 59		

STAR TRANSIT

To find the approximate time of transit of a star for any day of the month use the table above. All corrections are subtractive.

If the value taken from the table is greater than the time of transit for the first of the month, add 23h 56min to the time of transit before subtracting the correction.

Example: What time will Schedar (No 3) transit the Meridian on October 30?

	h min
Transit on Oct 1	00 02
	+23 56
	23 58
Corr for Oct 30	−01 54
Transit on Oct 30	22 04

OCTOBER DIARY

d h	
1 12	Uranus 3° S of Moon
4 07	Pallas 1.0° S of Moon
4 09	Jupiter 0.2° N of Moon
6 16	Mars 0.9° N of Regulus
7 01	Saturn 1.8° N of Moon
9 16	Aldebaran 0.4° S of Moon
11 11	Neptune stationary
15 17	Regulus 0.5° N of Moon
16 04	Mars 1.0° N of Moon
19 01	Uranus stationary
21 18	Mercury 7° S of Moon
23 19	Saturn at opposition
28 01	Juno in conjunction
28 03	Neptune 2° S of Moon
28 20	Uranus 2° S of Moon
30 04	Venus in conjunction
31 16	Jupiter 0.2° N of Moon

NOTES

Star declinations may be used as is for the whole month. To correct GHA for day, hour, minute, and second see page E 106; for further explanation of this table see page E 4.

OCTOBER 1998 SUN & ARIES GMT

Time	SUN GHA ° ′	Dec ° ′	ARIES GHA ° ′	Time	SUN GHA ° ′	Dec ° ′	ARIES GHA ° ′	Time	SUN GHA ° ′	Dec ° ′	ARIES GHA ° ′	Time
	Thursday, 1st October				**Tuesday, 6th October**				**Sunday, 11th October**			
00	182 31.8	S 3 01.3	9 31.5	00	182 55.3	S 4 57.2	14 27.2	00	183 16.4	S 6 51.5	19 22.9	00
02	212 32.2	3 03.3	39 36.4	02	212 55.7	4 59.1	44 32.1	02	213 16.8	6 53.4	49 27.8	02
04	242 32.7	3 05.2	69 41.3	04	242 56.1	5 01.0	74 37.0	04	243 17.1	6 55.3	79 32.7	04
06	272 33.1	3 07.2	99 46.3	06	272 56.4	5 03.0	104 41.9	06	273 17.4	6 57.2	109 37.6	06
08	302 33.5	3 09.1	129 51.2	08	302 56.8	5 04.9	134 46.9	08	303 17.7	6 59.1	139 42.6	08
10	332 33.9	3 11.0	159 56.1	10	332 57.2	5 06.8	164 51.8	10	333 18.1	7 01.0	169 47.5	10
12	2 34.3	3 13.0	190 01.0	12	2 57.6	5 08.7	194 56.7	12	3 18.4	7 02.8	199 52.4	12
14	32 34.7	3 14.9	220 06.0	14	32 57.9	5 10.6	225 01.7	14	33 18.7	7 04.7	229 57.4	14
16	62 35.1	3 16.8	250 10.9	16	62 58.3	5 12.5	255 06.6	16	63 19.0	7 06.6	260 02.3	16
18	92 35.5	3 18.8	280 15.8	18	92 58.7	5 14.5	285 11.5	18	93 19.3	7 08.5	290 07.2	18
20	122 35.9	3 20.7	310 20.8	20	122 59.0	5 16.4	315 16.4	20	123 19.7	7 10.4	320 12.1	20
22	152 36.3	S 3 22.7	340 25.7	22	152 59.4	S 5 18.3	345 21.4	22	153 20.0	S 7 12.3	350 17.1	22
	Friday, 2nd October				**Wednesday, 7th October**				**Monday, 12th October**			
00	182 36.7	S 3 24.6	10 30.6	00	182 59.8	S 5 20.2	15 26.3	00	183 20.3	S 7 14.1	20 22.0	00
02	212 37.1	3 26.5	40 35.5	02	213 00.1	5 22.1	45 31.2	02	213 20.6	7 16.0	50 26.9	02
04	242 37.5	3 28.5	70 40.5	04	243 00.5	5 24.0	75 36.2	04	243 20.9	7 17.9	80 31.8	04
06	272 37.9	3 30.4	100 45.4	06	273 00.9	5 26.0	105 41.1	06	273 21.2	7 19.8	110 36.8	06
08	302 38.3	3 32.3	130 50.3	08	303 01.2	5 27.9	135 46.0	08	303 21.6	7 21.7	140 41.7	08
10	332 38.7	3 34.3	160 55.3	10	333 01.6	5 29.8	165 50.9	10	333 21.9	7 23.5	170 46.6	10
12	2 39.1	3 36.2	191 00.2	12	3 01.9	5 31.7	195 55.9	12	3 22.2	7 25.4	200 51.6	12
14	32 39.5	3 38.1	221 05.1	14	33 02.3	5 33.6	226 00.8	14	33 22.5	7 27.3	230 56.5	14
16	62 39.9	3 40.1	251 10.0	16	63 02.7	5 35.5	256 05.7	16	63 22.8	7 29.2	261 01.4	16
18	92 40.3	3 42.0	281 15.0	18	93 03.0	5 37.4	286 10.7	18	93 23.1	7 31.0	291 06.3	18
20	122 40.7	3 44.0	311 19.9	20	123 03.4	5 39.3	316 15.6	20	123 23.4	7 32.9	321 11.3	20
22	152 41.1	S 3 45.9	341 24.8	22	153 03.7	S 5 41.2	346 20.5	22	153 23.7	S 7 34.8	351 16.2	22
	Saturday, 3rd October				**Thursday, 8th October**				**Tuesday, 13th October**			
00	182 41.5	S 3 47.8	11 29.7	00	183 04.1	S 5 43.2	16 25.4	00	183 24.0	S 7 36.7	21 21.1	00
02	212 41.9	3 49.8	41 34.7	02	213 04.5	5 45.1	46 30.4	02	213 24.3	7 38.5	51 26.1	02
04	242 42.3	3 51.7	71 39.6	04	243 04.8	5 47.0	76 35.3	04	243 24.7	7 40.4	81 31.0	04
06	272 42.7	3 53.6	101 44.5	06	273 05.2	5 48.9	106 40.2	06	273 25.0	7 42.3	111 35.9	06
08	302 43.0	3 55.6	131 49.5	08	303 05.5	5 50.8	136 45.1	08	303 25.3	7 44.1	141 40.8	08
10	332 43.4	3 57.5	161 54.4	10	333 05.9	5 52.7	166 50.1	10	333 25.6	7 46.0	171 45.8	10
12	2 43.8	3 59.4	191 59.3	12	3 06.2	5 54.6	196 55.0	12	3 25.9	7 47.9	201 50.7	12
14	32 44.2	4 01.3	222 04.2	14	33 06.6	5 56.5	226 59.9	14	33 26.2	7 49.7	231 55.6	14
16	62 44.6	4 03.3	252 09.2	16	63 06.9	5 58.4	257 04.9	16	63 26.5	7 51.6	262 00.6	16
18	92 45.0	4 05.2	282 14.1	18	93 07.3	6 00.3	287 09.8	18	93 26.8	7 53.5	292 05.5	18
20	122 45.4	4 07.1	312 19.0	20	123 07.6	6 02.2	317 14.7	20	123 27.1	7 55.3	322 10.4	20
22	152 45.8	S 4 09.1	342 24.0	22	153 08.0	S 6 04.1	347 19.6	22	153 27.4	S 7 57.2	352 15.3	22
	Sunday, 4th October				**Friday, 9th October**				**Wednesday, 14th October**			
00	182 46.2	S 4 11.0	12 28.9	00	183 08.3	S 6 06.0	17 24.6	00	183 27.6	S 7 59.1	22 20.3	00
02	212 46.6	4 12.9	42 33.8	02	213 08.7	6 07.9	47 29.5	02	213 27.9	8 00.9	52 25.2	02
04	242 46.9	4 14.9	72 38.7	04	243 09.0	6 09.8	77 34.4	04	243 28.2	8 02.8	82 30.1	04
06	272 47.3	4 16.8	102 43.7	06	273 09.4	6 11.7	107 39.4	06	273 28.5	8 04.7	112 35.1	06
08	302 47.7	4 18.7	132 48.6	08	303 09.7	6 13.6	137 44.3	08	303 28.8	8 06.5	142 40.0	08
10	332 48.1	4 20.6	162 53.5	10	333 10.1	6 15.5	167 49.2	10	333 29.1	8 08.4	172 44.9	10
12	2 48.5	4 22.6	192 58.5	12	3 10.4	6 17.4	197 54.1	12	3 29.4	8 10.2	202 49.8	12
14	32 48.9	4 24.5	223 03.4	14	33 10.7	6 19.3	227 59.1	14	33 29.7	8 12.1	232 54.8	14
16	62 49.3	4 26.4	253 08.3	16	63 11.1	6 21.2	258 04.0	16	63 30.0	8 14.0	262 59.7	16
18	92 49.6	4 28.4	283 13.2	18	93 11.4	6 23.1	288 08.9	18	93 30.3	8 15.8	293 04.6	18
20	122 50.0	4 30.3	313 18.2	20	123 11.8	6 25.0	318 13.9	20	123 30.6	8 17.7	323 09.6	20
22	152 50.4	S 4 32.2	343 23.1	22	153 12.1	S 6 26.9	348 18.8	22	153 30.8	S 8 19.5	353 14.5	22
	Monday, 5th October				**Saturday, 10th October**				**Thursday, 15th October**			
00	182 50.8	S 4 34.1	13 28.0	00	183 12.4	S 6 28.8	18 23.7	00	183 31.1	S 8 21.4	23 19.4	00
02	212 51.2	4 36.1	43 33.0	02	213 12.8	6 30.7	48 28.6	02	213 31.4	8 23.2	53 24.3	02
04	242 51.6	4 38.0	73 37.9	04	243 13.1	6 32.6	78 33.6	04	243 31.7	8 25.1	83 29.3	04
06	272 51.9	4 39.9	103 42.8	06	273 13.4	6 34.5	108 38.5	06	273 32.0	8 26.9	113 34.2	06
08	302 52.3	4 41.8	133 47.7	08	303 13.8	6 36.4	138 43.4	08	303 32.2	8 28.8	143 39.1	08
10	332 52.7	4 43.7	163 52.7	10	333 14.1	6 38.3	168 48.4	10	333 32.5	8 30.6	173 44.0	10
12	2 53.1	4 45.7	193 57.6	12	3 14.4	6 40.2	198 53.3	12	3 32.8	8 32.5	203 49.0	12
14	32 53.4	4 47.6	224 02.5	14	33 14.8	6 42.1	228 58.2	14	33 33.1	8 34.3	233 53.9	14
16	62 53.8	4 49.5	254 07.4	16	63 15.1	6 44.0	259 03.1	16	63 33.4	8 36.2	263 58.8	16
18	92 54.2	4 51.4	284 12.4	18	93 15.4	6 45.9	289 08.1	18	93 33.6	8 38.0	294 03.8	18
20	122 54.6	4 53.4	314 17.3	20	123 15.8	6 47.7	319 13.0	20	123 33.9	8 39.9	324 08.7	20
22	152 55.0	S 4 55.3	344 22.2	22	153 16.1	S 6 49.6	349 17.9	22	153 34.2	S 8 41.7	354 13.6	22

NOTES

Sun and Aries GHA corrections for additional hour, minutes, and seconds are on page E 106.
Correct sun declination by interpolation. For an example see page E 4.

OCTOBER 1998 — SUN & ARIES — GMT

Time	SUN GHA ° ′	Dec ° ′	ARIES GHA ° ′
Friday, 16th October			
00	183 34.4	S 8 43.6	24 18.5
02	213 34.7	8 45.4	54 23.5
04	243 35.0	8 47.3	84 28.4
06	273 35.3	8 49.1	114 33.3
08	303 35.5	8 50.9	144 38.3
10	333 35.8	8 52.8	174 43.2
12	3 36.1	8 54.6	204 48.1
14	33 36.3	8 56.5	234 53.0
16	63 36.6	8 58.3	264 58.0
18	93 36.9	9 00.1	295 02.9
20	123 37.1	9 02.0	325 07.8
22	153 37.4	S 9 03.8	355 12.8
Saturday, 17th October			
00	183 37.6	S 9 05.6	25 17.7
02	213 37.9	9 07.5	55 22.6
04	243 38.1	9 09.3	85 27.5
06	273 38.4	9 11.1	115 32.5
08	303 38.7	9 13.0	145 37.4
10	333 38.9	9 14.8	175 42.3
12	3 39.2	9 16.6	205 47.3
14	33 39.4	9 18.5	235 52.2
16	63 39.7	9 20.3	265 57.1
18	93 39.9	9 22.1	296 02.0
20	123 40.2	9 23.9	326 07.0
22	153 40.4	S 9 25.8	356 11.9
Sunday, 18th October			
00	183 40.7	S 9 27.6	26 16.8
02	213 40.9	9 29.4	56 21.7
04	243 41.2	9 31.2	86 26.7
06	273 41.4	9 33.0	116 31.6
08	303 41.6	9 34.9	146 36.5
10	333 41.9	9 36.7	176 41.5
12	3 42.1	9 38.5	206 46.4
14	33 42.4	9 40.3	236 51.3
16	63 42.6	9 42.1	266 56.2
18	93 42.8	9 43.9	297 01.2
20	123 43.1	9 45.8	327 06.1
22	153 43.3	S 9 47.6	357 11.0
Monday, 19th October			
00	183 43.5	S 9 49.4	27 16.0
02	213 43.8	9 51.2	57 20.9
04	243 44.0	9 53.0	87 25.8
06	273 44.2	9 54.8	117 30.7
08	303 44.5	9 56.6	147 35.7
10	333 44.7	9 58.4	177 40.6
12	3 44.9	10 00.2	207 45.5
14	33 45.2	10 02.0	237 50.5
16	63 45.4	10 03.8	267 55.4
18	93 45.6	10 05.6	298 00.3
20	123 45.8	10 07.4	328 05.2
22	153 46.1	S10 09.2	358 10.2
Tuesday, 20th October			
00	183 46.3	S10 11.0	28 15.1
02	213 46.5	10 12.8	58 20.0
04	243 46.7	10 14.6	88 24.9
06	273 46.9	10 16.4	118 29.9
08	303 47.1	10 18.2	148 34.8
10	333 47.4	10 20.0	178 39.7
12	3 47.6	10 21.8	208 44.7
14	33 47.8	10 23.6	238 49.6
16	63 48.0	10 25.4	268 54.5
18	93 48.2	10 27.2	298 59.4
20	123 48.4	10 29.0	329 04.4
22	153 48.6	S10 30.8	359 09.3

Time	SUN GHA ° ′	Dec ° ′	ARIES GHA ° ′
Wednesday, 21st October			
00	183 48.8	S10 32.6	29 14.2
02	213 49.0	10 34.3	59 19.2
04	243 49.3	10 36.1	89 24.1
06	273 49.5	10 37.9	119 29.0
08	303 49.7	10 39.7	149 33.9
10	333 49.9	10 41.5	179 38.9
12	3 50.1	10 43.3	209 43.8
14	33 50.3	10 45.0	239 48.7
16	63 50.5	10 46.8	269 53.7
18	93 50.7	10 48.6	299 58.6
20	123 50.9	10 50.4	330 03.5
22	153 51.1	S10 52.1	0 08.4
Thursday, 22nd October			
00	183 51.2	S10 53.9	30 13.4
02	213 51.4	10 55.7	60 18.3
04	243 51.6	10 57.5	90 23.2
06	273 51.8	10 59.2	120 28.2
08	303 52.0	11 01.0	150 33.1
10	333 52.2	11 02.8	180 38.0
12	3 52.4	11 04.5	210 42.9
14	33 52.6	11 06.3	240 47.9
16	63 52.8	11 08.1	270 52.8
18	93 52.9	11 09.8	300 57.7
20	123 53.1	11 11.6	331 02.7
22	153 53.3	S11 13.4	1 07.6
Friday, 23rd October			
00	183 53.5	S11 15.1	31 12.5
02	213 53.7	11 16.9	61 17.4
04	243 53.8	11 18.6	91 22.4
06	273 54.0	11 20.4	121 27.3
08	303 54.2	11 22.1	151 32.2
10	333 54.4	11 23.9	181 37.1
12	3 54.5	11 25.6	211 42.1
14	33 54.7	11 27.4	241 47.0
16	63 54.9	11 29.1	271 51.9
18	93 55.1	11 30.9	301 56.9
20	123 55.2	11 32.6	332 01.8
22	153 55.4	S11 34.4	2 06.7
Saturday, 24th October			
00	183 55.6	S11 36.1	32 11.6
02	213 55.7	11 37.9	62 16.6
04	243 55.9	11 39.6	92 21.5
06	273 56.0	11 41.4	122 26.4
08	303 56.2	11 43.1	152 31.4
10	333 56.4	11 44.8	182 36.3
12	3 56.5	11 46.6	212 41.2
14	33 56.7	11 48.3	242 46.1
16	63 56.8	11 50.1	272 51.1
18	93 57.0	11 51.8	302 56.0
20	123 57.1	11 53.5	333 00.9
22	153 57.3	S11 55.3	3 05.9
Sunday, 25th October			
00	183 57.5	S11 57.0	33 10.8
02	213 57.6	11 58.7	63 15.7
04	243 57.7	12 00.4	93 20.6
06	273 57.9	12 02.2	123 25.6
08	303 58.0	12 03.9	153 30.5
10	333 58.2	12 05.6	183 35.4
12	3 58.3	12 07.3	213 40.4
14	33 58.5	12 09.1	243 45.3
16	63 58.6	12 10.8	273 50.2
18	93 58.8	12 12.5	303 55.1
20	123 58.9	12 14.2	334 00.1
22	153 59.0	S12 15.9	4 05.0

Time	SUN GHA ° ′	Dec ° ′	ARIES GHA ° ′	Time
Monday, 26th October				
00	183 59.2	S12 17.7	34 09.9	00
02	213 59.3	12 19.4	64 14.9	02
04	243 59.4	12 21.1	94 19.8	04
06	273 59.6	12 22.8	124 24.7	06
08	303 59.7	12 24.5	154 29.6	08
10	333 59.8	12 26.2	184 34.6	10
12	4 00.0	12 27.9	214 39.5	12
14	34 00.1	12 29.6	244 44.4	14
16	64 00.2	12 31.3	274 49.4	16
18	94 00.3	12 33.0	304 54.3	18
20	124 00.5	12 34.7	334 59.2	20
22	154 00.6	S12 36.4	5 04.1	22
Tuesday, 27th October				
00	184 00.7	S12 38.1	35 09.1	00
02	214 00.8	12 39.8	65 14.0	02
04	244 01.0	12 41.5	95 18.9	04
06	274 01.1	12 43.2	125 23.8	06
08	304 01.2	12 44.9	155 28.8	08
10	334 01.3	12 46.6	185 33.7	10
12	4 01.4	12 48.3	215 38.6	12
14	34 01.5	12 50.0	245 43.6	14
16	64 01.6	12 51.7	275 48.5	16
18	94 01.8	12 53.4	305 53.4	18
20	124 01.9	12 55.0	335 58.3	20
22	154 02.0	S12 56.7	6 03.3	22
Wednesday, 28th October				
00	184 02.1	S12 58.4	36 08.2	00
02	214 02.2	13 00.1	66 13.1	02
04	244 02.3	13 01.8	96 18.1	04
06	274 02.4	13 03.4	126 23.0	06
08	304 02.5	13 05.1	156 27.9	08
10	334 02.6	13 06.8	186 32.8	10
12	4 02.7	13 08.5	216 37.8	12
14	34 02.8	13 10.1	246 42.7	14
16	64 02.9	13 11.8	276 47.6	16
18	94 03.0	13 13.5	306 52.6	18
20	124 03.1	13 15.2	336 57.5	20
22	154 03.2	S13 16.8	7 02.4	22
Thursday, 29th October				
00	184 03.3	S13 18.5	37 07.3	00
02	214 03.4	13 20.2	67 12.3	02
04	244 03.4	13 21.8	97 17.2	04
06	274 03.5	13 23.5	127 22.1	06
08	304 03.6	13 25.1	157 27.1	08
10	334 03.7	13 26.8	187 32.0	10
12	4 03.8	13 28.4	217 36.9	12
14	34 03.9	13 30.1	247 41.8	14
16	64 04.0	13 31.8	277 46.8	16
18	94 04.0	13 33.4	307 51.7	18
20	124 04.1	13 35.1	337 56.6	20
22	154 04.2	S13 36.7	8 01.6	22
Friday, 30th October				
00	184 04.3	S13 38.4	38 06.5	00
02	214 04.3	13 40.0	68 11.4	02
04	244 04.4	13 41.6	98 16.3	04
06	274 04.5	13 43.3	128 21.2	06
08	304 04.6	13 44.9	158 26.2	08
10	334 04.6	13 46.6	188 31.1	10
12	4 04.7	13 48.2	218 36.0	12
14	34 04.8	13 49.8	248 41.0	14
16	64 04.8	13 51.5	278 45.9	16
18	94 04.9	13 53.1	308 50.8	18
20	124 05.0	13 54.7	338 55.8	20
22	154 05.0	S13 56.4	9 00.7	22

Saturday, 31st October

Time	SUN GHA	Dec	ARIES GHA	Time	SUN GHA	Dec	ARIES GHA	Time	SUN GHA	Dec	ARIES GHA	Time
00	184 05.1	S13 58.0	39 05.6	08	304 05.3	S14 04.5	159 25.3	16	64 05.5	S14 11.0	279 45.0	16
02	214 05.1	13 59.6	69 10.5	10	334 05.4	14 06.1	189 30.3	18	94 05.6	14 12.6	309 50.0	18
04	244 05.2	14 01.3	99 15.5	12	4 05.4	14 07.7	219 35.2	20	124 05.6	14 14.2	339 54.9	20
06	274 05.2	S14 02.9	129 20.4	14	34 05.5	S14 09.4	249 40.1	22	154 05.6	S14 15.8	9 59.8	22

OCTOBER 1998 PLANETS 0h GMT

VENUS / JUPITER

VENUS Mer Pass h m	GHA ° ′	Mean Var/hr 14°+	Dec ° ′	Mean Var/hr ′	Day		JUPITER GHA ° ′	Mean Var/hr 15°+	Dec ° ′	Mean Var/hr ′	Mer Pass h m
11 25	**188 55.0**	59.6	**N 1 18.0**	1.2	**1**	**Th**	**17 03.5**	2.7	**S 4 56.9**	0.1	**22 48**
11 25	**188 45.6**	59.6	**N 0 48.1**	1.3	**2**	**Fri**	**18 09.4**	2.7	**S 4 59.7**	0.1	**22 43**
11 26	**188 36.2**	59.6	**N 0 18.1**	1.3	**3**	**Sat**	**19 15.2**	2.7	**S 5 02.5**	0.1	**22 39**
11 27	**188 26.8**	59.6	**S 0 11.9**	1.3	**4**	**SUN**	**20 20.9**	2.7	**S 5 05.2**	0.1	**22 34**
11 27	**188 17.4**	59.6	**S 0 42.0**	1.3	**5**	**Mon**	**21 26.5**	2.7	**S 5 07.8**	0.1	**22 30**
11 28	**188 07.9**	59.6	**S 1 12.0**	1.3	**6**	**Tu**	**22 32.0**	2.7	**S 5 10.4**	0.1	**22 26**
11 28	**187 58.5**	59.6	**S 1 42.1**	1.3	**7**	**Wed**	**23 37.4**	2.7	**S 5 13.0**	0.1	**22 21**
11 29	**187 49.0**	59.6	**S 2 12.1**	1.3	**8**	**Th**	**24 42.7**	2.7	**S 5 15.5**	0.1	**22 17**
11 30	**187 39.4**	59.6	**S 2 42.2**	1.3	**9**	**Fri**	**25 47.9**	2.7	**S 5 18.0**	0.1	**22 13**
11 30	**187 29.7**	59.6	**S 3 12.2**	1.2	**10**	**Sat**	**26 53.0**	2.7	**S 5 20.4**	0.1	**22 08**
11 31	**187 20.0**	59.6	**S 3 42.1**	1.2	**11**	**SUN**	**27 58.0**	2.7	**S 5 22.7**	0.1	**22 04**
11 32	**187 10.2**	59.6	**S 4 12.0**	1.2	**12**	**Mon**	**29 02.8**	2.7	**S 5 25.0**	0.1	**22 00**
11 32	**187 00.3**	59.6	**S 4 41.8**	1.2	**13**	**Tu**	**30 07.5**	2.7	**S 5 27.2**	0.1	**21 56**
11 33	**186 50.3**	59.6	**S 5 11.5**	1.2	**14**	**Wed**	**31 12.1**	2.7	**S 5 29.4**	0.1	**21 51**
11 34	**186 40.2**	59.6	**S 5 41.2**	1.2	**15**	**Th**	**32 16.5**	2.7	**S 5 31.5**	0.1	**21 47**
11 34	**186 30.0**	59.6	**S 6 10.7**	1.2	**16**	**Fri**	**33 20.8**	2.7	**S 5 33.5**	0.1	**21 43**
11 35	**186 19.6**	59.6	**S 6 40.1**	1.2	**17**	**Sat**	**34 25.0**	2.7	**S 5 35.5**	0.1	**21 38**
11 36	**186 09.1**	59.6	**S 7 09.3**	1.2	**18**	**SUN**	**35 29.0**	2.7	**S 5 37.4**	0.1	**21 34**
11 36	**185 58.4**	59.5	**S 7 38.4**	1.2	**19**	**Mon**	**36 32.9**	2.7	**S 5 39.2**	0.1	**21 30**
11 37	**185 47.5**	59.5	**S 8 07.4**	1.2	**20**	**Tu**	**37 36.6**	2.6	**S 5 41.0**	0.1	**21 26**
11 38	**185 36.5**	59.5	**S 8 36.1**	1.2	**21**	**Wed**	**38 40.2**	2.6	**S 5 42.6**	0.1	**21 22**
11 39	**185 25.3**	59.5	**S 9 04.7**	1.2	**22**	**Th**	**39 43.6**	2.6	**S 5 44.3**	0.1	**21 17**
11 39	**185 13.9**	59.5	**S 9 33.1**	1.2	**23**	**Fri**	**40 46.8**	2.6	**S 5 45.8**	0.1	**21 13**
11 40	**185 02.3**	59.5	**S10 01.2**	1.2	**24**	**Sat**	**41 49.9**	2.6	**S 5 47.3**	0.1	**21 09**
11 41	**184 50.5**	59.5	**S10 29.2**	1.1	**25**	**SUN**	**42 52.8**	2.6	**S 5 48.6**	0.1	**21 05**
11 42	**184 38.5**	59.5	**S10 56.8**	1.1	**26**	**Mon**	**43 55.5**	2.6	**S 5 49.9**	0.1	**21 01**
11 43	**184 26.3**	59.5	**S11 24.3**	1.1	**27**	**Tu**	**44 58.1**	2.6	**S 5 51.2**	0.0	**20 56**
11 43	**184 13.8**	59.5	**S11 51.4**	1.1	**28**	**Wed**	**46 00.5**	2.6	**S 5 52.3**	0.0	**20 52**
11 44	**184 01.1**	59.5	**S12 18.3**	1.1	**29**	**Th**	**47 02.7**	2.6	**S 5 53.4**	0.0	**20 48**
11 45	**183 48.1**	59.4	**S12 44.9**	1.1	**30**	**Fri**	**48 04.7**	2.6	**S 5 54.4**	0.0	**20 44**
11 46	**183 34.9**	59.4	**S13 11.1**	1.1	**31**	**Sat**	**49 06.6**	2.6	**S 5 55.3**	0.0	**20 40**

VENUS, Av. Mag. -3.9
S.H.A. October
5 175; 10 169; 15 163; 20 158; 25 152; 30 146.

JUPITER, Av. Mag.-2.8
S.H.A. October
5 8; 10 8; 15 9; 20 9; 25 10; 30 10.

MARS / SATURN

MARS Mer Pass h m	GHA ° ′	Mean Var/hr 15°+	Dec ° ′	Mean Var/hr ′	Day		SATURN GHA ° ′	Mean Var/hr 15°+	Dec ° ′	Mean Var/hr ′	Mer Pass h m
09 16	**220 50.0**	1.0	**N14 03.8**	0.5	**1**	**Th**	**338 50.3**	2.6	**N 9 33.2**	0.1	**01 24**
09 15	**221 13.1**	1.0	**N13 51.6**	0.5	**2**	**Fri**	**339 53.3**	2.6	**N 9 31.7**	0.1	**01 20**
09 13	**221 36.4**	1.0	**N13 39.3**	0.5	**3**	**Sat**	**340 56.4**	2.6	**N 9 30.2**	0.1	**01 16**
09 11	**221 59.8**	1.0	**N13 27.0**	0.5	**4**	**SUN**	**341 59.6**	2.6	**N 9 28.6**	0.1	**01 12**
09 10	**222 23.3**	1.0	**N13 14.6**	0.5	**5**	**Mon**	**343 02.8**	2.6	**N 9 27.0**	0.1	**01 08**
09 08	**222 46.9**	1.0	**N13 02.1**	0.5	**6**	**Tu**	**344 06.0**	2.6	**N 9 25.4**	0.1	**01 03**
09 07	**223 10.5**	1.0	**N12 49.6**	0.5	**7**	**Wed**	**345 09.3**	2.6	**N 9 23.8**	0.1	**00 59**
09 05	**223 34.3**	1.0	**N12 37.0**	0.5	**8**	**Th**	**346 12.7**	2.6	**N 9 22.2**	0.1	**00 55**
09 04	**223 58.2**	1.0	**N12 24.4**	0.5	**9**	**Fri**	**347 16.0**	2.6	**N 9 20.6**	0.1	**00 51**
09 02	**224 22.1**	1.0	**N12 11.7**	0.5	**10**	**Sat**	**348 19.4**	2.6	**N 9 19.0**	0.1	**00 47**
09 00	**224 46.2**	1.0	**N11 59.0**	0.5	**11**	**SUN**	**349 22.9**	2.6	**N 9 17.4**	0.1	**00 42**
08 59	**225 10.4**	1.0	**N11 46.2**	0.5	**12**	**Mon**	**350 26.4**	2.6	**N 9 15.7**	0.1	**00 38**
08 57	**225 34.6**	1.0	**N11 33.4**	0.5	**13**	**Tu**	**351 29.9**	2.6	**N 9 14.1**	0.1	**00 34**
08 55	**225 59.0**	1.0	**N11 20.5**	0.5	**14**	**Wed**	**352 33.4**	2.6	**N 9 12.4**	0.1	**00 30**
08 54	**226 23.4**	1.0	**N11 07.6**	0.5	**15**	**Th**	**353 37.0**	2.6	**N 9 10.8**	0.1	**00 25**
08 52	**226 47.9**	1.0	**N10 54.7**	0.5	**16**	**Fri**	**354 40.6**	2.7	**N 9 09.1**	0.1	**00 21**
08 51	**227 12.5**	1.0	**N10 41.7**	0.5	**17**	**Sat**	**355 44.2**	2.6	**N 9 07.4**	0.1	**00 17**
08 49	**227 37.3**	1.0	**N10 28.6**	0.5	**18**	**SUN**	**356 47.8**	2.7	**N 9 05.8**	0.1	**00 13**
08 47	**228 02.1**	1.0	**N10 15.6**	0.5	**19**	**Mon**	**357 51.5**	2.7	**N 9 04.1**	0.1	**00 09**
08 46	**228 27.0**	1.0	**N10 02.5**	0.6	**20**	**Tu**	**358 55.2**	2.6	**N 9 02.5**	0.1	**00 04**
08 44	**228 51.9**	1.0	**N 9 49.3**	0.5	**21**	**Wed**	**359 58.8**	2.7	**N 9 00.8**	0.1	**00 00**
08 42	**229 17.0**	1.1	**N 9 36.2**	0.6	**22**	**Th**	**1 02.5**	2.7	**N 8 59.1**	0.1	**23 52**
08 41	**229 42.2**	1.1	**N 9 23.0**	0.6	**23**	**Fri**	**2 06.2**	2.7	**N 8 57.5**	0.1	**23 47**
08 39	**230 07.5**	1.1	**N 9 09.8**	0.5	**24**	**Sat**	**3 09.9**	2.7	**N 8 55.8**	0.1	**23 43**
08 37	**230 32.8**	1.1	**N 8 56.6**	0.6	**25**	**SUN**	**4 13.6**	2.7	**N 8 54.2**	0.1	**23 39**
08 36	**230 58.3**	1.1	**N 8 43.3**	0.6	**26**	**Mon**	**5 17.3**	2.7	**N 8 52.5**	0.1	**23 35**
08 34	**231 23.8**	1.1	**N 8 30.0**	0.6	**27**	**Tu**	**6 21.0**	2.7	**N 8 50.9**	0.1	**23 30**
08 32	**231 49.4**	1.1	**N 8 16.7**	0.6	**28**	**Wed**	**7 24.7**	2.7	**N 8 49.3**	0.1	**23 26**
08 30	**232 15.1**	1.1	**N 8 03.4**	0.6	**29**	**Th**	**8 28.3**	2.7	**N 8 47.6**	0.1	**23 22**
08 29	**232 40.9**	1.1	**N 7 50.1**	0.6	**30**	**Fri**	**9 32.0**	2.7	**N 8 46.0**	0.1	**23 18**
08 27	**233 06.8**	1.1	**N 7 36.8**	0.6	**31**	**Sat**	**10 35.6**	2.7	**N 8 44.4**	0.1	**23 14**

MARS, Av. Mag. +1.6
S.H.A. October
5 209; 10 206; 15 203; 20 200; 25 197; 30 195.

NOTE: Planet corrections on pages E113-15

SATURN, Av. Mag. 0.0
S.H.A. October
5 330; 10 330; 15 330; 20 331; 25 331; 30 331.

OCTOBER 1998 MOON

Day	GMT hr	GHA ° ′	Mean Var/hr 14°+	Dec ° ′	Mean Var/hr ′
1 Th	**0**	65 00.1	27.9	**S 17 08.2**	5.2
	6	151 47.4	27.8	**S 16 36.9**	5.9
	12	238 34.1	27.7	**S 16 01.9**	6.5
	18	325 20.2	27.6	**S 15 23.2**	7.1
2 Fri	**0**	52 05.9	27.6	**S 14 40.8**	7.7
	6	138 51.1	27.4	**S 13 55.0**	8.3
	12	225 36.0	27.4	**S 13 05.7**	8.8
	18	312 20.5	27.4	**S 12 13.2**	9.3
3 Sat	**0**	39 04.7	27.3	**S 11 17.5**	9.8
	6	125 48.6	27.3	**S 10 19.0**	10.3
	12	212 32.2	27.2	**S 9 17.6**	10.7
	18	299 15.5	27.2	**S 8 13.7**	11.1
4 Sun	**0**	25 58.6	27.2	**S 7 07.5**	11.4
	6	112 41.2	27.1	**S 5 59.3**	11.7
	12	199 23.6	26.9	**S 4 49.3**	12.0
	18	286 05.4	26.9	**S 3 37.8**	12.1
5 Mon	**0**	12 46.9	26.8	**S 2 25.0**	12.2
	6	99 27.8	26.7	**S 1 11.5**	12.3
	12	186 08.2	26.7	**N 0 02.7**	12.4
	18	272 47.9	26.5	**N 1 17.0**	12.3
6 Tu	**0**	359 27.0	26.4	**N 2 31.2**	12.3
	6	86 05.4	26.2	**N 3 45.0**	12.2
	12	172 43.0	26.1	**N 4 57.9**	11.9
	18	259 19.8	26.0	**N 6 09.6**	11.7
7 Wed	**0**	345 55.8	25.9	**N 7 19.8**	11.4
	6	72 31.0	25.8	**N 8 28.2**	11.0
	12	159 05.4	25.6	**N 9 34.3**	10.6
	18	245 39.0	25.4	**N 10 38.0**	10.1
8 Th	**0**	332 11.9	25.3	**N 11 38.9**	9.6
	6	58 44.1	25.2	**N 12 36.7**	9.1
	12	145 15.7	25.2	**N 13 31.2**	8.4
	18	231 46.8	25.1	**N 14 22.1**	7.8
9 Fri	**0**	318 17.5	25.0	**N 15 09.2**	7.2
	6	44 48.0	25.0	**N 15 52.4**	6.4
	12	131 18.4	25.1	**N 16 31.5**	5.7
	18	217 48.9	25.1	**N 17 06.4**	5.0
10 Sat	**0**	304 19.6	25.2	**N 17 36.9**	4.3
	6	30 50.8	25.4	**N 18 03.0**	3.6
	12	117 22.7	25.5	**N 18 24.7**	2.8
	18	203 55.4	25.7	**N 18 41.9**	2.0
11 Sun	**0**	290 29.1	25.8	**N 18 54.7**	1.4
	6	17 04.1	26.0	**N 19 03.1**	0.6
	12	103 40.6	26.3	**N 19 07.1**	0.1
	18	190 18.6	26.7	**N 19 06.8**	0.8
12 Mon	**0**	276 58.4	27.0	**N 19 02.4**	1.5
	6	3 40.0	27.3	**N 18 53.9**	2.1
	12	90 23.7	27.7	**N 18 41.5**	2.8
	18	177 09.5	28.1	**N 18 25.3**	3.4
13 Tu	**0**	263 57.5	28.4	**N 18 05.4**	4.0
	6	350 47.8	28.8	**N 17 42.1**	4.5
	12	77 40.3	29.2	**N 17 15.4**	5.0
	18	164 35.2	29.6	**N 16 45.6**	5.5
14 Wed	**0**	251 32.4	30.0	**N 16 12.8**	6.0
	6	338 31.9	30.3	**N 15 37.2**	6.4
	12	65 33.7	30.7	**N 14 59.0**	6.8
	18	152 37.8	31.0	**N 14 18.3**	7.2
15 Th	**0**	239 43.9	31.4	**N 13 35.3**	7.5
	6	326 52.2	31.7	**N 12 50.2**	7.9
	12	54 02.5	32.1	**N 12 03.1**	8.2
	18	141 14.6	32.3	**N 11 14.2**	8.5
16 Fri	**0**	228 28.6	32.6	**N 10 23.7**	8.7
	6	315 44.2	32.9	**N 9 31.7**	8.9
	12	43 01.4	33.1	**N 8 38.4**	9.1
	18	130 20.1	33.3	**N 7 43.9**	9.2
17 Sat	**0**	217 40.1	33.5	**N 6 48.3**	9.4
	6	305 01.3	33.7	**N 5 51.9**	9.5
	12	32 23.6	33.8	**N 4 54.8**	9.7
	18	119 46.8	34.0	**N 3 57.0**	9.7
18 Sun	**0**	207 10.8	34.1	**N 2 58.8**	9.8
	6	294 35.5	34.2	**N 2 00.3**	9.8
	12	22 00.7	34.3	**N 1 01.5**	9.8
	18	109 26.4	34.3	**N 0 02.8**	9.8
19 Mon	**0**	196 52.3	34.4	**S 0 55.9**	9.8
	6	284 18.4	34.3	**S 1 54.4**	9.7
	12	11 44.6	34.4	**S 2 52.6**	9.6
	18	99 10.6	34.3	**S 3 50.2**	9.5
20 Tu	**0**	186 36.5	34.3	**S 4 47.3**	9.4
	6	274 02.1	34.2	**S 5 43.7**	9.3
	12	1 27.2	34.1	**S 6 39.2**	9.1
	18	88 51.8	34.0	**S 7 33.7**	8.9
21 Wed	**0**	176 15.8	33.9	**S 8 27.1**	8.7
	6	263 39.1	33.7	**S 9 19.2**	8.4
	12	351 01.5	33.6	**S 10 10.0**	8.2
	18	78 23.1	33.4	**S 10 59.3**	8.0
22 Th	**0**	165 43.6	33.2	**S 11 47.0**	7.7
	6	253 03.1	33.0	**S 12 32.9**	7.4
	12	340 21.5	32.9	**S 13 16.9**	7.0
	18	67 38.8	32.7	**S 13 59.0**	6.7
23 Fri	**0**	154 54.8	32.4	**S 14 38.9**	6.2
	6	242 09.5	32.2	**S 15 16.7**	5.9
	12	329 23.0	32.0	**S 15 52.0**	5.4
	18	56 35.1	31.8	**S 16 24.9**	5.0
24 Sat	**0**	143 45.9	31.6	**S 16 55.2**	4.6
	6	230 55.4	31.3	**S 17 22.9**	4.1
	12	318 03.6	31.1	**S 17 47.7**	3.7
	18	45 10.4	30.9	**S 18 09.6**	3.1
25 Sun	**0**	132 16.0	30.8	**S 18 28.6**	2.6
	6	219 20.4	30.5	**S 18 44.4**	2.1
	12	306 23.5	30.3	**S 18 57.1**	1.5
	18	33 25.5	30.2	**S 19 06.6**	1.0
26 Mon	**0**	120 26.3	29.9	**S 19 12.7**	0.4
	6	207 26.2	29.8	**S 19 15.4**	0.2
	12	294 25.0	29.7	**S 19 14.7**	0.7
	18	21 23.0	29.5	**S 19 10.5**	1.3
27 Tu	**0**	108 20.1	29.4	**S 19 02.7**	2.0
	6	195 16.5	29.3	**S 18 51.4**	2.6
	12	282 12.2	29.2	**S 18 36.5**	3.1
	18	9 07.3	29.1	**S 18 17.9**	3.7
28 Wed	**0**	96 01.8	29.0	**S 17 55.8**	4.3
	6	182 55.8	28.9	**S 17 30.1**	4.9
	12	269 49.4	28.8	**S 17 00.8**	5.5
	18	356 42.6	28.8	**S 16 28.0**	6.1
29 Th	**0**	83 35.5	28.7	**S 15 51.7**	6.7
	6	170 28.0	28.7	**S 15 12.0**	7.2
	12	257 20.2	28.7	**S 14 29.0**	7.8
	18	344 12.0	28.6	**S 13 42.6**	8.3
30 Fri	**0**	71 03.6	28.5	**S 12 53.1**	8.8
	6	157 54.8	28.5	**S 12 00.6**	9.3
	12	244 45.7	28.4	**S 11 05.1**	9.7
	18	331 36.2	28.4	**S 10 06.8**	10.1
31 Sat	**0**	58 26.2	28.2	**S 9 06.0**	10.5
	6	145 15.7	28.2	**S 8 02.7**	10.9
	12	232 04.6	28.0	**S 6 57.2**	11.2
	18	318 52.8	27.9	**S 5 49.6**	11.6

NOTE: Moon GHA corrections for additional hours, minutes, and seconds are on page E 108–110. Moon declination corrections are on page E 111. The mean var/hr value is needed for both corrections.

NOVEMBER 1998 SUN & MOON GMT

SUN

Yr	Mth	Week	Equation of Time 0h	Equation of Time 12h	Transit	Semi-diam	Lat 30°N Twi-light	Lat 30°N Sun-rise	Lat 30°N Sun-set	Lat 30°N Twi-light
			m s	m s	h m	′	h m	h m	h m	h m
305	**1**	**Sun**	-16 23	-16 24	**11 44**	16.1	05 49	06 14	17 13	17 38
306	**2**	**Mon**	-16 24	-16 25	**11 44**	16.1	05 50	06 14	17 12	17 37
307	**3**	**Tu**	-16 25	-16 25	**11 44**	16.2	05 50	06 15	17 12	17 36
308	**4**	**Wed**	-16 25	-16 25	**11 44**	16.2	05 51	06 16	17 11	17 36
309	**5**	**Th**	-16 25	-16 24	**11 44**	16.2	05 52	06 17	17 10	17 35
310	**6**	**Fri**	-16 23	-16 22	**11 44**	16.2	05 53	06 18	17 09	17 34
311	**7**	**Sat**	-16 21	-16 19	**11 44**	16.2	05 53	06 18	17 09	17 34
312	**8**	**Sun**	-16 18	-16 16	**11 44**	16.2	05 54	06 19	17 08	17 33
313	**9**	**Mon**	-16 13	-16 11	**11 44**	16.2	05 55	06 20	17 07	17 32
314	**10**	**Tu**	-16 09	-16 06	**11 44**	16.2	05 56	06 21	17 07	17 32
315	**11**	**Wed**	-16 03	-15 59	**11 44**	16.2	05 56	06 22	17 06	17 31
316	**12**	**Th**	-15 56	-15 52	**11 44**	16.2	05 57	06 22	17 06	17 31
317	**13**	**Fri**	-15 48	-15 44	**11 44**	16.2	05 58	06 23	17 05	17 30
318	**14**	**Sat**	-15 40	-15 36	**11 44**	16.2	05 59	06 24	17 05	17 30
319	**15**	**Sun**	-15 31	-15 26	**11 45**	16.2	05 59	06 25	17 04	17 29
320	**16**	**Mon**	-15 21	-15 15	**11 45**	16.2	06 00	06 26	17 04	17 29
321	**17**	**Tu**	-15 10	-15 04	**11 45**	16.2	06 01	06 26	17 03	17 29
322	**18**	**Wed**	-14 58	-14 52	**11 45**	16.2	06 02	06 27	17 03	17 28
323	**19**	**Th**	-14 45	-14 38	**11 45**	16.2	06 03	06 28	17 02	17 28
324	**20**	**Fri**	-14 32	-14 25	**11 46**	16.2	06 03	06 29	17 02	17 27
325	**21**	**Sat**	-14 17	-14 10	**11 46**	16.2	06 04	06 30	17 02	17 27
326	**22**	**Sun**	-14 02	-13 54	**11 46**	16.2	06 05	06 31	17 01	17 27
327	**23**	**Mon**	-13 46	-13 38	**11 46**	16.2	06 06	06 31	17 01	17 27
328	**24**	**Tu**	-13 30	-13 21	**11 47**	16.2	06 07	06 32	17 01	17 26
329	**25**	**Wed**	-13 12	-13 03	**11 47**	16.2	06 07	06 33	17 01	17 26
330	**26**	**Th**	-12 54	-12 45	**11 47**	16.2	06 08	06 34	17 00	17 26
331	**27**	**Fri**	-12 35	-12 25	**11 48**	16.2	06 09	06 35	17 00	17 26
332	**28**	**Sat**	-12 16	-12 05	**11 48**	16.2	06 10	06 36	17 00	17 26
333	**29**	**Sun**	-11 55	-11 45	**11 48**	16.2	06 10	06 36	17 00	17 26
334	**30**	**Mon**	-11 34	-11 23	**11 49**	16.2	06 11	06 37	17 00	17 26

Lat Corr to Sunrise, Sunset etc

Lat	Twi-light	Sun-rise	Sun-set	Twi-light
°	h m	h m	h m	h m
N60	+1 09	+1 32	−1 32	−1 09
55	+0 51	+1 06	−1 06	−0 51
50	+0 37	+0 47	−0 47	−0 37
45	+0 26	+0 32	−0 32	−0 26
N40	+0 16	+0 20	−0 20	−0 16
35	+0 08	+0 09	−0 09	−0 08
30	0 00	0 00	0 00	0 00
25	−0 07	−0 08	+0 08	+0 07
N20	−0 14	−0 16	+0 16	+0 14
15	−0 21	−0 23	+0 23	+0 21
10	−0 27	−0 30	+0 30	+0 27
N 5	−0 34	−0 37	+0 37	+0 34
0	−0 40	−0 44	+0 44	+0 40
S 5	−0 47	−0 50	+0 50	+0 47
10	−0 54	−0 57	+0 57	+0 54
15	−1 02	−1 04	+1 04	+1 02
S20	−1 10	−1 12	+1 12	+1 10
25	−1 19	−1 20	+1 20	+1 19
30	−1 29	−1 29	+1 29	+1 29
35	−1 41	−1 38	+1 38	+1 41
S40	−1 55	−1 50	+1 50	+1 55
45	−2 12	−2 03	+2 03	+2 12
50	−2 33	−2 20	+2 20	+2 33

NOTES

Latitude corrections are for mid-month.

Equation of time = mean time (LHA mean sun) minus apparent time (LHA true sun).

MOON

Yr	Mth	Week	Age	Transit (Upper)	Diff	Semi-diam	Hor Par	Lat 30°N Moon-rise	Lat 30°N Moon-set
			days	h m	m	′	′	h m	h m
305	**1**	**Sun**	12	**21 44**	**55**	16.5	60.5	15 35	02 53
306	**2**	**Mon**	13	**22 39**	**57**	16.6	61.1	16 18	03 59
307	**3**	**Tu**	14	**23 36**	**59**	16.7	61.4	17 02	05 07
308	**4**	**Wed**	15	**24 35**	**-**	16.7	61.4	17 50	06 17
309	**5**	**Th**	16	**00 35**	**60**	16.7	61.1	18 40	07 26
310	**6**	**Fri**	17	**01 35**	**60**	16.5	60.5	19 35	08 33
311	**7**	**Sat**	18	**02 35**	**59**	16.3	59.7	20 32	09 37
312	**8**	**Sun**	19	**03 34**	**56**	16.0	58.7	21 30	10 36
313	**9**	**Mon**	20	**04 30**	**54**	15.7	57.7	22 29	11 28
314	**10**	**Tu**	21	**05 24**	**50**	15.5	56.8	23 26	12 15
315	**11**	**Wed**	22	**06 14**	**47**	15.3	56.0	- -	12 57
316	**12**	**Th**	23	**07 01**	**44**	15.1	55.3	00 22	13 35
317	**13**	**Fri**	24	**07 45**	**44**	14.9	54.8	01 16	14 09
318	**14**	**Sat**	25	**08 29**	**42**	14.8	54.4	02 09	14 43
319	**15**	**Sun**	26	**09 11**	**42**	14.7	54.1	03 01	15 15
320	**16**	**Mon**	27	**09 53**	**42**	14.7	54.0	03 53	15 48
321	**17**	**Tu**	28	**10 35**	**44**	14.7	53.9	04 45	16 22
322	**18**	**Wed**	29	**11 19**	**45**	14.7	54.0	05 37	16 58
323	**19**	**Th**	00	**12 04**	**47**	14.8	54.2	06 29	17 36
324	**20**	**Fri**	01	**12 51**	**48**	14.8	54.4	07 22	18 18
325	**21**	**Sat**	02	**13 39**	**50**	14.9	54.7	08 14	19 03
326	**22**	**Sun**	03	**14 29**	**50**	15.0	55.1	09 05	19 52
327	**23**	**Mon**	04	**15 19**	**50**	15.1	55.6	09 55	20 44
328	**24**	**Tu**	05	**16 09**	**51**	15.3	56.1	10 42	21 40
329	**25**	**Wed**	06	**17 00**	**51**	15.5	56.8	11 26	22 37
330	**26**	**Th**	07	**17 51**	**50**	15.7	57.5	12 09	23 37
331	**27**	**Fri**	08	**18 41**	**51**	15.9	58.3	12 50	- -
332	**28**	**Sat**	09	**19 32**	**52**	16.1	59.1	13 30	00 38
333	**29**	**Sun**	10	**20 24**	**55**	16.3	59.9	14 10	01 41
334	**30**	**Mon**	11	**21 19**	**56**	16.5	60.5	14 52	02 45

Phases of the Moon

Phase	d	h	m
○ Full Moon	4	05	18
☽ Last Quarter	11	00	28
● New Moon	19	04	27
☾ First Quarter	27	00	23
Perigee	4	01	
Apogee	17	06	
On 0°	1	23	
Farthest N of 0°	8	00	
On 0°	15	01	
Farthest S of 0°	22	14	
On 0°	29	10	

NOTES

Diff equals daily change in transit time.

To correct moonrise, moonset, or transit times for latitude and/or longitude see page E 112. For further information about these tables see page E 4. For arc-to-time correction see page E 103.

NOVEMBER 1998 — STARS — 0h GMT November 1

No	Name	Mag	Transit h m	Dec ° ′	GHA ° ′	RA h m	SHA ° ′
	ARIES		**21 16**		**40 04.8**		
1	**Alpheratz**	2.2	**21 25**	N29 05.2	**37 59.7**	0 08	357 54.9
2	**Ankaa**	2.4	**21 42**	S 42 18.9	**33 31.4**	0 26	353 26.6
3	**Schedar**	2.5	**21 57**	N56 32.0	**29 57.7**	0 40	349 52.9
4	**Diphda**	2.2	**22 00**	S 17 59.7	**29 11.8**	0 44	349 07.0
5	**Achernar**	0.6	**22 54**	S 57 14.8	**15 39.5**	1 38	335 34.7
6	**POLARIS**	2.1	**23 48**	N89 15.5	**1 58.1**	2 32	321 53.3
7	**Hamal**	2.2	**23 23**	N23 27.4	**8 18.0**	2 07	328 13.2
8	**Acamar**	3.1	**0 18**	S 40 18.7	**355 31.3**	2 58	315 26.5
9	**Menkar**	2.8	**0 22**	N 4 05.1	**354 31.3**	3 02	314 26.5
10	**Mirfak**	1.9	**0 44**	N49 51.4	**349 00.8**	3 24	308 56.0
11	**Aldebaran**	1.1	**1 55**	N16 30.3	**331 06.8**	4 36	291 02.0
12	**Rigel**	0.3	**2 34**	S 8 12.2	**321 27.4**	5 14	281 22.6
13	**Capella**	0.2	**2 36**	N45 59.6	**320 55.5**	5 17	280 50.7
14	**Bellatrix**	1.7	**2 44**	N 6 20.8	**318 48.6**	5 25	278 43.8
15	**Elnath**	1.8	**2 45**	N28 36.3	**318 31.3**	5 26	278 26.5
16	**Alnilam**	1.8	**2 55**	S 1 12.2	**316 02.3**	5 36	275 57.5
17	**Betelgeuse**	0.1-1.2	**3 14**	N 7 24.3	**311 18.0**	5 55	271 13.2
18	**Canopus**	-0.9	**3 43**	S 52 41.6	**304 05.5**	6 24	264 00.7
19	**Sirius**	-1.6	**4 04**	S 16 42.8	**298 48.2**	6 45	258 43.4
20	**Adhara**	1.6	**4 18**	S 28 58.1	**295 25.9**	6 59	255 21.1
21	**Castor**	1.6	**4 53**	N31 53.3	**286 26.9**	7 35	246 22.1
22	**Procyon**	0.5	**4 58**	N 5 13.6	**285 16.2**	7 39	245 11.4
23	**Pollux**	1.2	**5 04**	N28 01.6	**283 46.1**	7 45	243 41.3
24	**Avior**	1.7	**5 41**	S 59 30.1	**274 27.2**	8 23	234 22.4
25	**Suhail**	2.2	**6 27**	S 43 25.4	**263 05.4**	9 08	223 00.6
26	**Miaplacidus**	1.8	**6 32**	S 69 42.4	**261 46.6**	9 13	221 41.8
27	**Alphard**	2.2	**6 46**	S 8 39.1	**258 11.9**	9 28	218 07.1
28	**Regulus**	1.3	**7 27**	N11 58.4	**248 00.4**	10 08	207 55.6
29	**Dubhe**	2.0	**8 22**	N61 45.2	**234 10.6**	11 04	194 05.8
30	**Denebola**	2.2	**9 07**	N14 34.8	**222 50.2**	11 49	182 45.4
31	**Gienah**	2.8	**9 34**	S 17 31.9	**216 08.9**	12 16	176 04.1
32	**Acrux**	1.1	**9 45**	S 63 05.2	**213 27.2**	12 27	173 22.4
33	**Gacrux**	1.6	**9 49**	S 57 06.1	**212 18.7**	12 31	172 13.9
34	**Mimosa**	1.5	**10 06**	S 59 40.6	**208 10.4**	12 48	168 05.6
35	**Alioth**	1.7	**10 12**	N55 57.9	**206 35.9**	12 54	166 31.1
36	**Spica**	1.2	**10 43**	S 11 09.1	**198 48.2**	13 25	158 43.4
37	**Alkaid**	1.9	**11 05**	N49 19.2	**193 13.1**	13 47	153 08.3
38	**Hadar**	0.9	**11 22**	S 60 21.8	**189 09.3**	14 04	149 04.5
39	**Menkent**	2.3	**11 24**	S 36 21.6	**188 26.1**	14 07	148 21.3
40	**Arcturus**	0.2	**11 33**	N19 11.5	**186 11.2**	14 16	146 06.4
41	**Rigil Kent.**	0.1	**11 57**	S 60 49.5	**180 12.6**	14 39	140 07.8
42	**Zuben'ubi**	2.9	**12 08**	S 16 02.0	**177 23.1**	14 51	137 18.3
43	**Kochab**	2.2	**12 08**	N74 09.7	**177 25.7**	14 51	137 20.9
44	**Alphecca**	2.3	**12 52**	N26 43.3	**166 25.8**	15 35	126 21.0
45	**Antares**	1.2	**13 47**	S 26 25.6	**152 45.3**	16 29	112 40.5
46	**Atria**	1.9	**14 06**	S 69 01.5	**147 57.8**	16 48	107 53.0
47	**Sabik**	2.6	**14 28**	S 15 43.2	**142 30.6**	17 10	102 25.8
48	**Shaula**	1.7	**14 51**	S 37 06.1	**136 42.4**	17 33	96 37.6
49	**Rasalhague**	2.1	**14 52**	N12 33.9	**136 22.0**	17 35	96 17.2
50	**Eltanin**	2.4	**15 14**	N51 29.7	**130 56.6**	17 57	90 51.8
51	**Kaus Aust.**	2.0	**15 41**	S 34 23.0	**124 03.9**	18 24	83 59.1
52	**Vega**	0.1	**15 54**	N38 47.3	**120 51.7**	18 37	80 46.9
53	**Nunki**	2.1	**16 12**	S 26 17.8	**116 17.3**	18 55	76 12.5
54	**Altair**	0.9	**17 08**	N 8 52.1	**102 24.2**	19 51	62 19.4
55	**Peacock**	2.1	**17 42**	S 56 44.5	**93 42.0**	20 26	53 37.2
56	**Deneb**	1.3	**17 58**	N45 16.9	**89 44.0**	20 41	49 39.2
57	**Enif**	2.5	**19 01**	N 9 52.3	**74 03.0**	21 44	33 58.2
58	**Al Na'ir**	2.2	**19 25**	S 46 58.2	**68 02.6**	22 08	27 57.8
59	**Fomalhaut**	1.3	**20 14**	S 29 37.8	**55 41.1**	22 58	15 36.3
60	**Markab**	2.6	**20 21**	N15 12.1	**53 54.2**	23 05	13 49.4

Star Transit Corr Table

Date	Corr h m	Date	Corr h m
1	0 00	17	–1 03
2	–0 04	18	–1 07
3	–0 08	19	–1 11
4	–0 12	20	–1 15
5	–0 16	21	–1 19
6	–0 20	22	–1 23
7	–0 24	23	–1 27
8	–0 28	24	–1 30
9	–0 31	25	–1 34
10	–0 35	26	–1 38
11	–0 39	27	–1 42
12	–0 43	28	–1 46
13	–0 47	29	–1 50
14	–0 51	30	–1 54
15	–0 55	31	–1 58
16	–0 59		

STAR TRANSIT

To find the approximate time of transit of a star for any day of the month use the table above. All corrections are subtractive.

If the value taken from the table is greater than the time of transit for the first of the month, add 23h 56min to the time of transit before subtracting the correction.

Example: What time will Acamar (No 8) transit the Meridian on November 30?

	h min
Transit on Nov 1	00 18
	+23 56
	24 14
Corr for Nov 30	–01 54
Transit on Nov 30	22 20

NOVEMBER DIARY

d h	
3 09	Saturn 1.7° N of Moon
6 02	Aldebaran 0.6° S of Moon
6 23	Pallas stationary
9 09	Mercury 1.9° N of Antares
11 09	Mercury greatest elong. E.(23°)
11 22	Regulus 0.3° N of Moon
13 18	Mars 0.5° S of Moon
14 01	Jupiter stationary
20 21	Mercury 7° S of Moon
21 14	Mercury stationary
24 09	Neptune 1.9° S of Moon
25 03	Uranus 2° S of Moon
28 01	Jupiter 0.6° N of Moon
30 08	Pluto in conjunction
30 17	Saturn 1.8° N of Moon

NOTES

Star declinations may be used as is for the whole month. To correct GHA for day, hour, minute, and second see page E 106; for further explanation of this table see page E 4.

NOVEMBER 1998 SUN & ARIES GMT

Sunday, 1st November

Time	SUN GHA ° ′	Dec ° ′	ARIES GHA ° ′
00	184 05.7	**S14 17.4**	40 04.8
02	214 05.7	**14 19.0**	70 09.7
04	244 05.8	**14 20.6**	100 14.6
06	274 05.8	**14 22.2**	130 19.5
08	304 05.9	**14 23.8**	160 24.5
10	334 05.9	**14 25.5**	190 29.4
12	4 05.9	**14 27.1**	220 34.3
14	34 06.0	**14 28.7**	250 39.3
16	64 06.0	**14 30.2**	280 44.2
18	94 06.0	**14 31.8**	310 49.1
20	124 06.1	**14 33.4**	340 54.0
22	154 06.1	**S14 35.0**	10 59.0

Monday, 2nd November

Time	SUN GHA ° ′	Dec ° ′	ARIES GHA ° ′
00	184 06.1	**S14 36.6**	41 03.9
02	214 06.1	**14 38.2**	71 08.8
04	244 06.2	**14 39.8**	101 13.7
06	274 06.2	**14 41.4**	131 18.7
08	304 06.2	**14 43.0**	161 23.6
10	334 06.2	**14 44.5**	191 28.5
12	4 06.2	**14 46.1**	221 33.5
14	34 06.3	**14 47.7**	251 38.4
16	64 06.3	**14 49.3**	281 43.3
18	94 06.3	**14 50.9**	311 48.2
20	124 06.3	**14 52.4**	341 53.2
22	154 06.3	**S14 54.0**	11 58.1

Tuesday, 3rd November

Time	SUN GHA ° ′	Dec ° ′	ARIES GHA ° ′
00	184 06.3	**S14 55.6**	42 03.0
02	214 06.3	**14 57.1**	72 08.0
04	244 06.4	**14 58.7**	102 12.9
06	274 06.4	**15 00.3**	132 17.8
08	304 06.4	**15 01.8**	162 22.7
10	334 06.4	**15 03.4**	192 27.7
12	4 06.4	**15 05.0**	222 32.6
14	34 06.4	**15 06.5**	252 37.5
16	64 06.4	**15 08.1**	282 42.5
18	94 06.4	**15 09.6**	312 47.4
20	124 06.4	**15 11.2**	342 52.3
22	154 06.4	**S15 12.7**	12 57.2

Wednesday, 4th November

Time	SUN GHA ° ′	Dec ° ′	ARIES GHA ° ′
00	184 06.4	**S15 14.3**	43 02.2
02	214 06.4	**15 15.8**	73 07.1
04	244 06.3	**15 17.4**	103 12.0
06	274 06.3	**15 18.9**	133 17.0
08	304 06.3	**15 20.5**	163 21.9
10	334 06.3	**15 22.0**	193 26.8
12	4 06.3	**15 23.6**	223 31.7
14	34 06.3	**15 25.1**	253 36.7
16	64 06.3	**15 26.6**	283 41.6
18	94 06.2	**15 28.2**	313 46.5
20	124 06.2	**15 29.7**	343 51.4
22	154 06.2	**S15 31.2**	13 56.4

Thursday, 5th November

Time	SUN GHA ° ′	Dec ° ′	ARIES GHA ° ′
00	184 06.2	**S15 32.8**	44 01.3
02	214 06.2	**15 34.3**	74 06.2
04	244 06.1	**15 35.8**	104 11.2
06	274 06.1	**15 37.3**	134 16.1
08	304 06.1	**15 38.9**	164 21.0
10	334 06.0	**15 40.4**	194 25.9
12	4 06.0	**15 41.9**	224 30.9
14	34 06.0	**15 43.4**	254 35.8
16	64 05.9	**15 44.9**	284 40.7
18	94 05.9	**15 46.4**	314 45.7
20	124 05.9	**15 47.9**	344 50.6
22	154 05.8	**S15 49.5**	14 55.5

Friday, 6th November

Time	SUN GHA ° ′	Dec ° ′	ARIES GHA ° ′
00	184 05.8	**S15 51.0**	45 00.4
02	214 05.7	**15 52.5**	75 05.4
04	244 05.7	**15 54.0**	105 10.3
06	274 05.7	**15 55.5**	135 15.2
08	304 05.6	**15 57.0**	165 20.2
10	334 05.6	**15 58.5**	195 25.1
12	4 05.5	**16 00.0**	225 30.0
14	34 05.5	**16 01.5**	255 34.9
16	64 05.4	**16 03.0**	285 39.9
18	94 05.4	**16 04.4**	315 44.8
20	124 05.3	**16 05.9**	345 49.7
22	154 05.2	**S16 07.4**	15 54.7

Saturday, 7th November

Time	SUN GHA ° ′	Dec ° ′	ARIES GHA ° ′
00	184 05.2	**S16 08.9**	45 59.6
02	214 05.1	**16 10.4**	76 04.5
04	244 05.1	**16 11.9**	106 09.4
06	274 05.0	**16 13.4**	136 14.4
08	304 04.9	**16 14.8**	166 19.3
10	334 04.9	**16 16.3**	196 24.2
12	4 04.8	**16 17.8**	226 29.2
14	34 04.7	**16 19.3**	256 34.1
16	64 04.7	**16 20.7**	286 39.0
18	94 04.6	**16 22.2**	316 43.9
20	124 04.5	**16 23.7**	346 48.9
22	154 04.5	**S16 25.1**	16 53.8

Sunday, 8th November

Time	SUN GHA ° ′	Dec ° ′	ARIES GHA ° ′
00	184 04.4	**S16 26.6**	46 58.7
02	214 04.3	**16 28.0**	77 03.7
04	244 04.2	**16 29.5**	107 08.6
06	274 04.1	**16 31.0**	137 13.5
08	304 04.1	**16 32.4**	167 18.4
10	334 04.0	**16 33.9**	197 23.4
12	4 03.9	**16 35.3**	227 28.3
14	34 03.8	**16 36.8**	257 33.2
16	64 03.7	**16 38.2**	287 38.2
18	94 03.6	**16 39.7**	317 43.1
20	124 03.5	**16 41.1**	347 48.0
22	154 03.4	**S16 42.5**	17 52.9

Monday, 9th November

Time	SUN GHA ° ′	Dec ° ′	ARIES GHA ° ′
00	184 03.3	**S16 44.0**	47 57.9
02	214 03.3	**16 45.4**	78 02.8
04	244 03.2	**16 46.9**	108 07.7
06	274 03.1	**16 48.3**	138 12.7
08	304 03.0	**16 49.7**	168 17.6
10	334 02.9	**16 51.2**	198 22.5
12	4 02.8	**16 52.6**	228 27.4
14	34 02.7	**16 54.0**	258 32.4
16	64 02.5	**16 55.4**	288 37.3
18	94 02.4	**16 56.9**	318 42.2
20	124 02.3	**16 58.3**	348 47.1
22	154 02.2	**S16 59.7**	18 52.1

Tuesday, 10th November

Time	SUN GHA ° ′	Dec ° ′	ARIES GHA ° ′
00	184 02.1	**S17 01.1**	48 57.0
02	214 02.0	**17 02.5**	79 01.9
04	244 01.9	**17 03.9**	109 06.9
06	274 01.8	**17 05.3**	139 11.8
08	304 01.6	**17 06.7**	169 16.7
10	334 01.5	**17 08.2**	199 21.6
12	4 01.4	**17 09.6**	229 26.6
14	34 01.3	**17 11.0**	259 31.5
16	64 01.2	**17 12.4**	289 36.4
18	94 01.0	**17 13.8**	319 41.4
20	124 00.9	**17 15.2**	349 46.3
22	154 00.8	**S17 16.5**	19 51.2

Wednesday, 11th November

Time	SUN GHA ° ′	Dec ° ′	ARIES GHA ° ′	Time
00	184 00.6	**S17 17.9**	49 56.1	00
02	214 00.5	**17 19.3**	80 01.1	02
04	244 00.4	**17 20.7**	110 06.0	04
06	274 00.2	**17 22.1**	140 10.9	06
08	304 00.1	**17 23.5**	170 15.9	08
10	334 00.0	**17 24.9**	200 20.8	10
12	3 59.8	**17 26.2**	230 25.7	12
14	33 59.7	**17 27.6**	260 30.6	14
16	63 59.6	**17 29.0**	290 35.6	16
18	93 59.4	**17 30.4**	320 40.5	18
20	123 59.3	**17 31.7**	350 45.4	20
22	153 59.1	**S17 33.1**	20 50.4	22

Thursday, 12th November

Time	SUN GHA ° ′	Dec ° ′	ARIES GHA ° ′	Time
00	183 59.0	**S17 34.5**	50 55.3	00
02	213 58.8	**17 35.8**	81 00.2	02
04	243 58.7	**17 37.2**	111 05.1	04
06	273 58.5	**17 38.6**	141 10.1	06
08	303 58.4	**17 39.9**	171 15.0	08
10	333 58.2	**17 41.3**	201 19.9	10
12	3 58.1	**17 42.6**	231 24.9	12
14	33 57.9	**17 44.0**	261 29.8	14
16	63 57.7	**17 45.3**	291 34.7	16
18	93 57.6	**17 46.7**	321 39.6	18
20	123 57.4	**17 48.0**	351 44.6	20
22	153 57.2	**S17 49.4**	21 49.5	22

Friday, 13th November

Time	SUN GHA ° ′	Dec ° ′	ARIES GHA ° ′	Time
00	183 57.1	**S17 50.7**	51 54.4	00
02	213 56.9	**17 52.0**	81 59.3	02
04	243 56.7	**17 53.4**	112 04.3	04
06	273 56.6	**17 54.7**	142 09.2	06
08	303 56.4	**17 56.0**	172 14.1	08
10	333 56.2	**17 57.4**	202 19.1	10
12	3 56.1	**17 58.7**	232 24.0	12
14	33 55.9	**18 00.0**	262 28.9	14
16	63 55.7	**18 01.4**	292 33.8	16
18	93 55.5	**18 02.7**	322 38.8	18
20	123 55.3	**18 04.0**	352 43.7	20
22	153 55.2	**S18 05.3**	22 48.6	22

Saturday, 14th November

Time	SUN GHA ° ′	Dec ° ′	ARIES GHA ° ′	Time
00	183 55.0	**S18 06.6**	52 53.6	00
02	213 54.8	**18 07.9**	82 58.5	02
04	243 54.6	**18 09.3**	113 03.4	04
06	273 54.4	**18 10.6**	143 08.3	06
08	303 54.2	**18 11.9**	173 13.3	08
10	333 54.0	**18 13.2**	203 18.2	10
12	3 53.8	**18 14.5**	233 23.1	12
14	33 53.6	**18 15.8**	263 28.1	14
16	63 53.4	**18 17.1**	293 33.0	16
18	93 53.2	**18 18.4**	323 37.9	18
20	123 53.0	**18 19.7**	353 42.8	20
22	153 52.8	**S18 21.0**	23 47.8	22

Sunday, 15th November

Time	SUN GHA ° ′	Dec ° ′	ARIES GHA ° ′	Time
00	183 52.6	**S18 22.2**	53 52.7	00
02	213 52.4	**18 23.5**	83 57.6	02
04	243 52.2	**18 24.8**	114 02.6	04
06	273 52.0	**18 26.1**	144 07.5	06
08	303 51.8	**18 27.4**	174 12.4	08
10	333 51.6	**18 28.6**	204 17.3	10
12	3 51.4	**18 29.9**	234 22.3	12
14	33 51.2	**18 31.2**	264 27.2	14
16	63 51.0	**18 32.5**	294 32.1	16
18	93 50.8	**18 33.7**	324 37.1	18
20	123 50.5	**18 35.0**	354 42.0	20
22	153 50.3	**S18 36.3**	24 46.9	22

NOTES

Sun and Aries GHA corrections for additional hour, minutes, and seconds are on page E 106.
Correct sun declination by interpolation. For an example see page E 4.

NOVEMBER 1998 SUN & ARIES GMT

Time	SUN GHA ° ′	Dec ° ′	ARIES GHA ° ′
Monday, 16th November			
00	183 50.1	S18 37.5	54 51.8
02	213 49.9	18 38.8	84 56.8
04	243 49.7	18 40.0	115 01.7
06	273 49.4	18 41.3	145 06.6
08	303 49.2	18 42.5	175 11.5
10	333 49.0	18 43.8	205 16.5
12	3 48.8	18 45.0	235 21.4
14	33 48.5	18 46.3	265 26.3
16	63 48.3	18 47.5	295 31.3
18	93 48.1	18 48.8	325 36.2
20	123 47.8	18 50.0	355 41.1
22	153 47.6	S18 51.3	25 46.0
Tuesday, 17th November			
00	183 47.4	S18 52.5	55 51.0
02	213 47.1	18 53.7	85 55.9
04	243 46.9	18 54.9	116 00.8
06	273 46.6	18 56.2	146 05.8
08	303 46.4	18 57.4	176 10.7
10	333 46.1	18 58.6	206 15.6
12	3 45.9	18 59.8	236 20.5
14	33 45.6	19 01.1	266 25.5
16	63 45.4	19 02.3	296 30.4
18	93 45.1	19 03.5	326 35.3
20	123 44.9	19 04.7	356 40.3
22	153 44.6	S19 05.9	26 45.2
Wednesday, 18th November			
00	183 44.4	S19 07.1	56 50.1
02	213 44.1	19 08.3	86 55.0
04	243 43.9	19 09.5	117 00.0
06	273 43.6	19 10.7	147 04.9
08	303 43.4	19 11.9	177 09.8
10	333 43.1	19 13.1	207 14.8
12	3 42.8	19 14.3	237 19.7
14	33 42.6	19 15.5	267 24.6
16	63 42.3	19 16.7	297 29.5
18	93 42.0	19 17.9	327 34.5
20	123 41.8	19 19.0	357 39.4
22	153 41.5	S19 20.2	27 44.3
Thursday, 19th November			
00	183 41.2	S19 21.4	57 49.2
02	213 40.9	19 22.6	87 54.2
04	243 40.7	19 23.7	117 59.1
06	273 40.4	19 24.9	148 04.0
08	303 40.1	19 26.1	178 09.0
10	333 39.8	19 27.2	208 13.9
12	3 39.6	19 28.4	238 18.8
14	33 39.3	19 29.6	268 23.7
16	63 39.0	19 30.7	298 28.7
18	93 38.7	19 31.9	328 33.6
20	123 38.4	19 33.0	358 38.5
22	153 38.1	S19 34.2	28 43.5
Friday, 20th November			
00	183 37.8	S19 35.3	58 48.4
02	213 37.5	19 36.5	88 53.3
04	243 37.3	19 37.6	118 58.2
06	273 37.0	19 38.8	149 03.2
08	303 36.7	19 39.9	179 08.1
10	333 36.4	19 41.0	209 13.0
12	3 36.1	19 42.2	239 18.0
14	33 35.8	19 43.3	269 22.9
16	63 35.5	19 44.4	299 27.8
18	93 35.2	19 45.5	329 32.7
20	123 34.9	19 46.7	359 37.7
22	153 34.6	S19 47.8	29 42.6

Time	SUN GHA ° ′	Dec ° ′	ARIES GHA ° ′
Saturday, 21st November			
00	183 34.3	S19 48.9	59 47.5
02	213 33.9	19 50.0	89 52.5
04	243 33.6	19 51.1	119 57.4
06	273 33.3	19 52.2	150 02.3
08	303 33.0	19 53.4	180 07.2
10	333 32.7	19 54.5	210 12.2
12	3 32.4	19 55.6	240 17.1
14	33 32.1	19 56.7	270 22.0
16	63 31.8	19 57.8	300 27.0
18	93 31.4	19 58.9	330 31.9
20	123 31.1	20 00.0	0 36.8
22	153 30.8	S20 01.0	30 41.7
Sunday, 22nd November			
00	183 30.5	S20 02.1	60 46.7
02	213 30.1	20 03.2	90 51.6
04	243 29.8	20 04.3	120 56.5
06	273 29.5	20 05.4	151 01.5
08	303 29.2	20 06.5	181 06.4
10	333 28.8	20 07.5	211 11.3
12	3 28.5	20 08.6	241 16.2
14	33 28.2	20 09.7	271 21.2
16	63 27.8	20 10.7	301 26.1
18	93 27.5	20 11.8	331 31.0
20	123 27.2	20 12.9	1 36.0
22	153 26.8	S20 13.9	31 40.9
Monday, 23rd November			
00	183 26.5	S20 15.0	61 45.8
02	213 26.1	20 16.0	91 50.7
04	243 25.8	20 17.1	121 55.7
06	273 25.5	20 18.1	152 00.6
08	303 25.1	20 19.2	182 05.5
10	333 24.8	20 20.2	212 10.5
12	3 24.4	20 21.3	242 15.4
14	33 24.1	20 22.3	272 20.3
16	63 23.7	20 23.3	302 25.2
18	93 23.4	20 24.4	332 30.2
20	123 23.0	20 25.4	2 35.1
22	153 22.7	S20 26.4	32 40.0
Tuesday, 24th November			
00	183 22.3	S20 27.5	62 44.9
02	213 22.0	20 28.5	92 49.9
04	243 21.6	20 29.5	122 54.8
06	273 21.2	20 30.5	152 59.7
08	303 20.9	20 31.5	183 04.7
10	333 20.5	20 32.6	213 09.6
12	3 20.2	20 33.6	243 14.5
14	33 19.8	20 34.6	273 19.4
16	63 19.4	20 35.6	303 24.4
18	93 19.1	20 36.6	333 29.3
20	123 18.7	20 37.6	3 34.2
22	153 18.3	S20 38.6	33 39.2
Wednesday, 25th November			
00	183 18.0	S20 39.6	63 44.1
02	213 17.6	20 40.6	93 49.0
04	243 17.2	20 41.5	123 53.9
06	273 16.8	20 42.5	153 58.9
08	303 16.5	20 43.5	184 03.8
10	333 16.1	20 44.5	214 08.7
12	3 15.7	20 45.5	244 13.7
14	33 15.3	20 46.5	274 18.6
16	63 14.9	20 47.4	304 23.5
18	93 14.6	20 48.4	334 28.4
20	123 14.2	20 49.4	4 33.4
22	153 13.8	S20 50.3	34 38.3

Time	SUN GHA ° ′	Dec ° ′	ARIES GHA ° ′	Time
Thursday, 26th November				
00	183 13.4	S20 51.3	64 43.2	00
02	213 13.0	20 52.2	94 48.2	02
04	243 12.6	20 53.2	124 53.1	04
06	273 12.2	20 54.2	154 58.0	06
08	303 11.9	20 55.1	185 02.9	08
10	333 11.5	20 56.1	215 07.9	10
12	3 11.1	20 57.0	245 12.8	12
14	33 10.7	20 57.9	275 17.7	14
16	63 10.3	20 58.9	305 22.7	16
18	93 09.9	20 59.8	335 27.6	18
20	123 09.5	21 00.8	5 32.5	20
22	153 09.1	S21 01.7	35 37.4	22
Friday, 27th November				
00	183 08.7	S21 02.6	65 42.4	00
02	213 08.3	21 03.5	95 47.3	02
04	243 07.9	21 04.5	125 52.2	04
06	273 07.5	21 05.4	155 57.2	06
08	303 07.1	21 06.3	186 02.1	08
10	333 06.7	21 07.2	216 07.0	10
12	3 06.2	21 08.1	246 11.9	12
14	33 05.8	21 09.0	276 16.9	14
16	63 05.4	21 09.9	306 21.8	16
18	93 05.0	21 10.9	336 26.7	18
20	123 04.6	21 11.8	6 31.6	20
22	153 04.2	S21 12.7	36 36.6	22
Saturday, 28th November				
00	183 03.8	S21 13.5	66 41.5	00
02	213 03.4	21 14.4	96 46.4	02
04	243 02.9	21 15.3	126 51.4	04
06	273 02.5	21 16.2	156 56.3	06
08	303 02.1	21 17.1	187 01.2	08
10	333 01.7	21 18.0	217 06.1	10
12	3 01.3	21 18.9	247 11.1	12
14	33 00.8	21 19.7	277 16.0	14
16	63 00.4	21 20.6	307 20.9	16
18	93 00.0	21 21.5	337 25.9	18
20	122 59.6	21 22.4	7 30.8	20
22	152 59.1	S21 23.2	37 35.7	22
Sunday, 29th November				
00	182 58.7	S21 24.1	67 40.6	00
02	212 58.3	21 24.9	97 45.6	02
04	242 57.8	21 25.8	127 50.5	04
06	272 57.4	21 26.6	157 55.4	06
08	302 57.0	21 27.5	188 00.4	08
10	332 56.5	21 28.3	218 05.3	10
12	2 56.1	21 29.2	248 10.2	12
14	32 55.7	21 30.0	278 15.1	14
16	62 55.2	21 30.9	308 20.1	16
18	92 54.8	21 31.7	338 25.0	18
20	122 54.3	21 32.5	8 29.9	20
22	152 53.9	S21 33.4	38 34.9	22
Monday, 30th November				
00	182 53.4	S21 34.2	68 39.8	00
02	212 53.0	21 35.0	98 44.7	02
04	242 52.6	21 35.9	128 49.6	04
06	272 52.1	21 36.7	158 54.6	06
08	302 51.7	21 37.5	188 59.5	08
10	332 51.2	21 38.3	219 04.4	10
12	2 50.8	21 39.1	249 09.3	12
14	32 50.3	21 39.9	279 14.3	14
16	62 49.9	21 40.7	309 19.2	16
18	92 49.4	21 41.5	339 24.1	18
20	122 48.9	21 42.3	9 29.1	20
22	152 48.5	S21 43.1	39 34.0	22

NOVEMBER 1998 PLANETS 0h GMT

VENUS						JUPITER				
Mer Pass h m	GHA ° ′	Mean Var/hr 14°+	Dec ° ′	Mean Var/hr ′	Day	GHA ° ′	Mean Var/hr 15°+	Dec ° ′	Mean Var/hr ′	Mer Pass h m
11 47	**183 21.5**	59.4	**S13 37.1**	1.1	**1 SUN**	**50 08.2**	2.6	**S 5 56.2**	0.0	**20 36**
11 48	**183 07.7**	59.4	**S14 02.7**	1.1	**2 Mon**	**51 09.7**	2.6	**S 5 56.9**	0.0	**20 32**
11 49	**182 53.7**	59.4	**S14 28.0**	1.0	**3 Tu**	**52 11.0**	2.5	**S 5 57.6**	0.0	**20 28**
11 50	**182 39.5**	59.4	**S14 52.9**	1.0	**4 Wed**	**53 12.2**	2.5	**S 5 58.2**	0.0	**20 24**
11 51	**182 24.9**	59.4	**S15 17.4**	1.0	**5 Th**	**54 13.1**	2.5	**S 5 58.7**	0.0	**20 20**
11 52	**182 10.0**	59.4	**S15 41.5**	1.0	**6 Fri**	**55 13.8**	2.5	**S 5 59.1**	0.0	**20 16**
11 53	**181 54.9**	59.4	**S16 05.3**	1.0	**7 Sat**	**56 14.4**	2.5	**S 5 59.5**	0.0	**20 12**
11 54	**181 39.4**	59.3	**S16 28.6**	1.0	**8 SUN**	**57 14.8**	2.5	**S 5 59.7**	0.0	**20 08**
11 55	**181 23.7**	59.3	**S16 51.5**	0.9	**9 Mon**	**58 15.0**	2.5	**S 5 59.9**	0.0	**20 04**
11 56	**181 07.6**	59.3	**S17 13.9**	0.9	**10 Tu**	**59 15.0**	2.5	**S 6 00.0**	0.0	**20 00**
11 57	**180 51.3**	59.3	**S17 35.9**	0.9	**11 Wed**	**60 14.8**	2.5	**S 6 00.1**	0.0	**19 56**
11 58	**180 34.6**	59.3	**S17 57.4**	0.9	**12 Th**	**61 14.4**	2.5	**S 6 00.0**	0.0	**19 52**
11 59	**180 17.6**	59.3	**S18 18.5**	0.9	**13 Fri**	**62 13.9**	2.5	**S 5 59.9**	0.0	**19 48**
12 01	**180 00.3**	59.3	**S18 39.0**	0.8	**14 Sat**	**63 13.1**	2.5	**S 5 59.6**	0.0	**19 44**
12 02	**179 42.7**	59.3	**S18 59.1**	0.8	**15 SUN**	**64 12.1**	2.5	**S 5 59.3**	0.0	**19 40**
12 03	**179 24.8**	59.2	**S19 18.6**	0.8	**16 Mon**	**65 11.0**	2.4	**S 5 58.9**	0.0	**19 36**
12 04	**179 06.6**	59.2	**S19 37.6**	0.8	**17 Tu**	**66 09.7**	2.4	**S 5 58.5**	0.0	**19 32**
12 05	**178 48.1**	59.2	**S19 56.0**	0.7	**18 Wed**	**67 08.2**	2.4	**S 5 57.9**	0.0	**19 28**
12 07	**178 29.3**	59.2	**S20 13.9**	0.7	**19 Th**	**68 06.4**	2.4	**S 5 57.3**	0.0	**19 24**
12 08	**178 10.2**	59.2	**S20 31.2**	0.7	**20 Fri**	**69 04.5**	2.4	**S 5 56.6**	0.0	**19 21**
12 09	**177 50.8**	59.2	**S20 48.0**	0.7	**21 Sat**	**70 02.4**	2.4	**S 5 55.8**	0.0	**19 17**
12 11	**177 31.1**	59.2	**S21 04.1**	0.6	**22 SUN**	**71 00.2**	2.4	**S 5 54.9**	0.0	**19 13**
12 12	**177 11.1**	59.2	**S21 19.6**	0.6	**23 Mon**	**71 57.7**	2.4	**S 5 54.0**	0.0	**19 09**
12 13	**176 50.8**	59.1	**S21 34.6**	0.6	**24 Tu**	**72 55.0**	2.4	**S 5 52.9**	0.0	**19 05**
12 15	**176 30.3**	59.1	**S21 48.9**	0.6	**25 Wed**	**73 52.2**	2.4	**S 5 51.8**	0.0	**19 02**
12 16	**176 09.5**	59.1	**S22 02.5**	0.5	**26 Th**	**74 49.1**	2.4	**S 5 50.6**	0.1	**18 58**
12 17	**175 48.5**	59.1	**S22 15.6**	0.5	**27 Fri**	**75 45.9**	2.4	**S 5 49.3**	0.1	**18 54**
12 19	**175 27.2**	59.1	**S22 27.9**	0.5	**28 Sat**	**76 42.5**	2.3	**S 5 48.0**	0.1	**18 50**
12 20	**175 05.8**	59.1	**S22 39.6**	0.5	**29 SUN**	**77 38.9**	2.3	**S 5 46.6**	0.1	**18 46**
12 22	**174 44.0**	59.1	**S22 50.7**	0.4	**30 Mon**	**78 35.1**	2.3	**S 5 45.1**	0.1	**18 43**

VENUS, Av. Mag. -3.9
S.H.A. November
5 138; 10 132; 15 126; 20 119; 25 113; 30 106.

JUPITER, Av. Mag.-2.6
S.H.A. November
5 10; 10 10; 15 10; 20 10; 25 10; 30 10.

MARS						SATURN				
Mer Pass h m	GHA ° ′	Mean Var/hr 15°+	Dec ° ′	Mean Var/hr ′	Day	GHA ° ′	Mean Var/hr 15°+	Dec ° ′	Mean Var/hr ′	Mer Pass h m
08 25	**233 32.7**	1.1	**N 7 23.4**	0.6	**1 SUN**	**11 39.2**	2.7	**N 8 42.8**	0.1	**23 09**
08 23	**233 58.8**	1.1	**N 7 10.1**	0.6	**2 Mon**	**12 42.9**	2.6	**N 8 41.3**	0.1	**23 05**
08 22	**234 24.9**	1.1	**N 6 56.7**	0.6	**3 Tu**	**13 46.4**	2.7	**N 8 39.7**	0.1	**23 01**
08 20	**234 51.1**	1.1	**N 6 43.3**	0.6	**4 Wed**	**14 50.0**	2.6	**N 8 38.1**	0.1	**22 57**
08 18	**235 17.4**	1.1	**N 6 29.9**	0.6	**5 Th**	**15 53.5**	2.6	**N 8 36.6**	0.1	**22 52**
08 16	**235 43.8**	1.1	**N 6 16.5**	0.6	**6 Fri**	**16 57.0**	2.6	**N 8 35.1**	0.1	**22 48**
08 15	**236 10.2**	1.1	**N 6 03.2**	0.6	**7 Sat**	**18 00.5**	2.6	**N 8 33.6**	0.1	**22 44**
08 13	**236 36.7**	1.1	**N 5 49.8**	0.6	**8 SUN**	**19 03.9**	2.6	**N 8 32.1**	0.1	**22 40**
08 11	**237 03.3**	1.1	**N 5 36.4**	0.6	**9 Mon**	**20 07.3**	2.6	**N 8 30.6**	0.1	**22 36**
08 09	**237 30.0**	1.1	**N 5 23.0**	0.6	**10 Tu**	**21 10.7**	2.6	**N 8 29.2**	0.1	**22 31**
08 08	**237 56.8**	1.1	**N 5 09.6**	0.6	**11 Wed**	**22 14.0**	2.6	**N 8 27.8**	0.1	**22 27**
08 06	**238 23.6**	1.1	**N 4 56.2**	0.6	**12 Th**	**23 17.3**	2.6	**N 8 26.4**	0.1	**22 23**
08 04	**238 50.5**	1.1	**N 4 42.9**	0.6	**13 Fri**	**24 20.5**	2.6	**N 8 25.0**	0.1	**22 19**
08 02	**239 17.5**	1.1	**N 4 29.5**	0.6	**14 Sat**	**25 23.7**	2.6	**N 8 23.6**	0.1	**22 15**
08 00	**239 44.5**	1.1	**N 4 16.2**	0.6	**15 SUN**	**26 26.9**	2.6	**N 8 22.3**	0.1	**22 10**
07 59	**240 11.7**	1.1	**N 4 02.9**	0.6	**16 Mon**	**27 30.0**	2.6	**N 8 21.0**	0.1	**22 06**
07 57	**240 38.9**	1.1	**N 3 49.5**	0.5	**17 Tu**	**28 33.0**	2.6	**N 8 19.7**	0.1	**22 02**
07 55	**241 06.2**	1.1	**N 3 36.3**	0.6	**18 Wed**	**29 36.0**	2.6	**N 8 18.4**	0.0	**21 58**
07 53	**241 33.6**	1.1	**N 3 23.0**	0.6	**19 Th**	**30 38.9**	2.6	**N 8 17.2**	0.0	**21 54**
07 51	**242 01.1**	1.1	**N 3 09.7**	0.5	**20 Fri**	**31 41.7**	2.6	**N 8 16.0**	0.0	**21 49**
07 49	**242 28.6**	1.2	**N 2 56.5**	0.5	**21 Sat**	**32 44.5**	2.6	**N 8 14.8**	0.0	**21 45**
07 48	**242 56.2**	1.2	**N 2 43.3**	0.6	**22 SUN**	**33 47.3**	2.6	**N 8 13.6**	0.0	**21 41**
07 46	**243 23.9**	1.2	**N 2 30.1**	0.5	**23 Mon**	**34 49.9**	2.6	**N 8 12.5**	0.0	**21 37**
07 44	**243 51.7**	1.2	**N 2 17.0**	0.6	**24 Tu**	**35 52.5**	2.6	**N 8 11.4**	0.0	**21 33**
07 42	**244 19.6**	1.2	**N 2 03.8**	0.5	**25 Wed**	**36 55.0**	2.6	**N 8 10.4**	0.0	**21 29**
07 40	**244 47.5**	1.2	**N 1 50.7**	0.5	**26 Th**	**37 57.5**	2.6	**N 8 09.3**	0.0	**21 24**
07 38	**245 15.6**	1.2	**N 1 37.7**	0.5	**27 Fri**	**38 59.8**	2.6	**N 8 08.4**	0.0	**21 20**
07 36	**245 43.7**	1.2	**N 1 24.7**	0.5	**28 Sat**	**40 02.1**	2.6	**N 8 07.4**	0.0	**21 16**
07 35	**246 11.9**	1.2	**N 1 11.7**	0.5	**29 SUN**	**41 04.3**	2.6	**N 8 06.5**	0.0	**21 12**
07 33	**246 40.1**	1.2	**N 0 58.7**	0.5	**30 Mon**	**42 06.5**	2.6	**N 8 05.6**	0.0	**21 08**

MARS, Av. Mag. +1.5
S.H.A. November
5 191; 10 189; 15 186; 20 183; 25 181; 30 178.

NOTE: Planet corrections on pages E113-15

SATURN, Av. Mag. +0.1
S.H.A. November
5 332; 10 332; 15 333; 20 333; 25 333; 30 333.

NOVEMBER 1998 MOON

Day	GMT hr	GHA ° ′	Mean Var/hr 14°+	Dec ° ′	Mean Var/hr ′
1 Sun	**0**	45 40.3	27.8	**S 4 40.3**	11.8
	6	132 27.0	27.6	**S 3 29.4**	12.1
	12	219 12.8	27.5	**S 2 17.2**	12.2
	18	305 57.5	27.3	**S 1 04.1**	12.3
2 Mon	**0**	32 41.1	27.1	**N 0 09.8**	12.4
	6	119 23.5	26.8	**N 1 24.0**	12.4
	12	206 04.6	26.6	**N 2 38.4**	12.4
	18	292 44.3	26.3	**N 3 52.4**	12.2
3 Tu	**0**	19 22.6	26.1	**N 5 05.8**	12.0
	6	105 59.5	25.9	**N 6 18.3**	11.8
	12	192 34.8	25.6	**N 7 29.4**	11.6
	18	279 08.5	25.4	**N 8 38.8**	11.2
4 Wed	**0**	5 40.8	25.1	**N 9 46.2**	10.8
	6	92 11.5	24.9	**N 10 51.1**	10.3
	12	178 40.8	24.6	**N 11 53.4**	9.9
	18	265 08.7	24.4	**N 12 52.5**	9.3
5 Th	**0**	351 35.4	24.3	**N 13 48.3**	8.7
	6	78 01.0	24.1	**N 14 40.4**	8.0
	12	164 25.6	23.9	**N 15 28.6**	7.2
	18	250 49.6	23.9	**N 16 12.6**	6.6
6 Fri	**0**	337 13.0	23.9	**N 16 52.3**	5.8
	6	63 36.2	23.9	**N 17 27.5**	5.1
	12	149 59.4	24.0	**N 17 58.0**	4.2
	18	236 22.8	24.0	**N 18 23.8**	3.4
7 Sat	**0**	322 46.9	24.2	**N 18 44.8**	2.6
	6	49 11.8	24.4	**N 19 00.9**	1.8
	12	135 37.9	24.6	**N 19 12.3**	1.0
	18	222 05.4	24.9	**N 19 18.9**	0.3
8 Sun	**0**	308 34.6	25.3	**N 19 20.8**	0.5
	6	35 05.8	25.6	**N 19 18.1**	1.2
	12	121 39.0	26.0	**N 19 11.0**	2.0
	18	208 14.6	26.4	**N 18 59.7**	2.6
9 Mon	**0**	294 52.8	26.8	**N 18 44.1**	3.3
	6	21 33.5	27.2	**N 18 24.7**	3.9
	12	108 16.9	27.7	**N 18 01.5**	4.5
	18	195 03.2	28.2	**N 17 34.7**	5.0
10 Tu	**0**	281 52.3	28.7	**N 17 04.6**	5.6
	6	8 44.2	29.2	**N 16 31.3**	6.1
	12	95 38.9	29.6	**N 15 55.1**	6.5
	18	182 36.5	30.1	**N 15 16.2**	7.0
11 Wed	**0**	269 36.8	30.5	**N 14 34.8**	7.3
	6	356 39.7	31.0	**N 13 51.0**	7.7
	12	83 45.2	31.4	**N 13 05.0**	8.0
	18	170 53.1	31.8	**N 12 17.2**	8.3
12 Th	**0**	258 03.3	32.1	**N 11 27.5**	8.5
	6	345 15.8	32.4	**N 10 36.2**	8.8
	12	72 30.3	32.8	**N 9 43.5**	9.1
	18	159 46.6	33.0	**N 8 49.5**	9.2
13 Fri	**0**	247 04.8	33.3	**N 7 54.4**	9.3
	6	334 24.5	33.5	**N 6 58.3**	9.5
	12	61 45.6	33.7	**N 6 01.3**	9.6
	18	149 08.0	33.9	**N 5 03.7**	9.7
14 Sat	**0**	236 31.6	34.1	**N 4 05.5**	9.8
	6	323 56.1	34.2	**N 3 06.8**	9.8
	12	51 21.3	34.3	**N 2 07.9**	9.9
	18	138 47.2	34.4	**N 1 08.8**	9.8
15 Sun	**0**	226 13.6	34.5	**N 0 09.6**	9.9
	6	313 40.3	34.5	**S 0 49.4**	9.8
	12	41 07.2	34.5	**S 1 48.3**	9.7
	18	128 34.2	34.5	**S 2 46.9**	9.7
16 Mon	**0**	216 01.0	34.5	**S 3 45.0**	9.6
	6	303 27.6	34.4	**S 4 42.6**	9.5
	12	30 53.8	34.3	**S 5 39.5**	9.3
	18	118 19.5	34.2	**S 6 35.5**	9.2

Day	GMT hr	GHA ° ′	Mean Var/hr 14°+	Dec ° ′	Mean Var/hr ′
17 Tu	**0**	205 44.6	34.0	**S 7 30.6**	9.0
	6	293 09.0	33.9	**S 8 24.7**	8.8
	12	20 32.5	33.7	**S 9 17.5**	8.6
	18	107 55.1	33.5	**S 10 09.1**	8.3
18 Wed	**0**	195 16.6	33.3	**S 10 59.2**	8.1
	6	282 37.0	33.2	**S 11 47.7**	7.7
	12	9 56.3	33.0	**S 12 34.4**	7.5
	18	97 14.3	32.7	**S 13 19.4**	7.1
19 Th	**0**	184 30.9	32.5	**S 14 02.3**	6.8
	6	271 46.3	32.4	**S 14 43.2**	6.4
	12	359 00.3	32.1	**S 15 21.8**	6.0
	18	86 12.8	31.8	**S 15 58.0**	5.6
20 Fri	**0**	173 24.0	31.6	**S 16 31.8**	5.1
	6	260 33.8	31.3	**S 17 02.9**	4.7
	12	347 42.2	31.2	**S 17 31.3**	4.2
	18	74 49.3	30.9	**S 17 56.8**	3.7
21 Sat	**0**	161 55.0	30.7	**S 18 19.4**	3.2
	6	248 59.6	30.5	**S 18 39.0**	2.7
	12	336 02.9	30.3	**S 18 55.4**	2.1
	18	63 05.2	30.2	**S 19 08.6**	1.6
22 Sun	**0**	150 06.4	30.0	**S 19 18.4**	1.0
	6	237 06.8	29.9	**S 19 24.9**	0.4
	12	324 06.3	29.8	**S 19 28.0**	0.1
	18	51 05.0	29.7	**S 19 27.6**	0.7
23 Mon	**0**	138 03.2	29.6	**S 19 23.6**	1.3
	6	225 00.8	29.5	**S 19 16.1**	1.9
	12	311 58.1	29.5	**S 19 05.1**	2.4
	18	38 55.0	29.4	**S 18 50.5**	3.1
24 Tu	**0**	125 51.7	29.4	**S 18 32.4**	3.7
	6	212 48.3	29.4	**S 18 10.7**	4.3
	12	299 44.7	29.4	**S 17 45.6**	4.8
	18	26 41.2	29.4	**S 17 17.0**	5.4
25 Wed	**0**	113 37.8	29.4	**S 16 45.0**	5.9
	6	200 34.4	29.5	**S 16 09.7**	6.5
	12	287 31.1	29.4	**S 15 31.1**	7.0
	18	14 28.0	29.5	**S 14 49.3**	7.5
26 Th	**0**	101 25.0	29.5	**S 14 04.5**	8.0
	6	188 22.1	29.6	**S 13 16.7**	8.5
	12	275 19.2	29.5	**S 12 26.1**	9.0
	18	2 16.4	29.5	**S 11 32.7**	9.4
27 Fri	**0**	89 13.5	29.5	**S 10 36.8**	9.8
	6	176 10.5	29.5	**S 9 38.4**	10.1
	12	263 07.3	29.4	**S 8 37.7**	10.5
	18	350 03.8	29.3	**S 7 34.9**	10.8
28 Sat	**0**	76 59.8	29.2	**S 6 30.1**	11.1
	6	163 55.4	29.1	**S 5 23.5**	11.4
	12	250 50.2	29.0	**S 4 15.4**	11.6
	18	337 44.3	28.9	**S 3 06.0**	11.8
29 Sun	**0**	64 37.4	28.7	**S 1 55.4**	12.0
	6	151 29.4	28.4	**S 0 43.9**	12.0
	12	238 20.2	28.2	**N 0 28.2**	12.0
	18	325 09.7	28.0	**N 1 40.6**	12.1
30 Mon	**0**	51 57.6	27.7	**N 2 53.2**	12.0
	6	138 43.9	27.5	**N 4 05.5**	12.0
	12	225 28.5	27.1	**N 5 17.3**	11.9
	18	312 11.2	26.8	**N 6 28.3**	11.6

NOTE: Moon GHA corrections for additional hours, minutes, and seconds are on page E 108–110. Moon declination corrections are on page E 111. The mean var/hr value is needed for both corrections.

DECEMBER 1998 SUN & MOON GMT

SUN

Yr	Day of Mth	Week	Equation of Time 0h	Equation of Time 12h	Transit	Semi-diam	Lat 30°N Twi-light	Lat 30°N Sun-rise	Lat 30°N Sun-set	Lat 30°N Twi-light
			m s	m s	h m	′	h m	h m	h m	h m
335	**1**	**Tu**	-11 13	-11 02	**11 49**	16.2	06 12	06 38	17 00	17 26
336	**2**	**Wed**	-10 50	-10 39	**11 49**	16.2	06 13	06 39	17 00	17 26
337	**3**	**Th**	-10 27	-10 16	**11 50**	16.3	06 13	06 40	17 00	17 26
338	**4**	**Fri**	-10 04	-09 52	**11 50**	16.3	06 14	06 40	17 00	17 26
339	**5**	**Sat**	-09 40	-09 28	**11 51**	16.3	06 15	06 41	17 00	17 26
340	**6**	**Sun**	-09 15	-09 03	**11 51**	16.3	06 16	06 42	17 00	17 26
341	**7**	**Mon**	-08 50	-08 37	**11 51**	16.3	06 16	06 43	17 00	17 26
342	**8**	**Tu**	-08 24	-08 11	**11 52**	16.3	06 17	06 43	17 00	17 26
343	**9**	**Wed**	-07 58	-07 45	**11 52**	16.3	06 18	06 44	17 00	17 27
344	**10**	**Th**	-07 31	-07 18	**11 53**	16.3	06 19	06 45	17 01	17 27
345	**11**	**Fri**	-07 04	-06 50	**11 53**	16.3	06 19	06 45	17 01	17 27
346	**12**	**Sat**	-06 36	-06 23	**11 54**	16.3	06 20	06 46	17 01	17 27
347	**13**	**Sun**	-06 08	-05 54	**11 54**	16.3	06 20	06 47	17 01	17 28
348	**14**	**Mon**	-05 40	-05 26	**11 55**	16.3	06 21	06 47	17 02	17 28
349	**15**	**Tu**	-05 12	-04 57	**11 55**	16.3	06 22	06 48	17 02	17 28
350	**16**	**Wed**	-04 43	-04 28	**11 56**	16.3	06 22	06 49	17 02	17 29
351	**17**	**Th**	-04 13	-03 59	**11 56**	16.3	06 23	06 49	17 03	17 29
352	**18**	**Fri**	-03 44	-03 29	**11 57**	16.3	06 24	06 50	17 03	17 29
353	**19**	**Sat**	-03 14	-03 00	**11 57**	16.3	06 24	06 50	17 04	17 30
354	**20**	**Sun**	-02 45	-02 30	**11 58**	16.3	06 25	06 51	17 04	17 30
355	**21**	**Mon**	-02 15	-02 00	**11 58**	16.3	06 25	06 52	17 04	17 31
356	**22**	**Tu**	-01 45	-01 30	**11 58**	16.3	06 26	06 52	17 05	17 31
357	**23**	**Wed**	-01 15	-01 00	**11 59**	16.3	06 26	06 53	17 05	17 32
358	**24**	**Th**	-00 45	-00 30	**11 59**	16.3	06 27	06 53	17 06	17 32
359	**25**	**Fri**	-00 15	00 00	**12 00**	16.3	06 27	06 53	17 07	17 33
360	**26**	**Sat**	+00 14	+00 29	**12 00**	16.3	06 27	06 54	17 07	17 34
361	**27**	**Sun**	+00 44	+00 59	**12 01**	16.3	06 28	06 54	17 08	17 34
362	**28**	**Mon**	+01 14	+01 28	**12 01**	16.3	06 28	06 55	17 08	17 35
363	**29**	**Tu**	+01 43	+01 58	**12 02**	16.3	06 29	06 55	17 09	17 35
364	**30**	**Wed**	+02 12	+02 26	**12 02**	16.3	06 29	06 55	17 10	17 36
365	**31**	**Th**	+02 41	+02 55	**12 03**	16.3	06 29	06 56	17 10	17 37

Lat Corr to Sunrise, Sunset etc

Lat	Twi-light	Sun-rise	Sun-set	Twi-light
°	h m	h m	h m	h m
N60	+1 38	+2 08	−2 08	−1 38
55	+1 11	+1 30	−1 30	−1 11
50	+0 51	+1 03	−1 03	−0 51
45	+0 36	+0 43	−0 43	−0 36
N40	+0 22	+0 26	−0 26	−0 22
35	+0 10	+0 12	−0 12	−0 10
30	0 00	0 00	0 00	0 00
25	−0 10	−0 11	+0 11	+0 10
N20	−0 19	−0 21	+0 21	+0 19
15	−0 27	−0 30	+0 30	+0 27
10	−0 36	−0 39	+0 39	+0 36
N 5	−0 44	−0 48	+0 48	+0 44
0	−0 53	−0 57	+0 57	+0 53
S 5	−1 02	−1 05	+1 05	+1 02
10	−1 11	−1 14	+1 14	+1 11
15	−1 21	−1 23	+1 23	+1 21
S20	−1 31	−1 33	+1 33	+1 31
25	−1 43	−1 43	+1 43	+1 43
30	−1 56	−1 55	+1 55	+1 56
35	−2 11	−2 08	+2 08	+2 11
S40	−2 29	−2 23	+2 23	+2 29
45	−2 52	−2 41	+2 41	+2 52
50	−3 21	−3 03	+3 03	+3 21

NOTES

Latitude corrections are for mid-month.

Equation of time = mean time (LHA mean sun) minus apparent time (LHA true sun).

MOON

Yr	Day of Mth	Week	Age	Transit (Upper)	Diff	Semi-diam	Hor Par	Lat 30°N Moon-rise	Lat 30°N Moon-set
			days	h m	m	′	′	h m	h m
335	**1**	**Tu**	12	**22 15**	**59**	16.6	61.0	15 36	03 52
336	**2**	**Wed**	13	**23 14**	**60**	16.6	61.1	16 24	05 00
337	**3**	**Th**	14	**24 14**	**-**	16.6	61.0	17 16	06 09
338	**4**	**Fri**	15	**00 14**	**61**	16.5	60.5	18 13	07 16
339	**5**	**Sat**	16	**01 15**	**60**	16.3	59.8	19 12	08 19
340	**6**	**Sun**	17	**02 15**	**57**	16.1	58.9	20 13	09 16
341	**7**	**Mon**	18	**03 12**	**53**	15.8	58.0	21 13	10 08
342	**8**	**Tu**	19	**04 05**	**50**	15.5	57.0	22 12	10 53
343	**9**	**Wed**	20	**04 55**	**47**	15.3	56.2	23 08	11 33
344	**10**	**Th**	21	**05 42**	**44**	15.1	55.4	- -	12 10
345	**11**	**Fri**	22	**06 26**	**43**	14.9	54.8	00 03	12 44
346	**12**	**Sat**	23	**07 09**	**42**	14.8	54.4	00 55	13 17
347	**13**	**Sun**	24	**07 51**	**42**	14.8	54.1	01 47	13 50
348	**14**	**Mon**	25	**08 33**	**43**	14.7	54.0	02 39	14 23
349	**15**	**Tu**	26	**09 16**	**45**	14.7	54.1	03 31	14 58
350	**16**	**Wed**	27	**10 01**	**46**	14.8	54.2	04 23	15 35
351	**17**	**Th**	28	**10 47**	**48**	14.8	54.5	05 16	16 15
352	**18**	**Fri**	29	**11 35**	**50**	14.9	54.8	06 09	17 00
353	**19**	**Sat**	01	**12 25**	**51**	15.0	55.2	07 01	17 48
354	**20**	**Sun**	02	**13 16**	**51**	15.2	55.6	07 52	18 40
355	**21**	**Mon**	03	**14 07**	**51**	15.3	56.1	08 41	19 35
356	**22**	**Tu**	04	**14 58**	**50**	15.4	56.6	09 27	20 32
357	**23**	**Wed**	05	**15 48**	**50**	15.6	57.1	10 10	21 31
358	**24**	**Th**	06	**16 38**	**50**	15.7	57.7	10 51	22 31
359	**25**	**Fri**	07	**17 28**	**50**	15.9	58.2	11 30	23 31
360	**26**	**Sat**	08	**18 18**	**51**	16.0	58.8	12 09	- -
361	**27**	**Sun**	09	**19 09**	**53**	16.2	59.3	12 48	00 33
362	**28**	**Mon**	10	**20 02**	**56**	16.3	59.8	13 29	01 36
363	**29**	**Tu**	11	**20 58**	**58**	16.4	60.1	14 13	02 41
364	**30**	**Wed**	12	**21 56**	**59**	16.4	60.3	15 02	03 47
365	**31**	**Th**	13	**22 55**	**61**	16.4	60.2	15 55	04 54

Phases of the Moon

		d	h	m
○	Full Moon	3	15	19
☽	Last Quarter	10	17	53
●	New Moon	18	22	42
☾	First Quarter	26	10	46

	d	h
Perigee	2	12
Apogee	14	17
Perigee	30	18
Farthest N of 0°	5	11
On 0°	12	09
Farthest S of 0°	19	21
On 0°	26	18

NOTES

Diff equals daily change in transit time.

To correct moonrise, moonset, or transit times for latitude and/or longitude see page E 112. For further information about these tables see page E 4. For arc-to-time correction see page E 103.

DECEMBER 1998 — STARS — 0h GMT December 1

No	Name	Mag	Transit h m	Dec ° ′	GHA ° ′	RA h m	SHA ° ′
	ARIES		**19 18**		**69 38.9**		
1	**Alpheratz**	2.2	**19 27**	N29 05.2	**67 33.9**	0 08	357 55.0
2	**Ankaa**	2.4	**19 44**	S 42 19.0	**63 05.6**	0 26	353 26.7
3	**Schedar**	2.5	**19 59**	N56 32.1	**59 32.0**	0 40	349 53.1
4	**Diphda**	2.2	**20 02**	S 17 59.7	**58 45.9**	0 44	349 07.0
5	**Achernar**	0.6	**20 56**	S 57 14.9	**45 13.8**	1 38	335 34.9
6	**POLARIS**	2.1	**21 50**	N89 15.7	**31 36.8**	2 32	321 57.9
7	**Hamal**	2.2	**21 25**	N23 27.4	**37 52.1**	2 07	328 13.2
8	**Acamar**	3.1	**22 16**	S 40 18.8	**25 05.4**	2 58	315 26.5
9	**Menkar**	2.8	**22 20**	N 4 05.0	**24 05.4**	3 02	314 26.5
10	**Mirfak**	1.9	**22 42**	N49 51.5	**18 34.9**	3 24	308 56.0
11	**Aldebaran**	1.1	**23 53**	N16 30.3	**0 40.8**	4 36	291 01.9
12	**Rigel**	0.3	**0 36**	S 8 12.3	**351 01.4**	5 15	281 22.5
13	**Capella**	0.2	**0 38**	N45 59.7	**350 29.4**	5 17	280 50.5
14	**Bellatrix**	1.7	**0 46**	N 6 20.8	**348 22.6**	5 25	278 43.7
15	**Elnath**	1.8	**0 48**	N28 36.3	**348 05.3**	5 26	278 26.4
16	**Alnilam**	1.8	**0 57**	S 1 12.3	**345 36.3**	5 36	275 57.4
17	**Betelgeuse**	0.1-1.2	**1 16**	N 7 24.3	**340 52.0**	5 55	271 13.1
18	**Canopus**	-0.9	**1 45**	S 52 41.8	**333 39.5**	6 24	264 00.6
19	**Sirius**	-1.6	**2 06**	S 16 42.9	**328 22.2**	6 45	258 43.3
20	**Adhara**	1.6	**2 20**	S 28 58.3	**324 59.9**	6 59	255 21.0
21	**Castor**	1.6	**2 55**	N31 53.3	**316 00.8**	7 35	246 21.9
22	**Procyon**	0.5	**3 00**	N 5 13.6	**314 50.1**	7 39	245 11.2
23	**Pollux**	1.2	**3 06**	N28 01.6	**313 20.0**	7 45	243 41.1
24	**Avior**	1.7	**3 43**	S 59 30.2	**304 00.9**	8 23	234 22.0
25	**Suhail**	2.2	**4 29**	S 43 25.5	**292 39.3**	9 08	223 00.4
26	**Miaplacidus**	1.8	**4 34**	S 69 42.5	**291 20.3**	9 13	221 41.4
27	**Alphard**	2.2	**4 48**	S 8 39.2	**287 45.8**	9 28	218 06.9
28	**Regulus**	1.3	**5 29**	N11 58.3	**277 34.2**	10 08	207 55.3
29	**Dubhe**	2.0	**6 24**	N61 45.1	**263 44.3**	11 04	194 05.4
30	**Denebola**	2.2	**7 09**	N14 34.7	**252 24.0**	11 49	182 45.1
31	**Gienah**	2.8	**7 36**	S 17 32.0	**245 42.8**	12 16	176 03.9
32	**Acrux**	1.1	**7 47**	S 63 05.2	**243 00.8**	12 27	173 21.9
33	**Gacrux**	1.6	**7 51**	S 57 06.1	**241 52.4**	12 31	172 13.5
34	**Mimosa**	1.5	**8 08**	S 59 40.6	**237 44.2**	12 48	168 05.3
35	**Alioth**	1.7	**8 14**	N55 57.8	**236 09.7**	12 54	166 30.8
36	**Spica**	1.2	**8 45**	S 11 09.2	**228 22.1**	13 25	158 43.2
37	**Alkaid**	1.9	**9 07**	N49 19.0	**222 47.0**	13 47	153 08.1
38	**Hadar**	0.9	**9 24**	S 60 21.7	**218 43.1**	14 04	149 04.2
39	**Menkent**	2.3	**9 26**	S 36 21.6	**218 00.0**	14 07	148 21.1
40	**Arcturus**	0.2	**9 35**	N19 11.3	**215 45.1**	14 16	146 06.2
41	**Rigil Kent.**	0.1	**9 59**	S 60 49.5	**209 46.5**	14 39	140 07.6
42	**Zuben'ubi**	2.9	**10 11**	S 16 02.0	**206 57.0**	14 51	137 18.1
43	**Kochab**	2.2	**10 10**	N74 09.5	**206 59.6**	14 51	137 20.7
44	**Alphecca**	2.3	**10 54**	N26 43.2	**195 59.8**	15 35	126 20.9
45	**Antares**	1.2	**11 49**	S 26 25.6	**182 19.3**	16 29	112 40.4
46	**Atria**	1.9	**12 08**	S 69 01.3	**177 31.8**	16 48	107 52.9
47	**Sabik**	2.6	**12 30**	S 15 43.2	**172 04.7**	17 10	102 25.8
48	**Shaula**	1.7	**12 53**	S 37 06.0	**166 16.5**	17 33	96 37.6
49	**Rasalhague**	2.1	**12 54**	N12 33.8	**165 56.1**	17 35	96 17.2
50	**Eltanin**	2.4	**13 16**	N51 29.5	**160 30.8**	17 57	90 51.9
51	**Kaus Aust.**	2.0	**13 43**	S 34 23.0	**153 38.0**	18 24	83 59.1
52	**Vega**	0.1	**13 56**	N38 47.2	**150 25.9**	18 37	80 47.0
53	**Nunki**	2.1	**14 14**	S 26 17.8	**145 51.5**	18 55	76 12.6
54	**Altair**	0.9	**15 10**	N 8 52.1	**131 58.3**	19 51	62 19.4
55	**Peacock**	2.1	**15 44**	S 56 44.4	**123 16.3**	20 26	53 37.4
56	**Deneb**	1.3	**16 00**	N45 16.8	**119 18.3**	20 41	49 39.4
57	**Enif**	2.5	**17 03**	N 9 52.3	**103 37.2**	21 44	33 58.3
58	**Al Na'ir**	2.2	**17 27**	S 46 58.2	**97 36.9**	22 08	27 58.0
59	**Fomalhaut**	1.3	**18 16**	S 29 37.9	**85 15.3**	22 58	15 36.4
60	**Markab**	2.6	**18 23**	N15 12.0	**83 28.4**	23 05	13 49.5

Star Transit Corr Table

Date	Corr h m	Date	Corr h m
1	0 00	17	–1 03
2	–0 04	18	–1 07
3	–0 08	19	–1 11
4	–0 12	20	–1 15
5	–0 16	21	–1 19
6	–0 20	22	–1 23
7	–0 24	23	–1 27
8	–0 28	24	–1 30
9	–0 31	25	–1 34
10	–0 35	26	–1 38
11	–0 39	27	–1 42
12	–0 43	28	–1 46
13	–0 47	29	–1 50
14	–0 51	30	–1 54
15	–0 55	31	–1 58
16	–0 59		

STAR TRANSIT

To find the approximate time of transit of a star for any day of the month use the table above. All corrections are subtractive.

If the value taken from the table is greater than the time of transit for the first of the month, add 23h 56min to the time of transit before subtracting the correction.

Example: What time will Rigel (No 12) transit the Meridian on December 30?

	h min
Transit on Dec 1	00 36
	+23 56
	24 32
Corr for Dec 30	–01 54
Transit on Dec 30	22 38

DECEMBER DIARY

d h	
1 15	Mercury in conjunction
3 06	Ceres 1.2° N of Moon
3 13	Aldebaran 0.6° S of Moon
9 06	Regulus 0.01° N of Moon
11 06	Mercury stationary
12 08	Mars 1.8° S of Moon
17 00	Mercury 3° S of Moon
20 04	Mercury greatest elong. W.(22°)
21 16	Neptune 1.7° S of Moon
22 02	Solstice
22 06	Mercury 7° N of Antares
22 11	Uranus 1.8° S of Moon
25 11	Jupiter 1.2° N of Moon
27 23	Saturn 2° N of Moon
30 16	Saturn stationary
30 23	Aldebaran 0.6° S of Moon

NOTES

Star declinations may be used as is for the whole month. To correct GHA for day, hour, minute, and second see page E 106; for further explanation of this table see page E 4.

DECEMBER 1998 — SUN & ARIES — GMT

Tuesday, 1st December

Time	SUN GHA ° ′	Dec ° ′	ARIES GHA ° ′
00	182 48.0	**S21 43.9**	69 38.9
02	212 47.6	**21 44.7**	99 43.8
04	242 47.1	**21 45.5**	129 48.8
06	272 46.7	**21 46.3**	159 53.7
08	302 46.2	**21 47.1**	189 58.6
10	332 45.7	**21 47.8**	220 03.6
12	2 45.3	**21 48.6**	250 08.5
14	32 44.8	**21 49.4**	280 13.4
16	62 44.3	**21 50.2**	310 18.3
18	92 43.9	**21 50.9**	340 23.3
20	122 43.4	**21 51.7**	10 28.2
22	152 42.9	**S21 52.5**	40 33.1

Wednesday, 2nd December

Time	SUN GHA ° ′	Dec ° ′	ARIES GHA ° ′
00	182 42.5	**S21 53.2**	70 38.1
02	212 42.0	**21 54.0**	100 43.0
04	242 41.5	**21 54.7**	130 47.9
06	272 41.0	**21 55.5**	160 52.8
08	302 40.6	**21 56.2**	190 57.8
10	332 40.1	**21 57.0**	221 02.7
12	2 39.6	**21 57.7**	251 07.6
14	32 39.1	**21 58.4**	281 12.6
16	62 38.7	**21 59.2**	311 17.5
18	92 38.2	**21 59.9**	341 22.4
20	122 37.7	**22 00.6**	11 27.3
22	152 37.2	**S22 01.4**	41 32.3

Thursday, 3rd December

Time	SUN GHA ° ′	Dec ° ′	ARIES GHA ° ′
00	182 36.7	**S22 02.1**	71 37.2
02	212 36.3	**22 02.8**	101 42.1
04	242 35.8	**22 03.5**	131 47.1
06	272 35.3	**22 04.2**	161 52.0
08	302 34.8	**22 05.0**	191 56.9
10	332 34.3	**22 05.7**	222 01.8
12	2 33.8	**22 06.4**	252 06.8
14	32 33.3	**22 07.1**	282 11.7
16	62 32.8	**22 07.8**	312 16.6
18	92 32.3	**22 08.5**	342 21.6
20	122 31.8	**22 09.2**	12 26.5
22	152 31.4	**S22 09.9**	42 31.4

Friday, 4th December

Time	SUN GHA ° ′	Dec ° ′	ARIES GHA ° ′
00	182 30.9	**S22 10.5**	72 36.3
02	212 30.4	**22 11.2**	102 41.3
04	242 29.9	**22 11.9**	132 46.2
06	272 29.4	**22 12.6**	162 51.1
08	302 28.9	**22 13.3**	192 56.1
10	332 28.4	**22 13.9**	223 01.0
12	2 27.9	**22 14.6**	253 05.9
14	32 27.4	**22 15.3**	283 10.8
16	62 26.9	**22 15.9**	313 15.8
18	92 26.3	**22 16.6**	343 20.7
20	122 25.8	**22 17.3**	13 25.6
22	152 25.3	**S22 17.9**	43 30.6

Saturday, 5th December

Time	SUN GHA ° ′	Dec ° ′	ARIES GHA ° ′
00	182 24.8	**S22 18.6**	73 35.5
02	212 24.3	**22 19.2**	103 40.4
04	242 23.8	**22 19.9**	133 45.3
06	272 23.3	**22 20.5**	163 50.3
08	302 22.8	**22 21.1**	193 55.2
10	332 22.3	**22 21.8**	224 00.1
12	2 21.8	**22 22.4**	254 05.0
14	32 21.2	**22 23.1**	284 10.0
16	62 20.7	**22 23.7**	314 14.9
18	92 20.2	**22 24.3**	344 19.8
20	122 19.7	**22 24.9**	14 24.8
22	152 19.2	**S22 25.5**	44 29.7

Sunday, 6th December

Time	SUN GHA ° ′	Dec ° ′	ARIES GHA ° ′
00	182 18.7	**S22 26.2**	74 34.6
02	212 18.1	**22 26.8**	104 39.5
04	242 17.6	**22 27.4**	134 44.5
06	272 17.1	**22 28.0**	164 49.4
08	302 16.6	**22 28.6**	194 54.3
10	332 16.0	**22 29.2**	224 59.3
12	2 15.5	**22 29.8**	255 04.2
14	32 15.0	**22 30.4**	285 09.1
16	62 14.5	**22 31.0**	315 14.0
18	92 13.9	**22 31.6**	345 19.0
20	122 13.4	**22 32.2**	15 23.9
22	152 12.9	**S22 32.7**	45 28.8

Monday, 7th December

Time	SUN GHA ° ′	Dec ° ′	ARIES GHA ° ′
00	182 12.4	**S22 33.3**	75 33.8
02	212 11.8	**22 33.9**	105 38.7
04	242 11.3	**22 34.5**	135 43.6
06	272 10.8	**22 35.0**	165 48.5
08	302 10.2	**22 35.6**	195 53.5
10	332 09.7	**22 36.2**	225 58.4
12	2 09.1	**22 36.7**	256 03.3
14	32 08.6	**22 37.3**	286 08.3
16	62 08.1	**22 37.8**	316 13.2
18	92 07.5	**22 38.4**	346 18.1
20	122 07.0	**22 38.9**	16 23.0
22	152 06.5	**S22 39.5**	46 28.0

Tuesday, 8th December

Time	SUN GHA ° ′	Dec ° ′	ARIES GHA ° ′
00	182 05.9	**S22 40.0**	76 32.9
02	212 05.4	**22 40.6**	106 37.8
04	242 04.8	**22 41.1**	136 42.8
06	272 04.3	**22 41.6**	166 47.7
08	302 03.7	**22 42.2**	196 52.6
10	332 03.2	**22 42.7**	226 57.5
12	2 02.6	**22 43.2**	257 02.5
14	32 02.1	**22 43.7**	287 07.4
16	62 01.5	**22 44.3**	317 12.3
18	92 01.0	**22 44.8**	347 17.3
20	122 00.5	**22 45.3**	17 22.2
22	151 59.9	**S22 45.8**	47 27.1

Wednesday, 9th December

Time	SUN GHA ° ′	Dec ° ′	ARIES GHA ° ′
00	181 59.3	**S22 46.3**	77 32.0
02	211 58.8	**22 46.8**	107 37.0
04	241 58.2	**22 47.3**	137 41.9
06	271 57.7	**22 47.8**	167 46.8
08	301 57.1	**22 48.3**	197 51.8
10	331 56.6	**22 48.8**	227 56.7
12	1 56.0	**22 49.3**	258 01.6
14	31 55.5	**22 49.8**	288 06.5
16	61 54.9	**22 50.2**	318 11.5
18	91 54.3	**22 50.7**	348 16.4
20	121 53.8	**22 51.2**	18 21.3
22	151 53.2	**S22 51.7**	48 26.2

Thursday, 10th December

Time	SUN GHA ° ′	Dec ° ′	ARIES GHA ° ′
00	181 52.7	**S22 52.1**	78 31.2
02	211 52.1	**22 52.6**	108 36.1
04	241 51.5	**22 53.1**	138 41.0
06	271 51.0	**22 53.5**	168 46.0
08	301 50.4	**22 54.0**	198 50.9
10	331 49.8	**22 54.4**	228 55.8
12	1 49.3	**22 54.9**	259 00.7
14	31 48.7	**22 55.3**	289 05.7
16	61 48.1	**22 55.8**	319 10.6
18	91 47.6	**22 56.2**	349 15.5
20	121 47.0	**22 56.6**	19 20.5
22	151 46.4	**S22 57.1**	49 25.4

Friday, 11th December

Time	SUN GHA ° ′	Dec ° ′	ARIES GHA ° ′
00	181 45.9	**S22 57.5**	79 30.3
02	211 45.3	**22 57.9**	109 35.2
04	241 44.7	**22 58.4**	139 40.2
06	271 44.2	**22 58.8**	169 45.1
08	301 43.6	**22 59.2**	199 50.0
10	331 43.0	**22 59.6**	229 55.0
12	1 42.4	**23 00.0**	259 59.9
14	31 41.9	**23 00.4**	290 04.8
16	61 41.3	**23 00.8**	320 09.7
18	91 40.7	**23 01.2**	350 14.7
20	121 40.1	**23 01.6**	20 19.6
22	151 39.5	**S23 02.0**	50 24.5

Saturday, 12th December

Time	SUN GHA ° ′	Dec ° ′	ARIES GHA ° ′
00	181 39.0	**S23 02.4**	80 29.5
02	211 38.4	**23 02.8**	110 34.4
04	241 37.8	**23 03.2**	140 39.3
06	271 37.2	**23 03.6**	170 44.2
08	301 36.6	**23 04.0**	200 49.2
10	331 36.1	**23 04.3**	230 54.1
12	1 35.5	**23 04.7**	260 59.0
14	31 34.9	**23 05.1**	291 04.0
16	61 34.3	**23 05.5**	321 08.9
18	91 33.7	**23 05.8**	351 13.8
20	121 33.1	**23 06.2**	21 18.7
22	151 32.6	**S23 06.5**	51 23.7

Sunday, 13th December

Time	SUN GHA ° ′	Dec ° ′	ARIES GHA ° ′
00	181 32.0	**S23 06.9**	81 28.6
02	211 31.4	**23 07.2**	111 33.5
04	241 30.8	**23 07.6**	141 38.4
06	271 30.2	**23 07.9**	171 43.4
08	301 29.6	**23 08.3**	201 48.3
10	331 29.0	**23 08.6**	231 53.2
12	1 28.4	**23 09.0**	261 58.2
14	31 27.9	**23 09.3**	292 03.1
16	61 27.3	**23 09.6**	322 08.0
18	91 26.7	**23 09.9**	352 12.9
20	121 26.1	**23 10.3**	22 17.9
22	151 25.5	**S23 10.6**	52 22.8

Monday, 14th December

Time	SUN GHA ° ′	Dec ° ′	ARIES GHA ° ′
00	181 24.9	**S23 10.9**	82 27.7
02	211 24.3	**23 11.2**	112 32.7
04	241 23.7	**23 11.5**	142 37.6
06	271 23.1	**23 11.8**	172 42.5
08	301 22.5	**23 12.1**	202 47.4
10	331 21.9	**23 12.4**	232 52.4
12	1 21.3	**23 12.7**	262 57.3
14	31 20.7	**23 13.0**	293 02.2
16	61 20.1	**23 13.3**	323 07.2
18	91 19.5	**23 13.6**	353 12.1
20	121 18.9	**23 13.9**	23 17.0
22	151 18.3	**S23 14.2**	53 21.9

Tuesday, 15th December

Time	SUN GHA ° ′	Dec ° ′	ARIES GHA ° ′
00	181 17.7	**S23 14.4**	83 26.9
02	211 17.1	**23 14.7**	113 31.8
04	241 16.5	**23 15.0**	143 36.7
06	271 15.9	**23 15.3**	173 41.7
08	301 15.3	**23 15.5**	203 46.6
10	331 14.7	**23 15.8**	233 51.5
12	1 14.1	**23 16.0**	263 56.4
14	31 13.5	**23 16.3**	294 01.4
16	61 12.9	**23 16.6**	324 06.3
18	91 12.3	**23 16.8**	354 11.2
20	121 11.7	**23 17.0**	24 16.2
22	151 11.1	**S23 17.3**	54 21.1

NOTES

Sun and Aries GHA corrections for additional hour, minutes, and seconds are on page E 106.
Correct sun declination by interpolation. For an example see page E 4.

DECEMBER 1998 SUN & ARIES GMT

Wednesday, 16th December

Time	SUN GHA ° ′	Dec ° ′	ARIES GHA ° ′
00	181 10.5	**S23 17.5**	84 26.0
02	211 09.9	**23 17.8**	114 30.9
04	241 09.3	**23 18.0**	144 35.9
06	271 08.7	**23 18.2**	174 40.8
08	301 08.1	**23 18.5**	204 45.7
10	331 07.5	**23 18.7**	234 50.6
12	1 06.8	**23 18.9**	264 55.6
14	31 06.2	**23 19.1**	295 00.5
16	61 05.6	**23 19.3**	325 05.4
18	91 05.0	**23 19.5**	355 10.4
20	121 04.4	**23 19.7**	25 15.3
22	151 03.8	**S23 19.9**	55 20.2

Thursday, 17th December

Time	SUN GHA ° ′	Dec ° ′	ARIES GHA ° ′
00	181 03.2	**S23 20.2**	85 25.1
02	211 02.6	**23 20.3**	115 30.1
04	241 02.0	**23 20.5**	145 35.0
06	271 01.4	**23 20.7**	175 39.9
08	301 00.7	**23 20.9**	205 44.9
10	331 00.1	**23 21.1**	235 49.8
12	0 59.5	**23 21.3**	265 54.7
14	30 58.9	**23 21.5**	295 59.6
16	60 58.3	**23 21.6**	326 04.6
18	90 57.7	**23 21.8**	356 09.5
20	120 57.1	**23 22.0**	26 14.4
22	150 56.5	**S23 22.1**	56 19.4

Friday, 18th December

Time	SUN GHA ° ′	Dec ° ′	ARIES GHA ° ′
00	180 55.8	**S23 22.3**	86 24.3
02	210 55.2	**23 22.5**	116 29.2
04	240 54.6	**23 22.6**	146 34.1
06	270 54.0	**23 22.8**	176 39.1
08	300 53.4	**23 22.9**	206 44.0
10	330 52.8	**23 23.1**	236 48.9
12	0 52.1	**23 23.2**	266 53.9
14	30 51.5	**23 23.3**	296 58.8
16	60 50.9	**23 23.5**	327 03.7
18	90 50.3	**23 23.6**	357 08.6
20	120 49.7	**23 23.7**	27 13.6
22	150 49.1	**S23 23.9**	57 18.5

Saturday, 19th December

Time	SUN GHA ° ′	Dec ° ′	ARIES GHA ° ′
00	180 48.4	**S23 24.0**	87 23.4
02	210 47.8	**23 24.1**	117 28.4
04	240 47.2	**23 24.2**	147 33.3
06	270 46.6	**23 24.3**	177 38.2
08	300 46.0	**23 24.4**	207 43.1
10	330 45.3	**23 24.6**	237 48.1
12	0 44.7	**23 24.7**	267 53.0
14	30 44.1	**23 24.8**	297 57.9
16	60 43.5	**23 24.9**	328 02.9
18	90 42.9	**23 24.9**	358 07.8
20	120 42.3	**23 25.0**	28 12.7
22	150 41.6	**S23 25.1**	58 17.6

Sunday, 20th December

Time	SUN GHA ° ′	Dec ° ′	ARIES GHA ° ′
00	180 41.0	**S23 25.2**	88 22.6
02	210 40.4	**23 25.3**	118 27.5
04	240 39.8	**23 25.4**	148 32.4
06	270 39.1	**23 25.4**	178 37.4
08	300 38.5	**23 25.5**	208 42.3
10	330 37.9	**23 25.6**	238 47.2
12	0 37.3	**23 25.6**	268 52.1
14	30 36.7	**23 25.7**	298 57.1
16	60 36.0	**23 25.8**	329 02.0
18	90 35.4	**23 25.8**	359 06.9
20	120 34.8	**23 25.9**	29 11.9
22	150 34.2	**S23 25.9**	59 16.8

Monday, 21st December

Time	SUN GHA ° ′	Dec ° ′	ARIES GHA ° ′
00	180 33.6	**S23 26.0**	89 21.7
02	210 32.9	**23 26.0**	119 26.6
04	240 32.3	**23 26.0**	149 31.6
06	270 31.7	**23 26.1**	179 36.5
08	300 31.1	**23 26.1**	209 41.4
10	330 30.4	**23 26.1**	239 46.4
12	0 29.8	**23 26.1**	269 51.3
14	30 29.2	**23 26.2**	299 56.2
16	60 28.6	**23 26.2**	330 01.1
18	90 27.9	**23 26.2**	0 06.1
20	120 27.3	**23 26.2**	30 11.0
22	150 26.7	**S23 26.2**	60 15.9

Tuesday, 22nd December

Time	SUN GHA ° ′	Dec ° ′	ARIES GHA ° ′
00	180 26.1	**S23 26.2**	90 20.8
02	210 25.5	**23 26.2**	120 25.8
04	240 24.8	**23 26.2**	150 30.7
06	270 24.2	**23 26.2**	180 35.6
08	300 23.6	**23 26.2**	210 40.6
10	330 23.0	**23 26.2**	240 45.5
12	0 22.3	**23 26.2**	270 50.4
14	30 21.7	**23 26.2**	300 55.3
16	60 21.1	**23 26.1**	331 00.3
18	90 20.5	**23 26.1**	1 05.2
20	120 19.8	**23 26.1**	31 10.1
22	150 19.2	**S23 26.1**	61 15.1

Wednesday, 23rd December

Time	SUN GHA ° ′	Dec ° ′	ARIES GHA ° ′
00	180 18.6	**S23 26.0**	91 20.0
02	210 18.0	**23 26.0**	121 24.9
04	240 17.4	**23 26.0**	151 29.8
06	270 16.7	**23 25.9**	181 34.8
08	300 16.1	**23 25.9**	211 39.7
10	330 15.5	**23 25.8**	241 44.6
12	0 14.9	**23 25.8**	271 49.6
14	30 14.2	**23 25.7**	301 54.5
16	60 13.6	**23 25.6**	331 59.4
18	90 13.0	**23 25.6**	2 04.3
20	120 12.4	**23 25.5**	32 09.3
22	150 11.7	**S23 25.4**	62 14.2

Thursday, 24th December

Time	SUN GHA ° ′	Dec ° ′	ARIES GHA ° ′
00	180 11.1	**S23 25.4**	92 19.1
02	210 10.5	**23 25.3**	122 24.1
04	240 09.9	**23 25.2**	152 29.0
06	270 09.3	**23 25.1**	182 33.9
08	300 08.6	**23 25.0**	212 38.8
10	330 08.0	**23 25.0**	242 43.8
12	0 07.4	**23 24.9**	272 48.7
14	30 06.8	**23 24.8**	302 53.6
16	60 06.1	**23 24.7**	332 58.6
18	90 05.5	**23 24.6**	3 03.5
20	120 04.9	**23 24.5**	33 08.4
22	150 04.3	**S23 24.3**	63 13.3

Friday, 25th December

Time	SUN GHA ° ′	Dec ° ′	ARIES GHA ° ′
00	180 03.7	**S23 24.2**	93 18.3
02	210 03.0	**23 24.1**	123 23.2
04	240 02.4	**23 24.0**	153 28.1
06	270 01.8	**23 23.9**	183 33.1
08	300 01.2	**23 23.7**	213 38.0
10	330 00.6	**23 23.6**	243 42.9
12	0 00.0	**23 23.5**	273 47.8
14	29 59.3	**23 23.3**	303 52.8
16	59 58.7	**23 23.2**	333 57.7
18	89 58.1	**23 23.1**	4 02.6
20	119 57.5	**23 22.9**	34 07.5
22	149 56.8	**S23 22.8**	64 12.5

Saturday, 26th December

Time	SUN GHA ° ′	Dec ° ′	ARIES GHA ° ′	Time
00	179 56.2	**S23 22.6**	94 17.4	**00**
02	209 55.6	**23 22.5**	124 22.3	**02**
04	239 55.0	**23 22.3**	154 27.3	**04**
06	269 54.4	**23 22.1**	184 32.2	**06**
08	299 53.7	**23 22.0**	214 37.1	**08**
10	329 53.1	**23 21.8**	244 42.0	**10**
12	359 52.5	**23 21.6**	274 47.0	**12**
14	29 51.9	**23 21.5**	304 51.9	**14**
16	59 51.3	**23 21.3**	334 56.8	**16**
18	89 50.7	**23 21.1**	5 01.8	**18**
20	119 50.0	**23 20.9**	35 06.7	**20**
22	149 49.4	**S23 20.7**	65 11.6	**22**

Sunday, 27th December

Time	SUN GHA ° ′	Dec ° ′	ARIES GHA ° ′	Time
00	179 48.8	**S23 20.5**	95 16.5	**00**
02	209 48.2	**23 20.4**	125 21.5	**02**
04	239 47.6	**23 20.2**	155 26.4	**04**
06	269 47.0	**23 20.0**	185 31.3	**06**
08	299 46.3	**23 19.7**	215 36.3	**08**
10	329 45.7	**23 19.5**	245 41.2	**10**
12	359 45.1	**23 19.3**	275 46.1	**12**
14	29 44.5	**23 19.1**	305 51.0	**14**
16	59 43.9	**23 18.9**	335 56.0	**16**
18	89 43.3	**23 18.7**	6 00.9	**18**
20	119 42.7	**23 18.5**	36 05.8	**20**
22	149 42.1	**S23 18.2**	66 10.8	**22**

Monday, 28th December

Time	SUN GHA ° ′	Dec ° ′	ARIES GHA ° ′	Time
00	179 41.4	**S23 18.0**	96 15.7	**00**
02	209 40.8	**23 17.8**	126 20.6	**02**
04	239 40.2	**23 17.5**	156 25.5	**04**
06	269 39.6	**23 17.3**	186 30.5	**06**
08	299 39.0	**23 17.0**	216 35.4	**08**
10	329 38.4	**23 16.8**	246 40.3	**10**
12	359 37.8	**23 16.6**	276 45.3	**12**
14	29 37.2	**23 16.3**	306 50.2	**14**
16	59 36.6	**23 16.0**	336 55.1	**16**
18	89 35.9	**23 15.8**	7 00.0	**18**
20	119 35.3	**23 15.5**	37 05.0	**20**
22	149 34.7	**S23 15.3**	67 09.9	**22**

Tuesday, 29th December

Time	SUN GHA ° ′	Dec ° ′	ARIES GHA ° ′	Time
00	179 34.1	**S23 15.0**	97 14.8	**00**
02	209 33.5	**23 14.7**	127 19.7	**02**
04	239 32.9	**23 14.4**	157 24.7	**04**
06	269 32.3	**23 14.2**	187 29.6	**06**
08	299 31.7	**23 13.9**	217 34.5	**08**
10	329 31.1	**23 13.6**	247 39.5	**10**
12	359 30.5	**23 13.3**	277 44.4	**12**
14	29 29.9	**23 13.0**	307 49.3	**14**
16	59 29.3	**23 12.7**	337 54.2	**16**
18	89 28.7	**23 12.4**	7 59.2	**18**
20	119 28.1	**23 12.1**	38 04.1	**20**
22	149 27.4	**S23 11.8**	68 09.0	**22**

Wednesday, 30th December

Time	SUN GHA ° ′	Dec ° ′	ARIES GHA ° ′	Time
00	179 26.8	**S23 11.5**	98 14.0	**00**
02	209 26.2	**23 11.2**	128 18.9	**02**
04	239 25.6	**23 10.9**	158 23.8	**04**
06	269 25.0	**23 10.6**	188 28.7	**06**
08	299 24.4	**23 10.2**	218 33.7	**08**
10	329 23.8	**23 09.9**	248 38.6	**10**
12	359 23.2	**23 09.6**	278 43.5	**12**
14	29 22.6	**23 09.3**	308 48.5	**14**
16	59 22.0	**23 08.9**	338 53.4	**16**
18	89 21.4	**23 08.6**	8 58.3	**18**
20	119 20.8	**23 08.2**	39 03.2	**20**
22	149 20.2	**S23 07.9**	69 08.2	**22**

Thursday, 31st December

Time	SUN GHA ° ′	Dec ° ′	ARIES GHA ° ′
00	179 19.6	**S23 07.6**	99 13.1
02	209 19.0	**23 07.2**	129 18.0
04	239 18.4	**23 06.9**	159 23.0
06	269 17.8	**S23 06.5**	189 27.9
08	299 17.3	**S23 06.1**	219 32.8
10	329 16.7	**23 05.8**	249 37.7
12	359 16.1	**23 05.4**	279 42.7
14	29 15.5	**S23 05.0**	309 47.6
16	59 14.9	**S23 04.7**	339 52.5
18	89 14.3	**23 04.3**	9 57.5
20	119 13.7	**23 03.9**	40 02.4
22	149 13.1	**S23 03.5**	70 07.3

DECEMBER 1998 PLANETS 0h GMT

VENUS						JUPITER				
Mer Pass h m	GHA ° ′	Mean Var/hr 14°+	Dec ° ′	Mean Var/hr ′	Day	GHA ° ′	Mean Var/hr 15°+	Dec ° ′	Mean Var/hr ′	Mer Pass h m
12 23	**174 22.1**	59.1	**S23 01.0**	0.4	**1 Tu**	**79 31.2**	2.3	**S 5 43.5**	0.1	**18 39**
12 25	**174 00.0**	59.1	**S23 10.6**	0.4	**2 Wed**	**80 27.1**	2.3	**S 5 41.8**	0.1	**18 35**
12 26	**173 37.7**	59.1	**S23 19.6**	0.3	**3 Th**	**81 22.8**	2.3	**S 5 40.1**	0.1	**18 32**
12 28	**173 15.2**	59.1	**S23 27.8**	0.3	**4 Fri**	**82 18.3**	2.3	**S 5 38.3**	0.1	**18 28**
12 29	**172 52.6**	59.0	**S23 35.4**	0.3	**5 Sat**	**83 13.6**	2.3	**S 5 36.4**	0.1	**18 24**
12 31	**172 29.8**	59.0	**S23 42.2**	0.3	**6 SUN**	**84 08.8**	2.3	**S 5 34.5**	0.1	**18 21**
12 32	**172 06.8**	59.0	**S23 48.3**	0.2	**7 Mon**	**85 03.8**	2.3	**S 5 32.5**	0.1	**18 17**
12 34	**171 43.8**	59.0	**S23 53.7**	0.2	**8 Tu**	**85 58.6**	2.3	**S 5 30.4**	0.1	**18 13**
12 35	**171 20.6**	59.0	**S23 58.3**	0.2	**9 Wed**	**86 53.3**	2.3	**S 5 28.2**	0.1	**18 10**
12 37	**170 57.4**	59.0	**S24 02.2**	0.1	**10 Th**	**87 47.7**	2.3	**S 5 26.0**	0.1	**18 06**
12 39	**170 34.1**	59.0	**S24 05.4**	0.1	**11 Fri**	**88 42.1**	2.3	**S 5 23.6**	0.1	**18 02**
12 40	**170 10.7**	59.0	**S24 07.8**	0.1	**12 Sat**	**89 36.2**	2.3	**S 5 21.3**	0.1	**17 59**
12 42	**169 47.3**	59.0	**S24 09.5**	0.0	**13 SUN**	**90 30.2**	2.2	**S 5 18.8**	0.1	**17 55**
12 43	**169 23.8**	59.0	**S24 10.4**	0.0	**14 Mon**	**91 24.0**	2.2	**S 5 16.3**	0.1	**17 52**
12 45	**169 00.4**	59.0	**S24 10.6**	0.0	**15 Tu**	**92 17.7**	2.2	**S 5 13.7**	0.1	**17 48**
12 46	**168 36.9**	59.0	**S24 10.0**	0.1	**16 Wed**	**93 11.2**	2.2	**S 5 11.1**	0.1	**17 45**
12 48	**168 13.5**	59.0	**S24 08.7**	0.1	**17 Th**	**94 04.6**	2.2	**S 5 08.4**	0.1	**17 41**
12 49	**167 50.1**	59.0	**S24 06.6**	0.1	**18 Fri**	**94 57.7**	2.2	**S 5 05.6**	0.1	**17 38**
12 51	**167 26.8**	59.0	**S24 03.8**	0.1	**19 Sat**	**95 50.8**	2.2	**S 5 02.7**	0.1	**17 34**
12 53	**167 03.5**	59.0	**S24 00.2**	0.2	**20 SUN**	**96 43.7**	2.2	**S 4 59.8**	0.1	**17 31**
12 54	**166 40.4**	59.0	**S23 55.9**	0.2	**21 Mon**	**97 36.4**	2.2	**S 4 56.8**	0.1	**17 27**
12 56	**166 17.3**	59.0	**S23 50.9**	0.2	**22 Tu**	**98 29.0**	2.2	**S 4 53.8**	0.1	**17 24**
12 57	**165 54.4**	59.1	**S23 45.1**	0.3	**23 Wed**	**99 21.4**	2.2	**S 4 50.7**	0.1	**17 20**
12 59	**165 31.6**	59.1	**S23 38.6**	0.3	**24 Th**	**100 13.7**	2.2	**S 4 47.5**	0.1	**17 17**
13 00	**165 09.0**	59.1	**S23 31.4**	0.3	**25 Fri**	**101 05.8**	2.2	**S 4 44.3**	0.1	**17 13**
13 02	**164 46.5**	59.1	**S23 23.4**	0.4	**26 Sat**	**101 57.8**	2.2	**S 4 41.0**	0.1	**17 10**
13 03	**164 24.2**	59.1	**S23 14.7**	0.4	**27 SUN**	**102 49.7**	2.2	**S 4 37.7**	0.1	**17 06**
13 05	**164 02.1**	59.1	**S23 05.3**	0.4	**28 Mon**	**103 41.4**	2.2	**S 4 34.3**	0.1	**17 03**
13 06	**163 40.3**	59.1	**S22 55.2**	0.4	**29 Tu**	**104 33.0**	2.1	**S 4 30.8**	0.1	**16 59**
13 08	**163 18.6**	59.1	**S22 44.5**	0.5	**30 Wed**	**105 24.5**	2.1	**S 4 27.3**	0.2	**16 56**
13 09	**162 57.2**	59.1	**S22 33.0**	0.5	**31 Th**	**106 15.8**	2.1	**S 4 23.7**	0.1	**16 53**

VENUS, Av. Mag. -3.9
S.H.A. December
5 99; 10 92; 15 86; 20 79; 25 72; 30 65.

JUPITER, Av. Mag.-2.4
S.H.A. December
5 10; 10 9; 15 9; 20 8; 25 8; 30 7.

MARS						SATURN				
Mer Pass h m	GHA ° ′	Mean Var/hr 15°+	Dec ° ′	Mean Var/hr ′	Day	GHA ° ′	Mean Var/hr 15°+	Dec ° ′	Mean Var/hr ′	Mer Pass h m
07 31	**247 08.5**	1.2	**N 0 45.8**	0.5	**1 Tu**	**43 08.5**	2.6	**N 8 04.7**	0.0	**21 04**
07 29	**247 36.9**	1.2	**N 0 32.9**	0.5	**2 Wed**	**44 10.5**	2.6	**N 8 03.9**	0.0	**21 00**
07 27	**248 05.4**	1.2	**N 0 20.1**	0.5	**3 Th**	**45 12.4**	2.6	**N 8 03.1**	0.0	**20 56**
07 25	**248 34.0**	1.2	**N 0 07.3**	0.5	**4 Fri**	**46 14.2**	2.6	**N 8 02.3**	0.0	**20 51**
07 23	**249 02.7**	1.2	**S 0 05.4**	0.5	**5 Sat**	**47 15.9**	2.6	**N 8 01.6**	0.0	**20 47**
07 21	**249 31.4**	1.2	**S 0 18.1**	0.5	**6 SUN**	**48 17.6**	2.6	**N 8 00.9**	0.0	**20 43**
07 19	**250 00.2**	1.2	**S 0 30.8**	0.5	**7 Mon**	**49 19.1**	2.6	**N 8 00.3**	0.0	**20 39**
07 17	**250 29.1**	1.2	**S 0 43.4**	0.5	**8 Tu**	**50 20.6**	2.6	**N 7 59.6**	0.0	**20 35**
07 16	**250 58.1**	1.2	**S 0 56.0**	0.5	**9 Wed**	**51 21.9**	2.6	**N 7 59.1**	0.0	**20 31**
07 14	**251 27.2**	1.2	**S 1 08.5**	0.5	**10 Th**	**52 23.2**	2.6	**N 7 58.5**	0.0	**20 27**
07 12	**251 56.3**	1.2	**S 1 20.9**	0.5	**11 Fri**	**53 24.4**	2.5	**N 7 58.0**	0.0	**20 23**
07 10	**252 25.6**	1.2	**S 1 33.3**	0.5	**12 Sat**	**54 25.5**	2.5	**N 7 57.6**	0.0	**20 19**
07 08	**252 54.9**	1.2	**S 1 45.6**	0.5	**13 SUN**	**55 26.5**	2.5	**N 7 57.1**	0.0	**20 15**
07 06	**253 24.3**	1.2	**S 1 57.9**	0.5	**14 Mon**	**56 27.4**	2.5	**N 7 56.8**	0.0	**20 11**
07 04	**253 53.9**	1.2	**S 2 10.1**	0.5	**15 Tu**	**57 28.2**	2.5	**N 7 56.4**	0.0	**20 07**
07 02	**254 23.5**	1.2	**S 2 22.3**	0.5	**16 Wed**	**58 28.9**	2.5	**N 7 56.1**	0.0	**20 03**
07 00	**254 53.2**	1.2	**S 2 34.4**	0.5	**17 Th**	**59 29.5**	2.5	**N 7 55.8**	0.0	**19 59**
06 58	**255 23.0**	1.2	**S 2 46.4**	0.5	**18 Fri**	**60 30.0**	2.5	**N 7 55.6**	0.0	**19 55**
06 56	**255 52.9**	1.3	**S 2 58.4**	0.5	**19 Sat**	**61 30.4**	2.5	**N 7 55.4**	0.0	**19 51**
06 54	**256 22.9**	1.3	**S 3 10.2**	0.5	**20 SUN**	**62 30.7**	2.5	**N 7 55.3**	0.0	**19 47**
06 52	**256 53.0**	1.3	**S 3 22.1**	0.5	**21 Mon**	**63 30.9**	2.5	**N 7 55.2**	0.0	**19 43**
06 50	**257 23.2**	1.3	**S 3 33.8**	0.5	**22 Tu**	**64 31.0**	2.5	**N 7 55.1**	0.0	**19 39**
06 48	**257 53.5**	1.3	**S 3 45.5**	0.5	**23 Wed**	**65 30.9**	2.5	**N 7 55.1**	0.0	**19 35**
06 46	**258 23.9**	1.3	**S 3 57.1**	0.5	**24 Th**	**66 30.8**	2.5	**N 7 55.1**	0.0	**19 31**
06 44	**258 54.4**	1.3	**S 4 08.6**	0.5	**25 Fri**	**67 30.6**	2.5	**N 7 55.2**	0.0	**19 27**
06 42	**259 25.1**	1.3	**S 4 20.0**	0.5	**26 Sat**	**68 30.3**	2.5	**N 7 55.3**	0.0	**19 23**
06 40	**259 55.8**	1.3	**S 4 31.4**	0.5	**27 SUN**	**69 29.9**	2.5	**N 7 55.4**	0.0	**19 19**
06 38	**260 26.7**	1.3	**S 4 42.7**	0.5	**28 Mon**	**70 29.3**	2.5	**N 7 55.6**	0.0	**19 15**
06 36	**260 57.7**	1.3	**S 4 53.8**	0.5	**29 Tu**	**71 28.7**	2.5	**N 7 55.8**	0.0	**19 11**
06 34	**261 28.8**	1.3	**S 5 05.0**	0.5	**30 Wed**	**72 28.0**	2.5	**N 7 56.1**	0.0	**19 07**
06 31	**262 00.0**	1.3	**S 5 16.0**	0.5	**31 Th**	**73 27.1**	2.5	**N 7 56.4**	0.0	**19 03**

MARS, Av. Mag. +1.2
S.H.A. December
5 175; 10 173; 15 170; 20 168; 25 166; 30 163.

NOTE: Planet corrections on pages E113-15

SATURN, Av. Mag. +0.3
S.H.A. December
5 334; 10 334; 15 334; 20 334; 25 334; 30 334.

DECEMBER 1998 — MOON

Day	GMT hr	GHA ° ′	Mean Var/hr 14°+	Dec ° ′	Mean Var/hr ′
1 Tu	0	38 52.0	26.4	N 7 38.2	11.3
	6	125 30.8	26.1	N 8 46.5	11.0
	12	212 07.6	25.7	N 9 53.1	10.7
	18	298 42.2	25.4	N10 57.5	10.3
2 Wed	0	25 14.8	25.1	N11 59.3	9.8
	6	111 45.4	24.7	N12 58.4	9.3
	12	198 14.1	24.5	N13 54.3	8.7
	18	284 40.9	24.1	N14 46.7	8.1
3 Th	0	11 06.0	23.9	N15 35.4	7.4
	6	97 29.7	23.7	N16 20.0	6.7
	12	183 52.1	23.6	N17 00.4	5.9
	18	270 13.5	23.4	N17 36.2	5.2
4 Fri	0	356 34.3	23.3	N18 07.4	4.4
	6	82 54.6	23.4	N18 33.8	3.5
	12	169 14.9	23.4	N18 55.3	2.7
	18	255 35.5	23.5	N19 11.8	1.9
5 Sat	0	341 56.7	23.7	N19 23.3	1.0
	6	68 18.9	24.0	N19 29.8	0.1
	12	154 42.5	24.2	N19 31.3	0.6
	18	241 07.7	24.5	N19 28.1	1.4
6 Sun	0	327 35.0	24.9	N19 20.1	2.2
	6	54 04.4	25.4	N19 07.5	2.9
	12	140 36.3	25.8	N18 50.6	3.6
	18	227 10.9	26.3	N18 29.4	4.3
7 Mon	0	313 48.4	26.8	N18 04.3	4.9
	6	40 28.8	27.3	N17 35.5	5.4
	12	127 12.3	27.8	N17 03.1	6.0
	18	213 58.9	28.4	N16 27.4	6.5
8 Tu	0	300 48.6	28.9	N15 48.7	6.9
	6	27 41.4	29.3	N15 07.2	7.4
	12	114 37.3	29.8	N14 23.2	7.7
	18	201 36.2	30.3	N13 36.7	8.1
9 Wed	0	288 37.9	30.8	N12 48.2	8.4
	6	15 42.5	31.2	N11 57.8	8.7
	12	102 49.7	31.6	N11 05.6	9.0
	18	189 59.4	32.1	N10 12.0	9.2
10 Th	0	277 11.5	32.4	N 9 17.0	9.4
	6	4 25.8	32.7	N 8 20.9	9.5
	12	91 42.2	33.1	N 7 23.8	9.7
	18	179 00.3	33.3	N 6 25.9	9.8
11 Fri	0	266 20.2	33.6	N 5 27.3	9.8
	6	353 41.5	33.8	N 4 28.2	9.9
	12	81 04.2	33.9	N 3 28.8	10.0
	18	168 28.0	34.2	N 2 29.0	10.0
12 Sat	0	255 52.7	34.3	N 1 29.1	10.0
	6	343 18.2	34.4	N 0 29.3	10.0
	12	70 44.3	34.5	S 0 30.5	9.9
	18	158 10.8	34.5	S 1 30.0	9.9
13 Sun	0	245 37.6	34.5	S 2 29.2	9.8
	6	333 04.4	34.5	S 3 27.9	9.7
	12	60 31.2	34.4	S 4 26.0	9.6
	18	147 57.7	34.4	S 5 23.5	9.4
14 Mon	0	235 23.9	34.3	S 6 20.1	9.2
	6	322 49.4	34.1	S 7 15.8	9.1
	12	50 14.3	34.0	S 8 10.5	8.9
	18	137 38.4	33.9	S 9 04.0	8.7
15 Tu	0	225 01.6	33.6	S 9 56.2	8.4
	6	312 23.7	33.4	S10 47.0	8.2
	12	39 44.6	33.2	S11 36.3	7.9
	18	127 04.2	33.0	S12 23.9	7.6
16 Wed	0	214 22.5	32.8	S13 09.7	7.3
	6	301 39.4	32.6	S13 53.6	6.9
	12	28 54.8	32.3	S14 35.4	6.6
	18	116 08.7	32.0	S15 15.1	6.2
17 Th	0	203 21.0	31.8	S15 52.4	5.7
	6	290 31.6	31.5	S16 27.3	5.4
	12	17 40.7	31.2	S16 59.6	4.9
	18	104 48.3	31.0	S17 29.1	4.4
18 Fri	0	191 54.2	30.7	S17 55.8	3.9
	6	278 58.7	30.5	S18 19.6	3.4
	12	6 01.7	30.3	S18 40.2	2.8
	18	93 03.3	30.1	S18 57.7	2.3
19 Sat	0	180 03.6	29.8	S19 11.9	1.8
	6	267 02.8	29.6	S19 22.7	1.2
	12	354 00.9	29.5	S19 30.1	0.6
	18	80 58.1	29.4	S19 33.9	0.0
20 Sun	0	167 54.4	29.3	S19 34.2	0.6
	6	254 50.1	29.1	S19 30.8	1.2
	12	341 45.3	29.1	S19 23.8	1.8
	18	68 40.1	29.0	S19 13.1	2.4
21 Mon	0	155 34.6	29.1	S18 58.8	3.1
	6	242 29.0	29.1	S18 40.9	3.6
	12	329 23.5	29.1	S18 19.3	4.3
	18	56 18.0	29.2	S17 54.2	4.8
22 Tu	0	143 12.9	29.2	S17 25.6	5.4
	6	230 08.0	29.3	S16 53.5	6.0
	12	317 03.5	29.3	S16 18.2	6.4
	18	43 59.5	29.4	S15 39.5	7.0
23 Wed	0	130 56.0	29.5	S14 57.8	7.5
	6	217 53.0	29.6	S14 13.0	8.0
	12	304 50.6	29.7	S13 25.4	8.4
	18	31 48.6	29.8	S12 35.0	8.8
24 Th	0	118 47.2	29.8	S11 42.0	9.3
	6	205 46.1	29.8	S10 46.6	9.6
	12	292 45.5	30.0	S 9 48.9	10.0
	18	19 45.1	29.9	S 8 49.0	10.4
25 Fri	0	106 44.9	30.0	S 7 47.2	10.6
	6	193 44.8	30.0	S 6 43.7	10.9
	12	280 44.6	29.9	S 5 38.6	11.1
	18	7 44.3	29.9	S 4 32.1	11.3
26 Sat	0	94 43.6	29.8	S 3 24.4	11.4
	6	181 42.5	29.7	S 2 15.7	11.5
	12	268 40.8	29.6	S 1 06.2	11.7
	18	355 38.3	29.5	N 0 03.8	11.7
27 Sun	0	82 34.9	29.2	N 1 14.1	11.7
	6	169 30.4	29.0	N 2 24.5	11.7
	12	256 24.6	28.8	N 3 34.7	11.7
	18	343 17.4	28.5	N 4 44.5	11.6
28 Mon	0	70 08.7	28.2	N 5 53.6	11.4
	6	156 58.3	27.9	N 7 01.8	11.2
	12	243 46.0	27.6	N 8 08.7	10.9
	18	330 31.9	27.3	N 9 14.1	10.6
29 Tu	0	57 15.7	26.9	N10 17.8	10.3
	6	143 57.4	26.5	N11 19.3	9.8
	12	230 37.0	26.2	N12 18.5	9.4
	18	317 14.5	25.9	N13 15.0	8.9
30 Wed	0	43 49.8	25.5	N14 08.5	8.4
	6	130 23.1	25.2	N14 58.8	7.7
	12	216 54.4	24.9	N15 45.5	7.1
	18	303 23.9	24.6	N16 28.5	6.4
31 Th	0	29 51.7	24.4	N17 07.4	5.7
	6	116 18.0	24.1	N17 42.1	5.0
	12	202 43.1	24.0	N18 12.4	4.3
	18	289 07.2	23.9	N18 38.1	3.4

NOTE: Moon GHA corrections for additional hours, minutes, and seconds are on page E 108–110. Moon declination corrections are on page E 111. The mean var/hr value is needed for both corrections.

POLARIS (POLE STAR) TABLE, 1998

LHA Aries	Q	LHA Aries	Q	LHA Aries	Q	LHA Aries	Q	LHA Aries	Q	LHA Aries	Q	LHA Aries	Q	LHA Aries	Q
° ′	′	° ′	′	° ′	′	° ′	′	° ′	′	° ′	′	° ′	′	° ′	′
358 33	- 35	**84 13**	- 30	**120 11**	- 5	**153 14**	+ 20	**213 47**	+ 45	**280 36**	+ 20	**313 49**	- 5	**349 20**	- 30
0 37	- 36	**85 57**	- 29	**121 28**	- 4	**154 41**	+ 21	**221 30**	+ 44	**282 03**	+ 19	**315 06**	- 6	**351 04**	- 31
2 47	- 37	**87 38**	- 28	**122 46**	- 3	**156 09**	+ 22	**230 28**	+ 43	**283 28**	+ 18	**316 24**	- 7	**352 51**	- 32
5 05	- 38	**89 16**	- 27	**124 03**	- 2	**157 37**	+ 23	**235 24**	+ 42	**284 52**	+ 17	**317 42**	- 8	**354 41**	- 33
7 31	- 39	**90 53**	- 26	**125 20**	- 1	**159 08**	+ 24	**239 16**	+ 41	**286 16**	+ 16	**319 00**	- 9	**356 35**	- 34
10 09	- 40	**92 27**	- 25	**126 37**	0	**160 40**	+ 25	**242 33**	+ 40	**287 39**	+ 15	**320 19**	- 10	**358 33**	- 35
13 03	- 41	**94 00**	- 24	**127 55**	+ 1	**162 13**	+ 26	**245 29**	+ 39	**289 01**	+ 14	**321 38**	- 11	**0 37**	- 36
16 18	- 42	**95 31**	- 23	**129 12**	+ 2	**163 48**	+ 27	**248 08**	+ 38	**290 22**	+ 13	**322 57**	- 12	**2 47**	- 37
20 07	- 43	**97 01**	- 22	**130 30**	+ 3	**165 25**	+ 28	**250 37**	+ 37	**291 43**	+ 12	**324 17**	- 13	**5 05**	- 38
24 59	- 44	**98 29**	- 21	**131 47**	+ 4	**167 05**	+ 29	**252 56**	+ 36	**293 03**	+ 11	**325 37**	- 14	**7 31**	- 39
33 50	- 45	**99 56**	- 20	**133 04**	+ 5	**168 47**	+ 30	**255 07**	+ 35	**294 23**	+ 10	**326 58**	- 15	**10 09**	- 40
41 27	- 44	**101 22**	- 19	**134 22**	+ 6	**170 31**	+ 31	**257 12**	+ 34	**295 42**	+ 9	**328 20**	- 16	**13 03**	- 41
50 18	- 43	**102 47**	- 18	**135 40**	+ 7	**172 19**	+ 32	**259 12**	+ 33	**297 01**	+ 8	**329 42**	- 17	**16 18**	- 42
55 10	- 42	**104 11**	- 17	**136 58**	+ 8	**174 10**	+ 33	**261 07**	+ 32	**298 19**	+ 7	**331 06**	- 18	**20 07**	- 43
58 59	- 41	**105 35**	- 16	**138 16**	+ 9	**176 05**	+ 34	**262 58**	+ 31	**299 37**	+ 6	**332 30**	- 19	**24 59**	- 44
62 14	- 40	**106 57**	- 15	**139 35**	+ 10	**178 05**	+ 35	**264 46**	+ 30	**300 55**	+ 5	**333 55**	- 20	**33 50**	- 45
65 08	- 39	**108 19**	- 14	**140 54**	+ 11	**180 10**	+ 36	**266 30**	+ 29	**302 13**	+ 4	**335 21**	- 21	**41 27**	- 44
67 46	- 38	**109 40**	- 13	**142 14**	+ 12	**182 21**	+ 37	**268 12**	+ 28	**303 30**	+ 3	**336 48**	- 22	**50 18**	- 43
70 12	- 37	**111 00**	- 12	**143 34**	+ 13	**184 40**	+ 38	**269 52**	+ 27	**304 47**	+ 2	**338 16**	- 23	**55 10**	- 42
72 30	- 36	**112 20**	- 11	**144 55**	+ 14	**187 09**	+ 39	**271 29**	+ 26	**306 05**	+ 1	**339 46**	- 24	**58 59**	- 41
74 40	- 35	**113 39**	- 10	**146 16**	+ 15	**189 48**	+ 40	**273 04**	+ 25	**307 22**	0	**341 17**	- 25	**62 14**	- 40
76 44	- 34	**114 58**	- 9	**147 38**	+ 16	**192 44**	+ 41	**274 37**	+ 24	**308 40**	- 1	**342 50**	- 26	**65 08**	- 39
78 42	- 33	**116 17**	- 8	**149 01**	+ 17	**196 01**	+ 42	**276 09**	+ 23	**309 57**	- 2	**344 24**	- 27	**67 46**	- 38
80 36	- 32	**117 35**	- 7	**150 25**	+ 18	**199 53**	+ 43	**277 40**	+ 22	**311 14**	- 3	**346 01**	- 28	**70 12**	- 37
82 26	- 31	**118 53**	- 6	**151 49**	+ 19	**204 49**	+ 44	**279 08**	+ 21	**312 31**	- 4	**347 39**	- 29	**72 30**	- 36
84 13		**120 11**		**153 14**		**213 47**		**280 36**		**313 49**		**349 20**		**74 40**	

NOTE: The table above can be used to find latitude directly from an observation of Polaris. GHA of Aries (from the monthly pages) – assumed West longitude (or + East) = LHA Aries. Apply correction Q to the observed sextant altitude (Ho) (which is the sextant altitude [Hs] corrected using the table on page E 108). In critical cases (when LHA Aries equals the number shown in the table), use the higher Q value.

AZIMUTH OF POLARIS, 1998

LHA Aries	Latitude (N) 0°	30°	50°	55°	60°	65°	70°	LHA Aries	Latitude (N) 0°	30°	50°	55°	60°	65°	70°
°	°	°	°	°	°	°	°	°	°	°	°	°	°	°	°
0	0.5	0.5	0.7	0.8	0.9	1.1	1.4	**180**	359.5	359.5	359.3	359.2	359.1	358.9	358.7
10	0.3	0.4	0.5	0.6	0.7	0.8	1.0	**190**	359.7	359.6	359.5	359.4	359.3	359.2	359.0
20	0.2	0.3	0.4	0.4	0.5	0.5	0.7	**200**	359.8	359.7	359.7	359.6	359.6	359.5	359.4
30	0.1	0.1	0.2	0.2	0.2	0.2	0.3	**210**	359.9	359.9	359.8	359.8	359.8	359.8	359.7
40	0.0	0.0	0.0	359.9	359.9	359.9	359.9	**220**	0.0	0.0	0.0	0.1	0.1	0.1	0.1
50	359.8	359.8	359.7	359.7	359.7	359.6	359.5	**230**	0.2	0.2	0.2	0.3	0.3	0.4	0.4
60	359.7	359.7	359.6	359.5	359.4	359.3	359.1	**240**	0.3	0.3	0.4	0.5	0.6	0.7	0.8
70	359.6	359.5	359.4	359.3	359.2	359.0	358.8	**250**	0.4	0.5	0.6	0.7	0.8	0.9	1.1
80	359.5	359.4	359.2	359.1	359.0	358.8	358.5	**260**	0.5	0.6	0.8	0.9	1.0	1.2	1.4
90	359.4	359.3	359.1	359.0	358.8	358.6	358.2	**270**	0.6	0.7	0.9	1.0	1.2	1.4	1.7
100	359.3	359.2	359.0	358.8	358.7	358.4	358.0	**280**	0.7	0.8	1.0	1.1	1.3	1.5	1.9
110	359.3	359.2	358.9	358.8	358.6	358.3	357.9	**290**	0.7	0.8	1.1	1.2	1.4	1.7	2.0
120	359.3	359.1	358.9	358.7	358.5	358.3	357.8	**300**	0.7	0.8	1.1	1.3	1.5	1.7	2.1
130	359.3	359.1	358.8	358.7	358.5	358.2	357.8	**310**	0.7	0.9	1.2	1.3	1.5	1.8	2.2
140	359.3	359.2	358.9	358.7	358.6	358.3	357.9	**320**	0.7	0.8	1.1	1.3	1.5	1.7	2.1
150	359.3	359.2	358.9	358.8	358.6	358.4	358.0	**330**	0.7	0.8	1.1	1.2	1.4	1.6	2.0
160	359.4	359.3	359.0	358.9	358.8	358.5	358.2	**340**	0.6	0.7	1.0	1.1	1.3	1.5	1.9
170	359.5	359.4	359.2	359.1	358.9	358.7	358.4	**350**	0.5	0.6	0.9	1.0	1.1	1.3	1.6
180	359.5	359.5	359.3	359.2	359.1	358.9	358.7	**360**	0.5	0.5	0.7	0.8	0.9	1.1	1.4

NOTE: When Cassiopeia is left (right), Polaris is west (east)

ARC-TO-TIME CONVERSION TABLE

Arc	Time	Arc	Time	Arc	Time	Arc	Time	Arc	Time	Arc	Time	Arc	Time	Arc	Time
°	h m	°	h m	°	h m	°	h m	°	h m	°	h m	′	m s	″ ′	s
0	0 00	**60**	4 00	**120**	8 00	**180**	12 00	**240**	16 00	**300**	20 00	**0**	0 00	**0=0.0**	0.00
1	0 04	**61**	4 04	**121**	8 04	**181**	12 04	**241**	16 04	**301**	20 04	**1**	0 04	**1**	0.07
2	0 08	**62**	4 08	**122**	8 08	**182**	12 08	**242**	16 08	**302**	20 08	**2**	0 08	**2**	0.13
3	0 12	**63**	4 12	**123**	8 12	**183**	12 12	**243**	16 12	**303**	20 12	**3**	0 12	**3**	0.20
4	0 16	**64**	4 16	**124**	8 16	**184**	12 16	**244**	16 16	**304**	20 16	**4**	0 16	**4**	0.27
5	0 20	**65**	4 20	**125**	8 20	**185**	12 20	**245**	16 20	**305**	20 20	**5**	0 20	**5**	0.33
6	0 24	**66**	4 24	**126**	8 24	**186**	12 24	**246**	16 24	**306**	20 24	**6**	0 24	**6=0.1**	0.40
7	0 28	**67**	4 28	**127**	8 28	**187**	12 28	**247**	16 28	**307**	20 28	**7**	0 28	**7**	0.47
8	0 32	**68**	4 32	**128**	8 32	**188**	12 32	**248**	16 32	**308**	20 32	**8**	0 32	**8**	0.53
9	0 36	**69**	4 36	**129**	8 36	**189**	12 36	**249**	16 36	**309**	20 36	**9**	0 36	**9**	0.60
10	0 40	**70**	4 40	**130**	8 40	**190**	12 40	**250**	16 40	**310**	20 40	**10**	0 40	**10**	0.67
11	0 44	**71**	4 44	**131**	8 44	**191**	12 44	**251**	16 44	**311**	20 44	**11**	0 44	**11**	0.73
12	0 48	**72**	4 48	**132**	8 48	**192**	12 48	**252**	16 48	**312**	20 48	**12**	0 48	**12=0.2**	0.80
13	0 52	**73**	4 52	**133**	8 52	**193**	12 52	**253**	16 52	**313**	20 52	**13**	0 52	**13**	0.87
14	0 56	**74**	4 56	**134**	8 56	**194**	12 56	**254**	16 56	**314**	20 56	**14**	0 56	**14**	0.93
15	1 00	**75**	5 00	**135**	9 00	**195**	13 00	**255**	17 00	**315**	21 00	**15**	1 00	**15**	1.00
16	1 04	**76**	5 04	**136**	9 04	**196**	13 04	**256**	17 04	**316**	21 04	**16**	1 04	**16**	1.07
17	1 08	**77**	5 08	**137**	9 08	**197**	13 08	**257**	17 08	**317**	21 08	**17**	1 08	**17**	1.13
18	1 12	**78**	5 12	**138**	9 12	**198**	13 12	**258**	17 12	**318**	21 12	**18**	1 12	**18=0.3**	1.20
19	1 16	**79**	5 16	**139**	9 16	**199**	13 16	**259**	17 16	**319**	21 16	**19**	1 16	**19**	1.27
20	1 20	**80**	5 20	**140**	9 20	**200**	13 20	**260**	17 20	**320**	21 20	**20**	1 20	**20**	1.33
21	1 24	**81**	5 24	**141**	9 24	**201**	13 24	**261**	17 24	**321**	21 24	**21**	1 24	**21**	1.40
22	1 28	**82**	5 28	**142**	9 28	**202**	13 28	**262**	17 28	**322**	21 28	**22**	1 28	**22**	1.47
23	1 32	**83**	5 32	**143**	9 32	**203**	13 32	**263**	17 32	**323**	21 32	**23**	1 32	**23**	1.53
24	1 36	**84**	5 36	**144**	9 36	**204**	13 36	**264**	17 36	**324**	21 36	**24**	1 36	**24=0.4**	1.60
25	1 40	**85**	5 40	**145**	9 40	**205**	13 40	**265**	17 40	**325**	21 40	**25**	1 40	**25**	1.67
26	1 44	**86**	5 44	**146**	9 44	**206**	13 44	**266**	17 44	**326**	21 44	**26**	1 44	**26**	1.73
27	1 48	**87**	5 48	**147**	9 48	**207**	13 48	**267**	17 48	**327**	21 48	**27**	1 48	**27**	1.80
28	1 52	**88**	5 52	**148**	9 52	**208**	13 52	**268**	17 52	**328**	21 52	**28**	1 52	**28**	1.87
29	1 56	**89**	5 56	**149**	9 56	**209**	13 56	**269**	17 56	**329**	21 56	**29**	1 56	**29**	1.93
30	2 00	**90**	6 00	**150**	10 00	**210**	14 00	**270**	18 00	**330**	22 00	**30**	2 00	**30=0.5**	2.00
31	2 04	**91**	6 04	**151**	10 04	**211**	14 04	**271**	18 04	**331**	22 04	**31**	2 04	**31**	2.07
32	2 08	**92**	6 08	**152**	10 08	**212**	14 08	**272**	18 08	**332**	22 08	**32**	2 08	**32**	2.13
33	2 12	**93**	6 12	**153**	10 12	**213**	14 12	**273**	18 12	**333**	22 12	**33**	2 12	**33**	2.20
34	2 16	**94**	6 16	**154**	10 16	**214**	14 16	**274**	18 16	**334**	22 16	**34**	2 16	**34**	2.27
35	2 20	**95**	6 20	**155**	10 20	**215**	14 20	**275**	18 20	**335**	22 20	**35**	2 20	**35**	2.33
36	2 24	**96**	6 24	**156**	10 24	**216**	14 24	**276**	18 24	**336**	22 24	**36**	2 24	**36=0.6**	2.40
37	2 28	**97**	6 28	**157**	10 28	**217**	14 28	**277**	18 28	**337**	22 28	**37**	2 28	**37**	2.47
38	2 32	**98**	6 32	**158**	10 32	**218**	14 32	**278**	18 32	**338**	22 32	**38**	2 32	**38**	2.53
39	2 36	**99**	6 36	**159**	10 36	**219**	14 36	**279**	18 36	**339**	22 36	**39**	2 36	**39**	2.60
40	2 40	**100**	6 40	**160**	10 40	**220**	14 40	**280**	18 40	**340**	22 40	**40**	2 40	**40**	2.67
41	2 44	**101**	6 44	**161**	10 44	**221**	14 44	**281**	18 44	**341**	22 44	**41**	2 44	**41**	2.73
42	2 48	**102**	6 48	**162**	10 48	**222**	14 48	**282**	18 48	**342**	22 48	**42**	2 48	**42=0.7**	2.80
43	2 52	**103**	6 52	**163**	10 52	**223**	14 52	**283**	18 52	**343**	22 52	**43**	2 52	**43**	2.87
44	2 56	**104**	6 56	**164**	10 56	**224**	14 56	**284**	18 56	**344**	22 56	**44**	2 56	**44**	2.93
45	3 00	**105**	7 00	**165**	11 00	**225**	15 00	**285**	19 00	**345**	23 00	**45**	3 00	**45**	3.00
46	3 04	**106**	7 04	**166**	11 04	**226**	15 04	**286**	19 04	**346**	23 04	**46**	3 04	**46**	3.07
47	3 08	**107**	7 08	**167**	11 08	**227**	15 08	**287**	19 08	**347**	23 08	**47**	3 08	**47**	3.13
48	3 12	**108**	7 12	**168**	11 12	**228**	15 12	**288**	19 12	**348**	23 12	**48**	3 12	**48=0.8**	3.20
49	3 16	**109**	7 16	**169**	11 16	**229**	15 16	**289**	19 16	**349**	23 16	**49**	3 16	**49**	3.27
50	3 20	**110**	7 20	**170**	11 20	**230**	15 20	**290**	19 20	**350**	23 20	**50**	3 20	**50**	3.33
51	3 24	**111**	7 24	**171**	11 24	**231**	15 24	**291**	19 24	**351**	23 24	**51**	3 24	**51**	3.40
52	3 28	**112**	7 28	**172**	11 28	**232**	15 28	**292**	19 28	**352**	23 28	**52**	3 28	**52**	3.47
53	3 32	**113**	7 32	**173**	11 32	**233**	15 32	**293**	19 32	**353**	23 32	**53**	3 32	**53**	3.53
54	3 36	**114**	7 36	**174**	11 36	**234**	15 36	**294**	19 36	**354**	23 36	**54**	3 36	**54=0.9**	3.60
55	3 40	**115**	7 40	**175**	11 40	**235**	15 40	**295**	19 40	**355**	23 40	**55**	3 40	**55**	3.67
56	3 44	**116**	7 44	**176**	11 44	**236**	15 44	**296**	19 44	**356**	23 44	**56**	3 44	**56**	3.73
57	3 48	**117**	7 48	**177**	11 48	**237**	15 48	**297**	19 48	**357**	23 48	**57**	3 48	**57**	3.80
58	3 52	**118**	7 52	**178**	11 52	**238**	15 52	**298**	19 52	**358**	23 52	**58**	3 52	**58**	3.87
59	3 56	**119**	7 56	**179**	11 56	**239**	15 56	**299**	19 56	**359**	23 56	**59**	3 56	**59**	3.93
60	4 00	**120**	8 00	**180**	12 00	**240**	16 00	**300**	20 00	**360**	24 00	**60**	4 00	**60=1.0**	4.00

REFRACTION, DIP, AND SUN PARALLAX TABLES

Mean Refraction
Subtract

App Alt ° ′	Refr ′	App Alt ° ′	Refr ′
0 00	34.9	**10 00**	5.3
10	32.8	**10**	5.2
20	30.9	**20**	5.1
30	29.1	**30**	5.0
40	27.4	**40**	5.0
50	25.8	**50**	4.9
1 00	24.4	**11 00**	4.8
10	23.1	**10**	4.7
20	21.9	**20**	4.7
30	20.9	**30**	4.6
40	19.9	**40**	4.5
50	19.0	**50**	4.5
2 00	18.1	**12 00**	4.4
10	17.4	**10**	4.4
20	16.7	**20**	4.3
30	16.0	**30**	4.2
40	15.4	**40**	4.2
50	14.8	**50**	4.1
3 00	14.2	**13 00**	4.1
10	13.7	**10**	4.0
20	13.3	**20**	4.0
30	12.8	**30**	3.9
40	12.4	**40**	3.9
50	12.0	**50**	3.8
4 00	11.7	**14 00**	3.8
10	11.3	**20**	3.7
20	11.0	**40**	3.6
30	10.7	**15 00**	3.5
40	10.4	**30**	3.4
50	10.1	**16 00**	3.3
5 00	9.8	**30**	3.2
10	9.5	**17 00**	3.1
20	9.3	**30**	3.0
30	9.0	**18 00**	2.9
40	8.8	**30**	2.9
50	8.6	**19 00**	2.8
6 00	8.4	**20 00**	2.6
10	8.2	**21 00**	2.5
20	8.0	**22 00**	2.4
30	7.8	**23 00**	2.3
40	7.7	**24 00**	2.2
50	7.5	**26 00**	2.0
7 00	7.3	**28 00**	1.8
10	7.2	**30 00**	1.7
20	7.0	**32 00**	1.5
30	6.9	**34 00**	1.4
40	6.8	**36 00**	1.3
50	6.6	**38 00**	1.2
8 00	6.5	**40 00**	1.1
10	6.4	**43 00**	1.0
20	6.3	**46 00**	0.9
30	6.1	**50 00**	0.8
40	6.0	**55 00**	0.7
50	5.9	**60 00**	0.6
9 00	5.8	**65 00**	0.5
10	5.7	**70 00**	0.4
20	5.6	**75 00**	0.3
30	5.5	**80 00**	0.2
40	5.4	**85 00**	0.1
50	5.4	**90 00**	0.0

Dip of Sea Horizon
Subtract

HE ft	Dip ′	HE m
2	1.5	**0.6**
3	1.8	**0.9**
4	2.0	**1.2**
5	2.2	**1.5**
6	2.4	**1.8**
7	2.6	**2.1**
8	2.8	**2.4**
9	2.9	**2.7**
10	3.1	**3.0**
11	3.3	**3.4**
12	3.4	**3.7**
13	3.5	**4.0**
14	3.7	**4.3**
15	3.8	**4.6**
16	3.9	**4.9**
17	4.0	**5.2**
18	4.2	**5.5**
19	4.3	**5.8**
20	4.4	**6.1**
22	4.6	**6.7**
24	4.8	**7.3**
26	5.0	**7.9**
28	5.2	**8.5**
30	5.4	**9.1**
32	5.5	**9.8**
34	5.7	**10.4**
36	5.9	**11.0**
38	6.0	**11.6**
40	6.2	**12.2**
42	6.4	**12.8**
44	6.5	**13.4**
46	6.7	**14.0**
48	6.8	**14.6**
50	6.9	**15.2**
52	7.1	**15.9**
54	7.2	**16.5**
56	7.3	**17.0**
58	7.5	**17.7**
60	7.6	**18.3**
65	7.9	**19.8**
70	8.2	**21.3**
80	8.8	**24.4**
90	9.3	**27.4**
100	9.8	**30.5**
120	10.7	**36.6**
140	11.6	**42.7**
160	12.4	**49.8**
180	13.2	**54.9**
200	13.7	**61.0**
220	14.5	**67.1**
240	15.2	**73.2**
260	15.8	**79.3**
280	16.4	**85.3**
300	17.0	**91.4**
350	18.3	**107**
400	19.6	**122**
450	20.8	**137**
500	21.9	**152**

Sun's Parallax in Altitude
Add

App Alt °	Parallax ′
0	0.15
5	0.15
10	0.14
15	0.14
20	0.14
25	0.13
30	0.12
40	0.11
50	0.10
60	0.08
70	0.06
80	0.03
90	0.00

NOTES

The tables on this page are primarily useful for correcting sun sights with sextant altitudes of less than 9°.

For more conventional sun sights, you will find the table on the opposite page much quicker.

To use this table, first correct your sextant altitude (Hs) for dip, and then use the resulting apparent altitude (Ha) to figure corrections for refraction and parallax. Finally, find the sun's semi-diameter from the first page of the appropriate monthly table. If you shot the lower limb of the sun, add it; if you shot the upper limb, subtract it.

SUN ALTITUDE TOTAL CORRECTION TABLE (LOWER LIMB)

Always add

	Height of Eye above Sea Level															
m	0.9	1.8	2.4	3.0	3.7	4.3	4.9	5.5	6.0	7.6	9.0	12	15	18	21	24
ft	3	6	8	10	12	14	16	18	20	25	30	40	50	60	70	80
Sextant Altitude (°)	′	′	′	′	′	′	′	′	′	′	′	′	′	′	′	′
9	8.6	8.0	7.6	7.2	6.9	6.6	6.4	6.2	5.9	5.4	4.9	4.1	3.4	2.7	2.1	1.5
10	9.1	8.5	8.1	7.9	7.5	7.2	7.0	6.7	6.6	6.0	5.5	4.7	3.9	3.3	2.7	2.1
11	9.6	9.0	8.6	8.3	8.0	7.7	7.4	7.2	7.0	6.4	6.0	5.2	4.4	3.7	3.1	2.5
12	10.0	9.4	9.0	8.7	8.4	8.1	7.8	7.6	7.4	6.8	6.4	5.6	4.8	4.1	3.5	2.9
13	10.3	9.7	9.3	9.0	8.7	8.4	8.2	7.9	7.7	7.2	6.7	5.9	5.2	4.5	3.9	3.3
14	10.6	10.0	9.6	9.3	9.0	8.7	8.5	8.2	8.0	7.5	7.0	6.2	5.5	4.8	4.2	3.6
15	10.9	10.2	9.9	9.5	9.2	9.0	8.7	8.5	8.2	7.7	7.2	6.4	5.7	5.0	4.4	3.8
16	11.1	10.5	10.1	9.7	9.5	9.2	8.9	8.7	8.5	7.9	7.5	6.7	5.9	5.2	4.6	4.1
17	11.3	10.7	10.3	10.0	9.7	9.4	9.1	8.9	8.7	8.2	7.7	6.9	6.1	5.5	4.9	4.3
18	11.5	10.8	10.5	10.1	9.9	9.6	9.3	9.1	8.9	8.3	7.9	7.0	6.3	5.6	5.0	4.5
19	11.6	11.0	10.6	10.3	10.0	9.7	9.5	9.2	9.0	8.5	8.0	7.2	6.5	5.8	5.2	4.6
20	11.8	11.2	10.8	10.4	10.2	9.9	9.6	9.4	9.2	8.6	8.2	7.4	6.6	5.9	5.3	4.8
21	11.9	11.3	10.9	10.6	10.3	10.0	9.8	9.5	9.3	8.8	8.3	7.5	6.8	6.1	5.5	4.9
22	12.0	11.4	11.0	10.7	10.4	10.1	9.9	9.7	9.4	8.9	8.4	7.6	6.9	6.2	5.6	5.0
23	12.1	11.5	11.1	10.8	10.5	10.2	10.0	9.8	9.5	9.0	8.5	7.7	7.0	6.3	5.7	5.1
24	12.2	11.6	11.2	10.9	10.6	10.3	10.1	9.9	9.6	9.1	8.6	7.8	7.1	6.4	5.8	5.2
25	12.3	11.7	11.3	11.0	10.7	10.4	10.2	10.0	9.7	9.2	8.7	7.9	7.2	6.5	5.9	5.3
26	12.4	11.8	11.4	11.1	10.8	10.5	10.3	10.1	9.8	9.3	8.8	8.0	7.3	6.6	6.0	5.4
27	12.5	11.9	11.5	11.2	10.9	10.6	10.4	10.1	9.9	9.4	8.9	8.1	7.4	6.7	6.1	5.5
28	12.6	12.0	11.6	11.3	11.0	10.7	10.4	10.2	10.0	9.5	9.0	8.2	7.4	6.8	6.2	5.6
30	12.7	12.1	11.7	11.4	11.1	10.8	10.6	10.4	10.1	9.6	9.1	8.3	7.6	6.9	6.3	5.7
32	12.9	12.2	11.9	11.5	11.2	11.0	10.7	10.5	10.2	9.7	9.3	8.4	7.7	7.0	6.4	5.8
34	13.0	12.3	12.0	11.6	11.3	11.1	10.8	10.6	10.3	9.8	9.4	8.5	7.8	7.1	6.5	5.9
36	13.1	12.4	12.1	11.7	11.4	11.2	10.9	10.7	10.4	9.9	9.5	8.6	7.9	7.2	6.6	6.0
38	13.2	12.5	12.1	11.8	11.5	11.2	11.0	10.8	10.5	10.0	9.5	8.7	8.0	7.3	6.7	6.1
40	13.3	12.6	12.2	11.9	11.6	11.3	11.1	10.8	10.6	10.1	9.6	8.8	8.1	7.4	6.8	6.2
42	13.4	12.7	12.3	12.0	11.7	11.4	11.2	10.9	10.7	10.2	9.7	8.9	8.2	7.5	6.9	6.3
44	13.4	12.7	12.4	12.0	11.7	11.5	11.2	11.0	10.7	10.2	9.8	8.9	8.2	7.5	6.9	6.3
46	13.5	12.8	12.4	12.1	11.8	11.5	11.3	11.0	10.8	10.3	9.8	9.0	8.3	7.6	7.0	6.4
48	13.6	12.9	12.5	12.2	11.9	11.6	11.3	11.1	10.9	10.4	9.9	9.1	8.3	7.7	7.1	6.4
50	13.6	12.9	12.5	12.2	11.9	11.6	11.4	11.1	10.9	10.4	9.9	9.1	8.4	7.7	7.1	6.5
52	13.6	13.0	12.6	12.3	12.0	11.7	11.4	11.2	11.0	10.5	10.0	9.2	8.4	7.8	7.2	6.5
54	13.7	13.0	12.6	12.3	12.0	11.7	11.5	11.3	11.0	10.5	10.0	9.2	8.5	7.8	7.2	6.6
56	13.7	13.1	12.7	12.4	12.1	11.8	11.5	11.3	11.1	10.6	10.1	9.3	8.5	7.9	7.3	6.7
58	13.8	13.1	12.7	12.4	12.1	11.8	11.6	11.3	11.1	10.6	10.1	9.3	8.6	7.9	7.3	6.8
60	13.8	13.1	12.8	12.4	12.1	11.9	11.6	11.4	11.1	10.6	10.2	9.3	8.6	7.9	7.3	6.8
62	13.9	13.2	12.8	12.5	12.2	11.9	11.7	11.4	11.2	10.7	10.2	9.4	8.7	8.0	7.4	6.8
64	13.9	13.2	12.8	12.5	12.2	11.9	11.7	11.5	11.2	10.7	10.2	9.4	8.7	8.0	7.4	6.9
66	14.0	13.2	12.9	12.5	12.3	12.0	11.7	11.5	11.3	10.7	10.3	9.5	8.7	8.1	7.5	7.0
70	14.1	13.3	12.9	12.6	12.3	12.0	11.8	11.6	11.3	10.8	10.3	9.5	8.8	8.1	7.5	7.0
80	14.2	13.5	13.1	12.8	12.5	12.2	11.9	11.7	11.5	11.0	10.5	9.7	8.9	8.3	7.7	7.1
90	14.3	13.6	13.2	12.9	12.6	12.3	12.1	11.9	11.6	11.1	10.6	9.8	9.1	8.4	7.8	7.2

Monthly Correction

Jan	Feb	Mar	Apr	May	Jun	Jul	Aug	Sep	Oct	Nov	Dec
+0.3′	+0.2′	+0.1′	0.0′	–0.1′	–0.2′	–0.2′	–0.2′	–0.1′	+0.1′	+0.2′	+0.3′

NOTES

For lower limb sun sights greater than 9°, this table combines the corrections for refraction, dip, and parallax together with a correction for the sun's semi-diameter. Simply add the one correction to obtain your observed altitude (Ho).

For slightly greater precision, also add or subtract the monthly correction, which compensates for slight changes in the sun's semi-diameter.

For upper limb sun sights, it is necessary to subtract twice the sun's semi-diameter (given on the monthly tables) from the sextant altitude before entering the table.

SUN, STAR, AND ARIES GHA CORRECTION TABLES

Always add

Hours, minutes, or seconds (α)

	SUN			STAR AND ARIES						
	α mins	1h+ α mins	α secs	Date GMT	Corr for Date	α hours	1h+ α mins	α mins	α secs	
	° ′	° ′	′		° ′	° ′	° ′	° ′	′	
0	0 00.0	15 00.0	0.0	**1st**	0 00.0	0 00.0	15 02.5	0 00.0	0.0	**0**
1	0 15.0	15 15.0	0.3	**2nd**	0 59.1	15 02.5	15 17.5	0 15.0	0.3	**1**
2	0 30.0	15 30.0	0.5	**3rd**	1 58.2	30 04.9	15 32.6	0 30.1	0.5	**2**
3	0 45.0	15 45.0	0.8	**4th**	2 57.3	45 07.4	15 47.6	0 45.1	0.8	**3**
4	1 00.0	16 00.0	1.0	**5th**	3 56.5	60 09.9	16 02.7	1 00.2	1.0	**4**
5	1 15.0	16 15.0	1.3	**6th**	4 55.6	75 12.3	16 17.7	1 15.2	1.3	**5**
6	1 30.0	16 30.0	1.5	**7th**	5 54.8	90 14.8	16 32.7	1 30.2	1.5	**6**
7	1 45.0	16 45.0	1.8	**8th**	6 54.0	105 17.2	16 47.8	1 45.3	1.8	**7**
8	2 00.0	17 00.0	2.0	**9th**	7 53.1	120 19.7	17 02.8	2 00.3	2.0	**8**
9	2 15.0	17 15.0	2.3	**10th**	8 52.2	135 22.2	17 17.9	2 15.4	2.3	**9**
10	2 30.0	17 30.0	2.5	**11th**	9 51.4	150 24.6	17 32.9	2 30.4	2.5	**10**
11	2 45.0	17 45.0	2.8	**12th**	10 50.5	165 27.1	17 48.0	2 45.5	2.8	**11**
12	3 00.0	18 00.0	3.0	**13th**	11 49.6	180 29.6	18 03.0	3 00.5	3.0	**12**
13	3 15.0	18 15.0	3.3	**14th**	12 48.8	195 32.0	18 18.0	3 15.5	3.3	**13**
14	3 30.0	18 30.0	3.5	**15th**	13 48.0	210 34.5	18 33.1	3 30.6	3.5	**14**
15	3 45.0	18 45.0	3.8	**16th**	14 47.1	225 37.0	18 48.1	3 45.6	3.8	**15**
16	4 00.0	19 00.0	4.0	**17th**	15 46.2	240 39.4	19 03.2	4 00.7	4.0	**16**
17	4 15.0	19 15.0	4.3	**18th**	16 45.3	255 41.9	19 18.2	4 15.7	4.3	**17**
18	4 30.0	19 30.0	4.5	**19th**	17 44.5	270 44.4	19 33.2	4 30.7	4.5	**18**
19	4 45.0	19 45.0	4.8	**20th**	18 43.6	285 46.8	19 48.3	4 45.8	4.8	**19**
20	5 00.0	20 00.0	5.0	**21st**	19 42.7	300 49.3	20 03.3	5 00.8	5.0	**20**
21	5 15.0	20 15.0	5.3	**22nd**	20 41.9	315 51.7	20 18.4	5 15.9	5.3	**21**
22	5 30.0	20 30.0	5.5	**23rd**	21 41.0	330 54.2	20 33.4	5 30.9	5.5	**22**
23	5 45.0	20 45.0	5.8	**24th**	22 40.1	345 56.7	20 48.4	5 45.9	5.8	**23**
24	6 00.0	21 00.0	6.0	**25th**	23 39.3	360 59.1	21 03.5	6 01.0	6.0	**24**
25	6 15.0	21 15.0	6.3	**26th**	24 38.4		21 18.5	6 16.0	6.3	**25**
26	6 30.0	21 30.0	6.5	**27th**	25 37.6		21 33.6	6 31.1	6.5	**26**
27	6 45.0	21 45.0	6.8	**28th**	26 36.7		21 48.6	6 46.1	6.8	**27**
28	7 00.0	22 00.0	7.0	**29th**	27 35.8		22 03.6	7 01.1	7.0	**28**
29	7 15.0	22 15.0	7.3	**30th**	28 35.0		22 18.7	7 16.2	7.3	**29**
30	7 30.0	22 30.0	7.5	**31st**	29 34.1		22 33.7	7 31.2	7.5	**30**
31	7 45.0	22 45.0	7.8				22 48.8	7 46.3	7.8	**31**
32	8 00.0	23 00.0	8.0				23 03.8	8 01.3	8.0	**32**
33	8 15.0	23 15.0	8.3				23 18.9	8 16.4	8.3	**33**
34	8 30.0	23 30.0	8.5				23 33.9	8 31.4	8.5	**34**
35	8 45.0	23 45.0	8.8				23 48.9	8 46.4	8.8	**35**
36	9 00.0	24 00.0	9.0				24 04.0	9 01.5	9.0	**36**
37	9 15.0	24 15.0	9.3				24 19.0	9 16.5	9.3	**37**
38	9 30.0	24 30.0	9.5				24 34.1	9 31.6	9.5	**38**
39	9 45.0	24 45.0	9.8				24 49.1	9 46.6	9.8	**39**
40	10 00.0	25 00.0	10.0				25 04.1	10 01.6	10.0	**40**
41	10 15.0	25 15.0	10.3				25 19.2	10 16.7	10.3	**41**
42	10 30.0	25 30.0	10.5				25 34.2	10 31.7	10.5	**42**
43	10 45.0	25 45.0	10.8				25 49.3	10 46.8	10.8	**43**
44	11 00.0	26 00.0	11.0				26 04.3	11 01.8	11.0	**44**
45	11 15.0	26 15.0	11.3				26 19.3	11 16.8	11.3	**45**
46	11 30.0	26 30.0	11.5				26 34.4	11 31.9	11.5	**46**
47	11 45.0	26 45.0	11.8				26 49.4	11 46.9	11.8	**47**
48	12 00.0	27 00.0	12.0				27 04.5	12 02.0	12.0	**48**
49	12 15.0	27 15.0	12.3				27 19.5	12 17.0	12.3	**49**
50	12 30.0	27 30.0	12.5				27 34.6	12 32.1	12.5	**50**
51	12 45.0	27 45.0	12.8				27 49.6	12 47.1	12.8	**51**
52	13 00.0	28 00.0	13.0				28 04.6	13 02.1	13.0	**52**
53	13 15.0	28 15.0	13.3				28 19.7	13 17.2	13.3	**53**
54	13 30.0	28 30.0	13.5				28 34.7	13 32.2	13.5	**54**
55	13 45.0	28 45.0	13.8				28 49.8	13 47.3	13.8	**55**
56	14 00.0	29 00.0	14.0				29 04.8	14 02.3	14.0	**56**
57	14 15.0	29 15.0	14.3				29 19.8	14 17.3	14.3	**57**
58	14 30.0	29 30.0	14.5				29 34.9	14 32.4	14.5	**58**
59	14 45.0	29 45.0	14.8				29 49.9	14 47.4	14.8	**59**
60	15 00.0	30 00.0	15.0				30 04.9	15 02.5	15.0	**60**

EXAMPLES

GHA sun, 15:13:25 GMT 1/12/98

from p. E 32	14 00	27 55.3
from 1H+ α min column	13	18 15.0
from α secs col.	25	6.3
GHA Sun		46 16.6

GHA Aries, 04:03:05 GMT 3/23/98

from p. E 45	04 00	238 19.1
from α min column	3	0 45.1
from α secs col.	5	1.3
GHA Aries		239 05.5

GHA Sirius, 23:58:42 GMT 9/26/98

from p. E 79		238 41.1
from date col	26th	24 38.4
from α hours col	23	345 56.7
from α min col	58	14 32.4
from α secs col.	42	10.5
GHA Sirius		623 59.1
(less 360°)		263 59.1

MOON ALTITUDE TOTAL CORRECTION TABLE

Add/subtract as indicated

Upper Limb — Add above line / Subtract below line

Hor Par / Sextant Altitude (°)	54′	55′	56′	57′	58′	59′	60′	61′
	′	′	′	′	′	′	′	′
10	23.4	24.0	24.6	25.5	26.0	26.7	27.5	28.3
12	23.8	24.6	25.2	26.0	26.5	27.2	28.0	28.7
14	24.0	24.8	25.4	26.1	26.7	27.5	28.3	29.0
16	24.0	24.8	25.5	26.1	26.7	27.5	28.3	28.8
18	23.8	24.6	25.2	26.0	26.5	27.3	28.0	28.6
20	23.6	24.2	25.0	25.5	26.2	27.0	27.5	28.2
22	23.2	23.8	24.6	25.0	25.7	26.5	27.0	27.8
24	22.7	23.2	24.0	24.5	25.3	25.8	26.5	27.0
26	22.0	22.6	23.4	24.0	24.5	25.0	25.7	26.5
28	21.4	22.0	22.6	23.3	23.8	24.5	25.0	25.5
30	20.6	21.2	21.8	22.3	23.0	23.5	24.3	24.7
32	19.8	20.2	21.0	21.3	22.0	22.5	23.2	23.7
34	19.0	19.4	20.0	20.5	21.0	21.5	22.2	22.7
36	18.0	18.4	19.0	19.5	20.0	20.5	21.0	21.7
38	16.8	17.4	17.8	18.5	19.0	19.5	20.0	20.4
40	15.8	16.2	16.8	17.3	17.7	18.2	18.8	19.2
42	14.7	15.2	15.6	16.0	16.5	17.0	17.5	18.0
44	13.5	13.8	14.2	14.6	15.0	15.5	16.0	16.5
46	12.0	12.6	13.0	13.4	13.8	14.2	14.5	15.0
48	10.5	11.2	11.6	12.0	12.4	12.8	13.2	13.5
50	9.3	10.0	10.2	10.6	11.0	11.3	11.7	12.0
52	8.0	8.4	8.6	9.2	9.5	9.7	10.0	10.5
54	6.7	6.8	7.2	7.5	7.8	8.2	8.5	8.7
56	5.2	5.5	5.6	6.0	6.3	6.5	7.0	7.0
58	3.7	3.7	4.2	4.5	4.5	5.0	5.0	5.5
60	2.0	2.2	2.5	2.7	3.0	3.2	3.5	3.5
62	+0.5	+0.7	+0.8	+1.0	+1.2	+1.5	+1.5	+1.7
64	−1.2	−1.0	−1.0	−0.8	−0.6	−0.5	−0.3	−0.1
66	3.0	2.8	2.6	2.5	2.4	2.3	2.0	2.0
68	4.5	4.5	4.4	4.3	4.2	4.0	4.0	4.0
70	6.3	6.2	6.2	6.1	6.0	6.0	5.8	5.8
72	8.0	8.0	8.0	8.0	8.0	8.0	7.8	7.8
74	9.7	9.7	9.7	9.7	9.7	9.7	9.7	9.7
76	11.5	11.5	11.5	11.5	11.6	11.7	11.7	11.7
78	13.5	13.5	13.5	13.6	13.6	13.7	13.7	13.7
80	15.4	15.4	15.4	15.5	15.6	15.7	15.7	16.0
82	17.0	17.0	17.2	17.3	17.5	17.7	17.8	18.0
84	18.8	19.0	19.2	19.3	19.5	19.7	19.9	20.0
86	20.8	21.0	21.0	21.2	21.5	21.7	22.0	22.0
88	22.6	22.8	23.0	23.2	23.4	23.7	24.0	24.2
90								

Lower Limb Add

Hor Par / Sextant Altitude (°)	54′	55′	56′	57′	58′	59′	60′	61′
	′	′	′	′	′	′	′	′
10	52.7	54.0	55.3	56.5	57.7	59.0	60.2	61.5
12	53.2	54.5	55.7	57.0	58.4	59.5	60.7	62.0
14	53.5	54.7	56.0	57.3	58.5	59.8	61.0	62.3
16	53.5	54.6	56.0	57.3	58.5	59.8	61.0	62.2
18	53.4	54.6	55.7	57.0	58.4	59.5	60.6	62.0
20	53.0	54.4	55.5	56.8	58.0	59.0	60.4	61.5
22	52.5	53.7	55.0	56.3	57.5	58.8	60.0	61.0
24	52.0	53.3	54.5	55.5	56.7	58.0	59.4	60.5
26	51.5	52.5	53.7	55.0	56.3	57.5	58.0	59.8
28	50.7	52.0	53.0	54.4	55.5	56.5	57.8	59.0
30	50.0	51.0	52.3	53.5	54.5	55.7	57.0	58.0
32	49.3	50.4	51.3	52.5	53.7	54.8	56.0	57.0
34	48.3	49.5	50.5	51.5	52.7	53.7	55.0	56.0
36	47.3	48.5	49.5	50.5	51.7	52.7	54.0	55.0
38	46.4	47.4	48.5	49.5	50.5	51.5	52.7	53.8
40	45.3	46.3	47.3	48.3	49.5	50.5	51.5	52.5
42	44.0	45.0	46.0	47.0	48.0	49.0	50.0	51.0
44	42.7	43.7	44.7	45.7	46.7	47.7	48.7	49.7
46	41.5	42.5	43.5	44.5	45.5	46.5	47.5	48.5
48	40.2	41.2	42.2	43.0	44.0	45.0	46.0	47.0
50	39.0	40.0	41.0	41.8	42.6	43.6	44.5	45.5
52	37.5	38.5	39.3	40.2	41.0	42.0	42.8	43.7
54	36.0	37.0	38.0	38.8	39.5	40.5	41.3	42.0
56	34.5	35.5	36.2	37.0	38.0	38.7	39.5	40.5
58	33.0	34.0	34.7	35.5	36.3	37.0	38.0	38.8
60	31.5	32.4	33.0	34.0	34.5	35.5	36.0	37.0
62	30.0	30.5	31.5	32.0	33.0	33.5	34.5	35.0
64	28.3	29.0	29.6	30.5	31.0	31.8	32.5	33.3
66	26.5	27.3	28.0	28.5	29.3	30.0	30.7	31.5
68	25.0	25.5	26.3	26.8	27.5	28.0	28.8	29.5
70	23.3	23.8	24.5	25.0	25.5	26.2	27.0	27.5
72	21.5	22.0	22.5	23.3	23.8	24.5	25.0	25.5
74	19.7	20.3	20.7	21.2	22.0	22.5	23.0	23.5
76	18.0	18.5	19.0	19.5	20.0	20.5	21.0	21.5
78	16.0	16.5	17.0	17.5	18.0	18.5	19.0	19.5
80	14.2	14.7	15.3	15.5	16.0	16.5	17.0	17.5
82	12.5	13.0	13.3	13.5	14.0	14.5	15.0	15.5
84	10.5	11.0	11.5	11.7	12.0	12.5	13.0	13.4
86	8.8	9.0	9.5	9.8	10.0	10.5	11.0	11.3
88	7.0	7.2	7.5	8.0	8.3	8.5	8.7	9.0
90								

HEIGHT OF EYE CORRECTION (MOON ONLY)

Add

Height of eye (m)	**0**	**1.5**	**3**	**4.6**	**6**	**7.6**	**9**	**10.7**	**12**	**14**	**15**	**17**	**18**	**20**	**21**	**23**	**24**	**26**	**30**
(ft)	**0**	**5**	**10**	**15**	**20**	**25**	**30**	**35**	**40**	**45**	**50**	**55**	**60**	**65**	**70**	**75**	**80**	**85**	**100**
Correction	9.8′	7.6′	6.7′	6.0′	5.5′	5.0′	4.5′	4.0′	3.5′	3.2′	3.0′	2.5′	2.3′	2.0′	1.7′	1.3′	1.0′	0.8′	0.0′

NOTE: This table is in two parts. The upper table corrects for semi-diameter, parallax, and refraction. The horizontal parallax figure required for the top line varies daily and can be found on the first of the monthly tables. The lower table is especially formulated to go with the upper table so that dip is always added.

STAR OR PLANET ALTITUDE TOTAL CORRECTION TABLE

Always subtract

	Height of Eye above Sea Level												
m	**1.5**	**3.0**	**4.6**	**6.0**	**7.6**	**9.0**	**10.7**	**12**	**13.7**	**15**	**16.8**	**18**	**21.3**
ft	**5**	**10**	**15**	**20**	**25**	**30**	**35**	**40**	**45**	**50**	**55**	**60**	**70**
Sextant Altitude (°)	′	′	′	′	′	′	′	′	′	′	′	′	′
9	8.0	8.9	9.6	10.3	10.7	11.2	11.6	12.0	12.4	12.8	13.1	13.5	14.1
10	7.4	8.4	9.1	9.7	10.2	10.6	11.1	11.5	11.8	12.2	12.5	12.9	13.5
11	7.0	7.9	8.6	9.2	9.7	10.2	10.6	11.0	11.4	11.8	12.0	12.4	13.0
12	6.6	7.5	8.2	8.8	9.3	9.8	10.2	10.6	11.0	11.4	11.6	12.0	12.6
13	6.2	7.2	7.9	8.4	9.0	9.4	9.9	10.3	10.6	11.0	11.3	11.6	12.3
14	5.9	6.9	7.6	8.1	8.6	9.2	9.6	10.0	10.3	10.7	11.0	11.3	12.0
15	5.7	6.6	7.3	7.9	8.4	8.9	9.3	9.7	10.1	10.4	10.8	11.1	11.7
16	5.5	6.4	7.1	7.7	8.2	8.7	9.1	9.5	9.9	10.2	10.5	10.9	11.5
17	5.3	6.2	6.9	7.5	8.0	8.5	8.9	9.3	9.7	10.0	10.3	10.7	11.3
18	5.1	6.0	6.7	7.3	7.8	8.3	8.7	9.1	9.5	9.8	10.2	10.5	11.1
19	4.9	5.8	6.5	7.1	7.6	8.1	8.5	8.9	9.3	9.7	10.0	10.3	11.0
20	4.8	5.7	6.4	7.0	7.5	8.0	8.4	8.8	9.2	9.6	9.9	10.2	10.8
25	4.2	5.1	5.8	6.4	6.9	7.4	7.8	8.2	8.6	9.0	9.3	9.6	10.2
30	3.8	4.7	5.4	6.0	6.5	7.0	7.4	7.8	8.2	8.6	8.9	9.2	9.8
35	3.5	4.4	5.1	5.7	6.3	6.7	7.2	7.6	7.9	8.3	8.6	8.9	9.5
40	3.3	4.2	4.9	5.5	6.0	6.5	6.9	7.3	7.7	8.1	8.4	8.7	9.3
50	3.0	3.9	4.6	5.2	5.7	6.2	6.6	7.0	7.4	7.7	8.1	8.4	9.0
60	2.7	3.6	4.4	4.9	5.5	5.9	6.4	6.8	7.1	7.5	7.8	8.1	8.8
70	2.5	3.4	4.1	4.7	5.3	5.7	6.2	6.6	6.9	7.3	7.6	7.9	8.6
80	2.3	3.3	4.0	4.6	5.1	5.5	6.0	6.4	6.7	7.1	7.4	7.8	8.4
90	2.2	3.1	3.8	4.4	4.9	5.4	5.8	6.2	6.6	6.9	7.3	7.6	8.2

NOTE: This table combines the corrections for refraction and dip in the tables on page E 104, but not for parallax or semi-diameter since these corrections are of no significance in the case of star or planet sights.

MOON GHA CORRECTION TABLE (HOURS)

Always add

Var/hr	**14°20′**	**14°20.5′**	**14°21′**	**14°21.5′**	**14°22′**	**14°22.5′**	**14°23′**	**14°23.5′**	**14°24′**	**14°24.5′**	**14°25′**	**14°25.5′**
	° ′	° ′	° ′	° ′	° ′	° ′	° ′	° ′	° ′	° ′	° ′	° ′
Hours 1	14 20.0	14 20.5	14 21.0	14 21.5	14 22.0	14 22.5	14 23.0	14 23.5	14 24.0	14 24.5	14 25.0	14 25.5
2	28 40.0	28 41.0	28 42.0	28 43.0	28 44.0	28 45.0	28 46.0	28 47.0	28 48.0	28 49.0	28 50.0	28 51.0
3	43 00.0	43 01.5	43 03.0	43 04.5	43 06.0	43 07.5	43 09.0	43 10.5	43 12.0	43 13.5	43 15.0	43 16.5
4	57 20.0	57 22.0	57 24.0	57 26.0	57 28.0	57 30.0	57 32.0	57 34.0	57 36.0	57 38.0	57 40.0	57 42.0
5	71 40.0	71 42.5	71 45.0	71 47.5	71 50.0	71 52.5	71 55.0	71 57.5	72 00.0	72 02.5	72 05.0	72 07.5

Var/hr	**14°26′**	**14°26.5′**	**14°27′**	**14°27.5′**	**14°28′**	**14°28.5′**	**14°29′**	**14°29.5′**	**14°30′**	**14°30.5′**	**14°31′**	**14°31.5′**
Hours 1	14 26.0	14 26.5	14 27.0	14 27.5	14 28.0	14°28.5	14 29.0	14 29.5	14 30.0	14 30.5	14 31.0	14 31.5
2	28 52.0	28 53.0	28 54.0	28 55.0	28 56.0	28 57.0	28 58.0	28 59.0	29 00.0	29 01.0	29 02.0	29 03.0
3	43 18.0	43 19.5	43 21.0	43 22.5	43 24.0	43 25.5	43 27.0	43 28.5	43 30.0	43 31.5	43 33.0	43 34.5
4	57 44.0	57 46.0	57 48.0	57 50.0	57 52.0	57 54.0	57 56.0	57 58.0	58 00.0	58 02.0	58 04.0	58 06.0
5	72 10.0	72 12.5	72 15.0	72 17.5	72 20.0	72 22.5	72 25.0	72 27.5	72 30.0	72 32.5	72 35.0	72 37.5

Var/hr	**14°32′**	**14°32.5′**	**14°33′**	**14°33.5′**	**14°34′**	**14°34.5′**	**14°35′**	**14°35.5′**	**14°36′**	**14°36.5′**	**14°37′**	**14°37.5′**
Hours 1	14 32.0	14 32.5	14 33.0	14 33.5	14 34.0	14 34.5	14 35.0	14 35.5	14 36.0	14 36.5	14 37.0	14 37.5
2	29 04.0	29 05.0	29 06.0	29 07.0	29 08.0	29 09.0	29 10.0	29 11.0	29 12.0	29 13.0	29 14.0	29 15.0
3	43 36.0	43 37.5	43 39.0	43 40.5	43 42.0	43 43.5	43 45.0	43 46.5	43 48.0	43 49.5	43 51.0	43 52.5
4	58 08.0	58 10.0	58 12.0	58 14.0	58 16.0	58 18.0	58 20.0	58 22.0	58 24.0	58 26.0	58 28.0	58 30.0
5	72 40.0	72 42.5	72 45.0	72 47.5	72 50.0	72 52.5	72 55.0	72 57.5	73 00.0	73 02.5	73 05.0	73 07.5

NOTE: The correction for moon GHA is taken in three parts. Using the figure for variation per hour given in the monthly tables, you get the correction for hours from the table above and the correction for minutes from the tables on the next two pages. Finally, take the correction for seconds from the right-hand side of whichever minute table you used.

MOON GHA CORRECTION TABLE (MINUTES)

Var/hr (Minutes)	14°20′	14°21′	14°22′	14°23′	14°24′	14°25′	14°26′	14°27′	14°28′	Diff 1′	Sec	Corr
	° ′	° ′	° ′	° ′	° ′	° ′	° ′	° ′	° ′	′		′
0	0 00.0	0 00.0	0 00.0	0 00.0	0 00.0	0 00.0	0 00.0	0 00.0	0 00.0	0.0	**0**	0.0
1	0 14.3	0 14.4	0 14.4	0 14.4	0 14.4	0 14.4	0 14.4	0 14.4	0 14.5	0.0	**1**	0.2
2	0 28.7	0 28.7	0 28.7	0 28.8	0 28.8	0 28.8	0 28.9	0 28.9	0 28.9	0.0	**2**	0.5
3	0 43.0	0 43.0	0 43.1	0 43.2	0 43.2	0 43.2	0 43.3	0 43.4	0 43.4	0.0	**3**	0.7
4	0 57.3	0 57.4	0 57.5	0 57.5	0 57.6	0 57.7	0 57.7	0 57.8	0 57.9	0.1	**4**	1.0
5	1 11.7	1 11.8	1 11.8	1 11.9	1 12.0	1 12.1	1 12.2	1 12.2	1 12.3	0.1	**5**	1.2
6	1 26.0	1 26.1	1 26.2	1 26.3	1 26.4	1 26.5	1 26.6	1 26.7	1 26.8	0.1	**6**	1.4
7	1 40.3	1 40.4	1 40.6	1 40.7	1 40.8	1 40.9	1 41.2	1 41.2	1 41.3	0.1	**7**	1.7
8	1 54.7	1 54.8	1 54.9	1 55.1	1 55.2	1 55.3	1 55.5	1 55.6	1 55.7	0.1	**8**	1.9
9	2 09.0	2 09.2	2 09.3	2 09.4	2 09.6	2 09.8	2 09.9	2 10.0	2 10.2	0.2	**9**	2.2
10	2 23.3	2 23.5	2 23.7	2 23.8	2 24.0	2 24.2	2 24.3	2 24.5	2 24.7	0.2	**10**	2.4
11	2 37.7	2 37.8	2 38.0	2 38.2	2 38.4	2 38.6	2 38.8	2 39.0	2 39.1	0.2	**11**	2.6
12	2 52.0	2 52.2	2 52.4	2 52.6	2 52.8	2 53.0	2 53.2	2 53.4	2 53.6	0.2	**12**	2.9
13	3 06.3	3 06.6	3 06.8	3 07.0	3 07.2	3 07.4	3 07.6	3 07.8	3 08.1	0.2	**13**	3.1
14	3 20.7	3 20.9	3 21.1	3 21.4	3 21.6	3 21.8	3 22.1	3 22.3	3 22.5	0.2	**14**	3.4
15	3 35.0	3 35.2	3 35.5	3 35.8	3 36.0	3 36.2	3 36.5	3 36.8	3 37.0	0.2	**15**	3.6
16	3 49.3	3 49.6	3 49.9	3 50.1	3 50.4	3 50.7	3 50.9	3 51.2	3 51.5	0.3	**16**	3.8
17	4 03.7	4 04.0	4 04.2	4 04.5	4 04.8	4 05.1	4 05.4	4 05.6	4 05.9	0.3	**17**	4.1
18	4 18.0	4 18.3	4 18.6	4 18.9	4 19.2	4 19.5	4 19.8	4 20.1	4 20.4	0.3	**18**	4.3
19	4 32.3	4 32.6	4 33.0	4 33.3	4 33.6	4 33.9	4 34.2	4 34.6	4 34.9	0.3	**19**	4.6
20	4 46.7	4 47.0	4 47.3	4 47.7	4 48.0	4 48.3	4 48.7	4 49.0	4 49.3	0.3	**20**	4.8
21	5 01.0	5 01.4	5 01.7	5 02.0	5 02.4	5 02.8	5 03.1	5 03.4	5 03.8	0.4	**21**	5.0
22	5 15.3	5 15.7	5 16.1	5 16.4	5 16.8	5 17.2	5 17.5	5 17.9	5 18.3	0.4	**22**	5.3
23	5 29.7	5 30.0	5 30.4	5 30.8	5 31.2	5 31.6	5 32.0	5 32.4	5 32.7	0.4	**23**	5.5
24	5 44.0	5 44.4	5 44.8	5 45.2	5 45.6	5 46.0	5 46.4	5 46.8	5 47.2	0.4	**24**	5.8
25	5 58.3	5 58.8	5 59.2	5 59.6	6 00.0	6 00.4	6 00.8	6 01.2	6 01.7	0.4	**25**	6.0
26	6 12.7	6 13.1	6 13.5	6 14.0	6 14.4	6 14.8	6 15.3	6 15.7	5 16.1	0.4	**26**	6.2
27	6 27.0	6 27.4	6 27.9	6 28.4	6 28.8	6 29.2	6 29.7	6 30.2	6 30.6	0.4	**27**	6.5
28	6 41.3	6 41.8	6 42.3	6 42.7	6 43.2	6 43.7	6 44.1	6 44.6	6 45.1	0.5	**28**	6.7
29	6 55.7	6 56.2	6 56.6	6 57.1	6 57.6	6 58.1	6 58.6	6 59.0	6 59.5	0.5	**29**	7.0
30	7 10.0	7 10.5	7 11.0	7 11.5	7 12.0	7 12.5	7 13.0	7 13.5	7 14.0	0.5	**30**	7.2
31	7 24.3	7 24.8	7 25.4	7 25.9	7 26.4	7 26.9	7 27.4	7 28.0	7 28.5	0.5	**31**	7.4
32	7 38.7	7 39.2	7 39.7	7 40.3	7 40.8	7 41.3	7 41.9	7 42.4	7 42.9	0.5	**32**	7.7
33	7 53.0	7 53.6	7 54.1	7 54.6	7 55.2	7 55.8	7 56.3	7 56.8	7 57.4	0.6	**33**	7.9
34	8 07.3	8 07.9	8 08.5	8 09.0	8 09.6	8 10.2	8 10.7	8 11.3	8 11.9	0.6	**34**	8.2
35	8 21.7	8 22.2	8 22.8	8 23.4	8 24.0	8 24.6	8 25.2	8 25.8	8 26.3	0.6	**35**	8.4
36	8 36.0	8 36.6	8 37.2	8 37.8	8 38.4	8 39.0	8 39.6	8 40.2	8 40.8	0.6	**36**	8.6
37	8 50.3	8 51.0	8 51.6	8 52.2	8 52.8	8 53.4	8 54.0	8 54.6	8 55.3	0.6	**37**	8.9
38	9 04.7	9 05.3	9 05.9	9 06.6	9 07.2	9 07.8	9 08.5	9 09.1	9 09.7	0.6	**38**	9.1
39	9 19.0	9 19.6	9 20.3	9 21.0	9 21.6	9 22.2	9 22.9	9 23.6	9 24.2	0.6	**39**	9.4
40	9 33.3	9 34.0	9 34.7	9 35.3	9 36.0	9 36.7	9 37.3	9 38.0	9 38.7	0.7	**40**	9.6
41	9 47.7	9 48.4	9 49.0	9 49.7	9 50.4	9 51.1	9 51.8	9 52.4	9 53.1	0.7	**41**	9.8
42	10 02.0	10 02.7	10 03.4	10 04.1	10 04.8	10 05.5	10 06.2	10 06.9	10 07.6	0.7	**42**	10.1
43	10 16.3	10 17.0	10 17.8	10 18.5	10 19.2	10 19.9	10 20.6	10 21.4	10 22.1	0.7	**43**	10.3
44	10 30.7	10 31.4	10 32.1	10 32.9	10 33.6	10 34.3	10 35.1	10 35.8	10 36.5	0.7	**44**	10.6
45	10 45.0	10 45.8	10 46.5	10 47.2	10 48.0	10 48.8	10 49.5	10 50.2	10 51.0	0.8	**45**	10.8
46	10 59.3	11 00.1	11 00.9	11 01.6	11 02.4	11 03.2	11 03.9	11 04.7	11 05.5	0.8	**46**	11.0
47	11 13.7	11 14.4	11 15.2	11 16.0	11 16.8	11 17.6	11 18.4	11 19.2	11 19.9	0.8	**47**	11.3
48	11 28.0	11 28.8	11 29.6	11 30.4	11 31.2	11 32.0	11 32.8	11 33.6	11 34.4	0.8	**48**	11.5
49	11 42.3	11 43.2	11 44.0	11 44.8	11 45.6	11 46.4	11 47.2	11 48.0	11 48.9	0.8	**49**	11.8
50	11 56.7	11 57.5	11 58.3	11 59.2	12 00.0	12 00.8	12 01.7	12 02.5	12 03.3	0.8	**50**	12.0
51	12 11.0	12 11.8	12 12.7	12 13.6	12 14.4	12 15.2	12 16.1	12 17.0	12 17.8	0.8	**51**	12.2
52	12 25.3	12 26.2	12 27.1	12 27.9	12 28.8	12 29.7	12 30.5	12 31.4	12 32.3	0.9	**52**	12.5
53	12 39.7	12 40.6	12 41.4	12 42.3	12 43.2	12 44.1	12 45.0	12 45.8	12 46.7	0.9	**53**	12.7
54	12 54.0	12 54.9	12 55.8	12 56.7	12 57.6	12 58.5	12 59.4	13 00.3	13 01.2	0.9	**54**	13.0
55	13 08.3	13 09.2	13 10.2	13 11.1	13 12.0	13 12.9	13 13.8	13 14.8	13 15.7	0.9	**55**	13.2
56	13 22.7	13 23.6	13 24.5	13 25.5	13 26.4	13 27.3	13 28.3	13 29.2	13 30.1	0.9	**56**	13.4
57	13 37.0	13 38.0	13 38.9	13 39.8	13 40.8	13 41.8	13 42.7	13 43.6	13 44.6	1.0	**57**	13.7
58	13 51.3	13 52.3	13 53.3	13 54.2	13 55.2	13 56.2	13 57.1	13 58.1	13 59.1	1.0	**58**	13.9
59	14 05.7	14 06.6	14 07.6	14 08.6	14 09.6	14 10.6	14 11.6	14 12.6	14 13.5	1.0	**59**	14.1
60	14 20.0	14 21.0	14 22.0	14 23.0	14 24.0	14 25.0	14 26.0	14 27.0	14 28.0	1.0	**60**	14.4

MOON GHA CORRECTION TABLE (MINUTES)

Var/hr (Minutes)	14°29′	14°30′	14°31′	14°32′	14°33′	14°34′	14°35′	14°36′	14°37′	Diff 1′	Sec	Corr
	° ′	° ′	° ′	° ′	° ′	° ′	° ′	° ′	° ′	′		′
0	0 00.0	0 00.0	0 00.0	0 00.0	0 00.0	0 00.0	0 00.0	0 00.0	0 00.0	0.0	**0**	0.0
1	0 14.5	0 14.5	0 14.5	0 14.6	0 14.6	0 14.6	0 14.6	0 14.6	0 14.6	0.0	**1**	0.2
2	0 29.0	0 29.0	0 29.0	0 29.1	0 29.1	0 29.1	0 29.2	0 29.2	0 29.2	0.0	**2**	0.5
3	0 43.4	0 43.5	0 43.6	0 43.6	0 43.6	0 43.7	0 43.8	0 43.8	0 43.8	0.0	**3**	0.7
4	0 57.9	0 58.0	0 58.1	0 58.1	0 58.2	0 58.3	0 58.3	0 58.4	0 58.5	0.1	**4**	1.0
5	1 12.4	1 12.5	1 12.6	1 12.7	1 12.8	1 12.8	1 12.9	1 13.0	1 13.1	0.1	**5**	1.2
6	1 26.9	1 27.0	1 27.1	1 27.2	1 27.3	1 27.4	1 27.5	1 27.6	1 27.7	0.1	**6**	1.5
7	1 41.4	1 41.5	1 41.6	1 41.7	1 41.8	1 42.0	1 42.1	1 42.2	1 42.3	0.1	**7**	1.7
8	1 55.9	1 56.0	1 56.1	1 56.3	1 56.4	1 56.5	1 56.7	1 56.8	1 56.9	0.1	**8**	1.9
9	2 10.4	2 10.5	2 10.6	2 10.8	2 11.0	2 11.1	2 11.2	2 11.4	2 11.6	0.2	**9**	2.2
10	2 24.8	2 25.0	2 25.2	2 25.2	2 25.5	2 25.7	2 25.8	2 26.0	2 26.2	0.2	**10**	2.4
11	2 39.3	2 39.5	2 39.7	2 39.9	2 40.0	2 40.2	2 40.4	2 40.6	2 40.8	0.2	**11**	2.7
12	2 53.8	2 54.0	2 54.2	2 54.4	2 54.6	2 54.8	2 55.0	2 55.2	2 55.4	0.2	**12**	2.9
13	3 08.3	3 08.5	3 08.7	3 08.9	3 09.2	3 09.4	3 09.6	3 09.8	3 10.0	0.2	**13**	3.2
14	3 22.8	3 23.0	3 23.2	3 23.5	3 23.7	3 23.9	3 24.2	3 24.4	3 24.6	0.2	**14**	3.4
15	3 37.2	3 37.5	3 37.8	3 38.0	3 38.2	3 38.5	3 38.8	3 39.0	3 39.2	0.2	**15**	3.6
16	3 51.7	3 52.0	3 52.3	3 52.5	3 52.8	3 53.1	3 53.3	3 53.6	3 53.9	0.3	**16**	3.9
17	4 06.2	4 06.5	4 06.8	4 07.1	4 07.4	4 07.6	4 07.9	4 08.2	4 08.5	0.3	**17**	4.1
18	4 20.7	4 21.0	4 21.3	4 21.6	4 21.9	4 22.2	4 22.5	4 22.8	4 23.1	0.3	**18**	4.4
19	4 35.2	4 35.5	4 35.8	4 36.1	4 36.4	4 36.8	4 37.1	4 37.4	4 37.7	0.3	**19**	4.6
20	4 49.7	4 50.0	4 50.3	4 50.7	4 51.0	4 51.3	4 51.7	4 52.0	4 52.3	0.3	**20**	4.9
21	5 04.2	5 04.5	5 04.8	5 05.2	5 05.6	5 05.9	5 06.2	5 06.6	5 07.0	0.4	**21**	5.1
22	5 18.6	5 19.0	5 19.4	5 19.7	5 20.1	5 20.5	5 20.8	5 21.2	5 21.6	0.4	**22**	5.3
23	5 33.1	5 33.5	5 33.9	5 34.3	5 34.6	5 35.0	5 35.4	5 35.8	5 36.2	0.4	**23**	5.6
24	5 47.6	5 48.0	5 48.4	5 48.8	5 49.2	5 49.6	5 50.0	5 50.4	5 50.8	0.4	**24**	5.8
25	6 02.1	6 02.5	6 02.9	6 03.3	6 03.8	6 04.2	6 04.6	6 05.0	6 05.4	0.4	**25**	6.1
26	6 16.6	6 17.0	6 17.4	6 17.9	6 18.3	6 18.7	6 19.2	6 19.6	6 20.0	0.4	**26**	6.3
27	6 31.0	6 31.5	6 32.0	6 32.4	6 32.8	6 33.3	6 33.8	6 34.2	6 34.6	0.4	**27**	6.5
28	6 45.5	6 46.0	6 46.5	6 46.9	6 47.4	6 47.9	6 48.3	6 48.8	6 49.3	0.5	**28**	6.8
29	7 00.0	7 00.5	7 01.0	7 01.5	7 02.0	7 02.4	7 02.9	7 03.4	7 03.9	0.5	**29**	7.0
30	7 14.5	7 15.0	7 15.5	7 16.0	7 16.5	7 17.0	7 17.5	7 18.0	7 18.5	0.5	**30**	7.3
31	7 29.0	7 29.5	7 30.0	7 30.5	7 31.0	7 31.6	7 32.1	7 32.6	7 33.1	0.5	**31**	7.5
32	7 43.5	7 44.0	7 44.5	7 45.1	7 45.6	7 46.1	7 46.7	7 47.2	7 47.7	0.5	**32**	7.8
33	7 58.0	7 58.5	7 59.0	7 59.6	8 00.2	8 00.7	8 01.2	8 01.8	8 02.4	0.6	**33**	8.0
34	8 12.4	8 13.0	8 13.6	8 14.1	8 14.7	8 15.3	8 15.8	8 16.4	8 17.0	0.6	**34**	8.2
35	8 26.9	8 27.5	8 28.1	8 28.7	8 29.2	8 29.8	8 30.4	8 31.0	8 31.6	0.6	**35**	8.5
36	8 41.4	8 42.0	8 42.6	8 43.2	8 43.8	8 44.4	8 45.0	8 45.6	8 46.2	0.6	**36**	8.7
37	8 55.9	8 56.5	8 57.1	8 57.7	8 58.4	8 59.0	8 59.6	9 00.2	9 00.8	0.6	**37**	9.0
38	9 10.4	9 11.0	9 11.6	9 12.3	9 12.9	9 13.5	9 14.2	9 14.8	9 15.4	0.6	**38**	9.2
39	9 24.8	9 25.5	9 26.2	9 26.8	9 27.4	9 28.1	9 28.8	9 29.4	9 30.0	0.6	**39**	9.5
40	9 39.3	9 40.0	9 40.7	9 41.3	9 42.0	9 42.7	9 43.3	9 44.0	9 44.7	0.7	**40**	9.7
41	9 53.8	9 54.5	9 55.2	9 55.9	9 56.6	9 57.2	9 57.9	9 58.6	9 59.3	0.7	**41**	9.9
42	10 08.3	10 09.0	10 09.7	10 10.4	10 11.1	10 11.8	10 12.5	10 13.2	10 13.9	0.7	**42**	10.2
43	10 22.8	10 23.5	10 24.2	10 24.9	10 25.6	10 26.4	10 27.1	10 27.8	10 28.5	0.7	**43**	10.4
44	10 37.3	10 38.0	10 38.7	10 39.5	10 40.2	10 40.9	10 41.7	10 42.2	10 43.1	0.7	**44**	10.7
45	10 51.8	10 52.5	10 53.2	10 54.0	10 54.8	10 55.5	10 56.2	10 57.0	10 57.8	0.8	**45**	10.9
46	11 06.2	11 07.0	11 07.8	11 08.5	11 09.3	11 10.1	11 10.8	11 11.6	11 12.4	0.8	**46**	11.2
47	11 20.7	11 21.5	11 22.3	11 23.1	11 23.8	11 24.6	11 25.4	11 26.2	11 27.0	0.8	**47**	11.4
48	11 35.2	11 36.0	11 36.8	11 37.6	11 38.4	11 39.2	11 40.0	11 40.8	11 41.6	0.8	**48**	11.6
49	11 49.7	11 50.5	11 51.3	11 52.1	11 53.0	11 53.8	11 54.6	11 55.4	11 56.2	0.8	**49**	11.9
50	12 04.2	12 05.0	12 05.8	12 06.7	12 07.5	12 08.3	12 09.2	12 10.0	12 10.8	0.8	**50**	12.1
51	12 18.6	12 19.5	12 20.4	12 21.2	12 22.0	12 22.9	12 23.8	12 24.6	12 25.4	0.8	**51**	12.4
52	12 33.1	12 34.0	12 34.9	12 35.7	12 36.6	12 37.5	12 38.3	12 39.2	12 40.1	0.9	**52**	12.6
53	12 47.6	12 48.5	12 49.4	12 50.3	12 51.2	12 52.0	12 52.9	12 53.8	12 54.7	0.9	**53**	12.9
54	13 02.1	13 03.0	13 03.9	13 04.8	13 05.7	13 06.6	13 07.5	13 08.4	13 09.3	0.9	**54**	13.1
55	13 16.6	13 17.5	13 18.4	13 19.3	13 20.2	13 21.2	13 22.1	13 23.0	13 23.9	0.9	**55**	13.3
56	13 31.1	13 32.0	13 32.9	13 33.9	13 34.8	13 35.7	13 36.7	13 37.6	13 38.5	0.9	**56**	13.6
57	13 45.6	13 46.5	13 47.4	13 48.4	13 49.4	13 50.3	13 51.2	13 52.2	13 53.2	1.0	**57**	13.9
58	14 00.0	14 01.0	14 02.0	14 02.9	14 03.9	14 04.9	14 05.8	14 06.8	14 07.8	1.0	**58**	14.1
59	14 14.5	14 15.5	14 16.5	14 17.5	14 18.4	14 19.4	14 20.4	14 21.4	14 22.4	1.0	**59**	14.3
60	14 29.0	14 30.0	14 31.0	14 32.0	14 33.0	14 34.0	14 35.0	14 36.0	14 37.0	1.0	**60**	14.6

MOON DECLINATION CORRECTION TABLE

Var/hr	0.0′	1.0′	2.0′	3.0′	4.0′	5.0′	6.0′	7.0′	8.0′	9.0′	10.0′	11.0′	12.0′	13.0′	14.0′	15.0′	16.0′	17.0′	18.0′
Minutes	′	′	′	′	′	′	′	′	′	′	′	′	′	′	′	′	′	′	′
0	0.0	0.0	0.0	0.0	0.0	0.0	0.0	0.0	0.0	0.0	0.0	0.0	0.0	0.0	0.0	0.0	0.0	0.0	0.0
1	0.0	0.0	0.0	0.0	0.1	0.1	0.1	0.1	0.1	0.2	0.2	0.2	0.2	0.2	0.2	0.2	0.3	0.3	0.3
2	0.0	0.0	0.1	0.1	0.1	0.2	0.2	0.2	0.3	0.3	0.3	0.4	0.4	0.4	0.5	0.5	0.5	0.6	0.6
3	0.0	0.0	0.1	0.2	0.2	0.2	0.3	0.4	0.4	0.4	0.5	0.6	0.6	0.6	0.7	0.8	0.8	0.8	0.9
4	0.0	0.1	0.1	0.2	0.3	0.3	0.4	0.5	0.5	0.6	0.7	0.7	0.8	0.9	0.9	1.0	1.1	1.1	1.2
5	0.0	0.1	0.2	0.2	0.3	0.4	0.5	0.6	0.7	0.8	0.8	0.9	1.0	1.1	1.2	1.2	1.3	1.4	1.5
6	0.0	0.1	0.2	0.3	0.4	0.5	0.6	0.7	0.8	0.9	1.0	1.1	1.2	1.3	1.4	1.5	1.6	1.7	1.8
7	0.0	0.1	0.2	0.4	0.5	0.6	0.7	0.8	0.9	1.0	1.2	1.3	1.4	1.5	1.6	1.8	1.9	2.0	2.1
8	0.0	0.1	0.3	0.4	0.5	0.7	0.8	0.9	1.1	1.2	1.3	1.5	1.6	1.7	1.9	2.0	2.1	2.3	2.4
9	0.0	0.2	0.3	0.4	0.6	0.8	0.9	1.0	1.2	1.4	1.5	1.6	1.8	2.0	2.1	2.2	2.4	2.6	2.7
10	0.0	0.2	0.3	0.5	0.7	0.8	1.0	1.2	1.3	1.5	1.7	1.8	2.0	2.2	2.3	2.5	2.7	2.8	3.0
11	0.0	0.2	0.4	0.6	0.7	0.9	1.1	1.3	1.5	1.6	1.8	2.0	2.2	2.4	2.6	2.8	2.9	3.1	3.3
12	0.0	0.2	0.4	0.6	0.8	1.0	1.2	1.4	1.6	1.8	2.0	2.2	2.4	2.6	2.8	3.0	3.2	3.4	3.6
13	0.0	0.2	0.4	0.6	0.9	1.1	1.3	1.5	1.7	2.0	2.2	2.4	2.6	2.8	3.0	3.2	3.5	3.7	3.9
14	0.0	0.2	0.5	0.7	0.9	1.2	1.4	1.6	1.9	2.1	2.3	2.6	2.8	3.0	3.3	3.5	3.7	4.0	4.2
15	0.0	0.2	0.5	0.8	1.0	1.2	1.5	1.8	2.0	2.2	2.5	2.8	3.0	3.2	3.5	3.8	4.0	4.2	4.5
16	0.0	0.3	0.5	0.8	1.1	1.3	1.6	1.9	2.1	2.4	2.7	2.9	3.2	3.5	3.7	4.0	4.3	4.5	4.8
17	0.0	0.3	0.6	0.8	1.1	1.4	1.7	2.0	2.3	2.6	2.8	3.1	3.4	3.7	4.0	4.2	4.5	4.8	5.1
18	0.0	0.3	0.6	0.9	1.2	1.5	1.8	2.1	2.4	2.7	3.0	3.3	3.6	3.9	4.2	4.5	4.8	5.1	5.4
19	0.0	0.3	0.6	1.0	1.3	1.6	1.9	2.2	2.5	2.8	3.2	3.5	3.8	4.1	4.4	4.8	5.1	5.4	5.7
20	0.0	0.3	0.7	1.0	1.3	1.7	2.0	2.3	2.7	3.0	3.3	3.7	4.0	4.3	4.7	5.0	5.3	5.7	6.0
21	0.0	0.4	0.7	1.0	1.4	1.8	2.1	2.4	2.8	3.2	3.5	3.8	4.2	4.6	4.9	5.2	5.6	6.0	6.3
22	0.0	0.4	0.7	1.1	1.5	1.8	2.2	2.6	2.9	3.3	3.7	4.0	4.4	4.8	5.1	5.5	5.9	6.2	6.6
23	0.0	0.4	0.8	1.2	1.5	1.9	2.3	2.7	3.1	3.4	3.8	4.2	4.6	5.0	5.4	5.8	6.1	6.5	6.9
24	0.0	0.4	0.8	1.2	1.6	2.0	2.4	2.8	3.2	3.6	4.0	4.4	4.8	5.2	5.6	6.0	6.4	6.8	7.2
25	0.0	0.4	0.8	1.2	1.7	2.1	2.5	2.9	3.3	3.8	4.2	4.6	5.0	5.4	5.8	6.2	6.7	7.1	7.5
26	0.0	0.4	0.9	1.3	1.7	2.2	2.6	3.0	3.5	3.9	4.3	4.8	5.2	5.6	6.1	6.5	6.9	7.4	7.8
27	0.0	0.4	0.9	1.3	1.8	2.2	2.7	3.2	3.6	4.0	4.5	5.0	5.4	5.8	6.3	6.8	7.2	7.6	8.1
28	0.0	0.5	0.9	1.4	1.9	2.3	2.8	3.3	3.7	4.2	4.7	5.1	5.6	6.1	6.5	7.0	7.5	7.9	8.4
29	0.0	0.5	1.0	1.4	1.9	2.4	2.9	3.4	3.9	4.4	4.8	5.3	5.8	6.3	6.8	7.2	7.7	8.2	8.7
30	0.0	0.5	1.0	1.5	2.0	2.5	3.0	3.5	4.0	4.5	5.0	5.5	6.0	6.5	7.0	7.5	8.0	8.5	9.0
31	0.0	0.5	1.0	1.6	2.1	2.6	3.1	3.6	4.1	4.6	5.2	5.7	6.2	6.7	7.2	7.8	8.3	8.8	9.3
32	0.0	0.5	1.1	1.6	2.1	2.7	3.2	3.7	4.3	4.8	5.3	5.9	6.4	6.9	7.5	8.0	8.5	9.1	9.6
33	0.0	0.6	1.1	1.6	2.2	2.8	3.3	3.8	4.4	5.0	5.5	6.0	6.6	7.2	7.7	8.2	8.8	9.4	9.9
34	0.0	0.6	1.1	1.7	2.3	2.8	3.4	4.0	4.5	5.1	5.7	6.2	6.8	7.4	7.9	8.5	9.1	9.6	10.2
35	0.0	0.6	1.2	1.8	2.3	2.9	3.5	4.1	4.7	5.2	5.8	6.4	7.0	7.6	8.2	8.8	9.3	9.9	10.5
36	0.0	0.6	1.2	1.8	2.4	3.0	3.6	4.2	4.8	5.4	6.0	6.6	7.2	7.8	8.4	9.0	9.6	10.2	10.8
37	0.0	0.6	1.2	1.8	2.5	3.1	3.7	4.3	4.9	5.6	6.2	6.8	7.4	8.0	8.6	9.2	9.9	10.5	11.1
38	0.0	0.6	1.3	1.9	2.5	3.2	3.8	4.4	5.1	5.7	6.3	7.0	7.6	8.2	8.9	9.5	10.1	10.8	11.4
39	0.0	0.6	1.3	2.0	2.6	3.2	3.9	4.6	5.2	5.8	6.5	7.2	7.8	8.4	9.1	9.8	10.4	11.0	11.7
40	0.0	0.7	1.3	2.0	2.7	3.3	4.0	4.7	5.3	6.0	6.7	7.3	8.0	8.7	9.3	10.0	10.7	11.3	12.0
41	0.0	0.7	1.4	2.0	2.7	3.4	4.1	4.8	5.5	6.2	6.8	7.5	8.2	8.9	9.6	10.2	10.9	11.6	12.3
42	0.0	0.7	1.4	2.1	2.8	3.5	4.2	4.9	5.6	6.3	7.0	7.7	8.4	9.1	9.8	10.5	11.2	11.9	12.6
43	0.0	0.7	1.4	2.2	2.9	3.6	4.3	5.0	5.7	6.4	7.2	7.9	8.6	9.3	10.0	10.8	11.5	12.2	12.9
44	0.0	0.7	1.5	2.2	2.9	3.7	4.4	5.1	5.9	6.6	7.3	8.1	8.8	9.5	10.3	11.0	11.7	12.5	13.2
45	0.0	0.7	1.5	2.2	3.0	3.8	4.5	5.2	6.0	6.8	7.5	8.2	9.0	9.8	10.5	11.2	12.0	12.8	13.5
46	0.0	0.8	1.5	2.3	3.1	3.8	4.6	5.4	6.1	6.9	7.7	8.4	9.2	10.0	10.7	11.5	12.3	13.0	13.8
47	0.0	0.8	1.6	2.4	3.1	3.9	4.7	5.5	6.3	7.0	7.8	8.6	9.4	10.2	11.0	11.8	12.5	13.3	14.1
48	0.0	0.8	1.6	2.4	3.2	4.0	4.8	5.6	6.4	7.2	8.0	8.8	9.6	10.4	11.2	12.0	12.8	13.6	14.4
49	0.0	0.8	1.6	2.4	3.3	4.1	4.9	5.7	6.5	7.4	8.2	9.0	9.8	10.6	11.4	12.2	13.1	13.9	14.7
50	0.0	0.8	1.7	2.5	3.3	4.2	5.0	5.8	6.7	7.5	8.3	9.2	10.0	10.8	11.7	12.5	13.3	14.2	15.0
51	0.0	0.8	1.7	2.6	3.4	4.2	5.1	6.0	6.8	7.6	8.5	9.4	10.2	11.0	11.9	12.8	13.6	14.4	15.3
52	0.0	0.9	1.7	2.6	3.5	4.3	5.2	6.1	6.9	7.8	8.7	9.5	10.4	11.3	12.1	13.0	13.9	14.7	15.6
53	0.0	0.9	1.8	2.6	3.5	4.4	5.3	6.2	7.1	8.0	8.8	9.7	10.6	11.5	12.4	13.2	14.1	15.0	15.9
54	0.0	0.9	1.8	2.7	3.6	4.5	5.4	6.3	7.2	8.1	9.0	9.9	10.8	11.7	12.6	13.5	14.4	15.3	16.2
55	0.0	0.9	1.8	2.8	3.7	4.6	5.5	6.4	7.3	8.2	9.2	10.1	11.0	11.9	12.8	13.8	14.7	15.6	16.5
56	0.0	0.9	1.9	2.8	3.7	4.7	5.6	6.5	7.5	8.4	9.3	10.3	11.2	12.1	13.1	14.0	14.9	15.9	16.8
57	0.0	1.0	1.9	2.8	3.8	4.8	5.7	6.6	7.6	8.6	9.5	10.4	11.4	12.4	13.3	14.2	15.2	16.2	17.1
58	0.0	1.0	1.9	2.9	3.9	4.8	5.8	6.8	7.7	8.7	9.7	10.6	11.6	12.6	13.5	14.5	15.5	16.4	17.4
59	0.0	1.0	2.0	3.0	3.9	4.9	5.9	6.9	7.9	8.8	9.8	10.8	11.8	12.8	13.8	14.8	15.7	16.7	17.7
60	0.0	1.0	2.0	3.0	4.0	5.0	6.0	7.0	8.0	9.0	10.0	11.0	12.0	13.0	14.0	15.0	16.0	17.0	18.0

NOTE: The table above gives a moon declination correction for minutes based on the Var/hr taken from the monthly table. The correction for hours is simply Var/hr multiplied by hours.

MOON MERIDIAN PASSAGE CORRECTION TABLE

Longitude (°E/W)	Daily Difference of Meridian Passage											
Minutes	**39**	**42**	**45**	**48**	**51**	**54**	**57**	**60**	**63**	**66**	**69**	
0	0	0	0	0	0	0	0	0	0	0	0	**0**
10	1	1	1	1	1	1	2	2	2	2	2	**10**
20	2	2	2	3	3	3	3	3	3	4	4	**20**
30	3	3	4	4	4	4	5	5	5	5	6	**30**
40	4	5	5	5	6	6	6	7	7	7	8	**40**
50	5	6	6	7	7	7	8	8	9	9	10	**50**
60	6	7	7	8	8	9	9	10	10	11	11	**60**
70	8	8	9	9	10	10	11	12	12	13	13	**70**
80	9	9	10	11	11	12	13	13	14	15	15	**80**
90	10	10	11	12	13	13	14	15	16	16	17	**90**
100	11	12	12	13	14	15	16	17	17	18	19	**100**
110	12	13	14	15	16	16	17	18	19	20	21	**110**
120	13	14	15	16	17	18	19	20	21	22	23	**120**
130	14	15	16	17	18	19	21	22	23	24	25	**130**
140	15	16	17	19	20	21	22	23	24	26	27	**140**
150	16	17	19	20	21	22	24	25	26	27	29	**150**
160	17	19	20	21	23	24	25	27	28	29	31	**160**
170	18	20	21	23	24	25	27	28	30	31	33	**170**
180	19	21	22	24	25	27	28	30	31	33	34	**180**

NOTE: Apply correction in **minutes** to the time of meridian passage given in the monthly tables. **Add** the correction if longitude is West; **subtract** if longitude is East.

MOON RISING AND SETTING CORRECTION TABLE

Latitude (°N) / **Dec (N/S)**	**2°**	**4°**	**6°**	**8°**	**10°**	**12°**	**14°**	**16°**	**18°**	**20°**	**22°**	**24°**	**26°**	**28°**	
	′	′	′	′	′	′	′	′	′	′	′	′	′	′	
0	5	9	14	19	23	28	33	38	43	49	54	60	65	72	**0**
2	4	9	13	17	22	26	31	36	41	46	51	56	62	67	**2**
4	4	8	12	16	21	25	29	34	38	43	47	52	58	63	**4**
6	4	8	11	15	19	23	27	31	35	40	44	49	54	59	**6**
8	3	7	11	14	18	21	25	29	33	37	41	45	50	54	**8**
10	3	6	10	13	16	20	23	27	30	34	38	42	46	50	**10**
12	3	6	9	12	15	18	21	24	27	31	34	38	42	46	**12**
14	3	5	8	11	13	16	19	22	25	28	31	34	37	41	**14**
16	2	5	7	9	12	14	17	19	22	25	27	30	33	36	**16**
18	2	4	6	8	10	12	15	17	19	21	24	26	29	32	**18**
20	2	3	5	7	9	10	12	14	16	18	20	22	25	27	**20**
22	1	3	4	6	7	8	10	12	13	15	16	18	20	22	**22**
24	1	2	3	4	5	6	8	9	10	11	13	14	15	17	**24**
26	1	1	2	3	4	4	5	6	7	8	8	9	10	11	**26**
28	0	1	1	1	2	2	3	3	3	4	4	5	5	6	**28**
30	0	0	0	0	0	0	0	0	0	0	0	0	0	0	**30**
32	0	1	1	2	2	2	3	3	4	4	5	5	6	6	**32**
34	1	2	2	3	4	5	6	6	7	8	9	10	11	13	**34**
36	1	2	4	5	6	7	9	10	11	13	14	16	18	19	**36**
38	2	3	5	7	8	10	12	14	16	18	20	22	24	27	**38**
40	2	4	6	8	11	13	15	18	20	23	25	28	31	34	**40**
42	3	5	8	10	13	16	19	22	25	28	31	35	39	43	**42**
44	3	6	9	13	16	19	23	26	30	34	38	42	47	52	**44**
46	4	7	11	15	19	23	27	31	35	40	45	50	56	62	**46**
48	4	9	13	17	22	26	31	36	41	47	53	59	66	73	**48**
50	5	10	15	20	25	30	36	42	48	54	61	69	77	86	**50**
52	6	11	17	23	29	35	41	48	55	63	71	79	89	100	**52**
54	6	13	19	26	33	40	47	55	63	72	81	92	103	117	**54**
56	7	15	22	29	37	45	54	63	72	82	93	106	120	137	**56**
58	8	16	25	33	42	51	61	71	82	94	107	122	140	162	**58**

NOTE: This table corrects the times of moonrise and moonset given for latitude 30°N in the monthly pages for your latitude. Find the moon's approximate declination from the moon table on the sixth page of the month. All corrections are in **minutes** and are to be applied as follows:

From latitude **0° to 30°N** with declination **N**: **Add** to moonrise, **subtract** from moonset
From latitude **0° to 30°N** with declination **S**: **Subtract** from moonrise, **add** to moonset
From latitude **30°N to 58°N** with declination **N**: **Subtract** from moonrise, **add** to moonset
From latitude **30°N to 58°N** with declination **S**: **Add** to moonrise, **subtract** from moonset

For greatest precision, also apply daily change (Diff) correction for your longitude from table at top of page.

PLANETS GHA CORRECTION TABLE (HOURS)

Always add

Var/hr	14°58.8′	14°59.0′	14°59.1′	14°59.3′	14°59.4′	14°59.6′	14°59.7′	14°59.9′	15°0.0′
Hours	° ′	° ′	° ′	° ′	° ′	° ′	° ′	° ′	° ′
0	0 00.0	0 00.0	0 00.0	0 00.0	0 00.0	0 00.0	0 00.0	0 00.0	0 00.0
1	14 58.8	14 59.0	14 59.1	14 59.3	14 59.4	14 59.6	14 59.7	14 59.9	15 00.0
2	29 57.6	29 58.0	29 58.2	29 58.6	29 58.8	29 59.2	29 59.4	29 59.8	30 00.0
3	44 56.4	44 57.0	44 57.3	44 57.9	44 58.2	44 58.8	44 59.1	44 59.7	45 00.0
4	59 55.2	59 56.0	59 56.4	59 57.2	59 57.6	59 58.4	59 58.8	59 59.6	60 00.0
5	74 54.0	74 55.0	74 55.5	74 56.5	74 57.0	74 58.0	74 58.5	74 59.5	75 00.0
6	89 52.8	89 54.0	89 54.6	89 55.8	89 56.4	89 57.6	89 58.2	89 59.4	90 00.0
7	104 51.6	104 53.0	104 53.7	104 55.1	104 55.8	104 57.2	104 57.9	104 59.3	105 00.0
8	119 50.4	119 52.0	119 52.8	119 54.4	119 55.2	119 56.8	119 57.6	119 59.2	120 00.0
9	134 49.2	134 51.0	134 51.9	134 53.7	134 54.6	134 56.4	134 57.3	134 59.1	135 00.0
10	149 48.0	149 50.0	149 51.0	149 53.0	149 54.0	149 56.0	149 57.0	149 59.0	150 00.0
11	164 46.8	164 49.0	164 50.1	164 52.3	164 53.4	164 55.6	164 56.7	164 58.9	165 00.0
12	179 45.6	179 48.0	179 49.2	179 51.6	179 52.8	179 55.2	179 56.4	179 58.8	180 00.0
13	194 44.4	194 47.0	194 48.3	194 50.9	194 52.2	194 54.8	194 56.1	194 58.7	195 00.0
14	209 43.2	209 46.0	209 47.4	209 50.2	209 51.6	209 54.4	209 55.8	209 58.6	210 00.0
15	224 42.0	224 45.0	224 46.5	224 49.5	224 51.0	224 54.0	224 55.5	224 58.5	225 00.0
16	239 40.8	239 44.0	239 45.6	239 48.8	239 50.4	239 53.6	239 55.2	239 58.4	240 00.0
17	254 39.6	254 43.0	254 44.7	254 48.1	254 49.8	254 53.2	254 54.9	254 58.3	255 00.0
18	269 38.4	269 42.0	269 43.8	269 47.4	269 49.2	269 52.8	269 54.6	269 58.2	270 00.0
19	284 37.2	284 41.0	284 42.9	284 46.7	284 48.6	284 52.4	284 54.3	284 58.1	285 00.0
20	299 36.0	299 40.0	299 42.0	299 46.0	299 48.0	299 52.0	299 54.0	299 58.0	300 00.0
21	314 34.8	314 39.0	314 41.1	314 45.3	314 47.4	314 51.6	314 53.7	314 57.9	315 00.0
22	329 33.6	329 38.0	329 40.2	329 44.6	329 46.8	329 51.2	329 53.4	329 57.8	330 00.0
23	344 32.4	344 37.0	344 39.3	344 43.7	344 46.2	344 50.8	344 53.1	344 57.7	345 00.0
24	359 31.2	359 36.0	359 38.4	359 43.2	359 45.6	359 50.4	359 52.8	359 57.6	0 00.0

Var/hr	15°00.2′	15°00.3′	15°00.5′	15°00.6′	15°00.8′	15°00.9′	15°01.1′	15°01.2′	15°01.4′
Hours	° ′	° ′	° ′	° ′	° ′	° ′	° ′	° ′	° ′
0	0 00.0	0 00.0	0 00.0	0 00.0	0 00.0	0 00.0	0 00.0	0 00.0	0 00.0
1	15 00.2	15 00.3	15 00.5	15 00.6	15 00.8	15 00.9	15 01.1	15 01.2	15 01.4
2	30 00.4	30 00.6	30 01.0	30 01.2	30 01.6	30 01.8	30 02.2	30 02.4	30 02.8
3	45 00.6	45 00.9	45 01.5	45 01.8	45 02.4	45 02.7	45 03.3	45 03.6	45 04.2
4	60 00.8	60 01.2	60 02.0	60 02.4	60 03.2	60 03.6	60 04.4	60 04.8	60 05.6
5	75 01.0	75 01.5	75 02.5	75 03.0	75 04.0	75 04.5	75 05.5	75 06.0	75 07.0
6	90 01.2	90 01.8	90 03.0	90 03.6	90 04.8	90 05.4	90 06.6	90 07.2	90 08.4
7	105 01.4	105 02.1	105 03.5	105 04.2	105 05.6	105 06.3	105 07.7	105 08.4	105 09.8
8	120 01.6	120 02.4	120 04.0	120 04.8	120 06.4	120 07.2	120 08.8	120 09.6	120 11.2
9	135 01.8	135 02.7	135 04.5	135 05.4	135 07.2	135 08.1	135 09.9	135 10.8	135 12.6
10	150 02.0	150 03.0	150 05.0	150 06.0	150 08.0	150 09.0	150 11.0	150 12.0	150 14.0
11	165 02.2	165 03.3	165 05.5	165 06.6	165 08.8	165 09.9	165 12.1	165 13.2	165 15.4
12	180 02.4	180 03.6	180 06.0	180 07.2	180 09.6	180 10.8	180 13.2	180 14.4	180 16.8
13	195 02.6	195 03.9	190 06.5	195 07.8	195 10.4	195 11.7	195 14.3	195 15.6	195 18.2
14	210 02.8	210 04.2	210 07.0	210 08.4	210 11.2	210 12.6	210 15.4	210 16.8	210 19.6
15	225 03.0	225 04.5	225 07.5	225 09.0	225 12.0	225 13.5	225 16.5	225 18.5	225 21.0
16	240 03.2	240 04.8	240 08.0	240 09.6	240 12.8	240 14.4	240 17.6	240 19.2	240 22.4
17	255 03.4	255 05.1	255 08.5	255 10.2	255 13.6	255 15.3	255 18.7	255 20.4	255 23.8
18	270 03.6	270 05.4	270 09.0	270 10.8	270 14.4	270 16.2	270 19.8	270 21.6	270 25.2
19	285 03.8	285 05.7	285 09.5	285 11.4	285 15.2	285 17.1	285 20.9	285 22.8	285 26.6
20	300 04.0	300 06.0	300 10.0	300 12.0	300 16.0	300 18.0	300 22.0	300 24.0	300 28.0
21	315 04.2	315 06.3	315 10.5	315 12.6	315 16.8	315 18.9	315 23.1	315 25.2	315 29.4
22	330 04.4	330 06.6	330 11.0	330 13.2	330 17.6	330 19.8	330 24.2	330 26.4	330 30.8
23	345 04.6	345 06.9	345 11.5	345 13.8	345 18.4	345 20.7	345 25.3	345 27.6	345 32.2
24	0 04.8	0 07.2	0 12.0	0 14.4	0 19.2	0 21.6	0 26.4	0 28.8	0 33.6

NOTE: The correction for planet GHA is taken in three parts. Using the daily figure for variation per hour given on the monthly page, you get the correction for hours from the table on this or the following page, and then the correction for minutes from the table on page E 115. Finally, take the correction for seconds from the right-hand side of the minute table.

PLANETS GHA CORRECTION TABLE (HOURS)

Always add

Var/hr	15°01.5′	15°01.7′	15°01.8′	15°02.0′	15°02.1′	15°02.3′	15°02.4′	15°02.6′	15°02.7′
Hours	° ′	° ′	° ′	° ′	° ′	° ′	° ′	° ′	° ′
0	0 00.0	0 00.0	0 00.0	0 00.0	0 00.0	0 00.0	0 00.0	0 00.0	0 00.0
1	15 01.5	15 01.7	15 01.8	15 02.0	15 02.1	15 02.3	15 02.4	15 02.6	15 02.7
2	30 03.0	30 03.4	30 03.6	30 04.0	30 04.2	30 04.6	30 04.8	30 05.2	30 05.4
3	45 04.5	45 05.1	45 05.4	45 06.0	45 06.3	45 06.9	45 07.2	45 07.8	45 08.1
4	60 06.0	60 06.8	60 07.2	60 08.0	60 08.4	60 09.2	60 09.6	60 10.4	60 10.8
5	75 07.5	75 08.5	75 09.0	75 10.0	75 10.5	75 11.5	75 12.0	75 13.0	75 13.5
6	90 09.0	90 10.2	90 10.8	90 12.0	90 12.6	90 13.8	90 14.4	90 15.6	90 16.2
7	105 10.5	105 11.9	105 12.6	105 14.0	105 14.7	105 16.1	105 16.8	105 18.2	105 18.9
8	120 12.0	120 13.6	120 14.4	120 16.0	120 16.8	120 18.4	120 19.2	120 20.8	120 21.6
9	135 13.5	135 15.3	135 16.2	135 18.0	135 18.9	135 20.7	135 21.6	135 23.4	135 24.3
10	150 15.0	150 17.0	150 18.0	150 20.0	150 21.0	150 23.0	150 24.0	150 26.0	150 27.0
11	165 16.5	165 18.7	165 19.8	165 22.0	165 23.1	165 25.8	165 26.4	165 28.6	165 29.7
12	180 18.0	180 20.4	180 21.6	180 24.0	180 25.2	180 27.6	180 28.8	180 31.2	180 32.4
13	195 19.5	195 22.1	195 23.4	195 26.0	195 27.3	195 29.9	195 31.2	195 33.8	195 35.1
14	210 21.0	210 23.8	210 25.2	210 28.0	210 29.4	210 32.2	210 33.6	210 36.4	210 37.8
15	225 22.5	225 25.5	225 27.0	225 30.0	225 31.5	225 34.5	225 36.0	225 39.0	225 40.5
16	240 24.0	240 27.2	240 28.8	240 32.0	240 33.6	240 36.8	240 38.4	240 41.6	240 43.2
17	255 25.5	255 28.9	255 30.6	255 34.0	255 35.7	255 39.1	255 40.8	255 44.8	255 45.9
18	270 27.0	270 30.6	270 32.4	270 36.0	270 37.8	270 41.4	270 43.2	270 46.8	270 48.6
19	285 28.5	285 32.3	285 34.2	285 38.0	285 39.9	285 43.7	285 45.6	285 49.4	285 51.3
20	300 30.0	300 34.0	300 36.0	300 40.0	300 42.0	300 46.0	300 48.0	300 52.0	300 54.0
21	315 31.5	315 35.7	315 37.8	315 42.0	315 44.1	315 48.3	315 50.4	315 54.6	315 56.7
22	330 33.0	330 37.4	330 39.6	330 44.0	330 46.2	330 50.6	330 52.8	330 57.2	330 59.4
23	345 34.5	345 39.1	345 41.4	345 46.0	345 48.3	345 52.9	345 55.2	345 59.8	346 02.1
24	0 36.0	0 40.8	0 43..2	0 48.0	0 50.4	0 55.2	0 57.6	1 02.4	1 04.8

Var/hr	15°02.9′	15°03.0′	15°03.2′	15°03.3′	15°03.5′	15°03.6′	15°03.8′	15°03.9′	15°04.1′
Hours	° ′	° ′	° ′	° ′	° ′	° ′	° ′	° ′	° ′
0	0 00.0	0 00.0	0 00.0	0 00.0	0 00.0	0 00.0	0 00.0	0 00.0	0 00.0
1	15 02.9	15 03.0	15 03.2	15 03.3	15 03.5	15 03.6	15 03.8	15 03.9	15 04.1
2	30 05.8	30 06.0	30 06.4	30 06.6	30 07.0	15 07.2	30 07.6	30 07.8	30 08.2
3	45 08.7	45 09.0	45 09.6	45 09.9	45 10.5	45 10.8	45 11.4	45 11.7	45 12.3
4	60 11.6	60 12.0	60 12.8	60 13.2	60 14.0	60 14.4	60 15.2	60 15.6	60 16.4
5	75 14.5	75 15.0	75 16.0	75 16.5	75 17.5	75 18.0	75 19.0	75 19.5	75 20.5
6	90 17.4	90 18.0	90 19.2	90 19.8	90 21.0	90 21.6	90 22.8	90 23.4	90 24.6
7	105 20.3	105 21.0	105 22.4	105 23.1	105 24.5	105 25.2	105 26.6	105 27.3	105 28.7
8	120 23.2	120 24.0	120 25.6	120 26.4	120 28.0	120 28.8	120 30.4	120 31.2	120 32.8
9	135 26.1	135 27.0	135 28.8	135 29.7	135 31.5	135 32.4	135 34.2	135 35.1	135 36.9
10	150 29.0	150 30.0	150 32.0	150 33.0	150 35.0	150 36.0	150 38.0	150 39.0	150 41.0
11	165 31.9	165 33.0	165 35.2	165 36.3	165 38.5	165 39.6	165 41.8	165 42.9	165 45.1
12	180 34.8	180 36.0	180 38.4	180 39.6	180 42.0	180 43.2	180 45.6	180 46.8	180 49.2
13	195 37.7	195 39.0	195 41.6	195 42.9	195 45.5	195 46.8	195 49.4	195 50.7	195 53.3
14	210 40.6	210 42.0	210 44.8	210 46.2	210 49.0	210 50.4	210 53.2	210 54.6	210 57.4
15	225 43.5	225 45.0	225 48.0	225 49.5	225 52.5	225 54.0	225 57.0	225 58.5	226 01.5
16	240 46.4	240 48.0	240 51.2	240 52.8	240 56.0	240 57.6	241 00.8	241 02.4	241 05.6
17	255 49.3	255 51.0	255 54.4	255 56.1	255 59.5	256 01.2	256 04.6	256 06.3	256 09.7
18	270 52.2	270 54.0	270 57.6	270 59.4	271 03.0	271 04.8	271 08.4	271 10.2	271 13.8
19	285 55.1	285 57.0	286 00.8	286 02.7	286 06.5	286 08.4	286 12.2	286 14.1	286 17.9
20	300 58.0	301 00.0	301 04.0	301 06.0	301 10.0	301 12.0	301 16.0	301 18.0	301 22.0
21	316 00.9	316 03.0	316 07.2	316 09.3	316 13.5	316 15.6	316 19.8	316 21.9	316 26.1
22	331 03.8	331 06.0	331 10.4	331 12.6	331 17.0	331 19.2	331 23.6	331 25.8	331 30.2
23	346 06.7	346 09.0	346 13.6	346 15.9	346 20.5	346 22.8	346 27.4	346 29.7	346 34.3
24	001 09.6	001 12.0	001 16.8	001 19.2	001 24.0	001 26.4	001 31.2	001 33.6	001 38.4

NOTE: The correction for planet GHA is taken in three parts. Using the daily figure for variation per hour given on the monthly page, you get the correction for hours from the table on this or the preceding page, and then the correction for minutes from the table on the opposite page. Finally, take the correction for seconds from the right-hand side of the minute table.

PLANETS GHA CORRECTION TABLE (MINUTES)

Always add

Var/hr	14°58.8′	14°59.4′	15°00.0′	15°00.6′	15°01.2′	15°01.8′	15°02.4′	15°03.0′	15°03.6′	Sec	
Minutes	° ′	° ′	° ′	° ′	° ′	° ′	° ′	° ′	° ′		′
0	0 00.0	0 00.0	0 00.0	0 00.0	0 00.0	0 00.0	0 00.0	0 00.0	0 00.0	**0**	0.0
1	0 15.0	0 15.0	0 15.0	0 15.0	0 15.0	0 15.0	0 15.0	0 15.0	0 15.1	**1**	0.3
2	0 30.0	0 30.0	0 30.0	0 30.0	0 30.0	0 30.1	0 30.1	0 30.1	0 30.1	**2**	0.5
3	0 44.9	0 45.0	0 45.0	0 45.0	0 45.1	0 45.1	0 45.1	0 45.1	0 45.2	**3**	0.8
4	0 59.9	1 00.0	1 00.0	1 00.0	1 00.1	1 00.1	1 00.2	1 00.2	1 00.2	**4**	1.0
5	1 14.9	1 14.9	1 15.0	1 15.0	1 15.1	1 15.2	1 15.2	1 15.2	1 15.3	**5**	1.3
6	1 29.9	1 29.9	1 30.0	1 30.1	1 30.1	1 30.2	1 30.2	1 30.3	1 30.4	**6**	1.5
7	1 44.9	1 44.9	1 45.0	1 45.1	1 45.1	1 45.2	1 45.3	1 45.3	1 45.4	**7**	1.8
8	1 59.8	1 59.9	2 00.0	2 00.1	2 00.2	2 00.2	2 00.3	2 00.4	2 00.5	**8**	2.0
9	2 14.8	2 14.9	2 15.0	2 15.1	2 15.2	2 15.3	2 15.4	2 15.4	2 15.5	**9**	2.3
10	2 29.8	2 29.9	2 30.0	2 30.1	2 30.2	2 30.3	2 30.4	2 30.5	2 30.6	**10**	2.5
11	2 44.8	2 44.9	2 45.0	2 45.1	2 45.2	2 45.3	2 45.4	2 45.5	2 45.7	**11**	2.8
12	2 59.8	2 59.9	3 00.0	3 00.1	3 00.2	3 00.4	3 00.5	3 00.6	3 00.7	**12**	3.0
13	3 14.7	3 14.9	3 15.0	3 15.1	3 15.3	3 15.4	3 15.5	3 15.6	3 15.8	**13**	3.3
14	3 29.7	3 29.9	3 30.0	3 30.1	3 30.3	3 30.4	3 30.6	3 30.7	3 30.8	**14**	3.5
15	3 44.7	3 44.8	3 45.0	3 45.2	3 45.3	3 45.4	3 45.6	3 45.7	3 45.9	**15**	3.8
16	3 59.7	3 59.8	4 00.0	4 00.2	4 00.3	4 00.5	4 00.6	4 00.8	4 01.0	**16**	4.0
17	4 14.7	4 14.8	4 15.0	4 15.2	4 15.3	4 15.5	4 15.7	4 15.8	4 16.0	**17**	4.3
18	4 29.6	4 29.8	4 30.0	4 30.2	4 30.4	4 30.5	4 30.7	4 30.9	4 31.1	**18**	4.5
19	4 44.6	4 44.8	4 45.0	4 45.2	4 45.4	4 45.6	4 45.8	4 45.9	4 46.1	**19**	4.8
20	4 59.6	4 59.8	5 00.0	5 00.2	5 00.4	5 00.6	5 00.8	5 01.0	5 01.2	**20**	5.0
21	5 14.6	5 14.8	5 15.0	5 15.2	5 15.4	5 15.6	5 15.8	5 16.0	5 16.3	**21**	5.3
22	5 29.6	5 29.8	5 30.0	5 30.2	5 30.4	5 30.7	5 30.9	5 31.1	5 31.3	**22**	5.5
23	5 44.5	5 44.8	5 45.0	5 45.2	5 45.5	5 45.7	5 45.9	5 46.1	5 46.4	**23**	5.8
24	5 59.5	5 59.8	6 00.0	6 00.2	6 00.5	6 00.7	6 01.0	6 01.2	6 01.4	**24**	6.0
25	6 14.5	6 14.8	6 15.0	6 15.2	6 15.5	6 15.7	6 16.0	6 16.2	6 16.5	**25**	6.3
26	6 29.5	6 29.7	6 30.0	6 30.3	6 30.5	6 30.8	6 31.0	6 31.3	6 31.6	**26**	6.5
27	6 04.5	6 44.7	6 45.0	6 45.3	6 45.5	6 45.8	6 46.1	6 46.3	6 46.6	**27**	6.8
28	6 59.4	6 59.7	7 00.0	7 00.3	7 00.6	7 00.8	7 01.1	7 01.4	7 01.7	**28**	7.0
29	7 14.4	7 14.7	7 15.0	7 15.3	7 15.6	7 15.9	7 16.2	7 16.4	7 16.7	**29**	7.3
30	7 29.4	7 29.7	7 30.0	7 30.3	7 30.6	7 30.9	7 31.2	7 31.5	7 31.8	**30**	7.5
31	7 44.4	7 44.7	7 45.0	7 45.3	7 45.6	7 45.9	7 46.2	7 46.5	7 46.9	**31**	7.8
32	7 59.4	7 59.7	8 00.0	8 00.3	8 00.6	8 01.0	8 01.3	8 01.6	8 01.9	**32**	8.0
33	8 14.3	8 14.7	8 15.0	8 15.3	8 15.7	8 16.0	8 16.3	8 16.6	8 17.0	**33**	8.3
34	8 29.3	8 29.7	8 30.0	8 30.3	8 30.7	8 31.0	8 31.4	8 31.7	8 32.0	**34**	8.5
35	8 44.3	8 44.6	8 45.0	8 45.5	8 45.7	8 46.0	8 46.4	8 46.7	8 47.1	**35**	8.8
36	8 59.3	8 59.6	9 00.0	9 00.4	9 00.7	9 01.1	9 01.4	9 01.8	9 02.2	**36**	9.0
37	9 14.3	9 14.6	9 15.0	9 15.4	9 15.7	9 16.1	9 16.5	9 16.8	9 17.2	**37**	9.3
38	9 29.2	9 29.6	9 30.0	9 30.4	9 30.8	9 31.1	9 31.5	9 31.9	9 32.3	**38**	9.5
39	9 44.2	9 44.6	9 45.0	9 45.4	9 45.8	9 46.2	9 46.6	9 46.9	9 47.3	**39**	9.8
40	9 59.2	9 59.6	10 00.0	10 00.4	10 00.8	10 01.2	10 01.6	10 02.0	10 02.4	**40**	10.0
41	10 14.2	10 14.6	10 15.0	10 15.4	10 15.8	10 16.2	10 16.6	10 17.0	10 17.5	**41**	10.3
42	10 29.2	10 29.6	10 30.0	10 30.4	10 30.8	10 31.3	10 31.7	10 32.1	10 32.5	**42**	10.5
43	10 44.1	10 44.6	10 45.0	10 45.4	10 45.9	10 46.3	10 46.7	10 47.1	10 47.6	**43**	10.8
44	10 59.1	10 59.6	11 00.0	11 00.4	11 00.9	11 01.3	11 01.8	11 02.2	11 02.6	**44**	11.0
45	11 14.1	11 14.6	11 15.0	11 15.4	11 15.9	11 16.3	11 16.8	11 17.2	11 17.7	**45**	11.3
46	11 29.1	11 29.5	11 30.0	11 30.5	11 30.9	11 31.4	11 31.8	11 32.3	11 32.8	**46**	11.5
47	11 44.1	11 44.5	11 45.0	11 45.5	11 45.9	11 46.4	11 46.9	11 47.3	11 47.8	**47**	11.8
48	11 59.0	11 59.5	12 00.0	12 00.5	12 01.0	12 01.4	12 01.9	12 02.4	12 02.9	**48**	12.0
49	12 14.0	12 14.5	12 15.0	12 15.5	12 16.0	12 16.5	12 17.0	12 17.4	12 17.9	**49**	12.3
50	12 29.0	12 29.5	12 30.0	12 30.5	12 31.0	12 31.5	12 32.0	12 32.5	12 33.0	**50**	12.5
51	12 44.0	12 44.5	12 45.0	12 45.5	12 46.0	12 46.5	12 47.0	12 47.5	12 48.1	**51**	12.8
52	12 59.0	12 59.5	13 00.0	13 00.5	13 01.0	13 01.6	13 02.1	13 02.6	13 03.1	**52**	13.1
53	13 13.9	13 14.5	13 15.0	13 15.5	13 16.1	13 16.6	13 17.1	13 17.6	13 18.2	**53**	13.3
54	13 28.9	13 29.5	13 30.0	13 30.5	13 31.1	13 31.6	13 32.2	13 32.7	13 33.2	**54**	13.5
55	13 43.9	13 44.4	13 45.0	13 45.6	13 46.1	13 46.6	13 47.2	13 47.7	13 48.3	**55**	13.8
56	13 58.9	13 59.4	14 00.0	14 00.6	14 01.1	14 01.7	14 02.2	14 02.8	14 03.4	**56**	14.0
57	14 13.9	14 14.4	14 15.0	14 15.6	14 16.1	14 16.7	14 17.3	14 17.8	14 18.4	**57**	14.3
58	14 28.8	14 29.4	14 30.0	14 30.6	14 31.2	14 31.7	14 32.3	14 32.9	14 33.5	**58**	14.5
59	14 43.8	14 44.4	14 45.0	14 45.6	14 46.2	14 46.8	14 47.4	14 47.9	14 48.5	**59**	14.8
60	14 58.8	14 59.4	15 00.0	15 00.6	15 01.2	15 01.8	15 02.4	15 03.0	15 03.6	**60**	15.0

PLANETS DECLINATION CORRECTION TABLE

Var/hr	0.0′	0.1′	0.2′	0.3′	0.4′	0.5′	0.6′	0.7′	0.8′	0.9′	1.0′	1.1′	1.2′	1.3′	1.4′	1.5′
h m	′	′	′	′	′	′	′	′	′	′	′	′	′	′	′	′
0 00	0.0	0.0	0.0	0.0	0.0	0.0	0.0	0.0	0.0	0.0	0.0	0.0	0.0	0.0	0.0	0.0
12	0.0	0.0	0.0	0.1	0.1	0.1	0.1	0.1	0.2	0.2	0.2	0.2	0.2	0.3	0.3	0.3
24	0.0	0.0	0.1	0.1	0.2	0.2	0.2	0.3	0.3	0.4	0.4	0.4	0.5	0.5	0.6	0.6
36	0.0	0.1	0.1	0.2	0.2	0.3	0.4	0.4	0.5	0.5	0.6	0.7	0.7	0.8	0.8	0.9
48	0.0	0.1	0.2	0.2	0.3	0.4	0.5	0.6	0.6	0.7	0.8	0.9	1.0	1.0	1.1	1.2
1 00	0.0	0.1	0.2	0.3	0.4	0.5	0.6	0.7	0.8	0.9	1.0	1.1	1.2	1.3	1.4	1.5
12	0.0	0.1	0.2	0.4	0.5	0.6	0.7	0.8	1.0	1.1	1.2	1.3	1.4	1.6	1.7	1.8
24	0.0	0.1	0.3	0.4	0.6	0.7	0.8	1.0	1.1	1.3	1.4	1.5	1.7	1.8	2.0	2.1
36	0.0	0.2	0.3	0.5	0.6	0.8	1.0	1.1	1.3	1.4	1.6	1.8	1.9	2.1	2.2	2.4
48	0.0	0.2	0.4	0.5	0.7	0.9	1.1	1.3	1.4	1.6	1.8	2.0	2.2	2.3	2.5	2.7
2 00	0.0	0.2	0.4	0.6	0.8	1.0	1.2	1.4	1.6	1.8	2.0	2.2	2.4	2.6	2.8	3.0
12	0.0	0.2	0.4	0.7	0.9	1.1	1.3	1.5	1.8	2.0	2.2	2.4	2.6	2.9	3.1	3.3
24	0.0	0.2	0.5	0.7	1.0	1.2	1.4	1.7	1.9	2.2	2.4	2.6	2.9	3.1	3.4	3.6
36	0.0	0.3	0.5	0.8	1.0	1.3	1.6	1.8	2.1	2.3	2.6	2.9	3.1	3.4	3.6	3.9
48	0.0	0.3	0.6	0.8	1.1	1.4	1.7	2.0	2.2	2.5	2.8	3.1	3.4	3.6	3.9	4.2
3 00	0.0	0.3	0.6	0.9	1.2	1.5	1.8	2.1	2.4	2.7	3.0	3.3	3.6	3.9	4.2	4.5
12	0.0	0.3	0.6	1.0	1.3	1.6	1.9	2.2	2.6	2.9	3.2	3.5	3.8	4.2	4.5	4.8
24	0.0	0.3	0.7	1.0	1.4	1.7	2.0	2.4	2.7	3.1	3.4	3.7	4.1	4.4	4.8	5.1
36	0.0	0.4	0.7	1.1	1.4	1.8	2.2	2.5	2.9	3.2	3.6	4.0	4.3	4.7	5.0	5.4
48	0.0	0.4	0.8	1.1	1.5	1.9	2.3	2.7	3.0	3.4	3.8	4.2	4.6	4.9	5.3	5.7
4 00	0.0	0.4	0.8	1.2	1.6	2.0	2.4	2.8	3.2	3.6	4.0	4.4	4.8	5.2	5.6	6.0
12	0.0	0.4	0.8	1.3	1.7	2.1	2.5	2.9	3.4	3.8	4.2	4.6	5.0	5.5	5.9	6.3
24	0.0	0.4	0.9	1.3	1.8	2.2	2.6	3.1	3.5	4.0	4.4	4.8	5.3	5.7	6.2	6.6
36	0.0	0.5	0.9	1.4	1.8	2.3	2.8	3.2	3.7	4.1	4.6	5.1	5.5	6.0	6.4	6.9
48	0.0	0.5	1.0	1.4	1.9	2.4	2.9	3.4	3.8	4.3	4.8	5.3	5.8	6.2	6.7	7.2
5 00	0.0	0.5	1.0	1.5	2.0	2.5	3.0	3.5	4.0	4.5	5.0	5.5	6.0	6.5	7.0	7.5
12	0.0	0.5	1.0	1.6	2.1	2.6	3.1	3.6	4.2	4.7	5.2	5.7	6.2	6.8	7.3	7.8
24	0.0	0.5	1.1	1.6	2.2	2.7	3.2	3.8	4.3	4.9	5.4	5.9	6.5	7.0	7.6	8.1
36	0.0	0.6	1.1	1.7	2.2	2.8	3.4	3.9	4.5	5.0	5.6	6.2	6.7	7.3	7.8	8.4
48	0.0	0.6	1.2	1.7	2.3	2.9	3.5	4.1	4.6	5.2	5.8	6.4	7.0	7.5	8.1	8.7
6 00	0.0	0.6	1.2	1.8	2.4	3.0	3.6	4.2	4.8	5.4	6.0	6.6	7.2	7.8	8.4	9.0
12	0.0	0.6	1.2	1.9	2.5	3.1	3.7	4.3	5.0	5.6	6.2	6.8	7.4	8.1	8.7	9.3
24	0.0	0.6	1.3	1.9	2.6	3.2	3.8	4.5	5.1	5.8	6.4	7.0	7.7	8.3	9.0	9.6
36	0.0	0.7	1.3	2.0	2.6	3.3	4.0	4.6	5.3	5.9	6.6	7.3	7.9	8.6	9.2	9.9
48	0.0	0.7	1.4	2.0	2.7	3.4	4.1	4.8	5.4	6.1	6.8	7.5	8.2	8.8	9.5	10.2
7 00	0.0	0.7	1.4	2.1	2.8	3.5	4.2	4.9	5.6	6.3	7.0	7.7	8.4	9.1	9.8	10.5
12	0.0	0.7	1.4	2.2	2.9	3.6	4.3	5.0	5.8	6.5	7.2	7.9	8.6	9.4	10.1	10.8
24	0.0	0.7	1.5	2.2	3.0	3.7	4.4	5.2	5.9	6.7	7.4	8.1	8.9	9.6	10.4	11.1
36	0.0	0.8	1.5	2.3	3.0	3.8	4.6	5.3	6.1	6.8	7.6	8.4	9.1	9.9	10.6	11.4
48	0.0	0.8	1.6	2.3	3.1	3.9	4.7	5.5	6.2	7.0	7.8	8.6	9.4	10.1	10.9	11.7
8 00	0.0	0.8	1.6	2.4	3.2	4.0	4.8	5.6	6.4	7.2	8.0	8.8	9.6	10.4	11.2	12.0
12	0.0	0.8	1.6	2.5	3.3	4.1	4.9	5.7	6.6	7.4	8.2	9.0	9.8	10.7	11.5	12.3
24	0.0	0.8	1.7	2.5	3.4	4.2	5.0	5.9	6.7	7.6	8.4	9.2	10.1	10.9	11.8	12.6
36	0.0	0.9	1.7	2.6	3.4	4.3	5.2	6.0	6.9	7.7	8.6	9.5	10.3	11.2	12.0	12.9
48	0.0	0.9	1.8	2.6	3.5	4.4	5.3	6.2	7.0	7.9	8.8	9.7	10.6	11.4	12.3	13.2
9 00	0.0	0.9	1.8	2.7	3.6	4.5	5.4	6.3	7.2	8.1	9.0	9.9	10.8	11.7	12.6	13.5
12	0.0	0.9	1.8	2.8	3.7	4.6	5.5	6.4	7.4	8.3	9.2	10.1	11.0	12.0	12.9	13.8
24	0.0	0.9	1.9	2.8	3.8	4.7	5.6	6.6	7.5	8.5	9.4	10.3	11.3	12.2	13.2	14.1
36	0.0	1.0	1.9	2.9	3.8	4.8	5.8	6.7	7.7	8.6	9.6	10.6	11.5	12.5	13.4	14.4
48	0.0	1.0	2.0	2.9	3.9	4.9	5.9	6.9	7.8	8.8	9.8	10.8	11.8	12.7	13.7	14.7
10 00	0.0	1.0	2.0	3.0	4.0	5.0	6.0	7.0	8.0	9.0	10.0	11.0	12.0	13.0	14.0	15.0
12	0.0	1.0	2.0	3.1	4.1	5.1	6.1	7.1	8.2	9.2	10.2	11.2	12.2	13.2	14.3	15.3
24	0.0	1.0	2.1	3.1	4.2	5.2	6.2	7.3	8.3	9.4	10.4	11.4	12.5	13.5	14.6	15.6
36	0.0	1.1	2.1	3.2	4.2	5.3	6.4	7.4	8.5	9.5	10.6	11.7	12.7	13.8	14.8	15.9
48	0.0	1.1	2.2	3.2	4.3	5.4	6.5	7.6	8.6	9.7	10.8	11.9	13.0	14.0	15.1	16.2
11 00	0.0	1.1	2.2	3.3	4.4	5.5	6.6	7.7	8.8	9.9	11.0	12.1	13.2	14.3	15.4	16.5
12	0.0	1.1	2.2	3.4	4.5	5.6	6.7	7.8	9.0	10.1	11.2	12.3	13.4	14.6	15.7	16.8
24	0.0	1.1	2.3	3.4	4.6	5.7	6.8	8.0	9.1	10.3	11.4	12.5	13.7	14.8	16.0	17.1
36	0.0	1.2	2.3	3.5	4.6	5.8	7.0	8.1	9.3	10.4	11.6	12.8	13.9	15.1	16.2	17.4
48	0.0	1.2	2.4	3.5	4.7	5.9	7.1	8.3	9.4	10.6	11.8	13.0	14.2	15.3	16.5	17.7
12 00	0.0	1.2	2.4	3.6	4.8	6.0	7.2	8.4	9.6	10.8	12.0	13.2	14.4	15.6	16.8	18.0

PLANETS DECLINATION CORRECTION TABLE

Var/hr	0.0′	0.1′	0.2′	0.3′	0.4′	0.5′	0.6′	0.7′	0.8′	0.9′	1.0′	1.1′	1.2′	1.3′	1.4′	1.5′
h m	′	′	′	′	′	′	′	′	′	′	′	′	′	′	′	′
12 00	0.0	1.2	2.4	3.6	4.8	6.0	7.2	8.4	9.6	10.8	12.0	13.2	14.4	15.6	16.8	18.0
12	0.0	1.2	2.4	3.7	4.9	6.1	7.3	8.5	9.8	11.0	12.2	13.4	14.6	15.9	17.1	18.3
24	0.0	1.2	2.5	3.7	5.0	6.2	7.4	8.7	9.9	11.2	12.4	13.6	14.9	16.1	17.4	18.6
36	0.0	1.3	2.5	3.8	5.0	6.3	7.6	8.8	10.1	11.3	12.6	13.9	15.1	16.4	17.6	18.9
48	0.0	1.3	2.6	3.8	5.1	6.4	7.7	9.0	10.2	11.5	12.8	14.1	15.4	16.6	17.9	19.2
13 00	0.0	1.3	2.6	3.9	5.2	6.5	7.8	9.1	10.4	11.7	13.0	14.3	15.6	16.9	18.2	19.5
12	0.0	1.3	2.6	4.0	5.3	6.6	7.9	9.2	10.6	11.9	13.2	14.5	15.8	17.2	18.5	19.8
24	0.0	1.3	2.7	4.0	5.4	6.7	8.0	9.4	10.7	12.1	13.4	14.7	16.1	17.4	18.8	20.1
36	0.0	1.4	2.7	4.1	5.4	6.8	8.2	9.5	10.9	12.2	13.6	15.0	16.3	17.7	19.0	20.4
48	0.0	1.4	2.8	4.1	5.5	6.9	8.3	9.7	11.0	12.4	13.8	15.2	16.6	17.9	19.3	20.7
14 00	0.0	1.4	2.8	4.2	5.6	7.0	8.4	9.8	11.2	12.6	14.0	15.4	16.8	18.2	19.6	21.0
12	0.0	1.4	2.8	4.3	5.7	7.1	8.5	9.9	11.4	12.8	14.2	15.6	17.0	18.5	19.9	21.3
24	0.0	1.4	2.9	4.3	5.8	7.2	8.6	10.1	11.5	13.0	14.4	15.8	17.3	18.7	20.2	21.6
36	0.0	1.5	2.9	4.4	5.8	7.3	8.8	10.2	11.7	13.1	14.6	16.1	17.5	19.0	20.4	21.9
48	0.0	1.5	3.0	4.4	5.9	7.4	8.9	10.4	11.8	13.3	14.8	16.3	17.8	19.2	20.7	22.2
15 00	0.0	1.5	3.0	4.5	6.0	7.5	9.0	10.5	12.0	13.5	15.0	16.5	18.0	19.5	21.0	22.5
12	0.0	1.5	3.0	4.6	6.1	7.6	9.1	10.6	12.2	13.7	15.2	16.7	18.2	19.8	21.3	22.8
24	0.0	1.5	3.1	4.6	6.2	7.7	9.2	10.8	12.3	13.9	15.4	16.9	18.5	20.0	21.6	23.1
36	0.0	1.6	3.1	4.7	6.2	7.8	9.4	10.9	12.5	14.0	15.6	17.2	18.7	20.3	21.8	23.4
48	0.0	1.6	3.2	4.7	6.3	7.9	9.5	11.1	12.6	14.2	15.8	17.4	19.0	20.5	22.1	23.7
16 00	0.0	1.6	3.2	4.8	6.4	8.0	9.6	11.2	12.8	14.4	16.0	17.6	19.2	20.8	22.4	24.0
12	0.0	1.6	3.2	4.9	6.5	8.1	9.7	11.3	13.0	14.6	16.2	17.8	19.4	21.1	22.7	24.3
24	0.0	1.6	3.3	4.9	6.6	8.2	9.8	11.5	13.1	14.8	16.4	18.0	19.7	21.3	23.0	24.6
36	0.0	1.7	3.3	5.0	6.6	8.3	10.0	11.6	13.3	14.9	16.6	18.3	19.9	21.6	23.2	24.9
48	0.0	1.7	3.4	5.0	6.7	8.4	10.1	11.8	13.4	15.1	16.8	18.5	20.2	21.8	23.5	25.2
17 00	0.0	1.7	3.4	5.1	6.8	8.5	10.2	11.9	13.6	15.3	17.0	18.7	20.4	22.1	23.8	25.5
12	0.0	1.7	3.4	5.2	6.9	8.6	10.3	12.0	13.8	15.5	17.2	18.9	20.6	22.4	24.1	25.8
24	0.0	1.7	3.5	5.2	7.0	8.7	10.4	12.2	13.9	15.7	17.4	19.1	20.9	22.6	24.4	26.1
36	0.0	1.8	3.5	5.3	7.0	8.8	10.6	12.3	14.1	15.8	17.6	19.4	21.1	22.9	24.6	26.4
48	0.0	1.8	3.6	5.3	7.1	8.9	10.7	12.5	14.2	16.0	17.8	19.6	21.4	23.1	24.9	26.7
18 00	0.0	1.8	3.6	5.4	7.2	9.0	10.8	12.6	14.4	16.2	18.0	19.8	21.6	23.4	25.2	27.0
12	0.0	1.8	3.6	5.5	7.3	9.1	10.9	12.7	14.6	16.4	18.2	20.0	21.8	23.7	25.5	27.3
24	0.0	1.8	3.7	5.5	7.4	9.2	11.0	12.9	14.7	16.6	18.4	20.2	22.1	23.9	25.8	27.6
36	0.0	1.9	3.7	5.6	7.4	9.3	11.2	13.0	14.9	16.7	18.6	20.5	22.3	24.2	26.0	27.9
48	0.0	1.9	3.8	5.6	7.5	9.4	11.3	13.2	15.0	16.9	18.8	20.7	22.6	24.4	26.3	28.2
19 00	0.0	1.9	3.8	5.7	7.6	9.5	11.4	13.3	15.2	17.1	19.0	20.9	22.8	24.7	26.6	28.5
12	0.0	1.9	3.8	5.8	7.7	9.6	11.5	13.4	15.4	17.3	19.2	21.1	23.0	25.0	26.9	28.8
24	0.0	1.9	3.9	5.8	7.8	9.7	11.6	13.6	15.5	17.5	19.4	21.3	23.3	25.2	27.2	29.1
36	0.0	2.0	3.9	5.9	7.8	9.8	11.8	13.7	15.7	17.6	19.6	21.6	23.5	25.5	27.4	29.4
48	0.0	2.0	4.0	5.9	7.9	9.9	11.9	13.9	15.8	17.8	19.8	21.8	23.8	25.7	27.7	29.7
20 00	0.0	2.0	4.0	6.0	8.0	10.0	12.0	14.0	16.0	18.0	20.0	22.0	24.0	26.0	28.0	30.0
12	0.0	2.0	4.0	6.1	8.1	10.1	12.1	14.1	16.2	18.2	20.2	22.2	24.2	26.3	28.3	30.3
24	0.0	2.0	4.1	6.1	8.2	10.2	12.2	14.3	16.3	18.4	20.4	22.4	24.5	26.5	28.6	30.6
36	0.0	2.1	4.1	6.2	8.2	10.3	12.4	14.4	16.5	18.5	20.6	22.7	24.7	26.8	28.8	30.9
48	0.0	2.1	4.2	6.2	8.3	10.4	12.5	14.6	16.6	18.7	20.8	22.9	25.0	27.0	29.1	31.2
21 00	0.0	2.1	4.2	6.3	8.4	10.5	12.6	14.7	16.8	18.9	21.0	23.1	25.2	27.3	29.4	31.5
12	0.0	2.1	4.2	6.4	8.5	10.6	12.7	14.8	17.0	19.1	21.2	23.3	25.4	27.6	29.7	31.8
24	0.0	2.1	4.3	6.4	8.6	10.7	12.8	15.0	17.1	19.3	21.4	23.5	25.7	27.8	30.0	32.1
36	0.0	2.2	4.3	6.5	8.6	10.8	13.0	15.1	17.3	19.4	21.6	23.8	25.9	28.1	30.2	32.4
48	0.0	2.2	4.4	6.5	8.7	10.9	13.1	15.3	17.4	19.6	21.8	24.0	26.2	28.3	30.5	32.7
22 00	0.0	2.2	4.4	6.6	8.8	11.0	13.2	15.4	17.6	19.8	22.0	24.2	26.4	28.6	30.8	33.0
12	0.0	2.2	4.4	6.7	8.9	11.1	13.3	15.5	17.8	20.0	22.2	24.4	26.6	28.9	31.1	33.3
24	0.0	2.2	4.5	6.7	9.0	11.2	13.4	15.7	17.9	20.2	22.4	24.6	26.9	29.1	31.4	33.6
36	0.0	2.3	4.5	6.8	9.0	11.3	13.6	15.8	18.1	20.3	22.6	24.9	27.1	29.4	31.6	33.9
48	0.0	2.3	4.6	6.8	9.1	11.4	13.7	16.0	18.2	20.5	22.8	25.1	27.4	29.6	31.9	34.2
23 00	0.0	2.3	4.6	6.9	9.2	11.5	13.8	16.1	18.4	20.7	23.0	25.3	27.6	29.9	32.2	34.5
12	0.0	2.3	4.6	7.0	9.3	11.6	13.9	16.2	18.6	20.9	23.2	25.5	27.8	30.2	32.5	34.8
24	0.0	2.3	4.7	7.0	9.4	11.7	14.0	16.4	18.7	21.1	23.4	25.7	28.1	30.4	32.8	35.1
36	0.0	2.4	4.7	7.1	9.4	11.8	14.2	16.5	18.9	21.2	23.6	26.0	28.3	30.7	33.0	35.4
48	0.0	2.4	4.8	7.1	9.5	11.9	14.3	16.7	19.0	21.4	23.8	26.2	28.6	30.9	33.3	35.7
24 00	0.0	2.4	4.8	7.2	9.6	12.0	14.4	16.8	19.2	21.6	24.0	26.4	28.8	31.2	33.6	36.0

ABC TABLES

A + if hour angle is listed at top
– if hour angle is listed at bottom

LHA	1° 359°	2° 358°	3° 357°	4° 356°	5° 355°	6° 354°	7° 353°	8° 352°	9° 351°	10° 350°	11° 349°	12° 348°	13° 347°	14° 346°	15° 345°
Latitude (°N/S)															
0	0.00	.000	.000	.000	.000	.000	.000	.000	.000	.000	.000	.000	.000	.000	.000
3	3.00	1.50	1.00	.749	.599	.499	.427	.373	.331	.297	.270	.247	.227	.210	.196
6	6.02	3.01	2.01	1.50	1.20	1.00	.856	.748	.664	.596	.541	.494	.455	.422	.392
9	9.07	4.54	3.02	2.27	1.81	1.51	1.29	1.13	1.00	.898	.815	.745	.686	.635	.591
12	12.2	6.09	4.06	3.04	2.43	2.02	1.73	1.51	1.34	1.21	1.09	1.00	.921	.853	.793
15	15.4	7.67	5.11	3.83	3.06	2.55	2.18	1.91	1.69	1.52	1.38	1.26	1.16	1.07	1.00
18	18.6	9.30	6.20	4.65	3.71	3.09	2.65	2.31	2.05	1.84	1.67	1.53	1.41	1.30	1.21
21	22.0	11.0	7.32	5.49	4.39	3.65	3.13	2.73	2.42	2.18	1.97	1.81	1.66	1.54	1.43
24	25.5	12.7	8.50	6.37	5.09	4.24	3.63	3.17	2.81	2.53	2.29	2.09	1.93	1.79	1.66
27	29.2	14.6	9.72	7.29	5.82	4.85	4.15	3.63	3.22	2.89	2.62	2.40	2.21	2.04	1.90
30	33.1	16.5	11.0	8.26	6.60	5.49	4.70	4.11	3.65	3.27	2.97	2.72	2.50	2.32	2.15
33	37.2	18.6	12.4	9.29	7.42	6.18	5.29	4.62	4.10	3.68	3.34	3.06	2.81	2.61	2.42
36	41.6	20.8	13.9	10.4	8.30	6.91	5.92	5.17	4.59	4.12	3.74	3.42	3.15	2.91	2.71
38	44.8	22.4	14.9	11.2	8.93	7.43	6.36	5.56	4.93	4.43	4.02	3.68	3.38	3.13	2.92
40	48.1	24.0	16.0	12.0	9.59	7.98	6.83	5.97	5.30	4.76	4.32	3.95	3.63	3.37	3.13
42	51.6	25.8	17.2	12.9	10.3	8.57	7.33	6.41	5.69	5.11	4.63	4.24	3.90	3.61	3.36
44	55.3	27.7	18.4	13.8	11.0	9.19	7.86	6.87	6.10	5.48	4.97	4.54	4.18	3.87	3.60
46	59.3	29.7	19.8	14.8	11.8	9.85	8.43	7.37	6.54	5.87	5.33	4.87	4.49	4.15	3.86
48	63.6	31.8	21.2	15.9	12.7	10.6	9.05	7.90	7.01	6.30	5.71	5.23	4.81	4.45	4.14
50	68.3	34.1	22.7	17.0	13.6	11.3	9.71	8.48	7.52	6.76	6.13	5.61	5.16	4.78	4.45
52	73.3	36.7	24.4	18.3	14.6	12.2	10.4	9.11	8.08	7.26	6.58	6.02	5.55	5.13	4.78
54	78.9	39.4	26.3	19.7	15.7	13.1	11.2	9.79	8.69	7.81	7.08	6.48	5.96	5.52	5.14
56	84.9	42.5	28.3	21.2	16.9	14.1	12.1	10.5	9.36	8.41	7.63	6.97	6.42	5.95	5.53
58	91.7	45.8	30.5	22.9	18.3	15.2	13.0	11.4	10.1	9.08	8.23	7.53	6.93	6.42	5.97
60	99.2	49.6	33.0	24.8	19.8	16.5	14.1	12.3	10.9	9.82	8.91	8.15	7.50	6.95	6.46
62	108	53.9	35.9	26.9	21.5	17.9	15.3	13.4	11.9	10.7	9.68	8.85	8.15	7.54	7.02
64	117	58.7	39.1	29.3	23.4	19.5	16.7	14.6	12.9	11.6	10.5	9.65	8.88	8.22	7.65
66	129	64.3	42.9	32.1	25.7	21.4	18.3	16.0	14.2	12.7	11.6	10.6	9.72	9.01	8.38
LHA	**179° 181°**	**178° 182°**	**177° 183°**	**176° 184°**	**175° 185°**	**174° 186°**	**173° 187°**	**172° 188°**	**171° 189°**	**170° 190°**	**169° 191°**	**168° 192°**	**167° 193°**	**166° 194°**	**165° 195°**

B – if latitude and declination have same name
+ if latitude and declination have different names

LHA	1° 359°	2° 358°	3° 357°	4° 356°	5° 355°	6° 354°	7° 353°	8° 352°	9° 351°	10° 350°	11° 349°	12° 348°	13° 347°	14° 346°	15° 345°
Declination (°N/S)															
0	0.00	.000	.000	.000	.000	.000	.000	.000	.000	.000	.000	.000	.000	.000	.000
3	3.00	1.50	1.00	.751	.601	.501	.430	.377	.335	.302	.275	.252	.233	.217	.202
6	6.02	3.01	2.01	1.51	1.21	1.01	.862	.755	.672	.605	.551	.506	.467	.434	.406
9	9.08	4.54	3.03	2.27	1.82	1.52	1.30	1.14	1.01	.912	.830	.762	.704	.655	.612
12	12.2	6.09	4.06	3.05	2.44	2.03	1.74	1.53	1.36	1.22	1.11	1.02	.945	.879	.821
15	15.4	7.68	5.12	3.84	3.07	2.56	2.20	1.93	1.71	1.54	1.40	1.29	1.19	1.11	1.04
18	18.6	9.31	6.21	4.66	3.73	3.11	2.67	2.33	2.08	1.87	1.70	1.56	1.44	1.34	1.26
21	22.0	11.0	7.33	5.50	4.40	3.67	3.15	2.76	2.45	2.21	2.01	1.85	1.71	1.59	1.48
24	25.5	12.8	8.51	6.38	5.11	4.26	3.65	3.20	2.85	2.56	2.33	2.14	1.98	1.84	1.72
27	29.2	14.6	9.74	7.30	5.85	4.87	4.18	3.66	3.26	2.93	2.67	2.45	2.27	2.11	1.97
30	33.1	16.5	11.0	8.28	6.62	5.52	4.74	4.15	3.69	3.32	3.03	2.78	2.57	2.39	2.23
33	37.2	18.6	12.4	9.31	7.45	6.21	5.33	4.67	4.15	3.74	3.40	3.12	2.89	2.68	2.51
36	41.6	20.8	13.9	10.4	8.34	6.95	5.96	5.22	4.64	4.18	3.81	3.49	3.23	3.00	2.81
38	44.8	22.4	14.9	11.2	8.96	7.47	6.41	5.61	4.99	4.50	4.09	3.76	3.47	3.23	3.02
40	48.1	24.0	16.0	12.0	9.63	8.03	6.89	6.03	5.36	4.83	4.40	4.04	3.73	3.47	3.24
42	51.6	25.8	17.2	12.9	10.3	8.61	7.39	6.47	5.76	5.19	4.72	4.33	4.00	3.72	3.48
44	55.3	27.7	18.5	13.8	11.1	9.24	7.92	6.94	6.17	5.56	5.06	4.64	4.29	3.99	3.73
46	59.3	29.7	19.8	14.8	11.9	9.91	8.50	7.44	6.62	5.96	5.43	4.98	4.60	4.28	4.00
48	63.6	31.8	21.2	15.9	12.7	10.6	9.11	7.98	7.10	6.40	5.82	5.34	4.94	4.59	4.29
50	68.3	34.1	22.8	17.1	13.7	11.4	9.78	8.56	7.62	6.86	6.25	5.73	5.30	4.93	4.60
52	73.3	36.7	24.5	18.3	14.7	12.2	10.5	9.20	8.18	7.37	6.71	6.16	5.69	5.29	4.95
54	78.9	39.4	26.3	19.7	15.8	13.2	11.3	9.89	8.80	7.93	7.21	6.62	6.12	5.69	5.32
56	84.9	42.5	28.3	21.3	17.0	14.2	12.2	10.7	9.48	8.54	7.77	7.13	6.59	6.13	5.73
58	91.7	45.9	30.6	22.9	18.4	15.3	13.1	11.5	10.2	9.22	8.39	7.70	7.11	6.62	6.18
60	99.2	49.6	33.1	24.8	19.9	16.6	14.2	12.5	11.1	9.97	9.08	8.33	7.70	7.16	6.69
62	108	53.9	35.9	27.0	21.6	18.0	15.4	13.5	12.0	10.8	9.86	9.05	8.36	7.77	7.27
LHA	**179° 181°**	**178° 182°**	**177° 183°**	**176° 184°**	**175° 185°**	**174° 186°**	**173° 187°**	**172° 188°**	**171° 189°**	**170° 190°**	**169° 191°**	**168° 192°**	**167° 193°**	**166° 194°**	**165° 195°**

ABC TABLES

A + if hour angle is listed at top
− if hour angle is listed at bottom

LHA	16° 344°	17° 343°	18° 342°	19° 341°	20° 340°	21° 339°	22° 338°	23° 337°	24° 336°	25° 335°	26° 334°	27° 333°	28° 332°	29° 331°	30° 330°
Latitude (°N/S) 0	.000	.000	.000	.000	.000	.000	.000	.000	.000	.000	.000	.000	.000	.000	.000
3	.183	.171	.161	.152	.144	.137	.130	.123	.118	.112	.107	.103	.099	.095	.091
6	.367	.344	.323	.305	.289	.274	.260	.248	.236	.225	.215	.206	.198	.190	.182
9	.552	.518	.487	.460	.435	.413	.392	.373	.356	.340	.325	.311	.298	.286	.274
12	.741	.695	.654	.617	.584	.554	.526	.501	.477	.456	.436	.417	.400	.383	.368
15	.934	.876	.825	.778	.736	.698	.663	.631	.602	.575	.549	.526	.504	.483	.464
18	1.13	1.06	1.00	.944	.893	.846	.804	.765	.730	.697	.666	.638	.611	.586	.563
21	1.34	1.26	1.18	1.11	1.05	1.00	.950	.904	.862	.823	.787	.753	.722	.693	.665
24	1.55	1.46	1.37	1.29	1.22	1.16	1.10	1.05	1.00	.955	.913	.874	.837	.803	.771
27	1.78	1.67	1.57	1.48	1.40	1.33	1.26	1.20	1.14	1.09	1.04	1.00	.958	.919	.883
30	2.01	1.89	1.78	1.68	1.59	1.50	1.43	1.36	1.30	1.24	1.18	1.13	1.09	1.04	1.00
33	2.26	2.12	2.00	1.89	1.78	1.69	1.61	1.53	1.46	1.39	1.33	1.27	1.22	1.17	1.12
36	2.53	2.38	2.24	2.11	2.00	1.89	1.80	1.71	1.63	1.56	1.49	1.43	1.37	1.31	1.26
38	2.72	2.56	2.40	2.27	2.15	2.04	1.93	1.84	1.75	1.68	1.60	1.53	1.47	1.41	1.35
40	2.93	2.74	2.58	2.44	2.31	2.19	2.08	1.98	1.88	1.80	1.72	1.65	1.58	1.51	1.45
42	3.14	2.95	2.77	2.61	2.47	2.35	2.23	2.12	2.02	1.93	1.85	1.77	1.69	1.62	1.56
44	3.37	3.16	2.97	2.80	2.65	2.52	2.39	2.28	2.17	2.07	1.98	1.90	1.82	1.74	1.67
46	3.61	3.39	3.19	3.01	2.85	2.70	2.56	2.44	2.33	2.22	2.12	2.03	1.95	1.87	1.79
48	3.87	3.63	3.42	3.23	3.05	2.89	2.75	2.62	2.49	2.38	2.28	2.18	2.09	2.00	1.92
50	4.16	3.90	3.67	3.46	3.27	3.10	2.95	2.81	2.68	2.56	2.44	2.34	2.24	2.15	2.06
52	4.46	4.19	3.94	3.72	3.52	3.33	3.17	3.02	2.87	2.74	2.62	2.51	2.41	2.31	2.22
54	4.80	4.50	4.24	4.00	3.78	3.59	3.41	3.24	3.09	2.95	2.82	2.70	2.59	2.48	2.38
56	5.17	4.85	4.56	4.31	4.07	3.86	3.67	3.49	3.33	3.18	3.04	2.91	2.79	2.67	2.57
58	5.58	5.23	4.93	4.65	4.40	4.17	3.96	3.77	3.59	3.43	3.28	3.14	3.01	2.89	2.77
60	6.04	5.67	5.33	5.03	4.76	4.51	4.29	4.08	3.89	3.71	3.55	3.40	3.26	3.12	3.00
62	6.56	6.15	5.79	5.46	5.17	4.90	4.65	4.43	4.22	4.03	3.86	3.69	3.54	3.39	3.26
64	7.15	6.71	6.31	5.95	5.63	5.34	5.07	4.83	4.61	4.40	4.20	4.02	3.86	3.70	3.55
66	7.83	7.35	6.91	6.52	6.17	5.85	5.56	5.29	5.04	4.82	4.61	4.41	4.22	4.05	3.89
LHA	164° 196°	163° 197°	162° 198°	161° 199°	160° 200°	159° 201°	158° 202°	157° 203°	156° 204°	155° 205°	154° 206°	153° 207°	152° 208°	151° 209°	150° 210°

B − if latitude and declination have same name
+ if latitude and declination have different names

LHA	16° 344°	17° 343°	18° 342°	19° 341°	20° 340°	21° 339°	22° 338°	23° 337°	24° 336°	25° 335°	26° 334°	27° 333°	28° 332°	29° 331°	30° 330°
Declination (°N/S) 0	.000	.000	.000	.000	.000	.000	.000	.000	.000	.000	.000	.000	.000	.000	.000
3	.190	.179	.170	.161	.153	.146	.140	.134	.129	.124	.120	.115	.112	.108	.105
6	.381	.359	.340	.323	.307	.293	.281	.269	.258	.249	.240	.232	.224	.217	.210
9	.575	.542	.513	.486	.463	.442	.423	.405	.389	.375	.361	.349	.337	.327	.317
12	.771	.727	.688	.653	.621	.593	.567	.544	.523	.503	.485	.468	.453	.438	.425
15	.972	.916	.867	.823	.783	.748	.715	.686	.659	.634	.611	.590	.571	.553	.536
18	1.18	1.11	1.05	.998	.950	.907	.867	.832	.799	.769	.741	.716	.692	.670	.650
21	1.39	1.31	1.24	1.18	1.12	1.07	1.02	.982	.944	.908	.876	.846	.818	.792	.768
24	1.62	1.52	1.44	1.37	1.30	1.24	1.19	1.14	1.09	1.05	1.02	.981	.948	.918	.890
27	1.85	1.74	1.65	1.57	1.49	1.42	1.36	1.30	1.25	1.21	1.16	1.12	1.09	1.05	1.02
30	2.09	1.97	1.87	1.77	1.69	1.61	1.54	1.48	1.42	1.37	1.32	1.27	1.23	1.19	1.15
33	2.36	2.22	2.10	1.99	1.90	1.81	1.73	1.66	1.60	1.54	1.48	1.43	1.38	1.34	1.30
36	2.64	2.48	2.32	2.23	2.12	2.03	1.94	1.86	1.79	1.72	1.66	1.60	1.55	1.50	1.45
38	2.83	2.67	2.53	2.40	2.28	2.18	2.09	2.00	1.92	1.85	1.78	1.72	1.66	1.61	1.56
40	3.04	2.87	2.72	2.58	2.45	2.34	2.24	2.15	2.06	1.99	1.91	1.85	1.79-	1.73	1.68
42	3.27	3.08	2.91	2.77	2.63	2.51	2.40	2.30	2.21	2.13	2.05	1.98	1.92	1.86	1.80
44	3.50	3.30	3.13	2.97	2.82	2.69	2.58	2.47	2.37	2.29	2.20	2.13	2.06	1.99	1.93
46	3.76	3.54	3.35	3.18	3.03	2.89	2.76	2.65	2.55	2.45	2.36	2.28	2.21	2.14	2.07
48	4.03	3.80	3.59	3.41	3.25	3.10	2.96	2.84	2.73	2.63	2.53	2.45	2.37	2.29	2.22
50	4.32	4.08	3.86	3.66	3.48	3.33	3.18	3.05	2.93	2.82	2.72	2.63	2.54	2.46	2.38
52	4.64	4.38	4.14	3.93	3.74	3.57	3.42	3.28	3.15	3.03	2.92	2.82	2.73	2.64	2.56
54	4.99	4.71	4.45	4.23	4.02	3.84	3.67	3.52	3.38	3.26	3.14	3.03	2.93	2.84	2.75
56	5.38	5.07	4.80	4.55	4.33	4.14	3.96	3.79	3.65	3.51	3.38	3.27	3.16	3.06	2.97
58	5.81	5.47	5.18	4.92	4.68	4.47	4.27	4.10	3.93	3.79	3.65	3.53	3.41	3.30	3.20
60	6.28	5.92	5.61	5.32	5.06	4.83	4.62	4.43	4.26	4.10	3.95	3.82	3.69	3.57	3.46
62	6.82	6.43	6.09	5.78	5.50	5.25	5.02	4.81	4.62	4.45	4.29	4.14	4.01	3.88	3.76
LHA	164° 196°	163° 197°	162° 198°	161° 199°	160° 200°	159° 201°	158° 202°	157° 203°	156° 204°	155° 205°	154° 206°	153° 207°	152° 208°	151° 209°	150° 210°

ABC TABLES

A

+ if hour angle is listed at top
− if hour angle is listed at bottom

Latitude (°N/S) / LHA	32° 328°	34° 326°	36° 324°	38° 322°	40° 320°	42° 318°	44° 316°	46° 314°	48° 312°	50° 310°	52° 308°	54° 306°	56° 304°	58° 302°	60° 300°
0	.000	.000	.000	.000	.000	.000	.000	.000	.000	.000	.000	.000	.000	.000	.000
3	.084	.078	.072	.067	.062	.058	.054	.051	.047	.044	.041	.038	.035	.033	.030
6	.168	.156	.145	.135	.125	.117	.109	.101	.095	.088	.082	.076	.071	.066	.061
9	.253	.235	.218	.203	.189	.176	.164	.153	.143	.133	.124	.115	.107	.099	.091
12	.340	.315	.293	.272	.253	.236	.220	.205	.191	.178	.166	.154	.143	.133	.123
15	.429	.397	.369	.343	.319	.298	.277	.259	.241	.225	.209	.195	.181	.167	.155
18	.520	.482	.447	.416	.387	.361	.336	.314	.293	.273	.254	.236	.219	.203	.188
21	.614	.569	.528	.491	.457	.426	.398	.371	.346	.322	.300	.279	.259	.240	.222
24	.713	.660	.613	.570	.531	.494	.461	.430	.401	.374	.348	.323	.300	.278	.257
27	.815	.755	.701	.652	.607	.566	.528	.492	.459	.428	.398	.370	.344	.318	.294
30	.924	.856	.795	.739	.688	.641	.598	.558	.520	.484	.451	.419	.389	.361	.333
33	1.04	.963	.894	.831	.774	.721	.672	.627	.585	.545	.507	.472	.438	.406	.375
36	1.16	1.08	1.00	.930	.866	.807	.752	.702	.654	.610	.568	.528	.490	.454	.419
38	1.25	1.16	1.08	1.00	.931	.868	.809	.754	.703	.656	.610	.568	.527	.488	.451
40	1.34	1.24	1.15	1.07	1.00	.932	.869	.810	.756	.704	.656	.610	.566	.524	.484
42	1.44	1.33	1.24	1.15	1.07	1.00	.932	.870	.811	.756	.703	.654	.607	.563	.520
44	1.55	1.43	1.33	1.24	1.15	1.07	1.00	.933	.870	.810	.754	.702	.651	.603	.558
46	1.66	1.54	1.43	1.33	1.23	1.15	1.07	1.00	.932	.869	.809	.752	.698	.647	.598
48	1.78	1.65	1.53	1.42	1.32	1.23	1.15	1.07	1.00	.932	.868	.807	.749	.694	.641
50	1.91	1.77	1.64	1.53	1.42	1.32	1.23	1.15	1.07	1.00	.931	.866	.804	.745	.688
52	2.05	1.90	1.76	1.64	1.53	1.42	1.33	1.24	1.15	1.07	1.00	.930	.863	.800	.739
54	2.20	2.04	1.89	1.76	1.64	1.53	1.43	1.33	1.24	1.15	1.08	1.00	.928	.860	.795
56	2.37	2.20	2.04	1.90	1.77	1.65	1.54	1.43	1.33	1.24	1.16	1.08	1.00	.926	.856
58	2.56	2.37	2.20	2.05	1.91	1.78	1.66	1.55	1.44	1.34	1.25	1.16	1.08	1.00	.924
60	2.77	2.57	2.38	2.22	2.06	1.92	1.79	1.67	1.56	1.45	1.35	1.26	1.17	1.08	1.00
62	3.01	2.79	2.59	2.41	2.24	2.09	1.95	1.82	1.69	1.58	1.47	1.37	1.27	1.18	1.09
64	3.28	3.04	2.82	2.62	2.44	2.28	2.12	1.98	1.85	1.72	1.60	1.49	1.38	1.28	1.18
66	3.59	3.33	3.09	2.87	2.68	2.49	2.33	2.17	2.02	1.88	1.75	1.63	1.52	1.40	1.30
LHA	**148° 212°**	**146° 214°**	**144° 216°**	**142° 218°**	**140° 220°**	**138° 222°**	**136° 224°**	**134° 226°**	**132° 228°**	**130° 230°**	**128° 232°**	**126° 234°**	**124° 236°**	**122° 238°**	**120° 240°**

B

− if latitude and declination have same name
+ if latitude and declination have different names

Declination (°N/S) / LHA	32° 328°	34° 326°	36° 324°	38° 322°	40° 320°	42° 318°	44° 316°	46° 314°	48° 312°	50° 310°	52° 308°	54° 306°	56° 304°	58° 302°	60° 300°
0	.000	.000	.000	.000	.000	.000	.000	.000	.000	.000	.000	.000	.000	.000	.000
3	.099	.094	.089	.085	.082	.078	.075	.073	.071	.068	.067	.065	.063	.062	.061
6	.198	.188	.179	.171	.164	.157	.151	.146	.141	.137	.133	.130	.127	.124	.121
9	.299	.283	.269	.257	.246	.237	.228	.220	.213	.207	.201	.196	.191	.187	.183
12	.401	.380	.362	.345	.331	.318	.306	.295	.286	.277	.270	.263	.256	.251	.245
15	.506	.479	.456	.435	.417	.400	.386	.372	.361	.350	.340	.331	.323	.316	.309
18	.613	.581	.553	.528	.505	.486	.468	.452	.437	.424	.412	.402	.392	.383	.375
21	.724	.686	.653	.623	.597	.574	.553	.534	.517	.501	.487	.474	.463	.453	.443
24	.840	.796	.757	.723	.693	.665	.641	.619	.599	.581	.565	.550	.537	.525	.514
27	.962	.911	.867	.828	.793	.761	.733	.708	.686	.665	.647	.630	.615	.601	.588
30	1.09	1.03	.982	.938	.898	.863	.831	.803	.777	.754	.733	.714	.696	.681	.667
33	1.23	1.16	1.11	1.05	1.01	.971	.935	.903	.874	.848	.824	.803	.783	.766	.750
36	1.37	1.30	1.24	1.18	1.13	1.09	1.05	1.01	.978	.948	.922	.898	.876	.857	.839
38	1.47	1.40	1.33	1.27	1.22	1.17	1.12	1.09	1.05	1.02	.991	.966	.942	.921	.902
40	1.58	1.50	1.43	1.36	1.31	1.25	1.21	1.17	1.13	1.10	1.06	1.04	1.01	.989	.969
42	1.70	1.61	1.53	1.46	1.40	1.35	1.30	1.25	1.21	1.18	1.14	1.11	1.09	1.06	1.04
44	1.82	1.73	1.64	1.57	1.50	1.44	1.39	1.34	1.30	1.26	1.23	1.19	1.16	1.14	1.12
46	1.95	1.85	1.76	1.68	1.61	1.55	1.49	1.44	1.39	1.35	1.31	1.28	1.25	1.22	1.20
48	2.10	1.99	1.89	1.80	1.73	1.66	1.60	1.54	1.49	1.45	1.41	1.37	1.34	1.31	1.28
50	2.25	2.13	2.03	1.94	1.85	1.78	1.72	1.66	1.60	1.56	1.51	1.47	1.44	1.41	1.38
52	2.42	2.29	2.18	2.08	1.99	1.91	1.84	1.78	1.72	1.67	1.62	1.58	1.54	1.51	1.48
54	2.60	2.46	2.34	2.24	2.14	2.06	1.98	1.91	1.85	1.80	1.75	1.70	1.66	1.62	1.59
56	2.80	2.65	2.52	2.41	2.31	2.22	2.13	2.06	2.00	1.94	1.88	1.83	1.79	1.75	1.71
58	3.02	2.86	2.72	2.60	2.49	2.39	2.30	2.22	2.15	2.09	2.03	1.98	1.93	1.89	1.85
60	3.27	3.10	2.95	2.81	2.69	2.59	2.49	2.41	2.33	2.26	2.20	2.14	2.09	2.04	2.00
62	3.55	3.36	3.20	3.05	2.93	2.81	2.71	2.61	2.53	2.46	2.39	2.32	2.27	2.22	2.17
LHA	**148° 212°**	**146° 214°**	**144° 216°**	**142° 218°**	**140° 220°**	**138° 222°**	**136° 224°**	**134° 226°**	**132° 228°**	**130° 230°**	**128° 232°**	**126° 234°**	**124° 236°**	**122° 238°**	**120° 240°**

ABC TABLES

A

\+ if hour angle is listed at top
− if hour angle is listed at bottom

LHA / Latitude (°N/S)	62° 298°	64° 296°	66° 294°	68° 292°	70° 290°	72° 288°	74° 286°	76° 284°	78° 282°	80° 280°	82° 278°	84° 276°	86° 274°	88° 272°	90° 270°
0	.000	.000	.000	.000	.000	.000	.000	.000	.000	.000	.000	.000	.000	.000	.000
3	.028	.026	.023	.021	.019	.017	.015	.013	.011	.009	.007	.006	.004	.002	.000
6	.056	.051	.047	.043	.038	.034	.030	.026	.022	.019	.015	.011	.007	.004	.000
9	.084	.077	.071	.064	.058	.051	.045	.039	.034	.028	.022	.017	.011	.006	.000
12	.113	.104	.065	.086	.077	.069	.061	.053	.045	.037	.030	.022	.015	.007	.000
15	.142	.131	.119	.108	.098	.087	.077	.067	.057	.047	.038	.028	.019	.009	.000
18	.173	.158	.145	.131	.118	.106	.093	.081	.069	.057	.046	.034	.023	.012	.000
21	.204	.187	.171	.155	.140	.125	.110	.096	.082	.068	.054	.040	.027	.013	.000
24	.237	.217	.198	.180	.162	.145	.128	.111	.095	.079	.063	.047	.031	.016	.000
27	.271	.249	.227	.206	.185	.166	.146	.127	.108	.090	.072	.054	.036	.018	.000
30	.307	.282	.257	.233	.210	.188	.166	.144	.123	.102	.081	.061	.040	.020	.000
33	.345	.317	.289	.262	.236	.211	.186	.162	.138	.115	.091	.068	.045	.023	.000
36	.386	.354	.323	.294	.264	.236	.208	.181	.154	.128	.102	.076	.051	.025	.000
38	.415	.381	.348	.316	.284	.254	.224	.195	.166	.138	.110	.082	.055	.027	.000
40	.446	.409	.374	.339	.305	.273	.241	.209	.178	.148	.118	.088	.059	.029	.000
42	.479	.439	.401	.364	.328	.293	.258	.224	.191	.159	.127	.095	.063	.031	.000
44	.513	.471	.430	.390	.351	.314	.277	.241	.205	.170	.136	.101	.068	.034	.000
46	.551	.505	.461	.418	.377	.336	.297	.258	.220	.183	.146	.109	.072	.036	.000
48	.591	.542	.494	.449	.404	.361	.318	.277	.236	.196	.156	.117	.078	.039	.000
50	.634	.581	.531	.481	.434	.387	.342	.297	.253	.210	.167	.125	.083	.042	.000
52	.681	.624	.570	.517	.466	.416	.367	.319	.272	.226	.180	.135	.090	.045	.000
54	.732	.671	.613	.556	.501	.447	.395	.343	.293	.243	.193	.145	.096	.048	.000
56	.788	.723	.660	.559	.540	.482	.425	.370	.315	.261	.208	.156	.104	.052	.000
58	.851	.781	.713	.647	.582	.520	.459	.399	.340	.282	.225	.168	.112	.056	.000
60	.921	.845	.771	.700	.630	.563	.497	.432	.368	.305	.243	.182	.121	.060	.000
62	1.00	.917	.837	.760	.685	.611	.539	.469	.400	.332	.264	.198	.132	.066	.000
64	1.09	1.00	.913	.828	.746	.666	.588	.511	.436	.362	.288	.215	.143	.072	.000
66	1.19	1.10	1.00	.907	.817	.730	.644	.560	.477	.396	.316	.236	.157	.078	.000
LHA	118° 242°	116° 244°	114° 246°	112° 248°	110° 250°	108° 252°	106° 254°	104° 256°	102° 258°	100° 260°	98° 262°	96° 264°	94° 266°	92° 268°	90° 270°

B

− if latitude and declination have same name
\+ if latitude and declination have different names

LHA / Declination (°N/S)	62° 298°	64° 296°	66° 294°	68° 292°	70° 290°	72° 288°	74° 286°	76° 284°	78° 282°	80° 280°	82° 278°	84° 276°	86° 274°	88° 272°	90° 270°
0	.000	.000	.000	.000	.000	.000	.000	.000	.000	.000	.000	.000	.000	.000	.000
3	.059	.058	.057	.057	.056	.055	.055	.054	.054	.053	.053	.053	.053	.052	.052
6	.119	.117	.115	.113	.112	.111	.109	.108	.107	.107	.106	.106	.105	.105	.105
9	.179	.176	.173	.171	.169	.167	.165	.163	.162	.161	.160	.159	.159	.158	.158
12	.241	.236	.233	.229	.226	.223	.221	.219	.217	.216	.215	.214	.213	.213	.213
15	.303	.298	.293	.289	.285	.282	.279	.276	.274	.272	.271	.269	.269	.268	.268
18	.368	.362	.356	.350	.346	.342	.338	.335	.332	.330	.328	.327	.326	.325	.325
21	.435	.427	.420	.414	.408	.404	.399	.396	.392	.390	.388	.386	.385	.384	.384
24	.504	.495	.487	.480	.474	.468	.463	.459	.455	.452	.450	.448	.446	.446	.445
27	.577	.567	.558	.550	.542	.536	.530	.525	.521	.517	.515	.512	.511	.510	.510
30	.654	.642	.632	.623	.614	.607	.601	.595	.590	.586	.583	.581	.579	.578	.577
33	.735	.723	.711	.700	.691	.683	.676	.669	.664	.659	.656	.653	.651	.650	.649
36	.823	.808	.795	.784	.773	.764	.756	.749	.743	.738	.734	.731	.728	.727	.727
38	.885	.869	.855	.843	.831	.821	.813	.805	.799	.793	.789	.786	.783	.782	.781
40	.950	.934	.919	.905	.893	.882	.873	.865	.858	.852	.847	.844	.841	.840	.839
42	1.02	1.00	.986	.971	.958	.947	.937	.928	.921	.914	.909	.905	.903	.901	.900
44	1.09	1.07	1.06	1.04	1.03	1.02	1.00	.995	.987	.981	.975	.971	.968	.966	.966
46	1.17	1.15	1.13	1.12	1.10	1.09	1.08	1.07	1.06	1.05	1.05	1.04	1.04	1.04	1.04
48	1.26	1.24	1.22	1.20	1.18	1.17	1.16	1.14	1.14	1.13	1.12	1.12	1.11	1.11	1.11
50	1.35	1.33	1.30	1.29	1.27	1.25	1.24	1.23	1.22	1.21	1.20	1.20	1.19	1.19	1.19
52	1.45	1.42	1.40	1.38	1.36	1.35	1.33	1.32	1.31	1.30	1.29	1.29	1.28	1.28	1.28
54	1.56	1.53	1.51	1.48	1.46	1.45	1.43	1.42	1.41	1.40	1.39	1.38	1.38	1.38	1.38
56	1.68	1.65	1.62	1.60	1.58	1.56	1.54	1.53	1.52	1.51	1.50	1.49	1.49	1.48	1.48
58	1.81	1.78	1.75	1.73	1.70	1.68	1.66	1.65	1.64	1.63	1.62	1.61	1.60	1.60	1.60
60	1.96	1.93	1.90	1.87	1.84	1.82	1.80	1.79	1.77	1.76	1.75	1.74	1.74	1.73	1.73
62	2.13	2.09	2.06	2.03	2.00	1.98	1.96	1.94	1.92	1.91	1.90	1.89	1.89	1.88	1.88
LHA	118° 242°	116° 244°	114° 246°	112° 248°	110° 250°	108° 252°	106° 254°	104° 256°	102° 258°	100° 260°	98° 262°	96° 264°	94° 266°	92° 268°	90° 270°

ABC TABLES

C C (correction) = A ± B

C	0.00	0.05	0.10	0.15	0.20	0.25	0.30	0.35	0.40	0.45	0.50	0.55	0.60	0.70
Lat (°)							Azimuth (°)							
0	90.0	87.1	84.3	81.5	78.7	76.0	73.3	70.7	68.2	65.8	63.4	61.2	59.0	55.0
10	90.0	87.2	84.4	81.6	78.9	76.2	73.5	71.0	68.5	66.1	63.8	61.6	59.4	55.4
20	90.0	87.3	84.6	82.0	79.4	76.8	74.3	71.8	69.4	67.1	64.8	62.7	60.6	56.7
24	90.0	87.4	84.8	82.2	79.6	77.1	74.7	72.3	69.9	67.7	65.5	63.3	61.3	57.4
28	90.0	87.5	85.0	82.5	80.0	77.6	75.2	72.8	70.5	68.3	66.2	64.1	62.1	58.3
30	90.0	87.5	85.1	82.6	80.2	77.8	75.4	73.1	70.9	68.7	66.6	64.5	62.5	58.8
32	90.0	87.6	85.2	82.8	80.4	78.0	75.7	73.5	71.3	69.1	67.0	65.0	63.0	59.3
34	90.0	87.6	85.3	82.9	80.6	78.3	76.0	73.8	71.7	69.5	67.5	65.5	63.6	59.9
36	90.0	87.7	85.4	83.1	80.8	78.6	76.4	74.2	72.1	70.0	68.0	66.0	64.1	60.5
38	90.0	87.7	85.5	83.3	81.0	78.9	76.7	74.6	72.5	70.5	68.5	66.6	64.7	61.1
40	90.0	87.8	85.6	83.4	81.3	79.2	77.1	75.0	73.0	71.0	69.0	67.2	65.3	61.8
42	90.0	87.9	85.7	83.6	81.5	79.5	77.4	75.4	73.4	71.5	69.6	67.8	66.0	62.5
44	90.0	87.9	85.9	83.8	81.8	79.8	77.8	75.9	73.9	72.1	70.2	68.4	66.7	63.3
46	90.0	88.0	86.0	84.1	82.1	80.1	78.2	76.3	74.5	72.6	70.8	69.1	67.4	64.1
48	90.0	88.1	86.2	84.3	82.4	80.5	78.6	76.8	75.0	73.2	71.5	69.8	68.1	64.9
50	90.0	88.2	86.3	84.5	82.7	80.9	79.1	77.3	75.6	73.9	72.2	70.5	68.9	65.8
52	90.0	88.2	86.5	84.7	83.0	81.2	79.5	77.8	76.2	74.5	72.9	71.3	69.7	66.7
54	90.0	88.3	86.6	85.0	83.3	81.6	80.0	78.4	76.8	75.2	73.6	72.1	70.6	67.6
56	90.0	88.4	86.8	85.2	83.6	82.0	80.5	78.9	77.4	75.9	74.4	72.9	71.5	68.6
58	90.0	88.5	87.0	85.5	84.0	82.5	81.0	79.5	78.0	76.6	75.2	73.8	72.4	69.6
60	90.0	88.6	87.1	85.7	84.3	82.9	81.5	80.1	78.7	77.3	76.0	74.6	73.3	70.7
62	90.0	88.7	87.3	86.0	84.6	83.3	82.0	80.7	79.4	78.1	76.8	75.5	74.3	71.8
64	90.0	88.7	87.5	86.2	85.0	83.7	82.5	81.3	80.1	78.8	77.6	76.4	75.3	72.9
66	90.0	88.8	87.7	86.5	85.3	84.2	83.0	81.9	80.8	79.6	78.5	77.4	76.3	74.1
68	90.0	88.9	87.9	86.8	85.7	84.6	83.6	82.5	81.5	80.4	79.4	78.4	77.3	75.3
	0.00	**0.05**	**0.10**	**0.15**	**0.20**	**0.25**	**0.30**	**0.35**	**0.40**	**0.45**	**0.50**	**0.55**	**0.60**	**0.70**

C	0.80	0.90	1.00	1.10	1.20	1.40	1.60	1.80	2.00	2.20	2.40	2.60	2.80
Lat (°)						Azimuth (°)							
0	51.3	48.0	45.0	42.3	39.8	35.5	32.0	29.1	26.6	24.4	22.6	21.0	19.7
10	51.8	48.4	45.4	42.7	40.2	36.0	32.4	29.4	26.9	24.8	22.9	21.3	19.9
20	53.1	49.8	46.8	44.1	41.6	37.2	33.6	30.6	28.0	25.8	23.9	22.3	20.8
24	53.8	50.6	47.6	44.9	42.4	38.0	34.4	31.3	28.7	26.5	24.5	22.8	21.4
28	54.8	51.5	48.6	45.8	43.3	39.0	35.3	32.2	29.5	27.2	25.3	23.5	22.0
30	55.3	52.1	49.1	46.4	43.9	39.5	35.8	32.7	30.0	27.7	25.7	23.9	22.4
32	55.8	52.7	49.7	47.0	44.5	40.1	36.4	33.2	30.5	28.2	26.2	24.4	22.8
34	56.4	53.3	50.3	47.6	45.1	40.7	37.0	33.8	31.1	28.7	26.7	24.9	23.3
36	57.1	53.9	51.0	48.3	45.8	41.4	37.7	34.5	31.7	29.3	27.3	25.4	23.8
38	57.8	54.7	51.8	49.1	46.6	42.2	38.4	35.2	32.4	30.0	27.9	26.0	24.4
40	58.5	55.4	52.5	49.9	47.4	43.0	39.2	36.0	33.1	30.7	28.5	26.7	25.0
42	59.3	56.2	53.4	50.7	48.3	43.9	40.1	36.8	33.9	31.5	29.3	27.4	25.7
44	60.1	57.1	54.3	51.6	49.2	44.8	41.0	37.7	34.8	32.3	30.1	28.1	26.4
46	60.9	58.0	55.2	52.6	50.2	45.8	42.0	38.7	35.7	33.2	31.0	29.0	27.2
48	61.8	58.9	56.2	53.6	51.2	46.9	43.0	39.7	36.8	34.2	31.9	29.9	28.1
50	62.8	60.0	57.3	54.7	52.4	48.0	44.2	40.8	37.9	35.3	33.0	30.9	29.1
52	63.8	61.0	58.4	55.9	53.5	49.2	45.4	42.1	39.1	36.4	34.1	32.0	30.1
54	64.8	62.1	59.6	57.1	54.8	50.6	46.8	43.4	40.4	37.7	35.3	33.2	31.3
56	65.9	63.3	60.8	58.4	56.1	51.9	48.2	44.8	41.8	39.1	36.7	34.5	32.6
58	67.0	64.5	62.1	59.8	57.5	53.4	49.7	46.4	43.4	40.6	38.2	36.0	34.0
60	68.2	65.8	63.4	61.2	59.0	55.0	51.3	48.0	45.0	42.3	39.8	37.6	35.5
62	69.4	67.1	64.9	62.7	60.6	56.7	53.1	49.8	46.8	44.1	41.6	39.3	37.3
64	70.7	68.5	66.3	64.3	62.3	58.5	55.0	51.7	48.8	46.0	43.5	41.3	39.2
66	72.0	69.9	67.9	65.9	64.0	60.3	56.9	53.8	50.9	48.2	45.7	43.4	41.3
68	73.3	71.4	69.5	67.6	65.8	62.3	59.1	56.0	53.2	50.5	48.0	45.8	43.6
	0.80	**0.90**	**1.00**	**1.10**	**1.20**	**1.40**	**1.60**	**1.80**	**2.00**	**2.20**	**2.40**	**2.60**	**2.80**

Naming the azimuth: If answer is **+**, azimuth is **South** in north latitudes and **North** in south latitudes; if answer is **–**, azimuth is **North** in north latitudes and **South** in south latitudes.
If hour angle is **less than 180°**, azimuth is **West**; if **more than 180°**, azimuth is **East**.

ABC TABLES

C

C (correction) = A ± B

C	3.20	3.60	4.00	4.50	5.00	6.00	7.00	8.00	9.00	10.0	15.0	20.0	40.0
Lat(°)						Azimuth (°)							
0	17.4	15.5	14.0	12.5	11.3	9.5	8.1	7.1	6.3	5.7	3.8	2.9	1.4
10	17.6	15.8	14.2	12.7	11.5	9.6	8.3	7.2	6.4	5.8	3.9	2.9	1.5
20	18.4	16.5	14.9	13.3	12.0	10.1	8.6	7.6	6.7	6.1	4.1	3.0	1.5
24	18.9	16.9	15.3	13.7	12.3	10.3	8.9	7.8	6.9	6.2	4.2	3.1	1.6
28	19.5	17.5	15.8	14.1	12.8	10.7	9.2	8.1	7.2	6.5	4.3	3.2	1.6
30	19.8	17.8	16.1	14.4	13.0	10.9	9.4	8.2	7.3	6.6	4.4	3.3	1.7
32	20.2	18.1	16.4	14.7	13.3	11.1	9.6	8.4	7.5	6.7	4.5	3.4	1.7
34	20.7	18.5	16.8	15.0	13.6	11.4	9.8	8.6	7.6	6.9	4.6	3.5	1.7
36	21.1	19.0	17.2	15.4	13.9	11.6	10.0	8.8	7.8	7.0	4.7	3.5	1.8
38	21.6	19.4	17.6	15.8	14.2	11.9	10.3	9.0	8.0	7.2	4.8	3.6	1.8
40	22.2	19.9	18.1	16.2	14.6	12.3	10.6	9.3	8.3	7.4	5.0	3.7	1.9
42	22.8	20.5	18.6	16.7	15.1	12.6	10.9	9.5	8.5	7.7	5.1	3.8	1.9
44	23.5	21.1	19.2	17.2	15.5	13.0	11.2	9.9	8.8	7.9	5.3	4.0	2.0
46	24.2	21.8	19.8	17.8	16.1	13.5	11.6	10.2	9.1	8.2	5.5	4.1	2.1
48	25.0	22.5	20.5	18.4	16.6	14.0	12.1	10.6	9.4	8.5	5.7	4.3	2.1
50	25.9	23.4	21.3	19.1	17.3	14.5	12.5	11.0	9.8	8.8	5.9	4.4	2.2
52	26.9	24.3	22.1	19.9	18.0	15.1	13.1	11.5	10.2	9.2	6.2	4.6	2.3
54	28.0	25.3	23.1	20.7	18.8	15.8	13.7	12.0	10.7	9.7	6.5	4.9	2.4
56	29.2	26.4	24.1	21.7	19.7	16.6	14.3	12.6	11.2	10.1	6.8	5.1	2.6
58	30.5	27.7	25.3	22.8	20.7	17.5	15.1	13.3	11.8	10.7	7.2	5.4	2.7
60	32.0	29.1	26.6	24.0	21.8	18.4	15.9	14.0	12.5	11.3	7.6	5.7	2.9
62	33.6	30.6	28.0	25.3	23.1	19.6	16.9	14.9	13.3	12.0	8.1	6.1	3.0
64	35.5	32.4	29.7	26.9	24.5	20.8	18.1	15.9	14.2	12.9	8.6	6.5	3.3
66	37.6	34.3	31.6	28.7	26.2	22.3	19.4	17.1	15.3	13.8	9.3	7.0	3.5
68	39.8	36.6	33.7	30.7	28.1	24.0	20.9	18.5	16.5	14.9	10.1	7.6	3.8
	3.20	**3.60**	**4.00**	**4.50**	**5.00**	**6.00**	**7.00**	**8.00**	**9.00**	**10.0**	**15.0**	**20.0**	**40.0**

Naming the azimuth: If answer is **+**, azimuth is **South** in north latitudes and **North** in south latitudes; if answer is **–**, azimuth is **North** in north latitudes and **South** in south latitudes.
If hour angle is **less than 180°**, azimuth is **West**; if **more than 180°**, azimuth is **East**.

VERSINES

′	0° Log	0° Nat	1° Log	1° Nat	2° Log	2° Nat	3° Log	3° Nat	4° Log	4° Nat	5° Log	5° Nat	6° Log	6° Nat	
0	∞	0.0000	1827	0002	6.7847	0006	1369	0014	7.3867	0024	5804	0038	7.7386	0.0055	60
1	2.6264	0.0000	1971	0002	6.7919	0006	1417	0014	7.3903	0025	5833	0038	7.7410	0.0055	59
2	3.2285	0.0000	2112	0002	6.7991	0006	1465	0014	7.3939	0025	5862	0039	7.7434	0.0055	58
3	3.5807	0.0000	2251	0002	6.8062	0006	1512	0014	7.3975	0025	5890	0039	7.7458	0.0056	57
4	3.8305	0.0000	2388	0002	6.8132	0007	1560	0014	7.4010	0025	5919	0039	7.7482	0.0056	56
5	4.0244	0.0000	2522	0002	6.8202	0007	1607	0014	7.4046	0025	5947	0039	7.7506	0.0056	55
6	4.1827	0.0000	2655	0002	6.8271	0007	1653	0015	7.4081	0026	5976	0040	7.7530	0.0057	54
7	4.3166	0.0000	2786	0002	6.8340	0007	1700	0015	7.4116	0026	6004	0040	7.7553	0.0057	53
8	4.4326	0.0000	2914	0002	6.8408	0007	1746	0015	7.4151	0026	6032	0040	7.7577	0.0057	52
9	4.5349	0.0000	3041	0002	6.8476	0007	1792	0015	7.4186	0026	6060	0040	7.7601	0.0058	51
10	4.6264	0.0000	3166	0002	6.8543	0007	1838	0015	7.4221	0026	6089	0041	7.7624	0.0058	50
11	4.7092	0.0000	3289	0002	6.8609	0007	1884	0015	7.4256	0027	6116	0041	7.7647	0.0058	49
12	4.7848	0.0000	3411	0002	6.8675	0007	1929	0016	7.4290	0027	6144	0041	7.7671	0.0058	48
13	4.8543	0.0000	3531	0002	6.8741	0007	1974	0016	7.4325	0027	6172	0041	7.7694	0.0059	47
14	4.9187	0.0000	3649	0002	6.8806	0008	2019	0016	7.4359	0027	6200	0042	7.7717	0.0059	46
15	4.9786	0.0000	3765	0002	6.8870	0008	2064	0016	7.4393	0027	6227	0042	7.7741	0.0059	45
16	5.0347	0.0000	3880	0002	6.8934	0008	2108	0016	7.4427	0028	6255	0042	7.7764	0.0060	44
17	5.0873	0.0000	3994	0003	6.8998	0008	2152	0016	7.4461	0028	6282	0042	7.7787	0.0060	43
18	5.1370	0.0000	4106	0003	6.9061	0008	2196	0017	7.4495	0028	6310	0043	7.7810	0.0060	42
19	5.1839	0.0000	4217	0003	6.9124	0008	2240	0017	7.4528	0028	6337	0043	7.7833	0.0061	41
20	5.2285	0.0000	4326	0003	6.9186	0008	2284	0017	7.4562	0029	6364	0043	7.7855	0.0061	40
21	5.2709	0.0000	4434	0003	6.9248	0008	2327	0017	7.4595	0029	6391	0044	7.7878	0.0061	39
22	5.3113	0.0000	4540	0003	6.9309	0009	2370	0017	7.4628	0029	6418	0044	7.7901	0.0062	38
23	5.3499	0.0000	4646	0003	6.9370	0009	2413	0017	7.4661	0029	6445	0044	7.7924	0.0062	37
24	5.3868	0.0000	4750	0003	6.9431	0009	2456	0018	7.4694	0029	6472	0044	7.7946	0.0062	36
25	5.4223	0.0000	4852	0003	6.9491	0009	2498	0018	7.4727	0030	6499	0045	7.7969	0.0063	35
26	5.4564	0.0000	4954	0003	6.9551	0009	2540	0018	7.4760	0030	6525	0045	7.7991	0.0063	34
27	5.4891	0.0000	5054	0003	6.9610	0009	2582	0018	7.4792	0030	6552	0045	7.8014	0.0063	33
28	5.5207	0.0000	5154	0003	6.9669	0009	2624	0018	7.4825	0030	6578	0045	7.8036	0.0064	32
29	5.5512	0.0000	5252	0003	6.9727	0009	2666	0018	7.4857	0031	6605	0046	7.8059	0.0064	31
30	5.5807	0.0000	5349	0003	6.9785	0010	2707	0019	7.4889	0031	6631	0046	7.8081	0.0064	30
31	5.6091	0.0000	5445	0004	6.9843	0010	2749	0019	7.4921	0031	6657	0046	7.8103	0.0065	29
32	5.6367	0.0000	5540	0004	6.9900	0010	2790	0019	7.4953	0031	6684	0047	7.8125	0.0065	28
33	5.6634	0.0000	5634	0004	6.9957	0010	2830	0019	7.4985	0032	6710	0047	7.8147	0.0065	27
34	5.6894	0.0000	5727	0004	7.0014	0010	2871	0019	7.5017	0032	6736	0047	7.8169	0.0066	26
35	5.7146	0.0001	5818	0004	7.0070	0010	2912	0020	7.5049	0032	6762	0047	7.8191	0.0066	25
36	5.7390	0.0001	5909	0004	7.0126	0010	2952	0020	7.5080	0032	6788	0048	7.8213	0.0066	24
37	5.7628	0.0001	5999	0004	7.0181	0010	2992	0020	7.5111	0032	6813	0048	7.8235	0.0067	23
38	5.7860	0.0001	6088	0004	7.0237	0011	3032	0020	7.5143	0033	6839	0048	7.8257	0.0067	22
39	5.8085	0.0001	6177	0004	7.0291	0011	3072	0020	7.5174	0033	6865	0049	7.8279	0.0067	21
40	5.8305	0.0001	6264	0004	7.0346	0011	3111	0020	7.5205	0033	6890	0049	7.8301	0.0068	20
41	5.8520	0.0001	6350	0004	7.0400	0011	3151	0021	7.5236	0033	6916	0049	7.8322	0.0068	19
42	5.8729	0.0001	6436	0004	7.0454	0011	3190	0021	7.5267	0034	6941	0049	7.8344	0.0068	18
43	5.8934	0.0001	6521	0004	7.0507	0011	3229	0021	7.5297	0034	6967	0050	7.8365	0.0069	17
44	5.9133	0.0001	6605	0005	7.0560	0011	3268	0021	7.5328	0034	6992	0050	7.8387	0.0069	16
45	5.9328	0.0001	6688	0005	7.0613	0012	3306	0021	7.5359	0034	7017	0050	7.8408	0.0069	15
46	5.9519	0.0001	6770	0005	7.0666	0012	3345	0022	7.5389	0035	7042	0051	7.8430	0.0070	14
47	5.9706	0.0001	6852	0005	7.0718	0012	3383	0022	7.5419	0035	7067	0051	7.8451	0.0070	13
48	5.9889	0.0001	6932	0005	7.0770	0012	3421	0022	7.5450	0035	7092	0051	7.8472	0.0070	12
49	6.0068	0.0001	7012	0005	7.0821	0012	3459	0022	7.5480	0035	7117	0051	7.8494	0.0071	11
50	6.0244	0.0001	7092	0005	7.0872	0012	3497	0022	7.5510	0036	7142	0052	7.8515	0.0071	10
51	6.0416	0.0001	7170	0005	7.0923	0012	3535	0023	7.5539	0036	7167	0052	7.8536	0.0071	9
52	6.0584	0.0001	7248	0005	7.0974	0013	3572	0023	7.5569	0036	7191	0052	7.8557	0.0072	8
53	6.0750	0.0001	7325	0005	7.1024	0013	3610	0023	7.5599	0036	7216	0053	7.8578	0.0072	7
54	6.0912	0.0001	7402	0005	7.1074	0013	3647	0023	7.5629	0037	7240	0053	7.8599	0.0072	6
55	6.1071	0.0001	7478	0006	7.1124	0013	3684	0023	7.5658	0037	7265	0053	7.8620	0.0073	5
56	6.1228	0.0001	7553	0006	7.1174	0013	3721	0024	7.5687	0037	7289	0054	7.8641	0.0073	4
57	6.1382	0.0001	7628	0006	7.1223	0013	3757	0024	7.5717	0037	7314	0054	7.8662	0.0073	3
58	6.1533	0.0001	7701	0006	7.1272	0013	3794	0024	7.5746	0038	7338	0054	7.8682	0.0074	2
59	6.1681	0.0001	7775	0006	7.1320	0014	3830	0024	7.5775	0038	7362	0054	7.8703	0.0074	1
60	6.1827	0.0002	7847	0006	7.1369	0014	3867	0024	7.5804	0038	7386	0055	7.8724	0.0075	0
′	359° Log	359° Nat	358° Log	358° Nat	357° Log	357° Nat	356° Log	356° Nat	355° Log	355° Nat	354° Log	354° Nat	353° Log	353° Nat	

NOTE: To save space, the leading digit has been dropped from every other log versine column. Scan the preceding or following log versine column for the necessary digit. The leading digit of natural versines is 0 or 1. Scan the first and last natural versine columns to determine the appropriate number.

VERSINES

′	7° Log	7° Nat	8° Log	8° Nat	9° Log	9° Nat	10° Log	10° Nat	11° Log	11° Nat	12° Log	12° Nat	13° Log	13° Nat	′
0	**7.8724**	0.0075	**9882**	0097	**8.0903**	0123	**1816**	0152	**8.2642**	0184	**3395**	0219	**8.4087**	0.0256	60
1	**7.8744**	0.0075	**9900**	0098	**8.0919**	0124	**1831**	0152	**8.2655**	0184	**3407**	0219	**8.4099**	0.0257	59
2	**7.8765**	0.0075	**9918**	0098	**8.0935**	0124	**1845**	0153	**8.2668**	0185	**3419**	0220	**8.4110**	0.0258	58
3	**7.8786**	0.0076	**9936**	0099	**8.0951**	0124	**1859**	0153	**8.2681**	0185	**3431**	0220	**8.4121**	0.0258	57
4	**7.8806**	0.0076	**9954**	0099	**8.0967**	0125	**1874**	0154	**8.2694**	0186	**3443**	0221	**8.4132**	0.0259	56
5	**7.8826**	0.0076	**9972**	0099	**8.0983**	0125	**1888**	0154	**8.2707**	0187	**3455**	0222	**8.4143**	0.0260	55
6	**7.8847**	0.0077	**9990**	0100	**8.0999**	0126	**1902**	0155	**8.2720**	0187	**3467**	0222	**8.4154**	0.0260	54
7	**7.8867**	0.0077	**0008**	0100	**8.1015**	0126	**1917**	0155	**8.2733**	0188	**3479**	0223	**8.4165**	0.0261	53
8	**7.8887**	0.0077	**0025**	0101	**8.1031**	0127	**1931**	0156	**8.2746**	0188	**3491**	0223	**8.4176**	0.0262	52
9	**7.8908**	0.0078	**0043**	0101	**8.1046**	0127	**1945**	0157	**8.2759**	0189	**3502**	0224	**8.4187**	0.0262	51
10	**7.8928**	0.0078	**0061**	0101	**8.1062**	0128	**1959**	0157	**8.2772**	0189	**3514**	0225	**8.4198**	0.0263	50
11	**7.8948**	0.0078	**0078**	0102	**8.1078**	0128	**1974**	0158	**8.2785**	0190	**3526**	0225	**8.4209**	0.0264	49
12	**7.8968**	0.0079	**0096**	0102	**8.1094**	0129	**1988**	0158	**8.2798**	0190	**3538**	0226	**8.4220**	0.0264	48
13	**7.8988**	0.0079	**0114**	0103	**8.1109**	0129	**2002**	0159	**8.2811**	0191	**3550**	0226	**8.4230**	0.0265	47
14	**7.9008**	0.0080	**0131**	0103	**8.1125**	0130	**2016**	0159	**8.2824**	0192	**3562**	0227	**8.4241**	0.0266	46
15	**7.9028**	0.0080	**0149**	0103	**8.1141**	0130	**2030**	0160	**8.2836**	0192	**3573**	0228	**8.4252**	0.0266	45
16	**7.9048**	0.0080	**0166**	0104	**8.1156**	0131	**2044**	0160	**8.2849**	0193	**3585**	0228	**8.4263**	0.0267	44
17	**7.9068**	0.0081	**0184**	0104	**8.1172**	0131	**2058**	0161	**8.2862**	0193	**3597**	0229	**8.4274**	0.0268	43
18	**7.9088**	0.0081	**0201**	0105	**8.1187**	0131	**2072**	0161	**8.2875**	0194	**3609**	0230	**8.4285**	0.0268	42
19	**7.9108**	0.0081	**0219**	0105	**8.1203**	0132	**2086**	0162	**8.2887**	0194	**3620**	0230	**8.4296**	0.0269	41
20	**7.9127**	0.0082	**0236**	0106	**8.1218**	0132	**2100**	0162	**8.2900**	0195	**3632**	0231	**8.4306**	0.0270	40
21	**7.9147**	0.0082	**0253**	0106	**8.1234**	0133	**2114**	0163	**8.2913**	0196	**3644**	0231	**8.4317**	0.0270	39
22	**7.9167**	0.0083	**0271**	0106	**8.1249**	0133	**2128**	0163	**8.2926**	0196	**3655**	0232	**8.4328**	0.0271	38
23	**7.9186**	0.0083	**0288**	0107	**8.1265**	0134	**2142**	0164	**8.2938**	0197	**3667**	0233	**8.4339**	0.0272	37
24	**7.9206**	0.0083	**0305**	0107	**8.1280**	0134	**2156**	0164	**8.2951**	0197	**3679**	0233	**8.4350**	0.0272	36
25	**7.9225**	0.0084	**0322**	0108	**8.1295**	0135	**2170**	0165	**8.2964**	0198	**3690**	0234	**8.4360**	0.0273	35
26	**7.9245**	0.0084	**0339**	0108	**8.1311**	0135	**2184**	0165	**8.2976**	0198	**3702**	0235	**8.4371**	0.0274	34
27	**7.9264**	0.0084	**0357**	0109	**8.1326**	0136	**2198**	0166	**8.2989**	0199	**3714**	0235	**8.4382**	0.0274	33
28	**7.9284**	0.0085	**0374**	0109	**8.1341**	0136	**2211**	0166	**8.3001**	0200	**3725**	0236	**8.4392**	0.0275	32
29	**7.9303**	0.0085	**0391**	0109	**8.1357**	0137	**2225**	0167	**8.3014**	0200	**3737**	0236	**8.4403**	0.0276	31
30	**7.9322**	0.0086	**0408**	0110	**8.1372**	0137	**2239**	0167	**8.3027**	0201	**3748**	0237	**8.4414**	0.0276	30
31	**7.9342**	0.0086	**0425**	0110	**8.1387**	0138	**2253**	0168	**8.3039**	0201	**3760**	0238	**8.4424**	0.0277	29
32	**7.9361**	0.0086	**0442**	0111	**8.1402**	0138	**2266**	0169	**8.3052**	0202	**3771**	0238	**8.4435**	0.0278	28
33	**7.9380**	0.0087	**0459**	0111	**8.1417**	0139	**2280**	0169	**8.3064**	0202	**3783**	0239	**8.4446**	0.0278	27
34	**7.9399**	0.0087	**0475**	0112	**8.1432**	0139	**2294**	0170	**8.3077**	0203	**3794**	0240	**8.4456**	0.0279	26
35	**7.9418**	0.0087	**0492**	0112	**8.1447**	0140	**2307**	0170	**8.3089**	0204	**3806**	0240	**8.4467**	0.0280	25
36	**7.9437**	0.0088	**0509**	0112	**8.1463**	0140	**2321**	0171	**8.3102**	0204	**3817**	0241	**8.4478**	0.0280	24
37	**7.9456**	0.0088	**0526**	0113	**8.1478**	0141	**2335**	0171	**8.3114**	0205	**3829**	0241	**8.4488**	0.0281	23
38	**7.9475**	0.0089	**0543**	0113	**8.1493**	0141	**2348**	0172	**8.3126**	0205	**3840**	0242	**8.4499**	0.0282	22
39	**7.9494**	0.0089	**0559**	0114	**8.1508**	0141	**2362**	0172	**8.3139**	0206	**3851**	0243	**8.4509**	0.0282	21
40	**7.9513**	0.0089	**0576**	0114	**8.1522**	0142	**2375**	0173	**8.3151**	0207	**3863**	0243	**8.4520**	0.0283	20
41	**7.9532**	0.0090	**0593**	0115	**8.1537**	0142	**2389**	0173	**8.3164**	0207	**3874**	0244	**8.4530**	0.0284	19
42	**7.9551**	0.0090	**0609**	0115	**8.1552**	0143	**2402**	0174	**8.3176**	0208	**3886**	0245	**8.4541**	0.0285	18
43	**7.9569**	0.0091	**0626**	0116	**8.1567**	0143	**2416**	0174	**8.3188**	0208	**3897**	0245	**8.4551**	0.0285	17
44	**7.9588**	0.0091	**0642**	0116	**8.1582**	0144	**2429**	0175	**8.3200**	0209	**3908**	0246	**8.4562**	0.0286	16
45	**7.9607**	0.0091	**0659**	0116	**8.1597**	0144	**2443**	0175	**8.3213**	0210	**3920**	0247	**8.4572**	0.0287	15
46	**7.9625**	0.0092	**0675**	0117	**8.1612**	0145	**2456**	0176	**8.3225**	0210	**3931**	0247	**8.4583**	0.0287	14
47	**7.9644**	0.0092	**0692**	0117	**8.1626**	0145	**2469**	0177	**8.3237**	0211	**3942**	0248	**8.4593**	0.0288	13
48	**7.9662**	0.0093	**0708**	0118	**8.1641**	0146	**2483**	0177	**8.3250**	0211	**3953**	0249	**8.4604**	0.0289	12
49	**7.9681**	0.0093	**0725**	0118	**8.1656**	0146	**2496**	0178	**8.3262**	0212	**3965**	0249	**8.4614**	0.0289	11
50	**7.9699**	0.0093	**0741**	0119	**8.1671**	0147	**2510**	0178	**8.3274**	0213	**3976**	0250	**8.4625**	0.0290	10
51	**7.9718**	0.0094	**0757**	0119	**8.1685**	0147	**2523**	0179	**8.3286**	0213	**3987**	0250	**8.4635**	0.0291	9
52	**7.9736**	0.0094	**0774**	0120	**8.1700**	0148	**2536**	0179	**8.3298**	0214	**3998**	0251	**8.4645**	0.0291	8
53	**7.9755**	0.0095	**0790**	0120	**8.1715**	0148	**2549**	0180	**8.3310**	0214	**4010**	0252	**8.4656**	0.0292	7
54	**7.9773**	0.0095	**0806**	0120	**8.1729**	0149	**2563**	0180	**8.3323**	0215	**4021**	0252	**8.4666**	0.0293	6
55	**7.9791**	0.0095	**0823**	0121	**8.1744**	0149	**2576**	0181	**8.3335**	0216	**4032**	0253	**8.4677**	0.0294	5
56	**7.9809**	0.0096	**0839**	0121	**8.1758**	0150	**2589**	0182	**8.3347**	0216	**4043**	0254	**8.4687**	0.0294	4
57	**7.9828**	0.0096	**0855**	0122	**8.1773**	0150	**2602**	0182	**8.3359**	0217	**4054**	0254	**8.4697**	0.0295	3
58	**7.9846**	0.0097	**0871**	0122	**8.1787**	0151	**2615**	0183	**8.3371**	0217	**4065**	0255	**8.4708**	0.0296	2
59	**7.9864**	0.0097	**0887**	0123	**8.1802**	0151	**2629**	0183	**8.3383**	0218	**4076**	0256	**8.4718**	0.0296	1
60	**7.9882**	0.0097	**0903**	0123	**8.1816**	0152	**2642**	0184	**8.3395**	0219	**4087**	0256	**8.4728**	0.0297	0
	Log	Nat	Log	Nat	Log	Nat	Log	Nat	Log	Nat	Log	Nat	Log	Nat	′
	352°		351°		350°		349°		348°		347°		346°		

NOTE: To save space, the leading digit has been dropped from every other log versine column. Scan the preceding or following log versine column for the necessary digit. The leading digit of natural versines is 0 or 1. Scan the first and last natural versine columns to determine the appropriate number.

VERSINES

E Ephemeris

VERSINES

	14°		15°		16°		17°		18°		19°		20°		
′	Log	Nat	Log	Nat	Log	Nat	Log	Nat	Log	Nat	Log	Nat	Log	Nat	
0	**8.4728**	0.0297	**5324**	0341	**8.5881**	0387	**6404**	0437	**8.6897**	0489	**7362**	0545	**8.7804**	0.0603	60
1	**8.4738**	0.0298	**5334**	0341	**8.5890**	0388	**6413**	0438	**8.6905**	0490	**7370**	0546	**8.7811**	0.0604	59
2	**8.4749**	0.0298	**5343**	0342	**8.5899**	0389	**6421**	0439	**8.6913**	0491	**7378**	0547	**8.7818**	0.0605	58
3	**8.4759**	0.0299	**5353**	0343	**8.5908**	0390	**6430**	0440	**8.6921**	0492	**7385**	0548	**8.7825**	0.0606	57
4	**8.4769**	0.0300	**5363**	0344	**8.5917**	0391	**6438**	0440	**8.6929**	0493	**7393**	0549	**8.7832**	0.0607	56
5	**8.4779**	0.0301	**5372**	0345	**8.5926**	0391	**6447**	0441	**8.6937**	0494	**7400**	0550	**8.7839**	0.0608	55
6	**8.4790**	0.0301	**5382**	0345	**8.5935**	0392	**6455**	0442	**8.6945**	0495	**7408**	0551	**8.7847**	0.0609	54
7	**8.4800**	0.0302	**5391**	0346	**8.5944**	0393	**6463**	0443	**8.6953**	0496	**7415**	0551	**8.7854**	0.0610	53
8	**8.4810**	0.0303	**5401**	0347	**8.5953**	0394	**6472**	0444	**8.6961**	0497	**7423**	0552	**8.7861**	0.0611	52
9	**8.4820**	0.0303	**5410**	0348	**8.5962**	0395	**6480**	0445	**8.6968**	0498	**7430**	0553	**8.7868**	0.0612	51
10	**8.4830**	0.0304	**5420**	0348	**8.5971**	0395	**6488**	0445	**8.6976**	0498	**7438**	0554	**8.7875**	0.0613	50
11	**8.4841**	0.0305	**5429**	0349	**8.5980**	0396	**6497**	0446	**8.6984**	0499	**7445**	0555	**8.7882**	0.0614	49
12	**8.4851**	0.0306	**5439**	0350	**8.5989**	0397	**6505**	0447	**8.6992**	0500	**7453**	0556	**8.7889**	0.0615	48
13	**8.4861**	0.0306	**5448**	0351	**8.5997**	0398	**6514**	0448	**8.7000**	0501	**7460**	0557	**8.7896**	0.0616	47
14	**8.4871**	0.0307	**5458**	0351	**8.6006**	0399	**6522**	0449	**8.7008**	0502	**7468**	0558	**8.7903**	0.0617	46
15	**8.4881**	0.0308	**5467**	0352	**8.6015**	0400	**6530**	0450	**8.7016**	0503	**7475**	0559	**8.7910**	0.0618	45
16	**8.4891**	0.0308	**5476**	0353	**8.6024**	0400	**6539**	0451	**8.7024**	0504	**7482**	0560	**8.7918**	0.0619	44
17	**8.4901**	0.0309	**5486**	0354	**8.6033**	0401	**6547**	0452	**8.7031**	0505	**7490**	0561	**8.7925**	0.0620	43
18	**8.4911**	0.0310	**5495**	0354	**8.6042**	0402	**6555**	0452	**8.7039**	0506	**7497**	0562	**8.7932**	0.0621	42
19	**8.4921**	0.0311	**5505**	0355	**8.6051**	0403	**6563**	0453	**8.7047**	0507	**7505**	0563	**8.7939**	0.0622	41
20	**8.4932**	0.0311	**5514**	0356	**8.6059**	0404	**6572**	0454	**8.7055**	0508	**7512**	0564	**8.7946**	0.0623	40
21	**8.4942**	0.0312	**5523**	0357	**8.6068**	0404	**6580**	0455	**8.7063**	0508	**7520**	0565	**8.7953**	0.0624	39
22	**8.4952**	0.0313	**5533**	0358	**8.6077**	0405	**6588**	0456	**8.7071**	0509	**7527**	0566	**8.7960**	0.0625	38
23	**8.4962**	0.0313	**5542**	0358	**8.6086**	0406	**6597**	0457	**8.7078**	0510	**7534**	0567	**8.7967**	0.0626	37
24	**8.4972**	0.0314	**5552**	0359	**8.6094**	0407	**6605**	0458	**8.7086**	0511	**7542**	0568	**8.7974**	0.0627	36
25	**8.4982**	0.0315	**5561**	0360	**8.6103**	0408	**6613**	0458	**8.7094**	0512	**7549**	0569	**8.7981**	0.0628	35
26	**8.4992**	0.0316	**5570**	0361	**8.6112**	0409	**6621**	0459	**8.7102**	0513	**7557**	0570	**8.7988**	0.0629	34
27	**8.5002**	0.0316	**5579**	0361	**8.6121**	0409	**6630**	0460	**8.7110**	0514	**7564**	0571	**8.7995**	0.0630	33
28	**8.5012**	0.0317	**5589**	0362	**8.6129**	0410	**6638**	0461	**8.7117**	0515	**7571**	0572	**8.8002**	0.0631	32
29	**8.5021**	0.0318	**5598**	0363	**8.6138**	0411	**6646**	0462	**8.7125**	0516	**7579**	0573	**8.8009**	0.0632	31
30	**8.5031**	0.0319	**5607**	0364	**8.6147**	0412	**6654**	0463	**8.7133**	0517	**7586**	0574	**8.8016**	0.0633	30
31	**8.5041**	0.0319	**5617**	0364	**8.6156**	0413	**6662**	0464	**8.7141**	0518	**7593**	0575	**8.8023**	0.0634	29
32	**8.5051**	0.0320	**5626**	0365	**8.6164**	0413	**6671**	0465	**8.7148**	0519	**7601**	0576	**8.8030**	0.0635	28
33	**8.5061**	0.0321	**5635**	0366	**8.6173**	0414	**6679**	0465	**8.7156**	0520	**7608**	0577	**8.8037**	0.0636	27
34	**8.5071**	0.0321	**5644**	0367	**8.6182**	0415	**6687**	0466	**8.7164**	0520	**7615**	0577	**8.8044**	0.0637	26
35	**8.5081**	0.0322	**5654**	0368	**8.6190**	0416	**6695**	0467	**8.7172**	0521	**7623**	0578	**8.8051**	0.0638	25
36	**8.5091**	0.0323	**5663**	0368	**8.6199**	0417	**6703**	0468	**8.7179**	0522	**7630**	0579	**8.8058**	0.0639	24
37	**8.5101**	0.0324	**5672**	0369	**8.6208**	0418	**6711**	0469	**8.7187**	0523	**7637**	0580	**8.8065**	0.0640	23
38	**8.5110**	0.0324	**5681**	0370	**8.6216**	0418	**6720**	0470	**8.7195**	0524	**7645**	0581	**8.8072**	0.0641	22
39	**8.5120**	0.0325	**5691**	0371	**8.6225**	0419	**6728**	0471	**8.7202**	0525	**7652**	0582	**8.8079**	0.0642	21
40	**8.5130**	0.0326	**5700**	0372	**8.6234**	0420	**6736**	0472	**8.7210**	0526	**7659**	0583	**8.8086**	0.0644	20
41	**8.5140**	0.0327	**5709**	0372	**8.6242**	0421	**6744**	0473	**8.7218**	0527	**7666**	0584	**8.8092**	0.0645	19
42	**8.5150**	0.0327	**5718**	0373	**8.6251**	0422	**6752**	0473	**8.7225**	0528	**7674**	0585	**8.8099**	0.0646	18
43	**8.5160**	0.0328	**5727**	0374	**8.6259**	0423	**6760**	0474	**8.7233**	0529	**7681**	0586	**8.8106**	0.0647	17
44	**8.5169**	0.0329	**5736**	0375	**8.6268**	0423	**6768**	0475	**8.7241**	0530	**7688**	0587	**8.8113**	0.0648	16
45	**8.5179**	0.0330	**5745**	0375	**8.6277**	0424	**6776**	0476	**8.7248**	0531	**7696**	0588	**8.8120**	0.0649	15
46	**8.5189**	0.0330	**5755**	0376	**8.6285**	0425	**6785**	0477	**8.7256**	0532	**7703**	0589	**8.8127**	0.0650	14
47	**8.5199**	0.0331	**5764**	0377	**8.6294**	0426	**6793**	0478	**8.7264**	0533	**7710**	0590	**8.8134**	0.0651	13
48	**8.5208**	0.0332	**5773**	0378	**8.6302**	0427	**6801**	0479	**8.7271**	0534	**7717**	0591	**8.8141**	0.0652	12
49	**8.5218**	0.0333	**5782**	0379	**8.6311**	0428	**6809**	0480	**8.7279**	0534	**7725**	0592	**8.8148**	0.0653	11
50	**8.5228**	0.0333	**5791**	0379	**8.6319**	0428	**6817**	0480	**8.7287**	0535	**7732**	0593	**8.8155**	0.0654	10
51	**8.5237**	0.0334	**5800**	0380	**8.6328**	0429	**6825**	0481	**8.7294**	0536	**7739**	0594	**8.8161**	0.0655	9
52	**8.5247**	0.0335	**5809**	0381	**8.6336**	0430	**6833**	0482	**8.7302**	0537	**7746**	0595	**8.8168**	0.0656	8
53	**8.5257**	0.0335	**5818**	0382	**8.6345**	0431	**6841**	0483	**8.7309**	0538	**7753**	0596	**8.8175**	0.0657	7
54	**8.5266**	0.0336	**5827**	0383	**8.6353**	0432	**6849**	0484	**8.7317**	0539	**7761**	0597	**8.8182**	0.0658	6
55	**8.5276**	0.0337	**5836**	0383	**8.6362**	0433	**6857**	0485	**8.7325**	0540	**7768**	0598	**8.8189**	0.0659	5
56	**8.5286**	0.0338	**5845**	0384	**8.6370**	0434	**6865**	0486	**8.7332**	0541	**7775**	0599	**8.8196**	0.0660	4
57	**8.5295**	0.0338	**5854**	0385	**8.6379**	0434	**6873**	0487	**8.7340**	0542	**7782**	0600	**8.8202**	0.0661	3
58	**8.5305**	0.0339	**5863**	0386	**8.6387**	0435	**6881**	0488	**8.7347**	0543	**7789**	0601	**8.8209**	0.0662	2
59	**8.5315**	0.0340	**5872**	0387	**8.6396**	0436	**6889**	0489	**8.7355**	0544	**7797**	0602	**8.8216**	0.0663	1
60	**8.5324**	0.0341	**5881**	0387	**8.6404**	0437	**6897**	0489	**8.7362**	0545	**7804**	0603	**8.8223**	0.0664	0
	Log	Nat	Log	Nat	Log	Nat	Log	Nat	Log	Nat	Log	Nat	Log	Nat	′
	345°		344°		343°		342°		341°		340°		339°		

NOTE To save space, the leading digit has been dropped from every other log versine column. Scan the preceding or following log versine column for the necessary digit. The leading digit of natural versines is 0 or 1. Scan the first and last natural versine columns to determine the appropriate number.

VERSINES

′	21° Log	21° Nat	22° Log	22° Nat	23° Log	23° Nat	24° Log	24° Nat	25° Log	25° Nat	26° Log	26° Nat	27° Log	27° Nat	
0	8.8223	0.0664	8622	0728	8.9003	0795	9368	0865	8.9717	0937	0052	1012	9.0374	0.1090	60
1	8.8230	0.0665	8629	0729	8.9010	0796	9374	0866	8.9723	0938	0058	1013	9.0379	0.1091	59
2	8.8237	0.0666	8635	0730	8.9016	0797	9380	0867	8.9728	0939	0063	1015	9.0385	0.1093	58
3	8.8243	0.0667	8642	0731	8.9022	0798	9386	0868	8.9734	0941	0068	1016	9.0390	0.1094	57
4	8.8250	0.0668	8648	0733	8.9028	0800	9392	0869	8.9740	0942	0074	1017	9.0395	0.1095	56
5	8.8257	0.0669	8655	0734	8.9034	0801	9398	0870	8.9745	0943	0079	1018	9.0400	0.1097	55
6	8.8264	0.0670	8661	0735	8.9041	0802	9403	0872	8.9751	0944	0085	1020	9.0406	0.1098	54
7	8.8271	0.0672	8668	0736	8.9047	0803	9409	0873	8.9757	0946	0090	1021	9.0411	0.1099	53
8	8.8277	0.0673	8674	0737	8.9053	0804	9415	0874	8.9762	0947	0096	1022	9.0416	0.1101	52
9	8.8284	0.0674	8681	0738	8.9059	0805	9421	0875	8.9768	0948	0101	1024	9.0421	0.1102	51
10	8.8291	0.0675	8687	0739	8.9065	0806	9427	0876	8.9774	0949	0107	1025	9.0426	0.1103	50
11	8.8298	0.0676	8693	0740	8.9071	0808	9433	0878	8.9779	0950	0112	1026	9.0432	0.1105	49
12	8.8304	0.0677	8700	0741	8.9078	0809	9439	0879	8.9785	0952	0117	1027	9.0437	0.1106	48
13	8.8311	0.0678	8706	0742	8.9084	0810	9445	0880	8.9791	0953	0123	1029	9.0442	0.1107	47
14	8.8318	0.0679	8713	0743	8.9090	0811	9451	0881	8.9796	0954	0128	1030	9.0447	0.1108	46
15	8.8325	0.0680	8719	0745	8.9096	0812	9457	0882	8.9802	0955	0134	1031	9.0453	0.1110	45
16	8.8331	0.0681	8726	0746	8.9102	0813	9462	0884	8.9808	0957	0139	1033	9.0458	0.1111	44
17	8.8338	0.0682	8732	0747	8.9108	0814	9468	0885	8.9813	0958	0145	1034	9.0463	0.1112	43
18	8.8345	0.0683	8738	0748	8.9114	0816	9474	0886	8.9819	0959	0150	1035	9.0468	0.1114	42
19	8.8351	0.0684	8745	0749	8.9121	0817	9480	0887	8.9825	0960	0155	1036	9.0473	0.1115	41
20	8.8358	0.0685	8751	0750	8.9127	0818	9486	0888	8.9830	0962	0161	1038	9.0479	0.1116	40
21	8.8365	0.0686	8758	0751	8.9133	0819	9492	0890	8.9836	0963	0166	1039	9.0484	0.1118	39
22	8.8372	0.0687	8764	0752	8.9139	0820	9498	0891	8.9841	0964	0172	1040	9.0489	0.1119	38
23	8.8378	0.0688	8770	0753	8.9145	0821	9503	0892	8.9847	0965	0177	1042	9.0494	0.1121	37
24	8.8385	0.0689	8777	0755	8.9151	0822	9509	0893	8.9853	0967	0182	1043	9.0499	0.1122	36
25	8.8392	0.0691	8783	0756	8.9157	0824	9515	0894	8.9858	0968	0188	1044	9.0505	0.1123	35
26	8.8398	0.0692	8790	0757	8.9163	0825	9521	0896	8.9864	0969	0193	1045	9.0510	0.1125	34
27	8.8405	0.0693	8796	0758	8.9169	0826	9527	0897	8.9869	0970	0198	1047	9.0515	0.1126	33
28	8.8412	0.0694	8802	0759	8.9175	0827	9533	0898	8.9875	0972	0204	1048	9.0520	0.1127	32
29	8.8418	0.0695	8809	0760	8.9182	0828	9538	0899	8.9881	0973	0209	1049	9.0525	0.1129	31
30	8.8425	0.0696	8815	0761	8.9188	0829	9544	0900	8.9886	0974	0215	1051	9.0530	0.1130	30
31	8.8432	0.0697	8821	0762	8.9194	0831	9550	0902	8.9892	0975	0220	1052	9.0536	0.1131	29
32	8.8438	0.0698	8828	0763	8.9200	0832	9556	0903	8.9897	0977	0225	1053	9.0541	0.1133	28
33	8.8445	0.0699	8834	0765	8.9206	0833	9562	0904	8.9903	0978	0231	1055	9.0546	0.1134	27
34	8.8452	0.0700	8840	0766	8.9212	0834	9568	0905	8.9909	0979	0236	1056	9.0551	0.1135	26
35	8.8458	0.0701	8847	0767	8.9218	0835	9573	0906	8.9914	0980	0241	1057	9.0556	0.1137	25
36	8.8465	0.0702	8853	0768	8.9224	0836	9579	0908	8.9920	0982	0247	1058	9.0561	0.1138	24
37	8.8471	0.0703	8859	0769	8.9230	0838	9585	0909	8.9925	0983	0252	1060	9.0566	0.1139	23
38	8.8478	0.0704	8866	0770	8.9236	0839	9591	0910	8.9931	0984	0257	1061	9.0572	0.1141	22
39	8.8485	0.0705	8872	0771	8.9242	0840	9596	0911	8.9936	0985	0263	1062	9.0577	0.1142	21
40	8.8491	0.0707	8878	0772	8.9248	0841	9602	0912	8.9942	0987	0268	1064	9.0582	0.1143	20
41	8.8498	0.0708	8885	0773	8.9254	0842	9608	0914	8.9947	0988	0273	1065	9.0587	0.1145	19
42	8.8504	0.0709	8891	0775	8.9260	0843	9614	0915	8.9953	0989	0279	1066	9.0592	0.1146	18
43	8.8511	0.0710	8897	0776	8.9266	0845	9620	0916	8.9959	0990	0284	1068	9.0597	0.1147	17
44	8.8518	0.0711	8903	0777	8.9272	0846	9625	0917	8.9964	0992	0289	1069	9.0602	0.1149	16
45	8.8524	0.0712	8910	0778	8.9278	0847	9631	0919	8.9970	0993	0295	1070	9.0607	0.1150	15
46	8.8531	0.0713	8916	0779	8.9284	0848	9637	0920	8.9975	0994	0300	1072	9.0613	0.1151	14
47	8.8537	0.0714	8922	0780	8.9290	0849	9643	0921	8.9981	0996	0305	1073	9.0618	0.1153	13
48	8.8544	0.0715	8929	0781	8.9296	0850	9648	0922	8.9986	0997	0311	1074	9.0623	0.1154	12
49	8.8550	0.0716	8935	0782	8.9302	0852	9654	0923	8.9992	0998	0316	1075	9.0628	0.1156	11
50	8.8557	0.0717	8941	0784	8.9308	0853	9660	0925	8.9997	0999	0321	1077	9.0633	0.1157	10
51	8.8564	0.0718	8947	0785	8.9314	0854	9666	0926	9.0003	1001	0327	1078	9.0638	0.1158	9
52	8.8570	0.0719	8954	0786	8.9320	0855	9671	0927	9.0008	1002	0332	1079	9.0643	0.1160	8
53	8.8577	0.0721	8960	0787	8.9326	0856	9677	0928	9.0014	1003	0337	1081	9.0648	0.1161	7
54	8.8583	0.0722	8966	0788	8.9332	0857	9683	0930	9.0019	1004	0342	1082	9.0653	0.1162	6
55	8.8590	0.0723	8972	0789	8.9338	0859	9688	0931	9.0025	1006	0348	1083	9.0658	0.1164	5
56	8.8596	0.0724	8979	0790	8.9344	0860	9694	0932	9.0030	1007	0353	1085	9.0664	0.1165	4
57	8.8603	0.0725	8985	0792	8.9350	0861	9700	0933	9.0036	1008	0358	1086	9.0669	0.1166	3
58	8.8609	0.0726	8991	0793	8.9356	0862	9706	0934	9.0041	1010	0363	1087	9.0674	0.1168	2
59	8.8616	0.0727	8997	0794	8.9362	0863	9711	0936	9.0047	1011	0369	1089	9.0679	0.1169	1
60	8.8622	0.0728	9003	0795	8.9368	0865	9717	0937	9.0052	1012	0374	1090	9.0684	0.1171	0
	Log	Nat	Log	Nat	Log	Nat	Log	Nat	Log	Nat	Log	Nat	Log	Nat	′
	338°		337°		336°		335°		334°		333°		332°		

NOTE: To save space, the leading digit has been dropped from every other log versine column. Scan the preceding or following log versine column for the necessary digit. The leading digit of natural versines is 0 or 1. Scan the first and last natural versine columns to determine the appropriate number.

VERSINES

′	28° Log	Nat	29° Log	Nat	30° Log	Nat	31° Log	Nat	32° Log	Nat	33° Log	Nat	34° Log	Nat	
0	**9.0684**	0.1171	**0982**	1254	**9.1270**	1340	**1548**	1428	**9.1817**	1520	**2077**	1613	**9.2329**	0.1710	60
1	**9.0689**	0.1172	**0987**	1255	**9.1275**	1341	**1553**	1430	**9.1821**	1521	**2081**	1615	**9.2333**	0.1711	59
2	**9.0694**	0.1173	**0992**	1257	**9.1280**	1343	**1557**	1431	**9.1826**	1523	**2086**	1616	**9.2337**	0.1713	58
3	**9.0699**	0.1175	**0997**	1258	**9.1284**	1344	**1562**	1433	**9.1830**	1524	**2090**	1618	**9.2341**	0.1715	57
4	**9.0704**	0.1176	**1002**	1259	**9.1289**	1346	**1566**	1434	**9.1835**	1526	**2094**	1620	**9.2346**	0.1716	56
5	**9.0709**	0.1177	**1007**	1261	**9.1294**	1347	**1571**	1436	**9.1839**	1527	**2098**	1621	**9.2350**	0.1718	55
6	**9.0714**	0.1179	**1012**	1262	**9.1298**	1348	**1576**	1437	**9.1843**	1529	**2103**	1623	**9.2354**	0.1719	54
7	**9.0719**	0.1180	**1016**	1264	**9.1303**	1350	**1580**	1439	**9.1848**	1530	**2107**	1624	**9.2358**	0.1721	53
8	**9.0724**	0.1181	**1021**	1265	**9.1308**	1351	**1585**	1440	**9.1852**	1532	**2111**	1626	**9.2362**	0.1723	52
9	**9.0729**	0.1183	**1026**	1267	**9.1313**	1353	**1589**	1442	**9.1857**	1533	**2115**	1628	**9.2366**	0.1724	51
10	**9.0734**	0.1184	**1031**	1268	**9.1317**	1354	**1594**	1443	**9.1861**	1535	**2120**	1629	**9.2370**	0.1726	50
11	**9.0739**	0.1186	**1036**	1269	**9.1322**	1356	**1598**	1445	**9.1865**	1537	**2124**	1631	**9.2374**	0.1728	49
12	**9.0744**	0.1187	**1041**	1271	**9.1327**	1357	**1603**	1446	**9.1870**	1538	**2128**	1632	**9.2378**	0.1729	48
13	**9.0749**	0.1188	**1046**	1272	**9.1331**	1359	**1607**	1448	**9.1874**	1540	**2132**	1634	**9.2383**	0.1731	47
14	**9.0754**	0.1190	**1050**	1274	**9.1336**	1360	**1612**	1449	**9.1879**	1541	**2137**	1636	**9.2387**	0.1732	46
15	**9.0759**	0.1191	**1055**	1275	**9.1341**	1362	**1616**	1451	**9.1883**	1543	**2141**	1637	**9.2391**	0.1734	45
16	**9.0764**	0.1192	**1060**	1276	**9.1345**	1363	**1621**	1452	**9.1887**	1544	**2145**	1639	**9.2395**	0.1736	44
17	**9.0769**	0.1194	**1065**	1278	**9.1350**	1365	**1625**	1454	**9.1892**	1546	**2149**	1640	**9.2399**	0.1737	43
18	**9.0775**	0.1195	**1070**	1279	**9.1355**	1366	**1630**	1455	**9.1896**	1547	**2154**	1642	**9.2403**	0.1739	42
19	**9.0780**	0.1197	**1075**	1281	**9.1359**	1368	**1634**	1457	**9.1900**	1549	**2158**	1644	**9.2407**	0.1741	41
20	**9.0785**	0.1198	**1079**	1282	**9.1364**	1369	**1639**	1458	**9.1905**	1550	**2162**	1645	**9.2411**	0.1742	40
21	**9.0790**	0.1199	**1084**	1284	**9.1369**	1370	**1643**	1460	**9.1909**	1552	**2166**	1647	**9.2415**	0.1744	39
22	**9.0795**	0.1201	**1089**	1285	**9.1373**	1372	**1648**	1461	**9.1913**	1554	**2170**	1648	**9.2419**	0.1746	38
23	**9.0800**	0.1202	**1094**	1286	**9.1378**	1373	**1652**	1463	**9.1918**	1555	**2175**	1650	**9.2423**	0.1747	37
24	**9.0805**	0.1204	**1099**	1288	**9.1383**	1375	**1657**	1464	**9.1922**	1557	**2179**	1652	**9.2428**	0.1749	36
25	**9.0810**	0.1205	**1104**	1289	**9.1387**	1376	**1661**	1466	**9.1926**	1558	**2183**	1653	**9.2432**	0.1751	35
26	**9.0814**	0.1206	**1108**	1291	**9.1392**	1378	**1666**	1468	**9.1931**	1560	**2187**	1655	**9.2436**	0.1752	34
27	**9.0819**	0.1208	**1113**	1292	**9.1397**	1379	**1670**	1469	**9.1935**	1561	**2191**	1656	**9.2440**	0.1754	33
28	**9.0824**	0.1209	**1118**	1294	**9.1401**	1381	**1675**	1471	**9.1939**	1563	**2196**	1658	**9.2444**	0.1755	32
29	**9.0829**	0.1210	**1123**	1295	**9.1406**	1382	**1679**	1472	**9.1944**	1565	**2200**	1660	**9.2448**	0.1757	31
30	**9.0834**	0.1212	**1128**	1296	**9.1410**	1384	**1684**	1474	**9.1948**	1566	**2204**	1661	**9.2452**	0.1759	30
31	**9.0839**	0.1213	**1132**	1298	**9.1415**	1385	**1688**	1475	**9.1952**	1568	**2208**	1663	**9.2456**	0.1760	29
32	**9.0844**	0.1215	**1137**	1299	**9.1420**	1387	**1693**	1477	**9.1957**	1569	**2212**	1664	**9.2460**	0.1762	28
33	**9.0849**	0.1216	**1142**	1301	**9.1424**	1388	**1697**	1478	**9.1961**	1571	**2217**	1666	**9.2464**	0.1764	27
34	**9.0854**	0.1217	**1147**	1302	**9.1429**	1390	**1702**	1480	**9.1965**	1572	**2221**	1668	**9.2468**	0.1765	26
35	**9.0859**	0.1219	**1151**	1304	**9.1434**	1391	**1706**	1481	**9.1970**	1574	**2225**	1669	**9.2472**	0.1767	25
36	**9.0864**	0.1220	**1156**	1305	**9.1438**	1393	**1711**	1483	**9.1974**	1575	**2229**	1671	**9.2476**	0.1769	24
37	**9.0869**	0.1222	**1161**	1306	**9.1443**	1394	**1715**	1484	**9.1978**	1577	**2233**	1672	**9.2480**	0.1770	23
38	**9.0874**	0.1223	**1166**	1308	**9.1447**	1396	**1720**	1486	**9.1983**	1579	**2238**	1674	**9.2484**	0.1772	22
39	**9.0879**	0.1224	**1171**	1309	**9.1452**	1397	**1724**	1487	**9.1987**	1580	**2242**	1676	**9.2489**	0.1774	21
40	**9.0884**	0.1226	**1175**	1311	**9.1457**	1399	**1728**	1489	**9.1991**	1582	**2246**	1677	**9.2493**	0.1775	20
41	**9.0889**	0.1227	**1180**	1312	**9.1461**	1400	**1733**	1490	**9.1996**	1583	**2250**	1679	**9.2497**	0.1777	19
42	**9.0894**	0.1229	**1185**	1314	**9.1466**	1401	**1737**	1492	**9.2000**	1585	**2254**	1680	**9.2501**	0.1779	18
43	**9.0899**	0.1230	**1190**	1315	**9.1470**	1403	**1742**	1493	**9.2004**	1586	**2258**	1682	**9.2505**	0.1780	17
44	**9.0904**	0.1231	**1194**	1317	**9.1475**	1404	**1746**	1495	**9.2009**	1588	**2263**	1684	**9.2509**	0.1782	16
45	**9.0909**	0.1233	**1199**	1318	**9.1480**	1406	**1751**	1496	**9.2013**	1590	**2267**	1685	**9.2513**	0.1784	15
46	**9.0914**	0.1234	**1204**	1319	**9.1484**	1407	**1755**	1498	**9.2017**	1591	**2271**	1687	**9.2517**	0.1785	14
47	**9.0919**	0.1236	**1209**	1321	**9.1489**	1409	**1760**	1500	**9.2021**	1593	**2275**	1689	**9.2521**	0.1787	13
48	**9.0923**	0.1237	**1213**	1322	**9.1493**	1410	**1764**	1501	**9.2026**	1594	**2279**	1690	**9.2525**	0.1789	12
49	**9.0928**	0.1238	**1218**	1324	**9.1498**	1412	**1768**	1503	**9.2030**	1596	**2283**	1692	**9.2529**	0.1790	11
50	**9.0933**	0.1240	**1223**	1325	**9.1503**	1413	**1773**	1504	**9.2034**	1597	**2288**	1693	**9.2533**	0.1792	10
51	**9.0938**	0.1241	**1228**	1327	**9.1507**	1415	**1777**	1506	**9.2039**	1599	**2292**	1695	**9.2537**	0.1793	9
52	**9.0943**	0.1243	**1232**	1328	**9.1512**	1416	**1782**	1507	**9.2043**	1601	**2296**	1697	**9.2541**	0.1795	8
53	**9.0948**	0.1244	**1237**	1330	**9.1516**	1418	**1786**	1509	**9.2047**	1602	**2300**	1698	**9.2545**	0.1797	7
54	**9.0953**	0.1245	**1242**	1331	**9.1521**	1419	**1791**	1510	**9.2052**	1604	**2304**	1700	**9.2549**	0.1798	6
55	**9.0958**	0.1247	**1247**	1332	**9.1525**	1421	**1795**	1512	**9.2056**	1605	**2308**	1701	**9.2553**	0.1800	5
56	**9.0963**	0.1248	**1251**	1334	**9.1530**	1422	**1799**	1513	**9.2060**	1607	**2312**	1703	**9.2557**	0.1802	4
57	**9.0968**	0.1250	**1256**	1335	**9.1535**	1424	**1804**	1515	**9.2064**	1609	**2317**	1705	**9.2561**	0.1803	3
58	**9.0973**	0.1251	**1261**	1337	**9.1539**	1425	**1808**	1516	**9.2069**	1610	**2321**	1706	**9.2565**	0.1805	2
59	**9.0977**	0.1252	**1266**	1338	**9.1544**	1427	**1813**	1518	**9.2073**	1612	**2325**	1708	**9.2569**	0.1807	1
60	**9.0982**	0.1254	**1270**	1340	**9.1548**	1428	**1817**	1520	**9.2077**	1613	**2329**	1710	**9.2573**	0.1808	0
	Log	Nat	Log	Nat	Log	Nat	Log	Nat	Log	Nat	Log	Nat	Log	Nat	′
	331°		330°		329°		328°		327°		326°		325°		

NOTE: To save space, the leading digit has been dropped from every other log versine column. Scan the preceding or following log versine column for the necessary digit. The leading digit of natural versines is 0 or 1. Scan the first and last natural versine columns to determine the appropriate number.

VERSINES

	35°		36°		37°		38°		39°		40°		41°		
′	Log	Nat	Log	Nat	Log	Nat	Log	Nat	Log	Nat	Log	Nat	Log	Nat	
0	**9.2573**	0.1808	**2810**	1910	**9.3040**	2014	**3263**	2120	**9.3480**	2229	**3691**	2340	**9.3897**	0.2453	60
1	**9.2577**	0.1810	**2814**	1912	**9.3044**	2015	**3267**	2122	**9.3484**	2230	**3695**	2341	**9.3900**	0.2455	59
2	**9.2581**	0.1812	**2818**	1913	**9.3047**	2017	**3270**	2123	**9.3487**	2232	**3698**	2343	**9.3904**	0.2457	58
3	**9.2585**	0.1813	**2822**	1915	**9.3051**	2019	**3274**	2125	**9.3491**	2234	**3702**	2345	**9.3907**	0.2459	57
4	**9.2589**	0.1815	**2825**	1917	**9.3055**	2021	**3278**	2127	**9.3494**	2236	**3705**	2347	**9.3910**	0.2461	56
5	**9.2593**	0.1817	**2829**	1918	**9.3059**	2022	**3281**	2129	**9.3498**	2238	**3709**	2349	**9.3914**	0.2462	55
6	**9.2597**	0.1819	**2833**	1920	**9.3062**	2024	**3285**	2131	**9.3502**	2240	**3712**	2351	**9.3917**	0.2464	54
7	**9.2601**	0.1820	**2837**	1922	**9.3066**	2026	**3289**	2132	**9.3505**	2241	**3716**	2353	**9.3920**	0.2466	53
8	**9.2605**	0.1822	**2841**	1924	**9.3070**	2028	**3292**	2134	**9.3509**	2243	**3719**	2355	**9.3924**	0.2468	52
9	**9.2609**	0.1824	**2845**	1925	**9.3074**	2029	**3296**	2136	**9.3512**	2245	**3723**	2356	**9.3927**	0.2470	51
10	**9.2613**	0.1825	**2849**	1927	**9.3077**	2031	**3300**	2138	**9.3516**	2247	**3726**	2358	**9.3931**	0.2472	50
11	**9.2617**	0.1827	**2853**	1929	**9.3081**	2033	**3303**	2140	**9.3519**	2249	**3729**	2360	**9.3934**	0.2474	49
12	**9.2621**	0.1829	**2856**	1930	**9.3085**	2035	**3307**	2141	**9.3523**	2251	**3733**	2362	**9.3937**	0.2476	48
13	**9.2625**	0.1830	**2860**	1932	**9.3089**	2036	**3311**	2143	**9.3526**	2252	**3736**	2364	**9.3941**	0.2478	47
14	**9.2629**	0.1832	**2864**	1934	**9.3093**	2038	**3314**	2145	**9.3530**	2254	**3740**	2366	**9.3944**	0.2480	46
15	**9.2633**	0.1834	**2868**	1936	**9.3096**	2040	**3318**	2147	**9.3534**	2256	**3743**	2368	**9.3947**	0.2482	45
16	**9.2637**	0.1835	**2872**	1937	**9.3100**	2042	**3322**	2149	**9.3537**	2258	**3747**	2370	**9.3951**	0.2484	44
17	**9.2641**	0.1837	**2876**	1939	**9.3104**	2044	**3325**	2150	**9.3541**	2260	**3750**	2371	**9.3954**	0.2485	43
18	**9.2645**	0.1839	**2880**	1941	**9.3107**	2045	**3329**	2152	**9.3544**	2262	**3754**	2373	**9.3957**	0.2487	42
19	**9.2649**	0.1840	**2883**	1942	**9.3111**	2047	**3333**	2154	**9.3548**	2263	**3757**	2375	**9.3961**	0.2489	41
20	**9.2653**	0.1842	**2887**	1944	**9.3115**	2049	**3336**	2156	**9.3551**	2265	**3760**	2377	**9.3964**	0.2491	40
21	**9.2657**	0.1844	**2891**	1946	**9.3119**	2051	**3340**	2158	**9.3555**	2267	**3764**	2379	**9.3967**	0.2493	39
22	**9.2661**	0.1845	**2895**	1948	**9.3122**	2052	**3343**	2159	**9.3558**	2269	**3767**	2381	**9.3971**	0.2495	38
23	**9.2665**	0.1847	**2899**	1949	**9.3126**	2054	**3347**	2161	**9.3562**	2271	**3771**	2383	**9.3974**	0.2497	37
24	**9.2669**	0.1849	**2903**	1951	**9.3130**	2056	**3351**	2163	**9.3565**	2273	**3774**	2385	**9.3977**	0.2499	36
25	**9.2673**	0.1850	**2907**	1953	**9.3134**	2058	**3354**	2165	**9.3569**	2275	**3778**	2387	**9.3981**	0.2501	35
26	**9.2677**	0.1852	**2910**	1955	**9.3137**	2059	**3358**	2167	**9.3572**	2276	**3781**	2388	**9.3984**	0.2503	34
27	**9.2681**	0.1854	**2914**	1956	**9.3141**	2061	**3362**	2168	**9.3576**	2278	**3784**	2390	**9.3987**	0.2505	33
28	**9.2685**	0.1855	**2918**	1958	**9.3145**	2063	**3365**	2170	**9.3579**	2280	**3788**	2392	**9.3991**	0.2507	32
29	**9.2688**	0.1857	**2922**	1960	**9.3149**	2065	**3369**	2172	**9.3583**	2282	**3791**	2394	**9.3994**	0.2509	31
30	**9.2692**	0.1859	**2926**	1961	**9.3152**	2066	**3372**	2174	**9.3586**	2284	**3795**	2396	**9.3998**	0.2510	30
31	**9.2696**	0.1861	**2930**	1963	**9.3156**	2068	**3376**	2176	**9.3590**	2286	**3798**	2398	**9.4001**	0.2512	29
32	**9.2700**	0.1862	**2933**	1965	**9.3160**	2070	**3380**	2178	**9.3594**	2287	**3802**	2400	**9.4004**	0.2514	28
33	**9.2704**	0.1864	**2937**	1967	**9.3163**	2072	**3383**	2179	**9.3597**	2289	**3805**	2402	**9.4008**	0.2516	27
34	**9.2708**	0.1866	**2941**	1968	**9.3167**	2074	**3387**	2181	**9.3601**	2291	**3808**	2404	**9.4011**	0.2518	26
35	**9.2712**	0.1867	**2945**	1970	**9.3171**	2075	**3390**	2183	**9.3604**	2293	**3812**	2405	**9.4014**	0.2520	25
36	**9.2716**	0.1869	**2949**	1972	**9.3175**	2077	**3394**	2185	**9.3608**	2295	**3815**	2407	**9.4017**	0.2522	24
37	**9.2720**	0.1871	**2953**	1974	**9.3178**	2079	**3398**	2187	**9.3611**	2297	**3819**	2409	**9.4021**	0.2524	23
38	**9.2724**	0.1872	**2956**	1975	**9.3182**	2081	**3401**	2188	**9.3615**	2299	**3822**	2411	**9.4024**	0.2526	22
39	**9.2728**	0.1874	**2960**	1977	**9.3186**	2082	**3405**	2190	**9.3618**	2300	**3826**	2413	**9.4027**	0.2528	21
40	**9.2732**	0.1876	**2964**	1979	**9.3189**	2084	**3409**	2192	**9.3622**	2302	**3829**	2415	**9.4031**	0.2530	20
41	**9.2736**	0.1877	**2968**	1981	**9.3193**	2086	**3412**	2194	**9.3625**	2304	**3832**	2417	**9.4034**	0.2532	19
42	**9.2740**	0.1879	**2972**	1982	**9.3197**	2088	**3416**	2196	**9.3629**	2306	**3836**	2419	**9.4037**	0.2534	18
43	**9.2744**	0.1881	**2975**	1984	**9.3201**	2090	**3419**	2198	**9.3632**	2308	**3839**	2421	**9.4041**	0.2536	17
44	**9.2747**	0.1883	**2979**	1986	**9.3204**	2091	**3423**	2199	**9.3636**	2310	**3843**	2422	**9.4044**	0.2537	16
45	**9.2751**	0.1884	**2983**	1987	**9.3208**	2093	**3427**	2201	**9.3639**	2312	**3846**	2424	**9.4047**	0.2539	15
46	**9.2755**	0.1886	**2987**	1989	**9.3212**	2095	**3430**	2203	**9.3643**	2313	**3849**	2426	**9.4051**	0.2541	14
47	**9.2759**	0.1888	**2991**	1991	**9.3215**	2097	**3434**	2205	**9.3646**	2315	**3853**	2428	**9.4054**	0.2543	13
48	**9.2763**	0.1889	**2994**	1993	**9.3219**	2098	**3437**	2207	**9.3650**	2317	**3856**	2430	**9.4057**	0.2545	12
49	**9.2767**	0.1891	**2998**	1994	**9.3223**	2100	**3441**	2208	**9.3653**	2319	**3860**	2432	**9.4061**	0.2547	11
50	**9.2771**	0.1893	**3002**	1996	**9.3226**	2102	**3444**	2210	**9.3657**	2321	**3863**	2434	**9.4064**	0.2549	10
51	**9.2775**	0.1894	**3006**	1998	**9.3230**	2104	**3448**	2212	**9.3660**	2323	**3866**	2436	**9.4067**	0.2551	9
52	**9.2779**	0.1896	**3010**	2000	**9.3234**	2106	**3452**	2214	**9.3664**	2325	**3870**	2438	**9.4071**	0.2553	8
53	**9.2783**	0.1898	**3013**	2001	**9.3237**	2107	**3455**	2216	**9.3667**	2326	**3873**	2440	**9.4074**	0.2555	7
54	**9.2787**	0.1900	**3017**	2003	**9.3241**	2109	**3459**	2218	**9.3670**	2328	**3877**	2441	**9.4077**	0.2557	6
55	**9.2790**	0.1901	**3021**	2005	**9.3245**	2111	**3462**	2219	**9.3674**	2330	**3880**	2443	**9.4080**	0.2559	5
56	**9.2794**	0.1903	**3025**	2007	**9.3248**	2113	**3466**	2221	**9.3677**	2332	**3883**	2445	**9.4084**	0.2561	4
57	**9.2798**	0.1905	**3028**	2008	**9.3252**	2115	**3469**	2223	**9.3681**	2334	**3887**	2447	**9.4087**	0.2563	3
58	**9.2802**	0.1906	**3032**	2010	**9.3256**	2116	**3473**	2225	**9.3684**	2336	**3890**	2449	**9.4090**	0.2565	2
59	**9.2806**	0.1908	**3036**	2012	**9.3259**	2118	**3477**	2227	**9.3688**	2338	**3893**	2451	**9.4094**	0.2567	1
60	**9.2810**	0.1910	**3040**	2014	**9.3263**	2120	**3480**	2229	**9.3691**	2340	**3897**	2453	**9.4097**	0.2569	0
	Log	Nat	Log	Nat	Log	Nat	Log	Nat	Log	Nat	Log	Nat	Log	Nat	′
	324°		323°		322°		321°		320°		319°		318°		

NOTE: To save space, the leading digit has been dropped from every other log versine column. Scan the preceding or following log versine column for the necessary digit. The leading digit of natural versines is 0 or 1. Scan the first and last natural versine columns to determine the appropriate

VERSINES

	42°		43°		44°		45°		46°		47°		48°		
′	Log	Nat	Log	Nat	Log	Nat	Log	Nat	Log	Nat	Log	Nat	Log	Nat	
0	**9.4097**	0.2569	**4292**	2686	**9.4482**	2807	**4667**	2929	**9.4848**	3053	**5024**	3180	**9.5197**	0.3309	60
1	**9.4100**	0.2570	**4295**	2688	**9.4485**	2809	**4670**	2931	**9.4851**	3056	**5027**	3182	**9.5199**	0.3311	59
2	**9.4103**	0.2572	**4298**	2690	**9.4488**	2811	**4673**	2933	**9.4854**	3058	**5030**	3184	**9.5202**	0.3313	58
3	**9.4107**	0.2574	**4301**	2692	**9.4491**	2813	**4676**	2935	**9.4857**	3060	**5033**	3186	**9.5205**	0.3315	57
4	**9.4110**	0.2576	**4305**	2694	**9.4494**	2815	**4679**	2937	**9.4860**	3062	**5036**	3189	**9.5208**	0.3317	56
5	**9.4113**	0.2578	**4308**	2696	**9.4497**	2817	**4682**	2939	**9.4863**	3064	**5039**	3191	**9.5211**	0.3320	55
6	**9.4117**	0.2580	**4311**	2698	**9.4501**	2819	**4685**	2941	**9.4866**	3066	**5042**	3193	**9.5214**	0.3322	54
7	**9.4120**	0.2582	**4314**	2700	**9.4504**	2821	**4688**	2943	**9.4869**	3068	**5045**	3195	**9.5216**	0.3324	53
8	**9.4123**	0.2584	**4317**	2702	**9.4507**	2823	**4691**	2945	**9.4872**	3070	**5048**	3197	**9.5219**	0.3326	52
9	**9.4126**	0.2586	**4321**	2704	**9.4510**	2825	**4694**	2947	**9.4875**	3072	**5050**	3199	**9.5222**	0.3328	51
10	**9.4130**	0.2588	**4324**	2706	**9.4513**	2827	**4698**	2950	**9.4878**	3074	**5053**	3201	**9.5225**	0.3330	50
11	**9.4133**	0.2590	**4327**	2708	**9.4516**	2829	**4701**	2952	**9.4881**	3076	**5056**	3203	**9.5228**	0.3333	49
12	**9.4136**	0.2592	**4330**	2710	**9.4519**	2831	**4704**	2954	**9.4883**	3079	**5059**	3206	**9.5231**	0.3335	48
13	**9.4140**	0.2594	**4333**	2712	**9.4522**	2833	**4707**	2956	**9.4886**	3081	**5062**	3208	**9.5233**	0.3337	47
14	**9.4143**	0.2596	**4337**	2714	**9.4525**	2835	**4710**	2958	**9.4889**	3083	**5065**	3210	**9.5236**	0.3339	46
15	**9.4146**	0.2598	**4340**	2716	**9.4529**	2837	**4713**	2960	**9.4892**	3085	**5068**	3212	**9.5239**	0.3341	45
16	**9.4149**	0.2600	**4343**	2718	**9.4532**	2839	**4716**	2962	**9.4895**	3087	**5071**	3214	**9.5242**	0.3343	44
17	**9.4153**	0.2602	**4346**	2720	**9.4535**	2841	**4719**	2964	**9.4898**	3089	**5074**	3216	**9.5245**	0.3346	43
18	**9.4156**	0.2604	**4349**	2722	**9.4538**	2843	**4722**	2966	**9.4901**	3091	**5076**	3218	**9.5247**	0.3348	42
19	**9.4159**	0.2606	**4352**	2724	**9.4541**	2845	**4725**	2968	**9.4904**	3093	**5079**	3221	**9.5250**	0.3350	41
20	**9.4162**	0.2608	**4356**	2726	**9.4544**	2847	**4728**	2970	**9.4907**	3095	**5082**	3223	**9.5253**	0.3352	40
21	**9.4166**	0.2610	**4359**	2728	**9.4547**	2849	**4731**	2972	**9.4910**	3097	**5085**	3225	**9.5256**	0.3354	39
22	**9.4169**	0.2612	**4362**	2730	**9.4550**	2851	**4734**	2974	**9.4913**	3100	**5088**	3227	**9.5259**	0.3356	38
23	**9.4172**	0.2613	**4365**	2732	**9.4553**	2853	**4737**	2976	**9.4916**	3102	**5091**	3229	**9.5262**	0.3359	37
24	**9.4175**	0.2615	**4368**	2734	**9.4556**	2855	**4740**	2978	**9.4919**	3104	**5094**	3231	**9.5264**	0.3361	36
25	**9.4179**	0.2617	**4372**	2736	**9.4560**	2857	**4743**	2981	**9.4922**	3106	**5097**	3233	**9.5267**	0.3363	35
26	**9.4182**	0.2619	**4375**	2738	**9.4563**	2859	**4746**	2983	**9.4925**	3108	**5099**	3236	**9.5270**	0.3365	34
27	**9.4185**	0.2621	**4378**	2740	**9.4566**	2861	**4749**	2985	**9.4928**	3110	**5102**	3238	**9.5273**	0.3367	33
28	**9.4188**	0.2623	**4381**	2742	**9.4569**	2863	**4752**	2987	**9.4931**	3112	**5105**	3240	**9.5276**	0.3369	32
29	**9.4192**	0.2625	**4384**	2744	**9.4572**	2865	**4755**	2989	**9.4934**	3114	**5108**	3242	**9.5278**	0.3372	31
30	**9.4195**	0.2627	**4387**	2746	**9.4575**	2867	**4758**	2991	**9.4937**	3116	**5111**	3244	**9.5281**	0.3374	30
31	**9.4198**	0.2629	**4391**	2748	**9.4578**	2870	**4761**	2993	**9.4940**	3119	**5114**	3246	**9.5284**	0.3376	29
32	**9.4201**	0.2631	**4394**	2750	**9.4581**	2872	**4764**	2995	**9.4942**	3121	**5117**	3248	**9.5287**	0.3378	28
33	**9.4205**	0.2633	**4397**	2752	**9.4584**	2874	**4767**	2997	**9.4945**	3123	**5120**	3251	**9.5290**	0.3380	27
34	**9.4208**	0.2635	**4400**	2754	**9.4587**	2876	**4770**	2999	**9.4948**	3125	**5122**	3253	**9.5292**	0.3383	26
35	**9.4211**	0.2637	**4403**	2756	**9.4590**	2878	**4773**	3001	**9.4951**	3127	**5125**	3255	**9.5295**	0.3385	25
36	**9.4214**	0.2639	**4406**	2758	**9.4594**	2880	**4776**	3003	**9.4954**	3129	**5128**	3257	**9.5298**	0.3387	24
37	**9.4218**	0.2641	**4410**	2760	**9.4597**	2882	**4779**	3005	**9.4957**	3131	**5131**	3259	**9.5301**	0.3389	23
38	**9.4221**	0.2643	**4413**	2762	**9.4600**	2884	**4782**	3008	**9.4960**	3133	**5134**	3261	**9.5304**	0.3391	22
39	**9.4224**	0.2645	**4416**	2764	**9.4603**	2886	**4785**	3010	**9.4963**	3135	**5137**	3263	**9.5306**	0.3393	21
40	**9.4227**	0.2647	**4419**	2766	**9.4606**	2888	**4788**	3012	**9.4966**	3138	**5140**	3266	**9.5309**	0.3396	20
41	**9.4231**	0.2649	**4422**	2768	**9.4609**	2890	**4791**	3014	**9.4969**	3140	**5142**	3268	**9.5312**	0.3398	19
42	**9.4234**	0.2651	**4425**	2770	**9.4612**	2892	**4794**	3016	**9.4972**	3142	**5145**	3270	**9.5315**	0.3400	18
43	**9.4237**	0.2653	**4428**	2772	**9.4615**	2894	**4797**	3018	**9.4975**	3144	**5148**	3272	**9.5318**	0.3402	17
44	**9.4240**	0.2655	**4432**	2774	**9.4618**	2896	**4800**	3020	**9.4978**	3146	**5151**	3274	**9.5320**	0.3404	16
45	**9.4244**	0.2657	**4435**	2776	**9.4621**	2898	**4803**	3022	**9.4981**	3148	**5154**	3276	**9.5323**	0.3407	15
46	**9.4247**	0.2659	**4438**	2778	**9.4624**	2900	**4806**	3024	**9.4984**	3150	**5157**	3278	**9.5326**	0.3409	14
47	**9.4250**	0.2661	**4441**	2780	**9.4627**	2902	**4809**	3026	**9.4986**	3152	**5160**	3281	**9.5329**	0.3411	13
48	**9.4253**	0.2663	**4444**	2782	**9.4630**	2904	**4812**	3028	**9.4989**	3155	**5162**	3283	**9.5331**	0.3413	12
49	**9.4256**	0.2665	**4447**	2784	**9.4633**	2906	**4815**	3030	**9.4992**	3157	**5165**	3285	**9.5334**	0.3415	11
50	**9.4260**	0.2667	**4450**	2786	**9.4637**	2908	**4818**	3033	**9.4995**	3159	**5168**	3287	**9.5337**	0.3417	10
51	**9.4263**	0.2669	**4454**	2788	**9.4640**	2910	**4821**	3035	**9.4998**	3161	**5171**	3289	**9.5340**	0.3420	9
52	**9.4266**	0.2671	**4457**	2790	**9.4643**	2912	**4824**	3037	**9.5001**	3163	**5174**	3291	**9.5343**	0.3422	8
53	**9.4269**	0.2673	**4460**	2792	**9.4646**	2915	**4827**	3039	**9.5004**	3165	**5177**	3294	**9.5345**	0.3424	7
54	**9.4273**	0.2675	**4463**	2794	**9.4649**	2917	**4830**	3041	**9.5007**	3167	**5180**	3296	**9.5348**	0.3426	6
55	**9.4276**	0.2677	**4466**	2797	**9.4652**	2919	**4833**	3043	**9.5010**	3169	**5182**	3298	**9.5351**	0.3428	5
56	**9.4279**	0.2679	**4469**	2799	**9.4655**	2921	**4836**	3045	**9.5013**	3172	**5185**	3300	**9.5354**	0.3431	4
57	**9.4282**	0.2681	**4472**	2801	**9.4658**	2923	**4839**	3047	**9.5016**	3174	**5188**	3302	**9.5357**	0.3433	3
58	**9.4285**	0.2682	**4476**	2803	**9.4661**	2925	**4842**	3049	**9.5018**	3176	**5191**	3304	**9.5359**	0.3435	2
59	**9.4289**	0.2684	**4479**	2805	**9.4664**	2927	**4845**	3051	**9.5021**	3178	**5194**	3307	**9.5362**	0.3437	1
60	**9.4292**	0.2686	**4482**	2807	**9.4667**	2929	**4848**	3053	**9.5024**	3180	**5197**	3309	**9.5365**	0.3439	0
	Log	Nat	Log	Nat	Log	Nat	Log	Nat	Log	Nat	Log	Nat	Log	Nat	′
	317°		316°		315°		314°		313°		312°		311°		

NOTE: To save space, the leading digit has been dropped from every other log versine column. Scan the preceding or following log versine column for the necessary digit. The leading digit of natural versines is 0 or 1. Scan the first and last natural versine columns to determine the appropriate number.

VERSINES

	49°		50°		51°		52°		53°		54°		55°		
′	Log	Nat	Log	Nat	Log	Nat	Log	Nat	Log	Nat	Log	Nat	Log	Nat	
0	**9.5365**	0.3439	**5529**	3572	**9.5690**	3707	**5847**	3843	**9.6001**	3982	**6151**	4122	**9.6298**	0.4264	60
1	**9.5368**	0.3442	**5532**	3574	**9.5693**	3709	**5850**	3846	**9.6003**	3984	**6154**	4125	**9.6301**	0.4267	59
2	**9.5370**	0.3444	**5535**	3577	**9.5695**	3711	**5852**	3848	**9.6006**	3986	**6156**	4127	**9.6303**	0.4269	58
3	**9.5373**	0.3446	**5537**	3579	**9.5698**	3714	**5855**	3850	**9.6008**	3989	**6159**	4129	**9.6306**	0.4271	57
4	**9.5376**	0.3448	**5540**	3581	**9.5701**	3716	**5857**	3853	**9.6011**	3991	**6161**	4132	**9.6308**	0.4274	56
5	**9.5379**	0.3450	**5543**	3583	**9.5703**	3718	**5860**	3855	**9.6014**	3993	**6164**	4134	**9.6311**	0.4276	55
6	**9.5381**	0.3453	**5546**	3586	**9.5706**	3720	**5863**	3857	**9.6016**	3996	**6166**	4136	**9.6313**	0.4279	54
7	**9.5384**	0.3455	**5548**	3588	**9.5709**	3723	**5865**	3859	**9.6019**	3998	**6169**	4139	**9.6315**	0.4281	53
8	**9.5387**	0.3457	**5551**	3590	**9.5711**	3725	**5868**	3862	**9.6021**	4000	**6171**	4141	**9.6318**	0.4283	52
9	**9.5390**	0.3459	**5554**	3592	**9.5714**	3727	**5870**	3864	**9.6024**	4003	**6174**	4143	**9.6320**	0.4286	51
10	**9.5393**	0.3461	**5556**	3594	**9.5716**	3729	**5873**	3866	**9.6026**	4005	**6176**	4146	**9.6323**	0.4288	50
11	**9.5395**	0.3464	**5559**	3597	**9.5719**	3732	**5876**	3869	**9.6029**	4007	**6178**	4148	**9.6325**	0.4290	49
12	**9.5398**	0.3466	**5562**	3599	**9.5722**	3734	**5878**	3871	**9.6031**	4010	**6181**	4150	**9.6327**	0.4293	48
13	**9.5401**	0.3468	**5564**	3601	**9.5724**	3736	**5881**	3873	**9.6034**	4012	**6183**	4153	**9.6330**	0.4295	47
14	**9.5404**	0.3470	**5567**	3603	**9.5727**	3738	**5883**	3876	**9.6036**	4014	**6186**	4155	**9.6332**	0.4298	46
15	**9.5406**	0.3472	**5570**	3606	**9.5730**	3741	**5886**	3878	**9.6039**	4017	**6188**	4158	**9.6335**	0.4300	45
16	**9.5409**	0.3475	**5572**	3608	**9.5732**	3743	**5888**	3880	**9.6041**	4019	**6191**	4160	**9.6337**	0.4302	44
17	**9.5412**	0.3477	**5575**	3610	**9.5735**	3745	**5891**	3882	**9.6044**	4021	**6193**	4162	**9.6340**	0.4305	43
18	**9.5415**	0.3479	**5578**	3612	**9.5738**	3748	**5894**	3885	**9.6046**	4024	**6196**	4165	**9.6342**	0.4307	42
19	**9.5417**	0.3481	**5581**	3615	**9.5740**	3750	**5896**	3887	**9.6049**	4026	**6198**	4167	**9.6344**	0.4310	41
20	**9.5420**	0.3483	**5583**	3617	**9.5743**	3752	**5899**	3889	**9.6051**	4028	**6201**	4169	**9.6347**	0.4312	40
21	**9.5423**	0.3486	**5586**	3619	**9.5745**	3754	**5901**	3892	**9.6054**	4031	**6203**	4172	**9.6349**	0.4314	39
22	**9.5426**	0.3488	**5589**	3621	**9.5748**	3757	**5904**	3894	**9.6056**	4033	**6206**	4174	**9.6352**	0.4317	38
23	**9.5428**	0.3490	**5591**	3624	**9.5751**	3759	**5906**	3896	**9.6059**	4035	**6208**	4176	**9.6354**	0.4319	37
24	**9.5431**	0.3492	**5594**	3626	**9.5753**	3761	**5909**	3899	**9.6061**	4038	**6210**	4179	**9.6356**	0.4322	36
25	**9.5434**	0.3494	**5597**	3628	**9.5756**	3763	**5912**	3901	**9.6064**	4040	**6213**	4181	**9.6359**	0.4324	35
26	**9.5437**	0.3497	**5599**	3630	**9.5759**	3766	**5914**	3903	**9.6066**	4042	**6215**	4184	**9.6361**	0.4326	34
27	**9.5439**	0.3499	**5602**	3632	**9.5761**	3768	**5917**	3905	**9.6069**	4045	**6218**	4186	**9.6364**	0.4329	33
28	**9.5442**	0.3501	**5605**	3635	**9.5764**	3770	**5919**	3908	**9.6071**	4047	**6220**	4188	**9.6366**	0.4331	32
29	**9.5445**	0.3503	**5607**	3637	**9.5766**	3773	**5922**	3910	**9.6074**	4049	**6223**	4191	**9.6368**	0.4334	31
30	**9.5448**	0.3506	**5610**	3639	**9.5769**	3775	**5924**	3912	**9.6076**	4052	**6225**	4193	**9.6371**	0.4336	30
31	**9.5450**	0.3508	**5613**	3641	**9.5772**	3777	**5927**	3915	**9.6079**	4054	**6228**	4195	**9.6373**	0.4338	29
32	**9.5453**	0.3510	**5615**	3644	**9.5774**	3779	**5930**	3917	**9.6081**	4056	**6230**	4198	**9.6376**	0.4341	28
33	**9.5456**	0.3512	**5618**	3646	**9.5777**	3782	**5932**	3919	**9.6084**	4059	**6233**	4200	**9.6378**	0.4343	27
34	**9.5458**	0.3514	**5621**	3648	**9.5779**	3784	**5935**	3922	**9.6086**	4061	**6235**	4202	**9.6380**	0.4346	26
35	**9.5461**	0.3517	**5623**	3650	**9.5782**	3786	**5937**	3924	**9.6089**	4063	**6237**	4205	**9.6383**	0.4348	25
36	**9.5464**	0.3519	**5626**	3653	**9.5785**	3789	**5940**	3926	**9.6091**	4066	**6240**	4207	**9.6385**	0.4350	24
37	**9.5467**	0.3521	**5629**	3655	**9.5787**	3791	**5942**	3929	**9.6094**	4068	**6242**	4210	**9.6388**	0.4353	23
38	**9.5469**	0.3523	**5631**	3657	**9.5790**	3793	**5945**	3931	**9.6096**	4070	**6245**	4212	**9.6390**	0.4355	22
39	**9.5472**	0.3525	**5634**	3659	**9.5793**	3795	**5947**	3933	**9.6099**	4073	**6247**	4214	**9.6392**	0.4358	21
40	**9.5475**	0.3528	**5637**	3662	**9.5795**	3798	**5950**	3935	**9.6101**	4075	**6250**	4217	**9.6395**	0.4360	20
41	**9.5478**	0.3530	**5639**	3664	**9.5798**	3800	**5953**	3938	**9.6104**	4078	**6252**	4219	**9.6397**	0.4362	19
42	**9.5480**	0.3532	**5642**	3666	**9.5800**	3802	**5955**	3940	**9.6106**	4080	**6255**	4221	**9.6400**	0.4365	18
43	**9.5483**	0.3534	**5645**	3668	**9.5803**	3804	**5958**	3942	**9.6109**	4082	**6257**	4224	**9.6402**	0.4367	17
44	**9.5486**	0.3537	**5647**	3671	**9.5806**	3807	**5960**	3945	**9.6111**	4085	**6259**	4226	**9.6404**	0.4370	16
45	**9.5489**	0.3539	**5650**	3673	**9.5808**	3809	**5963**	3947	**9.6114**	4087	**6262**	4229	**9.6407**	0.4372	15
46	**9.5491**	0.3541	**5653**	3675	**9.5811**	3811	**5965**	3949	**9.6116**	4089	**6264**	4231	**9.6409**	0.4374	14
47	**9.5494**	0.3543	**5655**	3677	**9.5813**	3814	**5968**	3952	**9.6119**	4092	**6267**	4233	**9.6412**	0.4377	13
48	**9.5497**	0.3545	**5658**	3680	**9.5816**	3816	**5970**	3954	**9.6121**	4094	**6269**	4236	**9.6414**	0.4379	12
49	**9.5499**	0.3548	**5661**	3682	**9.5819**	3818	**5973**	3956	**9.6124**	4096	**6272**	4238	**9.6416**	0.4382	11
50	**9.5502**	0.3550	**5663**	3684	**9.5821**	3820	**5975**	3959	**9.6126**	4099	**6274**	4240	**9.6419**	0.4384	10
51	**9.5505**	0.3552	**5666**	3686	**9.5824**	3823	**5978**	3961	**9.6129**	4101	**6277**	4243	**9.6421**	0.4386	9
52	**9.5508**	0.3554	**5669**	3689	**9.5826**	3825	**5981**	3963	**9.6131**	4103	**6279**	4245	**9.6423**	0.4389	8
53	**9.5510**	0.3557	**5671**	3691	**9.5829**	3827	**5983**	3966	**9.6134**	4106	**6281**	4248	**9.6426**	0.4391	7
54	**9.5513**	0.3559	**5674**	3693	**9.5832**	3830	**5986**	3968	**9.6136**	4108	**6284**	4250	**9.6428**	0.4394	6
55	**9.5516**	0.3561	**5677**	3695	**9.5834**	3832	**5988**	3970	**9.6139**	4110	**6286**	4252	**9.6431**	0.4396	5
56	**9.5518**	0.3563	**5679**	3698	**9.5837**	3834	**5991**	3973	**9.6141**	4113	**6289**	4255	**9.6433**	0.4398	4
57	**9.5521**	0.3565	**5682**	3700	**9.5839**	3837	**5993**	3975	**9.6144**	4115	**6291**	4257	**9.6435**	0.4401	3
58	**9.5524**	0.3568	**5685**	3702	**9.5842**	3839	**5996**	3977	**9.6146**	4117	**6294**	4259	**9.6438**	0.4403	2
59	**9.5527**	0.3570	**5687**	3705	**9.5845**	3841	**5998**	3980	**9.6149**	4120	**6296**	4262	**9.6440**	0.4406	1
60	**9.5529**	0.3572	**5690**	3707	**9.5847**	3843	**6001**	3982	**9.6151**	4122	**6298**	4264	**9.6442**	0.4408	0
	Log	Nat	Log	Nat	Log	Nat	Log	Nat	Log	Nat	Log	Nat	Log	Nat	′
	310°		309°		308°		307°		306°		305°		304°		

NOTE: To save space, the leading digit has been dropped from every other log versine column. Scan the preceding or following log versine column for the necessary digit. The leading digit of natural versines is 0 or 1. Scan the first and last natural versine columns to determine the appropriate number.

VERSINES

′	56°		57°		58°		59°		60°		61°		62°		
′	Log	Nat	Log	Nat	Log	Nat	Log	Nat	Log	Nat	Log	Nat	Log	Nat	
0	**9.6442**	0.4408	**6584**	4554	**9.6722**	4701	**6857**	4850	**9.6990**	5000	**7120**	5152	**9.7247**	0.5305	**60**
1	**9.6445**	0.4410	**6586**	4556	**9.6724**	4703	**6859**	4852	**9.6992**	5003	**7122**	5154	**9.7249**	0.5308	**59**
2	**9.6447**	0.4413	**6588**	4558	**9.6726**	4706	**6862**	4855	**9.6994**	5005	**7124**	5157	**9.7251**	0.5310	**58**
3	**9.6450**	0.4415	**6591**	4561	**9.6729**	4708	**6864**	4857	**9.6996**	5008	**7126**	5160	**9.7253**	0.5313	**57**
4	**9.6452**	0.4418	**6593**	4563	**9.6731**	4711	**6866**	4860	**9.6998**	5010	**7128**	5162	**9.7255**	0.5316	**56**
5	**9.6454**	0.4420	**6595**	4566	**9.6733**	4713	**6868**	4862	**9.7001**	5013	**7130**	5165	**9.7258**	0.5318	**55**
6	**9.6457**	0.4423	**6598**	4568	**9.6735**	4716	**6870**	4865	**9.7003**	5015	**7133**	5167	**9.7260**	0.5321	**54**
7	**9.6459**	0.4425	**6600**	4571	**9.6738**	4718	**6873**	4867	**9.7005**	5018	**7135**	5170	**9.7262**	0.5323	**53**
8	**9.6461**	0.4427	**6602**	4573	**9.6740**	4721	**6875**	4870	**9.7007**	5020	**7137**	5172	**9.7264**	0.5326	**52**
9	**9.6464**	0.4430	**6604**	4576	**9.6742**	4723	**6877**	4872	**9.7009**	5023	**7139**	5175	**9.7266**	0.5328	**51**
10	**9.6466**	0.4432	**6607**	4578	**9.6744**	4725	**6879**	4875	**9.7012**	5025	**7141**	5177	**9.7268**	0.5331	**50**
11	**9.6469**	0.4435	**6609**	4580	**9.6747**	4728	**6882**	4877	**9.7014**	5028	**7143**	5180	**9.7270**	0.5334	**49**
12	**9.6471**	0.4437	**6611**	4583	**9.6749**	4730	**6884**	4880	**9.7016**	5030	**7145**	5182	**9.7272**	0.5336	**48**
13	**9.6473**	0.4439	**6614**	4585	**9.6751**	4733	**6886**	4882	**9.7018**	5033	**7147**	5185	**9.7274**	0.5339	**47**
14	**9.6476**	0.4442	**6616**	4588	**9.6754**	4735	**6888**	4885	**9.7020**	5035	**7150**	5188	**9.7276**	0.5341	**46**
15	**9.6478**	0.4444	**6618**	4590	**9.6756**	4738	**6890**	4887	**9.7022**	5038	**7152**	5190	**9.7279**	0.5344	**45**
16	**9.6480**	0.4447	**6621**	4593	**9.6758**	4740	**6893**	4890	**9.7025**	5040	**7154**	5193	**9.7281**	0.5346	**44**
17	**9.6483**	0.4449	**6623**	4595	**9.6760**	4743	**6895**	4892	**9.7027**	5043	**7156**	5195	**9.7283**	0.5349	**43**
18	**9.6485**	0.4452	**6625**	4598	**9.6763**	4745	**6897**	4895	**9.7029**	5045	**7158**	5198	**9.7285**	0.5352	**42**
19	**9.6487**	0.4454	**6628**	4600	**9.6765**	4748	**6899**	4897	**9.7031**	5048	**7160**	5200	**9.7287**	0.5354	**41**
20	**9.6490**	0.4456	**6630**	4602	**9.6767**	4750	**6902**	4900	**9.7033**	5050	**7162**	5203	**9.7289**	0.5357	**40**
21	**9.6492**	0.4459	**6632**	4605	**9.6769**	4753	**6904**	4902	**9.7035**	5053	**7165**	5205	**9.7291**	0.5359	**39**
22	**9.6495**	0.4461	**6635**	4607	**9.6772**	4755	**6906**	4905	**9.7038**	5056	**7167**	5208	**9.7293**	0.5362	**38**
23	**9.6497**	0.4464	**6637**	4610	**9.6774**	4758	**6908**	4907	**9.7040**	5058	**7169**	5211	**9.7295**	0.5364	**37**
24	**9.6499**	0.4466	**6639**	4612	**9.6776**	4760	**6910**	4910	**9.7042**	5061	**7171**	5213	**9.7297**	0.5367	**36**
25	**9.6502**	0.4469	**6641**	4615	**9.6778**	4763	**6913**	4912	**9.7044**	5063	**7173**	5216	**9.7299**	0.5370	**35**
26	**9.6504**	0.4471	**6644**	4617	**9.6781**	4765	**6915**	4915	**9.7046**	5066	**7175**	5218	**9.7302**	0.5372	**34**
27	**9.6506**	0.4473	**6646**	4620	**9.6783**	4768	**6917**	4917	**9.7049**	5068	**7177**	5221	**9.7304**	0.5375	**33**
28	**9.6509**	0.4476	**6648**	4622	**9.6785**	4770	**6919**	4920	**9.7051**	5071	**7179**	5223	**9.7306**	0.5377	**32**
29	**9.6511**	0.4478	**6651**	4625	**9.6787**	4773	**6922**	4922	**9.7053**	5073	**7182**	5226	**9.7308**	0.5380	**31**
30	**9.6513**	0.4481	**6653**	4627	**9.6790**	4775	**6924**	4925	**9.7055**	5076	**7184**	5228	**9.7310**	0.5383	**30**
31	**9.6516**	0.4483	**6655**	4629	**9.6792**	4777	**6926**	4927	**9.7057**	5078	**7186**	5231	**9.7312**	0.5385	**29**
32	**9.6518**	0.4485	**6658**	4632	**9.6794**	4780	**6928**	4930	**9.7059**	5081	**7188**	5234	**9.7314**	0.5388	**28**
33	**9.6520**	0.4488	**6660**	4634	**9.6797**	4782	**6930**	4932	**9.7062**	5083	**7190**	5236	**9.7316**	0.5390	**27**
34	**9.6523**	0.4490	**6662**	4637	**9.6799**	4785	**6933**	4935	**9.7064**	5086	**7192**	5239	**9.7318**	0.5393	**26**
35	**9.6525**	0.4493	**6665**	4639	**9.6801**	4787	**6935**	4937	**9.7066**	5088	**7194**	5241	**9.7320**	0.5395	**25**
36	**9.6527**	0.4495	**6667**	4642	**9.6803**	4790	**6937**	4940	**9.7068**	5091	**7196**	5244	**9.7322**	0.5398	**24**
37	**9.6530**	0.4498	**6669**	4644	**9.6806**	4792	**6939**	4942	**9.7070**	5093	**7199**	5246	**9.7324**	0.5401	**23**
38	**9.6532**	0.4500	**6671**	4647	**9.6808**	4795	**6941**	4945	**9.7072**	5096	**7201**	5249	**9.7326**	0.5403	**22**
39	**9.6535**	0.4502	**6674**	4649	**9.6810**	4797	**6944**	4947	**9.7074**	5099	**7203**	5251	**9.7329**	0.5406	**21**
40	**9.6537**	0.4505	**6676**	4652	**9.6812**	4800	**6946**	4950	**9.7077**	5101	**7205**	5254	**9.7331**	0.5408	**20**
41	**9.6539**	0.4507	**6678**	4654	**9.6815**	4802	**6948**	4952	**9.7079**	5104	**7207**	5257	**9.7333**	0.5411	**19**
42	**9.6542**	0.4510	**6681**	4656	**9.6817**	4805	**6950**	4955	**9.7081**	5106	**7209**	5259	**9.7335**	0.5414	**18**
43	**9.6544**	0.4512	**6683**	4659	**9.6819**	4807	**6952**	4957	**9.7083**	5109	**7211**	5262	**9.7337**	0.5416	**17**
44	**9.6546**	0.4515	**6685**	4661	**9.6821**	4810	**6955**	4960	**9.7085**	5111	**7213**	5264	**9.7339**	0.5419	**16**
45	**9.6549**	0.4517	**6687**	4664	**9.6823**	4812	**6957**	4962	**9.7087**	5114	**7215**	5267	**9.7341**	0.5421	**15**
46	**9.6551**	0.4520	**6690**	4666	**9.6826**	4815	**6959**	4965	**9.7090**	5116	**7218**	5269	**9.7343**	0.5424	**14**
47	**9.6553**	0.4522	**6692**	4669	**9.6828**	4817	**6961**	4967	**9.7092**	5119	**7220**	5272	**9.7345**	0.5426	**13**
48	**9.6556**	0.4524	**6694**	4671	**9.6830**	4820	**6963**	4970	**9.7094**	5121	**7222**	5274	**9.7347**	0.5429	**12**
49	**9.6558**	0.4527	**6697**	4674	**9.6832**	4822	**6966**	4972	**9.7096**	5124	**7224**	5277	**9.7349**	0.5432	**11**
50	**9.6560**	0.4529	**6699**	4676	**9.6835**	4825	**6968**	4975	**9.7098**	5126	**7226**	5280	**9.7351**	0.5434	**10**
51	**9.6563**	0.4532	**6701**	4679	**9.6837**	4827	**6970**	4977	**9.7100**	5129	**7228**	5282	**9.7353**	0.5437	**9**
52	**9.6565**	0.4534	**6703**	4681	**9.6839**	4830	**6972**	4980	**9.7102**	5132	**7230**	5285	**9.7355**	0.5439	**8**
53	**9.6567**	0.4537	**6706**	4684	**9.6841**	4832	**6974**	4982	**9.7105**	5134	**7232**	5287	**9.7358**	0.5442	**7**
54	**9.6570**	0.4539	**6708**	4686	**9.6844**	4835	**6977**	4985	**9.7107**	5137	**7234**	5290	**9.7360**	0.5445	**6**
55	**9.6572**	0.4541	**6710**	4688	**9.6846**	4837	**6979**	4987	**9.7109**	5139	**7237**	5292	**9.7362**	0.5447	**5**
56	**9.6574**	0.4544	**6713**	4691	**9.6848**	4840	**6981**	4990	**9.7111**	5142	**7239**	5295	**9.7364**	0.5450	**4**
57	**9.6577**	0.4546	**6715**	4693	**9.6850**	4842	**6983**	4992	**9.7113**	5144	**7241**	5298	**9.7366**	0.5452	**3**
58	**9.6579**	0.4549	**6717**	4696	**9.6853**	4845	**6985**	4995	**9.7115**	5147	**7243**	5300	**9.7368**	0.5455	**2**
59	**9.6581**	0.4551	**6719**	4698	**9.6855**	4847	**6988**	4997	**9.7118**	5149	**7245**	5303	**9.7370**	0.5458	**1**
60	**9.6584**	0.4554	**6722**	4701	**9.6857**	4850	**6990**	5000	**9.7120**	5152	**7247**	5305	**9.7372**	0.5460	**0**
	Log	Nat	Log	Nat	Log	Nat	Log	Nat	Log	Nat	Log	Nat	Log	Nat	′
	303°		302°		301°		300°		299°		298°		297°		

NOTE: To save space, the leading digit has been dropped from every other log versine column. Scan the preceding or following log versine column for the necessary digit. The leading digit of natural versines is 0 or 1. Scan the first and last natural versine columns to determine the appropriate number.

VERSINES

′	63° Log	63° Nat	64° Log	64° Nat	65° Log	65° Nat	66° Log	66° Nat	67° Log	67° Nat	68° Log	68° Nat	69° Log	69° Nat	
0	9.7372	0.5460	7494	5616	9.7615	5774	7732	5933	9.7848	6093	7962	6254	9.8073	0.6416	60
4	9.7380	0.5470	7503	5627	9.7623	5784	7740	5943	9.7856	6103	7969	6265	9.8080	0.6427	56
8	9.7388	0.5481	7511	5637	9.7630	5795	7748	5954	9.7863	6114	7976	6276	9.8088	0.6438	52
12	9.7397	0.5491	7519	5648	9.7638	5805	7756	5965	9.7871	6125	7984	6286	9.8095	0.6449	48
16	9.7405	0.5502	7527	5658	9.7646	5816	7764	5975	9.7879	6136	7991	6297	9.8102	0.6460	44
20	9.7413	0.5512	7535	5669	9.7654	5827	7771	5986	9.7886	6146	7999	6308	9.8110	0.6471	40
24	9.7421	0.5522	7543	5679	9.7662	5837	7779	5997	9.7894	6157	8006	6319	9.8117	0.6482	36
28	9.7429	0.5533	7551	5690	9.7670	5848	7787	6007	9.7901	6168	8014	6330	9.8124	0.6492	32
32	9.7438	0.5543	7559	5700	9.7678	5858	7794	6018	9.7909	6179	8021	6340	9.8131	0.6503	28
36	9.7446	0.5554	7567	5711	9.7686	5869	7802	6029	9.7916	6189	8029	6351	9.8139	0.6514	24
40	9.7454	0.5564	7575	5721	9.7693	5880	7810	6039	9.7924	6200	8036	6362	9.8146	0.6525	20
44	9.7462	0.5575	7583	5732	9.7701	5890	7817	6050	9.7931	6211	8043	6373	9.8153	0.6536	16
48	9.7470	0.5585	7591	5742	9.7709	5901	7825	6061	9.7939	6222	8051	6384	9.8160	0.6547	12
52	9.7478	0.5595	7599	5753	9.7717	5911	7833	6071	9.7947	6232	8058	6395	9.8168	0.6558	8
56	9.7486	0.5606	7607	5763	9.7725	5922	7840	6082	9.7954	6243	8066	6405	9.8175	0.6569	4
60	9.7494	0.5616	7615	5774	9.7732	5933	7848	6093	9.7962	6254	8073	6416	9.8182	0.6580	0
	Log	Nat	Log	Nat	Log	Nat	Log	Nat	Log	Nat	Log	Nat	Log	Nat	′
	296°		295°		294°		293°		292°		291°		290°		

′	70° Log	70° Nat	71° Log	71° Nat	72° Log	72° Nat	73° Log	73° Nat	74° Log	74° Nat	75° Log	75° Nat	76° Log	76° Nat	
0	9.8182	0.6580	8289	6744	9.8395	6910	8498	7076	9.8600	7244	8699	7412	9.8797	0.7581	60
4	9.8189	0.6591	8296	6755	9.8402	6921	8505	7087	9.8606	7255	8706	7423	9.8804	0.7592	56
8	9.8197	0.6602	8304	6766	9.8409	6932	8512	7099	9.8613	7266	8712	7434	9.8810	0.7603	52
12	9.8204	0.6613	8311	6777	9.8416	6943	8519	7110	9.8620	7277	8719	7446	9.8817	0.7615	48
16	9.8211	0.6624	8318	6788	9.8422	6954	8525	7121	9.8626	7288	8726	7457	9.8823	0.7626	44
20	9.8218	0.6635	8325	6799	9.8429	6965	8532	7132	9.8633	7300	8732	7468	9.8829	0.7637	40
24	9.8225	0.6645	8332	6810	9.8436	6976	8539	7143	9.8640	7311	8739	7479	9.8836	0.7649	36
28	9.8232	0.6656	8339	6821	9.8443	6987	8546	7154	9.8646	7322	8745	7491	9.8842	0.7660	32
32	9.8240	0.6667	8346	6832	9.8450	6998	8552	7165	9.8653	7333	8752	7502	9.8849	0.7671	28
36	9.8247	0.6678	8353	6844	9.8457	7010	8559	7177	9.8660	7344	8758	7513	9.8855	0.7683	24
40	9.8254	0.6689	8360	6855	9.8464	7021	8566	7188	9.8666	7356	8765	7524	9.8861	0.7694	20
44	9.8261	0.6700	8367	6866	9.8471	7032	8573	7199	9.8673	7367	8771	7536	9.8868	0.7705	16
48	9.8268	0.6711	8374	6877	9.8478	7043	8579	7210	9.8679	7378	8778	7547	9.8874	0.7716	12
52	9.8275	0.6722	8381	6888	9.8484	7054	8586	7221	9.8686	7389	8784	7558	9.8881	0.7728	8
56	9.8282	0.6733	8388	6899	9.8491	7065	8593	7232	9.8693	7401	8791	7569	9.8887	0.7739	4
60	9.8289	0.6744	8395	6910	9.8498	7076	8600	7244	9.8699	7412	8797	7581	9.8893	0.7750	0
	Log	Nat	Log	Nat	Log	Nat	Log	Nat	Log	Nat	Log	Nat	Log	Nat	′
	289°		288°		287°		286°		285°		284°		283°		

′	77° Log	77° Nat	78° Log	78° Nat	79° Log	79° Nat	80° Log	80° Nat	81° Log	81° Nat	82° Log	82° Nat	83° Log	83° Nat	
0	9.8893	0.7750	8988	7921	9.9081	8092	9172	8264	9.9261	8436	9349	8608	9.9436	0.8781	60
4	9.8900	0.7762	8994	7932	9.9087	8103	9178	8275	9.9267	8447	9355	8620	9.9441	0.8793	56
8	9.8906	0.7773	9000	7944	9.9093	8115	9184	8286	9.9273	8459	9361	8631	9.9447	0.8804	52
12	9.8912	0.7785	9006	7955	9.9099	8126	9190	8298	9.9279	8470	9367	8643	9.9453	0.8816	48
16	9.8919	0.7796	9013	7966	9.9105	8138	9196	8309	9.9285	8482	9372	8654	9.9458	0.8828	44
20	9.8925	0.7807	9019	7978	9.9111	8149	9202	8321	9.9291	8493	9378	8666	9.9464	0.8839	40
24	9.8931	0.7819	9025	7989	9.9117	8160	9208	8332	9.9297	8505	9384	8677	9.9470	0.8851	36
28	9.8938	0.7830	9031	8001	9.9123	8172	9214	8344	9.9302	8516	9390	8689	9.9475	0.8862	32
32	9.8944	0.7841	9037	8012	9.9129	8183	9220	8355	9.9308	8528	9395	8701	9.9481	0.8874	28
36	9.8950	0.7853	9044	8023	9.9135	8195	9226	8367	9.9314	8539	9401	8712	9.9487	0.8885	24
40	9.8956	0.7864	9050	8035	9.9141	8206	9232	8378	9.9320	8551	9407	8724	9.9492	0.8897	20
44	9.8963	0.7875	9056	8046	9.9148	8218	9237	8390	9.9326	8562	9413	8735	9.9498	0.8908	16
48	9.8969	0.7887	9062	8058	9.9154	8229	9243	8401	9.9332	8574	9418	8747	9.9504	0.8920	12
52	9.8975	0.7898	9068	8069	9.9160	8241	9249	8413	9.9338	8585	9424	8758	9.9509	0.8932	8
56	9.8981	0.7910	9074	8080	9.9166	8252	9255	8424	9.9343	8597	9430	8770	9.9515	0.8943	4
60	9.8988	0.7921	9081	8092	9.9172	8264	9261	8436	9.9349	8608	9436	8781	9.9521	0.8955	0
	Log	Nat	Log	Nat	Log	Nat	Log	Nat	Log	Nat	Log	Nat	Log	Nat	′
	282°		281°		280°		279°		278°		277°		276°		

NOTE: To save space, the leading digit has been dropped from every other log versine column. Scan the preceding or following log versine column for the necessary digit. The leading digit of natural versines is 0 or 1. Scan the first and last natural versine columns to determine the appropriate number.

VERSINES

	84°		85°		86°		87°		88°		89°		90°		
′	Log	Nat	Log	Nat	Log	Nat	Log	Nat	Log	Nat	Log	Nat	Log	Nat	
0	9.9521	0.8955	9604	9128	9.9686	9302	9767	9477	9.9846	9651	9924	9825	0.0000	1.0000	60
4	9.9526	0.8966	9609	9140	9.9691	9314	9772	9488	9.9851	9663	9929	9837	0.0005	1.0012	56
8	9.9532	0.8978	9615	9152	9.9697	9326	9777	9500	9.9856	9674	9934	9849	0.0010	1.0023	52
12	9.9537	0.8989	9620	9163	9.9702	9337	9782	9512	9.9861	9686	9939	9860	0.0015	1.0035	48
16	9.9543	0.9001	9626	9175	9.9708	9349	9788	9523	9.9867	9698	9944	9872	0.0020	1.0047	44
20	9.9548	0.9013	9631	9186	9.9713	9360	9793	9535	9.9872	9709	9949	9884	0.0025	1.0058	40
24	9.9554	0.9024	9637	9198	9.9718	9372	9798	9546	9.9877	9721	9954	9895	0.0030	1.0070	36
28	9.9560	0.9036	9642	9210	9.9724	9384	9804	9558	9.9882	9732	9959	9907	0.0035	1.0081	32
32	9.9565	0.9047	9648	9221	9.9729	9395	9809	9570	9.9887	9744	9964	9919	0.0040	1.0093	28
36	9.9571	0.9059	9653	9233	9.9734	9407	9814	9581	9.9893	9756	9970	9930	0.0045	1.0105	24
40	9.9576	0.9071	9659	9244	9.9740	9419	9819	9593	9.9898	9767	9975	9942	0.0050	1.0116	20
44	9.9582	0.9082	9664	9256	9.9745	9430	9825	9604	9.9903	9779	9980	9953	0.0055	1.0128	16
48	9.9587	0.9094	9670	9268	9.9751	9442	9830	9616	9.9908	9791	9985	9965	0.0060	1.0140	12
52	9.9593	0.9105	9675	9279	9.9756	9453	9835	9628	9.9913	9802	9990	9977	0.0065	1.0151	8
56	9.9598	0.9117	9681	9291	9.9761	9465	9840	9639	9.9918	9814	9995	9988	0.0070	1.0163	4
60	9.9604	0.9128	9686	9302	9.9767	9477	9846	9651	9.9924	9825	0000	0000	0.0075	1.0175	0
	Log	Nat	Log	Nat	Log	Nat	Log	Nat	Log	Nat	Log	Nat	Log	Nat	′
	275°		274°		273°		272°		271°		270°		269°		

	91°		92°		93°		94°		95°		96°		97°		
′	Log	Nat	Log	Nat	Log	Nat	Log	Nat	Log	Nat	Log	Nat	Log	Nat	
0	0.0075	1.0175	0149	0349	0.0222	0523	0293	0698	0.0363	0872	0432	1045	0.0499	1.1219	60
4	0.0080	1.0186	0154	0361	0.0226	0535	0298	0709	0.0368	0883	0436	1057	0.0504	1.1230	56
8	0.0085	1.0198	0159	0372	0.0231	0547	0302	0721	0.0372	0895	0441	1068	0.0508	1.1242	52
12	0.0090	1.0209	0164	0384	0.0236	0558	0307	0732	0.0377	0906	0445	1080	0.0513	1.1253	48
16	0.0095	1.0221	0168	0396	0.0241	0570	0312	0744	0.0381	0918	0450	1092	0.0517	1.1265	44
20	0.0100	1.0233	0173	0407	0.0245	0581	0316	0756	0.0386	0929	0454	1103	0.0522	1.1276	40
24	0.0105	1.0244	0178	0419	0.0250	0593	0321	0767	0.0391	0941	0459	1115	0.0526	1.1288	36
28	0.0110	1.0256	0183	0430	0.0255	0605	0326	0779	0.0395	0953	0463	1126	0.0531	1.1299	32
32	0.0115	1.0268	0188	0442	0.0260	0616	0330	0790	0.0400	0964	0468	1138	0.0535	1.1311	28
36	0.0120	1.0279	0193	0454	0.0264	0628	0335	0802	0.0404	0976	0473	1149	0.0539	1.1323	24
40	0.0125	1.0291	0197	0465	0.0269	0640	0340	0814	0.0409	0987	0477	1161	0.0544	1.1334	20
44	0.0129	1.0302	0202	0477	0.0274	0651	0344	0825	0.0414	0999	0481	1172	0.0548	1.1346	16
48	0.0134	1.0314	0207	0488	0.0279	0663	0349	0837	0.0418	1011	0486	1184	0.0553	1.1357	12
52	0.0139	1.0326	0212	0500	0.0283	0674	0354	0848	0.0423	1022	0490	1196	0.0557	1.1369	8
56	0.0144	1.0337	0217	0512	0.0288	0686	0358	0860	0.0427	1034	0495	1207	0.0562	1.1380	4
60	0.0149	1.0349	0222	0523	0.0293	0698	0363	0872	0.0432	1045	0499	1219	0.0566	1.1392	0
	Log	Nat	Log	Nat	Log	Nat	Log	Nat	Log	Nat	Log	Nat	Log	Nat	′
	268°		267°		266°		265°		264°		263°		262°		

	98°		99°		100°		101°		102°		103°		104°		
′	Log	Nat	Log	Nat	Log	Nat	Log	Nat	Log	Nat	Log	Nat	Log	Nat	
0	0.0566	1.1392	0631	1564	0.0695	1736	0758	1908	0.0820	2079	0881	2250	0.0941	1.2419	60
4	0.0570	1.1403	0636	1576	0.0700	1748	0763	1920	0.0824	2090	0885	2261	0.0945	1.2431	56
8	0.0575	1.1415	0640	1587	0.0704	1759	0767	1931	0.0829	2102	0889	2272	0.0949	1.2442	52
12	0.0579	1.1426	0644	1599	0.0708	1771	0771	1942	0.0833	2113	0893	2284	0.0953	1.2453	48
16	0.0583	1.1438	0648	1610	0.0712	1782	0775	1954	0.0837	2125	0897	2295	0.0957	1.2464	44
20	0.0588	1.1449	0653	1622	0.0717	1794	0779	1965	0.0841	2136	0901	2306	0.0961	1.2476	40
24	0.0592	1.1461	0657	1633	0.0721	1805	0783	1977	0.0845	2147	0905	2317	0.0965	1.2487	36
28	0.0597	1.1472	0661	1645	0.0725	1817	0787	1988	0.0849	2159	0909	2329	0.0968	1.2498	32
32	0.0601	1.1484	0666	1656	0.0729	1828	0792	1999	0.0853	2170	0913	2340	0.0972	1.2509	28
36	0.0605	1.1495	0670	1668	0.0733	1840	0796	2011	0.0857	2181	0917	2351	0.0976	1.2521	24
40	0.0610	1.1507	0674	1679	0.0738	1851	0800	2022	0.0861	2193	0921	2363	0.0980	1.2532	20
44	0.0614	1.1518	0678	1691	0.0742	1862	0804	2034	0.0865	2204	0925	2374	0.0984	1.2543	16
48	0.0618	1.1530	0683	1702	0.0746	1874	0808	2045	0.0869	2215	0929	2385	0.0988	1.2554	12
52	0.0623	1.1541	0687	1714	0.0750	1885	0812	2056	0.0873	2227	0933	2397	0.0992	1.2566	8
56	0.0627	1.1553	0691	1725	0.0754	1897	0816	2068	0.0877	2238	0937	2408	0.0996	1.2577	4
60	0.0631	1.1564	0695	1736	0.0758	1908	0820	2079	0.0881	2250	0941	2419	0.1000	1.2588	0
	Log	Nat	Log	Nat	Log	Nat	Log	Nat	Log	Nat	Log	Nat	Log	Nat	′
	261°		260°		259°		258°		257°		256°		255°		

NOTE: To save space, the leading digit has been dropped from every other log versine column. Scan the preceding or following log versine column for the necessary digit. The leading digit of natural versines is 0 or 1. Scan the first and last natural versine columns to determine the appropriate number.

VERSINES

	105°		106°		107°		108°		109°		110°		111°		
′	Log	Nat	Log	Nat	Log	Nat	Log	Nat	Log	Nat	Log	Nat	Log	Nat	
0	**0.1000**	1.2588	**1057**	2756	**0.1114**	2924	**1169**	3090	**0.1224**	3256	**1278**	3420	**0.1330**	1.3584	60
4	**0.1004**	1.2599	**1061**	2768	**0.1118**	2935	**1173**	3101	**0.1228**	3267	**1281**	3431	**0.1334**	1.3595	56
8	**0.1007**	1.2611	**1065**	2779	**0.1121**	2946	**1177**	3112	**0.1231**	3278	**1285**	3442	**0.1337**	1.3605	52
12	**0.1011**	1.2622	**1069**	2790	**0.1125**	2957	**1180**	3123	**0.1235**	3289	**1288**	3453	**0.1341**	1.3616	48
16	**0.1015**	1.2633	**1072**	2801	**0.1129**	2968	**1184**	3134	**0.1238**	3300	**1292**	3464	**0.1344**	1.3627	44
20	**0.1019**	1.2644	**1076**	2812	**0.1133**	2979	**1188**	3145	**0.1242**	3311	**1295**	3475	**0.1347**	1.3638	40
24	**0.1023**	1.2656	**1080**	2823	**0.1136**	2990	**1191**	3156	**0.1246**	3322	**1299**	3486	**0.1351**	1.3649	36
28	**0.1027**	1.2667	**1084**	2835	**0.1140**	3002	**1195**	3168	**0.1249**	3333	**1302**	3497	**0.1354**	1.3660	32
32	**0.1031**	1.2678	**1088**	2846	**0.1144**	3013	**1199**	3179	**0.1253**	3344	**1306**	3508	**0.1358**	1.3670	28
36	**0.1034**	1.2689	**1091**	2857	**0.1147**	3024	**1202**	3190	**0.1256**	3355	**1309**	3518	**0.1361**	1.3681	24
40	**0.1038**	1.2700	**1095**	2868	**0.1151**	3035	**1206**	3201	**0.1260**	3365	**1313**	3529	**0.1365**	1.3692	20
44	**0.1042**	1.2712	**1099**	2879	**0.1155**	3046	**1210**	3212	**0.1263**	3376	**1316**	3540	**0.1368**	1.3703	16
48	**0.1046**	1.2723	**1103**	2890	**0.1158**	3057	**1213**	3223	**0.1267**	3387	**1320**	3551	**0.1372**	1.3714	12
52	**0.1050**	1.2734	**1106**	2901	**0.1162**	3068	**1217**	3234	**0.1271**	3398	**1323**	3562	**0.1375**	1.3724	8
56	**0.1053**	1.2745	**1110**	2913	**0.1166**	3079	**1220**	3245	**0.1274**	3409	**1327**	3573	**0.1378**	1.3735	4
60	**0.1057**	1.2756	**1114**	2924	**0.1169**	3090	**1224**	3256	**0.1278**	3420	**1330**	3584	**0.1382**	1.3746	0
	Log	Nat	Log	Nat	Log	Nat	Log	Nat	Log	Nat	Log	Nat	Log	Nat	′
	254°		253°		252°		251°		250°		249°		248°		

	112°		113°		114°		115°		116°		117°		118°		
′	Log	Nat	Log	Nat	Log	Nat	Log	Nat	Log	Nat	Log	Nat	Log	Nat	
0	**0.1382**	1.3746	**1432**	3907	**0.1482**	4067	**1531**	4226	**0.1579**	4384	**1626**	4540	**0.1672**	1.4695	60
4	**0.1385**	1.3757	**1436**	3918	**0.1485**	4078	**1534**	4237	**0.1582**	4394	**1629**	4550	**0.1675**	1.4705	56
8	**0.1389**	1.3768	**1439**	3929	**0.1489**	4089	**1537**	4247	**0.1585**	4405	**1632**	4561	**0.1678**	1.4715	52
12	**0.1392**	1.3778	**1442**	3939	**0.1492**	4099	**1541**	4258	**0.1588**	4415	**1635**	4571	**0.1681**	1.4726	48
16	**0.1395**	1.3789	**1446**	3950	**0.1495**	4110	**1544**	4268	**0.1591**	4425	**1638**	4581	**0.1684**	1.4736	44
20	**0.1399**	1.3800	**1449**	3961	**0.1498**	4120	**1547**	4279	**0.1594**	4436	**1641**	4592	**0.1687**	1.4746	40
24	**0.1402**	1.3811	**1452**	3971	**0.1502**	4131	**1550**	4289	**0.1598**	4446	**1644**	4602	**0.1690**	1.4756	36
28	**0.1406**	1.3821	**1456**	3982	**0.1505**	4142	**1553**	4300	**0.1601**	4457	**1647**	4612	**0.1693**	1.4766	32
32	**0.1409**	1.3832	**1459**	3993	**0.1508**	4152	**1557**	4310	**0.1604**	4467	**1650**	4623	**0.1696**	1.4777	28
36	**0.1412**	1.3843	**1462**	4003	**0.1511**	4163	**1560**	4321	**0.1607**	4478	**1653**	4633	**0.1699**	1.4787	24
40	**0.1416**	1.3854	**1466**	4014	**0.1515**	4173	**1563**	4331	**0.1610**	4488	**1656**	4643	**0.1702**	1.4797	20
44	**0.1419**	1.3864	**1469**	4025	**0.1518**	4184	**1566**	4342	**0.1613**	4498	**1659**	4654	**0.1705**	1.4807	16
48	**0.1422**	1.3875	**1472**	4035	**0.1521**	4195	**1569**	4352	**0.1616**	4509	**1662**	4664	**0.1708**	1.4818	12
52	**0.1426**	1.3886	**1476**	4046	**0.1524**	4205	**1572**	4363	**0.1619**	4519	**1666**	4674	**0.1711**	1.4828	8
56	**0.1429**	1.3897	**1479**	4057	**0.1528**	4216	**1576**	4373	**0.1623**	4530	**1669**	4684	**0.1714**	1.4838	4
60	**0.1432**	1.3907	**1482**	4067	**0.1531**	4226	**1579**	4384	**0.1626**	4540	**1672**	4695	**0.1717**	1.4848	0
	Log	Nat	Log	Nat	Log	Nat	Log	Nat	Log	Nat	Log	Nat	Log	Nat	′
	247°		246°		245°		244°		243°		242°		241°		

	119°		120°		121°		122°		123°		124°		125°		
′	Log	Nat	Log	Nat	Log	Nat	Log	Nat	Log	Nat	Log	Nat	Log	Nat	
0	**0.1717**	1.4848	**1761**	5000	**0.1804**	5150	**1847**	5299	**0.1888**	5446	**1929**	5592	**0.1969**	1.5736	60
4	**0.1720**	1.4858	**1764**	5010	**0.1807**	5160	**1849**	5309	**0.1891**	5456	**1932**	5602	**0.1972**	1.5745	56
8	**0.1723**	1.4868	**1767**	5020	**0.1810**	5170	**1852**	5319	**0.1894**	5466	**1934**	5611	**0.1974**	1.5755	52
12	**0.1726**	1.4879	**1770**	5030	**0.1813**	5180	**1855**	5329	**0.1896**	5476	**1937**	5621	**0.1977**	1.5764	48
16	**0.1729**	1.4889	**1773**	5040	**0.1816**	5190	**1858**	5339	**0.1899**	5485	**1940**	5630	**0.1979**	1.5774	44
20	**0.1732**	1.4899	**1775**	5050	**0.1818**	5200	**1861**	5348	**0.1902**	5495	**1942**	5640	**0.1982**	1.5783	40
24	**0.1734**	1.4909	**1778**	5060	**0.1821**	5210	**1863**	5358	**0.1905**	5505	**1945**	5650	**0.1985**	1.5793	36
28	**0.1737**	1.4919	**1781**	5070	**0.1824**	5220	**1866**	5368	**0.1907**	5515	**1948**	5659	**0.1987**	1.5802	32
32	**0.1740**	1.4929	**1784**	5080	**0.1827**	5230	**1869**	5378	**0.1910**	5524	**1950**	5669	**0.1990**	1.5812	28
36	**0.1743**	1.4939	**1787**	5090	**0.1830**	5240	**1872**	5388	**0.1913**	5534	**1953**	5678	**0.1992**	1.5821	24
40	**0.1746**	1.4950	**1790**	5100	**0.1833**	5250	**1875**	5398	**0.1916**	5544	**1956**	5688	**0.1995**	1.5831	20
44	**0.1749**	1.4960	**1793**	5110	**0.1835**	5260	**1877**	5407	**0.1918**	5553	**1958**	5698	**0.1998**	1.5840	16
48	**0.1752**	1.4970	**1796**	5120	**0.1838**	5270	**1880**	5417	**0.1921**	5563	**1961**	5707	**0.2000**	1.5850	12
52	**0.1755**	1.4980	**1799**	5130	**0.1841**	5279	**1883**	5427	**0.1924**	5573	**1964**	5717	**0.2003**	1.5859	8
56	**0.1758**	1.4990	**1801**	5140	**0.1844**	5289	**1886**	5437	**0.1926**	5582	**1966**	5726	**0.2005**	1.5868	4
60	**0.1761**	1.5000	**1804**	5150	**0.1847**	5299	**1888**	5446	**0.1929**	5592	**1969**	5736	**0.2008**	1.5878	0
	Log	Nat	Log	Nat	Log	Nat	Log	Nat	Log	Nat	Log	Nat	Log	Nat	′
	240°		239°		238°		237°		236°		235°		234°		

NOTE: To save space, the leading digit has been dropped from every other log versine column. Scan the preceding or following log versine column for the necessary digit. The leading digit of natural versines is 0 or 1. Scan the first and last natural versine columns to determine the appropriate number.

VERSINES

	126°		127°		128°		129°		130°		131°		132°		
′	Log	Nat	Log	Nat	Log	Nat	Log	Nat	Log	Nat	Log	Nat	Log	Nat	
0	**0.2008**	1.5878	**2046**	6018	**0.2084**	6157	**2120**	6293	**0.2156**	6428	**2191**	6561	**0.2225**	1.6691	60
6	**0.2012**	1.5892	**2050**	6032	**0.2087**	6170	**2124**	6307	**0.2159**	6441	**2194**	6574	**0.2228**	1.6704	54
12	**0.2016**	1.5906	**2054**	6046	**0.2091**	6184	**2127**	6320	**0.2163**	6455	**2198**	6587	**0.2232**	1.6717	48
18	**0.2019**	1.5920	**2057**	6060	**0.2095**	6198	**2131**	6334	**0.2166**	6468	**2201**	6600	**0.2235**	1.6730	42
24	**0.2023**	1.5934	**2061**	6074	**0.2098**	6211	**2134**	6347	**0.2170**	6481	**2205**	6613	**0.2238**	1.6743	36
30	**0.2027**	1.5948	**2065**	6088	**0.2102**	6225	**2138**	6361	**0.2173**	6494	**2208**	6626	**0.2242**	1.6756	30
36	**0.2031**	1.5962	**2069**	6101	**0.2106**	6239	**2142**	6374	**0.2177**	6508	**2211**	6639	**0.2245**	1.6769	24
42	**0.2035**	1.5976	**2072**	6115	**0.2109**	6252	**2145**	6388	**0.2180**	6521	**2215**	6652	**0.2248**	1.6782	18
48	**0.2039**	1.5990	**2076**	6129	**0.2113**	6266	**2149**	6401	**0.2184**	6534	**2218**	6665	**0.2252**	1.6794	12
54	**0.2042**	1.6004	**2080**	6143	**0.2116**	6280	**2152**	6414	**0.2187**	6547	**2222**	6678	**0.2255**	1.6807	6
60	**0.2046**	1.6018	**2084**	6157	**0.2120**	6293	**2156**	6428	**0.2191**	6561	**2225**	6691	**0.2258**	1.6820	0
	Log	Nat	Log	Nat	Log	Nat	Log	Nat	Log	Nat	Log	Nat	Log	Nat	′
	233°		232°		231°		230°		229°		228°		227°		

	133°		134°		135°		136°		137°		138°		139°		
′	Log	Nat	Log	Nat	Log	Nat	Log	Nat	Log	Nat	Log	Nat	Log	Nat	
0	**0.2258**	1.6820	**2291**	6947	**0.2323**	7071	**2354**	7193	**0.2384**	7314	**2413**	7431	**0.2442**	1.7547	60
6	**0.2262**	1.6833	**2294**	6959	**0.2326**	7083	**2357**	7206	**0.2387**	7325	**2416**	7443	**0.2445**	1.7559	54
12	**0.2265**	1.6845	**2297**	6972	**0.2329**	7096	**2360**	7218	**0.2390**	7337	**2419**	7455	**0.2448**	1.7570	48
18	**0.2268**	1.6858	**2300**	6984	**0.2332**	7108	**2363**	7230	**0.2393**	7349	**2422**	7466	**0.2451**	1.7581	42
24	**0.2271**	1.6871	**2304**	6997	**0.2335**	7120	**2366**	7242	**0.2396**	7361	**2425**	7478	**0.2453**	1.7593	36
30	**0.2275**	1.6884	**2307**	7009	**0.2338**	7133	**2369**	7254	**0.2399**	7373	**2428**	7490	**0.2456**	1.7604	30
36	**0.2278**	1.6896	**2310**	7022	**0.2341**	7145	**2372**	7266	**0.2402**	7385	**2431**	7501	**0.2459**	1.7615	24
42	**0.2281**	1.6909	**2313**	7034	**0.2344**	7157	**2375**	7278	**0.2405**	7396	**2434**	7513	**0.2462**	1.7627	18
48	**0.2284**	1.6921	**2316**	7046	**0.2347**	7169	**2378**	7290	**0.2408**	7408	**2436**	7524	**0.2464**	1.7638	12
54	**0.2288**	1.6934	**2319**	7059	**0.2351**	7181	**2381**	7302	**0.2410**	7420	**2439**	7536	**0.2467**	1.7649	6
60	**0.2291**	1.6947	**2323**	7071	**0.2354**	7193	**2384**	7314	**0.2413**	7431	**2442**	7547	**0.2470**	1.7660	0
	Log	Nat	Log	Nat	Log	Nat	Log	Nat	Log	Nat	Log	Nat	Log	Nat	′
	226°		225°		224°		223°		222°		221°		220°		

	140°		141°		142°		143°		144°		145°		146°		
′	Log	Nat	Log	Nat	Log	Nat	Log	Nat	Log	Nat	Log	Nat	Log	Nat	
0	**0.2470**	1.7660	**2497**	7771	**0.2524**	7880	**2549**	7986	**0.2574**	8090	**2599**	8192	**0.2622**	1.8290	60
6	**0.2473**	1.7672	**2500**	7782	**0.2526**	7891	**2552**	7997	**0.2577**	8100	**2601**	8202	**0.2625**	1.8300	54
12	**0.2476**	1.7683	**2503**	7793	**0.2529**	7902	**2554**	8007	**0.2579**	8111	**2603**	8211	**0.2627**	1.8310	48
18	**0.2478**	1.7694	**2505**	7804	**0.2531**	7912	**2557**	8018	**0.2582**	8121	**2606**	8221	**0.2629**	1.8320	42
24	**0.2481**	1.7705	**2508**	7815	**0.2534**	7923	**2560**	8028	**0.2584**	8131	**2608**	8231	**0.2631**	1.8329	36
30	**0.2484**	1.7716	**2511**	7826	**0.2537**	7934	**2562**	8039	**0.2587**	8141	**2611**	8241	**0.2634**	1.8339	30
36	**0.2486**	1.7727	**2513**	7837	**0.2539**	7944	**2565**	8049	**0.2589**	8151	**2613**	8251	**0.2636**	1.8348	24
42	**0.2489**	1.7738	**2516**	7848	**0.2542**	7955	**2567**	8059	**0.2591**	8161	**2615**	8261	**0.2638**	1.8358	18
48	**0.2492**	1.7749	**2518**	7859	**0.2544**	7965	**2569**	8070	**0.2594**	8171	**2618**	8271	**0.2641**	1.8368	12
54	**0.2495**	1.7760	**2521**	7869	**0.2547**	7976	**2572**	8080	**0.2596**	8181	**2620**	8281	**0.2643**	1.8377	6
60	**0.2497**	1.7771	**2524**	7880	**0.2549**	7986	**2574**	8090	**0.2599**	8192	**2622**	8290	**0.2645**	1.8387	0
	Log	Nat	Log	Nat	Log	Nat	Log	Nat	Log	Nat	Log	Nat	Log	Nat	′
	219°		218°		217°		216°		215°		214°		213°		

	147°		148°		149°		150°		151°		152°		153°		
′	Log	Nat	Log	Nat	Log	Nat	Log	Nat	Log	Nat	Log	Nat	Log	Nat	
0	**0.2645**	1.8387	**2667**	8480	**0.2689**	8572	**2709**	8660	**0.2729**	8746	**2748**	8829	**0.2767**	1.8910	60
6	**0.2647**	1.8396	**2669**	8490	**0.2691**	8581	**2711**	8669	**0.2731**	8755	**2750**	8838	**0.2769**	1.8918	54
12	**0.2650**	1.8406	**2671**	8499	**0.2693**	8590	**2713**	8678	**0.2733**	8763	**2752**	8846	**0.2771**	1.8926	48
18	**0.2652**	1.8415	**2674**	8508	**0.2695**	8599	**2715**	8686	**0.2735**	8771	**2754**	8854	**0.2772**	1.8934	42
24	**0.2654**	1.8425	**2676**	8517	**0.2697**	8607	**2717**	8695	**0.2737**	8780	**2756**	8862	**0.2774**	1.8942	36
30	**0.2656**	1.8434	**2678**	8526	**0.2699**	8616	**2719**	8704	**0.2739**	8788	**2758**	8870	**0.2776**	1.8949	30
36	**0.2658**	1.8443	**2680**	8536	**0.2701**	8625	**2721**	8712	**0.2741**	8796	**2760**	8878	**0.2778**	1.8957	24
42	**0.2661**	1.8453	**2682**	8545	**0.2703**	8634	**2723**	8721	**0.2743**	8805	**2761**	8886	**0.2779**	1.8965	18
48	**0.2663**	1.8462	**2684**	8554	**0.2705**	8643	**2725**	8729	**0.2745**	8813	**2763**	8894	**0.2781**	1.8973	12
54	**0.2665**	1.8471	**2686**	8563	**0.2707**	8652	**2727**	8738	**0.2746**	8821	**2765**	8902	**0.2783**	1.8980	6
60	**0.2667**	1.8480	**2689**	8572	**0.2709**	8660	**2729**	8746	**0.2748**	8829	**2767**	8910	**0.2785**	1.8988	0
	Log	Nat	Log	Nat	Log	Nat	Log	Nat	Log	Nat	Log	Nat	Log	Nat	′
	212°		211°		210°		209°		208°		207°		206°		

NOTE: To save space, the leading digit has been dropped from every other log versine column. Scan the preceding or following log versine column for the necessary digit. The leading digit of natural versines is 0 or 1. Scan the first and last natural versine columns to determine the appropriate number.

VERSINES

	154°		155°		156°		157°		158°		159°		160°		
′	Log	Nat	Log	Nat	Log	Nat	Log	Nat	Log	Nat	Log	Nat	Log	Nat	
0	**0.2785**	1.8988	**2802**	9063	**0.2818**	9135	**2834**	9205	**0.2849**	9272	**2864**	9336	**0.2877**	1.9397	60
6	**0.2787**	1.8996	**2804**	9070	**0.2820**	9143	**2836**	9212	**0.2851**	9278	**2865**	9342	**0.2879**	1.9403	54
12	**0.2788**	1.9003	**2805**	9078	**0.2822**	9150	**2837**	9219	**0.2852**	9285	**2866**	9348	**0.2880**	1.9409	48
18	**0.2790**	1.9011	**2807**	9085	**0.2823**	9157	**2839**	9225	**0.2854**	9291	**2868**	9354	**0.2881**	1.9415	42
24	**0.2792**	1.9018	**2809**	9092	**0.2825**	9164	**2840**	9232	**0.2855**	9298	**2869**	9361	**0.2883**	1.9421	36
30	**0.2793**	1.9026	**2810**	9100	**0.2826**	9171	**2842**	9239	**0.2857**	9304	**2871**	9367	**0.2884**	1.9426	30
36	**0.2795**	1.9033	**2812**	9107	**0.2828**	9178	**2843**	9245	**0.2858**	9311	**2872**	9373	**0.2885**	1.9432	24
42	**0.2797**	1.9041	**2814**	9114	**0.2829**	9184	**2845**	9252	**0.2859**	9317	**2873**	9379	**0.2887**	1.9438	18
48	**0.2799**	1.9048	**2815**	9121	**0.2831**	9191	**2846**	9259	**0.2861**	9323	**2875**	9385	**0.2888**	1.9444	12
54	**0.2800**	1.9056	**2817**	9128	**0.2833**	9198	**2848**	9265	**0.2862**	9330	**2876**	9391	**0.2889**	1.9449	6
60	**0.2802**	1.9063	**2818**	9135	**0.2834**	9205	**2849**	9272	**0.2864**	9336	**2877**	9397	**0.2890**	1.9455	0
	Log	Nat	Log	Nat	Log	Nat	Log	Nat	Log	Nat	Log	Nat	Log	Nat	′
	205°		204°		203°		202°		201°		200°		199°		

	161°		162°		163°		164°		165°		166°		167°		
′	Log	Nat	Log	Nat	Log	Nat	Log	Nat	Log	Nat	Log	Nat	Log	Nat	
0	**0.2890**	1.9455	**2903**	9511	**0.2914**	9563	**2925**	9613	**0.2936**	9659	**2945**	9703	**0.2954**	1.9744	60
6	**0.2892**	1.9461	**2904**	9516	**0.2915**	9568	**2926**	9617	**0.2937**	9664	**2946**	9707	**0.2955**	1.9748	54
12	**0.2893**	1.9466	**2905**	9521	**0.2917**	9573	**2927**	9622	**0.2938**	9668	**2947**	9711	**0.2956**	1.9751	48
18	**0.2894**	1.9472	**2906**	9527	**0.2918**	9578	**2929**	9627	**0.2939**	9673	**2948**	9715	**0.2957**	1.9755	42
24	**0.2895**	1.9478	**2907**	9532	**0.2919**	9583	**2930**	9632	**0.2940**	9677	**2949**	9720	**0.2958**	1.9759	36
30	**0.2897**	1.9483	**2909**	9537	**0.2920**	9588	**2931**	9636	**0.2941**	9681	**2950**	9724	**0.2959**	1.9763	30
36	**0.2898**	1.9489	**2910**	9542	**0.2921**	9593	**2932**	9641	**0.2942**	9686	**2951**	9728	**0.2959**	1.9767	24
42	**0.2899**	1.9494	**2911**	9548	**0.2922**	9598	**2933**	9646	**0.2942**	9690	**2952**	9732	**0.2960**	1.9770	18
48	**0.2900**	1.9500	**2912**	9553	**0.2923**	9603	**2934**	9650	**0.2943**	9694	**2953**	9736	**0.2961**	1.9774	12
54	**0.2901**	1.9505	**2913**	9558	**0.2924**	9608	**2935**	9655	**0.2944**	9699	**2953**	9740	**0.2962**	1.9778	6
60	**0.2903**	1.9511	**2914**	9563	**0.2925**	9613	**2936**	9659	**0.2945**	9703	**2954**	9744	**0.2963**	1.9781	0
	Log	Nat	Log	Nat	Log	Nat	Log	Nat	Log	Nat	Log	Nat	Log	Nat	′
	198°		197°		196°		195°		194°		193°		192°		

	168°		169°		170°		171°		172°		173°		174°		
′	Log	Nat	Log	Nat	Log	Nat	Log	Nat	Log	Nat	Log	Nat	Log	Nat	
0	**0.2963**	1.9781	**2970**	9816	**0.2977**	9848	**2983**	9877	**0.2989**	9903	**2994**	9925	**0.2998**	1.9945	60
6	**0.2963**	1.9785	**2971**	9820	**0.2978**	9851	**2984**	9880	**0.2990**	9905	**2995**	9928	**0.2999**	1.9947	54
12	**0.2964**	1.9789	**2972**	9823	**0.2978**	9854	**2985**	9882	**0.2990**	9907	**2995**	9930	**0.2999**	1.9949	48
18	**0.2965**	1.9792	**2972**	9826	**0.2979**	9857	**2985**	9885	**0.2991**	9910	**2995**	9932	**0.3000**	1.9951	42
24	**0.2966**	1.9796	**2973**	9829	**0.2980**	9860	**2986**	9888	**0.2991**	9912	**2996**	9934	**0.3000**	1.9952	36
30	**0.2966**	1.9799	**2974**	9833	**0.2980**	9863	**2986**	9890	**0.2992**	9914	**2996**	9936	**0.3000**	1.9954	30
36	**0.2967**	1.9803	**2974**	9836	**0.2981**	9866	**2987**	9893	**0.2992**	9917	**2997**	9938	**0.3001**	1.9956	24
42	**0.2968**	1.9806	**2975**	9839	**0.2982**	9869	**2987**	9895	**0.2993**	9919	**2997**	9940	**0.3001**	1.9957	18
48	**0.2969**	1.9810	**2976**	9842	**0.2982**	9871	**2988**	9898	**0.2993**	9921	**2998**	9942	**0.3001**	1.9959	12
54	**0.2969**	1.9813	**2977**	9845	**0.2983**	9874	**2989**	9900	**0.2994**	9923	**2998**	9943	**0.3002**	1.9960	6
60	**0.2970**	1.9816	**2977**	9848	**0.2983**	9877	**2989**	9903	**0.2994**	9925	**2998**	9945	**0.3002**	1.9962	0
	Log	Nat	Log	Nat	Log	Nat	Log	Nat	Log	Nat	Log	Nat	Log	Nat	′
	191°		190°		189°		188°		187°		186°		185°		

	175°		176°		177°		178°		179°		
′	Log	Nat	Log	Nat	Log	Nat	Log	Nat	Log	Nat	
0	**0.3002**	1.9962	**3005**	9976	**0.3007**	9986	**3009**	9994	**0.3010**	9998	60
6	**0.3002**	1.9963	**3005**	9977	**0.3008**	9987	**3009**	9995	**0.3010**	9999	54
12	**0.3003**	1.9965	**3006**	9978	**0.3008**	9988	**3009**	9995	**0.3010**	9999	48
18	**0.3003**	1.9966	**3006**	9979	**0.3008**	9989	**3009**	9996	**0.3010**	9999	42
24	**0.3003**	1.9968	**3006**	9980	**0.3008**	9990	**3009**	9996	**0.3010**	9999	36
30	**0.3004**	1.9969	**3006**	9981	**0.3008**	9990	**3010**	9997	**0.3010**	0000	30
36	**0.3004**	1.9971	**3006**	9982	**0.3008**	9991	**3010**	9997	**0.3010**	0000	24
42	**0.3004**	1.9972	**3007**	9983	**0.3009**	9992	**3010**	9997	**0.3010**	0000	18
48	**0.3004**	1.9973	**3007**	9984	**0.3009**	9993	**3010**	9998	**0.3010**	0000	12
54	**0.3005**	1.9974	**3007**	9985	**0.3009**	9993	**3010**	9998	**0.3010**	0000	6
60	**0.3005**	1.9976	**3007**	9986	**0.3009**	9994	**3010**	9998	**0.3010**	0000	0
	Log	Nat	Log	Nat	Log	Nat	Log	Nat	Log	Nat	′
	184°		183°		182°		181°		180°		

NOTE: To save space, the leading digit has been dropped from every other log versine column. Scan the preceding or following log versine column for the necessary digit. The leading digit of natural versines is 0 or 1. Scan the first and last natural versine columns to determine the appropriate number.

LOG COSINES

′	0°	1°	2°	3°	4°	5°	6°	7°	8°	9°	10°	11°	12°	13°	14°	
0	0.0000	9999	9997	9994	9989	9.9983	9976	9968	9958	9.9946	9934	9919	9904	9887	9.9869	**60**
1	0.0000	9999	9997	9994	9989	9.9983	9976	9967	9957	9.9946	9933	9919	9904	9887	9.9869	**59**
2	0.0000	9999	9997	9994	9989	9.9983	9976	9967	9957	9.9946	9933	9919	9904	9887	9.9868	**58**
3	0.0000	9999	9997	9994	9989	9.9983	9976	9967	9957	9.9946	9933	9919	9903	9886	9.9868	**57**
4	0.0000	9999	9997	9994	9989	9.9983	9976	9967	9957	9.9945	9933	9918	9903	9886	9.9868	**56**
5	0.0000	9999	9997	9994	9989	9.9983	9975	9967	9957	9.9945	9932	9918	9903	9886	9.9867	**55**
6	0.0000	9999	9997	9994	9989	9.9983	9975	9967	9956	9.9945	9932	9918	9902	9885	9.9867	**54**
7	0.0000	9999	9997	9994	9989	9.9983	9975	9966	9956	9.9945	9932	9918	9902	9885	9.9867	**53**
8	0.0000	9999	9997	9994	9989	9.9983	9975	9966	9956	9.9945	9932	9917	9902	9885	9.9867	**52**
9	0.0000	9999	9997	9993	9989	9.9982	9975	9966	9956	9.9944	9931	9917	9902	9885	9.9866	**51**
10	0.0000	9999	9997	9993	9989	9.9982	9975	9966	9956	9.9944	9931	9917	9901	9884	9.9866	**50**
11	0.0000	9999	9997	9993	9988	9.9982	9975	9966	9956	9.9944	9931	9917	9901	9884	9.9866	**49**
12	0.0000	9999	9997	9993	9988	9.9982	9975	9966	9955	9.9944	9931	9916	9901	9884	9.9865	**48**
13	0.0000	9999	9997	9993	9988	9.9982	9974	9965	9955	9.9944	9931	9916	9901	9883	9.9865	**47**
14	0.0000	9999	9997	9993	9988	9.9982	9974	9965	9955	9.9943	9930	9916	9900	9883	9.9865	**46**
15	0.0000	9999	9997	9993	9988	9.9982	9974	9965	9955	9.9943	9930	9916	9900	9883	9.9864	**45**
16	0.0000	9999	9997	9993	9988	9.9982	9974	9965	9955	9.9943	9930	9915	9900	9883	9.9864	**44**
17	0.0000	9999	9997	9993	9988	9.9982	9974	9965	9954	9.9943	9930	9915	9899	9882	9.9864	**43**
18	0.0000	9999	9996	9993	9988	9.9981	9974	9965	9954	9.9943	9929	9915	9899	9882	9.9863	**42**
19	0.0000	9999	9996	9993	9988	9.9981	9974	9964	9954	9.9942	9929	9915	9899	9882	9.9863	**41**
20	0.0000	9999	9996	9993	9988	9.9981	9973	9964	9954	9.9942	9929	9914	9899	9881	9.9863	**40**
21	0.0000	9999	9996	9993	9987	9.9981	9973	9964	9954	9.9942	9929	9914	9898	9881	9.9862	**39**
22	0.0000	9999	9996	9992	9987	9.9981	9973	9964	9954	9.9942	9929	9914	9898	9881	9.9862	**38**
23	0.0000	9999	9996	9992	9987	9.9981	9973	9964	9953	9.9941	9928	9914	9898	9880	9.9862	**37**
24	0.0000	9999	9996	9992	9987	9.9981	9973	9964	9953	9.9941	9928	9913	9897	9880	9.9861	**36**
25	0.0000	9999	9996	9992	9987	9.9981	9973	9964	9953	9.9941	9928	9913	9897	9880	9.9861	**35**
26	0.0000	9999	9996	9992	9987	9.9980	9973	9963	9953	9.9941	9928	9913	9897	9880	9.9861	**34**
27	0.0000	9999	9996	9992	9987	9.9980	9972	9963	9953	9.9941	9927	9913	9897	9879	9.9860	**33**
28	0.0000	9999	9996	9992	9987	9.9980	9972	9963	9952	9.9940	9927	9912	9896	9879	9.9860	**32**
29	0.0000	9999	9996	9992	9987	9.9980	9972	9963	9952	9.9940	9927	9912	9896	9879	9.9860	**31**
30	0.0000	9999	9996	9992	9987	9.9980	9972	9963	9952	9.9940	9927	9912	9896	9878	9.9859	**30**
31	0.0000	9998	9996	9992	9986	9.9980	9972	9963	9952	9.9940	9926	9912	9896	9878	9.9859	**29**
32	0.0000	9998	9996	9992	9986	9.9980	9972	9962	9952	9.9940	9926	9911	9895	9878	9.9859	**28**
33	0.0000	9998	9996	9992	9986	9.9980	9972	9962	9951	9.9939	9926	9911	9895	9877	9.9858	**27**
34	0.0000	9998	9996	9992	9986	9.9979	9971	9962	9951	9.9939	9926	9911	9895	9877	9.9858	**26**
35	0.0000	9998	9996	9992	9986	9.9979	9971	9962	9951	9.9939	9925	9911	9894	9877	9.9858	**25**
36	0.0000	9998	9996	9991	9986	9.9979	9971	9962	9951	9.9939	9925	9910	9894	9876	9.9857	**24**
37	0.0000	9998	9995	9991	9986	9.9979	9971	9962	9951	9.9939	9925	9910	9894	9876	9.9857	**23**
38	0.0000	9998	9995	9991	9986	9.9979	9971	9961	9951	9.9938	9925	9910	9894	9876	9.9857	**22**
39	0.0000	9998	9995	9991	9986	9.9979	9971	9961	9950	9.9938	9925	9910	9893	9876	9.9856	**21**
40	0.0000	9998	9995	9991	9986	9.9979	9971	9961	9950	9.9938	9924	9909	9893	9875	9.9856	**20**
41	0.0000	9998	9995	9991	9985	9.9979	9970	9961	9950	9.9938	9924	9909	9893	9875	9.9856	**19**
42	0.0000	9998	9995	9991	9985	9.9978	9970	9961	9950	9.9937	9924	9909	9892	9875	9.9855	**18**
43	0.0000	9998	9995	9991	9985	9.9978	9970	9960	9950	9.9937	9924	9909	9892	9874	9.9855	**17**
44	0.0000	9998	9995	9991	9985	9.9978	9970	9960	9949	9.9937	9923	9908	9892	9874	9.9855	**16**
45	0.0000	9998	9995	9991	9985	9.9978	9970	9960	9949	9.9937	9923	9908	9892	9874	9.9854	**15**
46	0.0000	9998	9995	9991	9985	9.9978	9970	9960	9949	9.9937	9923	9908	9891	9873	9.9854	**14**
47	0.0000	9998	9995	9991	9985	9.9978	9969	9960	9949	9.9936	9923	9908	9891	9873	9.9854	**13**
48	0.0000	9998	9995	9990	9985	9.9978	9969	9960	9949	9.9936	9922	9907	9891	9873	9.9853	**12**
49	0.0000	9998	9995	9990	9985	9.9978	9969	9959	9948	9.9936	9922	9907	9890	9872	9.9853	**11**
50	0.0000	9998	9995	9990	9985	9.9977	9969	9959	9948	9.9936	9922	9907	9890	9872	9.9853	**10**
51	0.0000	9998	9995	9990	9984	9.9977	9969	9959	9948	9.9936	9922	9906	9890	9872	9.9852	**9**
52	0.0000	9998	9995	9990	9984	9.9977	9969	9959	9948	9.9935	9921	9906	9890	9872	9.9852	**8**
53	9.9999	9998	9994	9990	9984	9.9977	9969	9959	9948	9.9935	9921	9906	9889	9871	9.9852	**7**
54	9.9999	9998	9994	9990	9984	9.9977	9968	9959	9947	9.9935	9921	9906	9889	9871	9.9851	**6**
55	9.9999	9998	9994	9990	9984	9.9977	9968	9958	9947	9.9935	9921	9905	9889	9871	9.9851	**5**
56	9.9999	9998	9994	9990	9984	9.9977	9968	9958	9947	9.9934	9920	9905	9888	9870	9.9851	**4**
57	9.9999	9997	9994	9990	9984	9.9977	9968	9958	9947	9.9934	9920	9905	9888	9870	9.9850	**3**
58	9.9999	9997	9994	9990	9984	9.9976	9968	9958	9947	9.9934	9920	9905	9888	9870	9.9850	**2**
59	9.9999	9997	9994	9989	9984	9.9976	9968	9958	9946	9.9934	9920	9904	9888	9869	9.9850	**1**
60	9.9999	9997	9994	9989	9983	9.9976	9968	9958	9946	9.9934	9919	9904	9887	9869	9.9849	**0**
	89°	**88°**	**87°**	**86°**	**85°**	**84°**	**83°**	**82°**	**81°**	**80°**	**79°**	**78°**	**77°**	**76°**	**75°**	**′**

LOG SINES

LOG COSINES

′	15°	16°	17°	18°	19°	20°	21°	22°	23°	24°	25°	26°	27°	28°	29°	
0	9.9849	9828	9806	9782	9757	9.9730	9702	9672	9640	9.9607	9573	9537	9499	9459	9.9418	**60**
1	9.9849	9828	9806	9782	9756	9.9729	9701	9671	9640	9.9607	9572	9536	9498	9459	9.9417	**59**
2	9.9849	9828	9805	9781	9756	9.9729	9701	9671	9639	9.9606	9572	9535	9498	9458	9.9417	**58**
3	9.9848	9827	9805	9781	9755	9.9728	9700	9670	9639	9.9606	9571	9535	9497	9457	9.9416	**57**
4	9.9848	9827	9804	9780	9755	9.9728	9700	9670	9638	9.9605	9570	9534	9496	9457	9.9415	**56**
5	9.9848	9827	9804	9780	9755	9.9728	9699	9669	9638	9.9604	9570	9534	9496	9456	9.9415	**55**
6	9.9847	9826	9804	9780	9754	9.9727	9699	9669	9637	9.9604	9569	9533	9495	9455	9.9414	**54**
7	9.9847	9826	9803	9779	9754	9.9727	9698	9668	9636	9.9603	9569	9532	9494	9455	9.9413	**53**
8	9.9847	9826	9803	9779	9753	9.9726	9698	9668	9636	9.9603	9568	9532	9494	9454	9.9413	**52**
9	9.9846	9825	9802	9778	9753	9.9726	9697	9667	9635	9.9602	9567	9531	9493	9453	9.9412	**51**
10	9.9846	9825	9802	9778	9752	9.9725	9697	9667	9635	9.9602	9567	9530	9492	9453	9.9411	**50**
11	9.9846	9824	9802	9778	9752	9.9725	9696	9666	9634	9.9601	9566	9530	9492	9452	9.9410	**49**
12	9.9845	9824	9801	9777	9751	9.9724	9696	9666	9634	9.9601	9566	9529	9491	9451	9.9410	**48**
13	9.9845	9824	9801	9777	9751	9.9724	9695	9665	9633	9.9600	9565	9529	9490	9451	9.9409	**47**
14	9.9845	9823	9801	9776	9751	9.9723	9695	9664	9633	9.9599	9564	9528	9490	9450	9.9408	**46**
15	9.9844	9823	9800	9776	9750	9.9723	9694	9664	9632	9.9599	9564	9527	9489	9449	9.9408	**45**
16	9.9844	9823	9800	9775	9750	9.9722	9694	9663	9632	9.9598	9563	9527	9488	9449	9.9407	**44**
17	9.9844	9822	9799	9775	9749	9.9722	9693	9663	9631	9.9598	9563	9526	9488	9448	9.9406	**43**
18	9.9843	9822	9799	9775	9749	9.9722	9693	9662	9631	9.9597	9562	9525	9487	9447	9.9406	**42**
19	9.9843	9821	9799	9774	9748	9.9721	9692	9662	9630	9.9597	9561	9525	9486	9447	9.9405	**41**
20	9.9843	9821	9798	9774	9748	9.9721	9692	9661	9629	9.9596	9561	9524	9486	9446	9.9404	**40**
21	9.9842	9821	9798	9773	9747	9.9720	9691	9661	9629	9.9595	9560	9524	9485	9445	9.9403	**39**
22	9.9842	9820	9797	9773	9747	9.9720	9691	9660	9628	9.9595	9560	9523	9485	9444	9.9403	**38**
23	9.9842	9820	9797	9773	9747	9.9719	9690	9660	9628	9.9594	9559	9522	9484	9444	9.9402	**37**
24	9.9841	9820	9797	9772	9746	9.9719	9690	9659	9627	9.9594	9558	9522	9483	9443	9.9401	**36**
25	9.9841	9819	9796	9772	9746	9.9718	9689	9659	9627	9.9593	9558	9521	9483	9442	9.9401	**35**
26	9.9841	9819	9796	9771	9745	9.9718	9689	9658	9626	9.9593	9557	9520	9482	9442	9.9400	**34**
27	9.9840	9818	9795	9771	9745	9.9717	9688	9658	9626	9.9592	9557	9520	9481	9441	9.9399	**33**
28	9.9840	9818	9795	9770	9744	9.9717	9688	9657	9625	9.9591	9556	9519	9481	9440	9.9398	**32**
29	9.9839	9818	9795	9770	9744	9.9716	9687	9657	9625	9.9591	9555	9519	9480	9440	9.9398	**31**
30	9.9839	9817	9794	9770	9743	9.9716	9687	9656	9624	9.9590	9555	9518	9479	9439	9.9397	**30**
31	9.9839	9817	9794	9769	9743	9.9715	9686	9656	9623	9.9590	9554	9517	9479	9438	9.9396	**29**
32	9.9838	9817	9793	9769	9743	9.9715	9686	9655	9623	9.9589	9554	9517	9478	9438	9.9396	**28**
33	9.9838	9816	9793	9768	9742	9.9714	9685	9655	9622	9.9589	9553	9516	9477	9437	9.9395	**27**
34	9.9838	9816	9793	9768	9742	9.9714	9685	9654	9622	9.9588	9552	9515	9477	9436	9.9394	**26**
35	9.9837	9815	9792	9767	9741	9.9714	9684	9654	9621	9.9587	9552	9515	9476	9436	9.9393	**25**
36	9.9837	9815	9792	9767	9741	9.9713	9684	9653	9621	9.9587	9551	9514	9475	9435	9.9393	**24**
37	9.9837	9815	9791	9767	9740	9.9713	9683	9652	9620	9.9586	9551	9513	9475	9434	9.9392	**23**
38	9.9836	9814	9791	9766	9740	9.9712	9683	9652	9620	9.9586	9550	9513	9474	9433	9.9391	**22**
39	9.9836	9814	9791	9766	9739	9.9712	9682	9651	9619	9.9585	9549	9512	9473	9433	9.9391	**21**
40	9.9836	9814	9790	9765	9739	9.9711	9682	9651	9618	9.9584	9549	9512	9473	9432	9.9390	**20**
41	9.9835	9813	9790	9765	9739	9.9711	9681	9650	9618	9.9584	9548	9511	9472	9431	9.9389	**19**
42	9.9835	9813	9789	9764	9738	9.9710	9681	9650	9617	9.9583	9548	9510	9471	9431	9.9388	**18**
43	9.9835	9812	9789	9764	9738	9.9710	9680	9649	9617	9.9583	9547	9510	9471	9430	9.9388	**17**
44	9.9834	9812	9789	9764	9737	9.9709	9680	9649	9616	9.9582	9546	9509	9470	9429	9.9387	**16**
45	9.9834	9812	9788	9763	9737	9.9709	9679	9648	9616	9.9582	9546	9508	9469	9429	9.9386	**15**
46	9.9833	9811	9788	9763	9736	9.9708	9679	9648	9615	9.9581	9545	9508	9469	9428	9.9385	**14**
47	9.9833	9811	9787	9762	9736	9.9708	9678	9647	9615	9.9580	9545	9507	9468	9427	9.9385	**13**
48	9.9833	9811	9787	9762	9735	9.9707	9678	9647	9614	9.9580	9544	9506	9467	9427	9.9384	**12**
49	9.9832	9810	9787	9761	9735	9.9707	9677	9646	9613	9.9579	9543	9506	9467	9426	9.9383	**11**
50	9.9832	9810	9786	9761	9734	9.9706	9677	9646	9613	9.9579	9543	9505	9466	9425	9.9383	**10**
51	9.9832	9809	9786	9761	9734	9.9706	9676	9645	9612	9.9578	9542	9505	9465	9424	9.9382	**9**
52	9.9831	9809	9785	9760	9734	9.9705	9676	9645	9612	9.9577	9542	9504	9465	9424	9.9381	**8**
53	9.9831	9809	9785	9760	9733	9.9705	9675	9644	9611	9.9577	9541	9503	9464	9423	9.9380	**7**
54	9.9831	9808	9785	9759	9733	9.9704	9675	9643	9611	9.9576	9540	9503	9463	9422	9.9380	**6**
55	9.9830	9808	9784	9759	9732	9.9704	9674	9643	9610	9.9576	9540	9502	9463	9422	9.9379	**5**
56	9.9830	9808	9784	9758	9732	9.9703	9674	9642	9610	9.9575	9539	9501	9462	9421	9.9378	**4**
57	9.9830	9807	9783	9758	9731	9.9703	9673	9642	9609	9.9575	9538	9501	9461	9420	9.9377	**3**
58	9.9829	9807	9783	9758	9731	9.9702	9673	9641	9608	9.9574	9538	9500	9461	9420	9.9377	**2**
59	9.9829	9806	9782	9757	9730	9.9702	9672	9641	9608	9.9573	9537	9499	9460	9419	9.9376	**1**
60	9.9828	9806	9782	9757	9730	9.9702	9672	9640	9607	9.9573	9537	9499	9459	9418	9.9375	**0**
	74°	**73°**	**72°**	**71°**	**70°**	**69°**	**68°**	**67°**	**66°**	**65°**	**64°**	**63°**	**62°**	**61°**	**60°**	**′**

LOG SINES

LOG COSINES

′	30°	31°	32°	33°	34°	35°	36°	37°	38°	39°	40°	41°	42°	43°	44°	
0	9.9375	9331	9284	9236	9186	9.9134	9080	9023	8965	9.8905	8843	8778	8711	8641	9.8569	**60**
1	9.9375	9330	9283	9235	9185	9.9133	9079	9023	8964	9.8904	8841	8777	8710	8640	9.8568	**59**
2	9.9374	9329	9283	9234	9184	9.9132	9078	9022	8963	9.8903	8840	8776	8708	8639	9.8567	**58**
3	9.9373	9328	9282	9233	9183	9.9131	9077	9021	8962	9.8902	8839	8775	8707	8638	9.8566	**57**
4	9.9372	9328	9281	9233	9182	9.9130	9076	9020	8961	9.8901	8838	8773	8706	8637	9.8564	**56**
5	9.9372	9327	9280	9232	9181	9.9129	9075	9019	8960	9.8900	8837	8772	8705	8635	9.8563	**55**
6	9.9371	9326	9279	9231	9181	9.9128	9074	9018	8959	9.8899	8836	8771	8704	8634	9.8562	**54**
7	9.9370	9325	9279	9230	9180	9.9127	9073	9017	8958	9.8898	8835	8770	8703	8633	9.8561	**53**
8	9.9369	9325	9278	9229	9179	9.9127	9072	9016	8957	9.8897	8834	8769	8702	8632	9.8560	**52**
9	9.9369	9324	9277	9229	9178	9.9126	9071	9015	8956	9.8896	8833	8768	8700	8631	9.8558	**51**
10	9.9368	9323	9276	9228	9177	9.9125	9070	9014	8955	9.8895	8832	8767	8699	8629	9.8557	**50**
11	9.9367	9322	9275	9227	9176	9.9124	9069	9013	8954	9.8894	8831	8766	8698	8628	9.8556	**49**
12	9.9367	9322	9275	9226	9175	9.9123	9069	9012	8953	9.8893	8830	8765	8697	8627	9.8555	**48**
13	9.9366	9321	9274	9225	9175	9.9122	9068	9011	8952	9.8892	8829	8763	8696	8626	9.8553	**47**
14	9.9365	9320	9273	9224	9174	9.9121	9067	9010	8951	9.8891	8828	8762	8695	8625	9.8552	**46**
15	9.9364	9319	9272	9224	9173	9.9120	9066	9009	8950	9.8890	8827	8761	8694	8624	9.8551	**45**
16	9.9364	9318	9272	9223	9172	9.9119	9065	9008	8949	9.8889	8825	8760	8692	8622	9.8550	**44**
17	9.9363	9318	9271	9222	9171	9.9119	9064	9007	8948	9.8888	8824	8759	8691	8621	9.8549	**43**
18	9.9362	9317	9270	9221	9170	9.9118	9063	9006	8947	9.8887	8823	8758	8690	8620	9.8547	**42**
19	9.9361	9316	9269	9220	9169	9.9117	9062	9005	8946	9.8885	8822	8757	8689	8619	9.8546	**41**
20	9.9361	9315	9268	9219	9169	9.9116	9061	9004	8945	9.8884	8821	8756	8688	8618	9.8545	**40**
21	9.9360	9315	9268	9219	9168	9.9115	9060	9003	8944	9.8883	8820	8755	8687	8616	9.8544	**39**
22	9.9359	9314	9267	9218	9167	9.9114	9059	9002	8943	9.8882	8819	8753	8686	8615	9.8542	**38**
23	9.9358	9313	9266	9217	9166	9.9113	9058	9001	8942	9.8881	8818	8752	8684	8614	9.8541	**37**
24	9.9358	9312	9265	9216	9165	9.9112	9057	9000	8941	9.8880	8817	8751	8683	8613	9.8540	**36**
25	9.9357	9312	9264	9215	9164	9.9111	9056	9000	8940	9.8879	8816	8750	8682	8612	9.8539	**35**
26	9.9356	9311	9264	9214	9163	9.9110	9056	8999	8939	9.8878	8815	8749	8681	8610	9.8537	**34**
27	9.9355	9310	9263	9214	9163	9.9110	9055	8998	8938	9.8877	8814	8748	8680	8609	9.8536	**33**
28	9.9355	9309	9262	9213	9162	9.9109	9054	8997	8937	9.8876	8813	8747	8679	8608	9.8535	**32**
29	9.9354	9308	9261	9212	9161	9.9108	9053	8996	8936	9.8875	8812	8746	8677	8607	9.8534	**31**
30	9.9353	9308	9260	9211	9160	9.9107	9052	8995	8935	9.8874	8810	8745	8676	8606	9.8532	**30**
31	9.9352	9307	9259	9210	9159	9.9106	9051	8994	8934	9.8873	8809	8743	8675	8604	9.8531	**29**
32	9.9352	9306	9259	9209	9158	9.9105	9050	8993	8933	9.8872	8808	8742	8674	8603	9.8530	**28**
33	9.9351	9305	9258	9209	9157	9.9104	9049	8992	8932	9.8871	8807	8741	8673	8602	9.8529	**27**
34	9.9350	9305	9257	9208	9156	9.9103	9048	8991	8931	9.8870	8806	8740	8672	8601	9.8527	**26**
35	9.9349	9304	9256	9207	9156	9.9102	9047	8990	8930	9.8869	8805	8739	8671	8600	9.8526	**25**
36	9.9349	9303	9255	9206	9155	9.9101	9046	8989	8929	9.8868	8804	8738	8669	8598	9.8525	**24**
37	9.9348	9302	9255	9205	9154	9.9101	9045	8988	8928	9.8867	8803	8737	8668	8597	9.8524	**23**
38	9.9347	9301	9254	9204	9153	9.9100	9044	8987	8927	9.8866	8802	8736	8667	8596	9.8522	**22**
39	9.9346	9301	9253	9204	9152	9.9099	9043	8986	8926	9.8865	8801	8734	8666	8595	9.8521	**21**
40	9.9346	9300	9252	9203	9151	9.9098	9042	8985	8925	9.8864	8800	8733	8665	8594	9.8520	**20**
41	9.9345	9299	9251	9202	9150	9.9097	9041	8984	8924	9.8863	8799	8732	8664	8592	9.8519	**19**
42	9.9344	9298	9251	9201	9149	9.9096	9041	8983	8923	9.8862	8797	8731	8662	8591	9.8517	**18**
43	9.9343	9298	9250	9200	9149	9.9095	9040	8982	8922	9.8860	8796	8730	8661	8590	9.8516	**17**
44	9.9343	9297	9249	9199	9148	9.9094	9039	8981	8921	9.8859	8795	8729	8660	8589	9.8515	**16**
45	9.9342	9296	9248	9198	9147	9.9093	9038	8980	8920	9.8858	8794	8728	8659	8588	9.8514	**15**
46	9.9341	9295	9247	9198	9146	9.9092	9037	8979	8919	9.8857	8793	8727	8658	8586	9.8512	**14**
47	9.9340	9294	9247	9197	9145	9.9091	9036	8978	8918	9.8856	8792	8725	8657	8585	9.8511	**13**
48	9.9340	9294	9246	9196	9144	9.9091	9035	8977	8917	9.8855	8791	8724	8655	8584	9.8510	**12**
49	9.9339	9293	9245	9195	9143	9.9090	9034	8976	8916	9.8854	8790	8723	8654	8583	9.8509	**11**
50	9.9338	9292	9244	9194	9142	9.9089	9033	8975	8915	9.8853	8789	8722	8653	8582	9.8507	**10**
51	9.9337	9291	9243	9193	9142	9.9088	9032	8974	8914	9.8852	8788	8721	8652	8580	9.8506	**9**
52	9.9337	9291	9242	9193	9141	9.9087	9031	8973	8913	9.8851	8787	8720	8651	8579	9.8505	**8**
53	9.9336	9290	9242	9192	9140	9.9086	9030	8972	8912	9.8850	8785	8719	8650	8578	9.8504	**7**
54	9.9335	9289	9241	9191	9139	9.9085	9029	8971	8911	9.8849	8784	8718	8648	8577	9.8502	**6**
55	9.9334	9288	9240	9190	9138	9.9084	9028	8970	8910	9.8848	8783	8716	8647	8575	9.8501	**5**
56	9.9334	9287	9239	9189	9137	9.9083	9027	8969	8909	9.8847	8782	8715	8646	8574	9.8500	**4**
57	9.9333	9287	9238	9188	9136	9.9082	9026	8968	8908	9.8846	8781	8714	8645	8573	9.8499	**3**
58	9.9332	9286	9238	9187	9135	9.9081	9025	8967	8907	9.8845	8780	8713	8644	8572	9.8497	**2**
59	9.9331	9285	9237	9187	9135	9.9080	9024	8966	8906	9.8844	8779	8712	8642	8571	9.8496	**1**
60	9.9331	9284	9236	9186	9134	9.9080	9023	8965	8905	9.8843	8778	8711	8641	8569	9.8495	**0**
	59°	**58°**	**57°**	**56°**	**55°**	**54°**	**53°**	**52°**	**51°**	**50°**	**49°**	**48°**	**47°**	**46°**	**45°**	**′**

LOG SINES

LOG COSINES

′	45°	46°	47°	48°	49°	50°	51°	52°	53°	54°	55°	56°	57°	58°	59°	
0	9.8495	8418	8338	8255	8169	9.8081	7989	7893	7795	9.7692	7586	7476	7361	7242	9.7118	60
1	9.8494	8416	8336	8254	8168	9.8079	7987	7892	7793	9.7690	7584	7474	7359	7240	9.7116	59
2	9.8492	8415	8335	8252	8167	9.8078	7986	7890	7791	9.7689	7582	7472	7357	7238	9.7114	58
3	9.8491	8414	8334	8251	8165	9.8076	7984	7889	7790	9.7687	7580	7470	7355	7236	9.7112	57
4	9.8490	8412	8332	8249	8164	9.8075	7982	7887	7788	9.7685	7579	7468	7353	7234	9.7110	56
5	9.8489	8411	8331	8248	8162	9.8073	7981	7885	7786	9.7683	7577	7466	7351	7232	9.7108	55
6	9.8487	8410	8330	8247	8161	9.8072	7979	7884	7785	9.7682	7575	7464	7349	7230	9.7106	54
7	9.8486	8409	8328	8245	8159	9.8070	7978	7882	7783	9.7680	7573	7462	7347	7228	9.7104	53
8	9.8485	8407	8327	8244	8158	9.8069	7976	7880	7781	9.7678	7571	7461	7345	7226	9.7102	52
9	9.8483	8406	8326	8242	8156	9.8067	7975	7879	7780	9.7676	7570	7459	7344	7224	9.7099	51
10	9.8482	8405	8324	8241	8155	9.8066	7973	7877	7778	9.7675	7568	7457	7342	7222	9.7097	50
11	9.8481	8403	8323	8240	8153	9.8064	7972	7876	7776	9.7673	7566	7455	7340	7220	9.7095	49
12	9.8480	8402	8322	8238	8152	9.8063	7970	7874	7774	9.7671	7564	7453	7338	7218	9.7093	48
13	9.8478	8401	8320	8237	8150	9.8061	7968	7872	7773	9.7669	7562	7451	7336	7216	9.7091	47
14	9.8477	8399	8319	8235	8149	9.8060	7967	7871	7771	9.7668	7561	7449	7334	7214	9.7089	46
15	9.8476	8398	8317	8234	8148	9.8058	7965	7869	7769	9.7666	7559	7447	7332	7212	9.7087	45
16	9.8475	8397	8316	8233	8146	9.8056	7964	7867	7768	9.7664	7557	7445	7330	7210	9.7085	44
17	9.8473	8395	8315	8231	8145	9.8055	7962	7866	7766	9.7662	7555	7444	7328	7208	9.7082	43
18	9.8472	8394	8313	8230	8143	9.8053	7960	7864	7764	9.7661	7553	7442	7326	7205	9.7080	42
19	9.8471	8393	8312	8228	8142	9.8052	7959	7863	7763	9.7659	7551	7440	7324	7203	9.7078	41
20	9.8469	8391	8311	8227	8140	9.8050	7957	7861	7761	9.7657	7550	7438	7322	7201	9.7076	40
21	9.8468	8390	8309	8225	8139	9.8049	7956	7859	7759	9.7655	7548	7436	7320	7199	9.7074	39
22	9.8467	8389	8308	8224	8137	9.8047	7954	7858	7758	9.7654	7546	7434	7318	7197	9.7072	38
23	9.8466	8387	8306	8223	8136	9.8046	7953	7856	7756	9.7652	7544	7432	7316	7195	9.7070	37
24	9.8464	8386	8305	8221	8134	9.8044	7951	7854	7754	9.7650	7542	7430	7314	7193	9.7068	36
25	9.8463	8385	8304	8220	8133	9.8043	7949	7853	7752	9.7648	7540	7428	7312	7191	9.7065	35
26	9.8462	8383	8302	8218	8131	9.8041	7948	7851	7751	9.7647	7539	7427	7310	7189	9.7063	34
27	9.8460	8382	8301	8217	8130	9.8040	7946	7849	7749	9.7645	7537	7425	7308	7187	9.7061	33
28	9.8459	8381	8300	8215	8128	9.8038	7945	7848	7747	9.7643	7535	7423	7306	7185	9.7059	32
29	9.8458	8379	8298	8214	8127	9.8037	7943	7846	7746	9.7641	7533	7421	7304	7183	9.7057	31
30	9.8457	8378	8297	8213	8125	9.8035	7941	7844	7744	9.7640	7531	7419	7302	7181	9.7055	30
31	9.8455	8377	8295	8211	8124	9.8034	7940	7843	7742	9.7638	7529	7417	7300	7179	9.7053	29
32	9.8454	8375	8294	8210	8122	9.8032	7938	7841	7740	9.7636	7528	7415	7298	7177	9.7050	28
33	9.8453	8374	8293	8208	8121	9.8031	7937	7840	7739	9.7634	7526	7413	7296	7175	9.7048	27
34	9.8451	8373	8291	8207	8120	9.8029	7935	7838	7737	9.7632	7524	7411	7294	7173	9.7046	26
35	9.8450	8371	8290	8205	8118	9.8027	7934	7836	7735	9.7631	7522	7409	7292	7171	9.7044	25
36	9.8449	8370	8289	8204	8117	9.8026	7932	7835	7734	9.7629	7520	7407	7290	7168	9.7042	24
37	9.8448	8369	8287	8203	8115	9.8024	7930	7833	7732	9.7627	7518	7406	7288	7166	9.7040	23
38	9.8446	8367	8286	8201	8114	9.8023	7929	7831	7730	9.7625	7517	7404	7286	7164	9.7037	22
39	9.8445	8366	8284	8200	8112	9.8021	7927	7830	7728	9.7624	7515	7402	7284	7162	9.7035	21
40	9.8444	8365	8283	8198	8111	9.8020	7926	7828	7727	9.7622	7513	7400	7282	7160	9.7033	20
41	9.8442	8363	8282	8197	8109	9.8018	7924	7826	7725	9.7620	7511	7398	7280	7158	9.7031	19
42	9.8441	8362	8280	8195	8108	9.8017	7922	7825	7723	9.7618	7509	7396	7278	7156	9.7029	18
43	9.8440	8361	8279	8194	8106	9.8015	7921	7823	7722	9.7616	7507	7394	7276	7154	9.7027	17
44	9.8439	8359	8277	8193	8105	9.8014	7919	7821	7720	9.7615	7505	7392	7274	7152	9.7025	16
45	9.8437	8358	8276	8191	8103	9.8012	7918	7820	7718	9.7613	7504	7390	7272	7150	9.7022	15
46	9.8436	8357	8275	8190	8102	9.8010	7916	7818	7716	9.7611	7502	7388	7270	7148	9.7020	14
47	9.8435	8355	8273	8188	8100	9.8009	7914	7816	7715	9.7609	7500	7386	7268	7146	9.7018	13
48	9.8433	8354	8272	8187	8099	9.8007	7913	7815	7713	9.7607	7498	7384	7266	7144	9.7016	12
49	9.8432	8353	8270	8185	8097	9.8006	7911	7813	7711	9.7606	7496	7382	7264	7141	9.7014	11
50	9.8431	8351	8269	8184	8096	9.8004	7910	7811	7710	9.7604	7494	7380	7262	7139	9.7012	10
51	9.8429	8350	8268	8182	8094	9.8003	7908	7810	7708	9.7602	7492	7379	7260	7137	9.7009	9
52	9.8428	8349	8266	8181	8093	9.8001	7906	7808	7706	9.7600	7491	7377	7258	7135	9.7007	8
53	9.8427	8347	8265	8180	8091	9.8000	7905	7806	7704	9.7599	7489	7375	7256	7133	9.7005	7
54	9.8426	8346	8264	8178	8090	9.7998	7903	7805	7703	9.7597	7487	7373	7254	7131	9.7003	6
55	9.8424	8345	8262	8177	8088	9.7997	7901	7803	7701	9.7595	7485	7371	7252	7129	9.7001	5
56	9.8423	8343	8261	8175	8087	9.7995	7900	7801	7699	9.7593	7483	7369	7250	7127	9.6998	4
57	9.8422	8342	8259	8174	8085	9.7993	7898	7800	7697	9.7591	7481	7367	7248	7125	9.6996	3
58	9.8420	8341	8258	8172	8084	9.7992	7897	7798	7696	9.7590	7479	7365	7246	7123	9.6994	2
59	9.8419	8339	8257	8171	8082	9.7990	7895	7796	7694	9.7588	7477	7363	7244	7120	9.6992	1
60	9.8418	8338	8255	8169	8081	9.7989	7893	7795	7692	9.7586	7476	7361	7242	7118	9.6990	0
	44°	43°	42°	41°	40°	39°	38°	37°	36°	35°	34°	33°	32°	31°	30°	′

LOG SINES

LOG COSINES

′	60°	61°	62°	63°	64°	65°	66°	67°	68°	69°	70°	71°	72°	73°	74°	
0	9.6990	6856	6716	6570	6418	9.6259	6093	5919	5736	9.5543	5341	5126	4900	4659	9.4403	**60**
1	9.6988	6853	6714	6568	6416	9.6257	6090	5916	5733	9.5540	5337	5123	4896	4655	9.4399	**59**
2	9.6985	6851	6711	6566	6413	9.6254	6087	5913	5729	9.5537	5334	5119	4892	4651	9.4395	**58**
3	9.6983	6849	6709	6563	6411	9.6251	6085	5910	5726	9.5533	5330	5115	4888	4647	9.4390	**57**
4	9.6981	6847	6707	6561	6408	9.6249	6082	5907	5723	9.5530	5327	5112	4884	4643	9.4386	**56**
5	9.6979	6844	6704	6558	6405	9.6246	6079	5904	5720	9.5527	5323	5108	4880	4639	9.4381	**55**
6	9.6977	6842	6702	6556	6403	9.6243	6076	5901	5717	9.5524	5320	5104	4876	4634	9.4377	**54**
7	9.6974	6840	6699	6553	6400	9.6240	6073	5898	5714	9.5520	5316	5101	4873	4630	9.4372	**53**
8	9.6972	6837	6697	6551	6398	9.6238	6070	5895	5711	9.5517	5313	5097	4869	4626	9.4368	**52**
9	9.6970	6835	6695	6548	6395	9.6235	6068	5892	5708	9.5514	5309	5093	4865	4622	9.4364	**51**
10	9.6968	6833	6692	6546	6392	9.6232	6065	5889	5704	9.5510	5306	5090	4861	4618	9.4359	**50**
11	9.6966	6831	6690	6543	6390	9.6230	6062	5886	5701	9.5507	5302	5086	4857	4614	9.4355	**49**
12	9.6963	6828	6687	6541	6387	9.6227	6059	5883	5698	9.5504	5299	5082	4853	4609	9.4350	**48**
13	9.6961	6826	6685	6538	6385	9.6224	6056	5880	5695	9.5500	5295	5078	4849	4605	9.4346	**47**
14	9.6959	6824	6683	6536	6382	9.6221	6053	5877	5692	9.5497	5292	5075	4845	4601	9.4341	**46**
15	9.6957	6821	6680	6533	6379	9.6219	6050	5874	5689	9.5494	5288	5071	4841	4597	9.4337	**45**
16	9.6955	6819	6678	6531	6377	9.6216	6047	5871	5685	9.5490	5285	5067	4837	4593	9.4332	**44**
17	9.6952	6817	6675	6528	6374	9.6213	6045	5868	5682	9.5487	5281	5064	4833	4588	9.4328	**43**
18	9.6950	6814	6673	6526	6371	9.6210	6042	5865	5679	9.5484	5278	5060	4829	4584	9.4323	**42**
19	9.6948	6812	6671	6523	6369	9.6208	6039	5862	5676	9.5480	5274	5056	4825	4580	9.4319	**41**
20	9.6946	6810	6668	6521	6366	9.6205	6036	5859	5673	9.5477	5270	5052	4821	4576	9.4314	**40**
21	9.6943	6808	6666	6518	6364	9.6202	6033	5856	5670	9.5474	5267	5049	4817	4572	9.4310	**39**
22	9.6941	6805	6663	6515	6361	9.6199	6030	5853	5666	9.5470	5263	5045	4813	4567	9.4305	**38**
23	9.6939	6803	6661	6513	6358	9.6197	6027	5850	5663	9.5467	5260	5041	4809	4563	9.4301	**37**
24	9.6937	6801	6659	6510	6356	9.6194	6024	5847	5660	9.5463	5256	5037	4805	4559	9.4296	**36**
25	9.6935	6798	6656	6508	6353	9.6191	6021	5844	5657	9.5460	5253	5034	4801	4555	9.4292	**35**
26	9.6932	6796	6654	6505	6350	9.6188	6019	5841	5654	9.5457	5249	5030	4797	4550	9.4287	**34**
27	9.6930	6794	6651	6503	6348	9.6186	6016	5838	5650	9.5453	5246	5026	4793	4546	9.4283	**33**
28	9.6928	6791	6649	6500	6345	9.6183	6013	5834	5647	9.5450	5242	5022	4789	4542	9.4278	**32**
29	9.6926	6789	6646	6498	6342	9.6180	6010	5831	5644	9.5447	5239	5019	4785	4538	9.4274	**31**
30	9.6923	6787	6644	6495	6340	9.6177	6007	5828	5641	9.5443	5235	5015	4781	4533	9.4269	**30**
31	9.6921	6784	6642	6493	6337	9.6175	6004	5825	5638	9.5440	5231	5011	4777	4529	9.4264	**29**
32	9.6919	6782	6639	6490	6335	9.6172	6001	5822	5634	9.5437	5228	5007	4773	4525	9.4260	**28**
33	9.6917	6780	6637	6488	6332	9.6169	5998	5819	5631	9.5433	5224	5003	4769	4521	9.4255	**27**
34	9.6914	6777	6634	6485	6329	9.6166	5995	5816	5628	9.5430	5221	5000	4765	4516	9.4251	**26**
35	9.6912	6775	6632	6483	6327	9.6163	5992	5813	5625	9.5426	5217	4996	4761	4512	9.4246	**25**
36	9.6910	6773	6629	6480	6324	9.6161	5990	5810	5621	9.5423	5213	4992	4757	4508	9.4242	**24**
37	9.6908	6770	6627	6477	6321	9.6158	5987	5807	5618	9.5420	5210	4988	4753	4503	9.4237	**23**
38	9.6905	6768	6625	6475	6319	9.6155	5984	5804	5615	9.5416	5206	4984	4749	4499	9.4232	**22**
39	9.6903	6766	6622	6472	6316	9.6152	5981	5801	5612	9.5413	5203	4981	4745	4495	9.4228	**21**
40	9.6901	6763	6620	6470	6313	9.6149	5978	5798	5609	9.5409	5199	4977	4741	4491	9.4223	**20**
41	9.6899	6761	6617	6467	6311	9.6147	5975	5795	5605	9.5406	5196	4973	4737	4486	9.4219	**19**
42	9.6896	6759	6615	6465	6308	9.6144	5972	5792	5602	9.5403	5192	4969	4733	4482	9.4214	**18**
43	9.6894	6756	6612	6462	6305	9.6141	5969	5789	5599	9.5399	5188	4965	4729	4478	9.4209	**17**
44	9.6892	6754	6610	6460	6303	9.6138	5966	5785	5596	9.5396	5185	4962	4725	4473	9.4205	**16**
45	9.6890	6752	6607	6457	6300	9.6135	5963	5782	5592	9.5392	5181	4958	4721	4469	9.4200	**15**
46	9.6887	6749	6605	6454	6297	9.6133	5960	5779	5589	9.5389	5177	4954	4717	4465	9.4195	**14**
47	9.6885	6747	6603	6452	6295	9.6130	5957	5776	5586	9.5385	5174	4950	4713	4460	9.4191	**13**
48	9.6883	6744	6600	6449	6292	9.6127	5954	5773	5583	9.5382	5170	4946	4709	4456	9.4186	**12**
49	9.6881	6742	6598	6447	6289	9.6124	5951	5770	5579	9.5379	5167	4942	4705	4452	9.4182	**11**
50	9.6878	6740	6595	6444	6286	9.6121	5948	5767	5576	9.5375	5163	4939	4700	4447	9.4177	**10**
51	9.6876	6737	6593	6442	6284	9.6119	5945	5764	5573	9.5372	5159	4935	4696	4443	9.4172	**9**
52	9.6874	6735	6590	6439	6281	9.6116	5943	5761	5570	9.5368	5156	4931	4692	4438	9.4168	**8**
53	9.6872	6733	6588	6437	6278	9.6113	5940	5758	5566	9.5365	5152	4927	4688	4434	9.4163	**7**
54	9.6869	6730	6585	6434	6276	9.6110	5937	5754	5563	9.5361	5148	4923	4684	4430	9.4158	**6**
55	9.6867	6728	6583	6431	6273	9.6107	5934	5751	5560	9.5358	5145	4919	4680	4425	9.4153	**5**
56	9.6865	6726	6580	6429	6270	9.6104	5931	5748	5556	9.5354	5141	4915	4676	4421	9.4149	**4**
57	9.6863	6723	6578	6426	6268	9.6102	5928	5745	5553	9.5351	5137	4911	4672	4417	9.4144	**3**
58	9.6860	6721	6575	6424	6265	9.6099	5925	5742	5550	9.5347	5134	4908	4668	4412	9.4139	**2**
59	9.6858	6718	6573	6421	6262	9.6096	5922	5739	5547	9.5344	5130	4904	4663	4408	9.4135	**1**
60	9.6856	6716	6570	6418	6259	9.6093	5919	5736	5543	9.5341	5126	4900	4659	4403	9.4130	**0**
	29°	**28°**	**27°**	**26°**	**25°**	**24°**	**23°**	**22°**	**21°**	**20°**	**19°**	**18°**	**17°**	**16°**	**15°**	**′**

LOG SINES

LOG COSINES

′	75°	76°	77°	78°	79°	80°	81°	82°	83°	84°	85°	86°	87°	88°	89°	
0	9.4130	3837	3521	3179	2806	9.2397	1943	1436	0859	9.0192	9403	8436	7188	5428	8.2419	**60**
1	9.4125	3832	3515	3173	2799	9.2390	1935	1427	0849	9.0180	9388	8418	7164	5392	8.2346	**59**
2	9.4121	3827	3510	3167	2793	9.2382	1927	1418	0838	9.0168	9374	8400	7140	5355	8.2271	**58**
3	9.4116	3822	3504	3161	2786	9.2375	1919	1409	0828	9.0156	9359	8381	7115	5318	8.2196	**57**
4	9.4111	3816	3499	3155	2780	9.2368	1911	1399	0818	9.0144	9345	8363	7090	5281	8.2119	**56**
5	9.4106	3811	3493	3149	2773	9.2361	1903	1390	0807	9.0132	9330	8345	7066	5243	8.2041	**55**
6	9.4102	3806	3488	3143	2767	9.2354	1895	1381	0797	9.0120	9315	8326	7041	5206	8.1961	**54**
7	9.4097	3801	3482	3137	2760	9.2346	1887	1372	0786	9.0107	9301	8307	7016	5167	8.1880	**53**
8	9.4092	3796	3477	3131	2754	9.2339	1879	1363	0776	9.0095	9286	8289	6991	5129	8.1797	**52**
9	9.4087	3791	3471	3125	2747	9.2332	1871	1354	0765	9.0083	9271	8270	6965	5090	8.1713	**51**
10	9.4083	3786	3466	3119	2740	9.2324	1863	1345	0755	9.0070	9256	8251	6940	5050	8.1627	**50**
11	9.4078	3781	3460	3113	2734	9.2317	1855	1336	0744	9.0058	9241	8232	6914	5011	8.1539	**49**
12	9.4073	3775	3455	3107	2727	9.2310	1847	1326	0734	9.0046	9226	8213	6889	4971	8.1450	**48**
13	9.4068	3770	3449	3101	2721	9.2303	1838	1317	0723	9.0033	9211	8194	6863	4930	8.1358	**47**
14	9.4063	3765	3444	3095	2714	9.2295	1830	1308	0712	9.0021	9196	8175	6837	4890	8.1265	**46**
15	9.4059	3760	3438	3089	2707	9.2288	1822	1299	0702	9.0008	9181	8156	6810	4848	8.1169	**45**
16	9.4054	3755	3432	3083	2701	9.2281	1814	1289	0691	8.9996	9166	8137	6784	4807	8.1072	**44**
17	9.4049	3750	3427	3077	2694	9.2273	1806	1280	0680	8.9983	9150	8117	6758	4765	8.0972	**43**
18	9.4044	3745	3421	3070	2687	9.2266	1797	1271	0670	8.9970	9135	8098	6731	4723	8.0870	**42**
19	9.4039	3739	3416	3064	2681	9.2258	1789	1261	0659	8.9958	9119	8078	6704	4680	8.0765	**41**
20	9.4035	3734	3410	3058	2674	9.2251	1781	1252	0648	8.9945	9104	8059	6677	4637	8.0658	**40**
21	9.4030	3729	3404	3052	2667	9.2244	1772	1242	0637	8.9932	9089	8039	6650	4593	8.0548	**39**
22	9.4025	3724	3399	3046	2661	9.2236	1764	1233	0626	8.9919	9073	8019	6622	4549	8.0435	**38**
23	9.4020	3719	3393	3040	2654	9.2229	1756	1224	0616	8.9907	9057	7999	6595	4504	8.0319	**37**
24	9.4015	3713	3387	3034	2647	9.2221	1747	1214	0605	8.9894	9042	7979	6567	4459	8.0200	**36**
25	9.4010	3708	3382	3027	2640	9.2214	1739	1205	0594	8.9881	9026	7959	6539	4414	8.0078	**35**
26	9.4006	3703	3376	3021	2634	9.2206	1731	1195	0583	8.9868	9010	7939	6511	4368	7.9952	**34**
27	9.4001	3698	3370	3015	2627	9.2199	1722	1186	0572	8.9855	8994	7918	6483	4322	7.9822	**33**
28	9.3996	3692	3365	3009	2620	9.2191	1714	1176	0561	8.9842	8978	7898	6454	4275	7.9689	**32**
29	9.3991	3687	3359	3003	2613	9.2184	1705	1167	0550	8.9829	8962	7877	6426	4227	7.9551	**31**
30	9.3986	3682	3353	2997	2606	9.2176	1697	1157	0539	8.9816	8946	7857	6397	4179	7.9408	**30**
31	9.3981	3677	3348	2990	2600	9.2169	1689	1147	0527	8.9803	8930	7836	6368	4131	7.9261	**29**
32	9.3976	3671	3342	2984	2593	9.2161	1680	1138	0516	8.9789	8914	7815	6339	4082	7.9109	**28**
33	9.3971	3666	3336	2978	2586	9.2153	1672	1128	0505	8.9776	8898	7794	6309	4032	7.8951	**27**
34	9.3966	3661	3331	2972	2579	9.2146	1663	1118	0494	8.9763	8882	7773	6279	3982	7.8787	**26**
35	9.3962	3655	3325	2965	2572	9.2138	1655	1109	0483	8.9750	8865	7752	6250	3931	7.8617	**25**
36	9.3957	3650	3319	2959	2565	9.2131	1646	1099	0472	8.9736	8849	7731	6220	3880	7.8439	**24**
37	9.3952	3645	3313	2953	2558	9.2123	1637	1089	0460	8.9723	8833	7710	6189	3828	7.8255	**23**
38	9.3947	3640	3308	2947	2551	9.2115	1629	1080	0449	8.9710	8816	7688	6159	3775	7.8061	**22**
39	9.3942	3634	3302	2940	2545	9.2108	1620	1070	0438	8.9696	8799	7667	6128	3722	7.7859	**21**
40	9.3937	3629	3296	2934	2538	9.2100	1612	1060	0426	8.9683	8783	7645	6097	3668	7.7648	**20**
41	9.3932	3624	3290	2928	2531	9.2092	1603	1050	0415	8.9669	8766	7623	6066	3613	7.7425	**19**
42	9.3927	3618	3284	2921	2524	9.2085	1594	1040	0403	8.9655	8749	7602	6035	3558	7.7190	**18**
43	9.3922	3613	3279	2915	2517	9.2077	1586	1030	0392	8.9642	8733	7580	6003	3502	7.6942	**17**
44	9.3917	3608	3273	2909	2510	9.2069	1577	1020	0380	8.9628	8716	7557	5972	3445	7.6678	**16**
45	9.3912	3602	3267	2902	2503	9.2061	1568	1011	0369	8.9614	8689	7535	5939	3388	7.6398	**15**
46	9.3907	3597	3261	2896	2496	9.2054	1560	1001	0357	8.9601	8682	7513	5907	3329	7.6099	**14**
47	9.3902	3591	3255	2890	2489	9.2046	1551	0991	0346	8.9587	8665	7491	5875	3270	7.5777	**13**
48	9.3897	3586	3250	2883	2482	9.2038	1542	0981	0334	8.9573	8647	7468	5842	3210	7.5429	**12**
49	9.3892	3581	3244	2877	2475	9.2030	1533	0971	0323	8.9559	8630	7445	5809	3150	7.5051	**11**
50	9.3887	3575	3238	2870	2468	9.2022	1525	0961	0311	8.9545	8613	7423	5776	3088	7.4637	**10**
51	9.3882	3570	3232	2864	2461	9.2015	1516	0951	0299	8.9531	8595	7400	5742	3025	7.4180	**9**
52	9.3877	3564	3226	2858	2454	9.2007	1507	0940	0287	8.9517	8578	7377	5708	2962	7.3668	**8**
53	9.3872	3559	3220	2851	2447	9.1999	1498	0930	0276	8.9503	8560	7354	5674	2898	7.3088	**7**
54	9.3867	3554	3214	2845	2439	9.1991	1489	0920	0264	8.9489	8543	7330	5640	2832	7.2419	**6**
55	9.3862	3548	3208	2838	2432	9.1983	1480	0910	0252	8.9475	8525	7307	5605	2766	7.1627	**5**
56	9.3857	3543	3202	2832	2425	9.1975	1471	0900	0240	8.9460	8508	7283	5571	2699	7.0658	**4**
57	9.3852	3537	3197	2825	2418	9.1967	1462	0890	0228	8.9446	8490	7260	5535	2630	6.9409	**3**
58	9.3847	3532	3191	2819	2411	9.1959	1453	0879	0216	8.9432	8472	7236	5500	2561	6.7648	**2**
59	9.3842	3526	3185	2812	2404	9.1951	1445	0869	0204	8.9417	8454	7212	5464	2490	6.4637	**1**
60	9.3837	3521	3179	2806	2397	9.1943	1436	0859	0192	8.9403	8436	7188	5428	2419	−∞	**0**
	14°	**13°**	**12°**	**11°**	**10°**	**9°**	**8°**	**7°**	**6°**	**5°**	**4°**	**3°**	**2°**	**1°**	**0°**	**′**

LOG SINES

LOG COSINES/SINES

E Ephemeris

NOTES

NOTES

SUBJECT INDEX

SUBJECT Index I

Place Index

A

PLACE Index I

D

PLACE Index I

E

F

G

H

I

J

M

N

PLACE Index I

O

P

Q

R

S

PLACE

Index

I

T

Index to Advertisers